Grob's Basic Electronics

2024 Release

Mitchel E. Schultz

Minnesota State College Southeast

Mc Graw Hill

GROB'S BASIC ELECTRONICS

Published by McGraw Hill LLC, 1325 Avenue of the Americas, New York, NY 10019.

This book is printed on acid-free paper.

2 3 4 5 6 7 8 9 CPI 29 28 27 26

ISBN 978-1-266-83498-1
MHID 1-266-83498-2

mheducation.com/highered

Dedication

This book is dedicated to all of the students I have had the honor of teaching over the span of my career. Your passion and level of commitment to learning has truly been inspiring.

Brief Contents

Contents

Chapter 16 Capacitance 492

Chapter 17 Capacitive Reactance 532

Chapter 18 Capacitive Circuits 554

Chapter 19 Inductance 580

Chapter 24 Complex Numbers for AC Circuits 740

Chapter 25 Resonance 770

Chapter 26 Filters 806

Chapter 27 Three-Phase AC Power Systems 850

Preface

The 2024 release of ***Grob's Basic Electronics*** provides students and instructors with complete and comprehensive coverage of the fundamentals of electricity and electronics. The book is written for beginning students who have little or no experience and/or knowledge about the field of electronics. A basic understanding of algebra and trigonometry is helpful since several algebraic equations and right-angle trigonometry problems appear throughout the text.

The opening material in the book, titled **"Introduction to Powers of 10,"** prepares students to work with numbers expressed in scientific and engineering notation as well as with the most common metric prefixes encountered in electronics. Students learn how to add, subtract, multiply, divide, square, and take the square root of numbers expressed in any form of powers of 10 notation.

Chapters 1 through 12 cover the basics of atomic structure, voltage, current, resistance, the resistor color code, Ohm's law, power, series circuits, parallel circuits, series-parallel (combination) circuits, voltage and current dividers, analog and digital meters, Kirchhoff's laws, network theorems, wire resistance, switches, insulators, primary and secondary cells, battery types, internal resistance, and maximum transfer of power. The first 12 chapters are considered DC chapters because the voltages and currents used in analyzing the circuits in these chapters are strictly DC.

Chapters 13 through 27 cover the basics of magnetism, electromagnetism, relays, alternating voltage and current, capacitance, capacitor types, capacitive reactance, capacitive circuits, inductance, transformers, inductive reactance, inductive circuits, *RC* and *L/R* time constants, real power, apparent power, power factor, complex numbers, resonance, filters, and three-phase AC power systems. Chapters 13–27 are considered the AC chapters since the voltages and currents used in analyzing the circuits in these chapters are primarily AC.

Chapters 28 through 34 cover the basics of electronic devices, which include semiconductor physics, diode characteristics, diode testing, half-wave and full-wave rectifier circuits, the capacitor input filter, light-emitting diodes (LEDs), zener diodes, bipolar junction transistors, transistor biasing techniques, the common-emitter, common-collector, and common-base amplifiers, JFET and MOSFET characteristics, JFET amplifiers, MOSFET amplifiers, class A, class B and class C amplifiers, diacs, SCRs, triacs, UJTs, op-amp characteristics, inverting amplifiers, noninverting amplifiers, and nonlinear op-amp circuits. ***These seven additional chapters covering electronic devices may qualify this text for those who want to use it for DC fundamentals, AC fundamentals, as well as electronic devices.***

Appendixes **A** through **H** serve as a resource for students seeking additional information on topics that may or may not be covered in the main part of the text. Appendix A provides a comprehensive list of electrical quantities and their symbols. It also includes a listing of the most popular multiple and submultiple units encountered in electronics as well as a listing of all the Greek letter symbols and their uses. Appendix B provides students with a comprehensive overview of solder and the soldering process. Appendix C provides a list of preferred values for resistors. The list of preferred values shows the multiple and submultiple values available for a specified tolerance. Appendix D provides a complete listing of electronic components and their respective schematic symbols. Appendix E provides students with an introduction on how to use an oscilloscope. Both analog and digital scopes are covered. Appendix F provides an extensive overview on the use of **Multisim**, which is an interactive circuit simulation software package that allows students to create and test electronic circuits. Appendix F introduces students to the main features of Multisim that directly relate to their study of DC circuits, AC circuits, and electronic devices. Appendix G provides thorough coverage of the damaging effects of electrostatic discharge (ESD). It also discusses the proper techniques and procedures to follow to prevent ESD from damaging sensitive electronic components and assemblies. Appendix H, which is new to the 2024 release, provides a comprehensive overview on the use of the breadboard for constructing electronic circuits in the laboratory.

What's New in the 2024 release of *Grob's Basic Electronics?*

New to the 2024 release of Grob's Basic Electronics is expanded coverage of *work and energy* in Chapter 3, Ohm's Law. The difference in the meaning of these two terms is clearly explained in the 2024 release. The definition of the term *power* is also more clearly explained.

The *GOOD TO KNOW* feature appearing in the margins of the page has also been updated and expanded in the 2024 release. New and interesting information has been added that adds merit and value to the material covered within a given chapter.

Several of the *Introductions* that appear at the beginning of each chapter have also been revised. Most notably, important applications for the material covered in that chapter have now been included.

Appendix H, called *Introduction to Breadboarding*, has also been added to the 2024 release of Grob's Basic Electronics. This appendix explains the layout of a typical prototype board, or breadboard as it is often called. It explains how the rows, columns, and buses are all interconnected (or not) with each other. And finally, this appendix illustrates how electronic circuits are constructed on the breadboard when doing laboratory experiments.

Ancillary Package

The following supplements are available to support *Grob's Basic Electronics*, 2024 Release.

Problems Manual for Use with Grob's Basic Electronics

This book, written by Mitchel E. Schultz, provides students and instructors with hundreds of additional practice problems for self-study, homework assignments, tests, and review. The book is organized to correlate with the first 27 chapters of the textbook, including the Introduction to Powers of 10 chapter. Each chapter contains a number of solved illustrative problems demonstrating step-by-step how representative problems on a particular topic are solved. Following the solved problems are sets of problems for the students to solve. Included at the end of each chapter is a brief true/false self-test. The *Problems Manual* is a must-have for students requiring additional practice in solving both DC and AC circuits. It is important to note that this book can be used as a supplement with any textbook covering DC and AC circuit theory.

Experiments Manual for Grob's Basic Electronics

This lab manual provides students and instructors with easy-to-follow laboratory experiments. The experiments range from an introduction to laboratory equipment to experiments dealing with operational amplifiers. All experiments have been student tested to ensure their effectiveness. The lab book is organized to correlate with the topics covered in the text, by chapter.

All experiments have a Multisim activity that is to be done prior to the actual physical lab activity. Multisim files are part of the Instructor's Resources on Connect. This prepares students to work with circuit simulation software, and also to do "pre-lab" preparation before doing a physical lab exercise. Multisim coverage also reflects the widespread use of circuit simulation software in today's electronics industries.

Digital Resources

Curated, Auto-Graded, Question Banks - with Instant Feedback & Insightful reporting

Assignable as high stakes assessment or unlimited practice problems. These algorithmic, auto-graded banks save faculty time while providing instant feedback to students. Instructors can edit these questions and/or add their own questions to these banks.

STEM Prep - Adaptive Pre-Requisite Assignments

Ensure each student is prepared for your course. STEM Prep modules provide a refresher on pre-requisite content. Adaptive learning technology provides an individualized learning path for each student. Instructors can assign the prep modules before the semester begins, during the early weeks of the semester, or as remedial work for struggling students.

SmartBook - Adaptive Reading Assignments

An adaptive reading assignment that can be assigned as pre-lecture homework to expose students to the terminology and general concepts in the text. It shows where students struggled, allowing instructors and students to spend time where they need it most.

Application-Based Activities in McGraw Hill Connect®

Prepare students for the real world with Application-Based Activities in Connect. These highly interactive, assignable exercises boost engagement and provide a safe space to apply concepts learned to real-world, course-specific problems. Each Application-Based Activity involves the application of multiple concepts, providing the ability to synthesize information and use critical thinking skills to solve realistic scenarios.

Polling

Every learner has unique needs. Uncover where and when you're needed with the new Polling tool in McGraw Hill Connect®! Polling allows you to discover where students are in real time. Engage students and help them create connections with your course content while gaining valuable insight during lectures. Leverage polling data to deliver personalized instruction when and where it is needed most.

Writing Assignment

Available within McGraw Hill Connect®, the Writing Assignment tool delivers a learning experience to help students improve written communication skills and conceptual understanding. Assign, monitor, grade, and provide feedback on writing more efficiently and effectively.

Test Builder in Connect

Available within McGraw Hill Connect®, Test Builder is a cloud-based tool that enables instructors to format tests that can be printed, administered within a Learning Management System, or exported as a Word document. Test Builder offers a modern, streamlined interface for easy content configuration that matches course needs, without requiring a download.

Test Builder allows you to:

- access all test bank content from a particular title.
- easily pinpoint the most relevant content through robust filtering options.
- manipulate the order of questions or scramble questions and/or answers.
- pin questions to a specific location within a test.
- determine your preferred treatment of algorithmic questions.
- choose the layout and spacing.
- add instructions and configure default settings.

Test Builder provides a secure interface for better protection of content and allows for just-in-time updates to flow directly into assessments.

ReadAnywhere® App

Available for iOS and Android smartphones or tablets, give users access to McGraw Hill tools including the eBook and SmartBook® or Adaptive Learning Assignments in McGraw Hill Connect®. Students can take notes, highlight, and complete assignments offline—all their work will sync when connected to Wi-Fi. Students log in with their Connect username and password to start learning—anytime, anywhere!

Proctorio

Remote Proctoring & Browser-Locking Capabilities

Remote proctoring and browser-locking capabilities, hosted by Proctorio within Connect, provide control of the assessment environment by enabling security options and verifying the identity of the student.

Seamlessly integrated within Connect, these services allow instructors to control the assessment experience by verifying identification, restricting browser activity, and monitoring student actions.

Instant and detailed reporting gives instructors an at-a-glance view of potential academic integrity concerns, thereby avoiding personal bias and supporting evidence-based claims.

OLC-Aligned Courses

Implementing High-Quality Instruction and Assessment through Preconfigured Courseware

In consultation with the Online Learning Consortium (OLC) and our certified Faculty Consultants, McGraw Hill has created preconfigured courseware using OLC's quality scorecard to align with best practices in online course delivery. This turnkey courseware contains a combination of formative assessments, summative assessments, homework, and application activities, and can easily be customized to meet an individual instructor's needs and desired course outcomes. For more information, visit https://www.mheducation.com/highered/olc.

Evergreen

Content and technology are ever-changing, and it is important that you can keep your course up-to-date with the latest information and assessments. That's why we want to deliver the most current and relevant content for your course, hassle-free.

Schultz, Grob's Basics Electronics is moving to an Evergreen delivery model, which means it has content, tools, and technology that is updated and relevant, with updates delivered directly to your existing McGraw Hill Connect® course. Engage students and freshen up assignments with up-to-date coverage of select topics and assessments, all without having to switch editions or build a new course.

Reflecting the Diverse World Around Us

McGraw Hill believes in unlocking the potential of every learner at every stage of life. To accomplish that, we are dedicated to creating products that reflect, and are accessible to, all the diverse, global customers we serve. Within McGraw Hill, we foster a culture of belonging, and we work with partners who share our commitment to equity, inclusion, and diversity in all forms. In McGraw Hill Higher Education, this includes, but is not limited to, the following:

- Refreshing and implementing inclusive content guidelines around topics including generalizations and stereotypes, gender, abilities/disabilities, race/ethnicity, sexual orientation, diversity of names, and age.
- Enhancing best practices in assessment creation to eliminate cultural, cognitive, and affective bias.
- Maintaining and continually updating a robust photo library of diverse images that reflect our student populations.
- Including more diverse voices in the development and review of our content.
- Strengthening art guidelines to improve accessibility by ensuring meaningful text and images are distinguishable and perceivable by users with limited color vision and moderately low vision.

Acknowledgments

The 2024 release of *Grob's Basic Electronics* would not have been possible without the help of some very dedicated people. I would like to thank the highly professional staff of McGraw Hill Higher Education, especially Theresa Collins and Beth Bettcher. Thank you for your patience and understanding during the preparation of this manuscript.

Reviewers

Chuck Ammond
Muskegon Community College
Michael Black
Northeastern Technical College
David Bohnenstiehl
Southwestern Illinois College
Dorina Cornea-Hasegan
Portland Community College
Ben Ficklin
San Jacinto College
Matthew Gannota
Suffolk County Community College
Lyle Gross
Kaskaskia College
Darryl J Gust
Ivy Tech Community College of Indiana
Dr. Ali Hariri
Schoolcraft College
Alex Heisler
Wichita Technical Institute
Jonathan Hymer
Mt. San Antonio College
Saliman Isa
South Carolina State University
Timothy Kidd
University of Northern Iowa
James E Leopard
Suffolk County Community College
Randy Libros
Community College of Philadelphia
Gregory Martinez
Wichita Technical Institute
Erie Morales
Seton Hall University
Samuel Navarro
Los Medanos College
Bryan Schroder
Northcentral Technical College
Jeff Schults
Los Medanos College
Jeffrey L. Schwartz
Queensborough Community College
Scott Segalewitz
University of Dayton
Mohammad Shanehsaz
Brookdale Community College
Tom Stanonis
Oakton College/Northeastern Illinois University
Jose M. Suchite
Bates Technical College

Mitchel E. Schultz

Mc Graw Hill connect®

A complete course platform

Connect enables you to build deeper connections with your students through cohesive digital content and tools, creating engaging learning experiences. We are committed to providing you with the right resources and tools to support all your students along their personal learning journeys.

65%
Less Time Grading

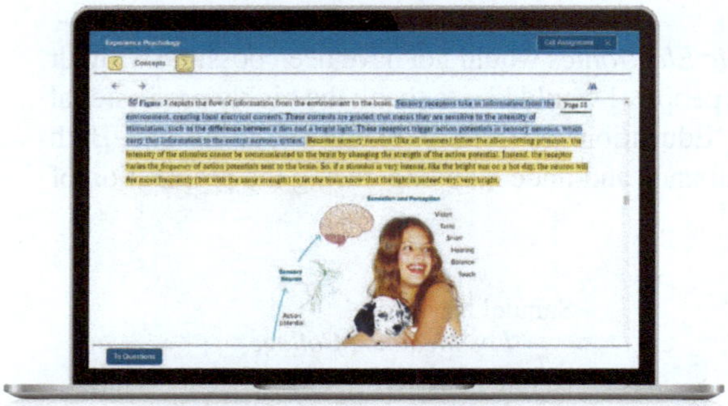

Laptop: Getty Images; Woman/dog: George Doyle/Getty Images

Every learner is unique

In Connect, instructors can assign an adaptive reading experience with SmartBook® 2.0. Rooted in advanced learning science principles, SmartBook 2.0 delivers each student a personalized experience, focusing students on their learning gaps, ensuring that the time they spend studying is time well spent.
mheducation.com/highered/connect/smartbook

Study anytime, anywhere

Encourage your students to download the free ReadAnywhere® app so they can access their online eBook, SmartBook® 2.0, or Adaptive Learning Assignments when it's convenient, even when they're offline. And since the app automatically syncs with their Connect account, all of their work is available every time they open it. Find out more at **mheducation.com/readanywhere**

"I really liked this app—it made it easy to study when you don't have your textbook in front of you."

Jordan Cunningham, a student at *Eastern Washington University*

Effective tools for efficient studying

Connect is designed to help students be more productive with simple, flexible, intuitive tools that maximize study time and meet students' individual learning needs. Get learning that works for everyone with Connect.

SUPPORT AT every step

Education for all

McGraw Hill works directly with Accessibility Services departments and faculty to meet the learning needs of all students. Please contact your Accessibility Services Office, and ask them to email **accessibility@mheducation.com**, or visit **mheducation.com/about/accessibility** for more information.

Affordable solutions, added value

Make technology work for you with LMS integration for single sign-on access, mobile access to the digital textbook, and reports to quickly show you how each of your students is doing. And with our Inclusive Access program, you can provide all these tools at the lowest available market price to your students. Ask your McGraw Hill representative for more information.

Solutions for your challenges

A product isn't a solution. Real solutions are affordable, reliable, and come with training and ongoing support when you need it and how you want it. Visit **supportateverystep.com** for videos and resources both you and your students can use throughout the term.

Updated and relevant content

EVERGREEN

Our new Evergreen delivery model provides the most current and relevant content for your course, hassle-free. Content, tools, and technology updates are delivered directly to your existing McGraw Hill Connect® course. Engage students and freshen up assignments with up-to-date coverage of select topics and assessments, all without having to switch editions or build a new course.

Before you read . . .

Chapter Introductions briefly outline the main chapter topics and concepts.

Chapter Outlines guide you through the material in the chapter ahead. The outlines breakdown the individual topics covered, and each outline is tied to a main heading to emphasize important topics throughout the chapter.

I

Introduction to Powers of 10

The electrical quantities you will encounter while working in the field of electronics are often extremely small or extremely large. For example, it is not at all uncommon to work with extremely small decimal numbers such as 0.000000000056 or extremely large numbers such as 1,296,000,000. To enable us to work conveniently with both very small and very large numbers, powers of 10 notation is used. With powers of 10 notation, any number, no matter how small or

Chapter Outline

- 1–1 Negative and Positive Polarities
- 1–2 Electrons and Protons in the Atom
- 1–3 Structure of the Atom
- 1–4 The Coulomb Unit of Electric Charge
- 1–5 The Volt Unit of Potential Difference
- 1–6 Charge in Motion Is Current
- 1–7 Resistance Is Opposition to Current
- 1–8 The Closed Circuit
- 1–9 The Direction of Current
- 1–10 Direct Current (DC) and Alternating Current (AC)
- 1–11 Sources of Electricity
- 1–12 The Digital Multimeter

Chapter Objectives

After studying this chapter, you should be able to

- *List* the two basic particles of electric charge.
- *Describe* the basic structure of the atom.
- *Define* the terms *conductor, insulator,* and *semiconductor* and give examples of each term.
- *Define* the coulomb unit of electric charge.
- *Define* potential difference and list its unit of measure.
- *Define* current and list its unit of measure.
- *Describe* the difference between voltage and current.
- *Define* resistance and conductance and list the unit of each.
- *List* three important characteristics of an electric circuit.
- *Define* the difference between electron flow and conventional current.
- *Describe* the difference between direct and alternating current.

Chapter Objectives organize and highlight the key concepts covered within the chapter text.

Important Terms

alternating current (AC)	conductor	electron valence	ohm
ampere	conventional current	element	potential difference
atom	coulomb	free electron	proton
atomic number	current	insulator	resistance
circuit	dielectric	ion	semiconductor
compound	direct current (DC)	molecule	siemens
conductance	electron	neutron	static electricity
	electron flow	nucleus	volt

Important Terms help students identify key words at the beginning of each chapter. They are defined in the text, at the end of the chapter, and in the glossary.

While you read . . .

Pioneers in Electronics offer background information on the scientists and engineers whose theories and discoveries were instrumental in the development of electronics.

Good to Know boxes provide additional information in the margins of the text.

Section Self-Reviews allow students to check their understanding of the material just presented. They are located at the end of each section within a chapter, with answers at the end of the chapter.

Bettmann/Getty Images

PIONEERS IN ELECTRONICS

French natural philosopher *Charles-Augustin Coulomb (1736–1806)* developed a method for measuring the force of attraction and repulsion between two electrically charged spheres. Coulomb established the law of inverse squares and defined the basic unit of charge quantity, the coulomb.

GOOD TO KNOW

As an aid for determining the added charge ($\pm Q$) to a neutral dielectric, use the following equation:

$$\pm Q = \frac{\text{Number of electrons added or removed}}{6.25 \times 10^{18}\ \text{electrons/C}}$$

Figure 1-5 Physical force between electric charges. (*a*) Opposite charges attract. (*b*) Two negative charges repel each other. (*c*) Two positive charges repel.

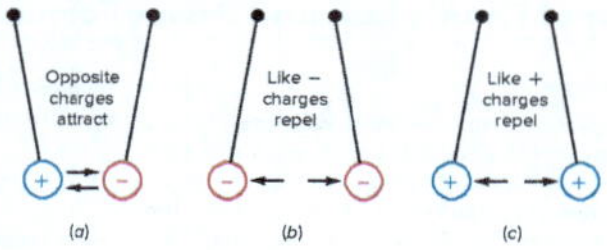

in Fig. 1-5*b*, and two positive charges of the same value repel each other in Fig. 1-5*c*.

Polarity of a Charge

An electric charge must have either negative or positive polarity, labeled $-Q$ or $+Q$, with an excess of either electrons or protons. A neutral condition is considered zero charge. On this basis, consider the following examples, remembering that the electron is the basic particle of charge and the proton has exactly the same amount, although of opposite polarity.

Example 1-1

A neutral dielectric has 12.5×10^{18} electrons added to it. What is its charge in coulombs?

ANSWER This number of electrons is double the charge of 1 C. Therefore, $-Q = 2$ C.

GOOD TO KNOW

Electricity is a form of energy, where energy refers to the ability to do work. More specifically, electrical energy refers to the energy associated with electric charges.

1-1 Self-Review

Answers at the end of the chapter.

a. **Is the charge of an electron positive or negative?**
b. **Is the charge of a proton positive or negative?**
c. **Is it true or false that the neutral condition means equal positive and negative charges?**

1-2 Electrons and Protons in the Atom

Although there are any number of possible methods by which electrons and protons might be grouped, they assemble in specific atomic combinations for a stable arrangement. (An **atom** is the smallest particle of the basic elements which forms

Examples throughout the text expand on key concepts and offer students a deeper understanding of complex material.

Multisim Icons, identify circuits for which there is a Multisim activity. Multisim files can be found on the Instructor Resources section for Connect.

MultiSim **Figure 1-8** Chemical cell as a voltage source. (*a*) Voltage output is the potential difference between the two terminals. (*b*) Schematic symbol of any DC voltage source with constant polarity. Longer line indicates positive side.

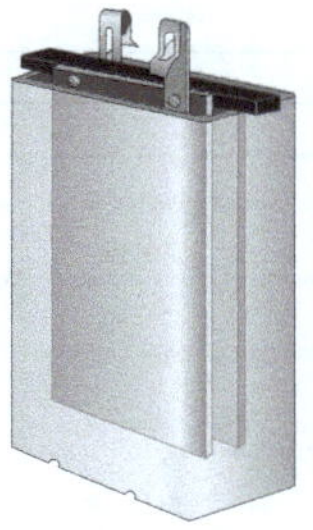

(*a*)

then, is a voltage source, or a source of electromotive force (emf). The schematic symbol for a battery or DC voltage source is shown in Fig. 1-8*b*.

Sometimes the symbol E is used for emf, but the standard symbol V represents any potential difference. This applies either to the voltage generated by a source or to the voltage drop across a passive component such as a resistor.

It may be helpful to think of voltage as an electrical pressure or force. The higher the voltage, the more electrical pressure or force. The electrical pressure of voltage is in the form of the attraction and repulsion of an electric charge, such as an electron.

The general equation for any voltage can be stated as

$$V = \frac{W}{Q} \qquad \textbf{(1–1)}$$

where V is the voltage in volts, W is the work or energy in joules, and Q is the charge in coulombs.

Let's take a look at an example.

Example 1-5

What is the output voltage of a battery that supplies 3.6 J of energy to move 0.5 C of charge?

ANSWER Use equation 1-1.

$$V = \frac{W}{Q}$$

After you've read . . .

Application of Ohm's Law and Power Formulas

HOME APPLIANCES

Every electrical appliance in our home has a ***nameplate*** attached to it. The nameplate provides important information about the appliance such as its make and model, its electrical specifications and the Underwriters Laboratories (UL) listing mark. The nameplate is usually located on the bottom or rear-side of the appliance. The electrical specifications listed are usually its power and voltage ratings. The voltage rating is the voltage at which the appliance is designed to operate. The power rating is the power dissipation of the appliance when the rated voltage is applied. With the rated voltage and power ratings listed on the nameplate, we can calculate the current drawn from the appliance when it's being used. To calculate the current (I) simply divide the power rating (P) in watts by the voltage rating (V) in volts. As an example, suppose you want to know how much current your toaster draws when it's toasting your bread. To find the answer you will probably need to turn your toaster upside rating of 120 V and a power rating of 850 W, the current drawn by the toaster is calculated as follows:

$$I = \frac{P}{V} = \frac{850\ W}{120\ V} = 7.083\ A$$

Some appliances in our homes have a voltage rating of 240 V rather than 120 V. These are typically the appliances with very high power ratings. Some examples include: electric stoves, electric clothes dryers, electric water heaters, and air conditioning units. These appliances may have power ratings as high as 7.2 kW or more. The reason the higher power appliances have a higher voltage rating is simple. At twice the voltage you only need half the current to obtain the desired power. With half as much current, the size of the conductors connecting the appliance to the power line can be kept much smaller. This is important because a smaller diameter wire costs less and is physically much easier to handle.

Real-world **applications** bring to life the concepts covered in a specific chapter.

Each chapter concludes with a **Summary,** a comprehensive recap of the major points and takeaways.

Summary

- Electricity is present in all matter in the form of electrons and protons.
- The electron is the basic particle of negative charge, and the proton is the basic particle of positive charge.
- A conductor is a material in which electrons can move easily from one atom to the next. Current flows freely or easily in a conductor.
- An insulator is a material in which electrons tend to stay in their own orbit. Insulators prevent or block the flow of current. Another name for insulator is dielectric.
- The atomic number of an element gives the number of protons in the
- One coulomb (C) of charge is a quantity of electricity corresponding to 6.25×10^{18} electrons or protons. The symbol for charge is Q.
- Potential difference or voltage is an electrical pressure or force that exists between two points. The unit of potential difference is the volt (V). $1\ V = \frac{1\ J}{1\ C}$ In general, $V = \frac{W}{Q}$.
- Current is the rate of movement of electric charge. The symbol for current is I, and the basic unit of measure is the ampere (A). $1\ A = \frac{1\ C}{1\ s}$ In general, $I = \frac{Q}{T}$.
- An electric circuit is a closed path for current flow. A voltage must be connected across a circuit to produce current flow. In the external circuit outside the voltage source, electrons flow from the negative terminal toward the positive terminal.
- A motion of positive charges, in the opposite direction of electron flow, is considered conventional current.
- Voltage can exist without current, but current cannot exist without voltage.
- Direct current has just one direction because a DC voltage source has

Related Formulas are a quick, easy way to locate the important formulas from the chapter.

Related Formulas

$1\ C = 6.25 \times 10^{18}$ electrons

$V = \frac{W}{Q}$

$I = Q/T$

$Q = I \times T$

$R = 1/G$

$G = 1/R$

Multiple-Choice Self-Tests at the end of every chapter allow for quick learning assessment.

Self-Test

Answers at the back of the book.

1. **The most basic particle of negative charge is the**
   a. coulomb.
   b. electron.
   c. proton.
   d. neutron.

2. **The coulomb is a unit of**
   a. electric charge.
   b. potential difference.
   c. current.
   d. voltage.

4. **The electron valence of a neutral copper atom is**
   a. +1.
   b. 0.
   c. ±4.
   d. −1.

5. **The unit of potential difference is the**
   a. volt.
   b. ampere.
   c. siemens.
   d. coulomb.

7. **In a metal conductor, such as a copper wire,**
   a. positive ions are the moving charges that provide current.
   b. free electrons are the moving charges that provide current.
   c. there are no free electrons.
   d. none of the above.

8. **A 100-Ω resistor has a conductance, *G*, of**
   a. 0.01 S.
   b. 0.1 S.
   c. 0.001 S.

Essay Questions

1. Name two good conductors, two good insulators, and two semiconductors.
2. In a metal conductor, what is a free electron?
3. What is the smallest unit of a compound with the same chemical characteristics?
4. Define the term ion.
5. How does the resistance of a conductor compare to that of an insulator?
6. Explain why potential difference is necessary to produce current in a circuit.
7. List three important characteristics of an electric circuit.
8. Describe the difference between an open circuit and a short circuit.
9. Is the power line voltage available in our homes a DC or an AC voltage?
10. What is the mathematical relationship between resistance and conductance?
11. Briefly describe the electric field of a static charge.

The **Essay Questions** at the end of each chapter are great ways to spark classroom discussion, and they make great homework assignments.

End-of-Chapter Problems, organized by chapter section, provide another opportunity for students to check their understanding, and for instructors to hone in on key concepts.

Critical Thinking Problems for each chapter provide students with more challenging problems, allowing them to polish critical skills needed on the job.

Problems

SECTION 1–4 THE COULOMB UNIT OF ELECTRIC CHARGE

1–1 If 31.25 × 10^{18} electrons are removed from a neutral dielectric, how much charge is stored in coulombs?

1–2 If 18.75 × 10^{18} electrons are added to a neutral dielectric, how much charge is stored in coulombs?

1–3 A dielectric with a positive charge of + 5 C has 18.75 × 10^{18} electrons added to it. What is the net charge of the dielectric in coulombs?

1–4 If 93.75 × 10^{18} electrons are removed from a neutral dielectric, how much charge is stored in coulombs?

1–5 If 37.5 × 10^{18} electrons are added to a neutral dielectric, how much charge is stored in coulombs?

SECTION 1–5 THE VOLT UNIT OF POTENTIAL DIFFERENCE

1–6 What is the output voltage of a battery if 10 J of energy is expended in moving 1.25 C of charge?

1–7 What is the output voltage of a battery if 6 J of energy is expended in moving 1 C of charge?

1–8 What is the output voltage of a battery if 12 J of energy is expended in moving 1 C of charge?

1–9 How much is the potential difference between two points if 0.5 J of energy is required to move 0.4 C of charge between the two points?

1–10 How much energy is expended, in joules, if a voltage of 12 V moves 1.25 C of charge between two points?

SECTION 1–6 CHARGE IN MOTION IS CURRENT

1–11 A charge of 2 C moves past a given point every 0.5 s. How much is the current?

1–12 A charge of 1 C moves past a given point every 0.1 s. How much is the current?

1–13 A charge of 0.05 C moves past a given point every 0.1 s. How much is the current?

1–14 A charge of 6 C moves past a given point every 0.3 s. How much is the current?

1–15 A charge of 0.1 C moves past a given point every 0.01 s. How much is the current?

1–16 If a current of 1.5 A charges a dielectric for 5 s, how much charge is stored in the dielectric?

1–17 If a current of 500 mA charges a dielectric for 2 s, how much charge is stored in the dielectric?

1–18 If a current of 200 μA charges a dielectric for 20 s, how much charge is stored in the dielectric?

SECTION 1–7 RESISTANCE IS OPPOSITION TO CURRENT

1–19 Calculate the resistance value in ohms for the following conductance values: (a) 0.001 S (b) 0.01 S (c) 0.1 S (d) 1 S.

1–20 Calculate the resistance value in ohms for the following conductance values: (a) 0.002 S (b) 0.004 S (c) 0.00833 S (d) 0.25 S.

1–21 Calculate the conductance value in siemens for each of the following resistance values: (a) 200 Ω (b) 100 Ω (c) 50 Ω (d) 25 Ω.

1–22 Calculate the conductance value in siemens for each of the following resistance values: (a) 1 Ω (b) 10 k Ω (c) 40 Ω (d) 0.5 Ω.

Critical Thinking

1–23 Suppose that 1000 electrons are removed from a neutral dielectric. How much charge, in coulombs, is stored in the dielectric?

1–24 How long will it take an insulator that has a charge of + 5 C to charge to + 30 C if the charging current is 2 A?

1–25 Assume that 6.25 × 10^{15} electrons flow past a given point in a conductor every 10 s. Calculate the current I in amperes.

1–26 The conductance of a wire at 100°C is one-tenth its value at 25°C. If the wire resistance equals 10 Ω at 25°C calculate the resistance of the wire at 100°C.

Laboratory Application Assignment

In your first lab application assignment you will use a DMM to measure the voltage, current, and resistance in Fig. 1–22. Refer to Section 1–12, "The Digital Multimeter," if necessary.

Equipment: Obtain the following items from your instructor.

- Variable dc power supply
- 1-kΩ, ½-W resistor
- DMM
- Connecting leads

Measuring Voltage

Set the DMM to measure DC voltage. Be sure the meter leads are inserted into the correct jacks (red lead in the VΩ jack and the black lead in the COM jack). Also, be sure the voltmeter range exceeds the voltage being measured. Connect the DMM test leads to the variable DC power supply as shown in Fig. 1–22*a*. Adjust the variable DC power supply voltage to any value between 5 and 15 V. Record your measured voltage. V = __________ Note: Keep the power supply voltage set to this value when measuring the current in Fig. 1–22*c*.

Measuring Resistance

Disconnect the meter leads from the power supply terminals. Set the DMM to measure resistance. Keep the meter leads in the same jacks you used for measuring voltage. Connect the DMM test leads to the leads of the 1 kΩ resistor, as shown in Fig. 1–22*b*. Record your measured resistance.
R = __________ (The measured resistance will most likely be displayed as a decimal fraction in kΩ.)

Measuring Current

Set the DMM to measure DC current. Also, move the red test lead to the appropriate jack for measuring small DC currents (usually labeled mA). Turn off the variable DC power supply. Connect the red test lead of the DMM to the positive (+) terminal of the variable DC power supply as shown in Fig. 1–22*c*. Also, connect the black test lead of the DMM to one lead of the 1 kΩ resistor as shown. Finally, connect the other lead of the resistor to the negative (–) terminal of the variable DC power supply. Turn on the variable DC power supply. Record your measured current.
I = __________

Figure 1–22 Measuring electrical quantities. (*a*) Measuring voltage. (*b*) Measuring resistance. (*c*) Measuring current.

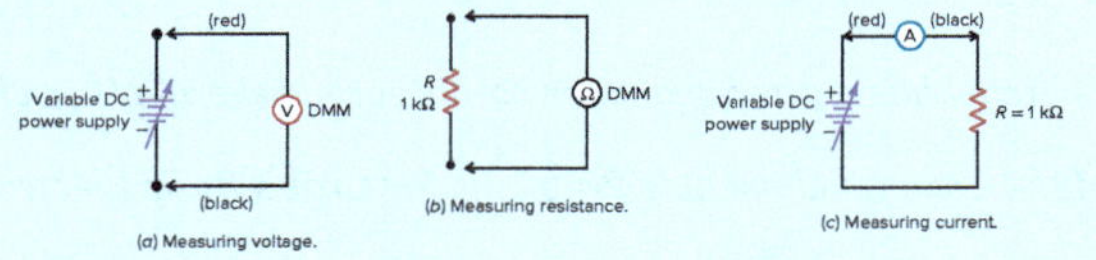

Laboratory Application Assignments, reinforce one or more of the chapter's main topics by asking students to build and test circuits in a laboratory environment.

Troubleshooting Challenge

Table 4–1 shows voltage measurements taken in Fig. 4–50. The first row shows the normal values that exist when the circuit is operating properly. Rows 2 to 15 are voltage measurements taken when one component in the circuit has failed. For each row, identify which component is defective and determine the type of defect that has occurred in the component.

Figure 4–50 Circuit diagram for Troubleshooting Challenge. Normal values for V_1, V_2, V_3, V_4, and V_5 are shown on schematic.

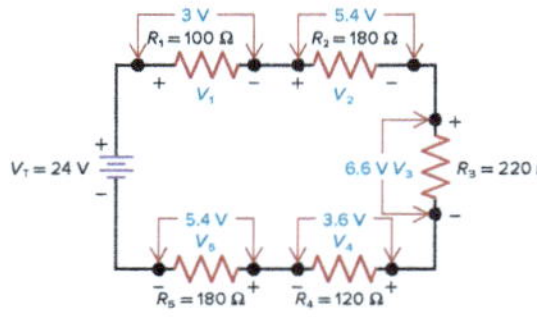

Troubleshooting Challenges appear in selected chapters to give students a feel for troubleshooting real circuits, again providing real-world applications of chapter content.

About the Author

Mitchel E. Schultz is an adjunct instructor at Minnesota State College Southeast (MSCS) in Winona, MN, where he has taught electronics for the past 3 years. Prior to teaching at MSCS, he taught electronics for 31years at Western Technical College in La Crosse, WI and for 8 years at Riverland Community College in Austin, Minnesota. He has also provided training for a variety of different electronic industries over the past 42 years.

Before he began teaching, Mitchel worked for several years as an electronic technician. His primary work experience was in the field of electronic communication, which included designing, testing, and troubleshooting rf communications systems. Mitchel graduated in 1978 from MSCS, where he earned an Associate's Degree in Electronics Technology. He also attended Winona State University, Mankato State University, and the University of Minnesota. He is an ISCET Certified Electronics Technician and also holds his Extra Class Amateur Radio License.

Mitchel has authored and/or co-authored several other electronic textbooks which include Problems for Grob's Basic Electronics, Electricity: Principles and Applications, Electric Circuits: A Text and Software Problems Manual, Electronic Devices: A Text and Software Problems Manual, Basic Mathematics for Electricity and Electronics, and Shaum's Outline of Theory and Problems of Electronic Communication.

Electric Shock—Dangers, Precautions, and First Aid

Electricity is a form of energy that provides an endless number of useful functions in our daily lives. However, no matter how useful electricity may be, it can also be very dangerous. Perhaps the greatest danger is from an electric shock. If a person comes into contact with a "live" conductor or circuit, it only takes a small amount of current through the human body to paralyze the victim, making it impossible for him or her to let go. A current in excessive of about $\frac{1}{100}$ of an Ampere (A), which is the basic unit of current, is about all it takes. If the current approaches $\frac{1}{10}$ of an Ampere, or more, the shock can be fatal. The danger of electric shock increases with higher voltages because a higher voltage can produce more current through the skin and internal organs. Lower voltages, such as those associated with AA or AAA batteries, for example, can be handled with little or no danger because the resistance of human skin is normally high enough to keep the current well below the threshold of sensation. However, when a person's skin is moist or cut, the resistance to the flow of current decreases drastically. When this happens, even moderate voltages can produce an electric shock. Therefore, safe practices must always be followed when working in and around electric circuits to avoid accidental electric shock, fires, and explosions.

Guideline of Safe Practices

The following is a list of ***safe practices*** that will help protect you and your fellow classmates while performing experiments in the laboratory. These same rules apply to those individuals working in industry. It is a good idea to review these safe practices from time to time so that you are reminded of their importance.

1. Never work on electrical equipment and/or machinery if you are under the influence of either drugs or alcohol.
2. Never work on electrical equipment and/or machinery if the lighting is poor or insufficient.
3. Never work on electrical equipment and/or machinery if your shoes and/or clothing are wet.
4. Wear rubber-soled shoes or stand on an insulated mat when working on electrical equipment.
5. If possible, never work alone.
6. Avoid wearing any metal objects such as bracelets, rings, necklaces, etc., when working in and around electric circuits.
7. Never assume that the power applied to a circuit is off! Either unplug the equipment you are working on or use a known-good meter to check for power.
8. Measure voltages with one hand in your pocket or behind your back when possible.
9. Do not remove safety grounds on three-prong power plugs and never use AC adapters to defeat the ground connection on any electrical equipment.
10. Power cords should always be checked before use. If the insulation is cracked or cut, they should not be used until they are properly repaired.

Figure S-1 Safety glasses are required when soldering and/or de-soldering.

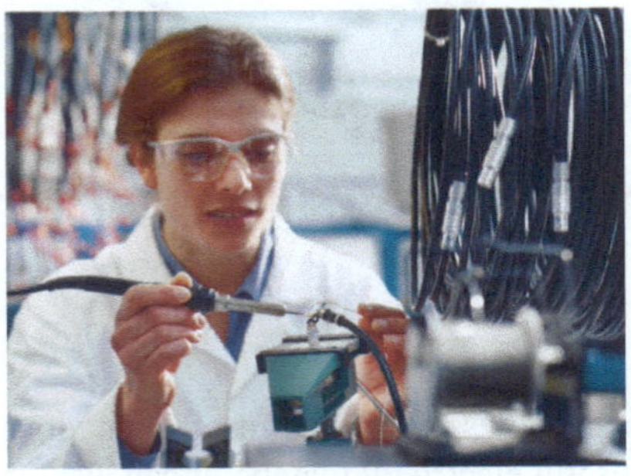

Monty Rakusen/Cultura/Getty Images

Figure S-2 Lockout-tagout (LOTO) procedure used in industry.

alacatr/iStock/Getty Images

11. Wear eye protection (safety glasses or goggles) when appropriate, especially when soldering, de-soldering, or clipping wires and/or wire leads. See Fig. S-1.
12. Avoid having liquids such as water, coffee, or soda at your workstation or around electrical equipment or machinery in general. Liquids are excellent conductors of electricity and therefore increase the risk of electric shock.
13. When possible, use a **lockout-tagout** (LOTO) procedure when working on electrical equipment and/or machinery. See Fig. S-2. Lockout-tagout or lock and tag is a safety procedure that is used in laboratory, industrial, and research settings. LOTO ensures that dangerous machines are properly shut off and are not able to be started up again prior to the completion of maintenance or repair work.
14. Some components, like capacitors, can store a lethal electric charge and should be fully discharged before repairing and/or replacing components or modules in an electronic system.
15. Never override any safety devices, such as an **interlock** switch, when working on electrical equipment and/or machinery. An **interlock** switch (usually a micro switch) is a switch that shuts off power to components (motors and lamps, for example) if a machine is opened. The purpose of an interlock switch is to prevent injury if someone inadvertently attempts to open a machine while it is still powered up and running. An interlock switch is sometimes called a "safety switch."
16. Keep your work station or work area organized and clean. A cluttered, disorganized work area is hazardous.
17. Be sure to secure loose-fitting clothing and ties when working near rotating machinery.
18. Make sure you know the proper procedures and potential safety hazards before working on any equipment, either electrical or mechanical.
19. Keep all tools and test equipment in good working condition. Be sure to regularly inspect the insulation on the handles of tools as well as the insulation on both test leads and insulated probes.
20. In a laboratory or industrial setting, be sure you know the location of the circuit breaker panel or main power-off switch so you can turn off power quickly, if necessary.
21. Know the location and operation of all fire alarms and fire extinguishers.
22. Know the location of all emergency exits.
23. Do not indulge in horseplay or practical jokes in the laboratory.
24. Know the location of the first-aid kit. If an accident should occur, notify your instructor and/or supervisor immediately.
25. Take a careful and deliberate approach to each task while working in the lab.

First Aid

The danger from an electric shock depends on

- The type of current (AC or DC)
- The amount of voltage present
- How the current traveled through the person's body
- The person's overall health
- How quickly the person receives treatment

An electric shock may cause minor to severe burns or leave no visible mark at all. Either way, an electric current can cause internal damage, including cardiac arrest or other injuries. Even a small amount of electricity through the body can be fatal under certain circumstances.

A person who has suffered an injury from contact with electricity should ALWAYS see a doctor!

If you are trying to help a person who is suffering from an electric shock, follow these guidelines:

- Do not touch them if they are still in contact with the source of electricity.
- Stay at least 20 feet away from any high-voltage wires until the power has been turned off.
- Don't move the person unless they are in immediate danger.

Call 911 if the injured person experiences the following:

- Injuries from a high-voltage wire or lightning
- Severe burns
- Confusion
- Loss of consciousness
- Breathing difficulty
- Cardiac arrest
- Muscle pain and contractions
- Heart rhythm problems
- Seizures

Take these actions immediately while waiting for medical help:

- If possible, turn off the electricity. If not, use a dry, nonconducting object (cardboard, plastic, or wood) to move the source away from you and the injured person.
- If there are no signs of circulation (breathing, coughing, or movement) begin CPR.
- Keep the injured person from becoming chilled.
- Cover any burned areas with a sterile gauze bandage, if possible. Otherwise use a clean cloth. Don't use a towel or blanket, as the loose fibers can stick to the burns.

Grob's Basic Electronics

I

Introduction to Powers of 10

The electrical quantities you will encounter while working in the field of electronics are often extremely small or extremely large. For example, it is not at all uncommon to work with extremely small decimal numbers such as 0.000000000056 or extremely large numbers such as 1,296,000,000. To enable us to work conveniently with both very small and very large numbers, powers of 10 notation is used. With powers of 10 notation, any number, no matter how small or large, can be expressed as a decimal number multiplied by a power of 10. A power of 10 is an exponent written above and to the right of 10, which is called the base. The power of 10 indicates how many times the base is to be multiplied by itself. For example, 10^3 means $10 \times 10 \times 10$ and 10^6 means $10 \times 10 \times 10 \times 10 \times 10 \times 10$. In electronics, the base 10 is common because multiples of 10 are used in the metric system of units.

Scientific and engineering notation are two common forms of powers of 10 notation. In electronics, engineering notation is generally more common than scientific notation because it ties in directly with the metric prefixes so often used. When a number is written in standard form without using any form of powers of 10 notation, it is said to be written in decimal notation (sometimes referred to as floating decimal notation). When selecting a calculator for solving problems in electronics, be sure to choose one that can display the answers in decimal, scientific, and engineering notation. ■

Chapter Outline

Chapter Objectives

After studying this chapter, you should be able to

- *Express* any number in scientific or engineering notation.
- *List* the metric prefixes and their corresponding powers of 10.
- *Change* a power of 10 in engineering notation to its corresponding metric prefix.
- *Convert* between metric prefixes.
- *Add* and *subtract* numbers expressed in powers of 10 notation.
- *Multiply* and *divide* numbers expressed in powers of 10 notation.
- *Determine* the reciprocal of a power of 10.
- *Find* the square of a number expressed in powers of 10 notation.
- *Find* the square root of a number expressed in powers of 10 notation.
- *Enter* numbers written in scientific and engineering notation into your calculator.

Important Terms

decimal notation
engineering notation
metric prefixes
powers of 10
scientific notation

Table I–1 Powers of 10

$1{,}000{,}000{,}000 = 10^9$
$100{,}000{,}000 = 10^8$
$10{,}000{,}000 = 10^7$
$1{,}000{,}000 = 10^6$
$100{,}000 = 10^5$
$10{,}000 = 10^4$
$1{,}000 = 10^3$
$100 = 10^2$
$10 = 10^1$
$1 = 10^0$
$0.1 = 10^{-1}$
$0.01 = 10^{-2}$
$0.001 = 10^{-3}$
$0.0001 = 10^{-4}$
$0.00001 = 10^{-5}$
$0.000001 = 10^{-6}$
$0.0000001 = 10^{-7}$
$0.00000001 = 10^{-8}$
$0.000000001 = 10^{-9}$
$0.0000000001 = 10^{-10}$
$0.00000000001 = 10^{-11}$
$0.000000000001 = 10^{-12}$

I–1 Scientific Notation

Before jumping directly into **scientific notation**, let's take a closer look at **powers of 10**. A power of 10 is an exponent of the base 10 and can be either positive or negative.

Base ⟶ 10^X ⟵ Exponent

Positive powers of 10 are used to indicate numbers greater than 1, whereas negative powers of 10 are used to indicate numbers less than 1. Table I–1 shows the powers of 10 ranging from 10^{-12} to 10^9 and their equivalent decimal values. In electronics, you will seldom work with powers of 10 outside this range. From Table I–1, notice that $10^0 = 1$ and that $10^1 = 10$. In the case of $10^0 = 1$, it is important to realize that any number raised to the zero power equals 1. In the case of $10^1 = 10$, it is important to note that any number written without a power is assumed to have a power of 1.

Expressing a Number in Scientific Notation

The procedure for using any form of powers of 10 notation is to write the original number as two separate factors. Scientific notation is a form of powers of 10 notation in which a number is expressed as a number between 1 and 10 times a power of 10. The power of 10 is used to place the decimal point correctly. The power of 10 indicates the number of places by which the decimal point has been moved to the left or right in the original number. If the decimal point is moved to the left in the original number, then the power of 10 will increase or become more positive. Conversely, if the decimal point is moved to the right in the original number then the power of 10 will decrease or become more negative. Let's take a look at an example.

Example I–1

Express the following numbers in scientific notation: (a) 3900 (b) 0.0000056.

ANSWER (a) To express 3900 in scientific notation, write the number as a number between 1 and 10, which is 3.9 in this case, times a power of 10. To do this, the decimal point must be shifted three places to the left. The number of places by which the decimal point is shifted to the left indicates the positive power of 10. Therefore, $3900 = 3.9 \times 10^3$ in scientific notation.

(b) To express 0.0000056 in scientific notation, write the number as a number between 1 and 10, which is 5.6 in this case, times a power of 10. To do this, the decimal point must be shifted six places to the right. The number of places by which the decimal point is shifted to the right indicates the negative power of 10. Therefore, $0.0000056 = 5.6 \times 10^{-6}$ in scientific notation.

When expressing a number in scientific notation, remember the following rules:

Rule 1:	Express the number as a number between 1 and 10 times a power of 10.

Rule 2:	If the decimal point is moved to the left in the original number, make the power of 10 positive. If the decimal point is moved to the right in the original number, make the power of 10 negative.

Rule 3:	The power of 10 always equals the number of places by which the decimal point has been shifted to the left or right in the original number.

Let's try some other examples.

Example I-2

Express the following numbers in scientific notation: (a) 235,000 (b) 364,000,000 (c) 0.000756 (d) 0.0000000000016.

ANSWER (a) To express the number 235,000 in scientific notation, move the decimal point five places to the left, which gives us a number of 2.35. Next, multiply this number by 10^5. Notice that the power of 10 is a positive 5 because the decimal point was shifted five places to the left in the original number. Therefore, $235{,}000 = 2.35 \times 10^5$ in scientific notation.

(b) To express 364,000,000 in scientific notation, move the decimal point eight places to the left, which gives us a number of 3.64. Next, multiply this number by 10^8. Notice that the power of 10 is a positive 8 because the decimal point was shifted eight places to the left in the original number. Therefore, $364{,}000{,}000 = 3.64 \times 10^8$ in scientific notation.

(c) To express 0.000756 in scientific notation, move the decimal point four places to the right, which gives us a number of 7.56. Next, multiply this number by 10^{-4}. Notice that the power of 10 is a negative 4 because the decimal point was shifted four places to the right in the original number. Therefore, $0.000756 = 7.56 \times 10^{-4}$.

(d) To express 0.0000000000016 in scientific notation, move the decimal point 13 places to the right, which gives us a number of 1.6. Next, multiply this number by 10^{-13}. Notice that the power of 10 is a negative 13 because the decimal point was shifted thirteen places to the right in the original number. Therefore, $0.0000000000016 = 1.6 \times 10^{-13}$ in scientific notation.

Decimal Notation

Numbers written in standard form without using any form of powers of 10 notation are said to be written in **decimal notation,** sometimes called floating decimal notation. In some cases, it may be necessary to change a number written in scientific notation into decimal notation. When converting from scientific to decimal notation, observe the following rules.

Rule 4:	If the exponent or power of 10 is positive, move the decimal point to the right, the same number of places as the exponent.

Rule 5:	If the exponent or power of 10 is negative, move the decimal point to the left, the same number of places as the exponent.

Example I-3

Convert the following numbers written in scientific notation into decimal notation: (a) 4.75×10^2 (b) 6.8×10^{-5}.

ANSWER (a) To convert 4.75×10^2 into decimal notation, the decimal point must be shifted two places to the right. The decimal point is shifted to the right because the power of 10, which is 2 in this case, is positive. Therefore, $4.75 \times 10^2 = 475$ in decimal notation.

(b) To convert 6.8×10^{-5} into decimal notation, the decimal point must be shifted five places to the left. The decimal point is shifted to the left because the power of 10, which is −5 in this case, is negative. Therefore, $6.8 \times 10^{-5} = 0.000068$ in decimal notation.

I–1 Self-Review

Answers at the end of the chapter.

a. Are positive or negative powers of 10 used to indicate numbers less than 1?

b. Are positive or negative powers of 10 used to indicate numbers greater than 1?

c. $10^0 = 1$. (True/False)

d. Express the following numbers in scientific notation: (a) 13,500 (b) 0.00825 (c) 95,600,000 (d) 0.104.

e. Convert the following numbers written in scientific notation into decimal notation: (a) 4.6×10^{-7} (b) 3.33×10^3 (c) 5.4×10^8 (d) 2.54×10^{-2}.

I–2 Engineering Notation and Metric Prefixes

Engineering notation is another form of powers of 10 notation. Engineering notation is similar to scientific notation except that in engineering notation, the powers of 10 are always multiples of 3 such as 10^{-12}, 10^{-9}, 10^{-6}, 10^{-3}, 10^3, 10^6, 10^9, 10^{12}, etc. More specifically, a number expressed in engineering notation is always expressed as a number between 1 and 1000 times a power of 10 which is a multiple of 3.

Example I-4

Express the following numbers in engineering notation: (a) 27,000 (b) 0.00047.

ANSWER (a) To express the number 27,000 in engineering notation, it must be written as a number between 1 and 1000 times a power of 10 which is a multiple of 3. It is often helpful to begin by expressing the number in scientific notation: $27{,}000 = 2.7 \times 10^4$. Next, examine the power of 10 to see if it should be increased to 10^6 or decreased to 10^3. If the power of 10 is increased to 10^6, then the decimal point in the number 2.7 would have to be shifted two places to the left.

Because 0.027 is not a number between 1 and 1000, the answer of 0.027×10^6 is not representative of engineering notation. If the power of 10 were decreased to 10^3, however, then the decimal point in the number 2.7 would have to be shifted one place to the right and the answer would be 27×10^3, which is representative of engineering notation. In summary, $27{,}000 = 2.7 \times 10^4 = 27 \times 10^3$ in engineering notation.

(b) To express the number 0.00047 in engineering notation, it must be written as a number between 1 and 1000 times a power of 10 which is a multiple of 3. Begin by expressing the number in scientific notation: $0.00047 = 4.7 \times 10^{-4}$. Next, examine the power of 10 to see if it should be increased to 10^{-3} or decreased to 10^{-6}. If the power of 10 were increased to 10^{-3}, then the decimal point in the number 4.7 would have to be shifted one place to the left. Because 0.47 is not a number between 1 and 1000, the answer 0.47×10^{-3} is not representative of engineering notation. If the power of 10 were decreased to 10^{-6}, however, then the decimal point in the number 4.7 would have to be shifted two places to the right and the answer would be 470×10^{-6} which is representative of engineering notation. In summary, $0.00047 = 4.7 \times 10^{-4} = 470 \times 10^{-6}$ in engineering notation.

When expressing a number in engineering notation, remember the following rules:

Rule 6: Express the original number in scientific notation first. If the power of 10 is a multiple of 3, the number appears the same in both scientific and engineering notation.

Rule 7: If the original number expressed in scientific notation does not use a power of 10 which is a multiple of 3, the power of 10 must either be increased or decreased until it is a multiple of 3. The decimal point in the numerical part of the expression must be adjusted accordingly to compensate for the change in the power of 10.

Rule 8: Each time the power of 10 is increased by 1, the decimal point in the numerical part of the expression must be moved one place to the left. Each time the power of 10 is decreased by 1, the decimal point in the numerical part of the expression must be moved one place to the right.

You know that a quantity is expressed in engineering notation when the original number is written as a number between 1 and 1000 times a power of 10 which is a multiple of 3.

Metric Prefixes

The **metric prefixes** represent those powers of 10 that are multiples of 3. In the field of electronics, engineering notation is much more common than scientific notation because most values of voltage, current, resistance, power, and so on are specified in terms of the metric prefixes. Once a number is expressed in engineering notation, its power of 10 can be replaced directly with its corresponding metric prefix. Table I–2 lists the most common metric prefixes and their corresponding powers of 10. Notice that uppercase letters are used for the abbreviations of the prefixes involving positive powers of 10, whereas lowercase letters are used for negative powers of 10. There is one exception to the rule however; the lowercase letter "k" is used for kilo

GOOD TO KNOW

The uppercase letter K is not used as the abbreviation for the metric prefix kilo because its use is reserved for the kelvin unit of absolute temperature.

Table I–2	Metric Prefixes	
Power of 10	**Prefix**	**Abbreviation**
10^{12}	tera	T
10^{9}	giga	G
10^{6}	mega	M
10^{3}	kilo	k
10^{-3}	milli	m
10^{-6}	micro	μ
10^{-9}	nano	n
10^{-12}	pico	p

corresponding to 10^3. Because the metric prefixes are used so often in electronics, it is common practice to express the value of a given quantity in engineering notation first so that the power of 10, which is a multiple of 3, can be replaced directly with its corresponding metric prefix. For example, a resistor whose value is 33,000 Ω can be expressed in engineering notation as 33×10^3 Ω. In Table I–2, we see that the metric prefix kilo (k) corresponds to 10^3. Therefore, 33,000 Ω or 33×10^3 Ω can be expressed as 33 kΩ. (Note that the unit of resistance is the ohm abbreviated Ω.) As another example, a current of 0.0000075 A can be expressed in engineering notation as 7.5×10^{-6} A. In Table I–2, we see that the metric prefix micro (μ) corresponds to 10^{-6}. Therefore, 0.0000075 A or 7.5×10^{-6} A can be expressed as 7.5 μA. (The unit of current is the ampere, abbreviated A.)

In general, when using metric prefixes to express the value of a given quantity, write the original number in engineering notation first and then substitute the appropriate metric prefix corresponding to the power of 10 involved. As this technique shows, metric prefixes are direct substitutes for the powers of 10 used in engineering notation.

Table I–3 lists many of the electrical quantities that you will encounter in your study of electronics. For each electrical quantity listed in Table I–3, take special note of the unit and symbol shown. In the examples and problems that follow, we will use several numerical values with various symbols and units from this table. Let's take a look at a few examples.

Table I–3	Electrical Quantities with Their Units and Symbols	
Quantity	**Unit**	**Symbol**
Current	Ampere (A)	I
Voltage	Volt (V)	V
Resistance	Ohm (Ω)	R
Frequency	Hertz (Hz)	f
Capacitance	Farad (F)	C
Inductance	Henry (H)	L
Power	Watt (W)	P

Example I-5

Express the resistance of 1,000,000 Ω using the appropriate metric prefix from Table I–2.

ANSWER First, express 1,000,000 Ω in engineering notation: 1,000,000 Ω = 1.0×10^6 Ω. Next, replace 10^6 with its corresponding metric prefix. Because the metric prefix mega (M) corresponds to 10^6, the value of 1,000,000 Ω can be expressed as 1 MΩ. In summary, 1,000,000 Ω = 1.0×10^6 Ω = 1 MΩ.

Example I-6

Express the voltage value of 0.015 V using the appropriate metric prefix from Table I–2.

ANSWER First, express 0.015 V in engineering notation: 0.015 V = 15×10^{-3} V. Next, replace 10^{-3} with its corresponding metric prefix. Because the metric prefix milli (m) corresponds to 10^{-3}, the value 0.015 V can be expressed as 15 mV. In summary, 0.015 V = 15×10^{-3} V = 15 mV.

Example I-7

Express the power value of 250 W using the appropriate metric prefix from Table I–2.

ANSWER In this case, it is not necessary or desirable to use any of the metric prefixes listed in Table I–2. The reason is that 250 W cannot be expressed as a number between 1 and 1000 times a power of 10 which is a multiple of 3. In other words, 250 W cannot be expressed in engineering notation. The closest we can come is 0.25×10^3 W, which is not representative of engineering notation. Although 10^3 can be replaced with the metric prefix kilo (k), it is usually preferable to express the power as 250 W and not as 0.25 kW.

In summary, whenever the value of a quantity lies between 1 and 1000, only the basic unit of measure should be used for the answer. As another example, 75 V should be expressed as 75 V and not as 0.075 kV or 75,000 mV, and so forth.

I–2 Self-Review

Answers at the end of the chapter.

a. **Express the following numbers in engineering notation: (a) 36,000,000 (b) 0.085 (c) 39,300 (d) 0.000093.**
b. **List the metric prefixes for each of the powers of 10 listed: (a) 10^{-9} (b) 10^6 (c) 10^{-12} (d) 10^3 (e) 10^4.**
c. **Express the following values using the appropriate metric prefixes: (a) 0.000010 A (b) 2,200,000 Ω (c) 0.000000045 V (d) 5600 Ω (e) 18 W.**

1–3 Converting between Metric Prefixes

As you have seen in the previous section, metric prefixes can be substituted for powers of 10 that are multiples of 3. This is true even when the value of the original quantity is not expressed in proper engineering notation. For example, a capacitance value of 0.047×10^{-6} F could be expressed as 0.047 μF. Also, a frequency of 1510×10^{3} Hz could be expressed as 1510 kHz. Furthermore, the values of like quantities in a given circuit may be specified using different metric prefixes such as 22 kΩ and 1.5 MΩ or 0.001 μF and 3300 pF, as examples. In some cases, therefore, it may be necessary or desirable to convert from one metric prefix to another when combining values. Converting from one metric prefix to another is actually a change in the power of 10. When the power of 10 is changed, however, care must be taken to make sure that the numerical part of the expression is also changed so that the value of the original number remains the same. When converting from one metric prefix to another, observe the following rule:

Rule 9: When converting from a larger metric prefix to a smaller one, increase the numerical part of the expression by the same factor by which the metric prefix has been decreased. Conversely, when converting from a smaller metric prefix to a larger one, decrease the numerical part of the expression by the same factor by which the metric prefix has been increased.

Example 1-8

Make the following conversions: (a) convert 25 mA to μA (b) convert 2700 kΩ to ΩM.

ANSWER (a) To convert 25 mA to μA, recall that the metric prefix milli (m) corresponds to 10^{-3} and that metric prefix micro (μ) corresponds to 10^{-6}. Since 10^{-6} is less than 10^{-3} by a factor of 1000 (10^{3}), the numerical part of the expression must be increased by a factor of 1000 (10^{3}). Therefore, 25 mA = 25×10^{-3} A = $25{,}000 \times 10^{-6}$ A = 25,000 μA.

(b) To convert 2700 kΩ to MΩ, recall that the metric prefix kilo (k) corresponds to 10^{3} and that the metric prefix mega (M) corresponds to 10^{6}. Since 10^{6} is larger than 10^{3} by a factor of 1000 (10^{3}), the numerical part of the expression must be decreased by a factor of 1000 (10^{3}). Therefore, 2700 kΩ = 2700×10^{3} Ω = 2.7×10^{6} Ω = 2.7 MΩ.

1–3 Self-Review

Answers at the end of the chapter.

a. **Converting from one metric prefix to another is actually a change in the power of 10. (True/False)**

b. **Make the following conversions: (a) convert 2.2 MΩ to kΩ (b) convert 47,000 pF to nF (c) convert 2500 μA to mA (d) convert 6.25 mW to μW.**

I–4 Addition and Subtraction Involving Powers of 10 Notation

When adding or subtracting numbers expressed in powers of 10 notation, observe the following rule:

Rule 10: Before numbers expressed in powers of 10 notation can be added or subtracted, both terms must be expressed using the same power of 10. When both terms have the same power of 10, just add or subtract the numerical parts of each term and multiply the sum or difference by the power of 10 common to both terms. Express the final answer in the desired form of powers of 10 notation.

Let's take a look at a couple of examples.

Example I-9

Add 170×10^3 and 23×10^4. Express the final answer in scientific notation.

ANSWER First, express both terms using either 10^3 or 10^4 as the common power of 10. Either one can be used. In this example, we will use 10^3 as the common power of 10 for both terms. Rewriting 23×10^4 using 10^3 as the power of 10 gives us 230×10^3. Notice that because the power of 10 was decreased by a factor of 10, the numerical part of the expression was increased by a factor of 10. Next, add the numerical parts of each term and multiply the sum by 10^3 which is the power of 10 common to both terms. This gives us $(170 + 230) \times 10^3$ or 400×10^3. Expressing the final answer in scientific notation gives us 4.0×10^5. In summary, $(170 \times 10^3) + (23 \times 10^4) = (170 \times 10^3) + (230 \times 10^3) = (170 + 230) \times 10^3 = 400 \times 10^3 = 4.0 \times 10^5$.

Example I-10

Subtract 250×10^3 from 1.5×10^6. Express the final answer in scientific notation.

ANSWER First, express both terms using either 10^3 or 10^6 as the common power of 10. Again, either one can be used. In this example, we will use 10^6 as the common power of 10 for both terms. Rewriting 250×10^3 using 10^6 as the power of 10 gives us 0.25×10^6. Notice that because the power of 10 was increased by a factor 1000 (10^3), the numerical part of the expression was decreased by a factor of 1000 (10^3). Next, subtract 0.25 from 1.5 and multiply the difference by 10^6, which is the power of 10 common to both terms. This gives us $(1.5 - 0.25) \times 10^6$ or 1.25×10^6. Notice that the final answer is already in scientific notation. In summary, $(1.5 \times 10^6) - (250 \times 10^3) = (1.5 \times 10^6) - (0.25 \times 10^6) = (1.5 - 0.25) \times 10^6 = 1.25 \times 10^6$.

I–4 Self-Review

Answers at the end of the chapter.

a. **Add the following terms expressed in powers of 10 notation. Express the answers in scientific notation. (a) $(470 \times 10^4) + (55 \times 10^6)$ (b) $(3.5 \times 10^{-2}) + (1500 \times 10^{-5})$.**

b. **Subtract the following terms expressed in powers of 10 notation. Express the answers in scientific notation. (a) $(65 \times 10^4) -$ (200×10^3) (b) $(850 \times 10^{-3}) - (3500 \times 10^{-4})$.**

I–5 Multiplication and Division Involving Powers of 10 Notation

When multiplying or dividing numbers expressed in powers of 10 notation, observe the following rules:

Rule 11: When multiplying numbers expressed in powers of 10 notation, multiply the numerical parts and powers of 10 separately. When multiplying powers of 10, simply add the exponents to obtain the new power of 10. Express the final answer in the desired form of powers of 10 notation.

Rule 12: When dividing numbers expressed in powers of 10 notation, divide the numerical parts and powers of 10 separately. When dividing powers of 10, subtract the power of 10 in the denominator from the power of 10 in the numerator. Express the final answer in the desired form of powers of 10 notation.

Let's take a look at a few examples.

Example I-11

Multiply (3×10^6) by (150×10^2). Express the final answer in scientific notation.

ANSWER First, multiply 3×150 to obtain 450. Next, multiply 10^6 by 10^2 to obtain $10^6 \times 10^2 = 10^{6+2} = 10^8$. To review, $(3 \times 10^6) \times (150 \times 10^2) =$ $(3 \times 150) \times (10^6 \times 10^2) = 450 \times 10^{6+2} = 450 \times 10^8$. The final answer expressed in scientific notation is 4.5×10^{10}.

Example I-12

Divide (5.0×10^7) by (2.0×10^4). Express the final answer in scientific notation.

ANSWER First, divide 5 by 2 to obtain 2.5. Next, divide 10^7 by 10^4 to obtain $10^{7-4} = 10^3$. To review, $\frac{5.0 \times 10^7}{2.0 \times 10^4} = \frac{5}{2} \times \frac{10^7}{10^4} = 2.5 \times 10^3$.
Notice that the final answer is already in scientific notation.

I–5 Self-Review

Answers at the end of the chapter.

a. **Multiply the following numbers expressed in powers of 10 notation. Express your answers in scientific notation. (a) $(3.3 \times 10^{-2}) \times (4.0 \times 10^{-3})$ (b) $(2.7 \times 10^{2}) \times (3 \times 10^{-5})$.**

b. **Divide the following numbers expressed in powers of 10 notation. Express your answers in scientific notation. (a) $(7.5 \times 10^{8}) \div (3.0 \times 10^{4})$ (b) $(15 \times 10^{-6}) \div (5 \times 10^{-3})$.**

I–6 Reciprocals with Powers of 10

Taking the reciprocal of a power of 10 is really just a special case of division using powers of 10 because 1 in the numerator can be written as 10^0 since $10^0 = 1$. With zero as the power of 10 in the numerator, taking the reciprocal results in a sign change for the power of 10 in the denominator. Let's take a look at an example to clarify this point.

Example I-13

Find the reciprocals for the following powers of 10: (a) 10^5 (b) 10^{-3}.

ANSWER (a) $\frac{1}{10^5} = \frac{10^0}{10^5} = 10^{0-5} = 10^{-5}$; therefore, $\frac{1}{10^5} = 10^{-5}$.

(b) $\frac{1}{10^{-3}} = \frac{10^0}{10^{-3}} = 10^{0-(-3)} = 10^3$; therefore, $\frac{1}{10^{-3}} = 10^3$.

Notice that in both (a) and (b), the power of 10 in the denominator is subtracted from zero, which is the power of 10 in the numerator.

Here's a simple rule for reciprocals of powers of 10.

Rule 13: When taking the reciprocal of a power of 10, simply change the sign of the exponent or power of 10.

Negative Powers of 10

Recall that a power of 10 indicates how many times the base, 10, is to be multiplied by itself. For example, $10^4 = 10 \times 10 \times 10 \times 10$. But you might ask how this definition fits with negative powers of 10. The answer is that negative powers of 10 are just reciprocals of positive powers of 10. For example, $10^{-4} = \frac{1}{10^4} = \frac{1}{10 \times 10 \times 10 \times 10}$.

I–6 Self-Review

Answers at the end of the chapter.

a. **Take the reciprocals of each of the powers of 10 listed. (a) 10^{-4} (b) 10^{9} (c) 10^{-18} (d) 10^{0}.**

I–7 Squaring Numbers Expressed in Powers of 10 Notation

When squaring a number expressed in powers of 10 notation, observe the following rule:

Rule 14: To square a number expressed in powers of 10 notation, square the numerical part of the expression and double the power of 10. Express the answer in the desired form of powers of 10 notation.

Example I–14

Square 3.0×10^4. Express the answer in scientific notation.

ANSWER First, square 3.0 to obtain 9.0. Next, square 10^4 to obtain $(10^4)^2 = 10^8$. Therefore, $(3.0 \times 10^4)^2 = 9.0 \times 10^8$.

I–7 Self-Review

Answers at the end of the chapter.

a. Obtain the following answers and express them in scientific notation. (a) $(4.0 \times 10^{-2})^2$ (b) $(6.0 \times 10^5)^2$ (c) $(2.0 \times 10^{-3})^2$.

I–8 Square Roots of Numbers Expressed in Powers of 10 Notation

When taking the square root of a number expressed in powers of 10 notation, observe the following rule:

Rule 15: To find the square root of a number expressed in powers of 10 notation, take the square root of the numerical part of the expression and divide the power of 10 by 2. Express the answer in the desired form of powers of 10 notation.

Example I–15

Find the square root of 4×10^6. Express the answer in scientific notation.

ANSWER $\sqrt{4 \times 10^6} = \sqrt{4} \times \sqrt{10^6} = 2 \times 10^3$

Notice that the answer is already in scientific notation.

Example I–16

Find the square root of 90×10^5. Express the answer in scientific notation.

ANSWER The problem can be simplified if we increase the power of 10 from 10^5 to 10^6 and decrease the numerical part of the expression from 90 to 9. This gives us $\sqrt{90 \times 10^5} = \sqrt{9 \times 10^6} = \sqrt{9} \times \sqrt{10^6} = 3.0 \times 10^3$. Again, the answer is already in scientific notation.

Figure I–1 Scientific calculator (Sharp EL–531 X).

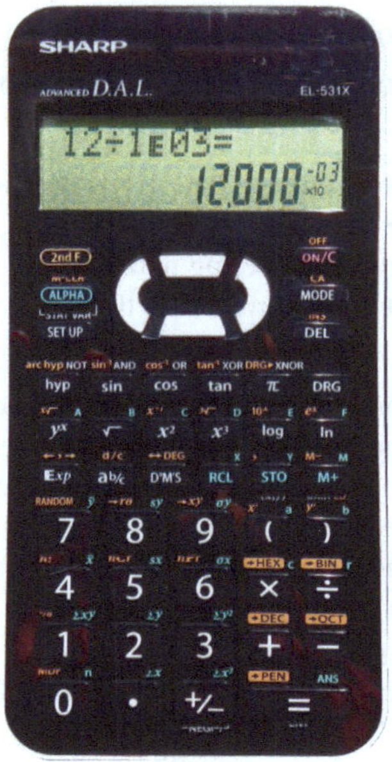

Sarah Schultz Photography

I–8 Self-Review

Answers at the end of the chapter.

a. **Obtain the following answers and express them in scientific notation. (a) $\sqrt{36 \times 10^4}$ (b) $\sqrt{160 \times 10^{-5}}$ (c) $\sqrt{25 \times 10^{-8}}$.**

I–9 The Scientific Calculator

Throughout your study of electronics, you will make several calculations involving numerical values that are expressed in decimal, scientific, or engineering notation. In most cases, you will want to use a scientific calculator to aid you in your calculations. Be sure to select a calculator that can perform all of the mathematical functions and operations that you will encounter in your study of electronics. Also, make sure the calculator you select can store and retrieve mathematical results from one or more memory locations. If the school or industry responsible for your training does not recommend or mandate a specific calculator, be sure to ask your instructor or supervisor for his or her recommendation on which calculator to buy. And finally, once you have purchased your calculator, carefully read the instructions that are included with it. At first, you may not understand many of your calculators functions and features, but as you progress in your studies, you will become more familiar with them. Figure I–1 shows an example of a typical scientific calculator.

Entering and Displaying Values

Scientific calculators typically have four notation systems for displaying calculation results: *floating decimal notation, fixed decimal notation (FIX), scientific notation (SCI)*, and *engineering notation (ENG)*. The calculator display typically shows the current notation system being used. When the FIX, SCI, or ENG symbol is displayed, the number of digits to the right of the decimal point can usually be set to any value from 0 to 9. With floating decimal notation, however, there is no set number of digits displayed for any given answer. For the examples that follow, assume that the calculator has been set to display three digits to the right of the decimal point.

Most scientific calculators have a key labeled [EXP], [EE], or [×10^] for entering the exponents associated with scientific and engineering notation. When entering a number expressed in any form of powers of 10 notation, always enter the numerical part of the expression first, followed by the exponent or power of 10. Use the *change sign* [+/−] key for entering negative exponents or for changing the sign of an existing exponent. To illustrate an example, the keystrokes involved in entering the number 25×10^{-6} would be as follows:

[2] [5] [EXP] [+/−] [6]

(Some calculators require that you press the [+/−] key after the exponent is entered.)

GOOD TO KNOW

When entering the number 25×10^{-6}, *do not* press the multiplication (×) key and then enter the number 10 prior to pressing the [EXP] key. If you do, the number you are intending to enter (25×10^{-6}) will be larger than it should be by a factor of 10. Since pressing the [EXP] key is equivalent to entering $\times 10^{00}$, you do not have to duplicate these steps! If you enter × 10 prior to pressing the [EXP] key, this is what you have actually entered: $25 \times 10 \times 10^{-6}$ which is equivalent to 250.000×10^{-6}.

It must be understood that pressing the [EXP] key is the same as entering $\times 10^{00}$. After the [EXP] key is pressed, the exponent in 10^{00} can be changed to any desired value, which is 10^{-06} in this case. Most calculators will display the value just entered as either

25.000×10^{-06} or 25E-06

For *25E-06*, the base 10 is implied by the uppercase letter *E*.

Most students, like yourself, are very comfortable with decimal notation because you have been exposed to it your entire life. In contrast, this chapter may be your first exposure to engineering notation. As a result, you may be tempted to enter and display all the values in decimal notation rather than engineering notation. For example, you may find yourself entering 47 kΩ as

[4] [7] [0] [0] [0] (decimal notation)

instead of

[4] [7] [EXP] [3] (engineering notation)

Entering and displaying values in decimal notation is a bad habit to get into for two reasons:

1. Very small and very large values cannot be entered in decimal notation because most calculators have only an 8- or 10-digit display.
2. Mentally converting between decimal and engineering notation is cumbersome and time-consuming, not to mention the fact that this practice is prone to error.

The main argument against using decimal notation is that most calculations encountered in electronics involve the use of the metric prefixes and hence engineering notation. By entering and displaying all values in engineering notation, you will be forced to learn the metric prefixes and their corresponding powers of 10.

When entering and displaying values in engineering notation remember that:

10^{-12} = pico (p)	10^{3} = kilo (k)
10^{-9} = nano (n)	10^{6} = mega (M)
10^{-6} = micro (μ)	10^{9} = giga (G)
10^{-3} = milli (m)	10^{12} = tera (T)

Example I-17

Show the keystrokes for multiplying 40×10^{-3} by 5×10^{6}.

ANSWER The keystrokes would be as follows:

[4] [0] [EXP] [+/−] [3] [×] [5] [EXP] [6] [=]

In engineering notation, the answer would be displayed as either

200.000×10^{03} or 200E03

As mentioned earlier, take the time to read the instruction manual for your calculator and keep it with you for future reference. I guarantee you, it will come in handy!

I–9 Self-Review

Answers at the end of the chapter.

a. When using a scientific calculator for the calculations encountered in electronics, decimal notation is the preferred notation system when entering and displaying values. (True/False)

b. Which key on a scientific calculator is used to enter the exponents associated with scientific and engineering notation?

Summary

- A power of 10 is an exponent that is written above and to the right of 10, which is called the base.
- A power of 10 indicates how many times the base, 10, is to be multiplied by itself.
- Positive powers of 10 indicate numbers greater than 1 and negative powers of 10 indicate numbers less than 1. Also, $10^0 = 1$ and $10^1 = 10$.
- Powers of 10 notation is a convenient method for expressing very small or very large numbers as a decimal number multiplied by a power of 10.
- Scientific and engineering notation are two forms of powers of 10 notation.
- A number expressed in scientific notation is always expressed as a number between 1 and 10 times a power of 10.
- A number expressed in engineering notation is always expressed as a number between 1 and 1000 times a power of 10 which is a multiple of 3.
- Decimal notation refers to those numbers that are written in standard form without any form of powers of 10 notation.
- Metric prefixes are letter symbols used to replace the powers of 10 that are multiples of 3. Refer to Table I–2 for a complete listing of the metric prefixes and their corresponding powers of 10.
- Converting from one metric prefix to another is a change in the power of 10 used to express a given quantity.
- Before numbers expressed in powers of 10 notation can be added or subtracted, both terms must have the same power of 10. When both terms have the same power of 10, just add or subtract the numerical parts of the expression and multiply the sum or difference by the power of 10 common to both terms.
- When multiplying numbers expressed in powers of 10 notation, multiply the numerical parts and powers of 10 separately. When multiplying powers of 10, simply add the exponents.
- When dividing numbers expressed in powers of 10 notation, divide the numerical parts and powers of 10 separately. When dividing powers of 10, simply subtract the power of 10 in the denominator from the power of 10 in the numerator.
- Taking the reciprocal of a power of 10 is the same as changing the sign of the exponent.
- To square a number expressed in powers of 10 notation, square the numerical part of the expression and double the power of 10.
- To take the square root of a number expressed in powers of 10 notation, take the square root of the numerical part of the expression and divide the power of 10 by 2.
- On a scientific calculator, the [EXP], [EE], or [x10^] key is used for entering the exponents associated with scientific and engineering notation.

Important Terms

Decimal notation — numbers that are written in standard form without using powers of 10 notation.

Engineering notation — a form of powers of 10 notation in which a number is expressed as a number between 1 and 1000 times a power of 10 that is a multiple of 3.

Metric prefixes — letter symbols used to replace the powers of 10 that are multiples of 3.

Powers of 10 — a numerical representation consisting of a base of 10 and an exponent; the base 10 raised to a power.

Scientific notation — a form of powers of 10 notation in which a number is expressed as a number between 1 and 10 times a power of 10.

Self-Test

Answers at the back of the book.

1. 10^4 means the same thing as

a. 10,000.
b. 10×4.
c. $10 \times 10 \times 10 \times 10$.
d. both a and c.

2. Negative powers of 10

a. indicate numbers less than 1.
b. are not used with engineering notation.
c. indicate numbers greater than 1.
d. are used only with scientific notation.

3. A number expressed in scientific notation is always expressed as a number between

a. 1 and 1000 times a power of 10 which is a multiple of 3.
b. 1 and 10 times a power of 10.
c. 1 and 100 times a power of 10.
d. 0 and 1 times a power of 10.

4. A number expressed in engineering notation is always expressed as a number between

a. 1 and 10 times a power of 10 that is a multiple of 3.
b. 1 and 10 times a power of 10.
c. 1 and 1000 times a power of 10 that is a multiple of 3.
d. 0 and 1 times a power of 10 that is a multiple of 3.

5. **10^0 equals**
   a. 0.
   b. 10.
   c. 1.
   d. none of the above.

6. **Metric prefixes are used only with those powers of 10 that are**
   a. multiples of 3.
   b. negative.
   c. associated with scientific notation.
   d. both a and b.

7. **40×10^{-3} A is the same as**
   a. 40 mA.
   b. 40 μA.
   c. 40 kA.
   d. 40 MA.

8. **3.9 MΩ is the same as**
   a. 3.9×10^3 Ω.
   b. 3.9×10^6 Ω.
   c. 3,900 kΩ.
   d. both b and c.

9. **A number written in standard form without any form of powers of 10 notation is said to be written in**
   a. scientific notation.
   b. decimal notation.
   c. engineering notation.
   d. metric prefix notation.

10. **The metric prefix pico (p) corresponds to**
    a. 10^{12}.
    b. 10^{-9}.
    c. 10^{-12}.
    d. 10^{-6}.

11. **Positive powers of 10**
    a. indicate numbers less than 1.
    b. are not used with engineering notation.
    c. indicate numbers greater than 1.
    d. are used only with scientific notation.

12. **10^1 equals**
    a. 0.
    b. 10.
    c. 1.
    d. none of the above.

13. **In engineering notation, the number 0.000452 is expressed as**
    a. 452×10^{-6}.
    b. 4.52×10^{-4}.
    c. 4.52×10^{-6}.
    d. 0.452×10^{-3}.

14. **$(40 \times 10^2) + (5.0 \times 10^3)$ equals**
    a. 90×10^3.
    b. 9.0×10^2.
    c. 20×10^5.
    d. 9.0×10^3.

15. **When dividing powers of 10**
    a. subtract the power of 10 in the numerator from the power of 10 in the denominator.
    b. change the sign of the power of 10 in the numerator.
    c. subtract the power of 10 in the denominator from the power of 10 in the numerator.
    d. add the exponents.

16. **When multiplying powers of 10**
    a. subtract the exponents.
    b. add the exponents.
    c. multiply the exponents.
    d. none of the above.

17. **10,000 μV is the same as**
    a. 0.01 mV.
    b. 10 kV.
    c. 10 mV.
    d. 0.0001 V.

18. **$\sqrt{81 \times 10^6}$ equals**
    a. 9×10^3.
    b. 9×10^6.
    c. 9×10^2.
    d. 81×10^3.

19. **$(4.0 \times 10^3)^2$ equals**
    a. 16×10^5.
    b. 1.6×10^7.
    c. 4.0×10^5.
    d. 16×10^1.

20. **The number 220×10^3 is the same as**
    a. 2.2×10^5.
    b. 220,000.
    c. 2200.
    d. both a and b.

Essay Questions

1. For 10^7, which is the base and which is the exponent?

2. Define: (a) scientific notation (b) engineering notation (c) decimal notation.

3. In electronics, why is engineering notation more common than scientific notation?

4. List the metric prefixes for each of the following powers of 10: (a) 10^{-3} (b) 10^3 (c) 10^{-6} (d) 10^6 (e) 10^{-9} (f) 10^9 (g) 10^{-12} (h) 10^{12}.

5. List the units and symbols for each of the following quantities: (a) frequency (b) voltage (c) power (d) resistance (e) capacitance (f) inductance (g) current.

Problems

SECTION I-1 SCIENTIFIC NOTATION

Express each of the following numbers in scientific notation:

I–1 3,500,000

I–2 678

I–3 160,000,000

I–4 0.00055

I–5 0.150

I–6 0.00000000000942

I–7 2270

I–8 42,100

I–9 0.033

I–10 0.000006

I–11 77,700,000

I–12 100

I–13 87

I–14 0.0018

I–15 0.000000095

I–16 18,200

I–17 640,000

I–18 0.011

I–19 0.00000000175

I–20 3,200,000,000,000

Convert each of the following numbers expressed in scientific notation into decimal notation.

I–21 1.65×10^{-4}

I–22 5.6×10^{5}

I–23 8.63×10^{2}

I–24 3.15×10^{-3}

I–25 1.7×10^{-9}

I–26 4.65×10^{6}

I–27 1.66×10^{3}

I–28 2.5×10^{-2}

I–29 3.3×10^{-12}

I–30 9.21×10^{4}

SECTION I-2 ENGINEERING NOTATION AND METRIC PREFIXES

Express each of the following numbers in engineering notation:

I–31 5500

I–32 0.0055

I–33 6,200,000

I–34 150,000

I–35 99,000

I–36 0.01

I–37 0.00075

I–38 0.55

I–39 10,000,000

I–40 0.0000000032

I–41 0.000068

I–42 92,000,000,000

I–43 270,000

I–44 0.000000000018

I–45 0.000000450

I–46 0.00010

I–47 2,570,000,000,000

I–48 20,000

I–49 0.000070

I–50 2500

Express the following values using the metric prefixes from Table I–2. (Note: The metric prefix associated with each answer must coincide with engineering notation.)

I–51 1000 W

I–52 10,000 Ω

I–53 0.035 V

I–54 0.000050 A

I–55 0.000001 F

I–56 1,570,000 Hz

I–57 2,200,000 Ω

I–58 162,000 V

I–59 1,250,000,000 Hz

I–60 0.00000000033 F

I–61 0.00025 A

I–62 0.000000000061 F

I–63 0.5 W

I–64 2200 Ω

I–65 180,000 Ω

I–66 240 V

I–67 4.7 Ω

I–68 0.001 H

I–69 0.00005 W

I–70 0.0000000001 A

SECTION I-3 CONVERTING BETWEEN METRIC PREFIXES

Make the following conversions:

I–71 55,000 μA = ___________ mA

I–72 10 nF = ___________ pF

I–73 6800 pF = ___________ μF

I–74 1.49 MHz = ___________ kHz

I–75 22,000 nF = ___________ μF

I–76 1500 μH = ___________ mH

I–77 1.5 MΩ = ___________ kΩ

I–78 2.2 GHz = ___________ MHz

I–79 0.039 MΩ = ___________ kΩ

I–80 5600 kΩ = ___________ MΩ

I–81 7500 μA = ___________ mA

I–82 1 mA = ___________ μA

I–83 100 kW = ___________ W

I–84 50 MW = ___________ kW

I–85 4700 pF = ___________ nF

I–86 560 nF = ___________ μF

I–87 1296 MHz = ___________ GHz

I–88 50 mH = ___________ μH

I–89 7.5 μF = ___________ pF

I–90 220,000 MΩ = ___________ GΩ

SECTION I-4 ADDITION AND SUBTRACTION INVOLVING POWERS OF 10 NOTATION

Add the following numbers and express your answers in scientific notation:

I–91 $(25 \times 10^3) + (5.0 \times 10^4)$

I–92 $(4500 \times 10^3) + (5.0 \times 10^6)$

I–93 $(90 \times 10^{-12}) + (0.5 \times 10^{-9})$

I–94 $(15 \times 10^{-3}) + (100 \times 10^{-4})$

I–95 $(150 \times 10^{-6}) + (2.0 \times 10^{-3})$

I–96 $(150 \times 10^0) + (0.05 \times 10^3)$

Subtract the following numbers and express your answers in scientific notation:

I–97 $(100 \times 10^6) - (0.5 \times 10^8)$

I–98 $(20 \times 10^{-3}) - (5000 \times 10^{-6})$

I–99 $(180 \times 10^{-4}) - (3.5 \times 10^{-3})$

I–100 $(7.5 \times 10^2) - (0.25 \times 10^3)$

I–101 $(5.0 \times 10^4) - (240 \times 10^2)$

I–102 $(475 \times 10^{-5}) - (1500 \times 10^{-7})$

SECTION I-5 MULTIPLICATION AND DIVISION INVOLVING POWERS OF 10 NOTATION

Multiply the following numbers and express your answers in scientific notation:

I–103 $(6.0 \times 10^3) \times (3.0 \times 10^2)$

I–104 $(4.0 \times 10^{-9}) \times (2.5 \times 10^6)$

I–105 $(50 \times 10^4) \times (6.0 \times 10^3)$

I–106 $(2.2 \times 10^{-2}) \times (6.5 \times 10^0)$

I–107 $(5.0 \times 10^{-5}) \times (2.0 \times 10^{-1})$

I–108 $(100 \times 10^{-3}) \times (50 \times 10^{-6})$

Divide the following numbers and express your answers in scientific notation:

I–109 $(100 \times 10^5) \div (4.0 \times 10^2)$

I–110 $(90 \times 10^{-9}) \div (3.0 \times 10^{-5})$

I–111 $(5.0 \times 10^6) \div (40 \times 10^3)$

I–112 $(750 \times 10^{-7}) \div (3.0 \times 10^{-4})$

I–113 $(55 \times 10^9) \div (11 \times 10^2)$

I–114 $(220 \times 10^3) \div (2.0 \times 10^7)$

SECTION I-6 RECIPROCALS WITH POWERS OF 10

Find the reciprocal for each power of 10 listed.

I–115 10^4

I–116 10^{-4}

I–117 10^1

I–118 10^{-8}

I–119 10^{-7}

I–120 10^{-13}

I–121 10^{15}

I–122 10^{18}

SECTION I-7 SQUARING NUMBERS EXPRESSED IN POWERS OF 10 NOTATION

Express the following answers in scientific notation:

I–123 $(5.0 \times 10^3)^2$

I–124 $(2.5 \times 10^{-7})^2$

I–125 $(90 \times 10^{4})^2$

I–126 $(7.0 \times 10^{5})^2$

I–127 $(12 \times 10^{-9})^2$

I–128 $(800 \times 10^{-12})^2$

SECTION I-8 SQUARE ROOTS OF NUMBERS EXPRESSED IN POWERS OF 10 NOTATION

Express the following answers in scientific notation:

I–129 $\sqrt{40 \times 10^{-5}}$

I–130 $\sqrt{50 \times 10^{4}}$

I–131 $\sqrt{36 \times 10^{-12}}$

I–132 $\sqrt{49 \times 10^{-3}}$

I–133 $\sqrt{150 \times 10^{-5}}$

I–134 $\sqrt{35 \times 10^{-6}}$

SECTION I-9 THE SCIENTIFIC CALCULATOR

Show the keystrokes on a scientific calculator for entering the following math problems. Display all answers in engineering notation.

I–135 $(15 \times 10^{-3}) \times (1.2 \times 10^{3})$

I–136 $60 \div (1.5 \times 10^{3})$

I–137 $12 \div (10 \times 10^{3})$

I–138 $(5 \times 10^{-3}) \times (120 \times 10^{3})$

I–139 $(6.5 \times 10^{4}) + (25 \times 10^{3})$

I–140 $(2.5 \times 10^{-4}) - (50 \times 10^{-6})$

Answers to Self-Reviews

I–1
a. negative powers of 10
b. positive powers of 10
c. true
d. (a) 1.35×10^{4}
(b) 8.25×10^{-3}
(c) 9.56×10^{7}
(d) 1.04×10^{-1}
e. (a) 0.00000046 (b) 3330
(c) 540,000,000 (d) 0.0254

I–2
a. (a) 36×10^{6} (b) 85×10^{-3}
(c) 39.3×10^{3} (d) 93×10^{-6}
b. (a) nano (n) (b) mega (M)
(c) pico (p) (d) kilo (k)
(e) none
c. (a) 10 μA (b) 2.2 MΩ
(c) 45 nV (d) 5.6 kΩ
(e) 18 W

I–3
a. true
b. (a) 2.2 MΩ = 2200 kΩ
(b) 47,000 pF = 47 nF
(c) 2500 μA = 2.5 mA
(d) 6.25 mW = 6250 μW

I–4
a. (a) 5.97×10^{7} (b) 5.0×10^{-2}
b. (a) 4.5×10^{5} (b) 5.0×10^{-1}

I–5
a. (a) 1.32×10^{-4} (b) 8.1×10^{-3}
b. (a) 2.5×10^{4} (b) 3.0×10^{-3}

I–6
a. (a) 10^{4} (b) 10^{-9}
(c) 10^{18} (d) 10^{0}

I–7
a. (a) 1.6×10^{-3} (b) 3.6×10^{11}
(c) 4.0×10^{-6}

I–8
a. (a) 6.0×10^{2} (b) 4.0×10^{-2}
(c) 5.0×10^{-4}

I–9
a. false
b. the [EXP], [EE], or [×10^] key

Design credits: Chapter Opener Background Image, and Multisim: Stockbyte/Getty Images

chapter 1

Electricity

We see applications of electricity all around us, especially in the electronic products we own and operate every day. For example, we depend on electricity for lighting, heating, and air conditioning and for the operation of our vehicles, cell phones, appliances, computers, and home entertainment systems, to name a few. The applications of electricity are extensive and almost limitless to the imagination. One thing is for sure, without electricity our lives would be much different and much more challenging.

Although there are many applications of electricity, electricity itself can be explained in terms of electric charge, voltage, and current. In this chapter, you will be introduced to the basic concepts of electricity, which include a discussion of the following topics: basic atomic structure, the coulomb unit of electric charge, the volt unit of potential difference, the ampere unit of current, and the ohm unit of resistance. You will also be introduced to conductors, semiconductors, insulators, and the basic characteristics of an electric circuit. ■

Chapter Outline

Chapter Objectives

After studying this chapter, you should be able to

- *List* the two basic particles of electric charge.
- *Describe* the basic structure of the atom.
- *Define* the terms *conductor*, *insulator*, and *semiconductor* and give examples of each term.
- *Define* the coulomb unit of electric charge.
- *Define* potential difference and list its unit of measure.
- *Define* current and list its unit of measure.
- *Describe* the difference between voltage and current.
- *Define* resistance and conductance and list the unit of each.
- *List* three important characteristics of an electric circuit.
- *Define* the difference between electron flow and conventional current.
- *Describe* the difference between direct and alternating current.

Important Terms

alternating current (AC)
ampere
atom
atomic number
circuit
compound
conductance
conductor
conventional current
coulomb
current
dielectric
direct current (DC)
electron
electron flow
electron valence
element
free electron
insulator
ion
molecule
neutron
nucleus
ohm
potential difference
proton
resistance
semiconductor
siemens
static electricity
volt

1–1 Negative and Positive Polarities

We see the effects of electricity in a battery, static charge, lightning, radio, television, and many other applications. What do they all have in common that is electrical in nature? The answer is basic particles of electric charge with opposite *polarities*. All the materials we know, including solids, liquids, and gases, contain two basic particles of electric charge: the ***electron*** and the ***proton***. An electron is the smallest amount of electric charge having the characteristic called *negative polarity*. The proton is a basic particle with *positive polarity*.

GOOD TO KNOW

A battery is a device that converts chemical energy into electrical energy.

Figure 1–1 Positive and negative polarities for the voltage output of a typical battery.

Cindy Schroeder/McGraw Hill

The negative and positive polarities indicate two opposite characteristics that seem to be fundamental in all physical applications. Just as magnets have north and south poles, electric charges have the opposite polarities labeled negative and positive. The opposing characteristics provide a method of balancing one against the other to explain different physical effects.

It is the arrangement of electrons and protons as basic particles of electricity that determines the electrical characteristics of all substances. For example, this paper has electrons and protons in it. There is no evidence of electricity, though, because the number of electrons equals the number of protons. In that case, the opposite electrical forces cancel, making the paper electrically neutral. The neutral condition means that opposing forces are exactly balanced, without any net effect either way.

When we want to use the electrical forces associated with the negative and positive charges in all matter, work must be done to separate the electrons and protons. Changing the balance of forces produces evidence of electricity. A battery, for instance, can do electrical work because its chemical energy separates electric charges to produce an excess number of electrons at its negative terminal and an excess number of protons at its positive terminal. With separate and opposite charges at the two terminals, electric energy can be supplied to a **circuit** connected to the battery. Figure 1–1 shows a battery with its negative (−) and positive (+) terminals marked to emphasize the two opposite polarities.

1–1 Self-Review

Answers at the end of the chapter.

a. Is the charge of an electron positive or negative?

b. Is the charge of a proton positive or negative?

c. Is it true or false that the neutral condition means equal positive and negative charges?

GOOD TO KNOW

Electricity is a form of energy, where energy refers to the ability to do work. More specifically, electrical energy refers to the energy associated with electric charges.

1–2 Electrons and Protons in the Atom

Although there are any number of possible methods by which electrons and protons might be grouped, they assemble in specific atomic combinations for a stable arrangement. (An **atom** is the smallest particle of the basic elements which forms the physical substances we know as solids, liquids, and gases.) Each stable combination of electrons and protons makes one particular type of atom. For example, Fig. 1–2 illustrates the electron and proton structure of one atom of the gas, hydrogen. This atom consists of a central mass called the ***nucleus*** and one electron outside. The proton in the nucleus makes it the massive and stable part of the atom because a proton is 1840 times heavier than an electron.

Figure 1–2 Electron and proton in hydrogen (H) atom.

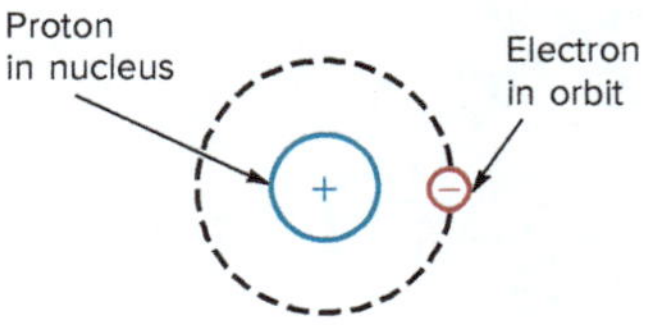

In Fig. 1–2, the one electron in the hydrogen atom is shown in an orbital ring or shell around the nucleus. To account for the electrical stability of the atom, we can consider the electron as spinning around the nucleus, as planets revolve around the sun. Then the electrical force attracting the electrons toward the nucleus is balanced by the mechanical force outward on the rotating electron. As a result, the electron stays in its orbit around the nucleus.

Figure 1–3 Atomic structure showing the nucleus and its orbital rings of electrons. (*a*) Carbon (C) atom has six orbital electrons to balance six protons in nucleus. (*b*) Copper (Cu) atom has 29 protons in nucleus and 29 orbital electrons.

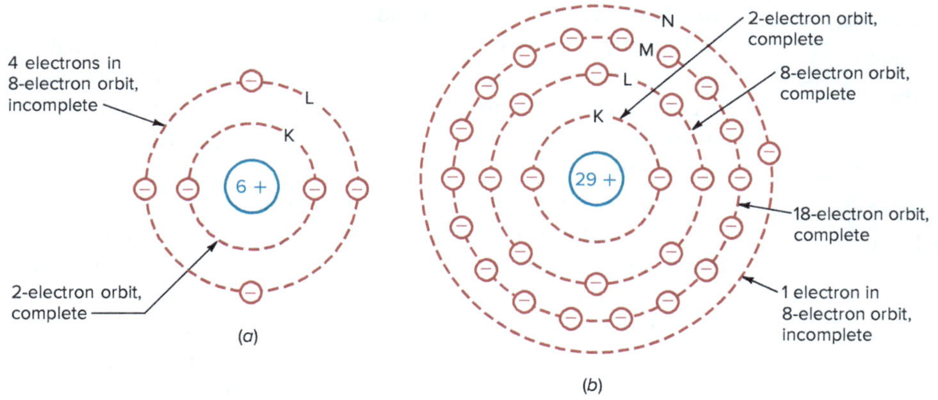

In an atom that has more electrons and protons than hydrogen, all protons are in the nucleus, and all the electrons are in one or more outside rings. For example, the carbon atom illustrated in Fig. 1–3*a* has six protons in the nucleus and six electrons in two outside rings. The total number of electrons in the outside rings must equal the number of protons in the nucleus in a neutral atom.

The distribution of electrons in the orbital rings determines the atom's electrical stability. Especially important is the number of electrons in the ring farthest from the nucleus. This outermost ring requires eight electrons for stability, except when there is only one ring, which has a maximum of two electrons.

In the carbon atom in Fig. 1–3*a*, with six electrons, there are just two electrons in the first ring because two is its maximum number. The remaining four electrons are in the second ring, which can have a maximum of eight electrons.

As another example, the copper atom in Fig. 1–3*b* has only one electron in the last ring, which can include eight electrons. Therefore, the outside ring of the copper atom is less stable than the outside ring of the carbon atom.

When many atoms are close together in a copper wire, the outermost orbital electron of each copper atom can easily break free from its home or parent atom. These electrons then can migrate easily from one atom to another at random. Such electrons that can move freely from one atom to the next are called *free electrons*. This freedom accounts for the ability of copper to conduct electricity very easily. It is the movement of free electrons that provides electric **current** in a metal **conductor**.

GOOD TO KNOW

Conductors have many free electrons whereas insulators have very few or none at all.

The net effect in the wire itself without any applied voltage, however, is zero because of the random motion of the free electrons. When voltage is applied, it forces all the free electrons to move in the same direction to produce **electron flow**, which is an electric current.

Conductors, Insulators, and Semiconductors

When electrons can move easily from atom to atom in a material, the material is a *conductor*. In general, all metals are good conductors, with silver the best and copper second. Their atomic structure allows free movement of the outermost orbital electrons. Copper wire is generally used for practical conductors because it costs much less than silver. The purpose of using conductors is to allow electric current to flow with minimum opposition.

GOOD TO KNOW

Even the best insulators release an occasional electron now and then. These electrons allow a minute current to flow through the insulator when a voltage is present. For all practical purposes, however, the current is so small that it can be considered zero.

The wire conductor is used only to deliver current produced by the voltage source to a device that needs the current to function. As an example, a bulb lights only when current flows through the filament.

A material with atoms in which the electrons tend to stay in their own orbits is an ***insulator*** because it cannot conduct electricity very easily. However, insulators can hold or store electricity better than conductors. An insulating material, such as glass, plastic, rubber, paper, air, or mica, is also called a ***dielectric,*** meaning it can store electric charge. It is interesting to note that most insulators are not pure elements. They are typically made up of two or more elements that join together to form a new substance.

Insulators are useful when it is necessary to prevent or block the flow of current. In addition, for applications requiring the storage of electric charge, as in capacitors, a dielectric material must be used because a good conductor cannot store any charge.

Carbon can be considered a **semiconductor**, conducting less than metal conductors but more than insulators. In the same group are germanium and silicon, which are commonly used for transistors and other semiconductor components. Practically all transistors are made of silicon.

Elements

The combinations of electrons and protons forming stable atomic structures result in different kinds of elementary substances having specific characteristics. A few familiar examples are the elements hydrogen, oxygen, carbon, copper, and iron. An ***element*** is defined as a substance that cannot be decomposed any further by chemical action. The atom is the smallest particle of an element that still has the same characteristics as the element. *Atom* is a Greek word meaning a "particle too small to be subdivided." As an example of the fact that atoms are too small to be visible, a particle of carbon the size of a pinpoint contains many billions of atoms. The electrons and protons within the atom are even smaller.

Table 1–1 lists some more examples of elements. These are just a few out of a total of 112. Notice how the elements are grouped. The metals listed across the top row are all good conductors of electricity. Each has an atomic structure with an unstable outside ring that allows many free electrons.

Table 1–1 Examples of the Chemical Elements

Group	Element	Symbol	Atomic Number	Electron Valence
Metal conductors, in order of conductance	Silver	Ag	47	+1
	Copper	Cu	29	+1*
	Gold	Au	79	+1*
	Aluminum	Al	13	+3
	Iron	Fe	26	+2*
Semiconductors	Carbon	C	6	±4
	Silicon	Si	14	±4
	Germanium	Ge	32	±4
Active gases	Hydrogen	H	1	±1
	Oxygen	O	8	−2
Inert gases	Helium	He	2	0
	Neon	Ne	10	0

* Some metals have more than one valence number in forming chemical compounds. Examples are cuprous or cupric copper, ferrous or ferric iron, and aurous or auric gold.

Semiconductors have four electrons in the outermost ring. This means that they neither gain nor lose electrons but share them with similar atoms. The reason is that four is exactly halfway to the stable condition of eight electrons in the outside ring.

The inert gas neon has a complete outside ring of eight electrons, which makes it chemically inactive. Remember that eight electrons in the outside ring is a stable structure.

Molecules and Compounds

A group of two or more atoms forms a **molecule**. For instance, two atoms of hydrogen (H) form a hydrogen molecule (H_2). When hydrogen unites chemically with oxygen, the result is water (H_2O), which is a **compound**. A compound, then, consists of two or more elements. The molecule is the smallest unit of a compound with the same chemical characteristics. We can have molecules for either elements or compounds. However, atoms exist only for elements.

■ *1–2 Self-Review*

Answers at the end of the chapter.

a. Which have more free electrons: conductors or insulators?
b. Which is the best conductor: silver, carbon, or iron?
c. Which is a semiconductor: copper, silicon, or neon?

1–3 Structure of the Atom

Our present planetary model of the atom was proposed by Niels Bohr in 1913. His contribution was joining the new ideas of a nuclear atom developed by Lord Rutherford (1871–1937) with the quantum theory of radiation developed by Max Planck (1858–1947) and Albert Einstein (1879–1955).

As illustrated in Figs. 1–2 and 1–3, the nucleus contains protons for all the positive charge in the atom. The number of protons in the nucleus is equal to the number of planetary electrons. Then the positive and negative charges are balanced because the proton and electron have equal and opposite charges. The orbits for the planetary electrons are also called *shells* or *energy levels*.

Atomic Number

This gives the number of protons or electrons required in the atom for each element. For the hydrogen atom in Fig. 1–2, the **atomic number** is one, which means that the nucleus has one proton balanced by one orbital electron. Similarly, the carbon atom in Fig. 1–3 with atomic number six has six protons in the nucleus and six orbital electrons. The copper atom has 29 protons and 29 electrons because its atomic number is 29. The atomic number listed for each of the elements in Table 1–1 indicates the atomic structure.

Orbital Rings

The planetary electrons are in successive shells or rings called K, L, M, N, O, P, and Q at increasing distances outward from the nucleus. Each shell has a maximum number of electrons for stability. As indicated in Table 1–2, these stable shells correspond to inert gases, such as helium and neon.

The K shell, closest to the nucleus, is stable with two electrons, corresponding to the atomic structure for the inert gas, helium. Once the stable number of electrons has filled a shell, it cannot take any more electrons. The atomic structure with all its shells filled to the maximum number for stability corresponds to an inert gas.

Table 1–2	Shells of Orbital Electrons in the Atom	
Shell	**Maximum Electrons**	**Inert Gas**
K	2	Helium
L	8	Neon
M	8 (up to calcium) or 18	Argon
N	8, 18, or 32	Krypton
O	8 or 18	Xenon
P	8 or 18	Radon
Q	8	—

Elements with a higher atomic number have more planetary electrons. These are in successive shells, tending to form the structure of the next inert gas in the periodic table. (The periodic table is a very useful grouping of all elements according to their chemical properties.) After the K shell has been filled with two electrons, the L shell can take up to eight electrons. Ten electrons filling the K and L shells is the atomic structure for the inert gas, neon.

The maximum number of electrons in the remaining shells can be 8, 18, or 32 for different elements, depending on their place in the periodic table. The maximum for an outermost shell, though, is always eight.

To illustrate these rules, we can use the copper atom in Fig. 1–3*b* as an example. There are 29 protons in the nucleus balanced by 29 planetary electrons. This number of electrons fills the K shell with two electrons, corresponding to the helium atom, and the L shell with eight electrons. The 10 electrons in these two shells correspond to the neon atom, which has an atomic number of 10. The remaining 19 electrons for the copper atom then fill the M shell with 18 electrons and one electron in the outermost N shell. These values can be summarized as follows:

K shell = 2 electrons
L shell = 8 electrons
M shell = 18 electrons
N shell = 1 electron
Total = 29 electrons

GOOD TO KNOW

Each orbital ring of electrons corresponds to a different energy level; the larger the orbit, the higher the energy level of the orbiting electrons.

For most elements, we can use the rule that the maximum number of electrons in a filled inner shell equals $2n^2$, where n is the shell number in sequential order outward from the nucleus. Then the maximum number of electrons in the first shell is $2 \times 1 = 2$; for the second shell $2 \times 2^2 = 8$, for the third shell $2 \times 3^2 = 18$, and for the fourth shell $2 \times 4^2 = 32$. These values apply only to an inner shell that is filled with its maximum number of electrons.

GOOD TO KNOW

The outermost shell of orbiting electrons is called the valence shell, and the electrons in this shell are called valence electrons.

Electron Valence

This value is the number of electrons in an incomplete outermost shell (valence shell). A completed outer shell has a valence of zero. Copper, for instance, has a valence of one, as there is one electron in the last shell, after the inner shells have been completed with their stable number. Similarly, hydrogen has a valence of one, and carbon has a valence of four. The number of outer electrons is considered positive valence because these electrons are in addition to the stable shells.

Except for H and He, the goal of valence is eight for all atoms, as each tends to form the stable structure of eight electrons in the outside ring. For this reason, valence can also be considered the number of electrons in the outside ring needed to make eight. This value is the negative valence. As examples, the valence of copper can be considered +1 or −7; carbon has the valence of ±4. The inert gases have zero valence because they all have complete outer shells.

The valence indicates how easily the atom can gain or lose electrons. For instance, atoms with a valence of +1 can lose this one outside electron, especially to atoms with a valence of +7 or −1, which need one electron to complete the outside shell with eight electrons.

Subshells

Although not shown in the illustrations, all shells except K are divided into subshells. This subdivision accounts for different types of orbits in the same shell. For instance, electrons in one subshell may have elliptical orbits, and other electrons in the same main shell have circular orbits. The subshells indicate magnetic properties of the atom.

Particles in the Nucleus

A stable nucleus (i.e., one that is not radioactive) contains protons and neutrons. The **neutron** is electrically neutral (it has no net charge). Its mass is almost the same as that of a proton.

A proton has the positive charge of a hydrogen nucleus. The charge is the same as that of an orbital electron but of opposite polarity. There are no electrons in the nucleus. Table 1–3 lists the charge and mass for these three basic particles in all atoms. The C in the charge column is for **coulombs**.

GOOD TO KNOW

Not all atoms of the same element have the same mass number. The reason is that the number of neutrons contained in the nucleus is not the same for all atoms of a given element. Since the neutron has no net electrical charge, how many of them are contained in the nucleus of an atom has no effect on the chemical and electrical characteristics of the atom. Atoms of the same element but with different atomic masses are called isotopes. Almost every element has two or more isotopes.

1–3 Self-Review

Answers at the end of the chapter.

a. **An element with 14 protons and 14 electrons has what atomic number?**

b. **What is the electron valence of an element with an atomic number of 3?**

c. **Except for H and He, what is the goal of valence for all atoms?**

Table 1–3 Stable Particles in the Atom

Particle	Charge	Mass
Electron, in orbital shells	0.16×10^{-18} C, negative	9.108×10^{-28} g
Proton, in nucleus	0.16×10^{-18} C, positive	1.672×10^{-24} g
Neutron, in nucleus	None	1.675×10^{-24} g

GOOD TO KNOW

The generation of static electricity is also referred to as the triboelectric effect.

1–4 The Coulomb Unit of Electric Charge

If you rub a hard rubber pen or comb on a sheet of paper, the rubber will attract a corner of the paper if it is free to move easily. The paper and rubber then give evidence of a static electric charge. The work of rubbing resulted in separating electrons and protons to produce a charge of excess electrons on the surface of the rubber and a charge of excess protons on the paper.

Because paper and rubber are dielectric materials, they hold their extra electrons or protons. As a result, the paper and rubber are no longer neutral, but each has an electric charge. The resultant electric charges provide the force of attraction between the rubber and the paper. This mechanical force of attraction or repulsion between charges is the fundamental method by which electricity makes itself evident.

Any charge is an example of ***static electricity*** because the electrons or protons are not in motion. There are many examples. When you walk across a wool rug, your body becomes charged with an excess of electrons. Similarly, silk, fur, and glass can be rubbed to produce a static charge. This effect is more evident in dry weather because a moist dielectric does not hold its charge so well. Also, plastic materials can be charged easily, which is why thin, lightweight plastics seem to stick to everything.

The charge of many billions of electrons or protons is necessary for common applications of electricity. Therefore, it is convenient to define a practical unit called the *coulomb* (C) as equal to the charge of 6.25×10^{18} electrons or protons stored in a dielectric (see Fig. 1–4). The analysis of static charges and their forces is called *electrostatics*.

Figure 1–4 The coulomb (C) unit of electric charge. (*a*) Quantity of 6.25×10^{18} excess electrons for a negative charge of 1 C. (*b*) Same amount of protons for a positive charge of 1 C, caused by removing electrons from neutral atoms.

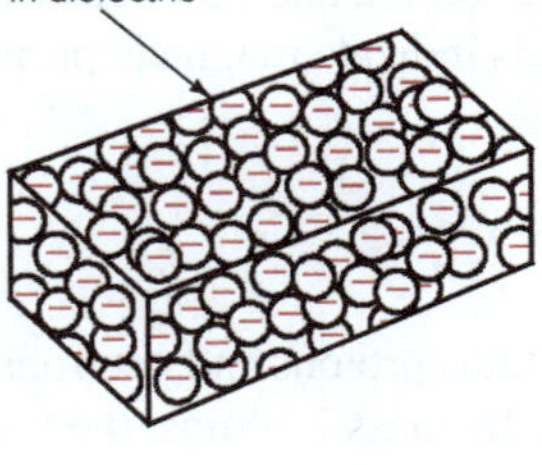

(*a*)

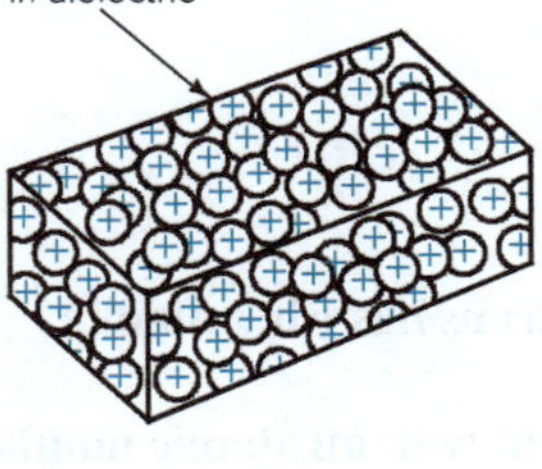

(*b*)

The symbol for electric charge is Q or q, standing for quantity. For instance, a charge of 6.25×10^{18} electrons is stated as $Q = 1$ C. This unit is named after Charles A. Coulomb (1736–1806), a French physicist, who measured the force between charges.

Negative and Positive Polarities

Historically, negative polarity has been assigned to the static charge produced on rubber, amber, and resinous materials in general. Positive polarity refers to the static charge produced on glass and other vitreous materials. On this basis, the electrons in all atoms are basic particles of negative charge because their polarity is the same as the charge on rubber. Protons have positive charge because the polarity is the same as the charge on glass.

Charges of Opposite Polarity Attract

GOOD TO KNOW

Unlike charges attract and like charges repel.

If two small charged bodies of light weight are mounted so that they are free to move easily and are placed close to each other, one can be attracted to the other when the two charges have opposite polarity (Fig. 1–5*a*). In terms of electrons and protons, they tend to be attracted to each other by the force of attraction between opposite charges. Furthermore, the weight of an electron is only about $\frac{1}{1840}$ the weight of a proton. As a result, the force of attraction tends to make electrons move to protons.

Charges of the Same Polarity Repel

In Fig. 1–5*b* and *c*, it is shown that when the two bodies have an equal amount of charge with the same polarity, they repel each other. The two negative charges repel

PIONEERS IN ELECTRONICS

French natural philosopher *Charles-Augustin Coulomb (1736–1806)* developed a method for measuring the force of attraction and repulsion between two electrically charged spheres. Coulomb established the law of inverse squares and defined the basic unit of charge quantity, the coulomb.

GOOD TO KNOW

As an aid for determining the added charge ($\pm Q$) to a neutral dielectric, use the following equation:

$$\pm Q = \frac{\text{Number of electrons added or removed}}{6.25 \times 10^{18}\ \text{electrons/C}}$$

Figure 1-5 Physical force between electric charges. (*a*) Opposite charges attract. (*b*) Two negative charges repel each other. (*c*) Two positive charges repel.

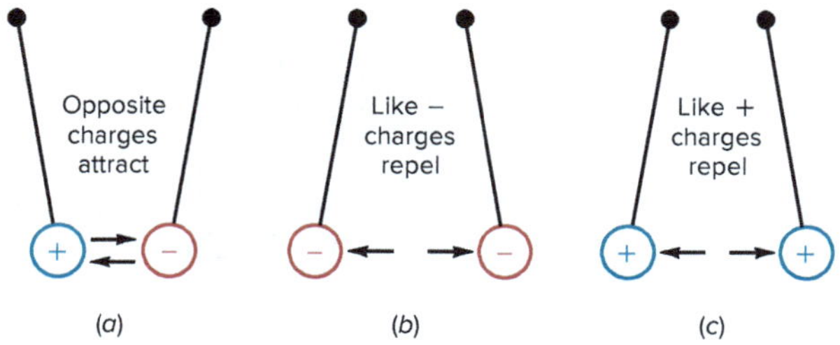

in Fig. 1–5*b*, and two positive charges of the same value repel each other in Fig. 1–5*c*.

Polarity of a Charge

An electric charge must have either negative or positive polarity, labeled $-Q$ or $+Q$, with an excess of either electrons or protons. A neutral condition is considered zero charge. On this basis, consider the following examples, remembering that the electron is the basic particle of charge and the proton has exactly the same amount, although of opposite polarity.

Example 1-1

A neutral dielectric has 12.5×10^{18} electrons added to it. What is its charge in coulombs?

ANSWER This number of electrons is double the charge of 1 C. Therefore, $-Q = 2$ C.

Example 1-2

A dielectric has a positive charge of 12.5×10^{18} protons. What is its charge in coulombs?

ANSWER This is the same amount of charge as in Example 1 but positive. Therefore, $+Q = 2$ C.

Example 1-3

A dielectric with $+Q$ of 2 C has 12.5×10^{18} electrons added. What is its charge then?

ANSWER The 2 C of negative charge added by the electrons cancels the 2 C of positive charge, making the dielectric neutral, for $Q = 0$.

Example 1-4

A neutral dielectric has 12.5×10^{18} electrons removed. What is its charge?

ANSWER The 2 C of electron charge removed allows an excess of 12.5×10^{18} protons. Since the proton and electron have exactly the same amount of charge, now the dielectric has a positive charge of $+Q = 2$ C.

Note that we generally consider that the electrons move, rather than heavier protons. However, a loss of a given number of electrons is equivalent to a gain of the same number of protons.

Charge of an Electron

The charge of a single electron, designated Q_e, is 0.16×10^{-18} C. This value is the reciprocal of 6.25×10^{18} electrons, which is the number of electrons in 1 coulomb of charge. Expressed mathematically,

$$-Q_e = 0.16 \times 10^{-18} \text{ C}$$

($-Q_e$ denotes that the charge of the electron is negative.)

It is important to note that the charge of a single proton, designated Q_P, is also equal to 0.16×10^{-18} C. However, its polarity is positive instead of negative.

In some cases, the charge of a single electron or proton will be expressed in scientific notation. In this case, $-Q_e = 1.6 \times 10^{-19}$ C. It is for convenience only that Q_e or Q_P is sometimes expressed as 0.16×10^{-18} C instead of 1.6×10^{-19} C. The convenience lies in the fact that 0.16 is the reciprocal of 6.25 and 10^{-18} is the reciprocal of 10^{18}.

Figure 1–6 Arrows indicate electric field around a stationary charge Q.

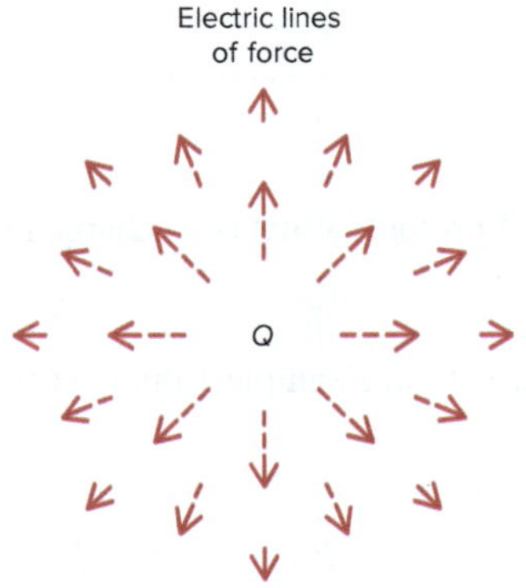

The Electric Field of a Static Charge

The ability of an electric charge to attract or repel another charge is a physical force. To help visualize this effect, lines of force are used, as shown in Fig. 1–6. All the lines form the electric field. The lines and the field are imaginary, since they cannot be seen. Just as the field of the force of gravity is not visible, however, the resulting physical effects prove that the field is there.

Each line of force, as shown in Fig. 1–6, is directed outward to indicate repulsion of another charge in the field with the same polarity as Q, either positive or negative. The lines are shorter farther away from Q to indicate that the force decreases inversely as the square of the distance. The larger the charge, the greater the force. These relations describe Coulomb's law of electrostatics.

1–4 Self-Review

Answers at the end of the chapter.

a. **How many electron charges are there in the practical unit of 1 coulomb?**

b. **How much is the charge in coulombs for a surplus of 18.75×10^{18} electrons?**

c. **Do opposite electric charges attract or repel each other?**

1–5 The Volt Unit of Potential Difference

Potential refers to the possibility of doing work. Any charge has the potential to do the work of moving another charge by either attraction or repulsion. When we consider two unlike charges, they have a *difference of potential*.

A charge is the result of work done in separating electrons and protons. Because of the separation, stress and strain are associated with opposite charges, since normally they would be balancing each other to produce a neutral condition. We could consider that the accumulated electrons are drawn tight and are straining themselves to be attracted toward protons to return to the neutral condition. Similarly, the work of producing the charge causes a condition of stress in the protons, which are trying to attract electrons and return to the neutral condition. Because of these forces, the charge of electrons or protons has potential because it is ready to give back the work put into producing the charge. The force between charges is in the electric field.

Potential between Different Charges

When one charge is different from the other, there must be a difference of potential between them. For instance, consider a positive charge of 3 C, shown at the right in Fig. 1–7*a*. The charge has a certain amount of potential, corresponding to the amount of work this charge can do. The work to be done is moving some electrons, as illustrated.

Assume that a charge of 1 C can move three electrons. Then the charge of +3 C can attract nine electrons toward the right. However, the charge of +1 C at the opposite side can attract three electrons toward the left. The net result, then, is that six electrons can be moved toward the right to the more positive charge.

In Fig. 1–7*b*, one charge is 2 C, and the other charge is neutral with 0 C. For the difference of 2 C, again 2×3 or 6 electrons can be attracted to the positive side.

In Fig. 1–7*c*, the difference between the charges is still 2 C. The +1 C attracts three electrons to the right side. The −1 C repels three electrons to the right side also. This effect is really the same as attracting six electrons.

Figure 1–7 The amount of work required to move electrons between two charges depends on their difference of potential. This potential difference (PD) is equivalent for the examples in (*a*), (*b*), and (*c*).

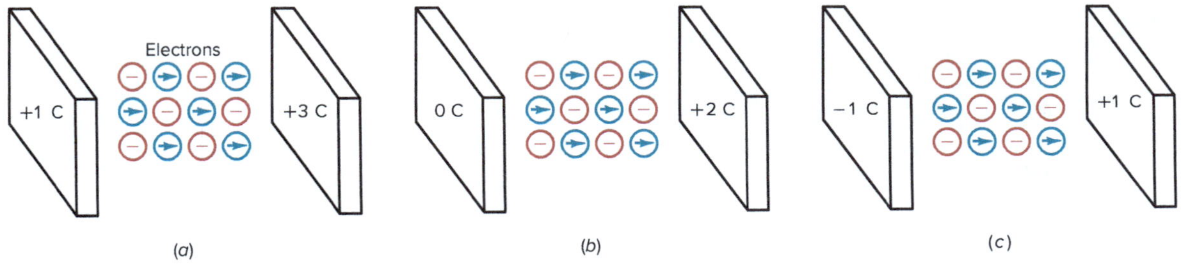

PIONEERS IN ELECTRONICS

In 1796, Italian physicist *Alessandro Volta (1745–1827)* developed the first chemical battery, which provided the first practical source of electricity.

Therefore, the net number of electrons moved in the direction of the more positive charge depends on the difference of potential between the two charges. This **potential difference** is the same for all three cases, as shown in Fig. 1–7. Potential difference is often abbreviated PD.

The only case without any potential difference between charges occurs when they both have the same polarity and are equal in amount. Then the repelling and attracting forces cancel, and no work can be done in moving electrons between the two identical charges.

The Volt Unit

The ***volt*** *unit* of potential difference is named after Alessandro Volta (1745–1827). Fundamentally, the volt is a measure of the amount of work or energy needed to move an electric charge. By definition, when 0.7376 foot-pound (ft·lb) of work is required to move 6.25×10^{18} electrons between two points, the potential difference between those two points is one volt. (Note that 6.25×10^{18} electrons make up one coulomb of charge.) The metric unit of work or energy is the joule (J). One joule is the same amount of work or energy as 0.7376 ft·lb. Therefore, we can say that the potential difference between two points is one volt when one joule of energy is required to move one coulomb of electric charge between those two points. Expressed as a formula, $1\text{ V} = \frac{1\text{ J}}{1\text{ C}}$.

In electronics, potential difference is commonly referred to as voltage, with the symbol *V*. Remember, voltage is the potential difference between two points and that two terminals are necessary for a potential difference to exist. A potential difference cannot exist at only one point!

Consider the 2.2-V lead-acid cell in Fig. 1–8*a*. Its output of 2.2 V means that this is the amount of potential difference between the two terminals. The lead-acid cell, then, is a voltage source, or a source of electromotive force (emf). The schematic symbol for a battery or DC voltage source is shown in Fig. 1–8*b*.

Sometimes the symbol *E* is used for emf, but the standard symbol *V* represents any potential difference. This applies either to the voltage generated by a source or to the voltage drop across a passive component such as a resistor.

It may be helpful to think of voltage as an electrical pressure or force. The higher the voltage, the more electrical pressure or force. The electrical pressure of voltage is in the form of the attraction and repulsion of an electric charge, such as an electron.

The general equation for any voltage can be stated as

$$V = \frac{W}{Q} \tag{1–1}$$

where *V* is the voltage in volts, *W* is the work or energy in joules, and *Q* is the charge in coulombs.

Let's take a look at an example.

Example 1-5

What is the output voltage of a battery that supplies 3.6 J of energy to move 0.5 C of charge?

ANSWER Use equation 1–1.

$$\begin{aligned} V &= \frac{W}{Q} \\ &= \frac{3.6\text{ J}}{0.5\text{ C}} \\ &= 7.2\text{ V} \end{aligned}$$

MultiSim **Figure 1–8** Chemical cell as a voltage source. (*a*) Voltage output is the potential difference between the two terminals. (*b*) Schematic symbol of any DC voltage source with constant polarity. Longer line indicates positive side.

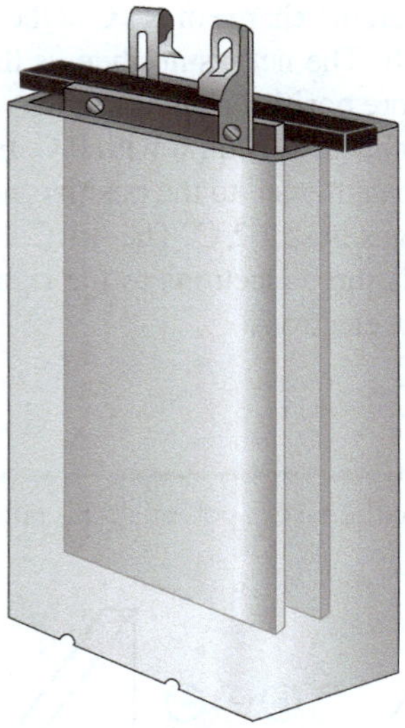

(*a*)

(*b*)

1–5 Self-Review

Answers at the end of the chapter.

a. **How much potential difference is there between two identical charges?**

b. **If 27 J of energy is expended in moving 3 C of charge between two points, how much voltage is there between those two points?**

1–6 Charge in Motion Is Current

When the potential difference between two charges forces a third charge to move, the charge in motion is an *electric current.* To produce current, therefore, charge must be moved by a potential difference.

In solid materials, such as copper wire, free electrons are charges that can be forced to move with relative ease by a potential difference, since they require relatively little work to be moved. As illustrated in Fig. 1–9, if a potential difference is connected across two ends of a copper wire, the applied voltage forces the free electrons to move. This current is a drift of electrons, from the point of negative charge at one end, moving through the wire, and returning to the positive charge at the other end.

To illustrate the drift of free electrons through the copper wire in Fig. 1–9, each **free electron** in the middle row is numbered, corresponding to a copper atom to which the free electron once belonged. The free electron at the left is labeled S to indicate that it comes from the negative charge of the source of potential difference. The free electron S is repelled from the negative charge $-Q$ at the left and is attracted by the positive charge $+Q$ at the right. Therefore, the potential difference of the voltage source can make the free electron S move toward the free electron from atom 1. Now the free electron from atom 1 is repelled toward the free electron from atom 2. In this way, there is a drift of free electrons moving from left to right from one atom to the next in the copper wire. The final result is that the free electron labeled 8 at the extreme right in Fig. 1–9 moves out from the wire to return to the positive charge of the voltage source.

Considering this case of moving electrons, note that the electron returning to the positive side of the voltage source is not the electron labeled S that left the negative side. All electrons are the same, however, and have the same charge. Therefore, the drift of free electrons resulted in the charge of one electron moving through the wire. This charge in motion is the current. With more electrons drifting through the wire, the charge of many electrons moves, resulting in more current.

MultiSim **Figure 1–9** Potential difference across the two ends of a copper wire conductor causes a drift of free electrons throughout the wire to produce electric current.

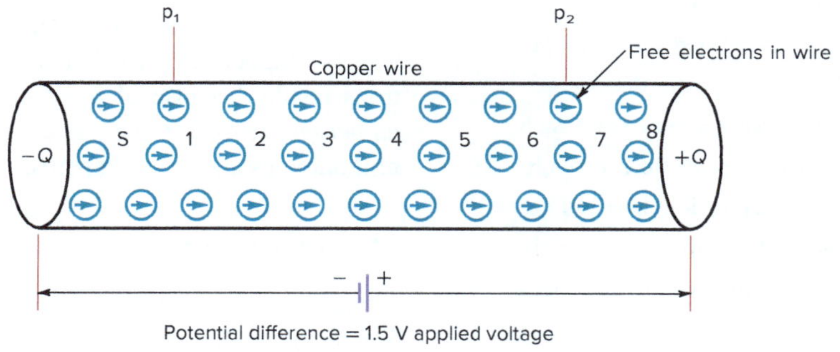

The current is a continuous flow of electrons. Only the electrons move, not the potential difference. For ordinary applications, where the wires are not long lines, the potential difference produces current instantaneously through the entire length of wire. Furthermore, the current must be the same at all points of the wire at any time.

Potential Difference Is Necessary to Produce Current

The number of free electrons that can be forced to drift through a wire to produce a moving charge depends upon the amount of potential difference across the wire. With more applied voltage, the forces of attraction and repulsion can make more free electrons drift, producing more charge in motion. A larger amount of charge moving during a given period of time means a higher value of current. Less applied voltage across the same wire results in a smaller amount of charge in motion, which is a smaller value of current. With zero potential difference across the wire, there is no current.

Two cases of zero potential difference and no current can be considered to emphasize that potential difference is needed to produce current. Assume that the copper wire is by itself, not connected to any voltage source, so that there is no potential difference across the wire. The free electrons in the wire can move from atom to atom, but this motion is random, without any organized drift through the wire. If the wire is considered as a whole, from one end to the other, the current is zero.

As another example, suppose that the two ends of the wire have the same potential. Then free electrons cannot move to either end because both ends have the same force and there is no current through the wire. A practical example of this case of zero potential difference would be to connect both ends of the wire to just one terminal of a battery. Each end of the wire would have the same potential, and there would be no current. The conclusion, therefore, is that two connections to two points at different potentials are needed to produce a current.

GOOD TO KNOW

In most electronic circuits, the current, *I*, is only a small fraction of an ampere. A typical value of current in an electronic circuit is 0.01 A, which is the same as 10 mA.

The Ampere of Current

Since current is the movement of charge, the unit for stating the amount of current is defined in rate of flow of charge. When the charge moves at the rate of 6.25×10^{18} electrons flowing past a given point per second, the value of the current is one ***ampere*** (A). This is the same as one coulomb of charge per second. The *ampere unit* of current is named after André M. Ampère (1775–1836).

Referring back to Fig. 1–9, note that if 6.25×10^{18} free electrons move past p_1 in 1 second (s), the current is 1 A. Similarly, the current is 1 A at p_2 because the electron drift is the same throughout the wire. If twice as many electrons moved past either point in 1 s, the current would be 2 A.

The symbol for current is *I* or *i* for intensity, since the current is a measure of how intense or concentrated the electron flow is. Two amperes of current in a copper wire is a higher intensity than one ampere; a greater concentration of moving electrons results because of more electrons in motion. Sometimes current is called *amperage*. However, the current in electronic circuits is usually in smaller units, milliamperes and microamperes.

Pixtal/Age Fotostock

PIONEERS IN ELECTRONICS

The unit of measure for current, the ampere (A), was named for French physicist *André Marie Ampère (1775–1836)*. Ampère discovered that two parallel wires attract each other when currents flow through them in the same direction and repel each other when currents are made to flow in opposite directions.

How Current Differs from Charge

Charge is a quantity of electricity accumulated in a dielectric, which is an insulator. The charge is static electricity, at rest, without any motion. When the charge moves, usually in a conductor, the current I indicates the intensity of the electricity in motion. This characteristic is a fundamental definition of current:

$$I = \frac{Q}{T} \qquad \textbf{(1–2)}$$

where I is the current in amperes, Q is in coulombs, and time T is in seconds. It does not matter whether the moving charge is positive or negative. The only question is how much charge moves and what its rate of motion is.

In terms of practical units,

$$1\text{ A} = \frac{1\text{ C}}{1\text{ s}} \qquad \textbf{(1–3)}$$

One ampere of current results when one coulomb of charge moves past a given point in 1 s. In summary, Q represents a specific amount or quantity of electric charge, whereas the current I represents the rate at which the electric charge, such as electrons, is moving. The difference between electric charge and current is similar to the difference between miles and miles per hour.

Example 1-6

The charge of 12 C moves past a given point every second. How much is the intensity of charge flow?

ANSWER

$$I = \frac{Q}{T} = \frac{12\text{ C}}{1\text{ s}}$$

$$I = 12\text{ A}$$

Example 1-7

The charge of 5 C moves past a given point in 1 s. How much is the current?

ANSWER

$$I = \frac{Q}{T} = \frac{5\text{ C}}{1\text{ s}}$$

$$I = 5\text{ A}$$

The fundamental definition of current can also be used to consider the charge as equal to the product of the current multiplied by the time. Or

$$Q = I \times T \qquad \textbf{(1–4)}$$

In terms of practical units,

$$1\text{ C} = 1\text{ A} \times 1\text{ s} \qquad \textbf{(1–5)}$$

One coulomb of charge results when one ampere of current accumulates charge during one second. The charge is generally accumulated in the dielectric of a capacitor or at the electrodes of a battery.

For instance, we can have a dielectric connected to conductors with a current of 0.4 A. If the current can deposit electrons for 0.2 s, the accumulated charge in the dielectric will be

$$Q = I \times T = 0.4 \text{ A} \times 0.2 \text{ s}$$
$$Q = 0.08 \text{ C}$$

The formulas $Q = IT$ for charge and $I = Q/T$ for current illustrate the fundamental nature of Q as an accumulation of static charge in an insulator, whereas I measures the intensity of moving charges in a conductor. Furthermore, current I is different from voltage V. You can have V without I, but you cannot have current without an applied voltage.

The General Nature of Current

The moving charges that provide current in metal conductors such as copper wire are the free electrons of the copper atoms. In this case, the moving charges have negative polarity. The direction of motion between two terminals for this *electron current*, therefore, is toward the more positive end. It is important to note, however, that there are examples of positive charges in motion. Common applications include current in liquids, gases, and semiconductors. For the current resulting from the motion of positive charges, its direction is opposite from the direction of electron flow. Whether negative or positive charges move, though, the current is still defined fundamentally as Q/T. Also, note that the current is provided by free charges, which are easily moved by an applied voltage.

Magnetic Field Associated with an Electric Current

Every current-carrying conductor has an associated magnetic field with magnetic field lines that extend outward from the conductor. The magnetic field lines are circular and exist in a plane perpendicular to the direction of current. Also, the strength of the magnetic field (the number of magnetic field lines) is proportional to the amount of current. In general, a magnetic field is always associated with electric charges in motion. In fact, even the orbiting electrons within an atom generate a magnetic field.

The magnetic field produced by an electric current is the basis for several electromagnetic applications in electronics. For example, relays, speakers, transformers, and coils in general are practical applications of electromagnetism. More details of electromagnetism and magnetic fields are covered in Chapter 14, "Electromagnetism."

1–6 Self-Review

Answers at the end of the chapter.

a. The flow of 2 C/s of electron charges is how many amperes of current?
b. The symbol for current is *I* for intensity. (True/False)
c. How much is the current with zero potential difference?

Bettmann/Getty Images

PIONEERS IN ELECTRONICS

The unit of measure for resistance, the ohm, was named for German physicist *Georg Simon Ohm (1789–1854)*. Ohm is also known for his development of Ohm's law: $I = \frac{V}{R}$.

1–7 Resistance Is Opposition to Current

The fact that a wire conducting current can become hot is evidence that the work done by the applied voltage in producing current must be accomplished against some form of opposition. This opposition, which limits the amount of current that can be produced by the applied voltage, is called ***resistance***. Conductors have very little resistance; insulators have a large amount of resistance.

The atoms of a copper wire have a large number of free electrons, which can be moved easily by a potential difference. Therefore, the copper wire has little opposition to the flow of free electrons when voltage is applied, corresponding to low resistance.

Figure 1–10 (*a*) Wire-wound type of resistor with cement coating for insulation. (*b*) Schematic symbol for any type of fixed resistor.

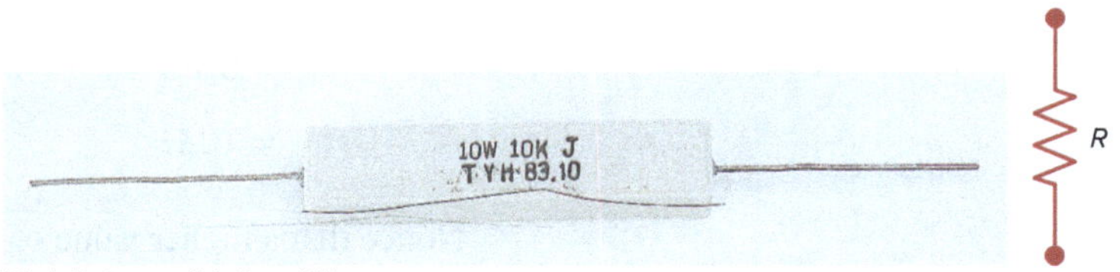

Mark Steinmetz/McGraw Hill

(*a*) (*b*)

Carbon, however, has fewer free electrons than copper. When the same amount of voltage is applied to carbon as to copper, fewer electrons will flow. Just as much current can be produced in carbon by applying more voltage. For the same current, though, the higher applied voltage means that more work is necessary, causing more heat. Carbon opposes the current more than copper, therefore, and has higher resistance.

The Ohm

The practical unit of resistance is the ***ohm***. A resistance that develops 0.24 calorie of heat when one ampere of current flows through it for one second has one ohm of opposition. As an example of a low resistance, a good conductor such as copper wire can have a resistance of 0.01 Ω for a 1-ft length. The resistance-wire heating element in a 600-W 120-V toaster has a resistance of 24 Ω, and the tungsten filament in a 100-W 120-V lightbulb has a resistance of 144 Ω. The ohm unit is named after Georg Simon Ohm (1789–1854), a German physicist.

Figure 1–10*a* shows a wire-wound resistor. Resistors are also made with powdered carbon. They can be manufactured with values from a few ohms to millions of ohms.

The symbol for resistance is *R*. The abbreviation used for the ohm unit is the Greek letter *omega*, written as Ω. In diagrams, resistance is indicated by a zigzag line, as shown by *R* in Fig. 1–10*b*.

Conductance

The opposite of resistance is ***conductance***. The lower the resistance, the higher the conductance. Its symbol is *G*, and the unit is the ***siemens*** **(S)**, named after Ernst von Siemens (1816–1892), a German inventor. (The old unit name for conductance is *mho*, which is *ohm* spelled backward.)

Specifically, *G* is the reciprocal of *R*, or $G = \frac{1}{R}$. Also, $R = \frac{1}{G}$.

Example 1-8

Calculate the resistance for the following conductance values: (a) 0.05 S (b) 0.1 S

ANSWER

$$\text{(a) } R = \frac{1}{G}$$

$$= \frac{1}{0.05\ \text{S}}$$

$$= 20\ \Omega$$

$$\text{(b) } R = \frac{1}{G}$$

$$= \frac{1}{0.1 \text{ S}}$$

$$= 10 \ \Omega$$

Notice that a higher value of conductance corresponds to a lower value of resistance.

Example 1-9

Calculate the conductance for the following resistance values: (a) 1 kΩ (b) 5 kΩ.

ANSWER

$$\text{(a) } G = \frac{1}{R}$$

$$= \frac{1}{1000 \ \Omega}$$

$$= 0.001 \text{ S or } 1 \text{ mS}$$

$$\text{(b) } G = \frac{1}{R}$$

$$= \frac{1}{5000 \ \Omega}$$

$$= 0.0002 \text{ S or } 200 \ \mu\text{S}$$

Notice that a higher value of resistance corresponds to a lower value of conductance.

Figure 1–11 Example of an electric circuit with a battery as a voltage source connected to a lightbulb as a resistance. (*a*) Wiring diagram of the closed path for current. (*b*) Schematic diagram of the circuit.

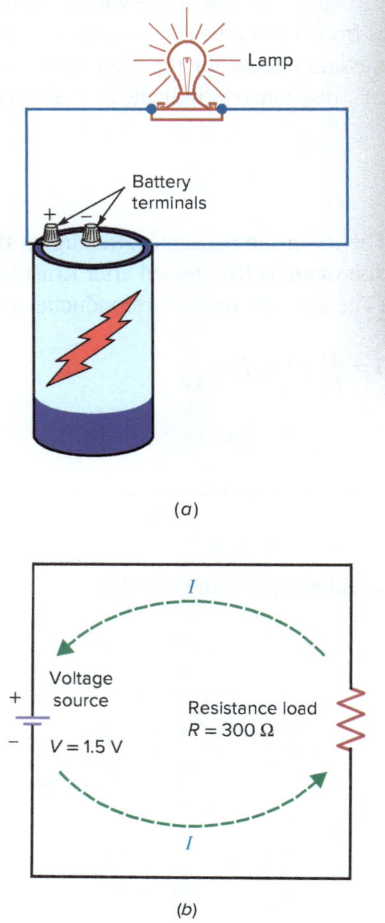

1–7 Self-Review

Answers at the end of the chapter.

a. Which has more resistance, carbon or copper?

b. With the same voltage applied, which resistance will allow more current, 4.7 Ω or 5000 Ω?

c. What is the conductance value in siemens units for a 10-Ω *R*?

1–8 The Closed Circuit

In applications requiring current, the components are arranged in the form of a *circuit*, as shown in Fig. 1–11. A circuit can be defined as a complete circular path for current flow. The purpose of this circuit is to light the incandescent bulb. The bulb lights when the tungsten-filament wire inside is white hot, producing an incandescent glow.

The tungsten filament cannot produce current by itself. A source of potential difference is necessary. Since the battery produces a potential difference of 1.5 V across its two output terminals, this voltage is connected across the filament of the bulb by the two wires so that the applied voltage can produce current through the filament.

In Fig. 1–11*b*, the schematic diagram of the circuit is shown. Here the components are represented by shorthand symbols. Note the symbols for the battery and resistance. The connecting wires are shown simply as straight lines because their resistance is small enough to be neglected. A resistance of less than 0.01 Ω for the wire is practically zero compared with the 300-Ω resistance of the bulb. If the resistance of the wire must be considered, the schematic diagram includes it as additional resistance in the same current path.

Note that the schematic diagram does not look like the physical layout of the circuit. The schematic shows only the symbols for the components and their electrical connections.

Any electric circuit has three important characteristics:

1. There must be a source of potential difference. Without the applied voltage, current cannot flow.
2. There must be a complete path for current flow, from one side of the applied voltage source, through the external circuit, and returning to the other side of the voltage source.
3. The current path normally has resistance. The resistance is in the circuit either to generate heat or limit the amount of current.

How the Voltage Is Different from the Current

It is the current that moves through the circuit. The potential difference (PD) does not move.

In Fig. 1–11, the voltage across the filament resistance makes electrons flow from one side to the other. While the current is flowing around the circuit, however, the potential difference remains across the filament to do the work of moving electrons through the resistance of the filament.

Figure 1–12 Comparison of voltage (*V*) across a resistance and the current (*I*) through *R*.

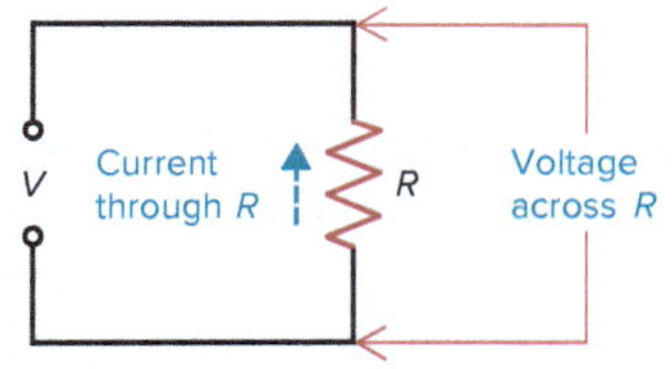

The circuit is redrawn in Fig. 1–12 to emphasize the comparison between *V* and *I*. The voltage is the potential difference across the two ends of the resistance. If you want to measure the PD, just connect the two leads of a voltmeter across the resistor. However, the current is the intensity of the electron flow past any one point in the circuit. Measuring the current is not as easy. You would have to break open the path at any point and then insert the current meter to complete the circuit.

The word *across* is used with voltage because it is the potential difference between two points. There cannot be a PD at one point. However, current can be considered at one point, as the motion of charges through that point.

To illustrate the difference between *V* and *I* in another way, suppose that the circuit in Fig. 1–11 is opened by disconnecting the bulb. Now no current can flow because there is no closed path. Still, the battery has its potential difference. If you measure across the two terminals, the voltmeter will read 1.5 V even though the current is zero. This is like a battery sitting on a store shelf. Even though the battery is not producing current in a circuit, it still has a voltage output between its two terminals. This brings us to a very important conclusion: **Voltage can exist without current, but current cannot exist without voltage.**

The Voltage Source Maintains the Current

As current flows in a circuit, electrons leave the negative terminal of the cell or battery in Fig. 1–11, and the same number of free electrons in the conductor are returned to the positive terminal. As electrons are lost from the negative charge and gained by the positive charge, the two charges tend to neutralize each other. The chemical action inside the battery, however, continuously separates electrons and protons to maintain the negative and positive charges on the outside terminals that provide the potential difference. Otherwise, the current would neutralize the charges, resulting in no potential difference, and the current would stop. Therefore, the battery keeps the current flowing by maintaining the potential difference across the circuit. The battery is the voltage source for the circuit.

The Circuit Is a Load on the Voltage Source

We can consider the circuit as a means whereby the energy of the voltage source is carried by the current through the filament of the bulb, where the electric energy is used in producing heat energy. On this basis, the battery is the *source* in the circuit, since its voltage output represents the potential energy to be used. The part of the circuit connected to the voltage source is the *load resistance*, since it determines how much work the source will supply. In this case, the bulb's filament is the load resistance on the battery.

The current that flows through the load resistance is the *load current*. Note that a lower value of ohms for the load resistance corresponds to a higher load current. Unless noted otherwise, the term *load* by itself can be assumed generally to mean the load current. Therefore, a heavy or big load electrically means a high current load, corresponding to a large amount of work supplied by the source.

GOOD TO KNOW

The actual load for a power source could be a motor, fan, lightbulb, or heating element, as examples.

In summary, we can say that the closed circuit, normal circuit, or just a circuit is a closed path that has V to produce I with R to limit the amount of current. The circuit provides a means of using the energy of the battery as a voltage source. The battery has its potential difference V with or without the circuit. However, the battery alone is not doing any work in producing load current. The bulb alone has resistance, but without current, the bulb does not light. With the circuit, the voltage source is used to produce current to light the bulb.

Open Circuit

When any part of the path is open or broken, the circuit is incomplete because there is no conducting path. The *open circuit* can be in the connecting wires or in the bulb's filament as the load resistance. The resistance of an open circuit is infinitely high. The result is no current in an open circuit.

Short Circuit

In this case, the voltage source has a closed path across its terminals, but the resistance is practically zero. The result is too much current in a *short circuit*. Usually, the short circuit is a bypass around the load resistance. For instance, a short across the tungsten filament of a bulb produces too much current in the connecting wires but no current through the bulb. Then the bulb is shorted out. The bulb is not damaged, but the connecting wires can become hot enough to burn unless the line has a fuse as a safety precaution against too much current.

GOOD TO KNOW

A short circuit and open circuit are opposites of each other. The resistance of a short circuit is approximately 0 Ω and the resistance of an open circuit is considered to be infinite (∞) Ω.

1–8 Self-Review

Answers at the end of the chapter.

Answer true or false for the circuit, shown in Fig. 1–11.

a. The bulb has a PD of 1.5 V across its filament only when connected to the voltage source.

b. The battery has a PD of 1.5 V across its terminals only when connected to the bulb.

c. The battery by itself, without the wires and the bulb, has a PD of 1.5 V.

1–9 The Direction of Current

Just as a voltage source has polarity, current has a direction. The reference is with respect to the positive and negative terminals of the voltage source. The direction of the current depends on whether we consider the flow of negative electrons or the motion of positive charges in the opposite direction.

Electron Flow

As shown in Fig. 1–13*a*, the direction of electron drift for the current *I* is out from the negative side of the voltage source. Current *I* flows through the external circuit with *R* and returns to the positive side of *V*. Note that this direction from the negative terminal applies to the external circuit connected to the output terminals of the voltage source. *Electron flow* is also shown in Fig. 1–13*c* with reversed polarity for *V*.

Inside the battery, the electrons move to the negative terminal because this is how the voltage source produces its potential difference. The battery is doing the work of separating charges, accumulating electrons at the negative terminal and protons at the positive terminal. Then the potential difference across the two output terminals can do the work of moving electrons around the external circuit. For the circuit outside the voltage source, however, the direction of the electron flow is from a point of negative potential to a point of positive potential.

Conventional Current

A motion of positive charges, in the opposite direction from electron flow, is considered ***conventional current***. This direction is generally used for analyzing circuits in electrical engineering. The reason is based on some traditional definitions in the science of physics. By the definitions of force and work with positive values, a

Figure 1–13 Direction of *I* in a closed circuit, shown for electron flow and conventional current. The circuit works the same way no matter which direction you consider. (*a*) Electron flow indicated with dashed arrow in diagram. (*b*) Conventional current indicated with solid arrow. (*c*) Electron flow as in (*a*) but with reversed polarity of voltage source. (*d*) Conventional *I* as in (*b*) but reversed polarity for *V*.

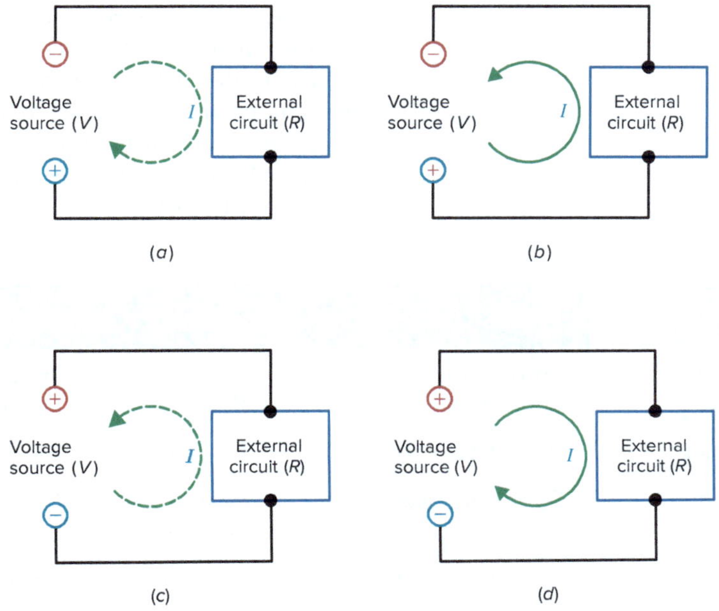

positive potential is considered above a negative potential. Then conventional current corresponds to a motion of positive charges "falling downhill" from a positive to a negative potential. The conventional current, therefore, is in the direction of positive charges in motion. An example is shown in Fig. 1–13*b*. The conventional *I* is out from the positive side of the voltage source, flows through the external circuit, and returns to the negative side of *V*. Conventional current is also shown in Fig. 1–13*d*, with the voltage source in reverse polarity.

GOOD TO KNOW

Electrical engineers usually analyze electronic circuits using conventional current flow, whereas electronic technicians usually use electron flow. Both directions of current flow produce the same results. Which one to use is mainly a matter of personal preference.

Examples of Mobile Positive Charges

An **ion** is an atom that has either lost or gained one or more valence electrons to become electrically charged. For example, a positive ion is created when a neutral atom loses one or more valence electrons, and thus becomes positively charged. Similarly, a negative ion is created when a neutral atom gains one or more valence electrons and thus becomes negatively charged. Depending on the number of valence electrons that have been added or removed, the charge of an ion may equal the charge of one electron (Q_e), two electrons ($2\ Q_e$), three electrons ($3\ Q_e$), and so on. Ions can be produced by applying voltage to liquids and gases to ionize the atoms. These ions are mobile charges that can provide an electric current. Positive or negative ions are much less mobile than electrons, however, because an ion includes a complex atom with its nucleus.

An example of positive charges in motion for conventional current, therefore, is the current of positive ions in either liquids or gases. This type of current is referred to as ionization current. The positive ions in a liquid or gas flow in the direction of conventional current because they are repelled by the positive terminal of the voltage source and attracted to the negative terminal. Therefore, the mobile positive ions flow from the positive side of the voltage source to the negative side.

Another example of a mobile positive charge is the hole. Holes exist in semiconductor materials such as silicon and germanium. A hole possesses the same amount of charge as an electron but instead has positive polarity. Although the details of the hole charge are beyond the scope of this discussion, you should be aware that in semiconductors, the movement of hole charges are in the direction of conventional current.

It is important to note that protons themselves are not mobile positive charges because they are tightly bound in the nucleus of the atom and cannot be released except by nuclear forces. Therefore, a current of positive charges is a flow of either positive ions in liquids and gases or positive holes in semiconductors. Table 1–4 summarizes different types of electric charge that can provide current in a circuit.

In this book, the current is considered as electron flow in the applications where electrons are the moving charges. A dotted or dashed arrow, as in Fig. 1–13*a* and *c*, is used to indicate the direction of electron flow for *I*. In Fig. 1–13*b* and *d*, the solid arrow means the direction of conventional current. These arrows are used for the unidirectional current in DC circuits. For AC circuits, the direction of current can be considered either way because *I* reverses direction every half-cycle with the reversals in polarity for *V*.

Table 1–4 Types of Electric Charges for Current

Types of Charges	Amount of Charge	Polarity	Types of Current	Applications
Electron	$Q_e = 0.16 \times 10^{-18}$ C	Negative	Electron flow	In wire conductors
Ion	Q_e or multiples of Q_e	Positive or negative	Ion current	In liquids and gases
Hole	$Q_e = 0.16 \times 10^{-18}$ C	Positive	Hole current	In p-type semiconductors

1–9 Self-Review

Answers at the end of the chapter.

a. Is electron flow out from the positive or negative terminal of the voltage source?

b. Does conventional current return to the positive or negative terminal of the voltage source?

c. Is it true or false that electron flow and conventional current are in opposite directions?

Figure 1–14 Steady DC voltage of fixed polarity, such as the output of a battery. Note schematic symbol at left.

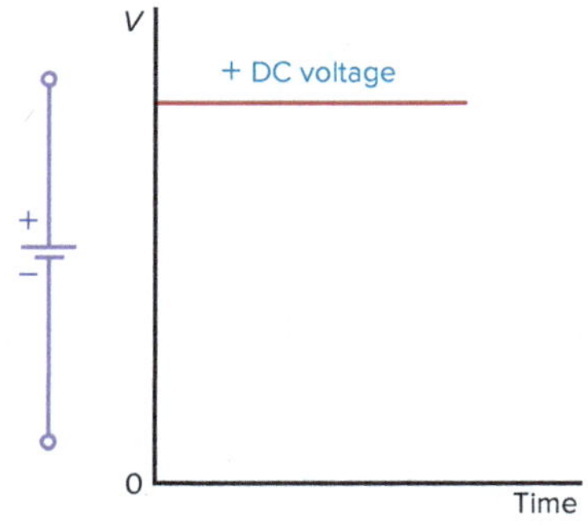

Figure 1–15 Sine wave AC voltage with alternating polarity, such as from an AC generator. Note schematic symbol at left. The AC line voltage in your home has this waveform.

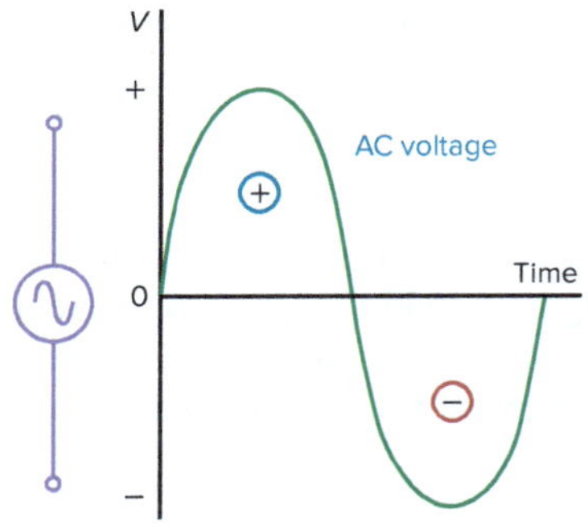

1–10 Direct Current (DC) and Alternating Current (AC)

The electron flow illustrated for the circuit with a bulb in Fig. 1–11 is ***direct current (DC)*** because it has just one direction. The reason for the unidirectional current is that the battery maintains the same polarity of output voltage across its two terminals.

The flow of charges in one direction and the fixed polarity of applied voltage are the characteristics of a DC circuit. The current can be a flow of positive charges, rather than electrons, but the conventional direction of current does not change the fact that the charges are moving only one way.

Furthermore, the DC voltage source can change the amount of its output voltage but, with the same polarity, direct current still flows only in one direction. This type of source provides a fluctuating or pulsating DC voltage. A battery is a steady DC voltage source because it has fixed polarity and its output voltage is a steady value.

An alternating voltage source periodically reverses or alternates in polarity. The resulting ***alternating current (AC)***, therefore, periodically reverses in direction. In terms of electron flow, the current always flows from the negative terminal of the voltage source, through the circuit, and back to the positive terminal, but when the generator alternates in polarity, the current must reverse its direction. The 60-cycle AC power line used in most homes is a common example. This frequency means that the voltage polarity and current direction go through 60 cycles of reversal per second.

The unit for 1 cycle per second is 1 hertz (Hz). Therefore, 60 cycles per second is a frequency of 60 Hz.

The details of AC circuits are explained in Chapter 15. Direct-current circuits are analyzed first because they usually are simpler. However, the principles of DC circuits also apply to AC circuits. Both types are important because most electronic circuits include AC voltages and DC voltages. A comparison of DC and AC voltages and their waveforms is illustrated in Figs. 1–14 and 1–15. Their uses are compared in Table 1–5.

Table 1–5 Comparison of DC Voltage and AC Voltage

DC Voltage	AC Voltage
Magnitude remains constant or steady with fixed polarity.	Varies in magnitude and reverses in polarity.
Steady DC voltage cannot be stepped up or down by a transformer.	Varying AC voltage can be stepped up or down with a transformer for electric power distribution.
Schematic symbol for DC voltage source. 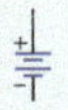	Schematic symbol for sine wave AC voltage source. 
The type of voltage available at the terminals of a battary.	The type of voltage available at the output of a rotary generator such as an alternator.
Heating effect is the same for direct or alternating current	

1–10 Self-Review

Answers at the end of the chapter.

a. When the polarity of the applied voltage reverses, the direction of current flow also reverses. (True/False)

b. A battery is a DC voltage source because it cannot reverse the polarity across its output terminals. (True/False)

1–11 Sources of Electricity

There are electrons and protons in the atoms of all materials, but to do useful work, the charges must be separated to produce a *potential difference* that can make current flow. Some of the more common methods of providing electrical effects are listed here.

Static Electricity by Friction

In this method, electrons in an insulator can be separated by the work of rubbing to produce opposite charges that remain in the dielectric. Examples of how *static electricity* can be generated include combing your hair, walking across a carpeted room, or sliding two pieces of plastic across each other. An *electrostatic discharge (ESD)* occurs when one of the charged objects comes into contact with another dissimilarly charged object. The electrostatic discharge is in the form of a spark. The current from the discharge lasts for only a very short time but can be very large.

Conversion of Chemical Energy

Wet or dry cells and batteries are the applications. Here a chemical reaction produces opposite charges on two dissimilar metals, which serve as the negative and positive terminals.

Electromagnetism

Electricity and magnetism are closely related. Any moving charge has an associated *magnetic field;* also, any changing magnetic field can produce current. A motor is an example showing how current can react with a magnetic field to produce motion; a generator produces voltage by means of a conductor rotating in a magnetic field.

Photoelectricity

Some materials are photoelectric, that is, they can emit electrons when light strikes the surface. The element cesium is often used as a source of *photoelectrons*. Also, photovoltaic cells or solar cells use silicon to generate output voltage from the light input. In another effect, the resistance of the element selenium changes with light. When this is combined with a fixed voltage source, wide variations between *dark current* and *light current* can be produced. Such characteristics are the basis of many photoelectric devices, including photoelectric cells and phototransistors as examples.

1–11 Self-Review

Answers at the end of the chapter.

a. The excess charges at the negative terminal of a battery are ______.

b. Any moving charge has an associated ______.

c. An electrostatic discharge (ESD) is in the form of a(n) ______.

Figure 1–16 Typical digital multimeters (DMMs) (*a*) Handheld DMM (*b*) Benchtop DMM.

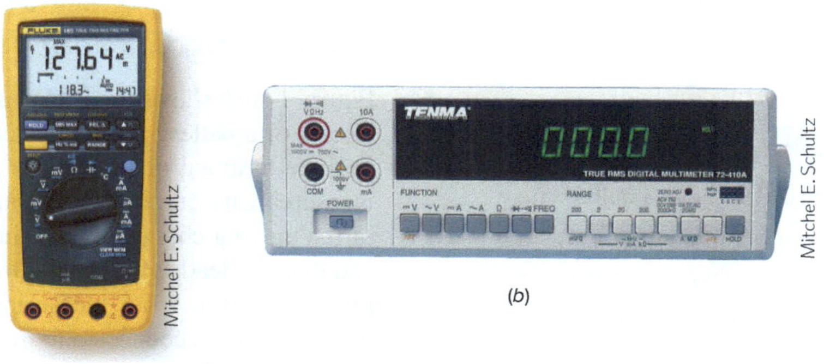

(*a*) (*b*)

1–12 The Digital Multimeter

As an electronics technician, you can expect to encounter many situations where it will be necessary to measure the voltage, current, or resistance in a circuit. When this is the case, a technician will most likely use a digital multimeter (DMM) to make these measurements. A DMM may be either a handheld or benchtop unit. Both types are shown in Fig. 1–16. All digital meters have numerical readouts that display the value of voltage, current, or resistance being measured.

Measuring Voltage

Figure 1–17*a* shows a typical DMM measuring the voltage across the terminals of a battery. To measure any voltage, the meter leads are connected directly across the two points where the potential difference or voltage exists. For DC voltages, the red lead of the meter is normally connected to the positive (+) side of the potential difference, whereas the black lead is normally connected to the negative (−) side.

Figure 1–17 DMM measurements (*a*) Measuring voltage (*b*) Measuring current (*c*) Measuring resistance.

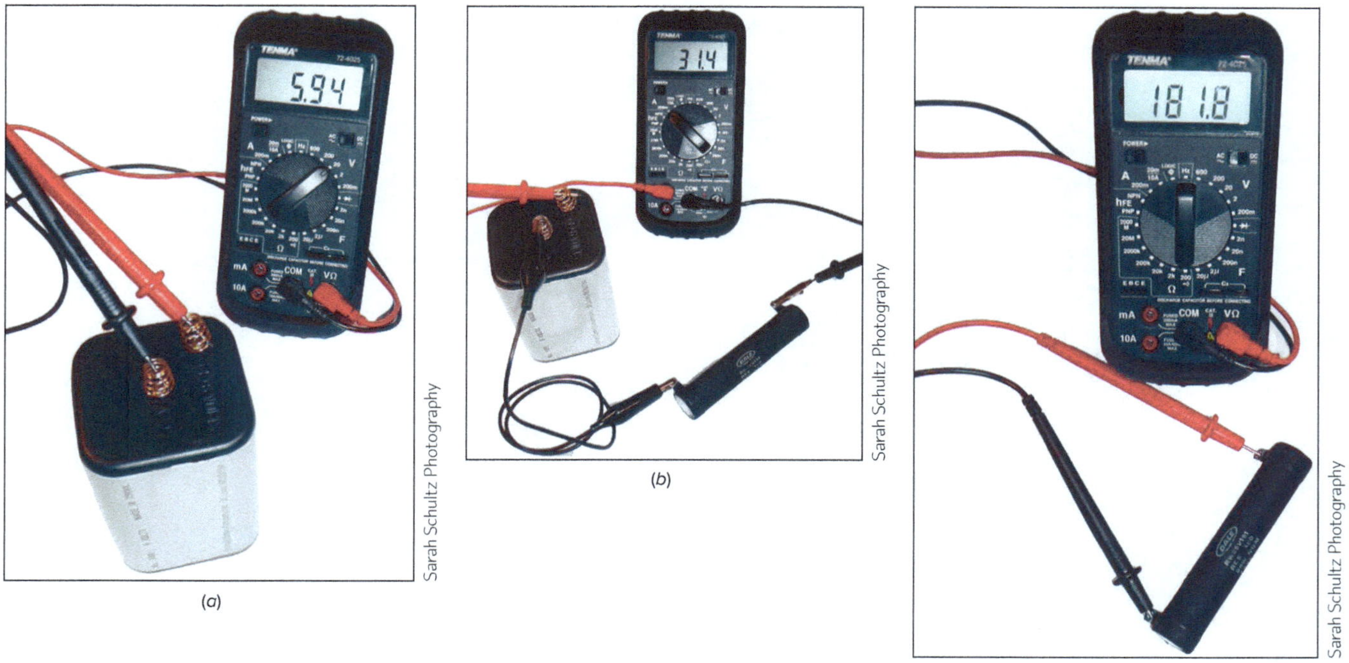

(*a*) (*b*) (*c*)

When measuring an alternating (AC) voltage, the orientation of the meter leads does not matter since the voltage periodically reverses polarity anyway.

Measuring Current

Figure 1–17*b* shows the DMM measuring the current in a simple DC circuit consisting of a battery and a resistor. Notice that the meter is connected between the positive terminal of the battery and the right lead of the resistor. Unlike voltage measurements, current measurements must be made by placing the meter in the path of the moving charges. To do this, the circuit must be broken open at some point, and then the leads of the meter must be connected across the open points to recomplete the circuit. When measuring the current in a DC circuit, the black lead of the meter should be connected to the point that traces directly back to the negative side of the potential difference. Likewise, the red lead of the meter should be connected to the point that traces directly back to the positive side of the potential difference. When measuring AC currents, the orientation of the meter leads is unimportant.

GOOD TO KNOW

Most DMMs have a built-in fuse that serves to protect the meter from becoming damaged when the measured current is excessively high. Excessive current could result from improperly connecting the meter into the circuit when measuring current.

Measuring Resistance

Figure 1–17*c* shows the DMM measuring the ohmic value of a single resistor. Note that the orientation of the meter leads is unimportant when measuring resistance. What is important is that no voltage is present across the resistance being measured, otherwise the meter could be damaged. Also, make sure that no other components are connected across the resistance being measured. If there are, the measurement will probably be both inaccurate and misleading.

1–12 Self-Review

Answers at the end of the chapter.

a. When using a DMM to measure voltage, place the meter leads directly across the two points of potential difference. (True/False)

b. When using a DMM to measure current, break open the circuit first and then insert the meter across the open points. (True/False)

c. When using a DMM to measure the value of a single resistor, the orientation of the meter leads is extremely important. (True/False)

Application in Understanding Alternative and Renewable Energy

When energy is generated in a way that does not deplete our natural resources or harm the environment, it is referred to as *alternative energy*. Alternative energy encompasses all of the different types of energy sources that do not burn fossil fuels. The most popular sources of alternative energy in use today include solar, wind, biomass, hydroelectric, and geothermal. The increased use of these alternative energy sources has reduced our dependency on fossil fuels considerably. The main benefit, of course, is that these alternative energy sources produce little or no pollution.

Another term often encountered is *renewable energy*. Renewable energy refers to energy that is generated from natural resources and is replenished naturally, meaning it does not run out or become exhausted over time. It is important to note that not all alternative energy sources are renewable. Nuclear power, for example, is an alternative energy source but it is not considered renewable because it is a finite resource that will eventually run out. Let's take a closer look at solar and wind energy, which are perhaps the two most popular types of alternative and/or renewable energy sources used for generating electricity.

SOLAR ENERGY

Solar energy, which is renewable, is the energy provided by the sun in the form of solar radiation. Solar energy is a clean, reliable form of renewable energy because it does not emit any greenhouse gases or air pollutants. When we think of solar energy, we usually think of the technology associated with harnessing the sun's energy and making it usable. **Photovoltaic (PV) cells** and **PV systems** in general convert light energy (solar energy) directly into electricity. The electricity is in the form of a DC voltage and current. Figure 1–18 shows an example of a basic PV cell, also called a solar cell. The PV cell is made from silicon, which is a semiconductor. When incoming light strikes the solar cell, it produces a small potential difference or voltage within the cell. (The theory behind the generation of this voltage is beyond the scope of this discussion and therefore will not be covered here.) Photovoltaic cells are usually connected in series and/or parallel to produce higher voltages, currents, and power levels. When this is done, the arrangement of solar cells is called a **PV module**. PV modules can be grouped together to form much larger **photovoltaic (PV) arrays**. The progression from PV cell to PV module to PV array is illustrated in Fig. 1-19.

Photovoltaic cells are used to power calculators, wrist watches, and decorative outdoor lighting, to name just a few of the more popular applications. Moderately sized PV systems, consisting of PV modules, are commonly used in applications where electronic equipment is located in remote areas. Some examples include providing power for communications equipment located on a mountaintop, powering warning signs along a highway or interstate, or for pumping water in remote areas of a farm or ranch. Many portable and/or remote applications of solar energy require power even when the sun is not shining. In these applications, a battery is part of the PV system and is continuously charged when the sun is shining. Much larger PV systems are now being used by electric utility companies to generate electricity for our homes and industries. There are endless applications of solar energy in use today, way too many to list or identify here. As shown in Fig. 1–20 it is

Figure 1–18 Photovoltaic cell, also called a solar cell.

Figure 1–19 PV cells can be connected in series and/or parallel to form a PV module. Similarly, PV modules can be arranged to form much larger PV arrays.

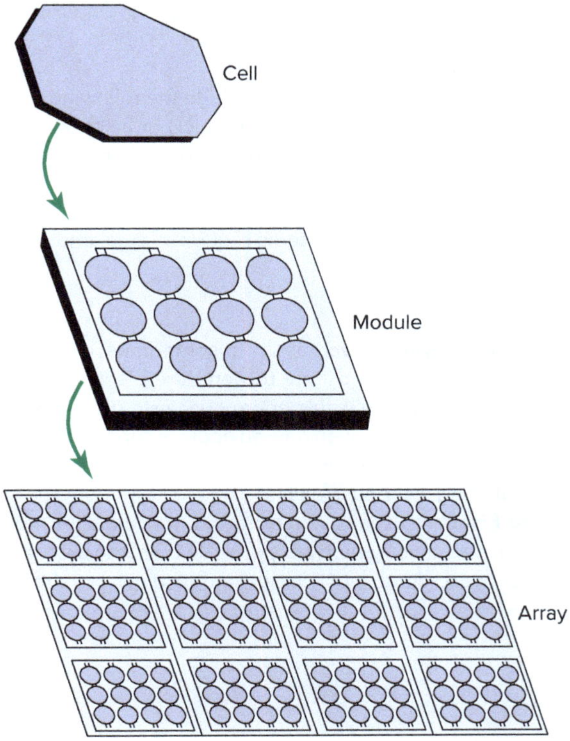

Figure 1–20 Solar panels mounted on the roof of a house.

becoming commonplace for home owners to install solar panels on the roofs of their homes. In doing so, the home owner can significantly reduce the amount of electricity they need to purchase from their local electric utility company. The electrical energy generated by roof-mounted solar panels is determined by their power rating as well as their efficiency. Also, the tilt angle and azimuth of the roof are other important factors in determining the amount of electricity that can be generated.

There are basically two types of PV systems, ***grid-connected systems*** and ***stand-alone systems***. The main difference between these two systems is that one is connected to the utility grid and the other is not. Grid-connected systems are designed to operate in conjunction with an electric ***utility grid***. The utility grid is a network of cables through which electricity is transported from power stations to homes and schools, as well as to businesses and industries.

The main benefits or advantages of PV systems include the following:

- They are clean, safe, and quiet to operate.
- They are highly reliable and thus require little or no maintenance.
- They are very cost-effective, especially in remote locations.
- They provide independence from the power grid and/or back-up power during power outages.
- It is a renewable source of energy.

The main limitations of PV systems include the following:

- They usually cannot operate continuously or all of the time.
- They are not well suited for energy intensive applications such as heating.
- They are relatively expensive to buy and install.
- Utility or large-scale PV systems require a large amount of land and/or space.

WIND ENERGY

Wind energy, or ***wind power*** as it is sometimes called, is a technology that uses wind flow to generate electricity. Wind energy is considered renewable energy because no matter how much of it is used on any given day, an endless supply is available for the future. Like solar energy, wind energy does not emit any air pollutants or greenhouse gases.

Wind turbines are used to convert wind flow, which is motion energy, into mechanical energy or power. In turn, the mechanical energy is converted into electricity. Wind turbines are similar to the propeller blades on an airplane because they turn or spin in the moving air. The turning blades are connected to a rotating shaft on an AC generator. The rotating shaft connects to a large coil of wire, which is forced to rotate through a strong magnetic field within the generator. The rotating coil of wire moving through the magnetic field generates electricity in the form of an alternating voltage and current (AC). (There are also wind turbines that generate DC voltages and currents but they are less common). Figure 1–21 shows a typical wind turbine used for generating large amounts of electricity. Wind turbines are often grouped together into a single wind power plant, also known as a **wind farm**. These wind farms generate bulk electrical power. Electricity from these wind turbines is fed into the utility grid and distributed to customers, just as with conventional power plants. Wind turbines are available in a wide variety of different sizes and capacity (power) ratings. In the United States, the average capacity of a ***utility-scale*** wind turbine falls somewhere in the range of about 2 to 3 megawatts (MW). However, some high-end utility-scale wind turbines have capacities approaching 10 MW. On a much smaller scale, single wind turbines typically have a capacity of 50 kilowatts (kW) or less. These smaller wind turbines are used for powering homes and telecommunications equipment and to pump water, to name just a few of the more popular applications. Wind farms have relatively little impact on the environment compared to conventional power plants, but some concern does exist over the noise produced by the turbine blades and the negative visual impact on the landscape.

Figure 1–21 Wind turbine.

Image Source/Getty Images

Summary

- Electricity is present in all matter in the form of electrons and protons.
- The electron is the basic particle of negative charge, and the proton is the basic particle of positive charge.
- A conductor is a material in which electrons can move easily from one atom to the next. Current flows freely or easily in a conductor.
- An insulator is a material in which electrons tend to stay in their own orbit. Insulators prevent or block the flow of current. Another name for insulator is dielectric.
- The atomic number of an element gives the number of protons in the nucleus of the atom, balanced by an equal number of orbital electrons.
- Electron valence refers to the number of electrons in the outermost shell of an atom. Except for H and He, the goal of valence is eight for all atoms.
- Charges of opposite polarity attract, and charges of like polarity repel.
- One coulomb (C) of charge is a quantity of electricity corresponding to 6.25×10^{18} electrons or protons. The symbol for charge is Q.
- Potential difference or voltage is an electrical pressure or force that exists between two points. The unit of potential difference is the volt (V). $1\text{ V} = \frac{1\text{ J}}{1\text{ C}}$ In general, $V = \frac{W}{Q}$.
- Current is the rate of movement of electric charge. The symbol for current is I, and the basic unit of measure is the ampere (A). $1\text{ A} = \frac{1\text{ C}}{1\text{ s}}$ In general, $I = \frac{Q}{T}$.
- Resistance is the opposition to the flow of current. The symbol for resistance is R, and the basic unit of measure is the ohm (Ω).
- Conductance is the reciprocal of resistance. The symbol for conductance is G, and the basic unit of measure is the siemens (S).
- $R = 1/G$ and $G = 1/R$.
- An electric circuit is a closed path for current flow. A voltage must be connected across a circuit to produce current flow. In the external circuit outside the voltage source, electrons flow from the negative terminal toward the positive terminal.
- A motion of positive charges, in the opposite direction of electron flow, is considered conventional current.
- Voltage can exist without current, but current cannot exist without voltage.
- Direct current has just one direction because a DC voltage source has fixed polarity. Alternating current periodically reverses in direction as the AC voltage source periodically reverses in polarity.
- Table 1–6 summarizes the main features of electric circuits.
- A digital multimeter is used to measure the voltage, current, or resistance in a circuit.

Table 1–6 Electrical Characteristics

Characteristic	Symbol	Unit	Description
Electric Charge	Q or q*	Coulomb (C)	Quantity of electrons or protons; $Q = I \times T$
Current	I or i*	Ampere (A)	Charge in motion; $I = Q/T$
Voltage	V or v*,†	Volt (V)	Potential difference between two unlike charges; makes charge move to produce I
Resistance	R or r‡	Ohm (Ω)	Opposition that reduces amount of current; $R = 1/G$
Conductance	G or g‡	Siemens (S)	Reciprocal of R, or $G = 1/R$

* Small letter q, i, or v is used for an instantaneous value of a varying charge, current, or voltage.
† E or e is sometimes used for a generated emf, but the standard symbol for any potential difference is V or v in the international system of units (SI).
‡ Small letter r or g is used for internal resistance or conductance of transistors.

Important Terms

Alternating current (AC) — a current that periodically reverses in direction as the alternating voltage periodically reverses in polarity.

Ampere — the basic unit of current. $1\text{ A} = \frac{1\text{ C}}{1\text{ s}}$

Atom — the smallest particle of an element that still has the same characteristics as the element.

Atomic number — the number of protons, balanced by an equal number of electrons, in an atom.

Circuit — a path for current flow.

Compound — a combination of two or more elements.

Conductance — the reciprocal of resistance.

Conductor — any material that allows the free movement of electric charges, such as electrons, to provide an electric current.

Conventional current — the direction of current flow associated with positive charges in motion. The current flow direction is from a positive to a negative potential, which is in the opposite direction of electron flow.

Coulomb — the basic unit of electric charge. 1 C = 6.25×10^{18} electrons or protons.

Current — a movement of electric charges around a closed path or circuit.

Dielectric — another name for insulator.

Direct current (DC) — a current flow that has just one direction.

Electron — the most basic particle of negative charge.

Electron flow — the movement of electrons that provides current in a circuit. The current flow direction is from a negative to a positive potential, which is in the opposite direction of conventional current.

Electron valence — the number of electrons in an incomplete outermost shell of an atom.

Element — a substance that cannot be decomposed any further by chemical action.

Free electron — an electron that can move freely from one atom to the next.

Insulator — a material with atoms in which the electrons tend to stay in their own orbits.

Ion — an atom that has either gained or lost one or more valence electrons to become electrically charged.

Molecule — the smallest unit of a compound with the same chemical characteristics.

Neutron — a particle contained in the nucleus of an atom that is electrically neutral.

Nucleus — the massive, stable part of the atom that contains both protons and neutrons.

Ohm — the unit of resistance.

Potential difference — a property associated with two unlike charges in close proximity to each other.

Proton — the most basic particle of positive charge.

Resistance — the opposition to the flow of current in an electric circuit.

Semiconductor — a material that is neither a good conductor nor a good insulator.

Siemens — the unit of conductance.

Static electricity — any charge, positive or negative, that is stationary or not in motion.

Volt — the unit of potential difference or voltage. $1\text{ V} = \frac{1\text{ J}}{1\text{ C}}$.

Related Formulas

$1\text{ C} = 6.25 \times 10^{18}$ electrons

$V = \frac{W}{Q}$

$I = Q/T$

$Q = I \times T$

$R = 1/G$

$G = 1/R$

Self-Test

Answers at the back of the book.

1. The most basic particle of negative charge is the

a. coulomb.

b. electron.

c. proton.

d. neutron.

2. The coulomb is a unit of

a. electric charge.

b. potential difference.

c. current.

d. voltage.

3. Which of the following is not a good conductor?

a. copper.

b. silver.

c. glass.

d. gold.

4. The electron valence of a neutral copper atom is

a. +1.

b. 0.

c. ±4.

d. −1.

5. The unit of potential difference is the

a. volt.

b. ampere.

c. siemens.

d. coulomb.

6. Which of the following statements is true?

a. Unlike charges repel each other.

b. Like charges repel each other.

c. Unlike charges attract each other.

d. Both b and c.

7. In a metal conductor, such as a copper wire,

a. positive ions are the moving charges that provide current.

b. free electrons are the moving charges that provide current.

c. there are no free electrons.

d. none of the above.

8. A 100-Ω resistor has a conductance, *G*, of

a. 0.01 S.

b. 0.1 S.

c. 0.001 S.

d. 1 S.

9. The most basic particle of positive charge is the

a. coulomb.

b. electron.

c. proton.

d. neutron.

10. If a neutral atom loses one of its valence electrons, it becomes a(n)

a. negative ion.
b. electrically charged atom.
c. positive ion.
d. both b and c.

11. The unit of electric current is the

a. volt.
b. ampere.
c. coulomb.
d. siemens.

12. A semiconductor, such as silicon, has an electron valence of

a. ±4.
b. +1.
c. −7.
d. 0.

13. Which of the following statements is true?

a. Current can exist without voltage.
b. Voltage can exist without current.
c. Current can flow through an open circuit.
d. Both b and c.

14. The unit of resistance is the

a. volt.
b. coulomb.
c. siemens.
d. ohm.

15. Except for hydrogen (H) and helium (He) the goal of valence for an atom is

a. 6.
b. 1.
c. 8.
d. 4.

16. One ampere of current corresponds to

a. $\frac{1\text{ C}}{1\text{ s}}$.
b. $\frac{1\text{ J}}{1\text{ C}}$.
c. 6.25×10^{18} electrons.
d. 0.16×10^{-18} C/s.

17. Conventional current is considered

a. the motion of negative charges in the opposite direction of electron flow.
b. the motion of positive charges in the same direction as electron flow.
c. the motion of positive charges in the opposite direction of electron flow.
d. none of the above.

18. When using a DMM to measure the value of a resistor

a. make sure that the resistor is in a circuit where voltage is present.
b. make sure there is no voltage present across the resistor.
c. make sure there is no other component connected across the leads of the resistor.
d. both b and c.

19. In a circuit, the opposition to the flow of current is called

a. conductance.
b. resistance.
c. voltage.
d. current.

20. Aluminum, with an atomic number of 13, has

a. 13 valence electrons.
b. 3 valence electrons.
c. 13 protons in its nucleus.
d. both b and c.

21. The nucleus of an atom is made up of

a. electrons and neutrons.
b. ions.
c. neutrons and protons.
d. electrons only.

22. How much charge is accumulated in a dielectric that is charged by a 4-A current for 5 seconds?

a. 16 C.
b. 20 C.
c. 1.25 C.
d. 0.8 C.

23. A charge of 6 C moves past a given point every 0.25 second. How much is the current flow in amperes?

a. 24 A.
b. 2.4 A.
c. 1.5 A.
d. 12 A.

24. What is the output voltage of a battery that expends 12 J of energy in moving 1.5 C of charge?

a. 18 V.
b. 6 V.
c. 125 mV.
d. 8 V.

25. Which of the following statements is false?

a. The resistance of an open circuit is practically zero.
b. The resistance of a short circuit is practically zero.
c. The resistance of an open circuit is infinitely high.
d. There is no current in an open circuit.

Essay Questions

1. Name two good conductors, two good insulators, and two semiconductors.
2. In a metal conductor, what is a free electron?
3. What is the smallest unit of a compound with the same chemical characteristics?
4. Define the term ion.
5. How does the resistance of a conductor compare to that of an insulator?
6. Explain why potential difference is necessary to produce current in a circuit.
7. List three important characteristics of an electric circuit.
8. Describe the difference between an open circuit and a short circuit.
9. Is the power line voltage available in our homes a DC or an AC voltage?
10. What is the mathematical relationship between resistance and conductance?
11. Briefly describe the electric field of a static charge.

12. List at least two examples that show how static electricity can be generated.

13. What is another name for an insulator?

14. List the particles in the nucleus of an atom.

15. Explain the difference between electron flow and conventional current.

16. Define −3 C of charge and compare it to a charge of +3 C.

17. Why protons are not considered a source of moving charges for current flow?

18. Write the formulas for each of the following statements: (a) current is the time rate of change of charge. (b) charge is current accumulated over a period of time.

19. Briefly define each of the following: (a) 1 coulomb (b) 1 volt (c) 1 ampere (d) 1 ohm.

20. Describe the difference between direct and alternating current.

Problems

SECTION 1–4 THE COULOMB UNIT OF ELECTRIC CHARGE

1–1 If 31.25×10^{18} electrons are removed from a neutral dielectric, how much charge is stored in coulombs?

1–2 If 18.75×10^{18} electrons are added to a neutral dielectric, how much charge is stored in coulombs?

1–3 A dielectric with a positive charge of + 5 C has 18.75×10^{18} electrons added to it. What is the net charge of the dielectric in coulombs?

1–4 If 93.75×10^{18} electrons are removed from a neutral dielectric, how much charge is stored in coulombs?

1–5 If 37.5×10^{18} electrons are added to a neutral dielectric, how much charge is stored in coulombs?

SECTION 1–5 THE VOLT UNIT OF POTENTIAL DIFFERENCE

1–6 What is the output voltage of a battery if 10 J of energy is expended in moving 1.25 C of charge?

1–7 What is the output voltage of a battery if 6 J of energy is expended in moving 1 C of charge?

1–8 What is the output voltage of a battery if 12 J of energy is expended in moving 1 C of charge?

1–9 How much is the potential difference between two points if 0.5 J of energy is required to move 0.4 C of charge between the two points?

1–10 How much energy is expended, in joules, if a voltage of 12 V moves 1.25 C of charge between two points?

SECTION 1–6 CHARGE IN MOTION IS CURRENT

1–11 A charge of 2 C moves past a given point every 0.5 s. How much is the current?

1–12 A charge of 1 C moves past a given point every 0.1 s. How much is the current?

1–13 A charge of 0.05 C moves past a given point every 0.1 s. How much is the current?

1–14 A charge of 6 C moves past a given point every 0.3 s. How much is the current?

1–15 A charge of 0.1 C moves past a given point every 0.01 s. How much is the current?

1–16 If a current of 1.5 A charges a dielectric for 5 s, how much charge is stored in the dielectric?

1–17 If a current of 500 mA charges a dielectric for 2 s, how much charge is stored in the dielectric?

1–18 If a current of 200 μA charges a dielectric for 20 s, how much charge is stored in the dielectric?

SECTION 1–7 RESISTANCE IS OPPOSITION TO CURRENT

1–19 Calculate the resistance value in ohms for the following conductance values: (a) 0.001 S (b) 0.01 S (c) 0.1 S (d) 1 S.

1–20 Calculate the resistance value in ohms for the following conductance values: (a) 0.002 S (b) 0.004 S (c) 0.00833 S (d) 0.25 S.

1–21 Calculate the conductance value in siemens for each of the following resistance values: (a) 200 Ω (b) 100 Ω (c) 50 Ω (d) 25 Ω.

1–22 Calculate the conductance value in siemens for each of the following resistance values: (a) 1 Ω (b) 10 kΩ (c) 40 Ω (d) 0.5 Ω.

Critical Thinking

1–23 Suppose that 1000 electrons are removed from a neutral dielectric. How much charge, in coulombs, is stored in the dielectric?

1–24 How long will it take an insulator that has a charge of + 5 C to charge to + 30 C if the charging current is 2 A?

1–25 Assume that 6.25×10^{15} electrons flow past a given point in a conductor every 10 s. Calculate the current I in amperes.

1–26 The conductance of a wire at 100°C is one-tenth its value at 25°C. If the wire resistance equals 10 Ω at 25°C calculate the resistance of the wire at 100°C.

Answers to Self-Reviews

1–1 **a.** negative **b.** positive **c.** true

1–2 **a.** conductors **b.** silver **c.** silicon

1–3 **a.** 14 **b.** 1 **c.** 8

1–4 **a.** 6.25×10^{18} **b.** $-Q = 3$ C **c.** attract

1–5 **a.** zero **b.** 9 V

1–6 **a.** 2 A **b.** true **c.** zero

1–7 **a.** carbon **b.** 4.7 Ω **c.** $^1/_{10}$ S or 0.1 S

1–8 **a.** true **b.** false **c.** true

1–9 **a.** negative **b.** negative **c.** true

1–10 **a.** true **b.** true

1–11 **a.** electrons **b.** magnetic field **c.** spark

1–12 **a.** true **b.** true **c.** false

Laboratory Application Assignment

In your first lab application assignment you will use a DMM to measure the voltage, current, and resistance in Fig. 1–22. Refer to Section 1–12, "The Digital Multimeter," if necessary.

Equipment: Obtain the following items from your instructor.

- Variable dc power supply
- 1-k**Ω**, ½-W resistor
- DMM
- Connecting leads

Measuring Voltage

Set the DMM to measure DC voltage. Be sure the meter leads are inserted into the correct jacks (red lead in the V**Ω** jack and the black lead in the COM jack). Also, be sure the voltmeter range exceeds the voltage being measured. Connect the DMM test leads to the variable DC power supply as shown in Fig. 1–22*a*. Adjust the variable DC power supply voltage to any value between 5 and 15 V. Record your measured voltage. $V =$ __________ Note: Keep the power supply voltage set to this value when measuring the current in Fig. 1–22*c*.

Measuring Resistance

Disconnect the meter leads from the power supply terminals. Set the DMM to measure resistance. Keep the meter leads in the same jacks you used for measuring voltage. Connect the DMM test leads to the leads of the 1 k**Ω** resistor, as shown in Fig. 1–22*b*. Record your measured resistance. $R =$ __________ (The measured resistance will most likely be displayed as a decimal fraction in k**Ω**.)

Measuring Current

Set the DMM to measure DC current. Also, move the red test lead to the appropriate jack for measuring small DC currents (usually labeled mA). Turn off the variable DC power supply. Connect the red test lead of the DMM to the positive (+) terminal of the variable DC power supply as shown in Fig. 1–22*c*. Also, connect the black test lead of the DMM to one lead of the 1 k**Ω** resistor as shown. Finally, connect the other lead of the resistor to the negative (−) terminal of the variable DC power supply. Turn on the variable DC power supply. Record your measured current.
$I =$ __________

Figure 1–22 Measuring electrical quantities. (*a*) Measuring voltage. (*b*) Measuring resistance. (*c*) Measuring current.

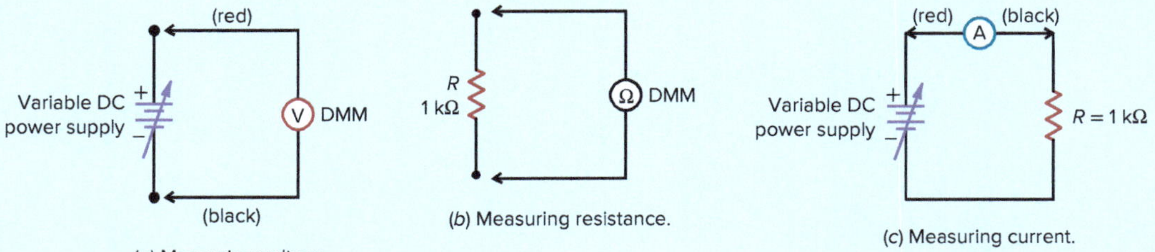

(*a*) Measuring voltage. (*b*) Measuring resistance. (*c*) Measuring current.

Design credits: Chapter Opener Background Image, and Multisim: Stockbyte/Getty Images

chapter 2

Resistors

Resistors are used in a wide variety of applications in all types of electronic circuits. Their main function in any circuit is to limit the amount of current or to produce a desired drop in voltage. Resistors are manufactured in a wide variety of shapes and sizes and have ohmic values ranging from a fraction of an ohm to several megohms. Resistors can be either fixed or variable. The power or wattage rating of a resistor is determined mainly by its physical size. There is, however, no direct correlation between the physical size of a resistor and its resistance value.

In this chapter, you will be presented with an in-depth discussion of the following resistor topics: resistor types, resistor color coding, potentiometers and rheostats, power ratings, and resistor troubles.

Chapter Outline

Chapter Objectives

After studying this chapter, you should be able to

- *List* several different types of resistors and describe the characteristics of each type.
- *Interpret* the resistor color code to determine the resistance and tolerance of a resistor.
- *Explain* the difference between a potentiometer and a rheostat.
- *Explain* the significance of a resistor's power rating.
- *List* the most common troubles with resistors.
- *Explain* the precautions that must be observed when measuring resistance with an ohmmeter.

Important Terms

carbon-composition resistor
carbon-film resistor
color coding
decade resistance box
derating curve
metal-film resistor
negative temperature coefficient (NTC)
positive temperature coefficient (PTC)
potentiometer
rheostat
surface-mount resistor
taper
thermistor
tolerance
wire-wound resistor
zero-ohm resistor
zero-power resistance

Figure 2–1 Carbon-composition resistor. (*a*) Internal construction. Length is about ¼ in. without leads for ¼-W power rating. Color stripes give *R* in ohms. Tinned leads have coating of solder. (*b*) Resistors mounted on printed-circuit (PC) board.

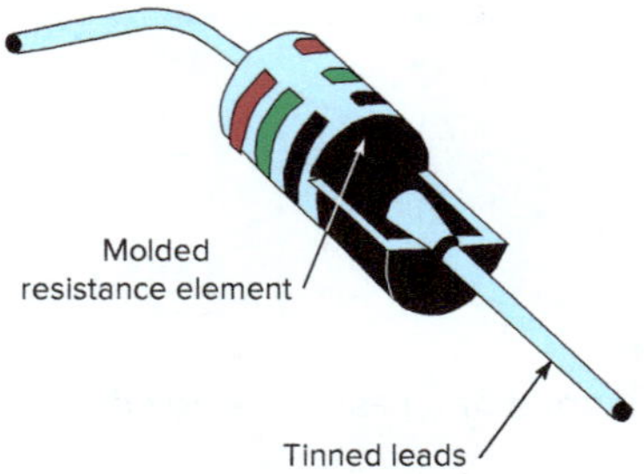

(*a*)

Igor Batenev/Shutterstock

(*b*)

2–1 Types of Resistors

The two main characteristics of a resistor are its resistance *R* in ohms and its power rating in watts (*W*). Resistors are available in a very wide range of *R* values, from a fraction of an ohm to many kilohms (kΩ) and megohms (MΩ). One kilohm is 1000 Ω and one megohm is 1,000,000 Ω. The power rating for resistors may be as high as several hundred watts or as low as 1⁄10 W.

The *R* is the resistance value required to provide the desired current or voltage. Also important is the wattage rating because it specifies the maximum power the resistor can dissipate without excessive heat. *Dissipation* means that the power is wasted, since the resultant heat is not used. Too much heat can make the resistor burn. The wattage rating of the resistor is generally more than the actual power dissipation, as a safety factor.

Most common in electronic equipment are carbon resistors with a power rating of 1 W or less. The construction is illustrated in Fig. 2–1*a*. The leads extending out from the resistor body can be inserted through the holes on a printed-circuit (PC) board for mounting as shown in Fig. 2–1*b*. The resistors on a PC board are often inserted automatically by machine. Note that resistors are not polarity-sensitive devices. This means that it does not matter which way the leads of a resistor are connected in a circuit.

Resistors with higher *R* values usually have lower wattage ratings because they have less current. As an example, a common value is 1 MΩ at ¼ W, for a resistor only ¼ in. long. The lower the power rating, the smaller the actual size of the resistor. However, the resistance value is not related to physical size. Figure 2–2 shows several carbon resistors with the same physical size but different resistance values. The different color bands on each resistor indicate a different ohmic value. The carbon resistors in Fig. 2–2 each have a power rating of ½ W, which is based on their physical size.

Wire-Wound Resistors

In this construction, a special type of wire called *resistance wire* is wrapped around an insulating core. The length of wire and its specific resistivity determine the *R* of the unit. Types of resistance wire include tungsten and manganin, as explained in

Figure 2–2 Carbon resistors with same physical size but different resistance values. The physical size indicates a power rating of ½ W.

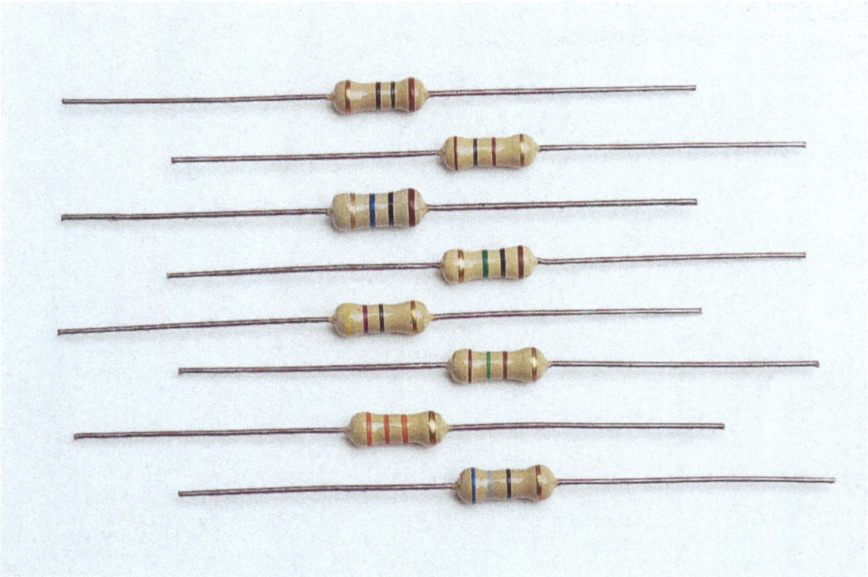

Mark Steinmetz/McGraw Hill

Figure 2–3 Large wire-wound resistors with 50-W power rating. (*a*) Fixed *R*, length of 5 in. (*b*) Variable *R*, diameter of 3 in.

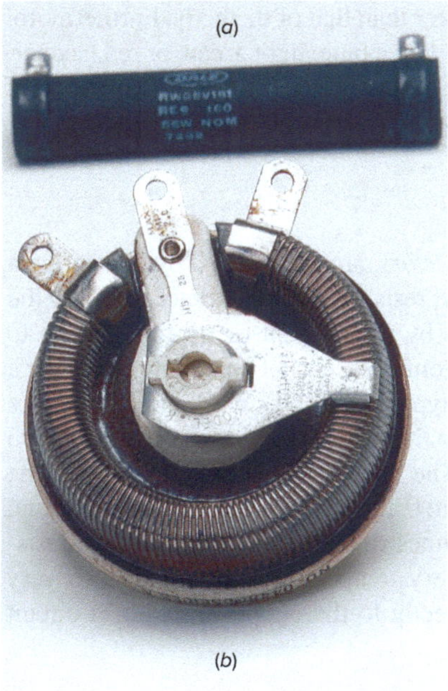

Cindy Schroeder/McGraw Hill

Chapter 11, "Conductors and Insulators." The insulated core is commonly porcelain, cement, or just plain pressed paper. Bare wire is used, but the entire unit is generally encased in an insulating material. Typical fixed and variable **wire-wound resistors** are shown in Fig. 2–3.

Since they are generally used for high-current applications with low resistance and appreciable power, *wire-wound resistors* are available in wattage ratings from 1 W up to 100 W or more. The resistance can be less than 1 Ω up to several thousand ohms. For 2 W or less, carbon resistors are preferable because they are generally smaller and cost less.

In addition, wire-wound resistors are used where accurate, stable resistance values are necessary. Examples are precision resistors for the function of an ammeter shunt or a precision **potentiometer** to adjust for an exact amount of *R*.

Carbon-Composition Resistors

These resistors are made of finely divided carbon or graphite mixed with a powdered insulating material as a binder in the proportions needed for the desired *R* value. As shown in Fig. 2–1*a*, the resistor element is enclosed in a plastic case for insulation and mechanical strength. Joined to the two ends of the carbon resistance element are metal caps with leads of tinned copper wire for soldering the connections into a circuit. These are called *axial leads* because they come straight out from the ends. **Carbon-composition resistors** normally have a brown body and are cylindrical.

Carbon-composition resistors are commonly available in *R* values of 1 Ω to 20 MΩ. Examples are 10 Ω, 220 Ω, 4.7 kΩ, and 68 kΩ. The power rating is generally 1⁄10, 1⁄8, 1⁄4, 1⁄2, 1, or 2 W.

Film-Type Resistors

There are two kinds of film-type resistors: *carbon-film* and ***metal-film resistors.*** The **carbon-film resistor,** whose construction is shown in Fig. 2–4, is made by depositing a thin layer of carbon on an insulated substrate. The carbon film is then cut in the form of a spiral to form the resistive element. The resistance value is controlled by varying the proportion of carbon to insulator. Compared to carbon-composition resistors, carbon-film resistors have the following advantages: tighter **tolerances,** less sensitivity to temperature changes and aging, and they generate less noise internally.

Metal-film resistors are constructed in a manner similar to the carbon-film type. However, in a metal-film resistor, a thin film of metal is sprayed onto a ceramic substrate and then cut in the form of a spiral. The construction of a metal-film resistor is shown in Fig. 2–5. The length, thickness, and width of the metal spiral determine the exact resistance value. Metal-film resistors offer more precise *R* values

GOOD TO KNOW

Resistor networks are composed of film-type resistors housed in either dual in-line packages (DIP) or single in-line packages (SIP). In each resistor network, the resistors are uniquely configured to meet the specific needs of the application. They are used in a variety of different applications but are especially common in digital electronic circuits.

Figure 2–4 Construction of carbon-film resistor.

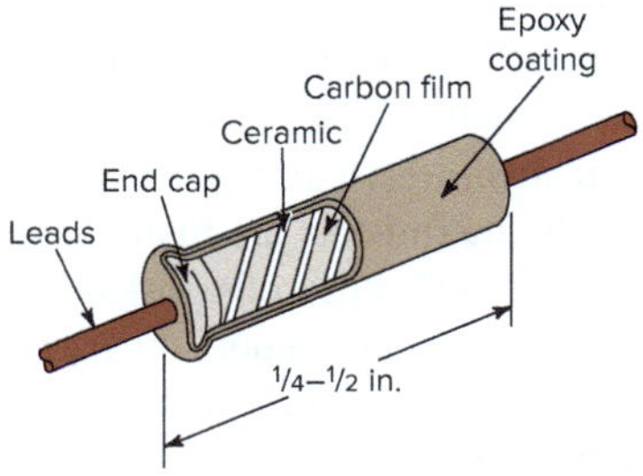

Figure 2–5 Construction of metal-film resistor.

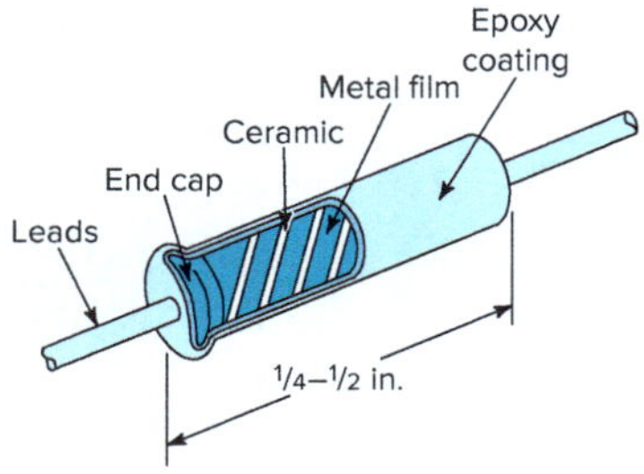

than carbon-film resistors. Like carbon-film resistors, metal-film resistors are affected very little by temperature changes and aging. They also generate very little noise internally. In overall performance, metal-film resistors are the best, carbon-film resistors are next, and carbon-composition resistors are last. Both carbon- and metal-film resistors can be distinguished from carbon-composition resistors by the fact that the diameter of the ends is a little larger than that of the body. Furthermore, metal-film resistors are almost always coated with a blue, light green, or red lacquer, which provides electrical, mechanical, and climate protection. The body color of carbon-film resistors is usually tan.

Figure 2-6 Typical chip resistors.

Mark Steinmetz/McGraw Hill

Surface-Mount Resistors

Surface-mount resistors, also called *chip resistors,* are constructed by depositing a thick carbon film on a ceramic base. The exact resistance value is determined by the composition of the carbon itself, as well as by the amount of trimming done to the carbon deposit. The resistance can vary from a fraction of an ohm to well over a million ohms. Power dissipation ratings are typically ⅛ to ¼ W. Figure 2–6 shows typical chip resistors. Electrical connection to the resistive element is made via two leadless solder end electrodes (terminals). The end electrodes are C-shaped. The physical dimensions of a ⅛-W chip resistor are 0.125 in. long by 0.063 in. wide and approximately 0.028 in. thick. This is many times smaller than a conventional resistor having axial leads. Chip resistors are very temperature-stable and also very rugged. The end electrodes are soldered directly to the copper traces of a circuit board, hence the name *surface-mount.*

GOOD TO KNOW

Another type of resistor is the varistor. Varistors are voltage-dependant resistors. This means that their resistance is dependant on the voltage across them.

Fusible Resistors

This type is a wire-wound resistor made to burn open easily when the power rating is exceeded. It then serves the dual functions of a fuse and a resistor to limit the current.

Thermistors

A ***thermistor*** is a thermally sensitive resistor whose resistance value changes with changes in its operating temperature. Because of the self-heating effect of current in a thermistor, the device changes resistance with changes in current. Thermistors, which are essentially semiconductors, exhibit either a ***positive temperature coefficient (PTC)*** or a ***negative temperature coefficient (NTC).*** If a thermistor has a PTC, its resistance increases as the operating temperature increases. Conversely, if a thermistor has an NTC, its resistance decreases as its operating temperature increases. How much the resistance changes with changes in operating temperature depends on the size and construction of the thermistor. Note that the resistance does not undergo instantaneous changes with changes in current or ambient temperature. A certain time interval, determined by the thermal mass (size) of the thermistor, is required for the resistance change. A thermistor with a small mass will change more rapidly than one with a large mass. Carbon- and metal-film resistors are different: their resistance does not change appreciably with changes in operating temperature.

Figure 2-7 (*a*) Thermistor schematic symbol. (*b*) Typical thermistor shapes and sizes.

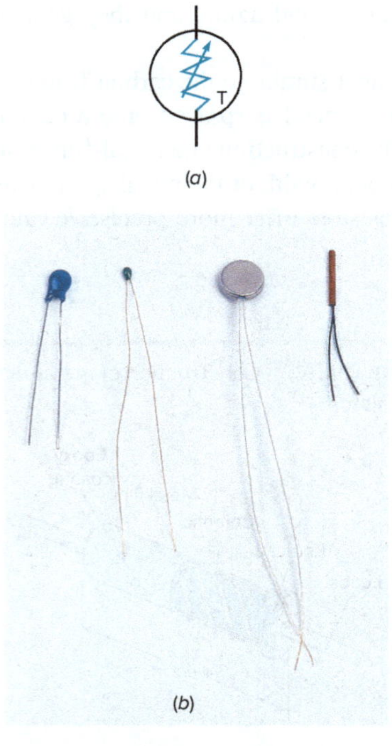

Cindy Schroeder/McGraw Hill

Figure 2–7*a* shows the standard schematic symbol for a thermistor. Notice the arrow through the resistor symbol and the letter *T* within the circle. The arrow indicates that the resistance is variable as the temperature *T* changes. As shown in Fig. 2–7*b*, thermistors are manufactured in a wide variety of shapes and sizes. The shapes include beads, rods, disks, and washers.

Thermistors are frequently used in electronic circuits in which it is desired to provide temperature measurement, temperature control, and temperature compensation.

2–1 Self-Review

Answers at the end of the chapter.

a. **An *R* of 10 Ω with a 25-W rating would most likely be a wire-wound resistor. (True/False)**

b. **A resistance of 10,000 Ω is the same as a resistance of 10 kΩ. (True/False)**

c. **Which is more temperature stable, a carbon-composition or a metal-film resistor?**

d. **Which is larger, a 1000-Ω, ½-W or a 1000-Ω, 1-W carbon-film resistor?**

e. **What happens to the resistance of an NTC thermistor when its operating temperature increases?**

GOOD TO KNOW

Because color-coded resistors are encountered so frequently in electronic circuits, it is highly recommended that you memorize the resistor color code. It may be helpful to develop a jingle (using mnemonics) to help memorize the color code. You may want to ask your instructor to share some of the color-code jingles that have been passed down over the years.

2–2 Resistor Color Coding

Because carbon resistors are small, they are *color-coded* to mark their *R* value in ohms. The basis of this system is the use of colors for numerical values, as listed in Table 2–1. In memorizing the colors, note that the darkest colors, black and brown, are for the lowest numbers, zero and one, whereas white is for nine. The **color coding** is standardized by the Electronic Industries Alliance (EIA).

Resistance Color Stripes

The use of colored bands or stripes is the most common system for color-coding resistors, as shown in Fig. 2–8. The colored bands or stripes completely encircle the body of the resistor and are usually crowded toward one end. Reading from the left to right, the first band closest to the edge gives the first digit in the numerical value of *R*. The next band indicates the second digit. The third band is the decimal multiplier, which tells us how many zeros to add after the first two digits.

Figure 2–8 How to read color stripes on carbon resistors for *R* in ohms.

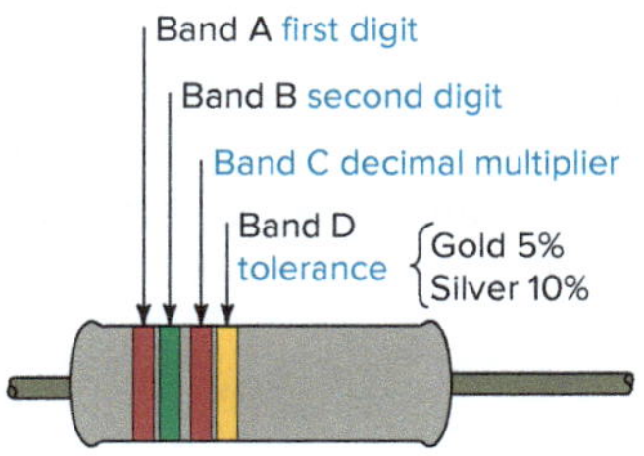

GOOD TO KNOW

The third colored band or stripe from the left is called the decimal multiplier. It is called the decimal multiplier because the first two digits are multiplied by 10^x, where x is the number or digit corresponding to the color used in the third band. For example, if the third color band is orange, the decimal multiplier is 10^3 or 1000.

Table 2–1 Color Code

Color	Numerical Value
Black	0
Brown	1
Red	2
Orange	3
Yellow	4
Green	5
Blue	6
Violet	7
Gray	8
White	9

MultiSim **Figure 2–9** Examples of color-coded *R* values with percent tolerance.

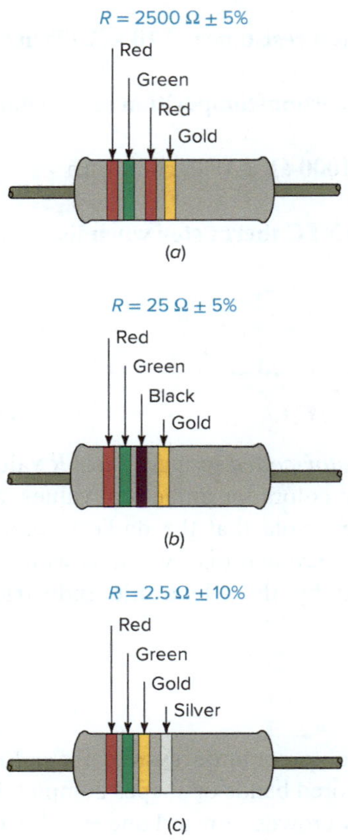

In Fig. 2–9*a*, the first stripe is red for 2 and the next stripe is green for 5. The red colored decimal multiplier in the third stripe means add two zeros to 25, or multiply the number 25 by 10^2. The result can be illustrated as follows:

Red	Green		Red	
↓	↓		↓	
2	5	×	100	= 2500

Therefore, this *R* value is 2500 Ω or 2.5 kΩ.

The example in Fig. 2–9*b* illustrates that black for the third stripe just means "do not add any zeros to the first two digits" or multiply 25 by 10. Since this resistor has red, green, and black stripes, the *R* value is 25 Ω.

Resistors under 10 Ω

For these values, the third stripe is either gold or silver, indicating a fractional decimal multiplier. When the third stripe is gold, multiply the first two digits by 0.1. In Fig. 2–9*c*, the *R* value is

$$25 \times 0.1 = 2.5\ \Omega.$$

Silver means a multiplier of 0.01. If the third band in Fig. 2–9*c* were silver, the *R* value would be

$$25 \times 0.01 = 0.25\ \Omega.$$

It is important to realize that the gold and silver colors represent fractional decimal multipliers only when they appear in the third stripe. Gold and silver are used most often however as a fourth stripe to indicate how accurate the *R* value is. The colors gold and silver will never appear in the first two color stripes.

Resistor Tolerance

The amount by which the actual *R* can differ from the color-coded value is the *tolerance,* usually given in percent. For instance, a 2000-Ω resistor with ±10% tolerance can have resistance 10% above or below the coded value. This *R*, therefore, is between 1800 and 2200 Ω. The calculations are as follows:

$$10\% \text{ of } 2000 \text{ is } 0.1 \times 2000 = 200.$$

For +10%, the value is

$$2000 + 200 = 2200\ \Omega.$$

For −10%, the value is

$$2000 - 200 = 1800\ \Omega.$$

As illustrated in Fig. 2–8, silver in the fourth band indicates a tolerance of ±10%, gold indicates ±5%. If there is no color band for tolerance, it is ±20%. One of the main disadvantages of carbon-composition resistors is that they tend to drift out of tolerance over time. This is due to the fact that the compressed particles of carbon relax over time, thus causing an increase in resistance. Also, as compared to other resistor types, carbon-composition resistors have sloppier tolerances. For these reasons, carbon-composition resistors are seldom used anymore. However, as an electronics technician, you may encounter carbon-composition resistors when working on older equipment.

Five-Band Color Code

Precision resistors (typically metal-film resistors) often use a five-band color code rather than the four-band code, as shown in Fig. 2–8. The purpose is to obtain more precise R values. With the five-band code, the first three color stripes indicate the first three digits, followed by the decimal multiplier in the fourth stripe and the tolerance in the fifth stripe. In the fifth stripe, the colors brown, red, green, blue, and violet represent the following tolerances:

Brown	±1%
Red	±2%
Green	±0.5%
Blue	±0.25%
Violet	±0.1%

Example 2–1

What is the resistance indicated by the five-band color code in Fig. 2–10? Also, what ohmic range is permissible for the specified tolerance?

ANSWER The first stripe is orange for the number 3, the second stripe is blue for the number 6, and the third stripe is green for the number 5. Therefore, the first three digits of the resistance are 3, 6, and 5, respectively. The fourth stripe, which is the multiplier, is black, which means add no zeros. The fifth stripe, which indicates the resistor tolerance, is green for ±0.5%. Therefore, $R = 365\ \Omega \pm 0.5\%$. The permissible ohmic range is calculated as $365 \times 0.005 = \pm 1.825\ \Omega$, or 363.175 to 366.825 Ω.

MultiSim **Figure 2–10** Five-band code.

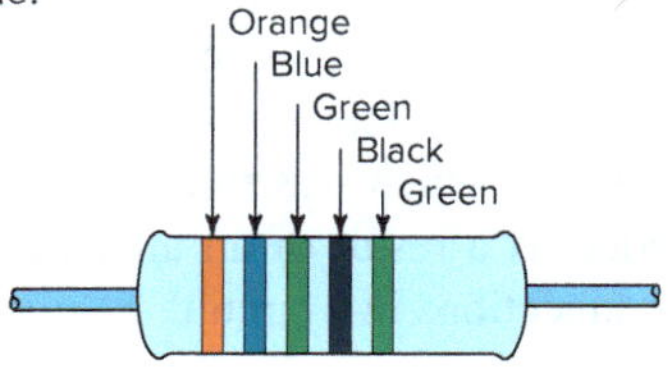

Wire-Wound-Resistor Marking

Usually, wire-wound resistors are big enough to have the R value printed on the insulating case. The tolerance is generally ±5% except for precision resistors, which have a tolerance of ±1% or less.

Some small wire-wound resistors may be color-coded with stripes, however, like carbon resistors. In this case, the first stripe is double the width of the others to indicate a wire-wound resistor. Wire-wound resistors that are color-coded generally have a power rating of 4 W or less.

Preferred Resistance Values

To minimize the problem of manufacturing different R values for an almost unlimited variety of circuits, specific values are made in large quantities so that they are cheaper and more easily available than unusual sizes. For resistors of ±10% the *preferred values* are 10, 12, 15, 18, 22, 27, 33, 39, 47, 56, 68, and 82 with their decimal multiples. As examples, 47, 470, 4700, and 47,000 are preferred values. In this way, there is a preferred value available within 10% of any R value needed in a circuit. See Appendix C for a listing of preferred resistance values for tolerances of ±20%, ±10%, and ±5%.

Figure 2–11 A zero-ohm resistor is indicated by a single black color band around the body of the resistor.

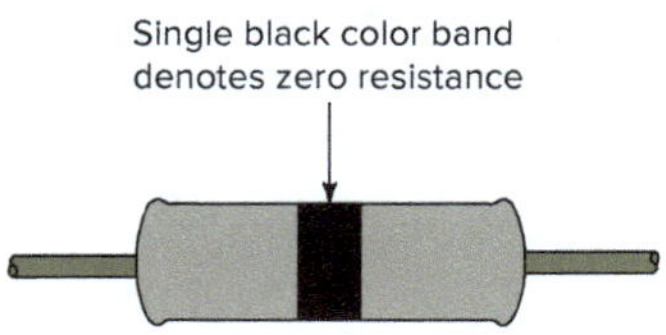

Zero-Ohm Resistors

Believe it or not, there is such a thing as a ***zero-ohm resistor.*** In fact, zero-ohm resistors are quite common. The zero-ohm value is denoted by the use of a single black band around the center of the resistor body, as shown in Fig. 2–11. Zero-ohm resistors are available in ⅛- or ¼-W sizes. The actual resistance of a so-called ⅛-W zero-ohm resistor is about 0.004 Ω, whereas a ¼-W zero-ohm resistor has a resistance of approximately 0.003 Ω.

Figure 2–12 Typical chip resistor coding system.

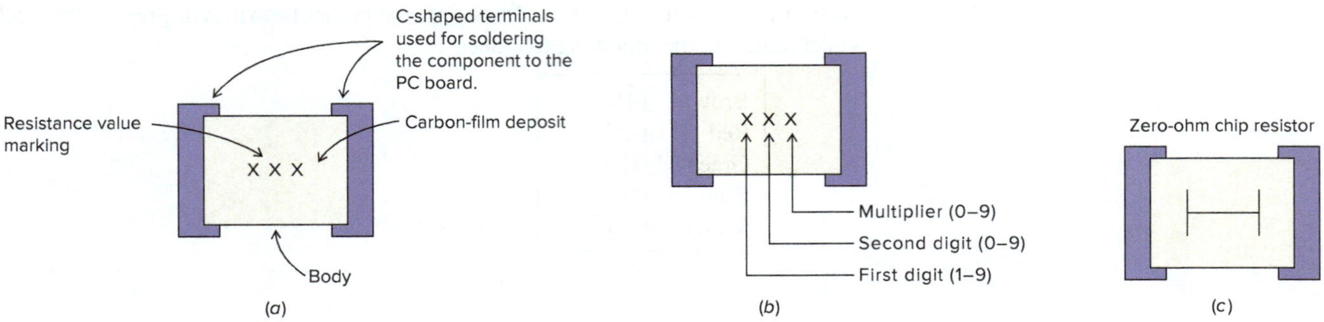

But why are zero-ohm resistors used in the first place? The reason is that for most printed-circuit boards, the components are inserted by automatic insertion machines (robots) rather than by human hands. In some instances, it may be necessary to short two points on the printed-circuit board, in which case a piece of wire has to be placed between the two points. Because the robot can handle only components such as resistors, and not wires, zero-ohm resistors are used. Before zero-ohm resistors were developed, jumpers had to be installed by hand, which was time-consuming and expensive. Zero-ohm resistors may be needed as a result of an after-the-fact design change that requires new point-to-point connections in a circuit.

Chip Resistor Coding System

The chip resistor, shown in Fig. 2–12*a*, has the following identifiable features:

1. A dark film on one side only (usually black, but may also be dark gray or green)
2. Two C-shaped terminals at each end of the resistor, used for soldering.
3. A three- or four-digit number on the dark film side of the resistor.

The resistance value of a chip resistor is determined from the three-digit number printed on the film or body side of the component. The three digits provide the same information as the first three color stripes on a four-band resistor. This is shown in Fig. 2–12*b*. The first two digits indicate the first two numbers in the numerical value of the resistance; the third digit indicates the multiplier. If a four-digit number is used, the first three digits indicate the first three numbers in the numerical value of the resistance, and the fourth digit indicates the multiplier. The letter *R* is used to signify a decimal point for values between 1 and 10 ohms as in $2R7 = 2.7\ \Omega$. Figure 2–12*c* shows the symbol used to denote a zero-ohm chip resistor. Chip resistors are typically available in tolerances of $\pm 1\%$ and $\pm 5\%$. It is important to note, however, that the tolerance of a chip resistor is not indicated by the three- or four-digit code.

GOOD TO KNOW

Even though chip resistors use a three- or four-digit code to indicate their resistance value in ohms, the digits may be too small to read with the naked eye. In other words, it may be necessary to use a magnifying device to read the value on the chip resistor.

Example 2–2

Determine the resistance of the chip resistor in Fig. 2–13.

ANSWER The first two digits are 5 and 6, giving 56 as the first two numbers in the resistance value. The third digit, 2, is the multiplier, which means add 2 zeros to 56 for a resistance of 5600 Ω or 5.6 kΩ.

Figure 2–13 Chip resistor with number coding.

Figure 2-14 Construction of variable carbon resistance control. Diameter is ¾ in. (*a*) External view. (*b*) Internal view of circular resistance element.

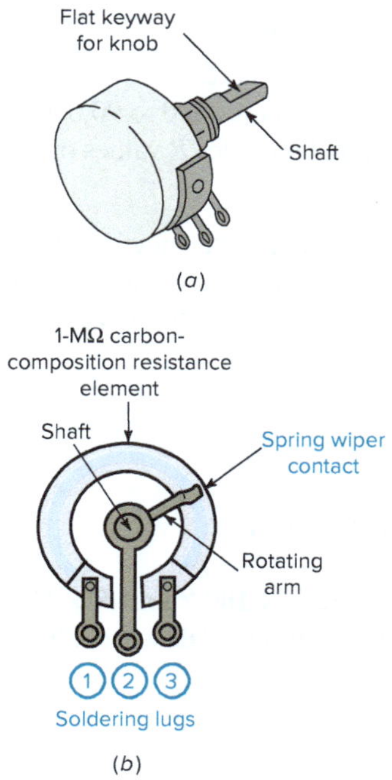

Thermistor Values

Thermistors are normally rated by the value of their resistance at a reference temperature *T* of 25°C. The value of *R* at 25°C is most often referred to as the **zero-power resistance** and is designated R_0. The term *zero-power resistance* refers to the resistance of the thermistor with zero-power dissipation. Thermistors normally do not have a code or marking system to indicate their resistance value in ohms. In rare cases, however, a three-dot code is used to indicate the value of R_0. In this case, the first and second dots indicate the first two significant digits, and the third dot is the multiplier. The colors used are the same as those for carbon resistors.

2–2 Self-Review

Answers at the end of the chapter.

a. **Give the color for the number 4.**
b. **What tolerance does a silver stripe represent?**
c. **Give the multiplier for red in the third stripe.**
d. **Give *R* and the tolerance for a resistor coded with yellow, violet, brown, and gold stripes.**
e. **Assume that the chip resistor in Fig. 2–13 is marked 333. What is its resistance value in ohms?**

2–3 Variable Resistors

Variable resistors can be wire-wound, as in Fig. 2–3*b*, or carbon type, illustrated in Fig. 2–14. Inside the metal case of Fig. 2–14*a*, the control has a circular disk, shown in Fig. 2–14*b*, that is the carbon-composition resistance element. It can be a thin coating on pressed paper or a molded carbon disk. Joined to the two ends are the external soldering-lug terminals 1 and 3. The middle terminal is connected to the variable arm that contacts the resistor element by a metal spring wiper. As the shaft of the control is turned, the variable arm moves the wiper to make contact at different points on the resistor element. The same idea applies to the slide control in Fig. 2–15, except that the resistor element is straight instead of circular.

When the contact moves closer to one end, the *R* decreases between this terminal and the variable arm. Between the two ends, however, *R* is not variable but always has the maximum resistance of the control.

Carbon controls are available with a total *R* from 1000 Ω to 5 MΩ, approximately. Their power rating is usually ½ to 2 W.

GOOD TO KNOW

The generic symbol for a variable resistor is:

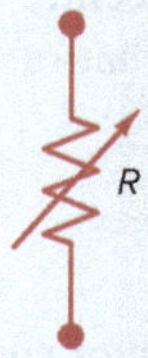

Tapered Controls

The way *R* varies with shaft rotation is called the ***taper*** of the control. With a linear taper, a one-half rotation changes *R* by one-half the maximum value. Similarly, all values of *R* change in direct proportion to rotation. For a nonlinear taper, though, *R* can change more gradually at one end with bigger changes at the opposite end. This effect is accomplished by different densities of carbon in the resistance element. For a volume control, its audio taper allows smaller changes in *R* at low settings. Then it is easier to make changes without having the volume too loud or too low.

Figure 2-15 Slide control for variable *R*. Length is 2 in.

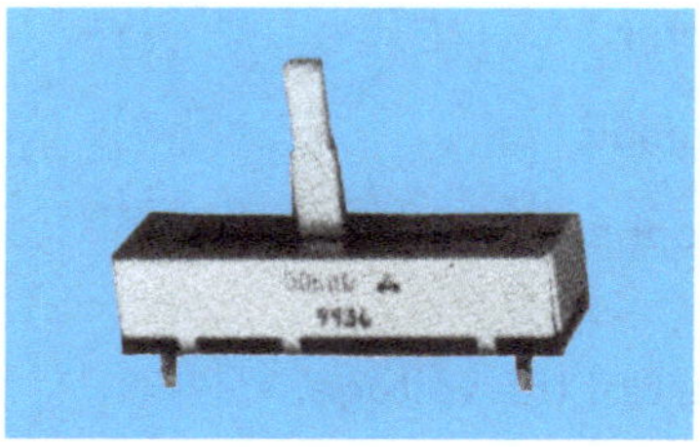

Mark Steinmetz/McGraw Hill

Decade Resistance Box

As shown in Fig. 2–16, the ***decade resistance box*** is a convenient unit for providing any one *R* within a wide range of values. It can be considered test equipment for

Figure 2–16 Decade resistance box for a wide range of *R* values.

Mark Steinmetz/McGraw Hill

trying different *R* values in a circuit. Inside the box are six series strings of resistors, with one string for each dial switch.

The first dial connects in an *R* of 0 to 9 Ω. It is the *units* or $R \times 1$ dial.
The second dial has units of 10 from 0 to 90 Ω. It is the *tens* or $R \times 10$ dial.
The hundreds or $R \times 100$ dial has an *R* of 0 to 900 Ω.
The thousands or $R \times 1$ k dial has an *R* of 0 to 9000 Ω.
The ten-thousands or $R \times 10$ k dial provides *R* values of 0 to 90,000 Ω.
The one-hundred-thousands or $R \times 100$ k dial provides *R* values of 0 to 900,000 Ω.

The six dial sections are connected internally so that their values add to one another. Then any value from 0 to 999,999 Ω can be obtained. Note the exact values that are possible. As an example, when all six dials are on 2, the total *R* equals $2 + 20 + 200 + 2000 + 20{,}000 + 200{,}000 = 222{,}222\ \Omega$.

2–3 Self-Review

Answers at the end of the chapter.

a. In Fig. 2–14, which terminal provides variable *R*?
b. Is an audio taper linear or nonlinear?
c. In Fig. 2–16, how much is the total *R* if the $R \times 100$ k and $R \times 10$ k dials are set to 4 and 7, respectively, and all other dials are set to zero?

2–4 Rheostats and Potentiometers

Rheostats and potentiometers are variable resistances, either carbon or wire-wound, used to vary the amount of current or voltage in a circuit. The controls can be used in either direct current (DC) or alternating current (AC) applications.

A *rheostat* is a variable *R* with two terminals connected in series with a load. The purpose is to vary the amount of current.

A *potentiometer,* generally called a *pot* for short, has three terminals. The fixed maximum *R* across the two ends is connected across a voltage source. Then the variable arm is used to vary the voltage division between the center terminal and the ends. This function of a potentiometer is compared with that of a rheostat in Table 2–2.

Rheostat Circuit

The function of the rheostat R_2 in Fig. 2–17 is to vary the amount of current through R_1. For instance, R_1 can be a small lightbulb that requires a specified value of current, *I*. Therefore, the two terminals of the rheostat R_2 are connected in series with R_1 and the voltage source to vary the total resistance R_T in the circuit. When R_T changes, the current *I* changes, as read by the meter.

GOOD TO KNOW

The resistance value of a potentiometer usually appears on the back side of its metal or plastic enclosure. In some cases, the tolerance is also indicated.

Table 2–2 Potentiometers and Rheostats

Rheostat	Potentiometer
Two terminals	Three terminals
In series with load and voltage source	Ends are connected across voltage source
Varies the current, *I*	Taps off part of the voltage, *V*

Figure 2–17 Rheostat connected in a series circuit to vary the current I. Symbol for current meter is A for amperes.

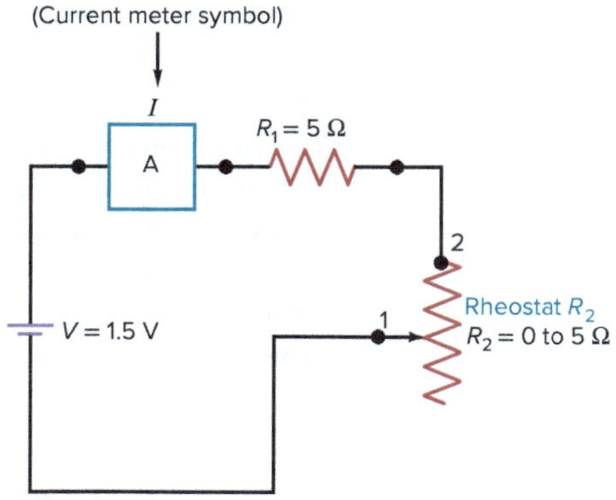

Figure 2–18 Potentiometer connected across voltage source to function as a voltage divider. (*a*) Wiring diagram. (*b*) Schematic diagram.

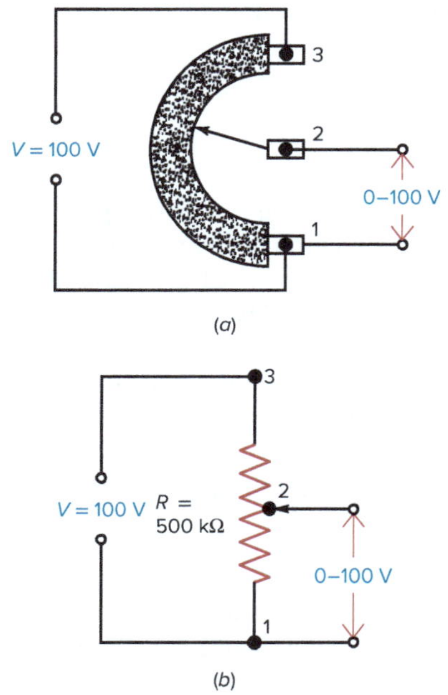

Figure 2–19 Small potentiometers and trimmers often used for variable controls in electronic circuits. Terminal leads are formed for insertion into a PC board.

In Fig. 2–17, R_1 is 5 Ω and the rheostat R_2 varies from 0 to 5 Ω. With R_2 at its maximum of 5 Ω, then the total resistance, R_T equals $5 + 5 = 10$ Ω.

When R_2 is at its minimum value of 0 Ω R_T equals 5 Ω. As a result, varying the rheostat changes the circuit resistance to vary the current through R_1. I increases as R decreases.

It is important that the rheostat have a wattage rating high enough for the maximum I when R is minimum. Rheostats are often wire-wound variable resistors used to control relatively large values of current in low-resistance circuits for AC power applications.

Potentiometer Circuit

The purpose of the circuit in Fig. 2–18 is to tap off a variable part of the 100 V from the source. Consider this circuit in two parts:

1. The applied voltage V is connected across the two end terminals of the potentiometer.
2. The variable voltage V is between the variable arm and an end terminal.

Two pairs of connections to the three terminals are necessary, with one terminal common to the input and output. One pair connects the source voltage V to the end terminals 1 and 3. The other pair of connections is between the variable arm at the center terminal and one end. This end has double connections for input and output. The other end has only an input connection.

When the variable arm is at the middle value of the 500-kΩ R in Fig. 2–18, the 50 V is tapped off between terminals 2 and 1 as one-half the 100-V input. The other 50 V is between terminals 2 and 3. However, this voltage is not used for the output.

As the control is turned up to move the variable arm closer to terminal 3, more of the input voltage is available between 2 and 1. With the control at its maximum R, the voltage between 2 and 1 is the entire 100 V. Actually, terminal 2 is then the same as 3.

When the variable arm is at minimum R, rotated to terminal 1, the output between 2 and 1 is zero. Now all the applied voltage is across 2 and 3 with no output for the variable arm. It is important to note that the source voltage is not short-circuited. The reason is that the maximum R of the potentiometer is always across the applied V, regardless of where the variable arm is set. Typical examples of small potentiometers used in electronic circuits are shown in Fig. 2–19.

Potentiometer Used as a Rheostat

Commercial rheostats are generally wire-wound, high-wattage resistors for power applications. However, a small, low-wattage rheostat is often needed in electronic circuits. One example is a continuous tone control in a receiver. The control requires the variable series resistance of a rheostat but dissipates very little power.

A method of wiring a potentiometer as a rheostat is to connect just one end of the control and the variable arm, using only two terminals. The third terminal is open, or floating, not connected to anything.

Another method is to wire the unused terminal to the center terminal. When the variable arm is rotated, different amounts of resistance are short-circuited. This method is preferable because there is no floating resistance.

Either end of the potentiometer can be used for the rheostat. The direction of increasing R with shaft rotation reverses, though, for connections at opposite ends. Also, the taper is reversed on a nonlinear control.

The resistance of a potentiometer is sometimes marked on the enclosure that houses the resistance element. The marked value indicates the resistance between the outside terminals.

2–4 Self-Review

Answers at the end of the chapter.

a. How many circuit connections to a potentiometer are needed?
b. How many circuit connections to a rheostat are needed?
c. In Fig. 2–18, with a 500-kΩ linear potentiometer, how much is the output voltage with 400 kΩ between terminals 1 and 2?

2–5 Power Rating of Resistors

In addition to having the required ohms value, a resistor should have a wattage rating high enough to dissipate the power produced by the current flowing through the resistance without becoming too hot. Carbon resistors in normal operation often become warm, but they should not get so hot that they "sweat" beads of liquid on the insulating case. Wire-wound resistors operate at very high temperatures; a typical value is 300°C for the maximum temperature. If a resistor becomes too hot because of excessive power dissipation, it can change appreciably in resistance value or burn open.

The power rating is a physical property that depends on the resistor construction, especially physical size. Note the following:

1. A larger physical size indicates a higher power rating.
2. Higher wattage resistors can operate at higher temperatures.
3. Wire-wound resistors are larger and have higher wattage ratings than carbon resistors.

For approximate sizes, a 2-W carbon resistor is about 1 in. long with a ¼-in. diameter; a ¼-W resistor is about 0.25 in. long with a diameter of 0.1 in.

For both types, a higher power rating allows a higher voltage rating. This rating gives the highest voltage that may be applied across the resistor without internal arcing. As examples for carbon resistors, the maximum voltage is 500 V for a 1-W rating, 350 V for ½-W, 250 V for ¼-W, and 150 V for ⅛-W. In wire-wound resistors, excessive voltage can produce an arc between turns; in carbon-composition resistors, the arc is between carbon granules.

GOOD TO KNOW

Many electronic products available today have built-in cooling fans to help reduce the buildup of heat inside of the equipment cabinet. This keeps components operating at cooler temperatures thus extending their life expectancy.

Power Derating Curve

When a carbon resistor is mounted on a PC board close to other resistors and components, all of which are producing heat and enclosed in a confined space, the ambient temperature can rise appreciably above 25°C. When carbon resistors are operated at ambient temperatures of 70°C or less, the commercial power rating, indicated by the physical size, remains valid. However, for ambient temperatures greater than 70°C, the power rating must be reduced or derated. This is shown in Fig. 2–20.

Figure 2-20 Resistor power derating curve.

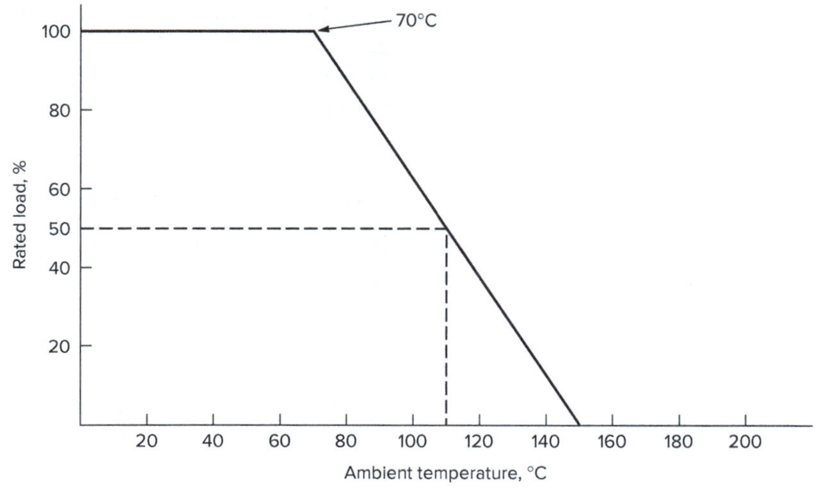

Notice that for ambient temperatures up to 70°C, the commercial power rating is the same (100%) as that determined by the resistor's physical size. Note, however, that above 70°C, the power rating decreases linearly.

For example, at an ambient temperature of 110°C, the power rating must be reduced to 50% of its rated value. This means that a 1-kΩ, ½-W resistor operating at 110°C can safely dissipate only ¼ W of power. Therefore, the physical size of the resistor must be increased if it is to safely dissipate ½ W at 110°C. In this case, a 1-kΩ, 1-W resistor would be necessary.

The curve in Fig. 2–20, called a *power* ***derating curve,*** is supplied by the resistor manufacturer. For a ½-W carbon resistor, the power derating curve corresponds to a 6.25 mW reduction in the power rating for each degree Celsius rise in temperature above 70°C. This corresponds to a derate factor of 6.25 mW/°C.

Shelf Life

Resistors keep their characteristics almost indefinitely when not used. Without any current in a circuit to heat the resistor, it has practically no change with age. The shelf life of resistors is therefore usually no problem.

2–5 Self-Review

Answers at the end of the chapter.

a. The power rating of a resistor is mainly determined by its physical size. (True/False)

b. The power rating of a carbon resistor is not affected by the ambient temperature in which it operates. (True/False)

2–6 Resistor Troubles

The most common trouble in resistors is an open. When the open resistor is a series component, there is no current in the entire series path.

Noisy Controls

In applications such as volume and tone controls, carbon controls are preferred because the smoother change in resistance results in less noise when the variable arm is rotated. With use, however, the resistance element becomes worn by the wiper contact, making the *control noisy*. When a volume or tone control makes a scratchy noise as the shaft is rotated, it indicates either a dirty or worn-out resistance element. If the control is just dirty, it can be cleaned by spraying the resistance element with a special contact cleaner. If the resistance element is worn out, the control must be replaced.

Checking Resistors with an Ohmmeter

Resistance is measured with an ohmmeter. The ohmmeter has its own voltage source so that it is always used without any external power applied to the resistance being measured. Separate the resistance from its circuit by disconnecting one lead of the resistor. Then connect the ohmmeter leads across the resistance to be measured.

An open resistor reads infinitely high ohms. For some reason, infinite ohms is often confused with zero ohms. Remember, though, that infinite ohms means an open circuit. The current is zero, but the resistance is infinitely high. Furthermore, it is practically impossible for a resistor to become short-circuited in itself. The resistor may be short-circuited by some other part of the circuit. However, the construction of resistors is such that the trouble they develop is an open circuit with infinitely high ohms.

The ohmmeter must have an ohms scale capable of reading the resistance value, or the resistor cannot be checked. In checking a 10-MΩ resistor, for instance, if the highest R the ohmmeter can read is 1 MΩ, it will indicate infinite resistance, even if the resistor has its normal value of 10 MΩ. An ohms scale of 100 MΩ or more should be used for checking such high resistances.

To check resistors of less than 10 Ω, a low-ohms scale of about 100 Ω or less is necessary. Center scale should be 6 Ω or less. Otherwise, the ohmmeter will read a normally low resistance value as zero ohms.

When checking resistance in a circuit, it is important to be sure there are no parallel resistance paths. Otherwise, the measured resistance can be much lower than the actual resistor value, as illustrated in Fig. 2–21*a*. Here, the ohmmeter reads the resistance of R_2 in parallel with R_1. To check across R_2 alone, one end is disconnected, as shown in Fig. 2–21*b*.

For very high resistances, it is important not to touch the ohmmeter leads. There is no danger of shock, but the body resistance of about 50,000 Ω as a parallel path will lower the ohmmeter reading.

GOOD TO KNOW

When measuring the value of a resistor in an electronic circuit, make absolutely sure that the power is off in the circuit being tested. Failure to do so could result in damage to the meter!

Figure 2–21 Parallel R_1 can lower the ohmmeter reading for testing R_2. (*a*) The two resistances R_1 and R_2 are in parallel. (*b*) R_2 is isolated by disconnecting one end of R_1.

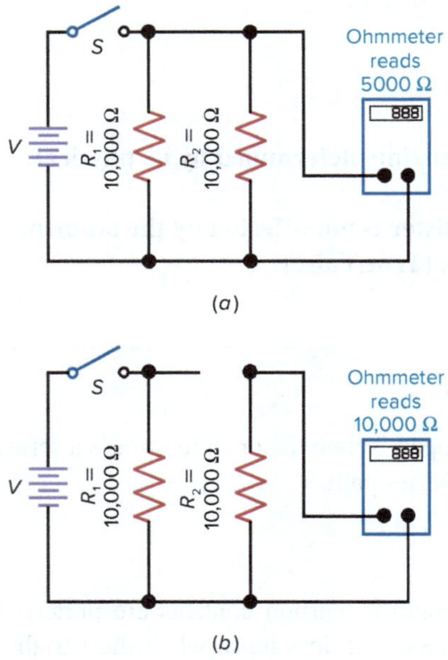

Changed Value of *R*

In many cases, the value of a carbon-composition resistor can exceed its allowed tolerance; this is caused by normal resistor heating over a long period of time. In most instances, the value change is seen as an increase in R. This is known as *aging*. As you know, carbon-film and metal-film resistors age very little. A surface-mount resistor should never be rubbed or scraped because this will remove some of the carbon deposit and change its resistance.

2–6 Self-Review

Answers at the end of the chapter.

a. What is the ohmmeter reading for a short circuit?
b. What is the ohmmeter reading for an open resistor?
c. Which has a higher *R*, an open or a short circuit?
d. Which is more likely to change in *R* value after many years of use, a metal-film or a carbon-composition resistor?

Application in Understanding Varistors and Surge Protectors

A voltage spike, sometimes called a transient voltage or power surge, is a short-duration voltage pulse that lasts for only a very short time, typically 1 to 30 microseconds. Voltage spikes or power surges can originate from a variety of different sources. For example, during a thunderstorm, nearby lightning strikes can produce voltage spikes as high as 100 kV, or more, in overhead power lines. These short-duration voltage spikes travel down the power line conductors and into our homes, schools, and industries. Voltage spikes or power surges can also originate from within our home, or other type of facility, when motors in air conditioners or refrigerators kick on or off, for example. These voltage spikes can reach values as high as 1 kV or more. Regardless of the source of the voltage spike, however, if the electronic devices and appliances plugged into the 120 or 240 V AC power line are not properly protected, they can be partially damaged or completely destroyed. To protect these devices or appliances, they should be plugged into a ***surge protector***, like the one shown in Fig. 2–22. As you can see, the surge protector looks like a standard outlet strip with a power cord and multiple outlets along its length. The surge protectors power cord is plugged directly into the 120 V AC power line and the electronic devices and/or appliances are plugged into one or more of the available outlets. If a high-energy voltage spike appears on the incoming 120 V AC power line, the surge protector limits the voltage to some predetermined maximum value such as 330 V. This voltage-limiting feature protects the electronic devices and appliances that are plugged into the outlet strip.

Figure 2–22 Surge Protector

Sarah Schultz Photography

VARISTOR CHARACTERISTICS

To protect against voltage spikes or power surges, surge protectors use a special type of resistor known as a ***varistor***. A varistor is a ***voltage-dependent***, nonlinear resistor, whose resistance value depends on its voltage. The current in a varistor varies as a power of the voltage, and for a particular varistor it may increase by many orders of magnitude when the voltage is doubled. Varistors are not polarity-sensitive components, which means they have identical characteristics for either polarity of voltage. It is also worth noting that ***metal oxide varistors*** have a negative temperature coefficient. Varistors are typically manufactured using metal oxides, such as zinc oxide, and as a result, ***metal oxide varistors*** are commonly referred to as **MOVs**. It is important to note that the terms varistor and MOV may be used interchangeably. Figure 2–23a shows the schematic symbol of an MOV and Fig. 2–23b shows a typical volt-ampere characteristic. Because the volt-ampere characteristic is identical for both polarities of voltage, MOVs are classified as being both bilateral and symmetrical. In Fig. 2–23b, notice that the current is practically zero until we reach the varistor's rated voltage. The varistor's rated voltage, sometimes called the ***nominal voltage***, V_{NOM}, is the minimum voltage at which the varistor begins conducting. The rated or nominal voltage is

Figure 2–23a Schematic symbol of an MOV.

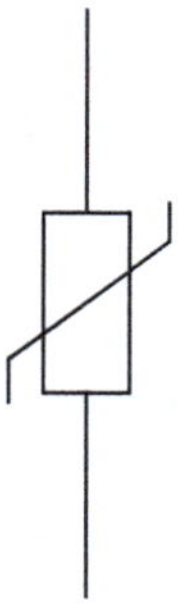

Figure 2–23b Volt-Ampere Characteristic of a typical MOV.

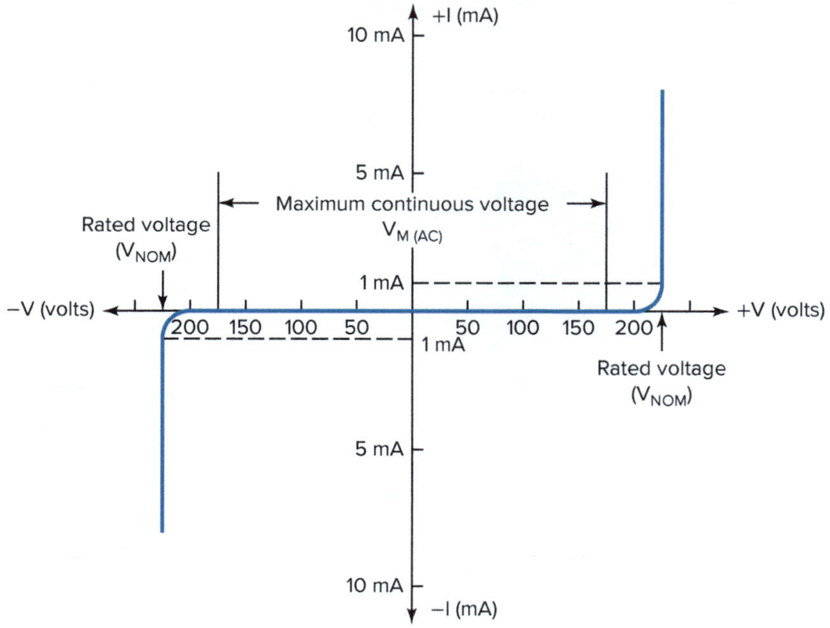

typically specified at a current of 1 mA. Below the rated voltage, the MOV exhibits a very high resistance and the MOV can be considered to be in its **"OFF"** or nonconducting state. In the absence of any voltage spikes or power surges, this is the normal region of operation for the MOV. In fact, an important MOV rating is its ***maximum continuous voltage*** rating, which is the maximum voltage that can continuously be applied across the MOV without conducting. This voltage rating is usually symbolized as $V_{M(AC)}$ or $V_{M(DC)}$. When the voltage across the MOV increases slightly above its rated voltage, however, the resistance of the MOV decreases significantly and the current rises rapidly. It is important to note, however, that the voltage across the MOV remains relatively constant even as the current through the MOV continues to increase. Perhaps the most important rating of an MOV is its ***clamping voltage,*** designated V_C. The clamping voltage, V_C, is the maximum guaranteed voltage across the MOV at a specified peak current, I_{PK}. When the MOV is conducting and the current is extremely high, the energy contained within the pulse is dissipated as heat in the MOV. The life of an MOV is determined by the number, duration, and intensity of power surges that it experiences over its lifetime.

If an MOV is exposed to a transient voltage whose amplitude is significantly greater than its clamping voltage, its resistance decreases drastically. The result is a sharp increase in current through the MOV. After a very short delay (usually less than 1 ns) the voltage across the MOV returns to its rated clamping voltage, V_C, thus limiting the transient voltage to a safe level. The energy contained in the transient voltage or power surge is absorbed by the MOV, thus protecting the intended circuitry.

MOVs come in a variety of different shapes and sizes, but the most common packaging style is the disc type shown in Fig. 2–24. The varistor voltage rating (the nominal voltage, V_{NOM}) is determined by several different factors, but perhaps the biggest determining factor is its thickness. Typically, the thicker the disc wafer, the higher the voltage rating.

The following is a summary of the most important MOV ratings.

1. **Nominal voltage, V_{NOM}**—The rated varistor voltage at a specified current of 1 (mA). This is considered the minimum varistor voltage at which the MOV begins conducting.
2. **Clamping voltage, V_C**—The maximum guaranteed varistor voltage at a specified peak current.
3. **Maximum continuous voltage, $V_{M(AC)}$**—The maximum voltage that can be continuously applied across the MOV without resulting in any considerable current. The maximum continuous voltage is always less than the nominal voltage, V_{NOM}.
4. **Maximum non-repetitive surge current**—The maximum peak surge current allowable through the MOV.
5. **Energy rating**—The maximum energy, in joules (J), the MOV is capable of dissipating under transient conditions without causing device failure.
6. **Response time**—The time it takes for an MOV to start reacting to a voltage spike once the magnitude of the voltage spike exceeds the rated clamping voltage, V_C.
7. **Operating ambient temperature range**—The range of operating temperatures over which the MOV can operate under specified ratings.

Figure 2–24 Metal-Oxide Varistor (MOV).

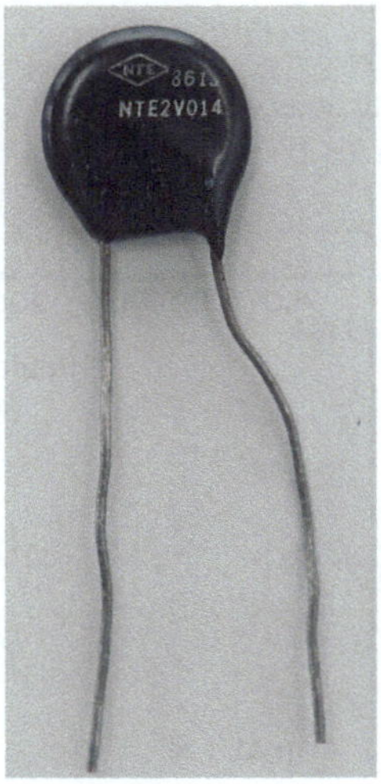

SURGE PROTECTOR RATINGS

Now let's take a look at the ratings typically associated with commercially available surge protectors. The surge protector in Fig. 2–22 has an ***energy absorption/dissipation rating*** of 1150 Joules, a ***voltage protection*** (clamping voltage) rating of 400 V, and a ***response time*** rating less than 1 ns. (These are the specifications for this model.) The energy absorption/dissipation rating, listed as 1150 Joules, indicates the amount of energy the surge protector can handle before it fails. The higher this rating is, the better the protection. The voltage protection rating, listed as 400 V, specifies the maximum voltage the surge protector will allow across the 120 V AC power line in the event of a voltage spike or power surge. A lower voltage rating indicates better protection. There are three levels or ratings of voltage protection listed by ***Underwriters Laboratory (UL)***: 330 V, 400 V, and 500 V. The response time rating, listed as being less than 1 ns, indicates how much of a delay there will be in responding to a voltage spike or power surge. The longer the response time, the longer the electronic devices plugged into the power strip will be exposed to the incoming power surge. Obviously, the shorter the response time, the better the protection.

The schematic diagram of a typical surge protector is shown in Fig. 2–25. The power cord has three conductors: ***hot (H), neutral (N),*** and ***ground (G).*** Inside the surge protector, MOVs are connected between hot and neutral (H-N), hot and ground

Figure 2–25 Schematic diagram of a surge protector.

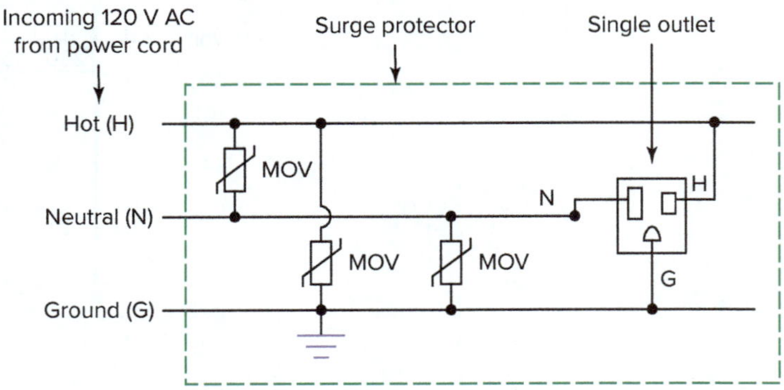

(H-G), and neutral and ground (N-G). Three MOVs are needed because voltage spikes may be present on any or all of the incoming power line conductors.

IMPORTANT TIPS

When purchasing a surge protector, look for one that has two indicator lights: ***Protected*** and ***Grounded***. If the ***Protected*** light-emitting diode (LED) is illuminated, it tells you that the voltage-limiting components (MOVs) are still functioning properly. If the ***Grounded*** LED is illuminated, it tells you that the 120 V AC receptacle, to which the surge protector is plugged into, is properly grounded. If the ***Protected*** light goes off, it indicates that the voltage-limiting components (MOVs) have been destroyed by a power surge. In this case, the surge protector has served its purpose and it should be replaced. Some surge protectors, like the one shown in Fig. 2–22, automatically shut down if a catastrophic power surge occurs, or if the MOVs have absorbed as much energy (in joules) as they are capable of absorbing. It is important to note, however, that most surge protectors will still function normally as a power strip even if the MOVs have been destroyed. Without some kind of indicator light, there is no way of knowing if the surge protector feature of the power strip is still functioning properly. If the ***Grounded*** LED goes off, it indicates that there is a ground-wiring problem in the 120 V AC outlet. Without a proper ground, the surge protector is useless. If the ***Grounded*** LED is off, contact an electrician and have him or her properly ground the 120 V AC outlet. It is important to note that the surge protector will still provide 120 V AC between hot (H) and neutral (N) even if the ***Grounded*** LED is not illuminated.

One final point: The surge protector shown in Fig. 2–22 has a lighted ***ON/OFF*** switch with a built-in circuit breaker. There are two reasons why the circuit breaker may trip: either the devices plugged into the surge protector are drawing more current than the surge protector is rated for (which is 15 A) or there has been a catastrophic power surge. In either case, if the circuit breaker trips, wait for a short time and then turn the switch from ***OFF*** to ***ON*** to reset it. As mentioned previously, if the ***Protected*** LED does not come back on, the surge protector has served its purpose and must be replaced.

Summary

- The most common types of resistors include carbon-composition, carbon-film, metal-film, wire-wound, and surface-mount or chip resistors. Carbon-film and metal-film resistors are better than carbon-composition resistors because they have tighter tolerances, are less affected by temperature and aging, and generate less noise internally.
- A thermistor is a thermally sensitive resistor whose resistance value changes with temperature. If the resistance of a thermistor increases with temperature, it is said to have a positive temperature coefficient (PTC). If the resistance of a thermistor decreases with temperature, it is said to have a negative temperature coefficient (NTC).
- Wire-wound resistors are typically used in high-current applications. Wire-wound resistors are available with wattage ratings of about 1 to 100 W.
- Resistors are usually color-coded to indicate their resistance value in ohms. Either a four-band or a five-band code is used. The five-band code is used for more precise *R* values. Chip resistors use a three- or four-digit code to indicate their resistance value.
- Zero-ohm resistors are used with automatic insertion machines when it is desired to short two points on a printed-circuit board. Zero-ohm resistors are available in ⅛- or ¼-W ratings.
- A potentiometer is a variable resistor with three terminals. It is used to vary the voltage in a circuit. A rheostat is a variable resistor with two terminals. It is used to vary the current in a circuit.
- The physical size of a resistor determines its wattage rating: the larger the physical size, the larger the wattage rating. There is no correlation between a resistor's physical size and its resistance value.
- The most common trouble in resistors is an open. An ohmmeter across the leads of an open resistor will read infinite, assuming there is no other parallel path across the resistor.

Important Terms

Carbon-composition resistor — a type of resistor made of finely divided carbon mixed with a powdered insulating material in the correct proportion to obtain the desired resistance value.

Carbon-film resistor — a type of resistor whose construction consists of a thin spiral layer of carbon on an insulated substrate.

Color coding — a scheme using colored bands or stripes around the body of a resistor to indicate the ohmic value and tolerance of a resistor.

Decade resistance box — a variable resistance box whose resistance value can be varied in 1-Ω, 10-Ω, 100-Ω, 1000-Ω, 10,000-Ω, or 100,000-Ω steps.

Derating curve — a graph showing how the power rating of a resistor decreases as its operating temperature increases.

Metal-film resistor — a type of resistor whose construction consists of a thin spiral film of metal on a ceramic substrate.

Negative temperature coefficient (NTC) — a characteristic of a thermistor indicating that its resistance decreases with an increase in operating temperature.

Positive temperature coefficient (PTC) — a characteristic of a thermistor indicating that its resistance increases with an increase in operating temperature.

Potentiometer — a three-terminal variable resistor used to vary the voltage between the center terminal and one of the outside terminals.

Rheostat — a two-terminal variable resistor used to vary the amount of current in a circuit.

Surface-mount resistor — a type of resistor constructed by depositing a thick carbon film on a ceramic base. (A surface-mount resistor is many times smaller than a conventional resistor and has no leads that extend out from the body itself.)

Taper — a word describing the way the resistance of a potentiometer or rheostat varies with the rotation of its shaft.

Thermistor — a resistor whose resistance value changes with changes in its operating temperature.

Tolerance — the maximum allowable percent difference between the measured and coded values of resistance.

Wire-wound resistor — a type of resistor whose construction consists of resistance wire wrapped on an insulating core.

Zero-ohm resistor — a resistor whose ohmic value is approximately zero ohms.

Zero-power resistance — the resistance of a thermistor with zero-power dissipation, designated R_0.

Self-Test

Answers at the back of the book.

1. A carbon composition resistor having only three color stripes has a tolerance of

a. ±5%.
b. ±20%.
c. ±10%.
d. ±100%.

2. A resistor with a power rating of 25 W is most likely a

a. carbon-composition resistor.
b. metal-film resistor.
c. surface-mount resistor.
d. wire-wound resistor.

3. When checked with an ohmmeter, an open resistor measures

a. infinite resistance.
b. its color-coded value.
c. zero resistance.
d. less than its color-coded value.

4. **One precaution to observe when checking resistors with an ohmmeter is to**
   a. check high resistances on the lowest ohms range.
   b. check low resistances on the highest ohms range.
   c. disconnect all parallel paths.
   d. make sure your fingers are touching each test lead.

5. **A chip resistor is marked 394. Its resistance value is**
   a. 39.4 Ω.
   b. 394 Ω.
   c. 390,000 Ω.
   d. 39,000 Ω.

6. **A carbon-film resistor is color-coded with red, violet, black, and gold stripes. What are its resistance and tolerance?**
   a. 27 Ω ± 5%.
   b. 270 Ω ± 5%.
   c. 270 Ω ± 10%.
   d. 27 Ω ± 10%.

7. **A potentiometer is a**
   a. three-terminal device used to vary the voltage in a circuit.
   b. two-terminal device used to vary the current in a circuit.
   c. fixed resistor.
   d. two-terminal device used to vary the voltage in a circuit.

8. **A metal-film resistor is color-coded with brown, green, red, brown, and blue stripes. What are its resistance and tolerance?**
   a. 1500 Ω ± 1.25%.
   b. 152 Ω ± 1%.
   c. 1521 Ω ± 0.5%.
   d. 1520 Ω ± 0.25%.

9. **Which of the following resistors has the smallest physical size?**
   a. wire-wound resistors.
   b. carbon-composition resistors.
   c. surface-mount resistors.
   d. potentiometers.

10. **Which of the following statements is true?**
    a. Resistors always have axial leads.
    b. Resistors are always made from carbon.
    c. There is no correlation between the physical size of a resistor and its resistance value.
    d. The shelf life of a resistor is about one year.

11. **If a thermistor has a negative temperature coefficient (NTC), its resistance**
    a. increases with an increase in operating temperature.
    b. decreases with a decrease in operating temperature.
    c. decreases with an increase in operating temperature.
    d. is unaffected by its operating temperature.

12. **With the four-band resistor color code, gold in the third stripe corresponds to a**
    a. fractional multiplier of 0.01.
    b. fractional multiplier of 0.1.
    c. decimal multiplier of 10.
    d. resistor tolerance of ±10%.

13. **Which of the following axial-lead resistor types usually has a blue, light green, or red body?**
    a. wire-wound resistors.
    b. carbon-composition resistors.
    c. carbon-film resistors.
    d. metal-film resistors.

14. **A surface-mount resistor has a coded value of 4R7. This indicates a resistance of**
    a. 4.7 Ω.
    b. 4.7 kΩ.
    c. 4.7 MΩ.
    d. none of the above.

15. **Reading from left to right, the colored bands on a resistor are yellow, violet, brown and gold. If the resistor measures 513 Ω with an ohmmeter, it is**
    a. well within tolerance.
    b. out of tolerance.
    c. right on the money.
    d. close enough to be considered within tolerance.

Essay Questions

1. List five different types of fixed resistors.
2. List the advantages of using a metal-film resistor versus a carbon-composition resistor.
3. Draw the schematic symbols for a (*a*) fixed resistor (*b*) potentiometer (*c*) rheostat (*d*) thermistor.
4. How can a technician identify a wire-wound resistor that is color-coded?
5. Explain an application using a decade resistance box.
6. List the differences between a potentiometer and a rheostat.
7. For resistors using the four-band code, what are the values for gold and silver as fractional decimal multipliers in the third band?
8. Briefly describe how you would check to see whether a 1-MΩ resistor is open or not. Give two precautions to make sure the test is not misleading.
9. Define the term "zero-power resistance" as it relates to thermistors.
10. Explain how the ambient temperature affects the power rating of a carbon resistor.

Problems

Answers to odd-numbered problems at the back of the book.

SECTION 2–2 RESISTOR COLOR CODING

2–1 Indicate the resistance and tolerance for each resistor shown in Fig. 2–26.

2–2 Indicate the resistance and tolerance for each resistor shown in Fig. 2–27.

2–3 Indicate the resistance for each chip resistor shown in Fig. 2–28.

Figure 2–26 Resistors for Prob. 2–1.

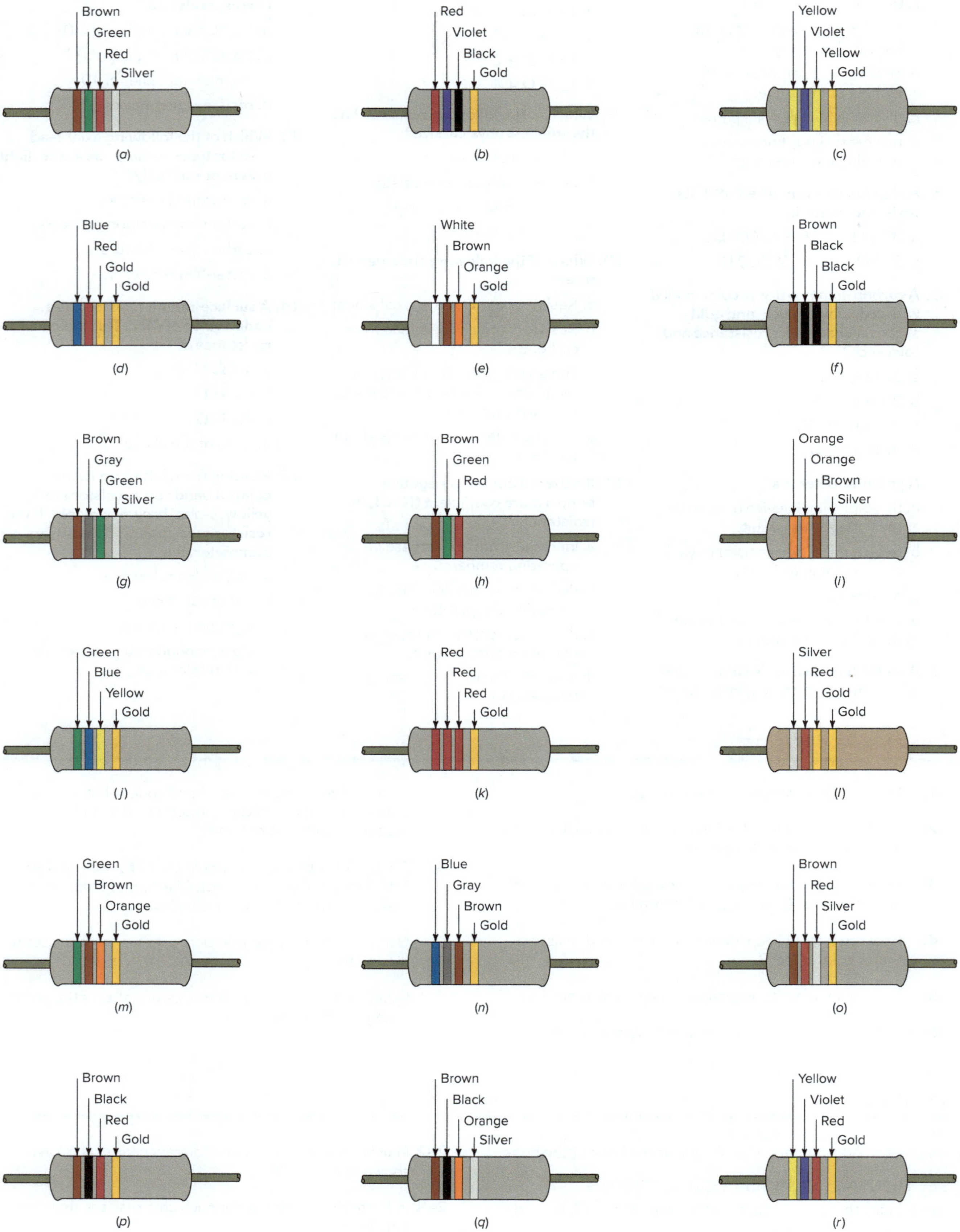

Figure 2-27 Resistors for Prob. 2–2.

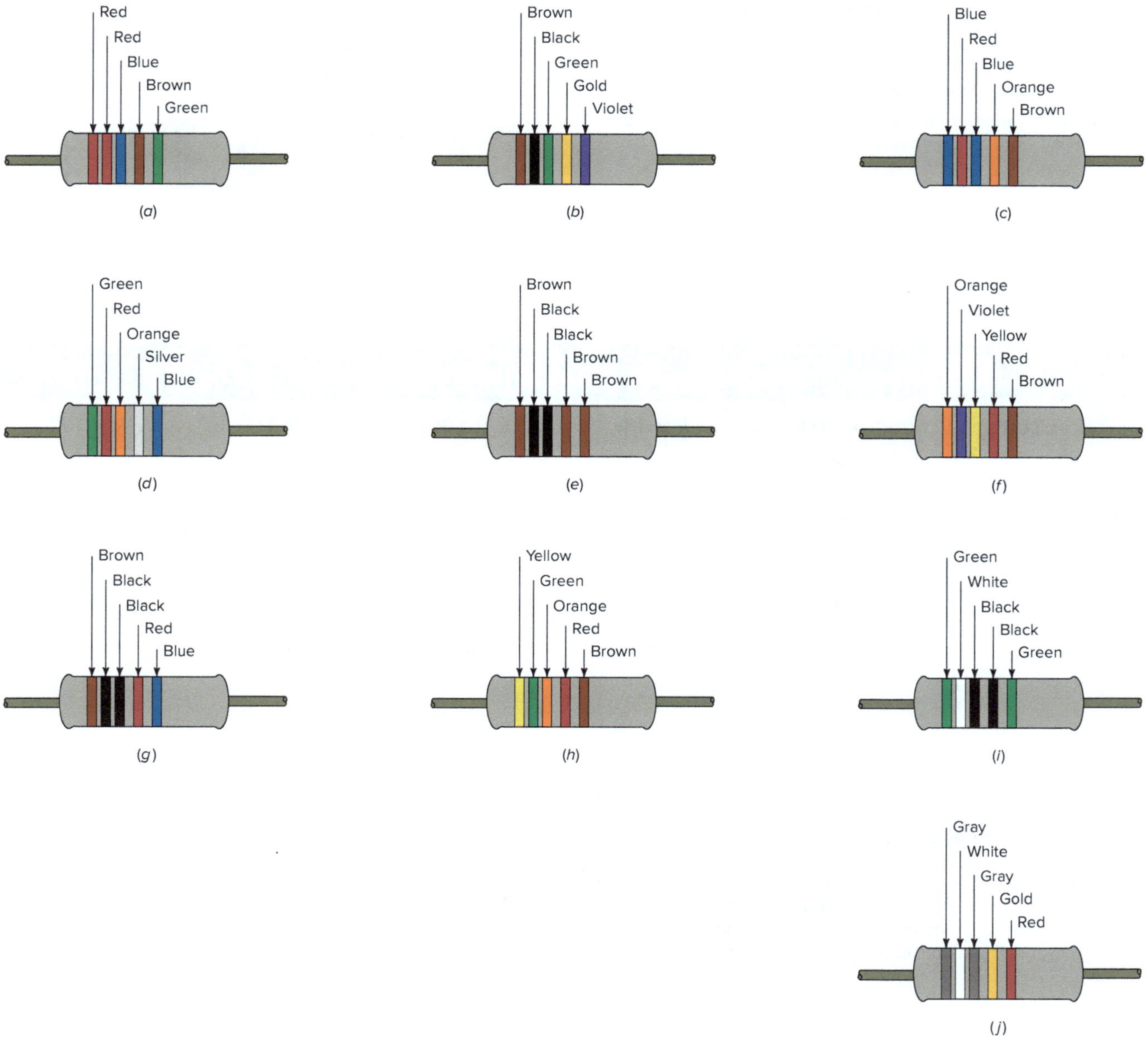

Figure 2-28 Chip resistors for Prob. 2–3.

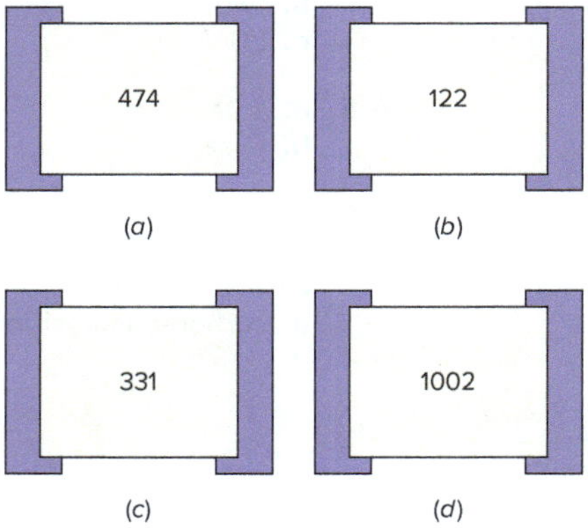

2–4 Calculate the permissible ohmic range of a resistor whose resistance value and tolerance are (*a*) 3.9 kΩ ± 5% (*b*) 100 Ω ± 10% (*c*) 120 kΩ ± 2% (*d*) 2.2 Ω ± 5% (*e*) 75 Ω ± 1%.

2–5 Using the four-band code, indicate the colors of the bands for each of the following resistors: (*a*) 10 kΩ ± 5% (*b*) 2.7 Ω ± 5% (*c*) 5.6 kΩ ± 10% (*d*) 1.5 MΩ ± 5% (*e*) 0.22 Ω ± 5%.

2–6 Using the five-band code, indicate the colors of the bands for each of the following resistors: (*a*) 110 Ω ± 1% (*b*) 34 kΩ ± 0.5% (*c*) 82.5 kΩ ± 2% (*d*) 62.6 Ω ± 1% (*e*) 105 kΩ ± 0.1%.

SECTION 2–3 VARIABLE RESISTORS

2–7 Refer to Fig. 2–29 at the right. Indicate the total resistance R_T for each of the different dial settings in Table 2–3.

SECTION 2–4 RHEOSTATS AND POTENTIOMETERS

2–8 Show two different ways to wire a potentiometer so that it will work as a rheostat.

Figure 2–29 Decade resistance box.

Mark Steinmetz/McGraw Hill

Table 2–3 Decade Resistance Box Dial Settings for Problem 2–7

	$R \times 100$ k	$R \times 10$ k	$R \times 1$ k	$R \times 100$	$R \times 10$	$R \times 1$
(*a*)	6	8	0	2	2	5
(*b*)	0	0	8	2	5	0
(*c*)	0	1	8	5	0	3
(*d*)	2	7	5	0	6	0
(*e*)	0	6	2	9	8	4

Critical Thinking

2–9 A manufacturer of carbon-film resistors specifies a maximum working voltage of 250 V for all its ¹⁄₄-W resistors. Exceeding 250 V causes internal arcing within the resistor. Above what minimum resistance will the maximum working voltage be exceeded before its ¹⁄₄-W power dissipation rating is exceeded? **Hint**: The maximum voltage that produces the rated power dissipation can be calculated as $V_{max} = \sqrt{P \times R}$.

2–10 What is the power rating of a ¹⁄₂-W carbon resistor if it is used at an ambient temperature of 120°C?

Answers to Self-Reviews

2–1
a. true
b. true
c. metal-film
d. 1000-Ω, 1-W
e. *R* decreases

2–2
a. yellow
b. ±10%
c. 100
d. 470 Ω ± 5%
e. 33,000 Ω or 33 kΩ

2–3
a. terminal 2
b. nonlinear
c. 470,000 Ω or 470 kΩ

2–4
a. four connections to three terminals
b. two
c. 80 V

2–5
a. true
b. false

2–6
a. 0 Ω
b. infinite ohms
c. open circuit
d. carbon-composition resistor

Laboratory Application Assignment

In this lab application assignment, you will examine the four-band resistor color code. You will also use a digital multimeter (DMM) to measure the resistance values of both fixed and variable resistors.

Equipment: Obtain the following items from your instructor.

- An assortment of carbon-film resistors
- 1-MΩ carbon-film resistor (any wattage rating)
- 10-kΩ potentiometer (any wattage rating)
- DMM

Resistor Color Code

Obtain five different carbon-film resistors from your instructor. In the space provided below, indicate the color of each band and its corresponding color coded value. The coded value should include both the resistance value and tolerance. Finally, measure and record the value of each of the five resistors.

	First Band	Second Band	Third Band	Fourth Band	Coded Value	Measured Value
R_1	______	______	______	______	______	______
R_2	______	______	______	______	______	______
R_3	______	______	______	______	______	______
R_4	______	______	______	______	______	______
R_5	______	______	______	______	______	______

Is the measured value of each resistor within its specified tolerance? __________If not, indentify which resistors are out of tolerance.______________

Resistance Measurement Precautions

Now let's use a DMM to measure the value of a 1-MΩ resistor.

Measure the value of the 1-MΩ resistor without allowing your fingers to touch its leads. Record your measured value. $R =$ ______

Remeasure the value of the 1-MΩ resistor with both of your fingers firmly grasping the resistor leads. Record your measured value. $R =$ ______

Are the measured values different? ______ If so, which measurement is incorrect and why? ______________

____

____

Potentiometer

Locate the 10-kΩ potentiometer and position it, as shown in Fig. 2–30.

Measure and record the resistance across terminals 1 and 3. $R =$ ________ Rotate the shaft of the potentiometer back and forth. Does the resistance vary? ________

Connect the DMM to terminals 1 and 2 of the potentiometer. Does the resistance increase or decrease with clockwise shaft rotation? ____________

Connect the DMM to terminals 2 and 3 of the potentiometer. Does the resistance increase or decrease with clockwise shaft rotation? ______________

____

Rotate the shaft of the potentiometer to its midway position. Measure and record the resistance across terminals 1 and 2. $R =$ ______ Measure and record the resistance across terminals 2 and 3. $R =$ ______ Do the sum of these resistance values add to equal the resistance across terminals 1 and 3? ______________

____

Rheostat

In this step, you will convert a potentiometer into a rheostat.

Connect a jumper across terminals 1 and 2. Connect your DMM to terminals 1 and 3. Explain how the resistance varies with clockwise shaft rotation. ______________

____

Remove the jumper across terminals 1 and 2, and place it across terminals 2 and 3. With your DMM connected across terminals 1 and 3, explain how the resistance varies with clockwise shaft rotation.______________

____

Figure 2–30 Potentiometer. (*a*) Component body. (*b*) Schematic symbol.

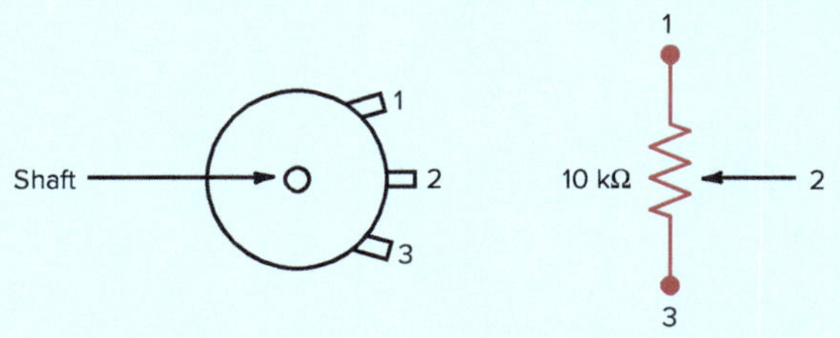

Design credits: Chapter Opener Background Image, and Multisim: Stockbyte/Getty Images

chapter

3

Ohm's Law

The mathematical relationship between voltage, current, and resistance was discovered in 1826 by Georg Simon Ohm. The relationship, known as Ohm's law, is the basic foundation for all circuit analysis in electronics. Ohm's law, which is the basis of this chapter, states that the amount of current, I, is directly proportional to the voltage, V, and inversely proportional to the resistance, R. Expressed mathematically, Ohm's law is stated as

$$I = \frac{V}{R}$$

Besides the coverage of Ohm's law, this chapter also introduces you to the concept of work, energy, and power. Power can be defined as the time rate of doing work or using energy. The symbol for power is P and the unit is the watt. All the mathematical relationships that exist between V, I, R, and P are covered in this chapter.

In addition to Ohm's law and power, this chapter also discusses electric shock and open- and short-circuit troubles. ■

Chapter Outline

Chapter Objectives

After studying this chapter, you should be able to

- *List* the three forms of Ohm's law.
- *Use* Ohm's law to calculate the current, voltage, or resistance in a circuit.
- *List* the multiple and submultiple units of voltage, current, and resistance.
- *Explain* the linear relationship between V and I when R is constant.
- *Explain* the difference between a linear and a nonlinear resistance.
- *Explain* the inverse relation between I and R when V is constant.
- *Explain* the difference between work, energy, and power and list the units of each.
- *Calculate* the power in a circuit when the voltage and current, current and resistance, or voltage and resistance are known.
- *Determine* the required resistance and appropriate wattage rating of a resistor.
- *Identify* the shock hazards associated with working with electricity.
- *Explain* the difference between an open circuit and short circuit.

Important Terms

ampere
electron volt (eV)
energy
horsepower (hp)
inverse relation
joule
kilowatt-hour (kWh)
linear proportion
linear resistance
maximum working voltage rating
nonlinear resistance
ohm
open circuit
power
short circuit
volt
volt-ampere characteristic
watt
work

GOOD TO KNOW

A power supply is a piece of test equipment that can provide a variable DC output voltage to electronic circuits under test. A red-colored jack provides connection to the positive (+) side of the DC voltage whereas a black-colored jack provides connection to the negative (−) side.

3–1 The Current $I = V/R$

If we keep the same resistance in a circuit but vary the voltage, the current will vary. The circuit in Fig. 3–1 demonstrates this idea. The applied voltage V can be varied from 0 to 12 V, as an example. The bulb has a 12-V filament, which requires this much voltage for its normal current to light with normal intensity. The meter I indicates the amount of current in the circuit for the bulb.

With 12 V applied, the bulb lights, indicating normal current. When V is reduced to 10 V, there is less light because of less I. As V decreases, the bulb becomes dimmer. For zero volts applied, there is no current and the bulb cannot light. In summary, the changing brilliance of the bulb shows that the current varies with the changes in applied voltage.

For the general case of any V and R, Ohm's law is

$$I = \frac{V}{R} \qquad \textbf{(3–1)}$$

where I is the amount of current through the resistance R connected across the source of potential difference V. With **volts** as the practical unit for V and **ohms** for R, the amount of current I is in **amperes.** Therefore,

$$\text{Amperes} = \frac{\text{volts}}{\text{ohms}}$$

This formula states simply to divide the voltage across R by the ohms of resistance between the two points of potential difference to calculate the amperes of current through R. In Fig. 3–2, for instance, with 6 V applied across a 3-Ω resistance, by Ohm's law, the amount of current I equals $6/3$ or 2 A.

High Voltage but Low Current

It is important to realize that with high voltage, the current can have a low value when there is a very high resistance in the circuit. For example, 1000 V applied across 1,000,000 Ω results in a current of only $1/1000$ A. By Ohm's law,

$$I = \frac{V}{R}$$

$$= \frac{1000\text{ V}}{1{,}000{,}000\ \Omega} = \frac{1}{1000}$$

$$I = 0.001\text{ A}$$

The practical fact is that high-voltage circuits usually do have small values of current in electronic equipment. Otherwise, tremendous amounts of **power** would be necessary.

Figure 3–1 Increasing the applied voltage V produces more current I to light the bulb with more intensity.

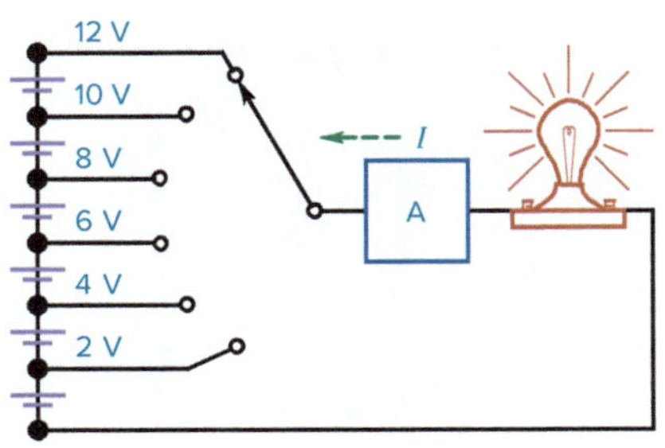

MultiSim **Figure 3–2** Example of using Ohm's law. (*a*) Wiring diagram of a circuit with a 6-V battery for *V* applied across a load *R*. (*b*) Schematic diagram of the circuit with values for *I* and *R* calculated by Ohm's law.

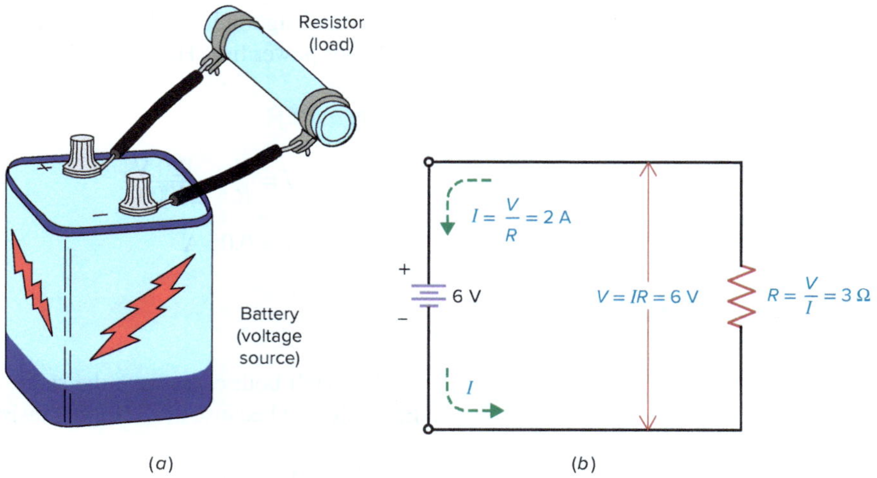

Low Voltage but High Current

At the opposite extreme, a low value of voltage in a very low resistance circuit can produce a very high current. A 6-V battery connected across a resistance of 0.01 Ω produces 600 A of current:

$$I = \frac{V}{R}$$

$$= \frac{6\text{ V}}{0.01\ \Omega}$$

$$I = 600\text{ A}$$

Less *I* with More *R*

Note also the values of *I* in the following two examples.

CALCULATOR

To do a division problem like *V*/*R* in Example 3–1 on the calculator, punch in the number 120 for the numerator, then press the (÷) key for division before punching in 8 for the denominator. Finally, press the (=) key for the answer of 15 on the display. The numerator must be punched in first.

Example 3–1

A heater with a resistance of 8 Ω is connected across the 120-V power line. How much is the current *I*?

ANSWER

$$I = \frac{V}{R} = \frac{120\text{ V}}{8\ \Omega}$$

$$I = 15\text{ A}$$

Example 3–2

A small lightbulb with a resistance of 2400 Ω is connected across the same 120-V power line. How much is current *I*?

ANSWER

$$I = \frac{V}{R} = \frac{120\text{ V}}{2400\ \Omega}$$

$$I = 0.05\text{ A}$$

Although both cases have the same 120 V applied, the current is much less in Example 3–2 because of the higher resistance.

Typical *V* and *I*

Transistors and integrated circuits generally operate with a DC supply of 3.3, 5, 6, 9, 12, 15, 24, or 50 V. The current is usually in millionths or thousandths of one ampere up to about 5 A.

3–1 Self-Review

Answers at the end of the chapter.

- **a. Calculate *I* for 24 V applied across 8 Ω.**
- **b. Calculate *I* for 12 V applied across 8 Ω.**
- **c. Calculate *I* for 24 V applied across 12 Ω.**
- **d. Calculate *I* for 6 V applied across 1 Ω.**

3–2 The Voltage *V* = *IR*

Referring back to Fig. 3–2, the voltage across *R* must be the same as the source *V* because the resistance is connected directly across the battery. The numerical value of this *V* is equal to the product *I* × *R*. For instance, the *IR* voltage in Fig. 3–2 is 2 A × 3 Ω, which equals the 6 V of the applied voltage. The formula is

$$V = IR \qquad \textbf{(3–2)}$$

CALCULATOR

To do a multiplication problem like *I* × *R* in Example 3–3 on the calculator, punch in the factor 2.5, then press the ⊗ key for multiplication before punching in 12 for the other factor. Finally, press the ⊜ key for the answer of 30 on the display. The factors can be multiplied in any order.

Example 3–3

If a 12-Ω resistor is carrying a current of 2.5 A, how much is its voltage?

ANSWER

$$V = IR$$

$$= 2.5\text{ A} \times 12\ \Omega$$

$$= 30\text{ V}$$

With I in ampere units and R in ohms, their product V is in volts. In fact, this must be so because the I value equal to V/R is the amount that allows the IR product to be the same as the voltage across R.

Besides the numerical calculations possible with the IR formula, it is useful to consider that the IR product means voltage. Whenever there is current through a resistance, it must have a potential difference across its two ends equal to the IR product. If there were no potential difference, no electrons could flow to produce the current.

3–2 Self-Review

Answers at the end of the chapter.

a. Calculate V for 0.002 A through 1000 Ω.
b. Calculate V for 0.004 A through 1000 Ω.
c. Calculate V for 0.002 A through 2000 Ω.

3–3 The Resistance $R = V/I$

GOOD TO KNOW

Since R and G are reciprocals of each other, the conductance, G, of a circuit can be calculated as $G = \frac{I}{V}$.

As the third and final version of Ohm's law, the three factors V, I, and R are related by the formula

$$R = \frac{V}{I} \qquad \textbf{(3–3)}$$

In Fig. 3–2, R is 3 Ω because 6 V applied across the resistance produces 2 A through it. Whenever V and I are known, the resistance can be calculated as the voltage across R divided by the current through it.

Physically, a resistance can be considered some material whose elements have an atomic structure that allows free electrons to drift through it with more or less force applied. Electrically, though, a more practical way of considering resistance is simply as a V/I ratio. Anything that allows 1 A of current with 10 V applied has a resistance of 10 Ω. This V/I ratio of 10 Ω is its characteristic. If the voltage is doubled to 20 V, the current will also double to 2 A, providing the same V/I ratio of a 10-Ω resistance.

Furthermore, we do not need to know the physical construction of a resistance to analyze its effect in a circuit, so long as we know its V/I ratio. This idea is illustrated in Fig. 3–3. Here, a box with some unknown material in it is connected in a circuit where we can measure the 12 V applied across the box and the 3 A of current through it. The resistance is 12V/3A, or 4 Ω. There may be liquid, gas, metal, powder, or any other material in the box; but electrically the box is just a 4-Ω resistance because its V/I ratio is 4.

Figure 3–3 The resistance R of any component is its V/I ratio.

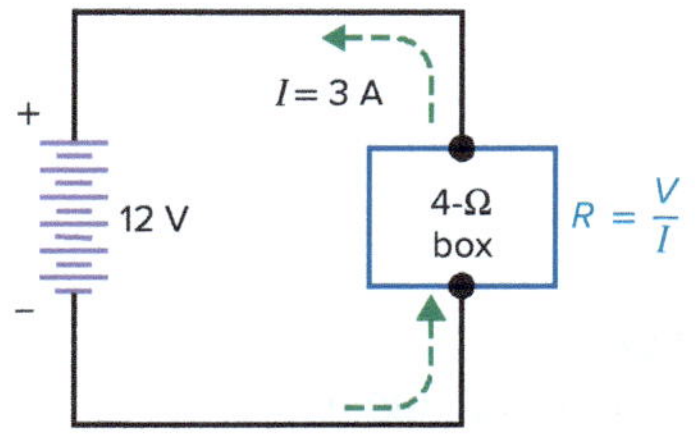

Example 3–4

How much is the resistance of a lightbulb if it draws 0.16 A from a 12-V battery?

ANSWER

$$R = \frac{V}{I}$$

$$= \frac{12\text{ V}}{0.16\text{ A}}$$

$$= 75\ \Omega$$

3–3 Self-Review

Answers at the end of the chapter.

a. **Calculate *R* for 12 V with 0.003 A.**
b. **Calculate *R* for 12 V with 0.006 A.**
c. **Calculate *R* for 12 V with 0.001 A.**

3–4 Practical Units

The three forms of Ohm's law can be used to define the practical units of current, potential difference, and resistance as follows:

$$1 \text{ ampere} = \frac{1 \text{ volt}}{1 \text{ ohm}}$$

$$1 \text{ volt} = 1 \text{ ampere} \times 1 \text{ ohm}$$

$$1 \text{ ohm} = \frac{1 \text{ volt}}{1 \text{ ampere}}$$

Figure 3–4 A circle diagram to help in memorizing the Ohm's law formulas $V = IR$, $I = V/R$, and $R = V/I$. The V is always at the top.

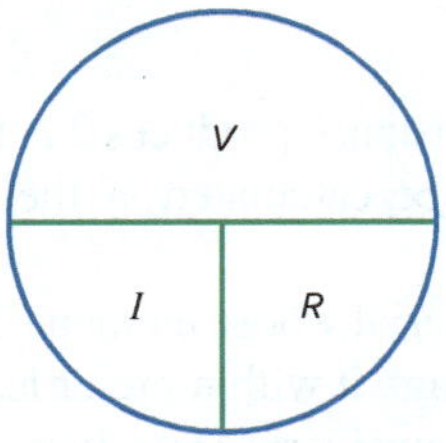

One **ampere** is the amount of current through a one-ohm resistance that has one volt of potential difference applied across it.

One **volt** is the potential difference across a one-ohm resistance that has one ampere of current through it.

One **ohm** is the amount of opposition in a resistance that has a V/I ratio of 1, allowing one ampere of current with one volt applied.

In summary, the circle diagram in Fig. 3–4 for $V = IR$ can be helpful in using Ohm's law. Put your finger on the unknown quantity and the desired formula remains. The three possibilities are

Cover V and you have IR.
Cover I and you have V/R.
Cover R and you have V/I.

3–4 Self-Review

Answers at the end of the chapter.

a. **Calculate *V* for 0.007 A through 5000 Ω.**
b. **Calculate the amount of *I* for 12,000 V across 6,000,000 Ω.**
c. **Calculate *R* for 8 V with 0.004 A.**

3–5 Multiple and Submultiple Units

The basic units—ampere, volt, and ohm—are practical values in most electric power circuits, but in many electronics applications, these units are either too small or too big. As examples, resistances can be a few million ohms, the output of a high-voltage power supply can be several kilovolts (kV), and the current in transistors and integrated circuits is generally thousandths or millionths of an ampere.

In such cases, it is often helpful to use multiples and submultiples of the basic units. These multiple and submultiple values are based on the metric system of units discussed earlier. The common conversions for V, I, and R are summarized at the end of this chapter, but a complete listing of all metric prefixes is in Table A–2 in Appendix A. Chapter I, Introduction to Powers of 10, provides extensive coverage of the use of metric prefixes in the study of electricity and electronics.

Example 3–5

MultiSim

The I of 8 mA flows through a 5-kΩ R. How much is the IR voltage?

ANSWER

$$V = IR = 8 \times 10^{-3} \times 5 \times 10^{3} = 8 \times 5$$
$$V = 40 \text{ V}$$

In general, milliamperes multiplied by kilohms results in volts for the answer, as 10^{-3} and 10^{3} cancel.

Example 3–6

MultiSim

How much current is produced by 60 V across 12 kΩ?

ANSWER

$$I = \frac{V}{R} = \frac{60}{12 \times 10^{3}}$$
$$= 5 \times 10^{-3} = 5 \text{ mA}$$

Note that volts across kilohms produces milliamperes of current. Similarly, volts across megohms produces microamperes.

In summary, common combinations to calculate the current I are

$$\frac{\text{V}}{\text{k}\Omega} = \text{mA} \quad \text{and} \quad \frac{\text{V}}{\text{M}\Omega} = \mu\text{A}$$

Also, common combinations to calculate IR voltage are

$$\text{mA} \times \text{k}\Omega = \text{V}$$
$$\mu\text{A} \times \text{M}\Omega = \text{V}$$

These relationships occur often in electronic circuits because the current is generally in units of milliamperes or microamperes. A useful relationship to remember is that 1 mA is equal to 1000 μA.

3–5 Self-Review

Answers at the end of the chapter.

- **a. Change the following to basic units with powers of 10 instead of metric prefixes: 6 mA, 5 kΩ, and 3 μA.**
- **b. Change the following powers of 10 to units with metric prefixes: 6×10^{-3} A, 5×10^{3} Ω, and 3×10^{-6} A.**
- **c. Which is larger, 2 mA or 20 μA?**
- **d. How much current flows in a 560-kΩ resistor if the voltage is 70 V?**

3–6 The Linear Proportion between V and I

The Ohm's law formula $I = V/R$ states that V and I are directly proportional for any one value of R. This relation between V and I can be analyzed by using a fixed resistance of 2 Ω for R_L, as in Fig. 3–5. Then when V is varied, the meter shows I values directly proportional to V. For instance, with 12 V, I equals 6 A; for 10 V, the current is 5 A; an 8-V potential difference produces 4 A.

All the values of V and I are listed in the table in Fig. 3–5*b* and plotted in the graph in Fig. 3–5*c*. The I values are one-half the V values because R is 2 Ω. However, I is zero with zero volts applied.

Plotting the Graph

The voltage values for V are marked on the horizontal axis, called the *x axis* or *abscissa*. The current values I are on the vertical axis, called the *y axis* or *ordinate*.

Because the values for V and I depend on each other, they are variable factors. The independent variable here is V because we assign values of voltage and note the resulting current. Generally, the independent variable is plotted on the x axis, which is why the V values are shown here horizontally and the I values are on the ordinate.

The two scales need not be the same. The only requirement is that equal distances on each scale represent equal changes in magnitude. On the x axis here, 2-V steps are chosen, whereas the y axis has 1-A scale divisions. The zero point at the origin is the reference.

The plotted points in the graph show the values in the table. For instance, the lowest point is 2 V horizontally from the origin, and 1 A up. Similarly, the next point is at the intersection of the 4-V mark and the 2-A mark.

A line joining these plotted points includes all values of I, for any value of V, with R constant at 2 Ω. This also applies to values not listed in the table. For instance, if we take the value of 7 V up to the straight line and over to the I axis, the graph shows 3.5 A for I.

MultiSim **Figure 3–5** Experiment to show that I increases in direct proportion to V with the same R. (*a*) Circuit with variable V but constant R. (*b*) Table of increasing I for higher V. (*c*) Graph of V and I values. This is a linear volt-ampere characteristic. It shows a direct proportion between V and I.

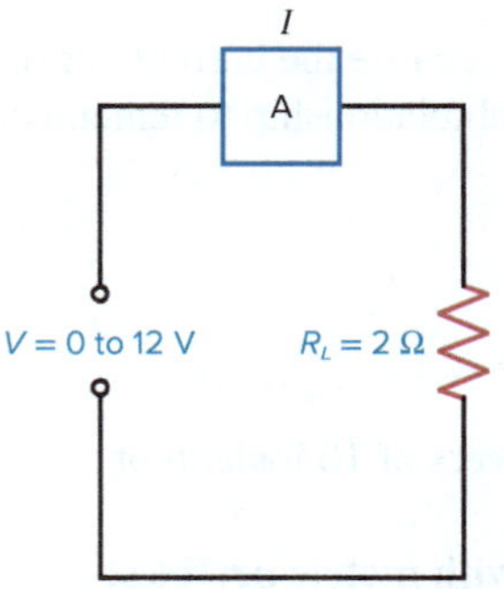

(*a*)

Volts V	Ohms Ω	Amperes A
0	2	0
2	2	1
4	2	2
6	2	3
8	2	4
10	2	5
12	2	6

(*b*)

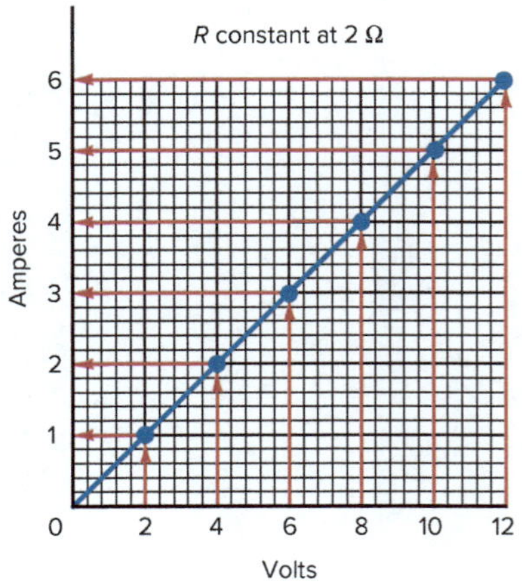

(*c*)

GOOD TO KNOW

In Fig. 3–5*c*, the slope of the straight line increases as *R* decreases. Conversely, the slope decreases as *R* increases. For any value of *R*, the slope of the straight line can be calculated as $\Delta I/\Delta V$ or $\frac{1}{R}$.

Volt-Ampere Characteristic

The graph in Fig. 3–5*c* is called the ***volt-ampere characteristic*** of *R*. It shows how much current the resistor allows for different voltages. Multiple and submultiple units of *V* and *I* can be used, though. For transistors, the units of *I* are often milliamperes or microamperes.

Linear Resistance

The *straight-line (linear) graph* in Fig. 3–5 shows that *R* is a linear resistor. A **linear resistance** has a constant value of ohms. Its *R* does not change with the applied voltage. Then *V* and *I* are directly proportional. Doubling the value of *V* from 4 to 8 V results in twice the current, from 2 to 4 A. Similarly, three or four times the value of *V* will produce three or four times *I*, for a proportional increase in current.

Nonlinear Resistance

This type of resistance has a nonlinear volt-ampere characteristic. As an example, the resistance of the tungsten filament in a lightbulb is nonlinear. The reason is that *R* increases with more current as the filament becomes hotter. Increasing the applied voltage does produce more current, but *I* does not increase in the same proportion as the increase in *V*. Another example of a nonlinear resistor is a thermistor.

Inverse Relation between *I* and *R*

Whether *R* is linear or not, the current *I* is less for more *R* with the voltage constant. This is an **inverse relation,** that is, *I* goes down as *R* goes up. Remember that in the formula $I = V/R$, the resistance is in the denominator. A higher value of *R* actually lowers the value of the complete fraction.

The inverse relation between *I* and *R* can be illustrated by using a fixed value of 12 V for the voltage, *V* in Fig. 3–6. Then when *R* is varied, the meter shows decreasing values of *I* for increasing values of *R*. For instance, in Fig. 3–6*a* when *R* equals 2 Ω, *I* equals 6 A; when *R* equals 4 Ω, *I* equals 3 A and when *R* equals 6 Ω, *I* equals 2 A. All the values of *I* and *R* are listed Table 3–6*b* and plotted in the graph of Fig. 3–6*c*.

Figure 3–6 The current (*I*) is inversely proportional to the resistance *R*, (*a*) Circuit with variable *R* but constant *V*, (*b*) Table of decreasing *I* for higher *R* values. (*c*) Graph of the *I* and R values. This graph shows that the relationship between *I* and *R* is not linear.

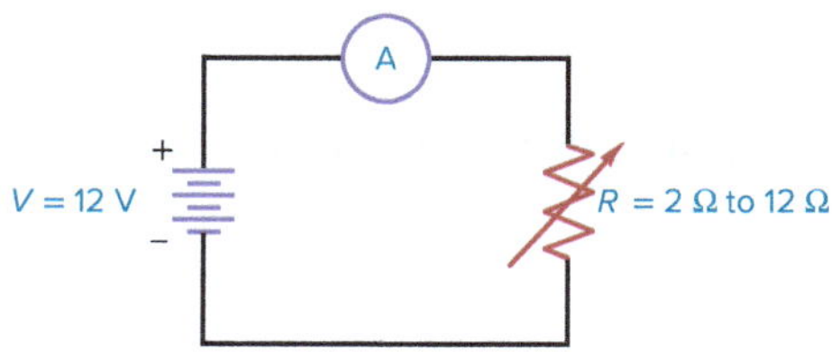

(*a*)

Ohms Ω	Volts V	Amperes A
2 Ω	12 V	6 A
4 Ω	12 V	3 A
6 Ω	12 V	2 A
8 Ω	12 V	1.5 A
10 Ω	12 V	1.2 A
12 Ω	12 V	1 A

(*b*)

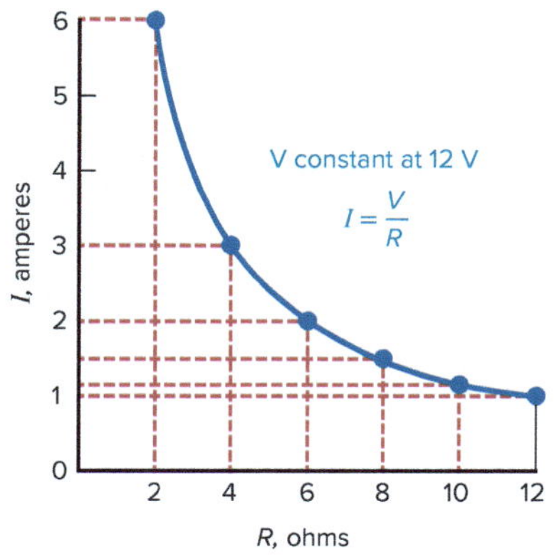

(*c*)

GOOD TO KNOW

When the voltage, V, is constant, the relationship between I and R is not linear. This means that equal changes in R do not produce equal changes in I. A graph of I versus R with V constant is called a hyperbola.

Notice that the relation between I and R is not linear. The table in Fig. 3–6*b* and the graph in Fig. 3–6*c* show that equal increases in the resistance, R do not produce equal decreases in the current, I. As a numerical example, when R is increased from 2 Ω to 4 Ω, I decreases from 6 A to 3A which is a 3 A change in current. However, when R increases from 4 Ω to 6 Ω, I decreases from 3 A to 2 A which is only a 1 A change. The graph of I versus R in Fig. 3–6*c* is called a hyperbola corresponding to the reciprocal relation $Y = 1/X$. This graph shows that the value of I drops sharply at first and then decreases more gradually as the value of R is increased.

3–6 Self-Review

Answers at the end of the chapter.

a. In Fig. 3–5*c* are the values of I on the y or x axis?

b. In Fig. 3–5*c* is R linear or nonlinear?

c. If the voltage across a 5-Ω resistor increases from 10 V to 20 V, what happens to I?

d. The voltage across a 5-Ω resistor is 10 V. If R is doubled to 10 Ω, what happens to I?

Bettmann/Getty Images

PIONEERS IN ELECTRONICS

The unit of electric power, the watt, is named for Scottish inventor and engineer *James Watt (1736–1819)*. One watt equals one joule of energy transferred in one second.

3–7 Electric Power

The unit of electric *power* is the ***watt*** **(W),** named after James Watt (1736–1819). One watt of power equals the work done in one second by one volt of potential difference in moving one coulomb of charge.

Remember that one coulomb per second is an ampere. Therefore, power in watts equals the product of volts times amperes.

$$\text{Power in watts} = \text{volts} \times \text{amperes}$$

$$P = V \times I \qquad \textbf{(3–4)}$$

When a 6-V battery produces 2 A in a circuit, for example, the battery is generating 12 W of power.

The power formula can be used in three ways:

$$P = V \times I$$

$$I = P \div V \quad \text{or} \quad \frac{P}{V}$$

$$V = P \div I \quad \text{or} \quad \frac{P}{I}$$

Which formula to use depends on whether you want to calculate P, I, or V. Note the following examples.

Example 3–7

A toaster takes 10 A from the 120-V power line. How much power is used?

ANSWER

$$P = V \times I = 120\ \text{V} \times 10\ \text{A}$$

$$P = 1200\ \text{W} \quad \text{or} \quad 1.2\ \text{kW}$$

Example 3–8

How much current flows in the filament of a 300-W bulb connected to the 120-V power line?

ANSWER

$$I = \frac{P}{V} = \frac{300\ \text{W}}{120\ \text{V}}$$

$$I = 2.5\ \text{A}$$

Example 3–9

How much current flows in the filament of a 60-W bulb connected to the 120-V power line?

ANSWER

$$I = \frac{P}{V} = \frac{60\ \text{W}}{120\ \text{V}}$$

$$I = 0.5\ \text{A} \quad \text{or} \quad 500\ \text{mA}$$

Note that the lower wattage bulb uses less current.

GOOD TO KNOW

Since $V = \frac{W}{Q}$ then $W = V \times Q$.
Therefore, $P = \frac{V \times Q}{T}$ or $P = V \times \frac{Q}{T}$.
Since $I = \frac{Q}{T}$ then $P = V \times I$.

Work, Energy, and Power

Work and **energy** are closely related terms that use the same symbol and have the same units. However, there is a significant difference between work and energy and therefore they are not interchangeable terms. Work can be defined as a force moving through a distance whereas energy can be defined as the capacity or ability to do work. It is interesting to note that energy can exist all by itself whether or not any work is being done. Work, on the other hand, requires energy.

Whenever a force is applied to an object, causing the object to be moved or displaced, work is done by the force. Energy is required to do the work of moving the object. As an example of work, if you move 100 lb a distance of 10 ft, the work is 100 lb × 10 ft or 1000 ft·lb, regardless of how fast or how slowly the work is done. Note that the unit of work is foot-pounds, without any reference to time.

However, power equals the work divided by the time it takes to do the work. If it takes 1 s, the power in this example is 1000 ft·lb/s; if the work takes 2 s, the power is 1000 ft·lb in 2 s, or 500 ft·lb/s.

Similarly, electric power is the time rate at which electric charge is forced to move by voltage. This is why power in watts is the product of volts and amperes. The voltage states the amount of work per unit of charge; the current value includes the rate at which the charge is moved.

Watts and Horsepower Units

A further example of how electric power corresponds to mechanical power is the fact that

$$746\ \text{W} = 1\ \text{hp} = 550\ \text{ft·lb/s}$$

This relation can be remembered more easily as 1 hp equals approximately ¾ kilowatt (kW). One kilowatt = 1000 W.

Bettmann/Getty Images

PIONEERS IN ELECTRONICS

The SI unit of measure for electrical energy is the joule. Named for English physicist *James Prescott Joule (1818–1889)*, one joule (J) is equal to one volt-coulomb.

GOOD TO KNOW

The joule (J) and the kilowatt-hour (kWh) are units of electrical energy.

Practical Units of Power and Work

Starting with the watt, we can develop several other important units. The fundamental principle to remember is that power is the time rate of doing work or using energy, whereas work is power used during a period of time. The formulas are

$$\text{Power} = \frac{\text{work}}{\text{time}} \qquad \textbf{(3–5)}$$

and

$$\text{Work} = \text{power} \times \text{time} \qquad \textbf{(3–6)}$$

With the watt unit for power, one watt used during one second equals the work of one **joule.** Or one watt equals one joule per second. Therefore, 1 W = 1 J/s. The **joule** is the base unit of work or energy.

To summarize these practical definitions,

$$1 \text{ joule} = 1 \text{ watt} \cdot \text{second}$$
$$1 \text{ watt} = 1 \text{ joule/second}$$

In terms of charge and current,

$$1 \text{ joule} = 1 \text{ volt} \cdot \text{coulomb}$$
$$1 \text{ watt} = 1 \text{ volt} \cdot \text{ampere}$$

Remember that the ampere unit includes time in the denominator, since the formula is 1 ampere = 1 coulomb/second.

Electron Volt (eV)

This unit of work can be used for an individual electron, rather than the large quantity of electrons in a coulomb. An electron is charge, and the volt is potential difference. Therefore, 1 eV is the amount of work required to move an electron between two points that have a potential difference of one volt.

The number of electrons in one coulomb for the joule unit equals 6.25×10^{18}. Also, the work of one joule is a volt-coulomb. Therefore, the number of **electron volts** equal to one joule must be 6.25×10^{18}. As a formula,

$$1 \text{ J} = 6.25 \times 10^{18} \text{ eV}$$

Either the electron volt or the joule unit of work is the product of charge times voltage, but the watt unit of power is the product of voltage times current. The division by time to convert work to power corresponds to the division by time that converts charge to current.

kilowatt-Hour Unit of Electrical Energy

The **kilowatt-hour (kWh)** is the unit most commonly used for large amounts of electrical work or energy. The number of kilowatt-hours is calculated simply as the product of the power in kilowatts and by the time in hours during which the power is used. As an example, if a lightbulb uses 300 W or 0.3 kW for 4 hours (h), the amount of energy is 0.3 × 4, which equals 1.2 kWh.

We pay for electricity in kilowatt-hours of energy. The power-line voltage is constant at 120 V. However, more appliances and lightbulbs require more current because they all add in the main line to increase the power.

Suppose that the total load current in the main line equals 20 A. Then the power in watts from the 120-V line is

$$P = 120 \text{ V} \times 20 \text{ A}$$
$$P = 2400 \text{ W} \quad \text{or} \quad 2.4 \text{ kW}$$

If this power is used for 5 h, then the energy or work supplied equals $2.4 \times 5 = 12$ kWh. If the cost of electricity is 12¢/kWh, then 12 kWh of electricity will cost $0.12 \times 12 = \$1.44$. This charge is for a 20-A load current from the 120-V line during the time of 5 h.

As a streamlined approach to calculating energy costs, follow the steps listed below.

1. Calculate the total power in kilowatts (kW).
2. Convert the number of watts (W) to kilowatts (kW) if necessary. $\left(\#\text{kW} = \#\text{W} \times \dfrac{1\text{ kW}}{1000\text{ W}}\right)$
3. Calculate the total number of hours (h) the power is used. $\left(\#\text{h} = \#\text{days} \times \dfrac{24\text{ h}}{\text{day}}\right)$
4. Multiply the number of kW by the number of hours (h) to get the number of kilowatt-hours (kWh).
5. Multiply the number of kWh by the cost/kWh to get the cost of energy consumption. $\left(\text{Cost (\$)} = \#\text{kWh} \times \dfrac{\text{cost (\$)}}{\text{kWh}}\right)$

Example 3–10

Assuming that the cost of electricity is 12 ¢/kWh, how much will it cost to light a 100-W lightbulb for 30 days?

ANSWER The first step in solving this problem is to express 100 W as 0.1 kW. The next step is to find the total number of hours in 30 days. Since there are 24 hours in a day, the total number of hours the light is on is calculated as

$$\text{Total hours} = \frac{24\text{ h}}{\text{day}} \times 30\text{ days} = 720\text{ h}$$

Next, calculate the number of kWh as

$$\begin{aligned}\text{kWh} &= \text{kW} \times \text{h}\\ &= 0.1\text{ kW} \times 720\text{ h}\\ &= 72\text{ kWh}\end{aligned}$$

And finally, determine the cost. (Note that 12¢ = \$0.12.)

$$\begin{aligned}\text{Cost} &= \text{kWh} \times \frac{\text{cost}}{\text{kWh}}\\ &= 72\text{ kWh} \times \frac{\$0.12}{\text{kWh}}\\ &= \$8.64\end{aligned}$$

3–7 Self-Review

Answers at the end of the chapter.

a. **An electric heater takes 15 A from the 120-V power line. Calculate the amount of power used.**
b. **How much is the load current for a 100-W bulb connected to the 120-V power line?**
c. **How many watts is the power of 200 J/s equal to?**
d. **How much will it cost to operate a 300-W lightbulb for 48 h if the cost of electricity is 15¢/kWh?**

GOOD TO KNOW

The power dissipated by a resistance is proportional to I^2. In other words, if the current, I, carried by a resistor is doubled, the power dissipation in the resistance increases by a factor of 4.

GOOD TO KNOW

The power dissipated by a resistance is proportional to V^2. In other words, if the voltage, V, across a resistor is doubled, the power dissipation in the resistance increases by a factor of 4.

GOOD TO KNOW

In the distribution of electric power, the power transmission lines often use a very high voltage such as 160 kV or more. With such a high voltage, the current carried by the transmission lines can be kept low to transmit the desired power from one location to another. The reduction in I with the much higher V considerably reduces the I^2R power losses in the transmission line conductors.

3–8 Power Dissipation in Resistance

When current flows in a resistance, heat is produced because friction between the moving free electrons and the atoms obstructs the path of electron flow. The heat is evidence that power is used in producing current. This is how a fuse opens, as heat resulting from excessive current melts the metal link in the fuse.

The power is generated by the source of applied voltage and consumed in the resistance as heat. As much power as the resistance dissipates in heat must be supplied by the voltage source; otherwise, it cannot maintain the potential difference required to produce the current.

The correspondence between electric power and heat is indicated by the fact that 1 W used during 1 s is equivalent to 0.24 calorie of heat energy. The electric energy converted to heat is considered dissipated or used up because the calories of heat cannot be returned to the circuit as electric energy.

Since power is dissipated in the resistance of a circuit, it is convenient to express the power in terms of the resistance R. The formula $P = V \times I$ can be rearranged as follows:

Substituting IR for V,

$$P = V \times I = IR \times I$$

$$P = I^2R \qquad \textbf{(3–7)}$$

This is a common form of the power formula because of the heat produced by current in a resistance.

For another form, substitute V/R for I. Then

$$P = V \times I = V \times \frac{V}{R}$$

$$P = \frac{V^2}{R} \qquad \textbf{(3–8)}$$

In all the formulas, V is the voltage across R in ohms, producing the current I in amperes, for power in watts.

Any one of the three formulas (3–4), (3–7), and (3–8) can be used to calculate the power dissipated in a resistance. The one to be used is a matter of convenience, depending on which factors are known.

In Fig. 3–7, for example, the power dissipated with 2 A through the resistance and 12 V across it is 2 × 12 = 24 W.

Or, calculating in terms of just the current and resistance, the power is the product of 2 squared, or 4, times 6, which equals 24 W.

Using the voltage and resistance, the power can be calculated as 12 squared, or 144, divided by 6, which also equals 24 W.

No matter which formula is used, 24 W of power is dissipated as heat. This amount of power must be generated continuously by the battery to maintain the potential difference of 12 V that produces the 2-A current against the opposition of 6 Ω.

MultiSim **Figure 3–7** Calculating the electric power in a circuit as $P = V \times I$, $P = I^2R$, or $P = V^2/R$.

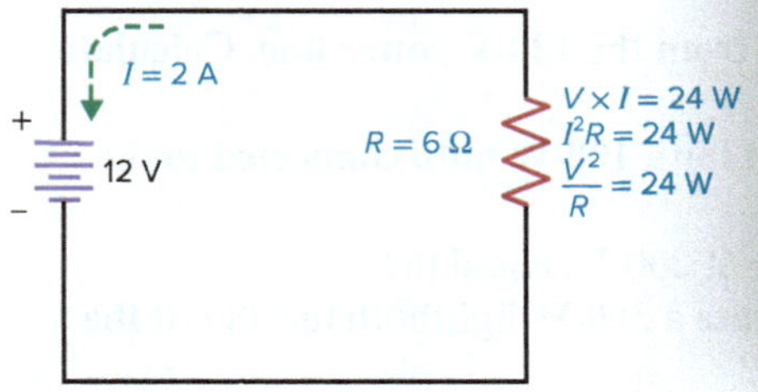

Example 3–11

MultiSim

Calculate the power in a circuit where the source of 100 V produces 2 A in a 50-Ω R.

ANSWER

$$P = I^2R = 2 \times 2 \times 50 = 4 \times 50$$

$$P = 200 \text{ W}$$

This means that the source delivers 200 W of power to the resistance and the resistance dissipates 200 W as heat.

CALCULATOR

To use the calculator for a problem like Example 3–12, in which I must be squared for $I^2 \times R$, use the following procedure:

- Punch in the value of 4 for I.
- Press the key marked (x^2) for the square of 4 equal to 16 on the display.
- Next, press the multiplication (×) key.
- Punch in the value of 25 for R.
- Finally, press the (=) key for the answer of 400 on the display.

Be sure to square only the I value before multiplying by the R value.

GOOD TO KNOW

In order to reduce the power dissipation in a given resistance by one-half, the voltage V and the current I must be reduced to 70.7% of their original values. To reduce the power dissipation by one-half for a given voltage V, the resistance R must be doubled. However, to reduce the power dissipation by one-half for a given current I, the resistance R must be cut in half.

GOOD TO KNOW

Every home appliance has a power rating and a voltage rating. To calculate the current drawn by an appliance, simply divide the power rating by the voltage rating.

Example 3–12

MultiSim

Calculate the power in a circuit in which the same source of 100 V produces 4 A in a 25-Ω R.

ANSWER

$$P = I^2R = 4^2 \times 25 = 16 \times 25$$
$$P = 400 \text{ W}$$

Note the higher power in Example 3–12 because of more I, even though R is less than that in Example 3–11.

In some applications, electric power dissipation is desirable because the component must produce heat to do its job. For instance, a 600-W toaster must dissipate this amount of power to produce the necessary amount of heat. Similarly, a 300-W lightbulb must dissipate this power to make the filament white-hot so that it will have the incandescent glow that furnishes the light. In other applications, however, the heat may be just an undesirable by-product of the need to provide current through the resistance in a circuit. In any case, though, whenever there is current I in a resistance R, it dissipates the amount of power P equal to I^2R.

Components that use the power dissipated in their resistance, such as lightbulbs and toasters, are generally rated in terms of power. The power rating is given at normal applied voltage, which is usually the 120 V of the power line. For instance, a 600-W, 120-V toaster has this rating because it dissipates 600 W in the resistance of the heating element when connected across 120 V.

Note this interesting point about the power relations. The lower the source voltage, the higher the current required for the same power. The reason is that $P = V \times I$. For instance, an electric heater rated at 240 W from a 120-V power line takes 240 W/120 V = 2 A of current from the source. However, the same 240 W from a 12-V source, as in a car or boat, requires 240 W/12 V = 20 A. More current must be supplied by a source with lower voltage, to provide a specified amount of power.

3–8 Self-Review

Answers at the end of the chapter.

a. **Current I is 2 A in a 5-Ω R. Calculate P.**
b. **Voltage V is 10 V across a 5-Ω R. Calculate P.**
c. **Resistance R has 10 V with 2 A. Calculate the values for P and R.**

3–9 Power Formulas

To calculate I or R for components rated in terms of power at a specified voltage, it may be convenient to use the power formulas in different forms. There are three basic power formulas, but each can be in three forms for nine combinations.

$$P = VI \qquad P = I^2R \qquad P = \frac{V^2}{R}$$

$$\text{or} \quad I = \frac{P}{V} \quad \text{or} \quad R = \frac{P}{I^2} \quad \text{or} \quad R = \frac{V^2}{P}$$

$$\text{or} \quad V = \frac{P}{I} \quad \text{or} \quad I = \sqrt{\frac{P}{R}} \quad \text{or} \quad V = \sqrt{PR}$$

CALCULATOR

To use the calculator for a problem like Example 3–14 that involves a square and division for V^2/R, use the following procedure:

- Punch in the V value of 120.
- Press the key marked (x^2) for the square of 120, equal to 14,400 on the display.
- Next, press the division (÷) key.
- Punch in the value of 600 for R.
- Finally, press the (=) key for the answer of 24 on the display. Be sure to square only the numerator before dividing.

CALCULATOR

For Example 3–15 with a square root and division, be sure to divide first, so that the square root is taken for the quotient, as follows:

- Punch in the P of 600.
- Press the division (÷) key.
- Punch in 24 for R.
- Press the (=) key for the quotient of 25.

Then press the (√) key for the square root. This key may be a second function of the same key for squares. If so, press the key marked (2nd F) or (SHIFT) before pressing the (√) key. As a result, the square root equal to 5 appears on the display. You do not need the (=) key for this answer. In general, the (=) key is pressed only for the multiplication, division, addition, and subtraction operations.

Example 3–13

How much current is needed for a 600-W, 120-V toaster?

ANSWER

$$I = \frac{P}{V} = \frac{600}{120}$$

$$I = 5\ \text{A}$$

Example 3–14

How much is the resistance of a 600-W, 120-V toaster?

ANSWER

$$R = \frac{V^2}{P} = \frac{(120)^2}{600} = \frac{14{,}400}{600}$$

$$R = 24\ \Omega$$

Example 3–15

How much current is needed for a 24-Ω R that dissipates 600 W?

ANSWER

$$I = \sqrt{\frac{P}{R}} = \sqrt{\frac{600\ \text{W}}{24\ \Omega}} = \sqrt{25}$$

$$I = 5\ \text{A}$$

Note that all these formulas are based on Ohm's law $V = IR$ and the power formula $P = VI$. The following example with a 300-W bulb also illustrates this idea. Refer to Fig. 3–8. The bulb is connected across the 120-V line. Its 300-W filament requires a current of 2.5 A, equal to P/V. These calculations are

$$I = \frac{P}{V} = \frac{300\ \text{W}}{120\ \text{V}} = 2.5\ \text{A}$$

The proof is that the VI product is 120 × 2.5, which equals 300 W.

Figure 3–8 All formulas are based on Ohm's law.

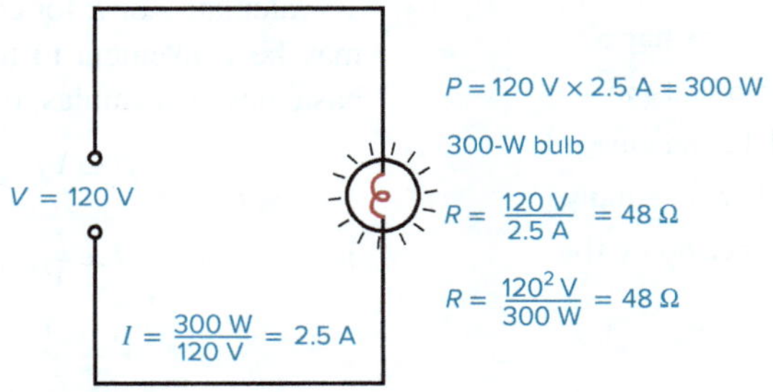

Furthermore, the resistance of the filament, equal to V/I, is 48 Ω. These calculations are

$$R = \frac{V}{I} = \frac{120\ \text{V}}{2.5\ \text{A}} = 48\ \Omega$$

If we use the power formula $R = V^2/P$, the answer is the same 48 Ω. These calculations are

$$R = \frac{V^2}{P} = \frac{120^2}{300}$$

$$R = \frac{14{,}400}{300} = 48\ \Omega$$

In any case, when this bulb is connected across 120 V so that it can dissipate its rated power, the bulb draws 2.5 A from the power line and the resistance of the white-hot filament is 48 Ω.

3–9 Self-Review

Answers at the end of the chapter.

a. How much is the *R* of a 100-W, 120-V lightbulb?
b. How much power is dissipated by a 2-Ω *R* with 10 V across it?
c. Calculate *P* for 2 A of *I* through a 2-Ω resistor.

3–10 Choosing a Resistor for a Circuit

When choosing a resistor for a circuit, first determine the required resistance value as $R = \frac{V}{I}$. Next, calculate the amount of power dissipated by the resistor using any one of the power formulas. Then, select a wattage rating for the resistor that will provide a reasonable amount of cushion between the actual power dissipation and the power rating of the resistor. Ideally, the power dissipation in a resistor should never be more than 50% of its power rating, which is a safety factor of 2. A safety factor of 2 allows the resistor to operate at a cooler temperature and thus last longer without breaking down from excessive heat. In practice, however, as long as the safety factor is reasonably close to 2, the resistor will not overheat.

Example 3–16

Determine the required resistance and appropriate wattage rating of a resistor to meet the following requirements: The resistor must have a 30-V *IR* drop when its current is 20 mA. The resistors available have the following wattage ratings: ⅛, ¼ , ½ , 1, and 2 W.

ANSWER First, calculate the required resistance.

$$\begin{aligned} R &= \frac{V}{I} \\ &= \frac{30\ \text{V}}{20\ \text{mA}} \\ &= 1.5\ \text{k}\Omega \end{aligned}$$

Next, calculate the power dissipated by the resistor using the formula $P = I^2R$.

$$\begin{aligned} P &= I^2R \\ &= (20\ \text{mA})^2 \times 1.5\ \text{k}\Omega \\ &= 0.6\ \text{W} \quad \text{or} \quad 600\ \text{mW} \end{aligned}$$

Now, select a suitable wattage rating for the resistor. In this example, a 1-W rating provides a safety factor that is reasonably close to 2. A resistor with a higher wattage rating could be used if there is space available for it to be mounted. In summary, a 1.5-kΩ, 1-W resistor will safely dissipate 600 mW of power while providing an *IR* voltage of 30 V when the current is 20 mA.

Maximum Working Voltage Rating

The **maximum working voltage rating** of a resistor is the maximum allowable voltage that the resistor can safely withstand without internal arcing. The higher the wattage rating of the resistor, the higher the maximum working voltage rating. For carbon-film resistors, the following voltage ratings are typical:

⅛ W – 150 V
¼ W – 250 V
½ W – 350 V
1 W – 500 V

It is interesting to note that with very large resistance values, the maximum working voltage rating may actually be exceeded before the power rating is exceeded. For example, a 1 MΩ, ¼ W carbon-film resistor with a maximum working voltage rating of 250 V, does not dissipate ¼ W of power until its voltage equals 500 V. Since 500 V exceeds its 250 V rating, internal arcing will occur within the resistor. Therefore, 250 V rather than 500 V is the maximum voltage that can safely be applied across this resistor. With 250 V across the 1-MΩ resistor, the actual power dissipation is 1⁄16 W which is only one-fourth its power rating.

For any resistor, the maximum voltage that produces the rated power dissipation is calculated as

$$V_{max} = \sqrt{P_{rating} \times R}$$

Exceeding V_{max} causes the resistor's power dissipation to exceed its power rating. Except for very large resistance values, the maximum working voltage rating is usually much larger than the maximum voltage that produces the rated power dissipation.

Example 3–17

Determine the required resistance and appropriate wattage rating of a carbon-film resistor to meet the following requirements: The resistor must have a 225-V *IR* drop when its current is 150 μA. The resistors available have the following wattage ratings: ⅛, ¼ , ½ , 1, and 2 W.

ANSWER First, calculate the required resistance.

$$R = \frac{V}{I}$$

$$= \frac{225\ \text{V}}{150\ \mu\text{A}}$$

$$= 1.5\ \text{M}\Omega$$

Next, calculate the power dissipated by the resistor using the formula $P = I^2R$.

$$P = I^2R$$
$$= (150\ \mu A)^2 \times 1.5\ M\Omega$$
$$= 33.75\ mW$$

Now, select a suitable wattage rating for the resistor.

In this application, a ⅛-W (125 mW) resistor could be considered because it will provide a safety factor of nearly 4. However, a ⅛-W resistor could not be used because its maximum working voltage rating is only 150 V and the resistor must be able to withstand a voltage of 225 V. Therefore, a higher wattage rating must be chosen just because it will have a higher maximum working voltage rating. In this application, a ½-W resistor would be a reasonable choice because it has a 350-V rating. A ¼-W resistor provides a 250-V rating which is only 25 V more than the actual voltage across the resistor. It's a good idea to play it safe and go with the higher voltage rating offered by the ½-W resistor. In summary, a 1.5-MΩ, ½-W resistor will safely dissipate 33.75 mW of power as well as withstand a voltage of 225 V.

3–10 Self-Review

Answers at the end of the chapter.

a. What is the maximum voltage that a 10-kΩ, ¼-W resistor can safely handle without exceeding its power rating? If the resistor has a 250-V maximum working voltage rating, is this rating being exceeded?

b. Determine the required resistance and appropriate wattage rating of a carbon-film resistor for the following conditions: the *IR* voltage must equal 100 V when the current is 100 μA. The available wattage ratings for the resistor are ⅛, ¼, ½, 1, and 2 W.

3–11 Electric Shock

While working on electric circuits, there is often the possibility of receiving an electric shock by touching the "live" conductors when the power is on. The shock is a sudden involuntary contraction of the muscles, with a feeling of pain, caused by current through the body. If severe enough, the shock can be fatal. Safety first, therefore, should always be the rule.

The greatest shock hazard is from high-voltage circuits that can supply appreciable amounts of power. The resistance of the human body is also an important factor. If you hold a conducting wire in each hand, the resistance of the body across the conductors is about 10,000 to 50,000 Ω. Holding the conductors tighter lowers the resistance. If you hold only one conductor, your resistance is much higher. It follows that the higher the body resistance, the smaller the current that can flow through you.

A safety tip, therefore, is to work with only one of your hands if the power is on. Place the other hand behind your back or in your pocket. Therefore, if a live circuit is touched with only one hand, the current will normally not flow directly through the heart. Also, keep yourself insulated from earth ground when working on power-line circuits, since one side of the power line is connected to earth ground. The final and best safety rule is to work on circuits with the power disconnected if at all possible and make resistance tests.

Note that it is current through the body, not through the circuit, which causes the electric shock. This is why high-voltage circuits are most important, since sufficient potential difference can produce a dangerous amount of current through the

Figure 3–9 Physiological effects of electric current.

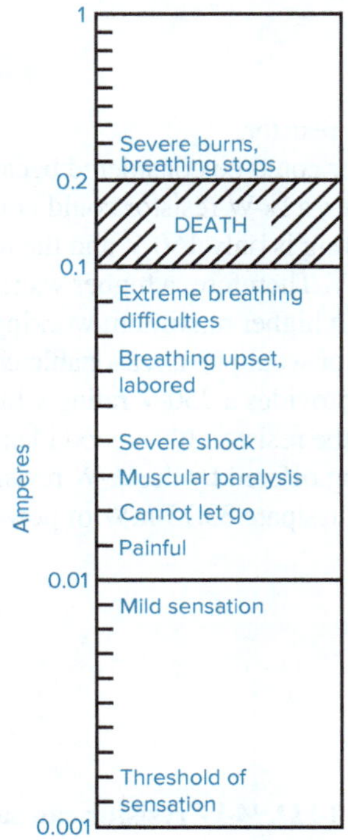

relatively high resistance of the body. For instance, 500 V across a body resistance of 25,000 Ω produces 0.02 A, or 20 mA, which can be fatal. As little as 1 mA through the body can cause an electric shock. The chart shown in Fig. 3–9 is a visual representation of the physiological effects of an electric current on the human body. As the chart shows, the threshold of sensation occurs when the current through the body is only slightly above 0.001 A or 1 mA. Slightly above 10 mA, the sensation of current through the body becomes painful and the person can no longer let go or free him or herself from the circuit. When the current through the body exceeds approximately 100 mA, the result is usually death.

In addition to high voltage, the other important consideration in how dangerous the shock can be is the amount of power the source can supply. A current of 0.02 A through 25,000 Ω means that the body resistance dissipates 10 W. If the source cannot supply 10 W, its output voltage drops with the excessive current load. Then the current is reduced to the amount corresponding to the amount of power the source can produce.

In summary, then, the greatest danger is from a source having an output of more than about 30 V with enough power to maintain the load current through the body when it is connected across the applied voltage. In general, components that can supply high power are physically big because of the need for dissipating heat.

3–11 Self-Review

Answers at the end of the chapter.

a. The potential difference of 120 V is more dangerous than 12 V for electric shock. (True/False)

b. Resistance in a circuit should be measured with its power off. (True/False)

3–12 Open-Circuit and Short-Circuit Troubles

Ohm's law is useful for calculating *I*, *V*, and *R* in a closed circuit with normal values. However, an **open circuit** or a **short circuit** causes trouble that can be summarized as follows: An open circuit (Fig. 3–10) has zero *I* because *R* is infinitely high. It does not matter how much the *V* is. A short circuit has zero *R*, which causes excessively high *I* in the short-circuit path because of no resistance (Fig. 3–11).

In Fig. 3–10*a*, the circuit is normal with *I* of 2 A produced by 10 V applied across *R* of 5 Ω. However, the resistor is shown open in Fig. 3–10*b*. Then the path for current has infinitely high resistance and there is no current in any part of the circuit. The trouble can be caused by an internal open in the resistor or a break in the wire conductors.

In Fig. 3–11*a*, the same normal circuit is shown with *I* of 2 A. In Fig. 3–11*b*, however, there is a short-circuit path across *R* with zero resistance. The result is excessively high current in the short-circuit path, including the wire conductors. It may be surprising, but there is no current in the resistor itself because all the current is in the zero-resistance path around it.

Figure 3–10 Effect of an open circuit. (*a*) Normal circuit with current of 2 A for 10 V across 5 Ω. (*b*) Open circuit with no current and infinitely high resistance.

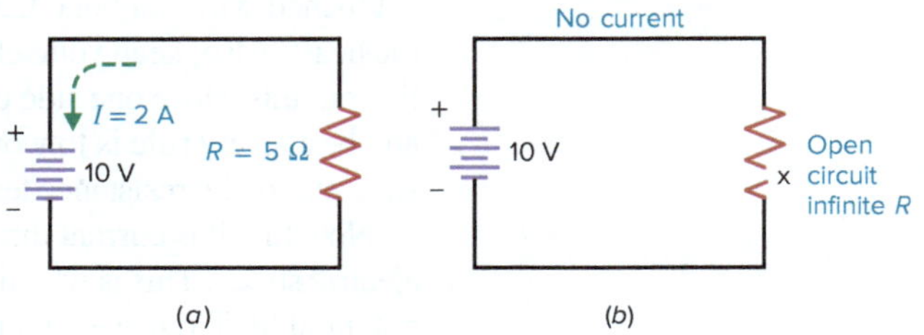

Figure 3-11 Effect of a short circuit. (*a*) Normal circuit with current of 2 A for 10 V across 5 Ω. (*b*) Short circuit with zero resistance and excessively high current.

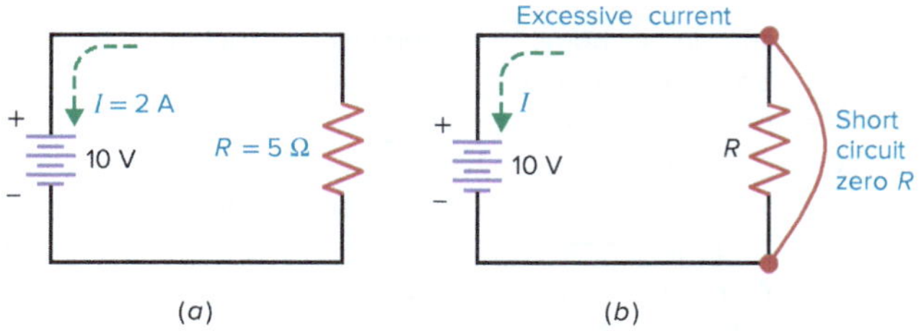

Theoretically, the amount of current could be infinitely high with no *R*, but the voltage source can supply only a limited amount of *I* before it loses its ability to provide voltage output. The wire conductors may become hot enough to burn open, which would open the circuit. Also, if there is any fuse in the circuit, it will open because of the excessive current produced by the short circuit.

Note that the resistor itself is not likely to develop a short circuit because of the nature of its construction. However, the wire conductors may touch, or some other component in a circuit connected across the resistor may become short-circuited.

3–12 Self-Review

Answers at the end of the chapter.

a. **An open circuit has zero current. (True/False)**

b. **A short circuit has excessive current. (True/False)**

c. **An open circuit and a short circuit have opposite effects on resistance and current. (True/False)**

Eric Tormey/Alamy Stock Photo

HOME APPLIANCES

Every electrical appliance in our home has a ***nameplate*** attached to it. The nameplate provides important information about the appliance such as its make and model, its electrical specifications and the Underwriters Laboratories (UL) listing mark. The nameplate is usually located on the bottom or rear-side of the appliance. The electrical specifications listed are usually its power and voltage ratings. The voltage rating is the voltage at which the appliance is designed to operate. The power rating is the power dissipation of the appliance when the rated voltage is applied. With the rated voltage and power ratings listed on the nameplate, we can calculate the current drawn from the appliance when it's being used. To calculate the current (I) simply divide the power rating (P) in watts by the voltage rating (V) in volts. As an example, suppose you want to know how much current your toaster draws when it's toasting your bread. To find the answer you will probably need to turn your toaster upside down to locate its nameplate. For example, the toaster in Fig. 3–12*a* has the nameplate shown in Fig. 3–12*b*. With a voltage rating of 120 V and a power rating of 850 W, the current drawn by the toaster is calculated as follows;

$$I = \frac{P}{V} = \frac{850\ W}{120\ V} = 7.083\ \text{A}$$

Some appliances in our homes have a voltage rating of 240 V rather than 120 V. These are typically the appliances with very high power ratings. Some examples include: electric stoves, electric clothes dryers, electric water heaters, and air conditioning units. These appliances may have power ratings as high as 7.2 kW or more. The reason the higher power appliances have a higher voltage rating is simple. At twice the voltage you only need half the current to obtain the desired power. With half as much current, the size of the conductors connecting the appliance to the power line can be kept much smaller. This is important because a smaller diameter wire costs less and is physically much easier to handle.

Table 3–1 lists the electrical specifications of some common household appliances.

Figure 3–12 Identifying electrical specifications of household appliances. (*a*) Toaster (*b*) Nameplate on bottom of toaster

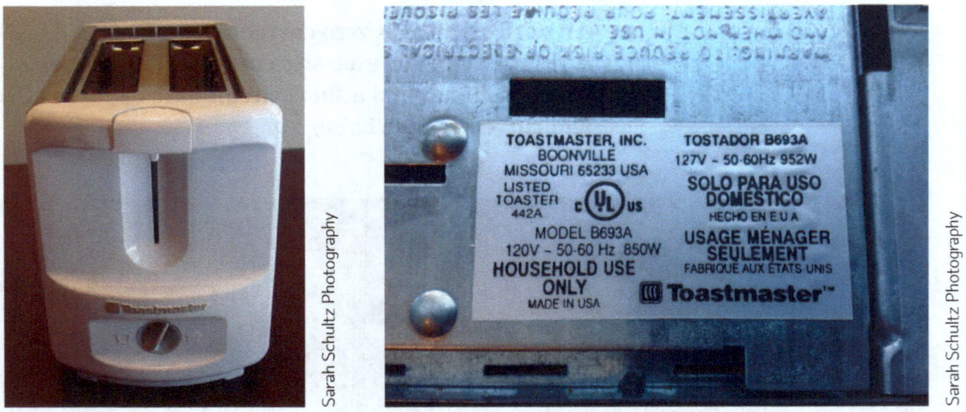

Table 3–1 Electrical Specifications of Common Household Appliances

Appliance	Power Rating	Voltage Rating
Clothes Iron	1.2 kW	120 V
Toaster	800 – 1500 W	120 V
Microwave	600 – 1500 W	120 V
Waffle Iron	1.2 kW	120 V
Coffee Maker	800 – 1200 W	120 V
Blow Dryer	1 kW	120 V
Flat Screen TV	100 – 150 W	120 V
Dishwasher	1200 – 2400 W	120 V
Water Heater	4.5 – 5.5 kW	240 V
Electric Stove	9.6 kW	240 V

Summary

- The three forms of Ohm's law are $I = V/R$, $V = IR$, and $R = V/I$.
- One ampere is the amount of current produced by one volt of potential difference across one ohm of resistance. This current of 1 A is the same as 1 C/s.
- With R constant, the amount of I increases in direct proportion as V increases. This linear relation between V and I is shown by the graph in Fig. 3–5.
- With V constant, the current I decreases as R increases. This is an inverse relation. The inverse relation between I and R is shown in Fig. 3–6.
- Power is the time rate of doing work or using energy. The unit is the watt. One watt equals 1 V × 1 A. Also, watts = joules per second.
- The unit of work or energy is the joule. One joule equals 1 W × 1 s. A larger unit of electrical energy is the kilowatt-hour (kWh). 1 kWh = 1 kW × 1 h.
- The most common multiples and submultiples of the practical units are listed in Table 3–2.
- Voltage applied across your body can produce a dangerous electric shock. Whenever possible, shut off the power and make resistance tests. If the power must be on, use only one hand when making measurements. Place your other hand behind your back or in your pocket.
- Table 3–3 summarizes the practical units of electricity.
- An open circuit has no current and infinitely high R. A short circuit has zero resistance and excessively high current.

Table 3–2 Summary of Conversion Factors

Prefix	Symbol	Relation to Basic Unit	Examples
mega	M	1,000,000 or 1×10^6	5 MΩ (megohms) = 5,000,000 ohms = 5×10^6 ohms
kilo	k	1000 or 1×10^3	18 kV (kilovolts) = 18,000 volts = 18×10^3 volts
milli	m	0.001 or 1×10^{-3}	48 mA (milliamperes) = 48×10^{-3} amperes = 0.048 amperes
micro	μ	0.000 001 or 1×10^{-6}	15 μV (microvolts) = 15×10^{-6} volts = 0.000 015 volts

Table 3–3 Summary of Practical Units of Electricity

Coulomb	Ampere	Volt	Watt	Ohm	Siemens
6.25×10^{18} electrons	$\frac{\text{coulomb}}{\text{second}}$	$\frac{\text{joule}}{\text{coulomb}}$	$\frac{\text{joule}}{\text{second}}$	$\frac{\text{volt}}{\text{ampere}}$	$\frac{\text{ampere}}{\text{volt}}$

Important Terms

Ampere — the basic unit of current. $1\text{ A} = \frac{1\text{ V}}{1\ \Omega}$.

Electron volt (eV) — a small unit of work or energy that represents the amount of work required to move a single electron between two points having a potential difference of 1 volt.

Energy — the ability or capacity to do work.

Horsepower (hp) — a unit of mechanical power corresponding to 550 ft·lb/s. In terms of electric power, 1 hp = 746 W.

Inverse relation — a relation in which the quotient of a fraction decreases as the value in the denominator increases with the numerator constant. In the equation $I = \frac{V}{R}$, I and R are inversely related because I decreases as R increases with V constant.

Joule — a practical unit of work or energy. 1 J = 1 W · 1 s.

Kilowatt-hour (kWh) — a large unit of electrical energy corresponding to 1 kW · 1 h.

Linear proportion — a relation between two quantities which shows how equal changes in one quantity produce equal changes in the other. In the equation $I = \frac{V}{R}$, I and V are directly proportional because equal changes in V produce equal changes in I with R constant.

Linear resistance — a resistance with a constant value of ohms.

Maximum working voltage rating — the maximum allowable voltage that a resistor can safely withstand without internal arcing.

Nonlinear resistance — a resistance whose value changes as a result of current producing power dissipation and heat in the resistance.

Ohm — the basic unit of resistance. $1\ \Omega = \frac{1\text{ V}}{1\text{ A}}$.

Open circuit — a broken or incomplete current path with infinitely high resistance.

Power — the time rate of doing work. $\text{Power} = \frac{\text{Work}}{\text{Time}}$.

Short circuit — a very low resistance path around or across a component such as a resistor. A short circuit with very low R can have excessively high current.

Volt — the basic unit of potential difference or voltage. 1 V = 1 A · 1 Ω

Volt ampere characteristic — a graph showing how much current a resistor allows for different voltages.

Watt — the basic unit of electric power. $1\text{ W} = \frac{1\text{ J}}{\text{s}}$.

Work — a force acting through a distance causing the displacement or movement of an object.

Related Formulas

$I = \frac{V}{R}$

$V = I \times R$

$R = \frac{V}{I}$

$P = V \times I$

$I = \frac{P}{V}$

$V = \frac{P}{I}$

1 hp = 746 W

1 J = 1 W × 1 s

$1\text{ W} = \frac{1\text{ J}}{1\text{ s}}$

1 J = 1 V × 1 C

$1\text{ J} = 6.25 \times 10^{18}\text{ eV}$

$P = I^2R$

$I = \sqrt{\frac{P}{R}}$

$R = \frac{P}{I^2}$

$P = \frac{V^2}{R}$

$V = \sqrt{PR}$

$R = \frac{V^2}{P}$

Self-Test

Answers at the back of the book.

1. With 24 V across a 1-kΩ resistor, the current, *I*, equals
a. 0.24 A.
b. 2.4 mA.
c. 24 mA.
d. 24 μA.

2. With 30 μA of current in a 120-kΩ resistor, the voltage, *V*, equals
a. 360 mV.
b. 3.6 kV.
c. 0.036 V.
d. 3.6 V.

3. How much is the resistance in a circuit if 15 V of potential difference produces 500 μA of current?
a. 30 kΩ.
b. 3 MΩ.
c. 300 kΩ.
d. 3 kΩ.

4. **A current of 1000 μA equals**
   a. 1 A.
   b. 1 mA.
   c. 0.01 A.
   d. none of the above.

5. **One horsepower equals**
   a. 746 W.
   b. 550 ft·lb/s.
   c. approximately ¾ kW.
   d. all of the above.

6. **With *R* constant**
   a. *I* and *P* are inversely related.
   b. *V* and *I* are directly proportional.
   c. *V* and *I* are inversely proportional.
   d. none of the above.

7. **One watt of power equals**
   a. 1 V × 1 A.
   b. $\frac{1\ \text{J}}{\text{s}}$
   c. $\frac{1\ \text{C}}{\text{s}}$
   d. both a and b.

8. **A 10-Ω resistor dissipates 1 W of power when connected to a DC voltage source. If the value of DC voltage is doubled, the resistor will dissipate**
   a. 1 W.
   b. 2 W.
   c. 4 W.
   d. 10 W.

9. **If the voltage across a variable resistance is held constant, the current, *I*, is**
   a. inversely proportional to resistance.
   b. directly proportional to resistance.
   c. the same for all values of resistance.
   d. both a and b.

10. **A resistor must provide a voltage drop of 27 V when the current is 10 mA. Which of the following resistors will provide the required resistance and appropriate wattage rating?**
   a. 2.7 kΩ, ⅛ W.
   b. 270 Ω, ½ W.
   c. 2.7 kΩ, ½ W
   d. 2.7 kΩ, ¼ W

11. **The resistance of an open circuit is**
   a. approximately 0 Ω.
   b. infinitely high.
   c. very low.
   d. none of the above.

12. **The current in an open circuit is**
   a. normally very high because the resistance of an open circuit is 0 Ω.
   b. usually high enough to blow the circuit fuse.
   c. zero.
   d. slightly below normal.

13. **Which of the following safety rules should be observed while working on a live electric circuit?**
   a. Keep yourself well insulated from earth ground.
   b. When making measurements in a live circuit place one hand behind your back or in your pocket.
   c. Make resistance measurements only in a live circuit.
   d. Both a and b.

14. **How much current does a 75-W lightbulb draw from the 120-V power line?**
   a. 625 mA.
   b. 1.6 A.
   c. 160 mA.
   d. 62.5 mA.

15. **The resistance of a short circuit is**
   a. infinitely high.
   b. very high.
   c. usually above 1 kΩ.
   d. approximately zero.

16. **Which of the following is considered a linear resistance?**
   a. lightbulb.
   b. thermistor.
   c. 1-kΩ, ½-W carbon-film resistor.
   d. both a and b.

17. **How much will it cost to operate a 4-kW air-conditioner for 12 hours if the cost of electricity is 12¢/kWh?**
   a. $5.76.
   b. 57.6¢.
   c. $576.
   d. $4.80.

18. **What is the maximum voltage a 150-Ω, ⅛-W resistor can safely handle without exceeding its power rating? (Assume no power rating safety factor.)**
   a. 18.75 V.
   b. 4.33 V.
   c. 6.1 V.
   d. 150 V.

19. **Which of the following voltages provides the greatest danger in terms of electric shock?**
   a. 12 V.
   b. 10,000 mV.
   c. 120 V.
   d. 9 V.

20. **If a short circuit is placed across the leads of a resistor, the current in the resistor itself would be**
   a. zero.
   b. much higher than normal.
   c. the same as normal.
   d. excessively high.

Essay Questions

1. State the three forms of Ohm's law relating *V*, *I*, and *R*.
2. (a) Why does higher applied voltage with the same resistance result in more current? (b) Why does more resistance with the same applied voltage result in less current?
3. Calculate the resistance of a 300-W bulb connected across the 120-V power line, using two different methods to arrive at the same answer.
4. State which unit in each of the following pairs is larger: (a) volt or kilovolt; (b) ampere or milliampere; (c) ohm or megohm; (d) volt or microvolt; (e) siemens or microsiemens; (f) electron volt or joule; (g) watt or kilowatt; (h) kilowatt-hour or joule; (i) volt or millivolt; (j) megohm or kilohm.
5. State two safety precautions to follow when working on electric circuits.

6. Explain the difference between work and energy.

7. Referring back to the resistor shown in Fig. 1–10 in Chapter 1, suppose that it is not marked. How could you determine its resistance by Ohm's law? Show your calculations that result in the V/I ratio of 10 kΩ. However, do not exceed the power rating of 10 W.

8. Give three formulas for electric power.

9. What is the difference between work and power? Give two units for each.

10. Prove that 1 kWh is equal to 3.6×10^6 J.

11. Give the metric prefixes for 10^{-6}, 10^{-3}, 10^3, and 10^6.

12. Which two units in Table 3–3 are reciprocals of each other?

13. A circuit has a constant R of 5000 Ω, and V is varied from 0 to 50 V in 10-V steps. Make a table listing the values of I for each value of V. Then draw a graph plotting these values of milliamperes vs. volts. (This graph should be like Fig. 3–5c.)

14. Give the voltage and power rating for at least two types of electrical equipment.

15. Which uses more current from the 120-V power line, a 600-W toaster or a 300-W lightbulb?

16. Give a definition for a short circuit and for an open circuit.

17. Compare the R of zero ohms and infinite ohms.

18. Derive the formula $P = I^2R$ from $P = IV$ by using an Ohm's law formula.

19. Explain why a thermistor is a nonlinear resistance.

20. What is meant by the maximum working voltage rating of a resistor?

21. Why do resistors often have a safety factor of 2 in regard to their power rating?

Problems

SECTION 3–1 THE CURRENT $I = V/R$

In Probs. 3–1 to 3–5, solve for the current, I, when V and R are known. As a visual aid, it may be helpful to insert the values of V and R into Fig. 3–13 when solving for I.

3-1 MultiSim a. $V = 10$ V, $R = 5$ Ω, $I = ?$
b. $V = 9$ V, $R = 3$ Ω, $I = ?$
c. $V = 24$ V, $R = 3$ Ω, $I = ?$
d. $V = 36$ V, $R = 9$ Ω, $I = ?$

3-2 MultiSim a. $V = 18$ V, $R = 3$ Ω, $I = ?$
b. $V = 16$ V, $R = 16$ Ω, $I = ?$
c. $V = 90$ V, $R = 450$ Ω, $I = ?$
d. $V = 12$ V, $R = 30$ Ω, $I = ?$

3-3 MultiSim a. $V = 15$ V, $R = 3{,}000$ Ω, $I = ?$
b. $V = 120$ V, $R = 6{,}000$ Ω, $I = ?$
c. $V = 27$ V, $R = 9{,}000$ Ω, $I = ?$
d. $V = 150$ V, $R = 10{,}000$ Ω, $I = ?$

3-4 If a 100-Ω resistor is connected across the terminals of a 12-V battery, how much is the current, I?

3-5 If one branch of a 120-V power line is protected by a 20-A fuse, will the fuse carry an 8-Ω load?

SECTION 3–2 THE VOLTAGE $V = IR$

In Probs. 3–6 to 3–10, solve for the voltage, V, when I and R are known. As a visual aid, it may be helpful to insert the values of I and R into Fig. 3–14 when solving for V.

3-6 MultiSim a. $I = 2$ A, $R = 5$ Ω, $V = ?$
b. $I = 6$ A, $R = 8$ Ω, $V = ?$
c. $I = 9$ A, $R = 20$ Ω, $V = ?$
d. $I = 4$ A, $R = 15$ Ω, $V = ?$

3-7 MultiSim a. $I = 5$ A, $R = 10$ Ω, $V = ?$
b. $I = 10$ A, $R = 3$ Ω, $V = ?$
c. $I = 4$ A, $R = 2.5$ Ω, $V = ?$
d. $I = 1.5$ A, $R = 5$ Ω, $V = ?$

Figure 3–13 Figure for Probs. 3–1 to 3–5.

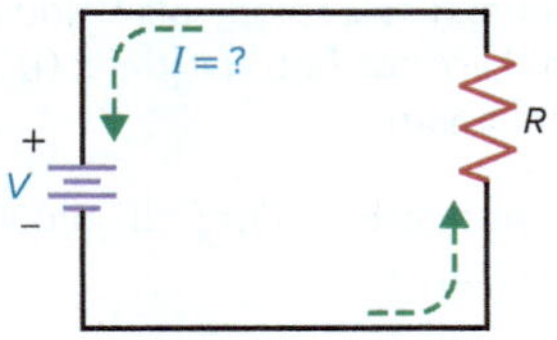

Figure 3–14 Figure for Probs. 3–6 to 3–10.

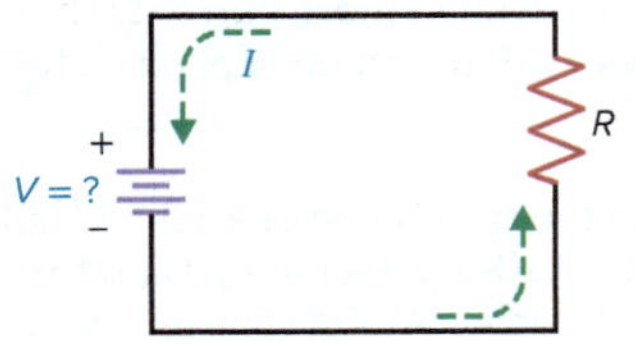

Figure 3–15 Figure for Probs. 3–11 to 3–15.

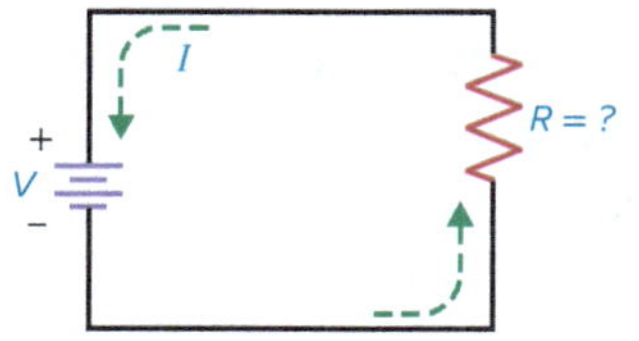

3–8 MultiSim a. $I = 0.05\text{ A}, R = 1200\ \Omega, V = ?$

b. $I = 0.2\text{ A}, R = 470\ \Omega, V = ?$

c. $I = 0.01\text{ A}, R = 15{,}000\ \Omega, V = ?$

d. $I = 0.006\text{ A}, R = 2200\ \Omega, V = ?$

3–9 How much voltage is developed across a 1000-Ω resistor if it has a current of 0.01 A?

3–10 A lightbulb drawing 1.25 A of current has a resistance of 96 Ω. How much is the voltage across the lightbulb?

SECTION 3–3 THE RESISTANCE $R = V/I$

In Probs. 3–11 to 3–15, solve for the resistance, R, when V and I are known. As a visual aid, it may be helpful to insert the values of V and I into Fig. 3–15 when solving for R.

3–11 a. $V = 14\text{ V}, I = 2\text{ A}, R = ?$

b. $V = 25\text{ V}, I = 5\text{ A}, R = ?$

c. $V = 6\text{ V}, I = 1.5\text{ A}, R = ?$

d. $V = 24\text{ V}, I = 4\text{ A}, R = ?$

3–12 a. $V = 36\text{ V}, I = 9\text{ A}, R = ?$

b. $V = 45\text{ V}, I = 5\text{ A}, R = ?$

c. $V = 100\text{ V}, I = 2\text{ A}, R = ?$

d. $V = 240\text{ V}, I = 20\text{ A}, R = ?$

3–13 a. $V = 12\text{ V}, I = 0.002\text{ A}, R = ?$

b. $V = 16\text{ V}, I = 0.08\text{ A}, R = ?$

c. $V = 50\text{ V}, I = 0.02\text{ A}, R = ?$

d. $V = 45\text{ V}, I = 0.009\text{ A}, R = ?$

3–14 How much is the resistance of a motor if it draws 2 A of current from the 120-V power line?

3–15 If a CD player draws 1.6 A of current from a 13.6-V_{DC} source, how much is its resistance?

SECTION 3–5 MULTIPLE AND SUBMULTIPLE UNITS

In Probs. 3–16 to 3–20, solve for the unknowns listed. As a visual aid, it may be helpful to insert the known values of I, V, or R into Figs. 3–13, 3–14, or 3–15 when solving for the unknown quantity.

3–16 a. $V = 10\text{ V}, R = 100\text{ k}\Omega, I = ?$

b. $V = 15\text{ V}, R = 2\text{ k}\Omega, I = ?$

c. $I = 200\ \mu\text{A}, R = 3.3\text{ M}\Omega, V = ?$

d. $V = 5.4\text{ V}, I = 2\text{ mA}, R = ?$

Figure 3–16 Circuit diagram for Prob. 3–21.

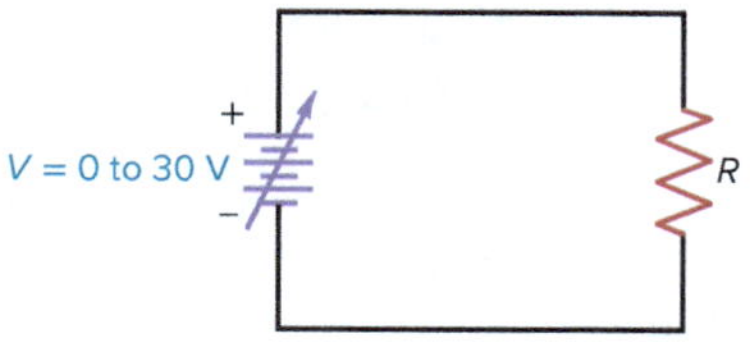

3–17 a. $V = 120\text{ V}, R = 1.5\text{ k}\Omega, I = ?$

b. $I = 50\ \mu\text{A}, R = 390\text{ k}\Omega, V = ?$

c. $I = 2.5\text{ mA}, R = 1.2\text{ k}\Omega, V = ?$

d. $V = 99\text{ V}, I = 3\text{ mA}, R = ?$

3–18 a. $V = 24\text{ V}, I = 800\ \mu\text{A}, R = ?$

b. $V = 160\text{ mV}, I = 8\ \mu\text{A}, R = ?$

c. $V = 13.5\text{ V}, R = 300\ \Omega, I = ?$

d. $I = 30\text{ mA}, R = 1.8\text{ k}\Omega, V = ?$

3–19 How much is the current, I, in a 470-kΩ resistor if its voltage is 23.5 V?

3–20 How much voltage will be dropped across a 40-kΩ resistance whose current is 250 μA?

SECTION 3–6 THE LINEAR PROPORTION BETWEEN V AND I

3–21 Refer to Fig. 3–16. Draw a graph of the I and V values if (*a*) $R = 2.5\ \Omega$; (*b*) $R = 5\ \Omega$; (*c*) $R = 10\ \Omega$. In each case, the voltage source is to be varied in 5-V steps from 0 to 30 V.

3–22 Refer to Fig. 3–17. Draw a graph of the I and R values when R is varied in 1 kΩ steps from 1 kΩ to 10 kΩ. (V is constant at 10 V.)

SECTION 3–7 ELECTRIC POWER

In Probs. 3–23 to 3–31, solve for the unknowns listed.

3–23 a. $V = 120\text{ V}, I = 12.5\text{ A}, P = ?$

b. $V = 120\text{ V}, I = 625\text{ mA}, P = ?$

c. $P = 1.2\text{ kW}, V = 120\text{ V}, I = ?$

d. $P = 100\text{ W}, I = 8.33\text{ A}, V = ?$

3–24 a. $V = 24\text{ V}, I = 25\text{ mA}, P = ?$

b. $P = 6\text{ W}, V = 12\text{ V}, I = ?$

c. $P = 10\text{ W}, I = 100\text{ mA}, V = ?$

d. $P = 50\text{ W}, V = 9\text{ V}, I = ?$

Figure 3–17 Circuit diagram for Prob. 3–22.

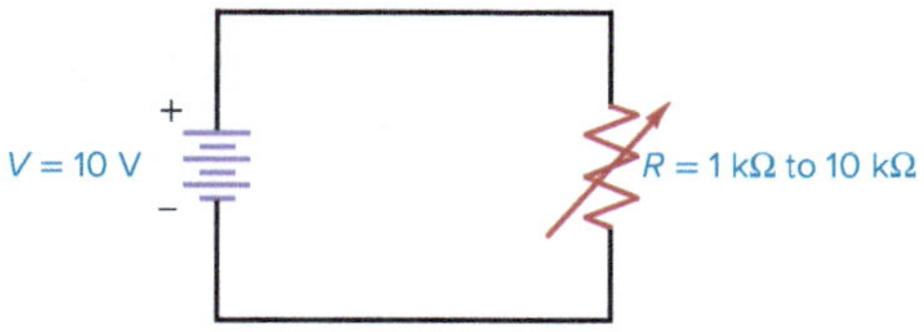

3–25 a. $V = 15.81$ V, $P = 500$ mW, $I = ?$
b. $P = 100$ mW, $V = 50$ V, $I = ?$
c. $V = 75$ mV, $I = 2$ mA, $P = ?$
d. $P = 20$ mW, $I = 100$ μA, $V = ?$

3–26 How much current do each of the following lightbulbs draw from the 120-V power line?
a. 60-W bulb
b. 75-W bulb
c. 100-W bulb
d. 300-W bulb

3–27 How much is the output voltage of a power supply if it supplies 75 W of power while delivering a current of 5 A?

3–28 How much power is consumed by a 12-V incandescent lamp if it draws 150 mA of current when lit?

3–29 How much will it cost to operate a 1500-W quartz heater for 48 h if the cost of electricity is 10¢/kWh?

3–30 How much does it cost to light a 300-W lightbulb for 30 days if the cost of electricity is 12¢/kWh?

3–31 How much will it cost to run an electric motor for 10 days if the motor draws 15 A of current from the 240-V power line? The cost of electricity is 7.5¢/kWh.

SECTION 3–8 POWER DISSIPATION IN RESISTANCE

In Probs. 3–32 to 3–38, solve for the power, *P*, dissipated by the resistance, *R*.

3–32 a. $I = 1$ A, $R = 100$ Ω, $P = ?$
b. $I = 20$ mA, $R = 1$ kΩ, $P = ?$
c. $V = 5$ V, $R = 150$ Ω, $P = ?$
d. $V = 22.36$ V, $R = 1$ kΩ, $P = ?$

3–33 a. $I = 300$ μA, $R = 22$ kΩ, $P = ?$
b. $I = 50$ mA, $R = 270$ Ω, $P = ?$
c. $V = 70$ V, $R = 200$ kΩ, $P = ?$
d. $V = 8$ V, $R = 50$ Ω, $P = ?$

3–34 a. $I = 40$ mA, $R = 10$ kΩ, $P = ?$
b. $I = 3.33$ A, $R = 20$ Ω, $P = ?$
c. $V = 100$ mV, $R = 10$ Ω, $P = ?$
d. $V = 1$ kV, $R = 10$ MΩ, $P = ?$

3–35 How much power is dissipated by a 5.6-kΩ resistor whose current is 9.45 mA?

3–36 How much power is dissipated by a 50-Ω load if the voltage across the load is 100 V?

3–37 How much power is dissipated by a 600-Ω load if the voltage across the load is 36 V?

3–38 How much power is dissipated by an 8-Ω load if the current in the load is 200 mA?

SECTION 3–9 POWER FORMULAS

In Probs. 3–39 to 3–51, solve for the unknowns listed.

3–39 a. $P = 250$ mW, $R = 10$ kΩ, $I = ?$
b. $P = 100$ W, $V = 120$ V, $R = ?$
c. $P = 125$ mW, $I = 20$ mA, $R = ?$
d. $P = 1$ kW, $R = 50$ Ω, $V = ?$

3–40 a. $P = 500$ μW, $V = 10$ V, $R = ?$
b. $P = 150$ mW, $I = 25$ mA, $R = ?$
c. $P = 300$ W, $R = 100$ Ω, $V = ?$
d. $P = 500$ mW, $R = 3.3$ kΩ, $I = ?$

3–41 a. $P = 50$ W, $R = 40$ Ω, $V = ?$
b. $P = 2$ W, $R = 2$ kΩ, $V = ?$
c. $P = 50$ mW, $V = 500$ V, $I = ?$
d. $P = 50$ mW, $R = 312.5$ kΩ, $I = ?$

3–42 Calculate the maximum current that a 1-kΩ, 1-W carbon resistor can safely handle without exceeding its power rating.

3–43 Calculate the maximum current that a 22-kΩ, ⅛-W resistor can safely handle without exceeding its power rating.

3–44 What is the hot resistance of a 60-W, 120-V lightbulb?

3–45 A 50-Ω load dissipates 200 W of power. How much voltage is across the load?

3–46 Calculate the maximum voltage that a 390-Ω, ½-W resistor can safely handle without exceeding its power rating.

3–47 What is the resistance of a device that dissipates 1.2 kW of power when its current is 10 A?

3–48 How much current does a 960-W coffeemaker draw from the 120-V power line?

3–49 How much voltage is across a resistor if it dissipates 2 W of power when the current is 40 mA?

3–50 If a 4-Ω speaker dissipates 15 W of power, how much voltage is across the speaker?

3–51 What is the resistance of a 20-W, 12-V halogen lamp?

SECTION 3–10 CHOOSING A RESISTOR FOR A CIRCUIT

In Probs. 3–52 to 3–60, determine the required resistance and appropriate wattage rating of a carbon-film resistor for the specific requirements listed. For all problems, assume that the following wattage ratings are available: ⅛ W, ¼ W, ½ W, 1 W, and 2 W. (Assume the maximum working voltage ratings listed on page 94.)

3–52 Required values of *V* and *I* are 54 V and 2 mA.

3–53 Required values of *V* and *I* are 12 V and 10 mA.

3–54 Required values of *V* and *I* are 390 V and 1 mA.

3–55 Required values of *V* and *I* are 36 V and 18 mA.

3–56 Required values of *V* and *I* are 340 V and 500 μA.

3–57 Required values of *V* and *I* are 3 V and 20 mA.

3–58 Required values of *V* and *I* are 33 V and 18.33 mA.

3–59 Required values of *V* and *I* are 264 V and 120 μA.

3–60 Required values of *V* and *I* are 9.8 V and 1.75 mA.

Critical Thinking

3–61 The percent efficiency of a motor can be calculated as

$$\% \text{ efficiency} = \frac{\text{power out}}{\text{power in}} \times 100$$

where power out represents horsepower (hp). Calculate the current drawn by a 5-hp, 240-V motor that is 72% efficient.

3–62 A ½-hp, 120-V motor draws 4.67 A when it is running. Calculate the motor's efficiency.

3–63 A ¾-hp motor with an efficiency of 75% runs 20% of the time during a 30-day period. If the cost of electricity is 7¢/kWh, how much will it cost the user?

3–64 An appliance uses 14.4×10^6 J of energy for 1 day. How much will this cost the user if the cost of electricity is 12¢/kWh?

3–65 A certain 1-kΩ resistor has a power rating of ½ W for temperatures up to 70°C. Above 70°C, however, the power rating must be reduced by a factor of 6.25 mW/°C. Calculate the maximum current that the resistor can allow at 120°C without exceeding its power dissipation rating at this temperature.

Keith Eng 2007

Answers to Self-Reviews

3-1 **a.** 3 A **b.** 1.5 A **c.** 2 A **d.** 6 A

3-2 **a.** 2 V **b.** 4 V **c.** 4 V

3-3 **a.** 4000 Ω **b.** 2000 Ω **c.** 12,000 Ω

3-4 **a.** 35 V **b.** 0.002 A **c.** 2000 Ω

3-5 **a.** See Prob. **b** **b.** See Prob. **a** **c.** 2 mA **d.** 125 μA

3-6 **a.** *y* axis **b.** linear **c.** *I* doubles from 2 A to 4 A **d.** *I* is halved from 2 A to 1 A

3-7 **a.** 1.8 kW **b.** 0.83 A **c.** 200 W **d.** $2.16

3-8 **a.** 20 W **b.** 20 W **c.** 20 W and 5 Ω

3-9 **a.** 144 Ω **b.** 50 W **c.** 8 W

3-10 **a.** 50 V; no **b.** $R = 1$ MΩ, $P_{rating} = 1/8$ W

3-11 **a.** true **b.** true

3-12 **a.** true **b.** true **c.** true

Laboratory Application Assignment

In this lab application assignment, you will examine the difference between a linear and nonlinear resistance. Recall from your reading that a linear resistance has a constant value of ohms. Conversely, a nonlinear resistance has an ohmic value that varies with different amounts of voltage and current.

Equipment: Obtain the following items from your instructor.

- Variable dc power supply
- DMM
- 330-Ω and 1 kΩ carbon-film resistors (½-W)
- 12-V incandescent bulb

Linear Resistance

Measure and record the value of the 330-Ω carbon-film resistor. $R =$ ___________

Connect the circuit in Fig. 3–18. Measure and record the current, *I*, with the voltage, *V*, set to 3 V. $I =$ ___________

Increase the voltage to 6 V and remeasure the current, *I*. $I =$ ___________

Increase the voltage one more time to 12 V, and remeasure the current, *I*. $I =$ ___________

For each value of voltage and current, calculate the resistance value as $R = V/I$. When $V = 3$ V, $R =$ _________. When $V = 6$ V, $R =$ _________. When $V = 12$ V, $R =$ _________. Does *R* remain the same even though *V* and *I* are changing? ___________

Nonlinear Resistance

Measure and record the cold resistance of the 12-V incandescent bulb. $R =$ ___________

In Fig. 3–18 replace the 330-Ω carbon-film resistor with the 12-V incandescent bulb.

Measure and record the current, *I*, with the voltage, *V*, set to 3 V. $I =$ ___________

Increase the voltage to 6 V, and remeasure the current, *I*. $I =$ ___________

Increase the voltage one more time to 12 V, and remeasure the current, *I*. $I =$ ___________

Calculate the resistance of the bulb as $R = V/I$ for each value of applied voltage. When $V = 3$ V, $R =$ _________. When $V = 6$ V, $R =$ ___________. When $V = 12$ V, $R =$ ___________

Does *R* remain constant as the values of voltage and current increase? _________

Does the bulb's resistance increase or decrease as *V* and *I* increase? _________

Calculating Power

(330-Ω resistor)

Calculate the power dissipated by the 330-Ω resistor with $V = 3$ V. $P =$ ____________W

Calculate the power dissipated by the 330-Ω resistor with $V = 6$ V. $P =$ ____________W

Calculate the power dissipated by the 330-Ω resistor with $V = 12$ V. $P =$ ____________W

What happens to the power dissipation each time the voltage, V, is doubled?

____

(12-V incandescent bulb)

Calculate the power dissipated by the 12-V incandescent bulb with $V = 3$ V. $P =$ ____________W

Calculate the power dissipated by the 12-V incandescent bulb with $V = 6$ V. $P =$ ____________W

Calculate the power dissipated by the 12-V incandescent bulb with $V = 12$ V. $P =$ ____________W

What happens to the power dissipation each time the voltage, V, is doubled?

____

How do the changes in power dissipation when V is doubled compare to the results obtained with the 330-Ω resistor? _____

____

Volt-Ampere Characteristic of *R*

In Fig. 3–18, replace the 12-V incandescent bulb with a 1 kΩ resistor. Measure the current, I as the voltage, V is increased in 1 V increments from 0 to 12 V. Create a table to record your answers. From the data recorded in your table, draw a graph of the V and I values. This graph represents the volt-ampere characteristic of the 1 kΩ R.

Reconnect the 12-V incandescent bulb in Fig. 3–18. Measure the current, I as the voltage, V is increased in 1 V increments from 0 to 12 V. Create a second table to record your answers. From the data recorded in your table, draw a graph of the V and I values. This graph represents the volt-ampere characteristic of the 12-V incandescent bulb.

How does the volt-ampere characteristic of the 1 kΩ resistor compare to that of the 12-V incandescent bulb? ____________

____

____

Does the 12-V incandescent bulb have a linear or nonlinear resistance? ______________________________

Explain your answer. ______________________________

____

Figure 3–18

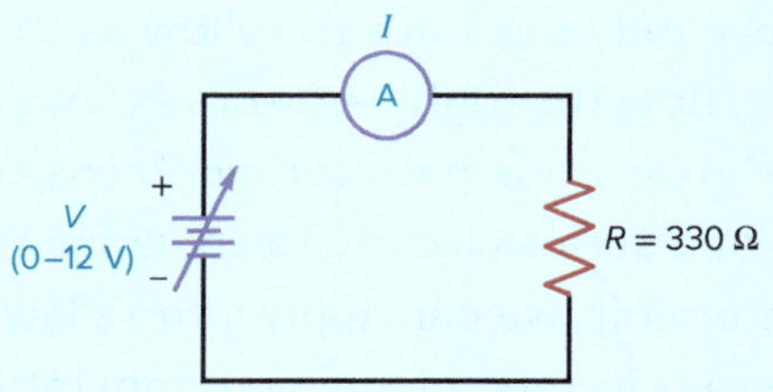

Design credits: Chapter Opener Background Image, and Multisim: Stockbyte/Getty Images

chapter

4

Series Circuits

A series circuit is any circuit that provides only one path for current flow. An example of a series circuit is shown in Fig. 4–1. Here two resistors are connected end to end with their opposite ends connected across the terminals of a voltage source. Figure 4–1*a* shows the pictorial wiring diagram, and Fig. 4–1*b* shows the schematic diagram. The small dots in Fig. 4–1*b* represent free electrons. Notice that the free electrons have only one path to follow as they leave the negative terminal of the voltage source, flow through resistors R_2 and R_1, and return to the positive terminal. Since there is only one path for electrons to follow, the current, I, must be the same in all parts of a series circuit. To solve for the values of voltage, current, or resistance in a series circuit, we can apply Ohm's law. This chapter covers all of the characteristics of series circuits, including important information about how to troubleshoot a series circuit containing a defective component. As you will learn, there are many applications for series circuits. One such application featured at the end of this chapter is holiday lights. Other applications include motor speed controls and lamp intensity controls to name a couple of the more common ones. ■

Chapter Outline

Chapter Objectives

After studying this chapter, you should be able to

- *Explain* why the current is the same in all parts of a series circuit.
- *Calculate* the total resistance of a series circuit.
- *Calculate* the current in a series circuit.
- *Determine* the individual resistor voltage drops in a series circuit.
- *Apply* Kirchhoff's voltage law to series circuits.
- *Determine* the polarity of a resistor's *IR* voltage drop.
- *Calculate* the total power dissipated in a series circuit.
- *Determine* the net voltage of series-aiding and series-opposing voltage sources.
- *Solve* for the voltage, current, resistance, and power in a series circuit having random unknowns.
- *Define* the terms earth ground and chassis ground.
- *Calculate* the voltage at a given point with respect to ground in a series circuit.
- *Describe* the effect of an open in a series circuit.
- *Describe* the effect of a short in a series circuit.
- *Troubleshoot* series circuits containing opens and shorts.

Important Terms

chassis ground
double subscript notation
earth ground
Kirchhoff's voltage law (KVL)
series-aiding voltages
series components
series-opposing voltages
series string
troubleshooting
voltage drop
voltage polarity

4–1 Why *I* Is the Same in All Parts of a Series Circuit

An electric current is a movement of charges between two points, produced by the applied voltage. When components are connected in successive order, as shown in Fig. 4–1, they form a series circuit. The resistors R_1 and R_2 are in series with each other and the battery.

In Fig. 4–2*a*, the battery supplies the potential difference that forces free electrons to drift from the negative terminal at A, toward B, through the connecting wires and resistances R_3, R_2, and R_1, back to the positive battery terminal at J. At the negative battery terminal, its negative charge repels electrons. Therefore, free electrons in the atoms of the wire at this terminal are repelled from A toward B. Similarly, free electrons at point B can then repel adjacent electrons, producing an electron drift toward C and away from the negative battery terminal.

At the same time, the positive charge of the positive battery terminal attracts free electrons, causing electrons to drift toward I and J. As a result, the free electrons in R_1, R_2, and R_3, are forced to drift toward the positive terminal.

The positive terminal of the battery attracts electrons just as much as the negative side of the battery repels electrons. Therefore, the motion of free electrons in the circuit starts at the same time and at the same speed in all parts of the circuit.

The electrons returning to the positive battery terminal are not the same electrons as those leaving the negative terminal. Free electrons in the wire are forced to move to the positive terminal because of the potential difference of the battery.

The free electrons moving away from one point are continuously replaced by free electrons flowing from an adjacent point in the series circuit. All electrons have the same speed as those leaving the battery. In all parts of the circuit, therefore, the electron drift is the same. An equal number of electrons move at one time with the same speed. That is why the current is the same in all parts of the series circuit.

In Fig. 4–2*b*, when the current is 2 A, for example, this is the value of the current through R_1, R_2, R_3, and the battery at the same instant. Not only is the amount of current the same throughout, but the current in all parts of a series circuit cannot differ in any way because there is just one current path for the entire circuit. Figure 4–2*c* shows how to assemble axial-lead resistors on a lab prototype board to form a series circuit.

Figure 4–1 A series circuit. (*a*) Pictorial wiring diagram. (*b*) Schematic diagram.

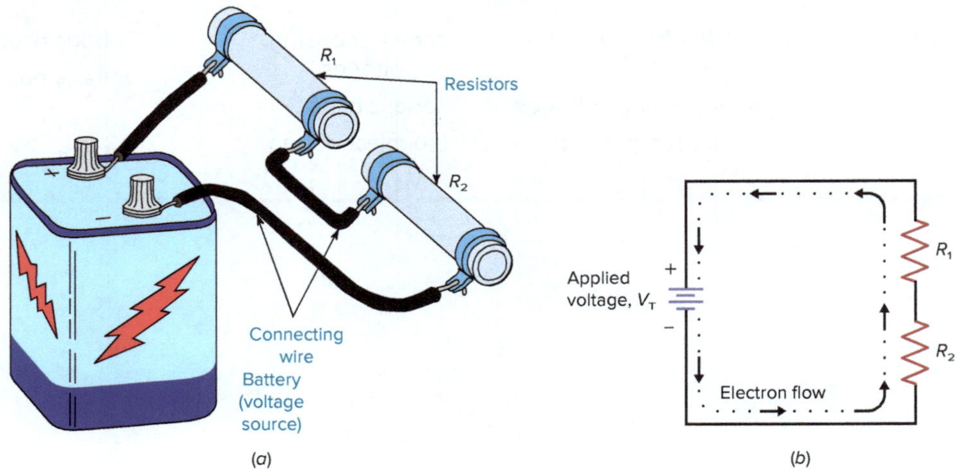

MultiSim **Figure 4–2** There is only one current through R_1, R_2, and R_3 in series. (*a*) Electron drift is the same in all parts of a series circuit. (*b*) Current I is the same at all points in a series circuit. (*c*) A series circuit assembled on a lab prototype board, using axial-lead resistors.

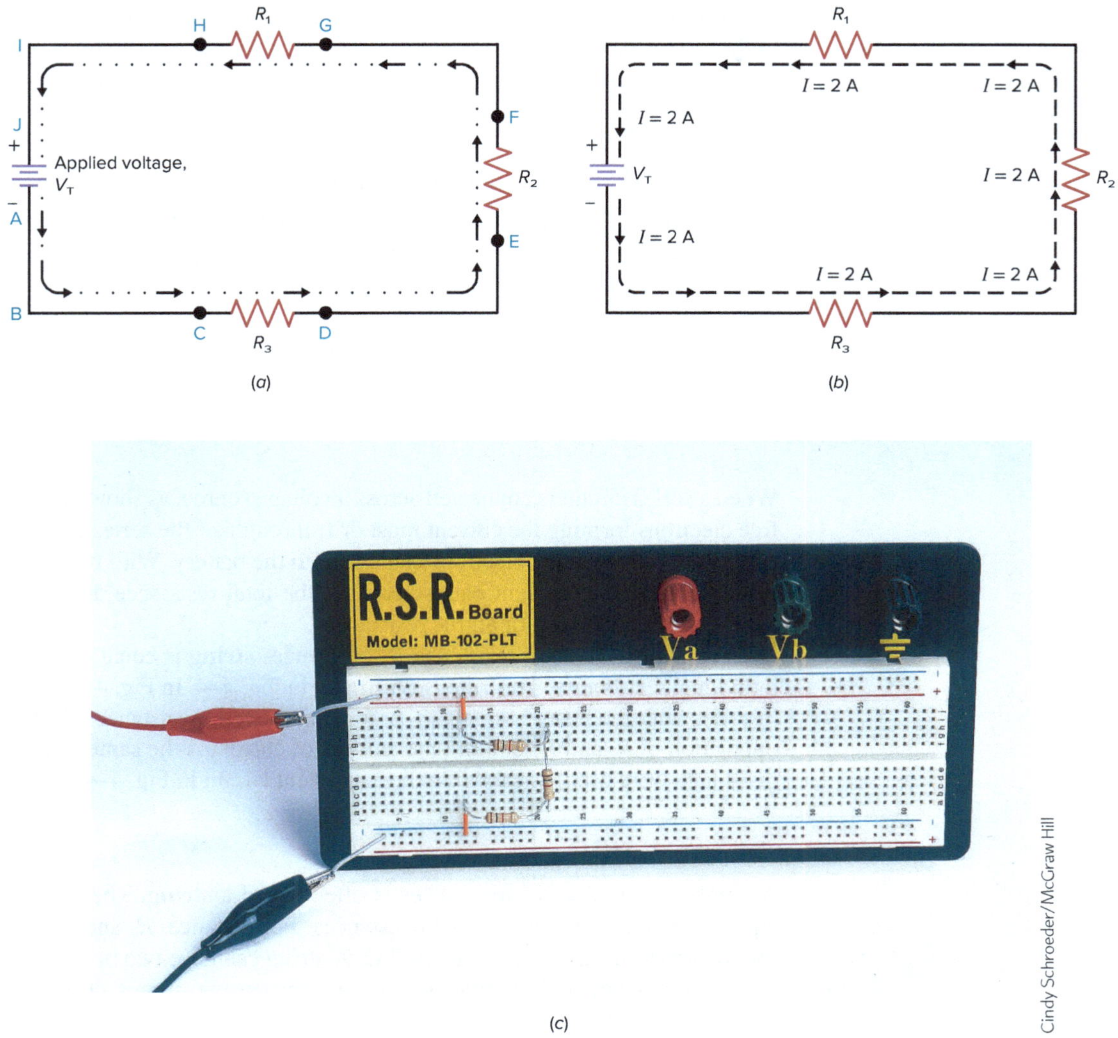

Cindy Schroeder/McGraw Hill

The order in which components are connected in series does not affect the current. In Fig. 4–3*b*, resistances R_1 and R_2 are connected in reverse order compared with Fig. 4–3*a*, but in both cases they are in series. The current through each is the same because there is only one path for the electron flow. Similarly, R_3, R_4, and R_5 are in series and have the same current for the connections shown in Fig. 4–3*c*, *d*, and *e*. Furthermore, the resistances need not be equal.

Figure 4–3 Examples of series connections: R_1 and R_2 are in series in both (*a*) and (*b*); also, R_3, R_4, and R_5 are in series in (*c*), (*d*), and (*e*).

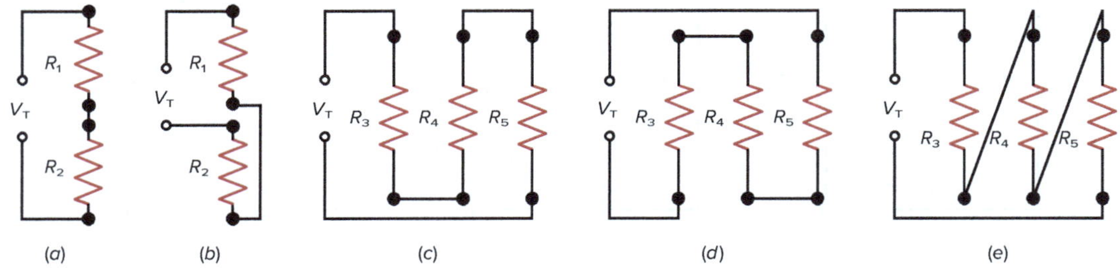

The question of whether a component is first, second, or last in a series circuit has no meaning in terms of current. The reason is that I is the same amount at the same time in all **series components.**

In fact, series components can be defined as those in the same current path. The path is from one side of the voltage source, through the series components, and back to the other side of the applied voltage. However, the series path must not have any point at which the current can branch off to another path in parallel.

4–1 Self-Review

Answers at the end of the chapter.

a. In Fig. 4–2*b*, name five parts that have the I of 2 A.

b. In Fig. 4–3*e*, when I in R_5 is 5 A, then I in R_3 is _______ A.

c. In Fig. 4–4*b*, how much is the I in R_2?

4–2 Total *R* Equals the Sum of All Series Resistances

When a series circuit is connected across a voltage source, as shown in Fig. 4–3, the free electrons forming the current must drift through all the series resistances. This path is the only way the electrons can return to the battery. With two or more resistances in the same current path, therefore, the total resistance across the voltage source is the opposition of all the resistances.

Specifically, the total resistance R_T of a **series string** is equal to the sum of the individual resistances. This rule is illustrated in Fig. 4–4. In Fig. 4–4*b*, 2 Ω is added in series with the 3 Ω of Fig. 4–4*a*, producing the total resistance of 5 Ω. The total opposition of R_1 and R_2 limiting the amount of current is the same as though a 5-Ω resistance were used, as shown in the equivalent circuit in Fig. 4–4*c*.

Series String

A combination of series resistances is often called a **string.** The string resistance equals the sum of the individual resistances. For instance, R_1 and R_2 in Fig. 4–4*b* form a series string having an R_T of 5 Ω. A string can have two or more resistors.

By Ohm's law, the amount of current between two points in a circuit equals the potential difference divided by the resistance between these points. Because the entire string is connected across the voltage source, the current equals the voltage applied across the entire string divided by the total series resistance of the string. Between points A and B in Fig. 4–4, for example, 10 V is applied across 5 Ω in Fig. 4–4*b* and *c* to produce 2 A. This current flows through R_1 and R_2 in one series path.

MultiSim **Figure 4–4** Series resistances are added for the total R_T. (*a*) R_1 alone is 3 Ω. (*b*) R_1 and R_2 in series total 5 Ω. (*c*) The R_T of 5 Ω is the same as one resistance of 5 Ω between points A and B.

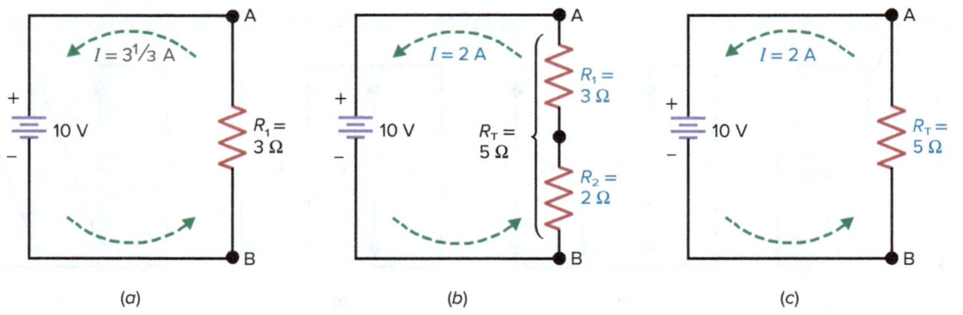

GOOD TO KNOW

When equal resistances are connected in series, the total resistance, $R_T = NR$, where R represents the value of a single resistance and N represents the number of identical resistors connected in series.

Series Resistance Formula

In summary, the *total resistance* of a series string equals the sum of the individual resistances. The formula is

$$R_T = R_1 + R_2 + R_3 + \cdots + \text{etc.} \qquad \textbf{(4–1)}$$

where R_T is the total resistance and R_1, R_2, and R_3 are individual series resistances.

This formula applies to any number of resistances, whether equal or not, as long as they are in the same series string. Note that R_T is the resistance to use in calculating the current in a series string. Then Ohm's law is

$$I = \frac{V_T}{R_T} \qquad \textbf{(4–2)}$$

where R_T is the sum of all the resistances, V_T is the voltage applied across the total resistance, and I is the current in all parts of the string.

Note that adding series resistance reduces the current. In Fig. 4–4*a*, the 3-Ω R_1 allows 10 V to produce 3⅓ A. However, I is reduced to 2 A when the 2-Ω R_2 is added for a total series resistance of 5 Ω opposing the 10-V source.

Example 4–1

MultiSim

Two resistances R_1 and R_2 of 5 Ω each and R_3 of 10 Ω are in series. How much is R_T?

ANSWER

$$R_T = R_1 + R_2 + R_3 = 5 + 5 + 10$$

$$R_T = 20\ \Omega$$

Example 4–2

MultiSim

With 80 V applied across the series string of Example 4–1, how much is the current in R_3?

ANSWER

$$I = \frac{V_T}{R_T} = \frac{80\text{ V}}{20\ \Omega}$$

$$I = 4\text{ A}$$

This 4-A current is the same in R_3, R_2, R_1, or any part of the series circuit.

4–2 Self-Review

Answers at the end of the chapter.

a. An applied voltage of 10 V is across a 5-kΩ resistor, R_1. How much is the current?

b. A 2-kΩ R_2 and 3-kΩ R_3 are added in series with R_1 in part a. Calculate R_T.

c. Calculate I in R_1, R_2, and R_3.

GOOD TO KNOW

When current flows through a resistance, electrical energy is converted into heat energy. The heat is caused by the collisions of the free electrons within the material carrying the current. When one electron collides with another, heat is given off. Therefore, the colliding electron gives up some of its acquired energy as it moves through the material. Since $V = \frac{W}{Q}$, it makes sense that the voltage (with respect to some point in the circuit) must be different on opposite ends of the resistor since the electrons entering and leaving the resistor have different energy levels. (Note that the same number of electrons enter and exit the resistor so Q is constant in the formula $V = \frac{W}{Q}$.)

4–3 Series *IR* Voltage Drops

With current I through a resistance, by Ohm's law, the voltage across R is equal to $I \times R$. This rule is illustrated in Fig. 4–5 for a string of two resistors. In this circuit, I is 1 A because the applied V_T of 10 V is across the total R_T of 10 Ω, equal to the 4-Ω R_1 plus the 6-Ω R_2. Then I is 10 V/10 Ω = 1 A.

For each IR voltage in Fig. 4–5, multiply each R by the 1 A of current in the series circuit. Then

$$V_1 = IR_1 = 1\text{ A} \times 4\ \Omega = 4\text{ V}$$

$$V_2 = IR_2 = 1\text{ A} \times 6\ \Omega = 6\text{ V}$$

The V_1 of 4 V is across the 4 Ω of R_1. Also, the V_2 of 6 V is across the 6 Ω of R_2. The two voltages V_1 and V_2 are in series.

The IR voltage across each resistance is called an *IR drop*, or a ***voltage drop***, because it reduces the potential difference available for the remaining resistances in the series circuit. Note that the symbols V_1 and V_2 are used for the voltage drops across each resistor to distinguish them from the source V_T applied across both resistors.

In Fig. 4–5, the V_T of 10 V is applied across the total series resistance of R_1 and R_2. However, because of the IR voltage drop of 4 V across R_1, the potential difference across R_2 is only 6 V. The positive potential drops from 10 V at point A, with respect to the common reference point at C, down to 6 V at point B with reference to point C. The potential difference of 6 V between B and the reference at C is the voltage across R_2.

Similarly, there is an IR voltage drop of 6 V across R_2. The positive potential drops from 6 V at point B with respect to point C, down to 0 V at point C with respect to itself. The potential difference between any two points on the return line to the battery must be zero because the wire has practically zero resistance and therefore no IR drop.

Note that voltage must be applied by a source of *potential difference* such as the battery to produce current and have an IR voltage drop across the resistance. With no current through a resistor, the resistor has only resistance. There is no potential difference across the two ends of the resistor.

The IR drop of 4 V across R_1 in Fig. 4–5 represents that part of the applied voltage used to produce the current of 1 A through the 4-Ω resistance. Also, the IR drop across R_2 is 6 V because this much voltage allows 1 A in the 6-Ω resistance. The IR drop is more in R_2 because more potential difference is necessary to produce the same amount of current in the higher resistance. For series circuits, in general, the highest R has the largest IR voltage drop across it.

MultiSim **Figure 4–5** An example of IR voltage drops V_1 and V_2 in a series circuit.

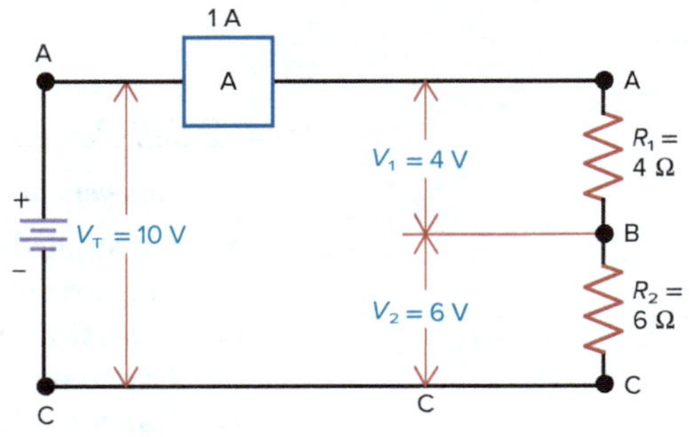

Figure 4–6 Circuit for Example 4–3.

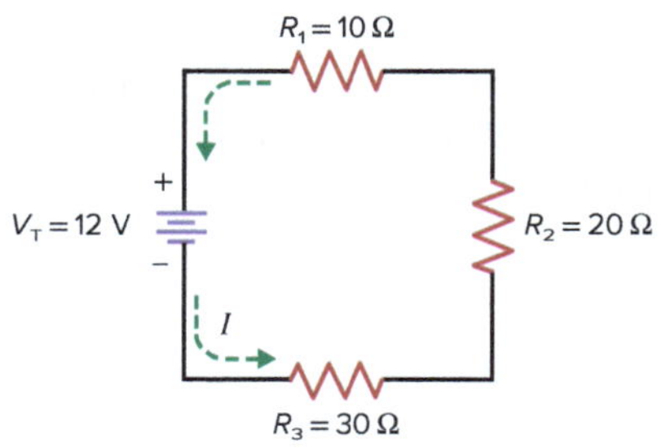

Example 4–3

In Fig. 4–6, solve for R_T, I, and the individual resistor voltage drops.

ANSWER

First, find R_T by adding the individual resistance values.

$$\begin{aligned} R_T &= R_1 + R_2 + R_3 \\ &= 10\ \Omega + 20\ \Omega + 30\ \Omega \\ &= 60\ \Omega \end{aligned}$$

Next, solve for the current, I.

$$\begin{aligned} I &= \frac{V_T}{R_T} \\ &= \frac{12\ V}{60\ \Omega} \\ &= 200\ \text{mA} \end{aligned}$$

Now we can solve for the individual resistor voltage drops.

$$\begin{aligned} V_1 &= I \times R_1 \\ &= 200\ \text{mA} \times 10\ \Omega \\ &= 2\ \text{V} \end{aligned}$$

$$\begin{aligned} V_2 &= I \times R_2 \\ &= 200\ \text{mA} \times 20\ \Omega \\ &= 4\ \text{V} \end{aligned}$$

$$\begin{aligned} V_3 &= I \times R_3 \\ &= 200\ \text{mA} \times 30\ \Omega \\ &= 6\ \text{V} \end{aligned}$$

Notice that the individual voltage drops are proportional to the series resistance values. For example, because R_3 is three times larger than R_1, V_3 will be three times larger than V_1. With the same current through all the resistors, the largest resistance must have the largest voltage drop.

4–3 Self-Review

Answers at the end of the chapter.

Refer to Fig. 4–5.

a. **How much is the sum of V_1 and V_2?**
b. **Calculate I as V_T/R_T.**
c. **How much is I through R_1?**
d. **How much is I through R_2?**

4–4 Kirchhoff's Voltage Law (KVL)

Kirchhoff's voltage law states that the sum of all resistor voltage drops in a series circuit equals the **applied voltage.** Expressed as an equation, Kirchhoff's voltage law is

$$V_T = V_1 + V_2 + V_3 + \cdots + \text{etc.} \qquad \textbf{(4–3)}$$

where V_T is the applied voltage and V_1, V_2, V_3 . . . are the individual *IR* voltage drops.

Example 4–4

A voltage source produces an *IR* drop of 40 V across a 20-Ω R_1, 60 V across a 30-Ω R_2, and 180 V across a 90-Ω R_3, all in series. According to Kirchhoff's voltage law, how much is the applied voltage V_T?

ANSWER

$$V_T = 40\text{ V} + 60\text{ V} + 180\text{ V}$$

$$V_T = 280\text{ V}$$

Note that the *IR* drop across each *R* results from the same current of 2 A, produced by 280 V across the total R_T of 140 Ω.

Example 4–5

An applied V_T of 120 V produces *IR* drops across two series resistors R_1 and R_2. If the voltage drop across R_1 is 40 V, how much is the voltage drop across R_2?

ANSWER Since V_1 and V_2 must total 120 V and V_1 is 40 V, the voltage drop across R_2 must be the difference between 120 V and 40 V, or

$$V_2 = V_T - V_1 = 120\text{ V} - 40\text{ V}$$

$$V_2 = 80\text{ V}$$

It is logical that V_T is the sum of the series *IR* drops. The current *I* is the same in all series components. For this reason, the total of all series voltages V_T is needed to produce the same *I* in the total of all series resistances R_T as the *I* that each resistor voltage produces in its *R*.

A practical application of voltages in a series circuit is illustrated in Fig. 4–7. In this circuit, two 120-V lightbulbs are operated from a 240-V line. If one bulb were

Figure 4–7 Series string of two 120-V lightbulbs operating from a 240-V line. (*a*) Wiring diagram. (*b*) Schematic diagram.

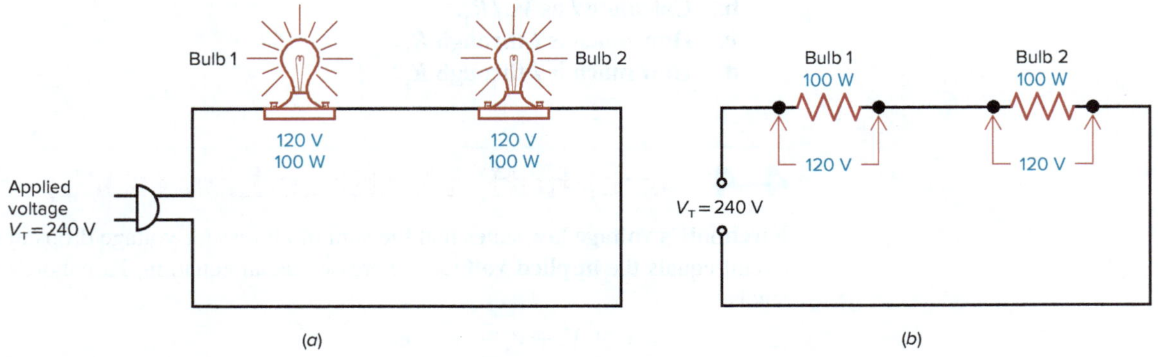

connected to 240 V, the filament would burn out. With the two bulbs in series, however, each has 120 V for proper operation. The two 120-V drops across the bulbs in series add to equal the applied voltage of 240 V.

Note: A more detailed explanation of Kirchhoff's voltage law is provided in Chapter 9 (Sec. 9–2).

4–4 Self-Review

Answers at the end of the chapter.

a. A series circuit has *IR* drops of 10, 20, and 30 V. How much is the applied voltage V_T of the source?

b. A 100-*V* source is applied across R_1 and R_2 in series. If V_1 is 25 V, how much is V_2?

c. A 120-*V* source is applied across three equal resistances in series. How much is the voltage drop across each individual resistor?

4–5 Polarity of *IR* Voltage Drops

When a voltage drop exists across a resistance, one end must be either more positive or more negative than the other end. Otherwise, without a potential difference no current could flow through the resistance to produce the voltage drop. The ***polarity*** of this *IR* voltage drop can be associated with the direction of *I* through *R*. In brief, electrons flow into the negative side of the *IR* voltage and out the positive side (see Fig. 4–8*a*).

If we want to consider conventional current, with positive charges moving in the opposite direction from electron flow, the rule is reversed for the positive charges. See Fig. 4–8*b*. Here the positive charges for *I* are moving into the positive side of the *IR* voltage.

However, for either electron flow or conventional current, the actual polarity of the *IR* drop is the same. In both *a* and *b* of Fig. 4–8, the top end of *R* in the diagrams is positive since this is the positive terminal of the source producing the current. After all, the resistor does not know which direction of current we are thinking of.

A series circuit with two *IR* voltage drops is shown in Fig. 4–9. We can analyze these polarities in terms of electron flow. The electrons move from the negative terminal of the source V_T through R_2 from point C to D. Electrons move into C and out from D. Therefore, C is the negative side of the voltage drop across R_2. Similarly, for the *IR* voltage drop across R_1, point E is the negative side, compared with point F.

A more fundamental way to consider the polarity of *IR* voltage drops in a circuit is the fact that between any two points the one nearer to the positive terminal of the voltage source is more positive; also, the point nearer to the negative terminal of the applied voltage is more negative. A point nearer the terminal means that there is less resistance in its path.

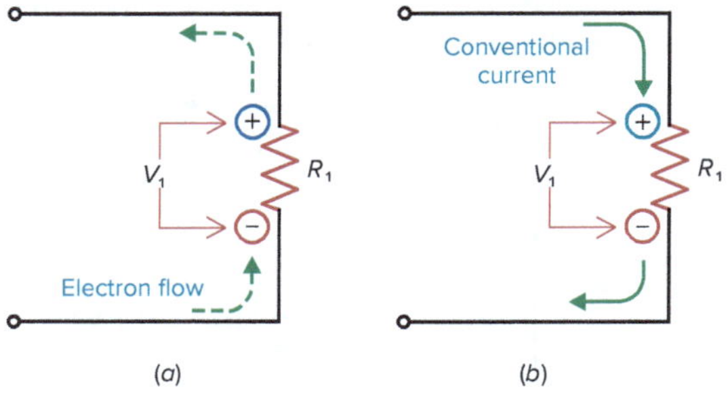

Figure 4–8 The polarity of *IR* voltage drops. (*a*) Electrons flow into the negative side of V_1 across R_1. (*b*) The same polarity of V_1 with positive charges into the positive side.

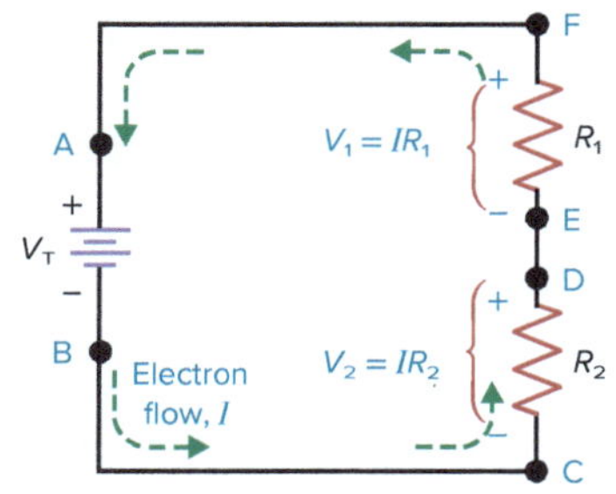

Figure 4–9 Example of two *IR* voltage drops in series. Electron flow shown for direction of *I*.

In Fig. 4–9, point C is nearer to the negative battery terminal than point D. The reason is that C has no resistance to B, whereas the path from D to B includes the resistance of R_2. Similarly, point F is nearer to the positive battery terminal than point E, which makes F more positive than E.

Notice that points D and E in Fig. 4–9 are marked with both plus and minus polarities. The plus polarity at D indicates that it is more positive than C. This polarity, however, is shown just for the voltage across R_2. Point D cannot be more positive than points F and A. The positive terminal of the applied voltage must be the most positive point because the battery is generating the *positive potential* for the entire circuit.

Similarly, points B and C must have the most negative potential in the entire string, since point B is the negative terminal of the applied voltage. In fact, the plus polarity marked at D means only that this end of R_2 is less negative than C by the amount of voltage drop across R_2.

Consider the potential difference between E and D in Fig. 4–9, which is only a piece of wire. This voltage is zero because there is no resistance between these two points. Without any resistance here, the current cannot produce the *IR* drop necessary for a difference in potential. Points E and D are, therefore, the same electrically since they have the same potential.

When we go around the external circuit from the negative terminal of V_T, with electron flow, the voltage drops are drops in *negative potential*. For the opposite direction, starting from the positive terminal of V_T, the voltage drops are drops in positive potential. Either way, the voltage drop of each series *R* is its proportional part of the V_T needed for the one value of current in all resistances.

4–5 Self-Review

Answers at the end of the chapter.

Refer to Fig. 4–9.

a. Which point in the circuit is the most negative?
b. Which point in the circuit is the most positive?
c. Which is more negative, point D or F?

GOOD TO KNOW

In a series circuit, the resistor with the largest resistance value dissipates the most power because $P = I^2R$, where *I* is the same in all the series resistances.

4–6 Total Power in a Series Circuit

The power needed to produce current in each series resistor is used up in the form of heat. Therefore, the *total power* used is the sum of the individual values of power dissipated in each part of the circuit. As a formula,

$$P_T = P_1 + P_2 + P_3 + \cdots + \text{etc.} \qquad (4\text{–}4)$$

As an example, in Fig. 4–10, R_1 dissipates 40 W for P_1, equal to 20 V × 2 A for the *VI* product. Or, the P_1 calculated as I^2R is (2 × 2) × 10 = 40 W. Also, P_1 is V^2/R, or (20 × 20)/10 = 40 W.

Figure 4–10 The sum of the individual powers P_1 and P_2 used in each resistance equals the total power P_T produced by the source.

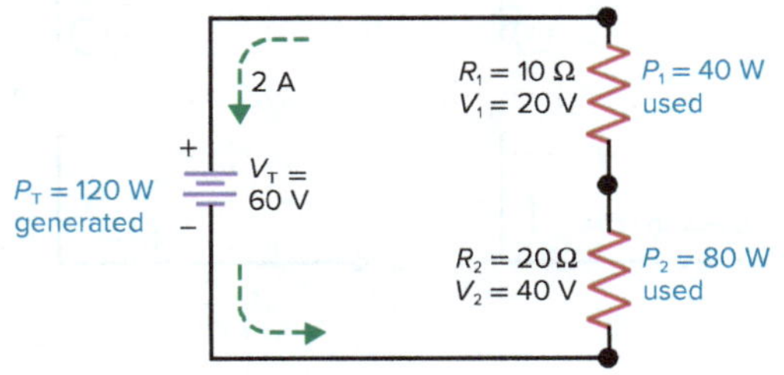

Similarly, P_2 for R_2 is 80 W. This value is 40×2 for VI, $(2 \times 2) \times 20$ for I^2R, or $(40 \times 40)/20$ for V^2/R. P_2 must be more than P_1 because R_2 is more than R_1 with the same current.

The total power dissipated by R_1 and R_2, then, is $40 + 80 = 120$ W. This power is generated by the source of applied voltage.

The total power can also be calculated as $V_T \times I$. The reason is that V_T is the sum of all series voltages and I is the same in all series components. In this case, then, $P_T = V_T \times I = 60 \times 2 = 120$ W.

The total power here is 120 W, calculated either from the total voltage or from the sum of P_1 and P_2. This is the amount of power produced by the battery. The voltage source produces this power, equal to the amount used by the resistors.

4–6 Self-Review

Answers at the end of the chapter.

a. Each of three equal resistances dissipates 2 W. How much P_T is supplied by the source?

b. A 1-kΩ R_1 and 40-kΩ R_2 are in series with a 50-V source. Which R dissipates more power?

4–7 Series-Aiding and Series-Opposing Voltages

Series-aiding voltages are connected with polarities that allow current in the same direction. In Fig. 4–11*a*, the 6 V of V_1 alone could produce a 3-A electron flow from the negative terminal, with the 2-Ω R. Also, the 8 V of V_2 could produce 4 A in the same direction. The total I then is 7 A.

Instead of adding the currents, however, the voltages V_1 and V_2 can be added, for a V_T of $6 + 8 = 14$ V. This 14 V produces 7 A in all parts of the series circuit with a resistance of 2 Ω. Then I is $14/2 = 7$ A.

Voltages are connected series-aiding when the plus terminal of one is connected to the negative terminal of the next. They can be added for a total equivalent voltage. This idea applies in the same way to voltage sources, such as batteries, and to voltage drops across resistances. Any number of voltages can be added, as long as they are connected with series-aiding polarities.

Series-opposing voltages are subtracted, as shown in Fig. 4–11*b*. Notice here that the positive terminals of V_1 and V_2 are connected. Subtract the smaller from the larger value, and give the net V the polarity of the larger voltage. In this example, V_T is $8 - 6 = 2$ V. The polarity of V_T is the same as V_2 because its voltage is higher than V_1.

Figure 4–11 Example of voltage sources V_1 and V_2 in series. (*a*) Note the connections for series-aiding polarities. Here $8\text{ V} + 6\text{ V} = 14$ V for the total V_T. (*b*) Connections for series-opposing polarities. Now $8\text{ V} - 6\text{ V} = 2$ V for V_T.

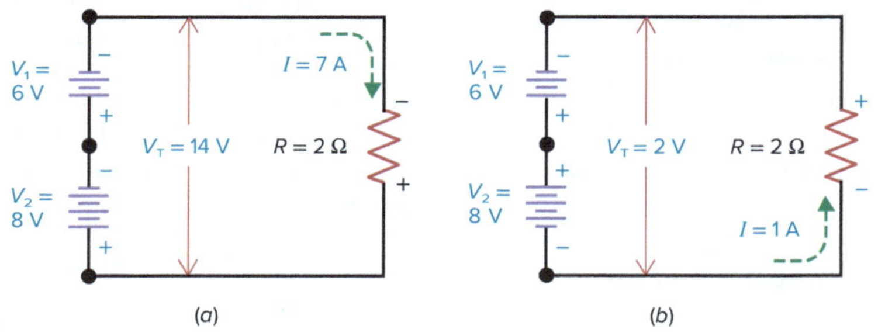

If two series-opposing voltages are equal, the net voltage will be zero. In effect, one voltage balances out the other. The current I also is zero, without any net potential difference.

4–7 Self-Review

Answers at the end of the chapter.

a. Voltage V_1 of 40 V is series-aiding with V_2 of 60 V. How much is V_T?
b. The same V_1 and V_2 are connected series-opposing. How much is V_T?

4–8 Analyzing Series Circuits with Random Unknowns

Figure 4–12 Analyzing a series circuit to find I, V_1, V_2, P_1, and P_2. See text for solution.

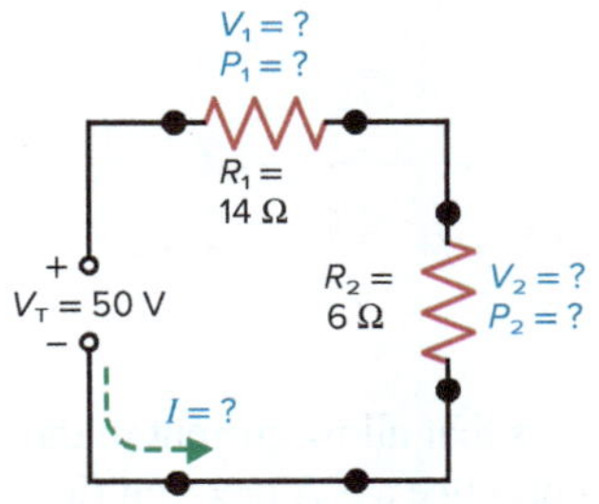

Refer to Fig. 4–12. Suppose that the source V_T of 50 V is known, with a 14-Ω R_1 and 6-Ω R_2. The problem is to find R_T, I, the individual voltage drops V_1 and V_2 across each resistor, and the power dissipated.

We must know the total resistance R_T to calculate I because the total applied voltage V_T is given. This V_T is applied across the total resistance R_T. In this example, R_T is 14 + 6 = 20 Ω.

Now I can be calculated as V_T/R_T, or 50/20, which equals 2.5 A. This 2.5-A I flows through R_1 and R_2.

The individual voltage drops are

$$V_1 = IR_1 = 2.5 \times 14 = 35 \text{ V}$$

$$V_2 = IR_2 = 2.5 \times 6 = 15 \text{ V}$$

Note that V_1 and V_2 total 50 V, equal to the applied V_T.

The calculations to find the power dissipated in each resistor are as follows:

$$P_1 = V_1 \times I = 35 \times 2.5 = 87.5 \text{ W}$$

$$P_2 = V_2 \times I = 15 \times 2.5 = 37.5 \text{ W}$$

These two values of dissipated power total 125 W. The power generated by the source equals $V_T \times I$ or 50 × 2.5, which is also 125 W.

GOOD TO KNOW

Solving a series circuit with random unknowns is similar to solving a crossword puzzle. Random clues are given for solving some of the values in the circuit, and then all of the clues are pieced together for the entire solution to the problem.

General Methods for Series Circuits

For other types of problems with series circuits, it is useful to remember the following:

1. When you know the I for one component, use this value for I in all components, for the current is the same in all parts of a series circuit.
2. To calculate I, the total V_T can be divided by the total R_T, or an individual IR drop can be divided by its R. For instance, the current in Fig. 4–12 could be calculated as V_2/R_2 or 15/6, which equals the same 2.5 A for I. However, do not mix a total value for the entire circuit with an individual value for only part of the circuit.
3. When you know the individual voltage drops around the circuit, these can be added to equal the applied V_T. This also means that a known voltage drop can be subtracted from the total V_T to find the remaining voltage drop.

Figure 4–13 Find the resistance of R_3. See text for the analysis of this series circuit.

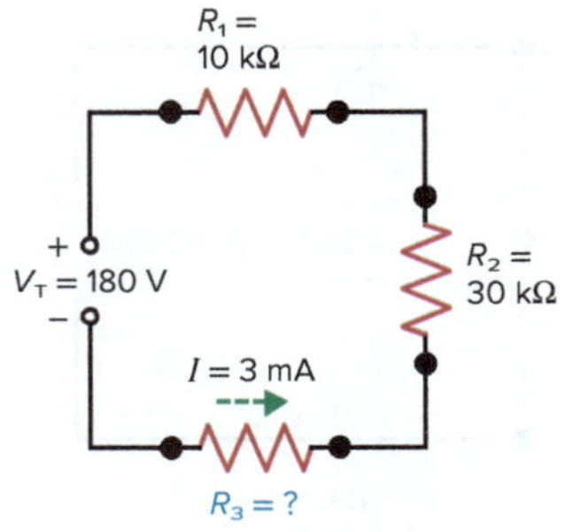

These principles are illustrated by the problem in Fig. 4–13. In this circuit, R_1 and R_2 are known but not R_3. However, the current through R_3 is given as 3 mA.

With just this information, all values in this circuit can be calculated. The I of 3 mA is the same in all three series resistances. Therefore,

$$V_1 = 3 \text{ mA} \times 10 \text{ K}\Omega = 30 \text{ V}$$

$$V_2 = 3 \text{ mA} \times 30 \text{ K}\Omega = 90 \text{ V}$$

The sum of V_1 and V_2 is 30 + 90 = 120 V. This 120 V plus V_3 must total 180 V. Therefore, V_3 is 180 − 120 = 60 V.

With 60 V for V_3, equal to IR_3, then R_3 must be 60/0.003, equal to 20,000 Ω or 20 kΩ. The total circuit resistance is 60 kΩ, which results in the current of 3 mA with 180 V applied, as specified in the circuit.

Another way of doing this problem is to find R_T first. The equation $I = V_T/R_T$ can be inverted to calculate R_T.

$$R_T = \frac{V_T}{I}$$

With a 3-mA I and 180 V for V_T, the value of R_T must be 180 V/3 mA = 60 kΩ. Then R_3 is 60 kΩ − 40 kΩ = 20 kΩ.

The power dissipated in each resistance is 90 mW in R_1, 270 mW in R_2, and 180 mW in R_3. The total power is 90 + 270 + 180 = 540 mW.

Series Voltage-Dropping Resistors

Figure 4–14 Example of a series voltage-dropping resistor R_S used to drop V_T of 12.6 V to 9 V for R_L. See text for calculations.

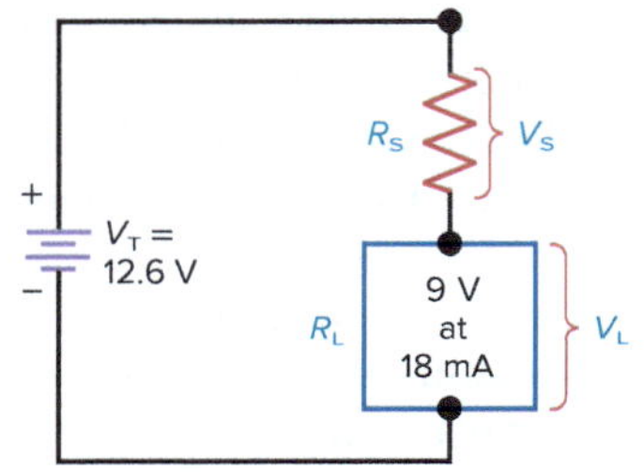

A common application of series circuits is to use a resistance to drop the voltage from the source V_T to a lower value, as in Fig. 4–14. The load R_L here represents a radio that operates normally with a 9-V battery. When the radio is on, the DC load current with 9 V applied is 18 mA. Therefore, the requirements are 9 V at 18 mA as the load.

To operate this radio from 12.6 V, the voltage-dropping resistor R_S is inserted in series to provide a voltage drop V_S that will make V_L equal to 9 V. The required voltage drop for V_S is the difference between V_L and the higher V_T. As a formula,

$$V_S = V_T - V_L = 12.6 - 9 = 3.6 \text{ V}$$

Furthermore, this voltage drop of 3.6 V must be provided with a current of 18 mA, for the current is the same through R_S and R_L. To calculate R_S,

$$R_S = \frac{3.6 \text{ V}}{18 \text{ mA}} = 0.2 \text{ k}\Omega = 200\ \Omega$$

Circuit with Voltage Sources in Series

See Fig. 4–15. Note that V_1 and V_2 are series-opposing, with + to + through R_1. Their net effect, then, is 0 V. Therefore, V_T consists only of V_3, equal to 4.5 V. The total R is 2 + 1 + 2 = 5 kΩ for R_T. Finally, I is V_T/R_T or 4.5 V/5 KΩ, which is equal to 0.9 mA, or 900 μA.

4–8 Self-Review

Answers at the end of the chapter.

Refer to Fig. 4–13.

a. **Calculate V_1 across R_1.**
b. **Calculate V_2 across R_2.**
c. **How much is V_3?**

Figure 4–15 Finding the I for this series circuit with three voltage sources. See text for solution.

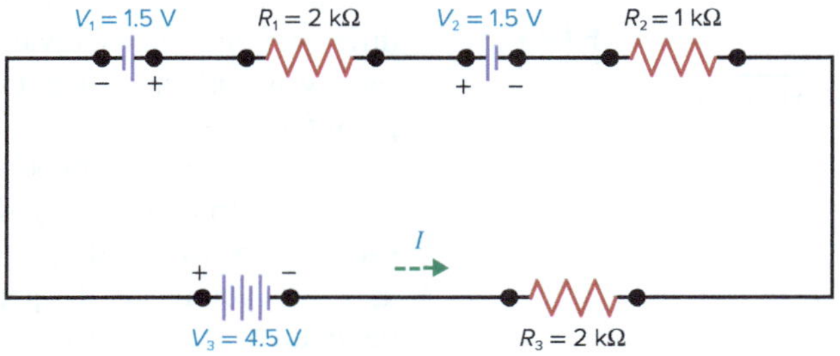

4–9 Ground Connections in Electrical and Electronic Systems

In most electrical and electronic systems, one side of the voltage source is connected to ground. For example, one side of the voltage source of the 120-V_{AC} power line in residential wiring is connected directly to ***earth ground.*** The reason for doing this is to reduce the possibility of electric shock. The connection to earth ground is usually made by driving copper rods into the ground and connecting the ground wire of the electrical system to these rods. The schematic symbol used for earth ground is shown in Fig. 4–16. In electronic circuits, however, not all ground connections are necessarily earth ground connections. The pitchfork-like symbol shown in Fig. 4–16 is considered by many people to be the most appropriate symbol for a metal chassis or copper foil ground on printed-circuit boards. This ***chassis ground*** *symbol* represents a common return path for current and may or may not be connected to an actual earth ground. Another ground symbol, common ground, is shown in Fig. 4–16. This is just another symbol used to represent a common return path for current in a circuit. In all cases, ground is assumed to be at a potential of 0 V, regardless of which symbol is shown. Some schematic diagrams may use two or all three of the ground symbols shown in Fig. 4–16. In this type of circuit, each ground represents a common return path for only those circuits using the same ground symbol. When more than one type of ground symbol is shown on a schematic diagram, it is important to realize that each one is electrically isolated from the other. The term *electrically isolated* means that the resistance between each ground or common point is infinite ohms.

Figure 4–16 Ground symbols.

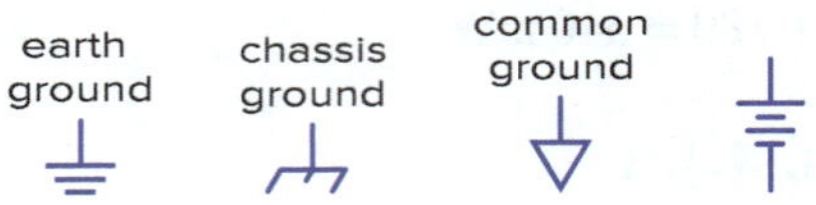

Although standards defining the use of each ground symbol in Fig. 4–16 have been set, the use of these symbols in the electronics industry seems to be inconsistent with their definitions. In other words, a schematic may show the earth ground symbol, even though it is a chassis ground connection. Regardless of the symbol used, the main thing to remember is that the symbol represents a common return path for current in a given circuit. In this text, the earth ground symbol shown in Fig. 4–16 has been arbitrarily chosen as the symbol representing a common return path for current.

Figure 4–17 shows a series circuit employing the earth ground symbol. Since each ground symbol represents the same electrical potential of 0 V, the negative terminal of V_T and the bottom end of R_3 are actually connected to the same point electrically. Electrons leaving the bottom of V_T flow through the common return path represented by the ground symbol and return to the bottom of R_3, as shown in the figure. One of the main reasons for using ground connections in electronic circuits is to simplify the wiring.

MultiSim **Figure 4–17** Series circuit using earth ground symbol to represent common return path for current.

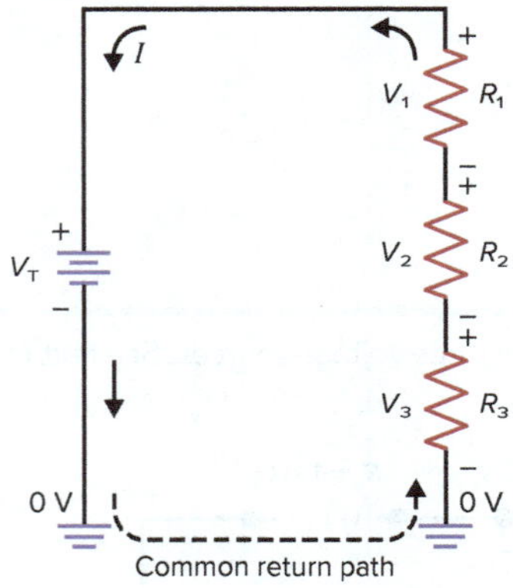

Voltages Measured with Respect to Ground

When a circuit has a ground as a common return, we generally measure the voltages with respect to this ground. The circuit in Fig. 4–18*a* is called a *voltage divider*. Let us consider this circuit without any ground, and then analyze the effect of grounding different points on the divider. It is important to realize that this circuit operates the same way with or without the ground. The only factor that changes is the reference point for measuring the voltages.

In Fig. 4–18*a*, the three 10-Ω resistances R_1, R_2, and R_3 divide the 30-V source equally. Then each voltage drop is 30/3 = 10 V for V_1, V_2, and V_3. The polarity of each resistor voltage drop is positive at the top and negative at the bottom, the same as V_T. As you recall, the polarity of a resistor's voltage drop is determined by the direction of current flow.

MultiSim **Figure 4–18** An example of calculating DC voltages measured with respect to ground. (*a*) Series circuit with no ground connection. (*b*) Negative side of V_T grounded to make all voltages positive with respect to ground. (*c*) Positive and negative voltages with respect to ground at point B. (*d*) Positive side of V_T grounded; all voltages are negative to ground.

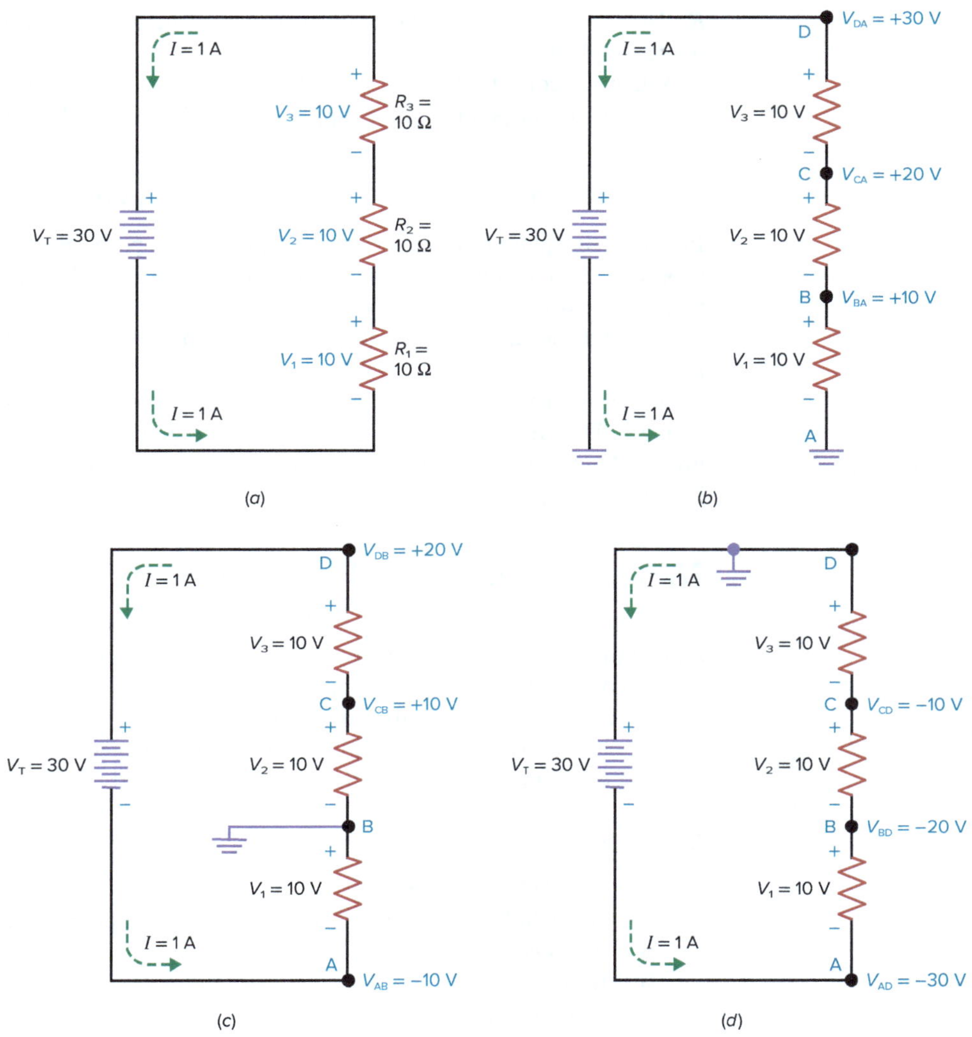

GOOD TO KNOW

It is commonplace for schematics to show only the value and polarity of a voltage source with respect to ground. In other words, the battery or DC voltage source symbol is usually not shown to represent the applied voltage.

If we want to consider the current, I is 30/30 = 1 A. Each IR drop is $1 \times 10 = 10$ V for V_1, V_2, and V_3.

Positive Voltages to Negative Ground

In Fig. 4–18*b*, the negative side of V_T is grounded and the bottom end of R_1 is also grounded to complete the circuit. The ground is at point A. Note that the individual voltages V_1, V_2, and V_3 are still 10 V each. Also, the current is still 1 A. The direction of current is also the same, from the negative side of V_T, through the common ground, to the bottom end of R_1. The only effect of the ground here is to provide a conducting path from one side of the source to one side of the load.

With the ground in Fig. 4–18*b*, though, it is useful to consider the voltages with respect to ground. In other words, the ground at point A will now be the reference for all voltages. When a voltage is indicated for only one point in a circuit, generally the other point is assumed to be ground. We must have two points for a potential difference.

Let us consider the voltages at points B, C, and D. The voltage at B to ground is V_{BA}. This ***double subscript notation*** indicates that we measure at B with respect to A. In general, the first letter indicates the point of measurement and the second letter is the reference point.

Then V_{BA} is +10 V. The positive sign is used here to emphasize the polarity. The value of 10 V for V_{BA} is the same as V_1 across R_1 because points B and A are across R_1. However, V_1 as the voltage across R_1 cannot be given any polarity without a reference point.

When we consider the voltage at C, then, V_{CA} is +20 V. This voltage equals $V_1 + V_2$. Also, for point D at the top, V_{DA} is +30 V for $V_1 + V_2 + V_3$.

Positive and Negative Voltages to a Grounded Tap

In Fig. 4–18*c*, point B in the divider is grounded. The purpose is to have the divider supply negative and positive voltages with respect to ground. The negative voltage here is V_{AB}, which equals −10 V. This value is the same 10 V as V_1, but V_{AB} is the voltage at the negative end A with respect to the positive end B. The other voltages in the divider are $V_{CB} = +10$ V and $V_{DB} = +20$ V.

We can consider the ground at B as a dividing point for positive and negative voltages. For all points toward the positive side of V_T, any voltage is positive to ground. Going the other way, at all points toward the negative side of V_T, any voltage is negative to ground.

GOOD TO KNOW

The ground connections in Fig. 4–18*b*, 4–18*c*, and 4–18*d* will not change the resistance, current, or voltage values from those shown in the original circuit of Fig. 4–18*a*. The only thing that changes is the reference point for making voltage measurements.

Negative Voltages to Positive Ground

In Fig. 4–18*d*, point D at the top of the divider is grounded, which is the same as grounding the positive side of the source V_T. The voltage source here is *inverted*, compared with Fig. 4–18*b*, as the opposite side is grounded. In Fig. 4–18*d*, all voltages on the divider are negative to ground. Here, $V_{CD} = -10$ V, $V_{BD} = -20$ V, and $V_{AD} = -30$ V. Any point in the circuit must be more negative than the positive terminal of the source, even when this terminal is grounded.

4–9 Self-Review

Answers at the end of the chapter.

Refer to Fig. 4–18*c* and give the voltage and polarity for

a. A to ground.
b. B to ground.
c. D to ground.
d. V_{DA} across V_T.

4–10 Troubleshooting: Opens and Shorts in Series Circuits

In many cases, electronic technicians are required to repair a piece of equipment that is no longer operating properly. The technician is expected to troubleshoot the equipment and restore it to its original operating condition. To *troubleshoot* means "to diagnose or analyze." For example, a technician may diagnose a failed electronic circuit by using a digital multimeter (DMM) to make voltage, current, and resistance measurements. Once the defective component has been located, it is removed and replaced with a good one. But here is one very important point that needs to be made about **troubleshooting**: To troubleshoot a defective circuit, you must understand how the circuit is supposed to work in the first place. Without this knowledge, your troubleshooting efforts could be nothing more than guesswork. What we will do next is analyze the effects of both opens and shorts in series circuits.

GOOD TO KNOW

Everyone who works with electronic hardware must be able to troubleshoot electronic equipment. The service technician, the production tester, the custom engineer, and the engineering designer all do some troubleshooting as part of their jobs.

Figure 4–19 Effect of an open in a series circuit. (*a*) Normal circuit with 1-A series current. (*b*) Open path between points P_1 and P_2 results in zero current in all parts of the circuit.

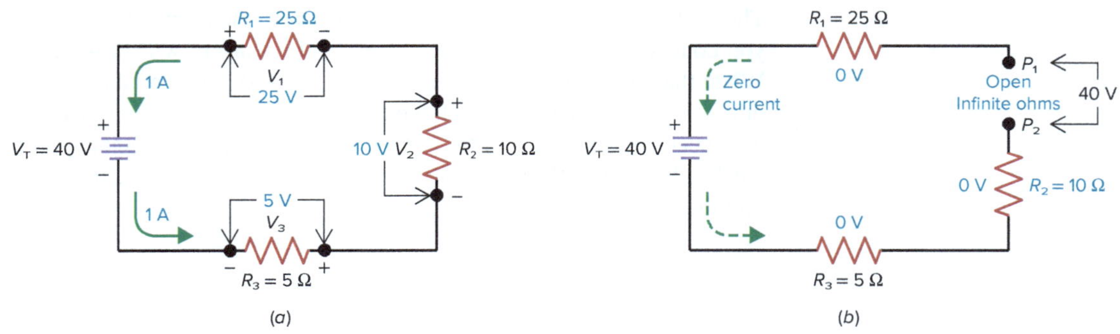

The Effect of an Open in a Series Circuit

An *open circuit* is a break in the current path. The resistance of an open circuit is extremely high because the air between the open points is a very good insulator. Air can have billions of ohms of resistance. For a series circuit, a break in the current path means zero current in all components.

Figure 4–19*a* shows a series circuit that is operating normally. With 40 V of applied voltage and 40 Ω of total resistance, the series current is 40 V/40 Ω = 1 A. This produces the following *IR* voltage drops across R_1, R_2, and R_3: $V_1 = 1\text{ A} \times 25\ \Omega = 25$ V, $V_2 = 1\text{ A} \times 10\ \Omega = 10$ V, and $V_3 = 1\text{ A} \times 5\ \Omega = 5$ V.

Now consider the effect of an open circuit between points P_1 and P_2 in Fig. 4–19*b*. Because there is practically infinite resistance between the open points, the current in the entire series circuit is zero. With zero current throughout the series circuit, each resistor's *IR* voltage will be 0 V even though the applied voltage is still 40 V. To calculate V_1, V_2, and V_3 in Fig. 4–19*b*, simply use 0 A for *I*. Then, $V_1 = 0\text{ A} \times 25\ \Omega = 0$ V, $V_2 = 0\text{ A} \times 10\ \Omega = 0$ V, and $V_3 = 0\text{ A} \times 5\ \Omega = 0$ V. But how much voltage is across points P_1 and P_2? The answer is 40 V. This might surprise you, but here's the proof: Let's assume that the resistance between P_1 and P_2 is $40 \times 10^9\ \Omega$, which is 40 GΩ (40 gigohms). Since the total resistance of a series circuit equals the sum of the series resistances, R_T is the sum of 25 Ω, 15 Ω, 10 Ω, and 40 GΩ. Since the 40 GΩ of resistance between P_1 and P_2 is so much larger than the other resistances, it is essentially the total resistance of the series circuit. Then the series current *I* is calculated as $40\text{ V}/40\text{ G}\Omega = 1 \times 10^{-9}\text{ A} = 1$ nA. For all practical purposes, the current *I* is zero. This is the value of current in the entire series circuit. This small current produces about 0 V across R_1, R_2, and R_3, but across the open points P_1 and P_2, where the resistance is high, the voltage is calculated as $V_{open} = 1 \times 10^{-9}\text{ A} \times 40 \times 10^9\ \Omega = 40$ V.

In summary, here is the effect of an open in a series circuit:

1. The current *I* is zero in all components.
2. The voltage drop across each good component is 0 V.
3. The voltage across the open points equals the applied voltage.

The Applied Voltage V_T Is Still Present with Zero Current

The open circuit in Fig. 4–19*b* is another example of how voltage and current are different. There is no current with the open circuit because there is no complete path for current flow between the two battery terminals. However, the battery still has its potential difference of 40 V across the positive and negative terminals. In other words, the applied voltage V_T is still present with or without current in the external circuit. If you measure V_T with a voltmeter, it will measure 40 V regardless of whether the circuit is closed, as in Fig. 4–19*a*, or open, as in Fig. 4–19*b*.

The same idea applies to the 120-V_{AC} voltage from the power line in our homes. The 120 V potential difference is available from the terminals of the wall outlet. If you connect a lamp or appliance to the outlet, current will flow in those circuits. When there is nothing connected, though, the 120 V potential is still present at the outlet. If you accidentally touch the metal terminals of the outlet when nothing else is connected, you will get an electric shock. The power company is maintaining the 120 V at the outlets as a source to produce current in any circuit that is plugged into the outlet.

Example 4–6

Figure 4–20 Series circuit for Example 4–6.

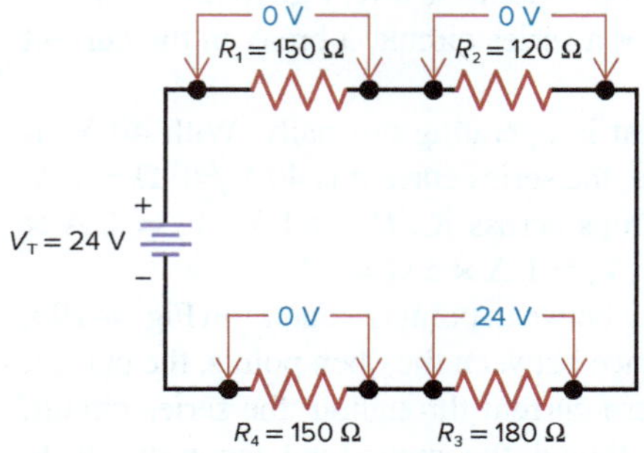

Assume that the series circuit in Fig. 4–20 has failed. A technician troubleshooting the circuit used a voltmeter to record the following resistor voltage drops:

$$V_1 = 0\ \text{V}$$
$$V_2 = 0\ \text{V}$$
$$V_3 = 24\ \text{V}$$
$$V_4 = 0\ \text{V}$$

Based on these voltmeter readings, which component is defective and what type of defect is it? (Assume that only one component is defective.)

ANSWER To help understand which component is defective, let's calculate what the values of V_1, V_2, V_3, and V_4 are supposed to be. Begin by calculating R_T and I.

$$R_T = R_1 + R_2 + R_3 + R_4$$
$$= 150\ \Omega + 120\ \Omega + 180\ \Omega + 150\ \Omega$$
$$R_T = 600\ \Omega$$

$$I = \frac{V_T}{R_T}$$
$$= \frac{24\ \text{V}}{600\ \Omega}$$
$$I = 40\ \text{mA}$$

Next,

$$V_1 = I \times R_1$$
$$= 40\ \text{mA} \times 150\ \Omega$$
$$V_1 = 6\ \text{V}$$
$$V_2 = I \times R_2$$
$$= 40\ \text{mA} \times 120\ \Omega$$
$$V_2 = 4.8\ \text{V}$$
$$V_3 = I \times R_3$$
$$= 40\ \text{mA} \times 180\ \Omega$$
$$V_3 = 7.2\ \text{V}$$
$$V_4 = I \times R_4$$
$$= 40\ \text{mA} \times 150\ \Omega$$
$$V_4 = 6\ \text{V}$$

Next, compare the calculated values with those measured in Fig. 4–20. When the circuit is operating normally, V_1, V_2, and V_4 should measure 6 V, 4.8 V, and 6 V, respectively. Instead, the measurements made in Fig. 4–20 show that each of these voltages is 0 V. This indicates that the current I in the circuit must be zero, caused by an open somewhere in the circuit. The reason that V_1, V_2, and V_4 are 0 V is simple: $V = I \times R$. If $I = 0$ A, then each good resistor must have a voltage drop of 0 V. The measured value of V_3 is 24 V, which is considerably higher than its calculated value of 7.2 V. Because V_3 is dropping the full value of the applied voltage, it must be open. The reason the open R_3 will drop the full 24 V is that it has billions of ohms of resistance and, in a series circuit, the largest resistance drops the most voltage. Since the open resistance of R_3 is so much higher than the values of R_1, R_2, and R_4, it will drop the full 24 V of applied voltage.

The Effect of a Short in a Series Circuit

A *short circuit* is an extremely low resistance path for current flow. The resistance of a short is assumed to be 0 Ω. This is in contrast to an open, which is assumed to have a resistance of infinite ohms. Let's reconsider the circuit in Fig. 4–19 with R_2 shorted. The circuit is redrawn for your convenience in Fig. 4–21. Recall from Fig. 4–19*a* that the normal values of V_1, V_2, and V_3 are 25 V, 10 V, and 5 V, respectively. With the 10-Ω R_2 shorted, the total resistance R_T will decrease from 40 Ω to 30 Ω. This will cause the series current to increase from 1 A to 1.33 A. This is calculated as 40 V/30 Ω = 1.33 A. The increase in current will cause the voltage drop across resistors R_1 and R_3 to increase from their normal values. The new voltage drops across R_1 and R_3 with R_2 shorted are calculated as follows:

$$V_1 = I \times R_1 = 1.33\text{ A} \times 25\ \Omega \qquad V_3 = I \times R_3 = 1.33\text{ A} \times 5\ \Omega$$

$$V_1 = 33.3\text{ V} \qquad V_3 = 6.67\text{ V}$$

The voltage drop across the shorted R_2 is 0 V because the short across R_2 effectively makes its resistance value 0 Ω. Then,

$$V_2 = I \times R_2 = 1.33\text{ A} \times 0\ \Omega$$

$$V_2 = 0\text{ V}$$

In summary, here is the effect of a short in a series circuit:

1. The current I increases above its normal value.
2. The voltage drop across each good component increases.
3. The voltage drop across the shorted component drops to 0 V.

Figure 4–21 Series circuit of Fig. 4–19 with R_2 shorted.

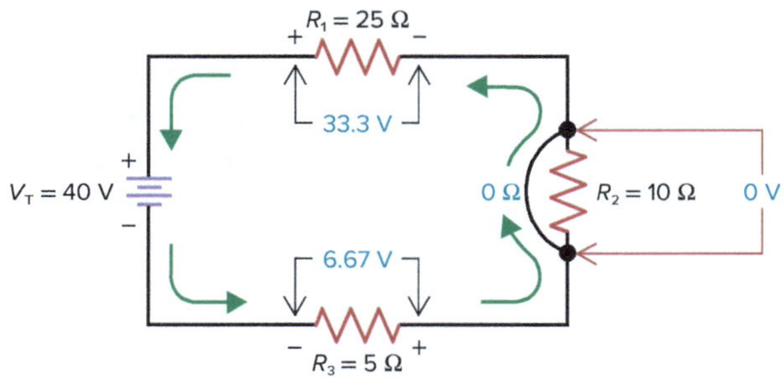

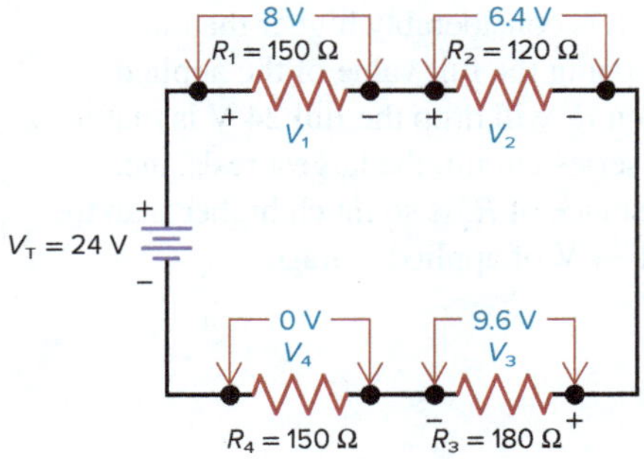

Figure 4–22 Series circuit for Example 4–7.

Example 4–7

Assume that the series circuit in Fig. 4–22 has failed. A technician troubleshooting the circuit used a voltmeter to record the following resistor voltage drops:

$$V_1 = 8\ \text{V}$$
$$V_2 = 6.4\ \text{V}$$
$$V_3 = 9.6\ \text{V}$$
$$V_4 = 0\ \text{V}$$

Based on the voltmeter readings, which component is defective and what type of defect is it? (Assume that only one component is defective.)

ANSWER This is the same circuit used in Example 4–6. Therefore, the normal values for V_1, V_2, V_3, and V_4 are 6 V, 4.8 V, 7.2 V, and 6 V, respectively. Comparing the calculated values with those measured in Fig. 4–22 reveals that V_1, V_2, and V_3 have increased from their normal values. This indicates that the current has increased, this is why we have a larger voltage drop across these resistors. The measured value of 0 V for V_4 shows a significant drop from its normal value of 6 V. The only way this resistor can have 0 V, when all other resistors show an increase in voltage, is if R_4 is shorted. Then $V_4 = I \times R_4 = I \times 0\ \Omega = 0$ V.

General Rules for Troubleshooting Series Circuits

When troubleshooting a series circuit containing three or more resistors, remember this important rule: The defective component will have a voltage drop that will change in the opposite direction as compared to the good components. In other words, in a series circuit containing an open, all the good components will have a voltage decrease from their normal value to 0 V. The defective component will have a voltage increase from its normal value to the full applied voltage. Likewise, in a series circuit containing a short, all good components will have a voltage increase from their normal values and the defective component's voltage drop will decrease from its normal value to 0 V. The point to be made here is simple: The component whose voltage changes in the opposite direction of the other components is the defective component. In the case of an open resistor, the voltage drop increases to the value of the applied voltage and all other resistor voltages decrease to 0 V. In the case of a short, all good components show their voltage drops increasing, whereas the shorted component shows a voltage decrease to 0 V. This same general rule applies to a series circuit that has components whose resistances have increased or decreased from their normal values but are neither open or shorted.

4–10 Self-Review

Answers at the end of the chapter.

a. **In Fig. 4–20, how much voltage is across R_1 if it is open?**
b. **In Fig. 4–20, how much voltage is across R_2 if it is shorted?**
c. **In Fig. 4–20, does the voltage across R_3 increase, decrease, or stay the same if the value of R_1 increases?**

Application of Series Circuits

HOLIDAY LIGHTS

During the holiday season, it is common to see strings of lights decorating trees and the outside of homes and businesses. Some strings of lights are on continuously while others may blink on and off rapidly. When examining a string of holiday lights for the first time, it may not be apparent that the bulbs are all connected in series. This is because the string of lights is interwoven between two other conductors making it difficult to figure out the exact wiring configuration. A typical set of holiday lights is shown in Figs. 4–23*a* and 4–23*b*. Notice that there is only one wire entering the socket of each bulb and only one wire leaving. This indicates the bulbs must be wired in series. Figure 4–23*c* shows the wiring diagram for a string of holiday lights with 50 bulbs. Notice that one end has a male plug so the string of lights can be plugged into the 120 V AC power line. The other end has a 120 V receptacle so that another string of lights can be plugged in (added) to the first one. The top and bottom conductors that run from the plug on one end to the receptacle on the other end serve as nothing more than an extension cord.

Now let's focus on the string of lights (bulbs) between the top and bottom conductors in Fig. 4–23*c*. Notice that all 50 bulbs are connected one after another in series. Furthermore, notice that one end of the series string is connected to the top conductor on the plug-end while the other end is connected to the bottom conductor on the receptacle-end. Therefore, the string of 50 bulbs is connected directly across the 120 V AC power line. Since the bulbs are all identical, the applied voltage of 120 V divides evenly among them. Since there are 50 bulbs, the voltage across each bulb equals 120 V/50 or 2.4 V. If the entire string of bulbs draws 166.7 mA of current, the power dissipated by each bulb is calculated as 2.4 V × 166.7 mA = 400 mW. The power dissipated by the entire string of bulbs equals 400 mW/bulb × 50 bulbs = 20 W. The values of voltage, current, and power listed here are very typical for a string of holiday lights with 50 incandescent bulbs.

If any bulb in Fig. 4–23*b* is removed from its socket, the path for current is broken and all of the bulbs go out (dark). This means that the voltage across each of the remaining bulbs is now 0 V and 120 V is across the connecting terminals of the empty bulb socket. (Remember, the voltage across an open in a series circuit equals the applied voltage.) But something very unusual happens when a bulb burns out while it's still in its socket. The bulb that burns out doesn't light anymore but all of the remaining bulbs continue to light, although a little brighter than before. To understand how this is possible, let's examine Fig. 4–23*d* which shows the internal construction of one of the bulbs. Besides the bulbs filament, notice the metal fuse link wrapped around the base of the bulbs filament wires. The fuse link, sometimes referred to as a shunt, is a very thin insulated wire. The insulation surrounding the metal fuse link is a very thin oxide coating. When the bulb's filament burns open, the full 120 V of applied voltage exists across the bulb terminals and the oxide coating surrounding the fuse link breaks down. This exposes the metal fuse link which in turn shorts the filament wires at the base of the bulb. The metal fuse link is actually spot welded to the wires at the base of the bulb's filament. (Under normal circumstances, the insulation prevents the metal fuse link from shorting the base of the bulb's filament.) Due to the

Figure 4–23a Holiday lights.

Sarah Schultz Photography

Figure 4–23b Close-up view of holiday lights.

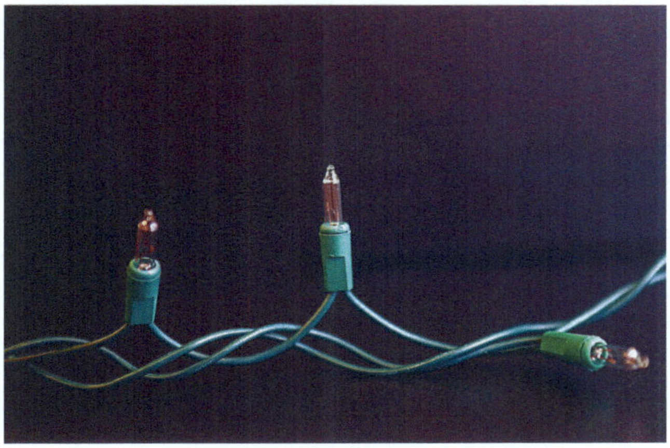

Sarah Schultz Photography

Figure 4–23c Wiring diagram for the lights in Fig. 4–23b.

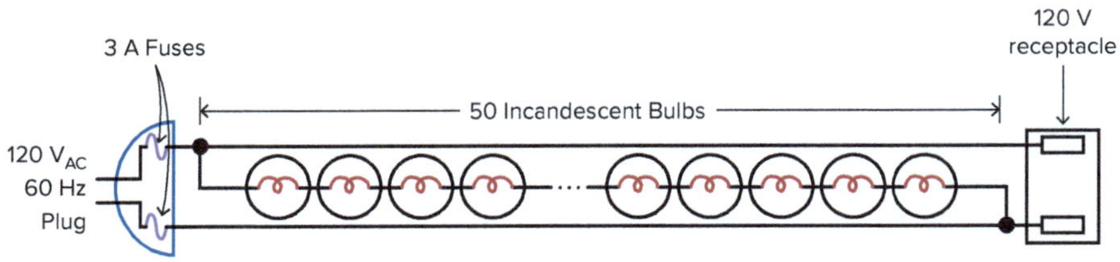

Figure 4–23d Internal construction of an incandescent bulb used with holiday lights.

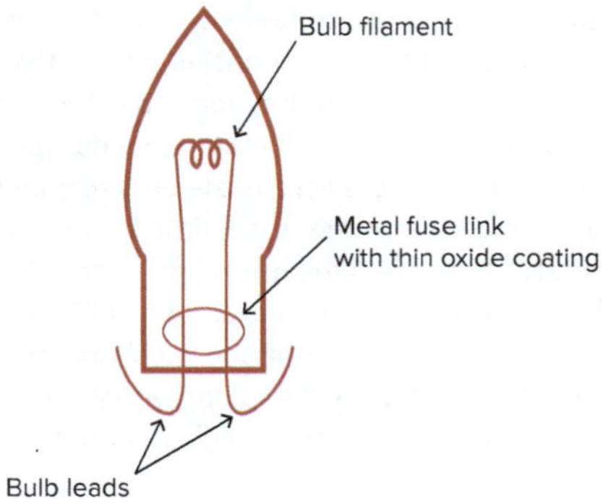

Figure 4–23e Flasher unit. Note the red dye on the tip of the bulb.

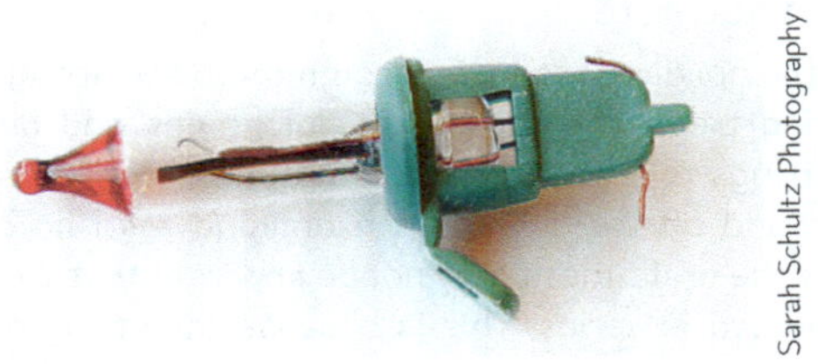

Figure 4–23f Internal construction of flasher unit.

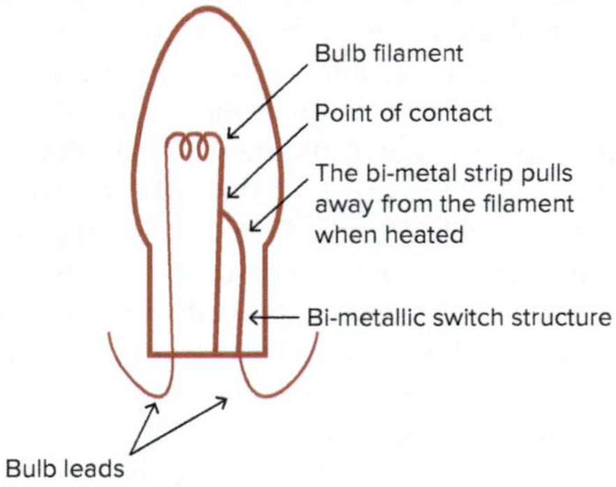

very low resistance of the metal fuse link, the voltage across the shorted filament is nearly 0 V. This causes the voltage across each of the remaining bulbs to increase slightly making them glow a little brighter. If too many bulbs burn out, the remaining good bulbs become stressed due to the increased voltage drop and power dissipation. In fact, if too many bulbs burn out it is likely they all will. This is due to the excessive power dissipation in the remaining good bulbs. This is why a string of holiday lights should be inspected regularly for burned out bulbs. When you see a bad bulb, replace it. If you don't, you'll risk the chance of all bulbs burning out (in rapid succession) due to excessive power dissipation.

It is important to realize that the total resistance of the string of lights continues to decrease as more and more bulbs burn out. This, in turn, causes the series current to increase above its normal value of 166.7 mA. The in-line fuses shown in Fig. 4–23*c* (inside of the male plug) are typically 3 A fuses.

FLASHER UNIT

Some holiday lights blink on and off rapidly. These strings of lights replace one of the series connected bulbs with a special flasher unit. The flasher unit looks like all the other bulbs except that the tip of the bulb is dipped in a dye so that it is easy to identify which bulb is the flasher unit. The flasher unit is designed to open and close as the bulb heats up and cools down. Figure 4–23*e* shows a picture of a typical flasher unit, whereas Fig. 4–23*f* shows its internal construction. The flasher unit incorporates a ***bi-metallic thermal switch***, which is nothing more than two dissimilar strips of metal that bend when heated. When the heat produced by the bulb's filament reaches some predetermined level, the bi-metallic thermal switch opens. When the bulb cools off, it closes again. Since the flasher unit is in series with all of the bulbs in the entire series string, all of the lights blink on and off rapidly. Because strings of holiday lights are often used outdoors, the outside temperature may affect how fast the lights blink on and off.

Here's one final point. The flasher unit does not have a built-in metal fuse link like the other bulbs. Therefore, if a flasher unit has an open filament, none of the bulbs in the string will light. In this case, the flasher unit must be replaced.

Chapter 5, ***Parallel Circuits***, examines the effects of connecting additional strings of lights to the first set. It also examines how a string of 100 holiday lights is wired.

Summary

- There is only one current, I, in a series circuit: $I = V_T/R_T$, where V_T is the voltage applied across the total series resistance R_T. This I is the same in all the series components.
- The total resistance R_T of a series string is the sum of the individual resistances.
- Kirchhoff's voltage law states that the applied voltage V_T equals the sum of the IR voltage drops in a series circuit.
- The negative side of an IR voltage drop is where electrons flow in, attracted to the positive side at the opposite end.
- The sum of the individual values of power used in the individual resistances equals the total power supplied by the source.
- Series-aiding voltages are added; series-opposing voltages are subtracted.
- An open circuit results in no current in all parts of the series circuit.
- For an open in a series circuit, the voltage across the two open terminals is equal to the applied voltage, and the voltage across the remaining components is 0 V.
- A short in a series circuit causes the current to increase above its normal value. The voltage drop across the shorted component decreases to 0 V, and the voltage drop across the remaining components increases.

Important Terms

Chassis ground — a common return path for current in a circuit. The common return path is often a direct connection to a metal chassis or frame or perhaps a copper foil trace on a printed-circuit board. The symbol for chassis ground is .

Double subscript notation — a notational system that identifies the points in the circuit where a voltage measurement is to be taken, that is, V_{AG}. The first letter in the subscript indicates the point in the circuit where the measurement is to be taken, and the second letter indicates the point of reference.

Earth ground — a direct connection to the earth usually made by driving copper rods into the earth and then connecting the ground wire of an electrical system to this point. The earth ground connection can serve as a common return path for the current in a circuit. The symbol for earth ground is 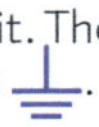.

Kirchhoff's voltage law (KVL) — a law stating that the sum of the voltage drops in a series circuit must equal the applied voltage.

Series-aiding voltages — voltage sources that are connected so that the polarities of the individual sources aid each other in producing current in the same direction in the circuit.

Series components — components that are connected in the same current path.

Series-opposing voltages — voltage sources that are connected so that the polarities of the individual sources will oppose each other in producing current flow in the circuit.

Series string — a combination of series resistances.

Troubleshooting — a term that refers to diagnosing or analyzing a faulty electronic circuit.

Voltage drop — a voltage across a resistor equal to the product of the current, I, and the resistance, R.

Voltage polarity — a term to describe the positive and negative ends of a potential difference across a component such as a resistor.

Related Formulas

$$R_T = R_1 + R_2 + R_3 + \cdots + \text{etc.}$$

$$R_T = \frac{V_T}{I}$$

$$I = \frac{V_T}{R_T}$$

$$V_R = I \times R$$

$$V_T = V_1 + V_2 + V_3 + \cdots + \text{etc.}$$

$$P_T = P_1 + P_2 + P_3 + \cdots + \text{etc.}$$

Self-Test

Answers at the back of the book.

1. Three resistors in series have individual values of 120 Ω, 680 Ω, and 1.2 kΩ. How much is the total resistance, R_T?

a. 1.8 kΩ.

b. 20 kΩ.

c. 2 kΩ.

d. none of the above.

2. In a series circuit, the current, I, is

a. different in each resistor.

b. the same everywhere.

c. the highest near the positive and negative terminals of the voltage source.

d. different at all points along the circuit.

3. In a series circuit, the largest resistance has

a. the largest voltage drop.

b. the smallest voltage drop.

c. more current than the other resistors.

d. both a and c.

4. The polarity of a resistor's voltage drop is determined by

a. the direction of current through the resistor.

b. how large the resistance is.

c. how close the resistor is to the voltage source.

d. how far away the resistor is from the voltage source.

5. A 10-Ω and 15-Ω resistor are in series across a DC voltage source. If the 10-Ω resistor has a voltage drop of 12 V, how much is the applied voltage?

a. 18 V.

b. 12 V.

c. 30 V.

d. It cannot be determined.

6. How much is the voltage across a shorted component in a series circuit?

a. The full applied voltage, V_T.

b. The voltage is slightly higher than normal.

c. 0 V.

d. It cannot be determined.

7. How much is the voltage across an open component in a series circuit?

a. The full applied voltage, V_T.

b. The voltage is slightly lower than normal.

c. 0 V.

d. It cannot be determined.

8. A voltage of 120 V is applied across two resistors, R_1 and R_2, in series. If the voltage across R_2 equals 90 V, how much is the voltage across R_1?

a. 90 V.

b. 30 V.

c. 120 V.

d. It cannot be determined.

9. If two series-opposing voltages each have a voltage of 9 V, the net or total voltage is

a. 0 V.

b. 18 V.

c. 9 V.

d. none of the above.

10. On a schematic diagram, what does the chassis ground symbol represent?

a. hot spots on the chassis.

b. the locations in the circuit where electrons accumulate.

c. a common return path for current in one or more circuits.

d. none of the above.

11. The notation, V_{BG}, means

a. the voltage at point G with respect to point B.

b. the voltage at point B with respect to point G.

c. the battery (b) or generator (G) voltage.

d. none of the above.

12. If a resistor in a series circuit is shorted, the series current, I,

a. decreases.

b. stays the same.

c. increases.

d. drops to zero.

13. A 6-V and 9-V source are connected in a series-aiding configuration. How much is the net or total voltage?

a. −3 V.

b. +3 V.

c. 0 V.

d. 15 V.

14. A 56-Ω and 82-Ω resistor are in series with an unknown resistor. If the total resistance of the series combination is 200 Ω, what is the value of the unknown resistor?

a. 138 Ω.

b. 62 Ω.

c. 26 Ω.

d. It cannot be determined.

15. How much is the total resistance, R_T, of a series circuit if one of the resistors is open?

a. infinite (∞) Ω.

b. 0 Ω.

c. R_T is much lower than normal.

d. none of the above.

16. If a resistor in a series circuit becomes open, how much is the voltage across each of the remaining resistors that are still good?

a. Each good resistor has the full value of applied voltage.

b. The applied voltage is split evenly among the good resistors.

c. 0 V.

d. It cannot be determined.

17. A 5-Ω and 10-Ω resistor are connected in series across a DC voltage source. Which resistor will dissipate more power?

a. the 5-Ω resistor.

b. the 10-Ω resistor.

c. It depends on how much the current is.

d. They will both dissipate the same amount of power.

18. Which of the following equations can be used to determine the total power in a series circuit?

a. $P_T = P_1 + P_2 + P_3 + \cdots +$ etc.

b. $P_T = V_T \times I$.

c. $P_T = I^2 R_T$.

d. all of the above.

19. Using electron flow, the polarity of a resistor's voltage drop is

a. positive on the side where electrons enter and negative on the side where they leave.

b. negative on the side were electrons enter and positive on the side where they leave.

c. opposite to that obtained with conventional current flow.

d. both b and c.

20. The schematic symbol for earth ground is

a. .

b. .

c. .

d. .

Essay Questions

1. Show how to connect two resistances in series across a DC voltage source.

2. State three rules for the current, voltage, and resistance in a series circuit.

3. For a given amount of current, why does a higher resistance have a larger voltage drop across it?

4. Two 300-W, 120-V lightbulbs are connected in series across a 240-V line. If the filament of one bulb burns open, will the other bulb light? Why? With the open circuit, how much is the voltage across the source and across each bulb?

5. Prove that if $V_T = V_1 + V_2 + V_3$, then $R_T = R_1 + R_2 + R_3$.

6. State briefly a rule for determining the polarity of the voltage drop across each resistor in a series circuit.

7. State briefly a rule to determine when voltages are series-aiding.

8. In a series string, why does the largest R dissipate the most power?

9. Give one application of series circuits.

10. In Fig. 4–18, explain why R_T, I, V_1, V_2, and V_3 are not affected by the placement of the ground at different points in the circuit.

Problems

SECTION 4–1 WHY *I* IS THE SAME IN ALL PARTS OF A SERIES CIRCUIT

4–1 MultiSim In Fig. 4–24, how much is the current, I, at each of the following points?

a. Point A
b. Point B
c. Point C
d. Point D
e. Point E
f. Point F

Figure 4–24

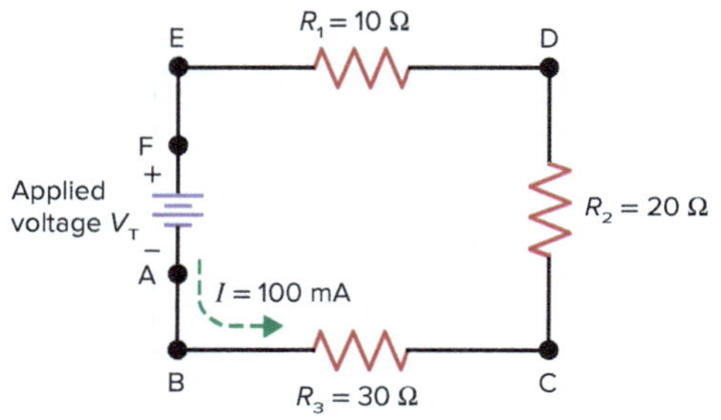

4–2 In Fig. 4–24, how much is the current, I, through each of the following resistors?

a. R_1
b. R_2
c. R_3

4–3 If R_1 and R_3 are interchanged in Fig. 4–24, how much is the current, I, in the circuit?

SECTION 4–2 TOTAL *R* EQUALS THE SUM OF ALL SERIES RESISTANCES

4–4 MultiSim In Fig. 4–25, solve for R_T and I.

4–5 MultiSim Recalculate the values for R_T and I in Fig. 4–25 if $R_1 = 220\ \Omega$ and $R_2 = 680\ \Omega$.

Figure 4–25

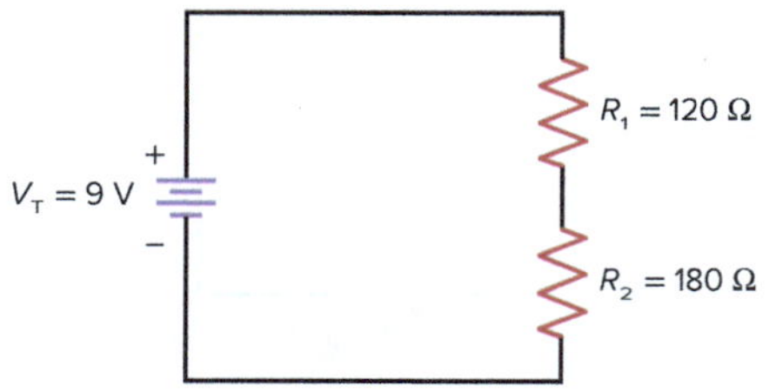

Figure 4–26

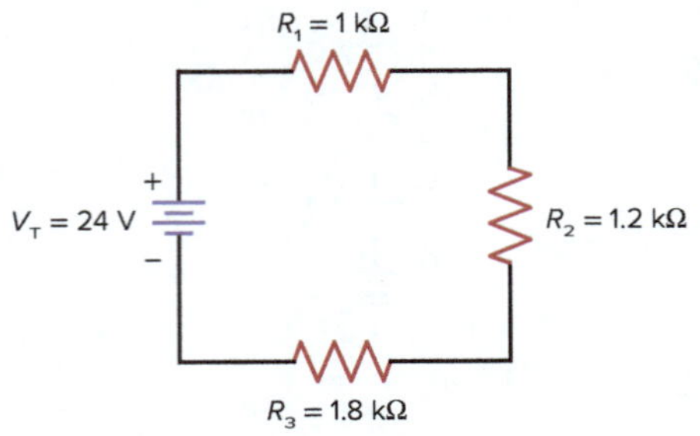

4–6 MultiSim In Fig. 4–26, solve for R_T and I.

4–7 MultiSim What are the new values for R_T and I in Fig. 4–26 if a 2-k**Ω** resistor, R_4, is added to the series circuit?

4–8 In Fig. 4–27, solve for R_T and I.

Figure 4–27

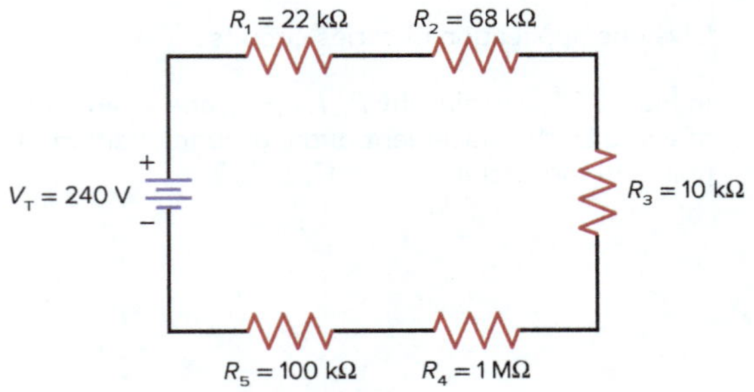

4–9 Recalculate the values for R_T and I in Fig. 4–27 if R_4 is changed to 100 k**Ω**.

SECTION 4–3 SERIES *IR* VOLTAGE DROPS

4–10 MultiSim In Fig. 4–25, find the voltage drops across R_1 and R_2.

4–11 MultiSim In Fig. 4–26, find the voltage drops across R_1, R_2, and R_3.

4–12 In Fig. 4–27, find the voltage drops across R_1, R_2, R_3, R_4, and R_5.

4–13 In Fig. 4–28, solve for R_T, I, V_1, V_2, and V_3.

Figure 4–28

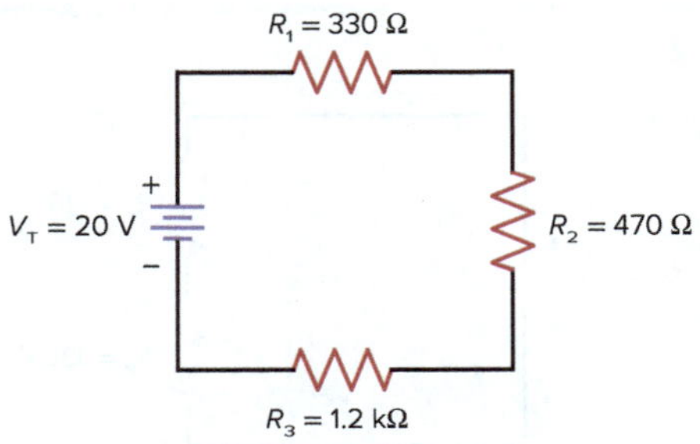

4–14 In Fig. 4–28, recalculate the values for R_T, I, V_1, V_2, and V_3 if V_T is increased to 60 V.

4–15 In Fig. 4–29, solve for R_T, I, V_1, V_2, V_3, and V_4.

Figure 4–29

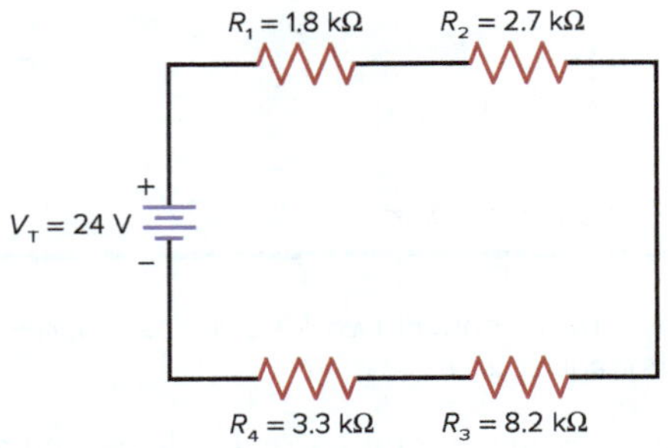

SECTION 4–4 KIRCHHOFF'S VOLTAGE LAW (KVL)

4–16 Using Kirchhoff's voltage law, determine the value of the applied voltage, V_T, in Fig. 4–30.

Figure 4–30

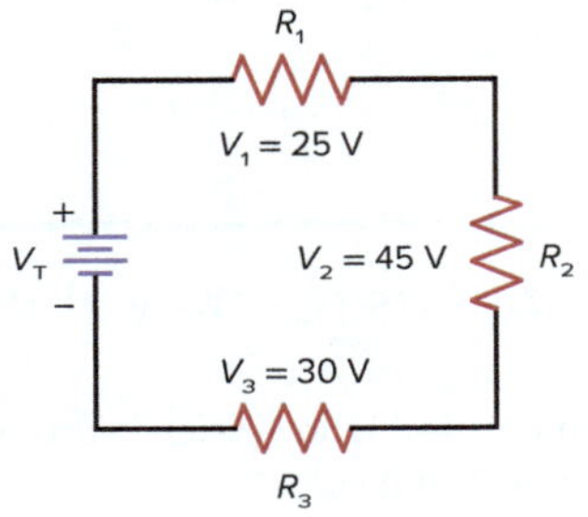

4–17 If $V_1 = 2\ V$, $V_2 = 6\ V$, and $V_3 = 7\ V$ in Fig. 4–30, how much is V_T?

4–18 Determine the voltage, V_2, in Fig. 4–31.

Figure 4–31

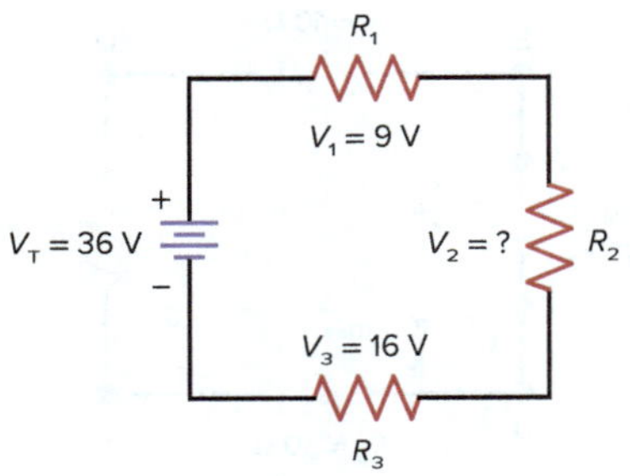

4–19 In Fig. 4–32, solve for the individual resistor voltage drops. Then, using Kirchhoff's voltage law, find V_T.

Figure 4–32

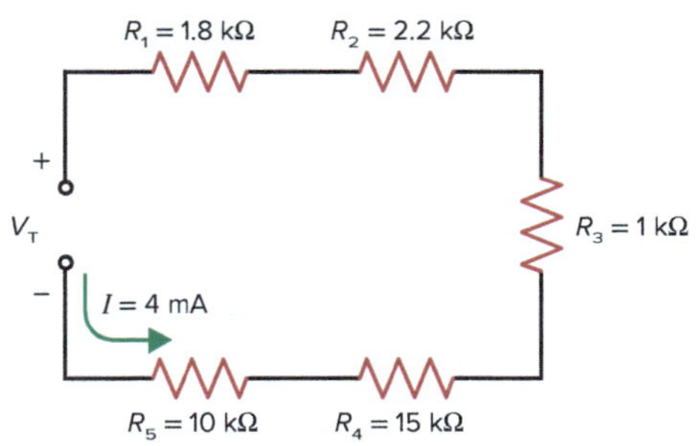

4–20 An applied voltage of 15 V is connected across resistors R_1 and R_2 in series. If $V_2 = 3$ V, how much is V_1?

SECTION 4–5 POLARITY OF *IR* VOLTAGE DROPS

4–21 In Fig. 4–33,

a. Solve for R_T, I, V_1, V_2, and V_3.

b. Indicate the direction of electron flow through R_1, R_2, and R_3.

c. Write the values of V_1, V_2, and V_3 next to resistors R_1, R_2, and R_3.

d. Indicate the polarity of each resistor voltage drop.

Figure 4–33

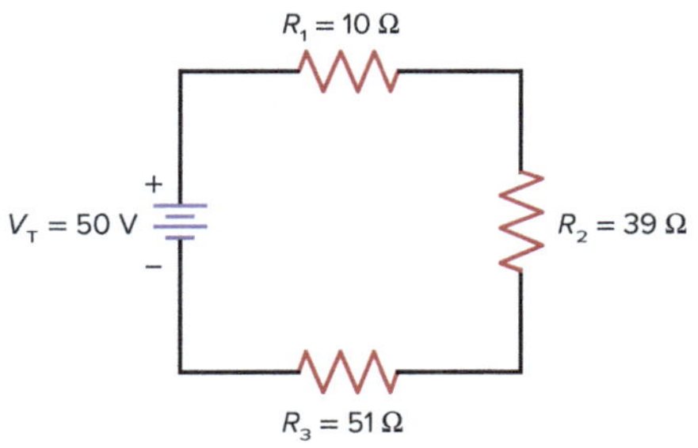

4–22 In Fig. 4–33, indicate the polarity for each resistor voltage drop using conventional current flow. Are the polarities opposite to those obtained with electron flow or are they the same?

4–23 If the polarity of V_T is reversed in Fig. 4–33, what happens to the polarity of the resistor voltage drops and why?

SECTION 4–6 TOTAL POWER IN A SERIES CIRCUIT

4–24 In Fig. 4–25, calculate P_1, P_2, and P_T.

4–25 In Fig. 4–26, calculate P_1, P_2, P_3, and P_T.

4–26 In Fig. 4–27, calculate P_1, P_2, P_3, P_4, P_5, and P_T.

4–27 In Fig. 4–28, calculate P_1, P_2, P_3, and P_T.

4–28 In Fig. 4–29, calculate P_1, P_2, P_3, P_4, and P_T.

SECTION 4–7 SERIES–AIDING AND SERIES–OPPOSING VOLTAGES

4–29 In Fig. 4–34,

a. How much is the net or total voltage, V_T across R_1?

b. How much is the current, I, in the circuit?

c. What is the direction of electron flow through R_1?

Figure 4–34

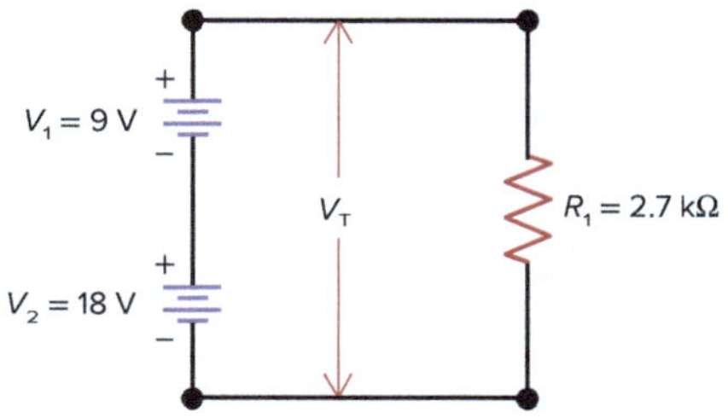

4–30 In Fig. 4–35,

a. How much is the net or total voltage, V_T, across R_1?

b. How much is the current, I, in the circuit?

c. What is the direction of electron flow through R_1?

Figure 4–35

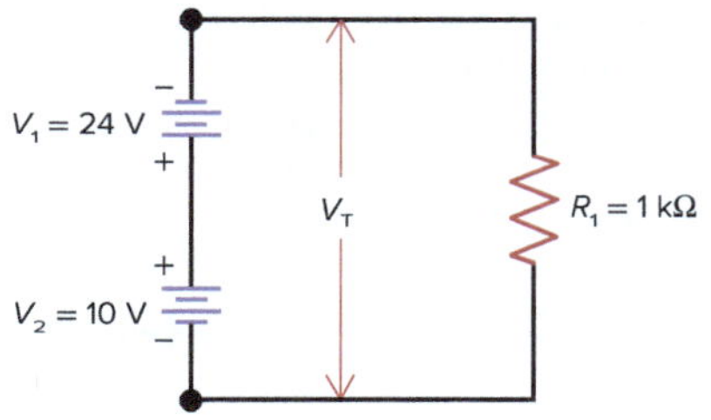

4–31 In Fig. 4–35, assume that V_2 is increased to 30 V. What is

a. The net or total voltage, V_T, across R_1?

b. The current, I, in the circuit?

c. The direction of electron flow through R_1?

4–32 In Fig. 4–36,

a. How much is the net or total voltage, V_T, across R_1?

b. How much is the current, I, in the circuit?

c. What is the direction of electron flow through R_1?

Figure 4–36

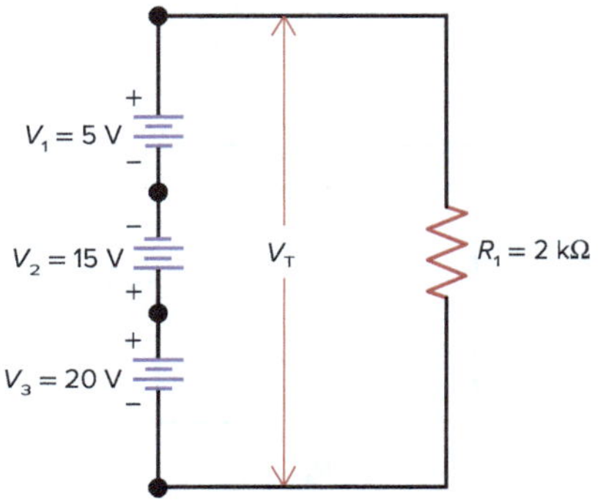

4–33 In Fig. 4–37,

a. How much is the net or total voltage, V_T, across R_1 and R_2 in series?
b. How much is the current, I, in the circuit?
c. What is the direction of electron flow through R_1 and R_2?
d. Calculate the voltage drops across R_1 and R_2.

Figure 4–37

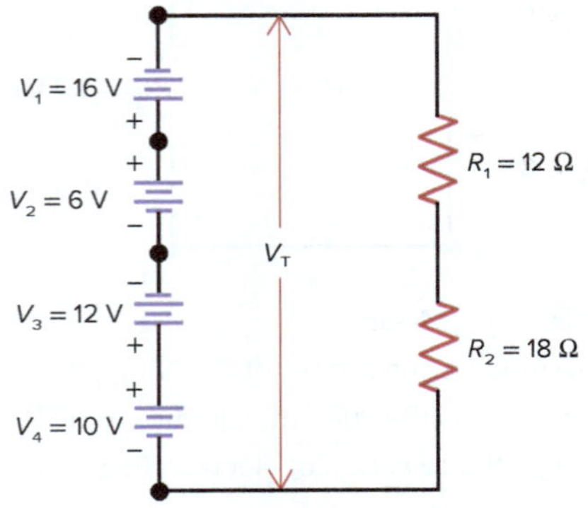

SECTION 4–8 ANALYZING SERIES CIRCUITS WITH RANDOM UNKNOWNS

4–34 In Fig. 4–38, calculate the value for the series resistor, R_s that will allow a 12-V, 150-mA CD player to be operated from a 30-V supply.

Figure 4–38

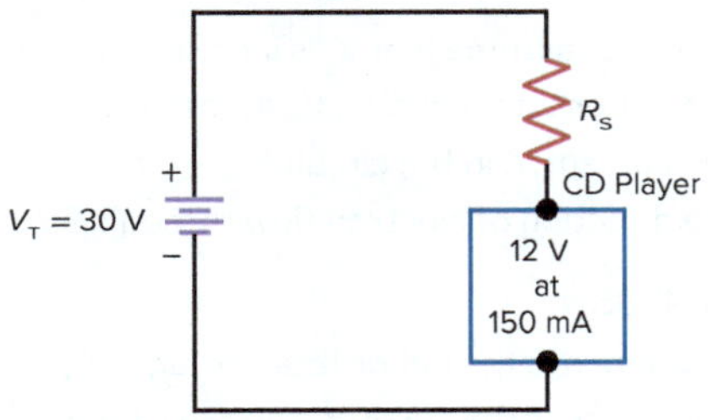

4–35 In Fig. 4–39, solve for I, V_1, V_2, V_3, V_T, R_3, P_T, P_2, and P_3.

Figure 4–39

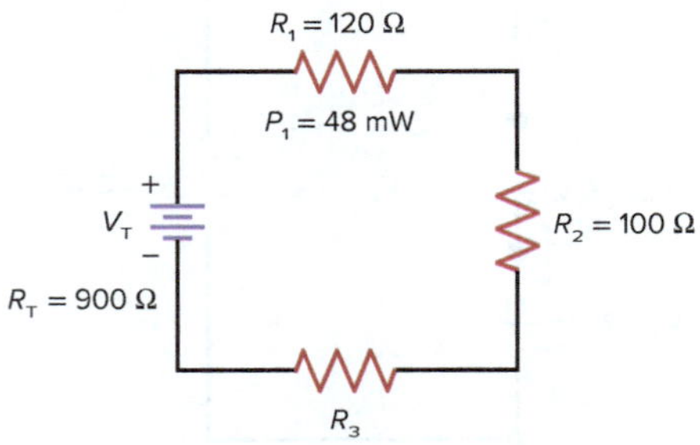

4–36 In Fig. 4–40, solve for R_T, V_1, V_3, V_4, R_2, R_3, P_T, P_1, P_2, P_3, and P_4.

Figure 4–40

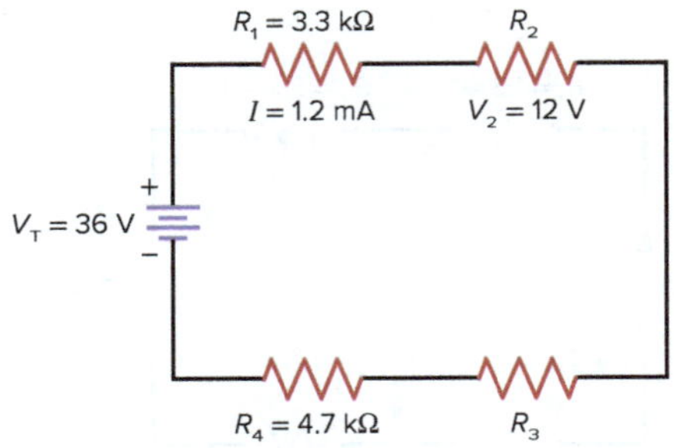

4–37 In Fig. 4–41, solve for I, R_T, V_T, V_2, V_3, V_4, R_4, P_1, P_2, P_3, and P_4.

Figure 4–41

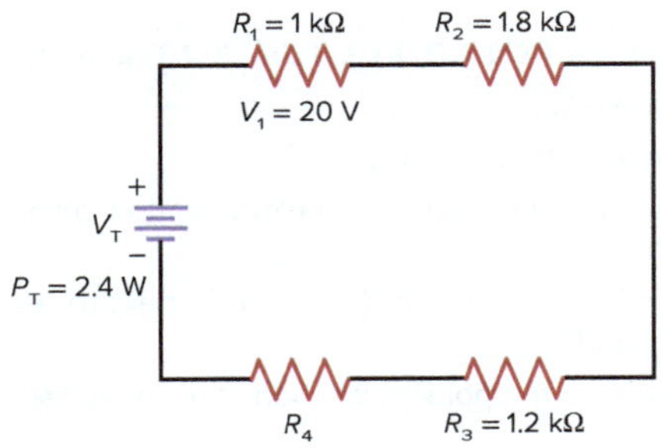

4–38 In Fig. 4–42, solve for I, R_T, R_2, V_2, V_3, P_1, P_2, P_3, and P_T.

Figure 4–42

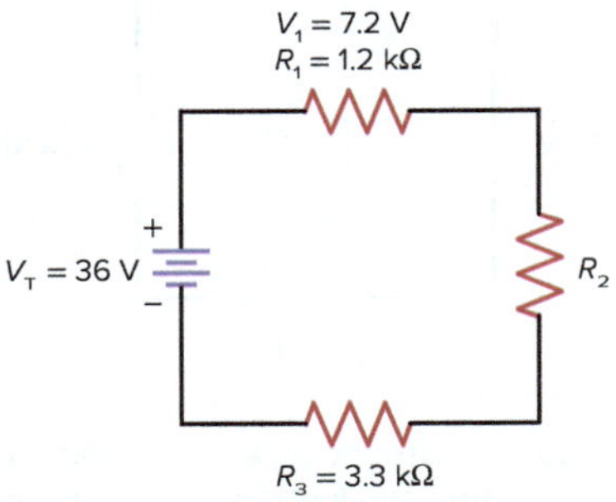

4–39 In Fig. 4–43, solve for R_3, I, V_T, V_1, V_2, V_3, V_4, P_1, P_3, P_4, and P_T.

Figure 4–43

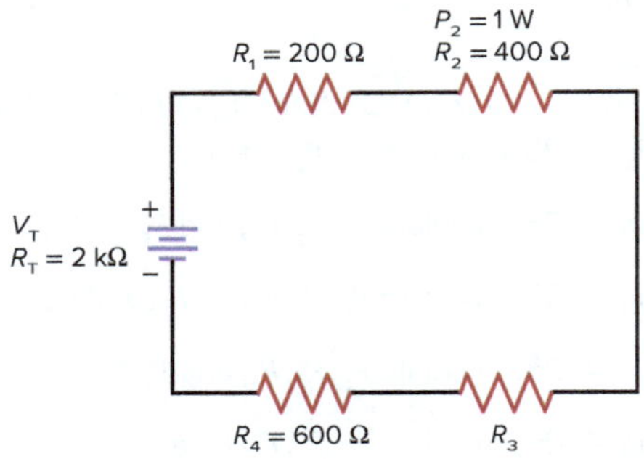

4–40 A 120-Ω resistor is in series with an unknown resistor. The voltage drop across the unknown resistor is 12 V and the power dissipated by the 120-Ω resistor is 4.8 W. Calculate the value of the unknown resistor.

4–41 A 1.5-kΩ resistor is in series with an unknown resistance. The applied voltage, V_T, equals 36 V and the series current is 14.4 mA. Calculate the value of the unknown resistor.

4–42 How much resistance must be added in series with a 6.3-V, 150-mA lightbulb if the bulb is to be operated from a 12-V source?

4–43 A 1-kΩ and 1.5-kΩ resistor are in series. If the total power dissipated by the resistors is 250 mW, how much is the applied voltage, V_T?

4–44 A 22-Ω resistor is in series with a 12-V motor that is drawing 150 mA of current. How much is the applied voltage, V_T?

SECTION 4–9 GROUND CONNECTIONS IN ELECTRICAL AND ELECTRONIC SYSTEMS

4–45 In Fig. 4–44, solve for V_{AG}, V_{BG}, and V_{CG}.

Figure 4–44

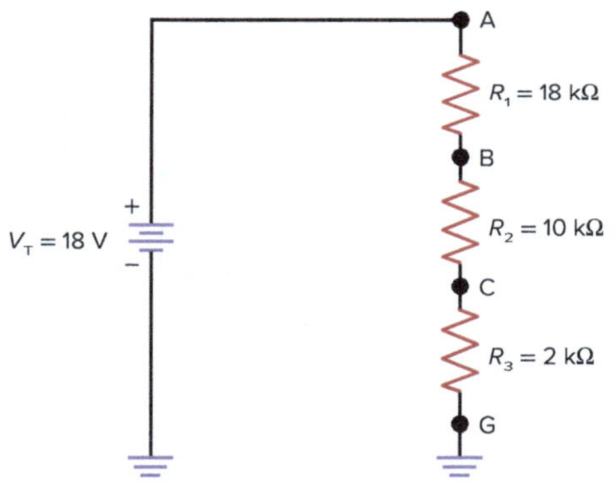

4–46 In Fig. 4–45, solve for V_{AG}, V_{BG}, and V_{CG}.

Figure 4–45

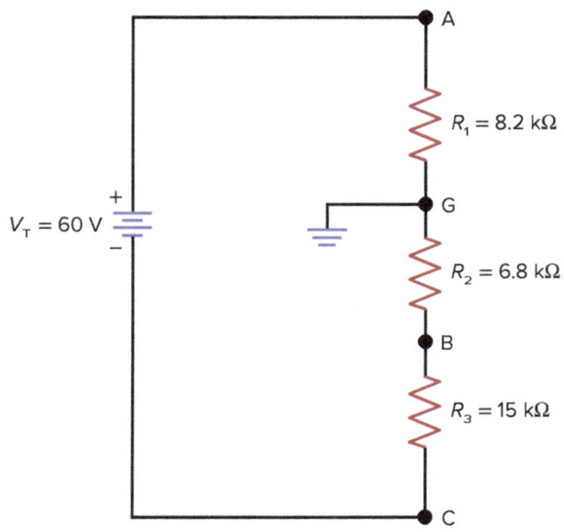

4–47 In Fig. 4–46, solve for V_{AG}, V_{BG}, V_{CG}, and V_{DG}.

Figure 4–46

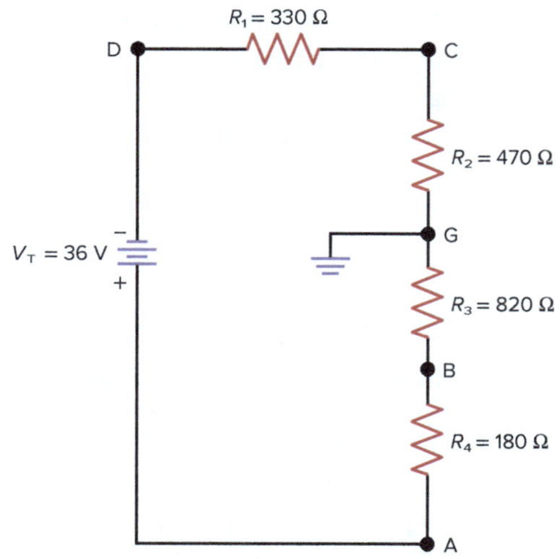

4–48 In Fig. 4–47, solve for V_{AG}, V_{BG}, and V_{CG}.

Figure 4–47

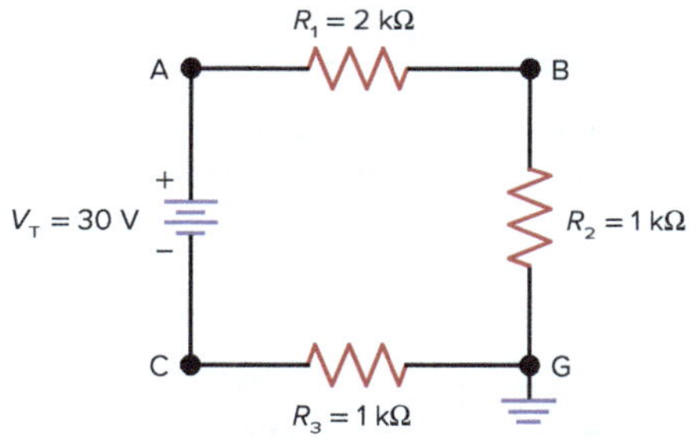

SECTION 4–10 TROUBLESHOOTING: OPENS AND SHORTS IN SERIES CIRCUITS

4–49 In Fig. 4–48, solve for R_T, I, V_1, V_2, and V_3.

Figure 4–48

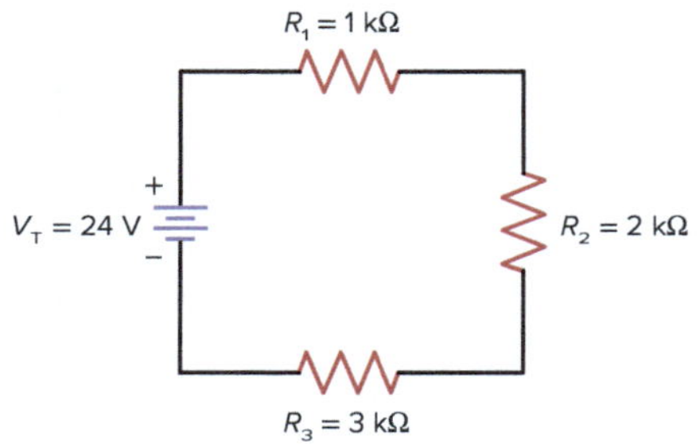

4–50 In Fig. 4–48, assume R_1 becomes open. How much is
a. the total resistance, R_T?
b. the series current, I?
c. the voltage across each resistor, R_1, R_2, and R_3?

4–51 In Fig. 4–48, assume R_3 shorts. How much is
a. the total resistance, R_T?
b. the series current, I?
c. the voltage across each resistor, R_1, R_2, and R_3?

4–52 In Fig. 4–48, assume that the value of R_2 has increased but is not open. What happens to
a. the total resistance, R_T?
b. the series current, I?
c. the voltage drop across R_2?
d. the voltage drops across R_1 and R_3?

Critical Thinking

4–53 Three resistors in series have a total resistance R_T of 2.7 kΩ. If R_2 is twice the value of R_1 and R_3 is three times the value of R_2, what are the values of R_1, R_2, and R_3?

4–54 Three resistors in series have an R_T of 7 kΩ. If R_3 is 2.2 times larger than R_1 and 1.5 times larger than R_2, what are the values of R_1, R_2, and R_3?

4–55 A 100-Ω, ⅛-W resistor is in series with a 330-Ω, ½-W resistor. What is the maximum series current this circuit can handle without exceeding the wattage rating of either resistor?

4–56 A 1.5-kΩ, ½-W resistor is in series with a 470-Ω, ¼-W resistor. What is the maximum voltage that can be applied to this series circuit without exceeding the wattage rating of either resistor?

4–57 Refer to Fig. 4–49. Select values for R_1 and V_T so that when R_2 varies from 1 kΩ to 0 Ω, the series current varies from 1 to 5 mA. V_T and R_1 are to have fixed or constant values.

Figure 4–49 Circuit diagram for Critical Thinking Prob. 4–57.

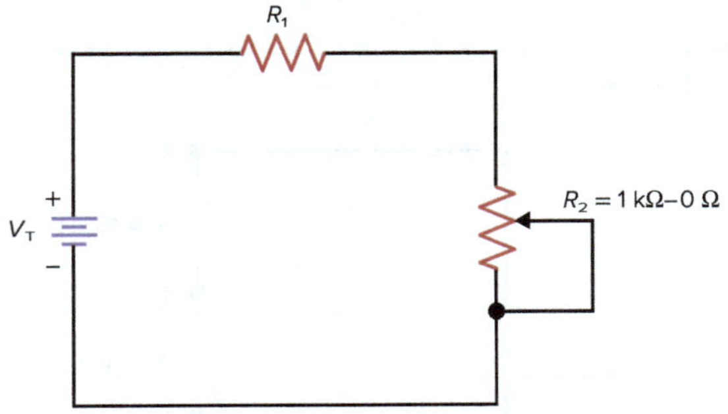

Troubleshooting Challenge

Table 4–1 shows voltage measurements taken in Fig. 4–50. The first row shows the normal values that exist when the circuit is operating properly. Rows 2 to 15 are voltage measurements taken when one component in the circuit has failed. For each row, identify which component is defective and determine the type of defect that has occurred in the component.

Figure 4–50 Circuit diagram for Troubleshooting Challenge. Normal values for V_1, V_2, V_3, V_4, and V_5 are shown on schematic.

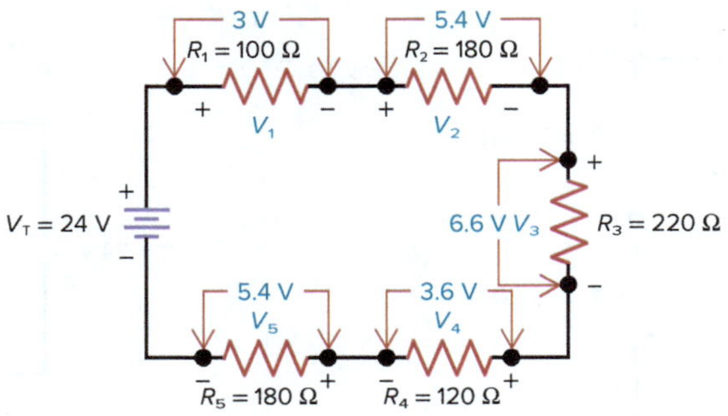

Table 4–1 Table for Troubleshooting Challenge

	V_1	V_2	V_3	V_4	V_5	Defective Component
			VOLTS			
1 Normal values	3	5.4	6.6	3.6	5.4	None
2 Trouble 1	0	24	0	0	0	
3 Trouble 2	4.14	7.45	0	4.96	7.45	
4 Trouble 3	3.53	6.35	7.76	0	6.35	
5 Trouble 4	24	0	0	0	0	
6 Trouble 5	0	6.17	7.54	4.11	6.17	
7 Trouble 6	0	0	0	24	0	
8 Trouble 7	3.87	0	8.52	4.64	6.97	
9 Trouble 8	0	0	24	0	0	
10 Trouble 9	0	0	0	0	24	
11 Trouble 10	2.4	4.32	5.28	2.88	9.12	
12 Trouble 11	4	7.2	0.8	4.8	7.2	
13 Trouble 12	3.87	6.97	8.52	4.64	0	
14 Trouble 13	15.6	2.16	2.64	1.44	2.16	
15 Trouble 14	3.43	6.17	7.55	0.68	6.17	

Answers to Self-Reviews

4–1 **a.** R_1, R_2, R_3, V_T, and the wires
b. 5 A
c. 2 A

4–2 **a.** 2 mA
b. 10 kΩ
c. 1 mA

4–3 **a.** 10 V
b. 1 A
c. 1 A
d. 1 A

4–4 **a.** 60 V
b. 75 V
c. 40 V

4–5 **a.** point B or C
b. point A or F
c. point D

4–6 **a.** 6 W
b. R_2

4–7 **a.** 100 V
b. 20 V

4–8 **a.** 30 V
b. 90 V
c. 60 V

4–9 **a.** −10 V
b. 0 V
c. +20 V
d. +30 V

4–10 **a.** 24 V
b. 0 V
c. decreases

Laboratory Application Assignment

In this lab application assignment, you will examine the characteristics of a simple series circuit. You will also troubleshoot a series with open and shorted resistors.

Equipment: Obtain the following items from your instructor.

- Variable dc power supply
- Assortment of carbon-film resistors
- DMM

Series Circuit Characteristics

Examine the series circuit in Fig. 4–51. Calculate and record the following values:

R_T = ______, I = ______, V_1 = ______, V_2 = ______, V_3 = ______

Figure 4–51

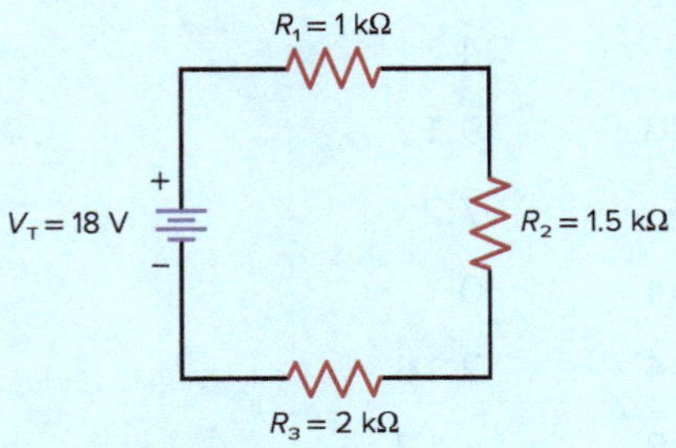

Construct the series circuit in Fig. 4–51. Measure and record the following values.

(Note that the power supply connections must be removed to measure R_T.)

R_T = ______, I = ______, V_1 = ______, V_2 = ______, V_3 = ______

How does the ratio V_2/V_1 compare to the ratio R_2/R_1? ______

How does the ratio V_3/V_1 compare to the ratio R_3/R_1? ______

Add the measured voltages V_1, V_2, and V_3. Record your answer.

______

How does this value compare to the value of V_T? ______

Does the sum of the resistor voltage drops satisfy KVL? ______

Using measured values, prove that the current, I, is the same in all parts of a series circuit. Show your calculations. ______

______

In Fig. 4–51, which series resistor dissipates the most amount of power? ______

Which resistor dissipates the least amount of power? ______

Troubleshooting a Series Circuit

In the following troubleshooting exercise, a 1 Ω resistor will be used to simulate a short circuit whereas a 1 MΩ resistor will be used to simulate an open circuit.

In Fig. 4–51, replace resistor R_2 with a 1 Ω resistor. Next, measure and record the current, I and the voltages V_1, V_2, and V_3.

I = ______, V_1 = ______, V_2 = ______, V_3 = ______

Did the current, I, increase or decrease with R_2 shorted? ______ Explain why. ______

______

______

Did the voltage drops across R_1 and R_3 increase or decrease with R_2 shorted? ______
Explain why. ______

______

Did the voltage drop across R_2 increase or decrease with R_2 shorted? ______ Explain why. ______

______

______

Next, change R_2 to a 1 MΩ resistor. Measure and record the current, I and the voltages V_1, V_2, and V_3.
I = ______, V_1 = ______, V_2 = ______, V_3 = ______

Did the current, I increase or decrease with R_2 open? ______
Explain why. ______

______

______

Did the voltage drops across R_1 and R_3 increase or decrease with R_2 open? ______
Explain why. ______

______

______

Did the voltage drop across R_2 increase or decrease with R_2 open? ______ Explain why.

______

______

Design credits: Chapter Opener Background Image, and Multisim: Stockbyte/Getty Images

chapter 5

Parallel Circuits

A parallel circuit is any circuit that provides one common voltage across all components. Each component across the voltage source provides a separate path or branch for current flow. The individual branch currents are calculated as $\frac{V_A}{R}$ where V_A is the applied voltage and R is the individual branch resistance. The total current, I_T, supplied by the applied voltage, must equal the sum of all individual branch currents.

The equivalent resistance of a parallel circuit equals the applied voltage, V_A, divided by the total current, I_T. The term equivalent resistance refers to a single resistance that would draw the same amount of current as all the parallel connected branches. The equivalent resistance of a parallel circuit is designated R_{EQ}. There are multiple applications for parallel circuits with the most prevalent being residential house wiring. Another application is the electrical system in our vehicles where the radio, heater, lights, and so on are all connected in parallel. Parallel circuits are also embedded in more complex electrical and electronic equipment. In this case, several different circuits are connected to one common voltage source with each circuit operating totally independent of each other.

This chapter covers all the characteristics of parallel circuits, including important information about how to troubleshoot a parallel circuit with a defective component. ■

Chapter Outline

Chapter Objectives

After studying this chapter, you should be able to

- *Explain* why voltage is the same across all the branches in a parallel circuit.
- *Calculate* the individual branch currents in a parallel circuit.
- *Calculate* the total current in a parallel circuit using Kirchhoff's current law.
- *Calculate* the equivalent resistance of two or more resistors in parallel.
- *Explain* why the equivalent resistance of a parallel circuit is always less than the smallest branch resistance.
- *Calculate* the total conductance of a parallel circuit.
- *Calculate* the total power in a parallel circuit.
- *Solve* for the voltage, current, power, and resistance in a parallel circuit having random unknowns.
- *Describe* the effects of an open and short in a parallel circuit.
- *Troubleshoot* parallel circuits containing opens and shorts.

Important Terms

equivalent resistance, R_{EQ}
Kirchhoff's current law (KCL)
main line
parallel bank
reciprocal resistance formula

5–1 The Applied Voltage V_A Is the Same across Parallel Branches

GOOD TO KNOW

Components can be connected in parallel even if they are not connected to a voltage source.

A parallel circuit is formed when two or more components are connected across a voltage source, as shown in Fig. 5–1. In this figure, R_1 and R_2 are in parallel with each other and a 1.5-V battery. In Fig. 5–1*b*, points A, B, C, and E are equivalent to a direct connection at the positive terminal of the battery because the connecting wires have practically no resistance. Similarly, points H, G, D, and F are the same as a direct connection at the negative battery terminal. Since R_1 and R_2 are directly connected across the two terminals of the battery, both resistances must have the same potential difference as the battery. It follows that the voltage is the same across components connected in parallel. The parallel circuit arrangement is used, therefore, to connect components that require the same voltage.

A common application of parallel circuits is typical house wiring to the power line, with many lights and appliances connected across the 120-V source (Fig. 5–2). The wall receptacle has a potential difference of 120 V across each pair of terminals. Therefore, any resistance connected to an outlet has an applied voltage of 120 V. The lightbulb is connected to one outlet and the toaster to another outlet, but both have the same applied voltage of 120 V. Therefore, each operates independently of any other appliance, with all the individual branch circuits connected across the 120-V line.

MultiSim **Figure 5–1** Example of a parallel circuit with two resistors. (*a*) Wiring diagram. (*b*) Schematic diagram.

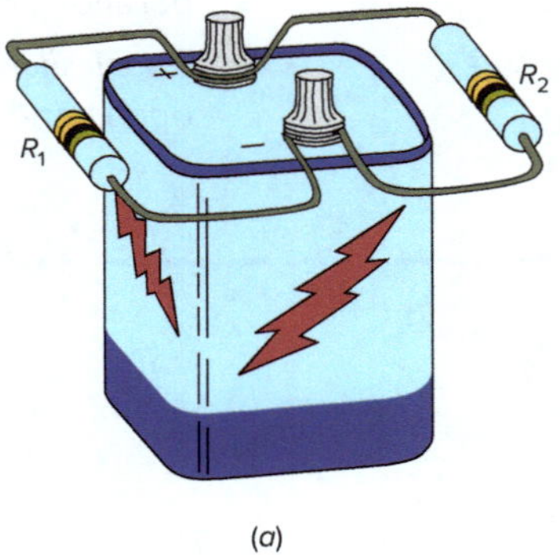

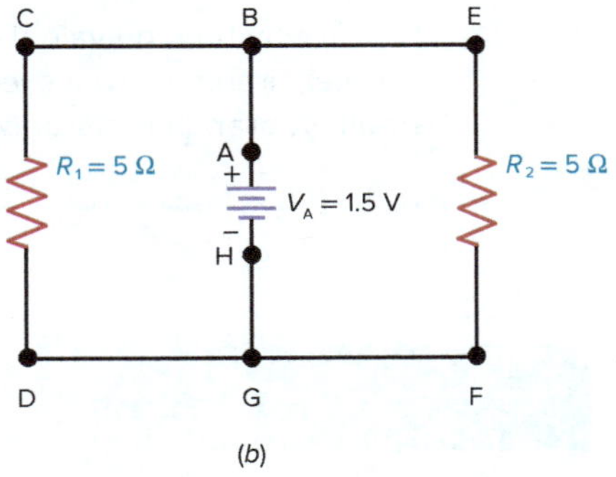

Figure 5–2 Lightbulb and toaster connected in parallel with the 120-V line. (*a*) Wiring diagram. (*b*) Schematic diagram.

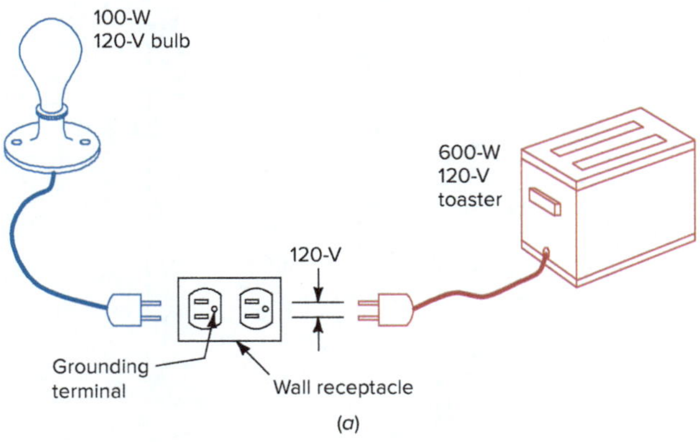

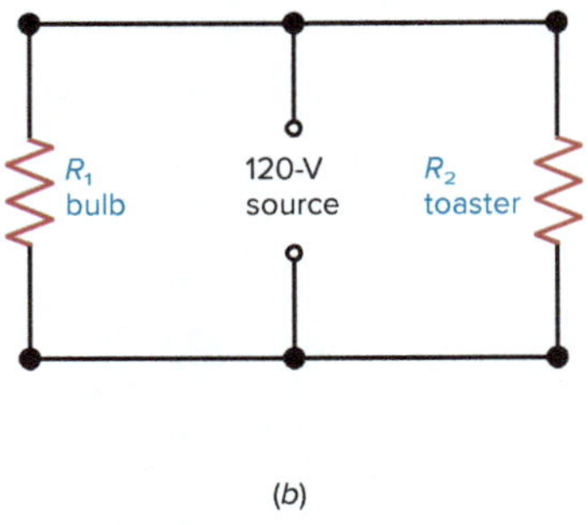

5–1 Self-Review

Answers at the end of the chapter.

a. **In Fig. 5–1, how much is the common voltage across R_1 and R_2?**
b. **In Fig. 5–2, how much is the common voltage across the bulb and the toaster?**
c. **How many parallel branch circuits are connected across the voltage source in Figs. 5–1 and 5–2?**

5–2 Each Branch I Equals V_A/R

GOOD TO KNOW

In a parallel circuit, the branch with the lowest resistance always has the most current. This must be true because each branch current is calculated as $\frac{V_A}{R}$ where V_A is the same across all branches.

In applying Ohm's law, it is important to note that the current equals the voltage applied across the circuit divided by the resistance between the two points where that voltage is applied. In Fig. 5–3*a*, 10 V is applied across the 5 Ω of R_2, resulting in the current of 2 A between points E and F through R_2. The battery voltage is also applied across the parallel resistance of R_1, applying 10 V across 10 Ω. Through R_1, therefore, the current is 1 A between points C and D. The current has a different value through R_1, with the same applied voltage, because the resistance is different. These values are calculated as follows:

$$I_1 = \frac{V_A}{R_1} = \frac{10}{10} = 1\text{ A}$$

$$I_2 = \frac{V_A}{R_2} = \frac{10}{5} = 2\text{ A}$$

Figure 5–3*b* shows how to assemble axial-lead resistors on a lab prototype board to form a parallel circuit.

Just as in a circuit with one resistance, any branch that has less R allows more I. If R_1 and R_2 were equal, however, the two branch currents would have the same value. For instance, in Fig. 5–1*b* each branch has its own current equal to $1.5\text{ V}/5\ \Omega = 0.3\text{ A}$.

The I can be different in parallel circuits that have different R because V is the same across all the branches. Any voltage source generates a potential difference across its two terminals. This voltage does not move. Only I flows around the circuit. The source voltage is available to make electrons move around any closed path connected to the terminals of the source. The amount of I in each separate path depends on the amount of R in each branch.

Figure 5–3 Parallel circuit. (*a*) The current in each parallel branch equals the applied voltage V_A divided by each branch resistance R. (*b*) Axial-lead resistors assembled on a lab prototype board, forming a parallel circuit.

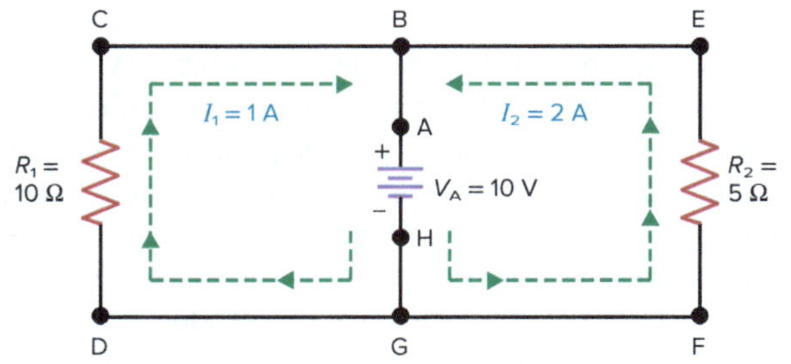

(*a*)

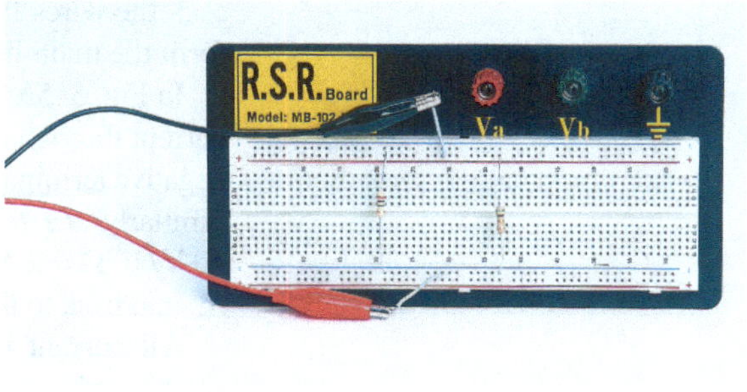

Cindy Schroeder/McGraw Hill

(*b*)

Example 5-1

MultiSim **Figure 5-4** Circuit for Example 5-1.

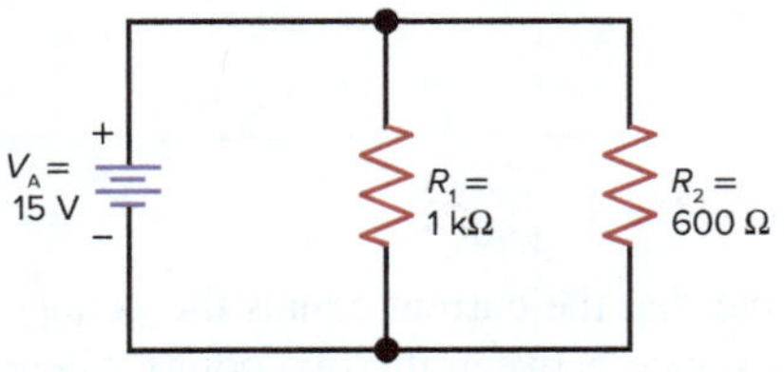

In Fig. 5-4, solve for the branch currents I_1 and I_2.

ANSWER The applied voltage, V_A, of 15 V is across both resistors R_1 and R_2. Therefore, the branch currents are calculated as $\frac{V_A}{R}$, where V_A is the applied voltage and R is the individual branch resistance.

$$I_1 = \frac{V_A}{R_1} = \frac{15\ \text{V}}{1\ \text{k}\Omega} = 15\ \text{mA}$$

$$I_2 = \frac{V_A}{R_2} = \frac{15\ \text{V}}{600\ \Omega} = 25\ \text{mA}$$

5-2 Self-Review

Answers at the end of the chapter.

Refer to Fig. 5-3.

a. How much is the voltage across R_1?
b. How much is I_1 through R_1?
c. How much is the voltage across R_2?
d. How much is I_2 through R_2?

5-3 Kirchhoff's Current Law (KCL)

Components to be connected in parallel are usually wired directly across each other, with the entire parallel combination connected to the voltage source, as illustrated in Fig. 5-5. This circuit is equivalent to wiring each parallel branch directly to the voltage source, as shown in Fig. 5-1, when the connecting wires have essentially zero resistance.

The advantage of having only one pair of connecting leads to the source for all the parallel branches is that usually less wire is necessary. The pair of leads connecting all the branches to the terminals of the voltage source is the **main line.** In Fig. 5-5, the wires from G to A on the negative side and from B to F in the return path form the main line.

In Fig. 5-5*b*, with 20 Ω of resistance for R_1 connected across the 20-V battery, the current through R_1 must be 20 V/20 Ω = 1 A. This current is electron flow from the negative terminal of the source, through R_1, and back to the positive battery terminal. Similarly, the R_2 branch of 10 Ω across the battery has its own branch current of 20 V/10 Ω = 2 A. This current flows from the negative terminal of the source, through R_2, and back to the positive terminal, since it is a separate path for electron flow.

All current in the circuit, however, must come from one side of the voltage source and return to the opposite side for a complete path. In the main line, therefore, the amount of current is equal to the total of the branch currents.

Figure 5–5 The current in the main line equals the sum of the branch currents. Note that from G to A at the bottom of this diagram is the negative side of the main line, and from B to F at the top is the positive side. (*a*) Wiring diagram. Arrows inside the lines indicate current in the main line for R_1; arrows outside indicate current for R_2. (*b*) Schematic diagram. I_T is the total line current for both R_1 and R_2.

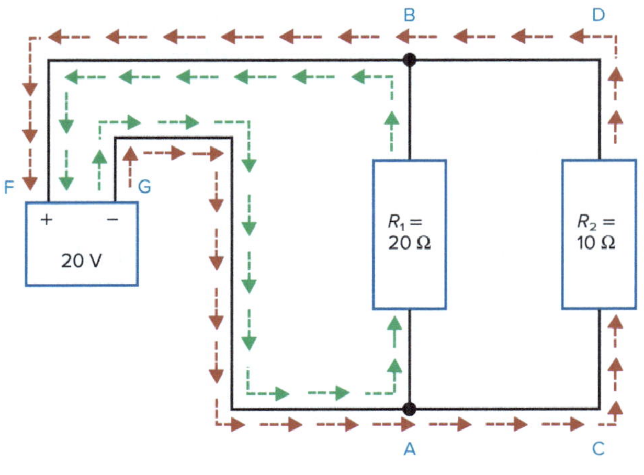

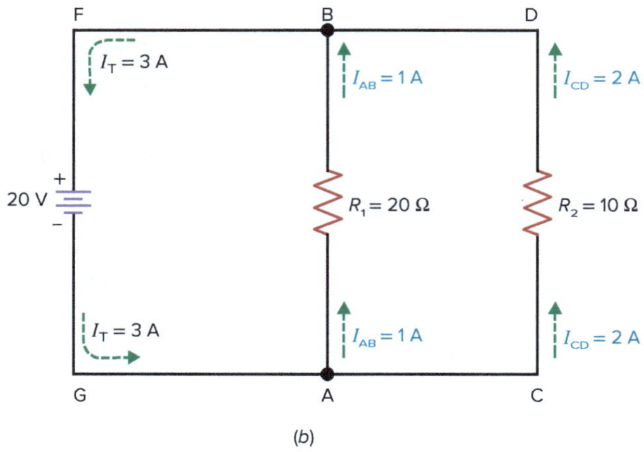

GOOD TO KNOW

As more branches are added to a parallel circuit, the total current, I_T, increases.

For example, in Fig. 5–5*b*, the total current in the line from point G to point A is 3 A. The total current at branch point A subdivides into its component branch currents for each of the branch resistances. Through the path of R_1 from A to B the current is 1 A. The other branch path ACDB through R_2 has a current of 2 A. At the branch point B, the electron flow from both parallel branches combines, so that the current in the main-line return path from B to F has the same value of 3 A as in the other side of the main line.

Kirchhoff's current law (KCL) states that the total current I_T in the main line of a parallel circuit equals the sum of the individual branch currents. Expressed as an equation, Kirchhoff's current law is

$$I_T = I_1 + I_2 + I_3 + \cdots + \text{etc.} \qquad \textbf{(5–1)}$$

where I_T is the total current and I_1, I_2, I_3 . . . are the individual branch currents. Kirchhoff's current law applies to any number of **parallel branches,** whether the resistances in the branches are equal or unequal.

Example 5-2

MultiSim

An R_1 of 20 Ω, an R_2 of 40 Ω, and an R_3 of 60 Ω are connected in parallel across the 120-V power line. Using Kirchhoff's current law, determine the total current I_T.

ANSWER Current I_1 for the R_1 branch is 120/20 or 6 A. Similarly, I_2 is 120/40 or 3 A, and I_3 is 120/60 or 2 A. The total current in the main line is

$$I_T = I_1 + I_2 + I_3 = 6 + 3 + 2$$
$$I_T = 11 \text{ A}$$

Example 5-3

Two branches R_1 and R_2 across the 120-V power line draw a total line current I_T of 15 A. The R_1 branch takes 10 A. How much is the current I_2 in the R_2 branch?

ANSWER $I_2 = I_T - I_1 = 15 - 10$

$I_2 = 5 \text{ A}$

With two branch currents, one must equal the difference between I_T and the other branch current.

Example 5-4

Three parallel branch currents are 0.1 A, 500 mA, and 800 μA. Using Kirchhoff's current law, calculate I_T.

ANSWER All values must be in the same units to be added. In this case, all units will be converted to milliamperes: 0.1 A = 100 mA and 800 μA = 0.8 mA. Applying Kirchhoff's current law

$$I_T = 100 + 500 + 0.8$$
$$I_T = 600.8 \text{ mA}$$

You can convert the currents to A, mA, or μA units, as long as the same unit is used for adding all currents.

5–3 Self-Review

Answers at the end of the chapter.

a. **Branch currents in a parallel circuit are 1 A for I_1, 2 A for I_2, and 3 A for I_3. How much is I_T?**
b. **Assume I_T = 6 A for three branch currents; I_1 is 1 A, and I_2 is 2 A. How much is I_3?**
c. **Branch currents in a parallel circuit are 1 A for I_1 and 200 mA for I_2. How much is I_T?**

5–4 Resistances in Parallel

The combined equivalent resistance across the main line in a parallel circuit can be found by Ohm's law: *Divide the common voltage across the parallel resistances by the total current of all the branches.* Referring to Fig. 5–6*a*, note that the parallel resistance of R_1 with R_2, indicated by the **equivalent resistance R_{EQ}**, is the opposition to the total current in the main line. In this example, V_A/I_T is 60 V/3 A = 20 Ω for R_{EQ}.

GOOD TO KNOW

The statement "current always takes the path of least resistance" is not always true. If it were, all the current in a parallel circuit would flow in the lowest branch resistance only.

MultiSim **Figure 5–6** Resistances in parallel. (*a*) Combination of R_1 and R_2 is the equivalent resistance R_{EQ} for the main line. (*b*) Equivalent circuit showing R_{EQ} drawing the same 3-A I_T as the parallel combination of R_1 and R_2 in (*a*).

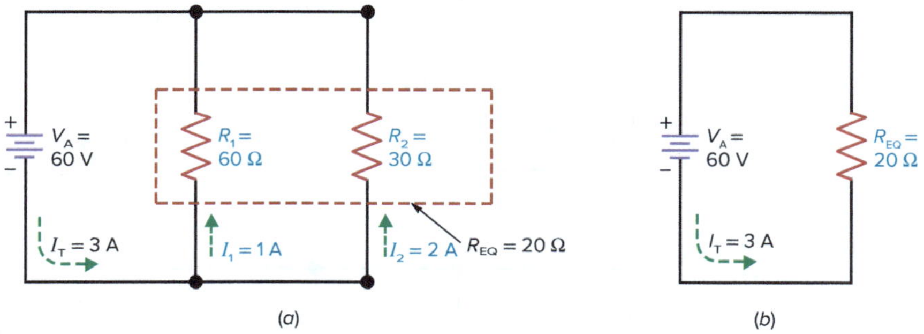

The total load connected to the source voltage is the same as though one equivalent resistance of 20 Ω were connected across the main line. This is illustrated by the equivalent circuit in Fig. 5–6*b*. For any number of parallel resistances of any value, use the following equation,

$$R_{EQ} = \frac{V_A}{I_T} \qquad \textbf{(5–2)}$$

where I_T is the sum of all the branch currents and R_{EQ} is the equivalent resistance of all parallel branches across the applied voltage source V_A.

The first step in solving for R_{EQ} is to add all the parallel branch currents to find the I_T being delivered by the voltage source. The voltage source thinks that it is connected to a single resistance whose value allows I_T to flow in the circuit according to Ohm's law. This single resistance is R_{EQ}. An illustrative example of a circuit with two parallel branches will be used to show how R_{EQ} is calculated.

Example 5-5

Two branches, each with a 5-A current, are connected across a 90-V source. How much is the equivalent resistance R_{EQ}?

ANSWER The total line current I_T is 5 + 5 = 10 A. Then,

$$R_{EQ} = \frac{V_A}{I_T} = \frac{90}{10}$$

$$R_{EQ} = 9\ \Omega$$

GOOD TO KNOW

Assume two resistors are connected in parallel. If one of the two resistors has a value 10 or more times larger than the other, the equivalent resistance, R_{EQ} is approximately equal to the value of the smaller resistor.

Parallel Bank

A combination of parallel branches is often called a **bank.** In Fig. 5–6, the bank consists of the 60-Ω R_1 and 30-Ω R_2 in parallel. Their combined parallel resistance R_{EQ} is the bank resistance, equal to 20 Ω in this example. A bank can have two or more parallel resistors.

When a circuit has more current with the same applied voltage, this greater value of I corresponds to less R because of their inverse relation. Therefore, the

Figure 5–7 How adding parallel branches of resistors increases I_T but decreases R_{EQ}. (*a*) One resistor. (*b*) Two branches. (*c*) Three branches. (*d*) Equivalent circuit of the three branches in (*c*).

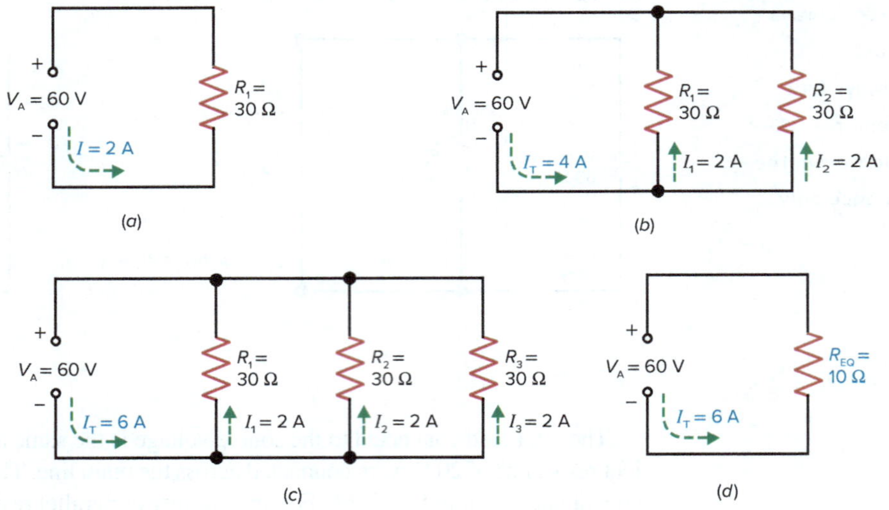

CALCULATOR

When using the calculator to find a reciprocal such as $1/R$, choose either of two methods. Either divide the number 1 by the value of R, or use the reciprocal key labeled $1/x$. As an example, to find the reciprocal of $R = 20\ \Omega$ by division:

- First punch in the number 1 on the key pad.
- Then press the division ÷ key.
- Punch in 20 for the value of R.
- Finally, press the equal = key for the quotient of 0.05 on the display.
- To use the reciprocal key, first punch in 20 for R. Then press the $1/x$ key. This may be a second function on some calculators, requiring that you push the 2ndF or SHIFT key before pressing $1/x$. The reciprocal equal to 0.05 is displayed without the need for the = key.

combination of parallel resistances R_{EQ} for the bank is always less than the smallest individual branch resistance. The reason is that I_T must be more than any one branch current.

Why R_{EQ} Is Less than Any Branch R

It may seem unusual at first that putting more resistance into a circuit lowers the equivalent resistance. This feature of parallel circuits is illustrated in Fig. 5–7. Note that equal resistances of 30 Ω each are added across the source voltage, one branch at a time. The circuit in Fig. 5–7*a* has just R_1, which allows 2 A with 60 V applied. In Fig. 5–7*b*, the R_2 branch is added across the same V_A. This branch also has 2 A. Now the parallel circuit has a 4-A total line current because of $I_1 + I_2$. Then the third branch, which also takes 2 A for I_3, is added in Fig. 5–7*c*. The combined circuit with three branches, therefore, requires a total load current of 6 A, which is supplied by the voltage source.

The combined resistance across the source, then, is V_A/I_T, which is 60/6, or 10 Ω. This equivalent resistance R_{EQ}, representing the entire load on the voltage source, is shown in Fig. 5–7*d*. More resistance branches reduce the combined resistance of the parallel circuit because more current is required from the same voltage source.

Reciprocal Resistance Formula

We can derive the **reciprocal resistance formula** from the fact that I_T is the sum of all the branch currents, or,

$$I_T = I_1 + I_2 + I_3 + \cdots + \text{etc.}$$

However, $I_T = V/R_{EQ}$. Also, each $I = V/R$. Substituting V/R_{EQ} for I_T on the left side of the formula and V/R for each branch I on the right side, the result is

$$\frac{V}{R_{EQ}} = \frac{V}{R_1} + \frac{V}{R_2} + \frac{V}{R_3} + \cdots + \text{etc.}$$

Dividing by V because the voltage is the same across all the resistances gives us

$$\frac{1}{R_{EQ}} = \frac{1}{R_1} + \frac{1}{R_2} + \frac{1}{R_3} + \cdots + \text{etc.}$$

MultiSim **Figure 5–8** Two methods of combining parallel resistances to find R_{EQ}. (*a*) Using the reciprocal resistance formula to calculate R_{EQ} as 4 Ω. (*b*) Using the total line current method with an assumed line voltage of 20 V gives the same 4 Ω for R_{EQ}.

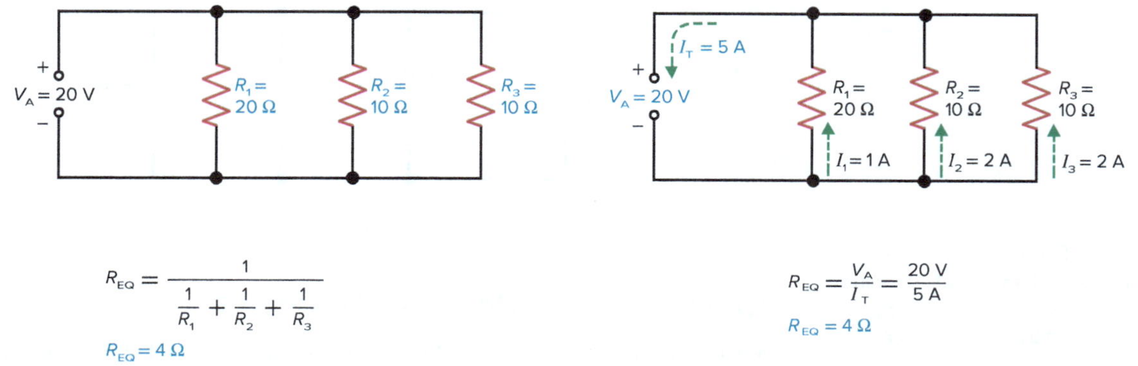

(*a*) (*b*)

Next, solve for R_{EQ}.

$$R_{EQ} = \frac{1}{{}^1\!/\!_{R_1} + {}^1\!/\!_{R_2} + {}^1\!/\!_{R_3} + \cdots + \text{etc.}} \tag{5–3}$$

This reciprocal formula applies to any number of parallel resistances of any value. Using the values in Fig. 5–8*a* as an example,

$$R_{EQ} = \frac{1}{{}^1\!/\!_{20} + {}^1\!/\!_{10} + {}^1\!/\!_{10}} = 4\ \Omega$$

Total-Current Method

It may be easier to work without fractions. Figure 5–8*b* shows how this same problem can be calculated in terms of total current instead of by the reciprocal formula. Although the applied voltage is not always known, any convenient value can be assumed because it cancels in the calculations. It is usually simplest to assume an applied voltage of the same numerical value as the highest resistance. Then one assumed branch current will automatically be 1 A and the other branch currents will be more, eliminating fractions less than 1 in the calculations.

In Fig. 5–8*b*, the highest branch R is 20 Ω. Therefore, assume 20 V for the applied voltage. Then the branch currents are 1 A in R_1, 2 A in R_2, and 2 A in R_3. Their sum is 1 + 2 + 2 = 5 A for I_T. The combined resistance R_{EQ} across the main line is V_A/I_T, or 20 V/5 A = 4 Ω. This is the same value calculated with the reciprocal resistance formula.

Special Case of Equal *R* in All Branches

If R is equal in all branches, the combined R_{EQ} equals the value of one branch resistance divided by the number of branches.

$$R_{EQ} = \frac{R}{n}$$

where R is the resistance in one branch and n is the number of branches.

This rule is illustrated in Fig. 5–9, where three 60-kΩ resistances in parallel equal 20 kΩ.

The rule applies to any number of parallel resistances, but they must all be equal. As another example, five 60-Ω resistances in parallel have the combined resistance of 60/5, or 12 Ω. A common application is two equal resistors wired in a **parallel bank** for R_{EQ} equal to one-half R.

Figure 5–9 For the special case of all branches having the same resistance, just divide R by the number of branches to find R_{EQ}. Here, $R_{EQ} = 60\ \text{k}\Omega/3 = 20\ \text{k}\Omega$.

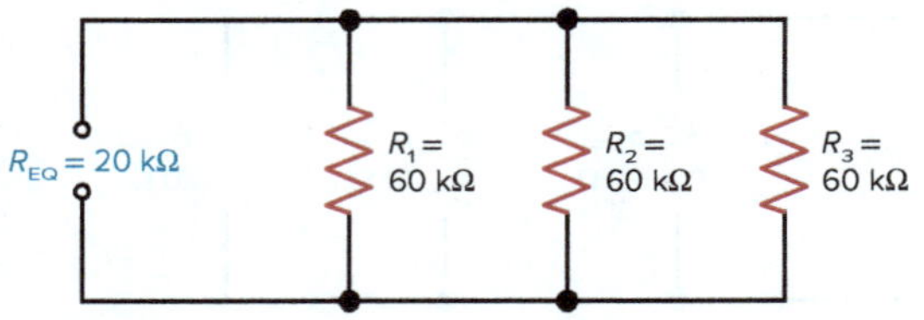

$$R_{EQ} = \frac{\text{value of one resistance}}{\text{number of resistances}} = \frac{60\ \text{k}\Omega}{3}$$

Figure 5–10 For the special case of only two branch resistances of any values R_{EQ} equals their product divided by the sum. Here, $R_{EQ} = 2400/100 = 24\ \Omega$.

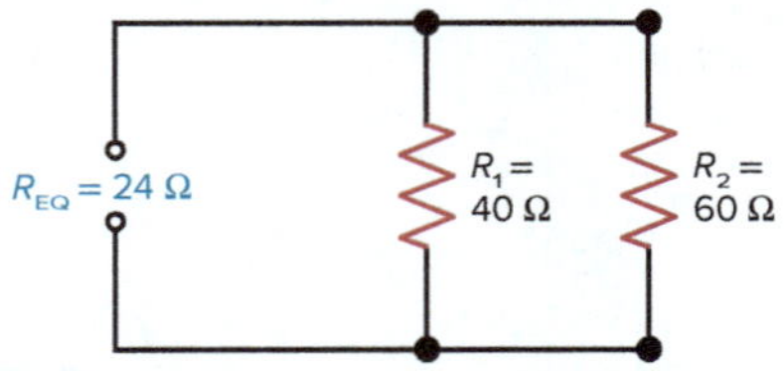

$$R_{EQ} = \frac{R_1 \times R_2}{R_1 + R_2} = \frac{2400}{100}$$

Special Case of Only Two Branches

When there are two parallel resistances and they are not equal, it is usually quicker to calculate the combined resistance by the method shown in Fig. 5–10. This rule states that the combination of two parallel resistances is their product divided by their sum.

$$R_{EQ} = \frac{R_1 \times R_2}{R_1 + R_2} \qquad \textbf{(5–4)}$$

where R_{EQ} is in the same units as all the individual resistances. For example, in Fig. 5–10,

$$R_{EQ} = \frac{R_1 \times R_2}{R_1 + R_2} = \frac{40 \times 60}{40 + 60} = \frac{2400}{100}$$

$$R_{EQ} = 24\ \Omega$$

Each R can have any value, but there must be only two resistances.

Short-Cut Calculations

Figure 5–11 shows how these special rules can help reduce parallel branches to a simpler equivalent circuit. In Fig. 5–11*a*, the 60-Ω R_1 and R_4 are equal and in parallel. Therefore, they are equivalent to the 30-Ω R_{14} in Fig. 5–11*b*. Similarly, the 20-Ω R_2 and R_3 are equivalent to the 10 Ω of R_{23}. The circuit in Fig. 5–11*a* is equivalent to the simpler circuit in Fig. 5–11*b* with just the two parallel resistances of 30 and 10 Ω.

Figure 5–11 An example of parallel resistance calculations with four branches. (*a*) Original circuit. (*b*) Resistors combined into two branches. (*c*) Equivalent circuit reduces to one R_{EQ} for all the branches.

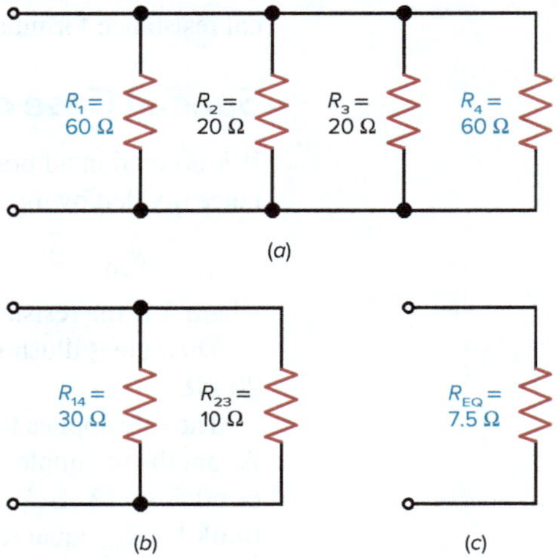

CALCULATOR

Formula (5–4) states a product over a sum. When using a calculator, group the R values in parentheses before dividing. The reason is that the division bar is a mathematical sign of grouping for terms to be added or subtracted. You must add $R_1 + R_2$ before dividing. By grouping R_1 and R_2 within parentheses, the addition will be done first before the division. The complete process is as follows.

Multiply the R values in the numerator. Press the divide ÷ key and then the left (or opening) parenthesis (key. Add the R values, $R_1 + R_2$, and press the right (or closing) parenthesis) key. Then press the equal = key for R_{EQ} on the display. Using the values in Fig. 5–10 as an example, multiply 40 × 60; press divide ÷ and left parenthesis (then 40 + 60 and the right parenthesis). Finally, press = to display 24 as the answer.

Finally, the combined resistance for these two equals their product divided by their sum, which is 300/40 or 7.5 Ω, as shown in Fig. 5–11*c*. This value of R_{EQ} in Fig. 5–11*c* is equivalent to the combination of the four branches in Fig. 5–11*a*. If you connect a voltage source across either circuit, the current in the main line will be the same for both cases.

The order of connections for parallel resistances does not matter in determining R_{EQ}. There is no question as to which is first or last because they are all across the same voltage source and receive their current at the same time.

GOOD TO KNOW

For more than two resistors connected in parallel, the value of an unknown resistance can be calculated using the following formula:

$$R_x = \frac{1}{1/R_{EQ} - 1/R_1 - 1/R_2 - \cdots \text{etc.}}$$

Finding an Unknown Branch Resistance

In some cases with two parallel resistors, it is useful to be able to determine what size R_X to connect in parallel with a known R to obtain a required value of R_{EQ}. Then the factors can be transposed as follows:

$$R_X = \frac{R \times R_{EQ}}{R - R_{EQ}} \qquad \textbf{(5–5)}$$

This formula is just another way of writing Formula (5–4).

Example 5-6

What R_X in parallel with 40 Ω will provide an R_{EQ} of 24 Ω?

ANSWER $R_X = \frac{R \times R_{EQ}}{R - R_{EQ}} = \frac{40 \times 24}{40 - 24} = \frac{960}{16}$

$R_X = 60\ \Omega$

This problem corresponds to the circuit shown before in Fig. 5–10.

Note that Formula (5–5) for R_X has a product over a difference. The R_{EQ} is subtracted because it is the smallest R. Remember that both Formulas (5–4) and (5–5) can be used with only two parallel branches.

Example 5-7

What R in parallel with 50 kΩ will provide an R_{EQ} of 25 kΩ?

ANSWER $R = 50\ \text{k}\Omega$

Two equal resistances in parallel have R_{EQ} equal to one-half R.

5–4 Self-Review

Answers at the end of the chapter.

a. Find R_{EQ} for three 4.7-MΩ resistances in parallel.
b. Find R_{EQ} for 3 MΩ in parallel with 2 MΩ.
c. Find R_{EQ} for two parallel 20-Ω resistances in parallel with 10 Ω.

5–5 Conductances in Parallel

Since conductance G is equal to $1/R$, the reciprocal resistance Formula (5–3) can be stated for conductance as $R_{EQ} = \frac{1}{G_T}$ where G_T is calculated as

$$G_T = G_1 + G_2 + G_3 + \cdots + \text{etc.} \qquad \textbf{(5–6)}$$

With R in ohms, G is in siemens. For the example in Fig. 5–12, G_1 is $1/20 = 0.05$, G_2 is $1/5 = 0.2$, and G_3 is $1/2 = 0.5$. Then

$$G_T = 0.05 + 0.2 + 0.5 = 0.75 \text{ S}$$

Notice that adding the conductances does not require reciprocals. Each value of G is the reciprocal of R.

The reason why *parallel conductances* are added directly can be illustrated by assuming a 1-V source across all branches. Then calculating the values of $1/R$ for the conductances gives the same values as calculating the branch currents. These values are added for the total I_T or G_T.

Working with G may be more convenient than working with R in parallel circuits, since it avoids the use of the reciprocal formula for R_{EQ}. Each branch current is directly proportional to its conductance. This idea corresponds to the fact that each voltage drop in series circuits is directly proportional to each of the series resistances. An example of the currents for parallel conductances is shown in Fig. 5–13. Note that the branch with G of 4 S has twice as much current as the 2-S branches because the branch conductance is doubled.

5–5 Self-Review

Answers at the end of the chapter.

a. **If G_1 is 2 S and G_2 in parallel is 4 S, calculate G_T.**

b. **If G_1 is 0.05 μS, G_2 is 0.2 μS, and G_3 is 0.5 μS, all in parallel, find G_T and its equivalent R_{EQ}.**

c. **If G_T is 4 μS for a parallel circuit, how much is R_{EQ}?**

Figure 5–12 Conductances G_1, G_2, and G_3 in parallel are added for the total G_T.

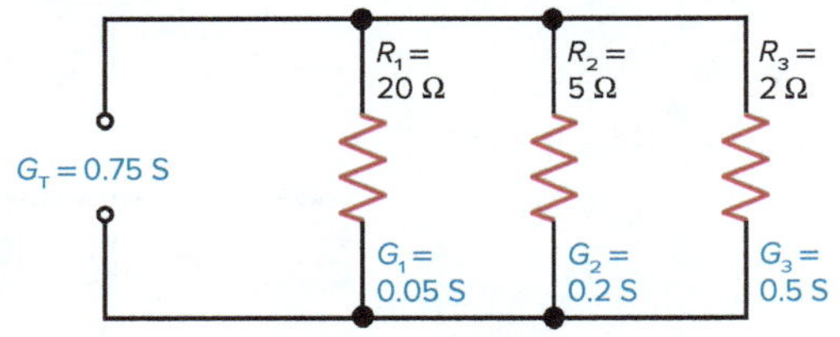

Figure 5–13 Example of how parallel branch currents are directly proportional to each branch conductance G.

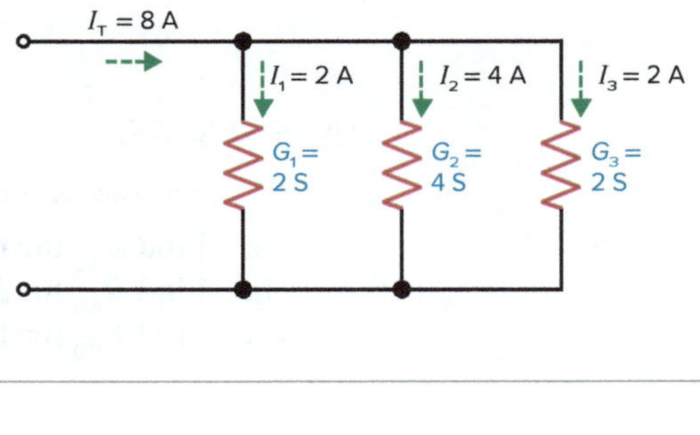

5–6 Total Power in Parallel Circuits

Since the power dissipated in the branch resistances must come from the voltage source, the **total power** equals the sum of the individual values of power in each branch. This rule is illustrated in Fig. 5–14. We can also use this circuit as an example of applying the rules of current, voltage, and resistance for a parallel circuit.

The applied 10 V is across the 10-Ω R_1 and 5-Ω R_2 in Fig. 5–14. The branch current I_1 then is V_A/R_1 or 10/10, which equals 1 A. Similarly, I_2 is 10/5, or 2 A. The total I_T is 1 + 2 = 3 A. If we want to find R_{EQ}, it equals V_A/I_T or 10/3, which is 3 ⅓ Ω.

The power dissipated in each branch R is $V_A \times I$. In the R_1 branch, I_1 is 10/10 = 1 A. Then P_1 is $V_A \times I_1$ or 10 × 1 = 10 W.

For the R_2 branch, I_2 is 10/5 = 2 A. Then P_2 is $V_A \times I_2$ or 10 × 2 = 20 W.

Adding P_1 and P_2, the answer is 10 + 20 = 30 W. This P_T is the total power dissipated in both branches.

This value of 30 W for P_T is also the total power supplied by the voltage source by means of its total line current I_T. With this method, the total power is $V_A \times I_T$ or 10 × 3 = 30 W for P_T. The 30 W of power supplied by the voltage source is dissipated or used up in the branch resistances.

It is interesting to note that in a parallel circuit, the smallest branch resistance will always dissipate the most power. Since $P = \frac{V^2}{R}$ and V is the same across all parallel branches, a smaller value of R in the denominator will result in a larger amount of power dissipation.

Also, note that in both parallel and series circuits, the sum of the individual values of power dissipated in the circuit equals the total power generated by the source. This can be stated as a formula

$$P_T = P_1 + P_2 + P_3 + \cdots + \text{etc.} \qquad \textbf{(5–7)}$$

The series or parallel connections can alter the distribution of voltage or current, but power is the rate at which energy is supplied. The circuit arrangement cannot change the fact that all the energy in the circuit comes from the source.

5–6 Self-Review

Answers at the end of the chapter.

a. Two parallel branches each have 2 A at 120 V. How much is P_T?

b. Three parallel branches of 10, 20, and 30 Ω have 60 V applied. How much is P_T?

c. Two parallel branches dissipate a power of 15 W each. How much is P_T?

Figure 5–14 The sum of the power values P_1 and P_2 used in each branch equals the total power P_T produced by the source.

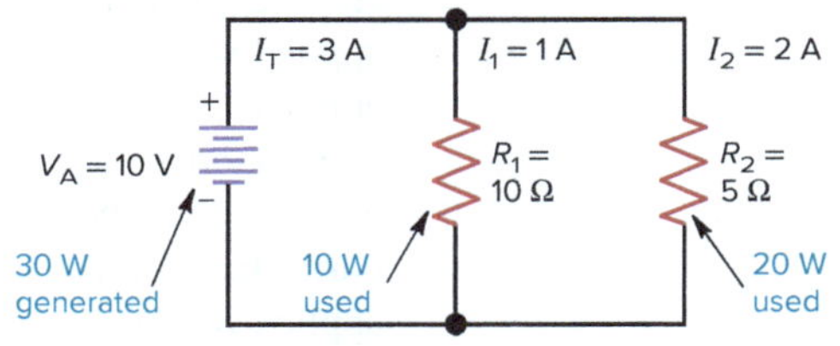

Figure 5–15 Analyzing a parallel circuit. What are the values for V_A and R_3? See solution in text.

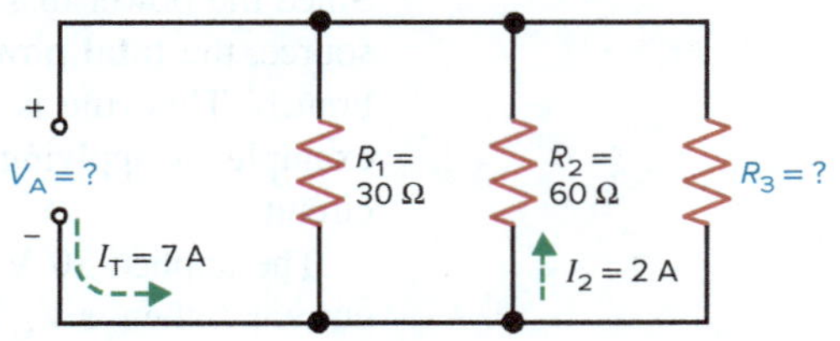

5–7 Analyzing Parallel Circuits with Random Unknowns

For many types of problems with parallel circuits, it is useful to remember the following points.

1. When you know the voltage across one branch, this voltage is across all the branches. There can be only one voltage across branch points with the same potential difference.
2. If you know I_T and one of the branch currents I_1, you can find I_2 by subtracting I_1 from I_T.

The circuit in Fig. 5–15 illustrates these points. The problem is to find the applied voltage V_A and the value of R_3. Of the three branch resistances, only R_1 and R_2 are known. However, since I_2 is given as 2 A, the I_2R_2 voltage must be $2 \times 60 = 120$ V.

Although the applied voltage is not given, this must also be 120 V. The voltage across all the parallel branches is the same 120 V that is across the R_2 branch.

Now I_1 can be calculated as V_A/R_1. This is $120/30 = 4$ A for I_1.

Current I_T is given as 7 A. The two branches take $2 + 4 = 6$ A. The third branch current through R_3 must be $7 - 6 = 1$ A for I_3.

Now R_3 can be calculated as V_A/I_3. This is $120/1 = 120$ Ω for R_3.

5–7 Self-Review

Answers at the end of the chapter.

Refer to Fig. 5–15.

a. How much is V_2 across R_2?
b. How much is I_1 through R_1?
c. How much is I_T?

5–8 Troubleshooting: Opens and Shorts in Parallel Circuits

In a parallel circuit, the effect of an open or a short is much different from that in a series circuit. For example, if one branch of a parallel circuit opens, the other branch currents remain the same. The reason is that the other branches still have the same applied voltage even though one branch has effectively been removed from the circuit. Also, if one branch of a parallel circuit becomes shorted, all branches are effectively shorted. The result is excessive current in the shorted branch and zero current in all other branches. In most cases, a fuse will be placed in the main line that will burn open (blow) when its current rating is exceeded. When the fuse blows, the applied voltage is removed from each of the parallel-connected branches. The effects of opens and shorts are examined in more detail in the following paragraphs.

The Effect of an Open in a Parallel Circuit

An open in any circuit is an infinite resistance that results in no current. However, in parallel circuits there is a difference between an open circuit in the main line and an open circuit in a parallel branch. These two cases are illustrated in Fig. 5–16. In Fig. 5–16*a*, the open circuit in the main line prevents any electron flow in the line to all the branches. The current is zero in every branch, therefore, and none of the bulbs can light.

However, in Fig. 5–16*b* the open is in the branch circuit for bulb 1. The **open branch** circuit has no current, then, and this bulb cannot light. The current in all the other parallel branches is normal, because each is connected to the voltage source. Therefore, the other bulbs light.

These circuits show the advantage of wiring components in parallel. An open in one component opens only one branch, whereas the other parallel branches have their normal voltage and current.

The Effect of a Short in a Parallel Circuit

A **short circuit** has practically zero resistance. Its effect, therefore, is to allow excessive current in the shorted circuit. Consider the example in Fig. 5–17. In Fig. 5–17*a*, the circuit is normal, with 1 A in each branch and 2 A for the total line current. However, suppose that the conducting wire at point G accidentally makes contact with the wire at point H, as shown in Fig. 5–17*b*. Since the wire is an excellent conductor, the short circuit results in practically zero resistance between points G and H. These two points are connected directly across the voltage source. Since the short circuit provides practically no opposition to current, the applied voltage could produce an infinitely high value of current through this current path.

Figure 5–16 Effect of an open in a parallel circuit. (*a*) Open path in the main line—no current and no light for all bulbs. (*b*) Open path in any branch—bulb for that branch does not light, but the other two bulbs operate normally.

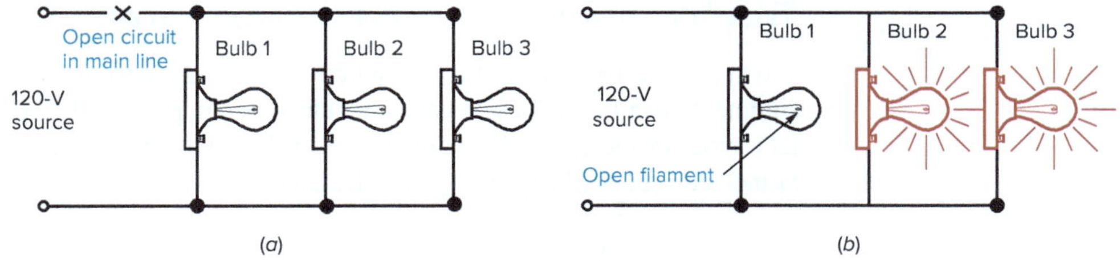

Figure 5–17 Effect of a short circuit across parallel branches. (*a*) Normal circuit. (*b*) Short circuit across points G and H shorts out all the branches.

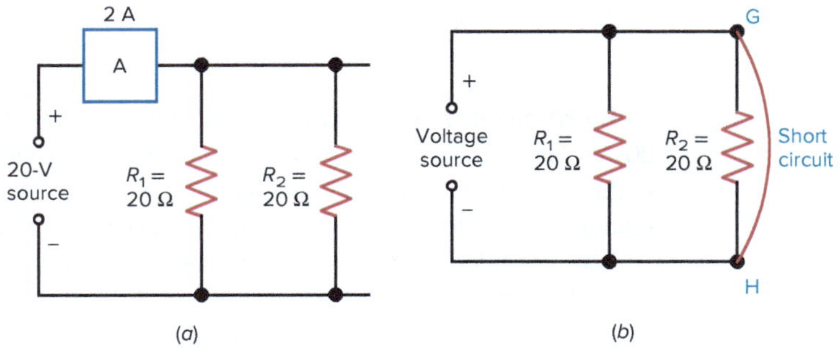

The Short-Circuit Current

Practically, the amount of current is limited by the small resistance of the wire. Also, the source usually cannot maintain its output voltage while supplying much more than its rated load current. Still, the amount of current can be dangerously high. For instance, the short-circuit current might be more than 100 A instead of the normal line current of 2 A in Fig. 5–17*a*. Because of the short circuit, excessive current flows in the voltage source, in the line to the short circuit at point H, through the short circuit, and in the line returning to the source from G. Because of the large amount of current, the wires can become hot enough to ignite and burn the insulation covering the wire. There should be a fuse that would open if there is too much current in the main line because of a short circuit across any of the branches.

The Short-Circuited Components Have No Current

For the short circuit in Fig. 5–17*b*, the I is 0 A in the parallel resistors R_1 and R_2. The reason is that the short circuit is a parallel path with practically zero resistance. Then all the current flows in this path, bypassing the resistors R_1 and R_2. Therefore, R_1 and R_2 are short-circuited or *shorted out* of the circuit. They cannot function without their normal current. If they were filament resistances of lightbulbs or heaters, they would not light without any current.

The short-circuited components are not damaged, however. They do not even have any current passing through them. Assuming that the short circuit has not damaged the voltage source and the wiring for the circuit, the components can operate again when the circuit is restored to normal by removing the short circuit.

All Parallel Branches Are Short-Circuited

If there were only one R in Fig. 5–17 or any number of parallel components, they would all be shorted out by the short circuit across points G and H. Therefore, a short circuit across one branch in a parallel circuit shorts out all parallel branches.

This idea also applies to a short circuit across the voltage source in any type of circuit. Then the entire circuit is shorted out.

Troubleshooting Procedures for Parallel Circuits

When a component fails in a parallel circuit, voltage, current, and resistance measurements can be made to locate the defective component. To begin our analysis, let's refer to the parallel circuit in Fig. 5–18*a*, which is normal. The individual branch currents I_1, I_2, I_3, and I_4 are calculated as follows:

$$I_1 = \frac{120\ \text{V}}{20\ \Omega} = 6\ \text{A}$$

$$I_2 = \frac{120\ \text{V}}{15\ \Omega} = 8\ \text{A}$$

$$I_3 = \frac{120\ \text{V}}{30\ \Omega} = 4\ \text{A}$$

$$I_4 = \frac{120\ \text{V}}{60\ \Omega} = 2\ \text{A}$$

By Kirchhoff's current law, the total current I_T equals 6 A + 8 A + 4 A + 2 A = 20 A. The total current I_T of 20 A is indicated by the ammeter M_1, which is placed in the main line between points J and K. The fuse F_1 between points A and B in the main line can safely carry 20 A, since its maximum rated current is 25 A, as shown.

Now consider the effect of an open branch between points D and I in Fig. 5–18*b*. With R_2 open, the branch current I_2 is 0 A. Also, the ammeter M_1 shows a total current I_T of 12 A, which is 8 A less than its normal value. This makes sense because I_2

Figure 5–18 Parallel circuit for troubleshooting analysis. (*a*) Normal circuit values; (*b*) circuit values with branch R_2 open; (*c*) circuit values with an open between points D and E; (*d*) circuit showing the effects of a shorted branch.

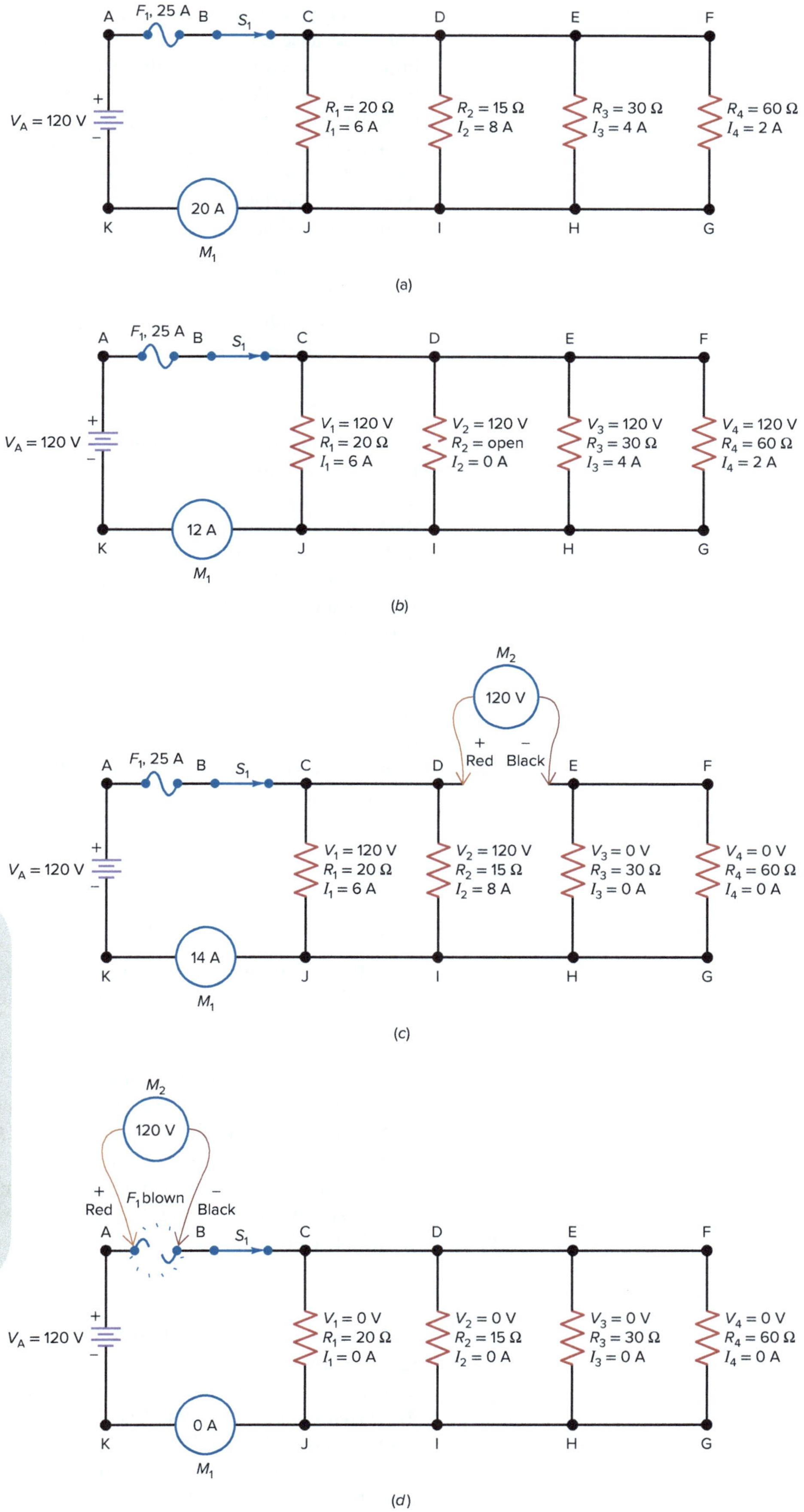

GOOD TO KNOW

A fuse is a safety device that serves to protect the circuit components and wiring in the event of a short circuit. Excessive current melts the fuse element which blows the fuse. With the fuse blown, there is no voltage across any of the parallel connected branches.

is normally 8 A. Notice that with R_2 open, all other branch currents remain the same. This is because each branch is still connected to the applied voltage of 120 V. It is important to realize that voltage measurements across the individual branches would not help determine which branch is open because even the open branch between points D and I will measure 120 V.

In most cases, the components in a parallel circuit provide a visual indication of failure. If a lamp burns open, it doesn't light. If a motor opens, it stops running. In these cases, the defective component is easy to spot.

In summary, here is the effect of an open branch in a parallel circuit:

1. The current in the open branch drops to 0 A.
2. The total current I_T decreases by an amount equal to the value normally drawn by the now open branch.
3. The current in all the remaining branches remains the same.
4. The applied voltage remains present across all branches whether they are open or not.

Next, let's consider the effect of an open between two branch points such as points D and E in Fig. 5–18*c*. With an open between these two points, the current through branch resistors R_3 and R_4 will be 0 A. Since $I_3 = 4$ A and $I_4 = 2$ A normally, the total current indicated by M_1 will drop from 20 A to 14 A as shown. The reason that I_3 and I_4 are now 0 A is that the applied voltage has effectively been removed from these two branches. If a voltmeter were placed across either points E and H or F and G, it would read 0 V. A voltmeter placed across points D and E would measure 120 V, however. This is indicated by the voltmeter M_2 as shown. The reason M_2 measures 120 V between points D and E is explained as follows: Notice that the positive (red) lead of M_2 is connected through S_1 and F_1 to the positive side of the applied voltage. Also, the negative (black) lead of M_2 is connected to the top of resistors R_3 and R_4. Since the voltage across R_3 and R_4 is 0 V, the negative lead of M_2 is in effect connected to the negative side of the applied voltage. In other words, M_2 is effectively connected directly across the 120-V source.

Example 5–8

In Fig. 5–18*a*, suppose that the ammeter M_1 reads 16 A instead of 20 A as it should. What could be wrong with the circuit?

ANSWER Notice that the current I_3 is supposed to be 4 A. If R_3 is open, this explains why M_1 reads a current that is 4 A less than its normal value. To confirm that R_3 is open; open S_1 and disconnect the top lead of R_3 from point E. Next place an ammeter between the top of R_3 and point E. Now, close S_1. If I_3 measures 0 A, you know that R_3 is open. If I_3 measures 4 A, you know that one of the other branches is drawing less current than it should. In this case, the next step would be to measure each of the remaining branch currents to find the defective component.

Consider the circuit in Fig. 5–18*d*. Notice that the fuse F_1 is blown and the ammeter M_1 reads 0 A. Notice also that the voltage across each branch measures 0 V and the voltage across the blown fuse measures 120 V as indicated by the voltmeter M_2. What could cause this? The most likely answer is that one of the parallel-connected branches has become short-circuited. This would cause the total current to rise well above the 25-A current rating of the fuse, thus causing it to blow. But how do we go about finding out which branch is shorted? There are at least three different approaches. Here's the first one: Start by opening switch S_1 and replacing the bad fuse. Next, with S_1 still open, disconnect all but one of the four parallel branches. For example, disconnect branch resistors R_1, R_2, and R_3 along the top (at points C, D, and E). With R_4 still connected, close S_1. If the fuse blows, you know R_4 is shorted! If the fuse does not blow, with only R_4 connected, open S_1 and reconnect R_3 to point E. Then, close S_1 and see if the fuse blows.

Repeat this procedure with branch resistors R_1 and R_2 until the shorted branch is identified. The shorted branch will blow the fuse when it is reconnected at the top (along points C, D, E, and F) with S_1 closed. Although this troubleshooting procedure is effective in locating the shorted branch, another fuse has been blown and this will cost you or the customer money.

Here's another approach to finding the shorted branch. Open S_1 and replace the bad fuse. Next, measure the resistance of each branch separately. It is important to remember that when you make resistance measurements in a parallel circuit, one end of each branch must be disconnected from the circuit so that the rest of the circuit does not affect the individual branch measurement. The branch that measures 0 Ω is obviously the shorted branch. With this approach, another fuse will not get blown.

Here is yet another approach that could be used to locate the shorted branch in Fig. 5–18*d*. With S_1 open, place an ohmmeter across points C and J. With a shorted branch, the ohmmeter will measure 0 Ω. To determine which branch is shorted, remove one branch at a time until the ohmmeter shows a value other than 0 Ω. The shorted component is located when removal of a given branch causes the ohmmeter to show a normal resistance.

In summary, here is the effect of a shorted branch in a parallel circuit:

1. The fuse in the main line will blow, resulting in zero current in the main line as well as in each parallel-connected branch.
2. The voltage across each branch will equal 0 V, and the voltage across the blown fuse will equal the applied voltage.
3. With power removed from the circuit, an ohmmeter will measure 0 Ω across all the branches.

Before leaving the topic of troubleshooting parallel circuits, one more point should be made about the fuse F_1 and the switch S_1 in Fig. 5–18*a*: The resistance of a good fuse and the resistance across the closed contacts of a switch are practically 0 Ω. Therefore, the voltage drop across a good fuse or a closed switch is approximately 0 V. This can be proven with Ohm's law, since $V = I \times R$. If $R = 0\ \Omega$, then $V = I \times 0\ \Omega = 0$ V. When a fuse blows or a switch opens, the resistance increases to such a high value that it is considered infinite. When used in the main line of a parallel circuit, the voltage across an open switch or a blown fuse is the same as the applied voltage. One way to reason this out logically is to treat all the parallel branches as a single equivalent resistance R_{EQ} in series with the switch and fuse. The result is a simple series circuit. Then, if either the fuse or the switch opens, apply the rules of an open to a series circuit. As you recall from your study of series circuits, the voltage across an open equals the applied voltage.

5–8 Self-Review

Answers at the end of the chapter.

a. In Fig. 5–16*b*, how much voltage is across bulb 1?

b. In Fig. 5–17*b*, how much is the resistance across points G and H?

c. In Fig. 5–18*a*, how much current will M_1 show if the wire between points C and D is removed?

d. With reference to Question *c*, how much voltage would be measured across R_4? Across points C and D?

e. In Fig. 5–18*a*, how much voltage will be measured across points A and B, assuming the fuse is blown?

Application of Parallel Circuits

HOLIDAY LIGHTS (CONTINUED)

In Chapter 4, ***Series Circuits***, we examined a string of holiday lights with 50 incandescent bulbs. In this chapter, we will examine the effect of connecting one or more additional sets of lights to the first one. Figure 5–19*a* shows a second set of lights plugged into the receptacle-end of the first set. Each set or string consists of 50 incandescent bulbs. Careful examination of the wiring diagram shows that the individual strings of lights are connected in parallel. Figure 5–19*b* shows the equivalent circuit. Because the two sets of lights are in parallel, the total current, I_T, equals the sum of the individual branch currents. Recall from Chapter 4 that the current drawn by a string of 50 bulbs is 166.7 mA. Therefore, the total current for two sets of lights is 333.3 mA. If a third set of lights is added, as in Fig. 5–19*c*, the total current, I_T, would be 500 mA or 0.5A. Figure 5–19*d* shows the equivalent circuit for Fig. 5–19*c*. Since the in-line fuses are rated at 3 A, more than three sets of lights can be connected together. Connecting too many sets of lights, however, will cause the fuses in the first set of lights (at the far left) to blow.

In Fig. 5–19*d*, it is important to note that the fuses in the last string of lights, at the far right, only carry a current of 166.7 mA. The fuses in the middle string of lights carry a current of 333.3 mA. The fuses in the first string, at the far left, carry 500 mA, or 0.5 A. Notice that the total current, I_T, is equal to the sum of the individual branch currents. In this case,

$$I_T = 166.7\text{ mA} + 166.7\text{ mA} + 166.7\text{ mA}$$
$$= 500\text{ mA or } 0.5\text{ A}$$

Figure 5–20 shows the wiring diagram for a string of holiday lights with 100 incandescent bulbs. Notice that all 100 bulbs are connected as one long string of lights. Notice, however, that at the middle of the string there is a wire running to the bottom conductor. There is also a wire that connects the top conductor on the plug-end to the top conductor on the receptacle-end. These two wires provide a parallel connection for two strings of lights, each with 50 bulbs. Because there are two 50-bulb strings in parallel, the total current, I_T, is 333.3 mA. Because the in-line fuses are rated at 3 A, several sets of lights can be connected together. If too many sets of lights are connected, the 3 A fuses in the first string will blow. (The first string is the string that is plugged directly into the 120 V_{AC} outlet.)

TROUBLESHOOTING

It is important to realize that if an entire string of lights is out (dark) the other strings will still light as normal. In Fig. 5–19*c*, for

Figure 5–19a Two sets of holiday lights connected together.

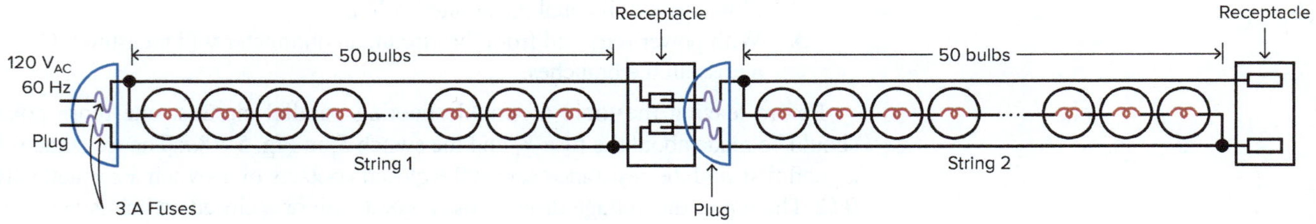

Figure 5–19b Equivalent circuit of Fig. 5–19*a* showing the individual branch currents and total current.

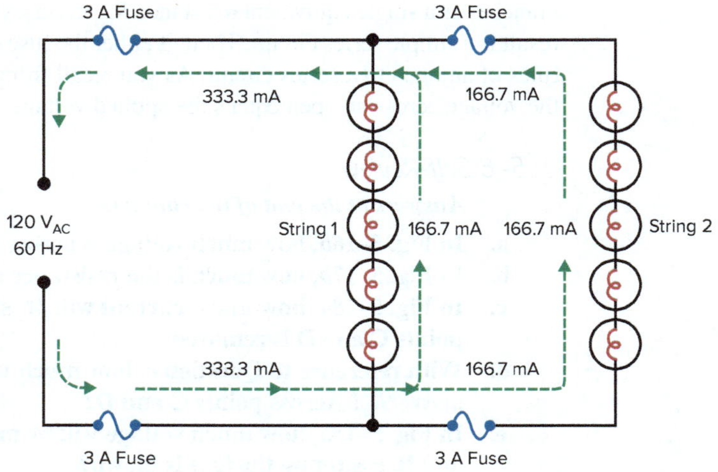

Figure 5–19c Three sets of holiday lights connected together.

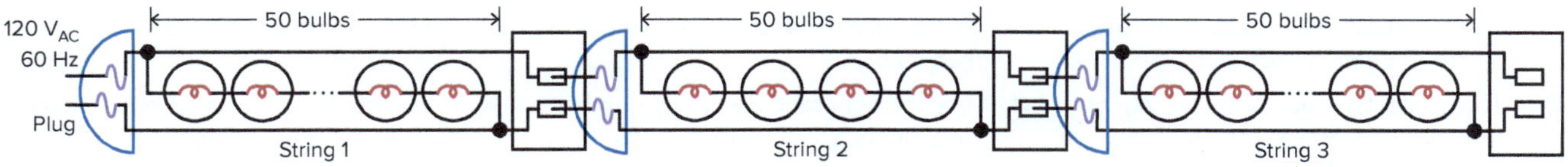

Figure 5–19d Equivalent circuit for Fig. 5–19*c* showing the individual branch currents and the total current.

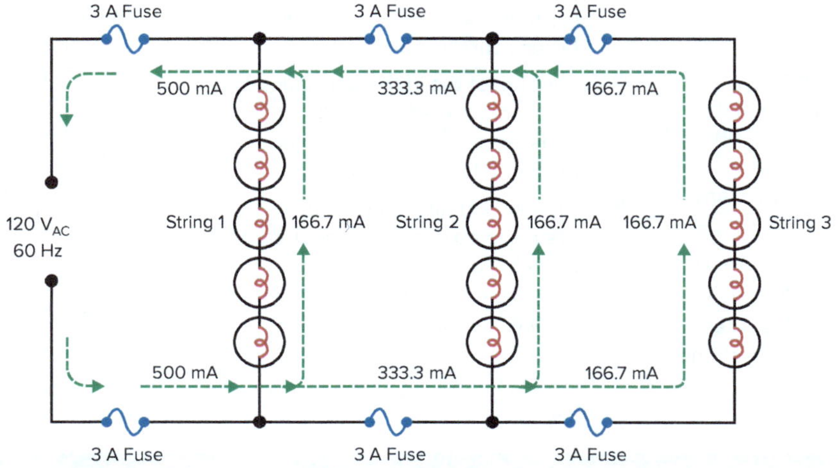

example, if the middle string of lights goes out (dark), the first and last sets will still light. The reason why is that each string of lights is in parallel with the other. If one branch of a parallel circuit opens, the remaining branches are unaffected. This is because the remaining sets of lights are still connected across the 120 V power line.

Also, it may seem unusual that one-half of the string of lights in Fig. 5–20 could be dark while the other half lights up as normal. This is actually a very common problem with a string of holiday lights with 100 bulbs. The reason this is possible, however, is that the wiring configuration creates a parallel circuit with two separate branches. In this case, each branch consists of 50 series connected bulbs. If one branch opens, the other half still lights as normal.

If too many strings of lights are connected together, the in-line fuses in the first string will blow. This, of course, does not mean that all of the individual sets of lights connected to the first set are bad. It only means that the fuses blew in the first string because the total current exceeded the fuses' current rating. To the inexperienced decorator, this can be frustrating and confusing.

Here's one final point. When multiple sets of blinking holiday lights are connected together, it is important to realize that each individual set of lights needs its own separate flasher unit. Otherwise, only the sets of lights having a flasher unit will blink. This is also true with the string of 100 bulbs, as shown in Fig. 5–20. Each string of 50 bulbs must have its own separate flasher unit. Remember, each string of lights is in parallel with the 120 V power line.

Figure 5–20 String of holiday lights with 100 incandescent bulbs. Wiring connection shows two 50 bulb strings in parallel.

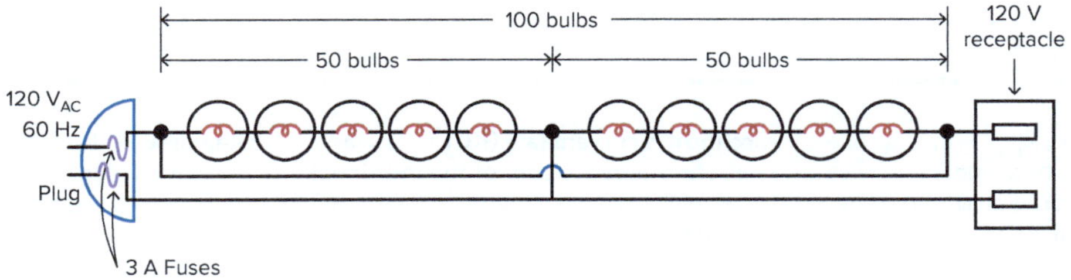

Summary

- There is only one voltage V_A across all components in parallel.
- The current in each branch I_b equals the voltage V_A across the branch divided by the branch resistance R_b, or $I_b = V_A/R_b$.
- Kirchhoff's current law states that the total current I_T in a parallel circuit equals the sum of the individual branch currents. Expressed as an equation, Kirchhoff's current law is $I_T = I_1 + I_2 + I_3 + \cdots + \text{etc.}$
- The equivalent resistance R_{EQ} of parallel branches is less than the smallest branch resistance, since all the branches must take more current from the source than any one branch.
- For only *two* parallel resistances of any value, $R_{EQ} = R_1R_2/(R_1 + R_2)$.
- For any number of *equal* parallel resistances, R_{EQ} is the value of one resistance divided by the number of resistances.
- For the general case of any number of branches, calculate R_{EQ} as V_A/I_T or use the reciprocal resistance formula:

$$R_{EQ} = \frac{1}{1/R_1 + 1/R_2 + 1/R_3 + \cdots + \text{etc.}}$$

- For any number of conductances in parallel, their values are added for G_T, in the same way as parallel branch currents are added.
- The sum of the individual values of power dissipated in parallel resistances equals the total power produced by the source.
- An open circuit in one branch results in no current through that branch, but the other branches can have their normal current. However, an open circuit in the main line results in no current for any of the branches.
- A short circuit has zero resistance, resulting in excessive current. When one branch is short-circuited, all parallel paths are also short-circuited. The entire current is in the short circuit and no current is in the short-circuited branches.
- The voltage across a good fuse and the voltage across a closed switch are approximately 0 V. When the fuse in the main line of a parallel circuit opens, the voltage across the fuse equals the full applied voltage. Likewise, when the switch in the main line of a parallel circuit opens, the voltage across the open switch equals the full applied voltage.
- Table 5–1 compares series and parallel circuits.

Table 5–1 Comparison of Series and Parallel Circuits

Series Circuit	Parallel Circuit
Current the same in all the components	Voltage the same across all the branches
V across each series *R* is $I \times R$	*I* in each branch *R* is V/R
$V_T = V_1 + V_2 + V_3 + \cdots + \text{etc.}$	$I_T = I_1 + I_2 + I_3 + \cdots + \text{etc.}$
$R_T = R_1 + R_2 + R_3 + \cdots + \text{etc.}$	$G_T = G_1 + G_2 + G_3 + \cdots + \text{etc.}$
R_T must be more than the largest individual *R*	R_{EQ} must be less than the smallest branch *R*
$P_1 = P_1 + P_2 + P_3 + \cdots + \text{etc.}$	$P_T = P_1 + P_2 + P_3 + \cdots + \text{etc.}$
Applied voltage is divided into *IR* voltage drops	Main-line current is divided into branch currents
The largest *IR* drop is across the largest series *R*	The largest branch *I* is in the smallest parallel *R*
Open in one component causes entire circuit to be open	Open in one branch does not prevent *I* in other branches

Important Terms

Equivalent resistance, R_{EQ} — in a parallel circuit, this refers to a single resistance that would draw the same amount of current as all of the parallel connected branches.

Kirchhoff's current law (KCL) — a law stating that the sum of the individual branch currents in a parallel circuit must equal the total current, I_T.

Main line — the pair of leads connecting all individual branches in a parallel circuit to the terminals of the applied voltage, V_A. The main line carries the total current, I_T, flowing to and from the terminals of the voltage source.

Parallel bank — a combination of parallel-connected branches.

Reciprocal resistance formula — a formula stating that the equivalent resistance, R_{EQ}, of a parallel circuit equals the reciprocal of the sum of the reciprocals of the individual branch resistances.

Related Formulas

$$I_1 = \frac{V_A}{R_1}, I_2 = \frac{V_A}{R_2}, I_3 = \frac{V_A}{R_3}$$

$$I_T = I_1 + I_2 + I_3 + \cdots + \text{etc.}$$

$$R_{EQ} = \frac{V_A}{I_T}$$

$$R_{EQ} = \frac{1}{1/R_1 + 1/R_2 + 1/R_3 + \cdots + \text{etc.}}$$

$$R_{EQ} = \frac{R}{n} \text{ } (R_{EQ} \text{ for equal branch resistances})$$

$$R_{EQ} = \frac{R_1 \times R_2}{R_1 + R_2} \text{ } (R_{EQ} \text{ for only two branch resistances})$$

$$R_X = \frac{R \times R_{EQ}}{R - R_{EQ}}$$

$$G_T = G_1 + G_2 + G_3 + \cdots + \text{etc.}$$

$$P_T = P_1 + P_2 + P_3 + \cdots + \text{etc.}$$

Self-Test

Answers at the back of the book.

1. A 120-kΩ resistor, R_1, and a 180-kΩ resistor, R_2, are in parallel. How much is the equivalent resistance, R_{EQ}?

a. 72 kΩ.

b. 300 kΩ.

c. 360 kΩ.

d. 90 kΩ.

2. A 100-Ω resistor, R_1, and a 300-Ω resistor, R_2, are in parallel across a DC voltage source. Which resistor dissipates more power?

a. The 300-Ω resistor.

b. Both resistors dissipate the same amount of power.

c. The 100-Ω resistor.

d. It cannot be determined.

3. Three 18-Ω resistors are in parallel. How much is the equivalent resistance, R_{EQ}?

a. 54 Ω.

b. 6 Ω.

c. 9 Ω.

d. none of the above.

4. Which of the following statements about parallel circuits is false?

a. The voltage is the same across all the branches in a parallel circuit.

b. The equivalent resistance, R_{EQ}, of a parallel circuit is always smaller than the smallest branch resistance.

c. In a parallel circuit, the total current, I_T, in the main line equals the sum of the individual branch currents.

d. The equivalent resistance, R_{EQ}, of a parallel circuit decreases when one or more parallel branches are removed from the circuit.

5. Two resistors, R_1 and R_2, are in parallel with each other and a DC voltage source. If the total current, I_T, in the main line equals 6 A and I_2 through R_2 is 4 A, how much is I_1 through R_1?

a. 6 A.

b. 2 A.

c. 4 A.

d. It cannot be determined.

6. How much resistance must be connected in parallel with a 360-Ω resistor to obtain an equivalent resistance, R_{EQ}, of 120 Ω?

a. 360 Ω.

b. 480 Ω.

c. 1.8 kΩ.

d. 180 Ω.

7. If one branch of a parallel circuit becomes open,

a. all remaining branch currents increase.

b. the voltage across the open branch will be 0 V.

c. the remaining branch currents do not change in value.

d. the equivalent resistance of the circuit decreases.

8. If a 10-Ω R_1, 40-Ω R_2, and 8-Ω R_3 are in parallel, calculate the total conductance, G_T, of the circuit.

a. 250 mS.

b. 58 S.

c. 4 Ω.

d. 0.25 μS.

9. Which of the following formulas can be used to determine the total power, P_T, dissipated by a parallel circuit.

a. $P_T = V_A \times I_T$.

b. $P_T = P_1 + P_2 + P_3 + \cdots + \text{etc.}$

c. $P_T = \frac{V_A^2}{R_{EQ}}$.

d. all of the above.

10. A 20-Ω R_1, 50-Ω R_2, and 100-Ω R_3 are connected in parallel. If R_2 is short-circuited, what is the equivalent resistance, R_{EQ}, of the circuit?

a. approximately 0 Ω.

b. infinite (∞) Ω.

c. 12.5 Ω.

d. It cannot be determined.

11. If the fuse in the main line of a parallel circuit opens,

a. the voltage across each branch will be 0 V.

b. the current in each branch will be zero.

c. the current in each branch will increase to offset the decrease in total current.

d. both a and b.

12. A 100-Ω R_1 and a 150-Ω R_2 are in parallel. If the current, I_1, through R_1 is 24 mA, how much is the total current, I_T?

a. 16 mA.

b. 40 mA.

c. 9.6 mA.

d. It cannot be determined.

13. A 2.2-kΩ R_1 is in parallel with a 3.3-kΩ R_2. If these two resistors carry a total current of 7.5 mA, how much is the applied voltage, V_A?

a. 16.5 V.

b. 24.75 V.

c. 9.9 V.

d. 41.25 V.

14. How many 120-Ω resistors must be connected in parallel to obtain an equivalent resistance, R_{EQ}, of 15 Ω?

a. 15.

b. 8.

c. 12.

d. 6.

15. A 220-Ω R_1, 2.2-kΩ R_2, and 200-Ω R_3 are connected across 15 V of applied voltage. What happens to R_{EQ} if the applied voltage is doubled to 30 V?

a. R_{EQ} doubles.

b. R_{EQ} cuts in half.

c. R_{EQ} does not change.

d. R_{EQ} increases but is not double its original value.

16. If one branch of a parallel circuit opens, the total current, I_T,

a. does not change.

b. decreases.

c. increases.

d. goes to zero.

17. In a normally operating parallel circuit, the individual branch currents are

a. independent of each other.

b. not affected by the value of the applied voltage.

c. larger than the total current, I_T.

d. none of the above.

18. If the total conductance, G_T, of a parallel circuit is 200 μS, how much is R_{EQ}?

a. 500 Ω.

b. 200 kΩ.

c. 5 kΩ.

d. 500 kΩ.

19. If one branch of a parallel circuit is short-circuited,

a. the fuse in the main line will blow.

b. the voltage across the short-circuited branch will measure the full value of applied voltage.

c. all the remaining branches are effectively short-circuited as well.

d. both a and c.

20. Two lightbulbs in parallel with the 120-V power line are rated at 60 W and 100 W, respectively. What is the equivalent resistance, R_{EQ}, of the bulbs when they are lit?

a. 144 Ω.

b. 90 Ω.

c. 213.3 Ω.

d. It cannot be determined.

Essay Questions

1. Draw a wiring diagram showing three resistances connected in parallel across a battery. Indicate each branch and the main line.

2. State two rules for the voltage and current values in a parallel circuit.

3. Explain briefly why the current is the same in both sides of the main line that connects the voltage source to the parallel branches.

4. **(a)** Show how to connect three equal resistances for a combined equivalent resistance one-third the value of one resistance. **(b)** Show how to connect three equal resistances for a combined equivalent resistance three times the value of one resistance.

5. Why can the current in parallel branches be different when they all have the same applied voltage?

6. Why does the current increase in the voltage source as more parallel branches are added to the circuit?

7. Show how the formula

$$R_{EQ} = R_1R_2/(R_1 + R_2)$$

is derived from the reciprocal formula

$$\frac{1}{R_{EQ}} = \frac{1}{R_1} + \frac{1}{R_2}$$

8. Redraw Fig. 5–17 with five parallel resistors R_1 to R_5 and explain why they all would be shorted out with a short circuit across R_3.

9. State briefly why the total power equals the sum of the individual values of power, whether a series circuit or a parallel circuit is used.

10. Explain why an open in the main line disables all the branches, but an open in one branch affects only that branch current.

11. Give two differences between an open circuit and a short circuit.

12. List as many differences as you can in comparing series circuits with parallel circuits.

13. Why are household appliances connected to the 120-V power line in parallel instead of in series?

14. Give one advantage and one disadvantage of parallel connections.

15. A 5-Ω and a 10-Ω resistor are in parallel across a DC voltage source. Which resistor will dissipate more power? Provide proof with your answer.

Problems

SECTION 5–1 THE APPLIED VOLTAGE V_A IS THE SAME ACROSS PARALLEL BRANCHES

5–1 MultiSim In Fig. 5–21, how much voltage is across points

a. A and B?

b. C and D?

c. E and F?

d. G and H?

Figure 5–21

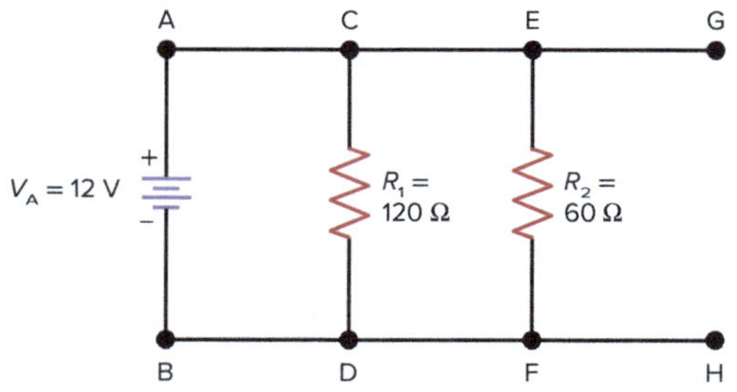

5–2 In Fig. 5–21, how much voltage is across

a. the terminals of the voltage source?

b. R_1?

c. R_2?

5–3 In Fig. 5–21, how much voltage will be measured across points C and D if R_1 is removed from the circuit?

SECTION 5–2 EACH BRANCH *I* EQUALS $\frac{V_A}{R}$

5–4 In Fig. 5–21, solve for the branch currents, I_1 and I_2.

5–5 In Fig. 5–21, explain why I_2 is double the value of I_1.

5–6 In Fig. 5–21, assume a 10-Ω resistor, R_3, is added across points G and H.

a. Calculate the branch current, I_3.

b. Explain how the branch currents, I_1 and I_2 are affected by the addition of R_3.

5–7 In Fig. 5–22, solve for the branch currents I_1, I_2, and I_3.

Figure 5–22

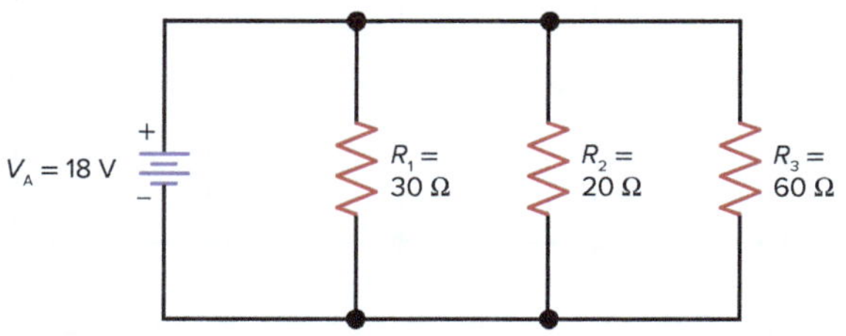

5–8 In Fig. 5–22, do the branch currents I_1 and I_3 remain the same if R_2 is removed from the circuit? Explain your answer.

5–9 In Fig. 5–23, solve for the branch currents I_1, I_2, I_3, and I_4.

Figure 5–23

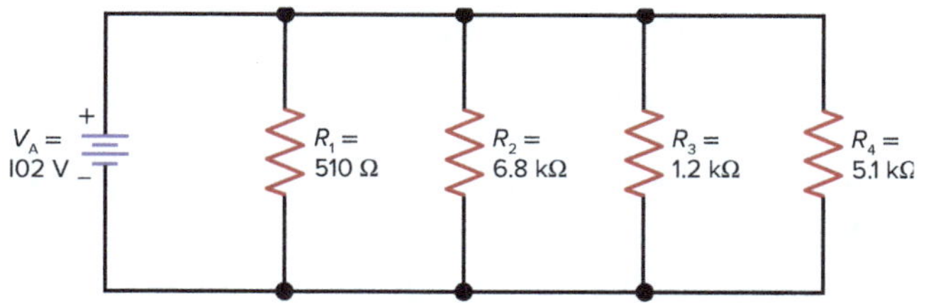

5–10 Recalculate the values for I_1, I_2, I_3, and I_4 in Fig. 5–23 if the applied voltage, V_A, is reduced to 51 V.

SECTION 5–3 KIRCHHOFF'S CURRENT LAW (KCL)

5–11 MultiSim In Fig. 5–21, solve for the total current, I_T.

5–12 MultiSim Figure 5–21 re-solve for the total current, I_T, if a 10-Ω resistor, R_3, is added across points G and H.

5–13 In Fig. 5–22, solve for the total current, I_T.

5–14 In Fig. 5–22, re-solve for the total current, I_T, if R_2 is removed from the circuit.

5–15 In Fig. 5–23, solve for the total current, I_T.

5–16 In Fig. 5–23, re-solve for the total current, I_T, if V_A is reduced to 51 V.

5–17 In Fig. 5–24, solve for I_1, I_2, I_3, and I_T.

Figure 5–24

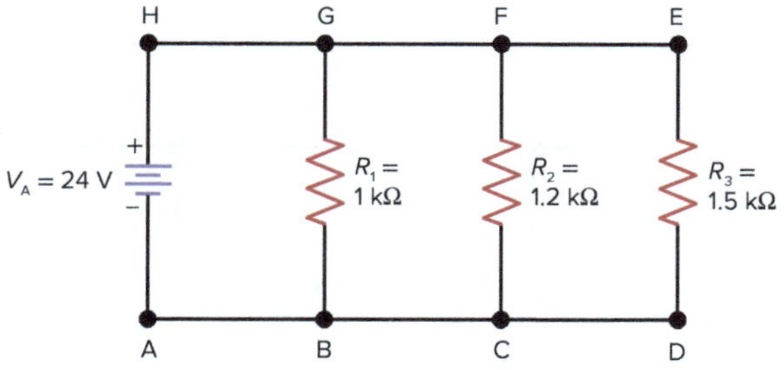

5–18 In Fig. 5–24, how much is the current in the wire between points

a. A and B?

b. B and C?

c. C and D?

d. E and F?

e. F and G?

f. G and H?

5–19 In Fig. 5–24, assume that a 100-Ω resistor, R_4, is added to the right of resistor, R_3. How much is the current in the wire between points

a. A and B?

b. B and C?

c. C and D?

d. E and F?

e. F and G?

f. G and H?

5–20 In Fig. 5–25, solve for I_1, I_2, I_3, and I_T.

Figure 5–25

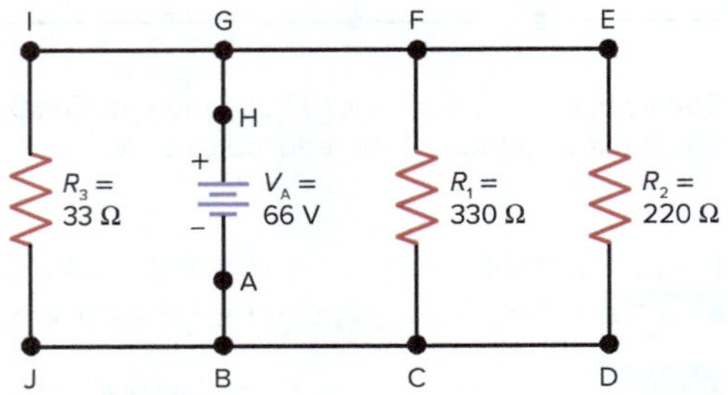

5–21 In Fig. 5–25, how much is the current in the wire between points

a. A and B?

b. B and C?

c. C and D?

d. E and F?

e. F and G?

f. G and H?

g. G and I?

h. B and J?

5–22 In Fig. 5–26, apply Kirchhoff's current law to solve for the unknown current, I_3.

Figure 5–26

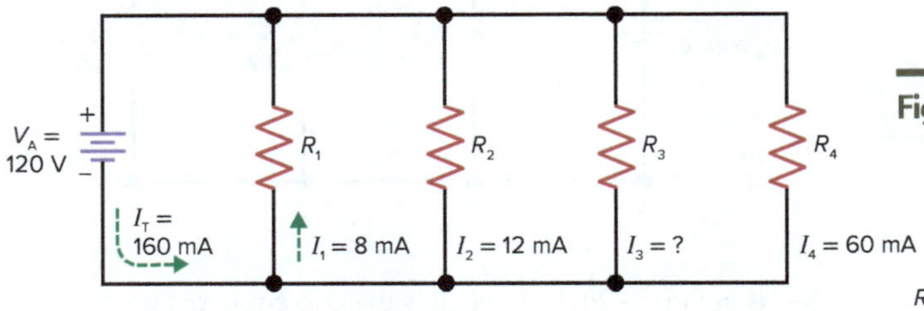

5–23 Two resistors R_1 and R_2 are in parallel with each other and a DC voltage source. How much is I_2 through R_2 if I_T = 150 mA and I_1 through R_1 is 60 mA?

SECTION 5–4 RESISTANCES IN PARALLEL

5–24 In Fig. 5–21, solve for R_{EQ}.

5–25 In Fig. 5–21, re-solve for R_{EQ} if a 10-Ω resistor, R_3 is added across points G and H.

5–26 In Fig. 5–22, solve for R_{EQ}.

5–27 In Fig. 5–22, re-solve for R_{EQ} if R_2 is removed from the circuit.

5–28 In Fig. 5–23, solve for R_{EQ}.

5–29 In Fig. 5–23, re-solve for R_{EQ} if V_A is reduced to 51 V.

5–30 In Fig. 5–24, solve for R_{EQ}.

5–31 In Fig. 5–25, solve for R_{EQ}.

5–32 In Fig. 5–26, solve for R_{EQ}.

5–33 MultiSim In Fig. 5–27, how much is R_{EQ} if R_1 = 100 Ω and R_2 = 25 Ω?

Figure 5–27

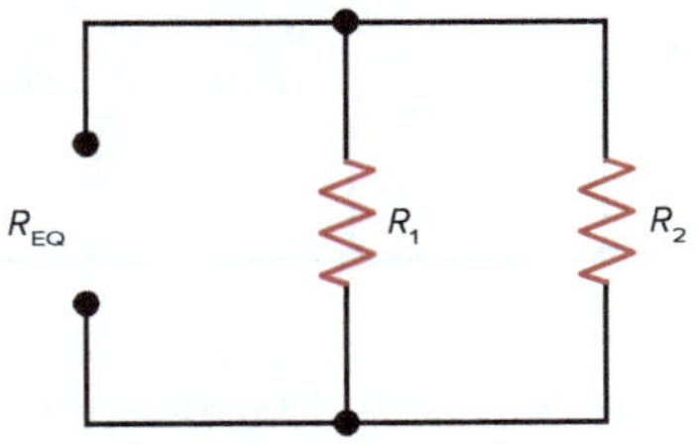

5–34 MultiSim In Fig. 5–27, how much is R_{EQ} if R_1 = 1.5 MΩ and R_2 = 1 MΩ?

5–35 MultiSim In Fig. 5–27, how much is R_{EQ} if R_1 = 2.2 kΩ and R_2 = 220 Ω?

5–36 In Fig. 5–27, how much is R_{EQ} if $R_1 = R_2$ = 10 kΩ?

5–37 In Fig. 5–27, how much resistance, R_2, must be connected in parallel with a 750 Ω R_1 to obtain an R_{EQ} of 500 Ω?

5–38 In Fig. 5–27, how much resistance, R_1, must be connected in parallel with a 6.8 kΩ R_2 to obtain an R_{EQ} of 1.02 kΩ?

5–39 How much is R_{EQ} in Fig. 5–28 if R_1 = 1 kΩ, R_2 = 4 kΩ, R_3 = 200 Ω, and R_4 = 240 Ω?

Figure 5–28

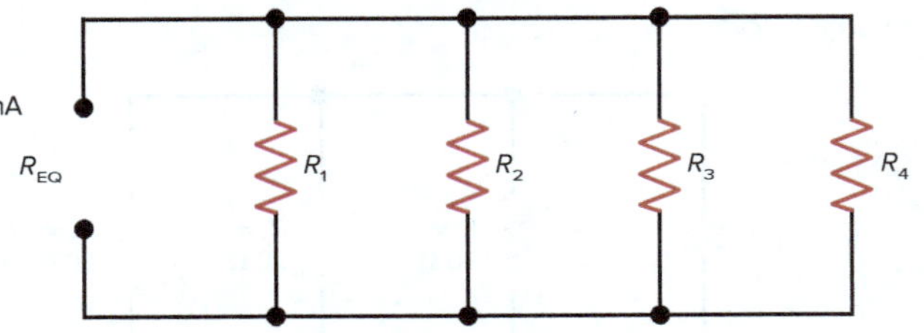

5–40 How much is R_{EQ} in Fig. 5–28 if R_1 = 5.6 kΩ, R_2 = 4.7 kΩ, R_3 = 8.2 kΩ, and R_4 = 2.7 kΩ?

5–41 MultiSim How much is R_{EQ} in Fig. 5–28 if R_1 = 1.5 kΩ, R_2 = 1 kΩ, R_3 = 1.8 kΩ, and R_4 = 150 Ω?

5–42 How much is R_{EQ} in Fig. 5–28 if $R_1 = R_2 = R_3 = R_4 =$ 2.2 kΩ?

5–43 A technician is using an ohmmeter to measure a variety of different resistor values. Assume the technician has a body resistance of 750 kΩ. How much resistance will the ohmmeter read if the fingers of the technician touch the leads of the ohmmeter when measuring the following resistors:

a. 270 Ω

b. 390 kΩ

c. 2.2 MΩ.

d. 1.5 kΩ.

e. 10 kΩ

SECTION 5–5 CONDUCTANCES IN PARALLEL

5–44 In Fig. 5–29, solve for G_1, G_2, G_3, G_T, and R_{EQ}.

Figure 5–29

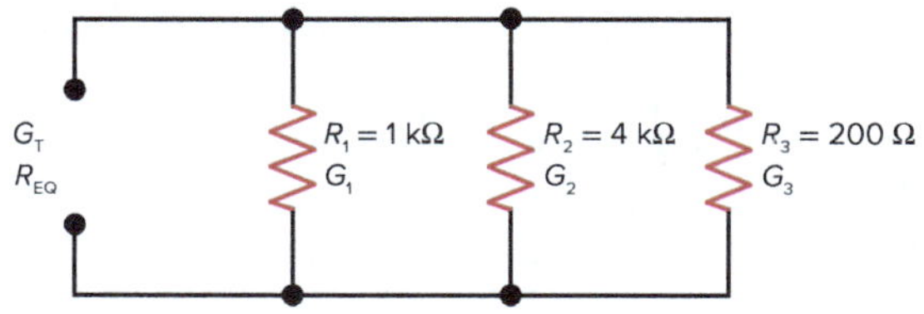

5–45 In Fig. 5–30, solve for G_1, G_2, G_3, G_4, G_T, and R_{EQ}.

Figure 5–30

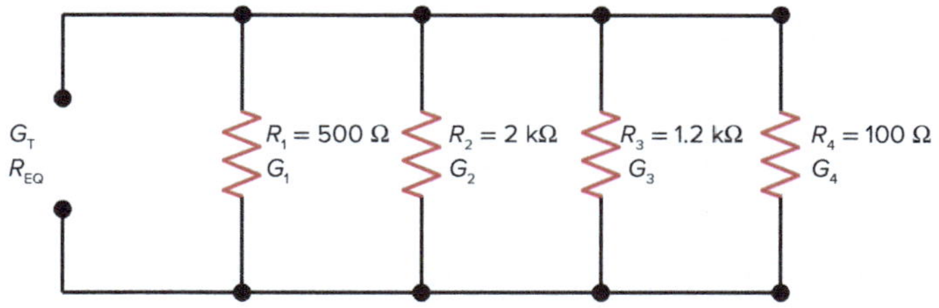

5–46 Find the total conductance, G_T for the following branch conductances; $G_1 = 1$ mS, $G_2 = 200$ μS, and $G_3 = 1.8$ mS. How much is R_{EQ}?

5–47 Find the total conductance, G_T for the following branch conductances; $G_1 = 100$ mS, $G_2 = 66.67$ mS, $G_3 = 250$ mS, and $G_4 = 83.33$ mS. How much is R_{EQ}?

SECTION 5–6 TOTAL POWER IN PARALLEL CIRCUITS

5–48 In Fig. 5–22, solve for P_1, P_2, P_3, and P_T.

5–49 In Fig. 5–23, solve for P_1, P_2, P_3, P_4, and P_T.

5–50 In Fig. 5–24, solve for P_1, P_2, P_3, and P_T.

5–51 In Fig. 5–25, solve for P_1, P_2, P_3, and P_T.

5–52 In Fig. 5–26, solve for P_1, P_2, P_3, P_4, and P_T.

SECTION 5–7 ANALYZING PARALLEL CIRCUITS WITH RANDOM UNKNOWNS

5–53 In Fig. 5–31, solve for V_A, R_1, I_2, R_{EQ}, P_1, P_2, and P_T.

Figure 5–31

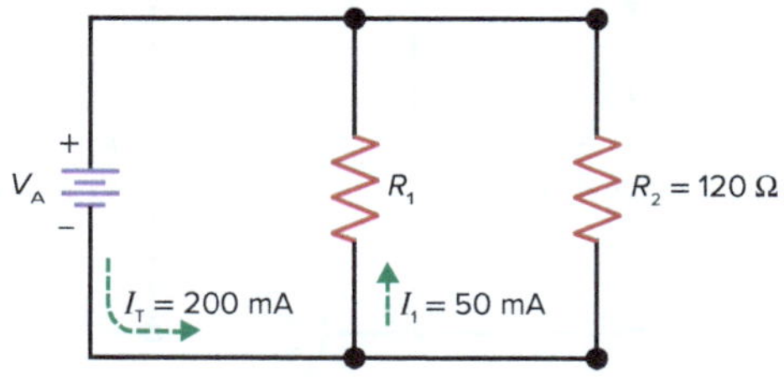

5–54 In Fig. 5–32, solve for V_A, I_1, I_2, R_2, I_T, P_2, and P_T.

Figure 5–32

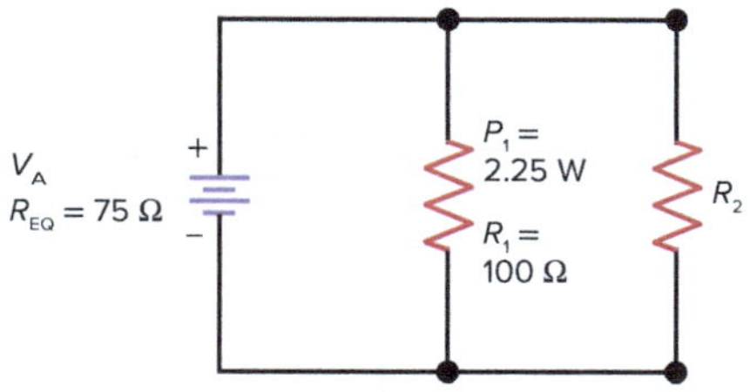

5–55 In Fig. 5–33, solve for R_3, V_A, I_1, I_2, I_T, P_1, P_2, P_3, and P_T.

Figure 5–33

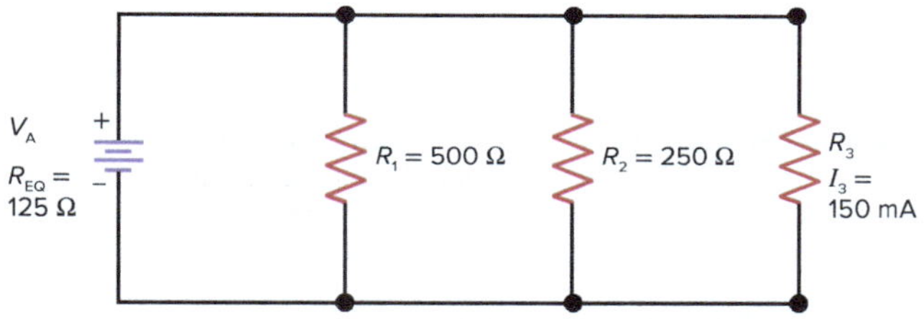

5–56 In Fig. 5–34, solve for I_T, I_1, I_2, R_1, R_2, R_3, P_2, P_3, and P_T.

Figure 5–34

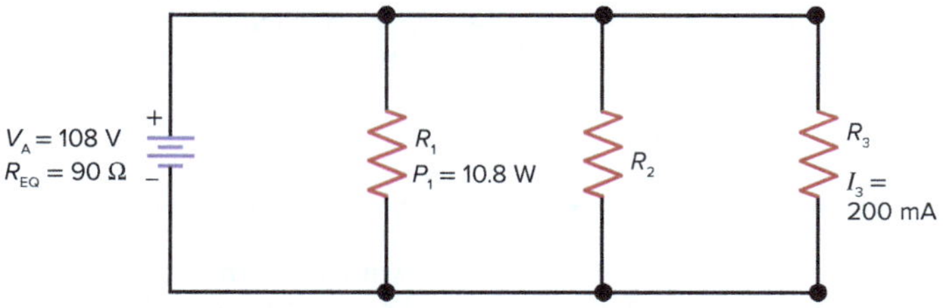

5–57 In Fig. 5–35, solve for I_T, I_1, I_2, I_4, R_3, R_4, P_1, P_2, P_3, P_4, and P_T.

Figure 5–35

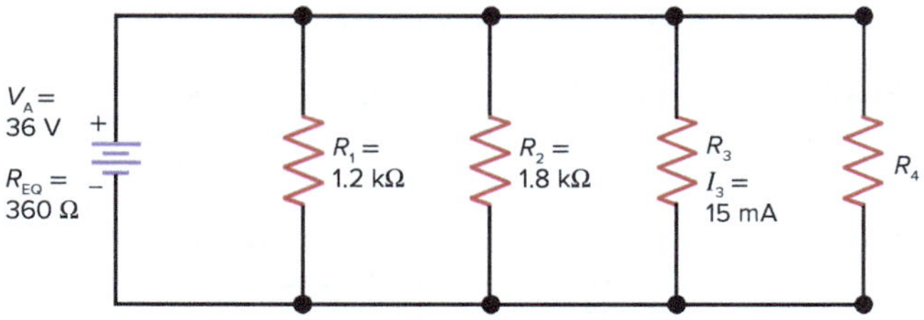

5–58 In Fig. 5–36, solve for V_A, I_1, I_2, R_2, R_3, I_4, and R_{EQ}.

Figure 5–36

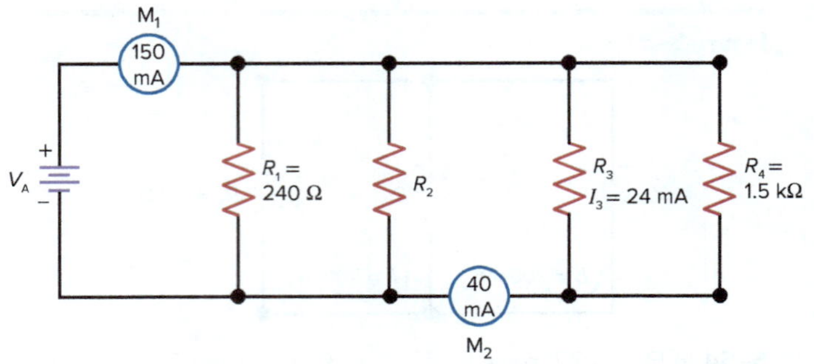

5–59 In Fig. 5–37, solve for V_A, I_1, I_2, I_4, R_1, R_3, and R_{EQ}.

Figure 5–37

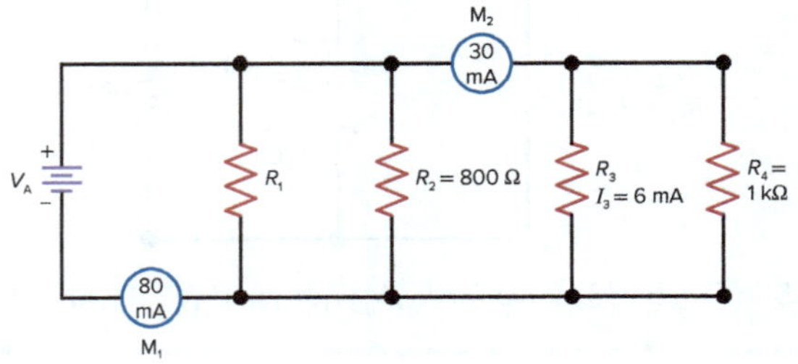

SECTION 5–8 TROUBLESHOOTING: OPENS AND SHORTS IN PARALLEL CIRCUITS

5–60 Figure 5–38 shows a parallel circuit with its normal operating voltages and currents. Notice that the fuse in the main line has a 25-A rating. What happens to the circuit components and their voltages and currents if

a. the appliance in Branch 3 shorts?

b. the motor in Branch 2 burns out and becomes an open?

c. the wire between points C and E develops an open?

d. the motor in Branch 2 develops a problem and begins drawing 16 A of current?

Figure 5–38

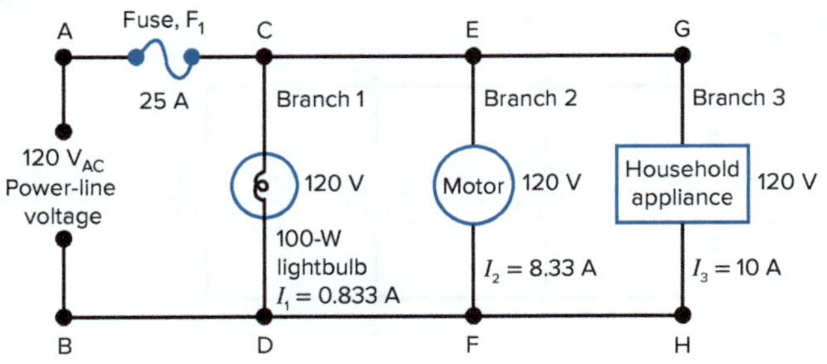

Critical Thinking

5–61 A 180-Ω, ¼-W resistor is in parallel with 1-kΩ, ½-W and 12-kΩ, 2-W resistors. What is the maximum total current, I_T, that this parallel combination can have before the wattage rating of any resistor is exceeded?

5–62 A 470-Ω, ⅛-W resistor is in parallel with 1-kΩ, ¼-W and 1.5-kΩ, ½-W resistors. What is the maximum voltage, V, that can be applied to this circuit without exceeding the wattage rating of any resistor?

5–63 Three resistors in parallel have a combined equivalent resistance R_{EQ} of 1 kΩ. If R_2 is twice the value of R_3 and three times the value of R_1, what are the values for R_1, R_2, and R_3?

5–64 Three resistors in parallel have a combined equivalent resistance R_{EQ} of 4 Ω. If the conductance, G_1, is one-fourth that of G_2 and one-fifth that of G_3, what are the values of R_1, R_2, and R_3?

5–65 A voltage source is connected in parallel across four resistors R_1, R_2, R_3, and R_4. The currents are labeled I_1, I_2, I_3, and I_4, respectively. If $I_2 = 2I_1$, $I_3 = 2I_2$, and $I_4 = 2I_3$, calculate the values for R_1, R_2, R_3, and R_4 if $R_{EQ} = 1$ kΩ.

Troubleshooting Challenge

Figure 5–39 shows a parallel circuit with its normal operating voltages and currents. Notice the placement of the meters M_1, M_2, and M_3 in the circuit. M_1 measures the total current I_T, M_2 measures the applied voltage V_A, and M_3 measures the current between points C and D. The following problems deal with troubleshooting the parallel circuit in Fig. 5–39.

5–66 If M_1 measures 2.8 A, M_2 measures 36 V, and M_3 measures 1.8 A, which component has most likely failed? How is the component defective?

5–67 If M_1 measures 2.5 A, M_2 measures 36 V, and M_3 measures 0 A, what is most likely wrong? How could you isolate the trouble by making voltage measurements?

5–68 If M_1 measures 3.3 A, M_2 measures 36 V, and M_3 measures 1.8 A, which component has most likely failed? How is the component defective?

5–69 If the fuse F_1 is blown, (a) How much current will be measured by M_1 and M_3? (b) How much voltage will be measured by M_2? (c) How much voltage will be measured across the blown fuse? (d) What is most likely to have caused the blown fuse? (e) Using resistance measurements, outline a procedure for finding the defective component.

5–70 If M_1 and M_3 measure 0 A but M_2 measures 36 V, what is most likely wrong? How could you isolate the trouble by making voltage measurements?

Figure 5–39 Circuit diagram for troubleshooting challenge. Normal values for I_1, I_2, I_3, and I_4 are shown on schematic.

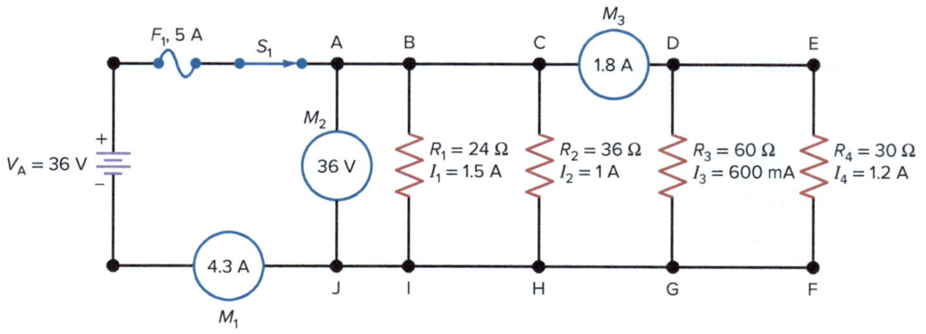

5–71 If the fuse F_1 has blown because of a shorted branch, how much resistance would be measured across points B and I? Without using resistance measurements, how could the shorted branch be identified?

5–72 If the wire connecting points F and G opens, (a) How much current will M_3 show? (b) How much voltage would be measured across R_4? (c) How much voltage would be measured across points D and E? (d) How much voltage would be measured across points F and G?

5–73 Assuming that the circuit is operating normally, how much voltage would be measured across, (a) the fuse F_1; (b) the switch S_1?

5–74 If the branch resistor R_3 opens, (a) How much voltage would be measured across R_3? (b) How much current would be indicated by M_1 and M_3?

5–75 If the wire between points B and C breaks open, (a) How much current will be measured by M_1 and M_3? (b) How much voltage would be measured across points B and C? (c) How much voltage will be measured across points C and H?

Answers to Self-Reviews

5–1 **a.** 1.5 V
b. 120 V
c. two each

5–2 **a.** 10 V
b. 1 A
c. 10 V
d. 2 A

5–3 **a.** 6 A
b. 3 A
c. 1.2 A

5–4 **a.** 1.57 MΩ
b. 1.2 MΩ
c. 5 Ω

5–5 **a.** 6 S
b. 0.75 μS, 1.33 MΩ
c. 0.25 MΩ

5–6 **a.** 480 W
b. 660 W
c. 30 W

5–7 **a.** 120 V
b. 4 A
c. 7 A

5–8 **a.** 120 V
b. 0 Ω
c. 6 A
d. 0 V, 120 V
e. 120 V

Laboratory Application Assignment

In this lab application assignment, you will examine the characteristics of a simple parallel circuit. You will also calculate and measure the equivalent resistance, R_{EQ}, of parallel connected resistors.

Equipment: Obtain the following items from your instructor.

- Variable dc power supply
- Assortment of carbon-film resistors
- DMM

Parallel Circuit Characteristics

Examine the parallel circuit in Fig. 5–40. Calculate and record the following values:

$I_1 =$ _____, $I_2 =$ _____, $I_3 =$ _____, $I_T =$ _____, $R_{EQ} =$ _____

Construct the parallel circuit in Fig. 5–40. Measure and record the following values. (Note that the power supply connections must be removed to measure R_{EQ}.)

$I_1 =$ _____, $I_2 =$ _____, $I_3 =$ _____, $I_T =$ _____, $R_{EQ} =$ _____

How does the ratio I_1/I_2 compare to the ratio R_2/R_1? __________

What is unique about comparing these ratios? __________

Add the measured branch currents I_1, I_2, and I_3. Record your answer. __________

How does this value compare to the measured value of I_T? __________

Does the sum of these individual branch currents satisfy KCL? __________

In Fig. 5–40, which branch resistance dissipates the most power? __________

Which branch resistance dissipates the least amount of power? __________

Parallel Circuit Analysis

In Fig. 5–40, add another 1.2 kΩ resistor, R_4, to the right of resistor R_3. Measure and record the total current, I_T. $I_T =$ _____ Next, calculate the equivalent resistance, R_{EQ}, using the equation $R_{EQ} = V_A/I_T$. $R_{EQ} =$ __________ Did the value of R_{EQ} increase or decrease from its original value? __________

Explain why. __________

Now remove both R_3 and R_4 from the circuit. Measure and record the total current, I_T. Recalculate the equivalent resistance, R_{EQ} as V_A/I_T. $R_{EQ} =$ _____ Did R_{EQ} increase or decrease from its original value? ________ Explain why. ______

Connect a 3.3 kΩ resistor, R_1, and a 330 Ω resistor, R_2, in parallel as shown in Fig. 5–41. Calculate and record the equivalent resistance, R_{EQ}, using equation 5–4 (product over the sum). $R_{EQ} =$ __________ Using a digital multimeter (DMM), measure and record the value of R_{EQ}. $R_{EQ} =$ __________

Connect another 150 Ω resistor, R_3, in parallel with R_1 and R_2 in Fig. 5–41. Calculate and record the equivalent resistance, R_{EQ}, using equation 5–3 (the reciprocal formula). $R_{EQ} =$ __________ Using a DMM, measure and record the value of R_{EQ}. $R_{EQ} =$ __________

Connect four 1 kΩ resistors in parallel as shown Fig. 5–42. Calculate and record the equivalent resistance, R_{EQ}. $R_{EQ} =$ __________ Using a DMM, measure and record the value of R_{EQ}. $R_{EQ} =$ __________

When calculating the value of R_{EQ} for equal resistances in parallel, which formula is the easiest to use? __________

Examine the circuit in Fig. 5–43. Determine what value of resistance, R_X, in parallel with a 1.2 kΩ R will provide an R_{EQ} of 720 Ω. (Use equation 5–5) $R_X =$ __________ Connect your calculated value of R_X across the 1.2 kΩ R in Fig. 5–43. Using a DMM, measure and record the value of R_{EQ}. $R_{EQ} =$ __________

Connect a 1 kΩ resistor, R_1, and a 10 Ω resistor, R_2, in parallel as in Fig. 5–44. Measure and record the value of R_{EQ}. $R_{EQ} =$ ________ Remove the 10 Ω R_2 from the circuit and replace it with a 100 kΩ resistor. Re-measure R_{EQ}. $R_{EQ} =$ __________. Explain the significance of the measurements in Fig. 5–44. __________

Figure 5–40

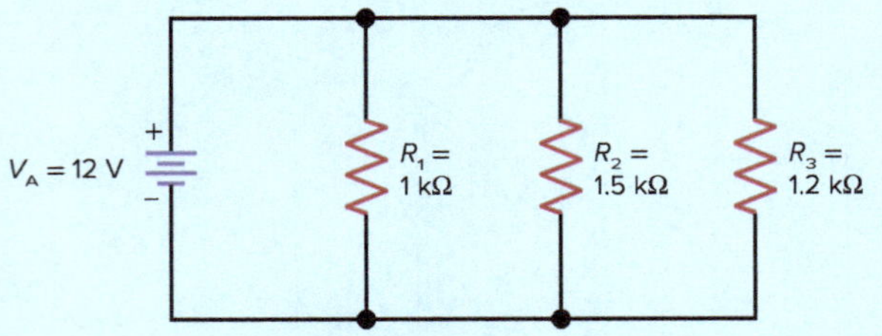

Figure 5–41

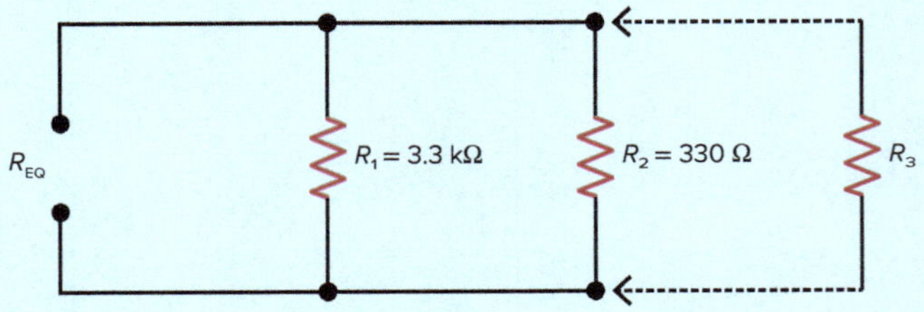

Figure 5–42

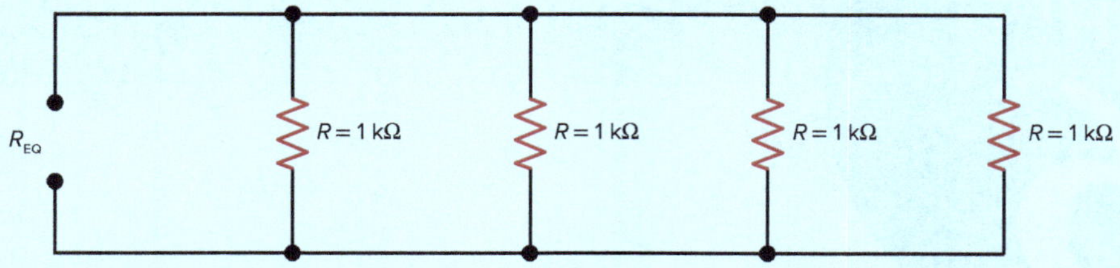

Figure 5–43

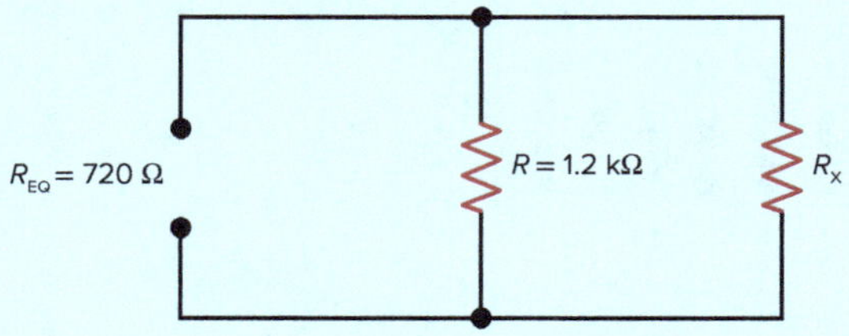

Figure 5–44

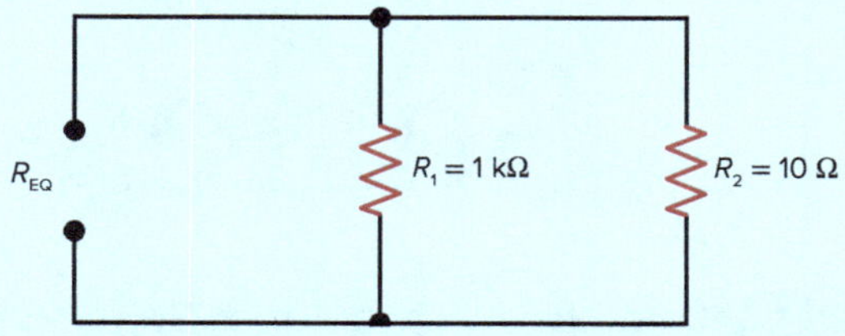

Design credits: Chapter Opener Background Image, and Multisim: Stockbyte/Getty Images

chapter

6

Series-Parallel Circuits

A series-parallel circuit, also called a combination circuit, is any circuit that combines both series and parallel connections. Although many applications exist for series or parallel circuits alone, most electronic circuits are actually a combination of the two. In general, series-parallel or combination circuits are used when it is necessary to obtain different voltage and current values from a single supply voltage, V_T. When analyzing combination circuits, the individual laws of series and parallel circuits can be applied to produce a much simpler overall circuit.

In this chapter, you will be presented with several different series-parallel combinations. For each type of combination circuit shown, you will learn how to solve for the unknown values of voltage, current, and resistance. You will also learn about a special circuit called the Wheatstone bridge. As you will see, this circuit has several very interesting applications in electronics. And finally, you will learn how to troubleshoot a series-parallel circuit containing both open and shorted components. ■

Chapter Outline

6–1 Finding R_T for Series-Parallel Resistances

6–2 Resistance Strings in Parallel

6–3 Resistance Banks in Series

6–4 Resistance Banks and Strings in Series-Parallel

6–5 Analyzing Series-Parallel Circuits with Random Unknowns

6–6 The Wheatstone Bridge

6–7 Troubleshooting: Opens and Shorts in Series-Parallel Circuits

Chapter Objectives

After studying this chapter, you should be able to

- *Determine* the total resistance of a series-parallel circuit.
- *Calculate* the voltage, current, resistance, and power in a series-parallel circuit.
- *Calculate* the voltage, current, resistance, and power in a series-parallel circuit having random unknowns.
- *Explain* how a Wheatstone bridge can be used to determine the value of an unknown resistor.
- *List* other applications of balanced bridge circuits.
- *Describe* the effects of opens and shorts in series-parallel circuits.
- *Troubleshoot* series-parallel circuits containing opens and shorts.

Important Terms

balanced bridge
banks in series
ratio arm
standard resistor
strings in parallel
Wheatstone bridge

6–1 Finding R_T for Series-Parallel Resistances

In Fig. 6–1, R_1 is in series with R_2. Also, R_3 is in parallel with R_4. However, R_2 is *not* in series with either R_3 or R_4. The reason is that the current through R_2 is equal to the sum of the branch currents I_3 and I_4 flowing into and away from point A (see Fig. 6–1*b*). As a result, the current through R_3 must be less than the current through R_2. Therefore, R_2 and R_3 cannot be in series because they do not have the same current. For the same reason, R_4 also cannot be in series with R_2. However, because the current in R_1 and R_2 is the same as the current flowing to and from the terminals of the voltage source, R_1, R_2, and V_T are in series.

The wiring is shown in Fig. 6–1*a* and the schematic diagram in Fig. 6–1*b*. To find R_T, we add the series resistances and combine the parallel resistances.

GOOD TO KNOW

Most electronic circuitry consists of a combination of series and parallel connections.

In Fig. 6–1*c*, the 0.5-kΩ R_1 and 0.5-kΩ R_2 in series total 1 kΩ for R_{1-2}. The calculations are

$$0.5\ \text{k}\Omega + 0.5\ \text{k}\Omega = 1\ \text{k}\Omega$$

Also, the 1-kΩ R_3 in parallel with the 1-kΩ R_4 can be combined, for an equivalent resistance of 0.5 kΩ for R_{3-4}, as in Fig. 6–1*d*. The calculations are

$$\frac{1\ \text{k}\Omega}{2} = 0.5\ \text{k}\Omega$$

MultiSim **Figure 6–1** Example of a series-parallel circuit. (*a*) Wiring of a series-parallel circuit. (*b*) Schematic diagram of a series-parallel circuit. (*c*) Schematic with R_1 and R_2 in series added for R_{1-2}. (*d*) Schematic with R_3 and R_4 in parallel combined for R_{3-4}. (*e*) Axial-lead resistors assembled on a lab prototype board to form the series-parallel circuit shown in part *c*.

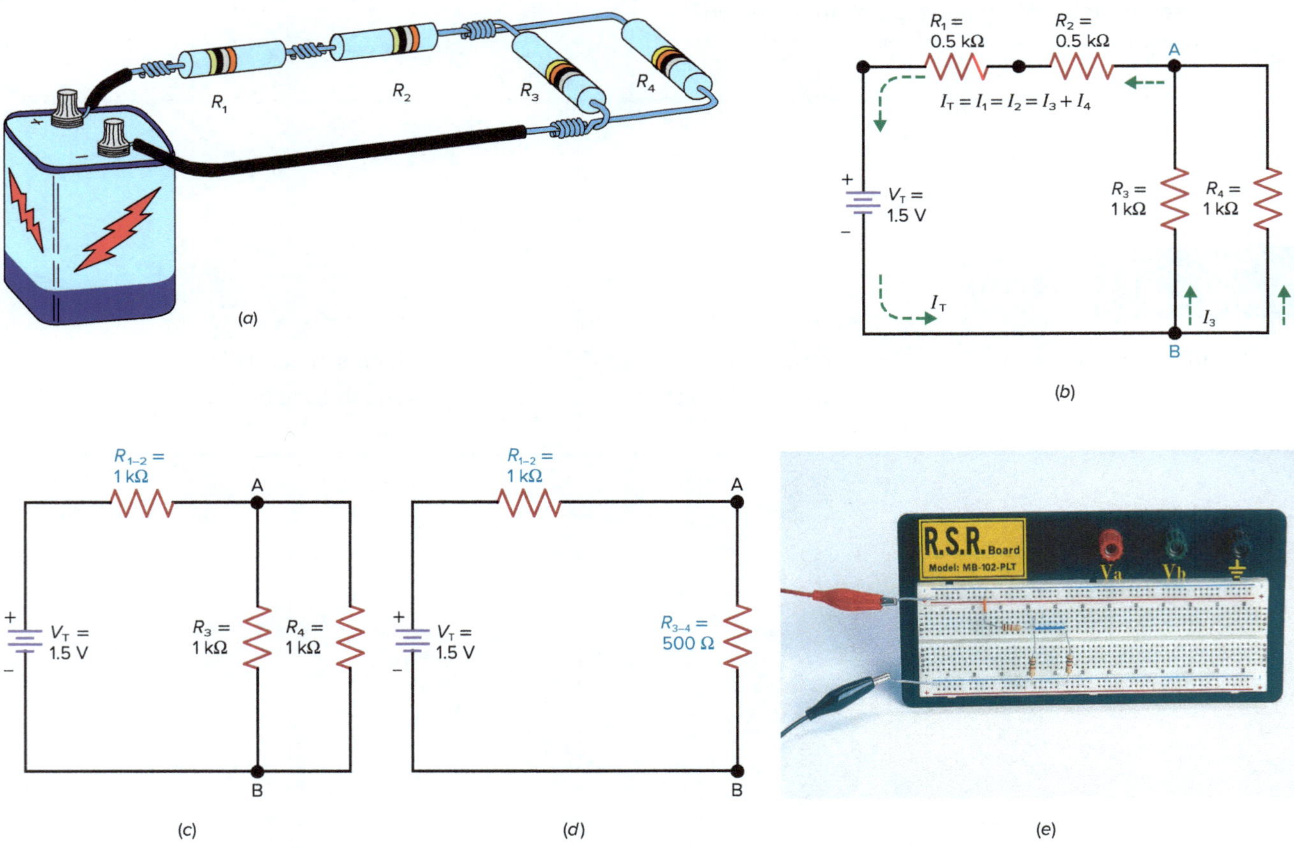

Cindy Schroeder/McGraw Hill

This parallel R_{3-4} combination of 0.5 kΩ is then added to the series R_{1-2} combination for the final R_T value of 1.5 kΩ. The calculations are

$$0.5\text{ k}\Omega + 1\text{ k}\Omega = 1.5\text{ k}\Omega$$

The 1.5 kΩ is the R_T of the entire circuit connected across the V_T of 1.5 V.

With R_T known to be 1.5 kΩ, we can find I_T in the main line produced by 1.5 V. Then

$$I_T = \frac{V_T}{R_T} = \frac{1.5\text{ V}}{1.5\text{ k}\Omega} = 1\text{ mA}$$

This 1-mA I_T is the current through resistors R_1 and R_2 in Fig. 6–1*a* and *b* or R_{1-2} in Fig. 6–1*c*.

At branch point B, at the bottom of the diagram in Fig. 6–1*b*, the 1 mA of electron flow for I_T divides into two branch currents for R_3 and R_4. Since these two branch resistances are equal, I_T divides into two equal parts of 0.5 mA each. At branch point A at the top of the diagram, the two 0.5-mA branch currents combine to equal the 1-mA I_T in the main line, returning to the source V_T.

Figure 6–1*e* shows axial-lead resistors assembled on a lab prototype board to form the series-parallel circuit shown in part *c*.

6–1 Self-Review

Answers at the end of the chapter.

Refer to Fig. 6–1*b*.

a. Calculate the series *R* of R_1 and R_2.

b. Calculate the parallel *R* of R_3 and R_4.

c. Calculate R_T across the source V_T.

6–2 Resistance Strings in Parallel

More details about the voltages and currents in a series-parallel circuit are illustrated in Fig. 6–2, which shows two identical series **strings in parallel.** Suppose that four 120-V, 100-W lightbulbs are to be wired with a voltage source that produces 240 V. Each bulb needs 120 V for normal brilliance. If the bulbs were connected directly across the source, each would have the applied voltage of 240 V. This would cause excessive current in all the bulbs that could result in burned-out filaments.

If the four bulbs were connected in series, each would have a potential difference of 60 V, or one-fourth the applied voltage. With too low a voltage, there would be insufficient current for normal operation, and the bulbs would not operate at normal brilliance.

Figure 6–2 Two identical series strings in parallel. All bulbs have a 120-V, 100-W rating. (*a*) Wiring diagram. (*b*) Schematic diagram.

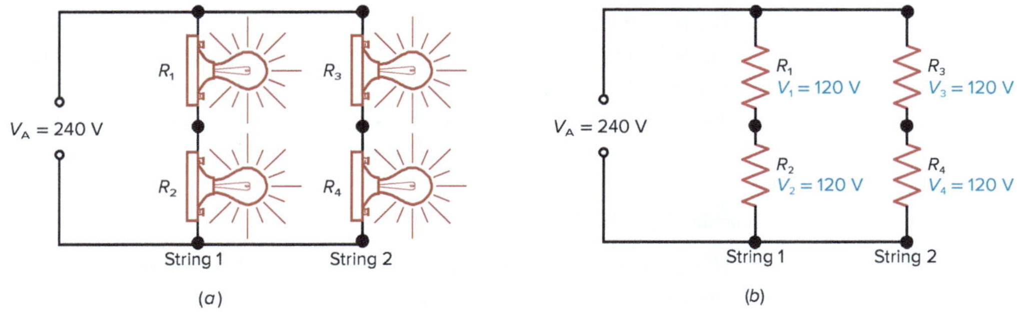

Figure 6–3 Series string in parallel with another branch. (*a*) Schematic diagram. (*b*) Equivalent circuit.

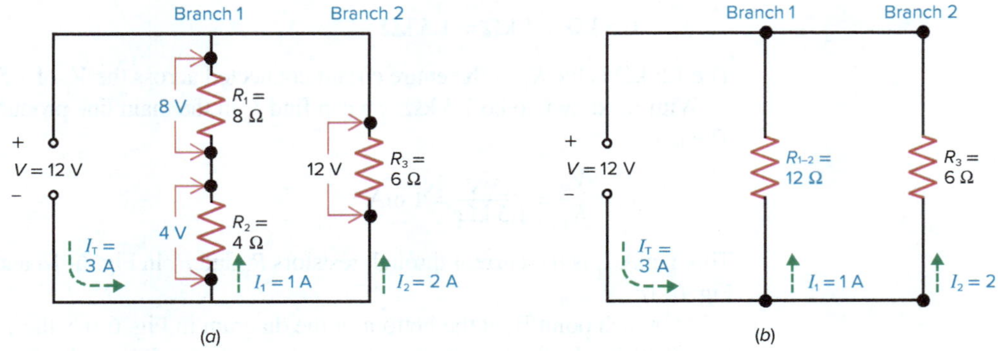

GOOD TO KNOW

When a parallel branch contains series resistors, both resistors have the same current but the individual resistor voltage drops will be less than the voltage applied across the entire branch. The individual resistor voltage drops add, however, to equal the voltage applied across the branch.

However, two bulbs in series across the 240-V line provide 120 V for each filament, which is the normal operating voltage. Therefore, the four bulbs are wired in strings of two in series, with the two strings in parallel across the 240-V source. Both strings have 240 V applied. In each string, two series bulbs divide the 240 V equally to provide the required 120 V for normal operation.

Another example is illustrated in Fig. 6–3. This circuit has just two parallel branches. One branch includes R_1 in series with R_2. The other branch has just the one resistance R_3. Ohm's law can be applied to each branch.

Branch Currents I_1 and I_2

In Fig. 6–3*a*, each branch current equals the voltage applied across the branch divided by the total resistance in the branch. In branch 1, R_1 and R_2 total $8 + 4 = 12\ \Omega$. With 12 V applied, this branch current I_1 is $12/12 = 1$ A. Branch 2 has only the 6-Ω R_3. Then I_2 in this branch is $12/6 = 2$ A.

Series Voltage Drops in a Branch

For any one resistance in a string, the current in the string multiplied by the resistance equals the *IR* voltage drop across that particular resistance. Also, the sum of the series *IR* drops in the string equals the voltage across the entire string.

Branch 1 is a string with R_1 and R_2 in series. The I_1R_1 drop equals 8 V, whereas the I_1R_2 drop is 4 V. These drops of 8 and 4 V add to equal the 12 V applied. The voltage across the R_3 branch is also the same 12 V.

Calculating I_T

The total line current equals the sum of the branch currents for all parallel strings. Here I_T is 3 A, equal to the sum of 1 A in branch 1 and 2 A in branch 2.

Calculating R_T

The resistance of the total series-parallel circuit across the voltage source equals the applied voltage divided by the total line current. In Fig. 6–3*a*, $R_T = 12\ \text{V}/3\ \text{A}$, or 4 Ω. This resistance can also be calculated as 12 Ω in parallel with 6 Ω. Figure 6–3*b* shows the equivalent circuit. Using the product divided by the sum formula, $72/18 = 4\ \Omega$ for the equivalent combined R_T.

Applying Ohm's Law

There can be any number of parallel strings and more than two series resistances in a string. Still, Ohm's law can be used in the same way for the series and parallel parts of the circuit. The series parts have the same current. The parallel parts have the same voltage. Remember that for V/R the R must include all the resistance across the two terminals of V.

6–2 Self-Review

Answers at the end of the chapter.

Refer to Fig. 6–3*a*.

a. **How much is the voltage across R_3?**
b. **If I in R_2 were 6 A, what would I in R_1 be?**
c. **If the source voltage were 18 V, what would V_3 be across R_3?**

6–3 Resistance Banks in Series

GOOD TO KNOW

When a parallel bank exists in a series path, both resistors have the same voltage but the individual branch currents are less than the series current. The branch currents add, however, to equal the series current entering and leaving the parallel bank.

In Fig. 6–4*a*, the group of parallel resistances R_2 and R_3 is a bank. This is in series with R_1 because the total current of the bank must go through R_1.

The circuit here has R_2 and R_3 in parallel in one bank so that these two resistances will have the same potential difference of 20 V across them. The source applies 24 V, but there is a 4-V drop across R_1.

The two series voltage drops of 4 V across R_1 and 20 V across the bank add to equal the applied voltage of 24 V. The purpose of a circuit like this is to provide the same voltage for two or more resistances in a bank, where the bank voltage must be less than the applied voltage by the amount of the IR drop across any series resistance.

To find the resistance of the entire circuit, combine the parallel resistances in each bank and add the series resistance. As shown in Fig. 6–4*b*, the two 10-Ω resistances, R_2 and R_3 in parallel, are equivalent to 5 Ω. Since the bank resistance of 5 Ω is in series with 1 Ω for R_1, the total resistance is 6 Ω across the 24-V source. Therefore, the main-line current is 24 V/6 Ω, which equals 4 A.

The total line current of 4 A divides into two parts of 2 A each in the parallel resistances R_2 and R_3. Note that each branch current equals the bank voltage divided by the branch resistance. For this bank, $20/10 = 2$ A for each branch.

The branch currents, I_2 and I_3, are combined in the main line to provide the total 4 A in R_1. This is the same total current flowing in the main line, in the source, into the bank, and out of the bank.

There can be more than two parallel resistances in a bank and any number of **banks in series.** Still, Ohm's law can be applied in the same way to the series and parallel parts of the circuit. The general procedure for circuits of this type is to find the equivalent resistance of each bank and then add all series resistances.

Figure 6–4 Parallel bank of R_2 and R_3 in series with R_1. (*a*) Original circuit. (*b*) Equivalent circuit.

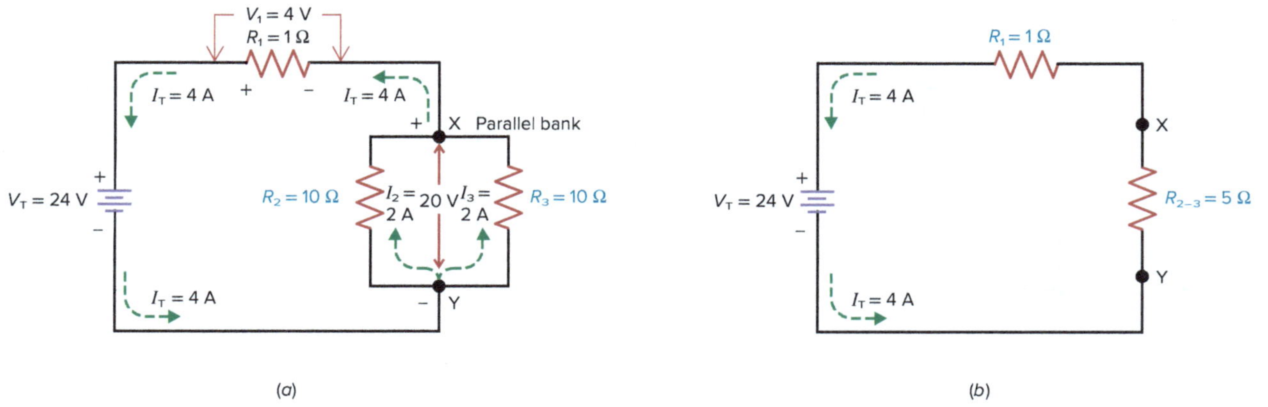

6–3 Self-Review

Answers at the end of the chapter.

Refer to Fig. 6–4*a*.

a. **If V_2 across R_2 were 40 V, what would V_3 across R_3 be?**

b. **If I in R_2 were 4 A, with 4 A in R_3, what would I in R_1 be?**

c. **How much is V_1 across R_1 in Fig. 6–4*b*?**

6–4 Resistance Banks and Strings in Series-Parallel

In the solution of such circuits, the most important fact to know is which components are in series with each other and which parts of the circuit are parallel branches. The series components must be in one current path without any branch points. A branch point such as point A or B in Fig. 6–5 is common to two or more current paths. For instance, R_1 and R_6 are *not* in series with each other. They do not have the same current because the current through R_1 equals the sum of the branch currents, I_5 and I_6, flowing into and away from point A. Similarly, R_5 is not in series with R_2 because of the branch point B.

To find the currents and voltages in Fig. 6–5, first find R_T to calculate the main-line current I_T as V_T/R_T. In calculating R_T, start reducing the branch farthest from the source and work toward the applied voltage. The reason for following this order is that you cannot tell how much resistance is in series with R_1 and R_2 until the parallel branches are reduced to their equivalent resistance. If no source voltage is shown, R_T can still be calculated from the outside in toward the open terminals where a source would be connected.

MultiSim **Figure 6–5** Reducing a series-parallel circuit to an equivalent series circuit to find the R_T. (*a*) Actual circuit. (*b*) R_3 and R_4 in parallel combined for the equivalent R_7. (*c*) R_7 and R_6 in series added for R_{13}. (*d*) R_{13} and R_5 in parallel combined for R_{18}. (*e*) The R_{18}, R_1, and R_2 in series are added for the total resistance of 50 Ω for R_T.

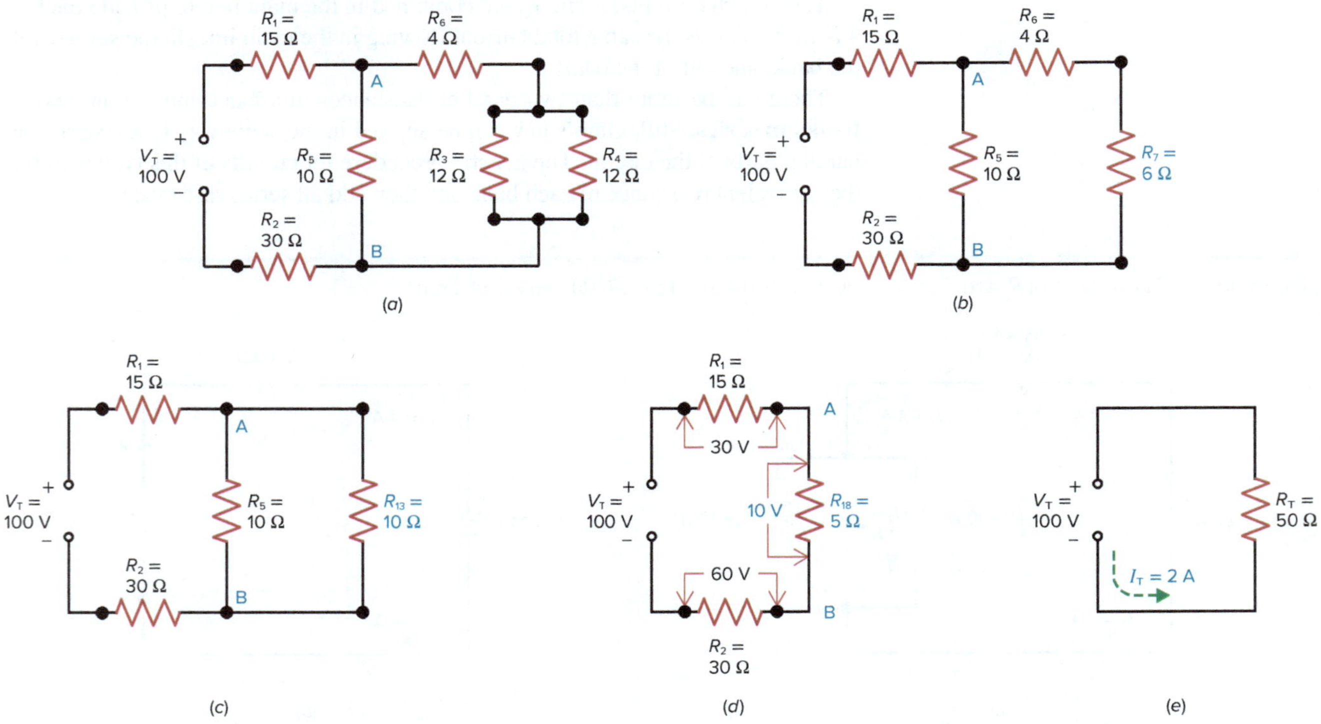

To calculate R_T in Fig. 6–5, the steps are as follows:

1. The bank of the 12-Ω R_3 and 12-Ω R_4 in parallel in Fig. 6–5*a* is equal to the 6-Ω R_7 in Fig. 6–5*b*.
2. The 6-Ω R_7 and 4-Ω R_6 in series in the same current path total 10 Ω for R_{13} in Fig. 6–5*c*.
3. The 10-Ω R_{13} is in parallel with the 10-Ω R_5, across the branch points A and B. Their equivalent resistance, then, is the 5-Ω R_{18} in Fig. 6–5*d*.
4. Now the circuit in Fig. 6–5*d* has just the 15-Ω R_1, 5-Ω R_{18}, and 30-Ω R_2 in series. These resistances total 50 Ω for R_T, as shown in Fig. 6–5*e*.
5. With a 50-Ω R_T across the 100-V source, the line current I_T is equal to 100/50 = 2 A.

GOOD TO KNOW

An equation for the total resistance, R_T, in Fig. 6–5a would be:

$$R_T = R_1 + [R_5 \parallel (R_6 + R_3 \parallel R_4)] + R_2$$

The vertical bars (||) represent a parallel connection.

To see the individual currents and voltages, we can use the I_T of 2 A for the equivalent circuit in Fig. 6–5*d*. Now we work from the source *V* out toward the branches. The reason is that I_T can be used to find the voltage drops in the main line. The *IR* voltage drops here are

$$V_1 = I_T R_1 = 2 \times 15 = 30 \text{ V}$$
$$V_{18} = I_T R_{18} = 2 \times 5 = 10 \text{ V}$$
$$V_2 = I_T R_2 = 2 \times 30 = 60 \text{ V}$$

The 10-V drop across R_{18} is actually the potential difference between branch points A and B. This means 10 V across R_5 and R_{13} in Fig. 6–5*c*. The 10 V produces 1 A in the 10-Ω R_5 branch. The same 10 V is also across the R_{13} branch.

Remember that the R_{13} branch is actually the string of R_6 in series with the R_3–R_4 bank. Since this branch resistance is 10 Ω with 10 V across it, the branch current here is 1 A. The 1 A through the 4 Ω of R_6 produces a voltage drop of 4 V. The remaining 6-V *IR* drop is across the R_3–R_4 bank. With 6 V across the 12-Ω R_3, its current is ½ A; the current is also ½ A in R_4.

Tracing all the current paths from the voltage source in Fig. 6–5*a*, the main-line current, I_T, through R_1 and R_2 is 2 A. The 2-A I_T flowing into point B subdivides into two separate branch currents: 1 A of the 2-A I_T flows up through resistor, R_5. The other 1 A flows into the branch containing resistors R_3, R_4, and R_6. Because resistors R_3 and R_4 are in parallel, the 1-A branch current subdivides further into ½ A for I_3 and ½ A for I_4. The currents I_3 and I_4 recombine to flow up through resistor R_6. At the branch point A, I_5 and I_6 combine resulting in the 2-A total current, I_T, flowing through R_1 back to the positive terminal of the voltage source.

6–4 Self-Review

Answers at the end of the chapter.

Refer to Fig. 6–5*a*.

a. **Which *R* is in series with R_2?**
b. **Which *R* is in parallel with R_3?**
c. **Which *R* is in series with the R_3 R_4 bank?**

6–5 Analyzing Series-Parallel Circuits with Random Unknowns

The circuits in Figs. 6–6 to 6–9 will be solved now. The following principles are illustrated:

1. With parallel strings across the main line, the branch currents and I_T can be found without R_T (see Figs. 6–6 and 6–7).
2. When parallel strings have series resistance in the main line, R_T must be calculated to find I_T, assuming no branch currents are known (see Fig. 6–9).

Figure 6–6 Finding all currents and voltages by calculating the branch currents first. See text for solution.

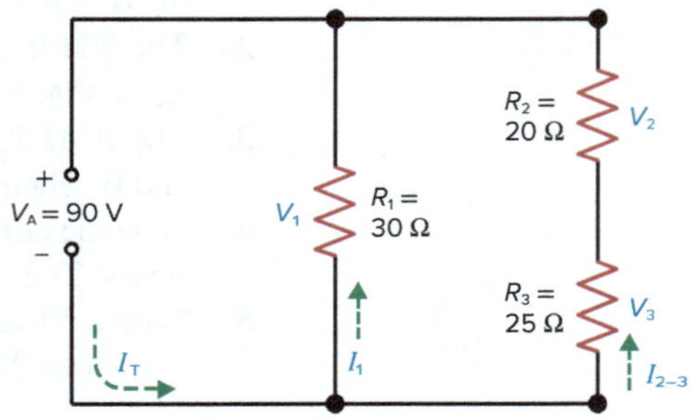

3. The source voltage is applied across the R_T of the entire circuit, producing an I_T that flows only in the main line.
4. Any individual series R has its own IR drop that must be less than the total V_T. In addition, any individual branch current must be less than I_T.

Solution for Figure 6–6

The problem here is to calculate the branch currents I_1 and I_{2-3}, total line current I_T, and the voltage drops V_1, V_2, and V_3. This order will be used for the calculations because we can find the branch currents from the 90 V across the known branch resistances.

In the 30-Ω branch of R_1, the branch current is $90/30 = 3$ A for I_1. The other branch resistance, with a 20-Ω R_2 and a 25-Ω R_3, totals 45 Ω. This branch current then is $90/45 = 2$ A for I_{2-3}. In the main line, I_T is 3 A + 2 A, which is equal to 5 A.

For the branch voltages, V_1 must be the same as V_A, equal to 90 V, or $V_1 = I_1 R_1$, which is $3 \times 30 = 90$ V.

In the other branch, the 2-A I_{2-3} flows through the 20-Ω R_2 and the 25-Ω R_3. Therefore, V_2 is $2 \times 20 = 40$ V. Also, V_3 is $2 \times 25 = 50$ V. Note that these 40-V and 50-V series IR drops in one branch add to equal the 90-V source.

If we want to know R_T, it can be calculated as V_A/I_T. Then 90 V/5 A equals 18 Ω. Or R_T can be calculated by combining the branch resistances of 30 Ω in parallel with 45 Ω. Then, using the product-divided-by-sum formula, R_T is $(30 \times 45)/(30 + 45)$ or 1350/75, which equals the same value of 18 Ω for R_T.

Solution for Figure 6–7

To find the applied voltage first, the I_1 branch current is given. This 3-A current through the 10-Ω R_1 produces a 30-V drop V_1 across R_1. The same 3-A current through the 20-Ω R_2 produces 60 V for V_2 across R_2. The 30-V and 60-V drops are in series

Figure 6–7 Finding the applied voltage V_A and then V_4 and R_4 from I_2 and the branch voltages. See text for calculations.

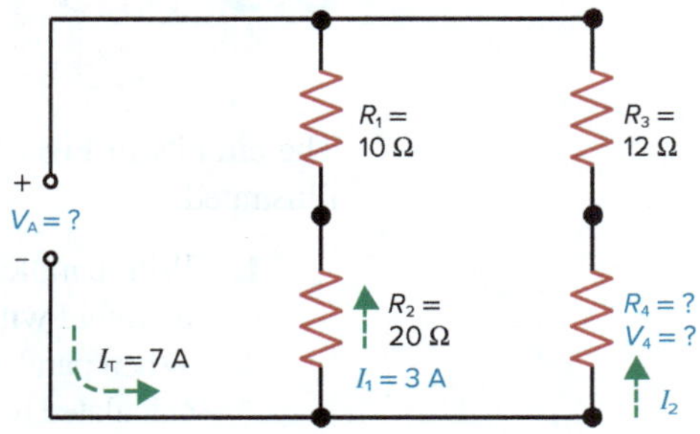

with each other across the applied voltage. Therefore, V_A equals the sum of 30 + 60, or 90 V. This 90 V is also across the other branch combining R_3 and R_4 in series.

The other branch current I_2 in Fig. 6–7 must be 4 A, equal to the 7-A I_T minus the 3-A I_1. With 4 A for I_2, the voltage drop across the 12-Ω R_3 equals 48 V for V_3. Then the voltage across R_4 is 90 − 48, or 42 V for V_4, as the sum of V_3 and V_4 must equal the applied 90 V.

Finally, with 42 V across R_4 and 4 A through it, this resistance equals 42/4, or 10.5 Ω. Note that 10.5 Ω for R_4 added to the 12 Ω of R_3 equals 22.5 Ω, which allows 90/22.5 or a 4-A branch current for I_2.

Figure 6–8 Finding R_2 in the parallel bank and its I_2. See text for solution.

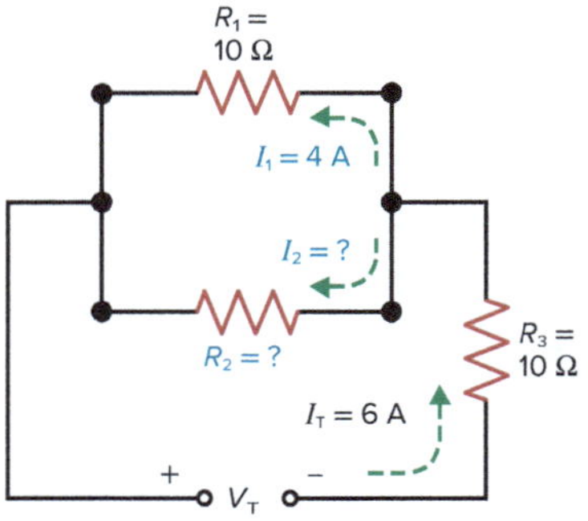

Solution for Figure 6–8

The division of branch currents also applies to Fig. 6–8, but the main principle here is that the voltage must be the same across R_1 and R_2 in parallel. For the branch currents, I_2 is 2 A, equal to the 6-A I_T minus the 4-A I_1. The voltage across the 10-Ω R_1 is 4 × 10, or 40 V. This same voltage is also across R_2. With 40 V across R_2 and 2 A through it, R_2 equals 40/2 or 20 Ω.

We can also find V_T in Fig. 6–8 from R_1, R_2, and R_3. The 6-A I_T through R_3 produces a voltage drop of 60 V for V_3. Also, the voltage across the parallel bank with R_1 and R_2 has been calculated as 40 V. This 40 V across the bank in series with 60 V across R_3 totals 100 V for the applied voltage.

Solution for Figure 6–9

To find all currents and voltage drops, we need R_T to calculate I_T through R_6 in the main line. Combining resistances for R_T, we start with R_1 and R_2 and work in toward the source. Add the 8-Ω R_1 and 8-Ω R_2 in series with each other for 16 Ω. This 16 Ω combined with the 16-Ω R_3 in parallel equals 8 Ω between points C and D. Add this 8 Ω to the series 12-Ω R_4 for 20 Ω. This 20 Ω with the parallel 20-Ω R_5 equals 10 Ω between points A and B. Add this 10 Ω in series with the 10-Ω R_6, to make R_T of 20 Ω for the entire series-parallel circuit.

Current I_T in the main line is V_T/R_T, or 80/20, which equals 4 A. This 4-A I_T flows through the 10-Ω R_6, producing a 40-V *IR* drop for V_6.

Now that we know I_T and V_6 in the main line, we use these values to calculate all other voltages and currents. Start from the main line, where we know the current, and work outward from the source. To find V_5, the *IR* drop of 40 V for V_6 in the main line is subtracted from the source voltage. The reason is that V_5 and V_6 must add to equal the 80 V of V_T. Then V_5 is 80 − 40 = 40 V.

Voltages V_5 and V_6 happen to be equal at 40 V each. They split the 80 V in half because the 10-Ω R_6 equals the combined resistance of 10 Ω between branch points A and B.

With V_5 known to be 40 V, then I_5 through the 20-Ω R_5 is 40/20 = 2 A. Since I_5 is 2 A and I_T is 4 A, I_4 must be 2 A also, equal to the difference between I_T and I_5. The current flowing into point A equals the sum of the branch currents I_4 and I_5.

Figure 6–9 Finding all currents and voltages by calculating R_T and then I_T to find V_6 across R_6 in the main line.

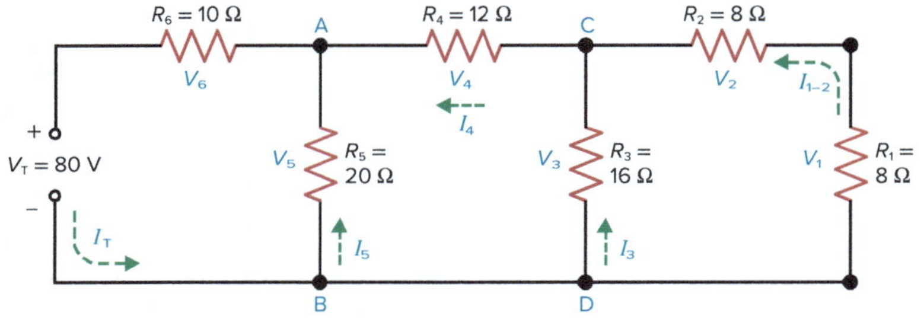

The 2-A I_4 through the 12-Ω R_4 produces an IR drop equal to 2 × 12 = 24 V for V_4. Note now that V_4 and V_3 must add to equal V_5. The reason is that both V_5 and the path with V_4 and V_3 are across the same two points AB or AD. Since the potential difference across any two points is the same regardless of the path, $V_5 = V_4 + V_3$. To find V_3 now, we can subtract the 24 V of V_4 from the 40 V of V_5. Then 40 − 24 = 16 V for V_3.

With 16 V for V_3 across the 16-Ω R_3, its current I_3 is 1 A. Also, I_{1-2} in the branch with R_1 and R_2 is equal to 1 A. The 2-A I_4 consists of the sum of the branch currents, I_3 and I_{1-2}, flowing into point C.

Finally, with 1A through the 8-Ω R_2 and 8-Ω R_1, their voltage drops are $V_2 = 8$ V and $V_1 = 8$ V. Note that the 8 V of V_1 in series with the 8 V of V_2 add to equal the 16-V potential difference V_3 between points C and D.

All answers for the solution of Fig. 6–9 are summarized below:

$R_T = 20\ \Omega$	$I_T = 4$ A	$V_6 = 40$ V
$V_5 = 40$ V	$I_5 = 2$ A	$I_4 = 2$ A
$V_4 = 24$ V	$V_3 = 16$ V	$I_3 = 1$ A
$I_{1-2} = 1$ A	$V_2 = 8$ V	$V_1 = 8$ V

6–5 Self-Review

Answers at the end of the chapter.

a. In Fig. 6–6, which *R* is in series with R_2?
b. In Fig. 6–6, which *R* is across V_A?
c. In Fig. 6–7, how much is I_2?
d. In Fig. 6–8, how much is V_3?

6–6 The Wheatstone Bridge

GOOD TO KNOW

With the availability of accurate, high-resolution DMMs, it is no longer practical to use a Wheatstone bridge to determine the value of an unknown resistor. However, the concepts of the Wheatstone bridge are covered here as a basic building block for balanced bridge circuits in general, which do find wide-spread use in electronics.

A **Wheatstone* bridge** is a circuit that is used to determine the value of an unknown resistance. A typical Wheatstone bridge is shown in Fig. 6–10. Notice that four resistors are configured in a diamond-like arrangement, which is typically how the Wheatstone bridge is drawn. In Fig. 6–10, the applied voltage V_T is connected to terminals A and B, which are considered the input terminals to the Wheatstone bridge. A very sensitive zero-centered current meter M_1, called a *galvonometer*, is connected between terminals C and D, which are considered the output terminals.

As shown in Fig. 6–10, the unknown resistor R_X is placed in the same branch as a variable **standard resistor** R_S. It is important to note that the standard resistor R_S is a precision resistance variable from 0–9999 Ω in 1-Ω steps. In the other branch, resistors R_1 and R_2 make up what is known as the **ratio arm.** Resistors R_1 and R_2 are also precision resistors having very tight resistance tolerances. To determine the value of an unknown resistance R_X, adjust the standard resistor R_S until the current in M_1 reads exactly 0 μA. With zero current in M_1 the Wheatstone bridge is said to be balanced.

Figure 6–10 Wheatstone bridge.

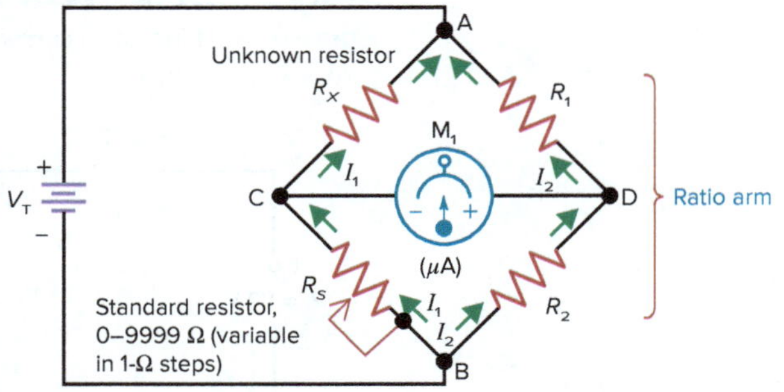

* Sir Charles Wheatstone (1802–1875), English physicist and inventor.

But how does the balanced condition provide the value of the unknown resistance R_X? Good question. With zero current in M_1, the voltage division among resistors R_X and R_S is identical to the voltage division among the ratio arm resistors R_1 and R_2. When the voltage division in the R_X–R_S branch is identical to the voltage division in the R_1–R_2 branch, the potential difference between points C and D will equal 0 V. With a potential difference of 0 V across points C and D, the current in M_1 will read 0 μA, which is the balanced condition. At balance, the equal voltage ratios can be stated as

$$\frac{I_1R_X}{I_1R_S} = \frac{I_2R_1}{I_2R_2}$$

Since I_1 and I_2 cancel in the equation, this yields

$$\frac{R_X}{R_S} = \frac{R_1}{R_2}$$

Solving for R_X gives us

$$R_X = R_S \times \frac{R_1}{R_2} \quad \textbf{(6–1)}$$

GOOD TO KNOW

A proportion states that two ratios are equal. As shown, the ratio of R_1 to R_2 is equal to the ratio of R_X to R_S when the bridge is balanced.

The ratio arm R_1/R_2 can be varied in most cases, typically in multiples of 10, such as 100/1, 10/1, 1/1, 1/10, and 1/100. However, the bridge is still balanced by varying the standard resistor R_S. The placement accuracy of the measurement of R_X is determined by the R_1/R_2 ratio. For example, if $R_1/R_2 = 1/10$, the value of R_X is accurate to within ±0.1 Ω. Likewise, if $R_1/R_2 = 1/100$, the value of R_X will be accurate to within ±0.01 Ω. The R_1/R_2 ratio also determines the maximum unknown resistance that can be measured. Expressed as an equation.

$$R_{X(\max)} = R_{S(\max)} \times \frac{R_1}{R_2} \quad \textbf{(6–2)}$$

Example 6–1

In Fig. 6–11, the current in M_1 reads 0 μA with the standard resistor R_S adjusted to 5642 Ω. What is the value of the unknown resistor R_X?

ANSWER Using Formula (6–1), R_X is calculated as follows:

$$R_X = R_S \times \frac{R_1}{R_2}$$

$$= 5642\ \Omega \times \frac{1\ \text{k}\Omega}{10\ \text{k}\Omega}$$

$$R_X = 564.2\ \Omega$$

MultiSim **Figure 6–11** Wheatstone bridge. See Examples 6–1 and 6–2.

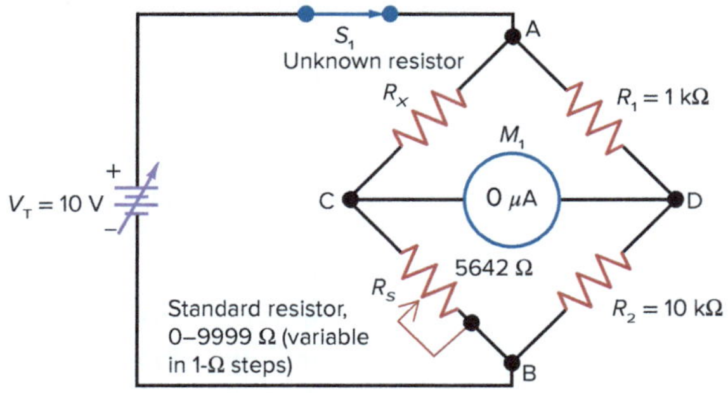

Example 6–2

In Fig. 6–11, what is the maximum unknown resistance R_X that can be measured for the ratio arm values shown?

ANSWER

$$R_{X(\text{max})} = R_{S(\text{max})} \times \frac{R_1}{R_2}$$

$$= 9999\ \Omega \times \frac{1\ \text{k}\Omega}{10\ \text{k}\Omega}$$

$$R_{X(\text{max})} = 999.9\ \Omega$$

If R_X is larger than 999.9 Ω, the bridge cannot be balanced because the voltage division will be greater than ⅒ in this branch. In other words, the current in M_1 cannot be adjusted to 0 μA. To measure an unknown resistance whose value is greater than 999.9 Ω, you would need to change the ratio arm fraction to $^1/_1$, $^{10}/_1$, or something higher.

Note that when the Wheatstone bridge is balanced, it can be analyzed simply as two series strings in parallel. The reason is that when the current through M_1 is zero, the path between points C and D is effectively open. When current flows through M_1, however, the bridge circuit must be analyzed by other methods described in Chaps. 9 and 10.

Other Balanced Bridge Applications

There are many other applications in electronics for **balanced bridge** circuits. For example, a variety of sensors are used in bridge circuits for detecting changes in pressure, flow, light, temperature, and so on. These sensors are used as one of the resistors in a bridge circuit. Furthermore, the bridge can be balanced or zeroed at some desired reference level of pressure, flow, light, or temperature. Then, when the condition being sensed changes, the bridge becomes unbalanced and causes a voltage to appear at the output terminals (C and D). This output voltage is then fed to the input of an amplifier or other device that modifies the condition being monitored, thus bringing the system back to its original preset level.

Consider the temperature control circuit in Fig. 6–12. In this circuit, a variable resistor R_3 is in the same branch as a negative temperature coefficient (NTC) thermistor whose resistance value at 25°C (R_0) equals 5 kΩ as shown. Assume that R_3 is adjusted to provide balance when the ambient (surrounding) temperature T_A equals 25°C. Remember, when the bridge is balanced, the output voltage across terminals C and D is 0 V. This voltage is fed to the input of an amplifier as shown. With 0 V into the amplifier, 0 V comes out of the amplifier.

Now let's consider what happens when the ambient temperature T_A increases above 25°C, say to 30°C. The increase in temperature causes the resistance of the thermistor to decrease, since it has an NTC. With a decrease in the thermistor's resistance, the voltage at point C decreases. However, the voltage at point D does not change because R_1 and R_2 are ordinary resistors. The result is that the output voltage V_{CD} goes negative. This negative voltage is fed into the amplifier, which in turn produces a positive output voltage. The positive output voltage from the amplifier turns on a cooling fan or air-conditioning unit. The air-conditioning unit remains on until the ambient temperature decreases to its original value of 25°C. As the temperature drops back to 25°C, the resistance of the thermistor increases to its

Figure 6–12 Temperature control circuit using the balanced bridge concept.

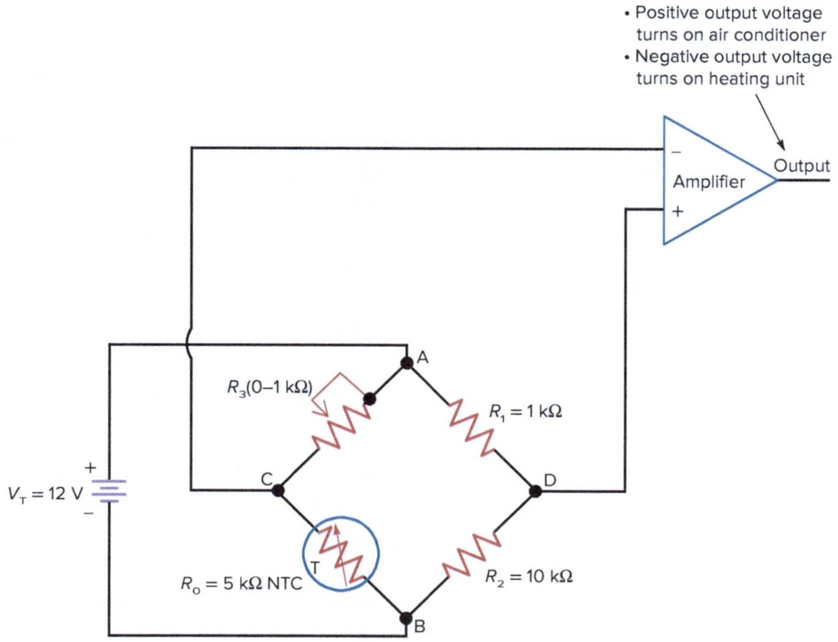

original value, thus causing the voltage V_{CD} to return to 0 V. This shuts off the air conditioner.

Next, let's consider what happens when the ambient temperature T_A decreases below 25°C, say to 20°C. The decrease in temperature causes the resistance of the thermistor to increase, thus making the voltage at point C more positive. The result is that V_{CD} goes positive. This positive voltage is fed into the amplifier, which in turn produces a negative output voltage. The negative output voltage from the amplifier turns on a heating unit, which remains on until the ambient temperature returns to its original value of 25°C. Although the details of the temperature-control circuit in Fig. 6–12 are rather vague, you should get the idea of how a balanced bridge circuit containing a thermistor could be used to control the temperature in a room. There are almost unlimited applications for balanced bridge circuits in electronics.

6–6 Self-Review

Answers at the end of the chapter.

a. **In Fig. 6–10, which terminals are the input terminals? Which terminals are the output terminals?**

b. **With reference to Fig. 6–10, how much current flows in M_1 when the bridge is balanced?**

c. **In Fig. 6–11, assume R_1 = 100 Ω and R_2 = 10 kΩ. If the bridge is balanced by adjusting R_S to 7135 Ω, what is the value of R_X?**

d. **With reference to question c, what is the maximum unknown resistance that can be measured for the circuit values given?**

e. **With reference to Fig. 6–12, to what value must the resistor R_3 be adjusted to provide 0 V output at 25°C?**

6–7 Troubleshooting: Opens and Shorts in Series-Parallel Circuits

GOOD TO KNOW

A short within a series-parallel circuit does not necessarily mean that the circuit will draw excessive current from the voltage source.

A short circuit has practically zero resistance. Its effect, therefore, is to allow excessive current. An open circuit has the opposite effect because an open circuit has infinitely high resistance with practically zero current. Furthermore, in series-parallel circuits, an open or short circuit in one path changes the circuit for the other resistances. For example, in Fig. 6–13, the series-parallel circuit in Fig. 6–13*a* becomes a series circuit with only R_1 when there is a short circuit between terminals A and B. As an example of an open circuit, the series-parallel circuit in Fig. 6–14*a* becomes a series circuit with just R_1 and R_2 when there is an open circuit between terminals C and D.

Effect of a Short Circuit

We can solve the series-parallel circuit in Fig. 6–13*a* to see the effect of the short circuit. For the normal circuit with S_1 open, R_2 and R_3 are in parallel. Although R_3 is drawn horizontally, both ends are across R_2. The switch S_1 has no effect as a parallel branch here because it is open.

The combined resistance of the 80-Ω R_2 in parallel with the 80-Ω R_3 is equivalent to 40 Ω. This 40 Ω for the bank resistance is in series with the 10-Ω R_1. Then R_T is $40 + 10 = 50$ Ω.

In the main line, I_T is $100/50 = 2$ A. Then V_1 across the 10-Ω R_1 in the main line is $2 \times 10 = 20$ V. The remaining 80 V is across R_2 and R_3 as a parallel bank. As a result, $V_2 = 80$ V and $V_3 = 80$ V.

Now consider the effect of closing switch S_1. A closed switch has zero resistance. Not only is R_2 short-circuited, but R_3 in the bank with R_2 is also short-circuited. The closed switch short-circuits everything connected between terminals A and B. The result is the series circuit shown in Fig. 6–13*b*.

Now the 10-Ω R_1 is the only opposition to current. I equals V/R_1, which is $100/10 = 10$ A. This 10 A flows through the closed switch, through R_1, and back to the positive terminal of the voltage source. With 10 A through R_1, instead of its normal 2 A,

Figure 6–13 Effect of a short circuit with series-parallel connections. (*a*) Normal circuit with S_1 open. (*b*) Circuit with short between points A and B when S_1 is closed; now R_2 and R_3 are short-circuited.

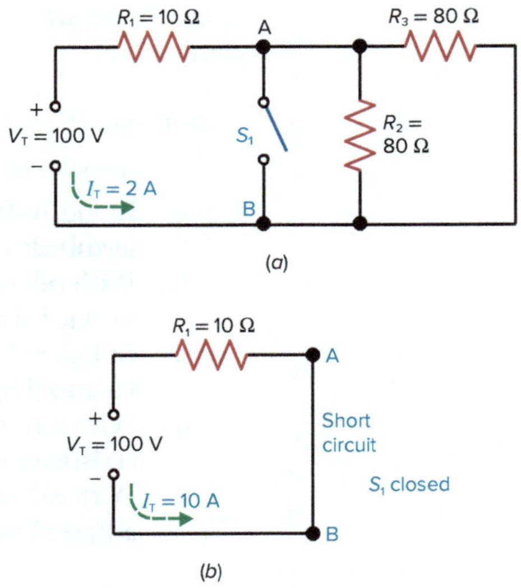

Figure 6–14 Effect of an open path in a series-parallel circuit. (*a*) Normal circuit with S_2 closed. (*b*) Series circuit with R_1 and R_2 when S_2 is open. Now R_3 in the open path has no current and zero *IR* voltage drop.

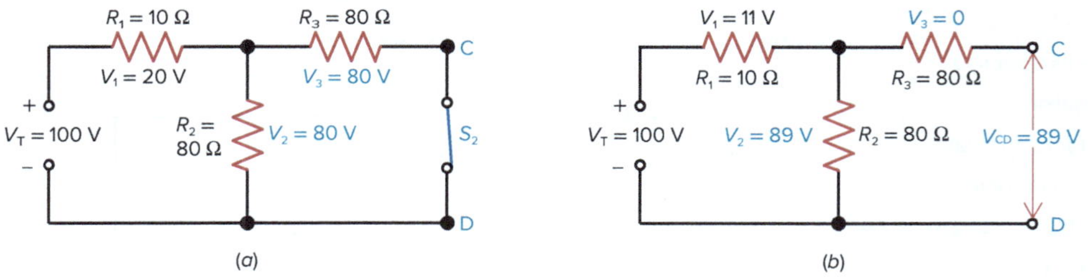

the excessive current can cause excessive heat in R_1. There is no current through R_2 and R_3, as they are short-circuited out of the path for current.

Effect of an Open Circuit

Figure 6–14*a* shows the same series-parallel circuit as Fig. 6–13*a*, except that switch S_2 is used now to connect R_3 in parallel with R_2. With S_2 closed for normal operation, all currents and voltages have the values calculated for the series-parallel circuit. However, let us consider the effect of opening S_2, as shown in Fig. 6–14*b*. An open switch has infinitely high resistance. Now there is an open circuit between terminals C and D. Furthermore, because R_3 is in the open path, its 80 Ω cannot be considered in parallel with R_2.

The circuit with S_2 open in Fig. 6–14*b* is really the same as having only R_1 and R_2 in series with the 100-V source. The open path with R_3 has no effect as a parallel branch because no current flows through R_3.

We can consider R_1 and R_2 in series as a voltage divider, where each *IR* drop is proportional to its resistance. The total series *R* is 80 + 10 = 90 Ω. The 10-Ω R_1 is 10/90 or ⅑ of the total *R* and the applied V_T. Then V_1 is ⅑ × 100 V = 11 V and V_2 is $\frac{8}{9}$ × 100 V = 89 V, approximately. The 11-V drop for V_1 and 89-V drop for V_2 add to equal the 100 V of the applied voltage.

Note that V_3 is zero. Without any current through R_3, it cannot have any voltage drop.

Furthermore, the voltage across the open terminals C and D is the same 89 V as the potential difference V_2 across R_2. Since there is no voltage drop across R_3, terminal C has the same potential as the top terminal of R_2. Terminal D is directly connected to the bottom end of resistor R_2. Therefore, the potential difference from terminal C to terminal D is the same 89 V that appears across resistor R_2.

Troubleshooting Procedures for Series-Parallel Circuits

The procedure for troubleshooting series-parallel circuits containing opens and shorts is a combination of the procedures used to troubleshoot individual series and parallel circuits. Figure 6–15*a* shows a series-parallel circuit with its normal operating voltages and currents. Across points A and B, the equivalent resistance R_{EQ} of R_2 and R_3 in parallel is calculated as

$$R_{EQ} = \frac{R_2 \times R_3}{R_2 + R_3}$$

$$= \frac{100\ \Omega \times 150\ \Omega}{100\ \Omega + 150\ \Omega}$$

$$R_{EQ} = 60\ \Omega$$

GOOD TO KNOW

Troubleshooting, a defective electronic circuit, is like being a crime scene investigator. A technician gathers facts and data about the defective electronic circuit by making voltage, current, and resistance measurements. These measurements are then carefully analyzed to pinpoint the defective electronic component.

Figure 6–15 Series-parallel circuit for troubleshooting analysis. (*a*) Normal circuit voltages and currents; (*b*) circuit voltages with R_3 open between points A and B; (*c*) circuit voltages with R_2 or R_3 shorted between points A and B.

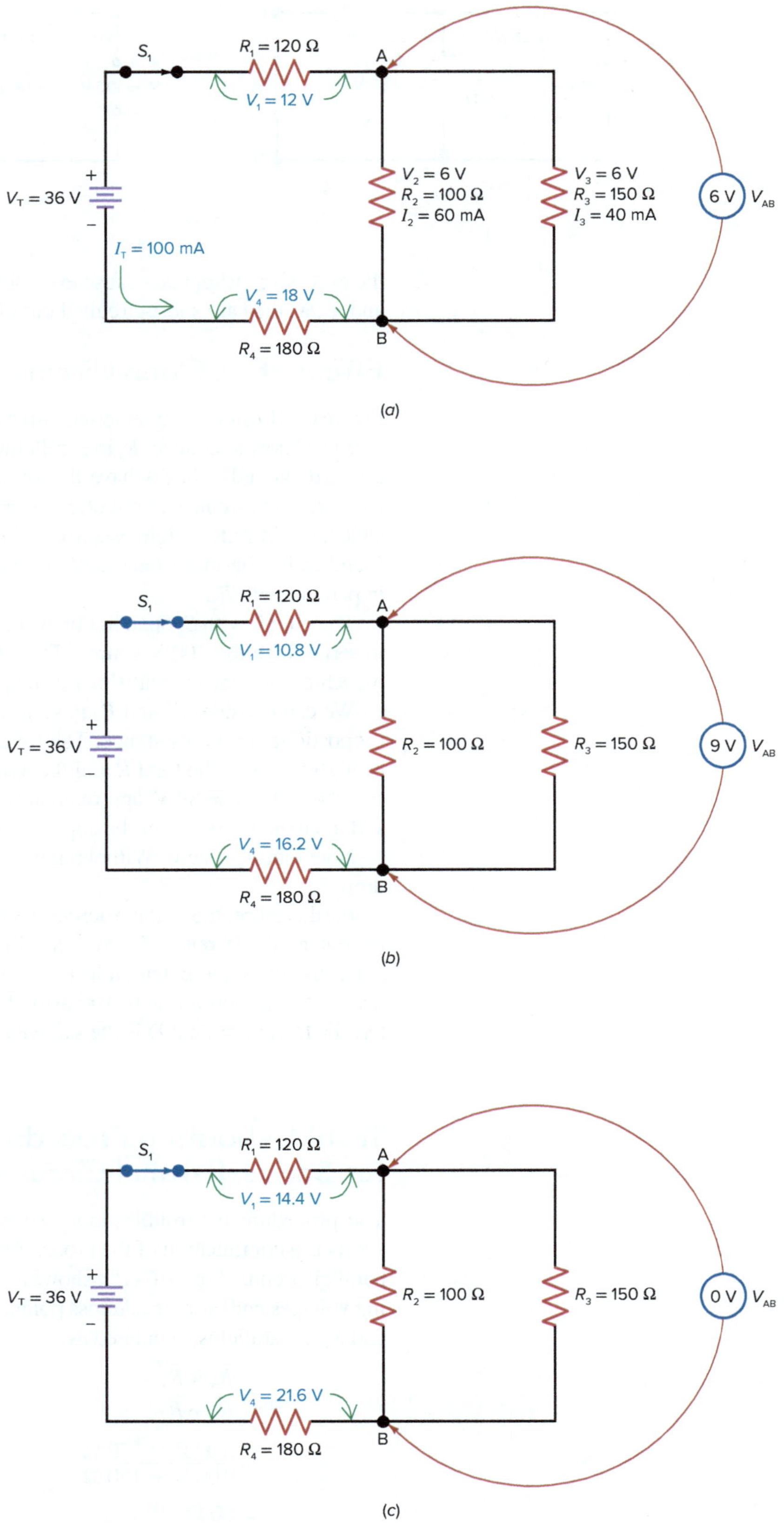

Since R_2 and R_3 are in parallel across points A and B, this equivalent resistance is designated R_{AB}. Therefore, $R_{AB} = 60\ \Omega$. The total resistance, R_T, is

$$R_T = R_1 + R_{AB} + R_4$$
$$= 120\ \Omega + 60\ \Omega + 180\ \Omega$$
$$R_T = 360\ \Omega$$

The total current, I_T, is

$$I_T = \frac{V_T}{R_T}$$
$$= \frac{36\ \text{V}}{360\ \Omega}$$
$$I_T = 100\ \text{mA}$$

The voltage drops across the individual resistors are calculated as

$$V_1 = I_T \times R_1$$
$$= 100\ \text{mA} \times 120\ \Omega$$
$$V_1 = 12\ \text{V}$$
$$V_2 = V_3 = V_{AB} = I_T \times R_{AB}$$
$$= 100\ \text{mA} \times 60\ \Omega$$
$$= 6\ \text{V}$$
$$V_4 = I_T \times R_4$$
$$= 100\ \text{mA} \times 180\ \Omega$$
$$= 18\ \text{V}$$

The current in resistors R_2 and R_3 across points A and B can be found as follows:

$$I_2 = \frac{V_{AB}}{R_2}$$
$$= \frac{6\ \text{V}}{100\ \Omega}$$
$$I_2 = 60\ \text{mA}$$
$$I_3 = \frac{V_{AB}}{R_3}$$
$$= \frac{6\ \text{V}}{150\ \Omega}$$
$$I_3 = 40\ \text{mA}$$

Example 6–3

Assume that the series-parallel circuit in Fig. 6–15*a* has failed. A technician troubleshooting the circuit has measured the following voltages:

$$V_1 = 10.8\ \text{V}$$
$$V_{AB} = 9\ \text{V}$$
$$V_4 = 16.2\ \text{V}$$

These voltage readings are shown in Fig. 6–15*b*. Based on the voltmeter readings shown, which component is defective and what type of defect does it have?

ANSWER If we consider the resistance between points A and B as a single resistance, the circuit can be analyzed as if it were a simple series circuit. Notice that V_1 and V_4 have decreased from their normal values of 12 V and 18 V, respectively, whereas the voltage V_{AB} across R_2 and R_3 has increased from 6 V to 9 V.

Recall that in a series circuit containing three or more components, the voltage across the defective component changes in a direction that is opposite to the direction of the change in voltage across the good components. Since the voltages V_1 and V_4 have decreased and the voltage V_{AB} has increased, the defective component must be either R_2 or R_3 across points A and B.

The increase in voltage across points A and B tells us that the resistance between points A and B must have increased. The increase in the resistance R_{AB} could be the result of an open in either or R_2 or R_3.

But how do we know which resistor is open? At least three approaches may be used to find this out. One approach would be to calculate the resistance across points A and B. To do this, find the total current in either R_1 or R_4. Let's find I_T in R_1.

$$I_T = \frac{V_1}{R_1}$$

$$= \frac{10.8\ \text{V}}{120\ \Omega}$$

$$I_T = 90\ \text{mA}$$

Next, divide the measured voltage V_{AB} by I_T to find R_{AB}.

$$R_{AB} = \frac{V_{AB}}{I_T}$$

$$= \frac{9\ \text{V}}{90\ \text{mA}}$$

$$R_{AB} = 100\ \Omega$$

Notice that the value of R_{AB} is the same as that of R_2. This means, of course, that R_3 must be open.

Another approach to finding which resistor is open would be to open the switch S_1 and measure the resistance across points A and B. This measurement would show that the resistance R_{AB} equals 100 Ω, again indicating that the resistor R_3 must be open.

The only other approach to determine which resistor is open would be to measure the currents I_2 and I_3 with the switch S_1 closed. In Fig. 6–15*b*, the current I_2 would measure 90 mA, whereas the current I_3 would measure 0 mA. With $I_3 = 0$ mA, R_3 must be open.

Example 6–4

Assume that the series-parallel circuit in Fig. 6–15*a* has failed. A technician troubleshooting the circuit has measured the following voltages:

$$V_1 = 14.4\ \text{V}$$

$$V_{AB} = 0\ \text{V}$$

$$V_4 = 21.6\ \text{V}$$

These voltage readings are shown in Fig. 6–15*c*. Based on the voltmeter readings shown, which component is defective and what type of defect does it have?

ANSWER Since the voltages V_1 and V_4 have both increased, and the voltage V_{AB} has decreased, the defective component must be either R_2 or R_3 across points A and B. Because the voltage V_{AB} is 0 V, either R_2 or R_3 must be shorted.

But how can we find out which resistor is shorted? One way would be to measure the currents I_2 and I_3. The shorted component is the one with all the current.

Another way to find out which resistor is shorted would be to open the switch S_1 and measure the resistance across points A and B. Disconnect one lead of either R_2 or R_3 from point A while observing the ohmmeter. If removing the top lead of R_3 from point A still shows a reading of 0 Ω, then you know that R_2 must be shorted. Similarly, if removing the top lead of R_2 from point A (with R_3 still connected at point A) still produces a reading of 0 Ω, then you know that R_3 is shorted.

6–7 Self-Review

Answers at the end of the chapter.

a. **In Fig. 6–13, the short circuit increases I_T from 2 A to what value?**
b. **In Fig. 6–14, the open branch reduces I_T from 2 A to what value?**
c. **In Fig. 6–15*a*, what is the voltage across points A and B if R_4 shorts?**
d. **In Fig. 6–15*a*, what is the voltage V_{AB} if R_1 opens?**

Summary

- In circuits combining series and parallel connections, the components in one current path without any branch points are in series; the parts of the circuit connected across the same two branch points are in parallel.
- To calculate R_T in a series-parallel circuit with R in the main line, combine resistances from the outside back toward the source.
- When the potential is the same at the two ends of a resistance, its voltage is zero. If no current flows through a resistance, it cannot have any *IR* voltage drop.
- A Wheatstone bridge circuit has two input terminals and two output terminals. When balanced, the Wheatstone bridge can be analyzed simply as two series strings in parallel. The Wheatstone bridge finds many uses in applications where comparison measurements are needed.
- The procedure for troubleshooting series-parallel circuits is a combination of the procedures used to troubleshoot series and parallel circuits.

Important Terms

Balanced bridge — a circuit consisting of two series strings in parallel. The balanced condition occurs when the voltage ratio in each series string is identical. The output from the bridge is taken between the centers of each series string. When the voltage ratios in each series string are identical, the output voltage is zero, and the bridge circuit is said to be balanced.

Banks in series — parallel resistor banks that are connected in series with each other.

Ratio arm — accurate, stable resistors in one leg of a Wheatstone bridge or bridge circuit in general. The ratio arm fraction, R_1/R_2, can be varied in most cases, typically in multiples of 10. The ratio arm fraction in a Wheatstone bridge determines two things: the placement accuracy of the measurement of an unknown resistor, R_X, and the maximum unknown resistance, $R_{X(\text{max})}$, that can be measured.

Standard resistor — a variable resistor in one leg of a Wheatstone bridge that is varied to provide equal voltage ratios in both series strings of the bridge. With equal voltage ratios in each series string, the bridge is said to be balanced.

Strings in parallel — series resistor strings that are connected in parallel with each other.

Wheatstone bridge — a balanced bridge circuit that can be used to find the value of an unknown resistor.

Related Formulas

$$R_X = R_S \times \frac{R_1}{R_2}$$

$$R_{X(\text{max})} = R_{S(\text{max})} \times \frac{R_1}{R_2}$$

Self-Test

Answers at the back of the book.

QUESTIONS 1–12 REFER TO FIG. 6–16.

1. In Fig. 6–16,

a. R_1 and R_2 are in series.
b. R_3 and R_4 are in series.
c. R_1 and R_4 are in series.
d. R_2 and R_4 are in series.

2. In Fig. 6–16,

a. R_2, R_3, and V_T are in parallel.
b. R_2 and R_3 are in parallel.
c. R_2 and R_3 are in series.
d. R_1 and R_4 are in parallel.

3. In Fig. 6–16, the total resistance, R_T, equals

a. 1.6 kΩ.
b. 3.88 kΩ.
c. 10 kΩ.
d. none of the above.

4. In Fig. 6–16, the total current, I_T, equals

a. 6.19 mA.
b. 150 mA.
c. 15 mA.
d. 25 mA.

5. In Fig. 6–16, how much voltage is across points A and B?

a. 12 V.
b. 18 V.
c. 13.8 V.
d. 10.8 V.

6. In Fig. 6–16, how much is I_2 through R_2?

a. 9 mA.
b. 15 mA.
c. 6 mA.
d. 10.8 mA.

Figure 6–16

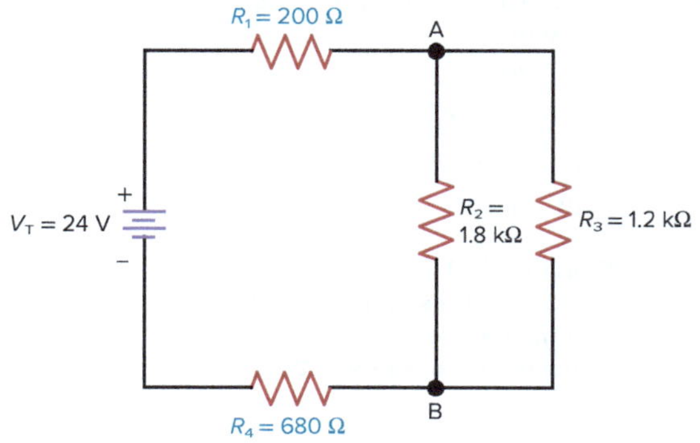

7. In Fig. 6–16, how much is I_3 through R_3?

a. 9 mA.

b. 15 mA.

c. 6 mA.

d. 45 mA.

8. If R_4 shorts in Fig. 6–16, the voltage, V_{AB}

a. increases.

b. decreases.

c. stays the same.

d. increases to 24 V.

9. If R_2 becomes open in Fig. 6–16,

a. the voltage across points A and B will decrease.

b. the resistors R_1, R_3, and R_4 will be in series.

c. the total resistance, R_T, will decrease.

d. the voltage across points A and B will measure 24 V.

10. If R_1 opens in Fig. 6–16,

a. the voltage across R_1 will measure 0 V.

b. the voltage across R_4 will measure 0 V.

c. the voltage across points A and B will measure 0 V.

d. both b and c.

11. If R_3 becomes open in Fig. 6–16, what happens to the voltage across points A and B?

a. It decreases.

b. It increases.

c. It stays the same.

d. none of the above.

12. If R_2 shorts in Fig. 6–16,

a. the voltage, V_{AB}, decreases to 0 V.

b. the total current, I_T, flows through R_3.

c. the current, I_3, in R_3 is zero.

d. both a and c.

QUESTIONS 13–20 REFER TO FIG. 6–17.

13. In Fig. 6–17, how much voltage exists between terminals C and D when the bridge is balanced?

a. 0 V.

b. 10.9 V.

c. 2.18 V.

d. 12 V.

14. In Fig. 6–17, assume that the current in M_1 is zero when R_S is adjusted to 55,943 Ω. What is the value of the unknown resistor, R_X?

a. 55,943 Ω.

b. 559.43 Ω.

c. 5,594.3 Ω.

d. 10 kΩ.

15. In Fig. 6–17, assume that the bridge is balanced when R_S is adjusted to 15,000 Ω. How much is the total current, I_T, flowing to and from the terminals of the voltage source, V_T?

a. zero.

b. approximately 727.27 μA.

c. approximately 1.09 mA.

d. approximately 1.82 mA.

16. In Fig. 6–17, what is the maximum unknown resistor, $R_{X(max)}$, that can be measured for the resistor values shown in the ratio arm?

a. 99.99 Ω.

b. 9,999.9 Ω.

c. 99,999 Ω.

d. 999,999 Ω.

17. In Fig. 6–17, the ratio R_1/R_2 determines

a. the placement accuracy of the measurement of R_X.

b. the maximum unknown resistor, $R_{X(max)}$, that can be measured.

c. the amount of voltage available across terminals A and B.

d. both a and b.

Figure 6–17

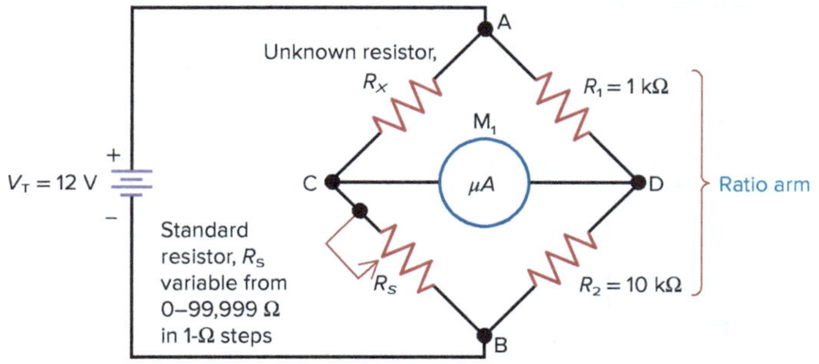

18. In Fig. 6–17, assume that the standard resistor, R_S, has been adjusted so that the current in M_1 is exactly 0 μA. How much voltage exists at terminal C with respect to terminal B?

a. 1.1 V.

b. 0 V.

c. 10.9 V.

d. none of the above.

19. In Fig. 6–17, assume that the ratio arm resistors, R_1 and R_2, are interchanged. What is the value of the unknown resistor, R_X, if R_S equals 33,950 Ω when the bridge is balanced?

a. 339.5 kΩ.

b. 3.395 kΩ.

c. 33,950 Ω.

d. none of the above.

20. In Fig. 6–17, assume that the standard resistor, R_S, cannot be adjusted high enough to provide a balanced condition. What modification must be made to the circuit?

a. Change the ratio arm fraction R_1/R_2, from $^1/_{10}$ to $^1/_{100}$ or something less.

b. Change the ratio arm fraction, R_1/R_2 from $^1/_{10}$ to $^1/_1$, $^{10}/_1$ or something greater.

c. Reverse the polarity of the applied voltage, V_T.

d. None of the above.

Essay Questions

1. In a series-parallel circuit, how can you tell which resistances are in series and which are in parallel?

2. Draw a schematic diagram showing two resistances in a parallel bank that is in series with one resistance.

3. Draw a diagram showing how to connect three resistances of equal value so that the combined resistance will be 1½ times the resistance of one unit.

4. Draw a diagram showing two strings in parallel across a voltage source, where each string has three series resistances.

5. Explain why components are connected in series-parallel, showing a circuit as an example of your explanation.

6. Give two differences between a short circuit and an open circuit.

7. Explain the difference between voltage division and current division.

8. In Fig. 6–12, assume that the thermistor has a positive temperature coefficient (PTC). Explain what happens to the voltage V_{CD} (a) if the ambient temperature decreases; (b) if the ambient temperature increases.

9. Draw a circuit with nine 40-V, 100-W bulbs connected to a 120-V source.

10. (a) Two 10-Ω resistors are in series with a 100-V source. If a third 10-Ω R is added in series, explain why I will decrease. (b) The same two 10-Ω resistors are in parallel with the 100-V source. If a third 10-Ω R is added in parallel, explain why I_T will increase.

Problems

SECTION 6–1 FINDING R_T FOR SERIES-PARALLEL RESISTANCES

6–1 In Fig. 6–18, identify which components are in series and which ones are in parallel.

6–2 In Fig. 6–18,

a. How much is the total resistance of just R_1 and R_2?

b. What is the equivalent resistance of R_3 and R_4 across points A and B?

c. How much is the total resistance, R_T, of the entire circuit?

d. How much is the total current, I_T, in the circuit?

e. How much current flows into point B?

f. How much current flows away from point A?

Figure 6–18

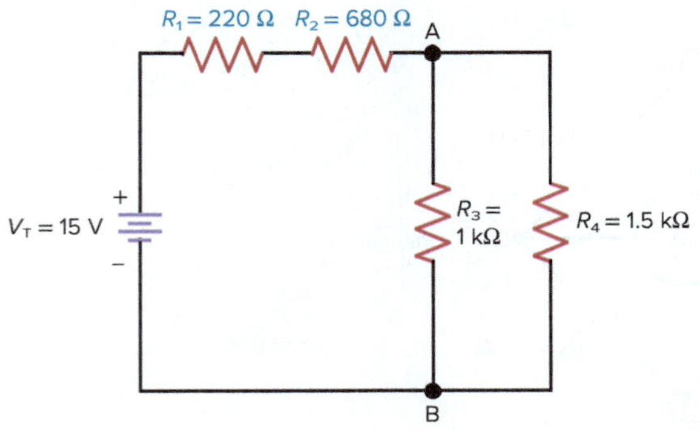

6–3 MultiSim In Fig. 6–18, solve for the following: I_1, I_2, V_1, V_2, V_3, V_4, I_3, and I_4.

6–4 In Fig. 6–19, identify which components are in series and which ones are in parallel.

6–5 In Fig. 6–19,

a. What is the equivalent resistance of R_2 and R_3 across points A and B?

b. How much is the total resistance, R_T of the entire circuit?

c. How much is the total current, I_T, in the circuit?

d. How much current flows into point B and away from point A?

6–6 MultiSim In Fig. 6–19, solve for I_1, V_1, V_2, V_3, I_2, and I_3.

Figure 6–19

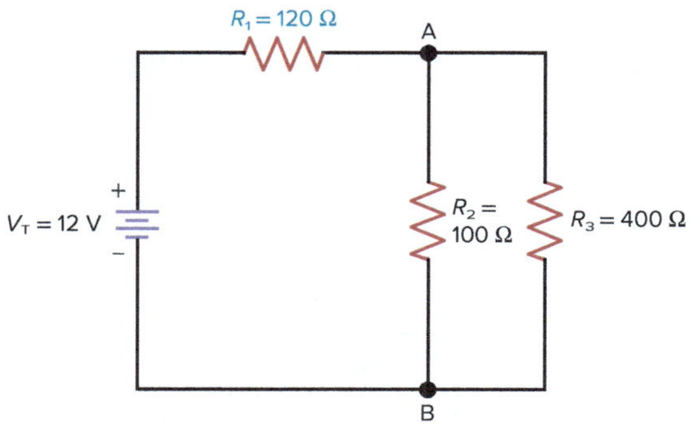

6–7 In Fig. 6–19, solve for P_1, P_2, P_3, and P_T.

6–8 In Fig. 6–20, identify which components are in series and which ones are in parallel.

Figure 6–20

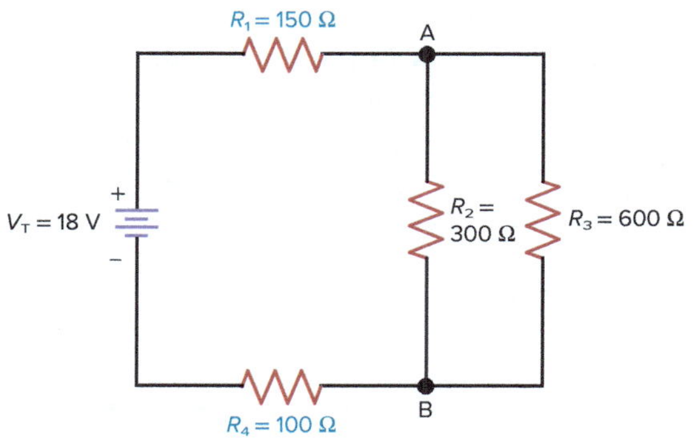

6–9 In Fig. 6–20,

a. How much is the total resistance of just R_1 and R_4?

b. What is the equivalent resistance of R_2 and R_3 across points A and B?

c. How much is the total resistance, R_T, of the entire circuit?

d. How much is the total current, I_T, in the circuit?

e. How much current flows into point B and away from point A?

6–10 In Fig. 6–20, solve for the following: I_1, V_1, V_2, V_3, I_2, I_3, I_4, V_4, P_1, P_2, P_3, P_4, and P_T.

SECTION 6–2 RESISTANCE STRINGS IN PARALLEL

6–11 In Fig. 6–21,

a. What is the total resistance of branch 1?

b. What is the resistance of branch 2?

c. How much are the branch currents I_1 and I_2?

d. How much is the total current, I_T, in the circuit?

e. How much is the total resistance, R_T, of the entire circuit?

f. What are the values of V_1, V_2, and V_3?

Figure 6–21

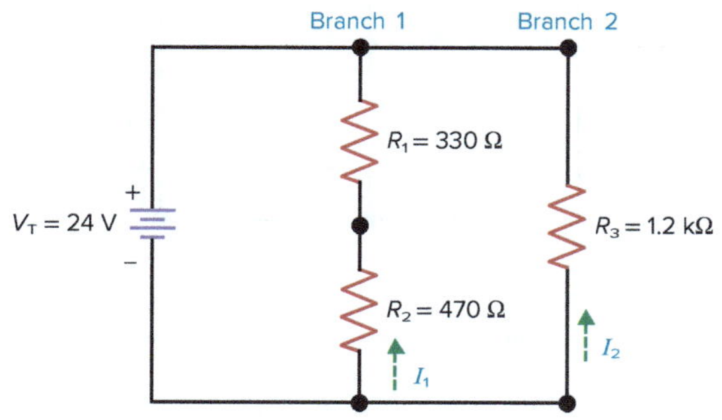

6–12 In Fig. 6–22,

a. What is the total resistance of branch 1?

b. What is the total resistance of branch 2?

c. How much are the branch currents I_1 and I_2?

d. How much is the total current, I_T, in the circuit?

e. How much is the total resistance, R_T, of the entire circuit?

f. What are the values of V_1, V_2, V_3, and V_4?

Figure 6–22

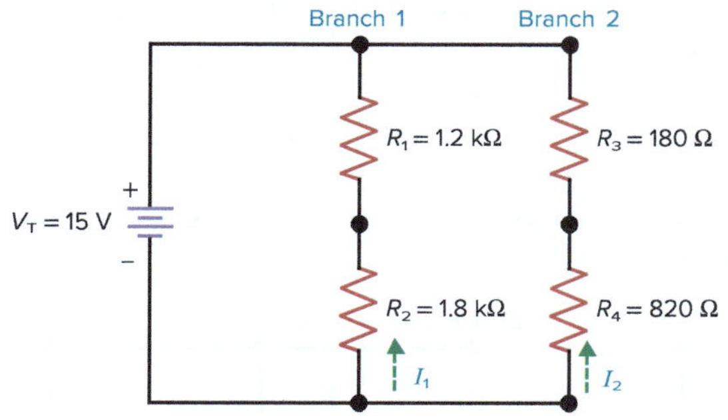

6–13 In Fig. 6–23,

a. What is the total resistance of branch 1?

b. What is the total resistance of branch 2?

c. How much are the branch currents I_1 and I_2?

d. How much is the total current, I_T, in the circuit?

e. How much is the total resistance, R_T, of the entire circuit?

f. What are the values of V_1, V_2, V_3, and V_4?

6–14 In Fig. 6–24, solve for

a. branch currents I_1, I_2 and the total current, I_T.

b. R_T.

c. V_1, V_2, V_3, and V_4.

d. P_1, P_2, P_3, P_4 and P_T.

Figure 6–23

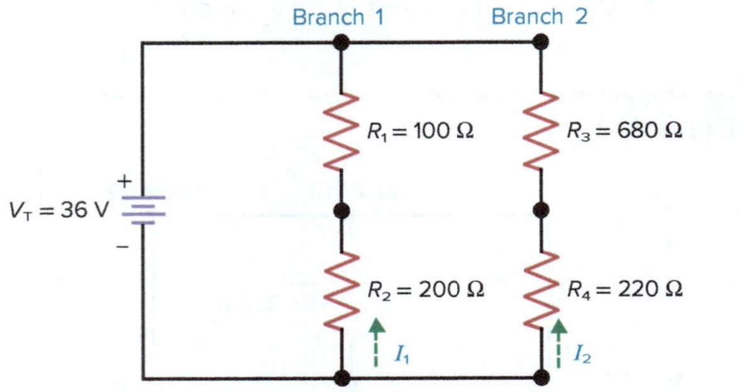

Figure 6–24

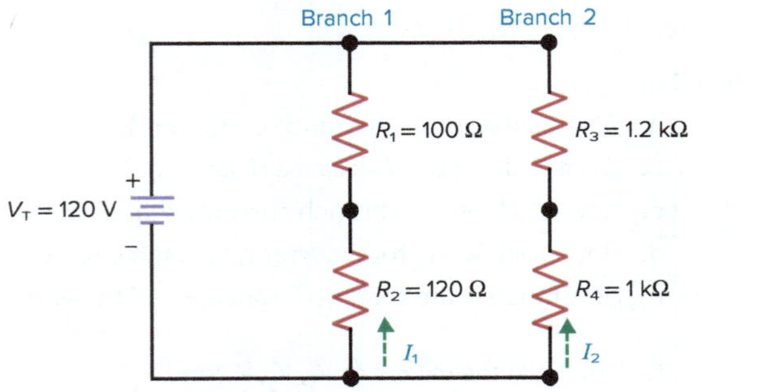

6–15 In Fig. 6–25, solve for

a. branch currents I_1, I_2, I_3 and the total current, I_T.

b. R_T.

c. V_1, V_2, V_3, V_4, and V_5.

Figure 6–25

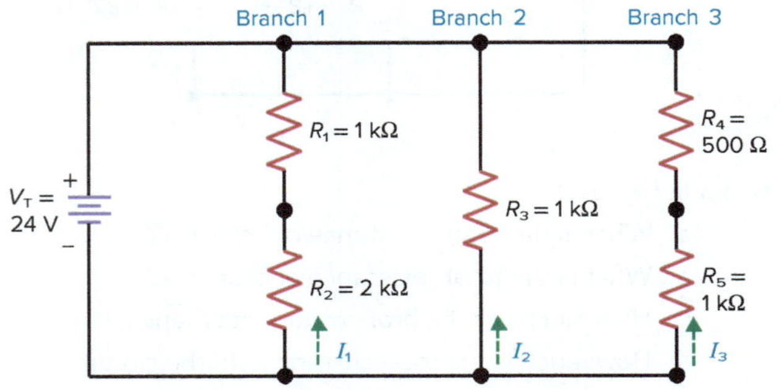

SECTION 6–3 RESISTANCE BANKS IN SERIES

6–16 In Fig. 6–26,

a. What is the equivalent resistance of R_1 and R_2 in parallel across points A and B?

b. What is the total resistance, R_T, of the circuit?

c. What is the total current, I_T, in the circuit?

d. How much voltage exists across points A and B?

e. How much voltage is dropped across R_3?

f. Solve for I_1 and I_2.

g. How much current flows into point B and away from point A?

Figure 6–26

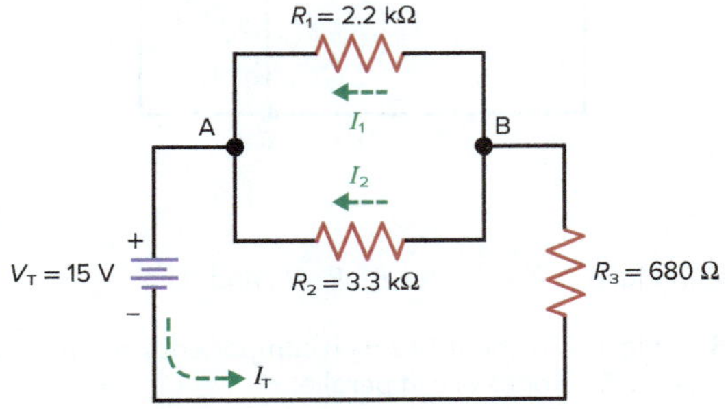

6–17 In Fig. 6–27,

a. What is the equivalent resistance of R_2 and R_3 in parallel across points A and B?

b. What is the total resistance, R_T, of the circuit?

c. What is the total current, I_T, in the circuit?

d. How much voltage exists across points A and B?

e. How much voltage is dropped across R_1?

f. Solve for I_2 and I_3.

g. How much current flows into point B and away from point A?

Figure 6–27

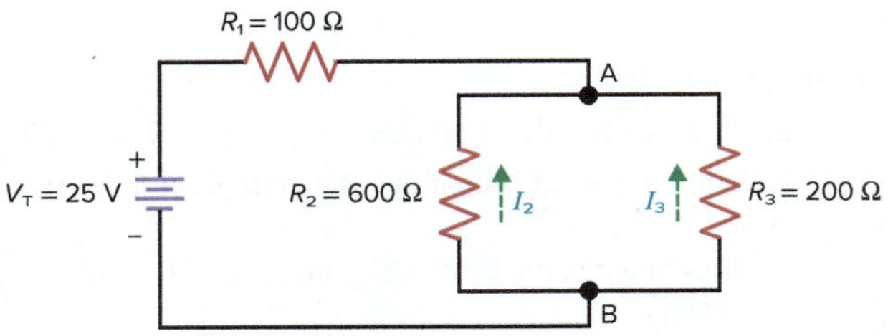

6–18 In Fig. 6–28, solve for

a. R_T.

b. I_T.

c. V_1, V_2, and V_3.

d. I_1 and I_2.

6–19 In Fig. 6–29, solve for

a. R_T.

b. I_T.

c. V_1, V_2, and V_3.

d. I_2 and I_3.

Figure 6–28

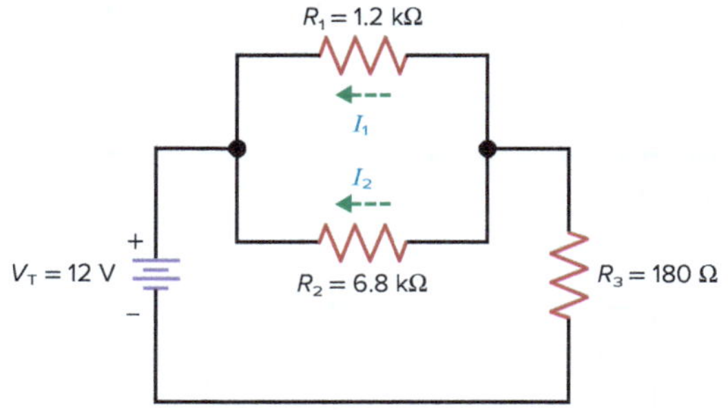

Figure 6–29

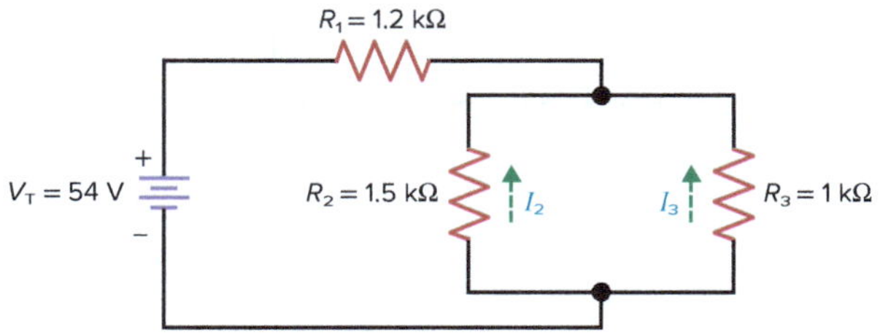

SECTION 6–4 RESISTANCE BANKS AND STRINGS IN SERIES-PARALLEL

6–20 In Fig. 6–30, solve for R_T, I_T, V_1, V_2, V_3, V_4, I_1, I_2, I_3, and I_4.

Figure 6–30

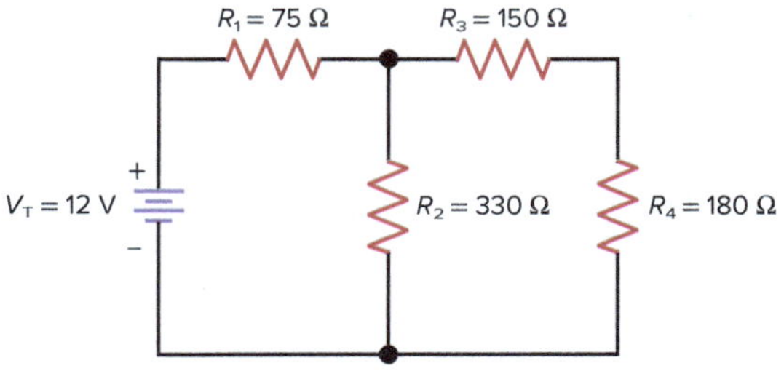

6–21 In Fig. 6–31, solve for R_T, I_T, V_1, V_2, V_3, V_4, V_5, I_1, I_2, I_3, I_4, and I_5.

Figure 6–31

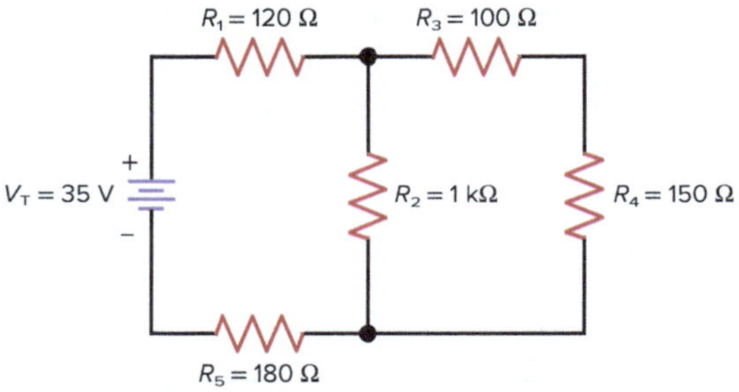

6–22 In Fig. 6–32, solve for R_T, I_T, V_1, V_2, V_3, V_4, V_5, V_6, I_1, I_2, I_3, I_4, I_5, and I_6.

Figure 6–32

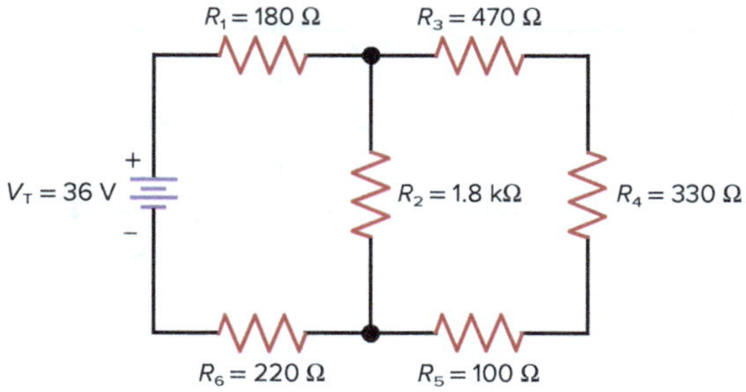

6–23 In Fig. 6–33, solve for R_T, I_T, V_1, V_2, V_3, V_4, V_5, I_1, I_2, I_3, I_4, and I_5.

Figure 6–33

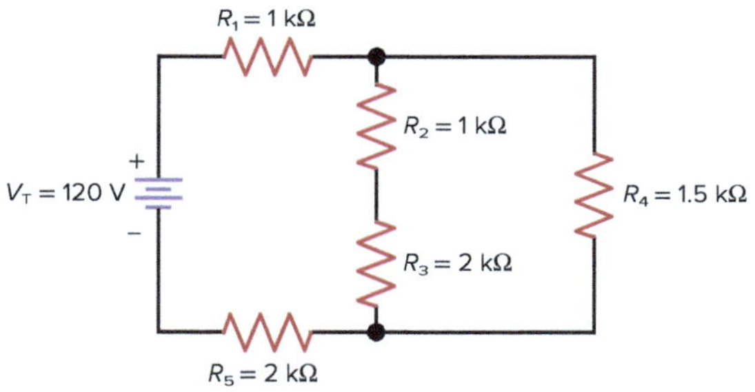

6–24 In Fig. 6–34, solve for R_T, I_T, V_1, V_2, V_3, V_4, V_5, V_6, I_1, I_2, I_3, I_4, I_5, and I_6.

Figure 6–34

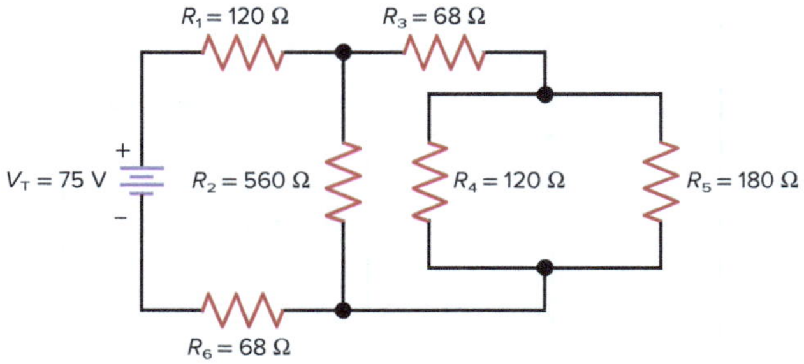

6–25 In Fig. 6–35, solve for R_T, I_T, V_1, V_2, V_3, V_4, V_5, V_6, V_7, I_1, I_2, I_3, I_4, I_5, I_6, and I_7.

Figure 6–35

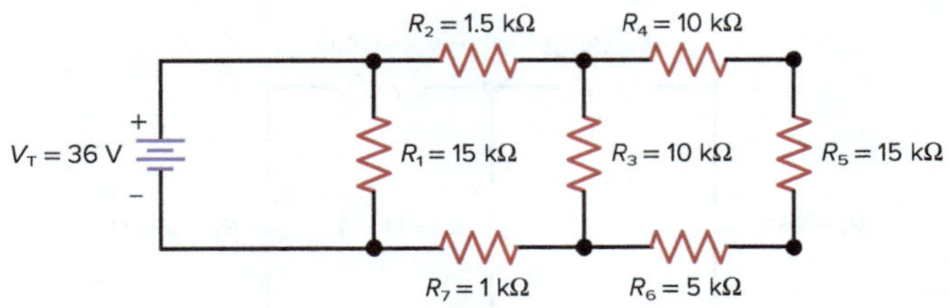

6–26 In Fig. 6–36, solve for R_T, I_1, I_2, I_3, V_1, V_2, V_3, and the voltage, V_{AB}.

Figure 6–36

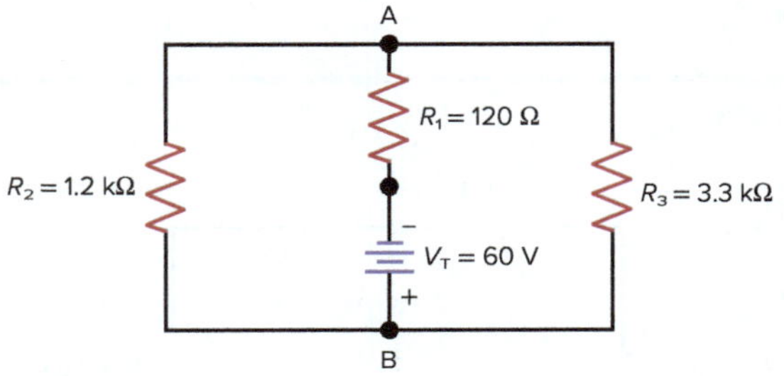

6–27 In Fig. 6–37, solve for R_T, I_T, V_1, V_2, V_3, V_4, V_5, V_6, I_1, I_2, I_3, I_4, I_5, and I_6.

Figure 6–37

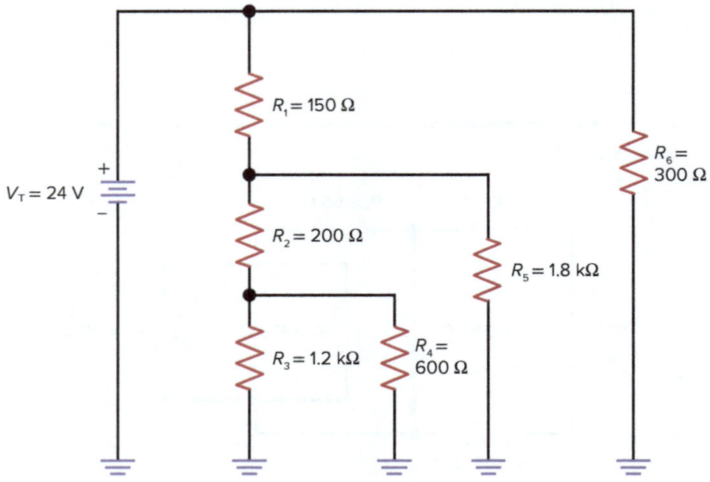

SECTION 6–5 ANALYZING SERIES-PARALLEL CIRCUITS WITH RANDOM UNKNOWNS

6–28 In Fig. 6–38, solve for V_1, V_2, V_3, I_2, R_3, R_T, and V_T.

Figure 6–38

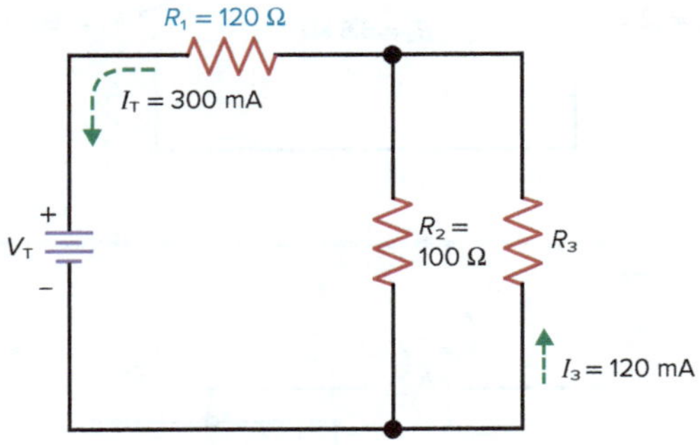

6–29 In Fig. 6–39, solve for R_T, I_T, V_T, V_1, V_2, V_4, I_2, and I_3.

Figure 6–39

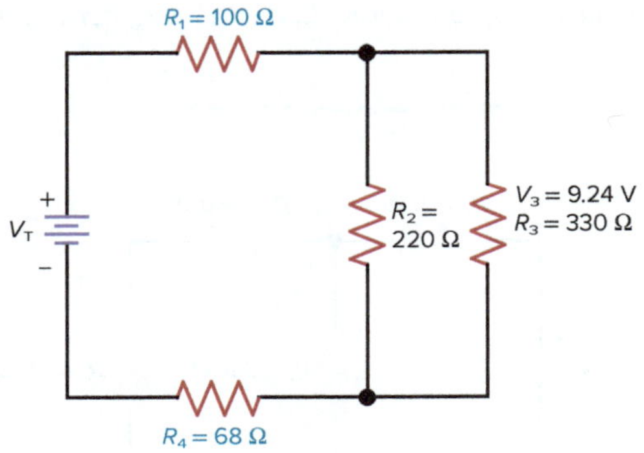

6–30 In Fig. 6–40, solve for R_2, V_1, V_2, V_3, I_3, I_T, and V_T.

Figure 6–40

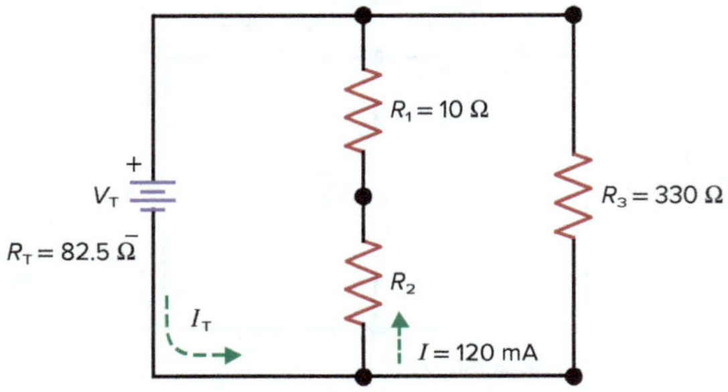

6–31 In Fig. 6–41, solve for R_T, I_T, V_T, V_1, V_2, V_3, V_4, V_5, V_6, I_2, I_3, I_4, and I_5.

Figure 6–41

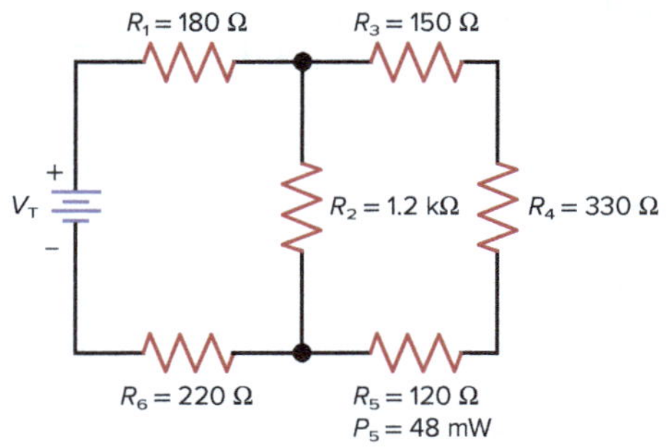

6–32 In Fig. 6–42, solve for R_T, I_T, V_1, V_2, V_3, V_4, V_5, V_6, I_1, I_2, I_3, I_4, I_5, and I_6.

Figure 6–42

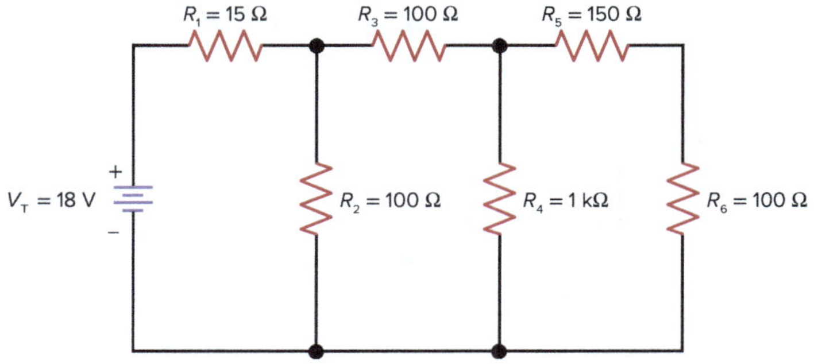

6–33 In Fig. 6–43, solve for I_T, R_T, R_2, V_2, V_3, V_4, V_5, V_6, I_2, I_3, I_4, I_5, and I_6.

Figure 6–43

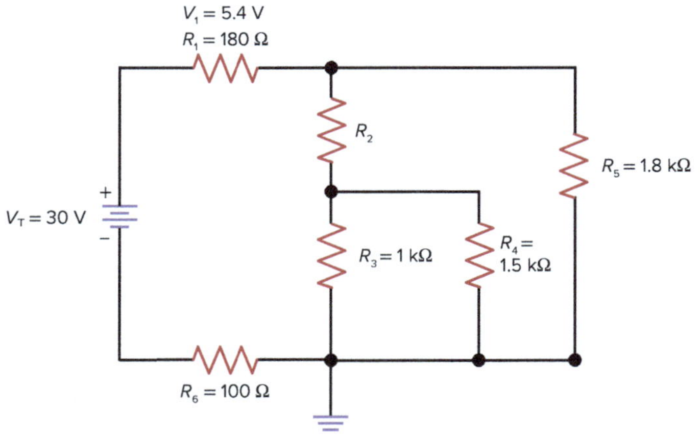

SECTION 6–6 THE WHEATSTONE BRIDGE

Probs. 34–38 refer to Fig. 6–44.

6–34 In Fig. 6–44,

a. How much current flows through M_1 when the Wheatstone bridge is balanced?

b. How much voltage exists between points C and D when the bridge is balanced?

Figure 6–44

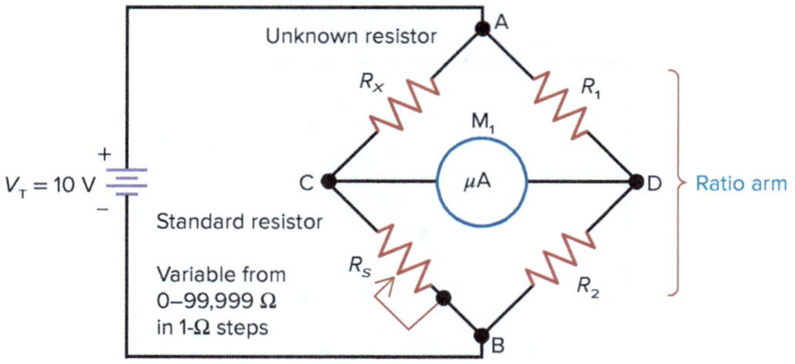

6–35 In Fig. 6–44, assume that the bridge is balanced when $R_1 = 1\ \text{k}\Omega$, $R_2 = 5\ \text{k}\Omega$, and $R_S = 34{,}080\ \Omega$. Determine

a. the value of the unknown resistor, R_X.

b. the voltages V_{CB} and V_{DB}.

c. the total current, I_T, flowing to and from the voltage source, V_T.

6–36 In reference to Prob. 6–35, which direction (C to D or D to C) will electrons flow through M_1 if

a. R_S is reduced in value?

b. R_S is increased in value?

6–37 In Fig. 6–44, calculate the maximum unknown resistor, $R_{X(max)}$, that can be measured for the following ratio arm values:

a. $\dfrac{R_1}{R_2} = \dfrac{1}{1000}$

b. $\dfrac{R_1}{R_2} = \dfrac{1}{100}$

c. $\dfrac{R_1}{R_2} = \dfrac{1}{10}$

d. $\dfrac{R_1}{R_2} = \dfrac{1}{1}$

e. $\dfrac{R_1}{R_2} = \dfrac{10}{1}$

f. $\dfrac{R_1}{R_2} = \dfrac{100}{1}$

6–38 Assume that the same unknown resistor, R_X, is measured using different ratio arm fractions in Fig. 6–44. In each case, the standard resistor, R_S, is adjusted to provide the balanced condition. The values for each measurement are

a. $R_S = 123\ \Omega$ and $\dfrac{R_1}{R_2} = \dfrac{1}{1}$.

b. $R_S = 1232\ \Omega$ and $\dfrac{R_1}{R_2} = \dfrac{1}{10}$.

c. $R_S = 12{,}317\ \Omega$ and $\dfrac{R_1}{R_2} = \dfrac{1}{100}$.

Calculate the value of the unknown resistor, R_X, for each measurement. Which ratio arm fraction provides the greatest accuracy?

PROBLEMS 39–41 REFER TO FIG. 6–45.

6–39 In Fig. 6–45, to what value must R_3 be adjusted to provide zero volts across terminals C and D when the ambient temperature, T_A, is 25°C? (Note: R_0 is the resistance of the thermistor at an ambient temperature, T_A, of 25°C.)

Figure 6–45

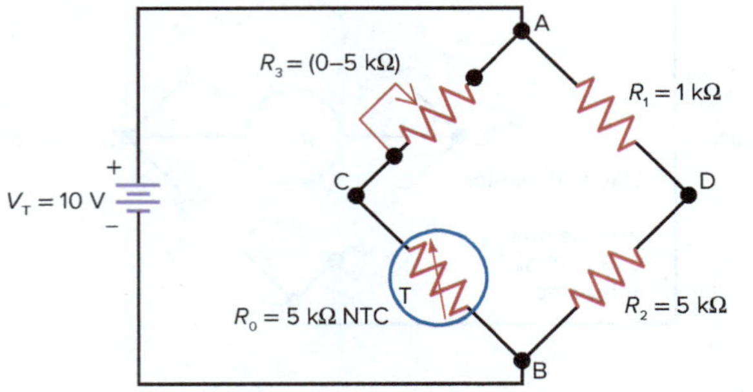

6–40 In Fig. 6–45, assume that R_S is adjusted to provide zero volts across terminals C and D at an ambient temperature, T_A, of 25°C. What happens to the polarity of the output voltage, V_{CD}, when

a. the ambient temperature, T_A, increases above 25°C?

b. the ambient temperature, T_A, decreases below 25°C?

6–41 In Fig. 6–45, assume that R_3 has been adjusted to 850 Ω to provide zero volts across the output terminals C and D. Determine

a. the resistance of the thermistor.

b. whether the ambient temperature, T_A, has increased or decreased from 25°C.

SECTION 6–7 TROUBLESHOOTING: OPENS AND SHORTS IN SERIES-PARALLEL CIRCUITS

Figure 6–46 shows a series-parallel circuit with its normal operating voltages and currents.

6–42 MultiSim In Fig. 6–46, determine the voltages V_1, V_{AB}, V_3, V_{CD}, and V_5 for each of the following component troubles:

a. R_4 is open.

b. R_2 is shorted.

c. R_3 is open.

d. R_4 is shorted.

e. R_2 is open.

f. R_1 is open.

g. R_1 is shorted.

Figure 6–46

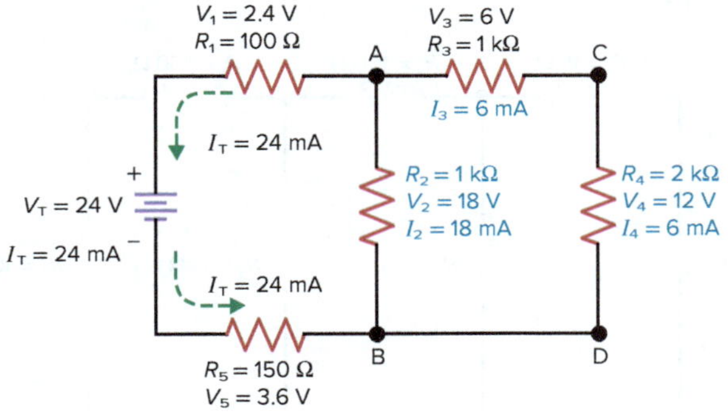

Critical Thinking

6–43 In Fig. 6–47, bulbs *A* and *B* each have an operating voltage of 28 V. If the wattage ratings for bulbs *A* and *B* are 1.12 W and 2.8 W, respectively, calculate (a) the required resistance of R_1; (b) the recommended wattage rating of R_1; (c) the total resistance R_T.

Figure 6–47 Circuit diagram for Critical Thinking Prob. 6–43.

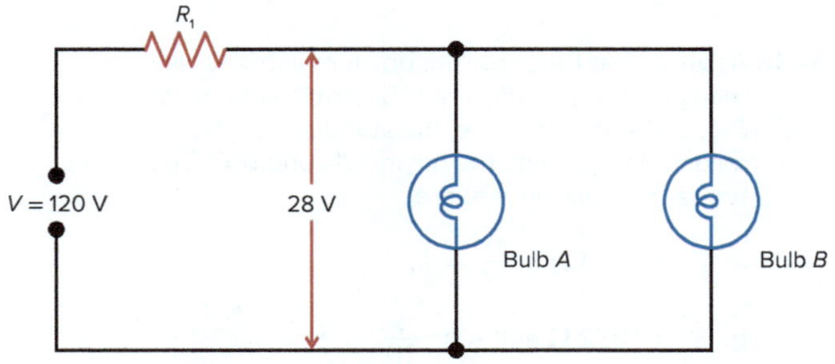

Figure 6–48 Circuit diagram for Critical Thinking Prob. 6–44.

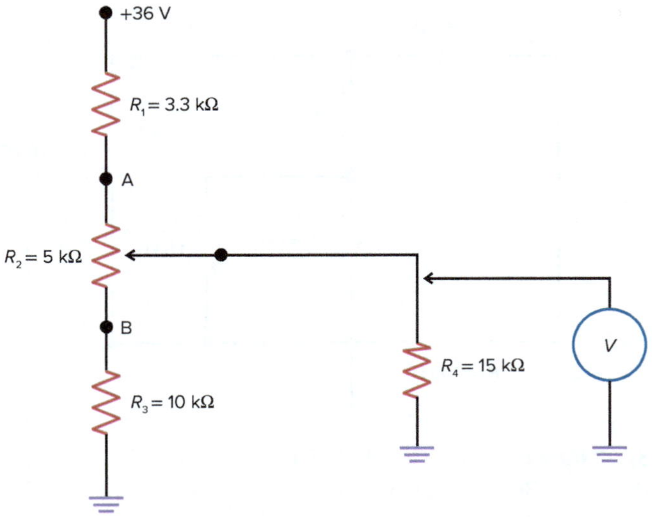

6–44 Refer to Fig. 6–48. How much voltage will be indicated by the voltmeter when the wiper arm of the linear potentiometer R_2 is set (a) to point A; (b) to point B; (c) midway between points A and B?

6–45 Explain how the temperature control circuit in Fig. 6–12 will be affected if the polarity of the applied voltage V_T is reversed.

Troubleshooting Challenge

Table 6–1 shows voltage measurements taken in Fig. 6–49. The first row shows the values that exist when the circuit is operating normally. Rows 2 to 13 are voltage measurements taken when one component in the circuit has failed. For each row in Table 6–1, identify which component is defective and determine the type of defect that has occurred in the component.

Figure 6–49 Circuit diagram for troubleshooting challenge. Normal operating voltages are shown.

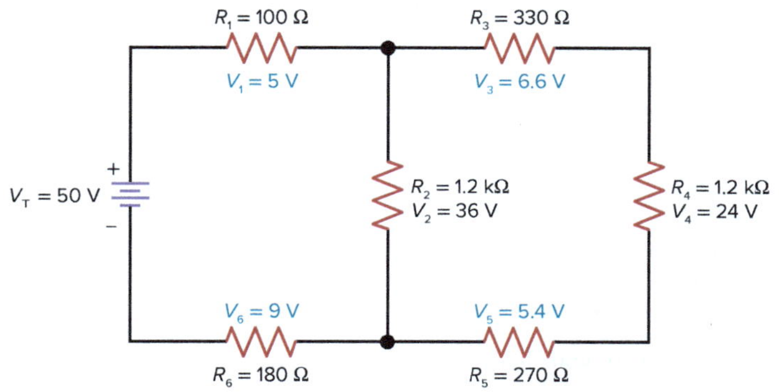

Table 6–1 Voltage Measurements for Troubleshooting Challenge

	V_1	V_2	V_3	V_4	V_5	V_6	Defective Component
	VOLTS						
1 Normal values	5	36	6.6	24	5.4	9	None
2 Trouble 1	0	0	0	0	0	50	
3 Trouble 2	17.86	0	0	0	0	32.14	
4 Trouble 3	3.38	40.54	0	40.54	0	6.08	
5 Trouble 4	3.38	40.54	0	0	40.54	6.08	
6 Trouble 5	6.1	43.9	8.05	29.27	6.59	0	
7 Trouble 6	2.4	43.27	7.93	28.85	6.49	4.33	
8 Trouble 7	50	0	0	0	0	0	
9 Trouble 8	7.35	29.41	16.18	0	13.24	13.24	
10 Trouble 9	0	40	7.33	26.67	6	10	
11 Trouble 10	3.38	40.54	40.54	0	0	6.08	
12 Trouble 11	5.32	35.11	0	28.67	6.45	9.57	
13 Trouble 12	5.25	35.3	7.61	27.7	0	9.45	

Answers to Self-Reviews

6–1 a. 1 kΩ
b. 0.5 kΩ
c. 1.5 kΩ

6–2 a. 12 V
b. 6 A
c. 18 V

6–3 a. 40 V
b. 8 A
c. 4 V

6–4 a. R_1
b. R_4
c. R_6

6–5 a. R_3
b. R_1
c. 4 A
d. 60 V

6–6 a. A and B are input; C and D are output.
b. Zero
c. 71.35 Ω
d. 99.99 Ω
e. 500 Ω

6–7 a. 10 A
b. 1.1 A
c. 12 V
d. 0 V

Laboratory Application Assignment

In this lab application assignment, you will examine three different series-parallel circuits. You will also troubleshoot a series-parallel circuit containing both shorts and opens.

Equipment: Obtain the following items from your instructor:

- Variable dc power supply
- Assortment of carbon-film resistors
- DMM

Series-Parallel Circuit Characteristics

Examine the series-parallel circuit in Fig. 6–50. Calculate and record the following values:

R_T = _____, I_T = _____, V_1 = _____, V_2 = _____,
V_3 = _____, V_4 = _____,
V_{AB} = _____ I_2 = _____, I_3 = _____

Construct the series-parallel circuit in Fig. 6–50. Measure and record the following values. (Note that the power supply connections must be removed to measure R_T.)

R_T = _____, I_T = _____, V_1 = _____, V_2 = _____,
V_3 = _____, V_4 = _____,
V_{AB} = _____ I_2 = _____, I_3 = _____

In Fig. 6–50, identify which components are in series and which components are in parallel. _____

Do your measured values of voltage and current support your answers? _____

Does the current entering point B equal the current leaving point A? _____

Figure 6–50

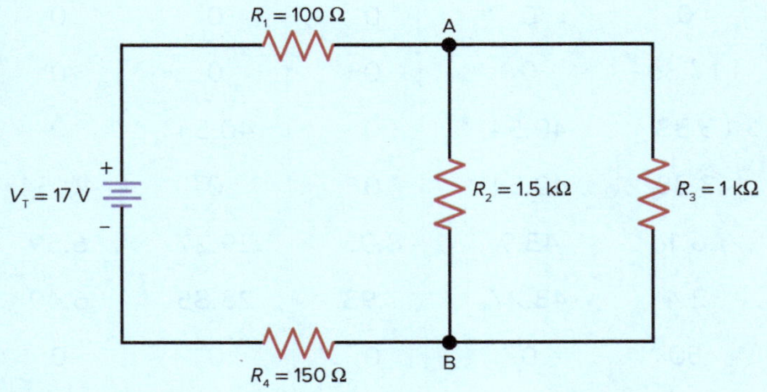

Add the measured values of V_1, V_{AB}, and V_4. Record your answer.

_____

How does this value compare to the value of V_T? Does the sum of these voltages satisfy KVL?

_____

Examine the series-parallel circuit in Fig. 6–51. Calculate and record the branch currents, I_1 and I_2 and the total current, I_T.
I_1 = _____ I_2 = _____ I_T = _____ Next, calculate and record the individual resistor voltage drops V_1, V_2, V_3 and V_4.
V_1 = _____, V_2 = _____, V_3 = _____,
V_4 _____ And finally, calculate and record the total

resistance, R_T. R_T = ________ Construct the circuit in Fig. 6–51. Measure and record the branch currents, I_1 and I_2 and the total current, I_T. I_1 = ________ I_2 = ________ I_T = ________ Next, measure and record the individual resistor voltage drops V_1, V_2, V_3 and V_4. V_1 = ________, V_2 = ________, V_3 = ________, V_4 ________ Finally, measure and record the total resistance R_T. (Note that the power supply connections must be removed to measure R_T.) R_T = ________

Do the measured values of V_1 and V_2 add to equal the applied voltage, V_T? ________ Do the measured values of V_3 and V_4 add to equal V_T? ________ Do the measured values of I_1 and I_2 add to equal the total current, I_T? ________

Examine the series-parallel circuit in Fig. 6–52. Calculate and record the following values:
R_T = ________ I_T = ________ V_1 = ________ V_{AB} = ________
I_2 = ________ I_3 = ________

Construct the circuit in Fig. 6–52. Measure and record the branch currents, I_2 and I_3, and the total current, I_T. I_2 = ________ I_3 = ________ I_T = ________ Next, measure and record the voltages V_1, and V_{AB}. V_1 = ________, V_{AB} = ________ Finally, measure and record the total resistance R_T. (Note that the power supply connections must be removed to measure R_T.) R_T = ________ Do the measured values of I_2 and I_3 add to equal the total current, I_T? ________ Do the measured values of V_1 and V_{AB} add to equal V_T? ________

Figure 6–51

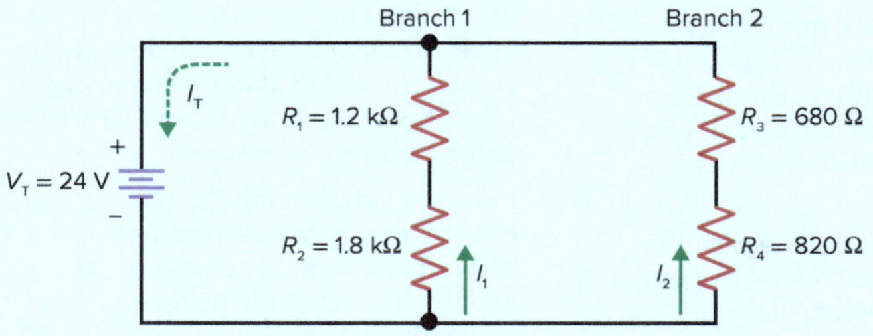

Figure 6–52

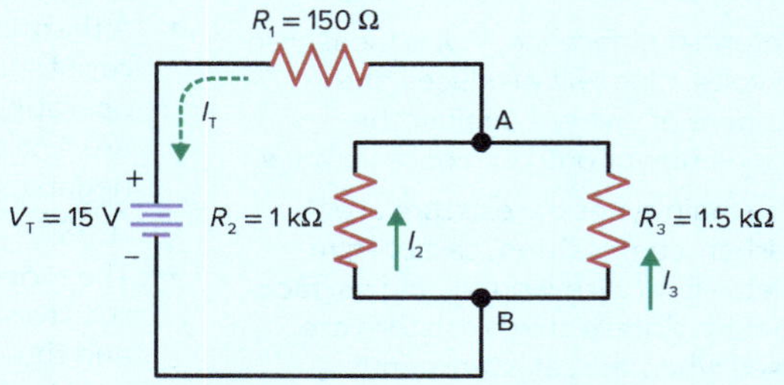

Series-Parallel Circuit Troubleshooting

In this troubleshooting assignment, you will insert faults into the series-parallel circuit of Fig. 6–50. However, you will not be asked to calculate every voltage and current for every possible defect. All you will be asked to do is to insert the fault specified in Table 6–2 and record the measured values for V_1, V_{AB}, V_4, I_T, I_2, and I_3. To simulate a short, replace the original resistor with a 1-Ω resistor. To simulate an open, replace the original resistor with a 1-MΩ resistor. Although you already know which component is defective, this exercise gives you practical hands-on experience in analyzing the effects of opens and shorts in series-parallel circuits. Let's get started.

Refer to Fig. 6–50. Recall the values of V_1, V_{AB}, V_4, I_T, I_2, and I_3, calculated earlier. Write each value next to its respective resistor on the schematic diagram. Next, use a DMM to measure V_1, V_{AB}, V_4, I_T, I_2, and I_3. Record these values in Table 6–2 in the first row labeled "Normal." Next, open R_1 (replace it with a 1-MΩ resistor) and measure V_1, V_{AB}, V_4, I_T, I_2, and I_3. Record these values in the second row of the table. Repeat this procedure for each fault listed in Table 6–2.

Table 6–2 Series-Parallel Circuit Troubleshooting

V_1	V_{AB}	V_4	I_T	I_2	I_3	Circuit Fault
						Normal
						R_1 open
						R_1 shorted
						R_2 open
						R_2 shorted
						R_3 open
						R_3 shorted
						R_4 open
						R_4 shorted

Cumulative Review Summary (Chapters 1–6)

- The electron is the most basic particle of negative electricity; the proton is the most basic particle of positive electricity. Both have the same charge, but they have opposite polarities.
- A quantity of electrons is a negative charge; a deficiency of electrons is a positive charge. Like charges repel each other; unlike charges attract.
- Charge, *Q*, is measured in coulombs; 6.25×10^{18} electrons equals one coulomb of charge. Charge in motion is current. One coulomb of charge flowing past a given point each second equals one ampere of current.
- Potential difference, PD, is measured in volts. One volt produces one ampere of current against the opposition of one ohm of resistance.
- The main types of resistors are carbon-composition, carbon-film, metal-film, wire-wound, and surface-mount. Wire-wound resistors are used when the resistance must dissipate a lot of power, such as 5 W or more. Carbon-film and metal-film resistors are better than the older carbon-composition type because they have tighter tolerances, are less sensitive to temperature changes and aging, and generate less noise internally.
- Resistors having a power rating less than 2 W are often color-coded to indicate their resistance value. To review the color code, refer to Table 2–1 and Fig. 2–8.
- Surface-mount resistors (also called chip resistors) typically use a three-digit number printed on the body to indicate the resistance value in ohms. The first two digits indicate the first two digits in the numerical value of the resistance; the third digit is the multiplier. For example, a surface-mount resistor which is marked 103 has a resistance value of 10,000 Ω or 10 kΩ.
- A potentiometer is a variable resistor that has three terminals. It is used as a variable voltage divider. A rheostat is a variable resistor that has only two terminals. It is used to vary the current in a circuit.
- A thermistor is a resistor whose resistance changes with changes in operating temperature. Thermistors are available with either a positive or a negative temperature coefficient (NTC).
- The most common trouble with resistors is that they develop opens and thus have infinitely high resistance.
- The three forms of Ohm's law are $I = V/R$, $V = IR$, and $R = V/I$.
- The three power formulas are $P = VI$, $P = I^2R$, and $P = V^2/R$.
- The most common multiple and submultiples of the practical units are mega (M) for 10^6, micro (μ) for 10^{-6}, kilo (k) for 10^3, and milli (m) for 10^{-3}.
- For series resistances: (1) the current is the same in all resistances; (2) *IR* drops can be different with unequal resistances; (3) the applied voltage equals the sum of the series *IR* drops; (4) the total resistance equals the sum of the individual resistances; (5) an open circuit in one resistance results in no current through the entire series circuit.
- For parallel resistances: (1) the voltage is the same across all resistances; (2) the branch currents can be different with unequal resistances; (3) the total line current equals the sum of the parallel branch currents; (4) the combined equivalent resistance, R_{EQ}, of parallel branches is less than the smallest resistance as determined by Formula (5–3); (5) an open circuit in one branch does not create an open in the other branches; (6) a short circuit across one branch short-circuits all branches.
- In series-parallel circuits, the resistances in one current path without any branch points are in series; all rules of series resistances apply. The resistances across the same two branch points are in parallel; all rules of parallel resistances apply.
- A Wheatstone bridge has two input terminals and two output terminals. When the bridge is balanced, the voltage across the output terminals is 0 V. When the bridge is unbalanced, however, the output voltage may be either positive or negative. Balanced bridge circuits find many useful applications in electronics.

Cumulative Self-Test

Answers at the back of the book.

1. A carbon resistor is color-coded with brown, green, red, and gold stripes from left to right. Its value is
(*a*) 1500 Ω ±5%;
(*b*) 6800 Ω ±5%;
(*c*) 10,000 Ω ±10%;
(*d*) 500,000 Ω ±5%.

2. A metal-film resistor is color-coded with orange, orange, orange, red, and green stripes, reading from left to right. Its value is
(*a*) 3.3 kΩ ±5%;
(*b*) 333 kΩ ±5%;
(*c*) 33.3 kΩ ±0.5%;
(*d*) 333 Ω ±0.5%.

3. With 30 V applied across two equal resistors in series, 10 mA of current flows. Typical values for each resistor to be used here are
(*a*) 10 Ω, 10 W;
(*b*) 1500 Ω, ½ W;
(*c*) 3000 Ω, 10 W;
(*d*) 30 MΩ, 2 W.

4. In which of the following circuits will the voltage source produce the most current?
(*a*) 10 V across a 10-Ω resistance;
(*b*) 10 V across two 10-Ω resistances in series;
(*c*) 10 V across two 10-Ω resistances in parallel;
(*d*) 1000 V across a 1-MΩ resistance.

5. Three 120-V, 100-W bulbs are in parallel across a 120 V power line. If one bulb burns open
(*a*) the other two bulbs cannot light;
(*b*) all three bulbs light;
(*c*) the other two bulbs can light;
(*d*) there is excessive current in the main line.

6. A circuit allows 1 mA of current to flow with 1 V applied. The conductance of the circuit equals
   (*a*) 0.002 Ω;
   (*b*) 0.005 μS;
   (*c*) 1000 μS;
   (*d*) 1 S.

7. If 2 A of current is allowed to accumulate charge for 5s, the resultant charge equals
   (*a*) 2 C;
   (*b*) 10 C;
   (*c*) 5 A;
   (*d*) 10 A.

8. A potential difference applied across a 1-MΩ resistor produces 1 mA of current. The applied voltage equals
   (*a*) 1 μV;
   (*b*) 1 mV;
   (*c*) 1 kV;
   (*d*) 1,000,000 V.

9. A string of two 1000-Ω resistances is in series with a parallel bank of two 1000-Ω resistances. The total resistance of the series-parallel circuit equals
   (*a*) 250 Ω;
   (*b*) 2500 Ω;
   (*c*) 3000 Ω;
   (*d*) 4000 Ω.

10. In the circuit of question 9, one of the resistances in the series string opens. Then the current in the parallel bank
   (*a*) increases slightly in both branches;
   (*b*) equals zero in one branch but is maximum in the other branch;
   (*c*) is maximum in both branches;
   (*d*) equals zero in both branches.

11. With 100 V applied across a 10,000-Ω resistance, the power dissipation equals
   (*a*) 1 mW;
   (*b*) 1 W;
   (*c*) 100 W;
   (*d*) 1 kW.

12. A source of 10 V is applied across R_1, R_2, and R_3 in series, producing 1 A in the series circuit. R_1 equals 6 Ω and R_2 equals 2 Ω. Therefore, R_3 equals
   (*a*) 2 Ω;
   (*b*) 4 Ω;
   (*c*) 10 Ω;
   (*d*) 12 Ω.

13. A 5-V source and a 3-V source are connected with series-opposing polarities. The combined voltage across both sources equals
   (*a*) 5 V;
   (*b*) 3 V;
   (*c*) 2 V;
   (*d*) 8 V.

14. In a circuit with three parallel branches, if one branch opens, the main-line current will be
   (*a*) more;
   (*b*) less;
   (*c*) the same;
   (*d*) infinite.

15. A 10-Ω R_1 and a 20-Ω R_2 are in series with a 30-V source. If R_1 opens, the voltage drop across R_2 will be
   (*a*) zero;
   (*b*) 20 V;
   (*c*) 30 V;
   (*d*) infinite.

16. A voltage V_1 of 40 V is connected series-opposing with V_2 of 50 V. The total voltage across both components is
   (*a*) 10 V;
   (*b*) 40 V;
   (*c*) 50 V;
   (*d*) 90 V.

17. Two series voltage drops V_1 and V_2 total 100 V for V_T. When V_1 is 60 V, then V_2 must equal
   (*a*) 40 V;
   (*b*) 60 V;
   (*c*) 100 V;
   (*d*) 160 V.

18. Two parallel branch currents I_1 and I_2 total 100 mA for I_T. When I_1 is 60 mA, then I_2 must equal
   (*a*) 40 mA;
   (*b*) 60 mA;
   (*c*) 100 mA;
   (*d*) 160 mA.

19. A surface-mount resistor is marked 224. Its resistance is
   (*a*) 224 Ω;
   (*b*) 220 kΩ;
   (*c*) 224 kΩ;
   (*d*) 22 kΩ.

20. If a variable voltage is connected across a fixed resistance,
   (*a*) *I* and *V* will vary in direct proportion;
   (*b*) *I* and *V* will be inversely proportional;
   (*c*) *I* will remain constant as *V* is varied;
   (*d*) none of the above.

21. If a fixed value of voltage is connected across a variable resistance,
   (*a*) *I* will vary in direct proportion to *R*;
   (*b*) *I* will be inversely proportional to *R*;
   (*c*) *I* will remain constant as *R* is varied;
   (*d*) none of the above.

chapter 7

Voltage Dividers and Current Dividers

Any series circuit is a voltage divider in which the individual resistor voltage drops are proportional to the series resistance values. Similarly, any parallel circuit is a current divider in which the individual branch currents are inversely proportional to the branch resistance values. When parallel-connected loads are added to a series circuit, the circuit becomes a loaded voltage divider. In fact, a loaded voltage divider is just a practical application of a series-parallel circuit.

In a series circuit, it is possible to find the individual resistor voltage drops without knowing the value of the series current. Likewise, it is possible to find the individual branch currents in a parallel circuit without knowing the value of the applied voltage. In this chapter, you will learn how to solve for the voltages in a series circuit and the currents in a parallel circuit using special formulas that provide shortcuts in the calculations. You will also learn how to design a loaded voltage divider that provides different load voltages and currents from a single supply voltage, V_T. ■

Chapter Outline

Chapter Objectives

After studying this chapter, you should be able to

- *Use* the voltage divider rule to calculate the voltage drops in an unloaded voltage divider.
- *Explain* why resistor voltage drops are proportional to the series resistance values in a series circuit.
- *Use* the current divider rule to calculate the branch currents in a parallel circuit.
- *Explain* why the branch currents are inversely proportional to the branch resistances in a parallel circuit.
- *Define* what is meant by the term *loaded voltage divider.*
- *Calculate* the voltage, current, and power values in a loaded voltage divider.

Important Terms

bleeder current
current divider
Current divider rule (CDR)
load currents
loaded voltage
voltage divider
Voltage divider rule (VDR)
voltage taps

GOOD TO KNOW

In the formula $V_R = IR$, substitute $\frac{V_T}{R_T}$ for I. This gives $V_R = \frac{V_T}{R_T} \times R$. Interchanging V_T and R gives $V_R = \frac{R}{R_T} \times V_T$, which is Formula (7–1).

7–1 Series Voltage Dividers

The current is the same in all resistances in a series circuit. Also, the voltage drops equal the product of I times R. Therefore, the IR voltages are proportional to the series resistances. A higher resistance has a greater IR voltage than a lower resistance in the same series circuit; equal resistances have the same amount of IR drop. If R_1 is double R_2, then V_1 will be double V_2.

The series string can be considered a ***voltage divider***. Each resistance provides an IR drop V_R equal to its proportional part of the applied voltage. Stated as a formula,

$$V_R = \frac{R}{R_T} \times V_T \qquad \textbf{(7–1)}$$

Formula (7–1) is called the **voltage divider rule (VDR)** because it allows us to calculate the voltage drops in a series circuit without knowing the value of the current, I.

Example 7-1

Three 50-kΩ resistors R_1, R_2, and R_3 are in series across an applied voltage of 180 V. How much is the IR voltage drop across each resistor?

ANSWER The voltage drop across each R is 60 V. Since R_1, R_2, and R_3 are equal, each resistor has one-third the total resistance of the circuit and one-third the total applied voltage. Using the formula,

$$V_R = \frac{R}{R_T} \times V_T = \frac{50\ \text{k}\Omega}{150\ \text{k}\Omega} \times 180\ \text{V}$$

$$= \frac{1}{3} \times 180\ \text{V}$$

$$= 60\ \text{V}$$

Note that R and R_T must be in the same units for the proportion. Then V is in the same units as V_T.

MultiSim **Figure 7-1** Series string of resistors as a proportional voltage divider. Each V_R is R/R_T fraction of the total source voltage V_T.

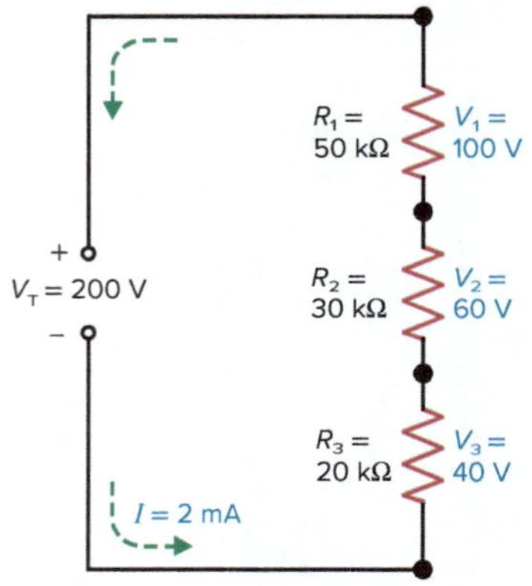

Typical Circuit

Figure 7–1 illustrates another example of a proportional voltage divider. Let the problem be to find the voltage across R_3. We can either calculate this voltage V_3 as IR_3 or determine its proportional part of the total applied voltage V_T. The answer is the same both ways. Note that R_T is 20 + 30 + 50 = 100 kΩ.

Proportional Voltage Method

Using Formula (7–1), V_3 equals 20/100 of the 200 V applied for V_T because R_3 is 20 kΩ and R_T is 100 kΩ. Then V_3 is 20/100 of 200 or ⅕ of 200, which is equal to 40 V. The calculations are

$$V_3 = \frac{R_3}{R_T} \times V_T = \frac{20}{100} \times 200\ \text{V}$$

$$V_3 = 40\ \text{V}$$

CALCULATOR

To do a problem like this on the calculator, you can divide R_3 by R_T first and then multiply by V_T. For the values here, to find V_3, the procedure can be as follows:

Punch in the number 20 for R_3.

Push the ⊕ key, then 100 for R_T and press the ⊗ key for the quotient of 0.2 on the display.

Next, punch in 200 for V_T and press the ⊜ key to display 40 as the answer for V_3.

As another method, you can multiply R_3 by V_T first and then divide by R_T. The answers will be the same for either method.

In the same way, V_2 is 60 V. The calculations are

$$V_2 = \frac{R_2}{R_T} \times V_T = \frac{30}{100} \times 200 \text{ V}$$

$$V_2 = 60 \text{ V}$$

Also, V_1 is 100 V. The calculations are

$$V_1 = \frac{R_1}{R_T} \times V_T = \frac{50}{100} \times 200 \text{ V}$$

$$V_1 = 100 \text{ V}$$

The sum of V_1, V_2, and V_3 in series is $100 + 60 + 40 = 200$ V, which is equal to V_T.

Method of *IR* Drops

If we want to solve for the current in Fig. 7–1, $I = V_T/R_T$ or 200 V/100 kΩ = 2 mA. This I flows through R_1, R_2, and R_3 in series. The IR drops are

$$V_1 = I \times R_1 = 2 \text{ mA} \times 50 \text{ k}\Omega = 100 \text{ V}$$
$$V_2 = I \times R_2 = 2 \text{ mA} \times 30 \text{ k}\Omega = 60 \text{ V}$$
$$V_3 = I \times R_3 = 2 \text{ mA} \times 20 \text{ k}\Omega = 40 \text{ V}$$

These voltages are the same values calculated by Formula (7–1) for proportional voltage dividers.

Two Voltage Drops in Series

For this case, it is not necessary to calculate both voltages. After you find one, subtract it from V_T to find the other.

As an example, assume that V_T is 48 V across two series resistors R_1 and R_2. If V_1 is 18 V, then V_2 must be $48 - 18 = 30$ V.

The Largest Series *R* Has the Most *V*

The fact that series voltage drops are proportional to the resistances means that a very small R in series with a much larger R has a negligible IR drop. An example is shown in Fig. 7–2*a*. Here the 1 kΩ of R_1 is in series with the much larger 999 kΩ of R_2. The V_T is 1000 V.

The voltages across R_1 and R_2 in Fig. 7–2*a* can be calculated using the voltage divider rule. Note that R_T is $1 + 999 = 1000$ kΩ.

$$V_1 = \frac{R_1}{R_T} \times V_T = \frac{1}{1000} \times 1000 \text{ V} = 1 \text{ V}$$

$$V_2 = \frac{R_2}{R_T} \times V_T = \frac{999}{1000} \times 1000 \text{ V} = 999 \text{ V}$$

The 999 V across R_2 is practically the entire applied voltage. Also, the very high series resistance dissipates almost all the power.

The current of 1 mA through R_1 and R_2 in Fig. 7–2*a* is determined almost entirely by the 999 kΩ of R_2. The I for R_T is 1000 V/1000 kΩ, which equals 1 mA. However, the 999 kΩ of R_2 alone would allow 1.001 mA of current, which differs very little from the original I of 1 mA.

Voltage Taps in a Series Voltage Divider

Consider the series voltage divider with **voltage taps** in Fig. 7–2*b*, where different voltages are available from the tap points A, B, and C. Note that the total resistance R_T is 20 kΩ, which can be found by adding the individual series resistance values. The voltage at each tap point is measured with respect to ground. The voltage at tap

Figure 7–2 (*a*) Example of a very small R_1 in series with a large R_2; V_2 is almost equal to the whole V_T. (*b*) Series voltage divider with voltage taps.

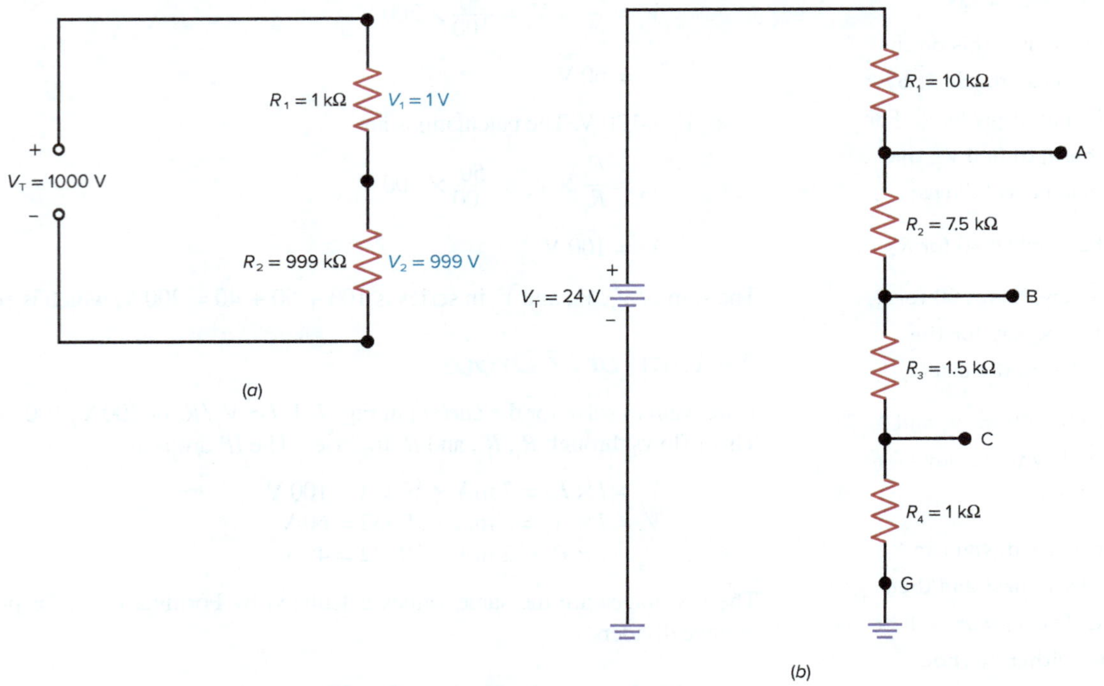

point C, designated V_{CG}, is the same as the voltage across R_4. The calculations for V_{CG} are as follows:

$$V_{CG} = \frac{R_4}{R_T} \times V_T$$

$$= \frac{1\ \text{k}\Omega}{20\ \text{k}\Omega} \times 24\ \text{V}$$

$$V_{CG} = 1.2\ \text{V}$$

The voltage at tap point B, designated V_{BG}, is the sum of the voltages across R_3 and R_4. The calculations for V_{BG} are

$$V_{BG} = \frac{R_3 + R_4}{R_T} \times V_T$$

$$= \frac{1.5\ \text{k}\Omega + 1\ \text{k}\Omega}{20\ \text{k}\Omega} \times 24\ \text{V}$$

$$V_{BG} = 3\ \text{V}$$

The voltage at tap point A, designated V_{AG}, is the sum of the voltages across R_2, R_3, and R_4. The calculations are

$$V_{AG} = \frac{R_2 + R_3 + R_4}{R_T} \times V_T$$

$$= \frac{7.5\ \text{k}\Omega + 1.5\ \text{k}\Omega + 1\ \text{k}\Omega}{20\ \text{k}\Omega} \times 24\ \text{V}$$

$$V_{AG} = 12\ \text{V}$$

Notice that the voltage V_{AG} equals 12 V, which is one-half of the applied voltage V_T. This makes sense, since $R_2 + R_3 + R_4$ make up 50% of the total resistance R_T. Similarly, since $R_3 + R_4$ make up 12.5% of the total resistance, the voltage V_{BG} will also be 12.5% of the applied voltage, which is 3 V in this case. The same analogy applies to V_{CG}.

Each tap voltage is positive because the negative terminal of the voltage source is grounded.

Potentiometer as a Variable Voltage Divider

Figure 7–3 shows a voltage divider where one of the resistors is a 10 kΩ potentiometer. Recall from Chapter 2, ***Resistors***, that the resistance across the outside terminals of a potentiometer is fixed or constant. Also, recall that the resistance between the wiper arm and either outside terminal varies as the shaft of the potentiometer is rotated. In Fig. 7–3, the outside terminals of the potentiometer are labeled ***A*** and ***B***, whereas the wiper arm is labeled as terminal ***C***. Since the resistance across terminals ***A*** and ***B*** does not vary as the wiper is moved up and down, the total resistance, R_T, is calculated as

$$R_T = R_1 + R_{AB}$$
$$= 15\ \text{k}\Omega + 10\ \text{k}\Omega$$
$$= 25\ \text{k}\Omega$$

With an applied voltage, V_T, of 25 V, the current, I, is calculated as:

$$I = \frac{V_T}{R_T}$$
$$= \frac{25\ \text{V}}{25\ \text{k}\Omega}$$
$$= 1\ \text{mA}$$

It is important to realize that the current, I, does not vary as the wiper of the potentiometer is moved up and down. The reason I doesn't vary is because the resistance, R_{AB}, has a fixed value of 10 kΩ.

The voltage across terminals ***A*** and ***B*** of the potentiometer can be calculated using the voltage divider rule:

$$V_{AB} = \frac{R_{AB}}{R_T} \times V_T$$
$$= \frac{10\ \text{k}\Omega}{25\ \text{k}\Omega} \times 25\ \text{V}$$
$$= 10\ \text{V}$$

With 10 V across the outside terminals of the potentiometer, the voltage at terminal C, with respect to ground, is adjustable over a range of 0 to 10 V. Assuming the potentiometer has a linear taper, the voltage at the wiper arm varies linearly as the shaft of the potentiometer is rotated through its range. For example, if the wiper is positioned midway between terminals ***A*** and ***B***, the voltage from terminal ***C*** to ground is 5 V.

Figure 7–3 Variable voltage divider.

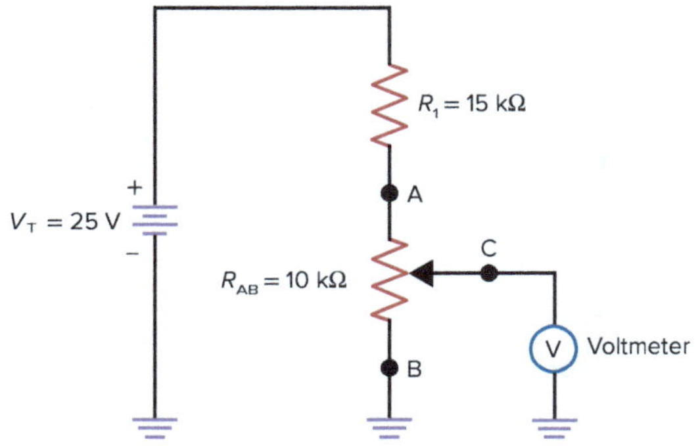

Advantage of the Voltage Divider Method

Using Formula (7–1), we can find the proportional voltage drops from V_T and the series resistances without knowing the amount of I. For odd values of R, calculating the I may be more troublesome than finding the proportional voltages directly. Also, in many cases, we can approximate the voltage division without the need for any written calculations.

7–1 Self-Review

Answers at the end of the chapter.

Refer to Fig. 7–1 for a to c.

a. How much is R_T?

b. What fraction of the applied voltage is V_3?

c. If each resistance is doubled in value, how much is V_1?

d. In Fig. 7–2*b*, how much is the voltage V_{BG} if resistors R_2 and R_3 are interchanged?

e. In Fig. 7–3, what is the adjustable voltage range at terminal C if the applied voltage, V_T, is changed to 15 V?

7–2 Current Divider with Two Parallel Resistances

It is often necessary to find the individual branch currents in a bank from the resistances and I_T, but without knowing the voltage across the bank. This problem can be solved by using the fact that currents divide inversely as the branch resistances. An example is shown in the ***current divider*** in Fig. 7–4. The formulas for the two branch currents are as follows:

$$I_1 = \frac{R_2}{R_1 + R_2} \times I_T$$

$$I_2 = \frac{R_1}{R_1 + R_2} \times I_T \qquad \textbf{(7–2)}$$

Formula 7–2 is called the **current divider rule (CDR)** because it allows us to calculate the branch currents in a parallel circuit without knowing the value of the applied voltage, V_A.

Notice that the formula for each branch I has the opposite R in the numerator. The reason is that each branch current is inversely proportional to the branch resistance.

CALCULATOR

To use the calculator for a problem like this with current division between two branch resistances, as in Formula (7–2), there are several points to note. The numerator has the R of the branch opposite from the desired I. In adding R_1 and R_2, the parentheses (parens) keys (and) should be used. The reason is that both terms in the denominator must be added before the division. The procedure for calculating I_1 in Fig. 7–4 can be as follows:

Punch in 4 for R_2.

Press the ÷ key followed by the opening parens key (.

Punch in 2 + 4 for R_1 and R_2 followed by the closing parens key). The sum of 6 will be displayed.

Press × and 30, then press = to display the answer 20 for I_1.

Figure 7–4 Current divider with two branch resistances. Each branch I is inversely proportional to its R. The smaller R has more I.

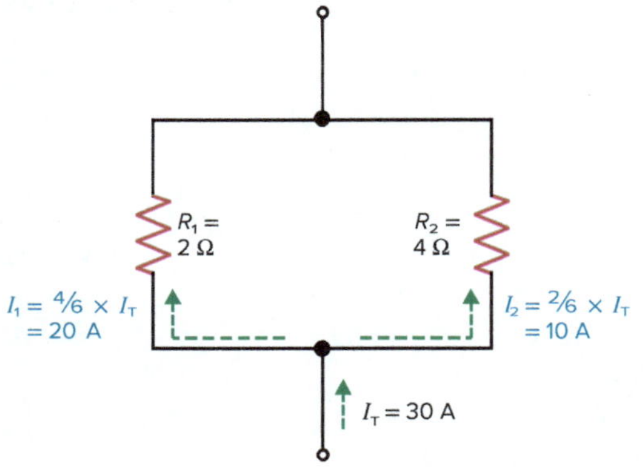

The denominator is the same in both formulas, equal to the sum of the two branch resistances.

To calculate the currents in Fig. 7–4, with a 30-A I_T, a 2-Ω R_1, and a 4-Ω R_2, use the current divider rule:

$$I_1 = \frac{4}{2+4} \times 30$$
$$= \frac{4}{6} \times 30$$
$$I_1 = 20 \text{ A}$$

For the other branch,

$$I_2 = \frac{2}{2+4} \times 30$$
$$= \frac{2}{6} \times 30$$
$$I_2 = 10 \text{ A}$$

With all the resistances in the same units, the branch currents are in units of I_T.

In fact, it is not necessary to calculate both currents. After one I is calculated, the other can be found by subtracting from I_T.

Notice that the division of branch currents in a parallel bank is opposite from the voltage division of resistance in a series string. With series resistances, a higher resistance develops a higher IR voltage proportional to its R; with parallel branches, a lower resistance takes more branch current, equal to V/R.

In Fig. 7–4, the 20-A I_1 is double the 10-A I_2 because the 2-Ω R_1 is one-half the 4-Ω R_2. This is an inverse relationship between I and R.

The inverse relation between I and R in a parallel bank means that a very large R has little effect with a much smaller R in parallel. As an example, Fig. 7–5 shows a 999-kΩ R_2 in parallel with a 1-kΩ R_1 dividing the I_T of 1000 mA. The branch currents are calculated as follows:

$$I_1 = \frac{999}{1000} \times 1000 \text{ mA}$$
$$= 999 \text{ mA}$$
$$I_2 = \frac{1}{1000} \times 1000 \text{ mA}$$
$$= 1 \text{ mA}$$

The 999 mA for I_1 is almost the entire line current of 1000 mA because R_1 is so small compared with R_2. Also, the smallest branch R dissipates the most power because it has the most I.

The current divider rule, Formula (7–2), can be used only for two branch resistances. The reason is the inverse relation between each branch I and its R. In comparison, the voltage divider rule, Formula (7–1), can be used for any number of series resistances because of the direct proportion between each voltage drop V and its R.

For more branches, it is possible to combine the branches to work with only two divided currents at a time. However, a better method is to use parallel conductances, because I and G are directly proportional, as explained in the next section.

Figure 7–5 Example of a very large R_2 in parallel with a small R_1. For the branch currents, the small R_1 has almost the entire total line current, I_T.

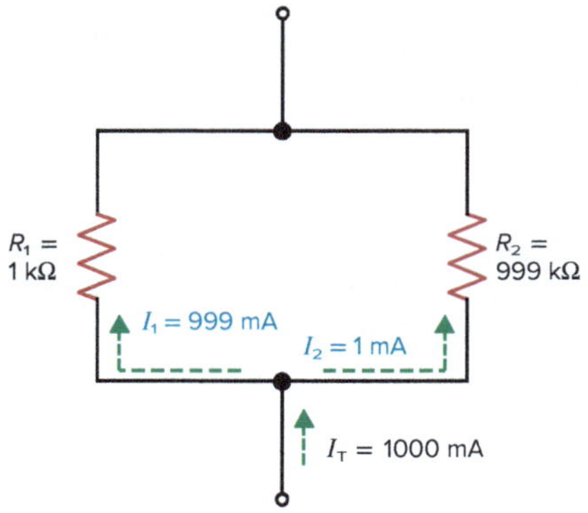

GOOD TO KNOW

For any number of resistances in parallel, the individual branch currents can be calculated as $I_R = \frac{R_{EQ}}{R} \times I_T$, where I_R and R represent the individual branch current and resistance, respectively. This is another form of the current divider rule (CDR).

7–2 Self-Review

Answers at the end of the chapter.

Refer to Fig. 7–4.

a. What is the ratio of R_2 to R_1?

b. What is the ratio of I_2 to I_1?

7–3 Current Division by Parallel Conductances

Remember that the conductance G is $1/R$. Therefore, conductance and current are directly proportional. More conductance allows more current, for the same V. With any number of parallel branches, each branch current is

$$I = \frac{G}{G_T} \times I_T \qquad (7\text{–}3)$$

where G is the conductance of one branch and G_T is the sum of all the parallel conductances. The unit for G is the siemens (S).

Note that Formula (7–3), for dividing branch currents in proportion to G, has the same form as Formula (7–1) for dividing series voltages in proportion to R. The reason is that both formulas specify a direct proportion.

GOOD TO KNOW

Formula 7-3 is yet another form of the current divider rule (CDR).

Two Branches

As an example of using Formula (7–3), we can go back to Fig. 7–4 and find the branch currents with G instead of R. For the 2 Ω of R_1, the G_1 is ½ = 0.5 S. The 4 Ω of R_2 has G_2 of ¼ = 0.25 S. Then G_T is 0.5 + 0.25 = 0.75 S.

The I_T is 30 A in Fig. 7–4. For the branch currents,

$$I_1 = \frac{G_1}{G_T} \times I_T = \frac{0.50}{0.75} \times 30\text{ A}$$
$$I_1 = 20\text{ A}$$

This 20 A is the same I_1 calculated before.

For the other branch, I_2 is 30 − 20 = 10 A. Also, I_2 can be calculated as 0.25/0.75 or ⅓ of I_T for the same 10-A value.

CALCULATOR

To calculate I_1 with a calculator, use the same procedure described before with Formula (7–1) for voltage dividers. Using Formula (7–3) for a proportional current divider with conductance, first divide G by G_T and then multiply the answer by I_T. The key strokes will be as follows: G (÷) G_T (×) I_T (=). Remember that both Formulas (7–1) and (7–3) are mathematically similar as a direct proportion.

Three Branches

A circuit with three branch currents is shown in Fig. 7–6. We can find G for the 10-Ω R_1, 2-Ω R_2, and 5-Ω R_3 as follows.

$$G_1 = \frac{1}{R_1} = \frac{1}{10\ \Omega} = 0.1\text{ S}$$
$$G_2 = \frac{1}{R_2} = \frac{1}{2\ \Omega} = 0.5\text{ S}$$
$$G_3 = \frac{1}{R_3} = \frac{1}{5\ \Omega} = 0.2\text{ S}$$

MultiSim **Figure 7–6** Current divider with branch conductances G_1, G_2, and G_3, each equal to $1/R$. Note that S is the siemens unit for conductance. With conductance values, each branch I is directly proportional to the branch G.

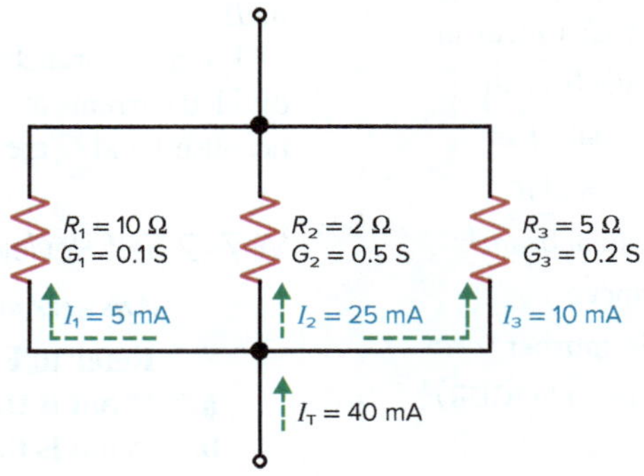

Remember that the siemens (S) unit is the reciprocal of the ohm (Ω) unit. The total conductance then is

$$G_T = G_1 + G_2 + G_3$$
$$= 0.1 + 0.5 + 0.2$$
$$G_T = 0.8 \text{ S}$$

The I_T is 40 mA in Fig. 7–6. To calculate the branch currents with Formula (7–3),

$$I_1 = 0.1/0.8 \times 40 \text{ mA} = 5 \text{ mA}$$
$$I_2 = 0.5/0.8 \times 40 \text{ mA} = 25 \text{ mA}$$
$$I_3 = 0.2/0.8 \times 40 \text{ mA} = 10 \text{ mA}$$

The sum is 5 + 25 + 10 = 40 mA for I_T.

Although three branches are shown here, Formula (7–3) can be used to find the currents for any number of parallel conductances because of the direct proportion between *I* and *G*. The method of conductances is usually easier to use than the method of resistances for three or more branches.

7–3 Self-Review

Answers at the end of the chapter.

Refer to Fig. 7–6.

a. What is the ratio of G_3 to G_1?

b. What is the ratio of I_3 to I_1?

7–4 Series Voltage Divider with Parallel Load Current

The voltage dividers shown so far illustrate just a series string without any branch currents. However, a voltage divider is often used to tap off part of the applied voltage for a load that needs less voltage than V_T. Then the added load is a parallel branch across part of the divider, as shown in Fig. 7–7. This example shows how the **loaded voltage** at the tap F is reduced below the potential it would have without the branch current for R_L.

MultiSim **Figure 7–7** Effect of a parallel load in part of a series voltage divider. (*a*) R_1 and R_2 in series without any branch current. (*b*) Reduced voltage across R_2 and its parallel load R_L. (*c*) Equivalent circuit of the loaded voltage divider.

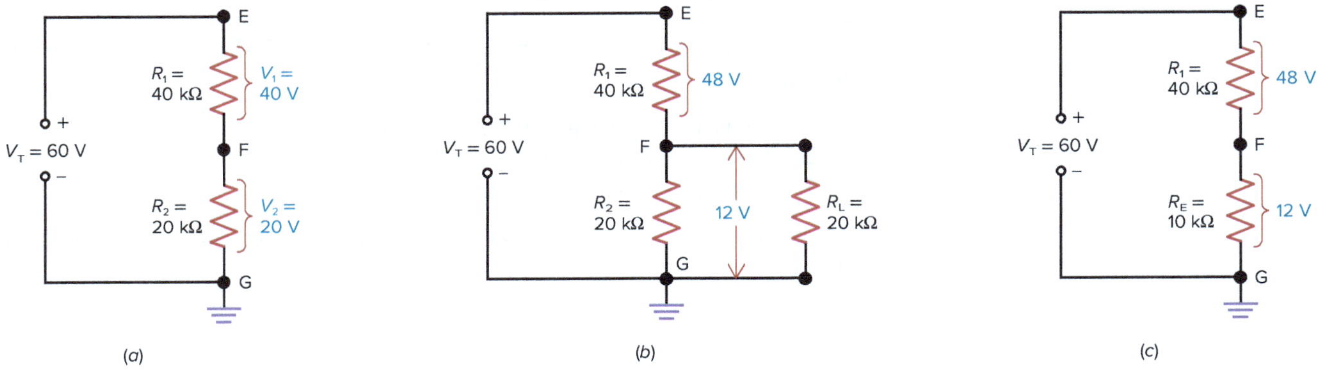

Why the Loaded Voltage Decreases

We can start with Fig. 7–7*a*, which shows an R_1–R_2 voltage divider alone. Resistances R_1 and R_2 in series simply form a proportional divider across the 60-V source for V_T.

For the resistances, R_1 is 40 kΩ and R_2 is 20 kΩ, making R_T equal to 60 kΩ. Also, the current $I = V_T/R_T$, or 60 V/60 kΩ = 1 mA. For the divided voltages in Fig. 7–7*a*,

$$V_1 = \frac{40}{60} \times 60 \text{ V} = 40 \text{ V}$$

$$V_2 = \frac{20}{60} \times 60 \text{ V} = 20 \text{ V}$$

Note that $V_1 + V_2$ is 40 + 20 = 60 V, which is the total applied voltage.

However, in Fig. 7–7*b,* the 20-kΩ branch of R_L changes the equivalent resistance at tap F to ground. This change in the proportions of R changes the voltage division. Now the resistance from F to G is 10 kΩ, equal to the 20-kΩ R_2 and R_L in parallel. This equivalent bank resistance is shown as the 10-kΩ R_E in Fig. 7–7*c*.

Resistance R_1 is still the same 40 kΩ because it has no parallel branch. The new R_T for the divider in Fig. 7–7*c* is 40 kΩ +10 kΩ = 50 kΩ. As a result, V_E from F to G in Fig. 7–6*c* becomes

$$V_E = \frac{R_E}{R_T} \times V_T = \frac{10}{50} \times 60 \text{ V}$$

$$V_E = 12 \text{ V}$$

Therefore, the voltage across the parallel R_2 and R_L in Fig. 7–7*b* is reduced to 12 V. This voltage is at the tap F for R_L.

Note that V_1 across R_1 increases to 48 V in Fig. 7–7*c*. Now V_1 is 40/50 × 60 V = 48 V. The V_1 increases here because there is more current through R_1.

The sum of $V_1 + V_E$ in Fig. 7–7*c* is 12 + 48 = 60 V. The *IR* drops still add to equal the applied voltage.

Path of Current for R_L

All current in the circuit must come from the source V_T. Trace the electron flow for R_L. It starts from the negative side of V_T, through R_L, to the tap at F, and returns through R_1 in the divider to the positive side of V_T. This current I_L goes through R_1 but not R_2.

Bleeder Current

In addition, both R_1 and R_2 have their own current from the source. This current through all the resistances in the divider is called the **bleeder current** I_B. The electron flow for I_B is from the negative side of V_T, through R_2 and R_1, and back to the positive side of V_T.

In summary, then, for the three resistances in Fig. 7–7*b*, note the following currents:

1. Resistance R_L has just its load current I_L.
2. Resistance R_2 has only the bleeder current I_B.
3. Resistance R_1 has both I_L and I_B.

Note that only R_1 is in the path for both the bleeder current and the load current.

7–4 Self-Review

Answers at the end of the chapter.

Refer to Fig. 7–7.

a. What is the proportion of R_2/R_T in Fig. 7–7*a*?

b. What is the proportion of R_E/R_T in Fig. 7–7*c*?

7–5 Design of a Loaded Voltage Divider

These principles can be applied to the design of a loaded voltage divider, like the one shown in Fig. 7–8. This type of circuit is used when it is necessary or desired to obtain different load voltage and load current values from a single supply voltage. The loads D, E, and F represent individual circuits within a complex electrical system. Each of these loads requires different voltage and current values for normal operation.

Note the load specifications in Fig. 7–8. Load F needs 18 V from point F to chassis ground. When the 18 V is supplied by this part of the divider, a 36-mA branch current will flow through the load. Similarly, 40 V is needed at tap E for 54 mA of I_E in load E. Also, 100 V is available at D with a load current I_D of 180 mA. The total load current here is 36 + 54 + 180 = 270 mA.

In addition, the bleeder current I_B through the entire divider is generally specified at about 10% of the total load current. For the example here, I_B is taken as 30 mA to make a total line current I_T of 270 + 30 = 300 mA from the voltage source. Remember that the 30-mA I_B flows through R_1, R_2, and R_3.

The design problem in Fig. 7–8 is to find the values of R_1, R_2, and R_3 needed to provide the specified voltages. Each R is calculated as its ratio of V/I. However, the question is what are the correct values of V and I to use for each part of the divider.

GOOD TO KNOW

A loaded voltage divider is just a practical application of a series-parallel circuit.

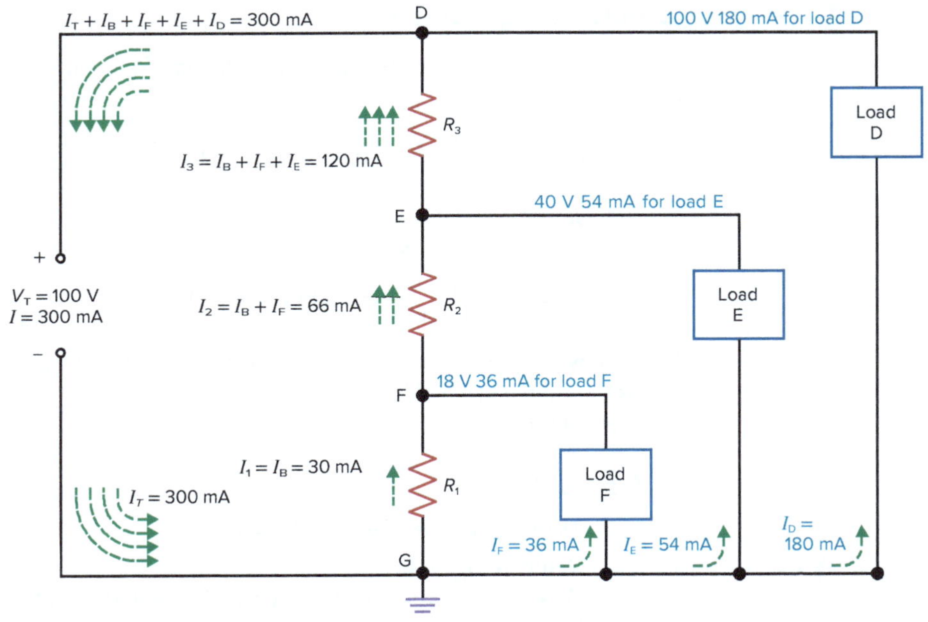

Figure 7–8 Voltage divider for different voltages and currents from the source V_T. See text for design calculations to find the values of R_1, R_2, and R_3.

Table 7–1	Design Values for Voltage Divider		
For Figure 7–8	Current, mA	Voltage, V	Resistance, Ω
R_1	30	18	600
R_2	66	22	333
R_3	120	60	500

Find the Current in Each *R*

We start with R_1 because its current is only the 30-mA bleeder current I_B. No load current flows through R_1. Therefore, I_1 through R_1 equals 30 mA.

The 36-mA current I_F for load F returns to the source through R_2 and R_3. Considering just R_2 now, its current is the I_F load current and the 30-mA bleeder current I_B. Therefore, I_2 through R_2 is 36 + 30 = 66 mA.

The 54-mA current I_E for load E returns to the source through R_3 alone. However, R_3 also has the 36-mA I_F and the 30-mA I_B. Therefore, I_3 through R_3 is 54 + 36 + 30 = 120 mA. The values for I_1, I_2, and I_3 are given in Table 7–1.

Note that the load current I_D for load D at the top of the diagram does not flow through R_3 or any of the resistors in the divider. However, the I_D of 180 mA is the main load current through the source of applied voltage. The 120 mA of bleeder and **load currents** plus the 180-mA I_D load add to equal 300 mA for I_T in the main line of the power supply.

Calculate the Voltage across Each *R*

The voltages at the taps in Fig. 7–8 give the potential to chassis ground. But we need the voltage across the two ends of each *R*. For R_1, the voltage V_1 is the indicated 18 V to ground because one end of R_1 is grounded. However, the voltage across R_2 is the difference between the 40-V potential at point E and the 18 V at F. Therefore, V_2 is 40 − 18 = 22 V. Similarly, V_3 is calculated as 100 V at point D minus the 40 V at E, or, V_3 is 100 − 40 = 60 V. These values for V_1, V_2, and V_3 are summarized in Table 7–1.

Calculating Each *R*

Now we can calculate the resistance of R_1, R_2, and R_3 as each V/I ratio. For the values listed in Table 7–1,

$$R_1 = \frac{V_1}{I_1} = \frac{18\text{ V}}{30\text{ mA}} = 0.6\text{ k}\Omega = 600\ \Omega$$

$$R_2 = \frac{V_2}{I_2} = \frac{22\text{ V}}{66\text{ mA}} = 0.333\text{ k}\Omega = 333\ \Omega$$

$$R_3 = \frac{V_3}{I_3} = \frac{60\text{ V}}{120\text{ mA}} = 0.5\text{ k}\Omega = 500\ \Omega$$

When these values are used for R_1, R_2, and R_3 and connected in a voltage divider across the source of 100 V, as in Fig. 7–8, each load will have the specified voltage at its rated current.

7–5 Self-Review

Answers at the end of the chapter.

Refer to Fig. 7–8.

a. How much is the bleeder current I_B through R_1, R_2, and R_3?
b. How much is the voltage for load E at tap E to ground?
c. How much is V_2 across R_2?
d. If load D opens, how much voltage will be measured at tap F to ground?

Summary

- In a series circuit, V_T is divided into IR voltage drops proportional to the resistances. Each $V_R = (R/R_T) \times V_T$, for any number of series resistances. The largest series R has the largest voltage drop.
- In a parallel circuit, I_T is divided into branch currents. Each I is inversely proportional to the branch R. The inverse division of branch currents is given by Formula (7–2) for only two resistances. The smaller branch R has the larger branch current.
- For any number of parallel branches, I_T is divided into branch currents directly proportional to each conductance G. Each $I = (G/G_T) \times I_T$.
- A series voltage divider is often tapped for a parallel load, as in Fig. 7–7. Then the voltage at the tap is reduced because of the load current.
- The design of a loaded voltage divider, as shown in Fig. 7–8, involves calculating each R. Find the I and potential difference V for each R. Then $R = V/I$.

Important Terms

Bleeder current — the current that flows through all resistors in a loaded voltage divider. The bleeder current, designated I_B, is generally specified at about 10% of the total load current.

Current divider — any parallel circuit is a current divider in which the individual branch currents are inversely proportional to the branch resistance values. With respect to conductances, the individual branch currents are directly proportional to the branch conductance values.

Current divider rule (CDR) — a formula that allows us to calculate the individual branch currents in a parallel circuit without knowing the value of the applied voltage. See Formulas (7–2) and (7–3).

Load currents — the currents drawn by the electronic devices and/or components connected as loads in a loaded voltage divider.

Loaded voltage — the voltage at a point in a series voltage divider where a parallel load has been connected.

Voltage divider — any series circuit is a voltage divider in which the individual resistor voltage drops are proportional to the series resistance values.

Voltage taps — the points in a series voltage divider that provide different voltages with respect to ground.

Voltage divider rule (VDR) — a formula that allows us to calculate the voltage drops in a series circuit without knowing the current, I. See Formula (7–1).

Related Formulas

$$V_R = \frac{R}{R_T} \times V_T$$

For two resistors in parallel: $I_1 = \dfrac{R_2}{R_1 + R_2} \times I_T$ $\qquad I_2 = \dfrac{R_1}{R_1 + R_2} \times I_T$

$$I = \frac{G}{G_T} \times I_T$$

Self-Test

Answers at the back of the book.

1. In a series circuit, the individual resistor voltage drops are

a. inversely proportional to the series resistance values.

b. proportional to the series resistance values.

c. unrelated to the series resistance values.

d. none of the above

2. In a parallel circuit, the individual branch currents are

a. not related to the branch resistance values.

b. directly proportional to the branch resistance values.

c. inversely proportional to the branch resistance values.

d. none of the above

3. Three resistors R_1, R_2, and R_3 are connected in series across an applied voltage, V_T, of 24 V. If R_2 is one-third the value of R_T, how much is V_2?

a. 8 V.

b. 16 V.

c. 4 V.

d. It cannot be determined.

4. **Two resistors R_1 and R_2 are in parallel. If R_1 is twice the value of R_2, how much is I_2 in R_2 if I_T equals 6 A?**
   a. 1 A.
   b. 2 A.
   c. 3 A.
   d. 4 A.

5. **Two resistors R_1 and R_2 are in parallel. If the conductance, G_1, of R_1 is twice the value of the conductance, G_2 of R_2, how much is I_2 if $I_T = 6$ A?**
   a. 1 A.
   b. 2 A.
   c. 3 A.
   d. 4 A.

PROBLEMS 6–10 REFER TO FIG. 7–9.

Figure 7–9

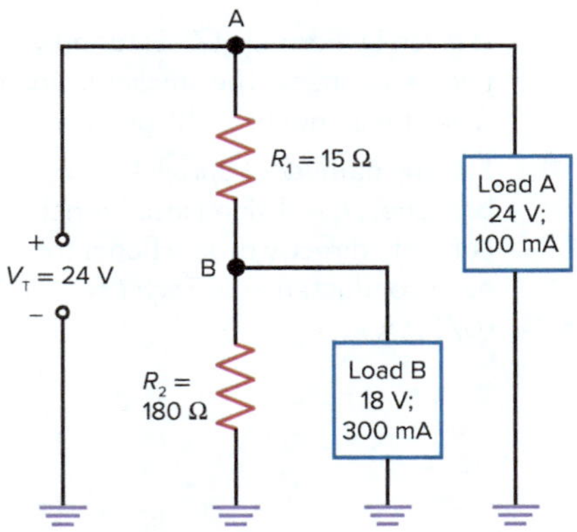

6. **In Fig. 7–9, how much is I_1 in R_1?**
   a. 400 mA.
   b. 300 mA.
   c. 100 mA.
   d. 500 mA.

7. **In Fig. 7–9, how much is the bleeder current, I_B?**
   a. 500 mA
   b. 400 mA
   c. 100 mA
   d. 300 mA

8. **In Fig. 7–9, how much is the total current, I_T?**
   a. 500 mA
   b. 400 mA
   c. 100 mA
   d. 300 mA

9. **In Fig. 7–9, what is the voltage, V_{BG}, if load B becomes open?**
   a. 18 V
   b. 19.2 V
   c. 6 V
   d. 22.15 V

10. **In Fig. 7–9, what happens to the voltage, V_{BG}, if load A becomes open?**
   a. It increases.
   b. It decreases.
   c. It remains the same.
   d. It cannot be determined.

Essay Questions

1. Define *series voltage divider.*
2. Define *parallel current divider.*
3. Give two differences between a series voltage divider and a parallel current divider.
4. Give three differences between Formula (7–2) for branch resistances and Formula (7–3) for branch conductances.
5. Define *bleeder current.*
6. What is the main difference between the circuits in Fig. 7–7*a* and *b*?
7. Referring to Fig. 7–1, why is V_1 series-aiding with V_2 and V_3 but in series opposition to V_T? Show the polarity of each *IR* drop.
8. Show the derivation of Formula (7–2) for each branch current in a parallel bank of two resistances. [Hint: The voltage across the bank is $I_T \times R_{EQ}$ and R_{EQ} is $R_1R_2/(R_1 + R_2)$.]

Problems

SECTION 7–1 SERIES VOLTAGE DIVIDERS

7–1 A 100 Ω R_1 is in series with a 200-Ω R_2 and a 300-Ω R_3. The applied voltage, V_T, is 18 V. Calculate V_1, V_2, and V_3.

7–2 A10- kΩ R_1 is in series with a 12- kΩ R_2, a 4.7- kΩ R_3, and a 3.3- kΩ R_4. The applied voltage, V_T, is 36 V. Calculate V_1, V_2, V_3, and V_4.

7–3 MultiSim In Fig. 7–10, calculate V_1, V_2, and V_3.

7–4 MultiSim In Fig. 7–10, recalculate V_1, V_2, and V_3 if $R_1 = 10\ \Omega$, $R_2 = 12\ \Omega$, $R_3 = 18\ \Omega$ and $V_T = 20$ V.

7–5 In Fig. 7–11, calculate V_1, V_2, and V_3. Note that resistor R_2 is three times the value of R_1 and resistor R_3 is two times the value of R_2.

Figure 7–10

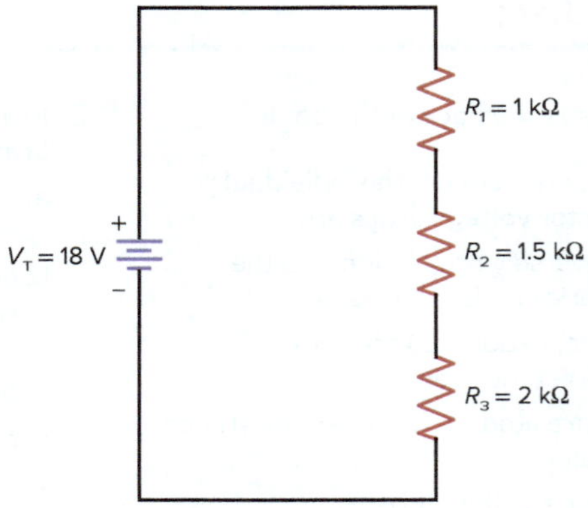

Figure 7–11

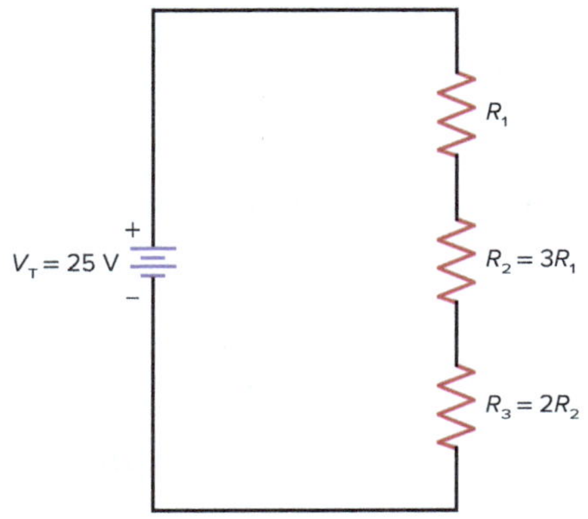

7-6 In Fig. 7–12, calculate

a. V_1, V_2, and V_3.

b. V_{AG}, V_{BG}, and V_{CG}.

Figure 7–12

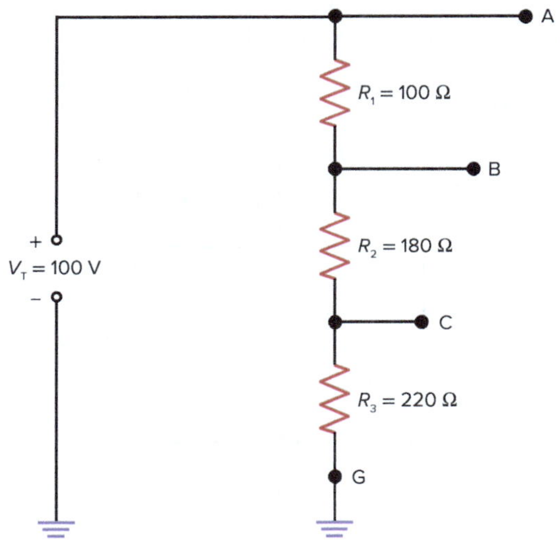

7-7 In Fig. 7–12, change R_1, R_2, R_3, and the applied voltage, V_T, to the following values: $R_1 = 9\ \text{k}\Omega$, $R_2 = 900\ \Omega$, $R_3 = 100\ \Omega$, and $V_T = 10$ V. Then, recalculate

a. V_1, V_2, and V_3.

b. V_{AG}, V_{BG}, and V_{CG}.

7-8 In Fig. 7–13, solve for

a. V_1, V_2, V_3, and V_4.

b. V_{AG}, V_{BG}, V_{CG}, and V_{DG}.

Figure 7–13

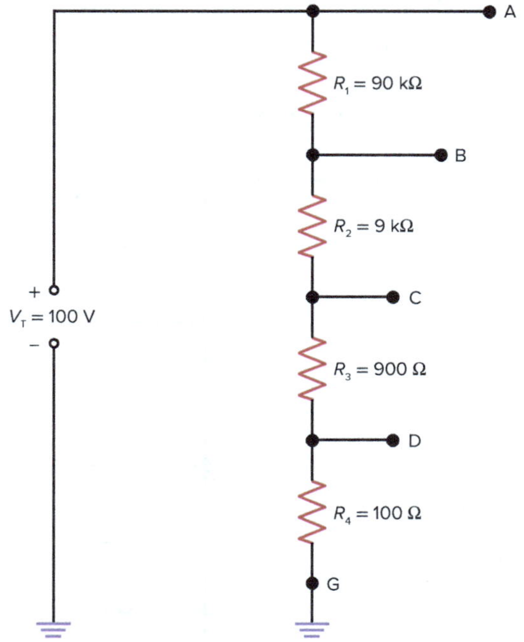

7-9 In Fig. 7–14, solve for

a. V_1, V_2, V_3, and V_4.

b. V_{AG}, V_{BG}, V_{CG}, and V_{DG}.

Figure 7–14

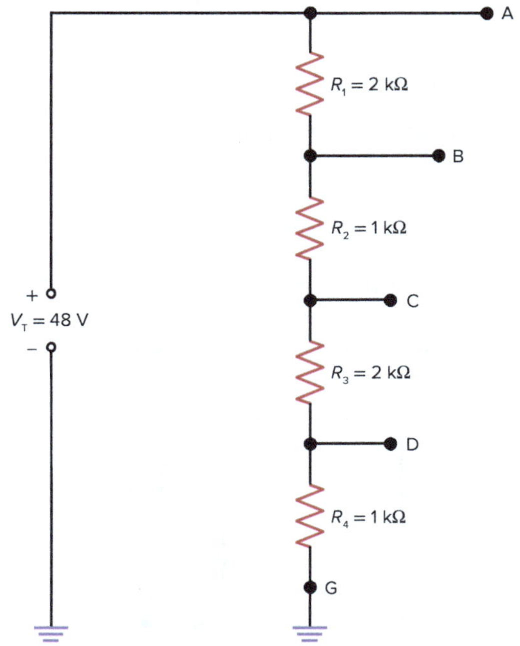

7-10 In Fig. 7–15, solve for

a. V_1, V_2, V_3, V_4, and V_5.

b. V_{AG}, V_{BG}, V_{CG}, and V_{DG}.

Figure 7–15

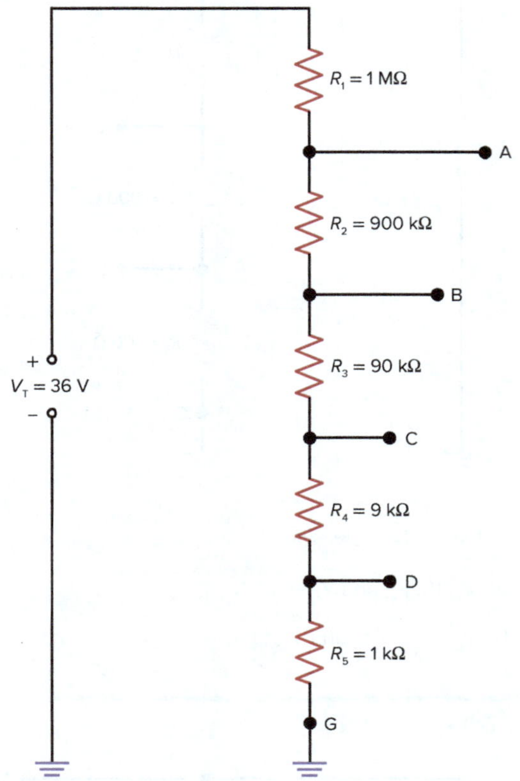

7-11 In Fig. 7–16, solve for

a. The total resistance, R_T.

b. The series current, I.

c. The voltages V_1 and V_{AB}.

d. The voltage range available at the wiper arm C.

Figure 7–16

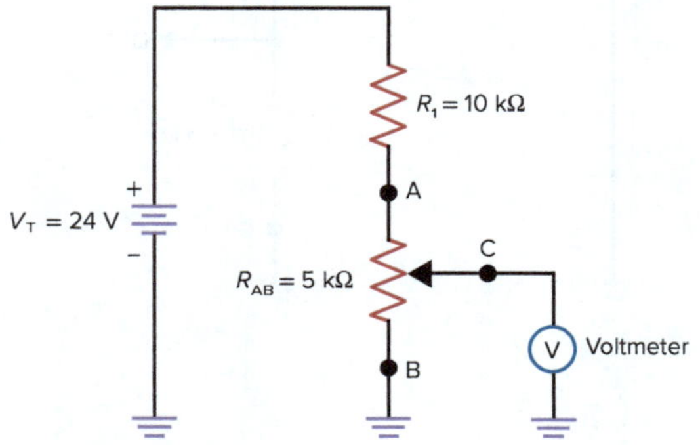

7-12 In Fig. 7–17, solve for

a. The total resistance, R_T.

b. The series current, I.

c. The voltages V_1 and V_{AB}.

d. The voltage range available at the wiper arm C.

Figure 7–17

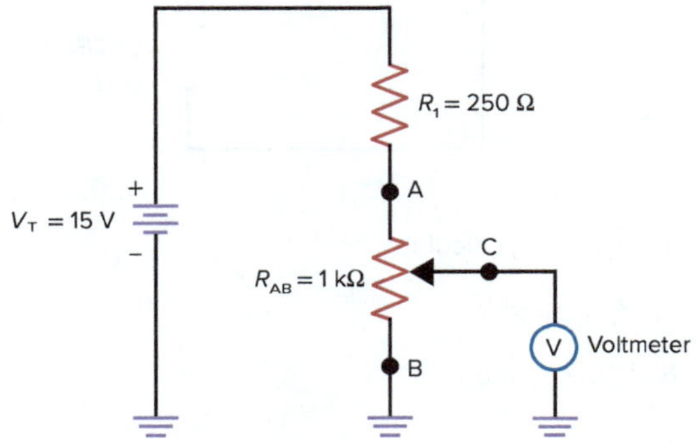

SECTION 7–2 CURRENT DIVIDER WITH TWO PARALLEL RESISTANCES

7-13 In Fig. 7–18, solve for I_1 and I_2.

Figure 7–18

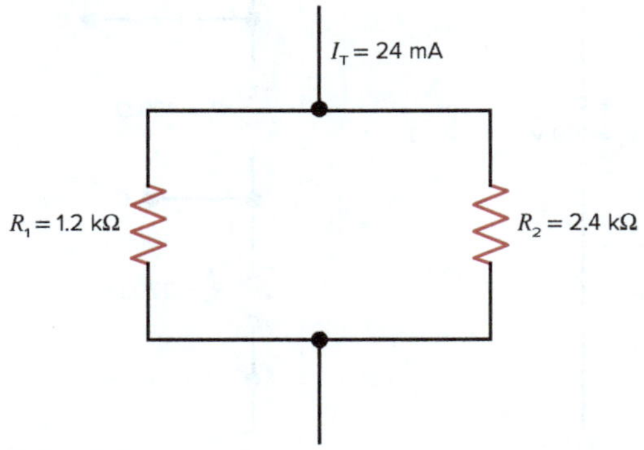

7-14 MultiSim In Fig. 7–19, solve for I_1 and I_2.

Figure 7–19

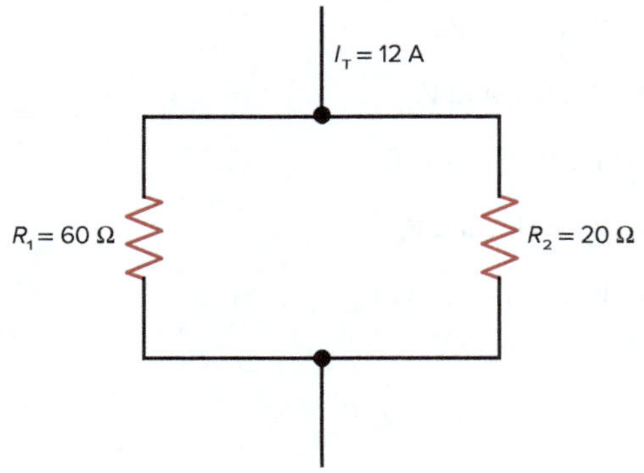

7-15 In Fig. 7–20, solve for I_1 and I_2.

Figure 7-20

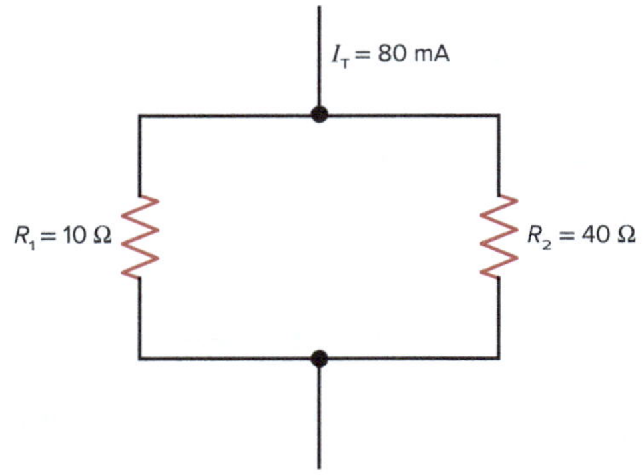

7-16 MultiSim In Fig. 7–21, solve for I_1 and I_2.

Figure 7-21

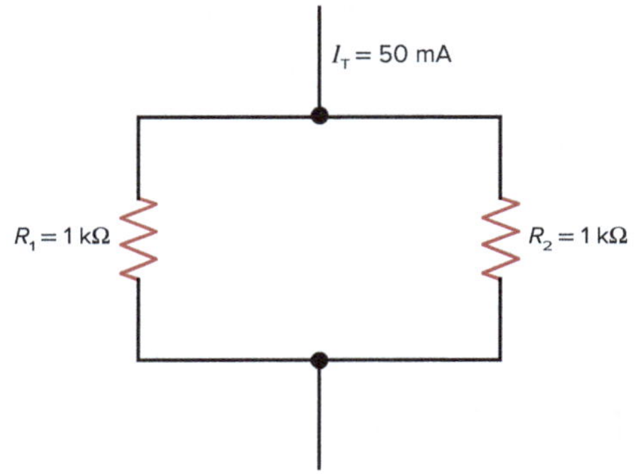

7-17 In Fig. 7–22, solve for I_1 and I_2.

Figure 7-22

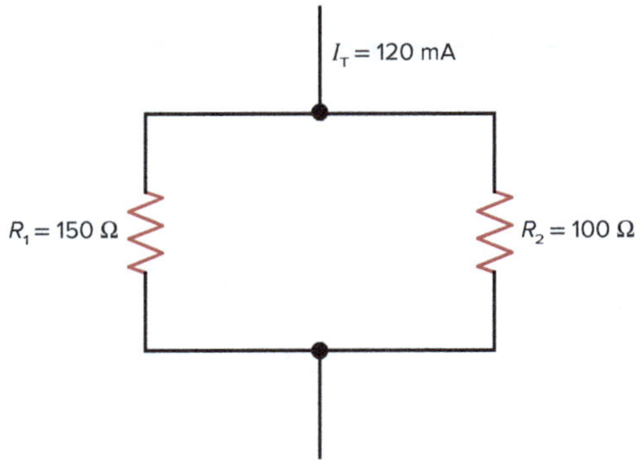

SECTION 7-3 CURRENT DIVISION BY PARALLEL CONDUCTANCES

7-18 In Fig. 7–23, solve for I_1, I_2, and I_3.

Figure 7-23

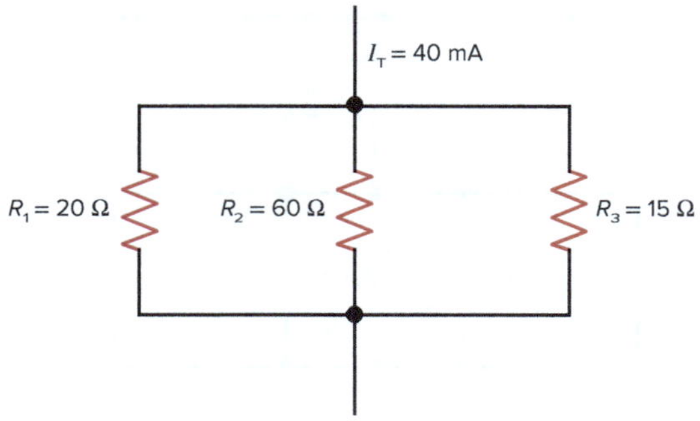

7-19 In Fig. 7–24, solve for I_1, I_2, and I_3.

Figure 7-24

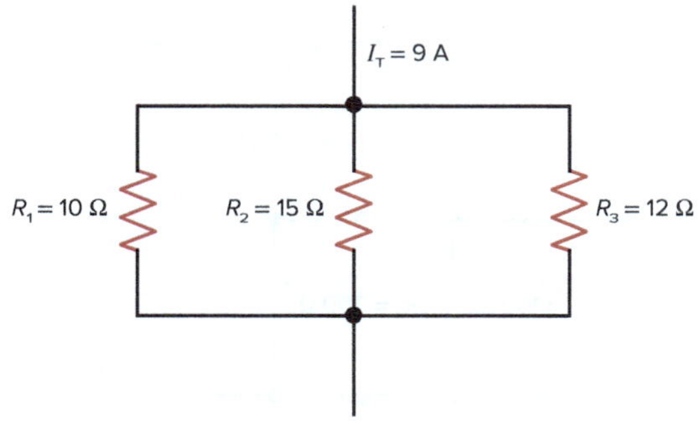

7-20 In Fig. 7–25, solve for I_1, I_2, and I_3.

Figure 7-25

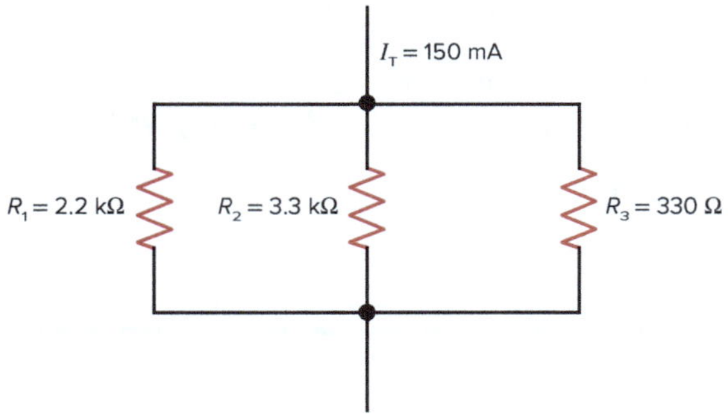

7-21 In Fig. 7–26, solve for I_1, I_2, and I_3.

Figure 7-26

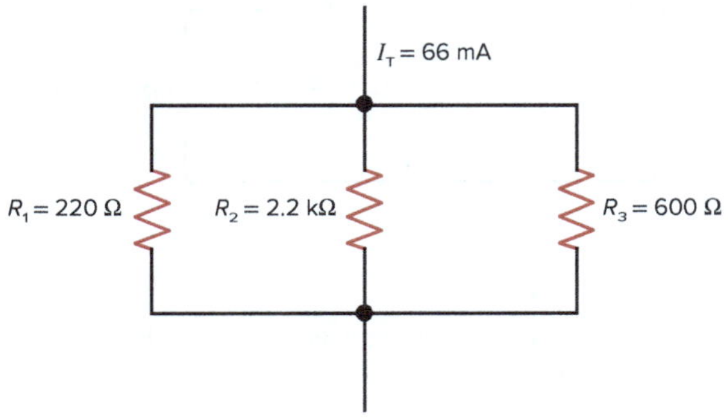

7–22 In Fig. 7–27, solve for I_1, I_2, I_3, and I_4.

Figure 7–27

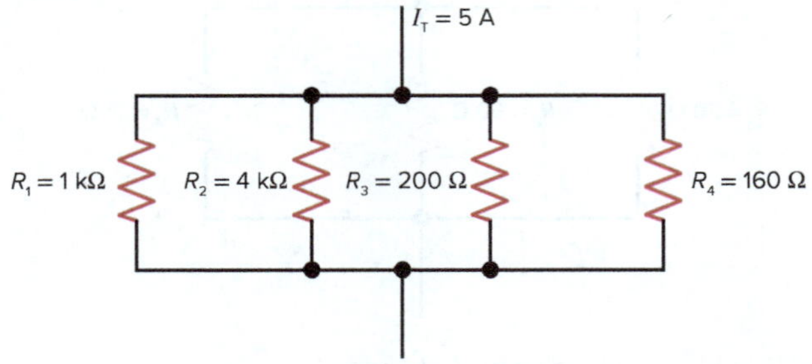

7–23 In Fig. 7–28, solve for I_1, I_2, I_3, and I_4.

Figure 7–28

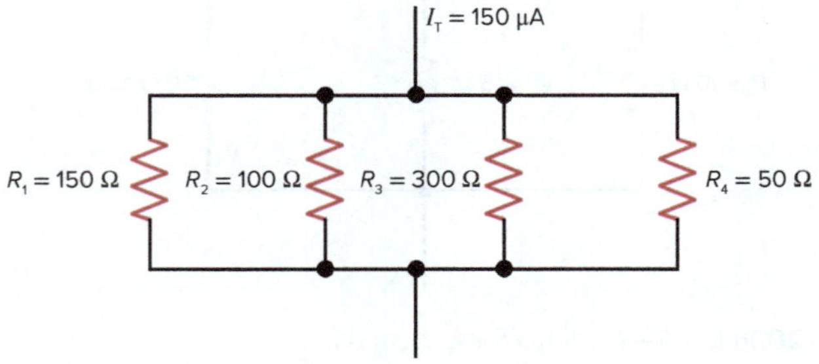

SECTION 7–4 SERIES VOLTAGE DIVIDER WITH PARALLEL LOAD CURRENT

7–24 In Fig. 7–29, calculate I_1, I_2, I_L, V_{BG}, and V_{AG} with

a. S_1 open.

b. S_1 closed.

Figure 7–29

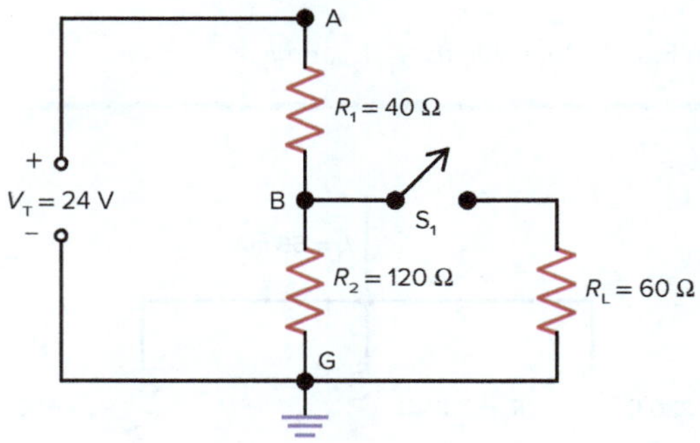

7–25 In Fig. 7–29, explain why the voltage, V_{BG}, decreases when the switch, S_1 is closed.

7–26 In Fig. 7–30, calculate I_1, I_2, I_L, V_{BG}, and V_{AG} with

a. S_1 open.

b. S_1 closed.

Figure 7–30

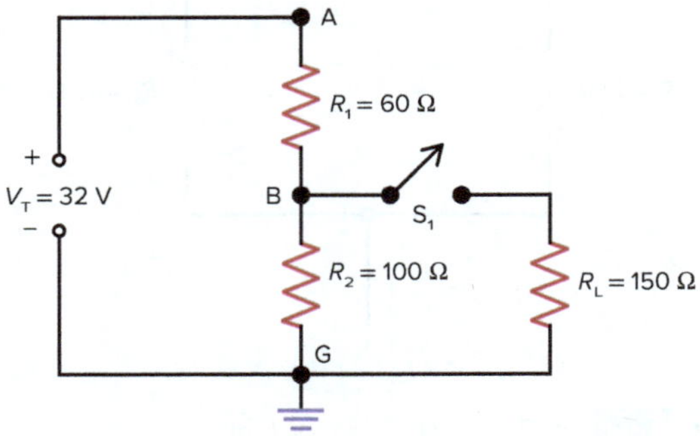

7–27 With S_1 closed in Fig. 7–30, which resistor has only the bleeder current, I_B, flowing through it?

7–28 In Fig. 7–31, calculate I_1, I_2, I_L, V_{BG}, and V_{AG} with

a. S_1 open.

b. S_1 closed.

Figure 7–31

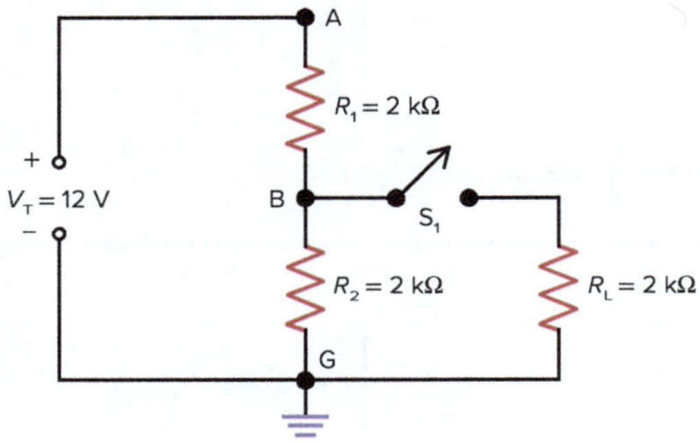

SECTION 7–5 DESIGN OF A LOADED VOLTAGE DIVIDER

7–29 If the bleeder current, I_B, is 10% of the total load current in Fig. 7–32, solve for

a. I_1, I_2, I_3, and I_T.

b. V_1, V_2, and V_3.

c. R_1, R_2, and R_3.

d. The power dissipated by R_1, R_2, and R_3.

Figure 7–32

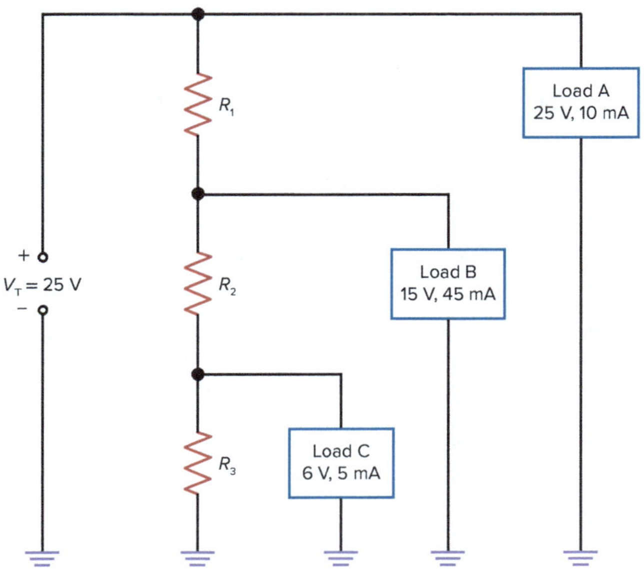

7–30 If the bleeder current, I_B, is 10% of the total load current in Fig. 7–33, solve for

a. I_1, I_2, I_3, and I_T.

b. V_1, V_2, and V_3.

c. R_1, R_2, and R_3.

d. The power dissipated by R_1, R_2, and R_3.

Figure 7–33

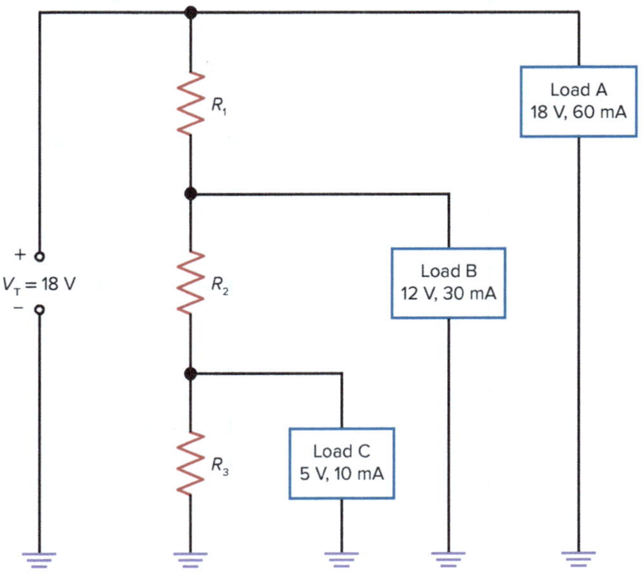

7–31 If the bleeder current, I_B, is 6 mA in Fig. 7–34, solve for

a. I_1, I_2, I_3, and I_T.

b. V_1, V_2, and V_3.

c. R_1, R_2, and R_3.

d. The power dissipated by R_1, R_2, and R_3.

Figure 7–34

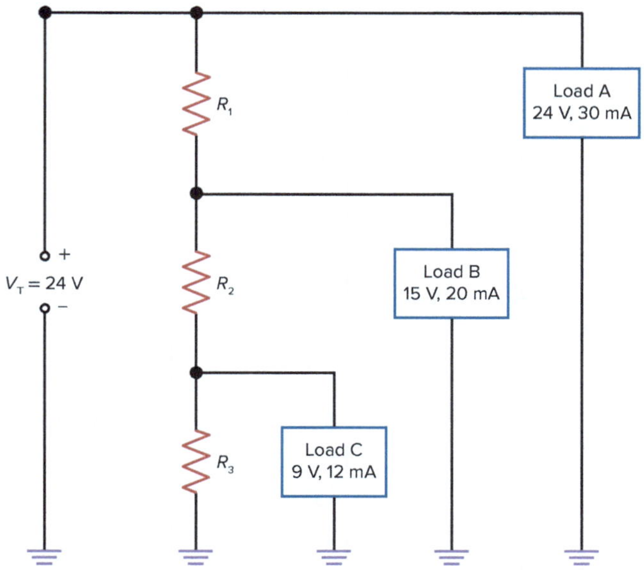

7–32 If the bleeder current, I_B, is 15 mA in Fig. 7–35, solve for

a. I_1, I_2, I_3, and I_T.

b. V_1, V_2, and V_3.

c. R_1, R_2, and R_3.

d. The power dissipated by R_1, R_2, and R_3.

Figure 7–35

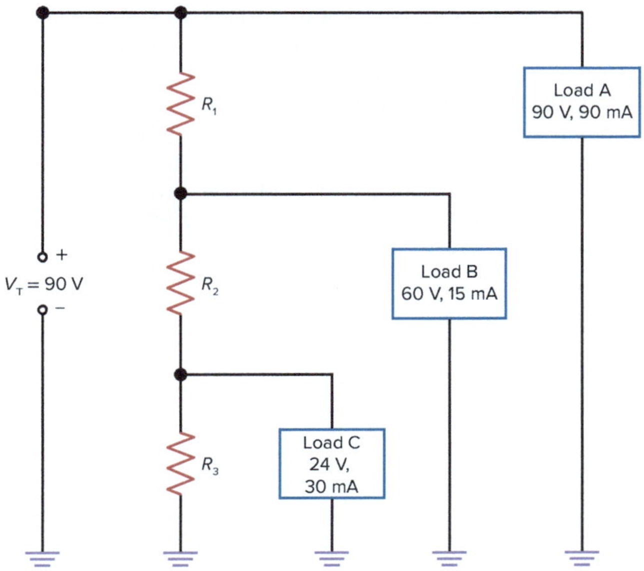

Critical Thinking

7–33 Refer to Fig. 7–36. Select values for R_1 and R_3 that will allow the output voltage to vary between 6 V and 15 V.

Figure 7–36

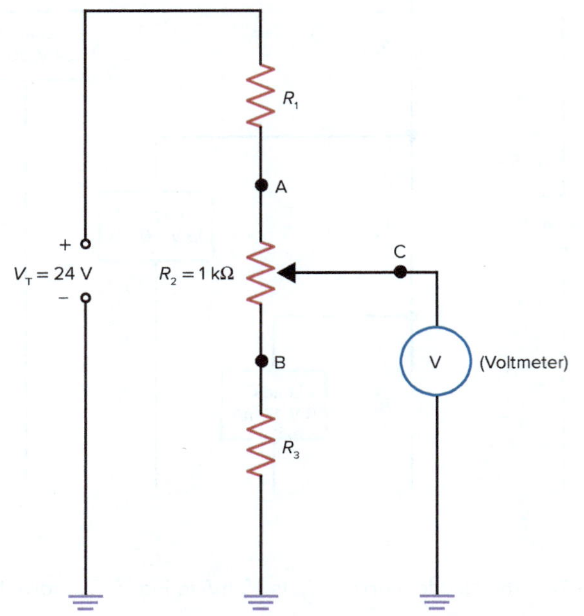

7–34 Design a loaded voltage divider, using a 25-V supply, to meet the following requirements: load A = 25 V, 10 mA; load B = 15 V, 45 mA; load C = 6 V, 5 mA; I_B = 10% of total load current. Draw the schematic diagram including all values.

7–35 Design a loaded voltage divider, using a 24-V supply, to meet the following requirements: load A = 18 V, 10 mA; load B = 12 V, 30 mA; load C = 5 V, 6 mA; I_B = 10% of total load current. Draw the schematic diagram including all values.

7–36 Design a loaded voltage divider, using a 25-V supply, to meet the following requirements: load A = 20 V, 25 mA; load B = 12 V, 10 mA; load C = −5 V, 10 mA; total current I_T = 40 mA. Draw the schematic diagram including all values.

Troubleshooting Challenge

Table 7–2 shows voltage measurements taken in Fig. 7–37. The first row shows the normal values when the circuit is operating properly. Rows 2 to 9 are voltage measurements taken when one component in the circuit has failed. For each row, identify which component is defective and determine the type of defect that has occurred in the component.

Table 7–3 shows voltage measurements taken in Fig. 7–38. The first row shows the normal values when the circuit is operating properly. Rows 2 to 9 are voltage measurements taken when one component in the circuit has failed. For each row, identify which component is defective and determine the type of defect that has occurred in the component.

Figure 7–37 Series voltage divider for Troubleshooting Challenge.

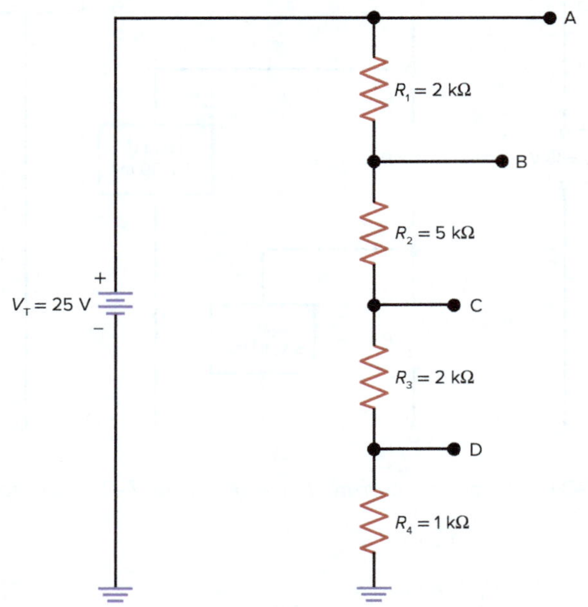

Table 7–2	Voltage Measurements Taken in Figure 7–37				
	V_{AG}	V_{BG}	V_{CG}	V_{DG}	Defective Component
		VOLTS			
1 Normal values	25	20	7.5	2.5	None
2 Trouble 1	25	25	0	0	
3 Trouble 2	25	0	0	0	
4 Trouble 3	25	18.75	3.125	3.125	
5 Trouble 4	25	25	25	25	
6 Trouble 5	25	15	15	5	
7 Trouble 6	25	25	9.375	3.125	
8 Trouble 7	25	25	25	0	
9 Trouble 8	25	19.4	5.56	0	

Table 7–3	Voltage Measurements Taken in Figure 7–38					
	V_A	V_B	V_C	V_D	Comments	Defective Component
			VOLTS			
1 Normal values	36	24	15	6	—	None
2 Trouble 1	36	32	0	0	—	
3 Trouble 2	36	16.94	0	0	R_2 Warm	
4 Trouble 3	36	25.52	18.23	0	—	
5 Trouble 4	36	0	0	0	R_1 Hot	
6 Trouble 5	36	0	0	0	—	
7 Trouble 6	36	27.87	23.23	9.29	—	
8 Trouble 7	36	23.25	13.4	0	—	
9 Trouble 8	36	36	22.5	9	—	

Figure 7–38 Loaded voltage divider for Troubleshooting Challenge.

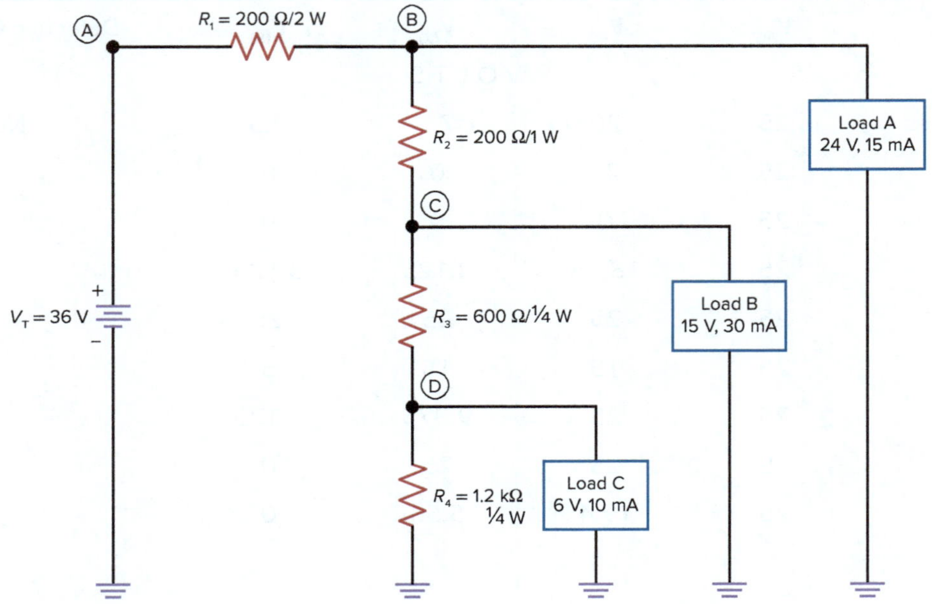

Answers to Self-Reviews

7–1 **a.** 100 kΩ
b. $(^2/_{10}) \times V_T$
c. 100 V
d. 10.2 V
e. 0.6 V

7–2 **a.** 2 to 1
b. 1 to 2

7–3 **a.** 2 to 1
b. 2 to 1

7–4 **a.** ⅓
b. ⅕

7–5 **a.** 30 mA
b. 40 V
c. 22 V
d. 18 V

Laboratory Application Assignment

In this lab application assignment, you will examine unloaded voltage dividers, current dividers, and loaded voltage dividers. You will also be presented with a challenging design problem involving loaded voltage dividers.

Equipment: Obtain the following items from your instructor.

- Variable dc power supply
- Assortment of carbon-film resistors
- DMM

Unloaded Voltage Divider

Examine the unloaded voltage divider in Fig. 7–39. Using Formula (7–1), calculate and record the following voltages:
V_{AG} = ____________, V_{BG} = ____________, V_{CG} = ____________, V_{DG} = ________________

Construct the voltage divider in Fig. 7–39. Measure and record the following voltages:
V_{AG} = ____________, V_{BG} = ____________, V_{CG} = ____________, V_{DG} = ____________

How does the ratio R_4/R_T compare to the ratio V_{DG}/V_T?
____

How does the ratio $(R_3 + R_4)/R_T$ compare to the ratio V_{CG}/V_T?
____

How does the ratio $(R_2 + R_3 + R_4)/R_T$ compare to the ratio V_{BG}/V_T? ____________

Figure 7–39

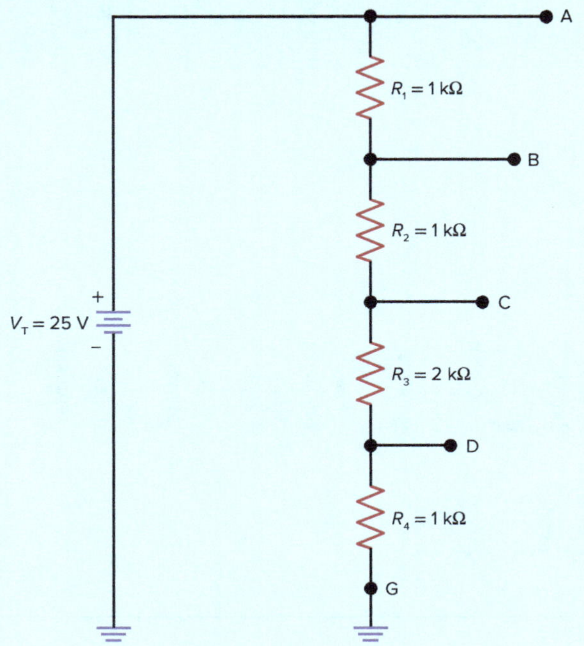

Current Divider

Examine the current divider in Fig. 7–40. Using Formula (7–2), calculate and record the currents, I_1 and I_2:

I_1 = ______________, I_2 = ______________

Construct the current divider in Fig. 7–40. Adjust the DC voltage source until the total current, I_T (as indicated by the DMM) measures exactly 15 mA. Now move the DMM to measure the individual branch currents, I_1 and I_2. Record your measured values.

I_1 = ______________, I_2 = ______________

How does the ratio I_1/I_2 compare to the ratio R_2/R_1?______________

What is unique about comparing these ratios?______________

Figure 7–40

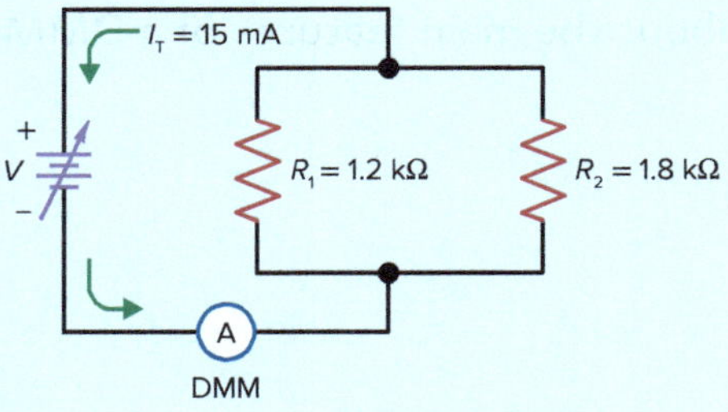

Series Voltage Divider with Parallel Load Current

In Fig. 7–41, calculate I_1, I_2, I_L, V_{BG}, and V_{AC} with S_1 open.

I_1 = ______________ I_2 = ______________ I_L = ______________
V_{BG} = ______________ V_{AC} = ______________

Next, calculate I_1, I_2, I_L, V_{BG}, and V_{AC} with S_1 closed.

I_1 = ______________ I_2 = ______________ I_L = ______________
V_{BG} = ______________ V_{AC} = ______________

Construct the circuit in Fig. 7–41. With S_1 open, measure and record the following values: I_1, I_2, I_L, V_{BG}, and V_{AC}.

I_1 = ______________ I_2 = ______________ I_L = ______________
V_{BG} = ______________ V_{AC} = ______________

Next, close S_1 and re-measure I_1, I_2, I_L, V_{BG}, and V_{AC}.

I_1 = ______________ I_2 = ______________ I_L = ______________
V_{BG} = ______________ V_{AC} = ______________

Did the voltage V_{BG} increase, decrease, or stay the same when S_1 was closed?______________

Why?______________

____

Did the voltage V_{AC} increase, decrease, or stay the same when S_1 was closed? ______________

Why?______________

____

Figure 7–41

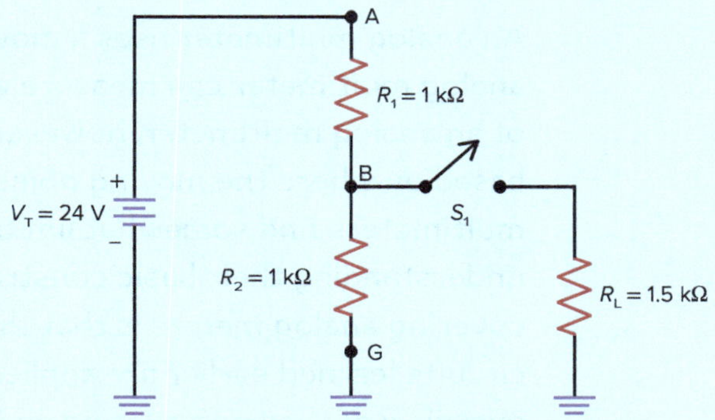

Design Challenge

Using a 24-V supply, design and build a loaded voltage divider to meet the following requirements:

Load A = 24 V @ 15 mA
Load B = 15 V @ 25 mA
Load C = 9 V @ 7.5 mA

I_B should equal approximately 10% of the total load current
Note: Use standard resistors for the actual loads.

chapter

8

Analog and Digital Multimeters

The digital multimeter (DMM) is the most common measuring instrument used by electronic technicians. Its main application is for measuring voltage, current, and resistance. However, most DMMs available today can also test diodes and measure capacitance values. A DMM uses a numeric display to indicate the value of the measured quantity.

An analog multimeter uses a moving pointer and a printed scale. Like a DMM, an analog multimeter can measure voltage, current, and resistance. One disadvantage of an analog multimeter, however, is that the meter reading must be interpreted based on where the moving pointer rests along the printed scale. Although analog multimeters find somewhat limited use in electronics, there is great value in understanding their basic construction and operation. One of the main reasons for covering analog meters is that the concepts of series, parallel, and series-parallel circuits learned earlier are applied. In this chapter, you will be provided with a basic overview of the construction and operation of an analog meter as well as a concept called voltmeter loading. You will also learn about the main features of a DMM. ■

Chapter Outline

Chapter Objectives

After studying this chapter, you should be able to

- *Explain* the difference between analog and digital meters.
- *Explain* the construction and operation of a moving-coil meter.
- *Calculate* the value of shunt resistance required to extend the current range of a basic moving-coil meter.
- *Calculate* the value of multiplier resistance required to make a basic moving-coil meter capable of measuring voltage.
- *Explain* the ohms-per-volt rating of a voltmeter.
- *Explain* what is meant by *voltmeter loading*.
- *Explain* how a basic moving-coil meter can be used with a battery to construct an ohmmeter.
- *List* the main features of a digital multimeter.

Important Terms

amp-clamp probe
analog multimeter
back-off ohmmeter scale
continuity testing
digital multimeter (DMM)
loading effect
multiplier resistor
ohms-per-volt (Ω/V) rating
shunt resistor
zero-ohms adjustment

8–1 Moving-Coil Meter

Figure 8–1 shows two types of multimeters used by electronic technicians. Figure 8–1*a* shows an analog volt-ohm-milliammeter (VOM), and Fig. 8–1*b* shows a DMM. Both types are capable of measuring voltage, current, and resistance.

A moving-coil meter movement, as shown in Fig. 8–2, is generally used in an analog VOM. The construction consists of a coil of fine wire wound on a drum mounted between the poles of a permanent magnet. When direct current flows in the coil, the magnetic field of the current reacts with the magnetic field of the permanent magnet.* The resultant force turns the drum with its pointer, winding up the restoring spring. When the current is removed, the pointer returns to zero. The amount of deflection indicates the amount of current in the coil. When the polarity is connected correctly, the pointer will read up-scale, to the right; the incorrect polarity forces the pointer off-scale, to the left. (It is interesting to note that the moving-coil arrangement is often called a D'Arsonval movement, after its inventor who patented this meter movement in 1881.)

The pointer deflection is directly proportional to the amount of current in the coil. If 100 μA is the current needed for full-scale deflection, 50 μA in the coil will produce half-scale deflection. The accuracy of the moving-coil meter mechanism is 0.1–2%.

Values of I_M

The full-scale deflection current I_M is the amount needed to deflect the pointer all the way to the right to the last mark on the printed scale. Typical values of I_M are from about 10 μA to 30 mA. In a VOM, the I_M is typically either 50 μA or 1 mA.

Figure 8–1 Typical multimeters used for measuring *V*, *I*, and *R*. (*a*) Analog VOM. (*b*) DMM.

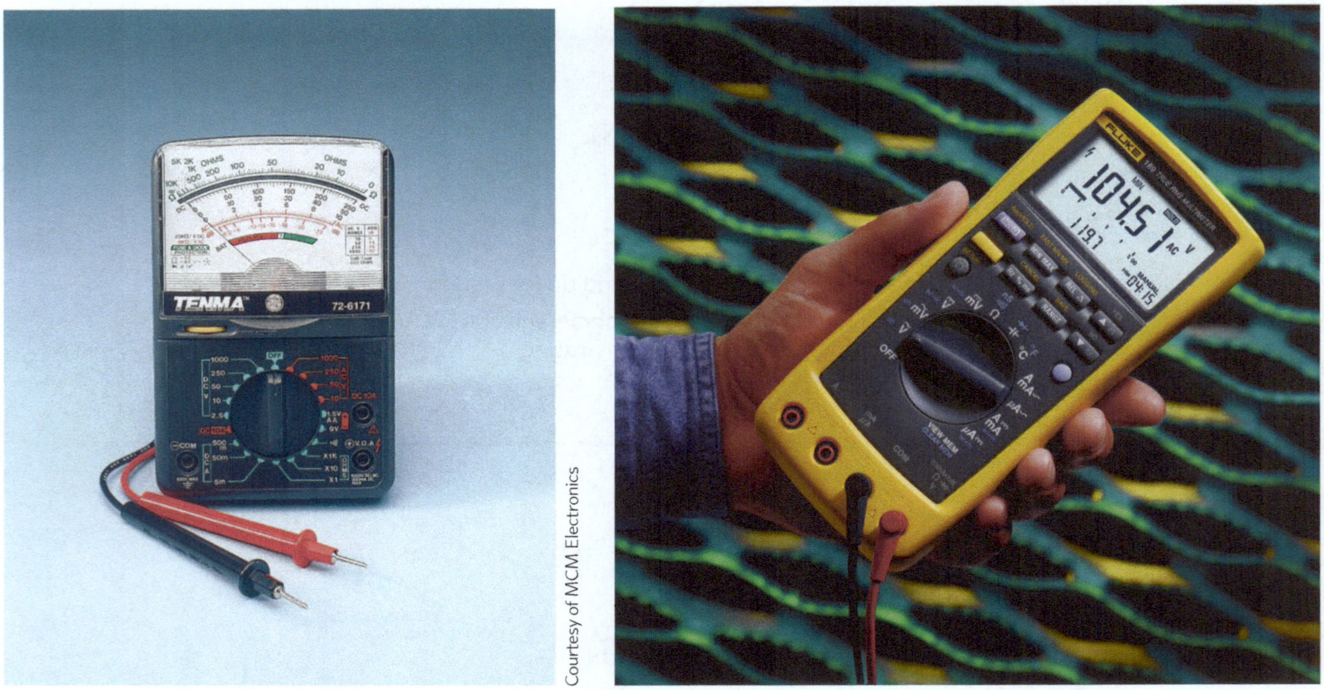

Courtesy of MCM Electronics

Courtesy of Fluke Corporation, Reproduced with Permission

(*a*) (*b*)

* For more details on the interaction between two magnetic fields, see Chapter 14, "Electromagnetism."

> **GOOD TO KNOW**
>
> Exceeding the value of I_M will cause the meter's pointer to deflect beyond the last mark on the right-hand side of the printed scale. When the moving pointer hits the right pointer stop, the meter is said to be pegged.

Figure 8–2 Construction of a moving-coil meter. The diameter of the coil can be ½ to 1 in.

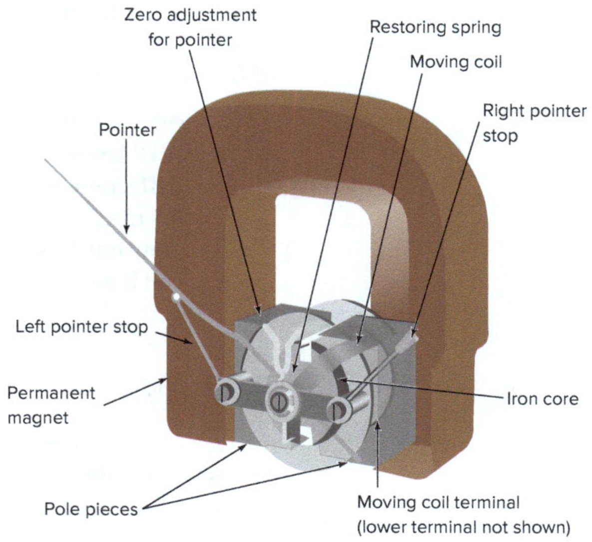

Refer to the analog VOM in Fig. 8–1*a*. The mirror along the scale is used to eliminate reading errors. The meter is read when the pointer and its mirror reflection appear as one. This eliminates the optical error called *parallax* caused by looking at the meter from the side.

Values of r_M

This is the internal resistance of the wire of the moving coil. Typical values range from 1.2 Ω for a 30-mA movement to 2000 Ω for a 50-μA movement. A movement with a smaller I_M has a higher r_M because many more turns of fine wire are needed. An average value of r_M for a 1-mA movement is about 50 Ω. Figure 8–3 provides a

Figure 8–3 Close-up view of a D'Arsonval moving-coil meter movement.

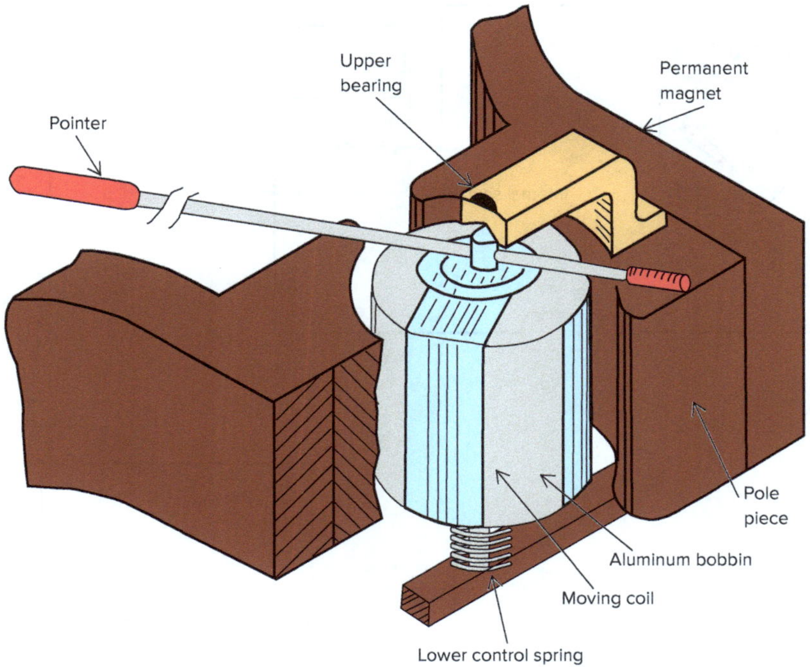

close-up view of the basic components contained within a D'Arsonval meter movement. Notice that the moving coil is wound around a drum which rotates when direct current flows through the wire of the moving coil.

8–1 Self-Review

Answers at the end of the chapter.

a. **A D'Arsonval movement has an I_M value of 1 mA. How much is the deflection of the meter pointer if the current in the moving coil is 0.5 mA?**

b. **How much is the deflection in question a if the current in the moving coil is zero?**

8–2 Meter Shunts

A *meter shunt* is a precision resistor connected across the meter movement for the purpose of shunting, or bypassing, a specific fraction of the circuit's current around the meter movement. The combination then provides a current meter with an extended range. The shunts are usually inside the meter case. However, the schematic symbol for the current meter usually does not show the shunt.

In current measurements, the parallel bank of the movement with its shunt is connected as a current meter in series in the circuit (Fig. 8–4). Note that the scale of a meter with an internal shunt is calibrated to take into account the current through both the shunt and the meter movement. Therefore, the scale reads total circuit current.

Resistance of the Meter Shunt

In Fig. 8–4*b*, the 1-mA meter movement has a resistance of 50 Ω, which is the resistance of the moving coil r_M. To double the range, the shunt resistance R_S is made

MultiSim **Figure 8–4** Example of meter shunt R_S in bypassing current around the meter movement to extend the range from 1 to 2 mA. (*a*) Wiring diagram. (*b*) Schematic diagram showing the effect of the shunt. With $R_S = r_M$ the current range is doubled. (*c*) Circuit with 2-mA meter to read the current.

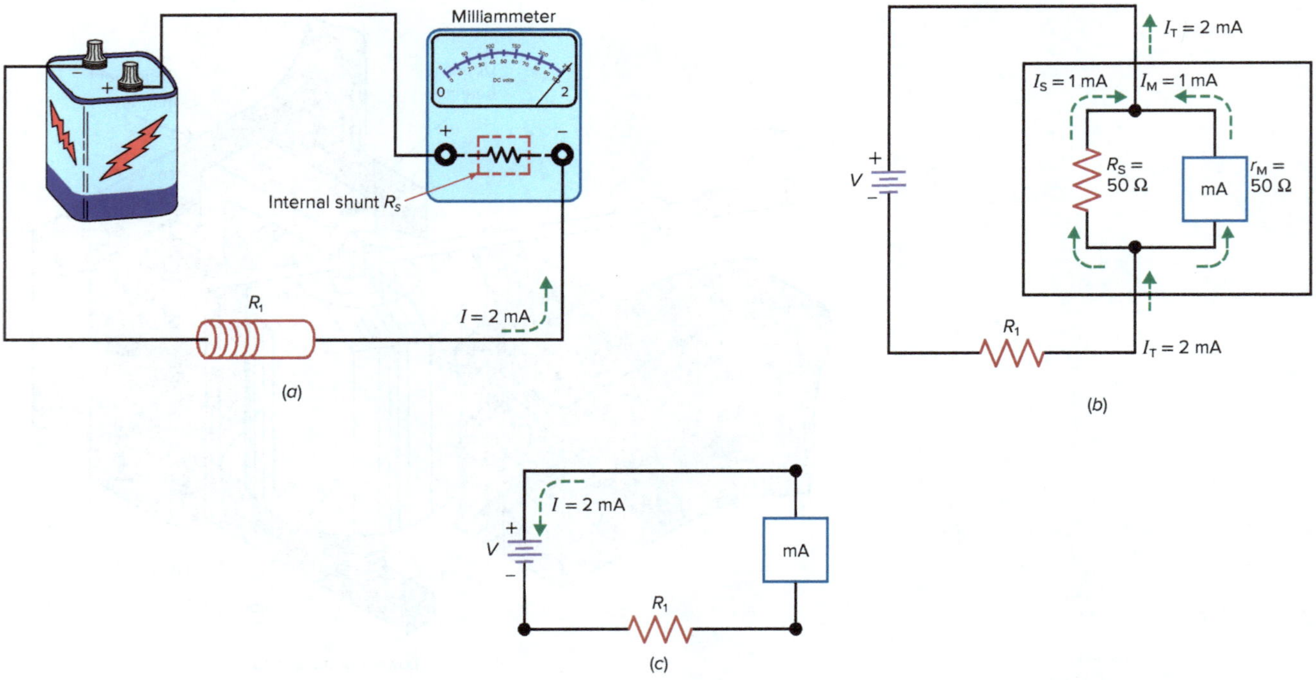

Figure 8–5 Calculating the resistance of a meter shunt. R_S is equal to V_M/I_S. See text for calculations.

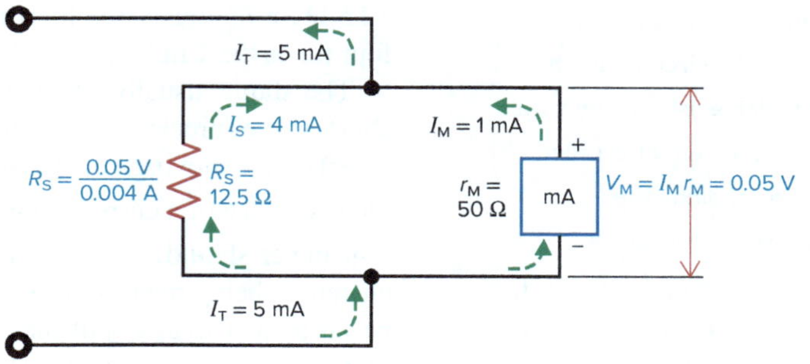

equal to the 50 Ω of the meter movement. When the meter is connected in series in a circuit where the current is 2 mA, this total current into one terminal of the meter divides equally between the shunt and the meter movement. At the opposite meter terminal, these two branch currents combine to provide the 2 mA of the circuit current.

Inside the meter, the current is 1 mA through the shunt and 1 mA through the moving coil. Since it is a 1-mA meter movement, this current produces full-scale deflection. The scale is doubled, however, reading 2 mA, to account for the additional 1 mA through the shunt. Therefore, the scale reading indicates total current at the meter terminals, not just coil current. The movement with its shunt, then, is a 2-mA meter. Its internal resistance is 50 × ½ = 25 Ω.

Another example is shown in Fig. 8–5. In general, the shunt resistance for any range can be calculated with Ohm's law from the formula

$$R_S = \frac{V_M}{I_S} \qquad \textbf{(8–1)}$$

where R_S is the resistance of the shunt and I_S is the current through it.

Voltage V_M is equal to $I_M \times r_M$. This is the voltage across both the shunt and the meter movement, which are in parallel.

Calculating I_S

The current through the shunt alone is the difference between the total current I_T and the current I_M through the meter movement or

$$I_S = I_T - I_M \qquad \textbf{(8–2)}$$

Use the values of current for full-scale deflection, as these are known. In Fig. 8–5,

$$I_S = 5 - 1 = 4\text{ mA}, \quad \text{or} \quad 0.004\text{ A}$$

Calculating R_S

The complete procedure for using the formula $R_S = V_M/I_S$ can be as follows:

1. Find V_M. Calculate this for full-scale deflection as $I_M \times r_M$. In Fig. 8–5, with a 1-mA full-scale current through the 50-Ω meter movement,

   $$V_M = 0.001 \times 50 = 0.05\text{ V} \quad \text{or} \quad 50\text{ mV}$$

2. Find I_S. For the values that are shown in Fig. 8–5,

   $$I_S = 5 - 1 = 4\text{ mA} = 0.004\text{ A} \quad \text{or} \quad 4\text{ mA}$$

3. Divide V_M by I_S to find R_S. For the final result,

   $$R_S = 0.05/0.004 = 12.5\ \Omega$$

GOOD TO KNOW

Memorize these formulas:

$V_M = I_M \times r_M$

$I_S = I_T - I_M$

$R_S = \frac{V_M}{I_S}$

GOOD TO KNOW

An alternate formula for calculating the shunt resistance is

$R_S = \frac{I_M}{I_T - I_M} \times r_M$

This shunt enables the 1-mA meter movement to be used for an extended range from 0 to 5 mA.

Note that R_S and r_M are inversely proportional to their full-scale currents. The 12.5 Ω for R_S equals one-fourth the 50 Ω of r_M because the shunt current of 4 mA is four times the 1 mA through the meter movement for full-scale deflection.

The shunts usually are precision wire-wound resistors. For very low values, a short wire of precise size can be used.

Since the moving-coil resistance, r_M, is in parallel with the shunt resistance, R_S, the resistance of a current meter can be calculated as $R_M = \frac{R_S \times r_M}{R_S + r_M}$. In general, a current meter should have very low resistance compared with the circuit where the current is being measured. As a general rule, the current meter's resistance should be no greater than 1⁄100 of the circuit resistance. The higher the current range of a meter, the lower its shunt resistance, R_S, and in turn the overall meter resistance, R_M.

GOOD TO KNOW

When the insertion of a current meter reduces the circuit current below that which exists without the meter present, the effect is called current meter loading. The correction formula for current meter loading is: $I = I_M(R_M/R_{CKT} + 1)$ where I is the actual current flowing in the circuit without the meter present, R_{CKT} is the resistance of the circuit without the meter present, R_M is the resistance of the current meter, and I_M is the actual value of current indicted by the meter when it is in the circuit. Current meter loading does not occur very, often but when it does it usually occurs in low-voltage, low-resistance circuits.

Example 8–1

A shunt extends the range of a 50-μA meter movement to 1 mA. How much is the current through the shunt at full-scale deflection?

ANSWER All currents must be in the same units for Formula (8–2). To avoid fractions, use 1000 μA for the 1-mA I_T. Then

$$I_S = I_T - I_M = 1000\ \mu A - 50\ \mu A$$

$$I_S = 950\ \mu A$$

GOOD TO KNOW

Most general-purpose analog VOMs have an accuracy of about ±2%. The percent accuracy is based on the full-scale reading of the meter. This means that for readings less than full-scale, the percent error is greater than the rated value. For example, a meter rated at ±2% on the 100-V range is accurate to within ±2 V for any voltage measured on this range; for example, if 10 V is being measured on the 100-V range, the reading could be anywhere from 8 to 12 V, which is an error of ±20%. Therefore, it is always important to make voltage measurements on the lowest possible range setting!

Example 8–2

A 50-μA meter movement has an r_M of 1000 Ω. What R_S is needed to extend the range to 500 μA?

ANSWER The shunt current I_S is 500 − 50, or 450 μA. Then

$$R_S = \frac{V_M}{I_S}$$

$$= \frac{50 \times 10^{-6}\ A \times 10^3\ \Omega}{450 \times 10^{-6}\ A} = \frac{50{,}000}{450}$$

$$R_S = 111.1\ \Omega$$

8–2 Self-Review

Answers at the end of the chapter.

A 50-μA meter movement with a 900-Ω r_M has a shunt R_S for the range of 500 μA.

- **a.** **How much is I_S?**
- **b.** **How much is V_M?**
- **c.** **What is the size of R_S?**

8–3 Voltmeters

Although a meter movement responds only to current in the moving coil, it is commonly used for measuring voltage by the addition of a high resistance in series with the meter movement (Fig. 8–6). The series resistance must be much higher than the coil resistance to limit the current through the coil. The combination of the meter movement with this added series resistance then forms a voltmeter. The series resistor, called a *multiplier*, is usually connected inside the voltmeter case.

Since a voltmeter has high resistance, it must be connected in parallel to measure the potential difference across two points in a circuit. Otherwise, the high-resistance multiplier would add so much series resistance that the current in the circuit would be reduced to a very low value. Connected in parallel, though, the high resistance of the voltmeter is an advantage. The higher the voltmeter resistance, the smaller the effect of its parallel connection on the circuit being tested.

The circuit is not opened to connect the voltmeter in parallel. Because of this convenience, it is common practice to make voltmeter tests in troubleshooting. The voltage measurements apply the same way to an *IR* drop or a generated emf.

The correct polarity must be observed in using a DC voltmeter. Connect the negative voltmeter lead to the negative side of the potential difference being measured and the positive lead to the positive side.

Multiplier Resistance

Figure 8–6 illustrates how the meter movement and its multiplier R_1 form a voltmeter. With 10 V applied by the battery in Fig. 8–6*a*, there must be 10,000 Ω of resistance to limit the current to 1 mA for full-scale deflection of the meter movement. Since the meter movement has a 50-Ω resistance, 9950 Ω is added in series, resulting in a 10,000-Ω total resistance. Then I is 10 V/10 kΩ = 1 mA.

MultiSim **Figure 8–6** Multiplier resistor R_1 added in series with meter movement to form a voltmeter. (*a*) Resistance of R_1 allows 1 mA for full-scale deflection in 1-mA movement with 10 V applied. (*b*) Internal multiplier R_1 forms a voltmeter. The test leads can be connected across a potential difference to measure 0 to 10 V. (*c*) 10-V scale of voltmeter and corresponding 1-mA scale of meter movement.

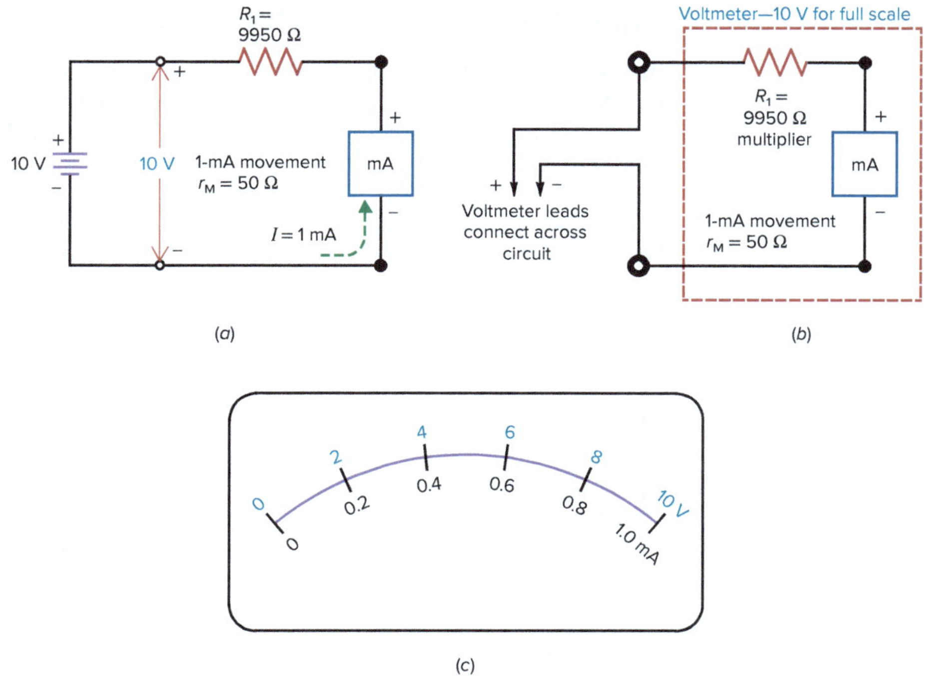

With 1 mA in the meter movement, the full-scale deflection can be calibrated as 10 V on the meter scale, as long as the 9950-Ω multiplier is included in series with the meter movement. It doesn't matter to which side of the meter movement the multiplier is connected.

If the battery is taken away, as shown in Fig. 8–6*b*, the meter movement with its multiplier forms a voltmeter that can indicate a potential difference of 0 to 10 V applied across its terminals. When the voltmeter leads are connected across a potential difference of 10 V in a DC circuit, the resulting 1-mA current through the meter movement produces full-scale deflection, and the reading is 10 V. In Fig. 8–6*c*, the 10-V scale is shown corresponding to the 1-mA range of the meter movement.

If the voltmeter is connected across a 5-V potential difference, the current in the meter movement is ½ mA, the deflection is one-half of full scale, and the reading is 5 V. Zero voltage across the terminals means no current in the meter movement, and the voltmeter reads zero. In summary, then, any potential difference up to 10 V, whether an *IR* voltage drop or a generated emf, can be applied across the meter terminals. The meter will indicate less than 10 V in the same ratio that the meter current is less than 1 mA.

The resistance of a multiplier can be calculated from the formula

$$R_{mult} = \frac{\text{Full-scale } V}{\text{Full-scale } I} - r_M \qquad \textbf{(8–3)}$$

Applying this formula to the example of R_1 in Fig. 8–6 gives

$$R_{mult} = \frac{10\text{ V}}{0.001\text{ A}} - 50\ \Omega = 10{,}000 - 50$$

$$R_{mult} = 9950\ \Omega \quad \text{or} \quad 9.95\text{ k}\Omega$$

We can take another example for the same 10-V scale but with a 50-μA meter movement, which is commonly used. Now the multiplier resistance is much higher, though, because less *I* is needed for full-scale deflection. Let the resistance of the 50-μA movement be 2000 Ω. Then

$$R_{mult} = \frac{10\text{ V}}{0.000\ 050\text{ A}} - 2000\ \Omega = 200{,}000 - 2000$$

$$R_{mult} = 198{,}000\ \Omega \quad \text{or} \quad 198\text{ k}\Omega$$

Typical Multiple Voltmeter Circuit

An example of a voltmeter with multiple voltage ranges is shown in Fig. 8–7. Resistance R_1 is the series multiplier for the lowest voltage range of 2.5 V. When higher resistance is needed for the higher ranges, the switch adds the required series resistors.

Figure 8–7 A typical voltmeter circuit with multiplier resistors for different ranges.

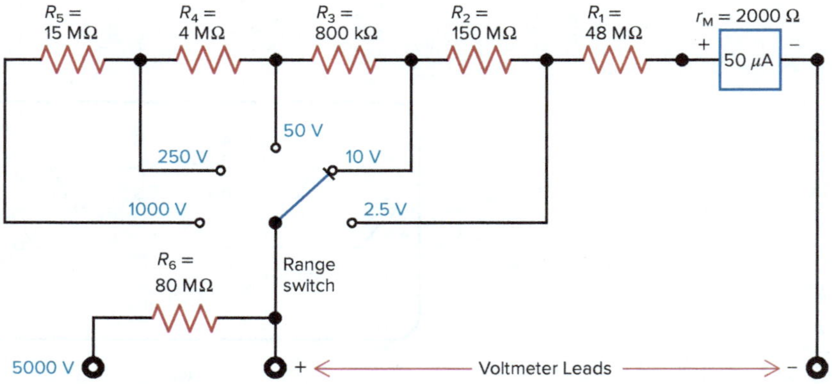

The meter in Fig. 8–7 requires 50 μA for full-scale deflection. For the 2.5-V range, a series resistance of $2.5/(50 \times 10^{-6})$, or 50,000 Ω, is needed. Since r_M is 2000 Ω, the value of R_1 is 50,000 − 2000, which equals 48,000 Ω or 48 kΩ.

For the 10-V range, a resistance of $10/(50 \times 10^{-6})$, or 200,000 Ω, is needed. Since $R_1 + r_M$ provides 50,000 Ω, R_2 is made 150,000 Ω, for a total of 200,000 Ω series resistance on the 10-V range. Similarly, additional resistors are switched in to increase the multiplier resistance for the higher voltage ranges. Note the separate jack and extra multiplier R_6 on the highest range for 5000 V. This method of adding series multipliers for higher voltage ranges is the circuit generally used in commercial multimeters.

GOOD TO KNOW

The voltmeter resistance of an analog VOM can be measured by connecting the leads of a DMM to the leads of the analog VOM. With the DMM set to measure resistance and the VOM set to measure voltage, the DMM will indicate the voltmeter resistance, R_V. The higher the voltmeter range, the higher the resistance.

Voltmeter Resistance

The high resistance of a voltmeter with a multiplier is essentially the value of the multiplier resistance. Since the multiplier is changed for each range, the voltmeter resistance changes.

Table 8–1 shows how the voltmeter resistance increases for higher ranges. The middle column lists the total internal resistance R_V, including R_{mult} and r_M, for the voltmeter circuit in Fig. 8–7. With a 50-μA meter movement, R_V increases from 50 kΩ on the 2.5-V range to 20 MΩ on the 1000-V range. Note that R_V has these values on each range whether you read full scale or not.

GOOD TO KNOW

$\frac{\Omega}{V}$ rating $= \frac{r_M}{V_M}$

or

$\frac{\Omega}{V}$ rating $= \frac{1}{I_M}$

Also,

$R_V = \frac{\Omega}{V}$ rating $\times V_{range}$

Ohms-per-Volt Rating

To indicate the voltmeter's resistance independently of the range, analog voltmeters are generally rated in ohms of resistance needed for 1 V of deflection. This value is the **ohms-per-volt rating** of the voltmeter. As an example, see the last column in Table 8–1. The values in the top row show that this meter needs 50,000-Ω R_V for 2.5 V of full-scale deflection. The resistance per 1 V of deflection then is 50,000/2.5, which equals 20,000 Ω/V.

The ohms-per-volt value is the same for all ranges because this characteristic is determined by the full-scale current I_M of the meter movement. To calculate the ohms-per-volt rating, take the reciprocal of I_M in ampere units. For example, a 1-mA meter movement results in 1/0.001 or 1000 Ω/V; a 50-μA meter movement allows 20,000 Ω/V, and a 20-μA meter movement allows 50,000 Ω/V. The ohms-per-volt rating is also called the *sensitivity* of the voltmeter.

A higher ohms-per-volt rating means higher voltmeter resistance R_V. R_V can be calculated as the product of the ohms-per-volt rating and the full-scale voltage of

Table 8–1 A Voltmeter Using a 50-μA Meter Movement

Full-Scale Voltage V_F	$R_V = R_{mult} + r_M$	Ohms per Volt $= R_V/V_F$
2.5	50 kΩ	20,000 Ω/V
10	200 kΩ	20,000 Ω/V
50	1 MΩ	20,000 Ω/V
250	5 MΩ	20,000 Ω/V
1000	20 MΩ	20,000 Ω/V

each range. For instance, across the second row in Table 8–1, on the 10-V range with a 20,000 Ω/V rating,

$$R_V = 10\text{ V} \times \frac{20{,}000\ \Omega}{\text{volt}}$$

$$R_V = 200{,}000\ \Omega$$

Usually the ohms-per-volt rating of a voltmeter is printed on the meter face.

8–3 Self-Review

Answers at the end of the chapter.

Refer to Fig. 8–7.

a. Calculate the voltmeter resistance R_V on the 2.5-V range.

b. Calculate the voltmeter resistance R_V on the 50-V range.

c. Is the voltmeter multiplier resistor in series or parallel with the meter movement?

d. Is a voltmeter connected in series or parallel with the potential difference to be measured?

e. How much is the total R of a voltmeter with a sensitivity of 20,000 Ω/V on the 25-V scale?

8–4 Loading Effect of a Voltmeter

When the voltmeter resistance is not high enough, connecting it across a circuit can reduce the measured voltage, compared with the voltage present without the voltmeter. This effect is called *loading down* the circuit, since the measured voltage decreases because of the additional load current for the meter.

Loading Effect

Voltmeter loading can be appreciable in high-resistance circuits, as shown in Fig. 8–8. In Fig. 8–8*a*, without the voltmeter, R_1 and R_2 form a voltage divider across the applied voltage of 120 V. The two equal resistances of 100 kΩ each divide the applied voltage equally, with 60 V across each.

When the voltmeter in Fig. 8–8*b* is connected across R_2 to measure its potential difference, however, the voltage division changes. The voltmeter resistance R_V of 100 kΩ is the value for a 1000-ohms-per-volt meter on the 100-V range. Now the voltmeter in parallel with R_2 draws additional current, and the equivalent resistance

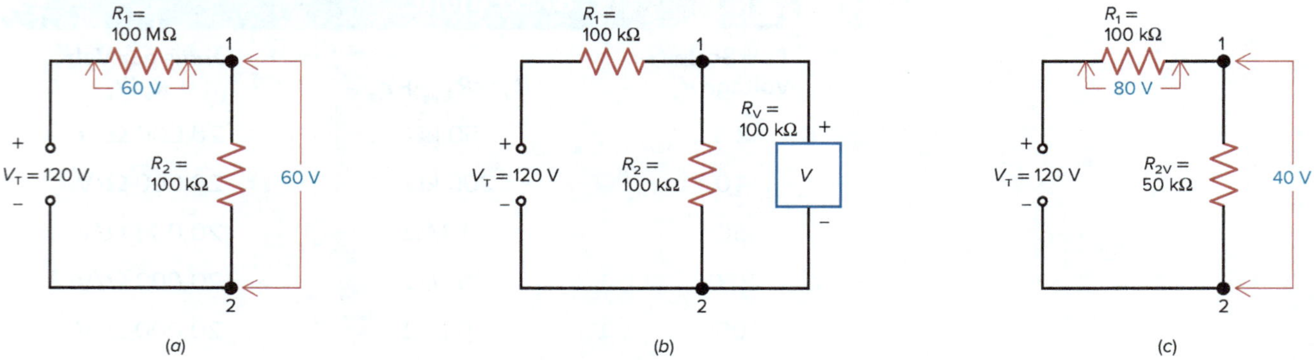

MultiSim **Figure 8–8** How the loading effect of the voltmeter can reduce the voltage reading. (*a*) High-resistance series circuit without voltmeter. (*b*) Connecting voltmeter across one of the series resistances. (*c*) Reduced *R* and *V* between points 1 and 2 caused by the voltmeter as a parallel branch across R_2. The R_{2V} is the equivalent of R_2 and R_V in parallel.

GOOD TO KNOW

The loading effect of a voltmeter is a primary concern when measuring voltages in high-resistance circuits. As a technician troubleshooting a high-resistance circuit, you must be able to determine whether a lower than normal voltage reading is the result of a defective component or the result of voltmeter loading.

Figure 8–9 Negligible loading effect with a high-resistance voltmeter. (*a*) High-resistance series circuit without voltmeter, as shown in Fig. 8–8*a*. (*b*) Same voltages in circuit with voltmeter connected because R_V is so high.

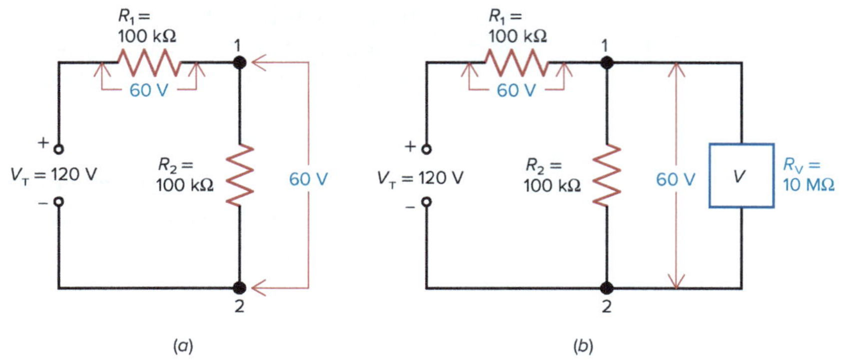

between the measured points 1 and 2 is reduced from 100,000 to 50,000 Ω. This resistance is one-third the total circuit resistance, and the measured voltage across points 1 and 2 drops to 40 V, as shown in Fig. 8–8*c*.

As additional current drawn by the voltmeter flows through the series resistance R_1, this voltage goes up to 80 V.

Similarly, if the voltmeter were connected across R_1, this voltage would go down to 40 V, with the voltage across R_2 rising to 80 V. When the voltmeter is disconnected, the circuit returns to the condition in Fig. 8–8*a*, with 60 V across both R_1 and R_2.

The **loading effect** is minimized by using a voltmeter with a resistance much greater than the resistance across which the voltage is measured. As shown in Fig. 8–9, with a voltmeter resistance of 10 MΩ, the loading effect is negligible. Because R_V is so high, it does not change the voltage division in the circuit. The 10 MΩ of the meter in parallel with the 100,000 Ω for R_2 results in an equivalent resistance practically equal to 100,000 Ω.

With multiple ranges on a VOM, the voltmeter resistance changes with the range selected. Higher ranges require more multiplier resistance, increasing the voltmeter resistance for less loading. As examples, a 20,000-ohms-per-volt meter on the 250-V range has an internal resistance R_V of 20,000 × 250, or 5 MΩ. However, on the 2.5-V range, the same meter has an R_V of 20,000 × 2.5, which is only 50,000 Ω.

On any one range, though, the voltmeter resistance is constant whether you read full-scale or less than full-scale deflection. The reason is that the multiplier resistance set by the range switch is the same for any reading on that range.

Correction for Loading Effect

The following formula can be used:

Actual reading + correction

$$V = \underset{\downarrow}{V_M} + \underset{\downarrow}{\frac{R_1R_2}{R_V(R_1 + R_2)}V_M} \qquad \textbf{(8–4)}$$

Voltage V is the corrected reading the voltmeter would show if it had infinitely high resistance. Voltage V_M is the actual voltage reading. Resistances R_1 and R_2 are the voltage-dividing resistances in the circuit without the voltmeter resistance R_V. For example, in Fig. 8–8,

$$\text{V} = 40\text{ V} + \frac{100\text{ k}\Omega \times 100\text{ k}\Omega}{100\text{ k}\Omega \times 200\text{ k}\Omega} \times 40\text{ V} = 40 + \frac{1}{2} \times 40 = 40 + 20$$

$$V = 60\text{ V}$$

The loading effect of a voltmeter causes too low a voltage reading because R_V is too low as a parallel resistance. This corresponds to the case of a current meter reading too low because R_M is too high as a series resistance. Both of these effects illustrate the general problem of trying to make any measurement without changing the circuit being measured.

Note that the DMM has practically no loading effect as a voltmeter. The input resistance is usually 10 MΩ or 20 MΩ, the same on all ranges.

8–4 Self-Review

Answers at the end of the chapter.

With the voltmeter across R_2 in Fig. 8–8*b*, what is the value for

a. V_1?
b. V_2?

8–5 Ohmmeters

An ohmmeter consists of an internal battery, the meter movement, and a current-limiting resistance, as illustrated in Fig. 8–10. For measuring resistance, the ohmmeter leads are connected across the external resistance to be measured. Power in the circuit being tested must be off. Then only the ohmmeter battery produces current for deflecting the meter movement. Since the amount of current through the meter depends on the external resistance, the scale can be calibrated in ohms.

The amount of deflection on the ohms scale indicates the measured resistance directly. The ohmmeter reads up-scale regardless of the polarity of the leads because the polarity of the internal battery determines the direction of current through the meter movement.

Series Ohmmeter Circuit

In Fig. 8–10*a*, the circuit has 1500 Ω for $(R_1 + r_M)$. Then the 1.5-V cell produces 1 mA, deflecting the moving coil full scale. When these components are enclosed in a case, as shown in Fig. 8–10*b*, the series circuit forms an ohmmeter. Note that M indicates the meter movement.

Figure 8–10 How meter movement M can be used as an ohmmeter with a 1.5-V battery. (*a*) Equivalent closed circuit with R_1 and the battery when ohmmeter leads are short-circuited for zero ohms of external *R*. (*b*) Internal ohmmeter circuit with test leads open, ready to measure an external resistance.

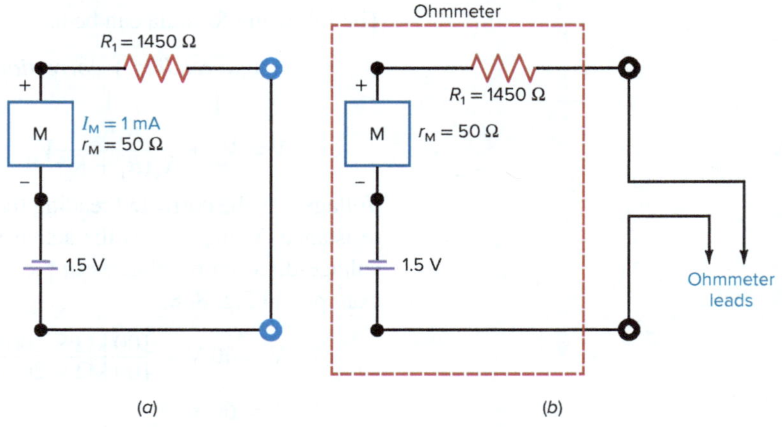

Table 8–2	Calibration of Ohmmeter in Figure 8–10				
External R_X, Ω	Internal $R_i = R_1 + r_M$, Ω	$R_T = R_X + R_i$, Ω	$I = V/R_T$, mA	Deflection	Scale Reading, Ω
0	1500	1500	1	Full scale	0
750	1500	2250	⅔ = 0.67	⅔ scale	750
1500	1500	3000	½ = 0.5	½ scale	1500
3000	1500	4500	⅓ = 0.33	⅓ scale	3000
150,000	1500	151,500	0.01	1⁄100 scale	150,000
500,000	1500	501,500	0	None	∞

If the leads are short-circuited together or connected across a short circuit, as in Fig. 8–10*a*, 1 mA flows. The meter movement is deflected full scale to the right. This ohmmeter reading is 0 Ω.

When the ohmmeter leads are open, not touching each other, the current is zero. The ohmmeter indicates infinitely high resistance or an open circuit across its terminals.

Therefore, the meter face can be marked zero ohms at the right for full-scale deflection and infinite ohms at the left for no deflection. In-between values of resistance result when less than 1 mA flows through the meter movement. The corresponding deflection on the ohm scale indicates how much resistance is across the ohmmeter terminals.

Back-Off Ohmmeter Scale

Table 8–2 and Fig. 8–11 illustrate the calibration of an ohmmeter scale in terms of meter current. The current equals V/R_T. Voltage V is the fixed applied voltage of 1.5 V supplied by the internal battery. Resistance R_T is the total resistance of R_X and the ohmmeter's internal resistance. Note that R_X is the external resistance to be measured.

The ohmmeter's internal resistance R_i is constant at 50 + 1450, or 1500 Ω here. If R_X also equals 1500 Ω, for example, R_T equals 3000 Ω. The current then is 1.5 V/3000 Ω, or 0.5 mA, resulting in half-scale deflection for the 1-mA movement. Therefore, the center of the ohm scale is marked for 1500 Ω. Similarly, the amount of current and meter deflection can be calculated for any value of the external resistance R_X.

Figure 8–11 Back-off ohmmeter scale with *R* readings increasing from right to left. (*a*) Series ohmmeter circuit for the unknown external resistor R_X to be measured. (*b*) Ohm scale has higher *R* readings to the left of the scale as more R_X decreases I_M. The *R* and *I* values are listed in Table 8–2.

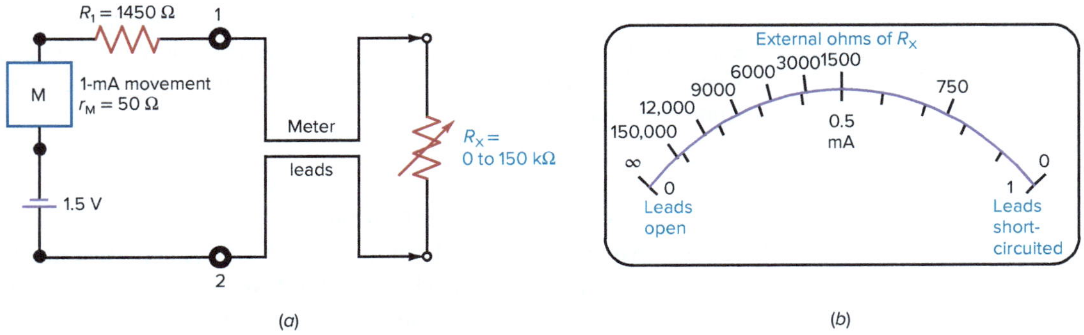

Note that the ohm scale increases from right to left. This arrangement is called a *back-off scale*, with ohm values increasing to the left as the current backs off from full-scale deflection. The back-off scale is a characteristic of any ohmmeter where the internal battery is in series with the meter movement. Then more external R_X decreases the meter current.

A **back-off ohmmeter scale** is expanded at the right near zero ohms and crowded at the left near infinite ohms. This nonlinear scale results from the relation $I = V/R$ with V constant at 1.5 V. Recall that with V constant, I and R are inversely related.

The highest resistance that can be indicated by the ohmmeter is about 100 times its total internal resistance. Therefore, the infinity mark on the ohms scale, or the "lazy eight" symbol ∞ for infinity, is only relative. It just means that the measured resistance is infinitely greater than the ohmmeter resistance.

It is important to note that the ohmmeter circuit in Fig. 8–11 is not entirely practical. The reason is that if the battery voltage, V_b, is not exactly 1.5 V, the ohmmeter scale will not be calibrated correctly. Also, a single ohmmeter range is not practical when it is necessary to measure very small or large resistance values. Without going into the circuit detail, you should be aware of the fact that commercially available ohmmeters are designed to provide multiple ohmmeter ranges as well as compensation for a change in the battery voltage, V_b.

Multiple Ohmmeter Ranges

Figure 8–12 Multiple ohmmeter ranges with just one ohm scale. The ohm reading is multiplied by the factor set on the range switch.

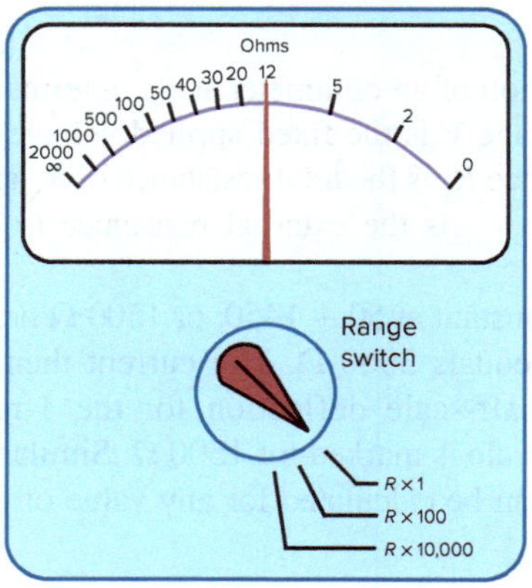

Commercial multimeters provide for resistance measurements from less than 1 Ω up to many megohms in several ranges. The range switch in Fig. 8–12 shows the multiplying factors for the ohm scale. On the $R \times 1$ range, for low-resistance measurements, read the ohm scale directly. In the example here, the pointer indicates 12 Ω. When the range switch is on $R \times 100$, multiply the scale reading by 100; this reading would then be 12 × 100 or 1200 Ω. On the $R \times 10{,}000$ range, the pointer would indicate 120,000 Ω.

A multiplying factor, instead of full-scale resistance, is given for each ohm range because the highest resistance is infinite on all ohm ranges. This method for ohms should not be confused with full-scale values for voltage ranges. For the ohmmeter ranges, always multiply the scale reading by the $R \times$ factor. On voltage ranges, you may have to multiply or divide the scale reading to match the full-scale voltage with the value on the range switch.

Zero-Ohms Adjustment

To compensate for lower voltage output as the internal battery ages, an ohmmeter includes a variable resistor to calibrate the ohm scale. A back-off ohmmeter is always adjusted for zero ohms. With the test leads short-circuited, vary the ZERO OHMS control on the front panel of the meter until the pointer is exactly on zero at the right edge of the ohm scale. Then the ohm readings are correct for the entire scale.

This type of ohmmeter must be zeroed again every time you change the range because the internal circuit changes.

When the adjustment cannot deflect the pointer all the way to zero at the right edge, it usually means that the battery voltage is too low and it must be replaced. Usually, this trouble shows up first on the $R \times 1$ range, which takes the most current from the battery.

8–5 Self-Review

Answers at the end of the chapter.

a. An ohmmeter reads 40 Ω on the $R \times 10$ range. How much is R_X?

b. A voltmeter reads 40 on the 300-V scale, but with the range switch on 30 V. How much is the measured voltage?

GOOD TO KNOW

When measuring an unknown value of current or voltage, always set the meter to its highest range and work your way down. The meter could be damaged if a very large value of current or voltage is attempted to be measured on the meter's lowest range setting.

8–6 Multimeters

Multimeters are also called *multitesters*, and they are used to measure voltage, current, or resistance. Table 8–3 compares the features of the main types of multimeters: first, the VOM in Fig. 8–13, and next the DMM in Fig. 8–14. The DMM is explained in more detail in the next section.

Beside its digital readout, an advantage of the DMM is its high input resistance R_V as a DC voltmeter. The R_V is usually 10 MΩ, the same on all ranges, which is high enough to prevent any loading effect by the voltmeter in most circuits. Some types have an R_V of 22 MΩ. Many modern DMMs are autoranging; that is, the internal circuitry selects the proper range for the meter and indicates the range as a readout.

Table 8–3 VOM Compared to DMM

VOM	DMM
Analog pointer reading	Digital readout
DC voltmeter R_V changes with range	R_V is 10 or 22 MΩ, the same on all ranges
Zero-ohms adjustment changed for each range	No zero-ohms adjustment
Ohm ranges up to $R \times 10{,}000$ Ω, as a multiplying factor	Ohm ranges up to 20 MΩ; each range is the maximum

Figure 8–13 Analog VOM that combines a function selector and range switch.

Avesun/Alamy Stock Photo

Figure 8–14 Portable digital multimeter.

Reproduced by permission of Tektronix, Inc.

For either a VOM or a DMM, it is important to have a low-voltage DC scale with resolution good enough to read 0.2 V or less. The range of 0.2 to 0.6 V, or 200 to 600 mV, is needed for measuring DC bias voltages in transistor circuits.

Low-Power Ohms (LPΩ)

Another feature needed for transistor measurements is an ohmmeter that does not have enough battery voltage to bias a semiconductor junction into the ON or conducting state. The limit is 0.2 V or less. The purpose is to prevent any parallel conduction paths in the transistor amplifier circuit that can lower the ohmmeter reading.

Decibel Scale

Most **analog multimeters** have an AC voltage scale calibrated in decibel (dB) units, for measuring AC signals. The decibel is a logarithmic unit used for comparison of power levels or voltage levels. The mark of 0 dB on the scale indicates the reference level, which is usually 0.775 V for 1 mW across 600 Ω. Positive decibel values above the zero mark indicate AC voltages above the reference of 0.775 V; negative decibel values are less than the reference level.

Amp-Clamp Probe

The problem of opening a circuit to measure *I* can be eliminated by using a probe with a clamp that fits around the current-carrying wire. The strength of the magnetic field is used to detect the amount of current carried by the conductor. An example is shown in Fig. 8–15. The clamp probe is generally used for measuring AC amperes such as those associated with the 120 or 240 V 60 Hz AC power line voltage.

Figure 8–15 DMM with amp-clamp accessory.

Courtesy of Fluke Corporation, Reproduced with Permission

High-Voltage Probe

An accessory probe can be used with a multimeter to measure DC voltages up to 30 kV. This probe is often referred to as a *high-voltage probe*. One application is measuring the high voltage of 20 to 30 kV at the anode of the color picture tube in a television receiver. The probe is just an external multiplier resistance for the DC voltmeter. The required R for a 30-kV probe is 580 MΩ with a 20-kΩ/V meter on the 1000-V range.

8–6 Self-Review

Answers at the end of the chapter.

a. How much is R_V on the 1-V range for a VOM with a sensitivity of 20 kΩ/V?

b. If R_V is 10 MΩ for a DMM on the 100-V range, how much is R_V on the 200-mV range?

c. The low-power ohm (LPΩ) function does not require an internal battery. (True/False)

Figure 8–16 Typical digital multimeter.

Courtesy of Fluke Corporation, Reproduced with Permission

8–7 Digital Multimeter (DMM)

The **digital multimeter** has become a very popular test instrument because the digital value of the measurement is displayed automatically with decimal point, polarity, and the unit for V, A, or Ω. Digital meters are generally easier to use because they eliminate the human error that often occurs in reading different scales on an analog meter with a pointer. Examples of the portable DMM are shown in Figs. 8–14 and 8–16.

The basis of the DMM operation is an analog-to-digital (A/D) converter circuit. It converts analog voltage values at the input to an equivalent binary form. These values are processed by digital circuits to be shown on a liquid-crystal display (LCD) as decimal values.

Voltage Measurements

The A/D converter requires a specific range of voltage input; typical values are −200 mV to +200 mV. For DMM input voltages that are higher, the voltage is divided down. When the input voltage is too low, it is increased by a DC amplifier circuit. The measured voltage can then be compared to a fixed reference voltage in the meter by a comparator circuit. Actually, all functions in the DMM depend on the voltage measurements by the converter and comparator circuits.

The input resistance of the DMM is in the range of 10 to 20 MΩ, shunted by 50 pF of capacitance. This R is high enough to eliminate the problem of voltmeter loading in most transistor circuits. Not only does the DMM have high input resistance, but the R is the same on all ranges.

With AC measurements, the AC input is converted to DC voltage for the A/D converter. The DMM has an internal diode rectifier that serves as an AC converter.

R Measurement

As an ohmmeter, the internal battery supplies I through the measured R for an IR drop measured by the DMM. The battery is usually the small 9-V type commonly used in portable equipment. A wide range of R values can be measured from a fraction of an ohm to more than 30 MΩ. Remember that power must be off in the circuit being tested with an ohmmeter.

A DMM ohmmeter usually has an open-circuit voltage across the meter leads, which is much too low to turn on a semiconductor junction. The result is low-power ohms (LPΩ) operation.

I Measurements

To measure current, internal resistors provide a proportional *IR* voltage. The display shows the *I* values. Note that the DMM still must be connected as a series component in the circuit when current is measured.

Diode Test

The DMM usually has a setting for testing semiconductor diodes, either silicon or germanium. Current is supplied by the DMM for the diode to test the voltage across its junction. Normal values are 0.7 V for silicon and 0.3 V for germanium. A short-circuited junction will read 0 V. The voltage across an open diode reads much too high. Most diodes are silicon. The diode test range can also be used for testing the continuity between two points in a circuit or across the ends of a wire or test lead. If continuity exists, an audible tone is produced by the meter. If there is no tone, it indicates the absence of continuity, which is an open circuit.

Resolution

This term for a DMM specifies how many places can be used to display the digits 0 to 9, regardless of the decimal point. For example, 9.99 V is a three-digit display; 9.999 V would be a four-digit display. Most portable units, however, compromise with a 3½-digit display. This means that the fourth digit at the left for the most significant place can only be a 1. If not, then the display has three digits. As examples, a 3½-digit display can show 19.99 V, but 29.99 V would be read as 30.0 V. Note that better resolution with more digits can be obtained with more expensive meters, especially the larger DMM units for bench mounting. Actually, 3½-digit resolution is enough for practically all measurements made in troubleshooting electronic equipment.

Range Overload

The DMM selector switch has specific ranges. Any value higher than the selected range is an overload. An indicator on the display warns that the value shown is not correct. Then a higher range is selected. Some units have an *autorange function* that shifts the meter automatically to a higher range as soon as an overload is indicated.

Typical DMM

The unit in Fig. 8–16 can be used as an example. On the front panel, the two jacks at the bottom right are for the test leads. The lower jack is the common lead, used for all measurements. Above is the jack for the "hot" lead, usually red, used for the measurements of *V* and *R* either DC or AC values. The two jacks at the bottom left side are for the red lead when measuring either DC or AC *I*.

Consider each function of the large selector switch at the center in Fig. 8–16. The first position, after the switch is turned clockwise from the OFF position, is used to measure AC volts, as indicated by the sine wave. No ranges are given as this meter has an autorange function. In operation, the meter has the ranges of 600 mV, 6 V, 60 V, 600 V, and, as a maximum, 1000 V.

If the autorange function is not desired, press the range button below the display to hold the range. Each touch of the button will change the range. Hold the button down to return to autorange operation.

The next two positions on the function switch are for DC volts. Polarity can be either positive or negative as indicated by the solid and dashed lines above the *V*. The ranges of DC voltages that can be measured are 6, 60, 600, and 1000 V as a maximum. For very low DC voltages, the mV switch setting should be used. Values below 600 mV can be measured on this range.

For an ohmmeter, the function switch is set to the position with the Ω symbol. The ohm values are from 0 to 50 MΩ in six ranges. Remember that power must be off in the circuit being measured, or the ohmmeter will read the wrong value. (Worse yet, the meter could be damaged.)

Next on the function switch is the position for testing semiconductor diodes, as shown by the diode symbol. Maximum diode test voltage is 2.4 V. This switch position is also used when it is desired to test for continuity between two points. The curved lines next to the diode symbol indicate that the meter produces an audible tone if continuity exists.

The last two positions on the function switch are for current measurements. The jacks at the lower left are used for larger or smaller current values.

In measuring AC values, either for V or I, the frequency range of the meter is limited to 45 to 1000 Hz, approximately. For amplitudes at higher frequencies, such as rf measurements, special meters are necessary. However, this meter can be used for V and I at the 60-Hz power-line frequency and the 400-Hz test frequency often used for audio equipment.

Analog Display

The bar at the bottom of the display in Fig. 8–16 is used only to show the relative magnitude of the input compared to the full-scale value of the range in use. This function is convenient when adjusting a circuit for a peak value or a minimum (null). The operation is comparable to watching the needle on a VOM for either a maximum or a null adjustment.

8–7 Self-Review

Answers at the end of the chapter.

a. **The typical resistance of a DMM voltmeter is 10 MΩ. (True/False)**

b. **A DMM voltmeter with 3½-digit resolution can display the value of 14.59 V. (True/False)**

8–8 Meter Applications

Table 8–4 summarizes the main points to remember when using a voltmeter, ohmmeter, or milliammeter. These rules apply whether the meter is a single unit or one function on a multimeter. The voltage and current tests also apply to either DC or AC circuits.

To avoid excessive current through the meter, it is good practice to start on a high range when measuring an unknown value of voltage or current. It is very important not to make the mistake of connecting a current meter in parallel because **usually this mistake ruins the meter.** The mistake of connecting a voltmeter in series does not damage the meter, but the reading will be wrong.

Table 8–4 Direct-Current Meters

Voltmeter	Milliammeter or Ammeter	Ohmmeter
Power on in circuit	Power on in circuit	Power off in circuit
Connect in parallel	Connect in series	Connect in parallel
High internal R	Low internal R	Has internal battery
Has internal series multipliers; higher R for higher ranges	Has internal shunts; lower resistance for higher current ranges	Higher battery voltage and more sensitive meter for higher ohm ranges

Figure 8–17 How to insert a current meter in different parts of a series-parallel circuit to read the desired current I. At point A, B, or C, the meter reads I_T; at D or E, the meter reads I_2; at F or G, the meter reads I_3.

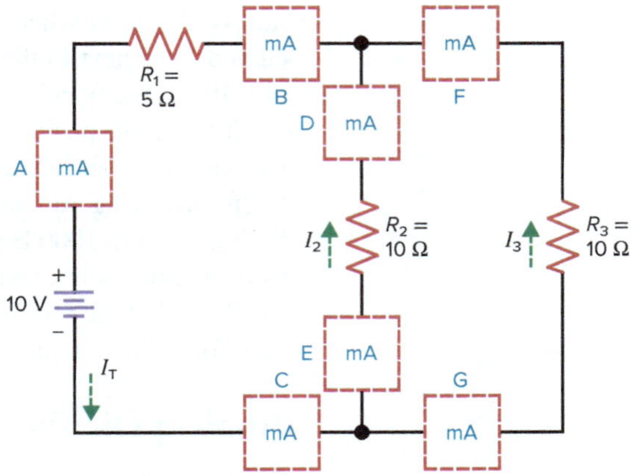

If the ohmmeter is connected to a circuit in which power is on, the meter can be damaged, beside giving the wrong reading. An ohmmeter has its own internal battery, and the power must be off in the circuit being tested. When R is tested with an ohmmeter, it may be necessary to disconnect one end of R from the circuit to eliminate parallel paths.

Connecting a Current Meter in the Circuit

In a series-parallel circuit, the current meter must be inserted in a branch to read branch current. In the main line, the meter reads the total current. These different connections are illustrated in Fig. 8–17. The meters are shown by dashed lines to illustrate the different points at which a meter could be connected to read the respective currents.

If the circuit is opened at point A to insert the meter in series in the main line here, the meter will read total line current I_T through R_1. A meter at B or C will read the same line current.

To read the branch current through R_2, this R must be disconnected from its junction with the main line at either end. A meter inserted at D or E, therefore, will read the R_2 branch current I_2. Similarly, a meter at F or G will read the R_3 branch current I_3.

Calculating *I* from Measured Voltage

Figure 8–18 With 15 V measured across a known R of 15 Ω, the I can be calculated as V/R or 15 V/15 Ω = 1 A.

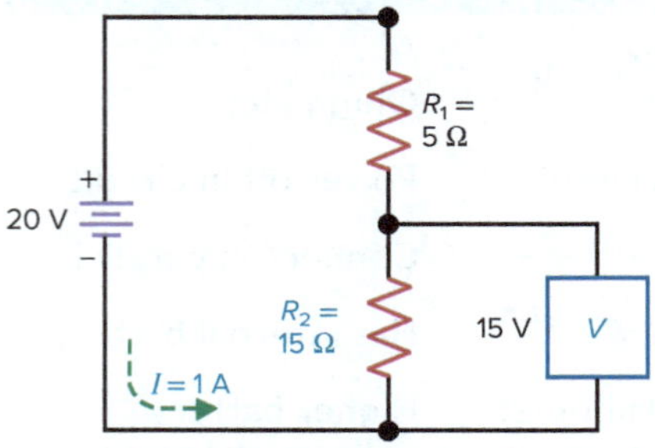

The inconvenience of opening the circuit to measure current can often be eliminated by the use of Ohm's law. The voltage and resistance can be measured without opening the circuit, and the current calculated as V/R. In the example in Fig. 8–18, when the voltage across R_2 is 15 V and its resistance is 15 Ω, the current through R_2 must be 1 A. When values are checked during troubleshooting, if the voltage and resistance are normal, so is the current.

This technique can also be convenient for determining I in low-resistance circuits where the resistance of a microammeter may be too high. Instead of measuring I, measure V and R and calculate I as V/R.

Furthermore, if necessary, we can insert a known resistance R_S in series in the circuit, temporarily, just to measure V_S. Then I is calculated as V_S/R_S. The resistance of R_S, however, must be small enough to have little effect on R_T and I in the series circuit.

This technique is often used with oscilloscopes to produce a voltage waveform of IR which has the same waveform as the current in a resistor. The oscilloscope must be connected as a voltmeter because of its high input resistance.

Figure 8–19 Voltage tests to localize an open circuit. (*a*) Normal circuit with voltages to chassis ground. (*b*) Reading of 0 V at point D shows R_3 is open.

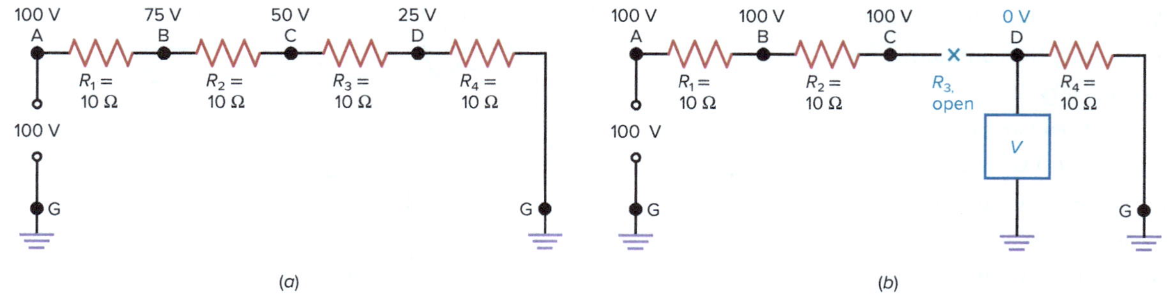

Checking Fuses

Turn the power off or remove the fuse from the circuit to check with an ohmmeter. A good fuse reads 0 Ω. A blown fuse is open, which reads infinity on the ohmmeter.

A fuse can also be checked with the power on in the circuit by using a voltmeter. Connect the voltmeter across the two terminals of the fuse. A good fuse reads 0 V because there is practically no *IR* drop. With an open fuse, though, the voltmeter reading is equal to the full value of the applied voltage. Having the full applied voltage seems to be a good idea, but it should not be across the fuse.

Voltage Tests for an Open Circuit

Figure 8–19 shows four equal resistors in series with a 100-V source. A ground return is shown here because voltage measurements are usually made with respect to chassis or earth ground. Normally, each resistor would have an *IR* drop of 25 V. Then, at point B, the voltmeter to ground should read 100 − 25 = 75 V. Also, the voltage at C should be 50 V, with 25 V at D, as shown in Fig. 8–19*a*.

However, the circuit in Fig. 8–19*b* has an open in R_3 toward the end of the series string of voltages to ground. Now when you measure at B, the reading is 100 V, equal to the applied voltage. This full voltage at B shows that the series circuit is open without any *IR* drop across R_1. The question is, however, which *R* has the open? Continue the voltage measurements to ground until you find 0 V. In this example, the open is in R_3 between the 100 V at C and 0 V at D.

The points that read the full applied voltage have a path back to the source of voltage. The first point that reads 0 V has no path back to the high side of the source. Therefore, the open circuit must be between points C and D in Fig. 8–19*b*.

8–8 Self-Review

Answers at the end of the chapter.

a. Which type of meter requires an internal battery?
b. How much is the normal voltage across a good fuse?
c. How much is the voltage across R_1 in Fig. 8–19*a*?
d. How much is the voltage across R_1 in Fig. 8–19*b*?

8–9 Checking Continuity with the Ohmmeter

A wire conductor that is continuous without a break has practically zero ohms of resistance. Therefore, the ohmmeter can be useful in testing for continuity. This test should be done on the lowest ohm range. There are many applications. A wire

GOOD TO KNOW

Most DMMs have a switch setting for testing the continuity between two points. If the resistance between two points is less than about 200 Ω, an audible tone will be heard. The audible tone provides immediate feedback to indicate whether or not there is continuity between the two points being measured. See Section 8–7, Digital Multimeter (DMM).

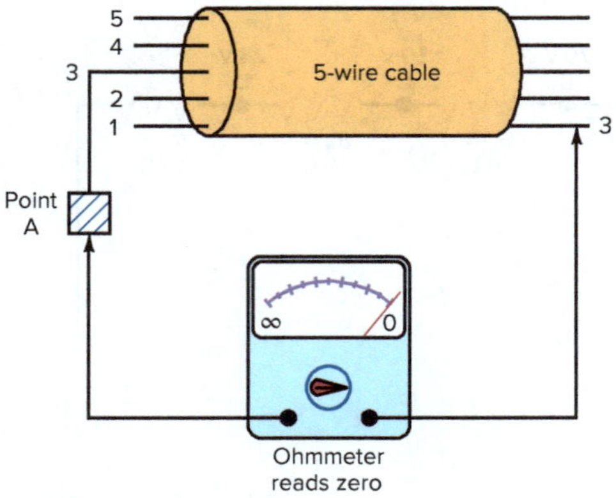

Figure 8–20 Continuity testing from point A to wire 3 shows that this wire is connected.

conductor can have an internal break which is not visible because of the insulated cover, or the wire can have a bad connection at the terminal. Checking for zero ohms between any two points along the conductor tests continuity. A break in the conducting path is evident from a reading of infinite resistance, showing an open circuit.

As another application of checking continuity, suppose that a cable of wires is harnessed together, as illustrated in Fig. 8–20, where the individual wires cannot be seen, but it is desired to find the conductor that connects to terminal A. This is done by checking continuity for each conductor to point A. The wire that has zero ohms to A is the one connected to this terminal. Often the individual wires are color-coded, but it may be necessary to check the continuity of each lead.

An additional technique that can be helpful is illustrated in Fig. 8–21. Here it is desired to check the continuity of the two-wire line, but its ends are too far apart for the ohmmeter leads to reach. The two conductors are temporarily short-circuited at one end, however, so that the continuity of both wires can be checked at the other end.

In summary, the ohmmeter is helpful in checking the continuity of any wire conductor. This includes resistance-wire heating elements, such as the wires in a toaster or the filament of an incandescent bulb. Their cold resistance is normally just a few ohms. Infinite resistance means that the wire element is open. Similarly, a good fuse has practically zero resistance. A burned-out fuse has infinite resistance; that is, it is open. Any coil for a transformer, solenoid, or motor will also have infinite resistance if the winding is open.

8–9 Self-Review

Answers at the end of the chapter.

a. On a back-off ohmmeter, is zero ohms at the left or the right end of the scale?

b. What is the ohmmeter reading for an open circuit?

Figure 8–21 Temporary short circuit at one end of a long two-wire line to check continuity from the opposite end.

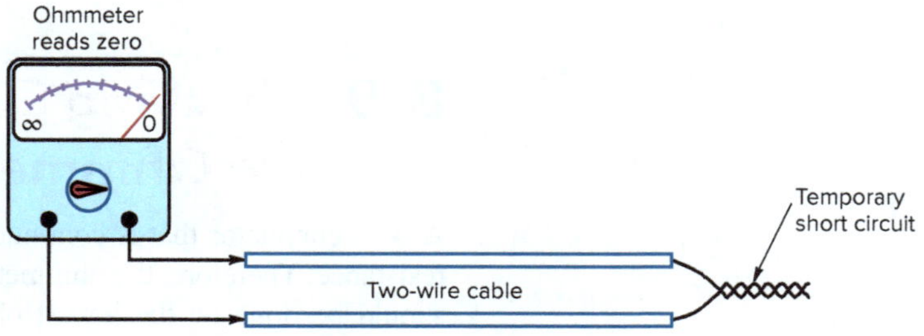

Application in Understanding Clamp-On Ammeters

Of the three electrical quantities—voltage, current, and resistance—measuring current is probably the most inconvenient and/or difficult of the three. The inconvenience or difficulty lies in the fact that the circuit must be opened and the meter must be placed in the path of the moving electric charges. Besides being difficult and/or inconvenient, this method of measuring current isn't very safe, especially when measuring currents in high-voltage AC circuits. Therefore, when measuring the current drawn by an electronic appliance or similar device, it is much easier, and safer, to use a clamp-on ammeter, such as the one shown in Fig. 8–22. In this figure, each control, key, and feature is identified so that their function and/or purpose can be described here.

METER CONTROLS, KEYS, AND FEATURES

Clamp head—The clamp head is used for clamping around a conductor when measuring current, either AC or DC.

Protective barrier—This is a physical border between the controls of the meter and the clamp head. For safety, make sure your hands and fingers are below this point when making current and/or non-contact voltage measurements.

Clamp head trigger—Pull the trigger to open the clamp head and release the trigger to close it.

Function select switch—This control can be rotated to select the desired electrical quantity to be measured.

Non-contact voltage (NCV) indicator—This feature is used to detect the presence of an AC voltage without the need for making any physical contact with live conductors. With the **function select** switch set to the NCV position, place the front end of the clamp head close to the voltage source. The NCV indicator will flash and the meter will beep when voltage is present.

HOLD/backlight key—This key is used to hold a measurement on the display for a delayed reading when making measurements in hard-to-get-at places. It is also used for turning the backlight of the display either ON or Off. Just press and hold this button for 2 seconds to either turn ON or turn OFF the display's backlight.

ZERO key—This key is used to zero the display before measuring DC amps and capacitance values.

SELECT key—This key selects the function mode such as ACV/DCV, ACA/DCA, and ohms (Ω), continuity, Diode test, and capacitance.

LCD display—The LCD display indicates the electrical quantity being measured and its corresponding numerical value.

Input jacks—The red and black test leads are inserted into these jacks when it is necessary to measure electrical quantities that cannot be detected or measured with the clamp head.

MEASURING CURRENT WITH A CLAMP-ON AMMETER

Recall from Chapter 1, "Electricity," that any current-carrying conductor has an associated magnetic field. The strength of the magnetic field is proportional to the current carried by the conductor. In other words, the greater the current, the stronger the magnetic field. The **clamp head**, which is made of ferrite material, detects the strength of an AC current's magnetic field and produces a voltage that is proportional to the magnitude of the current. This voltage is fed to an internal metering circuit, which is calibrated to display the measured value of current. The clamp head also contains a ***Hall effect transducer***, which allows it to detect the strength of DC currents.

MEASURING AC CURRENT

A stand-alone clamp-on ammeter, like the one shown in Fig. 8–22, is only capable of measuring the current in a single-wire conductor. Placing the clamp head around two current-carrying conductors, such as an extension cord, results in a reading of zero (or a false reading) because the opposing magnetic fields in each conductor cancel each other. To measure the amount of AC current in a single conductor, move the **Function select** switch to the desired range setting based on the anticipated value of current to be measured. Using the **SELECT** key select AC rather than DC amps. Next, press the trigger and open the jaws of the clamp head and position it so that the wire is inside of the jaws. Now release the trigger so that the jaws clamp around the current-carrying conductor. This procedure is illustrated in Figs. 8–23a and b. The measured value of current appears on the LCD display. For the most accurate reading, it is important to position the conductor in the center of the jaws. (There are special markings on the clamp head for centering the conductor.)

AC LINE SPLITTER

To measure the AC current drawn by an appliance, or similar device, an ***AC line splitter*** must be used. A typical AC line splitter is shown in Fig. 8–24a. Every AC line splitter has a plug-end and a receptacle-end. What the AC line splitter does is separate (split) the hot and neutral conductors of the power cord so that the clamp-on ammeter can be used to measure the current in just one conductor. The AC line splitter is plugged into the 120 V

Figure 8–22 Clamp-on Ammeter

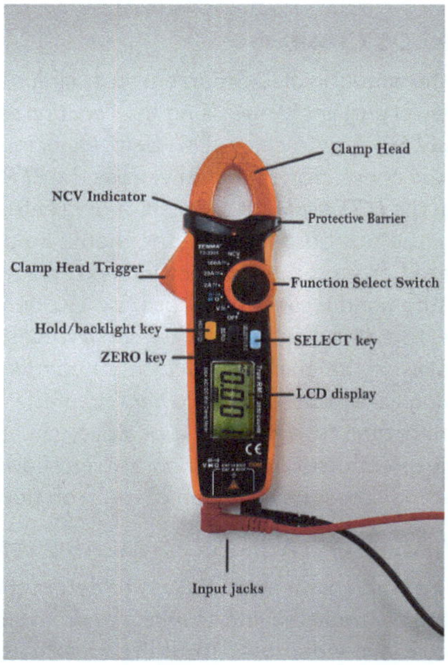

Figure 8–23a Press the trigger to open the jaws of the clamp head.

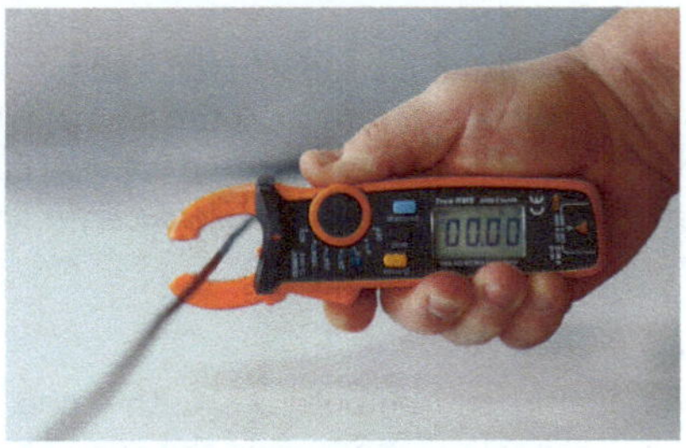

Sarah Schultz Photography

Figure 8–23b Release the trigger so that the jaws of the clamp head are around the current-carrying conductor.

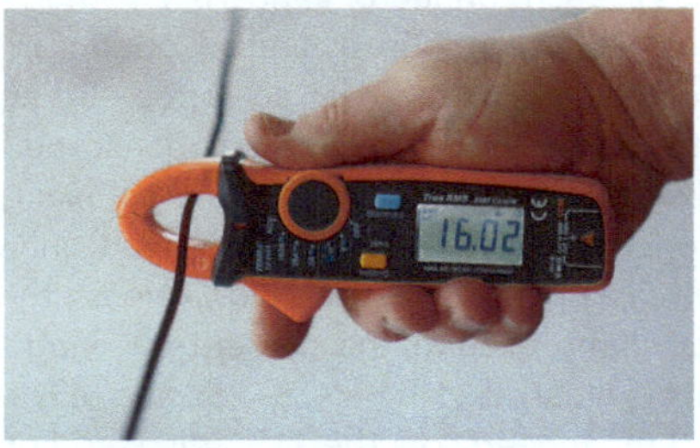

Sarah Schultz Photography

AC outlet, and the electronic appliance or device is plugged into the line splitter's receptacle-end. Next, the clamp-on ammeter is clamped around either the ×1 or ×10 square as shown in Fig. 8–24b. Clamping around the ×1 square gives the exact value of current drawn by the electronic device plugged into the AC line splitter. Clamping around the ×10 square multiplies the reading by a factor of 10. The ×10 square allows for measuring lower current values more accurately. Always remember, however, that the value of current displayed by the clamp-on ammeter must be divided by 10 when clamping around the ×10 square. This results in approximately the same value of current as when the meter is clamped around the ×1 square. Notice the two holes at the receptacle-end of the AC line splitter. These holes provide access to the hot and neutral conductors so that the AC power line voltage can be measured at this point.

MEASURING DC CURRENT

To measure the amount of DC current in a single conductor, select the desired current range using the ***Function select*** switch and use the ***SELECT*** key to select DC. Before placing the clamp head around the conductor, repeatedly press the ***ZERO*** key until the reading on the ***LCD display*** shows zero. For greatest accuracy when zeroing, the position of the meter should be the same as when it is placed around the conductor. Next, place the clamp head around the conductor to measure the DC current. If the reading is positive, it indicates that the current is flowing from the positive side of the clamp head to the negative side. (The polarity markings are located at the lower end of the clamp head jaws.) It is important to note that a positive value of current represents the direction of conventional current flow, whereas a negative value indicates the direction of electron flow.

NON-CONTACT VOLTAGE (NCV) MEASUREMENT

This feature is a convenient and safe way to detect the presence of an AC voltage without having to make any physical contact with live electrical conductors. To utilize this feature, simply

Figure 8–24a AC line splitter

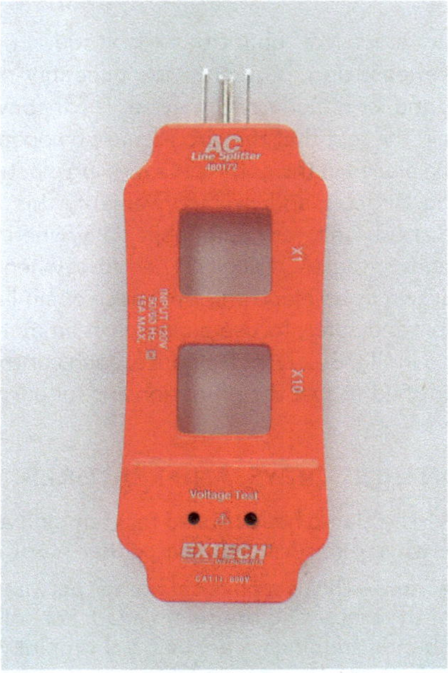

Sarah Schultz Photography

Figure 8–24b Clamping around the ×1 square of the line-splitter.

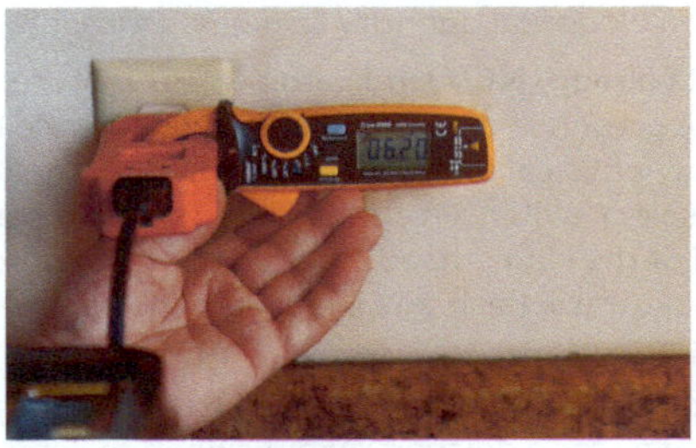

Sarah Schultz Photography

move the **FUNCTION select** switch to the **NCV** position. Next, place the clamp head close to the suspected live conductor. If an AC voltage is present, the **NCV** LED indicator flashes and the meter beeps repeatedly. Also, the LCD display shows one or more dash marks depending on the magnitude of the voltage. The higher the voltage, the greater the number of dash marks, the faster the **NCV** LED indicator flashes and the faster the meter beeps. (It is important to note that the NCV measurement is actually detecting the intensity of the electric field associated with the AC voltage. This meter is designed to detect four different levels of field intensity.)

As an example of how to use this feature refer to Fig. 8–25. In this example, the NCV feature is being used to determine if there is voltage present at a typical 120 V AC outlet. Notice the clamp head is placed in close proximity to the face of the outlet. In fact, it's OK if the clamp head touches the face of the outlet. In this case, the **NCV** indicator LED flashes and the meter beeps repeatedly, indicating that voltage is indeed present at the outlet.

ADDITIONAL METER FEATURES

To measure voltage, resistance, or capacitance or to test diodes, the red and black test leads must be inserted into the input jacks

Figure 8–25 NCV measurement

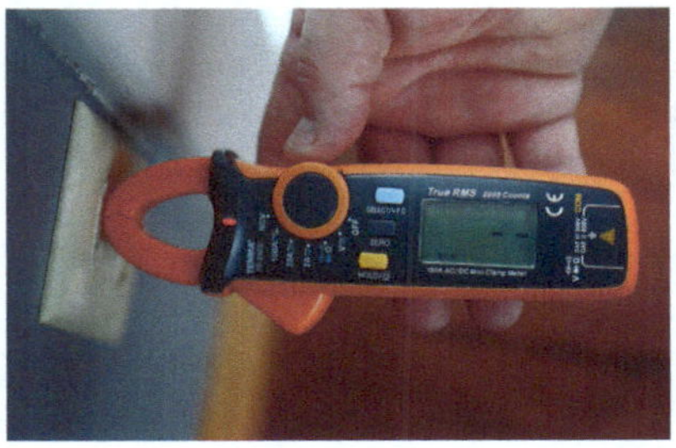

Sarah Schultz Photography

at the bottom of the meter. (This meter is ***not*** capable of making in-circuit current measurements like a typical DMM.) When making measurements, the meter's auto-range feature selects the correct range. It is important to note that the voltmeter portion of the meter has an input resistance of 10 MΩ. Also, when measuring continuity, the meter will produce a continuous tone if the resistance is less than 10 Ω.

To save energy and/or battery life, this meter has an automatic ***power-off feature*** that causes it to turn off automatically after 15 minutes if the meter's **FUNCTION select** switch is not moved. As a warning, one minute before the meter turns off, the meter will beep five times. A long, continuous beep will be made just before the meter actually turns off. To turn the meter back on, the **FUNCTION select** switch must be rotated back to the off position and then back to the desired setting. An easier way to turn the meter back on (or to wake it up) is to simply push any one of the three keys on the meter. The automatic **power-off** feature can also be disabled. Here's how to do it. With the **FUNCTION select** switch in the off position, press and hold the **SELECT** key and then rotate the **FUNCTION select** switch to the desired setting. Several continuous beeps will be heard when the meter first turns on. This indicates the automatic **power-off** feature has been disabled. With the automatic **power-off** feature disabled, the meter will put out five continuous beeps every 15 minutes (with five more beeps exactly 1 minute later) to remind the user that the meter has been left on. To reenable the automatic **power-off** feature, simply turn the meter off and then turn it back on.

Summary

- Direct current in a moving-coil meter deflects the coil in proportion to the amount of current.
- A current meter is a low-resistance meter connected in series to read the amount of current in a circuit.
- A meter shunt R_S in parallel with the meter movement extends the range of a current meter [see Formula (8–1)].
- A voltmeter consists of a meter movement in series with a high-resistance multiplier. The voltmeter with its multiplier is connected across two points to measure the potential difference in volts. The multiplier R can be calculated from Formula (8–3).
- The ohms-per-volt rating of a voltmeter with series multipliers specifies the sensitivity on all voltage ranges. It equals the reciprocal of the full-scale deflection current of the meter. A typical value is 20,000 Ω/V for a voltmeter using a 50-μA movement. The higher the ohms-per-volt rating, the better.
- Voltmeter resistance R_V is higher for higher ranges because of higher-resistance multipliers. Multiply the ohms-per-volt rating by the voltage range to calculate the R_V for each range.
- An ohmmeter consists of an internal battery in series with the meter movement. Power must be off in the circuit being checked with an ohmmeter. The series ohmmeter has a back-off scale with zero ohms at the right edge and infinity at the left. Adjust for zero ohms with the leads short-circuited each time the ohms range is changed.
- The VOM is a portable multimeter that measures volts, ohms, and milliamperes.
- The digital multimeter generally has an input resistance of 10 MΩ on all voltage ranges.
- In checking wire conductors, the ohmmeter reads 0 Ω or very low R for normal continuity and infinite ohms for an open.

Important Terms

Amp-clamp probe — a meter that can measure AC currents, generally from the 60-Hz AC power line, without breaking open the circuit. The probe of the meter is actually a clamp that fits around the current-carrying conductor.

Analog multimeter — a test instrument that is used to measure voltage, current, and resistance. An analog multimeter uses a moving pointer and a printed scale to display the value of the measured quantity.

Back-off ohmmeter scale — an ohmmeter scale that shows zero ohms (0 Ω) for full-scale deflection and infinite ohms (∞ Ω) for no deflection. As the name implies, the ohms of resistance increase from right to left on the scale as the pointer backs off from full-scale deflection.

Continuity testing — a resistance measurement that determines whether or not there is zero ohms of resistance (approximately) between two points, such as across the ends of a wire conductor.

Digital multimeter (DMM) — a popular test instrument that is used to measure voltage, current, and resistance. A DMM uses a numeric display to indicate directly the value of the measured quantity.

Loading effect — a term that describes the reduction in measured voltage when using a voltmeter to measure the voltage in a circuit. The term may also be applied to describe the reduction in current when using a current meter to measure the current in a circuit. The loading effect of a voltmeter occurs when the voltmeter resistance is not high enough. Conversely, the loading effect of a current meter occurs when the resistance of the current meter is too high.

Multiplier resistor — a large resistance in series with a moving-coil meter movement which allows the meter to measure voltages in a circuit.

Ohms-per-volt (Ω/V) rating — a voltmeter rating that specifies the ohms of resistance needed per 1 V of deflection. The Ω/V rating = $1/I_M$ or R_V/V_{range}.

Shunt resistor — a resistor placed in parallel with a basic moving-coil meter movement to extend the current range beyond the I_M value of the meter movement.

Zero-ohms adjustment — a control on an analog VOM that is adjusted for zero ohms with the ohmmeter leads shorted. This adjustment should be made each time the ohmmeter range is changed so that the ohmmeter scale remains calibrated.

Related Formulas

$$R_S = \frac{V_M}{I_S}$$

$$I_S = I_T - I_M$$

$$R_{mult} = \frac{\text{Full-scale } V}{\text{Full-scale } I} - r_M$$

$$V = V_M + \frac{R_1 R_2}{R_V(R_1 + R_2)} V_M$$

Self-Test

Answers at the back of the book.

1. **For a moving-coil meter movement, I_M is**
   a. the amount of current needed in the moving coil to produce full-scale deflection of the meter's pointer.
   b. the value of current flowing in the moving coil for any amount of pointer deflection.
   c. the amount of current required in the moving coil to produce half-scale deflection of the meter's pointer.
   d. none of the above.

2. **For an analog VOM with a mirror along the printed scale,**
   a. the pointer deflection will be magnified by the mirror when measuring small values of voltage, current, and resistance.
   b. the meter should always be read by looking at the meter from the side.
   c. the meter is read when the pointer and its mirror reflection appear as one.
   d. both a and b.

3. **A current meter should have a**
   a. very high internal resistance.
   b. very low internal resistance.
   c. infinitely high internal resistance.
   d. none of the above.

4. **A voltmeter should have a**
   a. resistance of about 0 Ω.
   b. very low resistance.
   c. very high internal resistance.
   d. none of the above.

5. **Voltmeter loading is usually a problem when measuring voltages in**
   a. parallel circuits.
   b. low-resistance circuits.
   c. a series circuit with low-resistance values.
   d. high-resistance circuits.

6. **To double the current range of a 50-μA, 2-kΩ moving-coil meter movement, the shunt resistance, R_S, should be**
   a. 2 kΩ.
   b. 1 kΩ.
   c. 18 kΩ.
   d. 50 kΩ.

7. **A voltmeter using a 20-μA meter movement has an Ω/V rating of**
   a. $\frac{20\ \text{k}\Omega}{\text{V}}$.
   b. $\frac{50\ \text{k}\Omega}{\text{V}}$.
   c. $\frac{1\ \text{k}\Omega}{\text{V}}$.
   d. $\frac{10\ \text{M}\Omega}{\text{V}}$.

8. **As the current range of an analog meter is increased, the overall meter resistance, R_M,**
   a. decreases.
   b. increases.
   c. stays the same.
   d. none of the above.

9. **As the voltage range of an analog VOM is increased, the total voltmeter resistance, R_V,**
   a. decreases.
   b. increases.
   c. stays the same.
   d. none of the above.

10. **An analog VOM has an Ω/V rating of 10 kΩ/V. What is the voltmeter resistance, R_V, if the voltmeter is set to the 25-V range?**
    a. 10 kΩ.
    b. 10 MΩ.
    c. 25 kΩ.
    d. 250 kΩ.

11. **What shunt resistance, R_S, is needed to make a 100-μA, 1-kΩ meter movement capable of measuring currents from 0 to 5 mA?**
    a. 25 Ω.
    b. 10.2 Ω.
    c. 20.41 Ω.
    d. 1 kΩ.

12. **For a 30-V range, a 50-μA, 2-kΩ meter movement needs a multiplier resistor of**
    a. 58 kΩ.
    b. 598 kΩ.
    c. 10 MΩ.
    d. 600 kΩ.

13. **When set to any of the voltage ranges, a typical DMM has an input resistance of**
    a. about 0 Ω.
    b. 20 kΩ.
    c. 10 MΩ.
    d. 1 kΩ.

14. **When using an ohmmeter to measure resistance in a circuit,**
    a. the power in the circuit being tested must be off.
    b. the power in the circuit being tested must be on.
    c. the power in the circuit being tested may be on or off.
    d. the power in the circuit being tested should be turned on after the leads are connected.

15. **Which of the following voltages cannot be displayed by a DMM with a 3½-digit display?**
    a. 7.64 V.
    b. 13.5 V.
    c. 19.98 V.
    d. 29.98 V.

16. **What type of meter can be used to measure AC currents without breaking open the circuit?**
    a. An analog VOM.
    b. An amp-clamp probe.
    c. A DMM.
    d. There isn't such a meter.

17. **Which of the following measurements is usually the most inconvenient and time-consuming when troubleshooting?**
    a. resistance measurements.
    b. DC voltage measurements.
    c. current measurements.
    d. AC voltage measurements.

18. **An analog ohmmeter reads 18 on the $R \times 10$ k range. What is the value of the measured resistance?**
    a. 180 kΩ.
    b. 18 kΩ.
    c. 18 Ω.
    d. 180 Ω.

19. Which meter has a higher resistance, a DMM with 10 MΩ of resistance on all DC voltage ranges or an analog VOM with a 50 kΩ/V rating set to the 250-V range?

a. the DMM.

b. the analog VOM.

c. They both have the same resistance.

d. It cannot be determined.

20. When using an ohmmeter to measure the continuity of a wire, the resistance should measure

a. about 0 Ω if the wire is good.

b. infinity if the wire is broken (open).

c. very high resistance if the wire is good.

d. both a and b.

Essay Questions

1. (a) Why is a milliammeter connected in series in a circuit? (b) Why should the milliammeter have low resistance?

2. (a) Why is a voltmeter connected in parallel in a circuit? (b) Why should the voltmeter have high resistance?

3. A circuit has a battery across two resistances in series. (a) Draw a diagram showing how to connect a milliammeter in the correct polarity to read current through the junction of the two resistances. (b) Draw a diagram showing how to connect a voltmeter in the correct polarity to read the voltage across one resistance.

4. Explain briefly why a meter shunt equal to the resistance of the moving coil doubles the current range.

5. Describe how to adjust the zero ohms control on a back-off ohmmeter.

6. What is meant by a 3½-digit display on a DMM?

7. Give two advantages of the DMM in Fig. 8–14 compared with the conventional VOM in Fig. 8–13.

8. What does the zero ohms control in the circuit of a back-off ohmmeter do?

9. State two precautions to be observed when you use a milliammeter.

10. State two precautions to be observed when you use an ohmmeter.

11. The resistance of a voltmeter R_V is 300 kΩ on the 300-V range when measuring 300 V. Why is R_V still 300 kΩ when measuring 250 V on the same range?

12. Give a typical value of voltmeter resistance for a DMM.

13. Would you rather use a DMM or VOM in troubleshooting? Why?

Problems

SECTION 8–2 METER SHUNTS

8–1 Calculate the value of the shunt resistance, R_S, needed to extend the range of the meter movement in Fig. 8–26 to (a) 2 mA; (b) 10 mA; (c) 25 mA; (d) 100 mA.

Figure 8–26

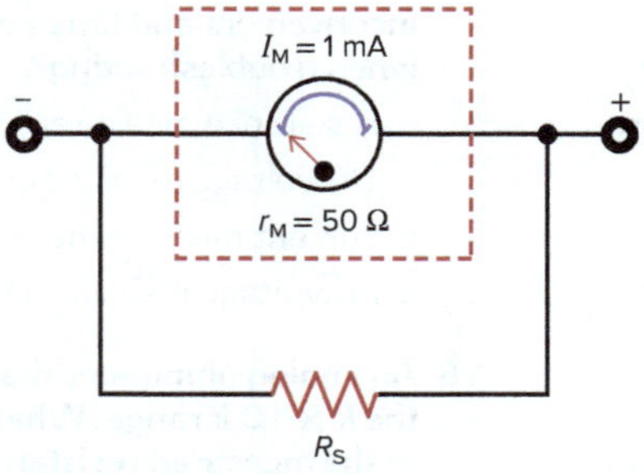

8–2 What is the resistance, R_M, of the meter (R_S in parallel with r_M) for each of the current ranges listed in Prob. 8–1?

8–3 Calculate the value of the shunt resistance, R_S, needed to extend the range of the meter movement in Fig. 8–27 to (a) 100 μA; (b) 1 mA; (c) 5 mA; (d) 10 mA; (e) 50 mA; and (f) 100 mA.

Figure 8–27

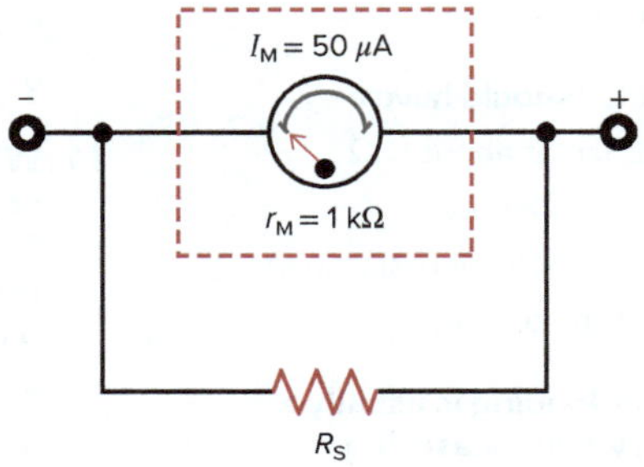

8–4 What is the resistance, R_M, of the meter (R_S in parallel with r_M) for each of the current ranges listed in Prob. 8–3?

8–5 Refer to Fig. 8–28. (a) Calculate the values for the separate shunt resistances, R_{S1}, R_{S2}, and R_{S3}. (b) Calculate the resistance, R_M, of the meter (R_S in parallel with r_M) for each setting of the range switch.

8–6 Repeat Prob. 8–5 if the meter movement has the following characteristics: $I_M = 250$ μA, $r_M = 2$ kΩ.

Figure 8–28

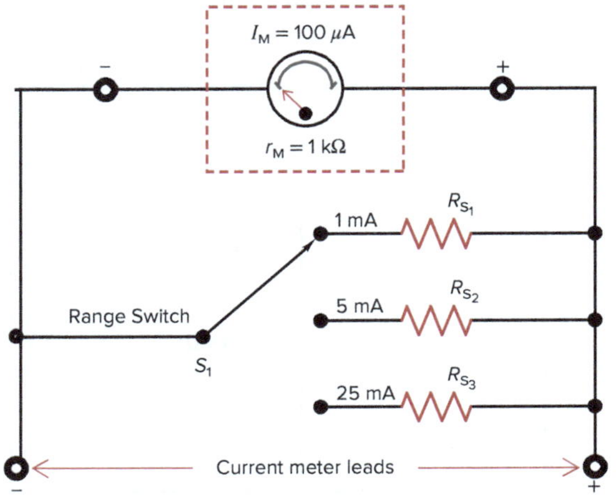

8–7 Why is it desirable for a current meter to have very low internal resistance?

SECTION 8–3 VOLTMETERS

8–8 Calculate the required multiplier resistance, R_{mult}, in Fig. 8–29 for each of the following voltage ranges: (a) 1 V; (b) 5 V; (c) 10 V; (d) 50 V; (e) 100 V; and (f) 500 V.

Figure 8–29

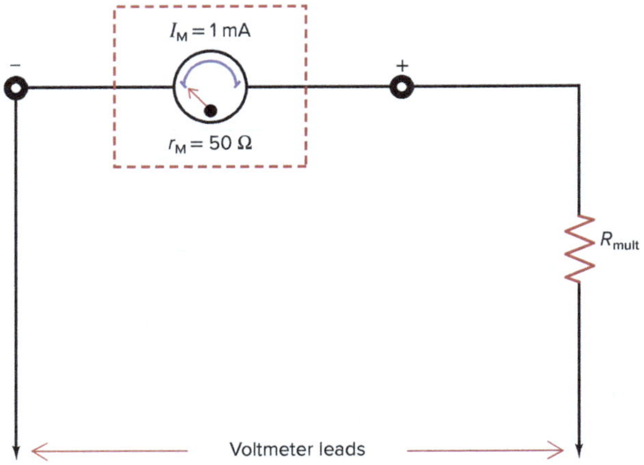

8–9 What is the Ω/V rating of the voltmeter in Prob. 8–8?

8–10 Calculate the required multiplier resistance, R_{mult}, in Fig. 8–30 for each of the following voltage ranges: (a) 3 V; (b) 10 V; (c) 30 V; (d) 100 V; and (e) 300 V.

Figure 8–30

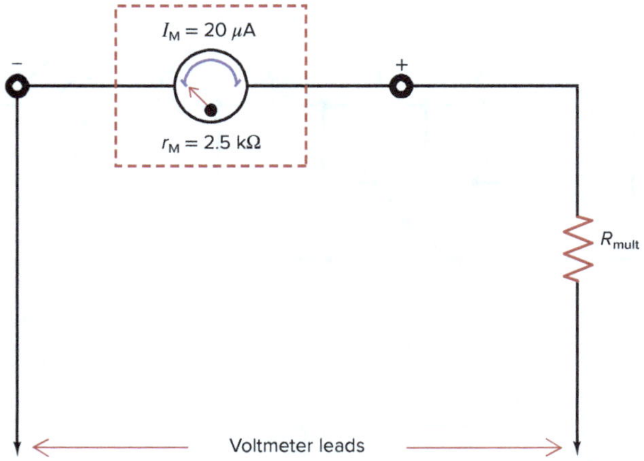

8–11 What is the Ω/V rating of the voltmeter in Prob. 8–10?

8–12 Refer to Fig. 8–31. (a) Calculate the values for the multiplier resistors R_1, R_2, R_3, R_4, R_5, and R_6. (b) Calculate the total voltmeter resistance, R_V, for each setting of the range switch. (c) Determine the Ω/V rating of the voltmeter.

Figure 8–31

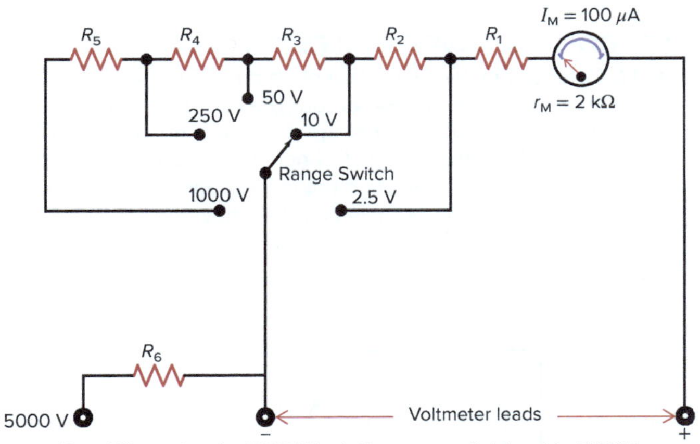

Note: When using the 5000-V jack, the range switch is set to 1000 V.

8–13 Refer to Fig. 8–32. (a) Calculate the values for the multiplier resistors R_1, R_2, R_3, R_4, R_5, and R_6. (b) Calculate the total voltmeter resistance, R_V, for each setting of the range switch. (c) Determine the Ω/V rating of the voltmeter.

Figure 8–32

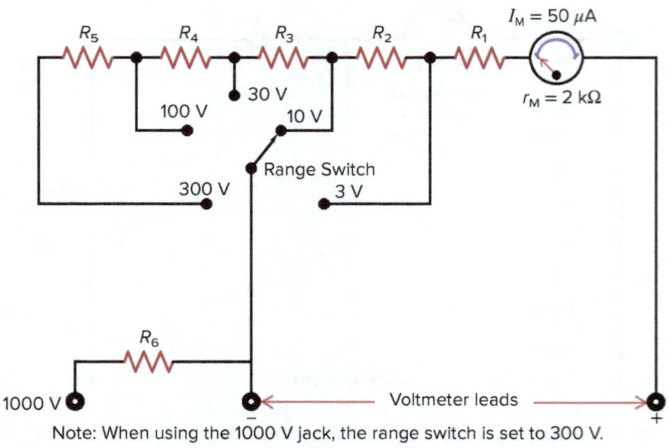

8–14 A certain voltmeter has an Ω/V rating of 25 kΩ/V. Calculate the total voltmeter resistance, R_V, for the following voltmeter ranges: (a) 2.5 V; (b) 10 V; (c) 25 V; (d) 100 V; (e) 250 V; (f) 1000 V; and (g) 5000 V.

8–15 Calculate the Ω/V rating of a voltmeter that uses a meter movement with an I_M value of (a) 1 mA; (b) 100 μA; (c) 50 μA; and (d) 10 μA.

SECTION 8–4 LOADING EFFECT OF A VOLTMETER

8–16 Refer to Fig. 8–33. (a) Calculate the DC voltage that should exist across R_2 without the voltmeter present. (b) Calculate the DC voltage that would be measured across R_2 using a 10 kΩ/V analog voltmeter set to the 10-V range. (c) Calculate the DC voltage that would be measured across R_2 using a DMM having an R_V of 10 MΩ on all DC voltage ranges.

Figure 8–33

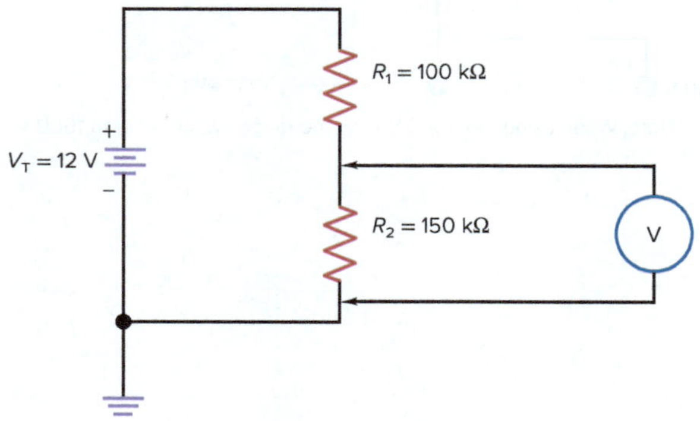

8–17 Repeat Prob. 8–16 if $R_1 = 1\ k\Omega$ and $R_2 = 1.5\ k\Omega$.

8–18 Refer to Fig. 8–34. (a) Calculate the DC voltage that should exist across R_2 without the voltmeter present. (b) Calculate the DC voltage that would be measured across R_2 using a 100 kΩ/V analog voltmeter set to the 10-V range. (c) Calculate the DC voltage that would be measured across R_2 using a DMM with an R_V of 10 MΩ on all DC voltage ranges.

Figure 8–34

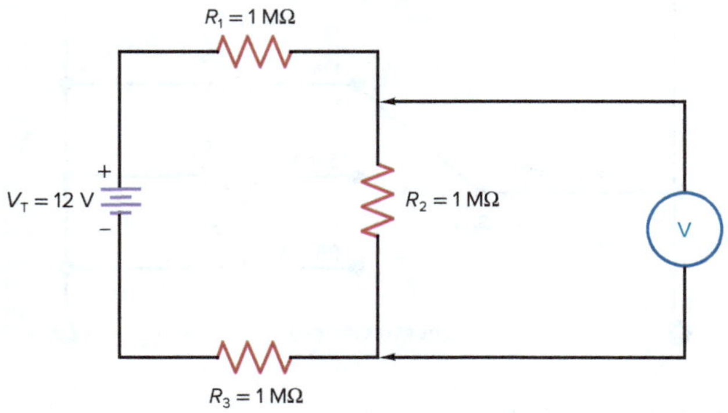

8–19 In Prob. 8–18, which voltmeter produced a greater loading effect? Why?

8–20 In Fig. 8–35, determine (a) the voltmeter resistance, R_V, and (b) the corrected voltmeter reading using Formula (8–4).

Figure 8–35

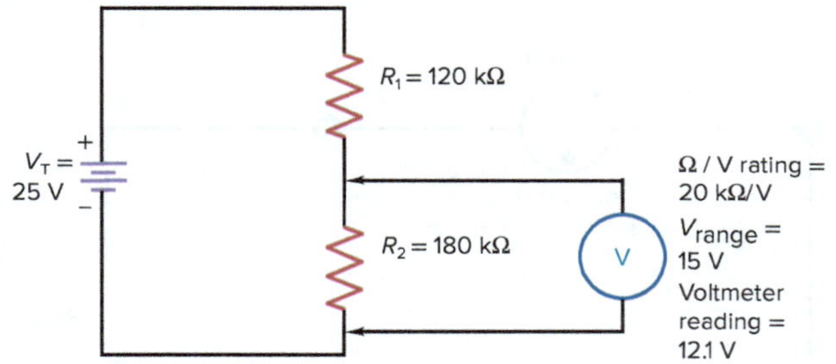

SECTION 8–5 OHMMETERS

8–21 Figure 8–36 shows a series ohmmeter and its corresponding meter face. How much is the external resistance, R_X, across the ohmmeter leads for (a) full-scale deflection; (b) three-fourths full-scale deflection; (c) one-half-scale deflection; (d) one-fourth full-scale deflection; and (e) no deflection?

8–22 In Fig. 8–36, how much is the external resistance, R_X, for (a) four-fifths full-scale deflection; (b) two-thirds full-scale deflection; (c) three-fifths full-scale deflection; (d) two-fifths full-scale deflection; (e) one-third full-scale deflection; and (f) one-fifth full-scale deflection.

8–23 If the resistance values in Probs. 8–21 and 8–22 were plotted on the scale of the meter face in Fig. 8–36, would the scale be linear or nonlinear? Why?

8–24 For the series ohmmeter in Fig. 8–36, is the orientation of the ohmmeter leads important when measuring the value of a resistor?

Figure 8–36

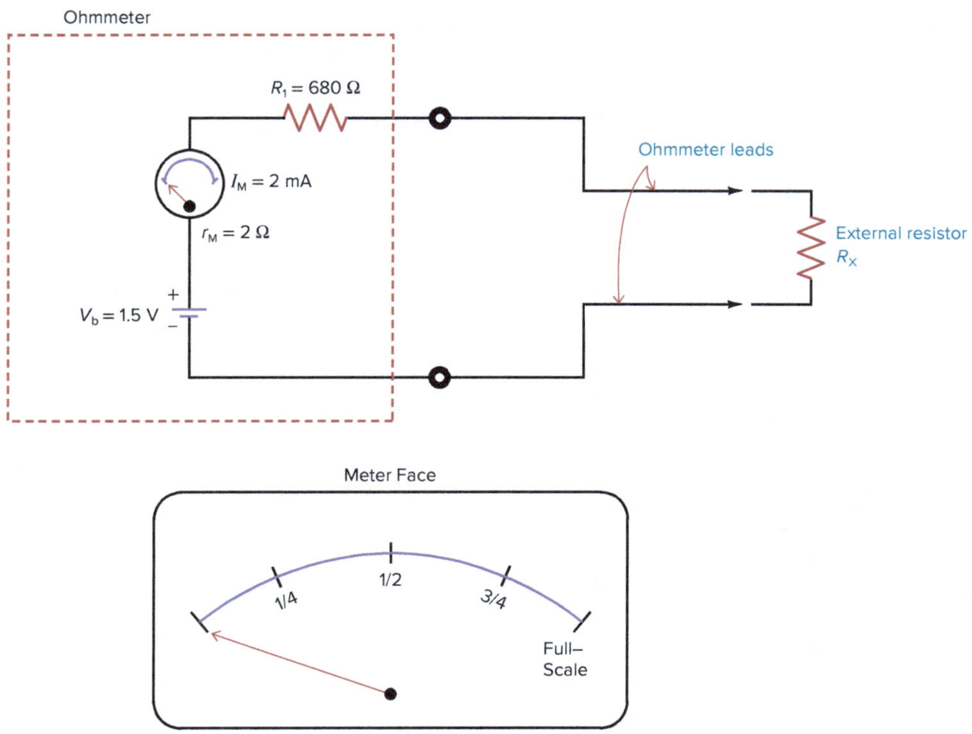

8–25 Why is the ohmmeter scale in Fig. 8–36 referred to as a back-off ohmmeter scale?

8–26 An analog ohmmeter has five range settings: $R \times 1$, $R \times 10$, $R \times 100$, $R \times 1$ k and $R \times 10$ k. Determine the measured resistance for each ohmmeter reading listed below.

Ohmmeter Reading	Range Setting	Measured Resistance
2.4	$R \times 100$	?
100	$R \times 10$	?
8.6	$R \times 10$ k	?
5.5	$R \times 1$	?
100	$R \times 1$ k	?
30	$R \times 100$	?
500	$R \times 10$ kΩ	?

8–27 Analog multimeters have a zero-ohm adjustment control for the ohmmeter portion of the meter. What purpose does it serve and how is it used?

SECTION 8–8 METER APPLICATIONS

8–28 On what range should you measure an unknown value of voltage or current? Why?

8–29 What might happen to an ohmmeter if it is connected across a resistor in a live circuit?

8–30 Why is one lead of a resistor disconnected from the circuit when measuring its resistance value?

8–31 Is a current meter connected in series or in parallel? Why?

8–32 How can the inconvenience of opening a circuit to measure current be eliminated in most cases?

8–33 What is the resistance of a

a. good fuse?

b. blown fuse?

8–34 In Fig. 8–37, list the voltages at points A, B, C, and D (with respect to ground) for each of the following situations:

a. All resistors normal.

b. R_1 open.

c. R_2 open.

d. R_3 open.

e. R_4 open.

Figure 8–37

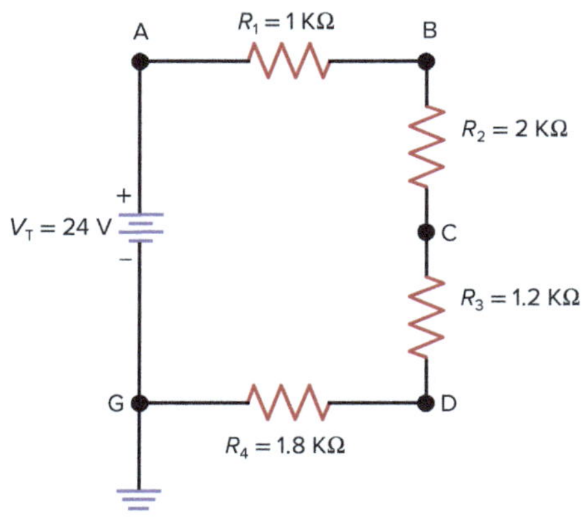

Critical Thinking

8–35 Figure 8–38 shows a universal-shunt current meter. Calculate the values for R_1, R_2, and R_3 that will provide current ranges of 2, 10, and 50 mA.

Figure 8–38 Circuit diagram for Critical Thinking Prob. 8–35.

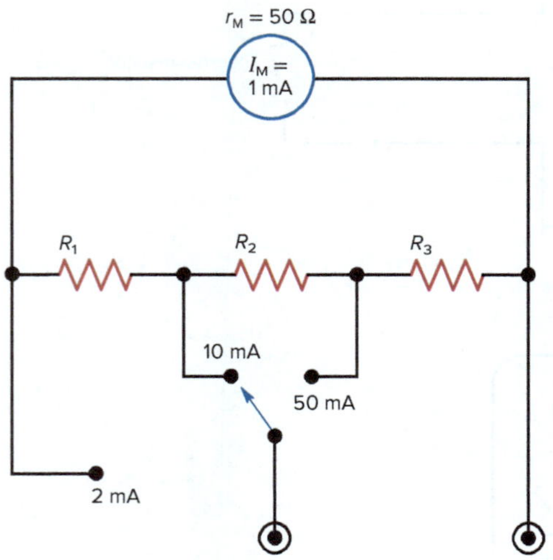

8–36 Design a series ohmmeter using a 2-kΩ, 50-μA meter movement and a 1.5-V battery. The center-scale ohm reading is to be 150 Ω.

8–37 The voltmeter across R_2 in Fig. 8–39 shows 20 V. If the voltmeter is set to the 30-V range, calculate the Ω/V rating of the meter.

Figure 8–39 Circuit diagram for Critical Thinking Prob. 8–37.

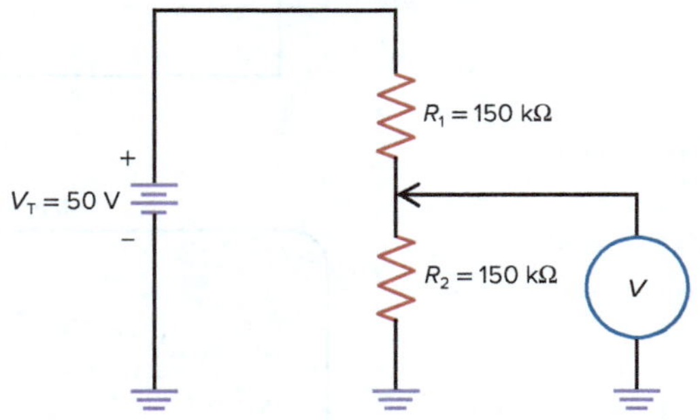

Answers to Self-Reviews

8–1 **a.** ½ scale
b. There is no deflection.

8–2 **a.** 450 μA
b. 0.045 V
c. 100 Ω

8–3 **a.** 50 kΩ
b. 1 MΩ
c. series
d. parallel
e. 500 kΩ

8–4 **a.** 80 V
b. 40 V

8–5 **a.** 400 Ω
b. 4 V

8–6 **a.** 20 kΩ
b. 10 MΩ
c. false

8–7 **a.** true
b. true

8–8 **a.** ohmmeter
b. 0 V
c. 25 V
d. 0 V

8–9 **a.** right edge
b. ∞ ohms

Laboratory Application Assignment

In this lab application assignment, you will examine the concept of voltmeter loading. As you will learn, voltmeter loading can be a problem when measuring voltages in high-resistance circuits but is usually not a problem when measuring voltages in low-resistance circuits. This lab application assignment also proves that a DMM produces less of a loading effect than an analog VOM.

Equipment: Obtain the following items from your instructor.

- Variable dc power supply
- Assortment of carbon-film resistors
- Simpson 260 analog VOM or equivalent (Ω/V rating = 20 kΩ/V)
- DMM

Voltmeter Loading

Calculate the voltmeter resistance, R_V, of the analog VOM when set to the 10-V range (R_V = Ω/V rating × V_{range} setting); R_V = ________. Next, connect the leads of your DMM to the leads of the analog VOM. With your DMM set to measure resistance and the analog VOM set to the 10-V range, measure and record the voltmeter resistance, R_V: R_V = ________

Examine the circuit in Fig. 8–40. Calculate and record the voltage across R_2: V_2 = ________

Figure 8–40

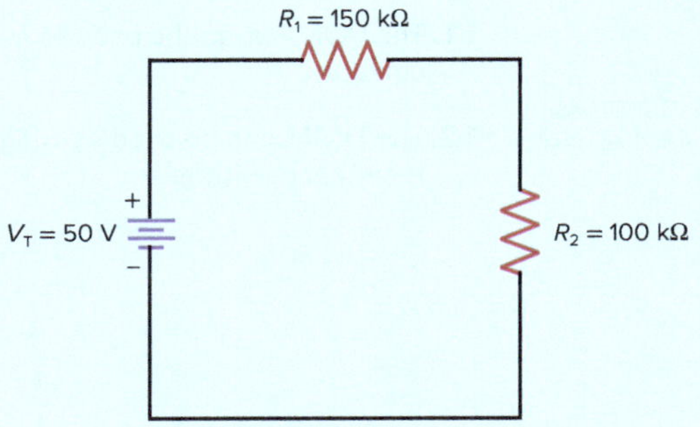

Construct the series circuit in Fig. 8–40. With your analog VOM set to the 10-V range, measure and record the voltage across R_2: V_2 = ________

How does your measured value compare to your calculated value? ________

Remeasure the voltage across R_2 using your DMM: V_2 = ________
How does this value compare to your calculated value? ________

Which meter loaded the circuit more, the analog VOM or the DMM? ________

How do you know? ________

What is the voltmeter resistance of your DMM on all voltage ranges? ________

Examine the circuit in Fig. 8–41. Calculate and record the voltage across R_2: V_2 = ________

Figure 8–41

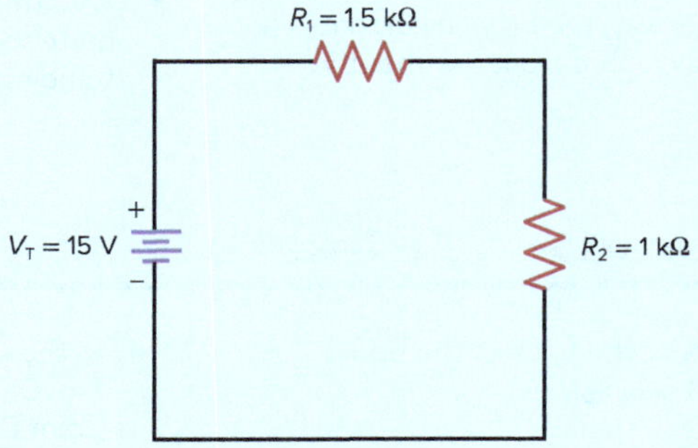

Construct the series circuit in Fig. 8–41. With your analog VOM set to the 10-V range, measure and record the voltage across R_2: V_2 = ________

How does your measured value compare to your calculated value? ________

Remeasure the voltage across R_2 using your DMM: V_2 = ________

How does this value compare to your calculated value? ________

Did the analog VOM load the circuit when measuring the voltage, V_2, in Fig. 8–41? ________

If not, why? ________

In Fig. 8–40, draw the equivalent circuit with the analog VOM connected to measure the voltage drop across R_2. Do the same in Fig. 8–41. Describe the differences in each of the equivalent circuits. ________

In general, describe why voltmeter loading is more likely to be a problem when measuring voltages in high-resistance circuits rather than low-resistance circuits. ________

Cumulative Review Summary (Chapters 7 and 8)

- In a series voltage divider, the *IR* drop across each resistance is proportional to its *R*. A larger *R* has a larger voltage drop. Each $V_R = (R/R_T) \times V_T$. In this way, the series voltage drops can be calculated from V_T without *I*.
- In a parallel current divider, each branch current is inversely related to its *R*. A smaller *R* has more branch current. For only two resistances, we can use the inverse relation

$$I_1 = [R_2/(R_1 + R_2)] \times I_T$$

- In this way, the branch currents can be calculated from I_T without *V*.
- In a parallel current divider, each branch current is directly proportional to its conductance *G*. A larger *G* has more branch current. For any number of parallel resistances, each branch $I = (G/G_T) \times I_T$.
- A milliammeter or ammeter is a low-resistance meter connected in series in a circuit to measure current.
- Different current ranges are obtained by meter shunts in parallel with the meter.
- A voltmeter is a high-resistance meter connected across the voltage to be measured.
- Different voltage ranges are obtained by multipliers in series with the meter.
- An ohmmeter has an internal battery to indicate the resistance of a component across its two terminals with external power off.
- In making resistance tests, remember that $R = 0\ \Omega$ for continuity or a short circuit, but the resistance of an open circuit is infinitely high.
- Figure 8–1 shows a VOM and DMM. Both types can be used for voltage, current, and resistance measurements.

Cumulative Self-Test

Answers at the back of the book.
Answer True or False.

1. The internal *R* of a milliammeter must be low to have minimum effect on *I* in the circuit.
2. The internal *R* of a voltmeter must be high to have minimum current through the meter.
3. Power must be off when checking resistance in a circuit because the ohmmeter has its own internal battery.
4. In the series voltage divider in Fig. 8–19, the normal voltage from point B to ground is 75 V.
5. In Fig. 8–19, the normal voltage across R_1, between A and B, is 75 V.
6. The highest ohm range is best for checking continuity with an ohmmeter.
7. With four equal resistors in a series voltage divider with V_T of 44.4 V, each *IR* drop is 11.1 V.
8. With four equal resistors in parallel with I_T of 44.4 mA, each branch current is 11.1 mA.
9. Series voltage drops divide V_T in direct proportion to each series *R*.
10. Parallel currents divide I_T in direct proportion to each branch *R*.
11. The VOM cannot be used to measure current.
12. The DMM can be used as a high-resistance voltmeter.

Design credits: Chapter Opener Background Image, and Multisim: Stockbyte/Getty Images

chapter

9

Kirchhoff's Laws

Many types of circuits have components that are not in series, in parallel, or in series-parallel. For example, a circuit may have two voltages applied in different branches. Another example is an unbalanced bridge circuit. When the rules of series and parallel circuits cannot be applied, more general methods of analysis become necessary. These methods include the application of Kirchhoff's laws, as described in this chapter.

All circuits can be solved by Kirchhoff's laws because the laws do not depend on series or parallel connections. Although Kirchhoff's voltage and current laws were introduced briefly in Chaps. 4 and 5, respectively, this chapter takes a more in-depth approach to using Kirchhoff's laws for circuit analysis.

Kirchhoff's voltage and current laws were stated in 1847 by German physicist Gustav R. Kirchhoff. As stated in Chaps. 4 and 5, Kirchhoff's voltage law and Kirchhoff's current law are abbreviated KVL and KCL, respectfully. Both laws are expressed here.

KVL: The algebraic sum of the voltage sources and *IR* voltage drops in any closed path must total zero.

KCL: At any point in a circuit, the algebraic sum of the currents directed into and out of a point must total zero.

These are the most precise statements of Kirchhoff's voltage and current laws. As you will see in this chapter, these statements do not conflict with the more general statements of Kirchhoff's laws used in earlier chapters. ■

Chapter Outline

Chapter Objectives

After studying this chapter, you should be able to

- *State* Kirchhoff's current law.
- *State* Kirchhoff's voltage law.
- *Use* the method of branch currents to solve for all voltages and currents in a circuit containing two or more voltage sources in different branches.
- *Use* node-voltage analysis to solve for the unknown voltages and currents in a circuit containing two or more voltage sources in different branches.
- *Use* the method of mesh currents to solve for the unknown voltages and currents in a circuit containing two or more voltage sources in different branches.

Important Terms

Kirchhoff's current law (KCL)
Kirchhoff's voltage law (KVL)
loop
loop equation
mesh
mesh current
node
principal node

GOOD TO KNOW

The currents entering and leaving point P in Fig. 9–1 can be compared to the flow of water in a pipeline. The total of all waters entering point P must equal the total of all waters leaving point P.

Figure 9–1 Current I_C out from point P equals 5 A + 3 A into P.

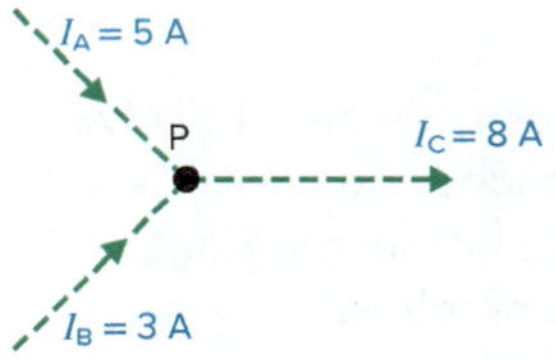

9–1 Kirchhoff's Current Law (KCL)

The algebraic sum of the currents entering and leaving any point in a circuit must equal zero. Or stated another way, *the algebraic sum of the currents into any point of the circuit must equal the algebraic sum of the currents out of that point.* Otherwise, charge would accumulate at the point, instead of having a conducting path. An *algebraic sum* means combining positive and negative values.

Algebraic Signs

In using Kirchhoff's laws to solve circuits, it is necessary to adopt conventions that determine the algebraic signs for current and voltage terms. A convenient system for currents is to *consider all currents into a branch point as positive and all currents directed away from that point as negative.*

For example, in Fig. 9–1, we can write the currents as

$$I_A + I_B - I_C = 0$$

or

$$5\text{ A} + 3\text{ A} - 8\text{ A} = 0$$

Currents I_A and I_B are positive terms because these currents flow into P, but I_C, directed out, is negative.

Current Equations

For a circuit application, refer to point C at the top of the diagram in Fig. 9–2. The 6-A I_T into point C divides into the 2-A I_3 and 4-A $I_{4\text{-}5}$, both directed out. Note that $I_{4\text{-}5}$ is the current through R_4 and R_5. The algebraic equation is

$$I_T - I_3 - I_{4\text{-}5} = 0$$

Substituting the values for these currents,

$$6\text{ A} - 2\text{ A} - 4\text{ A} = 0$$

For the opposite directions, refer to point D at the bottom of Fig. 9–2. Here the branch currents into D combine to equal the main-line current I_T returning to

MultiSim **Figure 9–2** Series-parallel circuit illustrating Kirchhoff's laws. See text for voltage and current equations.

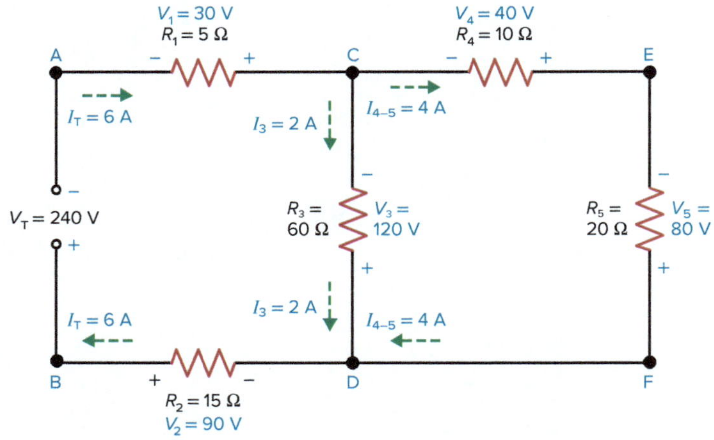

the voltage source. Now I_T is directed out from D with I_3 and $I_{4\text{-}5}$ directed in. The algebraic equation is

$$I_3 + I_{4\text{-}5} - I_T = 0$$

or

$$2\text{ A} + 4\text{ A} - 6\text{ A} = 0$$

$I_{in} = I_{out}$

Note that at either point C or point D in Fig. 9–2, the sum of the 2-A and 4-A branch currents must equal the 6-A total line current. Therefore, **Kirchhoff's current law** can also be stated as $\boldsymbol{I_{in} = I_{out}}$. For Fig. 9–2, the equations of current can be written:

At point C: $6\text{ A} = 2\text{ A} + 4\text{ A}$
At point D: $2\text{ A} + 4\text{ A} = 6\text{ A}$

Kirchhoff's current law is the basis for the practical rule in parallel circuits that the total line current must equal the sum of the branch currents.

Example 9–1

In Fig. 9–3, apply Kirchhoff's current law to solve for the unknown current, I_3.

Figure 9–3

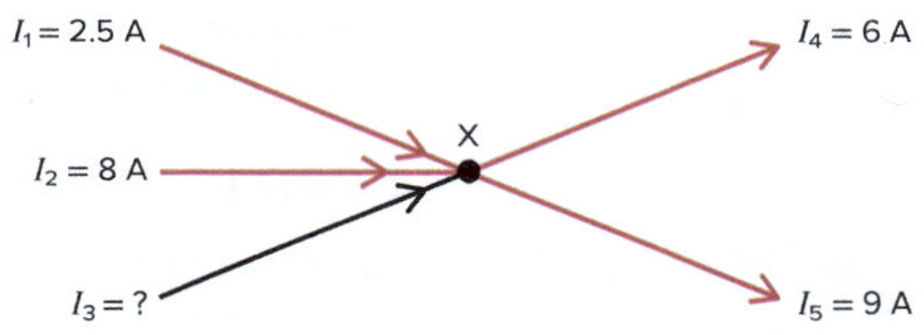

ANSWER In Fig. 9–3, the currents I_1, I_2, and I_3 flowing into point X are considered positive, whereas the currents I_4 and I_5 flowing away from point X are considered negative. Expressing the currents as an equation gives us

$$I_1 + I_2 + I_3 - I_4 - I_5 = 0$$

or

$$I_1 + I_2 + I_3 = I_4 + I_5$$

Inserting the values from Fig. 9–3,

$$2.5\text{ A} + 8\text{ A} + I_3 = 6\text{ A} + 9\text{ A}$$

Solving for I_3 gives us

$$I_3 = 6\text{ A} + 9\text{ A} - 2.5\text{ A} - 8\text{ A}$$
$$= 4.5\text{ A}$$

GOOD TO KNOW

Kirchhoff's voltage and current laws can be applied to all types of electronic circuits, not just those containing DC voltage sources and resistors. For example, KVL and KCL can be applied when analyzing circuits containing diodes, transistors, op-amps, etc.

■ *9–1 Self-Review*

Answers at the end of the chapter.

a. With a 1-A I_1, 2-A I_2, and 3-A I_3 into a point, how much is I_{out}?

b. If I_1 into a point is 3 A and I_3 out of that point is 7 A, how much is I_2 into that point?

9–2 Kirchhoff's Voltage Law (KVL)

The algebraic sum of the voltages around any closed path is zero. If you start from any point at one potential and come back to the same point and the same potential, the difference of potential must be zero.

Algebraic Signs

In determining the algebraic signs for voltage terms in a KVL equation, first mark the polarity of each voltage, as shown in Fig. 9–2. A convenient system is to *go around any closed path and consider any voltage whose negative terminal is reached first as a negative term and any voltage whose positive terminal is reached first as a positive term.* This method applies to *IR* voltage drops and voltage sources. The direction can be clockwise or counterclockwise.

Remember that electrons flowing into a resistor make that end negative with respect to the other end. For a voltage source, the direction of electrons returning to the positive terminal is the normal direction for electron flow, which means that the source should be a positive term in the voltage equation.

When you go around the closed path and come back to the starting point, the algebraic sum of all the voltage terms must be zero. There cannot be any potential difference for one point.

If you do not come back to the start, then the algebraic sum is the voltage between the start and finish points.

You can follow any closed path because the voltage between any two points in a circuit is the same regardless of the path used in determining the potential difference.

Loop Equations

Any closed path is called a ***loop***. A **loop equation** specifies the voltages around the loop.

Figure 9–2 has three loops. The outside loop, starting from point A at the top, through CEFDB, and back to A, includes the voltage drops V_1, V_4, V_5, and V_2 and the source V_T.

The inside loop ACDBA includes V_1, V_3, V_2, and V_T. The other inside loop, CEFDC with V_4, V_5, and V_3, does not include the voltage source.

Consider the voltage equation for the inside loop with V_T. In the clockwise direction starting from point A, the algebraic sum of the voltages is

$$-V_1 - V_3 - V_2 + V_T = 0$$

or

$$-30\ \text{V} - 120\ \text{V} - 90\ \text{V} + 240\ \text{V} = 0$$

Voltages V_1, V_3, and V_2 have negative signs, because the negative terminal for each of these voltages is reached first. However, the source V_T is a positive term because its plus terminal is reached first, going in the same direction.

For the opposite direction, going counterclockwise in the same loop from point B at the bottom, V_2, V_3, and V_1 have positive values and V_T is negative. Then

$$V_2 + V_3 + V_1 - V_T = 0$$

or

$$90\ \text{V} + 120\ \text{V} + 30\ \text{V} - 240\ \text{V} = 0$$

Bettmann/Getty Images

PIONEERS IN ELECTRONICS

German physicist *Gustav Kirchhoff (1824–1887)* is best known for his statement of two basic laws of the behavior of current and voltage. Developed in 1847, these laws enable scientists to understand, and therefore evaluate the behavior of networks.

When we transpose the negative term of −240 V, the equation becomes

$$90\text{ V} + 120\text{ V} + 30\text{ V} = 240\text{ V}$$

This equation states that the sum of the voltage drops equals the applied voltage.

$\Sigma V = V_T$

The Greek letter Σ means "sum of." In either direction, for any loop, the sum of the *IR* voltage drops must equal the applied voltage V_T. In Fig. 9–2, for the inside loop with the source V_T, going counterclockwise from point B,

$$90\text{ V} + 120\text{ V} + 30\text{ V} = 240\text{ V}$$

This system does not contradict the rule for algebraic signs. If 240 V were on the left side of the equation, this term would have a negative sign.

Stating a loop equation as $\Sigma V = V_T$ eliminates the step of transposing the negative terms from one side to the other to make them positive. In this form, the loop equations show that **Kirchhoff's voltage law** is the basis for the practical rule in series circuits that the sum of the voltage drops must equal the applied voltage.

When a loop does not have any voltage source, the algebraic sum of the *IR* voltage drops alone must total zero. For instance, in Fig. 9–2, for the loop CEFDC without the source V_T, going clockwise from point C, the loop equation of voltages is

$$\begin{aligned} -V_4 - V_5 + V_3 &= 0 \\ -40\text{ V} - 80\text{ V} + 120\text{ V} &= 0 \\ 0 &= 0 \end{aligned}$$

Notice that V_3 is positive now because its plus terminal is reached first by going clockwise from D to C in this loop.

Example 9–2

In Fig. 9–4*a*, apply Kirchhoff's voltage law to solve for the voltages V_{AG} and V_{BG}.

ANSWER In Fig. 9–4*a*, the voltage sources V_1 and V_2 are connected in a series-aiding fashion since they both force electrons to flow through the circuit in the same direction. The earth ground connection at the junction of V_1 and V_2 is used simply for a point of reference. The circuit is solved as follows:

$$\begin{aligned} V_T &= V_1 + V_2 \\ &= 18\text{ V} + 18\text{ V} = 36\text{ V} \\ R_T &= R_1 + R_2 + R_3 \\ &= 120\ \Omega + 100\ \Omega + 180\ \Omega = 400\ \Omega \end{aligned}$$

$$\begin{aligned} I &= \frac{V_T}{R_T} \\ &= \frac{36\text{ V}}{400\ \Omega} = 90\text{ mA} \end{aligned}$$

$$\begin{aligned} V_{R_1} &= I \times R_1 \\ &= 90\text{ mA} \times 120\ \Omega = 10.8\text{ V} \end{aligned}$$

$$\begin{aligned} V_{R_2} &= I \times R_2 \\ &= 90\text{ mA} \times 100\ \Omega = 9\text{ V} \end{aligned}$$

$$\begin{aligned} V_{R_3} &= I \times R_3 \\ &= 90\text{ mA} \times 180\ \Omega = 16.2\text{ V} \end{aligned}$$

Figure 9–4

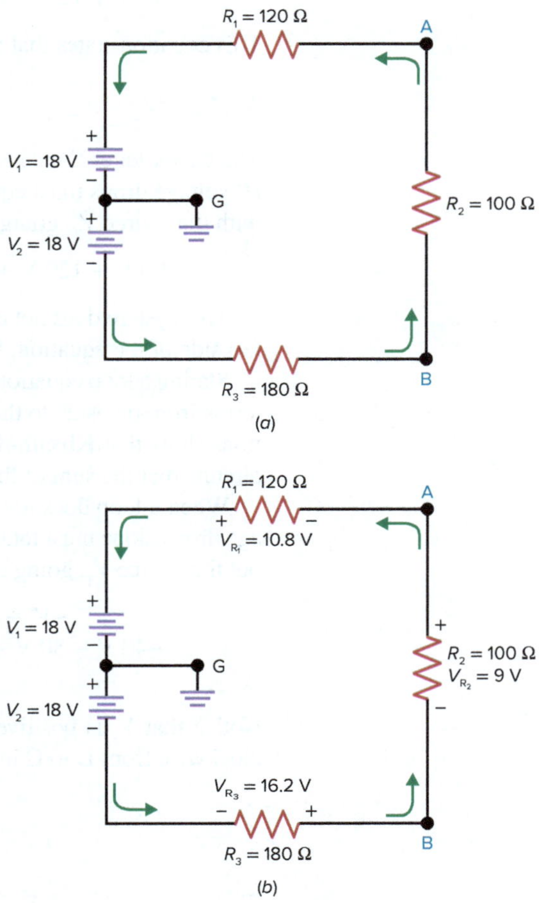

Figure 9–4*b* shows the voltage drops across each resistor. Notice that the polarity of each resistor voltage drop is negative at the end where the electrons enter the resistor and positive at the end where they leave.

Next, we can apply Kirchhoff's voltage law to determine if we have solved the circuit correctly. If we go counterclockwise (CCW) around the loop, starting and ending at the positive (+) terminal of V_1, we should obtain an algebraic sum of 0 V. The loop equation is written as

$$V_1 + V_2 - V_{R_3} - V_{R_2} - V_{R_1} = 0$$

Notice that the voltage sources V_1 and V_2 are considered positive terms in the equation because their positive (+) terminals were reached first when going around the loop. Similarly, the voltage drops V_{R_1}, V_{R_2}, and V_{R_3} are considered negative terms because the negative (−) end of each resistor's voltage drop is encountered first when going around the loop.

Substituting the values from Fig. 9–4*b* gives us

$$18\text{ V} + 18\text{ V} - 16.2\text{ V} - 9\text{ V} - 10.8\text{ V} = 0$$

It is important to realize that the sum of the resistor voltage drops must equal the applied voltage, V_T, which equals $V_1 + V_2$ or 36 V in this case. Expressed as an equation,

$$\begin{aligned} V_T &= V_{R_1} + V_{R_2} + V_{R_3} \\ &= 10.8\text{ V} + 9\text{ V} + 16.2\text{ V} \\ &= 36\text{ V} \end{aligned}$$

It is now possible to solve for the voltages V_{AG} and V_{BG} by applying Kirchhoff's voltage law. To do so, simply add the voltages algebraically between the start and finish points which are points A and G for V_{AG} and points B and G for V_{BG}. Using the values from Fig. 9–4*b*,

$$\begin{aligned} V_{AG} &= -V_{R_1} + V_1 \qquad \text{(CCW from A to G)} \\ &= -10.8\text{ V} + 18\text{ V} \\ &= 7.2\text{ V} \end{aligned}$$

Going clockwise (CW) from A to G produces the same result.

$$\begin{aligned} V_{AG} &= V_{R_2} + V_{R_3} - V_2 \qquad \text{(CW from A to G)} \\ &= 9\text{ V} + 16.2\text{ V} - 18\text{ V} \\ &= 7.2\text{ V} \end{aligned}$$

Since there are fewer voltages to add going counterclockwise from point A, it is the recommended solution for V_{AG}.

The voltage, V_{BG}, is found by using the same technique.

$$\begin{aligned} V_{BG} &= V_{R_3} - V_2 \qquad \text{(CW from B to G)} \\ &= 16.2\text{ V} - 18\text{ V} \\ &= -1.8\text{ V} \end{aligned}$$

Going around the loop in the other direction gives us

$$\begin{aligned} V_{BG} &= -V_{R_2} - V_{R_1} + V_1 \qquad \text{(CCW from B to G)} \\ &= -9\text{ V} - 10.8\text{ V} + 18\text{ V} \\ &= -1.8\text{ V} \end{aligned}$$

Since there are fewer voltages to add going clockwise from point B, it is the recommended solution for V_{BG}.

9–2 Self-Review

Answers at the end of the chapter.

Refer to Fig. 9–2.

a. For partial loop CEFD, what is the total voltage across CD with −40 V for V_4 and −80 V for V_5?

b. For loop CEFDC, what is the total voltage with −40 V for V_4, −80 V for V_5, and including 120 V for V_3?

9–3 Method of Branch Currents

Now we can use Kirchhoff's laws to analyze the circuit in Fig. 9–5. The problem is to find the currents and voltages for the three resistors.

First, indicate current directions and mark the voltage polarity across each resistor consistent with the assumed current. Remember that electron flow in a resistor produces negative polarity where the current enters. In Fig. 9–5, we assume that the source V_1 produces electron flow from left to right through R_1, and V_2 produces electron flow from right to left through R_2.

The three different currents in R_1, R_2, and R_3 are indicated as I_1, I_2, and I_3. However, three unknowns would require three equations for the solution. From Kirchhoff's current law, $I_3 = I_1 + I_2$, as the current out of point C must equal the current in. The current through R_3, therefore, can be specified as $I_1 + I_2$.

Figure 9–5 Application of Kirchhoff's laws to a circuit with two sources in different branches. See text for solution by finding the branch currents.

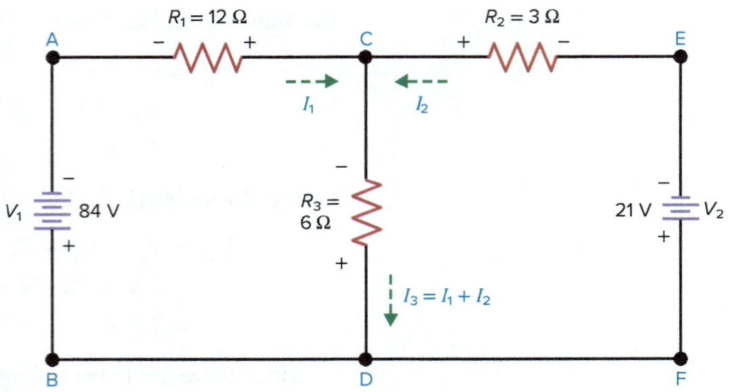

GOOD TO KNOW

In Fig. 9–5, $I_3 = I_1 + I_2$ as shown. The current $I_1 = I_3 - I_2$ and the current $I_2 = I_3 - I_1$.

With two unknowns, two independent equations are needed to solve for I_1 and I_2. These equations are obtained by writing two Kirchhoff's voltage law equations around two loops. There are three loops in Fig. 9–5, the outside loop and two inside loops, but we need only two. The inside loops are used for the solution here.

Writing the Loop Equations

For the loop with V_1, start at point B, at the bottom left, and go clockwise through V_1, V_{R_1}, and V_{R_3}. This equation for loop 1 is

$$84 - V_{R_1} - V_{R_3} = 0$$

For the loop with V_2, start at point F, at the lower right, and go counterclockwise through V_2, V_{R_2}, and V_{R_3}. This equation for loop 2 is

$$21 - V_{R_2} - V_{R_3} = 0$$

Using the known values of R_1, R_2, and R_3 to specify the *IR* voltage drops,

$$V_{R_1} = I_1R_1 = I_1 \times 12 = 12I_1$$
$$V_{R_2} = I_2R_2 = I_2 \times 3 = 3I_2$$
$$V_{R_3} = (I_1 + I_2)R_3 = 6(I_1 + I_2)$$

Substituting these values in the voltage equation for loop 1,

$$84 - 12I_1 - 6(I_1 + I_2) = 0$$

Also, in loop 2,

$$21 - 3I_2 - 6(I_1 + I_2) = 0$$

Multiplying $(I_1 + I_2)$ by 6 and combining terms and transposing, the two equations are

$$-18I_1 - 6I_2 = -84$$
$$-6I_1 - 9I_2 = -21$$

Divide the top equation by −6 and the bottom equation by −3 to reduce the equations to their simplest terms and to have all positive terms. The two equations in their simplest form then become

$$3I_1 + I_2 = 14$$
$$2I_1 + 3I_2 = 7$$

Solving for the Currents

These two equations with the two unknowns I_1 and I_2 contain the solution of the network. Note that the equations include every resistance in the circuit. Currents I_1 and I_2 can be calculated by any of the methods for the solution of simultaneous equations. Using the method of elimination, multiply the top equation by 3 to make the I_2 terms the same in both equations. Then

$$9I_1 + 3I_2 = 42$$
$$2I_1 + 3I_2 = 7$$

Subtract the bottom equation from the top equation, term by term, to eliminate I_2. Then, since the I_2 term becomes zero,

$$7I_1 = 35$$
$$I_1 = 5 \text{ A}$$

The 5-A I_1 is the current through R_1. Its direction is from A to C, as assumed, because the answer for I_1 is positive.

To calculate I_2, substitute 5 for I_1 in either of the two loop equations. Using the bottom equation for the substitution,

$$2(5) + 3I_2 = 7$$
$$3I_2 = 7 - 10$$
$$3I_2 = -3$$
$$I_2 = -1 \text{ A}$$

The negative sign for I_2 means that this current is opposite to the assumed direction. Therefore, I_2 flows through R_2 from C to E instead of from E to C as was previously assumed.

Why the Solution for I_2 Is Negative

In Fig. 9–5, I_2 was assumed to flow from E to C through R_2 because V_2 produces electron flow in this direction. However, the other voltage source V_1 produces electron flow through R_2 in the opposite direction from point C to E. This solution of −1 A for I_2 shows that the current through R_2 produced by V_1 is more than the current produced by V_2. The net result is 1 A through R_2 from C to E.

The actual direction of I_2 is shown in Fig. 9–6 with all the values for the solution of this circuit. Notice that the polarity of V_{R_2} is reversed from the assumed polarity

Figure 9–6 Solution of circuit in Fig. 9–5 with all currents and voltages.

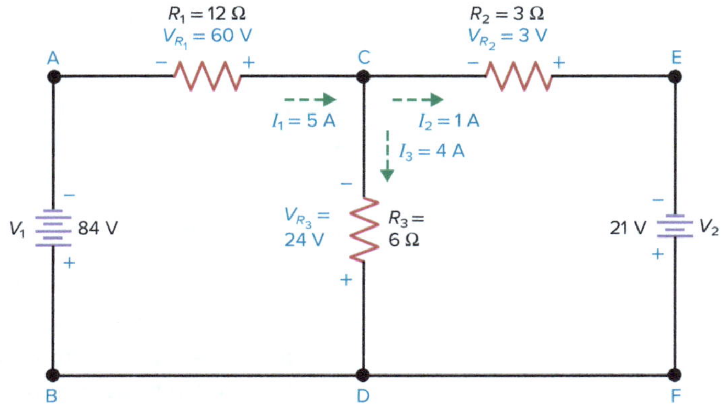

in Fig. 9–5. Since the net electron flow through R_2 is actually from C to E, the end of R_2 at C is the negative end. However, the polarity of V_2 is the same in both diagrams because it is a voltage source that generates its own polarity.

To calculate I_3 through R_3,

$$I_3 = I_1 + I_2 = 5 + (-1)$$
$$I_3 = 4\ \text{A}$$

The 4 A for I_3 is in the assumed direction from C to D. Although the negative sign for I_2 means only a reversed direction, its algebraic value of −1 must be used for substitution in the algebraic equations written for the assumed direction.

Calculating the Voltages

With all the currents known, the voltage across each resistor can be calculated as follows:

$$V_{R_1} = I_1R_1 = 5 \times 12 = 60\ \text{V}$$
$$V_{R_2} = I_2R_2 = 1 \times 3 = 3\ \text{V}$$
$$V_{R_3} = I_3R_3 = 4 \times 6 = 24\ \text{V}$$

All currents are taken as positive, in the correct direction, to calculate the voltages. Then the polarity of each *IR* drop is determined from the actual direction of current, with electron flow into the negative end (see Fig. 9–6). Notice that V_{R_3} and V_{R_2} have opposing polarities in loop 2. Then the sum of +3 V and −24 V equals the −21 V of V_2.

Checking the Solution

As a summary of all the answers for this problem, Fig. 9–6 shows the network with all currents and voltages. The polarity of each *V* is marked from the known directions. In checking the answers, we can see whether Kirchhoff's current and voltage laws are satisfied:

At point C: 5 A = 4 A + 1 A

At point D: 4 A + 1 A = 5 A

Around the loop with V_1 clockwise from B,

$$84\ \text{V} - 60\ \text{V} - 24\ \text{V} = 0$$

Around the loop with V_2 counterclockwise from F,

$$21\ \text{V} + 3\ \text{V} - 24\ \text{V} = 0$$

Note that the circuit has been solved using only the two Kirchhoff's Laws without any of the special rules for series and parallel circuits. Any circuit can be solved by applying Kirchhoff's laws for the voltages around a loop and the currents at a branch point.

9–3 Self-Review

Answers at the end of the chapter.

Refer to Fig. 9–6.

a. How much is the voltage around the partial loop CEFD?

b. How much is the voltage around loop CEFDC?

9–4 Node-Voltage Analysis

In the method of branch currents, these currents are used for specifying the voltage drops around the loops. Then loop equations are written to satisfy Kirchhoff's voltage law. Solving the loop equations, we can calculate the unknown branch currents.

Another method uses voltage drops to specify the currents at a branch point, also called a ***node.*** Then node equations of currents are written to satisfy Kirchhoff's current law. Solving the node equations, we can calculate the unknown node voltages. This method of node-voltage analysis is often shorter than the method of branch currents.

A node is simply a common connection for two or more components. A ***principal node*** has three or more connections. In effect, a principal node is a junction or branch point where currents can divide or combine. Therefore, we can always write an equation of currents at a principal node. In Fig. 9–7, points N and G are principal nodes.

However, one node must be the reference for specifying the voltage at any other node. In Fig. 9–7, point G connected to chassis ground is the reference node. Therefore, we need to write only one current equation for the other node N. In general, the number of current equations required to solve a circuit is one less than the number of principal nodes.

Writing the Node Equations

The circuit of Fig. 9–5, earlier solved by the method of branch currents, is redrawn in Fig. 9–7 to be solved now by node-voltage analysis. The problem here is to find the node voltage V_N from N to G. Once this voltage is known, all other voltages and currents can be determined.

The currents in and out of node N are specified as follows: I_1 is the only current through the 12-Ω R_1. Therefore, I_1 is V_{R_1}/R_1 or $V_{R_1}/12\ \Omega$. Similarly, I_2 is $V_{R_2}/3\ \Omega$. Finally, I_3 is $V_{R_3}/6\ \Omega$.

Note that V_{R_3} is the node voltage V_N that we are to calculate. Therefore, I_3 can also be stated as $V_N/6\ \Omega$. The equation of currents at node N is

$$I_1 + I_2 = I_3$$

or

$$\frac{V_{R_1}}{12} + \frac{V_{R_2}}{3} = \frac{V_N}{6}$$

Figure 9–7 Method of node-voltage analysis for the same circuit as in Fig. 9–5. See text for solution by finding V_N across R_3 from the principal node N to ground.

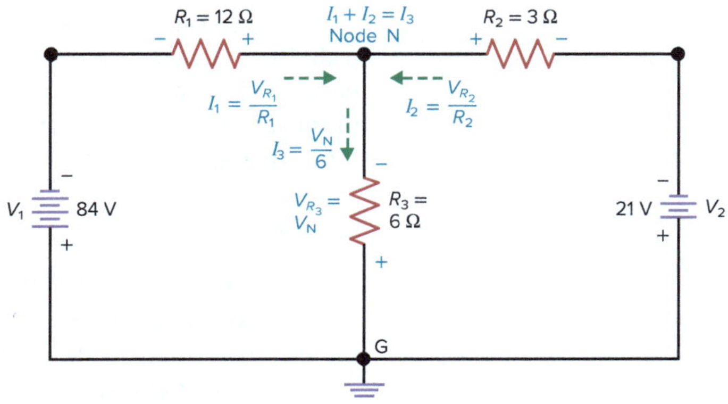

GOOD TO KNOW

Many people prefer to use node-voltage analysis rather than the method of branch currents when solving complex circuits. The reason is that node-voltage analysis does not always require the use of simultaneous equations as does the method of branch currents.

There are three unknowns here, but V_{R_1} and V_{R_2} can be specified in terms of V_N and the known values of V_1 and V_2. We can use Kirchhoff's voltage law because the applied voltage V must equal the algebraic sum of the voltage drops. For the loop with V_1 of 84 V,

$$V_{R_1} + V_N = 84 \quad \text{or} \quad V_{R_1} = 84 - V_N$$

For the loop with V_2 of 21 V,

$$V_{R_2} + V_N = 21 \quad \text{or} \quad V_{R_2} = 21 - V_N$$

Now substitute these values of V_{R_1} and V_{R_2} in the equation of currents:

$$I_1 + I_2 = I_3$$

$$\frac{V_{R_1}}{R_1} + \frac{V_{R_2}}{R_2} = \frac{V_{R_3}}{R_2}$$

Using the value of each V in terms of V_N,

$$\frac{84 - V_N}{12} + \frac{21 - V_N}{3} = \frac{V_N}{6}$$

This equation has only the one unknown, V_N. Clearing fractions by multiplying each term by 12, the equation is

$$(84 - V_N) + 4(21 - V_N) = 2\,V_N$$
$$84 - V_N + 84 - 4\,V_N = 2\,V_N$$
$$-7\,V_N = -168$$
$$V_N = 24\text{ V}$$

This answer of 24 V for V_N is the same as that calculated for V_{R_3} by the method of branch currents. The positive value means that the direction of I_3 is correct, making V_N negative at the top of R_3 in Fig. 9–7.

Calculating All Voltages and Currents

The reason for finding the voltage at a node, rather than some other voltage, is the fact that a node voltage must be common to two loops. As a result, the node voltage can be used for calculating all voltages in the loops. In Fig. 9–7, with a V_N of 24 V, then V_{R_1} must be 84 − 24 = 60 V. Also, I_1 is 60 V/12 Ω, which equals 5 A.

To find V_{R_2}, it must be 21–24, which equals −3 V. The negative answer means that I_2 is opposite to the assumed direction and the polarity of V_{R_2} is the reverse of the signs shown across R_2 in Fig. 9–7. The correct directions are shown in the solution for the circuit in Fig. 9–6. The magnitude of I_2 is 3 V/3 Ω, which equals 1 A.

The following comparisons can be helpful in using node equations and loop equations. A node equation applies Kirchhoff's current law to the currents in and out of a node. However, the currents are specified as V/R so that the equation of currents can be solved to find a node voltage.

A loop equation applies Kirchhoff's voltage law to the voltages around a closed path. However, the voltages are specified as IR so that the equation of voltages can be solved to find a loop current. This procedure with voltage equations is used for the method of branch currents explained before with Fig. 9–5 and for the method of **mesh currents** to be described next in Fig. 9–8.

9–4 Self-Review

Answers at the end of the chapter.

a. How many principal nodes does Fig. 9–7 have?

b. How many node equations are necessary to solve a circuit with three principal nodes?

9–5 Method of Mesh Currents

A **mesh** is the simplest possible closed path. The circuit in Fig. 9–8 has two meshes, ACDBA and CEFDC. The outside path ACEFDBA is a loop but not a mesh. Each mesh is like a single window frame. There is only one path without any branches.

A mesh current is assumed to flow around a mesh without dividing. In Fig. 9–8, the mesh current I_A flows through V_1, R_1, and R_3; mesh current I_B flows through V_2, R_2, and R_3. A resistance common to two meshes, such as R_3, has two mesh currents, which are I_A and I_B here.

The fact that a mesh current does not divide at a branch point is the difference between mesh currents and branch currents. A mesh current is an assumed current, and a branch current is the actual current. However, when the mesh currents are known, all individual currents and voltages can be determined.

For example, Fig. 9–8, which is the same circuit as Fig. 9–5, will now be solved by using the assumed mesh currents I_A and I_B. The mesh equations are

$$18I_A - 6I_B = 84 \text{ V} \qquad \text{in mesh A}$$
$$-6I_A + 9I_B = -21 \text{ V} \qquad \text{in mesh B}$$

Writing the Mesh Equations

The number of meshes equals the number of mesh currents, which is the number of equations required. Here two equations are used for I_A and I_B in the two meshes.

The assumed current is usually taken in the same direction around each mesh to be consistent. Generally, the clockwise direction is used, as shown for I_A and I_B in Fig. 9–8.

In each mesh equation, the algebraic sum of the voltage drops equals the applied voltage.

The voltage drops are added going around a mesh in the same direction as its mesh current. Any voltage drop in a mesh produced by its own mesh current is considered positive because it is added in the direction of the mesh current.

Since all the voltage drops of a mesh current in its own mesh must have the same positive sign, they can be written collectively as one voltage drop by adding all resistances in the mesh. For instance, in the first equation, for mesh A, the total resistance equals 12 + 6, or 18 Ω. Therefore, the voltage drop for I_A is $18I_A$ in mesh A.

GOOD TO KNOW

In Fig. 9–8, R_3 is often referred to as a mutual resistance because it is shared by two mesh currents.

Figure 9–8 The same circuit as Fig. 9–5 analyzed as two meshes. See text for solution by calculating the assumed mesh currents I_A and I_B.

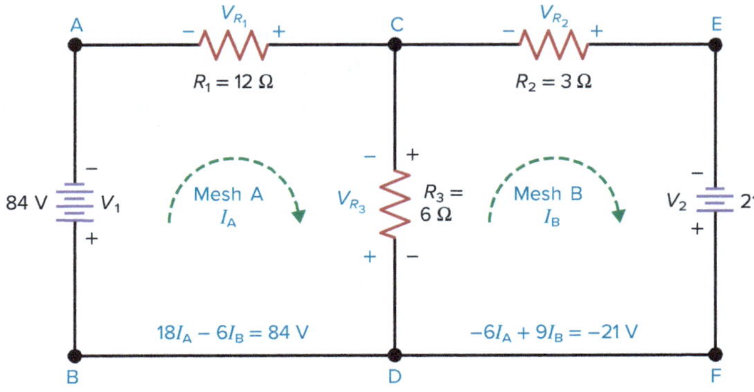

In the second equation, for mesh B, the total resistance is 3 + 6, or 9 Ω, making the total voltage drop $9I_B$ for I_B in mesh B. You can add all resistances in a mesh for one R_T because they can be considered in series for the assumed mesh current.

Any resistance common to two meshes has two opposite mesh currents. In Fig. 9–8, I_A flows down and I_B is up through the common R_3, with both currents clockwise. As a result, a common resistance has two opposing voltage drops. One voltage is positive for the current of the mesh whose equation is being written. The opposing voltage is negative for the current of the adjacent mesh.

In mesh A, the common 6-Ω R_3 has opposing voltages $6I_A$ and $-6I_B$. The $6I_A$ of R_3 adds to the $12I_A$ of R_1 for the total positive voltage drop of $18I_A$ in mesh A. With the opposing voltage of $-6I_B$, then the equation for mesh A is $18I_A - 6I_B = 84$ V.

The same idea applies to mesh B. However, now the voltage $6I_B$ is positive because the equation is for mesh B. The $-6I_A$ voltage is negative here because I_A is for the adjacent mesh. The $6I_B$ adds to the $3I_B$ of R_2 for the total positive voltage drop of $9I_B$ in mesh B. With the opposing voltage of $-6I_A$, the equation for mesh B then is $-6I_A + 9I_B = -21$ V.

The algebraic sign of the source voltage in a mesh depends on its polarity. When the assumed mesh current flows into the positive terminal, as for V_1 in Fig. 9–8, it is considered positive for the right-hand side of the mesh equation. This direction of electron flow produces voltage drops that must add to equal the applied voltage.

With the mesh current into the negative terminal, as for V_2 in Fig. 9–8, it is considered negative. This is why V_2 is −21 V in the equation for mesh B. Then V_2 is actually a load for the larger applied voltage of V_1, instead of V_2 being the source. When a mesh has no source voltage, the algebraic sum of the voltage drops must equal zero.

These rules for the voltage source mean that the direction of electron flow is assumed for the mesh currents. Then electron flow is used to determine the polarity of the voltage drops. Note that considering the voltage source as a positive value with electron flow into the positive terminal corresponds to the normal flow of electron charges. If the solution for a mesh current comes out negative, the actual current for the mesh must be in the direction opposite from the assumed current flow.

Solving the Mesh Equations to Find the Mesh Currents

The two equations for the two meshes in Fig. 9–8 are

$$18I_A - 6I_B = 84$$
$$-6I_A + 9I_B = -21$$

These equations have the same coefficients as the voltage equations written for the branch currents, but the signs are different because the directions of the assumed mesh currents are not the same as those of the branch currents.

The solution will give the same answers for either method, but you must be consistent in algebraic signs. Use either the rules for meshes with mesh currents or the rules for loops with branch currents, but do not mix the two methods.

To eliminate I_B and solve for I_A, divide the first equation by 2 and the second equation by 3. Then

$$9I_A - 3I_B = 42$$
$$-2I_A + 3I_B = -7$$

Add the equations, term by term, to eliminate I_B. Then

$$7I_A = 35$$
$$I_A = 5 \text{ A}$$

To calculate I_B, substitute 5 for I_A in the second equation:

$$-2(5) + 3I_B = -7$$
$$3I_B = -7 + 10 = 3$$
$$I_B = 1 \text{ A}$$

The positive solutions mean that the electron flow for both I_A and I_B is actually clockwise, as assumed.

Finding the Branch Currents and Voltage Drops

Referring to Fig. 9–8, the 5-A I_A is the only current through R_1. Therefore, I_A and I_1 are the same. Then V_{R_1} across the 12-Ω R_1 is 5 × 12, or 60 V. The polarity of V_{R_1} is marked negative at the left, with the electron flow into this side.

Similarly, the 1-A I_B is the only current through R_2. The direction of this electron flow through R_2 is from left to right. Note that this value of 1 A for I_B clockwise is the same as −1 A for I_2, assumed in the opposite direction in Fig. 9–3. Then V_{R_2} across the 3-Ω R_2 is 1 × 3 or 3 V, with the left side negative.

The current I_3 through R_3, common to both meshes, consists of I_A and I_B. Then I_3 is 5 − 1 or 4 A. The currents are subtracted because I_A and I_B are in opposing directions through R_3. When all the mesh currents are taken one way, they will always be in opposite directions through any resistance common to two meshes.

The direction of the net 4-A I_3 through R_3 is downward, the same as I_A, because it is more than I_B. Then, V_{R_3} across the is 6-Ω R_3 is 4 × 6 = 24 V, with the top negative.

The Set of Mesh Equations

The system for algebraic signs of the voltages in mesh equations is different from the method used with branch currents, but the end result is the same. The advantage of mesh currents is the pattern of algebraic signs for the voltages, without the need for tracing any branch currents. This feature is especially helpful in a more elaborate circuit, such as that in Fig. 9–9, that has three meshes. We can use Fig. 9–9 for more practice in writing mesh equations, without doing the numerical work of solving a set of three equations. Each R is 2 Ω.

In Fig. 9–9, the mesh currents are shown with solid arrows to indicate conventional current, which is a common way of analyzing these circuits. Also, the voltage sources V_1 and V_2 have the positive terminal at the top in the diagram. When the direction of conventional current is used, it is important to note that the voltage source is a positive value with mesh current into the negative terminal. This method corresponds to the normal flow of positive charges with conventional current.

For the three mesh equations in Fig. 9–9,

In mesh A:	$6I_A - 2I_B + 0 = 12$
In mesh B:	$-2I_A + 8I_B - 2I_C = 0$
In mesh C:	$0 - 2I_B + 6I_C = -8$

MultiSim **Figure 9–9** A circuit with three meshes. Each R is 2 Ω. See text for mesh equations.

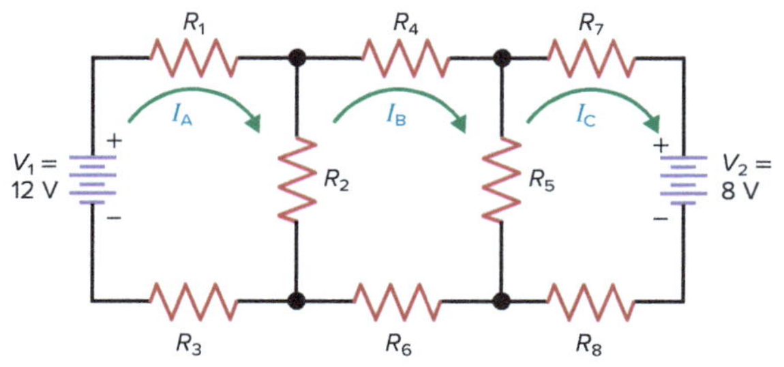

The zero term in equations A and C represents a missing mesh current. Only mesh B has all three mesh currents. However, note that mesh B has a zero term for the voltage source because it is the only mesh with only *IR* drops.

In summary, the only positive *IR* voltage in a mesh is for the R_T of each mesh current in its own mesh. All other voltage drops for any adjacent mesh current across a common resistance are always negative. This procedure for assigning algebraic signs to the voltage drops is the same whether the source voltage in the mesh is positive or negative. It also applies even if there is no voltage source in the mesh.

9–5 Self-Review

Answers at the end of the chapter.

a. A network with four mesh currents needs four mesh equations for a solution. (True/False)

b. An *R* common to two meshes has opposing mesh currents. (True/False)

Summary

- Kirchhoff's voltage law states that the algebraic sum of all voltages around any closed path must equal zero. Stated another way, the sum of the voltage drops equals the applied voltage.
- Kirchhoff's current law states that the algebraic sum of all currents directed in and out of any point in a circuit must equal zero. Stated another way, the current into a point equals the current out of that point.
- A closed path is a loop. The method of using algebraic equations for the voltages around the loops to calculate the branch currents is illustrated in Fig. 9–5.
- A principal node is a branch point where currents divide or combine. The method of using algebraic equations for the currents at a node to calculate the node voltage is illustrated in Fig. 9–7.
- A mesh is the simplest possible loop. A mesh current is assumed to flow around the mesh without branching. The method of using algebraic equations for the voltages around the meshes to calculate the mesh currents is illustrated in Fig. 9–8.

Important Terms

Kirchhoff's current law (KCL) — Kirchhoff's current law states that the algebraic sum of the currents entering and leaving any point in a circuit must equal zero.

Kirchhoff's voltage law (KVL) — Kirchhoff's voltage law states that the algebraic sum of the voltages around any closed path must equal zero.

Loop — another name for a closed path in a circuit.

Loop equation — an equation that specifies the voltages around a loop.

Mesh — the simplest possible closed path within a circuit.

Mesh current — a current that is assumed to flow around a mesh without dividing.

Node — a common connection for two or more components in a circuit where currents can combine or divide.

Principal node — a common connection for three or more components in a circuit where currents can combine or divide.

Self-Test

Answers at the back of the book.

1. Kirchhoff's current law states that

a. the algebraic sum of the currents flowing into any point in a circuit must equal zero.

b. the algebraic sum of the currents entering and leaving any point in a circuit must equal zero.

c. the algebraic sum of the currents flowing away from any point in a circuit must equal zero.

d. the algebraic sum of the currents around any closed path must equal zero.

2. When applying Kirchhoff's current law,

a. consider all currents flowing into a branch point positive and all the currents directed away from that point negative.

b. consider all currents flowing into a branch point negative and all the currents directed away from that point positive.

c. remember that the total of all the currents entering a branch point must always be greater than the sum of the currents leaving that point.

d. the algebraic sum of the currents entering and leaving a branch point does not necessarily have to be zero.

3. If a 10-A I_1 and a 3-A I_2 flow into point X, how much current must flow away from point X?

a. 7 A.

b. 30 A.

c. 13 A.

d. It cannot be determined.

4. Three currents I_1, I_2, and I_3 flow into point X, whereas current I_4 flows away from point X. If $I_1 = 2.5$ A, $I_3 = 6$ A, and $I_4 = 18$ A, how much is current I_2?

a. 21.5 A.

b. 14.5 A.

c. 26.5 A.

d. 9.5 A.

5. When applying Kirchhoff's voltage law, a closed path is commonly referred to as a

a. node.

b. principal node.

c. loop.

d. branch point.

6. Kirchhoff's voltage law states that

a. the algebraic sum of the voltage sources and *IR* voltage drops in any closed path must total zero.

b. the algebraic sum of the voltage sources and *IR* voltage drops around any closed path can never equal zero.

c. the algebraic sum of all the currents flowing around any closed loop must equal zero.

d. none of the above.

7. When applying Kirchhoff's voltage law,

a. consider any voltage whose positive terminal is reached first as negative and any voltage whose negative terminal is reached first as positive.

b. always consider all voltage sources as positive and all resistor voltage drops as negative.

c. consider any voltage whose negative terminal is reached first as negative and any voltage whose positive terminal is reached first as positive.

d. always consider all resistor voltage drops as positive and all voltage sources as negative.

8. The algebraic sum of +40 V and −30 V is

a. −10 V.

b. +10 V.

c. +70 V.

d. −70 V.

9. A principal node is

a. a closed path or loop where the algebraic sum of the voltages must equal zero.

b. the simplest possible closed path around a circuit.

c. a junction where branch currents can combine or divide.

d. none of the above.

10. How many equations are necessary to solve a circuit with two principal nodes?

a. 3.

b. 2.

c. 4.

d. 1.

11. The difference between a mesh current and a branch current is

a. a mesh current is an assumed current and a branch current is an actual current.

b. the direction of the currents themselves.

c. a mesh current does not divide at a branch point.

d. both a and c.

12. Using the method of mesh currents, any resistance common to two meshes has

a. two opposing mesh currents.

b. one common mesh current.

c. zero current.

d. none of the above.

13. The fact that the sum of the resistor voltage drops equals the applied voltage in a series circuit is the basis for

a. Kirchhoff's current law.

b. node-voltage analysis.

c. Kirchhoff's voltage law.

d. the method of mesh currents.

14. The fact that the sum of the individual branch currents equals the total current in a parallel circuit is the basis for

a. Kirchhoff's current law.

b. node-voltage analysis.

c. Kirchhoff's voltage law.

d. the method of mesh currents.

15. If you do not go completely around the loop when applying Kirchhoff's voltage law, then

a. the algebraic sum of the voltages will always be positive.

b. the algebraic sum is the voltage between the start and finish points.

c. the algebraic sum of the voltages will always be negative.

d. the algebraic sum of the voltages cannot be determined.

Essay Questions

1. State Kirchhoff's current law in two ways.

2. State Kirchhoff's voltage law in two ways.

3. What is the difference between a loop and a mesh?

4. What is the difference between a branch current and a mesh current?

5. Define *principal node*.

6. Define *node voltage*.

7. Use the values in Fig. 9–6 to show that the algebraic sum is zero for all voltages around the outside loop ACEFDBA.

8. Use the values in Fig. 9–6 to show that the algebraic sum is zero for all the currents into and out of node C and node D.

Problems

SECTION 9–1 KIRCHHOFF'S CURRENT LAW

9–1 If a 5-A I_1 and a 10-A I_2 flow into point X, how much is the current, I_3, directed away from that point?

9–2 Applying Kirchhoff's current law, write an equation for the currents directed into and out of point X in Prob. 9–1.

9–3 In Fig. 9–10, solve for the unknown current, I_3.

9–4 In Fig. 9–11, solve for the following unknown currents: I_3, I_5, and I_8.

9–5 Apply Kirchhoff's current law in Fig. 9–11 by writing an equation for the currents directed into and out of the following points:

a. Point X

b. Point Y

c. Point Z

Figure 9–10

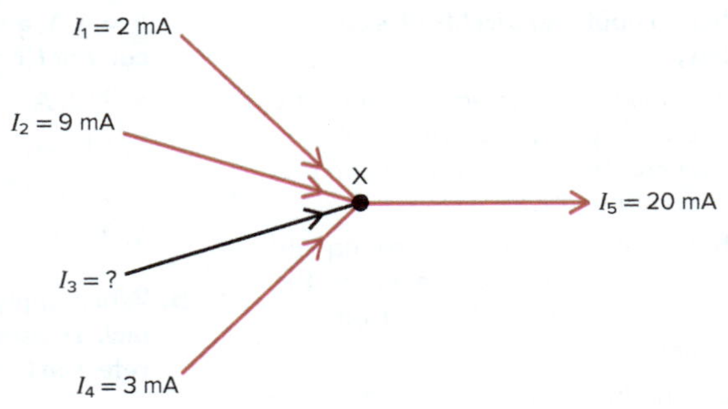

Figure 9–11

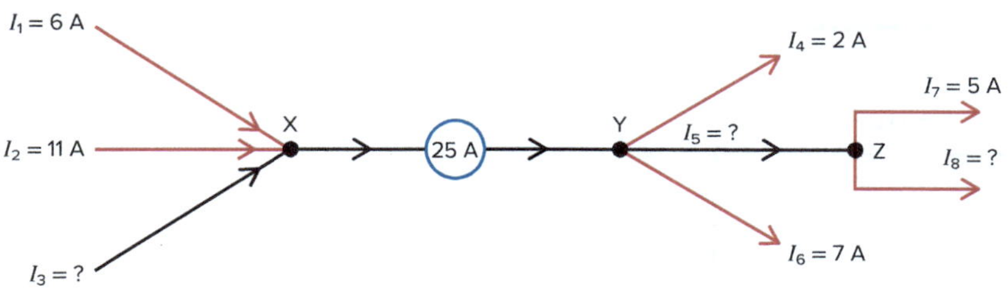

SECTION 9–2 KIRCHOFF'S VOLTAGE LAW

9–6 MultiSim In Fig. 9–12,

a. Write a KVL equation for the loop CEFDC going clockwise from point C.

b. Write a KVL equation for the loop ACDBA going clockwise from point A.

c. Write a KVL equation for the loop ACEFDBA going clockwise from point A.

Figure 9–12

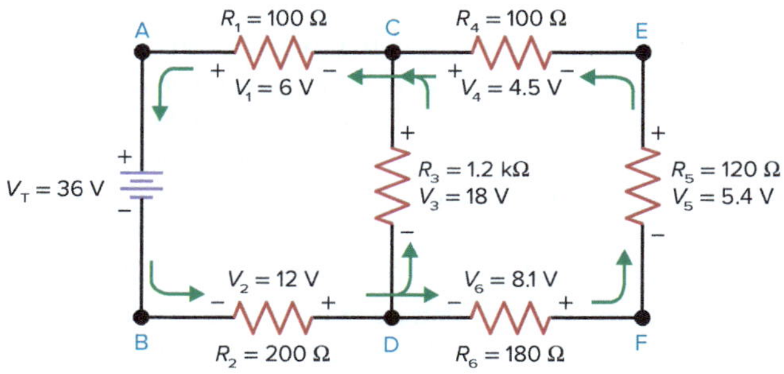

9–7 In Fig. 9–12,

a. Determine the voltage for the partial loop CEFD going clockwise from point C. How does your answer compare to the voltage drop across R_3?

b. Determine the voltage for the partial loop ACDB going clockwise from point A. How does your answer compare to the value of the applied voltage, V_T, across points A and B?

c. Determine the voltage for the partial loop ACEFDB going clockwise from point A. How does your answer compare to the value of the applied voltage, V_T, across points A and B?

d. Determine the voltage for the partial loop CDFE going counterclockwise from point C. How does your answer compare to the voltage drop across R_4?

9–8 In Fig. 9–13, solve for the voltages V_{AG} and V_{BG}. Indicate the proper polarity for each voltage.

9–9 In Fig. 9–14, solve for the voltages V_{AG} and V_{BG}. Indicate the proper polarity for each voltage.

9–10 In Fig. 9–15, solve for the voltages V_{AG} and V_{BG}. Indicate the proper polarity for each voltage.

Figure 9–13

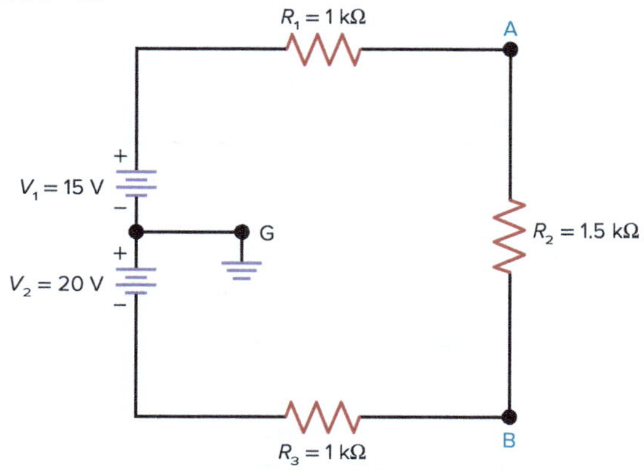

Figure 9–14

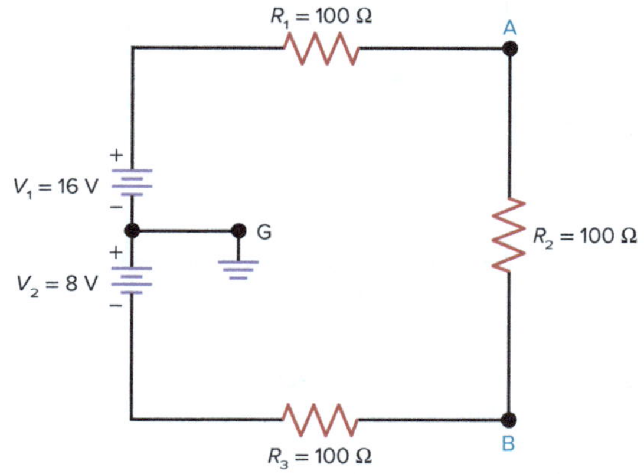

Figure 9–15

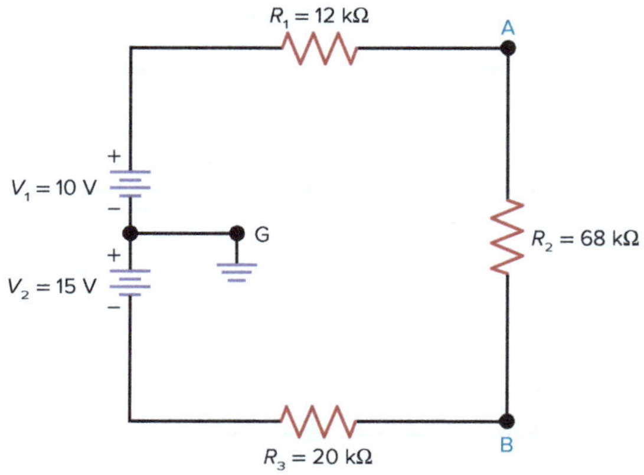

9–11 In Fig. 9–16, solve for the voltages V_{AG}, V_{BG}, V_{CG}, V_{DG}, and V_{AD}. Indicate the proper polarity for each voltage.

Figure 9–16

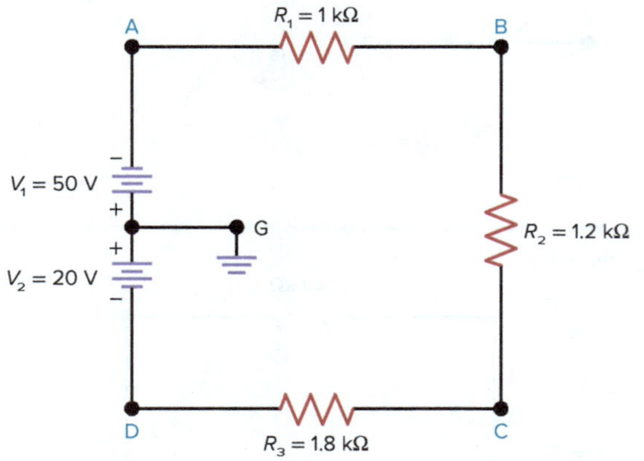

9–12 In Fig. 9–17, solve for the voltages V_{AG}, V_{BG}, and V_{CG}. Indicate the proper polarity for each voltage.

Figure 9–17

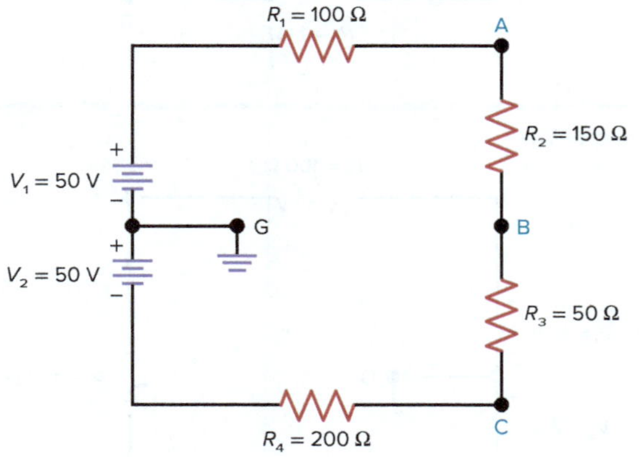

9–13 In Fig. 9–18, write a KVL equation for the loop ABDCA going counterclockwise from point A.

9–14 In Fig. 9–18, write a KVL equation for the loop EFDCE going clockwise from point E.

9–15 In Fig. 9–18, write a KVL equation for the loop ACEFDBA going clockwise from point A.

Figure 9–18

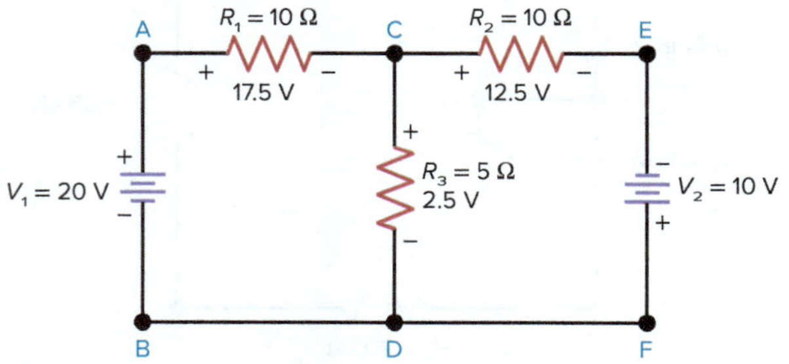

SECTION 9–3 METHOD OF BRANCH CURRENTS

9–16 Using the method of branch currents, solve for the unknown values of voltage and current in Fig. 9–19. To do this, complete steps a through m. The assumed direction of all currents is shown in the figure.

Figure 9–19

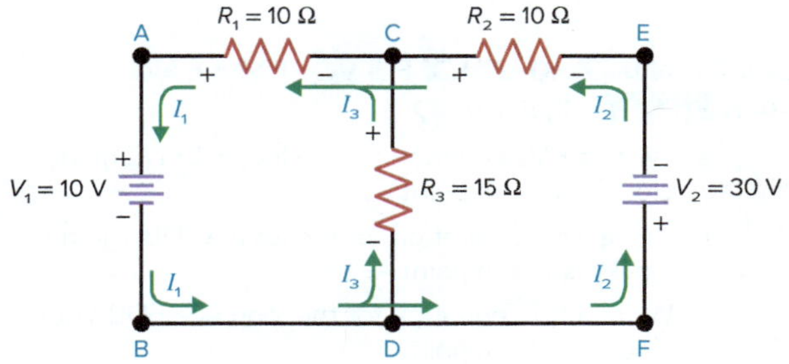

a. Using Kirchhoff's current law, write an equation for the currents I_1, I_2, and I_3 at point C.

b. Specify the current I_3 in terms of I_1 and I_2.

c. Write a KVL equation for the loop ABDCA, going counterclockwise from point A, using the terms V_1, V_{R_1} and V_{R_3}. This loop will be called Loop 1.

d. Write a KVL equation for the loop FECDF, going counterclockwise from point F, using the terms V_2, V_{R_2}, and V_{R_3}. This loop will be called Loop 2.

e. Specify each resistor voltage drop as an *IR* product using actual resistor values for R_1, R_2, and R_3.

f. Rewrite the KVL equation for Loop 1 in step c using the *IR* voltage values for V_{R_1} and V_{R_3} specified in step e.

g. Rewrite the KVL equation for Loop 2 in step d using the *IR* voltage values for V_{R_2} and V_{R_3} specified in step e.

h. Reduce the Loop 1 and Loop 2 equations in steps f and g to their simplest possible form.

i. Solve for currents I_1 and I_2 using any of the methods for the solution of simultaneous equations. Next, solve for I_3.

j. In Fig. 9–19, were the assumed directions of all currents correct? How do you know?

k. Using the actual values of I_1, I_2, and I_3, calculate the individual resistor voltage drops.

l. Rewrite the KVL loop equations for both Loops 1 and 2 using actual voltage values. Go counterclockwise around both loops when adding voltages. (Be sure that the resistor voltage drops all have the correct polarity based on the actual directions for I_1, I_2, and I_3.)

m. Based on the actual directions for I_1, I_2, and I_3, write a KCL equation for the currents at point C.

9–17 Repeat Prob. 9–16 for Fig. 9–20.

Figure 9–20

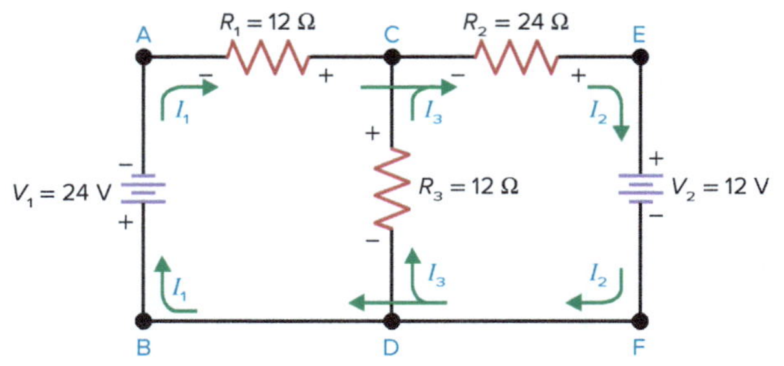

SECTION 9–4 NODE-VOLTAGE ANALYSIS

9–18 Using the method of node-voltage analysis, solve for all unknown values of voltage and current in Fig. 9–21. To do this, complete steps a through l. The assumed direction of all currents is shown in the figure.

Figure 9–21

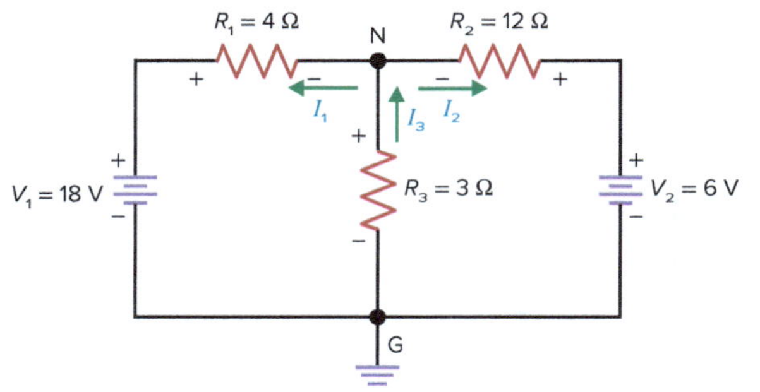

a. Using Kirchhoff's current law, write an equation for the currents I_1, I_2, and I_3 at the node-point, N.

b. Express the KCL equation in step a in terms of V_R/R and V_N/R.

c. Write a KVL equation for the loop containing V_1, V_{R_1}, and V_{R_3}.

d. Write a KVL equation for the loop containing V_2, V_{R_2}, and V_{R_3}.

e. Specify V_{R_1} and V_{R_2} in terms of V_N and the known values of V_1 and V_2.

f. Using the values for V_{R_1} and V_{R_2} from step e, write a KCL equation for the currents at the node-point N.

g. Solve for the node voltage, V_N.

h. Using the equations from step e, solve for the voltage drops V_{R_1} and V_{R_2}.

i. In Fig. 9–21, were all the assumed directions of current correct? How do you know?

j. Calculate the currents I_1, I_2, and I_3.

k. Rewrite the KVL loop equation for both inner loops using actual voltage values. (Be sure that the resistor voltage drops all have the correct polarity based on the final answers for V_{R_1} and V_{R_2}.)

l. Based on the actual values for I_1, I_2, and I_3, write a KCL equation for the currents at the node-point, N.

9–19 Repeat Prob. 9–18 for the circuit in Fig. 9–22.

Figure 9–22

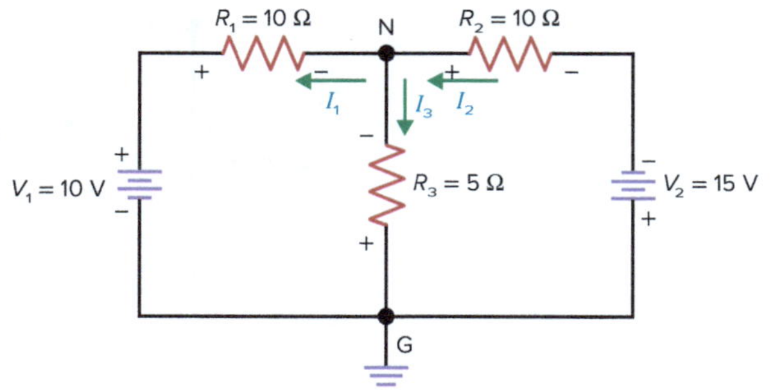

SECTION 9–5 METHOD OF MESH CURRENTS

9–20 Using the method of mesh currents, solve for all unknown values of voltage and current in Fig. 9–23. To do this, complete steps a through m.

Figure 9–23

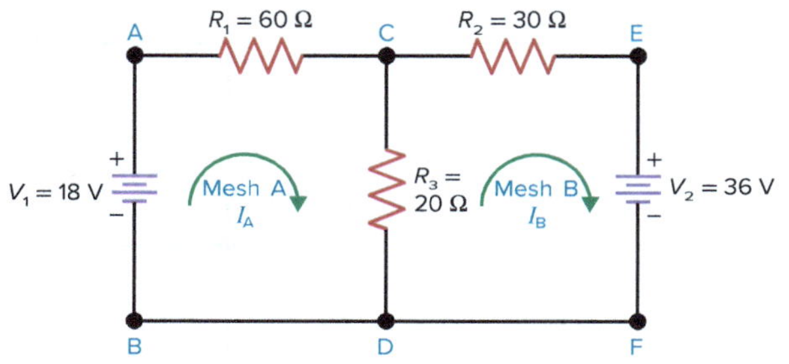

a. Identify the components through which the mesh current, I_A, flows.

b. Identify the components through which the mesh current, I_B, flows.

c. Which component has opposing mesh currents?

d. Write the mesh equation for mesh A.

e. Write the mesh equation for mesh B.

f. Solve for currents I_A and I_B using any of the methods for the solution of simultaneous equations.

g. Determine the values of currents I_1, I_2, and I_3.

h. Are the assumed directions of the mesh A and mesh B currents correct? How do you know?

i. What is the direction of the current, I_3, through R_3?

j. Solve for the voltage drops V_{R_1}, V_{R_2}, and V_{R_3}.

k. Using the final solutions for V_{R_1}, V_{R_2}, and V_{R_3}, write a KVL equation for the loop ACDBA going clockwise from point A.

l. Using the final solutions for V_{R_1}, V_{R_2}, and V_{R_3}, write a KVL equation for the loop EFDCE going clockwise from point E.

m. Using the final solutions (and directions) for I_1, I_2, and I_3, write a KCL equation for the currents at point C.

9–21 Repeat Prob. 9–20 for Fig. 9–24.

Figure 9–24

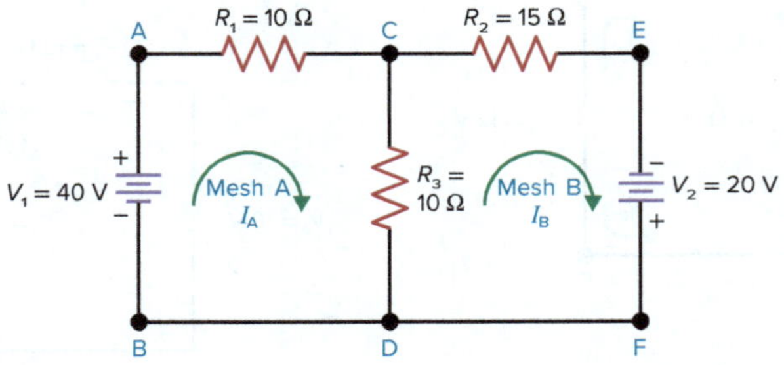

Critical Thinking

9–22 In Fig. 9–25, determine the values for R_1 and R_3 which will allow the output voltage to vary between −5 V and +5 V.

9–23 Refer to Fig. 9–9. If all resistances are 10 **Ω**, calculate (a) I_A, I_B, and I_C; (b) I_1, I_2, I_3, I_4, I_5, I_6, I_7, and I_8.

Figure 9–25 Circuit diagram for Critical Thinking Prob. 9–22.

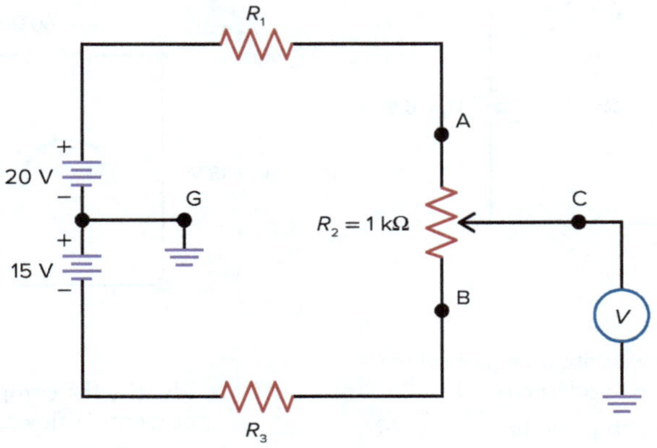

Answers to Self-Reviews

9–1 **a.** 6 A
b. 4 A

9–2 **a.** −120 V
b. 0 V

9–3 **a.** −24 V
b. 0 V

9–4 **a.** two
b. two

9–5 **a.** true
b. true

Laboratory Application Assignment

In this lab application assignment, you will examine Kirchhoff's voltage and current laws (KVL and KCL). You will actually apply both KVL and KCL in a simple series-parallel circuit. You will also apply KVL when solving for the voltages in a circuit containing series-aiding voltage sources.

Equipment: Obtain the following items from your instructor.

- Dual-output variable DC power supply
- Assortment of carbon-film resistors
- DMM

Applying KCL and KVL

Examine the circuit in Fig. 9–26. Calculate and record the following circuit values:
V_1 = ________, V_2 = ________, V_3 = ________,
V_4 = ________, V_5 = ________,
I_T = ________, I_3 = ________, I_{4-5} = ________

In Fig. 9–26, indicate the direction of all currents and the polarities of all resistor voltage drops.

Figure 9–26

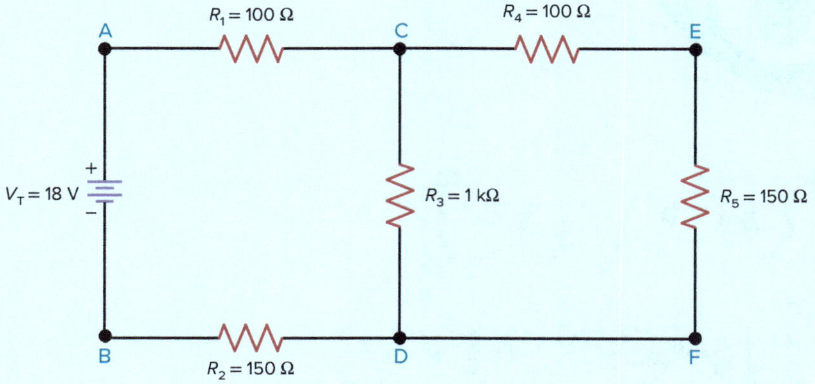

Construct the circuit in Fig. 9–26. Measure and record the following circuit values:

V_1 = ________, V_2 = ________, V_3 = ________,
V_4 = ________, V_5 = ________,
I_T = ________, I_3 = ________, I_{4-5} = ________

Write the measured values of voltage and current next to their respective resistors in Fig. 9–26.

Using measured values, write a KCL equation for the currents entering and leaving point C. ________________ Do the same for the currents entering and leaving point D. ________________ Do these circuit values satisfy KCL? ________

Using measured values, write a KVL equation for the voltages in the loop ACDBA. Go clockwise around the loop beginning at point A. ________________

Do these values satisfy KVL? ________

Using measured values, write a KVL equation for the voltages in the loop CEFDC. Go clockwise around the loop beginning at point C. ________________ Do these values satisfy KVL? ________

Beginning at point C and going clockwise, add the measured voltages in the partial loop CEFD. ________________
Is this value equal to the voltage across R_3? ________

Finally, using measured values, write a KVL equation for the voltages in the outside loop ACEFDBA. Go clockwise around the loop beginning at point A. ________________ Do these values satisfy KVL? ________

Examine the circuit in Fig. 9–27. Calculate and record the individual resistor voltage drops V_{R_1}, V_{R_2}, and V_{R_3}.

V_{R_1} = ________, V_{R_2} = ________, V_{R_3} = ________

Indicate the direction of current and the polarity of each resistor voltage drop. Next, apply KVL and solve for the voltages V_{AG} and V_{BG}. Record your answers.

V_{AG} = ________, V_{BG} = ________

Construct the circuit in Fig. 9–27. Before turning on the power, however, have your instructor check the circuit to make sure the power supplies are wired correctly.

Measure and record the voltages V_{AG} and V_{BG}. V_{AG} = ________, V_{BG} = ________

Do the measured voltages match your calculated values? ________

Figure 9–27

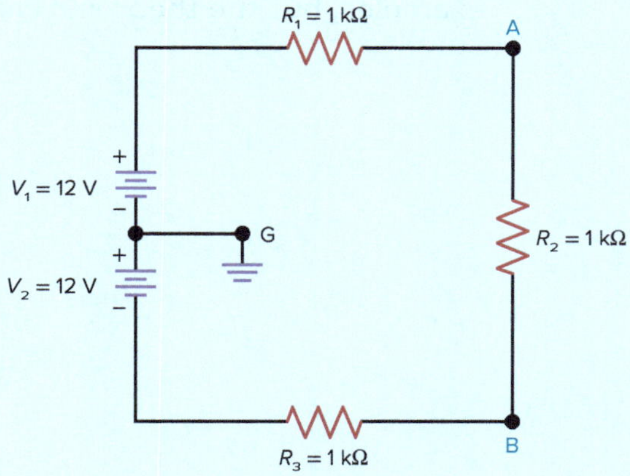

Design credits: Chapter Opener Background Image, and Multisim: Stockbyte/Getty Images

chapter

10

Network Theorems

A network is a combination of components, such as resistances and voltage sources, interconnected to achieve a particular end result. However, networks generally need more than the rules of series and parallel circuits for analysis. Kirchhoff's laws can always be applied for any circuit connections. The network theorems, though, usually provide shorter methods for solving a circuit.

Some theorems enable us to convert a network into a simpler circuit, equivalent to the original. Then the equivalent circuit can be solved by the rules of series and parallel circuits. Other theorems enable us to convert a given circuit into a form that permits easier solutions.

Only the applications are given here, although all network theorems can be derived from Kirchhoff's laws. Note that resistance networks with batteries are shown as examples, but the theorems can also be applied to AC networks.

Chapter Outline

Chapter Objectives

After studying this chapter, you should be able to

- *Apply* the superposition theorem to find the voltage across two points in a circuit containing more than one voltage source.
- *State* the requirements for applying the superposition theorem.
- *Determine* the Thevenin and Norton equivalent circuits with respect to any pair of terminals in a complex network.
- *Apply* Thevenin's and Norton's theorems in solving for an unknown voltage or current.
- *Convert* a Thevenin equivalent circuit to a Norton equivalent circuit and vice versa.
- *Apply* Millman's theorem to find the common voltage across any number of parallel branches.
- *Simplify* the analysis of a bridge circuit by using delta to wye conversion formulas.

Important Terms

active components
bilateral components
current source
linear component
Millman's theorem
Norton's theorem
passive components
superposition theorem
Thevenin's theorem
voltage source

10–1 Superposition Theorem

The **superposition theorem** is very useful because it extends the use of Ohm's law to circuits that have more than one source. In brief, we can calculate the effect of one source at a time and then superimpose the results of all sources. As a definition, the superposition theorem states: *In a network with two or more sources, the current or voltage for any component is the algebraic sum of the effects produced by each source acting separately.*

To use one source at a time, all other sources are "killed" temporarily. This means disabling the source so that it cannot generate voltage or current without changing the resistance of the circuit. A **voltage source** such as a battery is killed by assuming a short circuit across its potential difference. The internal resistance remains.

Voltage Divider with Two Sources

The problem in Fig. 10–1 is to find the voltage at P to chassis ground for the circuit in Fig. 10–1*a*. The method is to calculate the voltage at P contributed by each source separately, as in Fig. 10–1*b* and *c*, and then superimpose these voltages.

To find the effect of V_1 first, short-circuit V_2 as shown in Fig. 10–1*b*. Note that the bottom of R_1 then becomes connected to chassis ground because of the short circuit across V_2. As a result, R_2 and R_1 form a series voltage divider for the V_1 source.

Furthermore, the voltage across R_1 becomes the same as the voltage from P to ground. To find this V_{R_1} across R_1 as the contribution of the V_1 source, we use the voltage divider formula:

$$V_{R_1} = \frac{R_1}{R_1 + R_2} \times V_1 = \frac{60\ \text{k}\Omega}{60\ \text{k}\Omega + 30\ \text{k}\Omega} \times 24\ \text{V}$$

$$= \frac{60}{90} \times 24\ \text{V}$$

$$V_{R_1} = 16\ \text{V}$$

Next find the effect of V_2 alone, with V_1 short-circuited, as shown in Fig. 10–1*c*. Then point A at the top of R_2 becomes grounded. R_1 and R_2 form a series voltage divider again, but here the R_2 voltage is the voltage at P to ground.

With one side of R_2 grounded and the other side to point P, V_{R_2} is the voltage to calculate. Again we have a series divider, but this time for the negative voltage V_2.

MultiSim **Figure 10–1** Superposition theorem applied to a voltage divider with two sources V_1 and V_2. (*a*) Actual circuit with +13 V from point P to chassis ground. (*b*) V_1 alone producing +16 V at P. (*c*) V_2 alone producing −3 V at P.

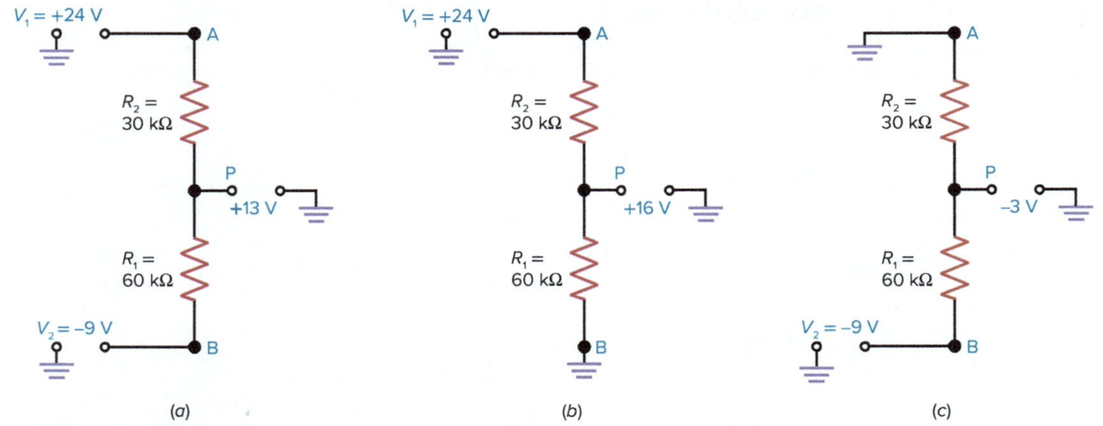

Using the voltage divider formula for V_{R_2} as the contribution of V_2 to the voltage at P,

$$V_{R_2} = \frac{R_2}{R_1 + R_2} \times V_2 = \frac{30\ \text{k}\Omega}{30\ \text{k}\Omega + 60\ \text{k}\Omega} \times -9\ \text{V}$$

$$= \frac{30}{90} \times -9\ \text{V}$$

$$V_{R_2} = -3\ \text{V}$$

This voltage is negative at P because V_2 is negative.

Finally, the total voltage at P is

$$V_P = V_{R_1} + V_{R_2} = 16 - 3$$

$$V_P = 13\ \text{V}$$

This algebraic sum is positive for the net V_P because the positive V_1 is larger than the negative V_2.

By superposition, therefore, this problem was reduced to two series voltage dividers. The same procedure can be used with more than two sources. Also, each voltage divider can have any number of series resistances. Note that in this case we were dealing with ideal voltage sources, that is, sources with zero internal resistance. If the source did have internal resistance, it would have been added in series with R_1 and R_2.

GOOD TO KNOW

When applying the superposition theorem to a DC network, it is important to realize that the power dissipated by a resistor in the network is not equal to the sum of the power dissipation values produced by each source acting separately. The reason is that power is not linearly related to either voltage or current. Recall that $P = \frac{V^2}{R}$ and $P = I^2R$.

Requirements for Superposition

All components must be linear and bilateral to superimpose currents and voltages. *Linear* means that the current is proportional to the applied voltage. Then the currents calculated for different source voltages can be superimposed.

Bilateral means that the current is the same amount for opposite polarities of the source voltage. Then the values for opposite directions of current can be combined algebraically. Networks with resistors, capacitors, and air-core inductors are generally linear and bilateral. These are also ***passive components***, that is, components that do not amplify or rectify. ***Active components***—such as transistors, semiconductor diodes, and electron tubes—are never bilateral and often are not linear.

Note: The superposition theorem can also be applied when solving for the individual voltage and current values in a complex electrical circuit. For expanded coverage of the superposition theorem, refer to Chapter 10 in the Problems Manual for *Grob's Basic Electronics, 13e.*

10–1 Self-Review

Answers at the end of the chapter.

a. In Fig. 10–1*b*, which *R* is shown grounded at one end?
b. In Fig. 10–1*c*, which *R* is shown grounded at one end?

10–2 Thevenin's Theorem

Named after M. L. Thevenin, a French engineer, **Thevenin's theorem** is very useful in simplifying the process of solving for the unknown values of voltage and current in a network. By Thevenin's theorem, many sources and components, no matter how they are interconnected, can be represented by an equivalent series circuit with respect to any pair of terminals in the network. In Fig. 10–2, imagine that the block at the left contains a network connected to terminals A and B. Thevenin's theorem states that the *entire* network connected to A and B can be replaced by a single voltage source V_{TH} in series with a single resistance R_{TH}, connected to the same two terminals.

Voltage V_{TH} is the open-circuit voltage across terminals A and B. This means finding the voltage that the network produces across the two terminals with an open circuit between A and B. The polarity of V_{TH} is such that it will produce current from A to B in the same direction as in the original network.

GOOD TO KNOW

Of all the different theorems covered in this chapter, Thevenin's theorem is by far the most widely used.

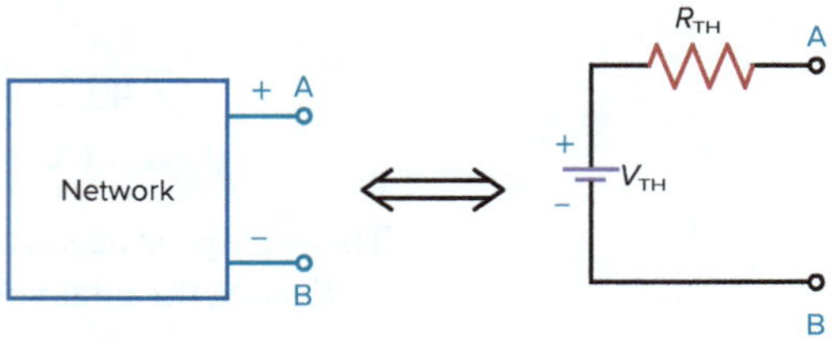

Figure 10–2 Any network in the block at the left can be reduced to the Thevenin equivalent series circuit at the right.

Resistance R_{TH} is the open-circuit resistance across terminals A and B, but with all the sources killed. This means finding the resistance looking back into the network from terminals A and B. Although the terminals are open, an ohmmeter across AB would read the value of R_{TH} as the resistance of the remaining paths in the network without any sources operating.

Thevenizing a Circuit

As an example, refer to Fig. 10–3*a*, where we want to find the voltage V_L across the 2-Ω R_L and its current I_L. To use Thevenin's theorem, mentally disconnect R_L. The two open ends then become terminals A and B. Now we find the Thevenin equivalent of the remainder of the circuit that is still connected to A and B. In general, open the part of the circuit to be analyzed and "thevenize" the remainder of the circuit connected to the two open terminals.

Our only problem now is to find the value of the open-circuit voltage V_{TH} across AB and the equivalent resistance R_{TH}. The Thevenin equivalent always consists of a single voltage source in series with a single resistance, as shown in Fig. 10–3*d*.

The effect of opening R_L is shown in Fig. 10–3*b*. As a result, the 3-Ω R_1 and 6-Ω R_2 form a series voltage divider without R_L.

Furthermore, the voltage across R_2 now is the same as the open-circuit voltage across terminals A and B. Therefore V_{R_2} with R_L open is V_{AB}. This is the V_{TH} we need for the Thevenin equivalent circuit. Using the voltage divider formula,

$$V_{R_2} = \frac{6}{9} \times 36\text{ V} = 24\text{ V}$$

$$V_{R_2} = V_{AB} = V_{TH} = 24\text{ V}$$

This voltage is positive at terminal A.

To find R_{TH}, the 2-Ω R_L is still disconnected. However, now the source V is short-circuited. So the circuit looks like Fig. 10–3*c*. The 3-Ω R_1 is now in parallel with the 6-Ω R_2 because both are connected across the same two points. This combined resistance is the product over the sum of R_1 and R_2.

$$R_{TH} = \frac{18}{9} = 2\ \Omega$$

Again, we assume an ideal voltage source whose internal resistance is zero.

As shown in Fig. 10–3*d*, the Thevenin circuit to the left of terminals A and B then consists of the equivalent voltage V_{TH}, equal to 24 V, in series with the equivalent series resistance R_{TH}, equal to 2 Ω. This Thevenin equivalent applies for any value of R_L because R_L was disconnected. We are actually thevenizing the circuit that feeds the open AB terminals.

To find V_L and I_L, we can finally reconnect R_L to terminals A and B of the Thevenin equivalent circuit, as shown in Fig. 10–3*e*. Then R_L is in series with R_{TH} and V_{TH}. Using the voltage divider formula for the 2-Ω R_{TH} and 2-Ω R_L, $V_L = 1/2 \times 24\text{ V} = 12\text{ V}$. To find I_L as V_L/R_L, the value is 12 V/2 Ω, which equals 6 A.

MultiSim **Figure 10–3** Application of Thevenin's theorem. (*a*) Actual circuit with terminals A and B across R_L. (*b*) Disconnect R_L to find that V_{AB} is 24 V. (*c*) Short-circuit *V* to find that R_{AB} is 2 Ω. (*d*) Thevenin equivalent circuit. (*e*) Reconnect R_L at terminals A and B to find that V_L is 12 V.

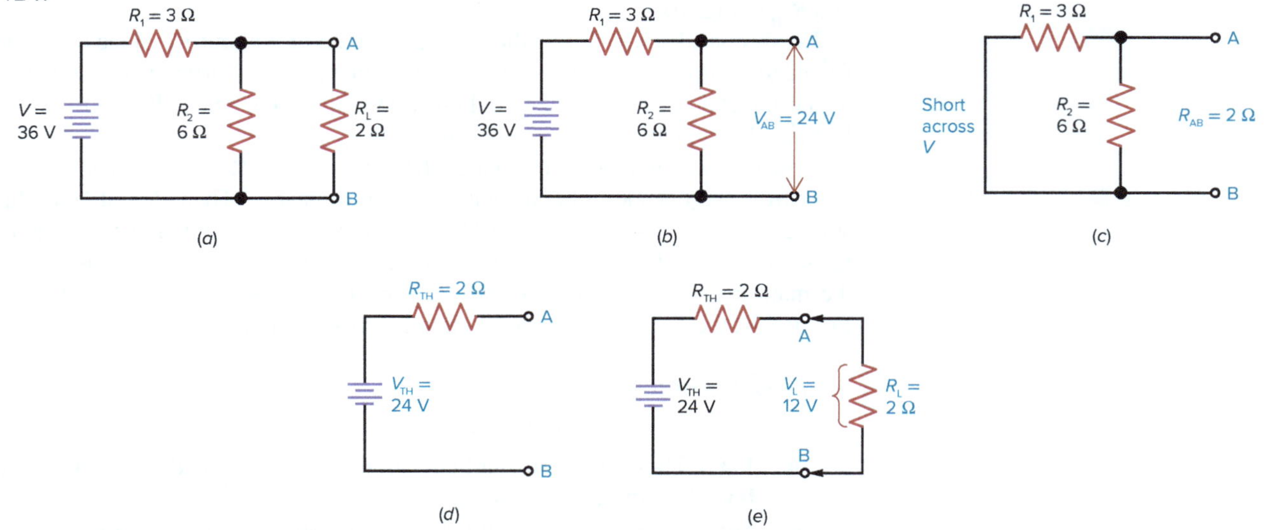

These answers of 6 A for I_L and 12 V for V_L apply to R_L in both the original circuit in Fig. 10–3*a* and the equivalent circuit in Fig. 10–3*e*. Note that the 6-A I_L also flows through R_{TH}.

The same answers could be obtained by solving the series-parallel circuit in Fig. 10–3*a*, using Ohm's law. However, the advantage of thevenizing the circuit is that the effect of different values of R_L can be calculated easily. Suppose that R_L is changed to 4 Ω. In the Thevenin circuit, the new value of V_L would be $4/6 \times 24\text{ V} = 16\text{ V}$. The new I_L would be 16 V/4 Ω, which equals 4 A. If we used Ohm's law in the original circuit, a complete, new solution would be required each time R_L was changed.

GOOD TO KNOW

The Thevenin equivalent circuit driving terminals A and B does not change even though the value of R_L may change.

Looking Back from Terminals A and B

The way we look at the resistance of a series-parallel circuit depends on where the source is connected. In general, we calculate the total resistance from the outside terminals of the circuit in toward the source as the reference.

When the source is short-circuited for thevenizing a circuit, terminals A and B become the reference. Looking back from A and B to calculate R_{TH}, the situation becomes reversed from the way the circuit was viewed to determine V_{TH}.

For R_{TH}, imagine that a source could be connected across AB, and calculate the total resistance working from the outside in toward terminals A and B. Actually, an ohmmeter placed across terminals A and B would read this resistance.

This idea of reversing the reference is illustrated in Fig. 10–4. The circuit in Fig. 10–4*a* has terminals A and B open, ready to be thevenized. This circuit is similar to that in Fig. 10–3 but with the 4-Ω R_3 inserted between R_2 and terminal A. The

Figure 10–4 Thevenizing the circuit of Fig. 10–3 but with a 4-Ω R_3 in series with the A terminal. (*a*) V_{AB} is still 24 V. (*b*) Now the R_{AB} is $2 + 4 = 6$ Ω. (*c*) Thevenin equivalent circuit.

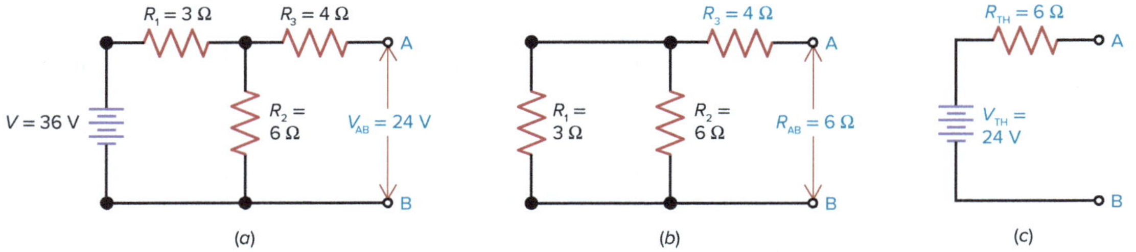

interesting point is that R_3 does not change the value of V_{AB} produced by the source V, but R_3 does increase the value of R_{TH}. When we look back from terminals A and B, the 4 Ω of R_3 is in series with 2 Ω to make R_{TH} 6 Ω, as shown for R_{AB} in Fig. 10–4*b* and R_{TH} in Fig. 10–4*c*.

Let us consider why V_{AB} is the same 24 V with or without R_3. Since R_3 is connected to the open terminal A, the source V cannot produce current in R_3. Therefore, R_3 has no IR drop. A voltmeter would read the same 24 V across R_2 and from A to B. Since V_{AB} equals 24 V, this is the value of V_{TH}.

Now consider why R_3 does change the value of R_{TH}. Remember that we must work from the outside in to calculate the total resistance. Then, A and B are like source terminals. As a result, the 3-Ω R_1 and 6-Ω R_2 are in parallel, for a combined resistance of 2 Ω. Furthermore, this 2 Ω is in series with the 4-Ω R_3 because R_3 is in the main line from terminals A and B. Then R_{TH} is $2 + 4 = 6$ Ω. As shown in Fig. 10–4*c*, the Thevenin equivalent circuit consists of $V_{TH} = 24$ V and $R_{TH} = 6$ Ω.

10–2 Self-Review

Answers at the end of the chapter.

a. For a Thevenin equivalent circuit, terminals A and B are open to find both V_{TH} and R_{TH}. (True/False)

b. For a Thevenin equivalent circuit, the source voltage is short-circuited only to find R_{TH}. (True/False)

10–3 Thevenizing a Circuit with Two Voltage Sources

The circuit in Fig. 10–5 has already been solved by Kirchhoff's laws, but we can use Thevenin's theorem to find the current I_3 through the middle resistance R_3. As shown in Fig. 10–5*a*, first mark the terminals A and B across R_3. In Fig. 10–5*b*, R_3 is disconnected. To calculate V_{TH}, find V_{AB} across the open terminals.

Figure 10–5 Thevenizing a circuit with two voltage sources V_1 and V_2. (*a*) Original circuit with terminals A and B across the middle resistor R_3. (*b*) Disconnect R_3 to find that V_{AB} is −33.6 V. (*c*) Short-circuit V_1 and V_2 to find that R_{AB} is 2.4 Ω. (*d*) Thevenin equivalent with R_L reconnected to terminals A and B.

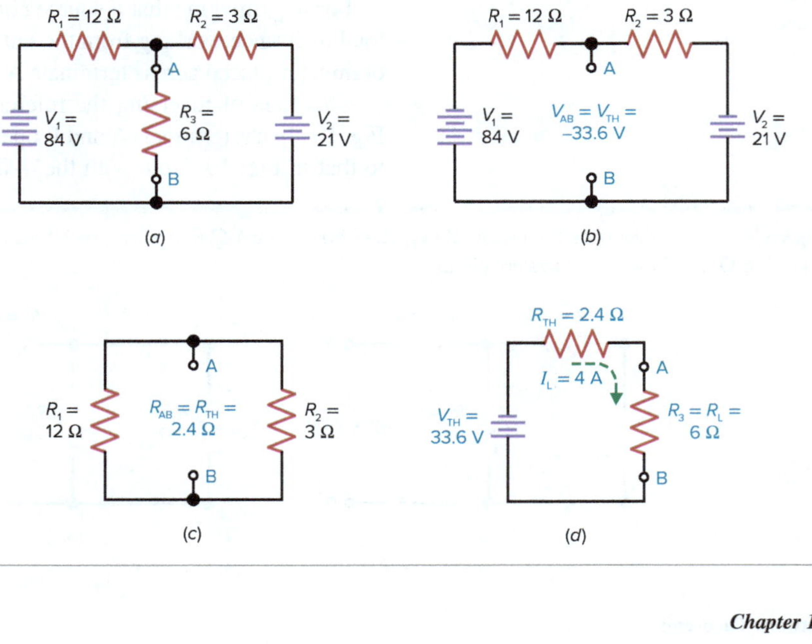

Superposition Method

With two sources, we can use superposition to calculate V_{AB}. First short-circuit V_2. Then the 84 V of V_1 is divided between R_1 and R_2. The voltage across R_2 is between terminals A and B. To calculate this divided voltage across R_2,

$$V_{R_2} = \frac{R_2}{R_{1-2}} \times V_1 = \frac{3}{15} \times (-84)$$
$$V_{R_2} = -16.8 \text{ V}$$

This is the only contribution of V_1 to V_{AB}. The polarity is negative at terminal A.

To find the voltage that V_2 produces between A and B, short-circuit V_1. Then the voltage across R_1 is connected from A to B. To calculate this divided voltage across R_1,

$$V_{R_1} = \frac{R_1}{R_{1-2}} \times V_2 = \frac{12}{15} \times (-21)$$
$$V_{R_1} = -16.8 \text{ V}$$

Both V_1 and V_2 produce -16.8 V across the AB terminals with the same polarity. Therefore, they are added.

The resultant value of $V_{AB} = -33.6$ V, shown in Fig. 10–5*b*, is the value of V_{TH}. The negative polarity means that terminal A is negative with respect to B.

To calculate R_{TH}, short-circuit the sources V_1 and V_2, as shown in Fig. 10–5*c*. Then the 12-Ω R_1 and 3-Ω R_2 are in parallel across terminals A and B. Their combined resistance is 36/15, or 2.4 Ω, which is the value of R_{TH}.

The final result is the Thevenin equivalent in Fig. 10–5*d* with an R_{TH} of 2.4 Ω and a V_{TH} of 33.6 V, negative toward terminal A.

To find the current through R_3, it is reconnected as a load resistance across terminals A and B. Then V_{TH} produces current through the total resistance of 2.4 Ω for R_{TH} and 6 Ω for R_3:

$$I_3 = \frac{V_{TH}}{R_{TH} + R_3} = \frac{33.6}{2.4 + 6} = \frac{33.6}{8.4} = 4 \text{ A}$$

This answer of 4 A for I_3 is the same value calculated before, using Kirchhoff's laws, as shown in Fig. 9–5.

It should be noted that this circuit can be solved by superposition alone, without using Thevenin's theorem, if R_3 is not disconnected. However, opening terminals A and B for the Thevenin equivalent simplifies the superposition, as the circuit then has only series voltage dividers without any parallel current paths. In general, a circuit can often be simplified by disconnecting a component to open terminals A and B for Thevenin's theorem.

GOOD TO KNOW

The polarity of V_{TH} is extremely critical because it allows us to determine the actual direction of I_3 through R_3.

10–3 Self-Review

Answers at the end of the chapter.

In the Thevenin equivalent circuit in Fig. 10–5*d*,

a. **How much is R_T?**
b. **How much is V_{R_L}?**

10–4 Thevenizing a Bridge Circuit

As another example of Thevenin's theorem, we can find the current through the 2-Ω R_L at the center of the bridge circuit in Fig. 10–6*a*. When R_L is disconnected to open terminals A and B, the result is as shown in Fig. 10–6*b*. Notice how the circuit has become simpler because of the open. Instead of the unbalanced bridge in

Figure 10–6 Thevenizing a bridge circuit. (*a*) Original circuit with terminals A and B across middle resistor R_L. (*b*) Disconnect R_L to find V_{AB} of −8 V. (*c*) With source *V* short-circuited, R_{AB} is 2 + 2.4 = 4.4 Ω. (*d*) Thevenin equivalent with R_L reconnected to terminals A and B.

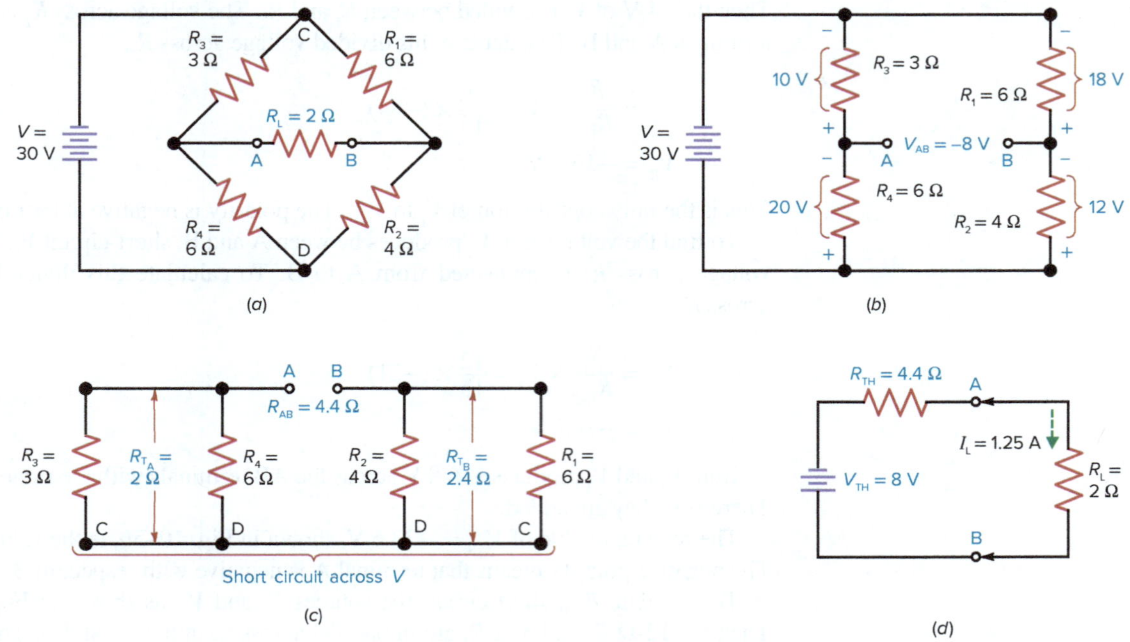

Fig. 10–6*a* which would require Kirchhoff's laws for a solution, the Thevenin equivalent in Fig. 10–6*b* consists of just two voltage dividers. Both the R_3–R_4 divider and the R_1–R_2 divider are across the same 30-V source.

Since the open terminal A is at the junction of R_3 and R_4, this divider can be used to find the potential at point A. Similarly, the potential at terminal B can be found from the R_1–R_2 divider. Then V_{AB} is the difference between the potentials at terminals A and B.

Note the voltages for the two dividers. In the divider with the 3-Ω R_3 and 6-Ω R_4, the bottom voltage V_{R_4} is 6⁄9 × 30 = 20 V. Then V_{R_3} at the top is 10 V because both must add up to equal the 30-V source. The polarities are marked negative at the top, the same as the source voltage *V*.

Similarly, in the divider with the 6-Ω R_1 and 4-Ω R_2, the bottom voltage V_{R_2} is 4⁄10 × 30 = 12 V. Then V_{R_1} at the top is 18 V because the two must add up to equal the 30-V source. The polarities are also negative at the top, the same as *V*.

Now we can determine the potentials at terminals A and B with respect to a common reference to find V_{AB}. Imagine that the positive side of the source *V* is connected to a chassis ground. Then we would use the bottom line in the diagram as our reference for voltages. Note that V_{R_4} at the bottom of the R_3–R_4 divider is the same as the potential of terminal A with respect to ground. This value is −20 V, with terminal A negative.

Similarly, V_{R_2} in the R_1–R_2 divider is the potential at B with respect to ground. This value is −12 V with terminal B negative. As a result, V_{AB} is the difference between the −20 V at A and the −12 V at B, both with respect to the common ground reference.

The potential difference V_{AB} then equals

$$V_{AB} = -20 - (-12) = -20 + 12 = -8 \text{ V}$$

Terminal A is 8 V more negative than B. Therefore, V_{TH} is 8 V, with the negative side toward terminal A, as shown in the Thevenin equivalent in Fig. 10–6*d*.

The potential difference V_{AB} can also be found as the difference between V_{R_3} and V_{R_1} in Fig. 10–6*b*. In this case, V_{R_3} is 10 V and V_{R_1} is 18 V, both positive with respect to the top line connected to the negative side of the source V. The potential difference between terminals A and B, then, is 10 − 18, which also equals −8 V. Note that V_{AB} must have the same value no matter which path is used to determine the voltage.

To find R_{TH}, the 30-V source is short-circuited while terminals A and B are still open. Then the circuit looks like Fig. 10–6*c*. Looking back from terminals A and B, the 3-Ω R_3 and 6-Ω R_4 are in parallel, for a combined resistance R_{T_A} of $^{18}/_{9}$ or 2 Ω. The reason is that R_3 and R_4 are joined at terminal A, while their opposite ends are connected by the short circuit across the source V. Similarly, the 6-Ω R_1 and 4-Ω R_2 are in parallel for a combined resistance R_{T_B} of $^{24}/_{10}$ = 2.4 Ω. Furthermore, the short circuit across the source now provides a path that connects R_{T_A} and R_{T_B} in series. The entire resistance is 2 + 2.4 = 4.4 Ω for R_{AB} or R_{TH}.

The Thevenin equivalent in Fig. 10–6*d* represents the bridge circuit feeding the open terminals A and B with 8 V for V_{TH} and 4.4 Ω for R_{TH}. Now connect the 2-Ω R_L to terminals A and B to calculate I_L. The current is

$$I_L = \frac{V_{TH}}{R_{TH} + R_L} = \frac{8}{4.4 + 2} = \frac{8}{6.4}$$

$$I_L = 1.25 \text{ A}$$

This 1.25 A is the current through the 2-Ω R_L at the center of the unbalanced bridge in Fig. 10–6*a*. Furthermore, the amount of I_L for any value of R_L in Fig. 10–6*a* can be calculated from the equivalent circuit in Fig. 10–6*d*.

Figure 10–7 General forms for a voltage source or current source connected to a load R_L across terminals A and B. (*a*) Voltage source V with series R. (*b*) Current source I with parallel R. (*c*) Current source I with parallel conductance G.

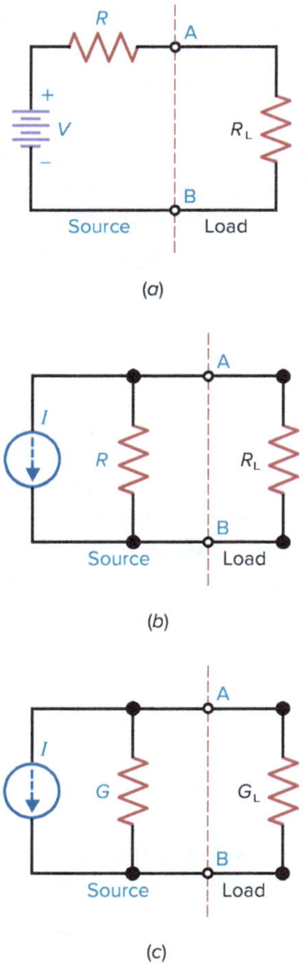

10–4 Self-Review

Answers at the end of the chapter.

In the Thevenin equivalent circuit in Fig. 10–6*d*,

a. **How much is R_T?**

b. **How much is V_{R_L}?**

10–5 Norton's Theorem

Named after E. L. Norton, an American scientist with Bell Telephone Laboratories, **Norton's theorem** is used to simplify a network in terms of currents instead of voltages. In many cases, analyzing the division of currents may be easier than voltage analysis. For current analysis, therefore, Norton's theorem can be used to reduce a network to a simple parallel circuit with a **current source**. The idea of a *current source* is that it supplies a total line current to be divided among parallel branches, corresponding to a *voltage source* applying a total voltage to be divided among series components. This comparison is illustrated in Fig. 10–7.

Example of a Current Source

A source of electric energy supplying voltage is often shown with a series resistance that represents the internal resistance of the source, as in Fig. 10–7*a*. This method corresponds to showing an actual voltage source, such as a battery for DC circuits. However, the source may also be represented as a current source with a parallel resistance, as in Fig. 10–7*b*. Just as a voltage source is rated at, say, 10 V, a current source may be rated at 2 A. For the purpose of analyzing parallel branches, the concept of a current source may be more convenient than the concept of a voltage source.

If the current I in Fig. 10–7*b* is a 2-A source, it supplies 2 A no matter what is connected across the output terminals A and B. Without anything connected across terminals A and B, all 2 A flows through the shunt R. When a load resistance R_L is connected across terminals A and B, then the 2-A I divides according to the current division rules for parallel branches.

GOOD TO KNOW

A current source symbol that uses a solid arrow indicates the direction of conventional current flow. A dashed or broken arrow indicates the direction of electron flow.

Figure 10–8 Any network in the block at the left can be reduced to the Norton equivalent parallel circuit at the right.

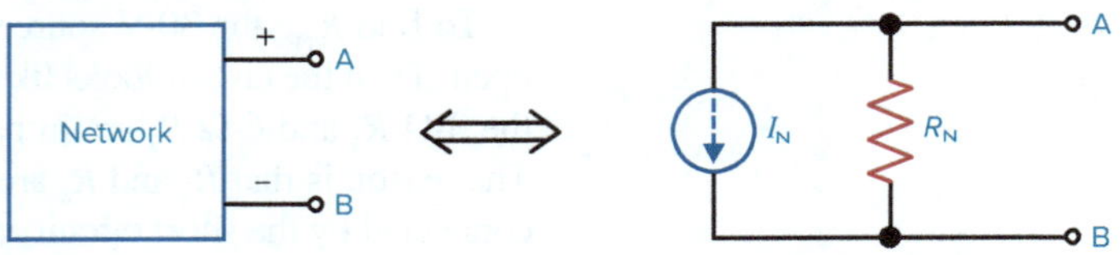

Remember that parallel currents divide inversely to branch resistances but directly with conductances. For this reason it may be preferable to consider the current source shunted by the conductance G, as shown in Fig. 10–7*c*. We can always convert between resistance and conductance because $1/R$ in ohms is equal to G in siemens.

The symbol for a current source is a circle with an arrow inside, as shown in Fig. 10–7*b* and *c*, to show the direction of current. This direction must be the same as the current produced by the polarity of the corresponding voltage source. Remember that a source produces electron flow out from the negative terminal.

An important difference between voltage and current sources is that a current source is killed by making it open, compared with short-circuiting a voltage source. Opening a current source kills its ability to supply current without affecting any parallel branches. A voltage source is short-circuited to kill its ability to supply voltage without affecting any series components.

The Norton Equivalent Circuit

As illustrated in Fig. 10–8, Norton's theorem states that the entire network connected to terminals A and B can be replaced by a single current source I_N in parallel with a single resistance R_N. The value of I_N is equal to the short-circuit current through the AB terminals. This means finding the current that the network would produce through A and B with a short circuit across these two terminals.

The value of R_N is the resistance looking back from open terminals A and B. These terminals are not short-circuited for R_N but are open, as in calculating R_{TH} for Thevenin's theorem. In fact, the single resistor is the same for both the Norton and Thevenin equivalent circuits. In the Norton case, this value of R_{AB} is R_N in parallel with the current source; in the Thevenin case, it is R_{TH} in series with the voltage source.

Nortonizing a Circuit

As an example, let us recalculate the current I_L in Fig. 10–9*a*, which was solved before by Thevenin's theorem. The first step in applying Norton's theorem is to imagine a short circuit across terminals A and B, as shown in Fig. 10–9*b*. How much current is flowing in the short circuit? Note that a short circuit across AB short-circuits R_L and the parallel R_2. Then the only resistance in the circuit is the 3-Ω R_1 in series with the 36-V source, as shown in Fig. 10–9*c*. The short-circuit current, therefore, is

$$I_N = \frac{36\ \text{V}}{3\ \Omega} = 12\ \text{A}$$

This 12-A I_N is the total current available from the current source in the Norton equivalent in Fig. 10–9*e*.

To find R_N, remove the short circuit across terminals A and B and consider the terminals open without R_L. Now the source V is considered short-circuited. As shown in Fig. 10–9*d*, the resistance seen looking back from terminals A and B is 6 Ω in parallel with 3 Ω, which equals 2 Ω for the value of R_N.

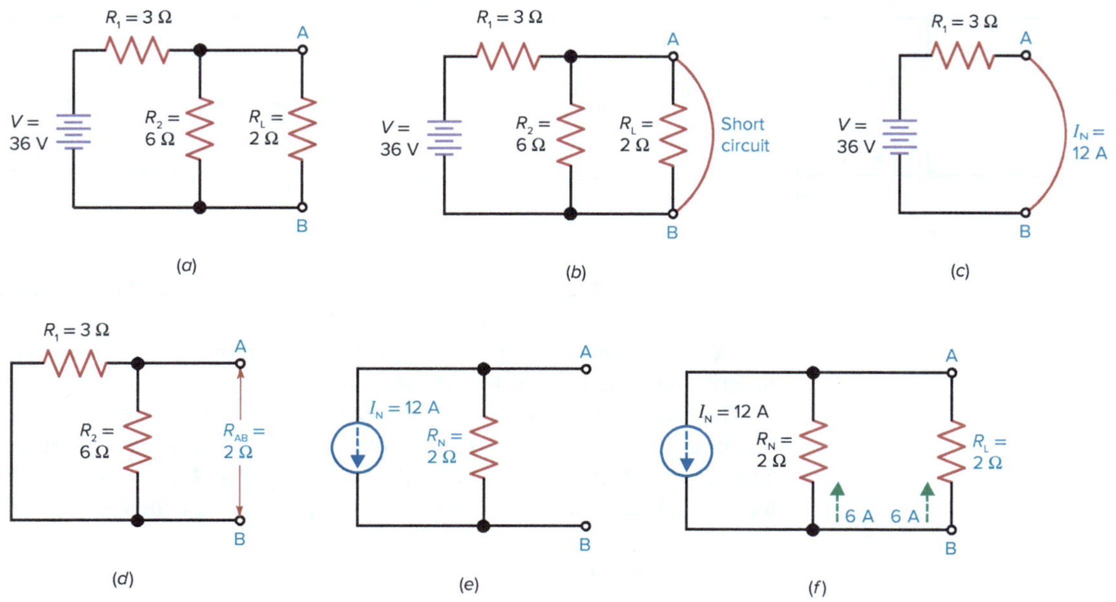

MultiSim **Figure 10–9** Same circuit as in Fig. 10–3, but solved by Norton's theorem. (*a*) Original circuit. (*b*) Short circuit across terminals A and B. (*c*) The short-circuit current I_N is $^{36}/_{3} = 12$ A. (*d*) Open terminals A and B but short-circuit V to find R_{AB} is 2 Ω, the same as R_{TH}. (*e*) Norton equivalent circuit. (*f*) R_L reconnected to terminals A and B to find that I_L is 6 A.

The resultant Norton equivalent is shown in Fig. 10–9*e*. It consists of a 12-A current source I_N shunted by the 2-Ω R_N. The arrow on the current source shows the direction of electron flow from terminal B to terminal A, as in the original circuit.

Finally, to calculate I_L, replace the 2-Ω R_L between terminals A and B, as shown in Fig. 10–9*f*. The current source still delivers 12 A, but now that current divides between the two branches of R_N and R_L. Since these two resistances are equal, the 12-A I_N divides into 6 A for each branch, and I_L is equal to 6 A. This value is the same current we calculated in Fig. 10–3, by Thevenin's theorem. Also, V_L can be calculated as $I_L R_L$, or 6 A × 2 Ω, which equals 12 V.

Looking at the Short-Circuit Current

In some cases, there may be a question of which current is I_N when terminals A and B are short-circuited. Imagine that a wire jumper is connected between A and B to short-circuit these terminals. Then I_N must be the current that flows in this wire between terminals A and B.

Remember that any components directly across these two terminals are also short-circuited by the wire jumper. Then these parallel paths have no effect. However, any components in series with terminal A or terminal B are in series with the wire jumper. Therefore, the short-circuit current I_N also flows through the series components.

An example of a resistor in series with the short circuit across terminals A and B is shown in Fig. 10–10. The idea here is that the short-circuit I_N is a branch current, not the main-line current. Refer to Fig. 10–10*a*. Here the short circuit connects R_3 across R_2. Also, the short-circuit current I_N is now the same as the current I_3 through R_3. Note that I_3 is only a branch current.

To calculate I_3, the circuit is solved by Ohm's law. The parallel combination of R_2 with R_3 equals $^{72}/_{18}$ or 4 Ω. The R_T is 4 + 4 = 8 Ω. As a result, the I_T from the source is 48 V/8 Ω = 6 A.

This I_T of 6 A in the main line divides into 4 A for R_2 and 2 A for R_3. The 2-A I_3 for R_3 flows through short-circuited terminals A and B. Therefore, this current of 2 A is the value of I_N.

Figure 10–10 Nortonizing a circuit where the short-circuit current I_N is a branch current. (*a*) I_N is 2 A through short-circuited terminals A and B and R_3. (*b*) $R_N = R_{AB} = 14.4\ \Omega$. (*c*) Norton equivalent circuit.

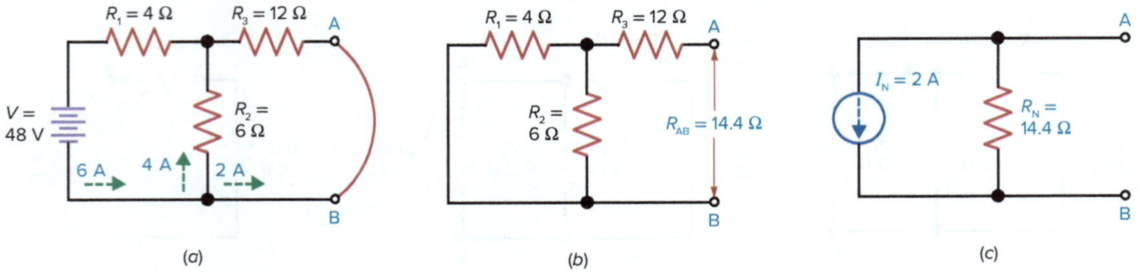

To find R_N in Fig. 10–10*b*, the short circuit is removed from terminals A and B. Now the source *V* is short-circuited. Looking back from open terminals A and B, the 4-Ω R_1 is in parallel with the 6-Ω R_2. This combination is $^{24}/_{10} = 2.4\ \Omega$. The 2.4 Ω is in series with the 12-Ω R_3 to make $R_{AB} = 2.4 + 12 = 14.4\ \Omega$.

The final Norton equivalent is shown in Fig. 10–10*c*. Current I_N is 2 A because this branch current in the original circuit is the current that flows through R_3 and short-circuited terminals A and B. Resistance R_N is 14.4 Ω looking back from open terminals A and B with the source *V* short-circuited the same way as for R_{TH}.

10–5 Self-Review

Answers at the end of the chapter.

a. For a Norton equivalent circuit, terminals A and B are short-circuited to find I_N. (True/False)

b. For a Norton equivalent circuit, terminals A and B are open to find R_N. (True/False)

10–6 Thevenin-Norton Conversions

Thevenin's theorem says that any network can be represented by a voltage source and series resistance, and Norton's theorem says that the same network can be represented by a current source and shunt resistance. It must be possible, therefore, to convert directly from a Thevenin form to a Norton form and vice versa. Such conversions are often useful.

Norton from Thevenin

Consider the Thevenin equivalent circuit in Fig. 10–11*a*. What is its Norton equivalent? Just apply Norton's theorem, the same as for any other circuit. The short-circuit current through terminals A and B is

$$I_N = \frac{V_{TH}}{R_{TH}} = \frac{15\ \text{V}}{3\ \Omega} = 5\ \text{A}$$

GOOD TO KNOW

An ideal current source is assumed to have an internal resistance of infinite ohms. Therefore, when calculating the Thevenin resistance, R_{TH}, it is only practical to consider a current source as an open circuit.

Figure 10–11 Thevenin equivalent circuit in (*a*) corresponds to the Norton equivalent in (*b*).

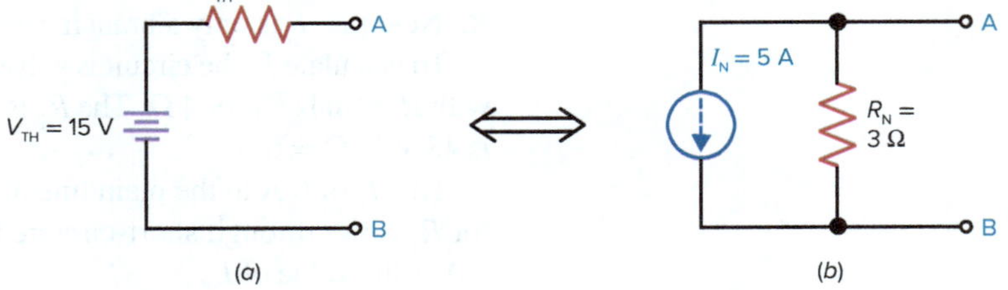

The resistance, looking back from open terminals A and B with the source V_{TH} short-circuited, is equal to the 3 Ω of R_{TH}. Therefore, the Norton equivalent consists of a current source that supplies the short-circuit current of 5 A, shunted by the same 3-Ω resistance that is in series in the Thevenin circuit. The results are shown in Fig. 10–11*b*.

Thevenin from Norton

For the opposite conversion, we can start with the Norton circuit of Fig. 10–11*b* and get back to the original Thevenin circuit. To do this, apply Thevenin's theorem, the same as for any other circuit. First, we find the Thevenin resistance by looking back from open terminals A and B. An important principle here, though, is that, although a voltage source is short-circuited to find R_{TH}, a current source is an open circuit. In general, a current source is killed by opening the path between its terminals. Therefore, we have just the 3-Ω R_N, in parallel with the infinite resistance of the open current source. The combined resistance then is 3 Ω.

In general, the resistance R_N always has the same value as R_{TH}. The only difference is that R_N is connected in parallel with I_N, but R_{TH} is in series with V_{TH}.

Now all that is required is to calculate the open-circuit voltage in Fig. 10–11*b* to find the equivalent V_{TH}. Note that with terminals A and B open, all current from the current source flows through the 3-Ω R_N. Then the open-circuit voltage across the terminals A and B is

$$I_N R_N = 5\text{ A} \times 3\ \Omega = 15\text{ V} = V_{TH}$$

As a result, we have the original Thevenin circuit, which consists of the 15-V source V_{TH} in series with the 3-Ω R_{TH}.

Conversion Formulas

In summary, the following formulas can be used for these conversions:

Thevenin from Norton:

$$R_{TH} = R_N$$
$$V_{TH} = I_N \times R_N$$

Norton from Thevenin:

$$R_N = R_{TH}$$
$$I_N = V_{TH}/R_{TH}$$

Another example of these conversions is shown in Fig. 10–12.

Figure 10–12 Example of Thevenin-Norton conversions. (*a*) Original circuit, the same as in Figs. 10–3*a* and 10–9*a*. (*b*) Thevenin equivalent. (*c*) Norton equivalent.

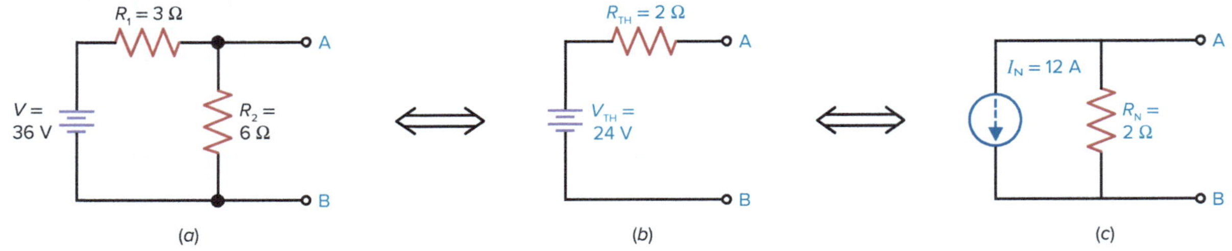

10–6 Self-Review

Answers at the end of the chapter.

a. In Thevenin-Norton conversions, resistances R_N and R_{TH} are equal. (True/False)

b. In Thevenin-Norton conversions, current I_N is V_{TH}/R_{TH}. (True/False)

c. In Thevenin-Norton conversions, voltage V_{TH} is $I_N \times R_N$. (True/False)

10–7 Conversion of Voltage and Current Sources

Norton conversion is a specific example of the general principle that any voltage source with its series resistance can be converted to an equivalent current source with the same resistance in parallel. In Fig. 10–13, the voltage source in Fig. 10–13*a* is equivalent to the current source in Fig. 10–13*b*. Just divide the source *V* by its series *R* to calculate the value of *I* for the equivalent current source shunted by the same *R*. Either source will supply the same current and voltage for any components connected across terminals A and B.

Conversion of voltage and current sources can often simplify circuits, especially those with two or more sources. Current sources are easier for parallel connections, where we can add or divide currents. Voltage sources are easier for series connections, where we can add or divide voltages.

Two Sources in Parallel Branches

In Fig. 10–14*a*, assume that the problem is to find I_3 through the middle resistor R_3. Note that V_1 with R_1 and V_2 with R_2 are branches in parallel with R_3. All three branches are connected across terminals A and B.

When we convert V_1 and V_2 to current sources in Fig. 10–14*b*, the circuit has all parallel branches. Current I_1 is $^{84}/_{12}$ or 7 A, and I_2 is $^{21}/_{3}$ which also happens to be 7 A. Current I_1 has its parallel *R* of 12 Ω, and I_2 has its parallel *R* of 3 Ω.

Furthermore, I_1 and I_2 can be combined for the one equivalent current source I_T in Fig. 10–14*c*. Since both sources produce current in the same direction through R_L, they are added for $I_T = 7 + 7 = 14$ A.

The shunt *R* for the 14-A combined source is the combined resistance of the 12-Ω R_1 and the 3-Ω R_2 in parallel. This *R* equals $^{36}/_{15}$ or 2.4 Ω, as shown in Fig. 10–14*c*.

To find I_L, we can use the current divider formula for the 6- and 2.4-Ω branches, into which the 14-A I_T from the current source was split. Then

$$I_L = \frac{2.4}{2.4 + 6} \times 14 = 4 \text{ A}$$

GOOD TO KNOW

Many electronic circuits contain both voltage and current sources.

Figure 10–13 The voltage source in (*a*) corresponds to the current source in (*b*).

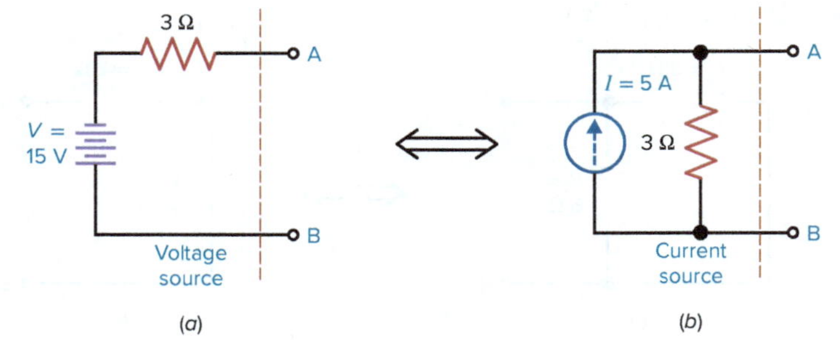

Figure 10–14 Converting two voltage sources in V_1 and V_2 in parallel branches to current sources I_1 and I_2 that can be combined. (*a*) Original circuit. (*b*) V_1 and V_2 converted to parallel current sources I_1 and I_2. (*c*) Equivalent circuit with one combined current source I_T.

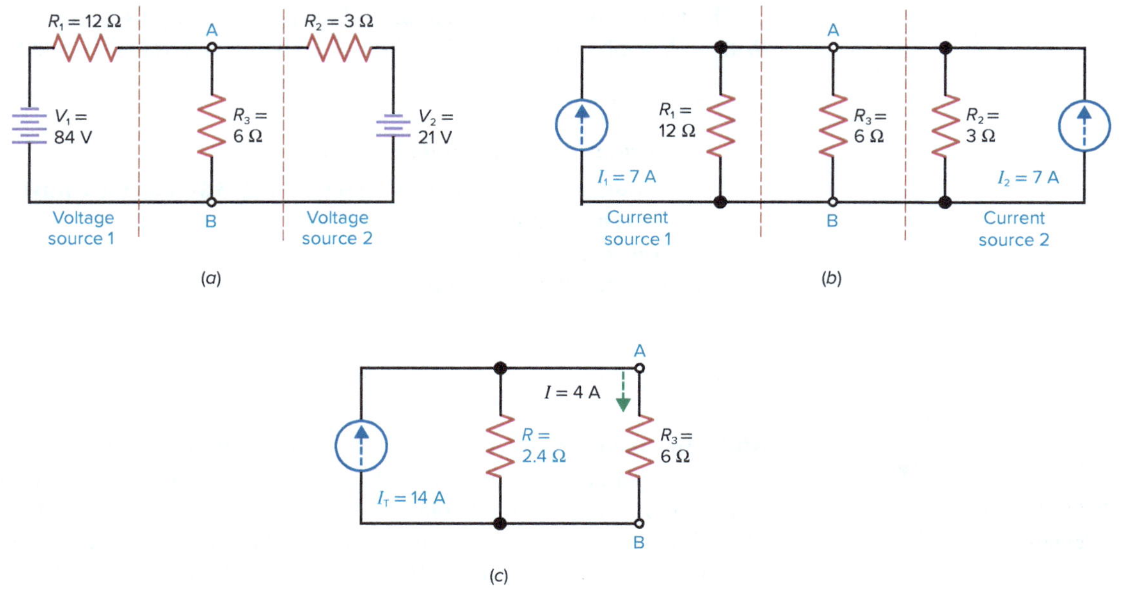

The voltage V_{R_3} across terminals A and B is $I_L R_L$, which equals $4 \times 6 = 24$ V. These are the same values calculated for V_{R_3} and I_3 by Kirchhoff's laws in Fig. 9–5 and by Thevenin's theorem in Fig. 10–5.

Two Sources in Series

Referring to Fig. 10–15, assume that the problem is to find the current I_L through the load resistance R_L between terminals A and B. This circuit has the two current sources I_1 and I_2 in series.

The problem here can be simplified by converting I_1 and I_2 to the series voltage sources V_1 and V_2, as shown in Fig. 10–15*b*. The 2-A I_1 with its shunt 4-Ω R_1 is equivalent to 4×2, or 8 V, for V_1 with a 4-Ω series resistance. Similarly, the 5-A I_2 with its shunt 2-Ω R_2 is equivalent to 5×2, or 10 V, for V_2 with a 2-Ω series resistance. The polarities of V_1 and V_2 produce electron flow in the same direction as I_1 and I_2.

The series voltages can now be combined as in Fig. 10–15*c*. The 8 V of V_1 and 10 V of V_2 are added because they are series-aiding, resulting in the total V_T of 18 V,

Figure 10–15 Converting two current sources I_1 and I_2 in series to voltage sources V_1 and V_2 that can be combined. (*a*) Original circuit. (*b*) I_1 and I_2 converted to series voltage sources V_1 and V_2. (*c*) Equivalent circuit with one combined voltage source V_T.

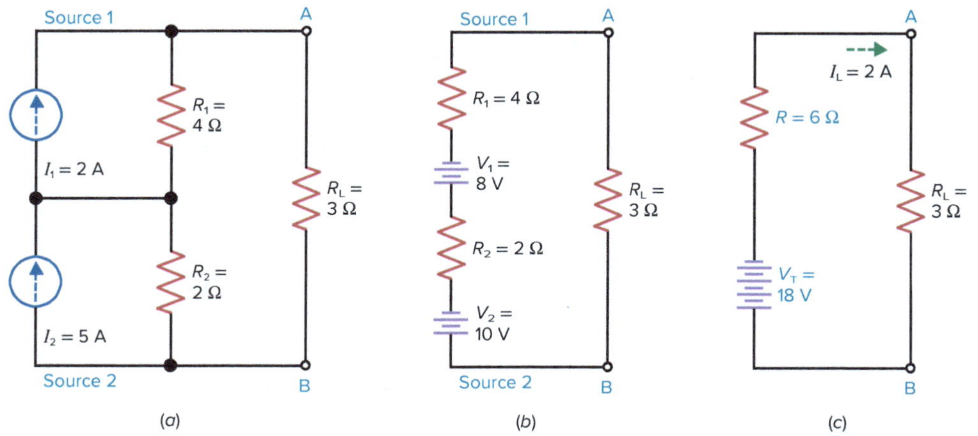

and, the resistances of 4 Ω for R_1 and 2 Ω for R_2 are added, for a combined R of 6 Ω. This is the series resistance of the 18-V source V_T connected across terminals A and B.

The total resistance of the circuit in Fig. 10–15*c* is R plus R_L, or 6 + 3 = 9 Ω. With 18 V applied, $I_L = 18/9 = 2$ A through R_L between terminals A and B.

10–7 Self-Review

Answers at the end of the chapter.

A voltage source has 21 V in series with 3 Ω. For the equivalent current source,

a. **How much is *I*?**
b. **How much is the shunt *R*?**

GOOD TO KNOW

When the values of I and G are known in any circuit, the voltage, V, can be calculated as $V = \frac{I}{G}$.

10–8 Millman's Theorem

Millman's theorem provides a shortcut for finding the common voltage across any number of parallel branches with different voltage sources. A typical example is shown in Fig. 10–16. For all branches, the ends at point Y are connected to chassis ground. Furthermore, the opposite ends of all branches are also connected to the common point X. The voltage V_{XY}, therefore, is the common voltage across all branches.

Finding the value of V_{XY} gives the net effect of all sources in determining the voltage at X with respect to chassis ground. To calculate this voltage,

$$V_{XY} = \frac{V_1/R_1 + V_2/R_2 + V_3/R_3}{1/R_1 + 1/R_2 + 1/R_3} \cdots \text{etc.} \tag{10–1}$$

This formula is derived by converting the voltage sources to current sources and combining the results. The numerator with V/R terms is the sum of the parallel current sources. The denominator with $1/R$ terms is the sum of the parallel conductances. The net V_{XY}, then, is the form of I/G or $I \times R$, which is in units of voltage.

Calculating V_{XY}

For the values in Fig. 10–16,

$$V_{XY} = \frac{32/4 + 0/2 - 8/4}{1/4 + 1/2 + 1/4}$$

$$= \frac{8 + 0 - 2}{1}$$

$$V_{XY} = 6 \text{ V}$$

Figure 10–16 Example of Millman's theorem to find V_{XY}, the common voltage across branches with separate voltage sources.

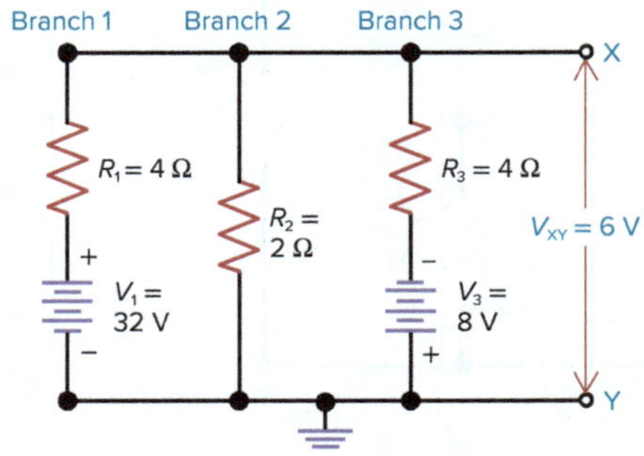

Figure 10–17 The same circuit as in Fig. 9–5 for Kirchhoff's laws, but shown with parallel branches to calculate V_{XY} by Millman's theorem.

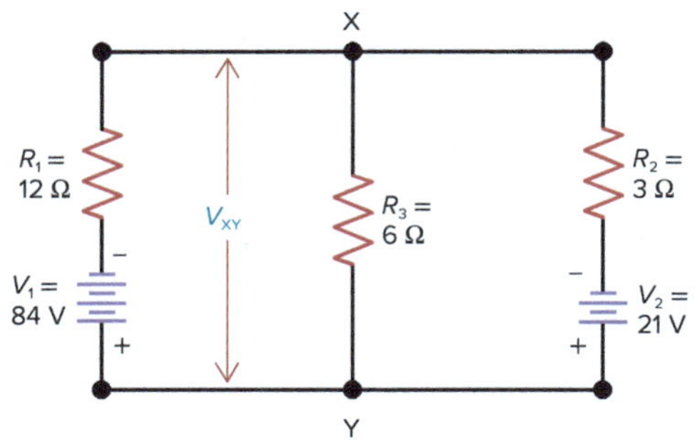

Note that in branch 3, V_3 is considered negative because it would make point X negative. However, all resistances are positive. The positive answer for V_{XY} means that point X is positive with respect to Y.

In branch 2, V_2 is zero because this branch has no voltage source. However, R_2 is still used in the denominator.

This method can be used for any number of branches, but all must be in parallel without any series resistances between branches. In a branch with several resistances, they can be combined as one R_T. When a branch has more than one voltage source, the voltages can be combined algebraically for one V_T.

Applications of Millman's Theorem

In many cases, a circuit can be redrawn to show the parallel branches and their common voltage V_{XY}. Then with V_{XY} known, the entire circuit can be analyzed quickly. For instance, Fig. 10–17 has been solved before by other methods. For Millman's theorem, the common voltage V_{XY} across all branches is the same as V_3 across R_3. This voltage is calculated with Formula (10–1), as follows:

$$V_{XY} = \frac{-84/12 + 0/6 - 21/3}{1/12 + 1/6 + 1/3} = \frac{-7 + 0 - 7}{7/12}$$

$$= \frac{-14}{7/12} = -14 \times \frac{12}{7}$$

$$V_{XY} = -24\text{ V} = V_3$$

The negative sign means that point X is the negative side of V_{XY}.

With V_3 known to be 24 V across the 6-Ω R_3, then I_3 must be $24/6 = 4$ A. Similarly, all voltages and currents in this circuit can then be calculated. (See Fig. 9–6 in Chapter 9.)

As another application, the example of superposition in Fig. 10–1 has been redrawn in Fig. 10–18 to show the parallel branches with a common voltage V_{XY} to be calculated by Millman's theorem. Then

$$V_{XY} = \frac{24\text{ V}/30\text{ k}\Omega - 9\text{ V}/60\text{ k}\Omega}{1/(30\text{ k}\Omega) + 1/(60\text{ k}\Omega)} = \frac{0.8\text{ mA} - 0.15\text{ mA}}{3/(60\text{ k}\Omega)}$$

$$= 0.65 \times \frac{60}{3} = \frac{39}{3}$$

$$V_{XY} = 13\text{ V} = V_P$$

The answer of 13 V from point P to ground, using Millman's theorem, is the same value calculated before by superposition.

Figure 10–18 Same circuit as in Fig. 10–1 for superposition, but shown with parallel branches to calculate V_{XY} by Millman's theorem.

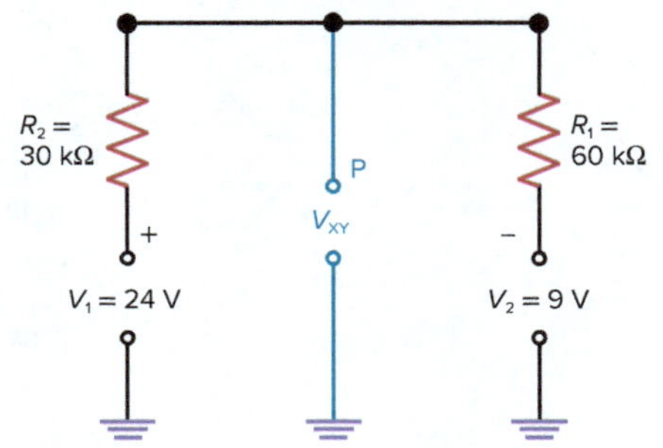

10–8 Self-Review

Answers at the end of the chapter.

For the example of Millman's theorem in Fig. 10–16,

a. **How much is V_{R_2}?**
b. **How much is V_{R_3}?**

10–9 T or Y and π or Δ Connections

The circuit in Fig. 10–19 is called a T (tee) or Y (wye) network, as suggested by the shape. They are different names for the same network; the only difference is that the R_2 and R_3 legs are shown at an angle in the Y.

The circuit in Fig. 10–20 is called a π (pi) or Δ (delta) network because the shape is similar to these Greek letters. Both forms are the same network. Actually, the network can have the R_A arm shown at the top or bottom, as long as it is connected between R_B and R_C. In Fig. 10–20, R_A is at the top, as an inverted delta, to look like the π network.

The circuits in Figs. 10–19 and 10–20 are passive networks without any energy sources. They are also three-terminal networks with two pairs of connections for

GOOD TO KNOW

You can expect to encounter delta (Δ) and wye (Y) networks when studying circuits involving three-phase AC power.

Figure 10–19 The form of a T or Y network.

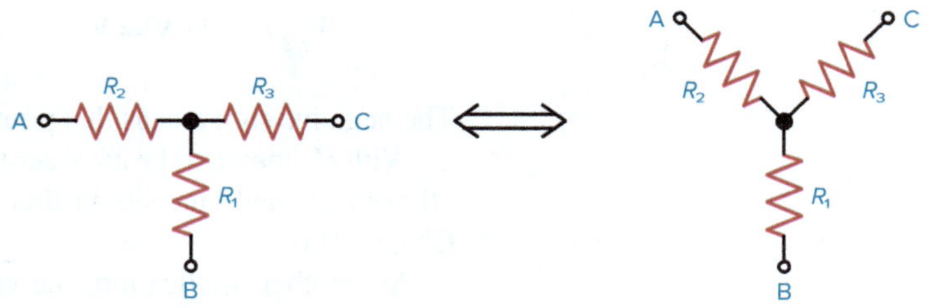

Figure 10–20 The form of a π or Δ network.

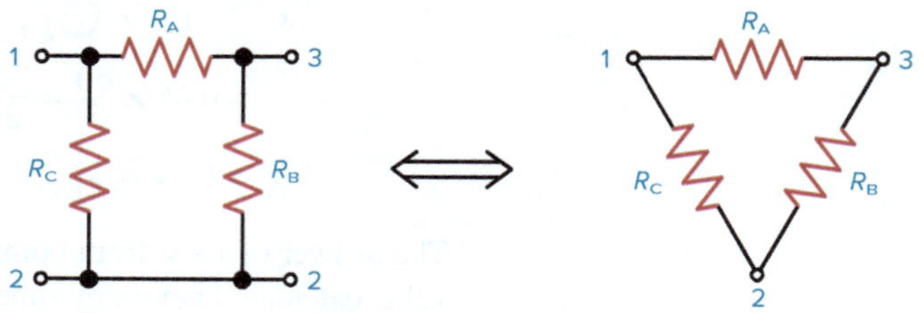

input and output voltages with one common. In Fig. 10–19, point B is the common terminal and point 2 is common in Fig. 10–20.

The Y and Δ forms are different ways to connect three resistors in a passive network. Note that resistors in the Y are labeled with subscripts 1, 2, and 3, whereas the Δ has subscripts A, B, and C to emphasize the different connections.

Conversion Formulas

In the analysis of networks, it is often helpful to convert a Δ to Y or vice versa. Either it may be difficult to visualize the circuit without the conversion, or the conversion makes the solution simpler. The formulas for these transformations are given here. All are derived from Kirchhoff's laws. Note that letters are used as subscripts for R_A, R_B, and R_C in the Δ, whereas the resistances are numbered R_1, R_2, and R_3 in the Y.

Conversions of Y to Δ or T to π:

$$R_A = \frac{R_1R_2 + R_2R_3 + R_3R_1}{R_1}$$

$$R_B = \frac{R_1R_2 + R_2R_3 + R_3R_1}{R_2} \qquad \textbf{(10–2)}$$

$$R_C = \frac{R_1R_2 + R_2R_3 + R_3R_1}{R_3}$$

or

$$R_\Delta = \frac{\Sigma \text{ all cross products in Y}}{\text{opposite } R \text{ in Y}}$$

These formulas can be used to convert a Y network to an equivalent Δ or a T network to π. Both networks will have the same resistance across any pair of terminals.

The three formulas have the same general form, indicated at the bottom as one basic rule. The symbol Σ is the Greek capital letter sigma, meaning "sum of."

For the opposite conversion,

Conversion of Δ to Y or π to T:

$$R_1 = \frac{R_BR_C}{R_A + R_B + R_C}$$

$$R_2 = \frac{R_CR_A}{R_A + R_B + R_C} \qquad \textbf{(10–3)}$$

$$R_3 = \frac{R_AR_B}{R_A + R_B + R_C}$$

or

$$R_Y = \frac{\text{product of two adjacent } R \text{ in } \Delta}{\Sigma \text{ all } R \text{ in } \Delta}$$

Figure 10–21 Conversion between Y and Δ networks. See text for conversion formulas.

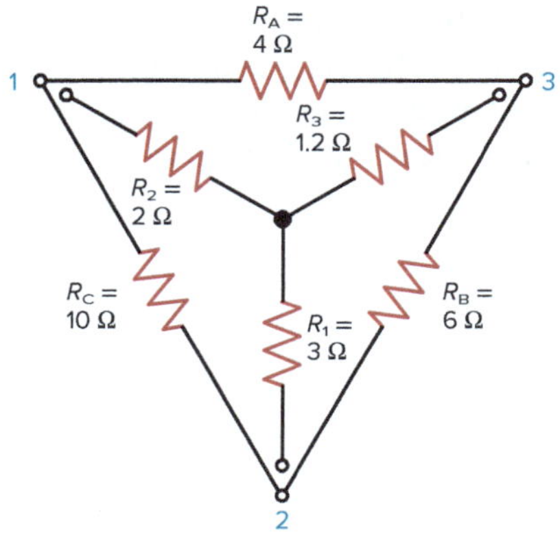

As an aid in using these formulas, the following scheme is useful. Place the Y inside the Δ, as shown in Fig. 10–21. Notice that the Δ has three closed sides, and the Y has three open arms. Also note how resistors can be considered opposite each other in the two networks. For instance, the open arm R_1 is opposite the closed side R_A, R_2 is opposite R_B, and R_3 is opposite R_C.

Furthermore, each resistor in an open arm has two adjacent resistors in the closed sides. For R_1, its adjacent resistors are R_B and R_C, also R_C and R_A are adjacent to R_2, and R_A and R_B are adjacent to R_3.

In the formulas for the Y-to-Δ conversion, each side of the delta is found by first taking all possible cross products of the arms of the wye, using two arms at a time. There are three such cross products. The sum of the three cross products is then divided by the opposite arm to find the value of each side of the delta. Note that the numerator remains the same, the sum of the three cross products. However, each side of the delta is calculated by dividing this sum by the opposite arm.

For the Δ-to-Y conversion, each arm of the wye is found by taking the product of the two adjacent sides in the delta and dividing by the sum of the three sides of the delta. The product of two adjacent resistors excludes the opposite resistor. The denominator for the sum of the three sides remains the same in the three formulas. However, each arm is calculated by dividing the sum into each cross product.

Example of Conversion

The values shown for the equivalent Y and Δ in Fig. 10–21 are calculated as follows: Starting with 4, 6, and 10 Ω for sides R_A, R_B, and R_C, respectively, in the delta, the corresponding arms in the wye are

$$R_1 = \frac{R_B R_C}{R_A + R_B + R_C} = \frac{6 \times 10}{4 + 6 + 10} = \frac{60}{20} = 3\ \Omega$$

$$R_2 = \frac{R_C R_A}{20} = \frac{10 \times 4}{20} = \frac{40}{20} = 2\ \Omega$$

$$R_3 = \frac{R_A R_B}{20} = \frac{4 \times 6}{20} = \frac{24}{20} = 1.2\ \Omega$$

As a check on these values, we can calculate the equivalent delta for this wye. Starting with values of 3, 2, and 1.2 Ω for R_1, R_2, and R_3, respectively, in the wye, the corresponding values in the delta are:

$$R_A = \frac{R_1 R_2 + R_2 R_3 + R_3 R_1}{R_1} = \frac{6 + 2.4 + 3.6}{3} = \frac{12}{3} = 4\ \Omega$$

$$R_B = \frac{12}{R_2} = \frac{12}{2} = 6\ \Omega$$

$$R_C = \frac{12}{R_3} = \frac{12}{1.2} = 10\ \Omega$$

These results show that the Y and Δ networks in Fig. 10–21 are equivalent to each other when they have the values obtained with the conversion formulas.

Note that the equivalent *R* values in the Y are less than those in the equivalent Δ network. The reason is that the Y has two legs between the terminals, whereas the Δ has only one.

Simplifying a Bridge Circuit

As an example of the use of such transformations, consider the bridge circuit of Fig. 10–22. The total current I_T from the battery is desired. Therefore, we must find the total resistance R_T.

One approach is to note that the bridge in Fig. 10–22*a* consists of two deltas connected between terminals P_1 and P_2. One of them can be replaced by an equivalent wye. We use the bottom delta with R_A across the top, in the same form as Fig. 10–21. We then replace this delta $R_A R_B R_C$ by an equivalent wye $R_1 R_2 R_3$, as shown in Fig. 10–22*b*. Using the conversion formulas,

$$R_1 = \frac{R_B R_C}{R_A + R_B + R_C} = \frac{24}{12} = 2\ \Omega$$

$$R_2 = \frac{R_C R_A}{12} = \frac{12}{12} = 1\ \Omega$$

$$R_3 = \frac{R_A R_B}{12} = \frac{8}{12} = \frac{2}{3}\ \Omega$$

Figure 10–22 Solving a bridge circuit by Δ-to-Y conversion. (*a*) Original circuit. (*b*) How the Y of $R_1R_2R_3$ corresponds to the Δ of $R_AR_BR_C$. (*c*) The Y substituted for the Δ network. The result is a series-parallel circuit with the same R_T as the original bridge circuit. (*d*) R_T is 4.5 Ω between points P_3 and P_4.

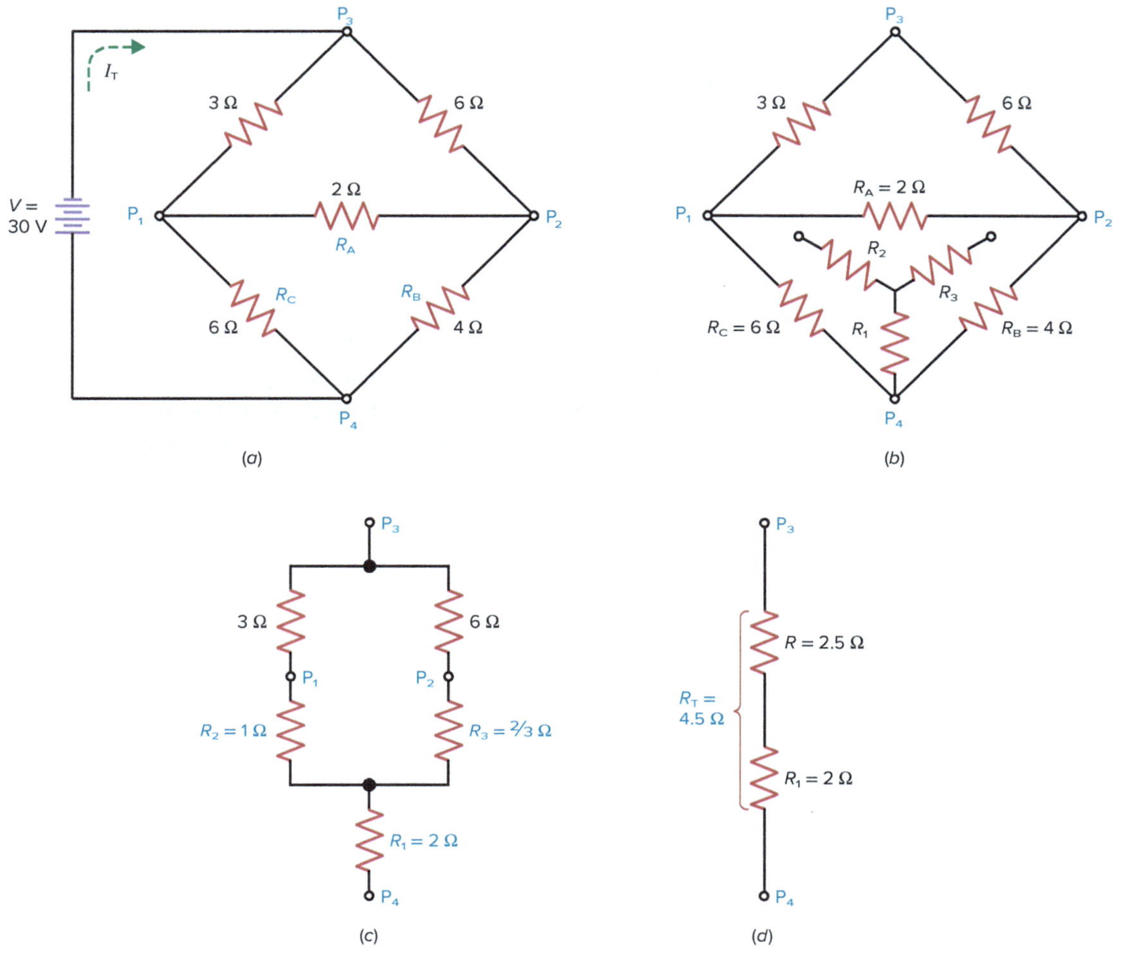

We next use these values for R_1, R_2, and R_3 in an equivalent wye to replace the original delta. Then the resistances form the series-parallel circuit shown in Fig. 10–22*c*. The combined resistance of the two parallel branches here is 4 × 6.67 divided by 10.67, which equals 2.5 Ω for *R*. Adding this 2.5 Ω to the series R_1 of 2 Ω, the total resistance is 4.5 Ω in Fig. 10–22*d*.

This 4.5 Ω is the R_T for the entire bridge circuit between P_3 and P_4 connected to source *V*. Then I_T is 30 V/4.5 Ω, which equals 6.67 A supplied by the source.

Another approach to finding R_T for the bridge circuit in Fig. 10–22*a* is to recognize that the bridge also consists of two T or Y networks between terminals P_3 and P_4. One of them can be transformed into an equivalent delta. The result is another series-parallel circuit but with the same R_T of 4.5 Ω.

Balanced Networks

When all the *R* values are equal in a network, it is balanced. Then the conversion is simplified, as *R* in the wye network is one-third the *R* in the equivalent delta. As an example, for $R_A = R_B = R_C$ equal to 6 Ω in the delta, the equivalent wye has $R_1 = R_2 = R_3$ equal to 6/3 or 2 Ω. Or, converting the other way, for 2-Ω *R* values in a balanced wye, the equivalent delta network has each *R* equal to 3 × 2 = 6 Ω. This example is illustrated in Fig. 10–23.

GOOD TO KNOW

In a balanced Network $R_Y = \dfrac{R_\Delta}{3}$ and $R_\Delta = 3R_Y$.

Figure 10–23 Equivalent balanced networks. (*a*) Delta form. (*b*) Wye form.

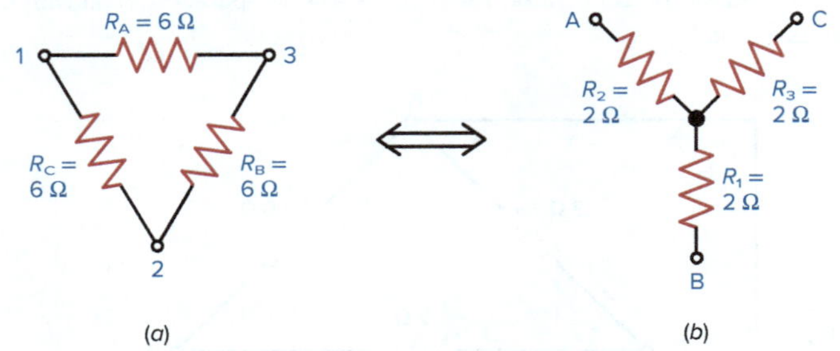

10–9 Self-Review

Answers at the end of the chapter.

In the standard form for conversion,

a. Which resistor in the Y is opposite R_A in the Δ?

b. Which two resistors in the Δ are adjacent to R_1, in the Y?

Summary

- *Superposition theorem*. In a linear, bilateral network having more than one source, the current and voltage in any part of the network can be found by adding algebraically the effect of each source separately. All other sources are temporarily killed by short-circuiting voltage sources and opening current sources.
- *Thevenin's theorem*. Any network with two open terminals A and B can be replaced by a single voltage source V_{TH} in series with a single resistance R_{TH} connected to terminals A and B. Voltage V_{TH} is the voltage produced by the network across terminals A and B. Resistance R_{TH} is the resistance across open terminals A and B with all voltage sources short-circuited.
- *Norton's theorem*. Any two-terminal network can be replaced by a single current source I_N in parallel with a single resistance R_N. The value of I_N is the current produced by the network through the short-circuited terminals. R_N is the resistance across the open terminals with all voltage sources short-circuited.
- *Millman's theorem*. The common voltage across parallel branches with different V sources can be determined with Formula (10–1).
- A voltage source V with its series R can be converted to an equivalent current source I with parallel R. Similarly, a current source I with a parallel R can be converted to a voltage source V with a series R. The value of I is V/R, or V is $I \times R$. The value of R is the same for both sources. However, R is in series with V but in parallel with I.
- The conversion between delta and wye networks is illustrated in Fig. 10–21. To convert from one network to the other, Formula (10–2) or (10–3) is used.

Important Terms

Active components — electronic components such as diodes and transistors that can rectify or amplify.

Bilateral components — electronic components that have the same current for opposite polarities of applied voltage.

Current source — a source that supplies a total current to be divided among parallel branches.

Linear component — an electronic component whose current is proportional to the applied voltage.

Millman's theorem — a theorem that provides a shortcut for finding the common voltage across any number of parallel branches with different voltage sources.

Norton's theorem — a theorem that states that an entire network connected to a pair of terminals can be replaced by a single current source, I_N, in parallel with a single resistance, R_N.

Passive components — electronic components that do not amplify or rectify.

Superposition theorem — a theorem that states that in a network with two or more sources, the current or voltage for any component is the algebraic sum of the effects produced by each source acting separately.

Thevenin's theorem — a theorem that states that an entire network connected to a pair of terminals can be replaced by a single voltage source, V_{TH}, in series with a single resistance, R_{TH}.

Voltage source — a source that supplies a total voltage to be divided among series components.

Related Formulas

Millman's theorem

$$V_{XY} = \frac{V_1/R_1 + V_2/R_2 + V_3/R_3}{1/R_1 + 1/R_2 + 1/R_3} \cdots \text{etc.}$$

Conversion of Y to Δ or T to π:

$$R_A = \frac{R_1R_2 + R_2R_3 + R_3R_1}{R_1}$$

$$R_B = \frac{R_1R_2 + R_2R_3 + R_3R_1}{R_2}$$

$$R_C = \frac{R_1R_2 + R_2R_3 + R_3R_1}{R_3}$$

or $$R_\Delta = \frac{\Sigma \text{ all cross products in Y}}{\text{opposite } R \text{ in Y}}$$

Conversion of Δ to Y or π to T.

$$R_1 = \frac{R_BR_C}{R_A + R_B + R_C}$$

$$R_2 = \frac{R_CR_A}{R_A + R_B + R_C}$$

$$R_3 = \frac{R_AR_B}{R_A + R_B + R_C}$$

or $$R_Y = \frac{\text{product of two adjacent } R \text{ in } \Delta}{\Sigma \text{ all } R \text{ in } \Delta}$$

Self-Test

Answers at the back of the book.

1. A resistor is an example of a(n)

a. bilateral component.
b. active component.
c. passive component.
d. both a and c.

2. To apply the superposition theorem, all components must be

a. the active type.
b. both linear and bilateral.
c. grounded.
d. both nonlinear and unidirectional.

3. When converting from a Norton equivalent circuit to a Thevenin equivalent circuit or vice versa,

a. R_N and R_{TH} have the same value.
b. R_N will always be larger than R_{TH}.
c. I_N is short-circuited to find V_{TH}.
d. V_{TH} is short-circuited to find I_N.

4. When solving for the Thevenin equivalent resistance, R_{TH},

a. all voltage sources must be opened.
b. all voltage sources must be short-circuited.
c. all voltage sources must be converted to current sources.
d. none of the above.

5. Thevenin's theorem states that an entire network connected to a pair of terminals can be replaced with

a. a single current source in parallel with a single resistance.
b. a single voltage source in parallel with a single resistance.
c. a single voltage source in series with a single resistance.
d. a single current source in series with a single resistance.

6. Norton's theorem states that an entire network connected to a pair of terminals can be replaced with

a. a single current source in parallel with a single resistance.
b. a single voltage source in parallel with a single resistance.
c. a single voltage source in series with a single resistance.
d. a single current source in series with a single resistance.

7. With respect to terminals A and B in a complex network, the Thevenin voltage, V_{TH}, is

a. the voltage across terminals A and B when they are short-circuited.
b. the open-circuit voltage across terminals A and B.
c. the same as the voltage applied to the complex network.
d. none of the above.

8. A Norton equivalent circuit consists of a 100-μA current source, I_N, in parallel with a 10-kΩ resistance, R_N. If this circuit is converted to a Thevenin equivalent circuit, how much is V_{TH}?

a. 1 kV.
b. 10 V.
c. 1 V.
d. It cannot be determined.

9. With respect to terminals A and B in a complex network, the Norton current, I_N, equals

a. the current flowing between terminals A and B when they are open.
b. the total current supplied by the applied voltage to the network.
c. zero when terminals A and B are short-circuited.
d. the current flowing between terminals A and B when they are short-circuited.

10. Which theorem provides a shortcut for finding the common voltage across any number of parallel branches with different voltage sources?

a. The superposition theorem.
b. Thevenin's theorem.
c. Norton's theorem.
d. Millman's theorem.

Essay Questions

1. State the superposition theorem, and discuss how to apply it.
2. State how to calculate V_{TH} and R_{TH} in Thevenin equivalent circuits.
3. State the method of calculating I_N and R_N for a Norton equivalent circuit.
4. How is a voltage source converted to a current source, and vice versa?
5. For what type of circuit is Millman's theorem used?
6. Draw a delta network and a wye network, and give the six formulas needed to convert from one to the other.

Problems

SECTION 10–1 SUPERPOSITION THEOREM

10–1 MultiSim In Fig. 10–24, use the superposition theorem to solve for the voltage, V_P, with respect to ground.

10–2 MultiSim In Fig. 10–25, use the superposition theorem to solve for the voltage, V_P, with respect to ground.

10–3 In Fig. 10–25, recalculate the voltage, V_P, if the resistors R_1 and R_2 are interchanged.

10–4 In Fig. 10–26, use the superposition theorem to solve for the voltage, V_{AB}.

Figure 10–24

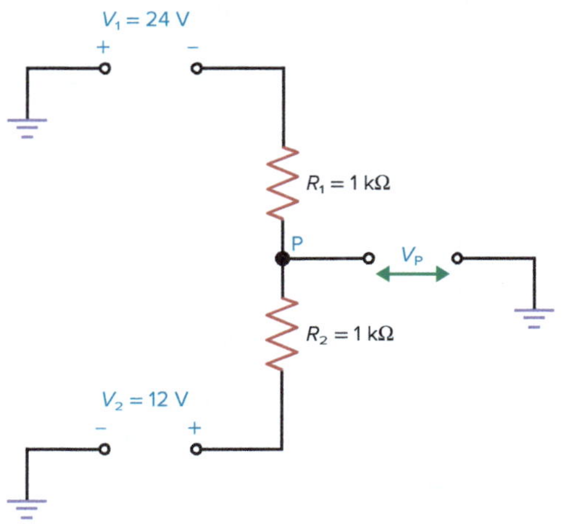

Figure 10–25

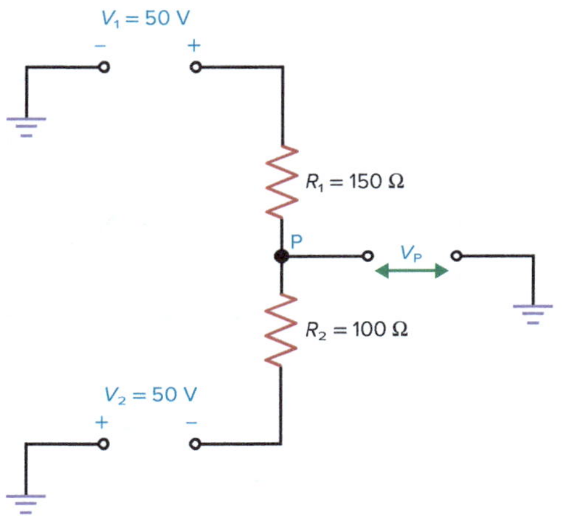

10–5 In Fig. 10–26, recalculate the voltage, V_{AB}, if the polarity of V_2 is reversed.

Figure 10–26

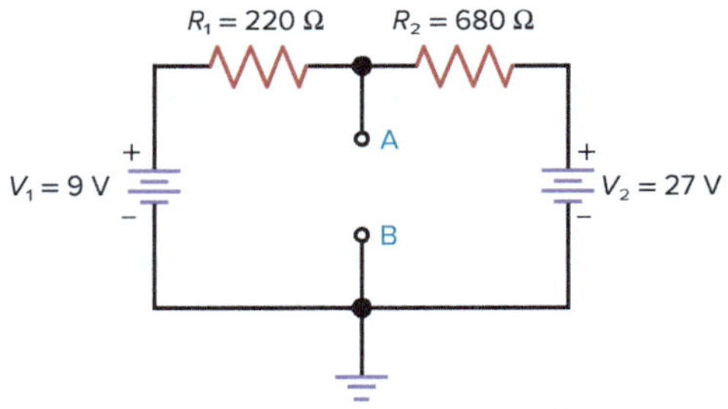

SECTION 10–2 THEVENIN'S THEOREM

10–6 MultiSim In Fig. 10–27, draw the Thevenin equivalent circuit with respect to terminals A and B (mentally remove R_L).

Figure 10–27

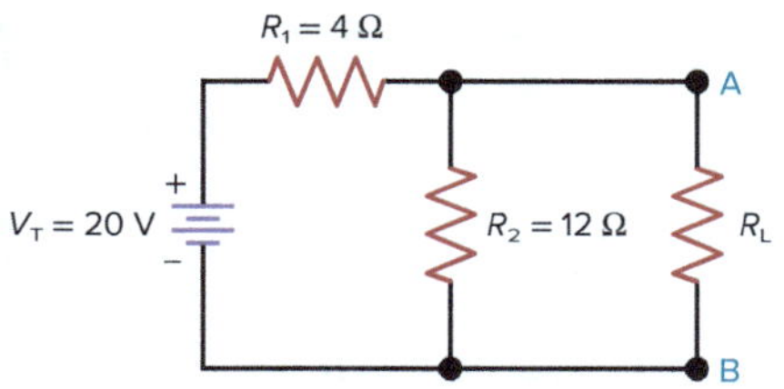

10–7 In Fig. 10–27, use the Thevenin equivalent circuit to calculate I_L and V_L for the following values of R_L: $R_L = 3\ \Omega$, $R_L = 6\ \Omega$, and $R_L = 12\ \Omega$.

10–8 MultiSim In Fig. 10–28, draw the Thevenin equivalent circuit with respect to terminals A and B (mentally remove R_L).

Figure 10–28

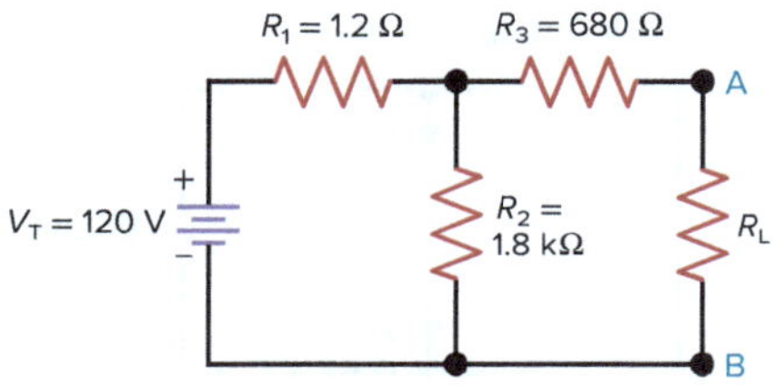

10–9 In Fig. 10–28, use the Thevenin equivalent circuit to calculate I_L and V_L for the following values of R_L: $R_L = 100\ \Omega$, $R_L = 1$ kΩ, and $R_L = 5.6$ kΩ.

10–10 In Fig. 10–29, draw the Thevenin equivalent circuit with respect to terminals A and B (mentally remove R_L).

Figure 10–29

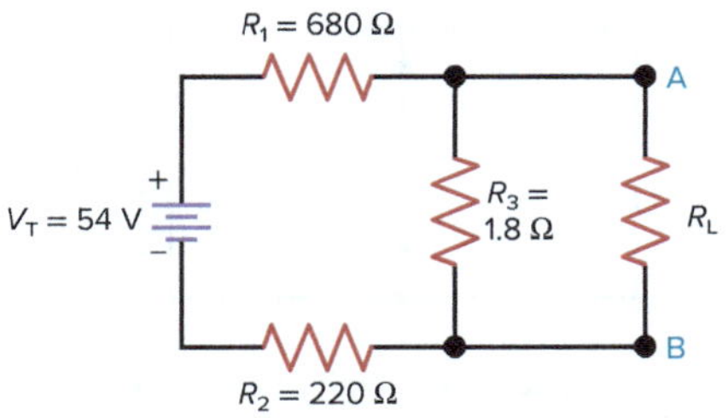

10–11 In Fig. 10–29, use the Thevenin equivalent circuit to calculate I_L and V_L for the following values of R_L: $R_L = 200\ \Omega$, $R_L = 1.2$ kΩ, and $R_L = 1.8$ k Ω.

10–12 In Fig. 10–30, draw the Thevenin equivalent circuit with respect to terminals A and B (mentally remove R_L).

Figure 10–30

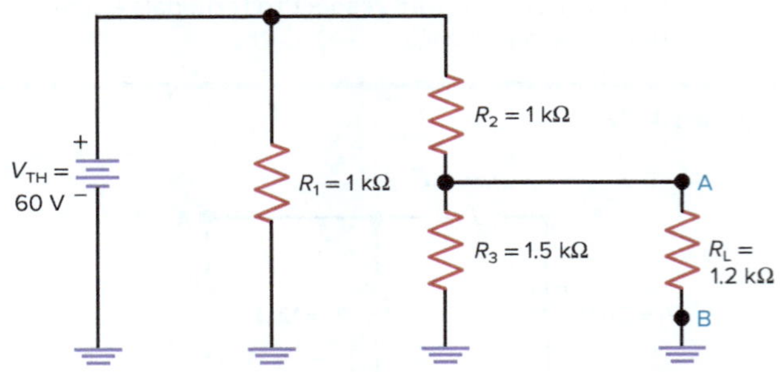

10–13 In Fig. 10–30, use the Thevenin equivalent circuit to solve for I_L and R_L.

SECTION 10–3 THEVENIZING A CIRCUIT WITH TWO VOLTAGE SOURCES

10–14 In Fig. 10–31, draw the Thevenin equivalent circuit with respect to terminals A and B (mentally remove R_3).

Figure 10–31

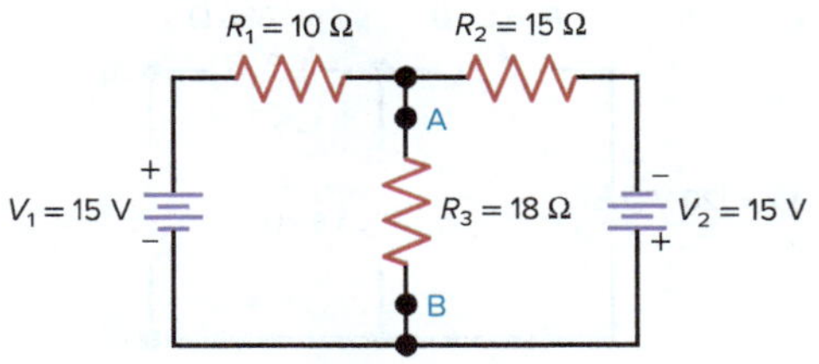

10–15 Using the Thevenin equivalent circuit for Fig. 10–31, calculate the values for I_3 and V_{R_3}.

10–16 In Fig. 10–32, draw the Thevenin equivalent circuit with respect to terminals A and B (mentally remove R_3).

Figure 10–32

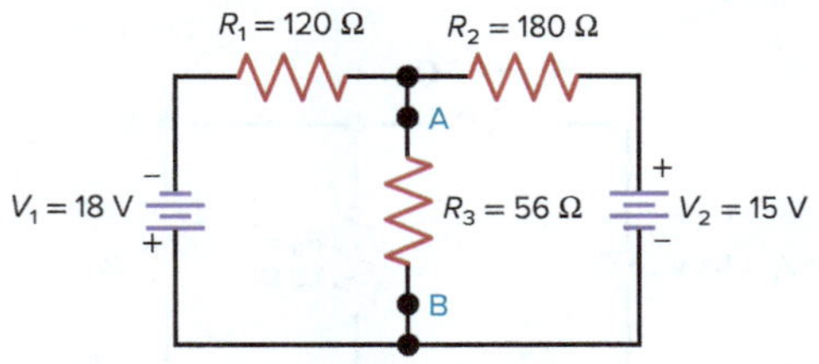

10–17 Using the Thevenin equivalent circuit for Fig. 10–32, calculate the values for I_3 and V_{R_3}.

10–18 With the polarity of V_1 reversed in Fig. 10–32, redraw the Thevenin equivalent circuit with respect to terminals A and B. Also, recalculate the new values for I_3 and V_{R_3}.

SECTION 10–4 THEVENIZING A BRIDGE CIRCUIT

10–19 In Fig. 10–33, draw the Thevenin equivalent circuit with respect to terminals A and B (mentally remove R_L).

Figure 10–33

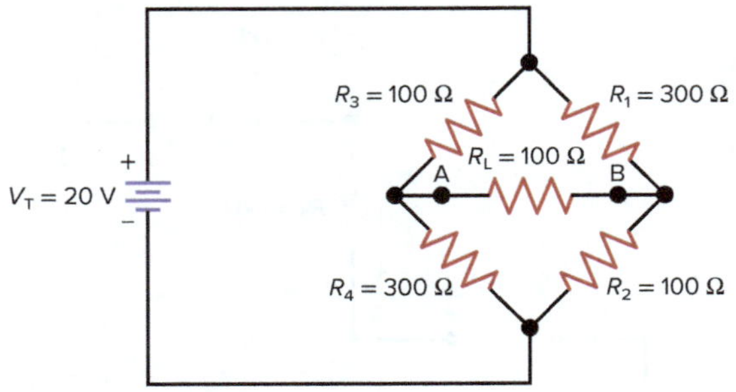

10–20 Using the Thevenin equivalent circuit for Fig. 10–33, calculate the values for I_L and V_L.

10–21 In Fig. 10–34, draw the Thevenin equivalent circuit with respect to terminals A and B (mentally remove R_L).

Figure 10–34

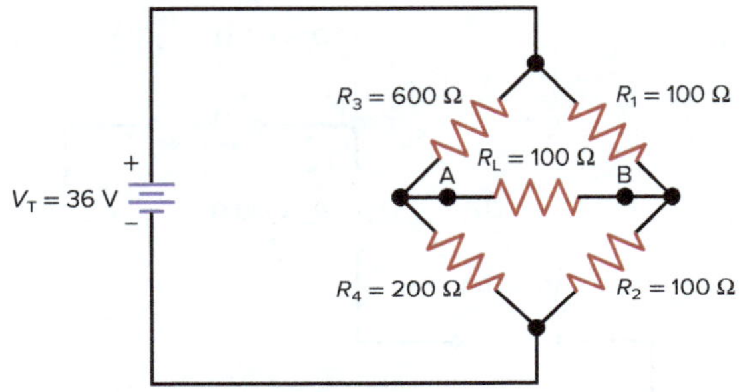

10–22 Using the Thevenin equivalent circuit for Fig. 10–34, calculate the values for I_L and V_L.

SECTION 10–5 NORTON'S THEOREM

10–23 MultiSim In Fig. 10–35, draw the Norton equivalent circuit with respect to terminals A and B (mentally remove R_L).

Figure 10–35

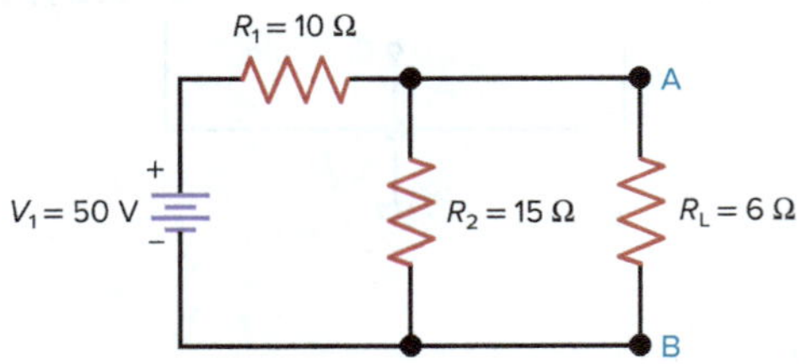

10–24 Using the Norton equivalent circuit for Fig. 10–35, calculate the values for I_L and V_L.

10–25 In Fig. 10–36, draw the Norton equivalent circuit with respect to terminals A and B (mentally remove R_L).

Figure 10–36

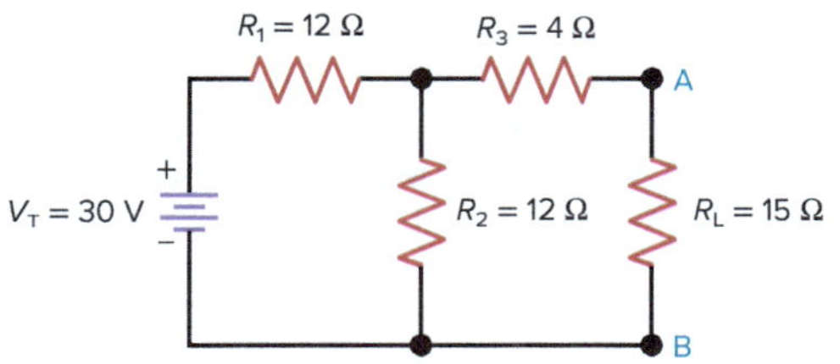

10–26 Using the Norton equivalent circuit for Fig. 10–36, calculate the values for I_L and V_L.

10–27 If R_3 is changed to 24 **Ω** in Fig. 10–36, redraw the Norton equivalent circuit with respect to terminals A and B. Also, recalculate the new values for I_L and V_L.

SECTION 10–6 THEVENIN-NORTON CONVERSIONS

10–28 Assume V_{TH} = 15 V and R_{TH} = 5 **Ω** for the Thevenin equivalent circuit in Fig. 10–37. What are the Norton equivalent values of I_N and R_N?

Figure 10–37

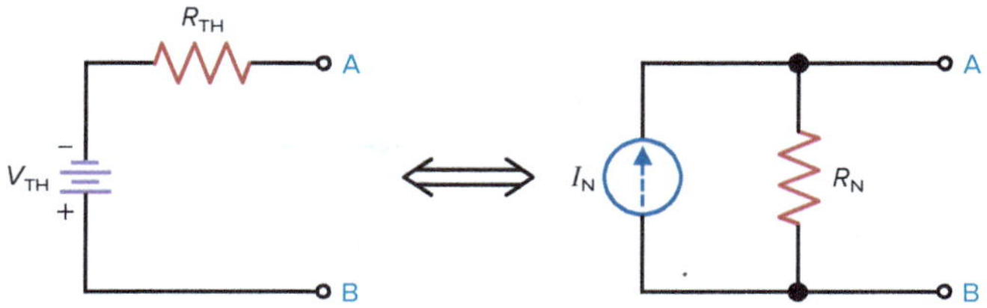

10–29 Assume I_N = 20 mA and R_N = 1.2 k**Ω** for the Norton equivalent circuit in Fig. 10–37. What are the Thevenin equivalent values of V_{TH} and R_{TH}?

10–30 Assume I_N = 5 mA and R_N = 1.5 k**Ω** for the Norton equivalent circuit in Fig. 10–37. What are the Thevenin equivalent values of V_{TH} and R_{TH}?

10–31 Assume V_{TH} = 36 V and R_{TH} = 1.2 k**Ω** for the Thevenin equivalent circuit in Fig. 10–37. What are the Norton equivalent values of I_N and R_N?

SECTION 10–7 CONVERSION OF VOLTAGE AND CURRENT SOURCES

10–32 In Fig. 10–38,

a. Convert voltage source 1 and voltage source 2 into equivalent current sources I_1 and I_2. Redraw the original circuit showing the current sources in place of V_1 and V_2.

b. Combine the current sources I_1 and I_2 into one equivalent current source, I_T. Draw the equivalent circuit.

c. Using the equivalent current source, I_T, calculate the values of I_3 and V_{R_3}.

Figure 10–38

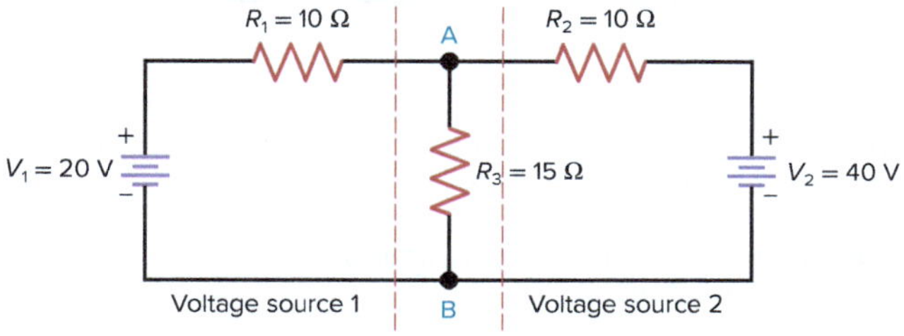

10–33 In Fig. 10–39,

a. Convert current source 1 and current source 2 into equivalent voltage sources V_1 and V_2. Redraw the original circuit showing the voltage sources in place of I_1 and I_2.

b. Combine the voltage sources V_1 and V_2 into one equivalent voltage source, V_T. Draw the equivalent circuit.

c. Using the equivalent voltage source, V_T, calculate the values of I_3 and V_{R_3}.

Figure 10–39

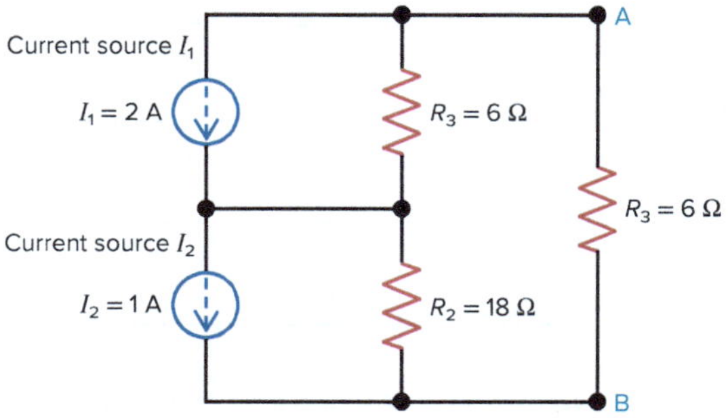

SECTION 10–8 MILLMAN'S THEOREM

10–34 In Fig. 10–40, apply Millman's theorem to solve the voltage, V_{XY}.

Figure 10–40

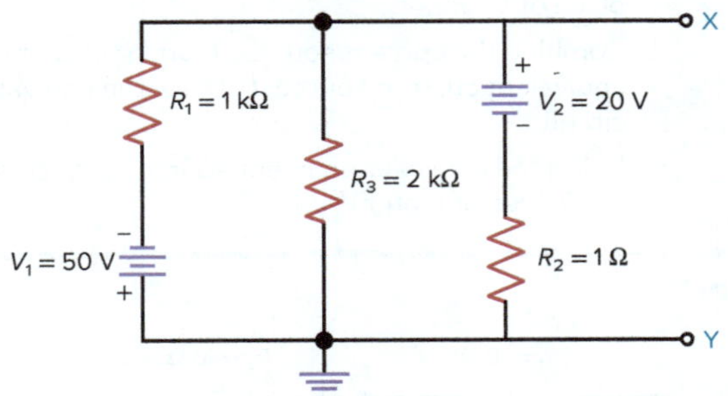

10–35 In Fig. 10–40, recalculate the voltage, V_{XY}, if the polarity of V_2 is reversed.

10–36 In Fig. 10–41, apply Millman's theorem to solve for the voltage, V_{XY}.

Figure 10–41

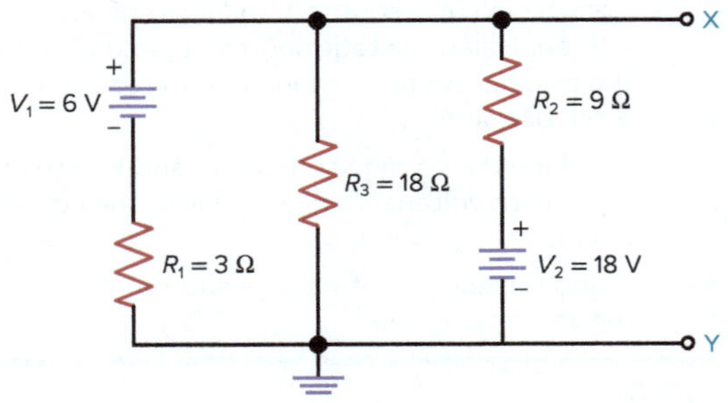

10–37 In Fig. 10–41, recalculate the voltage, V_{XY}, if the polarity of V_2 is reversed.

10–38 In Fig. 10–42, apply Millman's theorem to solve for the voltage, V_{XY}.

Figure 10–42

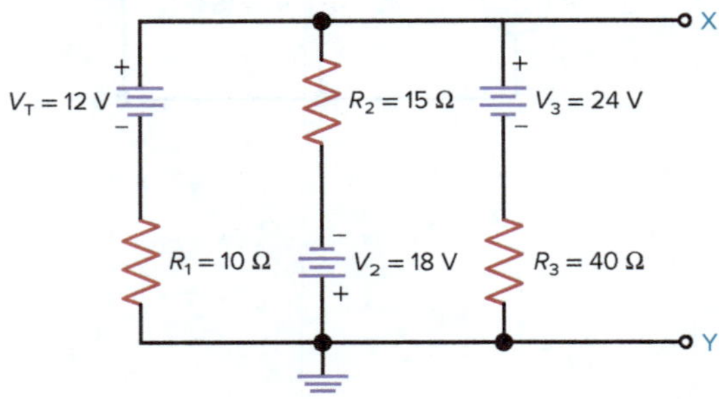

SECTION 10–9 T OR Y AND π OR Δ CONNECTIONS

10–39 Convert the T network in Fig. 10–43 to an equivalent π network.

Figure 10–43

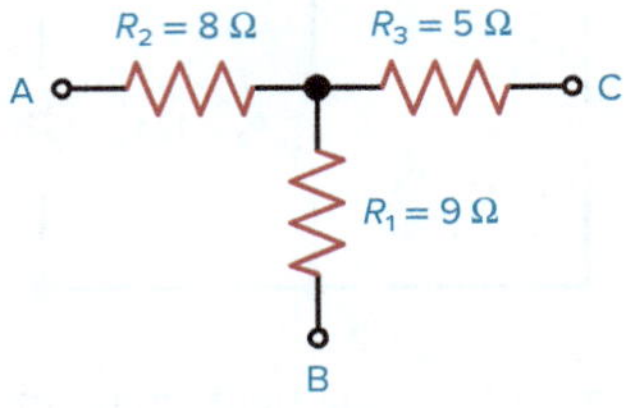

10–40 Convert the π network in Fig. 10–44 to an equivalent T network.

Figure 10–44

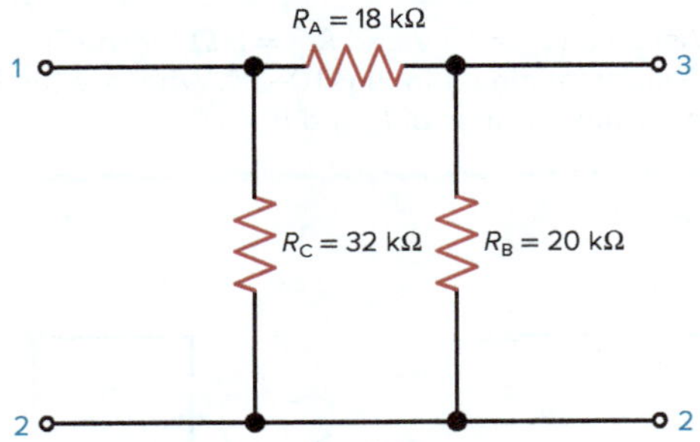

10–41 In Fig. 10–45, use delta-wye transformations to calculate both R_T and I_T.

Figure 10–45

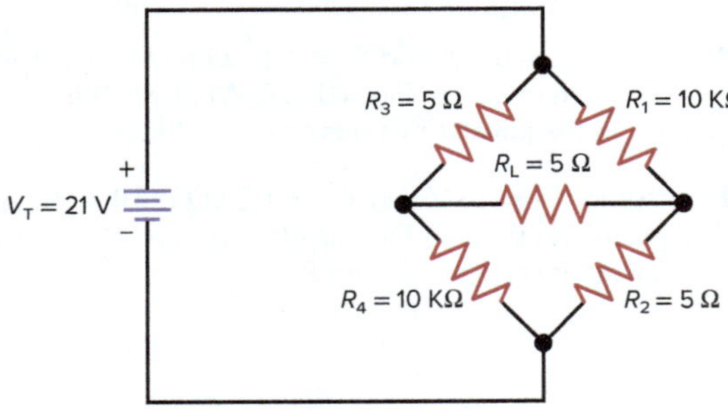

Critical Thinking

10–42 Thevenize the circuit driving terminals A and B in Fig. 10–46. Show the Thevenin equivalent circuit and calculate the values for I_L and V_L.

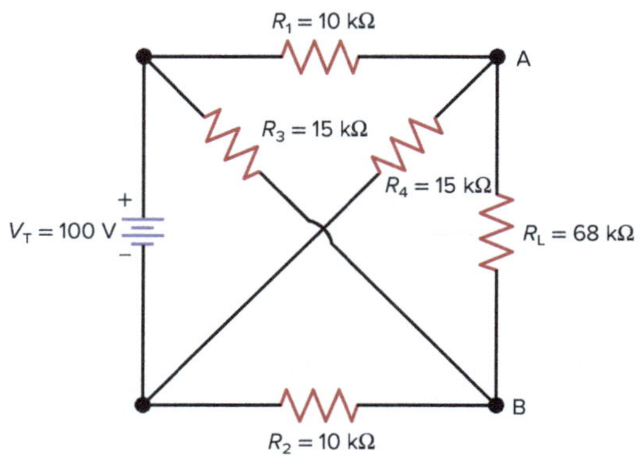

Figure 10–46 Circuit for Critical Thinking Problem 10–42.

10–43 In Fig. 10–47, use the superposition theorem to solve for I_L and V_L.

10–44 In Fig. 10–47, show the Thevenin equivalent circuit driving terminals A and B.

Figure 10–47 Circuit for Critical Thinking Probs. 10–43 and 10–44.

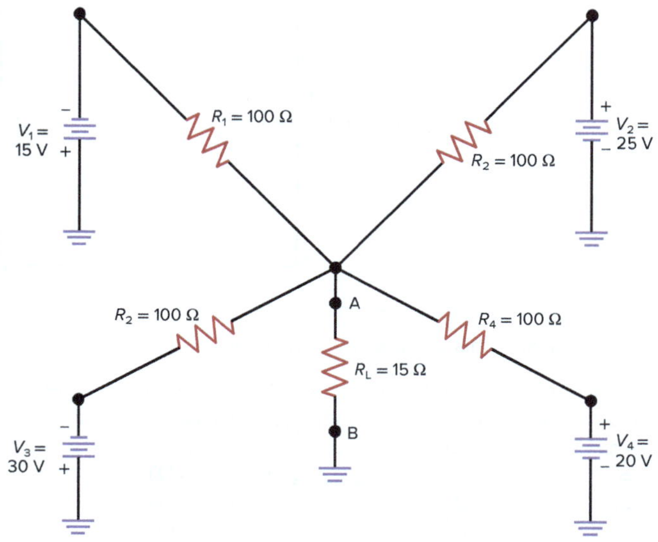

10–45 Refer to Fig. 10–34. Remove R_4 from the circuit, and show the Thevenin equivalent circuit driving the open terminals. Also, calculate the value of I_4 and V_{R_4}.

Answers to Self-Reviews

10–1 **a.** R_1
b. R_2

10–2 **a.** true
b. true

10–3 **a.** 8.4 Ω
b. 24 V

10–4 **a.** 6.4 Ω
b. 2.5 V

10–5 **a.** true
b. true

10–6 **a.** true
b. true
c. true

10–7 **a.** 7 A
b. 3 Ω

10–8 **a.** 6 V
b. 14 V

10–9 **a.** R_1
b. R_B and R_C

Laboratory Application Assignment

In this lab application assignment, you will apply Thevenin's theorem to solve for the unknown values of load voltage and load current in a circuit. You will begin by applying Thevenin's theorem to a relatively simple series-parallel circuit and then graduate to a more complex unbalanced bridge circuit.

Equipment: Obtain the following items from your instructor.

- Variable dc voltage source
- Assortment of carbon-film resistors
- DMM

Applying Ohm's Law

Examine the series-parallel circuit in Fig. 10–48. Using Ohm's law, calculate and record the current and voltage for the load resistor, R_L.

I_L = ____________, V_{RL} = ____________

Recalculate I_L and V_{RL} if R_L is changed to 1.8 kΩ. I_L = ____________, V_{RL} = ____________

Recalculate I_L and V_{RL} if R_L is changed to 2.7 kΩ. I_L = ____________, V_{RL} = ____________

Figure 10–48

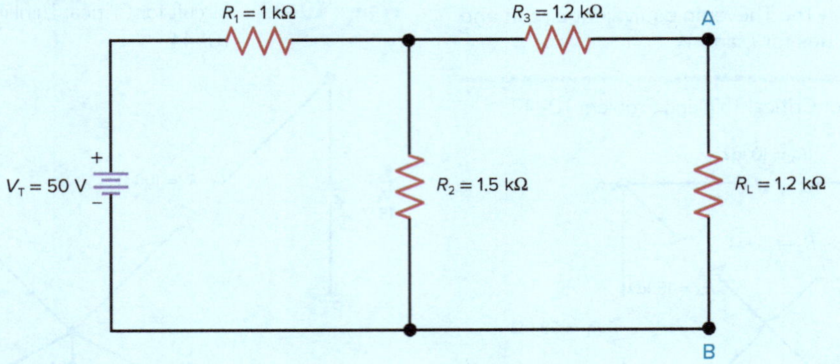

As you can see, this can become a very tedious task!

Construct the circuit in Fig. 10–48. Measure and record the load current, I_L, and load voltage, V_{RL}, for each of the different load resistance values.

I_L = __________, V_{RL} = __________(R_L = 1.2 kΩ)
I_L = __________, V_{RL} = __________(R_L = 1.8 kΩ)
I_L = __________, V_{RL} = __________(R_L = 2.7 kΩ)

Applying Thevenin's Theorem

Referring to Fig. 10–48, Thevenin's theorem states that the entire network connected to terminals A and B can be replaced with a single voltage source, V_{TH}, in series with a single resistance, R_{TH}. To find the values of V_{TH} and R_{TH}, proceed as follows. Mentally remove the load, R_L, from points A and B, and calculate the open-circuit voltage across these two points. This value is the Thevenin equivalent voltage, V_{TH}. Record this value in Fig. 10–49 as V_{TH} (calculated). Next, with the load, R_L, still removed, mentally short the voltage source, V_T, and calculate the resistance across the open terminals A and B. This value is the Thevenin equivalent resistance, R_{TH}. Record this value in Fig. 10–49 as R_{TH} (calculated).

Next, remove the load, R_L, in Fig. 10–48, and measure the open-circuit voltage across points A and B. Record this value in Fig. 10–49 as V_{TH} (measured). Next, short the voltage source, V_T, by removing the leads from the red and black power supply terminals and clipping them together. With V_T shorted, measure the resistance across terminals A and B. Record this value in Fig. 10–49 as R_{TH} (measured).

Using the calculated values of V_{TH} and R_{TH}, in Fig. 10–49, calculate and record the values of I_L and V_{RL} for each of the following load resistance values.

I_L = __________, V_{RL} = __________ (R_L = 1.2 kΩ)
I_L = __________, V_{RL} = __________ (R_L = 1.8 kΩ)
I_L = __________, V_{RL} = __________ (R_L = 2.7 kΩ)

How do these values compare to the calculated values in the original circuit of Fig. 10–48? ______________________________
____

Construct the Thevenin equivalent circuit in Fig. 10–49. Adjust both V_{TH} and R_{TH} to the measured values recorded in this figure. Now measure and record the values of I_L and V_{RL} for each of the following load resistance values.

I_L = __________, V_{RL} = __________ (R_L = 1.2 kΩ)
I_L = __________, V_{RL} = __________ (R_L = 1.8 kΩ)
I_L = __________, V_{RL} = __________ (R_L = 2.7 kΩ)

How do these values compare to the measured values in the original circuit of Fig. 10–48?______________________________
____

Figure 10–49

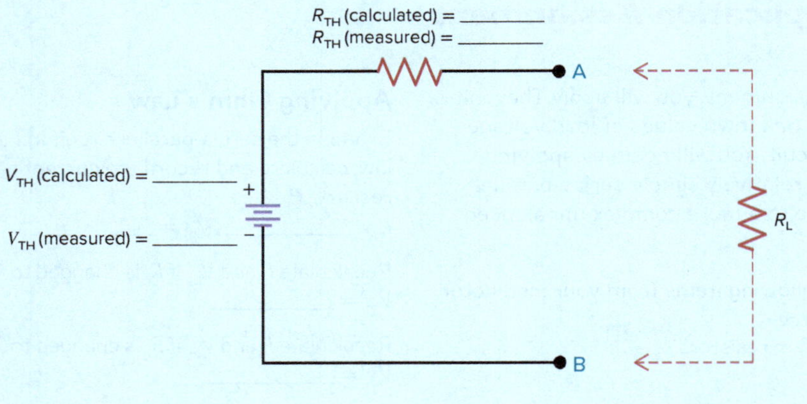

The magic in Thevenin's theorem lies in the fact that the Thevenin equivalent circuit driving terminals A and B remains the same regardless of the value of R_L. In the original circuit of Fig. 10–48, every time R_L was changed the entire circuit would have to be resolved. Not with Thevenin's theorem! Just plug the new value of R_L into the Thevenin equivalent circuit, and make one simple calculation.

Unbalanced Bridge Circuit

Refer to the unbalanced bridge circuit in Fig. 10–50. By applying Thevenin's theorem, determine the value of R_L that will provide a load current, I_L, of 1.2 mA. Show all your calculations as well as your Thevenin equivalent circuit.

Construct the original circuit in Fig. 10–50. For R_L, insert the value determined from your calculations. Finally, measure and record the value of I_L. I_L = ___________

Figure 10–50

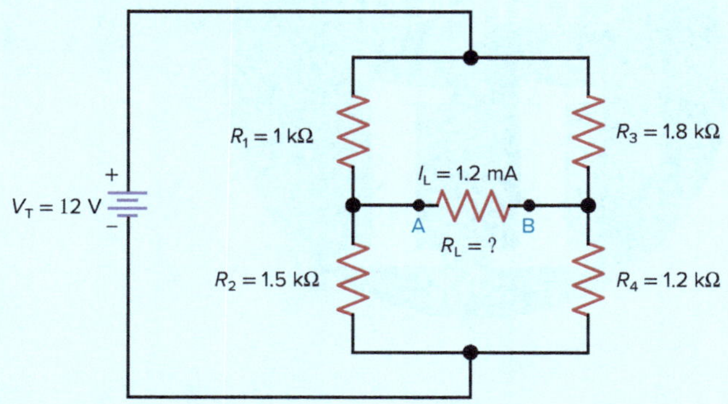

Cumulative Review Summary (Chapters 9–10)

- Methods of applying Kirchhoff's laws include
  (a) equations of voltages using the branch currents in the loops to specify the voltages.
  (b) equations of currents at a node using the node voltage to specify the node currents.
  (c) equations of voltages using assumed mesh currents to specify the voltages.
- Methods of reducing a network to a simple equivalent circuit include
  (a) the superposition theorem using one source at a time.
  (b) Thevenin's theorem to convert the network to a series circuit with one voltage source.
  (c) Norton's theorem to convert the network to a parallel circuit with one current source.
  (d) Millman's theorem to find the common voltage across parallel branches with different sources.
  (e) delta (Δ) wye (Y) conversions to transform a network into a series-parallel circuit.

Cumulative Self-Test

Answers at the back of the book.
Answer True or False.

1. In Fig. 9–5, V_3 can be found by using Kirchhoff's laws with either branch currents or mesh currents.
2. In Fig. 9–5, V_3 can be found by superposition, thevenizing, or using Millman's theorem.
3. In Fig. 10–6, I_L cannot be found by delta-wye conversion because R_L disappears in the transformation.
4. In Fig. 10–6, I_L can be calculated with Kirchhoff's laws, using mesh currents for three meshes.
5. With superposition, we can use Ohm's law for circuits that have more than one source.
6. A Thevenin equivalent is a parallel circuit.
7. A Norton equivalent is a series circuit.
8. Either a Thevenin or a Norton equivalent of a network will produce the same current in any load across terminals A and B.
9. A Thevenin-to-Norton conversion means converting a voltage source to a current source.
10. The volt unit is equal to (volts/ohms) ÷ siemens.
11. A node voltage is a voltage between current nodes.
12. A π network can be converted to an equivalent T network.
13. A 10-V source with 10-Ω series R will supply 5 V to a 10-Ω load R_L.
14. A 10-A source with 10-Ω parallel R will supply 5 A to a 10-Ω load R_L.
15. Current sources in parallel can be added when they supply current in the same direction through R_L.

Design credits: Chapter Opener Background Image, and Multisim: Stockbyte/Getty Images

chapter 11

Conductors and Insulators

If you think of a wire as a water pipe for electricity, then it makes sense that the larger the diameter of a wire, the more current it can carry. The smaller the diameter of a wire, the less current it can carry. For a given length of wire, therefore, the resistance, R, decreases as its diameter and cross-sectional area increase. For a given diameter and cross-sectional area, however, the resistance of a wire increases with length. In general, conductors offer very little opposition or resistance to the flow of current.

An insulator is any material that resists or prevents the flow of electric charge, such as electrons. The resistance of an insulator is very high, typically several hundreds of megohms or more. An insulator provides the equivalent of an open circuit with practically infinite resistance and almost zero current.

In this chapter, you will be introduced to a variety of topics that includes wire conductors, insulators, connectors, mechanical switches, and fuses. All of these topics relate to the discussion of conductors and insulators because they either pass or block the flow of current, depending on their condition or state. ■

Chapter Outline

Chapter Objectives

After studying this chapter, you should be able to

- *Explain* the main function of a conductor in an electric circuit.
- *Calculate* the cross-sectional area of round wire when the diameter is known.
- *List* the advantages of using stranded wire versus solid wire.
- *List* common types of connectors used with wire conductors.
- *Define* the terms *pole* and *throw* as they relate to switches.
- *Explain* how fast-acting and slow-blow fuses differ.
- *Calculate* the resistance of a wire conductor whose length, cross-sectional area, and specific resistance are known.
- *Explain* the meaning of *temperature coefficient of resistance*.
- *Explain* ion current and electron current.
- *Explain* why insulators are sometimes called *dielectrics*.
- *Explain* what is meant by the *corona effect*.

Important Terms

circuit breaker	fuse	specific resistance	throw
circular mil (cmil)	ionization current	switch	wire gage
corona effect	pole	temperature coefficient	
dielectric material	slow-blow fuse		

GOOD TO KNOW

In our homes, the higher-power appliances operate from 240 V rather than 120 V. The reason is that, at twice the voltage, only half the current is needed to obtain the required power. The reduction in I by one-half considerably reduces the I^2R power loss in the wire conductors that connect the load to the power line. With less current at a higher voltage, a smaller conductor size can be used which is an important advantage.

11–1 Function of the Conductor

In Fig. 11–1, the resistance of the two 10-ft lengths of copper-wire conductor is 0.08 Ω. This R is negligibly small compared with the 144-Ω R of the tungsten filament in the lightbulb. When the current of 0.833 A flows in the bulb and the series conductors, the IR voltage drop of the conductors is only 0.07 V, with 119.93 V across the bulb. Practically all the applied voltage is across the bulb filament. Since the bulb then has its rated voltage of 120 V, approximately, it will dissipate the rated power of 100 W and light with full brilliance.

The current in the wire conductors and the bulb is the same, since they are in series. However, the IR voltage drop in the conductor is practically zero because its R is almost zero.

Also, the I^2R power dissipated in the conductor is negligibly small, allowing the conductor to operate without becoming hot. Therefore, the conductor delivers energy from the source to the load with minimum loss by electron flow in the copper wires.

Although the resistance of wire conductors is very small, for some cases of high current, the resultant IR drop can be appreciable. For example, suppose that the 120-V power line is supplying 30 A of current to a load through two conductors, each of which has a resistance of 0.2 Ω. In this case, each conductor has an IR drop of 6 V, calculated as 30 A × 0.2 Ω = 6 V. With each conductor dropping 6 V, the load receives a voltage of only 108 V rather than the full 120 V. The lower-than-normal load voltage could result in the load not operating properly. Furthermore, the I^2R power dissipated in each conductor equals 180 W, calculated as $30^2 \times 0.2\ \Omega =$ 180 W. The I^2R power loss of 180 W in the conductors is considered excessively high.

11–1 Self-Review

Answers at the end of the chapter.

Refer to Fig. 11–1.

a. How much is *R* for the 20 ft of copper wire?

b. How much is the *IR* voltage drop for the wire conductors?

c. The *IR* voltage in question b is what percentage of the applied voltage?

MultiSim **Figure 11–1** The conductors should have minimum resistance to light the bulb with full brilliance. (*a*) Wiring diagram. (*b*) Schematic diagram. R_1 and R_2 represent the very small resistance of the wire conductors.

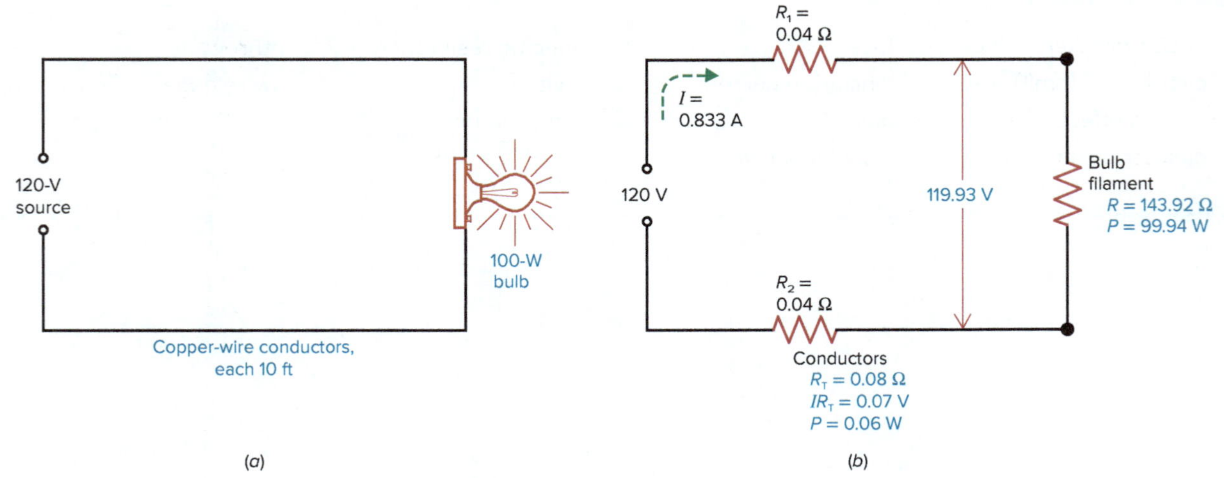

11–2 Standard Wire Gage Sizes

Table 11–1 lists the standard wire sizes in the system known as the American **Wire Gage** (AWG) or Brown and Sharpe (B&S) gage. The gage numbers specify the size of round wire in terms of its diameter and cross-sectional area. Note the following three points:

1. As the gage numbers increase from 1 to 40, the diameter and circular area decrease. Higher gage numbers indicate thinner wire sizes.
2. The circular area doubles for every three gage sizes. For example, No. 10 wire has approximately twice the area of No. 13 wire.
3. The higher the gage number and the thinner the wire, the greater the resistance of the wire for any given length.

Table 11–1 Copper-Wire Table

Gage No.	Diameter, Mils	Area, Circular Mils	Ohms per 1000 ft of Copper Wire at 25°C*	Gage No.	Diameter, Mils	Area, Circular Mils	Ohms per 1000 ft of Copper Wire at 25°C*
1	289.3	83,690	0.1264	21	28.46	810.1	13.05
2	257.6	66,370	0.1593	22	25.35	642.4	16.46
3	229.4	52,640	0.2009	23	22.57	509.5	20.76
4	204.3	41,740	0.2533	24	20.10	404.0	26.17
5	181.9	33,100	0.3195	25	17.90	320.4	33.00
6	162.0	26,250	0.4028	26	15.94	254.1	41.62
7	144.3	20,820	0.5080	27	14.20	201.5	52.48
8	128.5	16,510	0.6405	28	12.64	159.8	66.17
9	114.4	13,090	0.8077	29	11.26	126.7	83.44
10	101.9	10,380	1.018	30	10.03	100.5	105.2
11	90.74	8234	1.284	31	8.928	79.70	132.7
12	80.81	6530	1.619	32	7.950	63.21	167.3
13	71.96	5178	2.042	33	7.080	50.13	211.0
14	64.08	4107	2.575	34	6.305	39.75	266.0
15	57.07	3257	3.247	35	5.615	31.52	335.0
16	50.82	2583	4.094	36	5.000	25.00	423.0
17	45.26	2048	5.163	37	4.453	19.83	533.4
18	40.30	1624	6.510	38	3.965	15.72	672.6
19	35.89	1288	8.210	39	3.531	12.47	848.1
20	31.96	1022	10.35	40	3.145	9.88	1069

* 20° to 25°C or 68° to 77°F is considered average room temperature.

GOOD TO KNOW

There are gage sizes lower than No. 1 gage. They are No. 0, No. 00, No. 000, and No. 0000, with No. 0000 having the largest cross-sectional area. The wire sizes 0,00,000 and 0000 are usually represented as 1/0, 2/0, 3/0, and 4/0, respectively. The number after the slash is always a zero and is pronounced "aught." The Number before the slash indicates the wire size is X number of zeros. 1/0 is pronounced one aught, 2/0 is pronounced two aught, and so on.

Figure 11–2 Standard American Wire Gage for wire conductors.

Cindy Schroeder/McGraw Hill

In typical applications, hookup wire for electronic circuits with current of the order of milliamperes is generally about No. 22 gage. For this size, 0.5 to 1 A is the maximum current the wire can carry without excessive heating.

House wiring for circuits where the current is 5 to 15 A is usually No. 14 gage. Minimum sizes for house wiring are set by local electrical codes, which are usually guided by the National Electrical Code published by the National Fire Protection Association. A gage for measuring wire size is shown in Fig. 11–2.

Circular Mils

The cross-sectional area of round wire is measured in **circular mils**, abbreviated cmil. A mil is one-thousandth of an inch, or 0.001 in. One circular mil is the cross-sectional area of a wire with a diameter of 1 mil. The number of circular mils in any circular area is equal to the square of the diameter in mils or cmil = d^2(mils).

Example 11-1

What is the area in circular mils of a wire with a diameter of 0.005 in.?

ANSWER We must convert the diameter to mils. Since 0.005 in. equals 5 mil,

$$\text{Circular mil area} = (5\text{ mil})^2$$
$$\text{Area} = 25\text{ cmil}$$

Note that the circular mil is a unit of area, obtained by squaring the diameter, whereas the mil is a linear unit of length, equal to one-thousandth of an inch. Therefore, the circular-mil area increases as the square of the diameter. As illustrated in Fig. 11–3, doubling the diameter quadruples the area. Circular mils are convenient for round wire because the cross-section is specified without using the formula πr^2 or $\pi d^2/4$ for the area of a circle.

Figure 11–3 Cross-sectional area for round wire. Doubling the diameter increases the circular area by four times.

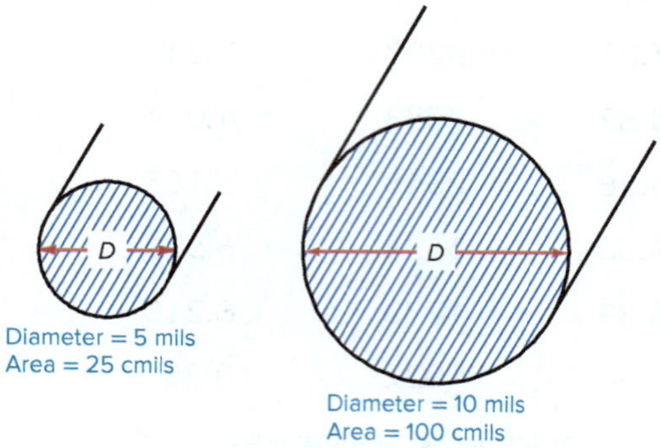

11–2 Self-Review

Answers at the end of the chapter.

a. **How much is *R* for 1 ft of No. 22 wire?**
b. **What is the cross-sectional area in circular mils for wire with a diameter of 0.025 in.?**
c. **What is the wire gage size in Fig. 11–1?**

Figure 11–4 Types of wire conductors. (*a*) Solid wire. (*b*) Stranded wire. (*c*) Braided wire for very low *R*. (*d*) Coaxial cable. Note braided wire for shielding the inner conductor. (*e*) Twin-lead cable.

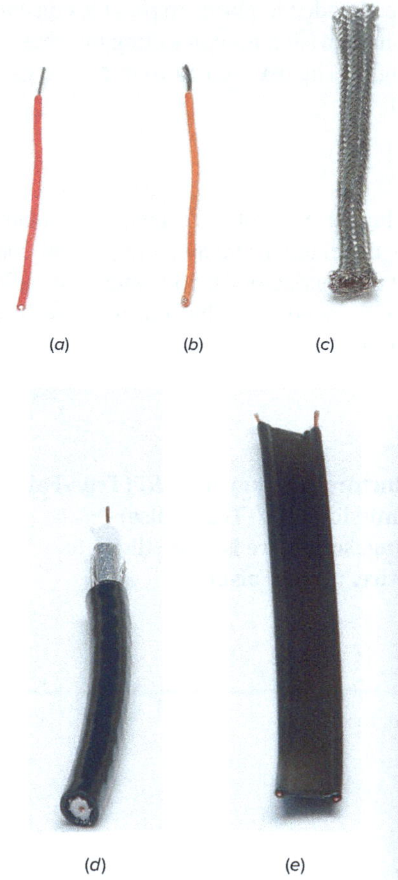

(a)–(e): Cindy Schroeder/McGraw Hill

11–3 Types of Wire Conductors

Most wire conductors are copper due to its low cost, although aluminum and silver are also used sometimes. The copper may be tinned with a thin coating of solder, which gives it a silvery appearance. Tinned wire is easier to solder for connections. The wire can be solid or stranded, as shown in Fig. 11–4*a* and *b*. Solid wire is made of only one conductor. If bent or flexed repeatedly, solid wire may break. Therefore, solid wire is used in places where bending and flexing is not encountered. House wiring is a good example of the use of solid wire. Stranded wire is made up of several individual strands put together in a braid. Some uses for stranded wire include telephone cords, extension cords, and speaker wire, to name a few.

Stranded wire is flexible, easier to handle, and less likely to develop an open break. Sizes for stranded wire are equivalent to the sum of the areas for the individual strands. For instance, two strands of No. 30 wire correspond to solid No. 27 wire.

Example 11-2

A stranded wire is made up of 16 individual strands of No. 27 gage wire. What is its equivalent gage size in solid wire?

ANSWER The equivalent gage size in solid wire is determined by the total circular area of all individual strands. Referring to Table 11–1, the circular area for No. 27 gage wire is 201.5 cmils. Since there are 16 individual strands, the total circular area is calculated as follows:

$$\text{Total cmil area} = 16 \text{ strands} \times \frac{201.5 \text{ cmils}}{\text{strand}}$$
$$= 3224 \text{ cmils}$$

Referring to Table 11–1, we see that the circular area of 3224 cmils corresponds very closely to the cmil area of No. 15 gage wire. Therefore, 16 strands of No. 27 gage wire is roughly equivalent to No. 15 gage solid wire.

Very thin wire, such as No. 30, often has an insulating coating of enamel or shellac. It may look like copper, but the coating must be scraped off the ends to make a good connection. This type of wire is used for small coils.

Heavier wires generally are in an insulating sleeve, which may be rubber or one of many plastic materials. General-purpose wire for connecting electronic components is generally plastic-coated hookup wire of No. 20 gage. Hookup wire that is bare should be enclosed in a hollow insulating sleeve called *spaghetti*.

The braided conductor in Fig. 11–4*c* is used for very low resistance. It is wide for low *R* and thin for flexibility, and the braiding provides many strands. A common application is a grounding connection, which must have very low *R*.

Transmission Lines

Constant spacing between two conductors through the entire length provides a transmission line. Common examples are the coaxial cable in Fig. 11–4*d* and the twin lead in Fig. 11–4*e*.

Coaxial cable with an outside diameter of ¼ in. is generally used for the signals in cable television. In construction, there is an inner solid wire, insulated from metallic braid that serves as the other conductor. The entire assembly is covered by an outer plastic jacket. In operation, the inner conductor has the desired signal voltage with respect to ground, and the metallic braid is connected to ground to shield the inner conductor against interference. Coaxial cable, therefore, is a shielded type of transmission line.

With twin-lead wire, two conductors are embedded in plastic to provide constant spacing. This type of line is commonly used in television for connecting the antenna to the receiver. In this application, the spacing is ⅝ in. between wires of No. 20 gage size, approximately. This line is not shielded.

Wire Cable

Two or more conductors in a common covering form a cable. Each wire is insulated from the others. Cables often consist of two, three, ten, or many more pairs of conductors, usually color-coded to help identify the conductors at both ends of a cable.

The ribbon cable in Fig. 11–5, has multiple conductors but not in pairs. This cable is used for multiple connections to a computer and associated equipment.

11–3 Self-Review

Answers at the end of the chapter.

a. The plastic coating on wire conductors has very high *R*. (True/False)

b. Coaxial cable is a shielded transmission line. (True/False)

c With repeated bending and flexing, solid wire is more likely to develop an open than stranded wire. (True/False)

Figure 11–5 Ribbon cable with multiple conductors.

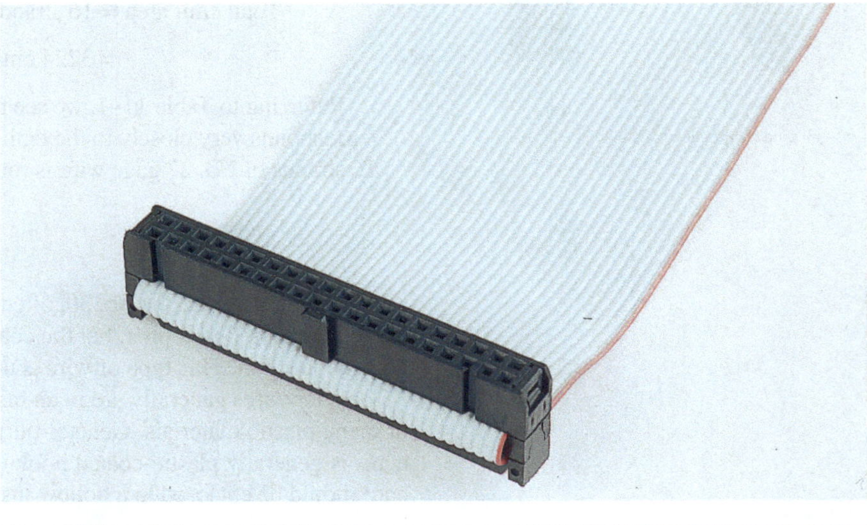

Mark Steinmetz/McGraw Hill

11–4 Connectors

> **GOOD TO KNOW**
>
> A lot of skill is required to properly attach a connector to a cable. Attaching a connector to a cable improperly (or poorly) can result in the malfunction of an entire system.

Refer to Fig. 11–6 for different types. The spade lug in Fig. 11–6*a* is often used for screw-type terminals. The alligator clip in Fig. 11–6*b* is convenient for a temporary connection. Alligator clips come in small and large sizes. The banana pins in Fig. 11–6*c* have spring-type sides that make a tight connection. The terminal strip in Fig. 11–6*d* provides a block for multiple solder connections.

The RCA-type plug in Fig. 11–6*e* is commonly used for shielded cables with audio equipment. The inner conductor of the shielded cable is connected to the center pin of the plug, and the cable braid is connected to the shield. Both connections must be soldered.

The phone plug in Fig. 11–6*f* is still used in many applications but usually in a smaller size. The ring is insulated from the sleeve to provide for two connections. There may be a separate tip, ring, and sleeve for three connections. The sleeve is usually the ground side.

The plug in Fig. 11–6*g* is called an *F connector*. It is universally used in cable television because of its convenience. The center conductor of the coaxial cable serves as the center pin of the plug, so that no soldering is needed. Also, the shield on the plug is press-fit onto the braid of the cable underneath the plastic jacket.

Figure 11–6*h* shows a multiple pin connector having many conductors. This type of connector is often used to connect the components of a computer system, such as the monitor and the keyboard, to the computer.

Figure 11–6*i* shows a spring-loaded metal hook as a grabber for a temporary connection to a circuit. This type of connector is often used with the test leads of a VOM or a DMM.

Figure 11–6 Common types of connectors for wire conductors. (*a*) Spade lug. (*b*) Alligator clip. (*c*) Double banana-pin plug. (*d*) Terminal strip. (*e*) RCA-type plug for audio cables. (*f*) Phone plug. (*g*) F-type plug for cable TV. (*h*) Multiple-pin connector plug. (*i*) Spring-loaded metal hook as grabber for temporary connection in testing circuits.

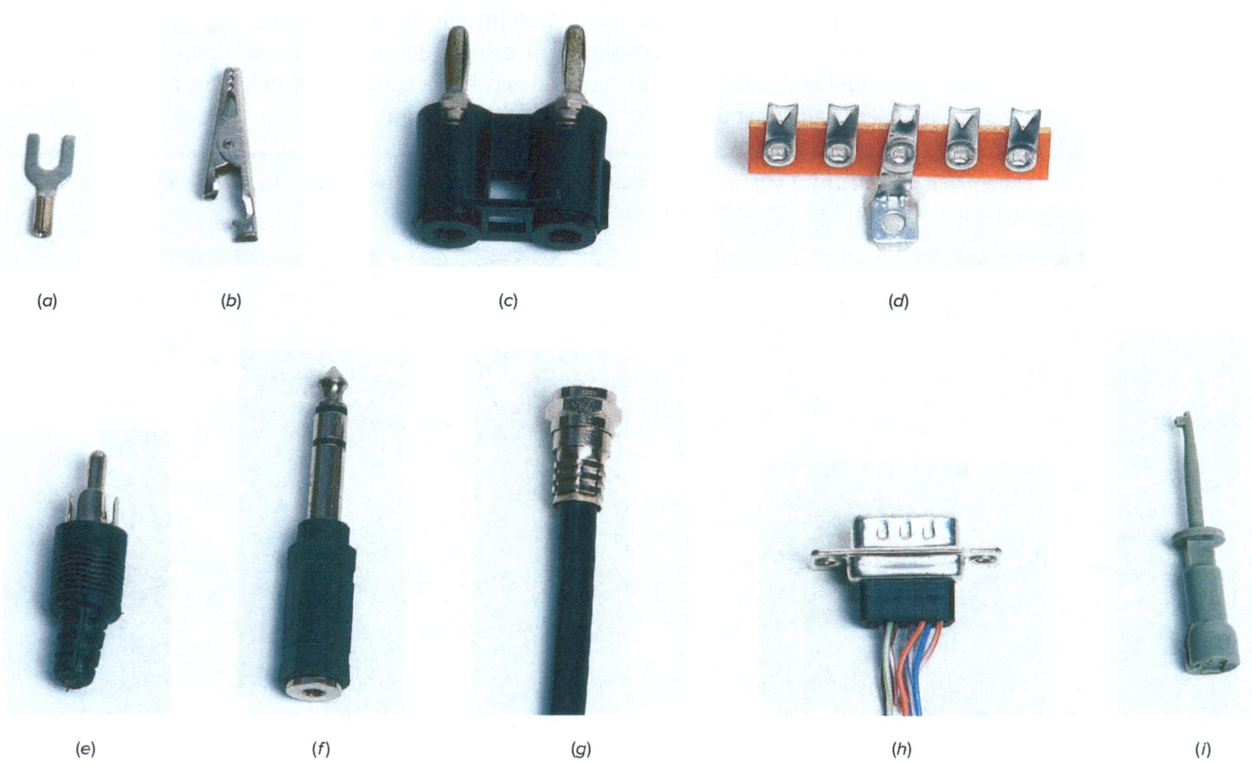

(a–i): Cindy Schroeder/McGraw Hill

11–4 Self-Review

Answers at the end of the chapter.

a. The RCA-type plug is commonly used for shielded cables with audio equipment. (True/False)
b. The F-type connector is used with coaxial cable. (True/False)
c. The F-type connector can also be used with twin-lead line. (True/False)

11–5 Printed Circuit Board

GOOD TO KNOW

Many printed-circuit boards are multilayer boards that contain several different layers of printed wiring. Printed-circuit boards that have four or five layers of printed wiring are not uncommon.

Most electronic components are mounted on a plastic or fiberglass insulating board with copper traces interconnecting the components. This is called a printed-circuit (PC) board. In some cases only one side of the PC board has the components, such as resistors, capacitors, coils, transistors, diodes, and integrated-circuits. The other side then has the conducting paths printed with silver or copper traces on the board, instead of using wires. On a double-sided board, the component side also has copper traces. Sockets, small metal eyelets, or holes in the board are used to connect the components to the wiring. Figure 11–7 shows a double-side PC board.

With a bright light on one side, you can see through to the opposite side to trace the connections. However, the circuit may be drawn on the PC board.

It is important not to use too much heat in soldering or desoldering. Otherwise the printed wiring can be lifted off the board. Use a small iron of about 25- to 30-W rating. When soldering semiconductor diodes and transistors, hold the lead with pliers or connect an alligator clip as a heat sink to conduct heat away from the semiconductor junction.

For desoldering, use a solder-sucker tool, with a soldering iron, to clean each terminal. Another method of removing solder is to use a copper-wire braid that is impregnated with rosin flux. This copper-wire braid, often called a desoldering braid, is excellent for attracting liquid or molten solder. Just put the desoldering braid on the solder joint and heat it until the solder runs up into the copper braid. The terminal must be clean enough to lift out the component easily without damaging the PC board. One advantage of using a desoldering braid versus a solder sucker tool is that the desoldering braid acts like a natural heat sink, thus reducing the risk of damaging the copper traces on the PC board.

Figure 11–7 Printed-circuit board (double-sided). (*a*) Top side of PC board with resistors, capacitors, transistors, and integrated circuits. (*b*) Bottom side of PC board with copper traces and additional components.

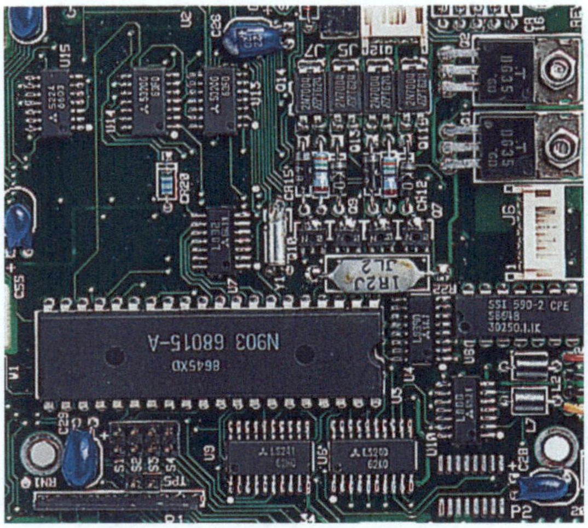

Mark Steinmetz/McGraw Hill

(*a*)

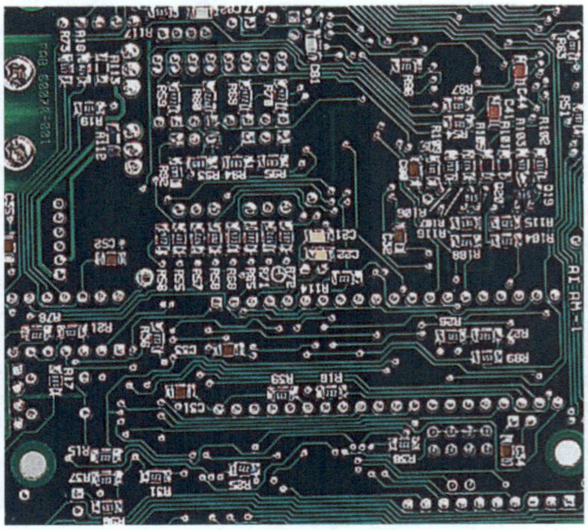

Mark Steinmetz/McGraw Hill

(*b*)

A small crack in the printed wiring acts like an open circuit preventing current flow. Cracks can be repaired by soldering a short length of bare wire over the open circuit. If a larger section of printed wiring is open, or if the board is cracked, you can bridge the open circuit with a length of hookup wire soldered at two convenient end terminals of the printed wiring. In many electronic industries, special kits are available for replacing damaged or open traces on PC boards.

11–5 Self-Review

Answers at the end of the chapter.

a. **Which is the best size of iron to use to solder on a PC board: 25, 100, or 150 W?**

b. **How much is the resistance of a printed-wire conductor with a crack in the middle?**

11–6 Switches

A **switch** is a component that allows us to control whether the current is ON or OFF in a circuit. A closed switch has practically zero resistance, whereas an open switch has nearly infinite resistance.

Figure 11–8 shows a switch in series with a voltage source and a lightbulb. With the switch closed, as in Fig. 11–8*a*, a complete path for current is provided and the light is ON. Since the switch has very low resistance when it is closed, all of the source voltage is across the load, with 0 V across the closed contacts of the switch. With the switch open, as in Fig. 11–8*b*, the path for current is interrupted and the bulb does not light. Since the switch has very high resistance when it is open, all of the source voltage is across the open switch contacts, with 0 V across the load.

GOOD TO KNOW

Switches can have more than two poles and two throws. For example, a single-pole switch might have 3, 4, 5, or more throws. Similarly, there are triple-pole, single-throw switches.

Switch Ratings

All switches have a current rating and a voltage rating. The current rating corresponds to the maximum allowable current that the switch can carry when closed. The current rating is based on the physical size of the switch contacts as well as the type of metal used for the contacts. Many switches have gold- or silver-plated contacts to ensure very low resistance when closed.

The voltage rating of a switch corresponds to the maximum voltage that can safely be applied across the open contacts without internal arcing. The voltage rating does not apply when the switch is closed, since the voltage drop across the closed switch contacts is practically zero.

MultiSim **Figure 11–8** A series switch used to open or close a circuit. (*a*) With the switch closed, current flows to light the bulb. The voltage drop across the closed switch is 0 V. (*b*) With the switch open, the light is OFF. The voltage drop across the open switch is 12 V.

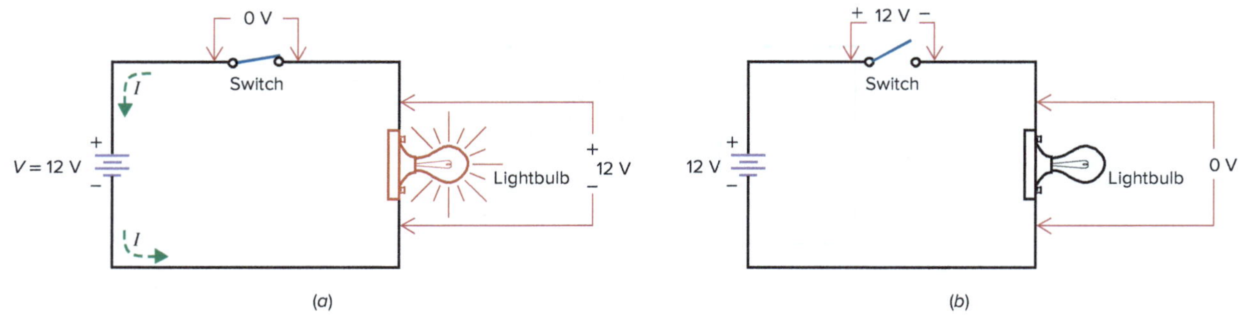

Figure 11–9 Switches. (*a*) Single-pole, single-throw. (*b*) Single-pole, double-throw. (*c*) Double-pole, single-throw. (*d*) Double-pole, double-throw.

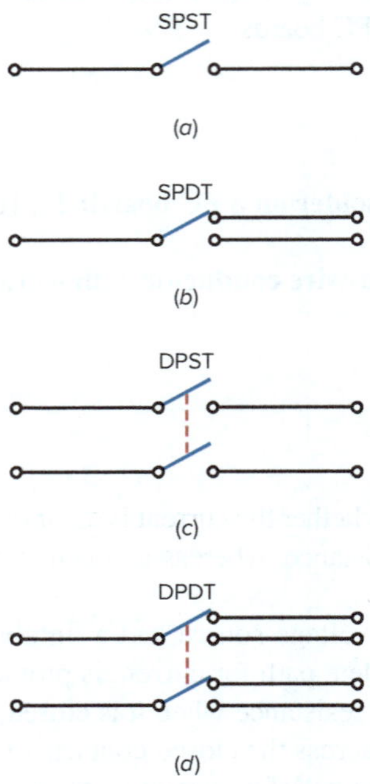

Switch Definitions

Toggle switches are usually described as having a certain number of **poles** and **throws**. For example, the switch in Fig. 11–8 is described as a *single-pole, single-throw (SPST)* switch. Other popular switch types include the *single-pole, double-throw (SPDT), double-pole, single-throw (DPST),* and *double-pole, double-throw (DPDT)*. The schematic symbols for each type are shown in Fig. 11–9. Notice that the SPST switch has two connecting terminals, whereas the SPDT has three, the DPST has four, and the DPDT has six.

The term *pole* is defined as the number of completely isolated circuits that can be controlled by the switch. The term *throw* is defined as the number of closed contact positions that exist per pole. The SPST switch in Fig. 11–8 can control the current in only one circuit, and there is only one closed contact position, hence the name single-pole, single-throw.

Figure 11–10 shows a variety of switch applications. In Fig. 11–10*a*, an SPDT switch is being used to switch a 12-V_{DC} source between one of two different loads. In Fig. 11–10*b*, a DPST switch is being used to control two completely separate circuits simultaneously. In Fig. 11–10*c*, a DPDT switch is being used to reverse the polarity of voltage across the terminals of a DC motor. (Reversing the polarity reverses the direction of the motor.) Note that the dashed lines shown between the poles in Fig. 11–10*b* and 11–10*c* indicate that both sets of contacts within the switch are opened and closed simultaneously.

Switch Types

Figure 11–11 shows a variety of toggle switches. Although the toggle switch is a very popular type of switch, several other types are found in electronic equipment. Additional types include push-button switches, rocker switches, slide switches, rotary switches, and DIP switches.

Push-button switches are often spring-loaded switches that are either normally open (NO) or normally closed (NC). Figure 11–12 shows the schematic symbols for both types. For the normally open switch in Fig. 11–12*a*, the switch contacts remain

Figure 11–10 Switch applications. (*a*) SPDT switch used to switch a 12-V source between one of two different loads. (*b*) DPST switch controlling two completely isolated circuits simultaneously. (*c*) DPDT switch used to reverse the polarity of voltage across a DC motor.

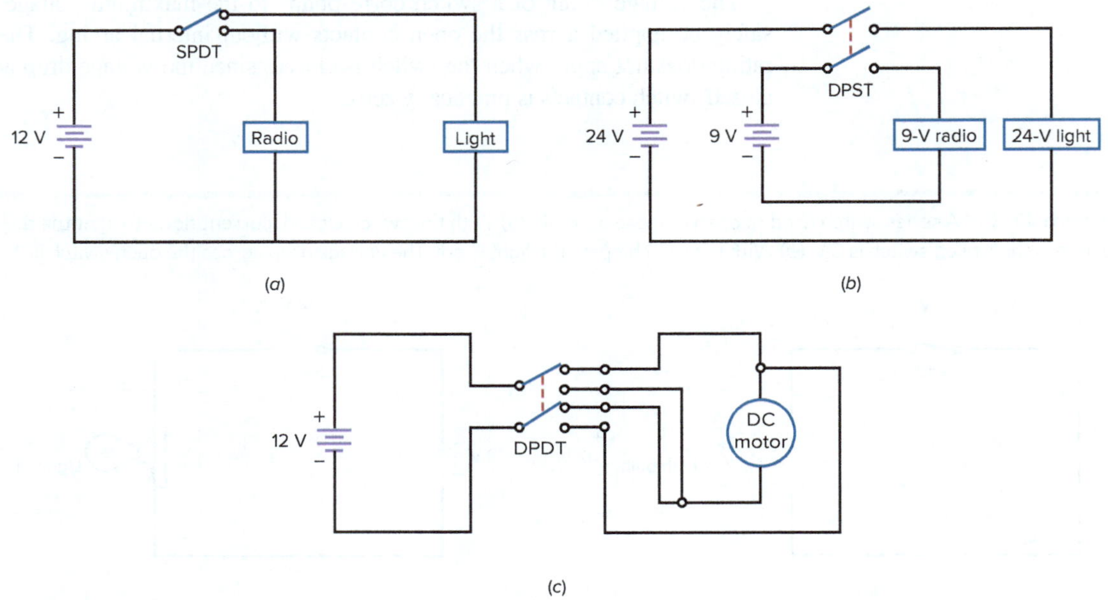

Figure 11–11 A variety of toggle switches.

Mark Steinmetz/McGraw Hill

Figure 11–12 Push-button switch schematic symbols. (*a*) Normally open push-button switch. (*b*) Normally closed push-button switch.

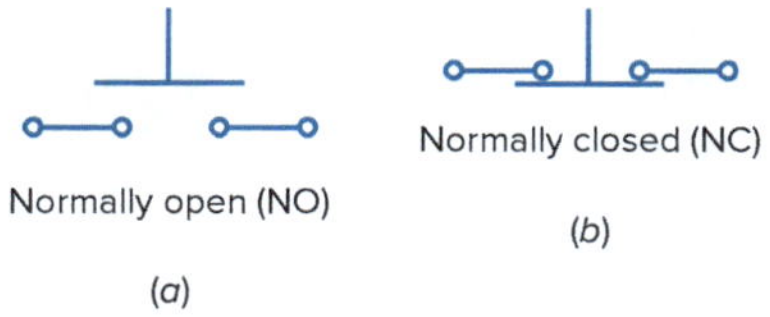

Figure 11–13 Typical push-button switch.

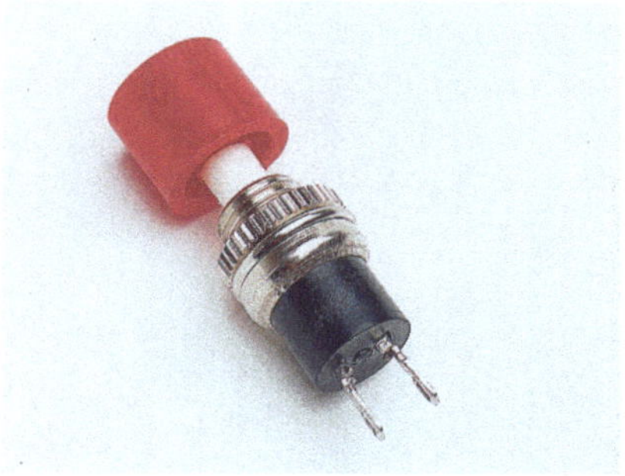

Mark Steinmetz/McGraw Hill

open until the push button is depressed. When the push button is depressed, the switch closes, allowing current to pass. The normally closed switch in Fig. 11–12*b* operates opposite the normally open switch in Fig. 11–12*a*. When the push button is depressed, the switch contacts open to interrupt current in the circuit. A typical push-button switch is shown in Fig. 11–13.

Figure 11–14 Dual-inline package (DIP) switch.

Mark Steinmetz/McGraw Hill

Figure 11–14 shows a DIP (dual-inline package) switch. It consists of eight miniature rocker switches, where each switch can be set separately. A DIP switch has pin connections that fit into a standard IC socket.

Figure 11–15 shows another type of switch known as a rotary switch. As shown, it consists of three wafers or decks mounted on a common shaft.

Figure 11–15 Rotary switch.

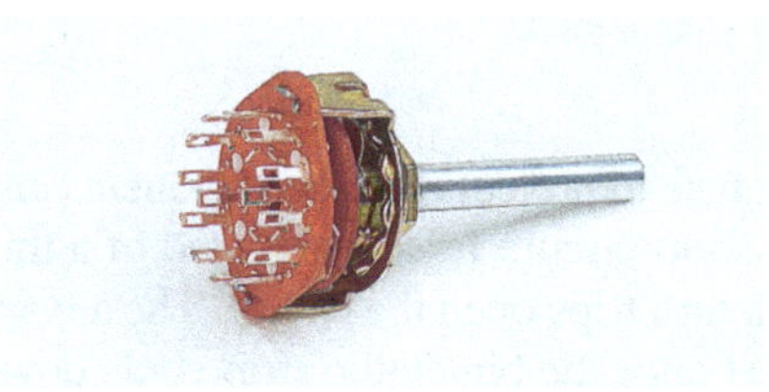

Mark Steinmetz/McGraw Hill

11–6 Self-Review

Answers at the end of the chapter.

a. How much is the *IR* voltage drop across a closed switch?

b. How many connections are there on an SPDT switch?

c. What is the resistance across the contacts of an open switch?

d. An SPST switch is rated at 10 A, 250 V. Should the switch be used to turn a 120-V, 1500-W heater ON and OFF?

11–7 Fuses

Many circuits have a **fuse** in series as a protection against an overload from a short circuit. Excessive current melts the fuse element, blowing the fuse and opening the series circuit. The purpose is to let the fuse blow before the components and wiring are damaged. The blown fuse can easily be replaced by a new one after the overload has been eliminated. A glass-cartridge fuse with holders is shown in Fig. 11–16. This is a type 3AG fuse with a diameter of ¼ in. and length of 1¼ in. *AG* is an abbreviation for "automobile glass," since that was one of the first applications of fuses in a glass holder to make the wire link visible. Aluminum, tin-coated copper, or nickel are the types of metal typically used as the fuse element. The schematic symbol for a fuse is , as shown in Fig. 11–18*a*.

Fuse Ratings

Fuses have three important ratings or characteristics that need to be considered. These three ratings are its ***current rating***, ***voltage rating***, and ***blowing characteristics***. These ratings apply to all fuses regardless of their size, shape, or style. The current rating specifies the maximum current the fuse can carry continuously without melting or burning open (blowing). The thinner the metal fuse element, the lower its current rating. The voltage rating specifies the maximum voltage a blown

Figure 11–16 *(a)* Glass-cartridge fuse. *(b)* Fuse holder. *(c)* Panel-mounted fuse holder.

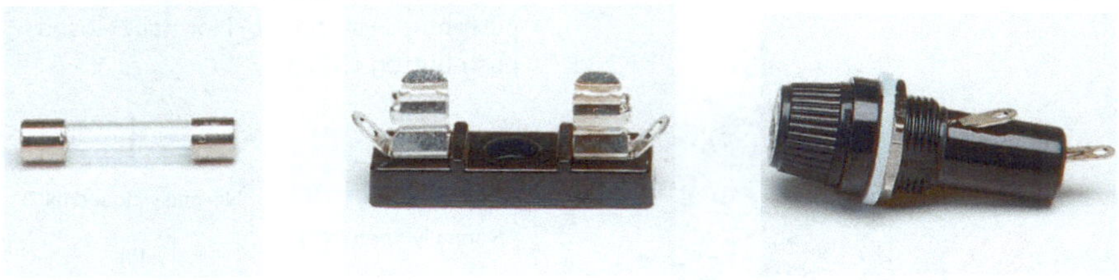

(a) (b) (c)

(a–c): Cindy Schroeder/McGraw Hill

fuse can withstand without internal arcing across the open fuse element. And finally, the blowing characteristic indicates how rapidly the fuse blows or opens when the current in the fuse exceeds its rated value by some specified amount. There are three blowing characteristics: ***fast-acting***, ***medium-acting***, and ***slow-acting***. (Slow-acting fuses are usually referred to as **slow-blow fuses**.) The blowing times of each type are compared in Fig. 11–17. Slow-blow fuses have a coiled construction and are designed to open only when there is a continued overload such as a short circuit. The purpose of the coiled construction is to prevent the fuse from blowing during a temporary current surge. Circuits with electric motors typically use slow-blow fuses because the initial starting current is much greater than when the motor is running at its normal speed. In most applications, however, fast-acting and medium-acting fuses are most common.

Fuses are available with current ratings ranging anywhere from 1/500 A to several hundred amperes. Also, fuses are available in a wide range of voltage ratings. The wide variety of different ratings is necessary to meet the vast number of applications in which they are used. The current and voltage ratings are typically listed somewhere on the fuse.

Circuit Breakers

A **circuit breaker** can be used in place of a fuse to protect circuit components and wiring against the high current caused by a short circuit. It is constructed of a thin bimetallic strip that expands with heat and in turn trips open the circuit. The advantage of a circuit breaker is that it can be reset once the bimetallic strip cools down

Figure 11–17 Chart showing percentage of rated current vs. blowing time for fuses.

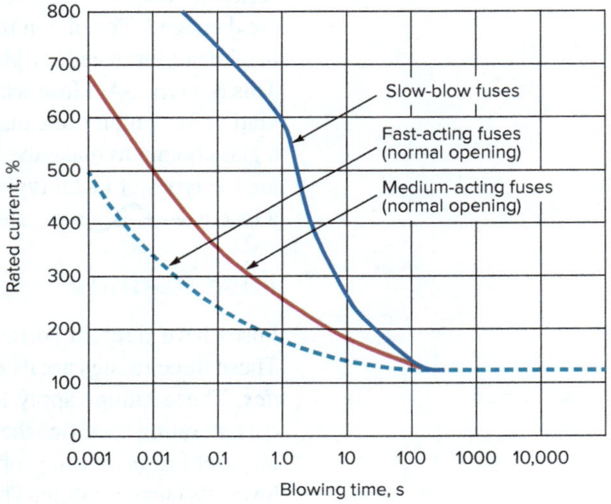

MultiSim **Figure 11–18** When a fuse opens, the applied voltage is across the fuse terminals. (*a*) Circuit closed with good fuse. Note schematic symbol for any type of fuse. (*b*) Fuse open. Voltage readings are explained in the text.

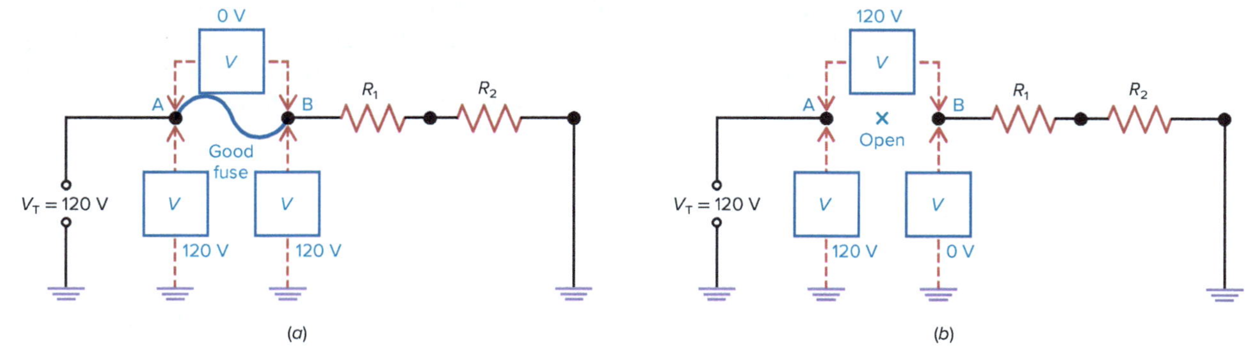

and the short circuit has been removed. Because they can be reset, almost all new residential house wiring is protected by circuit breakers rather than fuses. The schematic symbol for a circuit breaker is often shown as ⌒.

GOOD TO KNOW

For safety reasons, it is always best to remove a fuse from the circuit before measuring its resistance with an ohmmeter.

Testing Fuses

In glass fuses, you can usually see whether the wire element inside is burned open. When measured with an ohmmeter, a good fuse has practically zero resistance. An open fuse reads infinite ohms. Power must be off or the fuse must be out of the circuit to test a fuse with an ohmmeter.

When you test with a voltmeter, a good fuse has zero volts across its two terminals (Fig. 11–18*a*). If you read appreciable voltage across the fuse, this means that it is open. In fact, the full applied voltage is across the open fuse in a series circuit, as shown in Fig. 11–18*b*. This is why fuses also have a voltage rating, which gives the maximum voltage without arcing in the open fuse.

Referring to Fig. 11–18, notice the results when measuring the voltages to ground at the two fuse terminals. In Fig. 11–18*a*, the voltage is the same 120 V at both ends because there is no voltage drop across the good fuse. In Fig. 11–18*b*, however, terminal B reads 0 V because this end is disconnected from V_T by the open fuse. These tests apply to either DC or AC voltages.

11–7 Self-Review

Answers at the end of the chapter.

a. How much is the resistance of a good fuse?
b. How much is the *IR* voltage drop across a good fuse?

11–8 Wire Resistance

GOOD TO KNOW

The resistance of a wire conductor is directly proportional to its length and inversely proportional to its cross-sectional area.

The longer a wire, the higher its resistance. More work must be done to make electrons drift from one end to the other. However, the greater the diameter of the wire, the less the resistance, since there are more free electrons in the cross-sectional area. As a formula,

$$R = \rho \frac{l}{A} \qquad \textbf{(11–1)}$$

where R is the total resistance, l the length, A the cross-sectional area, and ρ the **specific resistance** or *resistivity* of the conductor. The factor ρ then enables the resistance of different materials to be compared according to their nature without regard to different lengths or areas. Higher values of ρ mean more resistance. Note that ρ is the Greek letter *rho*.

Specific Resistance

Table 11–2 lists resistance values for different metals having the standard wire size of a 1-ft length with a cross-sectional area of 1 cmil. This rating is the *specific resistance* of the metal, in circular-mil ohms per foot. Since silver, copper, gold, and aluminum are the best conductors, they have the lowest values of specific resistance. Tungsten and iron have much higher resistance.

Example 11-3

How much is the resistance of 100 ft of No. 20 gage copper wire?

ANSWER Note that from Table 11–1, the cross-sectional area for No. 20 gage wire is 1022 cmil; from Table 11–2, the ρ for copper is 10.4. Using Formula (11–1) gives

$$R = \rho \frac{l}{A}$$

$$= 10.4 \frac{\text{cmil} \cdot \Omega}{\text{ft}} \times \frac{100 \text{ ft}}{1022 \text{ cmil}}$$

$$R = 1.02\ \Omega$$

Table 11–2 Properties of Conducting Materials*

Material	Description and Symbol	Specific Resistance (ρ) at 20°C, cmil • Ω/ft	Temperature Coefficient per °C, α	Melting Point, °C
Aluminum	Element (Al)	17	0.004	660
Carbon	Element (C)	†	−0.0003	3000
Constantan	Alloy, 55% Cu, 45% Ni	295	0 (average)	1210
Copper	Element (Cu)	10.4	0.004	1083
Gold	Element (Au)	14	0.004	1063
Iron	Element (Fe)	58	0.006	1535
Manganin	Alloy, 84% Cu, 12% Mn, 4% Ni	270	0 (average)	910
Nichrome	Alloy 65% Ni, 23% Fe, 12% Cr	676	0.0002	1350
Nickel	Element (Ni)	52	0.005	1452
Silver	Element (Ag)	9.8	0.004	961
Steel	Alloy, 99.5% Fe, 0.5% C	100	0.003	1480
Tungsten	Element (W)	33.8	0.005	3370

* Listings approximate only, since precise values depend on exact composition of material.
† Carbon has about 2500 to 7500 times the resistance of copper. Graphite is a form of carbon.

All units cancel except the ohms for R. Note that 1.02 Ω for 100 ft is approximately one-tenth the resistance of 10.35 Ω for 1000 ft of No. 20 copper wire listed in Table 11–1, showing that the resistance is proportional to length. Also note that a wire that is three gage sizes higher has half the circular area and double the resistance for the same wire length.

Example 11-4

How much is the resistance of a 100-ft length of No. 23 gage copper wire?

ANSWER

$$R = \rho\frac{l}{A}$$

$$= 10.4\,\frac{\text{cmil} \cdot \Omega}{\text{ft}} \times \frac{100\text{ ft}}{509.5\text{ cmil}}$$

$$R = 2.04\ \Omega$$

Units of Ohm-Centimeters for ρ

Except for wire conductors, specific resistances are usually compared for the standard size of a 1-cm cube. Then ρ is specified in Ω • cm for the unit cross-sectional area of 1 cm^2.

As an example, pure germanium has $\rho = 55\ \Omega \cdot \text{cm}$, as listed in Table 11–3. This value means that R is 55 Ω for a cube with a cross-sectional area of 1 cm^2 and length of 1 cm.

For other sizes, use Formula (11–1) with l in cm and A in cm^2. Then all units of size cancel to give R in ohms.

Example 11-5

How much is the resistance of a slab of germanium 0.2 cm long with a cross-sectional area of 1 cm^2?

ANSWER

$$R = \rho\frac{l}{A}$$

$$= 55\ \Omega \cdot \text{cm} \times \frac{0.2\text{ cm}}{1\text{ cm}^2}$$

$$R = 11\ \Omega$$

The same size slab of silicon would have R of 11,000 Ω. Note from Table 11–3 that ρ is 1000 times more for silicon than for germanium.

Table 11–3	Comparison of Resistivities	
Material	**ρ, Ω · cm, at 25°C**	**Description**
Silver	1.6×10^{-6}	Conductor
Germanium	55	Semiconductor
Silicon	55,000	Semiconductor
Mica	2×10^{12}	Insulator

Types of Resistance Wire

For applications in heating elements, such as a toaster, an incandescent lightbulb, or a heater, it is necessary to use wire that has more resistance than good conductors like silver, copper, or aluminum. Higher resistance is preferable so that the required amount of I^2R power dissipated as heat in the wire can be obtained without excessive current. Typical materials for resistance wire are the elements tungsten, nickel, or iron and alloys* such as manganin, Nichrome, and constantan. These types are generally called *resistance wire* because R is greater than that of copper wire for the same length.

11–8 Self-Review

Answers at the end of the chapter.

a. Does Nichrome wire have less or more resistance than copper wire?

b. For 100 ft of No. 14 gage copper wire, R is 0.26 Ω. How much is R for 1000 ft?

11–9 Temperature Coefficient of Resistance

This factor with the symbol alpha (α) states how much the resistance changes for a change in temperature. A positive value for α means that R increases with temperature; with a negative α, R decreases; zero for α means that R is constant. Some typical values of α for metals and for carbon are listed in Table 11–2 in the fourth column.

Positive α

All metals in their pure form, such as copper and tungsten, have positive **temperature coefficients**. The α for tungsten, for example, is 0.005. Although α is not exactly constant, an increase in wire resistance caused by a rise in temperature can be calculated approximately from the formula

$$R_t = R_0 + R_0(\alpha \Delta t) \qquad \textbf{(11–2)}$$

where R_0 is the resistance at 20°C, R_t is the higher resistance at the higher temperature, and Δt is the temperature rise above 20°C.

* An *alloy* is a fusion of elements without chemical action between them. Metals are commonly alloyed to alter their physical characteristics.

Example 11-6

A tungsten wire has a 14-Ω R at 20°C. Calculate its resistance at 120°C.

ANSWER The temperature rise Δt here is 100°C; α is 0.005. Substituting in Formula (11–2),

$$R_t = 14 + 14(0.005 \times 100)$$
$$= 14 + 7$$
$$R_t = 21\ \Omega$$

The added resistance of 7 Ω increases the wire resistance by 50% because of the 100°C rise in temperature.

In practical terms, a positive α means that heat increases R in wire conductors. Then I is reduced for a specified applied voltage.

Negative α

Note that carbon has a negative temperature coefficient. In general, α is negative for all semiconductors, including germanium and silicon. Also, all electrolyte solutions, such as sulfuric acid and water, have a negative α.

A negative value of α means less resistance at higher temperatures. The resistance of semiconductor diodes and transistors, therefore, can be reduced appreciably when they become hot with normal load current.

Zero α

This means that R is constant with changes in temperature. The metal alloys constantan and manganin, for example, have a value of zero for α. They can be used for precision wire-wound resistors that do not change resistance when the temperature increases.

Hot Resistance

Because resistance wire is made of tungsten, Nichrome, iron, or nickel, there is usually a big difference in the amount of resistance the wire has when hot in normal operation and when cold without its normal load current. The reason is that the resistance increases with higher temperatures, since these materials have a positive temperature coefficient, as shown in Table 11–2.

As an example, the tungsten filament of a 100-W, 120-V incandescent bulb has a current of 0.833 A when the bulb lights with normal brilliance at its rated power, since $I = P/V$. By Ohm's law, the hot resistance is V/I, or $V/0.833$ A, which equals 144 Ω. If, however, the filament resistance is measured with an ohmmeter when the bulb is not lit, the cold resistance is only about 10 Ω.

The Nichrome heater elements in appliances and the tungsten heaters in vacuum tubes also become several hundred degrees hotter in normal operation. In these cases, only the cold resistance can be measured with an ohmmeter. The hot resistance must be calculated from voltage and current measurements with the normal value of load current. As a practical rule, the cold resistance is generally about one-tenth the hot resistance. In troubleshooting, however, the approach is usually just to check whether the heater element is open. Then it reads infinite ohms on the ohmmeter.

Superconductivity

The effect opposite to hot resistance occurs when cooling a metal down to very low temperatures to reduce its resistance. Near absolute zero, 0 K or −273°C, some metals abruptly lose practically all their resistance. As an example, when cooled by liquid helium, the metal tin becomes superconductive at 3.7 K. Tremendous currents can be produced, resulting in very strong electromagnetic fields. Such work at very low temperatures, near absolute zero, is called *cryogenics*.

New types of ceramic materials have been developed and are stimulating great interest in superconductivity because they provide zero resistance at temperatures much above absolute zero. One type is a ceramic pellet, with a 1-in. diameter, that includes yttrium, barium, copper, and oxygen atoms. The superconductivity occurs at a temperature of 93 K, equal to −160°C. This value is still far below room temperature, but the cooling can be done with liquid nitrogen, which is much cheaper than liquid helium. As research continues, it is likely that new materials will be discovered that are superconductive at even higher temperatures.

11–9 Self-Review

Answers at the end of the chapter.

a. Metal conductors have more R at higher temperatures. (True/False)
b. Tungsten can be used for resistance wire. (True/False)
c. A superconductive material has practically zero resistance. (True/False)

GOOD TO KNOW

A fluorescent light is a good example of an electric current in a gas.

11–10 Ion Current in Liquids and Gases

We usually think of metal wire for a conductor, but there are other possibilities. Liquids such as saltwater or dilute sulfuric acid can also allow the movement of electric charges. For gases, consider the neon glow lamp, in which neon serves as a conductor.

The mechanism may be different for conduction in metal wire, liquids, or gases, but in any case, the current is a motion of charges. Furthermore, either positive or negative charges can be the carriers that provide electric current. The amount of current is Q/T. For one coulomb of charge per second, the current is one ampere.

In solid materials such as metals, the atoms are not free to move among each other. Therefore, conduction of electricity must take place by the drift of free electrons. Each atom remains neutral, neither gaining nor losing charge, but the metals are good conductors because they have plenty of free electrons that can be forced to drift through the solid substance.

In liquids and gases, however, each atom can move freely among all the other atoms because the substance is not solid. As a result, the atoms can easily take on electrons or lose electrons, particularly the valence electrons in the outside shell. The result is an atom that is no longer electrically neutral. Adding one or more electrons produces a negative charge; the loss of one or more electrons results in a positive charge. The charged atoms are called *ions*. Such charged particles are commonly formed in liquids and gases.

Figure 11–19 Formation of ions. (*a*) Normal sodium (Na) atom. (*b*) Positively charged ion indicated as Na^+, missing one free electron.

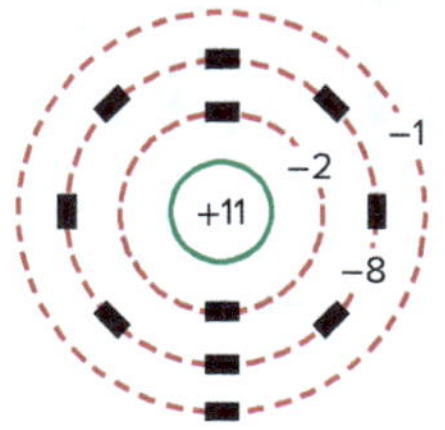

(*a*)

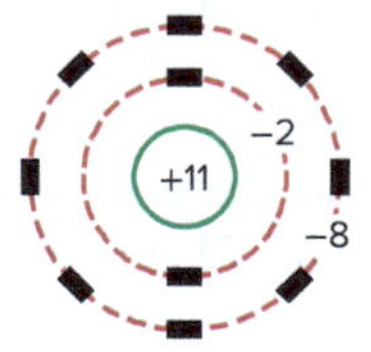

(*b*)

The Ion

An ion is an atom, or group of atoms, that has a net electric charge, either positive or negative, resulting from a loss or gain of electrons. In Fig. 11–19*a*, the sodium atom is neutral, with 11 positive charges in the nucleus balanced by 11 electrons in the outside shells. This atom has only one electron in the shell farthest from the nucleus. When the sodium is in solution, this one electron can easily leave the atom. The reason may be another atom close by that needs one electron to have a stable ring of eight electrons in its outside shell. Notice that if the sodium atom loses one valence electron, the atom will still have an outside ring of eight electrons, as shown in Fig. 11–19*b*. This sodium atom now is a positive ion, with a charge equal to one proton.

Current of Ions

Just as in electron flow, opposite ion charges are attracted to each other, and like charges repel. The resultant motion of ions provides electric current. In liquids and gases, therefore, conduction of electricity results mainly from the movement of ions. This motion of ion charges is called ***ionization current***. Since an ion includes the nucleus of the atom, the ion charge is much heavier than an electron charge and moves with less velocity. We can say that ion charges are less mobile than electron charges.

The direction of ionization current can be the same as that of electron flow or the opposite. When negative ions move, they are attracted to the positive terminal of an applied voltage in the same direction as electron flow. However, when positive ions move, this ionization current is in the opposite direction, toward the negative terminal of an applied voltage.

For either direction, though, the amount of ionization current is determined by the rate at which the charge moves. If 3 C of positive ion charges move past a given point per second, the current is 3 A, the same as 3 C of negative ions or 3 C of electron charges.

Ionization in Liquids

Ions are usually formed in liquids when salts or acids are dissolved in water. Saltwater is a good conductor because of ionization, but pure distilled water is an insulator. In addition, some metals immersed in acids or alkaline solutions ionize. Liquids that are good conductors because of ionization are called *electrolytes*. In general, electrolytes have a negative value of α, as more ionization at higher temperatures lowers the resistance.

Ionization in Gases

Gases have a minimum striking or ionization potential, which is the lowest applied voltage that will ionize the gas. Before ionization, the gas is an insulator, but the ionization current makes the ionized gas have a low resistance. An ionized gas usually glows. Argon, for instance, emits blue light when the gas is ionized. Ionized neon gas glows red. The amount of voltage needed to reach the striking potential varies with different gases and depends on the gas pressure. For example, a neon glow lamp for use as a night light ionizes at approximately 70 V.

Ionic Bonds

The sodium ion in Fig. 11–19*b* has a charge of 1+ because it is missing one electron. If such positive ions are placed near negative ions with a charge of 1−, there will be an electrical attraction to form an ionic bond.

Figure 11–20 Ionic bond between atoms of sodium (Na) and chlorine (Cl) to form a molecule of sodium chloride (NaCl).

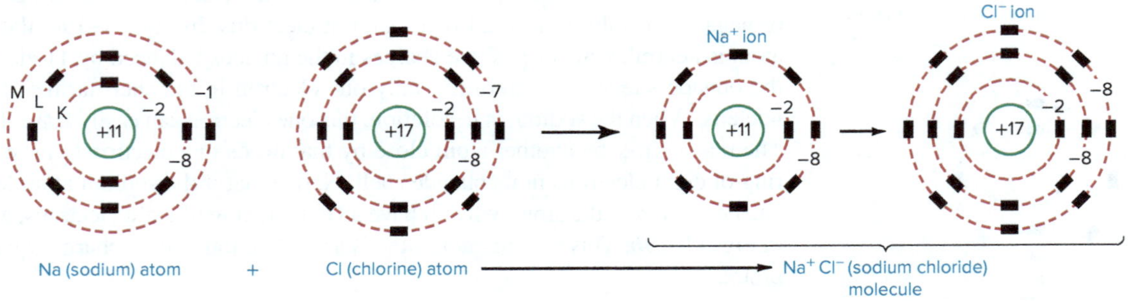

A common example is the combination of sodium (Na) ions and chlorine (Cl) ions to form table salt (NaCl), as shown in Fig. 11–20. Notice that the one outer electron of the Na atom can fit into the seven-electron shell of the Cl atom. When these two elements are combined, the Na atom gives up one electron to form a positive ion, with a stable L shell having eight electrons; also, the Cl atom adds this one electron to form a negative ion, with a stable M shell having eight electrons. The two opposite types of ions are bound in NaCl because of the strong attractive force between opposite charges close together.

The ions in NaCl can separate in water to make saltwater a conductor of electricity; pure water is not a conductor of electricity. When current flows in saltwater, then, the moving charges must be ions, another example of ionization current.

11–10 Self-Review

Answers at the end of the chapter.

a. How much is *I* for 2 C/s of positive ion charges?
b. Which have the greatest mobility: positive ions, negative ions, or electrons?
c. A dielectric material is a good conductor of electricity. (True/False)

11–11 Insulators

Substances that have very high resistance, of the order of many megohms, are classified as insulators. With such high resistance, an insulator cannot conduct appreciable current when voltage is applied. As a result, insulators can have either of two functions. One is to isolate conductors to eliminate conduction between them. The other is to store an electric charge when voltage is applied.

An insulator maintains its charge because electrons cannot flow to neutralize the charge. The insulators are commonly called ***dielectric materials***, which means that they can store a charge.

Among the best insulators, or dielectrics, are air, vacuum, rubber, wax, shellac, glass, mica, porcelain, oil, dry paper, textile fibers, and plastics such as Bakelite, Formica, and polystyrene. Pure water is a good insulator, but saltwater is not. Moist earth is a fairly good conductor, and dry, sandy earth is an insulator.

For any insulator, a high enough voltage can be applied to break down the internal structure of the material, forcing it to fail. This dielectric breakdown usually produces an arc. The intense heat of an electric arc, across the surface (tracking) or "through" an insulator (puncture), can permanently change the physical structure of the material, lowering its resistance and rendering it useless as an insulator. Table 11–4 compares several insulators in terms of dielectric strength, which is the voltage breakdown rating. The higher the dielectric strength, the

Table 11–4	Voltage Breakdown of Insulators		
Material	**Dielectric Strength, V/mil**	**Material**	**Dielectric Strength, V/mil**
Air or vacuum	20	Paraffin wax	200–300
Bakelite	300–550	Phenol, molded	300–700
Fiber	150–180	Polystyrene	500–760
Glass	335–2000	Porcelain	40–150
Mica	600–1500	Rubber, hard	450
Paper	1250	Shellac	900
Paraffin oil	380		

better the insulator, since it is less likely to break down at a high value of applied voltage. The breakdown voltages in Table 11–4 are approximate values for the standard thickness of 1 mil, or 0.001 in. More thickness allows a higher breakdown-voltage rating. Note that the value of 20 V/mil for air or vacuum is the same as 20 kV/in.

Insulator Discharge Current

An insulator in contact with a voltage source stores charge, producing a potential on the insulator. The charge tends to remain on the insulator, but it can be discharged by one of the following methods:

1. Conduction through a conducting path. For instance, a wire across the charged insulator provides a discharge path. Then the discharged dielectric has no potential.
2. Brush discharge. As an example, high voltage on a sharp pointed wire can discharge through the surrounding atmosphere by ionization of the air molecules. This may be visible in the dark as a bluish or reddish glow, called the ***corona effect***.
3. Spark discharge. This is a result of breakdown in the insulator because of a high potential difference that ruptures the dielectric. The current that flows across the insulator at the instant of breakdown causes the spark.

A corona is undesirable because it reduces the potential by brush discharge into the surrounding air. In addition, the corona often indicates the beginning of a spark discharge. A potential of the order of kilovolts is usually necessary for a corona because the breakdown voltage for air is approximately 20 kV/in. To reduce the corona effect, conductors that have high voltage should be smooth, rounded, and thick. This equalizes the potential difference from all the points on the conductor to the surrounding air. Any sharp point can have a more intense field, making it more susceptible to a corona and eventual spark discharge.

11–11 Self-Review

Answers at the end of the chapter.

a. Which has a higher voltage breakdown rating, air or mica?
b. Can 30 kV arc across an air gap of 1 in.?

GOOD TO KNOW

A mechanical switch can wear out by being opened and closed too many times. In most cases, a manufacturer will specify the number of times a switch can be opened or closed before wearing out.

11–12 Troubleshooting Hints for Wires and Connectors

For all types of electronic equipment, a common problem is an open circuit in the wire conductors, the connectors, and the switch contacts.

You can check continuity of conductors, both wires and printed wiring, with an ohmmeter. A good conductor reads 0 Ω for continuity. An open reads infinite ohms.

A connector can also be checked for continuity between the wire and the connector itself. Also, the connector may be tarnished, oxide coated, or rusted. Then it must be cleaned with either fine sandpaper or emery cloth. Sometimes, it helps just to pull out the plug and reinsert it to make sure of tight connections.

With a plug connector for cable, make sure the wires have continuity to the plug. Except for the F-type connector, most plugs require careful soldering to the center pin.

A switch with dirty or pitted contacts can produce intermittent operation. In most cases, the switch cannot be dissassembled for cleaning. Therefore, the switch must be replaced with a new one.

11–12 Self-Review

Answers at the end of the chapter.

a. Printed wiring cannot be checked for continuity with an ohmmeter. (True/False)

b. A tarnished or rusty connection has higher-than-normal resistance. (True/False)

Application of Standard Wire Gauge Sizes

In any electrical wire installation, such as residential house wiring, using the proper size (gauge) wire is critical. The required gauge of wire will depend on several factors which includes; the type of wire, the required length of the wire, the maximum current carried by the wire and the maximum allowable voltage drop across the wire. There are other factors to consider but they are of lesser importance and therefore not listed here. One very important wire rating to consider is its ***ampacity*** rating. The ampacity rating is the amount of current the wire can safely carry without becoming too hot. The I^2R power dissipation in the wire produces heat and if the power dissipation becomes excessive, the wire can get too hot and cause a fire. Table 11–5 lists several examples of the different gauges of wire used for the devices and appliances found in our homes. It should be noted that Table 11–5 is for copper (Cu) wire only. Notice that Table 11–5 lists the wires use or application first, followed by the typical wire size (gauge number) used for that application. At the far right, the ampacity rating is listed for each of the different gauge wires. This table provides only a general guideline and is in no way intended to reflect all of the criteria listed above in considering wire size.

NONMETALLIC-SHEATHED CABLE COLOR CODING SCHEME

Nonmetallic-sheathed cable (NM) is a type of covered electrical wire consisting of at least two insulated conductors and one bare conductor. This is the type of cable used for indoor residential house wiring. The outer sheath, or jacket, surrounds and protects the inner wire conductors and is made of a moisture-resistant, flame retardant, non metallic material. The most common type of indoor residential wiring is NM-B cable. The letter B (after NM) indicates an insulation heat rating of 194° Fahrenheit. This rating ensures that the wire conductors can operate at certain levels without overheating and thus melting the insulation surrounding the conductors. Overheating and melting the insulation could cause a fire and thus create a safety hazard. Most NM-B cable, made after 2001, is sheathed using different colors. The color of the outer sheath indicates the gauge number of the wires enclosed within the sheath. The four colors of NM-B cable used for indoor residential house wiring are; white, yellow, orange, and black. The different sheath colors and there corresponding wire gauge sizes are:

White – 14 gauge wire
Yellow – 12 gauge wire
Orange – 10 gauge wire
Black – 8 gauge wire and larger

It is important to note that the wire sizing guidelines specified in Table 11–5 are for NM-B cable, which is the type used for indoor residential wiring.

The color ***gray*** is also used in conjunction with residential electrical wiring but the letters UF-B rather than NM-B appear on the outer sheath. The letters UF stand for "Underground Feeder" indicating the cable can safely be buried underground. However, the color gray in no way indicates the gauge size of the wire conductors. Gray sheathed UF-B cable is available in a variety of different gauge sizes. The gray sheath is much different than the other colored sheaths, however, because it's actually molded around each conductor individually rather than just being a single sheath covering all of the conductors. This molded encasement protects and separates each wire within the sheath and prevents moisture and other external elements from corroding and deteriorating the wire conductors.

Having knowledge of the color coding scheme allows us to determine, at a glance, the gauge number of an NM-B cable.

Table 11–5 Electrical Wire Sizing Guidelines for NM-B Cable

Wire use	Wire Size	Rated Ampacity
Light Fixtures, Lamps, Lighting Runs and 120 V Receptacles	14 Gauge	15 Amperes
Kitchen Appliances, 120 V Air Conditioners, Sump Pumps, Some 120 V Receptacles	12 Gauge	20 Amperes
240 V Electric clothes dryers, Air Conditioners, Electric Water Heaters	10 Gauge	30 Amperes
Cook top Stoves	8 Gauge	45 Amperes
Electric Furnaces, Large Electric Heaters	6 Gauge	60 Amperes
Electric Furnaces, Large Electric Water Heaters and Sub Panels	4 Gauge	80 Amperes
Service Panels, Sub Panels	2 Gauge	100 Amperes
Service Entrance Panel	1/0 Gauge	150 Amperes
Service Entrance Panel	2/0 Gauge	200 Amperes

Note: Number 6 gauge wire and larger (no. 4, no. 2, etc.) is always made of stranded wire.

However, NM-B cable sold before 2001 may not use the color coding scheme described above. Therefore, always check the labeling printed on the outer sheath or jacket to be absolutely sure of the wires gauge size.

Figure 11-21

(a)

(b)

(c)

NM-B/UF-B SHEATHED CABLE OUTER JACKET LABELING

The labeling on the outer sheath or jacket of an NM-B or UF-B cable provides us with important information about the insulated wires concealed within the sheath. It tells us how many insulated conductors there are, the gauge of the wire, the type of insulation and the voltage rating of the cable. Figure 11-21*a* shows a white sheathed NM-B cable. The labeling on the outer sheath provides the following information; "Type NM-B, 14-2 with ground, 600 V". The letters "NM-B" tell us that this wire is suitable for indoor residential house wiring. Also, "14-2 with ground" tells us there are two insulated 14 gauge wires (black and white) plus one bare ground wire. The 600 V marking indicates the voltage rating of the cable. Figure 11-21*b* shows a yellow sheathed NM-B cable with the labeling "Type NM-B, 12-3 with ground, 600 V". Again, the letters "NM-B" tell us that this wire is suitable for indoor residential house wiring. Also "12-3 with ground" tell us that within the yellow sheath there are three insulated 12 gauge wires (black, red and white) plus one bare ground wire. The 600 V marking indicates the voltage rating of the cable. Notice that the white-colored sheath in Fig. 11-21*a* contains 14 gauge wires, whereas the yellow colored sheath in Fig. 11-21*b* contains 12 gauge wires. Figure 11-21*c* shows a gray UF-B cable with the markings 14-3, indicating it has three insulated conductors (black, red, and white) plus one bare ground wire.

MISCELLANEOUS APPLICATIONS OF INSULATED ELECTRICAL WIRE

The use of extension cords is commonplace in our homes, schools, businesses and industries. They are necessary when the power cord of an electrical device is not long enough to reach the nearest 120 V_{AC} outlet. Extension cords are made of stranded copper wire and come in a variety of different lengths and gauge sizes. The length and gauge size of an extension cord must be suitable for the application in which it is being used. If the extension cord is too long, or not of the appropriate gauge size, it can result in excessive voltage being dropped across the wire conductors. For example, when using extension cords in conjunction with power tools, be sure to select an extension cord with the proper gauge size wires. If the gauge size of the wire conductors is inadequate or if the extension cord is too long, it could result in either less than optimal performance of the power tool or the power tool not operating at all. The gauge size of an extension cord should have a current rating that equals or exceeds the current drawn by the load connected to it. Most extension cords available on the market today have both a current and a power (wattage) rating. The wattage rating indicates the highest wattage load that can be connected to the extension cord. When extension cords are connected together to extend the overall length, the current and power ratings of each individual extension cord no longer apply. This is because the total resistance of the wire conductors increases with length. In other words, the current and power ratings decrease when two or more extension cords are connected together. One more point. Extension cords are designed for either indoor or outdoor use. Extension cords designed for outdoor use have a heavier duty insulation jacket as compared to those designed for indoor use only. Although outdoor extension cords can always be used indoors, indoor extension cords should never be used outdoors! Also, never use an extension cord whose insulation is damaged or compromised. Exposed strands of copper wire present a safety hazard to those working in and around the area where the extension cord is being used.

Speaker wire is generally made of 16 or 18 gauge stranded copper wire. However, 12 or 14 gauge stranded wire is often required in home entertainment systems due to the high amount of power being delivered to the speakers. Holiday lights typically use number 22 gauge stranded copper wire. This is more than adequate because the amount of current carried by these conductors is typically less than three amperes.

Summary

- A conductor has very low resistance. All metals are good conductors; the best are silver, copper, and aluminum. Copper is generally used for wire conductors due to its lower cost.
- The sizes for copper wire are specified by the American Wire Gage. Higher gage numbers mean thinner wire. Typical sizes are No. 22 gage hookup wire for electronic circuits and No. 12 and No. 14 for house wiring.
- The cross-sectional area of round wire is measured in circular mils. One mil is 0.001 in. The area in circular mils equals the diameter in mils squared.
- The resistance *R* of a conductor can be found using the formula $R = \rho(l/A)$, where ρ is the specific resistance, l is the length of the conductor, and A is the cross-sectional area of the conductor. Wire resistance increases directly with length l, but decreases inversely with cross-sectional area A.
- The voltage drop across a closed switch in a series circuit is zero volts. When open, the switch has the applied voltage across it.
- A fuse protects circuit components and wiring against overload; excessive current melts the fuse element to open the entire series circuit. A good fuse has very low resistance and practically zero voltage across it.
- Ionization in liquids and gases produces atoms that are not electrically neutral. These are ions. Negative ions have an excess of electrons; positive ions have a deficiency of electrons. In liquids and gases, electric current is a result of the movement of ions.
- The resistance of pure metals increases with temperature. For semiconductors and liquid electrolytes, resistance decreases at higher temperatures.
- An insulator has very high resistance. Common insulating materials are air, vacuum, rubber, paper, glass, porcelain, shellac, and plastics. Insulators are also called *dielectrics*.
- Superconductors have practically no resistance.
- Common circuit troubles are an open in wire conductors; dirty contacts in switches; and dirt, oxides, and corrosion on connectors and terminals.

Important Terms

Circuit breaker — a device used to protect the components and wiring in a circuit in the event of a short circuit. It is constructed of a thin bimetallic strip that expands with heat and in turn trips open the circuit. The circuit breaker can be reset after the bimetallic strip cools down and the short circuit is removed.

Circular mil (cmil) — a unit that specifies the cross-sectional area of round wire. 1 cmil is the cross-sectional area of a wire with a diameter, *d*, of 1 mil, where 1 mil = 0.001 in. For round wire the cmil area is equal to the square of the diameter, *d*, in mils.

Corona effect — a visible blue or red glow caused by ionization of the air molecules when the high voltage on a sharp, pointed wire discharges into the atmosphere.

Dielectric material — another name used in conjunction with insulating materials. An insulator is commonly referred to as a dielectric because it can hold or store an electric charge.

Fuse — a device used to protect the components and wiring in a circuit in the event of a short circuit. The fuse element is made of either aluminum, tin-coated copper, or nickel. Excessive current melts the fuse element which blows the fuse.

Ionization current — a current from the movement of ion charges in a liquid or gas.

Pole — the number of completely isolated circuits that can be controlled by a switch.

Slow-blow fuse — a type of fuse that can handle a temporary surge current that exceeds the current rating of the fuse. This type of fuse has an element with a coiled construction and is designed to open only on a continued overload such as a short circuit.

Specific resistance — the resistance of a metal conductor whose cross-sectional area is 1 cmil and whose length is 1 ft. The specific resistance, designated ρ, is specified in cmil · Ω/ft.

Switch — a component that controls whether the current is on or off in a circuit.

Temperature coefficient — a factor that indicates how much the resistance of a material changes with temperature. A positive temperature coefficient means that the resistance increases with temperature, whereas a negative temperature coefficient means that the resistance decreases with temperature.

Throw — the number of closed contact positions that exist per pole on a switch.

Wire gage — a number assigned to a specific size of round wire in terms of its diameter and cross-sectional area. The American Wire Gage (AWG) system provides a table of all wire sizes that includes the gage size, the diameter, *d*, in mils and the area, *A*, in circular mils (cmils).

Related Formulas

$$R = \rho\frac{l}{A}$$

$$R_t = R_0 + R_0(\alpha\Delta t)$$

Self-Test

Answers at the back of the book.

1. A closed switch has a resistance of approximately

a. infinity.

b. zero ohms.

c. 1 MΩ.

d. none of the above.

2. An open fuse has a resistance that approaches

a. infinity.

b. zero ohms.

c. 1 to 2 Ω.

d. none of the above.

3. How many connecting terminals does an SPDT switch have?

a. 2.

b. 6.

c. 3.

d. 4.

4. The voltage drop across a closed switch equals

a. the applied voltage.

b. zero volts.

c. infinity.

d. none of the above.

5. For round wire, as the gage numbers increase from 1 to 40

a. the diameter and circular area increase.

b. the wire resistance decreases for a specific length and type.

c. the diameter increases but the circular area remains constant.

d. the diameter and circular area decrease.

6. The circular area of round wire, doubles for

a. every 2 gage sizes.

b. every 3 gage sizes.

c. each successive gage size.

d. every 10 gage sizes.

7. Which has more resistance, a 100-ft length of No. 12 gage copper wire or a 100-ft length of No. 12 gage aluminum wire?

a. The 100-ft length of No. 12 gage aluminum wire.

b. The 100-ft length of No. 12 gage copper wire.

c. They both have exactly the same resistance.

d. It cannot be determined.

8. In their pure form, all metals have a

a. negative temperature coefficient.

b. temperature coefficient of zero.

c. positive temperature coefficient.

d. very high resistance.

9. The current rating of a switch corresponds to the maximum current the switch can safely handle when it is

a. open.

b. either open or closed.

c. closed.

d. none of the above.

10. How much is the resistance of a 2000-ft length of No. 20 gage aluminum wire?

a. less than 1 Ω.

b. 20.35 Ω.

c. 3.33 kΩ.

d. 33.27 Ω.

11. How many completely isolated circuits can be controlled by a DPST switch?

a. 1.

b. 2.

c. 3.

d. 4.

12. Which of the following metals is the best conductor of electricity?

a. steel.

b. aluminum.

c. silver.

d. gold.

13. What is the area in circular mils (cmils) of a wire whose diameter, *d*, is 0.01 in.?

a. 0.001 cmil.

b. 10 cmil.

c. 1 cmil.

d. 100 cmil.

14. The term *pole* as it relates to switches is defined as

a. the number of completely isolated circuits that can be controlled by the switch.

b. the number of closed contact positions that the switch has.

c. the number of connecting terminals the switch has.

d. none of the above.

15. The motion of ion charges in a liquid or gas is called

a. the corona effect.

b. hole flow.

c. superconductivity.

d. ionization current.

Essay Questions

1. Name three good metal conductors in order of lowest to highest resistance. Describe at least one application.

2. Name four insulators. Give one application.

3. Name two semiconductors.

4. Name two types of resistance wire. Give one application.

5. What is meant by the "dielectric strength of an insulator"?

6. Why does ionization occur more readily in liquids and gases, compared with solid metals? Give an example of ionization current.

7. Define the following: ion, ionic bond, and electrolyte.

8. Draw a circuit with two bulbs, a battery, and an SPDT switch that determines which bulb lights.

9. Why is it not possible to measure the hot resistance of a filament with an ohmmeter?

10. Give one way in which negative ion charges are similar to electron charges and one way in which they are different.

11. Define the following abbreviations for switches: SPST, SPDT, DPST, DPDT, NO, and NC.

12. Give two common circuit troubles with conductors and connector plugs.

Problems

SECTION 11–1 FUNCTION OF THE CONDUCTOR

11–1 In Fig. 11–22, an 8-Ω heater is connected to the 120-V_{AC} power line by two 50-ft lengths of copper wire. If each 50-ft length of wire has a resistance of 0.08 Ω, then calculate the following:

a. The total length of copper wire that connects the 8-Ω heater to the 120-V_{AC} power line.
b. The total resistance, R_T, of the circuit.
c. The current, I, in the circuit.
d. The voltage drop across each 50-ft length of copper wire.
e. The voltage across the 8-Ω heater.
f. The I^2R power loss in each 50-ft length of copper wire.
g. The power dissipated by the 8-Ω heater.
h. The total power, P_T, supplied to the circuit by the 120-V_{AC} power line.
i. The percentage of the total power, P_T, dissipated by the 8-Ω heater.

Figure 11-22

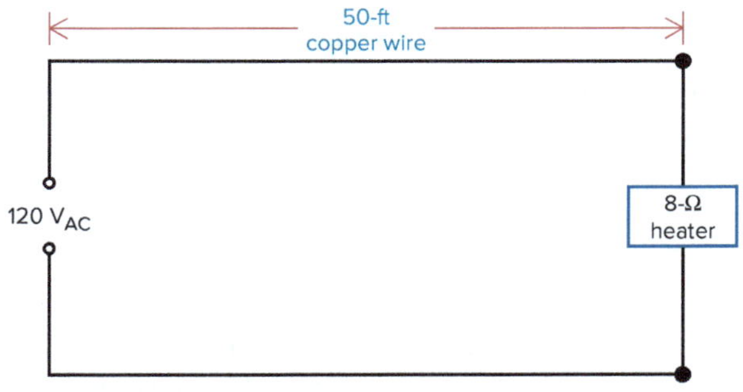

11–2 In Fig. 11–22, recalculate the values in steps a through i in Prob. 11–1 if the 8-Ω heater is replaced with a 24-Ω fan.

SECTION 11–2 STANDARD WIRE GAGE SIZES

11–3 Determine the area in circular mils for a wire if its diameter, d, equals

a. 0.005 in.
b. 0.021 in.
c. 0.032 in.
d. 0.05 in.
e. 0.1 in.
f. 0.2 in.

11–4 What is the approximate AWG size of a wire whose diameter, d, equals 0.072 in.?

11–5 Using Table 11–1, determine the resistance of a 1000-ft length of copper wire for the following gage sizes:

a. No. 10 gage.
b. No. 13 gage.
c. No. 16 gage.
d. No. 24 gage.

11–6 Which would you expect to have more resistance, a 1000-ft length of No. 14 gage copper wire or a 1000-ft length of No. 12 gage copper wire?

11–7 Which would you expect to have more resistance, a 1000-ft length of No. 23 gage copper wire or a 100-ft length of No. 23 gage copper wire?

SECTION 11–3 TYPES OF WIRE CONDUCTORS

11–8 A stranded wire consists of 41 strands of No. 30 gage copper wire. What is its equivalent gage size in solid wire?

11–9 If an extension cord is made up of 65 strands of No. 28 gage copper wire, what is its equivalent gage size in solid wire?

11–10 How many strands of No. 36 gage wire does it take to make a stranded wire whose equivalent gage size is No. 16?

11–11 What is the gage size of the individual strands in a No. 10 gage stranded wire if there are eight strands?

SECTION 11–6 SWITCHES

11–12 With the switch, S_1, closed in Fig. 11–23,

a. How much is the voltage across the switch?
b. How much is the voltage across the lamp?
c. Will the lamp light?
d. What is the current, I, in the circuit based on the specifications of the lamp?

Figure 11-23

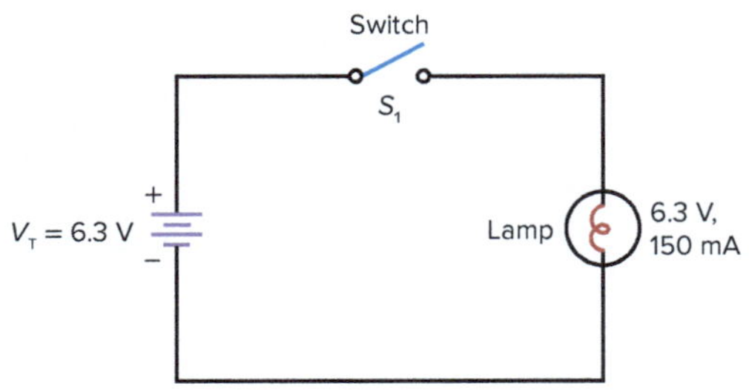

11–13 With the switch, S_1, open in Fig. 11–23,

a. How much is the voltage across the switch?

b. How much is the voltage across the lamp?

c. Will the lamp light?

d. What is the current, I, in the circuit based on the specifications of the lamp?

11–14 Draw a schematic diagram showing how an SPDT switch can be used to supply a resistive heating element with either 6 V or 12 V.

11–15 Draw a schematic diagram showing how a DPDT switch can be used to

a. allow a stereo receiver to switch between two different speakers.

b. reverse the polarity of voltage across a DC motor to reverse its direction.

11–16 An SPST switch is rated at 10 A/250 V. Can this switch be used to control a 120-V, 1000-W appliance?

SECTION 11–8 WIRE RESISTANCE

11–17 Calculate the resistance of the following conductors:

a. 250 ft of No. 20 gage copper wire.

b. 250 ft of No. 20 gage aluminum wire.

c. 250 ft of No. 20 gage steel wire.

11–18 Calculate the resistance of the following conductors:

a. 100 ft of No. 14 gage copper wire.

b. 200 ft of No. 14 gage copper wire.

11–19 Calculate the resistance of the following conductors:

a. 100 ft of No. 15 gage copper wire.

b. 100 ft of No. 18 gage copper wire.

11–20 How much is the resistance of a slab of silicon 0.1 cm long with a cross-sectional area of 1 cm^2?

11–21 What is the resistance for each conductor of a 50-ft extension cord made of No. 14 gage copper wire?

11–22 A 100-ft extension cord uses No. 14 gage copper wire for each of its conductors. If the extension cord is used to connect a 10-A load to the 120-V_{AC} power line, how much voltage is available at the load?

11–23 What is the smallest gage size copper wire that will limit the conductor voltage drop to 5 V when 120 V_{AC} is supplied to a 6-A load? The total line length of both conductors is 200 ft.

SECTION 11–9 TEMPERATURE COEFFICIENT OF RESISTANCE

11–24 A tungsten wire has a resistance, R, of 20 Ω at 20°C. Calculate its resistance at 70°C.

11–25 A steel wire has a resistance of 10 Ω at 20°C. Calculate its resistance at 100°C.

11–26 A nickel wire has a resistance of 150 Ω at 20°C. Calculate its resistance at 250°C.

11–27 An aluminum wire has a resistance of 100 Ω at 20°C. Calculate its resistance at 120°C.

11–28 The resistance of a Nichrome wire is 1 kΩ at 20°C. Calculate its resistance at 220°C.

11–29 The resistance of a steel wire is 10 Ω at 20°C. How much is its resistance at −30°C?

11–30 A No. 14 gage aluminum wire has 8.4 Ω of resistance at −20°C. What is its resistance at 50°C?

Critical Thinking

11–31 Use two switches having the appropriate number of poles and throws to build a partial decade resistance box. The resistance is to be adjustable in 1-Ω and 10-Ω steps from 0 to 99 Ω across two terminals identified as A and B. Draw the circuit showing all resistance values and switch connections.

11–32 Show how two SPDT switches can be wired to turn ON and OFF a light from two different locations. The voltage source is a 12-V battery.

Answers to Self-Reviews

11–1 **a.** 0.08 Ω
b. 0.07 V approx.
c. 0.06% approx.

11–2 **a.** 0.01646 Ω
b. 625 cmil
c. No. 16

11–3 **a.** true
b. true
c. true

11–4 **a.** true
b. true
c. false

11–5 **a.** 25 W
b. infinite ohms

11–6 **a.** zero
b. three
c. infinite
d. no

11–7 a. zero
b. zero

11–8 a. more
b. 2.6 Ω

11–9 a. true
b. true
c. true

11–10 a. 2 A
b. electrons
c. false

11–11 a. mica
b. yes

11–12 a. false
b. true

Laboratory Application Assignment

In this lab application assignment, you will learn how mechanical switches can be used to control or change the voltage and current in a circuit. This lab application assignment is different from the others in that you will not be given any schematic diagrams. It is up to you to draw the circuit diagram with the proper circuit connections based on the criteria specified.

Equipment: Obtain the following components from your instructor.

- Dual-output variable dc power supply
- Two 12-V incandescent lamps
- SPST, SPDT, and DPDT switches
- DMM
- 24-V_{DC} motor

Switching Circuits

1. In the space provided below, or on a separate sheet of paper, draw a schematic diagram showing how to use an SPDT switch to connect a 12-V incandescent lamp to either a 6-V or 12-V power supply. The brilliance of the bulb will be either dim or bright depending on the position of the switch. Build the circuit you have drawn, and verify that it functions properly. Have an instructor check your circuit.

2. In the space provided below, or on a separate sheet of paper, draw a schematic diagram showing how to use a DPDT switch to connect two 12-V incandescent lamps either in series or in parallel with a 12-V_{DC} power supply. The series or parallel connections must be controlled by the position of the switch. For one position, the 12-V incandescent lamps must be in series with the DC power supply. In this position, the lamps will be dim because each bulb will receive only 6 V. In the other position, the lamps must be in parallel with the DC power supply. In this position the bulbs will be much brighter because the full 12 V is across each bulb. Build the circuit you have drawn, and verify that it functions properly. Have an instructor check your circuit.

3. With the switches and DC voltage sources shown in Fig. 11–24, draw a schematic diagram that will control both the speed and direction of a 24-V_{DC} motor. Use the SPST switch to control whether the motor is on or off. The speed of the motor should be controlled using the SPDT switch and have two settings, ***low*** and ***high***. (The speed is determined by which voltage source is applied to the motor.) The direction of rotation is determined by the polarity of the voltage across the motor and should be controlled using the DPDT switch. When you have completed drawing the schematic diagram, construct the circuit and demonstrate its operation to an instructor.

Figure 11-24

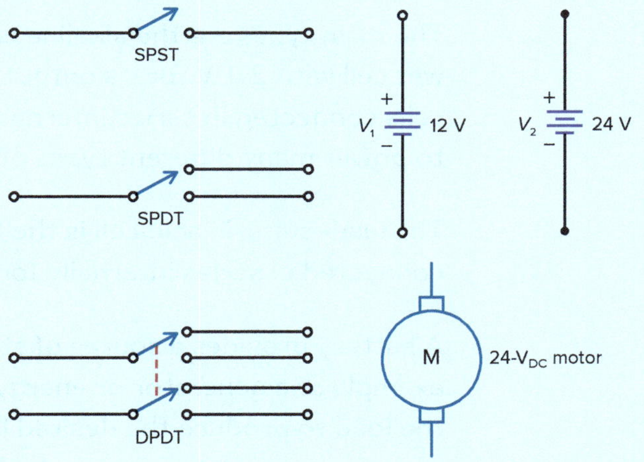

Design credits: Chapter Opener Background Image, and Multisim: Stockbyte/Getty Images

chapter

12

Batteries

A *battery* is a combination of individual voltaic cells that generate energy from an internal chemical reaction. Each cell consists of two different conducting materials as the electrodes that are immersed in an electrolyte. The chemical reaction between the electrodes and the electrolyte results in a separation of electric charges as ions and free electrons. Then the two electrodes have a difference of potential that provides a voltage output from the cell.

The main types are the alkaline cell with an output of 1.5 V and the lead–sulfuric acid wet cell with 2.1 V for its output. A common battery is the 9-V flat battery. It has six cells connected in series internally for an output of $6 \times 1.5 = 9$ V. Batteries are used to power many different types of portable electronic equipment.

The lead–sulfuric acid cell is the type used in most automobiles. Six cells are connected in series internally for a 12-V output.

A battery provides a source of steady DC voltage of fixed polarity and is a good example of a generator or energy source. The battery supplies voltage to a circuit as the load to produce the desired load current. An important factor is the internal resistance, r_i of the source, which affects the output voltage when a load is connected. A low r_i means that the source can maintain a constant output voltage for different values of load current. For the opposite case, a high r_i makes the output voltage drop, but a constant value of load current can be maintained.

Chapter Outline

Chapter Objectives

After studying this chapter, you should be able to

- *Explain* the difference between primary and secondary cells.
- *Define* what is meant by the *internal resistance of a cell*.
- *List* several different types of voltaic cells.
- *Explain* how cells can be connected to increase either the current capacity or voltage output of a battery.
- *Explain* why the terminal voltage of a battery drops with more load current.
- *Explain* the difference between voltage sources and current sources.
- *Explain* the concept of maximum power transfer.

Important Terms

ampere-hour (A•h) rating
battery
charging
constant-current generator
constant-voltage generator
discharging
float charging
fuel cell
hydrometer
internal resistance, r_i
open-circuit voltage
primary cell
secondary cell
specific gravity
storage cell
voltaic cell

GOOD TO KNOW

Most portable electronic equipment can be operated from either batteries or the 120-V_{AC} power line. When operated from the 120-V_{AC} power line, the 60-Hz alternating voltage is converted to the correct DC voltage by a special circuit known as a power supply.

12–1 Introduction to Batteries

We rely on batteries to power an almost unlimited number of electronic products available today. For example, batteries are used in cars, personal computers (PCs), handheld radios, laptops, cameras, MP3 players, and cell phones, to name just a few of the more common applications. Batteries are available in a wide variety of shapes and sizes and have many different voltage and current ratings. The different sizes and ratings are necessary to meet the needs of the vast number of applications. Regardless of the application, however, all batteries are made up of a combination of individual **voltaic cells**. Together, the cells provide a steady DC voltage at the output terminals of the **battery**. The voltage output and current rating of a battery are determined by several factors, including the type of elements used for the electrodes, the physical size of the electrodes, and the type of electrolyte.

As you know, some batteries become exhausted with use and cannot be recharged. Others can be recharged hundreds or even thousands of times before they are no longer able to produce or maintain the rated output voltage. Whether a battery is rechargeable or not is determined by the type of cells that make up the battery. There are two types of cells—**primary cells** and **secondary cells**.

Primary Cells

This type cannot be recharged. After it has delivered its rated capacity, the primary cell must be discarded because the internal chemical reaction cannot be restored. Figure 12–1 shows a variety of dry cells and batteries, all of which are of the primary type. In Table 12–1, several different cells are listed by name. Each of the cells is listed as either the primary or the secondary type. Notice the **open-circuit voltage** for each of the cell types listed.

Secondary Cells

This type can be recharged because the chemical action is reversible. When it supplies current to a load resistance, the cell is ***discharging*** because the current tends to neutralize the separated charges at the electrodes. For the opposite case, the current can be reversed to re-form the electrodes as the chemical action is reversed. This action is ***charging*** the cell. The charging current must be supplied by an external

Figure 12–1 Typical dry cells and batteries. These primary types cannot be recharged.

Mark Steinmetz/McGraw Hill

Table 12–1	Cell Types and Open-Circuit Voltage	
Cell Name	**Type**	**Nominal Open-Circuit* Voltage, V_{DC}**
Carbon-zinc	Primary	1.5
Zinc chloride	Primary	1.5
Manganese dioxide (alkaline)	Primary or secondary	1.5
Mercuric oxide	Primary	1.35
Silver oxide	Primary	1.5
Lithium	Primary	3.0
Lithium-ion	Secondary	3.7
Lead-acid	Secondary	2.1
Nickel-cadmium	Secondary	1.2
Nickel-metal-hydride	Secondary	1.2
Nickel-iron (Edison) cell	Secondary	1.2
Nickel-zinc	Secondary	1.6
Solar	Secondary	0.5

* Open-circuit *V* is the terminal voltage without a load.

GOOD TO KNOW

A battery is continually doing the work of separating the positive and negative charges within itself. This is true even when the battery is not in use. Because of this, batteries made up of primary cells can become exhausted before they ever leave the store. Therefore, always be sure to buy batteries that are fresh from the manufacturer.

DC voltage source, with the cell serving as a load resistance. The discharging and recharging is called *cycling* of the cell. Since a secondary cell can be recharged, it is also called a ***storage cell***. The most common type is the lead-acid cell generally used in automotive batteries (Fig. 12–2). In addition, the list in Table 12–1 indicates which are secondary cells.

Figure 12–2 Example of a 12-V auto battery using six lead-acid cells in series. This is a secondary type, which can be recharged.

Mark Steinmetz/McGraw Hill

Dry Cells

What we call a *dry cell* really has a moist electrolyte. However, the electrolyte cannot be spilled and the cell can operate in any position.

Sealed Rechargeable Cells

This type is a secondary cell that can be recharged, but it has a sealed electrolyte that cannot be refilled. These cells are capable of charge and discharge in any position.

12–1 Self-Review

Answers at the end of the chapter.

a. How much is the output voltage of a carbon-zinc cell?
b. How much is the output voltage of a lead-acid cell?
c. Which type can be recharged, a primary or a secondary cell?

12–2 The Voltaic Cell

When two different conducting materials are immersed in an electrolyte, as illustrated in Fig. 12–3*a*, the chemical action of forming a new solution results in the separation of charges. This device for converting chemical energy into electric energy is a voltaic cell. It is also called a *galvanic cell,* named after Luigi Galvani (1737–1798).

In Fig. 12–3*a*, the charged conductors in the electrolyte are the electrodes or plates of the cell. They are the terminals that connect the voltage output to an external circuit, as shown in Fig. 12–3*b*. Then the potential difference resulting from the separated charges enables the cell to function as a source of applied voltage. The voltage across the cell's terminals forces current to flow in the circuit to light the bulb.

Current Outside the Cell

Electrons from the negative terminal of the cell flow through the external circuit with R_L and return to the positive terminal. The chemical action in the cell separates charges continuously to maintain the terminal voltage that produces current in the circuit.

The current tends to neutralize the charges generated in the cell. For this reason, the process of producing load current is considered discharging of the cell. However, the internal chemical reaction continues to maintain the separation of charges that produces the output voltage.

Current Inside the Cell

The current through the electrolyte is a motion of ion charges. Notice in Fig. 12–3*b* that the current inside the cell flows from the positive terminal to the negative terminal. This action represents the work being done by the chemical reaction to generate the voltage across the output terminals.

The negative terminal in Fig. 12–3*a* is considered the anode of the cell because it forms positive ions in the electrolyte. The opposite terminal of the cell is its cathode.

Internal Resistance, r_i

Any practical voltage source has internal resistance, indicated as r_i, which limits the current it can deliver. For a chemical cell, as shown in Fig. 12–3, the r_i is mainly

Figure 12–3 How a voltaic cell converts chemical energy into electrical energy. (*a*) Electrodes or plates in liquid electrolyte solution. (*b*) Schematic of a circuit with a voltaic cell as a DC voltage source *V* to produce current in load R_L, which is the lightbulb.

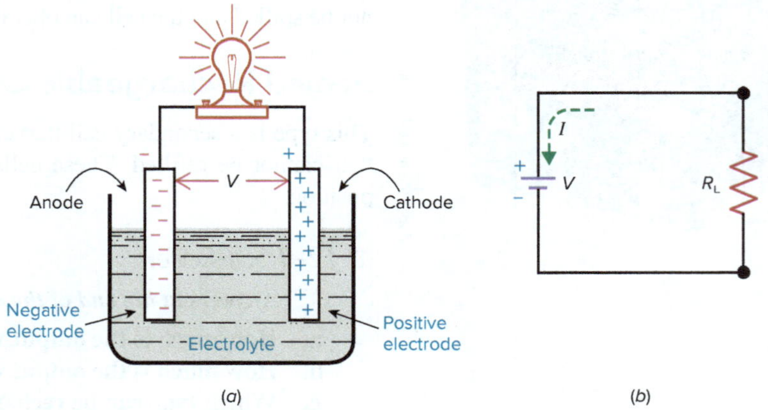

the resistance of the electrolyte. For a good cell, r_i is very low, with typical values less than 1 Ω. As the cell deteriorates, though, r_i increases, preventing the cell from producing its normal terminal voltage when load current is flowing because the internal voltage drop across r_i opposes the output terminal voltage. This is why you can often measure the normal voltage of a dry cell with a voltmeter, which drains very little current, but the terminal voltage drops when the load is connected.

The voltage output of a cell depends on the elements used for the electrodes and the electrolyte. The current rating depends mostly on the size. Larger batteries can supply more current. Dry cells are generally rated up to 250 mA, and the lead-acid wet cell can supply current up to 300 A or more. Note that a smaller r_i allows a higher current rating.

Electromotive Series

The fact that the voltage output of a cell depends on its elements can be seen from Table 12–2. This list, called the *electrochemical series* or *electromotive series,* gives the relative activity in forming ion charges for some of the chemical elements. The potential for each element is the voltage with respect to hydrogen as a zero reference. The difference between the potentials for two different elements indicates the voltage of an ideal cell using these electrodes. Note that other factors, such as the electrolyte, cost, stability, and long life, are important for the construction of commercial batteries.

12–2 Self-Review

Answers at the end of the chapter.

a. **The negative terminal of a chemical cell has a charge of excess electrons. (True/False)**
b. **The internal resistance of a cell limits the amount of output current. (True/False)**
c. **Two electrodes of the same metal provide the highest voltage output. (True/False)**

Table 12–2 Electromotive Series of Elements

Element	Potential, V
Lithium	−2.96
Magnesium	−2.40
Aluminum	−1.70
Zinc	−0.76
Cadmium	−0.40
Nickel	−0.23
Lead	−0.13
Hydrogen (reference)	0.00
Copper	+0.35
Mercury	+0.80
Silver	+0.80
Gold	+1.36

12–3 Common Types of Primary Cells

In this section, you will be introduced to several different types of primary cells in use today.

Carbon-Zinc

The carbon-zinc dry cell is a very common type because of its low cost. It is also called the *Leclanché cell,* named after its inventor. Examples are shown in Fig. 12–1, and Fig. 12–4 illustrates the internal construction of the D-size round cell. The voltage output of the carbon-zinc cell is 1.4 to 1.6 V, with a nominal value of 1.5 V. The suggested current range is up to 150 mA for the D size, which has a height of 2¼ in. and volume of 3.18 in.^{3} The C, AA, and AAA sizes are smaller, with lower current ratings.

The electrochemical system consists of a zinc anode and a manganese dioxide cathode in a moist electrolyte. The electrolyte is a combination of ammonium chloride and zinc chloride dissolved in water. For the round-cell construction, a carbon rod is used down the center, as shown in Fig. 12–4. The rod is chemically inert. However, it serves as a current collector for the positive terminal at the top. The path for current inside the cell includes the carbon rod as the positive terminal, the manganese dioxide, the electrolyte, and the zinc can which is the negative electrode. The carbon rod also prevents leakage of the electrolyte but is porous to allow the escape of gases which accumulate in the cell.

In operation of the cell, the ammonia releases hydrogen gas which collects around the carbon electrode. This reaction is called *polarization,* and it can reduce the voltage output. However, the manganese dioxide releases oxygen, which combines with the hydrogen to form water. The manganese dioxide functions as a *depolarizer.* Powdered carbon is also added to the depolarizer to improve conductivity and retain moisture.

Carbon-zinc dry cells are generally designed for an operating temperature of 70°F. Higher temperatures will enable the cell to provide greater output. However, temperatures of 125°F or more will cause rapid deterioration of the cell.

The chemical efficiency of the carbon-zinc cell increases with less current drain. Stated another way, the application should allow for the largest battery possible, within practical limits. In addition, performance of the cell is generally better with

Figure 12–4 Cutaway view of carbon-zinc dry cell. This is size D with a height of 2¼ in.

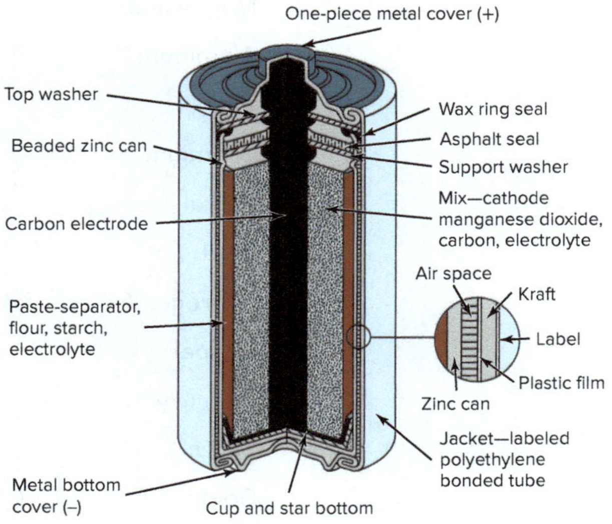

Figure 12–5 Construction of the alkaline cell.

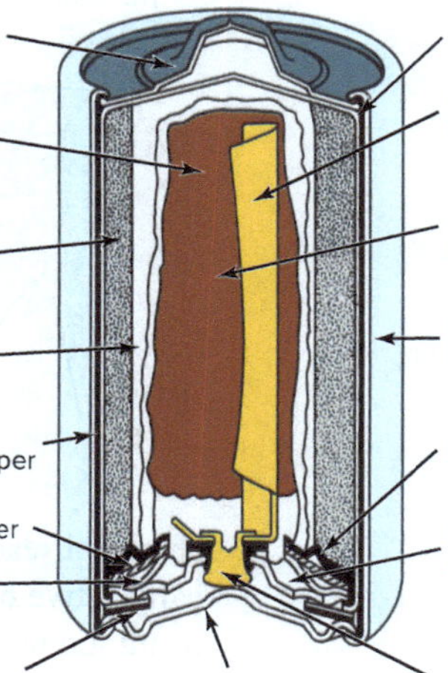

intermittent operation. The reason is that the cell can recuperate between discharges, probably by depolarization.

As an example of longer life with intermittent operation, a carbon-zinc D cell may operate for only a few hours with a continuous drain at its rated current. Yet the same cell could be used for a few months or even a year with intermittent operation of less than 1 hour at a time with smaller values of current.

Alkaline Cell

Another popular type is the manganese-zinc cell shown in Fig. 12–5, which has an alkaline electrolyte. It is available as either a primary or a secondary cell, but the primary type is more common.

The electrochemical system consists of a powdered zinc anode and a manganese dioxide cathode in an alkaline electrolyte. The electrolyte is potassium hydroxide, which is the main difference between the alkaline and Leclanché cells. Hydroxide compounds are alkaline with negative hydroxyl (OH) ions, whereas an acid electrolyte has positive hydrogen (H) ions. The voltage output from the alkaline cell is 1.5 V.

The alkaline cell has many applications because of its ability to work at high efficiency with continuous, high discharge rates. Compared with carbon-zinc cells, alkaline cells have a higher energy density and longer shelf life. Depending on the application, an alkaline cell can provide up to seven times the service life of a carbon-zinc cell.

The outstanding performance of the alkaline cell is due to its low internal resistance. Its r_i is low because of the dense cathode material, the large surface area of the anode in contact with the electrolyte, and the high conductivity of the electrolyte. In addition, alkaline cells perform satisfactorily at low temperatures.

Zinc Chloride Cells

This type is actually a modified carbon-zinc cell whose construction is illustrated in Fig. 12–4. However, the electrolyte contains only zinc chloride. The zinc chloride cell is often referred to as a *heavy-duty* type. It can normally deliver more current over a longer period of time than the Leclanché cell. Another difference is that the

Figure 12–6 Construction of miniature button type of primary cell. Diameter is 3/8 to 1 in. Note the chemical symbols AgO_2 for silver oxide, HgO for mercuric oxide, and MnO_2 for manganese dioxide.

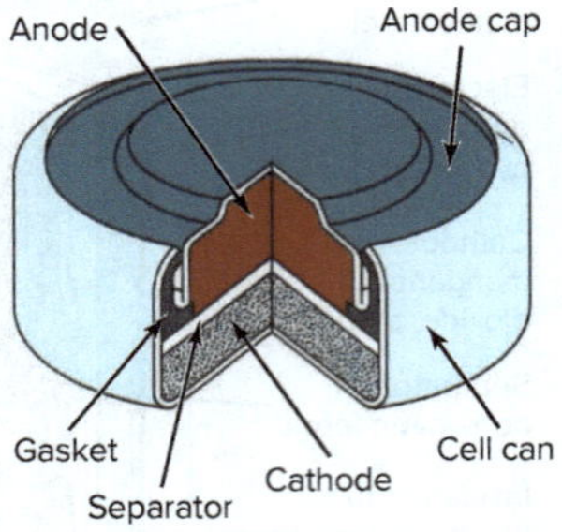

- Anodes are a gelled mixture of amalgamated zinc powder and electrolyte.
- Cathodes

Silver cells:	AgO_2, MnO_2, and conductor
Mercury cells:	HgO and conductor (may contain MnO_2)
Manganese dioxide cells:	MNO_2 and conductor

chemical reaction in the zinc chloride cell consumes water along with the chemically active materials, so that the cell is nearly dry at the end of its useful life. As a result, liquid leakage is not a problem.

Additional Types of Primary Cells

The miniature button construction shown in Fig. 12–6 is often used for the mercury cell and the silver oxide cell. The cell diameter is ⅜ to 1 in.

Mercury Cell

The electrochemical system consists of a zinc anode, a mercury compound for the cathode and an electrolyte of potassium or sodium hydroxide. Mercury cells are available as flat, round cylinders and miniature button shapes. Note, though, that some round mercury cells have the top button as the negative terminal and the bottom terminal positive. The open-circuit voltage is 1.35 V when the cathode is mercuric oxide (HgO) and 1.4 V or more with mercuric oxide/manganese dioxide. The 1.35-V type is more common.

The mercury cell is used where a relatively flat discharge characteristic is required with high current density. Its internal resistance is low and essentially constant. These cells perform well at elevated temperatures, up to 130°F continuously or 200°F for short periods. One drawback of the mercury cell is its relatively high cost compared with a carbon-zinc cell. Mercury cells are becoming increasingly unavailable due to the hazards associated with proper disposal after use.

Silver Oxide Cell

The electrochemical system consists of a zinc anode, a cathode of silver oxide (AgO_2) with small amounts of manganese dioxide, and an electrolyte of potassium or sodium hydroxide. It is commonly available in the miniature button shape shown in Fig. 12–6. The open-circuit voltage is 1.6 V, but the nominal output with a load is considered 1.5 V. Typical applications include hearing aids, cameras, and electronic watches, which use very little current.

Summary of the Most Common Types of Dry Cells

The most common types of dry cells include carbon-zinc, zinc chloride (heavy-duty), and manganese-zinc (alkaline). It should be noted that the alkaline cell is better for heavy-duty use than the zinc chloride type. They are available in the round, cylinder types, listed in Table 12–3, for the D, C, AA, and AAA sizes. The small button cells generally use either mercury or silver oxide. All these dry cells are the primary type and cannot be recharged. Each has an output of 1.5 V except for the 1.35-V mercury cell.

GOOD TO KNOW

It is common practice to store small nonrechargeable batteries in a cool place such as a refrigerator. The cooler storage temperature slows down the chemical activity within the battery, thus prolonging its life when not in use.

GOOD TO KNOW

Primary cells sometimes specify their capacity in watt-hours (W•h) which is determined by multiplying the primary cell's mA•h rating by its terminal voltage.

Table 12–3 Sizes for Popular Types of Dry Cells*

Size	Height, in.	Diameter, in.
D	2¼	1¼
C	1¾	1
AA	1⅞	9/16
AAA	1¾	⅜

* Cylinder shape shown in Fig. 12–1.

Any dry cell loses its ability to produce output voltage even when it is not being used. The shelf life is about 2 years for the alkaline type, but much less with the carbon-zinc cell, especially for small sizes and partially used cells. The reasons are self-discharge within the cell and loss of moisture from the electrolyte. Therefore, dry cells should be used fresh from the manufacturer. It is worth noting, however, that the shelf life of dry cells is steadily increasing due to recent advances in battery technology.

Note that shelf life can be extended by storing the cell at low temperatures, about 40 to 50°F. Even temperatures below freezing will not harm the cell. However, the cell should be allowed to return to normal room temperature before being used, preferably in its original packaging, to avoid condensation.

The alkaline type of dry cell is probably the most cost-efficient. It costs more but lasts much longer, besides having a longer shelf life. Compared with size-D batteries, the alkaline type can last about 10 times longer than the carbon-zinc type in continuous operation, or about seven times longer for typical intermittent operation. The zinc chloride heavy-duty type can last two or three times longer than the general-purpose carbon-zinc cell. For low-current applications of about 10 mA or less, however, there is not much difference in battery life.

Lithium Cell

The lithium cell is a relatively new primary cell. However, its high output voltage, long shelf life, low weight, and small volume make the lithium cell an excellent choice for special applications. The open-circuit output voltage is 3 V. Note the high potential of lithium in the electromotive list of elements shown before in Table 12–2. Figure 12–7 shows an example of a lithium battery with a 6-V output.

Figure 12–7 Lithium battery.

Mark Steinmetz/McGraw Hill

A lithium cell can provide at least 10 times more energy than the equivalent carbon-zinc cell. However, lithium is a very active chemical element. Many of the problems in construction have been solved, though, especially for small cells delivering low current. One interesting application is a lithium cell as the DC power source for a cardiac pacemaker. The long service life is important for this use.

Two forms of lithium cells are in widespread use, the lithium–sulfur dioxide ($LiSO_2$) type and the lithium–thionyl chloride type. Output is approximately 3 V.

In the $LiSO_2$ cell, the sulfur dioxide is kept in a liquid state by using a high-pressure container and an organic liquid solvent, usually methyl cyanide. One problem is safe encapsulation of toxic vapor if the container should be punctured or cracked. This problem can be significant for safe disposal of the cells when they are discarded after use.

The shelf life of the lithium cell, 10 years or more, is much longer than that of other types.

12–3 Self-Review

Answers at the end of the chapter.

a. Which has a longer shelf life, the alkaline or carbon-zinc cell?

b. For the same application, which will provide a longer service life, an alkaline or carbon-zinc dry cell?

c. Which size cell is larger, C or AA?

d. What type of cell is typically used in watches, hearing aids, cameras, etc.?

12–4 Lead-Acid Wet Cell

Figure 12–8 Common 12-V lead-acid battery used in automobiles.

Mark Steinmetz/McGraw Hill

Where high load current is necessary, the lead-acid cell is the type most commonly used. The electrolyte is a dilute solution of sulfuric acid (H_2SO_4). In the application of battery power to start the engine in an automobile, for example, the load current to the starter motor is typically 200 to 400 A. One cell has a nominal output of 2.1 V, but lead-acid cells are often used in a series combination of three for a 6-V battery and six for a 12-V battery. Examples are shown in Figs. 12–2 and 12–8.

The lead-acid type is a secondary cell or storage cell, which can be recharged. The charge and discharge cycle can be repeated many times to restore the output voltage, as long as the cell is in good physical condition. However, heat with excessive charge and discharge currents shortens the useful life to about 3 to 5 years for an automobile battery. The lead-acid type has a relatively high output voltage, which allows fewer cells for a specified battery voltage.

Construction

Inside a lead-acid battery, the positive and negative electrodes consist of a group of plates welded to a connecting strap. The plates are immersed in the electrolyte, consisting of eight parts of water to three parts of concentrated sulfuric acid. Each plate is a grid or framework, made of a lead-antimony alloy. This construction enables the active material, which is lead oxide, to be pasted into the grid. In manufacture of the cell, a forming charge produces the positive and negative electrodes. In the forming process, the active material in the positive plate is changed to lead peroxide (PbO_2). The negative electrode is spongy lead (Pb).

Automobile batteries are usually shipped dry from the manufacturer. The electrolyte is put in at installation, and then the battery is charged to form the plates. With maintenance-free batteries, little or no water need be added in normal service. Some types are sealed, except for a pressure vent, without provision for adding water.

Chemical Action

Sulfuric acid is a combination of hydrogen and sulfate ions. When the cell discharges, lead peroxide from the positive electrode combines with hydrogen ions to form water and with sulfate ions to form lead sulfate. The lead sulfate is also produced by combining lead on the negative plate with sulfate ions. Therefore, the net result of discharge is to produce more water, which dilutes the electrolyte, and to form lead sulfate on the plates.

As discharge continues, the sulfate fills the pores of the grids, retarding circulation of acid in the active material. Lead sulfate is the powder often seen on the outside terminals of old batteries. When the combination of weak electrolyte and sulfation on the plate lowers the output of the battery, charging is necessary.

On charge, the external DC source reverses the current in the battery. The reversed direction of ions flowing in the electrolyte results in a reversal of the chemical reactions. Now the lead sulfate on the positive plate reacts with water and sulfate ions to produce lead peroxide and sulfuric acid. This action re-forms the positive plate and makes the electrolyte stronger by adding sulfuric acid. At the same time, charging enables the lead sulfate on the negative plate to react with hydrogen ions; this also forms sulfuric acid while re-forming lead on the negative electrode.

As a result, the charging current can restore the cell to full output, with lead peroxide on the positive plates, spongy lead on the negative plate, and the required concentration of sulfuric acid in the electrolyte. The chemical equation for the lead-acid cell is

$$Pb + PbO_2 + 2H_2SO_4 \underset{\text{Discharge}}{\overset{\text{Charge}}{\rightleftarrows}} 2\,PbSO_4 + 2H_2O$$

On discharge, the Pb and PbO_2 combine with the SO_4 ions at the left side of the equation to form lead sulfate ($PbSO_4$) and water (H_2O) on the right side of the equation.

On charge, with reverse current through the electrolyte, the chemical action is reversed. Then the Pb ions from the lead sulfate on the right side of the equation re-form the lead and lead peroxide electrodes. Also, the SO_4 ions combine with H_2 ions from the water to produce more sulfuric acid on the left side of the equation.

Current Ratings

Lead-acid batteries are generally rated in terms of the amount of discharge current they can supply for a specified period of time. The output voltage should be maintained above a minimum level, which is 1.5 to 1.8 V per cell. A common rating is **ampere-hours** (A·h) based on a specific discharge time, which is often 8 h. Typical A·h ratings for automobile batteries are 100 to 300 A·h.

As an example, a 200-A·h battery can supply a load current of 200/8 or 25 A, based on an 8-h discharge. The battery can supply less current for a longer time or more current for a shorter time. Automobile batteries may be rated in "cold cranking amps" (CCAs), which is related to the job of starting the engine. The CCA rating specifies the amount of current, in amperes, the battery can deliver at 0°F for 30 seconds while maintaining an output voltage of 7.2 V for a 12-V battery. The higher the CCA rating, the greater the starting power of the battery.

Note that the ampere-hour unit specifies coulombs of charge. For instance, 200 A·h corresponds to 200 A × 3600 s (1 h = 3600 s). This equals 720,000 A·s, or coulombs. One ampere-second is equal to one coulomb. Then the charge equals 720,000 or 7.2×10^5 C. To put this much charge back into the battery would require 20 h with a charging current of 10 A.

The ratings for lead-acid batteries are given for a temperature range of 77 to 80°F. Higher temperatures increase the chemical reaction, but operation above 110°F shortens the battery life.

Low temperatures reduce the current capacity and voltage output. The ampere-hour capacity is reduced approximately 0.75% for each decrease of 1°F below the normal temperature rating. At 0°F, the available output is only 40% of the ampere-hour battery rating. In cold weather, therefore, it is very important to have an automobile battery up to full charge. In addition, the electrolyte freezes more easily when diluted by water in the discharged condition.

GOOD TO KNOW

Battery capacity, abbreviated C, is expressed in A·h or mA·h. Battery current, however, is usually described in units of C/T where T represents the discharge time in hours. For example, a lead-acid battery may have a 200 A·h rating at a C/10 discharge rate. This means the battery can supply 20 A of current $\left(\frac{200\text{ A·h}}{10\text{ h}}\right)$ for a period of 10 hours while maintaining a terminal voltage above some specified minimum value.

Specific Gravity

The state of discharge for a lead-acid cell is generally checked by measuring the **specific gravity** of the electrolyte. Specific gravity is a ratio comparing the weight of a substance with the weight of water. For instance, concentrated sulfuric acid is 1.835 times as heavy as water for the same volume. Therefore, its specific gravity equals 1.835. The specific gravity of water is 1, since it is the reference.

In a fully charged automotive cell, the mixture of sulfuric acid and water results in a specific gravity of 1.280 at room temperatures of 70 to 80°F. As the cell

Figure 12-9 Hydrometer to check specific gravity of lead-acid battery.

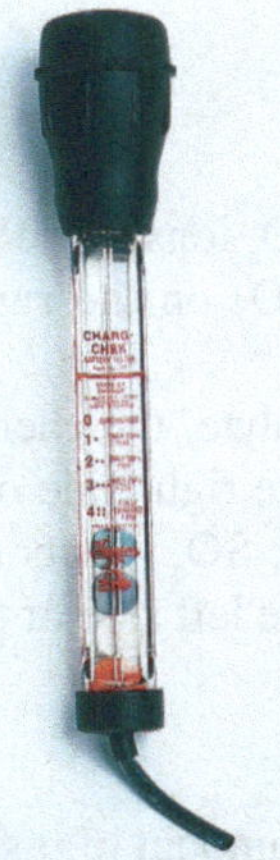

Cindy Schroeder/McGraw Hill

Figure 12-10 Reversed directions for charge and discharge currents of a battery. The r_i is internal resistance. (*a*) The V_B of the battery discharges to supply the load current for R_L. (*b*) The battery is the load resistance for V_G, which is an external source of charging voltage.

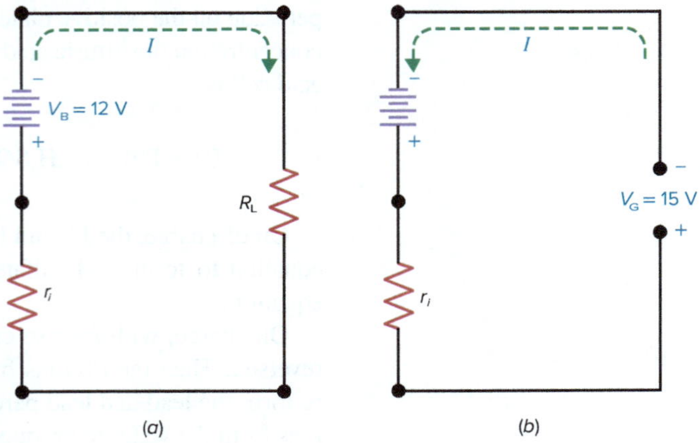

discharges, more water is formed, lowering the specific gravity. When the specific gravity is below about 1.145, the cell is considered completely discharged.

Specific-gravity readings are taken with a battery **hydrometer**, such as the one shown in Fig. 12–9. With this type of hydrometer, the state of charge of a cell within the battery is indicated by the number of floating disks. For example, one floating disk indicates the cell is at 25% of full charge. Two floating disks indicate the cell is at 50% of full charge. Similarly, three floating disks indicate 75% of full charge, whereas four floating disks indicate the cell is at 100% of full charge. The number of floating disks is directly correlated with the value of the specific gravity. As the specific gravity increases, more disks will float. Note that all cells within the battery must be tested for full charge.

The importance of the specific gravity can be seen from the fact that the open-circuit voltage of the lead-acid cell is approximately equal to

$$V = \text{specific gravity} + 0.84$$

For the specific gravity of 1.280, the voltage is $1.280 + 0.84 = 2.12$ V, as an example. These values are for a fully charged battery.

GOOD TO KNOW

Another way to test the condition of a lead-acid battery is to perform what is called a *load test*. This test involves drawing a load current equal to one-half the battery's CCA rating for a period of 15 seconds. At the end of the 15-second test interval, the battery voltage must be above some specified value that is primarily dependent on the battery's temperature. At 70°F, the battery voltage should be at or above 9.6 V after the 15-second time interval.

Charging the Lead-Acid Battery

The requirements are illustrated in Fig. 12–10. An external DC voltage source is necessary to produce current in one direction. Also, the charging voltage must be more than the battery emf. Approximately 2.5 V per cell is enough to overcome the cell emf so that the charging voltage can produce current opposite to the direction of the discharge current.

Note that the reversal of current is obtained by connecting the battery V_B and charging source V_G with + to + and – to –, as shown in Fig. 12–10*b*. The charging current is reversed because the battery effectively becomes a load resistance for V_G when it is higher than V_B. In this example, the net voltage available to produce a charging current is $15 - 12 = 3$ V.

A commercial charger for automobile batteries is shown in Fig. 12–11. This unit can also be used to test batteries and jump-start cars. The charger is essentially a DC power supply, rectifying input from an AC power line to provide DC output for charging batteries.

Float charging refers to a method in which the charger and the battery are always connected to each other to supply current to the load. In Fig. 12–12, the charger provides current for the load and the current necessary to keep the battery fully charged. The battery here is an auxiliary source for DC power.

Figure 12–11 Charger for auto batteries.

Drive Images/Alamy Stock Photo

Figure 12–12 Circuit for battery in float-charge application.

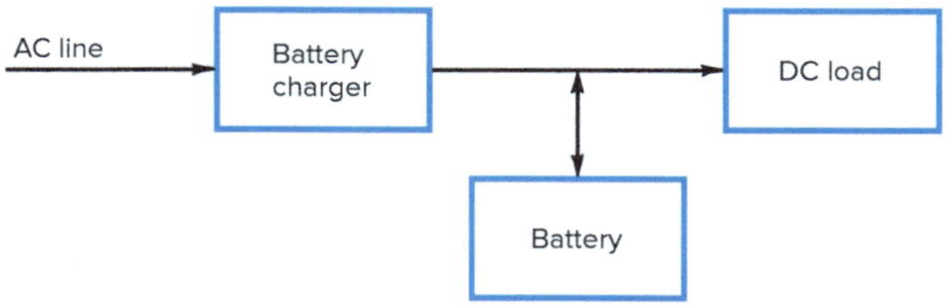

It may be of interest to note that an automobile battery is in a floating-charge circuit. The battery charger is an AC generator or alternator with rectifier diodes, driven by a belt from the engine. When you start the car, the battery supplies the cranking power. Once the engine is running, the alternator charges the battery. It is not necessary for the car to be moving. A voltage regulator is used in this system to maintain the output at approximately 13 to 15 V.

12–4 Self-Review

Answers at the end of the chapter.

a. How many lead-acid cells in series are needed for a 12-V battery?

b. A battery is rated for 120 A·h at an 8-h discharge rate at 77°F. How much discharge current can it supply for 8 h?

c. Which of the following is the specific gravity reading for a fully charged lead-acid cell: 1.080, 1.180, or 1.280?

12–5 Additional Types of Secondary Cells

A secondary cell is a storage cell that can be recharged by reversing the internal chemical reaction. A primary cell must be discarded after it has been completely discharged. The lead-acid cell is the most common type of storage cell. However, other types of secondary cells are available. Some of these are described next.

Nickel-Cadmium (NiCd) Cell

This type is popular because of its ability to deliver high current and to be cycled many times for recharging. Also, the cell can be stored for a long time, even when discharged, without any damage. The NiCd cell is available in both sealed and nonsealed designs, but the sealed construction shown in Fig. 12–13 is common. Nominal output voltage is 1.2 V per cell. Applications include portable power tools, alarm systems, and portable radio or video equipment.

The chemical equation for the NiCd cell can be written as follows:

$$2\,Ni(OH)_3 + Cd \underset{\text{Discharge}}{\overset{\text{Charge}}{\rightleftarrows}} 2\,Ni(OH)_2 + Cd(OH)_2$$

The electrolyte is potassium hydroxide (KOH), but it does not appear in the chemical equation because the function of this electrolyte is to act as a conductor for the transfer of hydroxyl (OH) ions. Therefore, unlike the lead-acid cell, the specific gravity of the electrolyte in the NiCd cell does not change with the state of charge.

The NiCd cell is a true storage cell with a reversible chemical reaction for recharging that can be cycled up to 1000 times. Maximum charging current is equal to the 10-h discharge rate. Note that a new NiCd battery may need charging before use. A disadvantage of NiCd batteries is that they can develop a memory whereby they won't accept full charge if they are routinely discharged to the same level and

Figure 12–13 Examples of nickel-cadmium cells. The output voltage for each is 1.2 V.

Mark Steinmetz/McGraw Hill

then charged. For this reason, it is a good idea to discharge them to different levels and occassionally to discharge them completely before recharging to erase the memory.

Nickel-Metal-Hydride (NiMH) Cell

Nickel-metal-hydride (NiMH) cells are currently finding widespread application in those high-end portable electrical and electronic products where battery performance parameters, notably run-time, are of major concern. NiMH cells are an extension of the proven, sealed, NiCd cells discussed previously. A NiMH cell has about 40% more capacity than a comparably sized NiCd cell, however. In other words, for a given weight and volume, a NiMH cell has a higher A•h rating than a NiCd cell.

With the exception of the negative electrode, NiMH cells use the same general types of components as a sealed NiCd cell. As a result, the nominal output voltage of a NiMH cell is 1.2 V, the same as the NiCd cell. In addition to having higher A•h ratings compared to NiCd cells, NiMH cells also do not suffer nearly as much from the memory effect. As a result of the advantages offered by the NiMH cell, they are finding widespread use in the power-tool market, where additional operating time and higher power are of major importance.

The disadvantage of NiMH cells versus NiCd cells is their higher cost. Also, NiMH cells self-discharge much more rapidly during storage or nonuse than NiCd cells. Furthermore, NiMH cells cannot be cycled as many times as their NiCd counterparts. NiMH cells are continually being improved, and it is foreseeable that they will overcome, at least to a large degree, the disadvantages listed here.

GOOD TO KNOW

Lithium-ion batteries are not available in standard cell sizes (AA, C, and D) like NiMH and NiCd batteries.

Lithium-Ion (Li-Ion) Cell

Lithium-ion cells (and batteries) are extremely popular and have found widespread use in today's consumer electronics market. They are commonly used in laptop computers, cell phones, handheld radios, and iPods to name a few of the more common applications. The electrodes of a lithium-ion cell are made of lightweight lithium and carbon. Since lithium is a highly reactive element, the energy density of lithium-ion batteries is very high. Their high energy density makes lithium-ion batteries significantly lighter than other rechargeable batteries of the same size. The nominal open-circuit output voltage of a single lithium-ion cell is approximately 3.7 V.

Unlike NiMH and NiCd cells, lithium-ion cells do not suffer from the memory effect. They also have a very low self-discharge rate of approximately 5% per month compared with approximately 30% or more per month with NiMH cells and 10% or more per month with NiCd cells. In addition, lithium-ion cells can handle several hundred charge-discharge cycles in their lifetime.

Lithium-ion batteries do have a few disadvantages, however. For one thing, they are more expensive than similar-capacity NiMH or NiCd batteries because the manufacturing process is much more complex. Also, they begin degrading the instant they leave the factory and last only about 2 or 3 years whether they are used or not. Another disadvantage is that higher temperatures cause them to degrade much more rapidly than normal.

Lithium-ion batteries are not as durable as NiMH and NiCd batteries either. In fact, they can be extremely dangerous if mistreated. For example, they may explode if they are overheated or charged to an excessively high voltage. Furthermore, they may be irreversibly damaged if discharged below a certain level. To avoid damaging a lithium-ion battery, a special circuit monitors the voltage output from the battery and shuts it down when it is discharged below a certain threshold level (typically 3 V) or charged above a certain limit (typically 4.2 V). This special circuitry makes lithium-ion batteries even more expensive than they already are.

Nickel-Iron (Edison) Cell

Developed by Thomas Edison, this cell was once used extensively in industrial truck and railway applications. However, it has been replaced almost entirely by the lead-acid battery. New methods of construction with less weight, though, are making this cell a possible alternative in some applications.

The Edison cell has a positive plate of nickel oxide, a negative plate of iron, and an electrolyte of potassium hydroxide in water with a small amount of lithium hydroxide added. The chemical reaction is reversible for recharging. Nominal output is 1.2 V per cell.

Nickel-Zinc Cell

This type of cell has been used in limited railway applications. There has been renewed interest in it for use in electric cars because of its high energy density. However, one drawback is its limited cycle life for recharging. Nominal output is 1.6 V per cell.

Fuel Cells

A **fuel cell** is an electrochemical device that converts hydrogen and oxygen into water and in the process produces electricity. A single fuel cell is a piece of plastic between a couple of pieces of carbon plates that are sandwiched between two end plates acting as electrodes. These plates have channels that distribute the fuel and oxygen. As long as the reactants—pure hydrogen and oxygen—are supplied to the fuel cell, it will continually produce electricity. A conventional battery has all of its chemicals stored inside and it converts the chemical energy into electrical energy. This means that a battery will eventually go dead or deteriorate to a point where it is no longer useful and must be discarded. Chemicals constantly flow into a fuel cell so it never goes dead. (This assumes of course that there is always a flow of chemicals into the cell.) Most fuel cells in use today use hydrogen and oxygen. However, research is being done on a new type of fuel cell that uses methanol and oxygen. This type of fuel cell is in the early stages of development, however. A fuel cell provides a DC voltage at its output and can be used to power motors, lights, and other electrical devices. Note that fuel cells are used extensively in the space program for DC power. Fuel cells are very efficient and can provide hundreds of kilowatts of power.

Several different types of fuel cells are available today, typically classified by the type of electrolyte that they use. The proton exchange membrane fuel cell (PEMFC) is one of the most promising. This is the type that is likely to be used to power cars, buses, and maybe even your house in the future.

Solar Cells

This type of cell converts the sun's light energy directly into electric energy. The cells are made of semiconductor materials, which generate voltage output with light input. Silicon, with an output of 0.5 V per cell, is mainly used now. Research is continuing, however, on other materials, such as cadmium sulfide and gallium arsenide, that might provide more output. In practice, the cells are arranged in modules that are assembled into a large solar array for the required power.

In most applications, the solar cells are used in combination with a lead-acid cell specifically designed for this use. When there is sunlight, the solar cells charge the battery and supply power to the load. When there is no light, the battery supplies the required power.

12–5 Self-Review

Answers at the end of the chapter.

a. **The NiCd cell is a primary type. (True/False)**
b. **The output of the NiCd cell is 1.2 V. (True/False)**
c. **NiMH cells cannot be cycled as many times as NiCd cells. (True/False)**
d. **The output of a solar cell is typically 0.5 V. (True/False)**

12–6 Series-Connected and Parallel-Connected Cells

An applied voltage higher than the voltage of one cell can be obtained by connecting cells in series. The total voltage available across the battery of cells is equal to the sum of the individual values for each cell. Parallel cells have the same voltage as one cell but have more current capacity. The combination of cells is called a *battery.*

Series Connections

Figure 12–14 shows series-aiding connections for three dry cells. Here the three 1.5-V cells in series provide a total battery voltage of 4.5 V. Notice that the two end terminals, A and B, are left open to serve as the plus and minus terminals of the battery. These terminals are used to connect the battery to the load circuit, as shown in Fig. 12–14*c*.

In the lead-acid battery in Fig. 12–2, short, heavy metal straps connect the cells in series. The current capacity of a battery with cells in series is the same as that for one cell because the same current flows through all series cells.

Parallel Connections

For more current capacity, the battery has cells in parallel, as shown in Fig. 12–15. All positive terminals are strapped together, as are all the negative terminals. Any point on the positive side can be the plus terminal of the battery, and any point on the negative side can be the negative terminal.

The parallel connection is equivalent to increasing the size of the electrodes and electrolyte, which increases the current capacity. The voltage output of the battery, however, is the same as that for one cell.

Identical cells in parallel supply equal parts of the load current. For example, with three identical parallel cells producing a load current of 300 mA, each cell has a

Figure 12–14 Cells connected in series for higher voltage. Current rating is the same as for one cell. (*a*) Wiring. (*b*) Schematic symbol for battery with three series cells. (*c*) Battery connected to load resistance R_L.

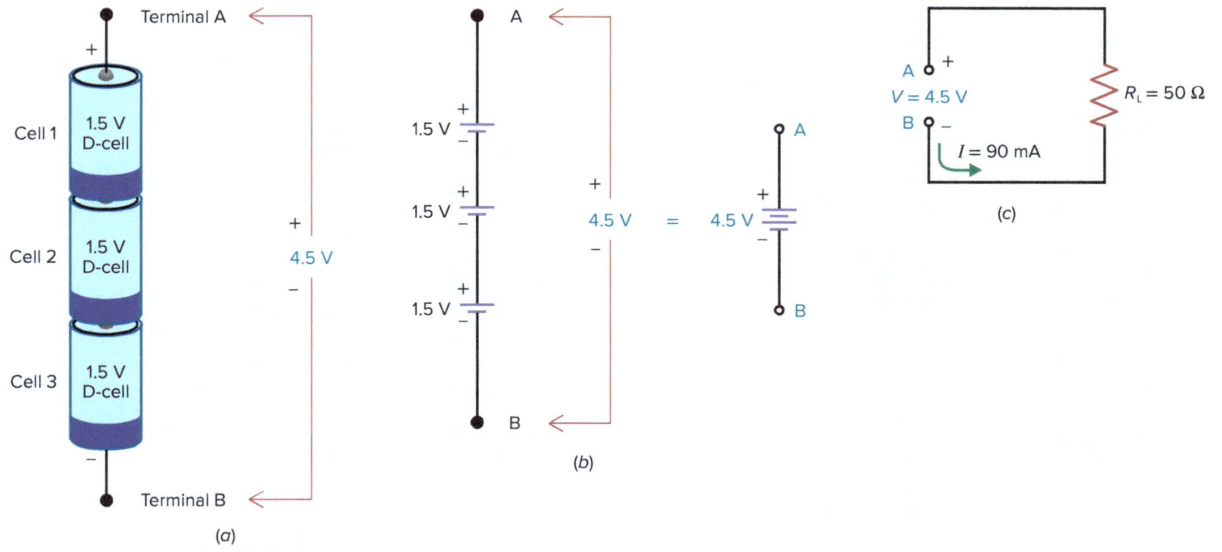

Figure 12–15 Cells connected in parallel for higher current rating. (*a*) Wiring. (*b*) Schematic symbol for battery with three parallel cells. (*c*) Battery connected to load resistance R_L.

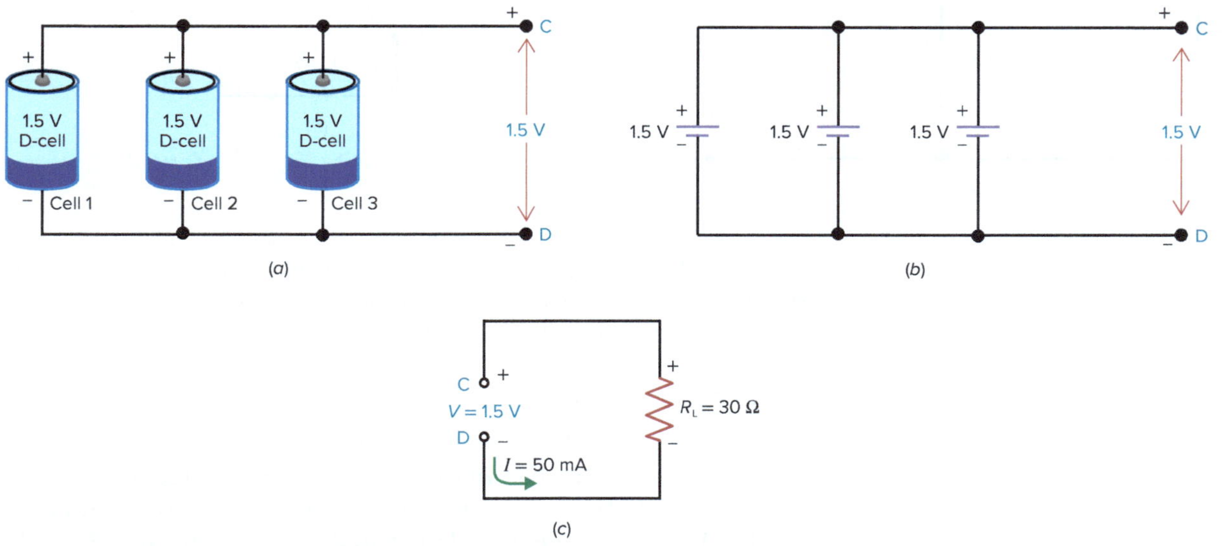

drain of 100 mA. Bad cells should not be connected in parallel with good cells, however, since the cells in good condition will supply more current, which may overload the good cells. In addition, a cell with lower output voltage will act as a load resistance, draining excessive current from the cells that have higher output voltage.

Series-Parallel Connections

To provide a higher output voltage and more current capacity, cells can be connected in series-parallel combinations. Figure 12–16 shows four D cells connected in series-parallel to form a battery that has a 3-V output with a current capacity of ½ A. Two of the 1.5-V cells in series provide 3 V total output voltage. This series string has a current capacity of ¼ A, however, assuming this current rating for one cell.

Multisim **Figure 12–16** Cells connected in series-parallel combinations. (*a*) Wiring two 3-V strings, each with two 1.5-V cells in series. (*b*) Wiring two 3-V strings in parallel. (*c*) Schematic symbol for the battery in (*b*) with output of 3 V. (*d*) Equivalent battery connected to load resistance R_L.

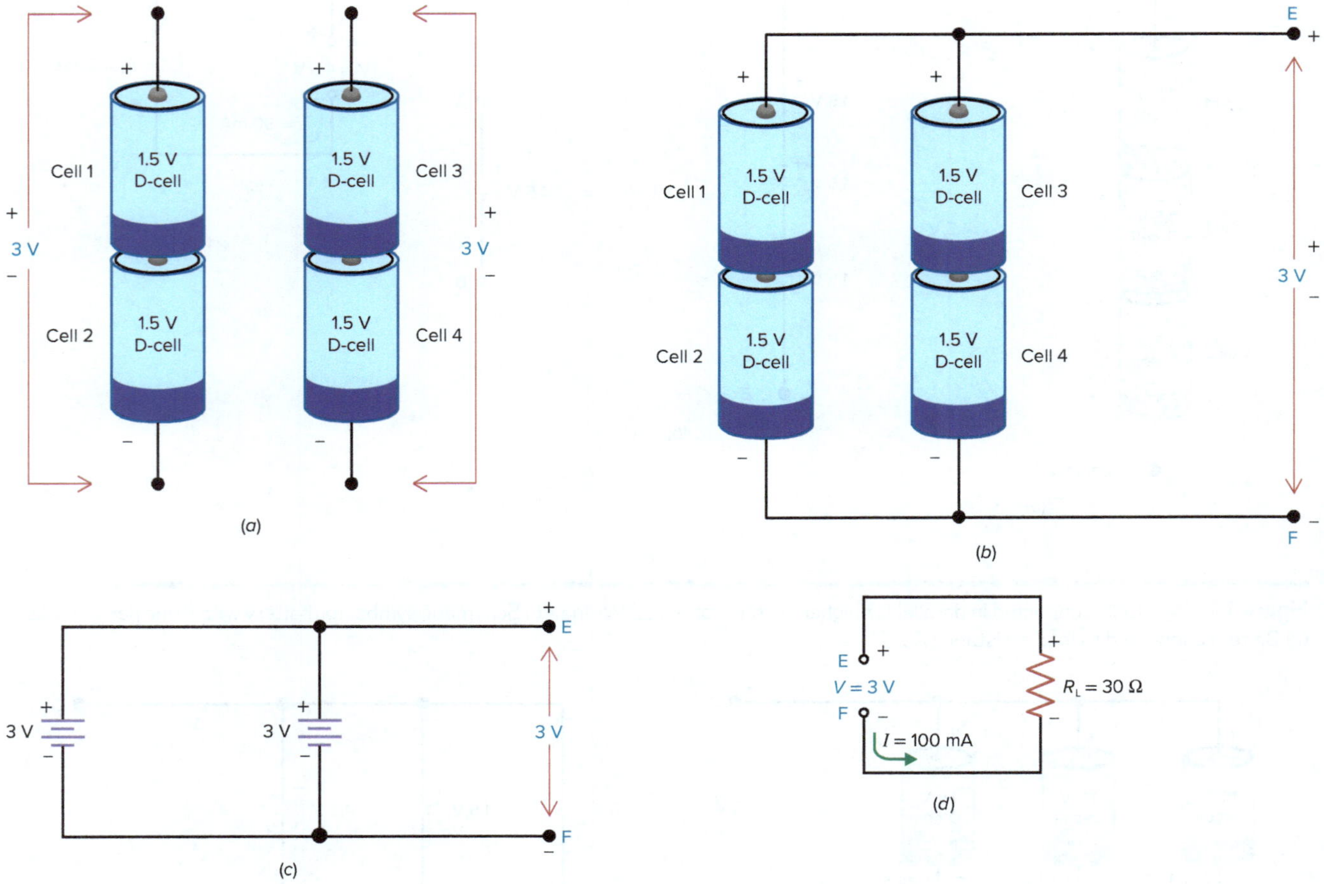

To double the current capacity, another string is connected in parallel. The two strings in parallel have the same 3-V output as one string, but with a current capacity of ½ A instead of the ¼ A for one string.

12–6 Self-Review

Answers at the end of the chapter.

a. How many carbon-zinc cells in series are required to obtain a 9-V_{DC} output? How many lead-acid cells are required to obtain 12.6 V_{DC}?

b. How many identical cells in parallel would be required to double the current rating of a single cell?

c. How many cells rated 1.5 V_{DC} 300 mA would be required in a series-parallel combination that would provide a rating of 900 mA at 6 V_{DC}?

12–7 Current Drain Depends on Load Resistance

It is important to note that the current rating of batteries, or any voltage source, is only a guide to typical values permissible for normal service life. The actual amount of current produced when the battery is connected to a load resistance is equal to $I = V/R$ by Ohm's law.

Figure 12–17 An example of how current drain from a battery used as a voltage source depends on the R of the load resistance. Different values of I are shown for the same V of 1.5 V. (*a*) V/R_1 equals I of 200 mA. (*b*) V/R_2 equals I of 10 mA. (*c*) V/R_3 equals I of 600 mA.

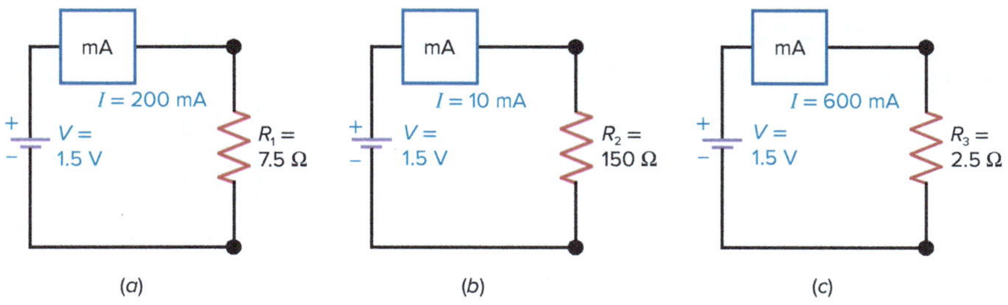

Figure 12–17 illustrates three different cases of using the applied voltage of 1.5 V from a dry cell. In Fig. 12–17*a*, the load resistance R_1 is 7.5 Ω. Then I is $1.5/7.5 = \frac{1}{5}$ A or 200 mA.

A No. 6 carbon-zinc cell with a 1500 mA·h rating could supply this load of 200 mA continuously for about 7.5 h at a temperature of 70°F before dropping to an end voltage of 1.2 V. If an end voltage of 1.0 V could be used, the same load would be served for a longer period of time.

In Fig. 12–17*b*, a larger load resistance R_2 is used. The value of 150 Ω limits the current to $1.5/150 = 0.01$ A or 10 mA. Again using the No. 6 carbon-zinc cell at 70°F, the load could be served continuously for 150 h with an end voltage of 1.2 V. The two principles here are

1. The cell delivers less current with higher resistance in the load circuit.
2. The cell can deliver a smaller load current for a longer time.

In Fig. 12–17*c*, the load resistance R_3 is reduced to 2.5 Ω. Then I is $1.5/2.5 = 0.6$ A or 600 mA. The No. 6 cell could serve this load continuously for only 2.5 h for an end voltage of 1.2 V. The cell could deliver even more load current, but for a shorter time. The relationship between current and time is not linear. For any one example, though, the amount of current is determined by the circuit, not by the current rating of the battery.

12–7 Self-Review

Answers at the end of the chapter.

a. A cell rated at 250 mA will produce this current for any value of R_L. (True/False)

b. A higher value of R_L allows the cell to operate at normal voltage for a longer time. (True/False)

12–8 Internal Resistance of a Generator

Any source that produces voltage output continuously is a generator. It may be a cell separating charges by chemical action or a rotary generator converting motion and magnetism into voltage output, for common examples. In any case, all generators have internal resistance, which is labeled r_i in Fig. 12–18.

The **internal resistance,** r_i, is important when a generator supplies load current because its internal voltage drop, Ir_i, subtracts from the generated emf, resulting in lower voltage across the output terminals. Physically, r_i may be the resistance of the wire in a rotary generator, or r_i is the resistance of the electrolyte between electrodes

Figure 12–18 Internal resistance r_i is in series with the generator voltage V_G. (*a*) Physical arrangement for a voltage cell. (*b*) Schematic symbol for r_i. (*c*) Equivalent circuit of r_i in series with V_G.

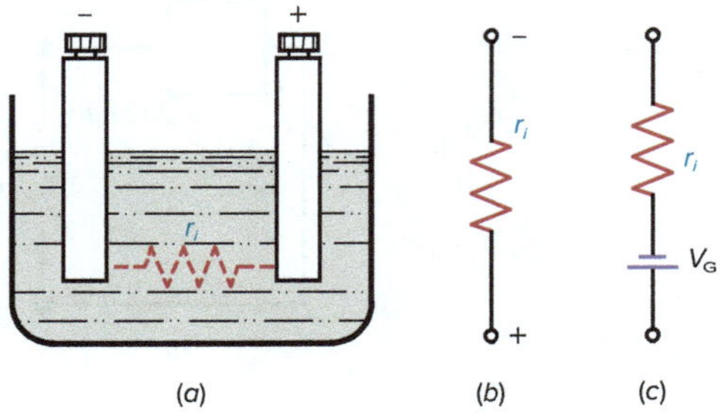

in a chemical cell. More generally, the internal resistance r_i is the opposition to load current inside the generator.

Since any current in the generator must flow through the internal resistance, r_i is in series with the generated voltage, as shown in Fig. 12–18*c*. It may be of interest to note that, with just one load resistance connected across a generator, they are in series with each other because R_L is in series with r_i.

If there is a short circuit across the generator, its r_i prevents the current from becoming infinitely high. As an example, if a 1.5-V cell is temporarily short-circuited, the short-circuit current I_{sc} could be about 15 A. Then r_i is V/I_{sc}, which equals 1.5/15, or 0.1 Ω for the internal resistance. These are typical values for a carbon-zinc D-size cell. (The value of r_i would be lower for a D-size alkaline cell.)

Why Terminal Voltage Drops with More Load Current

Figure 12–19 illustrates how the output of a 100-V source can drop to 90 V because of the internal 10-V drop across r_i. In Fig. 12–19*a*, the voltage across the output terminals is equal to the 100 V of V_G because there is no load current in an open circuit. With no current, the voltage drop across r_i is zero. Then the full generated voltage is available across the output terminals. This value is the generated emf, *open-circuit voltage,* or *no-load voltage.*

We cannot connect the test leads inside the source to measure V_G. However, measuring this no-load voltage without any load current provides a method of determining the internally generated emf. We can assume that the voltmeter draws practically no current because of its very high resistance.

MultiSim **Figure 12–19** Example of how an internal voltage drop decreases voltage at the output terminal of the generator. (*a*) Open-circuit voltage output equals V_G of 100 V because there is no load current. (*b*) Terminal voltage V_L between points A and B is reduced to 90 V because of 10-V drop across 100-Ω r_i with 0.1-A I_L.

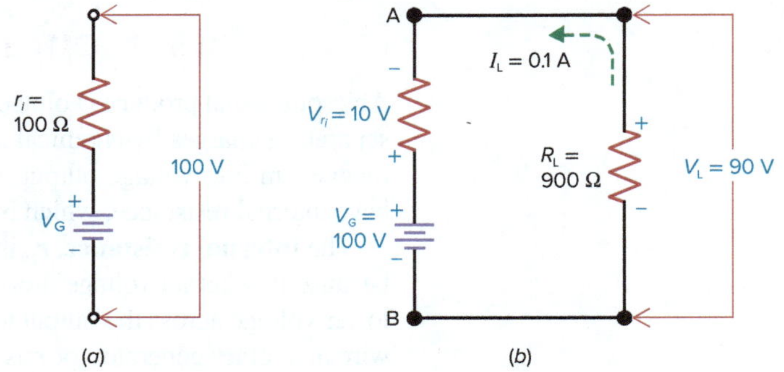

Figure 12–20 How terminal voltage V_L drops with more load current. The graph is plotted for values in Table 12–4.

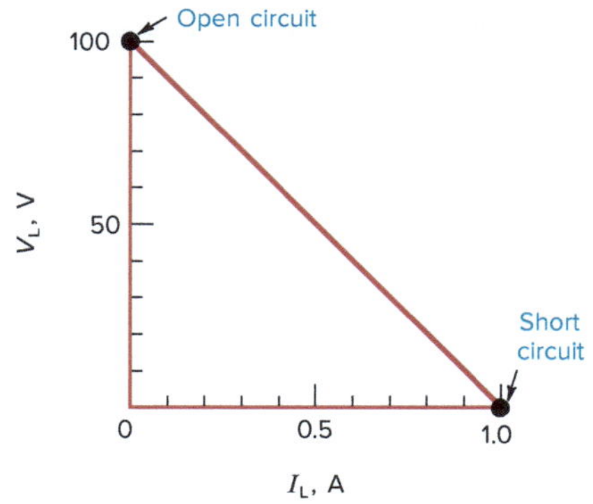

In Fig. 12–19*b* with a load, however, current of 0.1 A flows to produce a drop of 10 V across the 100 Ω of r_i. Note that R_T is 900 + 100 = 1000 Ω. Then I_L equals 100/1000, which is 0.1 A.

As a result, the voltage output V_L equals 100 − 10 = 90 V. This terminal voltage or load voltage is available across the output terminals when the generator is in a closed circuit with load current. The 10-V internal drop is subtracted from V_G because they are series-opposing voltages.

The graph in Fig. 12–20 shows how the terminal voltage V_L drops with increasing load current I_L. The reason is the greater internal voltage drop across r_i as shown by the calculated values listed in Table 12–4. For this example, V_G is 100 V and r_i is 100 Ω.

Across the top row, infinite ohms for R_L means an open circuit. Then I_L is zero, there is no internal drop V_i, and V_L is the same 100 V as V_G.

Across the bottom row, zero ohms for R_L means a short circuit. Then the short-circuit current of 1 A results in zero output voltage because the entire generator voltage is dropped across the internal resistance. Or we can say that with a short circuit of zero ohms across the load, the current is limited to V_G/r_i.

The lower the internal resistance of a generator, the better it is in producing full output voltage when supplying current for a load. For example, the very low r_i, about 0.01 Ω, for a 12-V lead-acid battery, is the reason it can supply high values of load current and maintain its output voltage.

For the opposite case, a higher r_i means that the terminal voltage of a generator is much less with load current. As an example, an old dry battery with r_i of 500 Ω would appear normal when measured by a voltmeter but be useless because of low voltage when normal load current flows in an actual circuit.

GOOD TO KNOW

The only true way to test the condition of a battery is to check the output voltage while drawing its rated output current.

How to Measure r_i

The internal resistance of any generator can be measured indirectly by determining how much the output voltage drops for a specified amount of load current. The difference between the no-load voltage and the load voltage is the amount of internal voltage drop $I_L r_i$. Dividing by I_L gives the value of r_i. As a formula,

$$r_i = \frac{V_{NL} - V_L}{I_L} \tag{12–1}$$

A convenient technique for measuring r_i is to use a variable load resistance R_L. Vary R_L until the load voltage is one-half the no-load voltage. This value of R_L is also the value of r_i, since they must be equal to divide the generator voltage equally. For the same 100-V generator with the 10-Ω r_i used in Example 12–1, if a 10-Ω R_L were used, the load voltage would be 50 V, equal to one-half the no-load voltage.

Table 12–4 How V_L Drops with More I_L (for Figure 12–20)

V_G, V	r_i, Ω	R_L, Ω	$R_T = R_L + r_i$, Ω	$I_L = V_G/R_T$, A	$V_i = I_L r_i$, V	$V_L = V_G - V_i$, V
100	100	∞	∞	0	0	100
100	100	900	1000	0.1	10	90
100	100	600	700	0.143	14.3	85.7
100	100	300	400	0.25	25	75
100	100	100	200	0.5	50	50
100	100	0	100	1.0	100	0

Example 12-1

Calculate r_i if the output of a generator drops from 100 V with zero load current to 80 V when $I_L = 2$ A.

ANSWER

$$r_i = \frac{100 - 80}{2}$$

$$= \frac{20}{2}$$

$$r_i = 10\ \Omega$$

GOOD TO KNOW

The internal resistance of a battery or generator can also be calculated as $r_i = \frac{V_{NL} - V_L}{V_L} \times R_L$. (This assumes, of course, that R_L is known.)

You can solve this circuit by Ohm's law to see that I_L is 5 A with 20 Ω for the combined R_T. Then the two voltage drops of 50 V each add to equal the 100 V of the generator.

12–8 Self-Review

Answers at the end of the chapter.

a. **For Formula (12–1), V_L must be more than V_{NL}. (True/False)**
b. **For Formula (12–1), when V_L is one-half V_{NL}, the r_i is equal to R_L. (True/False)**
c. **The generator's internal resistance, r_i, is in series with the load. (True/False)**
d. **More load current produces a larger voltage drop across r_i. (True/False)**

12–9 Constant-Voltage and Constant-Current Sources

A generator with very low internal resistance is considered a constant-voltage source. Then the output voltage remains essentially the same when the load current changes. This idea is illustrated in Fig. 12–21*a* for a 6-V lead-acid battery with an r_i of 0.005 Ω. If the load current varies over the wide range of 1 to 100 A, the internal Ir_i drop across 0.005 Ω is less than 0.5 V for any of these values.

Constant-Current Generator

It has very high resistance, compared with the external load resistance, resulting in constant current, although the output voltage varies.

The **constant-current generator**, shown in Fig. 12–22, has such high resistance, with an r_i of 0.9 MΩ, that it is the main factor determining how much current can be produced by V_G. Here R_L varies in a 3:1 range from 50 to 150 kΩ. Since the current is determined by the total resistance of R_L and r_i in series, however, I is essentially constant at 1.05 to 0.95 mA, or approximately 1 mA. This relatively constant I is shown by the graph in Fig. 12–22*b*.

MultiSim **Figure 12–21** Constant-voltage generator with low r_i. The V_L stays approximately the same 6 V as I varies with R_L. (*a*) Circuit. (*b*) Graph for V_L.

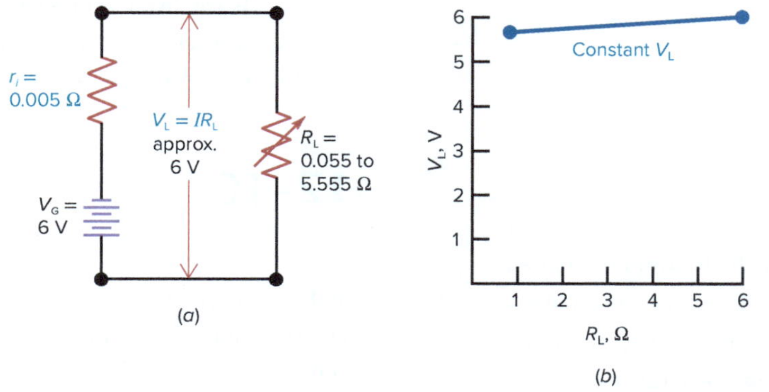

Figure 12–22 Constant-current generator with high r_i. The I stays approximately the same 1 mA as V_L varies with R_L. (*a*) Circuit. (*b*) Graph for I.

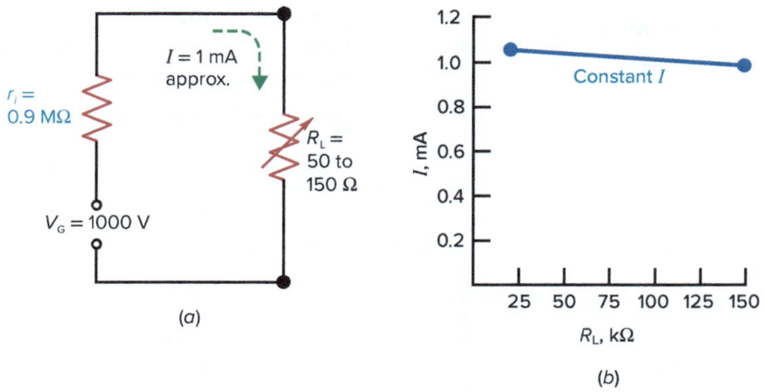

Note that the terminal voltage V_L varies in approximately the same 3:1 range as R_L. Also, the output voltage is much less than the generator voltage because of the high internal resistance compared with R_L. This is a necessary condition, however, in a circuit with a constant-current generator.

A common practice is to insert a series resistance to keep the current constant, as shown in Fig. 12–23*a*. Resistance R_1 must be very high compared with R_L. In this example, I_L is 50 μA with 50 V applied, and R_T is practically equal to the 1 MΩ of R_1. The value of R_L can vary over a range as great as 10:1 without changing R_T or I appreciably.

A circuit with an equivalent constant-current source is shown in Fig. 12–23*b*. Note the arrow symbol for a current source. As far as R_L is concerned, its terminals A and B can be considered as receiving either 50 V in series with 1 MΩ or 50 μA in a shunt with 1 MΩ.

Figure 12–23 Voltage source in (*a*) equivalent to current source in (*b*) for load resistance R_L across terminals A and B.

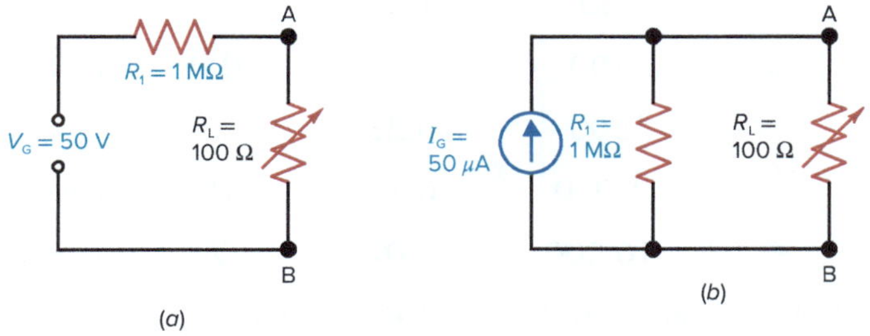

12–9 Self-Review

Answers at the end of the chapter.

Is the internal resistance high or low for

a. **a constant-voltage source?**

b. **a constant-current source?**

GOOD TO KNOW

To get the maximum radiated power from an antenna in a communications system, the radiation resistance of the antenna must match the output resistance of the radio transmitter. This is just one of many instances where it is crictical that $r_i = R_L$.

12–10 Matching a Load Resistance to the Generator r_i

In the diagram in Fig. 12–24, when R_L equals r_i, the load and generator are matched. The matching is significant because the generator then produces maximum power in R_L, as verified by the values listed in Table 12–5.

Maximum Power in R_L

When R_L is 100 Ω to match the 100 Ω of r_i, maximum power is transferred from the generator to the load. With higher resistance for R_L, the output voltage V_L is higher, but the current is reduced. Lower resistance for R_L allows more current, but V_L is less. When r_i and R_L both equal 100 Ω, this combination of current and voltage produces the maximum power of 100 W across R_L.

Figure 12–24 Circuit for varying R_L to match r_i. (*a*) Schematic diagram. (*b*) Equivalent voltage divider for voltage output across R_L. (*c*) Graph of power output P_L for different values of R_L. All values are listed in Table 12–5.

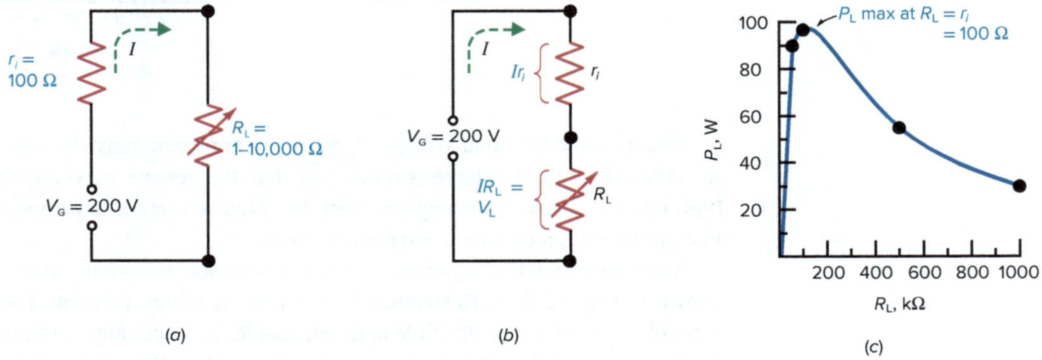

Table 12-5 Effect of Load Resistance on Generator Output*

	R_L, Ω	$I = V_G/R_T$, A	Ir_i, V	IR_L, V	P_L, W	P_i, W	P_T, W	Efficiency = P_L/P_T, %
	1	1.98	198	2	4	392	396	1
	50	1.33	133	67	89	178	267	33
$R_L = r_i \rightarrow$	100	1	100	100	100	100	200	50
	500	0.33	33	167	55	11	66	83
	1,000	0.18	18	180	32	3.24	35.24	91
	10,000	0.02	2	198	4	0.04	4.04	99

* Values calculated approximately for circuit in Fig. 12–24, with V_G = 200 V and r_i = 100 Ω.

With generators that have very low resistance, however, matching is often impractical. For example, if a 6-V lead-acid battery with a 0.003-Ω internal resistance were connected to a 0.003-Ω load resistance, the battery could be damaged by excessive current as high as 1000 A.

Maximum Voltage Across R_L

If maximum voltage, rather than power, is desired, the load should have as high a resistance as possible. Note that R_L and r_i form a voltage divider for the generator voltage, as illustrated in Fig. 12–24*b*. The values for IR_L listed in Table 12–5 show how the output voltage V_L increases with higher values of R_L.

Maximum Efficiency

Note also that the efficiency increases as R_L increases because there is less current, resulting in less power lost in r_i. When R_L equals r_i, the efficiency is only 50%, since one-half the total generated power is dissipated in r_i, the internal resistance of the generator. In conclusion, then, matching the load and generator resistances is desirable when the load requires maximum power rather than maximum voltage or efficiency, assuming that the match does not result in excessive current.

12–10 Self-Review

Answers at the end of the chapter.

a. When $R_L = r_i$, the P_L is maximum. (True/False)
b. The V_L is maximum when R_L is maximum. (True/False)

Application in Understanding Lead-Acid Battery Ratings, Charging, Testing, Storage, and Disposal

The battery in your vehicle supplies power to the starter and ignition system when the engine is initially turning over but is not yet running. In some vehicles, the battery also supplies the extra power necessary when the vehicle's electrical load exceeds the amount provided by the alternator. And finally, the battery acts a voltage stabilizer in the electrical system. In other words, the battery evens out the voltage spikes and prevents them from damaging other components in the electrical system.

When working with lead-acid batteries, it is helpful to have a good understanding of their ratings, how to properly charge them, how to test them, how to maintain them, and how to properly store and dispose of them.

IMPORTANT LEAD ACID BATTERY RATINGS

Cold Cranking Amperes (CCA) Rating: In our vehicles, a primary function of the battery is to provide power to crank the engine during starting. This process requires a large discharge of amperes from the battery over a short period of time. Since the ability of lead-acid batteries to deliver large amounts of current declines with colder temperatures, it is important to buy a battery that is capable of supplying enough current to start your vehicle when the temperature is very cold. This is when the CCA rating of your vehicle's battery becomes very important. For a lead-acid battery, the CCA rating is defined as the discharge load in amperes which a new, fully charged battery at 0°F can deliver for 30 seconds while maintaining a minimum battery voltage of 7.2 volts for a 12-V battery.

Cranking Amperes (CA) Rating: Lead-acid batteries also include a CA rating, which is defined as the discharge load in amperes that a new, fully charged battery at 32°F can deliver for 30 seconds while maintaining a minimum battery voltage of 7.2 volts for a 12-V battery. Notice that the CCA and CA ratings definitions are worded exactly the same, except that the temperatures are different. It should be noted that the CCA rating is the one typically used when comparing the cranking amp measurements of lead-acid batteries.

Hot Cranking Amperes (HCA) Rating: The HCA rating is defined as the amount of current a lead-acid battery can provide at 80°F for 30 seconds while maintaining a minimum battery voltage of 7.2 V for a 12-V battery.

Figure 12–25

Sarah Schultz Photography

Reserve Capacity (RC) Rating: The RC rating is defined as the number of minutes a lead acid-battery operating at 80°F can deliver 25 amperes of current while maintaining a minimum useable battery voltage of 10.5 V for a 12-V battery. This rating is sometimes referred to as the batteries' "staying power." This rating is becoming increasingly more important with lead-acid batteries because it tells us how long the battery can power the vehicle's electrical system in the event that the alternator should fail.

Figure 12-25 shows the top view of a typical lead-acid battery used in an automobile. Notice the battery has a CA rating of 725 and a CCA rating of 580. It is interesting to note that the CA rating of a lead-acid battery is typically about 1.25 times greater than its CCA rating.

CHARGING A LEAD-ACID BATTERY

A lead-acid battery should be charged if:

a. A hydrometer reading indicates the specific gravity of the electrolyte is at or below 1.225.

b. The open-circuit voltage measures less than 12.4 volts.

As a guideline, Table 12-6 shows the relationship between a battery's open-circuit voltage and its state of charge.

Table 12–6 Open-Circuit Voltage Test

Battery Voltage	Specific Gravity	State of Charge
12.6 V	1.265	100%
12.4 V	1.225	75%
12.2 V	1.190	50%
12.0 V	1.155	25%
11.8 V	1.120	0%

Here is a list of tips to follow when charging batteries:

1. Carefully read and follow the instructions that come with the charger you are using to avoid serious injury, property damage, and/or battery damage.
2. Unplug the charger before connecting or disconnecting a battery to avoid dangerous sparks which can cause the battery to explode.
3. Never leave a battery charge for more than 48 hours to avoid damaging the battery by overcharging. If gassing or spewing of electrolyte occurs, or if the battery case feels hot, reduce or temporarily halt charging to avoid damaging the battery.
4. Stop charging the battery when the hydrometer or open-circuit voltage readings recorded two hours apart indicate no increase. Further charging would be useless and may damage the battery and in turn shorten its life. If the battery will not take a full charge, it needs to be replaced.
5. Always leave the filler caps in place and make sure they are tight and secure to reduce the risk of battery explosion and serious injury!
6. Never attempt to charge a frozen battery! To avoid explosion and serious injury, allow it to warm to 60°F before charging.
7. ***Always wear safety glasses when working around batteries. Also, keep sparks, flames, and cigarettes away from batteries at all times. Batteries can explode!***

TESTING A BATTERY

Measuring the no-load voltage and/or checking the specific gravity of the electrolyte will indicate whether or not your battery needs charging, but these tests alone do not tell us if the battery can hold or maintain a charge. A battery can be fully charged but be so weak that it is unable to supply the current required to start your vehicle. The best way to determine the overall condition of a battery is to test it with a load tester. (A load tester is a commercially available piece of test equipment that is used for testing the condition of a battery.) Before running a load test, make sure the battery is at or near full charge. If it is not, the results of the load test will be inaccurate. The load tester should be set to draw a current from the battery equal to one-half the battery's CCA rating. The load test only lasts for 15 seconds. After 15 seconds, the battery's terminal voltage must be at or above 9.6 V. If the battery voltage is less than 9.6 V after 15 seconds, the battery needs to be replaced. If the battery voltage is greater than 9.6 V, the battery is good.

STORAGE

Batteries that are not in use must be cared for to extend both battery life and reliability:

1. Disconnect the cables of an unused battery to avoid self-discharge due to any electronic device(s) that may be drawing current when the vehicle is not turned on or running.
2. When placing a battery into storage, be sure the battery is fully charged. Also, be sure to maintain the battery at or above 75% of full charge while in storage. (See Table 12–6.) Also, check the state of charge every 90 days and recharge the battery if necessary.
3. Ideally, store the batteries in a cool, dry place with temperatures not below 32°F or above 80°F. Typically, batteries will self-discharge at faster rates at higher temperatures.

DISPOSAL

All rechargeable batteries must be disposed of properly, through approved recycling facilities. Lead-acid batteries are virtually 100% recyclable. If you are replacing an old lead-acid battery with a new one, be sure to inquire about where and how to dispose of your old battery. In most cases, the retailer where you purchase your new battery will take the old one off your hands at no extra expense. In most states, it is illegal to discard a battery by simply throwing it in the trash. Be environmentally conscientious and recycle your batteries!

Summary

- A voltaic cell consists of two different conductors as electrodes immersed in an electrolyte. The voltage output depends only on the chemicals in the cell. The current capacity increases with larger sizes. A primary cell cannot be recharged. A secondary or storage cell can be recharged.
- A battery is a group of cells in series or in parallel. With cells in series, the voltages add, but the current capacity is the same as that of one cell. With cells in parallel, the voltage output is the same as that of one cell, but the total current capacity is the sum of the individual values.
- The carbon-zinc dry cell is a common type of primary cell. Zinc is the negative electrode; carbon is the positive electrode. Its output voltage is approximately 1.5 V.
- The lead-acid cell is the most common form of storage battery. The positive electrode is lead peroxide; spongy lead is the negative electrode. Both are in a dilute solution of sulfuric acid as the electrolyte. The voltage output is approximately 2.1 V per cell.
- To charge a lead-acid battery, connect it to a DC voltage equal to approximately 2.5 V per cell. Connecting the positive terminal of the battery to the positive side of the charging source and the negative terminal to the negative side results in charging current through the battery.
- The nickel-cadmium cell is rechargeable and has an output of 1.2 V.
- A constant-voltage generator has very low internal resistance.
- A constant-current generator has very high internal resistance.
- Any generator has an internal resistance r_i. With load current I_L, the internal $I_L r_i$ drop reduces the voltage across the output terminals. When I_L makes the terminal voltage drop to one-half the no-load voltage, the external R_L equals the internal r_i.
- Matching a load to a generator means making the R_L equal to the generator's r_i. The result is maximum power delivered to the load from the generator.

Important Terms

Ampere-hour (A•h) rating — a common rating for batteries that indicates how much load current a battery can supply during a specified discharge time. For example, a battery with a 100 A•h rating can deliver 1 A for 100 h, 2 A for 50 h, 4 A for 25 h, etc.

Battery — a device containing a group of individual voltaic cells that provides a constant or steady DC voltage at its output terminals.

Charging — the process of reversing the current, and thus the chemical action, in a cell or battery to re-form the electrodes and the electrolyte.

Constant-current generator — a generator whose internal resistance is very high compared with the load resistance. Because its internal resistance is so high, it can supply constant current to a load whose resistance value varies over a wide range.

Constant-voltage generator — a generator whose internal resistance is very low compared with the load resistance. Because its internal resistance is so low, it can supply constant voltage to a load whose resistance value varies over a wide range.

Discharging — the process of neutralizing the separated charges on the electrodes of a cell or battery as a result of supplying current to a load resistance.

Float charging — a method of charging in which the charger and the battery are always connected to each other to supply current to the load. With this method, the charger provides the current for the load and the current necessary to keep the battery fully charged.

Fuel cell — an electrochemical device that converts hydrogen and oxygen into water and produces electricity. A fuel cell provides a steady DC output voltage that can power motors, lights, or other appliances. Unlike a regular battery, however, a fuel cell has chemicals constantly flowing into it so it never goes dead.

Hydrometer — a device used to check the state of charge of a cell within a lead-acid battery.

Internal resistance, r_i — the resistance inside a voltage source that limits the amount of current it can deliver to a load.

Open-circuit voltage — the voltage across the output terminals of a voltage source when no load is present.

Primary cell — a type of voltaic cell that cannot be recharged because the internal chemical reaction to restore the electrodes is not possible.

Secondary cell — a type of voltaic cell that can be recharged because the internal chemical reaction to restore the electrodes is possible.

Specific gravity — the ratio of the weight of a volume of a substance to that of water.

Storage cell — another name for a secondary cell.

Voltaic cell — a device that converts chemical energy into electric energy. The output voltage of a voltaic cell depends on the type of elements used for the electrodes and the type of electrolyte.

Related Formulas

$$r_i = \frac{V_{NL} - V_L}{I_L}$$

$$V = \text{Specific gravity} + 0.84$$

Self-Test

Answers at the back of the book.

1. Which of the following cells is not a primary cell?

a. carbon-zinc.

b. alkaline.

c. zinc chloride.

d. lead-acid.

2. The DC output voltage of a C-size alkaline cell is

a. 1.2 V.

b. 1.5 V.

c. 2.1 V.

d. about 3 V.

3. Which of the following cells is a secondary cell?

a. silver oxide.

b. lead-acid.

c. nickel-cadmium.

d. both b and c.

4. What happens to the internal resistance, r_i, of a voltaic cell as the cell deteriorates?

a. It increases.

b. It decreases.

c. It stays the same.

d. It usually disappears.

5. The DC output voltage of a lead-acid cell is

a. 1.35 V.

b. 1.5 V.

c. 2.1 V.

d. about 12 V.

6. Cells are connected in series to

a. increase the current capacity.

b. increase the voltage output.

c. decrease the voltage output.

d. decrease the internal resistance.

7. Cells are connected in parallel to

a. increase the current capacity.

b. increase the voltage output.

c. decrease the voltage output.

d. decrease the current capacity.

8. Five D-size alkaline cells in series have a combined voltage of

a. 1.5 V.

b. 5.0 V.

c. 7.5 V.

d. 11.0 V.

9. The main difference between a primary cell and a secondary cell is that

a. a primary cell can be recharged and a secondary cell cannot.

b. a secondary cell can be recharged and a primary cell cannot.

c. a primary cell has an unlimited shelf life and a secondary cell does not.

d. primary cells produce a DC voltage and secondary cells produce an AC voltage.

10. A constant-voltage source

a. has very high internal resistance.

b. supplies constant-current to any load resistance.

c. has very low internal resistance.

d. none of the above.

11. A constant-current source

a. has very low internal resistance.

b. supplies constant current to a wide range of load resistances.

c. has very high internal resistance.

d. both b and c.

12. The output voltage of a battery drops from 6.0 V with no load to 5.4 V with a load current of 50 mA. How much is the internal resistance, r_i?

a. 12 Ω.

b. 108 Ω.

c. 120 Ω.

d. It cannot be determined.

13. Maximum power is transferred from a generator to a load when

a. $R_L = r_i$.

b. R_L is maximum.

c. R_L is minimum.

d. R_L is 10 or more times the value of r_i.

14. What is the efficiency of power transfer for the matched load condition?

a. 100%.

b. 0%.

c. 50%.

d. It cannot be determined.

15. The internal resistance of a battery

a. cannot be measured with an ohmmeter.

b. can be measured with an ohmmeter.

c. can be measured indirectly by determining how much the output voltage drops for a given load current.

d. both a and c.

Essay Questions

1. What is the advantage of connecting cells in series?

2. (a) What is the advantage of connecting cells in parallel?
   (b) Why can the load be connected across any one of the parallel cells?

3. How many cells are necessary in a battery to double the voltage and current ratings of a single cell? Show the wiring diagram.

4. Draw a diagram showing two 12-V lead-acid batteries being charged by a 15-V source.

5. Why is a generator with very low internal resistance called a constant-voltage source?

6. Why does discharge current lower the specific gravity in a lead-acid cell?

7. Would you consider the lead-acid battery a constant-current source or a constant-voltage source? Why?

8. List five types of chemical cells, giving two features of each.

9. Referring to Fig. 12–21*b*, draw the corresponding graph that shows how I varies with R_L.

10. Referring to Fig. 12–22*b*, draw the corresponding graph that shows how V_L varies with R_L.

11. Referring to Fig. 12–24*c*, draw the corresponding graph that shows how V_L varies with R_L.

Problems

SECTION 12–6 SERIES-CONNECTED AND PARALLEL-CONNECTED CELLS

In Probs. 12–1 to 12–5, assume that each individual cell is identical and that the current capacity for each cell is not being exceeded for the load conditions presented.

12–1 In Fig. 12–26, solve for the load voltage, V_L, the load current, I_L, and the current supplied by each cell in the battery.

Figure 12–26

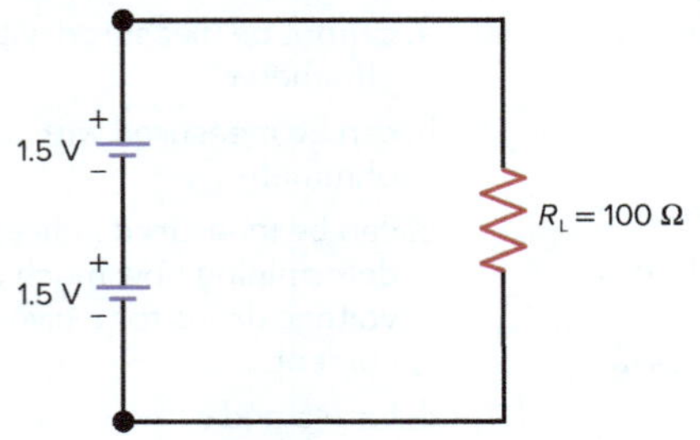

12–2 Repeat Prob. 12–1 for the circuit in Fig. 12–27.

Figure 12–27

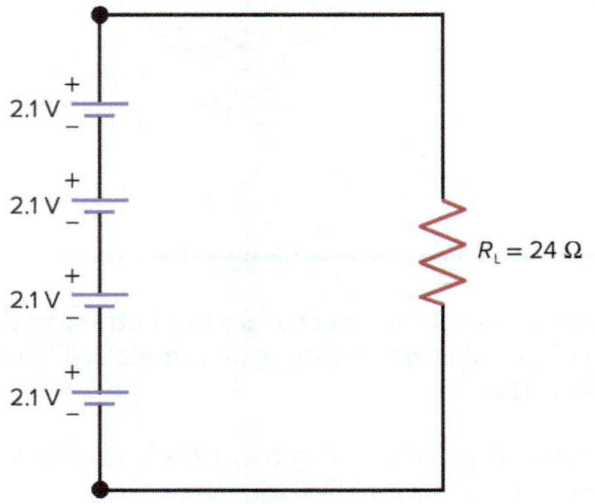

12–3 Repeat Prob. 12–1 for the circuit in Fig. 12–28.

Figure 12–28

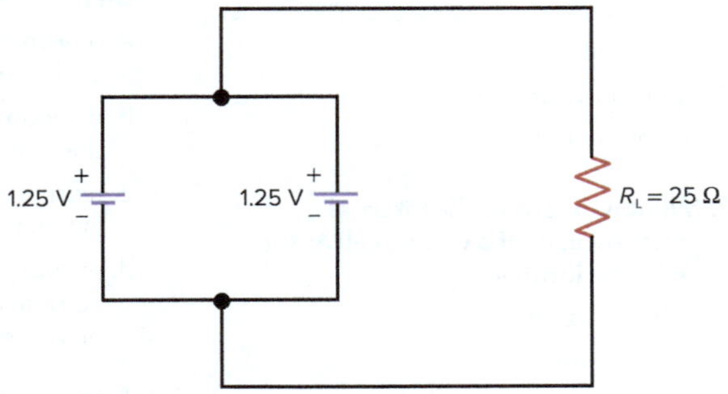

12–4 Repeat Prob. 12–1 for the circuit in Fig. 12–29.

Figure 12–29

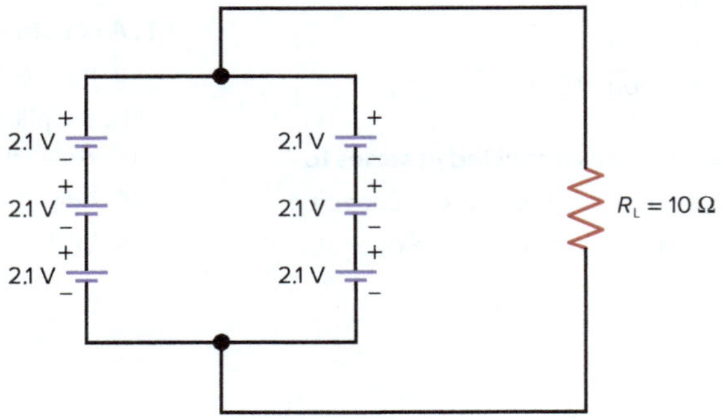

12–5 Repeat Prob. 12–1 for the circuit in Fig. 12–30.

Figure 12–30

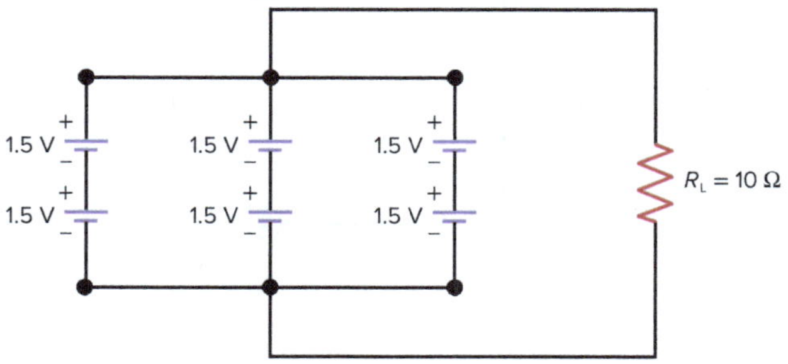

SECTION 12–8 INTERNAL RESISTANCE OF A GENERATOR

12–6 With no load, the output voltage of a battery is 9 V. If the output voltage drops to 8.5 V when supplying 50 mA of current to a load, how much is its internal resistance?

12–7 The output voltage of a battery drops from 6 V with no load to 5.2 V with a load current of 400 mA. Calculate the internal resistance, r_i.

12–8 MultiSim A 9-V battery has an internal resistance of 0.6 Ω. How much current flows from the 9-V battery in the event of a short circuit?

12–9 A 1.5-V "AA" alkaline cell develops a terminal voltage of 1.35 V while delivering 25 mA to a load resistance. Calculate r_i.

12–10 Refer to Fig. 12–31. With S_1 in position 1, $V = 50$ V. With S_1 in position 2, $V = 37.5$ V. Calculate r_i.

Figure 12–31

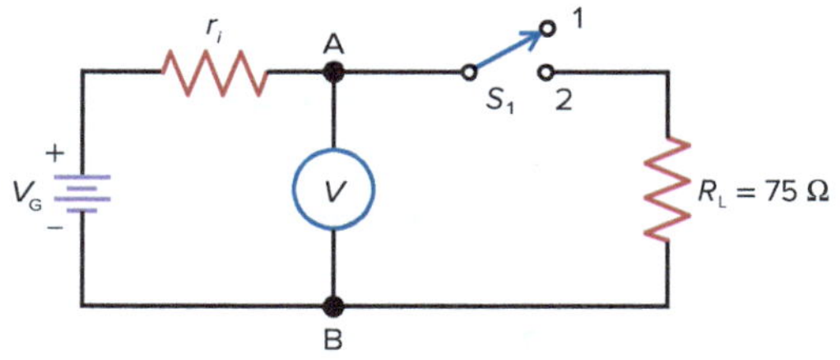

12–11 A generator has an open-circuit voltage of 18 V. Its terminal voltage drops to 15 V when a 75-Ω load is connected. Calculate r_i.

SECTION 12–9 CONSTANT-VOLTAGE AND CONSTANT-CURRENT SOURCES

12–12 Refer to Fig. 12–32. If $r_i = 0.01\ \Omega$, calculate I_L and V_L for the following values of load resistance:

a. $R_L = 1\ \Omega$.

b. $R_L = 5\ \Omega$.

c. $R_L = 10\ \Omega$.

d. $R_L = 100\ \Omega$.

Figure 12–32

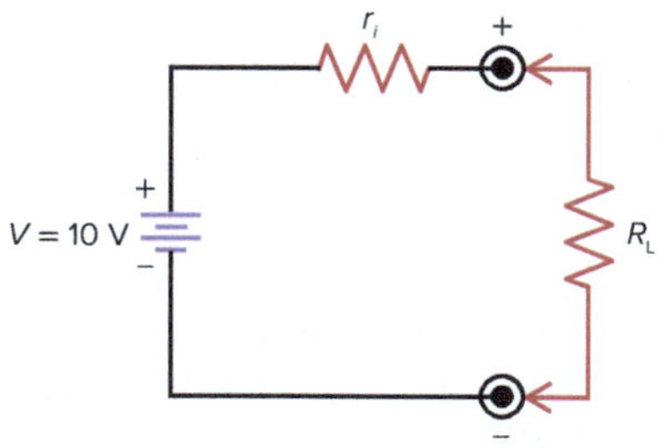

12–13 Refer to Fig. 12–32. If $r_i = 10$ MΩ, calculate I_L and V_L for the following values of load resistance:

a. $R_L = 0\ \Omega$.

b. $R_L = 100\ \Omega$.

c. $R_L = 1$ kΩ.

d. $R_L = 100$ kΩ.

12–14 Redraw the circuit in Fig. 12–33 using the symbol for a current source.

Figure 12–33

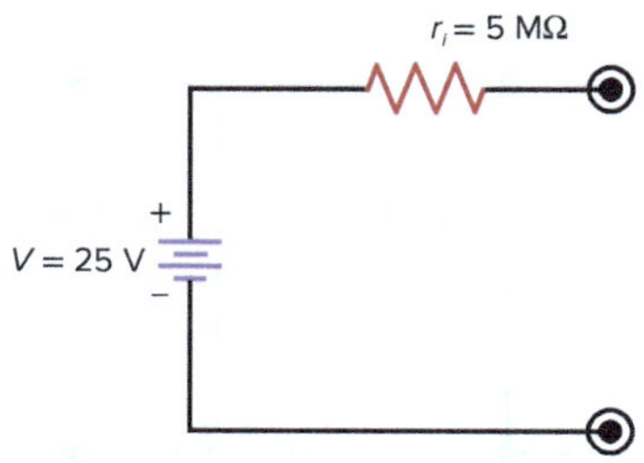

SECTION 12–10 MATCHING A LOAD RESISTANCE TO THE GENERATOR, r_i

12–15 Refer to Fig. 12–34. Calculate I_L, V_L, P_L, P_T, and percent efficiency for the following values of R_L:

a. $R_L = 10\ \Omega$.

b. $R_L = 25\ \Omega$.

c. $R_L = 50\ \Omega$.

d. $R_L = 75\ \Omega$.

e. $R_L = 100\ \Omega$.

Figure 12–34

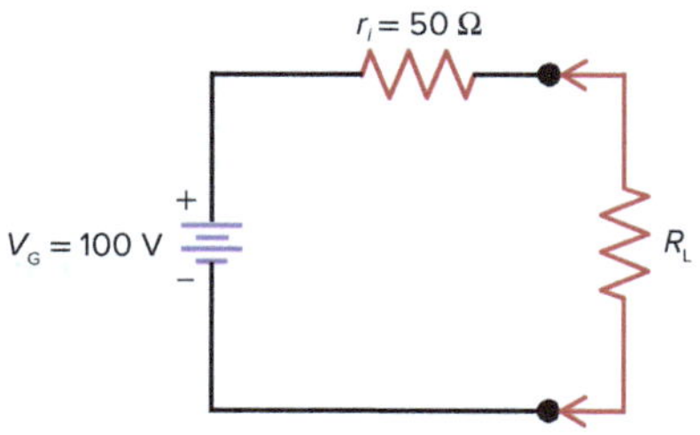

12–16 In Prob. 12–15, what value of R_L provides

a. the highest load voltage, V_L?
b. the smallest voltage drop across the 50-Ω r_i?
c. the maximum transfer of power?
d. the maximum efficiency?

12–17 In Fig. 12–35,

a. What value of R_L will provide maximum transfer of power from generator to load?
b. What is the load power for the matched load condition?
c. What percentage of the total power is delivered to R_L when $R_L = r_i$?

Figure 12–35

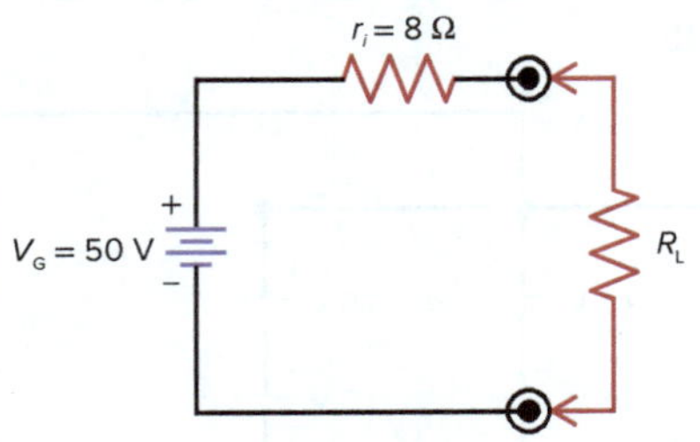

Critical Thinking

12–18 In Fig. 12–36, calculate (a) V_L, (b) I_L, and (c) the current supplied to R_L by each separate voltage source. Hint: Apply Millman's theorem.

12–19 In Fig. 12–37, calculate (a) the value of R_L for which the maximum transfer of power occurs; (b) the maximum power delivered to R_L.

Figure 12–36 Circuit diagram for Critical Thinking Prob. 12–18.

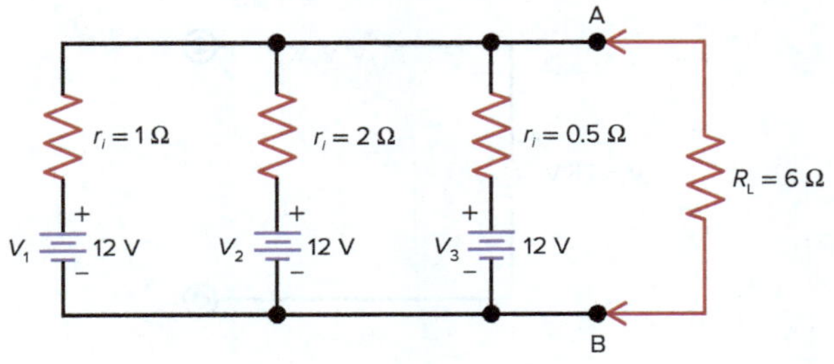

Figure 12–37 Circuit diagram for Critical Thinking Prob. 12–19.

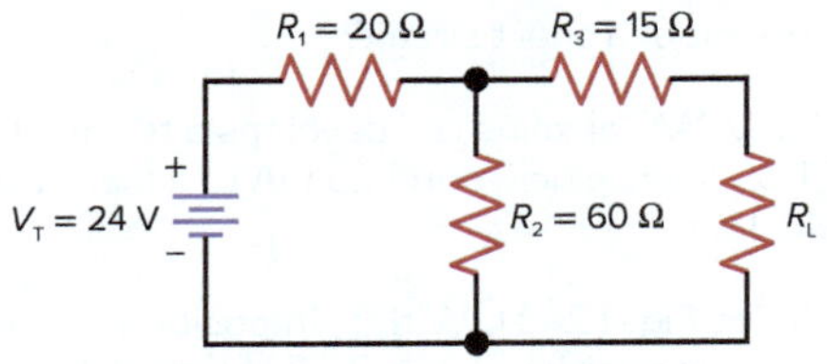

Answers to Self-Reviews

12–1 **a.** 1.5 V **b.** 2.1 V **c.** secondary

12–2 **a.** true **b.** true **c.** false

12–3 **a.** alkaline **b.** alkaline **c.** size C **d.** silver oxide

12–4 **a.** six **b.** 15 A **c.** 1.280

12–5 **a.** false **b.** true **c.** true **d.** true

12–6 **a.** six, six **b.** two **c.** twelve

12–7 **a.** false **b.** true

12–8 **a.** false **b.** true **c.** true **d.** true

12–9 **a.** low **b.** high

12–10 **a.** true **b.** true

Laboratory Application Assignment

In this lab application assignment, you will experimentally determine the internal resistance, r_i, of a DC voltage source. You will measure the no-load and full-load voltages and use Ohm's law to determine the load current.

Equipment: Obtain the following items from your instructor.

- Variable DC voltage source
- SPDT switch
- Assortment of carbon-film resistors
- DMM
- Black electrical tape

Internal Resistance, r_i

Have either your instructor or another student select a resistor whose value lies between 50 Ω and 500 Ω. You should not be allowed to see what its value is. The person who selected the resistor should cover it with black electrical tape so its value cannot be seen.

Construct the circuit in Fig. 12–38. The internal resistance, r_i is the resistor covered with black electrical tape. Note that the DMM is connected between points A and B.

With the switch in position 1, record the voltage indicated by the DMM. This value is the no-load voltage, V_{NL}. V_{NL} = ________

Move the switch to position 2, and record the voltage indicated by the DMM. This value is the full-load voltage, V_{FL}. V_{FL} = ______.

Calculate and record the load current, I_L. I_L = __________

Based on your values of V_{NL}, V_{FL}, and I_L, calculate and record the internal resistance, r_i. r_i = ________

Remove the resistor, r_i, from the circuit, and measure its value with a DMM. Record the measured value. r_i = __________

How does the measured value of r_i compare to the value determined experimentally? ______________________________

Can the internal resistance of a generator be measured directly with an ohmmeter? ____________ If not, why? ____________
____

Describe another experimental procedure that could be used to determine the internal resistance, r_i, in Fig. 12–38. __________
____
____

Figure 12–38

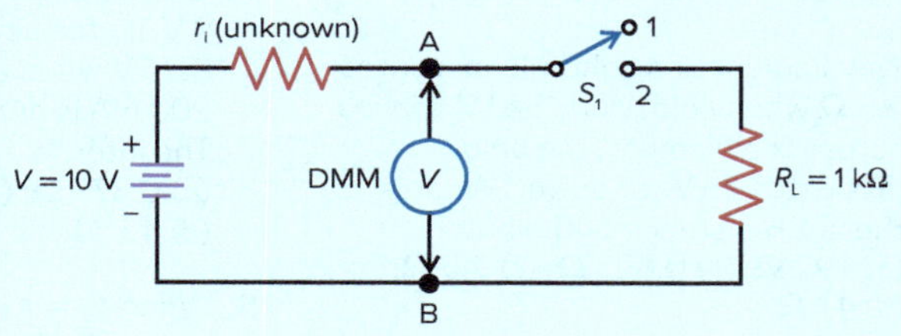

Cumulative Review Summary (Chapters 11–12)

- A conductor is a material whose resistance is very low. Some examples of good conductors are silver, copper, and aluminum; copper is generally used for wire. An insulator is a material whose resistance is very high. Some examples of good insulators include air, mica, rubber, porcelain, and plastics.
- The gage sizes for copper wires are listed in Table 11–1. As the gage sizes increase from 1 to 40, the diameter and circular area decrease. Higher gage numbers correspond to thinner wire.
- For switches, the term *pole* refers to the number of completely isolated circuits that can be controlled by the switch. The term *throw* refers to the number of closed-contact positions that exist per pole. A switch can have any number of poles and throws.
- A good fuse has very low resistance, with an *IR* voltage of practically zero. An open fuse has nearly infinite resistance. If an open fuse exists in a series circuit or in the main line of a parallel circuit, its voltage drop equals the applied voltage.
- The resistance of a wire is directly proportional to its length and inversely proportional to its cross-sectional area.
- All metals in their purest form have positive temperature coefficients, which means that their resistance increases with an increase in temperature. Carbon has a negative temperature coefficient, which means that its resistance decreases as the temperature increases.
- An ion is an atom that has either gained or lost electrons. A negative ion is an atom with more electrons than protons. Conversely, a positive ion is an atom with more protons than electrons. Ions can move to provide electric current in liquids and gases. The motion of ions is called *ionization current*.
- A battery is a combination of individual voltaic cells. A primary cell cannot be recharged, whereas a secondary cell can be recharged several times. The main types of cells for batteries include alkaline, silver oxide, nickel-cadmium, lithium, and lead-acid.
- With individual cells in series, the total battery voltage equals the sum of the individual cell voltages. This assumes that the cells are connected

in a series-aiding manner. The current rating of the series-aiding cells is the same as that for the cell with the lowest current rating.

- With individual cells in parallel, the voltage is the same as that across one cell. However, the current rating of the combination equals the sum of the individual current-rating values. Only cells that have the same voltage should be connected in parallel.
- All types of DC and AC generators have an internal resistance r_i. The value of r_i may be the resistance of the electrolyte in a battery or the wire in a rotary generator.
- When a generator supplies current to a load, the terminal voltage drops because some voltage is dropped across the internal resistance r_i.
- Matching a load to a generator means making R_L equal to r_i. When $R_L = r_i$, maximum power is delivered from the generator to the load.
- A constant-voltage source has very low internal r_i, whereas a constant-current source has very high internal r_i.

Cumulative Self-Test

Answers at the back of the book.

1. Which of the following is the best conductor of electricity? (*a*) carbon; (*b*) silicon; (*c*) rubber; (*d*) copper.

2. Which of the following wires has the largest cross-sectional area? (*a*) No. 28 gage; (*b*) No. 23 gage; (*c*) No. 12 gage; (*d*) No. 16 gage.

3. The filament of a lightbulb measures 2.5 Ω when cold. With 120 V applied across the filament, the bulb dissipates 75 W of power. What is the hot resistance of the bulb? (*a*) 192 Ω; (*b*) 0.625 Ω; (*c*) 2.5 Ω; (*d*) 47 Ω.

4. A DPST switch has how many terminal connections for soldering? (*a*) 3; (*b*) 1; (*c*) 4; (*d*) 6.

5. Which of the following materials has a negative temperature coefficient? (*a*) steel; (*b*) carbon; (*c*) tungsten; (*d*) Nichrome.

6. The *IR* voltage across a good fuse equals (*a*) the applied voltage; (*b*) one-half the applied voltage; (*c*) infinity; (*d*) zero.

7. A battery has a no-load voltage of 9 V. Its terminal voltage drops to 8.25 V when a load current of 200 mA is drawn from the battery. The internal resistance r_i equals (*a*) 0.375 Ω; (*b*) 3.75 Ω; (*c*) 41.25 Ω; (*d*) 4.5 Ω.

8. When $R_L = r_i$, (*a*) maximum voltage is across R_L; (*b*) maximum power is delivered to R_L; (*c*) the efficiency is 100%; (*d*) the minimum power is delivered to R_L.

9. A constant-current source has (*a*) very high internal resistance; (*b*) constant output voltage; (*c*) very low internal resistance; (*d*) output voltage that is always zero.

10. Cells can be connected in series-parallel to (*a*) increase the voltage above that of a single cell; (*b*) increase the current capacity above that of a single cell; (*c*) reduce the voltage and current rating below that of a single cell; (*d*) both (*a*) and (*b*).

Design credits: Chapter Opener Background Image, and Multisim: Stockbyte/Getty Images

chapter

13

Magnetism

The phenomenon known as magnetism was first discovered by the ancient Greeks in about 100 BC. Then it was observed that a peculiar stone had the property of attracting small fragments of iron to itself. The peculiar stone was called a lodestone, and the power of attraction it possessed was called magnetism. Any material possessing the property of magnetism is a magnet. Every magnet has both a north (N) pole and a south (S) pole. Just as “like” electric charges repel each other and “unlike” charges attract, “like” magnetic poles repel each other and “unlike” poles attract. The discovery of natural magnets led to the invention of the compass, which is a direction-finding device. Since the earth itself is a huge natural magnet, a freely suspended magnet will align itself with the North and South magnetic Poles of the earth.

Every magnet has invisible magnetic field lines that extend outward from the magnetic poles. The number of magnetic field lines and their concentration can be measured with special test equipment. In this chapter, you will be introduced to the basic units for magnetic fields. You will also learn about the different types of magnets and how magnetic materials are classified. ■

Chapter Outline

Chapter Objectives

After studying this chapter, you should be able to

- *Describe* the magnetic field surrounding a magnet.
- *Define* the units of *magnetic flux* and *flux density*.
- *Convert* between magnetic units.
- *Describe* how an iron bar is magnetized by induction.
- *Define* the term *relative permeability*.
- *Explain* the difference between a bar magnet and an electromagnet.
- *List* the three classifications of magnetic materials.
- *Explain* the electrical and magnetic properties of ferrites.
- *Describe* the Hall effect.

Important Terms

Curie temperature
diamagnetic
electromagnet
ferrite
ferromagnetic
flux density (B)
gauss (G)
Hall effect
induction
magnetic flux (ϕ)
maxwell (Mx)
paramagnetic
permanent magnet
relative permeability (μ_r)
tesla (T)
toroid
weber (Wb)

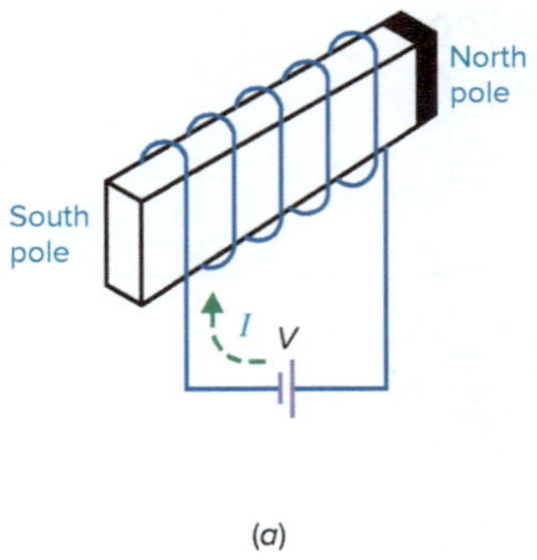

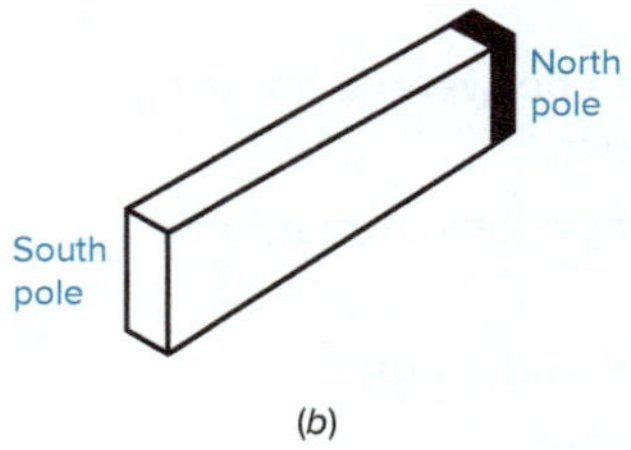

Figure 13–1 Poles of a magnet. (*a*) Electromagnet (EM) produced by current from a battery. (*b*) Permanent magnet (PM) without any external source of current.

13–1 The Magnetic Field

As shown in Figs. 13–1 and 13–2, the north and south poles of a magnet are the points of concentration of magnetic strength. The practical effects of this ferromagnetism result from the magnetic field of force between the two poles at opposite ends of the magnet. Although the magnetic field is invisible, evidence of its force can be seen when small iron filings are sprinkled on a glass or paper sheet placed over a bar magnet (Fig. 13–2*a*). Each iron filing becomes a small bar magnet. If the sheet is tapped gently to overcome friction so that the filings can move, they become aligned by the magnetic field.

Many filings cling to the ends of the magnet, showing that the magnetic field is strongest at the poles. The field exists in all directions but decreases in strength with increasing distance from the poles of the magnet.

Field Lines

To visualize the magnetic field without iron filings, we show the field as lines of force, as shown in Fig. 13–2*b*. The direction of the lines outside the magnet shows the path a north pole would follow in the field, repelled away from the north pole of the magnet and attracted to its south pole. Although we cannot actually have a unit north pole by itself, the field can be explored by noting how the north pole on a small compass needle moves.

The magnet can be considered the generator of an external magnetic field, provided by the two opposite magnetic poles at the ends. This idea corresponds to the two opposite terminals on a battery as the source of an external electric field provided by opposite charges.

Magnetic field lines are unaffected by nonmagnetic materials such as air, vacuum, paper, glass, wood, or plastics. When these materials are placed in the magnetic field of a magnet, the field lines are the same as though the material were not there.

However, the magnetic field lines become concentrated when a magnetic substance such as iron is placed in the field. Inside the iron, the field lines are more dense, compared with the field in air.

GOOD TO KNOW

Magnetic field lines never cross each other but instead push apart from each other.

North and South Magnetic Poles

The earth itself is a huge natural magnet, with its greatest strength at the North and South Poles. Because of the earth's magnetic poles, if a small bar magnet is suspended so that it can turn easily, one end will always point north. This end of the bar magnet is defined as the *north-seeking pole*, as shown in Fig. 13–3*a*. The opposite end is the *south-seeking pole*. When polarity is indicated on a magnet, the north-seeking end is the north pole (N) and the opposite end is the south pole (S).

Figure 13–2 Magnetic field of force around a bar magnet. (*a*) Field outlined by iron filings. (*b*) Field indicated by lines of force.

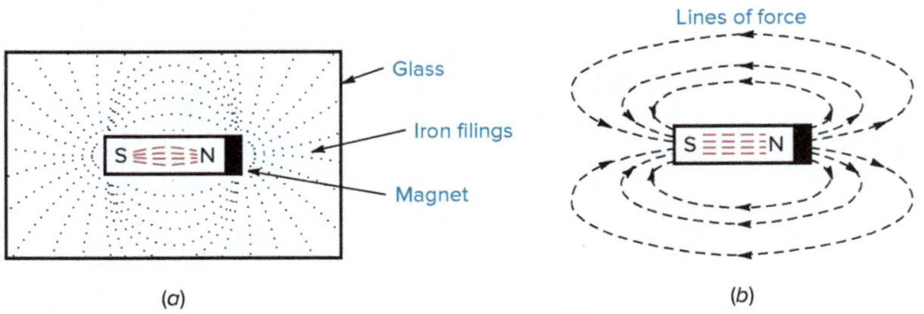

GOOD TO KNOW

Scientists estimate that on average, the earth undergoes a magnetic pole shift about every 300,000 years. This means, that over time, a compass needle will point in a direction other than the geographic north pole. In fact, it will eventually point south (S) instead of north (N). This shift is only temporary, however, since the magnetic poles always revert back to their original position.

Figure 13–3 Definition of north and south poles of a bar magnet. (*a*) North pole on bar magnet points to geographic North Pole of the earth. (*b*) Earth's magnetic field.

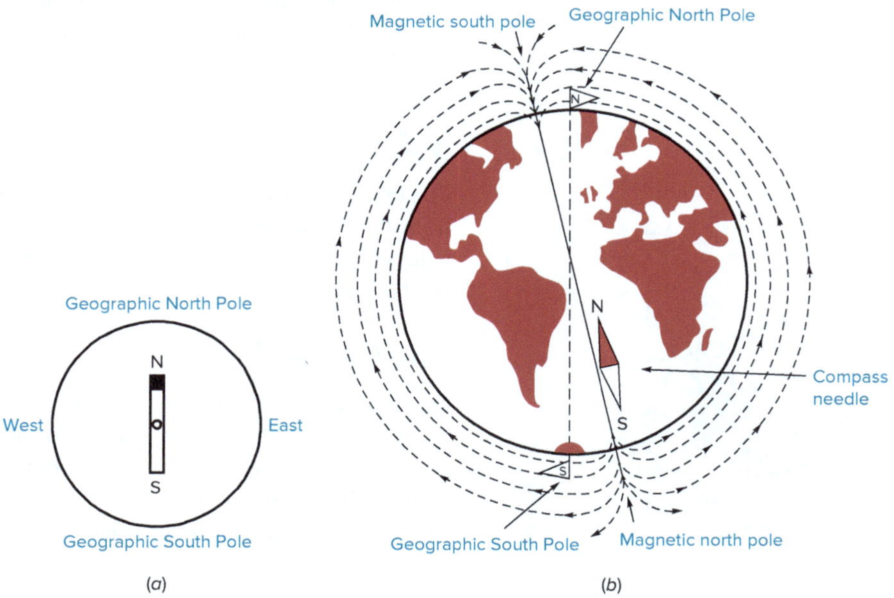

It is important to note that the earth's geographic North Pole has south magnetic polarity and the geographic South Pole has north magnetic polarity. This is shown in Fig. 13–3*b*.

Similar to the force between electric charges is the force between magnetic poles causing attraction of opposite poles and repulsion between similar poles:

1. A north pole (N) and a south pole (S) tend to attract each other.
2. A north pole (N) tends to repel another north pole (N), and a south pole (S) tends to repel another south pole (S).

These forces are illustrated by the fields of iron filings between opposite poles in Fig. 13–4*a* and between similar poles in Fig. 13–4*b*.

MultiSim **Figure 13–4** Magnetic field patterns produced by iron filings. (*a*) Field between opposite poles. The north and south poles could be reversed. (*b*) Field between similar poles. The two north poles could be south poles.

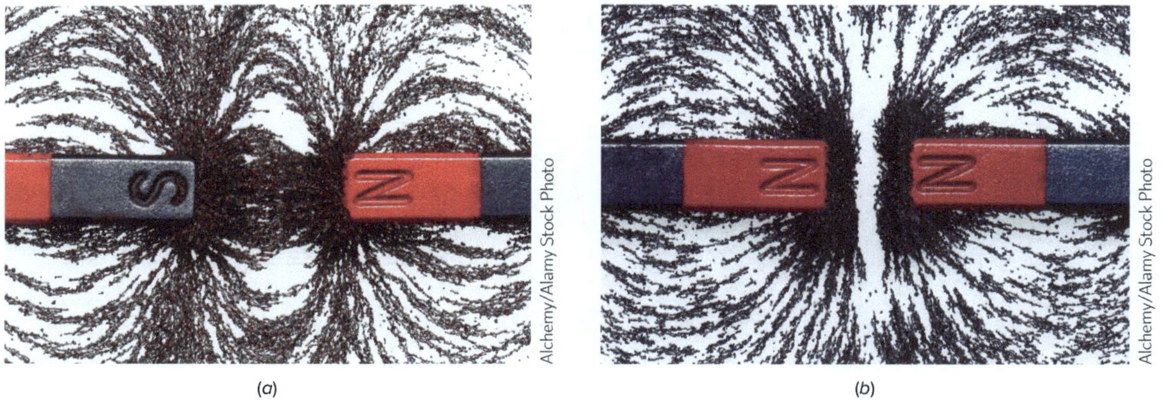

13–1 Self-Review

Answers at the end of the chapter.

a. On a magnet, the north-seeking pole is labeled *N*. (True/False)

b. Like poles have a force of repulsion. (True/False)

13–2 Magnetic Flux (ϕ)

The entire group of magnetic field lines, which can be considered flowing outward from the north pole of a magnet, is called *magnetic flux*. Its symbol is the Greek letter ϕ (phi). A strong magnetic field has more lines of force and more flux than a weak magnetic field.

The Maxwell

One **maxwell (Mx)** unit equals one magnetic field line. In Fig. 13–5, as an example, the flux illustrated is 6 Mx because there are six field lines flowing in or out for each pole. A 1-lb magnet can provide a **magnetic flux** ϕ of about 5000 Mx. This unit is named after James Clerk Maxwell (1831–1879), an important Scottish mathematical physicist, who contributed much to electrical and field theory.

The Weber

This is a larger unit of magnetic flux. One **weber (Wb)** equals 1×10^8 lines or maxwells. Since the weber is a large unit for typical fields, the microweber unit can be used. Then $1\ \mu\text{Wb} = 10^{-6}$ Wb. This unit is named after Wilhelm Weber (1804–1890), a German physicist.

To convert microwebers to lines or maxwells, multiply by the conversion factor 10^8 lines per weber, as follows:

$$1\ \mu\text{Wb} = 1 \times 10^{-6}\ \text{Wb} \times 10^8\ \frac{\text{lines}}{\text{Wb}}$$

$$= 1 \times 10^2\ \text{lines}$$

$$1\ \mu\text{Wb} = 100\ \text{lines or Mx}$$

Note that the conversion is arranged to make the weber units cancel, since we want maxwell units in the answer.

Even the microweber unit is larger than the maxwell unit. For the same 1-lb magnet, a magnetic flux of 5000 Mx corresponds to 50 μWb. The calculations for this conversion of units are

$$\frac{5000\ \text{Mx}}{100\ \text{Mx}/\mu\text{Wb}} = 50\ \mu\text{Wb}$$

Pixtal/Age Fotostock

PIONEERS IN ELECTRONICS

Physicist *James Clerk Maxwell (1831–1879)* unified scientific theories of electricity and magnetism into a unified theory of the electromagnetic field. In 1865, Maxwell proved that electromagnetic phenomena travel in waves at the speed of light. In 1873, he went on to state that light itself is an electromagnetic wave.

MultiSim **Figure 13–5** Total flux ϕ is six lines or 6 Mx. Flux density *B* at point P is two lines per square centimeter or 2 G.

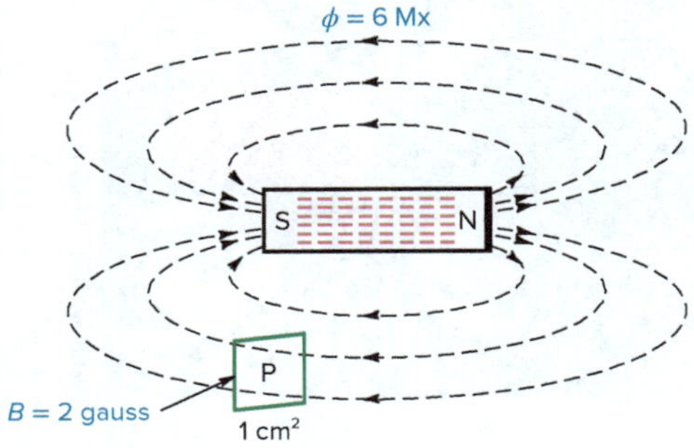

Note that the maxwell units cancel. Also, the $1/\mu$Wb becomes inverted from the denominator to μWb in the numerator.

Conversion between Units

Converting from maxwells (Mx) to webers (Wb) or vice versa is easier if you use the following conversion formulas:

$$\#\text{Wb} = \#\text{Mx} \times \frac{1\ \text{Wb}}{1 \times 10^8\ \text{Mx}}$$

$$\#\text{Mx} = \#\text{Wb} \times \frac{1 \times 10^8\ \text{Mx}}{1\ \text{Wb}}$$

Example 13–1

Make the following conversions: (a) 25,000 Mx to Wb; (b) 0.005 Wb to Mx.

ANSWER

$$\text{(a) } \#\text{Wb} = \#\text{Mx} \times \frac{1\ \text{Wb}}{1 \times 10^8\ \text{Mx}}$$

$$= 25{,}000\ \text{Mx} \times \frac{1\ \text{Wb}}{1 \times 10^8\ \text{Mx}}$$

$$\#\text{Wb} = 250 \times 10^{-6}\ \text{Wb or } 250\ \mu\text{Wb}$$

$$\text{(b) } \#\text{Mx} = \#\text{Wb} \times \frac{1 \times 10^8\ \text{Mx}}{1\ \text{Wb}}$$

$$= 0.005\ \text{Wb} \times \frac{1 \times 10^8\ \text{Mx}}{1\ \text{Wb}}$$

$$\#\text{Mx} = 5.0 \times 10^5\ \text{Mx}$$

PIONEERS IN ELECTRONICS

In 1833, *Carl Friedrich Gauss (1777–1855)* (shown in the photo) and *Wilhelm Eduard Weber (1804–1890)* set up the first telegraph in Göttingen, Germany. They used a mile-long double wire strung over rooftops to connect their observatory to their lab. They used a magnetometer to send first messages of words, then sentences. The telegraph didn't achieve widespread use until Samuel Morse improved the design in 1840. Gauss and Weber also explored Faraday's newly discovered induction of electricity. Gauss had such an impact on the new science of electricity and magnetism that the cgs unit of magnetic flux density, the gauss, is named for him.

Systems of Magnetic Units

The basic units in metric form can be defined in two ways:

1. The centimeter-gram-second system defines small units. This is the cgs system.
2. The meter-kilogram-second system is for larger units of a more practical size. This is the mks system.

Furthermore, the Système International (SI) units provide a worldwide standard in mks dimensions. They are practical values based on the ampere of current.

For magnetic flux ϕ, the maxwell (Mx) is a cgs unit, and the weber (Wb) is an mks or SI unit. The SI units are preferred for science and engineering, but the cgs units are still used in many practical applications of magnetism.

13–2 Self-Review

Answers at the end of the chapter.

The value of 2000 magnetic lines is how much flux in

a. maxwell units?
b. microweber units?

13–3 Flux Density (*B*)

As shown in Fig. 13–5, the *flux density* is the number of magnetic field lines per unit area of a section perpendicular to the direction of flux. As a formula,

$$B = \frac{\phi}{A} \qquad \textbf{(13–1)}$$

where ϕ is the flux through an area A and the flux density is B.

The Gauss

In the cgs system, this unit is one line per square centimeter, or 1 Mx/cm². As an example, in Fig. 13–5, the total flux ϕ is six lines, or 6 Mx. At point P in this field, however, the **flux density *B*** is 2 G because there are two lines per square centimeter. The flux density is higher close to the poles, where the flux lines are more crowded.

As an example of flux density, B for a 1-lb magnet would be 1000 G at the poles. This unit is named after Karl F. Gauss (1777–1855), a German mathematician.

Example 13–2

With a flux of 10,000 Mx through a perpendicular area of 5 cm², what is the flux density in gauss?

ANSWER

$$B = \frac{\phi}{A} = \frac{10{,}000\ \text{Mx}}{5\ \text{cm}^2} = 2000\ \frac{\text{Mx}}{\text{cm}^2}$$

$$B = 2000\ \text{G}$$

As typical values, B for the earth's magnetic field can be about 0.2 G; a large laboratory magnet produces B of 50,000 G. Since the gauss is so small, kilogauss units are often used, where 1 kG = 10^3 G.

The Tesla

In SI, the unit of flux density B is webers per square meter (Wb/m²). One weber per square meter is called a *tesla*, abbreviated T. This unit is named for Nikola Tesla (1856–1943), a Yugoslav-born American inventor in electricity and magnetism.

When converting between cgs and mks units, note that

$$1\ \text{m} = 100\ \text{cm} \quad \text{or} \quad 1 \times 10^2\ \text{cm}$$
$$1\ \text{m}^2 = 10{,}000\ \text{cm}^2 \quad \text{or} \quad 10^4\ \text{cm}^2$$

These conversions are from the larger m (meter) and m² (square meter) to the smaller units of cm (centimeter) and cm² (square centimeter). To go the opposite way,

$$1\ \text{cm} = 0.01\ \text{m} \quad \text{or} \quad 1 \times 10^{-2}\ \text{m}$$
$$1\ \text{cm}^2 = 0.0001\ \text{m}^2 \quad \text{or} \quad 1 \times 10^{-4}\ \text{m}^2$$

As an example, 5 cm² is equal to 0.0005 m² or 5×10^{-4} m². The calculations for the conversion are

$$5\ \text{cm}^2 \times \frac{0.0001\ \text{m}^2}{\text{cm}^2} = 0.0005\ \text{m}^2$$

In powers of 10, the conversion is

$$5 \text{ cm}^2 \times \frac{1 \times 10^{-4} \text{ m}^2}{\text{cm}^2} = 5 \times 10^{-4} \text{ m}^2$$

In both cases, note that the units of cm^2 cancel to leave m^2 as the desired unit.

Example 13–3

With a flux of 400 μWb through an area of 0.0005 m^2, what is the flux density B in tesla units?

ANSWER

$$B = \frac{\phi}{A} = \frac{400 \times 10^{-6} \text{ Wb}}{5 \times 10^{-4} \text{ m}^2}$$

$$= \frac{400}{5} \times 10^{-2}$$

$$= 80 \times 10^{-2} \text{ Wb/m}^2$$

$$B = 0.80 \text{ T}$$

The tesla is a larger unit than the gauss, as $1 \text{ T} = 1 \times 10^4 \text{ G}$.

For example, the flux density of 20,000 G is equal to 2 T. The calculations for this conversion are

$$\frac{20{,}000 \text{ G}}{1 \times 10^4 \text{ G/T}} = \frac{2 \times 10^4 \text{ T}}{1 \times 10^4} = 2 \text{ T}$$

Note that the G units cancel to leave T units for the desired answer. Also, the $1/\text{T}$ in the denominator becomes inverted to T units in the numerator.

Conversion between Units

Converting from **teslas (T)** to **gauss (G)**, or vice versa, is easier if you use the following conversion formulas:

$$\#\text{G} = \#\text{T} \times \frac{1 \times 10^4 \text{ G}}{1 \text{ T}}$$

$$\#\text{T} = \#\text{G} \times \frac{1 \text{ T}}{1 \times 10^4 \text{ G}}$$

Example 13–4

Make the following conversions: (a) 0.003 T to G; (b) 15,000 G to T.

ANSWER

$$\text{(a) } \#\text{G} = \#\text{T} \times \frac{1 \times 10^4 \text{ G}}{\text{T}}$$

$$= 0.003 \text{ T} \times \frac{1 \times 10^4 \text{ G}}{\text{T}}$$

$$\#\text{G} = 30 \text{ G}$$

$$(b)\ \#T = \#G \times \frac{1\ T}{1 \times 10^4\ G}$$

$$= 15{,}000\ G \times \frac{1\ T}{1 \times 10^4\ G}$$

$$\#T = 1.5\ T$$

Comparison of Flux and Flux Density

Remember that the flux ϕ includes total area, whereas the flux density B is for a specified unit area. The difference between ϕ and B is illustrated in Fig. 13–6 with cgs units. The total area A here is 9 cm^2, equal to 3 cm × 3 cm. For one unit box of 1 cm^2, 16 lines are shown. Therefore, the flux density B is 16 lines or maxwells per square centimeter, which equals 16 G. The total area includes nine of these boxes. Therefore, the total flux ϕ is 144 lines or maxwells, equal to 9 × 16 for $B \times A$.

For the opposite case, if the total flux ϕ is given as 144 lines or maxwells, the flux density is found by dividing 144 by 9 cm^2. This division of 144⁄9 equals 16 lines or maxwells per square centimeter, which is 16 G.

13–3 Self-Review

Answers at the end of the chapter.

a. **The ϕ is 9000 Mx through 3 cm^2. How much is B in gauss units?**

b. **How much is B in tesla units for ϕ of 90 μWb through 0.0003 m^2?**

Bettmann/Getty Images

PIONEERS IN ELECTRONICS

Nikola Tesla (1856–1943) had just immigrated to America from Eastern Europe in 1884 when he went to work for Thomas Edison. Tesla wanted to develop an AC induction motor, but Edison didn't want him working on this. Edison's DC power generation plants would be hurt by AC competition. So Tesla quit to work on his own, soon receiving patents for AC motors, dynamos, and systems of transmission. With the backing of industrialist George Westinghouse, Tesla's AC power was adopted in city after city. They even built an AC power plant using Niagara Falls. Tesla's AC power became the worldwide power standard.

13–4 Induction by the Magnetic Field

The electric or magnetic effect of one body on another without any physical contact between them is called ***induction***. For instance, a **permanent magnet** can induce an unmagnetized iron bar to become a magnet without the two touching. The iron bar then becomes a magnet, as shown in Fig. 13–7. What happens is that the magnetic lines of force generated by the permanent magnet make the internal molecular magnets in the iron bar line up in the same direction, instead of the random directions in

Figure 13–6 Comparison of total flux ϕ and flux density B. The total area of 9 cm^2 has 144 lines or 144 Mx. For 1 cm^2, the flux density is 144⁄9 = 16 G.

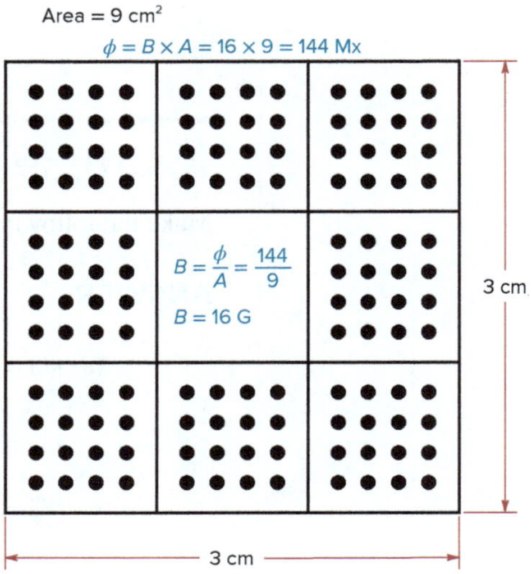

MultiSim **Figure 13–7** Magnetizing an iron bar by induction.

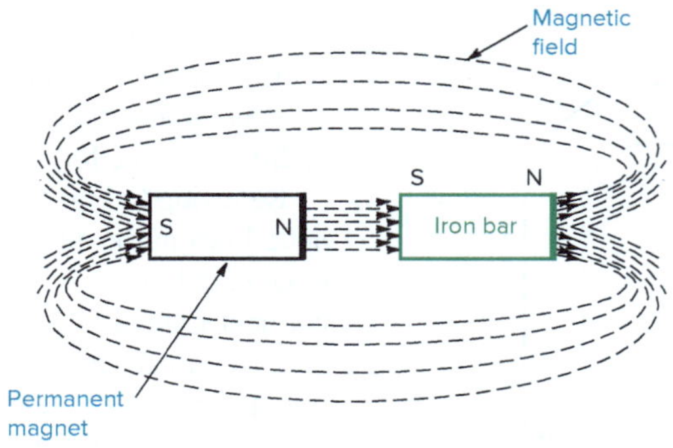

unmagnetized iron. The magnetized iron bar then has magnetic poles at the ends, as a result of magnetic induction.

Note that the induced poles in the iron have polarity opposite from the poles of the magnet. Since opposite poles attract, the iron bar will be attracted. Any magnet attracts to itself all magnetic materials by induction.

Although the two bars in Fig. 13–7 are not touching, the iron bar is in the magnetic flux of the permanent magnet. It is the invisible magnetic field that links the two magnets, enabling one to affect the other. Actually, this idea of magnetic flux extending outward from the magnetic poles is the basis for many inductive effects in AC circuits. More generally, the magnetic field between magnetic poles and the electric field between electric charges form the basis for wireless radio transmission and reception.

Polarity of Induced Poles

Note that the north pole of the permanent magnet in Fig. 13–7 induces an opposite south pole at this end of the iron bar. If the permanent magnet were reversed, its south pole would induce a north pole. The closest induced pole will always be of opposite polarity. This is the reason why either end of a magnet can attract another magnetic material to itself. No matter which pole is used, it will induce an opposite pole, and opposite poles are attracted.

Relative Permeability

Soft iron, as an example, is very effective in concentrating magnetic field lines by induction in the iron. This ability to concentrate magnetic flux is called *permeability*. Any material that is easily magnetized has high permeability, therefore, because the field lines are concentrated by induction.

Numerical values of permeability for different materials compared with air or vacuum can be assigned. For example, if the flux density in air is 1 G but an iron core in the same position in the same field has a flux density of 200 G, the **relative permeability** of the iron core equals $^{200}/_{1}$, or 200.

The symbol for relative permeability is μ_r (mu), where the subscript r indicates relative permeability. Typical values for μ_r are 100 to 9000 for iron and steel. There are no units because μ_r is a comparison of two flux densities and the units cancel. The symbol K_m may also be used for relative permeability to indicate this characteristic of a material for a magnetic field, corresponding to K_e for an electric field.

GOOD TO KNOW

The opposition to the production of magnetic flux is called reluctance. Reluctance is inversely related to permeability. The higher the permeability, the lower the reluctance and vice versa. It is important to note that there is no widely accepted unit for reluctance.

13–4 Self-Review

Answers at the end of the chapter.

a. Induced poles always have polarity opposite from the inducing poles. (True/False)

b. The relative permeability of air or vacuum is approximately 300. (True/False)

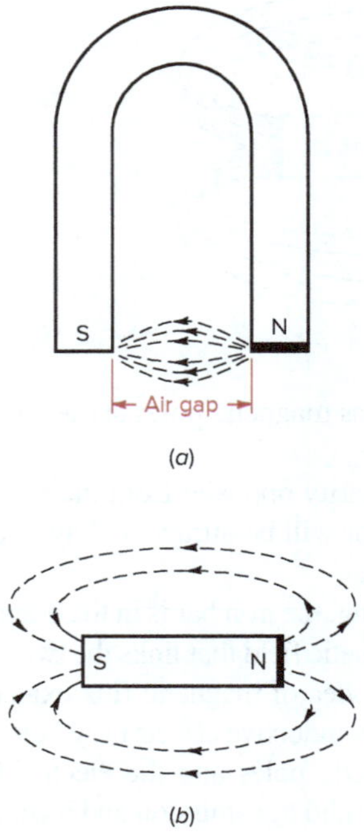

Figure 13–8 The horseshoe magnet in (*a*) has a smaller air gap than the bar magnet in (*b*).

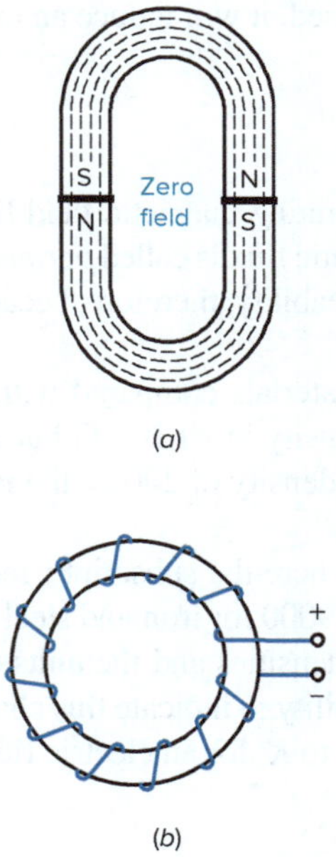

Figure 13–9 Examples of a closed magnetic ring without any air gap. (*a*) Two PM horseshoe magnets with opposite poles touching. (*b*) Toroid magnet.

13–5 Air Gap of a Magnet

As shown in Fig. 13–8, the air space between the poles of a magnet is its air gap. The shorter the air gap, the stronger the field in the gap for a given pole strength. Since air is not magnetic and cannot concentrate magnetic lines, a larger air gap provides additional space for the magnetic lines to spread out.

Referring to Fig. 13–8*a*, note that the horseshoe magnet has more crowded magnetic lines in the air gap, compared with the widely separated lines around the bar magnet in Fig. 13–8*b*. Actually, the horseshoe magnet can be considered a bar magnet bent around to place the opposite poles closer. Then the magnetic lines of the poles reinforce each other in the air gap. The purpose of a short air gap is to concentrate the magnetic field outside the magnet for maximum induction in a magnetic material placed in the gap.

Ring Magnet without Air Gap

When it is desired to concentrate magnetic lines within a magnet, however, the magnet can be formed as a closed magnetic loop. This method is illustrated in Fig. 13–9*a* by the two permanent horseshoe magnets placed in a closed loop with opposite poles touching. Since the loop has no open ends, there can be no air gap and no poles. The north and south poles of each magnet cancel as opposite poles touch.

Each magnet has its magnetic lines inside, plus the magnetic lines of the other magnet, but outside the magnets, the lines cancel because they are in opposite directions. The effect of the closed magnetic loop, therefore, is maximum concentration of magnetic lines in the magnet with minimum lines outside.

The same effect of a closed magnetic loop is obtained with the ***toroid*** or ring magnet in Fig. 13–9*b*, made in the form of a doughnut. Iron is often used for the core. This type of **electromagnet** has maximum strength in the iron ring and little flux outside. As a result, the toroidal magnet is less sensitive to induction from external magnetic fields and, conversely, has little magnetic effect outside the coil.

Note that, even if the winding is over only a small part of the ring, practically all the flux is in the iron core because its permeability is so much greater than that of air. The small part of the field in the air is called *leakage flux*.

Keeper for a Magnet

The principle of the closed magnetic ring is used to protect permanent magnets in storage. In Fig. 13–10*a*, four permanent-magnet bars are in a closed loop, while Fig. 13–10*b* shows a stacked pair. Additional even pairs can be stacked this way, with opposite poles touching. The closed loop in Fig. 13–10*c* shows one permanent horseshoe magnet with a soft-iron *keeper* across the air gap. The keeper maintains the strength of the permanent magnet as it becomes magnetized by induction to form a closed loop. Then any external magnetic field is concentrated in the closed loop without inducing opposite poles in the permanent magnet. If permanent magnets are not stored this way, the polarity can be reversed with induced poles produced by a strong external field from a DC source; an alternating field can demagnetize the magnet.

13–5 Self-Review

Answers at the end of the chapter.

a. A short air gap has a stronger field than a large air gap for the same magnetizing force. (True/False)

b. A toroid is made in the form of a doughnut. (True/False)

Figure 13–10 Storing permanent magnets in a closed loop, with opposite poles touching. (*a*) Four bar magnets. (*b*) Two bar magnets. (*c*) Horseshoe magnet with iron keeper across air gap.

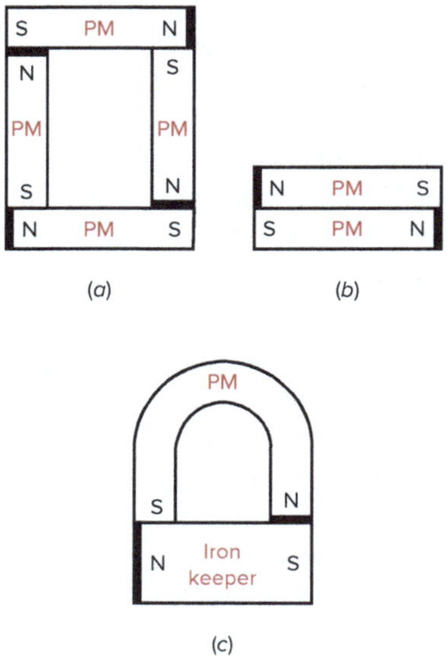

Figure 13–11 Electromagnet holding a nail when switch *S* is closed for current in the coil.

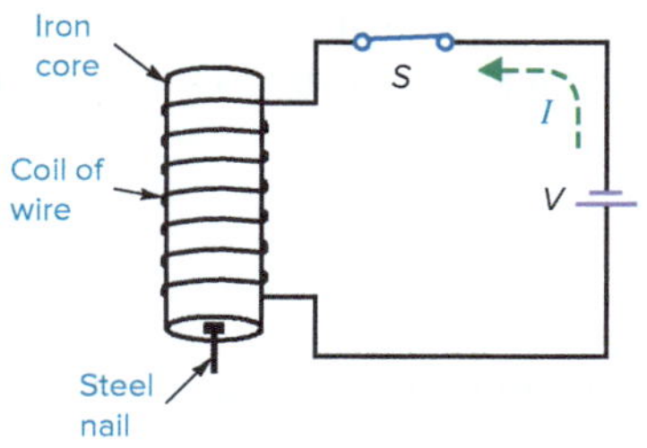

Figure 13–12 Example of a PM loudspeaker.

Mark Steinmetz/McGraw Hill

13–6 Types of Magnets

The two broad classes are permanent magnets and electromagnets. An electromagnet needs current from an external source to maintain its magnetic field. With a permanent magnet, not only is its magnetic field present without any external current, but the magnet can also maintain its strength indefinitely. Sharp mechanical shock as well as extreme heat, however, can cause demagnetization.

Electromagnets

Current in a wire conductor has an associated magnetic field. If the wire is wrapped in the form of a coil, as in Fig. 13–11, the current and its magnetic field become concentrated in a smaller space, resulting in a stronger field. With the length much greater than its width, the coil is called a *solenoid*. It acts like a bar magnet, with opposite poles at the ends.

More current and more turns make a stronger magnetic field. Also, the iron core concentrates magnetic lines inside the coil. Soft iron is generally used for the core because it is easily magnetized and demagnetized.

The coil in Fig. 13–11, with the switch closed and current in the coil, is an electromagnet that can pick up the steel nail shown. If the switch is opened, the magnetic field is reduced to zero, and the nail will drop off. This ability of an electromagnet to provide a strong magnetic force of attraction that can be turned on or off easily has many applications in lifting magnets, buzzers, bells or chimes, and relays. A *relay* is a switch with contacts that are opened or closed by an electromagnet.

Another common application is magnetic tape recording. The tape is coated with fine particles of iron oxide. The recording head is a coil that produces a magnetic field in proportion to the current. As the tape passes through the air gap of the head, small areas of the coating become magnetized by induction. On playback, the moving magnetic tape produces variations in electric current.

Permanent Magnets

These are made of hard magnetic materials, such as cobalt steel, magnetized by induction in the manufacturing process. A very strong field is needed for induction in these materials. When the magnetizing field is removed, however, residual induction makes the material a permanent magnet. A common PM material is *alnico*, a commercial alloy of aluminum, nickel, and iron, with cobalt, copper, and titanium added to produce about 12 grades. The Alnico V grade is often used for PM loudspeakers (Fig. 13–12). In this application, a typical size of PM slug for a steady magnetic field is a few ounces to about 5 lb, with a flux ϕ of 500 to 25,000 lines or maxwells. One advantage of a PM loudspeaker is that only two connecting leads are needed for the voice coil because the steady magnetic field of the PM slug is obtained without any field-coil winding.

Commercial permanent magnets will last indefinitely if they are not subjected to high temperatures, physical shock, or a strong demagnetizing field. If the magnet becomes hot, however, the molecular structure can be rearranged, resulting in loss of magnetism that is not recovered after cooling. The point at which a magnetic material loses its **ferromagnetic** properties is the ***Curie temperature***. For iron, this temperature is about 800°C, when the relative permeability drops to unity. A permanent magnet does not become exhausted with use because its magnetic properties are determined by the structure of the internal atoms and molecules.

Classification of Magnetic Materials

When we consider materials simply as either magnetic or nonmagnetic, this division is based on the strong magnetic properties of iron. However, weak magnetic

GOOD TO KNOW

A strong magnet will not attract a diamagnetic material at all, but it will slightly attract a paramagnetic material. Try doing this yourself for some of the paramagnetic and diamagnetic materials listed.

materials can be important in some applications. For this reason, a more exact classification includes the following three groups:

1. *Ferromagnetic materials.* These include iron, steel, nickel, cobalt, and commercial alloys such as alnico and Permalloy. They become strongly magnetized in the same direction as the magnetizing field, with high values of permeability from 50 to 5000. Permalloy has a μ_r of 100,000 but is easily saturated at relatively low values of flux density.
2. ***Paramagnetic** materials.* These include aluminum, platinum, manganese, and chromium. Their permeability is slightly more than 1. They become weakly magnetized in the same direction as the magnetizing field.
3. ***Diamagnetic** materials.* These include bismuth, antimony, copper, zinc, mercury, gold, and silver. Their permeability is less than 1. They become weakly magnetized but in the direction opposite from the magnetizing field.

The basis of all the magnetic effects is the magnetic field associated with electric charges in motion. Within the atom, the motion of its orbital electrons generates a magnetic field. There are two kinds of electron motion in the atom. First is the electron revolving in its orbit. This motion provides a diamagnetic effect. However, this magnetic effect is weak because thermal agitation at normal room temperature results in random directions of motion that neutralize each other.

More effective is the magnetic effect from the motion of each electron spinning on its own axis. The spinning electron serves as a tiny permanent magnet. Opposite spins provide opposite polarities. Two electrons spinning in opposite directions form a pair, neutralizing the magnetic fields. In the atoms of ferromagnetic materials, however, there are many unpaired electrons with spins in the same direction, resulting in a strong magnetic effect.

In terms of molecular structure, iron atoms are grouped in microscopically small arrangements called *domains*. Each domain is an elementary *dipole magnet*, with two opposite poles. In crystal form, the iron atoms have domains parallel to the axes of the crystal. Still, the domains can point in different directions because of the different axes. When the material becomes magnetized by an external magnetic field, though, the domains become aligned in the same direction. With PM materials, the alignment remains after the external field is removed.

13–6 Self-Review

Answers at the end of the chapter.

a. An electromagnet needs current to maintain its magnetic field. (True/False)
b. A relay coil is an electromagnet. (True/False)
c. Iron is a diamagnetic material. (True/False)

13–7 Ferrites

Ferrite is the name for nonmetallic materials that have the ferromagnetic properties of iron. Ferrites have very high permeability, like iron. However, a ferrite is a nonconducting ceramic material, whereas iron is a conductor. The permeability of ferrites is in the range of 50 to 3000. The specific resistance is $10^5\ \Omega\cdot$cm, which makes a ferrite an insulator.

A common application is a ferrite core, usually adjustable, in the coils of RF transformers. The ferrite core is much more efficient than iron when the current alternates at high frequency. The reason is that less I^2R power is lost by eddy currents in the core because of its very high resistance.

Figure 13–13 Ferrite bead equivalent to coil with 20 μH of inductance at 10 MHz.

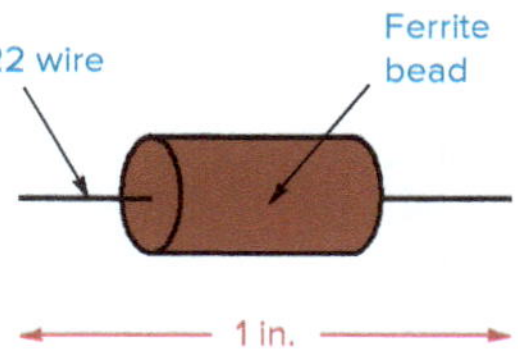

A ferrite core is used in small coils and transformers for signal frequencies up to 20 MHz, approximately. The high permeability means that the transformer can be very small. However, ferrites are easily saturated at low values of magnetizing current. This disadvantage means that ferrites are not used for power transformers.

Another application is in ferrite beads (Fig. 13–13). A bare wire is used as a string for one or more beads. The bead concentrates the magnetic field of the current in the wire. This construction serves as a simple, economical RF choke, instead of a coil. The purpose of the choke is to reduce the current just for an undesired radio frequency.

13–7 Self-Review

Answers at the end of the chapter.

a. **Which has more *R*, ferrites or soft iron?**
b. **Which has more I^2R losses, an insulator or a conductor?**

13–8 Magnetic Shielding

The idea of preventing one component from affecting another through their common electric or magnetic field is called *shielding*. Examples are the braided copper-wire shield around the inner conductor of a coaxial cable, a metal shield can that encloses an RF coil, or a shield of magnetic material enclosing a cathode-ray tube.

The problem in shielding is to prevent one component from inducing an effect in the shielded component. The shielding materials are always metals, but there is a difference between using good conductors with low resistance, such as copper and aluminum, and using good magnetic materials such as soft iron.

A good conductor is best for two shielding functions. One is to prevent induction of static electric charges. The other is to shield against the induction of a varying magnetic field. For static charges, the shield provides opposite induced charges, which prevent induction inside the shield. For a varying magnetic field, the shield has induced currents that oppose the inducing field. Then there is little net field strength to produce induction inside the shield.

The best shield for a steady magnetic field is a good magnetic material of high permeability. A steady field is produced by a permanent magnet, a coil with steady direct current, or the earth's magnetic field. A magnetic shield of high permeability concentrates the magnetic flux. Then there is little flux to induce poles in a component inside the shield. The shield can be considered a short circuit for the lines of magnetic flux.

Science & Society Picture Library/SSPL/Getty Images

PIONEERS IN ELECTRONICS

In 1879, *Edwin H. Hall* (1855–1938) was a graduate student at Johns Hopkins University when he discovered the Hall effect. When a wire carrying a current is placed in an applied magnetic field, a voltage across the wire is created that is proportional to the strength of the magnetic field. This effect is at the heart of a number of technologies such as antilock brake sensors and some computer keyboards.

13–8 Self-Review

Answers at the end of the chapter.

a. **A magnetic material with high permeability is a good shield for a steady magnetic field. (True/False)**
b. **A conductor is a good shield against a varying magnetic field. (True/False)**

13–9 The Hall Effect

In 1879, E. H. Hall observed that a small voltage is generated across a conductor carrying current in an external magnetic field. The Hall voltage was very small with typical conductors, and little use was made of this effect. However, with the development of semiconductors, larger values of Hall voltage can be generated. The

Figure 13–14 The Hall effect. Hall voltage V_H generated across the element is proportional to the perpendicular flux density *B*.

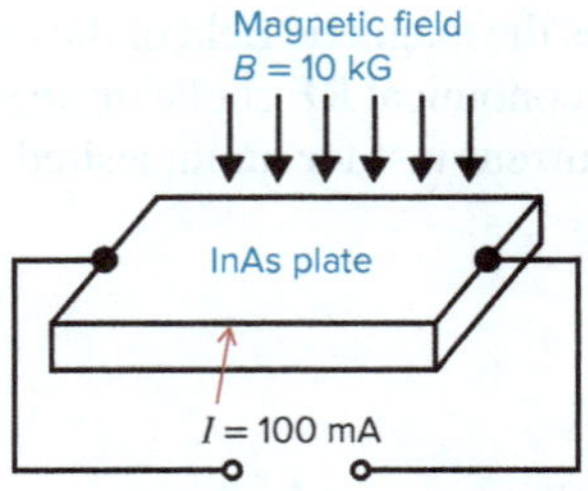

Figure 13–15 A gaussmeter to measure flux density, with a probe containing an indium arsenide element.

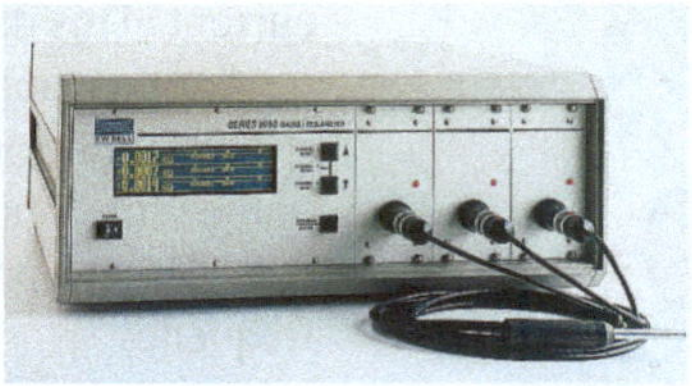

Courtesy of OECO, LLC

semiconductor material indium arsenide (InAs) is generally used. As illustrated in Fig. 13–14, the InAs element inserted in a magnetic field can generate 60 mV with *B* equal to 10 kG and an *I* of 100 mA. The applied flux must be perpendicular to the direction of the current. With current in the direction of the length of conductor, the generated voltage is developed across the width.

The amount of Hall voltage V_H is directly proportional to the value of flux density *B*. This means that values of *B* can be measured by V_H. As an example, the gaussmeter in Fig. 13–15 uses an InAs probe in the magnetic field to generate a proportional Hall voltage V_H. This value of V_H is then read by the meter, which is calibrated in gauss. The original calibration is made in terms of a reference magnet with a specified flux density.

13-9 Self-Review

Answers at the end of the chapter.

a. In Fig. 13–14, how much is the generated Hall voltage?

b. Does the gaussmeter in Fig. 13–15 measure flux or flux density?

Summary

- Iron, nickel, and cobalt are common examples of magnetic materials. Air, paper, wood, and plastics are nonmagnetic.
- The pole of a magnet that seeks the geographic North Pole of the earth is called a *north pole*; the opposite pole is a *south pole*.
- Opposite magnetic poles are attracted; similar poles repel.
- An electromagnet needs current from an external source to provide a magnetic field. Permanent magnets retain their magnetism indefinitely.
- Any magnet has an invisible field of force outside the magnet, indicated by magnetic field lines. Their direction is from the north to the south pole.
- The open ends of a magnet where it meets a nonmagnetic material provide magnetic poles. At opposite open ends, the poles have opposite polarity.
- A magnet with an air gap has opposite poles with magnetic lines of force across the gap. A closed magnetic ring has no poles.
- Magnetic induction enables the field of a magnet to induce magnetic poles in a magnetic material without touching.
- Permeability is the ability to concentrate magnetic flux. A good magnetic material has high permeability.
- Magnetic shielding means isolating a component from a magnetic field. The best shield against a steady magnetic field is a material with high permeability.
- The Hall voltage is a small voltage generated across the width of a conductor carrying current through its length, when magnetic flux is applied perpendicular to the current. This effect is generally used in the gaussmeter to measure flux density.
- Table 13–1 summarizes the units of magnetic flux ϕ and flux density B.

Table 13–1 Magnetic Flux ϕ and Flux Density B

Name	Symbol	cgs Units	mks or SI Units
Flux, or total lines	$\phi = B \times \text{area}$	1 maxwell (Mx) = 1 line	1 weber (Wb) = 10^8 Mx
Flux density, or lines per unit area	$B = \dfrac{\phi}{\text{area}}$	1 gauss (G) = $\dfrac{1 \text{ Mx}}{\text{cm}^2}$	1 tesla (T) = $\dfrac{1 \text{ Wb}}{\text{m}^2}$

Important Terms

Curie temperature — the temperature at which a magnetic material loses its ferromagnetic properties.

Diamagnetic — a classification of materials that become weakly magnetized but in the direction opposite to the magnetizing field. Diamagnetic materials have a permeability less than 1. Examples include antimony, bismuth, copper, gold, mercury, silver, and zinc.

Electromagnet — a magnet that requires an electric current flowing in the turns of a coil to create a magnetic field. With no current in the coil, there is no magnetic field.

Ferrite — a nonmetallic material that has the ferromagnetic properties of iron.

Ferromagnetic — a classification of materials that become strongly magnetized in the same direction as the magnetizing field. Ferromagnetic materials have high values of permeability in the range of 50 to 5000 or even higher. Examples include iron, steel, nickel, and cobalt.

Flux density (*B*) — the number of magnetic field lines per unit area of a section perpendicular to the direction of flux.

Gauss (G) — the cgs unit of flux density. $1 \text{ G} = \dfrac{1 \text{ Mx}}{\text{cm}^2}$.

Hall effect — the effect that describes a small voltage generated across the width of a conductor that is carrying current in an external magnetic field. To develop the Hall voltage, the current in the conductor and the external flux must be at right angles to each other.

Induction — the electric or magnetic effect of one body on another without any physical contact between them.

Magnetic flux (ϕ) — another name used to describe magnetic field lines.

Maxwell (Mx) — the cgs unit of magnetic flux. 1 Mx = 1 magnetic field line.

Paramagnetic — a classification of materials that become weakly magnetized in the same direction as the magnetizing field. Their permeability is slightly more than 1. Examples include aluminum, platinum, manganese, and chromium.

Permanent magnet — a hard magnetic material such as cobalt steel that is magnetized by induction in the manufacturing process. A permanent magnet retains its magnetic properties indefinitely as long as it is not subjected to very high temperatures, physical shock, or a strong demagnetizing field.

Relative permeability (μ_r) — the ability of a material to concentrate magnetic flux. Mathematically, relative permeability, designated μ_r, is the ratio of the flux density (*B*) in a material such as iron and the flux density, *B*, in air. There are no units for μ_r because it compares two flux densities and the units cancel.

Tesla (T) — the SI unit of flux density. $1 \text{ T} = \dfrac{1 \text{ Wb}}{\text{m}^2}$

Toroid — an electromagnet wound in the form of a doughnut. It has no magnetic poles and the maximum strength of the magnetic field is concentrated in its iron core.

Weber (Wb) — the SI unit of magnet flux. 1 Wb = 1×10^8 Mx or lines.

Related Formulas

$$B = \frac{\phi}{A}$$

$$\#Mx = \#Wb \times \frac{1 \times 10^8\ Mx}{1\ Wb}$$

$$\#Wb = \#Mx \times \frac{1\ Wb}{1 \times 10^8\ Mx}$$

$$\#G = \#T \times \frac{1 \times 10^4\ G}{1\ T}$$

$$\#T = \#G \times \frac{1\ T}{1 \times 10^4\ G}$$

Self-Test

Answers at the back of the book.

1. The maxwell (Mx) is a unit of
- a. flux density.
- b. permeability.
- c. magnetic flux.
- d. field intensity.

2. With bar magnets,
- a. like poles attract each other and unlike poles repel each other.
- b. unlike poles attract each other and like poles repel each other.
- c. there are no north or south poles on the ends of the magnet.
- d. none of the above.

3. The tesla (T) is a unit of
- a. flux density.
- b. magnetic flux.
- c. permeability.
- d. magnetomotive force.

4. 1 maxwell (Mx) is equal to
- a. 1×10^8 Wb.
- b. $\frac{1\ Wb}{m^2}$.
- c. 1×10^4 G.
- d. one magnetic field line.

5. 1 Wb is equal to
- a. 1×10^8 Mx.
- b. one magnetic field line.
- c. $\frac{1\ Mx}{cm^2}$.
- d. 1×10^4 kG.

6. The electric or magnetic effect of one body on another without any physical contact between them is called
- a. its permeability.
- b. induction.
- c. the Hall effect.
- d. hysteresis.

7. A commercial permanent magnet will last indefinitely if it is not subjected to
- a. a strong demagnetizing field.
- b. physical shock.
- c. high temperatures.
- d. all of the above.

8. What is the name for a nonmetallic material that has the ferromagnetic properties of iron?
- a. lodestone.
- b. toroid.
- c. ferrite.
- d. solenoid.

9. One tesla (T) is equal to
- a. $\frac{1\ Mx}{m^2}$.
- b. $\frac{1\ Mx}{cm^2}$.
- c. $\frac{1\ Wb}{m^2}$.
- d. $\frac{1\ Wb}{cm^2}$.

10. The ability of a material to concentrate magnetic flux is called its
- a. induction.
- b. permeability.
- c. Hall effect.
- d. diamagnetic.

11. If the north (N) pole of a permanent magnet is placed near a piece of soft iron, what is the polarity of the nearest induced pole?
- a. south (S) pole.
- b. north (N) pole.
- c. It could be either a north (N) or a south (S) pole.
- d. It cannot be determined.

12. A magnet that requires current in a coil to create the magnetic field is called a(n)
- a. permanent magnet.
- b. electromagnet.
- c. solenoid.
- d. both b and c.

13. The point at which a magnetic material loses its ferromagnetic properties is called the
- a. melting point.
- b. freezing point.
- c. Curie temperature.
- d. leakage point.

14. A material that becomes strongly magnetized in the same direction as the magnetizing field is classified as
- a. diamagnetic.
- b. ferromagnetic.
- c. paramagnetic.
- d. toroidal.

15. Which of the following materials are nonmagnetic?
- a. air.
- b. wood.
- c. glass.
- d. all of the above.

16. The gauss (G) is a unit of
- a. flux density.
- b. magnetic flux.
- c. permeability.
- d. none of the above.

17. One gauss (G) is equal to
- a. $\frac{1\ Mx}{m^2}$.
- b. $\frac{1\ Wb}{cm^2}$.
- c. $\frac{1\ Mx}{cm^2}$.
- d. $\frac{1\ Wb}{m}$.

18. 1 μWb equals
- a. 1×10^8 Mx.
- b. 10,000 Mx.
- c. 1×10^{-8} Mx.
- d. 100 Mx.

19. A toroid

a. is an electromagnet.

b. has no magnetic poles.

c. uses iron for the core around which the coil is wound.

d. all of the above.

20. When a small voltage is generated across the width of a conductor carrying current in an external magnetic field, the effect is called

a. the Doppler effect.

b. the Miller effect.

c. the Hall effect.

d. the Schultz effect.

21. The weber (Wb) is a unit of

a. magnetic flux.

b. flux density.

c. permeability.

d. none of the above.

22. The flux density in the iron core of an electromagnet is 0.25 T. When the iron core is removed, the flux density drops to 62.5×10^{-6} T. What is the relative permeability of the iron core?

a. $\mu_r = 4$.

b. $\mu_r = 250$.

c. $\mu_r = 4000$.

d. It cannot be determined.

23. What is the flux density, *B*, for a magnetic flux of 500 Mx through an area of 10 cm^2?

a. 50×10^{-3} T.

b. 50 G.

c. 5000 G.

d. both a and b.

24. The geographic North Pole of the earth has

a. no magnetic polarity.

b. south magnetic polarity.

c. north magnetic polarity.

d. none of the above.

25. With an electromagnet,

a. more current and more coil turns mean a stronger magnetic field.

b. less current and fewer coil turns mean a stronger magnetic field.

c. if there is no current in the coil, there is no magnetic field.

d. both a and c.

Essay Questions

1. Name two magnetic materials and three nonmagnetic materials.
2. Explain the difference between a permanent magnet and an electromagnet.
3. Draw a horseshoe magnet and its magnetic field. Label the magnetic poles, indicate the air gap, and show the direction of flux.
4. Define *relative permeability, shielding, induction,* and *Hall voltage*.
5. Give the symbols, cgs units, and SI units for magnetic flux and for flux density.
6. How are the north and south poles of a bar magnet determined with a magnetic compass?
7. Referring to Fig. 13–11, why can either end of the magnet pick up the nail?
8. What is the difference between flux ϕ and flux density *B*?

Problems

SECTION 13–2 MAGNETIC FLUX (Φ)

13–1 Define (a) the maxwell (Mx) unit of magnetic flux, ϕ; (b) the weber (Wb) unit of magnetic flux, ϕ.

13–2 Make the following conversions:

a. 0.001 Wb to Mx.

b. 0.05 Wb to Mx.

c. 15×10^{-4} Wb to Mx.

d. 1×10^{-8} Wb to Mx.

13–3 Make the following conversions:

a. 1000 Mx to Wb.

b. 10,000 Mx to Wb.

c. 1 Mx to Wb.

d. 100 Mx to Wb.

13–4 Make the following conversions:

a. 0.0002 Wb to Mx.

b. 5500 Mx to Wb.

c. 70 Mx to Wb.

d. 30×10^{-6} Wb to Mx.

13–5 Make the following conversions:

a. 0.00004 Wb to Mx.

b. 225 Mx to Wb.

c. 80,000 Mx to Wb.

d. 650×10^{-6} Wb to Mx.

13–6 A permanent magnet has a magnetic flux of 12,000 μWb. How many magnetic field lines does this correspond to?

13–7 An electromagnet produces a magnetic flux of 900 μWb. How many magnetic field lines does this correspond to?

13–8 A permanent magnet has a magnetic flux of 50,000 Mx. How many Webers (Wb) of magnetic flux does this correspond to?

SECTION 13–3 FLUX DENSITY (*B*)

13–9 Define (a) the gauss (G) unit of flux density, *B*; (b) the tesla (T) unit of flux density, *B*.

13–10 Make the following conversions:

a. 2.5 T to G.

b. 0.05 T to G.

c. 1×10^{-4} T to G.

d. 0.1 T to G.

13–11 Make the following conversions:

a. 4000 G to T.

b. 800,000 G to T.

c. 600 G to T.

d. 10,000 G to T.

13–12 Make the following conversions:

a. 0.004 T to G.

b. 1000 G to T.

c. 1×10^{5} G to T.

d. 10 T to G.

13–13 Make the following conversions:

a. 0.0905 T to G.

b. 100 T to G.

c. 75,000 G to T.

d. 1.75×10^{6} G to T.

13–14 With a flux of 250 Mx through an area of 2 cm^2, what is the flux density in gauss units?

13–15 A flux of 500 μWb exists in an area, *A*, of 0.01 m^2. What is the flux density in tesla units?

13–16 Calculate the flux density, in teslas, for a flux, ϕ, of 400 μWb in an area of 0.005 m^2.

13–17 Calculate the flux density in gauss units for a flux, ϕ, of 200 μWb in an area of 5×10^{-4} m^2.

13–18 With a magnetic flux, ϕ, of 30,000 Mx through a perpendicular area of 6 cm^2, what is the flux density in gauss units?

13–19 How much is the flux density in teslas for a flux, ϕ, of 160 μWb through an area of 0.0012 m^2?

13–20 With a flux, ϕ, of 2000 μWb through an area of 0.0004 m^2, what is the flux density in gauss units?

13–21 For a flux density of 30 kG at the north pole of a magnet through a cross-sectional area of 8 cm^2, how much is the total flux in maxwells?

13–22 The flux density in an iron core is 5×10^{-3} T. If the area of the core is 10 cm^2, calculate the total number of magnetic flux lines in the core.

13–23 The flux density in an iron core is 5 T. If the area of the core is 40 cm^2, calculate the magnetic flux in weber units.

13–24 The flux density in an iron core is 80 kG. If the area of the core is 0.2 m^2, calculate the magnetic flux in weber units.

13–25 If the flux density in 0.05 m^2 is 2000 G, how many magnetic field lines are there?

Critical Thinking

13–26 A flux ϕ of 25 μWb exists in an area of 0.25 in^2. What is the flux density *B* in (a) gauss units; (b) teslas?

13–27 At the north pole of an electromagnet, the flux density *B* equals 5 T. If the area *A* equals 0.125 in^2, determine the total number of flux lines ϕ in (a) maxwells; (b) webers.

Answers to Self-Reviews

13–1 **a.** true **b.** true

13–2 **a.** 2000 Mx **b.** 20 μWb

13–3 **a.** 3000 G **b.** 0.3 T

13–4 **a.** true **b.** false

13–5 **a.** true **b.** true

13–6 **a.** true **b.** true **c.** false

13–7 **a.** ferrites **b.** conductor

13–8 **a.** true **b.** true

13–9 **a.** 60 mV **b.** flux density

Laboratory Application Assignment

This lab application assignment is optional due to the fact that your laboratory may not have the items listed under "Equipment." If you do have the items listed, then you will be able to experimentally determine the pattern of magnetic field lines extending outward from the poles of a bar magnet.

Equipment: Obtain the following items from your instructor.

- Digital camera
- Compass
- Two bar magnets (If the pole ends are marked N and S cover them.)
- Iron filings
- Thin cardboard or sheet of glass (8½ × 11 in.)

Magnetic Field Pattern

Examine the compass assigned to you. How can you tell which end of the compass needle is the north (N) pole? ____________

____

Describe how the compass can be used to determine the pole polarities of the two unmarked bar magnets you have in your possession. ____________________________

____

____

Using the technique you described above, experimentally determine the pole polarities on each of the two bar magnets assigned to you.

Remove the coverings on each end of the bar magnets, and see if your technique was correct. Was it? __________

Place a thin piece of cardboard or thin sheet of glass over one of the bar magnets assigned to you. Sprinkle iron filings on the cardboard or glass directly above the bar magnet. Gently tap the cardboard or glass until you see a recognizable pattern. This is the pattern of a cross section of the lines of force surrounding the magnet. Take a picture of this pattern with a digital camera. Identify the position of the magnetic poles, and indicate the direction of the lines of force.

Now place both bar magnets (in line) under the cardboard or glass so that the N pole of one is adjacent to and about 1 in. away from the S pole of the other. Lightly sprinkle iron filings on the cardboard or glass directly above both bar magnets. Gently tap the cardboard or glass until you see a recognizable pattern. This is the pattern of a cross-section of the lines of force surrounding both magnets. Take a picture of this pattern with a digital camera. Identify the position of the magnetic poles, and indicate the direction of the lines of force.

Turn one of the two bar magnets completely around (180 degrees) so that the N pole of one magnet is adjacent to and about 1 in. away from the N pole of the other. Be sure the magnets are still in line. Lightly sprinkle iron filings on the cardboard or glass directly above both bar magnets. Gently tap the cardboard or glass until you see a recognizable pattern. This is the pattern of a cross section of the lines of force surrounding both magnets. Take a picture of this pattern with a digital camera. Identify the position of the magnetic poles, and indicate the direction of the lines of force.

Design credits: Chapter Opener Background Image, and Multisim: Stockbyte/Getty Images

chapter 14

Electromagnetism

A magnetic field is always associated with an electric current. Therefore, the units of measure for the strength or intensity of a magnetic field are based on the electric current that produces the field. For an electromagnet, the strength and intensity of the magnetic field depend on the amount of current flow and the number of coil turns in a given length. The electromagnet acts like a bar magnet with opposite magnetic poles at its ends.

When a conductor passes through a magnetic field, the work put into this action forces free electrons to move along the length of the conductor. The rate at which the conductor moves through the magnetic field and how many magnetic field lines are cut determines the amount of induced current and/or voltage. The effects of magnetism and electromagnetism play a very important role in a wide variety of electrical and electronic systems. In fact, motors, generators, transformers, speakers, relays, and all types of medical equipment in use today, depend on magnetism and its effects to function properly. In this chapter, you will learn about the units and laws of electromagnetism. You will also learn about an electromechanical device known as a relay. As you will see, a relay uses an electromagnet to open or close one or more sets of switching contacts.

Chapter Outline

Chapter Objectives

After studying this chapter, you should be able to

- *Define* the terms *magnetomotive force* and *field intensity* and list the units of each.
- *Explain* the *B-H* magnetization curve.
- *Define* the term *saturation* as it relates to a magnetic core.
- *Explain* what is meant by *magnetic hysteresis*.
- *Describe* the magnetic field of an electric current in a straight conductor.
- *Determine* the magnetic polarity of a solenoid using the left-hand rule.
- *Explain* the concept of motor action.
- *Explain* how an induced voltage can be developed across the ends of a conductor that passes through a magnetic field.
- *State* Lenz's law.
- *Apply* Faraday's law to *calculate* the induced voltage across a conductor being passed through a magnetic field.
- *Explain* the basic construction and operation of an electromechanical relay.
- *List and explain* important relay ratings.

Important Terms

ampere-turn (A·t)
ampere-turns/meter (A·t/m)
B-H magnetization curve
degaussing
Faraday's law
field intensity (*H*)
holding current
hysteresis
left-hand rule
Lenz's law
magnetomotive force (mmf)
motor action
pickup current
saturation

14–1 Ampere-Turns of Magnetomotive Force (mmf)

The strength of the magnetic field of a coil magnet depends on how much current flows in the turns of the coil. The more current, the stronger the magnetic field. Also, more turns in a specific length concentrate the field. The coil serves as a bar magnet with opposite poles at the ends, providing a magnetic field proportional to the **ampere-turns**. As a formula,

$$\text{Ampere-turns} = I \times N = \text{mmf} \tag{14–1}$$

where I is the current in amperes multiplied by the number of turns N. The quantity IN specifies the amount of *magnetizing force* or *magnetic potential,* which is the ***magnetomotive force* (*mmf*)**.

The practical unit is the ampere-turn. The SI abbreviation for ampere-turn is A, the same as for the ampere, since the number of turns in a coil usually is constant but the current can be varied. However, for clarity we shall use the abbreviation A·t.

As shown in Fig. 14–1, a solenoid with 5 turns and 2 amperes has the same magnetizing force as one with 10 turns and 1 ampere, as the product of the amperes and turns is 10 for both cases. With thinner wire, more turns can be placed in a given space. The amount of current is determined by the resistance of the wire and the source voltage. The number of ampere-turns necessary depends on the magnetic field strength required.

GOOD TO KNOW

The mmf of a coil is directly proportional to both the current, I, and the number of turns, N.

MultiSim **Figure 14–1** Two examples of equal ampere-turns for the same mmf. (*a*) IN is $2 \times 5 = 10$. (*b*) IN is $1 \times 10 = 10$.

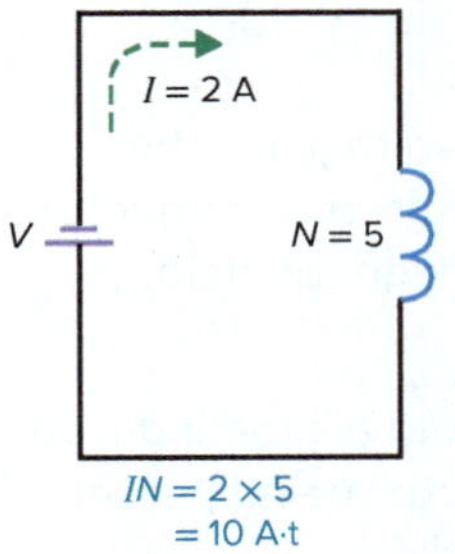

(*a*)

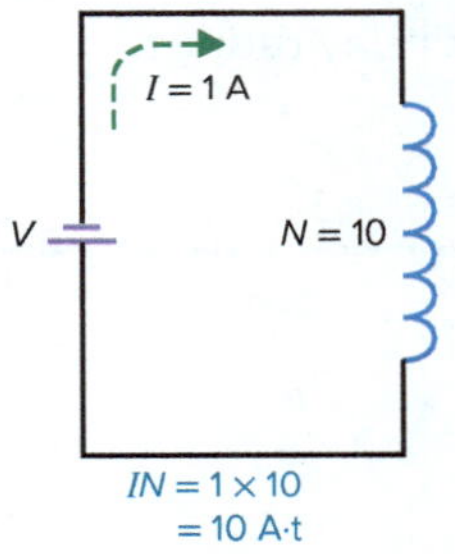

(*b*)

Example 14-1

Calculate the ampere-turns of mmf for a coil with 2000 turns and a 5-mA current.

ANSWER $\text{mmf} = I \times N = 2000 \times 5 \times 10^{-3}$

$= 10 \text{ A·t}$

Example 14-2

A coil with 4 A is to provide a magnetizing force of 600 A·t. How many turns are necessary?

ANSWER $N = \dfrac{\text{A·t}}{I} = \dfrac{600}{4}$

$= 150 \text{ turns}$

Example 14-3

A coil with 400 turns must provide 800 A·t of magnetizing force. How much current is necessary?

ANSWER $I = \frac{\text{A·t}}{\text{N}} = \frac{800}{400}$

$= 2\ \text{A}$

Example 14-4

The wire in a solenoid of 250 turns has a resistance of 3 Ω. (a) How much is the current when the coil is connected to a 6-V battery? (b) Calculate the ampere-turns of mmf.

ANSWER

a. $I = \frac{V}{R} = \frac{6\ \text{V}}{3\ \Omega}$

$= 2\ \text{A}$

b. $\text{mmf} = I \times N = 2\ \text{A} \times 250\ \text{t}$

$= 500\ \text{A·t}$

The ampere-turn, A·t, is an SI unit. It is calculated as *IN* with the current in amperes.

The cgs unit of mmf is the *gilbert,** abbreviated Gb. One ampere-turn equals 1.26 Gb. The number 1.26 is approximately $4\pi/10$, derived from the surface area of a sphere, which is $4\pi r^2$.

To convert *IN* to gilberts, multiply the ampere-turns by the constant conversion factor 1.26 Gb/1 A·t. As an example, 1000 A·t is the same mmf as 1260 Gb. The calculations are

$$1000\ \text{A·t} \times 1.26\ \frac{\text{Gb}}{1\ \text{A·t}} = 1260\ \text{Gb}$$

Note that the units of A·t cancel in the conversion.

14–1 Self-Review

Answers at the end of the chapter.

a. If the mmf is 243 A·t, and *I* is doubled from 2 to 4 A with the same number of turns, how much is the new mmf?

b. Convert 500 A·t to gilberts.

14–2 Field Intensity (*H*)

The ampere-turns of mmf specify the magnetizing force, but the intensity of the magnetic field depends on the length of the coil. At any point in space, a specific value of ampere-turns must produce less field intensity for a long coil than for a

* William Gilbert (1544–1603) was an English scientist who investigated the magnetism of the earth.

Figure 14–2 Relation between ampere-turns of mmf and the resultant field intensity H for different cores. Note that H = mmf/length. (*a*) Intensity H is 1000 A•t/m with an air core. (*b*) H = 1000 A•t/m in an iron core of the same length as the coil. (*c*) H is 1000/2 = 500 A•t/m in an iron core twice as long as the coil.

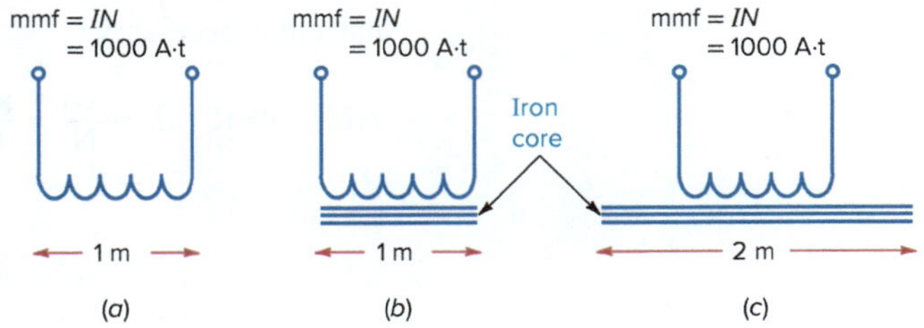

short coil that concentrates the same mmf. Specifically, the **field intensity *H*** in mks units is

$$H = \frac{\text{ampere-turns of mmf}}{l \text{ meters}} \qquad \textbf{(14–2)}$$

This formula is for a solenoid. The field intensity H is at the center of an air core. For an iron core, H is the intensity through the entire core. By means of units for H, the magnetic field intensity can be specified for either electromagnets or permanent magnets, since both provide the same kind of magnetic field.

The length in Formula (14–2) is between poles. In Fig. 14–2*a*, the length is 1 m between the poles at the ends of the coil. In Fig. 14–2*b*, l is also 1 m between the ends of the iron core. In Fig. 14–2*c*, though, l is 2 m between the poles at the ends of the iron core, although the winding is only 1 m long.

The examples in Fig. 14–2 illustrate the following comparisons:

1. In all three cases, the mmf is 1000 A•t for the same value of IN.
2. In Fig. 14–2*a* and *b*, H equals 1000 A•t/m. In *a*, this H is the intensity at the center of the air core; in *b*, this H is the intensity through the entire iron core.
3. In Fig. 14–2*c*, because l is 2 m, H is $^{1000}/_{2}$, or 500 A•t/m. This H is the intensity in the entire iron core.

Units for *H*

The field intensity is basically mmf per unit of length. In practical units, H is ampere-turns per meter. The cgs unit for H is the *oersted,** abbreviated Oe, which equals one gilbert of mmf per centimeter.

Conversion of Units

To convert SI units of A•t/m to cgs units of Oe, multiply by the conversion factor 0.0126 Oe per 1 A•t/m. As an example, 1000 A•t/m is the same H as 12.6 Oe. The calculations are

$$1000 \frac{\text{A·t}}{\text{m}} \times 0.0126 \frac{\text{Oe}}{1 \text{ A·t/m}} = 12.6 \text{ Oe}$$

Note that the units of A•t and m cancel. The m in the conversion factor becomes inverted to the numerator.

* H. C. Oersted (1777–1851), a Danish physicist, discovered electromagnetism.

Permeability (μ)

Whether we say H is 1000 A·t/m or 12.6 Oe, these units specify how much field intensity is available to produce magnetic flux. However, the amount of flux produced by H depends on the material in the field. A good magnetic material with high relative permeability can concentrate flux and produce a large value of flux density B for a specified H. These factors are related by the formula:

$$B = \mu \times H \qquad \textbf{(14–3)}$$

or

$$\mu = \frac{B}{H} \qquad \textbf{(14–4)}$$

Using SI units, B is the flux density in webers per square meter, or teslas; H is the field intensity in ampere-turns per meter. In the cgs system, the units are gauss for B and oersted for H. The factor μ is the absolute permeability, not referred to any other material, in units of B/H.

In the cgs system, the units of gauss for B and oersteds for H have been defined to give μ the value of 1 G/Oe, for vacuum, air, or space. This simplification means that B and H have the same numerical values in air and in vacuum. For instance, the field intensity H of 12.6 Oe produces a flux density of 12.6 G in air.

Furthermore, the values of relative permeability μ_r are the same as those for absolute permeability in B/H units in the cgs system. The reason is that μ is 1 for air or vacuum, used as the reference for the comparison. As an example, if μ_r for an iron sample is 600, the absolute μ is also 600 G/Oe.

In SI, however, the permeability of air or vacuum is not 1. This value is $4\pi \times 10^{-7}$, or 1.26×10^{-6}, with the symbol μ_0. Therefore, values of relative permeability μ_r must be multiplied by 1.26×10^{-6} for μ_0 to calculate μ as B/H in SI units.

For an example of $\mu_r = 100$, the SI value of μ can be calculated as follows:

$$\mu = \mu_r \times \mu_0$$
$$= 100 \times 1.26 \times 10^{-6} \frac{\text{T}}{\text{A·t/m}}$$
$$\mu = 126 \times 10^{-6} \frac{\text{T}}{\text{A·t/m}}$$

GOOD TO KNOW

The permeability of a material is similar in many respects to the conductivity in electric circuits.

Example 14-5

A magnetic material has a μ_r of 500. Calculate the absolute μ as B/H (a) in cgs units and (b) in SI units.

ANSWER

a. $\mu = \mu_r \times \mu_0$ in cgs units. Then

$$= 500 \times 1 \frac{\text{G}}{\text{Oe}}$$
$$= 500 \frac{\text{G}}{\text{Oe}}$$

b. $\mu = \mu_r \times \mu_0$ in SI units. Then

$$= 500 \times 1.26 \times 10^{-6} \frac{\text{T}}{\text{A·t/m}}$$
$$= 630 \times 10^{-6} \frac{\text{T}}{\text{A·t/m}}$$

Example 14-6

For this example of $\mu = 630 \times 10^{-6}$ in SI units, calculate the flux density B that will be produced by the field intensity H equal to 1000 A·t/m.

ANSWER $B = \mu H$

$$= \left(630 \times 10^{-6} \frac{\text{T}}{\text{A}\cdot\text{t/m}}\right)\left(1000 \frac{\text{A}\cdot\text{t}}{\text{m}}\right)$$

$$= 630 \times 10^{-3}\ \text{T}$$

$$= 0.63\ \text{T}$$

Note that the ampere-turns and meter units cancel, leaving only the tesla unit for the flux density B.

14–2 Self-Review

Answers at the end of the chapter.

a. What are the values of μ_r for air, vacuum, and space?

b. An iron core has 200 times more flux density than air for the same field intensity H. How much is μ_r?

c. An iron core produces 200 G of flux density for 1 Oe of field intensity H. How much is μ?

d. A coil with an mmf of 25 A·t is 0.1 m long. How much is the field intensity H?

e. Convert 500 $\frac{\text{A}\cdot\text{t}}{\text{m}}$ to oersted units.

14–3 *B-H* Magnetization Curve

The *B-H* curve in Fig. 14–3 is often used to show how much flux density B results from increasing the amount of field intensity H. This curve is for soft iron, plotted for the values in Table 14–1, but similar curves can be obtained for all magnetic materials.

GOOD TO KNOW

Manufacturers of magnetic materials usually provide information about the material's magnetic properties. For example, *B-H* curves and other data can usually be found in manuals provided by the manufacturer.

Calculating *H* and *B*

The values in Table 14–1 are calculated as follows:

1. The current I in the coil equals V/R. For a 10-Ω coil resistance with 20 V applied, I is 2 A, as listed in the top row of Table 14–1. Increasing values of V produce more current in the coil.
2. The ampere-turns IN of magnetizing force increase with more current. Since the turns are constant at 100, the values of IN increase from 200 for 2 A in the top row to 1000 for 10 A in the bottom row.
3. The field intensity H increases with higher IN. The values of H are in mks units of **ampere-turns per meter**. These values equal $IN/0.2$ because the length is 0.2 m. Therefore, each IN is divided by 0.2, or multiplied by 5, for the corresponding values of H. Since H increases in the same proportion as I, sometimes the horizontal axis on a *B-H* curve is given only in amperes, instead of in H units.

Figure 14–3 *B-H* magnetization curve for soft iron. No values are shown near zero, where μ may vary with previous magnetization.

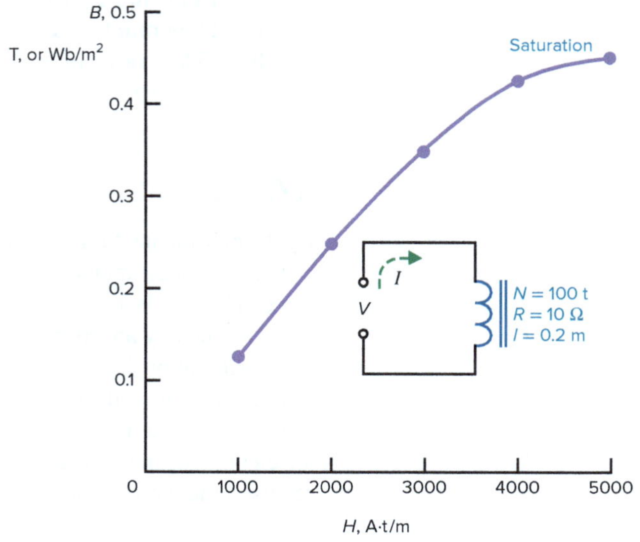

Table 14–1 *B-H* Values for Figure 14–3

V, Volts	*R*, Ω	*I* = *V*/*R*, Amperes	*N*, Turns	mmf, A·t	*l*, m	*H*, A·t/m	μ_r	$B = \mu \times H$, T
20	10	2	100	200	0.2	1000	100	0.126
40	10	4	100	400	0.2	2000	100	0.252
60	10	6	100	600	0.2	3000	100	0.378
80	10	8	100	800	0.2	4000	85	0.428
100	10	10	100	1000	0.2	5000	70	0.441

4. The flux density *B* depends on the field intensity *H* and the permeability of the iron. The values of *B* in the last column are obtained by multiplying $\mu \times H$. However, for SI units, the values of μ_r listed must be multiplied by 1.26×10^{-6} to obtain $\mu \times H$ in teslas.

Saturation

Note that the permeability decreases for the highest values of *H*. With less μ, the iron core cannot provide proportional increases in *B* for increasing values of *H*. In Fig. 14–3, for values of *H* above 4000 A·t/m, approximately, the values of *B* increase at a much slower rate, making the curve relatively flat at the top. The effect of little change in flux density when the field intensity increases is called ***saturation***.

Iron becomes saturated with magnetic lines of induction. After most of the molecular dipoles and the magnetic domains are aligned by the magnetizing force, very little additional induction can be produced. When the value of μ is specified for a magnetic material, it is usually the highest value before saturation.

■ *14–3 Self-Review*

Answers at the end of the chapter.

Refer to Fig. 14–3.

a. **How much is *B* in tesla units for 1500 A·t/m?**

b. **What value of *H* starts to produce saturation?**

14–4 Magnetic Hysteresis

Hysteresis means "lagging behind." With respect to the magnetic flux in an iron core of an electromagnet, the flux lags the increases or decreases in magnetizing force. Hysteresis results because the magnetic dipoles are not perfectly elastic. Once aligned by an external magnetizing force, the dipoles do not return exactly to their original positions when the force is removed. The effect is the same as if the dipoles were forced to move against internal friction between molecules. Furthermore, if the magnetizing force is reversed in direction by reversal of the current in an electromagnet, the flux produced in the opposite direction lags behind the reversed magnetizing force.

Hysteresis Loss

When the magnetizing force reverses thousands or millions of times per second, as with rapidly reversing alternating current, hysteresis can cause a considerable loss of energy. A large part of the magnetizing force is then used to overcome the internal friction of the molecular dipoles. The work done by the magnetizing force against this internal friction produces heat. This energy wasted in heat as the molecular dipoles lag the magnetizing force is called the *hysteresis loss.* For steel and other hard magnetic materials, hysteresis losses are much higher than in soft magnetic materials like iron.

When the magnetizing force varies at a slow rate, hysteresis losses can be considered negligible. An example is an electromagnet with direct current that is simply turned on and off or the magnetizing force of an alternating current that reverses 60 times per second or less. The faster the magnetizing force changes, however, the greater the hysteresis effect.

Hysteresis Loop

To show the hysteresis characteristics of a magnetic material, its values of flux density *B* are plotted for a periodically reversing magnetizing force. See Fig. 14–4. This curve is the hysteresis loop of the material. The larger the area enclosed by the curve, the greater the hysteresis loss. The hysteresis loop is actually a *B-H* curve with an AC magnetizing force.

Values of flux density *B* are indicated on the vertical axis. The units can be gauss or teslas.

The horizontal axis indicates values of field intensity *H.* On this axis, the units can be oersteds, ampere-turns per meter, ampere-turns, or magnetizing current because all factors are constant except *I.*

Opposite directions of current result in opposite directions of $+H$ and $-H$ for the field lines. Similarly, opposite polarities are indicated for flux density as $+B$ or $-B$.

The current starts from zero at the center, when the material is unmagnetized. Then positive *H* values increase *B* to saturation at $+B_{max}$. Next *H* decreases to zero, but *B* drops to the value B_R, instead of to zero, because of hysteresis. When *H* becomes negative, *B* drops to zero and continues to $-B_{max}$, which is saturation in the opposite direction from $+B_{max}$ because of the reversed magnetizing current.

Figure 14–4 Hysteresis loop for magnetic materials. This graph is a *B-H* curve like Fig. 14–3, but *H* alternates in polarity with alternating current.

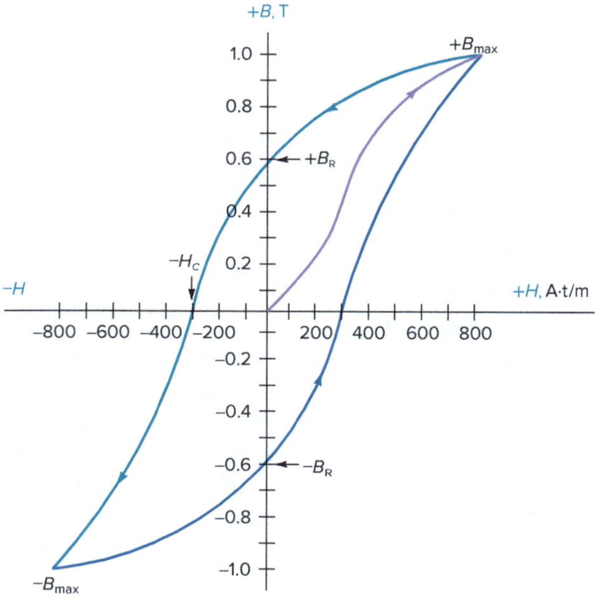

Then, as the $-H$ values decrease, the flux density is reduced to $-B_R$. Finally, the loop is completed; positive values of H produce saturation at B_{max} again. The curve does not return to the zero origin at the center because of hysteresis. As the magnetizing force periodically reverses, the values of flux density are repeated to trace out the hysteresis loop.

The value of either $+B_R$ or $-B_R$, which is the flux density remaining after the magnetizing force has been reduced to zero, is the *residual induction* of a magnetic material, also called its *retentivity*. In Fig. 14–4, the residual induction is 0.6 T, in either the positive or the negative direction.

The value of $-H_C$, which equals the magnetizing force that must be applied in the reverse direction to reduce the flux density to zero, is the *coercive force* of the material. In Fig. 14–4, the coercive force $-H_C$ is 300 A·t/m.

Demagnetization

To demagnetize a magnetic material completely, the residual induction B_R must be reduced to zero. This usually cannot be accomplished by a reversed DC magnetizing force because the material then would become magnetized with opposite polarity. The practical way is to magnetize and demagnetize the material with a continuously decreasing hysteresis loop. This can be done with a magnetic field produced by alternating current. Then, as the magnetic field and the material are moved away from each other or the current amplitude is reduced, the hysteresis loop becomes smaller and smaller. Finally, with the weakest field, the loop collapses practically to zero, resulting in zero residual induction. This method of demagnetization is also called ***degaussing***.

14–4 Self-Review

Answers at the end of the chapter.

a. Hysteresis loss increases with higher frequencies. (True/False)
b. Degaussing is done with alternating current. (True/False)

14–5 Magnetic Field around an Electric Current

In Fig. 14–5, the iron filings aligned in concentric rings around the conductor show the magnetic field of the current in the wire. The iron filings are dense next to the conductor, showing that the field is strongest at this point. Furthermore, the field strength decreases inversely as the square of the distance from the conductor. It is important to note the following two factors about the magnetic lines of force:

1. The magnetic lines are circular because the field is symmetrical with respect to the wire in the center.
2. The magnetic field with circular lines of force is in a plane perpendicular to the current in the wire.

From points C to D in the wire, the circular magnetic field is in the horizontal plane because the wire is vertical. Also, the vertical conductor between points EF and AB has the associated magnetic field in the horizontal plane. Where the conductor is horizontal, as from B to C and D to E, the magnetic field is in a vertical plane.

These two requirements of a circular magnetic field in a perpendicular plane apply to any charge in motion. Whether electron flow or motion of positive charges is considered, the associated magnetic field must be at right angles to the direction of current.

In addition, the current need not be in a wire conductor. As an example, the beam of moving electrons in the vacuum of a cathode-ray tube has an associated magnetic field. In all cases, the magnetic field has circular lines of force in a plane perpendicular to the direction of motion of the electric charges.

GOOD TO KNOW

The fact that an electric current has an associated magnetic field is the basis on which an amp-clamp probe operates. The amp-clamp probe has a pickup coil that senses the strength of the magnetic field set up by the current in the wire conductor. The deflection of the meter's pointer is proportional to the amount of current carried by the conductor. The meter is calibrated in amperes.

Clockwise and Counterclockwise Fields

With circular lines of force, the magnetic field would tend to move a magnetic pole in a circular path. Therefore, the direction of the lines must be considered either clockwise or counterclockwise. This idea is illustrated in Fig. 14–6, showing how a north pole would move in the circular field.

The directions are tested with a magnetic compass needle. When the compass is in front of the wire, the north pole on the needle points up. On the opposite side, the compass points down. If the compass were placed at the top, its needle would point toward the back of the wire; below the wire, the compass would point forward. When all these directions are combined, the result is the circular magnetic field shown with counterclockwise lines of force. (The counterclockwise direction of the magnetic field assumes that you are looking into the end of the wire, in the same direction as electron flow.)

Figure 14–5 How iron filings can be used to show the invisible magnetic field around the electric current in a wire conductor.

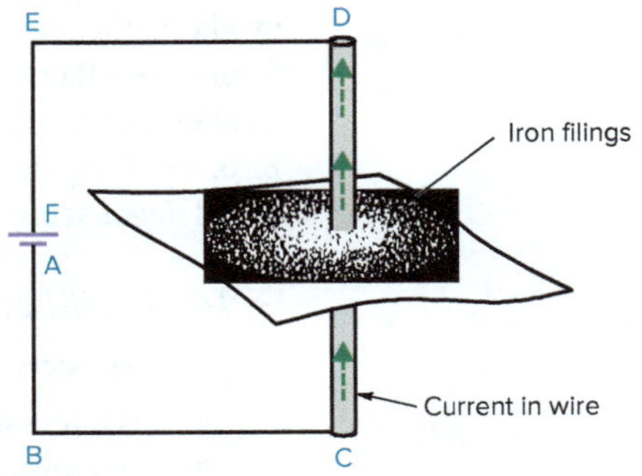

Figure 14–6 Rule for determining direction of circular field around a straight conductor. Field is counterclockwise for direction of electron flow shown here. Circular field is clockwise for reversed direction of electron flow.

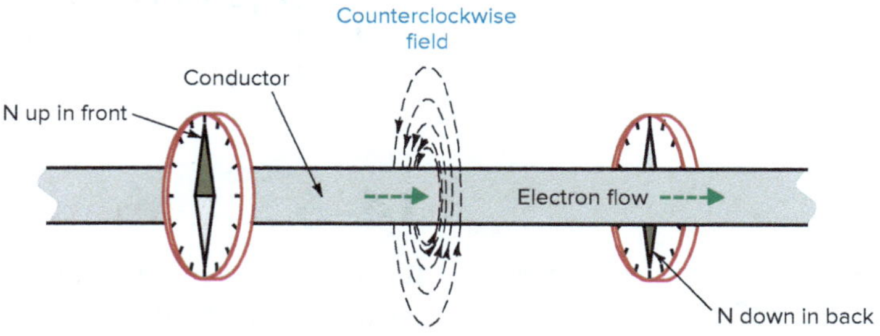

Instead of testing every conductor with a magnetic compass, however, we can use the following rule for straight conductors to determine the circular direction of the magnetic field: *If you grasp the conductor with your left hand so that the thumb points in the direction of electron flow, your fingers will encircle the conductor in the same direction as the circular magnetic field lines.* In Fig. 14–6, the direction of electron flow is from left to right. Facing this way, you can assume that the circular magnetic flux in a perpendicular plane has lines of force in the counterclockwise direction.

The opposite direction of electron flow produces a reversed field. Then the magnetic lines of force rotate clockwise. If the charges were moving from right to left in Fig. 14–6, the associated magnetic field would be in the opposite direction with clockwise lines of force.

Fields Aiding or Canceling

When the magnetic lines of two fields are in the same direction, the lines of force aid each other, making the field stronger. When magnetic lines are in opposite directions, the fields cancel.

In Fig. 14–7, the fields are shown for two conductors with opposite directions of electron flow. The dot in the middle of the field at the left indicates the tip of an arrowhead to show current up from the paper. The cross symbolizes the back of an arrow to indicate electron flow into the paper.

Notice that the magnetic lines *between the conductors* are in the same direction, although one field is clockwise and the other counterclockwise. Therefore, the fields aid here, making a stronger total field. On either side of the conductors, the two fields are opposite in direction and tend to cancel each other. The net result, then, is to strengthen the field in the space between the conductors.

MultiSim **Figure 14–7** Magnetic fields aiding between parallel conductors with opposite directions of current.

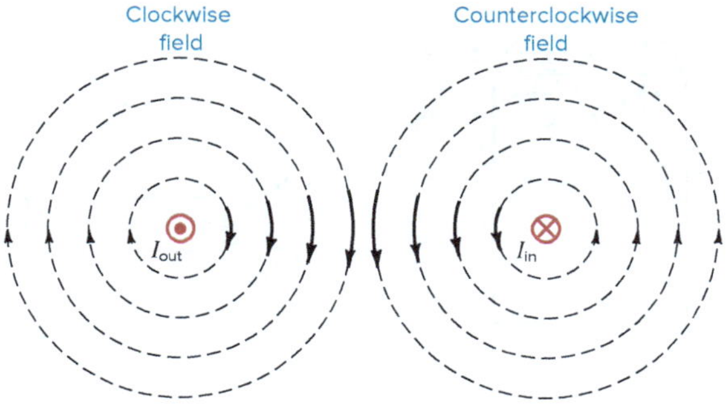

■ *14–5 Self-Review*

Answers at the end of the chapter.

a. Magnetic field lines around a conductor are circular in a perpendicular cross section of the conductor. (True/False)

b. In Fig. 14–7, the field is strongest between the conductors. (True/False)

14–6 Magnetic Polarity of a Coil

Figure 14–8 Magnetic poles of a current loop.

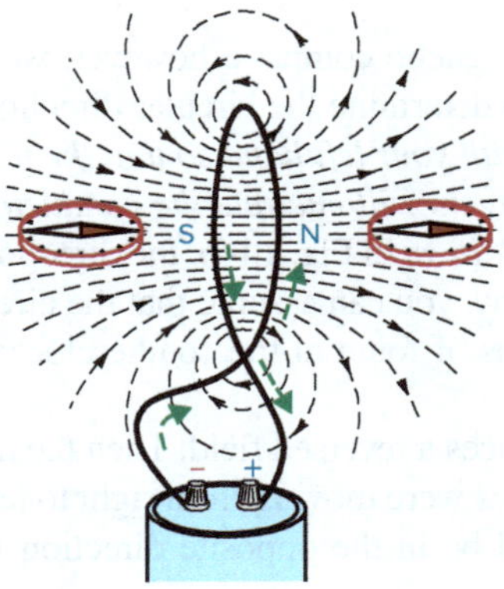

Bending a straight conductor into a loop, as shown in Fig. 14–8, has two effects. First, the magnetic field lines are more dense inside the loop. The total number of lines is the same as those for the straight conductor, but the lines inside the loop are concentrated in a smaller space. Furthermore, all lines inside the loop are aiding in the same direction. This makes the loop field effectively the same as a bar magnet with opposite poles at opposite faces of the loop.

Solenoid as a Bar Magnet

A coil of wire conductor with more than one turn is generally called a *solenoid.* An ideal solenoid, however, has a length much greater than its diameter. Like a single loop, the solenoid concentrates the magnetic field inside the coil and provides opposite magnetic poles at the ends. These effects are multiplied, however, by the number of turns as the magnetic field lines aid each other in the same direction inside the coil. Outside the coil, the field corresponds to a bar magnet with north and south poles at opposite ends, as illustrated in Fig. 14–9.

Magnetic Polarity

GOOD TO KNOW

The magnetic polarity of a solenoid can be verified with a compass.

To determine the magnetic polarity of a solenoid, use the ***left-hand rule*** illustrated in Fig. 14–10: *If the coil is grasped with the fingers of the left hand curled around the coil in the direction of electron flow, the thumb points to the north pole of the coil.* The left hand is used here because the current is electron flow.

The solenoid acts like a bar magnet, whether or not it has an iron core. Adding an iron core increases the flux density inside the coil. In addition, the field strength is uniform for the entire length of the core. The polarity is the same, however, for air-core and iron-core coils.

The magnetic polarity depends on the direction of current flow and the direction of winding. The current is determined by the connections to the voltage source. Electron flow is from the negative side of the voltage source, through the coil, and back to the positive terminal.

Figure 14–9 Magnetic poles of a solenoid. (*a*) Coil winding. (*b*) Equivalent bar magnet.

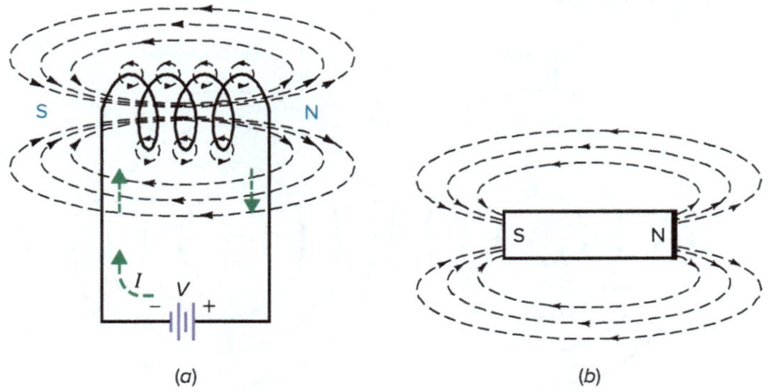

Figure 14–10 The left-hand rule for north pole of a coil with current *I*. The *I* is electron flow.

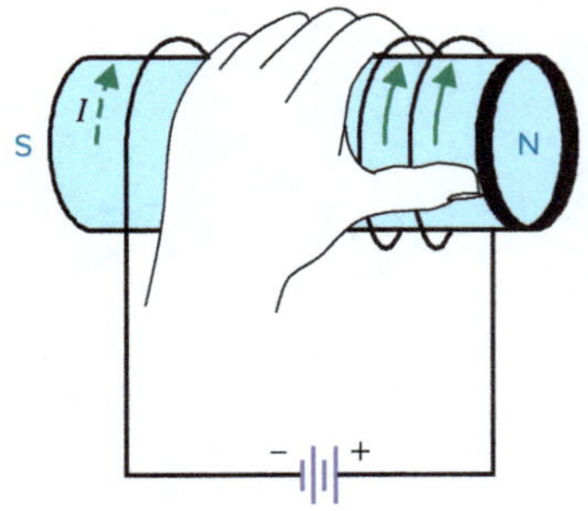

Figure 14–11 Examples for determining the magnetic polarity of a coil with direct current *I*. The *I* is electron flow. The polarities are reversed in (*a*) and (*b*) because the battery is reversed to reverse the direction of current. Also, (*d*) is the opposite of (*c*) because of the reversed winding.

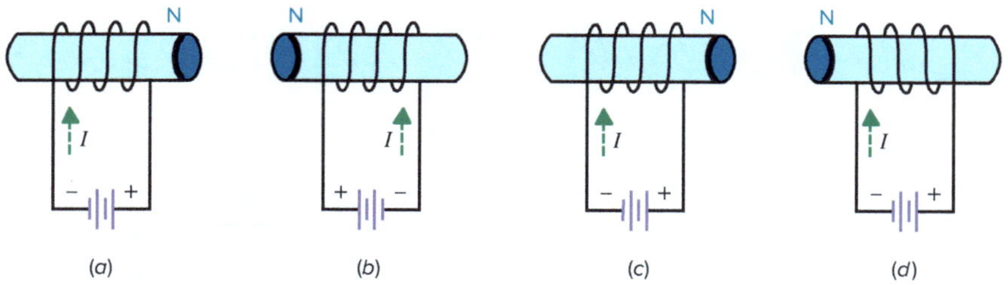

The direction of winding can be over and under, starting from one end of the coil, or under and over with respect to the same starting point. Reversing either the direction of winding or the direction of current reverses the magnetic poles of the solenoid. See Fig. 14–11. When both are reversed, though, the polarity is the same.

14–6 Self-Review

Answers at the end of the chapter.

a. **In Fig. 14–9, if the battery is reversed, will the north pole be at the left or the right?**

b. **If one end of a solenoid is a north pole, is the opposite end a north or a south pole?**

14–7 Motor Action between Two Magnetic Fields

The physical motion from the forces of magnetic fields is called ***motor action***. One example is the simple attraction or repulsion between bar magnets.

We know that like poles repel and unlike poles attract. It can also be considered that fields in the same direction repel and opposite fields attract.

Consider the repulsion between two north poles, illustrated in Fig. 14–12. Similar poles have fields in the same direction. Therefore, the similar fields of the two like poles repel each other.

A more fundamental reason for motor action, however, is the fact that the force in a magnetic field tends to produce motion from a stronger field toward a weaker field. In Fig. 14–12, note that the field intensity is greatest in the space between the two north poles. Here the field lines of similar poles in both magnets reinforce in the same direction. Farther away the field intensity is less, for essentially one magnet only. As a result, there is a difference in field strength, providing a net force that tends to produce motion. The direction of motion is always toward the weaker field.

To remember the directions, we can consider that the stronger field moves to the weaker field, tending to equalize field intensity. Otherwise, the motion would make the strong field stronger and the weak field weaker. This must be impossible because then the magnetic field would multiply its own strength without any work added.

Force on a Straight Conductor in a Magnetic Field

Current in a conductor has its associated magnetic field. When this conductor is placed in another magnetic field from a separate source, the two fields can react to produce motor action. The conductor must be perpendicular to the magnetic field,

Figure 14–12 Repulsion between similar poles of two bar magnets. The motion is from the stronger field to the weaker field.

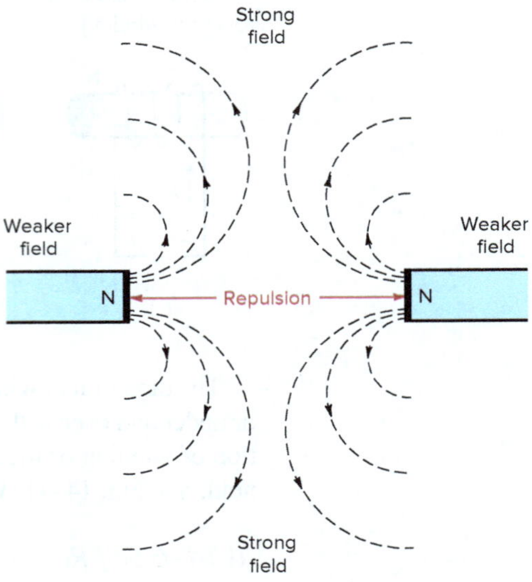

however, as shown in Fig. 14–13. This way, the perpendicular magnetic field produced by the current is in the same plane as the external magnetic field.

Unless the two fields are in the same plane, they cannot affect each other. In the same plane, however, lines of force in the same direction reinforce to make a stronger field, whereas lines in the opposite direction cancel and result in a weaker field.

To summarize these directions:

1. When the conductor is at 90°, or perpendicular to the external field, the reaction between the two magnetic fields is maximum.
2. When the conductor is at 0°, or parallel to the external field, there is no effect between them.
3. When the conductor is at an angle between 0 and 90°, only the perpendicular component is effective.

In Fig. 14–13, electrons flow in the wire conductor in the plane of the paper from the bottom to the top of the page. This flow provides the counterclockwise

GOOD TO KNOW

In Fig. 14–13, extend the thumb, forefinger, and middle finger of the right hand at right angles to each other. With the forefinger pointing in the direction of the magnetic flux and the middle finger in the direction of electron flow, the thumb points in the direction the conductor is moving. This is called the right-hand rule for motors.

MultiSim **Figure 14–13** Motor action of current in a straight conductor when it is in an external magnetic field. The H_1 is the circular field of the current. The H_M indicates field lines between the north and south poles of the external magnet.

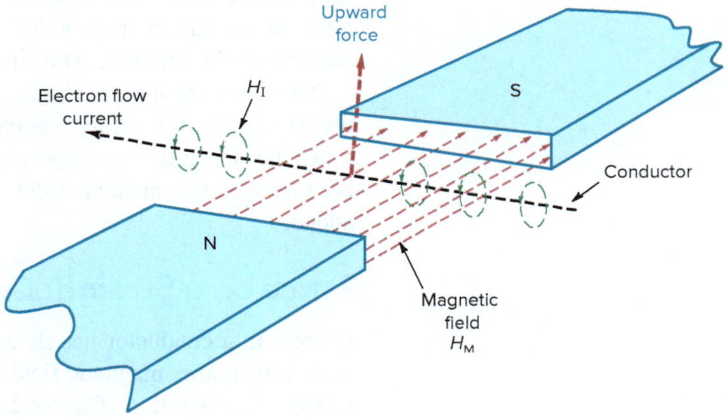

field H_I around the wire in a perpendicular plane cutting through the paper. The external field H_M has lines of force from left to right in the plane of the paper. Then lines of force in the two fields are parallel above and below the wire.

Below the conductor, its field lines are left to right in the same direction as the external field. Therefore, these lines reinforce to produce a stronger field. Above the conductor, the lines of the two fields are in opposite directions, causing a weaker field. As a result, the net force of the stronger field makes the conductor move upward out of the page toward the weaker field.

If electrons flow in the reverse direction in the conductor or if the external field is reversed, the motion action will be in the opposite direction. Reversing both the field and the current, however, results in the same direction of motion.

Rotation of a Conductor Loop in a Magnetic Field

When a loop of wire is in the magnetic field, opposite sides of the loop have current in opposite directions. Then the associated magnetic fields are opposite. The resulting forces are upward on one side of the loop and downward on the other side, making it rotate. This effect of a force in producing rotation is called *torque*.

The principle of motor action between magnetic fields producing rotational torque is the basis of all electric motors. The moving-coil meter described in Section 8–1 is a similar application. Since torque is proportional to current, the amount of rotation indicates how much current flows through the coil.

GOOD TO KNOW

In Fig. 14–14, extend the thumb, forefinger, and middle finger of the left hand at right angles to each other. With the forefinger pointing in the direction of magnetic flux, and the thumb in the direction the conductor is moving, the middle finger points in the direction of the induced current. This is called the left-hand generator rule.

14–7 Self-Review

Answers at the end of the chapter.

a. In Fig. 14–12, the field is strongest between the two north poles. (True/False)

b. In Fig. 14–13, if both the magnetic field and the current are reversed, the motion will still be upward. (True/False)

14–8 Induced Current

Just as electrons in motion provide an associated magnetic field, when magnetic flux moves, the motion of magnetic lines cutting across a conductor forces free electrons in the conductor to move, producing current. This action is called *induction* because there is no physical connection between the magnet and the conductor. The induced current is a result of generator action as the mechanical work put into moving the magnetic field is converted into electric energy when current flows in the conductor.

Referring to Fig. 14–14, let the conductor AB be placed at right angles to the flux in the air gap of the horseshoe magnet. Then, when the magnet is moved up or down, its flux cuts across the conductor. The action of magnetic flux cutting across the conductor generates current. The fact that current flows is indicated by the microammeter.

Figure 14–14 Induced current produced by magnetic flux cutting across a conductor. Direction of *I* here is for electron flow.

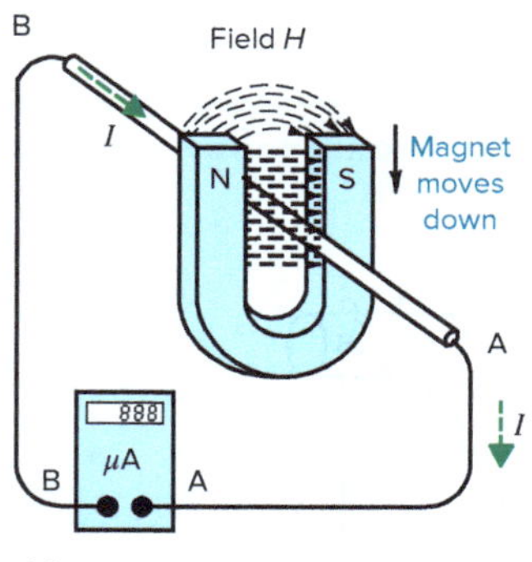

When the magnet is moved downward, current flows in the direction shown. If the magnet is moved upward, current will flow in the opposite direction. Without motion, there is no current.

Direction of Motion

Motion is necessary for the flux lines of the magnetic field to cut across the conductor. This cutting can be accomplished by motion of either the field or the conductor.

When the conductor is moved upward or downward, it cuts across the flux. The generator action is the same as moving the field, except that the relative motion is opposite. Moving the conductor upward, for instance, corresponds to moving the magnet downward.

Conductor Perpendicular to External Flux

To have electromagnetic induction, the conductor and the magnetic lines of flux must be perpendicular to each other. Then the motion makes the flux cut through the cross-sectional area of the conductor. As shown in Fig. 14–14, the conductor is at right angles to the lines of force in the field H.

The reason the conductor must be perpendicular is to make its induced current have an associated magnetic field in the same plane as the external flux. If the field of the induced current does not react with the external field, there can be no induced current.

How Induced Current Is Generated

The induced current can be considered the result of motor action between the external field H and the magnetic field of free electrons in every cross-sectional area of the wire. Without an external field, the free electrons move at random without any specific direction, and they have no net magnetic field. When the conductor is in the magnetic field H, there still is no induction without relative motion, since the magnetic fields for the free electrons are not disturbed. When the field or conductor moves, however, there must be a reaction opposing the motion. The reaction is a flow of free electrons resulting from motor action on the electrons.

Referring to Fig. 14–14, for example, the induced current must flow in the direction shown because the field is moved downward, pulling the magnet away from the conductor. The induced current of electrons then has a clockwise field; lines of force aid H above the conductor and cancel H below. When motor action between the two magnetic fields tends to move the conductor toward the weaker field, the conductor will be forced downward, staying with the magnet to oppose the work of pulling the magnet away from the conductor.

The effect of electromagnetic induction is increased when a coil is used for the conductor. Then the turns concentrate more conductor length in a smaller area. As illustrated in Fig. 14–15, moving the magnet into the coil enables the flux to cut across many turns of conductors.

Lenz's Law

Lenz's law is the basic principle for determining the direction of an induced voltage or current. Based on the principle of conservation of energy, the law simply states that the direction of the induced current must be such that its own magnetic field will oppose the action that produced the induced current.

Fine Art Images/Heritage Images/Getty Images

PIONEERS IN ELECTRONICS

Russian physicist *Heinrich Friedrich Emil Lenz (1804–1865)* demonstrated that an increase in temperature increases the resistance of a metal. In 1834, he formulated Lenz's law, which states that the current induced by a change flows so as to oppose the effect producing the change. In 1838, he demonstrated the "Peltier Effect" with reversing current by using bismuth-antimony rods.

Figure 14–15 Induced current produced by magnetic flux cutting across turns of wire in a coil. Direction of *I* here is for electron flow.

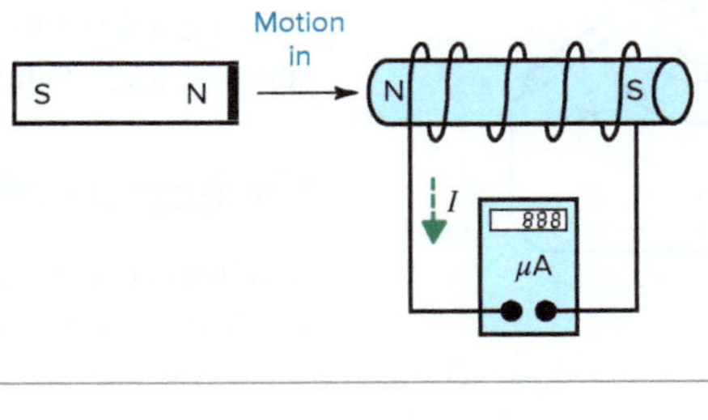

In Fig. 14–15, for example, the induced current has the direction that produces a north pole at the left to oppose the motion by repulsion of the north pole being moved in. This is why it takes some work to push the permanent magnet into the coil. The work expended in moving the permanent magnet is the source of energy for the current induced in the coil.

Using Lenz's law, we can start with the fact that the left end of the coil in Fig. 14–15 must be a north pole to oppose the motion. Then the direction of the induced current is determined by the left-hand rule for electron flow. If the fingers coil around the direction of electron flow shown, under and over the winding, the thumb will point to the left for the north pole.

For the opposite case, suppose that the north pole of the permanent magnet in Fig. 14–15 is moved away from the coil. Then the induced pole at the left end of the coil must be a south pole by Lenz's law. The induced south pole will attract the north pole to oppose the motion of the magnet being moved away. For a south pole at the left end of the coil, then, the electron flow will be reversed from the direction shown in Fig. 14–15. We could generate an alternating current in the coil by moving the magnet periodically in and out.

14–8 Self-Review

Answers at the end of the chapter.

Refer to Fig. 14–15.

a. If the north end of the magnet is moved away from the coil, will its left side be north or south?

b. If the south end of the magnet is moved in, will the left end of the coil be north or south?

c. Referring to Fig. 14–14, if the conductor is moved up, instead of moving the magnet down, will the induced current flow in the same direction?

14–9 Generating an Induced Voltage

Consider a magnetic flux cutting a conductor that is not in a closed circuit, as shown in Fig. 14–16. The motion of flux across the conductor forces free electrons to move, but in an open circuit, the displaced electrons produce opposite electric charges at the two open ends.

Figure 14–16 Voltage induced across open ends of conductor cut by magnetic flux.

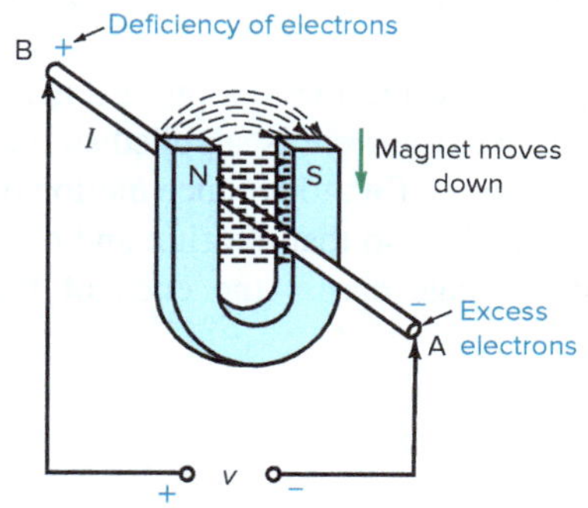

For the directions shown, free electrons in the conductor are forced to move to point A. Since the end is open, electrons accumulate here. Point A then develops a negative potential.

At the same time, point B loses electrons and becomes charged positively. The result is a potential difference across the two ends, provided by the separation of electric charges in the conductor.

The potential difference is an electromotive force (emf), generated by the work of cutting across the flux. You can measure this potential difference with a voltmeter. However, a conductor cannot store electric charge. Therefore, the voltage is present only while the motion of flux cutting across the conductor is producing the induced voltage.

Induced Voltage across a Coil

For a coil, as in Fig. 14–17*a*, the induced emf is increased by the number of turns. Each turn cut by flux adds to the induced voltage, since each turn cut forces free electrons to accumulate at the negative end of the coil with a deficiency of electrons at the positive end.

The polarity of the induced voltage follows from the direction of induced current. The end of the conductor to which the electrons go and at which they accumulate is

Figure 14–17 Voltage induced across coil cut by magnetic flux. (*a*) Motion of flux generating voltage across coil. (*b*) Induced voltage acts in series with the coil. (*c*) The induced voltage is a source that can produce current in an external load resistor R_L connected across the coil.

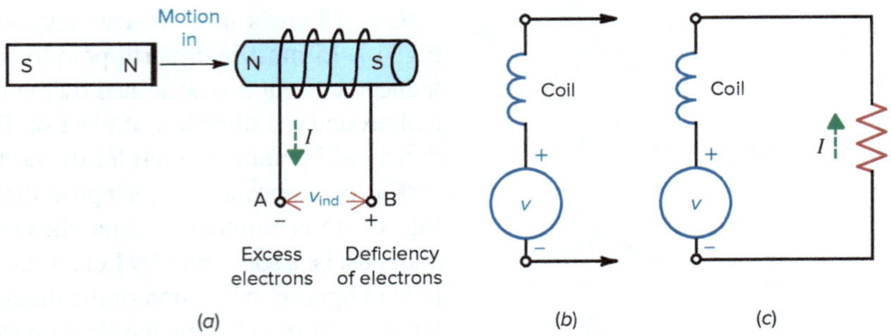

the negative side of the induced voltage. The opposite end, with a deficiency of electrons, is the positive side. The total emf across the coil is the sum of the induced voltages, since all the turns are in series.

Furthermore, the total induced voltage acts in series with the coil, as illustrated by the equivalent circuit in Fig. 14–17*b*, showing the induced voltage as a separate generator. This generator represents a voltage source with a potential difference resulting from the separation of charges produced by electromagnetic induction. The source v then can produce current in an external load circuit connected across the negative and positive terminals, as shown in Fig. 14–17*c*.

The induced voltage is in series with the coil because current produced by the generated emf must flow through all the turns. An induced voltage of 10 V, for example, with R_L equal to 5 Ω, results in a current of 2 A, which flows through the coil, the equivalent generator v, and the load resistance R_L.

The direction of current in Fig. 14–17*c* shows electron flow around the circuit. Outside the source v, the electrons move from its negative terminal, through R_L, and back to the positive terminal of v because of its potential difference.

Inside the generator, however, the electron flow is from the + terminal to the − terminal. This direction of electron flow results from the fact that the left end of the coil in Fig. 14–17*a* must be a north pole, by Lenz's law, to oppose the north pole being moved in.

Notice how motors and generators are similar in using the motion of a magnetic field, but with opposite applications. In a motor, current is supplied so that an associated magnetic field can react with the external flux to produce motion of the conductor. In a generator, motion must be supplied so that the flux and conductor can cut across each other to induce voltage across the ends of the conductor.

Faraday's Law of Induced Voltage

The voltage induced by magnetic flux cutting the turns of a coil depends on the number of turns and how fast the flux moves across the conductor. Either the flux or the conductor can move. Specifically, the amount of induced voltage is determined by the following three factors:

1. *Amount of flux.* The more magnetic lines of force that cut across the conductor, the higher the amount of induced voltage.
2. *Number of turns.* The more turns in a coil, the higher the induced voltage. The v_{ind} is the sum of all individual voltages generated in each turn in series.

3. *Time rate of cutting.* The faster the flux cuts a conductor, the higher the induced voltage. Then more lines of force cut the conductor within a specific period of time.

These factors are fundamental in many applications. Any conductor with current will have voltage induced in it by a change in current and its associated magnetic flux.

The amount of induced voltage can be calculated by **Faraday's law**:

$$v_{\text{ind}} = N\frac{d\phi\ (\text{webers})}{dt\ (\text{seconds})} \qquad \textbf{(14–5)}$$

where N is the number of turns and $d\phi/dt$ specifies how fast the flux ϕ cuts across the conductor. With $d\phi/dt$ in webers per second, the induced voltage is in volts.

As an example, suppose that magnetic flux cuts across 300 turns at the rate of 2 Wb/s.

To calculate the induced voltage,

$$\begin{aligned} v_{\text{ind}} &= N\frac{d\phi}{dt} \\ &= 300 \times 2 \\ v_{\text{ind}} &= 600\ \text{V} \end{aligned}$$

It is assumed that all flux links all turns, which is true for an iron core.

Rate of Change

The symbol d in $d\phi$ and dt is an abbreviation for *change*. The $d\phi$ means a change in the flux ϕ, and dt means a change in time. In mathematics, dt represents an infinitesimally small change in time, but in this book we are using the d to mean rate of change in general. The results are exactly the same for the practical changes used here because the rate of change is constant.

As an example, if the flux ϕ is 4 Wb one time but then changes to 6 Wb, the change in flux $d\phi$ is 2 Wb. The same idea applies to a decrease as well as an increase. If the flux changed from 6 to 4 Wb, $d\phi$ would still be 2 Wb. However, an increase is usually considered a change in the positive direction, with an upward slope, whereas a decrease has a negative slope downward.

Similarly, dt means a change in time. If we consider the flux at a time 2 s after the start and at a later time 3 s after the start, the change in time is 3 − 2, or 1 s for dt. Time always increases in the positive direction.

Combining the two factors of $d\phi$ and dt, we can say that for magnetic flux increasing by 2 Wb in 1 s, $d\phi/dt$ equals 2/1, or 2 Wb/s. This states the rate of change of the magnetic flux.

As another example, suppose that the flux increases by 2 Wb in 0.5 s. Then

$$\frac{d\phi}{dt} = \frac{2\ \text{Wb}}{0.5\ \text{s}} = 4\ \text{Wb/s}$$

Analysis of Induced Voltage as $N(d\phi/dt)$

This fundamental concept of voltage induced by a change in flux is illustrated by the graphs in Fig. 14–18, for the values listed in Table 14–2. The linear rise in Fig. 14–18*a* shows values of flux ϕ increasing at a uniform rate. In this case, the curve goes up 2 Wb for every 1-s interval. The slope of this curve, then, equal to $d\phi/dt$, is 2 Wb/s. Note that, although ϕ increases, the rate of change is constant because the linear rise has a constant slope.

For induced voltage, only the $d\phi/dt$ factor is important, not the actual value of flux. To emphasize this basic concept, the graph in Fig. 14–18*b* shows the $d\phi/dt$ values alone. This graph is a straight horizontal line for the constant value of 2 Wb/s.

Figure 14–18 Graphs of induced voltage produced by magnetic flux changes in a coil. (*a*) Linear increase of flux ϕ. (*b*) Constant rate of change for $d\phi/dt$ at 2 Wb/s. (*c*) Constant induced voltage of 600 V for a coil with 300 turns.

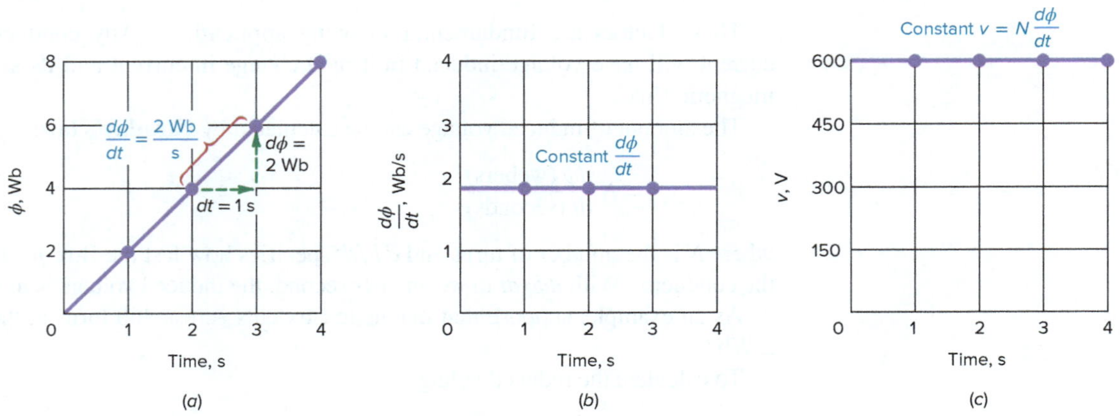

Table 14–2	Induced-Voltage Calculations for Figure 14–18					
ϕ, Wb	$d\phi$, Wb	t, s	dt, s	$d\phi/dt$, Wb/s	N, Turns	$N(d\phi/dt)$, V
2	2	1	1	2	300	600
4	2	2	1	2	300	600
6	2	3	1	2	300	600
8	2	4	1	2	300	600

The induced-voltage graph in Fig. 14–18*c* is also a straight horizontal line. Since $v_{ind} = N(d\phi/dt)$, the graph of induced voltage is the $d\phi/dt$ values multiplied by the number of turns. The result is a constant 600 V, with 300 turns cut by flux changing at the constant rate of 2 Wb/s.

The example illustrated here can be different in several ways without changing the basic fact that the induced voltage is equal to $N(d\phi/dt)$. First, the number of turns or the $d\phi/dt$ values can be greater or less than the values assumed here. More turns provide more induced voltage, whereas fewer turns mean less voltage. Similarly, a higher value for $d\phi/dt$ results in more induced voltage.

Note that two factors are included in $d\phi/dt$. Its value can be increased by a higher value of $d\phi$ or a smaller value of dt. As an example, the value of 2 Wb/s for $d\phi/dt$ can be doubled either by increasing $d\phi$ to 4 Wb or reducing dt to 0.5 s. Then $d\phi/dt$ is $4/1$ or $2/0.5$, which equals 4 Wb/s in either case. The same flux changing within a shorter time means a faster rate of flux cutting the conductor, resulting in a higher value of $d\phi/dt$ and more induced voltage.

For the opposite case, a smaller value of $d\phi/dt$, with less flux or a slower rate of change, results in a lower value of induced voltage. As $d\phi/dt$ decreases, the induced voltage will reverse polarity.

Finally, note that the $d\phi/dt$ graph in Fig. 14–18*b* has the constant value of 2 Wb/s because the flux is increasing at a linear rate. However, the flux need not have a uniform rate of change. Then the $d\phi/dt$ values will not be constant. In any case, though, the values of $d\phi/dt$ at all instants will determine the values of the induced voltage equal to $N(d\phi/dt)$.

Polarity of the Induced Voltage

The polarity is determined by Lenz's law. Any induced voltage has the polarity that opposes the change causing the induction. Sometimes this fact is indicated by using a negative sign for v_{ind} in Formula (14–5). However, the absolute polarity depends on whether the flux is increasing or decreasing, the method of winding, and which end of the coil is the reference.

When all these factors are considered, v_{ind} has polarity such that the current it produces and the associated magnetic field oppose the change in flux producing the induced voltage. If the external flux increases, the magnetic field of the induced current will be in the opposite direction. If the external field decreases, the magnetic field of the induced current will be in the same direction as the external field to oppose the change by sustaining the flux. In short, the induced voltage has polarity that opposes the change.

14–9 Self-Review

Answers at the end of the chapter.

a. The magnetic flux of 8 Wb changes to 10 Wb in 1 s. How much is $d\phi/dt$?

b. The flux of 8 μWb changes to 10 μWb in 1 μs. How much is $d\phi/dt$?

14–10 Relays

A *relay* is an electromechanical device that operates by electromagnetic induction. It uses either an AC- or a DC-actuated electromagnet to open or close one or more sets of contacts. Relay contacts that are open when the relay is not energized are called *normally open* (*NO*) contacts. Conversely, relay contacts that are closed when the relay is not energized are called *normally closed* (*NC*) contacts. Relay contacts are held in their resting or normal position either by a spring or by some type of gravity-actuated mechanism. In most cases, an adjustment of the spring tension is provided to set the restraining force on the normally open and normally closed contacts to some desired level based on predetermined circuit conditions.

Figure 14–19 Schematic symbols commonly used to represent relay contacts. (*a*) Symbols used to represent normally open contacts. (*b*) Symbols used to represent normally closed contacts.

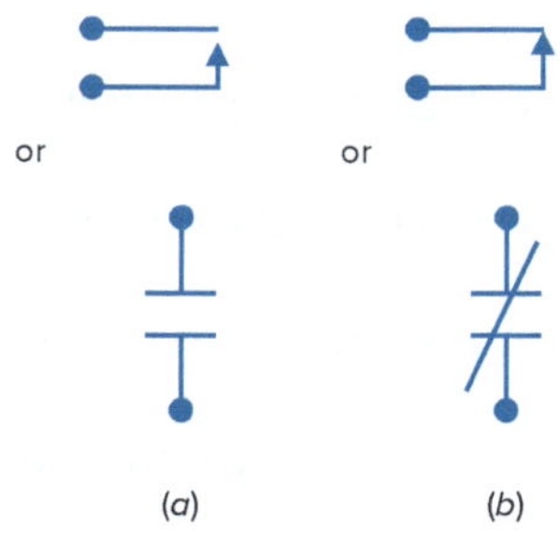

Figure 14–19 shows the schematic symbols that are commonly used to represent relay contacts. Figure 14–19*a* shows the symbols used to represent normally open contacts, and Fig. 14–19*b* shows the symbols used to represent normally closed contacts. When normally open contacts close, they are said to *make,* whereas when normally closed contacts open they are said to *break*. Like mechanical switches, the switching contacts of a relay can have any number of poles and throws.

Figure 14–20 shows the basic parts of an SPDT armature relay. Terminal connections 1 and 2 provide connection to the electromagnet (relay coil), and

Figure 14–20 Basic parts of an SPDT armature relay. Terminal connections 1 and 2 provide connection to the electromagnet, and terminal connections 3, 4, and 5 provide connections to the SPDT relay contacts which open or close when the relay is energized.

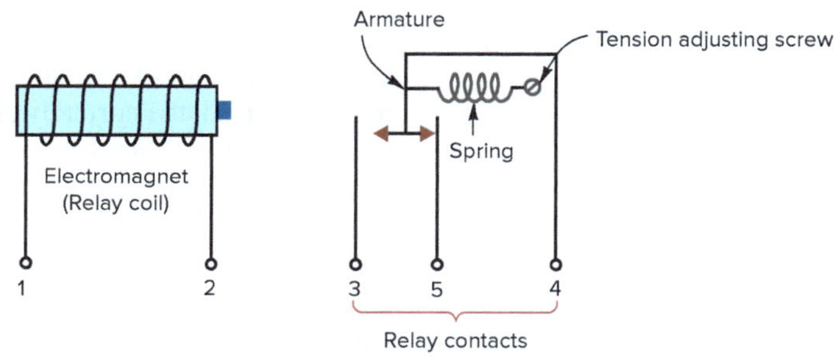

GOOD TO KNOW

Many relays are packaged in a hermetically sealed enclosure, which is a type of enclosure that isolates the relay, namely the switching contacts, from the environment.

Figure 14–21 Typical relay.

Mark Steinmetz/McGraw Hill

terminal connections 3, 4, and 5 provide connections to the SPDT relay contacts that open or close when the relay is energized. A relay is said to be *energized* when NO contacts close and NC contacts open. The movable arm of an electromechanical relay is called the *armature.* The armature is magnetic and has contacts that make or break with other contacts when the relay is energized. For example, when terminals 1 and 2 in Fig. 14–20 are connected to a DC source, current flows in the relay coil and an electromagnet is formed. If there is sufficient current in the relay coil, contacts 3 and 4 close (make) and contacts 4 and 5 open (break). The armature is attracted whether the electromagnet produces a north or a south pole on the end adjacent to the armature. Figure 14–21 is a photo of a typical relay.

Relay Specifications

Manufacturers of electromechanical relays always supply a specification sheet for each of their relays. The specification sheet contains voltage and current ratings for both the relay coil and its switch contacts. The specification sheet also includes information regarding the location of the relay coil and switching contact terminals. And finally, the specification sheet will indicate whether the relay can be energized from either an AC or a DC source. The following is an explanation of a relay's most important ratings.

Pickup voltage. The minimum amount of relay coil voltage necessary to energize or operate the relay.

Pickup current. The minimum amount of relay coil current necessary to energize or operate the relay.

Holding current. The minimum amount of current required to keep a relay energized or operating. (The holding current is less than the pickup current.)

Dropout voltage. The maximum relay coil voltage at which the relay is no longer energized.

Contact voltage rating. The maximum voltage the relay contacts can switch safely.

Contact current rating. The maximum current the relay contacts can switch safely.

Contact voltage drop. The voltage drop across the closed contacts of a relay when operating.

Insulation resistance. The resistance measured across the relay contacts in the open position.

Relay Applications

Figure 14–22 shows schematic diagrams for two relay systems. The diagram in Fig. 14–22*a* represents an open-circuit system. With the control switch S_1 open, the SPST relay contacts are open and the load is inoperative. Closing S_1 energizes the relay. This closes the NO relay contacts and makes the load operative.

Figure 14–22*b* represents a closed-circuit system. In this case, the relay is energized by the control switch S_1, which is closed during normal operation. With the relay energized, the normally closed relay contacts are open and the load is inoperative. When it is desired to operate the load, the control switch S_1 is opened. This returns the relay contacts to their normally closed position, thereby activating the load.

It is important to note that a relay can be energized using a low-voltage, low-power source. However, the relay contacts can be used to control a circuit whose load consumes much more power at a much higher voltage than the relay coil circuit.

MultiSim **Figure 14–22** Schematic diagrams for two relay systems. (*a*) Open-circuit system. (*b*) Closed-circuit system.

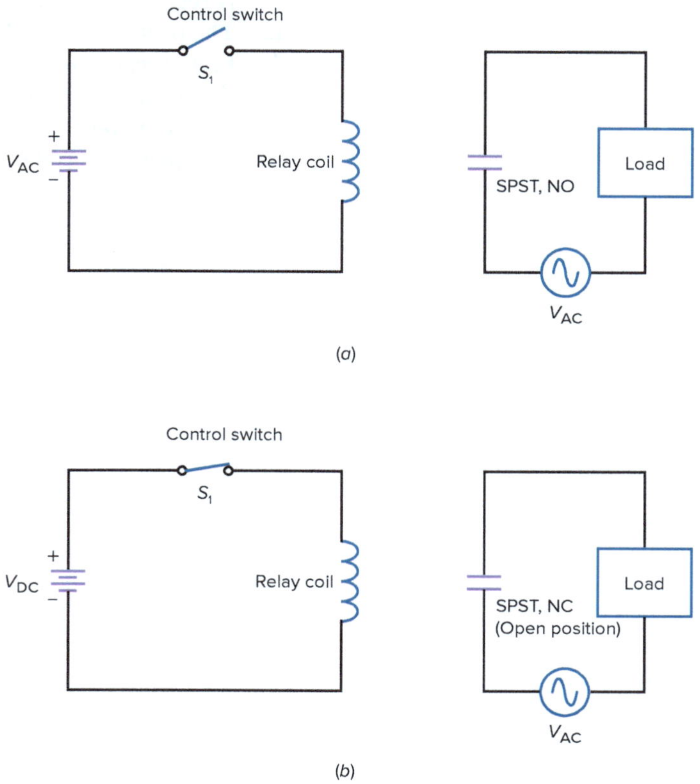

In fact, one of the main advantages of using a relay is its ability to switch or control very high power loads with a relatively low amount of input power. In remote-control applications, a relay can control a high power load a long distance away much more efficiently than a mechanical switch can. When a mechanical switch is used to control a high power load a long distance away, the I^2R power loss in the conductors carrying current to the load can become excessive. Critical Thinking Probs. 14–25 and 14–26 illustrate the advantages of using a relay to control a high power load a long distance away.

Common Relay Troubles

If a relay coil develops an open, the relay cannot be energized. The reason is simple. With an open relay coil, the current is zero and no magnetic field is set up by the electromagnet to attract the armature. An ohmmeter can be used to check for the proper relay coil resistance. An open relay coil measures infinite (∞) resistance. Since it is usually not practical to repair an open relay coil, the entire relay must be replaced.

A common problem with relays is dirty switch contacts. The switch contacts develop a thin carbon coating after extended use from arcing across the contact terminals when they are opened and closed. Dirty switch contacts usually produce intermittent operation of the load being controlled—for example, a motor. In some cases, the relay contacts may chatter (vibrate) if they are dirty.

One final point: The manufacturer of a relay usually indicates its life expectancy in terms of the number of times the relay can be energized (operated). A typical value is 5 million operations.

14–10 Self-Review

Answers at the end of the chapter.

a. **A relay is energized if NC contacts are opened. (True/False)**

b. **The pickup current is the minimum relay coil current required to keep a relay energized. (True/False)**

c. **The voltage drop across a set of closed relay contacts carrying 1 A of current is very low. (True/False)**

d. **An open relay coil measures 0 Ω with an ohmmeter. (True/False)**

Application in Understanding Solenoids

There are many useful devices in our homes and industries that operate on the principles of electromagnetism. One such device is a solenoid. A solenoid is an electromechanical device that converts electrical energy into mechanical energy (force and motion). In terms of its physical construction, a solenoid consists of a coil of wire and a movable iron core, called a plunger. The basic construction is shown in Fig. 14–23. The solenoid's coil is wound around a long nonmagnetic hollow cylinder. Slid inside the hollow cylinder are two iron cores, one which is fixed in position (stationary); the other is movable. As shown, the stationary and movable cores are attached to each other with a spring. With S_1 open in Fig. 14–24a, the current in the coil is zero. As a result, there is no associated magnetic field surrounding the coil to magnetize the iron cores. In this case, the spring forces the plunger out of the cylinder, as shown. However, when S_1 is closed, as shown in Fig. 14–24*b*, the current in the coil forms an electromagnet and both iron cores become magnetized. Notice that the magnetic polarity of both iron cores is the same. As a result, the ends of the iron cores attached to the spring have opposite magnetic polarity and thus attract each other. This attraction of opposite magnetic poles draws the plunger inside of the cylinder, causing the coil to compress. The plunger stays in this position as long as there is current in the coil. When S_1 is opened, however, the current drops to zero and the coil no longer serves as an electromagnet. Since soft iron has very low retentivity, the iron cores have very little residual magnetism after the DC current is removed from the coil. As a result, the iron cores are no longer magnetized and the spring forces the plunger back out of the cylinder.

In a solenoid, the force of the moving plunger can be used to open or close a valve or to move a lever to lock or unlock a door. It is common practice to use solenoids for controlling the locks on doors in offices, hotels, and automobiles. Figure 14–25 shows an example of a solenoid used for the automatic door locks in an automobile.

SOLENOID VALVES

A solenoid valve is an electromechanically operated device in which the operation or movement of a valve is controlled by opening or closing the path for current in the solenoid coil. In

Figure 14-23 Basic construction of a solenoid.

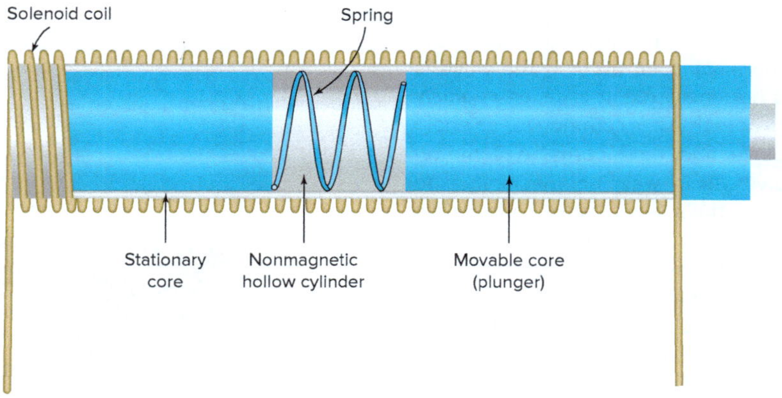

Figure 14-24 Operation of a solenoid. (*a*) With S_1 open, the current in the coil is zero and no electromagnet is formed. The iron cores are not magnetized, and the spring forces the plunger out of the cylinder.

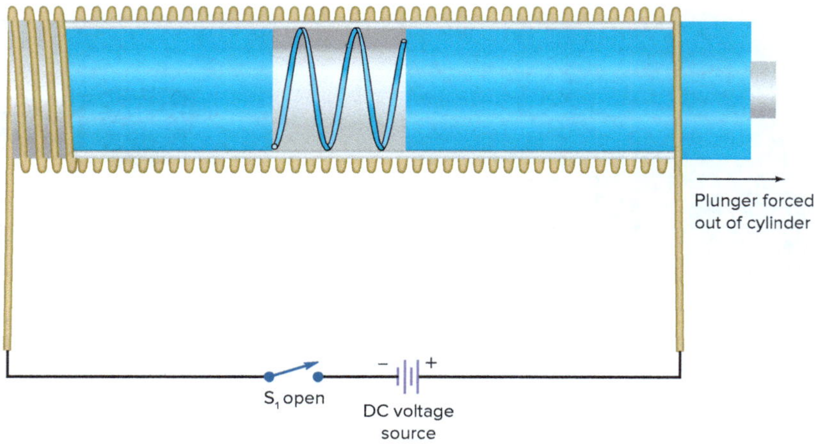

Figure 14-24 (*b*) With S_1 closed, the current in the coil forms an electromagnet. The electromagnet magnetizes the iron cores and draws the plunger into the cylinder.

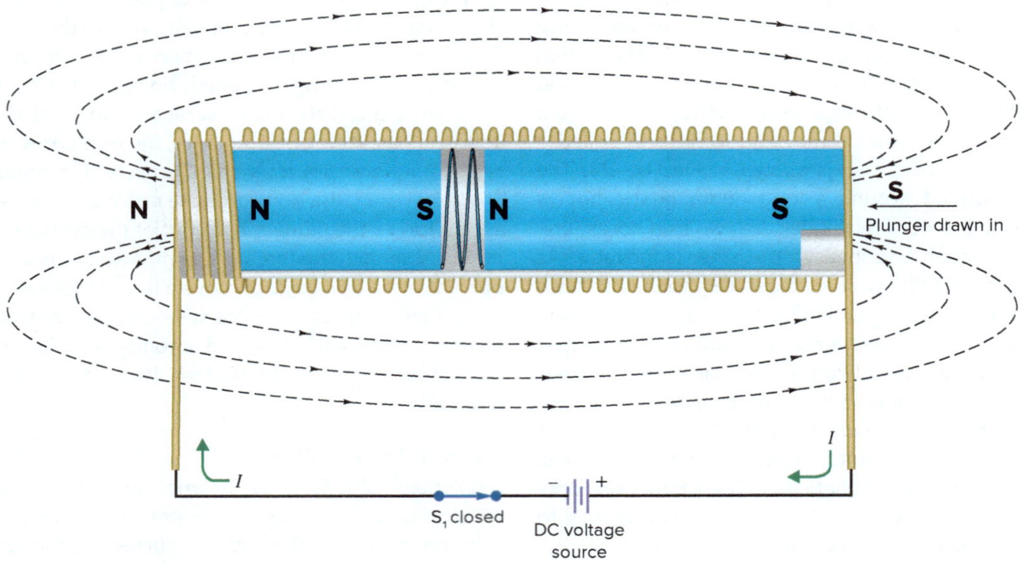

general, solenoid valves are used when it is necessary or desired to control the flow of fluids or gases in a system. For example, a solenoid valve may be used to open or close the pathway for fluids in a hose or pipe. The valve may be a moving plunger or a flap that twists or turns. If the valve is open when the solenoid is not energized, it is called ***normally open (NO)***. If the valve is closed when the solenoid is not energized, it is called ***normally closed (NC)***. It is important to note that it is common for solenoid valves to have multiple ports and fluid paths. A common application of solenoid valves is in washing machines and dishwashers. In this application, the solenoid valve controls the flow of water entering the machine. Other applications include using solenoid valves in fluid power systems, compressors and vacuum pumps, sprinkler and fire fighting systems, dental equipment, biomedical equipment, and boilers, to name a few. Because of the vast number of applications that exist, solenoid valves come in a wide variety of different shapes and sizes. Solenoid valves offer safe and fast switching, high reliability, relatively low power requirements, and a compact design.

Figure 14-25 Solenoid used for automobile door locks.

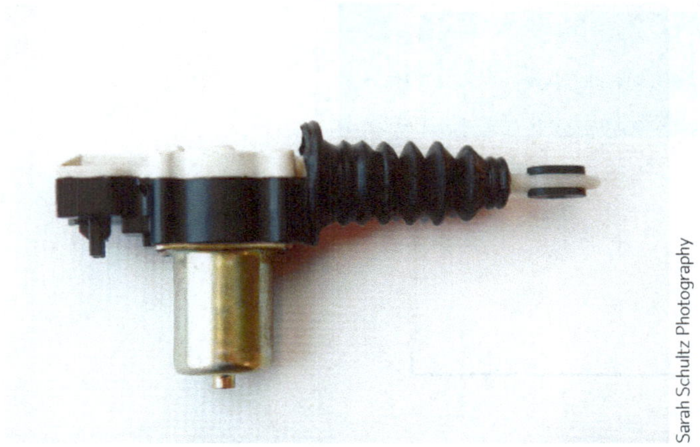

Figure 14–26 shows a solenoid valve used to control the flow of refrigerant in a commercial refrigerator. It should be noted that in many cases the coil assembly and valve body are two separate pieces.

Figure 14-26 Solendoid valve used to control the flow of refrigerant in a commercial refrigerator.

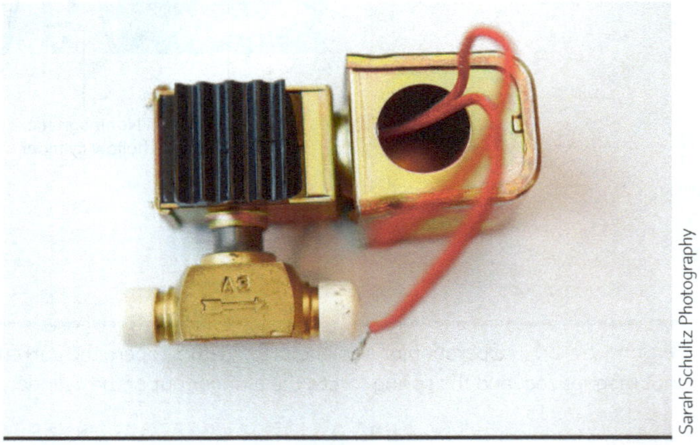

Summary

- For an electromagnet, the strength of the magnetic field depends on how much current flows in the turns of the coil. The coil serves as a bar magnet with opposite poles at the ends.
- The strength of an electromagnet is specified in SI units of ampere-turns. The product of amperes (A) and turns (t) indicates the magnetomotive force (mmf) of the coil. The cgs unit of mmf is the gilbert (Gb).
- The field intensity, H, of a coil specifies the mmf per unit length. The units of H are A·t/m and the oersted.
- The permeability of a material indicates its ability to concentrate magnetic flux.
- Demagnetization of a magnetic material is also known as degaussing.
- Current in a straight conductor has an associated magnetic field with circular lines of force in a plane perpendicular to the conductor. The direction of the circular field is counterclockwise when you look along the conductor in the direction of electron flow.
- The left-hand rule for determining the polarity of an electromagnet says that when your fingers curl around the turns in the direction of electron flow, the thumb points to the north pole.
- Motor action is the motion that results from the net force of two fields that can aid or cancel each other. The direction of the resultant force is always from the stronger field to the weaker field.
- Generator action refers to induced voltage. For N turns, $v_{ind} = N(d\phi/dt)$, where $d\phi/dt$ stands for the change in flux (ϕ) in time (t). The change is given in webers per second. There must be a change in flux to produce induced voltage.
- Lenz's law states that the direction of an induced current must be such that its own magnetic field will oppose the action that produced the induced current.
- The switching contacts of an electromechanical relay may be either normally open (NO) or normally closed (NC). The contacts are held in their normal or resting positions by springs or some gravity-actuated mechanism.
- The movable arm on a relay is called the *armature*. The armature is magnetic and has contacts that open or close with other contacts when the relay is energized.
- The pickup current of a relay is the minimum amount of relay coil current that will energize the relay. The holding current is the minimum relay coil current required to keep a relay energized.

Important Terms

Ampere-turn (A·t) — the SI unit of magnetomotive force (mmf).

Ampere-turns/meter (A·t/m) — the SI unit of field intensity, H.

***B-H* magnetization curve** — a graph showing how the flux density, B, in teslas, increases with the field intensity, H, in ampere-turns/meter.

Degaussing — a method of demagnetizing a material by using an alternating current. The method involves magnetizing and demagnetizing a material with a diminishing magnetic field until the material has practically zero residual induction.

Faraday's law — a law for determining the amount of induced voltage in a conductor. The amount of induced voltage, v_{ind}, is calculated as $v_{ind} = N\frac{d\phi}{dt}$.

Field intensity (*H*) — the amount of mmf per unit length. The units for field intensity are A·t/m and oersted.

Holding current — the minimum amount of relay coil current required to keep a relay energized or operating.

Hysteresis — hysteresis means lagging behind. With respect to the magnetic flux in an iron core of an electromagnet, the flux lags behind the increases and decreases in magnetizing force.

Left-hand rule — if a coil is grasped with the fingers of the left hand curled around the coil in the direction of electron flow, the thumb points to the north pole of the coil.

Lenz's law — Lenz's law states that the direction of the induced current in a conductor must be such that its own magnetic field will oppose the action that produced the induced current.

Magnetomotive force (mmf) — a measure of the strength of a magnetic field based on the amount of current flowing in the turns of a coil. The units of mmf are the ampere-turn (A·t) and the gilbert (Gb).

Motor action — a motion that results from the net force of two magnetic fields that can aid or cancel each other. The direction of the resultant force is always from a stronger field to a weaker field.

Pickup current — the minimum amount of relay coil current necessary to energize or operate a relay.

Saturation — the point in a magnetic material, such as an iron core, where further increases in field intensity produce no further increases in flux density.

Related Formulas

$$\text{Ampere-turns} = I \times N = \text{mmf}$$

$$H = \frac{\text{ampere-turns of mmf}}{l \text{ meters}}$$

$$B = \mu \times H$$

$$\mu = \frac{B}{H}$$

$$v_{ind} = N\frac{d\phi}{dt}$$

Self-Test

Answers at the back of the book.

1. A current of 20 mA flowing through a coil with 500 turns produces an mmf of

a. 100 A·t.

b. 1 A·t.

c. 10 A·t.

d. 7.93 A·t.

2. A coil with 1000 turns must provide an mmf of 50 A·t. The required current is

a. 5 mA.

b. 0.5 A.

c. 50 μA.

d. 50 mA.

3. The left-hand rule for solenoids states that

a. if the fingers of the left hand encircle the coil in the same direction as electron flow, the thumb points in the direction of the north pole.

b. if the thumb of the left hand points in the direction of current flow, the fingers point toward the north pole.

c. if the fingers of the left hand encircle the coil in the same direction as electron flow, the thumb points in the direction of the south pole.

d. if the thumb of the right hand points in the direction of electron flow, the fingers point in the direction of the north pole.

4. The physical motion resulting from the forces of two magnetic fields is called

a. Lenz's law.

b. motor action.

c. the left-hand rule for coils.

d. integration.

5. Motor action always tends to produce motion from

a. a stronger field toward a weaker field.

b. a weaker field toward a stronger field.

c. a north pole toward a south pole.

d. none of the above.

6. A conductor will have an induced current or voltage only when there is

a. a stationary magnetic field.

b. a stationary conductor.

c. relative motion between the wire and magnetic field.

d. both a and b.

7. The polarity of an induced voltage is determined by

a. motor action.

b. Lenz's law.

c. the number of turns in the coil.

d. the amount of current in the coil.

8. For a relay, the pickup current is defined as

a. the maximum current rating of the relay coil.

b. the minimum relay coil current required to keep a relay energized.

c. the minimum relay coil current required to energize a relay.

d. the minimum current in the switching contacts.

9. The moveable arm of an attraction-type relay is called the

a. contacts.

b. relay coil.

c. terminal.

d. armature.

10. For a conductor being moved through a magnetic field, the amount of induced voltage is determined by

a. the rate at which the conductor cuts the magnetic flux.

b. the number of magnetic flux lines that are cut by the conductor.

c. the time of day during which the conductor is moved through the field.

d. both a and b.

11. Degaussing is done with

a. strong permanent magnets.

b. alternating current.

c. static electricity.

d. direct current.

12. Hysteresis losses

a. increase with higher frequencies.

b. decrease with higher frequencies.

c. are greater with direct current.

d. increase with lower frequencies.

13. The saturation of an iron core occurs when

a. all of the molecular dipoles and magnetic domains are aligned by the magnetizing force.

b. the coil is way too long.

c. the flux density cannot be increased in the core when the field intensity is increased.

d. both a and c.

14. The unit of field intensity is the

a. oersted.

b. gilbert.

c. A·t/m.

d. both a and c.

15. For a single conductor carrying an alternating current, the associated magnetic field is

a. only on the top side.

b. parallel to the direction of current.

c. at right angles to the direction of current.

d. only on the bottom side.

16. A coil with 200 mA of current has an mmf of 80 A·t. How many turns does the coil have?

a. 4000 turns.

b. 400 turns.

c. 40 turns.

d. 16 turns.

17. The magnetic field surrounding a solenoid is

a. like that of a permanent magnet.

b. unable to develop north and south poles.

c. one without magnetic flux lines.

d. unlike that of a permanent magnet.

18. For a relay, the holding current is defined as

a. the maximum current the relay contacts can handle.

b. the minimum amount of relay coil current required to keep a relay energized.

c. the minimum amount of relay coil current required to energize a relay.

d. the maximum current required to operate a relay.

19. A vertical wire with electron flow into this page has an associated magnetic field which is

a. clockwise.

b. counterclockwise.

c. parallel to the wire.

d. none of the above.

20. How much is the induced voltage when a magnetic flux cuts across 150 turns at the rate of 5 Wb/s?

a. 7.5 kV.

b. 75 V.

c. 750 V.

d. 750 mV.

Essay Questions

1. State the rule for determining the magnetic polarity of a solenoid. (a) How can the polarity be reversed? (b) Why are there no magnetic poles when the current through the coil is zero?

2. Why does the motor action between two magnetic fields result in motion toward the weaker field?

3. Why does current in a conductor perpendicular to this page have a magnetic field in the plane of the paper?

4. Why must the conductor and the external field be perpendicular to each other to have motor action or to generate induced voltage?

5. Explain briefly how either motor action or generator action can be obtained with the same conductor in a magnetic field.

6. Assume that a conductor being cut by the flux of an expanding magnetic field has 10 V induced with the top end positive. Now analyze the effect of the following changes: (a) The magnetic flux continues to expand, but at a slower rate. How does this affect the amount of induced voltage and its polarity? (b) The magnetic flux is constant, neither increasing nor decreasing. How much is the induced voltage? (c) The magnetic flux contracts, cutting across the conductor with the opposite direction of motion. How does this affect the polarity of the induced voltage?

7. Redraw the graph in Fig. 14–18c for 500 turns with all other factors the same.

8. Redraw the circuit with the coil and battery in Fig. 14–10, showing two different ways to reverse the magnetic polarity.

9. Referring to Fig. 14–18, suppose that the flux decreases from 8 Wb to zero at the same rate as the increase. Tabulate all values as in Table 14–2 and draw the three graphs corresponding to those in Fig. 14–18.

10. Assume that you have a relay whose pickup and holding current values are unknown. Explain how you can determine their values experimentally.

11. List two factors that determine the strength of an electromagnet.

12. What is meant by magnetic hysteresis?

13. What is meant by the saturation of an iron core?

14. List the three main parts of a relay.

15. For a relay, what is meant by the terms normally open (NO) and normally closed (NC) contacts?

16. Define the following terms as they relate to relays: (a) pickup current, (b) holding current, and (c) dropout voltage.

17. List two common relay troubles.

Problems

SECTION 14–1 AMPERE-TURNS OF MAGNETOMOTIVE FORCE (mmf)

14–1 What is (a) the cgs unit of mmf? (b) the SI unit of mmf?

14–2 Calculate the ampere-turns of mmf for a coil with the following values:

a. $I = 10$ mA, $N = 150$ turns.

b. $I = 15$ mA, $N = 100$ turns.

c. $I = 2$ mA, $N = 5000$ turns.

d. $I = 100\ \mu$A, $N = 3000$ turns.

14–3 Calculate the ampere-turns of mmf for a coil with the following values:

a. $I = 5$ mA, $N = 4000$ turns.

b. $I = 40$ mA, $N = 50$ turns.

c. $I = 250$ mA, $N = 40$ turns.

d. $I = 600$ mA, $N = 300$ turns.

14–4 Calculate the current required in a coil to provide an mmf of 2 A·t if the number of turns equals

a. 50.

b. 500.

c. 100.

d. 2000.

14–5 Calculate the number of turns required in a coil to provide an mmf of 100 A·t if the current equals

a. $I = 100$ mA.

b. $I = 25$ mA.

c. $I = 40$ mA.

d. $I = 2$ A.

14–6 Convert the following values of mmf to gilberts (Gb):

a. 100 A•t.

b. 30 A•t.

c. 500 A•t.

14–7 Convert the following values of mmf to ampere-turns (A•t):

a. 126 Gb.

b. 37.8 Gb.

c. 630 Gb.

SECTION 14–2 FIELD INTENSITY (*H*)

14–8 What is (a) the cgs unit of field intensity? (b) the SI unit of field intensity?

14–9 Calculate the field intensity, H, in ampere-turns per meter, for each of the following cases:

a. mmf = 100 A•t, $l = 0.2$ m.

b. mmf = 25 A•t, $l = 0.25$ m.

c. mmf = 4 A•t, $l = 0.08$ m.

d. mmf = 20 A•t, $l = 0.1$ m.

14–10 Calculate the field intensity, H, in ampere-turns per meter, for each of the following cases:

a. $I = 40$ mA, $N = 500$ turns, $l = 0.2$ m.

b. $I = 100$ mA, $N = 1000$ turns, $l = 0.5$ m.

c. $I = 60$ mA, $N = 600$ turns, $l = 0.25$ m.

d. $I = 10$ mA, $N = 300$ turns, $l = 0.075$ m.

14–11 Convert the following values of field intensity to oersteds:

a. 50 A•t/m.

b. 150 A•t/m.

14–12 Convert the following values of field intensity to A•t/m.

a. 0.63 oersteds.

b. 1.89 oersteds.

14–13 Calculate the absolute permeability, μ, of a material if its relative permeability, μ_r, equals

a. 10.

b. 50.

c. 100.

d. 500.

e. 1000.

14–14 A coil with an iron core has a field intensity, H, of 50 A•t/m. If the relative permeability, μ_r, equals 300, calculate the flux density, B, in teslas.

14–15 Calculate the relative permeability, μ_r, of an iron core when a field intensity, H, of 750 A•t/m produces a flux density, B, of 0.126 T.

14–16 Calculate the field intensity, H, of an electromagnet if the flux density, B, equals 0.504 teslas and the relative permeability of the core is 200.

SECTION 14–3 *B-H* MAGNETIZATION CURVE

14–17 Referring to the *B-H* curve in Fig. 14–3, calculate the absolute permeability, μ, in SI units for the icon core at a field intensity, H, of

a. 3000 A•t/m.

b. 5000 A•t/m.

SECTION 14–9 GENERATING AN INDUCED VOLTAGE

14–18 A magnetic field cuts across a coil of 500 turns at the rate of 100 μWb/s. Calculate v_{ind}.

14–19 A magnetic field cuts across a coil of 400 turns at the rate of 0.02 Wb/s. Calculate v_{ind}.

14–20 A magnetic flux of 300 Mx cuts across a coil of 1500 turns in 200 μs. Calculate v_{ind}.

14–21 The magnetic flux surrounding a coil changes from 1000 to 6000 Mx in 5 μs. If the coil has 200 turns, how much is the induced voltage?

14–22 A coil has an induced voltage of 1 kV when the rate of flux change is 0.5 Wb/s. How many turns are in the coil?

Critical Thinking

14–23 Derive the value of 1.26×10^{-6} T/(A•t/m) for μ_0 from $\mu = B/H$.

14–24 What is the relative permeability (μ_r) of a piece of soft iron whose permeability (μ) equals 3.0×10^{-3} T/(A•t/m)?

14–25 Refer to Fig. 14–27*a*. Calculate (a) the total wire resistance R_w of the No. 12 gage copper wires; (b) the total resistance R_T of the circuit; (c) the voltage available across the load R_L; (d) the I^2R power loss in the wire conductors; (e) the load power P_L; (f) the total power P_T consumed by the circuit; (g) the percent efficiency of the system calculated as $(P_L/P_T) \times 100$.

14–26 Refer to Fig. 14–27*b*. Calculate (a) the total wire resistance R_w of the No. 20 gage copper wires; (b) the total resistance R_T of the relay coil circuit; (c) the voltage across the relay coil; (d) the I^2R power loss in the No. 20 gage copper wires in the relay coil circuit; (e) the total wire resistance R_w of the 10-ft length of No. 12 gage copper wires that connect the 16-Ω load R_L to the 240-V_{AC} power line; (f) the voltage available across the load R_L; (g) the I^2R power loss in the 10-ft length of the No. 12 gage copper wire; (h) the load power P_L; (i) the total power P_T consumed by the load side of the circuit; (j) the percent efficiency of the system calculated as $(P_L/P_T) \times 100$.

14–27 Explain the advantage of using a relay rather than an ordinary mechanical switch when controlling a high power load a long distance away. Use your solutions from Critical Thinking Probs. 14–25 and 14–26 to support your answer.

Figure 14-27 Circuit diagram for Critical Thinking Probs. 14–25 and 14–26. (*a*) Mechanical switch controlling a high power load a long distance away. (*b*) Relay controlling a high power load a long distance away.

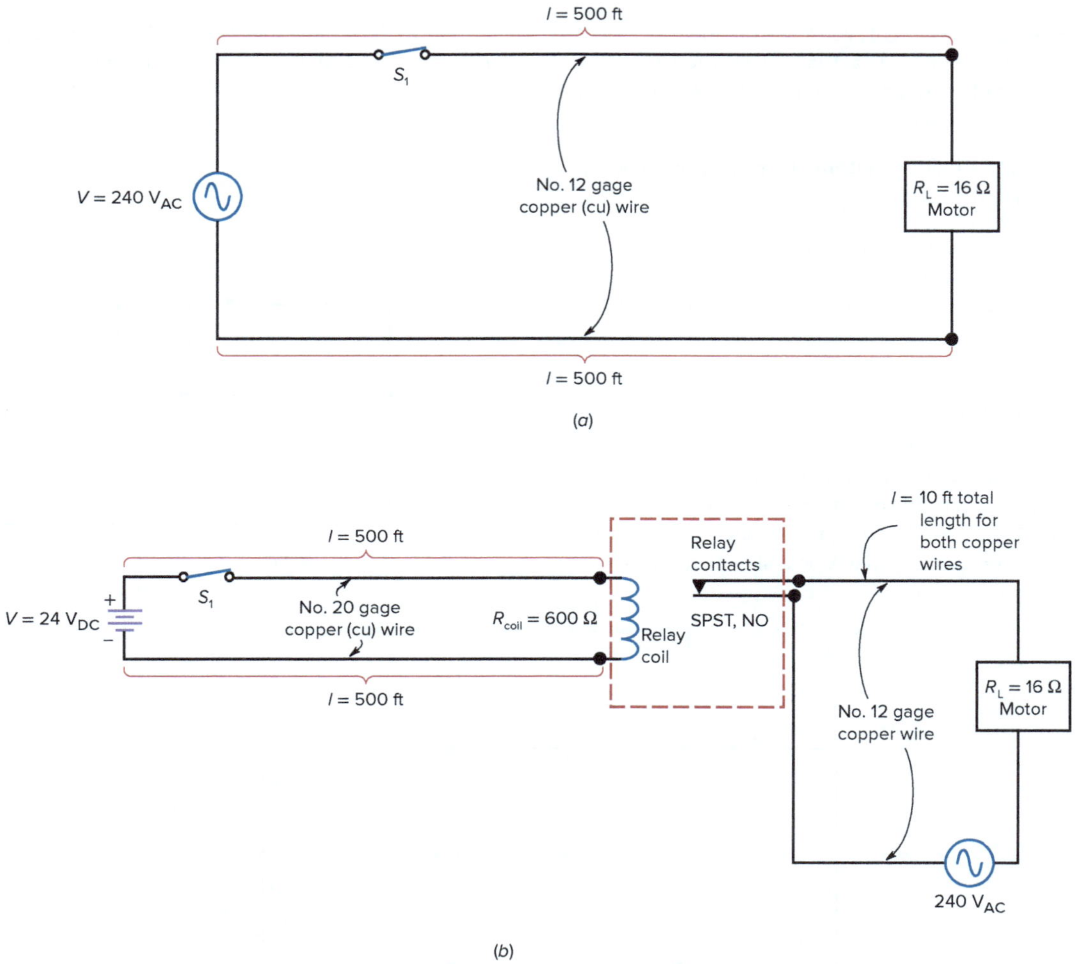

Answers to Self-Reviews

14–1 **a.** 486 A·t
b. 630 Gb

14–2 **a.** 1
b. 200
c. 200 G/Oe
d. $250 \frac{\text{A}\cdot\text{t}}{\text{m}}$
e. 6.3 oersteds

14–3 **a.** 0.189 T
b. 4000 A·t/m approx.

14–4 **a.** true
b. true

14–5 **a.** true
b. true

14–6 **a.** left
b. south

14–7 **a.** true
b. true

14–8 **a.** south
b. south
c. yes

14–9 **a.** 2 Wb/s
b. 2 Wb/s

14–10 **a.** true
b. false
c. true
d. false

Laboratory Application Assignment

In this lab application assignment, you will examine the operating characteristics of a DC actuated relay. You will also construct two different relay circuits that control which of two incandescent lamps are being lit.

Equipment: Obtain the following items from your instructor.

- Dual output variable dc power supply
- DMM
- 12-V_{DC} actuated relay with SPDT switching contacts
- Two 6.3-V and two 12-V incandescent lamps
- Normally open (NO) and normally closed (NC) push-button switches
- SPST switch

Relay Specifications

Examine the DC relay assigned to you. By inspection, locate the connecting terminals for the relay coil. Next, determine which terminal connects to the armature and which contact terminals are normally open (NO) and normally closed (NC). If the relay is in an enclosure, you will probably need to use an ohmmeter to determine this information. If the relay enclosure shows a diagram and has its terminal connections numbered, transfer these numbers to the relay circuits shown in Figs. 14–28 to 14–30.

Figure 14–28

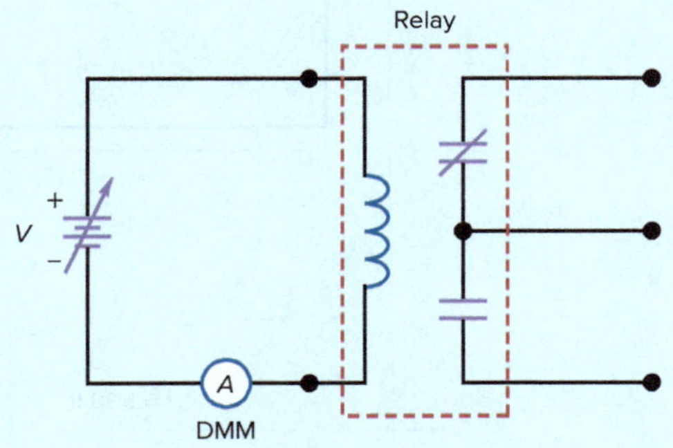

Figure 14–29

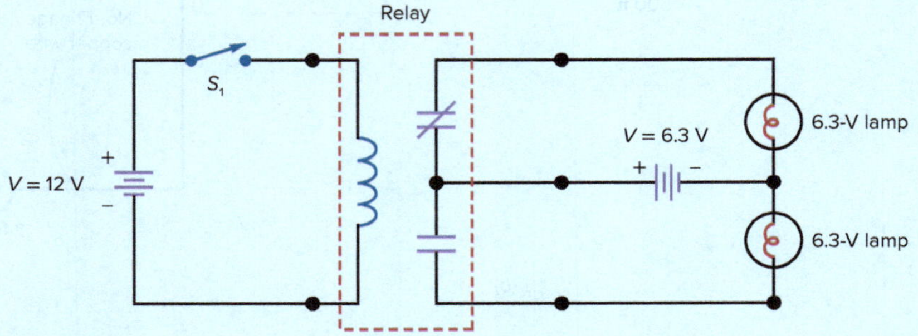

Figure 14–30

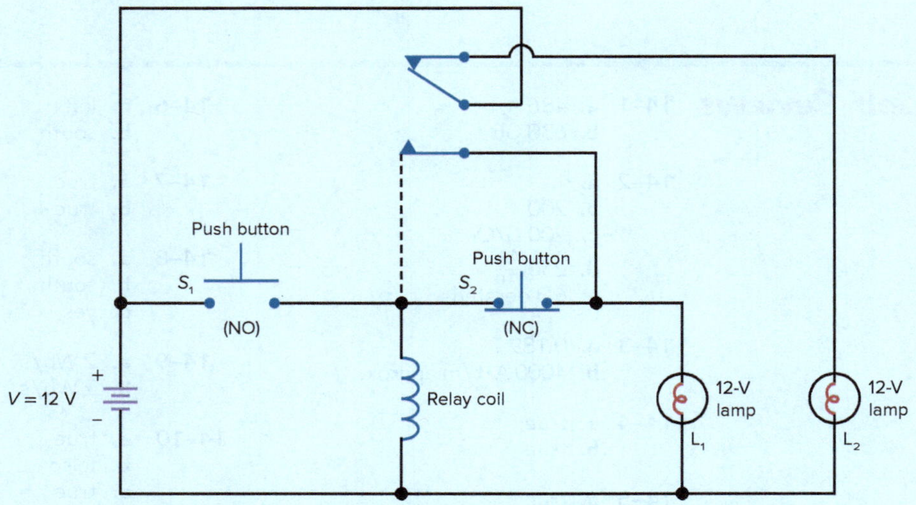

In this part of the lab, we will experimentally determine the pickup current and the holding current for your relay. Recall that the pickup current is defined as the minimum amount of relay coil current required to energize or operate a relay. The holding current is defined as the minimum amount of current required to keep a relay energized or operating.

Before connecting the relay to the circuit of Fig. 14–28, measure and record the resistance of the relay coil. R_{Coil} = ____

Connect the relay circuit shown in Fig. 14–28. Reduce the output from the DC power supply to 0 V. Slowly increase the voltage while monitoring the relay coil current indicated by the DMM. Carefully watch (or listen) for the relay contacts to open or close. When this happens, the relay is energized. Measure and record the minimum current that energizes the relay. This current is the relay's pickup current. I_{Pickup} = ______________

____

With the relay still energized, slowly decrease the output from the DC power supply while watching (or listening) for the relay contacts to open or close. Measure and record the minimum current that is still able to keep the relay energized or operating. This current is the relay's holding current, $I_{Holding}$ = ___________

Which current is larger, the pickup current or the holding current? ____

Is this what you expected? ______________________________

Relay Circuit 1

Construct the relay circuit shown in Fig. 14–29. Depending on the position of the switch S_1, one of the two 6.3-V bulbs should be on and the other should be off. Open and close S_1 several times to verify that the circuit is operating properly. Have an instructor check your circuit.

Relay Circuit 2

Examine the latching relay circuit in Fig. 14–30. When the relay is not energized, lamp L_2 will be lit and lamp L_1 will be off. Pressing the push-button switch S_1 energizes the relay. This turns lamp L_2 off and lamp L_1 on. To turn lamp L_2 on again and turn lamp L_1 off, press the push-button switch S_2 to deenergize the relay.

Construct the circuit in Fig. 14–30. (The relay switching contacts are shown in their deenergized state.) Press S_1 and S_2 in succession to verify that the circuit is operating normally. Have an instructor check your circuit.

Design credits: Chapter Opener Background Image, and Multisim: Stockbyte/Getty Images

chapter

15

Alternating Voltage and Current

In this chapter, and the chapters that follow, we will analyze circuits that operate using an alternating voltage source. Specifically, in this chapter we will analyze the characteristics and terminology associated with sinusoidal waveforms, although nonsinusoidal waveforms are also discussed. A sine wave alternating voltage is a voltage that continuously varies in amplitude and periodically reverses in polarity. Similarly, a sine wave alternating current is a current that continuously varies in amplitude and periodically reverses in direction. One cycle of alternating voltage includes two alternations in polarity. The number of cycles per second is the frequency, measured in hertz (Hz). One cycle per second equals 1 Hz. Every AC voltage, whether it's sinusoidal or not, has both amplitude variations and polarity reversals. The amplitude values and rate of polarity reversal, however, vary from one AC waveform to the next. An example of a sine wave alternating voltage is the 120 V AC power line voltage whose frequency is 60 Hz.

In this chapter, you will learn (a) how a sine wave alternating voltage is generated, (b) the different units of angular measure, (c) the different voltage and current values associated with a sine wave, and (d) what is meant by the frequency, period, wavelength, and phase angle of sinusoidal waveforms.

Chapter Outline

Chapter Objectives

After studying this chapter, you should be able to

- *Describe* how a sine wave of alternating voltage is generated.
- *Calculate* the instantaneous value of a sine wave of alternating voltage or current.
- *Define* the following values for a sine wave: peak, peak-to-peak, root-mean-square, and average.
- *Calculate* the rms, average, and peak-to-peak values of a sine wave when the peak value is known.
- *Define frequency* and *period* and list the units of each.
- *Calculate* the wavelength when the frequency is known.
- *Explain* the concept of phase angles.
- *Describe* the makeup of a nonsinusoidal waveform.
- *Define* the term *harmonics*.
- *Outline* the basics of residential house wiring.

Important Terms

alternation
average value
cycle
decade
effective value
form factor
frequency
generator
harmonic frequency
hertz (Hz)
motor
nonsinusoidal waveform
octave
peak value
period
phase angle
phasor
quadrature phase
radian
root-mean-square (rms) value
sine wave
wavelength

15–1 Alternating Current Applications

Figure 15–1 shows an AC voltage with the reversals between positive and negative polarities and the variations in amplitude. In Fig. 15–1*a*, the waveform shown simulates an AC voltage as it would appear on the screen of an oscilloscope, which is an important test instrument for AC voltages. The oscilloscope displays a graph of the AC voltage connected to its input terminals. The oscilloscope allows us to determine important information about the displayed waveform such as its amplitude, **period**, and **frequency**, all of which will be explained in greater detail later in this chapter. The details of how to use the oscilloscope for AC voltage measurements is explained in Appendix E.

In Fig. 15–1*b*, the graph of the AC waveform shows how the output from the **generator** in Fig. 15–1*c* varies with respect to time. Assume that this graph shows *V* at terminal 2 with respect to terminal 1. Then the voltage at terminal 1 corresponds to the zero axis in the graph as the reference level. At terminal 2, the output voltage has positive amplitude variations from zero up to the **peak value** and down to zero. All these voltage values are with respect to terminal 1. After a half-**cycle**, the voltage at terminal 2 becomes negative, still with respect to the other terminal. Then the same voltage variations are repeated at terminal 2, but they have negative polarity compared to the reference level. Note that if we take the voltage at terminal 1 with terminal 2 as the reference, the waveform in Fig. 15–1*b* would have the same shape but be inverted in polarity. The negative half-cycle would come first, but it does not matter which is first or second.

GOOD TO KNOW

An oscilloscope is used to view and measure AC waveforms. Most oscilloscopes can display two or more waveforms at the same time which is important for comparison purposes.

The characteristic of varying values is the reason that AC circuits have so many uses. For instance, a transformer can operate only with alternating current to step up or step down an AC voltage. The reason is that the changing current produces changes in its associated magnetic field. This application is just an example of inductance *L* in AC circuits, where the changing magnetic flux of a varying current can produce induced voltage. The details of inductance are explained in Chaps. 19, 20, and 21.

A similar but opposite effect in AC circuits is capacitance *C*. The capacitance is important with the changing electric field of a varying voltage. Just as *L* has an effect with alternating current, *C* has an effect that depends on alternating voltage. The details of capacitance are explained in Chaps. 16, 17, and 18.

MultiSim **Figure 15–1** Waveform of AC power-line voltage with frequency of 60 Hz. Two cycles are shown. (*a*) Oscilloscope display. (*b*) Details of waveform and alternating polarities. (*c*) Symbol for an AC voltage source.

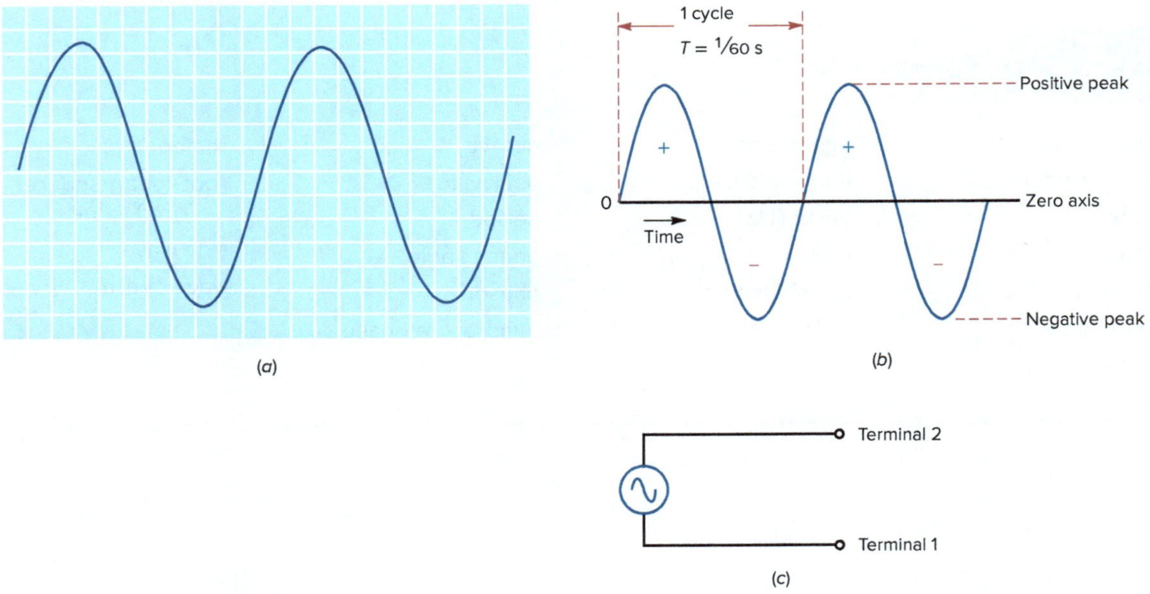

The L and C are additional factors, beside resistance R, in the operation of AC circuits. Note that R is the same for either a DC or an AC circuit. However, the effects of L and C depend on having an AC source. The rate at which the AC variations occur, which determines the frequency, allows a greater or lesser reaction by L and C. Therefore, the effect is different for different frequencies. One important application is a resonant circuit with L and C that is tuned to a particular frequency. The details of resonant circuits and tuning are examined in Chapter 25, "Resonance."

In general, electronic circuits are combinations of R, L, and C, with both direct current and alternating current. Audio, video, and radio signals are AC voltages and currents. However, amplifiers that use transistors and integrated circuits need DC voltages to conduct any current at all. The resulting output of an amplifier circuit, therefore, consists of direct current with a superimposed AC signal.

15–1 Self-Review

Answers at the end of the chapter.

a. An AC voltage varies in magnitude and reverses in polarity. (True/False)

b. A transformer can operate with either AC or a steady DC input. (True/False)

c. Inductance L and capacitance C are important factors in AC circuits. (True/False)

15–2 Alternating-Voltage Generator

A **sine wave** alternating voltage is a voltage that continuously varies in magnitude and periodically reverses in polarity. In Fig. 15–1, the variations up and down on the waveform show the changes in magnitude. The zero axis is a horizontal line across the center. Then voltages above the center have positive polarity, and values below center are negative.

Figure 15–2 shows how such a voltage waveform is produced by a rotary generator. The conductor loop rotates through the magnetic field to generate the induced AC voltage across its open terminals. The magnetic flux shown here is vertical, with lines of force in the plane of the paper.

In Fig. 15–2*a*, the loop is in its horizontal starting position in a plane perpendicular to the paper. When the loop rotates counterclockwise, the two longer conductors move around a circle. Note that in the flat position shown, the two long conductors of the loop move vertically up or down but parallel to the vertical flux lines. In this position, motion of the loop does not induce a voltage because the conductors are not cutting across the flux.

When the loop rotates through the upright position in Fig. 15–2*b*, however, the conductors cut across the flux, producing maximum induced voltage. The shorter connecting wires in the loop do not have any appreciable voltage induced in them.

Each of the longer conductors has opposite polarity of induced voltage because the conductor at the top is moving to the left while the bottom conductor is moving to the right. The amount of voltage varies from zero to maximum as the loop moves from a flat position to upright, where it can cut across the flux. Also, the polarity at the terminals of the loop reverses as the motion of each conductor reverses during each half-revolution.

With one revolution of the loop in a complete circle back to the starting position, therefore, the induced voltage provides a potential difference v across the loop, varying in the same way as the wave of voltage shown in Fig. 15–1. If the loop rotates at the speed of 60 revolutions per second, the AC voltage has a frequency of 60 Hz.

Figure 15–2 A loop rotating in a magnetic field to produce induced voltage with alternating polarities. (*a*) Loop conductors moving parallel to magnetic field results in zero voltage. (*b*) Loop conductors cutting across magnetic field produce maximum induced voltage.

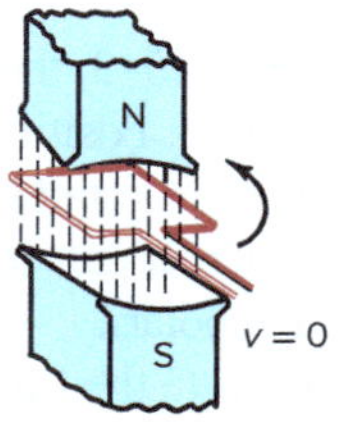

(*a*)

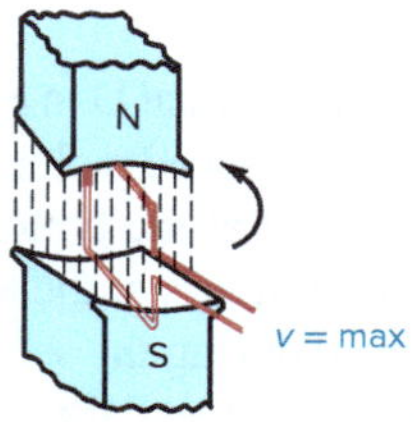

(*b*)

Figure 15–3 One cycle of alternating voltage generated by a rotating loop. The magnetic field, not shown here, is directed from top to bottom, as shown in Fig. 15–2.

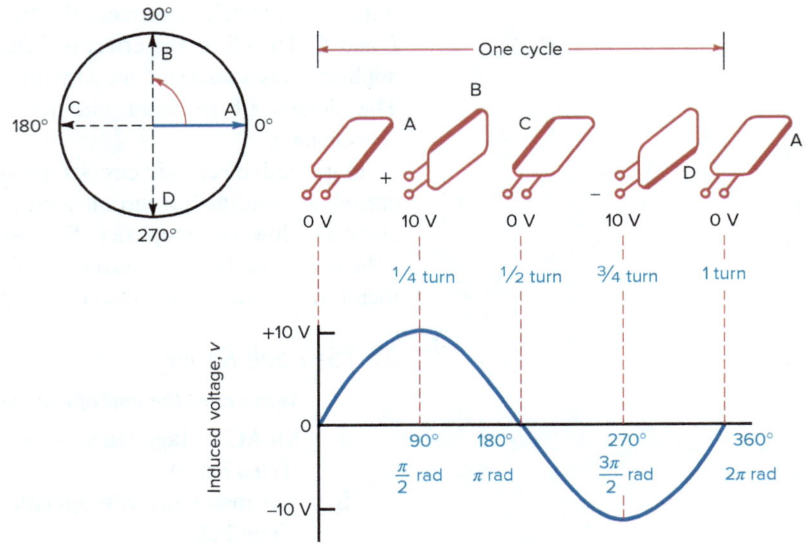

The Cycle

One complete revolution of the conductor loop around the circle is a *cycle*. In Fig. 15–3, the generator loop is shown in its position at each quarter-turn during one complete cycle. The corresponding wave of induced voltage also goes through one cycle. Although not shown, the magnetic field is from top to bottom of the page, as in Fig. 15–2.

At position A in Fig. 15–3, the loop is flat and moves parallel to the magnetic field, so that the induced voltage is zero. Counterclockwise rotation of the loop moves the dark conductor to the top at position B, where it cuts across the field to produce maximum induced voltage. The polarity of the induced voltage here makes the open end of the dark conductor positive. This conductor at the top is cutting across the flux from right to left. At the same time, the opposite conductor below is moving from left to right, causing its induced voltage to have opposite polarity. Therefore, maximum induced voltage is produced at this time across the two open ends of the loop. Now the top conductor is positive with respect to the bottom conductor.

In the graph of induced voltage values below the loop in Fig. 15–3, the polarity of the dark conductor is shown with respect to the other conductor. Positive voltage is shown above the zero axis in the graph. As the dark conductor rotates from its starting position parallel to the flux toward the top position, where it cuts maximum flux, more and more induced voltage is produced with positive polarity.

When the loop rotates through the next quarter-turn, it returns to the flat position shown in C, where it cannot cut across flux. Therefore, the induced voltage values shown in the graph decrease from the maximum value to zero at the half-turn, just as the voltage was zero at the start. The half-cycle of revolution is called an ***alternation***.

The next quarter-turn of the loop moves it to the position shown at D in Fig. 15–3, where the loop cuts across the flux again for maximum induced voltage. Note, however, that here the dark conductor is moving left to right at the bottom of the loop. This motion is reversed from the direction it had when it was at the top, moving right to left. Because the direction of motion is reversed during the second half-revolution, the induced voltage has opposite polarity with the dark conductor negative. This polarity is shown as negative voltage below the zero axis. The maximum value of induced voltage at the third quarter-turn is the same as at the first quarter-turn but with opposite polarity.

When the loop completes the last quarter-turn in the cycle, the induced voltage returns to zero as the loop returns to its flat position at A, the same as at the start. This cycle of values of induced voltage is repeated as the loop continues to rotate with one complete cycle of voltage values, as shown, for each circle of revolution.

Note that zero at the start and zero after the half-turn of an alternation are not the same. At the start, the voltage is zero because the loop is flat, but the dark conductor is moving upward in the direction that produces positive voltage. After one half-cycle, the voltage is zero with the loop flat, but the dark conductor is moving downward in the direction that produces negative voltage. After one complete cycle, the loop and its corresponding waveform of induced voltage are the same as at the start. *A cycle can be defined, therefore, as including the variations between two successive points having the same value and varying in the same direction.*

Angular Measure

Because the cycle of voltage in Fig. 15–3 corresponds to rotation of the conductor loop around a circle, it is convenient to consider parts of the cycle in angles. The most common unit of angular measure is the degree. One complete revolution of the conductor loop around the circle corresponds to 360 degrees. Therefore, one complete cycle includes 360°. One half-cycle, or one alternation, is 180° of revolution. A quarter-turn is 90°. The circle next to the loop positions in Fig. 15–3 illustrates the angular rotation of the dark conductor as it rotates counterclockwise from 0 to 90 to 180° for one half-cycle, then to 270°, and returning to 360° to complete the cycle.

GOOD TO KNOW

To convert from degrees to radians or vice versa, use the following conversion formulas:

$$\#deg = \#rad \times \frac{180^\circ}{\pi \text{ rad}}$$

$$\#rad = \#deg \times \frac{\pi \text{ rad}}{180^\circ}$$

Radian Measure

In angular measure, it is convenient to use a specific unit angle called the ***radian*** (abbreviated rad), which is an angle equal to 57.3°. Its convenience is due to the fact that a radian is the angular part of the circle that includes an arc equal to the radius r of the circle, as shown in Fig. 15–4. The circumference around the circle equals $2\pi r$. A circle includes 2π rad, then, as each radian angle includes one length r of the circumference. Therefore, one cycle equals 2π rad.

As shown in the graph in Fig. 15–3, divisions of the cycle can be indicated by angles in either degrees or radians. The comparison between degrees and radians can be summarized as follows:

Zero degrees is also zero radians
$360^\circ = 2\pi$ rad
$180^\circ = \frac{1}{2} \times 2\pi \text{ rad} = \pi$ rad
$90^\circ = \frac{1}{2} \times \pi \text{ rad} = \pi/2$ rad
$270^\circ = 180^\circ + 90^\circ$ or $\pi \text{ rad} + \pi/2 \text{ rad} = 3\pi/2$ rad

Figure 15–4 One radian (rad) is the angle equal to 57.3°. The complete circle of 360° includes 2π rad.

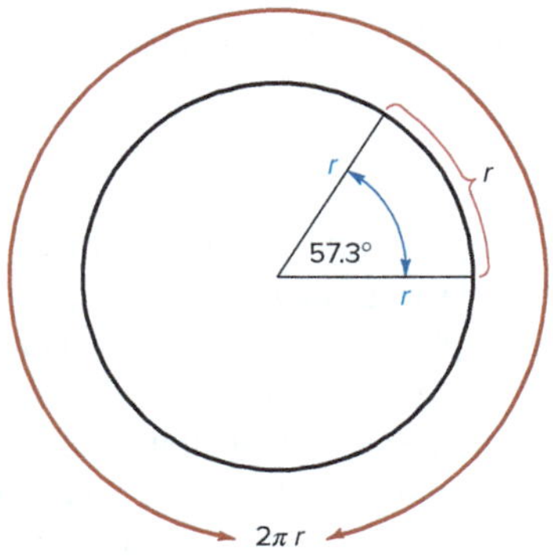

Circumference = 2π rad

The constant 2π in circular measure is numerically equal to 6.2832. This is double the value of 3.1416 for π. The Greek letter π (pi) is used to represent the ratio of the circumference to the diameter for any circle, which always has the numerical value of 3.1416. The fact that 2π rad is 360° can be shown as $2 \times 3.1416 \times 57.3^\circ = 360^\circ$ for a complete cycle.

15–2 Self-Review

Answers at the end of the chapter.

Refer to Fig. 15–3.

a. How much is the induced voltage at $\pi/2$ rad?
b. How many degrees are in a complete cycle?

15–3 The Sine Wave

The voltage waveform in Figs. 15–1 and 15–3 is called a *sine wave, sinusoidal wave*, or *sinusoid* because the amount of induced voltage is proportional to the sine of the angle of rotation in the circular motion producing the voltage. The sine is a trigonometric function of an angle; it is equal to the ratio of the opposite side to the hypotenuse in a right triangle. This numerical ratio increases from zero for 0° to a maximum value of 1 for 90° as the side opposite the angle becomes larger.

The voltage waveform produced by the circular motion of the loop is a sine wave because the induced voltage increases to a maximum at 90°, when the loop is vertical, in the same way that the sine of the angle of rotation increases to a maximum at 90°. The induced voltage and sine of the angle correspond for the full 360° of the cycle. Table 15–1 lists the numerical values of the sine for several important angles to illustrate the specific characteristics of a sine wave.

Notice that the sine wave reaches one-half its maximum value in 30°, which is only one-third of 90°. This fact means that the sine wave has a sharper slope of changing values when the wave is near the zero axis, compared with more gradual changes near the maximum value.

The instantaneous value of a sine-wave voltage for any angle of rotation is expressed by the formula

$$v = V_M \sin \theta \qquad \textbf{(15–1)}$$

where θ (Greek letter *theta*) is the angle, sin is the abbreviation for its sine, V_M is the maximum voltage value, and v is the instantaneous value of voltage at angle θ.

CALCULATOR

With the calculator, be sure it is set for degrees, not radians or grad units. To find the value of the sine function, just punch in the number for angle θ in degrees and push the (SIN) key to see the values of sin θ on the display.

Applying this procedure to Formula (15–1), find the value of sin θ and multiply by the peak value V_M. Specifically, for the first problem in Example 15–1 with V_M of 100 and θ of 30°, first punch in 30 on the calculator. Next press the (SIN) key to see 0.5 on the display, which is sin 30°. Then push the multiplication (×) key, punch in 100 for V_M, and press the (=) key for the final answer of 50. The same method is used for all the other values of angle θ.

Example 15–1

A sine wave of voltage varies from zero to a maximum of 100 V. How much is the voltage at the instant of 30° of the cycle? 45°? 90°? 270°?

ANSWER $v = V_M \sin \theta = 100 \sin \theta$

At 30°: $v = V_M \sin 30° = 100 \times 0.5 = 50$ V

At 45°: $v = V_M \sin 45° = 100 \times 0.707 = 70.7$ V

At 90°: $v = V_M \sin 90° = 100 \times 1 = 100$ V

At 270°: $v = V_M \sin 270° = 100 \times -1 = -100$ V

The value of −100 V at 270° is the same as that at 90° but with opposite polarity.

To do the problems in Example 15–1, you must either refer to a table of trigonometric functions or use a scientific calculator that has trig functions.

Between zero at 0° and maximum at 90°, the amplitudes of a sine wave increase exactly as the sine value of the angle of rotation. These values are for the first quadrant in the circle, that is, 0 to 90°. From 90 to 180° in the second quadrant, the values decrease as a mirror image of the first 90°. The values in the third and fourth quadrants, from 180 to 360°, are exactly the same as 0 to 180° but with opposite sign. At 360°, the waveform is back to 0° to repeat its values every 360°.

Table 15–1 Values in a Sine Wave

Angle θ			
Degrees	Radians	Sin θ	Loop Voltage
0	0	0	Zero
30	$\frac{\pi}{6}$	0.500	50% of maximum
45	$\frac{\pi}{4}$	0.707	70.7% of maximum
60	$\frac{\pi}{3}$	0.866	86.6% of maximum
90	$\frac{\pi}{2}$	1.000	Positive maximum value
180	π	0	Zero
270	$\frac{3\pi}{2}$	−1.000	Negative maximum value
360	2π	0	Zero

In summary, the characteristics of the sine-wave AC waveform are

1. The cycle includes 360° or 2π rad.
2. The polarity reverses each half-cycle.
3. The maximum values are at 90 and 270°.
4. The zero values are at 0 and 180°.
5. The waveform changes its values fastest when it crosses the zero axis.
6. The waveform changes its values slowest when it is at its maximum value. The values must stop increasing before they can decrease.

A perfect example of the sine-wave AC waveform is the 60-Hz power-line voltage in Fig. 15–1.

15–3 Self-Review

Answers at the end of the chapter.

A sine-wave voltage has a peak value of 170 V. What is its value at

a. **30°?**
b. **45°?**
c. **90°?**

15–4 Alternating Current

When a sine wave of alternating voltage is connected across a load resistance, the current that flows in the circuit is also a sine wave. In Fig. 15–5, let the sine-wave voltage at the left in the diagram be applied across R of 100 Ω. The resulting sine wave of alternating current is shown at the right in the diagram. Note that the frequency is the same for v and i.

During the first alternation of v in Fig. 15–5, terminal 1 is positive with respect to terminal 2. Since the direction of electron flow is from the negative side of v, through R, and back to the positive side of v, current flows in the direction indicated by arrow A for the first half-cycle. This direction is taken as the positive direction of current in the graph for i, corresponding to positive values of v.

The amount of current is equal to v/R. If several instantaneous values are taken, when v is zero, i is zero; when v is 50 V, i equals 50 V/100, or 0.5 A; when v is

Figure 15–5 A sine wave of alternating voltage applied across *R* produces a sine wave of alternating current in the circuit. (*a*) Waveform of applied voltage. (*b*) AC circuit. Note the symbol for sine-wave generator *V*. (*c*) Waveform of current in the circuit.

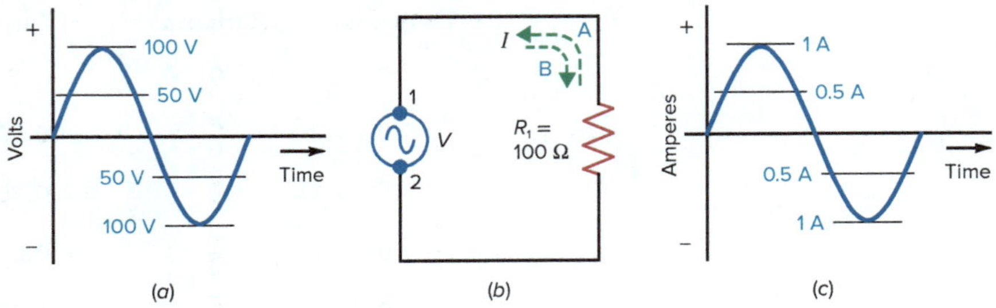

100 V, *i* equals 100 V/100, or 1 A. For all values of applied voltage with positive polarity, therefore, the current is in one direction, increasing to its maximum value and decreasing to zero, just like the voltage.

In the next half-cycle, the polarity of the alternating voltage reverses. Then terminal 1 is negative with respect to terminal 2. With reversed voltage polarity, current flows in the opposite direction. Electron flow is from terminal 1 of the voltage source, which is now the negative side, through *R*, and back to terminal 2. This direction of current, as indicated by arrow B in Fig. 15–5, is negative.

The negative values of *i* in the graph have the same numerical values as the positive values in the first half-cycle, corresponding to the reversed values of applied voltage. As a result, the alternating current in the circuit has sine-wave variations corresponding exactly to the sine-wave alternating voltage.

Only the waveforms for *v* and *i* can be compared. There is no comparison between relative values because the current and voltage are different quantities.

It is important to note that the negative half-cycle of applied voltage is just as useful as the positive half-cycle in producing current. The only difference is that the reversed polarity of voltage produces the opposite direction of current.

Furthermore, the negative half-cycle of current is just as effective as the positive values when heating the filament to light a bulb. With positive values, electrons flow through the filament in one direction. Negative values produce electron flow in the opposite direction. In both cases, electrons flow from the negative side of the voltage source, through the filament, and return to the positive side of the source. For either direction, the current heats the filament. The direction does not matter, since it is the motion of electrons against resistance that produces power dissipation. In short, resistance *R* has the same effect in reducing *I* for either direct current or alternating current.

15–4 Self-Review

Answers at the end of the chapter.

Refer to Fig. 15–5.

a. **When *v* is 70.7 V, how much is *i*?**

b. **How much is *i* at 30°?**

15–5 Voltage and Current Values for a Sine Wave

Since an alternating sine wave of voltage or current has many instantaneous values through the cycle, it is convenient to define specific magnitudes to compare one wave with another. The peak, average, and **root-mean-square (rms) values** can be

MultiSim **Figure 15–6** Definitions of important amplitude values for a sine wave of voltage or current.

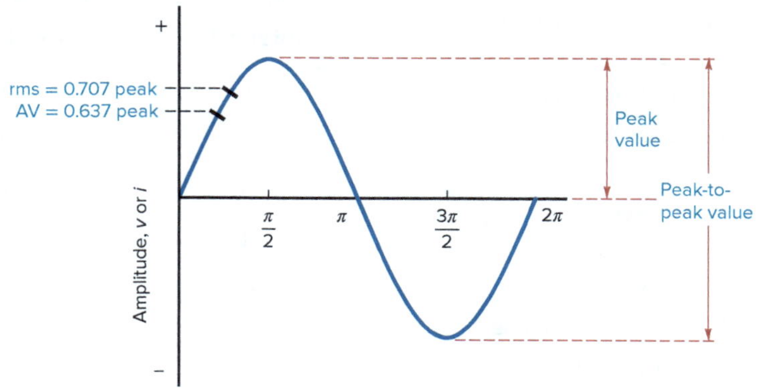

specified, as indicated in Fig. 15–6. These values can be used for either current or voltage.

Peak Value

This is the maximum value V_M or I_M. For example, specifying that a sine wave has a peak value of 170 V states the highest value the sine wave reaches. All other values during the cycle follow a sine wave. The peak value applies to either the positive or the negative peak.

To include both peak amplitudes, the *peak-to-peak (p-p) value* may be specified. For the same example, the peak-to-peak value is 340 V, double the peak value of 170 V, since the positive and negative peaks are symmetrical. Note that the two opposite peak values cannot occur at the same time. Furthermore, in some waveforms, the two peaks are not equal.

GOOD TO KNOW

The DC value of a waveform is its average value over one complete cycle. For a complete sine wave of alternating voltage, the average DC value is zero volts.

If just one alternation of the sine wave is considered, however, then the average value is calculated using Formula (15–2). Do not confuse the average value of one alternation with the waveform's DC value of 0 V.

Average Value

This is an arithmetic average of all values in a sine wave for one alternation, or half-cycle. The half-cycle is used for the average because over a full cycle the **average value** is zero, which is useless for comparison. If the sine values for all angles up to 180° for one alternation are added and then divided by the number of values, this average equals 0.637. These calculations are shown in Table 15–2.

Since the peak value of the sine function is 1 and the average equals 0.637, then

$$\text{Average value} = 0.637 \times \text{peak value} \qquad \textbf{(15–2)}$$

With a peak of 170 V, for example, the average value is 0.637 × 170 V, which equals approximately 108 V.

GOOD TO KNOW

When the amplitude of a sine wave is measured with an AC voltmeter or AC current meter, the value displayed by the meter will be the waveform's rms value.

Root-Mean-Square, or Effective, Value

The most common method of specifying the amount of a sine wave of voltage or current is by relating it to the DC voltage and current that will produce the same heating effect. This is called its *root-mean-square* value, abbreviated rms. The formula is

$$\text{rms value} = 0.707 \times \text{peak value} \qquad \textbf{(15–3)}$$

or

$$V_{rms} = 0.707V_{max}$$

GOOD TO KNOW

The rms value of an AC sine wave is its equivalent DC value for calculating power dissipation in a resistance. However, for any given sine wave, its own rms and DC values are different. For example, if the sine wave in Fig. 15–6 has a peak value of 10 V, then its rms value would be 7.07 V. However, the average or DC value of the waveform over one full cycle is 0 V.

Table 15–2 Derivation of Average and rms Values for a Sine-Wave Alternation

Interval	Angle θ	Sin θ	(Sin θ)^2
1	15°	0.26	0.07
2	30°	0.50	0.25
3	45°	0.71	0.50
4	60°	0.87	0.75
5	75°	0.97	0.93
6	90°	1.00	1.00
7*	105°	0.97	0.93
8	120°	0.87	0.75
9	135°	0.71	0.50
10	150°	0.50	0.25
11	165°	0.26	0.07
12	180°	0.00	0.00
	Total	7.62	6.00
	Average voltage: $\frac{7.62}{12} = 0.635^{\dagger}$	rms value: $\sqrt{6/12} = \sqrt{0.5} = 0.707$	

* For angles between 90 and 180°, *sin* θ = *sin* (180° − θ).

† More intervals and precise values are needed to get the exact average of 0.637.

and

$$I_{rms} = 0.707 I_{max}$$

With a peak of 170 V, for example, the rms value is 0.707 × 170, or 120 V, approximately. This is the voltage of the commercial AC power line, which is always given in rms value.

It is often necessary to convert from rms to peak value. This can be done by inverting Formula (15–3), as follows:

$$\text{Peak} = \frac{1}{0.707} \times \text{rms} = 1.414 \times \text{rms} \qquad \textbf{(15–4)}$$

or

$$V_{max} = 1.414 V_{rms}$$

and

$$I_{max} = 1.414 I_{rms}$$

Dividing by 0.707 is the same as multiplying by 1.414.

For example, commercial power-line voltage with an rms value of 120 V has a peak value of 120 × 1.414, which equals 170 V, approximately. Its peak-to-peak value is 2 × 170, or 340 V, which is double the peak value. As a formula,

$$\text{Peak-to-peak value} = 2.828 \times \text{rms value} \qquad \textbf{(15–5)}$$

GOOD TO KNOW

An oscilloscope can measure and display the peak and peak-to-peak values of an AC waveform.

Figure 15–7 Waveforms A and B have different amplitudes, but they are both sine waves.

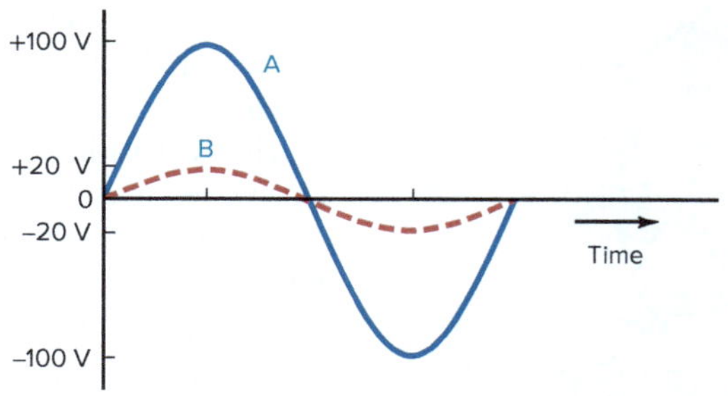

The factor 0.707 for the rms value is derived as the square root of the average (mean) of all the squares of the sine values. If we take the sine for each angle in the cycle, square each value, add all the squares, divide by the number of values added to obtain the average square, and then take the square root of this mean value, the answer is 0.707. These calculations are shown in Table 15–2 for one alternation from 0 to 180°. The results are the same for the opposite alternation.

The advantage of the rms value derived in terms of the squares of the voltage or current values is that it provides a measure based on the ability of the sine wave to produce power, which is I^2R or V^2/R. As a result, the rms value of an alternating sine wave corresponds to the same amount of direct current or voltage in heating power. An alternating voltage with an rms value of 120 V, for instance, is just as effective in heating the filament of a lightbulb as 120 V from a steady DC voltage source. For this reason, the rms value is also called the ***effective*** **value**.

Unless indicated otherwise, all sine-wave AC measurements are in rms values. The capital letters *V* and *I* are used, corresponding to the symbols for DC values. As an example, $V = 120$ V for AC power-line voltage.

The ratio of the rms to average values is the ***form factor***. For a sine wave, this ratio is $0.707/0.637 = 1.11$.

Note that sine waves can have different amplitudes but still follow the sinusoidal waveform. Figure 15–7 compares a low-amplitude voltage with a high-amplitude voltage. Although different in amplitude, they are both sine waves. In each wave, the rms value = 0.707 × peak value.

GOOD TO KNOW

The rms value of an AC waveform is sometimes referred to as its power producing value.

GOOD TO KNOW

$V_{rms} = V_{Average} \times 1.11$

15–5 Self-Review

Answers at the end of the chapter.

a. Convert 170 V peak to rms value.
b. Convert 10 V rms to peak value.
c. Convert 1 V rms to peak-to-peak value.

MultiSim **Figure 15–8** Number of cycles per second is the frequency in hertz (Hz) units. (*a*) $f = 1$ Hz. (*b*) $f = 4$ Hz.

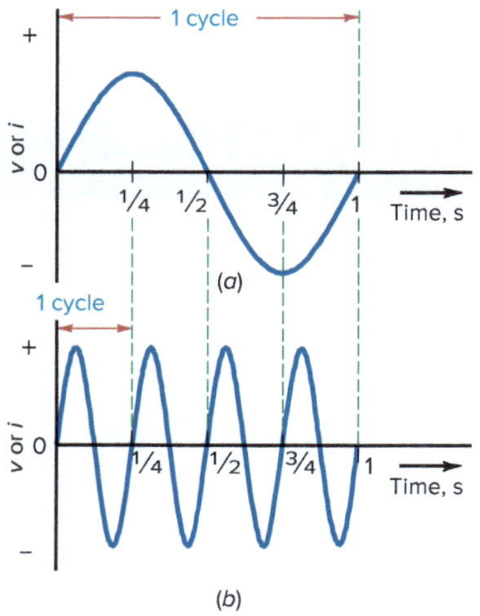

15–6 Frequency

The number of cycles per second is the *frequency*, with the symbol *f*. In Fig. 15–3, if the loop rotates through 60 complete revolutions, or cycles, during 1 s, the frequency of the generated voltage is 60 cps, or 60 Hz. You see only one cycle of the sine waveform, instead of 60 cycles, because the time interval shown here is $1/60$ s. Note that the factor of time is involved. More cycles per second means a higher frequency and less time for one cycle, as illustrated in Fig. 15–8. Then the changes in values are faster for higher frequencies.

Bettmann/Getty Images

PIONEERS IN ELECTRONICS

In 1887, German physicist *Heinrich Hertz (1857–1894)* proved that electricity could be transmitted in electromagnetic waves. In his honor, the hertz (Hz) is now the standard unit for the measurement of frequency. One Hz equals one complete cycle per second.

A complete cycle is measured between two successive points that have the same value and direction. In Fig. 15–8, the cycle is between successive points where the waveform is zero and ready to increase in the positive direction. Or the cycle can be measured between successive peaks.

On the time scale of 1 s, waveform *a* goes through one cycle; waveform *b* has much faster variations, with four complete cycles during 1 s. Both waveforms are sine waves, even though each has a different frequency.

In comparing sine waves, the amplitude has no relation to frequency. Two waveforms can have the same frequency with different amplitudes (Fig. 15–7), the same amplitude but different frequencies (Fig. 15–8), or different amplitudes and frequencies. The amplitude indicates the amount of voltage or current, and the frequency indicates the rate of change of amplitude variations in cycles per second.

Frequency Units

The unit called the ***hertz (Hz)***, named after Heinrich Hertz, is used for cycles per second. Then 60 cps = 60 Hz. All metric prefixes can be used. As examples

$$1 \text{ kilocycle per second} = 1 \times 10^3 \text{ Hz} = 1 \text{ kHz}$$
$$1 \text{ megacycle per second} = 1 \times 10^6 \text{ Hz} = 1 \text{ MHz}$$
$$1 \text{ gigacycle per second} = 1 \times 10^9 \text{ Hz} = 1 \text{ GHz}$$

Audio and Radio Frequencies

The entire frequency range of alternating voltage or current from 1 Hz to many megahertz can be considered in two broad groups: audio frequencies (af) and radio frequencies (rf). *Audio* is a Latin word meaning "I hear." The audio range includes frequencies that can be heard as sound waves by the human ear. This range of audible frequencies is approximately 16 to 16,000 Hz.

The higher the frequency, the higher the pitch or tone of the sound. High audio frequencies, about 3000 Hz and above, provide *treble* tone. Low audio frequencies, about 300 Hz and below, provide *bass* tone.

Loudness is determined by amplitude. The greater the amplitude of the af variation, the louder its corresponding sound.

Alternating current and voltage above the audio range provide rf variations, since electrical variations of high frequency can be transmitted by electromagnetic radio waves. Examples of frequency allocations are given in Table 15–3.

Sonic and Ultrasonic Frequencies

These terms refer to sound waves, which are variations in pressure generated by mechanical vibrations, rather than electrical variations. The velocity of sound waves

Table 15–3 Examples of Common Frequencies

Frequency	Use
60 Hz	AC power line (US)
50–15,000 Hz	Audio equipment
535–1605 kHz*	AM radio broadcast band
54–60 MHz	TV channel 2
88–108 MHz	FM radio broadcast band

* Expanded to 1705 kHz in 1991.

through dry air at 20°C equals 1130 ft/s. Sound waves above the audible range of frequencies are called *ultrasonic* waves. The range of frequencies for ultrasonic applications, therefore, is from 16,000 Hz up to several megahertz. Sound waves in the audible range of frequencies below 16,000 Hz can be considered *sonic* or sound frequencies. The term *audio* is reserved for electrical variations that can be heard when converted to sound waves.

15–6 Self-Review

Answers at the end of the chapter.

a. What is the frequency of the bottom waveform in Fig. 15–8?
b. Convert 1605 kHz to megahertz.

15–7 Period

The amount of time it takes for one cycle is called the *period*. Its symbol is T for time. With a frequency of 60 Hz, as an example, the time for one cycle is ¹⁄₆₀ s. Therefore, the period is ¹⁄₆₀ s. Frequency and period are reciprocals of each other:

$$T = \frac{1}{f} \quad \text{or} \quad f = \frac{1}{T} \qquad \textbf{(15–6)}$$

The higher the frequency, the shorter the period. In Fig. 15–8*a*, the period for the wave with a frequency of 1 Hz is 1 s, and the higher frequency wave of 4 Hz in Fig. 15–8*b* has a period of ¼ s for a complete cycle.

GOOD TO KNOW

An oscilloscope can measure the period and frequency of an AC waveform.

Units of Time

The second is the basic unit of time, but for higher frequencies and shorter periods, smaller units of time are convenient. Those used most often are:

$$T = 1 \text{ millisecond} = 1 \text{ ms} = 1 \times 10^{-3} \text{ s}$$
$$T = 1 \text{ microsecond} = 1\ \mu\text{s} = 1 \times 10^{-6} \text{ s}$$
$$T = 1 \text{ nanosecond} = 1 \text{ ns} = 1 \times 10^{-9} \text{ s}$$

These units of time for a period are reciprocals of the corresponding units for frequency. The reciprocal of frequency in kilohertz gives the period T in milliseconds; the reciprocal of megahertz is microseconds; the reciprocal of gigahertz is nanoseconds.

Example 15–2

An alternating current varies through one complete cycle in ¹⁄₁₀₀₀ s. Calculate the period and frequency.

ANSWER

$$T = \frac{1}{1000} \text{ s}$$

$$f = \frac{1}{T} = \frac{1}{1/1000}$$

$$= \frac{1000}{1} = 1000$$

$$= 1000 \text{ Hz or } 1 \text{ kHz}$$

Example 15–3

Calculate the period for the two frequencies of 1 MHz and 2 MHz.

ANSWER

a. For 1 MHz,

$$T = \frac{1}{f} = \frac{1}{1 \times 10^6}$$

$$= 1 \times 10^{-6}\ \text{s} = 1\ \mu\text{s}$$

b. For 2 MHz,

$$T = \frac{1}{f} = \frac{1}{2 \times 10^6}$$

$$= 0.5 \times 10^{-6}\ \text{s} = 0.5\ \mu\text{s}$$

To do these problems on a calculator, you need the reciprocal key, usually marked (1/x). Keep the powers of 10 separate and remember that the reciprocal has the same exponent with opposite sign. With f of 2×10^6, for $1/f$ just punch in 2 and then press (2ndF) and the (1/x) key to see 0.5 as the reciprocal. The 10^6 for f becomes 10^{-6} for T so that the answer is 0.5×10^{-6} s or 0.5 μs.

15–7 Self-Review

Answers at the end of the chapter.

a. $T = 1/400$ **s. Calculate** f**.**
b. $f = 400$ **Hz. Calculate** T**.**

15–8 Wavelength

When a periodic variation is considered with respect to distance, one cycle includes the ***wavelength***, which is the length of one complete wave or cycle (Fig. 15–9). For example, when a radio wave is transmitted, variations in the electromagnetic field travel through space. Also, with sound waves, the variations in air pressure corresponding to the sound wave move through air. In these applications, the distance traveled by the wave in one cycle is the wavelength. The wavelength depends upon the frequency of the variation and its velocity of transmission:

$$\lambda = \frac{\text{velocity}}{\text{frequency}} \qquad \textbf{(15–7)}$$

where λ (the Greek letter lambda) is the symbol for one complete wavelength.

Figure 15–9 Wavelength λ is the distance traveled by the wave in one cycle.

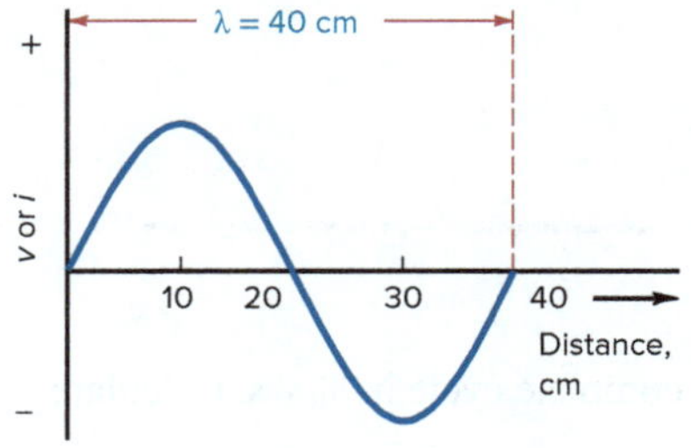

Wavelength of Radio Waves

The velocity of electromagnetic radio waves in air or vacuum is 186,000 mi/s, or 3×10^{10} cm/s, which is the speed of light. Therefore,

$$\lambda(\text{cm}) = \frac{3 \times 10^{10}\ \text{cm/s}}{f(\text{Hz})} \qquad \textbf{(15–8)}$$

Note that the higher the frequency, the shorter the wavelength. For instance, the short-wave radio broadcast band of 5.95 to 26.1 MHz includes frequencies higher than the standard AM radio broadcast band of 535 to 1605 kHz.

GOOD TO KNOW

The wavelength, λ, in meters is found using the following formula:

$$\lambda(\text{m}) = \frac{3 \times 10^8\ \text{m/s}}{f\ (\text{Hz})}$$

Example 15–4

Calculate λ for a radio wave with f of 30 GHz.

ANSWER

$$\lambda = \frac{3 \times 10^{10}\ \text{cm/s}}{30 \times 10^{9}\ \text{Hz}} = \frac{3}{30} \times 10\ \text{cm}$$
$$= 0.1 \times 10$$
$$= 1\ \text{cm}$$

Such short wavelengths are called *microwaves*. This range includes λ of 1 m or less for frequencies of 300 MHz or more.

Example 15–5

The length of a TV antenna is $\lambda/2$ for radio waves with f of 60 MHz. What is the antenna length in centimeters and feet?

ANSWER

a. $$\lambda = \frac{3 \times 10^{10}\ \text{cm/s}}{60 \times 10^{6}\ \text{Hz}} = \frac{1}{20} \times 10^{4}\ \text{cm}$$
$$= 0.05 \times 10^{4}$$
$$= 500\ \text{cm}$$

Then, $\lambda/2 = {}^{500}/_{2} = 250$ cm.

b. Since 2.54 cm = 1 in.,

$$\lambda/2 = \frac{250\ \text{cm}}{2.54\ \text{cm/in.}} = 98.4\ \text{in.}$$
$$= \frac{98.4\ \text{in}}{12\ \text{in./ft}} = 8.2\ \text{ft}$$

Example 15–6

For the 6-m band used in amateur radio, what is the corresponding frequency?

ANSWER The formula $\lambda = v/f$ can be inverted

$$f = \frac{v}{\lambda}$$

Then

$$f = \frac{3 \times 10\ \text{cm/s}}{6\ \text{m}} = \frac{3 \times 10^{10}\ \text{cm/s}}{6 \times 10^{2}\ \text{cm}}$$
$$= \frac{3}{6} \times 10^{8} = 0.5 \times 10^{8}\ \text{Hz}$$
$$= 50 \times 10^{6}\ \text{Hz} \quad \text{or} \quad 50\ \text{MHz}$$

Wavelength of Sound Waves

The velocity of sound waves is much lower than that of radio waves because sound waves result from mechanical vibrations rather than electrical variations. In average conditions, the velocity of sound waves in air equals 1130 ft/s. To calculate the wavelength, therefore,

$$\lambda = \frac{1130\ \text{ft/s}}{f\ \text{Hz}} \tag{15–9}$$

This formula can also be used for ultrasonic waves. Although their frequencies are too high to be audible, ultrasonic waves are still sound waves rather than radio waves.

Example 15–7

What is the wavelength of the sound waves produced by a loudspeaker at a frequency of 100 Hz?

ANSWER

$$\lambda = \frac{1130\ \text{ft/s}}{100\ \text{Hz}}$$

$$\lambda = 11.3\ \text{ft}$$

Example 15–8

For ultrasonic waves at a frequency of 34.44 kHz, calculate the wavelength in feet and in centimeters.

ANSWER

$$\lambda = \frac{1130}{34.44 \times 10^3}$$

$$= 32.8 \times 10^{-3}\ \text{ft}$$

$$= 0.0328\ \text{ft}$$

To convert to inches,

$$0.0328\ \text{ft} \times 12 = 0.3936\ \text{in.}$$

To convert to centimeters,

$$0.3936\ \text{in.} \times 2.54 = 1\ \text{cm} \quad \text{approximately}$$

Note that the 34.44-kHz sound waves in this example have the same wavelength (1 cm) as the 30-GHz radio waves in Example 15–4. The reason is that radio waves have a much higher velocity than sound waves.

15–8 Self-Review

Answers at the end of the chapter.

a. **The higher the frequency, the shorter the wavelength λ. (True/False)**
b. **The higher the frequency, the longer the period T. (True/False)**
c. **The velocity of propagation for radio waves in free space is 3×10^{10} cm/s. (True/False)**

15–9 Phase Angle

Referring back to Fig. 15–3, suppose that the generator started its cycle at point B, where maximum voltage output is produced, instead of starting at the point of zero output. If we compare the two cases, the two output voltage waves would be as shown in Fig. 15–10. Each is the same waveform of alternating voltage, but wave B starts at maximum, and wave A starts at zero. The complete cycle of wave B through 360° takes it back to the maximum value from which it started. Wave A starts and finishes its cycle at zero. With respect to time, therefore, wave B is ahead of wave A in values of generated voltage. The amount it leads in time equals one quarter-revolution, which is 90°. This angular difference is the **phase angle** between waves B and A. Wave B leads wave A by the phase angle of 90°.

The 90° phase angle between waves B and A is maintained throughout the complete cycle and in all successive cycles, as long as they both have the same frequency. At any instant, wave B has the value that A will have 90° later. For instance, at 180° wave A is at zero, but B is already at its negative maximum value, where wave A will be later at 270°.

To compare the phase angle between two waves, they must have the same frequency. Otherwise, the relative phase keeps changing. Also, they must have sine-wave variations because this is the only kind of waveform that is measured in angular units of time. The amplitudes can be different for the two waves, although they are shown the same here. We can compare the phases of two voltages, two currents, or a current with a voltage.

The 90° Phase Angle

The two waves in Fig. 15–10 represent a sine wave and a cosine wave 90° out of phase with each other. The 90° phase angle means that one has its maximum

GOOD TO KNOW

To compare the phase relationship between two AC sine waves, both sine waves must have exactly the same frequency.

Figure 15–10 Two sine-wave voltages 90° out of phase. (*a*) Wave B leads wave A by 90°. (*b*) Corresponding phasors V_B and V_A for the two sine-wave voltages with phase angle $\theta = 90°$. The right angle shows quadrature phase.

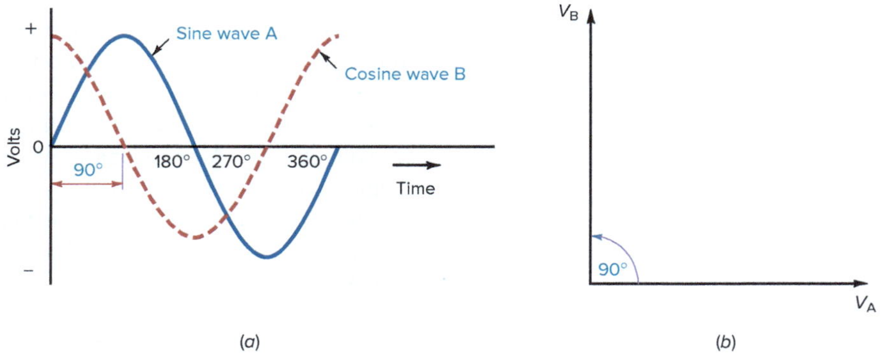

GOOD TO KNOW

The instantaneous value of a cosine-wave voltage for any angle of rotation is expressed by the formula

$$v = V_M \cos \theta.$$

amplitude when the other is at zero value. Wave A starts at zero, corresponding to the sine of 0°, has its peak amplitude at 90 and 270°, and is back to zero after one cycle of 360°. Wave B starts at its peak value, corresponding to the cosine of 0°, has its zero value at 90 and 270°, and is back to the peak value after one cycle of 360°.

However, wave B can also be considered a sine wave that starts 90° before wave A in time. This phase angle of 90° for current and voltage waveforms has many applications in sine-wave AC circuits with inductance or capacitance.

The sine and cosine waveforms have the same variations but displaced by 90°. Both waveforms are called *sinusoids*. The 90° angle is called ***quadrature phase***.

Phase-Angle Diagrams

To compare phases of alternating currents and voltages, it is much more convenient to use **phasor** diagrams corresponding to the voltage and current waveforms, as shown in Fig. 15–10*b*. The arrows here represent the phasor quantities corresponding to the generator voltage.

A phasor is a quantity that has magnitude and direction. The length of the arrow indicates the magnitude of the alternating voltage in rms, peak, or any AC value, as long as the same measure is used for all phasors. The angle of the arrow with respect to the horizontal axis indicates the phase angle.

The terms *phasor* and *vector* are used for a quantity that has direction, requiring an angle to specify the value completely. However, a vector quantity has direction in space, whereas a phasor quantity varies in time. As an example of a vector, a mechanical force can be represented by a vector arrow at a specific angle, with respect to either the horizontal or the vertical direction.

For phasor arrows, the angles shown represent differences in time. One sinusoid is chosen as the reference. Then the timing of the variations in another sinusoid can be compared to the reference by means of the angle between the phasor arrows.

The phasor corresponds to the entire cycle of voltage, but is shown only at one angle, such as the starting point, since the complete cycle is known to be a sine wave. Without the extra details of a whole cycle, phasors represent the alternating voltage or current in a compact form that is easier for comparing phase angles.

In Fig. 15–10*b*, for instance, the phasor V_A represents the voltage wave A with a phase angle of 0°. This angle can be considered the plane of the loop in the rotary generator where it starts with zero output voltage. The phasor V_B is vertical to show the phase angle of 90° for this voltage wave, corresponding to the vertical generator loop at the start of its cycle. The angle between the two phasors is the phase angle.

The symbol for a phase angle is θ (the Greek letter theta). In Fig. 15–10, as an example, $\theta = 90°$.

Phase–Angle Reference

The phase angle of one wave can be specified only with respect to another as reference. How the phasors are drawn to show the phase angle depends on which phase is chosen as the reference. Generally, the reference phasor is horizontal, corresponding to 0°. Two possibilities are shown in Fig. 15–11. In Fig. 15–11*a*, the voltage wave A or its phasor V_A is the reference. Then the phasor V_B is 90° counterclockwise. This method is standard practice, using counterclockwise rotation as the positive direction for angles. Also, a leading angle is positive. In this case, then, V_B is 90° counterclockwise from the reference V_A to show that wave B leads wave A by 90°.

However, wave B is shown as the reference in Fig. 15–11*b*. Now V_B is the horizontal phasor. To have the same phase angle, V_A must be 90° clockwise, or −90°

Figure 15–11 Leading and lagging phase angles for 90°. (*a*) When phasor V_A is the horizontal reference, phasor V_B leads by 90°. (*b*) When phasor V_B is the horizontal reference, phasor V_A lags by −90°.

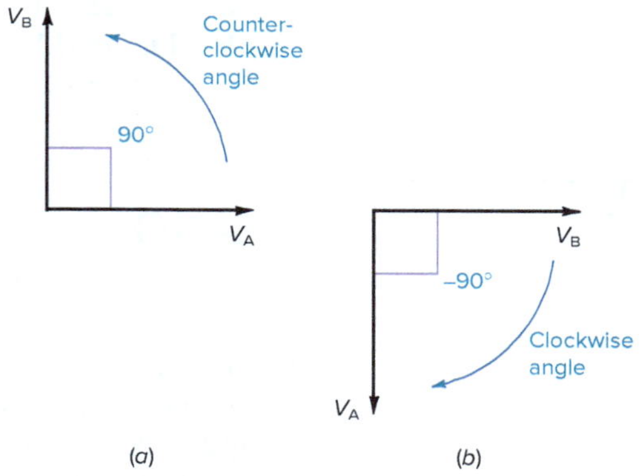

Figure 15–12 Phase angle of 60° is the time for 60/360 or 1/6 of the cycle. (*a*) Waveforms. (*b*) Phasor diagram.

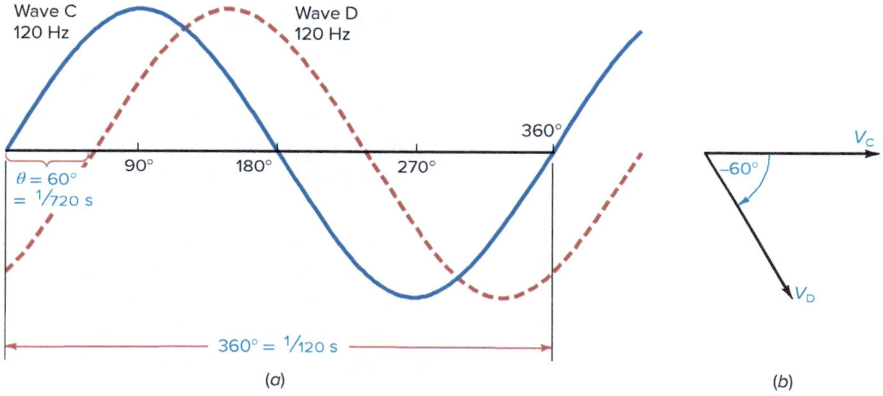

from V_B. This arrangement shows that negative angles, clockwise from the 0° reference, are used to show lagging phase angles. The reference determines whether the phase angle is considered leading or lagging in time.

The phase is not actually changed by the method of showing it. In Fig. 15–11, V_A and V_B are 90° out of phase, and V_B leads V_A by 90° in time. There is no fundamental difference whether we say V_B is ahead of V_A by +90° or V_A is behind V_B by −90°.

Two waves and their corresponding phasors can be out of phase by any angle, either less or more than 90°. For instance, a phase angle of 60° is shown in Fig. 15–12. For the waveforms in Fig. 15–12*a*, wave D is behind C by 60° in time. For the phasors in Fig. 15–12*b*, this lag is shown by the phase angle of −60°.

Figure 15–13 Two waveforms in phase, or the phase angle is 0°. (*a*) Waveforms. (*b*) Phasor diagram.

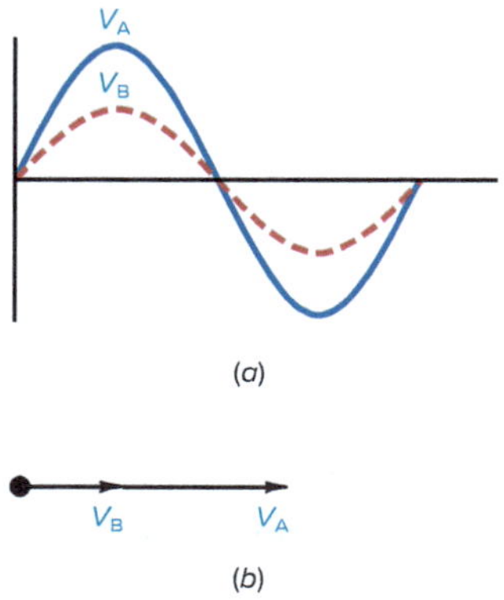

In-Phase Waveforms

A phase angle of 0° means that the two waves are in phase (Fig. 15–13).

Out-of-Phase Waveforms

An angle of 180° means opposite phase, or that the two waveforms are exactly out of phase (Fig. 15–14). Then the amplitudes are opposing.

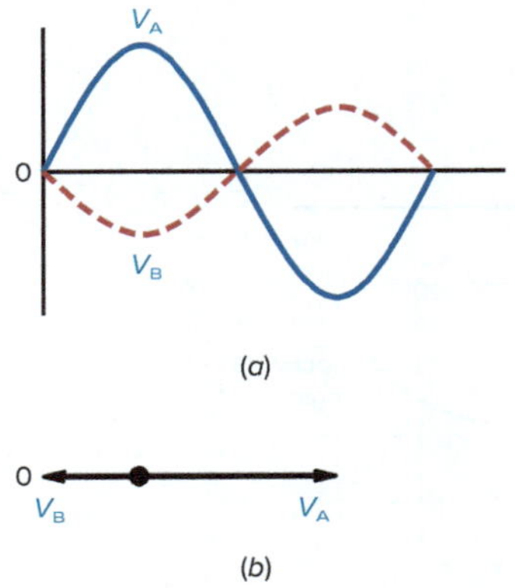

Figure 15–14 Two waveforms out of phase or in opposite phase with phase angle of 180°. (*a*) Waveforms. (*b*) Phasor diagram.

15–9 Self-Review

Answers at the end of the chapter.

Give the phase angle in

a. **Fig. 15–10.**
b. **Fig. 15–12.**
c. **Fig. 15–13.**

15–10 The Time Factor in Frequency and Phase

It is important to remember that the waveforms we are showing are graphs drawn on paper. The physical factors represented are variations in amplitude, usually on the vertical scale, with respect to equal intervals on the horizontal scale, which can represent either distance or time. To show wavelength, as in Fig. 15–9, the cycles of amplitude variations are plotted against distance or length. To show frequency, the cycles of amplitude variations are shown with respect to time in angular measure. The angle of 360° represents the time for one cycle, or the period T.

As an example of how frequency involves time, a waveform with stable frequency is actually used in electronic equipment as a clock reference for very small units of time. Assume a voltage waveform with the frequency of 10 MHz. The period T is 0.1 μs. Therefore, every cycle is repeated at 0.1-μs intervals. When each cycle of voltage variations is used to indicate time, then, the result is effectively a clock that measures 0.1-μs units. Even smaller units of time can be measured with higher frequencies. In everyday applications, an electric clock connected to the power line keeps correct time because it is controlled by the exact frequency of 60 Hz.

Furthermore, the phase angle between two waves of the same frequency indicates a specific difference in time. As an example, Fig. 15–12 shows a phase angle of 60°, with wave C leading wave D. Both have the same frequency of 120 Hz. The period T for each wave then is $\frac{1}{120}$ s. Since 60° is one-sixth of the complete cycle of 360°, this phase angle represents one-sixth of the complete period of $\frac{1}{120}$ s. If we multiply $\frac{1}{6} \times \frac{1}{120}$, the answer is $\frac{1}{720}$ s for the time corresponding to the phase angle of 60°. If we consider wave D lagging wave C by 60°, this lag is a time delay of $\frac{1}{720}$ s.

More generally, the time for a phase angle θ can be calculated as

$$t = \frac{\theta}{360} \times \frac{1}{f} \qquad \textbf{(15–10)}$$

where f is in Hz, θ is in degrees, and t is in seconds.

The formula gives the time of the phase angle as its proportional part of the total period of one cycle. For the example of θ equal to 60° with f at 120 Hz,

$$t = \frac{\theta}{360} \times \frac{1}{f}$$
$$= \frac{60}{360} \times \frac{1}{120} = \frac{1}{6} \times \frac{1}{120}$$
$$= \frac{1}{720} \text{ s}$$

GOOD TO KNOW

When t and f are known, θ can be calculated as $\theta = \frac{t}{T} \times 360°$ where $T = \frac{1}{f}$.

15–10 Self-Review

Answers at the end of the chapter.

a. **In Fig. 15–12, how much time corresponds to 180°?**
b. **For two waves with a frequency of 1 MHz, how much time is the phase angle of 36°?**

Figure 15–15 An AC circuit with resistance *R* alone. (*a*) Schematic diagram. (*b*) Waveforms.

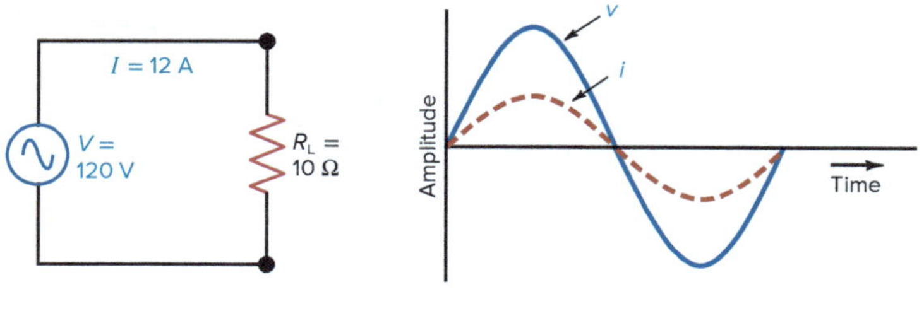

15–11 Alternating Current Circuits with Resistance

An AC circuit has an AC voltage source. Note the symbol in Fig. 15–15 used for any source of sine-wave alternating voltage. This voltage connected across an external load resistance produces alternating current of the same waveform, frequency, and phase as the applied voltage.

The amount of current equals V/R by Ohm's law. When V is an rms value, I is also an rms value. For any instantaneous value of V during the cycle, the value of I is for the corresponding instant.

In an AC circuit with only resistance, the current variations are in phase with the applied voltage, as shown in Fig. 15–15*b*. This in-phase relationship between V and I means that such an AC circuit can be analyzed by the same methods used for DC circuits, since there is no phase angle to consider. Circuit components that have R alone include resistors, the filaments of lightbulbs, and heating elements.

The calculations in AC circuits are generally in rms values, unless noted otherwise. In Fig. 15–15*a*, for example, the 120 V applied across the 10-Ω R_L produces rms current of 12 A. The calculations are

$$I = \frac{V}{R_L} = \frac{120\ \text{V}}{10\ \Omega} = 12\ \text{A}$$

Furthermore, the rms power dissipation is I^2R, or

$$P = 144 \times 10 = 1440\ \text{W}$$

GOOD TO KNOW

In AC circuits, power is always calculated using rms values of voltage and current.

Series AC Circuit with *R*

In Fig. 15–16, R_T is 30 Ω, equal to the sum of 10 Ω for R_1 plus 20 Ω for R_2. The current in the series circuit is

$$I = \frac{V_T}{R_T} = \frac{120\ \text{V}}{30\ \Omega} = 4\ \text{A}$$

The 4-A current is the same in all parts of the series circuit. This principle applies for either an AC or a DC source.

Next, we can calculate the series voltage drops in Fig. 15–16. With 4 A through the 10-Ω R_1, its IR voltage drop is

$$V_1 = I \times R_1 = 4\ \text{A} \times 10\ \Omega = 40\ \text{V}$$

The same 4 A through the 20-Ω R_2 produces an IR voltage drop of 80 V. The calculations are

$$V_2 = I \times R_2 = 4\ \text{A} \times 20\ \Omega = 80\ \text{V}$$

Note that the sum of 40 V for V_1 and 80 V for V_2 in series equals the 120 V applied.

Figure 15–16 Series AC circuit with resistance only.

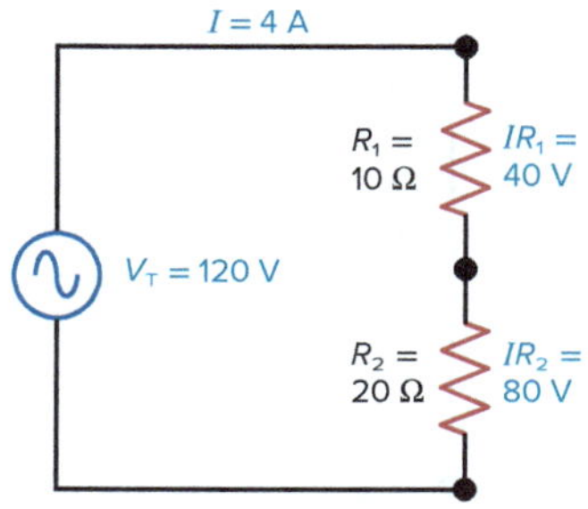

Figure 15–17 Parallel AC circuit with resistance only.

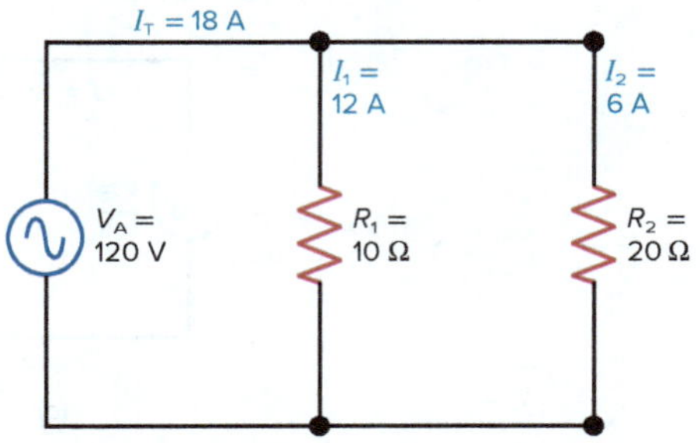

Parallel AC Circuit with *R*

In Fig. 15–17, the 10-Ω R_1 and 20-Ω R_2 are in parallel across the 120-V_{AC} source. Therefore, the voltage across the parallel branches is the same as the applied voltage.

Each branch current, then, is equal to 120 V divided by the branch resistance. The branch current for the 10-Ω R_1 is

$$I_1 = \frac{120\text{ V}}{10\ \Omega} = 12\text{ A}$$

The same 120 V is across the 20-Ω branch with R_2. Its branch current is

$$I_2 = \frac{120\text{ V}}{20\ \Omega} = 6\text{ A}$$

The total line current I_T is 12 + 6 = 18 A, or the sum of the branch currents.

Series–Parallel AC Circuit with *R*

See Fig. 15–18. The 20-Ω R_2 and 20-Ω R_3 are in parallel, for an equivalent bank resistance of $^{20}\!/_2$ or 10 Ω. This 10-Ω bank is in series with the 20-Ω R_1 in the main line and totals 30 Ω for R_T across the 120-V source. Therefore, the main line current produced by the 120-V source is

$$I_T = \frac{V_T}{R_T} = \frac{120\text{ V}}{30\ \Omega} = 4\text{ A}$$

The voltage drop across R_1 in the main line is calculated as

$$V_1 = I_T \times R_1 = 4\text{ A} \times 20\ \Omega = 80\text{ V}$$

Figure 15–18 Series-parallel AC circuit with resistance only.

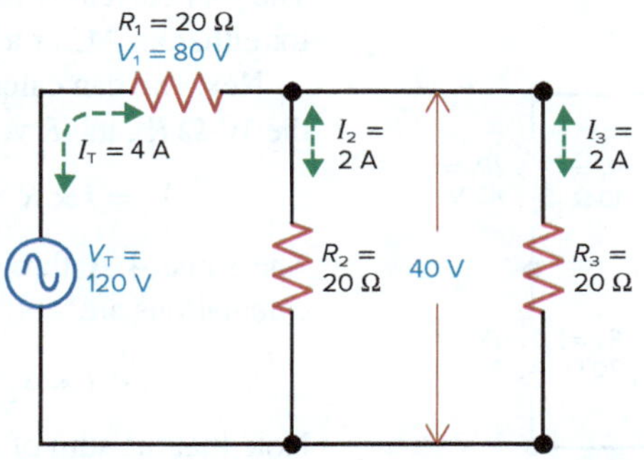

Subtracting this 80-V drop from the 120 V of the source, the remaining 40 V is across the bank of R_2 and R_3 in parallel. Since the branch resistances are equal, the 4-A I_T divides equally, 2 A in R_2 and 2 A in R_3. The branch currents can be calculated as

$$I_2 = \frac{40\ \text{V}}{20\ \Omega} = 2\ \text{A}$$

$$I_3 = \frac{40\ \text{V}}{20\ \Omega} = 2\ \text{A}$$

Note that the 2 A for I_2 and 2 A for I_3 in parallel branches add to equal the 4-A current in the main line.

15–11 Self-Review

Answers at the end of the chapter.

Calculate R_T in

a. **Fig. 15–16.**
b. **Fig. 15–17.**
c. **Fig. 15–18.**

15–12 Nonsinusoidal AC Waveforms

GOOD TO KNOW

A function generator is capable of generating both sinusoidal and nonsinusoidal type waveforms. The waveforms typically produced include sine waves, square waves, sawtooth waves, and triangular waves. The function generator has controls which allow you to adjust the frequency and amplitude of each type of waveform.

The sine wave is the basic waveform for AC variations for several reasons. This waveform is produced by a rotary generator; the output is proportional to the angle of rotation. In addition, electronic oscillator circuits with inductance and capacitance naturally produce sine-wave variations.

Because of its derivation from circular motion, any sine wave can be analyzed in terms of angular measure, either in degrees from 0 to 360° or in radians from 0 to 2π rad.

Another feature of a sine wave is its basic simplicity; the rate of change of the amplitude variations corresponds to a cosine wave that is similar but 90° out of phase. The sine wave is the only waveform that has this characteristic of a rate of change with the same waveform as the original changes in amplitude.

In many electronic applications, however, other waveshapes are important. Any waveform that is not a sine or cosine wave is a ***nonsinusoidal waveform***. Common examples are the square wave and sawtooth wave in Fig. 15–19.

For either voltage or current nonsinusoidal waveforms, there are important differences and similarities to consider. Note the following comparisons with sine waves.

1. In all cases, the cycle is measured between two points having the same amplitude and varying in the same direction. The period is the time for one cycle. In Fig. 15–19, T for any of the waveforms is 4 μs and the corresponding frequency is $1/T$, equal to 0.25 MHz.
2. Peak amplitude is measured from the zero axis to the maximum positive or negative value. However, peak-to-peak amplitude is better for measuring nonsinusoidal waveshapes because they can have unsymmetrical peaks, as in Fig. 15–19*d*. For all waveforms shown here, though, the peak-to-peak (p-p) amplitude is 20 V.
3. The rms value 0.707 of maximum applies only to sine waves because this factor is derived from the sine values in the angular measure used only for the sine waveform.
4. Phase angles apply only to sine waves because angular measure is used only for sine waves. Note that the horizontal axis for time is divided into angles for the sine wave in Fig. 15–19*a*, but there are no angles shown for the nonsinusoidal waveshapes.
5. All waveforms represent AC voltages. Positive values are shown above the zero axis, and negative values below the axis.

GOOD TO KNOW

Square wave and pulse-type wave forms are common in digital electronic circuitry. However, the waveforms encountered in these types of circuits usually vary between 0 V and some positive or negative value. In other words, the waveforms encountered in digital circuits are not true AC waveforms because they contain either positive or negative values but not both.

Figure 15–19 Comparison of sine wave with nonsinusoidal waveforms. Two cycles shown. (*a*) Sine wave. (*b*) Sawtooth wave. (*c*) Symmetrical square wave. (*d*) Unsymmetrical rectangular wave or pulse waveform.

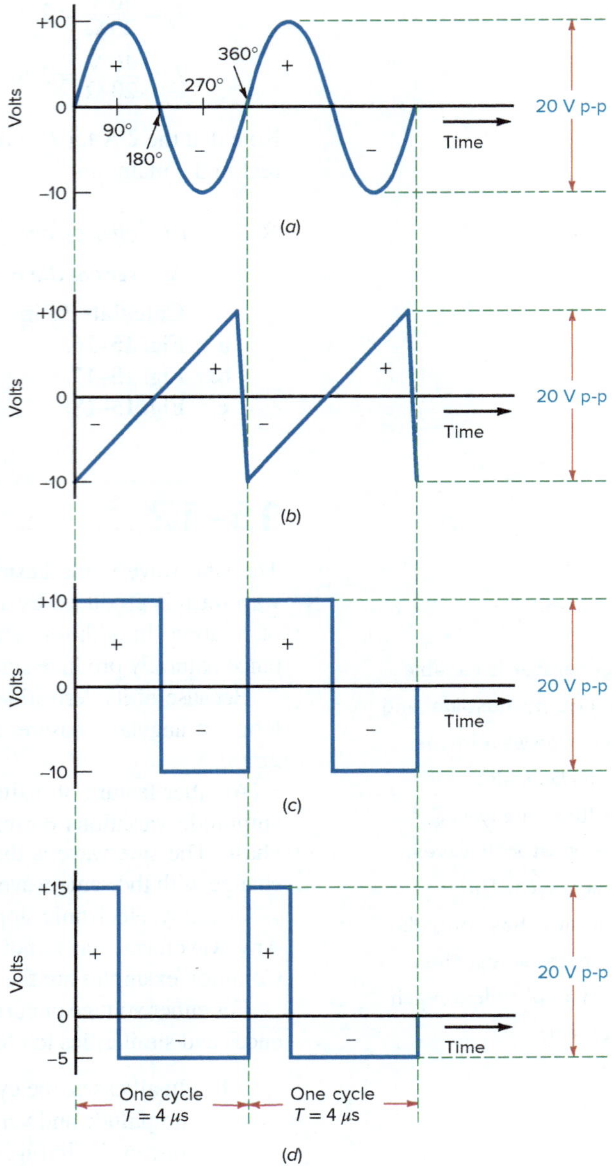

The sawtooth wave in Fig. 15–19*b* represents a voltage that slowly increases to its peak value with a uniform or linear rate of change and then drops sharply to its starting value. This waveform is also called a *ramp voltage*. It is also often referred to as a *time base* because of its constant rate of change.

Note that one complete cycle includes a slow rise and a fast drop in voltage. In this example, the period T for a complete cycle is 4 μs. Therefore, these sawtooth cycles are repeated at the frequency of 0.25 MHz. The sawtooth waveform of voltage or current is often used for horizontal deflection of the electron beam in the cathode-ray tube (CRT) for oscilloscopes and TV receivers.

The square wave in Fig. 15–19*c* represents a switching voltage. First, the 10-V peak is instantaneously applied in positive polarity. This voltage remains on for 2 μs, which is one half-cycle. Then the voltage is instantaneously switched to −10 V for another 2 μs. The complete cycle then takes 4 μs, and the frequency is 0.25 MHz.

The rectangular waveshape in Fig. 15–19*d* is similar, but the positive and negative half-cycles are not symmetrical either in amplitude or in time. However, the frequency is the same 0.25 MHz and the peak-to-peak amplitude is the same 20 V, as in all the waveshapes. This waveform shows pulses of voltage or current, repeated at a regular rate.

15–12 Self-Review

Answers at the end of the chapter.

a. In Fig. 15–19*c*, for how much time is the waveform at +10 V?
b. In Fig. 15–19*d*, what voltage is the positive peak amplitude?

15–13 Harmonic Frequencies

Consider a repetitive nonsinusoidal waveform, such as a 100-Hz square wave. Its fundamental rate of repetition is 100 Hz. Exact multiples of the fundamental frequency are called *harmonic frequencies*. The second harmonic is 200 Hz, the third harmonic is 300 Hz, etc. Even multiples are even harmonics, and odd multiples are odd harmonics.

Harmonics are useful in analyzing distorted sine waves or nonsinusoidal waveforms. Such waveforms consist of a pure sine wave at the fundamental frequency plus **harmonic frequency** components. For example, Fig. 15–20 illustrates how a square wave corresponds to a fundamental sine wave with odd harmonics. Typical audio waveforms include odd and even harmonics. The harmonic components make one source of sound different from another with the same fundamental frequency.

Figure 15–20 Fundamental and harmonic frequencies for an example of a 100-Hz square wave.

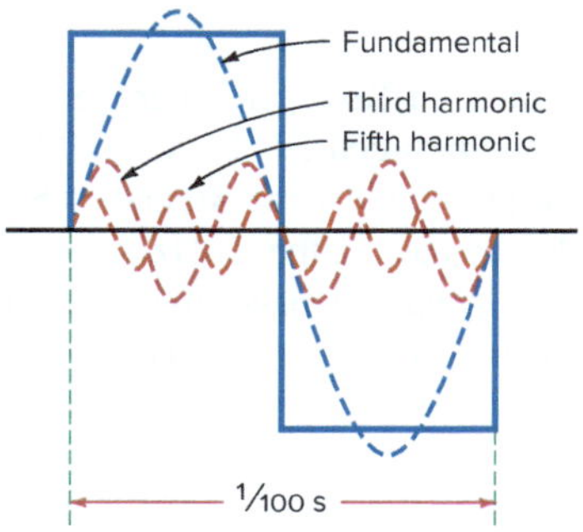

A common unit for frequency multiples is the ***octave***, which is a range of 2:1. Doubling the frequency range—from 100 to 200 Hz, from 200 to 400 Hz, and from 400 to 800 Hz, as examples—raises the frequency by one octave. The reason for this name is that an octave in music includes eight consecutive tones, for double the frequency. One-half the frequency is an octave lower.

Another unit for representing frequency multiples is the **decade**. A decade corresponds to a 10:1 range in frequencies such as 100 Hz to 1 kHz and 30 kHz to 300 kHz.

15–13 Self-Review

Answers at the end of the chapter.

a. What frequency is the fourth harmonic of 12 MHz?
b. Give the frequency one octave above 220 Hz.

15–14 The 60-Hz AC Power Line

Practically all homes in the United States are supplied alternating voltage between 115 and 125 V rms at a frequency of 60 Hz. This is a sine-wave voltage produced by a rotary generator. The electricity is distributed by high-voltage power lines from a generating station and reduced to the lower voltages used in the home. Here the incoming voltage is wired to all wall outlets and electrical equipment in parallel. The 120-V source of commercial electricity is the 60-Hz *power line* or the *mains*, indicating that it is the main line for all parallel branches.

Advantages

The incoming electric service to residences is normally given as 120 V rms. With an rms value of 120 V, the AC power is equivalent to 120-V_{DC} power in heating effect. If the value were higher, there would be more danger of a fatal electric shock. Lower voltages would be less efficient in supplying power.

Higher voltage can supply electric power with less I^2R loss, since the same power is produced with less I. Note that the I^2R power loss increases as the square of the current. For applications where large amounts of power are used, such as central air-conditioners and clothes dryers, a line voltage of 240 V is often used.

The advantage of AC over DC power is greater efficiency in distribution from the generating station. Alternating voltages can easily be stepped up by a transformer with very little loss, but a transformer cannot operate on direct current because it needs the varying magnetic field produced by an AC voltage.

With a transformer, the alternating voltage at the generating station can be stepped up to values as high as 500 kV for high-voltage distribution lines. These high-voltage lines supply large amounts of power with much less current and less I^2R loss, compared with a 120-V line. In the home, the lower voltage required is supplied by a step-down transformer. The step-up and step-down characteristics of a transformer refer to the ratio of voltages across the input and output connections.

The 60-Hz frequency is convenient for commercial AC power. Much lower frequencies would require much bigger transformers because larger windings would be necessary. Also, too low a frequency for alternating current in a lamp could cause the light to flicker. For the opposite case, too high a frequency results in excessive iron-core heating in the transformer because of eddy currents and hysteresis losses. Based on these factors, 60 Hz is the frequency of the AC power line in the United States. However, the frequency of the AC power mains in England and most European countries is 50 Hz.

The 60-Hz Frequency Reference

All power companies in the United States, except those in Texas, are interconnected in a grid that maintains the AC power-line frequency between 59.98 and 60.02 Hz. The frequency is compared with the time standard provided by the Bureau of Standards radio station WWV at Fort Collins, Colorado. As a result, the 60-Hz power-line frequency is maintained accurately to $\pm 0.033\%$. This accuracy makes the power-line voltage a good secondary standard for checking frequencies based on 60 Hz.

Residential Wiring

At the electrical service entrance (where power enters a house), most homes have the three-wire power lines illustrated in Fig. 15–21. The three wires, including the grounded neutral, can be used for either 240 or 120 V single phase. The 240 V at the residence is stepped down from the high-voltage distribution lines.

Note the color coding for the wiring in Fig. 15–21. The grounded neutral is white. Each high side can use any color except white or green, but usually black* or red is used. White is reserved for the neutral wire, and green or bare wire is reserved for grounding.

Figure 15–21 Three-wire, single-phase power lines that can provide either 240 or 120 V.

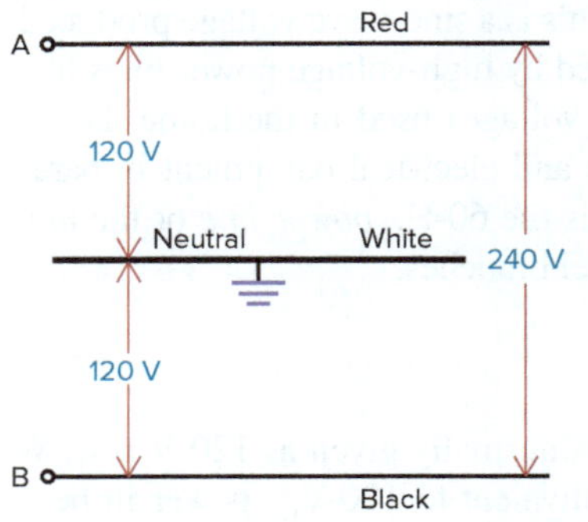

From either the red or black high side to the neutral, 120 V is available for separate branch circuits to the lights and outlets. Across the red and black wires, 240 V is available for high-power appliances. This three-wire service with a grounded neutral is called the *Edison system*.

The electrical service is commonly rated for 100 A. At 240 V, then, the power available is $100 \times 240 = 24{,}000$ W, or 24 kW.

The main wires to the service entrance are generally No. 2 gage or larger such as 1, 0, or 00. (Sizes 6 and heavier are always stranded wire.) The 120-V branch circuits, usually rated at 15 A or 20 A, use No. 12 or 14 gage wire. Each branch has its own fuse or circuit breaker. A main switch is usually included to cut off all power from the service entrance.

* Note that in electronic equipment, black is the color-coded wiring used for chassis-ground returns. However, in electric power work, black wire is used for high-side connections.

The neutral wire is grounded at the service entrance to a water pipe or a metal rod driven into the earth, which is *ground*. All 120-V branches must have one side connected to the grounded neutral. White wire is used for these connections. In addition, all metal boxes for outlets, switches, and lights must have a continuous ground to each other and to the neutral. The wire cable usually has a bare wire for grounding boxes.

Cables commonly used are armored sheath with the trade name BX and nonmetallic flexible cable with the trade name Romex. Each has two or more wires for the neutral, high-side connections, and grounding. Both cables contain an extra bare wire for grounding. Rules and regulations for residential wiring are governed by local electrical codes. These are usually based on the National Electrical Code (NEC) published by the National Fire Protection Association.

Grounding

In AC power distribution systems, grounding is the practice of connecting one side of the power line to earth or ground. The purpose is safety in two ways. First is protection against dangerous electric shock. Also, the power distribution lines are protected against excessively high voltage, particularly from lightning. If the system is struck by lightning, excessive current in the grounding system will energize a cutout device to deenergize the lines.

The grounding in the power distribution system means that it is especially important to have grounding for the electric wiring at the residence. For instance, suppose that an electric appliance such as a clothes dryer does not have its metal case grounded. An accidental short circuit in the equipment can connect the metal frame to the "hot" side of the AC power line. Then the frame has voltage with respect to earth ground. If somebody touches the frame and has a return to ground, the result is a dangerous electric shock. With the case grounded, however, the accidental short circuit blows the fuse or circuit breaker to cut off the power.

In normal operation, the electric circuits function the same way with or without the ground, but grounding is an important safety precaution. Figure 15–22 shows two types of plug connectors for the AC power line that help provide protection because they are polarized with respect to the ground connections. Although an AC voltage does not have any fixed polarity, the plugs ensure grounding of the chassis or frame of the equipment connected to the power line. In Fig. 15–22*a*, the plug has two blades for the 120-V line, but the wider blade will fit only the side of the outlet that is connected to the neutral wire. This wiring is standard practice. For the

GOOD TO KNOW

In residential house wiring, the bare or green ground wire and white neutral wire are grounded at the same point in the main entrance panel. Under normal circumstances, however, the current in the neutral wire is the same as the current in the black or red hot wire whereas the current in the ground wire is zero.

Figure 15–22 Plug connectors polarized for ground connection to an AC power line. (*a*) Wider blade connects to neutral. (*b*) Rounded pin connects to ground.

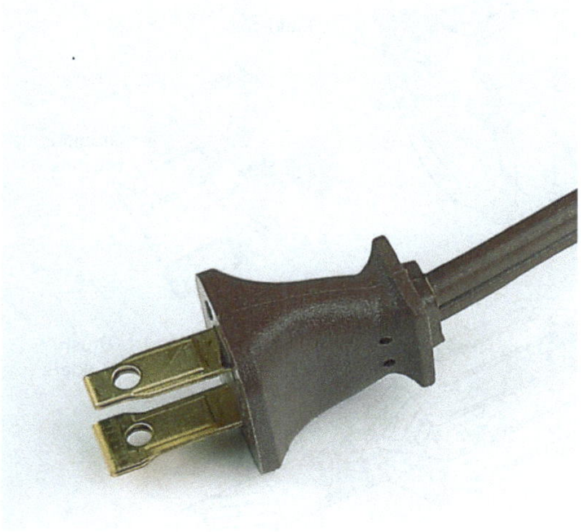

Sarah Schultz Photography

(*a*)

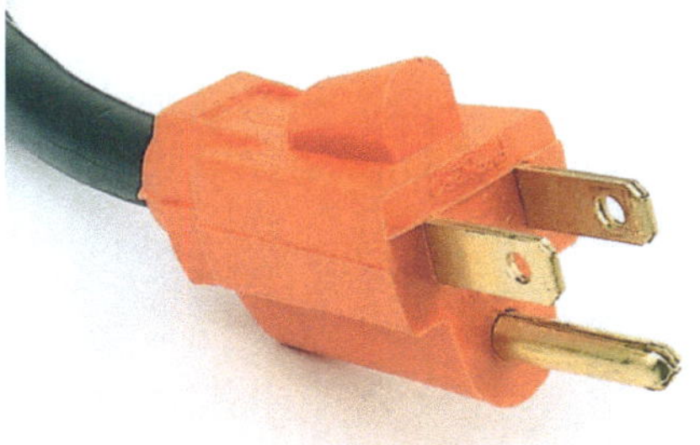

Sarah Schultz Photography

(*b*)

three-prong plug in Fig. 15–22*b*, the rounded pin is for a separate grounding wire, usually color-coded green.

It may be of interest to note that with high-fidelity audio equipment, the lack of proper grounding can cause a hum heard in the sound. The hum is usually not a safety problem, but it still is undesirable.

15–14 Self-Review

Answers at the end of the chapter.

a. The 120 V of the AC power line is a peak-to-peak value. (True/False)

b. The frequency of the AC power-line voltage is 60 Hz ± 0.033%. (True/False)

c. In Fig. 15–21, the voltage between the black and white wires is 120 V. (True/False)

d. The color code for grounding wires is green. (True/False)

15–15 Motors and Generators

A generator converts mechanical energy into electric energy; a **motor** does the opposite, converting electricity into rotary motion. The main parts in the assembly of motors and generators are essentially the same (Fig. 15–23).

Armature

In a generator, the armature connects to the external circuit to provide the generator output voltage. In a motor, it connects to the electrical source that drives the motor. The armature is often constructed in the form of a drum, using many conductor loops for increased output. In Fig. 15–23, the rotating armature is the *rotor* part of the assembly.

Field Winding

This electromagnet provides the flux cut by the rotor. In a motor, current for the field is produced by the same source that supplies the armature. In a generator,

Figure 15–23 Main parts of a DC motor.

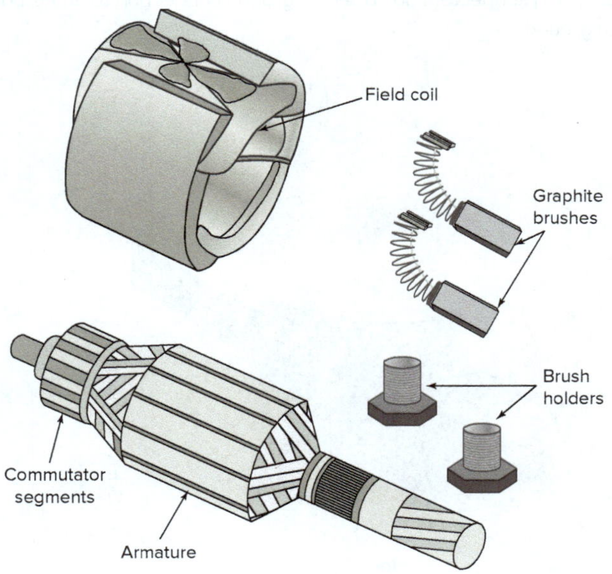

the field current may be obtained from a separate exciter source, or from its own armature output. Residual magnetism in the iron yoke of the field allows this *self-excited generator* to start.

The field coil may be connected in series with the armature, in parallel, or in a series-parallel *compound winding*. When the field winding is stationary, it is the *stator* part of the assembly.

Slip Rings

In an AC machine, two or more slip rings or *collector rings* connect the rotating loop to the stationary wire leads for the external circuit.

Brushes

These graphite connectors are spring-mounted to brush against the spinning rings on the rotor. The stationary external leads are connected to the brushes for connection to the rotating loop. Constant rubbing slowly wears down the brushes, and they must be replaced after they are worn.

Commutator

A DC machine has a commutator ring instead of slip rings. As shown in Fig. 15–23, the commutator ring has segments, one pair for each loop in the armature. Each of the commutator segments is insulated from the others by mica.

The commutator converts the AC machine to DC operation. In a generator, the commutator segments reverse the loop connections to the brushes every half-cycle to maintain a constant polarity of output voltage. For a DC motor, the commutator segments allow the DC source to produce torque in one direction.

Brushes are necessary with a commutator ring. The two stationary brushes contact opposite segments on the rotating commutator. Graphite brushes are used for very low resistance.

Alternating Current Induction Motor

This type, for alternating current only, does not have any brushes. The stator is connected directly to the AC source. Then alternating current in the stator winding induces current in the rotor without any physical connection between them. The magnetic field of the current induced in the rotor reacts with the stator field to produce rotation. Alternating current induction motors are economical and rugged without any troublesome brush arcing.

With a single-phase source, however, a starting torque must be provided for an AC induction motor. One method uses a starting capacitor in series with a separate starting coil. The capacitor supplies an out-of-phase current just for starting and then is switched out. Another method of starting uses shaded poles. A solid copper ring on the main field pole makes the magnetic field unsymmetrical to allow starting.

The rotor of an AC induction motor may be wire-wound or the squirrel-cage type. This rotor is constructed with a frame of metal bars.

Universal Motor

This type operates on either alternating or direct current because the field and armature are in series. Its construction is like that of a DC motor with the rotating armature connected to a commutator and brushes. The universal motor is commonly used for small machines such as portable drills and food mixers.

Alternators

Alternating current generators are alternators. For large power requirements, the alternator usually has a rotating field, and the armature is the stator.

15–15 Self-Review

Answers at the end of the chapter.

a. In Fig. 15–23, the commutator segments are on the armature. (True/False)

b. Motor brushes are made of graphite because of its very low resistance. (True/False)

c. A starting capacitor is used with DC motors that have small brushes. (True/False)

15–16 Three-Phase AC Power

In an alternator with three generator windings equally spaced around the circle, the windings produce output voltages 120° out of phase with each other. The three-phase output is illustrated by the sine-wave voltages in Fig. 15–24*a* and the corresponding phasors in Fig. 15–24*b*. The advantage of three-phase AC voltage is more efficient distribution of power. Also, AC induction motors are self-starting with three-phase alternating current. Finally, the AC ripple is easier to filter in the rectified output of a DC power supply.

In Fig. 15–25*a*, the three windings are in the form of a Y, also called *wye* or *star* connections. All three coils are joined at one end, and the opposite ends are for the output terminals A, B, and C. Note that any pair of terminals is across two coils in series. Each coil has 120 V. The voltage output across any two output terminals is $120 \times 1.73 = 208$ V, because of the 120° phase angle.

In Fig. 15–25*b*, the three windings are connected in the form of a *delta* (Δ). Any pair of terminals is across one generator winding. The output then is 120 V. However, the other coils are in a parallel branch. Therefore, the current capacity of the line is increased by the factor 1.73.

Figure 15–24 Three-phase alternating voltage or current with 120° between each phase. (*a*) Sine waves. (*b*) Phasor diagram.

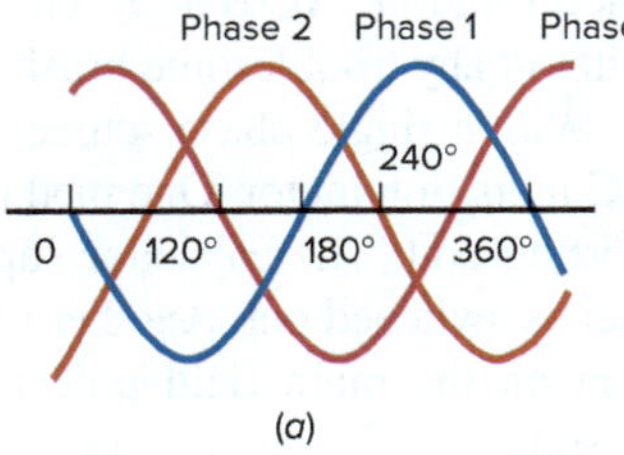

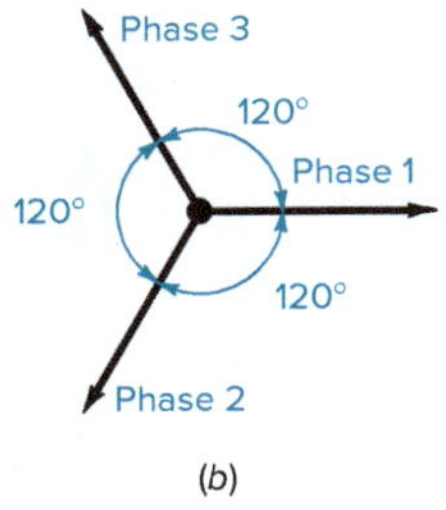

GOOD TO KNOW

Electrical utility companies generate and transmit their electrical power using the three-phase AC system rather than the single-phase AC system. The reason for this is that, by adding a third current-carrying conductor, the system's power handling capability is increased by approximately 58% over that of a single-phase system.

Figure 15–25 Types of connections for three-phase power. (*a*) Wye or Y. (*b*) Delta or Δ.

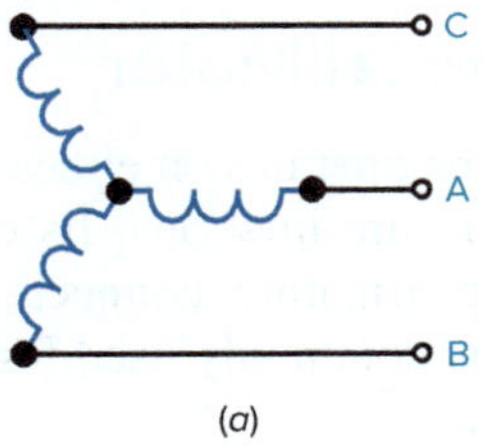

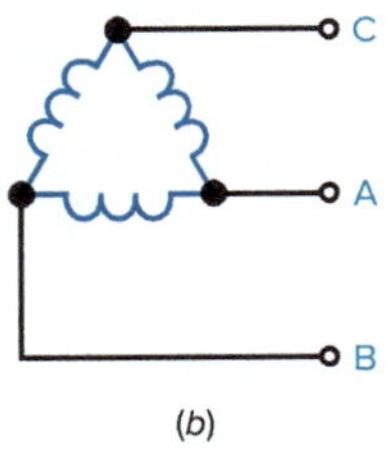

Figure 15–26 Y connections to a four-wire line with neutral.

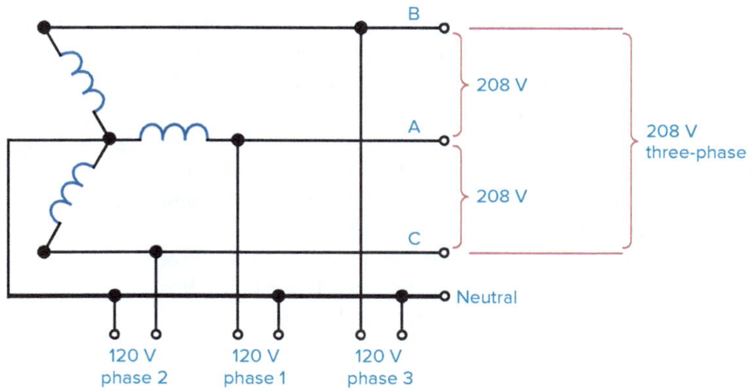

In Fig. 15–26, the center point of the Y is used for a fourth line, as the neutral wire in the three-phase power distribution system. This way, power is available at either 208 V three phase or 120 V single phase. Note that the three-phase voltage is 208 V, not the 240 V in the Edison single-phase system. From terminals A, B, or C to the neutral line in Fig. 15–26, the output is 120 V across one coil. This 120-V single-phase power is used in conventional lighting circuits. However, across terminals AB, BC, or CA, without the neutral, the output is 208 V for three-phase induction motors or other circuits that need three-phase power. Although illustrated here for the 120-V, 60-Hz power line, note that three-phase connections are commonly used for higher voltages.

More details of three-phase AC circuits are covered in Chapter 27, "Three-Phase AC Power Systems."

15–16 Self-Review

Answers at the end of the chapter.

a. What is the angle between three-phase voltages?

b. For the Y in Fig. 15–25*a*, how much is V_{AC} or V_{AB}?

Application in Understanding the 120-V Duplex Receptacle

In our homes, schools, businesses, and industries, most appliances and electrical devices are plugged into an electrical outlet that is hard-wired to the 120 V/60 Hz AC power line. By strict definition, the electrical outlet is called a duplex receptacle because there are two female receptacles in one enclosure. The duplex receptacle is wired directly to the 120-V_{AC} power line using the NM-B cable described in Chapter 11, ***Conductors and Insulators***. A standard 15-A duplex receptacle is shown in Fig. 15–27. The wide slot on the left-hand side is the neutral terminal, and the narrow slot on the right-hand side is the hot terminal. The U-shaped terminal directly below the wide and narrow slots is ground. On a 120-V duplex receptacle, the black hot wires are connected to the brass-colored screws on the same side of the receptacle as the narrow slots. The white neutral wires are connected to the silver-colored screws next to the wide slots. The bare copper ground wire (or green wire) is connected to the green hexagon screw at the bottom of the receptacle. It is important to note that the green hexagon screw is connected to all non-current-carrying metal parts on the receptacle. This is important because the ***National Electrical Code (NEC)*** specifies that all non-current-carrying metal parts on a 120-V duplex receptacle must be grounded.

VOLTAGE MEASUREMENTS

If the receptacle in Fig. 15–27 is wired correctly, a DMM should measure the following voltages:

Hot to Neutral—120 V_{AC}
Hot to Ground—120 V_{AC}
Neutral to Ground—0 V_{AC}

When measuring the voltages in a duplex receptacle, it is important to remember that the bare or green ground wire and the white neutral wire are both grounded to the same point in the ***service-entrance panel***.

RECEPTACLE TYPES

The 15-A duplex receptacle in Fig. 15–27 is usually wired with No. 14 gauge copper wire, and all circuits and wiring connected to this branch are protected by a 15-A circuit breaker or fuse. A 20-A duplex receptacle is shown in Fig. 15–28. Notice the horizontal slot branching off of the vertical neutral slot. Appliances, such as microwaves, often have a special 20-A plug that can only be plugged into a 20-A receptacle. All 20-A receptacles are wired with either No. 12 or No. 10 gauge copper wire and are protected by a 20-A circuit breaker or fuse. The larger diameter wire ensures that the wires will not overheat if a 20-A load is plugged into the receptacle. It is important to note that a standard 15-A plug will fit properly into a 20-A duplex receptacle but a 20-A plug will not fit into a standard 15-A receptacle.

Another type of receptacle specifically designed to help protect children from an electrical shock hazard is the ***tamper-resistant (TR)*** receptacle. These receptacles have spring-loaded shutters that close off the contact openings of the wide and narrow slots. When a proper plug is inserted into the receptacle, both springs are compressed and the shutters open, allowing the metal prongs of the plug to make contact with the hot, neutral, and ground terminals. Both springs must be compressed at the same time, otherwise the shutters do not open. Therefore, a child cannot insert an object into just one slot of the receptacle. A 15-A tamper-resistant receptacle looks identical to the one shown in Fig. 15–27. You have to look closely into the slots to see the shutters that block contact to the receptacle's metal terminals. Tamper-resistant receptacles are now required in all new and renovated dwellings and are making homes a safer place for children. It is important to note that the letters (TR) will always be stamped on a tamper-resistant receptacle.

Figure 15–27 Standard 15A duplex receptacle.

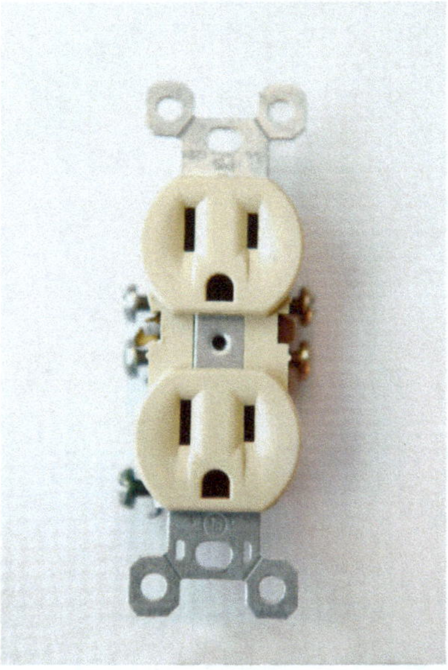

Sarah Schultz Photography

Figure 15–28 20 A duplex receptacle.

Sarah Schultz Photography

GROUND-FAULT CIRCUIT INTERRUPTER (GFCI)

Another type of receptacle, shown in Fig. 15–29, is the ***ground-fault circuit interrupter*** **(GFCI)**. According to the *NEC*, a "ground fault" is a conducting path, usually unintentional or accidental, between a hot or live electrical conductor and ground. If the conducting path is through a person's body, there is the potential for a severe electrical shock, which could result in death by electrocution. A current of 5 mA is considered the maximum current a person can be exposed to for a short period of time without risking serious injury. In general, however, a current of 5 mA, or more, through a person's body is considered dangerous. A current as low as 10 mA can cause involuntary muscle contractions that could prevent a person from letting go of a live conductor. GFCIs are specifically designed to prevent electrocution in the event that a person is the conducting path between a live conductor and ground. A GFCI is designed to ***trip*** or ***open*** if there is a current imbalance of 5 mA, or more, between the hot and neutral conductors. An imbalance in current between the hot and neutral conductors is most likely the result of a ground fault. When a GFCI trips, power is removed from each of its two receptacles. The response time of a GFCI, which is the length of time it takes to trip or open, is about 25 ms which is 1/40th of a second.

Figure 15–29 Ground-Fault Circuit Interrupter (GFCI).

Sarah Schultz Photography

The *NEC* requires the use of GFCIs in or around places such as bathrooms, unfinished basements, exterior outlets, spas, hot tubs, and swimming pools, to name a few of the most common locations. In general, GFCIs are used whenever an outlet is placed in or near an area where there is exposure to water or moisture. For additional safety, tamper resistant GFCIs are also available.

TEST AND RESET BUTTONS

Every GFCI has a ***TEST*** button and a ***RESET*** button. See Fig. 15–29. To make sure the GFCI is working properly, press the TEST button. If it is, the RESET button will pop out and power will be removed from the outlet. Push the RESET button back in to reset the GFCI. GFCIs should be tested at least once a month to make sure they are in proper working order.

It is important to note that pressing the TEST button simulates a ground fault. In fact, pressing the TEST button produces a leakage current of 5 mA between the hot conductor and ground. This simulates the current through a person's body should they be the reason for the ground fault.

Many of the newer GFCIs now include an LED, which provides a visual indication of its status. If the LED is green, the GFCI is in proper working order and you're good to go. If the RESET button is popped out and the LED is red (solid or blinking), the GFCI needs attention. It may only need to be reset. However, if the GFCI cannot be reset or if the LED remains red after it has been reset, the GFCI has failed and must be replaced. If the LED is not illuminated at all, the GFCI has either been tripped due to a ground fault or there is no power to the outlet. Try resetting the GFCI. If the GFCI resets and the LED is green, you're good to go. If the GFCI does not reset, the main circuit breaker may have tripped or the GFCI is faulty and needs to be replaced.

Summary

- Alternating voltage varies continuously in magnitude and periodically reverses in polarity. When alternating voltage is applied across a load resistance, the result is alternating current in the circuit.
- A complete set of values repeated periodically is one cycle of the AC waveform. The cycle can be measured from any one point on the wave to the next successive point having the same value and varying in the same direction. One cycle includes 360° in angular measure, or 2π rad.
- The rms value of a sine wave is 0.707 × peak value.
- The peak amplitude, at 90° and 270° in the cycle, is 1.414 × rms value.
- The peak-to-peak value is double the peak amplitude, or 2.828 × rms for a symmetrical AC waveform.
- The average value is 0.637 × peak value.
- The frequency equals the number of cycles per second. One cps is 1 Hz. The audio-frequency (af) range is 16 to 16,000 Hz. Higher frequencies up to 300,000 MHz are radio frequencies.
- The amount of time for one cycle is the period T. The period and frequency are reciprocals: $T = 1/f$, or $f = 1/T$. The higher the frequency, the shorter the period.
- Wavelength λ is the distance a wave travels in one cycle. The higher the frequency, the shorter the wavelength. The wavelength also depends on the velocity at which the wave travels: $\lambda = v/f$, where v is velocity of the wave and f is the frequency.
- Phase angle is the angular difference in time between corresponding values in the cycles for two waveforms of the same frequency.
- When one sine wave has its maximum value while the other is at zero, the two waves are 90° out of phase. Two waveforms with a zero phase angle between them are in phase; a 180° phase angle means opposite phase.
- The length of a phasor arrow indicates amplitude, and the angle corresponds to the phase. A leading phase is shown by counterclockwise angles.
- Sine-wave alternating voltage V applied across a load resistance R produces alternating current I in the circuit. The current has the same waveform, frequency, and phase as the applied voltage because of the resistive load. The amount of $I = V/R$.
- The sawtooth wave and square wave are two common examples of nonsinusoidal waveforms. The amplitudes of these waves are usually measured in peak-to-peak value.
- Harmonic frequencies are exact multiples of the fundamental frequency.
- The AC voltage used in residences range from 115 to 125 V rms with a frequency of 60 Hz. The nominal voltage is usually given as 120 V.
- For residential wiring, the three-wire, single-phase Edison system shown in Fig. 15–21 is used to provide either 120 or 240 V.
- In a motor, the rotating armature connects to the power line. The stator field coils provide the magnetic flux cut by the armature as it is forced to rotate. A generator has the opposite effect: it converts mechanical energy into electrical output.
- A DC motor has commutator segments contacted by graphite brushes for the external connections to the power source. An AC induction motor does not have brushes.
- In three-phase power, each phase angle is 120°. For the Y connections in Fig. 15–25*a*, each pair of output terminals has an output of 120 × 1.73 = 208 V. This voltage is known as the line-to-line voltage.

Important Terms

Alternation — one-half cycle of revolution of a conductor loop rotating through a magnetic field. This corresponds to one-half cycle of alternating voltage or current.

Average value — the arithmetic average of all values in a sine wave for one alternation. Average value = 0.637 × peak value.

Cycle — one complete revolution of a conductor loop rotating through a magnetic field. For any AC waveform, a cycle can be defined to include the variations between two successive points having the same value and varying in the same direction.

Decade — a unit for representing a 10:1 range in frequencies.

Effective value — another name for an rms value.

Form factor — the ratio of the rms to average values. For a sine wave, $\frac{\text{rms}}{\text{avg}} = 1.11$.

Frequency — the number of cycles a waveform completes each second.

Generator — a machine or device that converts mechanical energy into electrical energy.

Harmonic frequency — a frequency that is an exact multiple of the fundamental frequency.

Hertz (Hz) — the basic unit of frequency. 1 Hz = 1 cycle per second.

Motor — a machine or device that converts electrical energy into mechanical energy.

Nonsinusoidal wave form — any waveform that is not a sine wave or a cosine wave.

Octave — a unit for representing a 2:1 range in frequencies.

Peak value — the maximum amplitude of a sine wave.

Period — the amount of time it takes to complete one cycle of alternating voltage or current. The symbol for the period is T for time. The unit for T is the second (s).

Phase angle — the angular difference between two sinusoidal waveforms or phasors.

Phasor — a line representing the magnitude and direction of a quantity, such as voltage or current, with respect to time.

Quadrature phase — a phase angle of 90°.

Radian — an angle equal to approximately 57.3°.

Root-mean-square (rms) value — the value of a sine wave that corresponds to the same amount of direct current or voltage in heating power. Unless indicated otherwise, all sine-wave AC measurements are in rms values. rms value = 0.707 × peak value.

Sine wave — a waveform whose value is proportional to the sine of the angle of rotation in the circular motion producing the induced voltage or current.

Wavelength — the distance a waveform travels through space to complete one cycle.

Related Formulas

$v = V_M \sin \theta$

Average value = 0.637 × peak value

rms value = 0.707 × peak value

$\text{Peak} = \frac{1}{0.707} \times \text{rms} = 1.414 \times \text{rms}$

Peak-to-peak value = 2.828 × rms value

$T = \frac{1}{f}$ or $f = \frac{1}{T}$

$\lambda = \frac{\text{velocity}}{\text{frequency}}$

$\lambda\,(\text{cm}) = \frac{3 \times 10^{10}\ \text{cm/s}}{f\,(\text{Hz})}$ (radio wave)

$\lambda = \frac{1130\ \text{ft/s}}{f\,(\text{Hz})}$ (sound wave)

$t = \frac{\theta}{360} \times \frac{1}{f}$

Self-Test

Answers at the back of the book.

1. An alternating voltage is one that

a. varies continuously in magnitude.

b. reverses periodically in polarity.

c. never varies in amplitude.

d. both a and b.

2. One complete revolution of a conductor loop through a magnetic field is called a(n)

a. octave.

b. decade.

c. cycle.

d. alternation.

3. For a sine wave, one-half cycle is often called a(n)

a. alternation.

b. harmonic.

c. octave.

d. period.

4. One cycle includes

a. 180°.

b. 360°.

c. 2π rad.

d. both b and c.

5. In the United States, the frequency of the AC power-line voltage is

a. 120 Hz.

b. 60 Hz.

c. 50 Hz.

d. 100 Hz.

6. For a sine wave, the number of complete cycles per second is called the

a. period.

b. wavelength.

c. frequency.

d. phase angle.

7. A sine wave of alternating voltage has its maximum values at

a. 90° and 270°.

b. 0° and 180°.

c. 180° and 360°.

d. 30° and 150°.

8. To compare the phase angle between two waveforms, both must have

a. the same amplitude.

b. the same frequency.

c. different frequencies.

d. both a and b.

9. A 2-kHz sine wave has a period, T, of

a. 0.5 μs.

b. 50 μs.

c. 500 μs.

d. 2 ms.

10. If a sine wave has a period, T, of 40 μs, its frequency, f, equals

a. 25 kHz.

b. 250 Hz.

c. 40 kHz.

d. 2.5 kHz.

11. What is the wavelength of a radio wave whose frequency is 15 MHz?

a. 20 m.

b. 15 m.

c. 0.753 ft.

d. 2000 m.

12. The value of alternating current or voltage that has the same heating effect as a corresponding DC value is known as the

a. peak value.

b. average value.

c. rms value.

d. peak-to-peak value.

13. The wavelength of a 500-Hz sound wave is

a. 60 km.

b. 2.26 ft.

c. 4.52 ft.

d. 0.226 ft.

14. In residential house wiring, the hot wire is usually color-coded

a. white.

b. green.

c. black or red.

d. as a bare copper wire.

15. A sine wave with a peak value of 20 V has an rms value of

a. 28.28 V.

b. 14.14 V.

c. 12.74 V.

d. 56.6 V.

16. A sine wave whose rms voltage is 25.2 V has a peak value of approximately

a. 17.8 V.

b. 16 V.

c. 50.4 V.

d. 35.6 V.

17. The unit of frequency is the

a. hertz.

b. maxwell.

c. radian.

d. second.

18. For an AC waveform, the period, *T*, refers to

a. the number of complete cycles per second.

b. the length of time required to complete one cycle.

c. the time it takes for the waveform to reach its peak value.

d. none of the above.

19. The wavelength of a radio wave is

a. inversely proportional to its frequency.

b. directly proportional to its frequency.

c. inversely proportional to its amplitude.

d. unrelated to its frequency.

20. Exact multiples of the fundamental frequency are called

a. ultrasonic frequencies.

b. harmonic frequencies.

c. treble frequencies.

d. resonant frequencies.

21. Raising the frequency of 500 Hz by two octaves corresponds to a frequency of

a. 2 kHz.

b. 1 kHz.

c. 4 kHz.

d. 250 Hz.

22. In residential house wiring, the neutral wire is always color-coded

a. black.

b. bare copper.

c. green.

d. white.

23. The second harmonic of 7 MHz is

a. 3.5 MHz.

b. 28 MHz.

c. 14 MHz.

d. 7 MHz.

24. A sine wave has a peak voltage of 170 V. What is the instantaneous voltage at an angle of 45°?

a. 240 V.

b. 85 V.

c. 0 V.

d. 120 V.

25. Unless indicated otherwise, all sine-wave AC measurements are in

a. peak-to-peak values.

b. peak values.

c. rms values.

d. average values.

Essay Questions

1. (a) Define *alternating voltage*. (b) Define *alternating current*. (c) Why does AC voltage applied across a load resistance produce alternating current in the circuit?

2. (a) State two characteristics of a sine wave of voltage. (b) Why does the rms value of 0.707 × peak value apply just to sine waves?

3. Draw two cycles of an AC sawtooth voltage waveform with a peak-to-peak amplitude of 40 V. Do the same for a square wave.

4. Give the angle in degrees and radians for each of the following: one cycle, one half-cycle, one quarter-cycle, three quarter-cycles.

5. The peak value of a sine wave is 1 V. How much is its average value? rms value? Effective value? Peak-to-peak value?

6. State the following ranges in hertz: (a) audio frequencies; (b) radio frequencies; (c) standard AM radio broadcast band; (d) FM broadcast band; (e) VHF band; (f) microwave band.

7. Make a graph with two waves, one with a frequency of 500 kHz and the other with 1000 kHz. Mark the horizontal axis in time, and label each wave.

8. Draw the sine waves and phasor diagrams to show (a) two waves 180° out of phase; (b) two waves 90° out of phase.

9. Give the voltage value for the 60-Hz AC line voltage with an rms value of 120 V at each of the following times in a cycle: 0°, 30°, 45°, 90°, 180°, 270°, and 360°.

10. (a) The phase angle of 90° equals how many radians? (b) For two sine waves 90° out of phase with each other, compare their amplitudes at 0°, 90°, 180°, 270°, and 360°.

11. Tabulate the sine and cosine values every 30° from 0 to 360° and draw the corresponding sine wave and cosine wave.

12. Draw a graph of the values for $(\sin \theta)^2$ plotted against θ for every 30° from 0 to 360°.

13. Why is the wavelength of an ultrasonic wave at 34.44 kHz the same 1 cm as for the much higher frequency radio wave at 30 GHz?

14. Draw the sine waves and phasors to show wave V_1 leading wave V_2 by 45°.

15. Why are amplitudes for nonsinusoidal waveforms generally measured in peak-to-peak values, rather than rms or average value?

16. Define *harmonic frequencies*, giving numerical values.

17. Define *one octave*, with an example of numerical values.

18. Which do you consider more important for applications of alternating current—polarity reversals or variations in value?

19. Define the following parts in the assembly of motors: (a) armature rotor; (b) field stator; (c) collector rings; (d) commutator segments.

20. Show diagrams of Y and Δ connections for three-phase AC power.

Problems

SECTION 15–2 ALTERNATING-VOLTAGE GENERATOR

15–1 For a sine wave of alternating voltage, how many degrees are included in

a. ¼ cycle?

b. ½ cycle?

c. ¾ cycle?

d. 1 complete cycle?

15–2 For a sine wave of alternating voltage, how many radians are included in

a. ¼ cycle?

b. ½ cycle?

c. ¾ cycle?

d. 1 complete cycle?

15–3 At what angle does a sine wave of alternating voltage

a. reach its maximum positive value?

b. reach its maximum negative value?

c. cross the zero axis?

15–4 One radian corresponds to how many degrees?

SECTION 15–3 THE SINE WAVE

15–5 The peak value of a sine wave equals 20 V. Calculate the instantaneous voltage of the sine wave for the phase angles listed.

a. 30°.

b. 45°.

c. 60°.

d. 75°.

e. 120°.

f. 210°.

g. 300°.

15–6 The peak value of a sine wave equals 100 mV. Calculate the instantaneous voltage of the sine wave for the phase angles listed.

a. 15°.

b. 50°.

c. 90°.

d. 150°.

e. 180°.

f. 240°.

g. 330°.

15–7 A sine wave of alternating voltage has an instantaneous value of 45 V at an angle of 60°. Determine the peak value of the sine wave.

SECTION 15–4 ALTERNATING CURRENT

15–8 In Fig. 15–30, the sine wave of applied voltage has a peak or maximum value of 10 V, as shown. Calculate the instantaneous value of current for the phase angles listed.

a. 30°.

b. 60°.

c. 90°.

d. 120°.

e. 150°.

f. 180°.

g. 210°.

h. 240°.

i. 270°.

j. 300°.

k. 330°.

15–9 In Fig. 15–30, do electrons flow clockwise or counterclockwise in the circuit during

a. the positive alternation?

b. the negative alternation?

Note: During the positive alternation, terminal 1 is positive with respect to terminal 2.

Figure 15–30

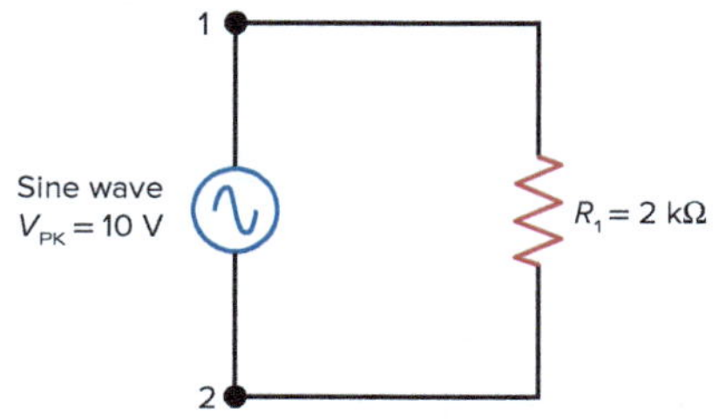

SECTION 15–5 VOLTAGE AND CURRENT VALUES FOR A SINE WAVE

15–10 If the sine wave in Fig. 15–31 has a peak value of 15 V, then calculate

a. the peak-to-peak value.

b. the rms value.

c. the average value.

Figure 15–31

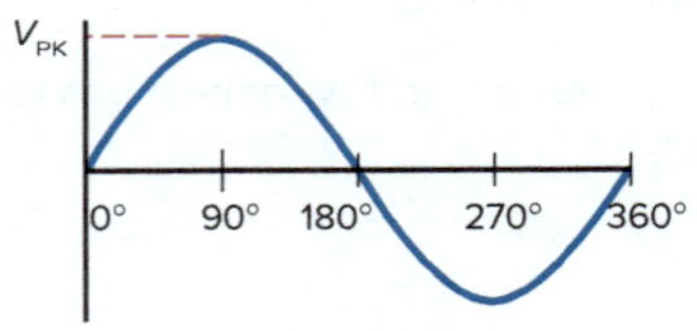

15–11 If the sine wave in Fig. 15–31 has a peak value of 50 V, then calculate

a. the peak-to-peak value.

b. the rms value.

c. the average value.

15–12 If the sine wave in Fig. 15–31 has an rms value of 60 V, then calculate

a. the peak value.

b. the peak-to-peak value.

c. the average value.

15–13 If the sine wave in Fig. 15–31 has an rms value of 40 V, then calculate

a. the peak value.

b. the peak-to-peak value.

c. the average value.

15–14 If the sine wave of alternating voltage in Fig. 15–32 has a peak value of 25 V, then calculate

a. the peak current value.

b. the peak-to-peak current value.

c. the rms current value.

d. the average current value.

Figure 15–32

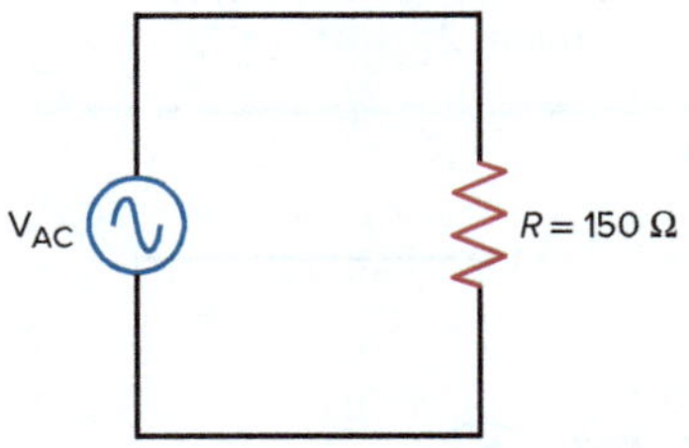

15–15 If the sine wave of alternating voltage in Fig. 15–32 has an rms value of 7.07 V, then calculate

a. the rms current value.

b. the peak current value.

c. the peak-to-peak current value.

d. the average current value.

15–16 Convert the following values into rms values:

a. 32 V peak.

b. 18 V peak-to-peak.

c. 90.09 V average.

d. 120 mA peak-to-peak.

15–17 Convert the following values into peak values:

a. 12 V rms.

b. 72 V average.

c. 50 V peak-to-peak.

d. 750 mV rms.

SECTION 15–6 FREQUENCY

15–18 What is the frequency, f, of a sine wave that completes

a. 10 cycles per second?

b. 500 cycles per second?

c. 50,000 cycles per second?

d. 2,000,000 cycles per second?

15–19 How many cycles per second (cps) do the following frequencies correspond to?

a. 2 kHz.

b. 15 MHz.

c. 10 kHz.

d. 5 GHz.

SECTION 15–7 PERIOD

15–20 Calculate the period, T, for the following sine wave frequencies:

a. 50 Hz.

b. 100 Hz.

c. 500 Hz.

d. 1 kHz.

15–21 Calculate the period, T, for the following sine wave frequencies:

a. 2 kHz.

b. 4 kHz.

c. 200 kHz.

d. 2 MHz.

15–22 Calculate the frequency, f, of a sine wave whose period, T, is

a. 40 μs.

b. 50 μs.

c. 2.5 ms.

d. 16.67 ms.

15–23 Calculate the frequency, f, of a sine wave whose period, T, is

a. 5 ms.

b. 10 μs.

c. 500 ns.

d. 33.33 μs.

15–24 For a 5-kHz sine wave, how long does it take for

a. ¼ cycle?

b. ½ cycle?

c. ¾ cycle?

d. 1 full cycle?

SECTION 15–8 WAVELENGTH

15–25 What is the velocity of an electromagnetic radio wave in

a. miles per second (mi/s)?

b. centimeters per sec (cm/s)?

c. meters per sec (m/s)?

15–26 What is the velocity in ft/s of a sound wave produced by mechanical vibrations?

15–27 What is the wavelength in cm of an electromagnetic radio wave whose frequency is

a. 3.75 MHz?

b. 7.5 MHz?

c. 15 MHz?

d. 20 MHz?

15–28 Convert the wavelengths in Prob. 15–27 into meters (m).

15–29 What is the wavelength in meters of an electromagnetic radio wave whose frequency is 150 MHz?

15–30 What is the wavelength in ft of a sound wave whose frequency is

a. 50 Hz?

b. 200 Hz?

c. 750 Hz?

d. 2 kHz?

e. 4 kHz?

f. 10 kHz?

15–31 What is the frequency of an electromagnetic radio wave whose wavelength is

a. 160 m?

b. 10 m?

c. 17 m?

d. 11 m?

15–32 What is the frequency of a sound wave whose wavelength is

a. 4.52 ft?

b. 1.13 ft?

c. 3.39 ft?

d. 0.226 ft?

SECTION 15–9 PHASE ANGLE

15–33 Describe the difference between a sine wave and a cosine wave.

15–34 Two voltage waveforms of the same amplitude, V_X and V_Y, are 45° out of phase with each other, with V_Y lagging V_X. Draw the phasors representing these voltage waveforms if

a. V_X is used as the reference phasor.

b. V_Y is used as the reference phasor.

SECTION 15–10 THE TIME FACTOR IN FREQUENCY AND PHASE

15–35 For two waveforms with a frequency of 1 kHz, how much time corresponds to a phase angle difference of

a. 30°?

b. 45°?

c. 60°?

d. 90°?

15–36 For two waveforms with a frequency of 50 kHz, how much time corresponds to a phase angle difference of

a. 15°?

b. 36°?

c. 60°?

d. 150°?

SECTION 15–11 ALTERNATING CURRENT CIRCUITS WITH RESISTANCE

15–37 In Fig. 15–33, solve for the following values: R_T, I, V_1, V_2, P_1, P_2, and P_T.

Figure 15–33

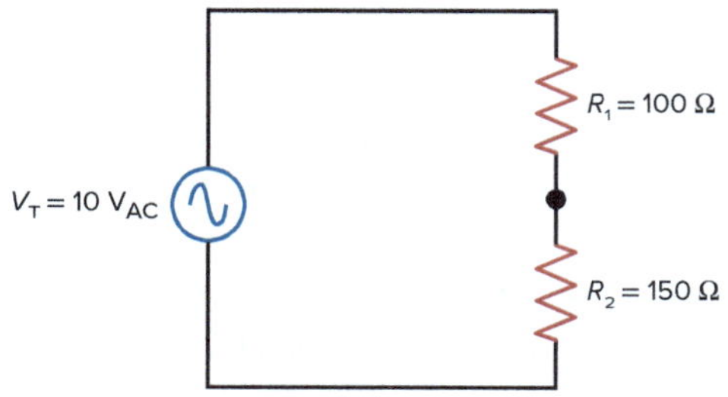

15–38 In Fig. 15–34, solve for the following values: I_1, I_2, I_T, R_{EQ}, P_1, P_2, and P_T.

Figure 15–34

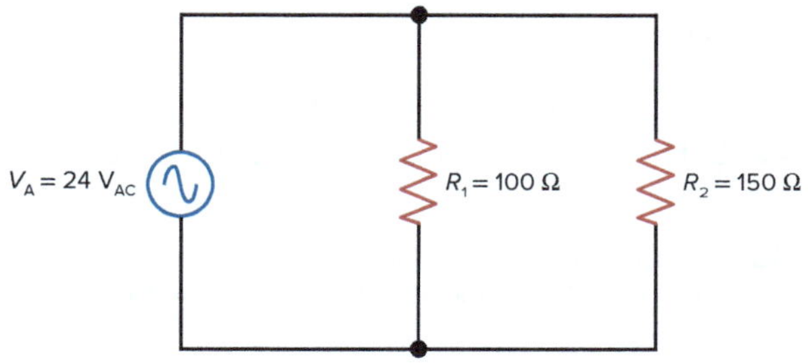

15–39 In Fig. 15–35, solve for the following values: R_T, I_T, V_1, V_2, V_3, I_2, I_3, P_1, P_2, P_3, and P_T.

Figure 15–35

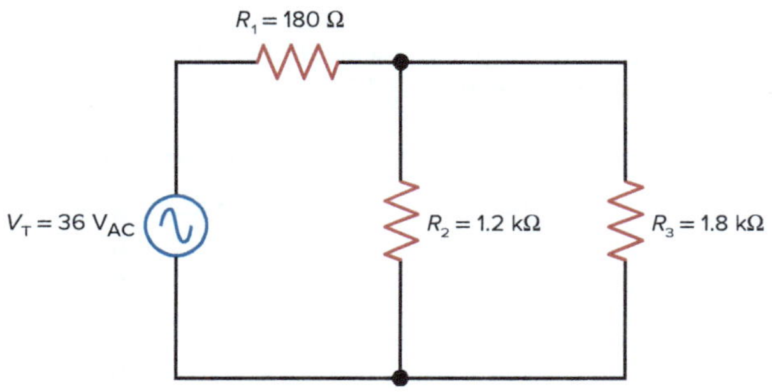

Figure 15–36

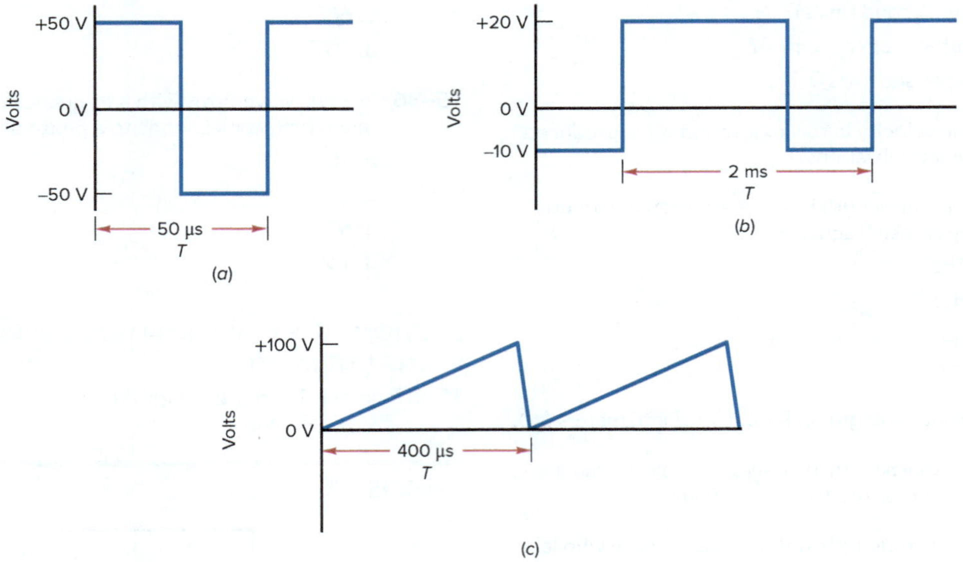

15–40 In Fig. 15–35, find the following values:
a. the peak-to-peak current through R_1.
b. the average voltage across R_2.
c. the peak voltage across R_3.
d. the average current through R_3.

SECTION 15–12 NONSINUSOIDAL AC WAVEFORMS

15–41 Determine the peak-to-peak voltage and frequency for the waveform in
a. Fig. 15–36*a*.
b. Fig. 15–36*b*.
c. Fig. 15–36*c*.

SECTION 15–13 HARMONIC FREQUENCIES

15–42 List the first four harmonics of a 3.8-MHz radio signal.

15–43 List the first seven harmonics of a 1-kHz sine wave. Label each harmonic as either an even or odd harmonic.

15–44 Raising the frequency of 250 Hz by one octave corresponds to what frequency?

15–45 Lowering the frequency of 3 kHz by two octaves corresponds to what frequency?

15–46 Raising the frequency of 300 Hz by three octaves corresponds to what frequency?

15–47 What is the frequency two decades above 1 kHz?

SECTION 15–14 THE 60-Hz AC POWER LINE

15–48 What is the frequency of the AC power line in most European countries?

15–49 What device or component is used to step up or step down an AC voltage in the distribution of AC power to our homes and industries?

15–50 What is the main reason for using extremely high voltages, such as 500 kV, on the distribution lines for AC power?

Critical Thinking

15–51 The electrical length of an antenna is to be one-half wavelength long at a frequency *f* of 7.2 MHz. Calculate the length of the antenna in (a) feet; (b) centimeters.

15–52 A transmission line has a length *l* of 7.5 m. What is its electrical wavelength at 10 MHz?

15–53 The total length of an antenna is 120 ft. At what frequency is the antenna one-half wavelength long?

15–54 A cosine wave of current has an instantaneous amplitude of 45 mA at $\theta = \pi/3$ rad. Calculate the waveform's instantaneous amplitude at $\theta = 3\pi/2$ rad.

Answers to Self-Reviews

15–1 **a.** true
b. false
c. true

15–2 **a.** 10 V
b. 360°

15–3 **a.** 85 V
b. 120 V
c. 170 V

15–4 **a.** 0.707 A
b. 0.5 A

15–5 **a.** 120 V rms **b.** 14.14 V peak **c.** 2.83 V p-p

15–6 **a.** 4 Hz **b.** 1.605 MHz

15–7 **a.** 400 Hz **b.** 1⁄400 s

15–8 **a.** true **b.** false **c.** true

15–9 **a.** 90° **b.** 60° **c.** 0°

15–10 **a.** 1⁄240 s **b.** 0.1 μs

15–11 **a.** 30 Ω **b.** 6.67 Ω **c.** 30 Ω

15–12 **a.** 2 μs **b.** 15 V

15–13 **a.** 48 MHz **b.** 440 Hz

15–14 **a.** false **b.** true **c.** true **d.** true

15–15 **a.** true **b.** true **c.** false

15–16 **a.** 120° **b.** 208 V

Laboratory Application Assignment

In this lab application assignment, you will use an oscilloscope to measure the amplitude, frequency, and period of a sine-wave AC voltage. You will also use a DMM to measure the voltage and current values in an AC circuit. As an aid in understanding the operation and use of the oscilloscope, refer to Appendix E. However, it is expected that your instructor will assist you with the operation of both the function generator and the oscilloscope when doing this experiment.

Equipment: Obtain the following items from your instructor.

- Function generator
- Assortment of carbon-film resistors
- Oscilloscope
- DMM

Using the Oscilloscope and Function Generator

Connect the channel 1 probe of the oscilloscope to the output of the function generator. Set the function generator to produce a sine-wave output. Next, while viewing the oscilloscope, adjust the function generator and oscilloscope controls to view one cycle of a 100-Hz, 8-V_{pp} sine wave. The displayed waveform should be similar to the one shown in Fig. 15–37. Have your instructor check the displayed waveform. If it is correct, proceed as follows.

What is the Volts/div. setting of the oscilloscope? Volts/div. = ____________ How many vertical divisions does the displayed waveform occupy? ____________
From this information, what is the measured peak-to-peak value of the displayed waveform? V_{pp} = ____________

What is the peak value of the displayed waveform? V_p = ____________ Using this value, calculate and record the waveform's rms value. V_{rms} = ____________ With your DMM

Figure 15–37

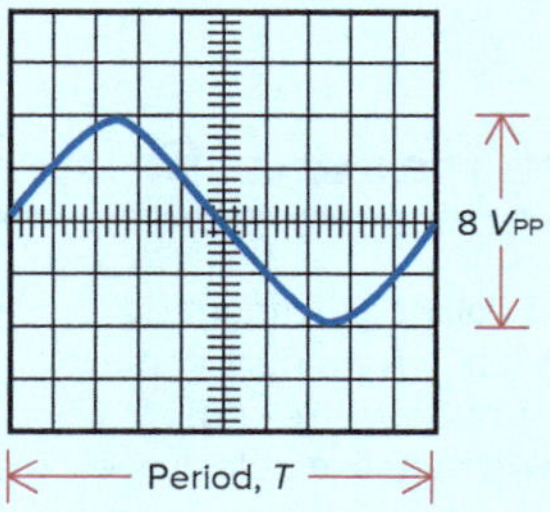

set to measure AC voltage, measure and record the rms voltage at the output of the function generator.
V_{rms} = ____________ How does this value compare to the calculated value? ____________

What is the Sec./div. setting of the oscilloscope? Sec./div. = ____________ How many horizontal divisions does one cycle occupy? ____________
From this information, what is the measured period, T, of the displayed waveform? T = ____________
Calculate and record the period, T, of the 100-Hz waveform as $T = 1/f$. T = ____________
How do the calculated and measured values compare? ____________

Adjust the Sec./div control of the oscilloscope until two cycles are displayed on the screen. What is the Sec./div. setting with two cycles displayed? Sec./div. = ____________

Have either your instructor or lab partner adjust the function generator controls to change the frequency and amplitude of the sine wave. Determine the period, T, frequency, f, and peak-to-peak value of the displayed waveform. Repeat this procedure several times until you become proficient in using the oscilloscope.

AC Circuit Measurements

Refer to Fig. 15–38. Calculate and record the following circuit values:

R_T = ______________, I_{rms} = ______________,
$V_{1(rms)}$ = ______________, $V_{2(rms)}$ = ______________

Construct the circuit in Fig. 15–38. Using your DMM, measure and record the following rms values: I_{rms} = ______________,
$V_{1(rms)}$ = ______________, $V_{2(rms)}$ = ______________

Using the oscilloscope, measure and record the following peak-to-peak values. (You will need to use both channels and the math mode to measure V_1.)

$V_{T(pp)}$ = ______________, $V_{1(pp)}$ = ______________,
$V_{2(pp)}$ = ______________

Convert the peak-to-peak values to rms values, and record your answers.

$V_{T(rms)}$ = ______________, $V_{1(rms)}$ = ______________,
$V_{2(rms)}$ = ______________

How do these values compare to the values measured with the DMM? ____

____

Figure 15–38

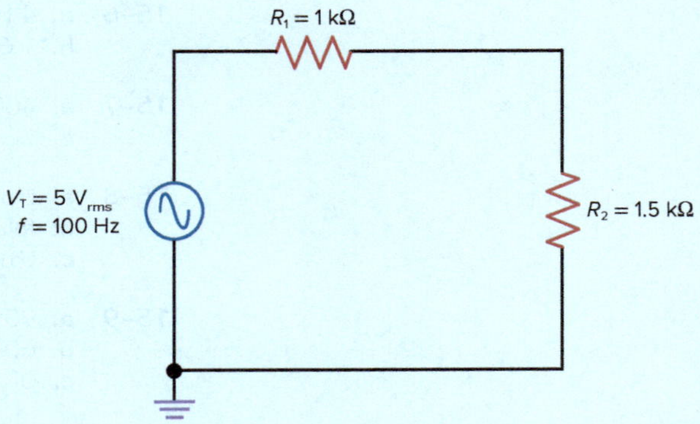

With your DMM set to measure DC voltage, measure and record the following values in Fig. 15–38:

$V_{T(DC)}$ = ______________, $V_{1(DC)}$ = ______________,
$V_{2(DC)}$ = ______________

Are these measurements what you expected? ______________

Cumulative Review Summary (Chapters 13–15)

- Iron, nickel, and cobalt are magnetic materials. Magnets have a north pole and a south pole at opposite ends. Opposite poles attract; like poles repel.
- A magnet has an invisible, external magnetic field. This magnetic flux is indicated by field lines. The direction of field lines outside the magnet is from a north pole to a south pole.
- An electromagnet has an iron core that becomes magnetized when current flows in the coil winding.
- Magnetic units are defined in Table 13–1.
- Continuous magnetization and demagnetization of an iron core by alternating current causes hysteresis losses that increase with higher frequencies.
- Current in a conductor has an associated magnetic field with circular lines of force in a plane perpendicular to the wire.
- Motor action results from the net force of two fields that can aid or cancel. The direction of the resultant force is from the stronger field to the weaker.
- The motion of flux cutting across a perpendicular conductor generates an induced voltage.
- Faraday's law of induced voltage states that $v = N\, d\phi/dt$.
- Lenz's law states that an induced voltage must have the polarity that opposes the change causing the induction.
- Alternating voltage varies in magnitude and reverses in polarity.
- One cycle includes the values between points having the same value and varying in the same direction. The cycle includes 360°, or 2π rad.
- Frequency f equals cycles per second (cps). One cps = 1 Hz.
- Period T is the time for one cycle. It equals $1/f$. When f is in cycles per second, T is in seconds.
- Wavelength λ is the distance a wave travels in one cycle. $\lambda = v/f$.
- The rms, or effective value, of a sine wave equals 0.707 × peak value. Or the peak value equals 1.414 × rms value. The average value equals 0.637 × peak value.
- Phase angle θ is the angular difference between corresponding values in the cycles for two sine waves of the same frequency. The angular difference can be expressed in time based on the frequency of the waves.
- Phasors, similar to vectors, indicate the amplitude and phase angle of alternating voltage or current. The length of the phasor is the amplitude, and the angle is the phase.
- The square wave and sawtooth wave are common examples of nonsinusoidal waveforms.
- Direct current motors generally use commutator segments with graphite brushes. Alternating current motors are usually the induction type without brushes.
- House wiring uses three-wire, single-phase power with a frequency of 60 Hz. The voltages for house wiring are 120 V to the grounded neutral and 240 V across the two high sides.
- Three-phase AC power has three legs 120° out of phase. A Y connection with 120 V across each phase has 208 V available across each two legs.

Cumulative Self-Test

Answers at the back of the book.

1. Which of the following statements is true? (*a*) Alnico is commonly used for electromagnets. (*b*) Paper cannot affect magnetic flux because it is not a magnetic material. (*c*) Iron is generally used for permanent magnets. (*d*) Ferrites have lower permeability than air or vacuum.

2. Hysteresis losses (*a*) are caused by high-frequency alternating current in a coil with an iron core; (*b*) generally increase with direct current in a coil; (*c*) are especially important for permanent magnets that have a steady magnetic field; (*d*) cannot be produced in an iron core because it is a conductor.

3. A magnetic flux of 25,000 lines through an area of 5 cm^2 results in (*a*) 5 lines of flux; (*b*) 5000 Mx of flux; (*c*) flux density of 5000 G; (*d*) flux density corresponding to 25,000 A.

4. If 10 V is applied across a relay coil with 100 turns having 2 Ω of resistance, the total force producing magnetic flux in the circuit is (*a*) 10 Mx; (*b*) 50 G; (*c*) 100 Oe; (*d*) 500 A·t.

5. The AC power-line voltage of 120 V rms has a peak value of (*a*) 100 V; (*b*) 170 V; (*c*) 240 V; (*d*) 338 V.

6. Which of the following can produce the most induced voltage? (*a*) 1-A direct current; (*b*) 50-A direct current; (*c*) 1-A 60-Hz alternating current; (*d*) 1-A 400-Hz alternating current.

7. Which of the following has the highest frequency? (*a*) $T = 1/1000$ s; (*b*) $T = 1/60$ s; (*c*) $T = 1$ s; (*d*) $T = 2$ s.

8. Two waves of the same frequency are opposite in phase when the phase angle between them is (*a*) 0°; (*b*) 90°; (*c*) 360°; (*d*) π rad.

9. A 120-V, 60-Hz power-line voltage is applied across a 120-Ω resistor. The current equals (*a*) 1 A, peak value; (*b*) 120 A, peak value; (*c*) 1 A, rms value; (*d*) 5 A, rms value.

10. When an alternating voltage reverses in polarity, the current it produces (*a*) reverses in direction; (*b*) has a steady DC value; (*c*) has a phase angle of 180°; (*d*) alternates at 1.4 times the frequency of the applied voltage.

Design credits: Chapter Opener Background Image, and Multisim: Stockbyte/Getty Images

chapter 16

Capacitance

Capacitance is the ability of a dielectric to hold or store an electric charge. The more charge stored for a given voltage, the higher the capacitance. The symbol for capacitance is C, and the unit is the farad (F), named after Michael Faraday.

A capacitor consists of two metal plates separated by an insulator (also called a dielectric). The metal plates make it possible to apply voltage across the dielectric. Different types of capacitors are manufactured for specific values of C. They are named according to the type of dielectric used by the capacitor. Common types are air, ceramic, mica, paper, plastic-film, and electrolytic capacitors. Capacitors used in electronic circuits are small and economical.

An important property of a capacitor is its ability to block a steady DC voltage while passing AC signals. The higher the frequency, the less the opposition to AC voltage.

Capacitors are a common source of trouble in electronic circuits because they can have either an open at the metal plates or a short circuit through the dielectric. These troubles are described here, including the method of checking a capacitor with an ohmmeter, even though a capacitor uses an insulator to separate the capacitor plates. ■

Chapter Outline

Chapter Objectives

After studying this chapter, you should be able to

- *Describe* how charge is stored in the dielectric of a capacitor.
- *Describe* how a capacitor charges and discharges.
- *Define* the farad unit of capacitance.
- *List* the physical factors affecting the capacitance of a capacitor.
- *List* several types of capacitors and the characteristics of each.
- *Explain* how an electrolytic capacitor is constructed.
- *Explain* how capacitors are coded.
- *Calculate* the total capacitance of parallel-connected capacitors.
- *Calculate* the equivalent capacitance of series-connected capacitors.
- *Calculate* the energy stored in a capacitor.
- *Define* the terms *leakage, dielectric absorption,* and *equivalent series resistance* as they relate to capacitors.
- *Describe* how an ohmmeter can be used to test a capacitor.

Important Terms

capacitance (C)
capacitor
charging
condenser
dielectric absorption
dielectric constant, K_ϵ
dielectric material
dielectric strength
discharging
electric field
equivalent series resistance (ESR)
farad (F) unit
ganged capacitors
leakage current
leakage resistance
microfarad (μF)
nanofarad (nF)
picofarad (pF)
relative permittivity, ϵ_r

16–1 How Charge Is Stored in a Capacitor

It is possible for **dielectric materials** such as air or paper to hold an electric charge because free electrons cannot flow through an insulator. However, the charge must be applied by some source. In Fig. 16–1*a*, the battery can charge the **capacitor** shown. With the dielectric contacting the two metal plates connected to the potential difference *V*, electrons from the voltage source accumulate on the side of the capacitor connected to the negative terminal of *V*. The opposite side of the capacitor connected to the positive terminal of *V* loses electrons.

As a result, the excess of electrons produces a negative charge on one side of the capacitor, and the opposite side has a positive charge. As an example, if 6.25×10^{18} electrons are accumulated, the negative charge equals 1 coulomb (C). The charge on only one plate need be considered because the number of electrons accumulated on one plate is exactly the same as the number taken from the opposite plate.

What the voltage source does is simply redistribute some electrons from one side of the capacitor to the other. This process is called ***charging*** the capacitor. Charging continues until the potential difference across the capacitor is equal to the applied voltage. Without any series resistance, the charging is instantaneous. Practically, however, there is always some series resistance. This charging current is transient, or temporary; it flows only until the capacitor is charged to the applied voltage. Then there is no current in the circuit.

The result is a device for storing charge in the dielectric. Storage means that the charge remains even after the voltage source is disconnected. The measure of how much charge can be stored is the **capacitance *C***. More charge stored for a given amount of applied voltage means more capacitance. Components made to provide a specified amount of capacitance are called *capacitors,* or by their old name ***condensers***.

Electrically, then, capacitance is the ability to store charge. A capacitor consists simply of two metal plates separated by an insulator. For example, Fig. 16–1*b* shows a variable capacitor using air for the dielectric between the metal plates. There are many types with different dielectric materials, including paper, mica, and ceramics, but the schematic symbols shown in Fig. 16–1*c* apply to all capacitors.

GOOD TO KNOW

In the automotive industry, capacitors are commonly referred to as condensers. In the electronics industry, however, capacitors are rarely, if ever, referred to as condensers.

Electric Field in the Dielectric

Any voltage has a field of electric lines of force between the opposite electric charges. The **electric field** corresponds to the magnetic lines of force of the

Figure 16–1 Capacitance stores the charge in the dielectric between two conductors. (*a*) Structure. (*b*) Air-dielectric variable capacitor. Length is 2 in. (*c*) Schematic symbols for fixed and variable capacitors.

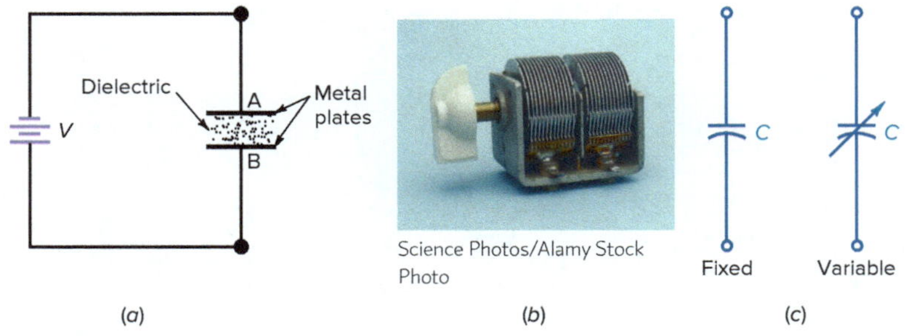

magnetic field associated with electric current. What a capacitor does is concentrate the electric field in the dielectric between the plates. This concentration corresponds to a magnetic field concentrated in the turns of a coil. The only function of the capacitor plates and wire conductors is to connect the voltage source *V* across the dielectric. Then the electric field is concentrated in the capacitor, instead of being spread out in all directions.

Electrostatic Induction

The capacitor has opposite charges because of electrostatic induction by the electric field. Electrons that accumulate on the negative side of the capacitor provide electric lines of force that repel electrons from the opposite side. When this side loses electrons, it becomes positively charged. The opposite charges induced by an electric field correspond to the opposite poles induced in magnetic materials by a magnetic field.

16–1 Self-Review

Answers at the end of the chapter.

a. In a capacitor, is the electric charge stored in the dielectric or on the metal plates?

b. What is the unit of capacitance?

16–2 Charging and Discharging a Capacitor

Charging and **discharging** are the two main effects of capacitors. Applied voltage puts charge in the capacitor. The accumulation of charge results in a buildup of potential difference across the capacitor plates. When the capacitor voltage equals the applied voltage, there is no more charging. The charge remains in the capacitor, with or without the applied voltage connected.

The capacitor discharges when a conducting path is provided across the plates, without any applied voltage. Actually, it is necessary only that the capacitor voltage be more than the applied voltage. Then the capacitor can serve as a voltage source, temporarily, to produce discharge current in the discharge path. The capacitor discharge continues until the capacitor voltage drops to zero or is equal to the applied voltage.

Applying the Charge

In Fig. 16–2*a,* the capacitor is neutral with no charge because it has not been connected to any source of applied voltage and there is no electrostatic field in the dielectric. Closing the switch in Fig. 16–2*b*, however, allows the negative battery terminal to repel free electrons in the conductor to plate A. At the same time, the positive terminal attracts free electrons from plate B. The side of the dielectric at plate A accumulates electrons because they cannot flow through the insulator, and plate B has an equal surplus of protons.

Remember that opposite charges have an associated potential difference, which is the voltage across the capacitor. The charging process continues until the capacitor voltage equals the battery voltage, which is 10 V in this example. Then no further charging is possible because the applied voltage cannot make free electrons flow in the conductors.

Note that the potential difference across the charged capacitor is 10 V between plates A and B. There is no potential difference from each plate to its battery terminal, however, which is why the capacitor stops charging.

MultiSim **Figure 16–2** Storing electric charge in a capacitance. (*a*) Capacitor without any charge. (*b*) Battery charges capacitor to applied voltage of 10 V. (*c*) Stored charge remains in the capacitor, providing 10 V without the battery. (*d*) Discharging the capacitor.

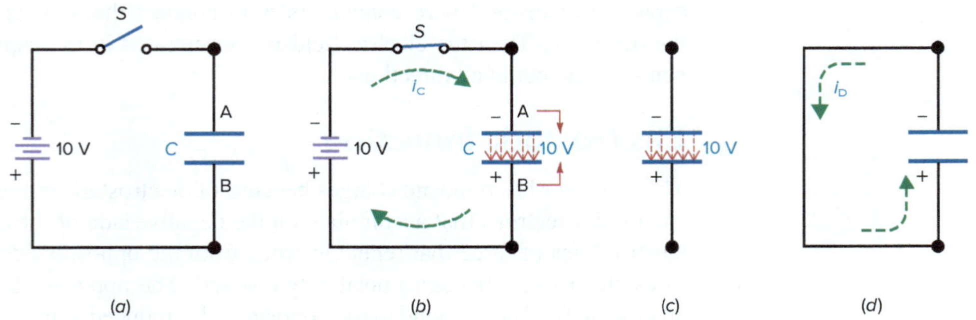

GOOD TO KNOW

In the word dielectric, "di" means opposing and "electric" means electric field.

Storing the Charge

The negative and positive charges on opposite plates of the capacitor have an associated electric field through the dielectric, as shown by the dotted lines in Fig. 16–2*b* and *c*. The direction of these electric lines of force is shown repelling electrons from plate B, making this side positive. The effect of the electric lines of force through the dielectric results in the storage of electric charge. The electric field distorts the molecular structure and polarizes it so that the dielectric is no longer neutral. The polarized dielectric has its own electric field that opposes the electric field produced by the positive and negative charges on the plates of the capacitor. The dielectric is actually stressed by the invisible force of the electric field. As evidence, the dielectric can be ruptured by a very intense field with high voltage across the capacitor.

The result of the electric field, then, is that the dielectric has charge supplied by the voltage source. Since the dielectric is an insulator that cannot conduct, the charge remains in the capacitor even after the voltage source is removed, as illustrated in Fig. 16–2*c*. You can now take this charged capacitor by itself out of the circuit, and it still has 10 V across the two terminals.

Discharging

The action of neutralizing the charge by connecting a conducting path across the dielectric is called *discharging* the capacitor. In Fig. 16–2*d*, the wire between plates A and B is a low-resistance path for discharge current. With the stored charge in the dielectric providing the potential difference, 10 V is available to produce discharge current. The negative plate repels electrons, which are attracted to the positive plate through the wire, until the positive and negative charges are neutralized. Then there is no net charge. The capacitor is completely discharged, the voltage across it equals zero, and there is no discharge current. Now the capacitor is in the same uncharged condition as in Fig. 16–2*a*. It can be charged again, however, by a source of applied voltage.

Nature of the Capacitance

A capacitor can store the amount of charge necessary to provide a potential difference equal to the charging voltage. If 100 V were applied in Fig. 16–2, the capacitor would charge to 100 V.

The capacitor charges to the applied voltage because it takes on more charge when the capacitor voltage is less. As soon as the capacitor voltage equals the applied voltage, no more charging current can flow. *Note that any charge or discharge current flows through the conducting wires to the plates but not through the dielectric.*

Oxford Science Archive/Print Collector/Getty Images

PIONEERS IN ELECTRONICS

The unit of measure for capacitance, the farad (F), was named for *Michael Faraday (1791–1867)*, an English chemist and physicist who discovered the principle of induction (1 F is the unit of capacitance that will store 1 coulomb [C] of charge when 1 volt [V] is applied).

GOOD TO KNOW

Capacitors are normally coded in either pF or μF units and rarely in nF units. However, most capacitance testers have capacitance ranges that are in nF units.

Charge and Discharge Currents

In Fig. 16–2*b*, i_C is in the opposite direction from i_D in Fig. 16–2*d*. In both cases, the current shown is electron flow. However, i_C is charging current to the capacitor and i_D is discharge current from the capacitor. The charge and discharge currents must always be in opposite directions. In Fig. 16–2*b*, the negative plate of C accumulates electrons from the voltage source. In Fig. 16–2*d*, the charged capacitor is a voltage source to produce electron flow around the discharge path.

More charge and discharge current result from a higher value of *C* for a given amount of voltage. Also, more *V* produces more charge and discharge current with a given amount of capacitance. However, the value of *C* does not change with the voltage because the amount of *C* depends on the physical construction of the capacitor.

16–2 Self-Review

Answers at the end of the chapter.

Refer to Fig. 16–2.

a. If the applied voltage were 14.5 V, how much would the voltage be across *C* after it is charged?

b. How much is the voltage across *C* after it is completely discharged?

c. Can a capacitor be charged again after it is discharged?

16–3 The Farad Unit of Capacitance

With more charging voltage, the electric field is stronger and more charge is stored in the dielectric. The amount of charge *Q* stored in the capacitance is therefore proportional to the applied voltage. Also, a larger capacitance can store more charge. These relations are summarized by the formula

$$Q = CV \text{ coulombs} \qquad \textbf{(16–1)}$$

where *Q* is the charge stored in the dielectric in coulombs (C), *V* is the voltage across the plates of the capacitor, and *C* is the capacitance in **farads**.

The *C* is a physical constant, indicating the capacitance in terms of the amount of charge that can be stored for a given amount of charging voltage. When one coulomb is stored in the dielectric with a potential difference of one volt, the capacitance is one *farad.*

Practical capacitors have sizes in millionths of a farad, or smaller. The reason is that typical capacitors store charge of microcoulombs or less. Therefore, the common units are

$$1 \text{ microfarad} = 1\ \mu\text{F} = 1 \times 10^{-6}\ \text{F}$$
$$1 \text{ nanofarad} = 1\ \text{nF} = 1 \times 10^{-9}\ \text{F}$$
$$1 \text{ picofarad} = 1\ \text{pF} = 1 \times 10^{-12}\ \text{F}$$

Although traditionally it has not been used, the **nanofarad** unit of capacitance is gaining acceptance in the electronics industry.

Example 16–1

How much charge is stored in a 2-μF capacitor connected across a 50-V supply?

ANSWER $Q = CV = 2 \times 10^{-6} \times 50$
$= 100 \times 10^{-6}\ \text{C}$

Example 16–2

How much charge is stored in a 40-μF capacitor connected across a 50-V supply?

ANSWER $Q = CV = 40 \times 10^{-6} \times 50$
$= 2000 \times 10^{-6}$ C

Note that the larger capacitor stores more charge for the same voltage, in accordance with the definition of capacitance as the ability to store charge.

The factors in $Q = CV$ can be inverted to

$$C = \frac{Q}{V} \tag{16–2}$$

or

$$V = \frac{Q}{C} \tag{16–3}$$

For all three formulas, the basic units are volts for V, coulombs for Q, and farads for C. Note that the formula $C = Q/V$ actually defines one farad of capacitance as one coulomb of charge stored for one volt of potential difference. The letter C (in italic type) is the symbol for capacitance. The same letter C (in roman type) is the abbreviation for the coulomb unit of charge. The difference between C and C will be made clearer in the examples that follow.

Example 16–3

A constant current of 2 μA charges a capacitor for 20 s. How much charge is stored? Remember $I = Q/t$ or $Q = I \times t$.

ANSWER $Q = I \times t$
$= 2 \times 10^{-6} \times 20$
$= 40 \times 10^{-6}$ or 40 μC

Example 16–4

The voltage across the charged capacitor in Example 16–3 is 20 V. Calculate C.

ANSWER $C = \frac{Q}{V} = \frac{40 \times 10^{-6}}{20} = 2 \times 10^{-6}$
$= 2\ \mu$F

Example 16–5

A constant current of 5 mA charges a 10-μF capacitor for 1 s. How much is the voltage across the capacitor?

ANSWER Find the stored charge first:

$$Q = I \times t = 5 \times 10^{-3} \times 1$$
$$= 5 \times 10^{-3}\ \text{C or } 5\ \text{mC}$$
$$V = \frac{Q}{C} = \frac{5 \times 10^{-3}}{10 \times 10^{-6}} = 0.5 \times 10^{3}$$
$$= 500\ \text{V}$$

Larger Plate Area Increases Capacitance

As illustrated in Fig. 16–3, when the area of each plate is doubled, the capacitance in Fig. 16–3*b* stores twice the charge of Fig. 16–3*a*. The potential difference in both cases is still 10 V. This voltage produces a given strength of electric field. A larger plate area, however, means that more of the dielectric surface can contact each plate, allowing more lines of force through the dielectric between the plates and less flux leakage outside the dielectric. Then the field can store more charge in the dielectric. The result of larger plate area is more charge stored for the same applied voltage, which means that the capacitance is larger.

Thinner Dielectric Increases Capacitance

As illustrated in Fig. 16–3*c*, when the distance between plates is reduced by one-half, the capacitance stores twice the charge of Fig. 16–3*a*. The potential difference is still 10 V, but its electric field has greater flux density in the thinner dielectric. Then the field between opposite plates can store more charge in the dielectric. With less distance between the plates, the stored charge is greater for the same applied voltage, which means that the capacitance is greater.

Figure 16–3 Increasing stored charge and capacitance by increasing the plate area and decreasing the distance between plates. (*a*) Capacitance of 1 μF. (*b*) A 2-μF capacitance with twice the plate area and the same distance. (*c*) A 2-μF capacitance with one-half the distance and the same plate area.

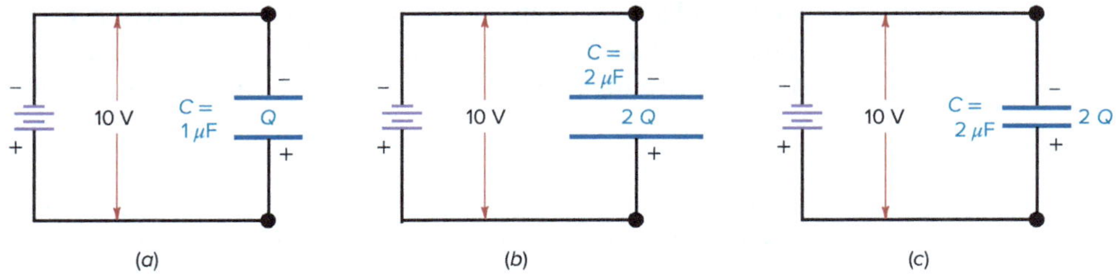

Table 16–1	Dielectric Materials*	
Material	Dielectric Constant K_ϵ	Dielectric Strength, V/mil**
Air or vacuum	1	20
Aluminum oxide	7	
Ceramics	80–1200	600–1250
Glass	8	335–2000
Mica	3–8	600–1500
Oil	2–5	275
Paper	2–6	1250
Plastic film	2–3	
Tantalum oxide	25	

* Exact values depend on the specific composition of different types.
** 1 mil equals one-thousandth of an inch or 0.001 in.

Dielectric Constant K_ϵ

This indicates the ability of an insulator to concentrate electric flux. Its numerical value is specified as the ratio of flux in the insulator compared with the flux in air or vacuum. The dielectric constant of air or vacuum is 1, since it is the reference.

Mica, for example, has an average dielectric constant of 6, which means that it can provide a density of electric flux six times as great as that of air or vacuum for the same applied voltage and equal size. Insulators generally have a **dielectric constant K_ϵ** greater than 1, as listed in Table 16–1. Higher values of K_ϵ allow greater values of capacitance.

Note that the aluminum oxide and tantalum oxide listed in Table 16–1 are used for the dielectric in electrolytic capacitors. Also, plastic film is often used instead of paper for the rolled-foil type of capacitor.

The dielectric constant for an insulator is actually its ***relative permittivity***. The symbol ϵ_r, or K_ϵ, indicates the ability to concentrate electric flux. This factor corresponds to relative permeability, with the symbol μ_r or K_m, for magnetic flux. Both ϵ_r and μ_r are pure numbers without units as they are just ratios.*

These physical factors for a parallel-plate capacitor are summarized by the formula

$$C = K_\epsilon \times \frac{A}{d} \times 8.85 \times 10^{-12}\ \text{F} \qquad \textbf{(16–4)}$$

where A is the area in square meters of either plate, d is the distance in meters between plates, K_ϵ is the dielectric constant, or relative permittivity, as listed in Table 16–1, and C is capacitance in farads. The constant factor 8.85×10^{-12} is the absolute permittivity of air or vacuum, in SI, since the farad is an SI unit.

GOOD TO KNOW

The capacitance of a capacitor is determined only by its physical construction and not by external circuit parameters such as frequency, voltage, etc.

* The absolute permittivity ϵ_0 is 8.854×10^{-12} F/m in SI units for electric flux in air or vacuum. This value corresponds to an absolute permeability μ_0 of $4\pi \times 10^{-7}$ H/m in SI units for magnetic flux in air or a vacuum.

Example 16–6

Calculate C for two plates, each with an area 2 m^2, separated by 1 cm, or 10^{-2} m, with a dielectric of air.

ANSWER Substituting in Formula (16–4),

$$C = 1 \times \frac{2}{10^{-2}} \times 8.85 \times 10^{-12}\ \text{F}$$
$$= 200 \times 8.85 \times 10^{-12}$$
$$= 1770 \times 10^{-12}\ \text{F or } 1770\ \text{pF}$$

This value means that the capacitor can store 1770×10^{-12} C of charge with 1 V. Note the relatively small capacitance, in picofarad units, with the extremely large plates of 2 m^2, which is really the size of a tabletop or a desktop.

If the dielectric used is paper with a dielectric constant of 6, then C will be six times greater. Also, if the spacing between plates is reduced by one-half to 0.5 cm, the capacitance will be doubled. Note that practical capacitors for electronic circuits are much smaller than this parallel-plate capacitor. They use a very thin dielectric with a high dielectric constant, and the plate area can be concentrated in a small space.

Dielectric Strength

Table 16–1 also lists breakdown-voltage ratings for typical dielectrics. ***Dielectric strength*** is the ability of a dielectric to withstand a potential difference without arcing across the insulator. This voltage rating is important because rupture of the insulator provides a conducting path through the dielectric. Then it cannot store charge because the capacitor has been short-circuited. Since the breakdown voltage increases with greater thickness, capacitors with higher voltage ratings have more distance between plates. This increased distance reduces the capacitance, however, all other factors remaining the same.

16–3 Self-Review

Answers at the end of the chapter.

a. A capacitor charged to 100 V has 1000 μC of charge. How much is C?
b. A mica capacitor and ceramic capacitor have the same physical dimensions. Which has more C?

16–4 Typical Capacitors

Commercial capacitors are generally classified according to the dielectric. Most common are air, mica, paper, plastic film, and ceramic capacitors, plus the electrolytic type. Electrolytic capacitors use a molecular-thin oxide film as the dielectric, resulting in large capacitance values in little space. These types are compared in Table 16–2 and discussed in the sections that follow.

Except for electrolytic capacitors, capacitors can be connected to a circuit without regard to polarity, since either side can be the more positive plate. Electrolytic capacitors are marked to indicate the side that must be connected to the positive or negative side of the circuit. *Note that the polarity of the charging source determines*

Table 16–2	Types of Capacitors		
Dielectric	**Construction**	**Capacitance**	**Breakdown, V**
Air	Meshed plates	10–400 pF	400 (0.02-in. air gap)
Ceramic	Tubular	0.5–1600 pF	500–20,000
	Disk	1 pF–1 μF	
Electrolytic	Aluminum	1–6800 μF	10–450
	Tantalum	0.047–330 μF	6–50
Mica	Stacked sheets	10–5000 pF	500–20,000
Paper	Rolled foil	0.001–1 μF	200–1600
Plastic film	Foil or metallized	100 pF–100 μF	50–600

the polarity of the capacitor voltage. Failure to observe the correct polarity can damage the dielectric and lead to the complete destruction of the capacitor.

Mica Capacitors

Thin mica sheets as the dielectric are stacked between tinfoil sections for the conducting plates to provide the required capacitance. Alternate strips of tinfoil are connected and brought out as one terminal for one set of plates, and the opposite terminal connects to the other set of interlaced plates. The construction is shown in Fig. 16–4*a*. The entire unit is generally in a molded Bakelite case. Mica capacitors are often used for small capacitance values of about 10 to 5000 pF; their length is ¾ in. or less with about ⅛-in. thickness. A typical mica capacitor is shown in Fig. 16–4*b*.

Paper Capacitors

In this construction shown in Fig. 16–5*a*, two rolls of tinfoil conductor separated by a paper dielectric are rolled into a compact cylinder. Each outside lead connects to its roll of tinfoil as a plate. The entire cylinder is generally placed in a cardboard container coated with wax or encased in plastic. Paper capacitors are often used for

Figure 16–4 Mica capacitor. (*a*) Physical construction. (*b*) Example of a mica capacitor.

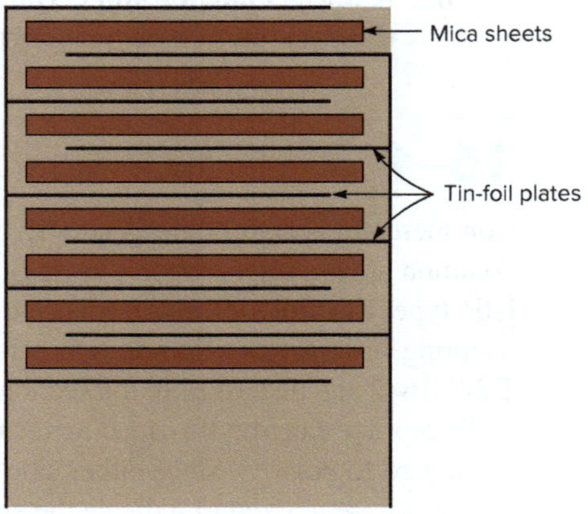

(*a*)

(*b*)

Figure 16–5 Paper capacitor. (*a*) Physical construction. (*b*) Example of a paper capacitor.

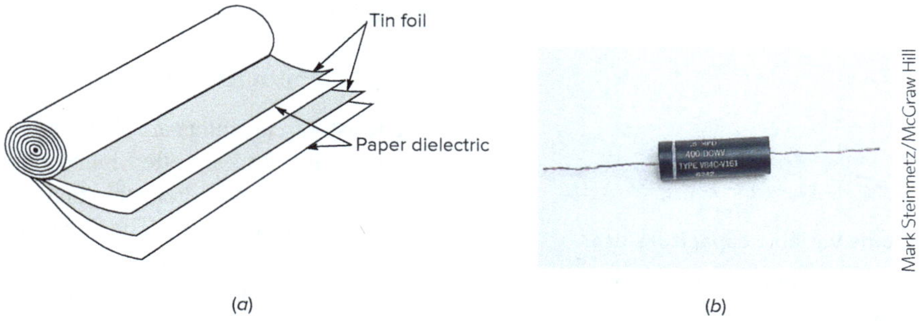

medium capacitance values of 0.001 to 1.0 μF, approximately. The size of a 0.05-μF capacitor is typically 1 in. long and 3⁄8-in. in diameter. A paper capacitor is shown in Fig. 16–5*b*.

A black or a white band at one end of a paper capacitor indicates the lead connected to the outside foil. This lead should be used for the ground or low-potential side of the circuit to take advantage of shielding by the outside foil. There is no required polarity, however, since the capacitance is the same no matter which side is grounded. Also note that in the schematic symbol for *C,* the curved line usually indicates the low-potential side of the capacitor.

Film Capacitors

Film capacitors are constructed much like paper capacitors except that the paper dielectric is replaced with a plastic film such as polypropylene, polystyrene, polycarbonate, or polyethelene terepthalate (Mylar). There are two main types of film capacitors: the foil type and the metallized type. The foil type uses sheets of metal foil, such as aluminum or tin, for its conductive plates. The metallized type is constructed by depositing (spraying) a thin layer of metal, such as aluminum or zinc, on the plastic film. The sprayed-on metal serves as the plates of the capacitor. The advantage of the metallized type over the foil type is that the metallized type is much smaller for a given capacitance value and breakdown voltage rating. The reason is that the metallized type has much thinner plates because they are sprayed on. Another advantage of the metallized type is that it is self-healing. This means that if the dielectric is punctured because its breakdown voltage rating is exceeded, the capacitor is not damaged permanently. Instead, the capacitor heals itself. This is not true of the foil type.

Film capacitors are very temperature-stable and are therefore used frequently in circuits that require very stable capacitance values. Some examples are radio-frequency oscillators and timer circuits. Film capacitors are available with values ranging from about 100 pF to 100 μF. Figure 16–6 shows a typical film capacitor.

Figure 16–6 Film capacitor.

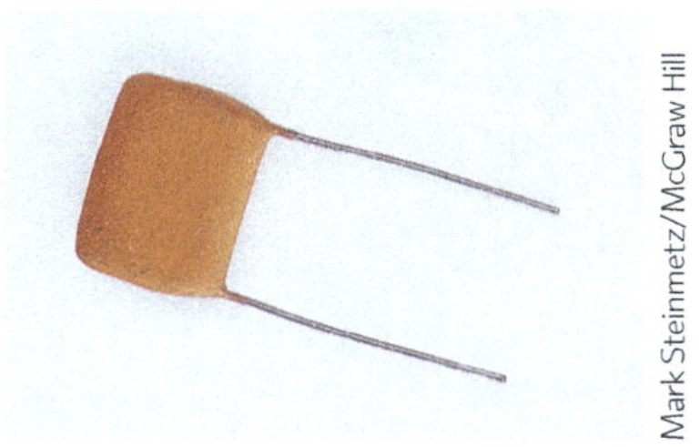

Ceramic Capacitors

The ceramic materials used in ceramic capacitors are made from earth fired under extreme heat. With titanium dioxide or one of several types of silicates, very high values of dielectric constant K_ϵ can be obtained. Most ceramic capacitors come in disk form, as shown in Fig. 16–7. In the disk form, silver is deposited on both sides of the ceramic dielectric to form the capacitor plates. Ceramic capacitors are available with values of 1 pF (or less) up to about 1 μF. The wide range of values is possible because the dielectric constant K_ϵ can be tailored to provide almost any desired value of capacitance.

Figure 16–7 Ceramic disk capacitor.

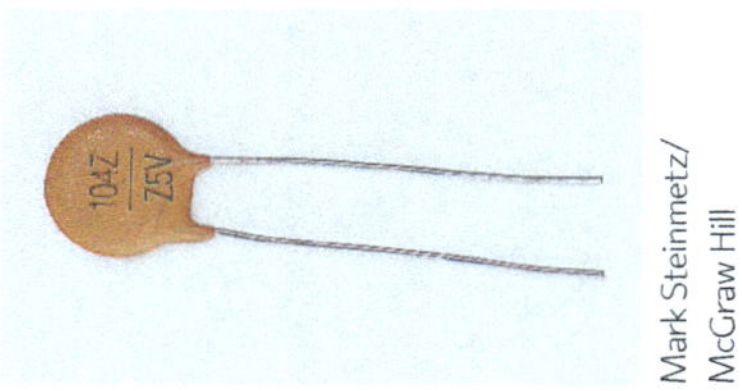

Note that ceramic capacitors are also available in forms other than disks. Some ceramic capacitors are available with axial leads and use a color code similar to that of a resistor.

Surface-Mount Capacitors

Like resistors, capacitors are also available as surface-mounted components. Surface-mounted capacitors are often called *chip capacitors.* Chip capacitors are constructed by placing a ceramic dielectric material between layers of conductive film which form the capacitor plates. The capacitance is determined by the dielectric constant K_e and the physical area of the plates. Chip capacitors are available in many sizes. A common size is 0.125 in. long by 0.063 in. wide in various thicknesses. Another common size is 0.080 in. long by 0.050 in. wide in various thicknesses. Figure 16–8 shows two sizes of chip capacitors. Like chip resistors, chip capacitors have their end electrodes soldered directly to the copper traces of the printed-circuit board. Chip capacitors are available with values ranging from a fraction of a picofarad up to several microfarads.

GOOD TO KNOW

Some variable capacitors use mica, ceramic, or plastic film as the dielectric. These types of capacitors are called trimmer capacitors. Trimmer capacitors typically have values less than 100 pF. Furthermore, trimmer capacitors are not adjusted very often; usually only during the alignment of equipment.

Variable Capacitors

Figure 16–1*b* shows a variable air capacitor. In this construction, the fixed metal plates connected together form the *stator.* The movable plates connected together on the shaft form the *rotor.* Capacitance is varied by rotating the shaft to make the rotor plates mesh with the stator plates. They do not touch, however, since air is the dielectric. Full mesh is maximum capacitance. Moving the rotor completely out of mesh provides minimum capacitance.

A common application is the tuning capacitor in radio receivers. When you tune to different stations, the capacitance varies as the rotor moves in or out of mesh. Combined with an inductance, the variable capacitance then tunes the receiver to a different resonant frequency for each station. Usually two or three capacitor sections are *ganged* on one common shaft.

Figure 16–8 Chip capacitors.

Mark Steinmetz/McGraw Hill

GOOD TO KNOW

To calculate the change in capacitance, ΔC, for a change in temperature, ΔT, use the following equation:

$$\Delta C = \frac{C}{10^6} \times \Delta T \times (\pm \text{ppm})$$

Temperature Coefficient

Ceramic capacitors are often used for temperature compensation to increase or decrease capacitance with a rise in temperature. The temperature coefficient is given in parts per million (ppm) per degree Celsius, with a reference of 25°C. As an example, a negative 750-ppm unit is stated as N750. A positive temperature coefficient of the same value would be stated as P750. Units that do not change in capacitance are labeled NPO.

Capacitance Tolerance

Ceramic disk capacitors for general applications usually have a tolerance of ±20%. For closer tolerances, mica or film capacitors are used. These have tolerance values of ±2 to 20% Silver-plated mica capacitors are available with a tolerance of ±1%.

The tolerance may be less on the minus side to make sure that there is enough capacitance, particularly with electrolytic capacitors, which have a wide tolerance. For instance, a 20-μF electrolytic with a tolerance of −10%, +50% may have a capacitance of 18 to 30 μF. However, the exact capacitance value is not critical in most applications of capacitors for filtering, AC coupling, and bypassing.

Voltage Rating of Capacitors

This rating specifies the maximum potential difference that can be applied across the plates without puncturing the dielectric. Usually the voltage rating is for temperatures up to about 60°C. Higher temperatures result in a lower voltage rating. Voltage ratings for general-purpose paper, mica, and ceramic capacitors are typically 200 to 500 V. Ceramic capacitors with ratings of 1 to 20 kV are also available.

Electrolytic capacitors are typically available in 16-, 35-, and 50-V ratings. For applications where a lower voltage rating is permissible, more capacitance can be obtained in a smaller size.

The potential difference across the capacitor depends on the applied voltage and is not necessarily equal to the voltage rating. A voltage rating higher than the potential difference applied across the capacitor provides a safety factor for long life in service. However, the actual capacitor voltage of electrolytic capacitors should be close to the rated voltage to produce the oxide film that provides the specified capacitance.

The voltage ratings are for DC voltage applied. The breakdown rating is lower for AC voltage because of the internal heat produced by continuous charge and discharge.

Capacitor Applications

In most electronic circuits, a capacitor has DC voltage applied, combined with a much smaller AC signal voltage. The usual function of the capacitor is to block the DC voltage but pass the AC signal voltage by means of the charge and discharge current. These applications include coupling, bypassing, and filtering of AC signals.

16–4 Self-Review

Answers at the end of the chapter.

a. **An electrolytic capacitor must be connected in the correct polarity. (True/False)**
b. **The potential difference across a capacitor is always equal to its maximum voltage rating. (True/False)**
c. **Ceramic and paper capacitors generally have less *C* than electrolytic capacitors. (True/False)**
d. **The letters NPO indicate zero temperature coefficient. (True/False)**

Figure 16–9 Construction of aluminum electrolytic capacitor. (*a*) Internal electrodes. (*b*) Foil rolled into cartridge. (*c*) Typical capacitor with multiple sections.

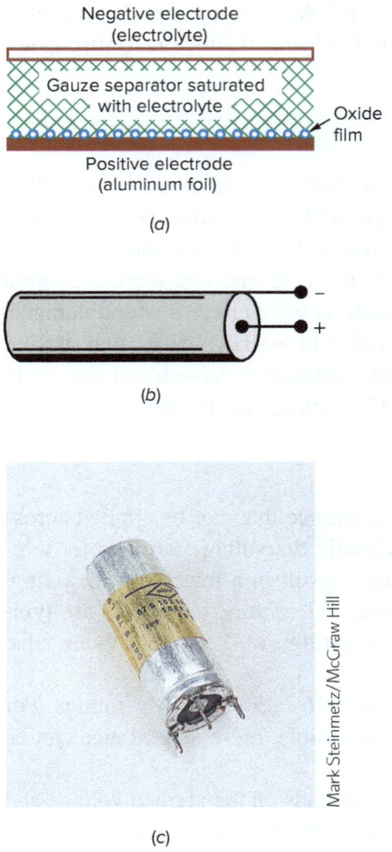

16–5 Electrolytic Capacitors

Electrolytic capacitors are commonly used for *C* values ranging from about 1 to 6800 μF because electrolytics provide the most capacitance in the smallest space with least cost.

Construction

Figure 16–9 shows the aluminum-foil type. The two aluminum electrodes are in an electrolyte of borax, phosphate, or carbonate. Between the two aluminum strips, absorbent gauze soaks up electrolyte to provide the required electrolysis that produces an oxide film. This type is considered a wet electrolytic, but it can be mounted in any position.

When DC voltage is applied to form the capacitance in manufacture, the electrolytic action accumulates a molecular-thin layer of aluminum oxide at the junction between the positive aluminum foil and the electrolyte. The oxide film is an insulator. As a result, capacitance is formed between the positive aluminum electrode and the electrolyte in the gauze separator. The negative aluminum electrode simply provides a connection to the electrolyte. Usually, the metal can itself is the negative terminal of the capacitor, as shown in Fig. 16–9*c*.

Because of the extremely thin dielectric film, very large *C* values can be obtained. The area is increased by using long strips of aluminum foil and gauze, which are rolled into a compact cylinder with very high capacitance. For example, an electrolytic capacitor the same size as a 0.1-μF paper capacitor, but rated at 10 V breakdown, may have 1000 μF of capacitance or more. Higher voltage ratings, up to 450 V, are available, with typical *C* values up to about 6800 μF. The very high *C* values usually have lower voltage ratings.

Polarity

Electrolytic capacitors are used in circuits that have a combination of DC voltage and AC voltage. The DC voltage maintains the required polarity across the electrolytic capacitor to form the oxide film. A common application is for electrolytic filter capacitors to eliminate the 60- or 120-Hz AC ripple in a DC power supply. Another use is for audio coupling capacitors in transistor amplifiers. In both applications, for filtering or coupling, electrolytics are needed for large *C* with a low-frequency AC component, whereas the circuit has a DC component for the required voltage polarity. Incidentally, the difference between filtering out an AC component or coupling it into a circuit is only a question of parallel or series connections. The filter capacitors for a power supply are typically 100 to 1000 μF. Audio capacitors are usually 10 to 47 μF.

If the electrolytic is connected in opposite polarity, the reversed electrolysis forms gas in the capacitor. It becomes hot and may explode. This is a possibility only with electrolytic capacitors.

Leakage Current

The disadvantage of electrolytics, in addition to the required polarization, is their relatively high **leakage current** compared with other capacitors, since the oxide film is not a perfect insulator. The problem with leakage current in a capacitor is that it allows part of the DC component to be coupled into the next circuit along with the AC component. In newer electrolytic capacitors, the leakage current is quite small. Section 16–10 takes a closer look at leakage current in capacitors.

Nonpolarized Electrolytics

This type is available for applications in circuits without any DC polarizing voltage, as in a 60-Hz AC power line. One application is the starting capacitor for

GOOD TO KNOW

The DC leakage current of an electrolytic capacitor is usually specified as a product of capacitance, voltage, and some decimal fraction. For example, the leakage current of an electrolytic may be specified as $I = 0.01\ CV + 3\ \mu A$ maximum. The DC leakage current is usually measured at 25°C with the rated DC working voltage applied.

Figure 16–10 Tantalum capacitors.

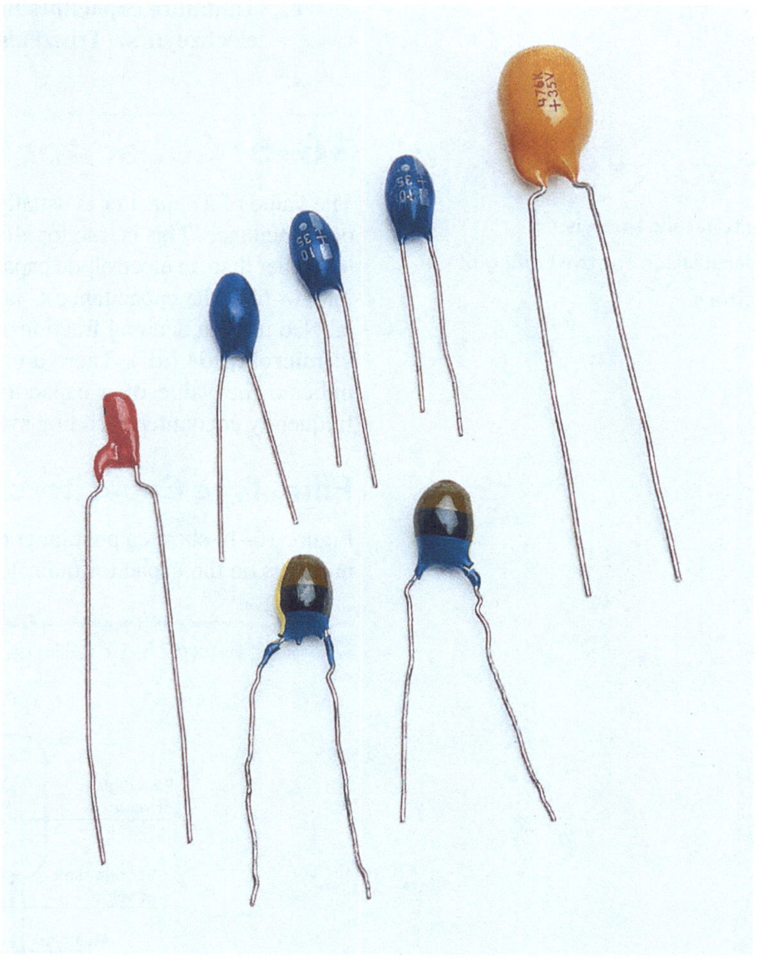

Mark Steinmetz/McGraw Hill

AC motors. A nonpolarized electrolytic actually contains two capacitors, connected internally in series-opposing polarity.

Tantalum Capacitors

This is another form of electrolytic capacitor, using tantalum (Ta) instead of aluminum. Titanium (Ti) is also used. Typical tantalum capacitors are shown in Fig. 16–10. They feature

1. Larger *C* in a smaller size
2. Longer shelf life
3. Less leakage current

However, tantalum electrolytics cost more than the aluminum type. Construction of tantalum capacitors include the wet-foil type and a solid chip or slug. The solid tantalum is processed in manufacture to have an oxide film as the dielectric. Referring back to Table 16–1, note that tantalum oxide has a dielectric constant of 25, compared with 7 for aluminum oxide.

16–5 Self-Review

Answers at the end of the chapter.

a. The rating of 1000 μF at 25 V is probably for an electrolytic capacitor. (True/False)

b. **Electrolytic capacitors allow more leakage current than mica capacitors. (True/False)**
c. **Tantalum capacitors have a longer shelf life than aluminum electrolytics. (True/False)**

GOOD TO KNOW

Unfortunately, there is no standardization for the coding of capacitors.

16–6 Capacitor Coding

The value of a capacitor is usually specified in either microfarad or picofarad units of capacitance. This is true for all types of capacitors. As a general rule, if a capacitor (other than an electrolytic capacitor) is marked using a whole number such as 33, 220, or 680, the capacitance *C* is in **picofarads (pF)**. Conversely, if a capacitor is labeled using a decimal fraction such as 0.1, 0.047, or 0.0082, the capacitance *C* is in **microfarads (μF)**. There are a variety of ways in which a manufacturer may indicate the value of a capacitor. What follows is an explanation of the most frequently encountered coding systems.

Film-Type Capacitors

Figure 16–11 shows a popular coding system for film-type capacitors. The first two numbers on the capacitor indicate the first two digits in the numerical value of the

MultiSim **Figure 16–11** Film capacitor coding system.

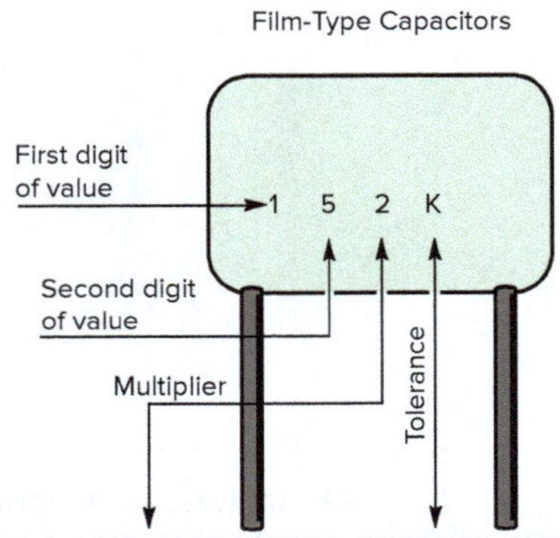

Multiplier		Tolerance of Capacitor		
For the Number	Multiplier	Letter	10 pF or Less	Over 10 pF
0	1	B	±0.1 pF	
1	10	C	±0.25 pF	
2	100	D	±0.5 pF	
3	1,000	F	±1.0 pF	±1%
4	10,000	G	±2.0 pF	±2%
5	100,000	H		±3%
		J		
8	0.01	K		±10%
9	0.1	M		±20%

Examples:
152K = 15 × A = 1500 pF or 0.0015 μF, ±10%
759J = 75 × 0.1 = 7.5 pF, ±5%

Note: The letter R may be used at times to signify a decimal point, as in 2R2 = 2.2 (pF or μF).

capacitance. The third number is the *multiplier,* indicating by what factor the first two digits must be multiplied. The letter at the far right indicates the capacitor's tolerance. In this coding system, the capacitance is always in picofarad units. The capacitor's breakdown voltage rating is usually printed on the body directly below the coded value of capacitance.

Example 16–7

Determine the value of capacitance for the film capacitors in Fig. 16–12*a* and *b*.

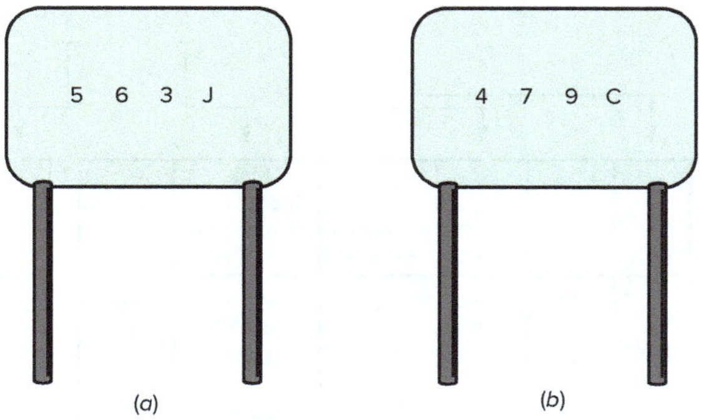

Figure 16–12 Film capacitors for Example 16–7.

ANSWER In Fig. 16–12*a*, the first two numbers are 5 and 6, respectively, for 56 as the first two digits in the numerical value of the capacitance. The third number, 3, indicates a multiplier of 1000, or 56 × 1000 = 56,000 pF. The letter J indicates a capacitor tolerance of ±5%.

In Fig. 16–12*b*, the first two numbers are 4 and 7, respectively, for 47 as the first two digits in the numerical value of the capacitance. The third number, 9, indicates a fractional multiplier of 0.1, or 47 × 0.1 = 4.7 pF. The letter C indicates a capacitor tolerance of ±0.25 pF.

Ceramic Disk Capacitors

Figure 16–13 shows how most ceramic disk capacitors are marked to indicate their capacitance. As you can see, the capacitance is expressed either as a whole number or as a decimal fraction. The type of coding system used depends on the manufacturer. Ceramic disk capacitors are often used for coupling and bypassing AC signals, where it is allowable to have a wide or lopsided tolerance.

Example 16–8

In Fig. 16–14, determine (a) the capacitance value and tolerance; (b) the temperature-range identification information.

ANSWER (a) Since the capacitance is expressed as a decimal fraction, its value is in microfarads. In this case, $C = 0.047\ \mu$F. The letter Z, to the right of 0.047, indicates a capacitor tolerance of +80%, −20%. Notice that the actual capacitance value can be as much as 80% above its coded value but only 20% below its coded value.

Figure 16–13 Ceramic disk capacitor coding system.

Ceramic Disk Capacitors

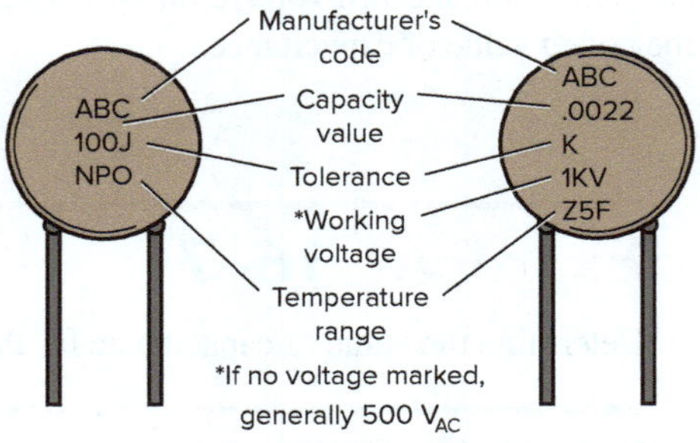

Typical Ceramic Disk Capacitor Markings

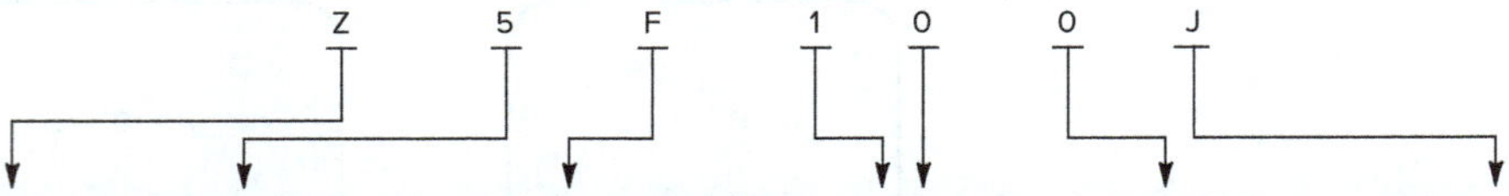

Low Temp.	Letter Symbol	High Temp.	Numerical Symbol	Max. Capacitance Change over Temp. Range	Letter Symbol
+10°C	Z	+45°C	2	+1.0%,	A
−30°C	Y	+65°C	4	±1.5%,	B
−55°C	X	+85°C	5	±1.1%,	C
		+105°C	6	±3.3%,	D
		+125°C	7	±4.7%,	E
				±7.5%,	F
				±10.0%	P
				±15.0%	R
				±22.0%	S
				+22%, −33%	T
				+22%, −56%	U
				+22%, −82%	V

Temperature Range Identification of Ceramic Disk Capacitors

1st & 2nd Fig. of Capacitance	Multiplier	Numerical Symbol	Tolerance on Capacitance	Letter Symbol
	1	0		
	10	1		
	100	2	±5%	J
	1,000	3	±10%	K
	10,000	4	±20%	M
	100,000	5	+100%, −0%	P
		—	+80%, −20%	Z
		—		
	0.01	8		
	0.1	9		

Capacity Value and Tolerance of Ceramic Disk Capacitors

Figure 16–14 Ceramic disk capacitor for Example 16–8.

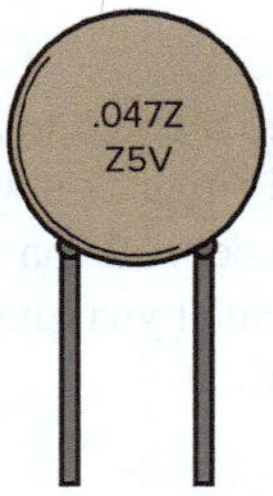

(b) The alphanumeric code, *Z*5V, printed below the capacitance value, provides additional capacitor information. Referring to Fig. 16–13, note that the letter Z and number 5 indicate the low and high temperatures of +10°C and +85°C, respectively. The letter V indicates that the maximum capacitance change over the specified temperature range (+10°C to +85°C) is +22%, −82%. For temperature changes less than the range indicated, the percent change in capacitance will be less than that indicated.

Chip Capacitors

Before determining the capacitance value of a chip capacitor, make sure it is a capacitor and not a resistor. Chip capacitors have the following identifiable features:

1. The body is one solid color, such as off-white, beige, gray, tan, or brown.
2. The end electrodes completely enclose the end of the part.

Figure 16–15 Chip capacitor coding system.

Value (33 Value Symbols)—Uppercase and Lowercase Letters					Multiplier
A-1.0	H-2.0	b-3.5	f-5.0	X-7.5	0 = × 1.0
B-1.1	J-2.2	P-3.6	T-5.1	t-8.0	1 = × 10
C-1.2	K-2.4	Q-3.9	U-5.6	Y-8.2	2 = × 100
D-1.3	a-2.5	d-4.0	m-6.0	y-9.0	3 = × 1,000
E-1.5	L-2.7	R-4.3	V-6.2	Z-9.1	4 = × 10,000
F-1.6	M-3.0	e-4.5	W-6.8		5 = × 100,000
G-1.8	N-3.3	S-4.7	n-7.0		etc.

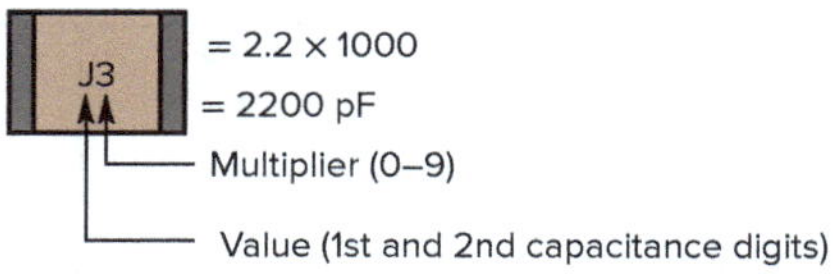

Three popular coding systems are used by manufacturers of chip capacitors. In all three systems, the values represented are in picofarads. One system, shown in Fig. 16–15, uses a two-place system in which a letter indicates the first and second digits of the capacitance value and a number indicates the multiplier (0 to 9). Thirty-three symbols are used to represent the two significant figures. The symbols used include 24 uppercase letters and 9 lowercase letters. In Fig. 16–15, note that J3 represents 2200 pF.

Another system, shown in Fig. 16–16, also uses two places. In this case, however, values below 100 pF are indicated using two numbers from which the capacitance value is read directly. Values above 100 pF are indicated by a letter and a number as before. In this system, only 24 uppercase letters are used. Also note that the alphanumeric codes in this system are 10 times higher than those in the system shown in Fig. 16–15.

Figure 16–17 shows yet another system, in which a single letter or number is used to designate the first two digits in the capacitance value. The multiplier is determined by the color of the letter. In the example shown, an orange-colored W represents a capacitance C of 4.7 pF.

Figure 16–16 Chip capacitor coding system.

Alternate Two-Place Code
• Values below 100 pF—Value read directly

• Values 100 pF and above—Letter/number code

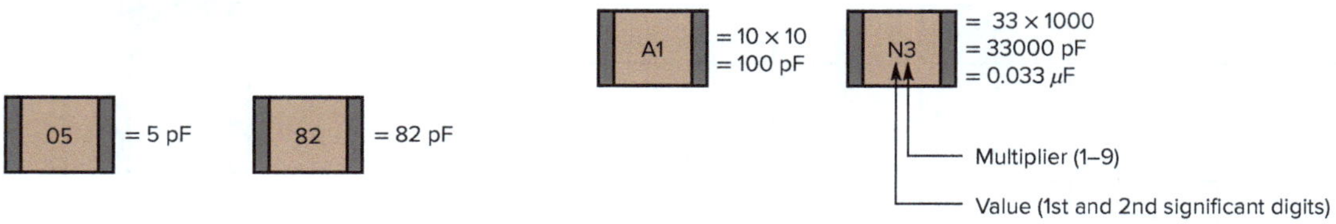

Value (24 Value Symbols)—Uppercase Letters Only					Multiplier
A-10	F-16	L-27	R-43	W-68	1 = × 10
B-11	G-18	M-30	S-47	X-75	2 = × 100
C-12	H-20	N-33	T-51	Y-82	3 = × 1,000
D-13	J-22	P-36	U-56	Z-91	4 = × 10,000
E-15	K-24	Q-39	V-62		5 = × 100,000 etc.

Figure 16–17 Chip capacitor coding system.

Standard Single-Place Code

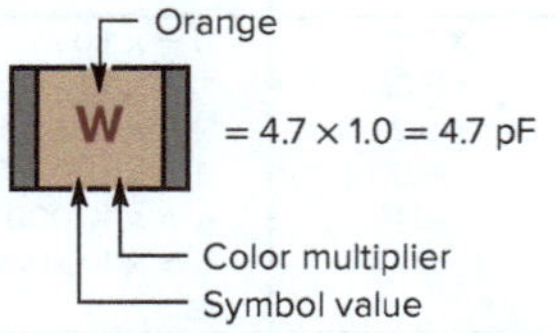

Examples: R (Green) = 3.3 × 100 = 330 pF
7 (Blue) = 8.2 × 1000 = 8200 pF

Value (24 Value Symbols)—Uppercase Letters and Numerals					Multiplier (Color)
A-1.0 B-1.1 C-1.2 D-1.3 E-1.5	H-1.6 I-1.8 J-2.0 K-2.2 L-2.4	N-2.7 O-3.0 R-3.3 S-3.6 T-3.9	V-4.3 W-4.7 X-5.1 Y-5.6 Z-6.2	3-6.8 4-7.5 7-8.2 9-9.1	Orange = × 1.0 Black = × 10 Green = × 100 Blue = × 1,000 Violet = × 10,000 Red = × 100,000

Note that other coding systems are used for chip capacitors; these systems are not covered here. However, the three coding systems shown in this section are the most common systems presently in use. Also note that some chip capacitors found on printed-circuit boards are not marked or coded. When this is the case, the only way to determine the capacitance value is to check it with a capacitance tester.

Tantalum Capacitors

Tantalum capacitors are frequently coded to indicate their capacitance in picofarads. Figure 16–18 shows how to interpret this system.

Figure 16–18 Tantalum capacitor coding system.

Color	Rated Voltage	Capacitance in Picofarads 1st Figure	2nd Figure	Multiplier
Black	4	0	0	—
Brown	6	1	1	—
Red	10	2	2	—
Orange	15	3	3	—
Yellow	20	4	4	10,000
Green	25	5	5	100,000
Blue	35	6	6	1,000,000
Violet	50	7	7	10,000,000
Gray	—	8	8	—
White	3	9	9	—

Dipped Tantalum Capacitors

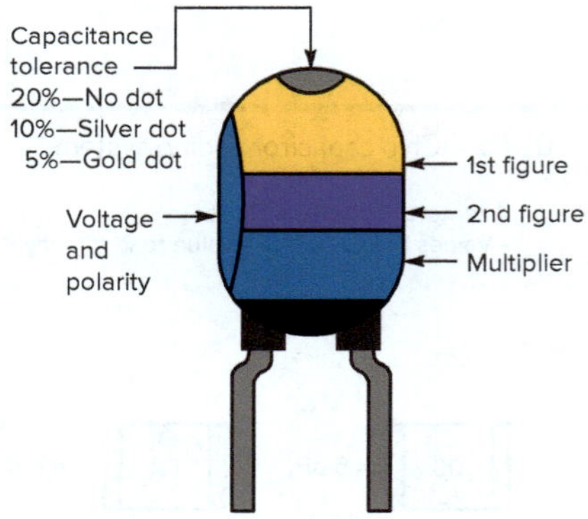

Figure 16–19 Tantalum capacitor for Example 16–10.

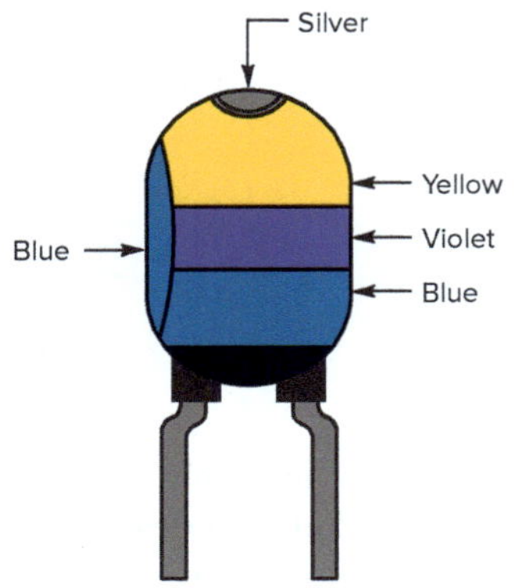

Example 16–9

For the tantalum capacitor shown in Fig. 16–19, determine the capacitance C in both pF and μF units. Also, determine the voltage rating and tolerance.

ANSWER Moving from top to bottom, the first two color bands are yellow and violet, which represent the digits 4 and 7, respectively. The third color band is blue, indicating a multiplier of 1,000,000. Therefore the capacitance C is $47 \times 1{,}000{,}000 = 47{,}000{,}000$ pF, or 47 μF. The blue color at the left indicates a voltage rating of 35 V. And, finally, the silver dot at the very top indicates a tolerance of $\pm 10\%$.

16–6 Self-Review

Answers at the end of the chapter.

a. **A ceramic disk capacitor that is marked .01 has a capacitance of 0.01 pF. (True/False)**
b. **A film capacitor that is marked 224 has a capacitance of 220,000 pF. (True/False)**
c. **A chip capacitor has a green letter E marked on it. Its capacitance is 150 pF. (True/False)**
d. **A ceramic disk capacitor is marked .001P. Its tolerance is +100%, −0%. (True/False)**

Figure 16–20 Capacitances in parallel.

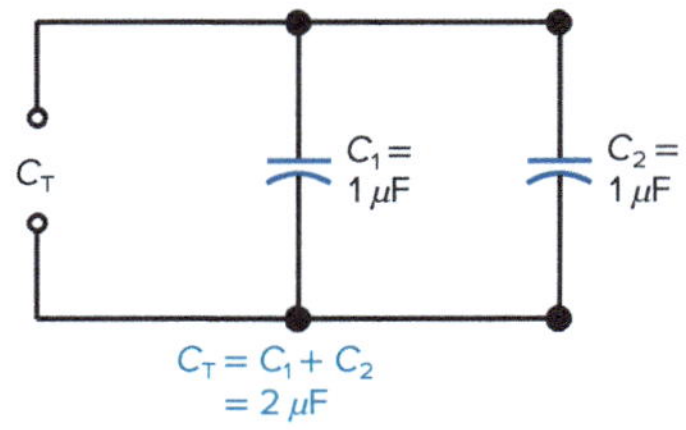

16–7 Parallel Capacitances

Connecting capacitances in parallel is equivalent to adding the plate areas. Therefore, the total capacitance is the sum of the individual capacitances. As illustrated in Fig. 16–20,

$$C_T = C_1 + C_2 + \cdots + \text{etc.} \qquad \textbf{(16–5)}$$

A 10-μF capacitor in parallel with a 5-μF capacitor, for example, provides a 15-μF capacitance for the parallel combination. The voltage is the same across the parallel capacitors. Note that adding parallel capacitances is opposite to inductances in parallel and resistances in parallel.

16–7 Self-Review

Answers at the end of the chapter.

a. **How much is C_T for 0.01 μF in parallel with 0.02 μF?**
b. **What C must be connected in parallel with 100 pF to make C_T 250 pF?**

16–8 Series Capacitances

Connecting capacitances in series is equivalent to increasing the thickness of the dielectric. Therefore, the combined capacitance is less than the smallest individual value. As shown in Fig. 16–21, the combined equivalent capacitance is calculated by the reciprocal formula:

Figure 16–21 Capacitances in series.

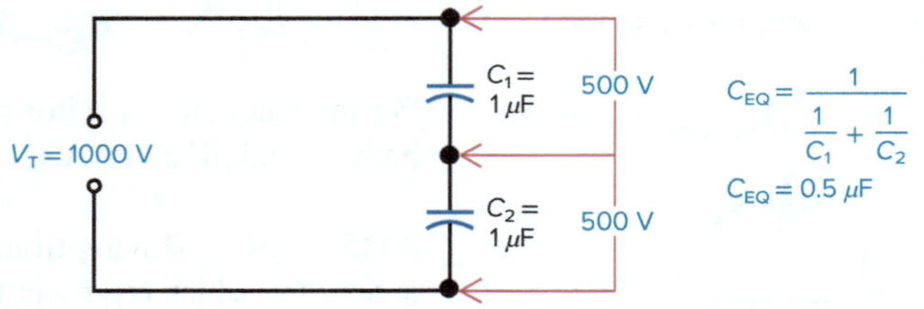

$$C_{EQ} = \frac{1}{\frac{1}{C_1} + \frac{1}{C_2} + \frac{1}{C_3} + \cdots + \text{etc.}} \qquad \textbf{(16–6)}$$

Any of the shortcut calculations for the reciprocal formula apply. For example, the combined capacitance of two equal capacitances of 10 μF in series is 5 μF.

Capacitors are used in series to provide a higher working voltage rating for the combination. For instance, each of three equal capacitances in series has one-third the applied voltage.

Division of Voltage across Unequal Capacitances

In series, the voltage across each C is inversely proportional to its capacitance, as illustrated in Fig. 16–22. The smaller capacitance has the larger proportion of the applied voltage. The reason is that the series capacitances all have the same charge because they are in one current path. With equal charge, a smaller capacitance has a greater potential difference.

We can consider the amount of charge in the series capacitors in Fig. 16–22. Let the charging current be 600 μA flowing for 1 s. The charge Q equals $I \times t$ or 600 μC. Both C_1 and C_2 have Q equal to 600 μC because they are in the same series path for charging current.

Although the charge is the same in C_1 and C_2, they have different voltages because of different capacitance values. For each capacitor, $V = Q/C$. For the two capacitors in Fig. 16–22, then,

$$V_1 = \frac{Q}{C_1} = \frac{600\ \mu\text{C}}{1\ \mu\text{F}} = 600\ \text{V}$$

$$V_2 = \frac{Q}{C_2} = \frac{600\ \mu\text{C}}{2\ \mu\text{F}} = 300\ \text{V}$$

GOOD TO KNOW

For two capacitors, C_1 and C_2 in series, the individual capacitor voltages can be calculated using the following equations:

$$V_{C_1} = \frac{C_2}{C_1 + C_2} \times V_T.$$

$$V_{C_2} = \frac{C_1}{C_1 + C_2} \times V_T.$$

Charging Current for Series Capacitances

The charging current is the same in all parts of the series path, including the junction between C_1 and C_2, even though this point is separated from the source voltage by two insulators. At the junction, the current is the result of electrons repelled by the negative plate of C_2 and attracted by the positive plate of C_1. The amount of current

Figure 16–22 With series capacitors, the smaller C has more voltage for the same charge.

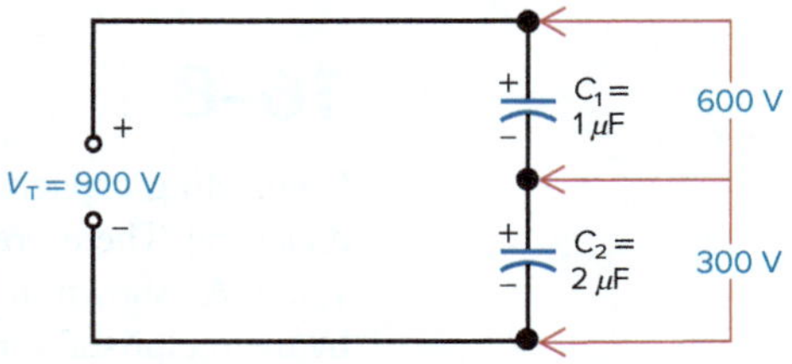

in the circuit is determined by the equivalent capacitance of C_1 and C_2 in series. In Fig. 16–22, the equivalent capacitance is ⅔ μF.

16–8 Self-Review

Answers at the end of the chapter.

a. How much is C_{EQ} for two 0.2-μF capacitors in series?
b. With 50 V applied across both, how much is V_C across each capacitor?
c. How much is C_{EQ} for 100 pF in series with 50 pF?

16–9 Energy Stored in Electrostatic Field of Capacitance

The electrostatic field of the charge stored in a dielectric has electric energy supplied by the voltage source that charges *C*. This energy is stored in the dielectric. The proof is the fact that the capacitance can produce discharge current when the voltage source is removed. The electric energy stored is

$$\text{Energy} = \mathscr{E} = \tfrac{1}{2}\,CV^2 \text{ (joules)} \qquad \textbf{(16–7)}$$

where *C* is the capacitance in farads, *V* is the voltage across the capacitor, and $\mathscr{E}$ is the electric energy in joules. For example, a 1-μF capacitor charged to 400 V has stored energy equal to

$$\mathscr{E} = \tfrac{1}{2}\,CV^2 = \frac{1 \times 10^{-6} \times (4 \times 10^2)^2}{2}$$

$$= \frac{1 \times 10^{-6} \times (16 \times 10^4)}{2} = 8 \times 10^{-2}$$

$$= 0.08 \text{ J}$$

This 0.08 J of energy is supplied by the voltage source that charges the capacitor to 400 V. When the charging circuit is opened, the stored energy remains as charge in the dielectric. With a closed path provided for discharge, the entire 0.08 J is available to produce discharge current. As the capacitor discharges, the energy is used in producing discharge current. When the capacitor is completely discharged, the stored energy is zero.

The stored energy is the reason that a charged capacitor can produce an electric shock, even when not connected in a circuit. When you touch the two leads of the charged capacitor, its voltage produces discharge current through your body. Stored energy greater than 1 J can be dangerous from a capacitor charged to a voltage high enough to produce an electric shock.

Example 16–10

The high-voltage circuit for a color picture tube can have 30 kV across 500 pF of *C*. Calculate the stored energy.

ANSWER

$$\mathscr{E} = \tfrac{1}{2}\,CV^2 = \frac{500 \times 10^{-12} \times (30 \times 10^3)^2}{2}$$

$$= 250 \times 10^{-12} \times 900 \times 10^6$$

$$= 225 \times 10^{-3}$$

$$= 0.225 \text{ J}$$

16–9 Self-Review

Answers at the end of the chapter.

a. The stored energy in *C* increases with more *V*. (True/False)
b. The stored energy decreases with less *C*. (True/False)

16–10 Measuring and Testing Capacitors

A *capacitance meter* is a piece of test equipment specifically designed to measure the capacitance value of capacitors. Although capacitance meters can be purchased as stand-alone units, many handheld and benchtop digital multimeters (DMMs) are capable of measuring a wide range of capacitance values. For example, the benchtop DMM shown in Fig. 16–23 has five capacitance ranges: 2 nF, 20 nF, 200 nF, 2000 nF, and 20 μF. To measure the value of a capacitor using this meter, insert the leads of the capacitor into the capacitance socket, labeled CX, located in the upper right-hand corner of the meter. Next, depress the CX (capacitance) button and select the desired capacitance range. The meter will display the measured capacitance value. For best accuracy, always select the lowest range setting that still displays the measured capacitance value. Note that the polarity markings next to the capacitance socket need to be observed when electrolytic capacitors are inserted. For nonelectrolytic capacitors, lead polarity does not matter. Before inserting any capacitor in the socket, it must be fully discharged to avoid damage to the meter.

Recall from Section 16–6 that capacitors are always coded in either microfarad or picofarad units but never in nanofarad units. Although this is standard industry practice, you will nevertheless encounter the nanofarad unit of capacitance when you use meters capable of measuring capacitance, such as that shown in Fig. 16–23. Therefore, it is important to know how to convert between the nanofarad unit and either microfarad or picofarad units. To convert from nanofarad units to picofarad units, simply move the decimal point three places to the right. For example, 33 nF $= 33 \times 10^{-9}$ F $= 33{,}000 \times 10^{-12}$ F $= 33{,}000$ pF. To convert from nanofarads to microfarads, move the decimal point three places to the left. For another example, 470 nF $= 470 \times 10^{-9}$ F $= 0.47 \times 10^{-6}$ F $= 0.47$ μF. When using meters having nanofarad capacitance ranges, you will need to make these conversions to compare the measured value of capacitance with the coded value.

Figure 16–23 Typical DMM with capacitance measurement capability.

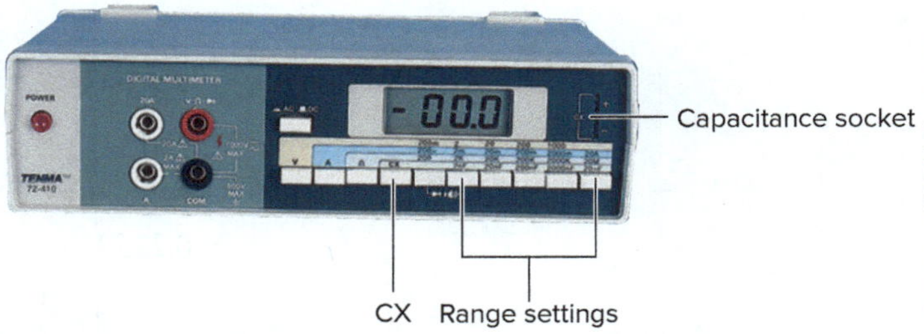

Courtesy of MCM Electronics

Example 16–11

Suppose a film capacitor, coded 393J, is being measured using the meter shown in Fig. 16–23. If the meter reads 37.6 on the 200-nF range, (a) What is the capacitance value in picofarad units? (b) Is the measured capacitance value within its specified tolerance?

ANSWER The capacitor code, 393J, corresponds to a capacitance value of 39,000 pF ±5%. (a) A reading of 37.6 on the 200-nF range corresponds to a capacitance of 37.6 nF. To convert 37.6 nF to picofarad units, move the decimal point three places to the right. This gives an answer of 37,600 pF. (b) The acceptable capacitance range is calculated as follows: 39,000 pF × 0.05 = ±1950 pF. Therefore, the measured value of capacitance can range anywhere from 37,050 pF to 40,950 pF and still be considered within tolerance. Note that in nanofarad units, this corresponds to a range of 37.05 to 40.95 nF. Since the measured value of 37.6 nF falls within this range, the measured capacitance value is within tolerance.

Figure 16–24 Leakage resistance R_ℓ of a capacitor.

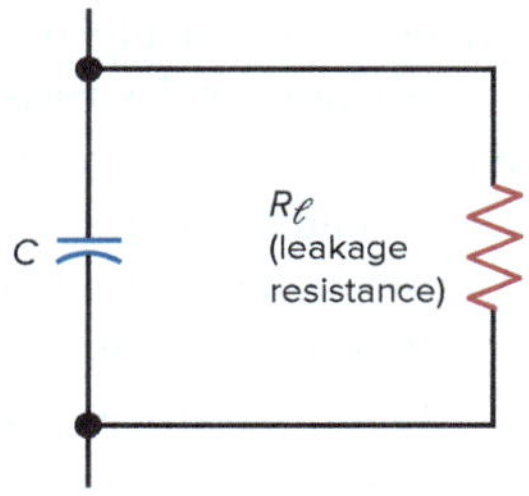

Leakage Resistance of a Capacitor

Consider a capacitor charged by a DC voltage source. After the charging voltage is removed, a perfect capacitor would hold its charge indefinitely. Because there is no such thing as a perfect insulator, however, the charge stored in the capacitor will eventually leak or bleed off, thus neutralizing the capacitor. There are three leakage paths through which the capacitor might discharge: (1) leakage through the dielectric, (2) leakage across the insulated case or body between the capacitor leads, and (3) leakage through the air surrounding the capacitor. For paper, film, mica, and ceramic, the leakage current is very slight, or inversely, the **leakage resistance** is very high. The combination of all leakage paths can be represented as a single parallel resistance R_ℓ across the capacitor plates, as shown in Fig. 16–24. For paper, film, mica, and ceramic capacitors, the leakage resistance R_ℓ is typically 100,000 MΩ or more. The leakage resistance is much less for larger capacitors such as electrolytics, however, with a typical value of R_ℓ ranging from about 500 kΩ up to 10 MΩ. In general, the larger the capacitance of a capacitor, the lower its leakage resistance. Note that the leakage current in capacitors is fairly temperature-sensitive. The higher the temperature, the greater the leakage current (because of lower leakage resistance).

The leakage resistance of a capacitor can be measured with a DMM or an analog ohmmeter, but this is not the best way to test a capacitor for leakage. The best way is to measure the leakage current in the capacitor while the rated working voltage is applied across the capacitor plates. A capacitor is much more likely to show leakage when the dielectric is under stress from the applied voltage. In fact, a capacitor may not show any leakage at all until the dielectric is under stress from the applied voltage. To measure the value of a capacitor and test it for leakage, technicians often use a capacitor-inductor analyzer like that shown in Fig. 16–25. This analyzer allows the user to apply the rated working voltage to the capacitor while testing for leakage. The amount of leakage acceptable depends on the type of capacitor. Most nonpolarized capacitors should have no leakage at all, whereas electrolytics will almost always show some. Pull-out charts showing the maximum allowable leakage for the most common electrolytic capacitors are usually provided with a capacitor-inductor analyzer.

Figure 16–25 Capacitor-inductor analyzer.

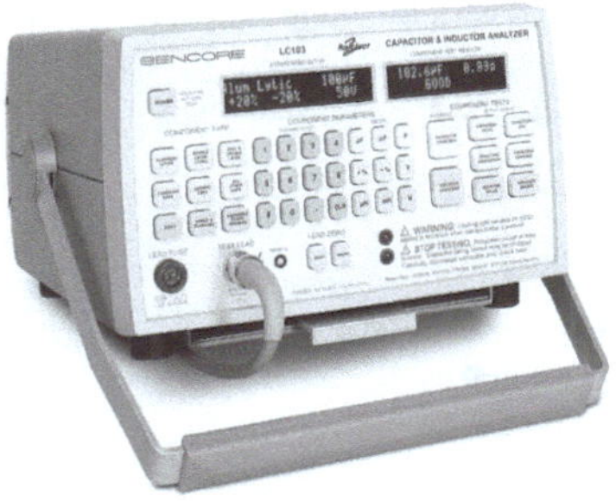

Dielectric Absorption

Dielectric absorption is the inability of a capacitor to completely discharge to zero. It is sometimes referred to as *battery action* or *capacitor memory* and is due to the dielectric of the capacitor retaining a charge after it is supposedly discharged. The effect of dielectric absorption is that it reduces the capacitance value of the capacitor. All capacitors have at least some dielectric absorption, but electrolytics have the highest amount. Dielectric absorption has an undesirable effect on circuit operation if it becomes excessive. The dielectric absorption of a capacitor can be checked using the capacitor-inductor analyzer in Fig. 16–25. Note that there is no way to test for dielectric absorption with an ohmmeter.

Equivalent Series Resistance (ESR)

With AC voltage applied to a capacitor, the continuous charge, discharge, and reverse charging action cannot be followed instantaneously in the dielectric. This corresponds to hysteresis in magnetic materials. With a high-frequency charging voltage applied to the capacitor, there may be a difference between the amount of AC voltage applied to the capacitor and the actual AC voltage across the dielectric. The difference, or loss, can be attributed to the effects of hysteresis in the dielectric. As you might expect, dielectric hysteresis losses increase with frequency.

All losses in a capacitor can be represented as a resistor either in series or in parallel with an ideal capacitor. For example, the losses from dielectric hysteresis can be represented as a single resistor in series with the capacitor as shown in Fig. 16–26*a*. The other resistor shown in series with the capacitor represents the resistance of the capacitor leads and plates. It also includes any resistance at the point where the capacitor leads are bonded to the metal plates. As before, the leakage resistance R_ℓ is shown directly in parallel with the capacitor. Collectively, the resistances shown in Fig. 16–26*a* can be lumped into one **equivalent series resistance (ESR)** as shown in Fig. 16–26*b*. This is an accurate and convenient way to represent all losses in a capacitor. Ideally, the ESR of a capacitor should be zero. For paper, film, ceramic, and mica capacitors, the ESR value is approximately zero. For electrolytics, however, the ESR may be several ohms or more depending on the way they are constructed. Note that ESR is most often a problem in capacitors used in high-frequency filtering applications. For example, most computers use switching power supplies to power the computer. These power supplies require capacitors for filtering high frequencies. In these applications, a high ESR interferes with the normal

Figure 16–26 Resistances representing losses in a capacitor. (*a*) Series and parallel resistance represents capacitor losses. (*b*) Equivalent series resistance (ESR) represents the total losses in a capacitor.

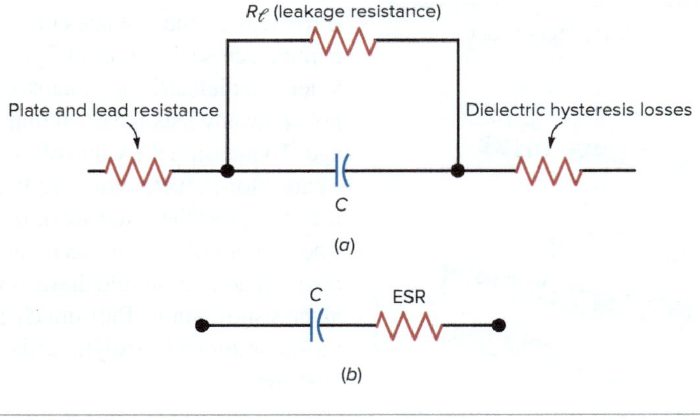

filtering action of the capacitor and therefore causes improper circuit operation. In some cases, the power dissipated by the ESR may cause the capacitor to overheat.

The ESR of a capacitor cannot be checked with an ohmmeter because the ESR is in series with the very high resistance of the dielectric. To check a capacitor for ESR, you must use a capacitor-inductor analyzer like that shown in Fig. 16–25. Pull-out charts showing the maximum allowable ESR for different types of capacitors are usually provided with the analyzer.

16–10 Self-Review

Answers at the end of the chapter.

a. A 150-nF capacitor is the same as a 0.15-μF capacitor. (True/False)
b. It is best to test a capacitor for leakage with the rated working voltage applied. (True/False)
c. Ideally, the ESR of an electrolytic capacitor should be infinite. (True/False)
d. Dielectric absorption in a capacitor can be detected with an ohmmeter. (True/False)

16–11 Troubles in Capacitors

Capacitors can become open or short-circuited. In either case, the capacitor is useless because it cannot store charge. A leaky capacitor is equivalent to a partial short circuit where the dielectric gradually loses its insulating properties under the stress of applied voltage, thus lowering its resistance. A good capacitor has very high resistance of the order of several megohms; a short-circuited capacitor has zero ohms resistance, or continuity; the resistance of a leaky capacitor is lower than normal. Capacitor-inductor analyzers, like that shown in Fig. 16–25, should be used to test a capacitor. However, if a capacitor-inductor analyzer is not available, an ohmmeter (preferably analog) may be able to identify the problem. What follows is a general procedure for testing capacitors using an analog ohmmeter.

Checking Capacitors with an Ohmmeter

A capacitor usually can be checked with an ohmmeter. The highest ohm range, such as $R \times 1\ \text{M}\Omega$ is preferable. Also, disconnect one side of the capacitor from the circuit to eliminate any parallel resistance paths that can lower the resistance. Keep your fingers off the connections, since body resistance lowers the reading.

As shown in Fig. 16–27, the ohmmeter leads are connected across the capacitor. For a good capacitor, the meter pointer moves quickly toward the low-resistance side of the scale and then slowly recedes toward infinity. When the pointer stops moving, the reading is the dielectric resistance of the capacitor, which is normally very high. For paper, film, mica, and ceramic capacitors, the resistance is usually so high that the needle of the meter rests on the infinity mark (∞). However, electrolytic capacitors will usually measure a much lower resistance of about 500 kΩ to 10 MΩ. In all cases, discharge the capacitor before checking with the ohmmeter.

Figure 16–27 Checking a capacitor with an ohmmeter. The *R* scale is shown right to left, as on a VOM. Use the highest ohms range. (*a*) Capacitor action as needle is moved by the charging current from the battery in the ohmmeter. (*b*) Practically infinite leakage resistance reading after the capacitor has been charged.

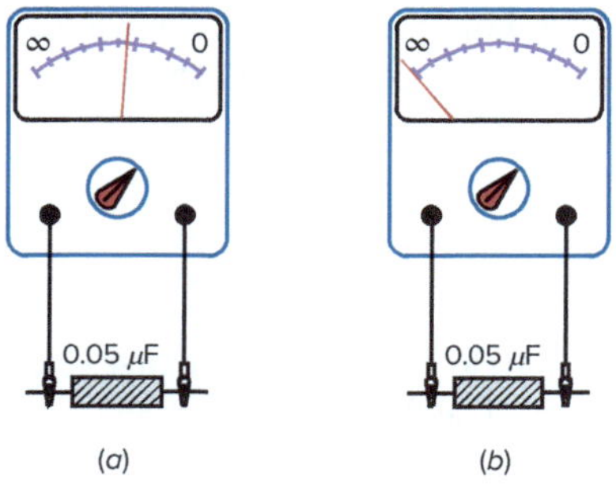

When the ohmmeter is initially connected, its battery charges the capacitor. This charging current is the reason the meter pointer moves away from infinity, since more current through the ohmmeter means less resistance. Maximum current flows at the first instant of charge. Then the charging current decreases as the capacitor voltage increases toward the applied voltage; therefore, the needle pointer slowly moves toward infinite resistance. Finally, the capacitor is completely charged to the ohmmeter battery voltage, the charging current is zero, and the ohmmeter reads just the small leakage current through the dielectric. This charging effect, called *capacitor action,* shows that the capacitor can store charge, indicating a normal capacitor. Note that both the rise and the fall of the meter readings are caused by charging. The capacitor discharges when the meter leads are reversed.

Ohmmeter Readings

Troubles in a capacitor are indicated as follows:

1. If an ohmmeter reading immediately goes practically to zero and stays there, the capacitor is short-circuited.
2. If a capacitor shows charging, but the final resistance reading is appreciably less than normal, the capacitor is leaky. Such capacitors are particularly troublesome in high-resistance circuits. When checking electrolytics, reverse the ohmmeter leads and take the higher of the two readings.
3. If a capacitor shows no charging action but reads very high resistance, it may be open. Some precautions must be remembered, however, since very high resistance is a normal condition for capacitors. Reverse the ohmmeter leads to discharge the capacitor, and check it again. In addition, remember that capacitance values of 100 pF, or less, normally have very little charging current for the low battery voltage of the ohmmeter.

Short-Circuited Capacitors

In normal service, capacitors can become short-circuited because the dielectric deteriorates with age, usually over a period of years under the stress of charging voltage, especially at higher temperatures. This effect is more common with paper and electrolytic capacitors. The capacitor may become leaky gradually, indicating a partial short circuit, or the dielectric may be punctured, causing a short circuit.

Open Capacitors

In addition to the possibility of an open connection in any type of capacitor, electrolytics develop high resistance in the electrolyte with age, particularly at high temperatures. After service of a few years, if the electrolyte dries up, the capacitor will be partially open. Much of the capacitor action is gone, and the capacitor should be replaced.

Leaky Capacitors

A leaky capacitor reads *R* less than normal with an ohmmeter. However, DC voltage tests are more definite. In a circuit, the DC voltage at one terminal of the capacitor should not affect the DC voltage at the other terminal.

Shelf Life

Except for electrolytics, capacitors do not deteriorate with age while stored, since there is no applied voltage. Electrolytic capacitors, however, like dry cells, should be used fresh from the manufacturer because the wet electrolyte may dry out over a period of time.

Capacitor Value Change

All capacitors can change value over time, but some are more prone to change than others. Ceramic capacitors often change value by 10 to 15% during the first year, as the ceramic material relaxes. Electrolytics change value from simply sitting because the electrolytic solution dries out.

Replacing Capacitors

Approximately the same *C* and *V* ratings should be used when installing a new capacitor. Except for tuning capacitors, the *C* value is usually not critical. Also, a higher voltage rating can be used. An important exception, however, is the electrolytic capacitor. Then the ratings should be close to the original values for two reasons. First, the specified voltage is needed to form the internal oxide film that provides the required capacitance. Also, too much *C* may allow excessive charging current in the circuit that charges the capacitor. Remember that electrolytics generally have large values of capacitance.

16–11 Self-Review

Answers at the end of the chapter.

a. **What is the ohmmeter reading for a shorted capacitor?**
b. **Does capacitor action with an ohmmeter show that the capacitor is good or bad?**
c. **Which type of capacitor is more likely to develop trouble, a mica or an electrolytic?**

Application in Understanding Supercapacitors

Supercapacitors, also called ultracapacitors, have capacitance values ranging from a fraction of a farad to several thousand farads. This is significantly more capacitance than an electrolytic capacitor is capable of providing. Because they have such large capacitance values, supercapacitors can store enormous amounts of electric charge and energy.

As you recall, a conventional capacitor consists of two metal plates separated by an insulator. The capacitance value is dependent on the area (A) of the plates, the distance (d) between the plates, and the dielectric constant (K_e) of the insulator. To increase the capacitance, C, it is necessary to increase the area of the plates, reduce the distance between the plates, and use an insulator with a high dielectric constant. But there are physical limitations with these three factors when trying to construct a capacitor with an extremely high value of capacitance. The construction of a supercapacitor is shown in Fig. 16-28. It consists of two metal plates coated with a highly porous carbon material that provides a huge surface area. These plates are immersed in a liquid electrolyte material that is highly ionized with both positive and negative ions. A separator between the plates isolates the charges on each plate. This construction gives the supercapacitor its alternate name: ***electrochemical double-layer capacitor (EDLC)***.

When a voltage is applied to the capacitor plates, the positive plate attracts negative ions from the electrolyte. At the same time, the negative plate attracts positive ions from the electrolyte. This produces two separately charged surfaces on either side of the separator, which is the equivalent of two capacitors in series. The charges on each plate are separated by a distance that is only the size of the ions in the electrolyte. This amounts to a plate separation of only a few angstroms. (Note: one angstrom equals 1×10^{-10} meters.) The extremely large surface area of the plates and the close spacing between them produces an extremely large value of capacitance. The only drawback to this construction is the very low breakdown voltage rating of the capacitor. A typical breakdown voltage rating is about 2.7 V. To increase the breakdown voltage rating, individual capacitors are connected in series with one another. This has the disadvantage, however, of reducing the overall capacitance. Also, a string of more than three series-connected capacitors requires voltage balancing to prevent any individual cell from having too much voltage.

Figure 16–28 Construction of a supercapacitor.

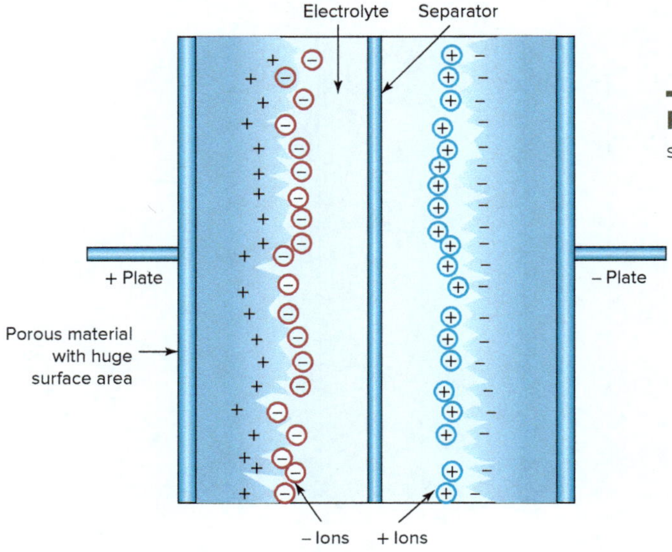

Supercapacitors store the most energy per unit volume or mass (energy density) of all capacitor types. As a result, supercapacitors bridge the gap between conventional capacitors and rechargeable batteries. They are ideal for applications involving frequent charge and discharge cycles where large amounts of current are supplied to a load for a short duration of time. It is important to note that supercapacitors are polarized and must therefore be connected with the proper polarity.

SUPERCAPACITOR APPLICATIONS

Supercapacitors store energy until it is needed. In many cases, they are used in conjunction with a battery to supply short-duration pulses of power. In some cases, they are used for backup power. It is common for a supercapacitor to be connected in parallel with a battery. In this case, the battery will charge the capacitor until the battery voltage and supercapacitor voltage are the same. If a load is connected across the two, current will be drawn from both the battery and the supercapacitor. However, if the load suddenly requires a large value of current for a very short duration of time, the supercapacitor will supply that current rather than the battery. This is because batteries have peak current limits, which are a function of their built-in internal resistance, r_i. The supercapacitor has an extremely low internal resistance, r_i, and is therefore able to supply the required amount of current during this short duration of time. As an example, supercapacitors are often used in conjunction with high-power car audio systems. When the high amplitude bass notes demand extremely high values of current, the battery alone cannot supply the required current. The supercapacitor, however, can easily supply the required high-amplitude current. It should be noted that the supercapacitor is located as close as possible to the DC input terminals of the amplifier. Another application of supercapacitors is in low-power consumer products. In this case, the supercapacitor provides voltage to a circuit or circuits if the battery dies or is being replaced. Examples are products with memories that may be erased or clocks that will lose their time setting if power is lost. Figure 16–29 shows two different supercapacitors used in conjunction with high-power car audio systems.

Figure 16–29 Supercapacitors used with high-power car audio systems.

Sarah Schultz Photography

Summary

- A capacitor consists of two metal plates separated by an insulator, or dielectric. Its ability to store charge is the capacitance C. Applying voltage to store charge is called *charging the capacitor;* short-circuiting the two leads or terminals of the capacitor to neutralize the charge is called *discharging the capacitor.* Schematic symbols for C are summarized in Fig. 16–30.

Figure 16–30 Schematic symbols for types of C. (*a*) Fixed type with air, paper, plastic film, mica, or ceramic dielectric. (*b*) Electrolytic type, which has polarity. (*c*) Variable. (*d*) Ganged variable capacitors on one shaft.

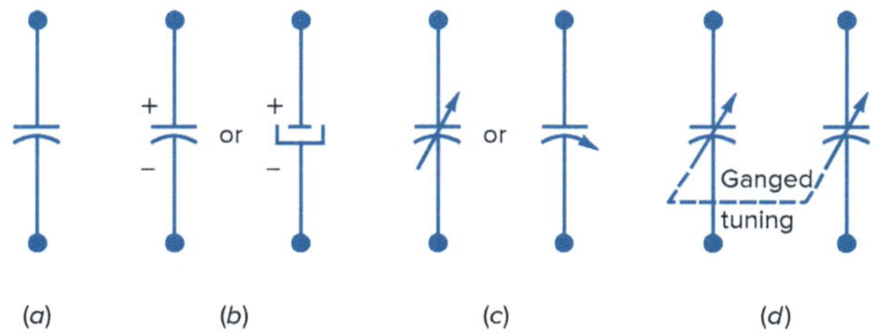

- The unit of capacitance is the farad. One farad of capacitance stores one coulomb of charge with one volt applied. Practical capacitors have much smaller capacitance values from 1 pF to 1000 μF. A capacitance of 1 pF is 1×10^{-12} F; 1 μF $= 1 \times 10^{-6}$ F; and 1 nF $= 1 \times 10^{-9}$ F.
- $Q = CV$, where Q is the charge in coulombs, C the capacitance in farads, and V is the potential difference across the capacitor in volts.
- Capacitance increases with larger plate area and less distance between plates.
- The ratio of charge stored in different insulators to the charge stored in air is the dielectric constant K_ϵ of the material. Air or vacuum has a dielectric constant of 1.
- The most common types of commercial capacitors are air, plastic film, paper, mica, ceramic, and electrolytic. Electrolytics are the only capacitors that require observing polarity when connecting to a circuit. The different types are compared in Table 16–2.
- Capacitors are coded to indicate their capacitance in either microfarads (μF) or picofarads (pF).
- For parallel capacitors, $C_T = C_1 + C_2 + C_3 + \cdots + \text{etc.}$
- For series capacitors, $C_{EQ} = \dfrac{1}{\dfrac{1}{C_1} + \dfrac{1}{C_2} + \dfrac{1}{C_3} + \cdots + \text{etc.}}$
- The electric field of a capacitance has stored energy $\mathscr{E} = {}^1\!/_2\, CV^2$, where V is in volts, C is in farads, and electric energy is in joules.
- When checked with an analog ohmmeter, a good capacitor shows charging current, and then the ohmmeter reading steadies at the leakage resistance. All types except electrolytics normally have very high leakage resistance such as 100,000 MΩ or more. Electrolytics have more leakage current; a typical leakage resistance is about 500 KΩ to 10 MΩ.

Important Terms

Capacitance (C) — the ability of a dielectric to hold or store an electric charge. The more charge stored for a given voltage, the greater the capacitance.

Capacitor — a component that can store electric charge. A capacitor consists of two metal plates separated by an insulator. Capacitors are named according to the type of dielectric used. Common capacitor types include air, ceramic, plastic film, mica, paper, and aluminum electrolytic.

Charging — increasing the amount of charge stored in a capacitor. The accumulation of stored charge results in a buildup of voltage across the capacitor.

Condenser — another (older) name for a capacitor.

Dielectric absorption — the inability of a capacitor to discharge completely to zero. Dielectric absorption is sometimes called battery action or capacitor memory.

Dielectric constant, K_ϵ — a factor that indicates the ability of an insulator to concentrate electric flux, also known as relative permittivity, ϵ_r.

Dielectric material — another name for an insulator.

Dielectric strength — the ability of a dielectric to withstand a potential difference without internal arcing.

Discharging — the action of neutralizing the charge stored in a capacitor by connecting a conducting path across the capacitor leads.

Electric field — the invisible lines of force between opposite electric charges.

Equivalent series resistance (ESR) — a resistance in series with an ideal capacitor that collectively represents all losses in a capacitor. Ideally, the ESR of a capacitor should be zero.

Farad (F) unit — the basic unit of capacitance. $1\ \text{F} = \dfrac{1\ \text{C}}{1\ \text{V}}$

Ganged capacitors — two or three capacitor sections on one common shaft that can be rotated.

Leakage current — the current that flows through the dielectric of a capacitor when voltage is applied across the capacitor plates.

Leakage resistance — a resistance in parallel with a capacitor that represents all leakage paths through which a capacitor can discharge.

Microfarad (μF) — a small unit of capacitance equal to 1×10^{-6} F.

Nanofarad (nF) — a small unit of capacitance equal to 1×10^{-9} F.

Picofarad (pF) — a small unit of capacitance equal to 1×10^{-12} F.

Relative permittivity, ϵ_r — a factor that indicates the ability of an insulator to concentrate electric flux, also known as the dielectric constant, K_ϵ.

Related Formulas

$Q = CV$ coulombs

$$C = \frac{Q}{V}$$

$$V = \frac{Q}{C}$$

$$C = K_\epsilon \times \frac{A}{d} \times 8.85 \times 10^{-12}\ \text{F}$$

$C_T = C_1 + C_2 + \cdots + \text{etc.}$ (parallel capacitors)

$$C_{EQ} = \frac{1}{\frac{1}{C_1} + \frac{1}{C_2} + \frac{1}{C_3} + \cdots + \text{etc.}}\ \text{(series capacitors)}$$

Energy $= \mathscr{E} = \frac{1}{2} CV^2$ joules

Self-Test

Answers at the back of the book.

1. In general, a capacitor is a component that can

a. pass a DC current.

b. store an electric charge.

c. act as a bar magnet.

d. step up or step down an AC voltage.

2. The basic unit of capacitance is the

a. farad.

b. henry.

c. tesla.

d. ohm.

3. Which of the following factors affect the capacitance of a capacitor?

a. the area, A, of the plates.

b. the distance, d, between the plates.

c. the type of dielectric used.

d. all of the above.

4. How much charge in coulombs is stored by a 50-μF capacitor with 20 V across its plates?

a. $Q = 100\ \mu$C.

b. $Q = 2.5\ \mu$C.

c. $Q = 1$ mC.

d. $Q = 1\ \mu$C.

5. A capacitor consists of

a. two insulators separated by a conductor.

b. a coil of wire wound on an iron core.

c. two metal plates separated by an insulator.

d. none of the above.

6. A capacitance of 82,000 pF is the same as

a. 0.082 μF.

b. 82 μF.

c. 82 nF.

d. both a and c.

7. A 47-μF capacitor has a stored charge of 2.35 mC. What is the voltage across the capacitor plates?

a. 50 V

b. 110 V approx.

c. 5 V

d. 100 V

8. Which of the following types of capacitors typically has the highest leakage current?

a. plastic-film.

b. electrolytic.

c. mica.

d. air-variable.

9. One of the main applications of a capacitor is to

a. block AC and pass DC.

b. block both DC and AC.

c. block DC and pass AC.

d. pass both DC and AC.

10. When checked with an ohmmeter, a shorted capacitor will measure

a. infinite ohms.

b. zero ohms.

c. somewhere in the range of 1 to 10 MΩ.

d. none of the above.

11. The equivalent capacitance, C_{EQ}, of a 10-μF and a 40-μF capacitor in series is

a. 50 μF.

b. 125 μF.

c. 8 μF.

d. 400 μF.

12. A 0.33-μF capacitor is in parallel with a 0.15-μF and a 220,000-pF capacitor. What is the total capacitance, C_T?
   a. 0.7 μF.
   b. 0.007 μF.
   c. 0.07 μF.
   d. 7 nF.

13. A 5-μF capacitor, C_1, and a 15-μF capacitor, C_2, are connected in series. If the charge stored in C_1 equals 90 μC, what is the voltage across the capacitor C_2?
   a. 18 V.
   b. 12 V.
   c. 9 V.
   d. 6 V.

14. A plastic-film capacitor, whose coded value is 333M, measures 0.025 μF when tested with a capacitor-inductor analyzer. The measured capacitance is
   a. well within tolerance.
   b. barely within tolerance.
   c. slightly out of tolerance.
   d. right on the money.

15. Capacitors are rarely coded in
   a. nanofarad units.
   b. microfarad units.
   c. picofarad units.
   d. both b and c.

16. Which type of capacitor could explode if the polarity of voltage across its plates is incorrect?
   a. air-variable.
   b. mica.
   c. ceramic disk.
   d. aluminum electrolytic.

17. The voltage rating of a capacitor is not affected by
   a. the area of the plates.
   b. the distance between the plates.
   c. the type of dielectric used.
   d. both b and c.

18. The leakage resistance of a capacitor is typically represented as a(n)
   a. resistance in series with the capacitor plates.
   b. electric field between the capacitor plates.
   c. resistance in parallel with the capacitor plates.
   d. closed switch across the dielectric material.

19. A 2200-μF capacitor with a voltage rating of 35 V is most likely a(n)
   a. electrolytic capacitor.
   b. air-variable capacitor.
   c. mica capacitor.
   d. paper capacitor.

20. A capacitor that can store 100 μC of charge with 10 V across its plates has a capacitance value of
   a. 0.01 μF.
   b. 10 μF.
   c. 10 nF.
   d. 100 mF.

21. Calculate the permissible capacitance range of a ceramic disk capacitor whose coded value is 0.068Z.
   a. 0.0544 μF to 0.1224 μF.
   b. 0.0136 μF to 0.0816 μF.
   c. 0.0136 μF to 0.1224 μF.
   d. 0.0544 pF to 0.1224 pF.

22. The equivalent series resistance (ESR) of a capacitor should ideally be
   a. infinite.
   b. as high as possible.
   c. around 100 kΩ.
   d. zero.

23. The charge and discharge current of a capacitor flows
   a. through the dielectric.
   b. to and from the capacitor plates.
   c. through the dielectric only until the capacitor is fully charged.
   d. straight through the dielectric from one plate to the other.

24. Capacitance increases with
   a. larger plate area and greater distance between the plates.
   b. smaller plate area and greater distance between the plates.
   c. larger plate area and less distance between the plates.
   d. higher values of applied voltage.

25. Two 0.02-μF, 500-V capacitors in series have an equivalent capacitance and breakdown voltage rating of
   a. 0.04 μF, 1 kV.
   b. 0.01 μF, 250 V.
   c. 0.01 μF, 500 V.
   d. 0.01 μF, 1 kV.

Essay Questions

1. Define *capacitance* with respect to physical structure and electrical function. Explain how a two-wire conductor has capacitance.

2. (a) What is meant by a dielectric material? (b) Name five common dielectric materials.

3. Explain briefly how to charge a capacitor. How is a charged capacitor discharged?

4. Define *1 F of capacitance*. Convert the following into farads using powers of 10: (a) 50 pF; (b) 0.001 μF; (c) 0.047 μF; (d) 0.01 μF; (e) 10 μF.

5. State the effect on capacitance of (a) larger plate area; (b) thinner dielectric; (c) higher value of dielectric constant.

6. Draw a diagram showing the fewest number of 400-V, 2-μF capacitors needed for a combination rated at 800 V with 2-μF total capacitance.

7. Suppose you are given two identical uncharged capacitors. One is charged to 50 V and connected across the other uncharged capacitor. Why will the voltage across both capacitors then be 25 V?

8. Describe briefly how you would check a 0.05-μF capacitor with an ohmmeter. State the ohmmeter indications when the capacitor is good, short-circuited, or open.

9. Define the following: (a) leakage resistance; (b) dielectric absorption; (c) equivalent series resistance.

10. Give two comparisons between the electric field in a capacitor and the magnetic field in a coil.

11. List three types of troubles in capacitors.

12. When a capacitor discharges, why is its discharge current in the opposite direction from the charging current?

13. Compare the features of aluminum and tantalum electrolytic capacitors.

14. Why can plastic film be used instead of paper for capacitors?

15. What two factors determine the breakdown voltage rating of a capacitor?

Problems

SECTION 16–3 THE FARAD UNIT OF CAPACITANCE

16–1 Calculate the amount of charge, Q, stored by a capacitor if
- a. $C = 10\ \mu$F and $V = 5$ V.
- b. $C = 1\ \mu$F and $V = 25$ V.
- c. $C = 0.01\ \mu$F and $V = 150$ V.
- d. $C = 0.22\ \mu$F and $V = 50$ V.
- e. $C = 680$ pF and $V = 200$ V.
- f. $C = 47$ pF and $V = 3$ kV.

16–2 How much charge, Q, is stored by a 0.05-μF capacitor if the voltage across the plates equals
- a. 10 V?
- b. 40 V?
- c. 300 V?
- d. 500 V?
- e. 1 kV?

16–3 How much voltage exists across the plates of a 200-μF capacitor if a constant current of 5 mA charges it for
- a. 100 ms?
- b. 250 ms?
- c. 0.5 s?
- d. 2 s?
- e. 3 s?

16–4 Determine the voltage, V, across a capacitor if
- a. $Q = 2.5\ \mu$C and $C = 0.01\ \mu$F.
- b. $Q = 49.5$ nC and $C = 330$ pF.
- c. $Q = 10$ mC and $C = 1{,}000\ \mu$F.
- d. $Q = 500\ \mu$C and $C = 0.5\ \mu$F.
- e. $Q = 188$ nC and $C = 0.0047\ \mu$F.
- f. $Q = 75$ nC and $C = 0.015\ \mu$F.

16–5 Determine the capacitance, C, of a capacitor if
- a. $Q = 15\ \mu$C and $V = 1$ V.
- b. $Q = 15\ \mu$C and $V = 30$ V.
- c. $Q = 100\ \mu$C and $V = 25$ V.
- d. $Q = 3.3\ \mu$C and $V = 15$ V.
- e. $Q = 0.12\ \mu$C and $V = 120$ V.
- f. $Q = 100\ \mu$C and $V = 2.5$ kV.

16–6 List the physical factors that affect the capacitance, C, of a capacitor.

16–7 Calculate the capacitance, C, of a capacitor for each set of physical characteristics listed.
- a. $A = 0.1\ \text{cm}^2, d = 0.005\ \text{cm}, K_\epsilon = 1$.
- b. $A = 0.05\ \text{cm}^2, d = 0.001\ \text{cm}, K_\epsilon = 500$.
- c. $A = 0.1\ \text{cm}^2, d = 1 \times 10^{-5}\ \text{cm}, K_\epsilon = 50$.
- d. $A = 1\ \text{cm}^2, d = 5 \times 10^{-6}\ \text{cm}, K_\epsilon = 6$.

SECTION 16–6 CAPACITOR CODING

16–8 Determine the capacitance and tolerance of each of the capacitors shown in Fig. 16–31.

Figure 16–31

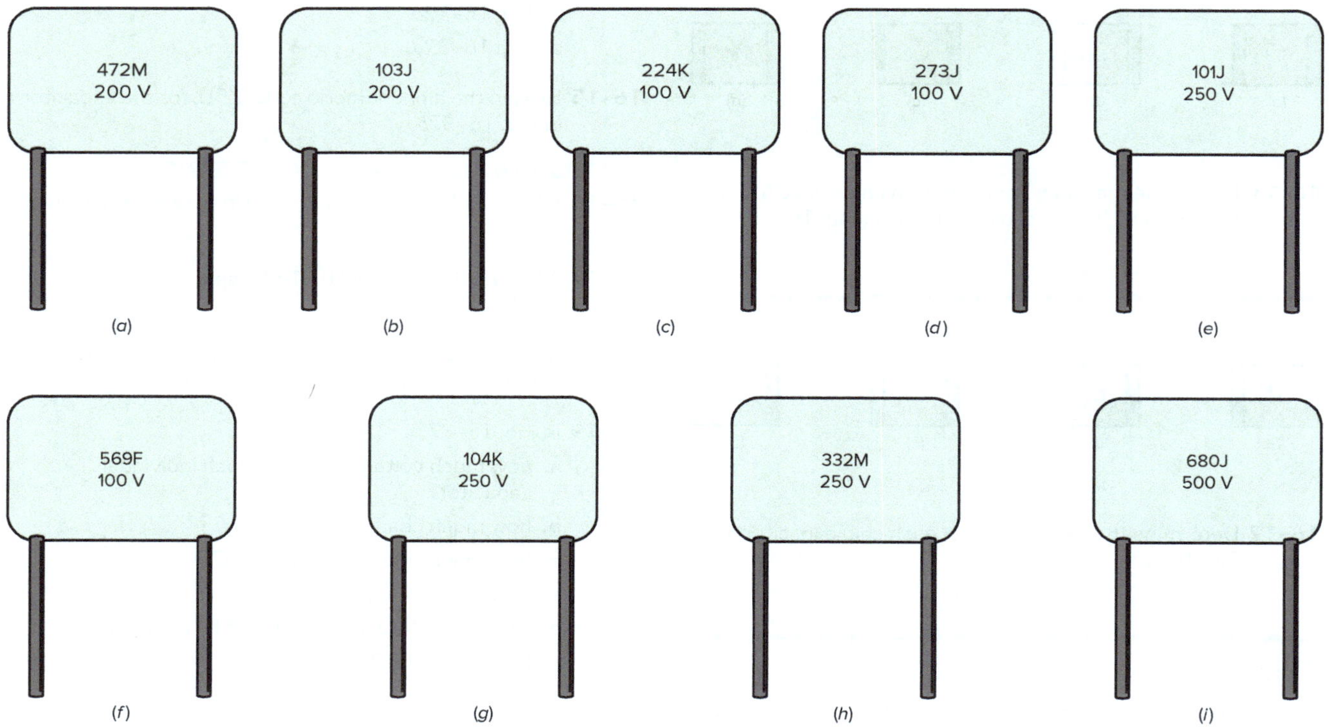

16–9 Determine the capacitance and tolerance of each of the capacitors shown in Fig. 16–32.

Figure 16–32

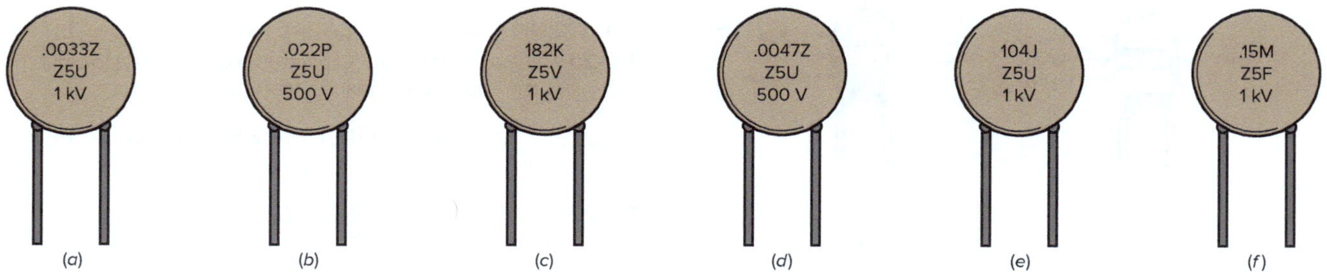

16–10 Determine the capacitance of each chip capacitor in Fig. 16–33. Use the coding scheme in Fig. 16–15.

Figure 16–33

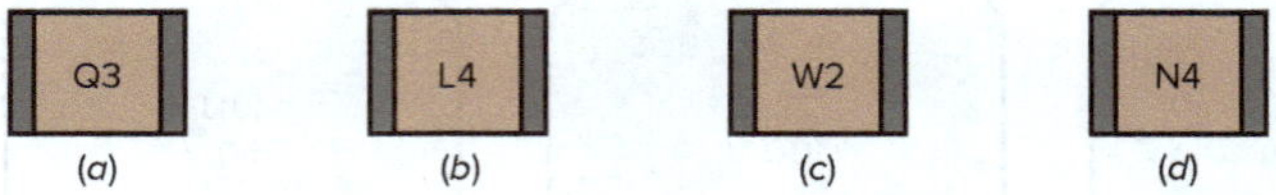

16–11 Determine the capacitance of each chip capacitor in Fig. 16–34. Use the coding scheme in Fig. 16–16.

Figure 16–34

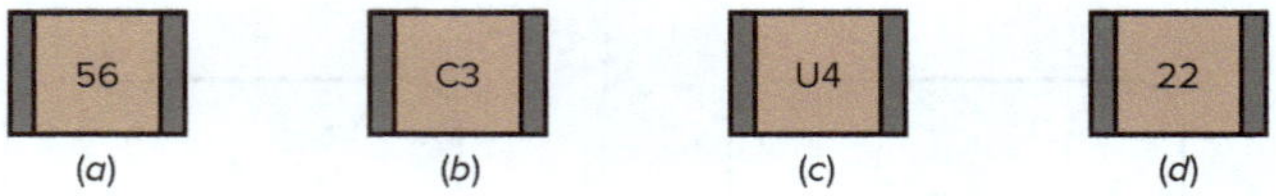

16–12 Determine the capacitance of each chip capacitor in Fig. 16–35.

Figure 16–35

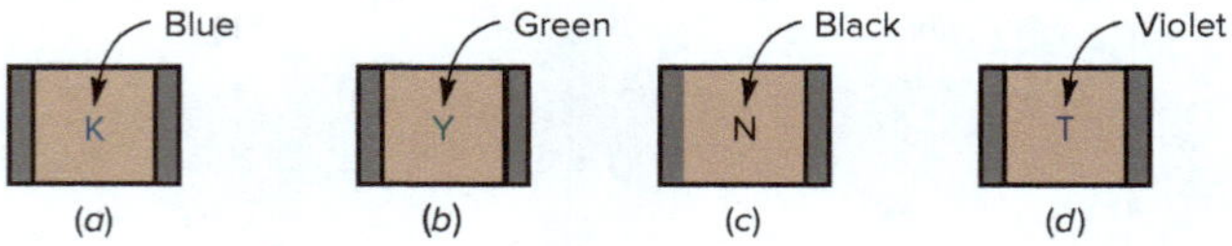

16–13 Determine the capacitance and tolerance of each capacitor in Fig. 16–36.

Figure 16–36

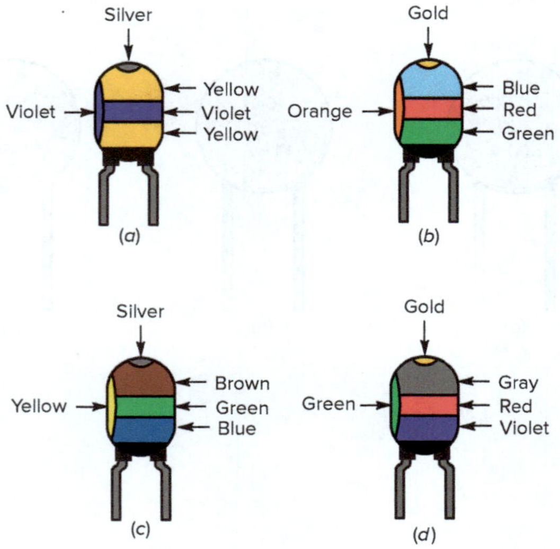

16–14 Determine the permissible capacitance range of the capacitors in

a. Fig. 16–31*a*.
b. Fig. 16–31*d*.
c. Fig. 16–31*f*.
d. Fig. 16–32*c*.
e. Fig. 16–32*d*.

16–15 Explain the alphanumeric code, Z5U, for the capacitor in Fig. 16–32*b*.

SECTION 16–7 PARALLEL CAPACITANCES

16–16 A 5-μF and 15-μF capacitor are in parallel. How much is C_T?

16–17 A 0.1-μF, 0.27-μF, and 0.01-μF capacitor are in parallel. How much is C_T?

16–18 A 150-pF, 330-pF, and 0.001-μF capacitor are in parallel. How much is C_T?

16–19 In Fig. 16–37,

a. how much voltage is across each individual capacitor?
b. how much charge is stored by C_1?
c. how much charge is stored by C_2?
d. how much charge is stored by C_3?
e. what is the total charge stored by all capacitors?
f. how much is C_T?

Figure 16–37

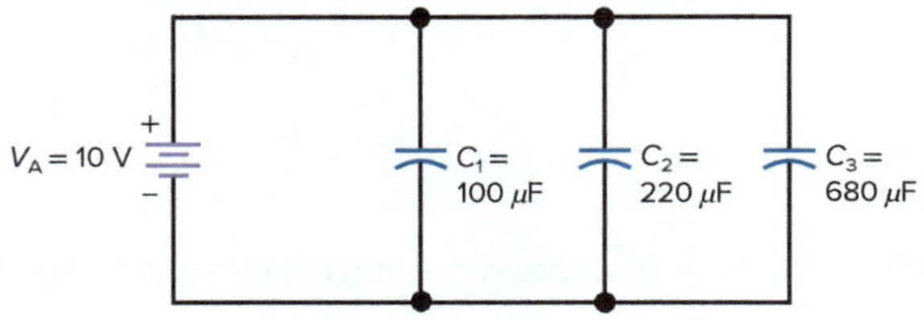

SECTION 16–8 SERIES CAPACITANCES

16–20 A 0.1-μF and 0.4-μF capacitor are in series. How much is the equivalent capacitance, C_{EQ}?

16–21 A 1500-pF and 0.001-μF capacitor are in series. How much is the equivalent capacitance, C_{EQ}?

16–22 A 0.082-μF, 0.047-μF, and 0.012 μF capacitor are in series. How much is the equivalent capacitance, C_{EQ}?

16–23 In Fig. 16–38, assume a charging current of 180 μA flows for 1 s. Solve for

a. C_{EQ}.
b. the charge stored by C_1, C_2, and C_3.
c. the voltage across C_1, C_2, and C_3.
d. the total charge stored by all capacitors.

Figure 16–38

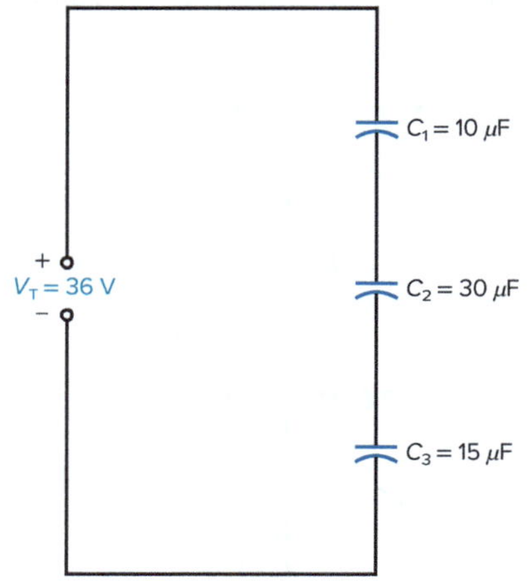

16–24 In Fig. 16–39, assume a charging current of 2.4 mA flows for 1 ms. Solve for

a. C_{EQ}.

b. the charge stored by C_1, C_2, and C_3.

c. the voltage across C_1, C_2, and C_3.

d. the total charge stored by all capacitors.

Figure 16–39

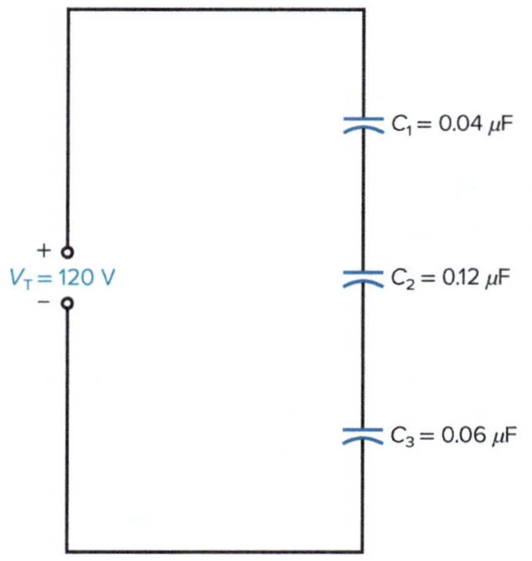

16–25 How much capacitance must be connected in series with a 120-pF capacitor to obtain an equivalent capacitance, C_{EQ}, of 100 pF.

SECTION 16–9 ENERGY STORED IN ELECTROSTATIC FIELD OF CAPACITANCE

16–26 How much energy is stored by a 100-μF capacitor which is charged to

a. 5 V?

b. 10 V?

c. 50 V?

16–27 How much energy is stored by a 0.027-μF capacitor which is charged to

a. 20 V?

b. 100 V?

c. 500 V?

16–28 Calculate the energy stored by each capacitor in Fig. 16–38.

SECTION 16–10 MEASURING AND TESTING CAPACITORS

16–29 Make the following conversions:

a. 0.047 μF to pF.

b. 0.0015 μF to pF.

c. 390,000 pF to μF.

d. 1000 pF to μF.

16–30 Make the following conversions:

a. 15 nF to pF.

b. 1 nF to pF.

c. 680 nF to pF.

d. 33,000 pF to nF.

e. 1,000,000 pF to nF.

f. 560,000 pF to nF.

16–31 A plastic-film capacitor has a coded value of 154K. If the measured value of capacitance is 0.160 μF, is the capacitance value within tolerance?

16–32 A ceramic disk capacitor is coded 102Z. If the measured value of capacitance is 680 pF, is the capacitance within tolerance?

16–33 A plastic-film capacitor has a coded value of 229B. If the measured value of capacitance is 2.05 pF, is the capacitance within tolerance?

SECTION 16–11 TROUBLES IN CAPACITORS

16–34 What is the ohmmeter reading for a(n)

a. shorted capacitor.

b. open capacitor.

c. leaky capacitor.

16–35 Describe the effect of connecting a 0.47-μF capacitor to the leads of an analog ohmmeter set to the $R \times 10$K range.

Critical Thinking

16–36 Three capacitors in series have a combined equivalent capacitance C_{EQ} of 1.6 nF. If $C_1 = 4C_2$ and $C_3 = 20C_1$, calculate the values for C_1, C_2, and C_3.

16–37 A 100-pF ceramic capacitor has a temperature coefficient T_C of N500. Calculate its capacitance at (a) 75°C; (b) 125°C; (c) −25°C.

16–38 (a) Calculate the energy stored by a 100-μF capacitor charged to 100 V. (b) If this capacitor is now connected across another 100-μF capacitor that is uncharged, calculate the total energy stored by both capacitors. (c) Is the energy stored by both capacitors in part (b) less than the energy stored by the single capacitor in part (a)? If yes, where did the energy go?

Answers to Self-Reviews

16–1 **a.** dielectric **b.** farad

16–2 **a.** 14.5 V **b.** 0 V **c.** yes

16–3 **a.** 10 μF **b.** ceramic

16–4 **a.** true **b.** false **c.** true **d.** true

16–5 **a.** true **b.** true **c.** true

16–6 **a.** false **b.** true **c.** true **d.** true

16–7 **a.** 0.03 μF **b.** 150 pF

16–8 **a.** 0.1 μF **b.** 25 V **c.** 33.3 pF

16–9 **a.** true **b.** true

16–10 **a.** true **b.** true **c.** false **d.** false

16–11 **a.** 0 Ω **b.** good **c.** electrolytic

Laboratory Application Assignment

In this lab application assignment, you will examine the coding systems used to indicate the capacitance and tolerance of a capacitor. You will also measure the value of a capacitor using either a Z meter or a DMM capable of measuring capacitance values. And finally, you will examine how capacitance values combine when connected in series and in parallel.

Equipment: Obtain the following items from your instructor.

- Assortment of plastic-film capacitors
- Z meter or DMM capable of measuring capacitance values

Measuring Capacitance

Obtain five plastic-film capacitors from your instructor. Make sure each capacitor has a different coded value. In the space provided below, indicate the coded value of each capacitor. Next, indicate the capacitance (in pF) corresponding to the coded value, including the tolerance. Finally, measure and record each capacitance value using either a Z meter or a DMM capable of measuring capacitance values. (If a measured value is displayed in nF or μF, convert it to pF.)

Coded Value	Capacitance Value	Measured Value
________	________	________
________	________	________
________	________	________
________	________	________
________	________	________

Is the measured value of any capacitor out of tolerance? ________ If so, which one(s)? ________

Series Capacitors

Connect a 0.1-μF capacitor in series with a 0.047-μF capacitor as shown in Fig. 16–40*a*. Calculate and record the equivalent capacitance, C_{EQ}, of this series combination. C_{EQ} = ________ Next, measure and record the equivalent capacitance, C_{EQ}, across terminals A and B. C_{EQ} = ________ Add another 0.022-μF capacitor, as shown in Fig. 16–40*b*. Calculate and record the equivalent capacitance, C_{EQ}, of this series combination. C_{EQ} = ________ Finally, measure and record the equivalent capacitance, C_{EQ}, across terminals A and B. C_{EQ} = ________

Figure 16–40

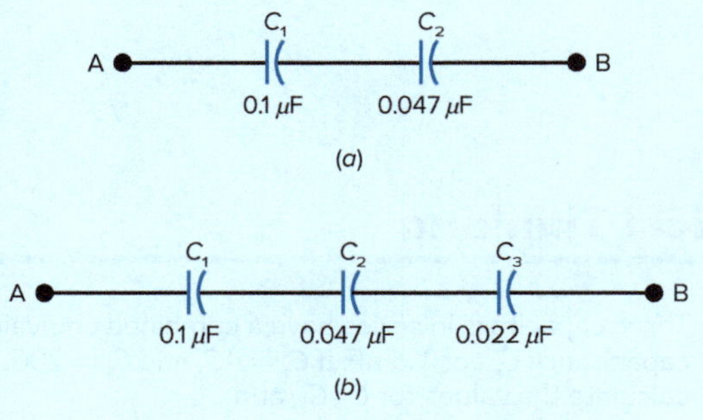

Parallel Capacitors

Connect a 0.1-μF capacitor in parallel with a 0.047-μF capacitor, as shown in Fig. 16–41*a*. Calculate and record the total capacitance, C_T, of this parallel combination. $C_T =$ __________ Next, measure and record the total capacitance, C_T, across terminals A and B. $C_T =$ __________ Add another 0.022-μF capacitor, as shown in Fig. 16–41*b*. Calculate and record the total capacitance, C_T, of this parallel combination. $C_T =$ __________ Finally, measure and record the total capacitance, C_T, across terminals A and B. $C_T =$ __________

Do capacitors in series combine the same way as resistors in parallel? __________

Do capacitors in parallel combine the same way as resistors in series? __________

Capacitor Leakage

Because there is no such thing as a perfect insulator, all capacitors have a small amount of current flowing through the dielectric. This current is called leakage current. For a good capacitor, this leakage current is usually insignificant and can therefore be ignored. Due to the very nature of their construction, electrolytic capacitors have a much higher leakage current than other types of capacitors. If a Z meter is available, have your instructor demonstrate how it can be used to measure the leakage current in a plastic-film and an electrolytic capacitor. When checking for leakage, be sure to apply the rated working voltage across the capacitor. Your instructor can also show you how to test the ESR value of a capacitor.

Figure 16–41

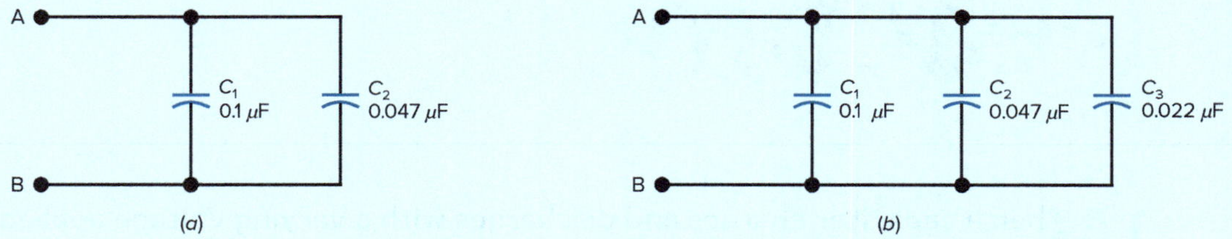

Design credits: Chapter Opener Background Image, and Multisim: Stockbyte/Getty Images

chapter 17

Capacitive Reactance

When a capacitor charges and discharges with a varying voltage applied, alternating current can flow. Although there cannot be any current through the dielectric, its charge and discharge current flows to and from the plates of the capacitor. If a sine-wave alternating voltage is applied across the capacitor, the amount of current, I, that flows depends on its capacitive reactance. The symbol for capacitive reactance is X_C, and its unit is the ohm. The X in X_C indicates reactance, whereas the subscript C specifies capacitive reactance.

The amount of X_C is a V/I ratio, but it can also be calculated as $X_C = 1/(2\pi fC)$. With the frequency, f in hertz (Hz), and the capacitance, C in farads (F), X_C is in units of ohms. The reciprocal relation in $1/(2\pi fC)$ means that X_C decreases for higher frequencies and more capacitance. This is because more charge and discharge current flows either with more capacitance or faster changes in the applied voltage. ■

Chapter Outline

Chapter Objectives

After studying this chapter, you should be able to

- *Explain* how alternating current can flow in a capacitive circuit.
- *Calculate* the capacitive reactance of a capacitor when the frequency and capacitance are known.
- *Calculate* the total capacitive reactance of series-connected capacitors.
- *Calculate* the equivalent capacitive reactance of parallel-connected capacitors.
- *Explain* how Ohm's law can be applied to capacitive reactance.
- *Calculate* the capacitive current when the capacitance and rate of voltage change are known.

Important Terms

capacitive reactance, X_C

charging current

discharge current

inversely proportional

phase angle

MultiSim **Figure 17–1** Current in a capacitive circuit. (*a*) The 4-μF capacitor allows enough current *I* to light the bulb brightly. (*b*) Less current with a smaller capacitor causes dim light. (*c*) The bulb cannot light with DC voltage applied because a capacitor blocks direct current.

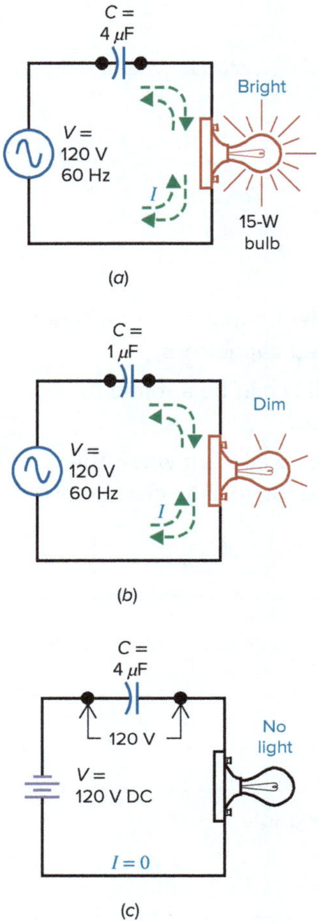

17–1 Alternating Current in a Capacitive Circuit

The fact that current flows with AC voltage applied is demonstrated in Fig. 17–1, where the bulb lights in Fig. 17–1*a* and *b* because of the capacitor charge and **discharge current**. There is no current through the dielectric, which is an insulator. While the capacitor is being charged by increasing applied voltage, however, the **charging current** flows in one direction in the conductors to the plates. While the capacitor is discharging, when the applied voltage decreases, the discharge current flows in the reverse direction. With alternating voltage applied, the capacitor alternately charges and discharges.

First the capacitor is charged in one polarity, and then it discharges; next the capacitor is charged in the opposite polarity, and then it discharges again. The cycles of charge and discharge current provide alternating current in the circuit at the same frequency as the applied voltage. This is the current that lights the bulb.

In Fig. 17–1*a*, the 4-μF capacitor provides enough alternating current (AC) to light the bulb brightly. In Fig. 17–1*b*, the 1-μF capacitor has less charge and discharge current because of the smaller capacitance, and the light is not so bright. Therefore, the smaller capacitor has more opposition to alternating current as less current flows with the same applied voltage; that is, it has more reactance for less capacitance.

In Fig. 17–1*c*, the steady DC voltage will charge the capacitor to 120 V. Because the applied voltage does not change, though, the capacitor will just stay charged. Since the potential difference of 120 V across the charged capacitor is a voltage drop opposing the applied voltage, no current can flow. Therefore, the bulb cannot light. The bulb may flicker on for an instant because charging current flows when voltage is applied, but this current is only temporary until the capacitor is charged. Then the capacitor has the applied voltage of 120 V, but there is zero voltage across the bulb.

As a result, the capacitor is said to *block* direct current or voltage. In other words, after the capacitor has been charged by a steady DC voltage, there is no current in the DC circuit. All the applied DC voltage is across the charged capacitor with zero voltage across any series resistance.

In summary, then, this demonstration shows the following points:

1. Alternating current flows in a capacitive circuit with AC voltage applied.
2. A smaller capacitance allows less current, which means more X_C with more ohms of opposition.
3. Lower frequencies for the applied voltage result in less current and more X_C. With a steady DC voltage source, which corresponds to a frequency of zero, the opposition of the capacitor is infinite and there is no current. In this case, the capacitor is effectively an open circuit.

These effects have almost unlimited applications in practical circuits because X_C depends on frequency. A very common use of a capacitor is to provide little opposition for AC voltage but to block any DC voltage. Another example is to use X_C for less opposition to a high-frequency alternating current, compared with lower frequencies.

GOOD TO KNOW

In a capacitor, the charge and discharge current flows to and from the plates but not through the dielectric.

Capacitive Current

The reason that a capacitor allows current to flow in an AC circuit is the alternate charge and discharge. If we insert an ammeter in the circuit, as shown in Fig. 17–2, the AC meter will read the amount of charge and discharge current. In this example, I_C is 0.12 A. This current is the same in the voltage source, the connecting leads, and

Figure 17–2 Capacitive reactance X_C is the ratio V_C/I_C.

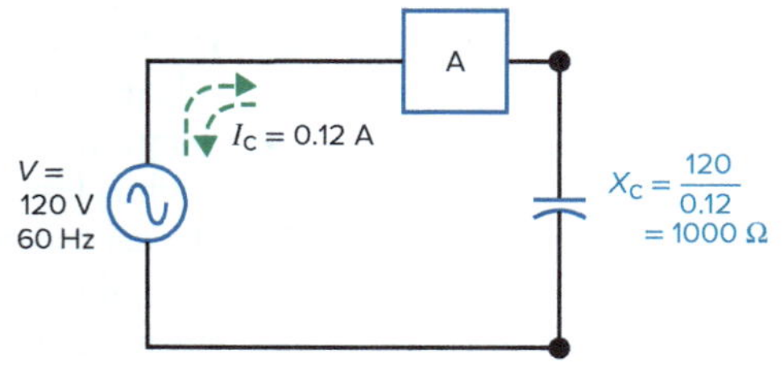

the plates of the capacitor. However, there is no current through the insulator between the plates of the capacitor.

Values for X_C

GOOD TO KNOW

Capacitive reactance, X_C, is a measure of a capacitor's opposition to the flow of sine-wave alternating current. The unit of X_C is the ohm (Ω).

When we consider the ratio of V_C/I_C for the ohms of opposition to the sine-wave current, this value is $^{120}\!/_{0.12}$, which equals 1000 Ω. This 1000 Ω is what we call X_C, to indicate how much current can be produced by sine-wave voltage applied to a capacitor. In terms of current, $X_C = V_C/I_C$. In terms of frequency and capacitance, $X_C = 1/(2\pi fC)$.

The X_C value depends on the amount of capacitance and the frequency of the applied voltage. If C in Fig. 17–2 were increased, it could take on more charge for more charging current and then produce more discharge current. Then X_C is less for more capacitance. Also, if the frequency in Fig. 17–2 were increased, the capacitor could charge and discharge faster to produce more current. This action also means that V_C/I_C would be less with more current for the same applied voltage. Therefore, X_C is less for higher frequencies. Reactance X_C can have almost any value from practically zero to almost infinite ohms.

17–1 Self-Review

Answers at the end of the chapter.

a. Which has more reactance, a 0.1- or a 0.5-μF capacitor, at the same frequency?

b. Which allows more charge and discharge current, a 0.1- or a 0.5-μF capacitor, at the same frequency?

17–2 The Amount of X_C Equals $1/(2\pi fC)$

The effects of frequency and capacitance are included in the formula for calculating ohms of reactance. The f is in hertz units and the C is in farads for X_C in ohms. As an example, we can calculate X_C for C of 2.65 μF and f of 60 Hz. Then

$$X_C = \frac{1}{2\pi fC} \quad (17\text{–}1)$$

$$= \frac{1}{2\pi \times 60 \times 2.65 \times 10^{-6}} = \frac{1}{6.28 \times 159 \times 10^{-6}}$$

$$= 0.00100 \times 10^{6}$$

$$= 1000\ \Omega$$

Note the following factors in the formula $X_C = \frac{1}{2\pi fC}$.

1. The constant factor 2π is always $2 \times 3.14 = 6.28$. It indicates the circular motion from which a sine wave is derived. Therefore, the

formula $X_C = \frac{1}{2\pi fC}$ applies only to sine-wave AC circuits. The 2π is actually 2π rad or 360° for a complete circle or cycle.

2. The frequency, *f*, is a time element. A higher frequency means that the voltage varies at a faster rate. A faster voltage change can produce more charge and discharge current for a given value of capacitance, *C*. The result is less X_C.
3. The capacitance, *C*, indicates the physical factors of the capacitor that determine how much charge and discharge current it can produce for a given change in voltage.
4. **Capacitive reactance, X_C,** is measured in ohms corresponding to the $\frac{V_C}{I_C}$ ratio for sine-wave AC circuits. The X_C value determines how much current *C* allows for a given value of applied voltage.

Example 17–1

How much is X_C for (a) 0.1 μF of *C* at 1400 Hz? (b) 1 μF of *C* at the same frequency?

ANSWER

a. $$X_C = \frac{1}{2\pi fC} = \frac{1}{6.28 \times 1400 \times 0.1 \times 10^{-6}}$$
$$= \frac{1}{6.28 \times 140 \times 10^{-6}} = 0.00114 \times 10^6$$
$$= 1140\ \Omega$$

b. At the same frequency, with 10 times more *C*, X_C is one-tenth or 1140⁄10, which equals 114 Ω.

Example 17–2

How much is the X_C of a 47-pF value of *C* at (a) 1 MHz? (b) 10 MHz?

ANSWER

a. $$X_C = \frac{1}{2\pi fC} = \frac{1}{6.28 \times 47 \times 10^{-12} \times 1 \times 10^6}$$
$$= \frac{1}{295.16 \times 10^{-6}} = 0.003388 \times 10^6$$
$$= 3388\ \Omega$$

b. At 10 times the frequency,
$$X_C = \frac{3388}{10} = 338\ \Omega.$$

CALCULATOR

When using Formula (17–1) with a calculator, probably the best method is to multiply all the factors in the denominator and then take the reciprocal of the total product. To save time, memorize 2π as $2 \times 3.14 = 6.28$. If your calculator does not have an (EXP) key, keep the powers of 10 separate. Remember that the negative sign of the exponent becomes positive in the reciprocal value. Specifically, for Example 17–1, the procedure can be as follows:

- Punch in 6.28 as the numbers for 2π.
- Press the (×) key and punch in the factor of 1400, then (×) and 0.1.
- Press the (=) key to see the total product of 879.2.
- While 879.2 is on the display, press the reciprocal key (1/x). This may require pushing the (2nd F) key first.
- The reciprocal value is 0.00114.
- The reciprocal of 10^{-6} in the denominator becomes 10^{6} in the numerator.
- For the final answer, then, move the decimal point six places to the right, as indicated by 10^{6}, for the final answer of 1140.

GOOD TO KNOW

The capacitive reactance, X_C, of a capacitor is infinite for DC. At the opposite extreme, the X_C of a capacitor can be approximately zero ohms at very high frequencies. When analyzing electronic circuits, therefore, capacitors are often treated as an open for DC and as a short for AC.

MultiSim **Figure 17–3** A table of values and a graph to show that capacitive reactance X_C decreases with higher values of C. Frequency is constant at 159 Hz.

X_C increases as C decreases

C, μF	$X_C{}^* = \frac{1}{2\pi fC}$, Ω
1.0	1000
0.5	2000
0.2	5000
0.1	10,000

*For f = 159 Hz

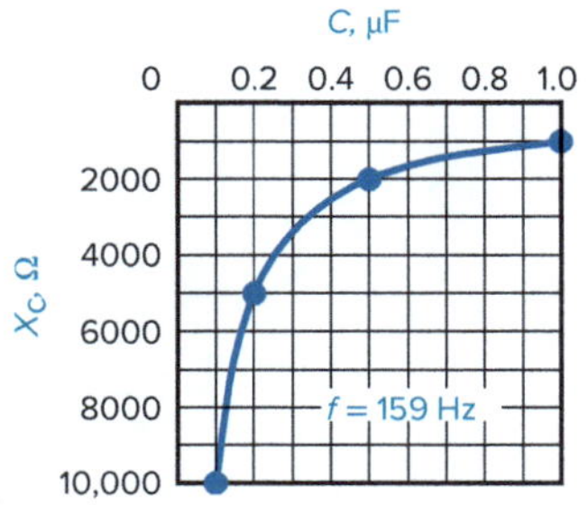

Note that X_C in Example 17–2b is one-tenth the value in Example 17–2a because f is 10 times greater.

X_C Is Inversely Proportional to Capacitance

This statement means that X_C increases as capacitance is decreased. In Fig. 17–3, when C is reduced by a factor of 1⁄10 from 1.0 to 0.1 μF, then X_C increases 10 times from 1000 to 10,000 Ω. Also, decreasing C by one-half from 0.2 to 0.1 μF doubles X_C from 5000 to 10,000 Ω.

This inverse relation between C and X_C is illustrated by the graph in Fig. 17–3. Note that values of X_C increase downward on the graph, indicating negative reactance that is opposite from inductive reactance. (Inductive reactance is covered in Chapter 20.) With C increasing to the right, the decreasing values of X_C approach the zero axis of the graph.

X_C Is Inversely Proportional to Frequency

Figure 17–4 illustrates the inverse relationship between X_C and f. With f increasing to the right in the graph from 0.1 to 1 MHz, the value of X_C for the 159-pF capacitor decreases from 10,000 to 1000 Ω as the X_C curve comes closer to the zero axis.

The graphs are nonlinear because of the inverse relation between X_C and f or C. At one end, the curves approach infinitely high reactance for zero capacitance or zero frequency. At the other end, the curves approach zero reactance for infinitely high capacitance or frequency.

Calculating C from Its Reactance

In some applications, it may be necessary to find the value of capacitance required for a desired amount of X_C. For this case, the reactance formula can be inverted to

$$C = \frac{1}{2\pi f X_C} \qquad \textbf{(17–2)}$$

Figure 17–4 A table of values and a graph to show that capacitive reactance X_C decreases with higher frequencies. C is constant at 159 pF.

X_C increases as f decreases

f, MHz	$X_C{}^* = \frac{1}{2\pi fC}$, Ω
1.0	1000
0.5	2000
0.2	5000
0.1	10,000

*For C = 159 pF

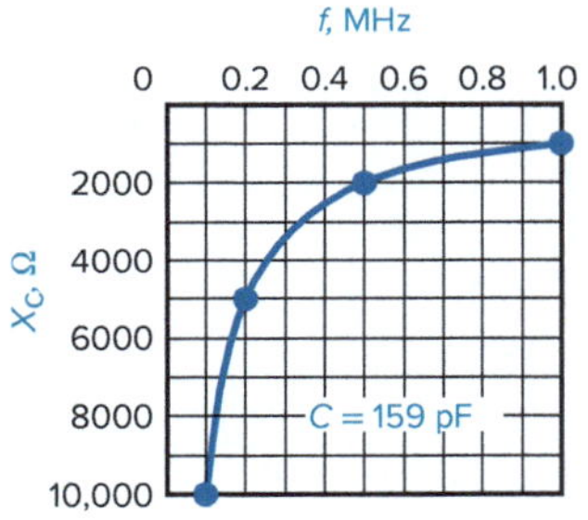

The value of 6.28 for 2π is still used. The only change from Formula (17–1) is that the C and X_C values are inverted between denominator and numerator on the left and right sides of the equation.

Example 17–3

What C is needed for X_C of 100 Ω at 3.4 MHz?

ANSWER

$$C = \frac{1}{2\pi f X_C} = \frac{1}{6.28 \times 3.4 \times 10^6 \times 100}$$

$$= \frac{1}{628 \times 3.4 \times 10^6}$$

$$= 0.000468 \times 10^{-6}\ \text{F} = 0.000468\ \mu\text{F} \quad \text{or} \quad 468\ \text{pF}$$

A practical size for this capacitor would be 470 pF. The application is to have low reactance at the specified frequency of 3.4 MHz.

Calculating Frequency from the Reactance

Another use is to find the frequency at which a capacitor has a specified amount of X_C. Again, the reactance formula can be inverted to the form shown in Formula (17–3).

$$f = \frac{1}{2\pi C X_C} \qquad \textbf{(17–3)}$$

The following example illustrates the use of this formula.

Example 17–4

At what frequency will a 10-μF capacitor have X_C equal to 100 Ω?

ANSWER

$$f = \frac{1}{2\pi C X_C} = \frac{1}{6.28 \times 10 \times 10^{-6} \times 100}$$

$$= \frac{1}{6280 \times 10^{-6}}$$

$$= 0.000159 \times 10^6$$

$$= 159\ \text{Hz}$$

This application is a capacitor for low reactance at audio frequencies.

Summary of X_C Formulas

Formula (17–1) is the basic form for calculating X_C when f and C are known values. As another possibility, the value of X_C can be measured as V_C/I_C.

With X_C known, the value of C can be calculated for a specified f by Formula (17–2), or f can be calculated with a known value of C by using Formula (17–3).

17–2 Self-Review

Answers at the end of the chapter.

The X_C for a capacitor is 400 Ω at 8 MHz.

a. **How much is X_C at 16 MHz?**

b. **How much is X_C at 4 MHz?**

c. **Is a smaller or larger C needed for less X_C?**

17–3 Series and Parallel Capacitive Reactances

Because capacitive reactance is an opposition in ohms, series or parallel reactances are combined in the same way as resistances. As shown in Fig. 17–5*a*, series capacitive reactances are added arithmetically.

Series capacitive reactance:

$$X_{C_T} = X_{C_1} + X_{C_2} + \cdots + \text{etc.} \qquad \textbf{(17–4)}$$

For parallel reactances, the combined equivalent reactance is calculated by the reciprocal formula, as shown in Fig. 17–5*b*.

Parallel capacitive reactance:

$$X_{C_{EQ}} = \frac{1}{\dfrac{1}{X_{C_1}} + \dfrac{1}{X_{C_2}} + \dfrac{1}{X_{C_3}} + \cdots + \text{etc.}} \qquad \textbf{(17–5)}$$

In Fig. 17–5*b*, the parallel combination of 100 and 200 Ω is 66⅔ Ω for $X_{C_{EQ}}$. The combined equivalent reactance is always less than the lowest branch reactance. Any shortcuts for combining parallel resistances also apply to parallel reactances.

Combining capacitive reactances is opposite to the way capacitances are combined. The two procedures are compatible, however, because capacitive reactance is **inversely proportional** to capacitance. The general case is that ohms of opposition add in series but combine by the reciprocal formula in parallel.

17-3 Self-Review

Answers at the end of the chapter.

a. **How much is X_{C_T} for a 200-Ω X_{C_1} in series with a 300-Ω X_{C_2}?**

b. **How much is $X_{C_{EQ}}$ for a 200-Ω X_{C_1} in parallel with a 300-Ω X_{C_2}?**

Figure 17–5 Reactances alone combine like resistances. (*a*) Addition of series reactances. (*b*) Two reactances in parallel equal their product divided by their sum.

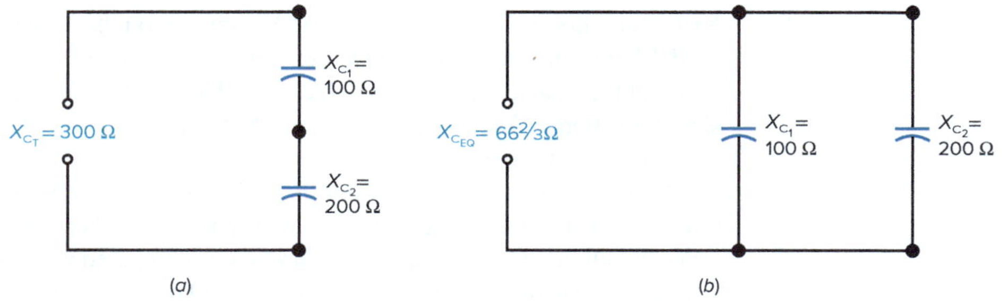

MultiSim **Figure 17–6** Example of circuit calculations with X_C. (*a*) With a single X_C, the $I = V/X_C$. (*b*) The sum of series voltage drops equals the applied voltage V_T. (*c*) The sum of parallel branch currents equals total line current I_T.

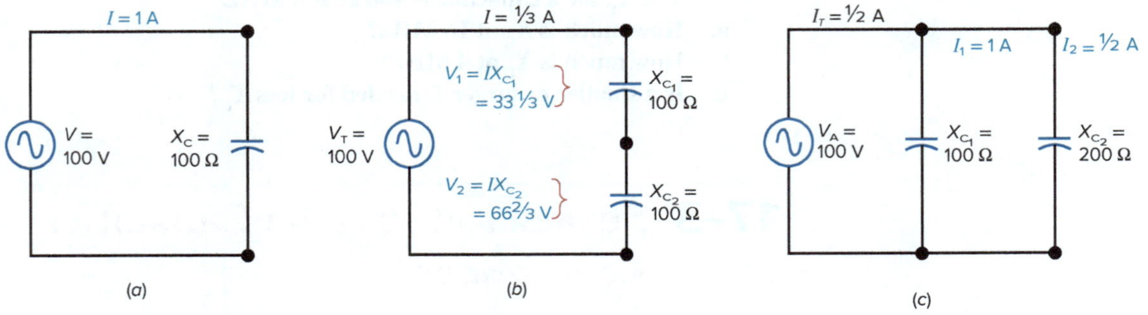

17–4 Ohm's Law Applied to X_C

The current in an AC circuit with X_C alone is equal to the applied voltage divided by the ohms of X_C. Three examples with X_C are illustrated in Fig. 17–6. In Fig. 17–6*a*, there is just one reactance of 100 Ω. The current I then is equal to V/X_C, or 100 V/100 Ω, which is 1 A.

For the series circuit in Fig. 17–6*b*, the total reactance, equal to the sum of the series reactances, is 300 Ω. Then the current is 100 V/300 Ω, which equals ⅓ A. Furthermore, the voltage across each reactance is equal to its IX_C product. The sum of these series voltage drops equals the applied voltage.

For the parallel circuit in Fig. 17–6*c*, each parallel reactance has its individual branch current, equal to the applied voltage divided by the branch reactance. The applied voltage is the same across both reactances, since all are in parallel. In addition, the total line current of 1½ A is equal to the sum of the individual branch currents of 1 and ½ A each. Because the applied voltage is an rms value, all the calculated currents and voltage drops in Fig. 17–6 are also rms values.

17–4 Self-Review

Answers at the end of the chapter.

a. **In Fig. 17–6*b*, how much is X_{C_T}?**
b. **In Fig. 17–6*c*, how much is $X_{C_{EQ}}$?**

17–5 Applications of Capacitive Reactance

The general use of X_C is to block direct current but provide low reactance for alternating current. In this way, a varying ac component can be separated from a steady direct current. Furthermore, a capacitor can have less reactance for alternating current of high frequencies, compared with lower frequencies.

Note the following difference in ohms of R and X_C. Ohms of R remain the same for DC circuits or AC circuits, whereas X_C depends on the frequency.

If 100 Ω is taken as a desired value of X_C, capacitor values can be calculated for different frequencies, as listed in Table 17–1. The C values indicate typical capacitor sizes for different frequency applications. Note that the required C becomes smaller for higher frequencies.

The 100 Ω of reactance for Table 17–1 is taken as a low X_C in common applications of C as a coupling capacitor, bypass capacitor, or filter capacitor for AC variations. For all these functions, the X_C must be low compared with the resistance in the circuit. Typical values of C, then, are 16 to 1600 pF for rf signals and 0.16 to 27 μF

Table 17–1 Capacitance Values for a Reactance of 100 Ω

C (Approx.)	Frequency	Remarks
27 μF	60 Hz	Power-line and low audio frequency
1.6 μF	1000 Hz	Audio frequency
0.16 μF	10,000 Hz	Audio frequency
1600 pF	1000 kHz (RF)	AM radio
160 pF	10 MHz (HF)	Short-wave radio
16 pF	100 MHz (VHF)	FM radio

for audio frequency (af) signals. The power-line frequency of 60 Hz, which is a low audio frequency, requires C values of about 27 μF or more.

17–5 Self-Review

Answers at the end of the chapter.

A capacitor C has 100 Ω X_C at 60 Hz.

a. How much is X_C at 120 Hz?

b. How much is X_C at 6 Hz?

17–6 Sine-Wave Charge and Discharge Current

In Fig. 17–7, sine-wave voltage applied across a capacitor produces alternating charge and discharge current. The action is considered for each quarter-cycle. Note that the voltage v_C across the capacitor is the same as the applied voltage v_A at all times because they are in parallel. The values of current i, however, depend on the charge and discharge of C. When v_A is increasing, it charges C to keep v_C at the same voltage as v_A; when v_A is decreasing, C discharges to maintain v_C at the same voltage as v_A. When v_A is not changing, there is no charge or discharge current.

During the first quarter-cycle in Fig. 17–7*a*, v_A is positive and increasing, charging C in the polarity shown. The electron flow is from the negative terminal of the source voltage, producing charging current in the direction indicated by the arrow for i. Next, when the applied voltage decreases during the second quarter-cycle, v_C also decreases by discharging. The discharge current is from the negative plate of C through the source and back to the positive plate. Note that the direction of discharge current in Fig. 17–7*b* is opposite that of the charge current in Fig. 17–7*a*.

Figure 17–7 Capacitive charge and discharge currents. (*a*) Voltage v_A increases positive to charge C. (*b*) The C discharges as v_A decreases. (*c*) Voltage v_A increases negative to charge C in opposite polarity. (*d*) The C discharges as reversed v_A decreases.

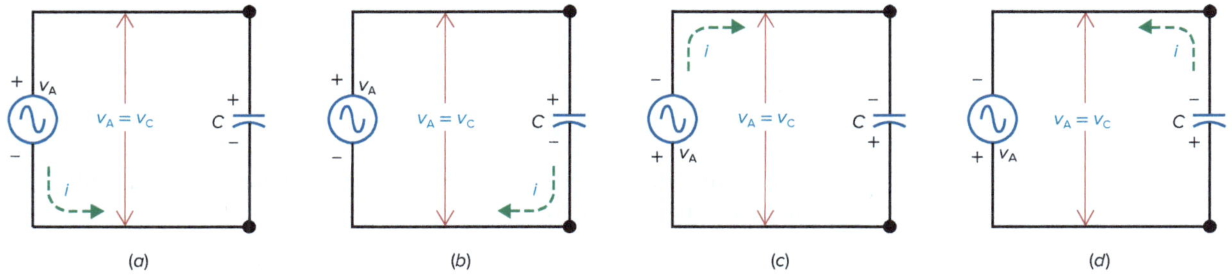

For the third quarter-cycle in Fig. 17–7*c*, the applied voltage v_A increases again but in the negative direction. Now *C* charges again but in reversed polarity. Here the charging current is in the direction opposite from the charge current in Fig. 17–7*a* but in the same direction as the discharge current in Fig. 17–7*b*. Finally, the negative applied voltage decreases during the final quarter-cycle in Fig. 17–7*d*. As a result, *C* discharges. This discharge current is opposite to the charge current in Fig. 17–7*c* but in the same direction as the charge current in Fig. 17–7*a*.

For the sine wave of applied voltage, therefore, the capacitor provides a cycle of alternating charge and discharge current. Notice that capacitive current flows for either charge or discharge, whenever the voltage changes, for either an increase or a decrease. Also, *i* and *v* have the same frequency.

Calculating the Values of i_C

GOOD TO KNOW

The rate of voltage change, $\frac{dv}{dt}$, for a sine wave is determined by both its amplitude and its frequency.

The greater the voltage change, the greater the amount of capacitive current. Furthermore, a larger capacitor can allow more charge current when the applied voltage increases and can produce more discharge current. Because of these factors the amount of capacitive current can be calculated as

$$i_C = C\frac{dv}{dt} \tag{17–6}$$

where *i* is in amperes, *C* is in farads, and dv/dt is in volts per second. As an example, suppose that the voltage across a 240-pF capacitor changes by 25 V in 1 μs. The amount of capacitive current then is

$$i_C = C\frac{dv}{dt} = 240 \times 10^{-12} \times \frac{25}{1 \times 10^{-6}}$$

$$= 240 \times 25 \times 10^{-6} = 6000 \times 10^{-6}$$

$$= 6 \times 10^{-3} \text{ A or 6 mA}$$

Notice how Formula (17–6) is similar to the capacitor charge formula $Q = CV$. When the voltage changes, this dv/dt factor produces a change in the charge *Q*. When the charge moves, this dq/dt change is the current i_C. Therefore, dq/dt or i_C is proportional to dv/dt. With the constant factor *C*, then, i_C becomes equal to $C(dv/dt)$.

By means of Formula (17–6), then, i_C can be calculated to find the instantaneous value of charge or discharge current when the voltage changes across a capacitor.

Example 17–5

Calculate the instantaneous value of charging current i_C produced by a 6-μF *C* when its potential difference is increased by 50 V in 1 s.

ANSWER

$$i_C = C\frac{dv}{dt} = 6 \times 10^{-6} \times \frac{50}{1}$$

$$= 300\ \mu\text{A}$$

Example 17-6

Calculate i_C for the same *C* as in Example 17–5 when its potential difference is decreased by 50 V in 1 s.

ANSWER For the same $C(dv/dt)$, i_C is the same 300 μA. However, this 300 μA is discharge current, which flows in the direction opposite from i_C on charge. If desired, the i_C for discharge current can be considered negative, or −300 μA.

Example 17–7

Calculate i_C produced by a 250-pF capacitor for a change of 50 V in 1 μs.

ANSWER

$$i_C = C\frac{dv}{dt}$$
$$= 250 \times 10^{-12} \times \frac{50}{1 \times 10^{-6}}$$
$$= 12{,}500 \times 10^{-6} \text{ A or } 12{,}500\ \mu\text{A or } 12.5 \text{ mA}$$

Notice that more i_C is produced in Example 17–7, although C is smaller than in Example 17–6, because dv/dt is a much faster voltage change.

Waveshapes of v_C and i_C

More details of capacitive circuits can be analyzed by plotting the values calculated in Table 17–2. Figure 17–8 shows the waveshapes representing these values.

Table 17–2 Values for $i_C = C(dv/dt)$ Cureves in Figure 17–8

Times		*dt*		*dv*,	*dv/dt*,	*C*,	$i_C = C(dv/dt)$,
θ	μs	θ	μs	V	V/μs	pF	mA
30°	2	30°	2	50	25	240	6
60°	4	30°	2	36.6	18.3	240	4.4
90°	6	30°	2	13.4	6.7	240	1.6
120°	8	30°	2	−13.4	−6.7	240	−1.6
150°	10	30°	2	−36.6	−18.3	240	−4.4
180°	12	30°	2	−50	−25	240	−6
210°	14	30°	2	−50	−25	240	−6
240°	16	30°	2	−36.6	−18.3	240	−4.4
270°	18	30°	2	−13.4	−6.7	240	−1.6
300°	20	30°	2	13.4	6.7	240	1.6
330°	22	30°	2	36.6	18.3	240	4.4
360°	24	30°	2	50	25	240	6

Figure 17–8 Waveshapes of capacitive circuits. (*a*) Waveshape of sine-wave voltage at top. (*b*) Changes in voltage below causing (*c*) current i_C charge and discharge waveshape. Values plotted are those given in Table 17–2.

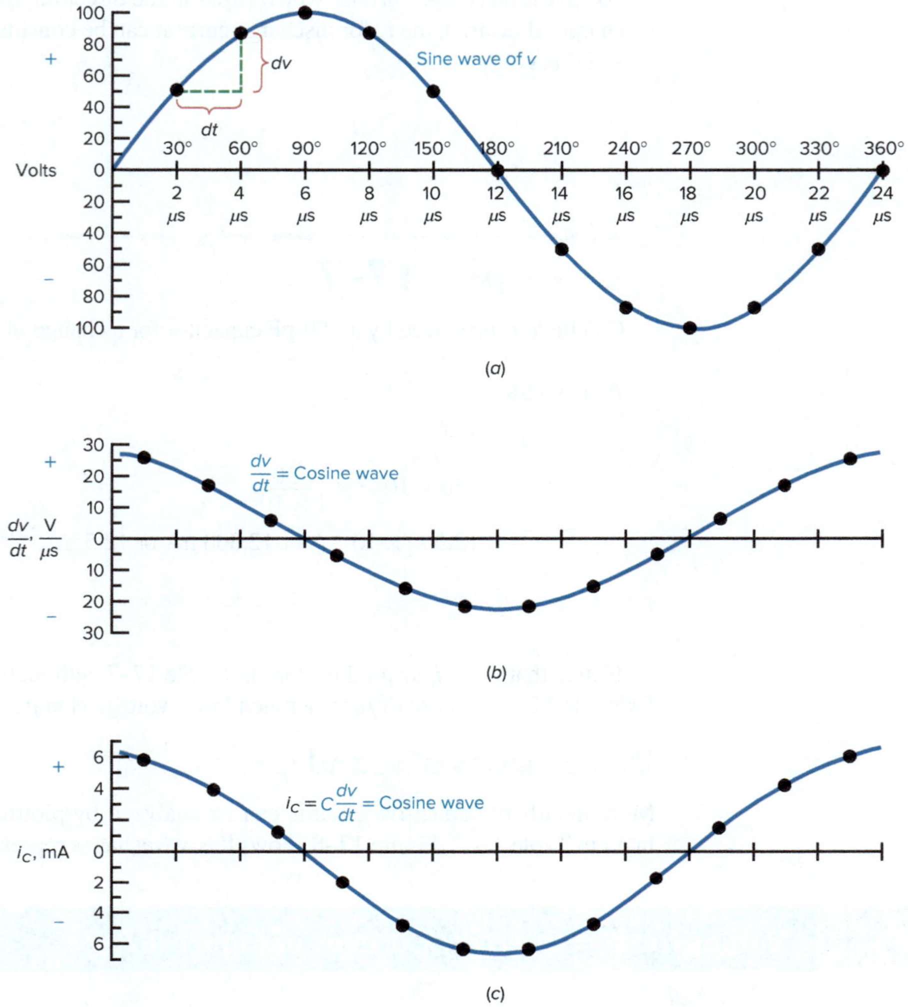

Figure 17–8*a* shows a sine wave of voltage v_C across a 240-pF capacitance *C*. Since the capacitive current i_C depends on the rate of change of voltage, rather than on the absolute value of *v*, the curve in Fig. 17–8*b* shows how much the voltage changes. In this curve, the dv/dt values are plotted for every 30° of the cycle.

Figure 17–8*c* shows the actual capacitive current i_C. This i_C curve is similar to the dv/dt curve because i_C equals the constant *C* multiplied by dv/dt.

90° Phase Angle

The i_C curve at the bottom of Fig. 17–8 has its zero values when the v_C curve at the top is at maximum. This comparison shows that the curves are 90° out of phase because i_C is a cosine wave of current for the sine wave of voltage v_C. The 90° phase difference results from the fact that i_C depends on the dv/dt rate of change, rather than on *v* itself. More details of this 90° **phase angle** for capacitance are explained in the next chapter.

For each of the curves, the period *T* is 24 μs. Therefore, the frequency is $1/T$ or $\frac{1}{24}$, which equals 41.67 kHz. Each curve has the same frequency, although there is a 90° phase difference between *i* and *v*.

Ohms of X_C

The ratio of v_C/i_C specifies the capacitive reactance in ohms. For this comparison, we use the actual value of v_C, which has a peak of 100 V. The rate-of-change factor is included in i_C. Although the peak of i_C at 6 mA is 90° ahead of the peak of v_C at 100 V, we can compare these two peak values. Then v_C/i_C is 100/0.006, which equals 16,667 Ω.

This X_C is only an approximate value because i_C cannot be determined exactly for the large dt changes every 30°. If we used smaller intervals of time, the peak i_C would be 6.28 mA with X_C then 15,900 Ω, the same as $1/(2\pi fC)$ with a 240-pF C and a frequency of 41.67 kHz.

17–6 Self-Review

Answers at the end of the chapter.

Refer to the curves in Fig. 17–8.

a. **At what angle does v have its maximum positive value?**
b. **At what angle does dv/dt have its maximum positive value?**
c. **What is the phase angle difference between v_C and i_C?**

Summary

- Capacitive reactance, X_C, is the opposition of a capacitance to the flow of sine-wave alternating current.
- X_C is measured in ohms because it limits the current to the value V/X_C. With V in volts and X_C in ohms, I is in amperes.
- $X_C = 1/(2\pi fC)$. With f in hertz and C in farads, X_C is in ohms.
- For the same value of capacitance, X_C decreases when the frequency increases.
- For the same frequency, X_C decreases when the capacitance *increases*.
- With X_C and f known, the capacitance $C = 1/(2\pi fX_C)$.
- With X_C and C known, the frequency $f = 1/(2\pi CX_C)$.
- The total X_C of capacitive reactances in series equals the sum of the individual values, as for series resistances. The series reactances have the same current. The voltage across each reactance is IX_C.
- The combined equivalent reactance of parallel capacitive reactances is calculated by the reciprocal formula, as for parallel resistances. Each branch current is V/X_C. The total line current is the sum of the individual branch currents.
- Table 17–3 summarizes the differences between C and X_C.

Table 17–3 Comparison of Capacitance and Capacitive Reactance

Capacitance	Capacitive Reactance
Symbol is C	Symbol is X_C
Measured in farad units	Measured in ohm units
Depends on construction of capacitor	Depends on frequency of sine-wave voltage
$C = i_C/(dv/dt)$ or Q/V	$X_C = v_C/i_C$ or $1/(2\pi fC)$

Important Terms

Capacitive reactance, X_C — a measure of a capacitor's opposition to the flow of sine-wave alternating current. X_C is measured in ohms. $X_C = \frac{1}{2\pi fC}$ or $X_C = \frac{V_C}{I_C}$. X_C applies only to sine-wave AC circuits.

Charging current — the current that flows to and from the plates of a capacitor as the charge stored by the capacitor increases.

Discharge current — the current that flows to and from the plates of a capacitor as the charge stored by the capacitor decreases. The discharge current of a capacitor is opposite in direction to the charging current.

Inversely proportional — the same as a reciprocal relation; as the value in the denominator increases the resultant quotient decreases. In the formula $X_C = \frac{1}{2\pi fC}$, X_C is inversely proportional to both f and C. This means that as f and C increase, X_C decreases.

Phase angle — the angular difference or displacement between two waveforms. For a capacitor, the charge and discharge current, i_C, reaches its maximum value 90° ahead of the capacitor voltage, v_C. As a result, the charge and discharge current, i_C, is said to lead the capacitor voltage, v_C, by a phase angle of 90°.

Related Formulas

$$X_C = \frac{1}{2\pi fC}$$

$$C = \frac{1}{2\pi fX_C}$$

$$f = \frac{1}{2\pi CX_C}$$

$$X_{C_T} = X_{C_1} + X_{C_2} + \cdots + \text{etc. (Series capacitive reactances)}$$

$$X_{C_{EQ}} = \frac{1}{\frac{1}{X_{C_1}} + \frac{1}{X_{C_2}} + \frac{1}{X_{C_3}} + \cdots + \text{etc.}} \text{ (Parallel capacitive reactances)}$$

$$i_C = C\frac{dv}{dt}$$

$$X_C = \frac{V_C}{I_C}$$

Self-Test

Answers at the back of the book.

1. The capacitive reactance, X_C, of a capacitor is

a. inversely proportional to frequency.

b. unaffected by frequency.

c. directly proportional to frequency.

d. directly proportional to capacitance.

2. The charge and discharge current of a capacitor flows

a. through the dielectric.

b. only when a DC voltage is applied.

c. to and from the plates.

d. both a and b.

3. For direct current (DC), a capacitor acts like a(n)

a. closed switch.

b. open.

c. short.

d. small resistance.

4. At the same frequency, a larger capacitance provides

a. more charge and discharge current.

b. less charge and discharge current.

c. less capacitive reactance, X_C.

d. both a and c.

5. How much is the capacitance, C, of a capacitor that draws 4.8 mA of current from a 12-V_{AC} generator? The frequency of the AC generator is 636.6 Hz.

a. 0.01 μF.

b. 0.1 μF.

c. 0.001 μF.

d. 100 pF.

6. At what frequency does a 0.015-μF capacitor have an X_C value of 2 kΩ?

a. 5.3 MHz.

b. 5.3 Hz.

c. 5.3 kHz.

d. 106 kHz.

7. What is the capacitive reactance, X_C, of a 330-pF capacitor at a frequency of 1 MHz?

a. 482 Ω.

b. 48.2 Ω.

c. 1 kΩ.

d. 482 MΩ.

8. What is the instantaneous value of charging current, i_C, of a 10-μF capacitor if the voltage across the capacitor plates changes at the rate of 250 V per second?

a. 250 μA.

b. 2.5 A.

c. 2.5 μA.

d. 2.5 mA.

9. For a capacitor, the charge and discharge current, i_C,

a. lags the capacitor voltage, v_C, by a phase angle of 90°.

b. leads the capacitor voltage, v_C, by a phase angle of 90°.

c. is in phase with the capacitor voltage, v_C.

d. none of the above.

10. Two 1-kΩ X_C values in series have a total capacitive reactance of

a. 1.414 kΩ.

b. 500 Ω.

c. 2 kΩ.

d. 707 Ω.

11. Two 5-kΩ X_C values in parallel have an equivalent capacitive reactance of

a. 7.07 kΩ.

b. 2.5 kΩ.

c. 10 kΩ.

d. 3.53 kΩ.

12. For any capacitor,

a. the stored charge increases with more capacitor voltage.

b. the charge and discharge currents are in opposite directions.

c. i_C leads v_C by 90°.

d. all of the above.

13. The unit of capacitive reactance, X_C, is the

a. ohm.

b. farad.

c. hertz.

d. radian.

14. The main difference between resistance, R, and capacitive reactance, X_C, is that

a. X_C is the same for both DC and AC, whereas R depends on frequency.

b. R is the same for both DC and AC, whereas X_C depends on frequency.

c. R is measured in ohms and X_C is measured in farads.

d. none of the above.

15. A very common use for a capacitor is to

a. block any DC voltage but provide very little opposition to an AC voltage.

b. block both DC and AC voltages.

c. pass both DC and AC voltages.

d. none of the above.

Essay Questions

1. Why is capacitive reactance measured in ohms? State two differences between capacitance and capacitive reactance.

2. Explain briefly why the bulb lights in Fig. 17–1*a* but not in *c*.

3. Explain briefly what is meant by two factors being inversely proportional. How does this apply to X_C and C? X_C and f?

4. In comparing X_C and R, give two differences and one similarity.

5. Why are the waves in Fig. 17–8*a* and *b* considered to be 90° out of phase, but the waves in Fig. 17–8*b* and *c* have the same phase?

6. Referring to Fig. 17–3, how does this graph show an inverse relation between X_C and C?

7. Referring to Fig. 17–4, how does this graph show an inverse relation between X_C and f?

8. Referring to Fig. 17–8, draw three similar curves but for a sine wave of voltage with a period $T = 12\ \mu s$ for the full cycle. Use the same C of 240 pF. Compare the value of X_C obtained as $1/(2\pi fC)$ and v_C/i_C.

9. (a) What is the relationship between charge q and current i? (b) How is this comparison similar to the relation between the two formulas $Q = CV$ and $i_C = C(dv/dt)$?

Problems

SECTION 17–1 ALTERNATING CURRENT IN A CAPACITIVE CIRCUIT

17–1 With the switch, S_1, closed in Fig. 17–9, how much is

a. the current, I, in the circuit?

b. the DC voltage across the 12-V lamp?

c. the DC voltage across the capacitor?

Figure 17-9

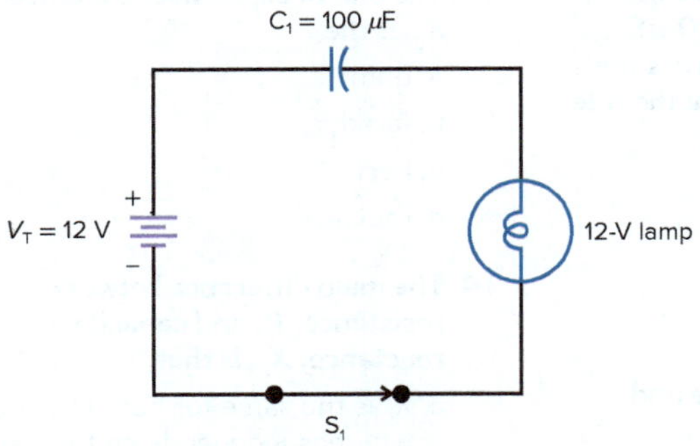

17–2 In Fig. 17–9, explain why the bulb will light for just an instant when S_1 is initially closed.

17–3 In Fig. 17–10, the capacitor and the lightbulb draw 400 mA from the 120-V_{AC} source. How much current flows

a. to and from the terminals of the 120-V_{AC} source?

b. through the lightbulb?

c. to and from the plates of the capacitor?

d. through the connecting wires?

e. through the dielectric of the capacitor?

Figure 17-10

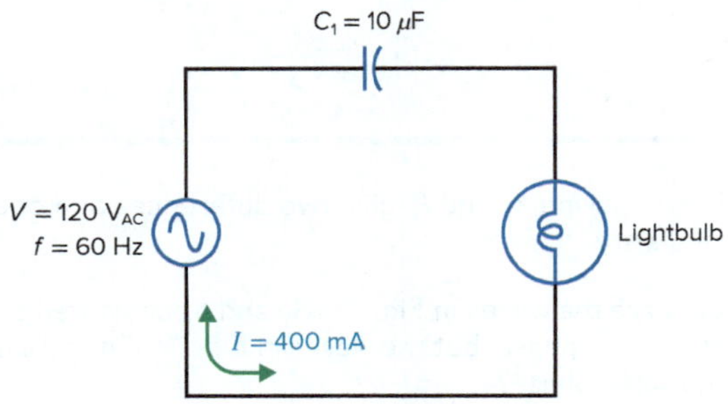

17–4 In Fig. 17–11, calculate the capacitive reactance, X_C, for the following values of V_{AC} and I?

a. $V_{AC} = 10$ V and $I = 20$ mA.

b. $V_{AC} = 24$ V and $I = 8$ mA.

c. $V_{AC} = 15$ V and $I = 300\ \mu A$.

d. $V_{AC} = 100$ V and $I = 50\ \mu A$.

Figure 17-11

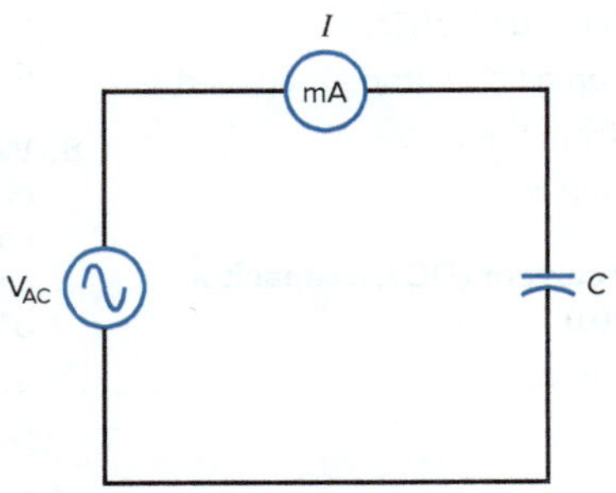

17–5 In Fig. 17–11, list three factors that can affect the amount of charge and discharge current flowing in the circuit.

SECTION 17–2 THE AMOUNT OF X_C EQUALS $\frac{1}{2\pi fC}$

17–6 Calculate the capacitive reactance, X_C, of a 0.1-μF capacitor at the following frequencies:

a. $f = 10$ Hz.

b. $f = 50$ Hz.

c. $f = 200$ Hz.

d. $f = 10$ kHz.

17–7 Calculate the capacitive reactance, X_C, of a 10-μF capacitor at the following frequencies:

a. $f = 60$ Hz.

b. $f = 120$ Hz.

c. $f = 500$ Hz.

d. $f = 1$ kHz.

17–8 What value of capacitance will provide an X_C of 1 kΩ at the following frequencies?

a. $f = 318.3$ Hz.

b. $f = 1.591$ kHz.

c. $f = 3.183$ kHz.

d. $f = 6.366$ kHz.

17–9 At what frequency will a 0.047-μF capacitor provide an X_C value of

a. 100 kΩ?

b. 5 kΩ?

c. 1.5 kΩ?

d. 50 Ω?

17–10 How much is the capacitance of a capacitor that draws 2 mA of current from a 10-V_{AC} generator whose frequency is 3.183 kHz?

17–11 At what frequency will a 820-pF capacitance have an X_C value of 250 Ω?

17–12 A 0.01-μF capacitor draws 50 mA of current when connected directly across a 50-V_{AC} source. What is the value of current drawn by the capacitor when

a. the frequency is doubled?

b. the frequency is decreased by one-half?

c. the capacitance is doubled to 0.02 μF?

d. the capacitance is reduced by one-half to 0.005 μF?

17–13 A capacitor has an X_C value of 10 kΩ at a given frequency. What is the new value of X_C when the frequency is

a. cut in half?

b. doubled?

c. quadrupled?

d. increased by a factor of 10?

17–14 Calculate the capacitive reactance, X_C, for the following capacitance and frequency values:

a. $C = 0.47\ \mu F, f = 1$ kHz.

b. $C = 100\ \mu F, f = 120$ Hz.

c. $C = 250$ pF, $f = 1$ MHz.

d. $C = 0.0022\ \mu F, f = 50$ kHz.

17–15 Determine the capacitance value for the following frequency and X_C values:

a. $X_C = 1\ k\Omega, f = 3.183$ kHz.

b. $X_C = 200\ \Omega, f = 63.66$ kHz.

c. $X_C = 25\ k\Omega, f = 1.592$ kHz.

d. $X_C = 1\ M\Omega, f = 100$ Hz.

17–16 Determine the frequency for the following capacitance and X_C values:

a. $C = 0.05\ \mu F, X_C = 4\ k\Omega$.

b. $C = 0.1\ \mu F, X_C = 1.591\ k\Omega$.

c. $C = 0.0082\ \mu F, X_C = 6.366\ k\Omega$.

d. $C = 50\ \mu F, X_C = 100\ \Omega$.

SECTION 17–3 SERIES AND PARALLEL CAPACITIVE REACTANCES

17–17 How much is the total capacitive reactance, X_{C_T}, for the following series capacitive reactances:

a. $X_{C_1} = 1\ k\Omega, X_{C_2} = 1.5\ k\Omega, X_{C_3} = 2.5\ k\Omega$.

b. $X_{C_1} = 500\ \Omega, X_{C_2} = 1\ k\Omega, X_{C_3} = 1.5\ k\Omega$.

c. $X_{C_1} = 20\ k\Omega, X_{C_2} = 10\ k\Omega, X_{C_3} = 120\ k\Omega$.

d. $X_{C_1} = 340\ \Omega, X_{C_2} = 570\ \Omega, X_{C_3} = 2.09\ k\Omega$.

17–18 What is the equivalent capacitive reactance, $X_{C_{EQ}}$, for the following parallel capacitive reactances:

a. $X_{C_1} = 100\ \Omega$ and $X_{C_2} = 400\ \Omega$.

b. $X_{C_1} = 1.2\ k\Omega$ and $X_{C_2} = 1.8\ k\Omega$.

c. $X_{C_1} = 15\ \Omega, X_{C_2} = 6\ \Omega, X_{C_3} = 10\ \Omega$.

d. $X_{C_1} = 2.5\ k\Omega, X_{C_2} = 10\ k\Omega, X_{C_3} = 2\ k\Omega, X_{C_4} = 1\ k\Omega$.

SECTION 17–4 OHM'S LAW APPLIED TO X_C

17–19 In Fig. 17–12, calculate the current, I.

Figure 17-12

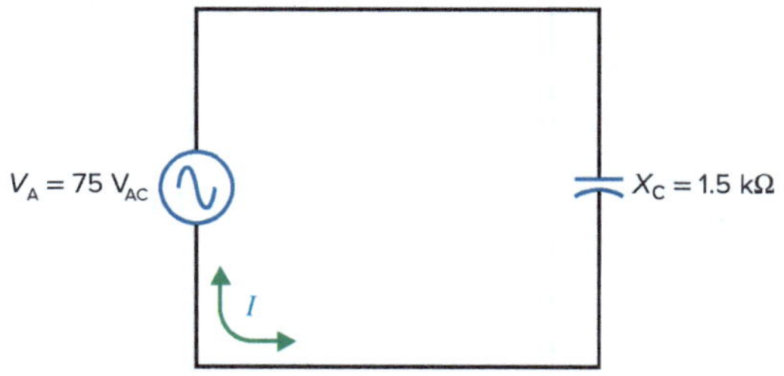

17–20 In Fig. 17–12, what happens to the current, I, when the frequency of the applied voltage

a. decreases?

b. increases?

17–21 In Fig. 17–13, solve for

a. X_{C_T}.

b. I.

c. V_{C_1}, V_{C_2}, and V_{C_3}.

Figure 17-13

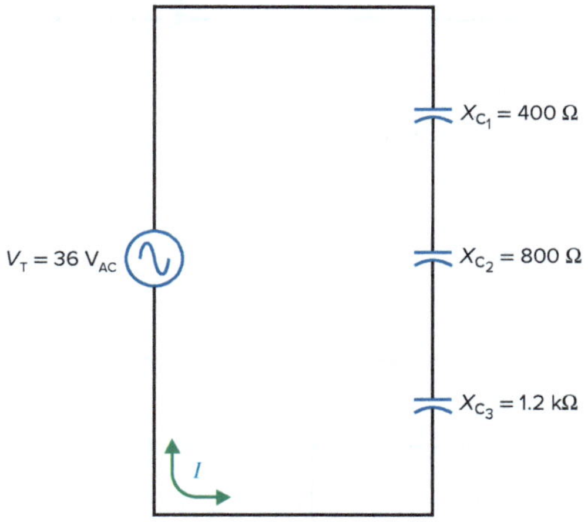

17–22 In Fig. 17–14, solve for

a. X_{C_1}, X_{C_2}, and X_{C_3}.

b. X_{C_T}.

c. I.

d. V_{C_1}, V_{C_2}, and V_{C_3}.

e. C_{EQ}.

17–23 In Fig. 17–13, solve for C_1, C_2, C_3, and C_{EQ} if the applied voltage has a frequency of 318.3 Hz.

17–24 In Fig. 17–15, solve for

a. I_{C_1}, I_{C_2}, and I_{C_3}.

b. I_T.

c. $X_{C_{EQ}}$.

Figure 17-14

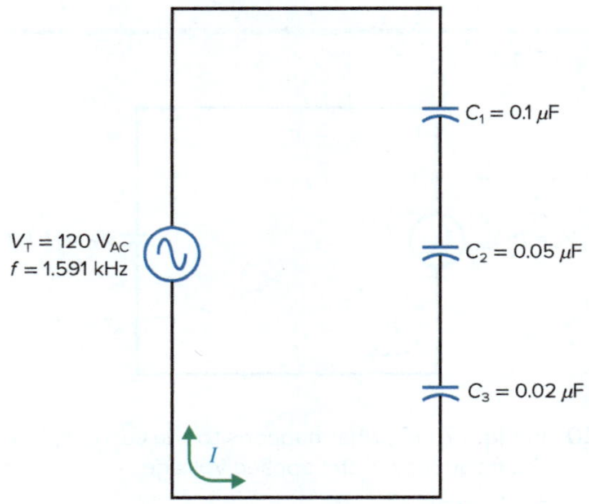

Figure 17-15

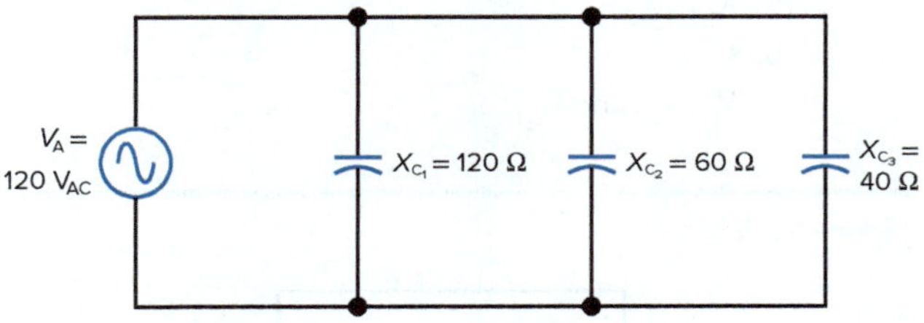

17–25 In Fig. 17–16, solve for

a. X_{C_1}, X_{C_2}, and X_{C_3}.

b. I_{C_1}, I_{C_2}, and I_{C_3}.

c. I_T.

d. $X_{C_{EQ}}$.

e. C_T.

Figure 17-16

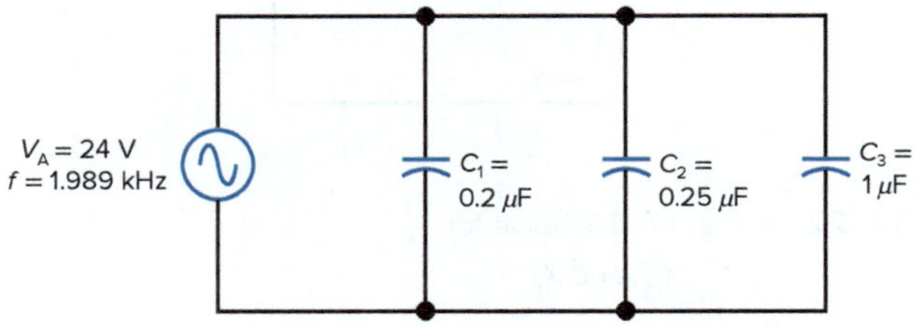

17–26 In Fig. 17–15, solve for C_1, C_2, C_3, and C_T if the frequency of the applied voltage is 6.366 kHz.

SECTION 17–5 APPLICATIONS OF CAPACITIVE REACTANCE

17–27 Calculate the value of capacitance, C, required to produce an X_C value of 500 Ω at the following frequencies:

a. $f = 100$ Hz.

b. $f = 2$ kHz.

c. $f = 50$ kHz.

d. $f = 10$ MHz.

SECTION 17–6 SINE-WAVE CHARGE AND DISCHARGE CURRENT

17–28 Calculate the instantaneous charging current, i_C, for a 0.33-μF capacitor if the voltage across the capacitor plates changes at the rate of 10 V/1 ms.

17–29 Calculate the instantaneous charging current, i_C, for a 0.01-μF capacitor if the voltage across the capacitor plates changes at the rate of

a. 100 V/s.

b. 100 V/ms.

c. 50 V/μs.

17–30 What is the instantaneous discharge current, i_C, for a 100-μF capacitor if the voltage across the capacitor plates decreases at the rate of

a. 10 V/s.

b. 1 V/ms.

c. 50 V/ms.

17–31 For a capacitor, what is the phase relationship between the charge and discharge current, i_C, and the capacitor voltage, v_C? Explain your answer.

17–32 A capacitor has a discharge current, i_C, of 15 mA when the voltage across its plates decreases at the rate of 150 V/μs. Calculate C.

17–33 What rate of voltage change, $\frac{dv}{dt}$, will produce a charging current of 25 mA in a 0.01-μF capacitor? Express your answer in volts per second.

Critical Thinking

17–34 Explain an experimental procedure for determining the value of an unmarked capacitor. (Assume that a capacitance meter is not available.)

17–35 In Fig. 17–17, calculate X_{C_T}, X_{C_1}, X_{C_2}, C_1, C_3, V_{C_1}, V_{C_2}, V_{C_3}, I_{C_2}, and I_{C_3}.

Figure 17-17 Circuit for Critical Thinking Prob. 17–35.

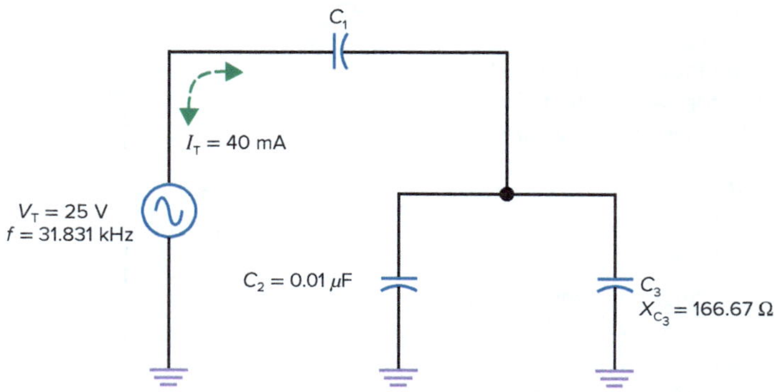

Answers to Self-Reviews

17–1 **a.** 0.1 μF
b. 0.5 μF

17–2 **a.** 200 Ω
b. 800 Ω
c. larger

17–3 **a.** 500 Ω
b. 120 Ω

17–4 **a.** 300 Ω
b. 66.7 Ω

17–5 **a.** 50 Ω
b. 1000 Ω

17–6 **a.** 90°
b. 0 or 360°
c. 90°

Laboratory Application Assignment

In this lab application assignment, you will examine how the capacitive reactance, X_C, of a capacitor decreases when the frequency, f, increases. You will also see that more capacitance, C, at a given frequency results in less capacitive reactance, X_C. Finally, you will observe how X_C values combine in series and in parallel.

Equipment: Obtain the following items from your instructor.

- Function generator
- Assortment of capacitors
- DMM

Capacitive Reactance, X_C

Refer to Fig. 17–18*a*. Calculate and record the value of X_C for each of the following frequencies listed below. Calculate X_C as $1/(2\pi fC)$.

X_C = ____________ @ f = 100 Hz
X_C = ____________ @ f = 200 Hz
X_C = ____________ @ f = 400 Hz

Connect the circuit in Fig. 17–18*a*. Set the voltage source to exactly 5 V_{rms}. For each of the following frequencies listed below, measure and record the current, I. (Use a DMM to measure I.) Next, calculate X_C as V/I.

I = ____________ @f = 100 Hz; X_C = ____________
I = ____________ @f = 200 Hz; X_C = ____________
I = ____________ @f = 400 Hz; X_C = ____________

How do the experimental values of X_C compare to those initially calculated?____________________________

Based on your experimental values, what happens to the value of X_C each time the frequency, f, is doubled? ____________

Figure 17-18

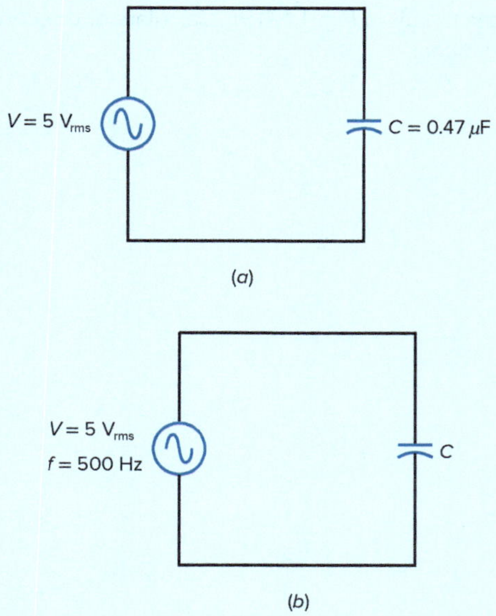

Is X_C proportional or inversely proportional to the frequency, f?

____

Refer to Fig. 17–18*b*. With the frequency, f, set to 500 Hz, calculate and record the value of X_C for each of the following capacitance values listed below. Calculate X_C as $1/(2\pi fC)$.

X_C = ________ when $C = 0.1\ \mu F$
X_C = ________ when $C = 0.22\ \mu F$
X_C = ________ when $C = 0.47\ \mu F$

Connect the circuit in Fig. 17–18*b*. Adjust the frequency of the function generator to exactly 500 Hz. For each of the following capacitance values listed below, measure and record the current, I. (Use a DMM to measure I.) Next, calculate X_C as V/I.

I = ________ when $C = 0.1\ \mu F$; X_C = ________
I = ________ when $C = 0.22\ \mu F$; X_C = ________
I = ________ when $C = 0.47\ \mu F$; X_C = ________

Is X_C proportional or inversely proportional to the value of capacitance? ________

Series Capacitive Reactances

Refer to the circuit in Fig. 17–19*a*. Calculate and record the following values:

X_{C_1} = ________, X_{C_2} = ________, X_{C_T} = ________, I = ________,
V_{C_1} = ________, V_{C_2} = ________

Do V_{C_1} and V_{C_2} add to equal V_T?

Construct the circuit in Fig. 17–19*a*. Set the frequency of the function generator to exactly 500 Hz. Next, using a DMM, measure and record the following values:

I = ________, V_{C_1} = ________, V_{C_2} = ________

Using the measured values of voltage and current, calculate the following values:

X_{C_1} = ________, X_{C_2} = ________, X_{C_T} = ________

Are the experimental values calculated here close to those initially calculated above? ________

Parallel Capacitive Reactances

Refer to the circuit in Fig. 17–19*b*. Calculate and record the following values:

Figure 17-19

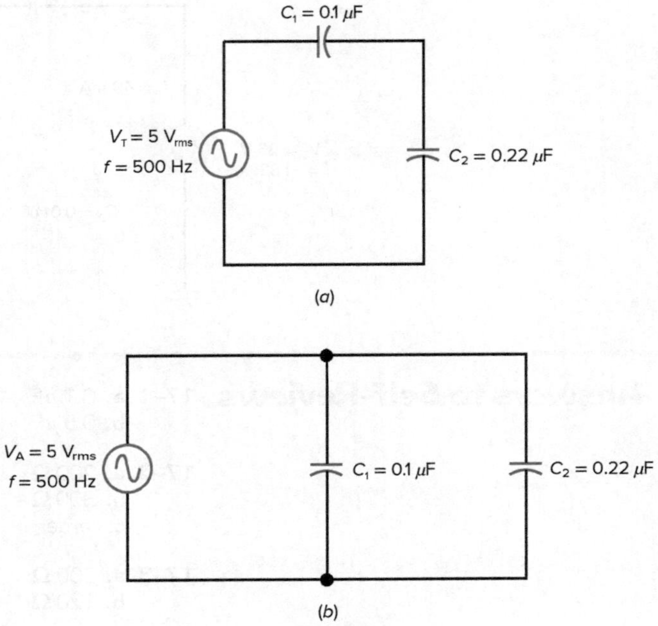

X_{C_1} = ________, X_{C_2} = ________, I_{C_1} = ________, I_{C_2} = ________,
I_T = ________, $X_{C_{EQ}}$ = ________

Do I_{C_1} and I_{C_2} add to equal I_T? ________

Construct the circuit in Fig. 17–19*b*. Set the frequency of the function generator to exactly 500 Hz. Next, using a DMM, measure and record the following values:

I_{C_1} = ________, I_{C_2} = ________, I_T = ________

Using the measured values of voltage and current, calculate the following values:

X_{C_1} = ________, X_{C_2} = ________, $X_{C_{EQ}}$ = ________

Are the experimental values calculated here close to those initially calculated above? ________

Design credits: Chapter Opener Background Image, and Multisim: Stockbyte/Getty Images

chapter 18

Capacitive Circuits

This chapter analyzes circuits that combine capacitive reactance X_C and resistance R. The main questions are, how do we combine the ohms of opposition, how much current flows, and what is the phase angle? Although both X_C and R are measured in ohms, they have different characteristics. Specifically, for a sine-wave AC voltage, X_C decreases with more capacitance and higher frequencies whereas R is the same for both DC and AC circuits. Furthermore, the phase angle for the voltage across X_C is at $-90°$ measured in the clockwise direction with i_C as the reference at $0°$.

In addition, the practical application of a coupling capacitor shows how a low value of X_C can be used to pass the desired AC signal variations, while blocking the steady DC level of a fluctuating DC voltage. In a coupling circuit with C and R in series, the AC component is across R for the output voltage, but the DC component is blocked from reaching the output by the capacitor, C.

Finally, the general case of capacitive charge and discharge current produced when the applied voltage changes is shown with nonsinusoidal voltage variations. In this case, we compare the waveshapes of v_C and i_C. Remember that the $-90°$ phase angle for an IX_C voltage applies only to sine waves. ■

Chapter Outline

Chapter Objectives

After studying this chapter, you should be able to

- *Explain* why the current leads the voltage by 90° for a capacitor.
- *Define* the term *impedance*.
- *Calculate* the total impedance and phase angle of a series *RC* circuit.
- *Describe* the operation and application of an *RC* phase-shifter circuit.
- *Calculate* the total current, equivalent impedance, and phase angle of a parallel *RC* circuit.
- *Explain* how a capacitor can couple some AC frequencies but not others.
- *Calculate* the individual capacitor voltage drops for capacitors in series.
- *Calculate* the capacitive current that flows with nonsinusoidal waveforms.

Important Terms

arctangent (arctan)
capacitive voltage divider
coupling capacitor, C_C
impedance, *Z*
phase angle, θ
phasor triangle
RC phase-shifter
tangent (tan)

18–1 Sine Wave v_C Lags i_C by 90°

For a sine wave of applied voltage, a capacitor provides a cycle of alternating charge and discharge current, as shown in Fig. 18–1*a*. In Fig. 18–1*b*, the waveshape of this charge and discharge current i_C is compared with the voltage v_C.

Examining the v_C and i_C Waveforms

In Fig. 18–1*b*, note that the instantaneous value of i_C is zero when v_C is at its maximum value. At either its positive or its negative peak, v_C is not changing. For one instant at both peaks, therefore, the voltage must have a static value before changing its direction. Then v is not changing and C is not charging or discharging. The result is zero current at this time.

Also note that i_C is maximum when v_C is zero. When v_C crosses the zero axis, i_C has its maximum value because then the voltage is changing most rapidly.

Therefore, i_C and v_C are 90° out of phase, since the maximum value of one corresponds to the zero value of the other; i_C leads v_C because i_C has its maximum value a quarter-cycle before the time that v_C reaches its peak. The phasors in Fig. 18–1*c* show i_C leading v_C by the counterclockwise angle of 90°. Here v_C is the horizontal phasor for the reference angle of 0°. In Fig. 18–1*d*, however, the current i_C is the horizontal phasor for reference. Since i_C must be 90° leading, v_C is shown lagging by the clockwise angle of −90°. In series circuits, the current i_C is the reference, and then the voltage v_C can be considered to lag i_C by 90°.

GOOD TO KNOW

For any capacitor, the 90° phase relationship between v_C and i_C exists for all values of capacitance or any frequency of sine-wave alternating voltage.

Why i_C Leads v_C by 90°

The 90° **phase angle** results because i_C depends on the rate of change of v_C. In other words, i_C has the phase of dv/dt, not the phase of v. As shown previously in Fig. 17–8 for a sine wave of v_C, the capacitive charge and discharge current is a cosine wave. This 90° phase between v_C and i_C is true in any sine-wave AC circuit, whether C is in series or parallel and whether C is alone or combined with other components. We can always say that for any X_C, its current and voltage are 90° out of phase.

Capacitive Current Is the Same in a Series Circuit

The leading phase angle of capacitive current is only with respect to the voltage across the capacitor, which does not change the fact that the current is the same in all parts of a series circuit. In Fig. 18–1*a*, for instance, the current in the generator, the connecting wires, and both plates of the capacitor must be the same because they are all in the same path.

Figure 18–1 Capacitive current i_C leads v_C by 90°. (*a*) Circuit with sine wave V_A across C. (*b*) Waveshapes of i_C 90° ahead of v_C. (*c*) Phasor diagram of i_C leading the horizontal reference v_C by a counterclockwise angle of 90°. (*d*) Phasor diagram with i_C as the reference phasor to show v_C lagging i_C by an angle of −90°.

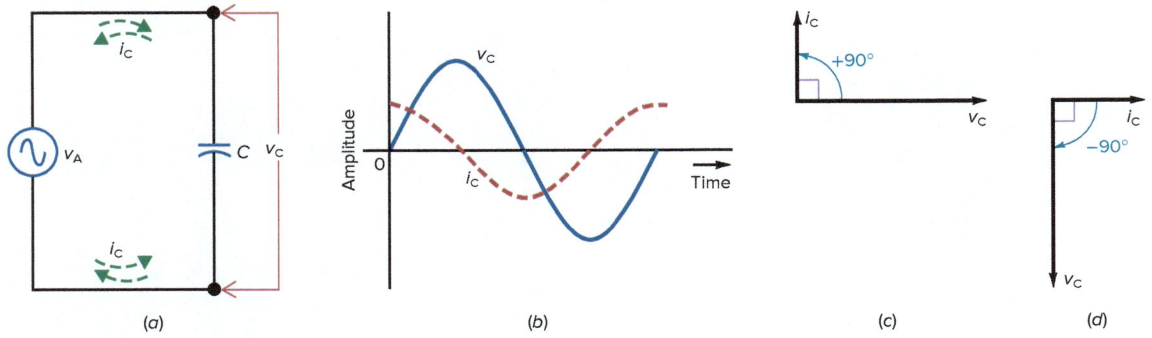

Capacitive Voltage Is the Same across Parallel Branches

In Fig. 18–1*a*, the voltage is the same across the generator and *C* because they are in parallel. There cannot be any lag or lead in time between these two parallel voltages. At any instant, whatever the voltage value is across the generator at that time, the voltage across *C* is the same. With respect to the series current, however, both v_A and v_C are 90° out of phase with i_C.

The Frequency Is the Same for v_C and i_C

Although v_C lags i_C by 90°, both waves have the same frequency. For example, if the frequency of the sine wave v_C in Fig. 18–1*b* is 100 Hz, this is also the frequency of i_C.

18–1 Self-Review

Answers at the end of the chapter.

Refer to Fig. 18–1.

a. What is the phase angle between v_A and v_C?
b. What is the phase angle between v_C and i_C?
c. Does v_C lead or lag i_C?

18–2 X_C and *R* in Series

When a capacitor and a resistor are connected in series, as shown in Fig. 18–2*a*, the current *I* is limited by both X_C and *R*. The current *I* is the same in both X_C and *R* since they are in series. However, each component has its own series voltage drop, equal to *IR* for the resistance and IX_C for the capacitive reactance.

Note the following points about a circuit that combines both X_C and *R* in series, like that in Fig. 18–2*a*.

1. The current is labeled *I*, rather than I_C, because *I* flows through all series components.
2. The voltage across X_C, labeled V_C, can be considered an IX_C voltage drop, just as we use V_R for an *IR* voltage drop.
3. The current *I* through X_C must lead V_C by 90° because this is the phase angle between the voltage and current for a capacitor.
4. The current *I* through *R* and its *IR* voltage drop are in phase. There is no reactance to sine-wave alternating current in any resistance. Therefore, *I* and *IR* have a phase angle of 0°.

It is important to note that the values of *I* and *V* may be in rms, peak, peak-to-peak, or instantaneous, as long as the same measure is applied to the entire circuit. Peak values will be used here for convenience in comparing waveforms.

Phase Comparisons

Note the following points about a circuit containing series resistance and reactance:

1. The voltage V_C is 90° out of phase with *I*.
2. However, V_R and *I* are in phase.
3. If *I* is used as the reference, V_C is 90° out of phase with V_R.

Specifically, V_C lags V_R by 90° just as the voltage V_C lags the current *I* by 90°. The phase relationships between *I*, V_R, V_C, and V_T are shown by the waveforms in Fig. 18–2*b*. Figure 18–2*c* shows the phasors representing *I*, V_R, and V_C.

MultiSim **Figure 18–2** Circuit with X_C and R in series. (*a*) Schematic diagram. (*b*) Waveforms of current and voltages. (*c*) Phasor diagram.

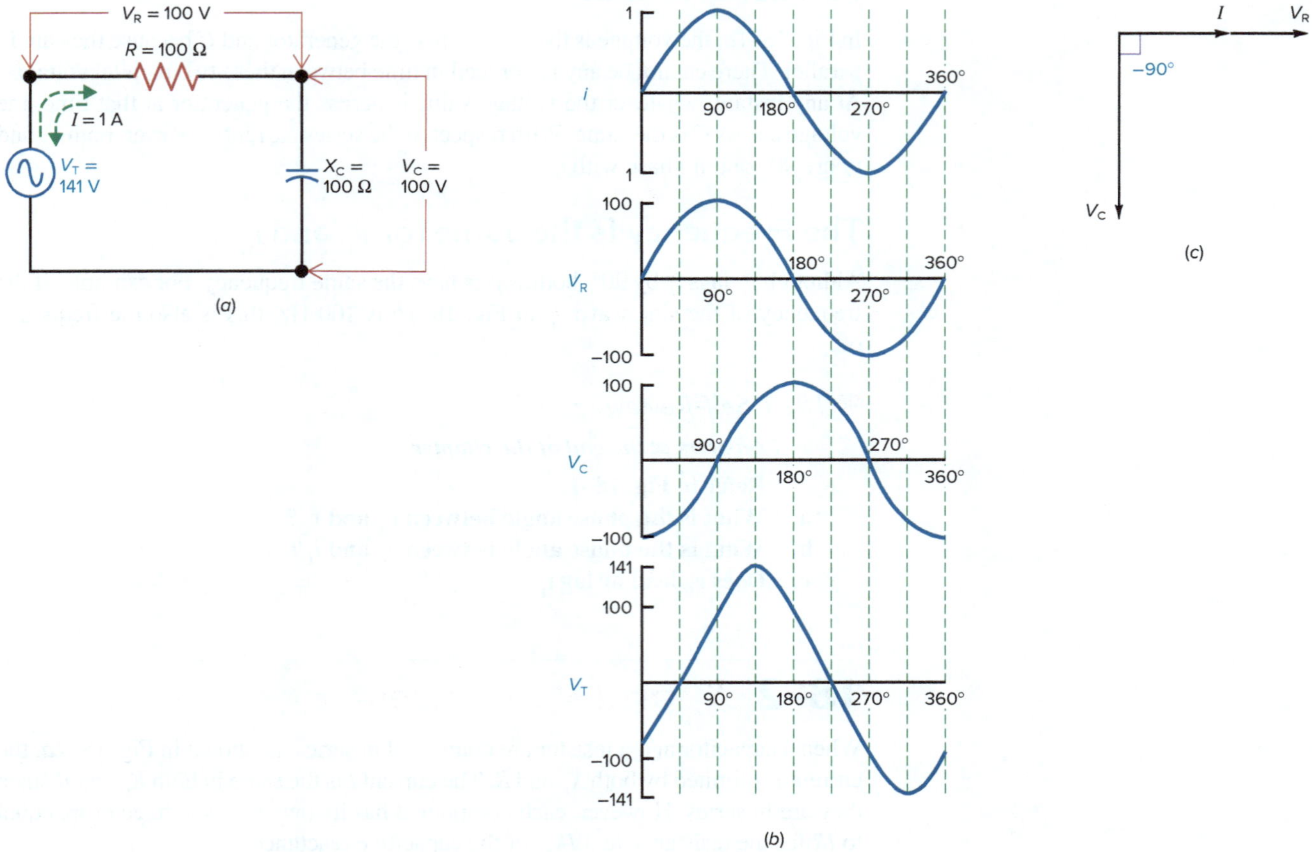

Figure 18–3 Addition of two voltages 90° out of phase. (*a*) Phasors for V_C and V_R are 90° out of phase. (*b*) Resultant of the two phasors is the hypotenuse of the right triangle for V_T.

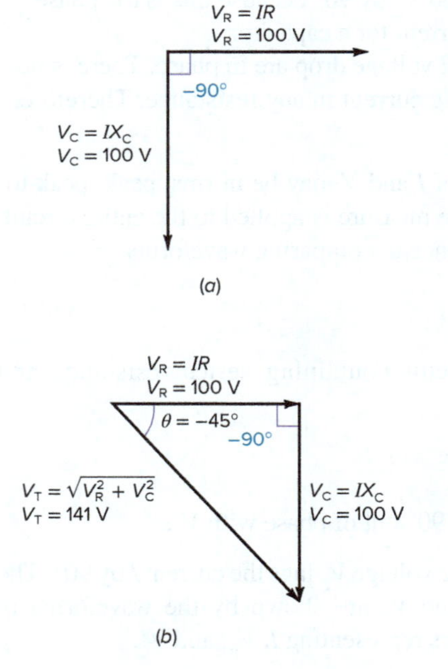

Combining V_R and V_C

As shown in Fig. 18–2*b*, when the voltage wave V_R is combined with the voltage wave V_C, the result is the voltage wave of the applied voltage V_T. The voltage drops, V_R and V_C, must add to equal the applied voltage V_T. The 100-V peak values for V_R and V_C total 141 V, however, instead of 200 V, because of the 90° phase difference.

Consider some instantaneous values in Fig. 18–2*b*, to see why the 100-V peak V_R and 100-V peak V_C cannot be added arithmetically. When V_R is at its maximum of 100 V, for instance, V_C is at zero. The total voltage V_T at this instant, then, is 100 V. Similarly, when V_C is at its maximum of 100 V, V_R is at zero and the total voltage V_T is again 100 V.

Actually, V_T reaches its maximum of 141 V when V_C and V_R are each at 70.7 V. When series voltage drops that are out of phase are combined, therefore, they cannot be added without taking the phase difference into account.

Phasor Voltage Triangle

Instead of combining waveforms that are out of phase, as shown in Fig. 18–2*b*, we can add them more quickly by using their equivalent phasors, as shown in Fig. 18–3. The phasors in Fig. 18–3*a* show the 90° phase angle without any addition. The method in Fig. 18–3*b* is to add the tail of one phasor to the arrowhead of the other, using the angle required to show their relative phase. Note that voltages V_R and V_C are at right angles to each other because they are 90° out of phase. Note also that the phasor for V_C is downward at an angle of −90° from the phasor for V_R. Here V_R is used as the reference phasor because it has the same phase as the series current I, which is the same everywhere in the circuit. The phasor V_T, extending from the tail of the V_R phasor to the arrowhead of the V_C phasor, represents the applied voltage V_T, which is the phasor sum of V_R and V_C. Since V_R and V_C form a right angle, the resultant phasor V_T is the hypotenuse of a right triangle. The hypotenuse is the side opposite the 90° angle.

From the geometry of a right triangle, the Pythagorean theorem states that the hypotenuse is equal to the square root of the sum of the squares of the sides. For the voltage triangle in Fig. 18–3*b*, therefore, the resultant is

$$V_T = \sqrt{V_R^2 + V_C^2} \qquad \textbf{(18–1)}$$

where V_T is the phasor sum of the two voltages V_R and V_C 90° out of phase.

This formula is for V_R and V_C when they are in series, since they are 90° out of phase. All voltages must be expressed in the same units. When V_T is an rms value, V_R and V_C must also be rms values. For the voltage triangle in Fig. 18–3*b*,

$$\begin{aligned} V_T &= \sqrt{100^2 + 100^2} = \sqrt{10{,}000 + 10{,}000} \\ &= \sqrt{20{,}000} \\ &= 141 \text{ V} \end{aligned}$$

CALCULATOR

To do a problem like this on a calculator, remember that the square root sign is a sign of grouping. All terms within the group must be added before you take the square root. Also, each term must be squared individually before adding for the sum. Specifically, for this problem:

- Punch in 100 and push the (x^2) key for 10,000 as the square.
- Next, punch the (+) key and then punch in 100 and (x^2). Press the (=) key. The display should read 20,000.
- Press (√) to read the answer of 141.421.

In some calculators, either the (x^2) or the (√) key must be preceded by the second function key (2ndF).

18–2 Self-Review

Answers at the end of the chapter.

a. **In a series circuit with X_C and R, what is the phase angle between I and V_R?**

b. **What is the phase angle between V_R and V_C?**

c. **In a series circuit with X_C and R, does the series current I lead or lag the applied voltage V_T?**

18–3 Impedance Z Triangle

A triangle of R and X_C in series corresponds to the voltage triangle, as shown in Fig. 18–4. It is similar to the voltage triangle in Fig. 18–3*b*, but the common factor I cancels because the series current I is the same in X_C and R. The resultant of the phasor addition of X_C and R is their total opposition in ohms, called ***impedance***, with the symbol Z_T. The Z takes into account the 90° phase relation between R and X_C.

For the impedance triangle of a series circuit with capacitive reactance X_C and resistance R,

$$Z_T = \sqrt{R^2 + X_C^2} \qquad \textbf{(18–2)}$$

where R, X_C, and Z_T are all in ohms. For the **phasor triangle** in Fig. 18–4,

$$\begin{aligned} Z_T &= \sqrt{100^2 + 100^2} = \sqrt{10{,}000 + 10{,}000} \\ &= \sqrt{20{,}000} \\ Z_T &= 141\ \Omega \end{aligned}$$

Figure 18–4 Addition of R and X_C 90° out of phase in a series RC circuit to find the total impedance Z_T.

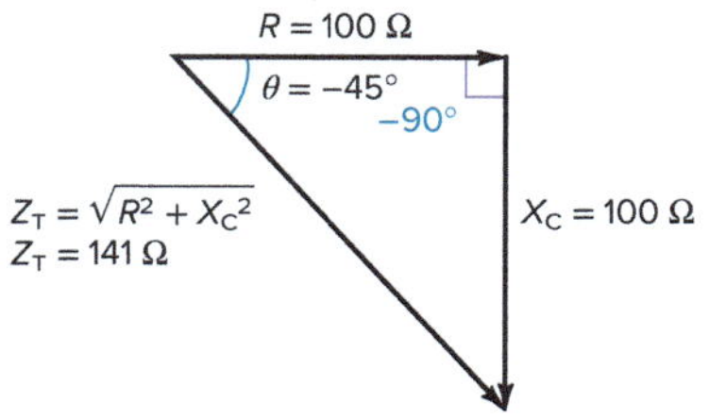

This is the total impedance Z_T in Fig. 18–2*a*.

Note that the applied voltage V_T of 141 V divided by the total impedance of 141 Ω results in 1 A of current in the series circuit. The IR voltage V_R is 1 A × 100 Ω or 100 V; the IX_C voltage is also 1 A × 100 Ω or 100 V. The series IR and IX_C voltage drops of 100 V each are added using phasors to equal the applied voltage V_T of 141 V. Finally, the applied voltage equals IZ_T or 1 A × 141 Ω, which is 141 V.

Summarizing the similar phasor triangles for voltage and ohms in a series RC circuit,

1. The phasor for R, IR, or V_R is used as a reference at 0°.
2. The phasor for X_C, IX_C, or V_C is at −90°.
3. The phasor for Z_T, IZ_T, or V_T has the phase angle θ of the complete circuit.

Phase Angle with Series X_C and R

The angle between the applied voltage V_T and the series current I is the phase angle of the circuit. Its symbol is θ (theta). In Fig. 18–3*b*, the phase angle between V_T and IR is −45°. Since IR and I have the same phase, the angle is also −45° between V_T and I.

In the corresponding impedance triangle in Fig. 18–4, the angle between Z_T and R is also equal to the phase angle. Therefore, the phase angle can be calculated from the impedance triangle of a series RC circuit by the formula

$$\tan \theta_Z = -\frac{X_C}{R} \qquad \textbf{(18–3)}$$

The **tangent (tan)** is a trigonometric function of an angle, equal to the ratio of the opposite side to the adjacent side of a triangle. In this impedance triangle, X_C is the opposite side and R is the adjacent side of the angle. We use the subscript Z for θ to show that θ_Z is found from the impedance triangle for a series circuit. To calculate this phase angle,

$$\tan \theta_Z = -\frac{X_C}{R} = -\frac{100}{100} = -1$$

The angle that has the tangent value of −1 is −45° in this example. The numerical values of the trigonometric functions can be found from a table or by using a scientific calculator. Note that the phase angle of −45° is halfway between 0° and −90° because R and X_C are equal.

CALCULATOR

To do the trigonometry in Example 18–1 with a calculator, keep in mind the following points:

- The ratio of $-X_C/R$ specifies the angle's tangent function as a numerical value, but this is not the angle θ in degrees. Finding X_C/R is a division problem.
- The angle θ itself is an inverse function of tan θ that is indicated as arctan θ or $\tan^{-1}\theta$. A scientific calculator can give the trigonometric functions directly from the value of an angle, or inversely show the angle from its trigonometric functions.
- As a check on your values, note that $\tan \theta = -1$, $\tan^{-1}$ (arctan θ) is −45°. Tangent values less than −1 must be for angles smaller than −45°; angles more than −45° must have tangent values higher than −1.

For the values in Example 18–1 specifically, punch in −40 for X_C, press the (÷) key, punch in 30 for R, and press the (=) key for the ratio of −1.33 on the display. This value is tan θ. Although it is on the display, push the (TAN⁻¹) key, and the answer of −53.1° will appear for the angle. Use of the (TAN⁻¹) key is usually preceded by pressing the second function key, (2ndF).

Example 18–1

If a 30-Ω R and a 40-Ω X_C are in series with 100 V applied, find the following: Z_T, I, V_R, V_C, and θ_Z. What is the phase angle between V_C and V_R with respect to I? Prove that the sum of the series voltage drops equals the applied voltage V_T.

ANSWER

$$Z_T = \sqrt{R^2 + X_C^2} = \sqrt{30^2 + 40^2}$$
$$= \sqrt{900 + 1600}$$
$$= \sqrt{2500}$$
$$= 50\ \Omega$$
$$I = \frac{V_T}{Z_T} = \frac{100\ \text{V}}{50\ \Omega} = 2\ \text{A}$$
$$V_R = IR = 2\ \text{A} \times 30\ \Omega = 60\ \text{V}$$
$$V_C = IX_C = 2\ \text{A} \times 40\ \Omega = 80\ \text{V}$$
$$\tan \theta_Z = -\frac{X_C}{R} = -\frac{40}{30} = -1.333$$
$$\theta_Z = -53.1°$$

Therefore, V_T lags I by 53.1°. Furthermore, I and V_R are in phase, and V_C lags I by 90°. Finally,

$$V_T = \sqrt{V_R^2 + V_C^2} = \sqrt{60^2 + 80^2} = \sqrt{3600 + 6400}$$
$$= \sqrt{10{,}000}$$
$$= 100\ \text{V}$$

Note that the phasor sum of the voltage drops equals the applied voltage V_T.

GOOD TO KNOW

For a series *RC* circuit, when $X_C \geq 10R$, $Z_T \cong X_C$. When $R \geq 10X_C$, $Z_T \cong R$.

Table 18–1 Series *R* and X_C Combinations

R, Ω	X_C, Ω	Z_T, Ω (Approx.)	Phase Angle θ_Z
1	10	$\sqrt{101} = 10$	−84.3°
10	10	$\sqrt{200} = 14$	−45°
10	1	$\sqrt{101} = 10$	−5.7°

Note: θ_Z is the phase angle of Z_T or V_T with respect to the reference phasor *I* in series circuits.

Series Combinations of X_C and *R*

In series, the higher the X_C compared with *R*, the more capacitive the circuit. There is more voltage drop across the capacitive reactance X_C, and the phase angle increases toward −90°. The series X_C always makes the series current *I* lead the applied voltage V_T. With all X_C and no *R*, the entire applied voltage V_T is across X_C and θ equals −90°.

Several combinations of X_C and *R* in series are listed in Table 18–1 with their resultant impedance values and phase angle. Note that a ratio of 10:1, or more, for X_C/R means that the circuit is practically all capacitive. The phase angle of −84.3° is almost −90°, and the total impedance Z_T is approximately equal to X_C. The voltage drop across X_C in the series circuit is then practically equal to the applied voltage V_T with almost none across *R*.

At the opposite extreme, when *R* is 10 times more than X_C, the series circuit is mainly resistive. The phase angle of −5.7° then means that the current is almost in phase with the applied voltage V_T; Z_T is approximately equal to *R*, and the voltage drop across *R* is practically equal to the applied voltage V_T with almost none across X_C.

When X_C and *R* equal each other, the resultant impedance Z_T is 1.41 times either one. The phase angle then is −45°, halfway between 0° for resistance alone and −90° for capacitive reactance alone.

18–3 Self-Review

Answers at the end of the chapter.

a. How much is Z_T for a 20-Ω *R* in series with a 20-Ω X_C?
b. How much is V_T for 20 V across *R* and 20 V across X_C in series?
c. What is the phase angle θ_Z of this circuit?

18–4 *RC* Phase-Shifter Circuit

GOOD TO KNOW

In Fig. 18–5*a*, another *RC* phase-shifting network could be added at the output of the first one to provide an even greater range in overall phase shift.

Figure 18–5 shows an application of X_C and *R* in series to provide a desired phase shift in the output V_R compared with the input V_T. The *R* can be varied up to 100 kΩ to change the phase angle. The *C* is 0.05 μF here for the 60-Hz AC power-line voltage, but a smaller *C* would be used for a higher frequency. The capacitor must have an appreciable value of reactance for the phase shift.

For the circuit in Fig. 18–5*a*, assume that *R* is set for 50 kΩ at its middle value. The reactance of the 0.05-μF capacitor at 60 Hz is approximately 53 kΩ. For these values of X_C and *R*, the phase angle of the circuit is −46.7°. This angle has a tangent of $-^{53}/_{50} = -1.06$.

The phasor triangle in Fig. 18–5*b* shows that *IR* or V_R is out of phase with V_T by the leading angle of 46.7°. Note that V_C is always 90° lagging V_R in a series circuit. The angle between V_C and V_T then becomes 90° − 46.7° = 43.3°.

MultiSim **Figure 18–5** An *RC* phase-shifter circuit. (*a*) Schematic diagram. (*b*) Phasor triangle with *IR*, or V_R, as the horizontal reference. V_R leads V_T by 46.7° with *R* set at 50 kΩ. (*c*) Phasors shown with V_T as the horizontal reference.

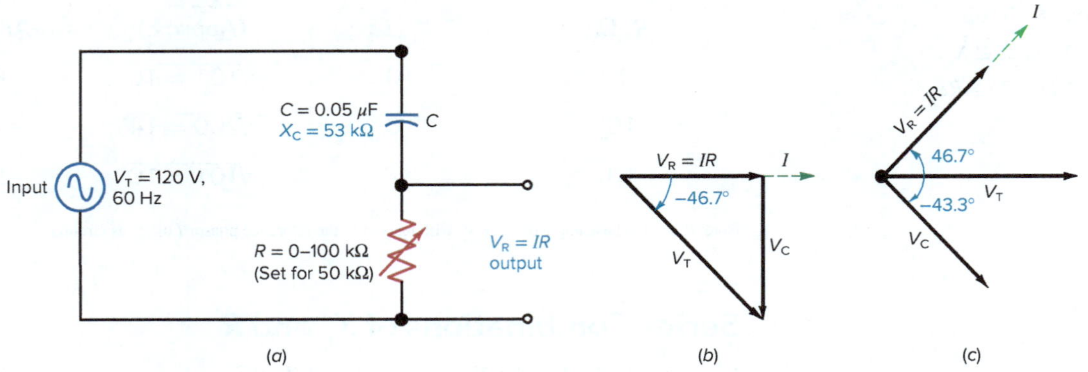

This circuit provides a phase-shifted voltage V_R at the output with respect to the input. For this reason, the phasors are redrawn in Fig. 18–5*c* to show the voltages with the input V_T as the horizontal reference. The conclusion, then, is that the output voltage across *R* leads the input V_T by 46.7°, whereas V_C lags V_T by 43.3°.

Now let *R* be varied for a higher value at 90 kΩ, while X_C stays the same. The phase angle becomes −30.5°. This angle has a tangent of $-53/90 = -0.59$. As a result, V_R leads V_T by 30.5°, and V_C lags V_T by 59.5°.

For the opposite case, let *R* be reduced to 10 kΩ. Then the phase angle becomes −79.3°. This angle has the tangent $-53/10 = -5.3$. Then V_R leads V_T by 79.3° and V_C lags V_T by 10.7°. Notice that the phase angle between V_R and V_T becomes larger as the series circuit becomes more capacitive with less resistance.

A practical application for this circuit is providing a voltage of variable phase to set the conduction time of semiconductors in power-control circuits. In this case, the output voltage is taken across the capacitor *C*. This provides a lagging phase angle with respect to the input voltage V_T. As *R* is varied from 0 Ω to 100 kΩ, the phase angle between V_C and V_T increases from 0° to about −62°. If *R* were changed so that it varied from 0 to 1 MΩ, the phase angle between V_C and V_T would vary between 0° and −90° approximately.

18–4 Self-Review

Answers at the end of the chapter.

In Fig. 18–5, give the phase angle between

a. V_R **and** V_T.
b. V_R **and** V_C.
c. V_C **and** V_T.

18–5 X_C and *R* in Parallel

For parallel circuits with X_C and *R*, the 90° phase angle must be considered for each of the branch currents. Remember that any series circuit has different voltage drops but one common current. A parallel circuit has different branch currents but one common voltage.

In the parallel circuit in Fig. 18–6*a*, the applied voltage V_A is the same across X_C, *R*, and the generator, since they are all in parallel. There cannot be any phase difference between these voltages. Each branch, however, has its own individual current. For the resistive branch, $I_R = V_A/R$; for the capacitive branch, $I_C = V_A/X_C$.

The resistive branch current I_R is in phase with the generator voltage V_A. The capacitive branch current I_C leads V_A, however, because the charge and discharge

Figure 18–6 Capacitive reactance X_C and R in parallel. (*a*) Circuit. (*b*) Waveforms of the applied voltage V_A, branch currents I_R and I_C, and total current I_T. (*c*) Phasor diagram.

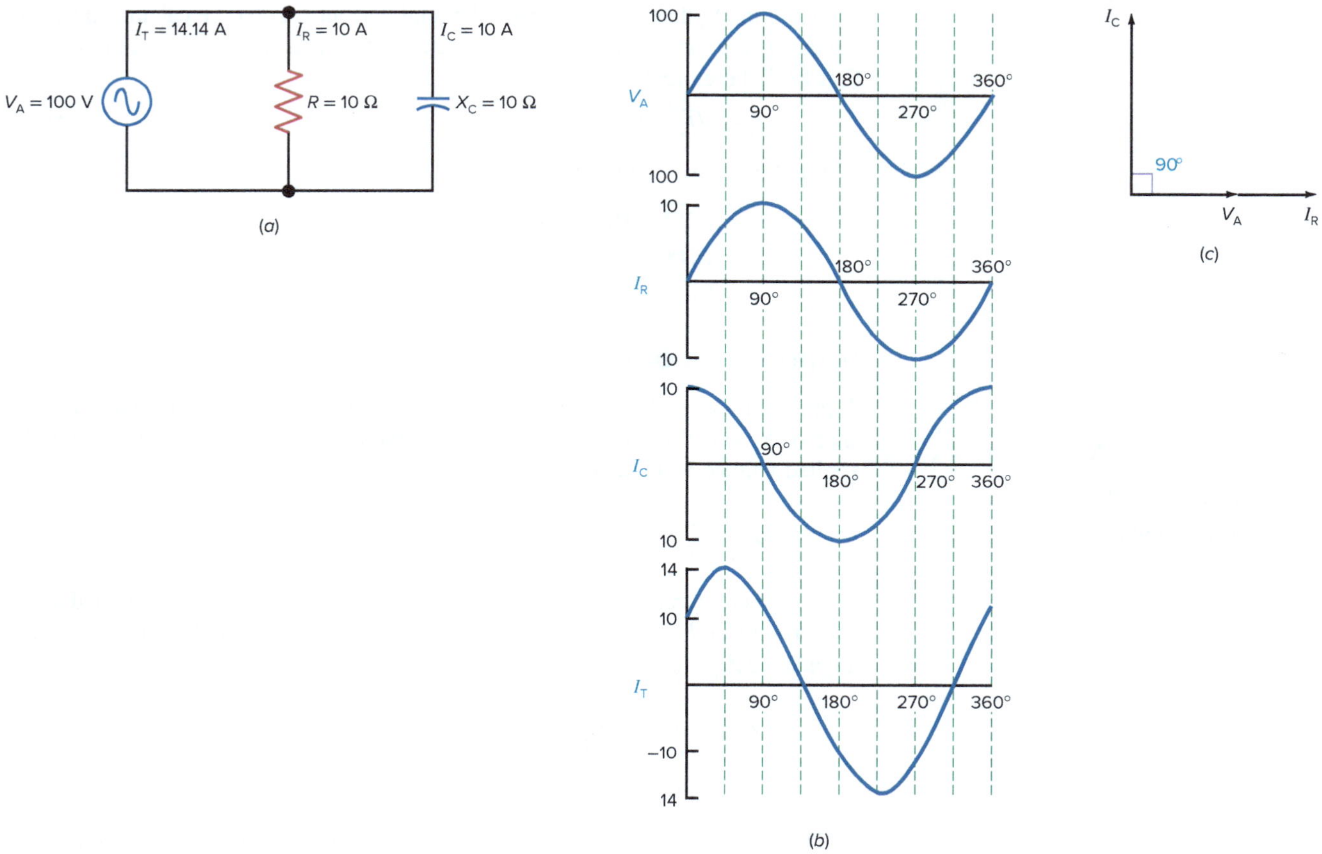

current of a capacitor leads the capacitor voltage by 90°. The waveforms for V_A, I_R, I_C, and I_T in Fig. 18–6*a* are shown in Fig. 18–6*b*. The individual branch currents I_R and I_C must add to equal the total current I_T. The 10-A peak values for I_R and I_C total 14.14 A, however, instead of 20 A, because of the 90° phase difference.

Consider some instantaneous values in Fig. 18–6*b* to see why the 10-A peak for I_R and 10-A peak for I_C cannot be added arithmetically. When I_C is at its maximum of 10 A, for instance, I_R is at zero. The total for I_T at this instant then is 10 A. Similarly, when I_R is at its maximum of 10 A, I_C is at zero and the total current I_T at this instant is also 10 A.

Actually, I_T has its maximum of 14.14 A when I_R and I_C are each 7.07 A. When branch currents that are out of phase are combined, therefore, they cannot be added without taking the phase difference into account.

Figure 18–6*c* shows the phasors representing V_A, I_R, and I_C. Notice that I_C leads V_A and I_R by 90°. In this case, the applied voltage V_A is used as the reference phasor since it is the same across both branches.

Figure 18–7 Phasor triangle of capacitive and resistive branch currents 90° out of phase in a parallel circuit to find the resultant I_T.

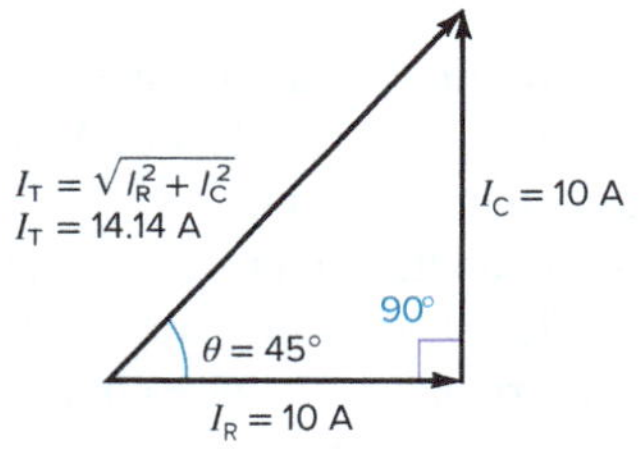

Phasor Current Triangle

Figure 18–7 shows the phasor current triangle for the parallel *RC* circuit in Fig. 18–6*a*. Note that the resistive branch current I_R is used as the reference phasor since V_A and I_R are in phase. The capacitive branch current I_C is drawn upward at an angle of +90° since I_C leads V_A and thus I_R by 90°. The sum of the I_R and I_C phasors is indicated by the phasor for I_T, which connects the tail of the I_R phasor to the tip of the I_C phasor. The I_T phasor is the hypotenuse of the right triangle. The phase angle between I_T and I_R represents the phase angle of the circuit. Peak values are shown here for convenience, but rms and peak-to-peak values could also be used.

Using the Pythagorean theorem, the total current I_T could be calculated by taking the square root of the sum of the squares of the sides. For the current triangle in Fig. 18–7 therefore, the resultant I_T is

$$I_T = \sqrt{I_R^2 + I_C^2} \qquad \textbf{(18–4)}$$

For the values in Fig. 18–6,

$$\begin{aligned} I_T &= \sqrt{10^2 + 10^2} = \sqrt{100 + 100} \\ &= \sqrt{200} \\ &= 14.14 \text{ A} \end{aligned}$$

Impedance of X_C and R in Parallel

GOOD TO KNOW

For R in parallel with X_C, Z_{EQ} can also be calculated as:

$$Z_{EQ} = \frac{X_C R}{\sqrt{R^2 + X_C^2}}$$

A practical approach to the problem of calculating the total or equivalent impedance of X_C and R in parallel is to calculate the total line current I_T and divide the applied voltage V_A by this value.

$$Z_{EQ} = \frac{V_A}{I_T} \qquad \textbf{(18–5)}$$

For the circuit in Fig. 18–6*a*, V_A is 100 V, and the total current I_T, obtained as the phasor sum of I_R and I_C, is 14.14 A. Therefore, we can calculate the equivalent impedance Z_{EQ} as

$$\begin{aligned} Z_{EQ} &= \frac{V_A}{I_T} = \frac{100 \text{ V}}{14.14 \text{ A}} \\ &= 7.07\ \Omega \end{aligned}$$

This impedance, the combined opposition in ohms across the generator, is equal to the 10-Ω resistance in parallel with the 10-Ω X_C.

Note that the impedance Z_{EQ} for equal values of R and X_C in parallel is not one-half but instead equals 70.7% of either one. Still, the value of Z_{EQ} will always be less than the lowest ohm value in the parallel branches.

For the general case of calculating the Z_{EQ} of X_C and R in parallel, any number can be assumed for the applied voltage V_A because, in the calculations for Z_{EQ} in terms of the branch currents, the value of V_A cancels. A good value to assume for V_A is the value of either R or X_C, whichever is the larger number. This way, there are no fractions smaller than that in the calculation of the branch currents.

Phase Angle in Parallel Circuits

GOOD TO KNOW

For parallel *RC* circuits, tan θ_I can also be calculated as $\tan \theta_I = \frac{R}{X_C}$.

In Fig. 18–7, the phase angle θ is 45° because R and X_C are equal, resulting in equal branch currents. The phase angle is between the total current I_T and the generator voltage V_A. However, V_A and I_R are in phase. Therefore, θ is also between I_T and I_R.

Using the tangent formula to find θ from the current triangle in Fig. 18–7 gives

$$\tan \theta_I = \frac{I_C}{I_R} \qquad \textbf{(18–6)}$$

The phase angle is positive because the I_C phasor is upward, leading V_A by 90°. This direction is opposite from the lagging phasor of series X_C. The effect of X_C is no different, however. Only the reference is changed for the phase angle.

Note that the phasor triangle of branch currents for parallel circuits gives θ_I as the angle of I_T with respect to the generator voltage V_A. This phase angle for I_T is labeled θ_I with respect to the applied voltage. For the phasor triangle of voltages in a series circuit, the phase angle for Z_T and V_T is labeled θ_Z with respect to the series current.

Example 18–2

A 30-mA I_R is in parallel with another branch current of 40 mA for I_C. The applied voltage V_A is 72 V. Calculate I_T, Z_{EQ}, and θ_I.

ANSWER This problem can be calculated in mA units for I and kΩ for Z without powers of 10.

$$I_T = \sqrt{I_R^2 + I_C^2} = \sqrt{(30)^2 + (40)^2}$$
$$= \sqrt{900 + 1600} = 2500$$
$$= 50 \text{ mA}$$
$$Z_{EQ} = \frac{V_A}{I_T} = \frac{72 \text{ V}}{50 \text{ mA}}$$
$$= 1.44 \text{ k}\Omega$$
$$\tan \theta_I = \frac{I_C}{I_R} = \frac{40}{30} = 1.333$$
$$= \arctan(1.333)$$
$$\theta_I = 53.1°$$

Parallel Combinations of X_C and R

In Table 18–2, when X_C is 10 times R, the parallel circuit is practically resistive because there is little leading capacitive current in the main line. The small value of I_C results from the high reactance of shunt X_C. Then the total impedance of the parallel circuit is approximately equal to the resistance, since the high value of X_C in a parallel branch has little effect. The phase angle of 5.7° is practically 0° because almost all of the line current is resistive.

As X_C becomes smaller, it provides more leading capacitive current in the main line. When X_C is 1⁄10 R, practically all of the line current is the I_C component. Then, the parallel circuit is practically all capacitive with a total impedance practically equal to X_C. The phase angle of 84.3° is almost 90° because the line current is mostly capacitive. Note that these conditions are opposite to the case of X_C and R in series. With X_C and R equal, their branch currents are equal and the phase angle is 45°.

GOOD TO KNOW

For a parallel *RC* circuit,
When $X_C \geq 10R$, $Z_{EQ} \cong R$.
When $R \geq 10X_C$, $Z_{EQ} \cong X_C$.

Table 18–2 Parallel Resistance and Capacitance Combinations*

R, Ω	X_C, Ω	I_R, A	I_C, A	I_T, A (Approx.)	Z_{EQ}, Ω (Approx.)	Phase Angle θ_I
1	10	10	1	$\sqrt{101} = 10$	1	5.7°
10	10	1	1	$\sqrt{2} = 1.4$	7.07	45°
10	1	1	10	$\sqrt{101} = 10$	1	84.3°

* V_A = 10 V. Note that θ_I is the phase angle of I_T with respect to the reference V_A in parallel circuits.

As additional comparisons between series and parallel *RC* circuits, remember that

1. The series voltage drops V_R and V_C have individual values that are 90° out of phase. Therefore, V_R and V_C are added by phasors to equal the applied voltage V_T. The negative phase angle $-\theta_Z$ is between V_T and the common series current *I*. More series X_C allows more V_C to make the circuit more capacitive with a larger negative phase angle for V_T with respect to *I*.
2. The parallel branch currents I_R and I_C have individual values that are 90° out of phase. Therefore, I_R and I_C are added by phasors to equal I_T, which is the main-line current. The positive phase angle θ_I is between the line current I_T and the common parallel voltage V_A. Less parallel X_C allows more I_C to make the circuit more capacitive with a larger positive phase angle for I_T with respect to V_A.

18–5 Self-Review

Answers at the end of the chapter.

a. How much is I_T for branch currents I_R of 2 A and I_C of 2 A?
b. Find the phase angle θ_I between I_T and V_A.

18–6 RF and AF Coupling Capacitors

In Fig. 18–8, C_C is used in the application of a **coupling capacitor**. Its low reactance allows developing practically all the AC signal voltage of the generator across *R*. Very little of the ac voltage is across C_C.

The coupling capacitor is used for this application because it provides more reactance at lower frequencies, resulting in less ac voltage coupled across *R* and more across C_C. For DC voltage, all voltage is across *C* with none across *R*, since the capacitor blocks direct current. As a result, the output signal voltage across *R* includes the desired higher frequencies but not direct current or very low frequencies. This application of C_C, therefore, is called *ac coupling*.

The dividing line for C_C to be a coupling capacitor at a specific frequency can be taken as X_C one-tenth or less of the series *R*. Then the series *RC* circuit is primarily resistive. Practically all the voltage drop of the ac generator is across *R*, with little across *C*. In addition, the phase angle is almost 0°.

Typical values of a coupling capacitor for audio or radio frequencies can be calculated if we assume a series resistance of 16,000 Ω. Then X_C must be 1600 Ω or less. Typical values for C_C are listed in Table 18–3. At 100 Hz, a coupling capacitor must be 1 μF to provide 1600 Ω of reactance. Higher frequencies allow a smaller value of C_C for a coupling capacitor having the same reactance. At 100 MHz in the VHF range, the required capacitance is only 1 pF.

Figure 18–8 Series circuit for *RC* coupling. Small X_C compared with *R* allows practically all the applied voltage to be developed across *R* for the output, with little across *C*.

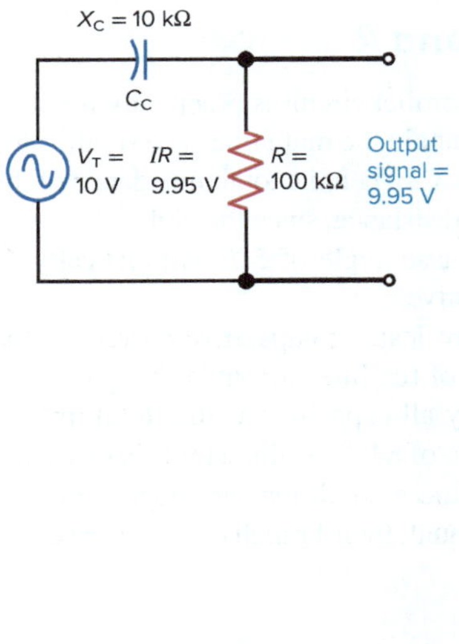

Table 18–3 Coupling Capacitors with a Reactance of 1600 Ω*

f	C_C	Remarks
100 Hz	1 μF	Low audio frequencies
1000 Hz	0.1 μF	Audio frequencies
10 kHz	0.01 μF	Audio frequencies
1000 kHz	100 pF	Radio frequencies
100 MHz	1 pF	Very high frequencies

* For an X_C one-tenth of a series *R* of 16,000 Ω

Note that the C_C values are calculated for each frequency as a lower limit. At higher frequencies, the same size C_C will have less reactance than one-tenth of R, which improves coupling.

Choosing a Coupling Capacitor for a Circuit

As an example of using these calculations, suppose that we have the problem of determining C_C for an audio amplifier. This application also illustrates the relatively large capacitance needed with low series resistance. The C is to be a coupling capacitor for audio frequencies of 50 Hz and up with a series R of 4000 Ω. Then the required X_C is $^{4000}/_{10}$ or 400 Ω. To find C at 50 Hz,

$$C = \frac{1}{2\pi f X_C} = \frac{1}{6.28 \times 50 \times 400}$$

$$= \frac{1}{125{,}600} = 0.0000079$$

$$= 7.9 \times 10^{-6} \quad \text{or} \quad 7.9\ \mu\text{F}$$

A 10-μF electrolytic capacitor would be a good choice for this application. The slightly higher capacitance value is better for coupling. The voltage rating should exceed the actual voltage across the capacitor in the circuit. Although electrolytic capacitors have a slight leakage current, they can be used for coupling capacitors in this application because of the low series resistance.

18–6 Self-Review

Answers at the end of the chapter.

a. **The X_C of a coupling capacitor is 70 Ω at 200 Hz. How much is its X_C at 400 Hz?**

b. **From Table 18–3, what C would be needed for 1600 Ω of X_C at 50 MHz?**

18–7 Capacitive Voltage Dividers

When capacitors are connected in series across a voltage source, the series capacitors serve as a voltage divider. Each capacitor has part of the applied voltage, and the sum of all the series voltage drops equals the source voltage.

The amount of voltage across each is inversely proportional to its capacitance. For instance, with 2 μF in series with 1 μF, the smaller capacitor has double the voltage of the larger capacitor. Assuming 120 V applied, one-third of this, or 40 V, is across the 2-μF capacitor, and two-thirds, or 80 V, is across the 1-μF capacitor.

The two series voltage drops of 40 and 80 V add to equal the applied voltage of 120 V. The phasor addition is the same as the arithmetic sum of the two voltages because they are in phase. When voltages are out of phase with each other, arithmetic addition is not possible and phasor addition becomes necessary.

AC Divider

With sine-wave alternating current, the voltage division between series capacitors can be calculated on the basis of reactance. In Fig. 18–9*a*, the total reactance is 120 Ω across the 120-V source. The current in the series circuit is 1 A. This current is the same for X_{C_1} and X_{C_2} in series. Therefore, the IX_C voltage across C_1 is 40 V with 80 V across C_2.

The voltage division is proportional to the series reactances, as it is to series resistances. However, reactance is inversely proportional to capacitance. As a result, the smaller capacitance has more reactance and a greater part of the applied voltage.

Figure 18–9 Series capacitors divide V_T inversely proportional to each C. The smaller C has more V. (*a*) An AC divider with more X_C for the smaller C. (*b*) A DC divider.

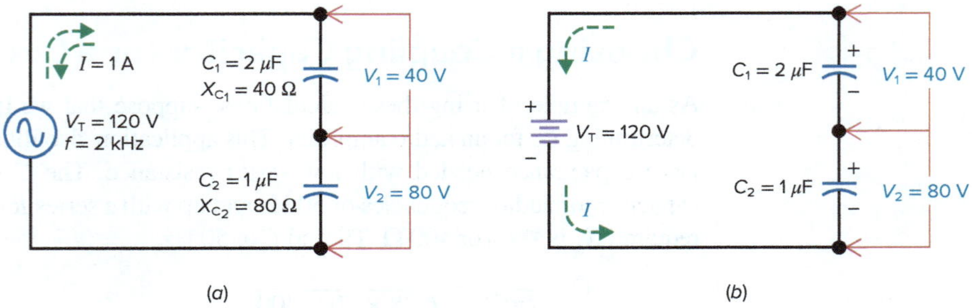

DC Divider

In Fig. 18–9*b*, both C_1 and C_2 will be charged by the battery. The voltage across the series combination of C_1 and C_2 must equal V_T. When charging current flows, electrons repelled from the negative battery terminal accumulate on the negative plate of C_2, repelling electrons from its positive plate. These electrons flow through the conductor to the negative plate of C_1. As the positive battery terminal attracts electrons, the charging current from the positive plate of C_1 returns to the positive side of the DC source. Then C_1 and C_2 become charged in the polarity shown.

Since C_1 and C_2 are in the same series path for charging current, both have the same amount of charge. However, the potential difference provided by the equal charges is inversely proportional to capacitance. The reason is that $Q = CV$, or $V = Q/C$. Therefore, the 1-μF capacitor has double the voltage of the 2-μF capacitor with the same charge in both.

If you measure across C_1 with a DC voltmeter, the meter reads 40 V. Across C_2, the DC voltage is 80 V. The measurement from the negative side of C_2 to the positive side of C_1 is the same as the applied battery voltage of 120 V.

If the meter is connected from the positive side of C_2 to the negative plate of C_1, however, the voltage is zero. These plates have the same potential because they are joined by a conductor of zero resistance.

The polarity marks at the junction between C_1 and C_2 indicate the voltage at this point with respect to the opposite plate of each capacitor. This junction is positive compared with the opposite plate of C_2 with a surplus of electrons. However, the same point is negative compared with the opposite plate of C_1, which has a deficiency of electrons.

In general, the following formula can be used for capacitances in series as a voltage divider:

$$V_C = \frac{C_{EQ}}{C} \times V_T \qquad \textbf{(18–7)}$$

Note that C_{EQ} is in the numerator, since it must be less than the smallest individual C with series capacitances. For the divider examples in Fig. 18–9*a* and *b*,

$$V_1 = \frac{C_{EQ}}{C_1} \times 120 = \frac{2/3}{2} \times 120 = 40 \text{ V}$$

$$V_2 = \frac{C_{EQ}}{C_2} \times 120 = \frac{2/3}{1} \times 120 = 80 \text{ V}$$

This method applies to series capacitances as dividers for either DC or AC voltage, as long as there is no series resistance. Note that the case of capacitive DC dividers also applies to pulse circuits. Furthermore, bleeder resistors may be used across each of the capacitors to ensure more exact division.

GOOD TO KNOW

Connecting a DC voltmeter across either C_1 or C_2 in Fig. 18–9*b* will cause the capacitor to discharge through the resistance of the meter. It is best to use a DMM with a high internal resistance so that the amount of discharge is minimal. The voltmeter reading should be taken immediately after it is connected across the capacitor.

18–7 Self-Review

Answers at the end of the chapter.

a. Capacitance C_1 of 10 pF and C_2 of 90 pF are across 20 kV. Calculate the amount of V_1 and V_2.

b. In Fig. 18–9*a*, how much is X_{C_T}?

18–8 The General Case of Capacitive Current i_C

The capacitive charge and discharge current i_C is always equal to $C(dv/dt)$. A sine wave of voltage variations for v_C produces a cosine wave of current i. This means that v_C and i_C have the same waveform, but they are 90° out of phase.

It is usually convenient to use X_C for calculations in sine-wave circuits. Since X_C is $1/(2\pi fC)$, the factors that determine the amount of charge and discharge current are included in f and C. Then I_C equals V_C/X_C. Or, if I_C is known, V_C can be calculated as $I_C \times X_C$.

With a nonsinusoidal waveform for voltage v_C, the concept of reactance cannot be used. Reactance X_C applies only to sine waves. Then i_C must be determined as $C(dv/dt)$. An example is illustrated in Fig. 18–10 to show the change of waveform here, instead of the change of phase angle in sine-wave circuits.

Note that the sawtooth waveform of voltage v_C corresponds to a rectangular waveform of current. The linear rise of the sawtooth wave produces a constant amount of charging current i_C because the rate of change is constant for the charging voltage. When the capacitor discharges, v_C drops sharply. Then the discharge current is in the direction opposite from the charge current. Also, the discharge current has a much larger value because of the faster rate of change in v_C.

GOOD TO KNOW

For a capacitor, the value of charge or discharge current is zero when dv/dt is zero. When dv/dt is a constant value, the charge or discharge current is also a constant value.

18–8 Self-Review

Answers at the end of the chapter.

a. In Fig. 18–10*a*, how much is dv/dt in V/s for the sawtooth rise from 0 to 90 V in 90 μs?

b. How much is the charge current i_C, as $C(dv/dt)$ for this dv/dt?

Figure 18–10 Waveshape of i_C equal to $C(dv/dt)$. (*a*) Sawtooth waveform of V_C. (*b*) Rectangular current waveform of i_C resulting from the uniform rate of change in the sawtooth waveform of voltage.

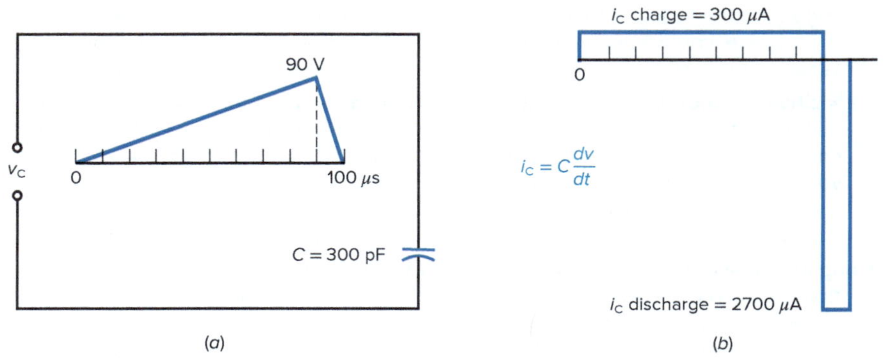

Summary

- In a sine-wave AC circuit, the voltage across a capacitance lags its charge and discharge current by 90°.
- Therefore, capacitive reactance X_C is a phasor quantity out of phase with its series resistance by −90° because $i_C = C(dv/dt)$. This fundamental fact is the basis of all the following relations.
- The combination of X_C and R in series is their total impedance Z_T. These three types of ohms of opposition to current are compared in Table 18–4.
- The opposite characteristics for series and parallel circuits with X_C and R are summarized in Table 18–5.
- Two or more capacitors in series across a voltage source serve as a voltage divider. The smallest C has the largest part of the applied voltage.
- A coupling capacitor has X_C less than its series resistance by a factor of one-tenth or less to provide practically all the AC applied voltage across R with little across C.
- In sine-wave circuits, $I_C = V_C/X_C$. Then I_C is out of phase with V_C by 90°.
- For a circuit with X_C and R in series, $\tan\theta_Z = -(X_C/R)$, and in parallel, $\tan\theta_I = I_C/I_R$. See Table 18–5.
- When the voltage is not a sine wave, $i_C = C(dv/dt)$. Then the waveshape of i_C is different from that of the voltage.

Table 18–4 Comparison of R, X_C, and Z

R	$X_C = 1/(2\pi fC)$	$Z_T = \sqrt{R^2 + X_C^2}$
Ohm unit	Ohm unit	Ohm unit
IR voltage in phase with I	IX_C voltage lags I_C by 90°	IZ_T is the applied voltage
Same ohm value for all f	Ohm value decreases for higher f	Becomes more resistive with more f Becomes more capacitive with less f

Table 18–5 Series and Parallel RC Circuits

X_C and R in Series	X_C and R in Parallel
I the same in X_C and R	V the same across X_C and R
$V_T = \sqrt{V_R^2 + V_C^2}$	$I_T = \sqrt{I_R^2 + I_C^2}$
$Z_T = \sqrt{R^2 + X_C^2}$	$Z_{EQ} = \dfrac{V_A}{I_T}$
V_C lags V_R by 90°	I_C leads I_R by 90°
$\tan\theta_Z = -\dfrac{X_C}{R}$; θ_Z increases as X_C increases, resulting in more V_C	$\tan\theta_I = \dfrac{I_C}{I_R}$; θ_I decreases as X_C increases, resulting in less I_C

Important Terms

Arctangent (arctan) — an inverse trigonometric function that specifies the angle, θ, corresponding to a given tangent (tan) value.

Capacitive voltage divider — a voltage divider that consists of series-connected capacitors. The amount of voltage across each capacitor is inversely proportional to its capacitance value.

Coupling capacitor, C_C — a capacitor that is selected to pass ac signals above a specified frequency from one point in a circuit to another. The dividing line for calculating the coupling capacitance, C_C, is to make X_C one-tenth the value of the series R. The value of C_C is calculated for a specified frequency as a lower limit.

Impedance, Z — the total opposition to the flow of current in a sine-wave AC circuit. In an RC circuit, the impedance, Z, takes into account the 90° phase relation between X_C and R. Impedance, Z, is measured in ohms.

Phase angle, θ — the angle between the generator voltage and current in a sine-wave ac circuit.

Phasor triangle — a right triangle that represents the phasor sum of two quantities 90° out of phase with each other.

***RC* phase-shifter** — an application of a series RC circuit in which the output across either R or C provides a desired phase shift with respect to the input voltage. RC phase-shifter circuits are commonly used to control the conduction angle of semiconductors in power-control circuits.

Tangent (tan) — a trigonometric function of an angle, equal to the ratio of the opposite side to the adjacent side of a right triangle.

Related Formulas

Series *RC* Circuits

$$V_T = \sqrt{V_R^2 + V_C^2}$$

$$Z_T = \sqrt{R^2 + X_C^2}$$

$$\tan \theta_Z = -\frac{X_C}{R}$$

Parallel *RC* Circuits

$$I_T = \sqrt{I_R^2 + I_C^2}$$

$$Z_{EQ} = \frac{V_A}{I_T}$$

$$\tan \theta_I = \frac{I_C}{I_R}$$

Series Capacitors

$$V_C = \frac{C_{EQ}}{C} \times V_T$$

Self-Test

Answers at the back of the book.

1. For a capacitor in a sine-wave AC circuit,

a. V_C lags i_C by 90°.

b. i_C leads V_C by 90°.

c. i_C and V_C have the same frequency.

d. all of the above.

2. In a series *RC* circuit,

a. V_C leads V_R by 90°.

b. V_C and I are in phase.

c. V_C lags V_R by 90°.

d. both b and c.

3. In a series *RC* circuit where $V_C = 15$ V and $V_R = 20$ V, how much is the total voltage, V_T?

a. 35 V.

b. 25 V.

c. 625 V.

d. 5 V.

4. A 10-Ω resistor is in parallel with a capacitive reactance of 10 Ω. The combined equivalent impedance, Z_{EQ}, of this combination is

a. 7.07 Ω.

b. 20 Ω.

c. 14.14 Ω.

d. 5 Ω.

5. In a parallel *RC* circuit,

a. I_C lags I_R by 90°.

b. I_R and I_C are in phase.

c. I_C leads I_R by 90°.

d. I_R leads I_C by 90°.

6. In a parallel *RC* circuit where $I_R = 8$ A and $I_C = 10$ A, how much is the total current, I_T?

a. 2 A.

b. 12.81 A.

c. 18 A.

d. 164 A.

7. In a series *RC* circuit where $R = X_C$, the phase angle, θ_Z, is

a. +45°.

b. −90°.

c. 0°.

d. −45°.

8. A 10-μF capacitor, C_1, and a 15-μF capacitor, C_2, are connected in series with a 12-V_{DC} source. How much voltage is across C_2?

a. 4.8 V.

b. 7.2 V.

c. 12 V.

d. 0 V.

9. The dividing line for a coupling capacitor at a specific frequency can be taken as

a. X_C 10 or more times the series resistance.

b. X_C equal to R.

c. X_C one-tenth or less the series resistance.

d. none of the above.

10. A 100-Ω resistance is in series with a capacitive reactance of 75 Ω. The total impedance, Z_T, is

a. 125 Ω.

b. 25 Ω.

c. 175 Ω.

d. 15.625 kΩ.

11. In a series *RC* circuit,

a. V_C and V_R are in phase.

b. V_T and I are always in phase.

c. V_R and I are in phase.

d. V_R leads I by 90°.

12. In a parallel *RC* circuit,

a. V_A and I_R are in phase.

b. V_A and I_C are in phase.

c. I_C and I_R are in phase.

d. V_A and I_R are 90° out of phase.

13. When the frequency of the applied voltage increases in a parallel *RC* circuit,

a. the phase angle, θ_I, increases.

b. Z_{EQ} increases.

c. Z_{EQ} decreases.

d. both a and c.

14. When the frequency of the applied voltage increases in a series *RC* circuit,

a. the phase angle, θ, becomes more negative.

b. Z_T increases.

c. Z_T decreases.

d. both a and c.

15. Capacitive reactance, X_C,

a. applies only to nonsinusoidal waveforms or DC.

b. applies only to sine waves.

c. applies to either sinusoidal or nonsinusoidal waveforms.

d. is directly proportional to frequency.

Essay Questions

1. (a) Why does a capacitor charge when the applied voltage increases? (b) Why does the capacitor discharge when the applied voltage decreases?

2. A sine wave of voltage V is applied across a capacitor C. (a) Draw the schematic diagram. (b) Draw the sine waves of voltage and current out of phase by 90°. (c) Draw a phasor diagram showing the phase angle of −90° between V and I.

3. Why will a circuit with R and X_C in series be less capacitive as the frequency of the applied voltage is increased?

4. Define the following: coupling capacitor, sawtooth voltage, capacitive voltage divider.

5. State two troubles possible in coupling capacitors, and describe briefly how you would check the capacitor with an ohmmeter.

6. Explain the function of R and C in an RC coupling circuit.

7. Explain briefly why a capacitor can block DC voltage.

8. What is the waveshape of i_C for a sine wave v_C?

9. Explain why the impedance Z_{EQ} of a parallel RC circuit decreases as the frequency increases.

10. Explain why θ_Z in a series RC circuit increases (becomes more negative) as frequency decreases.

Problems

SECTION 18–1 SINE WAVE v_C LAGS i_C BY 90°

18–1 In Fig. 18–11, what is the
a. peak value of the capacitor voltage, V_C?
b. peak value of the charge and discharge current, i_C?
c. frequency of the charge and discharge current?
d. phase relationship between V_C and i_C?

Figure 18–11

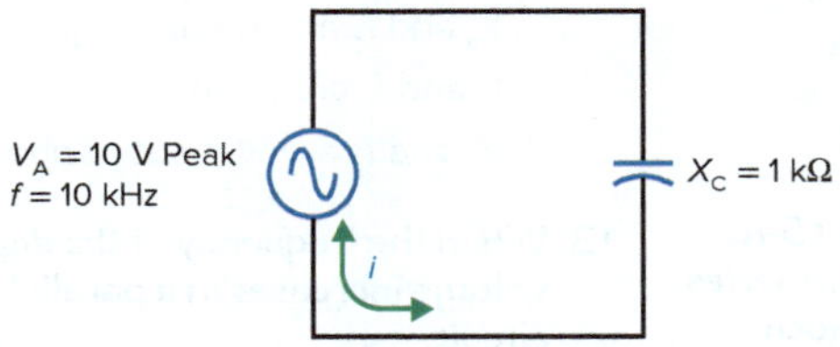

18–2 In Fig. 18–11, what is the value of the capacitor current, i_C, at the instant when V_C equals
a. its positive peak of +10 V?
b. 0 V?
c. its negative peak of −10 V?

18–3 In Fig. 18–11, draw the phasors representing V_C and i_C using
a. V_C as the reference phasor.
b. i_C as the reference phasor.

SECTION 18–2 X_C AND R IN SERIES

18–4 In Fig. 18–12, how much current, I, is flowing
a. through the 30-Ω resistor, R?
b. through the 40-Ω capacitive reactance, X_C?
c. to and from the terminals of the applied voltage, V_T?

Figure 18–12

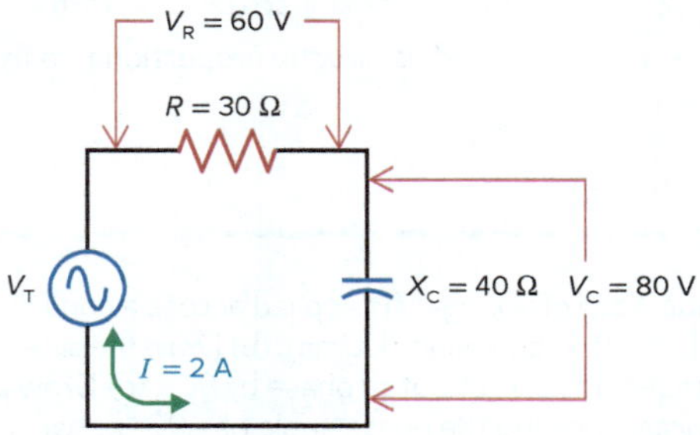

18–5 In Fig. 18–12, what is the phase relationship between
a. I and V_R?
b. I and V_C?
c. V_C and V_R?

18–6 In Fig. 18–12, how much is the applied voltage, V_T?

18–7 Draw the phasor voltage triangle for the circuit in Fig. 18–12. (Use V_R as the reference phasor.)

18–8 In Fig. 18–13, solve for
a. the resistor voltage, V_R.
b. the capacitor voltage, V_C.
c. the total voltage, V_T.

Figure 18–13

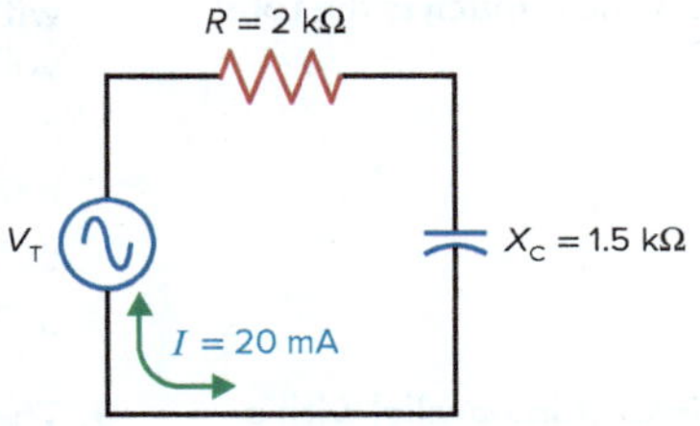

18–9 In Fig. 18–14, solve for
a. the resistor voltage, V_R.
b. the capacitor voltage, V_C.
c. the total voltage, V_T.

Figure 18–14

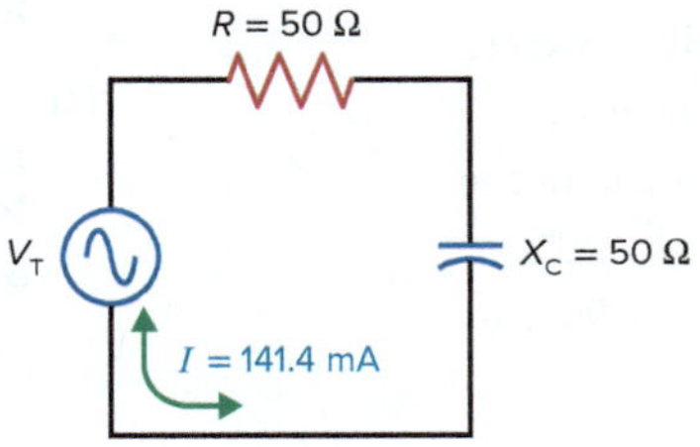

18–10 In a series RC circuit, solve for the applied voltage, V_T if
a. V_R = 40 V and V_C = 40 V.
b. V_R = 10 V and V_C = 5 V.
c. V_R = 48 V and V_C = 72 V.
d. V_R = 12 V and V_C = 18 V.

SECTION 18–3 IMPEDANCE *Z* TRIANGLE

18–11 In Fig. 18–15, solve for Z_T, I, V_C, V_R, and θ_Z.

Figure 18–15

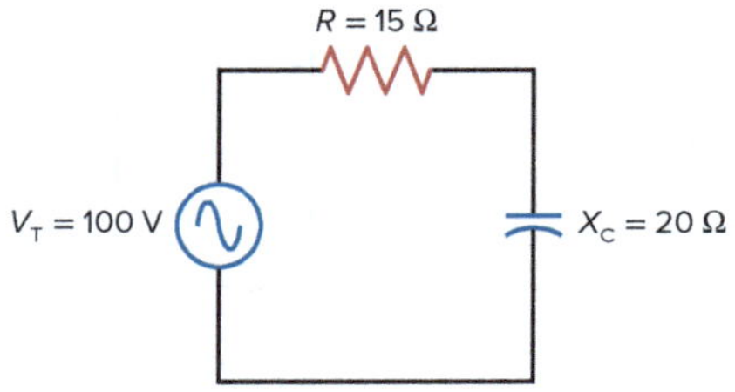

18–12 Draw the impedance triangle for the circuit in Fig. 18–15. (Use *R* as the reference phasor.)

18–13 In Fig. 18–16, solve for Z_T, I, V_C, V_R, and θ_Z.

Figure 18–16

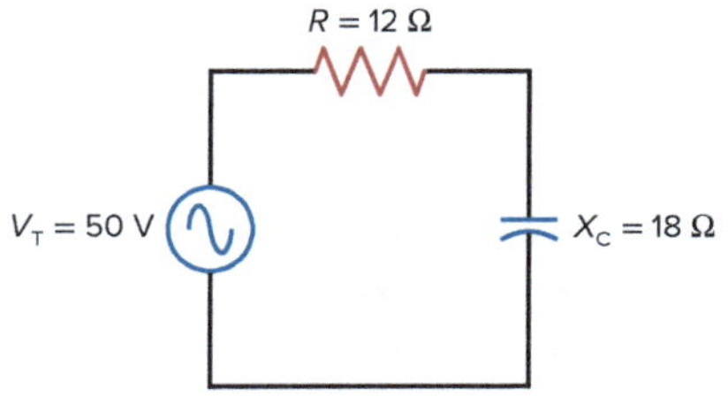

18–14 In Fig. 18–17, solve for Z_T, I, V_C, V_R, and θ_Z.

Figure 18–17

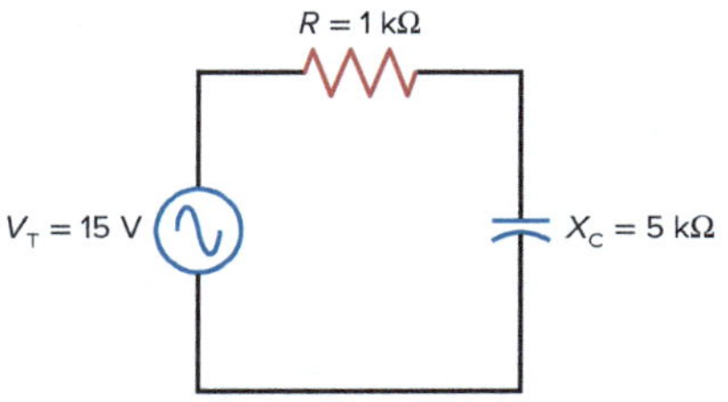

18–15 In Fig. 18–18, solve for Z_T, I, V_C, V_R, and θ_Z.

Figure 18–18

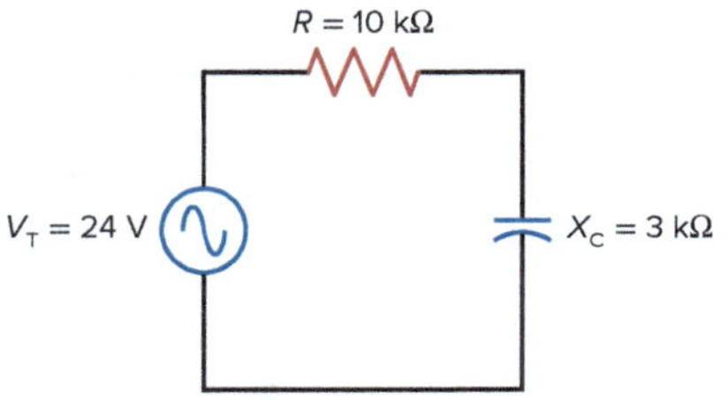

18–16 In Fig. 18–19, solve for Z_T, I, V_C, V_R, and θ_Z for the following circuit values:

a. $X_C = 30\ \Omega$, $R = 40\ \Omega$, and $V_T = 50$ V.

b. $X_C = 200\ \Omega$, $R = 200\ \Omega$, and $V_T = 56.56$ V.

c. $X_C = 10\ \Omega$, $R = 100\ \Omega$, and $V_T = 10$ V.

d. $X_C = 100\ \Omega$, $R = 10\ \Omega$, and $V_T = 10$ V.

Figure 18–19

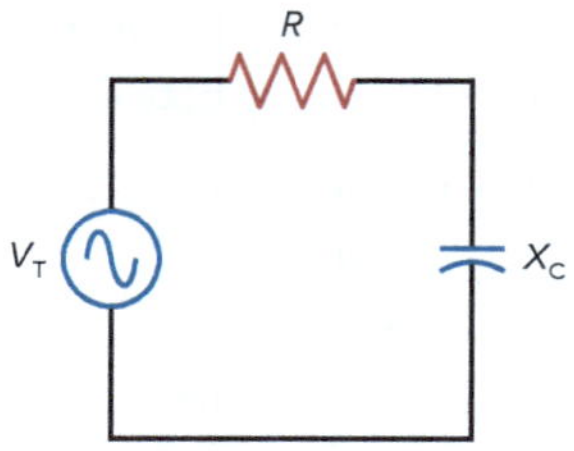

18–17 In Fig. 18–20, solve for X_C, Z_T, I, V_C, V_R, and θ_Z.

Figure 18–20

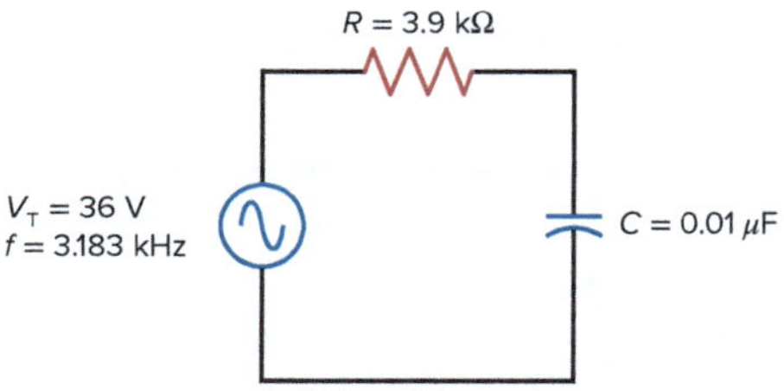

18–18 In Fig. 18–20, what happens to each of the following quantities if the frequency of the applied voltage increases?

a. X_C.

b. Z_T.

c. I.

d. V_C.

e. V_R.

f. θ_Z.

18–19 Repeat Prob. 18–18 if the frequency of the applied voltage decreases.

SECTION 18–4 *RC* PHASE-SHIFTER CIRCUIT

18–20 With *R* set to 50 kΩ in Fig. 18–21, solve for X_C, Z_T, I, V_R, V_C, and θ_Z.

Figure 18–21

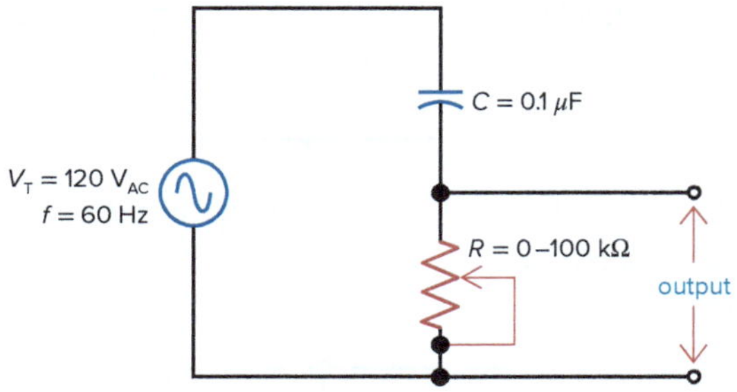

18–21 With *R* set to 50 kΩ in Fig. 18–21, what is the phase relationship between

a. V_T and V_R?

b V_T and V_C?

18–22 Draw the phasors for V_R, V_C, and V_T in Fig. 18–21 with *R* set at 50 kΩ. Use V_T as the reference phasor.

18–23 With R set at 1 kΩ in Fig. 18–21, solve for

a. Z_T, I, V_R, V_C, and θ_Z.

b. the phase relationship between V_T and V_R.

c. the phase relationship between V_T and V_C.

18–24 With R set at 100 kΩ in Fig. 18–21, solve for

a. Z_T, I, V_R, V_C, and θ_Z.

b. the phase relationship between V_T and V_R.

c. the phase relationship between V_T and V_C.

SECTION 18–5 X_C AND R IN PARALLEL

18–25 In Fig. 18–22, how much voltage is across

a. the 40-Ω resistor, R?

b. the 30-Ω capacitive reactance, X_C?

Figure 18–22

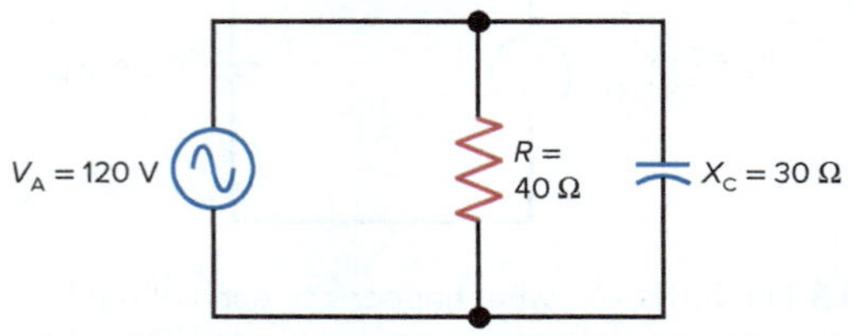

18–26 In Fig. 18–22, what is the phase relationship between

a. V_A and I_R?

b. V_A and I_C?

c. I_C and I_R?

18–27 In Fig. 18–22, solve for I_R, I_C, I_T, Z_{EQ}, and θ_I.

18–28 Draw the phasor current triangle for the circuit in Fig. 18–22. (Use I_R as the reference phasor.)

18–29 In Fig. 18–23, solve for I_R, I_C, I_T, Z_{EQ}, and θ_I.

Figure 18–23

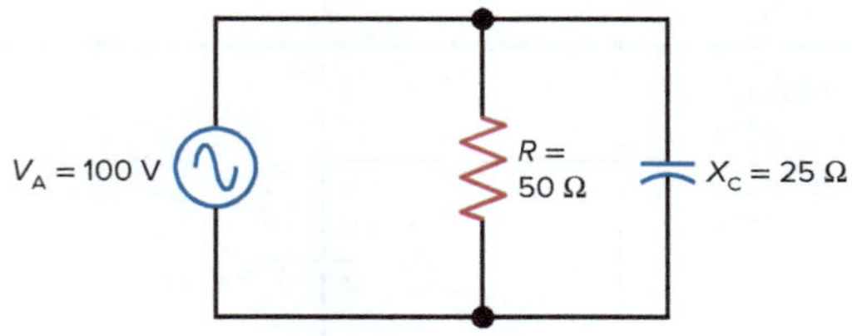

18–30 In Fig. 18–24, solve for I_R, I_C, I_T, Z_{EQ}, and θ_I.

Figure 18–24

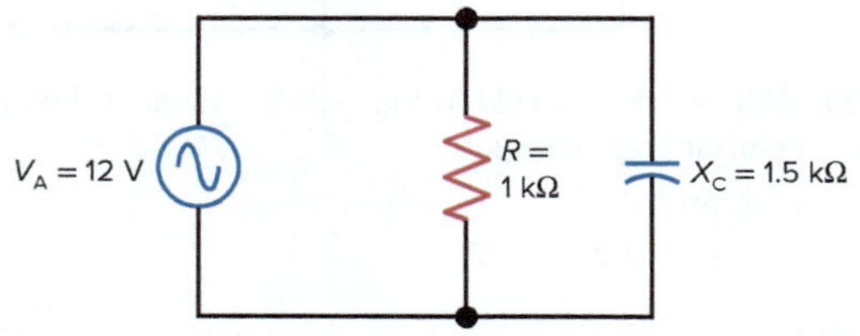

18–31 In Fig. 18–25, solve for I_R, I_C, I_T, Z_{EQ}, and θ_I.

Figure 18–25

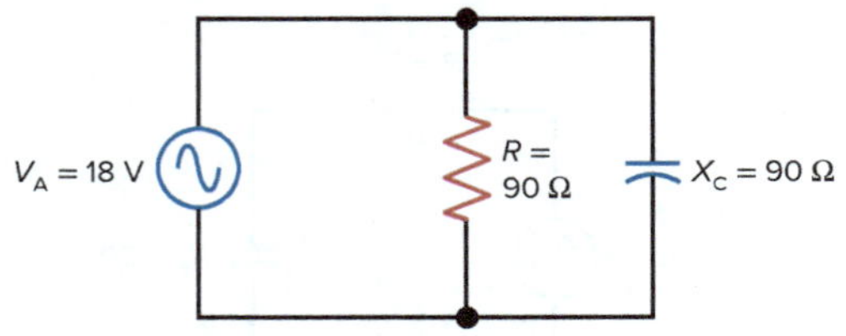

18–32 In Fig. 18–26, solve for I_R, I_C, I_T, Z_{EQ}, and θ_I.

Figure 18–26

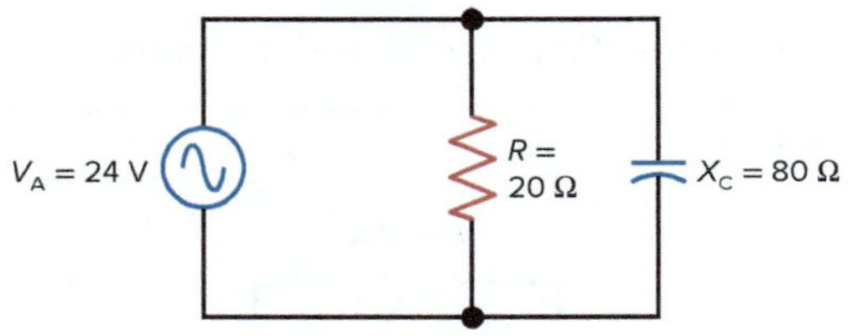

18–33 In Fig. 18–27, solve for I_R, I_C, I_T, Z_{EQ}, and θ_I for the following circuit values:

a. $R = 50\ \Omega$, $X_C = 50\ \Omega$, and $V_A = 50$ V.

b. $R = 10\ \Omega$, $X_C = 100\ \Omega$, and $V_A = 20$ V.

c. $R = 100\ \Omega$, $X_C = 10\ \Omega$, and $V_A = 20$ V.

Figure 18–27

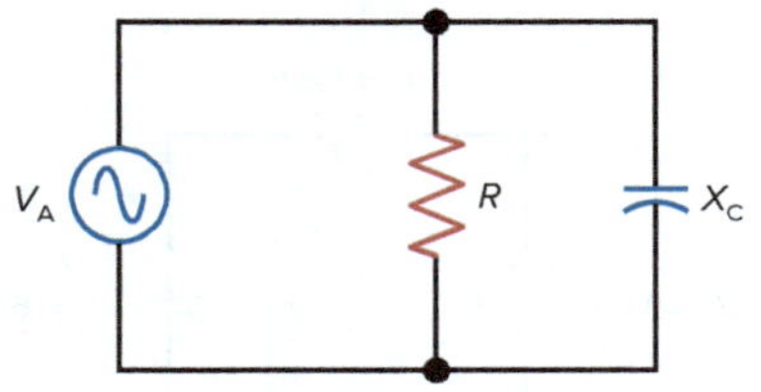

18–34 In Fig. 18–27, how much is Z_{EQ} if $R = 60\ \Omega$ and $X_C = 80\ \Omega$?

18–35 In Fig. 18–28, solve for X_C, I_R, I_C, I_T, Z_{EQ}, and θ_I.

Figure 18–28

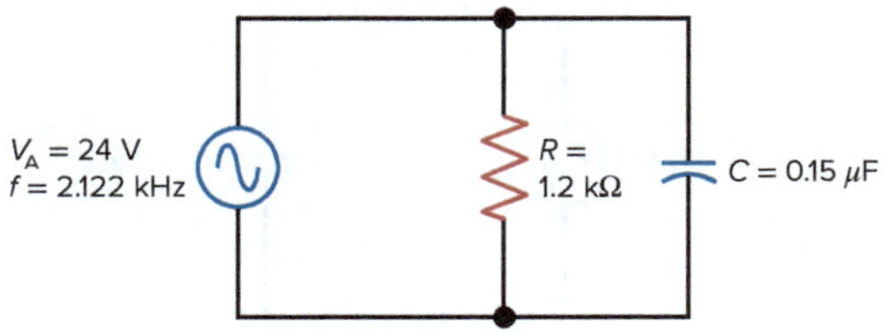

18–36 In Fig. 18–28, what happens to each of the following quantities if the frequency of the applied voltage increases?

a. I_R.

b. I_C.

c. I_T.

d. Z_{EQ}.

e. θ_I.

18–37 Repeat Prob. 18–36 if the frequency of the applied voltage decreases.

SECTION 18–6 RF AND AF COUPLING CAPACITORS

18–38 In Fig. 18–29, calculate the minimum coupling capacitance, C_C, in series with the 1-kΩ resistance, R, if the frequency of the applied voltage is

a. 159.1 Hz.

b. 1591 Hz.

c. 15.91 kHz.

Figure 18–29

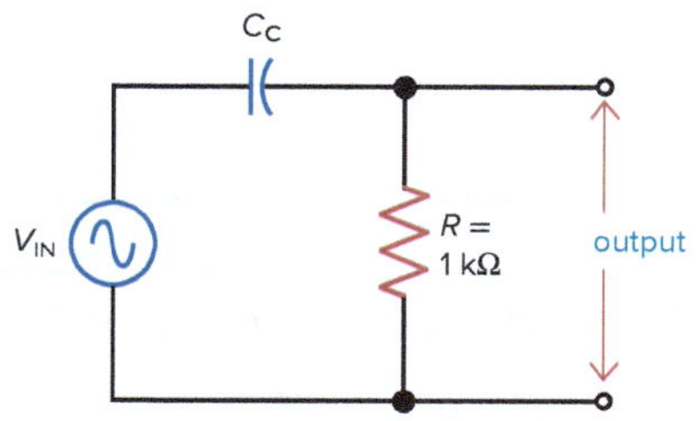

18–39 In Fig. 18–29, assume that $C_C = 0.047\mu$F and $R = 1$ kΩ, as shown. For these values, what is the lowest frequency of the applied voltage that will provide an X_C of 100 Ω? At this frequency, what is the phase angle, θ_Z?

SECTION 18–7 CAPACITIVE VOLTAGE DIVIDERS

18–40 In Fig. 18–30, calculate the following:

a. X_{C_1}, X_{C_2}, X_{C_3}, X_{C_4}, and X_{C_T}.

b. I.

c. V_{C_1}, V_{C_2}, V_{C_3}, and V_{C_4}.

Figure 18–30

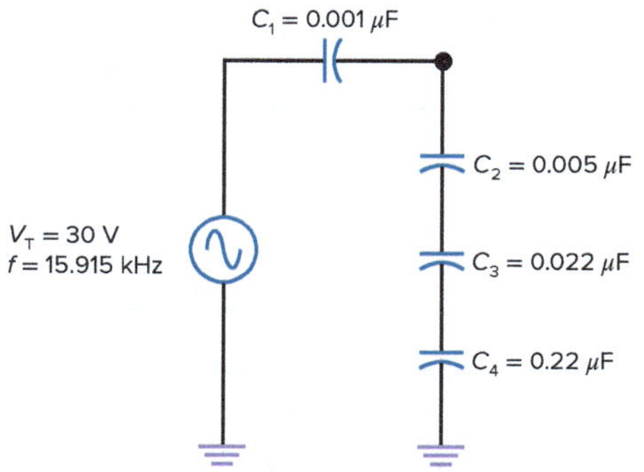

18–41 In Fig. 18–31, calculate V_{C_1}, V_{C_2}, and V_{C_3}.

Figure 18–31

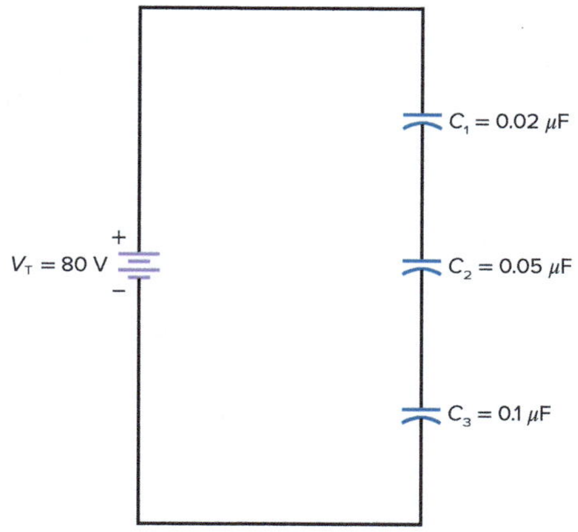

SECTION 18–8 THE GENERAL CASE OF CAPACITIVE CURRENT, i_C

18–42 For the waveshape of capacitor voltage, V_C, in Fig. 18–32, show the corresponding charge and discharge current, i_C, with values for a 200-pF capacitance.

Figure 18–32

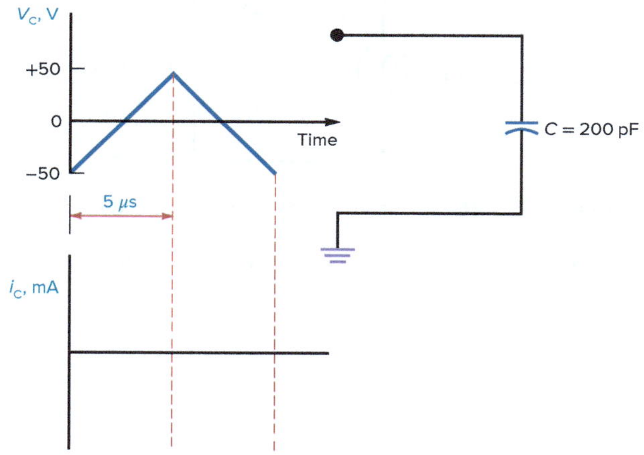

18–43 In Fig. 18–33, show the corresponding charge and discharge current for the waveshape of capacitor voltage shown.

Figure 18–33

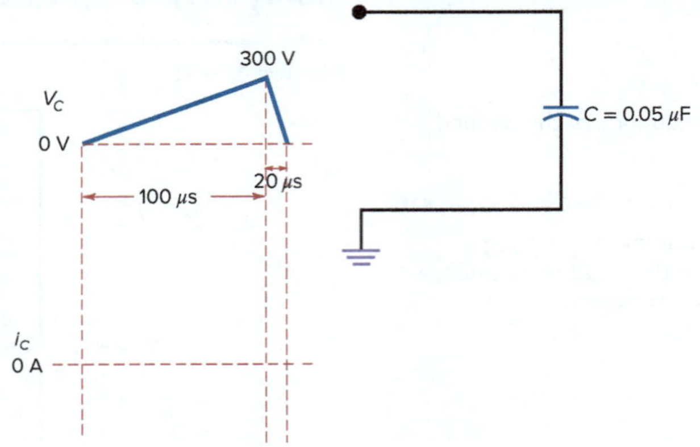

Critical Thinking

18–44 In Fig. 18–34, calculate X_C, Z_T, I, f, V_T, and V_R.

Figure 18–34 Circuit for Critical Thinking Prob. 18–44.

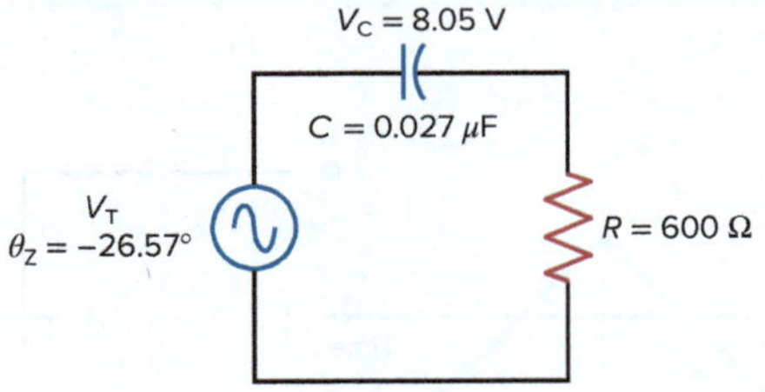

18–45 In Fig. 18–35, calculate I_C, I_R, V_A, X_C, C, and Z_{EQ}.

18–46 In Fig. 18–36, calculate I_C, I_R, I_T, X_C, R, and C.

Figure 18–35 Circuit for Critical Thinking Prob. 18–45.

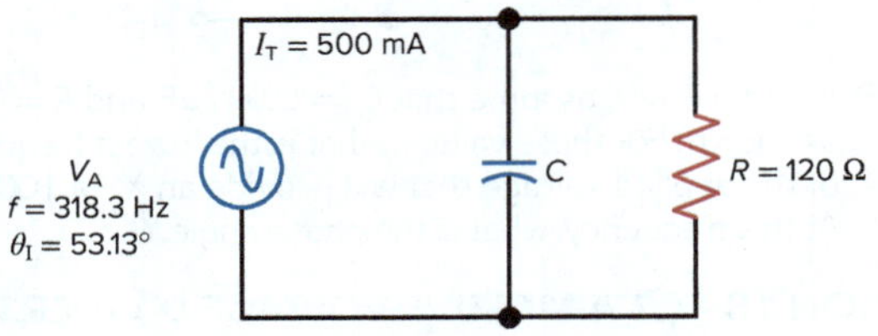

Figure 18–36 Circuit for Critical Thinking Prob. 18–46.

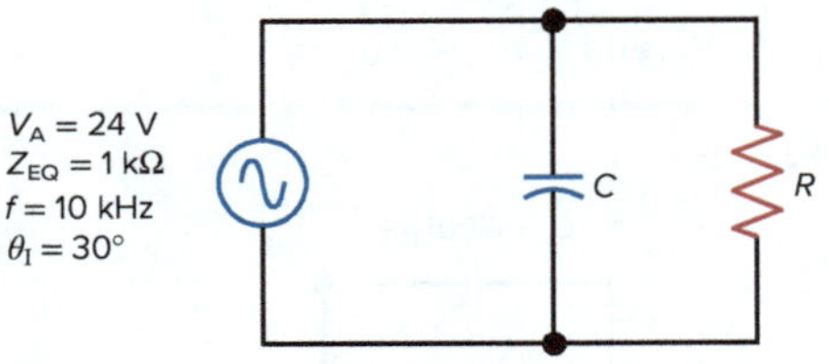

Answers to Self-Reviews

18–1 **a.** 0° **b.** 90° **c.** lag

18–2 **a.** 0° **b.** 90° **c.** lead

18–3 **a.** 28.28 Ω **b.** 28.28 V **c.** −45°

18–4 **a.** 46.7° **b.** 90° **c.** 43.3°

18–5 **a.** 2.828 A **b.** 45°

18–6 **a.** 35 Ω **b.** 2 pF

18–7 **a.** 18 kV; 2 kV **b.** 120 Ω

18–8 **a.** 1×10^6 V/s **b.** 300 μA

Laboratory Application Assignment

In this lab application assignment, you will examine both series and parallel *RC* circuits. In the series *RC* circuit, you will measure the individual component voltages as well the circuit current and phase angle. In the parallel *RC* circuit, you will measure the individual branch currents, the total current, and the circuit phase angle.

Equipment: Obtain the following items from your instructor.

- Function generator
- Oscilloscope
- Assortment of carbon-film resistors and plastic-film capacitors
- DMM

Series *RC* Circuit

Examine the series *RC* circuit in Fig. 18–37. Calculate and record the following circuit values:

X_C = ____________, Z_T = ____________, I = ____________,
V_C = ____________, V_R = ____________, θ_Z = ____________

Figure 18–37

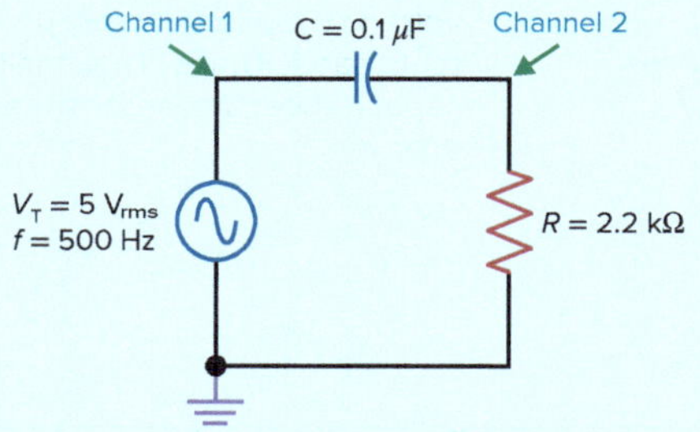

Construct the circuit in Fig. 18–37. Set the total voltage, V_T, to 5 V rms and the frequency, f, to 500 Hz. Using a DMM, measure and record the following circuit values:

I = ____________, V_C = ____________, V_R = ____________

Using the measured values of V_C and V_R, calculate the total voltage, V_T, as $V_T = \sqrt{V_R^2 + V_C^2}$. Does this value equal the applied voltage, V_T, of 5 V? __________. Using the measured values of voltage and current, calculate X_C as V_C/I and Z_T as V_T/I. X_C = ____________, Z_T = ____________. Using Formula (18–3), determine the phase angle, θ_Z. θ_Z = ____________. How do these values compare to those originally calculated? ____________

In the space provided below, draw the phasor voltage triangle, including the phase angle, θ_V for the circuit of Fig. 18–37. Use measured values for V_R, V_C, and V_T.

Ask your instructor for assistance in using the oscilloscope to measure the phase angle, θ, in Fig. 18–37. Note the connections designated for channels 1 and 2 in the figure.

Parallel *RC* Circuit

Examine the parallel *RC* circuit in Fig. 18–38*a*. Calculate and record the following circuit values:

X_C = ____________, I_C = ____________, I_R = ____________,
I_T = ____________, Z_{EQ} = ____________, θ_I = ____________

Construct the circuit in Fig. 18–38*a*. Set the applied voltage, V_A, to 5 V rms and the frequency, f, to 500 Hz. Using a DMM, measure and record the following circuit values:

I_C = ____________, I_R = ____________, I_T = ____________

Using the measured values of I_C and I_R, calculate the total current, I_T, as $I_T = \sqrt{I_R^2 + I_C^2}$.

Does this value agree with the measured value of total current? __________. Using the measured values of I_C and I_R, calculate the phase angle, θ_I, using Formula (18–6).
θ_I = __________. Also, calculate X_C as V_A/I_C and Z_{EQ} as V_A/I_T using measured values. X_C = ____________.
Z_{EQ} = __________. How do these values compare to those originally calculated in Fig. 18–38*a*? ____________

In the space provided below, draw the phasor current triangle, including the phase angle, θ_I, for the circuit of Fig. 18–38*a*. Use measured values for I_C, I_R, and I_T.

Ask your instructor for assistance in using the oscilloscope to measure the phase angle, θ_I, in Fig. 18–38*b*. Note the connections designated for channel 1 and channel 2 in the figure. (The voltage drop across the sensing resistor (R_{sense}) has the same phase as the total current, I_T.)

Figure 18–38

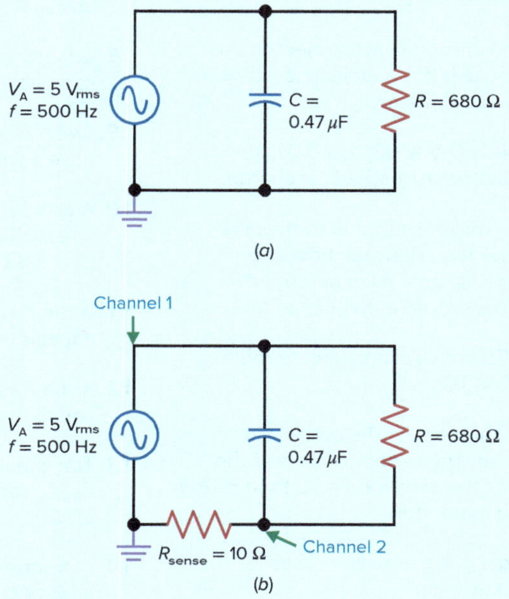

Cumulative Review Summary (Chapters 16–18)

- A capacitor consists of two metal plates separated by an insulator, which is a dielectric material. When voltage is applied to the metal plates, charge is stored in the dielectric. One coulomb of charge stored with 1 volt applied corresponds to 1 farad of capacitance *C*. The common units of capacitance are microfarads (1 μF $= 10^{-6}$ F) or picofarads (1 pF $= 10^{-12}$ F).
- Capacitance increases with plate area and larger values of dielectric constant but decreases with increased distance between plates.
- The most common types of capacitors are air, plastic film, paper, mica, ceramic disk, surface-mount (chip), and electrolytic. Electrolytics must be connected in the correct polarity. The capacitance coding systems for film, ceramic disk, and tantalum capacitors are illustrated in Figs. 16–11, 16–13, and 16–18, respectively. The capacitance coding systems used with chip capacitors are illustrated in Figs. 16–15, 16–16, and 16–17.
- The total capacitance of parallel capacitors is the sum of individual values; the combined equivalent capacitance of series capacitors is found by the reciprocal formula. These rules are opposite from the formulas used for resistors in series or parallel.
- When a good capacitor is checked with an analog ohmmeter, it shows charging current, and then the ohmmeter reads a very high value of ohms equal to the insulation resistance. A short-circuited capacitor reads zero ohms; an open capacitor does not show any charging current.
- $X_C = 1/(2\pi fC)$ Ω, where *f* is in hertz, *C* is in farads, and X_C is in ohms. The higher the frequency and the greater the capacitance, the smaller the X_C.
- The total X_C of capacitive reactances in series equals the sum of the individual values, just as for series resistances. The series reactances have the same current. The voltage across each X_C equals IX_C.
- With parallel capacitive reactances, the combined equivalent reactance is calculated using the reciprocal formula, as for parallel resistances. Each branch current equals V_A/X_C. The total current is the sum of the individual branch currents.
- A common application of X_C is in audio frequency (af) or radio frequency (rf) coupling capacitors, which have low reactance for higher frequencies but more reactance for lower frequencies.
- Reactance X_C is a phasor quantity in which the voltage across the capacitor lags 90° behind its charge and discharge current.
- In a series *RC* circuit, *R* and X_C are added by phasors because the voltage drops are 90° out of phase. Therefore, the total impedance $Z_T = \sqrt{R^2 + X_C^2}$; the current $I = V_T/Z_T$.
- For parallel *RC* circuits, the resistive and capacitive branch currents are added by phasors, $I_T = \sqrt{I_R^2 + I_C^2}$; the impedance $Z_{EQ} = V_A/I_T$.
- Capacitive charge or discharge current i_C is equal to $C(dv/dt)$ for any waveshape of v_C.
- For a series capacitor, the amount of voltage drop is inversely proportional to its capacitance. The smaller the capacitance, the larger the voltage drop.

Cumulative Self-Test

Answers at the back of the book.

Answer True or False.

1. A capacitor can store electric charge because it has a dielectric between two metal plates.
2. With 100-V applied, a 0.01-μF capacitor stores 1 μC of charge.
3. The smaller the capacitance, the higher the potential difference across it for a given amount of charge stored in the capacitor.
4. A 250-pF capacitance equals 250×10^{-12} F.
5. The thinner the dielectric, the greater the capacitance and the lower the breakdown voltage rating for a capacitor.
6. Larger plate area increases capacitance.
7. Capacitors in series provide less capacitance but a higher breakdown voltage rating for the combination.
8. Capacitors in parallel increase the total capacitance with the same voltage rating.
9. Two 0.01-μF capacitors in parallel have a total *C* of 0.005 μF.
10. A good 0.1-μF film capacitor will show charging current and read 500 MΩ or more on an ohmmeter.
11. If the capacitance is doubled, the capacitive reactance is halved.
12. If the frequency is doubled, the capacitive reactance is doubled.
13. The capacitive reactance of a 0.1-μF capacitor at 60 Hz is approximately 60 Ω.
14. In a series *RC* circuit, the voltage across X_C lags 90° behind the current.
15. The phase angle of a series *RC* circuit can be any angle between 0° and −90°, depending on the ratio of X_C to *R*.
16. In a parallel *RC* circuit, the voltage across X_C lags 90° behind its capacitive branch current.
17. If a parallel circuit has two resistances with 1 A in each branch, the total line current equals 1.414 A.
18. A 1000-Ω X_C in parallel with a 1000-Ω *R* has a combined *Z* of 707 Ω.
19. A 1000-Ω X_C in series with a 1000-Ω *R* has a total *Z* of 1414 Ω.
20. Neglecting its sign, the phase angle is 45° for both circuits in Probs. 18 and 19.
21. The total impedance of a 1-MΩ *R* in series with a 5-Ω X_C is approximately 1 MΩ with a phase angle of 0°.

22. The combined equivalent impedance of a 5-Ω R in parallel with a 1-MΩ X_C is approximately 5 Ω with a phase angle of 0°.

23. Both resistance and impedance are measured in ohms.

24. The impedance Z of an RC circuit can change with frequency because the circuit includes reactance.

25. Capacitors in series have the same charge and discharge current.

26. Capacitors in parallel have the same voltage.

27. The phasor combination of a 30-Ω R in series with a 40-Ω X_C equals 70 Ω of total impedance.

28. A film capacitor coded 103 has a value of 0.001 μF.

29. Capacitive current can be considered leading current in a series circuit.

30. In a series RC circuit, the higher the value of X_C, the greater its voltage drop compared with the IR drop.

31. Electrolytic capacitors typically have more leakage current than plastic-film capacitors.

32. A 0.04-μF capacitor in series with a 0.01-μF capacitor has a combined equivalent capacitance, C_{EQ} of 0.008 μF.

33. A shorted capacitor measures 0 Ω.

34. An open capacitor measures infinite ohms.

35. The X_C of a capacitor is inversely proportional to both f and C.

36. The equivalent series resistance, ESR, of a capacitor can be measured with an ohmmeter.

37. In an RC coupling circuit, the output is taken across C.

38. The equivalent impedance, Z_{EQ}, of a parallel RC circuit will decrease if the frequency of the applied voltage increases.

39. Electrolytic capacitors usually have lower breakdown voltage ratings than mica, plastic film, and ceramic capacitors.

40. A 10-μF and a 5-μF capacitor are in series with a DC voltage source. The 10-μF capacitor will have the larger voltage drop.

Design credits: Chapter Opener Background Image, and Multisim: Stockbyte/Getty Images

chapter 19

Inductance

Inductance is the ability of a conductor to produce an induced voltage in itself when the current varies. A long wire has more inductance than a short wire, since more conductor length cut by magnetic flux produces more induced voltage. Similarly, a coil has more inductance than the equivalent length of straight wire because the coil concentrates the magnetic flux. Components manufactured to have a definite value of inductance are coils of wire, called *inductors*. The symbol for inductance is L, and the unit is the henry (H).

The wire for a coil can be wound around a hollow, insulating tube, or the coil can be just the wire itself. This type is called an air-core coil because the magnetic field of the current in the coil is in air. With another basic type, the wire is wound on an iron core to concentrate the magnetic flux for more inductance.

Air-core coils are used in radio frequency (rf) circuits because higher frequencies need less L for the required inductive effect. Iron-core inductors are used in the audio-frequency (af) range, especially at the AC power-line frequency of 60 Hz and for lower frequencies in general. ■

Chapter Outline

Chapter Objectives

After studying this chapter, you should be able to

- *Explain* the concept of self-inductance.
- *Define* the henry unit of inductance *and define* mutual inductance.
- *Calculate* the inductance when the induced voltage and rate of current change are known.
- *List* the physical factors affecting the inductance of an inductor.
- *Calculate* the induced voltage across an inductor, given the inductance and rate of current change.
- *Explain* how induced voltage opposes a change in current.
- *Describe* how a transformer works and *list* important transformer ratings.
- *Calculate* the currents, voltages, and impedances of a transformer circuit.
- *Identify* the different types of transformer cores.
- *Calculate* the total inductance of series-connected inductors.
- *Calculate* the combined equivalent inductance of parallel-connected inductors.
- *List* some common troubles with inductors.

Important Terms

autotransformer
coefficient of coupling, k
counter emf (cemf)
eddy current
efficiency
ferrite core
henry (H)
impedance matching
inductance, L
leakage flux
Lenz's law
mutual inductance, L_M
phasing dots
reflected impedance
series-aiding
series-opposing
stray capacitance
stray inductance
transformer
turns ratio
Variac
volt-ampere (VA)

19–1 Induction by Alternating Current

Induced voltage is the result of flux cutting across a conductor. This action can be produced by physical motion of either the magnetic field or the conductor. When the current in a conductor varies in amplitude, however, the variations of current and its associated magnetic field are equivalent to motion of the flux. As the current increases in value, the magnetic field expands outward from the conductor. When the current decreases, the field collapses into the conductor. As the field expands and collapses with changes of current, the flux is effectively in motion. Therefore, a varying current can produce induced voltage without the need for motion of the conductor.

Figure 19–1 illustrates the changes in the magnetic field of a sine wave of alternating current. Since the alternating current varies in amplitude and reverses in direction, its magnetic field has the same variations. At point A, the current is zero and there is no flux. At B, the positive direction of current provides some field lines taken here in the counterclockwise direction. Point C has maximum current and maximum counterclockwise flux.

At D, there is less flux than at C. Now the field is collapsing because of reduced current. At E, with zero current, there is no magnetic flux. The field can be considered as having collapsed into the wire.

The next half-cycle of current allows the field to expand and collapse again, but the directions are reversed. When the flux expands at points F and G, the field lines are clockwise, corresponding to current in the negative direction. From G to H and I, this clockwise field collapses into the wire.

The result of an expanding and collapsing field, then, is the same as that of a field in motion. This moving flux cuts across the conductor that is providing the current, producing induced voltage in the wire itself. Furthermore, any other conductor in the field, whether or not carrying current, also is cut by the varying flux and has induced voltage.

It is important to note that induction by a varying current results from the change in current, not the current value itself. The current must change to provide motion of the flux. A steady direct current of 1000 A, as an example of a large current, cannot produce any induced voltage as long as the current value is constant. A current of 1 μA changing to 2 μA, however, does induce voltage. Also, the faster the current changes, the higher the induced voltage because when the flux moves at a higher speed, it can induce more voltage.

Since inductance is a measure of induced voltage, the amount of inductance has an important effect in any circuit in which the current changes. The inductance is an

GOOD TO KNOW

In Fig. 19–1, the rate of flux change is greatest when the alternating current passes through zero. Conversely, the rate of flux change is zero at the instant the alternating current reaches either of its two peak values.

Figure 19–1 The magnetic field of an alternating current is effectively in motion as it expands and contracts with current variations.

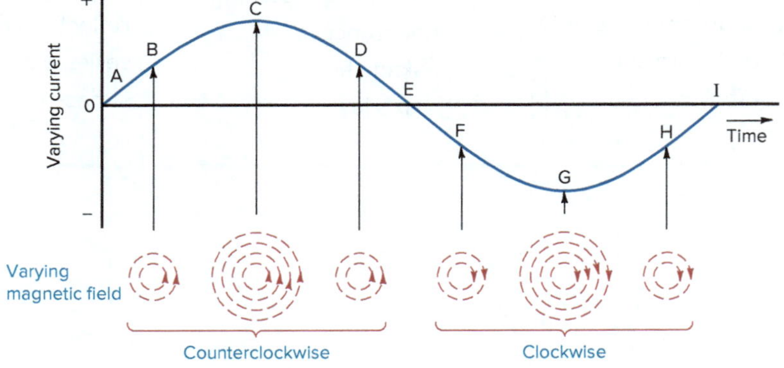

additional characteristic of a circuit beside its resistance. The characteristics of inductance are important in

1. *AC circuits*. Here the current is continuously changing and producing induced voltage. Lower frequencies of alternating current require more inductance to produce the same amount of induced voltage as a higher-frequency current. The current can have any waveform, as long as the amplitude is changing.
2. *DC circuits in which the current changes in value*. It is not necessary for the current to reverse direction. One example is a DC circuit turned on or off. When the direct current is changing between zero and its steady value, the inductance affects the circuit at the time of switching. This effect of a sudden change is called the circuit's *transient response*. A steady direct current that does not change in value is not affected by inductance, however, because there can be no induced voltage without a change in current.

PIONEERS IN ELECTRONICS

The work of American physicist *Joseph Henry (1797–1878)* provided the basis for much of electrical technology. His primary contributions were in the field of electromagnetism. Henry was the first to wind insulated wires around an iron core to obtain powerful electromagnets. Further, he found that if a single cell is used for a given magnet, the magnet should be wound with several coils of wire in parallel; but if a battery of many cells is used, the magnet winding should be a single long wire. Pictured is an early electromagnet, which was built by Joseph Henry. Through his studies, Henry found that self-inductance is greatly affected by the configuration of a circuit, especially the coiling of the wire. Among Henry's other credits are the invention of the electric motor and the development of the first electromagnetic telegraph, which formed the basis for the commercial telegraphic system. The unit of inductance, called the "henry," was named in his honor.

19–1 Self-Review

Answers at the end of the chapter.

a. For the same number of turns and frequency, which has more inductance, a coil with an iron core or one without an iron core?

b. In Fig. 19–1, are the changes of current faster at time B or C?

19–2 Self-Inductance *L*

The ability of a conductor to induce voltage in itself when the current changes is its *self-inductance* or simply *inductance*. The symbol for inductance is *L*, for linkages of the magnetic flux, and its unit is the ***henry*** **(H)**. This unit is named after Joseph Henry (1797–1878).

Definition of the Henry Unit

As illustrated in Fig. 19–2, 1 henry is the amount of inductance that allows one volt to be induced when the current changes at the rate of one ampere per second. The formula is

$$L = \frac{v_L}{di/dt} \qquad \textbf{(19–1)}$$

where v_L is in volts and di/dt is the current change in amperes per second.

Again the symbol *d* is used to indicate an infinitesimally small change in current with time. The factor di/dt for the current variation with respect to time specifies how fast the current's magnetic flux is cutting the conductor to produce v_L.

Figure 19–2 When a current change of 1 A/s induces 1 V across *L*, its inductance equals 1 H.

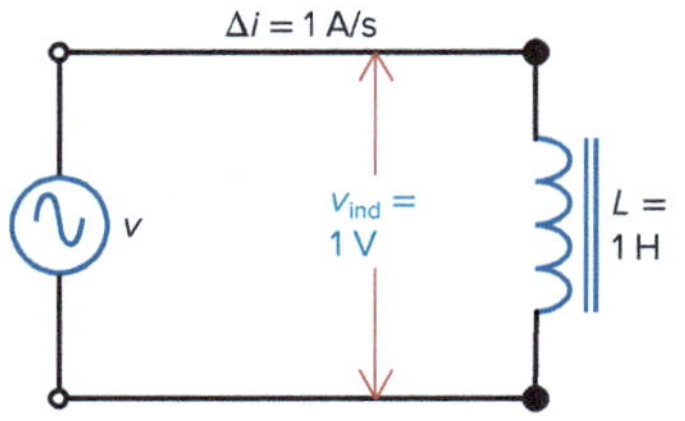

Example 19–1

The current in an inductor changes from 12 to 16 A in 1 s. How much is the *di*/*dt* rate of current change in amperes per second?

ANSWER The *di* is the difference between 16 and 12, or 4 A in 1 s. Then

$$\frac{di}{dt} = 4 \text{ A/s}$$

Example 19–2

The current in an inductor changes by 50 mA in 2 μs. How much is the *di*/*dt* rate of current change in amperes per second?

ANSWER

$$\frac{di}{dt} = \frac{50 \times 10^{-3}}{2 \times 10^{-6}} = 25 \times 10^{3}$$

$$= 25{,}000 \text{ A/s}$$

Example 19–3

How much is the inductance of a coil that induces 40 V when its current changes at the rate of 4 A/s?

ANSWER

$$L = \frac{V_L}{di/dt} = \frac{40}{4}$$

$$= 10 \text{ H}$$

Example 19–4

How much is the inductance of a coil that induces 1000 V when its current changes at the rate of 50 mA in 2 μs?

ANSWER For this example, the $1/dt$ factor in the denominator of Formula (19–1) can be inverted to the numerator.

$$L = \frac{v_L}{di/dt} = \frac{v_L \times dt}{di}$$
$$= \frac{1 \times 10^3 \times 2 \times 10^{-6}}{50 \times 10^{-3}}$$
$$= \frac{2 \times 10^{-3}}{50 \times 10^{-3}} = \frac{2}{50}$$
$$= 0.04 \text{ H or } 40 \text{ mH}$$

Notice that the smaller inductance in Example 19–4 produces much more v_L than the inductance in Example 19–3. The very fast current change in Example 19–4 is equivalent to 25,000 A/s.

Inductance of Coils

In terms of physical construction, the inductance depends on how a coil is wound. Note the following factors.

1. A greater number of turns N increases L because more voltage can be induced. L increases in proportion to N^2. Double the number of turns in the same area and length increases the inductance four times.
2. More area A enclosed by each turn increases L. This means that a coil with larger turns has more inductance. The L increases in direct proportion to A and as the square of the diameter of each turn.
3. The L increases with the permeability of the core. For an air core, μ_r is 1. With a magnetic core, L is increased by the μ_r factor because the magnetic flux is concentrated in the coil.
4. The L decreases with more length for the same number of turns because the magnetic field is less concentrated.

GOOD TO KNOW

The inductance of a coil with a magnetic core will vary with the amount of current (both DC and AC) that passes through the coil. Too much current will saturate the magnetic core, thus reducing its permeability, μ_r, and in turn the inductance, L. Most iron-core inductors (also known as chokes) have an inductance rating at a predetermined value of direct current. As an example, an iron-core choke may have the following rating: 8.5 H @ 50 mA.

These physical characteristics of a coil are illustrated in Fig. 19–3. For a long coil, where the length is at least 10 times the diameter, the inductance can be calculated from the formula

$$L = \mu_r \times \frac{N^2 \times A}{l} \times 1.26 \times 10^{-6} \text{ H} \qquad \textbf{(19–2)}$$

where L is in henrys, l is in meters, and A is in square meters. The constant factor 1.26×10^{-6} is the absolute permeability of air or vacuum in SI units to calculate L in henrys.

For the air-core coil in Fig. 19–3,

$$L = 1 \times \frac{10^4 \times 2 \times 10^{-4}}{0.2} \times 1.26 \times 10^{-6}$$
$$= 12.6 \times 10^{-6} \text{ H} = 12.6\ \mu\text{H}$$

Figure 19–3 Physical factors for inductance L of a coil. See text for calculating L.

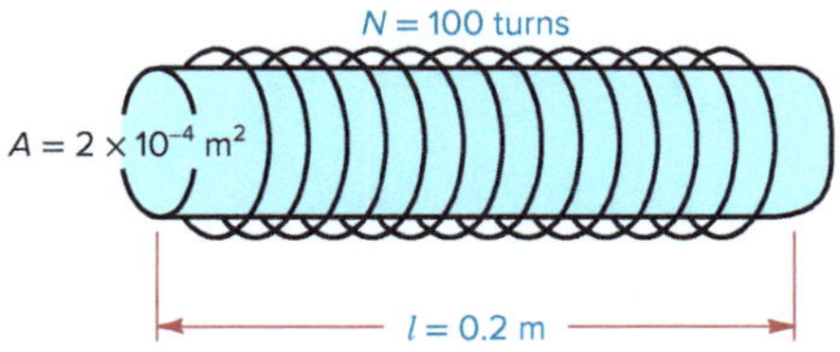

Figure 19–4 Typical inductors with symbols. (*a*) Air-core coil used as rf choke. Length is 2 in. (*b*) Iron-core coil used for 60 Hz. Height is 2 in.

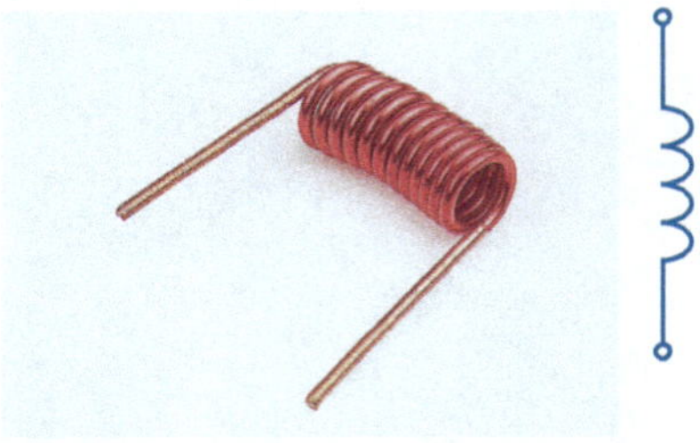

Mark Steinmetz/McGraw Hill

(*a*)

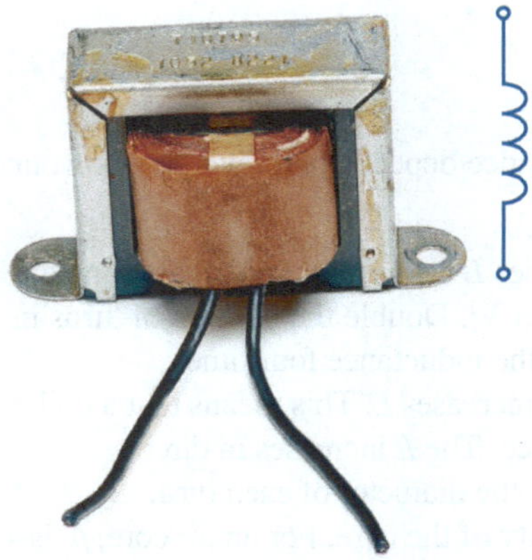

Cindy Schroeder/McGraw Hill

(*b*)

This value means that the coil can produce a self-induced voltage of 12.6 μV when its current changes at the rate of 1 A/s because $v_L = L(di/dt)$. Furthermore, if the coil has an iron core with $\mu_r = 100$, then L will be 100 times greater.

Typical Coil Inductance Values

Air-core coils for rf applications have L values in millihenrys (mH) and microhenrys (μH). A typical air-core rf inductor (called a *choke*) is shown with its schematic symbol in Fig. 19–4*a*. Note that

$$1 \text{ mH} = 1 \times 10^{-3} \text{ H}$$
$$1 \ \mu\text{H} = 1 \times 10^{-6} \text{ H}$$

For example, an rf coil for the radio broadcast band of 535 to 1605 kHz may have an **inductance *L*** of 250 μH, or 0.250 mH.

Iron-core inductors for the 60-Hz power line and for audio frequencies have inductance values of about 1 to 25 H. An iron-core choke is shown in Fig. 19–4*b*.

19–2 Self-Review

Answers at the end of the chapter.

a. A coil induces 2 V with *di/dt* of 1 A/s. How much is *L*?

b. A coil has *L* of 8 mH with 125 turns. If the number of turns is doubled, how much will *L* be?

19–3 Self-Induced Voltage v_L

The self-induced voltage across an inductance L produced by a change in current di/dt can be stated as

$$v_L = L\frac{di}{dt} \qquad \textbf{(19–3)}$$

where v_L is in volts, L is in henrys, and di/dt is in amperes per second. This formula is an inverted version of Formula (19–1), which defines inductance.

Actually, both versions are based on Formula (14–5): $v = N(d\phi/dt)$ for magnetism. This gives the voltage in terms of the amount of magnetic flux cut by a conductor per second. When the magnetic flux associated with the current varies the same as i, then Formula (19–3) gives the same results for calculating induced voltage. Also, remember that the induced voltage across the coil is actually the result of inducing electrons to move in the conductor, so that there is also an induced current. In using Formula (19–3) to calculate v_L, multiply L by the di/dt factor.

GOOD TO KNOW

A steady DC current cannot induce a voltage in a coil because the magnetic flux is stationary.

Example 19–5

How much is the self-induced voltage across a 4-H inductance produced by a current change of 12 A/s?

ANSWER

$$v_L = L\frac{di}{dt} = 4 \times 12$$
$$= 48 \text{ V}$$

Example 19–6

The current through a 200-mH L changes from 0 to 100 mA in 2 μs. How much is v_L?

ANSWER

$$v_L = L\frac{di}{dt}$$

$$= 200 \times 10^{-3} \times \frac{100 \times 10^{-3}}{2 \times 10^{-6}}$$

$$= 10{,}000 \text{ V or } 10 \text{ kV}$$

Note the high voltage induced in the 200-mH inductance because of the fast change in current.

The induced voltage is an actual voltage that can be measured, although v_L is produced only while the current is changing. When di/dt is present for only a short time, v_L is in the form of a voltage pulse. For a sine-wave current, which is always changing, v_L is a sinusoidal voltage 90° out of phase with i_L.

19–3 Self-Review

Answers at the end of the chapter.

a. If L is 2 H and di/dt is 1 A/s, how much is v_L?
b. For the same coil, the di/dt is increased to 100 A/s. How much is v_L?

19–4 How v_L Opposes a Change in Current

By **Lenz's law**, the induced voltage v_L must produce current with a magnetic field that opposes the change of current that induces v_L. The polarity of v_L, therefore, depends on the direction of the current variation *di*. When *di* increases, v_L has polarity that opposes the increase in current; when *di* decreases, v_L has opposite polarity to oppose the decrease in current.

In both cases, the change in current is opposed by the induced voltage. Otherwise, v_L could increase to an unlimited amount without the need to add any work. *Inductance, therefore, is the characteristic that opposes any change in current.* This is the reason that an induced voltage is often called a ***counter emf*** or *back emf.*

GOOD TO KNOW

Counter emf or back emf is also known as a bucking voltage.

More details of applying Lenz's law to determine the polarity of v_L in a circuit are shown in Fig. 19–5. Note the directions carefully. In Fig. 19–5*a*, the electron flow is into the top of the coil. This current is increasing. By Lenz's law, v_L must have the polarity needed to oppose the increase. The induced voltage shown with the top side negative opposes the increase in current. The reason is that this polarity of v_L can produce current in the opposite direction, from minus to plus in the external circuit. Note that for this opposing current, v_L is the generator. This action tends to keep the current from increasing.

In Fig. 19–5*b*, the source is still producing electron flow into the top of the coil, but *i* is decreasing because the source voltage is decreasing. By Lenz's law, v_L must have the polarity needed to oppose the decrease in current. The induced voltage shown with the top side positive now opposes the decrease. The reason is that this

Figure 19–5 Determining the polarity of v_L that opposes the change in *i*. (*a*) The *i* is increasing, and v_L has the polarity that produces an opposing current. (*b*) The *i* is decreasing, and v_L produces an aiding current. (*c*) The *i* is increasing but is flowing in the opposite direction. (*d*) The same direction of *i* as in (*c*) but with decreasing values.

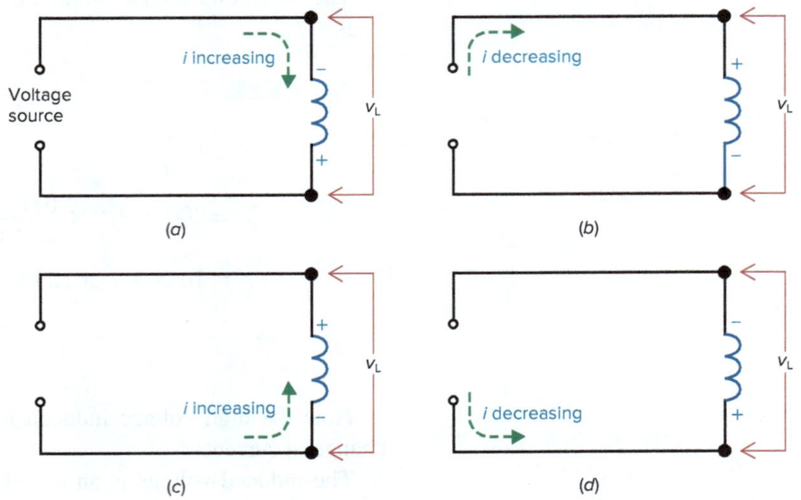

polarity of v_L can produce current in the same direction, tending to keep the current from decreasing.

In Fig. 19–5*c*, the voltage source reverses polarity to produce current in the opposite direction, with electron flow into the bottom of the coil. The current in this reversed direction is now increasing. The polarity of v_L must oppose the increase. As shown, now the bottom of the coil is made negative by v_L to produce current opposing the source current. Finally, in Fig. 19–5*d*, the reversed current is decreasing. This decrease is opposed by the polarity shown for v_L to keep the current flowing in the same direction as the source current.

Notice that the polarity of v_L reverses for either a reversal of direction for *i* or a reversal of change in *di* between increasing or decreasing values. When both the direction of the current and the direction of change are reversed, as in a comparison of Fig. 19–5*a* and *d*, the polarity of v_L remains unchanged.

Sometimes the formulas for induced voltage are written with minus signs to indicate that v_L opposes the change, as specified by Lenz's law. However, the negative sign is omitted here so that the actual polarity of the self-induced voltage can be determined in typical circuits.

In summary, Lenz's law states that the reaction v_L opposes its cause, which is the change in *i*. When *i* is increasing, v_L produces an opposing current. For the opposite case when *i* is decreasing, v_L produces an aiding current.

19–4 Self-Review

Answers at the end of the chapter.

a. In Fig. 19–5*a* and *b*, the v_L has opposite polarities. (True/False)
b. In Fig. 19–5*b* and *c*, the polarity of v_L is the same. (True/False)

19–5 Mutual Inductance L_M

When the current in an inductor changes, the varying flux can cut across any other inductor nearby, producing induced voltage in both inductors. In Fig. 19–6, the coil L_1 is connected to a generator that produces varying current in the turns. The winding L_2 is not connected to L_1, but the turns are linked by the magnetic field. A

GOOD TO KNOW

In Fig. 19–6, the induced voltage across L_2 can be determined if L_1, L_M, and V_{L_1} are known. The formula is $V_{L_2} = \frac{L_M}{L_1} \times V_{L_1}$.

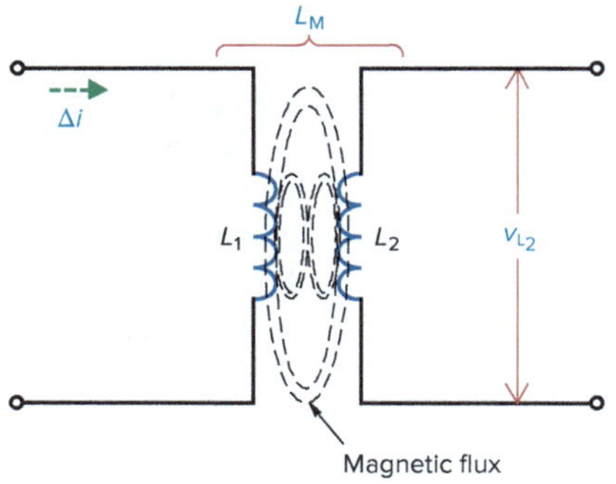

Figure 19-6 Mutual inductance L_M between L_1 and L_2 linked by magnetic flux.

varying current in L_1, therefore, induces voltage across L_1 and across L_2. If all flux of the current in L_1 links all turns of the coil L_2, each turn in L_2 will have the same amount of induced voltage as each turn in L_1. Furthermore, the induced voltage v_{L_2} can produce current in a load resistance connected across L_2.

When the induced voltage produces current in L_2, its varying magnetic field induces voltage in L_1. The two coils, L_1 and L_2, have **mutual inductance**, therefore, because current in one can induce voltage in the other.

The unit of mutual inductance is the henry, and the symbol is L_M. *Two coils have L_M of 1 H when a current change of 1 A/s in one coil induces 1 V in the other coil.*

The schematic symbol for two coils with mutual inductance is shown in Fig. 19–7*a* for an air core and in Fig. 19–7*b* for an iron core. Iron increases the mutual inductance, since it concentrates magnetic flux. Any magnetic lines that do not link the two coils result in ***leakage flux***.

Figure 19-7 Schematic symbols for two coils with mutual inductance. (*a*) Air core. (*b*) Iron core.

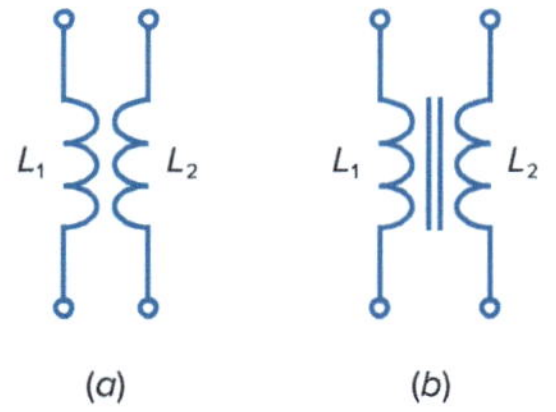

Coefficient of Coupling

The fraction of total flux from one coil linking another coil is the **coefficient of coupling *k*** between the two coils. As examples, if all the flux of L_1 in Fig. 19–6 links L_2, then k equals 1, or unity coupling; if half the flux of one coil links the other, k equals 0.5. Specifically, the coefficient of coupling is

$$k = \frac{\text{flux linkages between } L_1 \text{ and } L_2}{\text{flux produced by } L_1}$$

There are no units for k, because it is a ratio of two values of magnetic flux. The value of k is generally stated as a decimal fraction, like 0.5, rather than as a percent.

The coefficient of coupling is increased by placing the coils close together, possibly with one wound on top of the other, by placing them parallel rather than perpendicular to each other, or by winding the coils on a common iron core. Several examples are shown in Fig. 19–8.

Figure 19-8 Examples of coupling between two coils linked by L_M. (*a*) L_1 or L_2 on paper or plastic form with air core; k is 0.1. (*b*) L_1 wound over L_2 for tighter coupling; k is 0.3. (*c*) L_1 and L_2 on the same iron core; k is 1. (*d*) Zero coupling between perpendicular air-core coils.

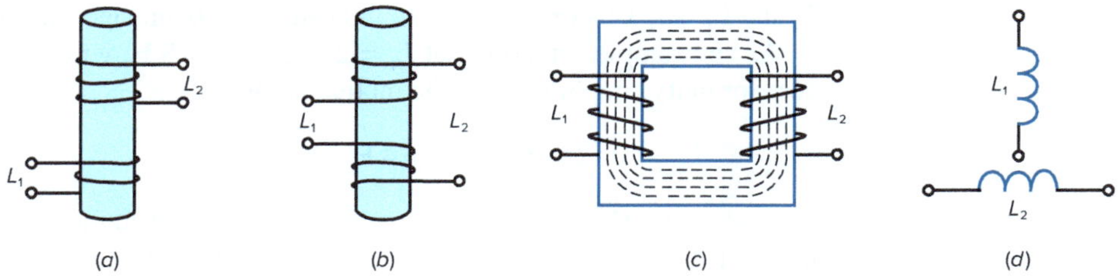

A high value of k, called *tight coupling*, allows the current in one coil to induce more voltage in the other coil. *Loose coupling*, with a low value of k, has the opposite effect. In the extreme case of zero coefficient of coupling, there is no mutual inductance. Two coils may be placed perpendicular to each other and far apart for essentially zero coupling to minimize interaction between the coils.

Air-core coils wound on one form have values of k equal to 0.05 to 0.3, approximately, corresponding to 5 to 30% linkage. Coils on a common iron core can be considered to have practically unity coupling, with k equal to 1. As shown in Fig. 19–8*c*, for both windings L_1 and L_2, practically all magnetic flux is in the common iron core. Mutual inductance is also called *mutual coupling.*

Example 19–7

A coil L_1 produces 80 μWb of magnetic flux. Of this total flux, 60 μWb are linked with L_2. How much is k between L_1 and L_2?

ANSWER

$$k = \frac{60\ \mu\text{Wb}}{80\ \mu\text{Wb}}$$

$$= 0.75$$

Example 19–8

A 10-H inductance L_1 on an iron core produces 4 Wb of magnetic flux. Another coil L_2 is on the same core. How much is k between L_1 and L_2?

ANSWER Unity or 1. All the coils on a common iron core have practically perfect coupling.

Calculating L_M

Mutual inductance increases with higher values for the primary and secondary inductances and tighter coupling:

$$L_M = k\sqrt{L_1 \times L_2} \qquad \textbf{(19–4)}$$

where L_1 and L_2 are the self-inductance values of the two coils, k is the coefficient of coupling, and L_M is the mutual inductance linking L_1 and L_2, in the same units as L_1 and L_2. The k factor is needed to indicate the flux linkages between the two coils.

As an example, suppose that $L_1 = 2$ H and $L_2 = 8$ H, with both coils on an iron core for unity coupling. Then the mutual inductance is

$$L_M = 1\sqrt{2 \times 8} = \sqrt{16} = 4\text{ H}$$

The value of 4 H for L_M in this example means that when the current changes at the rate of 1 A/s in either coil, it will induce 4 V in the other coil.

CALCULATOR

To do Example 19–9 on a calculator that does not have an (EXP) key, multiply $L_1 \times L_2$, take the square root of the product, and multiply by k. Keep the powers of 10 separate. Specifically, punch in 400 for L_1, push the (×) key, punch in 400 for L_2, and push the (=) key for the product, 16,000. Press the (√) key, which is sometimes the (2ndF) of the (x^2) key, to get 400. While it is on the display, push the (×) key, punch in 0.2, and press the (=) key for the answer of 80. For the powers of 10, $10^{-3} \times 10^{-3} = 10^{-6}$, and the square root is equal to 10^{-3} for the unit of millihenry in the answer.

For Example 19–10, the formula is L_M divided by $\sqrt{L_1 \times L_2}$. Specifically, punch in 40 for the value in the numerator, press the (÷) key, then the (( ) key, multiply 400 × 400, and press the ( )) key, followed by the (√) and (=) keys. The display will read 0.1. The powers of 10 cancel with 10^{-3} in the numerator and denominator. Also, there are no units for k, since the units of L cancel.

Example 19–9

Two 400-mH coils L_1 and L_2 have a coefficient of coupling k equal to 0.2. Calculate L_M.

ANSWER

$$\begin{aligned} L_M &= k\sqrt{L_1 \times L_2} \\ &= 0.2\sqrt{400 \times 10^{-3} \times 400 \times 10^{-3}} \\ &= 0.2 \times 400 \times 10^{-3} \\ &= 80 \times 10^{-3} \text{ H or } 80 \text{ mH} \end{aligned}$$

Example 19–10

If the two coils in Example 19–9 had a mutual inductance L_M of 40 mH, how much would k be?

ANSWER Formula (19–4) can be inverted to find k.

$$\begin{aligned} k &= \frac{L_M}{\sqrt{L_1 \times L_2}} \\ &= \frac{40 \times 10^{-3}}{\sqrt{400 \times 10^{-3} \times 400 \times 10^{-3}}} \\ &= \frac{40 \times 10^{-3}}{400 \times 10^{-3}} \\ &= 0.1 \end{aligned}$$

Notice that the same two coils have one-half the mutual inductance L_M because the coefficient of coupling k is 0.1 instead of 0.2.

19–5 Self-Review

Answers at the end of the chapter.

a. All flux from the current in L_1 links L_2. How much is the coefficient of coupling k?

b. Mutual inductance L_M is 9 mH with k of 0.2. If k is doubled to 0.4, how much will L_M be?

19–6 Transformers

The **transformer** is an important application of mutual inductance. As shown in Fig. 19–9, a transformer has a primary winding inductance L_P connected to a voltage source that produces alternating current, and the secondary winding inductance L_S is connected across the load resistance R_L. The purpose of the transformer is to transfer power from the primary, where the generator is connected, to the secondary, where the induced secondary voltage can produce current in the load resistance that is connected across L_S.

MultiSim **Figure 19–9** Iron-core transformer with a 1:10 turn ratio. Primary current I_P induces secondary voltage V_S, which produces current in secondary load R_L.

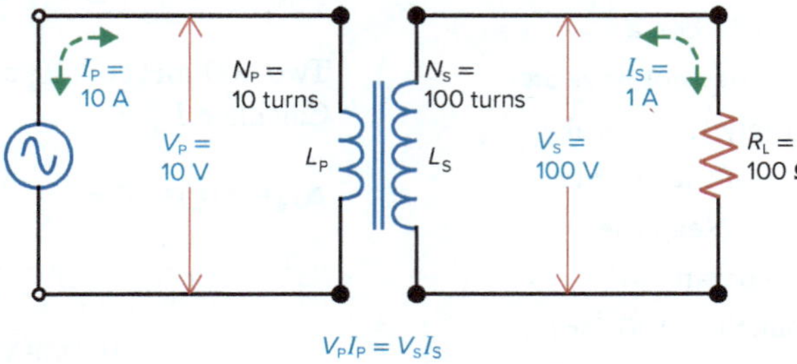

Figure 19–10 (*a*) Air-core rf transformer. Height is 2 in. (*b*) Color code and typical DC resistance of windings.

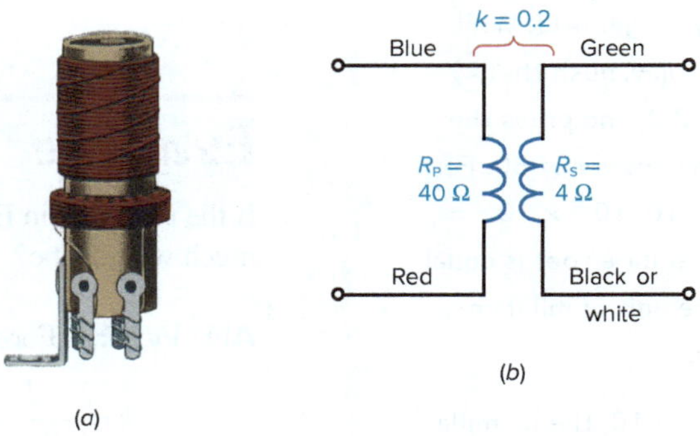

Although the primary and secondary are not physically connected to each other, power in the primary is coupled into the secondary by the magnetic field linking the two windings. The transformer is used to provide power for the load resistance R_L, instead of connecting R_L directly across the generator, whenever the load requires an AC voltage higher or lower than the generator voltage. By having more or fewer turns in L_S, compared with L_P, the transformer can step up or step down the generator voltage to provide the required amount of secondary voltage. Typical transformers are shown in Figs. 19–10 and 19–11. Note that a steady DC voltage cannot be stepped up or down by a transformer because a steady current cannot produce induced voltage.

Figure 19–11 Iron-core power transformer.

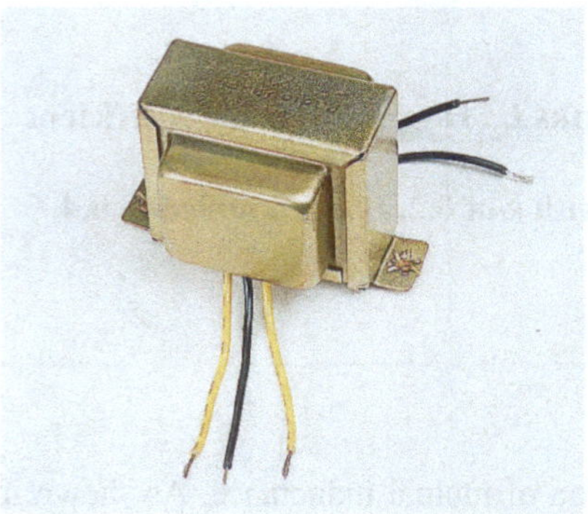

Mark Steinmetz/McGraw Hill

Turns Ratio

The ratio of the number of turns in the primary to the number in the secondary is the **turns ratio** of the transformer:

$$\text{Turns ratio} = \frac{N_P}{N_S} \qquad \textbf{(19–5)}$$

where N_P = number of turns in the primary and N_S = number of turns in the secondary. For example, 500 turns in the primary and 50 turns in the secondary provide a turns ratio of $^{500}/_{50}$, or 10:1, which is stated as "ten-to-one."

Voltage Ratio

With unity coupling between primary and secondary, the voltage induced in each turn of the secondary is the same as the self-induced voltage of each turn in the primary. Therefore, the voltage ratio is in the same proportion as the turns ratio:

$$\frac{V_P}{V_S} = \frac{N_P}{N_S} \qquad \textbf{(19–6)}$$

GOOD TO KNOW

The turns ratio N_P/N_S is sometimes represented by the lowercase letter a, where $a = N_P/N_S$.

When the secondary has more turns than the primary, the secondary voltage is higher than the primary voltage and the primary voltage is said to be stepped up. This principle is illustrated in Fig. 19–9 with a step-up ratio of $^{10}\!/_{100}$, or 1:10. When the secondary has fewer turns, the voltage is stepped down.

In either case, the ratio is in terms of the primary voltage, which may be stepped up or down in the secondary winding.

These calculations apply only to iron-core transformers with unity coupling. Air-core transformers for rf circuits (as shown in Fig. 19–10*a*) are generally tuned to resonance. In this case, the resonance factor is considered instead of the turns ratio.

Example 19–11

A power transformer has 100 turns for N_P and 600 turns for N_S. What is the turns ratio? How much is the secondary voltage V_S if the primary voltage V_P is 120 V?

ANSWER The turns ratio is $^{100}\!/_{600}$, or 1:6. Therefore, V_P is stepped up by the factor 6, making V_S equal to 6×120, or 720 V.

Example 19–12

A power transformer has 100 turns for N_P and 5 turns for N_S. What is the turns ratio? How much is the secondary voltage V_S with a primary voltage of 120 V?

ANSWER The turns ratio is $^{100}\!/_{5}$, or 20:1. The secondary voltage is stepped down by a factor of $^{1}\!/_{20}$, making V_S equal to $^{120}\!/_{20}$, or 6 V.

Secondary Current

By Ohm's law, the amount of secondary current equals the secondary voltage divided by the resistance in the secondary circuit. In Fig. 19–9, with a value of 100 Ω for R_L and negligible coil resistance assumed,

$$I_S = \frac{V_S}{R_L} = \frac{100\ \text{V}}{100\ \Omega} = 1\ \text{A}$$

Power in the Secondary

The power dissipated by R_L in the secondary is $I_S^2 \times R_L$ or $V_S \times I_S$, which equals 100 W in this example. The calculations are

$$P = I_S^2 \times R_L = 1 \times 100 = 100\ \text{W}$$
$$P = V_S \times I_S = 100 \times 1 = 100\ \text{W}$$

It is important to note that power used by the secondary load, such as R_L in Fig. 19–9, is supplied by the generator in the primary. How the load in the secondary draws power from the generator in the primary can be explained as follows.

With current in the secondary winding, its magnetic field opposes the varying flux of the primary current. The generator must then produce more primary current to maintain the self-induced voltage across L_P and the secondary voltage developed in L_S by mutual induction. If the secondary current doubles, for instance, because the load resistance is reduced by one-half, the primary current will also double in value to provide the required power for the secondary. Therefore, the effect of the secondary-load power on the generator is the same as though R_L were in the primary, except that the voltage for R_L in the secondary is stepped up or down by the turns ratio.

Current Ratio

With zero losses assumed for the transformer, the power in the secondary equals the power in the primary:

$$V_S I_S = V_P I_P \tag{19–7}$$

or

$$\frac{I_S}{I_P} = \frac{V_P}{V_S} \tag{19–8}$$

The current ratio is the inverse of the voltage ratio, that is, voltage step-up in the secondary means current step-down, and vice versa. The secondary does not generate power but takes it from the primary. Therefore, the current step-up or step-down is in terms of the secondary current I_S, which is determined by the load resistance across the secondary voltage. These points are illustrated by the following two examples.

Example 19–13

A transformer with a 1:6 turns ratio has 720 V across 7200 Ω in the secondary. (a) How much is I_S? (b) Calculate the value of I_P.

ANSWER

(a) $I_S = \frac{V_S}{R_L} = \frac{720\ \text{V}}{7200\ \Omega}$

$= 0.1\ \text{A}$

(b) With a turns ratio of 1:6, the current ratio is 6:1. Therefore,

$I_P = 6 \times I_S = 6 \times 0.1$

$= 0.6\ \text{A}$

Example 19–14

A transformer with a 20:1 voltage step-down ratio has 6 V across 0.6 Ω in the secondary. (a) How much is I_S? (b) How much is I_P?

ANSWER

(a) $I_S = \frac{V_S}{R_L} = \frac{6\ \text{V}}{0.6\ \Omega}$

$= 10\ \text{A}$

(b) $I_P = \frac{1}{20} \times I_S = \frac{1}{20} \times 10$

$= 0.5\ \text{A}$

Figure 19–12 Total power used by two secondary loads R_1 and R_2 is equal to the power supplied by the source in the primary.

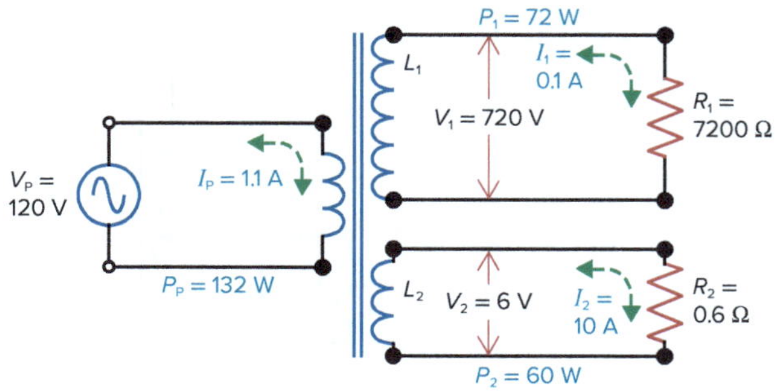

As an aid in these calculations, remember that the side with the higher voltage has the lower current. The primary and secondary V and I are in the same proportion as the number of turns in the primary and secondary.

Total Secondary Power Equals Primary Power

Figure 19–12 illustrates a power transformer with two secondary windings L_1 and L_2. There can be one, two, or more secondary windings with unity coupling to the primary as long as all the windings are on the same iron core. Each secondary winding has induced voltage in proportion to its turns ratio with the primary winding, which is connected across the 120 V source.

The secondary winding L_1 has a voltage step-up of 6:1, providing 720 V. The 7200-Ω load resistance R_1, across L_1, allows the 720 V to produce 0.1 A for I_1 in this secondary circuit. The power here is 720 V × 0.1 A = 72 W.

The other secondary winding L_2 provides voltage step-down with the ratio 20:1, resulting in 6 V across R_2. The 0.6-Ω load resistance in this circuit allows 10 A for I_2. Therefore, the power here is 6 V × 10 A, or 60 W. Since the windings have separate connections, each can have its individual values of voltage and current.

The total power used in the secondary circuits is supplied by the primary. In this example, the total secondary power is 132 W, equal to 72 W for P_1 and 60 W for P_2. The power supplied by the 120-V source in the primary then is 72 + 60 = 132 W.

The primary current I_P equals the primary power P_P divided by the primary voltage V_P. This is 132 W divided by 120 V, which equals 1.1 A for the primary current. The same value can be calculated as the sum of 0.6 A of primary current providing power for L_1 plus 0.5 A of primary current for L_2, resulting in the total of 1.1 A as the value of I_P.

This example shows how to analyze a loaded power transformer. The main idea is that the primary current depends on the secondary load. The calculations can be summarized as follows:

1. Calculate V_S from the turns ratio and V_P.
2. Use V_S to calculate I_S: $I_S = V_S/R_L$.
3. Use I_S to calculate P_S: $P_S = V_S \times I_S$.
4. Use P_S to find P_P: $P_P = P_S$.
5. Finally, I_P can be calculated: $I_P = P_P/V_P$.

With more than one secondary, calculate each I_S and P_S. Then add all P_S values for the total secondary power, which equals the primary power.

GOOD TO KNOW

Up until the invention of solid-state power control devices, autotransformers were used in theatrical stage lighting.

Autotransformers

As illustrated in Fig. 19–13, an **autotransformer** consists of one continuous coil with a tapped connection such as terminal 2 between the ends at terminals 1 and 3.

Figure 19–13 Autotransformer with tap at terminal 2 for 10 turns of the complete 60-turn winding. (*a*) V_p between terminals 1 and 2 stepped up across 1 and 3. (*b*) V_p between terminals 1 and 3 stepped down across 1 and 2.

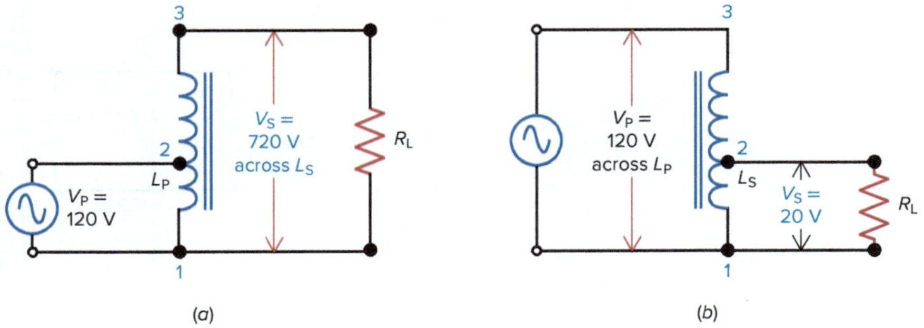

In Fig. 19–13*a*, the autotransformer steps up the generator voltage. Voltage V_P between 1 and 2 is connected across part of the total turns, and V_S is induced across all the turns. With six times the turns for the secondary voltage, V_S also is six times V_P.

In Fig. 19–13*b*, the autotransformer steps down the primary voltage connected across the entire coil. Then the secondary voltage is taken across less than the total turns.

The winding that connects to the voltage source to supply power is the primary, and the secondary is across the load resistance R_L. The turns ratio and voltage ratio apply the same way as in a conventional transformer having an isolated secondary winding.

Autotransformers are used often because they are compact and efficient and usually cost less since they have only one winding. Note that the autotransformer in Fig. 19–13 has only three leads, compared with four leads for the transformer in Fig. 19–9 with an isolated secondary.

Isolation of the Secondary

In a transformer with a separate winding for L_S, as shown in Fig. 19–9, the secondary load is not connected directly to the AC power line in the primary. This isolation is an advantage in reducing the chance of electric shock. With an autotransformer, as in Fig. 19–13, the secondary is not isolated. Another advantage of an isolated secondary is that any direct current in the primary is blocked from the secondary. Sometimes a transformer with a 1:1 turns ratio is used for isolation from the AC power line.

Transformer Efficiency

Efficiency is defined as the ratio of power out to power in. Stated as a formula,

$$\%\ \text{Efficiency} = \frac{P_{out}}{P_{in}} \times 100 \qquad \textbf{(19–9)}$$

For example, when the power out in watts equals one-half the power in, the efficiency is one-half, which equals 0.5 × 100% , or 50%. In a transformer, power out is secondary power, and power in is primary power.

Assuming zero losses in the transformer, power out equals power in and the efficiency is 100%. Actual power transformers, however, have an efficiency slightly less than 100%. The efficiency is approximately 80 to 90% for transformers that have high power ratings. Transformers for higher power are more efficient because they require heavier wire, which has less resistance. In a transformer that is less than 100% efficient, the primary supplies more than the secondary power. The primary power that is lost is dissipated as heat in the transformer, resulting from I^2R in the conductors and certain losses in the core material. The *R* of the primary winding is generally about 10 Ω or less for power transformers.

GOOD TO KNOW

As the load on the secondary of a transformer increases, the power losses in the transformer become a smaller and smaller percentage of the total power supplied to the primary. Thus, the more heavily a transformer is loaded, the higher its efficiency.

19–6 Self-Review

Answers at the end of the chapter.

a. **A transformer connected to the 120-V_{AC} power line has a turns ratio of 1:2. Calculate the stepped-up V_S.**

b. **A V_S of 240 V is connected across a 2400-Ω R_L. Calculate I_S.**

c. **An autotransformer has an isolated secondary. (True/False)**

d. **With more I_S for the secondary load, does the I_P increase or decrease?**

19–7 Transformer Ratings

Like other components, transformers have voltage, current, and power ratings that must not be exceeded. Exceeding any of these ratings will usually destroy the transformer. What follows is a brief description of the most important transformer ratings.

Voltage Ratings

Manufacturers of transformers always specify the voltage rating of the primary and secondary windings. Under no circumstances should the primary voltage rating be exceeded. In many cases, the rated primary and secondary voltages are printed on the transformer. For example, consider the transformer shown in Fig. 19–14*a*. Its rated primary voltage is 120 V, and its secondary voltage is specified as 12.6–0–12.6, which indicates that the secondary is center-tapped. The notation 12.6–0–12.6 indicates that 12.6 V is available between the center tap connection and either outside secondary lead. The total secondary voltage available is 2 × 12.6 V or 25.2 V. In Fig. 19–14*a*, the black leads coming out of the top of the transformer provide connection to the primary winding. The two yellow leads coming out of the bottom of the transformer provide connection to the outer leads of the secondary winding. The bottom middle black lead connects to the center tap on the secondary winding.

Note that manufacturers may specify the secondary voltages of a transformer differently. For example, the secondary in Fig. 19–14*a* may be specified as 25.2 V CT, where CT indicates a center-tapped secondary. Another way to specify the secondary voltage in Fig. 19–14*a* would be 12.6 V each side of center.

Regardless of how the secondary voltage of a transformer is specified, the rated value is always specified under full-load conditions with the rated primary voltage applied. A transformer is considered fully loaded when the rated current is drawn from the secondary. When unloaded, the secondary voltage will measure a value

MultiSim **Figure 19–14** Transformer with primary and secondary voltage ratings. (*a*) Top black leads are primary leads. Yellow and black leads on bottom are secondary leads. (*b*) Schematic symbol.

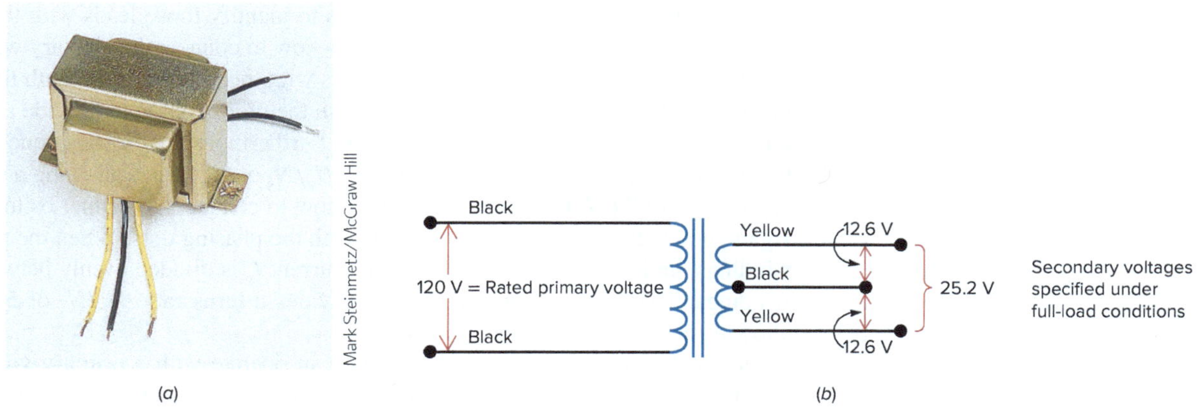

Figure 19–15 Transformer with multiple primary windings. (*a*) Phasing dots show primary leads with same instantaneous polarity. (*b*) Primary windings connected in series to work with a primary voltage of 240 V; N_P/N_S = 10:1. (*c*) Primary windings connected in parallel to work with a primary voltage of 120 V; N_P/N_S = 5:1.

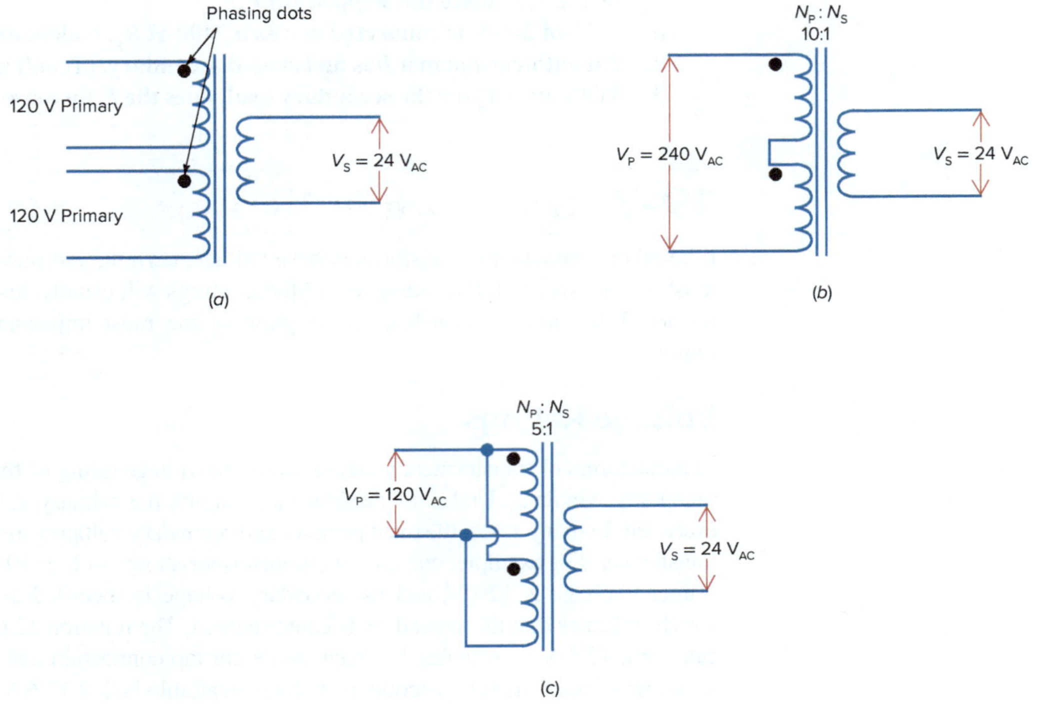

that is approximately 5 to 10% higher than its rated value. Let's use the transformer in Fig. 19–14*a* as an example. It has a rated secondary current of 2 A. If 120 V is connected to the primary and no load is connected to the secondary, each half of the secondary will measure somewhere between 13.2 and 13.9 V approximately. However, with the rated current of 2 A drawn from the secondary, each half of the secondary will measure approximately 12.6 V.

Figure 19–14*b* shows the schematic diagram for the transformer in Fig. 19–14*a*. Notice that the colors of each lead are identified for clarity.

As you already know, transformers can have more than one secondary winding. They can also have more than one primary winding. The purpose is to allow using the transformer with more than one value of primary voltage. Figure 19–15 shows a transformer with two separate primaries and a single secondary. This transformer can be wired to work with a primary voltage of either 120 or 240 V. For either value of primary voltage, the secondary voltage is 24 V. Figure 19–15*a* shows the individual primary windings with **phasing dots** to identify those leads with the same instantaneous polarity. Figure 19–15*b* shows how to connect the primary windings to 240 V. Notice the connections of the leads with the phasing dots. With this connection, each half of the primary voltage is in the proper phase to provide a **series-aiding** connection of the induced voltages. Furthermore, the series connection of the primary windings provides a turns ratio N_P/N_S of 10:1, thus allowing a secondary voltage of 24 V. Figure 19–15*c* shows how to connect the primaries to 120 V. Again, notice the connection of the leads with the phasing dots. When the primary windings are in parallel, the total primary current I_P is divided evenly between the windings. The parallel connection also provides a turns ratio N_P/N_S of 5:1, thus allowing a secondary voltage of 24 V.

Figure 19–16 shows a transformer that can operate with a primary voltage of either 120 or 440 V. In this case, only one of the primary windings is used with a

Figure 19–16 Transformer that has two primaries, which are used separately and never together.

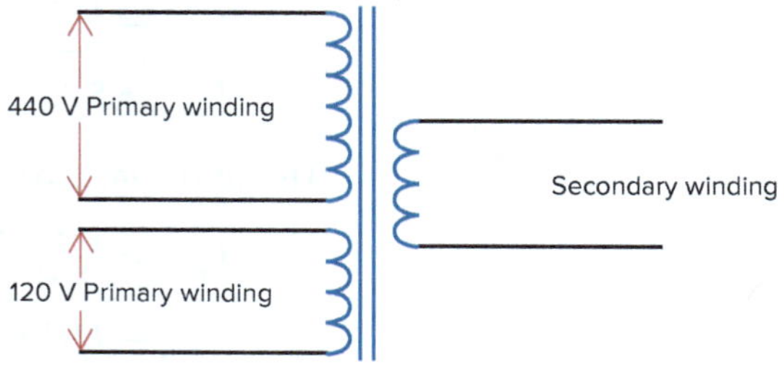

given primary voltage. For example, if 120 V is applied to the lower primary, the upper primary winding is not used. Conversely, if 440 V is applied to the upper primary, the lower primary winding is not used.

Current Ratings

Manufacturers of transformers usually specify current ratings only for the secondary windings. The reason is quite simple. If the secondary current is not exceeded, there is no possible way the primary current can be exceeded. If the secondary current exceeds its rated value, excessive I^2R losses will result in the secondary winding. This will cause the secondary, and perhaps the primary, to overheat, thus eventually destroying the transformer. The IR voltage drop across the secondary windings is the reason that the secondary voltage decreases as the load current increases.

Example 19–15

In Fig. 19–14*b*, calculate the primary current I_P if the secondary current I_S equals its rated value of 2 A.

ANSWER Rearrange Formula (19–8) and solve for the primary current I_P.

$$I_P = \frac{V_S}{V_P} \times I_S$$

$$= \frac{25.2 \text{ V}}{120 \text{ V}} \times 2 \text{ A}$$

$$= 0.42 \text{ A} \qquad \text{or} \qquad 420 \text{ mA}$$

Power Ratings

The power rating of a transformer is the amount of power the transformer can deliver to a resistive load. The power rating is specified in **volt-amperes (VA)** rather than watts (W) because the power is not actually dissipated by the transformer. The product VA is called *apparent power*, since it is the power that is *apparently* used by the transformer. The unit of apparent power is VA because the watt unit is reserved for the dissipation of power in a resistance.

Assume that a power transformer whose primary and secondary voltage ratings are 120 and 25 V, respectively, has a power rating of 125 VA. What does this mean? It means that the product of the transformer's primary, or secondary, voltage and current must not exceed 125 VA. If it does, the transformer will overheat and be destroyed. The maximum allowable secondary current for this transformer can be calculated as

$$I_{S(max)} = \frac{125\ \text{VA}}{25\ \text{V}}$$

$$I_{S(max)} = 5\ \text{A}$$

The maximum allowable primary current can be calculated as

$$I_{P(max)} = \frac{125\ \text{VA}}{120\ \text{V}}$$

$$I_{P(max)} = 1.04\text{A}$$

With multiple secondary windings, the VA rating of each individual secondary may be given without any mention of the primary VA rating. In this case, the sum of all secondary VA ratings must be divided by the rated primary voltage to determine the maximum allowable primary current.

In summary, you will never overload a transformer or exceed any of its maximum ratings if you obey two fundamental rules:

1. Never apply more than the rated voltage to the primary.
2. Never draw more than the rated current from the secondary.

Frequency Ratings

All transformers have a frequency rating that must be adhered to. Typical frequency ratings for power transformers are 50, 60, and 400 Hz. A power transformer with a frequency rating of 400 Hz cannot be used at 50 or 60 Hz because it will overheat. However, many power transformers are designed to operate at either 50 or 60 Hz because many types of equipment may be sold in both Europe and the United States, where the power-line frequencies are 50 and 60 Hz, respectively. Power transformers with a 400-Hz rating are often used in aircraft because these transformers are much smaller and lighter than 50- or 60-Hz transformers having the same power rating.

19–7 Self-Review

Answers at the end of the chapter.

a. The measured voltage across an unloaded secondary is usually 5 to 10% higher than its rated value. (True/False)

b. The current rating of a transformer is usually specified only for the secondary windings. (True/False)

c. A power rating of 300 VA for a transformer means that the transformer secondary must be able to dissipate this amount of power. (True/False)

19–8 Impedance Transformation

Transformers can be used to change or transform a secondary load impedance to a new value as seen by the primary. The secondary load impedance is said to be reflected back into the primary and is therefore called a ***reflected impedance***. The reflected impedance of the secondary may be stepped up or down in accordance with the square of the transformer turns ratio.

By manipulating the relationships between the currents, voltages, and turns ratio in a transformer, an equation for the reflected impedance can be developed. This relationship is

$$Z_P = \left(\frac{N_P}{N_S}\right)^2 \times Z_S \qquad \textbf{(19–10)}$$

where Z_P = primary impedance and Z_S = secondary impedance (see Fig. 19–17). If the turns ratio N_P/N_S is greater than 1, Z_S will be stepped up in value. Conversely, if

MultiSim **Figure 19–17** The secondary load impedance Z_S is reflected back into the primary as a new value that is proportional to the square of the turns ratio, N_P/N_S.

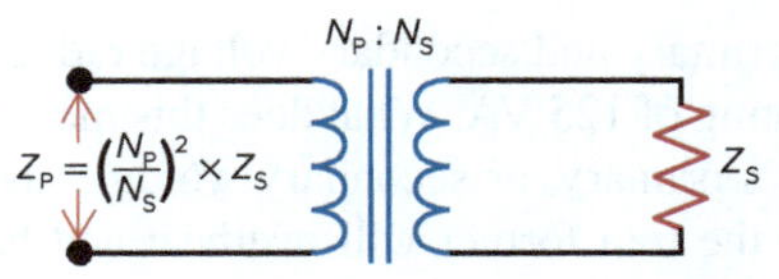

GOOD TO KNOW

If the secondary impedance, Z_S, of a transformer is capacitive or inductive in nature, the reflected impedance will also be capacitive or inductive in nature.

the turns ratio N_P/N_S is less than 1, Z_S will be stepped down in value. It should be noted that the term *impedance* is used rather loosely here, since the primary and secondary impedances may be purely resistive. In the discussions and examples that follow, Z_P and Z_S will be assumed to be purely resistive. The concept of reflected impedance has several practical applications in electronics.

To find the required turns ratio when the impedance ratio is known, rearrange Formula (19–10) as follows:

$$\frac{N_P}{N_S} = \sqrt{\frac{Z_P}{Z_S}} \qquad \textbf{(19–11)}$$

Example 19–16

Determine the primary impedance Z_P for the transformer circuit in Fig. 19–18.

Figure 19–18 Circuit for Example 19–16.

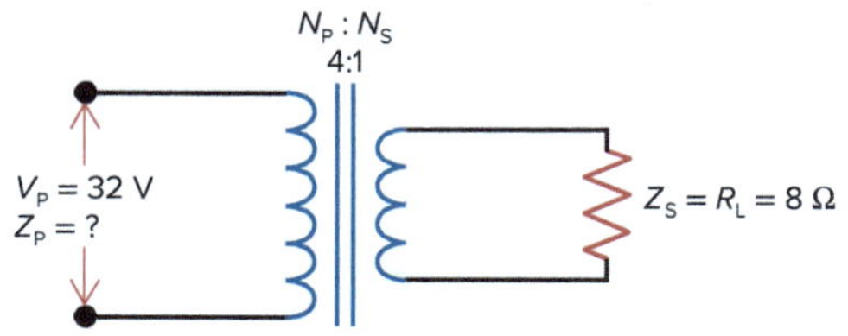

ANSWER Use Formula (19–10). Since $Z_S = R_L$, we have

$$Z_P = \left(\frac{N_P}{N_S}\right)^2 \times R_L$$
$$= \left(\frac{4}{1}\right)^2 \times 8\ \Omega$$
$$= 16 \times 8\ \Omega$$
$$= 128\ \Omega$$

The value of 128 Ω obtained for Z_P using Formula (19–10) can be verified as follows.

$$V_S = \frac{N_S}{N_P} \times V_P$$
$$= \frac{1}{4} \times 32\ \text{V}$$
$$= 8\ \text{V}$$
$$I_S = \frac{V_S}{R_L}$$
$$= \frac{8\ \text{V}}{8\ \Omega}$$
$$= 1\ \text{A}$$
$$I_P = \frac{V_S}{V_P} \times I_S$$
$$= \frac{8\ \text{V}}{32\ \text{V}} \times 1\ \text{A}$$
$$= 0.25\ \text{A}$$

And finally,

$$
\begin{aligned}
Z_P &= \frac{V_P}{I_P} \\
&= \frac{32\ \text{V}}{0.25\ \text{A}} \\
&= 128\ \Omega
\end{aligned}
$$

Example 19–17

In Fig. 19–19, calculate the turns ratio N_P/N_S that will produce a reflected primary impedance Z_P of (a) 75 Ω; (b) 600 Ω.

Figure 19–19 Circuit for Example 19–17.

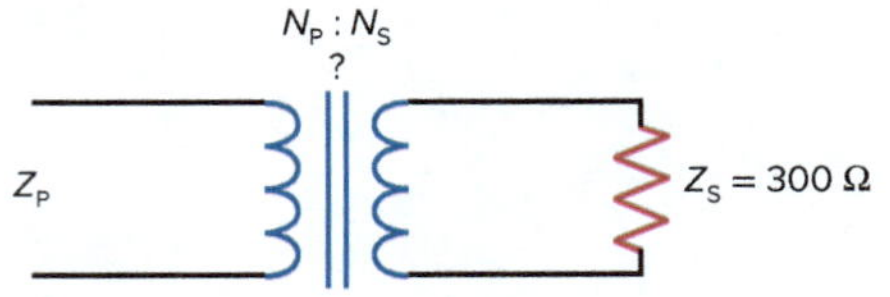

ANSWER (a) Use Formula (19–11).

$$
\begin{aligned}
\frac{N_P}{N_S} &= \sqrt{\frac{Z_P}{Z_S}} \\
&= \sqrt{\frac{75\ \Omega}{300\ \Omega}} \\
&= \sqrt{\frac{1}{4}} \\
&= \frac{1}{2}
\end{aligned}
$$

(b)

$$
\begin{aligned}
\frac{N_P}{N_S} &= \sqrt{\frac{Z_P}{Z_S}} \\
&= \sqrt{\frac{600\ \Omega}{300\ \Omega}} \\
&= \sqrt{\frac{2}{1}} \\
&= \frac{1.414}{1}
\end{aligned}
$$

Impedance Matching for Maximum Power Transfer

Transformers are used when it is necessary to achieve maximum transfer of power from a generator to a load when the generator and load impedances are not the same. This application of a transformer is called ***impedance matching***.

As an example, consider the amplifier and load in Fig. 19–20*a*. Notice that the internal resistance r_i of the amplifier is 200 Ω and the load R_L is 8 Ω. If the amplifier

Figure 19–20 Transferring power from an amplifier to a load R_L. (*a*) Amplifier has $r_i = 200\ \Omega$ and $R_L = 8\ \Omega$. (*b*) Connecting the amplifier directly to R_L. (*c*) Using a transformer to make the 8-Ω R_L appear like 200 Ω in the primary.

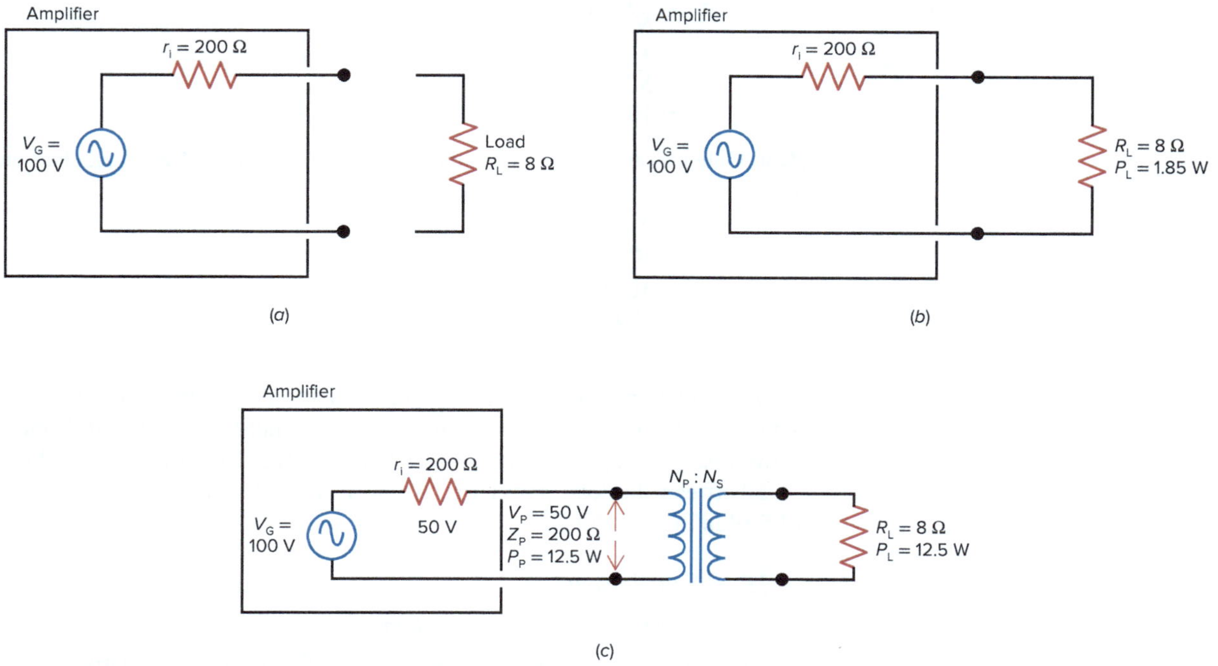

and load are connected directly, as shown in Fig. 19–20*b*, the load receives 1.85 W of power, which is calculated as

$$P_L = \left(\frac{V_G}{r_i + R_L}\right)^2 \times R_L$$
$$= \left(\frac{100\ \text{V}}{200\ \Omega + 8\ \Omega}\right)^2 \times 8\ \Omega$$
$$= 1.85\ \text{W}$$

To increase the power delivered to the load, a transformer can be used between the amplifier and load. This is shown in Fig. 19–20*c*. We know that to transfer maximum power from the amplifier to the load, R_L must be transformed to a value equaling 200 Ω in the primary. With Z_P equaling r_i, maximum power will be delivered from the amplifier to the primary. Since the primary power P_P must equal the secondary power P_S, maximum power will also be delivered to the load R_L. In Fig. 19–20*c*, the turns ratio that provides a Z_P of 200 Ω can be calculated as

$$\frac{N_P}{N_S} = \sqrt{\frac{Z_P}{Z_S}}$$
$$= \sqrt{\frac{200\ \Omega}{8\ \Omega}}$$
$$= \frac{5}{1}$$

With r_i and Z_P equal, the power delivered to the primary can be calculated as

$$P_L = \left(\frac{V_G}{r_i + Z_P}\right)^2 \times Z_P$$
$$= \left(\frac{100\ \text{V}}{400\ \Omega}\right)^2 \times 200\ \Omega$$
$$= 12.5\ \text{W}$$

Since $P_P = P_S$, the load R_L also receives 12.5 W of power. As proof, calculate the secondary voltage.

$$V_S = \frac{N_S}{N_P} \times V_P$$
$$= \frac{1}{5} \times 50 \text{ V}$$
$$= 10 \text{ V}$$

(Notice that V_P is ½ V_G, since r_i and Z_P divide V_G evenly.) Next, calculate the load power P_L.

$$P_L = \frac{V_S^2}{R_L}$$
$$= \frac{10^2 \text{ V}}{8\ \Omega}$$
$$= 12.5 \text{ W}$$

Notice how the transformer has been used as an impedance matching device to obtain the maximum transfer of power from the amplifier to the load. Compare the power dissipated by R_L in Fig. 19–20*b* to that in Fig. 19–20*c*. There is a big difference between the load power of 1.85 W in Fig. 19–20*b* and the load power of 12.5 W in Fig. 19–20*c*.

19–8 Self-Review

Answers at the end of the chapter.

a. The turns ratio of a transformer will not affect the primary impedance Z_P. (True/False)

b. When the turns ratio N_P/N_S is greater than 1, the primary impedance Z_P is less than the value of Z_S. (True/False)

c. If the turns ratio N_P/N_S of a transformer is ²⁄₁ and $Z_S = 50\ \Omega$, the primary impedance $Z_P = 200\ \Omega$. (True/False)

19–9 Core Losses

The fact that the magnetic core can become warm, or even hot, shows that some of the energy supplied to the coil is used up in the core as heat. The two main effects are eddy-current losses and hysteresis losses.

Eddy Currents

In any inductance with an iron core, alternating current induces voltage in the core itself. Since it is a conductor, the iron core has current produced by the induced voltage. This current is called an ***eddy current*** because it flows in a circular path through the cross-section of the core, as illustrated in Fig. 19–21.

Figure 19–21 Cross-sectional view of iron core showing eddy currents.

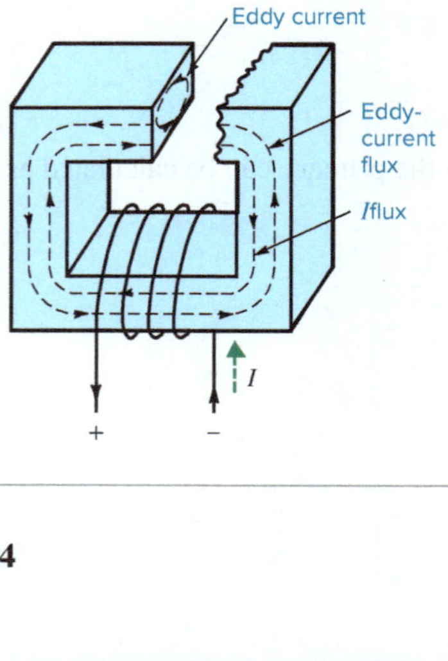

The eddy currents represent wasted power dissipated as heat in the core. Note in Fig. 19–21 that the eddy-current flux opposes the coil flux, so that more current is required in the coil to maintain its magnetic field. The higher the frequency of the alternating current in the inductance, the greater the eddy-current loss.

Eddy currents can be induced in any conductor near a coil with alternating current, not only in its core. For instance, a coil has eddy-current losses in a metal cover. In fact, the technique of induction heating is an application of heat resulting from induced eddy currents.

RF Shielding

The reason that a coil may have a metal cover, usually copper or aluminum, is to provide a shield against the varying flux of rf current. In this case, the shielding

Figure 19–22 Laminated iron core. (*a*) Shell-type construction. (*b*) E- and I-shaped laminations. (*c*) Symbol for iron core.

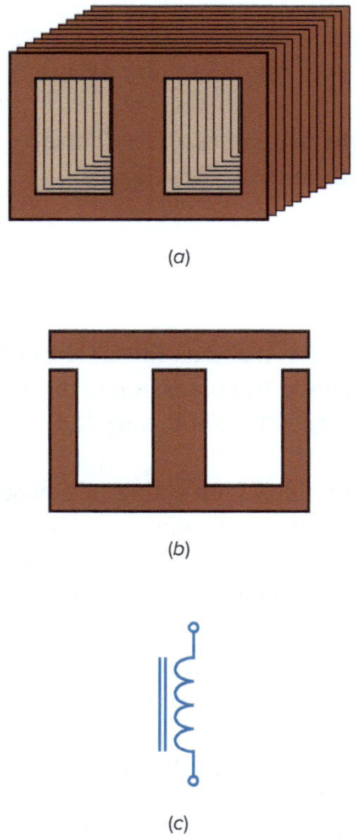

Figure 19–23 RF coils with ferrite core. Width of coil is ½ in. (*a*) Variable *L* from 1 to 3 mH. (*b*) Tuning coil.

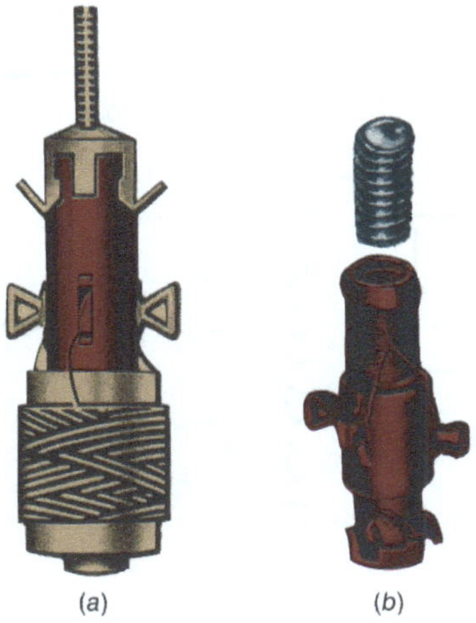

effect depends on using a good conductor for the eddy currents produced by the varying flux, rather than magnetic materials used for shielding against static magnetic flux.

The shield cover not only isolates the coil from external varying magnetic fields but also minimizes the effect of the coil's rf current for external circuits. The reason that the shield helps both ways is the same, as the induced eddy currents have a field that opposes the field that is inducing the current. Note that the clearance between the sides of the coil and the metal should be equal to or greater than the coil radius to minimize the effect of the shield in reducing the inductance.

Hysteresis Losses

Another loss factor present in magnetic cores is hysteresis, although hysteresis losses are not as great as eddy-current losses. The hysteresis losses result from the additional power needed to reverse the magnetic field in magnetic materials in the presence of alternating current. The greater the frequency, the more hysteresis losses.

Air-Core Coils

Note that air has practically no losses from eddy currents or hysteresis. However, the inductance for small coils with an air core is limited to low values in the microhenry or millihenry range.

19–9 Self-Review

Answers at the end of the chapter.

a. Which has greater eddy-current losses, an iron core or an air core?
b. Which produces more hysteresis losses, 60 Hz or 60 MHz ?

19–10 Types of Cores

To minimize losses while maintaining high flux density, the core can be made of laminated steel layers insulated from each other. Insulated powdered-iron granules and ferrite materials can also be used. These core types are illustrated in Figs. 19–22 and 19–23. The purpose is to reduce the amount of eddy currents. The type of steel itself can help reduce hysteresis losses.

Laminated Core

Figure 19–22*a* shows a shell-type core formed with a group of individual laminations. Each laminated section is insulated by a very thin coating of iron oxide, silicon steel, or varnish. The insulating material increases the resistance in the cross-section of the core to reduce the eddy currents but allows a low-reluctance path for high flux density around the core. Transformers for audio frequencies and 60-Hz power are generally made with a laminated iron core.

Powdered-Iron Core

Powdered iron is generally used to reduce eddy currents in the iron core of an inductance for radio frequencies. It consists of individual insulated granules pressed into one solid form called a *slug*.

Ferrite Core

Ferrites are synthetic ceramic materials that are ferromagnetic. They provide high values of flux density, like iron, but have the advantage of being insulators. Therefore, a **ferrite core** can be used for high frequencies with minimum eddy-current losses.

This core is usually a slug that can move in or out of the coil to vary *L*, as in Fig. 19–23*a*. In Fig. 19–23*b*, the core has a hole to fit a plastic alignment tool for tuning the coil. Maximum *L* results with the slug in the coil.

19–10 Self-Review

Answers at the end of the chapter.

a. An iron core provides a coefficient of coupling *k* of unity or 1. (True/False)

b. A laminated iron core reduces eddy-current losses. (True/False)

c. Ferrites have less eddy-current losses than iron. (True/False)

19–11 Variable Inductance

The inductance of a coil can be varied by one of the methods illustrated in Fig. 19–24. In Fig. 19–24*a*, more or fewer turns can be used by connection to one of the taps on the coil. Also, in Fig. 19–24*b*, a slider contacts the coil to vary the number of turns used. These methods are for large coils.

Figure 19–24*c* shows the schematic symbol for a coil with a slug of powdered iron or ferrite. The dotted lines indicate that the core is not solid iron. The arrow shows that the slug is variable. Usually, an arrow at the top means that the adjustment is at the top of the coil. An arrow at the bottom, pointing down, shows that the adjustment is at the bottom.

The symbol in Fig. 19–24*d* is a *variometer*, which is an arrangement for varying the position of one coil within the other. The total inductance of the series-aiding coils is minimum when they are perpendicular.

For any method of varying *L*, the coil with an arrow in Fig. 19–24*e* can be used. However, an adjustable slug is usually shown as in Fig. 19–24*c*.

A practical application of variable inductance is the ***Variac.*** The Variac is an autotransformer with a variable tap to change the turns ratio. The output voltage in the secondary can be varied from 0 to approximately 140 V, with input from the 120-V, 60-Hz power line. One use is to test equipment with voltage above or below the normal line voltage.

The Variac is plugged into the power line, and the equipment to be tested is plugged into the Variac. Note that the power rating of the Variac should be equal to or more than the power used by the equipment being tested. Figure 19–25 shows a Variac with an isolated output.

GOOD TO KNOW

A Variac is a common piece of test equipment used by technicians. It allows the technician to increase the AC voltage slowly while monitoring the operation of the equipment being repaired.

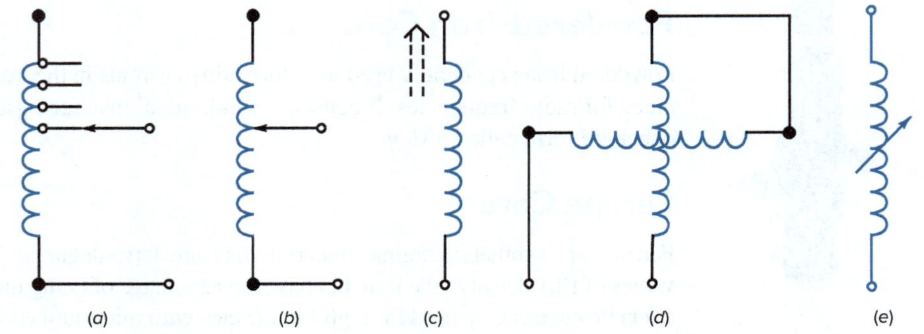

Figure 19–24 Methods of varying inductance. (*a*) Tapped coil. (*b*) Slider contact. (*c*) Adjustable slug. (*d*) Variometer. (*e*) Symbol for variable *L*.

Figure 19–25 Variac with isolated output.

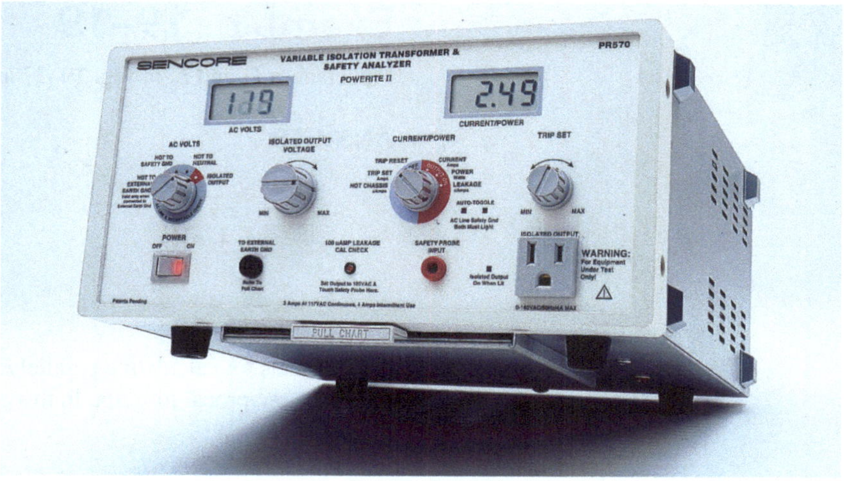

Courtesy of Sencore, Inc.

■ 19–11 *Self-Review*

Answers at the end of the chapter.

a. A Variac is a transformer with a variable secondary voltage. (True/False)

b. Figure 19–24*c* shows a ferrite or powdered-iron core. (True/False)

19–12 Inductances in Series or Parallel

Figure 19–26 Inductances L_1 and L_2 in series without mutual coupling.

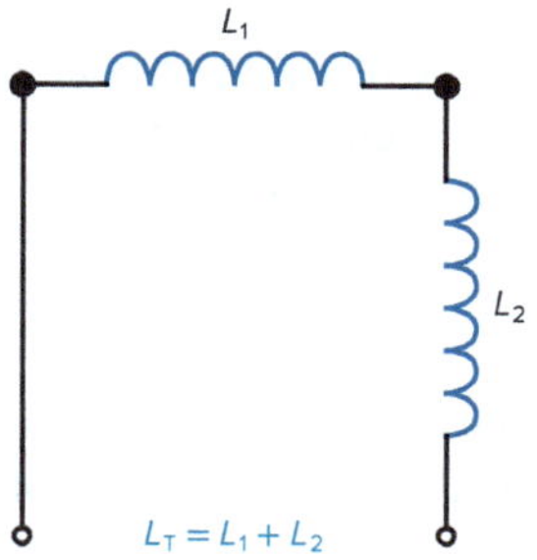

As shown in Fig. 19–26, the total inductance of coils connected in series is the sum of the individual *L* values, as for series *R*. Since the series coils have the same current, the total induced voltage is a result of the total number of turns. Therefore, total series inductance is,

$$L_T = L_1 + L_2 + L_3 + \cdots + \text{etc.} \qquad \textbf{(19–12)}$$

where L_T is in the same units of inductance as L_1, L_2, and L_3. This formula assumes no mutual induction between the coils.

Example 19–18

Inductance L_1 in Fig. 19–26 is 5 mH and L_2 is 10 mH. How much is L_T?

ANSWER $L_T = 5 \text{ mH} + 10 \text{ mH} = 15 \text{ mH}.$

Figure 19–27 Inductances L_1 and L_2 in parallel without mutual coupling.

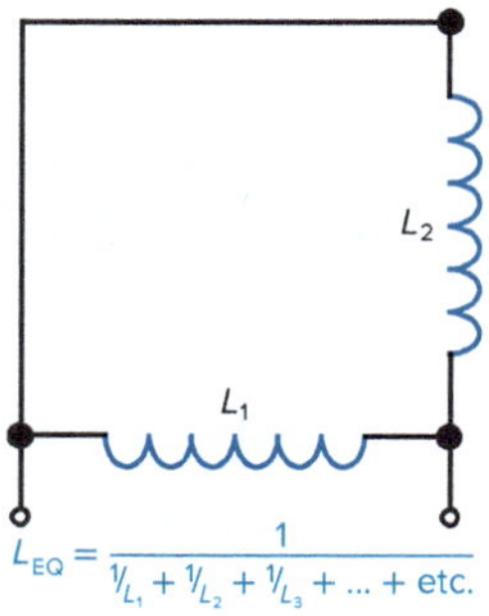

With coils connected in parallel, the combined equivalent inductance is calculated from the reciprocal formula

$$L_{EQ} = \frac{1}{\dfrac{1}{L_1} + \dfrac{1}{L_2} + \dfrac{1}{L_3} + \cdots + \text{etc.}} \qquad \textbf{(19–13)}$$

Again, no mutual induction is assumed, as illustrated in Fig. 19–27.

Example 19–19

Inductances L_1 and L_2 in Fig. 19–27 are each 8 mH. How much is L_{EQ}?

ANSWER

$$L_{EQ} = \frac{1}{1/8 + 1/8}$$
$$= 4 \text{ mH}$$

All shortcuts for calculating parallel R can be used with parallel L, since both are based on the reciprocal formula. In this example, L_{EQ} is $1/2 \times 8 = 4$ mH.

Series Coils with L_M

This depends on the amount of mutual coupling and on whether the coils are connected series-aiding or **series-opposing**. *Series-aiding* means that the common current produces the same direction of magnetic field for the two coils. The *series-opposing* connection results in opposite fields.

The coupling depends on the coil connections and direction of winding. Reversing either one reverses the field. Inductances L_1 and L_2 with the same direction of winding are connected series-aiding in Fig. 19–28*a*. However, they are series-opposing in Fig. 19–28*b* because L_1 is connected to the opposite end of L_2. To calculate the total inductance of two coils that are series-connected and have mutual inductance,

$$L_T = L_1 + L_2 \pm 2L_M \qquad \textbf{(19–14)}$$

The mutual inductance L_M is plus, increasing the total inductance, when the coils are series-aiding, or minus when they are series-opposing to reduce the total inductance.

Note the phasing dots above the coils in Fig. 19–28. Coils with phasing dots at the same end have the same direction of winding. When current enters the dotted ends for two coils, their fields are aiding and L_M has the same sense as L.

GOOD TO KNOW

It is quite complicated to determine the combined equivalent inductance, L_{EQ}, when mutual inductance exists between parallel-connected inductors.

How to Measure L_M

Formula (19–14) provides a method of determining the mutual inductance between two coils L_1 and L_2 of known inductance. First, the total inductance is measured for

Figure 19–28 Inductances L_1 and L_2 in series but with mutual coupling L_M. (*a*) Aiding magnetic fields. (*b*) Opposing magnetic fields.

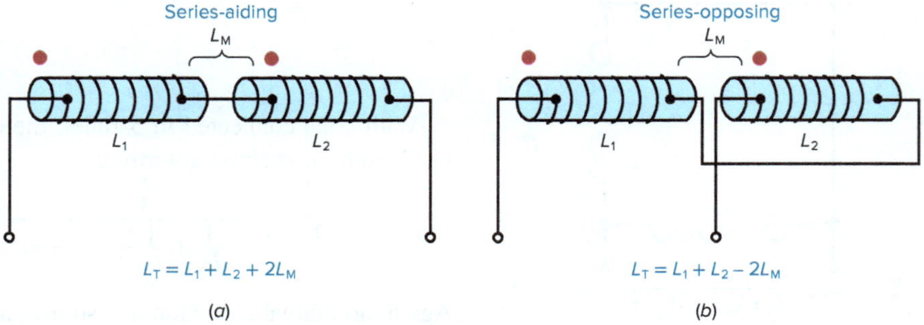

the series-aiding connection. Let this be L_{T_a}. Then the connections to one coil are reversed to measure the total inductance for the series-opposing coils. Let this be L_{T_o}. Then

$$L_M = \frac{L_{T_a} - L_{T_o}}{4} \tag{19–15}$$

When the mutual inductance is known, the coefficient of coupling k can be calculated from the fact that $L_M = k\sqrt{L_1L_2}$.

Example 19–20

Two series coils, each with an L of 250 μH, have a total inductance of 550 μH connected series-aiding and 450 μH series-opposing. (a) How much is the mutual inductance L_M between the two coils? (b) How much is the coupling coefficient k?

ANSWER

(a) $$L_M = \frac{L_{T_a} - L_{T_o}}{4}$$

$$= \frac{550 - 450}{4} = \frac{100}{4}$$

$$= 25\ \mu\text{H}$$

(b) $$L_M = k\sqrt{L_1L_2}, \quad \text{or}$$

$$k = \frac{L_M}{\sqrt{L_1L_2}} = \frac{25}{\sqrt{250 \times 250}}$$

$$= \frac{25}{250} = \frac{1}{10}$$

$$= 0.1$$

Coils may also be in parallel with mutual coupling. However, the inverse relations with parallel connections and the question of aiding or opposing fields make this case complicated. Actually, it would hardly ever be used.

19–12 Self-Review

Answers at the end of the chapter.

a. A 500-μH coil and a 1-mH coil are in series without L_M. Calculate L_T.
b. The same coils are in parallel without L_M. Calculate L_{EQ}.

19–13 Energy in a Magnetic Field of Inductance

The magnetic flux of the current in an inductance has electric energy supplied by the voltage source producing the current. The energy is stored in the magnetic field, since it can do the work of producing induced voltage when the flux moves. The amount of electric energy stored is

$$\text{Energy} = \mathscr{E} = \tfrac{1}{2}LI^2 \tag{19–16}$$

The factor of ½ gives the average result of I in producing energy. With L in henrys and I in amperes, the energy is in watt-seconds, or *joules*. For a 10-H L with a 3-A I, the electric energy stored in the magnetic field equals

$$\text{Energy} = \tfrac{1}{2}\,LI^2 = \frac{10 \times 9}{2} = 45 \text{ J}$$

This 45 J of energy is supplied by the voltage source that produces 3 A in the inductance. When the circuit is opened, the magnetic field collapses. The energy in the collapsing magnetic field is returned to the circuit in the form of induced voltage, which tends to keep the current flowing.

The entire 45 J is available for the work of inducing voltage, since no energy is dissipated by the magnetic field. With resistance in the circuit, however, the I^2R loss with induced current dissipates all energy after a period of time.

CALCULATOR

To do Example 19–21 on a calculator, first square the I, multiply by L, and divide by 2. Specifically, punch in 1.2 and push the x^2 key for 1.44. While this is on the display, push the × key, punch in 0.4, and push the = key for 0.576 on the display. Now push the ÷ key, punch in 2, and then push the = key to get 0.288 as the answer.

Example 19–21

A current of 1.2 A flows in a coil with an inductance of 0.4 H. How much energy is stored in the magnetic field?

ANSWER

$$\text{Energy} = \frac{LI^2}{2} = \frac{0.4 \times 1.44}{2}$$
$$= 0.288 \text{ J}$$

19–13 Self-Review

Answers at the end of the chapter.

a. What is the unit of electric energy stored in a magnetic field?

b. Does a 4-H coil store more or less energy than a 2-H coil for the same current?

19–14 Stray Capacitive and Inductive Effects

Stray capacitive and inductive effects can occur in all circuits with all types of components. A capacitor has a small amount of inductance in the conductors. A coil has some capacitance between windings. A resistor has a small amount of inductance and capacitance. After all, physically a capacitance is simply an insulator between two conductors having a difference of potential. An inductance is basically a conductor carrying current.

However, these stray effects are usually quite small, compared with the concentrated or lumped values of capacitance and inductance. Typical values of **stray capacitance** may be 1 to 10 pF, whereas **stray inductance** is usually a fraction of 1 μH. For very high radio frequencies, however, when small values of L and C must be used, the stray effects become important. As another example, any wire cable has capacitance between the conductors.

A practical case of problems caused by stray L and C is a long cable used for rf signals. If the cable is rolled in a coil to save space, a serious change in the electrical characteristics of the line will take place. Specifically, for twin-lead or coaxial cable feeding the antenna input to a television receiver, the line should not be coiled because the added L or C can affect the signal. Any excess line should be cut off, leaving the little slack that may be needed. This precaution is not so important with audio cables.

Stray Circuit Capacitance

The wiring and components in a circuit have capacitance to the metal chassis. This stray capacitance C_S is typically 5 to 10 pF. To reduce C_S, the wiring should be short with the leads and components placed high off the chassis. Sometimes, for very high frequencies, stray capacitance is included as part of the circuit design. Then changing the placement of components or wiring affects the circuit operation. Such critical *lead dress* is usually specified in the manufacturer's service notes.

Stray Inductance

Although practical inductors are generally made as coils, all conductors have inductance. The amount of L is $v_L/(di/dt)$, as with any inductance producing induced voltage when the current changes. The inductance of any wiring not included in the conventional inductors can be considered stray inductance. In most cases, stray inductance is very small; typical values are less than 1 μH. For high radio frequencies, though, even a small L can have an appreciable inductive effect.

One source of stray inductance is connecting leads. A wire 0.04 in. in diameter and 4 in. long has an L of approximately 0.1 μH. At low frequencies, this inductance is negligible. However, consider the case of rf current, where i varies from 0- to 20-mA peak value, in the short time of 0.025 μs, for a quarter-cycle of a 10-MHz sine wave. Then v_L equals 80 mV, which is an appreciable inductive effect. This is one reason that connecting leads must be very short in rf circuits.

As another example, wire-wound resistors can have appreciable inductance when wound as a straight coil. This is why carbon resistors are preferred for minimum stray inductance in rf circuits. However, noninductive wire-wound resistors can also be used. These are wound so that adjacent turns have current in opposite directions and the magnetic fields oppose each other to cancel the inductance. Another application of this technique is twisting a pair of connecting leads to reduce the inductive effect.

Inductance of a Capacitor

Capacitors with coiled construction, particularly paper and electrolytic capacitors, have some internal inductance. The larger the capacitor, the greater its series inductance. Mica and ceramic capacitors have very little inductance, however, which is why they are generally used for radio frequencies.

For use above audio frequencies, the rolled-foil type of capacitor must have noninductive construction. This means that the start and finish of the foil winding must not be the terminals of the capacitor. Instead, the foil windings are offset. Then one terminal can contact all layers of one foil at one edge, and the opposite edge of the other foil contacts the second terminal. Most rolled-foil capacitors, including the paper and film types, are constructed this way.

Figure 19–29 Equivalent circuit of an rf coil. (*a*) Distributed capacitance C_d between turns of wire. (*b*) Equivalent circuit.

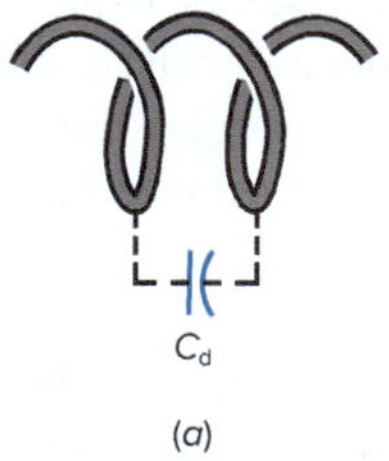

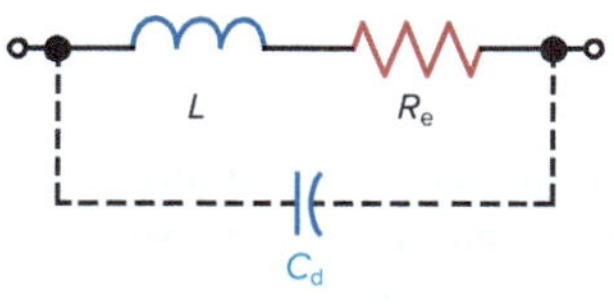

Distributed Capacitance of a Coil

As illustrated in Fig. 19–29, a coil has distributed capacitance C_d between turns. Note that each turn is a conductor separated from the next turn by an insulator, which is the definition of capacitance. Furthermore, the potential of each turn is different from the next, providing part of the total voltage as a potential difference to charge C_d. The result then is the equivalent circuit shown for an rf coil. The L is the inductance and R_e its internal effective AC resistance in series with L, and the total distributed capacitance C_d for all turns is across the entire coil.

Special methods for minimum C_d include *space-wound* coils, where the turns are spaced far apart; the honeycomb or *universal* winding, with the turns crossing each other at right angles; and the *bank winding*, with separate sections called *pies*. These windings are for rf coils. In audio and power transformers, a grounded conductor shield, called a *Faraday screen*, is often placed between windings to reduce capacitive coupling.

Reactive Effects in Resistors

Figure 19–30 High-frequency equivalent circuit of a resistor.

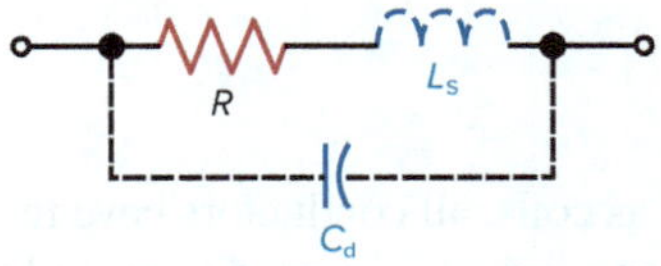

As illustrated by the high-frequency equivalent circuit in Fig. 19–30, a resistor can include a small amount of inductance and capacitance. The inductance of carbon-composition resistors is usually negligible. However, approximately 0.5 pF of capacitance across the ends may have an effect, particularly with large resistances used for high radio frequencies. Wire-wound resistors definitely have enough inductance to be evident at radio frequencies. However, special resistors are available with double windings in a noninductive method based on cancellation of opposing magnetic fields.

Capacitance of an Open Circuit

An open switch or a break in a conducting wire has capacitance C_O across the open. The reason is that the open consists of an insulator between two conductors. With a voltage source in the circuit, C_O charges to the applied voltage. Because of the small C_O, of the order of picofarads, the capacitance charges to the source voltage in a short time. This charging of C_O is the reason that an open series circuit has the applied voltage across the open terminals. After a momentary flow of charging current, C_O charges to the applied voltage and stores the charge needed to maintain this voltage.

■ *19–14 Self-Review*

Answers at the end of the chapter.

a. A two-wire cable has distributed *C* between the conductors. (True/False)

b. A coil has distributed *C* between the turns. (True/False)

c. Stray inductance and stray capacitance are most likely to be a problem at high frequencies. (True/False)

19–15 Measuring and Testing Inductors

Figure 19–31 Typical LCR meter.

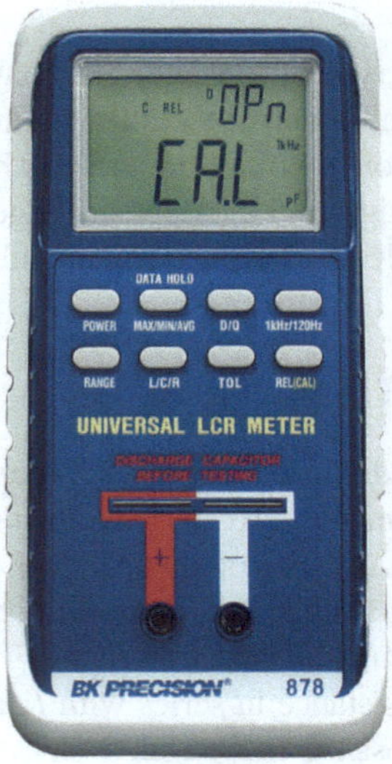

Courtesy of B&K Precision Corporation

Although many DMMs are capable of measuring the value of a capacitor, few are capable of measuring the value of an inductor. Therefore, when it is necessary to measure the value of an inductor, you may want to use a capacitor-inductor analyzer like that shown earlier in Chapter 16. The capacitor-inductor analyzer can also test the quality (*Q*) of the inductor by using something called a *ringing test*.

Another test instrument that is capable of measuring inductance *L*, capacitance *C*, and resistance *R*, is an LCR meter. A typical LCR meter is shown in Fig. 19–31. Although this is a handy piece of test equipment, most LCR meters are not capable of measuring anything except the value of a component. Note, however, that some LCR meters are capable of making a few additional tests besides measuring the component value.

Inductor Coding

Inductors may or may not be coded to indicate their inductance value in henrys (H), millihenrys (mH), or microhenrys (μH). Some very small inductors used in rf circuits may consist of five or six turns of bare wire and therefore cannot be coded. Larger inductors, such as chokes used in the audio-frequency range, normally have their inductance values printed on them. Some inductors use a coding system similar to that used with film capacitors. In this case, a three-digit code is used to indicate the inductance value in microhenrys. For example, an inductor may be coded 103; this is interpreted as follows: The first two digits (1 and 0) represent the first and second digits in the inductance value. The last digit (3), called the *multiplier digit*, tells how many zeros to add after the first two digits. In this case, 103 corresponds to an inductance of 10,000 μH. Some manufacturers put the multiplier digit first instead of last. For example, for an inductor coded 210, the second and third digits represent the first and second digits of the inductance value, and the first digit tells how many zeros to add. In this case, the code 210 corresponds to an inductance value of 1000 μH. Usually the three-digit codes include no tolerance rating. Sometimes inductors have their value printed on the body, sometimes they have colored stripes, and sometimes they are not coded at all. Confusing you say? Absolutely! Sometimes, the only sure way to determine the value of an inductor is to measure its value. Because no standardization is in place for the coding of inductors, no further coverage of the topic is provided here.

Troubles in Coils

Figure 19–32 An open coil reads infinite ohms when its continuity is checked with an ohmmeter.

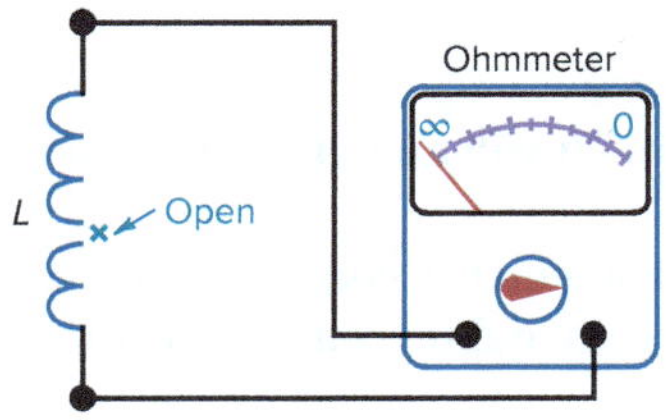

The most common trouble in coils is an open winding. As illustrated in Fig. 19–32, an ohmmeter connected across the coil reads infinite resistance for the open circuit. It does not matter whether the coil has an air core or an iron core. Since the coil is open, it cannot conduct current and therefore has no inductance because it cannot produce induced voltage. When the resistance is checked, the coil should be disconnected from the external circuit to eliminate any parallel paths that could affect the resistance readings.

Direct Current Resistance of a Coil

A coil has DC resistance equal to the resistance of the wire used in the winding. The amount of resistance is less with heavier wire and fewer turns. For rf coils with inductance values up to several millihenrys, requiring 10 to 100 turns of fine wire, the DC resistance is 1 to 20 Ω, approximately. Inductors for 60 Hz and audio frequencies with several hundred turns may have resistance values of 10 to 500 Ω, depending on the wire size.

As shown in Fig. 19–33, the DC resistance and inductance of a coil are in series, since the same current that induces voltage in the turns must overcome the resistance of the wire. Although resistance has no function in producing induced voltage, it is useful to know the DC coil resistance because if it is normal, usually the inductance can also be assumed to have its normal value.

GOOD TO KNOW

Two inductors can have identical inductance values even though the DC resistance of the wire used to wind the inductors is different.

Open Coil

An open winding has infinite resistance, as indicated by an ohmmeter reading. With a transformer that has four leads or more, check the resistance across the two leads for the primary, across the two leads for the secondary, and across any other pairs of leads for additional secondary windings. For an autotransformer with three leads, check the resistance from one lead to each of the other two.

When the open circuit is inside the winding, it is usually not practical to repair the coil, and the entire unit is replaced. In some cases, an open connection at the terminals can be resoldered.

Figure 19–33 The internal DC resistance r_i of a coil is in series with its inductance L.

Value Change

The value of an inductor can change over time because of core breakage, windings relaxing, or shorted turns. Note that a coil whose inductance value is changed may check okay with an ohmmeter. To check the value of an inductor, use either a capacitor-inductor analyzer or an LCR meter.

Open Primary Winding

When the primary of a transformer is open, no primary current can flow, and no voltage is induced in any of the secondary windings.

Open Secondary Winding

When the secondary of a transformer is open, it cannot supply power to any load resistance across the open winding. Furthermore, with no current in the secondary, the primary current is also practically zero, as though the primary winding were open. The only primary current needed is the small magnetizing current to sustain the field producing induced voltage across the secondary without any load. If the transformer has several secondary windings, however, an open winding in one secondary does not affect transformer operation for the secondary circuits that are normal.

GOOD TO KNOW

Except for autotransformers, the DC resistance between the primary and secondary windings should always measure infinite ohms.

Short across Secondary Winding

In this case, excessive primary current flows, as though it were short-circuited, often burning out the primary winding. The reason is that the large secondary current has a strong field that opposes the flux of the self-induced voltage across the primary, making it draw more current from the generator.

19–15 Self-Review

Answers at the end of the chapter.

a. **The normal *R* of a coil is 18 Ω. How much will an ohmmeter read if the coil is open?**
b. **The primary of a 1:3 step-up autotransformer is connected to a 120-V_{AC} power line. How much will the secondary voltage be if the primary is open?**
c. **Are the DC resistance and inductance of a coil in series or in parallel?**

Application in Understanding Isolation Transformers

Any transformer with separate primary and secondary windings is technically an ***isolation transformer***. With two separate windings, the resistance between the primary and secondary is infinitely high. In other words, there is no continuity between the two windings. Although the primary and secondary windings are not electrically connected to each other, the magnetic field linking the two windings transfers power from the AC voltage source in the primary to the load in the secondary. Ideally, the transformer itself does not dissipate any power. Realistically, however, a transformer always dissipates some power due to the I^2R losses in the windings as well as hysteresis and eddy-current losses in the core. The losses associated with a transformer are the reason why it can become warm during normal use or operation.

Although any transformer with separate primary and secondary windings is technically an isolation transformer, the name ***isolation transformer*** is usually reserved for iron-core transformers with a 1:1 turn's ratio. This is because the main function of this type of transformer is not to step up or step down the primary voltage but rather to isolate the earth ground connection of the primary's voltage source from the load in the secondary. Figure 19–34 shows the 120-V_{AC} power line voltage connected to the primary of an isolation transformer. Because the transformer has a 1:1 turn's ratio, the secondary voltage is also 120 V. In Fig. 19–34, notice that the upper primary lead is connected to the black hot wire of the 120-V_{AC} power line, whereas the bottom primary lead is connected to the white neutral wire, which is grounded. Although the voltage across both secondary leads measures 120 V, the voltage measured from either secondary lead to ground is 0 V. This is because the secondary voltage of an isolation transformer is not referenced to earth ground like the 120-V_{AC} power line is in the primary. If a person standing on earth ground accidentally comes into contact with the black hot wire of the 120-V_{AC} power line, as in Fig. 19–35*a*, he or she could receive a severe electric shock and possibly risk being electrocuted. In Fig. 19–35*b*, however, the secondary of the isolation transformer is not referenced to earth ground. Therefore, if a person accidentally comes into contact

Figure 19–34 The primary winding of an isolation transformer is connected to the 120-V_{AC} power line. Notice the earth ground connection of the white neutral wire. The primary voltage of 120 V is measured with respect to earth ground. The measured voltage across the secondary winding is also 120 V, but the voltage measured from either secondary lead to ground is 0 V.

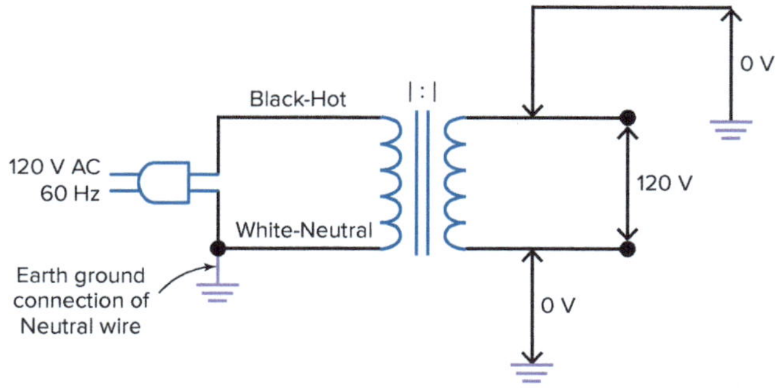

Figure 19–35 This illustration shows how an isolation transformer lessens the likelihood of an electric shock. (*a*) Accidently touching the hot wire of the 120-V_{AC} power line while in contact with earth ground results in a severe electric shock or even electrocution. (*b*) Touching only one secondary lead of an isolation transformer while in contact with earth ground does not result in an electric shock.

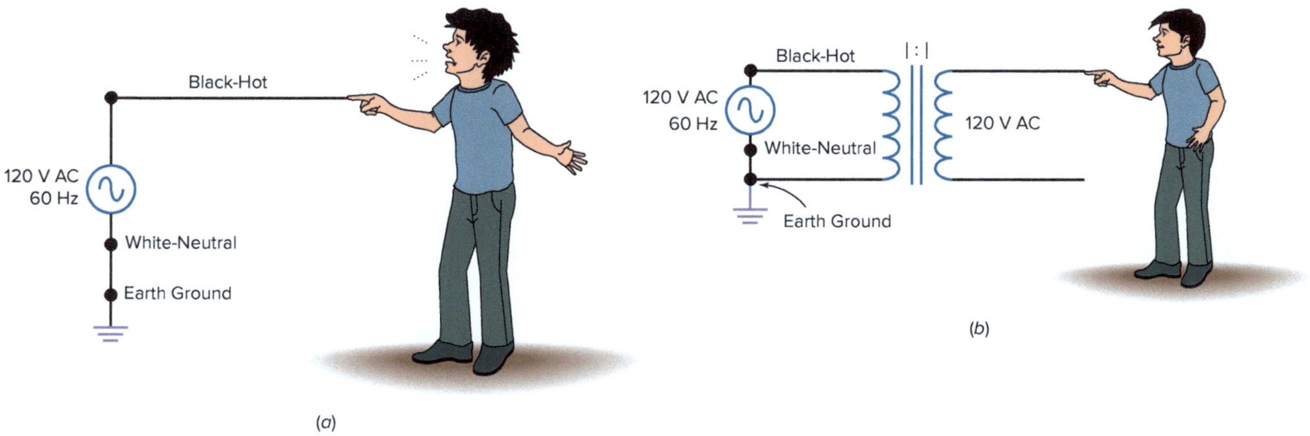

with just one lead of the transformer secondary, there is no path back to earth ground to complete the circuit and as a result, no electric shock.

If a technician is troubleshooting and/or servicing an electronic product and the product is plugged directly into the 120-V_{AC} power line, they are at a high risk of receiving an electric shock or being electrocuted if their body accidentally comes into contact with a live circuit. To reduce the risk of an electric shock, electronic technicians often use an isolation transformer when troubleshooting and/or servicing electronic equipment. The primary side of the isolation transformer is plugged directly into the 120-V_{AC} power line, and the electronic product being serviced is plugged into the secondary side of the transformer. If just one hand of the technician accidently touches a live circuit in the equipment plugged into the secondary, there is no chance of an electric shock because there is no path back to earth ground. It should be noted, however, that a technician can still receive a lethal electric shock even if an isolation transformer is used. For example, if both hands of a technician are across two points having a significant amount of potential difference, current will flow through the body between the two hands.

Therefore, to avoid an electric shock when making voltage measurements in a live circuit, it is always best to keep one hand in your pocket, or behind your back.

It should be noted that an isolation transformer is built using special insulation between the primary and secondary windings. As a result, isolation transformers can withstand a higher than normal voltage between the two windings without breaking down and conducting. Isolation transformers are also designed to minimize any capacitive coupling between the primary and secondary, which could cause AC currents to flow between the two windings.

It is also important to note that not all isolation transformers are designed to be used in conjunction with the 120-V_{AC} power line. Isolation transformers are also used in home theater systems, for example, to reduce any hum and noise associated with one or more ground loops in the system. In this case, the isolation transformer is much smaller and is designed to work at much higher frequencies than 60 Hz.

Summary

- Varying current induces voltage in a conductor, since the expanding and collapsing field of the current is equivalent to flux in motion.
- Lenz's law states that the induced voltage produces *I* that opposes the change in current causing the induction. Inductance, therefore, tends to keep the current from changing.
- The ability of a conductor to produce induced voltage across itself when the current varies is its self-inductance, or inductance. The symbol is *L*, and the unit of inductance is the henry. One henry of inductance allows 1 V to be induced when the current changes at the rate of 1 A/s. For smaller units, 1 mH = 1×10^{-3} H and 1 μH = 1×10^{-6} H.
- To calculate self-induced voltage, $v_L = L(di/dt)$, with *v* in volts, *L* in henrys, and di/dt in amperes per second.
- Mutual inductance is the ability of varying current in one conductor to induce voltage in another conductor nearby. Its symbol is L_M, measured in henrys. $L_M = k\sqrt{L_1L_2}$, where *k* is the coefficient of coupling between conductors.
- A transformer consists of two or more windings with mutual inductance. The primary winding connects to the source voltage; the load resistance is connected across the secondary winding. A separate winding is an isolated secondary.

The transformer is used to step up or step down AC voltage.

- An autotransformer is a tapped coil, used to step up or step down the primary voltage. There are three leads with one connection common to both the primary and the secondary.
- A transformer with an iron core has essentially unity coupling. Therefore, the voltage ratio is the same as the turns ratio: $V_P/V_S = N_P/N_S$.
- Assuming 100% efficiency for an iron-core power transformer, the power supplied to the primary equals the power used in the secondary.
- The voltage rating of a transformer's secondary is always specified under full load conditions with the rated primary voltage applied. The measured voltage across an unloaded secondary is usually 5 to 10% higher than its rated value.
- The current or power rating of a transformer is usually specified only for the secondary windings.
- Transformers can be used to reflect a secondary load impedance back into the primary as a new value that is either larger or smaller than its actual value. The primary impedance Z_P can be determined using Formula (19–10).
- The impedance transforming properties of a transformer make it possible to obtain maximum transfer of power from a generator to a load when the generator and load impedances are not equal. The required turns ratio can be determined using Formula (19–11).
- Eddy currents are induced in the iron core of an inductance, causing wasted power that heats the core. Eddy-current losses increase with higher frequencies of alternating current. To reduce eddy currents, the iron core is laminated. Powdered-iron and ferrite cores have minimum eddy-current losses at radio frequencies. Hysteresis also causes power loss.
- With no mutual coupling, series inductances are added like series resistances. The combined equivalent inductance of parallel connected inductances is calculated by the reciprocal formula, as for parallel resistances.
- The magnetic field of an inductance has stored energy $\mathscr{E} = \frac{1}{2} LI^2$. With *I* in amperes and *L* in henrys, energy $\mathscr{E}$ is in joules.
- In addition to its inductance, a coil has DC resistance equal to the resistance of the wire in the coil. An open coil has infinitely high resistance.
- An open primary in a transformer results in no induced voltage in any of the secondary windings.
- Figure 19–36 summarizes the main types of inductors, or coils, with their schematic symbols.

Figure 19–36 Summary of types of inductors. (*a*) Air-core coil. (*b*) Iron-core coil. (*c*) Adjustable ferrite core. (*d*) Air-core transformer. (*e*) Variable L_P and L_S. (*f*) Iron-core transformer. (*g*) Autotransformer.

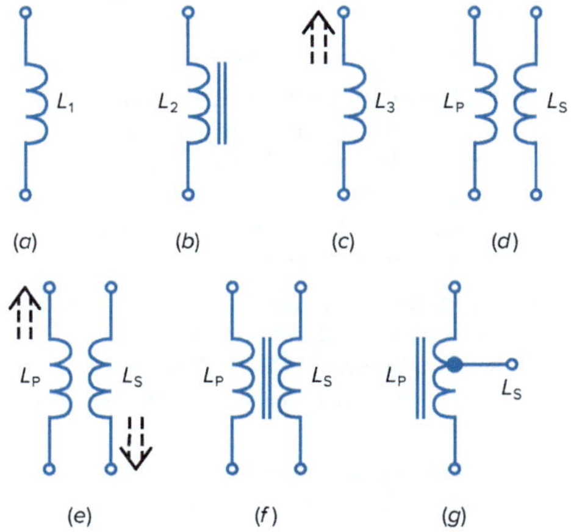

Table 19–1	Comparison of Capacitance and Inductance
Capacitance	**Inductance**
Symbol is *C*	Symbol is *L*
Unit is the farad (F)	Unit is the henry (H)
Needs dielectric as insulator	Needs conductor for circuit path
More plate area allows more *C*	More turns allow more *L*
Dielectric can concentrate electric field for more *C*	Core can concentrate magnetic field for more *L*
$C_{EQ} = \dfrac{1}{\frac{1}{C_1} + \frac{1}{C_2} + \cdots + \text{etc.}}$ in series	$L_T = L_1 + L_2$ in series
$C_T = C_1 + C_2$ in parallel	$L_{EQ} = \dfrac{1}{1/L_1 + 1/L_2}$ in parallel

- The characteristics of inductance and capacitance are compared in Table 19–1.
- Stray inductance can be considered the inductance of any wiring not included in conventional inductors. Stray capacitance can be considered the capacitance of any two conductors separated from each other by an insulator and not included in conventional capacitors.

Important Terms

Autotransformer — a transformer made of one continuous coil with a tapped connection between the end terminals. An autotransformer has only three leads and provides no isolation between the primary and secondary.

Coefficient of coupling, *k* — the fraction of total flux from one coil linking another coil nearby.

Counter emf (cemf) — a term used to describe the effect of an induced voltage in opposing a change in current.

Eddy current — the current that flows in a circular path through the cross section of the iron core in a transformer.

Efficiency — the ratio of power output to power input. In a transformer, power out is secondary power and power in is primary power.

Ferrite core — a type of core that has high flux density, like iron, but is an insulator. A ferrite core used in a coil has minimum eddy-current losses due to its high resistance.

Henry (H) — the basic unit of inductance. 1 H is the amount of inductance that produces 1 volt of induced voltage when the current varies at the rate of 1 A/s.

Impedance matching — an application of a transformer in which the secondary load resistance is stepped up or down to provide maximum transfer of power from the generator to the load.

Inductance, *L* — the ability of a conductor to produce an induced voltage in itself when the current changes. Inductance can also be defined as the characteristic that opposes any change in current.

Leakage flux — any magnetic field lines that do not link two coils that are close to each other.

Lenz's law — Lenz's law states that the polarity of an induced voltage must be such that it opposes the current that produces the induced voltage.

Mutual inductance, L_M — the ability of one coil to induce a voltage in another coil nearby. Two coils have a mutual inductance, L_M, of 1 H when a current change of 1 A/s in one coil induces 1 V in the other coil.

Phasing dots — dots on the primary and secondary leads of a transformer schematic symbol that identify those leads having the same instantaneous polarity.

Reflected impedance — the term used to describe the transformation of a secondary load resistance to a new value as seen by the primary.

Series-aiding — a connection of coils in which the coil current produces the same direction of magnetic field for both coils.

Series-opposing — a connection of coils in which the coil current produces opposing magnetic fields for each coil.

Stray capacitance — a very small capacitance that exists between any two conductors separated by an insulator. For example, the capacitance can be between two wires in a wiring harness or between a single wire and a metal chassis.

Stray inductance — the small inductance of any length of conductor or component lead. The effects of both stray inductance and stray capacitance are most noticeable at very high frequencies.

Transformer — a device that uses the concept of mutual inductance to step up or step down an alternating voltage.

Turns ratio — the ratio of the number of turns in the primary to the number of turns in the secondary of a transformer.

Variac — a piece of test equipment that provides a variable output voltage. The Variac is plugged into the 120-V_{AC} power line and the equipment under test is plugged into the Variac. Most Variacs available today have an isolated output.

Volt-ampere (VA) — the unit of apparent power that specifies the power rating of a transformer. The product VA is called the apparent power because it is the power that is apparently used by the transformer.

Related Formulas

$$L = \frac{V_L}{di/dt}$$

$$L = \mu_r \times \frac{N^2 \times A}{l} \times 1.26 \times 10^{-6}\ \text{H}$$

$$v_L = L\frac{di}{dt}$$

$$L_M = k\sqrt{L_1 \times L_2}$$

$$\text{Turns ratio} = \frac{N_P}{N_S}$$

$$\frac{V_P}{V_S} = \frac{N_P}{N_S}$$

$$V_S I_S = V_P I_P$$

$$\frac{I_S}{I_P} = \frac{V_P}{V_S}$$

$$\%\ \text{Efficiency} = \frac{P_{out}}{P_{in}} \times 100$$

$$Z_P = \left(\frac{N_P}{N_S}\right)^2 \times Z_S$$

$$\frac{N_P}{N_S} = \sqrt{\frac{Z_P}{Z_S}}$$

Series connection with no L_M

$$L_T = L_1 + L_2 + L_3 + \cdots + \text{etc.}$$

Series connection with L_M

$$L_T = L_1 + L_2 \pm 2L_M$$

Parallel connection (No L_M)

$$L_{EQ} = \frac{1}{\frac{1}{L_1} + \frac{1}{L_2} + \frac{1}{L_3} + \cdots + \text{etc.}}$$

$$L_M = \frac{L_{T_a} - L_{T_o}}{4}$$

$$\text{Energy} = \mathscr{E} = \tfrac{1}{2} LI^2$$

Self-Test

Answers at the back of the book.

1. **The unit of inductance is the**
   a. henry.
   b. farad.
   c. ohm.
   d. volt-ampere.

2. **The inductance, *L*, of an inductor is affected by**
   a. number of turns.
   b. area enclosed by each turn.
   c. permeability of the core.
   d. all of the above.

3. **A transformer cannot be used to**
   a. step up or down an AC voltage.
   b. step up or down a DC voltage.
   c. match impedances.
   d. transfer power from primary to secondary.

4. **The interaction between two inductors physically close together is called**
   a. counter emf.
   b. self-inductance.
   c. mutual inductance.
   d. hysteresis.

5. **If the secondary current in a step-down transformer increases, the primary current will**
   a. not change.
   b. increase.
   c. decrease.
   d. drop a little.

6. **Inductance can be defined as the characteristic that**
   a. opposes a change in current.
   b. opposes a change in voltage.
   c. aids or enhances any change in current.
   d. stores electric charge.

7. **If the number of turns in a coil is doubled in the same length and area, the inductance, *L*, will**
   a. double.
   b. quadruple.
   c. stay the same.
   d. be cut in half.

8. **An open coil has**
   a. zero resistance and zero inductance.
   b. infinite inductance and zero resistance.
   c. normal inductance but infinite resistance.
   d. infinite resistance and zero inductance.

9. **Two 10-H inductors are connected in series-aiding and have a mutual inductance, L_M, of 0.75 H. The total inductance, L_T, of this combination is**
   a. 18.5 H.
   b. 20.75 H.
   c. 21.5 H.
   d. 19.25 H.

10. **How much is the self-induced voltage, V_L, across a 100-mH inductor produced by a current change of 50,000 A/s?**
   a. 5 kV.
   b. 50 V.
   c. 5 mV.
   d. 500 kV.

11. The measured voltage across an unloaded secondary of a transformer is usually

a. the same as the rated secondary voltage.

b. 5 to 10% higher than the rated secondary voltage.

c. 50% higher than the rated secondary voltage.

d. 5 to 10% lower than the rated secondary voltage.

12. A laminated iron-core transformer has reduced eddy-current losses because

a. the laminations are stacked vertically.

b. more wire can be used with less DC resistance.

c. the magnetic flux is in the air gap of the core.

d. the laminations are insulated from each other.

13. How much is the inductance of a coil that induces 50 V when its current changes at the rate of 500 A/s?

a. 100 mH.

b. 1 H.

c. 100 μH.

d. 10 μH.

14. A 100-mH inductor is in parallel with a 150-mH and a 120-mH inductor. Assuming no mutual inductance between coils, how much is L_{EQ}?

a. 400 mH.

b. 370 mH.

c. 40 mH.

d. 80 mH.

15. A 400-μH coil is in series with a 1.2-mH coil without mutual inductance. How much is L_T?

a. 401.2 μH.

b. 300 μH.

c. 160 μH.

d. 1.6 mH.

16. A step-down transformer has a turns ratio, $\frac{N_P}{N_S}$, of 4:1. If the primary voltage, V_P, is 120 V_{AC}, how much is the secondary voltage, V_S?

a. 480 V_{AC}.

b. 120 V_{AC}.

c. 30 V_{AC}.

d. It cannot be determined.

17. If an iron-core transformer has a turns ratio, $\frac{N_P}{N_S}$, of 3:1 and Z_S = 16 Ω, how much is Z_P?

a. 48 Ω.

b. 144 Ω.

c. 1.78 Ω.

d. 288 Ω.

18. How much is the induced voltage, V_L, across a 5-H inductor carrying a steady DC current of 200 mA?

a. 0 V.

b. 1 V.

c. 100 kV.

d. 120 V_{AC}.

19. The secondary current, I_s, in an iron-core transformer equals 1.8 A. If the turns ratio, $\frac{N_P}{N_S}$, equals 3:1, how much is the primary current, I_P?

a. $I_P = 1.8$ A.

b. $I_P = 600$ mA.

c. $I_P = 5.4$ A.

d. none of the above.

20. For a coil, the DC resistance, r_i, and inductance, L, are

a. in parallel.

b. infinite.

c. the same thing.

d. in series.

Essay Questions

1. Define 1 H of self-inductance and 1 H of mutual inductance.

2. State Lenz's law in terms of induced voltage produced by varying current.

3. Refer to Fig. 19–5. Explain why the polarity of v_L is the same for the examples in Fig. 19–5*a* and *d*.

4. Make a schematic diagram showing the primary and secondary of an iron-core transformer with a 1:6 voltage step-up ratio (a) using an autotransformer; (b) using a transformer with isolated secondary winding. Then (c) with 100 turns in the primary, how many turns are in the secondary for both cases?

5. Define the following: coefficient of coupling, transformer efficiency, stray inductance, and eddy-current losses.

6. Why are eddy-current losses reduced with the following cores: (a) laminated; (b) powdered iron; (c) ferrite?

7. Why is a good conductor used for an rf shield?

8. Show two methods of providing a variable inductance.

9. (a) Why will the primary of a power transformer have excessive current if the secondary is short-circuited? (b) Why is there no voltage across the secondary if the primary is open?

10. (a) Describe briefly how to check a coil for an open winding with an ohmmeter. Which ohmmeter range should be used? (b) Which leads will be checked on an autotransformer with one secondary and a transformer with two isolated secondary windings?

11. Derive the formula $L_M = (L_{T_a} - L_{T_o})/4$ from the fact that $L_{T_a} = L_1 + L_2 + 2L_M$ and $L_{T_o} = L_1 + L_2 - 2L_M$.

12. Explain how a transformer with a 1:1 turns ratio and an isolated secondary can be used to reduce the chance of electric shock from the 120-V_{AC} power line.

13. Explain the terms *stray inductance* and *stray capacitance* and give an example of each.

Problems

SECTION 19–1 INDUCTION BY ALTERNATING CURRENT

19–1 Which can induce more voltage in a conductor, a steady DC current of 10 A or a small current change of 1 to 2 mA?

19–2 Examine the sine wave of alternating current in Fig. 19–1. Identify the points on the waveform (using the letters A–I) where the rate of current change, $\frac{di}{dt}$, is

a. greatest.

b. zero.

19–3 Which will induce more voltage across a conductor, a low-frequency alternating current or a high-frequency alternating current?

SECTION 19–2 SELF-INDUCTANCE *L*

19–4 Convert the following current changes, $\frac{di}{dt}$, to amperes per second:

a. 0 to 3 A in 2 s.

b. 0 to 50 mA in 5 μs.

c. 100 to 150 mA in 5 ms.

d. 150 to 100 mA in 20 μs.

e. 30 to 35 mA in 1 μs.

f. 80 to 96 mA in 0.4 μs.

g. 10 to 11 A in 1 s.

19–5 How much inductance, *L*, will be required to produce an induced voltage, V_L, of 15 V for each of the $\frac{di}{dt}$ values listed in Prob. 19–4?

19–6 How much is the inductance, *L*, of a coil that induces 75 V when the current changes at the rate of 2500 A/s?

19–7 How much is the inductance, *L*, of a coil that induces 20 V when the current changes at the rate of 400 A/s?

19–8 Calculate the inductance, *L*, for the following long coils: (Note: 1 m = 100 cm and 1 m^2 = 10,000 cm^2.)

a. air core, 20 turns, area 3.14 cm^2, length 25 cm.

b. same coil as step a with ferrite core having a μ_r of 5000.

c. air core, 200 turns, area 3.14 cm^2, length 25 cm.

d. air core, 20 turns, area 3.14 cm^2, length 50 cm.

e. iron core with μ_r of 2000, 100 turns, area 5 cm^2, length 10 cm.

19–9 Recalculate the inductance, *L*, in Prob. 19–8a if the number of turns is doubled to 40.

19–10 What is another name for an rf inductor?

SECTION 19–3 SELF-INDUCED VOLTAGE V_L

19–11 How much is the self-induced voltage across a 5-H inductance produced by a current change of 100 to 200 mA in 1 ms?

19–12 How much is the self-induced voltage across a 33-H inductance when the current changes at the rate of 1500 A/s?

19–13 Calculate the self-induced voltage across a 100-mH inductor for the following values of $\frac{di}{dt}$:

a. 100 A/s.

b. 200 A/s.

c. 50 A/s.

d. 1000 A/s.

SECTION 19–5 MUTUAL INDUCTANCE L_M

19–14 A coil, L_1, produces 200 μWb of magnetic flux. A nearby coil, L_2, is linked with L_1 by 50 μWb of magnetic flux. What is the coefficient of coupling, *k*, between L_1 and L_2?

19–15 A coil, L_1, produces 40 μWb of magnetic flux. A coil, L_2, nearby, is linked with L_1 by 30 μWb of magnetic flux. What is the value of *k*?

19–16 Two 50-mH coils, L_1 and L_2, have a coefficient of coupling, *k*, equal to 0.6. Calculate L_M.

19–17 Two inductors, L_1 and L_2, have a coefficient of coupling, *k*, equal to 0.5. L_1 = 100 mH and L_2 = 150 mH. Calculate L_M.

19–18 What is the assumed value of *k* for an iron-core transformer?

SECTION 19–6 TRANSFORMERS

19–19 In Fig. 19–37, solve for

a. the secondary voltage, V_S.

b. the secondary current, I_S.

c. the secondary power, P_{sec}.

d. the primary power, P_{pri}.

e. the primary current, I_P.

Figure 19–37

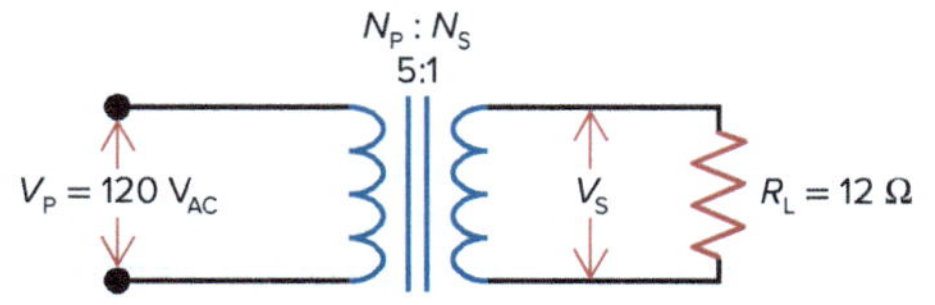

19–20 Repeat Prob. 19–19 if N_P:N_S = 10:1.

19–21 In Fig. 19–38, solve for

a. V_{S_1} (secondary 1 voltage).

b. V_{S_2} (secondary 2 voltage).

c. I_{S_1} (secondary 1 current).

d. I_{S_2} (secondary 2 current).

e. P_{Sec_1}.

f. P_{Sec_2}.

g. P_{pri}.

h. I_P.

Figure 19–38

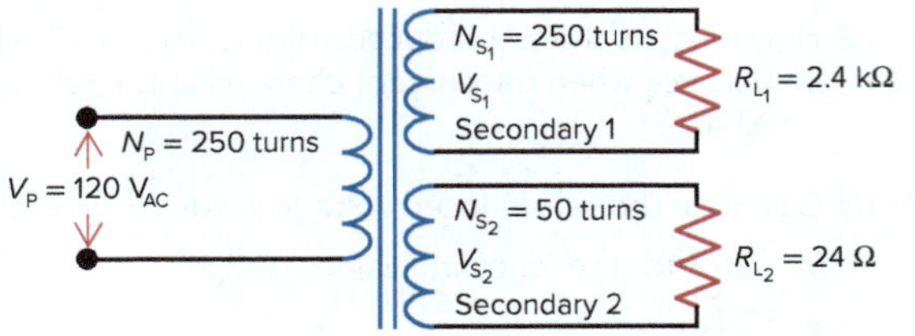

19–22 In Fig. 19–38, calculate the primary current, I_P, if R_{L_1} opens.

19–23 In Fig. 19–39, solve for

a. the turns ratio $\frac{N_P}{N_S}$.

b. the secondary current, I_S.

c. the primary current, I_P.

Figure 19–39

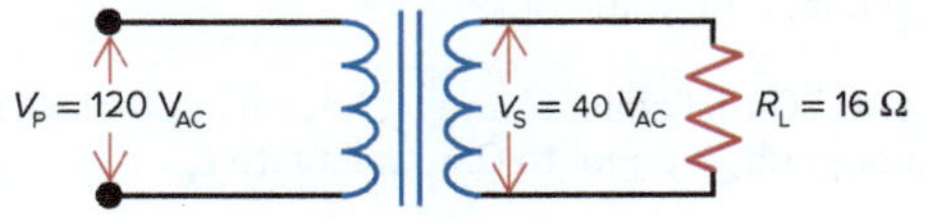

19–24 In Fig. 19–40, what turns ratio, $\frac{N_P}{N_S}$, is required to obtain a secondary voltage of

a. 60 V_{AC}?

b. 600 V_{AC}?

c. 420 V_{AC}?

d. 24 V_{AC}?

e. 12.6 V_{AC}?

Figure 19–40

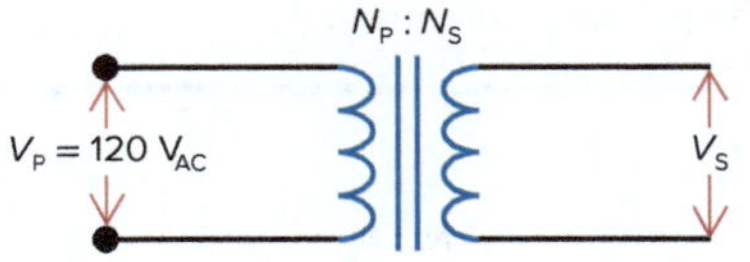

19–25 A transformer delivers 400 W to a load connected to its secondary. If the input power to the primary is 500 W, what is the efficiency of the transformer?

19–26 Explain the advantages of a transformer having an isolated secondary.

SECTION 19–7 TRANSFORMER RATINGS

19–27 How is the power rating of a transformer specified?

19–28 To avoid overloading a transformer, what two rules should be observed?

19–29 What is the purpose of phasing dots on the schematic symbol of a transformer?

19–30 Assume that a 6-Ω load is connected to the secondary in Fig. 19–15*b* and *c*. How much is the current in each individual primary winding in

a. Fig. 19–15*b*?

b. Fig. 19–15*c*?

19–31 Refer to Fig. 19–41. Calculate the following:

a. V_{Sec_1}.

b. V_{Sec_2}.

c. the maximum allowable current in secondary 1.

d. the maximum allowable current in secondary 2.

e. the maximum allowable primary current.

Figure 19–41

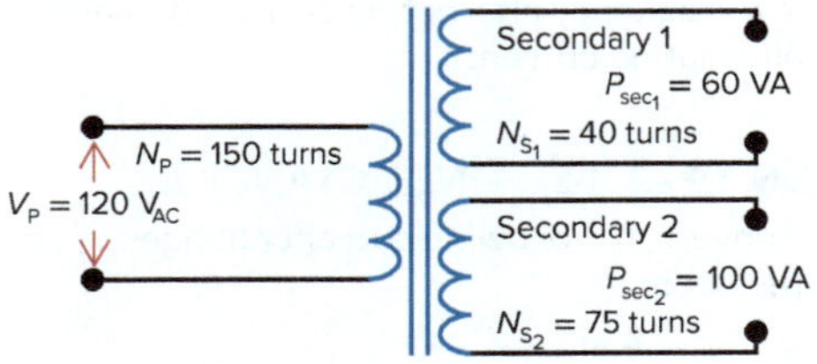

19–32 Refer to the transformer in Fig. 19–42. How much voltage would a DMM measure across the following secondary leads if the secondary current is 2 A?

a. V_{AB}.

b. V_{AC}.

c. V_{BC}.

Figure 19–42

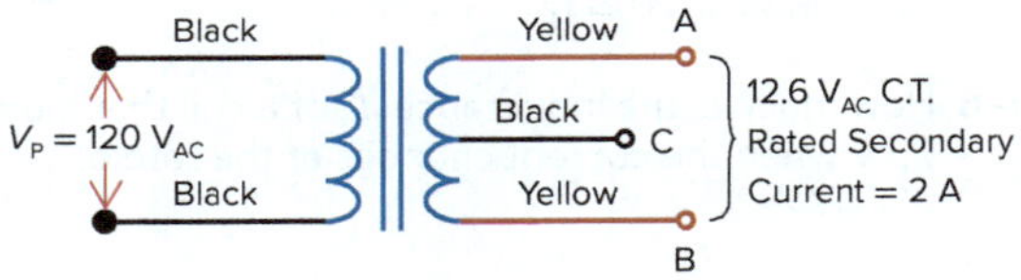

19–33 How much is the primary current, I_P, in Fig. 19–42 if the secondary current is 2 A?

19–34 Repeat Prob. 19–32 if the secondary is unloaded.

SECTION 19–8 IMPEDANCE TRANSFORMATION

19–35 In Fig. 19–43, calculate the primary impedance, Z_P, for a turns ratio $\frac{N_P}{N_S}$ of

a. 2:1.

b. 1:2.

c. 11.18:1.

d. 10:1.

e. 1:3.16.

Figure 19–43

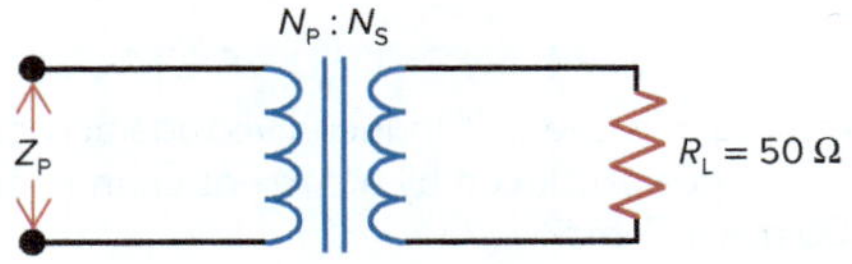

19–36 In Fig. 19–44, calculate the required turns ratio $\frac{N_P}{N_S}$ for

a. $Z_P = 10\ k\Omega$ and $R_L = 75\ \Omega$.
b. $Z_P = 100\ \Omega$ and $R_L = 25\ \Omega$.
c. $Z_P = 100\ \Omega$ and $R_L = 10\ k\Omega$.
d. $Z_P = 1\ k\Omega$ and $R_L = 200\ \Omega$.
e. $Z_P = 50\ \Omega$ and $R_L = 600\ \Omega$.
f. $Z_P = 200\ \Omega$ and $R_L = 10\ \Omega$.

Figure 19–44

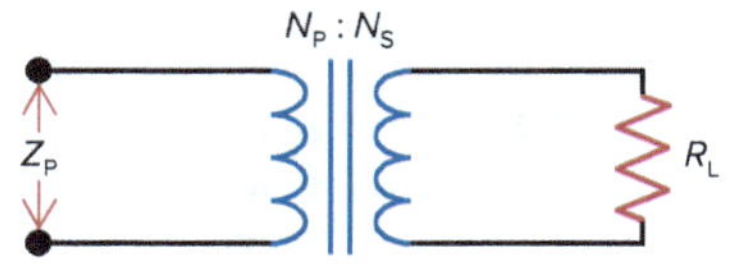

19–37 In Fig. 19–45, what turns ratio, $\frac{N_P}{N_S}$, will provide maximum transfer of power from the amplifier to the 4-Ω speaker?

Figure 19–45

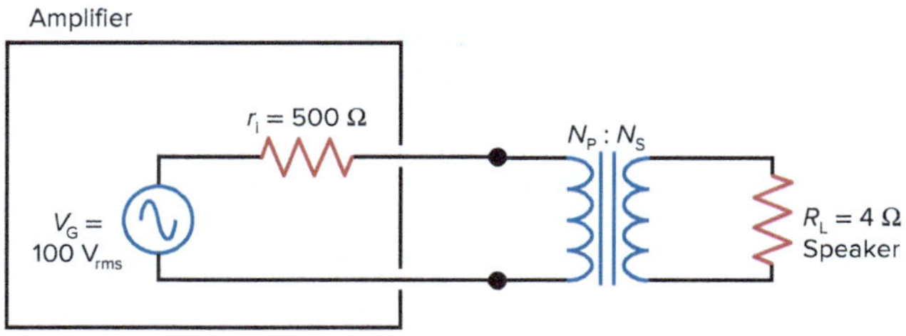

19–38 Using your answer from Prob. 19–37, calculate

a. the primary impedance, Z_P.
b. the power delivered to the 4-Ω speaker.
c. the primary power.

SECTION 19–12 INDUCTANCES IN SERIES OR PARALLEL

19–39 Calculate the total inductance, L_T, for the following combinations of series inductors. Assume no mutual induction.

a. $L_1 = 5$ mH and $L_2 = 15$ mH.
b. $L_1 = 12$ mH and $L_2 = 6$ mH.
c. $L_1 = 220\ \mu$H, $L_2 = 330\ \mu$H, and $L_3 = 450\ \mu$H.
d. $L_1 = 1$ mH, $L_2 = 500\ \mu$H, $L_3 = 2.5$ mH, and $L_4 = 6$ mH.

19–40 Assuming that the inductor combinations listed in Prob. 19–39 are in parallel rather than series, calculate the equivalent inductance, L_{EQ}. Assume no mutual induction.

19–41 A 100-mH and 300-mH inductor are connected in series-aiding and have a mutual inductance, L_M, of 130 mH. What is the total inductance, L_T?

19–42 If the inductors in Prob. 19–41 are connected in a series-opposing arrangement, how much is L_T?

19–43 A 20-mH and 40-mH inductor have a coefficient of coupling, k, of 0.4. Calculate L_T if the inductors are

a. series-aiding.
b. series-opposing.

19–44 Two 100-mH inductors in series have a total inductance, L_T, of 100 mH when connected in a series-opposing arrangement and 300 mH when connected in a series-aiding arrangement. Calculate

a. L_M.
b. k.

SECTION 19–13 ENERGY IN A MAGNETIC FIELD OF INDUCTANCE

19–45 Calculate the energy in joules stored by a magnetic field created by 90 mA of current in a 60-mH inductor.

19–46 Calculate the energy in joules stored by a magnetic field created by 200 mA in a 5-H inductor.

19–47 A current of 3 A flows in a coil with an inductance of 150 mH. How much energy is stored in the magnetic field?

Critical Thinking

19–48 Derive the formula:

$$Z_P = \left(\frac{N_P}{N_S}\right)^2 \times Z_S$$

19–49 Calculate the primary impedance Z_P in Fig. 19–46.

Figure 19–46 Circuit for Critical Thinking Prob. 19–49.

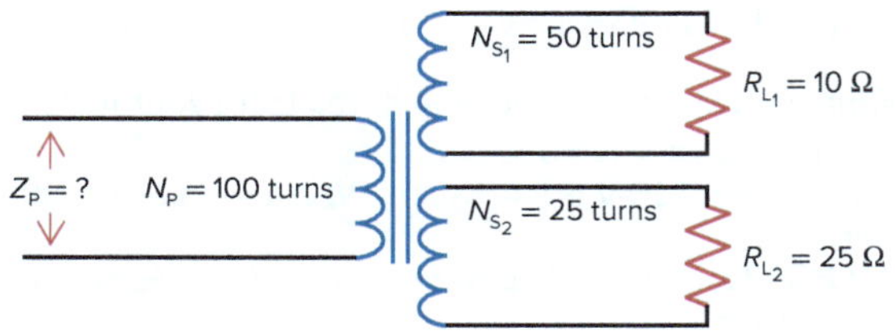

19–50 In Fig. 19–47, calculate the impedance Z_p across primary leads: (a) 1 and 3; (b) 1 and 2. (Note: Terminal 2 is a center-tap connection on the transformer primary. Also, the turns ratio of 4:1 is specified using leads 1 and 3 of the primary.)

19–51 Refer to Fig. 19–38. If the transformer has an efficiency of 80 percent, calculate the primary current I_p.

Figure 19–47 Circuit for Critical Thinking Prob. 19–50.

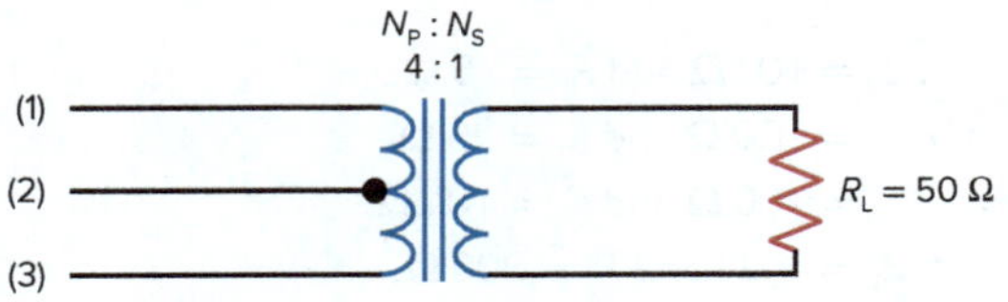

Answers to Self-Reviews

19–1 **a.** coil with an iron core **b.** time B

19–2 **a.** 2 H **b.** 32 mH

19–3 **a.** 2 V **b.** 200 V

19–4 **a.** true **b.** true

19–5 **a.** 1 **b.** 18 mH

19–6 **a.** 240 V **b.** 0.1 A **c.** false **d.** increase

19–7 **a.** true **b.** true **c.** false

19–8 **a.** false **b.** false **c.** true

19–9 **a.** iron core **b.** 60 MHz

19–10 **a.** true **b.** true **c.** true

19–11 **a.** true **b.** true

19–12 **a.** 1.5 mH **b.** 0.33 mH

19–13 **a.** joule **b.** more

19–14 **a.** true **b.** true **c.** true

19–15 **a.** infinite ohms **b.** 120 V **c.** series

Laboratory Application Assignment

In this lab application assignment, you will examine how a transformer can be used to step up or step down an AC voltage. You will measure the primary and secondary voltages as well as the primary and secondary currents for different values of load resistance connected to the secondary. From the measured values of voltage and current you will determine the primary and secondary power as well as the percent efficiency.

Equipment: Obtain the following items from your instructor.

- Isolation transformer and Variac
- Transformer: 120-V primary, 25.2-V, 2-A secondary with center tap
- 25-Ω, 50-Ω, and 100-Ω power resistors (50-W power rating)
- SPST switch
- 2 DMMs

Resistance Measurements

Examine the transformer supplied to you for this experiment. By inspection, determine the primary and secondary leads of the transformer and relate it to the schematic symbol, as shown in Fig. 19–48.

Figure 19–48

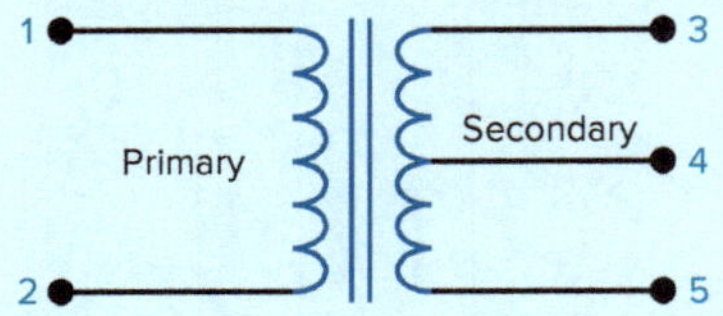

With a DMM, measure and record the resistance across each of the following transformer terminals. (Set the DMM to the lowest resistance range.)

$R_{1\text{-}2}$ = ________, $R_{3\text{-}4}$ = ________, $R_{4\text{-}5}$ = ________, $R_{3\text{-}5}$ = ________,
$R_{1\text{-}3}$ = ________, $R_{2\text{-}5}$ = ________, $R_{1\text{-}4}$ = ________, $R_{2\text{-}4}$ = ________

Which resistance measurements indicate isolation between the transformer windings? ________________

Effect of DC Voltage and Current

Connect a 10-V_{DC} supply to primary terminals 1 and 2 in Fig. 19–48. Next, measure and record the following DC voltages in the secondary:

$V_{3\text{-}4}$ = ________, $V_{4\text{-}5}$ = ________, $V_{3\text{-}5}$ = ________

Are these measured voltages what you expected? If so, why? ________________

Transformer Circuit

Caution: In this part of the lab you will be working with 120 V_{AC}. For your safety, you will need to use an isolation transformer. Plug the isolation transformer into the 120–V_{AC} outlet on your benchtop and in turn plug the Variac into the isolation transformer. Next adjust the Variac for an output of 120 V_{AC}. This is the voltage you will apply directly to the transformer primary.

Unloaded Secondary

Connect the circuit in Fig. 19–49. (Be sure the DMM in the primary is set to measure AC current.) Switch S_1 is open. With exactly 120 V_{AC} applied to the primary, measure and record the following secondary voltages:

$V_{3\text{-}4}$ = ________, $V_{4\text{-}5}$ = ________, $V_{3\text{-}5}$ = ________

Based on your measured values, calculate the transformer turns ratio from the primary (1 and 2) to secondary (3 and 5). (Recall that $N_P/N_S = V_P/V_S$.)

$V_{1\text{-}2}/V_{3\text{-}5}$ = ________ / ________

Does the full secondary voltage, $V_{3\text{-}5}$, measure higher than its rated value? ________________

Why? ________________

Figure 19–49

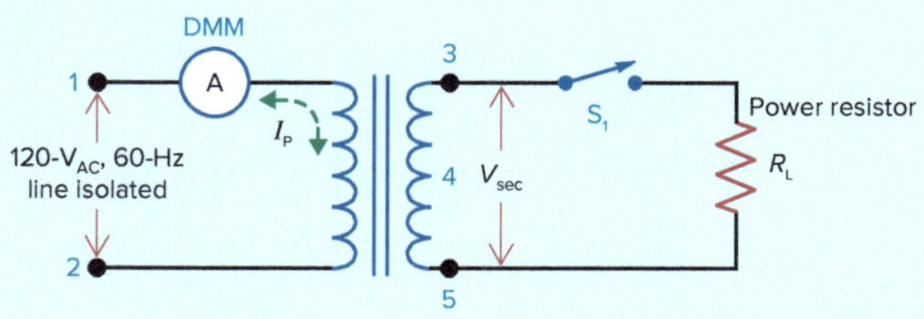

Record the primary current, I_P, indicated by the DMM.

I_P = ________________

$R_L = 100\ \Omega$

Close S_1. Measure and record the following values:

V_S = ________, I_P = ________

Make the following calculations based on the measured values of V_S and I_P.

Calculate I_S as V_S/R_L. I_S = ________

Calculate P_S as $V_S \times I_S$. P_S = ________

Calculate P_P as $V_P \times I_P$. P_P = ________

Calculate the % efficiency as $P_S/P_P \times 100$.
% efficiency = ________________

Repeat this procedure for each of the remaining load resistance values.

$R_L = 50\ \Omega$

Close S_1. Measure and record the following values:

V_S = ________, I_P = ________

Make the following calculations based on the measured values of V_S and I_P.

Calculate I_S as V_S/R_L. I_S = ________

Calculate P_S as $V_S \times I_S$. P_S = ________

Calculate P_P as $V_P \times I_P$. P_P = ________

Calculate the % efficiency as $P_S/P_P \times 100$.
% efficiency = ________________

$R_L = 25\ \Omega$

Close S_1. Measure and record the following values:

V_S = ________, I_P = ________

Make the following calculations based on the measured values of V_S and I_P.

Calculate I_S as V_S/R_L. I_S = ________

Calculate P_S as $V_S \times I_S$. P_S = ________

Calculate P_P as $V_P \times I_P$. P_P = ________

Calculate the % efficiency as $P_S/P_P \times 100$.
% efficiency = ________________

As the load resistance decreased in value, what happened to each of the following quantities?

I_S? ________, I_P? ________, P_S? ________, P_P? ________ % efficiency? ________

chapter 20

Inductive Reactance

When alternating current flows in an inductance L, the amount of current is much less than the DC resistance alone would allow. The reason is that the current variations induce a voltage across L that opposes the applied voltage. This additional opposition of an inductance to sine-wave alternating current is specified by the amount of its inductive reactance X_L. It is an opposition to current, measured in ohms. The X_L is the ohms of opposition, therefore, that an inductance L has for sine-wave current.

The amount of X_L equals $2\pi fL$ ohms, with f in hertz and L in henrys. Note that the opposition in ohms of X_L increases for higher frequencies and more inductance. The constant factor 2π indicates sine-wave variations.

The requirements for X_L correspond to what is needed to produce induced voltage. There must be variations in current and its associated magnetic flux. For a steady direct current without any changes in current, X_L is zero. However, with sine-wave alternating current, X_L is the best way to analyze the effect of L.

Chapter Outline

Chapter Objectives

After studying this chapter, you should be able to

- *Explain* how inductive reactance reduces the amount of alternating current.
- *Calculate* the inductive reactance of an inductor when the frequency and inductance are known.
- *Calculate* the total inductive reactance of series-connected inductors.
- *Calculate* the combined equivalent inductive reactance of parallel-connected inductors.
- *Explain* how Ohm's law can be applied to inductive reactance.
- *Describe* the waveshape of induced voltage produced by sine-wave alternating current.

Important Terms

inductive reactance, X_L

phase angle

proportional

MultiSim **Figure 20–1** Illustrating the effect of inductive reactance X_L in reducing the amount of sine-wave alternating current. (*a*) Bulb lights with 2.4 A. (*b*) Inserting an X_L of 1000 Ω reduces I to 0.12 A, and the bulb cannot light. (*c*) With direct current, the coil has no inductive reactance, and the bulb lights.

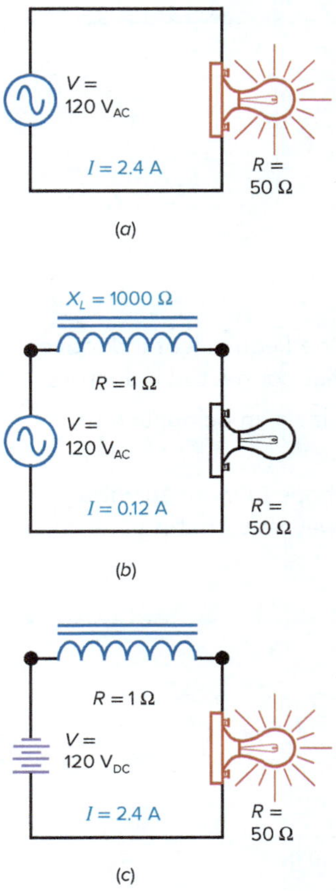

20–1 How X_L Reduces the Amount of I

Figure 20–1 illustrates the effect of X_L in reducing the alternating current for a lightbulb. The more ohms of X_L, the less current flows. When X_L reduces I to a very small value, the bulb cannot light.

In Fig. 20–1*a*, there is no inductance, and the AC voltage source produces a 2.4-A current to light the bulb with full brilliance. This 2.4-A I results from 120 V applied across the 50-Ω R of the bulb's filament.

In Fig. 20–1*b*, however, a coil is connected in series with the bulb. The coil has a DC resistance of only 1 Ω, which is negligible, but the reactance of the inductance is 1000 Ω. This 1000-Ω X_L is a measure of the coil's reaction to sine-wave current in producing a self-induced voltage that opposes the applied voltage and reduces the current. Now I is 120 V/1000 Ω, approximately, which equals 0.12 A. This I is not enough to light the bulb.

Although the DC resistance is only 1 Ω, the X_L of 1000 Ω for the coil limits the amount of alternating current to such a low value that the bulb cannot light. This X_L of 1000 Ω for a 60-Hz current can be obtained with an inductance L of approximately 2.65 H.

In Fig. 20–1*c*, the coil is also in series with the bulb, but the applied battery voltage produces a steady value of direct current. Without any current variations, the coil cannot induce any voltage and, therefore, it has no reactance. The amount of direct current, then, is practically the same as though the DC voltage source were connected directly across the bulb, and it lights with full brilliance. In this case, the coil is only a length of wire because there is no induced voltage without current variations. The DC resistance is the resistance of the wire in the coil.

In summary, we can draw the following conclusions:

1. An inductance can have appreciable X_L in AC circuits to reduce the amount of current. Furthermore, the higher the frequency of the alternating current, and the greater the inductance, the higher the X_L opposition.
2. There is no X_L for steady direct current. In this case, the coil is a resistance equal to the resistance of the wire.

These effects have almost unlimited applications in practical circuits. Consider how useful ohms of X_L can be for different kinds of current, compared with resistance, which always has the same ohms of opposition. One example is to use X_L where it is desired to have high ohms of opposition to alternating current but little opposition to direct current. Another example is to use X_L for more opposition to a high-frequency alternating current, compared with lower frequencies.

X_L Is an Inductive Effect

An inductance can have X_L to reduce the amount of alternating current because self-induced voltage is produced to oppose the applied voltage. In Fig. 20–2, V_L is the voltage across L, induced by the variations in sine-wave current produced by the applied voltage V_A.

MultiSim **Figure 20–2** The inductive reactance X_L equals the V_L/I_L ratio in ohms.

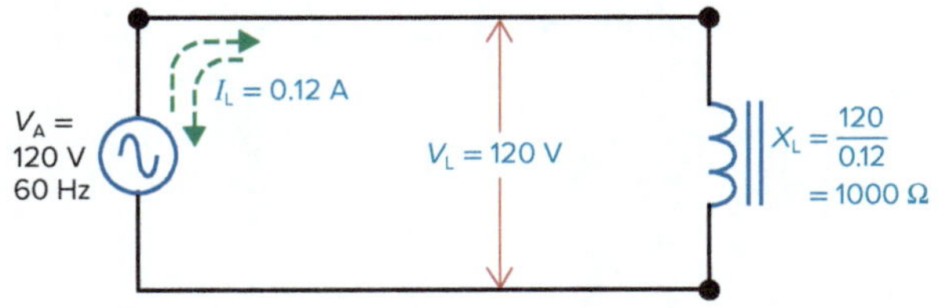

GOOD TO KNOW

The inductive reactance, X_L, of a coil is a result of the inductor's counter emf with a varying current. The polarity of the counter emf is always such that it opposes a change in current.

The two voltages V_A and V_L are the same because they are in parallel. However, the current I_L is the amount that allows the self-induced voltage V_L to be equal to V_A. In this example, I is 0.12 A. This value of a 60-Hz current in the inductance produces a V_L of 120 V.

The Inductive Reactance Is a *V/I* Ratio

The V/I ratio for the ohms of opposition to the sine-wave current is $^{120}/_{0.12}$, which equals 1000 Ω. This 1000 Ω is what we call X_L, to indicate how much current can be produced by sine-wave voltage across an inductance. The ohms of X_L can be almost any amount, but the 1000 Ω here is a typical example.

The Effect of *L* and *f* on X_L

The X_L value depends on the amount of inductance and on the frequency of the alternating current. If L in Fig. 20–2 were increased, it could induce the same 120 V for V_L with less current. Then the ratio of V_L/I_L would be greater, meaning more X_L for more inductance.

Also, if the frequency were increased in Fig. 20–2, the current variations would be faster with a higher frequency. Then the same L could produce the 120 V for V_L with less current. For this condition also, the V_L/I_L ratio would be greater because of the smaller current, indicating more X_L for a higher frequency.

■ *20–1 Self-Review*

Answers at the end of the chapter.

a. For the DC circuit in Fig. 20–1*c*, how much is X_L?
b. For the AC circuit in Fig. 20–1*b*, how much is the V/I ratio for X_L?

GOOD TO KNOW

Inductive reactance, X_L, is a measure of a coil's opposition to the flow of sine-wave alternating current. X_L is measured in ohms.

20–2 $X_L = 2\pi fL$

The formula $X_L = 2\pi fL$ includes the effects of frequency and inductance for calculating the inductive reactance. The frequency is in hertz, and L is in henrys for an X_L in ohms. As an example, we can calculate X_L for an inductance of 2.65 H at the frequency of 60 Hz:

$$\begin{aligned} X_L &= 2\pi fL \\ &= 6.28 \times 60 \times 2.65 \qquad \textbf{(20–1)} \\ X_L &= 1000\ \Omega \end{aligned}$$

Note the following factors in the formula $X_L = 2\pi fL$.

1. The constant factor 2π is always $2 \times 3.14 = 6.28$. It indicates the circular motion from which a sine wave is derived. Therefore, this formula applies only to sine-wave AC circuits. The 2π is actually 2π rad or 360° for a complete circle or cycle.
2. The frequency f is a time element. Higher frequency means that the current varies at a faster rate. A faster current change can produce more self-induced voltage across a given inductance. The result is more X_L.
3. The inductance L indicates the physical factors of the coil that determine how much voltage it can induce for a given current change.
4. **Inductive reactance** X_L is in ohms, corresponding to a V_L/I_L ratio for sine-wave AC circuits, to determine how much current L allows for a given applied voltage.

Stating X_L as V_L/I_L and as $2\pi fL$ are two ways of specifying the same value of ohms. The $2\pi fL$ formula gives the effect of L and f on the X_L. The V_L/I_L ratio gives the result of $2\pi fL$ in reducing the amount of I.

The formula $2\pi fL$ shows that X_L is **proportional** to frequency. When f is doubled, for instance, X_L is doubled. This linear increase in inductive reactance with frequency is illustrated in Fig. 20–3.

The reactance formula also shows that X_L is proportional to the inductance. When the value of henrys for L is doubled, the ohms of X_L is also doubled. This linear increase of inductive reactance with inductance is illustrated in Fig. 20–4.

MultiSim **Figure 20–3** Graph of values to show linear increase of X_L for higher frequencies. The L is constant at 0.32 H.

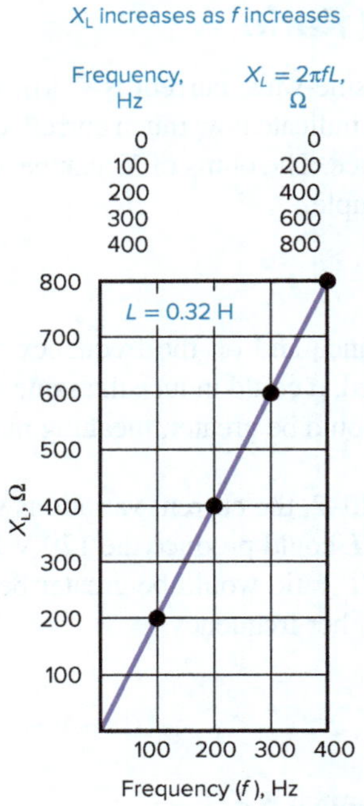

Figure 20–4 Graph of values to show linear increase of X_L for higher values of inductance L. The frequency is constant at 100 Hz.

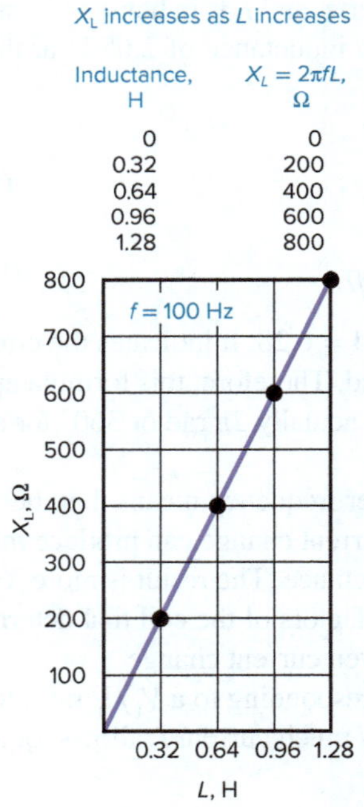

Example 20-1

How much is X_L of a 6-mH L at 41.67 kHz?

ANSWER

$$X_L = 2\pi fL$$
$$= 6.28 \times 41.67 \times 10^3 \times 6 \times 10^{-3}$$
$$= 1570\ \Omega$$

Example 20-2

Calculate the X_L of (a) a 10-H L at 60 Hz and (b) a 5-H L at 60 Hz.

ANSWER

(a) For a 10-H L,

$$X_L = 2\pi fL = 6.28 \times 60 \times 10$$
$$= 3768\ \Omega$$

(b) For a 5-H L,

$$X_L = \tfrac{1}{2} \times 3768 = 1884\ \Omega$$

Example 20-3

Calculate the X_L of 250-μH coil at (a) 1 MHz and (b) 10 MHz.

ANSWER

(a) At 1 MHz,

$$X_L = 2\pi fL = 6.28 \times 1 \times 10^6 \times 250 \times 10^{-6}$$
$$= 1570\ \Omega$$

(b) At 10 MHz,

$$X_L = 10 \times 1570 = 15{,}700\ \Omega$$

CALCULATOR

To do a problem like Example 20–1 with a calculator requires continued multiplication. Multiply all the factors and then press the (=) key only at the end. If the calculator does not have an (EXP) (exponential) function key, do the powers of 10 separately without the calculator. Specifically, for this example with $2\pi \times 6 \times 10^{-3} \times 41.67 \times 10^{3}$, the 10^3 and 10^{-3} cancel. Then calculate $2\pi \times 6 \times 41.67$ as factors. To save time in the calculation, 2π can be memorized as 6.28, since it occurs in many AC formulas. For the multiplication, punch in 6.28 for 2π and then push the (×) key, punch in 6 and push the (×) key again, punch in 41.67, and push the (=) key for the total product of 1570 as the final answer. It is not necessary to use the (=) key until the last step for the final product. The factors can be multiplied in any order.

The last two examples illustrate the fact that X_L is proportional to frequency and inductance. In Example 20–2*b*, X_L is one-half the value in Example 20–2*a* because the inductance is one-half. In Example 20–3*b*, the X_L is 10 times more than in Example 20–3*a* because the frequency is 10 times higher.

Finding L from X_L

Not only can X_L be calculated from f and L, but if any two factors are known, the third can be found. Very often X_L can be determined from voltage and current measurements. With the frequency known, L can be calculated as

$$L = \frac{X_L}{2\pi f} \qquad \textbf{(20–2)}$$

This formula has the factors inverted from Formula (20–1). Use the basic units with ohms for X_L and hertz for f to calculate L in henrys.

It should be noted that Formula (20–2) can also be stated as

$$L = \frac{1}{2\pi f} \times X_L$$

This form is easier to use with a calculator because $1/2\pi f$ can be found as a reciprocal value and then multiplied by X_L.

The following problems illustrate how to find X_L from V and I measurements and using X_L to determine L with Formula (20–2).

Example 20-4

A coil with negligible resistance has 62.8 V across it with 0.01 A of current. How much is X_L?

ANSWER

$$X_L = \frac{V_L}{I_L} = \frac{62.8\ \text{V}}{0.01\ \text{A}}$$

$$= 6280\ \Omega$$

Example 20-5

Calculate L of the coil in Example 20–4 when the frequency is 1000 Hz.

ANSWER

$$L = \frac{X_L}{2\pi f} = \frac{6280}{6.28 \times 1000}$$

$$= 1\ \text{H}$$

Example 20-6

Calculate L of a coil that has 15,700 Ω of X_L at 12 MHz.

ANSWER

$$L = \frac{X_L}{2\pi f} = \frac{1}{2\pi f} = X_L$$

$$= \frac{1}{6.28 \times 12 \times 10^6} \times 15{,}700$$

$$= 0.0133 \times 10^{-6} \times 15{,}700$$

$$= 208.8 \times 10^{-6}\ \text{H} \quad \text{or} \quad 208.8\ \mu\text{H}$$

GOOD TO KNOW

Although L can be determined when X_L and f are known, its value is determined strictly by its physical construction.

CALCULATOR

To do Example 20–6 with a calculator, first find the product $2\pi f$ and then take the reciprocal to multiply by 15,700. Note that with powers of 10 a reciprocal value has the sign reversed for the exponent. Specifically, 10^6 in the denominator here becomes 10^{-6} as the reciprocal. To multiply the factors, punch in 6.28 and then push the (×) key, punch in 12, and push the (=) key for the total product of 75.36. Take the reciprocal by using the (1/x) key, while the product is still on the display. This may require pushing the (2ndF) or "shift" key on the calculator. The reciprocal value is 0.0133. Now press the (×) key, punch in 15,700, and push the (=) key for the answer of 208.8×10^{-6}.

Finding f from X_L

For a third version of the inductive reactance formula,

$$f = \frac{X_L}{2\pi L} \tag{20–3}$$

Use the basic units of ohms for X_L and henrys for L to calculate the frequency in hertz.

Formula 20-3 can also be stated as

$$f = \frac{1}{2\pi L} \times X_L$$

This form is easier to use with a calculator. Find the reciprocal value and multiply by X_L, as explained before in Example 20–6.

Example 20-7

At what frequency will an inductance of 1 H have a reactance of 1000 Ω?

ANSWER

$$f = \frac{1}{2\pi L} \times X_L = \frac{1}{6.28 \times 1} \times 1000$$

$$= 0.159 \times 1000$$

$$= 159\ \text{Hz}$$

20–2 Self-Review

Answers at the end of the chapter.

Calculate X_L for the following:

a. **L is 1 H and f is 100 Hz.**
b. **L is 0.5 H and f is 100 Hz.**
c. **L is 1 H and f is 1000 Hz.**

Figure 20-5 Combining ohms of X_L for inductive reactances. (*a*) X_{L_1} and X_{L_2} in series. (*b*) X_{L_1} and X_{L_2} in parallel.

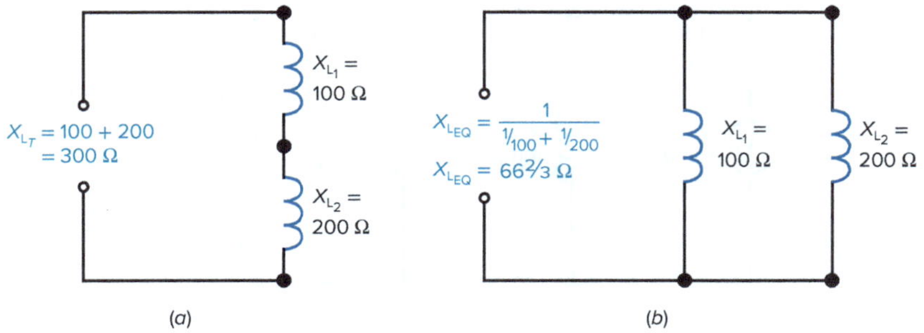

20–3 Series and Parallel Inductive Reactances

Since reactance is an opposition in ohms, the values of X_L in series or in parallel are combined the same way as ohms of resistance. With series reactances, the total is the sum of the individual values, as shown in Fig. 20–5*a*. For example, the series reactances of 100 and 200 Ω add to equal 300 Ω of X_L across both reactances. Therefore, in series,

$$X_{L_T} = X_{L_1} + X_{L_2} + X_{L_3} + \cdots + \text{etc.} \qquad \textbf{(20–4)}$$

The combined equivalent reactance of parallel connected reactances is calculated using the reciprocal formula. As shown in Fig. 20–5*b*, in parallel

$$X_{L_{EQ}} = \frac{1}{\frac{1}{X_{L_1}} + \frac{1}{X_{L_2}} + \frac{1}{X_{L_3}} + \cdots + \text{etc.}} \qquad \textbf{(20–5)}$$

The combined equivalent parallel reactance is always less than the lowest branch reactance. Any shortcuts for calculating parallel resistances also apply to parallel reactances. For instance, the combined equivalent reactance of two equal reactances in parallel is one-half either reactance.

20–3 Self-Review

Answers at the end of the chapter.

a. **An X_L of 200 Ω is in series with a 300-Ω X_L. How much is the total X_{L_T}?**

b. **An X_L of 200 Ω is in parallel with a 300-Ω X_L. How much is the combined $X_{L_{EQ}}$?**

20–4 Ohm's Law Applied to X_L

The amount of current in an AC circuit with only inductive reactance is equal to the applied voltage divided by X_L. Three examples are given in Fig. 20–6. No DC resistance is indicated, since it is assumed to be practically zero for the coils shown. In Fig. 20–6*a*, there is one reactance of 100 Ω. Then I equals V/X_L, or 100 V/100 Ω, which is 1 A.

In Fig. 20–6*b*, the total reactance is the sum of the two individual series reactances of 100 Ω each, for a total of 200 Ω. The current, calculated as V/X_{L_T}, then equals 100 V/200 Ω, which is 0.5 A. This current is the same in both series reactances. Therefore, the voltage across each reactance equals its IX_L product. This is 0.5 A × 100 Ω, or 50 V across each X_L.

Figure 20–6 Circuit calculations with V, I, and ohms of reactance X_L. (*a*) One reactance. (*b*) Two series reactances. (*c*) Two parallel reactances.

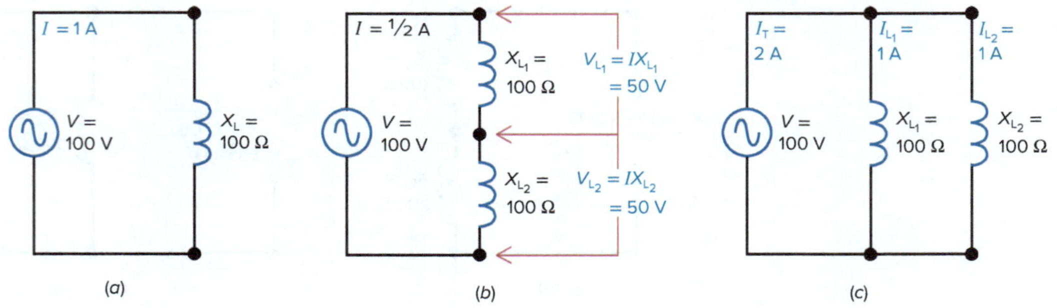

In Fig. 20–6*c*, each parallel reactance has its individual branch current, equal to the applied voltage divided by the branch reactance. Then each branch current equals 100 V/100 Ω, which is 1 A. The voltage is the same across both reactances, equal to the generator voltage, since they are all in parallel.

The total line current of 2 A is the sum of the two individual 1-A branch currents. With the rms value for the applied voltage, all calculated values of currents and voltage drops in Fig. 20–6 are also rms values.

20–4 Self-Review

Answers at the end of the chapter.

a. In Fig. 20–6*b*, how much is the *I* through both X_{L_1} and X_{L_2}?
b. In Fig. 20–6*c*, how much is the *V* across both X_{L_1} and X_{L_2}?

20–5 Applications of X_L for Different Frequencies

The general use of inductance is to provide minimum reactance for relatively low frequencies but more for higher frequencies. In this way, the current in an AC circuit can be reduced for higher frequencies because of more X_L. There are many circuits in which voltages of different frequencies are applied to produce current with different frequencies. Then, the general effect of X_L is to allow the most current for direct current and low frequencies, with less current for higher frequencies, as X_L increases.

Compare this frequency factor for ohms of X_L with ohms of resistance. The X_L increases with frequency, but R has the same effect in limiting direct current or alternating current of any frequency.

If 1000 Ω is taken as a suitable value of X_L for many applications, typical inductances can be calculated for different frequencies. These are listed in Table 20–1.

At 60 Hz, for example, the inductance L in the top row of Table 20–1 is 2.65 H for 1000 Ω of X_L. The calculations are

$$L = \frac{X_L}{2\pi f} = \frac{1000}{2\pi \times 60}$$
$$= \frac{1000}{377}$$
$$= 2.65 \text{ H}$$

For this case, the inductance has practically no reactance for direct current or for very low frequencies below 60 Hz. However, above 60 Hz, the inductive reactance increases to more than 1000 Ω.

Table 20–1	Values of Inductance L for X_L of 1000 Ω	
L* (Approx.)	Frequency	Remarks
2.65 H	60 Hz	Power-line frequency and low audio frequency
160 mH	1000 Hz	Medium audio frequency
16 mH	10,000 Hz	High audio frequency
160 μH	1000 kHz (RF)	In radio broadcast band
16 μH	10 MHz (HF)	In shortwave radio band
1.6 μH	100 MHz (VHF)	In FM broadcast band

* Calculated as $L = 1000/(2\pi f)$.

To summarize, the effects of increasing frequencies for this 2.65-H inductance are as follows:

Inductive reactance X_L is zero for 0 Hz which corresponds to a steady direct current.
Inductive reactance X_L is less than 1000 Ω for frequencies below 60 Hz.
Inductive reactance X_L equals 1000 Ω at 60 Hz.
Inductive reactance X_L is more than 1000 Ω for frequencies above 60 Hz.

GOOD TO KNOW

For the most part, inductor values below 1 μH are not practical for use in electronic circuits. For this reason, the nanohenry unit of inductance is practically unheard of.

Note that the smaller inductances at the bottom of the first column still have the same X_L of 1000 Ω as the frequency is increased. Typical rf coils, for instance, have an inductance value of the order of 100 to 300 μH. For the very high frequency (VHF) range, only several microhenrys of inductance are needed for an X_L of 1000 Ω.

It is necessary to use smaller inductance values as the frequency is increased because a coil that is too large can have excessive losses at high frequencies. With iron-core coils, particularly, the hysteresis and eddy-current losses increase with frequency.

20–5 Self-Review

Answers at the end of the chapter.

Refer to Table 20–1.

a. Which frequency requires the smallest L for 1000 Ω of X_L?

b. How much would X_L be for a 1.6-μH L at 200 MHz?

20–6 Waveshape of v_L Induced by Sine-Wave Current

More details of inductive circuits can be analyzed by means of the waveshapes in Fig. 20–7, plotted for the calculated values in Table 20–2. The top curve shows a sine wave of current i_L flowing through a 6-mH inductance L. Since induced voltage depends on the rate of change of current rather than on the absolute value of i, the curve in Fig. 20–7b shows how much the current changes. In this curve, the di/dt values are plotted for the current changes every 30° of the cycle. The bottom curve shows the actual induced voltage v_L. This v_L curve is similar to the di/dt curve because v_L equals the constant factor L multiplied by di/dt. Note that di/dt indicates infinitely small changes in i and t.

Figure 20–7 Waveshapes in inductive circuits. (*a*) Sine-wave current *i*; (*b*) changes in current with time *di*/*dt*; (*c*) induced voltage v_L.

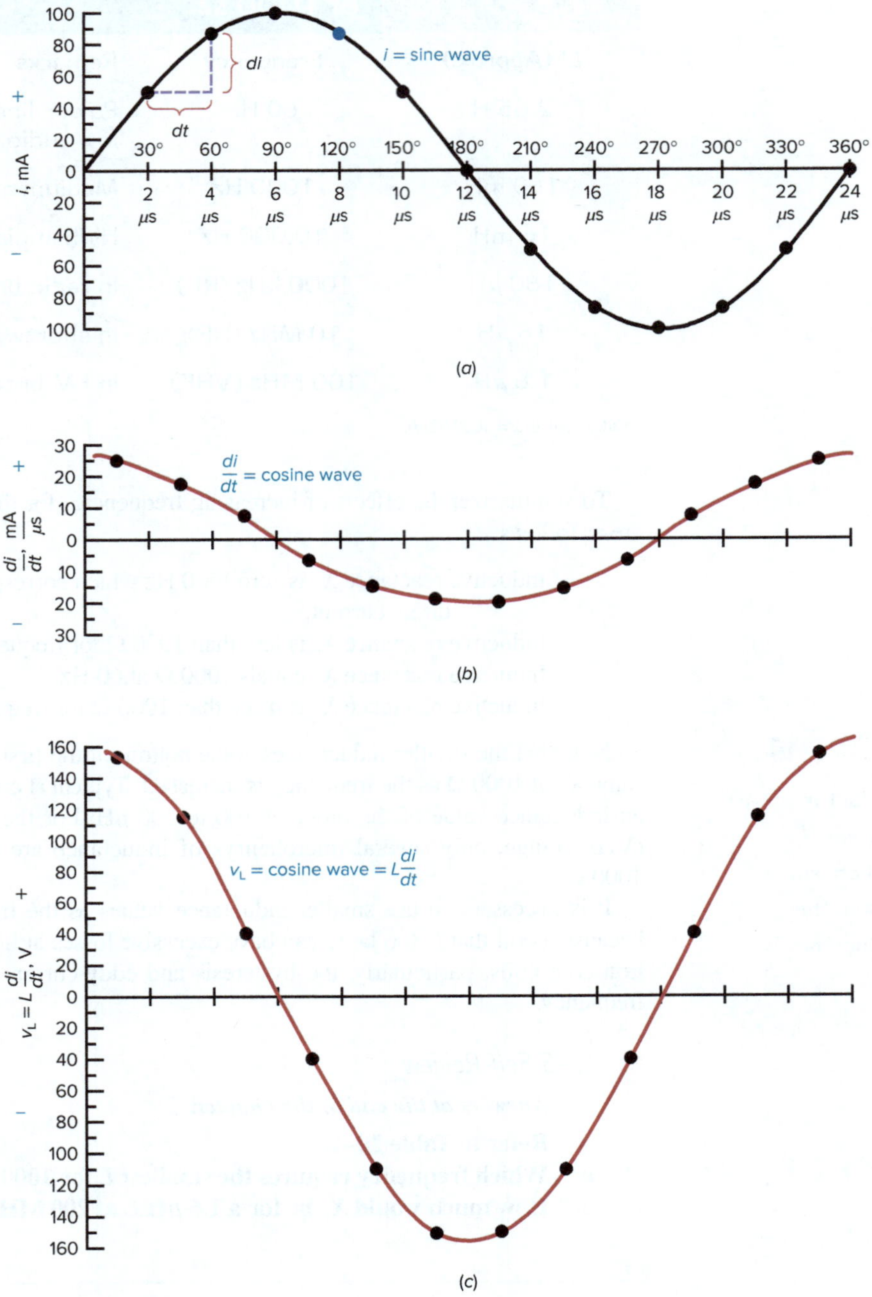

90° Phase Angle

The v_L curve at the bottom of Fig. 20–7 has its zero values when the i_L curve at the top is at maximum. This comparison shows that the curves are 90° out of phase. The v_L is a cosine wave of voltage for the sine wave of current i_L.

The 90° phase difference results from the fact that v_L depends on the *di*/*dt* rate of change, rather than on *i* itself. More details of this 90° **phase angle** between v_L and i_L for inductance are explained in the next chapter.

Table 20–2		Values for $v_L = L(di/dt)$ Curves in Figure 20–7					
Time		*dt*					
θ	μs	θ	μs	*di*, mA	*di/dt*, mA/μs	*L*, mH	$v_L = L(di/dt)$, V
30°	2	30°	2	50	25	6	150
60°	4	30°	2	36.6	18.3	6	109.8
90°	6	30°	2	13.4	6.7	6	40.2
120°	8	30°	2	−13.4	−6.7	6	−40.2
150°	10	30°	2	−36.6	−18.3	6	−109.8
180°	12	30°	2	−50	−25	6	−150
210°	14	30°	2	−50	−25	6	−150
240°	16	30°	2	−36.6	−18.3	6	−109.8
270°	18	30°	2	−13.4	−6.7	6	−40.2
300°	20	30°	2	13.4	6.7	6	40.2
330°	22	30°	2	36.6	18.3	6	109.8
360°	24	30°	2	50	25	6	150

Frequency

For each of the curves, the period T is 24 μs. Therefore, the frequency is $1/T$ or $\frac{1}{24}$ μs, which equals 41.67 kHz. Each curve has the same frequency.

Ohms of X_L

The ratio of v_L/i_L specifies the inductive reactance in ohms. For this comparison, we use the actual value of i_L, which has a peak value of 100 mA. The rate-of-change factor is included in the induced voltage v_L. Although the peak of v_L at 150 V is 90° before the peak of i_L at 100 mA, we can compare these two peak values. Then v_L/i_L is $^{150}/_{0.1}$, which equals 1500 Ω.

This X_L is only approximate because v_L cannot be determined exactly for the large *dt* changes every 30°. If we used smaller intervals of time, the peak v_L would be 157 V. Then X_L would be 1570 Ω, the same as $2\pi fL$ Ω with a 6-mH L and a frequency of 41.67 kHz. This is the same X_L problem as Example 20–1.

The Tabulated Values from 0° to 90°

The numerical values in Table 20–2 are calculated as follows: The *i* curve is a sine wave. This means that it rises to one-half its peak value in 30°, to 0.866 of the peak in 60°, and the peak value is at 90°.

In the *di/dt* curve, the changes in *i* are plotted. For the first 30°, the *di* is 50 mA; the *dt* change is 2 μs. Then *di/dt* is $^{50}/_{2}$ or 25 mA/μs. This point is plotted between 0° and 30° to indicate that 25 mA/μs is the rate of change of current for the 2-μs interval between 0° and 30°. If smaller intervals were used, the *di/dt* values could be determined more accurately.

During the next 2-μs interval from 30° to 60°, the current increases from 50 to 86.6 mA. The change of current during this time is 86.6 – 50, which equals 36.6 mA. The time is the same 2 μs for all the intervals. Then *di*/*dt* for the next plotted point is 36.6/2, or 18.3.

For the final 2-μs change before *i* reaches its peak at 100 mA, the *di* value is 100 – 86.6, or 13.4 mA, and the *di*/*dt* value is 6.7. All of these values are listed in Table 20–2.

Notice that the *di*/*dt* curve in Fig. 20–7*b* has its peak at the zero value of the *i* curve and the peak *i* values correspond to zero on the *di*/*dt* curves. These conditions result because the sine wave of *i* has its sharpest slope at the zero values. The rate of change is greatest when the *i* curve is going through the zero axis. The *i* curve flattens near the peaks and has a zero rate of change exactly at the peak. The curve must stop going up before it can come down. In summary, then, the *di*/*dt* curve and the *i* curve are 90° out of phase with each other.

The v_L curve follows the *di*/*dt* curve exactly, as $v_L = L(di/dt)$. The phase of the v_L curve is exactly the same as that of the *di*/*dt* curve, 90° out of phase with the *i* curve. For the first plotted point,

$$v_L = L\frac{di}{dt} = 6 \times 10^{-3} \times \frac{50 \times 10^{-3}}{2 \times 10^{-6}}$$

$$= 150 \text{ V}$$

The other v_L values are calculated the same way, multiplying the constant factor of 6 mH by the *di*/*dt* value for each 2-μs interval.

90° to 180°

In this quarter-cycle, the sine wave of *i* decreases from its peak of 100 mA at 90° to zero at 180°. This decrease is considered a negative value for *di*, as the slope is negative going downward. Physically, the decrease in current means that its associated magnetic flux is collapsing, compared with the expanding flux as the current increases. The opposite motion of the collapsing flux must make v_L of opposite polarity, compared with the induced voltage polarity for increasing flux. This is why the *di* values are negative from 90° to 180°. The *di*/*dt* values are also negative, and the v_L values are negative.

180° to 270°

In this quarter-cycle, the current increases in the reverse direction. If the magnetic flux is considered counterclockwise around the conductor with +*i* values, the flux is in the reversed clockwise direction with –*i* values. Any induced voltage produced by expanding flux in one direction will have opposite polarity from voltage induced by expanding flux in the opposite direction. This is why the *di* values are considered negative from 180° to 270°, as in the second quarter-cycle, compared with the positive *di* values from 0° to 90°. Actually, increasing negative values and decreasing positive values are changing in the same direction. This is why v_L is negative for both the second and third quarter-cycles.

270° to 360°

In the last quarter-cycle, the negative *i* values are decreasing. Now the effect on polarity is like two negatives making a positive. The current and its magnetic flux have the negative direction. But the flux is collapsing, which induces opposite voltage from increasing flux. Therefore, the *di* values from 270° to 360° are positive, as are the *di*/*dt* values and the induced voltages v_L.

The same action is repeated for each cycle of sine-wave current. Then the current i_L and the induced voltage v_L are 90° out of phase. The reason is that v_L depends on *di*/*dt*, not on *i* alone.

Figure 20–8 How a 90° phase angle for the V_L applies in a complex circuit with more than one inductance. The current I_1 lags V_{L_1} by 90°, I_2 lags V_{L_2} by 90°, and I_3 lags V_{L_3} by 90°.

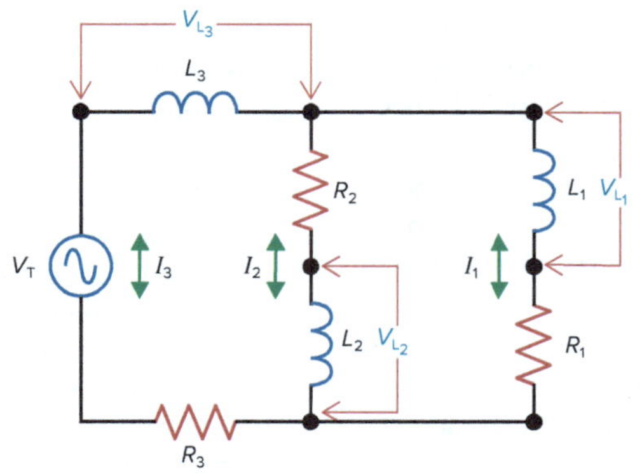

Application of the 90° Phase Angle in a Circuit

The phase angle of 90° between V_L and I will always apply for any L with sine-wave current. Remember, though, that the specific comparison is only between the induced voltage across any one coil and the current flowing in its turns. To emphasize this important principle, Fig. 20–8 shows an AC circuit with a few coils and resistors. The details of this complex circuit are not to be analyzed now. However, for each L in the circuit, the V_L is 90° out of phase with its I. The I lags V_L by 90°, or V_L leads I. For the three coils in Fig. 20–8,

Current I_1 lags V_{L_1} by 90°.
Current I_2 lags V_{L_2} by 90°.
Current I_3 lags V_{L_3} by 90°.
Note that I_3 is also I_T for the series-parallel circuit.

20–6 Self-Review

Answers at the end of the chapter.

Refer to Fig. 20–7.

a. **At what angle does i have its maximum positive value?**
b. **At what angle does v_L have its maximum positive value?**
c. **What is the phase angle difference between the waveforms for i and v_L?**

Summary

- Inductive reactance is the opposition of an inductance to the flow of sine-wave alternating current. The symbol for inductive reactance is X_L.
- Reactance X_L is measured in ohms because it limits the current to the value $I = V/X_L$. With V in volts and X_L in ohms, I is in amperes.
- $X_L = 2\pi fL$, where f is in hertz, L is in henrys, and X_L is in ohms.
- With a constant L, X_L increases proportionately with higher frequencies.
- At a constant frequency, X_L increases proportionately with higher inductances.
- With X_L and f known, the inductance $L = X_L/(2\pi f)$.
- With X_L and L known, the frequency $f = X_L/(2\pi L)$.
- The total X_L of reactances in series is the sum of the individual values, as for series resistances. Series reactances have the same current. The voltage across each inductive reactance is IX_L.
- The equivalent inductive reactance of parallel connected reactances is calculated using the reciprocal formula, as for parallel resistances. Each branch current is V/X_L. The total line current is the sum of the individual branch currents.
- Table 20–3 summarizes the differences between L and X_L.
- Table 20–4, compares X_L and R.
- Table 20–5, summarizes the differences between capacitive reactance and inductive reactance.

Table 20–3 Comparison of Inductance and Inductive Reactance

Inductance	Inductive Reactance
Symbol is L	Symbol is X_L
Measured in henry units	Measured in ohm units
Depends on construction of coil	Depends on frequency and inductance
$L = v_L/(di/dt)$, in H units	$X_L = v_L/i_L$ or $2\pi fL$, in Ω units

Table 20–4 Comparison of X_L and R

X_L	R
Ohm unit	Ohm unit
Increases for higher frequencies	Same for all frequencies
Current lags voltage by 90° ($\theta = 90°$)	Current in phase with voltage ($\theta = 0°$)

Table 20–5 Comparison of Capacitive and Inductive Reactances

X_C, Ω	X_L, Ω
Decreases with more capacitance C	Increases with more inductance L
Decreases with increase in frequency f	Increases with increase in frequency f
Allows less current at lower frequencies; blocks direct current	Allows more current at lower frequencies; passes direct current

Important Terms

Inductive reactance, X_L — a measure of an inductor's opposition to the flow of alternating current. X_L is measured in ohms and is calculated as $X_L = 2\pi fL$ or $X_L = \frac{V_L}{I_L}$. X_L applies only to sine-wave alternating current.

Phase angle — the angular difference or displacement between two waveforms. For an inductor, the induced voltage, v_L, reaches its maximum value 90° ahead of the inductor current, i_L. As a result, the induced voltage, v_L, across an inductor is said to lead the inductor current, i_L, by a phase angle of 90°.

Proportional — a mathematical term used to describe the relationship between two quantities. For example, in the formula $X_L = 2\pi fL$, X_L is directly proportional to both the frequency, f, and the inductance, L. The term proportional means that if either f or L is doubled X_L will double. Similarly, if either f or L is reduced by one-half, X_L will be reduced by one-half. In other words, X_L will increase or decrease in direct proportion to either f or L.

Related Formulas

$$X_L = 2\pi fL$$

$$L = \frac{X_L}{2\pi f}$$

$$f = \frac{X_L}{2\pi L}$$

$$X_L = \frac{V_L}{I_L}$$

$$X_{L_T} = X_{L_1} + X_{L_2} + X_{L_3} + \cdots + \text{etc. (Series inductors)}$$

$$X_{L_{EQ}} = \frac{1}{1/X_{L_1} + 1/X_{L_2} + 1/X_{L_3} + \cdots + \text{etc.}} \text{ (Parallel inductors)}$$

Self-Test

Answers at the back of the book.

1. The unit of inductive reactance, X_L, is the
a. henry.
b. ohm.
c. farad.
d. hertz.

2. The inductive reactance, X_L, of an inductor is
a. inversely proportional to frequency.
b. unaffected by frequency.
c. directly proportional to frequency.
d. inversely proportional to inductance.

3. For an inductor, the induced voltage, V_L,
a. leads the inductor current, i_L, by 90°.
b. lags the inductor current, i_L, by 90°.
c. is in phase with the inductor current, i_L.
d. none of the above.

4. For a steady DC current, the X_L of an inductor is
a. infinite.
b. extremely high.
c. usually about 10 kΩ.
d. 0 Ω.

5. What is the inductive reactance, X_L, of a 100-mH coil at a frequency of 3.183 kHz?
a. 2 kΩ.
b. 200 Ω.
c. 1 M Ω.
d. 4 Ω.

6. At what frequency does a 60-mH inductor have an X_L value of 1 kΩ?
a. 377 Hz.
b. 265 kHz.
c. 2.65 kHz.
d. 15.9 kHz.

7. What value of inductance will provide an X_L of 500 Ω at a frequency of 159.15 kHz?
a. 5 H.
b. 500 μH.
c. 500 mH.
d. 750 μH.

8. Two inductors, L_1 and L_2, are in series. If $X_{L_1} = 4$ kΩ and $X_{L_2} = 2$ kΩ, how much is X_{L_T}?
a. 6 kΩ.
b. 1.33 kΩ.
c. 4.47 kΩ.
d. 2 kΩ.

9. Two inductors, L_1 and L_2, are in parallel. If $X_{L_1} = 1$ kΩ and $X_{L_2} =$ 1 kΩ, how much is $X_{L_{EQ}}$?
a. 707 Ω.
b. 2 kΩ.
c. 1.414 kΩ.
d. 500 Ω.

10. How much is the inductance of a coil that draws 25 mA of current from a 24-V_{AC} source whose frequency is 1 kHz?
a. 63.7 μH.
b. 152.8 mH.
c. 6.37 H.
d. 15.28 mH.

Essay Questions

1. Explain briefly why X_L limits the amount of alternating current.

2. Give two differences and one similarity between X_L and R.

3. Explain why X_L increases with higher frequencies and more inductance.

4. Give two differences between the inductance L of a coil and its reactance X_L.

5. Why are the waves in Fig. 20–7*a* and *b* considered 90° out of phase, whereas the waves in Fig. 20–7*b* and *c* have the same phase?

6. Referring to Fig. 20–3, how does this graph show a linear proportion between X_L and frequency?

7. Referring to Fig. 20–4, how does this graph show a linear proportion between X_L and L?

8. Referring to Fig. 20–3, tabulate the values of L that would be needed for each frequency listed but for an X_L of 2000 Ω. (Do not include 0 Hz.)

9. (**a**) Draw the circuit for a 40-Ω R across a 120-V, 60-Hz source. (**b**) Draw the circuit for a 40-Ω X_L across a 120-V, 60-Hz source. (**c**) Why is I equal to 3 A for both circuits? (**d**) Give two differences between the circuits.

10. Why are coils for rf applications generally smaller than af coils?

Problems

SECTION 20–1 HOW X_L REDUCES THE AMOUNT OF I

20–1 How much is the inductive reactance, X_L, of a coil for a steady DC current?

20–2 List two factors that determine the amount of inductive reactance a coil will have.

20–3 In Fig. 20–9, how much DC current will be indicated by the ammeter, M_1, with S_1 in position 1?

Figure 20-9

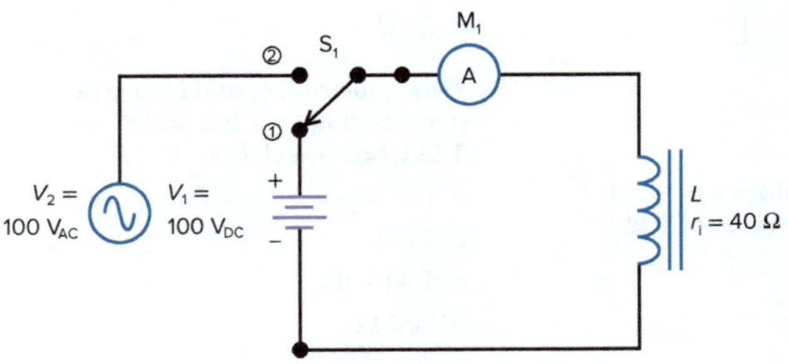

20–4 In Fig. 20–9, how much inductive reactance, X_L, does the coil have with S_1 in position 1? Explain your answer.

20–5 In Fig. 20–9, the ammeter, M_1, reads an AC current of 25 mA with S_1 in position 2.

a. Why is there less current in the circuit with S_1 in position 2 compared to position 1?

b. How much is the inductive reactance, X_L, of the coil? (Ignore the effect of the coil resistance, r_i.)

20–6 In Fig. 20–10, how much is the inductive reactance, X_L, for each of the following values of V_{AC} and I?

a. $V_{AC} = 10$ V and $I = 2$ mA.

b. $V_{AC} = 50$ V and $I = 20$ μA.

c. $V_{AC} = 12$ V and $I = 15$ mA.

d. $V_{AC} = 6$ V and $I = 40$ μA.

e. $V_{AC} = 120$ V and $I = 400$ mA.

SECTION 20–2 $X_L = 2\pi fL$

Figure 20-10

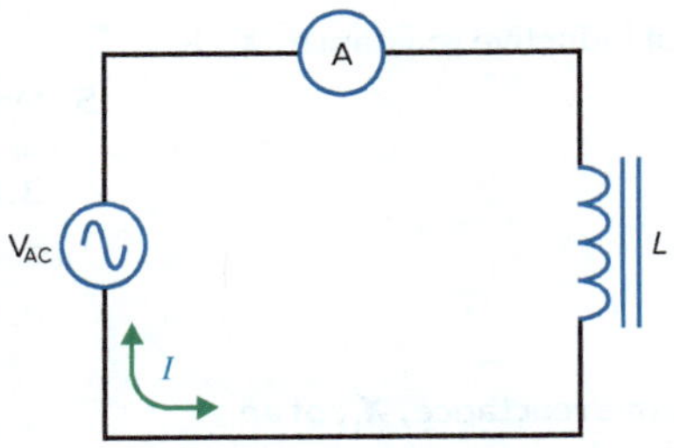

20–7 Calculate the inductive reactance, X_L, of a 100-mH inductor at the following frequencies:

a. $f = 60$ Hz.

b. $f = 120$ Hz.

c. $f = 1.592$ kHz.

d. $f = 10$ kHz.

20–8 Calculate the inductive reactance, X_L, of a 50-μH coil at the following frequencies:

a. $f = 60$ Hz.

b. $f = 10$ kHz.

c. $f = 500$ kHz.

d. $f = 3.8$ MHz.

20–9 What value of inductance, L, will provide an X_L value of 1 kΩ at the following frequencies?

a. $f = 318.3$ Hz.

b. $f = 1.591$ kHz.

c. $f = 5$ kHz.

d. $f = 6.36$ kHz.

20–10 At what frequency will a 30-mH inductor provide an X_L value of

a. 50 Ω?

b. 200 Ω?

c. 1 kΩ?

d. 40 kΩ?

20–11 How much is the inductance of a coil that draws 15 mA from a 24-V_{AC} source whose frequency is 1 kHz?

20–12 At what frequency will a stray inductance of 0.25 μH have an X_L value of 100 Ω?

20–13 A 25-mH coil draws 2 mA of current from a 10-V_{AC} source. What is the value of current drawn by the inductor when

a. the frequency is doubled?

b. the frequency is reduced by one-half?

c. the inductance is doubled to 50 mH?

d. the inductance is reduced by one-half to 12.5 mH?

20–14 A coil has an inductive reactance, X_L, of 10 kΩ at a given frequency. What is the value of X_L when the frequency is

a. cut in half?

b. doubled?

c. quadrupled?

d. increased by a factor 10?

20–15 Calculate the inductive reactance, X_L, for the following inductance and frequency values:

a. $L = 7$ H, $f = 60$ Hz.

b. $L = 25$ μH, $f = 7$ MHz.

c. $L = 500$ mH, $f = 318.31$ Hz.

d. $L = 1$ mH, $f = 159.2$ kHz.

20–16 Determine the inductance value for the following frequency and X_L values:

a. $X_L = 50$ Ω, $f = 15.91$ kHz.

b. $X_L = 2$ kΩ, $f = 5$ kHz.

c. $X_L = 10$ Ω, $f = 795.7$ kHz.

d. $X_L = 4$ kΩ, $f = 6$ kHz.

20–17 Determine the frequency for the following inductance and X_L values:

a. $L = 80$ mH, $X_L = 1$ kΩ.

b. $L = 60$ μH, $X_L = 200$ Ω.

c. $L = 5$ H, $X_L = 100$ kΩ.

d. $L = 150$ mH, $X_L = 7.5$ kΩ.

SECTION 20–3 SERIES AND PARALLEL INDUCTIVE REACTANCES

20–18 How much is the total inductive reactance, X_{L_T}, for the following series inductive reactances:

a. $X_{L_1} = 250$ Ω and $X_{L_2} = 1.5$ kΩ.

b. $X_{L_1} = 200$ Ω, $X_{L_2} = 400$ Ω and $X_{L_3} = 800$ Ω.

c. $X_{L_1} = 10$ kΩ, $X_{L_2} = 30$ kΩ and $X_{L_3} = 15$ kΩ.

d. $X_{L_1} = 1.8$ kΩ, $X_{L_2} = 2.2$ kΩ and $X_{L_3} = 1$ kΩ.

20–19 What is the equivalent inductive reactance, $X_{L_{EQ}}$, for the following parallel inductive reactances?

a. $X_{L_1} = 1.2$ kΩ and $X_{L_2} = 1.8$ kΩ.

b. $X_{L_1} = 1.5$ kΩ and $X_{L_2} = 1$ kΩ.

c. $X_{L_1} = 1.2$ kΩ, $X_{L_2} = 400$ Ω and $X_{L_3} = 300$ Ω.

d. $X_{L_1} = 1$ kΩ, $X_{L_2} = 4$ kΩ, $X_{L_3} = 800$ Ω, and $X_{L_4} = 200$ Ω.

SECTION 20–4 OHM'S LAW APPLIED TO X_L

20–20 In Fig. 20–11, calculate the current, I.

Figure 20–11

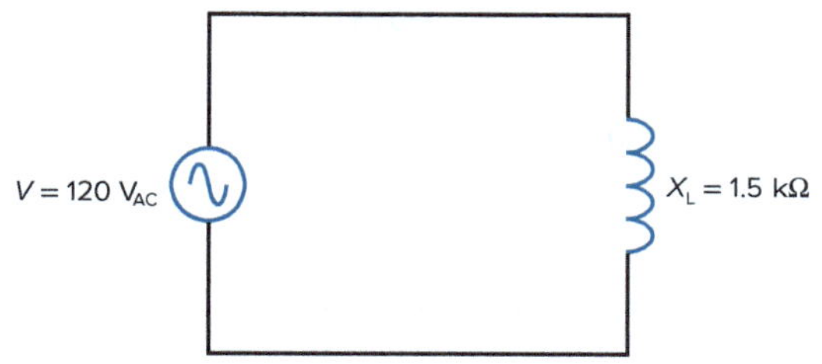

20–21 In Fig. 20–11, what happens to the current, I, when the frequency of the applied voltage

a. decreases?

b. increases?

20–22 In Fig. 20–12, solve for

a. X_{L_T}.

b. I.

c. V_{L_1}, V_{L_2}, and V_{L_3}.

Figure 20–12

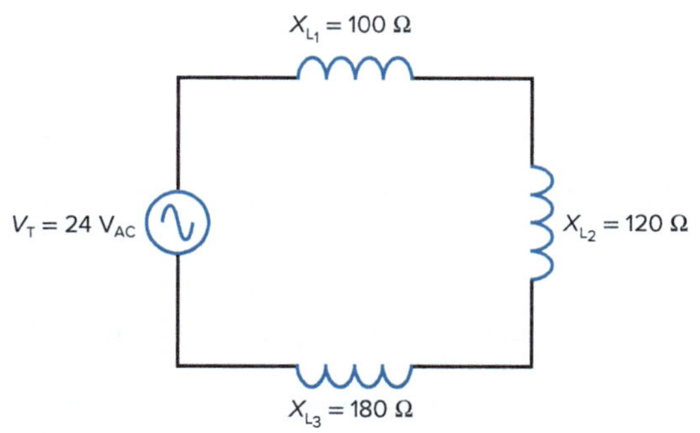

20–23 In Fig. 20–12, solve for L_1, L_2, L_3, and L_T if the frequency of the applied voltage is 1.591 kHz.

20–24 In Fig. 20–13, solve for

a. X_{L_1}, X_{L_2}, and X_{L_3}.

b. X_{L_T}.

c. I.

d. V_{L_1}, V_{L_2}, and V_{L_3}.

e. L_T.

Figure 20–13

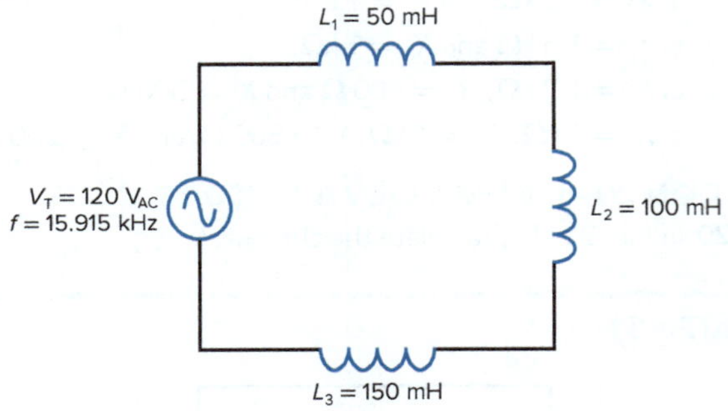

20–25 In Fig. 20–14, solve for

a. I_{L_1}, I_{L_2}, and I_{L_3}.

b. I_T.

c. $X_{L_{EQ}}$.

Figure 20–14

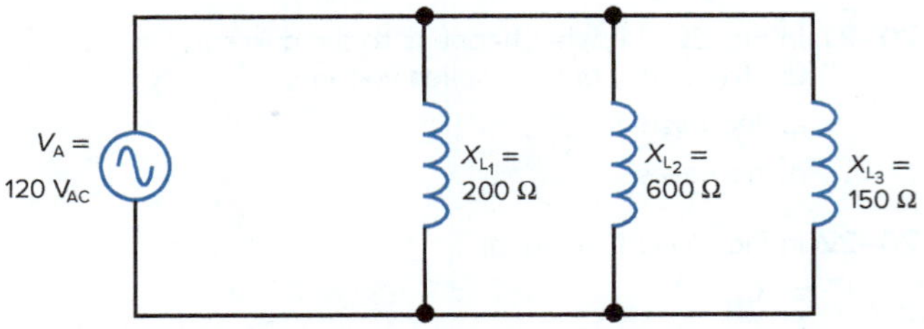

20–26 In Fig. 20–14, solve for L_1, L_2, L_3, and L_T if the frequency of the applied voltage is 6.366 kHz.

20–27 In Fig. 20–15, solve for

a. X_{L_1}, X_{L_2}, and X_{L_3}.

b. I_{L_1}, I_{L_2}, and I_{L_3}.

c. I_T.

d. $X_{L_{EQ}}$.

e. L_{EQ}.

Figure 20–15

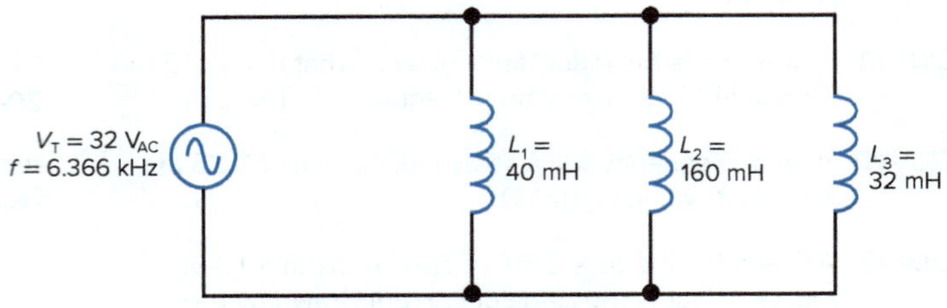

SECTION 20–5 APPLICATIONS OF X_L FOR DIFFERENT FREQUENCIES

20–28 Calculate the value of inductance, L, required to produce an X_L value of 500 Ω at the following frequencies:

a. f = 250 Hz.

b. f = 636.6 Hz.

c. f = 3.183 kHz.

d. f = 7.957 kHz.

SECTION 20–6 WAVESHAPE OF V_L INDUCED BY SINE-WAVE CURRENT

20–29 For an inductor, what is the phase relationship between the induced voltage, V_L, and the inductor current, i_L? Explain your answer.

20–30 For a sine wave of alternating current flowing through an inductor, at what angles in the cycle will the induced voltage be

a. maximum?

b. zero?

Critical Thinking

20–31 In Fig. 20–16, calculate L_1, L_2, L_3, L_T, X_{L_1}, X_{L_2}, X_{L_T}, V_{L_1}, V_{L_3}, I_{L_2}, and I_{L_3}.

20–32 Two inductors in series without L_M have a total inductance L_T of 120 μH. If $L_1/L_2 = 1/20$, what are the values for L_1 and L_2?

20–33 Three inductors in parallel have an equivalent inductance L_{EQ} of 7.5 mH. If $L_2 = 3\,L_3$ and $L_3 = 4\,L_1$, calculate L_1, L_2, and L_3.

Figure 20–16 Circuit for Critical Thinking Prob. 20–31.

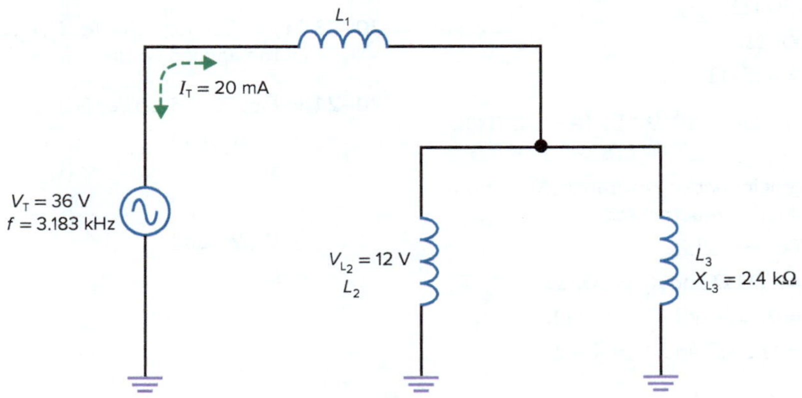

Answers to Self-Reviews

20–1 **a.** 0 Ω
b. 1000 Ω

20–2 **a.** 628 Ω
b. 314 Ω
c. 6280 Ω

20–3 **a.** 500 Ω
b. 120 Ω

20–4 **a.** 0.5 A
b. 100 V

20–5 **a.** 100 MHz
b. 2000 Ω

20–6 **a.** 90°
b. 0° or 360°
c. 90°

Laboratory Application Assignment

In this lab application assignment, you will examine how the inductive reactance, X_L, of an inductor increases when the frequency, f, increases. You will also see that more inductance, L, at a given frequency results in more inductive reactance, X_L. Finally, you will observe how X_L values combine in series and in parallel.

Equipment: Obtain the following items from your instructor.

- Function generator
- 33-mH and 100-mH inductors
- DMM

DC Resistance, r_i, of a Coil

With a DMM, measure and record the DC resistance of each inductor. Set the DMM to the lowest resistance range when measuring r_i.

r_i = ________ (33 mH)
r_i = ________ (100 mH)

Inductive Reactance, X_L

Refer to Fig. 20–17*a*. Calculate and record the value of X_L for each of the following frequencies listed below. Calculate X_L as $2\pi fL$.

X_L = ________ @ f = 500 Hz
X_L = ________ @ f = 1 kHz
X_L = ________ @ f = 2 kHz

Connect the circuit in Fig. 20–17*a*. Set the voltage source to exactly 5 Vrms. For each of the following frequencies listed below, measure and record the current, I. (Use a DMM to measure I.) Next, calculate X_L as V/I.

I = ________ @ f = 500 Hz; X_L = ________
I = ________ @ f = 1 kHz; X_L = ________
I = ________ @ f = 2 kHz; X_L = ________

How do the experimental values of X_L compare to the calculated values? ________________

Based on your experimental values, what happens to the value of X_L every time the frequency, f, doubles? ________________

Is X_L proportional or inversely proportional to the frequency, f?
____

Figure 20–17

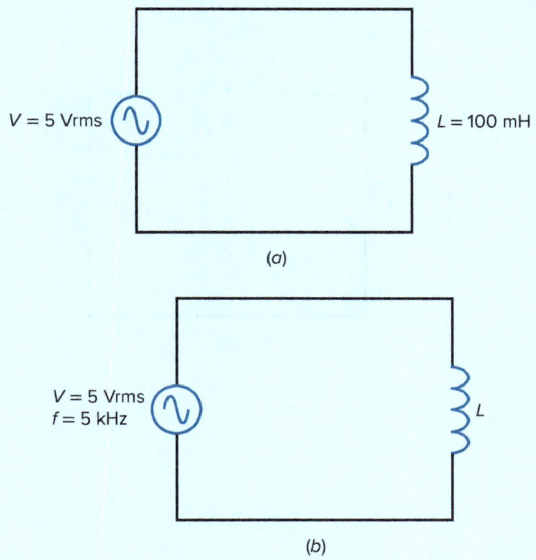

Refer to Fig. 20–17*b*. With the frequency, f, set to 5 kHz, calculate and record the value of X_L for each of the following inductance values listed below. Calculate X_L as $2\pi fL$.

X_L = ________ when L = 33 mH
X_L = ________ when L = 100 mH

Connect the circuit in Fig. 20–17*b*. Adjust the frequency, f, of the function generator to exactly 5 kHz. For each inductance value listed below, measure and record the current, I. (Use a DMM to measure I.) Next, calculate X_L as V/I.

I = ________ when L = 33 mH; X_L = ________
I = ________ when L = 100 mH; X_L = ________

Is X_L proportional or inversely proportional to the value of inductance? ________________

Did the DC resistance, r_i, of the inductors affect any of your measurements? ________________
If so, explain. ________________

Series Inductive Reactances

Refer to the circuit in Fig. 20–18*a*. Calculate and record the following values:

X_{L_1} = ________, X_{L_2} = ________, X_{L_T} = ________,
I = ________, V_{L_1} = ________, V_{L_2} = ________

Do V_{L_1} and V_{L_2} add to equal V_T? ________

Construct the circuit in Fig. 20–18*a*. Set the frequency of the function generator to exactly 5 kHz. Next, using a DMM, measure and record the following values:

I = ________, V_{L_1} = ________, V_{L_2} = ________

Using the measured values of voltage and current, calculate the following values:

X_{L_1} = ________, X_{L_2} = ________, X_{L_T} = ________

Are the experimental values calculated here close to those initially calculated? ________________

Parallel Inductive Reactances

Refer to the circuit in Fig. 20–18*b*. Calculate and record the following values:

X_{L_1} = ________, X_{L_2} = ________, I_{L_1} = ________,
I_{L_2} = ________, I_T = ________, $X_{L_{EQ}}$ = ________

Do I_{L_1} and I_{L_2} add to equal I_T? ________

Construct the circuit in Fig. 20–18*b*. Set the frequency of the function generator to exactly 5 kHz. Next, using a DMM, measure and record the following values:

I_{L_1} = ________, I_{L_2} = ________, I_T = ________

Using the measured values of voltage and current, calculate the following values:

X_{L_1} = ________, X_{L_2} = ________, $X_{L_{EQ}}$ = ________

Are the experimental values calculated here similar to those initially calculated? ________________

Figure 20–18

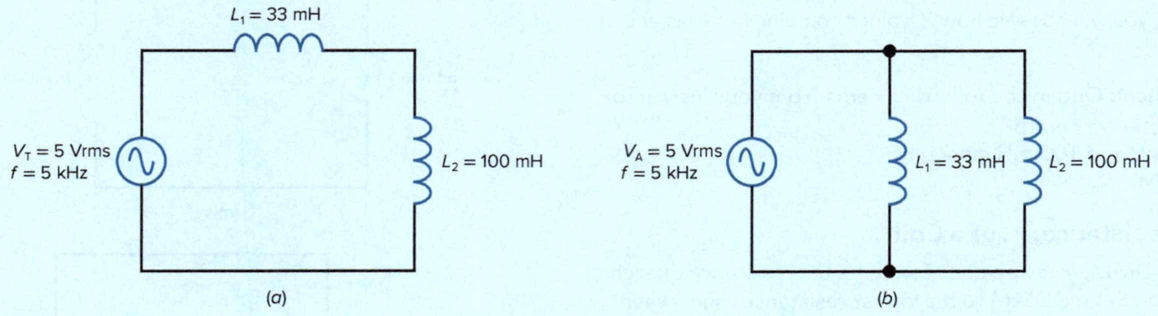

Design credits: Chapter Opener Background Image, and Multisim: Stockbyte/Getty Images

chapter

21

Inductive Circuits

This chapter analyzes circuits that combine inductive reactance X_L and resistance R. The main questions are, how do we combine the ohms of opposition, how much current flows, and what is the phase angle? Although X_L and R are both measured in ohms, they have different characteristics. Specifically, with sine-wave AC current, X_L increases with more inductance and higher frequencies, whereas R is the same for both DC or AC circuits. Furthermore, the phase angle for the voltage across X_L is at 90° with respect to the current through L.

In addition, the practical application of using a coil as a choke to reduce the current for a specific frequency is explained here. For a circuit with L and R in series, the X_L can be high for an undesired AC signal frequency, whereas R is the same for either direct current or alternating current.

Finally, the general case of induced voltage produced across L is shown with nonsinusoidal current variations. In this case, we compare the waveshapes of i_L and v_L instead of their phase. Remember that the 90° angle for an IX_L voltage applies only to sine waves.

With nonsinusoidal waveforms, such as pulses of current or voltage, the circuit can be analyzed in terms of its L/R time constant, as explained in Chapter 22. ■

Chapter Outline

Chapter Objectives

After studying this chapter, you should be able to

- *Explain* why the voltage leads the current by 90° for an inductor.
- *Calculate* the total impedance and phase angle of a series *RL* circuit.
- *Calculate* the total current, equivalent impedance, and phase angle of a parallel *RL* circuit.
- *Define* what is meant by the Q of a coil.
- *Explain* how an inductor can be used to pass some AC frequencies but block others.
- *Calculate* the induced voltage that is produced by a nonsinusoidal current.

Important Terms

AC effective resistance, R_e

arctangent (arctan)

choke

impedance, Z

phase angle, θ

phasor triangle

Q of a coil

skin effect

tangent (tan)

21–1 Sine Wave i_L Lags v_L by 90°

When sine-wave variations of current produce an induced voltage, the current lags its induced voltage by exactly 90°, as shown in Fig. 21–1. The inductive circuit in Fig. 21–1*a* has the current and voltage waveshapes shown in Fig. 21–1*b*. The phasors in Fig. 21–1*c* show the 90° phase angle between i_L and v_L. Therefore, we can say that i_L lags v_L by 90°, or v_L leads i_L by 90°.

This 90° phase relationship between i_L and v_L is true in any sine-wave AC circuit, whether *L* is in series or parallel and whether *L* is alone or combined with other components. We can always say that the voltage across any X_L is 90° out of phase with the current through it.

GOOD TO KNOW

A cosine wave has its maximum values at 0° and 180° and its minimum values at 90° and 270°.

Why the Phase Angle Is 90°

This results because v_L depends on the rate of change of i_L. As previously shown in Fig. 20–7 for a sine wave of i_L, the induced voltage is a cosine wave. In other words, v_L has the phase of di/dt, not the phase of *i*.

Why i_L Lags v_L

The 90° difference can be measured between any two points having the same value on the i_L and v_L waves. A convenient point is the positive peak value. Note that the i_L wave does not have its positive peak until 90° after the v_L wave. Therefore, i_L lags v_L by 90°. This 90° lag is in time. The time lag equals one quarter-cycle, which is one-quarter of the time for a complete cycle.

Inductive Current Is the Same in a Series Circuit

The time delay and resultant phase angle for the current in an inductance apply only with respect to the voltage across the inductance. This condition does not change the fact that the current is the same in all parts of a series circuit. In Fig. 21–1*a*, the current in the generator, the connecting wires, and *L* must be the same because they are in series. Whatever the current value is at any instant, it is the same in all series components. The time lag is between current and voltage.

Inductive Voltage Is the Same across Parallel Branches

In Fig. 21–1*a*, the voltage across the generator and the voltage across *L* are the same because they are in parallel. There cannot be any lag or lead in time between these two parallel voltages. Whatever the voltage value is across the generator at any instant, the voltage across *L* is the same. The parallel voltage v_A or v_L is 90° out of phase with the current.

MultiSim **Figure 21–1** (*a*) Circuit with inductance *L*. (*b*) Sine wave of i_L *lags* v_L by 90°. (*c*) Phasor diagram.

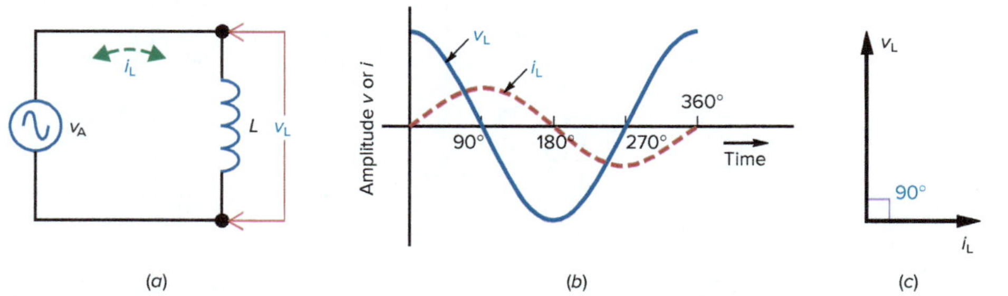

The voltage across L in this circuit is determined by the applied voltage, since they must be the same. The inductive effect here is to make the current have values that produce $L(di/dt)$ equal to the parallel voltage.

The Frequency Is the Same for i_L and v_L

Although i_L lags v_L by 90°, both waves have the same frequency. The i_L wave reaches its peak values 90° later than the v_L wave, but the complete cycles of variations are repeated at the same rate. As an example, if the frequency of the sine wave v_L in Fig. 21–1*b* is 100 Hz, this is also the frequency for i_L.

21–1 Self-Review

Answers at the end of the chapter.

Refer to Fig. 21–1.

a. What is the phase angle between v_A and v_L?
b. What is the phase angle between v_L and i_L?
c. Does i_L lead or lag v_L?

21–2 X_L and R in Series

When a coil has series resistance, the current is limited by both X_L and R. This current I is the same in X_L and R, since they are in series. Each has its own series voltage drop, equal to IR for the resistance and IX_L for the reactance.

Note the following points about a circuit that combines series X_L and R, as in Fig. 21–2:

1. The current is labeled I, rather than I_L, because I flows through all series components.
2. The voltage across X_L, labeled V_L, can be considered an IX_L voltage drop, just as we use V_R for an IR voltage drop.

MultiSim **Figure 21–2** Inductive reactance X_L and resistance R in series. (*a*) Circuit. (*b*) Waveforms of current and voltage. (*c*) Phasor diagram.

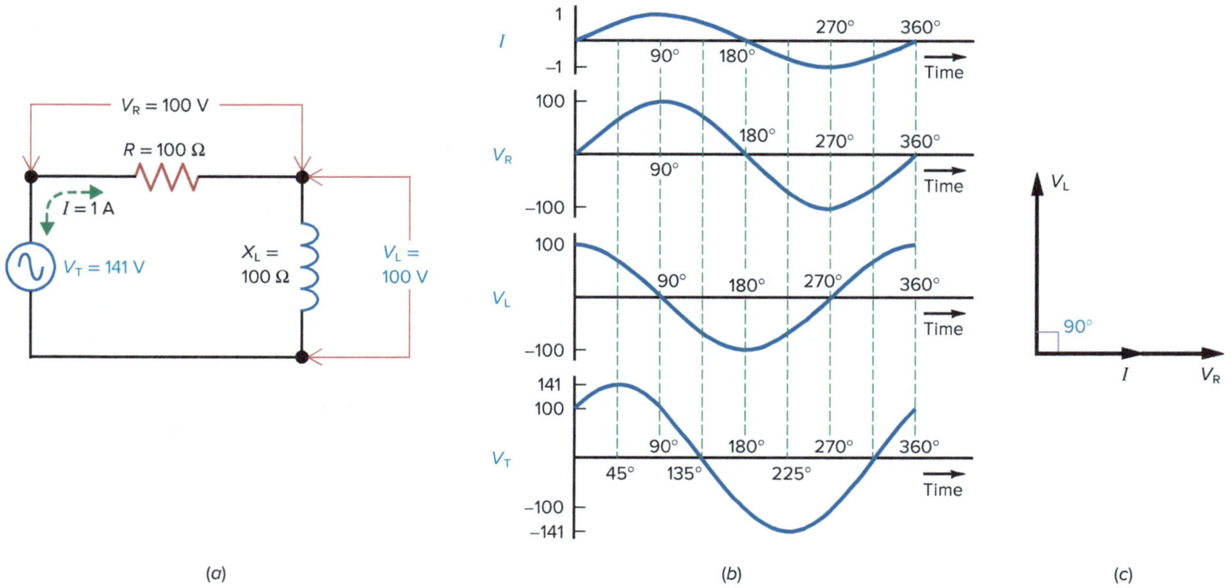

3. The current I through X_L must lag V_L by 90° because this is the phase angle between current through an inductance and its self-induced voltage.
4. The current I through R and its IR voltage drop are in phase. There is no reactance to sine-wave current in any resistance. Therefore, I and IR have a phase angle of 0°.

Resistance R can be either the internal resistance of the coil or an external series resistance. The I and V values may be rms, peak, or instantaneous, as long as the same measure is applied to all. Peak values are used here for convenience in comparing waveforms.

Phase Comparisons

Note the following:

1. Voltage V_L is 90° out of phase with I.
2. However, V_R and I are in phase.
3. If I is used as the reference, V_L is 90° out of phase with V_R.

Specifically, V_R lags V_L by 90°, just as the current I lags V_L. These phase relations are shown by the waveforms in Fig. 21–2*b* and the phasors in Fig. 21–2*c*.

Combining V_R and V_L

As shown in Fig. 21–2*b*, when the V_R voltage wave is combined with the V_L voltage wave, the result is the voltage wave for the applied generator voltage V_T. The voltage drops must add to equal the applied voltage. The 100-V peak values for V_R and for V_L total 141 V, however, instead of 200 V, because of the 90° phase difference.

Consider some instantaneous values to see why the 100-V peak V_R and 100-V peak V_L cannot be added arithmetically. When V_R is at its maximum value of 100 V, for instance, V_L is at zero. The total for V_T then is 100 V. Similarly, when V_L is at its maximum value of 100 V, then V_R is zero and the total V_T is also 100 V.

Actually, V_T has its maximum value of 141 V when V_L and V_R are each 70.7 V. When series voltage drops that are out of phase are combined, therefore, they cannot be added without taking the phase difference into account.

Phasor Voltage Triangle

Instead of combining waveforms that are out of phase, we can add them more quickly by using their equivalent phasors, as shown in Fig. 21–3. The phasors in Fig. 21–3*a* show only the 90° angle without any addition. The method in Fig. 21–3*b* is to add the tail of one phasor to the arrowhead of the other, using the angle required to show their relative phase. Voltages V_R and V_L are at right angles because they are 90° out of phase. The sum of the phasors is a resultant phasor from the start of one to the end of the other. Since the V_R and V_L phasors form a right angle, the resultant phasor is the hypotenuse of a right triangle. The hypotenuse is the side opposite the 90° angle.

From the geometry of a right triangle, the Pythagorean theorem states that the hypotenuse is equal to the square root of the sum of the squares of the sides. For the voltage triangle in Fig. 21–3*b*, therefore, the resultant is

$$V_T = \sqrt{V_R^2 + V_L^2} \qquad \textbf{(21–1)}$$

where V_T is the phasor sum of the two voltages V_R and V_L 90° out of phase.

CALCULATOR

To do a problem like this on the calculator, remember that the square root sign is a sign of grouping. All terms within the group must be added before you take the square root. Also, each term must be squared individually before adding for the sum. Specifically for this problem:

- Punch in 100 and push the (x^2) button for 10,000 as the square. Press (+).
- Next punch in 100 and (x^2). Press (=). The display should read 20,000.
- Press (√) to read the answer 141.421.

In some calculators, either the (x^2) or the (√) key must be preceded by the second function key (2ndF).

Figure 21–3 Addition of two voltages 90° out of phase. (*a*) Phasors for V_L and V_R are 90° out of phase. (*b*) Resultant of the two phasors is the hypotenuse of a right triangle for the value of V_T.

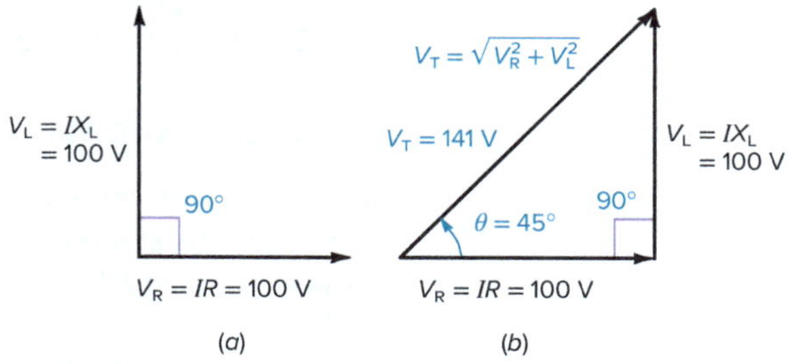

This formula is for V_R and V_L when they are in series, since then they are 90° out of phase. All voltages must be in the same units. When V_T is an rms value, V_R and V_L are also rms values. For the example in Fig. 21–3,

$$V_T = \sqrt{100^2 + 100^2} = \sqrt{10{,}000 + 10{,}000}$$
$$= \sqrt{20{,}000}$$
$$= 141 \text{ V}$$

21–2 Self-Review

Answers at the end of the chapter.

a. In a series circuit with X_L and R, what is the phase angle between I and V_R?

b. What is the phase angle between V_R and V_L?

21–3 Impedance Z Triangle

A triangle of R and X_L in series corresponds to a voltage triangle, as shown in Fig. 21–4. It is similar to the voltage triangle in Fig. 21–3, but the common factor I cancels because the current is the same in X_L and R. The resultant of the phasor addition of R and X_L is their total opposition in ohms, called ***impedance***, with the symbol Z_T.* The Z takes into account the 90° phase relation between R and X_L.

For the impedance triangle of a series circuit with reactance and resistance,

$$Z_T = \sqrt{R^2 + X_L^2} \quad \textbf{(21–2)}$$

where R, X_L, and Z_T are all in ohms. For the example in Fig. 21–4,

$$Z_T = \sqrt{100^2 + 100^2} = \sqrt{10{,}000 + 10{,}000}$$
$$= \sqrt{20{,}000}$$
$$= 141\ \Omega$$

Figure 21–4 Addition of R and X_L 90° out of phase in series circuit, to find the resultant impedance Z_T.

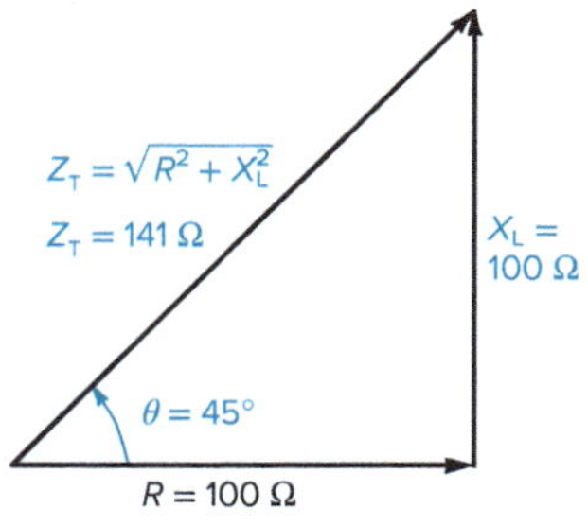

Note that the applied voltage of 141 V divided by the total impedance of 141 Ω results in 1 A of current in the series circuit. The IR voltage is 1 × 100, or 100 V; the IX_L voltage is also 1 × 100, or 100 V. The total of the series IR and IX_L drops of 100 V each, added by phasors, equals the applied voltage of 141 V. Finally, the applied voltage equals IZ, or 1 × 141, which is 141 V.

* Although Z_T is a passive component, we consider it a phasor here because it determines the phase angle of V and I.

Summarizing the similar **phasor triangles** for volts and ohms in a series circuit,

1. The phasor for R, IR, or V_R is used as a reference at 0°.
2. The phasor for X_L, IX_L, or V_L is at 90°.
3. The phasor for Z, IZ, or V_T has the **phase angle θ** of the complete circuit.

Phase Angle with Series X_L

The angle between the generator voltage and its current is the phase angle of the circuit. Its symbol is θ (theta). In Fig. 21–3, the phase angle between V_T and IR is 45°. Since IR and I have the same phase, the angle is also 45° between V_T and I.

In the corresponding impedance triangle in Fig. 21–4, the angle between Z_T and R is also equal to the phase angle. Therefore, the phase angle can be calculated from the impedance triangle of a series circuit by the formula

$$\tan \theta_Z = \frac{X_L}{R} \qquad \textbf{(21–3)}$$

The **tangent (tan)** is a trigonometric function of any angle, equal to the ratio of the opposite side to the adjacent side of a right triangle. In this impedance triangle, X_L is the opposite side and R is the adjacent side of the angle. We use the subscript z for θ to show that θ_Z is found from the impedance triangle for a series circuit. To calculate this phase angle,

$$\tan \theta_Z = \frac{X_L}{R} = \frac{100}{100} = 1$$

The angle whose tangent is equal to 1 is 45°. Therefore, the phase angle is 45° in this example. The numerical values of the trigonometric functions can be found from a table or from a scientific calculator.

Note that the phase angle of 45° is halfway between 0° and 90° because R and X_L are equal.

Example 21–1

If a 30-Ω R and a 40-Ω X_L are in series with 100 V applied, find the following: Z_T, I, V_R, V_L, and θ_Z. What is the phase angle between V_L and V_R with respect to I? Prove that the sum of the series voltage drops equals the applied voltage V_T.

ANSWER

$$Z_T = \sqrt{R^2 + X_L^2} = \sqrt{900 + 1600}$$
$$= \sqrt{2500}$$
$$= 50\ \Omega$$
$$I = \frac{V_T}{Z_T} = \frac{100}{50} = 2\ \text{A}$$
$$V_R = IR = 2 \times 30 = 60\ \text{V}$$
$$V_L = IX_L = 2 \times 40 = 80\ \text{V}$$
$$\tan \theta_Z = \frac{X_L}{R} = \frac{40}{30} = \frac{4}{3} = 1.33$$
$$\theta_Z = 53.1°$$

Therefore, I lags V_T by 53.1°. Furthermore, I and V_R are in phase, and I lags V_L by 90°. Finally,

$$V_T = \sqrt{V_R^2 + V_L^2} = \sqrt{60^2 + 80^2} = \sqrt{3600 + 6400}$$
$$= \sqrt{10{,}000}$$
$$= 100 \text{ V}$$

Note that the phasor sum of the voltage drops equals the applied voltage.

Series Combinations of X_L and R

In a series circuit, the higher the value of X_L compared with R, the more inductive the circuit. This means that there is more voltage drop across the inductive reactance and the phase angle increases toward 90°. The series current lags the applied generator voltage. With all X_L and no R, the entire applied voltage is across X_L, and θ_Z equals 90°.

Several combinations of X_L and R in series are listed in Table 21–1 with their resultant impedance and phase angles. Note that a ratio of 10:1 or more for X_L/R means that the circuit is practically all inductive. The phase angle of 84.3° is only slightly less than 90° for the ratio of 10:1, and the total impedance Z_T is approximately equal to X_L. The voltage drop across X_L in the series circuit will be practically equal to the applied voltage, with almost none across R.

At the opposite extreme, when R is 10 times as large as X_L, the series circuit is mainly resistive. The phase angle of 5.7°, then, means that the current is almost in phase with the applied voltage, the total impedance Z_T is approximately equal to R, and the voltage drop across R is practically equal to the applied voltage, with almost none across X_L.

When X_L and R equal each other, their resultant impedance Z_T is 1.41 times the value of either one. The phase angle then is 45°, halfway between 0° for resistance alone and 90° for inductive reactance alone.

21–3 Self-Review

Answers at the end of the chapter.

a. How much is Z_T for a 20-Ω R in series with a 20-Ω X_L?

b. How much is V_T for 20 V across R and 20 V across X_L in series?

c. What is the phase angle of the circuit in Questions a and b?

Table 21–1 Series R and X_L Combinations

R, Ω	X_L, Ω	Z_T, Ω (Approx.)	Impedance Angle θ_Z
1	10	$\sqrt{101} = 10$	84.3°
10	10	$\sqrt{200} = 14.1$	45°
10	1	$\sqrt{101} = 10$	5.7°

Note: θ_Z is the angle of Z_T with respect to the reference I in a series circuit.

CALCULATOR

To do the trigonometry in Example 21–1 with a calculator, there are several points to keep in mind:

- The ratio of X_L/R specifies the angle's tangent as a numerical value, but this is not the angle θ in degrees. Finding X_L/R is just a division problem.
- The angle θ itself is an *inverse function* of tan θ that is indicated as arctan θ or $\tan^{-1} \theta$. A scientific calculator can give the trigonometric functions directly from the value of the angle or inversely show the angle from its trig functions.
- As a check on your values, note that for $\tan \theta = 1$, $\tan^{-1} \theta$ is 45°. Tangent values less than 1 must be for angles smaller than 45°; angles more than 45° must have tangent values higher than 1.

For the values in Example 21–1, specifically, punch in 40 for X_L, push the (÷) key, punch in 30 for R, and push the (=) key for the ratio of 1.33 on the display. This value is tan θ. While it is on the display, push the (TAN^{-1}) key and the answer of 53.1° appears for angle θ. Use of the (TAN^{-1}) key is usually preceded by pressing the (2ndF) function key.

GOOD TO KNOW

For a series RL circuit, when $X_L \geq 10R$, $Z_T \cong X_L$. When $R \geq 10X_L$, $Z_T \cong R$.

21–4 X_L and R in Parallel

For parallel circuits with X_L and R, the 90° phase angle must be considered for each of the branch currents, instead of the voltage drops. Remember that any series circuit has different voltage drops but one common current. A parallel circuit has different branch currents but one common voltage.

In the parallel circuit in Fig. 21–5*a*, the applied voltage V_A is the same across X_L, R, and the generator, since they are all in parallel. There cannot be any phase difference between these voltages. Each branch, however, has its individual current. For the resistive branch, $I_R = V_A/R$; in the inductive branch, $I_L = V_A/X_L$.

The resistive branch current I_R is in phase with the generator voltage V_A. The inductive branch current I_L lags V_A, however, because the current in an inductance lags the voltage across it by 90°.

The total line current, therefore, consists of I_R and I_L, which are 90° out of phase with each other. The phasor sum of I_R and I_L equals the total line current I_T. These phase relations are shown by the waveforms in Fig. 21–5*b*, and the phasors in Fig. 21–5*c*. Either way, the phasor sum of 10 A for I_R and 10 A for I_L is equal to 14.14 A for I_T.

Both methods illustrate the general principle that quadrature components must be combined by phasor addition. The branch currents are added by phasors here because they are the factors that are 90° out of phase in a parallel circuit. This method is similar to combining voltage drops 90° out of phase in a series circuit.

Phasor Current Triangle

Note that the phasor diagram in Fig. 21–5*c* has the applied voltage V_A of the generator as the reference phasor because V_A is the same throughout the parallel circuit.

The phasor for I_L is down, compared with up for an X_L phasor. Here the parallel branch current I_L lags the parallel voltage reference V_A. In a series circuit, the X_L voltage leads the series current reference I. For this reason, the I_L phasor is shown with a negative 90° angle. The −90° means that the current I_L lags the reference phasor V_A.

MultiSim **Figure 21–5** Inductive reactance X_L and R in parallel. (*a*) Circuit. (*b*) Waveforms of applied voltage and branch currents. (*c*) Phasor diagram.

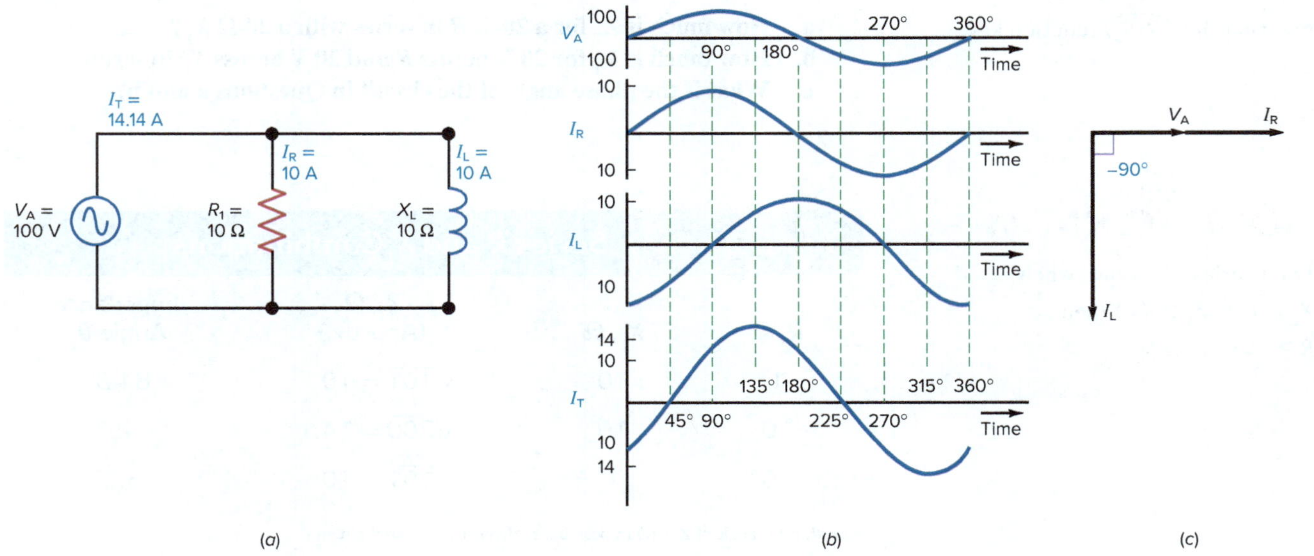

Figure 21–6 Phasor triangle of inductive and resistive branch currents 90° out of phase in a parallel circuit to find resultant I_T.

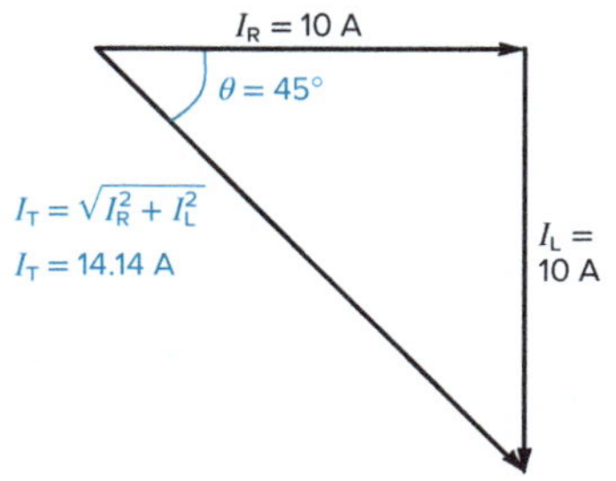

GOOD TO KNOW

For X_L in parallel with R, Z_{EQ} can also be calculated as

$$Z_{EQ} = \frac{X_L R}{\sqrt{R^2 + X_L^2}}.$$

The phasor addition of the branch currents in a parallel circuit can be calculated by the phasor triangle for currents shown in Fig. 21–6. Peak values are used for convenience in this example, but when the applied voltage is an rms value, the calculated currents are also in rms values. To calculate the total line current,

$$I_T = \sqrt{I_R^2 + I_L^2} \tag{21–4}$$

For the values in Fig. 21–6,

$$I_T = \sqrt{10^2 + 10^2} = \sqrt{100 + 100}$$
$$= \sqrt{200}$$
$$= 14.14 \text{ A}$$

Impedance of X_L and R in Parallel

A practical approach to the problem of calculating the equivalent impedance of X_L and R in parallel is to calculate the total line current I_T and divide this value into the applied voltage V_A:

$$Z_{EQ} = \frac{V_A}{I_T} \tag{21–5}$$

For example, in Fig. 21–5, V_A is 100 V and the resultant I_T, obtained as the phasor sum of the resistive and reactive branch currents, is equal to 14.14 A. Therefore, we calculate the equivalent impedance as

$$Z_{EQ} = \frac{V_A}{I_T} = \frac{100 \text{ V}}{14.14 \text{ A}}$$
$$= 7.07\ \Omega$$

This impedance is the combined opposition in ohms across the generator, equal to the resistance of 10 Ω in parallel with the reactance of 10 Ω.

Note that the impedance for equal values of R and X_L in parallel is not one-half but equals 70.7% of either one. Still, the combined value of ohms must be less than the lowest ohms value in the parallel branches.

For the general case of calculating the equivalent impedance of X_L and R in parallel, any number can be assumed for the applied voltage because the value of V_A cancels in the calculations for Z in terms of the branch currents. A good value to assume for V_A is the value of either R or X_L, whichever is the higher number. This way, there are no fractions smaller than 1 in the calculation of the branch currents.

Example 21–2

What is the total Z of a 600-Ω R in parallel with a 300-Ω X_L? Assume 600 V for the applied voltage.

ANSWER

$$I_R = \frac{600 \text{ V}}{600\ \Omega} = 1 \text{ A}$$
$$I_L = \frac{600 \text{ V}}{300\ \Omega} = 2 \text{ A}$$
$$I_T = \sqrt{I_R^2 + I_L^2}$$
$$= \sqrt{1 + 4} = \sqrt{5}$$
$$= 2.24 \text{ A}$$

Then, dividing the assumed value of 600 V for the applied voltage by the total line current gives

$$Z_{EQ} = \frac{V_A}{I_T} = \frac{600\text{ V}}{2.24\text{ A}}$$
$$= 268\ \Omega$$

The combined impedance of a 600-Ω R in parallel with a 300-Ω X_L is equal to 268 Ω, no matter how much the applied voltage is.

GOOD TO KNOW

For parallel RL circuits, $\tan\theta_I$ can also be calculated as

$\tan\theta_I = -\frac{R}{X_L}$.

Phase Angle with Parallel X_L and R

In a parallel circuit, the phase angle is between the line current I_T and the common voltage V_A applied across all branches. However, the resistive branch current I_R has the same phase as V_A. Therefore, the phase of I_R can be substituted for the phase of V_A. This is shown in Fig. 21–5*c*. The triangle of currents is shown in Fig. 21–6. To find θ_I from the branch currents, use the tangent formula:

$$\tan\theta_I = -\frac{I_L}{I_R} \quad \textbf{(21–6)}$$

We use the subscript I for θ to show that θ_I is found from the triangle of branch currents in a parallel circuit. In Fig. 21–6, θ_I is −45° because I_L and I_R are equal. Then $\tan\theta_I = -1$.

The negative sign is used for this current ratio because I_L is lagging at −90°, compared with I_R. The phase angle of −45° here means that I_T lags I_R and V_A by 45°.

Note that the phasor triangle of branch currents gives θ_I as the angle of I_T with respect to the generator voltage V_A. This phase angle for I_T is with respect to the applied voltage as the reference at 0°. For the phasor triangle of voltages in a series circuit, the phase angle θ_Z for Z_T and V_T is with respect to the series current as the reference phasor at 0°.

GOOD TO KNOW

For parallel RL circuits, when $X_L \geq 10R$, $Z_{EQ} \cong R$. When $R \geq 10X_L$, $Z_{EQ} \cong X_L$.

Parallel Combinations of X_L and R

Several combinations of X_L and R in parallel are listed in Table 21–2. When X_L is 10 times R, the parallel circuit is practically resistive because there is little inductive current in the line. The small value of I_L results from the high X_L. The total impedance of the parallel circuit is approximately equal to the resistance, then, since the high value of X_L in a parallel branch has little effect. The phase angle of −5.7° is practically 0° because almost all of the line current is resistive.

As X_L becomes smaller, it provides more inductive current in the main line. When X_L is 1⁄10 R, practically all of the line current is the I_L component. Then the

Table 21–2 Parallel Resistance and Inductance Combinations*

R, Ω	X_L, Ω	I_R, A	I_L, A	I_T, A (Approx.)	$Z_{EQ} = V_A/I_T$, Ω	Phase Angle θ_I
1	10	10	1	$\sqrt{101} = 10$	1	−5.7°
10	10	1	1	$\sqrt{2} = 1.4$	7.07	−45°
10	1	1	10	$\sqrt{101} = 10$	1	−84.3°

* V_A = 10 V. Note that θ_I is the angle of I_T with respect to the reference V_A in parallel circuits.

parallel circuit is practically all inductive, with a total impedance practically equal to X_L. The phase angle of −84.3° is almost −90° because the line current is mostly inductive. Note that these conditions are opposite from those of X_L and R in series.

When X_L and R are equal, their branch currents are equal and the phase angle is −45°. All these phase angles are negative for parallel I_L and I_R.

As additional comparisons between series and parallel circuits, remember that

1. The series voltage drops V_R and V_L have individual values that are 90° out of phase. Therefore, V_R and V_L are added by phasors to equal the applied voltage V_T. The phase angle θ_Z is between V_T and the common series current I. More series X_L allows more V_L to make the circuit more inductive, with a larger positive phase angle for V_T with respect to I.
2. The parallel branch currents I_R and I_L have individual values that are 90° out of phase. Therefore, I_R and I_L are added by phasors to equal I_T, which is the main-line current. The negative phase angle $-\theta_I$ is between the line current I_T and the common parallel voltage V_A. Less parallel X_L allows more I_L to make the circuit more inductive, with a larger negative phase angle for I_T with respect to V_A.

21–4 Self-Review

Answers at the end of the chapter.

a. How much is I_T for a branch current I_R of 2 A and I_L of 2 A?
b. Find the phase angle θ_I.

21–5 Q of a Coil

Figure 21–7 The Q of a coil depends on its inductive reactance X_L and resistance r_i.

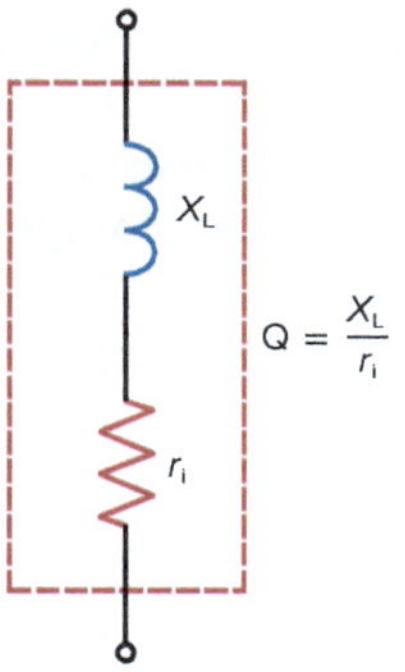

The ability of a coil to produce self-induced voltage is indicated by X_L, since it includes the factors of frequency and inductance. However, a coil has internal resistance equal to the resistance of the wire in the coil. This internal r_i of the coil reduces the current, which means less ability to produce induced voltage. Combining these two factors of X_L and r_i, the *quality* or *merit* of a coil is indicated by

$$Q = \frac{X_L}{r_i} = \frac{2\pi fL}{r_i} \qquad \textbf{(21–7)}$$

As shown in Fig. 21–7, the internal r_i is in series with X_L.

As an example, a coil with X_L of 500 Ω and r_i of 5 Ω has a Q of $^{500}/_5 = 100$. The Q is a numerical value without any units, since the ohms cancel in the ratio of reactance to resistance. This Q of 100 means that the X_L of the coil is 100 times more than its r_i.

The Q of coils may range in value from less than 10 for a low-Q coil up to 1000 for a very high Q. Radio-frequency (rf) coils generally have Qs of about 30 to 300.

At low frequencies, r_i is just the DC resistance of the wire in the coil. However, for rf coils, the losses increase with higher frequencies and the effective r_i increases. The increased resistance results from eddy currents and other losses.

Because of these losses, the ***Q* of a coil** does not increase without limit as X_L increases for higher frequencies. Generally, Q can increase by a factor of about 2 for higher frequencies, within the range for which the coil is designed. The highest Q for rf coils generally results from an inductance value that provides an X_L of about 1000 Ω at the operating frequency.

More fundamentally, Q is defined as the ratio of reactive power in the inductance to the real power dissipated in the resistance. Then

$$Q = \frac{P_L}{P_{r_i}} = \frac{I^2 X_L}{I^2 r_i} = \frac{X_L}{r_i} = \frac{2\pi fL}{r_i}$$

which is the same as Formula (21–7).

GOOD TO KNOW

The Q of a parallel RL circuit is calculated as $Q = \frac{R}{X_L}$, assuming the series resistance of the coil is negligible. The Q formula for parallel RL circuits is derived as follows: $Q = \frac{P_L}{P_R} = \frac{V_A^2/X_L}{V_A^2/R} = \frac{R}{X_L}$.

Note: R is the resistance in parallel with X_L.

Skin Effect

Radio-frequency current tends to flow at the surface of a conductor at very high frequencies (VHF), with little current in the solid core at the center. This **skin effect** results from the fact that current in the center of the wire encounters slightly more inductance because of the magnetic flux concentrated in the metal, compared with the edges, where part of the flux is in air. For this reason, conductors for VHF currents are often made of hollow tubing. The skin effect increases the effective resistance because a smaller cross-sectional area is used for the current path in the conductor.

AC Effective Resistance

When the power and current applied to a coil are measured for rf applied voltage, the I^2R loss corresponds to a much higher resistance than the DC resistance measured with an ohmmeter. This higher resistance is the **AC effective resistance R_e**. Although it is a result of high-frequency alternating current, R_e is not a reactance; R_e is a resistive component because it draws in-phase current from the AC voltage source.

The factors that make the R_e of a coil more than its DC resistance include skin effect, eddy currents, and hysteresis losses. Air-core coils have low losses but are limited to small values of inductance.

For a magnetic core in rf coils, a powdered-iron or ferrite slug is generally used. In a powdered-iron slug, the granules of iron are insulated from each other to reduce eddy currents. Ferrite materials have small eddy-current losses because they are insulators, although magnetic. A ferrite core is easily saturated. Therefore, its use must be limited to coils with low values of current. A common application is the ferrite-core antenna coil in Fig. 21–8.

To reduce the R_e for small rf coils, stranded wire can be made with separate strands insulated from each other and braided so that each strand is as much on the outer surface as all other strands. This is called *litzendraht* or *litz wire*.

As an example of the total effect of AC losses, assume that an air-core rf coil of 50-μH inductance has a DC resistance of 1 Ω measured with the battery in an ohmmeter. However, in an AC circuit with a 2-MHz current, the effective coil resistance R_e can increase to 12 Ω. The increased resistance reduces the Q of the coil.

Actually, the Q can be used to determine the effective AC resistance. Since Q is X_L/R_e, then R_e equals X_L/Q. For this 50-μH L at 2 MHz, its X_L, equal to $2\pi fL$, is 628 Ω.

Figure 21–8 Ferrite-coil antenna for a radio receiver.

Mark Steinmetz/McGraw Hill

The Q of the coil can be measured on a Q meter, which operates on the principle of resonance. Let the measured Q be 50. Then $R_e = 628/50$ equal to 12.6 Ω.

In general, the lower the internal resistance of a coil, the higher its Q.

Example 21–3

An air-core coil has an X_L of 700 Ω and an R_e of 2 Ω. Calculate the value of Q for this coil.

ANSWER

$$Q = \frac{X_L}{R_e} = \frac{700}{2}$$
$$= 350$$

Example 21–4

A 200-μH coil has a Q of 40 at 0.5 MHz. Find R_e.

ANSWER

$$R_e = \frac{X_L}{Q} = \frac{2\pi fL}{Q}$$
$$= \frac{2\pi \times 0.5 \times 10^6 \times 200 \times 10^{-6}}{40}$$
$$= \frac{628}{40}$$
$$= 15.7\ \Omega$$

The *Q* of a Capacitor

The quality Q of a capacitor in terms of minimum loss is often indicated by its power factor. The lower the numerical value of the power factor, the better the quality of the capacitor. Since the losses are in the dielectric, the power factor of the capacitor is essentially the power factor of the dielectric, independent of capacitance value or voltage rating. At radio frequencies, approximate values of power factor are 0.000 for air or vacuum, 0.0004 for mica, about 0.01 for paper, and 0.0001 to 0.03 for ceramics.

The reciprocal of the power factor can be considered the Q of the capacitor, similar to the idea of the Q of a coil. For instance, a power factor of 0.001 corresponds to a Q of 1000. A higher Q therefore means better quality for the capacitor. If the leakage resistance R_l is known, the Q can be calculated as $Q = 2\pi fR_lC$. Capacitors have Qs that are much higher than those of inductors. The Q of capacitors typically ranges into the thousands, depending on design.

21–5 Self-Review

Answers at the end of the chapter.

a. A 200-μH coil with an 8-Ω internal R_e has an X_L of 600 Ω. Calculate the Q.

b. A coil with a Q of 50 has a 500-Ω X_L at 4 MHz. Calculate its internal R_e.

GOOD TO KNOW

Always remember that an $\frac{X_L}{R}$ ratio of 10:1 is considered the dividing line for calculating the value of the choke inductance, *L*. This 10:1 ratio of $\frac{X_L}{R}$ should exist for the lowest frequency intended to be blocked from the output.

Figure 21–9 Coil used as a choke with X_L at least $10 \times R$. Note that *R* is an external resistor; V_L across *L* is practically all of the applied voltage with very little V_R. (*a*) Circuit with X_L and *R* in series. (*b*) Input and output voltages.

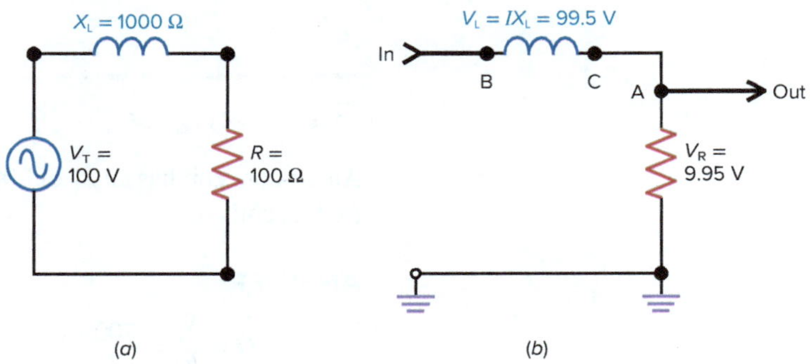

21–6 AF and RF Chokes

Inductance has the useful characteristic of providing more ohms of reactance at higher frequencies. Resistance has the same opposition at all frequencies and for direct current. The skin effect for *L* at VHF is not being considered here. These characteristics of *L* and *R* are applied to the circuit in Fig. 21–9 where X_L is much greater than *R* for the frequency of the AC source V_T. The result is that *L* has practically all the voltage drop in this series circuit with very little of the applied voltage across *R*.

The inductance *L* is used here as a ***choke***. Therefore, a choke is an inductance in series with an external *R* to prevent the AC signal voltage from developing any appreciable output across *R* at the frequency of the source.

The dividing line in calculations for a choke can be taken as X_L 10 or more times the series *R*. Then the circuit is primarily inductive. Practically all the AC voltage drop is across *L*, with little across *R*. This case also results in θ of practically 90°, but the phase angle is not related to the action of X_L as a choke.

Figure 21–9*b* illustrates how a choke is used to prevent AC voltage in the input from developing voltage in the output for the next circuit. Note that the output here is V_R from point A to earth ground. Practically all AC input voltage is across X_L between points B and C. However, this voltage is not coupled out because neither B nor C is grounded.

The desired output across *R* could be direct current from the input side without any AC component. Then X_L has no effect on the steady DC component. Practically all DC voltage would be across *R* for the output, but the AC voltage would be just across X_L. The same idea applies to passing an af signal through to *R*, while blocking an rf signal as IX_L across the choke because of more X_L at the higher frequency.

Calculations for a Choke

Typical values for audio or radio frequencies can be calculated if we assume a series resistance of 100 Ω as an example. Then X_L must be at least 1000 Ω. As listed in Table 21–3, at 100 Hz the relatively large inductance of 1.6 H provides 1000 Ω of X_L. Higher frequencies allow a smaller value of *L* for a choke with the same reactance. At 100 MHz in the VHF range, the choke is only 1.6 μH.

Some typical chokes are shown in Fig. 21–10. The iron-core choke in Fig. 21–10*a* is for audio frequencies. The air-core choke in Fig. 21–10*b* is for radio frequencies.

Table 21–3	Typical Chokes for a Reactance of 1000 Ω*	
f	***L***	**Remarks**
100 Hz	1.6 H	Low audio frequency
1000 Hz	0.16 H	Audio frequency
10 kHz	16 mH	Audio frequency
1000 kHz	0.16 mH	Radio frequency
100 MHz	1.6 μH	Very high radio frequency

* For an X_L that is 10 times a series R of 100 Ω.

Figure 21–10 Typical chokes. (*a*) Choke for 60 Hz with 8-H inductance and r_i of 350 Ω. Width is 2 in. (*b*) RF choke with 5 mH of inductance and r_i of 50 Ω. Length is 1 in. (*c*) Small rf choke encapsulated in plastic with leads for printed-circuit board; $L = 42\ \mu$H. Width is ¾ in.

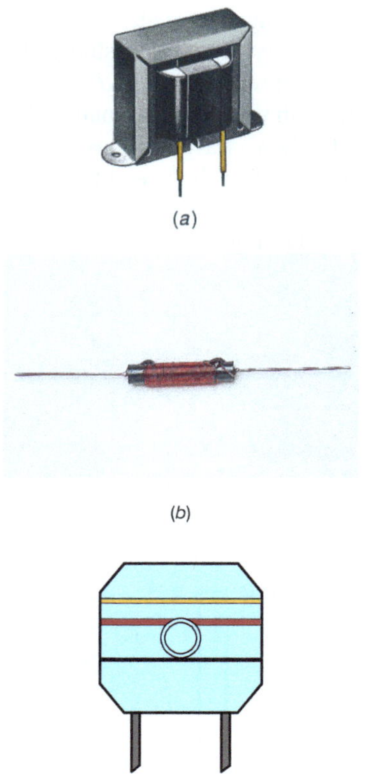

(*a*)

(*b*)

(*c*)

(*b*) Mark Steinmetz/McGraw Hill

The rf choke in Fig. 21–10*c* has color coding, which is often used for small coils. The color values are the same as for resistors, except that the values of *L* are given in microhenrys. As an example, a coil with yellow, red, and black stripes or dots is 42 μH.

Note that inductors are also available as surface-mount components. There are basically two body styles: completely encased and open. The encased body style looks like a thick capacitor with a black body. The open body style inductor is easy to identify because the coil is visible. The value of a surface-mount inductor, if marked, is usually represented using the same three-digit system used for resistors, with the value displayed in microhenrys (μH).

Choosing a Choke for a Circuit

As an example of using these calculations, suppose that we have the problem of determining what kind of coil to use as a choke for the following application. The *L* is to be an rf choke in series with an external *R* of 300 Ω, with a current of 90 mA and a frequency of 0.2 MHz. Then X_L must be at least 10 × 300 = 3000 Ω. At *f* of 0.2 MHz,

$$L = \frac{X_L}{2\pi f} = \frac{3000}{2\pi \times 0.2 \times 10^6} = \frac{3 \times 10^3}{1.256 \times 10^6}$$

$$= \frac{3}{1.256} \times 10^{-3}$$

$$= 2.4 \text{ mH}$$

A typical and easily available commercial size is 2.5 mH, with a current rating of 115 mA and an internal resistance of 20 Ω, similar to the rf choke in Fig. 21–10*b*. Note that the higher current rating is suitable. Also, the internal resistance is negligible compared with the external *R*. An inductance a little higher than the calculated value will provide more X_L, which is better for a choke.

21–6 Self-Review

Answers at the end of the chapter.

a. How much is the minimum X_L for a choke in series with *R* of 80 Ω?

b. If X_L is 800 Ω at 3 MHz, how much will X_L be at 6 MHz for the same coil?

21–7 The General Case of Inductive Voltage

The voltage across any inductance in any circuit is always equal to $L(di/dt)$. This formula gives the instantaneous values of v_L based on the self-induced voltage produced by a change in magnetic flux from a change in current.

A sine waveform of current i produces a cosine waveform for the induced voltage v_L, equal to $L(di/dt)$. This means that v_L has the same waveform as i, but v_L and i are 90° out of phase for sine-wave variations.

The inductive voltage can be calculated as IX_L in sine-wave AC circuits. Since X_L is $2\pi fL$, the factors that determine the induced voltage are included in the frequency and inductance. Usually, it is more convenient to work with IX_L for the inductive voltage in sine-wave AC circuits, instead of $L(di/dt)$.

However, with a nonsinusoidal current waveform, the concept of reactance cannot be used. The X_L applies only to sine waves. Then v_L must be calculated as $L(di/dt)$, which applies for any inductive voltage.

An example is illustrated in Fig. 21–11*a* for sawtooth current. The sawtooth rise is a uniform or linear increase of current from zero to 90 mA in this example. The sharp drop in current is from 90 mA to zero. Note that the rise is relatively slow; it takes 90 μs. This is nine times longer than the fast drop in 10 μs.

The complete period of one cycle of this sawtooth wave is 100 μs. A cycle includes the rise of i to the peak value and its drop back to the starting value.

The Slope of *i*

GOOD TO KNOW

For any inductor, the induced voltage, V_L, remains constant if the rate of current change is constant.

The slope of any curve is a measure of how much it changes vertically for each horizontal unit. In Fig. 21–11*a*, the increase in current has a constant slope. Here i increases 90 mA in 90 μs, or 10 mA for every 10 μs of time. Then di/dt is constant at 10 mA/10 μs for the entire rise time of the sawtooth waveform. Actually, di/dt is the slope of the i curve. The constant di/dt is why the v_L waveform has a constant value of voltage during the linear rise of i. Remember that the amount of induced voltage depends on the change in current with time.

The drop in i is also linear but much faster. During this time, the slope is 90 mA/10 μs for di/dt.

Figure 21–11 Rectangular waveshape of v_L produced by sawtooth current through inductance L. (*a*) Waveform of current i. (*b*) Induced voltage v_L equal to $L(di/dt)$.

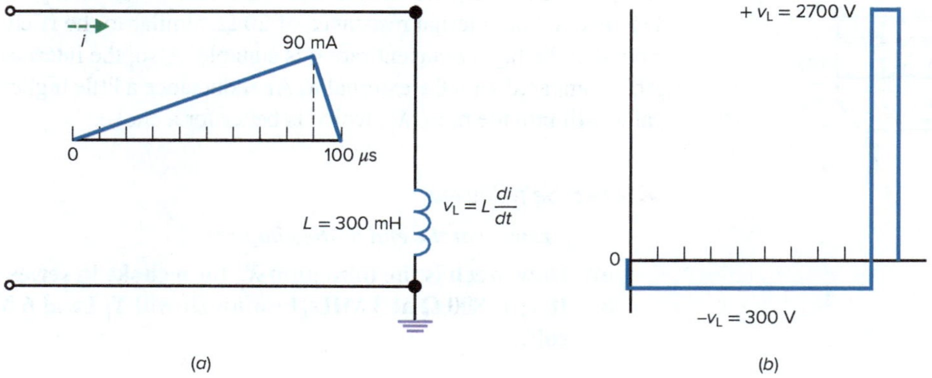

The Polarity of v_L

In Fig. 21–11, apply Lenz's law to indicate that v_L opposes the change in current. With electron flow into the top of L, the v_L is negative to oppose an increase in current. This polarity opposes the direction of electron flow shown for the current i produced by the source. For the rise time, then, the induced voltage here is labeled $-v_L$.

During the drop in current, the induced voltage has opposite polarity, which is labeled $+v_L$. These voltage polarities are for the top of L with respect to earth ground.

Calculations for v_L

The values of induced voltage across the 300-mH L are calculated as follows.

For the sawtooth rise

$$\begin{aligned}-v_L &= L\frac{di}{dt}\\ &= 300 \times 10^{-3} \times \frac{10 \times 10^{-3}}{10 \times 10^{-6}}\\ &= 300 \text{ V}\end{aligned}$$

For the sawtooth drop

$$\begin{aligned}+v_L &= L\frac{di}{dt}\\ &= 300 \times 10^{-3} \times \frac{90 \times 10^{-3}}{10 \times 10^{-6}}\\ &= 2700 \text{ V}\end{aligned}$$

The decrease in current produces nine times more voltage because the sharp drop in i is nine times faster than the relatively slow rise.

Remember that the di/dt factor can be very large, even with small currents, when the time is short. For instance, a current change of 1 mA in 1 μs is equivalent to the very high di/dt value of 1000 A/s.

An interesting feature of the inductive waveshapes in Fig. 21–11 is that they are the same as the capacitive waveshapes shown before in Fig. 18–10, but with current and voltage waveshapes interchanged. This comparison follows from the fact that both v_L and i_C depend on the rate of change. Then i_C is $C(dv/dt)$, and v_L is $L(di/dt)$.

It is important to note that v_L and i_L have different waveshapes with nonsinusoidal current. In this case, we compare the waveshapes instead of the phase angle. Common examples of nonsinusoidal waveshapes for either v or i are the sawtooth waveform, square wave, and rectangular pulses. For a sine wave, the $L(di/dt)$ effects result in a cosine wave, as shown before, in Fig. 20–7.

21–7 Self-Review

Answers at the end of the chapter.

Refer to Fig. 21–11.

a. **How much is di/dt in amperes per second for the sawtooth rise of i?**
b. **How much is di/dt in amperes per second for the drop in i?**

Summary

- In a sine-wave AC circuit, the current through an inductance lags 90° behind the voltage across the inductance because $v_L = L(di/dt)$. This fundamental fact is the basis of all the following relations.
- Therefore, inductive reactance X_L is a phasor quantity 90° out of phase with R. The phasor combination of X_L and R is their impedance Z_T.
- These three types of opposition to current are compared in Table 21–4.
- The phase angle θ is the angle between the applied voltage and its current.
- The opposite characteristics for series and parallel circuits with X_L and R are summarized in Table 21–5.
- The Q of a coil is X_L/r_i, where r_i is the coil's internal resistance.
- A choke is an inductance with X_L greater than the series R by a factor of 10 or more.
- In sine-wave circuits, $V_L = IX_L$. Then V_L is out of phase with I by an angle of 90°.
- For a circuit with X_L and R in series, $\tan \theta_Z = X_L/R$. When the components are in parallel, $\tan \theta_I = -(I_L/I_R)$. See Table 21–5.
- When the current is not a sine wave, $v_L = L(di/dt)$. Then the waveshape of V_L is different from the waveshape of i.
- Inductors are available as surface-mount components. Surface-mount inductors are available in both completely encased and open body styles.

Table 21–4 Comparison of R, X_L, and Z_T

R	$X_L = 2\pi f L$	$Z_T = \sqrt{R^2 + X_L^2}$
Ohm unit	Ohm unit	Ohm unit
IR voltage in phase with I	IX_L voltage leads I by 90°	IZ is applied voltage; it leads line I by $\theta°$
Same for all frequencies	Increases as frequency increases	Increases with X_L at higher frequencies

Table 21–5 Series and Parallel *RL* Circuits

X_L and R in Series	X_L and R in Parallel
I the same in X_L and R	V_A the same across X_L and R
$V_T = \sqrt{V_R^2 + V_L^2}$	$I_T = \sqrt{I_R^2 + I_L^2}$
$Z_T = \sqrt{R^2 + X_L^2}$	$Z_{EQ} = \dfrac{V_A}{I_T}$
V_L leads V_R by 90°	I_L lags I_R by 90°
$\tan \theta_Z = \dfrac{X_L}{R}$	$\tan \theta_I = -\dfrac{I_L}{I_R}$
The θ_Z increases with more X_L, which means more V_L, thus making the circuit more inductive	The $-\theta_I$ decreases with more X_L, which means less I_L, thus making the circuit less inductive

Important Terms

AC effective resistance, R_e — the resistance of a coil for higher-frequency alternating current. The value of R_e is more than the DC resistance of the coil because it includes the losses associated with high-frequency alternating current in a coil. These losses include skin effect, eddy currents, and hysteresis losses.

Arctangent (arctan) — an inverse trigonometric function that specifies the angle, θ, corresponding to a given tangent (tan) value.

Choke — a name used for a coil when its application is to appreciably reduce the amount of AC voltage that is developed across a series resistor, R. The dividing line for calculating the choke inductance is to make X_L 10 or more times larger than the series R at a specific frequency as a lower limit.

Impedance, Z — the total opposition to the flow of current in a sine-wave AC circuit. In an RL circuit, the impedance, Z, takes into account the 90° phase relation between X_L and R. Impedance is measured in ohms.

Phase angle, θ — the angle between the applied voltage and current in a sine-wave AC circuit.

Phasor triangle — a right triangle that represents the phasor sum of two quantities 90° out of phase with each other.

Q of a coil — the quality or figure of merit for a coil. More specifically, the Q of a coil is the ratio of reactive power in the inductance to the real power dissipated in the coil's resistance, $Q = \frac{X_L}{r_i}$.

Skin effect — a term used to describe current flowing on the outer surface of a conductor at VHF. The skin effect causes the effective resistance of a coil to increase at higher frequencies since the effect is the same as reducing the cmil area of the wire.

Tangent (tan) — a trigonometric function of an angle, equal to the ratio of the opposite side to the adjacent side of a right triangle.

Related Formulas

Series RL circuit

$$V_T = \sqrt{V_R^2 + V_L^2}$$

$$Z_T = \sqrt{R^2 + X_L^2}$$

$$\tan \theta_Z = \frac{X_L}{R}$$

Parallel RL circuit

$$I_T = \sqrt{I_R^2 + I_L^2}$$

$$Z_{EQ} = \frac{V_A}{I_T}$$

$$\tan \theta_I = -\frac{I_L}{I_R}$$

Q of a coil

$$Q = \frac{X_L}{r_i} = \frac{2\pi fL}{r_i}$$

Self-Test

Answers at the back of the book.

1. Inductive reactance, X_L,

a. applies only to nonsinusoidal waveforms or DC.

b. applies only to sine waves.

c. applies to either sinusoidal or nonsinusoidal waveforms.

d. is inversely proportional to frequency.

2. For an inductor in a sine-wave AC circuit,

a. V_L leads i_L by 90°.

b. V_L lags i_L by 90°.

c. V_L and i_L are in phase.

d. none of the above.

3. In a series RL circuit,

a. V_L lags V_R by 90°.

b. V_L leads V_R by 90°.

c. V_R and I are in phase.

d. both b and c.

4. In a series RL circuit where $V_L = 9$ V and $V_R = 12$ V, how much is the total voltage, V_T?

a. 21 V.

b. 225 V.

c. 15 V.

d. 3 V.

5. A 50-Ω resistor is in parallel with an inductive reactance, X_L, of 50 Ω. The combined equivalent impedance, Z_{EQ} of this combination is

a. 70.7 Ω.

b. 100 Ω.

c. 35.36 Ω.

d. 25 Ω.

6. In a parallel RL circuit,

a. I_L lags I_R by 90°.

b. I_L leads I_R by 90°.

c. I_L and I_R are in phase.

d. I_R lags I_L by 90°.

7. In a parallel RL circuit, where I_R = 1.2 A and I_L = 1.6 A, how much is the total current, I_T?

a. 2.8 A.

b. 2 A.

c. 4 A.

d. 400 mA.

8. In a series RL circuit where $X_L = R$, the phase angle, θ_Z, is

a. −45°.

b. 0°.

c. +90°.

d. +45°.

9. In a parallel RL circuit,

a. V_A and I_L are in phase.

b. I_L and I_R are in phase.

c. V_A and I_R are in phase.

d. V_A and I_R are 90° out of phase.

10. A 1-kΩ resistance is in series with an inductive reactance, X_L, of 2 kΩ. The total impedance, Z_T, is

a. 2.24 kΩ.

b. 3 kΩ.

c. 1 kΩ.

d. 5 MΩ.

11. When the frequency of the applied voltage decreases in a parallel RL circuit,

a. the phase angle, θ_I, becomes less negative.

b. Z_{EQ} increases.

c. Z_{EQ} decreases.

d. both a and b.

12. When the frequency of the applied voltage increases in a series RL circuit,

a. θ_Z increases.

b. Z_T decreases.

c. Z_T increases.

d. both a and c.

13. The dividing line for calculating the value of a choke inductance is to make

a. X_L 10 or more times larger than the series R.

b. X_L one-tenth or less than the series R.

c. X_L equal to R.

d. R 10 or more times larger than the series X_L.

14. The Q of a coil is affected by

a. frequency.

b. the resistance of the coil.

c. skin effect.

d. all of the above.

15. If the current through a 300-mH coil increases at the linear rate of 50 mA per 10 μs, how much is the induced voltage, V_L?

a. 1.5 V.

b. 1.5 kV.

c. This is impossible to determine because X_L is unknown.

d. This is impossible to determine because V_L also increases at a linear rate.

Essay Questions

1. What characteristic of the current in an inductance determines the amount of induced voltage? State briefly why.

2. Draw a schematic diagram showing an inductance connected across a sine-wave voltage source, and indicate the current and voltage that are 90° out of phase with one another.

3. Why is the voltage across a resistance in phase with the current through the resistance?

4. (a) Draw the sine waveforms for two voltages 90° out of phase, each with a peak value of 100 V. (b) Why does their phasor sum equal 141 V and not 200 V? (c) When will the sum of two 100-V drops in series equal 200 V?

5. (a) Define the phase angle of a sine-wave AC circuit. (b) State the formula for the phase angle in a circuit with X_L and R in series.

6. Define the following: (a) Q of a coil; (b) AC effective resistance; (c) rf choke; (d) sawtooth current.

7. Why do all waveshapes in Fig. 21–2*b* have the same frequency?

8. Describe how to check for an open coil with an ohmmeter.

9. Redraw the circuit and graph in Fig. 21–11 for a sawtooth current with a peak of 30 mA.

10. Why is the R_e of a coil considered resistance rather than reactance?

11. Why are rf chokes usually smaller than af chokes?

12. What is the waveshape of v_L for a sine wave i_L?

Problems

SECTION 21–1 SINE WAVE i_L LAGS v_L BY 90°

21–1 In Fig. 21–12, what is the

a. peak value of the inductor voltage, V_L?

b. peak value of the inductor current, i_L?

c. frequency of the inductor current, i_L?

d. phase relationship between V_L and i_L?

Figure 21–12

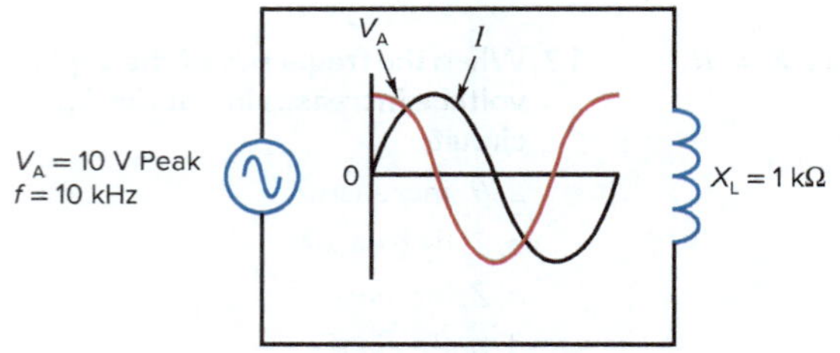

21–2 In Fig. 21–12, what is the value of the induced voltage, V_L, when i_L is at

a. 0 mA?

b. its positive peak of 10 mA?

c. its negative peak of 10 mA?

21–3 In Fig. 21–12, draw the phasors representing V_L and i_L using

a. i_L as the reference phasor.

b. V_L as the reference phasor.

SECTION 21–2 X_L AND R IN SERIES

21–4 In Fig. 21–13, how much current, I, is flowing

a. through the 15-Ω resistor, R?

b. through the 20-Ω inductive reactance, X_L?

c. to and from the terminals of the applied voltage, V_T?

Figure 21–13

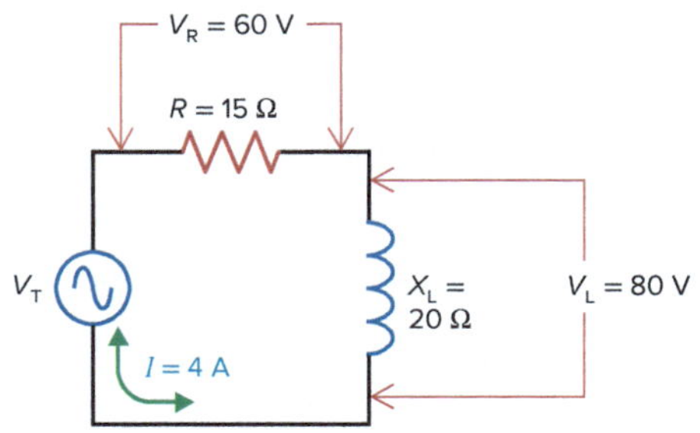

21–5 In Fig. 21–13, what is the phase relationship between

a. I and V_R?

b. I and V_L?

c. V_L and V_R?

21–6 In Fig. 21–13, how much is the applied voltage, V_T?

21–7 Draw the phasor voltage triangle for the circuit in Fig. 21–13. (Use V_R as the reference phasor.)

21–8 In Fig. 21–14, solve for

a. the resistor voltage, V_R.

b. the inductor voltage, V_L.

c. the total voltage, V_T.

Figure 21–14

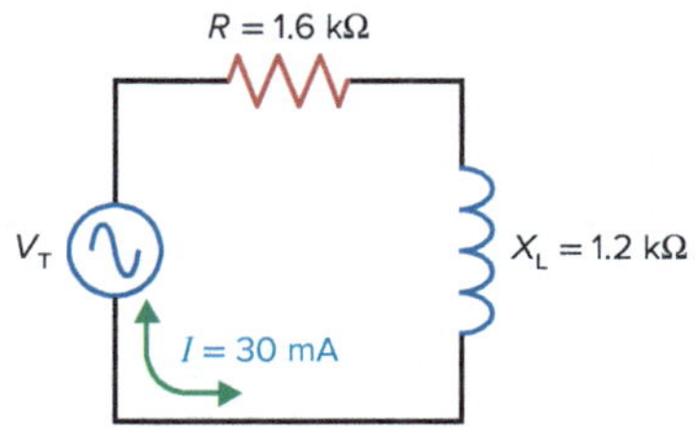

21–9 In Fig. 21–15, solve for

a. the resistor voltage, V_R.

b. the inductor voltage, V_L.

c. the total voltage, V_T.

Figure 21–15

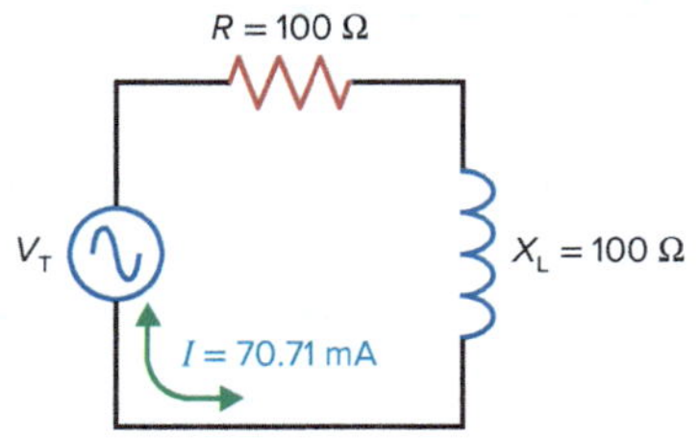

21–10 In a series *RL* circuit, solve for the applied voltage, V_T, if

a. $V_R = 12$ V and $V_L = 6$ V.

b. $V_R = 25$ V and $V_L = 40$ V.

c. $V_R = 9$ V and $V_L = 16$ V.

d. $V_R = 40$ V and $V_L = 40$ V.

SECTION 21–3 IMPEDANCE *Z* TRIANGLE

21–11 In Fig. 21–16, solve for Z_T, I, V_L, V_R, and θ_Z.

Figure 21–16

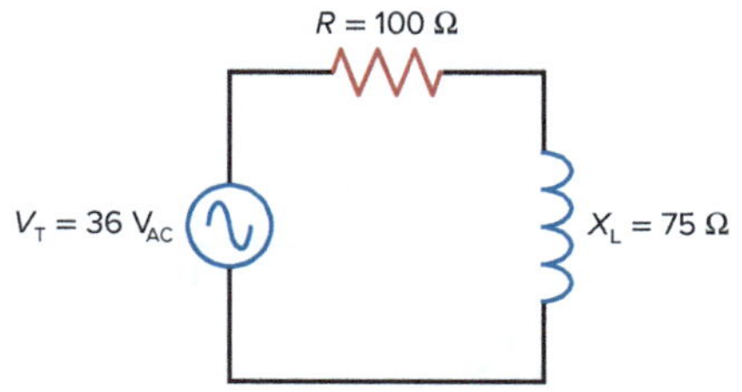

21–12 Draw the impedance triangle for the circuit in Fig. 21–16. (Use *R* as the reference phasor.)

21–13 In Fig. 21–17, solve for Z_T, I, V_L, V_R, and θ_Z.

Figure 21–17

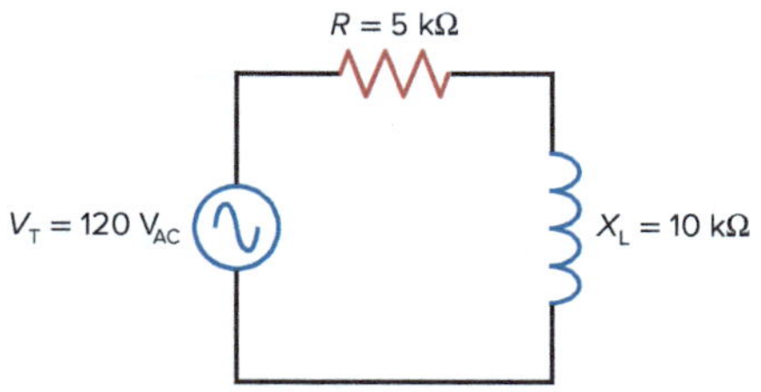

21–14 In Fig. 21–18, solve for Z_T, I, V_L, V_R, and θ_Z.

Figure 21–18

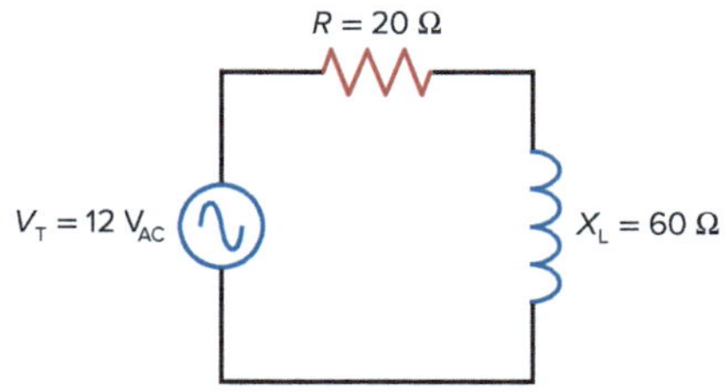

21–15 In Fig. 21–19, solve for Z_T, I, V_L, V_R, and θ_Z.

Figure 21–19

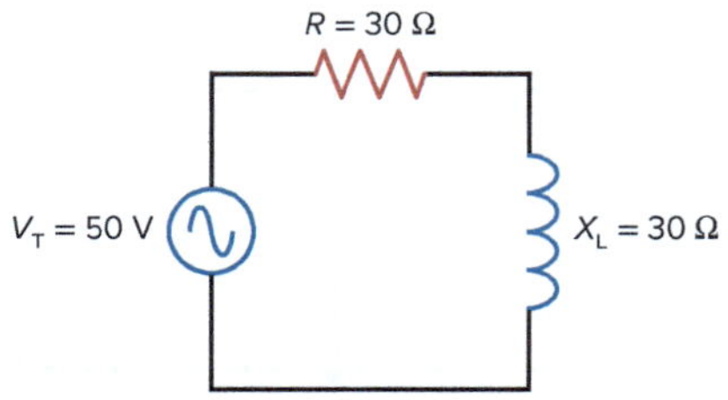

21–16 In Fig. 21–20, solve for Z_T, I, V_L, V_R, and θ_Z for the following circuit values:

a. $X_L = 30\ \Omega$, $R = 40\ \Omega$, and $V_T = 50$ V.

b. $X_L = 50\ \Omega$, $R = 50\ \Omega$, and $V_T = 141.4$ V.

c. $X_L = 10\ \Omega$, $R = 100\ \Omega$, and $V_T = 10$ V.

d. $X_L = 100\ \Omega$, $R = 10\ \Omega$, and $V_T = 10$ V.

Figure 21–20

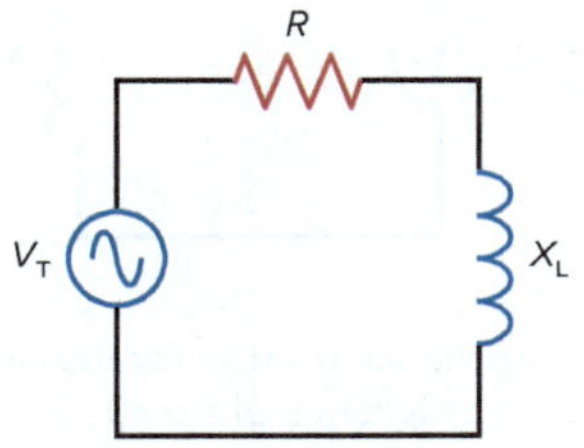

21–17 In Fig. 21–21, solve for X_L, Z_T, I, V_R, V_L, and θ_Z.

Figure 21–21

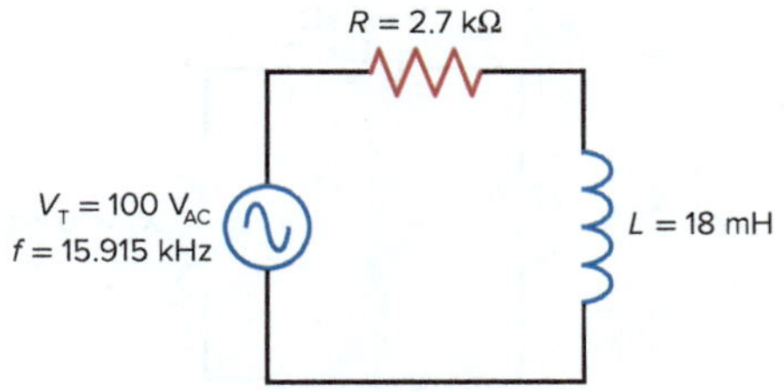

21–18 In Fig. 21–21, what happens to each of the following quantities if the frequency of the applied voltage increases?

a. X_L.

b. Z_T.

c. I.

d. V_R.

e. V_L.

f. θ_Z.

21–19 Repeat Prob. 21–18 if the frequency of the applied voltage decreases.

SECTION 21–4 X_L AND *R* IN PARALLEL

21–20 In Fig. 21–22, how much voltage is across

a. the 30-Ω resistor, *R*?

b. the 40-Ω inductive reactance, X_L?

Figure 21–22

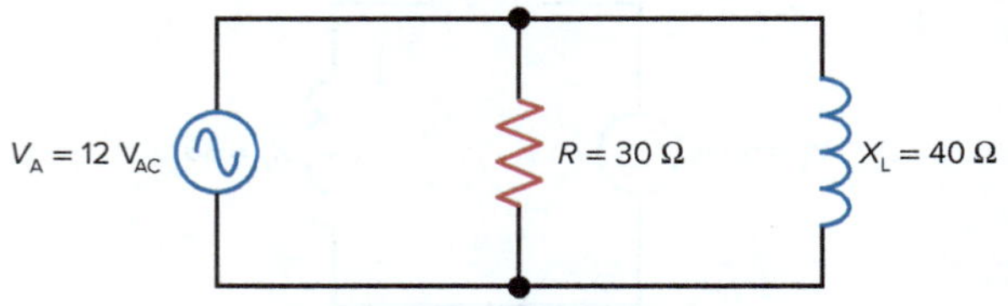

21–21 In Fig. 21–22, what is the phase relationship between

a. V_A and I_R?

b. V_A and I_L?

c. I_L and I_R?

21–22 In Fig. 21–22, solve for I_R, I_L, I_T, Z_{EQ}, and θ_I.

21–23 Draw the phasor current triangle for the circuit in Fig. 21–22. (Use I_R as the reference phasor.)

21–24 In Fig. 21–23, solve for I_R, I_L, I_T, Z_{EQ}, and θ_I.

Figure 21–23

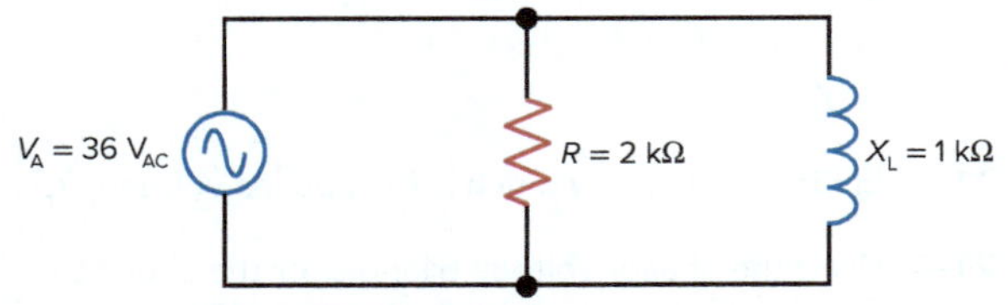

21–25 In Fig. 21–24, solve for I_R, I_L, I_T, Z_{EQ}, and θ_I.

Figure 21–24

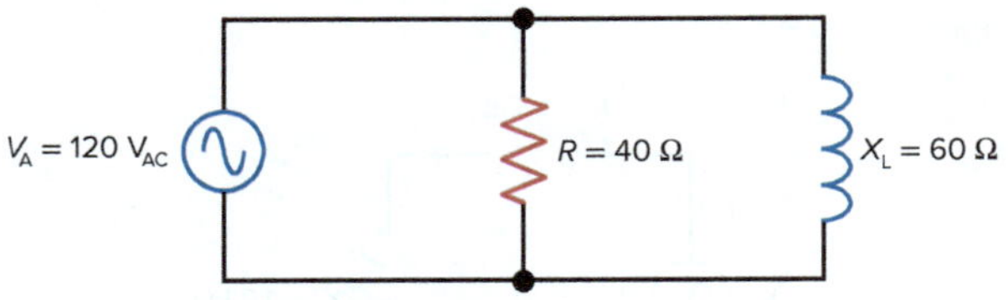

21–26 In Fig. 21–25, solve for I_R, I_L, I_T, Z_{EQ}, and θ_I.

Figure 21–25

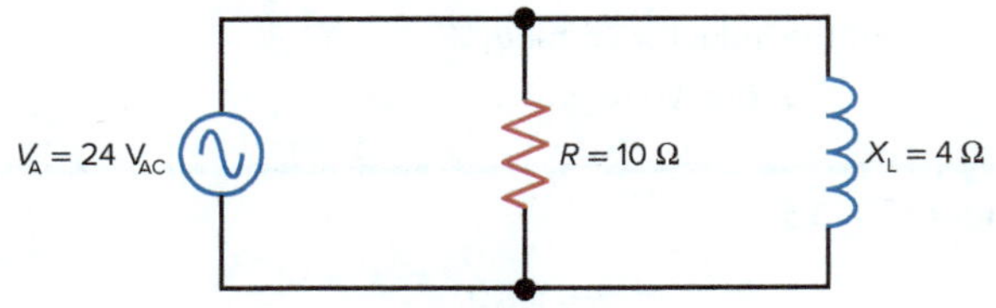

21–27 In Fig. 21–26, solve for I_R, I_L, I_T, Z_{EQ}, and θ_I.

Figure 21–26

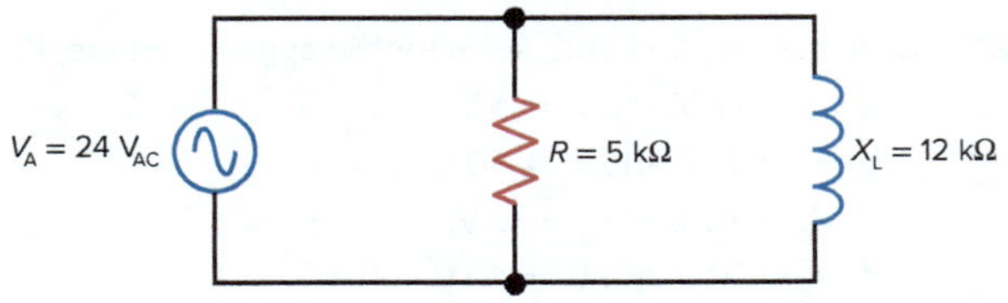

21–28 In Fig. 21–27, solve for I_R, I_L, I_T, Z_{EQ}, and θ_I for the following circuit values?

a. $R = 50\ \Omega$, $X_L = 50\ \Omega$, and $V_A = 50$ V.

b. $R = 10\ \Omega$, $X_L = 100\ \Omega$, and $V_A = 20$ V.

c. $R = 100\ \Omega$, $X_L = 10\ \Omega$, and $V_A = 20$ V.

Figure 21–27

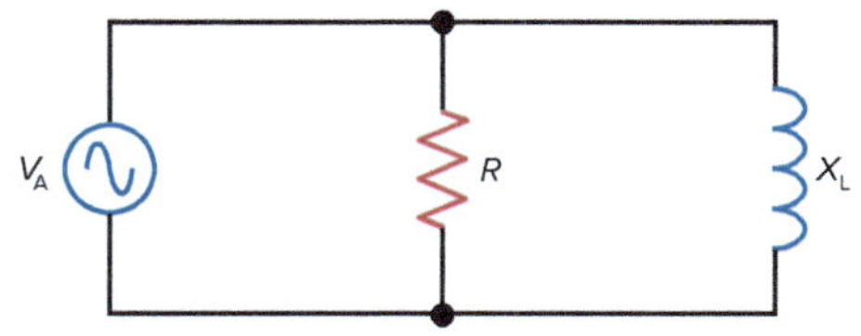

21–29 In Fig. 21–27, how much is Z_{EQ} if $R = 320\ \Omega$ and $X_L = 240\ \Omega$?

21–30 In Fig. 21–28, solve for X_L, I_R, I_L, I_T, Z_{EQ}, and θ_I.

Figure 21–28

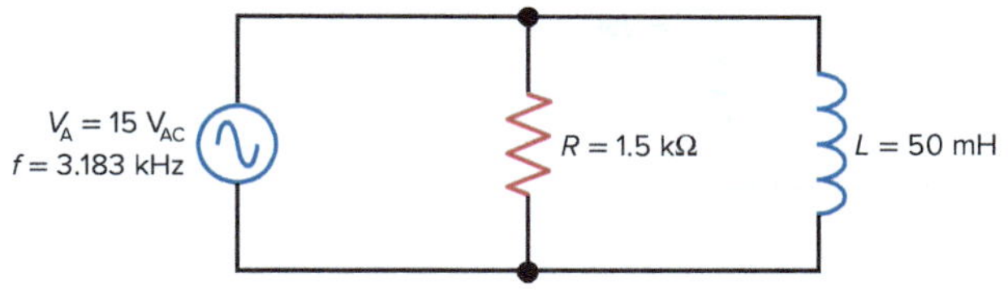

21–31 In Fig. 21–28, what happens to each of the following quantities if the frequency of the applied voltage increases?

a. I_R.

b. I_L.

c. I_T.

d. Z_{EQ}.

e. θ_I.

21–32 Repeat Prob. 21–31 if the frequency of the applied voltage decreases.

SECTION 21–5 *Q* OF A COIL

21–33 For the inductor shown in Fig. 21–29, calculate the Q for the following frequencies:

a. $f = 500$ Hz.

b. $f = 1$ kHz.

c. $f = 1.592$ kHz.

d. $f = 10$ kHz.

Figure 21–29

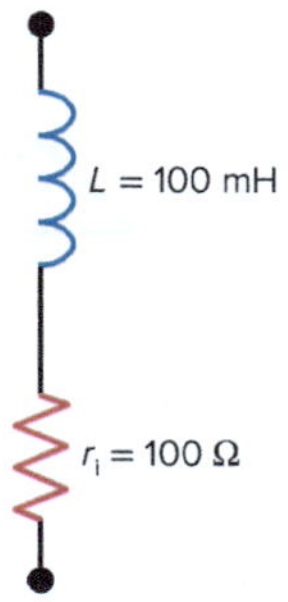

21–34 Why can't the Q of a coil increase without limit as the value of X_L increases for higher frequencies?

21–35 Calculate the AC effective resistance, R_e, of a 350-μH inductor whose Q equals 35 at 1.5 MHz.

21–36 Recalculate the value of R_e in Prob. 21–35 if the value of Q decreases to 25 at 5 MHz.

SECTION 21–6 AF AND RF CHOKES

21–37 In Fig. 21–30, calculate the required value of the choke inductance, L, at the following frequencies:

a. $f = 500$ Hz.

b. $f = 2.5$ kHz.

c. $f = 200$ kHz.

d. $f = 1$ MHz.

Figure 21–30

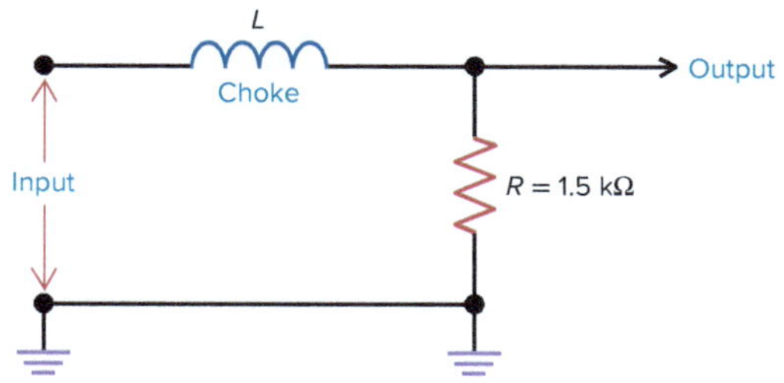

21–38 If $L = 50$ mH in Fig. 21–30, then what is the lowest frequency at which L will serve as a choke?

21–39 In Fig. 21–30 assume that the input voltage equals 10 V peak-to-peak for all frequencies. If $L = 150$ mH, then calculate V_{out} for the following frequencies:

a. 159.2 Hz.

b. 1.592 kHz.

c. 15.92 kHz.

SECTION 21–7 THE GENERAL CASE OF INDUCTIVE VOLTAGE

21–40 In Fig. 21–31, draw the waveform of induced voltage, V_L, across the 8-mH inductor for the triangular current waveform shown.

21–41 In Fig. 21–32, draw the waveform of induced voltage, V_L, across the 250-mH inductor for the sawtooth current waveform shown.

Figure 21–31

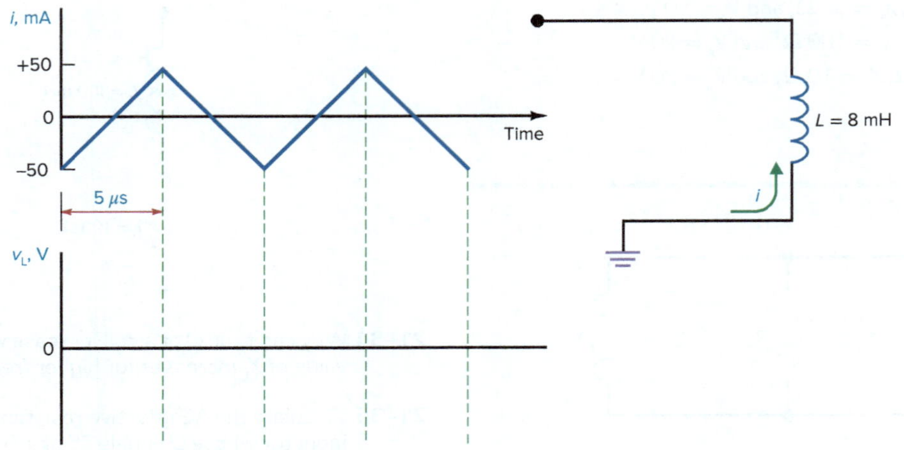

Figure 21–32

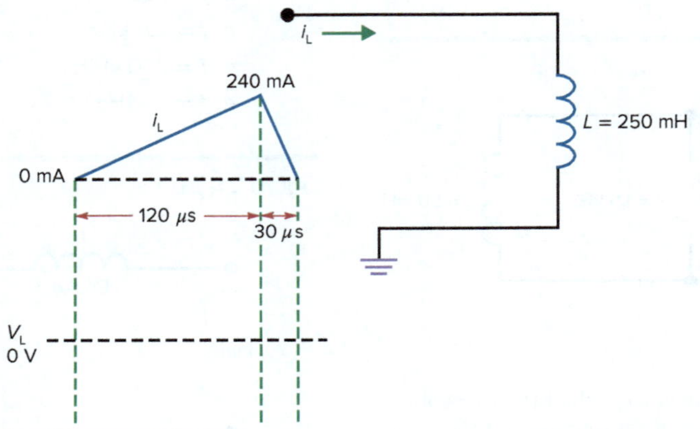

Critical Thinking

21–42 In Fig. 21–33, calculate X_L, R, L, I, V_L, and V_R.

21–43 In Fig. 21–34, calculate I_T, I_R, I_L, X_L, R, and L.

Figure 21–33 Circuit for Critical Thinking Prob. 21–42.

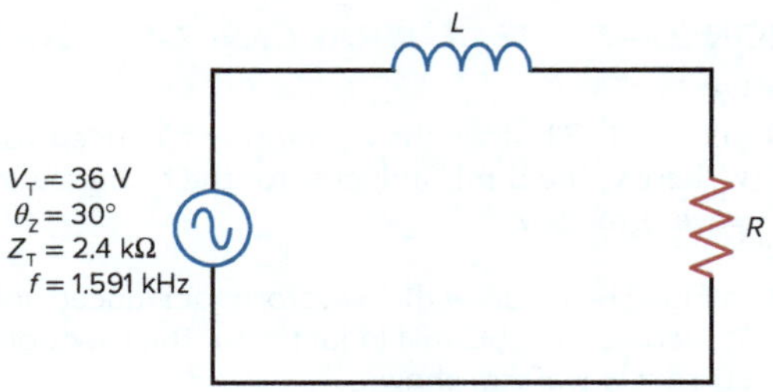

Figure 21–34 Circuit for Critical Thinking Prob. 21–43.

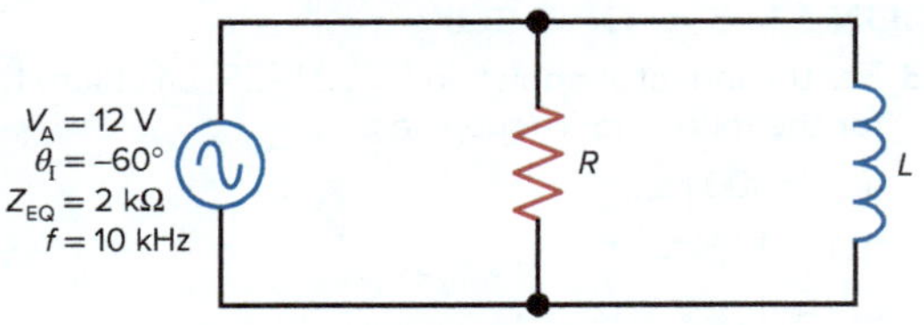

21–44 In Fig. 21–35, calculate V_R, V_{L_1}, X_{L_1}, X_{L_2}, I, Z_T, L_1, L_2, and θ_Z.

Figure 21–35 Circuit for Critical Thinking Prob. 21–44.

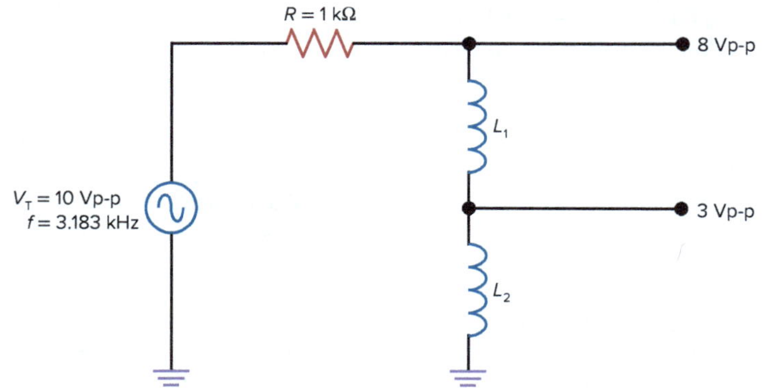

Answers to Self-Reviews

21–1 a. 0°
b. 90°
c. lag

21–2 a. 0°
b. 90°

21–3 a. 28.28 Ω
b. 28.28 V
c. 45°

21–4 a. 2.828 A
b. −45°

21–5 a. 75
b. 10 Ω

21–6 a. 800 Ω
b. 1600 Ω

21–7 a. 1000 A/s
b. 9000 A/s

Laboratory Application Assignment

In this lab application assignment, you will examine both series and parallel *RL* circuits. In the series *RL* circuit, you will measure the individual component voltages as well the circuit current and phase angle. In the parallel *RL* circuit, you will measure the individual branch currents, the total current, and the circuit phase angle.

Equipment: Obtain the following items from your instructor.

- Function generator
- Oscilloscope
- 10-Ω and 1-kΩ carbon-film resistors and a 100-mH inductor
- DMM

Series *RL* Circuit

Examine the series *RL* circuit in Fig. 21–36. Calculate and record the following circuit values:

X_L = ________, Z_T = ________, I = ________, V_L = ________, V_R = ________, θ_Z = ________

Figure 21–36

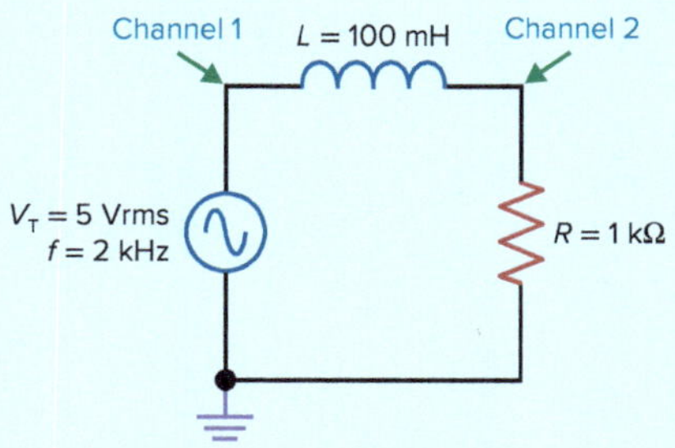

Construct the circuit in Fig. 21–36. Set the total voltage, V_T, to 5 Vrms and the frequency, f, to 2 kHz. Using a DMM, measure and record the following circuit values:

I = ________, V_L = ________, V_R = ________

Using the measured values of V_L and V_R, calculate the total voltage, V_T, as $V_T = \sqrt{V_R^2 + V_L^2}$. Does this value equal the applied voltage, V_T, of 5 V? ________ Using the measured values of voltage and current, calculate X_L as V_L/I and Z_T as V_T/I. X_L = ________, Z_T = ________ Using Formula (21–3), determine the phase angle, θ_Z. θ_Z = ________. How do these values compare to those originally calculated in Fig. 21–36. ________

In the space provided below, draw the phasor voltage triangle, including the phase angle, θ_V, for the circuit of Fig. 21–36. Use measured values for V_R, V_L, and V_T.

Ask your instructor for assistance in using the oscilloscope to measure the phase angle, θ_Z, in Fig. 21–36. Note the connections designated for channels 1 and 2 in the figure.

Parallel *RL* Circuit

Examine the parallel *RL* circuit in Fig. 21–37*a*. Calculate and record the following circuit values:
X_L = ________, I_L = ________, I_R = ________, I_T = ________, Z_{EQ} = ________, θ_I = ________

Construct the circuit in Fig. 21–37*a*. Set the applied voltage, V_A, to 5 Vrms and the frequency, *f*, to 2 kHz. Using a DMM, measure and record the following circuit values:
I_L = ________, I_R = ________, I_T = ________

Using the measured values of I_L and I_R, calculate the total current, I_T, as $I_T = \sqrt{I_R^2 + I_L^2}$. Does this value agree with the measured value of total current? ________ Using the measured values of I_L and I_R, calculate the phase angle, θ_I, using Formula (21–6). θ_I = ________ Also, calculate X_L as V_A/I_L and Z_{EQ} as V_A/I_T using measured values. X_L = ________, Z_{EQ} = ________. How do these values compare to those originally calculated in Fig. 21–37*a*? ________

In the space provided below, draw the phasor current triangle, including the phase angle, θ_I, for the circuit of Fig. 21–37*a*. Use measured values for I_L, I_R, and I_T.

Ask your instructor for assistance in using the oscilloscope to measure the phase angle, θ_I, in Fig. 21–37*b*. Note the connections designated for channels 1 and 2 in the figure. [The voltage drop across the sensing resistor (R_{sense}) has the same phase as the total current, I_T.]

Figure 21–37

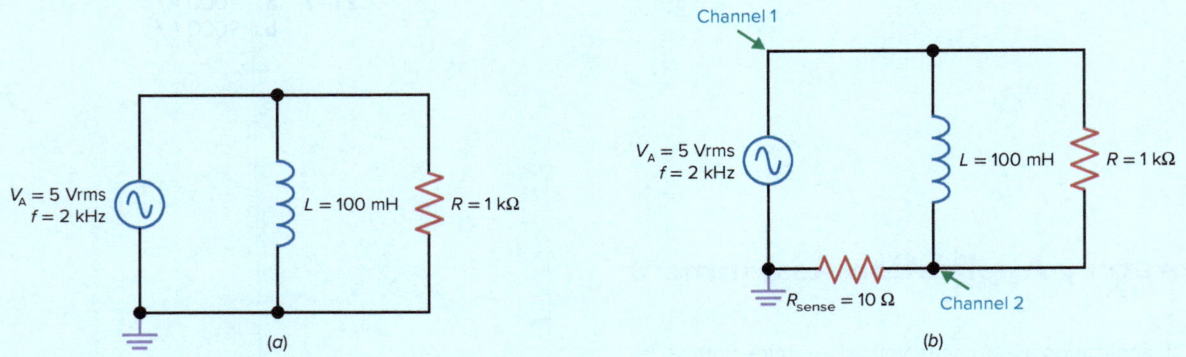

Design credits: Chapter Opener Background Image, and Multisim: Stockbyte/Getty Images

chapter

22

RC and *L/R* Time Constants

Many applications of inductance are for sine-wave AC circuits, but anytime the current changes, *L* has the effect of producing induced voltage. Examples of nonsinusoidal waveshapes include DC voltages that are switched on or off, square waves, sawtooth waves, and other rectangular type pulses. For capacitance, also, many applications are for sine-wave AC circuits, but whenever the voltage changes, *C* produces a charge or discharge current.

With nonsinusoidal voltage and current, the effect of *L* or *C* is to produce a change in waveshape. This effect can be analyzed by means of the time constant for capacitive and inductive circuits. The time constant is the time for a change of 63.2% in either the current through *L* or the voltage across *C*.

Many electronic circuits, such as digital circuits, have pulse inputs rather than sinusoidal inputs. In these instances, electronic technicians and engineers are more concerned with how a circuit responds over a given period of time to rapid changes in either voltage or current. How the time constant of the circuit relates to the pulse width and period of the input waveform determines the wave shapes of voltage and current created. Actually, *RC* circuits are more common than *RL* circuits because capacitors are smaller and more economical and do not have strong magnetic fields. ■

Chapter Outline

Chapter Objectives

After studying this chapter, you should be able to

- *Define* the term *transient response*.
- *Define* the term *time constant*.
- *Calculate* the time constant of a circuit containing resistance and inductance.
- *Explain* the effect of producing a high voltage when opening an *RL* circuit.
- *Calculate* the time constant of a circuit containing resistance and capacitance.
- *Explain* how capacitance opposes a change in voltage.
- *List* the criteria for proper differentiation and integration.
- *Explain* why a long time constant is required for an *RC* coupling circuit.
- *Use* the universal time constant graph to solve for voltage and current values in an *RC* or *RL* circuit that is charging or discharging.
- *Explain* the difference between time constants and reactance.

Important Terms

differentiator
integrator
long time constant
short time constant
steady-state value
time constant
transient response
universal time constant graph

22–1 Response of Resistance Alone

To emphasize the special features of L and C, the circuit in Fig. 22–1*a* illustrates how an ordinary resistive circuit behaves. When the switch is closed, the battery supplies 10 V across the 10-Ω R and the resultant I is 1 A. The graph in Fig. 22–1*b* shows that I changes from 0 to 1 A instantly when the switch is closed. If the applied voltage is changed to 5 V, the current will change instantly to 0.5 A. If the switch is opened, I will immediately drop to zero.

Resistance has only opposition to current; there is no reaction to a change because R has no concentrated magnetic field to oppose a change in I, like inductance, and no electric field to store charge that opposes a change in V, like capacitance.

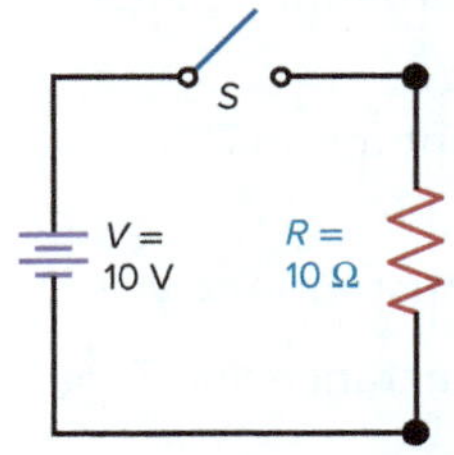

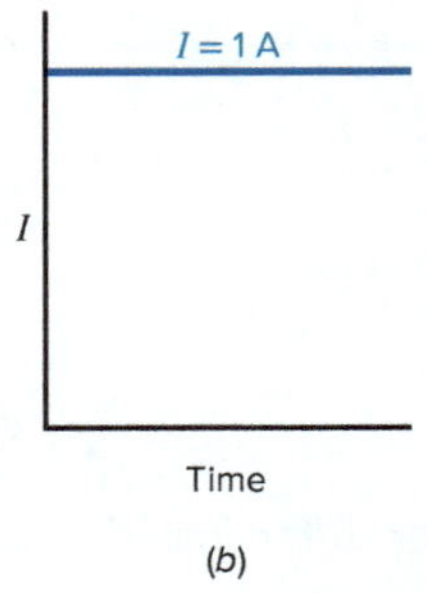

Figure 22–1 Response of circuit with R alone. When switch is closed, current I is 10 V/10 Ω = 1 A. (*a*) Circuit. (*b*) Graph of steady I.

22–1 Self-Review

Answers at the end of the chapter.

a. **Resistance R does not produce induced voltage for a change in I. (True/False)**

b. **Resistance R does not produce charge or discharge current for a change in V. (True/False)**

22–2 L/R Time Constant

Consider the circuit in Fig. 22–2, where L is in series with R. When S is closed, the current changes as I increases from zero. Eventually, I will reach the steady value of 1 A, equal to the battery voltage of 10 V divided by the circuit resistance of 10 Ω. While the current is building up from 0 to 1 A, however, I is changing and the inductance opposes the change. The action of the RL circuit during this time is its ***transient response***, which means that a temporary condition exists only until the steady-state current of 1 A is reached. Similarly, when S is opened, the transient response of the RL circuit opposes the decay of current toward the **steady-state value** of zero.

The transient response is measured in terms of the ratio L/R, which is the **time constant** of an inductive circuit. To calculate the time constant,

$$T = \frac{L}{R} \qquad \textbf{(22–1)}$$

MultiSim **Figure 22–2** Transient response of circuit with R and inductance L. When the switch is closed, I rises from zero to the steady-state value of 1 A. (*a*) Circuit with time constant L/R of 1 H/10 Ω = 0.1 s. (*b*) Graph of I during five time constants. Compare with graph in Fig. 22–1*b*.

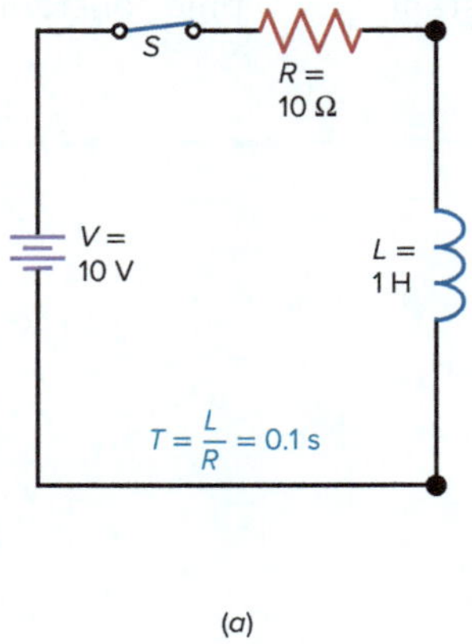

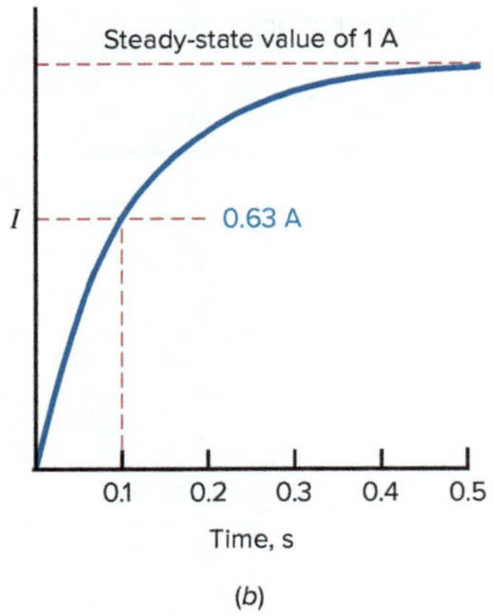

GOOD TO KNOW

Theoretically, the current, I, in Fig. 22–2 never reaches its steady-state value of 1 A with the switch closed.

GOOD TO KNOW

The time constant is often symbolized by τ, which is the greek letter tau. Therefore, Formula (22–1) often appears as $\tau = \frac{L}{R}$.

where T is the time constant in seconds, L is the inductance in henrys, and R is the resistance in ohms. The resistance in series with L is either the coil resistance, an external resistance, or both in series. In Fig. 22–2,

$$T = \frac{L}{R} = \frac{1}{10} = 0.1 \text{ s}$$

Specifically, the time constant is a measure of how long it takes the current to change by 63.2%, or approximately 63%. In Fig. 22–2, the current increases from 0 to 0.63 A, which is 63% of the steady-state value, in a period of 0.1 s, which is one time constant. In a period of five time constants, the current is practically equal to its steady-state value of 1 A.

The reason why L/R equals time can be illustrated as follows: Since induced voltage $V = L(di/dt)$, by transposing terms, L has the dimensions of $V \times T/I$. Dividing L by R results in $V \times T/IR$. As the IR and V factors cancel, T remains to indicate the dimension of time for the ratio L/R.

Example 22–1

What is the time constant of a 20-H coil having 100 Ω of series resistance?

ANSWER

$$T = \frac{L}{R} = \frac{20 \text{ H}}{100\ \Omega}$$
$$= 0.2 \text{ s}$$

Example 22–2

An applied DC voltage of 10 V will produce a steady-state current of 100 mA in the 100-Ω coil of Example 22–1. How much is the current after 0.2 s? After 1 s?

ANSWER Since 0.2 s is one time constant, I is 63% of 100 mA, which equals 63 mA. After five time constants, or 1 s (0.2 s × 5), the current will reach its steady-state value of 100 mA and remain at this value as long as the applied voltage stays at 10 V.

Example 22–3

If a 1-MΩ R is added in series with the coil of Example 22–1, how much will the time constant be for the higher resistance RL circuit?

ANSWER

$$T = \frac{L}{R} = \frac{20 \text{ H}}{1{,}000{,}000\ \Omega}$$
$$= 20 \times 10^{-6} \text{ s}$$
$$= 20\ \mu\text{s}$$

The L/R time constant becomes longer with larger values of L. More series R, however, makes the time constant shorter. With more series resistance, the circuit is less inductive and more resistive.

22–2 Self-Review

Answers at the end of the chapter.

a. Calculate the time constant for 2 H in series with 100 Ω.
b. Calculate the time constant for 2 H in series with 4000 Ω.

22–3 High Voltage Produced by Opening an *RL* Circuit

When an inductive circuit is opened, the time constant for current decay becomes very short because L/R becomes smaller with the high resistance of the open circuit. Then the current drops toward zero much faster than the rise of current when the switch is closed. The result is a high value of self-induced voltage V_L across a coil whenever an *RL* circuit is opened. This high voltage can be much greater than the applied voltage.

There is no gain in energy, though, because the high-voltage peak exists only for the short time the current is decreasing at a very fast rate at the start of the decay. Then, as I decays at a slower rate, the value of V_L is reduced. After the current has dropped to zero, there is no voltage across L.

This effect can be demonstrated by a neon bulb connected across a coil, as shown in Fig. 22–3. The neon bulb requires 90 V for ionization, at which time it glows. The source here is only 8 V, but when the switch is opened, the self-induced voltage is high enough to light the bulb for an instant. The sharp voltage pulse or spike is more than 90 V just after the switch is opened, when I drops very fast at the start of the decay in current.

Note that the 100-Ω R_1 is the internal resistance of the 2-H coil. This resistance is in series with L whether S is closed or open. The 4-kΩ R_2 across the switch is in the circuit only when S is opened, to have a specific resistance across the open switch. Since R_2 is much more than R_1, the L/R time constant is much shorter with the switch open.

Closing the Circuit

In Fig. 22–3*a*, the switch is closed to allow current in L and to store energy in the magnetic field. Since R_2 is short-circuited by the switch, the 100-Ω R_1 is the only

Figure 22–3 Demonstration of high voltage produced by opening inductive circuit. (*a*) With switch closed, 8 V applied cannot light the 90-V neon bulb. (*b*) When the switch is opened, the short L/R time constant results in high V_L, which lights the bulb.

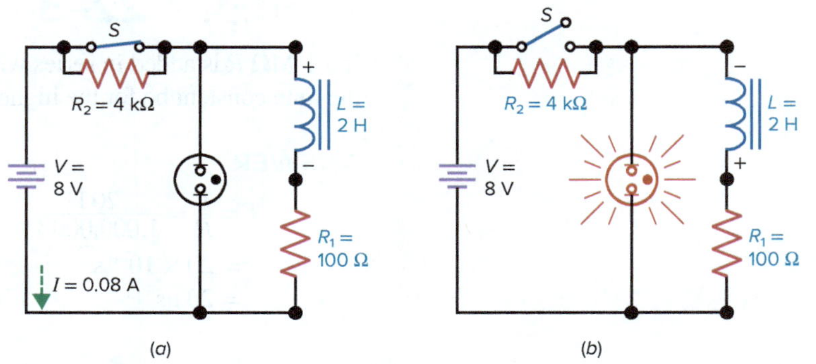

resistance. The steady-state I is $V/R_1 = 8/100 = 0.08$ A. This value of I is reached after five time constants.

One time constant is $L/R = 2/100 = 0.02$ s. Five time constants equal $5 \times 0.02 = 0.1$ s. Therefore, I is 0.08 A after 0.1 s, or 100 ms. The energy stored in the magnetic field is 64×10^{-4} J, equal to $\frac{1}{2} LI^2$.

Opening the Circuit

When the switch is opened in Fig. 22–3*b*, R_2 is in series with L, making the total resistance 4100 Ω, or approximately 4 kΩ. The result is a much shorter time constant for current decay. Then L/R is $2/4000$, or 0.5 ms. The current decays practically to zero in five time constants, or 2.5 ms.

This rapid drop in current results in a magnetic field collapsing at a fast rate, inducing a high voltage across L. The peak v_L in this example is 302 V. Then v_L serves as the voltage source for the bulb connected across the coil. As a result, the neon bulb becomes ionized, and it lights for an instant. One problem is arcing produced when an inductive circuit is opened. Arcing can destroy contact points and under certain conditions cause fires or explosions.

Calculating the Peak of v_L

The value of 320 V for the peak induced voltage when S is opened in Fig. 22–3 can be determined as follows: With the switch closed, I is 0.08 A in all parts of the series circuit. The instant S is opened, R_2 is added in series with L and R_1. The energy stored in the magnetic field maintains I at 0.08 A for an instant before the current decays. With 0.08 A in the 4-kΩ R_2, its potential difference is $0.08 \times 4000 = 320$ V. The collapsing magnetic field induces this 320-V pulse to allow an I of 0.08 A at the instant the switch is opened.

The *di/dt* for v_L

The required rate of change in current is 160 A/s for the v_L of 320 V induced by the L of 2 H. Since $v_L = L(di/dt)$, this formula can be transposed to specify di/dt as equal to v_L/L. Then di/dt corresponds to 320 V/2 H, or 160 A/s. This value is the actual di/dt at the start of the decay in current when the switch is opened in Fig. 22–3*b*, as a result of the **short time constant**.*

Applications of Inductive Voltage Pulses

There are many uses for the high voltage generated by opening an inductive circuit. One example is the high voltage produced for the ignition system in an automobile. Here the circuit of the battery in series with a high-inductance spark coil is opened by the breaker points of the distributor to produce the high voltage needed for each spark plug. When an inductive circuit is opened very rapidly, 10,000 V can easily be produced.

22–3 Self-Review

Answers at the end of the chapter.

a. **Is the L/R time constant longer or shorter in Fig. 22–3 when S is opened?**
b. **Which produces more v_L, a faster di/dt or a slower di/dt?**

---

* The di/dt value can be calculated from the slope at the start of decay, shown by the dashed line for curve *b* in Fig. 22–9.

22–4 *RC* Time Constant

The transient response of capacitive circuits is measured in terms of the product $R \times C$. To calculate the time constant,

$$T = R \times C \tag{22–2}$$

where R is in ohms, C is in farads, and T is in seconds. In Fig. 22–4, for example, with an R of 3 MΩ and a C of 1 μF,

$$T = 3 \times 10^6 \times 1 \times 10^{-6}$$
$$= 3 \text{ s}$$

Note that the 10^6 for megohms and the 10^{-6} for microfarads cancel. Therefore, multiplying the units of MΩ × μF gives the *RC* product in seconds.

Common combinations of units for the *RC* time constant are

$$\text{M}\Omega \times \mu\text{F} = \text{s}$$
$$\text{k}\Omega \times \mu\text{F} = \text{ms}$$
$$\text{M}\Omega \times \text{pF} = \mu\text{s}$$

The reason that the *RC* product is expressed in units of time can be illustrated as follows: $C = Q/V$. The charge Q is the product of $I \times T$. The factor V is IR. Therefore, RC is equivalent to $(R \times Q)/V$, or $(R \times IT)/IR$. Since I and R cancel, T remains to indicate the dimension of time.

GOOD TO KNOW

Another way to show Formula (22–2) is $\tau = RC$.

The Time Constant Indicates the Rate of Charge or Discharge

RC specifies the time it takes *C* to charge to 63% of the charging voltage. Similarly, *RC* specifies the time it takes *C* to discharge 63% of the way down to the value equal to 37% of the initial voltage across *C* at the start of discharge.

MultiSim **Figure 22–4** Details of how a capacitor charges and discharges in an *RC* circuit. (*a*) With S_1 closed, *C* charges through *R* to 63% of V_T in one *RC* time constant of 3 s and is almost completely charged in five time constants. (*b*) With S_1 opened to disconnect the battery and S_2 closed for *C* to discharge through *R*, V_C drops to 37% of its initial voltage in one time constant of 3 s and is almost completely discharged in five time constants.

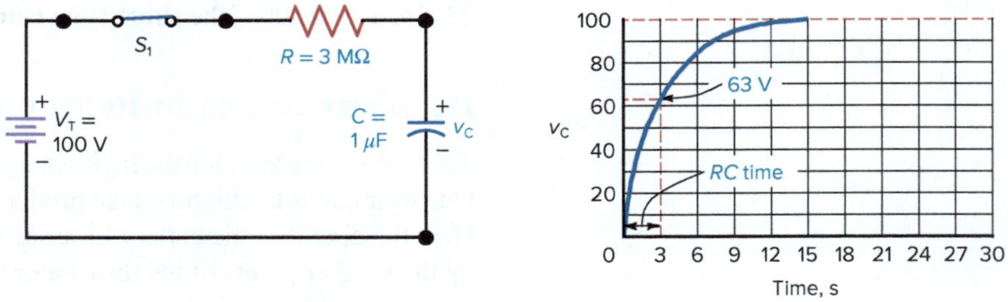

(*a*)

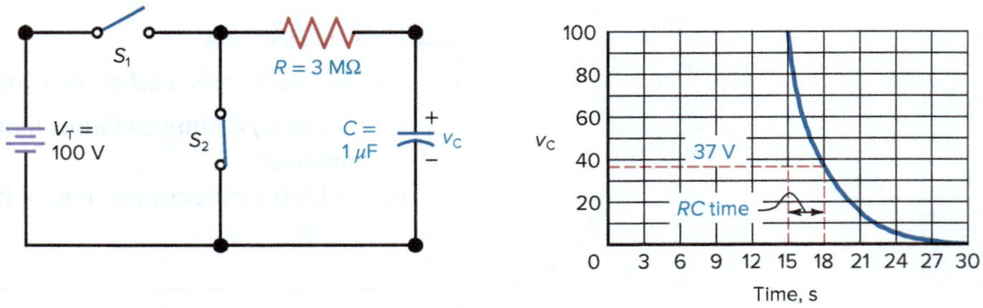

(*b*)

In Fig. 22–4*a*, for example, the time constant on charge is 3 s. Therefore, in 3 s, *C* charges to 63% of the 100 V applied, reaching 63 V in *RC* time. After five time constants, which is 15 s here, *C* is almost completely charged to the full 100 V applied. If *C* discharges after being charged to 100 V, then *C* will discharge down to 36.8 V or approximately 37 V in 3 s. After five time constants, *C* discharges to zero.

A shorter time constant allows the capacitor to charge or discharge faster. If the *RC* product in Fig. 22–4 is 1 s, then *C* will charge to 63 V in 1 s instead of 3 s. Also, v_C will reach the full applied voltage of 100 V in 5 s instead of 15 s. Charging to the same voltage in less time means a faster charge.

On discharge, the shorter time constant will allow *C* to discharge from 100 to 37 V in 1 s instead of 3 s. Also, v_C will be down to zero in 5 s instead of 15 s.

For the opposite case, a longer time constant means slower charge or discharge of the capacitor. More *R* or *C* results in a longer time constant.

RC Applications

Several examples are given here to illustrate how the time constant can be applied to *RC* circuits.

Example 22–4

What is the time constant of a 0.01-μF capacitor in series with a 1-MΩ resistance?

ANSWER

$$T = R \times C = 1 \times 10^6 \times 0.01 \times 10^{-6}$$
$$= 0.01 \text{ s}$$

The time constant in Example 22–4 is for charging or discharging, assuming the series resistance is the same for charge or discharge.

Example 22–5

With a DC voltage of 300 V applied, how much is the voltage across *C* in Example 22–4 after 0.01 s of charging? After 0.05 s? After 2 hours? After 2 days?

ANSWER Since 0.01 s is one time constant, the voltage across *C* then is 63% of 300 V, which equals 189 V. After five time constants, or 0.05 s, *C* will be charged practically to the applied voltage of 300 V. After 2 hours or 2 days, *C* will still be charged to 300 V if the applied voltage is still connected.

Example 22–6

If the capacitor in Example 22–5 is allowed to charge to 300 V and then discharged, how much is the capacitor voltage 0.01 s after the start of discharge? The series resistance is the same on discharge as on charge.

ANSWER In one time constant, *C* discharges to 37% of its initial voltage, or 0.37×300 V, which equals 111 V.

Example 22–7

Assume the capacitor in Example 22–5 is discharging after being charged to 200 V. How much will the voltage across *C* be 0.01 s after the beginning of discharge? The series resistance is the same on discharge as on charge.

ANSWER In one time constant, *C* discharges to 37% of its initial voltage, or 0.37×200, which equals 74 V.

Example 22–7 shows that the capacitor can charge or discharge from any voltage value. The rate at which it charges or discharges is determined by *RC*, counting from the time the charge or discharge starts.

Example 22–8

If a 1-MΩ resistance is added in series with the capacitor and resistor in Example 22–4, how much will the time constant be?

ANSWER Now the series resistance is 2 MΩ. Therefore, *RC* is 2×0.01, or 0.02 s.

The *RC* time constant becomes longer with larger values of *R* and *C*. More capacitance means that the capacitor can store more charge. Therefore, it takes longer to store the charge needed to provide a potential difference equal to 63% of the applied voltage. More resistance reduces the charging current, requiring more time to charge the capacitor.

Note that the *RC* time constant only specifies a rate. The actual amount of voltage across *C* depends on the amount of applied voltage as well as on the *RC* time constant.

A capacitor takes on charge whenever its voltage is less than the applied voltage. The charging continues at the *RC* rate until the capacitor is completely charged, or the voltage is disconnected.

A capacitor discharges whenever its voltage is more than the applied voltage. The discharge continues at the *RC* rate until the capacitor is completely discharged, the capacitor voltage equals the applied voltage, or the load is disconnected.

To summarize these two important principles:

1. Capacitor *C* charges when the net charging voltage is more than v_C.
2. Capacitor *C* discharges when v_C is more than the net charging voltage.

The net charging voltage equals the difference between v_C and the applied voltage.

22–4 Self-Review

Answers at the end of the chapter.

a. How much is the *RC* time constant for 470 pF in series with 2 MΩ on charge?

b. How much is the *RC* time constant for 470 pF in series with 1 kΩ on discharge?

22–5 *RC* Charge and Discharge Curves

In Fig. 22–4, the rise is shown in the *RC* charge curve because the charging is fastest at the start and then tapers off as *C* takes on additional charge at a slower rate. As *C* charges, its potential difference increases. Then the difference in voltage between V_T and v_C is reduced. Less potential difference reduces the current that puts the charge in *C*. The more *C* charges, the more slowly it takes on additional charge.

Similarly, on discharge, *C* loses its charge at a declining rate. At the start of discharge, v_C has its highest value and can produce maximum discharge current. As the discharge continues, v_C goes down and there is less discharge current. The more *C* discharges, the more slowly it loses the remainder of its charge.

Charge and Discharge Current

There is often the question of how current can flow in a capacitive circuit with a battery as the DC source. The answer is that current flows anytime there is a change in voltage. When V_T is connected, the applied voltage changes from zero. Then charging current flows to charge *C* to the applied voltage. After v_C equals V_T, there is no net charging voltage and *I* is zero.

Similarly, *C* can produce discharge current anytime v_C is greater than V_T. When V_T is disconnected, v_C can discharge down to zero, producing discharge current in the direction opposite from the charging current. After v_C equals zero, there is no current.

Capacitance Opposes Voltage Changes across Itself

GOOD TO KNOW

The voltage across a capacitor cannot change instantaneously.

This ability corresponds to the ability of inductance to oppose a change in current. When the applied voltage in an *RC* circuit increases, the voltage across the capacitance cannot increase until the charging current has stored enough charge in *C*. The increase in applied voltage is present across the resistance in series with *C* until the capacitor has charged to the higher applied voltage. When the applied voltage decreases, the voltage across the capacitor cannot go down immediately because the series resistance limits the discharge current.

The voltage across the capacitance in an *RC* circuit, therefore, cannot follow instantaneously the changes in applied voltage. As a result, the capacitance is able to oppose changes in voltage across itself. The instantaneous variations in V_T are present across the series resistance, however, since the series voltage drops must add to equal the applied voltage at all times.

22–5 Self-Review

Answers at the end of the chapter.

a. From the curve in Fig. 22–4*a*, how much is v_C after 3 s of charge?
b. From the curve in Fig. 22–4*b*, how much is v_C after 3 s of discharge?

22–6 High Current Produced by Short-Circuiting an *RC* Circuit

GOOD TO KNOW

Very large capacitors (1 farad or more) are sometimes placed across the terminals of a battery to improve its performance. The capacitor serves as a reservoir during instances of very high current draw from the battery. The capacitor is said to improve the transient response of the battery.

A capacitor can be charged slowly by a small charging current through a high resistance and then be discharged quickly through a low resistance to obtain a momentary surge, or pulse, of discharge current. This idea corresponds to the pulse of high voltage obtained by opening an inductive circuit.

The circuit in Fig. 22–5 illustrates the application of a battery-capacitor (BC) unit to fire a flashbulb for cameras. The flashbulb needs 5 A to ignite, but this is too much load current for the small 15-V battery, which has a rating of 30 mA for normal load current. Instead of using the bulb as a load for the battery, though, the 100-μF capacitor is charged by the battery through the 3-kΩ R in Fig. 22–5*a*, and then the capacitor is discharged through the bulb in Fig. 22–5*b*.

Charging the Capacitor

In Fig. 22–5*a*, S_1 is closed to charge C through the 3-kΩ R without the bulb. The time constant of the *RC* charging circuit is 0.3 s.

After five time constants, or 1.5 s, C is charged to the 15 V of the battery. The peak charging current, at the first instant of charge, is V/R or 15 V/3 kΩ, which equals 5 mA. This value is an easy load current for the battery.

Discharging the Capacitor

In Fig. 22–5*b*, v_C is 15 V without the battery. Now S_2 is closed, and C discharges through the 3-Ω resistance of the bulb. The time constant for discharge with the lower r of the bulb is $3 \times 100 \times 10^{-6}$, which equals 300 μs. At the first instant of discharge, when v_C is 15 V, the peak discharge current is $^{15}/_{3}$, which equals 5 A. This current is enough to fire the bulb.

Energy Stored in *C*

When the 100-μF C is charged to 15 V by the battery, the energy stored in the electric field is $CV^2/2$, which equals 0.01 J, approximately. This energy is available to maintain v_C at 15 V for an instant when the switch is closed. The result is the 5-A I

Figure 22-5 Demonstration of high current produced by discharging a charged capacitor through a low resistance. (*a*) When S_1 is closed, C charges to 15 V through 3 kΩ. (*b*) Without the battery, S_2 is closed to allow V_C to produce the peak discharge current of 5 A through the 3-Ω bulb. V_C in (*b*) is across the same C used in (*a*).

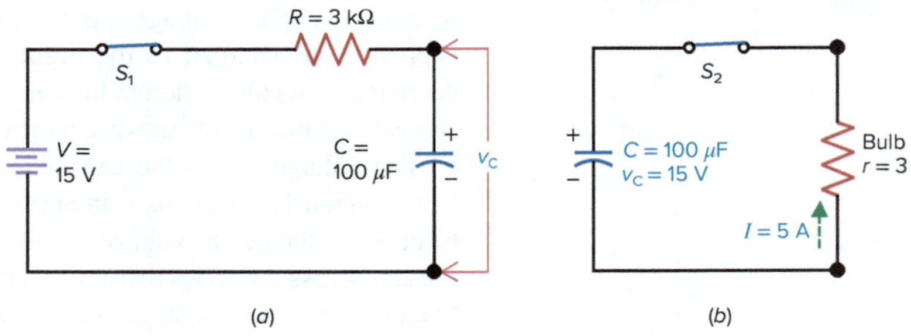

through the 3-Ω r of the bulb at the start of the decay. Then v_C and i_C drop to zero in five time constants.

The dv/dt for i_C

The required rate of change in voltage is 0.05×10^6 V/s for the discharge current i_C of 5 A produced by the C of 100 μF. Since $i_C = C(dv/dt)$, this formula can be transposed to specify dv/dt as equal to i_C/C. Then dv/dt corresponds to 5 A/100 μF, or 0.05×10^6 V/s. This value is the actual dv/dt at the start of discharge when the switch is closed in Fig. 22–5*b*. The dv/dt is high because of the short *RC* time constant.*

22–6 Self-Review

Answers at the end of the chapter.

a. **Is the *RC* time constant longer or shorter in Fig. 22–5*b* compared with Fig. 22–5*a*?**
b. **Which produces more i_C, a faster dv/dt or a slower dv/dt?**

22–7 *RC* Waveshapes

The voltage and current waveshapes in the *RC* circuit in Fig. 22–6 show when a capacitor is allowed to charge through a resistance for *RC* time and then discharge through the same resistance for the same amount of time. Note that this particular case is not typical of practical *RC* circuits, but the waveshapes show some useful details about the voltage and current for charging and discharging. The *RC* time constant here equals 0.1 s to simplify the calculations.

Square Wave of Applied Voltage

The idea of closing S_1 to apply 100 V and then opening it to disconnect V_T at a regular rate corresponds to a square wave of applied voltage, as shown by the waveform in Fig. 22–6*a*. When S_1 is closed for charge, S_2 is open; when S_1 is open, S_2 is closed for discharge. Here the voltage is on for the *RC* time of 0.1 s and off for the same time of 0.1 s. The period of the square wave is 0.2 s, and f is 1/0.2 s, which equals 5 Hz for the frequency.

Capacitor Voltage v_C

As shown in Fig. 22–6*b*, the capacitor charges to 63 V, equal to 63% of the charging voltage, in the *RC* time of 0.1 s. Then the capacitor discharges because the applied V_T drops to zero. As a result, v_C drops to 37% of 63 V, or 23.3 V in *RC* time.

The next charge cycle begins with v_C at 23.3 V. The net charging voltage now is $100 - 23.3 = 76.7$ V. The capacitor voltage increases by 63% of 76.7 V, or 48.3 V. When 48.3 V is added to 23.3 V, v_C rises to 71.6 V. On discharge, after 0.3 s, v_C drops to 37% of 71.6 V, or to 26.5 V.

Charge and Discharge Current

As shown in Fig. 22–6*c*, the current i has its positive peak at the start of charge and its negative peak at the start of discharge. On charge, i is calculated as the net charging voltage, which is $(V_T - v_C)$ divided by R. On discharge, i always equals v_C/R.

At the start of charge, i is maximum because the net charging voltage is maximum before C charges. Similarly, the peak i for discharge occurs at the start, when v_C is maximum before C discharges.

---

* See footnote on p. 681.

GOOD TO KNOW

The average DC value of the applied voltage is its average height over one full cycle. The DC value can be calculated as

$$V_{DC} = \frac{t_p}{\text{Prt}} \times V_p$$

where t_p = the length of time the pulse is on

Prt = the pulse repetition time, also known as the waveform's period

V_p = the peak value of the waveform

In Fig. 22–6*a*, V_{DC} = 50 V.

Figure 22–6 Waveshapes for the charge and discharge of an *RC* circuit in *RC* time. Circuit on top with S_1 and S_2 provides the square wave of applied voltage.

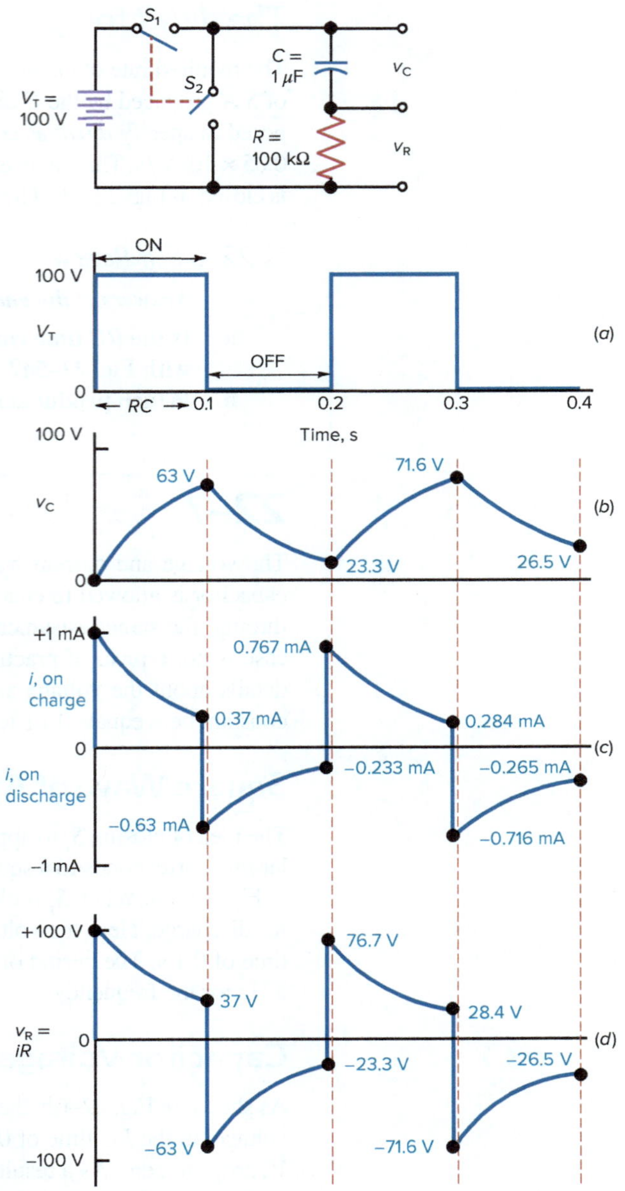

Note that *i* is an AC waveform around the zero axis, since the charge and discharge currents are in opposite directions. We are arbitrarily taking the charging current as positive values for *i*.

Resistor Voltage v_R

This waveshape in Fig. 22–6*d* follows the waveshape of current because v_R is $i \times R$. Because of the opposite directions of charge and discharge current, the *iR* waveshape is an AC voltage.

Note that on charge, v_R must always be equal to $V_T - v_C$ because of the series circuit.

On discharge, v_R has the same values as v_C because they are in parallel, without v_T. Then S_2 is closed to connect *R* across *C*.

Why the i_C Waveshape Is Important?

The v_C waveshape of capacitor voltage in Fig. 22–6 shows the charge and discharge directly, but the i_C waveshape is very interesting. First, the voltage waveshape across R is the same as the i_C waveshape. Also, whether C is charging or discharging, the i_C waveshape is the same except for the reversed polarity. We can see the i_C waveshape as the voltage across R. It generally is better to connect an oscilloscope for voltage waveshapes across R, especially with one side grounded.

Finally, we can tell what v_C is from the v_R waveshape. The reason is that at any instant, V_T must equal the sum of v_R and v_C. Therefore, v_C is equal to $V_T - v_R$, when V_T is charging C. When C is discharging, there is no V_T. Then v_R is the same as v_C.

22–7 Self-Review

Answers at the end of the chapter.

Refer to the waveforms in Fig. 22–6.

a. **When v_C is 63 V, how much is v_R?**
b. **When v_R is 76.7 V, how much is v_C?**

22–8 Long and Short Time Constants

Useful waveshapes can be obtained by using RC circuits with the required time constant. In practical applications, RC circuits are used more than RL circuits because almost any value of an RC time constant can be obtained easily. With coils, the internal series resistance cannot be short-circuited and the distributed capacitance often causes resonance effects.

GOOD TO KNOW

Inductors and resistors can also be used to obtain either a differentiated or integrated output. For an *RL* differentiator, the output is taken across the inductor. Conversely, for an *RL* integrator the output is taken across the resistor.

Long *RC* Time

Whether an RC time constant is long or short depends on the pulse width of the applied voltage. We can arbitrarily define a **long time constant** as at least five times longer than the pulse width, in time, for the applied voltage. As a result, C takes on very little charge. The time constant is too long for v_C to rise appreciably before the applied voltage drops to zero and C must discharge. On discharge also, with a long time constant, C discharges very little before the applied voltage rises to make C charge again.

Short *RC* Time

A short time constant is defined as no more than one-fifth the pulse width, in time, for the applied voltage V_T. Then V_T is applied for a period of at least five time constants, allowing C to become completely charged. After C is charged, v_C remains at the value of V_T while the voltage is applied. When V_T drops to zero, C discharges completely in five time constants and remains at zero while there is no applied voltage. On the next cycle, C charges and discharges completely again.

Differentiation

The voltage across R in an RC circuit is called a *differentiated output* because v_R can change instantaneously. A short time constant is always used for differentiating circuits to provide sharp pulses of v_R.

Integration

The voltage across C is called an *integrated output* because it must accumulate over a period of time. A medium or long time constant is always used for integrating circuits.

22–8 Self-Review

Answers at the end of the chapter.

a. Voltage V_T is on for 0.4 s and off for 0.4 s. *RC* is 6 ms for charge and discharge. Is this a long or short *RC* time constant?

b. Voltage V_T is on for 2 μs and off for 2 μs. *RC* is 6 ms for charge and discharge. Is this a long or short *RC* time constant?

22–9 Charge and Discharge with a Short *RC* Time Constant

GOOD TO KNOW

Differentiation and integration are mathematical terms used in calculus. Time constant circuits using capacitors, resistors, and inductors, can be used to produce waveforms that are approximated by these mathematical functions.

Usually, the time constant is made much shorter or longer than a factor of 5 to obtain better waveshapes. In Fig. 22–7, *RC* is 0.1 ms. The frequency of the square wave is 25 Hz, with a period of 0.04 s, or 40 ms. One-half this period is the time when V_T is applied. Therefore, the applied voltage is on for 20 ms and off for 20 ms. The *RC* time constant of 0.1 ms is shorter than the pulse width of 20 ms by a factor of 1⁄200. Note that the time axis of all waveshapes is calibrated in seconds for the period of V_T, not in *RC* time constants.

MultiSim **Figure 22–7** Charge and discharge of an *RC* circuit with a short time constant. Note that the waveshape of V_R in (*d*) has sharp voltage peaks for the leading and trailing edges of the square-wave applied voltage.

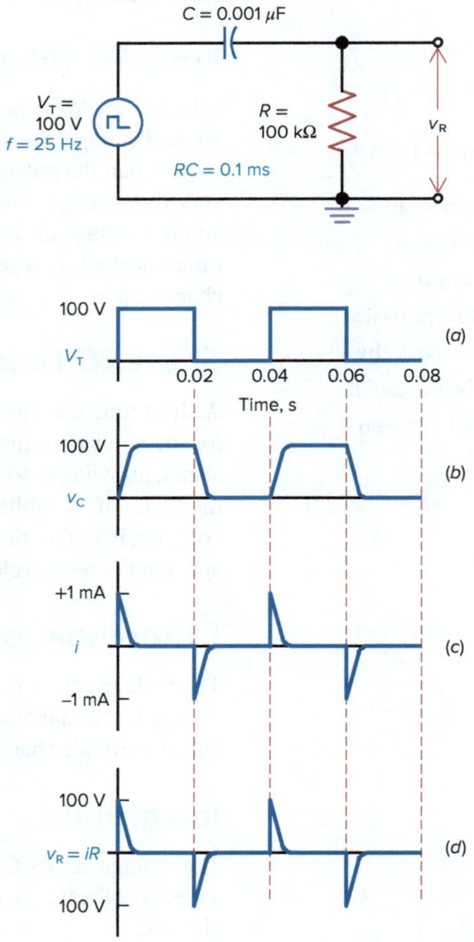

Square Wave of V_T Is across C

The waveshape of v_C in Fig. 22–7*b* is the same as the square wave of applied voltage because the short time constant allows C to charge or discharge completely very soon after V_T is applied or removed. The charge or discharge time of five time constants is much less than the pulse width.

Sharp Pulses of i

The waveshape of i shows sharp peaks for the charge or discharge current. Each current peak is $V_T/R = 1$ mA, decaying to zero in five RC time constants. These pulses coincide with the leading and trailing edges of the square wave of V_T.

Actually, the pulses are much sharper than shown. They are not to scale horizontally to indicate the charge and discharge action. Also, v_C is, in fact, a square wave, like the applied voltage, but with slightly rounded corners for the charge and discharge.

Sharp Pulses of v_R

The waveshape of voltage across the resistor follows the current waveshape because $v_R = iR$. Each current pulse of 1 mA across the 100-kΩ R results in a voltage pulse of 100 V.

More fundamentally, the peaks of v_R equal the applied voltage V_T before C charges. Then v_R drops to zero as v_C rises to the value of V_T.

On discharge, $v_R = v_C$, which is 100 V at the start of discharge. Then the pulse drops to zero in five time constants. The pulses of v_R in Fig. 22–7 are useful as timing pulses that match the edges of the square-wave applied voltage V_T. Either the positive or the negative pulses can be used.

The RC circuit in Fig. 22–7*a* is a good example of an RC **differentiator**. With the RC time constant much shorter than the pulse width of V_T, the voltage V_R follows instantaneously the changes in the applied voltage. Keep in mind that a differentiator must have a short time constant with respect to the pulse width of V_T to provide good differentiation. For best results, an RC differentiator should have a time constant which is one-tenth or less of the pulse width of V_T.

22–9 Self-Review

Answers at the end of the chapter.

Refer to Fig. 22–7.

a. Is the time constant here short or long?

b. Is the square wave of applied voltage across *C* or *R*?

22–10 Long Time Constant for an *RC* Coupling Circuit

The RC circuit in Fig. 22–8 is the same as that in Fig. 22–7, but now the RC time constant is long because of the higher frequency of the applied voltage. Specifically, the RC time of 0.1 ms is 200 times longer than the 0.5-μs pulse width of V_T with a frequency of 1 MHz. Note that the time axis is calibrated in microseconds for the period of V_T, not in RC time constants.

Very Little of V_T Is across C

The waveshape of v_C in Fig. 22–8*b* shows very little voltage rise because of the long time constant. During the 0.5 μs when V_T is applied, C charges to only 1⁄200 of the charging voltage. On discharge, also, v_C drops very little.

Figure 22-8 Charge and discharge of an *RC* circuit with a long time constant. Note that the waveshape of V_R in (*d*) has the same waveform as the applied voltage.

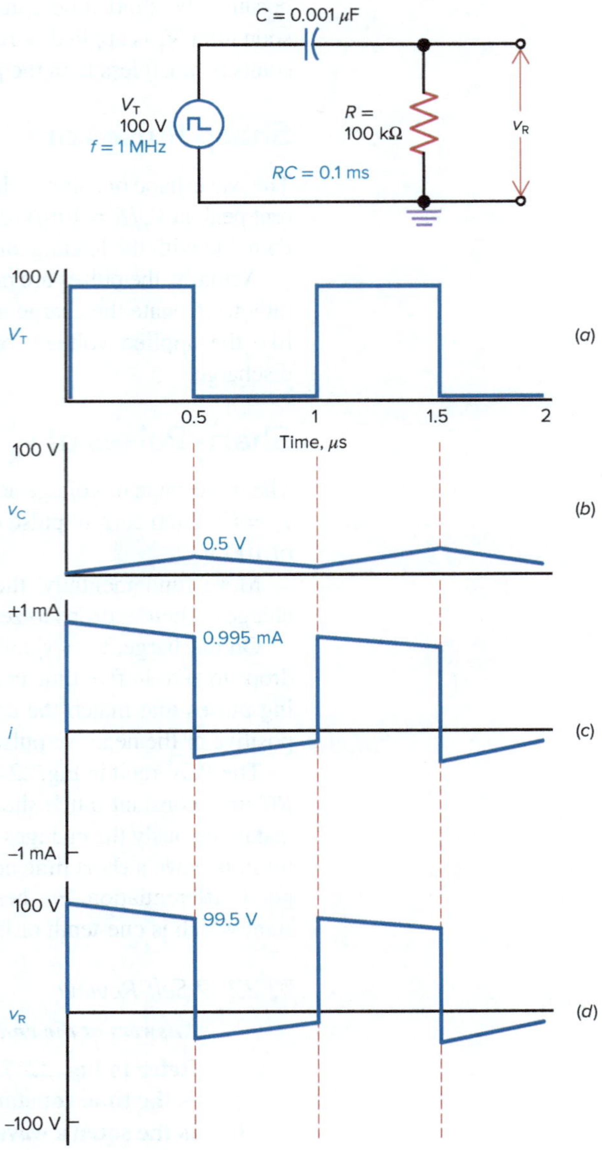

Square Wave of *i*

The waveshape of *i* stays close to the 1-mA peak at the start of charging. The reason is that v_C does not increase much, allowing V_T to maintain the charging current. On discharge, the reverse *i* for discharge current is very small because v_C is low.

Square Wave of V_T Is across *R*

The waveshape of v_R is the same square wave as *i* because $v_R = iR$. The waveshapes of *i* and v_R are essentially the same as the square-wave V_T applied. They are not shown to scale vertically to indicate the slight charge and discharge action.

Eventually, v_C will climb to the average DC value of 50 V, *i* will vary ±0.5 mA above and below zero, and v_R will vary ±50 V above and below zero. This application is an *RC* coupling circuit to block the average value of the varying DC voltage V_T as the capacitive voltage v_C, and v_R provides an ac voltage output having the same variations as V_T.

GOOD TO KNOW

An *RC* integrator can also be described as a circuit whose output is proportional to the charge being stored.

If the output is taken across C rather than R in Fig. 22–8*a*, the circuit is classified as an *RC* **integrator**. In Fig. 22–8*b*, it can be seen that C combines or integrates its original voltage with the new change in voltage. Eventually, however, the voltage across C will reach a steady-state value of 50 V after the input waveform has been applied for approximately five *RC* time constants. Keep in mind that an integrator must have a long time constant with respect to the pulse width of V_T to provide good integration. For best results, an *RC* integrator should have a time constant which is 10 or more times longer than the pulse width of V_T.

22–10 Self-Review

Answers at the end of the chapter.

Refer to Fig. 22–8.

a. Is the *RC* time constant here short or long?

b. Is the square wave of applied voltage across *R* or *C*?

22–11 Advanced Time Constant Analysis

We can determine transient voltage and current values for any amount of time with the curves in Fig. 22–9. The rising curve *a* shows how v_C builds up as C charges in an *RC* circuit; the same curve applies to i_L, increasing in the inductance for an *RL* circuit. The decreasing curve *b* shows how v_C drops as C discharges or i_L decays in an inductance.

Note that the horizontal axis is in units of time constants rather than absolute time. Suppose that the time constant of an *RC* circuit is 5 μs. Therefore, one *RC* time unit = 5 μs, two *RC* units = 10 μs, three *RC* units = 15 μs, four *RC* units = 20 μs, and five *RC* units = 25 μs.

As an example, to find v_C after 10 μs of charging, we can take the value of curve *a* in Fig. 22–9 at two *RC*. This point is at 86% amplitude. Therefore, we can say that in this *RC* circuit with a time constant of 5 μs, v_C charges to 86% of the applied V_T after 10 μs. Similarly, some important values that can be read from the curve are listed in Table 22–1.

Figure 22–9 Universal time constant chart for *RC* and *RL* circuits. The rise or fall changes by 63% in one time constant.

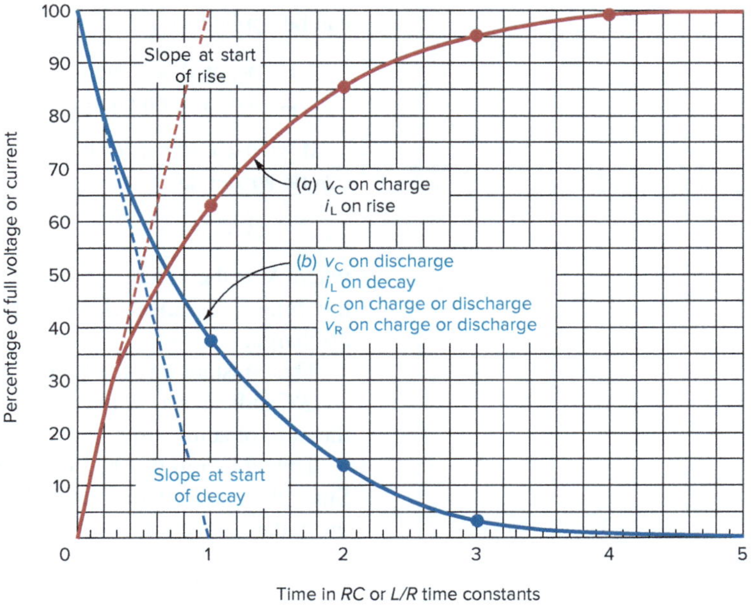

Table 22–1 Time Constant Factors

Factor	Amplitude
0.2 time constant	20%
0.5 time constant	40%
0.7 time constant	50%
1 time constant	63%
2 time constants	86%
3 time constants	96%
4 time constants	98%
5 time constants	99%

If we consider curve *a* in Fig. 22–9 as an *RC* charge curve, v_C adds 63% of the net charging voltage for each additional unit of one time constant, although it may not appear so. For instance, in the second interval of *RC* time, v_C adds 63% of the net charging voltage, which is 0.37 V_T. Then 0.63 × 0.37 equals 0.23, which is added to 0.63 to give 0.86, or 86%, as the total charge from the start.

Slope at $t = 0$

The curves in Fig. 22–9 can be considered approximately linear for the first 20% of change. In 0.1 time constant, for instance, the change in amplitude is 10%; in 0.2 time constant, the change is 20%. The dashed lines in Fig. 22–9 show that if this constant slope continued, the result would be 100% change in one time constant. This does not happen because the change is opposed by the energy stored in *L* and *C*. However, at the first instant of rise or decay, at $t = 0$, the change in v_C or i_L can be calculated from the dotted slope line.

Equation of the Decay Curve

The rising curve *a* in Fig. 22–9 may seem more interesting because it describes the buildup of v_C or i_L, but the decaying curve *b* is more useful. For *RC* circuits, curve *b* can be applied to

1. v_C on discharge
2. *i* and v_R on charge or discharge

If we use curve *b* for the voltage in *RC* circuits, the equation of this decay curve can be written as

$$v = V \times \epsilon^{-t/RC} \qquad \textbf{(22–3)}$$

where V is the voltage at the start of decay and *v* is the instantaneous voltage after the time *t*. Specifically, *v* can be v_R on charge and discharge or v_C only on discharge.

The constant ϵ is the base 2.718 for natural logarithms. The negative exponent $-t/RC$ indicates a declining exponential or logarithmic curve. The value of t/RC is the ratio of actual time of decline *t* to the *RC* time constant.

This equation can be converted to common logarithms for easier calculations. Since the natural base ϵ is 2.718, its logarithm to base 10 equals 0.434. Therefore, the equation becomes

$$v = \text{antilog}\left(\log V - 0.434 \times \frac{t}{RC}\right) \qquad \textbf{(22–4)}$$

Figure 22–10 How v_C and v_R add to equal the applied voltage v_T of 100 V. (*a*) Zero time at the start of charging. (*b*) After one *RC* time constant. (*c*) After two *RC* time constants. (*d*) After five or more *RC* time constants.

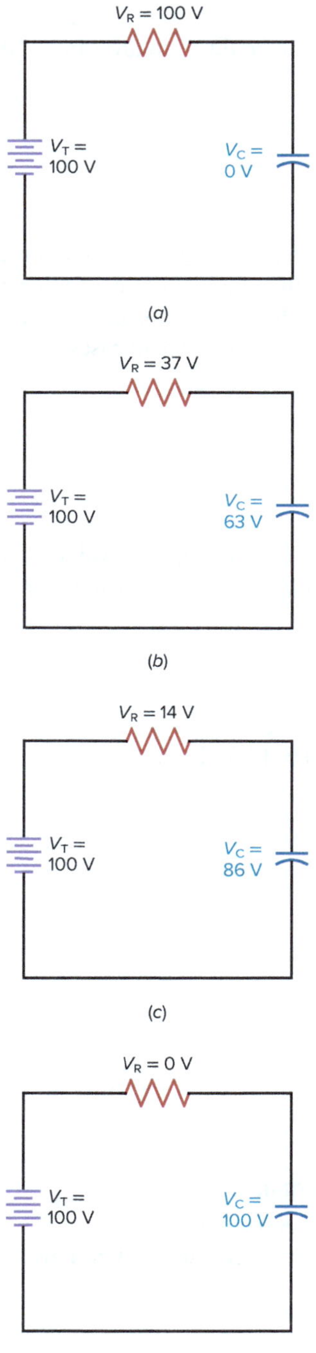

Calculations for v_R

As an example, let us calculate v_R dropping from 100 V, after *RC* time. Then the factor t/RC is 1. Substituting these values,

$$\begin{aligned} v_R &= \text{antilog}\,(\log 100 - 0.434 \times 1) \\ &= \text{antilog}\,(2 - 0.434) \\ &= \text{antilog}\,1.566 \\ &= 37\text{ V} \end{aligned}$$

All these logs are to base 10. Note that log 100 is taken first so that 0.434 can be subtracted from 2 before the antilog of the difference is found. The antilog of 1.566 is 37.

We can also use V_R to find V_C, which is $V_T - V_R$. Then 100 – 37 = 63 V for V_C. These answers agree with the fact that in one time constant, V_R drops 63% and V_C rises 63%.

Figure 22–10 illustrates how the voltages across *R* and *C* in series must add to equal the applied voltage V_T. The four examples with 100 V applied are

1. At time zero, at the start of charging, V_R is 100 V and V_C is 0 V. Then 100 + 0 = 100 V.
2. After one time constant, V_R is 37 V and V_C is 63 V. Then 37 + 63 = 100 V.
3. After two time constants, V_R is 14 V and V_C is 86 V. Then 14 + 86 = 100 V.
4. After five time constants, V_R is 0 V and V_C is 100 V, approximately. Then 0 + 100 = 100 V.

It should be emphasized that Formulas (22–3) and (22–4) can be used to calculate any decaying value on curve *b* in Fig. 22–9. These applications for an *RC* circuit include V_R on charge or discharge, *i* on charge or discharge, and V_C only on discharge. For an *RC* circuit in which *C* is charging, Formula (22–5) can be used to calculate the capacitor voltage v_C at any point along curve *a* in Fig. 22–9:

$$v_C = V(1 - \epsilon^{-t/RC}) \qquad \textbf{(22–5)}$$

In Formula (22–5), *V* represents the maximum voltage to which *C* can charge, whereas v_C is the instantaneous capacitor voltage after time *t*. Formula (22–5) is derived from the fact that v_C must equal $V_T - V_R$ while *C* is charging.

Example 22–9

An *RC* circuit has a time constant of 3 s. The capacitor is charged to 40 V. Then *C* is discharged. After 6 s of discharge, how much is V_R?

ANSWER Note that 6 s is twice the *RC* time of 3 s. Then $t/RC = 2$.

$$\begin{aligned} v_R &= \text{antilog}\,(\log 40 - 0.434 \times 2) \\ &= \text{antilog}\,(1.602 - 0.868) \\ &= \text{antilog}\,(0.734) \\ &= 5.42\text{ V} \end{aligned}$$

Note that in two *RC* time constants, the v_R is down to approximately 14% of its initial voltage, a drop of about 86%.

Calculations for t

Furthermore, Formula (22–4), can be transposed to find the time t for a specific voltage decay. Then

$$t = 2.3\ RC \log \frac{V}{v} \qquad \textbf{(22–6)}$$

where V is the higher voltage at the start and v is the lower voltage at the finish. The factor 2.3 is $1/0.434$.

As an example, let RC be 1 s. How long will it take for v_R to drop from 100 to 50 V? The required time for this decay is

$$\begin{aligned} t &= 2.3 \times 1 \times \log \frac{100}{50} = 2.3 \times 1 \times \log 2 \\ &= 2.3 \times 1 \times 0.3 \\ &= 0.7\ \text{s} \quad \text{approximately} \end{aligned}$$

This answer agrees with the fact that a drop of 50% takes 0.7 time constant. Formula (22–6) can also be used to calculate the time for any decay of v_C or v_R.

Formula (22–6) cannot be used for a rise in v_C. However, if you convert this rise to an equivalent drop in v_R, the calculated time is the same for both cases.

Example 22–10

An RC circuit has an R of 10 kΩ and a C of 0.05 μF. The applied voltage for charging is 36 V. (a) Calculate the time constant. (b) How long will it take C to charge to 24 V?

ANSWER

a. RC is 10 kΩ × 0.05 μF = 0.5 ms or 0.5×10^{-3} s.

b. The v_C rises to 24 V while v_R drops from 36 to 12 V. Then

$$\begin{aligned} t &= 2.3\ RC \log \frac{V}{v} \\ &= 2.3 \times 0.5 \times 10^{-3} \times \log \frac{36}{12} \\ &= 2.3 \times 0.5 \times 10^{-3} \times 0.477 \\ &= 0.549 \times 10^{-3}\ \text{s} \quad \text{or} \quad 0.549\ \text{ms} \end{aligned}$$

GOOD TO KNOW

When a capacitor charges from an initial voltage other than zero, the capacitor voltage can be determined at any time, t, with the use of the following equation:

$$v_C = (V_F - V_i)(1 - \epsilon^{-t/RC}) + V_i$$

where V_F and V_i represent the final and initial voltages, respectively. The quantity $(V_F - V_i)$ represents the net charging voltage.

22–11 Self-Review

Answers at the end of the chapter.

For the universal curves in Fig. 22–9,

a. Curve a applies to v_C on charge. (True/False)

b. Curve b applies to v_C on discharge. (True/False)

c. Curve b applies to v_R when C charges or discharges. (True/False)

22–12 Comparison of Reactance and Time Constant

The formula for capacitive reactance includes the factor of time in terms of frequency as $X_C = 1/(2\pi fC)$. Therefore, X_C and the RC time constant are both measures of the reaction of C to a change in voltage. The reactance X_C is a special case but a very important one that applies only to sine waves. The *RC* time constant can be applied to square waves and rectangular pulses.

Phase Angle of Reactance

The capacitive charge and discharge current i_C is always equal to $C(dv/dt)$. A sine wave of voltage variations for v_C produces a cosine wave of current i_C. This means that v_C and i_C are both sinusoids, but 90° out of phase.

In this case, it is usually more convenient to use X_C for calculations in sine-wave ac circuits to determine *Z*, *I*, and the phase angle θ. Then $I_C = V_C/X_C$. Moreover, if I_C is known, $V_C = I_C \times X_C$. The phase angle of the circuit depends on the amount of X_C compared with the resistance *R*.

Changes in Waveshape

With nonsinusoidal voltage applied, X_C cannot be used. Then i_C must be calculated as $C(dv/dt)$. In this comparison of i_C and v_C, their waveshapes can be different, instead of the change in phase angle for sine waves. The waveshapes of v_C and i_C depend on the *RC* time constant.

Coupling Capacitors

If we consider the application of a coupling capacitor, X_C must be one-tenth or less of its series *R* at the desired frequency. This condition is equivalent to having an *RC* time constant that is long compared with the period of one cycle. In terms of X_C, the *C* has little IX_C voltage, with practically all the applied voltage across the series *R*. In terms of a long *RC* time constant, *C* cannot take on much charge. Practically all the applied voltage is developed as $v_R = iR$ across the series resistance by the charge and discharge current. These comparisons are summarized in Table 22–2.

Inductive Circuits

Similar comparisons can be made between $X_L = 2\pi fL$ for sine waves and the L/R time constant. The voltage across any inductance is $v_L = L(di/dt)$. Sine-wave variations for i_L produce a cosine wave of voltage v_L, 90° out of phase.

In this case, X_L can be used to determine *Z*, *I*, and the phase angle θ. Then $I_L = V_L/X_L$. Furthermore, if I_L is known, $V_L = I_L \times X_L$. The phase angle of the circuit depends on the amount of X_L compared with *R*.

Table 22–2 Comparison of Reactance X_C and *RC* Time Constant

Sine-Wave Voltage	Nonsinusoidal Voltage
Examples are 60-Hz power line, af signal voltage, rf signal voltage	Examples are DC circuit turned on and off, square waves, rectangular pulses
Reactance $X_C = \frac{1}{2\pi fC}$	Time constant $T = RC$
Larger C results in smaller reactance X_C	Larger C results in longer time constant
Higher frequency results in smaller X_C	Shorter pulse width corresponds to longer time constant
$I_C = \frac{V_C}{X_C}$	$i_C = C\frac{dv}{dt}$
X_C makes I_C and V_C 90° out of phase	Waveshape changes between i_C and v_C

With nonsinusoidal voltage, however, X_L cannot be used. Then v_L must be calculated as $L(di/dt)$. In this comparison, i_L and v_L can have different waveshapes, depending on the L/R time constant.

Choke Coils

For this application, the idea is to have almost all the applied AC voltage across L. The condition of X_L being at least 10 times R corresponds to a long time constant. The high value of X_L means that practically all the applied AC voltage is across X_L as IX_L, with little IR voltage.

The long L/R time constant means that i_L cannot rise appreciably, resulting in little v_R voltage across the resistor. The waveform for i_L and v_R in an inductive circuit corresponds to v_C in a capacitive circuit.

When Do We Use the Time Constant?

In electronic circuits, the time constant is useful in analyzing the effect of L or C on the waveshape of nonsinusoidal voltages, particularly rectangular pulses. Another application is the transient response when a DC voltage is turned on or off. The 63% change in one time constant is a natural characteristic of v or i, where the magnitude of one is proportional to the rate of change of the other.

When Do We Use Reactance?

X_L and X_C are generally used for sine-wave V or I. We can determine Z, I, voltage drops, and phase angles. The phase angle of 90° is a natural characteristic of a cosine wave when its magnitude is proportional to the rate of change in a sine wave.

GOOD TO KNOW

Always remember that X_L and X_C are quantities that apply only to sinusoidal waveforms.

22–12 Self-Review

Answers at the end of the chapter.

a. Does an *RC* coupling circuit have a small or large X_C compared with *R*?

b. Does an *RC* coupling circuit have a long or short time constant for the frequency of the applied voltage?

Summary

- The transient response of an inductive circuit with nonsinusoidal current is indicated by the time constant L/R. With L in henrys and R in ohms, T is the time in seconds for the current i_L to change by 63%. In five time constants, i_L reaches the steady value of V_T/R.
- At the instant an inductive circuit is opened, high voltage is generated across L because of the fast current decay with a short time constant. The induced voltage $v_L = L(di/dt)$. The di is the change in i_L.
- The transient response of a capacitive circuit with nonsinusoidal voltage is indicated by the time constant RC. With C in farads and R in ohms, T is the time in seconds for the voltage across the capacitor v_C to change by 63%. In five time constants, v_C reaches the steady value of V_T.
- At the instant a charged capacitor is discharged through a low resistance, a high value of discharge current can be produced. The discharge current $i_C = C(dv/dt)$ can be large because of the fast discharge with a short time constant. The dv is the change in v_C.
- The waveshapes of v_C and i_L correspond, as both rise relatively slowly to the steady-state value.
- Also, i_C and v_L correspond because they are waveforms that can change instantaneously.
- The resistor voltage $v_R = iR$ for both RC and RL circuits.
- A short time constant is one-fifth or less of the pulse width, in time, for the applied voltage.
- A long time constant is greater than the pulse width, in time, for the applied voltage by a factor of 5 or more.
- An RC circuit with a short time constant produces sharp voltage spikes for v_R at the leading and trailing edges of a square wave of applied voltage. The waveshape of voltage V_T is across the capacitor as v_C. See Fig. 22–7.
- An RC circuit with a long time constant allows v_R to be essentially the same as the variations in applied voltage V_T, and the average DC value of V_T is blocked as v_C. See Fig. 22–8.
- The universal rise and decay curves in Fig. 22–9 can be used for current or voltage in RC and RL circuits for any time up to five time constants.
- A differentiator is a circuit whose output is proportional to the change in applied voltage.
- An integrator is a circuit whose output combines, or integrates, its original voltage with the new change in voltage.
- The concept of reactance is only useful for sine-wave AC circuits with L and C.
- The time constant method is used with L or C to analyze nonsinusoidal waveforms.

Important Terms

Differentiator — a circuit whose output is proportional to the change in applied voltage. To provide good differentiation, the time constant of a circuit must be short with respect to the pulse width of the applied voltage.

Integrator — a circuit whose output combines or integrates its original voltage with the new change in voltage. For best integration, the time constant of a circuit must be long with respect to the pulse width of the applied voltage.

Long time constant — a long time constant is arbitrarily defined as one that is five or more times longer than the pulse width of the applied voltage.

Short time constant — a short time constant is arbitrarily defined as one that is one-fifth or less the time of the pulse width of the applied voltage.

Steady-state value — the final condition of a circuit after it has passed through its initial transitional state.

Time constant — a measure of how long it takes for a 63.2% change to occur.

Transient response — a term to describe the transitional state of a circuit when power is first applied or removed.

Universal time constant graph — a graph that shows the percent change in voltage or current in an RC or RL circuit with respect to the number of time constants that have elapsed.

Related Formulas

$$T = \frac{L}{R}$$

$$T = R \times C$$

$$v = V \times \epsilon^{-t/RC}$$

$$v = \text{antilog}\left(\log V - 0.434 \times \frac{t}{RC}\right)$$

$$v_C = V(1 - \epsilon^{-t/RC})$$

$$t = 2.3\ RC \log \frac{V}{v}$$

Self-Test

Answers at the back of the book.

1. **What is the time constant of the circuit in Fig. 22–11 with S_1 closed?**
   a. 250 μs.
   b. 31.6 μs.
   c. 50 μs.
   d. 5 ms.

Figure 22–11

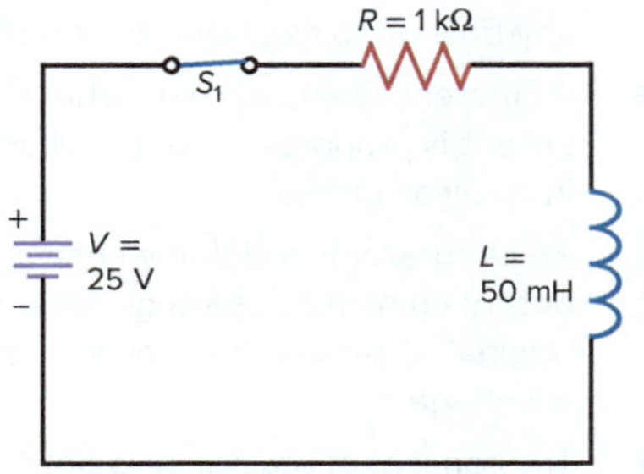

2. **With S_1 closed in Fig. 22–11, what is the eventual steady-state value of current?**
   a. 15.8 mA.
   b. 12.5 mA.
   c. 0 mA.
   d. 25 mA.

3. **In Fig. 22–11, how long does it take the current, I, to reach its steady-state value after S_1 is closed?**
   a. 50 μs.
   b. 250 μs.
   c. 500 μs.
   d. It cannot be determined.

4. **In Fig. 22–11, how much is the resistor voltage at the very first instant ($t = 0$ s) S_1 is closed?**
   a. 0 V.
   b. 25 V.
   c. 15.8 V.
   d. 9.2 V.

5. **In Fig. 22–11, what is the value of the resistor voltage exactly one time constant after S_1 is closed?**
   a. 15.8 V.
   b. 9.2 V.
   c. 6.32 V.
   d. 21.5 V.

6. **If a 2-MΩ resistor is placed across the switch, S_1, in Fig. 22–11, how much is the peak inductor voltage, V_L, when S_1 is opened?**
   a. 0 V.
   b. 25 V.
   c. 50 kV.
   d. It cannot be determined.

7. **In Fig. 22–11, what is the value of the current 35 μs after S_1 is closed?**
   a. approximately 20 mA.
   b. approximately 12.5 mA.
   c. 15.8 mA.
   d. 20 mA.

8. **With S_1 closed in Fig. 22–11, the length of one time constant could be increased by**
   a. decreasing L.
   b. decreasing R.
   c. increasing L.
   d. both b and c.

9. **In Fig. 22–11, what is the value of the inductor voltage five time constants after S_1 is closed?**
   a. 50 kV.
   b. 25 V.
   c. 0 V.
   d. 9.2 V.

10. **In Fig. 22–11, how much is the resistor voltage exactly 100 μs after S_1 is closed?**
    a. 12 V.
    b. 21.6 V.
    c. 3.4 V.
    d. 15.8 V.

11. **In Fig. 22–12, what is the time constant of the circuit with S_1 in Position 1?**
    a. 2 s.
    b. 5 s.
    c. 10 s.
    d. 1 s.

12. **In Fig. 22–12, what is the time constant of the circuit with S_1 in Position 2?**
    a. 2 s.
    b. 5 s.
    c. 10 s.
    d. 1 s.

Figure 22–12

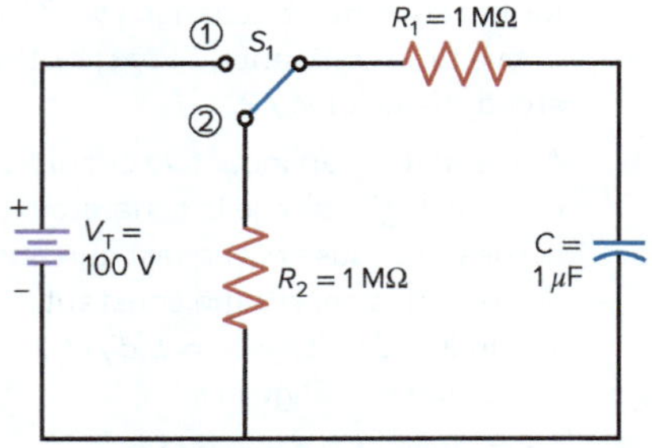

13. **In Fig. 22–12, how long will it take for the voltage across C to reach 100 V after S_1 is placed in Position 1?**
    a. 1 s.
    b. 2 s.
    c. 10 s.
    d. 5 s.

14. **In Fig. 22–12, how much voltage is across resistor, R_1, at the first instant the switch is moved from Position 2 to Position 1? (Assume that C was completely discharged with S_1 in Position 2.)**
    a. 100 V.
    b. 63.2 V.
    c. 0 V.
    d. 36.8 V.

15. **In Fig. 22–12, assume that C is fully charged to 100 V with S_1 in Position 1. How long will it take for C to discharge fully if S_1 is moved to Position 2?**
    a. 1 s.
    b. 5 s.
    c. 10 s.
    d. 2 s.

16. **In Fig. 22–12, assume that C is completely discharged while in Position 2. What is the voltage across C exactly 1 s after S_1 is moved to Position 1?**
    a. 50 v.
    b. 63.2 V.
    c. 36.8 V.
    d. 100 V.

17. In Fig. 22–12, assume that *C* is completely discharged while in Position 2. What is the voltage across R_1 exactly two time constants after S_1 is moved to Position 1?

a. 37 V.

b. 13.5 V.

c. 50 V.

d. 86 V.

18. In Fig. 22–12, what is the steady-state value of current with S_1 in Position 1?

a. 100 μA.

b. 50 μA.

c. 1 A.

d. 0 μA.

19. In Fig. 22–12, assume that *C* is fully charged to 100 V with S_1 in Position 1. What is the value of the capacitor voltage 3 s after S_1 is moved to Position 2?

a. 77.7 V.

b. 0 V.

c. 22.3 V.

d. 36.8 V.

20. In Fig. 22–12, assume that *C* is charging with S_1 in Position 1. At the instant the capacitor voltage reaches 75 V, S_1 is moved to Position 2. What is the approximate value of the capacitor voltage 0.7 time constant after S_1 is moved to Position 2?

a. 75 V.

b. 27.6 V.

c. 50 V.

d. 37.5 V.

21. For best results, an *RC* coupling circuit should have a

a. short time constant.

b. medium time constant.

c. long time constant.

d. zero time constant.

22. A differentiator is a circuit whose

a. output combines its original voltage with the new change in voltage.

b. output is always one-half of V_{in}.

c. time constant is long with the output across *C*.

d. output is proportional to the change in applied voltage.

23. An integrator is a circuit whose

a. output combines its original voltage with the new change in voltage.

b. output is always equal to V_{in}.

c. output is proportional to the change in applied voltage.

d. time constant is short with the output across *R*.

24. The time constant of an *RL* circuit is 47 μs. If $L = 4.7$ mH, calculate *R*.

a. $R = 10$ kΩ.

b. $R = 100$ Ω.

c. $R = 10$ MΩ.

d. $R = 1$ kΩ.

25. The time constant of an *RC* circuit is 300 μs. If $R = 1$ kΩ, calculate *C*.

a. $C = 0.33$ μF.

b. $C = 0.033$ μF.

c. $C = 3.3$ μF.

d. $C = 330$ pF.

Essay Questions

1. Give the formula, with units, for calculating the time constant of an *RL* circuit.

2. Give the formula, with units, for calculating the time constant of an *RC* circuit.

3. Redraw the *RL* circuit and graph in Fig. 22–2 for a 2 - H *L* and a 100-Ω *R*.

4. Redraw the graphs in Fig. 22–4 to fit the circuit in Fig. 22–5 with a 100-μF *C*. Use a 3000-Ω *R* for charge but a 3-Ω *R* for discharge.

5. List two comparisons of *RC* and *RL* circuits for nonsinusoidal voltage.

6. List two comparisons between *RC* circuits with nonsinusoidal voltage and sine-wave voltage applied.

7. Define the following: (a) a long time constant; (b) a short time constant; (c) an *RC* differentiating circuit; (d) an *RC* integrating circuit.

8. Redraw the horizontal time axis of the universal curve in Fig. 22–9, calibrated in absolute time units of milliseconds for an *RC* circuit with a time constant equal to 2.3 ms.

9. Redraw the circuit and graphs in Fig. 22–7 with everything the same except that *R* is 20 kΩ, making the *RC* time constant shorter.

10. Redraw the circuit and graphs in Fig. 22–8 with everything the same except that *R* is 500 kΩ, making the *RC* time constant longer.

11. Invert the equation $T = RC$, in two forms, to find *R* or *C* from the time constant.

12. Show three types of nonsinusoidal waveforms.

13. Give an application in electronic circuits for an *RC* circuit with a long time constant and with a short time constant.

14. Why can arcing voltage be a problem with coils used in switching circuits?

Problems

SECTION 22–1 RESPONSE OF RESISTANCE ALONE

22–1 In Fig. 22–13, how long does it take for the current, I, to reach its steady-state value after S_1 is closed?

Figure 22-13

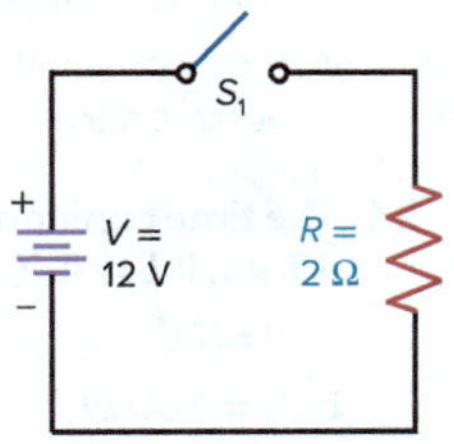

22–2 In Fig. 22–13, what is the current with S_1 closed?

22–3 Explain how the resistor in Fig. 22–13 reacts to the closing or opening of S_1.

SECTION 22–2 *L/R* TIME CONSTANT

22–4 In Fig. 22–14,

a. what is the time constant of the circuit with S_1 closed?

b. what is the eventual steady-state current with S_1 closed?

c. what is the value of the circuit current at the first instant S_1 is closed? ($t = 0$ s)

d. what is the value of the circuit current exactly one time constant after S_1 is closed?

e. how long after S_1 is closed will it take before the circuit current reaches its steady-state value?

Figure 22-14

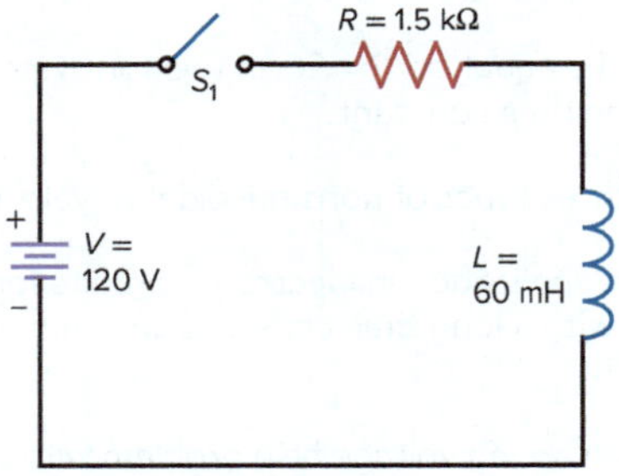

22–5 Repeat Prob. 22–4 if $L = 100$ mH and $R = 500\ \Omega$.

22–6 Calculate the time constant for an inductive circuit with the following values:

a. $L = 10$ H, $R = 1$ kΩ.

b. $L = 500$ mH, $R = 2$ kΩ.

c. $L = 250\ \mu$H, $R = 50\ \Omega$.

d. $L = 15$ mH, $R = 7.5$ kΩ.

22–7 List two ways to

a. increase the time constant of an inductive circuit.

b. decrease the time constant of an inductive circuit.

SECTION 22–3 HIGH VOLTAGE PRODUCED BY OPENING AN *RL* CIRCUIT

22–8 Assume that the switch, S_1, in Fig. 22–14 has been closed for more than five L/R time constants. If a 1-MΩ resistor is placed across the terminals of the switch, calculate

a. the approximate time constant of the circuit with S_1 open.

b. the peak inductor voltage, V_L, when S_1 is opened.

c. the di/dt value at the instant S_1 is opened.

d. how long it takes for the current to decay to zero after S_1 is opened (approximately).

22–9 Without a resistor across S_1 in Fig. 22–14, is it possible to calculate the time constant of the circuit with the switch open? Also, what effect will probably occur inside the switch when it is opened?

SECTION 22–4 *RC* TIME CONSTANT

22–10 In Fig. 22–15, what is the time constant of the circuit with the switch, S_1, in Position

a. 1?

b. 2?

Figure 22-15

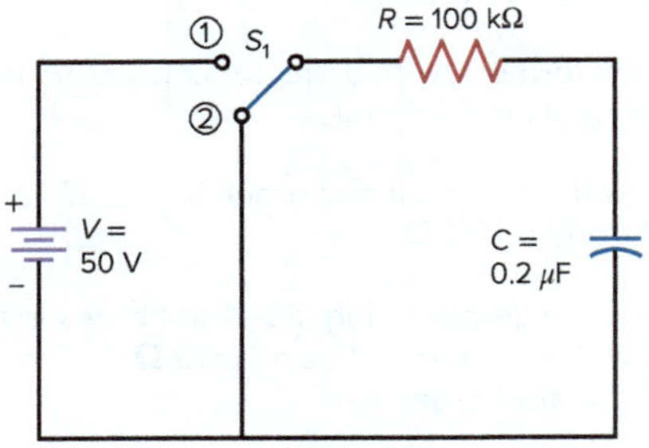

22–11 Assume that the capacitor in Fig. 22–15 is fully discharged with S_1 in Position 2. How much is the capacitor voltage, V_C,

a. exactly one time constant after S_1 is moved to Position 1?

b. five time constants after S_1 is moved to Position 1?

c. one week after S_1 is moved to Position 1?

22–12 Assume that the capacitor in Fig. 22–15 is fully charged with S_1 in Position 1. How much is the capacitor voltage, V_C,

a. exactly one time constant after S_1 is moved to Position 2?

b. five time constants after S_1 is moved to Position 2?

c. one week after S_1 is moved to Position 2?

22–13 Calculate the time constant of a capacitive circuit with the following values:

a. $R = 1\ \text{M}\Omega, C = 1\ \mu\text{F}$.

b. $R = 150\ \Omega, C = 0.01\ \mu\text{F}$.

c. $R = 330\ \text{k}\Omega, C = 270\ \text{pF}$.

d. $R = 5\ \text{k}\Omega, C = 40\ \mu\text{F}$.

22–14 List two ways to

a. increase the time constant of a capacitive circuit.

b. decrease the time constant of a capacitive circuit.

22–15 Assume that the capacitor in Fig. 22–15 is discharging from 50 V with S_1 in Position 2. At the instant the capacitor voltage reaches 25 V, S_1 is moved back to Position 1. What is

a. the net charging voltage at the first instant S_1 is put back in Position 1?

b. the value of the capacitor voltage exactly one time constant after S_1 is moved back to Position 1?

c. the value of the capacitor voltage five time constants after S_1 is moved back to Position 1?

22–16 Assume that the capacitor in Fig. 22–15 is charging from 0 V with S_1 in Position 1. At the instant the capacitor voltage reaches 35 V, S_1 is moved back to Position 2. What is

a. the value of the capacitor voltage exactly one time constant after S_1 is moved back to Position 2?

b. the value of the capacitor voltage five time constants after S_1 is moved back to Position 2?

SECTION 22–5 *RC* CHARGE AND DISCHARGE CURVES

22–17 Assume that the capacitor in Fig. 22–15 is fully discharged with S_1 in Position 2. What is

a. the value of the charging current at the first instant S_1 is moved to Position 1?

b. the value of the charging current five time constants after S_1 is moved to Position 1?

c. the value of the resistor voltage exactly one time constant after S_1 is moved to Position 1?

d. the value of the charging current exactly one time constant after S_1 is moved to Position 1?

22–18 Assume that the capacitor in Fig. 22–15 is fully charged to 50 V with S_1 in Position 1. What is the value of the discharge current

a. at the first instant S_1 is moved to Position 2?

b. exactly one time constant after S_1 is moved to Position 2?

c. five time constants after S_1 is moved to Position 2?

SECTION 22–6 HIGH CURRENT PRODUCED BY SHORT-CIRCUITING AN *RC* CIRCUIT

22–19 In Fig. 22–16, what is the *RC* time constant with S_1 in Position

a. 1?

b. 2?

Figure 22–16

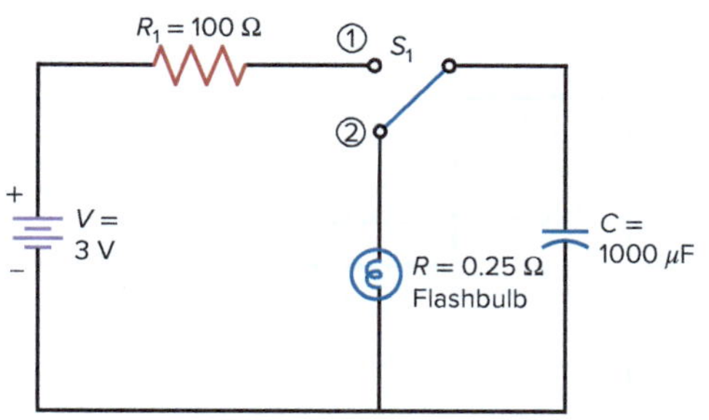

22–20 In Fig. 22–16, how long will it take the capacitor voltage to

a. reach 3 V after S_1 is moved to Position 1?

b. discharge to 0 V after S_1 is moved to Position 2?

22–21 Assume that the capacitor in Fig. 22–16 is fully discharged with S_1 in Position 2. At the first instant S_1 is moved to Position 1, how much is

a. the voltage across the capacitor?

b. the voltage across the resistor?

c. the initial charging current?

22–22 Assume that the capacitor in Fig. 22–16 is fully charged with S_1 in Position 1. At the first instant S_1 is moved to Position 2, what is

a. the DC voltage across the flashbulb?

b. the initial value of the discharge current?

c. the initial rate of voltage change, dv/dt?

22–23 How much energy is stored by the capacitor in Fig. 22–16 if it is fully charged to 3 V?

SECTION 22–7 *RC* WAVESHAPES

22–24 For the circuit in Fig. 22–17,

a. calculate the *RC* time constant.

b. draw the capacitor voltage waveform and include voltage values at times t_0, t_1, t_2, t_3, and t_4.

c. draw the resistor voltage waveform and include voltage values at times t_0, t_1, t_2, t_3, and t_4.

d. draw the charge and discharge current waveform and include current values at times t_0, t_1, t_2, t_3, and t_4.

Figure 22–17

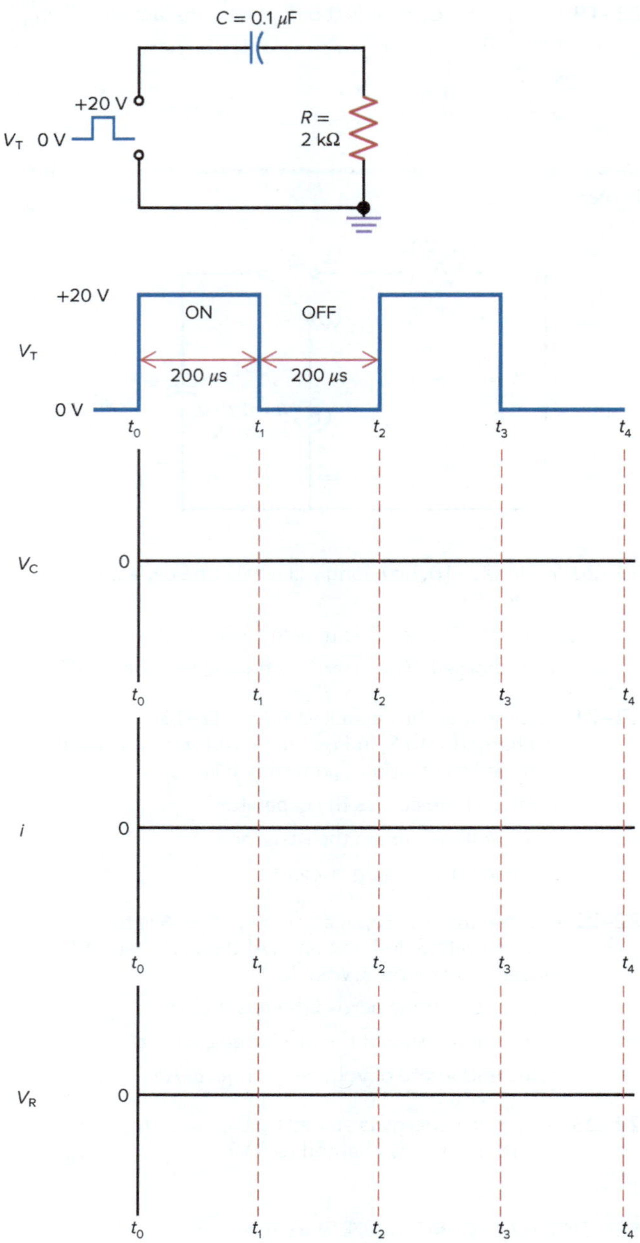

SECTION 22–8 LONG AND SHORT TIME CONSTANTS

22–25 In Fig. 22–17, is the time constant of the circuit considered long or short with respect to the pulse width of the applied voltage, V_T, if the resistance, R, is

a. increased to 10 kΩ?

b. decreased to 400 Ω?

22–26 For an *RC* circuit used as a differentiator,

a. across which component is the output taken?

b. should the time constant be long or short with respect to the pulse width of the applied voltage?

22–27 For an *RC* circuit used as an integrator,

a. across which component is the output taken?

b. should the time constant be long or short with respect to the pulse width of the applied voltage?

SECTION 22–9 CHARGE AND DISCHARGE WITH A SHORT *RC* TIME CONSTANT

22–28 For the circuit in Fig. 22–18,

a. calculate the *RC* time constant.

b. draw the capacitor voltage waveform and include voltage values at times t_0, t_1, t_2, t_3, and t_4.

c. draw the resistor voltage waveform and include voltage values at times t_0, t_1, t_2, t_3, and t_4.

d. specify the ratio of the pulse width of the applied voltage to the *RC* time constant.

Figure 22–18

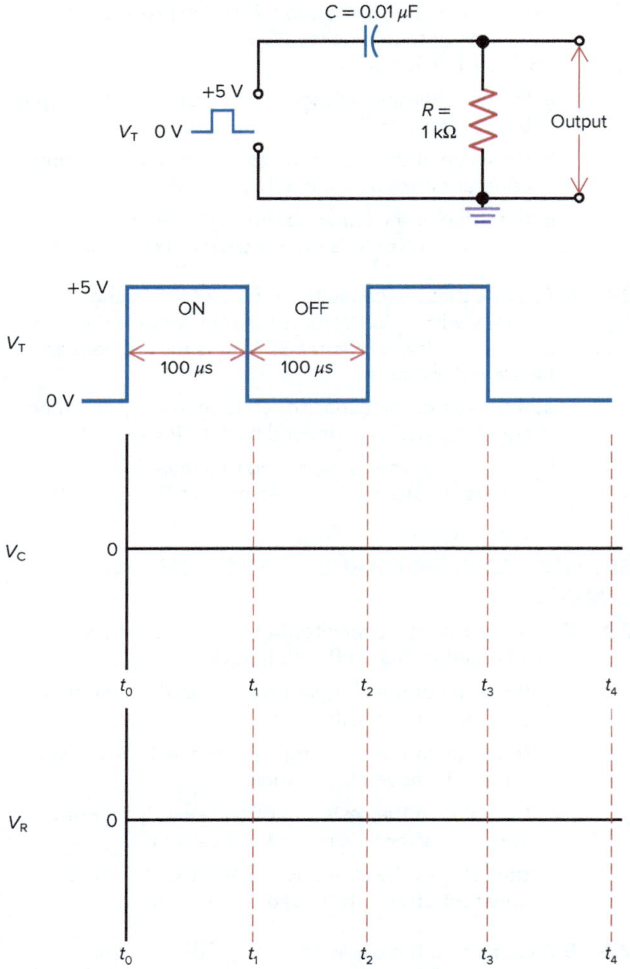

SECTION 22–10 LONG TIME CONSTANT FOR AN *RC* COUPLING CIRCUIT

22–29 Assume that the resistance, *R*, in Fig. 22–18 is increased to 100 kΩ but the frequency of the applied voltage, V_T, remains the same. Determine

a. the new *RC* time constant of the circuit.

b. the ratio of the pulse width of the applied voltage to the *RC* time constant.

c. the approximate capacitor and resistor voltage waveforms, assuming that the input voltage has been applied for longer than five *RC* time constants.

SECTION 22–11 ADVANCED TIME CONSTANT ANALYSIS

22–30 What is the time constant of the circuit in Fig. 22–19?

Figure 22–19

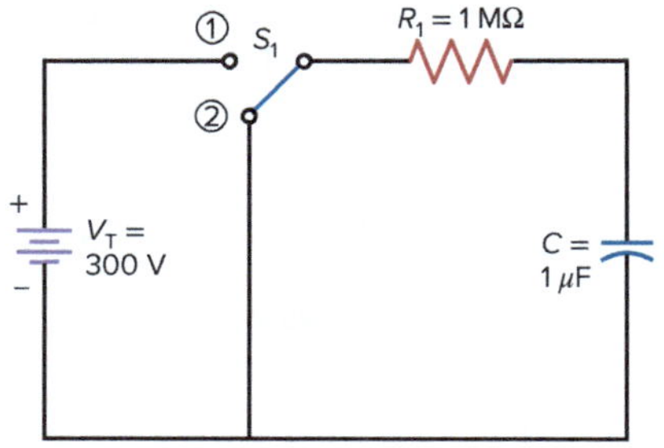

22–31 Assume that *C* in Fig. 22–19 is completely discharged with S_1 in Position 2. If S_1 is moved to Position 1, how much is the capacitor voltage at the following time intervals?

a. $t = 0$ s.

b. $t = 0.7$ s.

c. $t = 1$ s.

d. $t = 1.5$ s.

e. $t = 2$ s.

f. $t = 2.5$ s.

g. $t = 3.5$ s.

22–32 Assume that *C* in Fig. 22–19 is fully charged with S_1 in Position 1. If S_1 is moved to Position 2, how much is the resistor voltage at the following time intervals?

a. $t = 0$ s.

b. $t = 0.7$ s.

c. $t = 1$ s.

d. $t = 1.5$ s.

e. $t = 2$ s.

f. $t = 2.5$ s.

g. $t = 3.5$ s.

22–33 Assume that *C* in Fig. 22–19 is completely discharged with S_1 in Position 2. If S_1 is moved back to Position 1, how long will it take for the capacitor voltage to reach

a. 90 V?

b. 150 V?

c. 200 V?

d. 240 V?

e. 270 V?

22–34 Assume that the capacitor in Fig. 22–19 is discharging from 300 V with S_1 in Position 2. At the instant V_C reaches 150 V, S_1 is moved back to Position 1. What is the value of the capacitor voltage 1.25 s later?

22–35 What is the time constant of the circuit in Fig. 22–20?

Figure 22–20

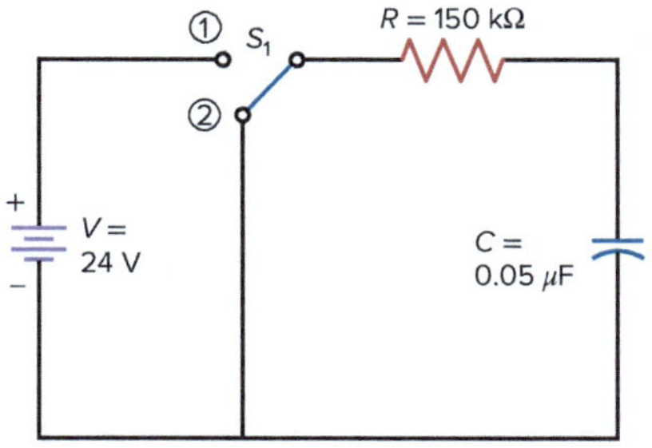

22–36 Assume that *C* in Fig. 22–20 is completely discharged with S_1 in Position 2. If S_1 is moved back to Position 1, how long will it take for the capacitor voltage to reach

a. 3 V?

b. 6 V?

c. 15 V?

d. 20 V?

22–37 Assume that *C* in Fig. 22–20 is completely discharged with S_1 in Position 2. If S_1 is moved back to Position 1, how much is the resistor voltage at the following time intervals?

a. $t = 0$ s.

b. $t = 4.5$ ms.

c. $t = 10$ ms.

d. $t = 15$ ms.

e. $t = 25$ ms.

22–38 Assume that *C* in Fig. 22–20 is fully charged with S_1 in Position 1. If S_1 is moved to Position 2, how long will it take the capacitor to discharge to

a. 4 V?

b. 8 V?

c. 12 V?

d. 18 V?

SECTION 22–12 COMPARISON OF REACTANCE AND TIME CONSTANT

22–39 When analyzing a sine-wave AC circuit containing resistance and capacitance or resistance and inductance, do we use the concepts involving reactance or time constants?

22–40 When analyzing a circuit having a square-wave input voltage, should we use the concepts involving reactance or time constants?

22–41 Should an *RC* coupling circuit have a long or short time constant with respect to the period of the AC input voltage?

Critical Thinking

22–42 Refer to Fig. 22–21. (*a*) If S_1 is closed long enough for the capacitor *C* to become fully charged, what voltage is across *C*? (b) With *C* fully charged, how long will it take *C* to discharge fully when S_1 is opened?

Figure 22–21 Circuit for Critical Thinking Probs. 22–42 and 22–43.

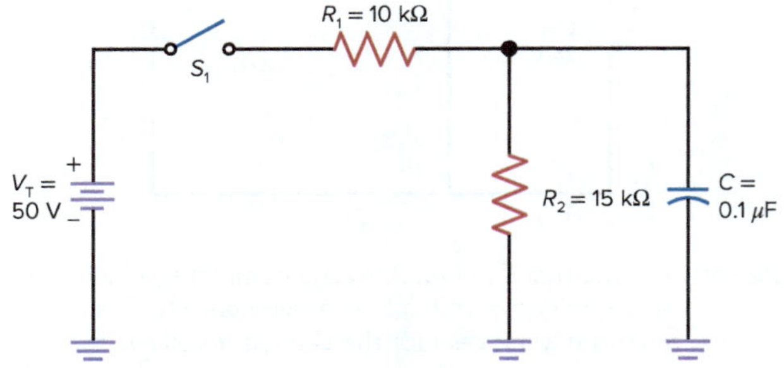

22–43 Refer to Fig. 22–21. (a) How long will it take *C* to fully charge after S_1 is closed? (b) What is V_C 1 ms after S_1 is initially closed? (c) What is V_C 415.8 μs after S_1 is initially closed? (d) What is V_C 1.5 ms after S_1 is initially closed?

22–44 Refer to Fig. 22–22. Assume that *C* is allowed to charge fully and then the polarity of V_T is suddenly reversed. What is the capacitor voltage v_C for the following time intervals after the reversal of V_T: (a) 0 s; (b) 6.93 ms; (c) 10 ms; (d) 15 ms; (e) 30 ms?

Figure 22–22 Circuit for Critical Thinking Prob. 22–44.

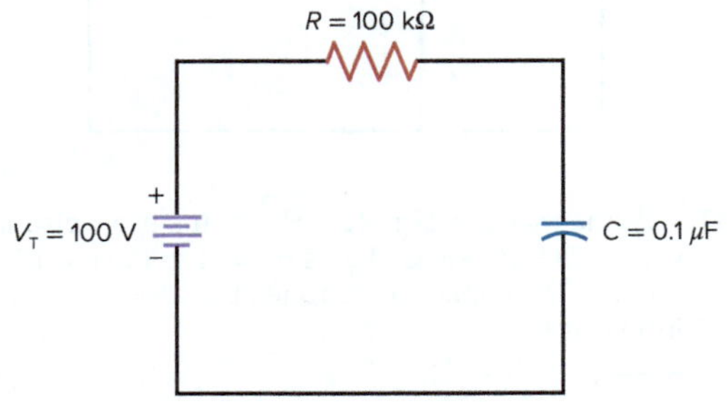

Answers to Self-Reviews

22–1 true
true

22–2 0.02 s
0.5 ms

22–3 shorter
faster

22–4 940 μs
470 ns

22–5 63 V
37 V

22–6 shorter
faster

22–7 37 V
23.3 V

22–8 short
long

22–9 short
across *C*

22–10 long
across *R*

22–11 true
true
true

22–12 small X_C
long time constantCritical

Laboratory Application Assignment

In this lab application assignment, you will examine *RC* differentiators and integrators. In each type of circuit, you will measure the resistor and capacitor voltage waveforms and draw them in the proper time relation with respect to the input voltage applied. For both the differentiator and integrator, pay close attention to how the *RC* time constant relates to the pulse time, t_p, of the applied voltage.

Equipment: Obtain the following items from your instructor.

- Function generator
- Oscilloscope
- 10-kΩ carbon-film resistor
- 0.01-μF and 0.1-μF capacitors
- DMM

RC Differentiator

In Fig. 22–23*a*, calculate and record the *RC* time constant. $RC =$ ___________ Is this value long or short with respect to the pulse time, t_p, of the applied voltage? ___________. Calculate and record the t_p/RC ratio: _____/_____ Will this ratio provide proper differentiation? ___________

Construct the *RC* differentiator in Fig. 22–23*a*. Connect channel 1 of your oscilloscope to the input side of the circuit, and leave it there. Set the channel 1 input coupling switch to DC. Next, adjust the DC offset, amplitude, and frequency controls of the function generator to produce the input waveform shown at the top of Fig. 22–23*b*. Have your instructor verify that the input waveform is indeed a 0- to +10-V square wave with a frequency of 500 Hz.

Connect channel 2 of your oscilloscope across the resistor, *R*, which is the output of the differentiator. Set the channel 2 input coupling switch to DC. Draw this waveform in Fig. 22–23*b* in the space allocated for V_R. Next, use the differential measurement capabilities of your oscilloscope to measure the voltage across the capacitor, *C*. Draw this waveform in Fig. 22–23*b* in the space allocated for V_C. Be certain that V_R and V_C are both drawn in proper time relation with respect to V_{in}. Include all voltage values for both the V_R and V_C waveforms.

Using your DMM, measure and record the DC value of the applied voltage, V_{IN}. $V_{IN(DC)} =$ ________
Next, measure and record the DC voltage across *R* and *C*.
$V_{C(DC)} =$ ________, $V_{R(DC)} =$ ________
What's significant about these DC voltage measurements?______
____

RC Integrator

In Fig. 22–24*a*, calculate and record the *RC* time constant. $RC =$ ___________ Is this value long or short with respect to the pulse time, t_p, of the applied voltage? __________. Calculate and record the t_p/RC ratio: ______ /______ Will this ratio provide proper integration? __________

Construct the *RC* integrator in Fig. 22–24*a*. Connect channel 1 of your oscilloscope to the input side of the circuit, and leave it there. Set the channel 1 input coupling switch to DC. Next, adjust the DC offset, amplitude, and frequency controls of the function generator to produce the input waveform shown at the top of Fig. 22–24*b*. Have your instructor verify that the input waveform is indeed a 0 - to +10-V square wave with a frequency of 5 kHz.

Connect channel 2 of your oscilloscope across the capacitor, *C*, which is the output of the integrator. Set the channel 2 input

Figure 22–23

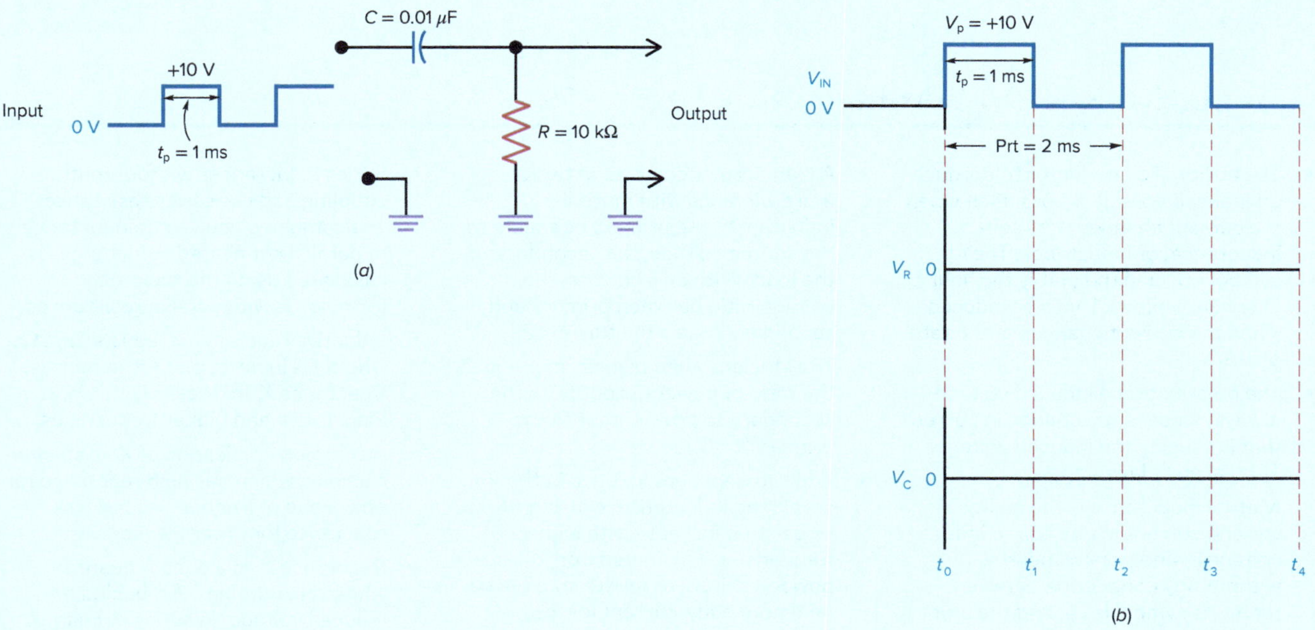

coupling switch to DC. Draw this waveform in Fig. 22–24*b* in the space allocated for V_C. Next, use the differential measurement capabilities of your oscilloscope to measure the voltage across the resistor, *R*. Draw this waveform in Fig. 22–24*b* in the space allocated for V_R. Be certain that V_C and V_R are both drawn in proper time relation with respect to V_{in}. Include all voltage values for both the V_C and V_R waveforms.

Was the capacitor voltage waveform difficult to view with the channel 2 input coupling switch set to DC?__________ Was it nearly a straight line centered around +5 V? ________________ Move the channel 2 input coupling switch to AC. Reduce the channel 2 volts/div. setting until the capacitor voltage waveform is recognizable as a triangular wave. Explain the displayed waveform.________________

____

Using your DMM, measure and record the DC value of the applied voltage, V_{IN}.

$V_{IN(DC)}$ = ________

Next, measure and record the DC voltage across *R* and *C*.

$V_{C(DC)}$ = ________, $V_{R(DC)}$ = ________

What's significant about these voltage measurements?________

____

____

Figure 22–24

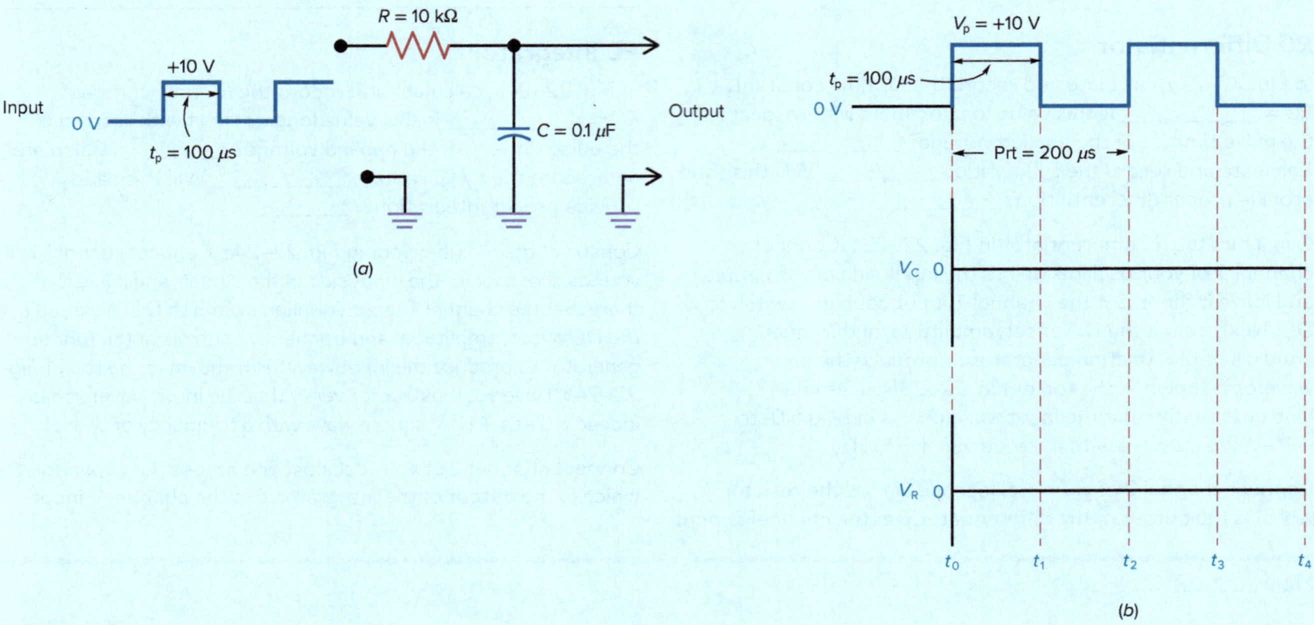

Cumulative Review Summary (Chapters 19–22)

- The ability of a conductor to produce an induced voltage across itself when the current changes is its self-inductance, or inductance. The symbol is *L*, and the unit is the henry. One henry allows 1 V to be induced when the current changes at the rate of 1 A/s.
- The polarity of the induced voltage always opposes the change in current that is causing the induced voltage. This is Lenz's law.
- Mutual inductance is the ability of varying current in one coil to induce voltage in another coil nearby, without any connection between them. Its symbol is L_M, and the unit is also the henry.
- A transformer consists of two or more windings with mutual inductance. The primary connects to the source voltage, the secondary to the load. With an iron core, the voltage ratio between primary and secondary equals the turns ratio.
- The efficiency of a transformer equals the ratio of power output from the secondary to power input to the primary × 100.
- Eddy currents are induced in the iron core of an inductance, causing I^2R losses that increase with higher frequencies. Laminated iron, powdered-iron, or ferrite cores have minimum eddy-current losses. Hysteresis also increases the losses.
- Series inductances without mutual coupling add like series resistances. The combined equivalent inductance of parallel connected inductances is calculated using the reciprocal formula, as with parallel resistances.
- Inductive reactance X_L equals $2\pi fL\ \Omega$, where *f* is in hertz and *L* is in henrys. Reactance X_L increases with more inductance and higher frequencies.
- A common application of X_L is an af or rf choke, which has high reactance for one group of frequencies but less reactance for lower frequencies.
- Reactance X_L is a phasor quantity whose current lags 90° behind its induced voltage. In series circuits, *R*

and X_L are added by phasors because their voltage drops are 90° out of phase. In parallel circuits, the resistive and inductive branch currents are 90° out of phase.

- Impedance Z, in ohms, is the total opposition of an AC circuit with resistance and reactance. For series circuits, $Z_T = \sqrt{R^2 + X_L^2}$ and $I = V_T/Z_T$. For Parallel circuits, $I_T = \sqrt{I_R^2 + I_L^2}$ and $Z_{EQ} = V_A/I_T$.
- The Q of a coil is X_L/r_i.
- Energy stored by an inductance is ½LI^2, where I is in amperes, L is in henrys, and the energy is in joules.
- The voltage across L is always equal to $L(di/dt)$ for any waveshape of current.
- The transient response of a circuit refers to the temporary condition that exists until the circuit's current or voltage reaches its steady-state value. The transient response of a circuit is measured in time constants, where one time constant is defined as the length of time during which a 63.2% change in current or voltage occurs.
- For an inductive circuit, one time constant is the time in seconds for the current to change by 63.2%. For inductive circuits, one time constant equals L/R, that is, $T = L/R$, where L is in henrys, R is in ohms, and T is in seconds. The current reaches its steady-state value after five L/R time constants have elapsed.
- In a capacitive circuit, one time constant is the time in seconds for the capacitor voltage to change by 63.2%. For capacitive circuits, one time constant equals RC, that is, $T = RC$, where R is in ohms, C is in farads, and T is in seconds. The capacitor voltage reaches its steady-state value after five RC time constants have elapsed.
- When the input voltage to an inductive or capacitive circuit is nonsinusoidal, time constants rather than reactances are used to determine the circuit's voltage and current values.
- Whether an L/R or RC time constant is considered short or long depends on its relationship to the pulse width of the applied voltage. In general, a short time constant is considered one that is one-fifth or less the time of the pulse width of the applied voltage. Conversely, a long time constant is generally considered one that is five or more times longer than the pulse width of the applied voltage.
- To calculate the voltage across a capacitor during charge, use curve *a* in Fig. 22–9 or use Formula (22–5). To calculate the voltage across a resistor during charge, use curve *b* in Fig. 22–9 or Formula (22–3). To calculate the voltage across a capacitor or resistor during discharge, use curve *b* in Fig. 22–9 or Formula (22–3).

Cumulative Self-Test

Answers at the back of the book.

1. A coil induces 200 mV when the current changes at the rate of 1 A/s. The inductance L is (*a*) 1 mH; (*b*) 2 mH; (*c*) 200 mH; (*d*) 100 mH.

2. Alternating current in an inductance produces maximum induced voltage when the current has its (*a*) maximum value; (*b*) maximum change in magnetic flux; (*c*) minimum change in magnetic flux; (*d*) rms value of 0.707 × peak.

3. An iron-core transformer connected to a 120-V, 60-Hz power line has a turns ratio of 1:20. The voltage across the secondary equals (*a*) 20 V; (*b*) 60 V; (*c*) 120 V; (*d*) 2400 V.

4. Two 250-mH chokes in series have a total inductance of (*a*) 60 mH; (*b*) 125 mH; (*c*) 250 mH; (*d*) 500 mH.

5. Which of the following will have minimum eddy-current losses? (*a*) Solid iron core; (*b*) laminated iron core; (*c*) powdered-iron core; (*d*) air core.

6. Which of the following will have maximum inductive reactance? (*a*) 2-H inductance at 60 Hz; (*b*) 2-mH inductance at 60 kHz; (*c*) 5-mH inductance at 60 kHz; (*d*) 5-mH inductance at 100 kHz.

7. A 100-Ω R is in series with 100 Ω of X_L. The total impedance Z equals (*a*) 70.7 Ω; (*b*) 100 Ω; (*c*) 141 Ω; (*d*) 200 Ω.

8. A 100-Ω R is in parallel with 100 Ω of X_L. The total impedance Z equals (*a*) 70.7 Ω; (*b*) 100 Ω; (*c*) 141 Ω; (*d*) 200 Ω.

9. If two waves have the frequency of 1000 Hz and one is at the maximum value when the other is at zero, the phase angle between them is (*a*) 0°; (*b*) 90°; (*c*) 180°; (*d*) 360°.

10. If an ohmmeter check on a 50-μH choke reads 3 Ω, the coil is probably (*a*) open; (*b*) defective; (*c*) normal; (*d*) partially open.

11. An inductive circuit with $L = 100$ mH and $R = 10$ kΩ has a time constant of (*a*) 1 μs; (*b*) 100 μs; (*c*) 10 μs; (*d*) 1000 μs.

12. A capacitive circuit with $R = 1.5$ kΩ and $C = 0.01$ μF has a time constant of (*a*) 15 μs; (*b*) 1.5 μs; (*c*) 150 μs; (*d*) 150 s.

13. With respect to the pulse width of the applied voltage, the time constant of an RC integrator should be (*a*) short; (*b*) the same as the pulse width of V_T; (*c*) long; (*d*) shorter than the pulse width of V_T.

14. With respect to the pulse width of the applied voltage, the time constant of an RC differentiator should be (*a*) long; (*b*) the same as the pulse width of V_T; (*c*) longer than the pulse width of V_T; (*d*) short.

15. The current rating of a transformer is usually specified for (*a*) the primary windings only; (*b*) the secondary windings only; (*c*) both the primary and secondary windings; (*d*) the core only.

16. The secondary of a transformer is connected to a 15-Ω resistor. If the turns ratio $N_P/N_S = 3{:}1$, the primary impedance Z_P equals (*a*) 135 Ω; (*b*) 45 Ω; (*c*) 5 Ω; (*d*) none of these.

Design credits: Chapter Opener Background Image, and Multisim: Stockbyte/Getty Images

chapter

23

Alternating Current Circuits

This chapter shows how to analyze sine-wave AC circuits that have R, X_L, and X_C. How do we combine these three types of ohms of opposition, how much current flows, and what is the phase angle? These questions are answered for both series and parallel circuits.

The problems are simplified by the fact that in series circuits X_L is at 90° and X_C is at −90°, which are opposite phase angles. Then all of one reactance can be canceled by part of the other reactance, resulting in only a single net reactance.

Similarly, in parallel circuits, I_L and I_C have opposite phase angles. These phasor currents oppose each other and result in a single net reactive line current.

Finally, the idea of how AC and DC power differ due to the presence of inductive and capacitive reactances in an AC circuit is explained. To fully understand the concepts of AC power you must understand the definitions of real power, apparent power, volt-ampere reactive, and power factor (PF). Also included in this chapter are the different types of AC current meters as well as the wattmeter. ■

Chapter Outline

Chapter Objectives

After studying this chapter, you should be able to

- *Explain* why opposite reactances in series cancel.
- *Determine* the total impedance and phase angle of a series circuit containing resistance, capacitance, and inductance.
- *Determine* the total current, equivalent impedance, and phase angle of a parallel circuit containing resistance, capacitance, and inductance.
- *Define* the terms *real power, apparent power, volt-ampere reactive,* and *power factor.*
- *Calculate* the power factor of a circuit.

Important Terms

apparent power
double-subscript notation
power factor (PF)
real power
volt-ampere (VA)
volt-ampere reactive (VAR)
wattmeter

GOOD TO KNOW

In a resistor, *V* and *I* are always in phase.

23–1 AC Circuits with Resistance but No Reactance

Combinations of series and parallel resistances are shown in Fig. 23–1. In Fig. 23–1*a* and *b*, all voltages and currents throughout the resistive circuit are in phase. There is no reactance to cause a lead or lag in either current or voltage.

Series Resistances

For the circuit in Fig. 23–1*a*, with two 50-Ω resistances in series across the 100-V source, the calculations are as follows:

$$R_T = R_1 + R_2 = 50 + 50 = 100\ \Omega$$

$$I = \frac{V_T}{R_T} = \frac{100}{100} = 1\text{ A}$$

$$V_1 = IR_1 = 1 \times 50 = 50\text{ V}$$

$$V_2 = IR_2 = 1 \times 50 = 50\text{ V}$$

Note that the series resistances R_1 and R_2 serve as a voltage divider, as in DC circuits. Each *R* has one-half the applied voltage for one-half the total series resistance.

The voltage drops V_1 and V_2 are both in phase with the series current *I*, which is the common reference. Also, *I* is in phase with the applied voltage V_T because there is no reactance.

Parallel Resistances

For the circuit in Fig. 23–1*b*, with two 50-Ω resistances in parallel across the 100-V source, the calculations are

$$I_1 = \frac{V_A}{R_1} = \frac{100}{50} = 2\text{ A}$$

$$I_2 = \frac{V_A}{R_2} = \frac{100}{50} = 2\text{ A}$$

$$I_T = I_1 + I_2 = 2 + 2 = 4\text{ A}$$

With a total current of 4 A in the main line from the 100-V source, the combined parallel resistance is 25 Ω. This R_{EQ} equals 100 V/4 A for the two 50-Ω branches.

Each branch current has the same phase as that of the applied voltage. Voltage V_A is the reference because it is common to both branches.

Figure 23–1 Alternating current circuits with resistance but no reactance. (*a*) Resistances R_1 and R_2 in series. (*b*) Resistances R_1 and R_2 in parallel.

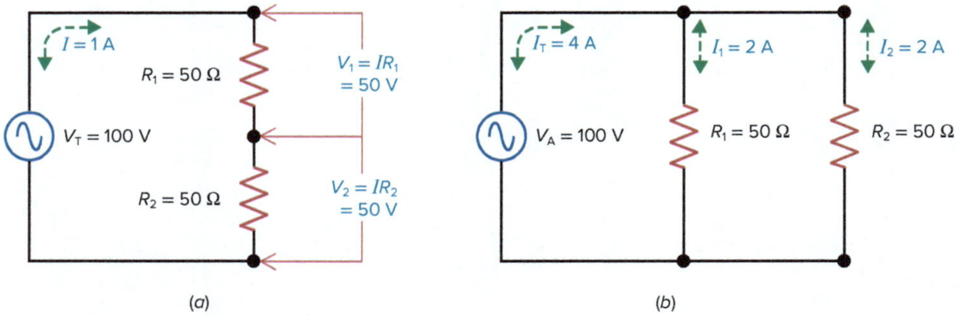

23–1 Self-Review

Answers at the end of the chapter.

a. In Fig. 23–1*a*, what is the phase angle between V_T and I?
b. In Fig. 23–1*b*, what is the phase angle between I_T and V_A?

23–2 Circuits with X_L Alone

The circuits with X_L in Figs. 23–2 and 23–3 correspond to the series and parallel circuits in Fig. 23–1, with ohms of X_L equal to R values. Since the applied voltage is the same, the values of current correspond because ohms of X_L are just as effective as ohms of R in limiting the current or producing a voltage drop.

Although X_L is a phasor quantity with a 90° phase angle, all ohms of opposition are the same kind of reactance in this example. Therefore, without any R or X_C, the series ohms of X_L can be combined directly. Similarly, the parallel I_L currents can be added.

GOOD TO KNOW

For series inductors, the voltage drops can be added algebraically, since each voltage drop has the same phase relationship with respect to the series current.

X_L Values in Series

For Fig. 23–2*a*, the calculations are

$$X_{L_T} = X_{L_1} + X_{L_2} = 50 + 50 = 100\ \Omega$$

$$I = \frac{V_T}{X_{L_T}} = \frac{100}{100} = 1\ \text{A}$$

$$V_1 = IX_{L_1} = 1 \times 50 = 50\ \text{V}$$

$$V_2 = IX_{L_2} = 1 \times 50 = 50\ \text{V}$$

Note that the two series voltage drops of 50 V each add to equal the total applied voltage of 100 V.

Figure 23–2 Series circuit with X_L alone. (*a*) Schematic diagram. (*b*) Phasor diagram of voltages and series current.

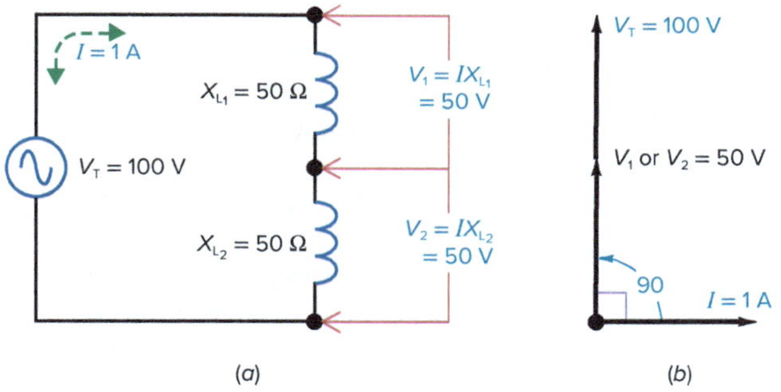

Figure 23–3 Parallel circuit with X_L alone. (*a*) Schematic diagram. (*b*) Phasor diagram of branch and total line currents and applied voltage.

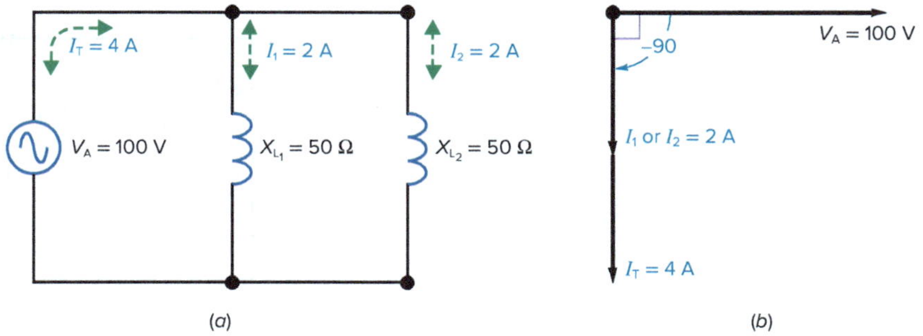

With regard to the phase angle for the inductive reactance, the voltage across any X_L always leads the current through it by 90°. In Fig. 23–2*b*, I is the reference phasor because it is common to all series components. Therefore, the voltage phasors for V_1 and V_2 across either reactance, or V_T across both reactances, are shown leading I by 90°.

I_L Values in Parallel

For 23–3*a*, the calculations are

$$I_1 = \frac{V_A}{X_{L_1}} = \frac{100}{50} = 2 \text{ A}$$

$$I_2 = \frac{V_A}{X_{L_2}} = \frac{100}{50} = 2 \text{ A}$$

$$I_T = I_1 + I_2 = 2 + 2 = 4 \text{ A}$$

These two branch currents can be added because both have the same phase. This angle is 90° lagging the voltage reference phasor, as shown in Fig. 23–3*b*.

Since the voltage V_A is common to the branches, this voltage is across X_{L_1} and X_{L_2}. Therefore, V_A is the reference phasor for parallel circuits.

Note that there is no fundamental change between Fig. 23–2*b*, which shows each X_L voltage leading its current by 90°, and Fig. 23–3*b*, showing each X_L current lagging its voltage by −90°. The phase angle between the inductive current and voltage is still the same 90°.

GOOD TO KNOW

For parallel inductors, the branch currents can be added algebraically since each branch current has the same phase relationship with respect to the applied voltage.

23–2 Self-Review

Answers at the end of the chapter.

a. In Fig. 23–2, what is the phase angle of V_T with respect to I?
b. In Fig. 23–3, what is the phase angle of I_T with respect to V_A?

23–3 Circuits with X_C Alone

Again, reactances are shown in Figs. 23–4 and 23–5 but with X_C values of 50 Ω. Since there is no R or X_L, the series ohms of X_C can be combined directly. Also, the parallel I_C currents can be added.

GOOD TO KNOW

For series capacitors, the voltage drops can be added algebraically to find V_T. Similarly, with parallel capacitors the branch currents can be added to find I_T.

X_C Values in Series

For Fig. 23–4*a*, the calculations for V_1 and V_2 are the same as before. These two series voltage drops of 50 V each add to equal the total applied voltage.

With regard to the phase angle for the capacitive reactance, the voltage across any X_C always lags its capacitive charge and discharge current I by 90°. For the

Figure 23–4 Series circuit with X_C alone. (*a*) Schematic diagram. (*b*) Phasor diagram of voltages and series current.

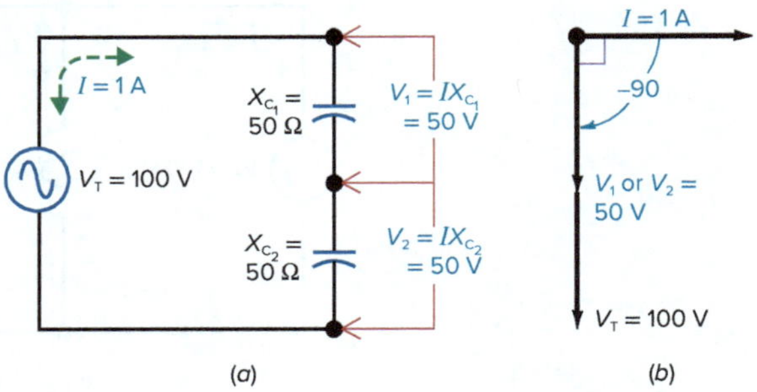

Figure 23–5 Parallel circuit with X_C alone. (*a*) Schematic diagram. (*b*) Phasor diagram of branch and total line currents and applied voltage.

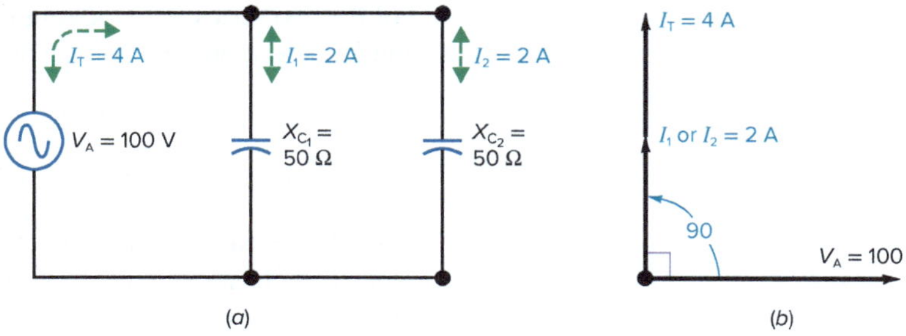

series circuit in Fig. 23–4, I is the reference phasor. The capacitive current leads by 90°, or we can say that each voltage lags I by −90°.

I_C Values in Parallel

For Fig. 23–5, V_A is the reference phasor. The calculations for I_1 and I_2 are the same as before. However, now each of the capacitive branch currents or the I_T leads V_A by 90°.

23–3 Self-Review

Answers at the end of the chapter.

a. In Fig. 23–4, what is the phase angle of V_T with respect to I?
b. In Fig. 23–5, what is the phase angle of I_T with respect to V_A?

23–4 Opposite Reactances Cancel

In a circuit with both X_L and X_C, the opposite phase angles enable one to offset the effect of the other. For X_L and X_C in series, the net reactance is the difference between the two series reactances, resulting in less reactance than in either one. In parallel circuits, the net reactive current is the difference between the I_L and I_C branch currents, resulting in less total line current than in either branch current.

X_L and X_C in Series

For the example in Fig. 23–6, the series combination of a 60-Ω X_L and a 40-Ω X_C in Fig. 23–6*a* and *b* is equivalent to the net reactance of the 20-Ω X_L shown in

MultiSim **Figure 23–6** When X_L and X_C are in series, their ohms of reactance subtract. (*a*) Series circuit with 60-Ω X_L and 40-Ω X_C. (*b*) Phasor diagram. (*c*) Equivalent circuit with net value of 20 Ω of X_L for the total reactance.

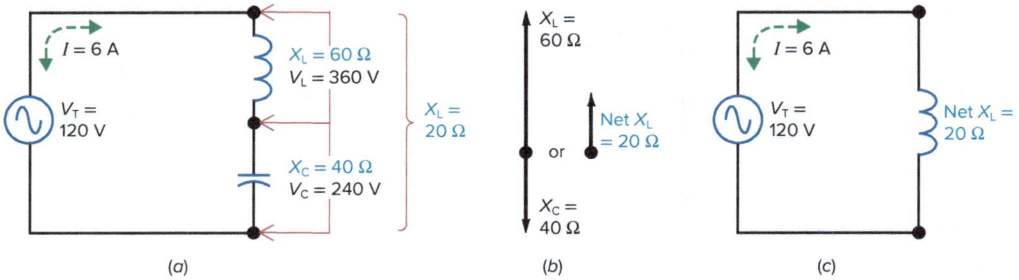

Fig. 23–6*c*. Then, with 20 Ω as the net reactance across the 120-V source, the current is 6 A. This current lags the applied voltage V_T by 90° because the net reactance is inductive.

For the two series reactances in Fig. 23–6*a*, the current is the same through both X_L and X_C. Therefore, the voltage drops can be calculated as

$$V_L \text{ or } IX_L = 6\text{ A} \times 60\ \Omega = 360\text{ V}$$

$$V_C \text{ or } IX_C = 6\text{ A} \times 40\ \Omega = 240\text{ V}$$

Note that each individual reactive voltage drop can be more than the applied voltage. The phasor sum of the series voltage drops still is 120 V, however, equal to the applied voltage because the IX_L and IX_C voltages are opposite. The IX_L voltage leads the series current by 90°; the IX_C voltage lags the same current by 90°. Therefore, IX_L and IX_C are 180° out of phase with each other, which means that they are of opposite polarity and offset each other. Then the total voltage across the two in series is 360 V minus 240 V, which equals the applied voltage of 120 V.

If the values in Fig. 23–6 were reversed, with an X_C of 60 Ω and an X_L of 40 Ω, the net reactance would be a 20-Ω X_C. The current would be 6 A again but with a lagging phase angle of −90° for the capacitive voltage. The IX_C voltage would then be greater at 360 V than an IX_L value of 240 V, but the difference would still equal the applied voltage of 120 V.

GOOD TO KNOW

For X_L and X_C in parallel, the equivalent impedance, Z_{EQ}, can be calculated as $Z_{EQ} = \frac{X_L \cdot X_C}{X}$ where X represents the difference between X_L and X_C.

X_L and X_C in Parallel

In Fig. 23–7, the 60-Ω X_L and 40-Ω X_C are in parallel across the 120-V source. Then the 60-Ω X_L branch current I_L is 2 A, and the 40-Ω X_C branch current I_C is 3 A. The X_C branch has more current because its reactance is less than X_L.

In terms of phase angle, I_L lags the parallel voltage V_A by 90°, and I_C leads the same voltage by 90°. Therefore, the opposite reactive branch currents are 180° out of phase with each other. The net line current then is the difference between 3 A for I_C and 2 A for I_L, which equals the net value of 1 A. This resultant current leads V_A by 90° because it is capacitive current.

If the values in Fig. 23–7 were reversed, with an X_C of 60 Ω and an X_L of 40 Ω, I_L would be larger. The I_L would then equal 3 A, with an I_C of 2 A. The net line current would be 1 A again but inductive with a net I_L.

23–4 Self-Review

Answers at the end of the chapter.

a. In Fig. 23–6, how much is the net X_L?
b. In Fig. 23–7, how much is the net I_C?

MultiSim **Figure 23–7** When X_L and X_C are in parallel, their branch currents subtract. (*a*) Parallel circuit with 3-A I_C and 2-A I_L. (*b*) Phasor diagram. (*c*) Equivalent circuit with net value of 1 A of I_C for the total line current.

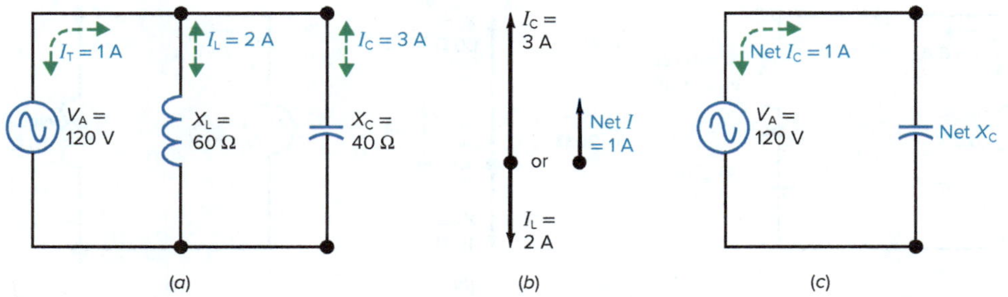

MultiSim **Figure 23–8** Impedance Z_T of series circuit with resistance and reactance. (*a*) Circuit with R, X_L, and X_C in series. (*b*) Equivalent circuit with one net reactance. (*c*) Phasor diagram. The voltage triangle of phasors is equivalent to an impedance triangle for series circuits.

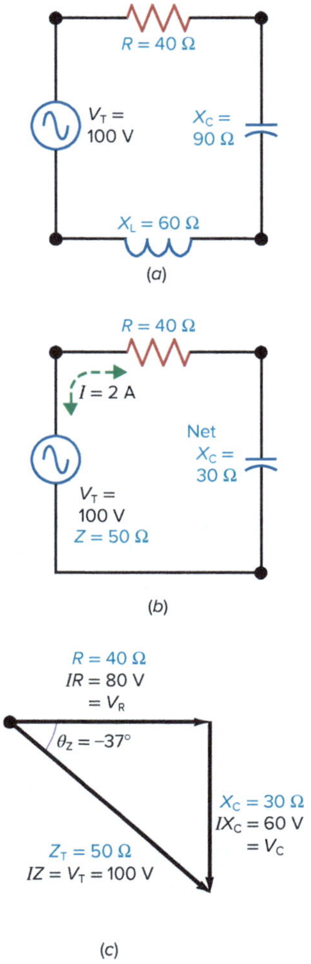

23–5 Series Reactance and Resistance

In the case of series reactance and resistance, the resistive and reactive effects must be combined by phasors. For series circuits, the ohms of opposition are added to find Z_T. First add all the series resistances for one total R. Also, combine all series reactances, adding all X_Ls and all X_Cs and finding the net X by subtraction. The result is one net reactance. It may be either capacitive or inductive, depending on which kind of reactance is larger. Then the total R and net X can be added by phasors to find the total ohms of opposition in the entire series circuit.

Magnitude of Z_T

After the total R and net reactance X are found, they can be combined by the formula

$$Z_T = \sqrt{R^2 + X^2} \tag{23–1}$$

The circuit's total impedance Z_T is the phasor sum of the series resistance and reactance. Whether the net X is at +90° for X_L or −90° for X_C does not matter in calculating the magnitude of Z_T.

An example is illustrated in Fig. 23–8. Here the net series reactance in Fig. 23–8*b* is a 30-Ω X_C. This value is equal to a 60-Ω X_L subtracted from a 90-Ω X_C, as shown in Fig. 23–8*a*. The net 30-Ω X_C in Fig. 23–8*b* is in series with a 40-Ω R. Therefore,

$$Z_T = \sqrt{R^2 + X^2} = \sqrt{(40)^2 + (30)^2} = \sqrt{1600 + 900}$$
$$= \sqrt{2500}$$
$$= 50\ \Omega$$

$I = V/Z_T$

The current is 100 V/50 Ω in this example, or 2 A. This value is the magnitude without considering the phase angle.

Series Voltage Drops

All series components have the same 2-A current. Therefore, the individual drops in Fig. 23–8*a* are

$$V_R = IR = 2 \times 40 = 80\ \text{V}$$
$$V_C = IX_C = 2 \times 90 = 180\ \text{V}$$
$$V_L = IX_L = 2 \times 60 = 120\ \text{V}$$

Since IX_C and IX_L are voltages of opposite polarity, the net reactive voltage is 180 V minus 120 V, which equals 60 V. The phasor sum of IR at 80 V and the net reactive voltage IX of 60 V equals the applied voltage V_T of 100 V.

Angle of Z_T

The impedance angle of the series circuit is the angle whose tangent equals X/R. This angle is negative for X_C but positive for X_L.

In this example, X is the net reactance of 30 Ω for X_C and R is 40 Ω. Then $\tan \theta_Z = -0.75$, and θ_Z is −37°, approximately.

The negative angle for Z indicates a net capacitive reactance for the series circuit. If the values of X_L and X_C were reversed, θ_Z would be +37°, instead of −37° because of the net X_L. However, the magnitude of Z would still be the same.

GOOD TO KNOW

For a series *RLC* circuit, Z_T can be less than either X_L or X_C alone. This is possible because X_L and X_C are 180° out of phase, thus producing a cancellation of some portion of reactance.

Example 23–1

A 27-Ω R is in series with 54 Ω of X_L and 27 Ω of X_C. The applied voltage V_T is 50 mV. Calculate Z_T, I, and θ_Z.

ANSWER The net X_L is 27 Ω. Then

$$Z_T = \sqrt{R^2 + X_L^2} = \sqrt{(27)^2 + (27)^2}$$
$$= \sqrt{729 + 729} = \sqrt{1458}$$
$$= 38.18\ \Omega$$
$$I = \frac{V_T}{Z_T} = \frac{50\text{ mV}}{38.18\ \Omega}$$
$$= 1.31\text{ mA}$$
$$\tan\theta_Z = X/R = \frac{27\ \Omega}{27\ \Omega}$$
$$= 1$$
$$\theta_Z = \arctan(1)$$
$$= 45°$$

In general, when the series resistance and reactance are equal, Z_T is 1.414 times either value. Here, Z_T is 1.414 × 27 = 38.18 Ω. Also, tan θ must be 1 and the angle is 45° for equal sides in a right triangle. To find Z_T on a calculator, see the procedure described on page 645 for the square root of the sum of two squares.

More Series Components

Figure 23–9 shows how to combine any number of series resistances and reactances. Here the total series R of 40 Ω is the sum of 30 Ω for R_1 and 10 Ω for R_2. Note that the order of connection does not matter, since the current is the same in all series components.

The total series X_C is 90 Ω, equal to the sum of 70 Ω for X_{C_1} and 20 Ω for X_{C_2}. Similarly, the total series X_L is 60 Ω. This value is equal to the sum of 30 Ω for X_{L_1} and 30 Ω for X_{L_2}.

The net reactance X equals 30 Ω, which is 90 Ω of X_C minus 60 Ω of X_L. Since X_C is larger than X_L, the net reactance is capacitive. The circuit in Fig. 23–9 is equivalent to Fig. 23–8, therefore, since a 40-Ω R is in series with a net X_C of 30 Ω.

Figure 23–9 Series circuit with more components than Fig. 23–8 but the same Z_T, I, and θ_Z.

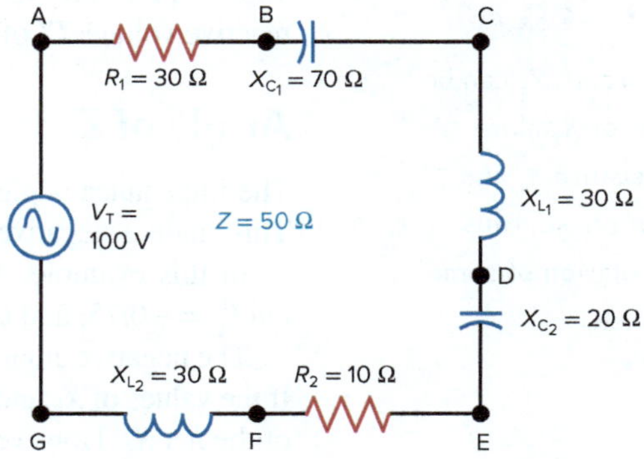

Double-Subscript Notation

This method for specifying AC and DC voltages is useful to indicate the polarity or phase. For instance, in Fig. 23–9 the voltage across R_2 can be taken as either V_{EF} or V_{FE}. With opposite subscripts, these two voltages are 180° out of phase. In using double subscripts, note that the first letter in the subscript is the point of measurement with respect to the second letter.

23–5 Self-Review

Answers at the end of the chapter.

a. **In Fig. 23–8, how much is the net reactance?**
b. **In Fig. 23–9, how much is the net reactance?**
c. **In Fig. 23–9, give the phase difference between V_{CD} and V_{DC}.**

23–6 Parallel Reactance and Resistance

In parallel circuits, the branch currents for resistance and reactance are added by phasors. Then the total line current is found by the formula

$$I_T = \sqrt{I_R^2 + I_X^2} \qquad \textbf{(23–2)}$$

Calculating I_T

As an example, Fig. 23–10*a*, shows a circuit with three branches. Since the voltage across all the parallel branches is the applied 100 V, the individual branch currents are

$$I_R = \frac{V_A}{R} = \frac{100\text{ V}}{25\ \Omega} = 4\text{ A}$$

$$I_L = \frac{V_A}{X_L} = \frac{100\text{ V}}{25\ \Omega} = 4\text{ A}$$

$$I_C = \frac{V_A}{X_C} = \frac{100\text{ V}}{100\ \Omega} = 1\text{ A}$$

Figure 23–10 Total line current I_T of parallel circuit with resistance and reactance. (*a*) Parallel branches with I_R, I_C, and I_L. (*b*) Equivalent circuit with net I_X. (*c*) Phasor diagram.

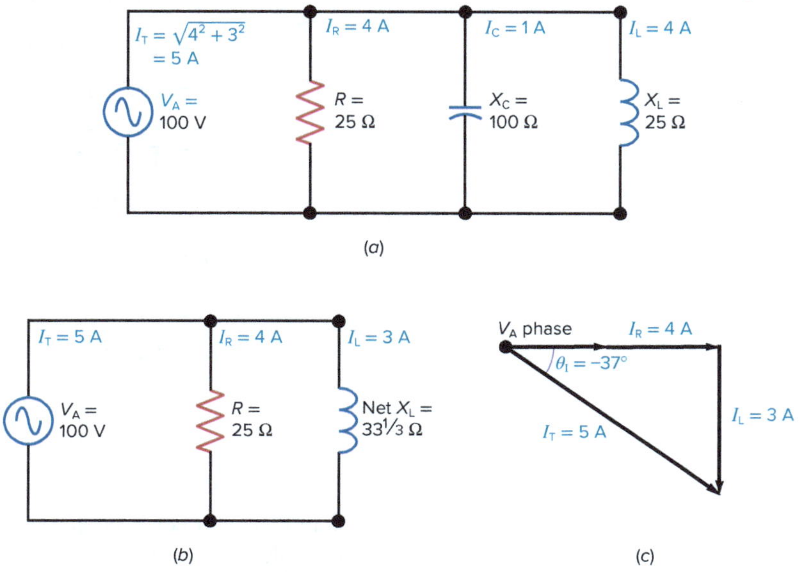

GOOD TO KNOW

For a parallel *RLC* circuit, I_T can be less than either I_L or I_C alone. The reason is that I_L and I_C are 180° out of phase, thus producing a cancellation of some portion of reactive branch current.

The net reactive branch current I_X is 3 A, then, equal to the difference between the 4-A I_L and the 1-A I_C, as shown in Fig. 23–10*b*.

The next step is to calculate I_T as the phasor sum of I_R and I_X. Then

$$I_T = \sqrt{I_R^2 + I_X^2} = \sqrt{4^2 + 3^2} = \sqrt{16 + 9} = \sqrt{25}$$
$$= 5 \text{ A}$$

The phasor diagram for I_T is shown in Fig. 23–10*c*.

$Z_{EQ} = V_A/I_T$

This gives the total impedance of a parallel circuit. In this example, Z_{EQ} is 100 V/5 A, which equals 20 Ω. This value is the equivalent impedance of all three branches in parallel across the source.

Phase Angle

The phase angle of the parallel circuit is found from the branch currents. Now θ is the angle whose tangent equals I_X/I_R.

For this example, I_X is the net inductive current of 3-A I_L. Also, I_R is 4 A. These phasors are shown in Fig. 23–10*c*. Then θ is a negative angle with a tangent of −0.75. This phase angle is approximately −37°.

The negative angle for I_T indicates lagging inductive current. The value of −37° is the phase angle of I_T with respect to the voltage reference V_A.

When Z_{EQ} is calculated as V_A/I_T for a parallel circuit, the phase angle is the same value as for I_T but with opposite sign. In this example, Z_{EQ} is 20 Ω with a phase angle of +37°, for an I_T of 5 A with an angle of −37°. We can consider that Z_{EQ} has the phase angle of the voltage source with respect to I_T.

Example 23–2

The following branch currents are supplied from a 50-mV source: $I_R = 1.8$ mA; $I_L = 2.8$ mA; $I_C = 1$ mA. Calculate I_T, Z_{EQ}, and θ_I.

ANSWER The net I_X is 1.8 mA. Then

$$I_T = \sqrt{I_R^2 + I_X^2} = \sqrt{(1.8)^2 + (1.8)^2}$$
$$= \sqrt{3.24 + 3.24} = \sqrt{6.48}$$
$$= 2.55 \text{ mA}$$
$$Z_{EQ} = \frac{V_A}{I_T} = \frac{50 \text{ mV}}{2.55 \text{ mA}}$$
$$= 19.61\ \Omega$$
$$\tan \theta_I = -\frac{I_L}{I_R} = -\frac{1.8 \text{ mA}}{1.8 \text{ mA}}$$
$$= -1$$
$$\theta_I = \arctan (1)$$
$$= -45°$$

Note that with equal branch currents, I_T is 1.414 × 1.8 = 2.55 mA. Also, the phase angle θ_I is negative for inductive branch current.

Figure 23–11 Parallel AC circuit with more components than Fig. 23–10 but the same values of I_T, Z_{EQ}, and θ.

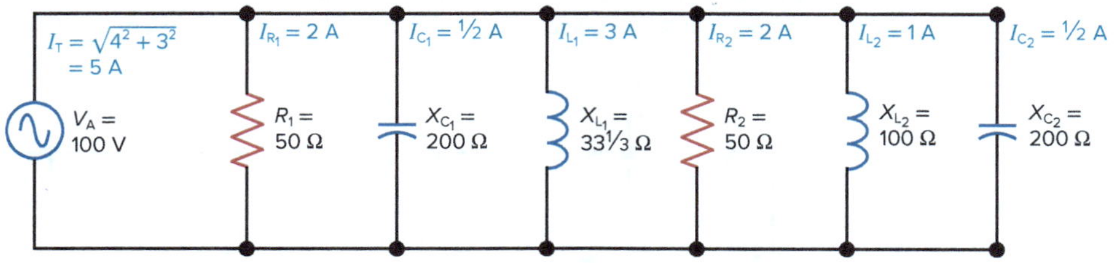

More Parallel Branches

Figure 23–11 shows how any number of parallel resistances and reactances can be combined. The total resistive branch current I_R of 4 A is the sum of 2 A each for the R_1 branch and the R_2 branch. Note that the order of connection does not matter, since the parallel branch currents add in the main line. Effectively, two 50-Ω resistances in parallel are equivalent to one 25-Ω resistance.

Similarly, the total inductive branch current I_L is 4 A, equal to 3 A for I_{L_1} and 1 A for I_{L_2}. Also, the total capacitive branch current I_C is 1 A, equal to ½ A each for I_{C_1} and I_{C_2}.

The net reactive branch current I_X is 3 A, then, equal to a 4-A I_L minus a 1-A I_C. Since I_L is larger, the net current is inductive.

Therefore, the circuit in Fig. 23–11 is equivalent to the circuit in Fig. 23–10. Both have a 4-A resistive current I_R and a 3-A net reactive current I_X. These values added by phasors make a total of 5 A for I_T in the main line.

23–6 Self-Review

Answers at the end of the chapter.

a. In Fig. 23–10, what is the net reactive branch current?
b. In Fig. 23–11, what is the net reactive branch current?

23–7 Series-Parallel Reactance and Resistance

Figure 23–12 shows how a series-parallel circuit can be reduced to a series circuit with just one reactance and one resistance. The method is straightforward as long as resistance and reactance are not combined in one parallel bank or series string.

Working backward toward the generator from the outside branch in Fig. 23–12*a*, we have an X_{L_1} and an X_{L_2} of 100 Ω each in series, which total 200 Ω. This string in Fig. 23–12*a* is equivalent to X_{L_5} in Fig. 23–12*b*.

In the other branch, the net reactance of X_{L_3} and X_C is equal to 600 Ω minus 400 Ω. This is equivalent to the 200 Ω of X_{L_4} in Fig. 23–12*b*. The X_{L_4} and X_{L_5} of 200 Ω each in parallel are combined for an X_L of 100 Ω.

In Fig. 23–12*c*, the 100-Ω X_L is in series with the 100-Ω R_{1-2}. This value is for R_1 and R_2 in parallel.

The triangle diagram for the equivalent circuit in Fig. 23–12*d* shows the total impedance Z of 141 Ω for a 100-Ω R in series with a 100-Ω X_L.

With a 141-Ω impedance across the applied V_T of 100 V, the current in the generator is 0.7 A. The phase angle θ is 45° for this circuit.*

* More complicated AC circuits with series-parallel impedances are analyzed with complex numbers, as explained in Chapter 24.

Figure 23–12 Reducing an AC series-parallel circuit with R, X_L, and X_C to a series circuit with one net resistance and one net reactance. (*a*) Actual circuit. (*b*) Simplified arrangement. (*c*) Equivalent series circuit. (*d*) Impedance triangle with phase angle.

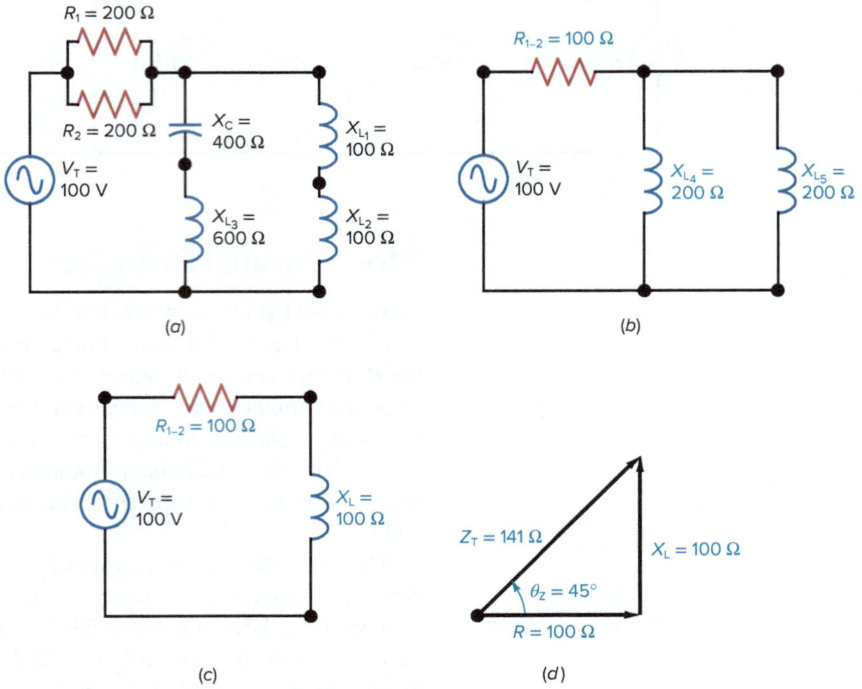

23–7 Self-Review

Answers at the end of the chapter.

Refer to Fig. 23–12.

a. **How much is $X_{L_1} + X_{L_2}$?**
b. **How much is $X_{L_3} - X_C$?**
c. **How much is X_{L_4} in parallel with X_{L_5}?**

23–8 Real Power

In an AC circuit with reactance, the current I supplied by the generator either leads or lags the generator voltage V. Then the product VI is not the **real power** produced by the generator, since the instantaneous voltage may have a high value while at the same time the current is near zero, or vice versa. The real power, in watts, however, can always be calculated as I^2R, where R is the total resistive component of the circuit, because current and voltage are in phase in a resistance. To find the corresponding value of power as VI, this product must be multiplied by the cosine of the phase angle θ. Then

$$\text{Real power} = P = I^2R \tag{23–3}$$

or

$$\text{Real power} = P = VI\cos\theta \tag{23–4}$$

where V and I are in rms values, and P, the real power, is in watts. Multiplying VI by the cosine of the phase angle provides the resistive component for real power equal to I^2R.

Figure 23–13 Real power, *P*, in a series circuit. (*a*) Schematic diagram. (*b*) Impedance triangle with phase angle.

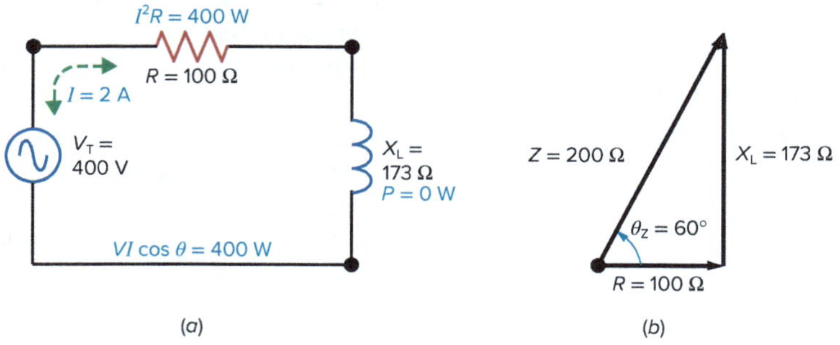

For example, the AC circuit in Fig. 23–13 has 2 A through a 100-Ω *R* in series with the X_L of 173 Ω. Therefore,

$$P = I^2R = 4 \times 100 = 400 \text{ W}$$

Furthermore, in this circuit, the phase angle is 60° with a cosine of 0.5. The applied voltage is 400 V. Therefore,

$$P = VI \cos\theta = 400 \times 2 \times 0.5 = 400 \text{ W}$$

In both examples, the real power is the same 400 W because this is the amount of power supplied by the generator and dissipated in the resistance. Either formula can be used for calculating the real power, depending on which is more convenient.

Real power, sometimes referred to as true power, can be considered resistive power that is dissipated as heat. A reactance does not dissipate power but stores energy in an electric or magnetic field.

Power Factor

Because it indicates the resistive component, cos *θ* is the **power factor** of the circuit, converting the *VI* product to real power. The power factor formulas are

For series circuits:

$$\text{Power factor} = PF = \cos\theta = \frac{R}{Z} \qquad \textbf{(23–5)}$$

For parallel circuits:

$$\text{Power factor} = \cos\theta = \frac{I_R}{I_T} \qquad \textbf{(23–6)}$$

GOOD TO KNOW

The formulas for power factor really tell us what fraction of Z_T or I_T is resistive.

In Fig. 23–13, as an example of a series circuit, we use *R* and *Z* for the calculations:

$$PF = \cos\theta = \frac{R}{Z} = \frac{100\ \Omega}{200\ \Omega} = 0.5$$

For the parallel circuit in Fig. 23–10, we use the resistive current I_R and the I_T:

$$PF = \cos\theta = \frac{I_R}{I_T} = \frac{4\text{ A}}{5\text{ A}} = 0.8$$

The power factor is not an angular measure but a numerical ratio with a value between 0 and 1, equal to the cosine of the phase angle.

With all resistance and zero reactance, *R* and *Z* are the same for a series circuit, or I_R and I_T are the same for a parallel circuit, and the ratio is 1. Therefore, unity power factor means a resistive circuit. At the opposite extreme, all reactance with zero resistance makes the power factor zero, which means that the circuit is all reactive. The power factor is frequently given in percent so that unity power factor is 100%. To convert from decimal *PF* to percent *PF*, just multiply by 100.

GOOD TO KNOW

The apparent power is the power that is "apparently" used by the circuit before the phase angle between V and I is considered.

Apparent Power

When V and I are out of phase because of reactance, the product of $V \times I$ is called ***apparent power***. The unit is ***volt-amperes (VA)*** instead of watts, since the watt is reserved for real power.

For the example in Fig. 23–13, with 400 V and the 2-A I, 60° out of phase, the apparent power is VI, or $400 \times 2 = 800$ VA. Note that apparent power is the VI product alone, without considering the power factor cos θ.

The power factor can be calculated as the ratio of real power to apparent power because this ratio equals cos θ. As an example, in Fig. 23–13, the real power is 400 W, and the apparent power is 800 VA. The ratio of $^{400}/_{800}$, then, is 0.5 for the power factor, the same as cos 60°.

GOOD TO KNOW

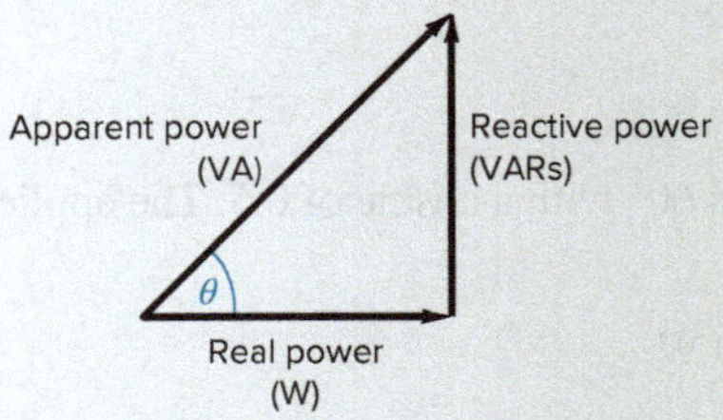

The power triangle shows the phase relationship between apparent power, real power, and reactive power.

The VAR

This is an abbreviation for **volt-ampere reactive**. Specifically, VARs are volt-amperes at the angle of 90°. VAR is also known as reactive power.

In general, for any phase angle θ between V and I, multiplying VI by sin θ gives the vertical component at 90° for the value of the VARs. In Fig. 23–13, the value of VI sin 60° is $800 \times 0.866 = 692.8$ VAR.

Note that the factor sin θ for the VARs gives the vertical or reactive component of the apparent power VI. However, multiplying VI by cos θ as the power factor gives the horizontal or resistive component of the real power.

Correcting the Power Factor

In commercial use, the power factor should be close to unity for efficient distribution of electric power. However, the inductive load of motors may result in a power factor of 0.7, as an example, for the phase angle of 45°. To correct for this lagging inductive component of the current in the main line, a capacitor can be connected across the line to draw leading current from the source. To bring the power factor up to 1.0, that is, unity PF, the value of capacitance is calculated to take the same amount of volt-amperes as the VARs of the load.

23–8 Self-Review

Answers at the end of the chapter.

a. What is the unit of real power?
b. What is the unit of apparent power?
c. Is I^2R real or apparent power?

23–9 AC Meters

The D'Arsonval moving-coil type of meter movement will not read if it is used in an AC circuit because the AC wave changes polarity too rapidly. Since the two opposite polarities cancel, an alternating current cannot deflect the meter movement either up-scale or down-scale. An AC meter must produce deflection of the meter pointer up-scale regardless of polarity. This deflection is accomplished by one of the following three methods for nonelectronic AC meters.

1. *Thermal type.* In this method, the heating effect of the current, which is independent of polarity, is used to provide meter deflection. Two examples are the thermocouple type and hot-wire meter.
2. *Electromagnetic type.* In this method, the relative magnetic polarity is maintained constant although the current reverses. Examples are the iron-vane meter, dynamometer, and **wattmeter**.

3. *Rectifier type.* The rectifier changes the AC input to DC output for the meter, which is usually a D'Arsonval movement. This type is the most common for AC voltmeters generally used for audio and radio frequencies.

All analog AC meters (meters with scales and pointers) have scales calibrated in rms values, unless noted otherwise on the meter.

A thermocouple consists of two dissimilar metals joined together at one end but open at the opposite side. Heat at the short-circuited junction produces a small DC voltage across the open ends, which are connected to a DC meter movement. In the hot-wire meter, current heats a wire to make it expand, and this motion is converted into meter deflection. Both types are used as AC meters for radio frequencies.

The iron-vane meter and dynamometer have very low sensitivity compared with a D'Arsonval movement. They are used in power circuits for either direct current or 60-Hz alternating current.

23–9 Self-Review

Answers at the end of the chapter.

a. **The iron-vane meter can read alternating current. (True/False)**
b. **The D'Arsonval meter movement works with direct current only. (True/False)**

Figure 23–14 Schematic of voltage and current coils of an analog wattmeter.

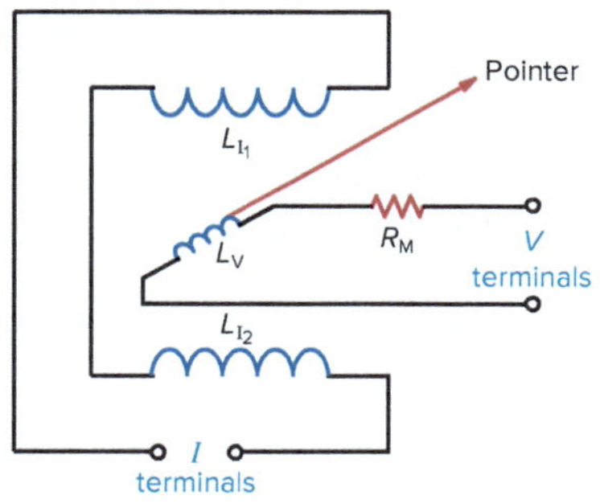

23–10 Wattmeters

The wattmeter uses fixed coils to measure current in a circuit, and the movable coil measures voltage (Fig. 23–14). The deflection, then, is proportional to power. Either DC power or real AC power can be read directly by the wattmeter.

In Fig. 23–14, the coils L_{I_1} and L_{I_2} in series are heavy stationary coils serving as an ammeter to measure current. The two *I* terminals are connected in one side of the line in series with the load. The movable coil L_V and its multiplier resistance R_M are used as a voltmeter with the *V* terminals connected across the line in parallel with the load. Then, the current in the fixed coils is proportional to *I*, and the current in the movable coil is proportional to *V*. As a result, the deflection is proportional to *V* and *I*.

Furthermore, it is the *VI* product for each instant that produces deflection. For instance, if the *V* value is high when the *I* value is low for a phase angle close to 90°, there will be little deflection. The meter deflection is proportional to the watts of real power, therefore, regardless of the power factor in AC circuits. The wattmeter is commonly used to measure power from the 60-Hz power line. For radio frequencies, however, power is generally measured in terms of heat transfer.

GOOD TO KNOW

Four circuit connections are needed when using the wattmeter in Fig. 23–14.

23–10 Self-Review

Answers at the end of the chapter.

a. **Does a wattmeter measure real or apparent power?**
b. **In Fig. 23–14, does the movable coil of a wattmeter measure *V* or *I*?**

23–11 Summary of Types of Ohms in AC Circuits

The differences in *R*, X_L, X_C, and Z_T are listed in Table 23–1, but the following general features should also be noted. Ohms of opposition limit the amount of current in DC circuits or AC circuits. Resistance *R* is the same for either case.

Table 23–1	Types of Ohms in AC Circuits			
	Resistance R, Ω	Inductive Reactance X_L, Ω	Capacitive Reactance X_C, Ω	Impedance Z_T, Ω
Definition	In-phase opposition to alternating or direct current	90° leading opposition to alternating current	90° lagging opposition to alternating current	Combination of resistance and reactance $Z_T = \sqrt{R^2 + X^2}$
Effect of frequency	Same for all frequencies	Increases with higher frequencies	Decreases at higher frequencies	X_L component increases, but X_C decreases at higher frequencies
Phase angle	0°	I_L lags V_L by 90°	I_C leads V_C by 90°	$\tan\theta_Z = \pm X/R$ in series, $\tan\theta_I = \pm I_X/I_R$ in parallel

However, AC circuits can have ohms of reactance because of the variations in alternating current or voltage. Reactance X_L is the reactance of an inductance with sine-wave changes in current. Reactance X_C is the reactance of a capacitor with sine-wave changes in voltage.

Both X_L and X_C are measured in ohms, like R, but reactance has a 90° phase angle, whereas the phase angle for resistance is 0°. A circuit with steady direct current cannot have any reactance.

Ohms of X_L or X_C are opposite because X_L has a phase angle of +90° and X_C has an angle of −90°. Any individual X_L or X_C always has a phase angle that is exactly 90°.

Ohms of impedance Z result from the phasor combination of resistance and reactance. In fact, Z can be considered the general form of any ohms of opposition in AC circuits.

Impedance can have any phase angle, depending on the relative amounts of R and X. When Z consists mostly of R with little reactance, the phase angle of Z is close to 0°. With R and X equal, the phase angle of Z is 45°. Whether the angle is positive or negative depends on whether the net reactance is inductive or capacitive. When Z consists mainly of X with little R, the phase angle of Z is close to 90°.

The phase angle is θ_Z for Z or V_T with respect to the common I in a series circuit. With parallel branch currents, θ_I is for I_T in the main line with respect to the common voltage.

23–11 *Self-Review*

Answers at the end of the chapter.

a. Which of the following does not change with frequency: Z, X_L, X_C, or R?
b. Which has lagging current: R, X_L, or X_C?
c. Which has leading current: R, X_L, or X_C?

23–12 Summary of Types of Phasors in AC Circuits

Phasors for ohms, volts, and amperes are shown in Fig. 23–15. Note the similarities and differences.

Figure 23–15 Summary of phasor relations in AC circuits. (*a*) Series *R* and X_L. (*b*) Series *R* and X_C. (*c*) Parallel branches with I_R and I_C. (*d*) Parallel branches with I_R and I_L.

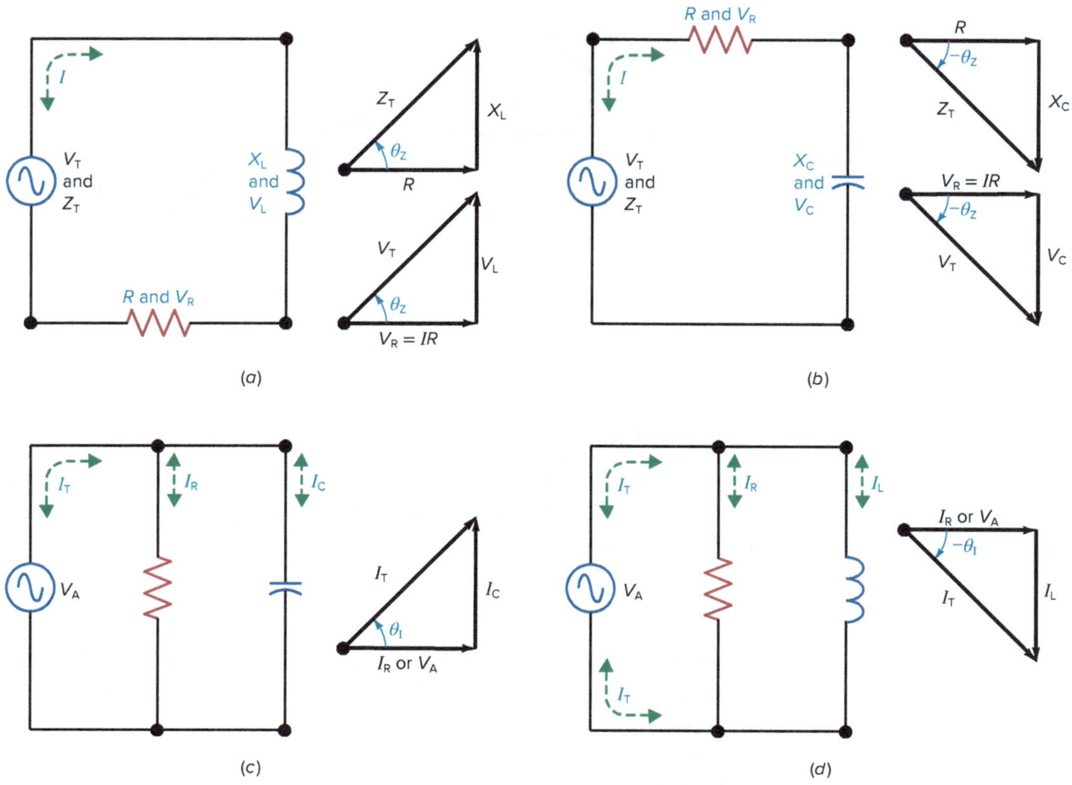

Series Components

In series circuits, ohms and voltage drops have similar phasors. The reason is the common *I* for all series components. Therefore,

V_R or *IR* has the same phase as *R*.

V_L or IX_L has the same phase as X_L.

V_C or IX_C has the same phase as X_C.

Resistance

The *R*, V_R, and I_R always have the same phase angle because there is no phase shift in a resistance. This applies to *R* in either a series or a parallel circuit.

Reactance

Reactances X_L and X_C are 90° phasors in opposite directions. The X_L or V_L has an angle of +90° with an upward phasor, and the X_C or V_C has an angle of −90° with a downward phasor.

Reactive Branch Currents

The phasor of a parallel branch current is opposite from its reactance. Therefore, I_C is upward at +90°, opposite from X_C downward at −90°. Also, I_L is downward at −90°, opposite from X_L upward at +90°.

In short, I_C and I_L are opposite each other, and both are opposite from their corresponding reactances.

Figure 23–16 Positive and negative values of θ as the phase angle for an AC circuit. (a) Positive θ with I leading V. (b) Negative θ with I lagging V.

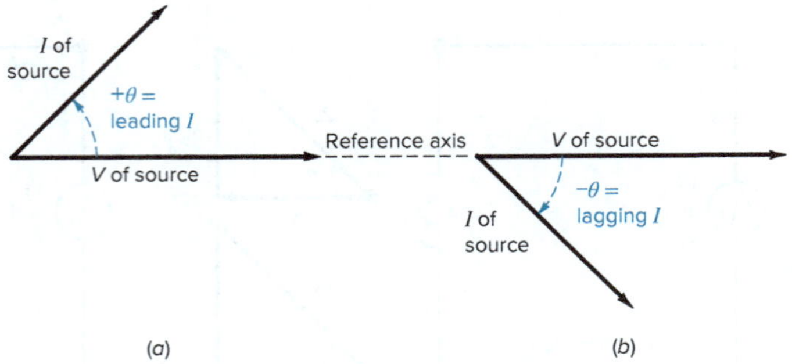

Angle θ_Z

The phasor resultant for ohms of reactance and resistance is Z. The phase angle θ for Z can be any angle between 0° and 90°. In a series circuit, θ_Z for Z is the same as θ for V_T with respect to the common current I.

Angle θ_I

The phasor resultant of branch currents is the total line current I_T. The phase angle of I_T can be any angle between 0° and 90°. In a parallel circuit, θ_I is the angle of I_T with respect to the applied voltage V_A.

Such phasor combinations are necessary in sine-wave AC circuits to take into account the effect of reactance. Phasors can be analyzed either graphically, as in Fig. 23–15, or by the shorter technique of complex numbers, with a j operator that corresponds to the 90° phasor. Complex numbers are explained in the next chapter.

Circuit Phase Angle θ

The phase angle for all types of sine-wave AC circuits is usually considered the angle between the current I from the source and its applied voltage as the reference. This angle can be labeled θ, without any subscript. No special identification is necessary because θ is the phase angle of the circuit. Then there are only the two possibilities, as shown in Fig. 23–16. In Fig. 23–16*a*, the θ is a counterclockwise angle for a positive value, which means that I leads V. The leading I is in a circuit with series X_C or with I_C in a parallel branch. In Fig. 23–16*b*, the phase angle is clockwise for $-\theta$, which means that I lags V. The lagging I is produced in a circuit with series X_L or with I_L in a parallel branch.

Note that, in general, θ is the same as θ_I in parallel branch currents. However, θ has a sign opposite from θ_Z with series reactances.

23–12 Self-Review

Answers at the end of the chapter.

a. Of the following phasors, which two are 180° apart: V_L, V_C, or V_R?
b. Of the following phasors, which two are out of phase by 90°: I_R, I_T, or I_L?

Application in Understanding Power Factor Correction

In any AC circuit with capacitive and/or inductive reactance, the current either leads or lags the applied voltage by some phase angle between 0 and 90°. In a purely resistive AC circuit, the phase angle, θ, is 0°. The power calculations in a purely resistive AC circuit are no different than those in DC circuits, except that in AC circuit's rms values of voltage and current are used. When the phase angle is between 0 and 90°, however, the power consumed by the circuit is more difficult to calculate.

CALCULATING POWER IN AC CIRCUITS

The three different types of power in AC circuits are real power, reactive power, and apparent power. Real power represents electrical energy that has been converted into another form, such as heat energy, light energy, or rotating mechanical energy. Reactive power represents the energy stored in the magnetic field of an inductor or the electric field of a capacitor. The apparent power represents the power that is "apparently" being consumed by the circuit before the phase angle is taken into account. The following symbols, units, and formulas show each type of power:

Real Power — The symbol is P and the unit is the watt (W). P is calculated as $P = I_R^2 R$, $P = \frac{V_R^2}{R}$, $P = V_R \times I_R$ or $P = V_A \times I_T \times \cos(\theta)$.

Reactive Power — The symbol is Q and the unit is the volt–ampere reactive (VAR). It is important to note that the reactive power in the inductor is designated Q_L, whereas the reactive power in the capacitor is designated Q_C. Q is calculated as $Q_L = I_L^2 X_L$ or $Q_C = I_C^2 X_C$, $Q_L = \frac{V_L^2}{X_L}$ or $Q_C = \frac{V_C^2}{X_C}$, $Q_L = V_L \times I_L$ or $Q_C = V_C \times I_C$ or $Q = V_A \times I_T \times \sin(\theta)$.

Apparent Power — The symbol is S and the unit is the volt–ampere (VA). S is calculated as $S = V_A \times I_T$, $S = I_T^2 Z$, $S = \frac{V_A^2}{Z}$ or $S = \sqrt{P^2 + Q^2}$.

The power triangle is shown in Fig. 23–17. The real or resistive power is at an angle of 0°, whereas the reactive power is at an angle of 90°. The apparent power, which is the hypotenuse of the right triangle, is the phasor sum of the real and reactive power. It should be noted that the reactive power, Q_L, in an inductor is at an angle of +90°, whereas the reactive power, Q_C, in a capacitor is at an angle of −90°.

Figure 23–17 Power triangle.

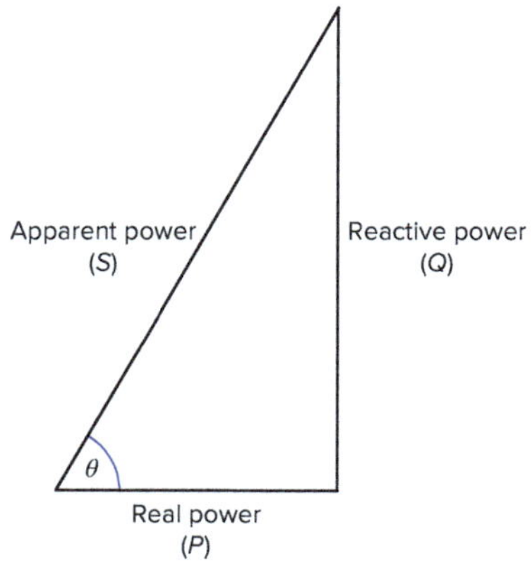

POWER FACTOR

In an AC circuit, the ratio of real to apparent power is called the power factor (usually abbreviated PF). The power factor is a numerical ratio with a value between 0 and 1. Expressed as a formula

$$\text{PF} = \frac{\textit{Real Power}}{\textit{Apparent Power}} \quad \text{or}$$

$$\text{PF} = \cos\theta$$

Basically, the power factor tells us what fraction of apparent power is actually real power.

CIRCUIT EXAMPLE

In Fig. 23–18

$$I_R = \frac{240\text{ V}}{3\Omega} = 80\text{ A}$$

$$I_L = \frac{240\ V}{4\Omega} = 60\text{ A}$$

$$I_T = \sqrt{80^2 + 60^2} = 100\text{ A}$$

$$\theta = \arctan -\frac{60}{80} = -36.87°$$

$$S = 240\text{ V} \times 100\text{ A} = 24\text{ kVA}$$

$$Q_L = 240\text{ V} \times 100\text{ A} \times \sin(-36.87°) = 14.4\text{ kVARs}$$

$$P = 240\text{ V} \times 100\text{ A} \times \cos(-36.87°) = 19.2\text{ kW}$$

$$\text{PF} = \frac{19.2\ kW}{24\ kVA} = 0.8$$

When utility companies deliver sine-wave AC power to their customers, it is desired to have the power factor as close to 1 as possible. This is because as the power factor decreases, the power generating station has to supply more power, which in turn means that the power lines will have a higher current. In other words, the lower the power factor, the higher the losses in the overall system. Because there are so many inductive loads, such as motors and transformers, the power factor will always be less than 1. How much less than 1 is determined by the number of inductive loads, which can vary from one location to the next or even with the time of day. Just remember, as the

Figure 23–18 A parallel RL circuit represents the inductive nature of a typical load encountered in electrical power distribution.

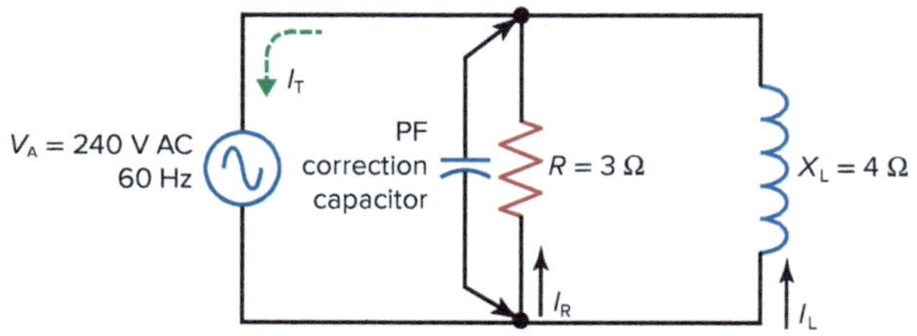

power factor decreases, the apparent power, total current, and overall power losses in the system increase. To say it another way, the ratio of delivered power to generated power decreases as the power factor decreases.

CORRECTING THE POWER FACTOR

To correct the power factor, a capacitor is added in parallel with the AC power line to offset the effects of the lagging current of the inductive loads. Factories, schools, and large facilities in general all use power factor correction capacitors to keep the power factor as close to 1 as possible. The power factor correction capacitor is always located at the point where the electric power lines enter the building. To increase the power factor to 1 in Fig. 23–18, a capacitor would need to draw the same amount of current as the inductive branch. Since X_L equals 4 Ω, X_C also needs to be 4 Ω. The value of capacitance that has an X_C of 4 Ω at 60 Hz is calculated as follows:

$$X_C = \frac{1}{2 \times \pi \times 60 \text{ Hz} \times 4 \,\Omega} = 663.14\ \mu\text{F}$$

With this amount of capacitance across the power line connections in Fig. 23–18, the reactive power (stored energy) in the capacitor will be identical but opposite to that of the inductor. The net reactive power is the difference between Q_L and Q_C, which is zero in this case.

Because the inductive and capacitive branch currents are equal and 180° out of phase, they completely cancel each other in the main line. Therefore, the only current flowing in the main line is the resistive branch current, I_R, of 80 A. This is a 20% reduction in the total current, I_T.

THE BOTTOM LINE

All utility companies have a way to switch in power factor correction capacitors to help improve the overall efficiency of power delivery. These capacitors may be located in substations or may actually be on power poles near transformers. By monitoring the phase angle of the power, the correct amount of capacitance can be switched in as needed. The inductive loading will vary over time, so the process of monitoring and correcting the power factor is a continuous one. The resulting energy savings of power factor correction is enormous!

It should be noted that it is not entirely practical to obtain a power factor of one or unity in the distribution of electrical power. This is because inductive loads are constantly being switched on and off throughout the day. This in turn results in a constantly changing power factor. In most cases, utility companies try to maintain a power factor of around 0.95. Utility companies usually charge a penalty fee to customers with power factors less than 0.95. Therefore, as a customer, there is great financial incentive in improving the power factor.

Summary

- In AC circuits with resistance alone, the circuit is analyzed the same way as DC circuits, generally with rms AC values. Without any reactance, the phase angle between V and I is zero.
- When capacitive reactances alone are combined, the X_C values are added in series and combined by the reciprocal formula in parallel, just like ohms of resistance. Similarly, ohms of X_L alone can be added in series or combined by the reciprocal formula in parallel, just like ohms of resistance.
- Since X_C and X_L are opposite reactances, they offset each other. In series, ohms of X_C and X_L can be subtracted. In parallel, the capacitive and inductive branch currents I_C and I_L can be subtracted.
- In AC circuits, R, X_L, and X_C can be reduced to one equivalent resistance and one net reactance.
- In series, the total R and net X at 90° are combined as $Z_T = \sqrt{R^2 + X^2}$. The phase angle of the series R and X is the angle with tangent $\pm X/R$. To find I, first we calculate Z_T and then divide into V_T.
- For parallel branches, the total I_R and net reactive I_X at 90° are combined as $I_T = \sqrt{I_R^2 + I_X^2}$. The phase angle of the parallel R and X is the angle with tangent $\pm I_X/I_R$. To find Z_{EQ} first we calculate I_T and then divide into V_A.
- The quantities R, X_L, X_C, and Z in AC circuits all are ohms of opposition. The differences with respect to frequency and phase angle are summarized in Table 23–1.
- The phase relations for resistance and reactance are summarized in Fig. 23–15.
- In AC circuits with reactance, the real power P in watts equals I^2R, or $VI \cos \theta$, where θ is the phase angle. The real power is the power dissipated as heat in resistance. Cos θ is the power factor of the circuit.
- The wattmeter measures real AC power or DC power.

Important Terms

Apparent power — the power that is apparently consumed by an AC circuit. Apparent power is calculated as $V \times I$ without considering the phase angle. The unit of apparent power is the volt-ampere (VA) since the watt unit is reserved for real power.

Double-subscript notation — a notational system used to specify the DC and AC voltages in a circuit. The first letter in the subscript indicates the point of measurement, whereas the second letter indicates the point of reference. With double-subscript notation, the polarity or phase of a voltage can be indicated.

Power factor (PF) — a numerical ratio between 0 and 1 that specifies the ratio of real to apparent power in an AC circuit. For any AC circuit, the power factor is equal to the cosine of the phase angle.

Real power — the actual power dissipated as heat in the resistance of an AC circuit. The unit of real power is the watt (W). Real power can be calculated as I^2R where R is the resistive component of the circuit or as $V \times I \times \cos \theta$ where θ is the phase angle of the circuit.

Volt-ampere (VA) — the unit of apparent power.

Volt-ampere reactive (VAR) — the volt-amperes at the angle of 90°.

Wattmeter — a test instrument used to measure the real power in watts.

Related Formulas

$Z_T = \sqrt{R^2 + X^2}$

$I_T = \sqrt{I_R^2 + I_X^2}$

Real power $= P = I^2R$

Real power $= P = VI \cos \theta$

Power factor $= PF = \cos \theta = \dfrac{R}{Z}$ (Series circuits)

Power factor $= PF = \cos \theta = \dfrac{I_R}{I_T}$ (Parallel circuits)

Self-Test

Answers at the back of the book.

1. In an AC circuit with only series resistances,

a. V_T and I are in phase.

b. $R_T = R_1 + R_2 + R_3 \cdots +$ etc.

c. each voltage drop is in phase with the series current.

d. all of the above.

2. In an AC circuit with only parallel inductors,

a. I_T lags V_A by 90°.

b. V_A lags I_T by 90°.

c. V_A and I_T are in phase.

d. none of the above.

3. A series circuit contains 150 Ω of X_L and 250 Ω of X_C. What is the net reactance?

a. 400 Ω, X_L.

b. 400 Ω, X_C.

c. 100 Ω, X_C.

d. 291.5 Ω, X_C.

4. What is the power factor (PF) of a purely resistive AC circuit?

a. 0.

b. 1.

c. 0.707.

d. Without values, it cannot be determined.

5. The unit of apparent power is the

a. volt-ampere (VA).

b. watt (W).

c. volt-ampere reactive (VAR).

d. joule (J).

6. A 15-Ω resistance is in series with 50 Ω of X_L and 30 Ω of X_C. If the applied voltage equals 50 V, how much real power is dissipated by the circuit?

a. 60 W.

b. 100 W.

c. 100 VA.

d. 4.16 W.

7. An AC circuit has a 100-Ω *R*, a 300-Ω X_L, and a 200-Ω X_C all in series. What is the phase angle of the circuit?

a. 78.7°.

b. 45°.

c. −90°.

d. 56.3°.

8. A 10-Ω resistor is in parallel with an X_L of 10 Ω. If the applied voltage is 120 V, what is the power factor of the circuit?

a. 0.

b. 0.5.

c. 1.

d. 0.707.

9. An AC circuit has an 80-Ω *R*, 20-Ω X_L, and a 40-Ω X_C in parallel. If the applied voltage is 24 Vac, what is the phase angle of the circuit?

a. −26.6°.

b. 45°.

c. −63.4°.

d. −51.3°.

10. In an AC circuit with only series capacitors,

a. V_T leads *I* by 90°.

b. V_T lags *I* by 90°.

c. each capacitor voltage drop leads *I* by 90°.

d. both a and c.

11. A 10-Ω is in parallel with a 15-Ω X_L. The applied voltage is 120 V_{AC}. How much is the apparent power in the circuit?

a. 2.4 kW.

b. 1.44 kVA.

c. 1.44 kW.

d. 1.73 kVA.

12. The unit of real power is the

a. watt (W).

b. volt-ampere (VA).

c. joule (J).

d. volt-ampere reactive (VAR).

13. In a parallel AC circuit with X_L and X_C,

a. I_L and I_C are 90° out of phase.

b. I_L and I_C are in phase.

c. I_L and I_C are 180° out of phase.

d. X_L and X_C are 90° out of phase.

14. In a series *RLC* circuit,

a. X_L and X_C are 180° out of phase.

b. I_L and I_C are 180° out of phase.

c. X_L and X_C are 90° out of phase.

d. X_L and X_C are in phase.

15. A parallel AC circuit with 120 V_{AC} applied has a total current, I_T, of 5 A. If the phase angle of the circuit is −53.13°, how much real power is dissipated by the circuit?

a. 600 VA.

b. 480 W.

c. 360 W.

d. 3.6 kVA.

Essay Questions

1. Why can series or parallel resistances be combined in AC circuits the same way as in DC circuits?

2. (a) Why do X_L and X_C reactances in series offset each other? (b) With X_L and X_C reactances in parallel, why can their branch currents be subtracted?

3. Give one difference in electrical characteristics comparing *R* and X_C, *R* and *Z*, X_C and *C*, X_L and *L*.

4. Name three types of AC meters.

5. Make a diagram showing a resistance R_1 in series with the load resistance R_L, with a wattmeter connected to measure the power in R_L.

6. Make a phasor diagram for the circuit in Fig. 23–8*a* showing the phase of the voltage drops *IR*, IX_C, and IX_L with respect to the reference phase of the common current *I*.

7. Explain briefly why the two opposite phasors at +90° for X_L and −90° for I_L both follow the principle that any self-induced voltage leads the current through the coil by 90°.

8. Why is it that a reactance phasor is always at exactly 90° but an impedance phasor can be less than 90°?

9. Why must the impedance of a series circuit be more than either its *X* or its *R*?

10. Why must I_T in a parallel circuit be more than either I_R or I_X?

11. Compare real power and apparent power.

12. Define *power factor*.

13. Make a phasor diagram showing the opposite direction of positive and negative angles.

14. In Fig. 23–15, which circuit has leading current with a positive phase angle *θ* where *I* from the source leads the *V* applied by the source?

Problems

SECTION 23–1 AC CIRCUITS WITH RESISTANCE BUT NO REACTANCE

23–1 In Fig. 23–19, solve for R_T, I, V_1, and V_2.

Figure 23–19

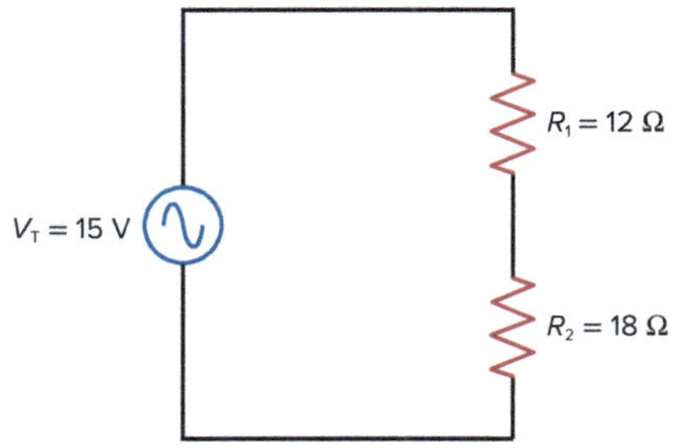

23–2 In Fig. 23–19, what is the phase relationship between

a. V_T and I?

b. V_1 and I?

c. V_2 and I?

23–3 In Fig. 23–20, solve for I_1, I_2, I_T, and R_{EQ}.

Figure 23–20

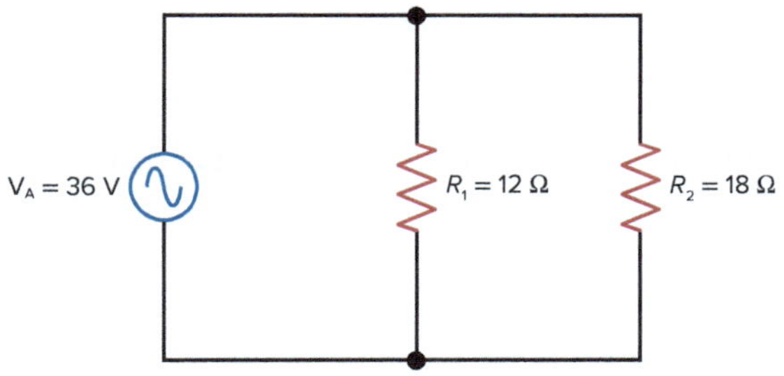

23–4 In Fig. 23–20, what is the phase relationship between

a. V_A and I_1?

b. V_A and I_2?

c. V_A and I_T?

SECTION 23–2 CIRCUITS WITH X_L ALONE

23–5 In Fig. 23–21, solve for X_{L_T}, I, V_1, and V_2.

Figure 23–21

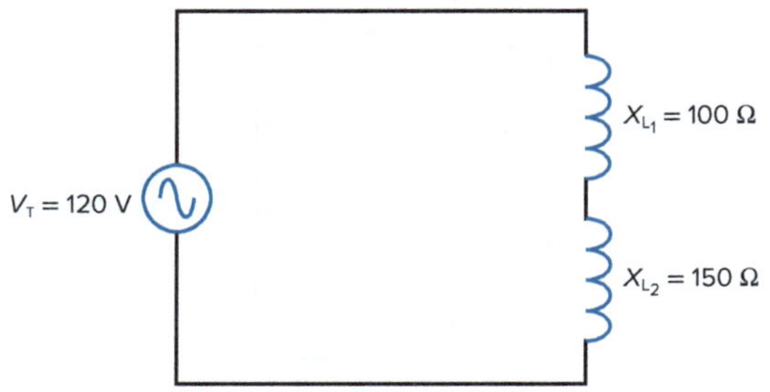

23–6 In Fig. 23–21, what is the phase relationship between

a. V_T and I?

b. V_1 and I?

c. V_2 and I?

23–7 In Fig. 23–22, solve for I_1, I_2, I_T, and $X_{L_{EQ}}$.

Figure 23–22

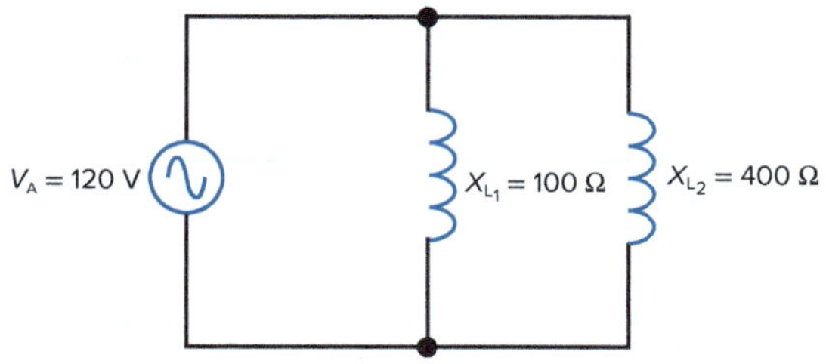

23–8 In Fig. 23–22, what is the phase relationship between

a. V_A and I_1?

b. V_A and I_2?

c. V_A and I_T?

SECTION 23–3 CIRCUITS WITH X_C ALONE

23–9 In Fig. 23–23, solve for X_{C_T}, I, V_1, and V_2.

Figure 23–23

23–10 In Fig. 23–23, what is the phase relationship between

a. V_T and I?

b. V_1 and I?

c. V_2 and I?

23–11 In Fig. 23–24, solve for I_1, I_2, I_T, and $X_{C_{EQ}}$.

Figure 23–24

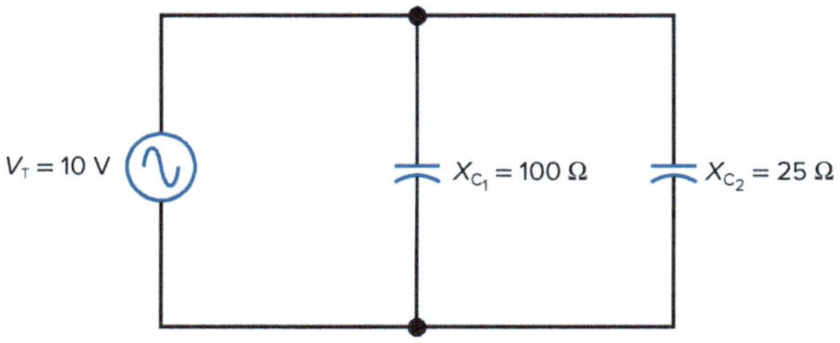

23–12 In Fig. 23–24, what is the phase relationship between

a. V_A and I_1?
b. V_A and I_2?
c. V_A and I_T?

SECTION 23–4 OPPOSITE REACTANCES CANCEL

23–13 In Fig. 23–25, solve for

a. the net reactance, X.
b. the current, I.
c. the inductor voltage, V_L.
d. the capacitor voltage, V_C.

Figure 23–25

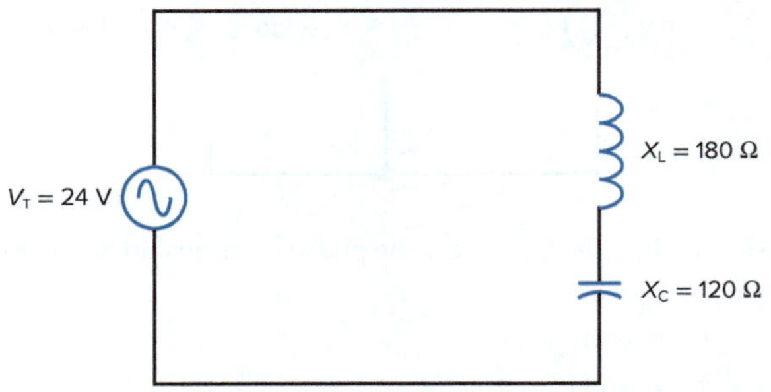

23–14 In Fig. 23–25, what is the phase relationship between

a. X_L and X_C?
b. V_L and I?
c. V_C and I?
d. V_T and I?
e. V_L and V_C?
f. V_T and I if the values of X_L and X_C are interchanged?

23–15 In Fig. 23–26, solve for

a. the inductive branch current, I_L.
b. the capacitive branch current, I_C.
c. the net line current, I_T.
d. the net reactance, X.

Figure 23–26

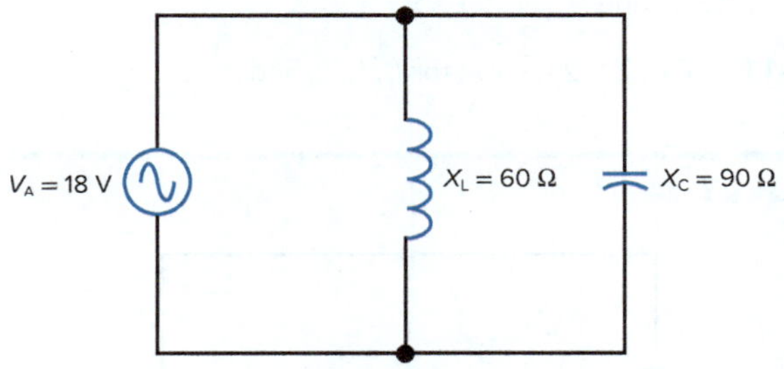

23–16 In Fig. 23–26, what is the phase relationship between

a. I_L and V_A?
b. I_C and V_A?
c. I_L and I_C?
d. V_A and I_T?
e. V_A and I_T if the values of X_L and X_C are interchanged?

SECTION 23–5 SERIES REACTANCE AND RESISTANCE

23–17 In Fig. 23–27, solve for

a. the net reactance, X.
b. Z_T.
c. I.
d. V_R.
e. V_L.
f. V_C.
g. θ_Z.

Figure 23–27

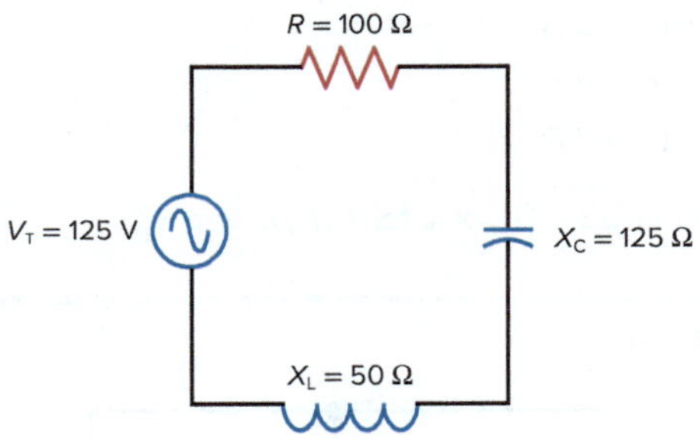

23–18 In Fig. 23–27, what is the phase relationship between

a. X_L and X_C?
b. V_L and I?
c. V_C and I?
d. V_R and I?
e. V_L and V_C?
f. V_T and I?
g. V_T and V_R?
h. V_T and I if the values of X_L and X_C are interchanged?

23–19 Repeat Prob. 23–17 for the circuit in Fig. 23–28.

Figure 23–28

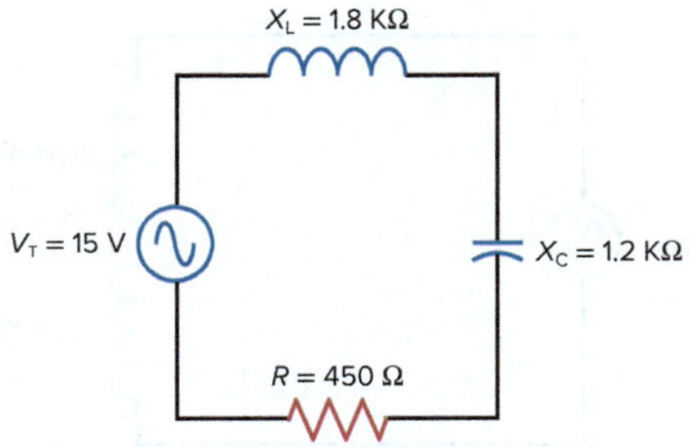

23–20 In Fig. 23–29, solve for

a. the net reactance, X.
b. Z_T.
c. I.
d. V_{R_1} and V_{R_2}.
e. V_{L_1} and V_{L_2}.
f. V_{C_1} and V_{C_2}.
g. θ_Z.

Figure 23–29

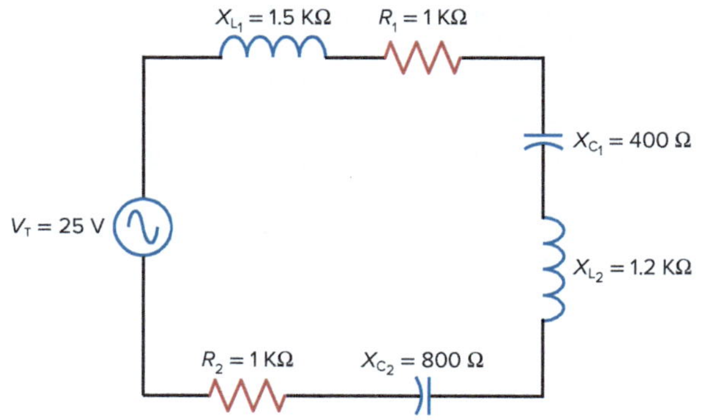

SECTION 23–6 PARALLEL REACTANCE AND RESISTANCE

23–21 In Fig. 23–30, solve for

a. I_R.
b. I_C.
c. I_L.
d. the net reactive branch current, I_X.
e. I_T.
f. Z_{EQ}.
g. θ_I.

Figure 23–30

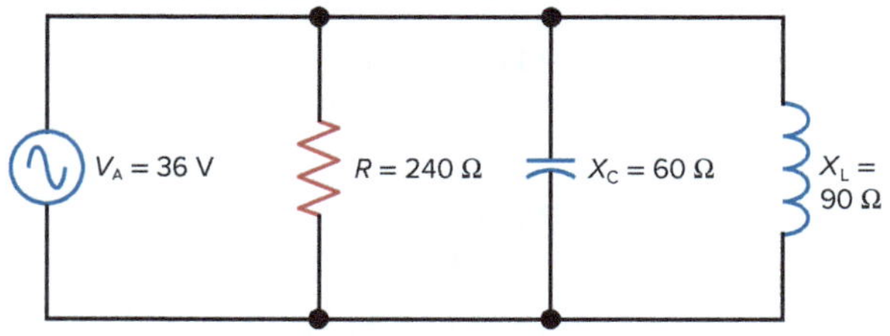

23–22 In Fig. 23–30, what is the phase relationship between

a. V_A and I_R?
b. V_A and I_C?
c. V_A and I_L?
d. I_L and I_C?
e. V_A and I_T?
f. V_A and I_T if the values of X_L and X_C are interchanged?

23–23 Repeat Prob. 23–21 for the circuit in Fig. 23–31.

Figure 23–31

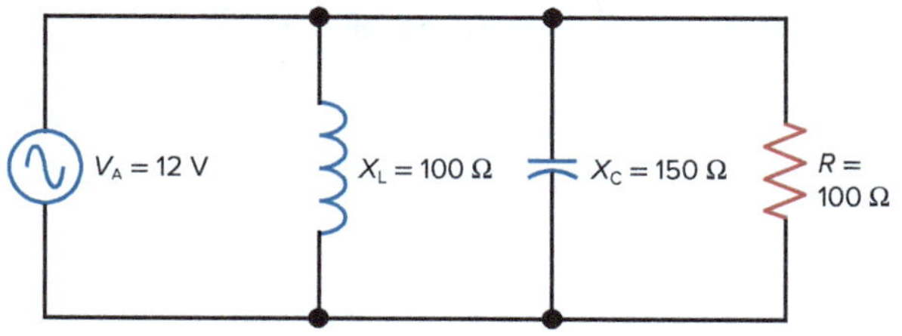

23–24 In Fig. 23–32, solve for

a. I_{R_1} and I_{R_2}.
b. I_{L_1} and I_{L_2}.
c. I_{C_1} and I_{C_2}.
d. the net reactive branch current, I_X.
e. I_T.
f. Z_{EQ}.
g. θ_I.

SECTION 23–7 SERIES-PARALLEL REACTANCE AND RESISTANCE

23–25 In Fig. 23–33, solve for

a. Z_T.
b. I_T.
c. V_{R_1}.
d. V_{C_1}, V_{C_2}, and V_{C_3}.
e. V_{L_1} and V_{L_2}.
f. θ_Z.

Figure 23–32

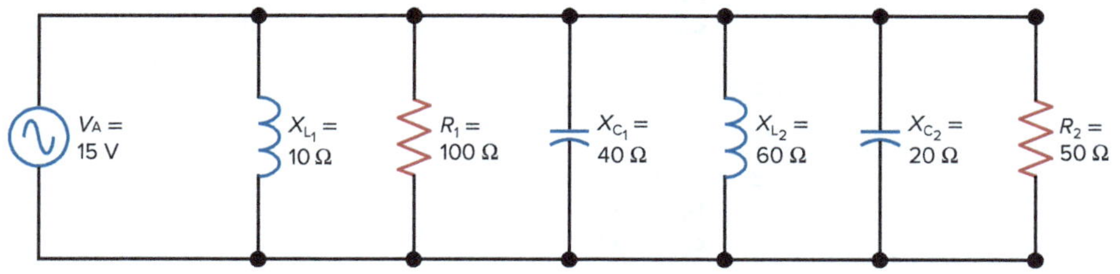

Figure 23–33

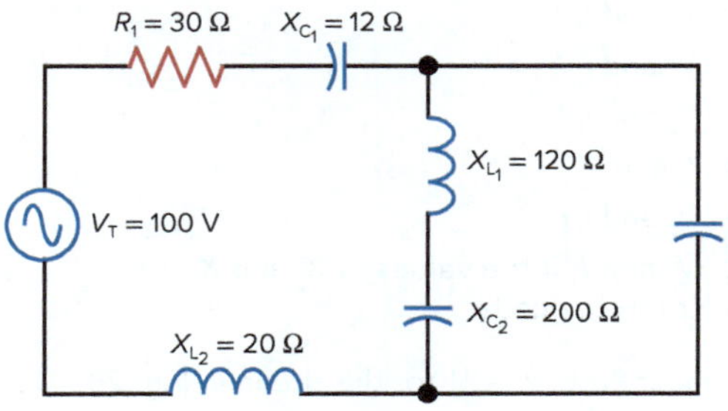

SECTION 23–8 REAL POWER

23–26 Determine the real power, apparent power, and power factor (PF) for each of the following circuits:

a. Fig. 23–19.

b. Fig. 23–21.

c. Fig. 23–24.

d. Fig. 23–26.

23–27 Determine the real power, apparent power, and power factor (PF) for each of the following circuits:

a. Fig. 23–27.

b. Fig. 23–28.

c. Fig. 23–30.

d. Fig. 23–31.

23–28 Calculate the real power, apparent power, and power factor for each of the following circuit conditions:

a. A parallel *RLC* circuit with $V_A = 120$ V, $I_T = 5$ A, and $\theta_I = -45°$.

b. A parallel *RLC* circuit with $V_A = 240$ V, $I_T = 18$ A, and $\theta_I = -26.56°$.

c. A parallel *RLC* circuit with $V_A = 100$ V, $I_T = 3$ A, and $\theta_I = -78°$.

d. A parallel *RLC* circuit with $V_A = 120$ V, $I_T = 8$ A, and $\theta_I = -56°$.

Critical Thinking

23–29 In Fig. 23–34, what value of *L* will produce a circuit power factor of 0.8?

23–30 In Fig. 23–35, what value of *C* in parallel with *R* and *L* will produce a power factor of 0.8?

Figure 23–34 Circuit for Critical Thinking Prob. 23–29.

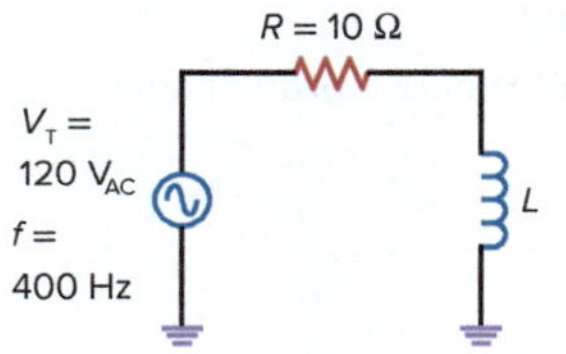

Figure 23–35 Circuit for Critical Thinking Prob. 23–30.

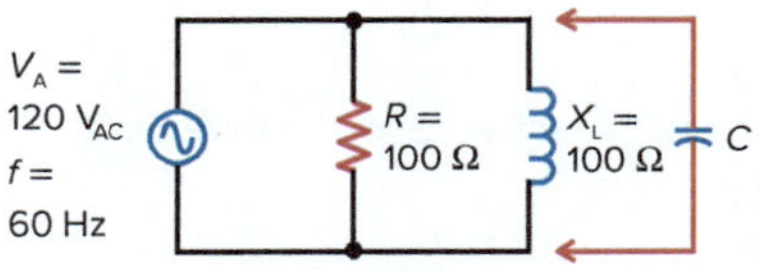

Answers to Self-Reviews

23–1 **a.** 0°
b. 0°

23–2 **a.** 90°
b. −90°

23–3 **a.** −90°
b. 90°

23–4 **a.** 20 Ω
b. 1 A

23–5 **a.** 30 Ω
b. 30 Ω
c. 180°

23–6 **a.** 3 A
b. 3 A

23–7 **a.** 200 Ω
b. 200 Ω
c. 100 Ω

23–8 **a.** watt
b. volt-ampere
c. real

23–9 **a.** true
b. true

23–10 **a.** real power
b. *V*

23–11 **a.** *R*
b. X_L
c. X_C

23–12 **a.** V_L and V_C
b. I_R and I_L

Laboratory Application Assignment

In this lab application assignment, you will examine the real power, apparent power, and power factor (PF) in a parallel *RL* circuit. You will also examine how a capacitor can be added in parallel to bring the power factor closer to 1 or unity. The procedure of adding a capacitor in parallel to raise the power factor is called power factor correction.

Note: *In this lab we will assume that the dc resistance, ri, of the inductor is negligible.*

Equipment: Obtain the following items from your instructor.

- Function generator
- Oscilloscope
- DMM
- 100-mH inductor
- 0.22-μF capacitor
- 10-Ω and 680-Ω resistors

Real Power, Apparent Power, and Power Factor

Examine the parallel *RL* circuit in Fig. 23–36*a*. (Ignore the 0.22-μF capacitor.) Calculate and record the following circuit values:

X_L = __________, I_L = __________, I_R = __________,
I_T = __________, Z_{EQ} = __________
θ_I = __________, real power = __________,
apparent power = __________, PF = __________

Construct the circuit in Fig. 23–36*a*. (Again, ignore the 0.22-μF capacitor.)
Adjust the applied voltage, V_A, to exactly 5 Vrms. With a DMM, measure and record the following circuit values:

I_L = __________, I_R = __________,
I_T = __________

Using the measured values of I_L and I_R, calculate the total current, I_T as $I_T = \sqrt{I_R^2 + I_L^2}$. Does this value agree with the measured value of total current? __________ If not, list one possible reason why. __________

Using the measured values of I_L and I_R, calculate the circuit's phase angle, θ_I. (Recall that in a parallel circuit, $\tan \theta_I = -I_L/I_R$.) θ_I = __________. Next, using measured values, determine the following: real power = __________, apparent power = __________, PF = __________ How do these experimental values compare to those initially calculated? __________

Power Factor Correction

Mentally connect the 0.22-μF capacitor in Fig. 23–36*a*. Calculate and record the following circuit values:

X_L = __________, X_C = __________, I_L = __________,
I_C = __________, I_X = __________
I_R = __________, I_T = __________, Z_{EQ} = __________,
θ_I = __________, apparent power = __________,
real power = __________, PF = __________

Construct the circuit in Fig. 23–36*a* including the 0.22-μF capacitor. Adjust the applied voltage, V_A, to exactly 5 Vrms. With a DMM, measure and record the following circuit values:

I_L = __________, I_C = __________,
I_R = __________, I_T = __________

Figure 23–36

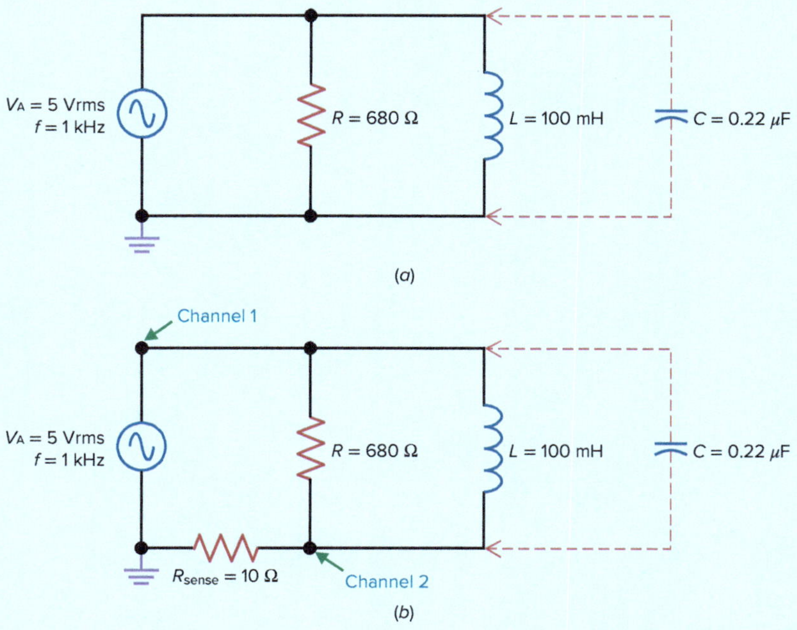

While measuring the total current, I_T, connect and disconnect the 0.22-μF capacitor several times. You should notice that I_T decreases when the capacitor is connected. Explain why this happens. ______________________________
____

Using measured values, calculate the net reactive current, I_X, as $I_L - I_C$ or $I_C - I_L$ depending on which current is larger.
I_X = ______________
Using the experimental value of I_X and the measured value of I_R, calculate the circuit's phase angle, θ_I. θ_I = ______________
Using measured values, calculate the following:
apparent power = __________, real power = __________,
PF = ______________

In Figure 23–36*a*

a. Did the apparent power increase, decrease, or stay the same when the capacitor was added? ______________
b. Did the phase angle, θ_I, become more negative, less negative, or did it stay the same when the capacitor was added? ______________
c. Did the real power increase, decrease, or stay the same when the capacitor was added? ______________
d. Did the power factor, PF, increase, decrease, or stay the same when the capacitor was added? ______________

Add a 10-Ω sensing resistor as shown in Fig. 23–36*b*. Next, connect the oscilloscope to measure the phase angle, θ_I, between V_A and I_T. Note the connections designated for channels 1 and 2 in the figure. While viewing the displayed waveforms on the oscilloscope, connect and disconnect the 0.22-μF capacitor several times. Explain what happens to the phase angle when the capacitor is connected. ______________
____

Design credits: Chapter Opener Background Image, and Multisim: Stockbyte/Getty Images

chapter

24

Complex Numbers for AC Circuits

Complex numbers refer to a numerical system that includes the phase angle of a quantity with its magnitude. Therefore, complex numbers are useful in AC circuits when the reactance of X_L or X_C makes it necessary to consider the phase angle. For instance, complex notation explains why θ_Z is negative with X_C and θ_I is negative with I_L

Any type of AC circuit can be analyzed with complex numbers. They are especially convenient for solving series-parallel circuits that have both resistance and reactance in one or more branches. Although graphical analysis with phasor arrows can be used, the method of complex numbers is probably the best way to analyze AC circuits with series-parallel impedances. ■

Chapter Outline

Chapter Objectives

After studying this chapter, you should be able to

- *Explain* the *j* operator.
- *Define* a complex number.
- *Add, subtract, multiply,* and *divide* complex numbers.
- *Explain* the difference between the rectangular and polar forms of a complex number.
- *Convert* a complex number from polar to rectangular form and vice versa.
- *Explain* how to use complex numbers to solve series and parallel AC circuits containing resistance, capacitance, and inductance.

Important Terms

Admittance, *Y*
Complex number
Imaginary number
j operator
Polar form
Real number
Rectangular form
Susceptance, *B*

24–1 Positive and Negative Numbers

Our common use of numbers as either positive or negative represents only two special cases. In their more general form, numbers have both quantity and phase angle. In Fig. 24–1, positive and negative numbers are shown corresponding to the phase angles of 0° and 180°, respectively.

For example, the numbers 2, 4, and 6 represent units along the horizontal or x axis, extending toward the right along the line of zero phase angle. Therefore, positive numbers represent units having the phase angle of 0°, or this phase angle corresponds to the factor of +1. To indicate 6 units with zero phase angle, then, 6 is multiplied by +1 as a factor for the positive number 6. The + sign is often omitted, as it is assumed unless indicated otherwise.

In the opposite direction, negative numbers correspond to 180°, or this phase angle corresponds to the factor of −1. Actually, −6 represents the same quantity as 6 but rotated through the phase angle of 180°. The angle of rotation is the *operator* for the number. The operator for −1 is 180°; the operator for +1 is 0°.

24–1 Self-Review

Answers at the end of the chapter.

a. What is the angle for the number +5?
b. What is the angle for the number −5?

24–2 The *j* Operator

The operator for a number can be any angle between 0° and 360°. Since the angle of 90° is important in AC circuits, the factor j is used to indicate 90°. See Fig. 24–2. Here, the number 5 means 5 units at 0°, the number −5 is at 180°, and $j5$ indicates the number 5 at the 90° angle.

The j is usually written before the number. The reason is that the j sign is a 90° operator, just as the + sign is a 0° operator and the − sign is a 180° operator. Any quantity at right angles to the zero axis, or 90° counterclockwise, is on the $+j$ axis.

GOOD TO KNOW

The induced voltage, V_L across an inductor is an example of a quantity that is represented on the $+j$ axis.

Figure 24–1 Positive and negative numbers.

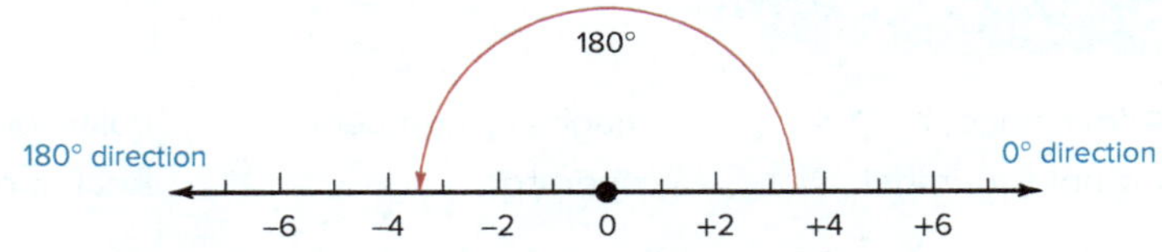

Figure 24–2 The j axis at 90° from the horizontal real axis.

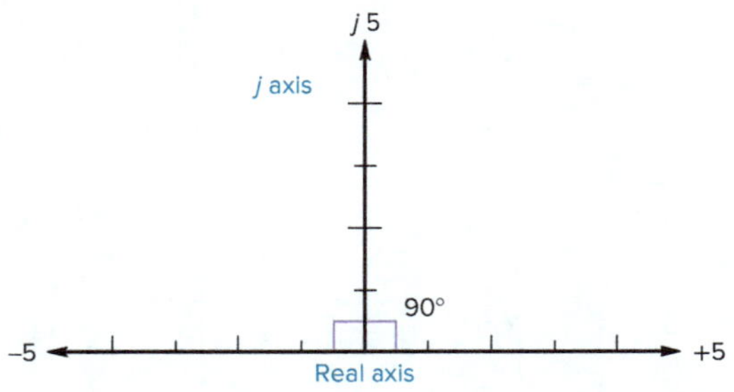

Figure 24–3 The j operator indicates 90° rotation from the real axis; the $-j$ operator is −90°; j^2 operation is 180° rotation back to the real axis in a negative direction.

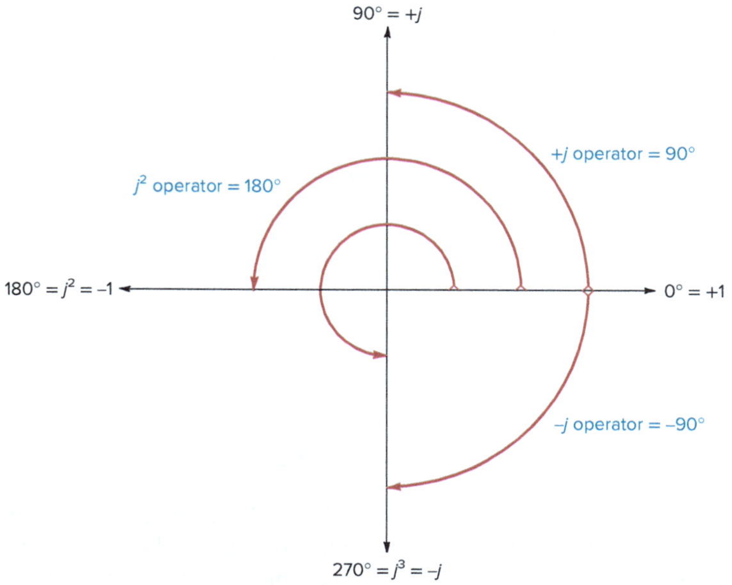

In mathematics, numbers on the horizontal axis are ***real numbers,*** including positive and negative values. Numbers on the j axis are called ***imaginary numbers*** because they are not on the real axis. In mathematics, the abbreviation i is used to indicate imaginary numbers. In electricity, however, j is used to avoid confusion with i as the symbol for current. Furthermore, there is nothing imaginary about electrical quantities on the j axis. An electric shock from j500 V is just as dangerous as 500 V positive or negative.

More features of the ***j* operator** are shown in Fig. 24–3. The angle of 180° corresponds to the j operation of 90° repeated twice. This angular rotation is indicated by the factor j^2. Note that the j operation multiplies itself, instead of adding.

Since j^2 means 180°, which corresponds to the factor of −1, we can say that j^2 is the same as −1. In short, the operator j^2 for a number means to multiply by −1. For instance, $j^2 8$ is −8.

Furthermore, the angle of 270° is the same as −90°, which corresponds to the operator $-j$. These characteristics of the j operator are summarized as follows:

GOOD TO KNOW

$j^2 = -1$

$\sqrt{j^2} = \sqrt{-1}$

and

$j = \sqrt{-1}$

$$0^\circ = 1$$
$$90^\circ = j$$
$$180^\circ = j^2 = -1$$
$$270^\circ = j^3 = j^2 \times j = -1 \times j = -j$$
$$360^\circ = \text{same as } 0^\circ$$

As examples, the number 4 or −4 represents 4 units on the real horizontal axis; $j4$ means 4 units with a leading phase angle of 90°; $-j4$ means 4 units with a lagging phase angle of −90°.

24–2 Self-Review

Answers at the end of the chapter.

a. What is the angle for the operator *j*?

b. What is the angle for the operator −*j*?

Figure 24–4 Phasors corresponding to real terms and imaginary (*j*) terms, in rectangular coordinates.

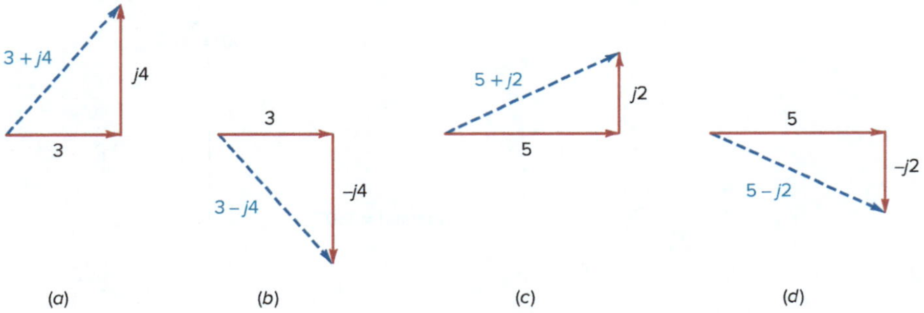

GOOD TO KNOW

When phasors are added, the tail of one phasor is placed at the arrowhead of the other. The resultant phasor extends from the tail of the first phasor to the arrowhead of the second.

24–3 Definition of a Complex Number

The combination of a real and an imaginary term is called a ***complex number***. Usually, the real number is written first. As an example, 3 + *j*4 is a complex number including 3 units on the real axis added to 4 units 90° out of phase on the *j* axis. Complex numbers must be added as phasors.

Phasors for complex numbers shown in Fig. 24–4 are typical examples. The +*j* phasor is up for 90°; the −*j* phasor is down for −90°. The phasors are shown with the end of one joined to the start of the next, to indicate addition. Graphically, the sum is the hypotenuse of the right triangle formed by the two phasors. Since a number like 3 + *j*4 specifies the phasors in rectangular coordinates, this system is the ***rectangular form*** of complex numbers.

Be careful to distinguish a number like *j*2, where 2 is a coefficient, from j^2, where 2 is the exponent. The number *j*2 means 2 units up on the *j* axis of 90°. However, j^2 is the operator of −1, which is on the real axis in the negative direction.

Another comparison to note is between *j*3 and j^3. The number *j*3 is 3 units up on the *j* axis, and j^3 is the same as the −*j* operator, which is down on the −90° axis.

Also note that either the real term or the *j* term can be the larger of the two. When the *j* term is larger, the angle is more than 45°; when the *j* term is smaller, the angle is less than 45°. If the *j* term and the real term are equal, the angle is 45°.

24–3 Self-Review

Answers at the end of the chapter.

a. **For 7 + *j*6, the 6 is at 90° leading the 7. (True/False)**
b. **For 7 − *j*6, the 6 is at −90° lagging the 7. (True/False)**

GOOD TO KNOW

Inductive reactance and capacitive reactance are sometimes indicated as $+jX_L$ and $-jX_C$, respectively. Conversely, inductive and capacitive branch currents are indicated as $-jI_L$ and $+jI_C$, respectively.

24–4 How Complex Numbers Are Applied to AC Circuits

Applications of complex numbers are a question of using a real term for 0°, +*j* for 90° and −*j* for −90°, to denote phase angles. Figure 24–5 illustrates the following rules:

An *angle of* 0° or a real number without any *j* operator is used for resistance *R*. For instance, 3 Ω of *R* is stated as 3 Ω.

An *angle of* 90° or +*j* is used for inductive reactance X_L. For instance, a 4-Ω X_L is *j*4 Ω. This rule always applies to X_L, whether it is in series or parallel with *R*. The

Figure 24–5 Rectangular form of complex numbers for impedances. (*a*) Reactance X_L is $+j$. (*b*) Reactance X_C is $-j$.

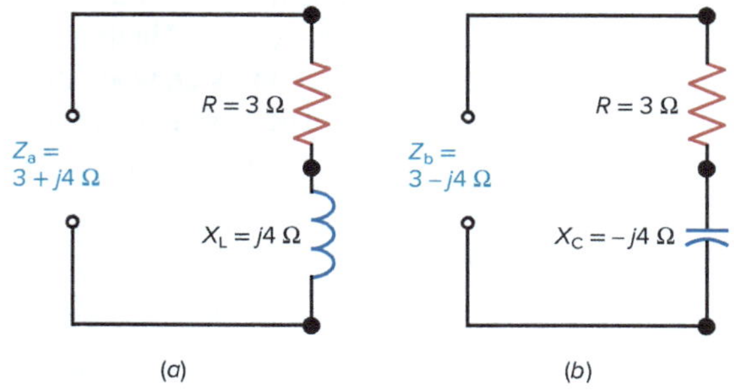

Figure 24–6 Rectangular form of complex numbers for branch currents. (*a*) Current I_L is $-j$. (*b*) Current I_C is $+j$.

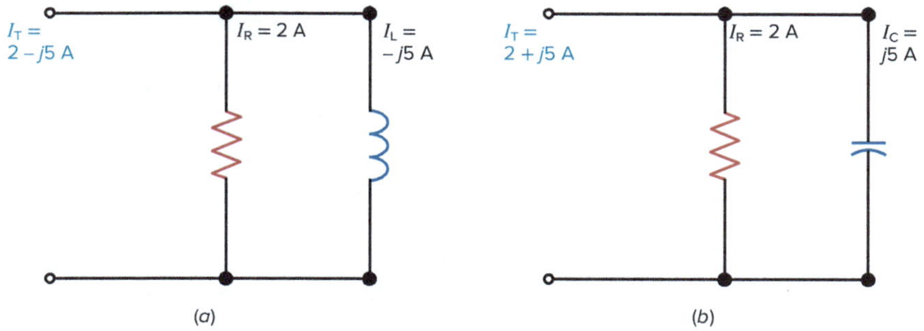

reason is the fact that IX_L represents voltage across an inductance, which always leads the current in the inductance 90°. The $+j$ is also used for V_L.

An *angle of* −90° or $-j$ is used for X_C. For instance, a 4-Ω X_C is $-j4$ Ω. This rule always applies to X_C, whether it is in series or parallel with R. The reason is that IX_C is the voltage across a capacitor, which always lags the capacitor's charge and discharge current by −90°. The $-j$ is also used for V_C.

With reactive branch currents, the sign for j is reversed, compared with reactive ohms, because of the opposite phase angle. In Fig. 24–6*a* and *b*, $-j$ is used for inductive branch current I_L, and $+j$ is used for capacitive branch current I_C.

24–4 Self-Review

Answers at the end of the chapter.

a. Write 3 kΩ of X_L using the j operator.
b. Write 5 mA of I_L using the j operator.

24–5 Impedance in Complex Form

The rectangular form of complex numbers is a convenient way to state the impedance of series resistance and reactance. In Fig. 24–5*a*, the impedance is $3 + j4$ because Z_a is the phasor sum of a 3-Ω R in series with $j4$ Ω for X_L. Similarly, Z_b is $3 - j4$ for a 3-Ω R in series with $-j4$ Ω for X_C. The minus sign in Z_b results from adding the negative term for $-j$, that is, $3 + (-j4) = 3 - j4$.

For a 4-kΩ R and a 2-kΩ X_L in series: $Z_T = 4000 + j2000$ Ω
For a 3-kΩ R and a 9-kΩ X_C in series: $Z_T = 3000 - j9000$ Ω

For $R = 0$ and a 7-Ω X_L in series: $Z_T = 0 + j7$ Ω
For a 12-Ω R and $X = 0$ in series: $Z_T = 12 + j0$

Note the general form of stating $Z = R \pm jX$. If one term is zero, substitute 0 for this term to keep Z in its general form. This procedure is not required, but there is usually less confusion when the same form is used for all types of Z.

The advantage of this method is that multiple impedances written as complex numbers can then be calculated as follows:

For series impedances:

$$Z_T = Z_1 + Z_2 + Z_3 + \cdots + \text{etc.} \tag{24–1}$$

For parallel impedances:

$$\frac{1}{Z_T} = \frac{1}{Z_1} + \frac{1}{Z_2} + \frac{1}{Z_3} + \cdots + \text{etc.} \tag{24–2}$$

For two parallel impedances:

$$Z_T = \frac{Z_1 \times Z_2}{Z_1 + Z_2} \tag{24–3}$$

Examples are shown in Fig. 24–7. The circuit in Fig. 24–7a is a series combination of resistances and reactances. Combining the real terms and j terms separately, $Z_T = 12 + j4$. The calculations are $3 + 9 = 12$ Ω for R and $j6$ added to $-j2$ equals $j4$ for the net X.

The parallel circuit in Fig. 24–7*b* shows that X_L is $+j$ and X_C is $-j$, even though they are in parallel branches, because they are reactances, not currents.

So far, these types of circuits can be analyzed with or without complex numbers. For the series-parallel circuit in Fig. 24–7*c*, however, the notation of complex numbers is necessary to state the complex impedance Z_T, consisting of branches with reactance and resistance in one or more of the branches. Impedance Z_T is stated here in its form as a complex impedance. To calculate Z_T, some of the rules described in the next section must be used for combining complex numbers.

24–5 Self-Review

Answers at the end of the chapter.

Write the following impedances in complex form:

a. X_L **of 7 Ω in series with** R **of 4 Ω.**
b. X_C **of 7 Ω in series with zero** R**.**

Figure 24–7 Reactance X_L is a $+j$ term and X_C is a $-j$ term whether in series or parallel. (*a*) Series circuit. (*b*) Parallel branches. (*c*) Complex branch impedances Z_1 and Z_2 in parallel.

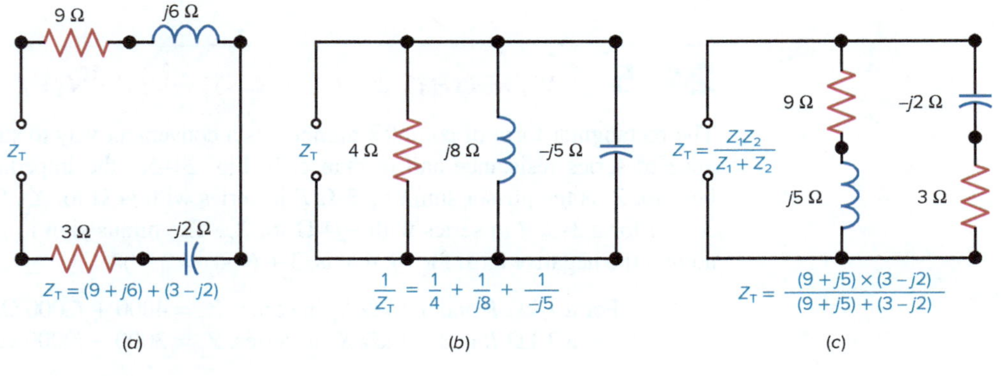

24–6 Operations with Complex Numbers

Real numbers and j terms cannot be combined directly because they are 90° out of phase. The following rules apply.

For Addition or Subtraction

Add or subtract the real and j terms separately:

$$(9 + j5) + (3 + j2) = 9 + 3 + j5 + j2$$
$$= 12 + j7$$
$$(9 + j5) + (3 - j2) = 9 + 3 + j5 - j2$$
$$= 12 + j3$$
$$(9 + j5) + (3 - j8) = 9 + 3 + j5 - j8$$
$$= 12 - j3$$

The answer should be in the form of $R \pm jX$, where R is the algebraic sum of all the real or resistive terms and X is the algebraic sum of all the imaginary or reactive terms.

To Multiply or Divide a *j* Term by a Real Number

Multiply or divide the numbers. The answer is still a j term. Note the algebraic signs in the following examples. If both factors have the same sign, either + or −, the answer is +; if one factor is negative, the answer is negative.

$$4 \times j3 = j12 \qquad j12 \div 4 = j3$$
$$j5 \times 6 = j30 \qquad j30 \div 6 = j5$$
$$j5 \times (-6) = -j30 \qquad -j30 \div (-6) = j5$$
$$-j5 \times 6 = -j30 \qquad -j30 \div 6 = -j5$$
$$-j5 \times (-6) = j30 \qquad j30 \div (-6) = -j5$$

To Multiply or Divide a Real Number by a Real Number

Just multiply or divide the real numbers, as in arithmetic. There is no j operation. The answer is still a real number.

To Multiply a *j* Term by a *j* Term

Multiply the numbers and the j coefficients to produce a j^2 term. The answer is a real term because j^2 is −1, which is on the real axis. Multiplying two j terms shifts the number 90° from the j axis to the real axis of 180°. As examples,

$$j4 \times j3 = j^2 12 = (-1)(12)$$
$$= -12$$
$$j4 \times (-j3) = -j^2 12 = -(-1)(12)$$
$$= 12$$

To Divide a *j* Term by a *j* Term

Divide the j coefficients to produce a real number; the j factors cancel. For instance:

$$j12 \div j4 = 3 \qquad -j12 \div j4 = -3$$
$$j30 \div j5 = 6 \qquad j30 \div (-j6) = -5$$
$$j15 \div j3 = 5 \qquad -j15 \div (-j3) = 5$$

GOOD TO KNOW

It is a lot less work to multiply and divide complex numbers that are in polar form as compared to complex numbers that are in rectangular form.

To Multiply Complex Numbers

Follow the rules of algebra for multiplying two factors, each having two terms:

$$\begin{aligned}(9 + j5) \times (3 - j2) &= 27 - j18 + j15 - j^2 10 \\ &= 27 - j3 - (-1)10 \\ &= 27 - j3 + 10 \\ &= 37 - j3\end{aligned}$$

Note that $-j^2 10$ equals $+10$ because the operator j^2 is -1 and $-(-1)10$ becomes $+10$.

To Divide Complex Numbers

This process becomes more involved because division of a real number by an imaginary number is not possible. Therefore, the denominator must first be converted to a real number without any j term.

Converting the denominator to a real number without any j term is called *rationalization* of the fraction. To do this, multiply both numerator and denominator by the *conjugate* of the denominator. Conjugate complex numbers have equal terms but opposite signs for the j term. For instance, $(1 + j2)$ has the conjugate $(1 - j2)$.

Rationalization is permissible because the value of a fraction is not changed when both numerator and denominator are multiplied by the same factor. This procedure is the same as multiplying by 1. In the following example of division with rationalization, the denominator $(1 + j2)$ has the conjugate $(1 - j2)$:

$$\begin{aligned}\frac{4 - j1}{1 + j2} &= \frac{4 - j1}{1 + j2} \times \frac{(1 - j2)}{(1 - j2)} = \frac{4 - j8 - j1 + j^2 2}{1 - j2 + j2 - j^2 4} \\ &= \frac{4 - j9 - 2}{1 + 4} \\ &= \frac{2 - j9}{5} \\ &= 0.4 - j1.8\end{aligned}$$

As a result of the rationalization, $4 - j1$ has been divided by $1 + j2$ to find the quotient that is equal to $0.4 - j1.8$.

Note that the product of a complex number and its conjugate always equals the sum of the squares of the numbers in each term. As another example, the product of $(2 + j3)$ and its conjugate $(2 - j3)$ must be $4 + 9$, which equals 13. Simple numerical examples of division and multiplication are given here because when the required calculations become too long, it is easier to divide and multiply complex numbers in **polar form**, as explained soon in Sec. 24–8.

24–6 Self-Review

Answers at the end of the chapter.

a. $\mathbf{(2 + j3) + (3 + j4) = ?}$
b. $\mathbf{(2 + j3) \times 2 = ?}$

GOOD TO KNOW

Learn how to use the polar-to-rectangular and rectangular-to-polar conversion keys on your calculator.

24–7 Magnitude and Angle of a Complex Number

In electrical terms, the complex impedance $(4 + j3)$ means 4 Ω of resistance and 3 Ω of inductive reactance with a leading phase angle of 90°. See Fig. 24–8*a*. The magnitude of Z is the resultant, equal to $\sqrt{16 + 9} = \sqrt{25} = 5\ \Omega$. Finding the square root of the sum of the squares is vector or phasor addition of two terms in quadrature, 90° out of phase.

Figure 24–8 Magnitude and angle of a complex number. (*a*) Rectangular form. (*b*) Polar form.

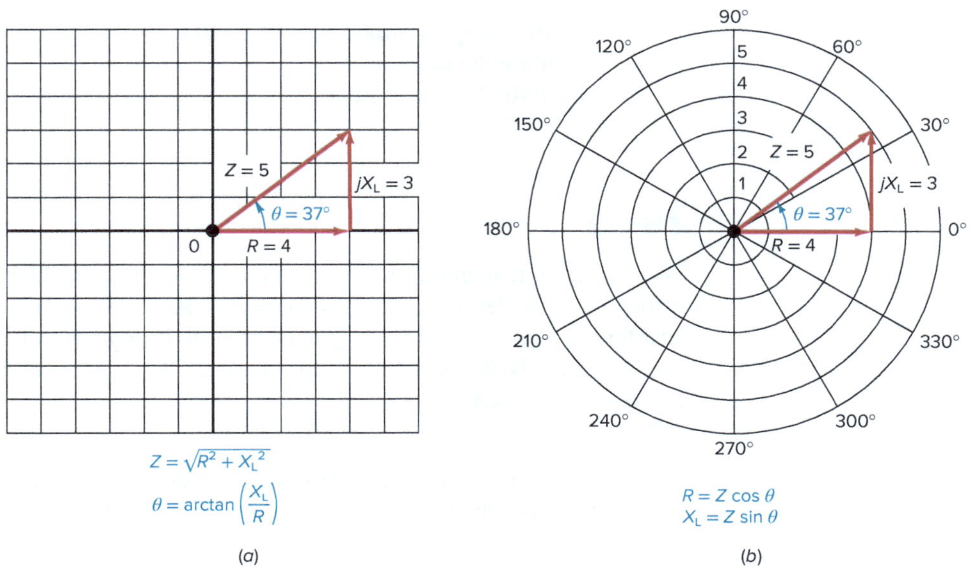

The phase angle of the resultant is the angle whose tangent is 0.75. This angle equals 37°. Therefore, $4 + j3 = 5\angle 37^\circ$.

When calculating the tangent ratio, note that the *j* term is the numerator and the real term is the denominator because the tangent of an angle is the ratio of the opposite side to the adjacent side. For a negative *j* term, the tangent is negative, which means a negative angle.

Note the following definitions: $(4 + j3)$ is the complex number in rectangular coordinates. The real term is 4. The imaginary term is $j3$. The resultant 5 is the magnitude, absolute value, or modulus of the complex number. Its phase angle or argument is 37°. The resultant value by itself can be written as $|5|$; the vertical lines indicate that it is the magnitude without the phase angle. The magnitude is the value a meter would read. For instance, with a current of $5\angle 37^\circ$ A in a circuit, an ammeter reads 5 A. As additional examples,

$$2 + j4 = \sqrt{4 + 16}\ \angle \arctan 2 = 4.47\angle 63^\circ$$

$$4 + j2 = \sqrt{16 + 4}\ \angle \arctan 0.5 = 4.47\angle 26.5^\circ$$

$$8 + j6 = \sqrt{64 + 36}\ \angle \arctan 0.75 = 10\angle 37^\circ$$

$$8 - j6 = \sqrt{64 + 36}\ \angle \arctan -0.75 = 10\angle -37^\circ$$

$$4 + j4 = \sqrt{16 + 16}\ \angle \arctan 1 = 5.66\angle 45^\circ$$

$$4 - j4 = \sqrt{16 + 16}\ \angle \arctan -1 = 5.66\angle -45^\circ$$

Note that arctan 0.75 in the third example means the angle with a tangent equal to 0.75. This value is ⁶⁄₈ or ¾ for the ratio of the opposite side to the adjacent side. The arctan can also be indicated as $\tan^{-1} 0.75$. In either case, this angle has 0.75 for its tangent, which makes the angle 36.87°.

Many scientific calculators have keys that can convert from rectangular coordinates to the magnitude–phase angle form (called *polar coordinates*) directly. See your calculator manual for the particular steps used. If your calculator does not have these keys, the problem can be done in two separate parts: (1) the magnitude as the square root of the sum of two squares and (2) the angle as the arctan equal to the *j* term divided by the real term.

CALCULATOR

Using the calculator for the magnitude in the first example, punch in 2 and then press the x^2 key for the square, equal to 4. Press + and then 4; press x^2, =, and √ in sequence. The display will show 4.47, which is the magnitude.

To find the angle from its tangent value, after the display is cleared, punch in 4 for the opposite side. Then press the ÷ key, punch in 2 for the adjacent side, and push the = key for the ratio of 2 as the tangent. With 2 as tan θ on the display, press the TAN^{-1} key, which is usually the second function of the TAN key. Then 63.4 appears on the display as the angle. Be sure that the calculator is set for degrees in the answer, not rad or grad units.

24–7 Self-Review

Answers at the end of the chapter.

For the complex impedance $10 + j10\ \Omega$,

a. **calculate the magnitude.**

b. **calculate the phase angle.**

24–8 Polar Form of Complex Numbers

Calculating the magnitude and phase angle of a complex number is actually converting to an angular form in polar coordinates. As shown in Fig. 24–8, the rectangular form $4 + j3$ is equal to $5\angle 37°$ in polar form. In polar coordinates, the distance from the center is the magnitude of the phasor Z. Its phase angle θ is counterclockwise from the 0° axis.

To convert any complex number to polar form,

1. Find the magnitude by phasor addition of the j term and real term.
2. Find the angle whose tangent is the j term divided by the real term.

As examples,

$$2 + j4 = 4.47\angle 63°$$
$$4 + j2 = 4.47\angle 26.5°$$
$$8 + j6 = 10\angle 37°$$
$$8 - j6 = 10\angle -37°$$
$$4 + j4 = 5.66\angle 45°$$
$$4 - j4 = 5.66\angle -45°$$

These examples are the same as those given before for finding the magnitude and phase angle of a complex number.

The magnitude in polar form must be more than either term in rectangular form, but less than their arithmetic sum. For instance, in $8 + j6 = 10\angle 37°$ the magnitude of 10 is more than 8 or 6 but less than their sum of 14.

Applied to AC circuits with resistance for the real term and reactance for the j term, then, the polar form of a complex number states the resultant impedance and its phase angle. Note the following cases for an impedance where either the resistance or the reactance is zero:

$$0 + j5 = 5\angle 90°$$
$$0 - j5 = 5\angle -90°$$
$$5 + j0 = 5\angle 0°$$

The polar form is much more convenient for multiplying or dividing complex numbers. The reason is that multiplication in polar form merely involves multiplying the magnitudes and adding the angles. Division involves dividing the magnitudes and subtracting the angles. The following rules apply.

For Multiplication

Multiply the magnitudes but add the angles algebraically:

$$24\angle 40° \times 2\angle 30° = 24 \times 2\angle 40° + 30° = 48\angle +70°$$
$$24\angle 40° \times (-2\angle 30°) = -48\angle +70°$$
$$12\angle -20° \times 3\angle -50° = 36\angle -70°$$
$$12\angle -20° \times 4\angle 5° = 48\angle -15°$$

When you multiply by a real number, just multiply the magnitudes:

$$4 \times 2\angle 30^\circ = 8\angle 30^\circ$$
$$4 \times 2\angle -30^\circ = 8\angle -30^\circ$$
$$-4 \times 2\angle 30^\circ = -8\angle 30^\circ$$
$$-4 \times (-2\angle 30^\circ) = 8\angle 30^\circ$$

This rule follows from the fact that a real number has an angle of 0°. When you add 0° to any angle, the sum equals the same angle.

For Division

Divide the magnitudes and subtract the angles algebraically:

$$24\angle 40^\circ \div 2\angle 30^\circ = 24 \div 2\angle 40^\circ - 30^\circ = 12\angle 10^\circ$$
$$12\angle 20^\circ \div 3\angle 50^\circ = 4\angle -30^\circ$$
$$12\angle -20^\circ \div 4\angle 50^\circ = 3\angle -70^\circ$$

To divide by a real number, just divide the magnitudes:

$$12\angle 30^\circ \div 2 = 6\angle 30^\circ$$
$$12\angle -30^\circ \div 2 = 6\angle -30^\circ$$

This rule is also a special case that follows from the fact that a real number has a phase angle of 0°. When you subtract 0° from any angle, the remainder equals the same angle.

For the opposite case, however, when you divide a real number by a complex number, the angle of the denominator changes its sign in the answer in the numerator. This rule still follows the procedure of subtracting angles for division, since a real number has a phase angle of 0°. As examples,

$$\frac{10}{5\angle 30^\circ} = \frac{10\angle 0^\circ}{5\angle 30^\circ} = 2\angle 0^\circ - 30^\circ$$
$$= 2\angle -30^\circ$$
$$\frac{10}{5\angle -30^\circ} = \frac{10\angle 0^\circ}{5\angle -30^\circ} = 2\angle 0^\circ - (-30^\circ)$$
$$= 2\angle +30^\circ$$

Stated another way, we can say that the reciprocal of an angle is the same angle but with opposite sign. Note that this operation is similar to working with powers of 10. Angles and powers of 10 follow the general rules of exponents.

24–8 Self-Review

Answers at the end of the chapter.

a. $\mathbf{6\angle 20^\circ \times 2\angle 30^\circ = ?}$
b. $\mathbf{6\angle 20^\circ \div 2\angle 30^\circ = ?}$

24–9 Converting Polar to Rectangular Form

Complex numbers in polar form are convenient for multiplication and division, but they cannot be added or subtracted if their angles are different because the real and imaginary parts that make up the magnitude are different. When complex numbers in polar form are to be added or subtracted, therefore, they must be converted into rectangular form.

CALCULATOR

Conversion to rectangular form can be done fast with a calculator. Again, some scientific calculators contain conversion keys that make going from polar coordinates to rectangular coordinates a simple four-key procedure. Check your calculator manual for the exact procedure. If your calculator does not have this capability, use the following routine. Punch in the value of the angle θ in degrees. Make sure that the correct sign is used and the calculator is set to handle angles in degrees. Find cos θ or sin θ, and multiply by the magnitude for each term. Remember to use cos θ for the real term and sin θ for the *j* term. For the example of $100\,\underline{/30^\circ}$, punch in the number 30 and press the (COS) key for 0.866 as cos θ. While it is on the display, press the (×) key, punch in 100, and press the (=) key for the answer of 86.6 as the real term. Clear the display for the next operation with sin θ. Punch in 30, push the (SIN) key for 0.5 as sin θ, press the (×) key, punch in 100, and push the (=) key for the answer of 50 as the *j* term.

Figure 24–9 Converting the polar form of $Z\underline{/\theta}$ to the rectangular form of $R \pm jX$. (*a*) The positive angle θ in the first quadrant has a $+j$ term. (*b*) The negative angle $-\theta$ in the fourth quadrant has a $-j$ term.

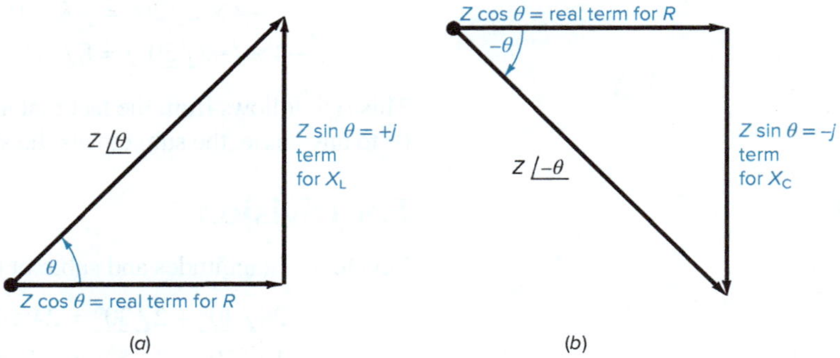

Consider the impedance $Z\underline{/\theta}$ in polar form. Its value is the hypotenuse of a right triangle with sides formed by the real term and *j* term in rectangular coordinates. See Fig. 24–9. Therefore, the polar form can be converted to rectangular form by finding the horizontal and vertical sides of the right triangle. Specifically,

$$\text{Real term for } R = Z\cos\theta$$
$$j \text{ term for } X = Z\sin\theta$$

In Fig. 24–9a, assume that $Z\underline{/\theta}$ in polar form is $5\underline{/37^\circ}$. The sine of 37° is 0.6 and its cosine is 0.8.

To convert to rectangular form,

$$R = Z\cos\theta = 5 \times 0.8 = 4$$
$$X = Z\sin\theta = 5 \times 0.6 = 3$$

Therefore,

$$5\underline{/37^\circ} = 4 + j3$$

This example is the same as the illustration in Fig. 24–8. The + sign for the *j* term means that it is X_L, not X_C.

In Fig. 24–9*b*, the values are the same, but the *j* term is negative when θ is negative. The negative angle has a negative *j* term because the opposite side is in the fourth quadrant, where the sine is negative. However, the real term is still positive because the cosine is positive.

Note that *R* for cos θ is the horizontal component, which is an adjacent side of the angle. The *X* for sin θ is the vertical component, which is opposite the angle. The $+X$ is X_L; the $-X$ is X_C.

These rules apply for angles in the first or fourth quadrant, from 0 to 90° or from 0 to −90°. As examples,

$$14.14\underline{/45^\circ} = 14.14\cos 45^\circ + 14.14\sin 45^\circ = 10 + j10$$
$$14.14\underline{/-45^\circ} = 14.14\cos(-45^\circ) + 14.14\sin(-45^\circ) = 10 + j(-10)$$
$$= 10 - j10$$
$$10\underline{/90^\circ} = 0 + j10$$
$$10\underline{/-90^\circ} = 0 - j10$$
$$100\underline{/30^\circ} = 86.6 + j50$$
$$100\underline{/-30^\circ} = 86.6 - j50$$
$$100\underline{/60^\circ} = 50 + j86.6$$
$$100\underline{/-60^\circ} = 50 - j86.6$$

When going from one form to the other, keep in mind whether the angle is smaller or greater than 45° and whether the *j* term is smaller or larger than the real term. For angles between 0 and 45°, the opposite side, which is the *j* term, must be smaller than the real term. For angles between 45 and 90°, the *j* term must be larger than the real term.

To summarize how complex numbers are used in AC circuits in rectangular and polar form:

1. For addition or subtraction, complex numbers must be in rectangular form. This procedure applies to the addition of impedances in a series circuit. If the series impedances are in rectangular form, combine all the real terms and the *j* terms separately. If the series impedances are in polar form, they must be converted to rectangular form to be added.
2. For multiplication and division, complex numbers are generally used in polar form because the calculations are faster. If the complex number is in rectangular form, convert to polar form. With the complex number available in both forms, you can quickly add or subtract in rectangular form and multiply or divide in polar form. Sample problems showing how to apply these methods in AC circuits are given in the following sections.

GOOD TO KNOW

It is important to note that complex numbers expressed in polar form can be added or subtracted if their phase angles are exactly the same.

24–9 Self-Review

Answers at the end of the chapter.

Convert to rectangular form.

a. $14.14\,\underline{/45^\circ}$.

b. $14.14\,\underline{/-45^\circ}$.

24–10 Complex Numbers in Series AC Circuits

Refer to Fig. 24–10. Although a circuit like this with only series resistances and reactances can be solved graphically with phasor arrows, the complex numbers show more details of the phase angles.

Z_T in Rectangular Form

The total Z_T in Fig. 24–10*a* is the sum of the impedances:

$$Z_T = 2 + j4 + 4 - j12$$
$$= 6 - j8$$

The total series impedance then is $6 - j8$. Actually, this amounts to adding all the series resistances for the real term and finding the algebraic sum of all series reactances for the *j* term.

Z_T in Polar Form

We can convert Z_T from rectangular to polar form as follows:

$$Z_T = 6 - j8$$
$$= \sqrt{36 + 64}\,\underline{/\text{arctan } -8/6}$$
$$= \sqrt{100}\,\underline{/\text{arctan } -1.33}$$
$$= 10\underline{/-53^\circ}\ \Omega$$

The angle of -53° for Z_T means that the applied voltage and the current are 53° out of phase. Specifically, this angle is θ_Z.

MultiSim **Figure 24–10** Complex numbers applied to series AC circuits. See text for analysis. (*a*) Circuit with series impedances. (*b*) Current and voltages in the circuit. (*c*) Phasor diagram of current and voltages.

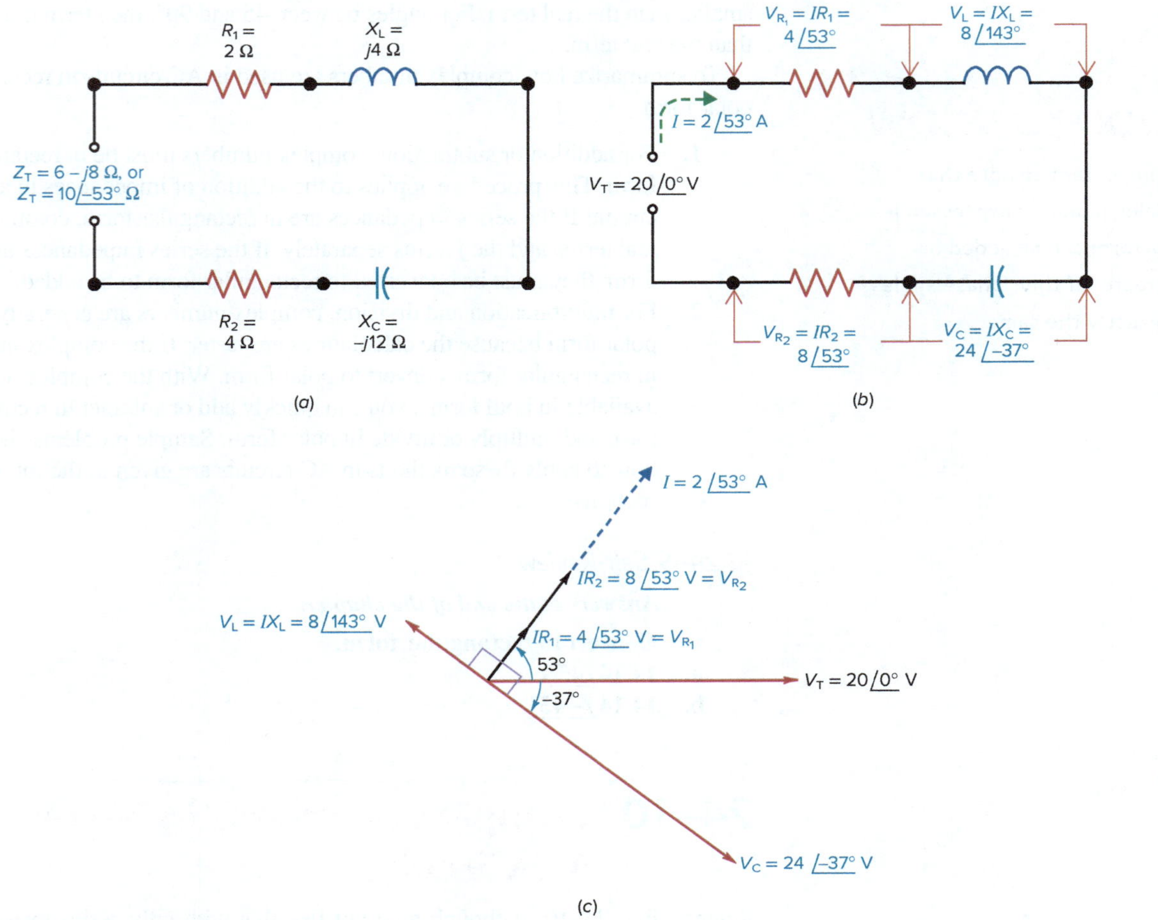

Calculating *I*

The reason for the polar form is to divide the applied voltage V_T by Z_T to calculate the current *I*. See Fig. 24–10*b*. Note that the V_T of 20 V is a real number without any *j* term. Therefore, the applied voltage is $20\angle 0^\circ$. This angle of 0° for V_T makes it the reference phase for the following calculations. We can find the current as

$$I = \frac{V_T}{Z_T} = \frac{20\angle 0^\circ}{10\angle -53^\circ} = 2\angle 0^\circ - (-53^\circ)$$
$$= 2\angle 53^\circ \text{ A}$$

Note that Z_T has a negative angle of −53° but the sign changes to +53° for *I* because of the division into a quantity with the angle of 0°. In general, the reciprocal of an angle in polar form is the same angle with opposite sign.

Phase Angle of the Circuit

The fact that *I* has an angle of +53° means that it leads V_T. The positive angle for *I* shows that the series circuit is capacitive with leading current. This angle is more than 45° because the net reactance is more than the total resistance, resulting in a tangent greater than 1.

Finding Each Voltage Drop

To calculate the voltage drops around the circuit, each resistance or reactance can be multiplied by I:

$$V_{R_1} = IR_1 = 2\underline{/53^\circ} \times 2\underline{/0^\circ} = 4\underline{/53^\circ}\ \text{V}$$
$$V_L = IX_L = 2\underline{/53^\circ} \times 4\underline{/90^\circ} = 8\underline{/143^\circ}\ \text{V}$$
$$V_C = IX_C = 2\underline{/53^\circ} \times 12\underline{/-90^\circ} = 24\underline{/-37^\circ}\ \text{V}$$
$$V_{R_2} = IR_2 = 2\underline{/53^\circ} \times 4\underline{/0^\circ} = 8\underline{/53^\circ}\ \text{V}$$

Phase Angle of Each Voltage

The phasors for these voltages are in Fig. 24–10*c*. They show the phase angles using the applied voltage V_T as the zero reference phase.

The angle of 53° for V_{R1} and V_{R2} shows that the voltage across a resistance has the same phase as I. These voltages lead V_T by 53° because of the leading current.

The angle of −37° for V_C means that it lags the generator voltage V_T by this much. However, this voltage across X_C still lags the current by 90°, which is the difference between 53° and −37°.

The angle of 143° for V_L in the second quadrant is still 90°, leading the current at 53° because 143° − 53° = 90°. With respect to the generator voltage V_T, though, the phase angle of V_L is 143°.

Total Voltage V_T Equals the Phasor Sum of the Series Voltage Drops

If we want to add the voltage drops around the circuit and find out whether they equal the applied voltage, each V must be converted to rectangular form. Then these values can be added. In rectangular form, then, the individual voltages are

$$V_{R_1} = 4\underline{/53^\circ} = 2.408 + j3.196\ \text{V}$$
$$V_L = 8\underline{/143^\circ} = -6.392 + j4.816\ \text{V}$$
$$V_C = 24\underline{/-37^\circ} = 19.176 - j14.448\ \text{V}$$
$$V_{R_2} = 8\underline{/53^\circ} = \underline{4.816 + j6.392\ \text{V}}$$
$$\text{Total } V = 20.008 - j0.044\ \text{V}$$

or converting to polar form,

$$V_T = 20\underline{/0^\circ}\text{V} \quad \text{approximately}$$

Note that for $8\underline{/143^\circ}$ in the second quadrant, the cosine is negative for a negative real term but the sine is positive for a positive j term.

24–10 Self-Review

Answers at the end of the chapter.

Refer to Fig. 24–10.

a. **What is the phase angle of I with reference to V_T?**
b. **What is the phase angle of V_L with reference to V_T?**
c. **What is the phase angle of V_L with reference to V_R?**

24–11 Complex Numbers in Parallel AC Circuits

A useful application is converting a parallel circuit to an equivalent series circuit. See Fig. 24–11, with a 10-Ω X_L in parallel with a 10-Ω R. In complex notation, R is

Figure 24–11 Complex numbers used for a parallel AC circuit to convert a parallel bank to an equivalent series impedance.

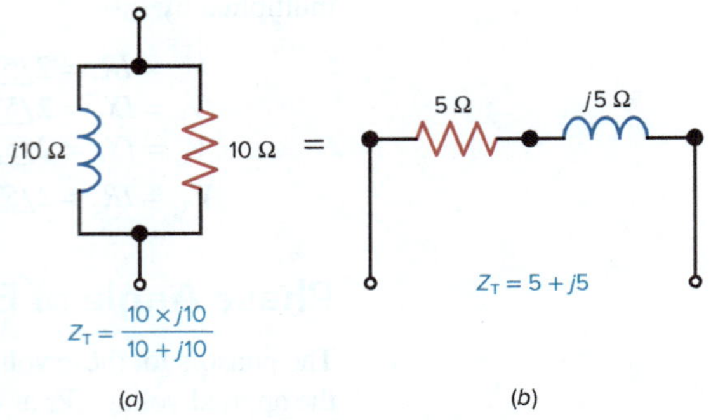

$10 + j0$ and X_L is $0 + j10$. Their combined parallel impedance Z_T equals the product divided by the sum. For Fig. 24–11*a*, then,

$$Z_T = \frac{(10 + j0) \times (0 + j10)}{(10 + j0) + (0 + j10)} = \frac{10 \times j10}{10 + j10} = \frac{j100}{10 + j10}$$
$$= \frac{j10}{1 + j1}$$

Converting to polar form for division,

$$Z_T = \frac{j100}{10 + j10} = \frac{100\angle 90^\circ}{14.14\angle +45^\circ} = 7.07\angle 45^\circ$$

Converting the Z_T of $7.07\angle 45^\circ$ into rectangular form to see its resistive and reactive components,

$$\begin{aligned}\text{Real term} &= 7.07 \cos 45^\circ \\ &= 7.07 \times 0.707 = 5 \\ j\ \text{term} &= 7.07 \sin 45^\circ \\ &= 7.07 \times 0.707 = 5\end{aligned}$$

Therefore,

$$Z_T = 7.07\angle 45^\circ \quad \text{in polar form}$$
$$Z_T = 5 + j5 \quad \text{in rectangular form}$$

The rectangular form of Z_T means that a 5-Ω *R* in series with a 5-Ω X_L is the equivalent of 10-Ω *R* in parallel with 10-Ω X_L, as shown in Fig. 24–11*b*.

Admittance *Y* and Susceptance *B*

In parallel circuits, it is usually easier to add branch currents than to combine reciprocal impedances. For this reason, branch conductance *G* is often used instead of branch resistance, where $G = 1/R$. Similarly, reciprocal terms can be defined for complex impedances. The two main types are ***admittance Y***, which is the reciprocal of impedance, and ***susceptance B***, which is the reciprocal of reactance. These reciprocals can be summarized as follows:

$$\text{Conductance} = G = \frac{1}{R}\,\text{S} \qquad \textbf{(24–4)}$$

$$\text{Susceptance} = B = \frac{1}{\pm X}\,\text{S} \qquad \textbf{(24–5)}$$

$$\text{Admittance} = Y = \frac{1}{Z}\,\text{S} \qquad \textbf{(24–6)}$$

With *R*, *X*, and *Z* in units of ohms, the reciprocals *G*, *B*, and *Y* are in siemens (S) units.

The phase angle for B or Y is the same as that of the current. Therefore, the sign is opposite from the angle of X or Z because of the reciprocal relation. An inductive branch has susceptance $-jB$, whereas a capacitive branch has susceptance $+jB$, with the same angle as a branch current.

For parallel branches of conductance and susceptance, the total admittance $Y_T = G \pm jB$. For the two branches in Fig. 24–11*a*, as an example, G is 0.1 and B is also 0.1.

In rectangular form,

$$Y_T = 0.1 - j0.1 \text{ S}$$

In polar form,

$$Y_T = 0.14 \underline{/-45^\circ} \text{ S}$$

This value for Y_T is the same as I_T with 1 V applied across Z_T of $7.07 \underline{/45^\circ}\Omega$.

As another example, suppose that a parallel circuit has 4 Ω for R in one branch and $-j4$ Ω for X_C in the other branch. In rectangular form, then, Y_T is $0.25 + j0.25$ S. Also, the polar form is $Y_T = 0.35 \underline{/45^\circ}$ S.

24–11 Self-Review

Answers at the end of the chapter.

a. **A Z of $3 + j4$ Ω is in parallel with an R of 2 Ω. State Z_T in rectangular form.**

b. **Do the same as in *a* for X_C instead of X_L.**

24–12 Combining Two Complex Branch Impedances

A common application is a circuit with two branches Z_1 and Z_2, where each is a complex impedance with both reactance and resistance. A circuit such as that in Fig. 24–12 can be solved only graphically or by complex numbers. Actually, using complex numbers is the shortest method.

The procedure here is to find Z_T as the product divided by the sum of Z_1 and Z_2. A good way to start is to state each branch impedance in both rectangular and polar forms. Then Z_1 and Z_2 are ready for addition, multiplication, and division. The solution of this circuit is as follows:

$$Z_1 = 6 + j8 = 10 \underline{/53^\circ}$$
$$Z_2 = 4 - j4 = 5.66 \underline{/-45^\circ}$$

The combined impedance is

$$Z_T = \frac{Z_1 \times Z_2}{Z_1 + Z_2}$$

MultiSim **Figure 24–12** Finding Z_T for any two complex impedances Z_1 and Z_2 in parallel. See text for solution.

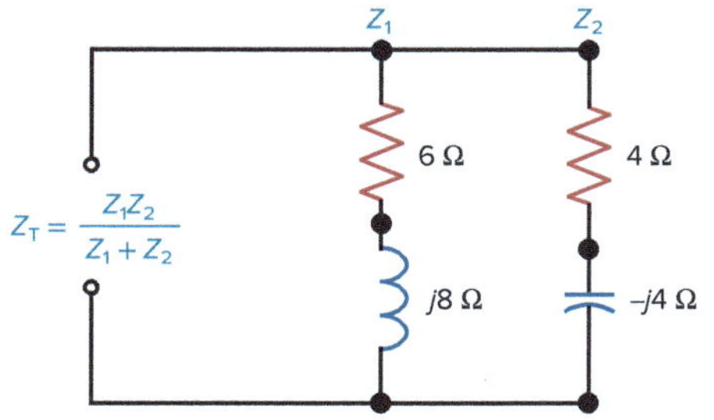

Use the polar form of Z_1 and Z_2 to multiply, but add in rectangular form:

$$Z_T = \frac{10\angle 53^\circ \times 5.66\angle -45^\circ}{6 + j8 + 4 - j4}$$
$$= \frac{56.6\angle 8^\circ}{10 + j4}$$

Converting the denominator to polar form for easier division,

$$10 + j4 = 10.8\angle 22^\circ$$

Then

$$Z_T = \frac{56.6\angle 8^\circ}{10.8\angle 22^\circ} = 5.24\angle -14^\circ\ \Omega$$

We can convert Z_T into rectangular form. The R component is $5.24 \times \cos(-14^\circ)$ or $5.24 \times 0.97 = 5.08$. Note that $\cos\theta$ is positive in the first and fourth quadrants. The j component equals $5.24 \times \sin(-14^\circ)$ or $5.24 \times (-0.242) = -1.27$. In rectangular form, then,

$$Z_T = 5.08 - j1.27$$

Therefore, this series-parallel circuit combination is equivalent to 5.08 Ω of R in series with 1.27 Ω of X. Notice that the minus j term means that the circuit is capacitive. This problem can also be done in rectangular form by rationalizing the fraction for Z_T.

24–12 Self-Review

Answers at the end of the chapter.

Refer to Fig. 24–12.

a. Add $(6 + j8) + (4 - j4)$ for the sum of Z_1 and Z_2.

b. Multiply $10\angle 53^\circ \times 5.66\angle -45^\circ$ for the product of Z_1 and Z_2.

24–13 Combining Complex Branch Currents

Figure 24–13 gives an example of finding I_T for two branch currents. The branch currents can just be added in rectangular form for the total I_T of parallel branches. This method corresponds to adding series impedances in rectangular form to find Z_T. The rectangular form is necessary for the addition of phasors.

Adding the branch currents in Fig. 24–13,

$$I_T = I_1 + I_2$$
$$= (6 + j6) + (3 - j4)$$
$$= 9 + j2\text{ A}$$

Figure 24–13 Finding I_T for two branch currents in parallel.

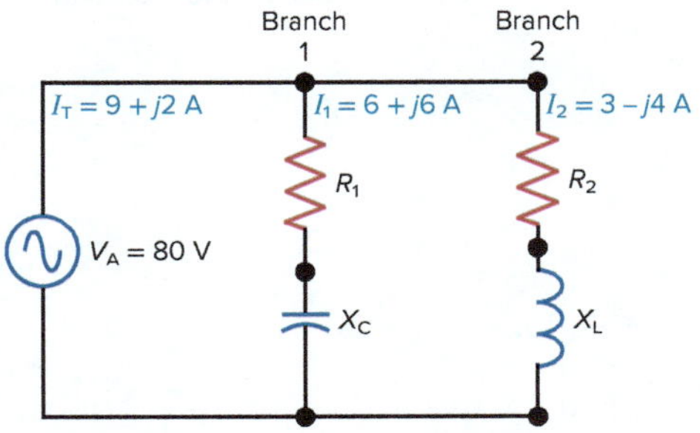

Note that I_1 has $+j$ for the $+90°$ of capacitive current, and I_2 has $-j$ for inductive current. These current phasors have signs opposite from their reactance phasors.

In polar form, the I_T of $9 + j2$ A is calculated as the phasor sum of the branch currents.

$$I_T = \sqrt{9^2 + 2^2} = \sqrt{85}$$
$$= 9.22 \text{ A}$$
$$\tan \theta = \frac{2}{9} = 0.222$$
$$\theta_1 = \arctan (0.22)$$
$$= 12.53°$$

Therefore, I_T is $9 + j2$ A in rectangular form or $9.22\angle 12.53°$ A in polar form. The complex currents for any number of branches can be added in rectangular form.

24–13 Self-Review

Answers at the end of the chapter.

a. **Find I_T in rectangular form for I_1 of $0 + j2$ A and I_2 of $4 + j3$ A.**

b. **Find I_T in rectangular form for I_1 of $6 + j7$ A and I_2 of $3 - j9$ A.**

24–14 Parallel Circuit with Three Complex Branches

Because the circuit in Fig. 24–14 has more than two complex impedances in parallel, use the method of branch currents. There will be several conversions between rectangular and polar form, since addition must be in rectangular form, but division is easier in polar form. The sequence of calculations is

1. Convert each branch impedance to polar form. This is necessary for dividing into the applied voltage V_A to calculate the individual branch currents. If V_A is not given, any convenient value can be assumed. Note that V_A has a phase angle of 0° because it is the reference.
2. Convert the individual branch currents from polar to rectangular form so that they can be added for the total line current. This step is necessary because the resistive and reactive components must be added separately.
3. Convert the total line current from rectangular to polar form for dividing into the applied voltage to calculate Z_T.
4. The total impedance can remain in polar form with its magnitude and phase angle or can be converted to rectangular form for its resistive and reactive components.

Figure 24–14 Finding Z_T for any three complex impedances in parallel. See text for solution by means of branch currents.

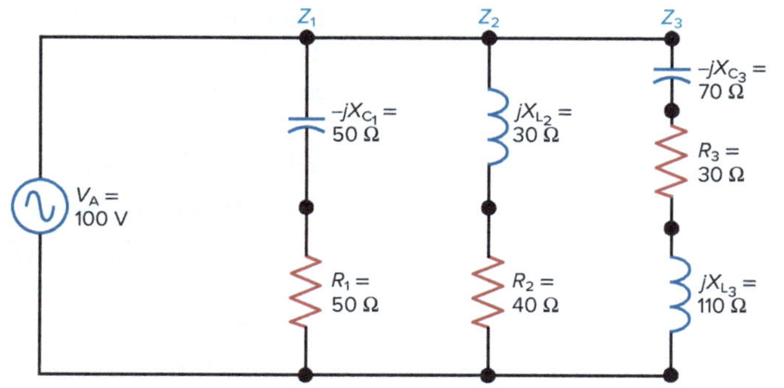

These steps are used in the following calculations to solve the circuit in Fig. 24–14. All the values are in A, V, or Ω units.

Branch Impedances

Each Z is converted from rectangular form to polar form:

$$Z_1 = 50 - j50 = 70.7\,\underline{/-45^\circ}$$
$$Z_2 = 40 + j30 = 50\,\underline{/+37^\circ}$$
$$Z_3 = 30 + j40 = 50\,\underline{/+53^\circ}$$

Branch Currents

Each I is calculated as V_A divided by Z in polar form:

$$I_1 = \frac{V_A}{Z_1} = \frac{100\,\underline{/0^\circ}}{70.7\,\underline{/-45^\circ}} = 1.414\,\underline{/+45^\circ} = 1 + j1$$

$$I_2 = \frac{V_A}{Z_2} = \frac{100\,\underline{/0^\circ}}{50\,\underline{/+37^\circ}} = 2.00\,\underline{/-37^\circ} = 1.6 - j1.2$$

$$I_3 = \frac{V_A}{Z_3} = \frac{100\,\underline{/0^\circ}}{50\,\underline{/+53^\circ}} = 2.00\,\underline{/-53^\circ} = 1.2 - j1.6$$

The polar form of each I is converted to rectangular form for addition of the branch currents.

Total Line Current

In rectangular form,

$$\begin{aligned} I_T &= I_1 + I_2 + I_3 \\ &= (1 + j1) + (1.6 - j1.2) + (1.2 - j1.6) \\ &= 1 + 1.6 + 1.2 + j1 - j1.2 - j1.6 \\ &= 3.8 - j1.8 \end{aligned}$$

Converting $3.8 - j1.8$ into polar form,

$$I_T = 4.2\,\underline{/-25.4^\circ}$$

Total Impedance

In polar form,

$$\begin{aligned} Z_T &= \frac{V_A}{I_T} = \frac{100\,\underline{/0^\circ}}{4.2\,\underline{/-25.4^\circ}} \\ &= 23.8\,\underline{/+25.4^\circ}\ \Omega \end{aligned}$$

Converting $23.8\,\underline{/+25.4^\circ}$ into rectangular form,

$$Z_T = 21.5 + j10.2\ \Omega$$

Therefore, the complex AC circuit in Fig. 24–14 is equivalent to the combination of 21.5 Ω of R in series with 10.2 Ω of X_L. The circuit is inductive.

This problem can also be done by combining Z_1 and Z_2 in parallel as $Z_1Z_2/(Z_1 + Z_2)$. Then combine this value with Z_3 in parallel to find the total Z_T of the three branches.

24–14 Self-Review

Answers at the end of the chapter.

Refer to Fig. 24–14.

a. State Z_2 in rectangular form for branch 2.
b. State Z_2 in polar form.
c. Find I_2.

Summary

- In complex numbers, resistance R is a real term and reactance is a j term. Thus, an 8-Ω R is 8; an 8-Ω X_L is $j8$; an 8-Ω X_C is $-j8$. The general form of a complex impedance with series resistance and reactance, then, is $Z_T = R \pm jX$, in rectangular form.
- The same notation can be used for series voltages where $V_T = V_R \pm jV_X$.
- For branch currents $I_T = I_R \pm jI_X$, but the reactive branch currents have signs opposite from impedances. Capacitive branch current is jI_C, and inductive branch current is $-jI_L$.
- The complex branch currents are added in rectangular form for any number of branches to find I_T.
- To convert from rectangular to polar form: $R \pm jX = Z_T \underline{/\theta}$. The angle is θ_Z. The magnitude of Z_T is $\sqrt{R^2 + X^2}$. Also, θ_Z is the angle with tan $= X/R$.
- To convert from polar to rectangular form, $Z_T \underline{/\theta_Z} = R \pm jX$, where R is Z_T cos θ_Z and the j term is Z_T sin θ_Z. A positive angle has a positive j term; a negative angle has a negative j term. Also, the angle is more than 45° for a j term larger than the real term; the angle is less than 45° for a j term smaller than the real term.
- The rectangular form must be used for addition or subtraction of complex numbers.
- The polar form is usually more convenient in multiplying and dividing complex numbers. For multiplication, multiply the magnitudes and add the angles; for division, divide the magnitudes and subtract the angles.
- To find the total impedance Z_T of a series circuit, add all resistances for the real term and find the algebraic sum of the reactances for the j term. The result is $Z_T = R \pm jX$. Then convert Z_T to polar form for dividing into the applied voltage to calculate the current.
- To find the total impedance Z_T of two complex branch impedances Z_1 and Z_2 in parallel, Z_T can be calculated as $Z_1Z_2/(Z_1 + Z_2)$.

Important Terms

Admittance, *Y*—the reciprocal of impedance or $Y = 1/Z$. The unit of admittance is siemens (S).

Complex number—the combination of a real and an imaginary term. Complex numbers can be expressed in either rectangular or polar form.

Imaginary number—a number on the j axis is called an imaginary number because it is not on the real axis. (Any quantity at right angles to the zero axis, or at 90°, is considered on the j axis.)

***j* operator**—the j operator indicates a phase angle of either plus or minus 90°. For example, $+j1$ kΩ indicates 1 kΩ of inductive reactance, X_L, on the $+j$ axis of 90°. Similarly, $-j1$ kΩ indicates 1 kΩ of capacitive reactance, X_C, on the $-j$ axis of $-90°$.

Polar form—the form of a complex number that specifies its magnitude and phase angle. The general form of a complex number expressed in polar form is $r\angle\theta$ where r is the magnitude of the resultant phasor and θ is the phase angle with respect to the horizontal or real axis.

Real number—any number on the horizontal axis, either positive (0°) or negative (180°).

Rectangular form—the form of a complex number that specifies the real and imaginary terms individually. The general form of a complex number specified in rectangular form is $a \pm jb$ where a is the real term and $\pm jb$ is the imaginary term at $\pm 90°$.

Susceptance, *B*—the reciprocal of reactance or $B = 1/\pm X$. The unit of susceptance is siemens (S).

Related Formulas

$Z_T = Z_1 + Z_2 + Z_3 + \cdots +$ etc. (Series impedances)

$\frac{1}{Z_T} = \frac{1}{Z_1} + \frac{1}{Z_2} + \frac{1}{Z_3} + \cdots +$ etc. (Parallel impedances)

$Z_T = \frac{Z_1Z_2}{Z_1 + Z_2}$ (Two parallel impedances)

Conductance $= G = \frac{1}{R}$ S

Susceptance $= B = \frac{1}{\pm X}$ S

Admittance $= Y = \frac{1}{Z}$ S

Self-Test

Answers at the back of the book.

1. **Numbers on the horizontal axis are called**
   a. imaginary numbers.
   b. conjugate numbers.
   c. real numbers.
   d. complex numbers.

2. **Numbers on the plus or minus *j* axis are called**
   a. imaginary numbers.
   b. conjugate numbers.
   c. real numbers.
   d. complex numbers.

3. **A value of $-j500\ \Omega$ represents**
   a. 500 Ω of inductive reactance.
   b. 500 Ω of capacitive reactance.
   c. 500 Ω of resistance.
   d. 500 Ω of conductance.

4. **An inductive reactance of 20 Ω can be expressed as**
   a. +20 Ω.
   b. $j^2 20\ \Omega$.
   c. $-j20\ \Omega$.
   d. $+j20\ \Omega$.

5. **A series AC circuit consists of 10 Ω of resistance and 15 Ω of inductive reactance. What is the impedance of this circuit when expressed in polar form?**
   a. $15\ \Omega + j10\ \Omega$.
   b. $18\angle 56.3°\ \Omega$.
   c. $18\angle 33.7°\ \Omega$.
   d. $25\angle 56.3°\ \Omega$.

6. **An AC circuit has an impedance, *Z*, of $50\angle -36.87°\ \Omega$. What is the impedance of this circuit when expressed in rectangular form?**
   a. $40\ \Omega - j30\ \Omega$.
   b. $40\ \Omega + j30\ \Omega$.
   c. $30\ \Omega + j40\ \Omega$.
   d. $30\ \Omega - j40\ \Omega$.

7. **When adding or subtracting complex numbers, all numbers must be in**
   a. polar form.
   b. scientific notation.
   c. rectangular form.
   d. none of the above.

8. **When multiplying complex numbers in polar form,**
   a. multiply the magnitudes and subtract the phase angles.
   b. multiply the magnitudes and add the phase angles.
   c. multiply the angles and add the magnitudes.
   d. multiply both the magnitudes and phase angles.

9. **When dividing complex numbers in polar form,**
   a. divide the magnitudes and subtract the phase angles.
   b. divide the magnitudes and add the phase angles.
   c. divide the phase angles and subtract the magnitudes.
   d. divide both the magnitudes and phase angles.

10. **What is the admittance, *Y*, of a parallel branch whose impedance is $200\angle -63.43°\ \Omega$?**
   a. $5\angle -63.43°$ mS.
   b. $5\angle -26.57°$ mS.
   c. $200\angle -63.43$ mS.
   d. $5\angle -63.43$ mS.

11. **In complex numbers, j^2 corresponds to**
   a. 180°.
   b. −1.
   c. −90°.
   d. both a and b.

12. **Susceptance, *B*, is**
   a. the reciprocal of impedance.
   b. the reciprocal of reactance.
   c. the reciprocal of resistance.
   d. the same as conductance.

13. **A branch current of $+j250$ mA represents**
   a. 250 mA of inductive current.
   b. 250 mA of resistive current.
   c. 250 mA of capacitive current.
   d. 250 mA of in-phase current.

14. **A parallel AC circuit has an admittance, Y_T, of 6 mS $+j8$ mS. What is the impedance, *Z*, in polar form?**
   a. $10\angle 45°$ kΩ.
   b. $100\angle 53.13°\ \Omega$.
   c. $100\angle -53.13°\ \Omega$.
   d. $14\angle 53.13°\ \Omega$.

15. **What is the resistance, *R*, of an AC circuit whose impedance, *Z*, is $300\angle 53.13°\ \Omega$?**
   a. 240 Ω.
   b. 180 Ω.
   c. 270 Ω.
   d. 60 Ω.

Essay Questions

1. Give the mathematical operator for the angles of 0°, 90°, 180°, 270°, and 360°.
2. Define the sine, cosine, and tangent functions of an angle.
3. Compare the following combinations: resistance *R* and conductance *G*; reactance *X* and susceptance *B*; impedance *Z* and admittance *Y*.
4. What are the units for admittance *Y* and susceptance *B*?
5. Why do Z_T and I_T for a circuit have angles with opposite signs?

Problems

SECTION 24–1 POSITIVE AND NEGATIVE NUMBERS

24–1 What is the phase angle for
- a. positive numbers on the horizontal or *x* axis?
- b. negative numbers on the horizontal or *x* axis?

24–2 What factor corresponds to a phase angle of
- a. 0°?
- b. 180°?

SECTION 24–2 THE *j* OPERATOR

24–3 What is the name of the axis at right angles to the real or horizontal axis?

24–4 What is the phase angle for numbers on the
- a. $+j$ axis?
- b. $-j$ axis?

24–5 What is the name given to numbers on the
- a. horizontal axis?
- b. *j* axis?

24–6 List the phase angle for each of the following factors:
- a. +1.
- b. −1.
- c. $+j$.
- d. $-j$.
- e. j^2.
- f. j^3.

24–7 What do the following numbers mean?
- a. $j25$.
- b. $-j36$.

SECTION 24–3 DEFINITION OF A COMPLEX NUMBER

24–8 What is the definition of a complex number?

24–9 In what form is the complex number 100 Ω + *j*400 Ω?

24–10 For the complex number $8 + j6$, identify the real and imaginary terms.

24–11 In each of the following examples, identify when the phase angle is less than 45°, greater than 45°, or equal to 45°:
- a. $3 + j5$.
- b. $180 + j60$.
- c. $40 - j40$.
- d. $100 - j120$.
- e. $40 + j30$.

SECTION 24–4 HOW COMPLEX NUMBERS ARE APPLIED TO AC CIRCUITS

24–12 Is a resistance value considered a real or imaginary number?

24–13 What is the phase angle of a positive real number?

24–14 Express the following quantities using the *j* operator.
- a. 50 Ω of X_L.
- b. 100 Ω of X_C.
- c. V_L of 25 V.
- d. V_C of 15 V.
- e. 4 A of I_L.
- f. 600 mA of I_C.

SECTION 24–5 IMPEDANCE IN COMPLEX FORM

24–15 Express the following impedances in rectangular form.
- a. 10 Ω of *R* in series with 20 Ω of X_L.
- b. 10 Ω of X_L in series with 15 Ω of *R*.
- c. 0 Ω of *R* in series with 1 kΩ of X_C.
- d. 1.5 kΩ of *R* in series with 2 kΩ of X_C.
- e. 150 Ω of *R* in series with 0 Ω of *X*.
- f. 75 Ω of *R* in series with 75 Ω of X_C.

24–16 In the following examples, combine the real terms and *j* terms separately, and express the resultant values in rectangular form.
- a. 40 Ω of *R* in series with 30 Ω of X_C and 60 Ω of X_L.
- b. 500 Ω of *R* in series with 150 Ω of X_L and 600 Ω of X_C.
- c. 1 kΩ of X_C in series with 2 kΩ of X_L, 3 kΩ of *R*, and another 2 kΩ of *R*.

SECTION 24–6 OPERATIONS WITH COMPLEX NUMBERS

24–17 Add the following complex numbers:
- a. $(6 + j9) + (9 + j6)$.
- b. $(25 + j10) + (15 - j30)$.
- c. $(0 + j100) + (200 + j50)$.
- d. $(50 - j40) + (40 - j10)$.
- e. $(12 + j0) + (24 - j48)$.

24–18 Multiply or divide the following *j* terms and real numbers:
- a. $j10 \times 5$.
- b. $-j60 \times (-4)$.
- c. $-j8 \times 9$.
- d. $j4 \times (-8)$.
- e. $j100 \div 20$.
- f. $-j600 \div 6$.
- g. $-j400 \div (-20)$.
- h. $j16 \div (-8)$.

24–19 Multiply or divide the following *j* terms.
- a. $j8 \times j9$.
- b. $-j12 \times j5$.
- c. $-j7 \times (-j4)$.
- d. $j3 \times j8$.
- e. $j12 \div j6$.
- f. $-j100 \div j8$.
- g. $-j250 \div (-j10)$.
- h. $j1000 \div (-j40)$.

24–20 Multiply the following complex numbers:

a. $(3 + j5) \times (4 + j3)$.
b. $(6 - j8) \times (8 + j6)$.
c. $(12 + j3) \times (5 + j9)$.
d. $(4 - j2) \times (8 - j12)$.

24–21 Divide the following complex numbers:

a. $(15 - j3) \div (10 + j4)$.
b. $(6 + j3) \div (24 - j8)$.
c. $(10 + j2) \div (20 - j4)$.
d. $(2 - j6) \div (4 - j4)$.

SECTION 24–7 MAGNITUDE AND ANGLE OF A COMPLEX NUMBER

24–22 Calculate the resultant magnitude and phase angle for each of the following complex numbers expressed in rectangular form:

a. $5 - j8$.
b. $10 + j15$.
c. $100 + j50$.
d. $20 - j35$.
e. $150 + j200$.
f. $75 - j75$.
g. $0 + j100$.
h. $100 + j0$.
i. $10 - j40$.
j. $2000 - j6000$.

SECTION 24–8 POLAR FORM OF COMPLEX NUMBERS

24–23 Convert the following complex numbers, written in rectangular form, into polar form:

a. $10 + j10$.
b. $8 - j10$.
c. $12 + j18$.
d. $140 - j55$.

24–24 Multiply the following complex numbers expressed in polar form:

a. $50\angle 30° \times 2\angle -65°$.
b. $3\angle -15° \times 5\angle -40°$.
c. $9\angle 20° \times 8\angle 30°$.
d. $15\angle -70° \times 4\angle 10°$.
e. $2 \times 150\angle -45°$.
f. $1000\angle -90° \times 0.5\angle 90°$.
g. $40\angle 25° \times 1.5$.

24–25 Divide the following complex numbers expressed in polar form:

a. $48\angle -80° \div 16\angle 45°$.
b. $120\angle 60° \div 24\angle -90°$.
c. $172\angle -45° \div 43\angle -45°$.
d. $210\angle 22° \div 45\angle -44°$.
e. $180\angle 75° \div 6$.
f. $750\angle 80° \div 30$.
g. $2500\angle 50° \div 200$.

SECTION 24–9 CONVERTING POLAR TO RECTANGULAR FORM

24–26 Convert the following numbers expressed in polar form into rectangular form:

a. $50\angle 45°$.
b. $100\angle 60°$.
c. $250\angle -53.13°$.
d. $1000\angle -30°$.
e. $12\angle 0°$.
f. $180\angle -78.5°$.
g. $5\angle 36.87°$.
h. $25\angle 15°$.
i. $45\angle 100°$.
j. $60\angle -90°$.
k. $100\angle 53.13°$.
l. $40\angle 90°$.

SECTION 24–10 COMPLEX NUMBERS IN SERIES AC CIRCUITS

24–27 In Fig. 24–15, state

a. Z_T in rectangular form.
b. Z_T in polar form.
c. I in polar form.
d. V_R in polar form.
e. V_L in polar form.
f. V_C in polar form.

Figure 24–15

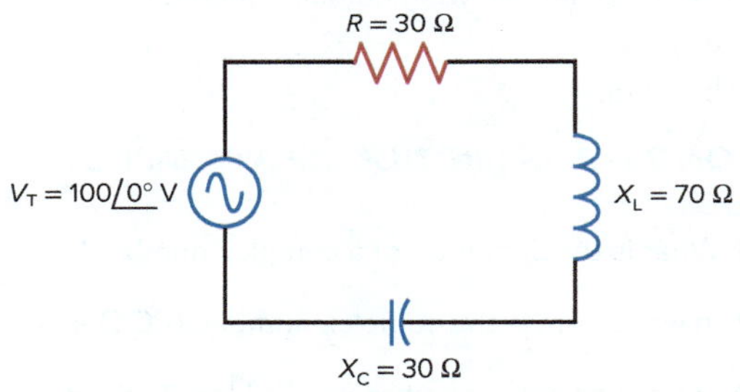

24–28 Repeat Prob. 24–27 for Fig. 24–16.

Figure 24–16

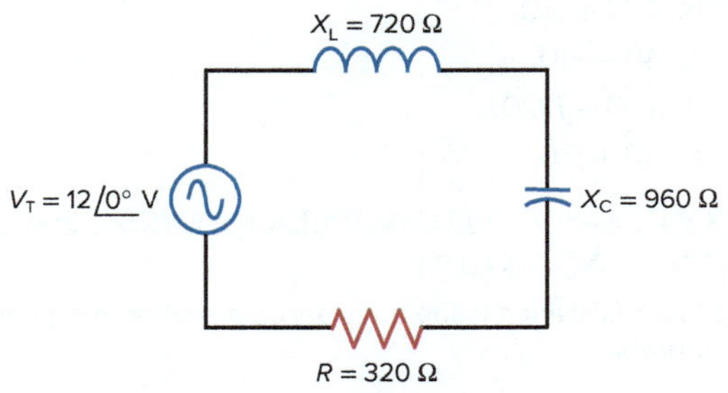

SECTION 24–11 COMPLEX NUMBERS IN PARALLEL AC CIRCUITS

24–29 In Fig. 24–17, state the total impedance, Z_T, in both polar and rectangular form. Use Formula 24–3) to solve for Z_T.

Figure 24–17

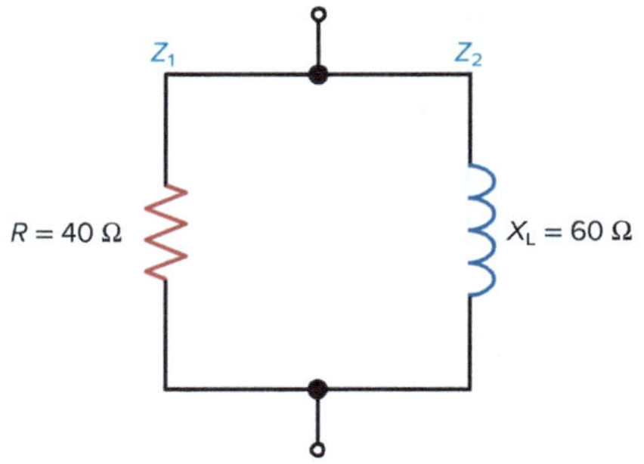

24–30 In Fig. 24–17, state the total admittance, Y_T, in both polar and rectangular form. Using the polar form of Y_T, solve for Z_T.

24–31 In Fig. 24–18, state the total admittance, Y_T, in both rectangular and polar form. Solve for Z_T from Y_T.

Figure 24–18

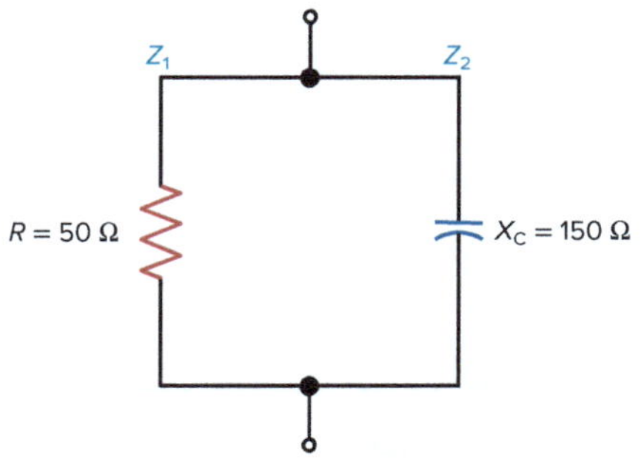

24–32 In Fig. 24–18, solve for the total impedance, Z_T, using Formula 24–3). State Z_T in both polar and rectangular form.

24–33 Draw the equivalent series circuit for the circuit in

a. Fig. 24–17.

b. Fig. 24–18.

SECTION 24–12 COMBINING TWO COMPLEX BRANCH IMPEDANCES

24–34 In Fig. 24–19,

a. state Z_1 in both rectangular and polar form.

b. state Z_2 in both rectangular and polar form.

c. state Z_T in both rectangular and polar form.

Figure 24–19

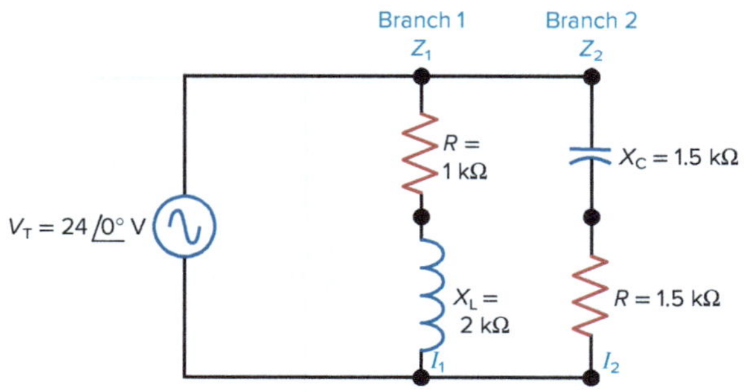

24–35 Repeat Prob. 24–34 for Fig. 24–20.

Figure 24–20

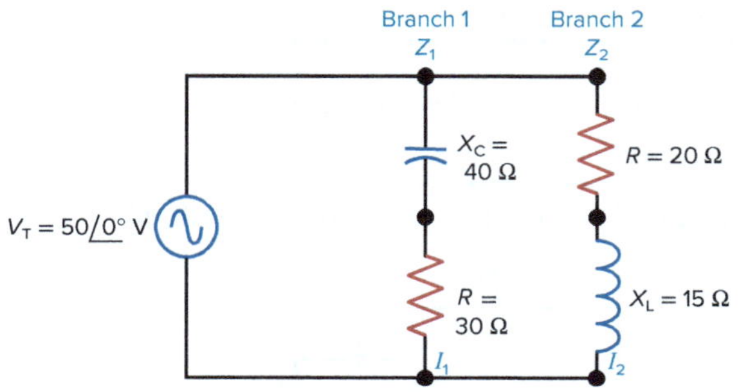

SECTION 24–13 COMBINING COMPLEX BRANCH CURRENTS

24–36 In Fig. 24–19,

a. state the branch current, I_1, in both polar and rectangular form.

b. state the branch current, I_2, in both polar and rectangular form.

c. state the total current, I_T, in both polar and rectangular form.

24–37 Repeat Prob. 24–36 for Fig. 24–20.

SECTION 24–14 PARALLEL CIRCUIT WITH THREE COMPLEX BRANCHES

24–38 In Fig. 24–21,

a. state Z_1 in polar form.

b. state Z_2 in polar form.

c. state Z_3 in polar form.

d. state I_1 in polar and rectangular form.

e. state I_2 in polar and rectangular form.

f. state I_3 in polar and rectangular form.

g. state I_T in polar and rectangular form.

h. state Z_T in polar and rectangular form.

Figure 24–21

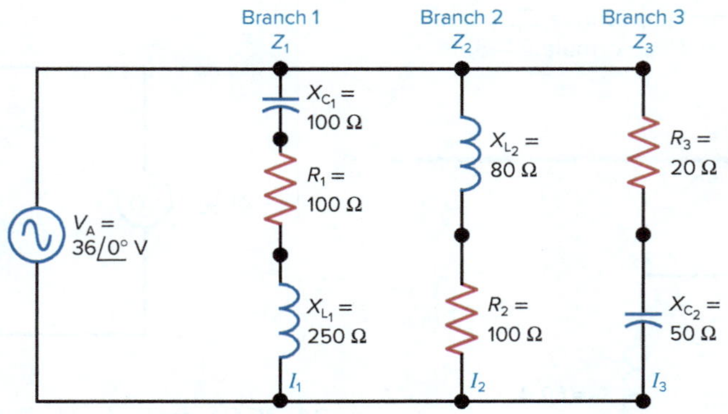

Critical Thinking

24–39 In Fig. 24–22, calculate the input voltage V_{in}, in polar form.

Figure 24–22 Circuit for Critical Thinking Prob. 24–39.

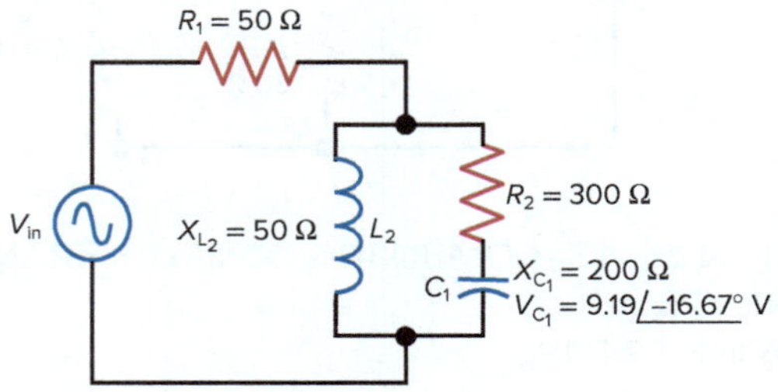

Answers to Self-Reviews

24–1 **a.** 0°
b. 180°

24–2 **a.** 90°
b. −90 or 270°

24–3 **a.** true
b. true

24–4 **a.** $j3$ kΩ
b. $-j5$ mA

24–5 **a.** $4 + j7$
b. $0 - j7$

24–6 **a.** $5 + j7$
b. $4 + j6$

24–7 **a.** 14.14 Ω
b. 45°

24–8 **a.** $12\angle 50°$
b. $3\angle -10°$

24–9 **a.** $10 + j10$
b. $10 - j10$

24–10 **a.** 53°
b. 143°
c. 90°

24–11 **a.** $(6 + j8)/(5 + j4)$
b. $(6 - j8)/(5 - j4)$

24–12 **a.** $10 + j4$
b. $56.6\angle 8°$

24–13 **a.** $4 + j5$ A
b. $9 - j2$ A

24–14 **a.** $40 + j30$
b. $50\angle 37°$ Ω
c. $2\angle -37°$ A

Laboratory Application Assignment

In this lab application assignment, you will examine how complex numbers can be used to solve an AC circuit containing both series and parallel impedances. More specifically, you will use complex numbers to solve for the magnitude and phase angle of the output voltage in a series-parallel *RC* network. Finally, you will build the *RC* network and confirm, through measurement, that your calculations are correct.

Equipment: Obtain the following items from your instructor.

- Function generator
- Oscilloscope
- Two 1-k**Ω** resistors and two 0.01-μF capacitors

Circuit Calculations

Examine the *RC* network in Fig. 24–23. Note the frequency, magnitude, and phase angle of the input voltage, V_{in}. With the use of complex numbers, calculate the magnitude and phase angle of the output voltage, V_{out}. Show all your work in the space provided below. Circle your final answer.

Hint: *Convert the parallel connection of* R_2 *and* C_2 *into an equivalent series circuit.*

Circuit Measurements

Construct the circuit in Fig. 24–23. Connect channel 1 of the oscilloscope to measure the input voltage and channel 2 to measure the output voltage. Set the amplitude of the input voltage to 10 $V_{p\text{-}p}$, and adjust the frequency to approximately 16 kHz. Measure and record the magnitude of the output voltage, V_{out}. V_{out} = ________

What is the ratio of V_{out}/V_{in} at 16 kHz?_____/_____

While viewing both V_{in} and V_{out} on the oscilloscope, measure and record the phase angle, θ, that exists between them. θ = ________

Adjust the frequency dial above and below 16 kHz. What happens to the magnitude of the output voltage as the frequency is increased and decreased from 16 kHz? ________

What happens to the phase relationship between V_{in} and V_{out} as the frequency is increased and decreased from 16 kHz? ________

Figure 24–23

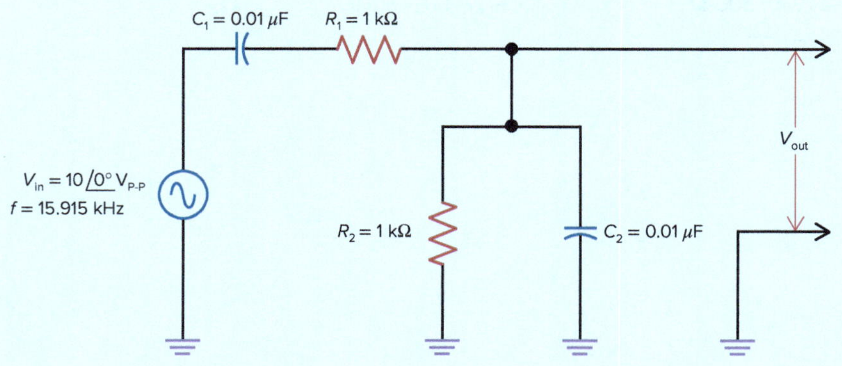

Cumulative Review Summary (Chapters 23–24)

- Reactances X_C and X_L are opposite. In series, the ohms of X_C and X_L cancel. In parallel, the branch currents I_C and I_L cancel.
- As a result, circuits with R, X_C, and X_L can be reduced to one net reactance X and one equivalent R.
- In series circuits, the net X is added with the total R by phasors for the impedance: $Z_T = \sqrt{R^2 + X^2}$. Then $I = V_T/Z_T$.
- For branch currents in parallel circuits, the net I_X is added with I_R by phasors for the total line current: $I_T = \sqrt{I_R^2 + I_X^2}$. Then $Z_{EQ} = V/I_T$.
- The characteristics of ohms of R, X_C, X_L, and Z in AC circuits are compared in Table 23–1.
- In AC circuits with reactance, the real power in watts equals I^2R. This value equals $VI \cos \theta$, where θ is the phase angle of the circuit and $\cos \theta$ is the power factor.
- The wattmeter uses an AC meter movement to read V and I at the same time, measuring watts of real power.
- In complex numbers, R is a real term at 0° and reactance is a $\pm j$ term at $\pm 90°$. In rectangular form, $Z_T = R \pm jX$. For example, 10 Ω of R in series with 10 Ω of X_L is $10 + j10$ Ω.
- The polar form of $10 + j10$ Ω is $14\angle 45°$ Ω. The angle of 45° is arctan X/R. The magnitude of 14 is $\sqrt{R^2 + X^2}$.
- The rectangular form of complex numbers must be used for addition and subtraction. Add or subtract the real terms and the j terms separately.
- The polar form of complex numbers is easier for multiplication and division. For multiplication, multiply the magnitudes and add the angles. For division, divide the magnitudes and subtract the angle of the divisor.
- In double-subscript notation for a voltage, such as V_{BE}, the first letter in the subscript is the point of measurement with respect to the second letter. So V_{BE} is the base voltage with respect to the emitter in a transistor.

Cumulative Self-Test

Answers at the back of the book.

Fill in the numerical answer.

1. An AC circuit with 100-Ω R_1 in series with 200-Ω R_2 has R_T of ________ Ω.
2. With 100-Ω X_{L_1} in series with 200-Ω X_{L_2}, the total X_L is ________ Ω.
3. For 200-Ω X_{C_1} in series with 100-Ω X_{C_2}, the total X_C is ________ Ω.
4. Two X_C branches of 500 Ω each in parallel have combined X_C of ________ Ω.
5. Two X_L branches of 500 Ω each in parallel have combined X_L of ________ Ω.
6. A 500-Ω X_L is in series with a 300-Ω X_C. The net X_L is ________ Ω.
7. For 500-Ω X_C in series with 300-Ω X_{L_1} the net X_C is ________ Ω.
8. A 10-Ω X_L is in series with a 10-Ω R. The total Z_T is ________ Ω.
9. With a 10-Ω X_C in series with a 10-Ω R, the total Z_T is ________ Ω.
10. With 14 V applied across 14-Ω Z_T, the I is ________ A.
11. For 10-Ω X_L and 10-Ω R in series, the phase angle θ is ________ degrees.
12. For 10-Ω X_C and 10-Ω R in series, the phase angle θ is ________ degrees.
13. A 10-Ω X_L and a 10-Ω R are in parallel across 10 V. The amount of each branch I is ________ A.
14. In Question 13, the total line current I_T equals ________ A.
15. In Questions 13 and 14, Z_T of the parallel branches equals ________ Ω.
16. With 120 V, an I of 10 A, and θ of 60°, a wattmeter reads ________ W.
17. The Z of $4 + j4$ Ω converted to polar form is ________ Ω.
18. The impedance value of $8\angle 40° / 2\angle 30°$ is equal to ________ Ω.

Answer True/False.

19. In an AC circuit with X_C and R in series, if the frequency is raised, the current will increase.
20. In an AC circuit with X_L and R in series, if the frequency is increased, the current will be reduced.
21. The volt-ampere is a unit of apparent power.
22. The polar form of complex numbers is best for adding impedance values.

Design credits: Chapter Opener Background Image, and Multisim: Stockbyte/Getty Images

chapter

25

Resonance

This chapter explains how X_L and X_C can be combined to favor one particular frequency, the resonant frequency to which the *LC* circuit is tuned. The resonance effect occurs when the inductive and capacitive reactances are equal. In a series RLC circuit, the impedance is minimum and the current is maximum at the resonant frequency. This is because the equal reactances cancel each other leaving only the resistance *R* to oppose the current. The opposite effect occurs in a parallel RLC circuit. In this case, the equal reactances result in equal reactive branch currents that cancel each other at the resonant frequency. Therefore, a parallel RLC circuit has minimal line current and maximum impedance at the resonant frequency.

In radio frequency (rf) circuits, the main application of resonance is for tuning to an AC signal of the desired frequency. Applications of resonance include tuning in communication receivers, transmitters, and electronic equipment in general.

Tuning by means of the resonant effect provides a practical application of selectivity. The resonant circuit can be operated to select a particular frequency for the output with many different frequencies at the input. ■

Chapter Outline

Chapter Objectives

After studying this chapter, you should be able to

- *Define* the term *resonance*.
- *List* four characteristics of a series resonant circuit.
- *List* three characteristics of a parallel resonant circuit.
- *Explain* how the resonant frequency formula is derived.
- *Calculate* the *Q* of a series or parallel resonant circuit.
- *Calculate* the equivalent impedance of a parallel resonant circuit.
- *Explain* what is meant by the *bandwidth* of a resonant circuit.
- *Calculate* the bandwidth of a series or parallel resonant circuit.
- *Explain* the effect of varying *L* or *C* in tuning an *LC* circuit.
- *Calculate* *L* or *C* for a resonant circuit.

Important Terms

antiresonance
bandwidth
damping
flywheel effect
half-power points
Q of a resonant circuit
resonant frequency
tank circuit
tuning

GOOD TO KNOW

A resonant *LC* circuit generally has hundreds or thousands of signals present at its input, but only one is selected to be present at its output. For example, the antenna for an FM receiver intercepts the signals of many different FM broadcast stations, but by tuning an *LC* circuit to resonance, the listener can select only the station he or she would like to listen to.

25–1 The Resonance Effect

Inductive reactance increases as the frequency is increased, but capacitive reactance decreases with higher frequencies. Because of these opposite characteristics, for any *LC* combination, there must be a frequency at which the X_L equals the X_C because one increases while the other decreases. This case of equal and opposite reactances is called *resonance,* and the AC circuit is then a *resonant circuit.*

Any *LC* circuit can be resonant. It all depends on the frequency. At the **resonant frequency**, an *LC* combination provides the resonance effect. Off the resonant frequency, either below or above, the *LC* combination is just another AC circuit.

The frequency at which the opposite reactances are equal is the *resonant frequency.* This frequency can be calculated as $f_r = 1/(2\pi\sqrt{LC})$, where *L* is the inductance in henrys, *C* is the capacitance in farads, and f_r is the resonant frequency in hertz that makes $X_L = X_C$.

In general, we can say that large values of *L* and *C* provide a relatively low resonant frequency. Smaller values of *L* and *C* allow higher values for f_r. The resonance effect is most useful for radio frequencies, where the required values of microhenrys for *L* and picofarads for *C* are easily obtained.

The most common application of resonance in rf circuits is called ***tuning***. In this use, the *LC* circuit provides maximum voltage output at the resonant frequency, compared with the amount of output at any other frequency either below or above resonance. This idea is illustrated in Fig. 25–1, where the *LC* circuit resonant at 1000 kHz magnifies the effect of this particular frequency. The result is maximum output at 1000 kHz, compared with lower or higher frequencies.

Tuning in radio and television receivers is an application of resonance. When you tune a radio to one station, the *LC* circuits are tuned to resonance for that particular carrier frequency. Also, when you tune a television receiver to a particular channel, the *LC* circuits are tuned to resonance for that station. There are almost unlimited uses for resonance in AC circuits.

25–1 Self-Review

Answers at the end of the chapter.

Refer to Fig. 25–1.

a. **Give the resonant frequency.**
b. **Give the frequency that has maximum output.**

25–2 Series Resonance

When the frequency of the applied voltage is 1000 kHz in the series AC circuit in Fig. 25–2*a*, the reactance of the 239-μH inductance equals 1500 Ω. At the same frequency, the reactance of the 106-pF capacitance also is 1500 Ω. Therefore, this

Figure 25–1 *LC* circuit resonant at f_r of 1000 kHz to provide maximum output at this frequency.

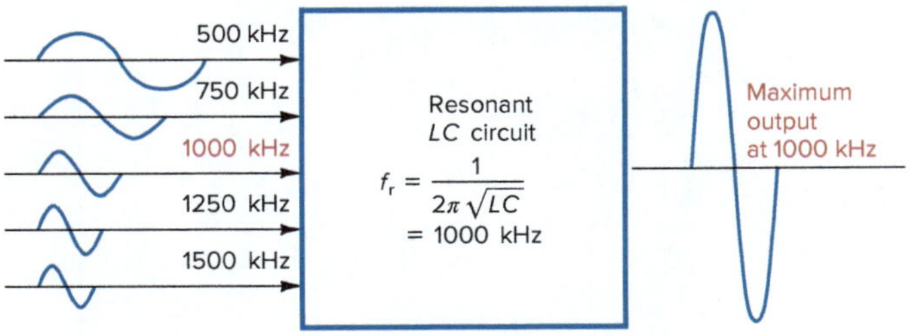

MultiSim **Figure 25–2** Series resonance. (*a*) Schematic diagram of series r_S, L, and C. (*b*) Graph to show that reactances X_C and X_L are equal and opposite at the resonant frequency f_r. Inductive reactance is shown up for jX_L and capacitive reactance is down for $-jX_C$.

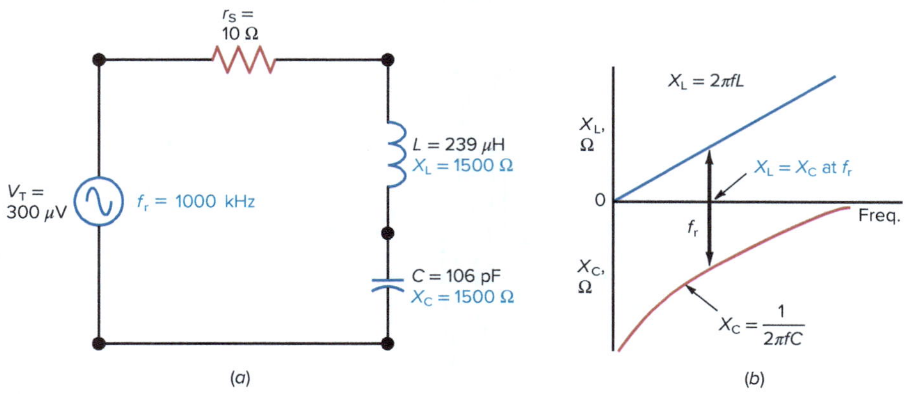

GOOD TO KNOW

The coil resistance, r_S, is usually more than just the DC resistance of the wire of the coil. At higher frequencies, skin effect causes r_S to be higher than the coil's DC resistance alone.

LC combination is resonant at 1000 kHz. This is f_r because the inductive reactance and capacitive reactance are equal at this frequency.

In a series AC circuit, inductive reactance leads by 90°, compared with the zero reference angle of the resistance, and capacitive reactance lags by 90°. Therefore, X_L and X_C are 180° out of phase. The opposite reactances cancel each other completely when they are equal.

Figure 25–2*b* shows X_L and X_C equal, resulting in a net reactance of zero ohms. The only opposition to current, then, is the coil resistance r_S, which limits how low the series resistance in the circuit can be. With zero reactance and just the low value of series resistance, the generator voltage produces the greatest amount of current in the series *LC* circuit at the resonant frequency. The series resistance should be as small as possible for a sharp increase in current at resonance.

Maximum Current at Series Resonance

The main characteristic of series resonance is the resonant rise of current to its maximum value of V_T/r_S at the resonant frequency. For the circuit in Fig. 25–2*a*, the maximum current at series resonance is 30 μA, equal to 300 μV/10 Ω. At any other frequency, either below or above the resonant frequency, there is less current in the circuit.

This resonant rise of current to 30 μA at 1000 kHz is shown in Fig. 25–3. In Fig. 25–3*a*, the amount of current is shown as the amplitude of individual cycles of the alternating current produced in the circuit by the AC generator voltage. Whether the amplitude of one AC cycle is considered in terms of peak, rms, or average value, the amount of current is greatest at the resonant frequency. In Fig. 25–3*b*, the current amplitudes are plotted on a graph for frequencies at and near the resonant frequency, producing a typical *response curve* for a series resonant circuit. The response curve in Fig. 25–3*b* can be considered an outline of the increasing and decreasing amplitudes of the individual cycles shown in Fig. 25–3*a*.

The response curve of the series resonant circuit shows that the current is small below resonance, rises to its maximum value at the resonant frequency, and then drops off to small values above resonance. To prove this fact, Table 25–1 lists the calculated values of impedance and current in the circuit of Fig. 25–2 at the resonant frequency of 1000 kHz and at two frequencies below and two frequencies above resonance.

Below resonance, at 600 kHz, X_C is more than X_L and there is appreciable net reactance, which limits the current to a relatively low value. At the higher frequency of 800 kHz, X_C decreases and X_L increases, making the two reactances closer to the same value. The net reactance is then smaller, allowing more current.

GOOD TO KNOW

For a series LC circuit at resonance, $Z_T = r_S$.

Figure 25-3 Graphs showing maximum current at resonance for the series circuit in Fig. 25–2. (*a*) Amplitudes of individual cycles. (*b*) Response curve to show the amount of I below and above resonance. Values of I are in Table 25–1.

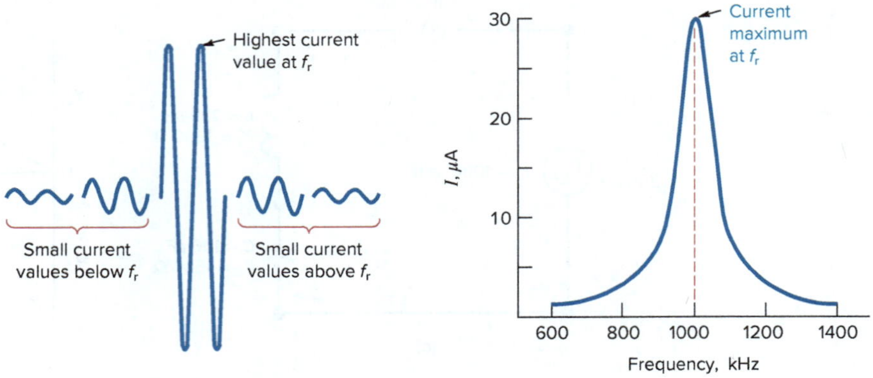

(*a*) (*b*)

At the resonant frequency, X_L and X_C are equal, the net reactance is zero, and the current has its maximum value equal to V_T/r_S.

Above resonance at 1200 and 1400 kHz, X_L is greater than X_C, providing net reactance that limits the current to values much smaller than at resonance. In summary,

1. Below the resonant frequency, X_L is small, but X_C has high values that limit the amount of current.
2. Above the resonant frequency, X_C is small, but X_L has high values that limit the amount of current.
3. At the resonant frequency, X_L equals X_C, and they cancel to allow maximum current.

Minimum Impedance at Series Resonance

Since reactances cancel at the resonant frequency, the impedance of the series circuit is minimum, equal to just the low value of series resistance. This minimum impedance at resonance is resistive, resulting in zero phase angle. At resonance, therefore, the resonant current is in phase with the generator voltage.

Table 25–1 Series-Resonance Calculations for the Circuit in Figure 25–2*

Frequency, kHz	$X_L = 2\pi fL$, Ω	$X_C = 1/(2\pi fC)$, Ω	Net Reactance, Ω		Z_T, Ω†	$I = V_T/Z_T$, μA†	$V_L = IX_L$, μV	$V_C = IX_C$, μV
			$X_C - X_L$	$X_L - X_C$				
600	900	2500	1600		1600	0.19	171	475
800	1200	1875	675		675	0.44	528	825
$f_r \rightarrow$ 1000	1500	1500	0	0	10	30	45,000	45,000
1200	1800	1250		550	550	0.55	990	688
1400	2100	1070		1030	1030	0.29	609	310

* $L = 239$ μH, $C = 106$ pF, $V_T = 300$ μV, $r_S = 10$ Ω.

† Z_T and I calculated without r_S when its resistance is very small compared with the net X_L or X_C. Z_T and I are resistive at f_r.

MultiSim **Figure 25–4** Series circuit selects frequency by producing maximum IX_C voltage output across C at resonance.

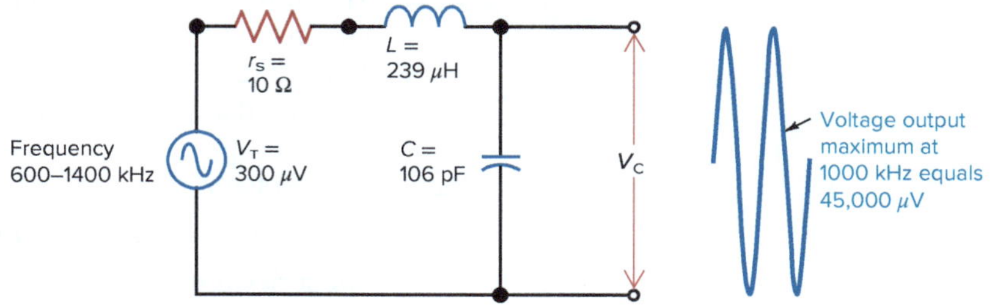

Resonant Rise in Voltage across Series *L* or *C*

The maximum current in a series *LC* circuit at resonance is useful because it produces maximum voltage across either X_L or X_C at the resonant frequency. As a result, the series resonant circuit can select one frequency by providing much more voltage output at the resonant frequency, compared with frequencies above and below resonance. Figure 25–4 illustrates the resonant rise in voltage across the capacitance in a series AC circuit. At the resonant frequency of 1000 kHz, the voltage across *C* rises to 45,000 μV, and the input voltage is only 300 μV.

In Table 25–1, the voltage across *C* is calculated as IX_C, and across *L* as IX_L. Below the resonant frequency, X_C has a higher value than at resonance, but the current is small. Similarly, above the resonant frequency, X_L is higher than at resonance, but the current has a low value because of inductive reactance. At resonance, although X_L and X_C cancel each other to allow maximum current, each reactance by itself has an appreciable value. Since the current is the same in all parts of a series circuit, the maximum current at resonance produces maximum voltage IX_C, across *C* and an equal IX_L voltage across *L* for the resonant frequency.

Although the voltage across X_C and X_L is reactive, it is an actual voltage that can be measured. In Fig. 25–5, the voltage drops around the series resonant circuit are 45,000 μV across *C*, 45,000 μV across *L*, and 300 μV across r_S. The voltage across the resistance is equal to and in phase with the generator voltage.

Across the series combination of both *L* and *C*, the voltage is zero because the two series voltage drops are equal and opposite. To use the resonant rise of voltage, therefore, the output must be connected across either *L* or *C* alone. We can consider the V_L and V_C voltages similar to the idea of two batteries connected in series opposition. Together, the resultant is zero for equal and opposite voltages, but each battery still has its own potential difference.

Figure 25–5 Voltage drops around series resonant circuit.

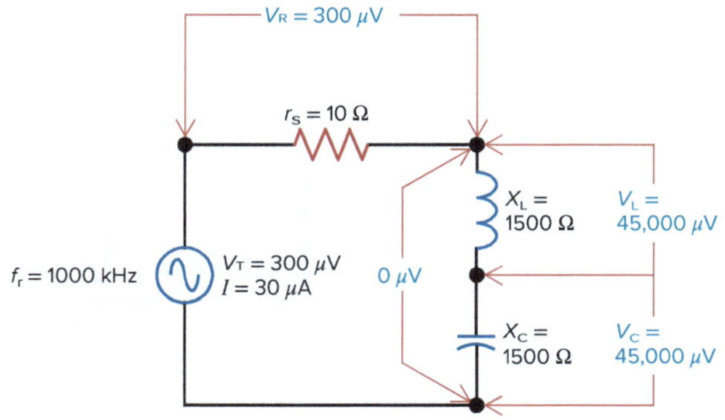

In summary, the main characteristics of a series resonant circuit are

1. The current I is maximum at the resonant frequency f_r.
2. The current I is in phase with the generator voltage, or the phase angle of the circuit is 0°.
3. The voltage is maximum across either L or C alone.
4. The impedance is minimum at f_r, equal only to the low r_S.

25–2 Self-Review

Answers at the end of the chapter.

For series resonance,

a. X_L **and** X_C **are maximum. (True/False)**
b. X_L **and** X_C **are equal. (True/False)**
c. current I **is maximum. (True/False)**

25–3 Parallel Resonance

When L and C are in parallel, as shown in Fig. 25–6, and X_L equals X_C, the reactive branch currents are equal and opposite at resonance. Then they cancel each other to produce minimum current in the main line. Since the line current is minimum, the impedance is maximum. These relations are based on r_S being very small compared with X_L at resonance. In this case, the branch currents are practically equal when X_L and X_C are equal.

MultiSim **Figure 25–6** Parallel resonant circuit. (*a*) Schematic diagram of L and C in parallel branches. (*b*) Response curve of I_T shows that the line current dips to a minimum at f_r. (*c*) Response curve of Z_{EQ} shows that it rises to a maximum at f_r.

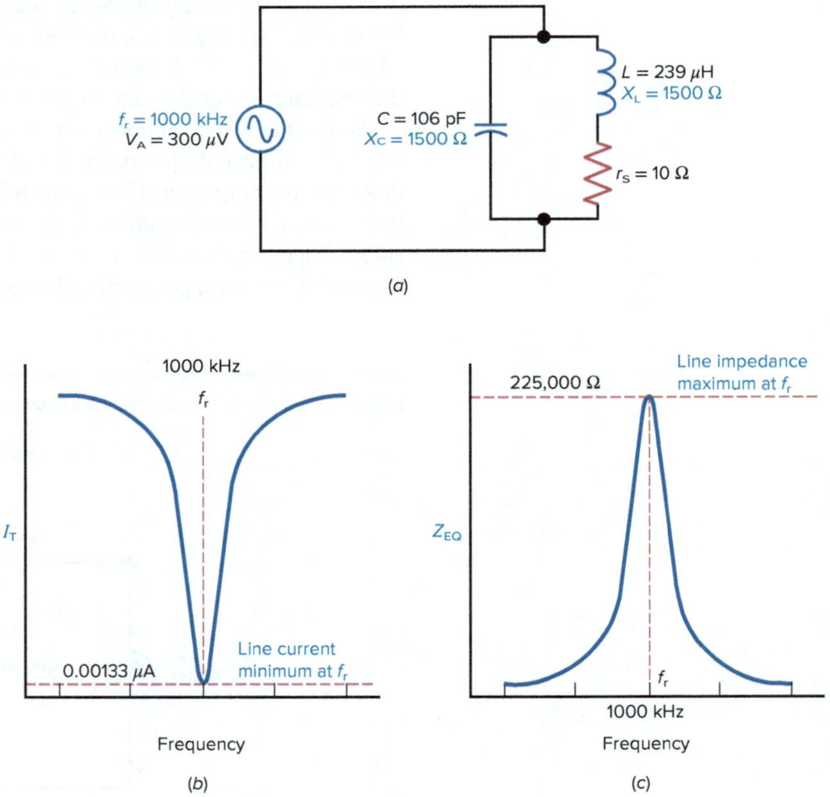

Table 25–2 Parallel-Resonance Calculations for the Circuit in Figure 25–6*

Frequency, kHz	$X_C = 1/(2\pi fC)$, Ω	$X_L = 2\pi fL$, Ω	$I_C = V/X_C$, μA	$I_L = V/X_L$, μA†	Net Reactive Line Current, μA		I_T, μA†	$Z_{EQ} = V_A/I_T$, Ω†
					$I_L - I_C$	$I_C - I_L$		
600	2500	900	0.12	0.33	0.21		0.21	1400
800	1875	1200	0.16	0.25	0.09		0.09	3333
$f_r \rightarrow$ 1000	1500	1500	0.20	0.20	0	0	0.00133	225,000‡
1200	1250	1800	0.24	0.17		0.07	0.07	3800
1400	1070	2100	0.28	0.14		0.14	0.14	2143

* $L = 239\ \mu$H, $C = 106$ pF, $V_T = 3000\ \mu$V, $r_S = 10\ \Omega$.

† Z_{EQ} and I calculated without r_S when its resistance is very small compared with the net X_L or X_C. Z_{EQ} and I are resistive at f_r.

‡ At resonance, Z_{EQ} is calculated by Formula (25–7). Z_{EQ} and I_T are resistive at f_r.

Minimum Line Current at Parallel Resonance

To show how the current in the main line dips to its minimum value when the parallel *LC* circuit is resonant, Table 25–2 lists the values of branch currents and the total line current for the circuit in Fig. 25–6.

With *L* and *C* the same as in the series circuit of Fig. 25–2, X_L and X_C have the same values at the same frequencies. Since *L*, *C*, and the generator are in parallel, the voltage applied across the branches equals the generator voltage of 300 μV. Therefore, each reactive branch current is calculated as 300 μV divided by the reactance of the branch.

The values in the top row of Table 25–2 are obtained as follows: At 600 kHz, the capacitive branch current equals 300 μV/2500 Ω, or 0.12 μA. The inductive branch current at this frequency is 300 μV/900 Ω, or 0.33 μA. Since this is a parallel AC circuit, the capacitive current leads by 90°, whereas the inductive current lags by 90°, compared with the reference angle of the generator voltage, which is applied across the parallel branches. Therefore, the opposite currents are 180° out of phase. The net current in the line, then, is the difference between 0.33 and 0.12, which equals 0.21 μA.

Following this procedure, the calculations show that as the frequency is increased toward resonance, the capacitive branch current increases because of the lower value of X_C and the inductive branch current decreases with higher values of X_L. As a result, there is less net line current as the two branch currents become more nearly equal.

At the resonant frequency of 1000 kHz, both reactances are 1500 Ω, and the reactive branch currents are both 0.20 μA, canceling each other completely.

Above the resonant frequency, there is more current in the capacitive branch than in the inductive branch, and the net line current increases above its minimum value at resonance.

The dip in I_T to its minimum value at f_r is shown by the graph in Fig. 25–6*b*. At parallel resonance, I_T is minimum and Z_{EQ} is maximum.

The in-phase current due to r_S in the inductive branch can be ignored off-resonance because it is so small compared with the reactive line current. At the resonant frequency when the reactive currents cancel, however, the resistive component is the entire line current. Its value at resonance equals 0.00133 μA in this example. This small resistive current is the minimum value of the line current at parallel resonance.

GOOD TO KNOW

The coil resistance, r_S, is the reason why the impedance of a parallel resonant circuit cannot reach infinity at the resonant frequency, f_r.

Maximum Line Impedance at Parallel Resonance

The minimum line current resulting from parallel resonance is useful because it corresponds to maximum impedance in the line across the generator. Therefore, an impedance that has a high value for just one frequency but a low impedance for other frequencies, either below or above resonance, can be obtained by using a parallel *LC* circuit resonant at the desired frequency. This is another method of selecting one frequency by resonance. The response curve in Fig. 25–6*c* shows how the impedance rises to a maximum for parallel resonance.

The main application of parallel resonance is the use of an *LC* tuned circuit as the load impedance Z_L in the output circuit of rf amplifiers. Because of the high impedance, then, the gain of the amplifier is maximum at f_r. The voltage gain of an amplifier is directly proportional to Z_L. The advantage of a resonant *LC* circuit is that *Z* is maximum only for an AC signal at the resonant frequency. Also, *L* has practically no DC resistance, which means practically no DC voltage drop.

Referring to Table 25–2, the total impedance of the parallel AC circuit is calculated as the generator voltage divided by the total line current. At 600 kHz, for example, Z_{EQ} equals 300 μV/0.21 μA, or 1400 Ω. At 800 kHz, the impedance is higher because there is less line current.

At the resonant frequency of 1000 kHz, the line current is at its minimum of 0.00133 μA. Then, the impedance is maximum and is equal to 300 μV/0.00133 μA, or 225,000 Ω.

Above 1000 kHz, the line current increases, and the impedance decreases from its maximum.

How the line current can be very low even though the reactive branch currents are appreciable is illustrated in Fig. 25–7. In Fig. 25–7*a*, the resistive component of the total line current is shown as though it were a separate branch drawing an amount of resistive current from the generator in the main line equal to the current resulting from the coil resistance. Each reactive branch current has its value equal to the generator voltage divided by the reactance. Since they are equal and of opposite phase, however, in any part of the circuit where both reactive currents are present, the net amount of electron flow in one direction at any instant corresponds to zero current. The graph in Fig. 25–7*b* shows how equal and opposite currents for I_L and I_C cancel.

If a meter is inserted in series with the main line to indicate total line current I_T, it dips sharply to the minimum value of line current at the resonant frequency. With minimum current in the line, the impedance across the line is maximum at the resonant frequency. The maximum impedance at parallel resonance corresponds to a

Figure 25–7 Distribution of currents in a parallel circuit at resonance. Resistive current shown as an equivalent branch for I_R. (*a*) Circuit with branch currents for *R*, *L*, and *C*. (*b*) Graph of equal and opposite reactive currents I_L and I_C.

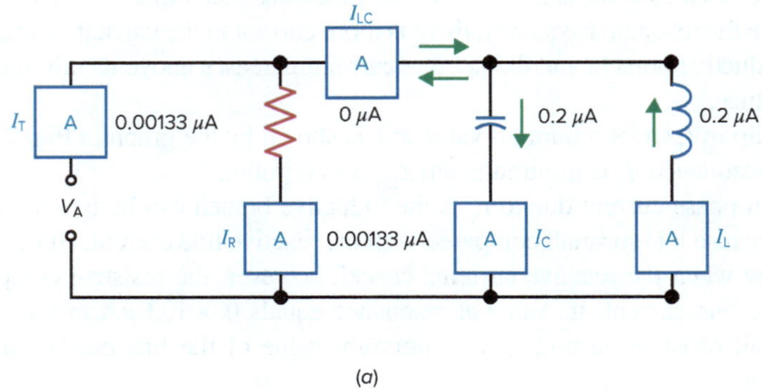

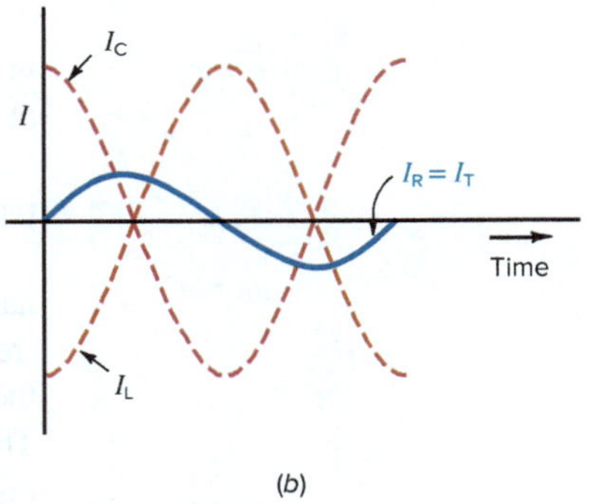

high value of resistance, without reactance, since the line current is then resistive with zero phase angle.

In summary, the main characteristics of a parallel resonant circuit are

1. The line current I_T is minimum at the resonant frequency.
2. The current I_T is in phase with the generator voltage V_A, or the phase angle of the circuit is 0°.
3. The impedance Z_{EQ}, equal to V_A/I_T, is maximum at f_r because of the minimum I_T.

The *LC* Tank Circuit

Note that the individual branch currents are appreciable at resonance, although I_T is minimum. For the example in Table 25–2, at f_r, either the I_L or the I_C equals 0.2 μA. This current is greater than the I_C values below f_r or the I_L values above f_r.

The branch currents cancel in the main line because I_C is at 90° with respect to the source V_A while I_L is at −90°, making them opposite with respect to each other.

However, inside the *LC* circuit, I_L and I_C do not cancel because they are in separate branches. Then I_L and I_C provide a circulating current in the *LC* circuit, which equals 0.2 μA in this example. For this reason, a parallel resonant *LC* circuit is often called a ***tank circuit***.

Because of the energy stored by *L* and *C*, the circulating tank current can provide full sine waves of current and voltage output when the input is only a pulse. The sine-wave output is always at the natural resonant frequency of the *LC* tank circuit. This ability of the *LC* circuit to supply complete sine waves is called the ***flywheel effect***. Also, the process of producing sine waves after a pulse of energy has been applied is called *ringing* of the *LC* circuit.

25–3 Self-Review

Answers at the end of the chapter.

For parallel resonance,

a. **currents I_L and I_C are maximum. (True/False)**
b. **currents I_L and I_C are equal. (True/False)**
c. **current I_T is minimum. (True/False)**

25–4 Resonant Frequency $f_r = 1/(2\pi\sqrt{LC})$

The formula for the resonant frequency is derived from $X_L = X_C$. Using f_r to indicate the resonant frequency in the formulas for X_L and X_C,

$$2\pi f_r L = \frac{1}{2\pi f_r C}$$

Inverting the factor f_r gives

$$2\pi L(f_r)^2 = \frac{1}{2\pi C}$$

Inverting the factor $2\pi L$ gives

$$(f_r)^2 = \frac{1}{(2\pi)^2 LC}$$

The square root of both sides is then

$$f_r = \frac{1}{2\pi\sqrt{LC}} \qquad \textbf{(25–1)}$$

CALCULATOR

To do this problem on a calculator, keep in mind the following points. If your calculator does not have an exponential (EXP) key, work with the powers of 10 separately, without the calculator. For multiplication, add the exponents. The square root has one-half the exponent, but be sure that the exponent is an even number before dividing by 2. The reciprocal has the same exponent but with opposite sign.

For the example just solved with 239 μH for L and 106 pF for C, first punch in 239, push the (×) key, punch in 106, and then press the (=) key for the product of 25,334. While this number is on the display, push the (√) key for 159.2. Keep this display, press the (×) key, punch in 6.28 for 2π, and push the (=) key for the total product of approximately 1000 in the denominator.

The powers of 10 in the denominator are $10^{-6} \times 10^{-12} = 10^{-18}$. Its square root is 10^{-9}.

While 1000 for the denominator is on the display, press the (1/x) key for the reciprocal, equal to 0.001. The reciprocal of 10^{-9} is 10^{9}. The answer for f_r then is 0.001×10^{9}, which equals 1×10^{6}.

where L is in henrys, C is in farads, and the resonant frequency f_r is in hertz (Hz). For example, to find the resonant frequency of the LC combination in Fig. 25–2, the values of 239×10^{-6} and 106×10^{-12} are substituted for L and C. Then,

$$f_r = \frac{1}{2\pi\sqrt{LC}} = \frac{1}{2\pi\sqrt{239 \times 10^{-6} \times 106 \times 10^{-12}}}$$

$$= \frac{1}{6.28\sqrt{25{,}334 \times 10^{-18}}} = \frac{1}{6.28 \times 159.2 \times 10^{-9}} = \frac{1}{1000 \times 10^{-9}}$$

$$= 1 \times 10^{6}\ \text{Hz} = 1\ \text{MHz} = 1000\ \text{kHz}$$

For any series or parallel LC circuit, the f_r equal to $1/(2\pi\sqrt{LC})$ is the resonant frequency that makes the inductive and capacitive reactances equal.

How the f_r Varies with L and C

It is important to note that higher values of L and C result in lower values of f_r. Either L or C, or both, can be varied. An LC circuit can be resonant at any frequency from a few hertz to many megahertz.

As examples, an LC combination with the relatively large values of an 8-H inductance and a 20-μF capacitance is resonant at the low audio frequency of 12.6 Hz. For a much higher frequency in the rf range, a small inductance of 2 μH will resonate with the small capacitance of 3 pF at an f_r of 64.9 MHz. These examples are solved in the next two problems for more practice with the resonant frequency formula. Such calculations are often used in practical applications of tuned circuits. Probably the most important feature of any LC combination is its resonant frequency, especially in rf circuits. The applications of resonance are mainly for radio frequencies.

Example 25-1

Calculate the resonant frequency for an 8-H inductance and a 20-μF capacitance.

ANSWER

$$f_r = \frac{1}{2\pi\sqrt{LC}}$$

$$= \frac{1}{2\pi\sqrt{8 \times 20 \times 10^{-6}}}$$

$$= \frac{1}{6.28\sqrt{160 \times 10^{-6}}}$$

$$= \frac{1}{6.28 \times 12.65 \times 10^{-3}}$$

$$= \frac{1}{79.44 \times 10^{-3}}$$

$$= 0.0126 \times 10^{3}$$

$$= 12.6\ \text{Hz} \qquad \text{(approx.)}$$

Example 25-2

Calculate the resonant frequency for a 2-μH inductance and a 3-pF capacitance.

ANSWER

$$f_r = \frac{1}{2\pi\sqrt{LC}}$$

$$= \frac{1}{2\pi\sqrt{2 \times 10^{-6} \times 3 \times 10^{-12}}}$$

$$= \frac{1}{6.28\sqrt{6 \times 10^{-18}}}$$

$$= \frac{1}{6.28 \times 2.45 \times 10^{-9}}$$

$$= \frac{1}{15.4 \times 10^{-9}} = 0.065 \times 10^{9}$$

$$= 65 \times 10^{6} \text{ Hz} = 65 \text{ MHz}$$

Specifically, because of the square root in the denominator of Formula (25–1), the f_r decreases inversely as the square root of L or C. For instance, if L or C is quadrupled, the f_r is reduced by one-half. The ½ is equal to the square root of ¼.

As a numerical example, suppose that f_r is 6 MHz with particular values of L and C. If either L or C is made four times larger, then f_r will be reduced to 3 MHz.

Or, to take the opposite case of doubling the frequency from 6 MHz to 12 MHz, the following can be done:

1. Use one-fourth the L with the same C.
2. Use one-fourth the C with the same L.
3. Reduce both L and C by one-half.
4. Use any new combination of L and C whose product will be one-fourth the original product of L and C.

LC Product Determines f_r

There are any number of LC combinations that can be resonant at one frequency. With more L, then less C can be used for the same f_r. Or less L can be used with more C. Table 25–3 lists five possible combinations of L and C resonant at 1000 kHz, as

Table 25–3 *LC* Combinations Resonant at 1000 kHz

L, μH	C, pF	$L \times C$ LC Product	X_L, Ω at 1000 kHz	X_C, Ω at 1000 kHz
23.9	1060	25,334	150	150
119.5	212	25,334	750	750
239	106	25,334	1500	1500
478	53	25,334	3000	3000
2390	10.6	25,334	15,000	15,000

an example of one f_r. The resonant frequency is the same 1000 kHz here for all five combinations. When either *L* or *C* is increased by a factor of 10 or 2, the other is decreased by the same factor, resulting in a constant value for the *LC* product.

The reactance at resonance changes with different combinations of *L* and *C*, but in all five cases, X_L and X_C are equal to each other at 1000 kHz. This is the resonant frequency determined by the value of the *LC* product in $f_r = 1/(2\pi\sqrt{LC})$.

Measuring *L* or *C* by Resonance

Of the three factors *L*, *C*, and f_r in the resonant-frequency formula, any one can be calculated when the other two are known. The resonant frequency of the *LC* combination can be found experimentally by determining the frequency that produces the resonant response in an *LC* combination. With a known value of either *L* or *C*, and the resonant frequency determined, the third factor can be calculated. This method is commonly used for measuring inductance or capacitance. A test instrument for this purpose is the *Q* meter, which also measures the *Q* of a coil.

Calculating *C* from f_r

The *C* can be taken out of the square root sign or radical in the resonance formula, as follows:

$$f_r = \frac{1}{2\pi\sqrt{LC}}$$

Squaring both sides to eliminate the radical gives

$$f_r^2 = \frac{1}{(2\pi)^2 LC}$$

Inverting C and f_r^2 gives

$$C = \frac{1}{4\pi^2 f_r^2 L} \qquad \textbf{(25–2)}$$

where f_r is in hertz, *C* is in farads, and *L* is in henrys.

Calculating *L* from f_r

Similarly, the resonance formula can be transposed to find *L*. Then

$$L = \frac{1}{4\pi^2 f_r^2 C} \qquad \textbf{(25–3)}$$

With Formula (25–3), *L* is determined by f_r with a known value of *C*. Similarly, *C* is determined from Formula (25–2) by f_r with a known value of *L*.

Example 25-3

What value of *C* resonates with a 239-μH *L* at 1000 kHz?

ANSWER

$$C = \frac{1}{4\pi^2 f_r^2 L}$$

$$= \frac{1}{4\pi^2 (1000 \times 10^3)^2 239 \times 10^{-6}}$$

$$= \frac{1}{39.48 \times 1 \times 10^6 \times 239}$$

$$= \frac{1}{9435.75 \times 10^6}$$

$$= 0.000106 \times 10^{-6}\ \text{F} = 106\ \text{pF}$$

Note that 39.48 is a constant for $4\pi^2$.

Example 25-4

What value of L resonates with a 106-pF C at 1000 kHz, equal to 1 MHz?

ANSWER

$$L = \frac{1}{4\pi^2 f_r^2 C}$$

$$= \frac{1}{39.48 \times 1 \times 10^{12} \times 106 \times 10^{-12}}$$

$$= \frac{1}{4184.88}$$

$$= 0.000239\ \text{H} = 239\ \mu\text{H}$$

Note that 10^{12} and 10^{-12} in the denominator cancel each other. Also, 1×10^{12} is the square of 1×10^6, or 1 MHz.

The values in Examples 25–3 and 25–4 are from the LC circuit illustrated in Fig. 25–2 for series resonance and Fig. 25–6 for parallel resonance.

25–4 Self-Review

Answers at the end of the chapter.

a. To increase f_r, should C be increased or decreased?

b. If C is increased from 100 to 400 pF, L must be decreased from 800 μH to what value for the same f_r?

c. Give the constant value for $4\pi^2$.

25–5 Q Magnification Factor of a Resonant Circuit

The quality, or *figure of merit*, of the resonant circuit, in sharpness of resonance, is indicated by the factor Q. In general, the higher the ratio of the reactance at resonance to the series resistance, the higher the Q and the sharper the resonance effect.

Q of Series Circuit

In a series resonant circuit, we can calculate Q from the following formula:

$$Q = \frac{X_L}{r_S} \qquad \textbf{(25–4)}$$

where Q is the figure of merit, X_L is the inductive reactance in ohms at the resonant frequency, and r_S is the resistance in ohms in series with X_L. For the series resonant circuit in Fig. 25–2,

$$Q = \frac{1500\ \Omega}{10\ \Omega} = 150$$

The Q is a numerical factor without any units, because it is a ratio of reactance to resistance and the ohms cancel. Since the series resistance limits the amount of current at resonance, the lower the resistance, the sharper the increase to maximum current at the resonant frequency, and the higher the Q. Also, a higher value of reactance at resonance allows the maximum current to produce higher voltage for the output.

The Q has the same value if it is calculated with X_C instead of X_L, since they are equal at resonance. However, the Q of the circuit is generally considered in terms of X_L because usually the coil has the series resistance of the circuit. In this case, the Q of the coil and the Q of the series resonant circuit are the same. If extra resistance is added, the Q of the circuit will be less than the Q of the coil. The highest possible Q for the circuit is the Q of the coil.

The value of 150 can be considered a high Q. Typical values are 50 to 250, approximately. Less than 10 is a low Q; more than 300 is a very high Q.

Higher L/C Ratio Can Provide Higher Q

As shown before in Table 25–3, different combinations of L and C can be resonant at the same frequency. However, the amount of reactance at resonance is different. More X_L can be obtained with a higher L and lower C for resonance, although X_L and X_C must be equal at the resonant frequency. Therefore, both X_L and X_C are higher with a higher L/C ratio for resonance.

More X_L can allow a higher Q if the AC resistance does not increase as much as the reactance. An approximate rule for typical rf coils is that maximum Q can be obtained when X_L is about 1000 Ω. In many cases, though, the minimum C is limited by stray capacitance in the circuit.

GOOD TO KNOW

Formula (25–5) is derived as follows: $V_L = I \times X_L$ where $I = \frac{V_{gen}}{r_S}$ at the resonant frequency, f_r. Therefore, $V_L = \frac{V_{gen}}{r_S} \times X_L$ or $V_L = \frac{X_L}{r_S} \times V_{gen}$. Since $Q = \frac{X_L}{r_S}$ we have $V_L = Q \times V_{gen}$.

Since $X_L = X_C$ at f_r, Formula (25–5) can be used to find both V_L and V_C.

Q Rise in Voltage across Series L or C

The Q of the resonant circuit can be considered a magnification factor that determines how much the voltage across L or C is increased by the resonant rise of current in a series circuit. Specifically, the voltage output at series resonance is Q times the generator voltage:

$$V_L = V_C = Q \times V_{gen} \qquad \textbf{(25–5)}$$

In Fig. 25–4, for example, the generator voltage is 300 μV and Q is 150. The resonant rise of voltage across either L or C then equals 300 μV × 150, or 45,000 μV. Note that this is the same value calculated in Table 25–1 for V_C or V_L at resonance.

How to Measure Q in a Series Resonant Circuit

The fundamental nature of Q for a series resonant circuit is seen from the fact that the Q can be determined experimentally by measuring the Q rise in voltage across either L or C and comparing this voltage with the generator voltage. As a formula,

$$Q = \frac{V_{out}}{V_{in}} \qquad \textbf{(25–6)}$$

where V_{out} is the AC voltage measured across the coil or capacitor and V_{in} is the generator voltage.

Referring to Fig. 25–5, suppose that you measure with an AC voltmeter across L or C and this voltage equals 45,000 μV at the resonant frequency. Also, measure the generator input of 300 μV. Then

$$Q = \frac{V_{out}}{V_{in}}$$

$$= \frac{45{,}000\ \mu\text{V}}{300\ \mu\text{V}}$$

$$= 150$$

This method is better than the X_L/r_S formula for determining Q because r_S is the AC resistance of the coil, which is not so easily measured. Remember that the coil's AC resistance can be more than double the DC resistance measured with an ohmmeter. In fact, measuring Q with Formula (25–6) makes it possible to calculate the AC resistance. These points are illustrated in the following examples.

Example 25-5

A series circuit resonant at 0.4 MHz develops 100 mV across a 250-μH L with a 2-mV input. Calculate Q.

ANSWER

$$Q = \frac{V_{out}}{V_{in}} = \frac{100\text{ mV}}{2\text{ mV}}$$

$$= 50$$

Example 25-6

What is the AC resistance of the coil in the preceding example?

ANSWER The Q of the coil is 50. We need to know the reactance of this 250-μH coil at the frequency of 0.4 MHz.Then,

$$X_L = 2\pi f\text{L} = 6.28 \times 0.4 \times 10^6 \times 250 \times 10^{-6}$$

$$= 628\ \Omega$$

$$\text{Also, } Q = \frac{X_L}{r_S} \quad \text{or} \quad r_S = \frac{X_L}{Q}$$

$$r_S = \frac{628\ \Omega}{50}$$

$$= 12.56\ \Omega$$

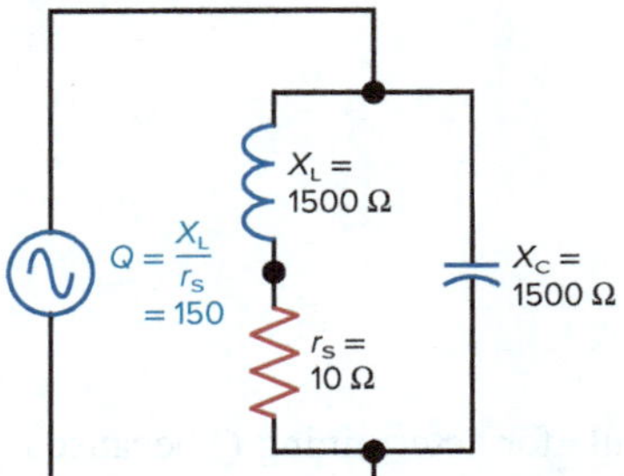

Figure 25–8 The Q of a parallel resonant circuit in terms of X_L and its series resistance r_S.

Q of Parallel Circuit

In a parallel resonant circuit where r_S is very small compared with X_L, the Q also equals X_L/r_S. Note that r_S is still the resistance of the coil in series with X_L (see Fig. 25–8). The Q of the coil determines the Q of the parallel circuit here because it is less than the Q of the capacitive branch. Capacitors used in tuned circuits generally have a very high Q because of their low losses. In Fig. 25–8, the Q is 1500 Ω/10 Ω, or 150, the same as the series resonant circuit with the same values.

This example assumes that the generator resistance is very high and that there is no other resistance branch shunting the tuned circuit. Then the Q of the parallel resonant circuit is the same as the Q of the coil. Actually, shunt resistance can lower the Q of a parallel resonant circuit, as analyzed in Section 25–10.

Q Rise in Impedance across a Parallel Resonant Circuit

For parallel resonance, the Q magnification factor determines by how much the impedance across the parallel LC circuit is increased because of the minimum line current. Specifically, the impedance across the parallel resonant circuit is Q times the inductive reactance at the resonant frequency:

$$Z_{EQ} = Q \times X_L \qquad \textbf{(25–7)}$$

Referring back to the parallel resonant circuit in Fig. 25–6 as an example, X_L is 1500 Ω and Q is 150. The result is a rise of impedance to the maximum value of 150 × 1500 Ω, or 225,000 Ω, at the resonant frequency.

Since the line current equals V_A/Z_{EQ}, the minimum line current is 300 μV/225,000 Ω, which equals 0.00133 μA.

At f_r, the minimum line current is $1/Q$ of either branch current. In Fig. 25–7, I_L or I_C is 0.2 μA and Q is 150. Therefore, I_T is $^{0.2}/_{150}$, or 0.00133 μA, which is the same answer as V_A/Z_{EQ}. Or, stated another way, the circulating tank current is Q times the minimum I_T.

GOOD TO KNOW

Another formula that can be used to calculate Z_{EQ} at f_r is

$Z_{EQ} = (1 + Q^2)r_S$, where $Q = \frac{X_L}{r_S}$.

How to Measure Z_{EQ} of a Parallel Resonant Circuit

Formula (25–7) for Z_{EQ} is also useful in its inverted form as $Q = Z_{EQ}/X_L$. We can measure Z_{EQ} by the method illustrated in Fig. 25–9. Then Q can be calculated from the value of Z_{EQ} and the inductive reactance of the coil.

Figure 25–9 How to measure Z_{EQ} of a parallel resonant circuit. Adjust R_1 to make its V_R equal to V_{LC}. Then $Z_{EQ} = R_1$.

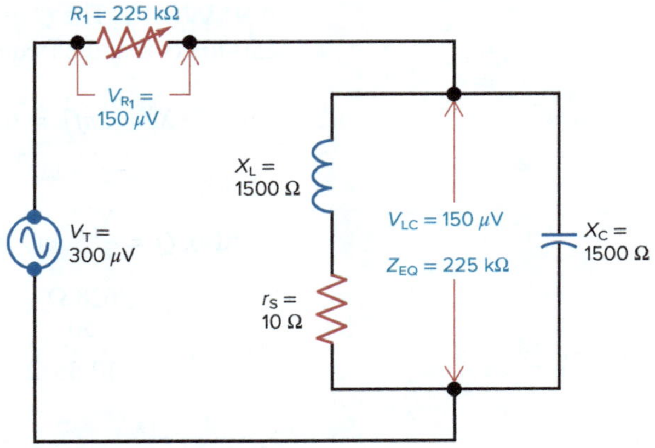

To measure Z_{EQ}, first tune the LC circuit to resonance. Then adjust R_1 in Fig. 25–9 to the resistance that makes its AC voltage equal to the AC voltage across the tuned circuit. With equal voltages, the Z_{EQ} must have the same value as R_1.

For the example here, which corresponds to the parallel resonance shown in Figs. 25–6 and 25–8, Z_{EQ} is equal to 225,000 Ω. This high value is a result of parallel resonance. The X_L is 1500 Ω. Therefore, to determine Q, the calculations are

$$Q = \frac{Z_{EQ}}{X_L} = \frac{225{,}000}{1500} = 150$$

Example 25-7

In Fig. 25–9, assume that with a 4-mVac input signal for V_T, the voltage across R_1 is 2 mV when R_1 is 225 kΩ. Determine Z_{EQ} and Q.

ANSWER Because they divide V_T equally, Z_{EQ} is 225 kΩ, the same as R_1. The amount of input voltage does not matter, as the voltage division determines the relative proportions between R_1 and Z_{EQ}. With 225 kΩ for Z_{EQ} and 1.5 kΩ for X_L, the Q is $^{225}/_{1.5}$, or $Q = 150$.

Example 25-8

A parallel LC circuit tuned to 200 kHz with a 350-μH L has a measured Z_{EQ} of 17,600 Ω. Calculate Q.

ANSWER First, calculate X_L as $2\pi fL$ at f_r:

$$X_L = 2\pi \times 200 \times 10^3 \times 350 \times 10^{-6} = 440\ \Omega$$

Then,

$$Q = \frac{Z_{EQ}}{X_L} = \frac{17{,}600}{440}$$
$$= 40$$

25–5 Self-Review

Answers at the end of the chapter.

a. In a series resonant circuit, V_L is 300 mV with an input of 3 mV. Calculate Q.

b. In a parallel resonant circuit, X_L is 500 Ω with a Q of 50, calculate Z_{EQ}.

25–6 Bandwidth of a Resonant Circuit

When we say that an LC circuit is resonant at one frequency, this is true for the maximum resonance effect. However, other frequencies close to f_r also are effective. For series resonance, frequencies just below and above f_r produce increased current, but a little less than the value at resonance. Similarly, for parallel resonance, frequencies close to f_r can provide high impedance, although a little less than the maximum Z_{EQ}.

Therefore, any resonant frequency has an associated band of frequencies that provide resonance effects. How wide the band is depends on the Q of the resonant circuit. Actually, it is practically impossible to have an LC circuit with a resonant effect at only one frequency. The width of the resonant band of frequencies centered around f_r is called the ***bandwidth*** of the tuned circuit.

Measurement of Bandwidth

The group of frequencies with a response 70.7% of maximum, or more, is generally considered the bandwidth of the tuned circuit, as shown in Fig. 25–10*b*. The resonant response here is increasing current for the series circuit in Fig. 25–10*a*. Therefore, the bandwidth is measured between the two frequencies f_1 and f_2 producing 70.7% of the maximum current at f_r.

For a parallel circuit, the resonant response is increasing impedance Z_{EQ}. Then the bandwidth is measured between the two frequencies allowing 70.7% of the maximum Z_{EQ} at f_r.

The bandwidth indicated on the response curve in Fig. 25–10*b* equals 20 kHz. This is the difference between f_2 at 60 kHz and f_1 at 40 kHz, both with 70.7% response.

Compared with the maximum current of 100 mA for f_r at 50 kHz, f_1 below resonance and f_2 above resonance each allows a rise to 70.7 mA. All frequencies in this band 20 kHz wide allow 70.7 mA, or more, as the resonant response in this example.

Bandwidth Equals f_r/Q

Sharp resonance with high Q means narrow bandwidth. The lower the Q, the broader the resonant response and the greater the bandwidth.

Also, the higher the resonant frequency, the greater the range of frequency values included in the bandwidth for a given sharpness of resonance. Therefore, the bandwidth of a resonant circuit depends on the factors f_r and Q. The formula is

$$f_2 - f_1 = \Delta f = \frac{f_r}{Q} \qquad \textbf{(25–8)}$$

where Δf is the total bandwidth in the same units as the resonant frequency f_r. The bandwidth Δf can also be abbreviated BW.

Figure 25–10 Bandwidth of a tuned *LC* circuit. (*a*) Series circuit with input of 0 to 100 kHz. (*b*) Response curve with bandwidth Δf equal to 20 kHz between f_1 and f_2.

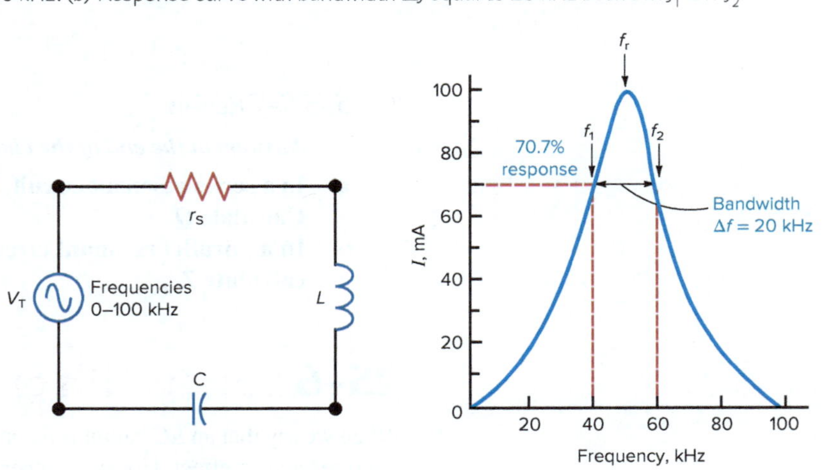

For example, a series circuit resonant at 800 kHz with a Q of 100 has a bandwidth of $^{800}/_{100}$, or 8 kHz. Then the I is 70.7% of maximum, or more, for all frequencies for a band 8 kHz wide. This frequency band is centered around 800 kHz, from 796 to 804 kHz.

With a parallel resonant circuit having a Q higher than 10, Formula (25–8) also can be used for calculating the bandwidth of frequencies that provide 70.7% or more of the maximum Z_{EQ}. However, the formula cannot be used for parallel resonant circuits with low Q, as the resonance curve then becomes unsymmetrical.

High *Q* Means Narrow Bandwidth

Th e effect for different values of Q is illustrated in Fig. 25–11. Note that a higher Q for the same resonant frequency results in less bandwidth. The slope is sharper for the sides or *skirts* of the response curve, in addition to its greater amplitude.

High Q is generally desirable for more output from the resonant circuit. However, it must have enough bandwidth to include the desired range of signal frequencies.

GOOD TO KNOW

Another formula that can be used to calculate the bandwidth of a series resonant circuit is $\Delta f = \frac{R}{2\pi L}$ where R represents the total series resistance of the circuit.

The Edge Frequencies

Both f_1 and f_2 are separated from f_r by one-half of the total bandwidth. For the top curve in Fig. 25–11, as an example, with a Q of 80, Δf is ±5 kHz centered around 800 kHz for f_r. To determine the edge frequencies,

$$f_1 = f_r - \frac{\Delta f}{2} = 800 - 5 = 795 \text{ kHz}$$

$$f_2 = f_r + \frac{\Delta f}{2} = 800 + 5 = 805 \text{ kHz}$$

These examples assume that the resonance curve is symmetrical. This is true for a high-Q parallel resonant circuit and a series resonant circuit with any Q.

Figure 25–11 Higher Q provides a sharper resonant response. Amplitude is I for series resonance or Z_{EQ} for parallel resonance. Bandwidth at half-power frequencies is Δf.

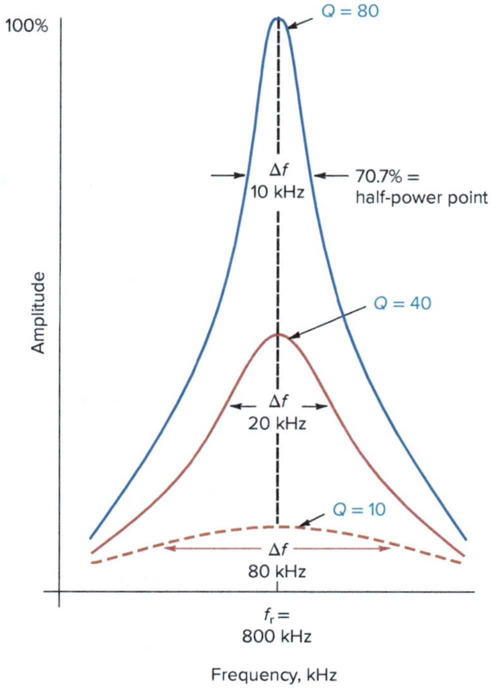

Example 25-9

An LC circuit resonant at 2000 kHz has a Q of 100. Find the total bandwidth Δf and the edge frequencies f_1 and f_2.

ANSWER

$$\Delta f = \frac{f_r}{Q} = \frac{2000 \text{ kHz}}{100} = 20 \text{ kHz}$$

$$f_1 = f_r - \frac{\Delta f}{2} = 2000 - 10 = 1990 \text{ kHz}$$

$$f_2 = f_r + \frac{\Delta f}{2} = 2000 + 10 = 2010 \text{ kHz}$$

Example 25-10

Repeat Example 25–9 for an f_r equal to 6000 kHz and the same Q of 100.

ANSWER

$$f = \frac{f_r}{Q} = \frac{6000 \text{ kHz}}{100} = 60 \text{ kHz}$$

$$f_1 = 6000 - 30 = 5970 \text{ kHz}$$

$$f_2 = 6000 + 30 = 6030 \text{ kHz}$$

Notice that Δf is three times as wide as Δf in Example 25–9 for the same Q because f_r is three times higher.

Half-Power Points

It is simply for convenience in calculations that the bandwidth is defined between the two frequencies having 70.7% response. At each of these frequencies, the net capacitive or inductive reactance equals the resistance. Then the total impedance of the series reactance and resistance is 1.4 times greater than R. With this much more impedance, the current is reduced to $1/1.414$, or 0.707, of its maximum value.

Furthermore, the relative current or voltage value of 70.7% corresponds to 50% in power, since power is I^2R or V^2/R and the square of 0.707 equals 0.50. Therefore, the bandwidth between frequencies having 70.7% response in current or voltage is also the bandwidth in terms of **half-power points**. Formula (25–8) is derived for Δf between the points with 70.7% response on the resonance curve.

Measuring Bandwidth to Calculate Q

The half-power frequencies f_1 and f_2 can be determined experimentally. For series resonance, find the two frequencies at which the current is 70.7% of maximum I, or for parallel resonance, find the two frequencies that make the impedance 70.7% of the maximum Z_{EQ}. The following method uses the circuit in Fig. 25–9 for measuring Z_{EQ}, but with different values to determine its bandwidth and Q:

1. Tune the circuit to resonance and determine its maximum Z_{EQ} at f_r. In this example, assume that Z_{EQ} is 10,000 Ω at the resonant frequency of 200 kHz.
2. Keep the same amount of input voltage, but change its frequency slightly below f_r to determine the frequency f_1 that results in a Z_1 equal to 70.7% of Z_{EQ}. The required value here is 0.707 × 10,000, or 7070 Ω, for Z_1 at f_1. Assume that this frequency f_1 is determined to be 195 kHz.
3. Similarly, find the frequency f_2 above f_r that results in the impedance Z_2 of 7070 Ω. Assume that f_2 is 205 kHz.
4. The total bandwidth between the half-power frequencies equals $f_2 - f_1$ or 200 − 195. Then the value of $\Delta f = 10$ kHz.
5. Then $Q = f_r/\Delta f$ or 200 kHz/10 kHz = 20 for the calculated value of Q.

In this way, measuring the bandwidth makes it possible to determine Q. With Δf and f_r, Q can be determined for either parallel or series resonance.

25–6 Self-Review

Answers at the end of the chapter.

a. **An *LC* circuit with f_r of 10 MHz has a Q of 40. Calculate the half-power bandwidth.**
b. **For an f_r of 500 kHz and bandwidth Δf of 10 kHz, calculate Q.**

25–7 Tuning

Tuning means obtaining resonance at different frequencies by varying either L or C. As illustrated in Fig. 25–12, the variable capacitance C can be adjusted to tune the series LC circuit to resonance at any one of the five different frequencies. Each of the voltages V_1 to V_5 indicates an AC input with a specific frequency. Which one is selected for maximum output is determined by the resonant frequency of the LC circuit.

When C is set to 424 pF, for example, the resonant frequency of the LC circuit is 500 kHz for f_{r1}. The input voltage whose frequency is 500 kHz then produces a resonant rise of current that results in maximum output voltage across C. At other frequencies, such as 707 kHz, the voltage output is less than the input. With C at 424 pF, therefore, the LC circuit tuned to 500 kHz selects this frequency by providing much more voltage output than other frequencies.

Suppose that we want maximum output for the AC input voltage that has the frequency of 707 kHz. Then C is set at 212 pF to make the LC circuit resonant at 707 kHz for f_{r2}. Similarly, the tuned circuit can resonate at a different frequency for each input voltage. In this way, the LC circuit is tuned to select the desired frequency.

Figure 25–12 Tuning a series *LC* circuit. (*a*) Input voltages at different frequencies. (*b*) Relative response for each frequency when *C* is varied (not to scale).

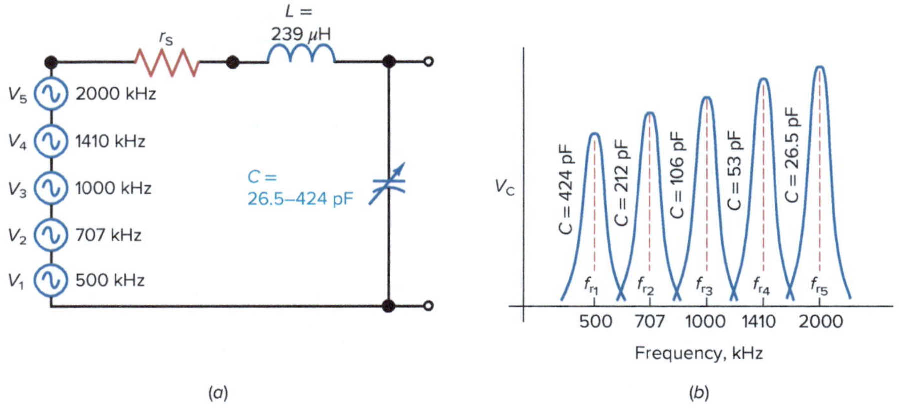

Table 25–4	Tuning *LC* Circuit by Varying *C*	
$L, \mu H$	C, pF	f_r, kHz
239	424	500
239	212	707
239	106	1000
239	53	1410
239	26.5	2000

GOOD TO KNOW

The tuning ratio, *TR*, for a capacitor is the ratio of its maximum capacitance to its minimum capacitance, or $TR = \frac{C_{max}}{C_{min}}$. For a given range of frequencies, the required capacitance tuning ratio can be calculated as $TR = \left(\frac{f_{r(max)}}{f_{r(min)}}\right)^2$.

The variable capacitance *C* can be set at the values listed in Table 25–4 to tune the *LC* circuit to different frequencies. Only five frequencies are listed here, but any one capacitance value between 26.5 and 424 pF can tune the 239-μH coil to resonance at any frequency in the range of 500 to 2000 kHz. Note that a parallel resonant circuit also can be tuned by varying *C* or *L*.

Tuning Ratio

When an *LC* circuit is tuned, the change in resonant frequency is inversely proportional to the square root of the change in *L* or *C*. Referring to Table 25–4, notice that when *C* is decreased to one-fourth, from 424 to 106 pF, the resonant frequency doubles from 500 to 1000 kHz, or the frequency is increased by the factor $1/\sqrt{1/4}$, which equals 2.

Suppose that we want to tune through the whole frequency range of 500 to 2000 kHz. This is a tuning ratio of 4:1 for the highest to the lowest frequency. Then the capacitance must be varied from 424 to 26.5 pF, which is a 16:1 capacitance ratio.

Radio Tuning Dial

Figure 25–13 illustrates a typical application of resonant circuits in tuning a receiver to the carrier frequency of a desired station in the AM broadcast band. The tuning is done by the air capacitor *C*, which can be varied from 360 pF with the plates completely in mesh to 40 pF out of mesh. The fixed plates form the *stator*, whereas the *rotor* has plates that move in and out.

Figure 25–13 Application of tuning an *LC* circuit through the AM radio band.

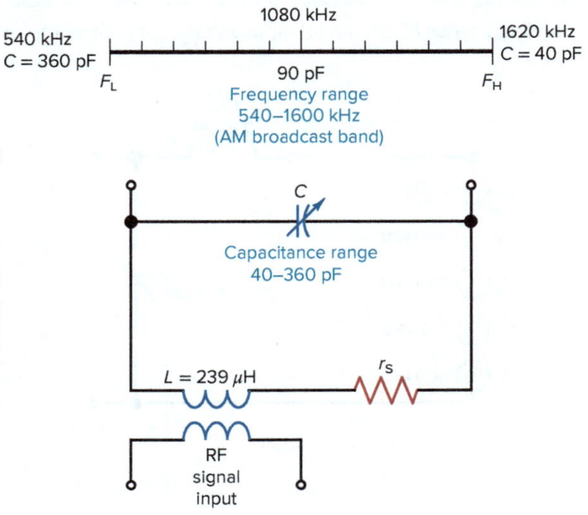

Note that the lowest frequency F_L at 540 kHz is tuned with the highest C at 360 pF. Resonance at the highest frequency F_H at 1620 kHz results from the lowest C at 40 pF.

The capacitance range of 40 to 360 pF tunes through the frequency range from 1620 kHz down to 540 kHz. Frequency F_L is one-third F_H because the maximum C is nine times the minimum C.

The same idea applies to tuning through the commercial FM broadcast band of 88 to 108 MHz with smaller values of L and C. Also, television receivers are tuned to a specific broadcast channel by resonance at the desired frequencies.

For electronic tuning, the C is varied by a *varactor*. This is a semiconductor diode that varies in capacitance when its voltage is changed.

25–7 Self-Review

Answers at the end of the chapter.

a. When a tuning capacitor is completely in mesh, is the *LC* circuit tuned to the highest or lowest frequency in the band?

b. A tuning ratio of 2:1 in frequency requires what ratio of variable *L* or *C*?

25–8 Mistuning

Suppose that a series LC circuit is tuned to 1000 kHz, but the frequency of the input voltage is 17 kHz, completely off-resonance. The circuit could provide a Q rise in output voltage for current having the frequency of 1000 kHz, but there is no input voltage and therefore no current at this frequency.

The input voltage produces current that has a frequency of 17 kHz. This frequency cannot produce a resonant rise in current, however, because the current is limited by the net reactance. When the frequency of the input voltage and the resonant frequency of the LC circuit are not the same, therefore, the mistuned circuit has very little output compared with the Q rise in voltage at resonance.

Similarly, when a parallel circuit is mistuned, it does not have a high value of impedance. Furthermore, the net reactance off-resonance makes the LC circuit either inductive or capacitive.

Series Circuit Off-Resonance

When the frequency of the input voltage is lower than the resonant frequency of a series LC circuit, the capacitive reactance is greater than the inductive reactance. As a result, there is more voltage across the capacitive reactance than across the inductive reactance. The series LC circuit is capacitive below resonance, therefore, with capacitive current leading the generator voltage.

Above the resonant frequency, the inductive reactance is greater than the capacitive reactance. As a result, the circuit is inductive above resonance with inductive current that lags the generator voltage. In both cases, there is much less output voltage than at resonance.

Parallel Circuit Off-Resonance

With a parallel LC circuit, the smaller amount of inductive reactance below resonance results in more inductive branch current than capacitive branch current. The net line current is inductive, therefore, making the parallel LC circuit inductive below resonance, as the line current lags the generator voltage.

Above the resonant frequency, the net line current is capacitive because of the higher value of capacitive branch current. In addition, the parallel LC circuit is capacitive with line current leading the generator voltage. In both cases, the total impedance of the parallel circuit is much less than the maximum impedance at resonance. Note that the capacitive and inductive effects off-resonance are opposite for series and parallel *LC* circuits.

25–8 Self-Review

Answers at the end of the chapter.

a. Is a series resonant circuit inductive or capacitive below resonance?
b. Is a parallel resonant circuit inductive or capacitive below resonance?

25–9 Analysis of Parallel Resonant Circuits

Parallel resonance is more complex than series resonance because the reactive branch currents are not exactly equal when X_L equals X_C. The reason is that the coil has its series resistance r_S in the X_L branch, whereas the capacitor has only X_C in its branch.

For high-*Q* circuits, we consider r_S negligible. In low-*Q* circuits, however, the inductive branch must be analyzed as a complex impedance with X_L and r_S in series. This impedance is in parallel with X_C, as shown in Fig. 25–14. The total impedance Z_{EQ} can then be calculated by using complex numbers, as explained in Chap. 24.

Figure 25–14 General method of calculating Z_{EQ} for a parallel resonant circuit as $(Z_1 \times Z_2)/(Z_1 + Z_2)$ with complex numbers.

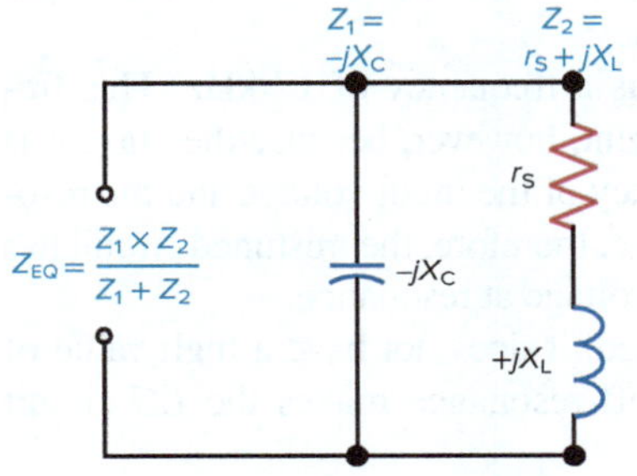

High-*Q* Circuit

We can apply the general method in Fig. 25–14 to the parallel resonant circuit shown before in Fig. 25–6 to see whether Z_{EQ} is 225,000 Ω. In this example, X_L and X_C are 1500 Ω and r_S is 10 Ω. The calculations are

$$Z_{EQ} = \frac{Z_1 \times Z_2}{Z_1 + Z_2} = \frac{-j1500 \times (j1500 + 10)}{-j1500 + j1500 + 10}$$

$$= \frac{-j^2 2.25 \times 10^6 - j15{,}000}{10} = -j^2 2.25 \times 10^5 - j1500$$

$$= 225{,}000 - j1500 = 225{,}000\underline{/0^\circ}\ \Omega$$

Note that $-j^2$ is +1. Also, the reactive $j1500$ Ω is negligible compared with the resistive 225,000 Ω. This answer for Z_{EQ} is the same as $Q \times X_L$, or 150 × 1500, because of the high *Q* with negligibly small r_S.

Low-*Q* Circuit

We can consider a *Q* less than 10 as low. For the same circuit in Fig. 25–6, if r_S is 300 Ω with an X_L of 1500 Ω, the *Q* will be $^{1500}/_{300}$, which equals 5. For this case of appreciable r_S, the branch currents cannot be equal when X_L and X_C are equal because then the inductive branch will have more impedance and less current.

With a low-*Q* circuit, Z_{EQ} must be calculated in terms of the branch impedances. For this example, the calculations are simpler with all impedances stated in kilohms:

$$Z_{EQ} = \frac{Z_1 \times Z_2}{Z_1 + Z_2} = \frac{-j1.5 \times (j1.5 + 0.3)}{-j1.5 + j1.5 + 0.3} = \frac{-j^2 2.25 - j0.45}{0.3}$$

$$= 7.5 - j1.5\ \Omega = 7.65\underline{/-11.3^\circ}\ \text{k}\Omega = 7650\underline{/-11.3^\circ}\ \Omega$$

The phase angle θ is not zero because the reactive branch currents are unequal, even though X_L and X_C are equal. The appreciable value of r_S in the X_L branch makes this branch current smaller than I_C in the X_C branch.

GOOD TO KNOW

When $Q < 10$ in a parallel resonant circuit, Z_{EQ} is maximum and I_T is minimum at a frequency slightly less than f_r.

Criteria for Parallel Resonance

The frequency f_r that makes $X_L = X_C$ is always $1/(2\pi\sqrt{LC})$. However, for low-Q circuits, f_r does not necessarily provide the desired resonance effect. The three main criteria for parallel resonance are

1. Zero phase angle and unity power factor.
2. Maximum impedance and minimum line current.
3. $X_L = X_C$. This is resonance at $f_r = 1/(2\pi\sqrt{LC})$.

These three effects do not occur at the same frequency in parallel circuits that have low Q. The condition for unity power factor is often called ***antiresonance*** in a parallel LC circuit to distinguish it from the case of equal X_L and X_C.

Note that when Q is 10 or higher, though, the parallel branch currents are practically equal when $X_L = X_C$. Then at $f_r = 1/(2\pi\sqrt{LC})$, the line current is minimum with zero phase angle, and the impedance is maximum.

For a series resonant circuit, there are no parallel branches to consider. Therefore, the current is maximum at exactly f_r, whether the Q is high or low.

25–9 Self-Review

Answers at the end of the chapter.

a. Is a Q of 8 a high or low value?

b. With this Q, will the I_L be more or less than I_C in the parallel branches when $X_L = X_C$?

25–10 Damping of Parallel Resonant Circuits

In Fig. 25–15*a*, the shunt R_P across L and C is a **damping** resistance because it lowers the Q of the tuned circuit. The R_P may represent the resistance of the external source driving the parallel resonant circuit, or R_P can be an actual resistor added for lower Q and greater bandwidth. Using the parallel R_P to reduce Q is better than

Figure 25–15 The Q of a parallel resonant circuit in terms of coil resistance r_S and parallel damping resistor R_P. See Formula (25–10) for calculating Q. (*a*) Parallel R_P but negligible r_S. (*b*) Series r_S but no R_P branch. (*c*) Both R_P and r_S.

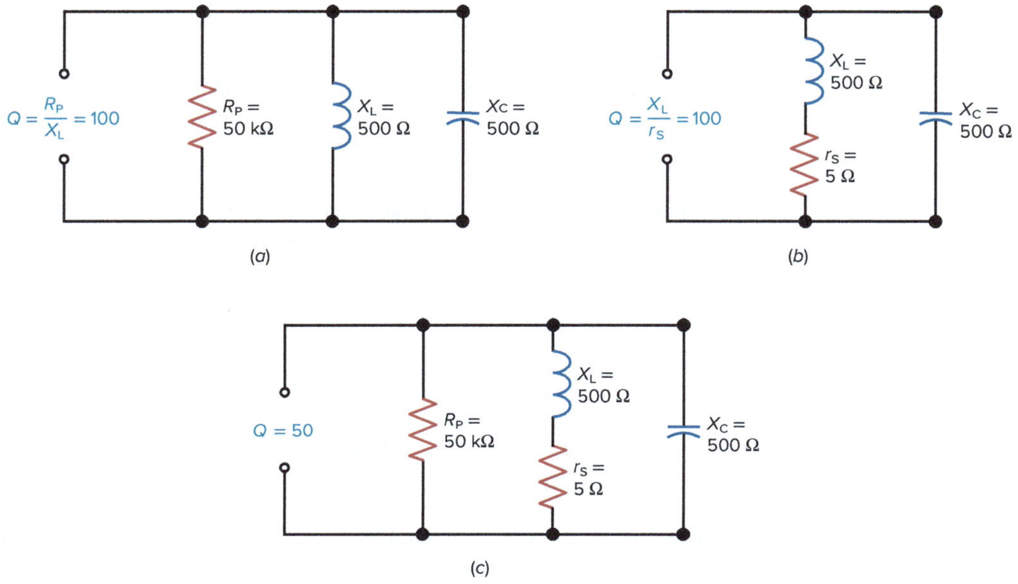

increasing the series resistance r_S because the resonant response is more symmetrical with shunt damping.

The effect of varying the parallel R_P is opposite from that of the series r_S. A lower value of R_P lowers the Q and reduces the sharpness of resonance. Remember that less resistance in a parallel branch allows more current. This resistive branch current cannot be canceled at resonance by the reactive currents. Therefore, the resonant dip to minimum line current is less sharp with more resistive line current. Specifically, when Q is determined by parallel resistance

$$Q = \frac{R_P}{X_L} \qquad \textbf{(25–9)}$$

This relationship with shunt R_P is the reciprocal of the Q formula with series r_S. Reducing R_P decreases Q, but reducing r_S increases Q. The damping can be done by series r_S, parallel R_P, or both.

Parallel R_P without r_S

In Fig. 25–15*a*, Q is determined only by the R_P, as no series r_S is shown. We can consider that r_S is zero or very small. Then the Q of the coil is infinite or high enough to be greater than the damped Q of the tuned circuit, by a factor of 10 or more. The Q of the damped resonant circuit here is $R_P/X_L = 50{,}000/500 = 100$.

Series r_S without R_P

In Fig. 25–15*b*, Q is determined only by the coil resistance r_S, as no shunt damping resistance is used. Then $Q = X_L/r_S = 500/5 = 100$. This is the Q of the coil, which is also the Q of the parallel resonant circuit without shunt damping.

Conversion of r_S or R_P

For the circuits in both Fig. 25–15*a* and *b*, Q is 100 because the 50,000-Ω R_P is equivalent to the 5-Ω r_S as a damping resistance. One value can be converted to the other. Specifically,

$$r_S = \frac{X_L^2}{R_P}$$

or

$$R_P = \frac{X_L^2}{r_S}$$

In this example, r_S equals $250{,}000/50{,}000 = 5\ \Omega$, or R_P is $250{,}000/5 = 50{,}000\ \Omega$.

Damping with Both r_S and R_P

Figure 25–15*c* shows the general case of damping where both r_S and R_P must be considered. Then the Q of the circuit can be calculated as

$$Q = \frac{X_L}{r_S + X_L^2/R_P} \qquad \textbf{(25–10)}$$

For the values in Fig. 25–15*c*,

$$Q = \frac{500}{5 + 250{,}000/50{,}000} = \frac{500}{5+5} = \frac{500}{10}$$
$$= 50$$

The Q is lower here compared with Fig. 25–15*a* or *b* because this circuit has both series and shunt damping.

Note that for an r_S of zero, Formula (25–10) can be inverted and simplified to $Q = R_P/X_L$. This is the same as Formula (25–9) for shunt damping alone.

For the opposite case where R_P is infinite, that is, an open circuit, Formula (25–10) reduces to X_L/r_S. This is the same as Formula (25–4) without shunt damping.

25–10 Self-Review

Answers at the end of the chapter.

a. A parallel resonant circuit has an X_L of 1000 Ω and an r_s of 20 Ω, without any shunt damping. Calculate Q.

b. A parallel resonant circuit has an X_L of 1000 Ω, negligible r_s, and shunt R_P of 50 kΩ. Calculate Q.

c. How much is Z_{EQ} at f_r for the circuits in Questions a and b?

25–11 Choosing *L* and *C* for a Resonant Circuit

The following example illustrates how resonance is an application of X_L and X_C. Suppose that we have the problem of determining the inductance and capacitance for a circuit resonant at 159 kHz. First, we need a known value for either L or C to calculate the other. Which one to choose depends on the application. In some cases, particularly at very high frequencies, C must be the minimum possible value, which might be about 10 pF. At medium frequencies, though, we can choose L for the general case when an X_L of 1000 Ω is desirable and can be obtained. Then the inductance of the required L, equal to $X_L/2\pi f$, is 0.001 H or 1 mH, for the inductive reactance of 1000 Ω.

For resonance at 159 kHz with a 1-mH L, the required C is 0.001 μF or 1000 pF. This value of C can be calculated for an X_C of 1000 Ω, equal to X_L at the f_r of 159 kHz, or from Formula (25–2). In either case, if you substitute 1×10^{-9} F for C and 1×10^{-3} H for L in the resonant frequency formula, f_r will be 159 kHz.

This combination is resonant at 159 kHz whether L and C are in series or parallel. In series, the resonant effect produces maximum current and maximum voltage across L or C at 159 kHz. The effect is desirable for the input circuit of an rf amplifier tuned to f_r because of the maximum signal. In parallel, the resonant effect at 159 kHz is minimum line current and maximum impedance across the generator. This effect is desirable for the output circuit of an rf amplifier, as the gain is maximum at f_r because of the high Z.

If we assume that the 1-mH coil used for L has an internal resistance of 20 Ω, the Q of the coil is 1000 Ω/20 Ω, which equals 50. This value is also the Q of the series resonant circuit. If there is no shunt damping resistance across the parallel LC circuit, its Q is also 50. With a Q of 50, the bandwidth of the resonant circuit is 159 kHz/50, which equals 3.18 kHz for Δf.

25–11 Self-Review

Answers at the end of the chapter.

a. What is f_r for 1000 pF of C and 1 mH of L?

b. What is f_r for 250 pF of C and 1 mH of L?

Summary

- Series and parallel resonance are compared in Table 25–5. The main difference is that series resonance produces maximum current and very low impedance at f_r, but with parallel resonance, the line current is minimum to provide very high impedance. Remember that these formulas for parallel resonance are very close approximations that can be used for circuits with a Q higher than 10. For series resonance, the formulas apply whether the Q is high or low.

Table 25–5 Comparison of Series and Parallel Resonance

Series Resonance	Parallel Resonance (high Q)
$f_r = \dfrac{1}{2\pi\sqrt{LC}}$	$f_r = \dfrac{1}{2\pi\sqrt{LC}}$
I maximum at f_r with θ of 0°	I_T minimum at f_r with θ of 0°
Impedance Z minimum at f_r	Impedance Z maximum at f_r
$Q = X_L/r_S$, or	$Q = X_L/r_S$, or
$Q = V_{out}/V_{in}$	$Q = Z_{max}/X_L$
Q rise in voltage $= Q \times V_{gen}$	Q rise in impedance $= Q \times X_L$
Bandwidth $\Delta f = f_r/Q$	Bandwidth $\Delta f = f_r/Q$
Circuit capacitive below f_r, but inductive above f_r	Circuit inductive below f_r, but capacitive above f_r
Needs low-resistance source for low r_S, high Q, and sharp tuning	Needs high-resistance source for high R_P, high Q, and sharp tuning
Source is inside LC circuit	Source is outside LC circuit

Important Terms

Antiresonance — a term to describe the condition of unity power factor in a parallel LC circuit. *Antiresonance* is used to distinguish it from the case of equal X_L and X_C values in a series LC circuit.

Bandwidth — the width of the resonant band of frequencies centered around the resonant frequency of an LC circuit.

Damping — a technique for reducing the Q of a resonant circuit to increase the bandwidth. For a parallel resonant circuit, damping is typically accomplished by adding a parallel resistor across the tank circuit.

Flywheel effect — the effect that reproduces complete sine waves in a parallel LC tank circuit when the input is only a pulse.

Half-power points — the frequencies above and below the resonant frequency that have a current or voltage value equal to 70.7% of its value at resonance.

***Q* of a resonant circuit** — a measure of the sharpness of a resonant circuit's response curve. The higher the ratio of the reactance at resonance to the series resistance, the higher the Q and the sharper the resonant effect.

Resonant frequency — the frequency at which the inductive reactance, X_L, and the capacitive reactance, X_C, of an LC circuit are equal.

Tank circuit — another name for a parallel resonant LC circuit.

Tuning — a means of obtaining resonance at different frequencies by varying either L or C in an LC circuit.

Related Formulas

$$f_r = \frac{1}{2\pi\sqrt{LC}}$$

$$C = \frac{1}{4\pi^2 f_r^2 L}$$

$$L = \frac{1}{4\pi^2 f_r^2 C}$$

$Q = \frac{X_L}{r_S}$ (series resonant circuit or parallel resonant circuit with no R_P)

$$V_L = V_C = Q \times V_{gen}$$

$$Q = \frac{V_{out}}{V_{in}}$$

$Z_{EQ} = Q \times X_L$ (parallel resonant circuit)

$$f_2 - f_1 = \Delta f = \frac{f_r}{Q}$$

$Q = \frac{R_P}{X_L}$ (Q for parallel resonant circuit without series r_S)

$Q = \frac{X_L}{r_S + X_L^2/R_P}$ (Q for parallel resonant circuit with both r_S and R_P)

Self-Test

Answers at the back of the book.

1. The resonant frequency of an *LC* circuit is the frequency where

a. $X_L = 0\ \Omega$ and $X_C = 0\ \Omega$.
b. $X_L = X_C$.
c. X_L and r_S of the coil are equal.
d. X_L and X_C are in phase.

2. The impedance of a series *LC* circuit at resonance is

a. maximum.
b. nearly infinite.
c. minimum.
d. both a and b.

3. The total line current, I_T, of a parallel *LC* circuit at resonance is

a. minimum.
b. maximum.
c. equal to I_L and I_C.
d. Q times larger than I_L or I_C.

4. The current at resonance in a series *LC* circuit is

a. zero.
b. minimum.
c. different in each component.
d. maximum.

5. The impedance of a parallel *LC* circuit at resonance is

a. zero.
b. maximum.
c. minimum.
d. equal to the r_S of the coil.

6. The phase angle of an *LC* circuit at resonance is

a. 0°.
b. +90°.
c. 180°.
d. −90°.

7. Below resonance, a series *LC* circuit appears

a. inductive.
b. resistive.
c. capacitive.
d. none of the above.

8. Above resonance, a parallel *LC* circuit appears

a. inductive.
b. resistive.
c. capacitive.
d. none of the above.

9. A parallel *LC* circuit has a resonant frequency of 3.75 MHz and a *Q* of 125. What is the bandwidth?

a. 15 kHz.
b. 30 kHz.
c. 60 kHz.
d. none of the above.

10. What is the resonant frequency of an *LC* circuit with the following values: $L = 100\ \mu$H and $C = 63.3$ pF?

a. $f_r = 1$ MHz.
b. $f_r = 8$ MHz.
c. $f_r = 2$ MHz.
d. $f_r = 20$ MHz.

11. What value of capacitance is needed to provide a resonant frequency of 1 MHz if *L* equals 50 μH?

a. 506.6 pF.
b. 506.6 μF.
c. 0.001 μF.
d. 0.0016 μF.

12. When either *L* or *C* is increased, the resonant frequency of an *LC* circuit

a. decreases.
b. increases.
c. doesn't change.
d. It cannot be determined.

13. A series *LC* circuit has a *Q* of 100 at resonance. If $V_{in} = 5\ mV_{p\text{-}p}$, how much is the voltage across *C*?

a. 50 $\mu V_{p\text{-}p}$.
b. 5 $mV_{p\text{-}p}$.
c. 50 $mV_{p\text{-}p}$.
d. 500 $mV_{p\text{-}p}$.

14. In a low *Q* parallel resonant circuit, when $X_L = X_C$,

a. $I_L = I_C$.
b. I_L is less than I_C.
c. I_C is less than I_L.
d. I_L is more than I_C.

15. To double the resonant frequency of an *LC* circuit with a fixed value of *L*, the capacitance, *C*, must be

a. doubled.
b. quadrupled.
c. reduced by one-half.
d. reduced by one-quarter.

16. A higher *Q* for a resonant circuit provides a

a. dampened response curve.
b. wider bandwidth.
c. narrower bandwidth.
d. none of the above.

17. The current at the resonant frequency of a series *LC* circuit is 10 $mA_{p\text{-}p}$. What is the value of current at the half-power points?

a. 7.07 $mA_{p\text{-}p}$.
b. 14.14 $mA_{p\text{-}p}$.
c. 5 $mA_{p\text{-}p}$.
d. 10 $mA_{p\text{-}p}$.

18. **The Q of a parallel resonant circuit can be lowered by**

a. placing a resistor in parallel with the tank.

b. adding more resistance in series with the coil.

c. decreasing the value of L or C.

d. both a and b.

19. **The ability of an LC circuit to supply complete sine waves when the input to the tank is only a pulse is called**

a. tuning.

b. the flywheel effect.

c. antiresonance.

d. its Q.

20. **Which of the following can provide a higher Q?**

a. a higher L/C ratio.

b. a lower L/C ratio.

c. more resistance in series with the coil.

d. either b or c.

Essay Questions

1. (a) State two characteristics of series resonance. (b) With a microammeter measuring current in the series LC circuit of Fig. 25–2, describe the meter readings for the different frequencies from 600 to 1400 kHz.

2. (a) State two characteristics of parallel resonance. (b) With a microammeter measuring current in the main line for the parallel LC circuit in Fig. 25–6*a*, describe the meter readings for frequencies from 600 to 1400 kHz.

3. State the Q formula for the following LC circuits: (a) series resonant; (b) parallel resonant, with series resistance r_S in the inductive branch; (c) parallel resonant with zero series resistance but shunt R_P.

4. Explain briefly why a parallel LC circuit is inductive but a series LC circuit is capacitive below f_r.

5. What is the effect on Q and bandwidth of a parallel resonant circuit if its shunt damping resistance is decreased from 50,000 to 10,000 Ω?

6. Describe briefly how you would use an AC meter to measure the bandwidth of a series resonant circuit to calculate the circuit Q.

7. Why is a low-resistance generator good for high Q in series resonance, but a high-resistance generator is needed for high Q in parallel resonance?

8. Referring to Fig. 25–13, why is it that the middle frequency of 1080 kHz does not correspond to the middle capacitance value of 200 pF?

9. (a) Give three criteria for parallel resonance. (b) Why is the antiresonant frequency f_a different from f_r with a low-Q circuit? (c) Why are they the same for a high-Q circuit?

10. Show how Formula (25–10) reduces to R_P/X_L when r_S is zero.

11. (a) Specify the edge frequencies f_1 and f_2 for each of the three response curves in Fig. 25–11. (b) Why does lower Q allow more bandwidth?

12. (a) Why does maximum Z for a parallel resonant circuit correspond to minimum line current? (b) Why does zero phase angle for a resonant circuit correspond to unity power factor?

13. Explain how manual tuning of an LC circuit can be done with a capacitor or a coil.

14. What is meant by *electronic tuning*?

15. Suppose it is desired to tune an LC circuit from 540 to 1600 kHz by varying either L or C. Explain how the bandwidth Δf is affected by (a) varying L to tune the LC circuit; (b) varying C to tune the LC circuit.

Problems

SECTION 25–1 THE RESONANCE EFFECT

25–1 Define what is meant by a resonant circuit.

25–2 What is the main application of resonance?

25–3 If an inductor in a resonant LC circuit has an X_L value of 1 kΩ, how much is X_C?

SECTION 25–2 SERIES RESONANCE

25–4 List the main characteristics of a series resonant circuit.

25–5 Figure 25–16 shows a series resonant circuit with the values of X_L and X_C at f_r. Calculate the

a. net reactance, X.

b. total impedance, Z_T.

c. current, I.

d. phase angle, θ.

e. voltage across L.

f. voltage across C.

g. voltage across r_S.

Figure 25–16

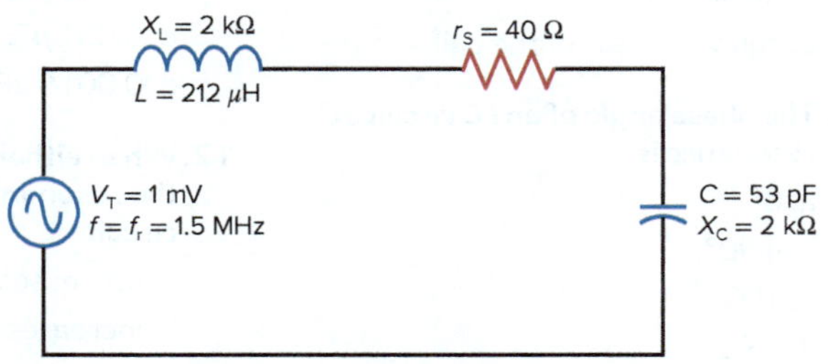

25–6 In Fig. 25–16, what happens to Z_T and I if the frequency of the applied voltage increases or decreases from the resonant frequency, f_r? Explain your answer.

25–7 In Fig. 25–16, why is the phase angle, θ, 0° at f_r?

25–8 In Fig. 25–16, assume that the frequency of the applied voltage increases slightly above f_r and $X_L = 2.02$ kΩ, and $X_C = 1.98$ kΩ. Calculate the

a. net reactance, X.

b. total impedance, Z_T.

c. current, I.

d. phase angle, θ.

e. voltage across L.

f. voltage across C.

g. voltage across r_S.

SECTION 25–3 PARALLEL RESONANCE

25–9 List the main characteristics of a parallel resonant circuit.

25–10 Figure 25–17 shows a parallel resonant circuit with the same values for X_L, X_C, and r_S as in Fig. 25–16. With an applied voltage of 10 V and a total line current, I_T, of 100 μA, calculate the

a. inductive current, I_L (ignore r_S).

b. capacitive current, I_C.

c. net reactive branch current, I_X.

d. equivalent impedance, Z_{EQ}, of the tank circuit.

Figure 25–17

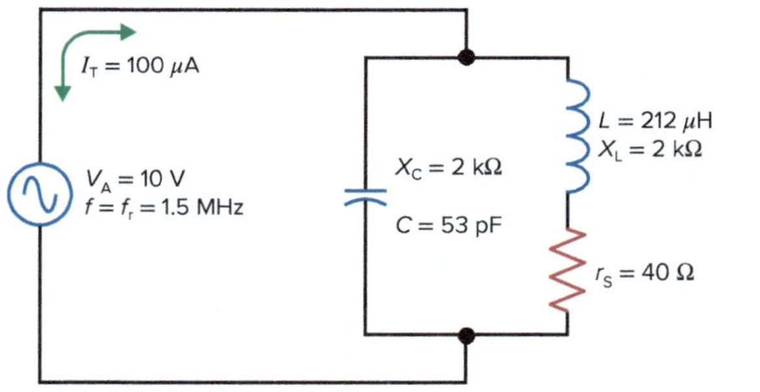

25–11 In Fig. 25–17, is X_L, X_C, or r_S responsible for the line current, I_T, of 100 μA at f_r?

25–12 In Fig. 25–17, what happens to Z_{EQ} and I_T as the frequency of the applied voltage increases or decreases from the resonant frequency, f_r? Explain your answer.

SECTION 25–4 RESONANT FREQUENCY $f_r = 1/(2\pi\sqrt{LC})$

25–13 Calculate the resonant frequency, f_r, of an LC circuit with the following values:

a. $L = 100$ μH and $C = 40.53$ pF.

b. $L = 250$ μH and $C = 633.25$ pF.

c. $L = 40$ μH and $C = 70.36$ pF.

d. $L = 50$ μH and $C = 20.26$ pF.

25–14 What value of inductance, L, must be connected in series with a 50-pF capacitance to obtain an f_r of 3.8 MHz

25–15 What value of capacitance, C, must be connected in parallel with a 100-μH inductance to obtain an f_r of 1.9 MHz?

25–16 In Fig. 25–18, what is the range of resonant frequencies as C is varied from 40 to 400 pF?

Figure 25–18

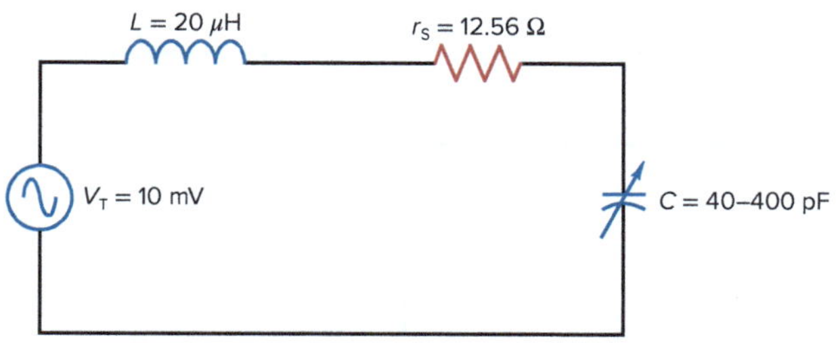

25–17 With C set at 50.67 pF in Fig. 25–18, solve for the following:

a. f_r.

b. X_L and X_C at f_r.

c. Z_T at f_r.

d. I at f_r.

e. V_L and V_C at f_r.

f. θ_Z at f_r.

25–18 In Fig. 25–18, what value of C will provide an f_r of 2.5 MHz?

25–19 If C is set at 360 pF in Fig. 25–18, what is the resonant frequency, f_r? What value of C will double the resonant frequency?

SECTION 25–5 *Q* MAGNIFICATION FACTOR OF A RESONANT CIRCUIT

25–20 What is the Q of the series resonant circuit in Fig. 25–18 with C set at

a. 202.7 pF?

b. 50.67pF?

25–21 A series resonant circuit has the following values: $L =$ 50 μH, $C = 506.6$ pF, $r_S = 3.14$ Ω, and $V_{in} = 10$ mV. Calculate the following:

a. f_r.

b. Q.

c. V_L and V_C.

25–22 In reference to Prob. 25–21, assume that L is doubled to 100 μH and C is reduced by one-half to 253.3 pF. If r_S and V_{in} remain the same, recalculate the following:

a. f_r.

b. Q.

c. V_L and V_C.

25–23 What is the Q of a series resonant circuit if the output voltage across the capacitor is 15 $V_{p\text{-}p}$ with an input voltage of 50 $mV_{p\text{-}p}$?

25–24 Explain why the Q of a resonant circuit cannot increase without limit as X_L increases for higher frequencies.

25–25 In Fig. 25–19, solve for the following:

a. f_r.
b. X_L and X_C at f_r.
c. I_L and I_C at f_r.
d. Q.
e. Z_{EQ} at f_r.
f. I_T.

Figure 25–19

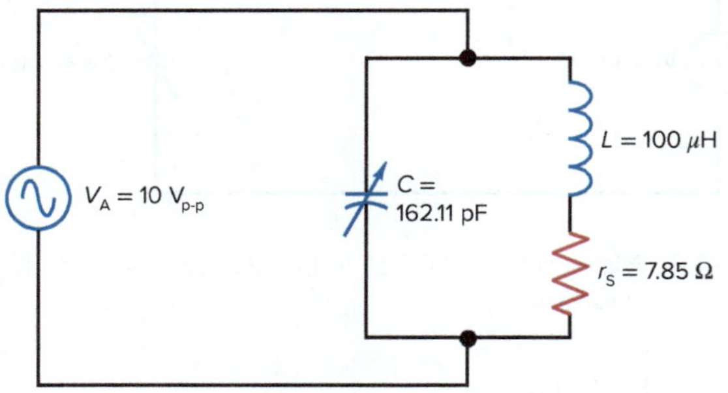

25–26 For the parallel resonant circuit in Fig. 25–17, prove that $Z_{EQ} = 100$ kΩ, as calculated earlier in step d of Prob. 25–10.

25–27 The equivalent impedance, Z_{EQ}, of an *LC* tank circuit measures 150 kΩ using the experimental technique shown in Fig. 25–9. If the resonant frequency is 2.5 MHz and $L = 50$ μH, calculate the Q of the resonant circuit.

SECTION 25–6 BANDWIDTH OF A RESONANT CIRCUIT

25–28 With *C* set at 50.67 pF in Fig. 25–18, calculate the following:

a. the bandwidth, Δf.
b. the edge frequencies f_1 and f_2.
c. the current, I, at f_r, f_1, and f_2.

25–29 In Fig. 25–19, calculate the following:

a. the bandwidth, Δf.
b. the edge frequencies f_1 and f_2.
c. the equivalent impedance, Z_{EQ} at f_r, f_1, and f_2.

25–30 Does a higher Q correspond to a wider or narrower bandwidth?

25–31 In Fig. 25–20, calculate the following:

a. f_r.
b. X_L and X_C at f_r.
c. Z_T at f_r.
d. I at f_r.
e. Q.
f. V_L and V_C at f_r.
g. θ_Z at f_r.
h. Δf, f_1, and f_2.
i. I at f_1 and f_2.

Figure 25–20

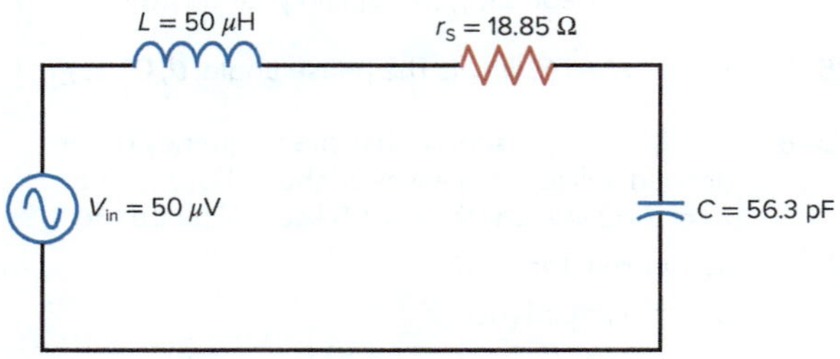

25–32 In Fig. 25–20, calculate the following values at the edge frequency, f_1:

a. X_L.
b. X_C.
c. the net reactance, X.
d. Z_T.
e. I.
f. θ_Z

25–33 Using the values from Probs. 25–31 and 25–32, compare Z_T and I at f_1 and f_r.

25–34 In Fig. 25–21, calculate the following:

a. f_r.
b. X_L and X_C at f_r.
c. I_L and I_C at f_r.
d. Q.
e. Z_{EQ}.
f. I_T.
g. θ_I.
h. Δf, f_1, and f_2.

Figure 25–21

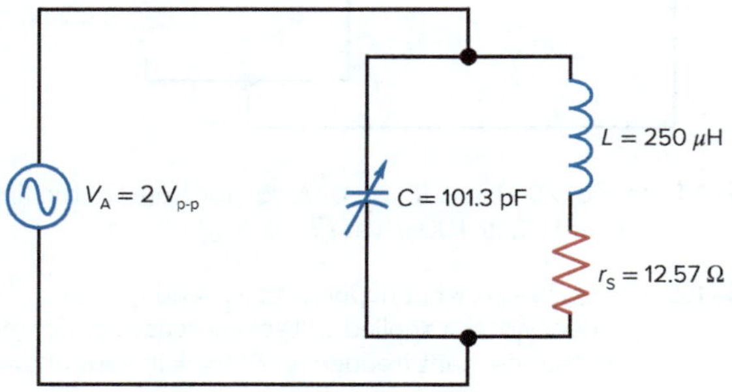

25–35 In Fig. 25–21, how much is Z_{EQ} at f_1 and f_2? How much is I_T?

SECTION 25–7 TUNING

25–36 In Fig. 25–18, calculate

a. the capacitance tuning ratio.
b. the ratio of the highest to lowest resonant frequency when *C* is varied from its lowest to its highest value.

25–37 In Fig. 25–18, is the voltage across the capacitor the same for all resonant frequencies across the tuning range? Why or why not?

SECTION 25–8 MISTUNING

25–38 Does a series *LC* circuit appear capacitive or inductive when the frequency of the input voltage is

a. lower than its resonant frequency?

b. higher than its resonant frequency?

25–39 Does a parallel *LC* circuit appear capacitive or inductive when the frequency of the input voltage is

a. lower than its resonant frequency?

b. higher than its resonant frequency?

SECTION 25–9 ANALYSIS OF PARALLEL RESONANT CIRCUITS

25–40 Is a *Q* of 5 considered a low or a high *Q*?

25–41 In Fig. 25–21, is the *Q* at f_r considered a low or a high *Q*?

25–42 If r_S is increased to 392.8 Ω in Fig. 25–21, calculate the following values:

a. f_r.

b. *Q*.

c. Z_{EQ}.

d. I_T.

25–43 In Prob. 25–42, calculate I_L and I_C at f_r. Are they equal? If not, why?

25–44 In reference to step c of Prob. 25–42, calculate Z_{EQ} as $Q \times X_L$. Does this answer agree with the original value obtained in Prob. 25–42, step c? If not, why?

25–45 In reference to Prob. 25–42, do you think Z_{EQ} is maximum at f_r or is Z_{EQ} maximum above or below f_r?

SECTION 25–10 DAMPING OF PARALLEL RESONANT CIRCUITS

25–46 In Fig. 25–19, calculate the *Q* and bandwidth, Δf, if a 100-kΩ resistor is placed in parallel with the tank circuit.

25–47 In Fig. 25–21, calculate the *Q* and bandwidth, Δf, if a 2-MΩ resistor is placed in parallel with the tank circuit.

25–48 In Fig. 25–19, convert the series resistance, r_S, to an equivalent parallel resistance, R_P.

25–49 Repeat Prob. 25–48 for Fig. 25–21.

SECTION 25–11 CHOOSING L AND C FOR A RESONANT CIRCUIT

25–50 Assume that it is desired to have an X_L value of 1.5 kΩ at the resonant frequency of 2 MHz. What are the required values of *L* and *C*?

Critical Thinking

25–51 Prove that

$$X_L = \sqrt{L/C}$$

for an *LC* circuit at f_r.

25–52 Suppose you are an engineer designing a coil to be used in a resonant *LC* circuit. Besides obtaining the required inductance *L*, your main concern is to reduce the skin effect to obtain as high a *Q* as possible for the *LC* circuit. List three design techniques that would reduce or minimize the skin effect in the coil windings.

Answers to Self-Reviews

25–1 **a.** 1000 kHz
b. 1000 kHz

25–2 **a.** false
b. true
c. true

25–3 **a.** false
b. true
c. true

25–4 **a.** decreased
b. 200 μH
c. 39.48

25–5 **a.** 100
b. 25 kΩ

25–6 **a.** 0.25 MHz
b. 50

25–7 **a.** lowest
b. 1:4

25–8 **a.** capacitive
b. inductive

25–9 **a.** low
b. less

25–10 **a.** 50
b. 50
c. 50 kΩ

25–11 **a.** 159 kHz
b. 318 kHz

Laboratory Application Assignment

In this lab application assignment, you will examine the resonant effect in both series and parallel *LC* circuits. You will construct a series *LC* circuit and see how the current, I, rises to its maximum value at resonance. You will also construct a parallel *LC* circuit and see how the impedance, *Z*, rises to its maximum value at resonance.

Equipment: Obtain the following items from your instructor.

- Function generator
- Oscilloscope
- 4.7-mH inductor
- 0.001-μF capacitor
- 47-Ω and 100 kΩ carbon-film resistors

Determining the Resonant Frequency, f_r, of a Series *LC* Circuit

Examine the series *LC* circuit in Fig. 25–22*a*. Calculate and record the resonant frequency, f_r. f_r = ______________

Construct the series *LC* circuit in Fig. 25–22*a*. Connect channel 1 of the oscilloscope to the input voltage, V_{in}, and channel 2 across the resistor, *R*, as shown. Set the input voltage to 1 $V_{p\text{-}p}$ and the frequency, *f*, to the value calculated for f_r. While viewing the resistor voltage, V_R, move the frequency dial back and forth. Measure and record the frequency that produces the maximum resistor voltage, V_R. This frequency is the resonant frequency, f_r. f_r = ______________

Record the measured value of V_R at f_r. (Make sure that V_{in} is still set to 1 $V_{p\text{-}p}$.) V_R = ______________ Does the fact that V_R is maximum at f_r prove that the series current, *I*, is also maximum at f_r? ______________ If your answer was yes, then explain why.____
____

Using the measured value of V_R, calculate and record the series current, *I*, at f_r [$I = V_{R(p\text{-}p)}/R$]. *I* = ______________ p-p Based on the fact that the current, *I*, decreases to 70.7% of its maximum value at the edge frequencies f_1 and f_2, experimentally determine the bandwidth, Δf, of the resonant circuit. (This may be difficult to do.) Δf = ______________ Calculate the *Q* of the circuit using the measured values of f_r and Δf. *Q* = ______________

Using the oscilloscope, measure and record the phase relationship between V_{in} and V_R at f_r. θ = ______________

Figure 25–22

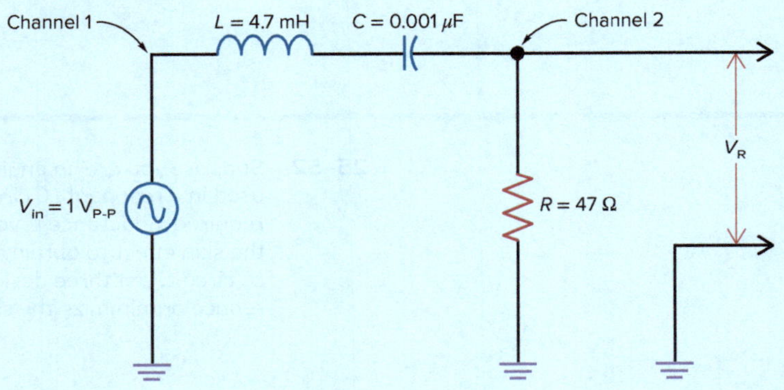

(*a*)

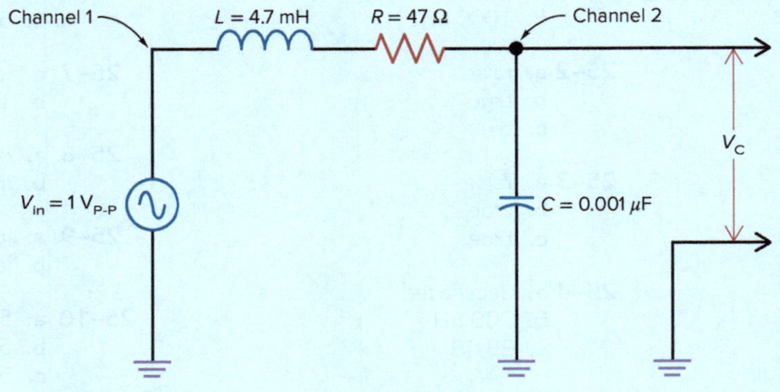

(*b*)

Does the series current have the same phase as the resistor voltage, V_R? ____________ If so, what is the phase relationship between V_{in} and I at f_r? $\theta =$ ____________

While adjusting the frequency dial above and below resonance, does the input voltage, V_{in}, dip at the resonant frequency? ____________ If your answer was yes, explain why this occurs. ____________

Rearrange the components as in Fig. 25–22*b*. Move the frequency dial back and forth until the capacitor voltage, V_C, is at its maximum value. The frequency at which V_C is maximum is the resonant frequency, f_r. This frequency should be very close to the frequency where V_R was maximum in Fig. 25–22*a*. Record the peak-to-peak capacitor voltage, V_C. $V_C =$ ____________ P-P
Is this value larger than the input voltage, V_{in}? ____________ If so, how is this possible? ____________

Determining the Resonant Frequency, f_r, of a Parallel Resonant Circuit

Examine the parallel *LC* circuit in Fig. 25–23. Calculate and record the resonant frequency, f_r. $f_r =$ ____________

Construct the parallel *LC* circuit in Fig. 25–23. Connect channel 1 of the oscilloscope to the input voltage, V_{in}, and channel 2 across the tank circuit. Set the input voltage to 10 $V_{p\text{-}p}$ and the frequency, f, to the value calculated for f_r. While viewing the tank voltage, V_{tank}, move the frequency dial back and forth. Measure and record the frequency that produces the maximum tank voltage, V_{tank}. This frequency is the resonant frequency, f_r. $f_r =$ ____________
Is the resonant frequency, f_r, the same as it was in Fig. 25–22? ____________

Does this prove that the resonant frequency is calculated the same way for both series and parallel *LC* circuits? ____________

In Fig. 25–23, why is the tank voltage, V_{tank}, maximum at f_r? ____________

From the measured values in Fig. 25–23, explain how the tank impedance, Z_{tank}, can be experimentally determined. ____________

What is the value of the tank impedance in Fig. 25–23?
$Z_{tank} =$ ____________

What is the measured phase relationship between V_{in} and V_{tank} at f_r? ____________

Does the input voltage, V_{in}, dip at the resonant frequency like it did in the series *LC* circuit of Fig. 25–22? ____________ If not, explain why. ____________

Figure 25–23

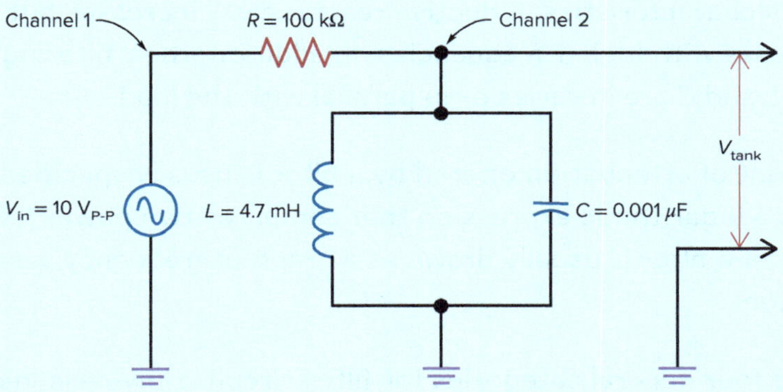

Design credits: Chapter Opener Background Image, and Multisim: Stockbyte/Getty Images

chapter 26

Filters

A filter separates different components that are mixed together. For instance, a mechanical filter can separate particles from liquid or small particles from large particles. An electrical filter can separate different frequency components.

Generally, inductors and capacitors are used for filtering because of their opposite frequency characteristics. Inductive reactance X_L increases but capacitive reactance X_C decreases with higher frequencies. In addition, their filtering action depends on whether L and C are in series or in parallel with the load.

The amount of attenuation offered by a filter is usually specified in decibels (dB). The decibel is a logarithmic expression that compares two power levels. The frequency response of a filter is usually drawn as a graph of frequency versus decibel attenuation.

There are four general categories for filter circuits: low-pass, high-pass, band-pass, and band-stop. Each type of filter serves to provide the desired filtering in a specific application. The most common filtering applications are separating audio from radio frequencies, or vice versa, and separating AC variations from the average DC level. There are many of these applications in electronic circuits.

Chapter Outline

Chapter Objectives

After studying this chapter, you should be able to

- *State* the difference between a low-pass and a high-pass filter.
- *Explain* what is meant by *pulsating direct current*.
- *Explain* how a transformer acts as a high-pass filter.
- *Explain* how an *RC* coupling circuit couples alternating current but blocks direct current.
- *Explain* the function of a bypass capacitor.
- *Calculate* the cutoff frequency, output voltage, and phase angle of basic *RL* and *RC* filters.
- *Explain* the operation of band-pass and band-stop filters.
- *Explain* why log-log graph paper or semilog graph paper is used to plot a frequency response.
- *Define* the term *decibel*.
- *Explain* how resonant circuits can be used as band-pass or band-stop filters.
- *Describe* the function of a power-line filter and a television antenna filter.

Important Terms

attenuation
band-pass filter
band-stop filter
bypass capacitor
crystal filter
cutoff frequency
decade
decibel (dB)
fluctuating DC
high-pass filter
low-pass filter
octave
pulsating DC

GOOD TO KNOW

The AC signal voltages present at the input to a filter usually span a wide range of frequencies.

26–1 Examples of Filtering

Electronic circuits often have currents of different frequencies corresponding to voltages of different frequencies because a source produces current with the same frequency as the applied voltage. As examples, the AC signal applied to an audio circuit can have high and low audio frequencies; an rf circuit can have a wide range of radio frequencies at its input; the audio detector in a radio has both radio frequencies and audio frequencies in the output. Finally, the rectifier in a power supply produces DC output with an AC ripple superimposed on the average DC level.

In such applications where the current has different frequency components, it is usually necessary either to favor or to reject one frequency or a band of frequencies. Then an electrical filter is used to separate higher or lower frequencies.

The electrical filter can pass the higher-frequency component to the load resistance, which is the case of a **high-pass filter**, or a **low-pass filter** can be used to favor the lower frequencies. In Fig. 26–1*a*, the high-pass filter allows 10 kHz to produce output, while rejecting or attenuating the lower frequency of 100 Hz. In Fig. 26–1*b*, the filtering action is reversed to pass the lower frequency of 100 Hz, while attenuating 10 kHz. These examples are meant for high and low audio frequencies.

For the case of audio frequencies mixed with radio frequencies, a low-pass filter allows the audio frequencies in the output, whereas a high-pass filter allows passing the radio frequencies to the output.

26–1 Self-Review

Answers at the end of the chapter.

A high-pass filter will pass which of the following:

a. 100 Hz or 500 kHz.

b. 60 Hz or a steady DC level.

26–2 Direct Current Combined with Alternating Current

Current that varies in amplitude but does not reverse in polarity is considered *pulsating* or *fluctuating* direct current. It is not a steady direct current because its value fluctuates. However, it is not alternating current because the polarity remains the same, either positive or negative. The same idea applies to voltages.

Figure 26–1 Function of electrical filters. (*a*) High-pass filter couples higher frequencies to the load. (*b*) Low-pass filter couples lower frequencies to the load.

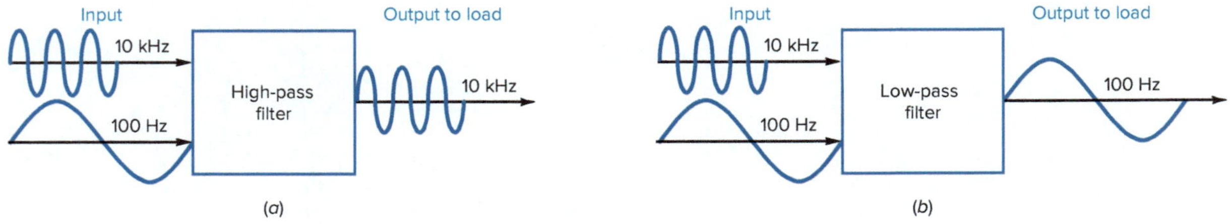

Figure 26–2 An example of a pulsating or fluctuating direct current and voltage. (*a*) Circuit. (*b*) Graph of voltage across R_L. This V equals V_B of the battery plus V_A of the AC source with a frequency of 1000 Hz.

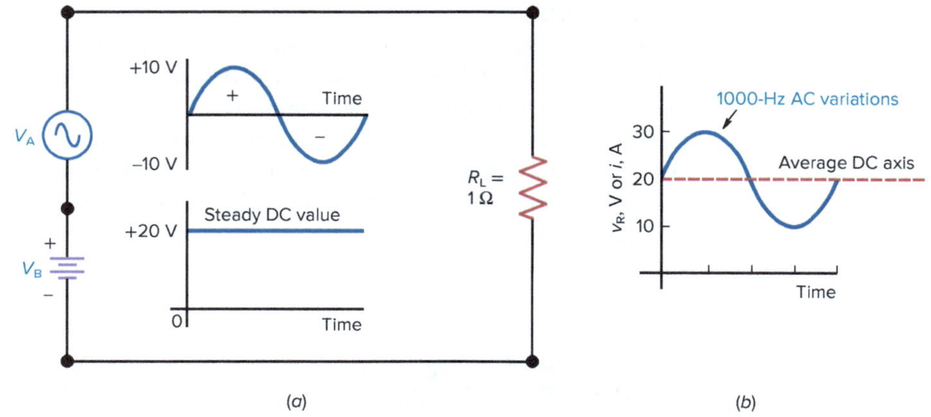

GOOD TO KNOW

In Fig. 26–2, both the DC and AC voltage sources deliver power to the load, R_L. The combined voltage of the AC and DC sources is called the effective voltage, designated V_{eff}. The effective voltage V_{eff} can be calculated as $V_{eff} = \sqrt{V_{DC}^2 + V_{rms}^2}$

Figure 26–2 illustrates how a circuit can have pulsating direct current or voltage. Here, the steady DC voltage of the battery V_B is in series with the AC voltage V_A. Since the two series generators add, the voltage across R_L is the sum of the two applied voltages, as shown by the waveshape of v_R in Fig. 26–2*b*.

If values are taken at opposite peaks of the AC variation, when V_A is at +10 V, it adds to the +20 V of the battery to provide +30 V across R_L; when the AC voltage is −10 V, it bucks the battery voltage of +20 V to provide +10 V across R_L. When the AC voltage is at zero, the voltage across R_L equals the battery voltage of +20 V.

The combined voltage v_R then consists of the AC variations fluctuating above and below the battery voltage as the axis, instead of the zero axis for AC voltage. The result is a **pulsating DC** voltage, since it is fluctuating but always has positive polarity with respect to zero.

The pulsating direct current i through R_L has the same waveform, fluctuating above and below the steady DC level of 20 A. The i and v values are the same because R_L is 1 Ω.

Another example is shown in Fig. 26–3. If a 100-Ω R_L is connected across 120 V, 60 Hz, as shown in Fig. 26–3*a*, the current in R_L will be V/R_L. This is an AC sine wave with an rms value of $^{120}\!/_{100}$ or 1.2 A.

Figure 26–3 A combination of AC and DC voltage to provide fluctuating DC voltage across R_L. (*a*) An AC source alone. (*b*) A DC source alone. (*c*) The AC source and DC source in series for the fluctuating voltage across R_L.

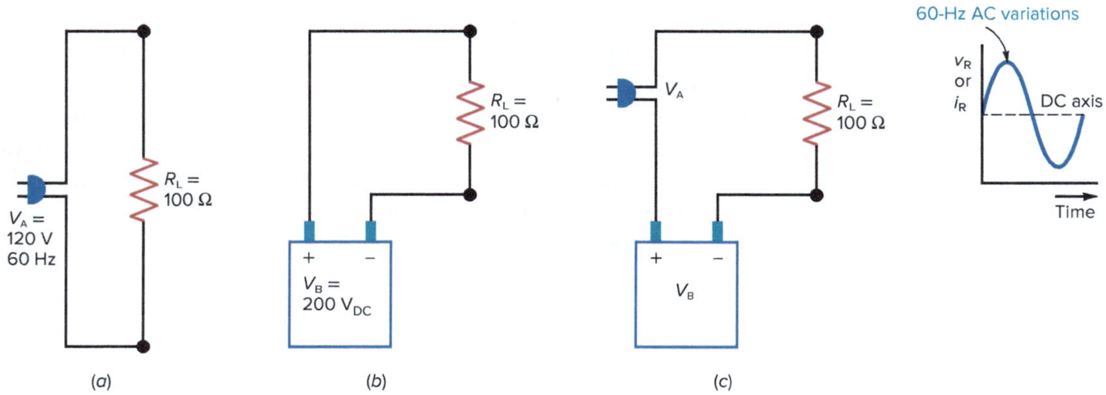

Also, if you connect the same R_L across the 200-V_{DC} source in Fig. 26–3*b*, instead of using the AC source, the steady direct current in R_L will be 200/100, or 2 A. The battery source voltage and its current are considered steady DC values because there are no variations.

However, suppose that the AC source V_A and DC source V_B are connected in series with R_L, as in Fig. 26–3*c*. What will happen to the current and voltage for R_L? Will V_A or V_B supply the current? The answer is that both sources will. Each voltage source produces current as though the other were not there, assuming the sources have negligibly small internal impedance. The result then is the **fluctuating DC** voltage or current shown, with the AC variations of V_A superimposed on the average DC level of V_B.

DC and AC Components

The pulsating DC voltage v_R in Fig. 26–3*c* is the original AC voltage V_A with its axis shifted to a DC level by the battery voltage V_B. In effect, a DC component has been inserted into the AC variations. This effect is called *DC insertion*.

Referring back to Fig. 26–2, if you measure across R_L with a DC voltmeter, it will read the DC level of 20 V. An Ac-coupled oscilloscope* will show only the peak-to-peak variations of ±10 V.

It is convenient, therefore, to consider the pulsating or fluctuating voltage and current in two parts. One is the steady DC component, which is the axis or average level of the variations; the other is the AC component, consisting of the variations above and below the DC axis. Here the DC level for V_T is +20 V, and the AC component equals 10 V peak or 7.07 V rms value. The AC component is also called Ac *ripple*.

Note that with respect to the DC level, the fluctuations represent alternating voltage or current that actually reverses in polarity. For example, the change of v_R from +20 to +10 V is a decrease in positive voltage compared with zero. However, compared with the DC level of +20 V, the value of +10 V is 10 V more negative than the axis.

GOOD TO KNOW

When a sine wave is symmetrical, its DC value can be calculated as $V_{DC} = \frac{V_{P\text{-}P}}{2} \pm V_{min}$ where V_{min} represents the negative-most peak of the AC sine wave.

Typical Examples of DC Level with AC Component

As a common application, transistors and ICs always have fluctuating DC voltage or current when used for amplifying an AC signal. The transistor or IC amplifier needs steady DC voltages to operate. The signal input is an AC variation, usually with a DC axis to establish the desired operating level. The amplified output is also an AC variation superimposed on a DC supply voltage that supplies the required power output. Therefore, the input and output circuits have fluctuating DC voltage.

The examples in Fig. 26–4 illustrate two possibilities in terms of polarities with respect to chassis ground. In Fig. 26–4*a*, the waveform is always positive, as in the previous examples. This example could apply to the collector voltage on an *npn* transistor amplifier. Note the specific values. The average DC axis is the steady DC level. The positive peak equals the DC level plus the peak AC value. The minimum point equals the DC level minus the peak AC value. The peak-to-peak value of the AC component and its rms value are the same as that of the AC signal alone. However, it is better to subtract the minimum from the maximum for the peak-to-peak value in case the waveform is unsymmetrical.

* See Appendix E for an explanation of how to use the oscilloscope.

Figure 26–4 Typical examples of a DC voltage with an AC component. (*a*) Positive fluctuating DC values because of a large positive DC component. (*b*) Negative fluctuating DC values because of a large negative DC component.

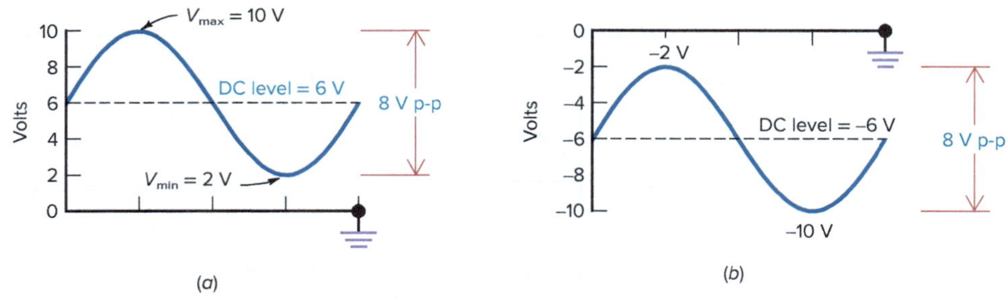

In Fig. 26–4*b*, all values are negative. Notice that here the positive peak of the AC component subtracts from the DC level because of opposite polarities. Now the negative peak adds to the negative DC level to provide a maximum point of negative voltage.

Separating the AC Component

In many applications, the circuit has pulsating DC voltage, but only the AC component is desired. Then the AC component can be passed to the load, while the steady DC component is blocked, either by transformer coupling or by capacitive coupling. A transformer with a separate secondary winding isolates or blocks steady direct current in the primary. A capacitor isolates or blocks a steady DC voltage.

26–2 Self-Review

Answers at the end of the chapter.

For the fluctuating DC waveform in Fig. 26–4*a*, specify the following voltages:

a. average DC level.
b. maximum and minimum values.
c. peak-to-peak of AC component.
d. peak and rms of AC component.

26–3 Transformer Coupling

Remember that a transformer produces induced secondary voltage just for variations in primary current. With pulsating direct current in the primary, the secondary has output voltage, therefore, only for the AC variations. The steady DC component in the primary has no effect in the secondary.

In Fig. 26–5, the pulsating DC voltage in the primary produces pulsating primary current. The DC axis corresponds to a steady value of primary current that has a constant magnetic field, but only when the field changes, can secondary voltage be induced. Therefore, only the fluctuations in the primary can produce output in the secondary. Since there is no output for the steady primary current, this DC level corresponds to the zero level for the AC output in the secondary.

When the primary current increases above the steady level, this increase produces one polarity for the secondary voltage as the field expands; when the primary current decreases below the steady level, the secondary voltage has reverse polarity as the field contracts. The result in the secondary is an AC variation having opposite polarities with respect to the zero level.

The phase of the AC secondary voltage may be as shown or 180° opposite, depending on the connections and direction of the windings. Also, the AC

Figure 26–5 Transformer coupling blocks the DC component. With fluctuating direct current in the primary L_P, only the AC component produces induced voltage in the secondary L_S.

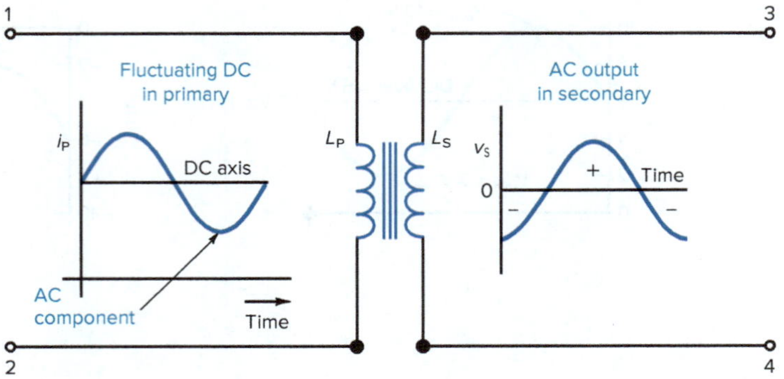

secondary output may be more or less than the AC component in the primary, depending on the turns ratio. This ability to isolate the steady DC component in the primary while providing AC output in the secondary applies to all iron-core and air-core transformers with a separate secondary winding.

26–3 Self-Review

Answers at the end of the chapter.

a. Is transformer coupling an example of a high-pass or low-pass filter?

b. In Fig. 26–5, what is the level of v_s for the average DC level of i_p?

26–4 Capacitive Coupling

Capacitive coupling is probably the most common type of coupling in amplifier circuits. The coupling connects the output of one circuit to the input of the next. The requirements are to include all frequencies in the desired signal, while rejecting undesired components. Usually, the DC component must be blocked from the input to AC amplifiers. The purpose is to maintain a specific DC level for the amplifier operation.

In Fig. 26–6, the pulsating DC voltage across input terminals 1 and 2 is applied to the *RC* coupling circuit. Capacitance C_C will charge to the steady DC level, which

Figure 26–6 The *RC* coupling blocks the DC component. With fluctuating DC voltage applied, only the AC component produces charge and discharge current for the output voltage across *R*.

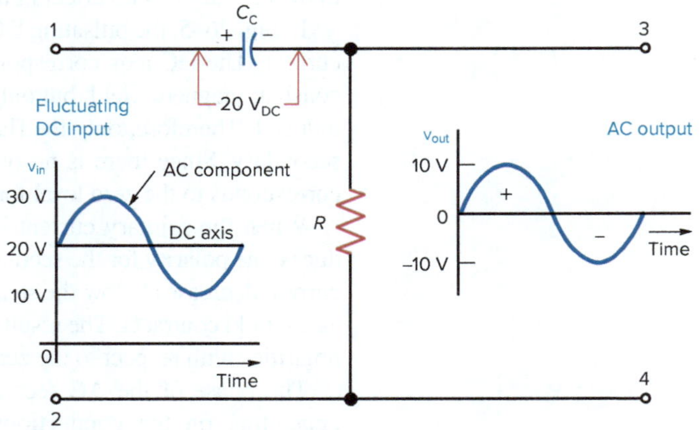

is the average charging voltage. The steady DC component is blocked, therefore, since it cannot produce voltage across *R*. However, the AC component is developed across *R*, between the output terminals 3 and 4, because the AC voltage allows *C* to produce charge and discharge current through *R*. Note that the zero axis of the AC voltage output corresponds to the average level of the pulsating DC voltage input.

The DC Component across *C*

GOOD TO KNOW

A DC voltage may be present at both the input and output sides of an *RC* coupling circuit. For example, the DC component on the input side may be the DC collector voltage of a transistor in one stage, whereas the DC component on the output side may be the DC base voltage of a transistor in another stage. The coupling capacitor isolates the DC voltages in each stage while still coupling the AC signal.

The voltage across C_C is the steady DC component of the input voltage because the variations of the AC component are symmetrical above and below the average level. Furthermore, the series resistance is the same for charge and discharge. As a result, any increase in charging voltage above the average level is counteracted by an equal discharge below the average.

In Fig. 26–6, for example, when v_{in} increases from 20 to 30 V, this effect on charging C_C is nullified by the discharge when v_{in} decreases from 20 to 10 V. At all times, however, v_{in} has a positive value that charges C_C in the polarity shown.

The net result is that only the average level is effective in charging C_C, since the variations from the axis neutralize each other. After a period of time, depending on the *RC* time constant, C_C will charge to the average value of the pulsating DC voltage applied, which is 20 V here.

The AC Component across *R*

Although C_C is charged to the average DC level, when the pulsating input voltage varies above and below this level, the charge and discharge current produces *IR* voltage corresponding to the fluctuations of the input. When v_{in} increases above the average level, C_C takes on charge, producing charging current through *R*. Even though the charging current may be too small to affect the voltage across C_C appreciably, the *IR* drop across a large value of resistance can be practically equal to the AC component of the input voltage. In summary, a long *RC* time constant is needed for good coupling.

If the polarity is considered, in Fig. 26–6, the charging current produced by an increase of v_{in} produces electron flow from the low side of *R* to the top, adding electrons to the negative side of C_C. The voltage at the top of *R* is then positive with respect to the line below.

When v_{in} decreases below the average level, *C* loses charge. The discharge current then is in the opposite direction through *R*. The result is negative polarity for the AC voltage output across *R*.

When the input voltage is at its average level, there is no charge or discharge current, resulting in zero voltage across *R*. The zero level in the AC voltage across *R* corresponds to the average level of the pulsating DC voltage applied to the *RC* circuit.

The end result is that with positive pulsating DC voltage applied, the values above the average produce the positive half-cycle of the AC voltage across *R*; the values below the average produce the negative half-cycle. Only this AC voltage across *R* is coupled to the next circuit, as terminals 3 and 4 provide the output from the *RC* coupling circuit.

It is important to note that there is practically no phase shift. This rule applies to all *RC* coupling circuits, since *R* must be 10 or more times X_C. Then the reactance is negligible compared with the series resistance, and the phase angle of less than 5.7° is practically zero.

Voltages around the *RC* Coupling Circuit

If you measure the fluctuating DC voltage across the input terminals 1 and 2 in Fig. 26–6 with a DC voltmeter, it will read the average DC level of 20 V. If you connect an ac-coupled oscilloscope across the same two points, it will show only the

Table 26–1	Typical Audio Frequency and Radio Frequency Coupling Capacitors*			
	Values of C_C			
Frequency	$R = 1.6$ kΩ	$R = 16$ kΩ	$R = 160$ kΩ	Frequency Band
100 Hz	10 μF	1 μF	0.1 μF	Audio frequency
1000 Hz	1 μF	0.1 μF	0.01 μF	Audio frequency
10 kHz	0.1 μF	0.01 μF	0.001 μF	Audio frequency
100 kHz	0.01 μF	0.001 μF	100 pF	Radio frequency
1 MHz	0.001 μF	100 pF	10 pF	Radio frequency
10 MHz	100 pF	10 pF	1 pF	Radio frequency
100 MHz	10 pF	1 pF	0.1 pF	Very high frequency

* For coupling circuit in Fig. 26–6; $X_{C_C} = \frac{1}{10} R$.

fluctuating AC component. These voltage variations have a peak value of 10 V, a peak-to-peak value of 20 V, or an rms value of $0.707 \times 10 = 7.07$ V.

Across points 1 and 3 for V_C in Fig. 26–6, a DC voltmeter reads the steady DC value of 20 V. An AC voltmeter across points 1 and 3 reads practically zero.

However, an AC voltmeter across the output R between points 3 and 4 will read the AC voltage of 7 V, approximately, for V_R. Furthermore, a DC voltmeter across R reads zero. The DC component of the input voltage is across C_C but is blocked from the output across R.

Typical Coupling Capacitors

Common values of rf and af coupling capacitors for different sizes of series R are listed in Table 26–1. In all the cases, the coupling capacitor blocks the steady DC component of the input voltage, and the AC component is passed to the resistance.

The size of C_C required depends on the frequency of the AC component. At each frequency listed at the left in Table 26–1, the values of capacitance in the horizontal row have an X_C equal to one-tenth the resistance value for each column. The R increases from 1.6 to 16 to 160 kΩ for the three columns, allowing smaller values of C_C. Typical audio coupling capacitors, then, are about 0.1 to 10 μF, depending on the lowest audio frequency to be coupled and the size of the series resistance. Typical rf coupling capacitors are about 1 to 100 pF.

Values of C_C more than about 1 μF are usually electrolytic capacitors, which must be connected in the correct polarity. These can be very small; many are ½ in. long with a low voltage rating of 6 to 25 V for transistor circuits. The small leakage current of electrolytic capacitors is not a serious problem in this application because of the low voltage and small series resistance of transistor coupling circuits.

26–4 Self-Review

Answers at the end of the chapter.

a. In Fig. 26–6, what is the level of v_{out} across R corresponding to the average DC level of v_{in}?

b. Which of the following is a typical audio coupling capacitor with a 1-kΩ R: 1 pF; 0.001 μF; or 5 μF?

26–5 Bypass Capacitors

A bypass is a path around a component. In circuits, the bypass is a parallel or shunt path. Capacitors are often used in parallel with resistance to bypass the AC component of a pulsating DC voltage. The result, then, is steady DC voltage across the *RC* parallel combination, if the bypass capacitance is large enough to have little reactance at the lowest frequency of the AC variations.

As illustrated in Fig. 26–7, the capacitance C_1 in parallel with R_1 is an AC **bypass capacitor** for R_1. For any frequency at which X_{C_1} is one-tenth of R_1, or less, the AC component is bypassed around R_1 through the low reactance in the shunt path. The result is practically zero AC voltage across the bypass capacitor because of its low reactance.

Since the voltage is the same across R_1 and C_1, because they are in parallel, there is also no AC voltage across R_1 for the frequency at which C_1 is a bypass capacitor. We can say that *R* is bypassed for the frequency at which X_C is one-tenth of *R*. The bypassing also applies to higher frequencies where X_C is less than one-tenth of *R*. Then the AC voltage across the bypass capacitor is even closer to zero because of its lower reactance.

Bypassing the AC Component of a Pulsating DC Voltage

The voltages in Fig. 26–7 are calculated by considering the effect of C_1 separately for V_{DC} and for V_{AC}. For direct current, C_1 is practically an open circuit. Then its reactance is so high compared with the 5000-Ω R_1 that X_{C_1} can be ignored as a parallel branch. Therefore, R_1 can be considered a voltage divider in series with R_2. Since R_1 and R_2 are equal, each has 5 V, equal to one-half V_{DC}. Although this DC voltage division depends on R_1 and R_2, the DC voltage across C_1 is the same 5 V as across its parallel R_1.

For the AC component of the applied voltage, however, the bypass capacitor has very low reactance. In fact, X_{C_1} must be one-tenth of R_1, or less. Then the 5000-Ω R_1 is so high compared with the low value of X_{C_1} that R_1 can be ignored as a parallel branch. Therefore, the 500-Ω X_{C_1} can be considered a voltage divider in series with R_2.

With an X_{C_1} of 500 Ω, this value in series with the 5000-Ω R_2 allows approximately one-eleventh of V_{AC} across C_1. This AC voltage, equal to 0.9 V here, is the same across R_1 and C_1 in parallel. The remainder of the AC applied voltage, approximately equal to 9.1 V, is across R_2. In summary, then, the bypass capacitor provides an AC short circuit across its shunt resistance, so that little or no AC voltage can be developed without affecting the DC voltages.

Measuring voltages around the circuit in Fig. 26–7, a DC voltmeter reads 5 V across R_1 and 5 V across R_2. An AC voltmeter across R_2 reads 9.1 V, which is almost

GOOD TO KNOW

The AC voltages shown in Fig. 26–7 are only approximate values. The actual values could be determined by applying the concepts learned in Chapter 24, "Complex Numbers for AC Circuits."

Figure 26–7 Low reactance of bypass capacitor C_1 short-circuits R_1 for an AC component of fluctuating DC input voltage.

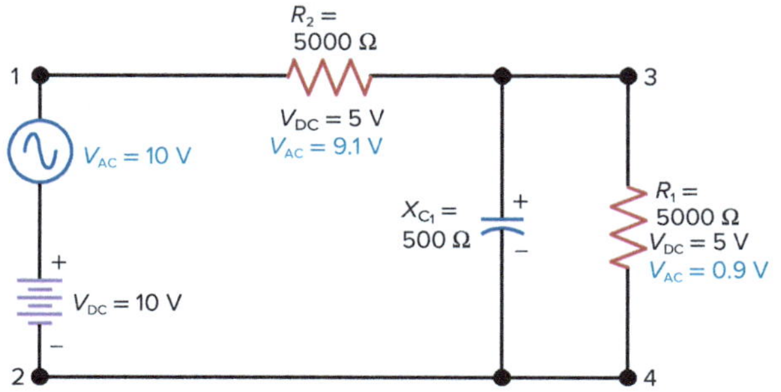

Table 26–2 Typical Audio Frequency and Radio Frequency Bypass Capacitors*

	Values of C			
Frequency	$R = 16\ k\Omega$	$R = 1.6\ k\Omega$	$R = 160\ \Omega$	Frequency Band
100 Hz	1 μF	10 μF	100 μF	Audio frequency
1000 Hz	0.1 μF	1 μF	10 μF	Audio frequency
10 kHz	0.01 μF	0.1 μF	1 μF	Audio frequency
100 kHz	0.001 μF	0.01 μF	0.1 μF	Radio frequency
1 MHz	100 pF	0.001 μF	0.01 μF	Radio frequency
10 MHz	10 pF	100 pF	0.001 μF	Radio frequency
100 MHz	1 pF	10 pF	100 pF	Very high frequency

* For *RC* bypass circuit in Fig. 26–7; $X_{C_1} = \frac{1}{10} R$.

all of the AC input voltage. Across the bypass capacitor C_1, the AC voltage is only 0.9 V.

Typical sizes of rf and af bypass capacitors are listed in Table 26–2. The values of C have been calculated at different frequencies for an X_C one-tenth the shunt resistance given in each column. The R decreases for the three columns, from 16 kΩ to 1.6 kΩ and 160 Ω. Note that smaller values of R require larger values of C for bypassing. Also, when X_C equals one-tenth of R at one frequency, X_C will be even less for higher frequencies, improving the bypassing action. Therefore, the size of bypass capacitors should be considered on the basis of the lowest frequency to be bypassed.

Note that the applications of coupling and bypassing for C are really the same, except that C_C is in series with R and the bypass C is in parallel with R. In both cases, X_C must be one-tenth or less of R. Then C_C couples the AC signal to R, or the shunt bypass short-circuits R for the AC signal.

Bypassing Radio Frequencies but Not Audio Frequencies

See Fig. 26–8. At the audio frequency of 1000 Hz, C_1 has a reactance of 1.6 MΩ. This reactance is so much higher than R_1 that the impedance of the parallel combination is essentially equal to the 16,000 Ω of R_1. Then R_1 and R_2 serve as a voltage divider for the applied af voltage of 10 V. Each of the equal resistances has one-half

Figure 26–8 Capacitor C_1 bypasses R_1 for radio frequencies but not for audio frequencies.

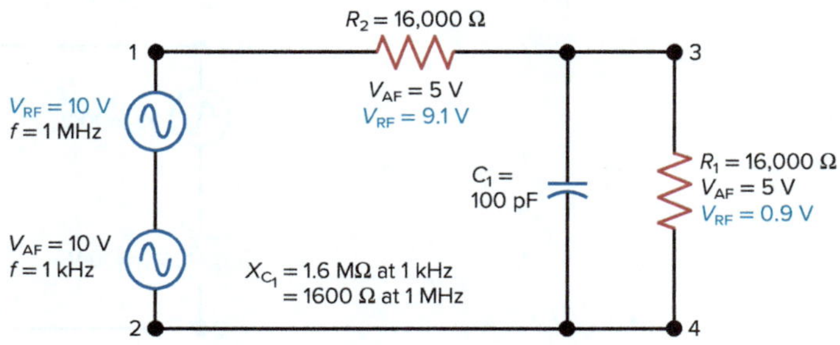

the applied voltage, equal to the 5 V across R_2 and 5 V across R_1. This 5 V at 1000 Hz is also present across C_1, since it is in parallel with R_1.

For the rf voltage at 1 MHz, however, the reactance of the bypass capacitor is only 1600 Ω. This is one-tenth of R_1. Then X_{C_1} and R_1 in parallel have a combined impedance approximately equal to 1600 Ω.

Now, with a 1600-Ω impedance for the R_1C_1 bank in series with the 16,000 Ω of R_2, the voltage across R_1 and C_1 is one-eleventh the applied rf voltage. Then there is 0.9 V across the lower impedance of R_1 and C_1 with 9.1 V across the larger resistance of R_2. As a result, the rf component of the applied voltage can be considered bypassed. The capacitor C_1 is the rf bypass across R_1.

26–5 Self-Review

Answers at the end of the chapter.

a. In Fig. 26–8, is C_1 an af or rf bypass?

b. Which of the following is a typical audio bypass capacitor across a 1-kΩ R: 1 pF; 0.001 μF; or 5 μF?

26–6 Filter Circuits

In terms of their function, filters can be classified as either low-pass or high-pass. A low-pass filter allows the lower-frequency components of the applied voltage to develop output voltage across the load resistance, whereas the higher-frequency components are attenuated, or reduced, in the output. A high-pass filter does the opposite, allowing the higher-frequency components of the applied voltage to develop voltage across the output load resistance.

An *RC* coupling circuit is an example of a high-pass filter because the AC component of the input voltage is developed across *R* while the DC voltage is blocked by the series capacitor. Furthermore, with higher frequencies in the AC component, more AC voltage is coupled. For the opposite case, a bypass capacitor is an example of a low-pass filter. The higher frequencies are bypassed, but the lower the frequency, the less the bypassing action. Then lower frequencies can develop output voltage across the shunt bypass capacitor.

To make the filtering more selective in terms of which frequencies are passed to produce output voltage across the load, filter circuits generally combine inductance and capacitance. Since inductive reactance increases with higher frequencies and capacitive reactance decreases, the two opposite effects improve the filtering action.

With combinations of *L* and *C*, filters are named to correspond to the circuit configuration. Most common types of filters are the *L*, *T*, and π. Any one of the three can function as either a low-pass filter or a high-pass filter.

The reactance X_L of either low-pass or high-pass filters with *L* and *C* increases with higher frequencies, while X_C decreases. The frequency characteristics of X_L and X_C cannot be changed. However, the circuit connections are opposite to reverse the filtering action.

In general, high-pass filters use

1. Coupling capacitance *C* in series with the load. Then X_C can be low for high frequencies to be passed to R_L, while low frequencies are blocked.
2. Choke inductance *L* in parallel across R_L. Then the shunt X_L can be high for high frequencies to prevent a short circuit across R_L, while low frequencies are bypassed.

The opposite characteristics for low-pass filters are

1. Inductance *L* in series with the load. The high X_L for high frequencies can serve as a choke, while low frequencies can be passed to R_L.

GOOD TO KNOW

An ideal filter would offer zero attenuation within the passband and infinite attenuation throughout the stop band.

2. Bypass capacitance C in parallel across R_L. Then high frequencies are bypassed by a small X_C, while low frequencies are not affected by the shunt path.

The ability of any filter to reduce the amplitude of undesired frequencies is called the ***attenuation*** of the filter. The frequency at which the attenuation reduces the output to 70.7% is the ***cutoff frequency***, usually designated f_c.

26–6 Self-Review

Answers at the end of the chapter.

a. Does high-pass filtering or low-pass filtering require series C?
b. Which filtering requires parallel C?

26–7 Low-Pass Filters

Figure 26–9 illustrates low-pass circuits from a single filter element with a shunt bypass capacitor in Fig. 26–9*a* or a series choke in *b*, to the more elaborate combinations of an inverted-L type filter in *c*, a T type in *d*, and a π type in *e* and *f*. With an applied input voltage having different frequency components, the low-pass filter action results in maximum low-frequency voltage across R_L, while most of the high-frequency voltage is developed across the series choke or resistance.

In Fig. 26–9*a*, the shunt capacitor C bypasses R_L at high frequencies. In Fig. 26–9*b*, the choke L acts as a voltage divider in series with R_L. Since L has maximum reactance for the highest frequencies, this component of the input voltage is developed across L with little across R_L. At lower frequencies, L has low reactance, and most of the input voltage can be developed across R_L.

In Fig. 26–9*c*, the use of both the series choke and the bypass capacitor improves the filtering by providing a sharper cutoff between the low frequencies that can develop voltage across R_L and the higher frequencies stopped from the load by producing maximum voltage across L. Similarly, the T-type circuit in Fig. 26–9*d* and the π-type circuits in *e* and *f* improve filtering.

Using the series resistance in Fig. 26–9*f* instead of a choke provides an economical π filter in less space.

Figure 26–9 Low-pass filter circuits. (*a*) Bypass capacitor C in parallel with R_L. (*b*) Choke L in series with R_L. (*c*) Inverted-L type with choke and bypass capacitor. (*d*) The T type with two chokes and one bypass capacitor. (*e*) The π type with one choke and bypass capacitors at both ends. (*f*) The π type with a series resistor instead of a choke.

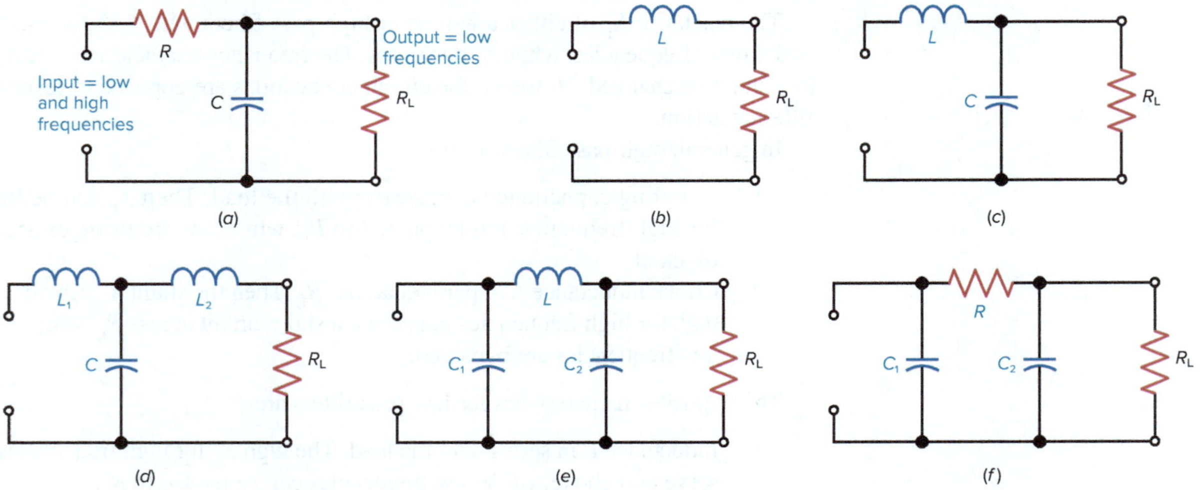

Figure 26–10 The response of a low-pass filter with cutoff at 15 kHz. The filter passes the audio signal but attenuates radio frequencies.

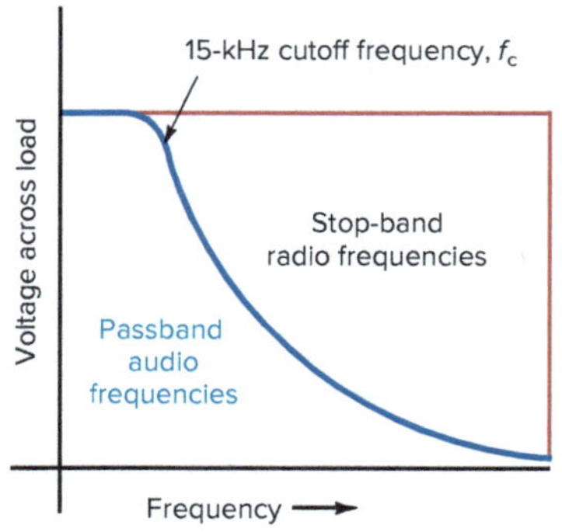

Passband and Stop Band

As illustrated in Fig. 26–10, a low-pass filter attenuates frequencies above the cutoff frequency f_c of 15 kHz in this example. Any component of the input voltage having a frequency lower than 15 kHz can produce output voltage across the load. These frequencies are in the *passband*. Frequencies of 15 kHz or more are in the *stop band*. The sharpness of filtering between the passband and the stop band depends on the type of circuit. In general, the more *L* and *C* components, the sharper the response of the filter. Therefore, π and T types are better filters than the L type and the bypass or choke alone.

The response curve in Fig. 26–10 is illustrated for the application of a low-pass filter attenuating rf voltages while passing audio frequencies to the load. This is necessary when the input voltage has rf and af components but only the audio voltage is desired for the af circuits that follow the filter.

A good example is filtering the audio output of the detector circuit in a radio receiver, after the rf-modulated carrier signal has been rectified. Another common application of low-pass filtering is separating the steady DC component of a pulsating DC input from the higher-frequency 60-Hz AC component, as in the pulsating DC output of the rectifier in a power supply.

Circuit Variations

The choice between the T-type filter with a series input choke and the π type with a shunt input capacitor depends on the internal resistance of the generator supplying input voltage to the filter. A low-resistance generator needs the T filter so that the choke can provide high series impedance for the bypass capacitor. Otherwise, the bypass capacitor must have extremely large values to short-circuit the low-resistance generator at high frequencies.

The π filter is more suitable with a high-resistance generator when the input capacitor can be effective as a bypass. For the same reasons, the L filter can have the shunt bypass either in the input for a high-resistance generator or across the output for a low-resistance generator.

In all filter circuits, the series choke can be connected either in the high side of the line, as shown in Fig. 26–9, or in series in the opposite side of the line, without having any effect on the filtering action. Also, the series components can be connected in both sides of the line for a *balanced filter* circuit.

Passive and Active Filters

All circuits here are passive filters, as they use only capacitors, inductors, and resistors, which are passive components. An active filter, however, uses an operational amplifier (op amp) on an IC chip, with *R* and *C*. The purpose is to eliminate the need for inductance *L*. This feature is important in filters for audio frequencies when large coils would be necessary.

26–7 Self-Review

Answers at the end of the chapter.

a. **Which diagrams in Fig. 26–9 show a π-type filter?**
b. **Does the response curve in Fig. 26–10 show low-pass or high-pass filtering?**

26–8 High-Pass Filters

As illustrated in Fig. 26–11, the high-pass filter passes to the load all frequencies higher than the cutoff frequency f_c, whereas lower frequencies cannot develop appreciable voltage across the load. The graph in Fig. 26–11*a* shows the response of

Figure 26–11 High-pass filters. (*a*) The response curve for an audio frequency filter cutting off at 50 Hz. (*b*) An *RC* coupling circuit. (*c*) Inverted-L type. (*d*) The T type. (*e*) The π type.

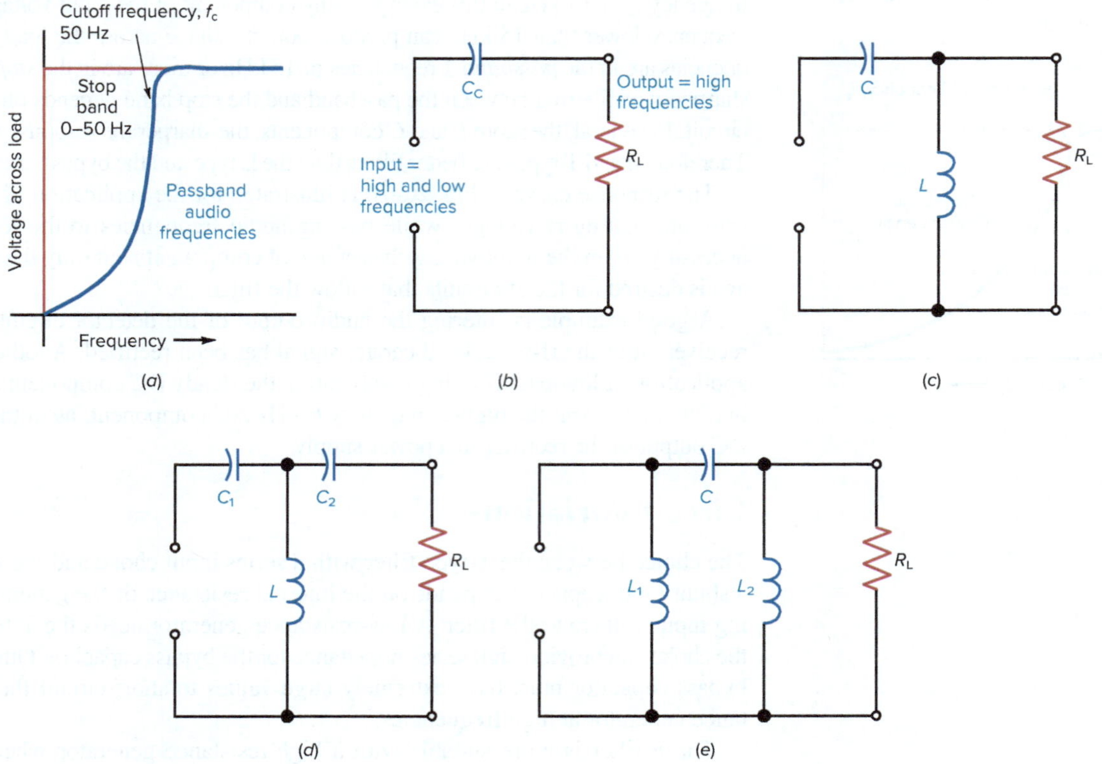

a high-pass filter with a stop band of 0 to 50 Hz. Above the cutoff frequency of 50 Hz, the higher audio frequencies in the passband can produce af voltage across the output load resistance.

The high-pass filtering action results from using C_C as a coupling capacitor in series with the load, as in Fig. 26–11*b*. The *L*, *T*, and π types use the inductance for a high-reactance choke across the line. In this way, the higher-frequency components of the input voltage can develop very little voltage across the series capacitance, allowing most of this voltage to be produced across R_L. The inductance across the line has higher reactance with increasing frequencies, allowing the shunt impedance to be no lower than the value of R_L.

For low frequencies, however, R_L is effectively short-circuited by the low inductive reactance across the line. Also, C_C has high reactance and develops most of the voltage at low frequencies, stopping these frequencies from developing voltage across the load.

26–8 Self-Review

Answers at the end of the chapter.

a. Which diagram in Fig. 26–11 shows a T-type filter?
b. Does the response curve in Fig. 26–11*a* show high-pass or low-pass filtering?

26–9 Analyzing Filter Circuits

Any low-pass or high-pass filter can be thought of as a frequency-dependent voltage divider, since the amount of output voltage is a function of frequency. Special formulas can be used to calculate the output voltage for any frequency of the applied

Figure 26–12 *RC* low-pass filter. (*a*) Circuit. (*b*) Graph of V_{out} versus frequency.

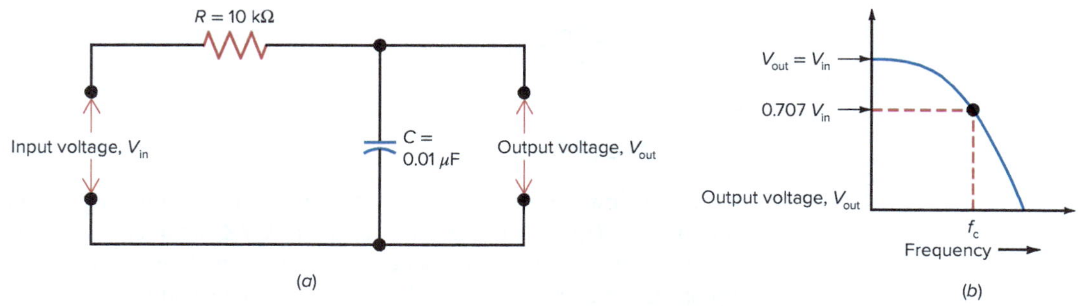

voltage. What follows is a more mathematical approach in analyzing the operation of the most basic low-pass and high-pass filter circuits.

RC Low-Pass Filter

Figure 26–12*a* shows a simple *RC* low-pass filter, and Fig. 26–12*b* shows how its output voltage V_{out} varies with frequency. Let's examine how the *RC* low-pass filter responds when $f = 0$ Hz (DC) and $f = \infty$ Hz. At $f = 0$ Hz, the capacitor *C* has infinite capacitive reactance X_C, calculated as

$$X_C = \frac{1}{2\pi fC}$$
$$= \frac{1}{2 \times \pi \times 0\text{ Hz} \times 0.01\ \mu\text{F}}$$
$$= \infty\ \Omega$$

Figure 26–13*a* shows the equivalent circuit for this condition. Notice that *C* appears as an open. Since all the input voltage appears across the open in a series circuit, V_{out} must equal V_{in} when $f = 0$ Hz.

At the other extreme, consider the circuit when the frequency *f* is very high or infinitely high. Then $X_C = 0\ \Omega$, calculated as

$$X_C = \frac{1}{2\pi fC}$$
$$= \frac{1}{2 \times \pi \times \infty\text{ Hz} \times 0.01\ \mu\text{F}}$$
$$= 0\ \Omega$$

Figure 26–13*b* shows the equivalent circuit for this condition. Notice that *C* appears as a short. Since the voltage across a short is zero, the output voltage for very high frequencies must be zero.

Figure 26–13 *RC* low-pass equivalent circuits. (*a*) Equivalent circuit for $f = 0$ Hz. (*b*) Equivalent circuit for very high frequencies, or $f = \infty$ Hz.

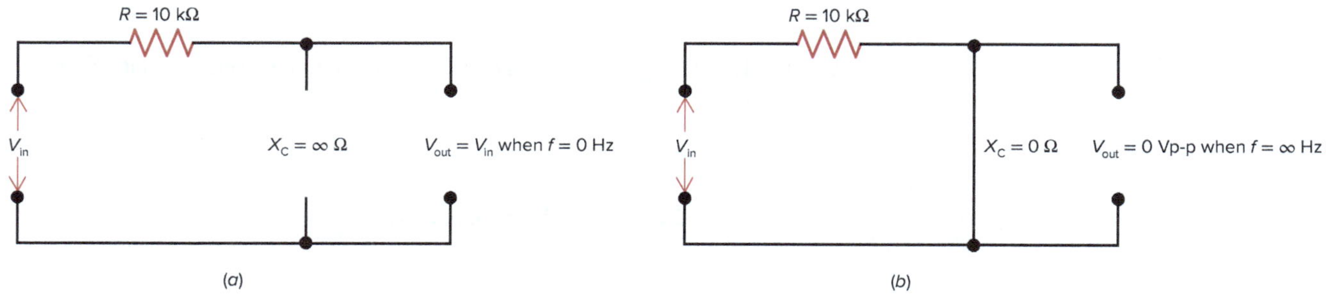

When the frequency of the input voltage is somewhere between zero and infinity, the output voltage can be determined by using Formula (26–1):

$$V_{out} = \frac{X_C}{Z_T} \times V_{in} \tag{26–1}$$

where

$$Z_T = \sqrt{R^2 + X_C^2}$$

At very low frequencies, where X_C approaches infinity, V_{out} is approximately equal to V_{in}. This is true because the ratio X_C/Z_T approaches one as X_C and Z_T become approximately the same value. At very high frequencies, where X_C approaches zero, the ratio X_C/Z_T becomes very small, and V_{out} is approximately zero.

With respect to the input voltage V_{in}, the phase angle θ of the output voltage V_{out} can be calculated as

$$\theta = \arctan\left(-\frac{R}{X_C}\right) \tag{26–2}$$

At very low frequencies, X_C is very large and θ is approximately 0°. At very high frequencies, however, X_C is nearly zero and θ approaches −90°.

The frequency where $X_C = R$ is the *cutoff frequency*, designated f_c. At f_c, the series current I is at 70.7% of its maximum value because the total impedance Z_T is 1.41 times larger than the resistance of R. The formula for the cutoff frequency f_c of an RC low-pass filter is derived as follows. Because $X_C = R$ at f_c,

$$\frac{1}{2\pi f_c C} = R$$

Solving for f_c gives

$$f_c = \frac{1}{2\pi RC} \tag{26–3}$$

The response curve in Fig. 26–12*b* shows that $V_{out} = 0.707\ V_{in}$ at the cutoff frequency f_c.

Example 26–1

In Fig. 26–12*a*, calculate (a) the cutoff frequency f_c; (b) V_{out} at f_c; (c) θ at f_c. (Assume $V_{in} = 10\ V_{p\text{-}p}$ for all frequencies.)

ANSWER

a. To calculate f_c, use Formula (26–3):

$$f_c = \frac{1}{2\pi RC}$$
$$= \frac{1}{2 \times \pi \times 10\ \text{k}\Omega \times 0.01\ \mu\text{F}}$$
$$= 1.592\ \text{kHz}$$

b. To calculate V_{out} at f_c, use Formula (26–1). First, however, calculate X_C and Z_T at f_c:

$$X_C = \frac{1}{2\pi f_c C}$$
$$= \frac{1}{2 \times \pi \times 1.592\ \text{kHz} \times 0.01\ \mu\text{F}}$$
$$= 10\ \text{k}\Omega$$

$$Z_T = \sqrt{R^2 + X_C^2}$$
$$= \sqrt{10^2\ \text{k}\Omega + 10^2\ \text{k}\Omega}$$
$$= 14.14\ \text{k}\Omega$$

Next,

$$V_{out} = \frac{X_C}{Z_T} \times V_{in}$$
$$= \frac{10\ \text{k}\Omega}{14.14\ \text{k}\Omega} \times 10\ \text{V}_{p\text{-}p}$$
$$= 7.07\ \text{V}_{p\text{-}p}$$

c. To calculate θ, use Formula (26–2):

$$\theta = \arctan\left(-\frac{R}{X_C}\right)$$
$$= \arctan\left(-\frac{10\ \text{k}\Omega}{10\ \text{k}\Omega}\right)$$
$$= \arctan(-1)$$
$$= -45°$$

The phase angle of $-45°$ tells us that V_{out} lags V_{in} by 45° at the cutoff frequency f_c.

GOOD TO KNOW

Another way to calculate the output voltage of an *RC* or *RL* low-pass filter is

$$V_{out} = \frac{V_{in}}{\sqrt{1 + (f/f_c)^2}}$$

where *f* represents any frequency.

RL Low-Pass Filter

Figure 26–14*a* shows a simple *RL* low-pass filter, and Fig. 26–14*b* shows how its output voltage V_{out} varies with frequency. For the analysis that follows, it is assumed that the coil's DC resistance r_s is negligible in comparison with the series resistance *R*.

Figure 26–15*a* shows the equivalent circuit when $f = 0$ Hz (DC). Notice that the inductor *L* acts as a short, since X_L must equal 0 Ω when $f = 0$ Hz. As a result, $V_{out} = V_{in}$ at very low frequencies and for direct current (0 Hz). At very high frequencies, X_L approaches infinity and the equivalent circuit appears as in Fig. 26–15*b*. Since *L* is basically equivalent to an open at very high frequencies, all of the input voltage will be dropped across *L* rather than *R*. Therefore, $V_{out} = 0\ \text{V}_{p\text{-}p}$ at very high frequencies.

To calculate the output voltage at any frequency in Fig. 26–14*a*, use Formula (26–4):

$$V_{out} = \frac{R}{Z_T} \times V_{in} \qquad \textbf{(26–4)}$$

where

$$Z_T = \sqrt{R^2 + X_L^2}$$

Figure 26–14 *RL* low-pass filter. (*a*) Circuit. (*b*) Graph of V_{out} versus frequency.

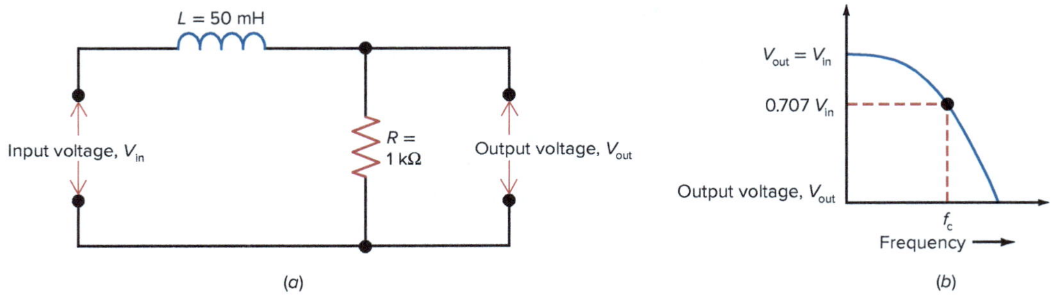

Figure 26–15 *RL* low-pass equivalent circuits. (*a*) Equivalent circuit for $f = 0$ Hz. (*b*) Equivalent circuit for very high frequencies, or $f = \infty$ Hz.

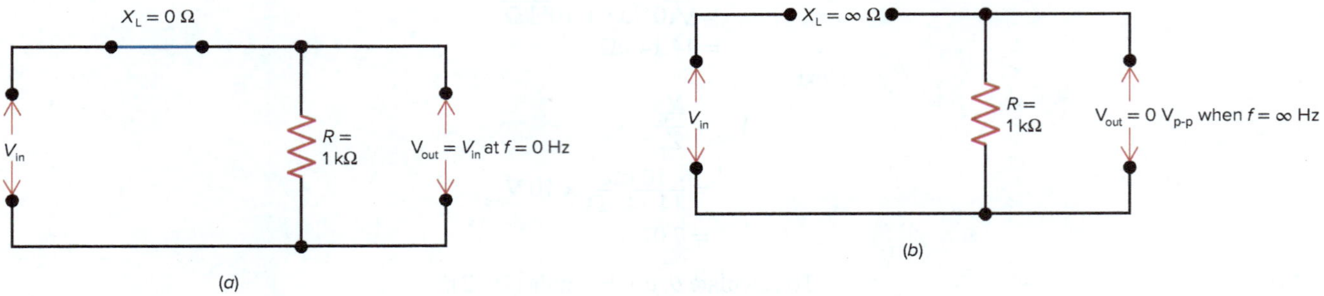

At very low frequencies, where X_L is very small, V_{out} is approximately equal to V_{in}. This is true because the ratio R/Z_T approaches one as Z_T and R become approximately the same value. At very high frequencies, the output voltage is approximately zero, because the ratio R/Z_T becomes very small as X_L and thus Z_T approach infinity.

The phase angle θ between V_{in} and V_{out} can be determined using Formula (26–5):

$$\theta = \arctan\left(-\frac{X_L}{R}\right) \qquad \textbf{(26–5)}$$

At very low frequencies, X_L approaches zero and θ is approximately 0°. At very high frequencies, X_L approaches infinity and θ is approximately −90°.

The frequency at which $X_L = R$ is the cutoff frequency f_c. At f_c, the series current I is at 70.7% of its maximum value, since $Z_T = 1.41R$ when $X_L = R$. The formula for the cutoff frequency of an *RL* low-pass filter is derived as follows. Since $X_L = R$ at f_c,

$$2\pi f_c L = R$$

Solving for f_c gives

$$f_c = \frac{R}{2\pi L} \qquad \textbf{(26–6)}$$

The response curve in Fig. 26–14*b* shows that $V_{out} = 0.707V_{in}$ at the cutoff frequency f_c.

Example 26–2

In Fig. 26–14*a*, calculate (a) the cutoff frequency f_c; (b) V_{out} at 1 kHz; (c) θ at 1 kHz. (Assume $V_{in} = 10\ V_{p\text{-}p}$ for all frequencies.)

ANSWER

a. To calculate f_c, use Formula (26–6):

$$f_c = \frac{R}{2\pi L}$$

$$= \frac{1\ \text{k}\Omega}{2 \times \pi \times 50\ \text{mH}}$$

$$= 3.183\ \text{kHz}$$

b. To calculate V_{out} at 1 kHz, use Formula (26–4). First, however, calculate X_L and Z_T at 1 kHz:

$$X_L = 2\pi fL$$
$$= 2 \times \pi \times 1 \text{ kHz} \times 50 \text{ mH}$$
$$= 314\ \Omega$$
$$Z_T = \sqrt{R^2 + X_L^2}$$
$$= \sqrt{1^2 \text{ k}\Omega + 314^2\ \Omega}$$
$$= 1.05 \text{ k}\Omega$$

Next,

$$V_{out} = \frac{R}{Z_T} \times V_{in}$$
$$= \frac{1 \text{ k}\Omega}{1.05 \text{ k}\Omega} \times 10\ V_{p\text{-}p}$$
$$= 9.52\ V_{p\text{-}p}$$

Notice that $V_{out} \cong V_{in}$, since 1 kHz is in the passband of the low-pass filter.

c. To calculate θ at 1 kHz, use Formula (26–5). Recall that $X_L = 314\ \Omega$ at 1 kHz:

$$\theta = \arctan\left(-\frac{X_L}{R}\right)$$
$$= \arctan\left(-\frac{314\ \Omega}{1 \text{ k}\Omega}\right)$$
$$= \arctan(-0.314)$$
$$= -17.4°$$

The phase angle of −17.4° tells us that V_{out} lags V_{in} by 17.4° at a frequency of 1 kHz.

RC High-Pass Filter

Figure 26–16*a* shows an *RC* high-pass filter. Notice that the output is taken across the resistor *R* rather than across the capacitor *C*. Figure 26–16*b* shows how the output voltage varies with frequency. To calculate the output voltage V_{out} at any frequency, use Formula (26–7):

$$V_{out} = \frac{R}{Z_T} \times V_{in} \qquad \textbf{(26–7)}$$

where

$$Z_T = \sqrt{R^2 + X_C^2}$$

GOOD TO KNOW

Another way to calculate the output voltage of an *RC* or *RL* high-pass filter is

$$V_{out} = \frac{V_{in}}{\sqrt{1 + (f_c/f)^2}}$$

where *f* represents any frequency.

MultiSim **Figure 26–16** *RC* high-pass filter. (*a*) Circuit. (*b*) Graph of V_{out} versus frequency.

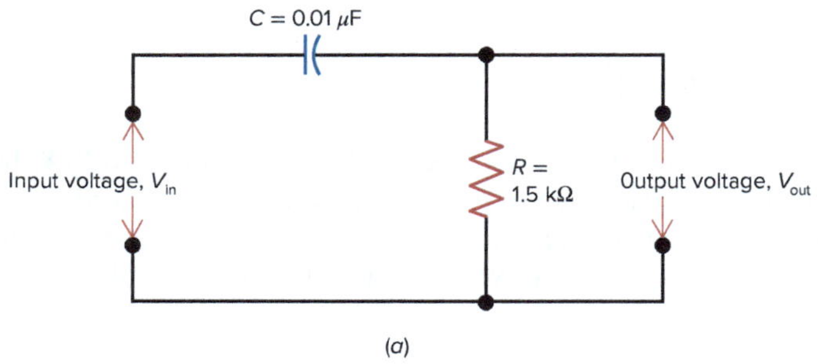

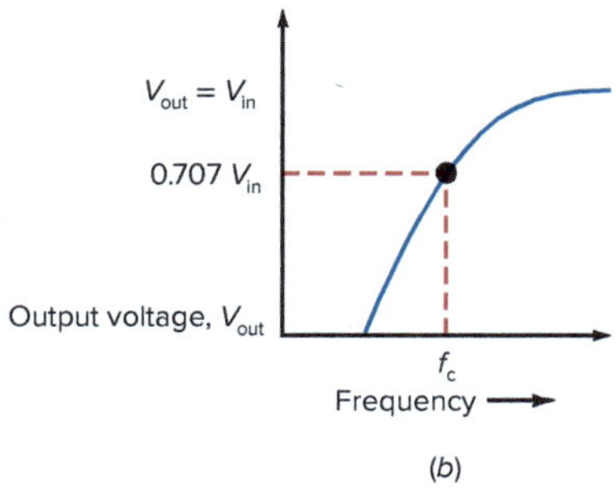

Figure 26–17 *RL* high-pass filter. (*a*) Circuit. (*b*) Graph of V_{out} versus frequency.

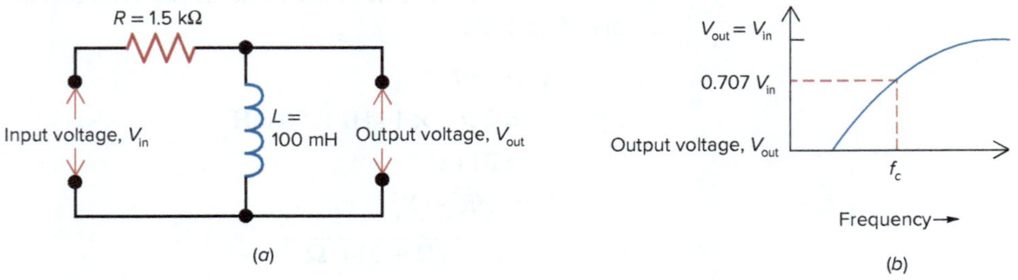

At very low frequencies, the output voltage approaches zero because the ratio R/Z_T becomes very small as X_C and thus Z_T approach infinity. At very high frequencies, V_{out} is approximately equal to V_{in}, because the ratio R/Z_T approaches one as Z_T and R become approximately the same value.

The phase angle of V_{out} with respect to V_{in} for an *RC* high-pass filter can be calculated using Formula (26–8):

$$\theta = \arctan\left(\frac{X_C}{R}\right) \qquad \textbf{(26–8)}$$

At very low frequencies where X_C is very large, θ is approximately 90°. At very high frequencies where X_C approaches zero, θ is approximately 0°.

To calculate the cutoff frequency f_c for an *RC* high-pass filter, use Formula (26–3). Although this formula is used to calculate f_c for an *RC* low-pass filter, it can also be used to calculate f_c for an *RC* high-pass filter. The reason is that, in both circuits, $X_C = R$ at the cutoff frequency. In Fig. 26–16*b*, notice that $V_{out} = 0.707\ V_{in}$ at f_c.

RL High-Pass Filter

An *RL* high-pass filter is shown in Fig. 26–17*a*, and its response curve is shown in Fig. 26–17*b*. In Fig. 26–17*a*, notice that the output is taken across the inductor *L* rather than across the resistance *R*.

To calculate the output voltage V_{out} at any frequency, use Formula (26–9):

$$V_{out} = \frac{X_L}{Z_T} \times V_{in} \qquad \textbf{(26–9)}$$

where

$$Z_T = \sqrt{R^2 + X_L^2}$$

At very low frequencies, where X_L is very small, V_{out} is approximately zero. At very high frequencies, $V_{out} = V_{in}$ because the ratio X_L/Z_T is approximately one.

The phase angle θ of the output voltage V_{out} with respect to the input voltage V_{in} is

$$\theta = \arctan\left(\frac{R}{X_L}\right) \qquad \textbf{(26–10)}$$

At very low frequencies, θ approaches 90° because the ratio R/X_L becomes very large when X_L approaches zero. At very high frequencies, θ approaches 0° because the ratio R/X_L becomes approximately zero as X_L approaches infinity. To calculate the cutoff frequency of an *RL* high-pass filter, use Formula (26–6).

Example 26-3

Calculate the cutoff frequency for (a) the *RC* high-pass filter in Fig. 26–16*a*; (b) the *RL* high-pass filter in Fig. 26–17*a*.

ANSWER

a. Use Formula (26–3):

$$f_c = \frac{1}{2\pi RC}$$

$$= \frac{1}{2 \times \pi \times 1.5\ \text{k}\Omega \times 0.01\ \mu\text{F}}$$

$$= 10.61\ \text{kHz}$$

b. Use Formula (26–6):

$$f_c = \frac{R}{2\pi L}$$

$$= \frac{1.5\ \text{k}\Omega}{2 \times \pi \times 100\ \text{mH}}$$

$$= 2.39\ \text{kHz}$$

RC Bandpass Filter

A high-pass filter can be combined with a low-pass filter when it is desired to pass only a certain band of frequencies. This type of filter is called a *bandpass filter*. Figure 26–18*a* shows an *RC* bandpass filter, and Fig. 26–18*b* shows how its output voltage varies with frequency. In Fig. 26–18*a*, R_1 and C_1 constitute the high-pass filter, and R_2 and C_2 constitute the low-pass filter. To ensure that the low-pass filter does not load the high-pass filter, R_2 is usually 10 or more times larger than the resistance of R_1. The cutoff frequency of the high-pass filter is designated f_{c_1}, and the cutoff frequency of the low-pass filter is designated f_{c_2}. These two frequencies can be found on the response curve in Fig. 26–18*b*. To calculate the values for f_{c_1} and f_{c_2}, use the formulas given earlier for individual *RC* low-pass and *RC* high-pass filter circuits.

MultiSim **Figure 26–18** *RC* bandpass filter. (*a*) Circuit. (*b*) Graph of V_{out} versus frequency.

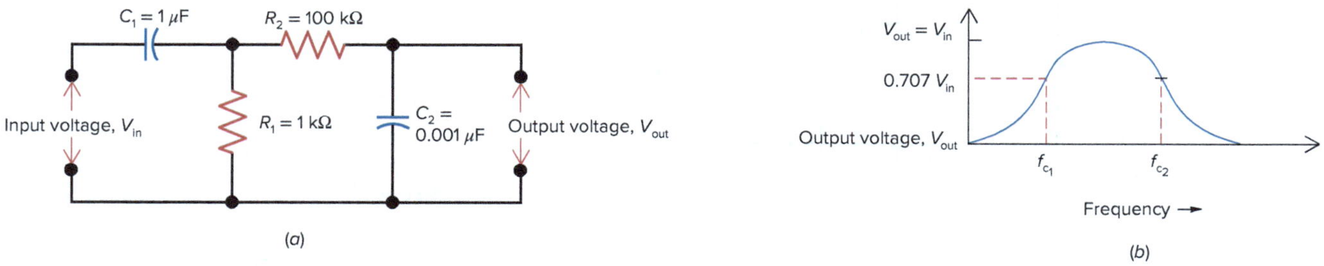

Example 26–4

In Fig. 26–18*a*, calculate the cutoff frequencies f_{c_1} and f_{c_2}.

ANSWER Calculate f_{c_1} for the high-pass filter consisting of R_1 and C_1:

$$f_{c_1} = \frac{1}{2\pi R_1 C_1}$$

$$= \frac{1}{2 \times \pi \times 1\ \text{k}\Omega \times 1\ \mu\text{F}}$$

$$= 159\ \text{Hz}$$

Next, calculate f_{c_2}.

$$f_{c_2} = \frac{1}{2\pi R_2 C_2}$$

$$= \frac{1}{2 \times \pi \times 100\ \text{k}\Omega \times 0.001\ \mu\text{F}}$$

$$= 1.59\ \text{kHz}$$

The frequencies below 159 Hz and above 1.59 kHz are severely attenuated, whereas those between 159 Hz and 1.59 kHz are effectively passed from the input to the output.

RC Band-Stop Filter

A high-pass filter can also be combined with a low-pass filter when it is desired to block or severely attenuate a certain band of frequencies. Such a filter is called a *band-stop* or *notch filter*. Figure 26–19*a* shows an *RC* **band-stop filter**, and Fig. 26–19*b* shows how its output voltage varies with frequency. In Fig. 26–19*a*, the components identified as $2R_1$ and $2C_1$ constitute the low-pass filter section, and the components identified as R_1 and C_1 constitute the high-pass filter section. Notice that the individual filters are in parallel. The frequency of maximum attenuation is called the *notch frequency*, identified as f_N in Fig. 26–19*b*. Notice that the maximum value of V_{out} below f_N is less than the maximum value of V_{out} above f_N. The reason for this is that the series resistances ($2R_1$) in the low-pass filter provide greater circuit losses than the series capacitors (C_1) in the high-pass filter.

Figure 26–19 Notch filter. (*a*) Circuit. (*b*) Graph of V_{out} versus frequency.

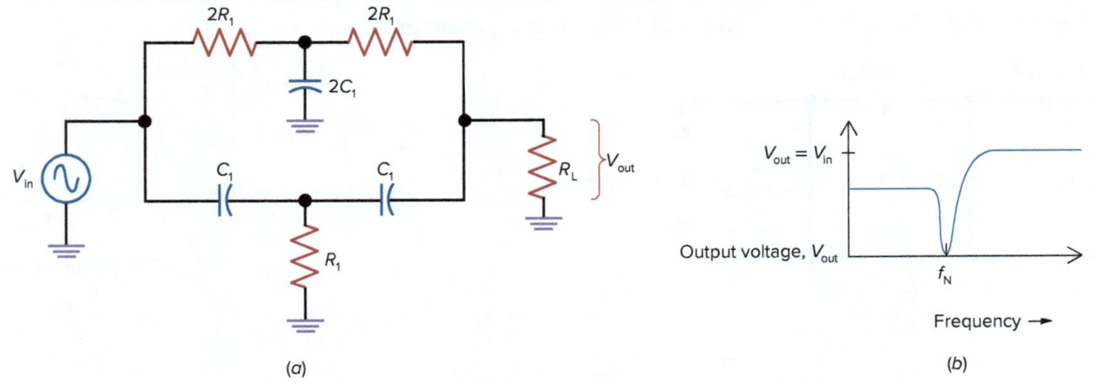

To calculate the notch frequency f_N in Fig. 26–19*a*, use Formula (26–11):

$$f_N = \frac{1}{4\pi R_1 C_1} \qquad \textbf{(26–11)}$$

Example 26-5

Calculate the notch frequency f_N in Fig. 26–19*a* if $R_1 = 1$ kΩ and $C_1 = 0.01$ μF. Also, calculate the required values for $2R_1$ and $2C_1$ in the low-pass filter.

ANSWER Use Formula (26–11):

$$f_N = \frac{1}{4\pi R_1 C_1}$$

$$= \frac{1}{4 \times \pi \times 1\ \text{k}\Omega \times 0.01\ \mu\text{F}}$$

$$= 7.96\ \text{kHz}$$

$$2R_1 = 2 \times 1\ \text{k}\Omega$$

$$= 2\ \text{k}\Omega$$

$$2C_1 = 2 \times 0.01\ \mu\text{F}$$

$$= 0.02\ \mu\text{F}$$

26–9 Self-Review

Answers at the end of the chapter.

a. **Increasing the capacitance C in Fig. 26–12*a* raises the cutoff frequency f_c. (True/False)**
b. **Decreasing the inductance L in Fig. 26–14*a* raises the cutoff frequency f_c. (True/False)**
c. **Increasing the value of C_2 in Fig. 26–18*a* reduces the passband. (True/False)**
d. **In Fig. 26–17*a*, V_{out} is approximately zero at very low frequencies. (True/False)**

26–10 Decibels and Frequency Response Curves

In analyzing filters, the **decibel (dB)** unit is often used to describe the amount of attenuation offered by the filter. In basic terms, the *decibel* is a logarithmic expression that compares two power levels. Expressed mathematically,

$$N_{dB} = 10 \log \frac{P_{out}}{P_{in}} \qquad \textbf{(26–12)}$$

where

N_{dB} = gain or loss in decibels
P_{in} = input power
P_{out} = output power

If the ratio P_{out}/P_{in} is greater than one, the N_{dB} value is positive, indicating an increase in power from input to output. If the ratio P_{out}/P_{in} is less than one, the N_{dB} value is negative, indicating a loss or reduction in power from input to output. A reduction in power, corresponding to a negative N_{dB} value, is referred to as *attenuation.*

Example 26–6

A certain amplifier has an input power of 1 W and an output power of 100 W. Calculate the dB power gain of the amplifier.

ANSWER Use Formula (26–12):

$$\begin{aligned} N_{dB} &= 10 \log \frac{P_{out}}{P_{in}} \\ &= 10 \log \frac{100\text{ W}}{1\text{ W}} \\ &= 10 \times 2 \\ &= 20\text{ dB} \end{aligned}$$

Example 26–7

The input power to a filter is 100 mW, and the output power is 5 mW. Calculate the attenuation, in decibels, offered by the filter.

ANSWER

$$\begin{aligned} N_{dB} &= 10 \log \frac{P_{out}}{P_{in}} \\ &= 10 \log \frac{5\text{ mW}}{100\text{ mW}} \\ &= 10 \times (-1.3) \\ &= -13\text{ dB} \end{aligned}$$

The power gain or loss in decibels can also be computed from a voltage ratio if the measurements are made across equal resistances.

$$N_{dB} = 20 \log \frac{V_{out}}{V_{in}} \qquad \textbf{(26–13)}$$

where

N_{dB} = gain or loss in decibels
V_{in} = input voltage
V_{out} = output voltage

Figure 26–20 *RC* low-pass filter.

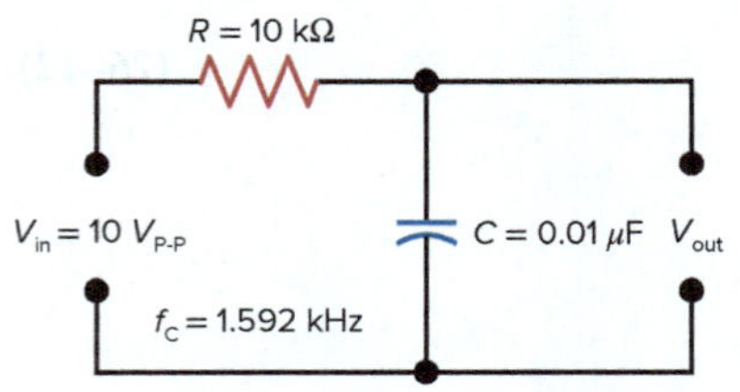

The N_{dB} values of the passive filters discussed in this chapter can never be positive because V_{out} can never be greater than V_{in}.

Consider the *RC* low-pass filter in Fig. 26–20. The cutoff frequency f_c for this circuit is 1.592 kHz, as determined by Formula (26–3). Recall that the formula for V_{out} at any frequency is

$$V_{out} = \frac{X_C}{Z_T} \times V_{in}$$

Dividing both sides of the equation by V_{in} gives

$$\frac{V_{out}}{V_{in}} = \frac{X_C}{Z_T}$$

Substituting X_C/Z_T for V_{out}/V_{in} in Formula (26–13) gives

$$N_{dB} = 20 \log \frac{X_C}{Z_T}$$

Example 26–8

In Fig. 26–20, calculate the attenuation, in decibels, at the following frequencies: (a) 0 Hz; (b) 1.592 kHz; (c) 15.92 kHz. (Assume that $V_{in} = 10\ V_{p\text{-}p}$ at all frequencies.)

ANSWER

a. At 0 Hz, $V_{out} = V_{in} = 10\ V_{p\text{-}p}$ because the capacitor C appears as an open. Therefore,

$$\begin{aligned} N_{dB} &= 20 \log \frac{V_{out}}{V_{in}} \\ &= 20 \log \frac{10\ V_{p\text{-}p}}{10\ V_{p\text{-}p}} \\ &= 20 \log 1 \\ &= 20 \times 0 \\ &= 0\ \text{dB} \end{aligned}$$

b. Since 1.592 kHz is the cutoff frequency f_c, V_{out} will be $0.707 \times V_{in}$ or $7.07\ V_{p\text{-}p}$. Therefore,

$$\begin{aligned} N_{dB} &= 20 \log \frac{V_{out}}{V_{in}} \\ &= 20 \log \frac{7.07\ V_{p\text{-}p}}{10\ V_{p\text{-}p}} \\ &= 20 \log 0.707 \\ &= 20 \times (-0.15) \\ &= -3\ \text{dB} \end{aligned}$$

c. To calculate N_{dB} at 15.92 kHz, X_C and Z_T must first be determined.

$$\begin{aligned} X_C &= \frac{1}{2\pi fC} \\ &= \frac{1}{2 \times \pi \times 15.92\ \text{kHz} \times 0.01\ \mu\text{F}} \\ &= 1\ \text{k}\Omega \end{aligned}$$

$$\begin{aligned} Z_T &= \sqrt{R^2 + X_C^2} \\ &= \sqrt{10^2\ \text{k}\Omega + 1^2\ \text{k}\Omega} \\ &= 10.05\ \text{k}\Omega \end{aligned}$$

Next,

$$\begin{aligned} N_{dB} &= 20 \log \frac{X_C}{Z_T} \\ &= 20 \log \frac{1\ \text{k}\Omega}{10.05\ \text{k}\Omega} \\ &= 20 \log 0.0995 \\ &= 20\,(-1) \\ &= -20\ \text{dB} \end{aligned}$$

GOOD TO KNOW

Some active filters provide a power gain for those frequencies in or near the passband. For example, an active filter may have a power gain of 20 dB in the passband. At the cutoff frequency, the power gain will be reduced to 17 dB.

Figure 26–21 *RC* and *RL* filter circuits, showing formulas for calculating decibel attenuation.

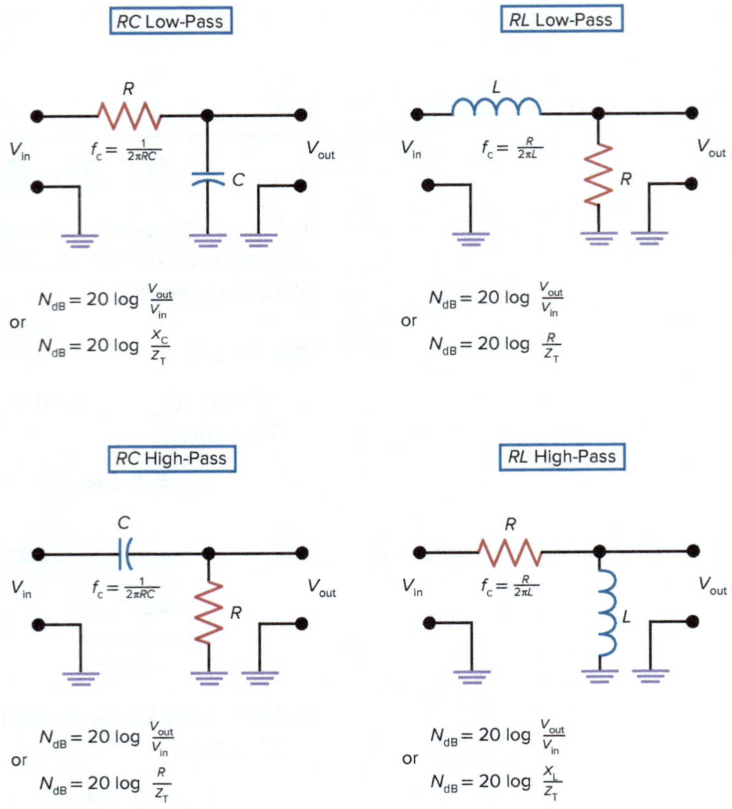

In Example 26–8, notice that N_{dB} is 0 dB at a frequency of 0 Hz, which is in the filter's passband. This may seem unusual, but the 0-dB value simply indicates that there is no attenuation at this frequency. For an ideal passive filter, $N_{dB} = 0$ dB in the passband. As another point of interest from Example 26–8, N_{dB} is −3 dB at the cutoff frequency of 1.592 kHz. Since $V_{out} = 0.707\ V_{in}$ at f_c for any passive filter, N_{dB} is always −3 dB at the cutoff frequency of a passive filter.

The N_{dB} value of loss can be determined for any filter if the values of V_{in} and V_{out} are known. Figure 26–21 shows the basic *RC* and *RL* low-pass and high-pass filters. The formula for calculating the N_{dB} attenuation is provided for each filter.

Frequency Response Curves

The frequency response of a filter is typically shown by plotting its gain (or loss) versus frequency on logarithmic graph paper. The two types of logarithmic graph paper are log-log and semilog. On *semilog graph paper*, the divisions along one axis are spaced logarithmically, and the other axis has conventional linear spacing between divisions. On *log-log graph paper*, both axes have logarithmic spacing between divisions. Logarithmic spacing results in a scale that expands the display of smaller values and compresses the display of larger values. On logarithmic graph paper, a 2-to-1 range of frequencies is called an ***octave***, and a 10-to-1 range of values is called a ***decade***.

One advantage of logarithmic spacing is that a larger range of values can be shown in one plot without losing resolution in the smaller values. For example, if frequencies between 10 Hz and 100 kHz were plotted on 100 divisions of linear graph paper, each division would represent approximately 1000 Hz and it would be impossible to plot values in the decade between 10 Hz and 100 Hz. On the other hand, by using logarithmic graph paper, the decade between 10 Hz and 100 Hz would occupy the same space on the graph as the decade between 10 kHz and 100 kHz.

Log-log or semilog graph paper is specified by the number of decades it contains. Each decade is a *graph cycle*. For example, 2-cycle by 4-cycle log-log paper has two decades on one axis and four on the other. The number of cycles must be adequate for the range of data plotted. For example, if the frequency response extends from 25 Hz to 40 kHz, 4 cycles are necessary to plot the frequencies corresponding to the decades 10 Hz to 100 Hz, 100 Hz to 1 kHz, 1 kHz to 10 kHz, and 10 kHz to 100 kHz. A typical sheet of log-log graph paper is shown in Fig. 26–22. Because there are three decades on the horizontal axis and five decades on the vertical axis, this graph paper is called 3-cycle by 5-cycle log-log paper. Notice that each octave corresponds to a 2-to-1 range in values and each decade corresponds to a 10-to-1 range in values. For clarity, several octaves and decades are shown in Fig. 26–22.

When semilog graph paper is used to plot a frequency response, the observed or calculated values of gain (or loss) must first be converted to decibels before plotting. On the other hand, since decibel voltage gain is a logarithmic function, the gain or loss values can be plotted on log-log paper without first converting to decibels.

RC Low-Pass Frequency Response Curve

Figure 26–23*a* shows an *RC* low-pass filter whose cutoff frequency f_c is 1.592 kHz as determined by Formula (26–3). Figure 26–23*b* shows its frequency response curve plotted on semilog graph paper. Notice there are 6 cycles on the horizontal axis, which spans a frequency range from 1 Hz to 1 MHz. Notice that the vertical axis specifies the N_{dB} loss, which is the amount of attenuation offered by the filter in decibels. Notice that $N_{dB} = -3$ dB at the cutoff frequency of 1.592 kHz. Above f_c, N_{dB} decreases at the rate of approximately 6 dB/octave, which is equivalent to a rate of 20 dB/decade.

Example 26–9

From the graph in Fig. 26–23*b*, what is the attenuation in decibels at (a) 100 Hz; (b) 10 kHz; (c) 50 kHz?

ANSWER

a. At $f = 100$ Hz, $N_{dB} = 0$ dB, as indicated by point A on the graph.
b. At $f = 10$ kHz, $N_{dB} = -16$ dB, as indicated by point B on the graph.
c. At $f = 50$ kHz, $N_{dB} = -30$ dB, as indicated by point C.

For filters such as the inverted *L*, *T*, or π type, the response curve rolloff is much steeper beyond the cutoff frequency f_c. For example, a low-pass filter with a series inductor and a shunt capacitor has a rolloff rate of 12 dB/octave or 40 dB/decade above the cutoff frequency f_c. To increase the rate of rolloff, more inductors and capacitors must be used in the filter design. Filters are available whose rolloff rates exceed 36 dB/octave.

Figure 26–22 Log-log graph paper. Notice that each octave corresponds to a 2-to-1 range of values and each decade corresponds to a 10-to-1 range of values.

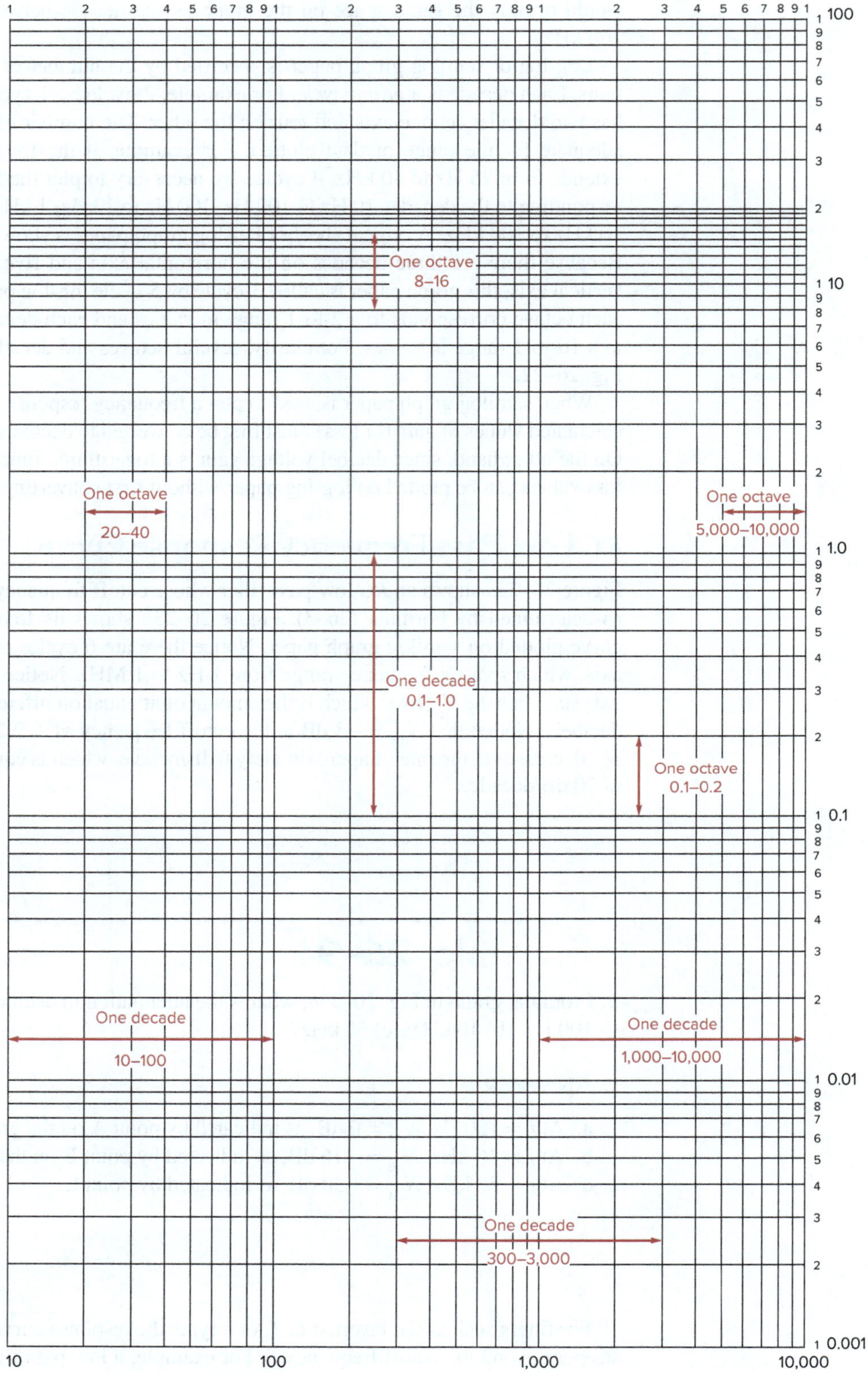

Figure 26–23 *RC* low-pass filter frequency response curve. (*a*) Circuit. (*b*) Frequency response curve.

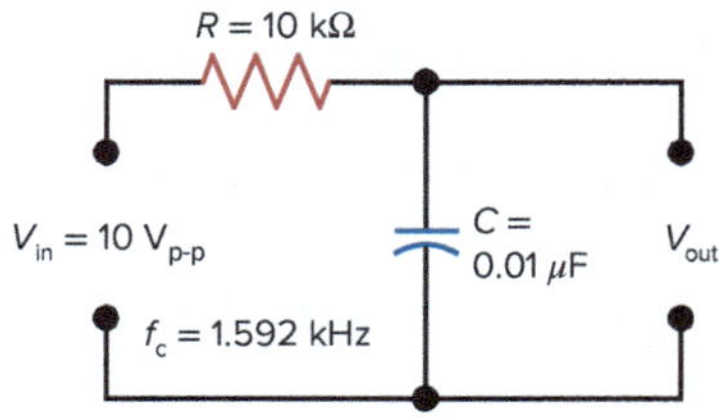

(*a*)

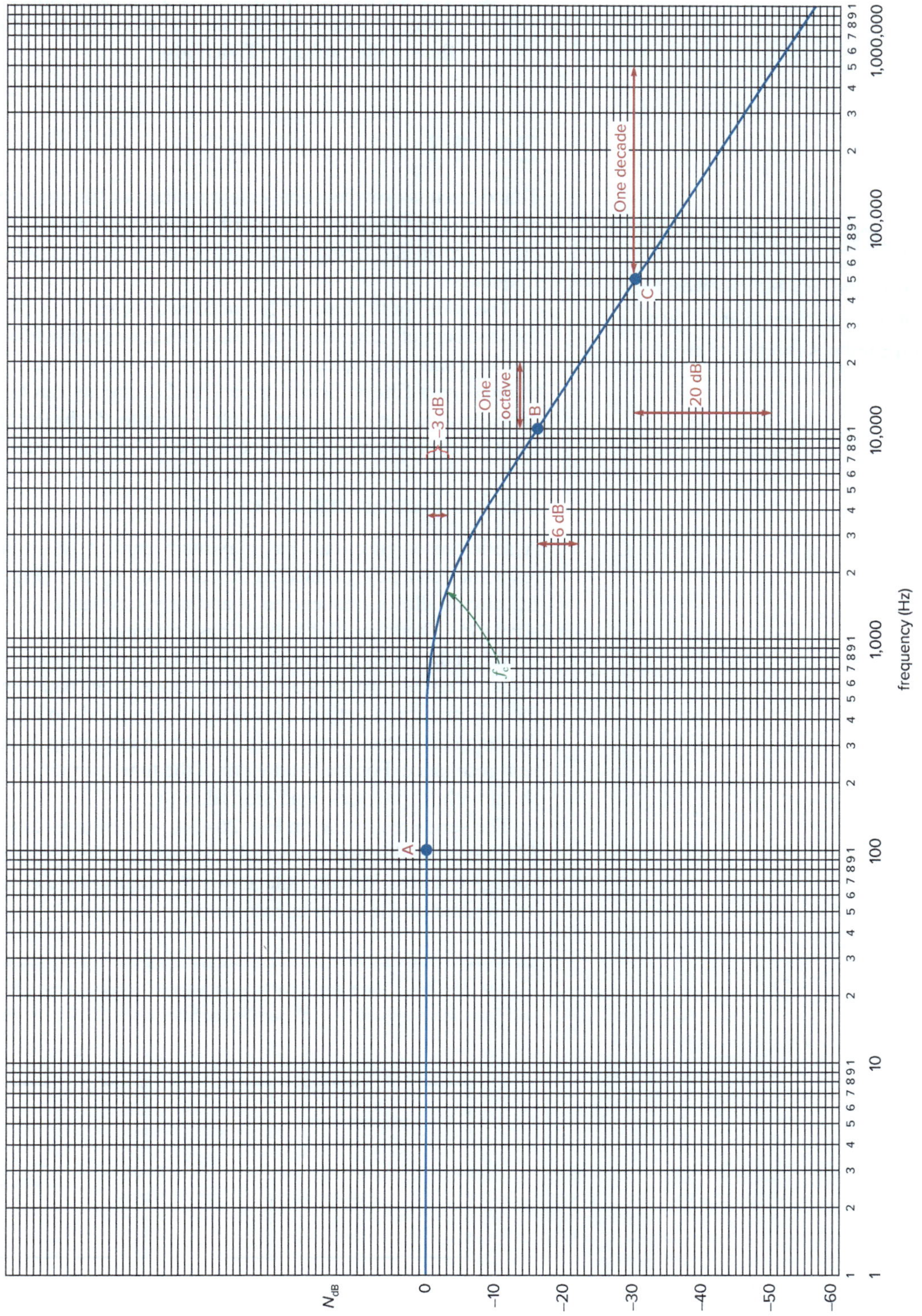

(*b*)

26–10 Self-Review

Answers at the end of the chapter.

a. At very low frequencies, a low-pass filter provides an attenuation of 0 dB. (True/False)

b. At the cutoff frequency, a low-pass filter has an N_{dB} loss of −3 dB. (True/False)

c. On logarithmic graph paper, one cycle is the same as one octave. (True/False)

d. The advantage of semilog and log-log graph paper is that a larger range of values can be shown in one plot without losing resolution in the smaller values. (True/False)

26–11 Resonant Filters

GOOD TO KNOW

RF transmitters and receivers contain several resonant filters, both band pass and band stop.

Tuned circuits provide a convenient method of filtering a band of radio frequencies because relatively small values of *L* and *C* are necessary for resonance. A tuned circuit provides filtering action by means of its maximum response at the resonant frequency.

The width of the band of frequencies affected by resonance depends on the *Q* of the tuned circuit; a higher *Q* provides a narrower bandwidth. Since resonance is effective for a band of frequencies below and above f_r, resonant filters are called *band-stop* or ***band-pass* filters**. Series or parallel *LC* circuits can be used for either function, depending on the connections with respect to R_L. In the application of a band-stop filter to suppress certain frequencies, the *LC* circuit is often called a *wavetrap.*

Series Resonance Filters

A series resonant circuit has maximum current and minimum impedance at the resonant frequency. Connected in series with R_L, as in Fig. 26–24*a*, the series-tuned *LC* circuit allows frequencies at and near resonance to produce maximum output across R_L. Therefore, this is band-pass filtering.

When the series *LC* circuit is connected across R_L as in Fig. 26–24*b*, however, the resonant circuit provides a low-impedance shunt path that short-circuits R_L. Then there is minimum output. This action corresponds to a shunt bypass capacitor, but the resonant circuit is more selective, short-circuiting R_L just for frequencies at and near resonance. For the bandwidth of the tuned circuit, the series resonant circuit in shunt with R_L provides band-stop filtering.

The series resistor R_S in Fig. 26–24*b* is used to isolate the low resistance of the *LC* filter from the input source. At the resonant frequency, practically all of the input voltage is across R_S with little across R_L because the *LC* tuned circuit then has very low resistance due to series resonance.

Figure 26–24 The filtering action of a series resonant circuit. (*a*) Band-pass filter when *L* and *C* are in series with R_L. (*b*) Band-stop filter when *LC* circuit is in shunt with R_L.

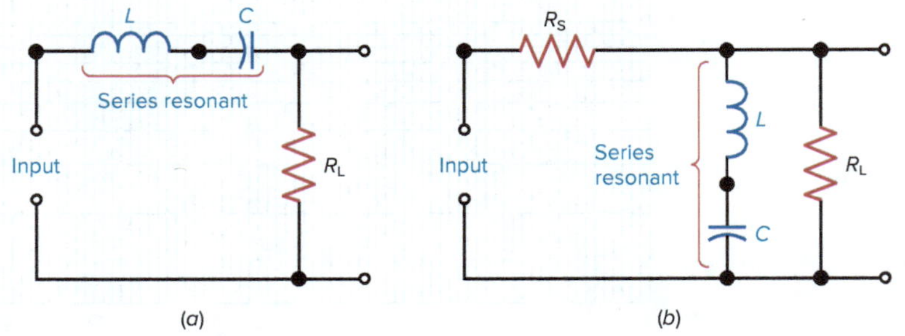

Figure 26–25 The filtering action of a parallel resonant circuit. (*a*) Band-stop filter when *LC* tank is in series with R_L. (*b*) Band-pass filter when *LC* tank is in shunt with R_L.

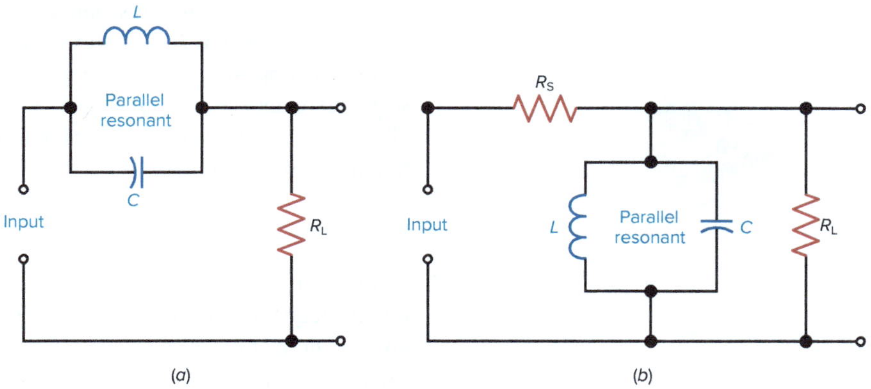

Parallel Resonance Filters

A parallel resonant circuit has maximum impedance at the resonant frequency. Connected in series with R_L, as in Fig. 26–25*a*, the parallel-tuned *LC* circuit provides maximum impedance in series with R_L at and near the resonant frequency. Then these frequencies produce maximum voltage across the *LC* circuit but minimum output voltage across R_L. This is a band-stop filter, therefore, for the bandwidth of the tuned circuit.

The parallel *LC* circuit connected across R_L, however, as in Fig. 26–25*b*, provides a band-pass filter. At resonance, the high impedance of the parallel *LC* circuit allows R_L to develop its output voltage. Below resonance, R_L is short-circuited by the low reactance of *L*; above resonance, R_L is short-circuited by the low reactance of *C*. For frequencies at or near resonance, though, R_L is shunted by high impedance, resulting in maximum output voltage.

The series resistor R_S in Fig. 26–25*b* is used to improve the filtering effect. Note that the parallel *LC* combination and R_S divide the input voltage. At the resonant frequency, though, the *LC* circuit has very high resistance for parallel resonance. Then most of the input voltage is across the *LC* circuit and R_L with little across R_S.

L-Type Resonant Filter

Series and parallel resonant circuits can be combined in *L*, *T*, or π sections for sharper discrimination of the frequencies to be filtered. Examples of an L-type filter are shown in Fig. 26–26.

Figure 26–26 Inverted-L filter with resonant circuits. (*a*) Band-stop filtering action. (*b*) Band-pass filtering action.

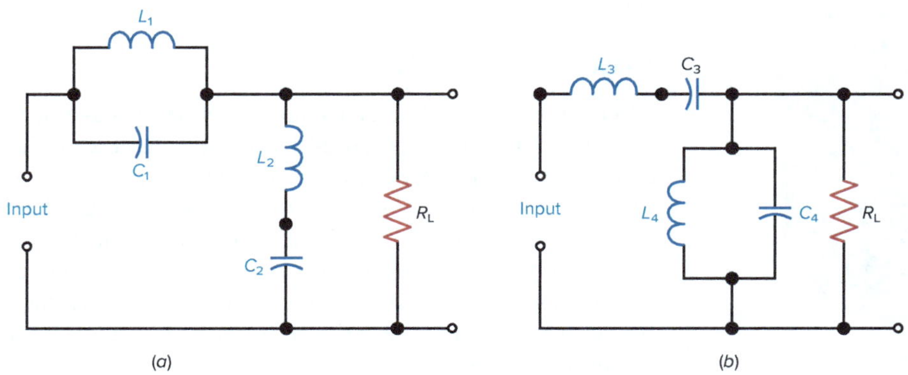

Figure 26–27 Quartz crystal. Size is ½ in. wide.

Mark Steinmetz/McGraw Hill

The circuit in Fig. 26–26*a* is a band-stop filter. The reason is that the parallel resonant L_1C_1 circuit is in series with the load, whereas the series resonant L_2C_2 circuit is in shunt with R_L. There is a dual effect as a voltage divider across the input source voltage. The high resistance of L_1C_1 reduces voltage output to the load. Also, the low resistance of L_2C_2 reduces the output voltage.

For the opposite effect, the circuit in Fig. 26–26*b* is a band-pass filter. Now the series resonant L_3C_3 circuit is in series with the load. Here the low resistance of L_3C_3 allows more output for R_L at resonance. Also, the high resistance of L_4C_4 allows maximum output voltage.

Crystal Filters

A thin slice of quartz provides a resonance effect by mechanical vibrations at a particular frequency, like an *LC* circuit. The quartz crystal can be made to vibrate by a voltage input or produce voltage output when it is compressed, expanded, or twisted. This characteristic of some crystals is known as the *piezoelectric effect.* As a result, crystals are often used in place of resonant circuits. In fact, the *Q* of a resonant crystal is much higher than that of *LC* circuits. However, the crystal has a specific frequency that cannot be varied because of its stability. Crystals are used for radio frequencies in the range of about 0.5 to 30 MHz. Figure 26–27 shows a crystal for the frequency of 3.579545 MHz for use in the color oscillator circuit of a television receiver. Note the exact frequency.

Special ceramic materials, such as lead titanate, can also be used as **crystal filters**. They have a piezoelectric effect like quartz crystals. Ceramic crystals are smaller and cost less, but they have a lower *Q* than quartz crystals.

GOOD TO KNOW

In some cases, rf interference cannot be eliminated by a power-line filter because it is radiated from the source. In this case, the rf energy induces voltage and currents into the circuit in addition to traveling through the 60-Hz AC power line.

26–11 Self-Review

Answers at the end of the chapter.

a. A parallel-resonant *LC* circuit in series with the load is a band-stop filter. (True/False)

b. A series resonant *LC* circuit in series with the load is a band-pass filter. (True/False)

c. Quartz crystals can be used as resonant filters. (True/False)

26–12 Interference Filters

Voltage or current not at the desired frequency represents interference. Usually, such interference can be eliminated by a filter. Some typical applications are (1) low-pass filter to eliminate rf interference from the 60-Hz power-line input to a receiver, (2) high-pass filter to eliminate rf interference from the signal picked up by a television receiving antenna, and (3) resonant filter to eliminate an interfering radio frequency from the desired rf signal. As noted earlier, the resonant band-stop filter is called a *wavetrap.*

Figure 26–28 Power-line filter unit. (*a*) Circuit of balanced L-type low-pass filter. (*b*) Filter unit.

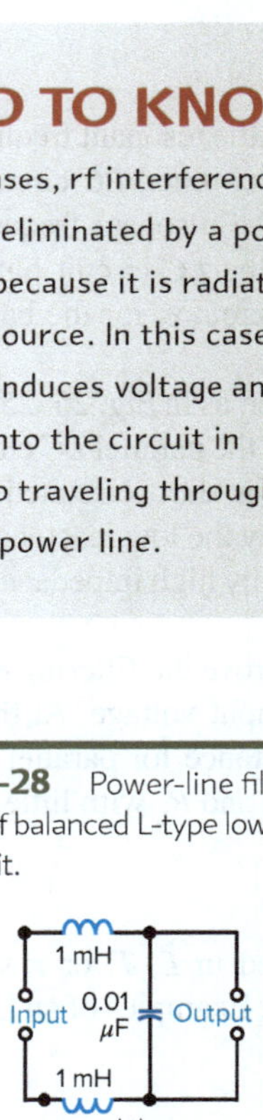

(*a*)

(*b*)

Mark Steinmetz/McGraw Hill

Power-Line Filter

Although the power line is a source of 60-Hz voltage, it is also a conductor of interfering rf currents produced by motors, fluorescent lighting circuits, and rf equipment. When a receiver is connected to the power line, the rf interference can produce noise and whistles in the receiver output. The filter, as shown in Fig. 26–28, can be used to minimize this interference. The filter is plugged into the wall outlet for 60-Hz power, and the receiver is plugged into the filter. An rf bypass capacitor across the line with two series rf chokes forms a low-pass balanced L-type filter. Using a choke in each side of the line makes the circuit balanced to ground.

The chokes provide high impedance for interfering rf current but not for 60 Hz, isolating the receiver input connections from rf interference in the power line. Also, the bypass capacitor short-circuits the receiver input for radio frequencies but not for 60 Hz. The unit then is a low-pass filter for 60-Hz power applied to the receiver while rejecting higher frequencies.

Television Antenna Filter

Figure 26–29 A television antenna filter to pass TV channel frequencies above 54 MHz but attenuate lower frequencies that can cause interference.

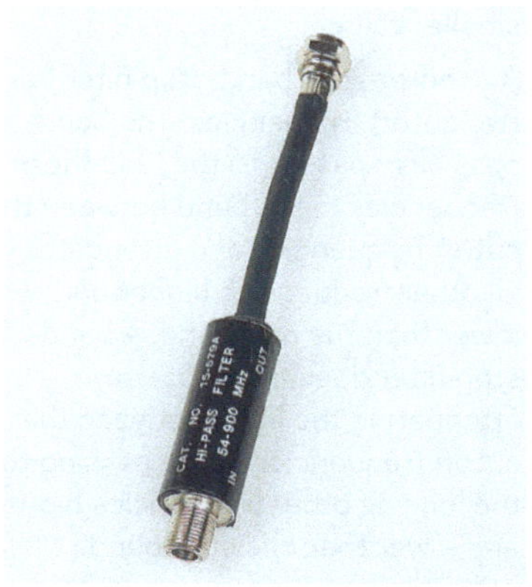

Mark Steinmetz/McGraw Hill

When a television receiver has interference in the picture resulting from radio frequencies below the television broadcast band that are picked up by the receiving antenna, this rf interference can be reduced by the high-pass filter shown in Fig. 26–29. The filter attenuates frequencies below 54 MHz, which is the lowest frequency for Channel 2.

At frequencies lower than 54 MHz, series capacitances provide increasing reactance with a larger voltage drop, whereas the shunt inductances have less reactance and short-circuit the load. Higher frequencies are passed to the load as the series capacitive reactance decreases and the shunt inductive reactance increases.

Connections to the filter unit are made at the receiver end of the line from the antenna. Either end of the filter is connected to the antenna terminals on the receiver with the opposite end connected to the antenna line.

26–12 Self-Review

Answers at the end of the chapter.

a. A wavetrap is a band-stop filter. (True/False)

b. The TV antenna filter in Fig. 26–29 is a high-pass filter with series capacitors. (True/False)

Summary

- A filter can separate high and low frequencies. With input of different frequencies, the high-pass filter allows the higher frequencies to produce output voltage across the load; a low-pass filter provides output voltage at lower frequencies.
- Pulsating or fluctuating direct current varies in amplitude but does not reverse its direction. Similarly, a pulsating or fluctuating DC voltage varies in amplitude but maintains one polarity, either positive or negative.
- Pulsating direct current or voltage consists of a steady DC level, equal to the average value, and an AC component that reverses in polarity with respect to the average level. The dc and AC can be separated by filters.
- An *RC* coupling circuit is a high-pass filter for pulsating direct current. Capacitance C_C blocks the steady DC voltage but passes the ac component.
- A transformer with an isolated secondary winding also is a high-pass filter. With pulsating direct current in the primary, only the AC component produces output voltage in the secondary.
- A bypass capacitor in parallel with *R* provides a low-pass filter.
- Combinations of *L*, *C*, and *R* can be arranged as L, T, or π filters for more selective filtering. All three arrangements can be used for either low-pass or high-pass action. See Figs. 26–9 and 26–11.
- In high-pass filters, the capacitance must be in series with the load as a coupling capacitor with shunt *R* or *L* across the line.
- For low-pass filters, the capacitance is across the line as a bypass capacitor, and *R* or *L* then must be in series with the load.
- The cutoff frequency f_c of a filter is the frequency at which the output voltage is reduced to 70.7% of its maximum value.
- For an *RC* low-pass or high-pass filter, $X_C = R$ at the cutoff frequency. Similarly, for an *RL* low-pass or high-pass filter, $X_L = R$ at the cutoff frequency. To calculate f_c for an *RC* low-pass or high-pass filter, use the formula $f_c = 1/(2\pi RC)$. To calculate f_c for an *RL* low-pass or high-pass filter, use the formula $f_c = R/2\pi L$.
- For an *RC* or *RL* filter, either low pass or high pass, the phase angle θ between V_{in} and V_{out} is approximately 0° in the passband. In the stop band, $\theta = \pm 90°$. The sign of θ depends on the type of filter.
- *RC* low-pass filters can be combined with *RC* high-pass filters when it is desired to either pass or block only a certain band of frequencies. These types of filters are called band-pass and band-stop filters, respectively.
- The decibel (dB) unit of measurement is used to compare two power levels. A passive filter has an attenuation of −3 dB at the cutoff frequency.
- Semilog and log-log graph paper are typically used to show the frequency response of a filter. On semilog graph paper, the vertical axis uses conventional linear spacing; the horizontal axis uses logarithmically spaced divisions.
- The advantage of using semilog or log-log graph paper is that a larger range of values can be shown in one plot without losing resolution in the smaller values.
- A band-pass or band-stop filter has two cutoff frequencies. The band-pass filter passes to the load those frequencies in the band between the cutoff frequencies and attenuates all other frequencies higher and lower than the passband. A band-stop filter does the opposite, attenuating the band between the cutoff frequencies, while passing to the load all other frequencies higher and lower than the stop band.
- Resonant circuits are generally used for band-pass or band-stop filtering with radio frequencies.
- For band-pass filtering, the series resonant *LC* circuit must be in series with the load, for minimum series opposition; the high impedance of parallel resonance is across the load.
- For band-stop filtering, the circuit is reversed, with the parallel resonant *LC* circuit in series with the load; the series resonant circuit is in shunt across the load.
- A wavetrap is an application of the resonant band-stop filter.

Important Terms

Attenuation — a reduction in signal amplitude.

Band-pass filter — a filter designed to pass only a specific band of frequencies from its input to its output.

Band-stop filter — a filter designed to block or severely attenuate only a specific band of frequencies.

Bypass capacitor — a capacitor that bypasses or shunts the AC component of a pulsating DC voltage around a component such as a resistor. The value of a bypass capacitor should be chosen so that its X_C value is one-tenth or less of the parallel resistance at the lowest frequency intended to be bypassed.

Crystal filter — a filter made of a crystalline material such as quartz. Crystal filters are often used in place of conventional *LC* circuits because their *Q* is so much higher.

Cutoff frequency — the frequency at which the attenuation of a filter reduces the output amplitude to 70.7% of its value in the passband.

Decade — a 10-to-1 range in frequencies.

Decibel (dB) — a logarithmic expression that compares two power levels.

Fluctuating DC — a DC voltage or current that varies in magnitude but does not reverse in polarity or direction. Another name for fluctuating DC is pulsating DC.

High-pass filter — a filter that allows the higher-frequency components of the applied voltage to develop appreciable output voltage while at the same time attenuating or eliminating the lower-frequency components.

Low-pass filter — a filter that allows the lower-frequency components of the applied voltage to develop appreciable output voltage while at the same time attenuating or eliminating the higher-frequency components.

Octave — a 2–to-1 range in frequencies.

Pulsating DC — a DC voltage or current that varies in magnitude but does not reverse in polarity or direction. Another name for pulsating DC is fluctuating DC

Related Formulas

RC Low-Pass Filters

$$V_{out} = \frac{X_C}{Z_T} \times V_{in}$$

$$\theta = \arctan\ (-R/X_C)$$

$$f_c = \frac{1}{2\pi RC}$$

RL Low-Pass Filters

$$V_{out} = \frac{R}{Z_T} \times V_{in}$$

$$\theta = \arctan\ (-X_L/R)$$

$$f_c = R/2\pi L$$

RC High-Pass Filters

$$V_{out} = \frac{R}{Z_T} \times V_{in}$$

$$\theta = \arctan\ (X_C/R)$$

$$f_c = 1/2\pi RC$$

RL High-Pass Filters

$$V_{out} = \frac{X_L}{Z_T} \times V_{in}$$

$$\theta = \arctan\ (R/X_L)$$

$$f_c = R/2\pi L$$

Notch Filter

$$f_N = 1/4\pi R_1 C_1$$

Decibels

$$N_{dB} = 10 \log \frac{P_{out}}{P_{in}}$$

$$N_{dB} = 20 \log \frac{V_{out}}{V_{in}}$$

Self-Test

Answers at the back of the book.

1. A voltage that varies in magnitude but does not reverse in polarity is called a(n)

a. alternating voltage.
b. steady DC voltage.
c. pulsating DC voltage.
d. none of the above.

2. The capacitor in an *RC* coupling circuit

a. blocks the steady DC component of the input voltage.
b. blocks the AC component of the input voltage.
c. appears like a short to a steady DC voltage.
d. will appear like an open to the AC component of the input voltage.

3. The value of a bypass capacitor should be chosen so that its X_C value is

a. 10 or more times the parallel resistance at the highest frequency to be bypassed.
b. one-tenth or less the parallel resistance at the lowest frequency to be bypassed.
c. one-tenth or less the parallel resistance at the highest frequency to be bypassed.
d. equal to the parallel resistance at the lowest frequency to be bypassed.

4. In an *RC* low-pass filter, the output is taken across the

a. resistor.
b. inductor.
c. capacitor.
d. none of the above.

5. On logarithmic graph paper, a 10-to-1 range of frequencies is called a(n)

a. octave.
b. decibel (dB).
c. harmonic.
d. decade.

6. The cutoff frequency, f_c, of a filter is the frequency at which the output voltage is

a. reduced to 50% of its maximum.
b. reduced to 70.7% of its maximum.
c. practically zero.
d. exactly equal to the input voltage.

7. The decibel attenuation of a passive filter at the cutoff frequency is

a. −3 dB.
b. 0 dB.
c. −20 dB.
d. −6 dB.

8. To increase the cutoff frequency of an *RL* high-pass filter, you can

a. decrease the value of *R*.
b. decrease the value of *L*.
c. increase the value of *R*.
d. both b and c.

9. An *RC* low-pass filter uses a 2.2-kΩ *R* and a 0.01-μF *C*. What is its cutoff frequency?

a. 3.5 MHz

b. 72.3 Hz.

c. 7.23 kHz.

d. 1.59 kHz.

10. For either an *RC* low-pass or high-pass filter,

a. $X_c = 0\ \Omega$ at the cutoff frequency.

b. $X_c = R$ at the cutoff frequency.

c. X_c is infinite at the cutoff frequency.

d. none of the above.

11. When a pulsating DC voltage is applied as an input to the primary of a transformer, the output from the secondary contains

a. only the steady DC component of the input signal.

b. a stepped up or down version of the pulsating DC voltage.

c. only the AC component of the input signal.

d. none of the above.

12. A power-line filter used to reduce rf interference is an example of a

a. low-pass filter.

b. high-pass filter.

c. notch filter.

d. band-pass filter.

13. On logarithmic graph paper, a 2-to-1 range of frequencies is called a(n)

a. decade.

b. decibel (dB).

c. harmonic.

d. octave.

14. What is the decibel (dB) attenuation of a filter with a 100-mV input and a 1-mV output at a given frequency?

a. −40 dB.

b. −20 dB.

c. −3 dB.

d. 0 dB.

15. In an *RL* high-pass filter, the output is taken across the

a. resistor.

b. inductor.

c. capacitor.

d. none of the above.

16. An *RL* high-pass filter uses a 60-mH *L* and a 1-kΩ *R*. What is its cutoff frequency?

a. 2.65 kHz.

b. 256 kHz.

c. 600 kHz.

d. 32 kHz.

17. A T-type low-pass filter consists of

a. series capacitors and a parallel inductor.

b. series inductors and a bypass capacitor.

c. series capacitors and a parallel resistor.

d. none of the above.

18. A π-type high-pass filter consists of

a. series inductors and parallel capacitors.

b. series inductors and a parallel resistor.

c. a series capacitor and parallel inductors.

d. none of the above.

19. When examining the frequency response curve of an *RC* low-pass filter, it can be seen that the rate of rolloff well above the cutoff frequency is

a. 6 dB/octave.

b. 6 dB/decade.

c. 20 dB/decade.

d. both a and c.

20. For signal frequencies in the passband, an *RC* high-pass filter has a phase angle of approximately

a. 45°.

b. 0°.

c. +90°.

d. −90°.

Essay Questions

1. What is the function of an electrical filter?

2. Give two examples where the voltage has different frequency components.

3. (a) What is meant by *pulsating* direct current or voltage? (b) What are the two components of a pulsating DC voltage? (c) How can you measure the value of each of the two components?

4. Define the function of the following filters in terms of output voltage across the load resistance: (a) High-pass filter. Why is an $R_C C_C$ coupling circuit an example? (b) Low-pass filter. Why is an $R_b C_b$ bypass circuit an example? (c) Band-pass filter. How does it differ from a coupling circuit? (d) Band-stop filter. How does it differ from a band-pass filter?

5. Draw circuit diagrams for the following filter types. No values are necessary. (a) T-type high-pass and T-type low-pass; (b) π-type low-pass, balanced with a filter reactance in both sides of the line.

6. Draw the circuit diagrams for L-type band-pass and L-type band-stop filters. How do these two circuits differ from each other?

7. Draw the response curve for each of the following filters: (a) low-pass cutting off at 20,000 Hz; (b) high-pass cutting off at 20 Hz; (c) band-pass for 20 to 20,000 Hz; (d) band-pass for 450 to 460 kHz.

8. Give one similarity and one difference in comparing a coupling capacitor and a bypass capacitor.

9. Give two differences between a low-pass filter and a high-pass filter.

10. Explain briefly why the power-line filter in Fig. 26–28 passes 60-Hz alternating current but not 1-MHz rf current.

11. Explain the advantage of using semilog and log-log graph paper for plotting a frequency response curve.

12. Can an *RC* band-stop filter be designed by interchanging the low-pass and high-pass filters in Fig. 26–18*a*?

Problems

SECTION 26–1 EXAMPLES OF FILTERING

26–1 Explain the basic function of a

a. low-pass filter.

b. high-pass filter.

SECTION 26–2 DIRECT CURRENT COMBINED WITH ALTERNATING CURRENT

26–2 For the values shown in Fig. 26–30,

a. draw the waveform of voltage that is present across the load, R_L. Indicate the average and peak values on the waveform.

b. draw the waveform of current that exists in the load, R_L. Indicate the average and peak values on the waveform.

Figure 26–30

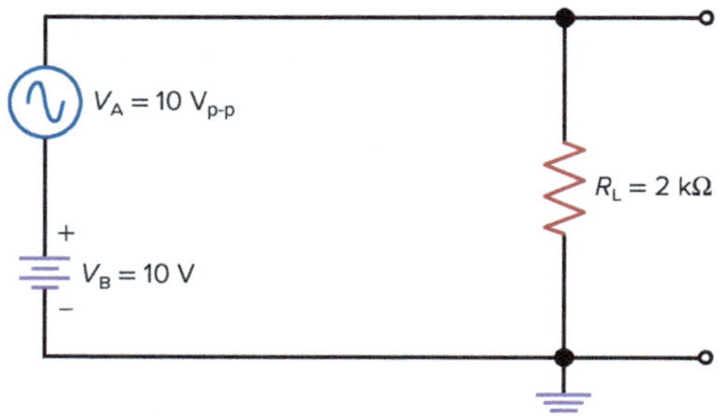

26–3 In Fig. 26–30, how much is the

a. average DC voltage across the load, R_L?

b. average DC current through the load, R_L?

26–4 In terms of the AC component in Fig. 26–30, how much is the

a. peak voltage?

b. peak-to-peak voltage?

c. rms voltage?

26–5 In Fig. 26–30, redraw the waveform of voltage present across the load, R_L, if the polarity of V_B is reversed. Indicate the average and peak values on the waveform.

SECTION 26–3 TRANSFORMER COUPLING

26–6 Figure 26–31 shows the application of transformer coupling. Notice that the transformer has a turns ratio of 1:1. How much is the

a. steady DC voltage in the primary?

b. steady DC voltage in the secondary?

c. peak-to-peak AC voltage in the primary?

d. peak-to-peak AC voltage in the secondary?

26–7 In Fig. 26–31, indicate the peak voltage values for the AC output in the secondary.

26–8 In Fig. 26–31, compare the average or DC value of the primary and secondary voltage waveforms. Explain any difference.

Figure 26–31

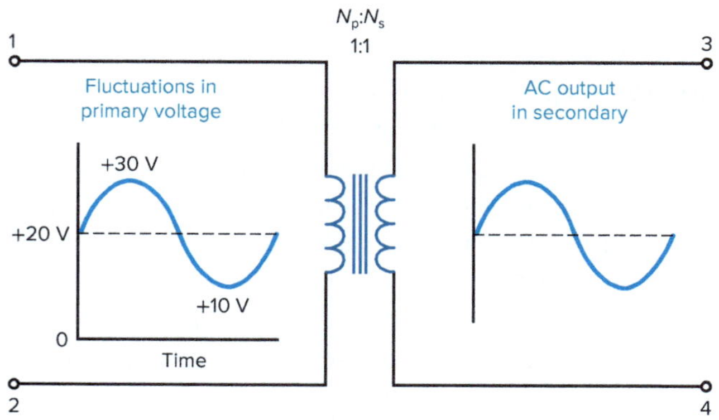

SECTION 26–4 CAPACITIVE COUPLING

26–9 In Fig. 26–32, determine the following:

a. the X_C value for C_C at $f = 10$ kHz.

b. the DC voltage across input terminals 1 and 2.

c. the DC voltage across the coupling capacitor, C_C.

d. the DC voltage across the resistor, R.

e. the peak-to-peak AC voltage across terminals 1 and 2.

f. the approximate peak-to-peak AC voltage across the coupling capacitor, C_C.

g. the peak-to-peak AC voltage across the resistor, R.

h. the rms voltage across the resistor, R.

Figure 26–32

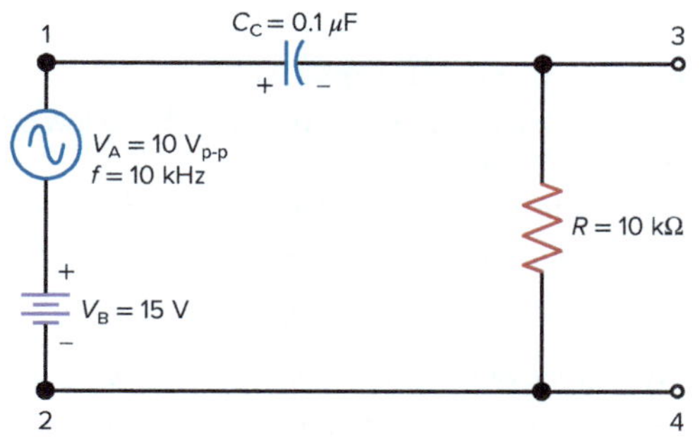

26–10 In Fig. 26–32, draw the voltage waveforms (including average and peak values) that exist across terminals

a. 1 and 2.

b. 3 and 4.

26–11 In Fig. 26–32, does C_C charge or discharge during the

a. positive alternation of V_A?

b. negative alternation of V_A?

26–12 In Fig. 26–32, is the positive or negative half-cycle of output voltage developed across R when C_C is

a. charging?

b. discharging?

26–13 In Fig. 26–32, what is the lowest frequency of V_A that will produce an X_C/R ratio of 1⁄10?

26–14 An *RC* coupling circuit is to be designed to couple frequencies above 500 Hz. If $R = 4.7$ kΩ, what is the minimum value for C_C?

SECTION 26–5 BYPASS CAPACITORS

26–15 In Fig. 26–33, determine the following:

a. the X_C value of C_1 at $f = 1$ MHz.

b. the DC voltage across input terminals 1 and 2.

c. the DC voltage across R_1.

d. the DC voltage across R_2.

e. the DC voltage across C_1.

f. the peak-to-peak AC voltage across terminals 1 and 2.

g. the approximate peak-to-peak AC voltage across terminals 3 and 4.

h. the approximate peak-to-peak AC voltage across R_1.

Figure 26–33

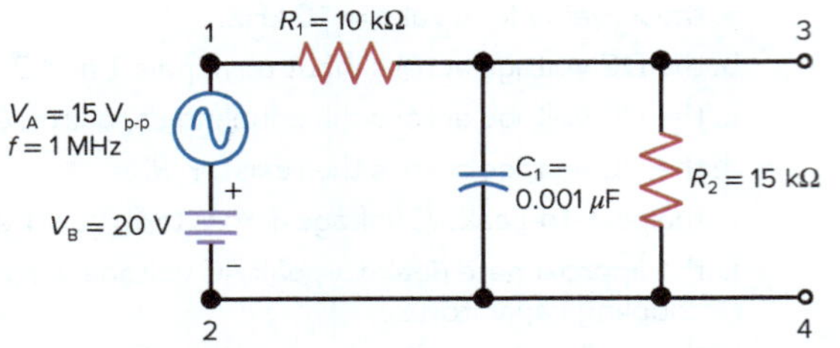

26–16 In Fig. 26–33, will the bypass capacitor, C_1, bypass R_2 if the frequency of V_A is 1 kHz?

26–17 What minimum value of capacitance will bypass a 1 kΩ resistor if the lowest frequency to be bypassed is 250 Hz?

SECTION 26–6 FILTER CIRCUITS

26–18 Classify each of the following as either a low-pass or high-pass filter:

a. transformer coupling (see Fig. 26–31).

b. *RC* coupling circuit (see Fig. 26–32).

c. bypass capacitor (see Fig. 26–33).

26–19 What type of filter, low-pass or high-pass, uses

a. series inductance and parallel capacitance?

b. series capacitance and parallel inductance?

26–20 Suppose that a low-pass filter has a cutoff frequency of 1 kHz. If the input voltage for a signal at this frequency is 30 mV, how much is the output voltage?

SECTION 26–7 LOW-PASS FILTERS

26–21 For a low-pass filter, define what is meant by the terms

a. passband.

b. stop band.

26–22 Assume that both the *RC* low-pass filter in Fig. 26–9*a* and the π-type filter in Fig. 26–9e have the same cutoff frequency, f_c. How do the filtering characteristics of these two filters differ?

SECTION 26–8 HIGH-PASS FILTERS

26–23 Do the terms passband and stop band apply to high-pass filters?

26–24 In Fig. 26–11, does the T-type filter provide sharper filtering than the RC filter? If so, why?

SECTION 26–9 ANALYZING FILTER CIRCUITS

26–25 Identify the filters in each of the following figures as either low-pass or high-pass:

a. Fig. 26–34

b. Fig. 26–35

c. Fig. 26–36

d. Fig. 26–37

Figure 26–34

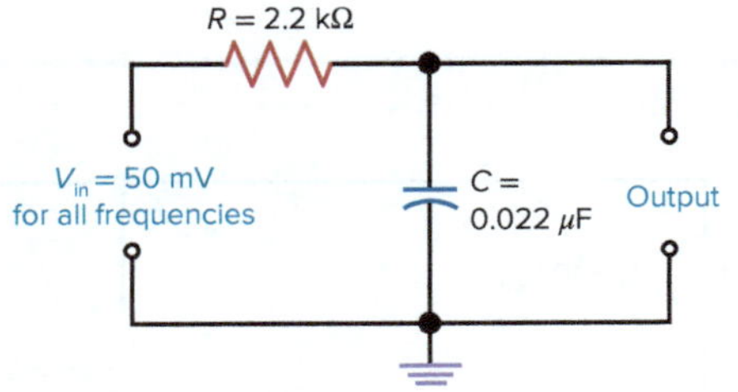

Figure 26–35

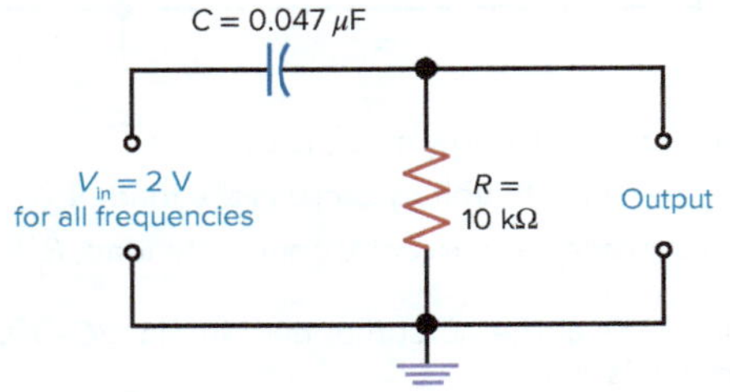

Figure 26–36

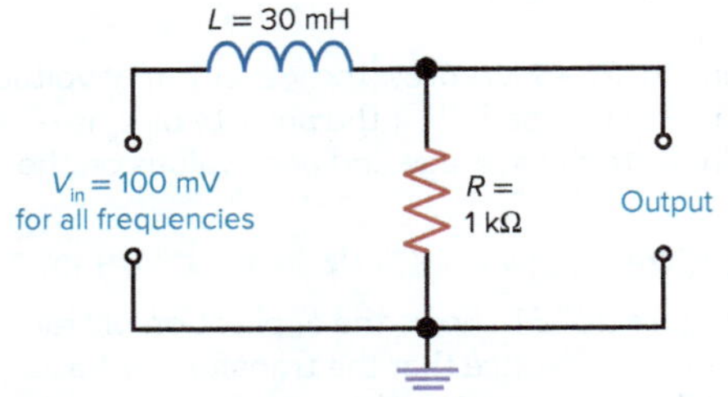

Figure 26–37

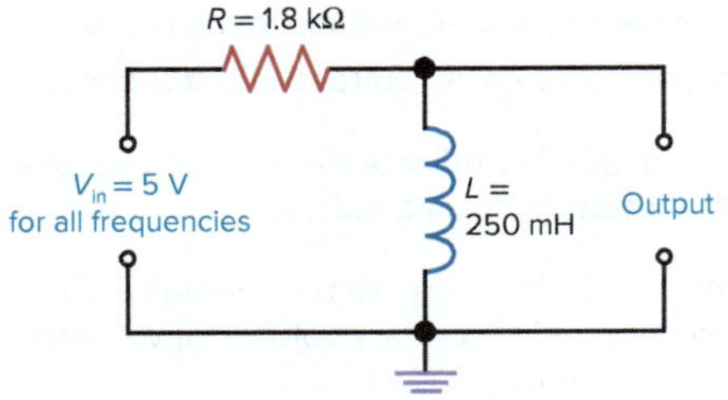

26–26 Calculate the cutoff frequency, f_c, for the filters in each of the following figures:

a. Fig. 26–34.

b. Fig. 26–35.

c. Fig. 26–36.

d. Fig. 26–37.

26–27 In Fig. 26–34, calculate the output voltage, V_{out}, and phase angle, θ, at the following frequencies:

a. 50 Hz.

b. 200 Hz.

c. 1 kHz.

d. f_c.

e. 10 kHz.

f. 20 kHz.

g. 100 kHz.

26–28 In Fig. 26–35, calculate the output voltage, V_{out}, and phase angle, θ, at the following frequencies:

a. 10 Hz.

b. 50 Hz.

c. 100 Hz.

d. f_c.

e. 1 kHz.

f. 20 kHz.

g. 500 kHz.

26–29 In Fig. 26–36, calculate the output voltage, V_{out}, and phase angle, θ, at the following frequencies:

a. 100 Hz.

b. 500 Hz.

c. 2 kHz.

d. f_c.

e. 15 kHz.

f. 30 kHz.

g. 100 kHz.

26–30 In Fig. 26–37, calculate the output voltage, V_{out}, and phase angle, θ, at the following frequencies:

a. 50 Hz.

b. 100 Hz.

c. 500 Hz.

d. f_c.

e. 3 kHz.

f. 10 kHz.

g. 25 kHz.

26–31 For the filters in Figs. 26–34 through 26–37, what is the ratio of V_{out}/V_{in} at the cutoff frequency?

26–32 Without regard to sign, what is the phase angle, θ, at the cutoff frequency for each of the filters in Figs. 26–34 through 26–37?

26–33 For a low-pass filter, what is the approximate phase angle, θ, for frequencies

a. well below the cutoff frequency?

b. well above the cutoff frequency?

26–34 Repeat Prob. 26–33 for a high-pass filter.

26–35 What type of filter is shown in Fig. 26–38?

Figure 26–38

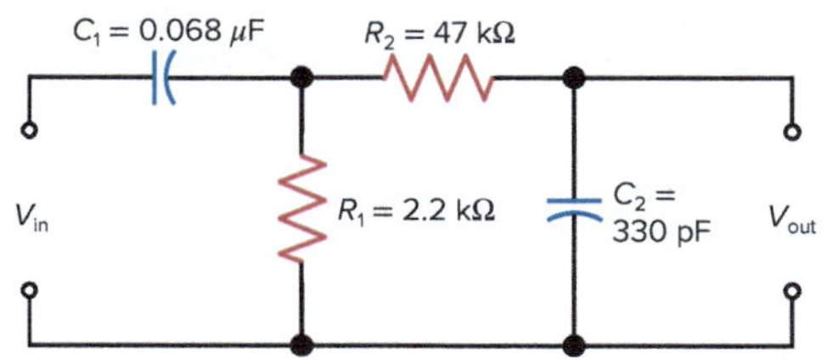

26–36 In Fig. 26–38, which components make up the

a. high-pass filter?

b. low-pass filter?

26–37 In Fig. 26–38, calculate

a. the cutoff frequency, f_{C_1}.

b. the cutoff frequency, f_{C_2}.

c. the bandwidth, $f_{C_2} - f_{C_1}$.

26–38 In Fig. 26–38, why is it important to make R_2 at least 10 times larger than R_1?

26–39 Calculate the notch frequency, f_N, in Fig. 26–39.

Figure 26–39

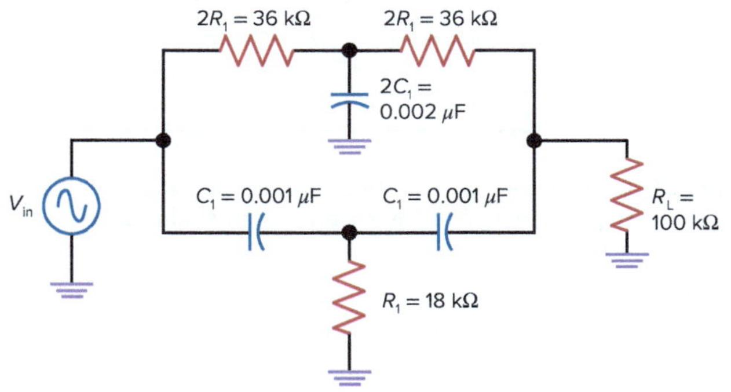

SECTION 26–10 DECIBELS AND FREQUENCY RESPONSE CURVES

26–40 Calculate the dB power gain of an amplifier for the following values of P_{in} and P_{out}:

a. P_{in} = 1 W, P_{out} = 2 W.

b. P_{in} = 1 W, P_{out} = 10 W.

c. P_{in} = 50 W, P_{out} = 1 kW.

d. P_{in} = 10 W, P_{out} = 400 W.

26–41 Calculate the dB attenuation of a filter for the following values of P_{in} and P_{out}:

a. P_{in} = 1 W, P_{out} = 500 mW.

b. P_{in} = 100 mW, P_{out} = 10 mW.

c. P_{in} = 5 W, P_{out} = 5 μW.

d. P_{in} = 10 W, P_{out} = 100 mW.

26–42 In Prob. 26–27, you calculated the output voltage for the *RC* filter in Fig. 26–34 at several different frequencies. For each frequency listed in Prob. 26–27, determine the dB attenuation offered by the filter.

26–43 In Prob. 26–28, you calculated the output voltage for the *RC* filter in Fig. 26–35 at several different frequencies. For each frequency listed in Prob. 26–28, determine the dB attenuation offered by the filter.

26–44 What is the rolloff rate of an *RC* low-pass filter for signal frequencies well beyond the cutoff frequency? Do the values calculated for the dB attenuation in Prob. 26–42 verify this rolloff rate?

SECTION 26–11 RESONANT FILTERS

26–45 What determines the width of the band of frequencies that are allowed to pass through a resonant band-pass filter?

26–46 Identify the following configurations as either band-pass or band-stop filters:

a. series *LC* circuit in series with R_L.

b. parallel *LC* circuit in series with R_L.

c. parallel *LC* circuit in parallel with R_L.

d. series *LC* circuit in parallel with R_L.

SECTION 26–12 INTERFERENCE FILTERS

26–47 To prevent the radiation of harmonic frequencies from a radio transmitter, a filter is placed between the transmitter and antenna. The filter must pass all frequencies below 30 MHz and severely attenuate all frequencies above 30 MHz. What type of filter must be used?

26–48 What type of filter should be used to eliminate an interfering signal having a very narrow range of frequencies?

Critical Thinking

26–49 In Fig. 26–40 calculate (a) the cutoff frequency f_c; (b) the output voltage at the cutoff frequency f_c; (c) the output voltage at 50 kHz.

Figure 26–40 Circuit for Critical Thinking Prob. 26–49.

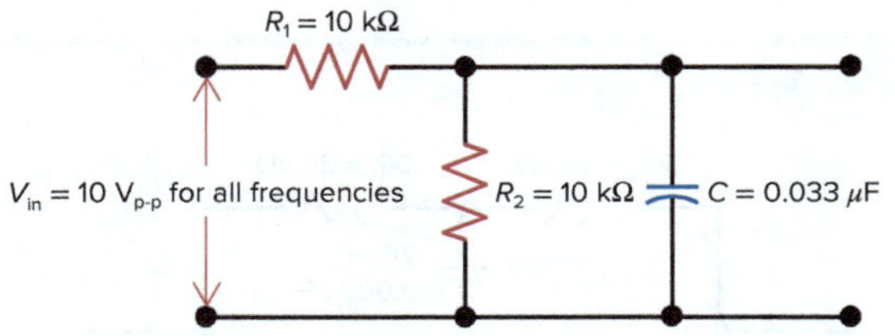

26–50 In Fig. 26–41, calculate the values of *L* and *C* required to provide an f_r of 1 MHz and a bandwidth Δf of 40 kHz.

Figure 26–41 Circuit for Critical Thinking Prob. 26–50.

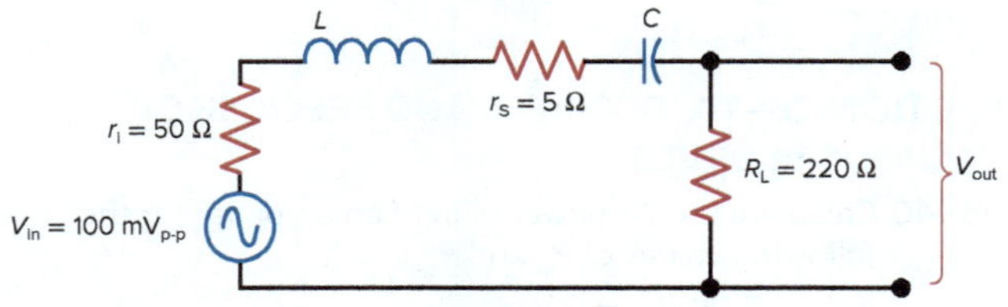

26–51 In Fig. 26–42 calculate the values of *L* and *C* required to provide an f_r of 1 MHz and a bandwidth Δf of 20 kHz.

Figure 26–42 Circuit for Critical Thinking Prob. 26–51.

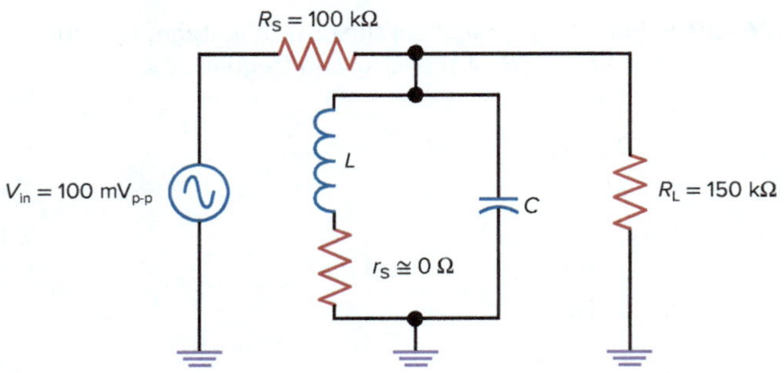

Answers to Self-Reviews

26–1 **a.** 500 kHz
b. 60 Hz

26–2 **a.** 6 V
b. 10 and 2 V
c. 8 V
d. 4 and 2.8 V

26–3 **a.** high-pass
b. 0 V

26–4 **a.** 0 V
b. 5 μF

26–5 **a.** RF
b. 5 μF

26–6 **a.** high-pass
b. low-pass

26–7 **a.** *e* and *f*
b. low-pass

26–8 **a.** *d*
b. high-pass

26–9 **a.** false
b. true
c. true
d. true

26–10 **a.** true
a. true
b. false
c. true

26–11 **a.** true
b. true
c. true

26–12 **a.** true
b. true

Laboratory Application Assignment

In this lab application assignment, you will examine an *RC* coupling circuit and an *RC* low-pass filter. In the *RC* coupling circuit, you will see how the series capacitor blocks the DC component of the input voltage but passes the AC component. In the *RC* low-pass filter, you will see how the low frequencies are passed from input to output with little or no attenuation but the higher frequencies are severely attenuated or blocked.

Equipment: Obtain the following items from your instructor.

- Function generator
- Oscilloscope
- DMM
- 0.1-μF and 0.22-μF capacitors
- 2.2-kΩ and 10-kΩ carbon-film resistors

RC Coupling Circuit

Examine the *RC* coupling circuit in Fig. 26–43*a*. Notice the input voltage is a pulsating DC voltage whose value remains entirely positive. The input waveform (across terminals 1 and 2) is shown in Fig. 26–43*b*. The output from the *RC* coupling circuit is taken across terminals 3 and 4, which is across the resistor, *R*.

Figure 26–43

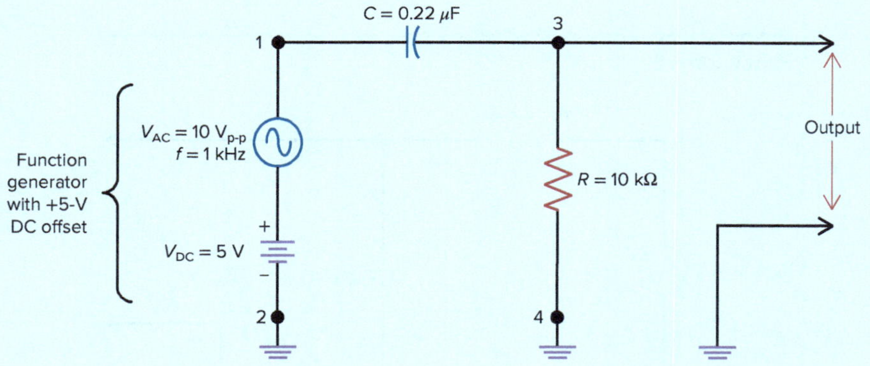

(*a*)

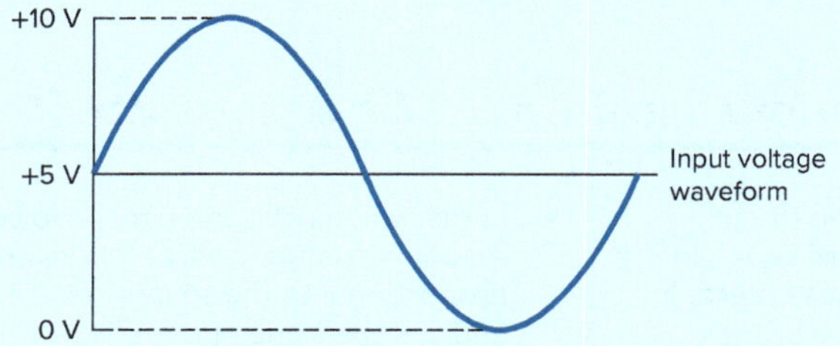

(*b*)

What value of DC voltage would you expect to measure across input terminals 1 and 2? $V_{in(DC)}$ = __________
What value of rms voltage would you expect to measure across input terminals 1 and 2? $V_{in(rms)}$ = __________
How much DC voltage would you expect to measure across the capacitor, *C*? $V_{C(DC)}$ = __________
How much DC voltage would you expect to measure across the resistor, *R*? $V_{R(DC)}$ = __________

Construct the circuit in Fig. 26–43*a*. With a DMM connected to the output of the function generator, adjust the DC offset control to obtain a DC value of $+5V_{DC}$. Also, while viewing the oscilloscope, adjust the amplitude and frequency controls of the function generator to obtain an output voltage of 10 $V_{p\text{-}p}$ with a frequency of 1 kHz. Have your instructor check your settings. Next, measure and record the following values in Fig. 26–43*a*:
$V_{in(DC)}$ = __________, $V_{in(rms)}$ = __________,
$V_{C(DC)}$ = __________, $V_{R(DC)}$ = __________
How do these values compare to those predicted?
__________ __________

In Fig. 26–43*a*, calculate X_C and Z_T at 1 kHz. X_C = __________,
Z_T = __________
Next, calculate and record the following rms values:
I = __________, V_C = __________, V_R = __________
Using your DMM, measure and record the following rms values:
V_C = __________, V_R = __________
How do your calculated and measured rms values compare?
____

Measure and record the peak-to-peak output voltage across *R* using the oscilloscope. V_{out} = __________ p-p. How does this value compare to the peak-to-peak value of input voltage? _____
____

RC Low-Pass Filter

Examine the *RC* low-pass filter in Fig. 26–44. Calculate and record the cutoff frequency, f_C. f_C = __________

Construct the *RC* low-pass filter in Fig. 26–44. The input voltage should be set to exactly 10 $V_{p\text{-}p}$ with no DC offset.

Measure and record the output voltage from the *RC* low-pass filter for each of the frequencies listed below. (Use the oscilloscope to measure the output voltage.) Next, calculate the decibel attenuation offered by the filter at each frequency.

f = 100 Hz $V_{out(p\text{-}p)}$ = __________ N_{dB} = __________
f = 250 Hz $V_{out(p\text{-}p)}$ = __________ N_{dB} = __________
f = 500 Hz $V_{out(p\text{-}p)}$ = __________ N_{dB} = __________
f_C (Calculated) $V_{out(p\text{-}p)}$ = __________ N_{dB} = __________
f = 10 kHz $V_{out(p\text{-}p)}$ = __________ N_{dB} = __________
f = 20 kHz $V_{out(p\text{-}p)}$ = __________ N_{dB} = __________
f = 100 kHz $V_{out(p\text{-}p)}$ = __________ N_{dB} = __________

Do the measured values of output voltage confirm that the circuit is a low-pass filter? __________

Rate of Rolloff

Based on your measured values in Fig. 26–44, what is the rate of rolloff when *f* is increased by one octave from 10 kHz to 20 kHz?
__________.

Based on your measured values in Fig. 26–44, what is the rate of rolloff when *f* is raised by one decade from 10 kHz to 100 kHz?
__________.

Figure 26–44

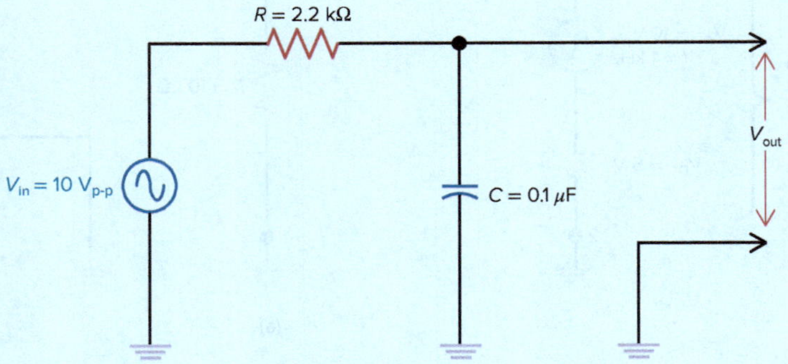

Cumulative Review Summary (Chapters 25–26)

- Resonance occurs when the reactances X_L and X_C are equal. In series, the net reactance is zero. In parallel, the net reactive branch current is zero. The specific frequency that makes $X_L = X_C$ is the resonant frequency $f_r = 1/(2\pi\sqrt{LC})$.
- Larger values of *L* and *C* mean lower resonant frequencies, as f_r is inversely proportional to the square root of *L* and *C*. If the value of *L* or *C* is quadrupled, for instance, f_r will decrease by one-half.
- For a series resonant *LC* circuit, the current is maximum. The voltage drop across the reactances is equal and opposite; the phase angle is zero. The reactive voltage at resonance is *Q* times greater than the applied voltage.

- For a parallel resonant *LC* circuit, the impedance is maximum with minimum line current, since the reactive branch currents cancel. The impedance at resonance is *Q* times the X_L value, but it is resistive with a phase angle of zero.
- The *Q* of the resonant circuit equals X_L/r_S for resistance in series with X_L or R_P/X_L for resistance in parallel with X_L.
- The bandwidth between half-power points is f_r/Q.
- A filter uses inductance and capacitance to separate high or low frequencies. A low-pass filter allows low frequencies to develop output voltage across the load; a high-pass filter does the same for high frequencies. Series inductance or shunt capacitance provides low-pass filtering; series capacitance or shunt inductance provides high-pass filtering.
- A fluctuating or pulsating DC is equivalent to an AC component varying in opposite directions around the average-value axis.
- An *RC* coupling circuit is effectively a high-pass filter for pulsating DC voltage, passing the AC component but blocking the DC component.
- A transformer with an isolated secondary is a high-pass filter for pulsating direct current, allowing alternating current in the secondary but no DC output level.
- A bypass capacitor in parallel with *R* is a low-pass filter, since its low reactance reduces the voltage across *R* at high frequencies.
- The main types of filter circuits are π, L, and T types. These can be high-pass or low-pass, depending on how *L* and *C* are connected.
- Resonant circuits can be used as band-pass or band-stop filters. For band-pass filtering, series resonant circuits are in series with the load or parallel resonant circuits are across the load. For band-stop filtering, parallel resonant circuits are in series with the load or series resonant circuits are across the load.
- A wavetrap is an application of a resonant band-stop filter.
- The cutoff frequency of a filter is the frequency at which the output voltage is reduced to 70.7% of its maximum value.
- The cutoff frequency of an *RC* low-pass or high-pass filter can be calculated from $f_c = 1/2\pi RC$. Similarly, the cutoff frequency of an *RL* low-pass or high-pass filter can be calculated from $f_c = R/2\pi L$.
- The decibel (dB) is a logarithmic expression that compares two power levels. In the passband, a passive filter provides 0 dB of attenuation. At the cutoff frequency, a passive filter provides attenuation of −3 dB.
- Semilog and log-log graph paper are typically used to show the frequency response of a filter. The advantage of using logarithmic graph paper is that a wide range of frequencies can be shown in one plot without losing resolution in the smaller values.

Cumulative Self-Test

Answers at the back of the book.

Fill in the numerical answer.

1. An *L* of 10 H and *C* of 40 μF has f_r of ________ Hz.

2. An *L* of 100 μH and *C* of 400 pF has f_r of ________ MHz.

3. In Question 2, if $C = 400$ pF and *L* is increased to 400 μH, the f_r decreases to ________ MHz.

4. In a series resonant circuit with 10 mV applied across a 1-Ω *R*, a 1000-Ω X_L, and a 1000-Ω X_C, at resonance, the current is ________ mA.

5. Imagine a parallel resonant circuit. It has a 1-Ω r_S in series with a 1000-Ω X_L in one branch and a 1000-Ω X_C in the other branch. With 10 mV applied, the voltage across X_C equals ________ mV.

6. In Question 5, the *Z* of the parallel resonant circuit equals ________ MΩ.

7. An *LC* circuit resonant at 500 kHz has a *Q* of 100. Its total bandwidth between half-power points equals ________ kHz.

8. A coupling capacitor for 40 to 15,000 Hz in series with a 0.5-MΩ resistor has a capacitance of ________ μF.

9. A bypass capacitor for 40 to 15,000 Hz in shunt with a 1000-Ω *R* has a capacitance of ________ μF.

10. A pulsating DC voltage varying in a symmetrical sine wave between 100 and 200 V has an average value of ________ V.

11. An *RC* low-pass filter has the following values: $R = 1$ kΩ, $C = 0.005$ μF. The cutoff frequency f_c is ________.

12. The input voltage to a filter is 10 $V_{p\text{-}p}$ and the output voltage is 100 $\mu V_{p\text{-}p}$. The amount of attenuation is ________ dB.

13. On logarithmic graph paper, a 2-to-1 range of values is called a(n) ________, and a 10-to-1 range of values is called a(n) ________.

14. At the cutoff frequency, the output voltage is reduced to ________ % of its maximum.

Answer True or False.

15. A series resonant circuit has low *I* and high *Z*.

16. A steady direct current in the primary of a transformer cannot produce any AC output voltage in the secondary.

17. A π-type filter with shunt capacitances is a low-pass filter.

18. An L-type filter with a parallel resonant *LC* circuit in series with the load is a band-stop filter.

19. A resonant circuit can be used as a band-stop filter.

20. In the passband, an *RC* low-pass filter provides approximately 0 dB of attenuation.

21. The frequency response of a filter is never shown on logarithmic graph paper.

Design credits: Chapter Opener Background Image, and Multisim: Stockbyte/Getty Images

chapter

27

Three-Phase AC Power Systems

A single-phase AC circuit can be defined as any circuit that uses only one AC voltage source. In a single-phase AC circuit, electrical power is transferred from the source to the load using only two conductors or wires. By contrast, a ***polyphase*** AC circuit is any circuit that uses two or more AC voltage sources, each having a different phase angle. (Poly means more than one.) The most common polyphase AC circuits are those with three AC voltage sources and either three or four conductors or wires. A polyphase AC circuit with three AC voltage sources is generally referred to as a ***three-phase*** AC circuit.

Although three-phase AC power was briefly introduced back in Chapter 15, "Alternating Voltage and Current," this chapter takes a much more in-depth approach in analyzing three-phase AC circuits. For example, in this chapter, you will learn how to calculate the voltage and current values in a three-phase AC circuit containing both wye (Y) and delta (Δ) circuit connections. You will also learn how to calculate the total power in a three-phase AC circuit with a balanced load. (A balanced load is one with three identical load impedances.) And finally, you will be able to list the advantages of three-phase AC power systems versus single-phase AC power systems. ■

Chapter Outline

Chapter Objectives

After studying this chapter, you should be able to

- *Describe* the difference between a single-phase and polyphase AC circuit.
- *Explain* the difference between wye (*Y*)- and delta (Δ)-connected three-phase AC generators.
- *Explain* the difference between the phase voltage, V_θ, and line voltage, V_L, in a wye (Y)-connected generator.
- *Explain* the mathematical relationship between V_L and V_θ in a wye (Y)-connected generator.
- *Explain* the difference between the phase current, I_θ, and line current, I_L, in a delta (Δ)-connected generator.
- *Explain* the mathematical relationship between I_L and I_θ in a delta (Δ)-connected generator.
- *Calculate* the voltage and current values in a three-phase AC circuit containing wye (Y) and delta (Δ) connections.
- *Calculate* power in a three-phase AC circuit.
- *List* the advantages of using three-phase AC power versus single-phase AC power.

Important Terms

balanced load
line current, I_L
line voltage, V_L
phase current, I_θ
phase voltage, V_θ
polyphase AC circuit
single-phase AC circuit
three-phase AC circuit

27–1 Three-Phase AC Generators

A generator with three separate windings, equally spaced around the circle, will produce three different output voltages 120° out of phase with each other. Such a generator is called a three-phase AC generator. The coil windings in a three-phase AC generator are shown in Fig. 27–1. This connection of the generator windings is called the wye (Y) or star connection. As shown in Fig. 27–1, all three generator windings are joined at one end, which is called the neutral (N) point. In this case, each coil winding has an induced voltage of 120 V with respect to the neutral point. The three-phase output at terminals **A**, **B**, and **C**, with respect to the neutral point, **N**, is represented by the waveforms in Fig. 27–2*a* and the phasors in Fig. 27–2*b*. The voltages V_{AN}, V_{BN}, and V_{CN} are commonly referred to as ***phase voltages*** and are generally designated as V_θ. When the output is taken across any two terminals (lines), such as **A** and **B**, **B** and **C**, or **C** and **A**, the output voltage is 208 V because of the 120° phase difference between the phase voltages. The output voltage across any two terminals or lines is called the *line voltage* and is designated V_L. As will be proved in Section 27–2, the **line voltage, V_L**, in a Y-connected three-phase AC generator is greater than the **phase voltage, V_θ**, by a factor of $\sqrt{3}$, which equals a factor of 1.732. In other words, in a Y-connected three-phase AC generator, $V_L = \sqrt{3}V_\theta$.

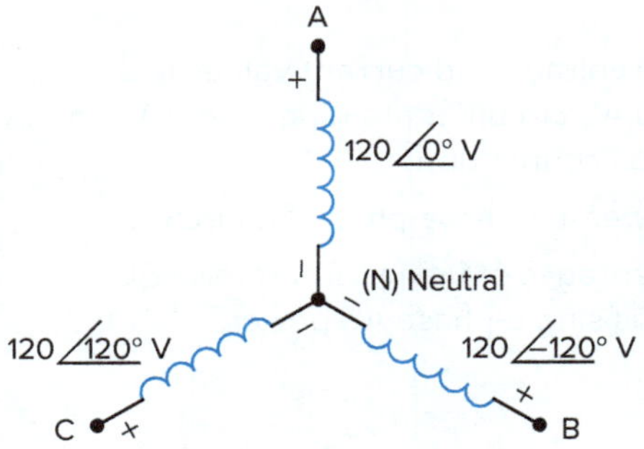

Figure 27–1 Coil windings in a wye (Y)-connected three-phase AC generator.

Figure 27–3 shows the coil windings in a delta (Δ)-connected three-phase AC generator. In the delta-connected generator, the phase voltage, V_θ, and the line voltage, V_L, are the same and equal to the voltage induced across each coil winding of the AC generator. In other words, $V_L = V_\theta$ in a delta-connected three-phase AC generator. It is important to note, however, that each available voltage, V_{AB}, V_{BC}, and V_{CA}, are still 120° out of phase with each other. Because of the 120° phase difference between each of the phase voltages, the currents in each of the coil windings are also 120° out of phase with each other. The currents in the individual coil windings are commonly called phase currents and are generally designated I_θ. The currents flowing out from terminals A, B, and C are referred to as line currents, designated I_L. Because each line current is a combination of two phase currents, the **line current**,

Figure 27–2 Waveforms and phasors representing the outputs at terminals A, B, and C with respect to N in Fig. 27–1. (*a*) Waveforms at terminals A, B, and C with respect to N. (*b*) Phasors representing the waveforms at terminals A, B, and C with respect to N.

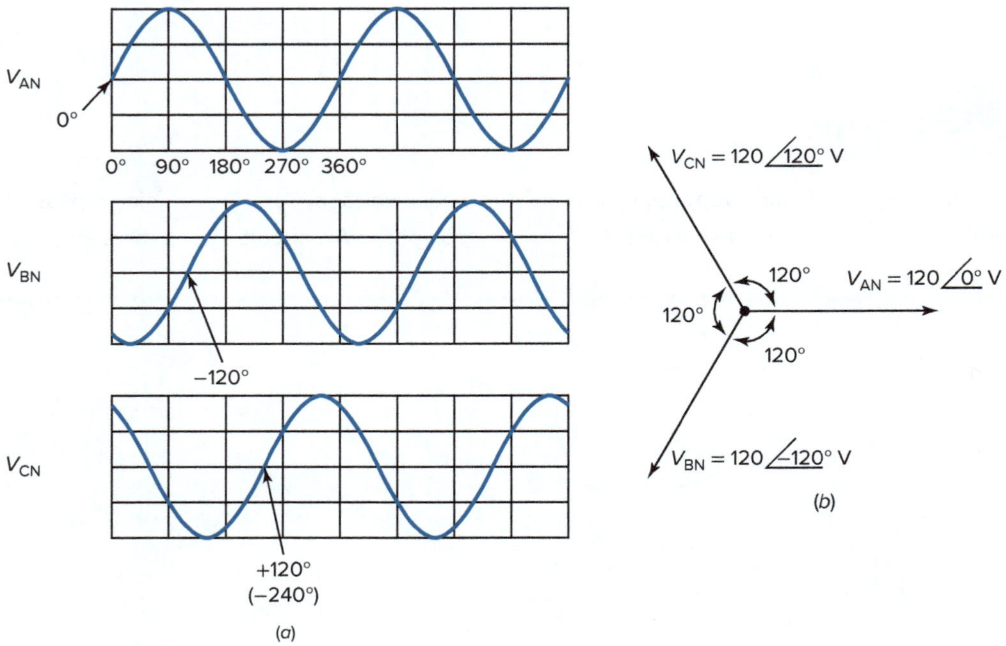

Figure 27–3 Coil windings in a delta-connected three-phase AC generator.

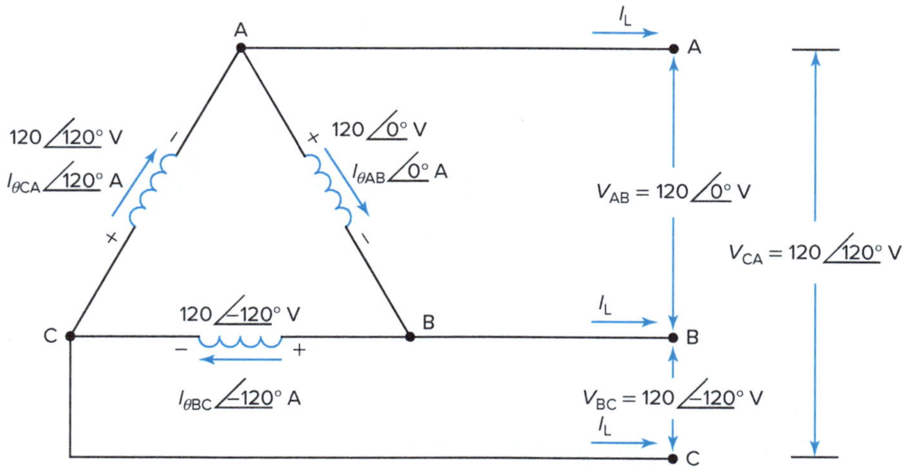

Figure 27–4 Phasors representing the phase relationship between each phase current in Fig. 27–3.

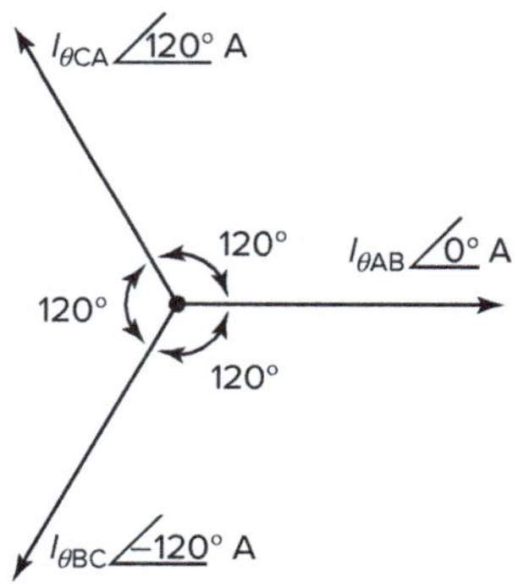

I_L, is greater than any individual phase current by a factor of $\sqrt{3}$, which equals a factor of 1.732. In other words, in a delta-connected three-phase AC generator, $I_L = \sqrt{3}I_\theta$. This equation will be proved in Section 27–3. Figure 27–4 shows the phase relationship between each phase current in a delta-connected three-phase AC generator.

It is important to note that in the electric utility industry, the term "phase" always refers to the wye (Y) configuration rather than the delta (Δ) configuration. However, in this chapter, we will use the term "phase voltage" and/or "phase current" when discussing the delta (Δ) configuration in order to assist in the explanation of the concepts.

27–1 Self-Review

Answers at the end of the chapter.

a. In a wye (Y)-connected three-phase AC generator, what is the phase relationship between each phase voltage?

b. In a delta (Δ)-connected three-phase AC generator, what is the phase relationship between each phase voltage?

27–2 The Wye (Y)-Connected Three-Phase Generator

In this section, we will examine the mathematical relationship between the phase voltage, V_θ, and the line voltage, V_L, in a wye (Y)-connected three-phase AC generator. Before we begin our analysis, however, let's revisit double subscript notation as it relates to AC voltages and currents. In Fig. 27–5, the phasor $V_{BN} = 120\angle -120°$ V as shown. In this case, V_{BN} represents the voltage measured at point B with respect to N in a wye (Y)-connected three-phase AC generator. (See Fig. 27–1.) But what if we reverse the reference point, and the point of measurement is V_{NB}? The answer is that V_{NB} equals the negative of the voltage V_{BN}. In the case of an AC voltage, this means that V_{BN} and V_{NB} are equal in magnitude but 180° out of phase with each other. Expressed mathematically, $V_{NB}\angle\theta = -V_{BN}\angle\theta$. In order to obtain the negative of any phasor, the following rules apply:

1. In polar form, add 180° to the phase angle.
2. In rectangular form, change the sign of both terms.

MultiSim **Figure 27–5** Phasors equal in magnitude but 180 degrees out of phase.

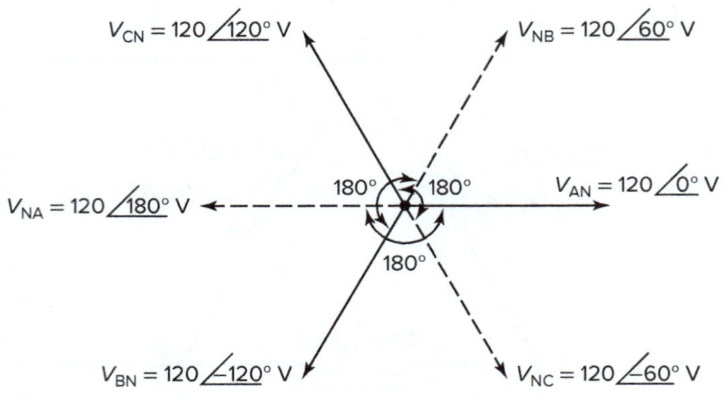

In Fig. 27–5, $V_{NB}\angle\theta = -V_{BN}\angle\theta$

$$= -120\angle -120°\text{ V}$$
$$= 120\angle -120° + 180°\text{ V}$$
$$V_{NB}\angle\theta = 120\angle 60°\text{ V}$$

Similarly, $V_{BN}\angle\theta = -V_{NB}\angle\theta$

$$= -120\angle 60°\text{ V}$$
$$= 120\angle 60° + 180°\text{ V}$$
$$V_{BN}\angle\theta = 120\angle 240°\text{ V or } 120\angle -120°\text{ V}$$

The voltage, V_{BN} of $120\angle -120°$ V in rectangular form equals

$$120\angle -120°\text{ V} = -60\text{ V} - j103.923\text{ V}$$

Applying rule number 2 to solve for $V_{NB}\angle\theta$ gives us

$$V_{NB}\angle\theta = -V_{BN}\angle\theta$$
$$= -(-60\text{ V} - j103.923\text{ V})$$
$$V_{NB}\angle\theta = 60\text{ V} + j103.923\text{ V} = 120\angle 60°\text{ V}$$

In Fig. 27–5, notice the phasors $V_{BN}\angle\theta$ and $V_{NB}\angle\theta$ are equal in magnitude but 180° out of phase with each other. Furthermore, the phasors $V_{NA}\angle\theta$ and $V_{NC}\angle\theta$ are also 180° out of phase with the phasors $V_{AN}\angle\theta$ and $V_{CN}\angle\theta$, respectively.

Calculating the Line Voltage, V_L

In a wye (Y)-connected generator, the voltage across any combination of output terminals such as **A** and **B**, **B** and **C**, or **C** and **A** is called the line voltage, V_L. To calculate the line voltage, V_L, across terminals A and B, refer to Fig. 27–6. As shown, the voltage, $V_{AN} = 120\angle 0°$ V and $V_{BN} = 120\angle -120°$ V. The line voltage can be determined by applying KVL.

$$V_{AB}\angle\theta = V_{AN}\angle\theta - V_{BN}\angle\theta$$

To solve for $V_{AB}\angle\theta$, state both $V_{AN}\angle\theta$ and $V_{BN}\angle\theta$ in rectangular form so their values can be combined.

$$V_{AN}\angle\theta = 120\angle 0°\text{ V} = 120\text{ V} + j0\text{ V}$$
$$V_{BN}\angle\theta = 120\angle -120°\text{ V} = -60\text{ V} - j103.923\text{ V}$$

Combining terms gives us the line voltage, V_L, across terminals A and B.

$$V_{AB}\angle\theta = (120\text{ V} + j0\text{ V}) - (-60\text{ V} - j103.923\text{ V})$$

Removing parenthesis, we have

$$V_{AB}\angle\theta = 120\text{ V} + j0\text{ V} + 60\text{ V} + j103.923\text{ V}$$
$$V_{AB}\angle\theta = 180\text{ V} + j103.923\text{ V}$$
$$V_{AB}\angle\theta = 207.85\angle 30^\circ\text{ V}$$

In a similar manner, the line voltages across terminals B and C (V_{BC}) and terminals C and A (V_{CA}) are calculated as follows:

$$V_{BC}\angle\theta = V_{BN}\angle\theta - V_{CN}\angle\theta$$

Stating $V_{BN}\angle\theta$ and $V_{CN}\angle\theta$ in rectangular form gives us

$$V_{BN}\angle\theta = 120\angle -120^\circ\text{ V} = -60\text{ V} - j103.923\text{ V}$$
$$V_{CN}\angle\theta = 120\angle 120^\circ\text{ V} = -60\text{ V} + j103.923\text{ V}$$

Inserting values for $V_{BC}\angle\theta$ gives us

$$V_{BC}\angle\theta = (-60\text{ V} - j103.923\text{ V}) - (-60\text{ V} + j103.923\text{ V})$$

Removing parenthesis, we have

$$V_{BC}\angle\theta = -60\text{ V} - j103.923\text{ V} + 60\text{ V} - j103.923\text{ V}$$
$$V_{BC}\angle\theta = 0\text{ V} - j207.85\text{ V}$$
$$V_{BC}\angle\theta = 207.85\angle -90^\circ\text{ V}$$

And finally, $V_{CA}\angle\theta = V_{CN}\angle\theta - V_{AN}\angle\theta$

Inserting values for V_{CA} gives us

$$V_{CA}\angle\theta = (-60\text{ V} + j103.923\text{ V}) - (120\text{ V} + j0\text{ V})$$

Removing parenthesis gives us

$$V_{CA}\angle\theta = -60\text{ V} + j103.923\text{ V} - 120\text{ V} - j0\text{ V}$$

This simplifies to

$$V_{CA}\angle\theta = -180\text{ V} + j103.923\text{ V}$$
$$V_{CA}\angle\theta = 207.85\angle 150^\circ\text{V}$$

In summary, the line voltages available in a wye (Y)-connected generator are

$$V_{AB}\angle\theta = 207.85\angle 30^\circ\text{ V}$$
$$V_{BC}\angle\theta = 207.85\angle -90^\circ\text{ V}$$
$$V_{CA}\angle\theta = 207.85\angle 150^\circ\text{ V}$$

Figure 27–7 shows the phase relationship between the phase voltages V_{AN}, V_{BN}, V_{CN} and the line voltages V_{AB}, V_{BC}, and V_{CA}. Note that each line voltage is 30° out of phase with the nearest phase voltage. Also, note that the line voltages are 120° out

Figure 27–6 Calculating the line voltage across terminals A and B.

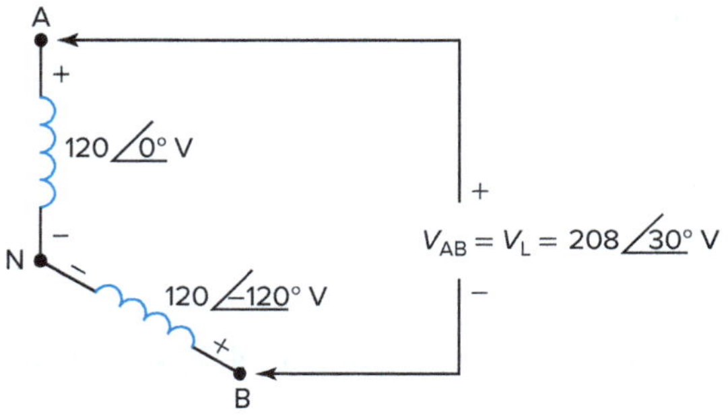

Figure 27–7 Phasors for phase and line voltages in a wye (Y)-connected three-phase AC generator.

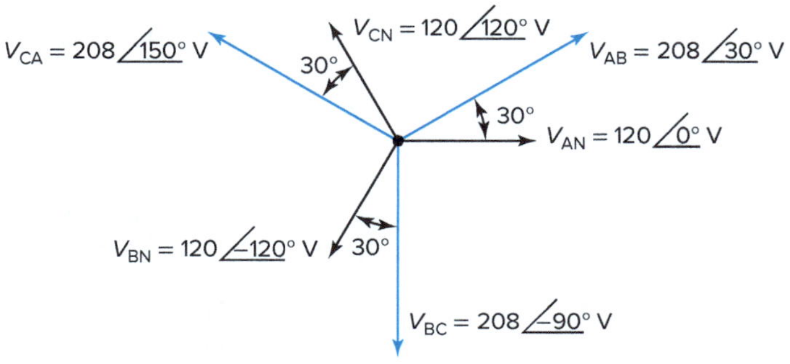

Figure 27–8 Combining phasors to prove Formula (27–1).

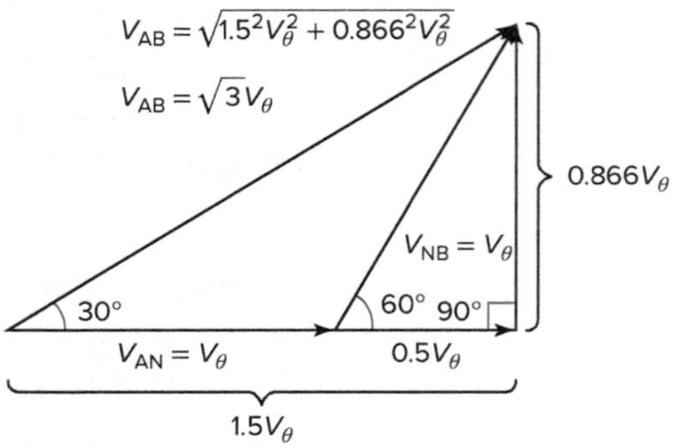

of phase with each other. In general, the line voltage, V_L, is related to the phase voltage, V_θ, as follows:

$$V_L = \sqrt{3}V_\theta \tag{27–1}$$

To prove Formula (27–1), refer to Fig. 27–8, which shows how the phasors V_{AN} and V_{NB} are combined for the resultant phasor, V_{AB}. Recall earlier from Fig. 27–6 that the line voltage V_{AB} was calculated as

$$V_{AB}\angle\theta = V_{AN}\angle\theta - V_{BN}\angle\theta$$

Since $-V_{BN}\angle\theta = V_{NB}\angle\theta$, then it is true that

$$V_{AB}\angle\theta = V_{AN}\angle\theta + V_{NB}\angle\theta$$

where $V_{AN}\angle\theta = 120\angle 0°$ V and $V_{NB}\angle\theta = 120\angle 60°$ V

(For mathematical purposes, both of these phase voltages are represented as V_θ in Fig. 27–8.)

In combining the phasors V_{AN} and V_{NB}, notice that the tail of the V_{NB} phasor is connected to the tip of the V_{AN} phasor. The resultant phasor, V_{AB}, then, is the phasor that connects the tail of the V_{AN} phasor to the tip of the V_{NB} phasor. If the V_{NB} (V_θ) phasor is converted into rectangular form for the rectangular coordinates of the inner right triangle, then

$$V_{NB}\angle\theta = V_\theta\angle 60° = 0.5V_\theta + j0.866V_\theta$$

Similarly, $V_{AN}\angle\theta = V_\theta\angle 0° = V_\theta + j0$ V

Now, if the rectangular coordinates of $V_{AN}\angle\theta$ and $V_{NB}\angle\theta$ are combined, we have $V_\theta + 0.5V_\theta = 1.5V_\theta$ for one side of the larger right triangle. The other side of the larger right triangle is $0.866V_\theta$ as shown. To solve for the hypotenuse, which is the phasor, V_{AB}, we have

$$V_{AB} = \sqrt{1.5^2V_\theta^2 + 0.866^2V_\theta^2}$$

Which simplifies to

$$V_{AB} = \sqrt{2.25V_\theta^2 + 0.75V_\theta^2}$$
$$V_{AB} = \sqrt{3V_\theta^2}$$
$$V_{AB} = \sqrt{3}V_\theta$$

The other line voltages are calculated the same way. In general, $V_L = \sqrt{3}V_\theta$ in a Y-connected generator.

Example 27-1

Figure 27–9 shows a Y-connected three-phase AC generator with a phase voltage, V_θ, of 277 V. How much is the line voltage, V_L?

Figure 27–9 Wye (Y)-connected three-phase AC generator.

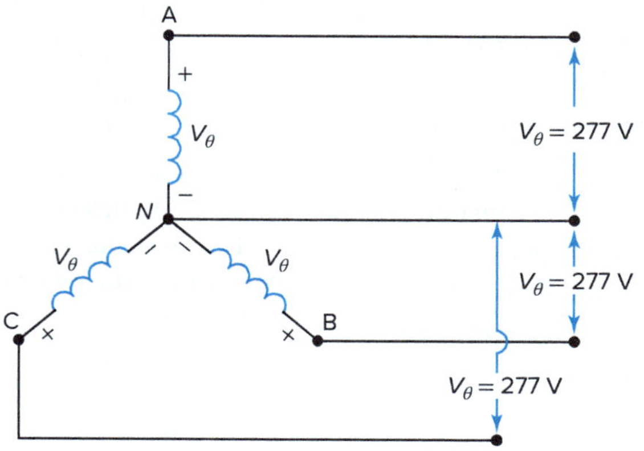

ANSWER The line voltage, V_L, can be determined by applying Formula (27–1).

$$\begin{aligned} V_L &= \sqrt{3}V_\theta \\ &= 1.732 \times 277 \text{ V} \\ &= 480 \text{ V} \end{aligned}$$

Example 27-2

What is the phase voltage, V_θ, in a Y-connected three-phase AC generator whose line voltage, V_L, equals 416 V.

ANSWER Rearrange Formula (27–1) to solve for the phase voltage, V_θ.

$$V_L = \sqrt{3}V_\theta \quad \text{so}$$

$$V_\theta = \frac{V_L}{\sqrt{3}}$$

$$\text{Inserting values: } V_\theta = \frac{416 \text{ V}}{1.732}$$

$$V_\theta = 240 \text{ V}$$

27–2 Self-Review

Answers at the end of the chapter.

a. If the phasor $V_{XY} = 240\angle 30°$ V, what is the value of the phasor V_{YX}?

b. What is the formula for calculating the line voltage, V_L, in a Y-connected three-phase AC generator?

GOOD TO KNOW

An advantage of the delta-connected generator is that of higher reliability. If one of the three generator windings should open, (which is called an open delta) the generator still produces the full voltage on all three phases. Also, the line currents remain the same even though one winding is open. The disadvantage of an open delta, however, is that the power capacity is reduced to 57.7% of its normal capacity with all three windings present.

27–3 The Delta (Δ)-Connected Three-Phase Generator

In this section, we will examine the mathematical relationship between the **phase current, I_θ**, and the line current, I_L, in a delta-connected three-phase AC generator. Recall that in a delta (Δ)-connected generator, the phase voltage, V_θ, and line voltage, V_L, are equivalent and equal to the voltage induced across each coil winding in the generator. Recall also that the phase voltages are 120° out of phase with each other. As shown in Fig. 27–10:

$$V_{AB}\angle\theta = V_\theta\angle 0^\circ \text{ V}$$
$$V_{BC}\angle\theta = V_\theta\angle -120^\circ \text{ V}$$
$$V_{CA}\angle\theta = V_\theta\angle 120^\circ \text{ V}$$

Because the voltage across any two lines is the same as an individual phase voltage, $V_\theta = V_L$ in a delta-connected three-phase AC generator.

Because of the 120° phase difference between each phase voltage, the phase currents are also 120° out of phase with each other. The phase angles of the phase currents in Fig. 27–10 are

$$I_{\theta AB}\angle 0^\circ \text{ A}$$
$$I_{\theta BC}\angle -120^\circ \text{ A}$$
$$I_{\theta CA}\angle 120^\circ \text{ A}$$

The letters used for the subscripts of each phase current indicate the assumed direction of current. The first letter indicates the point or terminal where current enters the phase winding, and the second letter indicates the point or terminal where current leaves the phase winding.

By examining Fig. 27–10, it is clear that the line current, I_L, flowing out from any terminal (I_{LA}, I_{LB}, and I_{LC}) is a combination of two phase currents. For our analysis here, let's assume that the magnitude of each phase current is $1\angle\theta$A. Applying KCL for the currents directed into and away from point A gives us the following equation:

$$I_{\theta CA} - I_{\theta AB} - I_{LA} = 0$$

or $$I_{LA} = I_{\theta CA} - I_{\theta AB}$$

Where $I_{\theta CA}$ and $I_{\theta AB}$ are stated as

$$I_{\theta CA} = 1\angle 120^\circ\text{A} = -0.5 \text{ A} + j0.866 \text{ A}$$
$$I_{\theta AB} = 1\angle 0^\circ\text{A} = 1 \text{ A} + j0 \text{ A}$$

Figure 27–10 Delta-connected three-phase AC generator showing the phase voltages, phase currents, and line currents.

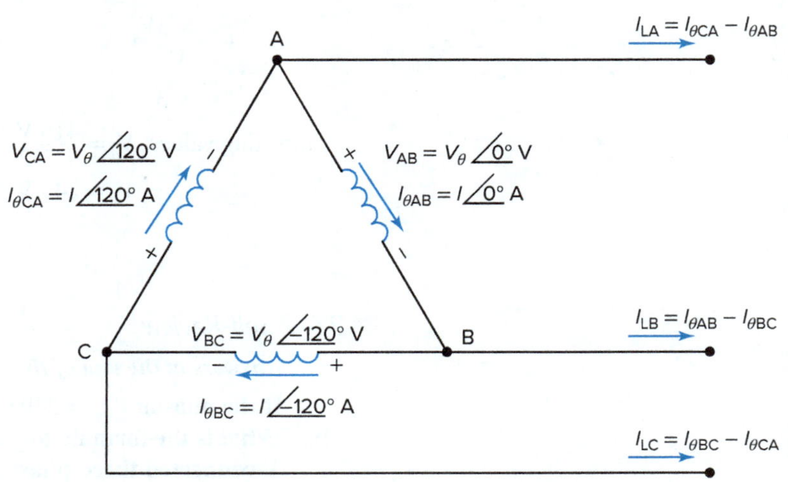

Inserting values into the equation for I_{LA} gives us

$$I_{LA} = (-0.5\text{ A} + j0.866\text{ A}) - (1\text{ A} + j0\text{ A})$$

Removing parenthesis, we have

$$I_{LA} = -0.5\text{ A} + j0.866\text{ A} - 1\text{ A} - j0\text{ A}$$

Which simplifies to

$$I_{LA} = -1.5\text{ A} + j0.866\text{ A} = 1.732\angle 150^\circ\text{ A}$$

Applying KCL for the currents directed into and away from point B gives us

$$I_{\theta AB} - I_{\theta BC} - I_{LB} = 0$$

or

$$I_{LB} = I_{\theta AB} - I_{\theta BC}$$

Where $I_{\theta AB}$ and $I_{\theta BC}$ are stated as

$$I_{\theta AB} = 1\angle 0^\circ\text{A} = 1\text{ A} + j0\text{ A}$$
$$I_{\theta BC} = 1\angle -120^\circ\text{A} = -0.5\text{ A} - j0.866\text{ A}$$

Inserting values into the equation for I_{LB} gives us

$$I_{LB} = (1\text{ A} + j0\text{ A}) - (-0.5\text{ A} - j0.866\text{ A})$$

Removing parenthesis, we have

$$I_{LB} = 1\text{ A} + j0\text{ A} + 0.5\text{ A} + j0.866\text{ A}$$

Which simplifies to

$$I_{LB} = 1.5\text{ A} + j0.866\text{ A} = 1.732\angle 30^\circ\text{ A}$$

And finally, applying KCL for the currents directed into and out of point C yields the following equation:

$$I_{\theta BC} - I_{\theta CA} - I_{LC} = 0$$

or

$$I_{LC} = I_{\theta BC} - I_{\theta CA}$$

where $I_{\theta BC}$ and $I_{\theta CA}$ are stated as

$$I_{\theta BC} = 1\angle -120^\circ\text{ A} = -0.5\text{ A} - j0.866\text{ A}$$
$$I_{\theta CA} = 1\angle 120^\circ\text{ A} = -0.5\text{ A} + j0.866\text{ A}$$

Inserting values into the equation for I_{LC} gives us

$$I_{LC} = (-0.5\text{ A} - j0.866\text{ A}) - (-0.5\text{ A} + j0.866\text{ A})$$

Removing parenthesis, we have

$$I_{LC} = -0.5\text{ A} - j0.866\text{ A} + 0.5\text{ A} - j0.866\text{ A}$$

Which simplifies to

$$I_{LC} = 0\text{ A} - j1.732\text{ A} = 1.732\angle -90^\circ\text{ A}$$

The phase relationship between the line currents and each phase current is shown in Fig. 27–11. The line currents are identified as I_{LA}, I_{LB}, and I_{LC}, where the second letter in the subscript denotes from which terminal (**A**, **B**, or **C**) the line current flows. Notice that each line current is 30° out of phase with the nearest phase current. In general, the line current, I_L, is related to the phase current, I_θ as

$$I_L = \sqrt{3}I_\theta \tag{27–2}$$

Figure 27–11 Phasors representing the phase currents and line currents in Fig. 27–10.

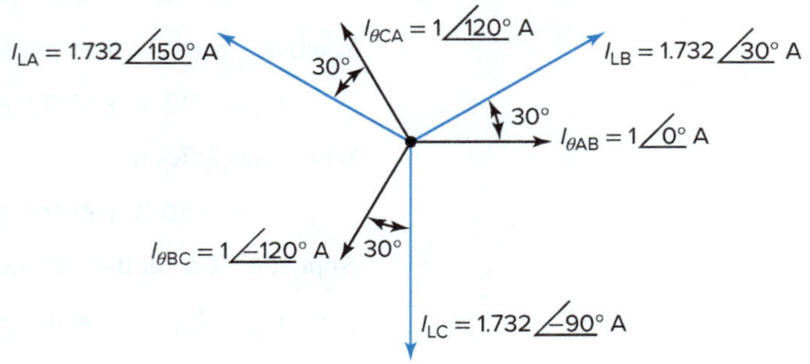

Figure 27–12 offers the proof for Formula (27–2). As you can see, this is the same technique used to verify Formula (27–1) for the line voltages in a Y-connected three-phase AC generator. Recall the KCL equation for the currents at point B in Fig. 27–10.

$$I_{LB} = I_{\theta AB} - I_{\theta BC}$$
$$\text{where } -I_{\theta BC} = I_{\theta CB} = 1\angle -120° + 180° \text{ A}$$
$$I_{\theta CB} = 1\angle 60° \text{ A}$$

Therefore, $I_{LB} = I_{\theta AB} + I_{\theta BC}$ for the currents at point B.

In Fig. 27–12, we see that the tail of the $I_{\theta CB}$ phasor is connected to the tip of the $I_{\theta AB}$ phasor. The resultant phasor I_{LB} is connected from the tail of the $I_{\theta AB}$ phasor to the tip of the $I_{\theta CB}$ phasor. Now, if the $I_{\theta CB}$ phasor is converted into rectangular form for the rectangular coordinates of the inner right triangle, then

$$I_{\theta CB} = 1\angle 60° \text{ A} = 0.5I_\theta + j0.866I_\theta$$

Similarly, $I_{\theta AB} = 1\angle 0° \text{ A} = I_\theta + j0I_\theta$

Now, if the rectangular coordinates for $I_{\theta AB}$ and $I_{\theta cb}$ are combined, we have $0.5I_\theta + I_\theta = 1.5I_\theta$ for one side of the larger right triangle in Fig. 27–12. The other side of the larger right triangle is $0.866I_\theta$, as shown. To solve for the hypotenuse, which is the phasor I_{LB}, we have

$$I_{LB} = \sqrt{1.5^2 I_\theta^2 + 0.866^2 I_\theta^2}$$

Which simplifies to

$$I_{LB} = \sqrt{3I_\theta^2}$$
$$I_{LB} = \sqrt{3}I_\theta$$

The other line currents are determined the same way. In general, $I_L = \sqrt{3}I_\theta$ for a delta-connected three-phase AC generator.

Figure 27–12 Combining phasors to prove Formula (27–2).

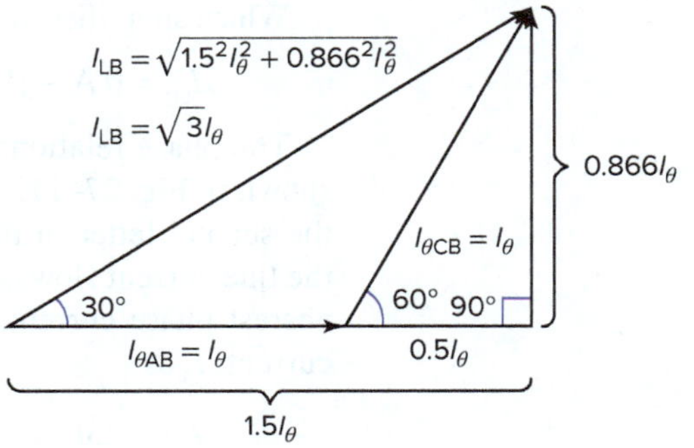

Example 27-3

In Fig. 27–10, how much is the line current, I_L, if each of the phase currents, I_θ, are 30 A?

ANSWER The line current, I_L, can be calculated using Formula (27–2).

$$\begin{aligned} I_L &= \sqrt{3} I_\theta \\ &= 1.732 \times 30 \text{ A} \\ &= 51.96 \text{ A} \end{aligned}$$

Example 27-4

In Fig. 27–10, how much is each phase current, I_θ, if the line currents, I_L, each equal 20 A?

ANSWER Rearrange Formula (27–2) to solve for the phase current, I_θ.

$$I_L = \sqrt{3} I_\theta \quad \text{so}$$

$$I_\theta = \frac{I_L}{\sqrt{3}}$$

Inserting values: $I_\theta = \dfrac{20 \text{ A}}{1.732}$

$$I_\theta = 11.55 \text{ A}$$

27–3 Self-Review

Answers at the end of the chapter.

a. **If the voltage across a coil winding in a delta-connected three-phase AC generator is 120 V, how much is the line voltage, V_L?**

b. **What is the formula for calculating the line current, I_L, in a delta-connected three-phase AC generator?**

27–4 Three-Phase Source/Load Configurations

The output from a wye (Y)- or delta (Δ)-connected three-phase AC generator may be connected to a load that is arranged in either a wye (Y) or delta (Δ) configuration. There are four possible system configurations, which include the following:

1. A wye (Y)-connected generator driving a wye (Y)-connected load. (Y-Y)
2. A wye (Y)-connected generator driving a delta (Δ)-connected load. (Y-Δ)
3. A delta (Δ)-connected generator driving a delta (Δ)-connected load. (Δ-Δ)
4. A delta (Δ)-connected generator driving a wye (Y)-connected load. (Δ-Y)

In any configuration, the loads can be purely resistive, purely reactive, or complex impedances that contain both resistance and reactance. If all of the impedances in a load are identical, the load is said to be ***balanced***. Conversely, if all of the impedances in a load are not identical, the load is said to be ***unbalanced***. Only **balanced loads** will be discussed in this chapter.

The Y-Y Configuration

Figure 27–13 shows a Y-connected generator driving a balanced Y-connected load. Since the neutral (N) point is connected between the source and the load, this is a three-phase four-wire system. As you will see, all other system configurations are strictly three-wire systems because there are only three connection points in a delta (Δ) configuration. Four-wire systems are useful when it is desired to have different load voltages. In a typical application, low-voltage loads such as lights and small appliances are connected between one of the lines (A, B or C) and neutral for operation at the phase voltage of 120 V. On the other hand, high-voltage loads, such as heaters, can be connected across two lines for operation at the line voltage of 208 V.

The Y-Y configuration in Fig. 27–13 is relatively simple to analyze. Close examination of the circuit connections reveals that each load impedance is in parallel with a phase voltage, V_θ, because of the common neutral (N) point. For a Y-Y configuration, therefore,

$$V_Z = V_\theta \tag{27–3}$$

Where V_Z is used to denote the load voltage to avoid confusion with the line voltage, V_L, available from a Y-connected generator. In Fig. 27–13:

$$V_Z = V_\theta = 120\angle\theta \text{ V}$$

To calculate the load currents in Fig. 27–13, just divide the phase voltage, V_θ, by the load impedance, Z.

$$I_Z = \frac{V_\theta}{Z} \tag{27–4}$$

where I_Z represents the current in the load impedance, Z.

In Fig. 27–13, each load impedance is identical and thus the load is said to be balanced. To calculate each load current, use Formula (27–4). In using

MultiSim **Figure 27–13** Y-Y configuration.

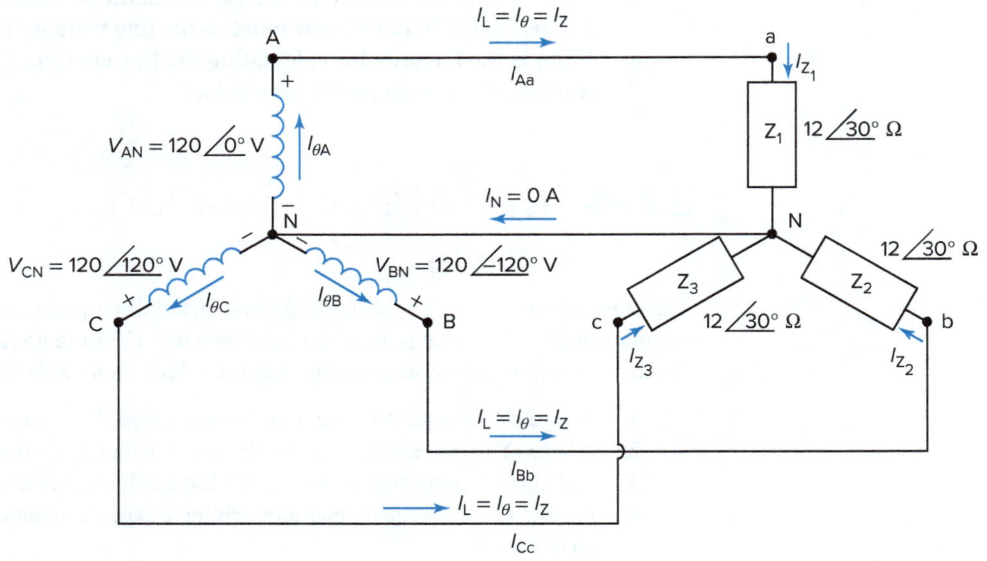

Formula (27–4), however, it is important to include the proper phase angle of each separate phase voltage. Across Z_1, $V_\theta = V_{AN} = 120\angle 0°$ V. Similarly, across Z_2, $V_\theta = V_{BN} = 120\angle -120°$ V and across Z_3, $V_\theta = V_{CN} = 120\angle 120°$ V. Now we can calculate each load current.

$$I_{Z1} = \frac{V_{AN}}{Z_1}$$
$$= \frac{120\angle 0°\ \text{V}}{12\angle 30°\ \Omega}$$
$$I_{Z1} = 10\angle -30°\ \text{A} = 8.66\ \text{A} - \text{j}5\ \text{A}$$

$$I_{Z2} = \frac{V_{BN}}{Z_2}$$
$$= \frac{120\angle -120°\ \text{V}}{12\angle 30°\ \Omega}$$
$$I_{Z2} = 10\angle -150°\ \text{A} = -8.66\ \text{A} - \text{j}5\ \text{A}$$

$$I_{Z3} = \frac{V_{CN}}{Z_3}$$
$$= \frac{120\angle 120°\ \text{V}}{12\angle 30°\ \Omega}$$
$$I_{Z3} = 10\angle 90°\ \text{A} = 0\ \text{A} + \text{j}10\ \text{A}$$

In Fig. 27–13, the current in the wires (lines) connecting points **A**, **B**, and **C** on the Y generator to points **a**, **b**, and **c** on the Y load is called the line current, designated I_L. As shown in the figure, the currents I_θ, I_L and I_Z are all equal within each phase of the Y-Y system. For a Y-Y configuration, therefore,

$$I_\theta = I_L = I_Z \qquad \textbf{(27–5)}$$

In Fig. 27–13 then, each line current equals $10\angle\theta$ A, where the phase angle $\angle\theta$ depends on which line current is being considered. In some cases, the individual line currents are identified using the connecting terminals between the source and load as subscripts. For example, in the line connecting terminals **A** and **a**, the line current is identified as I_{Aa}. Similarly, I_{Bb} and I_{Cc} can be used to identify the currents in the lines connecting terminals **B** and **b** and **C** and **c**, respectively. Also, each phase current, I_θ, can be identified by the terminal to which the phase winding connects. For example, $I_{\theta A}$ is the current in the phase winding connected to terminal **A**, and $I_{\theta B}$ and $I_{\theta C}$ are the currents in the phase windings connected to terminals **B** and **C**, respectively. In Fig. 27–13, therefore,

$$I_{\theta A} = I_{Aa} = I_{Z1} = 10\angle -30°\text{A}$$
$$I_{\theta B} = I_{Bb} = I_{Z2} = 10\angle -150°\text{A}$$
$$I_{\theta C} = I_{Cc} = I_{Z3} = 10\angle 90°\text{A}$$

The only current we haven't accounted for in Fig. 27–13 is the neutral current, I_N. By applying KCL at point N, we have the following equation of currents:

$$I_{Z1} + I_{Z2} + I_{Z3} - I_N = 0 \quad \text{or}$$
$$I_N = I_{Z1} + I_{Z2} + I_{Z3}$$

Inserting the rectangular form of each current, so values can be combined, gives us

$$I_N = (8.66\ \text{A} - \text{j}5\ \text{A}) + (-8.66\ \text{A} - \text{j}5\ \text{A}) + (0\ \text{A} + \text{j}10\ \text{A})$$

Removing parenthesis:

$$I_N = 8.66\ \text{A} - \text{j}5\ \text{A} - 8.66\ \text{A} - \text{j}5\ \text{A} + 0\ \text{A} + \text{j}10\ \text{A}$$

Which simplifies to

$$I_N = 0\ \text{A} + \text{j}0\ \text{A} = 0\angle 0°\ \text{A}$$

In a Y-Y system with a balanced load, the current in the neutral wire is always zero. In fact, because the neutral wire does not carry any current for the balanced load condition, it can be removed without having any effect on how the circuit operates. In a practical Y-Y system, however, the neutral wire is retained because a perfectly balanced load is difficult to achieve and almost impossible to maintain. In the case of an unbalanced load, the neutral wire will carry some current. The amount of current carried by the neutral wire depends on how much of an imbalance exists between each load impedance. In this chapter, we will deal exclusively with balanced loads.

The Y-Δ Configuration

Figure 27–14 shows a Y-connected generator driving a balanced, Δ-connected load. Notice that the neutral (N) point in the Y-connected generator is not connected to the load because there are only three connecting points on the delta. As a result, there are only three wires (lines) connecting the three-phase generator to the load.

In the Y-Δ configuration, each load impedance has the full line voltage across it. As you recall, in a Y-connected generator, the line voltage, V_L, is calculated as $V_L = \sqrt{3}V_\theta$. In Fig. 27–14, therefore, the load voltage, V_Z, is calculated as

$$V_Z = V_L = \sqrt{3}\,V_\theta \qquad \textbf{(27–6)}$$

Since $V_\theta = 120\text{ V}\underline{/\theta}\text{ V}$, we have

$$V_L = \sqrt{3} \times 120\underline{/\theta}\text{ V}$$
$$V_L \approx 208\underline{/\theta}\text{ V}$$

The phase angle of each line voltage was determined back in Section 27–2 and is displayed in Fig. 27–7. The phase angle between each line voltage is 120°, and each line voltage is 30° out of phase with the nearest phase voltage. In Fig. 27–14, the line voltage, V_{CA}, of $208\underline{/150°}$ V is across Z_1; the line voltage, V_{AB}, of $208\underline{/30°}$ V is

MultiSim **Figure 27–14** Y-Δ configuration.

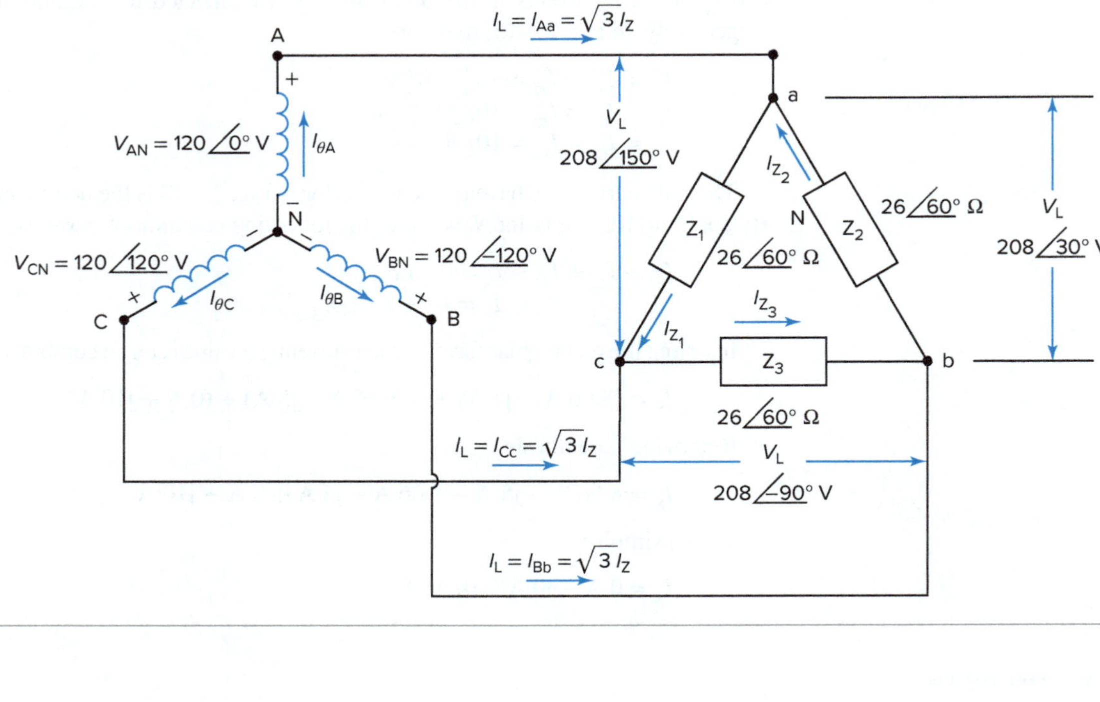

across Z_2; and the line voltage, V_{BC}, of $208\angle -90°$ V is across Z_3. The individual load currents are calculated as

$$I_{Z1} = \frac{V_{CA}}{Z_1}$$
$$= \frac{208\angle 150° \text{ V}}{26\angle 60° \ \Omega}$$
$$I_{Z1} = 8\angle 90° \text{A} = 0 \text{ A} + \text{j}8 \text{ A}$$
$$I_{Z2} = \frac{V_{AB}}{Z_2}$$
$$= \frac{208\angle 30° \text{ V}}{26\angle 60° \ \Omega}$$
$$I_{Z2} = 8\angle -30° \text{A} = 6.93 \text{ A} - \text{j}4 \text{ A}$$
$$I_{Z3} = \frac{V_{BC}}{Z_3}$$
$$= \frac{208\angle -90° \text{ V}}{26\angle 60° \ \Omega}$$
$$I_{Z3} = 8\angle -150° \text{A} = -6.93 \text{ A} - \text{j}4 \text{ A}$$

In Fig. 27–14, the line currents I_{Aa}, I_{Bb}, and I_{Cc} equal the corresponding phase currents $I_{\theta A}$, $I_{\theta B}$, and $I_{\theta C}$ in the Y-connected generator. That is, $I_L = I_\theta$. However, each line current divides into two load currents as shown. Using the assumed direction of current for I_{Z1}, I_{Z2} and I_{Z3} in Fig. 27–14, apply KCL at points **a**, **b**, and **c** to obtain the following equations for I_{Aa}, I_{Bb}, and I_{Cc}.

At point **a :** $I_{Aa} + I_{Z2} - I_{Z1} = 0$ or

$$I_{Aa} = I_{Z1} - I_{Z2}$$

At point **b :** $I_{Bb} + I_{Z3} - I_{Z2} = 0$ or

$$I_{Bb} = I_{Z2} - I_{Z3}$$

At point **c :** $I_{Cc} + I_{Z1} - I_{Z3} = 0$ or

$$I_{Cc} = I_{Z3} - I_{Z1}$$

Inserting values to solve for each line current gives us

$$I_{Aa} = I_{Z1} - I_{Z2}$$
$$= (0 \text{ A} + \text{j}8 \text{ A}) - (6.93 \text{ A} - \text{j}4 \text{ A})$$

Which simplifies to: $I_{Aa} = -6.93 \text{ A} + \text{j}12 \text{ A} = 13.86\angle 120° \text{ A}$

Next:

$$I_{Bb} = I_{Z2} - I_{Z3}$$
$$= (6.93 \text{ A} - \text{j}4 \text{ A}) - (-6.93 \text{ A} - \text{j}4 \text{ A})$$

Which simplifies to: $I_{Bb} = 13.86 \text{ A} + \text{j}0 \text{ A} = 13.86\angle 0° \text{ A}$

And finally:

$$I_{Cc} = I_{Z3} - I_{Z1} = (-6.93 \text{ A} - \text{j}4 \text{ A}) - (0 \text{ A} + \text{j}8 \text{ A})$$

Which simplifies to $I_{Cc} = -6.93 \text{ A} - \text{j}12 \text{ A} = 13.86\angle -120° \text{ A}$

In general,

$$I_L = \sqrt{3}I_Z$$
$$= 1.732 \times 8 \text{ A}$$
$$I_L = 13.86 \text{ A}$$

In Fig. 27–14, each line current is 30° out of phase with the nearest load current, I_Z. For example, I_{Aa} leads I_{Z1} by 30°, I_{Bb} leads I_{Z2} by 30°, and I_{Cc} leads I_{Z3} by 30°. Furthermore, it is important to note that the phasor sum of all the line currents is zero.

The Δ-Δ Configuration

Figure 27–15 shows a Δ-connected generator driving a balanced, Δ-connected load. In this configuration, the load voltage, line voltage, and phase voltage of a given phase are all equal.

$$V_Z = V_L = V_\theta \qquad \textbf{(27–7)}$$

Figure 27–15 Δ-Δ configuration.

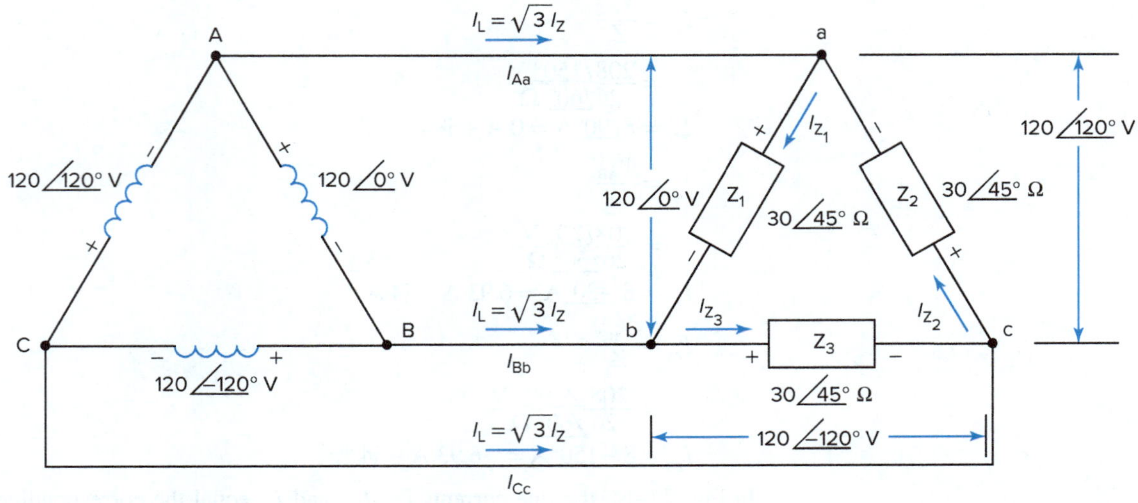

Since $V_Z = V_L = V_\theta$, each load current is calculated as

$$I_Z = \frac{V_\theta}{Z}$$

In Fig. 27–15, each load voltage is shown as

$$V_{Z1} = V_{AB} = 120\angle 0^\circ \text{ V}$$
$$V_{Z2} = V_{CA} = 120\angle 120^\circ \text{ V}$$
$$V_{Z3} = V_{BC} = 120\angle -120^\circ \text{ V}$$

Although the magnitude of each load voltage is the same, the load voltages are 120° out of phase with each other. The individual load currents are calculated as

$$I_{Z1} = \frac{V_{AB}}{Z_1}$$
$$= \frac{120\angle 0^\circ \text{ V}}{30\angle 45^\circ \ \Omega}$$
$$I_{Z1} = 4\angle -45^\circ \text{ A} = 2.828 \text{ A} - j2.828 \text{ A}$$
$$I_{Z2} = \frac{V_{CA}}{Z_2}$$
$$= \frac{120\angle 120^\circ \text{ V}}{30\angle 45^\circ \ \Omega}$$
$$I_{Z2} = 4\angle 75^\circ \text{ A} = 1.035 \text{ A} + j3.864 \text{ A}$$
$$I_{Z3} = \frac{V_{BC}}{Z_3}$$
$$= \frac{120\angle -120^\circ \text{ V}}{30\angle 45^\circ \ \Omega}$$
$$I_{Z3} = 4\angle -165^\circ \text{ A} = -3.864 \text{ A} - j1.035 \text{ A}$$

Next, apply KCL at points **a**, **b**, and **c** using the assumed direction of currents shown. Doing so yields the following equations for the line currents I_{Aa}, I_{Bb}, and I_{Cc}.

$$I_{Aa} = I_{Z1} - I_{Z2}$$
$$I_{Bb} = I_{Z3} - I_{Z1}$$
$$I_{Cc} = I_{Z2} - I_{Z3}$$

Inserting values to solve for each line current gives us

$$I_{Aa} = (2.828 \text{ A} - j2.828 \text{ A}) - (1.035 \text{ A} + j3.864 \text{ A})$$

Which simplifies to

$$I_{Aa} = 1.793 \text{ A} - j6.692 \text{ A} = 6.928\underline{/-75^\circ} \text{ A}$$

Next,

$$I_{Bb} = (-3.864 \text{ A} - j1.035 \text{ A}) - (2.828 \text{ A} - j2.828 \text{ A})$$

Which simplifies to

$$I_{Bb} = -6.692 \text{ A} + j1.793 \text{ A} = 6.928\underline{/165^\circ} \text{ A}$$

And finally

$$I_{Cc} = (1.035 \text{ A} + j3.864 \text{ A}) - (-3.864 \text{ A} - j1.035 \text{ A})$$

Which simplifies to

$$I_{Cc} = 4.899 \text{ A} + j4.899 \text{ A} = 6.928\underline{/45^\circ} \text{ A}$$

In general:

$$\begin{aligned} I_L &= \sqrt{3}I_Z \\ &= 1.732 \times 4 \text{ A} \\ I_L &= 6.928 \text{ A} \end{aligned}$$

In Fig. 27–15, each line current is 30° out of phase with the nearest load current, I_Z. For example, I_{Aa} lags I_{Z1} by 30°, I_{Bb} lags I_{Z3} by 30°, and I_{Cc} lags I_{Z2} by 30°. Furthermore, the line currents are 120° out of phase with each other.

The Δ-Y Configuration

Figure 27–16(a) shows a Δ-connected generator driving a balanced, Y-connected load. Close examination of the circuit connections shows that the current in each load is equal to its corresponding line current, I_L. That is, $I_L = I_Z$. If we can determine each line current, then each individual load voltage can be calculated as $V_Z = I_L Z$. Since the Y-connected load is balanced, it can be converted into an equivalent Δ network whose impedances are $Z_\Delta = 3Z_Y$. In this case, $Z_\Delta = 3 \times 120\underline{/45^\circ}\ \Omega = 360\underline{/45^\circ}\ \Omega$. The equivalent Δ-Δ connections are shown in Fig. 27–16(b). To calculate the line currents, we must first calculate the current in each leg of the Δ-connected load. In Fig. 27–16(b),

$$\begin{aligned} I_{ZA} &= \frac{V_{BC}}{Z_A} \\ &= \frac{480\underline{/-120^\circ} \text{ V}}{360\underline{/45^\circ}\ \Omega} \\ I_{ZA} &= 1.333\underline{/-165^\circ} \text{ A} = -1.288 \text{ A} - j345 \text{ mA} \end{aligned}$$

Also,

$$\begin{aligned} I_{ZB} &= \frac{V_{AB}}{Z_B} \\ &= \frac{480\underline{/0^\circ} \text{ V}}{360\underline{/45^\circ}\ \Omega} \\ I_{ZB} &= 1.333\underline{/-45^\circ} \text{ A} = 942.6 \text{ mA} - j942.6 \text{ mA} \end{aligned}$$

$$\begin{aligned} I_{ZC} &= \frac{V_{CA}}{Z_C} \\ &= \frac{480\underline{/120^\circ} \text{ V}}{360\underline{/45^\circ}\ \Omega} \\ I_{ZC} &= 1.333\underline{/75^\circ} \text{ A} = 345 \text{ mA} + j1.288 \text{ A} \end{aligned}$$

Figure 27–16 Δ-Y configuration. (*a*) Original circuit. (*b*) Converting the Y-connected load to an equivalent Δ-connected load to simplify analysis

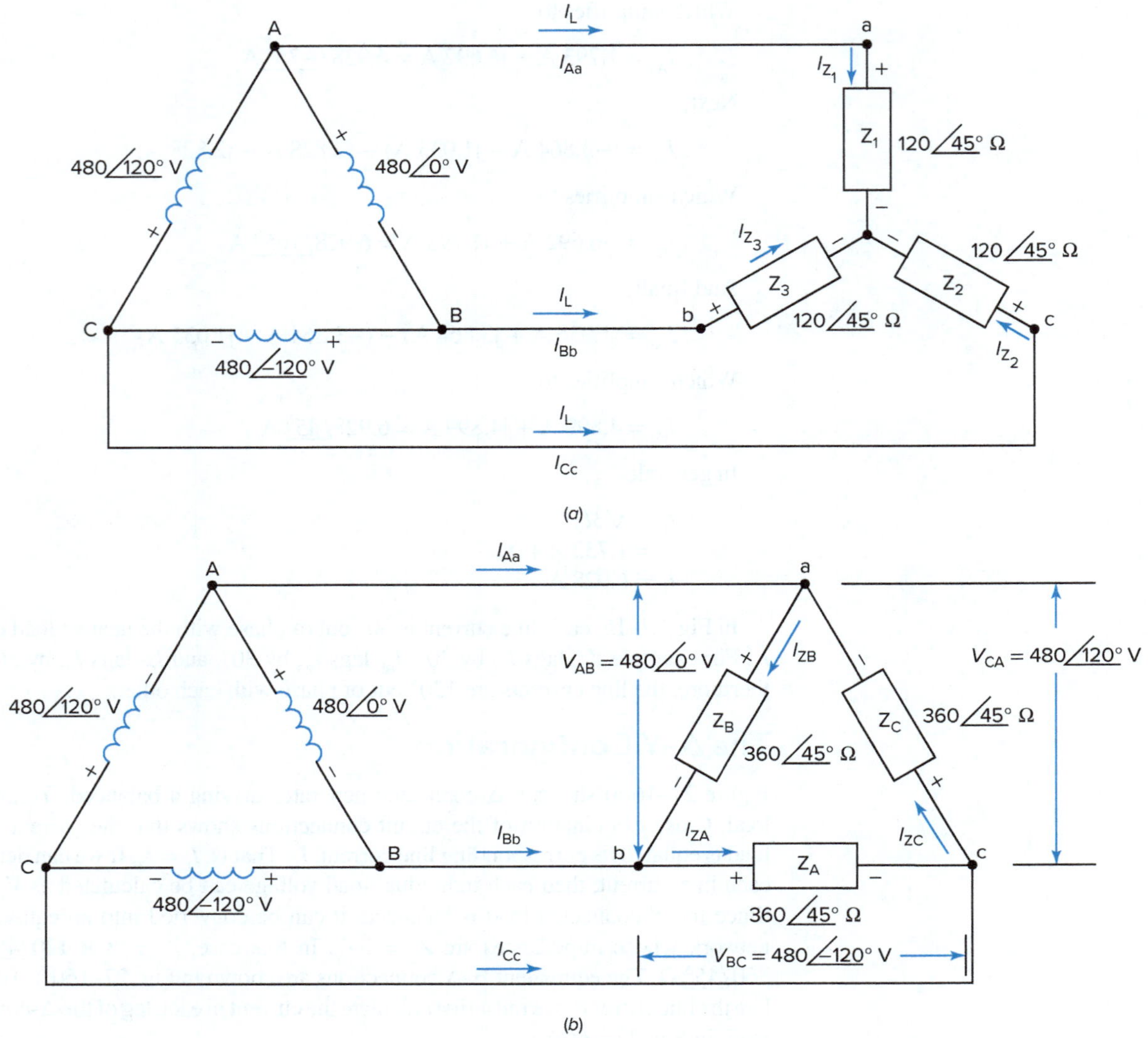

Next, apply KCL at points **a**, **b**, and **c** in Fig. 27–16(b) using the assumed direction of currents shown. Doing so yields the following equations for the line currents I_{Aa}, I_{Bb}, and I_{Cc}.

$$I_{Aa} = I_{ZB} - I_{ZC}$$
$$I_{Bb} = I_{ZA} - I_{ZB}$$
$$I_{Cc} = I_{ZC} - I_{ZA}$$

Inserting values to solve for I_{Aa}, I_{Bb}, and I_{Cc} gives us

$$I_{Aa} = (942.6\text{ mA} - j942.6\text{ mA}) - (345\text{ mA} + j1.288\text{ A})$$

Which simplifies to

$$I_{Aa} = 597.6\text{ mA} - j2.231\text{ A} = 2.31\underline{/-75^\circ}\text{ A}$$

Also,

$$I_{Bb} = (-1.288\text{ A} - j345\text{ mA}) - (942.6\text{ mA} - j942.6\text{ mA})$$

Which simplifies to

$$I_{Bb} = -2.231\text{ A} + j597.6\text{ mA} = 2.31\underline{/165^\circ}\text{ A}$$

And finally,

$$I_{Cc} = (345\ \text{mA} + j1.288\ \text{A}) - (-1.288\ \text{A} - j345\ \text{mA})$$

Which simplifies to

$$I_{Cc} = 1.633\ \text{A} + j1.633\ \text{A} = 2.31\underline{/45°}\ \text{A}$$

These line currents I_{Aa}, I_{Bb}, and I_{Cc} are the same in Fig. 27–16(a). Knowing the value of each line current allows us to calculate each load voltage. Refer to Fig. 28–16(a).

$$V_{Z1} = I_{Aa} \times Z_1$$
$$= 2.31\underline{/-75°}\ \text{A} \times 120\ \underline{/45°}\ \Omega$$
$$V_{Z1} = 277.2\underline{/-30°}\ \text{V}$$

Next,

$$V_{Z2} = I_{Cc} \times Z_2$$
$$= 2.31\ \underline{/45°}\ \text{A} \times 120\underline{/45°}\ \Omega$$
$$V_{Z2} = 277.2\underline{/90°}\ \text{V}$$

And finally,

$$V_{Z3} = I_{Bb} \times Z_3$$
$$= 2.31\underline{/165°}\ \text{A} \times 120\underline{/45°}\ \Omega$$
$$V_{Z3} = 277.2\ \underline{/210°}\ \text{V}$$

In general, the load voltage, V_Z in a Δ-Y configuration equals

$$V_Z = \frac{V_L}{\sqrt{3}} \qquad \textbf{(27–8)}$$

Where the line voltage, V_L, equals the induced voltage across each generator winding. In Fig. 27–16(a),

$$V_Z = \frac{480\underline{/\theta}\ \text{V}}{1.732} \quad \text{or}$$

$$V_Z \approx 277\ \text{V}$$

27–4 Self-Review

Answers at the end of the chapter.

a. **In a Y-Y configuration, how much current flows in the neutral wire if the load is balanced?**

b. **In the Y-Δ configuration, is the load voltage, V_Z, equal to the phase voltage, V_θ, or the line voltage, V_L?**

c. **In the Δ-Δ configuration, how is each line current, I_L, related to the load current, I_Z?**

d. **In the Δ-Y configuration, how is each load voltage, V_Z, related to the line voltage, V_L?**

27–5 Three-Phase Power Calculations

In a balanced Y- or Δ-connected load, each load impedance dissipates an equal amount of power, which can be calculated as

$$P_Z = V_Z I_Z \cos\theta \qquad \textbf{(27–9)}$$

where Cos θ is the power factor of the circuit.

The total power, P_T, dissipated by a balanced load is three times the power dissipated by an individual load.

$$P_T = 3P_Z \qquad \textbf{(27–10)}$$

The following voltage and current relationships exist for each of the four possible three-phase source/load configurations.

Y-Y Configuration

$$V_Z = V_\theta = \frac{V_L}{\sqrt{3}} \quad \text{or} \quad V_L = \sqrt{3}V_\theta \text{ and } I_Z = I_L$$

Y-Δ Configuration

$$V_Z = V_L \quad \text{and} \quad I_L = \sqrt{3}I_Z \text{ or } I_Z = \frac{I_L}{\sqrt{3}}$$

Δ-Δ Configuration

$$V_Z = V_L \quad \text{and} \quad I_L = \sqrt{3}I_Z \text{ or } I_Z = \frac{I_L}{\sqrt{3}}$$

Δ-Y Configuration

$$V_Z = \frac{V_L}{\sqrt{3}} \quad \text{and} \quad I_Z = I_L$$

If the foregoing expressions for V_Z and I_Z are substituted into Formula (27–10) for any of the four possible source/load configurations, we have

$$P_T = \sqrt{3}V_L I_L \text{ Cos } \theta \qquad \textbf{(27–11)}$$

As an example of how this formula is derived, let's use the expressions for V_Z and I_Z in the Y-Y configuration. Recall that the total power in a balanced load is calculated as

$$P_T = 3P_Z \quad \text{or}$$
$$P_T = 3V_Z I_Z \text{ Cos } \theta$$

Substituting for V_Z and I_Z, we have

$$P_T = 3\frac{V_L}{\sqrt{3}}I_L \text{ Cos } \theta$$

Since $3 = \sqrt{3} \times \sqrt{3}$, then $P_T = \sqrt{3} \times \sqrt{3}\frac{V_L}{\sqrt{3}}I_L \text{ Cos } \theta$

Which can be reduced to Formula (27–11), which is

$$P_T = \sqrt{3}V_L I_L \text{ Cos } \theta$$

About the Phase Angle θ

When using any of the power formulas, it is important to note that the phase angle, θ, is the phase angle of the impedance, Z. Since θ equals the phase angle between the applied voltage and current, then

$$\theta = \angle V_Z - \angle I_Z$$

Since $\angle I_Z = \angle V_Z - \angle Z$ then

$$\theta = \angle V_Z - (\angle V_Z - \angle Z) \quad \text{which simplifies to:}$$

$$\theta = \angle Z$$

Example 27-5

MultiSim

Calculate the total power, P_T, for the balanced load in (a) Fig. 27–13 and (b) Fig. 27–14.

ANSWER

(a) In Fig. 27–13, recall that $V_Z = V_\theta = 120$ V. Therefore, $V_L = \sqrt{3}V_\theta \approx 208$ V. Also, $I_Z = I_L = 10$ A and the phase angle, θ, of each load impedance is 30°. Using Formula (27–11), we have

$$P_T = \sqrt{3}V_L I_L \operatorname{Cos}\theta$$
$$= 1.732 \times 208 \text{ V} \times 10 \text{ A} \times \operatorname{Cos}(30°)$$
$$P_T = 3.12 \text{ kW}$$

(b) In Fig. 27–14, recall that $V_Z = V_L = 208$ V and $I_L = \sqrt{3}I_Z = 13.86$ A. Also, $\theta_Z = 60°$. Inserting values into Formula (27–11), we have

$$P_T = \sqrt{3}V_L I_L \operatorname{Cos}\theta$$
$$= 1.732 \times 208 \text{ V} \times 13.86 \text{ A} \times \operatorname{Cos}(60°)$$
$$= 2.497 \text{ kW}$$

Advantages of Three-Phase AC Power

Three-phase systems have several advantages over single-phase systems. In general, three-phase systems are more efficient and, in turn, more economical to implement and use than single-phase systems. The most notable advantages are listed here.

Economics of Power Transmission—One notable advantage to using a three-phase system is that for a given amount of load power, there is less current, and in turn, less I^2R power loss in the wires or lines connecting the source to the load. As you will see, for the same amount of transmitted power, this allows for the use of a smaller wire size (higher gauge number) with three-phase AC power systems. This, in turn, results in a significant cost savings because the total volume of wire needed to transmit a given amount of power is significantly reduced. Let's take a look at an example. Figure 27–17(a) shows a single-phase system in which a 1.8 kW load is connected to the 120 V power line through two copper conductors. The current drawn by the load is calculated as

$$I = \frac{P}{V}$$
$$= \frac{1.8 \text{ kW}}{120 \text{ V}}$$
$$I = 15 \text{ A}$$

The two conductors connected to the load must each carry a current of 15 A.

Now consider the three-phase circuit in Fig. 27–17(b) which uses a Δ-Δ configuration. Each load receives a voltage of 120 V because $V_Z = V_L = V_\theta$ in a Δ-Δ configuration. In the balanced Δ-connected load, each load dissipates 600 W of power

Figure 27–17 Economics of power transmission. (*a*) Single-phase system with 1.8 kW load. (*b*) Three-phase system with 3-600 W loads for a total of 1.8 kW.

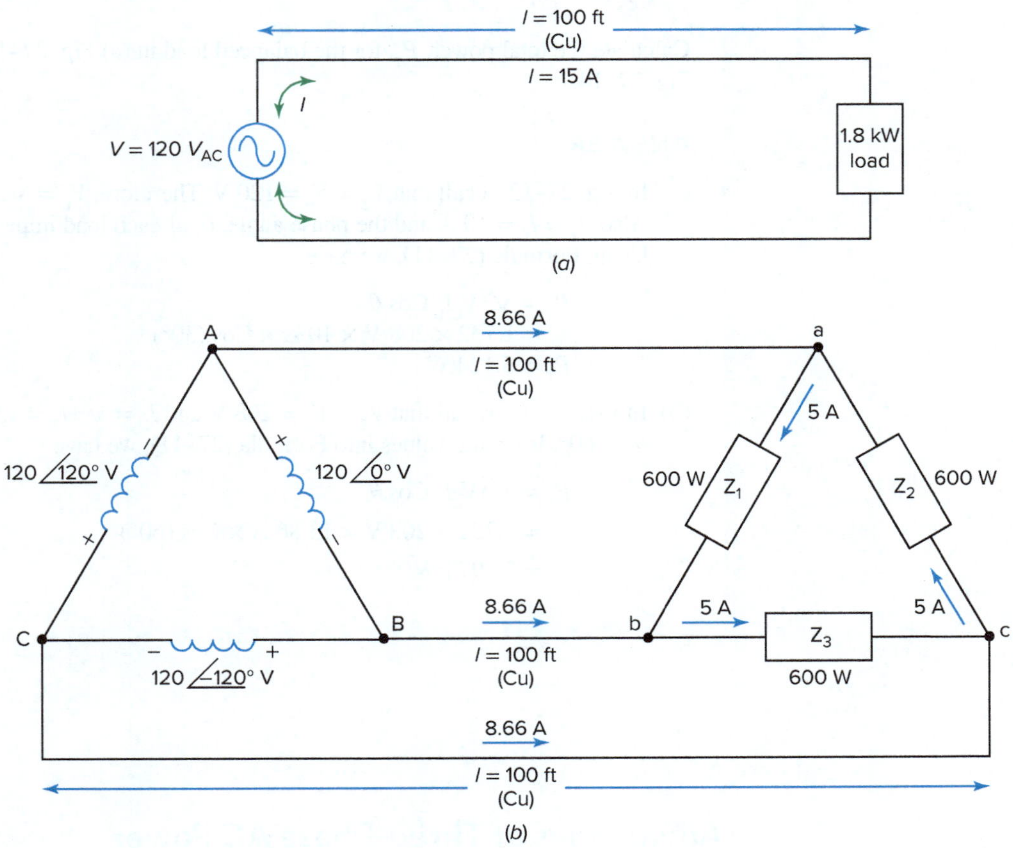

for a total power, P_T of 1.8 kW, as in Fig. 27–17(a). With 120 V across each load, each respective load current equals

$$I = \frac{P}{V}$$
$$= \frac{600\text{ W}}{120\text{ V}}$$
$$I = 5\text{ A}$$

Recall that in a Δ-Δ configuration, the line current, I_L is calculated as

$$I_L = \sqrt{3}I_Z$$
$$= 1.732 \times 5\text{ A}$$
$$I_L = 8.66\text{ A}$$

In Fig. 27–17(a) and (b), notice that the length of each conductor is 100 ft. If we go by the rule that 500 cmils can safely carry 1 A of current, then the load in Fig. 27–17(a) requires the use of a copper wire whose total cmil area equals

$$\text{cmil} = \frac{500\text{ cmil}}{\text{A}} \times 15\text{ A}$$
$$\text{cmil} = 7500\text{ cmil}$$

In Fig. 27–17(a), therefore, the total volume of copper wire, in units of cmil·ft, equals 7500 cmil × 200 ft = 1.5×10^6 cmil·ft.

In Fig. 27–17(b), the cmil area required for each conductor is

$$\text{cmil} = \frac{500\ \text{cmil}}{\text{A}} \times 8.66\ \text{A}$$
$$\text{cmil} = 4330\ \text{cmil}$$

In this case, the total volume of copper used by the conductors is calculated as 4330 cmil × 300 ft. = 1.299×10^6 cmil · ft. Note that the total volume of copper wire calculated in Fig. 27–17(b) is less than what was calculated in Fig. 27–17(a). In fact, the ratio of the two is

$$\frac{1.299 \times 10^6\ \text{cmil} \cdot \text{ft}}{1.5 \times 10^6\ \text{cmil} \cdot \text{ft}} = 0.866$$

This corresponds to approximately 13.4% less copper wire for the Δ-Δ three-phase configuration in Fig. 27–17(b) versus the single-phase system in Fig. 27–17(a). This cost savings is very significant when long distances and very large amounts of power are being transferred from one location to another.

Constant Power—In a **single-phase AC circuit**, the power decreases to zero twice during each cycle of alternating voltage. Therefore, in a single-phase AC circuit, the power delivered to the load pulsates or fluctuates between zero and some maximum value. In a balanced **three-phase AC circuit**, however, the instantaneous power delivered to the load is steady or constant. This is important when the load is a motor because it results in a uniform or constant torque on the shaft of the motor. This results in less variation and vibration, and in turn, less wear and tear on the motor during the life of its operation.

Power Generation—Three-phase AC power can be generated more economically and more efficiently than single-phase AC power. Also, three-phase AC power sources are capable of supplying power to both single-phase and three-phase loads when different load voltages are required in a given application.

Superiority of Three-Phase Motors—As compared to single-phase motors, three-phase motors are smaller in size and weight and have higher efficiencies and power factors. They also do not use or need a special starting capacitor as is required with single-phase motors. In general, three-phase motors are more economical to use as compared to single-phase motors.

It is important to note that a detailed explanation of some of these listed advantages is beyond the scope of this chapter. A more detailed analysis of the advantages listed would most likely be covered in more advanced subsequent power courses.

27–5 Self-Review

Answers at the end of the chapter.

a. **A Δ-Δ configuration has the following values: V_Z = 480 V, I_Z = 12 A, and θ_Z = 60°. Calculate the total power, P_T, dissipated by the load.**

b. **A Y-Y configuration has the following values: V_Z = 120 V, I_Z = 6 A, and θ_Z = 18°. Calculate the total power, P_T, dissipated by the load.**

Summary

- A single-phase AC circuit contains only one AC voltage source.
- A polyphase AC circuit contains two or more AC voltage sources that are out of phase with each other.
- A three-phase AC generator produces three different output voltages that are 120° out of phase with each other.
- In a Y-connected, three-phase AC generator, the voltage across each coil winding is called the phase voltage, designated V_θ. The output voltage across any two coil windings is called the line voltage and is designated V_L.
- In a Y-connected generator, $V_L = \sqrt{3}V_\theta$.
- In a Δ-connected generator, the phase voltage, V_θ, and line voltage, V_L, are the same. That is, $V_\theta = V_L$. Each phase voltage is still 120° out of phase with each other, however.
- The currents in the coil windings of a Δ-connected three-phase AC generator are called phase currents, designated, I_θ. The currents flowing out from the terminals (lines) of the Δ-connected generator are called line currents, designated I_L.
- In a Δ-connected three-phase AC generator, $I_L = \sqrt{3}I_\theta$.
- To obtain the negative of any phasor expressed in polar form, just add 180° to the phase angle, θ. In rectangular form, change the sign of both terms.
- In a Y-connected three-phase AC generator, each line voltage is 30° out of phase with the nearest phase voltage.
- In a Δ-connected three-phase AC generator, each line current is 30° out of phase with the nearest phase current.
- In the Y-Y configuration, the load voltage, V_Z, equals the phase voltage, V_θ. Also, the load current, I_Z, equals the line current, I_L, as well as the phase current, I_θ.
- In a Y-Y configuration with a balanced load, the current in the neutral wire is zero.
- In a Y-Δ configuration, the load voltage, V_Z, equals the line voltage, V_L, which is $V_L = \sqrt{3}V_\theta$. The line current, $I_L = \sqrt{3}I_Z$.
- In the Δ-Δ configuration, $V_Z = V_L = V_\theta$. The line current, $I_L = \sqrt{3}I_Z$.
- In the Δ-Y configuration, the load voltage, $V_Z = \frac{V_L}{\sqrt{3}}$. The line current, $I_L = I_Z$.
- In a balanced Y- or Δ-connected load, each load impedance dissipates an equal amount of power calculated as $P_Z = V_Z I_Z \text{ Cos } \theta$. The total power dissipated by a balanced load is calculated as $P_T = \sqrt{3}V_L I_L \text{ Cos } \theta$. The phase angle, θ, is the phase angle of the impedance, Z.
- The distribution of three-phase AC power is more efficient than the distribution of single-phase AC power. This is because, for the same load power, there is less I^2R power loss in the wires or lines connecting the source to the load. Because of this, a three-phase system can get by with using a smaller wire size, which results in considerable cost savings in terms of copper wire.
- Three-phase AC power can be generated more economically and more efficiently than single-phase AC power.

Important Terms

Balanced Load—A wye (Y)- or delta (Δ)-connected load in which each individual load impedance is identical.

Line Current, I_L—The current flowing in the wires (lines), which connects the terminals of a three-phase AC generator to the terminals of a Y- or Δ-connected load.

Line Voltage, V_L—The voltage measured across two terminals or lines (not including neutral) in a three-phase AC generator.

Phase Current, I_θ—The current flowing in a coil winding of a three-phase AC generator.

Phase Voltage, V_θ—The voltage directly across a coil winding in a three-phase AC generator.

Polyphase AC Circuit—A circuit containing two or more AC voltage sources, each with a different phase angle.

Single-Phase AC Circuit—A circuit containing only one AC voltage source.

Three-Phase AC Circuit—An AC circuit containing three separate AC voltage sources, which are 120° out of phase with each other.

Related Formulas

Y-Connected Generator

$$V_L = \sqrt{3}V_\theta$$

Δ-Connected Generator

$$I_L = \sqrt{3}I_\theta$$

Y-Y Configuration

$$V_Z = V_\theta$$

$$I_Z = \frac{V_Z}{Z}$$

$$I_\theta = I_L = I_Z$$

Y-Δ Configuration

$$V_Z = V_L = \sqrt{3}V_\theta$$

$$I_Z = \frac{V_Z}{Z}$$

$$I_L = \sqrt{3}I_Z$$

Δ-Δ Configuration

$$V_Z = V_L = V_\theta$$

$$I_Z = \frac{V_Z}{Z}$$

$$I_L = \sqrt{3}I_Z$$

Δ-Y Configuration

$V_Z = \frac{V_L}{\sqrt{3}}$ or $V_Z = I_L Z$

$I_Z = I_L$ where $I_Z = \frac{V_Z}{Z}$

Power Calculations

$P_Z = V_Z I_Z \text{ Cos } \theta$ (Individual Load)

$P_T = \sqrt{3} V_L I_L \text{ Cos } \theta$ (Total Load Power)

Self-Test

Answers at the back of the book.

1. **A single-phase AC circuit contains:**
   a. Three out-of-phase AC voltages.
   b. Only one AC voltage source.
   c. Two out-of-phase AC voltage sources.
   d. Three in-phase AC voltage sources.

2. **In a three-phase AC generator, each phase voltage is:**
   a. In-phase with each other.
   b. 30° out of phase with each other.
   c. 120° out of phase with each other.
   d. 180° out of phase with each other.

3. **In a four-wire Y-Y configuration, how much current flows in the neutral wire if the load is balanced?**
   a. $I_N = \sqrt{3} I_Z$.
   b. $I_N = I_Z$.
   c. $I_N = 0$ A.
   d. This is impossible to determine.

4. **If the phase voltage in a Y-connected three-phase generator is 3 kV, how much is the line voltage, V_L?**
   a. 5.2 kV.
   b. 3 kV.
   c. 1.732 kV.
   d. 4.24 kV.

5. **If the phasor $V_{XZ} = 50\angle{-40°}$ V, what is the value of the phasor V_{ZX}?**
   a. $50\angle{40°}$ V.
   b. $50\angle{-160°}$ V.
   c. $50\angle{180°}$ V.
   d. $50\angle{140°}$ V.

6. **In a Δ-Δ configuration, each load current equals 25 A. How much is each corresponding line current?**
   a. 14.43 A.
   b. 43.3 A.
   c. 0 A.
   d. 25 A.

7. **In a Y-Δ configuration, each load voltage is equal to**
   a. V_θ.
   b. $\frac{V_L}{\sqrt{3}}$.
   c. V_L.
   d. Both a and b.

8. **In a Y-connected three-phase generator, what is the phase relationship between each line voltage and its nearest phase voltage?**
   a. 30°.
   b. 120°.
   c. 60°.
   d. 180°.

9. **In a Δ-Y configuration, how much is each load voltage, V_Z, if the line voltage, V_L, is 208 V?**
   a. 208 V.
   b. 360 V.
   c. 147 V.
   d. 120 V.

10. **A Y-connected three-phase generator, whose phase voltage, V_θ, equals 120 V, is connected to a Y-connected load whose impedances are $Z_1 = Z_2 = Z_3 = 5\angle{-45°}\ \Omega$. How much is the total power, P_T, delivered to the load?**
    a. 3.53 kW.
    b. 6.11 kW.
    c. 2.04 kW.
    d. 10.6 kW.

11. **How much is each individual load current, I_Z in a Y-Δ configuration, if the line current, I_L, is 20 A?**
    a. 20 A.
    b. 11.55 A.
    c. 34.64 A.
    d. This is impossible to determine.

12. **In a Y-Y configuration, the load current, I_Z, equals:**
    a. I_θ.
    b. I_L.
    c. $\sqrt{3} I_\theta$.
    d. Both a and b.

13. **The distribution of three-phase AC power is more efficient than the distribution of single-phase AC power because**
    a. For a given amount of load power, each conductor has less current and thus less I^2R power loss.
    b. Less copper is required.
    c. More copper is required.
    d. Both a and b.

14. **In the Δ-Δ configuration:**
    a. $I_Z = I_L = I_\theta$.
    b. $V_Z = V_L = V_\theta$.
    c. $V_Z = \sqrt{3} V_\theta$.
    d. $I_Z = \sqrt{3} I_L$.

15. **In a four-wire Y-Y configuration, current will flow in the neutral wire only when**
    a. The load is unbalanced.
    b. The load is purely resistive.
    c. The load is balanced.
    d. The load is purely reactive.

Essay Questions

1. In a Y-connected three-phase AC generator, define the terms phase voltage and line voltage.

2. In a Δ-connected three-phase AC generator, define the terms phase current and line current.

3. In a Y-connected three-phase AC generator, what is the phase angle between
   a. Each phase voltage?
   b. Each line voltage?
   c. Each line voltage and the nearest phase voltage?

4. Explain what is meant by a balanced and unbalanced load.

5. Show how the formula $P_T = \sqrt{3}V_L I_L \text{ Cos } \theta$ is derived for each of the following circuit configurations:
   a. Y-Y.
   b. Y-Δ.
   c. Δ-Δ.
   d. Δ-Y.

Problems

SECTION 27–1 THREE-PHASE AC GENERATORS

27–1 What type of connection is shown for the generator coil windings in Fig. 27–18?

Figure 27–18

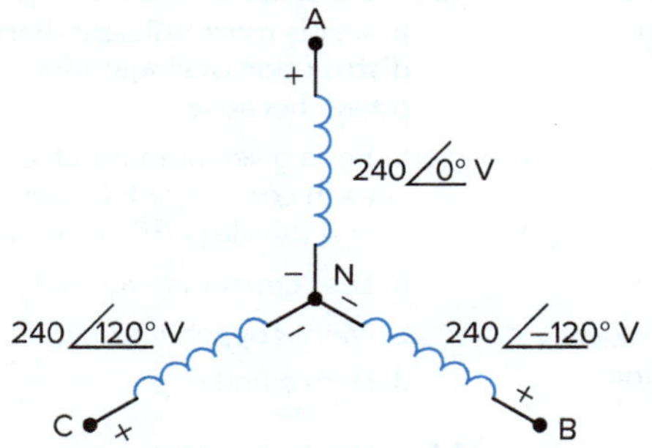

27–2 In Fig. 27–18, what is the phase angle between each of the three coil voltages?

27–3 In Fig. 27–18, what is the mathematical relationship between the line voltage, V_L, and the phase voltage, V_θ?

27–4 In Fig. 27–18, what is the magnitude of
a. Each phase voltage, V_θ?
b. The line voltage, V_L

27–5 What type of connection is shown for the generator coil windings in Fig. 27–19?

27–6 In Fig. 27–19, what is the magnitude of
a. Each phase voltage, V_θ?
b. The line voltage, V_L?

27–7 In Fig. 27–19, what is the phase angle between each
a. Phase voltage?
b. Phase current?

27–8 In Fig. 27–19, what is the mathematical relationship between the line current, I_L, and the phase current, I_θ? How much is the line current, I_L?

Figure 27–19

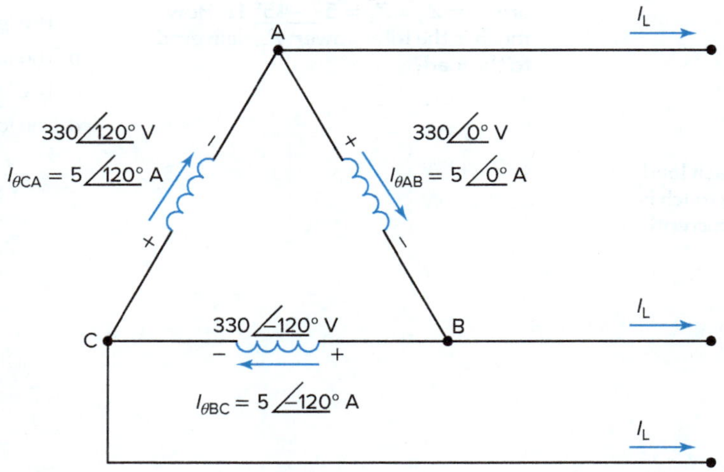

SECTION 27–2 THE WYE (Y)-CONNECTED THREE-PHASE GENERATOR

27–9 If the phasor $V_{AN} = 60\angle{-30^\circ}$ V, the phasor $V_{NA} = ?$

27–10 If the phasor $V_{BC} = 200\angle{0^\circ}$ V, the phasor $V_{CB} = ?$

27–11 If the phasor $V_{XZ} = 25\angle{60^\circ}$ V, the phasor $V_{ZX} = ?$

27–12 If the phasor $V_{CN} = -40$ V + j30 V, then the phasor $V_{NC} = ?$

27–13 Calculate the line voltage, V_L, in a Y-connected three-phase AC generator if the phase voltage, V_θ, equals

a. 45 V.

b. 100 V.

c. 2 kV.

27–14 Calculate the phase voltage, V_θ in a Y-connected three-phase AC generator, if the line voltage, V_L, equals:

a. 500 V.

b. 75 V.

c. 3.3 V.

SECTION 27–3 THE DELTA (Δ)-CONNECTED THREE-PHASE GENERATOR

27–15 Calculate the line current, I_L, in a Δ-connected three-phase AC generator if the phase current, I_θ, equals:

a. 3A.

b. 29 A.

c. 7.5 A.

27–16 Calculate the phase current, I_θ in a Δ-connected three-phase AC generator if the line current, I_L, equals:

a. 18 A.

b. 29 A.

c. 2.7 A.

SECTION 27–4 THREE-PHASE SOURCE/LOAD CONFIGURATIONS

27–17 In Fig. 27–20, solve for the following:

a. The phase voltages, V_{AN}, V_{BN}, and V_{CN}.

b. The line voltages, V_{AB}, V_{BC}, and V_{CA}.

c. The load currents, I_{Z1}, I_{Z2}, and I_{Z3}.

d. The line currents, I_{Aa}, I_{Bb}, and I_{Cc}.

e. The phase currents, $V_{\theta A}$, $V_{\theta B}$, and $V_{\theta C}$.

f. The neutral current, I_N.

Figure 27–20

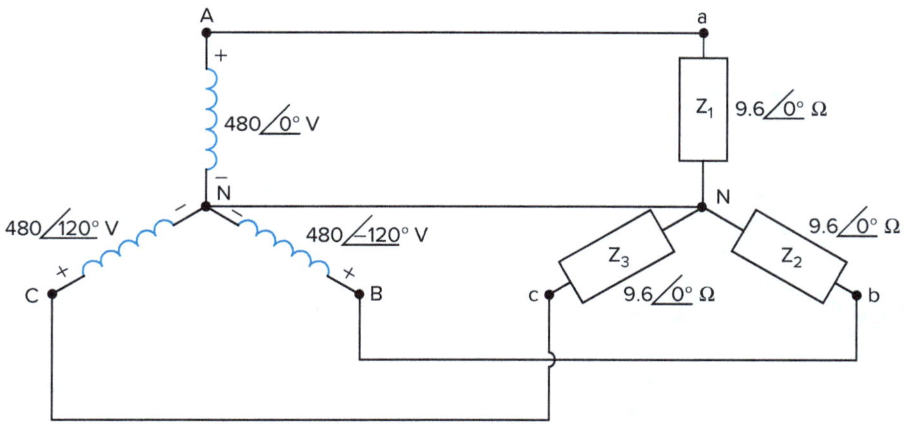

27–18 In Fig. 27–21, solve for the following:

a. The phase voltages, V_{AN}, V_{BN}, and V_{CN}.

b. The line voltages, V_{AB}, V_{BC}, and V_{CA}.

c. The load currents, I_{Z1}, I_{Z2}, and I_{Z3}. (Note the assumed direction of each load current.)

d. The line currents, I_{Aa}, I_{Bb}, and I_{Cc}.

e. The phase currents, $I_{\theta A}$, $I_{\theta B}$, and $I_{\theta C}$.

Figure 27–21

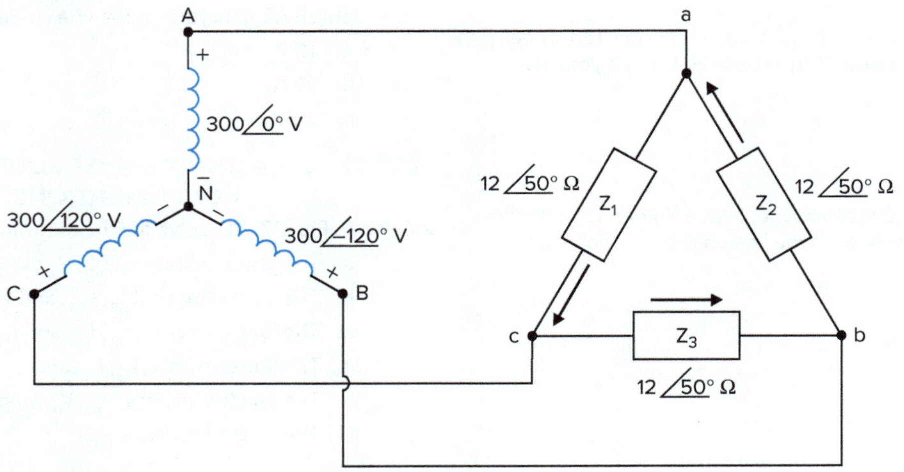

27–19 In Fig. 27–22, solve for the following:

a. The load voltages, V_{Z1}, V_{Z2}, and V_{Z3}.

b. The load currents, I_{Z1}, I_{Z2}, and I_{Z3}. (Note the assumed direction of each load current.)

c. The lin currents, I_{Aa}, I_{Bb}, and I_{Cc}.

Figure 27–22

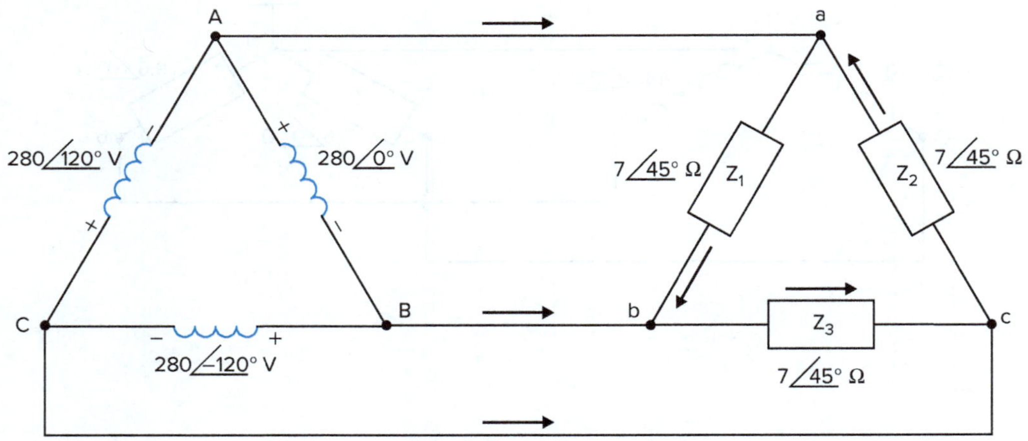

27–20 In Fig. 27–23, solve for the following:

a. The load voltages, V_{Z1}, V_{Z2}, and V_{Z3}.

b. The load currents, I_{Z1}, I_{Z2}, and I_{Z3}.

c. The line currents, I_{Aa}, I_{Bb}, and I_{Cc}.

Figure 27–23

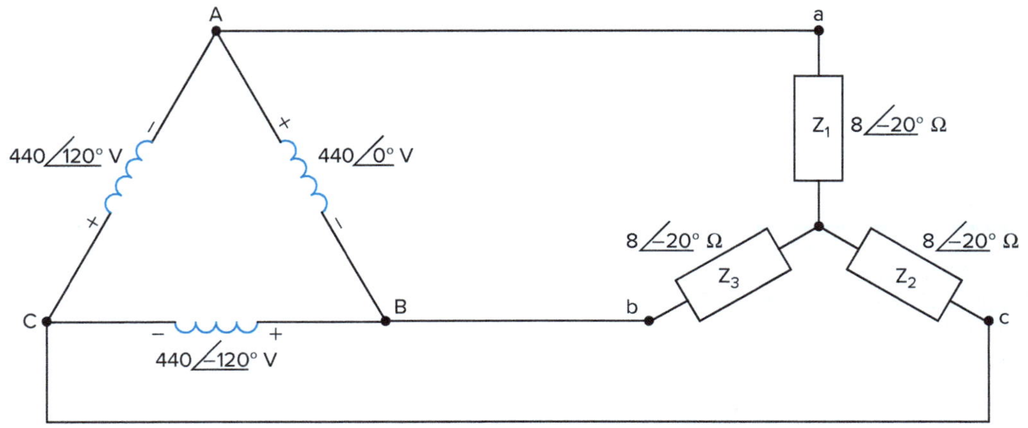

27–21 In Fig. 27–24, a Y-connected load is connected to a Y-connected generator whose line voltage, V_L, equals 480 V. Determine the magnitude of

a. Each load voltage, V_Z.

b. Each load current, I_Z.

c. Each line current, I_L.

d. The neutral current, I_N.

Figure 27–24

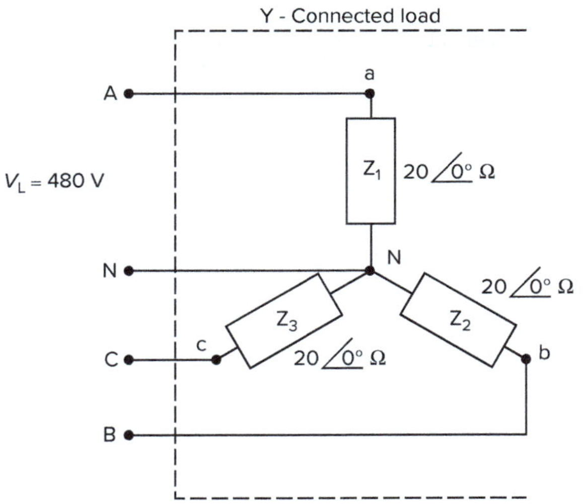

27–22 In Fig. 27–25, determine the magnitude of

a. Each load voltage, V_Z.

b. Each load current, I_Z.

c. Each line current, I_L.

Figure 27–25

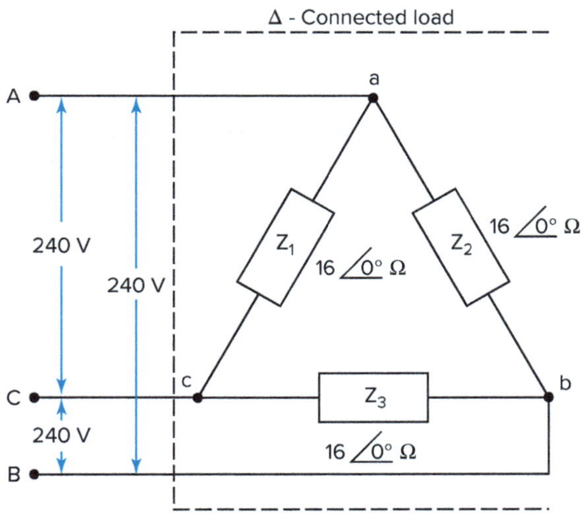

SECTION 27–5 THREE-PHASE POWER CALCULATIONS

27–23 Calculate the power dissipated in each individual load in the following figures:

a. 27–20.

b. 27–21.

c. 27–22.

d. 27–23.

e. 27–24.

f. 27–25.

27–24 Calculate the total power, P_T, dissipated by the balanced loads in the following figures:

a. 27–20.

b. 27–21.

c. 27–22.

d. 27–23.

e. 27–24.

f. 27–25.

Critical Thinking

27–25 In Fig. 27–20, assume an unbalanced load where $Z_1 = 10\underline{/0°}\ \Omega$, $Z_2 = 24\underline{/-60°}\ \Omega$, and $Z_3 = 15\underline{/60°}\ \Omega$. Solve for the following:

a. The load currents, I_{Z1}, I_{Z2}, and I_{Z3}.

b. The line currents, I_{Aa}, I_{Bb}, and I_{Cc}.

c. I_N.

d. P_T.

Answers to Self-Reviews

27–1 **a.** 120°
b. 120°

27–2 **a.** $240\underline{/210°}$ V
b. $V_L = \sqrt{3}V_\theta$

27–3 **a.** 120 V
b. $I_L = \sqrt{3}I_\theta$

27–4 **a.** Zero
b. Line voltage, V_L.
c. $I_L = \sqrt{3}I_Z$
d. $V_Z = \dfrac{V_L}{\sqrt{3}}$

27–5 **a.** $P_T = 8.64$ kW
b. $P_T = 2.056$ kW

Laboratory Application Assignment

Due to the high voltages and dangers associated with three-phase AC power, as well as the inconsistency of lab equipment from one school to the next, no Laboratory Application Assignment will be provided for this chapter.

Design credits: Chapter Opener Background Image, and Multisim: Stockbyte/Getty Images

chapter

28

Diodes and Diode Applications

A semiconductor is a material that is neither a good conductor nor a good insulator. In their purest form, semiconductors have few applications in electronics. However, when the characteristics of a pure semiconductor are altered through a process known as doping, many useful electronic devices can be developed. The most basic semiconductor device is the diode, which is a device that allows current to pass through it in only one direction. This characteristic of a diode has many useful applications in electronics. One of the most useful applications is converting an AC voltage into a DC voltage. When used for this purpose, diodes are typically referred to as rectifier diodes.

In this chapter, you will learn about the basic construction and operation of a semiconductor diode. You will learn how a diode can be turned on or off by applying the proper polarity of voltage across the diode terminals. You will also be introduced to half-wave and full-wave rectifiers which use diodes to convert an AC voltage into a DC voltage. And finally, you will be introduced to two special-purpose diodes, the light-emitting diode (LED) and the zener diode. ■

Chapter Outline

Chapter Objectives

After studying this chapter, you should be able to

- *Explain* the process of doping a semiconductor to produce both *n*- and *p*-type material.
- *Describe* the basic construction of a diode.
- *Draw* the schematic symbol of a diode and *identify* the anode and cathode terminals.
- *Describe* how to forward- and reverse-bias a diode.
- *Test* a diode with an analog VOM or DMM.
- *Explain* the operation of a half-wave and full-wave rectifier.
- *Calculate* the output voltage of half-wave and full-wave rectifiers.
- *Explain* the effect of a capacitor filter on the operation of half-wave and full-wave rectifiers.
- *List* the characteristics of a light-emitting diode (LED).
- *List* the forward- and reverse-bias characteristics of a zener diode.
- *Calculate* the voltage and current values in a loaded zener voltage regulator.

Important Terms

avalanche
barrier potential, V_B
bias
breakdown voltage, V_{BR}
bulk resistance, r_B
covalent bonding
depletion zone
diode
doping
electron-hole pair
extrinsic semiconductor
forward bias
full-wave rectifier
half-wave rectifier
hole
intrinsic semiconductor
leakage current
light-emitting diode (LED)
majority current carrier
minority current carrier
n-type semiconductor
p-type semiconductor
peak inverse voltage (PIV)
pentavalent atom
reverse bias
trivalent atom
valence electrons
zener current, I_z
zener diode

GOOD TO KNOW

The semiconductor element carbon (C) is mainly used in the production of resistors.

28–1 Semiconductor Materials

Semiconductors conduct less than metal conductors but more than insulators. Some common semiconductor materials are silicon (Si), germanium (Ge), and carbon (C). Silicon is the most widely used semiconductor material in the electronics industry. Almost all **diodes**, transistors, and ICs manufactured today are made from silicon.

Intrinsic semiconductors are semiconductors in their purest form. An example would be a semiconductor crystal with only silicon atoms. ***Extrinsic semiconductors*** are semiconductors with other atoms mixed in. These other atoms are called *impurity* atoms. The process of adding impurity atoms is called ***doping***. Doping alters the characteristics of the semiconductor, mainly its conductivity. The impurity atoms have either fewer than four **valence electrons** or more than four valence electrons. At room temperature (about 25°C), an intrinsic semiconductor acts more like an insulator than a conductor. The conductivity of an extrinsic semiconductor is greater than that of an intrinsic semiconductor. The level of conductivity is dependent mainly on the number of impurity atoms that have been added during the doping process.

Atomic Structure

Figure 28–1 Atomic structure of a silicon atom. (*a*) Atomic structure of a silicon atom showing the nucleus and its orbital rings of electrons. (*b*) Simplified drawing of a silicon atom. The core includes the nucleus and inner electrons.

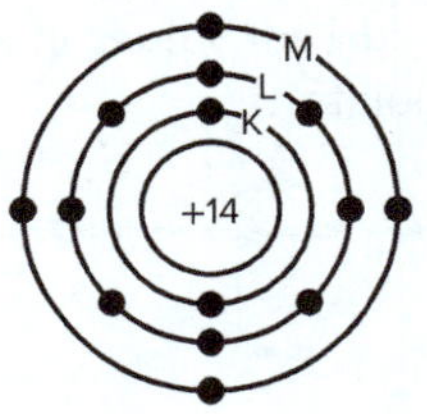

(*a*)

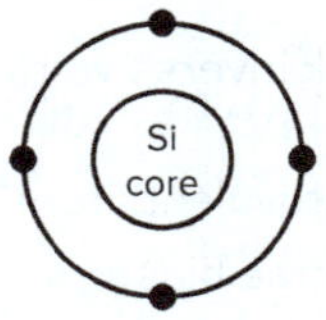

(*b*)

Figure 28–1 shows the atomic structure of a silicon atom. The atomic number of silicon is 14, meaning that there are 14 protons in its nucleus, balanced by 14 orbiting electrons. Notice in Fig. 28–1*a* that the first shell (K-shell) surrounding the nucleus has two electrons, the second shell (L-shell) has eight electrons, and the third shell (M-shell) has four electrons. The outermost ring of an atom is called the *valence ring*, and the electrons in this ring are called *valence electrons.* All semiconductors have four valence electrons. The number of valence electrons possessed by any atom determines its electrical conductivity. The number of valence electrons in an atom also determines how it will combine with other atoms. The best conductors have only one valence electron, whereas the best insulators have complete shells.

The simplified drawing of a silicon atom is shown in Fig. 28–1*b*. The core represents the nucleus and inner electrons. The outer four electrons represent the valence electrons of the silicon atom. As you will see, the nucleus and inner electrons are not that important when analyzing how atoms combine with each other; hence the reason for the simplified drawing. One more point: The Si core has a net charge of + 4 because it contains 14 protons and 10 inner electrons.

Forming a Crystal

When silicon atoms are grouped together, something very interesting happens. Each silicon atom shares its four valence electrons with other nearby atoms, thereby forming a solid crystalline structure. Each atom of the six inner silicon atoms in Fig. 28–2 has eight valence electrons as a result of the electron sharing, which is the amount required for maximum electrical stability. Notice in Fig. 28–2 that only the core and valence electrons are shown for each atom.

This sharing of valence electrons is called ***covalent bonding***. The covalent bonds between each silicon atom produce the solid crystalline structure.

GOOD TO KNOW

A hole and an electron each possess a charge of 0.16×10^{-18} C but of opposite polarity.

Thermally Generated Electron-Hole Pairs

All valence electrons of a silicon crystal at absolute zero (−273°C) remain locked in their respective covalent bonds. This means that no free electrons will be floating around in the silicon material. Above absolute zero, however, some valence electrons may gain enough energy from heat, radiation, or other sources to escape from their parent atoms. When an electron leaves its covalent bond, it becomes a free

Figure 28–2 Bonding diagram of a silicon crystal.

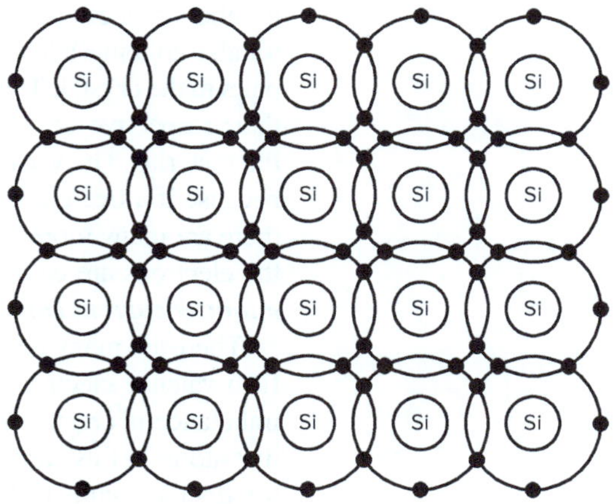

Figure 28–3 Thermal energy produces a free electron and a hole. This is often called an electron-hole pair.

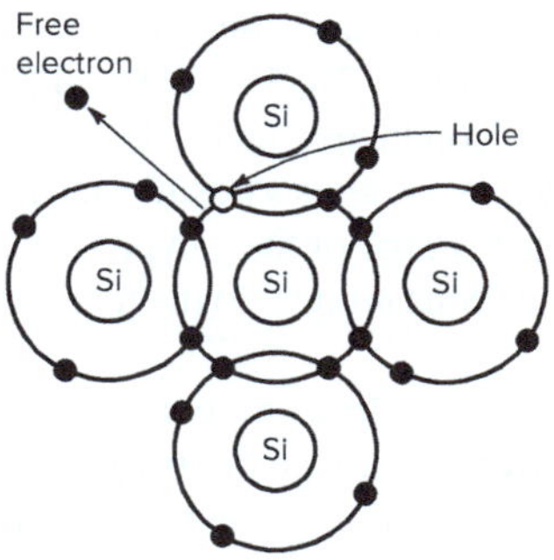

electron that can move freely in the material. This free electron also produces a vacancy or **hole** in the covalent bond structure that it left. Thermal energy is the main cause for the creation of an **electron-hole pair**, as shown in Fig. 28–3.

As the temperature increases, more thermally generated electron-hole pairs are created. In Fig. 28–3, the hole acts like a positive charge because it attracts a free electron passing through the crystal.

Note that a silicon semiconductor material has fewer thermally generated electron-hole pairs than a germanium crystal at the same temperature. This implies that a silicon crystal is more stable than a germanium crystal at higher temperatures. Its stability is the primary reason that silicon is the number one semiconductor material used in manufacturing diodes, transistors, and integrated circuits.

It is important to note that intrinsic semiconductor materials have only a few thermally generated electron-hole pairs at room temperature and therefore are still relatively good insulators.

Doping

As mentioned earlier, doping is a process that involves adding impurity atoms to an intrinsic semiconductor. Intrinsic semiconductors are of limited use in the field of electronics. Intrinsic semiconductor materials such as silicon or germanium are almost always doped with impurity atoms to increase their conductivity. An extrinsic semiconductor material, then, is one that has been doped with impurity atoms.

Figure 28–4 Doping a silicon crystal with a pentavalent impurity. Arsenic (As) is shown in this illustration, but other pentavalent impurities such as antimony (Sb) or phosphorous (P) could also be used.

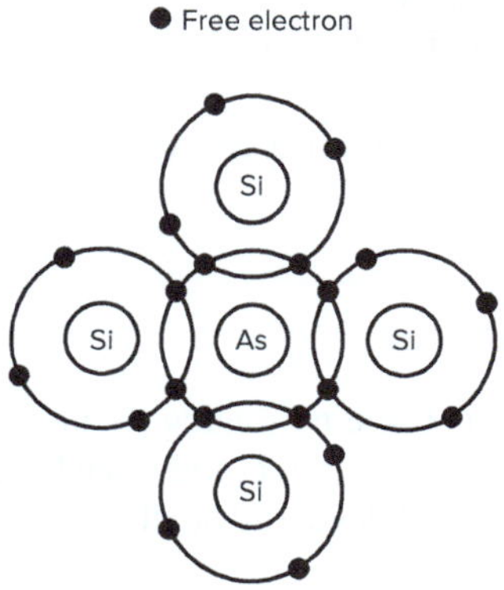

n-Type Semiconductors

A ***pentavalent atom*** is one that has five valence electrons. Some examples are antimony (Sb), arsenic (As), and phosphorous (P). A silicon crystal doped with a large number of pentavalent impurity atoms results in many free electrons in the material. This occurs because there is one electron at the location of each pentavalent atom that is not used in the covalent bond structure. Remember, only eight electrons can exist in the outermost ring of any atom. Therefore, one of the valence electrons in the pentavalent impurity atom is not needed in the covalent bond structure and can float through the material as a free electron. This is illustrated in Fig. 28–4. The free electron shown belongs to the arsenic atom, but since the covalent bond is already complete with eight valence electrons, the electron is extra, or not needed. When millions of pentavalent impurity atoms are added to an intrinsic silicon crystal, there are millions of free electrons that can float through the material. Since the electron is the basic particle of negative charge, we call this an ***n*-type semiconductor** material. The net charge of the *n*-type material is still neutral, however, since the total number of electrons is equal to the total number of protons. An *n*-type semiconductor also contains a few holes because thermal energy still creates a few electron-hole pairs in the crystal. The few valence electrons that do absorb enough energy to leave their respective covalent bonds increase further the number of free electrons in the material. The vacancies, or holes, created in the crystal act like positive charges because if a free electron passes by, it will be attracted to the hole and fill it. Since there are many more free electrons than holes in an *n*-type semiconductor material, the electrons are called the ***majority current carriers*** and the holes are called the ***minority current carriers***.

There are many positive ions in *n*-type semiconductor material because when the fifth valence electron of the pentavalent atom leaves its home or parent atom, an imbalance is created in the number of positive and negative charges that exist for that atom. In this case, the nucleus of the impurity atom will contain one more proton than its number of orbiting electrons. The positive ions are fixed charges in the crystal that are unable to move.

Figure 28–5 Doping a silicon crystal with a trivalent impurity. Aluminum (Al) is shown in this illustration, but other trivalent impurities such as boron (B) or gallium (Ga) could also be used.

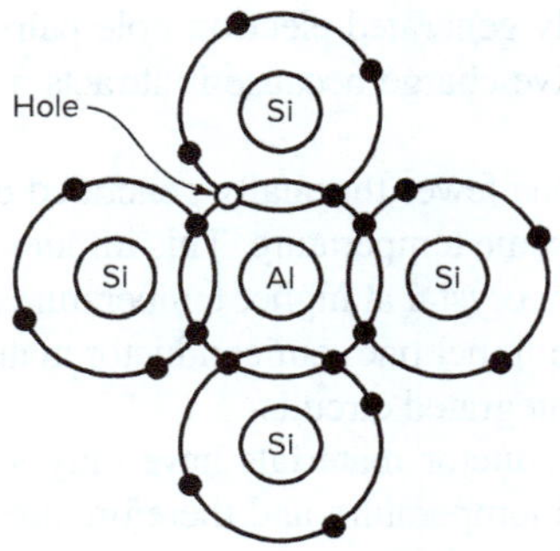

p-Type Semiconductors

A ***trivalent* atom** is one that has only three valence electrons. Some examples are aluminum (Al), boron (B), and gallium (Ga). A silicon crystal doped with a large number of trivalent impurity atoms results in many holes, or vacancies, in the covalent bond structure of the material. This happens because one more valence electron is needed at the location of each trivalent atom in the crystal to obtain the maximum electrical stability with eight electrons, as shown in Fig. 28–5. When millions of trivalent impurities are added to an intrinsic semiconductor material, millions of holes are created throughout the material. Since a hole exhibits a positive charge, we call this a ***p*-type semiconductor** material. The net charge of the *p*-type material is still neutral, however, since the total number of electrons is equal to the total number of protons.

A *p*-type semiconductor also contains a few free electrons because thermal energy still produces a few electron-hole pairs. Electrons are the minority carriers in *p*-type semiconductor material, whereas holes are the majority current carriers.

A *p*-type semiconductor material contains many negative ions because free electrons passing by may fill the holes in the covalent bond structure created by the trivalent impurity atoms. Thus, the trivalent impurity atom will have one more orbiting electron than it has protons in its nucleus, thereby creating a negative ion. The negative ions are fixed charges and are unable to move in the crystal.

Figure 28–6 *p*-*n* junction. (*a*) Basic construction of a diode showing the separate *p* and *n* regions. (*b*) Schematic symbol for a semiconductor diode showing the anode (A) and cathode (K) terminals.

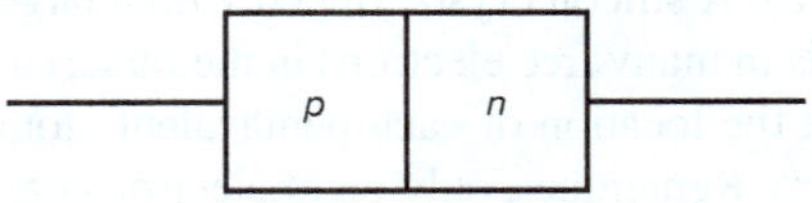

(*a*)

(*b*)

28–1 Self-Review

Answers at the end of the chapter.

a. What type of semiconductor material is created when a silicon crystal is doped with pentavalent impurity atoms?

b. What are the minority current carriers in a *p*-type semiconductor material?

c. Does a hole exhibit a positive, negative, or neutral charge?

28–2 The *p*-*n* Junction Diode

A popular semiconductor device called a *diode* is made by joining *p*- and *n*-type semiconductor materials, as shown in Fig. 28–6*a*. Notice that the doped regions meet to form a *p*-*n junction*. Diodes are unidirectional devices that allow current to flow through them in only one direction.

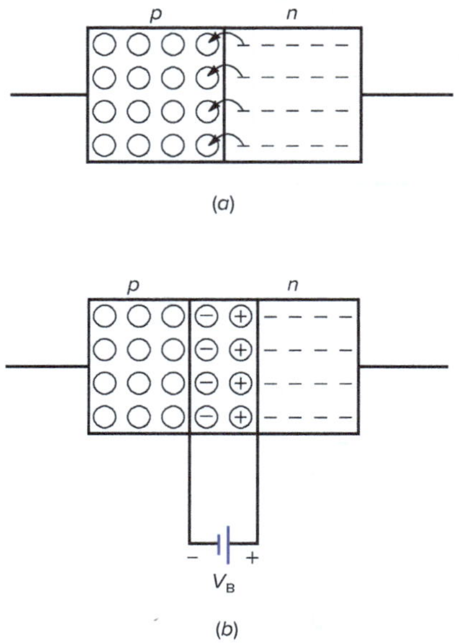

Figure 28–7 *p-n* junction. (*a*) *p-n* junction showing the electrons (–) in the *n* side and holes (O) in the *p* side. (*b*) Formation of depletion zone with positive ions at the left edge of the *n* material and negative ions at the right edge of the *p* material.

The schematic symbol for a semiconductor diode is shown in Fig. 28–6*b*. The *p* side of the diode is called the *anode* (A), whereas the *n* side of the diode is called the *cathode* (K).

Depletion Zone

Figure 28–7*a* shows a *p-n* junction with free electrons on the *n* side and holes on the *p* side. Notice that the free electrons are represented as dash (–) marks and the holes are represented as small circles (O).

At the instant the *p-n* junction is formed, free electrons on the *n* side migrate or diffuse across the junction to the *p* side. Once on the *p* side, the free electrons are minority current carriers. The lifetime of these free electrons is short, however, because they fall into holes shortly after crossing over to the *p* side. The important effect here is that when a free electron leaves the *n* side and falls into a hole on the *p* side, two ions are created: a positive ion on the *n* side and a negative ion on the *p* side (see Fig. 28–7*b*). As the process of diffusion continues, a **barrier potential**, V_B, is created and the diffusion of electrons from the *n* side to the *p* side stops. Electrons diffusing from the *n* side sense a large negative potential on the *p* side that repels them back to the *n* side. Likewise, holes from the *p* side are repelled back to the *p* side by the positive potential on the *n* side. The area where the positive and negative ions are located is called the ***depletion zone***. Other names commonly used are *depletion region* and *depletion layer.* The word *depletion* is used because the area has been depleted of all charge carriers. The positive and negative ions in the depletion zone are fixed in the crystalline structure and are therefore unable to move.

Barrier Potential, V_B

Ions create a potential difference at the *p-n* junction, as shown in Fig. 28–7*b*. This potential difference is called the *barrier potential* and is usually designated V_B. For silicon, the barrier potential at the *p-n* junction is approximately 0.7 V. For germanium, V_B is about 0.3 V. The barrier potential cannot be measured externally with a voltmeter, but it does exist at the *p-n* junction. The barrier potential stops the diffusion of current carriers.

Forward-Biased *p-n* Junction

The term ***bias*** is defined as a control voltage or current. Forward-biasing a diode allows current to flow easily through the diode. Figure 28–8*a* illustrates a *p-n* junction that is forward-biased.

In Fig. 28–8*a*, notice that the *n* material is connected to the negative terminal of the voltage source, *V*, and the *p* material is connected to the positive terminal of the voltage source, *V.* The voltage source, *V*, must be large enough to overcome the internal barrier potential V_B. The voltage source repels free electrons in the *n* side across the depletion zone and into the *p* side. Once on the *p* side, the free electron falls into a hole. The electron will then travel from hole to hole as it is attracted to the positive terminal of the voltage source, *V.* For every free electron entering the *n* side, one electron leaves the *p* side. Notice in Fig. 28–8*a* that if the *p-n* junction is made from silicon, the external voltage source must be 0.7 V or more to neutralize the effect of the internal barrier potential, V_B, and in turn produce current flow. (It should be noted that in a practical circuit, a resistance would be added in series with the diode to limit the current flow.)

Figure 28–8*b* shows the schematic symbol of a diode with the voltage source, *V*, connected to provide **forward bias**. Notice that forward bias exists when the anode, A is positive with respect to the cathode, K. Notice that electrons flow to the *n* side, against the arrow on the diode symbol. The arrow on the diode symbol points in the direction of conventional current flow. Either current direction works well when

GOOD TO KNOW

A colored band near one end of a diode is used to indicate the cathode lead.

Figure 28-8 Forward-biased *p-n* junction. (*a*) External voltage forces free electrons from the *n* side across the depletion zone to the *p* side where the electrons fall into a hole. Once on the *p* side, the electrons jump from hole to hole in the valence band. (*b*) Electron flow is against the arrow, whereas conventional current is in the same direction as the arrow.

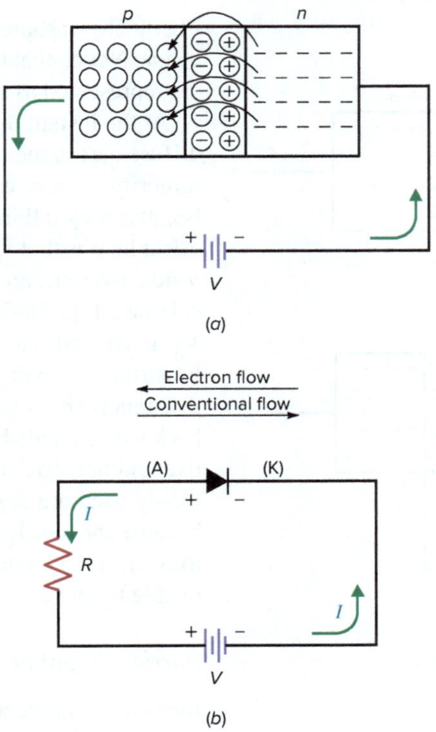

analyzing diode circuits. In this book, however, we will use electron flow when analyzing circuits containing diodes.

Reverse-Biased *p-n* Junctions

Figure 28–9*a* shows how to **reverse-bias** a *p-n* junction. Notice that the negative terminal of the voltage source, *V*, is connected to the *p*-type semiconductor material and that the positive terminal of the voltage source, *V*, is connected to the *n*-type semiconductor material. The effect is that charge carriers in both sections are pulled away from the junction. This increases the width of the depletion zone, as shown. Free electrons on the *n* side are attracted away from the junction because of the attraction of the positive terminal of the voltage source, *V*. Likewise, holes in the *p* side are attracted away from the junction because of the attraction by the negative terminal of the voltage source, *V*.

Figure 28–9*b* shows the schematic symbol of a diode with the voltage source, *V*, connected to provide reverse bias. The result of reverse bias is that the diode is in a nonconducting state and acts like an open switch, ideally with infinite resistance.

Leakage Current

Even a reverse-biased diode conducts a small amount of current, called ***leakage current***. The leakage current is mainly due to the minority current carriers in both sections of the diode. The minority current carriers are holes in the *n* side and free electrons in the *p* side. The minority current carriers exist as a result of thermal energy producing a few electron-hole pairs. Since temperature determines the number of electron-hole pairs generated, leakage current is mainly affected by

Figure 28–9 Reverse-biased *p-n* junction. (*a*) External voltage pulls majority current carriers away from the *p-n* junction. This widens the depletion zone. (*b*) Schematic symbol showing how a diode is reverse-biased with the external voltage, *V*.

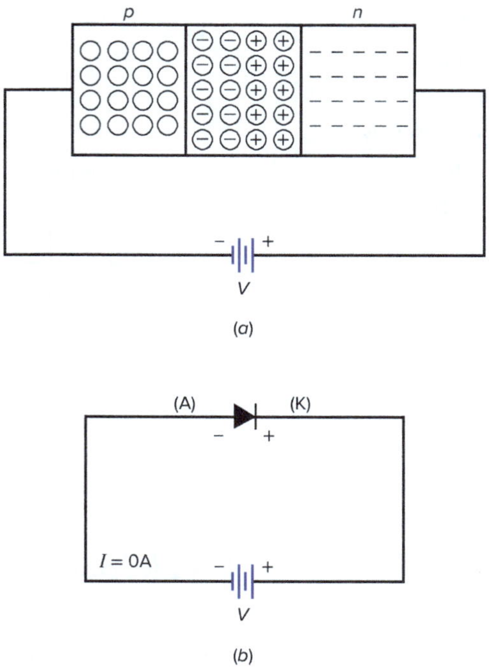

temperature. Any increase in the temperature of the diode increases the leakage current in the diode. These minority current carriers move in a direction that is opposite to the direction provided with forward bias.

28–2 Self-Review

Answers at the end of the chapter.

a. **What is the barrier potential, V_B, for a silicon diode?**
b. **The *p* side of a diode is called the (anode/cathode) and the *n* side is called the (anode/cathode).**
c. **To forward-bias a diode, the anode must be (positive/negative) with respect to its cathode.**
d. **A reverse-biased diode acts like a(n) (open/closed) switch.**

28–3 Volt-Ampere Characteristic Curve

Figure 28–10 is a graph of diode current versus diode voltage for a silicon diode. The graph includes the diode current for both forward- and reverse-bias voltages. The upper right quadrant of the graph represents the forward-bias condition. Notice that very little diode current flows when the forward voltage, V_F, is less than about 0.6 V. Beyond 0.6 V of forward bias, however, the diode current increases sharply. Notice that the forward voltage drop, V_F, remains relatively constant as I_F increases. A voltage of 0.7 V is the approximate value assumed for the barrier potential of a silicon *p-n* junction. The barrier potential of germanium diodes is approximately 0.3 V. Therefore, if the graph in Fig. 28–10 were for a germanium diode, the current would increase sharply for a forward voltage of about 0.3 V.

GOOD TO KNOW

For any diode, the forward voltage, V_F, decreases as the temperature of the diode increases. As a rough approximation, V_F decreases by about 2 mV for each degree Celsius rise in temperature.

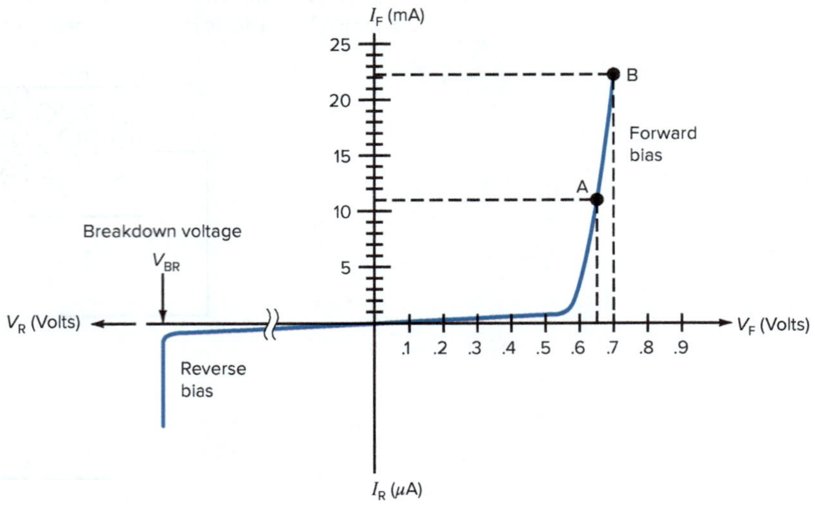

Figure 28–10 Volt-ampere characteristic curve of a silicon diode.

Breakdown Voltage, V_{BR}

The lower left quadrant of the graph in Fig. 28–10 represents the reverse-bias condition. Notice that only a very small current flows until the **breakdown voltage, V_{BR},** is reached. The current that flows prior to breakdown is mainly the result of thermally produced minority current carriers. As mentioned earlier, this current is called leakage current and is usually designated I_R. Leakage current increases mainly with temperature and is relatively independent of changes in reverse-bias voltage. The slight increase in reverse current, I_R, with increases in the reverse voltage, V_R, is a result of surface leakage current. Surface leakage current exists because there are many holes on the edges of a silicon crystal due to unfilled covalent bonds. These holes provide a path for a few electrons along the surfaces of the crystal.

Avalanche occurs when the reverse-bias voltage, V_R, becomes excessive. Thermally produced free electrons on the *p* side are accelerated by the voltage source to very high speeds as they move through the diode. These electrons collide with valence electrons in other orbits. These valence electrons are also set free and accelerated to very high speeds, thereby dislodging even more valence electrons. The process is cumulative; hence, we have an avalanche effect.

When the breakdown voltage, V_{BR}, is reached, the reverse current, I_R, increases sharply. Diodes should not be operated in the breakdown region. Most rectifier diodes have breakdown voltages exceeding 50 V.

GOOD TO KNOW

Exceeding the breakdown voltage of a diode does not necessarily mean that you will destroy the diode. As long as the product of reverse voltage and reverse current does not exceed the diode's power rating, the diode will recover fully.

DC Resistance of a Diode

Examine the forward-bias region of the graph shown in Fig. 28–10.

The graph of V_F versus I_F shows that a diode is a nonlinear device because the diode current, I_F, does not increase in direct proportion to the diode voltage, V_F. For example, the diode voltage does not have to be doubled to double the diode current. The DC resistance of a forward-biased diode can be calculated using Formula (28–1).

$$R_F = \frac{V_F}{I_F} \qquad \textbf{(28–1)}$$

where V_F is the forward voltage drop and I_F is the forward current.

Example 28–1

For the diode curve in Fig. 28–10, calculate the DC resistance, R_F, at points A and B.

ANSWER Using Formula (28–1), the calculations are

$$\text{Point A}: R_F = \frac{V_F}{I_F} = \frac{0.65\ \text{V}}{11\ \text{mA}} = 59.1\ \Omega$$

$$\text{Point B}: R_F = \frac{V_F}{I_F} = \frac{0.7\ \text{V}}{22.5\ \text{mA}} = 31.1\ \Omega$$

Notice that as the diode conducts more heavily, the forward resistance, R_F, decreases.

Using an Ohmmeter to Check a Diode

The condition of a semiconductor diode can be determined with an ohmmeter. When using an analog meter, check the resistance of the diode in one direction; then reverse the meter leads and measure the resistance of the diode in the other direction. If the diode is good, it should measure a very high resistance in one direction and a low resistance in the other direction. For a silicon diode, the ratio of reverse resistance, R_R, to forward resistance, R_F, should be very large, such as 1000:1 or more.

If the diode is shorted, it will measure a low resistance in both directions. If the diode is open, it will measure a high resistance in both directions.

A word of caution. When using analog ohmmeters to check a diode, do not use the $R \times 1$ range because the current forced through the diode by the meter may exceed the current rating of the diode. The $R \times 100$ range is usually the best range on which to check a diode.

GOOD TO KNOW

On the diode range, a DMM will supply a constant current to whatever device is connected to its leads. Most DMMs supply 1 mA of current on the diode range. Try connecting a 470-Ω, 1-kΩ, or 1.5-kΩ resistor across the leads of a DMM set to the diode range, and you will see that the meter will indicate the resistor voltage drop. For a 1-kΩ *R*, the voltage will read about 1 V. For a diode, the forward voltage will measure somewhere between 0.6 and 0.7 V.

Using a DMM to Check a Diode

Most digital multimeters (DMMs) cannot be used to measure the forward or reverse resistance of a diode junction. This is because the ohmmeter ranges in most digital multimeters do not provide the proper forward bias to turn on the diode being tested. Therefore, the resistance ranges on a DMM are often referred to as low power ohm (LPΩ) ranges.

Most digital multimeters provide a special range for testing diodes. This range is called the diode (→⊢) range. This is the only range setting on the DMM that can provide the proper amount of forward bias for the diode being tested. It is important to note that when the digital multimeter forward-biases the diode being tested, the digital display will indicate the forward voltage dropped across the diode rather than the forward resistance, R_F. A good silicon diode tested with a DMM should show a voltage between 0.6 and 0.7 V for one connection of the meter leads and an over-range condition for the opposite connection of the leads. An open diode will show an overrange condition for both connections of the meter leads, whereas a shorted diode will show a very low or zero reading for both connections of the meter leads.

28–3 Self-Review

Answers at the end of the chapter.

a. The leakage current in a reverse-biased diode is mainly due to the (majority/minority) current carriers in both sections of the diode.

b. What is the DC resistance of a diode if the forward current, I_F, is 30 mA when the forward voltage, V_F, is 0.66 V?

c. When checking a diode with an analog ohmmeter, is the diode good or bad if the ohmmeter reads a high resistance for both connections of the meter leads?

28–4 Diode Approximations

Three different diode approximations can be used when analyzing diode circuits. The one used depends on the desired accuracy of your circuit calculations.

First Approximation

GOOD TO KNOW

The first approximation is sometimes referred to as the ideal diode model.

The first approximation treats a forward-biased diode like a closed switch with a voltage drop of zero volts, as shown in Fig. 28–11*a*. Likewise, the first approximation treats a reverse-biased diode like an open switch with zero current, as shown in Fig. 28–11*b*. The graph in Fig. 28–11*c* indicates the ideal forward- and reverse-bias characteristics.

The first approximation of a diode is often used if only a rough idea is needed of what the circuit voltages and currents should be.

The first approximation is sometimes called the *ideal diode approximation.*

Second Approximation

GOOD TO KNOW

The second approximation is sometimes referred to as the practical diode model.

The second approximation treats a forward-biased diode like an ideal diode in series with a battery, as shown in Fig. 28–12*a*. For silicon diodes, the battery voltage is assumed to be 0.7 V, the same as the barrier potential, V_B, at a silicon *p-n* junction.

The second approximation of a reverse-biased diode is an open switch. See Fig. 28–12*b*.

The graph in Fig. 28–12*c* indicates the forward- and reverse-bias characteristics of the second approximation. Notice that the diode is considered off until the

Figure 28–11 First approximation of a diode. (*a*) Forward-biased diode treated like a closed switch. (*b*) Reverse-biased diode treated like an open switch. (*c*) Graph showing ideal forward and reverse characteristics.

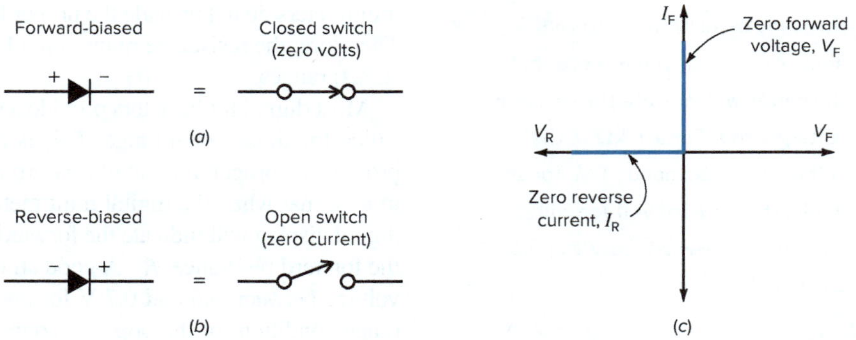

Figure 28–12 Second approximation of a diode. (*a*) Forward-biased diode treated like an ideal diode in series with a battery. (*b*) Reverse-biased diode treated like an open switch. (*c*) Graph showing forward and reverse characteristics.

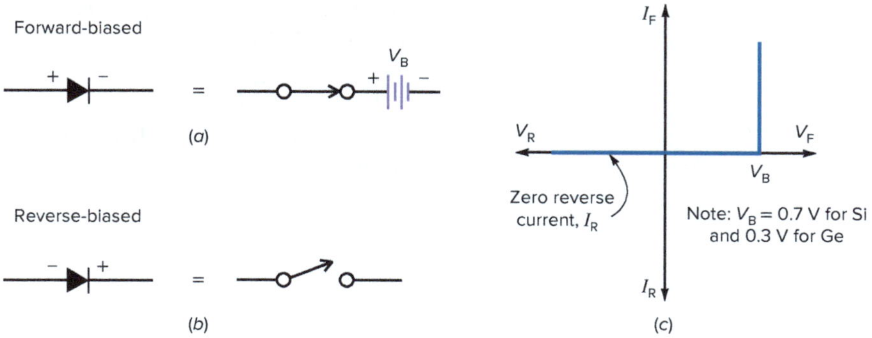

forward voltage, V_F, reaches 0.7 V. Also, the diode is assumed to drop 0.7 V for all currents that pass through it.

The second approximation is used if more accurate answers are needed for circuit calculations.

Third Approximation

GOOD TO KNOW

The third approximation is sometimes referred to as the complete diode model.

The third approximation of a diode includes the bulk resistance, designated r_B. The **bulk resistance, r_B**, is the resistance of the *p* and *n* materials. Its value is dependent on the doping level and the size of the *p* and *n* materials.

The third approximation of a forward-biased diode is shown in Fig. 28–13*a*. The total diode voltage drop using the third approximation is calculated using Formula (28–2).

$$V_F = V_B + I_F r_B \qquad \textbf{(28–2)}$$

The bulk resistance, r_B, causes the forward voltage across a diode to increase slightly with increases in the diode current.

Figure 28–13*b* shows the third approximation of a reverse-biased diode. The resistance across the open switch illustrates the high leakage resistance for the reverse-bias condition. Notice the small leakage current in the graph of Fig. 28–13*c* when the diode is reverse-biased. This is a result of the high resistance that exists when the diode is reverse-biased.

Figure 28–13 Third approximation of a diode. (*a*) Forward-biased diode including the barrier potential, V_B, and the bulk resistance, r_B. (*b*) Reverse-biased diode showing high resistance (not infinite) of the reverse-bias condition. (*c*) Graph showing forward and reverse characteristics.

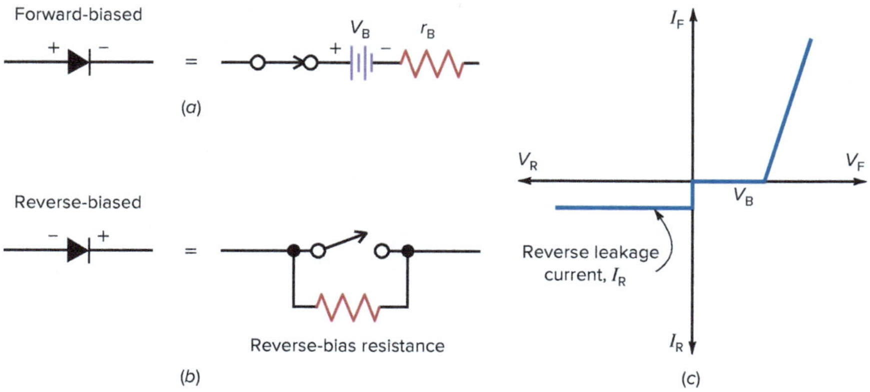

More on Bulk Resistance

The graph in Fig. 28–13*c* shows the forward- and reverse-bias characteristics included with the third approximation. Notice the slope of the diode curve when forward-biased. The value of the bulk resistance, r_B, can be determined by using Formula (28–3).

$$r_B = \frac{\Delta V}{\Delta I} \qquad \textbf{(28–3)}$$

where ΔV represents the change in diode voltage produced by the changes in diode current, ΔI.

Example 28–2

A silicon diode has a forward voltage drop of 1.1 V for a forward diode current, I_F, of 1 A. Calculate the bulk resistance, r_B.

ANSWER First, we can assume that the diode current, I_F, is zero when the forward voltage of the silicon diode is exactly 0.7 V. Then we can use Formula (28–3) as shown:

$$r_B = \frac{\Delta V}{\Delta I}$$

$$= \frac{1.1\ \text{V} - 0.7\ \text{V}}{1\ \text{A} - 0\ \text{A}}$$

$$= 0.4\ \Omega$$

Figure 28–14 Circuits used to illustrate the use of the first, second, and third diode approximations in calculating the circuits' voltage and current values. (*a*) Original circuit. (*b*) First approximation of a diode. (*c*) Second approximation of a diode. (*d*) Third approximation of a diode.

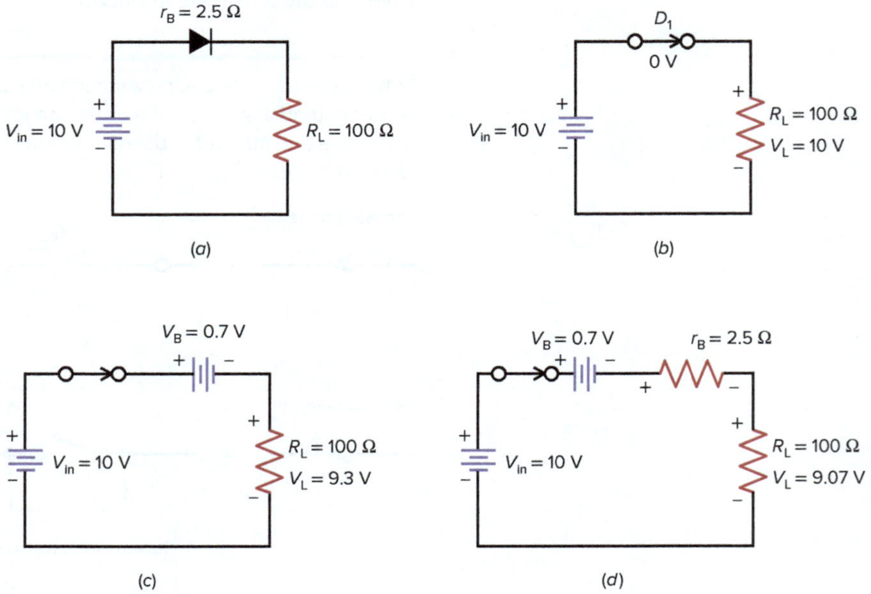

Example 28–3

In Fig. 28–14*a*, solve for the load voltage and current using the first, second, and third diode approximations.

ANSWER First approximation: see Fig. 28–14*b*

$$V_L = V_{in} = 10\ V_{DC}$$

$$I_L = \frac{V_L}{R_L} = \frac{10\ V}{100\ \Omega} = 100\ mA$$

Second approximation: see Fig. 28–14*c*

$$V_L = V_{in} - V_B = 10\ V - 0.7\ V = 9.3\ V$$

$$I_L = \frac{V_L}{R_L} = \frac{9.3\ V}{100\ \Omega} = 93\ mA$$

Third approximation: see Fig. 28–14*d*.

$$I_L = \frac{V_{in} - V_B}{R_L + r_B} = \frac{10\ V - 0.7\ V}{100\ \Omega + 2.5\ \Omega} = 90.73\ mA$$

$$V_L = I_L \times R_L = 90.73\ mA \times 100\ \Omega = 9.07\ V$$

28–4 Self-Review

Answers at the end of the chapter.

a. Which diode approximation includes the bulk resistance, r_B, when forward-biased?

b. Which diode approximation treats a forward-biased diode like a closed switch with a voltage drop of 0 V?

c. When forward-biased, which diode approximation takes into account the barrier potential of the diode but not the bulk resistance?

28–5 Diode Ratings

The following is a list of maximum ratings and electrical characteristics of semiconductor diodes.

GOOD TO KNOW

The part number for a diode will always begin with the prefix 1N, where the number 1 indicates one junction and the letter N indicates that the device is a semiconductor diode.

Breakdown Voltage Rating, V_{BR}

The reverse breakdown voltage rating is extremely important since the diode is usually destroyed if this rating is exceeded. The breakdown voltage, V_{BR}, is the voltage at which avalanche occurs.

This rating can be designated by any of the following: **peak inverse voltage (PIV)**; peak reverse voltage (PRV); breakdown voltage rating (V_{BR}); or peak reverse voltage maximum (V_{RRM}). There are other ways to designate the breakdown voltage rating, however; those most commonly used are listed here. Breakdown voltage ratings are maximum ratings and should never be exceeded.

Average Forward-Current Rating, I_O

This important rating indicates the maximum allowable average current that the diode can handle safely. The average forward-current rating is usually designated as I_O. Exceeding the diode's I_O rating will destroy the diode.

Maximum Forward-Surge Current Rating, I_{FSM}

The maximum forward-surge current (I_{FSM}) rating is the maximum instantaneous current the diode can handle safely from a single pulse. Diodes are often connected to large electrolytic capacitors in power supplies, as shown in the next section. When power is first applied, the initial charge current for the capacitor can be very high. Exceeding the I_{FSM} rating will destroy the diode.

Maximum Reverse Current, I_R

Almost all data sheets list at least one value of reverse current, I_R, for a specified amount of reverse-bias voltage. For example, the data sheet of a 1N4002 silicon diode specifies a typical I_R of 0.05 μA for a diode junction temperature, T_J, of 25°C and a reverse voltage, V_R, of 100 V. With these data, the reverse resistance, R_R, of the diode can be calculated:

$$
\begin{aligned}
R_R &= \frac{V_R}{I_R} \\
&= \frac{100 \text{ V}}{0.05 \ \mu\text{A}} \\
&= 2 \times 10^9 \ \Omega \quad \text{or} \quad 2 \text{ G}\Omega
\end{aligned}
$$

It should be emphasized that the maximum ratings of a diode should never be exceeded under any circumstances. If any maximum ratings are exceeded, there is a good chance the diode will fail and need to be replaced.

28–5 Self-Review

a. Which diode rating, if exceeded, causes the avalanche effect?

b. The maximum instantaneous current that a diode can safely handle from a single pulse is designated I_O. (True/False)

28–6 Rectifier Circuits

Most electronic equipment requires DC voltages to operate properly. Since most equipment is connected to the 120-V_{AC} power line, this AC voltage must somehow be converted to the required DC value. A circuit that converts the AC power-line voltage to the required DC value is called a *power supply*. The

GOOD TO KNOW

The rms value of a half-wave signal can be determined using the following formula: $V_{rms} = 1.57\ V_{avg}$, where $V_{avg} = V_{DC} = 0.318\ V_p$. Another formula that works is $V_{rms} = \frac{V_p}{2}$. For any waveform, the rms value corresponds to the equivalent DC value that will produce the same heating effect.

most important components in power supplies are *rectifier diodes*, which convert AC line voltage to DC voltage. Diodes are able to produce a DC output voltage because they are unidirectional devices allowing current to flow through them in only one direction. For the circuits that follow, assume that all diodes are silicon.

The Half-Wave Rectifier

The circuit shown in Fig. 28–15*a* is called a ***half-wave rectifier***. T_1 is a step-down transformer, which provides the secondary voltage, V_S, as shown in Fig. 28–15*b*. When the top of the transformer secondary voltage is positive, D_1 is forward-biased, producing current flow in the load, R_L. When the top of the secondary is negative, D_1 is reverse-biased and acts like an open switch. This results in zero current in the load, R_L. As a result of this action, the output voltage is a series of positive pulses, as shown in Fig. 28–15*c*.

Figure 28–15 Half-wave rectifier. (*a*) Circuit. (*b*) Secondary voltage, V_S. (*c*) Output waveform, V_{out}.

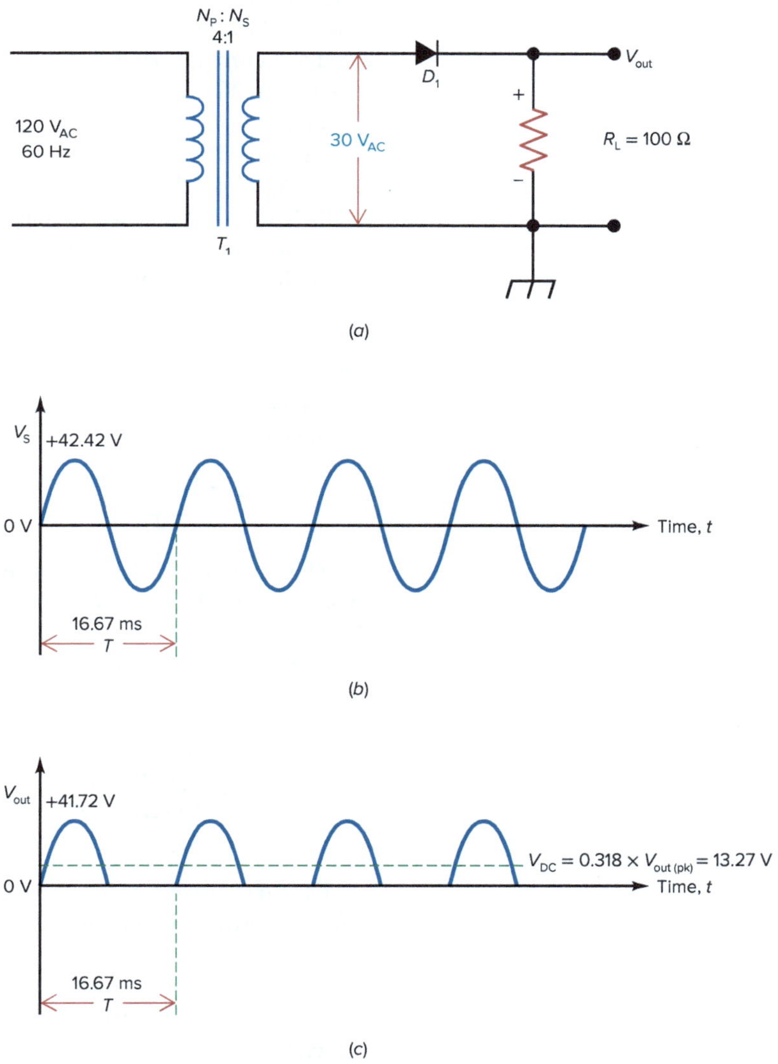

Transformer Calculations

The transformer in Fig. 28–15*a* has a turns ratio, $N_P:N_S$, of 4:1. Therefore, the root-mean-square (rms) secondary voltage is calculated as shown:

$$V_S = \frac{N_S}{N_P} \times V_P$$
$$= \frac{1}{4} \times 120\ V_{AC}$$
$$= 30\ V_{AC}$$

This means that if the secondary voltage is measured with an AC meter, it would read 30 V_{AC}.

To calculate the peak secondary voltage, we proceed as shown:

$$V_{S(pk)} = V_S \times 1.414$$
$$= 30\ V \times 1.414$$
$$= 42.42\ V$$

The peak-to-peak value of the secondary voltage equals $2 \times V_{S(pk)}$ or $2 \times 42.42\ V = 84.84\ V_{p-p}$. The values for the AC secondary voltage are shown in Fig. 28–15*b*.

Analyzing Circuit Operation

The output waveform for the half-wave rectifier of Fig. 28–15*a* is shown in Fig. 28–15*c*. Anytime the secondary voltage in Fig. 28–15*b* is positive, D_1 conducts and produces current flow in the load, R_L. Notice again that the output is a series of positive pulses. Notice also that when the secondary voltage in Fig. 28–15*b* is negative, the output voltage is zero.

Using the first approximation of a diode in Fig. 28–15*a*, the peak output voltage across R_L equals 42.42 V. Using the second approximation, the peak output voltage is 0.7 V less than the peak secondary voltage, which is 42.42 V − 0.7 V = 41.72 V.

The average or DC voltage at the output in Fig. 28–15*c* can be found using Formula (28–4) shown here:

$$V_{DC} = 0.318 \times V_{out(pk)} \qquad \textbf{(28–4)}$$

where $V_{out(pk)}$ is the peak value of the load voltage.

Using the second approximation of a diode, the DC output voltage in Fig. 28–15 is calculated as shown:

$$V_{DC} = 0.318 \times 41.72\ V$$
$$= 13.27\ V$$

This is the value of DC voltage that would be measured if a DC voltmeter were placed across the load, R_L. Notice also that this average value is depicted in Fig. 28–15*c*.

The DC load current is calculated as follows:

$$I_L = \frac{V_{DC}}{R_L}$$
$$= \frac{13.27\ V}{100\ \Omega}$$
$$= 132.7\ mA$$

For a half-wave rectifier, the DC load current and DC diode current are the same. This is expressed in Formula (28–5):

$$I_{diode} = I_{L(DC)} \qquad \textbf{(28–5)}$$

In Fig. 28–15, the DC current carried by the diode equals the load current, I_L, which is 132.7 mA.

Figure 28–16 Half-wave rectifier circuit showing D_1 reverse-biased during negative alternation of secondary voltage. D_1 must withstand the peak secondary voltage $V_{S(pk)}$ of 42.42 V.

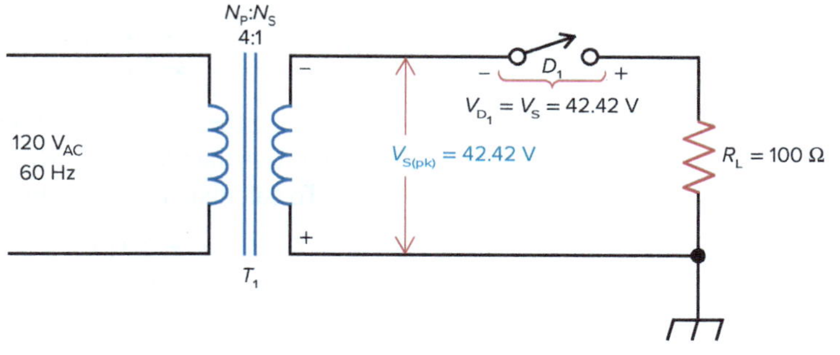

Frequency of the Output Waveform

By definition, a cycle includes the variations between two successive points having the same value and varying in the same direction. Since the frequency of the AC power line is 60 Hz, the period for one cycle is 16.67 ms, calculated as $1/f$, where $f = 60$ Hz. The period for one cycle of secondary voltage is shown in Fig. 28–15*b*. Notice directly below in Fig. 28–15*c* that one cycle of output voltage also repeats every 16.67 ms. Therefore, the frequency of the output waveform in a half-wave rectifier equals the input frequency applied to the rectifier. Expressed as a formula,

$$f_{out} = f_{in} \tag{28–6}$$

In Fig. 28–15, $f_{out} = f_{in} = 60$ Hz.

PIV

During the negative alternation of secondary voltage, the diode D_1 is off because it is reverse-biased. The equivalent circuit for this condition is shown in Fig. 28–16. When zero current flows during the negative alternation of secondary voltage, the output voltage is zero.

Notice that the secondary of T_1 is in series with D_1 and R_L. Remember from basic circuit theory that the voltage across an open in a simple series circuit equals the input voltage, which in this case is the transformer secondary voltage.

As shown in Fig. 28–16, D_1 must be able to withstand the peak value of secondary voltage, which is 42.42 V. The peak inverse voltage (PIV) rating of D_1 must be greater than the peak value of secondary voltage or the diode will break down and become damaged. For any unfiltered half-wave rectifier, the PIV for the diode always equals the peak value of the full secondary voltage.

Example 28–4

If the turns ratio $N_p:N_s = 3:1$ in Fig. 28–15*a*, calculate the following: V_S, V_{DC}, I_L, I_{diode}, PIV for D_1, and f_{out}.

ANSWER We begin by calculating the secondary voltage, V_S:

$$V_S = \frac{N_S}{N_P} \times V_P$$

$$= \frac{1}{3} \times 120\ V_{AC}$$

$$= 40\ V_{AC}$$

Next we calculate the peak value of the secondary voltage:

$$V_{S(pk)} = V_S \times 1.414$$
$$= 40\ V_{AC} \times 1.414$$
$$= 56.56\ V$$

Using the second approximation of a diode, the peak output voltage will be 0.7 V less than 55.56 V, which is 55.86 V.

To calculate the DC output voltage, we use Formula (28–4):

$$= 0.318 \times V_{out(pk)}$$
$$= 0.318 \times 55.86\ V$$
$$= 17.76\ V$$

The DC load current equals

$$I_L = \frac{V_{DC}}{R_L}$$
$$= \frac{17.76\ V}{100\ \Omega}$$
$$= 177.6\ mA$$

The DC diode current is calculated using Formula (28–5):

$$I_{diode} = I_L$$
$$= 177.6\ mA$$

Finally, the PIV for D1 equals the peak secondary voltage, which is 56.56 V. Also, the frequency of the output waveform equals 60 Hz.

Note that if it is desirable to obtain a negative output voltage in Fig. 28–15*a*, the diode, D_1, must be reversed.

The Full-Wave Rectifier

The circuit shown in Fig. 28–17*a* is called a **full-wave rectifier**. T_1 is a step-down transformer, which provides the secondary voltage as shown in Fig. 28–17*b* and *c*. When the top of the secondary is positive, D_1 is forward-biased, causing current to flow in the load, R_L. During this polarity of secondary voltage, D_2 is off because it is reverse-biased.

When the top of the secondary is negative, D_2 is forward-biased, causing current to flow in the load, R_L. During this polarity of secondary voltage, D_1 is off because it is reverse-biased. It is important to note that the direction of current through R_L is the same for both half-cycles of secondary voltage.

Transformer Calculations

The transformer used for the half-wave rectifier in Figs. 28–15 is again used for the full-wave rectifier in Fig. 28–17*a*. Notice, however, that for the full-wave rectifier shown in Fig. 28–17*a*, the transformer secondary is center-tapped.

In Fig. 28–17*a* each half of the secondary has a voltage of 15 V_{AC}, which is one-half the total secondary voltage, V_S, of 30 V_{AC}. The voltage for the top half of the secondary is designated V_1, whereas the voltage for the bottom half of the secondary is designated V_2. The secondary voltages V_1 and V_2 are shown in

GOOD TO KNOW

The rms value of a full-wave signal is $V_{rms} = 0.707\ V_P$ which is the same as V_{rms} for a full sine wave.

Figure 28–17 Full-wave rectifier with center tap in the transformer secondary. (*a*) Circuit. (*b*) Top half of secondary voltage, V_1. (*c*) Bottom half of secondary voltage, V_2. (*d*) Output voltage produced when D_1 conducts. (*e*) Output voltage produced when D_2 conducts. (*f*) Combined output voltage produced by D_1 and D_2 conducting during opposite alternations of secondary voltage.

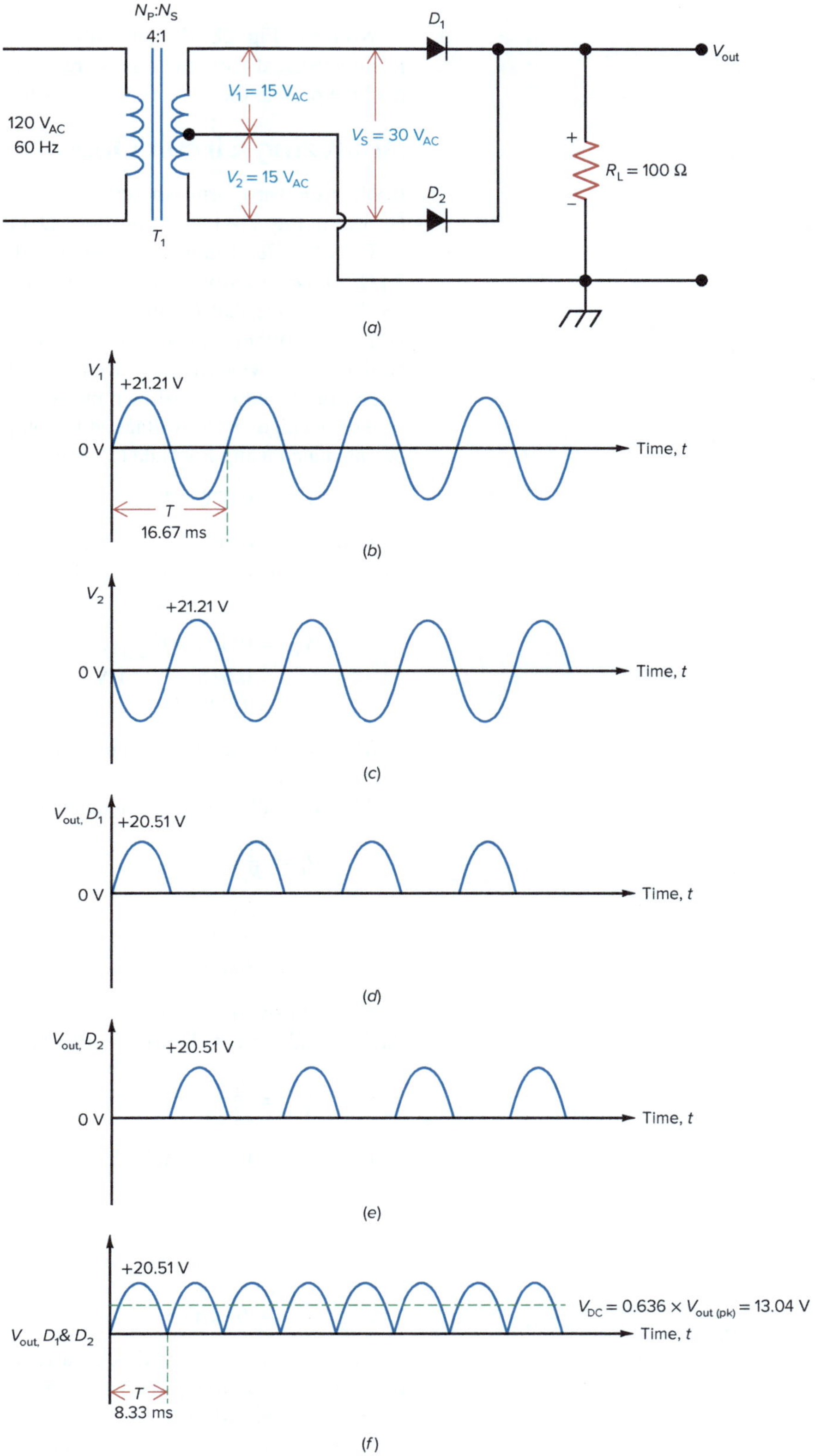

Fig. 28–17*b* and *c*, respectively. To calculate the peak voltage for V_1 and V_2, proceed as follows:

$$V_{1(pk)} = V_{2(pk)} = 1.414 \times \frac{V_S}{2}$$
$$= 1.414 \times 15\ V_{AC}$$
$$= 21.21\ V$$

Notice in Fig. 28–17*b* and *c* that V_1 are V_2 are 180° out of phase. V_1 reaches its positive peak at the same time V_2 reaches its negative peak. Likewise, V_2 reaches its positive peak at the same time V_1, reaches its negative peak.

Analyzing Circuit Operation

Look at the waveforms shown in Fig. 28–17. Whenever the secondary voltage, V_1 (shown in Fig. 28–17*b*), is positive, D_1 conducts and provides the output waveform in Fig. 28–17*d*. Likewise, whenever the secondary voltage, V_2 (shown in Fig. 28–17*c*), becomes positive, D_2 conducts and provides the output waveform in Fig. 28–17*e*. Notice that D_1 and D_2 conduct on opposite half-cycles of the secondary voltage, V_S. When V_1 is positive, D_2 is off. Likewise, when V_2 is positive, D_1 is off. Each diode provides a half-wave rectified waveform for the load, R_L. The combined effects of D_1 and D_2 are shown in Fig. 28–17*f*.

The average or DC voltage at the output of an unfiltered full-wave rectifier can be calculated using Formula (28–7):

$$V_{DC} = 0.636 \times V_{out(pk)} \qquad \textbf{(28–7)}$$

Using the second approximation of a diode, the peak load voltage across R_L = 21.21 V − 0.7 V = 20.51 V. The DC output voltage in Fig. 28–17*f* is calculated as follows:

$$V_{DC} = 0.636 \times V_{out(pk)}$$
$$= 0.636 \times 20.51\ V$$
$$= 13.04\ V$$

This is the value that would be measured if a DC voltmeter were placed across the load resistor, R_L.

The DC load current is calculated as follows:

$$I_L = \frac{V_{DC}}{R_L}$$
$$= \frac{13.04\ V}{100\ \Omega}$$
$$= 130.4\ mA$$

For a full-wave rectifier, the DC current carried by each diode equals one-half the DC load current. This is clearly expressed in Formula (28–8):

$$I_{diode} = \frac{I_L}{2} \qquad \textbf{(28–8)}$$

For Fig. 28–17*a*, each diode has a DC current calculated as follows:

$$I_{diode} = \frac{I_L}{2}$$
$$= \frac{130.4\ mA}{2}$$
$$= 65.2\ mA$$

The fact that the DC diode current is one-half the DC load current is best explained by examining the waveforms in Fig. 28–17*d* and *e*. If the peak value of output voltage is 20.51 V for each waveform, the peak load current at this instant is 205.1 mA, calculated as 20.51 V/100 Ω. Since the waveforms in Fig. 28–17*d* and *e* are each half-wave rectified waveforms, the average DC current passed by each diode is calculated as follows:

$$\begin{aligned} I_{DC} &= 0.318 \times I_{out(pk)} \\ &= 0.318 \times 205.1 \text{ mA} \\ &= 65.2 \text{ mA} \end{aligned}$$

This proves that each diode in a full-wave rectifier passes only half of the DC load current. Again, this is because each diode supplies its own half-wave rectified waveform to the load, R_L.

Frequency of the Output Waveform

Figure 28–17*b* shows that one cycle of secondary voltage has a period, *T*, of 16.67 ms, which equals 1/*f* where *f* = 60 Hz. Now look at the output waveform in Fig. 28–17*f*. Notice that one cycle is completed every 8.33 ms. Therefore, the frequency of the output waveform equals:

$$\begin{aligned} f_{out} &= \frac{1}{T} \\ &= \frac{1}{8.33 \text{ ms}} \\ &= 120 \text{ Hz} \end{aligned}$$

Therefore, for a full-wave rectifier, the following formula is true:

$$f_{out} = 2f_{in} \quad \textbf{(28–9)}$$

We must remember that the definition of a cycle states that it includes the variations between two successive points having the same value and varying in the same direction.

PIV

Figure 28–18 shows the diodes D_1 and D_2 represented using the second approximation. The circuit shows the voltages at the instant the top of the secondary reaches its positive peak of 42.42 V. To calculate the peak inverse voltage to which D_2 will be subjected, we use Kirchhoff's voltage law.

Starting at the anode (A) terminal of D_2 and going clockwise back to the cathode terminal,

$$\begin{aligned} V_{AK} &= -V_{S(pk)} + V_B \\ &= -42.42 \text{ V} + 0.7 \text{ V} \\ &= -41.72 \text{ V} \end{aligned}$$

Notice that this value is 0.7 V less than the peak value of the full secondary voltage.

Figure 28–18 Full-wave rectifier circuit showing D_2 reverse-biased during positive alternation of secondary voltage. Both diodes D_1 and D_2 must withstand a peak inverse voltage that is 0.7 V less than the peak value of the full secondary voltage.

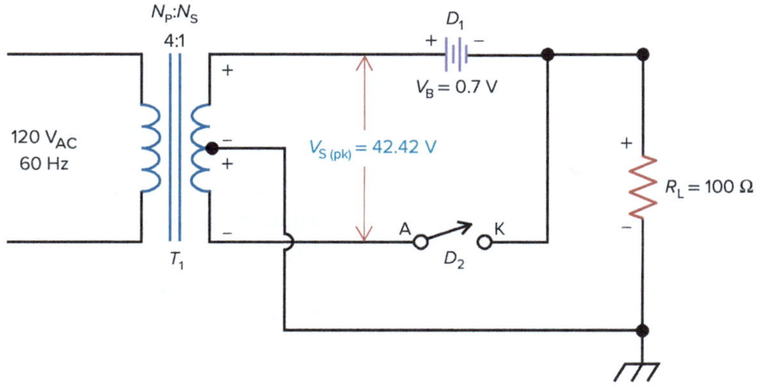

The same analogy can also be used to find the peak inverse voltage across the diode D_1 for the opposite polarity of the secondary voltage. Incidentally, the PIV for D_1 also equals −41.72 V. For any full-wave rectifier using a center-tapped transformer, the PIV for each diode will be 0.7 V less than the peak value of the full secondary voltage.

Example 28–5

If the turns ratio $N_P:N_S = 3:1$ in Fig. 28–17*a*, calculate the following: V_{DC}, I_L, I_{diode}, PIV for D_1, and f_{out}.

ANSWER Begin by calculating the total secondary voltage, V_S:

$$V_S = \frac{N_S}{N_P} \times V_P$$
$$= \frac{1}{3} \times 120\ V_{AC}$$
$$= 40\ V_{AC}$$

Next, we must realize that because of the secondary center tap, $V_1 = V_2 = \frac{V_S}{2} = 20\ V_{AC}$. To calculate the peak value for V_1 and V_2, we proceed as follows:

$$V_{1(pk)} = V_{2(pk)} = 1.414 \times \frac{V_S}{2}$$
$$= 1.414 \times 20\ V_{AC}$$
$$= 28.28\ V$$

Using the second approximation of a diode, the peak output voltage will be 0.7 V less than 28.28 V, which is 27.58 V.

To calculate the DC output voltage, we use Formula (28–7):

$$V_{DC} = 0.636 \times V_{out(pk)}$$
$$= 0.636 \times 27.58\ V$$
$$= 17.54\ V$$

The DC load current equals

$$I_L = \frac{V_{DC}}{R_L}$$
$$= \frac{17.54\ V}{100\ \Omega}$$
$$= 175.4\ mA$$

The DC diode current is calculated using Formula (28–8):

$$I_{diode} = \frac{I_L}{2}$$
$$= \frac{175.4\ mA}{2}$$
$$= 87.7\ mA$$

The PIV for D_1 and D_2 equals 55.86 V, which is 0.7 V less than the peak value of the full secondary voltage. The frequency of the output waveform is 120 Hz, the same as before.

If it is desirable to obtain a negative output voltage in Fig. 28–17*a*, the diodes D_1 and D_2 must be reversed.

GOOD TO KNOW

When a bridge rectifier, as opposed to a two-diode full-wave rectifier, is used, the same DC output voltage can be obtained with a transformer having a higher turns ratio, $N_P : N_S$. This means that with a bridge rectifier, fewer turns of wire are needed in the secondary of the transformer. Therefore, the transformer used with a bridge rectifier will be smaller, lighter, and will probably cost less. These benefits alone outweigh using four diodes instead of two in a conventional two-diode full-wave rectifier.

The Full-Wave Bridge Rectifier

The circuit shown in Fig. 28–19*a* is called a *full-wave bridge rectifier*. T_1 is a step-down transformer, which provides the secondary voltage shown in Fig. 28–19*b*. When the top of the secondary is positive, diodes D_2 and D_3 are forward-biased. This produces current flow in the load, R_L. For this polarity of secondary voltage, D_1 and D_4 are reverse-biased and do not conduct.

When the top of the secondary is negative, D_1 and D_4 are forward-biased, producing current flow in the load R_L. For this polarity of secondary voltage, D_2 and D_3 are reverse-biased and do not conduct. It is important to note that the direction of current through R_L is the same for both half-cycles of the secondary voltage. For the diode connections shown, the output voltage is positive.

Transformer Calculations

The transformer used for the half-wave rectifier in Fig. 28–15*a* is again used for the full-wave bridge rectifier in Fig. 28–19*a*. Remember that the rms value of secondary voltage was 30 V_{AC} for the turns ratio $N_P : N_S = 4 : 1$. Likewise, the peak value of the secondary voltage is 42.42 V.

Analyzing Circuit Operation

Look at the waveforms in Fig. 28–19*b* through *e*. When the secondary voltage, V_S, in Fig. 28–19*b* is positive, diodes D_2 and D_3 conduct, thus creating the output waveform shown in Fig. 28–19*c*. Likewise, when the top of the secondary is negative, diodes D_1 and D_4 conduct, giving the output waveform shown in Fig. 28–19*d*. Notice that the diode pairs D_2-D_3 and D_1-D_4 conduct on opposite half-cycles of the secondary voltage. The combined effects of D_2-D_3 and D_1-D_4 are shown in Fig. 28–19*e*. Each diode pair provides a half-wave rectified waveform for the load, R_L.

Using the second approximation of a diode, the peak load voltage across R_L equals 42.42 V − 1.4 V = 41.02 V. Notice that two diode voltage drops are subtracted from the peak secondary voltage. This is explained by the fact that when the diode pairs conduct, they are in series with the transformer secondary and the load, R_L.

The DC voltage at the output of an unfiltered full-wave bridge rectifier can be determined by using Formula (28–7). The calculations are

$$\begin{aligned} V_{DC} &= 0.636 \times V_{out(pk)} \\ &= 0.636 \times 41.02 \text{ V} \\ &= 26.09 \text{ V} \end{aligned}$$

The DC load current is calculated as follows:

$$\begin{aligned} I_L &= \frac{V_{DC}}{R_L} \\ &= \frac{26.09 \text{ V}}{100\ \Omega} \\ &= 260.9 \text{ mA} \end{aligned}$$

The DC diode current is calculated using Formula (28–8):

$$\begin{aligned} I_{diode} &= \frac{I_L}{2} \\ &= \frac{260.9 \text{ mA}}{2} \\ &= 130.4 \text{ mA} \end{aligned}$$

The DC diode current is one-half the DC load current because each diode pair in the bridge rectifier supplies a half-wave rectified waveform to the load, R_L.

MultiSim **Figure 28-19** Full-wave bridge rectifier. (*a*) Circuit. (*b*) Secondary voltage, V_S. (*c*) Output voltage produced when diodes D_2 and D_3 conduct. (*d*) Output voltage produced when diodes D_1 and D_4 conduct. (*e*) Combined output voltage.

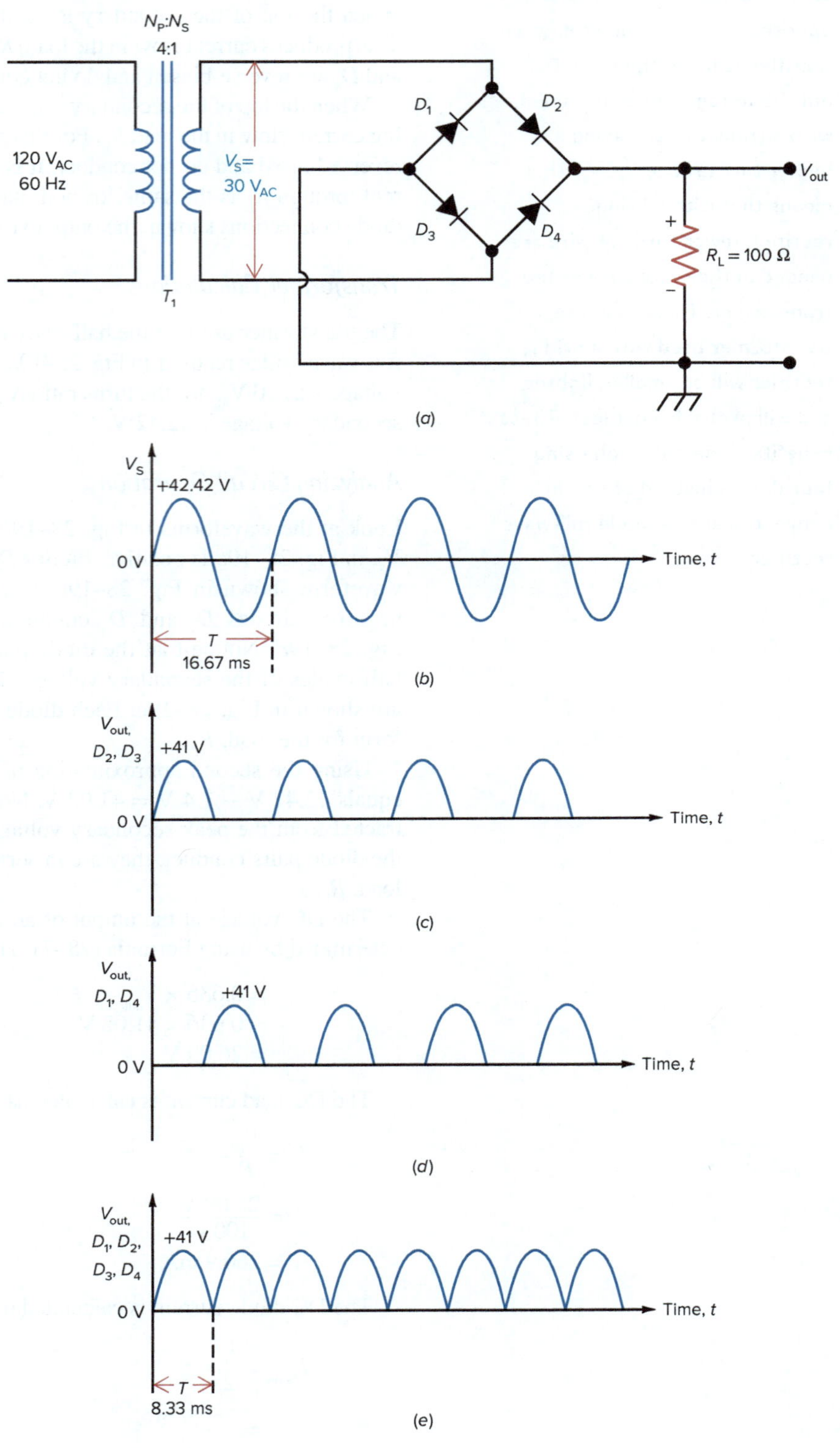

Figure 28–20 Full-wave bridge rectifier showing diodes D_1 and D_4 reverse-biased during positive alternation of secondary voltage. Each diode in the bridge must withstand a peak inverse voltage that is 0.7 V less than the peak value of the full secondary voltage.

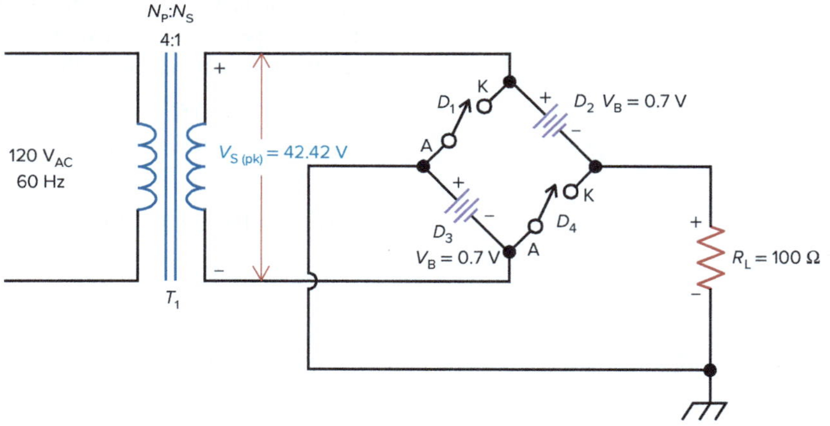

Frequency of the Output Waveform

Since the bridge rectifier in Fig. 28–19*a* provides a full-wave output, the frequency of the output waveform is found using Formula (28–9):

$$\begin{aligned} f_{out} &= 2f_{in} \\ &= 2 \times 60 \text{ Hz} \\ &= 120 \text{ Hz} \end{aligned}$$

PIV

Figure 28–20 shows the equivalent circuit of the bridge rectifier at the instant the secondary voltage reaches its maximum positive peak. Notice that diodes D_2 and D_3 are forward-biased and are replaced with their second approximation equivalent circuit. Notice also, that for this polarity of secondary voltage, diodes D_1 and D_4 are reverse-biased and are represented as open switches.

To calculate the peak inverse voltage (PIV) to which D_1 and D_4 will be subjected, use Kirchhoff's voltage law. For D_1, start at the anode (A) terminal and go clockwise around the loop back to its cathode (K) terminal. This gives

$$\begin{aligned} V_{AK}(D_1) &= V_B - V_{S(pk)} \\ &= 0.7 \text{ V} - 42.42 \text{ V} \\ &= -41.72 \text{ V} \end{aligned}$$

For D_4, start at its anode (A) terminal and go clockwise around the loop back to its cathode (K) terminal:

$$\begin{aligned} V_{AK}(D_4) &= -V_{S(pk)} + V_B \\ &= -42.42 \text{ V} + 0.7 \text{ V} \\ &= -41.72 \text{ V} \end{aligned}$$

Note that the PIV for each diode in a bridge rectifier will always be 0.7 V less than the peak value of the full secondary voltage.

Example 28–6

If the turns ratio $N_P : N_S = 3:1$ in Fig. 28–19*a*, calculate the following: V_{DC}, I_L, I_{diode}, PIV for each diode, and f_{out}.

ANSWER We already know from the previous examples that the secondary voltage equals 40 V_{AC} when the transformer turns ratio $N_P:N_S = 3:1$. Also, the peak secondary voltage equals 56.56 V. Subtracting two diode voltage drops gives us a peak load voltage of 56.56 V − 1.4 V = 55.16 V. To calculate the DC load voltage, use Formula (28–7):

$$\begin{aligned} V_{DC} &= 0.636 \times V_{out(pk)} \\ &= 0.636 \times 55.16\ \text{V} \\ &= 35.08\ \text{V} \end{aligned}$$

To calculate the DC load current, I_L, proceed as follows:

$$\begin{aligned} I_L &= \frac{V_{DC}}{R_L} \\ &= \frac{35.08\ \text{V}}{100\ \Omega} \\ &= 350.8\ \text{mA} \end{aligned}$$

The DC diode current is one-half the DC load current. This is calculated using Formula (28–8):

$$\begin{aligned} I_{diode} &= \frac{I_L}{2} \\ &= \frac{350.8\ \text{mA}}{2} \\ &= 175.4\ \text{mA} \end{aligned}$$

The *PIV* for each diode equals 56.56 V − 0.7 V = 55.86 V. Also, the frequency of the output waveform equals 120 Hz.

If it is desirable to obtain a negative output voltage in Fig. 28–19*a*, the diodes D_1, D_2, D_3, and D_4 must be reversed.

Capacitor Input Filter

The unfiltered output from a half-wave or full-wave rectifier is a pulsating DC voltage. For most applications, this DC voltage must be smoothed or filtered to be useful. One way to smooth out the pulsations in DC voltage is to connect a capacitor at the output of the rectifier. Figure 28–21*a* shows a half-wave rectifier with its output filtered by the capacitor, *C*. The filter capacitors used in this application are electrolytic capacitors with values typically larger than 100 μF.

Half-Wave Rectifier Filtering

When the top of the secondary goes positive initially in Fig. 28–21*a*, the diode, D_1, conducts and the capacitor, *C*, charges. Notice the time before t_0 in Fig. 28–21*b*. During this time, the capacitor voltage follows the positive-going secondary voltage. At time t_0, the voltage across *C* reaches its peak positive value.

Since $N_P:N_S = 8:1, V_S = \frac{1}{8} \times 120\ V_{AC} = 15\ V_{AC}$. To calculate the peak voltage to which *C* charges, we must first calculate the peak secondary voltage.

$$\begin{aligned} V_{S(pk)} &= V_S \times 1.414 \\ &= 15\ V_{AC} \times 1.414 \\ &= 21.21\ \text{V} \end{aligned}$$

Subtracting 0.7 V for the voltage drop across D_1 gives us a peak capacitor voltage of 20.51 V. This is the DC output voltage under ideal conditions.

Figure 28–21 Half-wave rectifier with capacitor input filter. (*a*) Circuit. (*b*) Output ripple voltage.

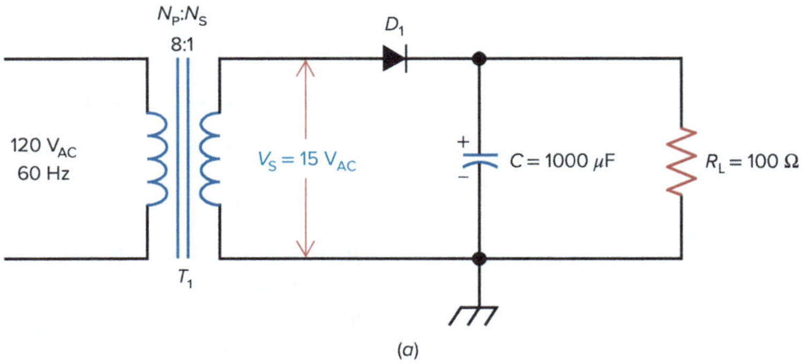

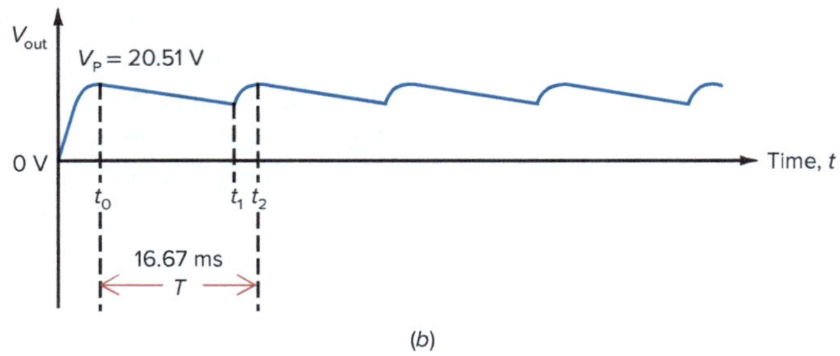

When the secondary voltage drops below its peak value of 21.21 V, D_1 is reverse-biased and the capacitor begins discharging through R_L. The discharge interval is between times t_0 and t_1 in Fig. 28–21*b*. At time t_1, the diode is once again forward-biased by the positive-going secondary voltage. This allows the capacitor to recharge to the peak value of 20.51 V at time t_2. Notice that the capacitor is charging only for the short time interval between t_1 and t_2. As a rough approximation, the capacitor is allowed to discharge for 16.67 ms in a half-wave rectifier, which is the period for one cycle of the input frequency of 60 Hz. The waveform of voltage in Fig. 28–21*b* is called the ripple voltage, designated V_{ripple}.

To calculate the peak-to-peak ripple voltage in Fig. 28–21*a*, we use Formula (28–10):

$$V_{\text{ripple}} = V_{\text{out(pk)}}\left(1 - \varepsilon^{\frac{-t}{R_L C}}\right) \qquad \textbf{(28–10)}$$

where t = discharge time for the filter capacitance, C and $R_L C$ = filtering time constant.

Inserting the values from Fig. 28–21 gives

$$\begin{aligned} V_{\text{ripple}} &= 20.51\text{ V}(1 - \varepsilon^{-0.167}) \\ &= 20.51\text{ V}(1 - 0.846) \\ &= 20.51\text{ V} \times 0.154 \\ &= 3.15\text{ V}_{\text{p-p}} \end{aligned}$$

Note that this peak-to-peak ripple voltage is undesirable. Ideally, we should have a steady DC voltage at the output equal to the peak value of 20.51 V. One way to reduce the ripple voltage is to increase the value of the filter capacitance, C.

A more accurate calculation for the DC voltage in Fig. 28–21 includes the ripple voltage, V_{ripple}. This is shown in Formula (28–11):

$$V_{\text{DC}} = V_{\text{out(pk)}} - \frac{V_{\text{ripple}}}{2} \qquad \textbf{(28–11)}$$

In Fig. 28–21, V_{DC} is calculated as shown:

$$V_{DC} = 20.51\text{ V} - \frac{V_{ripple}}{2}$$
$$= 20.51\text{ V} - 1.575\text{ V}$$
$$= 18.93\text{ V}$$

This is the approximate value that would be indicated by a DC voltmeter connected across the output.

Full-Wave Rectifier Filtering

Figure 28–22*a* shows a full-wave rectifier with its output filtered by the capacitor, *C*. When the top of the secondary is positive, D_1 conducts and charges *C* to the peak value of 20.51 V, equal to $V_{1(pk)} - 0.7$ V. When the bottom of the secondary is positive with respect to ground, D_2 conducts and recharges *C* to the peak value of 20.51 V. This is the DC output voltage under ideal conditions.

The difference between the half-wave rectifier in Fig. 28–21*a* and the full-wave rectifier in Fig. 28–22*a* is that *C* is charged twice as often in the full-wave rectifier. This also means that *C* has less time to discharge in the full-wave rectifier. In Fig. 28–22*b*, we see that the period for one cycle equals 8.33 ms.

The ripple voltage at the output of the full-wave rectifier in Fig. 28–22*a* can also be calculated using Formula (28–10). The calculations are

$$V_{ripple} = V_{out(pk)}\left(1 - \varepsilon^{\frac{-t}{R_L C}}\right)$$
$$= 20.51\text{ V }(1 - 0.92)$$
$$= 20.51\text{ V} \times 0.08$$
$$= 1.64\text{ V}_{p\text{-}p}$$

where $t = 8.33$ ms in the full-wave rectifier. Notice that the peak-to-peak ripple voltage is about one-half the value calculated earlier for the half-wave rectifier.

Figure 28–22 Full-wave rectifier with capacitor input filter. (*a*) Circuit. (*b*) Output ripple voltage.

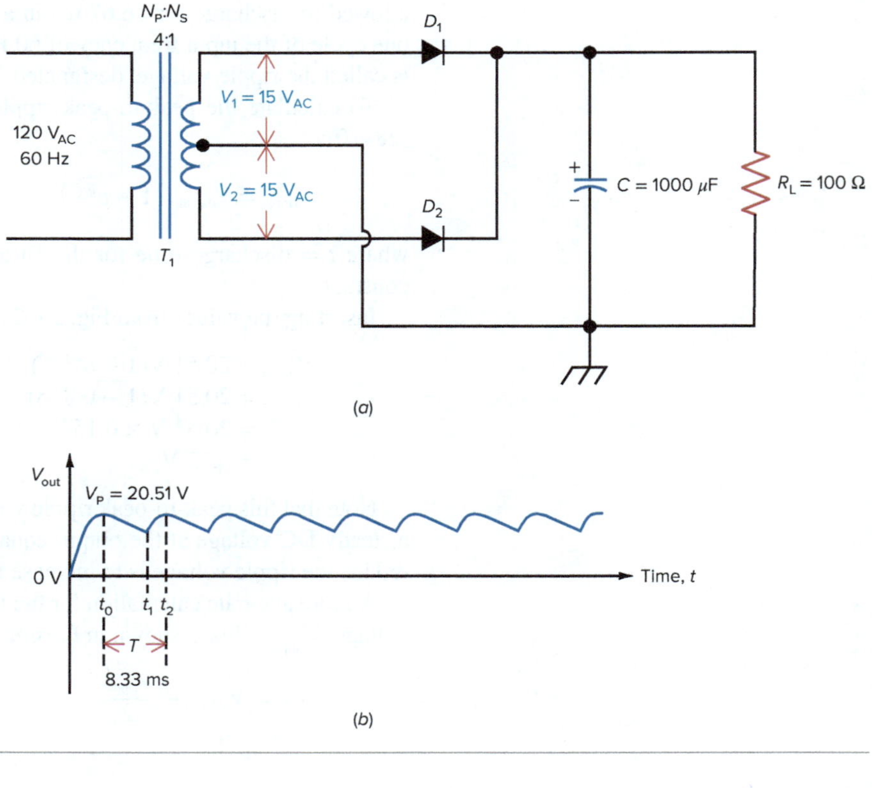

To get an accurate calculation of the DC voltage at the output, use Formula (28–11):

$$\begin{aligned} V_{DC} &= V_{out(pk)} - \frac{V_{ripple}}{2} \\ &= 20.51 \text{ V} - 0.82 \text{ V} \\ &= 19.69 \text{ V} \end{aligned}$$

Effect of Increasing the Load Current

If R_L is reduced in either Fig. 28–21 or 28–22, I_L and V_{ripple} will increase and V_{DC} will decrease. This is easily proven with a numerical example. Assume that R_L is decreased to 75 Ω in Fig. 28–22. This gives us an R_LC time constant of 75 Ω × 1000 μF = 75 ms. The ripple voltage is then calculated as follows:

$$\begin{aligned} V_{ripple} &= V_{out(pk)}\left(1 - \varepsilon^{\frac{-t}{R_LC}}\right) \\ &= 20.51 \text{ V } (1 - 0.8948) \\ &= 20.51 \text{ V} \times 0.105 \\ &= 2.15 \text{ V}_{p\text{-}p} \end{aligned}$$

Next, we calculate V_{DC} using Formula (28–11):

$$\begin{aligned} V_{DC} &= V_{out(pk)} - \left(\frac{V_{ripple}}{2}\right) \\ &= 20.51 \text{ V} - \frac{2.15 \text{ V}_{p\text{-}p}}{2} \\ &= 19.4 \text{ V} \end{aligned}$$

Notice that as the load resistance decreases from its previous value of 100 Ω to its present value of 75 Ω, V_{ripple} increases from 1.64 $V_{p\text{-}p}$ to 2.15 $V_{p\text{-}p}$. This causes V_{DC} to decrease from 19.69 V to 19.4 V when R_L decreases from 100 Ω to 75 Ω.

Example 28–7

Assume the transformer turns ratio $N_P{:}N_S = 4{:}1$ in Fig. 28–21*a* and 2:1 in Fig. 28–22*a*. Compare V_{ripple} and V_{DC} if $C = 500$ μF and $R_L = 250$ Ω.

ANSWER We begin with the half-wave rectifier in Fig. 28–21*a*. First we calculate the secondary voltage.

$$\begin{aligned} V_S &= \frac{N_S}{N_P} \times V_P \\ &= \frac{1}{4} \times 120 \text{ V}_{AC} \\ &= 30 \text{ V}_{AC} \end{aligned}$$

This gives us a peak secondary voltage of

$$\begin{aligned} V_{S(pk)} &= V_S \times 1.414 \\ &= 30 \text{ V}_{AC} \times 1.414 \\ &= 42.42 \text{ V} \end{aligned}$$

At the positive peak of secondary voltage, C charges to 0.7 V less than $V_{S(pk)}$. The calculations are

$$\begin{aligned} V_{out(pk)} &= V_{S(pk)} - V_B \\ &= 42.42 \text{ V} - 0.7 \text{ V} \\ &= 41.72 \text{ V} \end{aligned}$$

The peak-to-peak ripple voltage across C is calculated using Formula (28–10). Remember that for a half-wave rectifier, the discharge time t for the capacitor is approximately 16.67 ms. The calculations for V_{ripple} are

$$\begin{aligned} V_{\text{ripple}} &= V_{\text{out(pk)}}\left(1 - \varepsilon^{\frac{-t}{R_L C}}\right) \\ &= 41.72\ \text{V}\,(1 - \varepsilon^{-0.133}) \\ &= 41.72\ \text{V} \times 0.125 \\ &= 5.21\ \text{V}_{\text{p-p}} \end{aligned}$$

The DC voltage is calculated using Formula (28–11):

$$\begin{aligned} V_{\text{DC}} &= V_{\text{out(pk)}} - \frac{V_{\text{ripple}}}{2} \\ &= 41.72\ \text{V} - \frac{5.21\ \text{V}}{2} \\ &= 39.12\ \text{V} \end{aligned}$$

Next, calculate the values for the full-wave rectifier in Fig. 28–22*a*. The secondary voltage is calculated as follows:

$$\begin{aligned} V_{\text{S}} &= \frac{N_{\text{S}}}{N_{\text{P}}} \times V_{\text{P}} \\ &= \frac{1}{2} \times 120\ \text{V}_{\text{AC}} \\ &= 60\ \text{V}_{\text{AC}} \end{aligned}$$

To calculate V_1 and V_2, we divide V_{S} by 2. The calculations are

$$\begin{aligned} V_1 = V_2 &= \frac{V_{\text{S}}}{2} \\ &= \frac{60\ \text{V}_{\text{AC}}}{2} \\ &= 30\ \text{V}_{\text{AC}} \end{aligned}$$

The peak value for V_1 and V_2 is calculated as follows:

$$\begin{aligned} V_{1\text{(pk)}} = V_{2\text{(pk)}} &= 30\ \text{V}_{\text{AC}} \times 1.414 \\ &= 42.42\ \text{V} \end{aligned}$$

At the positive peak of secondary voltage, C charges to a value 0.7 V less than the peak value for V_1 or V_2. The calculations are

$$\begin{aligned} V_{\text{out(pk)}} &= \frac{V_{\text{S(pk)}}}{2} - 0.7\ \text{V} \\ &= 42.42\ \text{V} - 0.7\ \text{V} \\ &= 41.72\ \text{V} \end{aligned}$$

The peak-to-peak ripple voltage across C is calculated using Formula (28–10). Remember that the discharge time t for the capacitor in a full-wave rectifier is approximately 8.33 ms. The calculations are

$$\begin{aligned} V_{\text{ripple}} &= V_{\text{out(pk)}}\left(1 - \varepsilon^{\frac{-t}{R_L C}}\right) \\ &= 41.72\ \text{V}\,(1 - \varepsilon^{-0.066}) \\ &= 41.72\ \text{V} \times 0.0645 \\ &= 2.69\ \text{V}_{\text{p-p}} \end{aligned}$$

The DC voltage is calculated using Formula (28–11):

$$V_{\text{DC}} = V_{\text{out(pk)}} - \frac{V_{\text{ripple}}}{2}$$

$$= 41.72\text{ V} - \frac{2.69\ V_{p\text{-}p}}{2}$$

$$= 40.38\text{ V}$$

The peak-to-peak ripple voltage and DC voltage for the half-wave and full-wave rectifiers in this example are compared below:

	V_{ripple}	V_{DC}
Half-Wave	$5.21\ V_{p\text{-}p}$	$39.12\ V_{DC}$
Full-Wave	$2.69\ V_{p\text{-}p}$	$40.38\ V_{DC}$

Notice that the full-wave rectifier provides a larger DC output voltage with less ripple because the discharge time for a full-wave rectifier is one-half that of a half-wave rectifier.

Diode Currents with a Capacitor Input Filter

The diodes in a rectifier with a capacitor input filter conduct for less than 180° of the secondary voltage because the DC voltage on the filter capacitor holds the diodes off until the secondary voltage reaches a value high enough to provide the proper amount of forward bias. As a result, the diodes conduct for very short intervals of time. The average DC current in the diode of a half-wave rectifier, however, is still equal to the DC load current, I_L, given by Formula (28–5). Likewise, the DC diode current in a full-wave rectifier connected to a capacitor input filter still equals one-half the DC load current, I_L, as indicated in Formula (28–8).

When the power is first applied to a rectifier with a capacitor input filter, the diode current required to charge the capacitor can be extremely high because the large filter capacitor connected to the output of the rectifier is initially uncharged. The current flowing through the diode during this time is called the *surge* current. The surge current must be less than the diode's maximum forward surge current rating (I_{FSM}). After a few cycles of applied voltage, the capacitor achieves its full charge and the diode current is more normal, conducting only at or near the peak(s) of secondary voltage.

Peak Inverse Voltage with a Capacitor Input Filter

The peak inverse voltage across a nonconducting diode in either a full-wave rectifier or full-wave bridge rectifier with a capacitor input filter is still 0.7 V less than the peak value of the full secondary voltage.

The peak inverse voltage across the diode in a half-wave rectifier with a capacitor input filter, however, equals approximately two times the peak value of the full secondary voltage. This is because the filter capacitance remains charged to approximately the peak value during the negative alternation of the secondary voltage.

28–6 Self-Review

Answers at the end of the chapter.

a. The peak output voltage from an unfiltered half-wave rectifier is 38.5 V. How much is the DC output voltage?

b. The peak output voltage from an unfiltered full-wave rectifier is 18.9 V. How much is the DC output voltage?

c. The peak output voltage from an unfiltered full-wave bridge rectifier is 38.5 V. How much is the DC output voltage?

d. The peak output voltage from a full-wave rectifier is 10 V. If a 2200-μF capacitor is connected to the output of the rectifier, what is the approximate DC output voltage?

28–7 Special Diodes

Besides rectification, a semiconductor diode has many other useful applications. For example, semiconductor diodes can be manufactured to regulate voltage and emit different colors of light. This section introduces you to two special purpose diodes, the **light-emitting diode (LED)** and the **zener diode**.

GOOD TO KNOW

An LED is considered an optoelectronic device because it combines the concepts of optics and electronics.

Light-Emitting Diodes

When elements such as gallium, arsenic, and phosphorus are used in doping, a manufacturer can make diodes that emit different colors of light. These diodes are called *light-emitting diodes (LEDs)*. Some common LED colors are red, green, yellow, orange, and even infrared (invisible) light.

LEDs now operate in place of incandescent lamps in many cases. A semitransparent material is used with LEDs so that light can escape and be visible.

How It Happens

For any diode that is forward-biased, free electrons and holes combine at the junction. When free electrons from the *n* side cross over into the *p* side, they fall into a hole. When an electron falls, it releases energy. This energy is mainly heat or light. For the normal silicon diode, the light cannot escape because the device is not transparent. Because LEDs use a semitransparent material, however, light can escape to the surrounding environment. The color of the light emitted from the LED depends on the type of element used in the manufacture of the LED.

LED Characteristics

A light-emitting diode is represented using the schematic symbol shown in Fig. 28–23. The arrows pointing outward indicate the emitted light with forward bias. The internal barrier potential, V_B, for an LED is considerably higher than that of an ordinary silicon diode. Typical values of V_B for an LED range from approximately 1.5 to 2.5 V for currents between 10 and 50 mA. The exact amount of forward voltage drop varies with the color of the LED and also with the forward current through the LED. In most cases, the LED voltage drop can be assumed to be 2.0 V for all LED colors and all values of forward current. This is a convenient value to use in troubleshooting and design.

Figure 28–23 Schematic symbol of an LED.

It is important to note that as the forward current in an LED increases, so does the amount of emitted light. For smaller LEDs, a forward current of 20–25 mA produces adequate light for most applications.

Example 28–8

Calculate the LED current in Fig. 28–24*a*.

ANSWER The current through the LED can be found by dividing the resistor voltage by its resistance. Assume the LED has a voltage drop of 2.0 V. The calculations are

$$I_{\text{LED}} = \frac{V_{\text{in}} - V_{\text{LED}}}{R_S}$$

$$= \frac{24\text{ V} - 2\text{ V}}{2.2\text{ k}\Omega}$$

$$= 10\text{ mA}$$

Unless indicated otherwise assume that a forward-biased LED drops 2.0 V.

Figure 28–24 Circuit used for Example 28–8. (*a*) LED circuit. (*b*) Circuit drawn as it is often seen in commercial schematic diagrams. (*c*) D_1 connected across LED to protect the LED from negative voltages accidentally applied to the circuit.

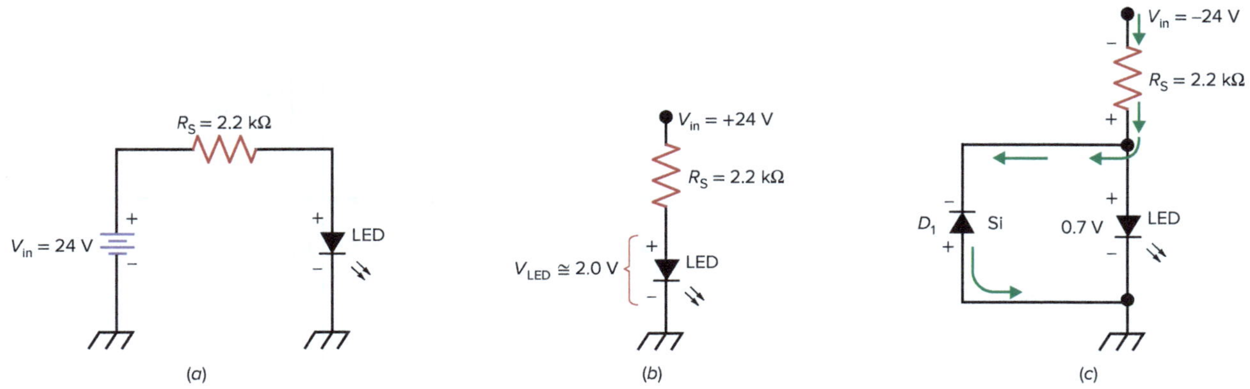

Figure 28–24*b* shows how the circuit of Fig. 28–24*a* is normally drawn in commercial schematics. It is common practice to show only the potential difference and its polarity with respect to chassis ground.

Example 28–9

In Fig. 28–24*b*, calculate the resistance, R_S, required to provide an LED current of 25 mA.

ANSWER As stated before, assume a forward voltage of 2.0 V for an LED. The value of the resistor, R_S, can be calculated by dividing the resistor voltage by the desired LED current of 25 mA. The calculations are

$$R_S = \frac{V_{in} - V_{LED}}{I_{LED}}$$

$$= \frac{24\ \text{V} - 2\ \text{V}}{25\ \text{mA}}$$

$$= 880\ \Omega$$

The nearest standard value for tolerances of ±5% is 910 Ω. This value will produce an LED current of nearly the desired value.

Breakdown Voltage Rating, V_{BR}

LEDs have a very low breakdown voltage rating. Typical values of V_{BR} range from 3 to 15 V. Because of the low value of breakdown voltage, accidentally applying even a small value of reverse voltage can destroy the LED or severely degrade its performance.

One way to protect an LED against excessive reverse voltage is to connect a silicon diode in parallel with the LED, as shown in Fig. 28–24*c*. The parallel connection ensures that the LED cannot accidentally receive a reverse-bias voltage greater than its breakdown voltage rating, V_{BR}. In this case, the LED has a maximum reverse voltage, V_R, equal to the forward voltage of −0.7 V across D_1.

Note that a negative voltage would never be intentionally applied to the circuit shown in Fig. 28–24*c*. The negative value of −24 V for V_{in} represents an accidental application of negative voltage caused by a fault in the power supply circuit that provides power for the LED.

GOOD TO KNOW

LEDs have replaced incandescent lamps in many applications because they have a lower operating voltage, a longer life, and faster ON-OFF switching.

Figure 28–25 Zener diode. (*a*) Schematic symbol. (*b*) Graph of a zener diode showing forward- and reverse-bias regions.

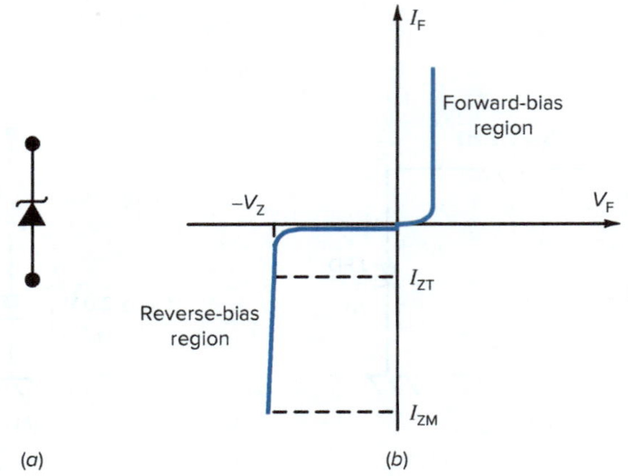

Zener Diodes and Their Characteristics

A *zener* diode is a special diode that has been optimized for operation in the breakdown region. These devices are unlike ordinary rectifier diodes, which are never intended to be operated at or near breakdown. Voltage regulation is the most common application of a zener diode. The zener diode is connected in parallel with the load of the power supply. The zener voltage remains constant despite load current variations. Figure 28–25*a* shows the schematic symbol of a zener diode.

> **GOOD TO KNOW**
>
> The voltage across a zener diode does not remain perfectly constant as the zener current varies. The reason is that all zeners have some value of zener impedance, designated R_Z. R_Z acts like a small resistance in series with the zener. The effect of R_Z is that small changes in V_Z will occur when I_Z varies.

Volt-Ampere Characteristic Curve

Figure 28–25*b* shows the volt-ampere characteristic curve for a typical silicon zener diode. In the forward region, the zener acts like an ordinary silicon rectifier diode with a forward voltage drop of about 0.7 V when conducting.

In the reverse-bias region, a small reverse leakage current flows until the breakdown voltage is reached. At this point, the reverse current through the zener increases sharply. The reverse current is called *zener current*, designated I_Z. Notice that the breakdown voltage, designated V_Z, remains nearly constant as the **zener current, I_Z**, increases. Because of this characteristic, a zener diode can be used in voltage regulation circuits, since the zener voltage, V_Z, remains constant even though the zener current, I_Z, varies over a wide range.

Most manufacturers specify the zener voltage, V_Z, at a specified test current designated I_{ZT}. For example, a 1N4742A zener diode has a rated zener voltage, V_Z, of 12.0 V for a test current, I_{ZT}, of 21 mA. The suffix A in the part number 1N4742A indicates a zener voltage tolerance of $\pm 5\%$.

Zener Ratings

An important zener rating is its power rating. In terms of power dissipation,

$$P_Z = V_Z I_Z \qquad \textbf{(28–12)}$$

where P_Z equals the power dissipated by the zener, V_Z equals the zener voltage, and I_Z equals the zener current.

For example, if a 12-V zener has 30 mA of current, its power dissipation, P_Z, is

$$\begin{aligned} P_Z &= V_Z I_Z \\ &= 12\text{ V} \times 30\text{ mA} \\ &= 360\text{ mW} \end{aligned}$$

The power dissipation in a zener diode must always be less than its power dissipation rating. The power rating of a zener is designated P_{ZM}. The maximum current that a zener can safely handle is given in Formula (28–13):

$$I_{ZM} = \frac{P_{ZM}}{V_Z} \qquad \textbf{(28–13)}$$

where V_Z equals the zener voltage, I_{ZM} equals the maximum-rated zener current, and P_{ZM} equals the power rating of the zener. I_{ZM} is shown on the graph in Fig. 28–25*b*. Exceeding the value of I_{ZM} will burn out the zener.

Example 28–10

Calculate the maximum-rated zener current for a 1-W, 10-V zener.

ANSWER Using Formula (28–13), the calculations are

$$I_{ZM} = \frac{P_{ZM}}{V_Z}$$
$$= \frac{1\ \text{W}}{10\ \text{V}}$$
$$= 100\ \text{mA}$$

For this zener diode, the current I_Z must never exceed 100 mA. If it does, the diode is likely to fail due to excessive power dissipation.

Zener Diode Applications

Figure 28–26 shows an unloaded voltage regulator that uses a 6.2-V zener diode. Notice that the zener diode is reverse-biased with the positive terminal of V_{in} connected to the cathode of the zener diode through the series limiting resistor, R_S.

The zener diode provides an output voltage of 6.2 V. The zener current is calculated by dividing the voltage across the series resistor, R_S, by the value of R_S. The calculations are

$$I_Z = \frac{V_{in} - V_Z}{R_S}$$
$$= \frac{25\ \text{V} - 6.2\ \text{V}}{1\ \text{k}\Omega}$$
$$= 18.8\ \text{mA}$$

Figure 28–26 Unloaded zener regulator with the output across the zener.

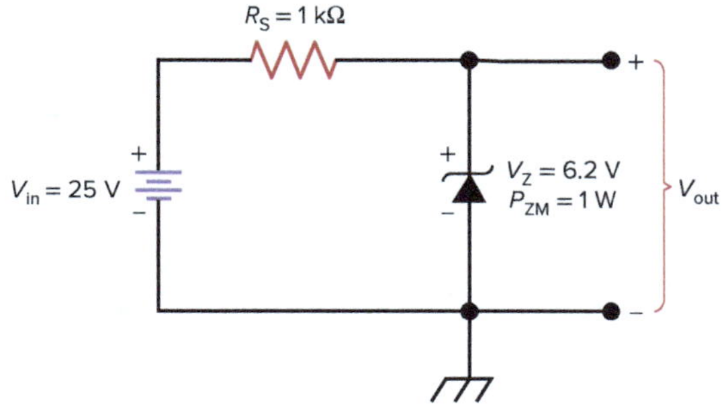

If the input voltage, V_{in}, varies, the zener current, I_Z, also varies. However, V_Z remains relatively constant. Any fluctuation in the zener voltage, V_Z, is due to the small change in the voltage drop across the zener impedance, R_Z.

Example 28–11

If $V_Z = 10$ V in Fig. 28–26, calculate I_Z.

ANSWER The calculations are

$$I_Z = \frac{V_{in} - V_Z}{R_S}$$
$$= \frac{25\text{ V} - 10\text{ V}}{1\text{ k}\Omega}$$
$$= 15\text{ mA}$$

GOOD TO KNOW

Zener diodes can also be used to regulate the voltage from a power supply whose output is negative. All you need to do is reverse the connection of the zener so that the cathode is grounded and the anode is connected to the negative side of the unregulated supply.

Loaded Zener Regulators

The unloaded voltage regulator shown in Fig. 28–26 has few applications in electronics. Usually, a load resistor is connected across the output, as shown in Fig. 28–27. This is a typical loaded voltage regulator. Since R_L is across the zener, the load voltage equals the zener voltage, or $V_L = V_Z$.

It is important to note in Fig. 28–27 that the voltage dropped across the series resistor, R_S, is $V_{in} - V_Z$. Thus, the current, I_S, through the series resistor is calculated as

$$I_S = \frac{V_{in} - V_Z}{R_S}$$
$$= \frac{15\text{ V} - 7.5\text{ V}}{100\ \Omega}$$
$$= 75\text{ mA}$$

The current I_L through the load resistor is calculated as

$$I_L = \frac{V_Z}{R_L}$$
$$= \frac{7.5\text{ V}}{150\ \Omega}$$
$$= 50\text{ mA}$$

MultiSim **Figure 28–27** Loaded zener regulator.

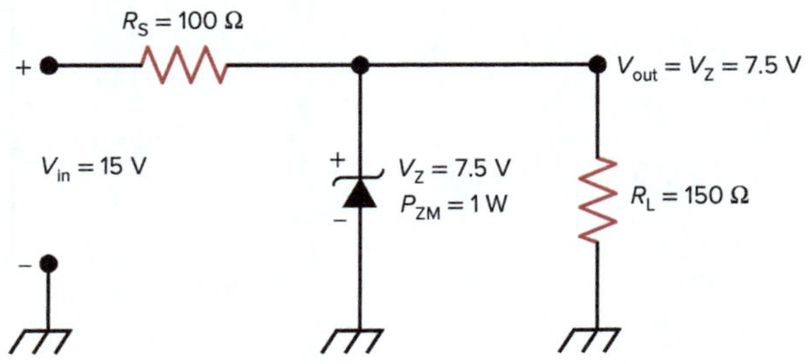

Note that the output voltage equals the zener voltage, V_Z.

Because the zener is in parallel with R_L, the series current, I_S, equals $I_Z + I_L$. This is expressed in Formula (28–14):

$$I_S = I_Z + I_L \qquad \textbf{(28–14)}$$

Formula (28–14) can also be arranged to solve for the zener current, I_Z:

$$I_Z = I_S - I_L$$

To calculate the currents in a loaded voltage regulator, I_S should be calculated first, then I_L, and, last, the zener current, I_Z. I_Z must be calculated indirectly because its DC resistance is not a fixed quantity. In Fig. 28–27, I_Z is found as follows:

$$\begin{aligned} I_Z &= I_S - I_L \\ &= 75 \text{ mA} - 50 \text{ mA} \\ &= 25 \text{ mA} \end{aligned}$$

Example 28–12

If R_L increases to 250 Ω in Fig. 28–27, calculate the following: I_S, I_L, I_Z, and P_Z.

ANSWER I_S remains constant at 75 mA even though R_L changes because V_{in}, V_Z, and R_S remain constant.

To calculate I_L, divide V_Z by the R_L value of 250 Ω:

$$\begin{aligned} I_L &= \frac{V_Z}{R_L} \\ &= \frac{7.5 \text{ V}}{250\ \Omega} \\ &= 30 \text{ mA} \end{aligned}$$

To calculate I_Z, proceed as follows:

$$\begin{aligned} I_Z &= I_S - I_L \\ &= 75 \text{ mA} - 30 \text{ mA} \\ &= 45 \text{ mA} \end{aligned}$$

P_Z is calculated as follows:

$$\begin{aligned} P_Z &= V_Z I_Z \\ &= 7.5 \text{ V} \times 45 \text{ mA} \\ &= 337.5 \text{ mW} \end{aligned}$$

Notice that increasing R_L from 150 Ω to 250 Ω in Fig. 28–27 causes I_L to decrease and I_Z to increase because I_S must remain constant at 75 mA.

Also, notice that the power dissipation in the zener is well below the power dissipation rating of 1 W in this example.

Example 28–13

In Fig. 28–28, calculate I_S, I_L, and I_Z for (a) $R_L = 200\ \Omega$; (b) $R_L = 500\ \Omega$.

ANSWER Begin by calculating the current, I_S, through resistor R_S:

$$I_S = \frac{V_{in} - V_Z}{R_S} = \frac{16\ \text{V} - 10\ \text{V}}{100\ \Omega} = 60\ \text{mA}$$

This current is the same for both load resistance values listed in (a) and (b). For $R_L = 200\ \Omega$ in Part (a), calculate I_L as follows:

$$I_L = \frac{V_Z}{R_L} = \frac{10\ \text{V}}{200\ \Omega} = 50\ \text{mA}$$

Next, solve for the zener current, I_Z. The calculations are

$$I_Z = I_S - I_L = 60\ \text{mA} - 50\ \text{mA} = 10\ \text{mA}$$

For $R_L = 500\ \Omega$ in Part (b), calculate I_L as follows:

$$I_L = \frac{V_Z}{R_L} = \frac{10\ \text{V}}{500\ \Omega} = 20\ \text{mA}$$

Next solve for I_Z as follows:

$$I_Z = I_S - I_L = 60\ \text{mA} - 20\ \text{mA} = 40\ \text{mA}$$

Compare the I_Z and I_L values for $R_L = 200\ \Omega$ and $R_L = 500\ \Omega$. Notice that when R_L increases from 200 Ω to 500 Ω, I_L decreases from 50 mA to 20 mA, which in turn causes I_Z to increase from 10 mA to 40 mA. Notice that the zener current, I_Z, increases by 30 mA, the same amount by which the load current, I_L, decreases.

When V_{in} is constant, I_Z and I_L will always have equal but opposite changes in value.

Figure 28–28 Loaded zener regulator with R_L adjustable. See Example 28–13.

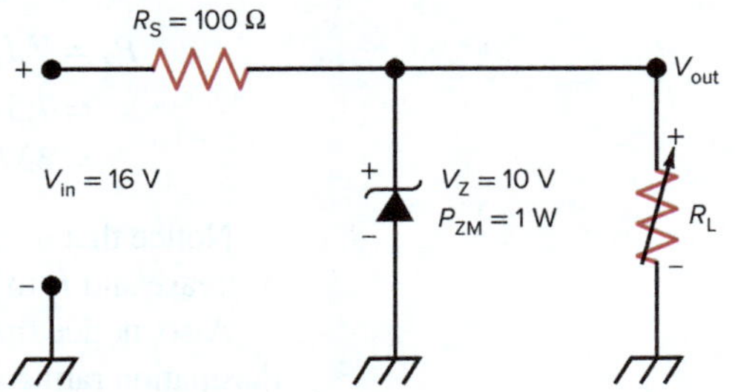

28–7 Self-Review

Answers at the end of the chapter.

a. The approximate voltage drop across a forward-biased LED is 2.0 V. (True/False)

b. A typical LED has a breakdown voltage rating around 1 kV. (True/False)

c. A zener diode is normally reverse-biased when used as a voltage regulator. (True/False)

Summary

- Semiconductor atoms have four valence electrons. Both germanium (Ge) and silicon (Si) are examples of semiconductor materials.
- A pure semiconductor material with only one type of atom is called an intrinsic semiconductor. An intrinsic semiconductor is neither a good conductor nor a good insulator.
- An extrinsic semiconductor is a semiconductor with impurity atoms added to it through a process known as doping. Doping increases the conductivity of a semiconductor material.
- *n*-type semiconductors have many free electrons as a result of adding pentavalent impurity atoms during the doping process. A *p*-type semiconductor has many holes or vacancies in its covalent bond structure as a result of adding trivalent impurity atoms during the doping process. A hole exhibits a positive charge.
- A diode is a unidirectional device that allows current to flow through it in only one direction.
- A diode is forward-biased by making its anode positive relative to its cathode. A diode is reverse-biased by making its anode negative relative to its cathode.
- A forward-biased diode has relatively low resistance, whereas a reverse-biased diode has very high resistance.
- When testing a silicon diode with an analog ohmmeter, the ratio of reverse resistance, R_R, to forward resistance, R_F, should be at least 1000 : 1.
- The first approximation treats a forward-biased diode like a closed switch with a voltage drop of 0 V. The second approximation includes the barrier potential, V_B, which is 0.7 V for a silicon diode.
- Both the first and second approximations of a diode treat a reverse-biased diode like an open switch with zero current.
- The third approximation of a diode includes both V_B and the bulk resistance when forward-biased. The bulk resistance, r_B, of a diode is the resistance of the *p* and *n* materials.
- The third approximation of a reverse-biased diode takes into account the reverse resistance, R_R.
- Half-wave, full-wave, and bridge rectifier circuits convert an AC voltage into a pulsating DC voltage.
- For an unfiltered half-wave rectifier, the DC output voltage is $0.318 \times V_{out(pk)}$. For an unfiltered full-wave rectifier, the DC output voltage is $0.636 \times V_{out(pk)}$.
- When a filter capacitor is connected to the output of a half-wave or full-wave rectifier, the DC output voltage is approximately equal to the peak output voltage from the rectifier.
- The ripple frequency at the output of a half-wave rectifier is the same as the frequency of the AC input voltage. The ripple frequency of a full-wave rectifier at the output is twice the frequency of the AC input voltage.
- A diode that emits light when forward-biased is called a light-emitting diode (LED). LEDs are doped with elements such as gallium, arsenic, and phosphorus because these elements emit different colors of light such as yellow, red, green, and orange.
- The voltage drop across a forward-biased LED ranges from about 1.5 to 2.5 V. When making calculations, an approximate voltage of 2.0 V can be assumed for a forward-biased LED. The breakdown voltage rating of an LED is typically 3 to 15 V.
- A zener diode is a special diode designed for operation in the breakdown region. The most common application of a zener diode is voltage regulation.

Important Terms

Avalanche—the effect that causes a sharp increase in reverse current, I_R, when the reverse-bias voltage across a diode becomes excessive.

Barrier potential, V_B—the potential difference at the *p-n* junction of a diode. V_B exists between the wall of positive and negative ions that are created as a result of free electrons diffusing from the *n* side of the diode to the *p* side.

Bias—a control voltage or current.

Breakdown voltage, V_{BR}—the reverse bias voltage at which the avalanche effect occurs. The avalanche effect causes the reverse current, I_R, to increase sharply.

Bulk resistance, r_B—the resistance of the *p* and *n* regions in a diode.

Covalent bonding—the sharing of valence electrons between neighboring atoms in a silicon crystal or other crystalline structure.

Depletion zone—the area at the *p-n* junction of a diode that is void or depleted of all charge carriers.

Diode—a unidirectional device that allows current to flow through it in only one direction.

Doping—the process of adding impurity atoms to a pure semiconductor material such as silicon.

Electron-hole pair—a free electron and a hole are created when a valence electron gains enough energy to leave its covalent bond in a silicon crystal.

Extrinsic semiconductor—a semiconductor that has been doped with impurity atoms to alter the characteristics of the material, mainly its conductivity.

Forward bias—the polarity of voltage across a diode that permits current to flow through it easily.

Full-wave rectifier—a circuit that provides an entirely positive or negative output voltage when an AC input voltage is applied. A full-wave rectifier provides an output for both the positive and negative alternations of the input voltage.

Half-wave rectifier—a circuit that provides an entirely positive or negative output voltage when an AC input voltage is applied. A half-wave rectifier provides an output for either the positive or negative alternation of the input voltage but not both.

Hole—the absence of a valence electron in a covalent bond structure.

Intrinsic semiconductor—a semiconductor material with only one type of atom.

Leakage current—the very small current that flows when a diode is reverse-biased. The leakage current is mainly due to the thermally generated minority carriers in both sections of the diode.

Light-emitting diode (LED)—a diode that emits a certain color light when forward-biased. The color of light emitted by the diode is determined by the type of material used in doping.

Majority current carrier—the dominant type of charge carrier in a doped semiconductor material. In an *n*-type semiconductor, free electrons are the majority current carriers, whereas in a *p*-type semiconductor, holes are the majority current carriers.

Minority current carrier—the type of charge carrier that appears sparsely throughout a doped semiconductor material. In an *n*-type semiconductor, holes are the minority current carriers, whereas free electrons are the minority current carriers in a *p*-type semiconductor.

***n*-type semiconductor**—a semiconductor that has been doped with pentavalent impurity atoms. The result is a large number of free electrons throughout the material. Since the electron is the basic particle of negative charge, the material is called *n*-type semiconductor material.

***p*-type semiconductor**—a semiconductor that has been doped with trivalent impurity atoms. The result is a large number of holes in the material. Since a hole exhibits a positive charge, the material is called *p*-type semiconductor material.

Peak inverse voltage (PIV)—the maximum instantaneous reverse-bias voltage across a diode.

Pentavalent atom—an atom with five valence electrons.

Reverse bias—the polarity of voltage across a diode that prevents the diode from conducting any current.

Trivalent atom—an atom with three valence electrons.

Valence electrons—the electrons in the outermost ring or shell of an atom.

Zener current, I_z—the name for the reverse current in a zener diode.

Zener diode—a diode that has been optimized for operation in the breakdown region.

Related Formulas

Diodes

$R_F = V_F/I_F$

$V_F = V_B + I_F r_B$

$r_B = \dfrac{\Delta V}{\Delta I}$

Half-Wave Rectifier (Unfiltered)

$V_{DC} = 0.318 \times V_{out(pk)}$

$I_{diode} = I_L$

$f_{out} = f_{in}$

Full-Wave Rectifier (Unfiltered)

$V_{DC} = 0.636 \times V_{out(pk)}$

$I_{diode} = I_L/2$

$f_{out} = 2f_{in}$

Rectifier with Capacitor Input Filter

$V_{ripple} = V_{out(pk)}\left(1 - \varepsilon^{\frac{-t}{R_L C}}\right)$

$V_{DC} = V_{out(pk)} - V_{ripple}/2$

Zener Diode

$P_Z = V_Z I_Z$

$I_{ZM} = P_{ZM}/V_Z$

Loaded Zener Regulator

$I_S = I_Z + I_L$

Self-Test

Answers at the back of the book.

1. A pure semiconductor is often referred to as a(n)

a. extrinsic semiconductor.

b. intrinsic semiconductor.

c. doped semiconductor.

d. none of the above.

2. An *n*-type semiconductor is a semiconductor that has been doped with

a. trivalent impurity atoms.

b. impurity atoms whose electron valence is +4.

c. pentavalent impurity atoms.

d. none of the above.

3. For a silicon diode, the barrier potential, V_B, is approximately

a. 0.7 V.

b. 0.3 V.

c. 2.0 V.

d. 6.8 V.

4. **In a *p*-type semiconductor, the majority current carriers are**
   a. free electrons.
   b. valence electrons.
   c. protons.
   d. holes.

5. **To forward-bias a diode,**
   a. the anode voltage must be positive with respect to its cathode.
   b. the anode voltage must be negative with respect to its cathode.
   c. the cathode voltage must be positive with respect to its anode.
   d. either a or b.

6. **A reverse-biased diode acts like a(n)**
   a. closed switch.
   b. open switch.
   c. small resistance.
   d. none of the above.

7. **The sharing of valence electrons in a silicon crystal is called**
   a. doping.
   b. the avalanche effect.
   c. covalent bonding.
   d. coupling.

8. **When used as a voltage regulator, a zener diode is normally**
   a. forward-biased.
   b. reverse-biased.
   c. not biased.
   d. none of the above.

9. **In an *n*-type semiconductor, the minority current carriers are**
   a. free electrons.
   b. protons.
   c. valence electrons.
   d. holes.

10. **A *p*-type semiconductor is a semiconductor doped with**
    a. trivalent impurity atoms.
    b. impurity atoms whose electron valence is +4.
    c. pentavalent impurity atoms.
    d. none of the above.

11. **To a first approximation, a forward-biased diode is treated like a(n)**
    a. open switch with infinite resistance.
    b. closed switch with a voltage drop of 0 V.
    c. closed switch in series with a battery voltage of 0.7 V.
    d. closed switch in series with a small resistance and a battery.

12. **What is the DC output voltage of an unfiltered half-wave rectifier whose peak output voltage is 9.8 V?**
    a. 6.23 V.
    b. 19.6 V.
    c. 9.8 V.
    d. 3.1 V.

13. **What is the frequency of the capacitor ripple voltage in a full-wave rectifier circuit if the frequency of the transformer secondary voltage is 60 Hz?**
    a. 60 Hz.
    b. 50 Hz.
    c. 120 Hz.
    d. It cannot be determined.

14. **In a full-wave rectifier, the DC load current equals 1 A. How much DC current is carried by each diode?**
    a. ½ A.
    b. 1 A.
    c. 2 A.
    d. 0 A.

15. **A 12-V zener diode has a 1-W power rating. What is the maximum-rated zener current?**
    a. 120 mA.
    b. 83.3 mA.
    c. 46.1 mA.
    d. 1 A.

16. **In a loaded zener regulator, the series resistor has a current, I_S, of 120 mA. If the load current, I_L, is 45 mA, how much is the zener current, I_Z?**
    a. 45 mA.
    b. 165 mA.
    c. 75 mA.
    d. It cannot be determined.

17. **The approximate voltage drop across a forward-biased LED is**
    a. 0.3 V.
    b. 0.7 V.
    c. 5.6 V.
    d. 2.0 V.

18. **The output from an unfiltered half-wave or full-wave rectifier is a**
    a. pulsating DC voltage.
    b. steady DC voltage.
    c. smooth DC voltage.
    d. none of the above.

19. **A diode is a**
    a. unidirectional device.
    b. linear device.
    c. nonlinear device.
    d. both a and c.

20. **What is the approximate DC output voltage from a filtered bridge rectifier whose peak output voltage is 30 V?**
    a. 19.1 V.
    b. 9.5 V.
    c. 30 V.
    d. none of the above.

Essay Questions

1. Explain why an *n*-type semiconductor material is electrically neutral and not negatively charged.
2. What are two other names for depletion zone.
3. Can a silicon diode be forward-biased if the anode voltage is negative? Explain your answer.
4. Give examples of when to use the first, second, and third diode approximations.
5. Explain why a bridge rectifier would be used instead of a two-diode full-wave rectifier.
6. Explain why the zener current and load current variations in a loaded zener regulator are equal but opposite.

Problems

SECTION 28–1 SEMICONDUCTOR MATERIALS

28–1 How many valence electrons does a silicon or germanium atom have?

28–2 What is it called when a silicon atom shares its four valence electrons with other nearby silicon atoms?

28–3 Define what is meant by an
a. intrinsic semiconductor.
b. extrinsic semiconductor.

28–4 In a semiconductor material, what is an electron-hole pair and how is it created?

28–5 What type of impurity atom is added during the doping process to create a(n)
a. *n*-type semiconductor material?
b. *p*-type semiconductor material?

28–6 What are the majority and minority current carriers in a(n)
a. *n*-type semiconductor?
b. *p*-type semiconductor?

SECTION 28–2 THE *p-n* JUNCTION DIODE

28–7 Why is a diode called a unidirectional device?

28–8 Which side of a diode, the *p* side or the *n* side, is called the
a. anode?
b. cathode?

28–9 How much is the barrier potential, V_B, for a
a. germanium diode?
b. silicon diode?

28–10 In a diode, what is the depletion zone and why is it given that name?

28–11 Describe the proper polarities for
a. forward-biasing a diode.
b. reverse-biasing a diode.

28–12 In a reverse-biased diode, what is the main cause of leakage current?

28–13 Does a reverse-biased diode resemble an open or closed switch?

SECTION 28–3 VOLT-AMPERE CHARACTERISTIC CURVE

28–14 How much current flows through a silicon diode for a forward-bias voltage less than 0.5 V?

28–15 In terms of forward bias, what is the most obvious difference between a silicon and germanium diode?

28–16 Is the leakage current in a diode mainly temperature or voltage dependent?

28–17 For a reverse-biased diode, what is meant by the breakdown voltage, V_{BR}?

28–18 Is a diode a linear or nonlinear device? Explain your answer.

28–19 Calculate the DC resistance of a diode for the following values of V_F and I_F:
a. $V_F = 0.5$ V, $I_F = 50$ μA.
b. $V_F = 0.55$ V, $I_F = 500$ μA.
c. $V_F = 0.6$ V, $I_F = 1$ mA.
d. $V_F = 0.625$ V, $I_F = 5$ mA.
e. $V_F = 0.65$ V, $I_F = 15$ mA.
f. $V_F = 0.68$ V, $I_F = 40$ mA.
g. $V_F = 0.7$ V, $I_F = 70$ mA.

28–20 From the values calculated in Prob. 28–19, what happens to the DC resistance of a diode as the forward bias increases?

28–21 Suppose an analog meter is used to test a diode. What should the meter read for both connections of the meter leads if the diode is
a. good?
b. shorted?
c. open?

28–22 When using an analog meter to test a diode, why should the R × 1 range be avoided?

28–23 Can a DMM set to measure resistance be used to test a diode? Why or why not?

28–24 Explain how a DMM can be used to test a diode.

SECTION 28–4 DIODE APPROXIMATIONS

28–25 Which diode approximation treats a forward-biased diode like a
a. closed switch in series with a battery?
b. closed switch with a voltage drop of 0 V?
c. closed switch in series with a battery and a resistor?

28–26 Which diode approximation
a. treats a reverse-biased diode like an open switch with zero current?
b. includes the high leakage resistance when the diode is reverse-biased?

28–27 Which diode approximation is used if only a rough approximation of the circuit's voltages and currents are needed?

28–28 What is the bulk resistance of a silicon diode for each of the following sets of values:
a. when $V_F = 0.8$ V, $I_F = 100$ mA and when $V_F = 0.72$ V, $I_F = 40$ mA?
b. when $V_F = 0.75$ V, $I_F = 60$ mA and when $V_F = 0.67$ V, $I_F = 12$ mA?
c. when $V_F = 1$ V, $I_F = 800$ mA and when $V_F = 0.7$ V, $I_F = 0$ mA?

28–29 In Fig. 28–29, solve for the load current, I_L, and the load voltage, V_L using
a. the first diode approximation.
b. the second diode approximation.
c. the third diode approximation.

Figure 28–29

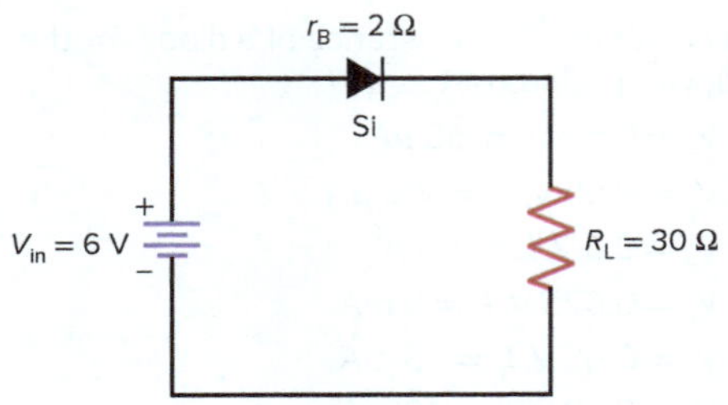

28–30 In Fig. 28–29, how much is the total diode drop when the third approximation is used to solve for V_L and I_L?

28–31 In Fig. 28–30, solve for the load current, I_L, and the load voltage, V_L, using

a. the first diode approximation.

b. the second diode approximation.

c. the third diode approximation.

Figure 28–30

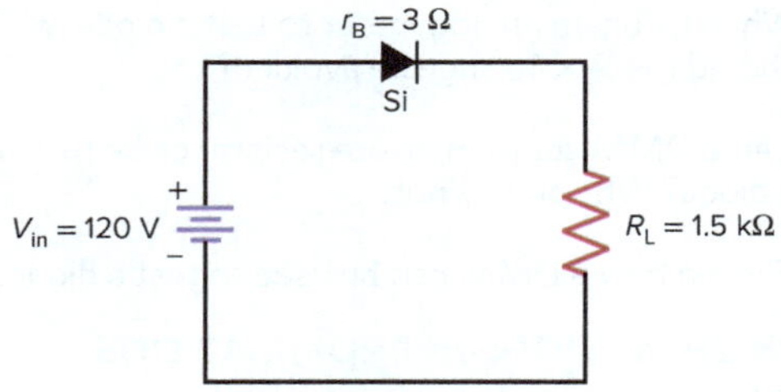

28–32 In Fig. 28–30, which diode approximations are not necessary when solving for I_L and V_L? Justify your answer.

SECTION 28–5 DIODE RATINGS

28–33 Which diode rating, if exceeded, causes the avalanche effect?

28–34 What is the designation for the average forward current rating?

28–35 What is the reverse resistance of a diode if $I_R = 0.01\ \mu$A when $V_R = 200$ V ?

SECTION 28–6 RECTIFIER CIRCUITS

28–36 What type of rectifier is shown in Fig. 28–31?

Figure 28–31

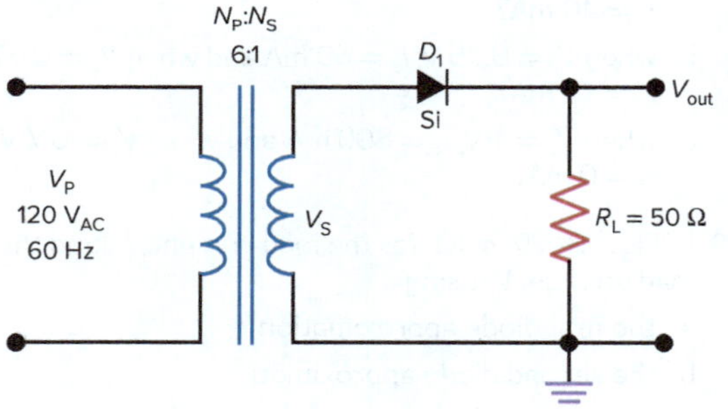

28–37 In Fig. 28–31, calculate the following (use the second diode approximation):

a. V_S.

b. $V_{out(pk)}$.

c. V_{DC}.

d. I_L.

e. I_{diode}.

f. PIV for D_1.

g. f_{out}.

28–38 Recalculate the values in Prob. 28–37 for a transformer turns ratio, N_P:N_S of 2.5:1.

28–39 What type of rectifier is shown in Fig. 28–32?

Figure 28–32

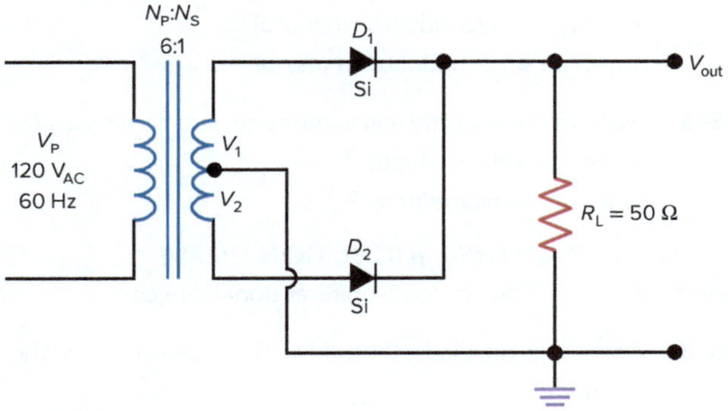

28–40 In Fig. 28–32, calculate the following (use the second diode approximation):

a. $V_{out(pk)}$.

b. V_{DC}.

c. I_L.

d. V_{diode}.

e. PIV for D_1 and D_2.

f. f_{out}.

28–41 Recalculate the values in Prob. 28–40 for a transformer turns ratio, N_P:N_S, of 2.5:1.

28–42 What type of rectifier is shown in Fig. 28–33?

Figure 28–33

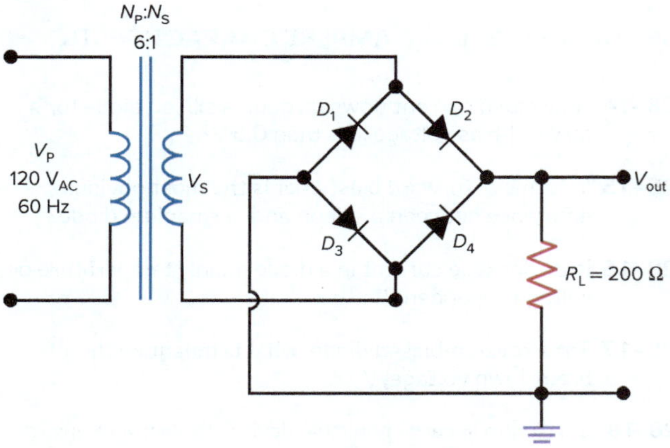

28–43 In Fig. 28–33, calculate the following (use the second diode approximation):

a. $V_{out(pk)}$.
b. V_{DC}.
c. I_L.
d. I_{diode}.
e. PIV for any diode.
f. f_{out}.

28–44 Recalculate the values in Prob. 28–43 for a transformer turns ratio, $N_P:N_S$, of 5:1.

28–45 If a 2200-μF capacitor is added to the output in Fig. 28–31, calculate the following:

a. V_{ripple}.
b. V_{DC}.
c. I_L.
d. PIV.

28–46 If a 1000-μF capacitor is added to the output in Fig. 28–32, calculate the following:

a. V_{ripple}.
b. V_{DC}.
c. I_L.
d. PIV.

28–47 If a 680-μF capacitor is added to the output in Fig. 28–33, calculate the following:

a. V_{ripple}.
b. V_{DC}.
c. I_L.
d. PIV.

SECTION 28–7 SPECIAL DIODES

28–48 In Fig. 28–34, calculate the LED current for each of the following values of R_S:

a. $R_S = 2.7\ \text{k}\Omega$.
b. $R_S = 1.5\ \text{k}\Omega$.
c. $R_S = 1\ \text{k}\Omega$.
d. $R_S = 510\ \Omega$.

Figure 28–34

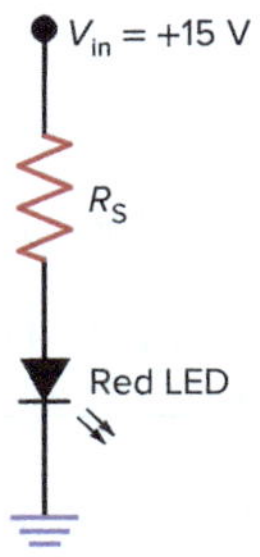

28–49 In Fig. 28–34, what value of R_S will provide an LED current of 20 mA?

28–50 Calculate the maximum-rated zener current, I_{ZM}, for the following ½-W zener diodes:

a. $V_Z = 5.6$ V.
b. $V_Z = 6.8$ V.
c. $V_Z = 10$ V.
d. $V_Z = 18$ V.

28–51 In Fig. 28–35, solve for the following:

a. I_S.
b. I_L.
c. I_Z.

Figure 28–35

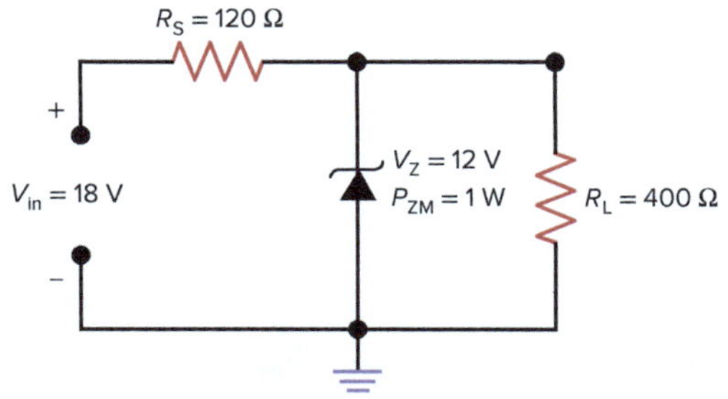

28–52 In Fig. 28–36, solve for the following:

a. I_S.
b. I_L.
c. I_Z.

Figure 28–36

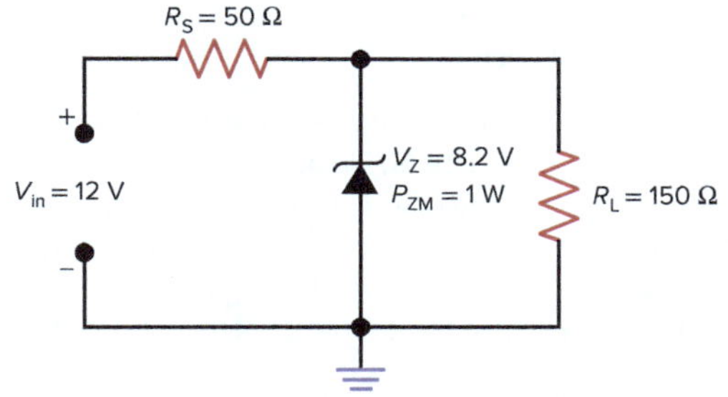

28–53 In Fig. 28–36, how much is I_Z if R_L opens?

Answers to Self-Reviews

28–1 **a.** *n*-type
b. electrons
c. positive

28–2 **a.** 0.7 V
b. anode, cathode
c. positive
d. open

28–3 **a.** minority
b. 22 Ω
c. bad

28–4 **a.** the third approximation
b. the first approximation
c. the second approximation

28–5 **a.** breakdown voltage rating, V_{BR}
b. false

28–6 **a.** 12.24 V
b. 12 V
c. 24.5 V
d. 10 V

28–7 **a.** true
b. false
c. true

Laboratory Application Assignment

In this lab application assignment, you will examine half-wave and full-wave rectifier circuits. In each circuit you will analyze the output voltage waveforms and values with and without a filter capacitor connected to the output.

Equipment: Obtain the following items from your instructor.

- Isolation transformer and Variac
- Transformer: 120-V primary, 25.2-V, 2-A secondary with center tap
- Two 1N4002 silicon diodes or equivalent
- 470-μF electrolytic capacitor
- 1 kΩ, ½-watt carbon-film resistor
- DMM and oscilloscope

Caution: In this lab, you will be working with 120 V_{AC}. For your safety, you will need to use an isolation transformer. Plug the isolation transformer into the 120-V_{AC} outlet on your benchtop and in turn plug a Variac into the isolation transformer. Next, adjust the Variac for an output of 120 V_{AC}. This is the voltage you will apply directly to the primary of the transformer.

Transformer Measurements

Connect the circuit in Fig. 28–37*a*. With exactly 120 V_{AC} applied to the primary, measure and record the following rms values of secondary voltage. (Use your DMM.) Note that V_1 and V_2 each represent the voltage measured from one side of the transformer secondary to the center tap, whereas V_S represents the full secondary voltage.

V_1 = ________, V_2 = ________, V_S = ________

Are these voltages approximately 10% higher than the rated values?________________________________

If yes, explain why. ________________________________

____

Use these measured values in all your calculations that follow.

Half-Wave Rectifier

Examine the half-wave rectifier in Fig. 28–37*a*. Calculate and record the following circuit values:

$V_{out(pk)}$ = ________, V_{DC} = ________, I_L = ________,
I_D = ________, f_{out} = ________

Connect channel 1 of your oscilloscope to the top of the transformer secondary and channel 2 across the load resistor, R_L. Set the channel 2 input coupling switch to DC. Adjust the sec./div. control of the oscilloscope to view at least two complete cycles of secondary voltage. Draw the channels 1 and 2 waveforms on the scope graticule provided in Fig. 28–38. Label each waveform. From your displayed waveforms, what is

a. The peak output voltage across the load resistor, R_L?
$V_{out(pk)}$ = ________
b. The period, *T*, and frequency, *f*, of the secondary voltage?
T = ________, *f* = ________
c. The period, *T*, and frequency, *f*, of the load voltage?
T = ________, *f* = ________

Next, measure and record the DC load voltage and current:
V_{DC} = ________, I_L = ________
Connect a 470-μF filter capacitor across R_L. (Observe polarity.)
Remeasure V_{DC} and I_L. V_{DC} = ____________, I_L = ____________
Did the filter capacitor increase the DC load voltage?________
If yes, why did this happen? ________________________________

____

Explain the waveform that is now displayed on channel 2 of your oscilloscope. ________________________________

____

Change the channel 2 input coupling switch to AC, and reduce the volts/div. setting. Measure and record the peak-to-peak ripple voltage. V_{ripple} = ________

Figure 28–37

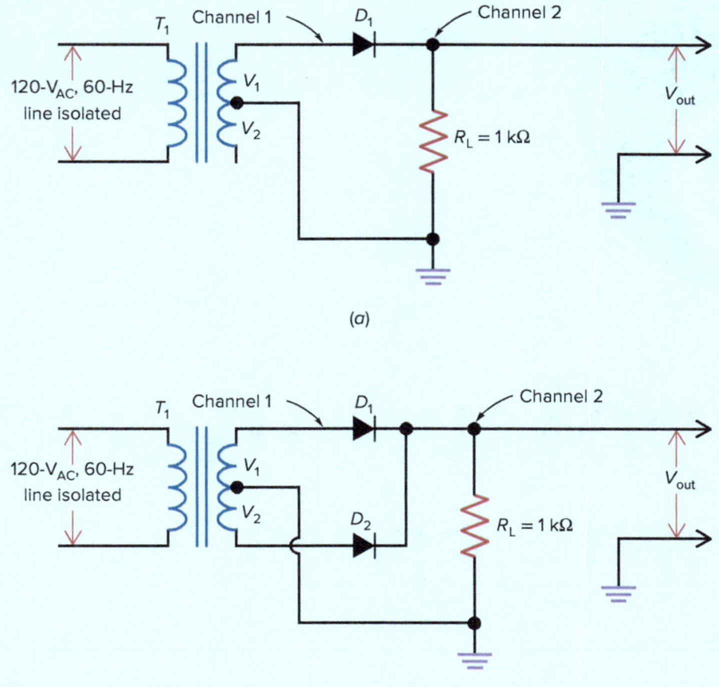

Full-Wave Rectifier

Examine the full-wave rectifier in Fig. 28–37*b*. Calculate and record the following circuit values:
$V_{out(pk)}$ = ________, V_{DC} = ________, I_L = ________,
I_D = ________, f_{out} = ________

Construct the full-wave rectifier in Fig. 28–37*b*. Connect channel 1 of your oscilloscope to the top of the transformer secondary and channel 2 across the load resistor, R_L. Set the channel 2 input coupling switch to DC. Adjust the sec./div. control of the oscilloscope to view at least two complete cycles of secondary voltage. Draw the channels 1 and 2 waveforms on the scope graticule provided in Fig. 28–39. Label each waveform. From your displayed waveforms, what is

a. The peak output voltage across the load resistor, R_L?
$V_{out(pk)}$ = ________

b. The period, T, and frequency, f, of the secondary voltage?
T = ________, f = ________

c. The period, T, and frequency, f, of the load voltage?
T = ________, f = ________

Next, measure and record the following DC values:
V_{DC} = ________, I_L = ________, I_{D_1} = ________,
I_{D_2} = ________

Connect a 470-μF filter capacitor across R_L. (Observe polarity.) Remeasure V_{DC} and I_L. V_{DC} = ________, I_L = ________

Did the filter capacitor increase the DC load voltage? ________

If yes, why did this happen? ____

Explain the waveform that is now displayed on channel 2 of your oscilloscope. ____

Change the channel 2 input coupling switch to AC, and reduce the volts /div. setting. Measure and record the peak-to-peak ripple voltage. V_{ripple} = ________ How does this value compare to what was measured in the half-wave rectifier? ____

Figure 28–38

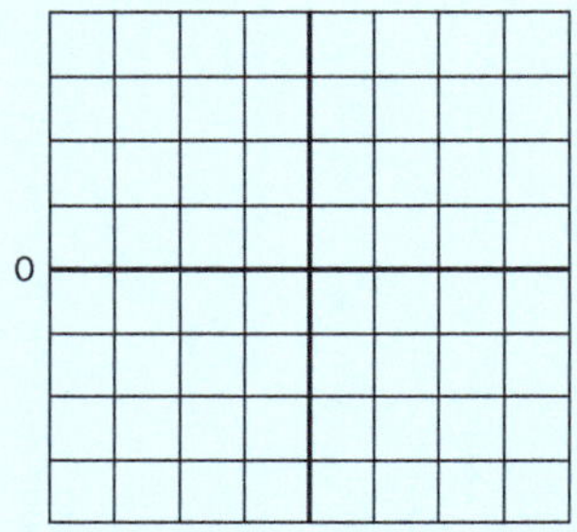

Figure 28–39

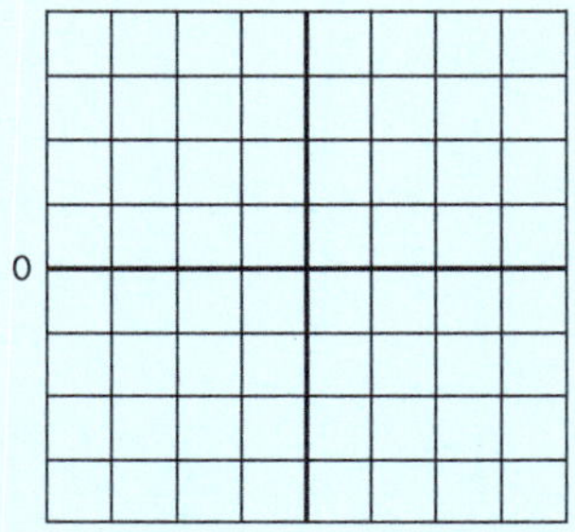

Design credits: Chapter Opener Background Image, and Multisim: Stockbyte/Getty Images

chapter

29

Bipolar Junction Transistors

Transistors are used when it is necessary to amplify voltage, current, and power. With a small signal applied to the transistor amplifier, the transistor and its associated circuitry can produce an amplified version of the input signal. The output signal can be hundreds or even thousands of times larger than the input. In computer circuits, the transistor can be used as an electronic switch. When used as a switch, the transistor is either ON or OFF. In this chapter, you will study basic transistor construction, the proper biasing arrangement, and the general characteristics of transistors. You will also learn about the most common ways to bias a transistor which include base bias, voltage divider bias, and emitter bias. ■

Chapter Outline

Chapter Objectives

After studying this chapter, you should be able to

- *List* the three doped regions of a transistor.
- *Explain* the role of each doped region in a transistor.
- *Identify* the schematic symbol of *npn* and *pnp* transistors.
- *Explain* how to properly bias the emitter-base and collector-base junctions of a transistor.
- *State* the mathematical relationship between the emitter, base, and collector currents in a transistor.
- *Define* α_{DC} and β_{DC}.
- *Define* the active, saturation, cutoff, and breakdown operating regions of a transistor.
- *Draw* the DC equivalent circuit of a transistor.
- *Calculate* the power dissipated by a transistor.
- *Explain* how to test a transistor with an analog ohmmeter and a DMM.
- *Calculate* the voltages and currents in a transistor circuit using base bias, voltage divider bias, and emitter bias.
- *Draw* the DC load line for a transistor circuit.
- *Locate* the *Q* point on the DC load line.

Important Terms

active region
base
breakdown region
collector
cutoff
DC alpha, α_{DC}
DC beta, β_{DC}
DC load line
derating factor
emitter
midpoint bias
Q point
saturation
transistor

29–1 Transistor Construction

A **transistor** has three doped regions, as shown in Fig. 29–1. Figure 29–1*a* shows an *npn* transistor, and Fig. 29–1*b* shows a *pnp* transistor. Notice that for both types, the **base** is a narrow region sandwiched between the larger **collector** and **emitter** regions. The emitter region of a transistor is heavily doped. Its job is to emit or inject current carriers into the base. The base region is very thin and lightly doped. Most of the current carriers injected into the base from the emitter do not flow out the base lead. Instead, most of the current carriers injected into the base pass on to the collector. The collector region is moderately doped and is the largest of all three regions. The collector region attracts the current carriers that are injected into the thin and lightly doped base region. Incidentally, the collector region is the largest of all the three regions because it must dissipate more heat than the emitter or base regions.

In *npn* transistors, the majority current carriers are free electrons in the emitter and collector, whereas the majority current carriers are holes in the base. The opposite is true in a *pnp* transistor where the majority current carriers are holes in the emitter and collector, and the majority current carriers are free electrons in the base.

Figure 29–2 shows the depletion layers in an unbiased *npn* transistor. The diffusion of electrons from both *n* regions into the *p*-type base causes a barrier potential, V_B, for both *p-n* junctions. The *p-n* junction at the left is the emitter-base junction; the *p-n* junction at the right is the collector-base junction. For silicon, the barrier potential for both the emitter-base (EB) and collector-base (CB) junctions equals approximately 0.7 V.

Notice in Fig. 29–2 that the EB depletion layer is narrower than the CB depletion layer. The reason for the different widths can be attributed to the doping level of the emitter and collector regions. With heavy doping in the emitter region, the penetration into the *n* material is minimal due to the availability of many free electrons. On the collector side, however, there are fewer free electrons available due to the more moderate doping level in this region. Therefore, the depletion layer must penetrate deeper into the collector region to set up the barrier potential, V_B, of 0.7 V. In Fig. 29–2, dash marks are used in the *n*-type emitter and collector to indicate the large number of free electrons in these regions. Small circles are used to indicate the holes in the *p*-type base region. (For an *npn* transistor, holes are the minority current carriers in the *n*-type emitter and collector regions, whereas free electrons are the minority current carriers in the *p*-type base.)

GOOD TO KNOW

On December 23, 1947, Walter H. Brattain and John Bardeen demonstrated the amplifying action of the first transistor at the Bell telephone laboratories. The first transistor was called a point-contact transistor, which was the predecessor to the junction transistor invented by Schockley.

Figure 29–1 Transistor construction showing the three doped regions. (*a*) *npn* transistor. (*b*) *pnp* transistor.

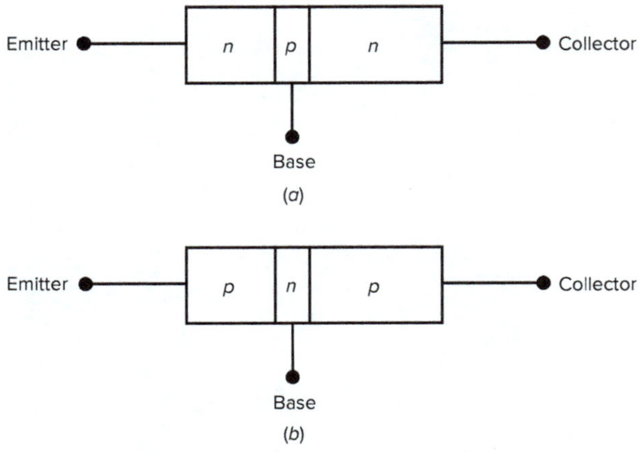

Figure 29-2 Depletion layers in an *npn* transistor.

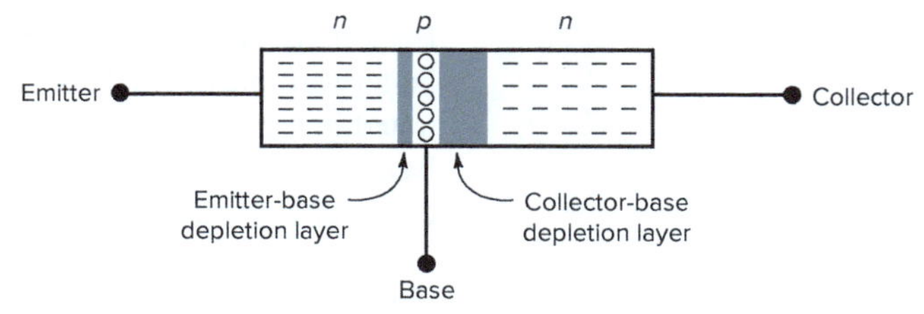

Figure 29-3 Schematic symbols for transistors. (*a*) *npn* transistor. (*b*) *pnp* transistor.

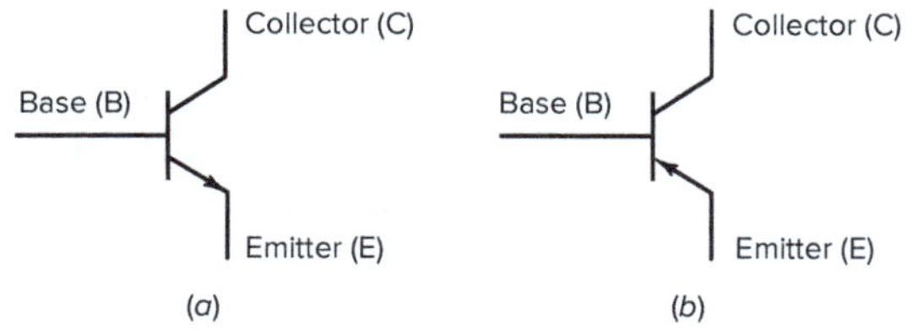

Schematic Symbols

Figure 29–3 shows the schematic symbols for both the *npn* and *pnp* transistors. Notice the arrow on the emitter lead for both types. For the *npn* transistor in Fig. 29–3*a*, the arrow on the emitter lead points outward, and in the *pnp* transistor of Fig. 29–3*b*, the arrow on the emitter lead points inward.

The *npn* and *pnp* transistors are not different in terms of their ability to amplify voltage, current, or power. Each type, however, does require different polarities of operating voltages. For example, the collector-emitter voltage, V_{CE}, of an *npn* transistor must be positive, and the collector-emitter voltage, V_{CE}, must be negative for the *pnp* type.

In summary, it is important to note the following points about the construction of a transistor:

1. *The emitter region is heavily doped.* Its job is to emit or inject current carriers into the base region. For *npn* transistors, the *n*-type emitter injects free electrons into the base. For *pnp* transistors, the *p*-type emitter injects holes into the base.
2. *The base is very thin and lightly doped.* Most of the current carriers injected into the base region cross over into the collector side and do not flow out the base lead.
3. *The collector region is moderately doped.* It is also the largest region within the transistor. Its function is to collect or attract current carriers injected into the base region.

29–1 Self-Review

Answers at the end of the chapter.

a. Which region in a transistor is the most heavily doped?
b. Which region in a transistor is the largest?
c. Which region in a transistor is very thin and lightly doped?
d. Which lead of a transistor schematic symbol has an arrow on it?

29–2 Proper Transistor Biasing

For a transistor to function properly as an amplifier, the emitter-base junction must be forward-biased, and the collector-base junction must be reverse-biased, as illustrated in Fig. 29–4*a*. Notice the common connection for the voltage sources at the base lead of the transistor. The emitter-base supply voltage is designated V_{EE} and the collector-base supply voltage is designated V_{CC}.

Transistor Currents

Figure 29–4*b* shows the emitter current, I_E, the base current, I_B, and the collector current, I_C. Electrons in the *n*-type emitter are repelled into the base by the negative terminal of the emitter supply voltage, V_{EE}. Since the base is very thin and lightly doped, only a few electrons combine with holes in the base. The small current flowing out of the base lead (which is the base current, I_B) is called recombination current because free electrons injected into the base must fall into a hole before they can flow out the base lead.

Notice in Fig. 29–4*b* that most of the emitter-injected electrons pass through the base region and into the collector region. The reason is twofold. First, only a few holes are available for recombination in the base. Second, the positive collector-base voltage attracts the free electrons in the *p*-type base over to the collector side before they can recombine with holes in the base. In most transistors, the collector current, I_C, is nearly identical to the emitter current, I_E. This is equivalent to saying that the recombination current, I_B, is very small.

Only a small voltage is needed to create an electric field strong enough in the collector-base junction to collect almost all the free electrons injected into the base. After the collector-base voltage reaches a certain level, increasing it further will have little or no effect on the number of free electrons entering the collector. As a matter of fact, after the collector-base voltage is slightly above zero, full current is

Figure 29–4 Transistor biasing for the common-base connection. (*a*) Proper biasing for an *npn* transistor. The EB junction is forward-biased by the emitter supply voltage, V_{EE}. V_{CC} reverse-biases the CB junction. (*b*) Currents in a transistor.

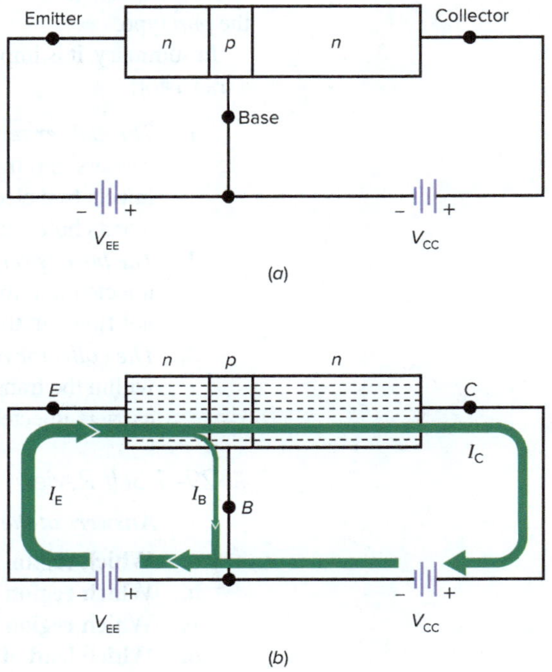

obtained in the collector. If the voltage across the collector-base junction is too large, however, the breakdown voltage may be exceeded, which could destroy the transistor.

Notice the relative size of the current arrows shown in Fig. 29–4*b*. The currents are illustrated in this manner to emphasize their relationship with each other. The currents in a transistor are related as shown in Formulas (29–1), (29–2), and (29–3).

$$I_E = I_B + I_C \tag{29–1}$$

$$I_C = I_E - I_B \tag{29–2}$$

$$I_B = I_E - I_C \tag{29–3}$$

Example 29-1

A transistor has the following currents: $I_B = 20$ mA and $I_C = 4.98$ A. Calculate I_E.

ANSWER Using Formula (29–1), the calculations are

$$\begin{aligned} I_E &= I_B + I_C \\ &= 20 \text{ mA} + 4.98 \text{ A} \\ &= 0.02 \text{ A} + 4.98 \text{ A} \\ &= 5 \text{ A} \end{aligned}$$

Example 29-2

A transistor has the following currents: $I_E = 100$ mA, $I_B = 1.96$ mA. Calculate I_C.

ANSWER Using Formula (29–2), the calculations are

$$\begin{aligned} I_C &= I_E - I_B \\ &= 100 \text{ mA} - 1.96 \text{ mA} \\ &= 98.04 \text{ mA} \end{aligned}$$

Example 29-3

A transistor has the following currents: $I_E = 50$ mA, $I_C = 49$ mA. Calculate I_B.

ANSWER Using Formula (29–3), the calculations are

$$\begin{aligned} I_B &= I_E - I_C \\ &= 50 \text{ mA} - 49 \text{ mA} \\ &= 1 \text{ mA} \end{aligned}$$

DC Alpha

The circuit shown in Fig. 29–4 is called a *common-base (CB)* connection because the base lead is common to both the input and output sides of the circuit. A characteristic that describes how closely the emitter and collector currents are in a common base circuit is called the ***DC alpha,*** designated α_{DC}. This is expressed in Formula (29–4).

$$\alpha_{DC} = \frac{I_C}{I_E} \qquad \textbf{(29–4)}$$

In most cases, the DC alpha is 0.99 or greater. The thinner and more lightly doped the base, the closer alpha is to one, or unity. In most discussions, the DC alpha is so close to one that we ignore the small difference that exists.

Example 29-4

A transistor has the following currents: $I_E = 15$ mA, $I_B = 60\ \mu$A. Calculate α_{DC}.

ANSWER First calculate I_C using Formula (29–2). The calculations are

$$\begin{aligned} I_C &= I_E - I_B \\ &= 15 \text{ mA} - 60\ \mu\text{A} \\ &= 15 \text{ mA} - 0.06 \text{ mA} \\ &= 14.94 \text{ mA} \end{aligned}$$

Next, use Formula (29–4) to calculate α_{DC}:

$$\begin{aligned} \alpha_{DC} &= \frac{I_C}{I_E} \\ &= \frac{14.94 \text{ mA}}{15 \text{ mA}} \\ &= 0.996 \end{aligned}$$

DC Beta

Figure 29–5 shows another way to connect external voltages to the *npn* transistor. V_{BB} provides the forward bias for the base-emitter junction, and V_{CC} provides the reverse bias for the collector-base junction. This connection is called the common-emitter (CE) connection since the emitter lead is common to both the input and

GOOD TO KNOW

The symbol h_{FE} is also used to indicate the DC beta of a transistor. The symbol h_{FE} represents the forward current transfer ratio in the common-emitter configuration. The symbol h_{FE} is a hybrid (h) parameter symbol. The h-parameter system in use today is the most common way of specifying transistor parameters.

Figure 29–5 Transistor biasing for the common-emitter connection.

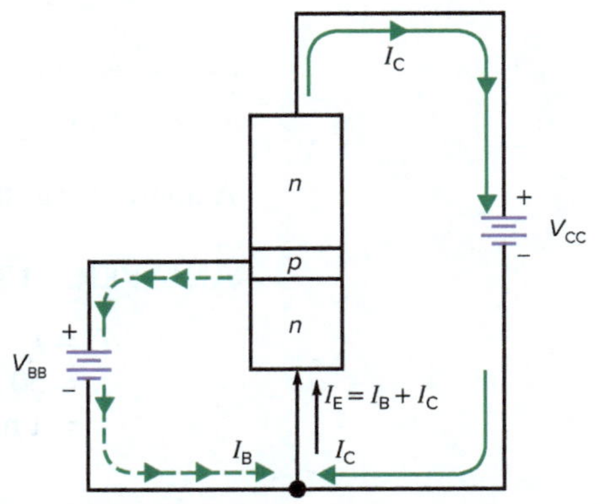

output sides of the circuit. Notice the arrows indicating the direction of the transistor currents I_E, I_C, and I_B.

The DC current gain of a transistor in the common-emitter connection is called the ***DC beta,*** usually designated, β_{DC}. The DC beta is expressed in Formula (29–5).

$$\beta_{DC} = \frac{I_C}{I_B} \tag{29–5}$$

Example 29-5

A transistor has the following currents: $I_C = 10$ mA, $I_B = 50$ μA. Calculate β_{DC}.

ANSWER Using Formula (29–5), the calculations are

$$\beta_{DC} = \frac{I_C}{I_B}$$
$$= \frac{10 \text{ mA}}{50\ \mu\text{A}}$$
$$= 200$$

Example 29-6

A transistor has $\beta_{DC} = 150$ and $I_B = 75$ μA. Calculate I_C.

ANSWER Begin with

$$\beta_{DC} = \frac{I_C}{I_B}$$

Next, rearrange Formula (29–5) to solve for I_C:

$$I_C = \beta_{DC} \times I_B$$
$$= 150 \times 75\ \mu\text{A}$$
$$= 11.25 \text{ mA}$$

Relating β_{DC} and α_{DC}

If β_{DC} is known, α_{DC} can be found by using Formula (29–6):

$$\alpha_{DC} = \frac{\beta_{DC}}{1 + \beta_{DC}} \tag{29–6}$$

Likewise, if α_{DC} is known, β_{DC} can be found by using Formula (29–7):

$$\beta_{DC} = \frac{\alpha_{DC}}{1 - \alpha_{DC}} \tag{29–7}$$

These formulas are derived from Formulas (29–1), (29–2), and (29–3).

Example 29-7

A transistor has $\beta_{DC} = 100$. Calculate α_{DC}.

ANSWER Using Formula (29–6), the calculations are

$$\alpha_{DC} = \frac{\beta_{DC}}{1 + \beta_{DC}}$$

$$= \frac{100}{1 + 100}$$

$$= 0.99$$

Example 29-8

A transistor has $\alpha_{DC} = 0.995$. Calculate β_{DC}.

ANSWER Using Formula (29–7), the calculations are

$$\beta_{DC} = \frac{\alpha_{DC}}{1 - \alpha_{DC}}$$

$$= \frac{0.995}{1 - 0.995}$$

$$= 199$$

In Example 29–7, notice how close α_{DC} is to one. For this reason, it can usually be assumed to have a value of one, or unity, in most transitor circuit analyses.

29–2 Self-Review

Answers at the end of the chapter.

a. For normal transistor operation, the EB junction is (forward/reverse)-biased and the CB junction is (forward/reverse)-biased.

b. A transistor has an emitter current of 15 mA and a collector current of 14.88 mA. How much is the base current?

c. In Question b, how much is the DC alpha?

d. A transistor has a collector current of 90 mA and a base current of 500 μA. What is the transistor's DC beta?

29–3 Transistor Operating Regions

Figure 29–6*a* shows an *npn* transistor in a CE connection. Notice that the base supply voltage, V_{BB}, and the collector supply voltage, V_{CC}, are variable. Notice also that a base resistor, R_B, is used to control the amount of base current, I_B. With a fixed value for R_B, V_{BB} can be adjusted to produce the desired value of base current.

Figure 29–6 Common-emitter connection. (a) Circuit. (b) Graph of I_C versus V_{CE} for different base current values.

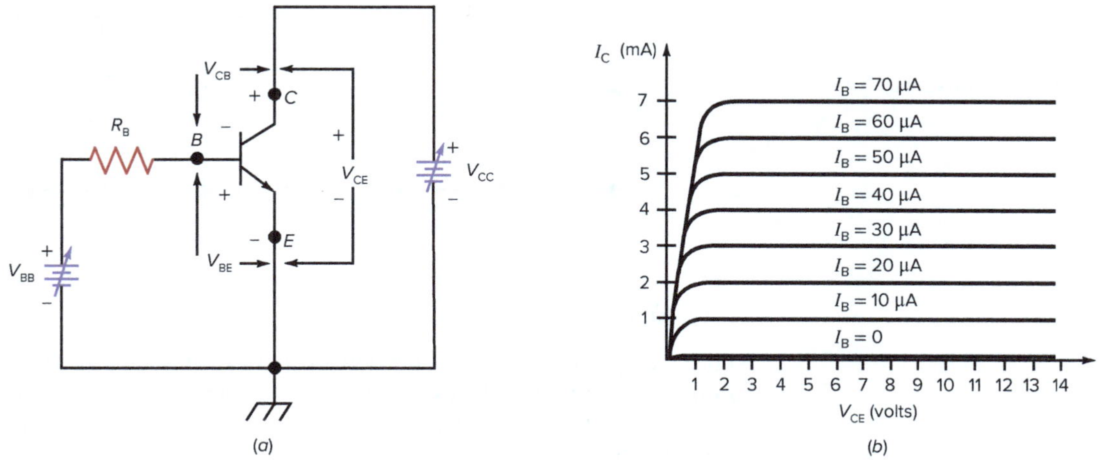

GOOD TO KNOW

When displayed on a curve tracer, the collector curves in Fig. 29–6*b* actually have a slight upward slope. This is the result of the base region becoming slightly narrower as V_{CE} increases. (As V_{CE} increases, the CB depletion layer widens, thus narrowing the base.) With a smaller base region, there are fewer holes available for recombination. Since each curve represents a constant base current, the effect is seen as an increase in collector current.

GOOD TO KNOW

When a transistor is saturated, further increases in base current produce no further increases in collector current.

Transistor Voltages and Currents

In Fig. 29–6*a*, V_{BB} can be adjusted to provide a wide range of base and collector current values. Assume that V_{BB} has been adjusted to produce an I_B of 50 μA. If $\beta_{DC} = 100$, then I_C is

$$I_C = \beta_{DC} \times I_B$$
$$= 100 \times 50\ \mu A$$
$$= 5\ mA$$

As long as the collector-base junction remains reverse-biased, I_C remains at 5 mA. This is true regardless of the actual voltage between the collector and base.

In Fig. 29–6, V_{CC} can be varied from a few tenths of a volt to several volts without having any effect on the collector current, I_C! This is true provided the collector-base breakdown voltage rating of the transistor is not exceeded. If V_{BB} is increased to provide a base current, I_B, of 100 μA, then $I_C = 100 \times 100\ \mu A = 10$ mA. Again, if V_{CC} is varied from a few tenths of a volt to several volts, I_C remains constant.

In Fig. 29–6*a*, notice that $V_{CE} = V_{CB} + V_{BE}$. When V_{CB} is a few tenths of a volt above zero, the collector-base diode is reverse-biased and $I_C = I_B \times \beta_{DC}$. This means that I_C is controlled solely by the base current, I_B, and **not** by the collector supply voltage V_{CC}.

Saturation Region

Figure 29–6*b* shows the action of the transistor for several different base currents. As can be seen, when V_{CE} is zero, I_C is zero because the collector-base function is not reverse-biased when $V_{CE} = 0$. Without a positive voltage at the collector, it cannot attract electrons from the base. When V_{CE} increases from zero, however, I_C increases linearly. The vertical portion of the curves near the origin is called the ***saturation region***. When a transistor is saturated, the collector current, I_C, is not controlled solely by the base current, I_B.

Breakdown Region

When the collector-base voltage is too large, the collector-base diode breaks down, causing a large, undesired collector current to flow. This is the ***breakdown region***. This area of operation should always be avoided in transistor circuits. This region is not shown in Fig. 29–6*b* because it is assumed that breakdown will not occur when the circuit is designed properly.

GOOD TO KNOW

A bipolar junction transistor is frequently used as a constant current source.

Figure 29–7 DC equivalent circuit of a transistor operating in the active region.

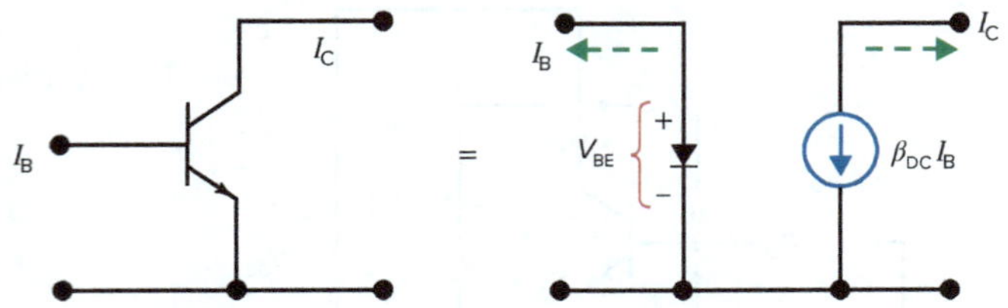

Cutoff Region

Notice the $I_B = 0$ curve nearest the horizontal axis in Fig. 29–6*b*. This is called the ***cutoff** region* because only a small collector current, I_C, flows. For silicon transistors, this current is very small and is therefore usually ignored. A transistor is said to be cut off when its collector current, I_C, is zero.

Active Region

The **active region** of a transistor is where the collector curves are nearly horizontal. When a transistor operates in the active region, the collector current, I_C, is greater than the base current, I_B, by a factor of beta or $I_C = \beta_{DC} \times I_B$. In the active region, the collector circuit acts like a current source.

DC Equivalent of a Transistor

Figure 29–7 shows the DC equivalent circuit of a transistor operating in the active region. Notice that the base-emitter junction acts like a forward-biased diode with a current, I_B. Usually, the second approximation of a diode is used, rather than the first or third. If the transistor is silicon, assume that V_{BE} equals 0.7 V.

Notice also that the collector circuit in Fig. 29–7 is replaced with a current source. The collector current source has an output current equal to $\beta_{DC} \times I_B$. Ideally, the current source has infinite internal impedance. With the CE connection in Fig. 29–7, we note that the collector current, I_C, is controlled only by the base current, I_B, assuming β_{DC} is a fixed quantity. When I_B changes, I_C still equals $\beta_{DC} I_B$. In Fig. 29–7, the arrow in the current source symbol points in the direction of conventional current flow. Of course, electron flow is in the opposite direction, indicated by the dashed arrows for I_B and I_C.

29–3 Self-Review

Answers at the end of the chapter.

a. In what operating region does the collector of a transistor act like a current source?

b. In what region is a transistor operating if the collector current is zero?

c. When a transistor is operating in its active region, is I_C controlled by I_B or V_{CC}?

29–4 Transistor Ratings

A transistor, like any other device, has limitations on its operations. These limitations are specified in the manufacturer's data sheet. In all the cases, the maximum ratings are given for collector-base voltage, collector-emitter voltage, emitter-base voltage, collector current, and power dissipation.

Power Dissipation Rating, $P_{d(max)}$

The product of V_{CE} and I_C gives the power dissipation, P_d, of the transistor. This is shown in Formula (29–8):

$$P_d = V_{CE} \times I_C \qquad \textbf{(29–8)}$$

The product $V_{CE} \times I_C$ must not exceed the maximum power dissipation rating, $P_{d(max)}$, of the transistor. Formula (29–8) can also be rearranged to solve for the maximum allowable collector current, I_C, for a specified value of collector-emitter voltage, V_{CE}.

Example 29-9

In Fig. 29–6*a* calculate P_d if $V_{CC} = 10$ V and $I_B = 50\ \mu$A. Assume $\beta_{DC} = 100$.

ANSWER First calculate the collector current, I_C. The calculations are

$$\begin{aligned} I_C &= \beta_{DC} \times I_B \\ &= 100 \times 50\ \mu\text{A} \\ &= 5\ \text{mA} \end{aligned}$$

Next, from Fig. 29–6*a*, $V_{CC} = V_{CE} = 10$ V. Therefore, the power dissipation, P_d can be calculated using Formula (29–8).

$$\begin{aligned} P_d &= V_{CE} \times I_C \\ &= 10\ \text{V} \times 5\ \text{mA} \\ &= 50\ \text{mW} \end{aligned}$$

The transistor must have a power rating higher than 50 mW to avoid becoming damaged. Incidentally, note that small signal transistors have power dissipation ratings of less than 0.5 W, whereas power transistors have P_d ratings greater than 0.5 W.

Example 29-10

The transistor in Fig. 29–6*a* has a power rating of 0.5 W. If $V_{CE} = 20$ V, calculate the maximum allowable collector current, I_C, that can exist without exceeding the transistor's power rating.

ANSWER Begin by rearranging Formula (29–8) to solve for the collector current, I_C:

$$I_{C(max)} = \frac{P_{d(max)}}{V_{CE}}$$

Inserting the values of $V_{CE} = 20$ V, and $P_{d(max)} = 0.5$ W gives

$$I_{C(max)} = \frac{0.5\ \text{W}}{20\ \text{V}}$$

$$= 25\ \text{mA}$$

Derating Factor

The power dissipation rating of a transistor is usually given at 25°C. At higher temperatures, the power dissipation rating is less. For most silicon transistors, the maximum allowable junction temperature is between 150° and 200°C. Temperatures higher than this will destroy the transistor. This is why a manufacturer must specify a maximum power rating for the transistor. The transistor's power dissipation rating must be kept to less than its rated value so that the junction temperature will not reach destructive levels. Manufacturers usually supply **derating factors** for determining the power dissipation rating at any temperature above 25°C. The derating factor is specified in W/°C. For example, if a transistor has a derating factor of 2 mW/°C then for each 1°C rise in junction temperature, the power rating of the transistor is reduced by 2 mW.

Example 29-11

Assume that a transistor has a power rating $P_{d(max)}$ of 350 mW at an ambient temperature T_A of 25°C. The derating factor is 2.8 mW/°C. Calculate the power rating at 50°C.

ANSWER First, calculate the change in temperature. $\Delta T = 50°C - 25°C = 25°C$. Next, multiply the change in temperature by the derating factor of 2.8 mW/°C. Finally, subtract this answer from 350 mW to get the new P_d rating. The calculations are

$$\Delta P_d = \Delta T \times \text{derating factor}$$

$$= (50° - 25°C) \times \left(\frac{2.8\text{ mW}}{°C}\right)$$

$$= 25°C \times \frac{2.8\text{ mW}}{°C}$$

$$= 70\text{ mW}$$

It is important to note that ΔT is the change in temperature and ΔP_d is the reduction in the P_d rating. Therefore, the power dissipation rating at 50°C is

$$350\text{ mW} - 70\text{ mW} = 280\text{ mW}$$

For larger power transistors the derating factor is usually specified for the case temperature, T_C, rather than the ambient temperature, T_A.

Breakdown Voltage Ratings

A data sheet lists the breakdown voltage ratings for the emitter-base, collector-base, and collector-emitter junctions. For example, the data sheet of a 2N3904 small signal transistor has the following breakdown voltage ratings:

$$V_{CBO} = 60\ V_{DC}$$
$$V_{CEO} = 40\ V_{DC}$$
$$V_{EBO} = 6.0\ V_{DC}$$

The first two letters in the subscript indicate the two transistor terminals for which the voltage rating applies, and the third letter indicates the condition of the unmentioned terminal. The first voltage, V_{CBO}, indicates the maximum allowable collector-to-base voltage with the emitter terminal open. The second voltage, V_{CEO}, is the maximum allowable collector-emitter voltage with the base open. The voltage rating, V_{EBO}, is the maximum allowable emitter-base voltage with the collector open. Exceeding any one of these voltage ratings can destroy the transistor.

29–4 Self-Review

Answers at the end of the chapter.

a. **How much power is dissipated by a transistor if V_{CE} = 15 V and I_C = 300 mA?**

b. **A transistor has a power rating of 1 W at 25°C. If the derate factor is 4 mW/°C, what is the transistor's power rating at 125°C?**

29–5 Checking a Transistor with an Ohmmeter

An analog ohmmeter can be used to check a transistor because the emitter-base and collector-base junctions are *p-n* junctions. This is illustrated in Fig. 29–8 where the *npn* transistor is replaced by its diode equivalent circuit.

To check the base-emitter junction of an *npn* transistor, first connect the ohmmeter as shown in Fig. 29–9*a*, and then reverse the ohmmeter leads, as shown in Fig. 29–9*b*. In Fig. 29–9*a*, the resistance indicated by the ohmmeter should be low since the base-emitter junction is forward-biased. In Fig. 29–9*b*, the resistance indicated by the ohmmeter should read high because the base-emitter junction is reverse-biased. For a good *p-n* junction made of silicon, the ratio R_R/R_F should be equal to or greater than 1000:1.

To check the collector-base junction, repeat the process described for the base-emitter junction. For clarity, the ohmmeter connections are shown in Fig. 29–10. Notice that in Fig. 29–10*a*, the ohmmeter reads a low resistance because the collector-base junction is forward-biased. Conversely, in Fig. 29–10*b*, the ohmmeter reads a high resistance because the collector-base junction is reverse-biased. Although not

Figure 29–8 An *npn* transistor and its diode equivalent.

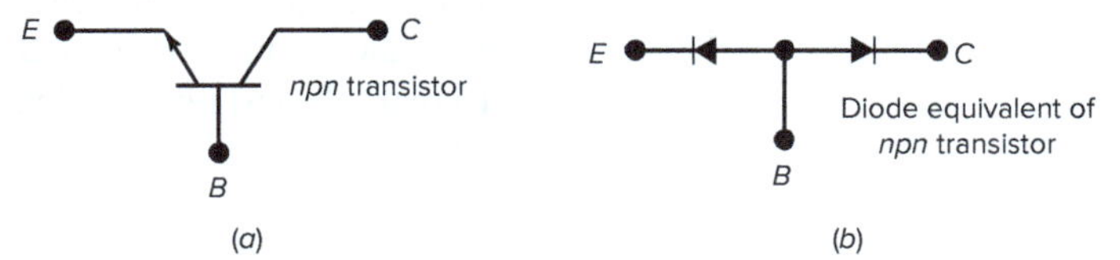

Figure 29–9 Testing the base-emitter junction of a transistor with an analog ohmmeter. (*a*) Low resistance is measured because the analog meter forward-biases the base-emitter junction. (*b*) High resistance is measured because the analog meter reverse-biases the base-emitter junction.

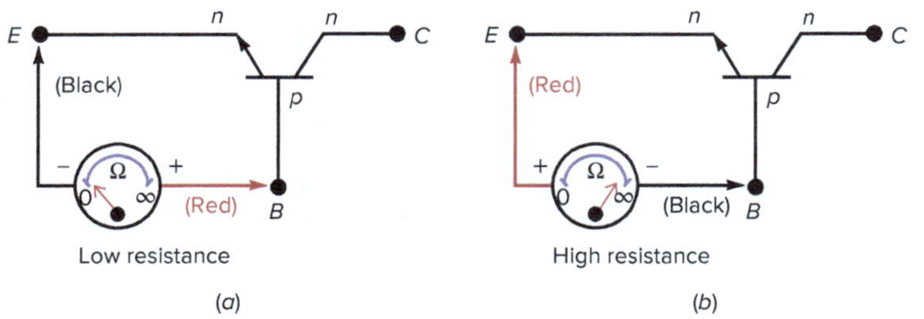

Figure 29–10 Testing the collector-base junction of a transistor with an analog ohmmeter. (*a*) Low resistance is measured because the analog meter forward-biases the collector-base junction. (*b*) High resistance is measured because the analog meter reverse-biases the collector-base junction.

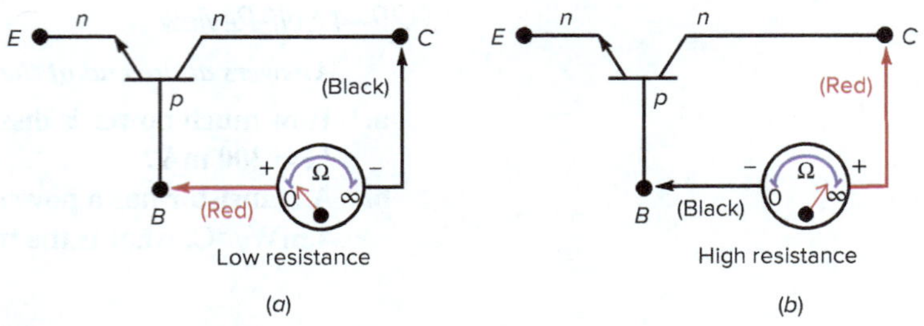

shown, the resistance measured between the collector and emitter should read high or infinite for both connections of the meter leads.

Shorted and Open Junctions

For either Fig. 29–9 or Fig. 29–10, low resistance across the junction in both directions implies that the emitter-base or collector-base junctions are shorted. If the ohmmeter indicates high resistance in both directions, then the junctions are open. In both cases, the transistor is defective and must be replaced.

Checking for Proper Transistor Action

An analog ohmmeter can also check to see whether the transistor functions properly as an amplifier, as shown in Fig. 29–11. Notice in Fig. 29–11*a* that the ohmmeter leads are connected so that the collector is positive with respect to the emitter. For this connection, the ohmmeter reads high or infinite (∞) resistance. Reversing the ohmmeter leads should not change the reading indicated by the ohmmeter.

Next, connect a resistor between the collector and base, as shown in Fig. 29–11*b*. This connection provides a positive voltage at the base with respect to the emitter, thereby forward-biasing the base-emitter junction. (Remember, an analog ohmmeter uses an internal battery.) Also, the collector is made positive with respect to its emitter, which is the required polarity for an *npn* transistor. This causes the ohmmeter to read approximately midscale because the transistor has collector current flowing through the ohmmeter. This action implies that the forward-biased base-emitter junction has turned on the transistor, causing a collector current, I_C, to flow.

To make the test illustrated in Fig. 29–11, it is important to use an R value no less than 10 kΩ and no greater than 100 kΩ. Also, it is recommended that the $R \times 10$ range

Figure 29–11 Checking for transistor action. (*a*) Ohmmeter reads high resistance. (*b*) Connecting *R* between collector and base biases the transistor so that it conducts.

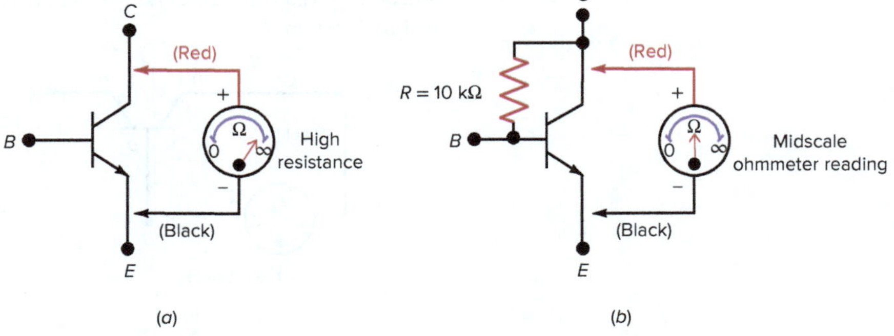

of the ohmmeter be used. These limits will produce the most reliable results. This test is rather crude, but it does provide a unique way of testing for transistor action.

All of the transistor tests shown in this section should be done only with the transistor out of the circuit. Also, it is recommended that the analog meter be set to the $R \times 10$ or $R \times 100$ ranges. This will protect the transistor against excessive currents and voltages that could exist if the transistor is checked on the lowest and highest ohm ranges of the ohmmeter.

Checking a Transistor with a Digital Multimeter (DMM)

A transistor can also be checked with a DMM. The difference, however, is that the ohm ranges are typically not capable of forward-biasing a silicon or germanium *p-n* junction. As mentioned in Chapter 27, these ranges are typically designated as LPΩ (low power ohm). The LPΩ ranges of a DMM are useful when it is necessary to measure resistances in a transistor circuit where the forward-biasing of a *p-n* junction could cause an undesired parallel path across the resistance being measured. In fact, if a typical DMM is used, the ohmmeter tests indicated in Figs. 29–9 and 29–10 could not be performed because the DMM would show an overrange condition on the display for both connections of the ohmmeter. This has fooled many technicians into believing that one or both junctions in the transistor being tested are bad.

When using a DMM to check the diode junctions in a transistor, the diode range (→+) must be used. However, the meter will show the forward voltage dropped across the *p-n* junction being tested rather than the actual value of forward or reverse resistance. For a forward-biased, emitter-base, or collector-base, silicon *p-n* junction, the DMM usually indicates a voltage between 0.6 and 0.7 V. For the reverse-bias condition, the meter indicates an overrange condition.

Note that the transistor action test illustrated in Fig. 29–11 cannot be performed with the DMM.

GOOD TO KNOW

Many DMMs available today are capable of measuring the DC beta (β_{DC}) of a bipolar junction transistor—either *npn* or *pnp*. The leads of the transistor are inserted into a small socket and a switch selects between *npn* and *pnp* for the measurement of β_{DC}.

29–5 Self-Review

Answers at the end of the chapter.

a. When tested with an analog ohmmeter, the base-emitter junction of a transistor should measure high resistance for one polarity of the meter leads and low resistance for the other polarity. (True/False)

b. A DMM can measure the forward and reverse resistance of a diode. (True/False)

29–6 Transistor Biasing Techniques

For a transistor to function properly as an amplifier, an external DC supply voltage (or voltages) must be applied to produce the desired collector current, I_C. Recall from Chapter 28 that the term *bias* is defined as a control voltage or current. Transistors must be biased correctly to produce the desired circuit voltages and currents. Several biasing techniques exist; the most common are discussed in this section. They include base bias, voltage divider bias, and emitter bias.

Base Bias

Figure 29–12*a* shows the simplest way to bias a transistor, called *base bias*. V_{BB} is the base supply voltage, which is used to forward-bias the base-emitter junction. R_B is used to provide the desired value of base current, I_B. V_{CC} is the collector supply voltage, which provides the reverse-bias voltage required for the collector-base junction of the transistor. The collector resistor, R_C, provides the desired voltage in the collector circuit. Figure 29–12*b* shows the DC equivalent circuit. For silicon

GOOD TO KNOW

In Fig. 29–12*a*, I_C is totally independent of V_{CC} and R_C. This is true only if the transistor is operating in its active region, however.

Figure 29–12 Base bias. (*a*) Circuit. (*b*) DC equivalent circuit.

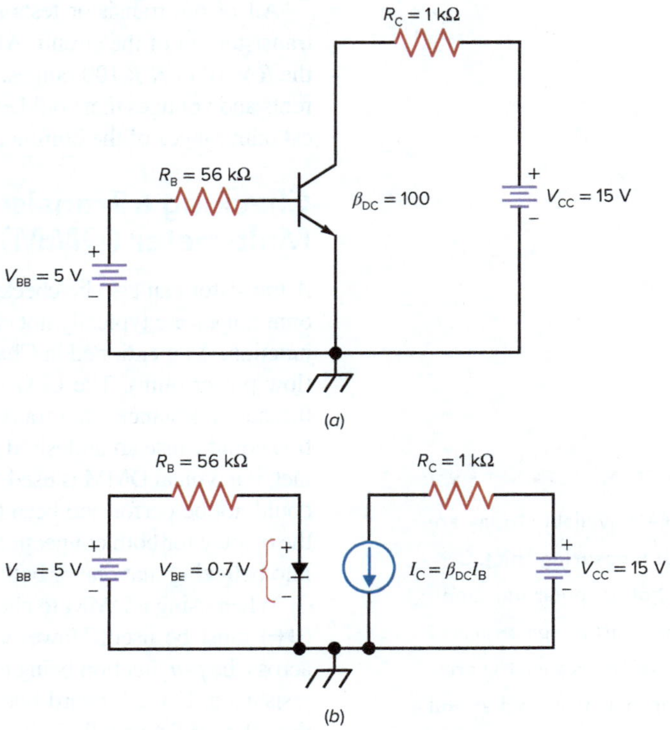

transistors, V_{BE} equals 0.7 V. Notice that the collector circuit is represented as a current source whose value is dependent only on the values of β_{DC} and I_B. Collector supply voltage variations have little or no effect on the collector current, I_C.

In Fig. 29–12*a*, the base current I_B can be found by dividing the voltage drop across R_B by the value of R_B. This is shown in Formula (29–9).

$$I_B = \frac{V_{BB} - V_{BE}}{R_B} \qquad \textbf{(29–9)}$$

Since the transistor is silicon, V_{BE} equals 0.7 V. Therefore, I_B is calculated as follows:

$$I_B = \frac{V_{BB} - V_{BE}}{R_B}$$
$$= \frac{5\ V - 0.7\ V}{56\ k\Omega}$$
$$= 76.78\ \mu A$$

The collector current, I_C, can be calculated next:

$$I_C = \beta_{DC} \times I_B$$
$$= 100 \times 76.78\ \mu A$$
$$\approx 7.68\ mA$$

With I_C known, the collector-emitter voltage, V_{CE}, can be found. This is shown in Formula (29–10):

$$V_{CE} = V_{CC} - I_C R_C \qquad \textbf{(29–10)}$$
$$= 15\ V - (7.68\ mA \times 1\ k\Omega)$$
$$= 15\ V - 7.68\ V$$
$$= 7.32\ V$$

Figure 29–13 Base bias using a single power supply.

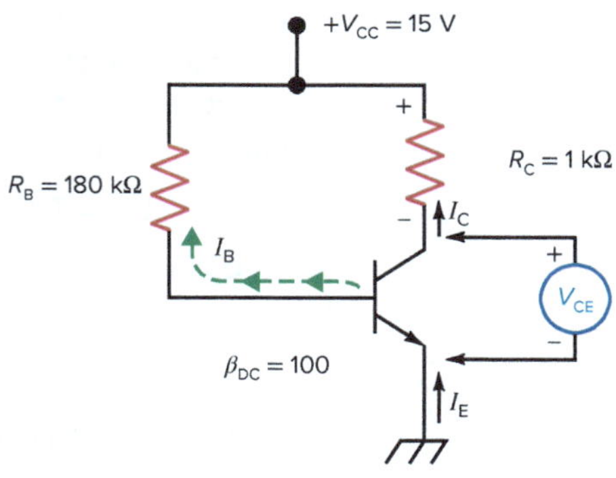

Notice that the calculation for V_{CE} involves subtracting the voltage drop across R_C from the collector supply voltage, V_{CC}. Increasing I_C reduces V_{CE} because of the increased voltage drop across the collector resistor, R_C. Conversely, decreasing I_C increases V_{CE} due to the reduction of the $I_C R_C$ voltage drop.

Base Bias with a Single Supply

In most cases, a single voltage source provides the base bias for a transistor. One example is shown in Fig. 29–13. Notice that the base supply voltage, V_{BB}, has been omitted and R_B is connected to the positive (+) terminal of V_{CC}. For this circuit, I_B is calculated using Formula (29–11):

$$I_B = \frac{V_{CC} - V_{BE}}{R_B} \qquad \textbf{(29–11)}$$

For the circuit values given, I_B, I_C, and V_{CE} are calculated as follows:

$$I_B = \frac{15\text{ V} - 0.7\text{ V}}{180\text{ k}\Omega\text{V}}$$
$$= \frac{14.3\text{ V}}{180\text{ k}\Omega}$$
$$= 79.44\ \mu\text{A}$$

$$I_C = \beta_{DC} \times I_B$$
$$= 100 \times 79.44\ \mu\text{A}$$
$$= 7.94\text{ mA}$$

Next, calculate V_{CE} using Formula (29–10). The calculations are

$$V_{CE} = V_{CC} - I_C R_C$$
$$= 15\text{ V} - (7.94\text{ mA} \times 1\text{ k}\Omega)$$
$$= 15\text{ V} - 7.94\text{ V}$$
$$= 7.06\text{ V}$$

Figure 29–13 shows a more practical way to provide base bias because only one power supply is required.

Figure 29–14 DC load line.

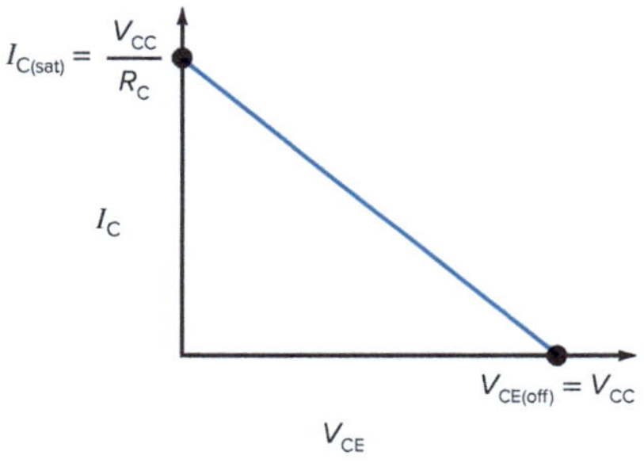

DC Load Line

The **DC load line** is a graph that allows us to determine all possible combinations of I_C and V_{CE} for a given amplifier. For every value of collector current, I_C, the corresponding value of V_{CE} can be found by examining the DC load line. A sample DC load line is shown in Fig. 29–14.

The endpoints of the DC load line are labeled $I_{C(sat)}$ and $V_{CE(off)}$. $I_{C(sat)}$ represents the collector current, I_C, when the transistor is saturated. $V_{CE(off)}$ represents the collector-emitter voltage with $I_C = 0$ for the cutoff condition.

When a transistor is saturated, treat the collector-emitter region like a short since V_{CE} equals zero for this condition. With V_{CE} equal to zero, the voltage across R_C must equal V_{CC}. This is expressed as

$$V_{CC} = I_C R_C$$

Rearranging, this gives Formula (29–12):

$$I_{C(sat)} = \frac{V_{CC}}{R_C} \qquad \textbf{(29–12)}$$

This is an ideal value for $I_{C(sat)}$, with V_{CE} assumed to be zero. When a transistor is saturated, note the following points:

1. Further increases in I_B produce no further increases in I_C.
2. The collector circuit no longer acts like a current source because V_{CE} is approximately zero and the collector-base junction of the transistor is not properly reverse-biased.

When the transistor is cut off, visualize the collector-emitter region as an open circuit because $I_C = 0$. With zero collector current, the $I_C R_C$ voltage drop is zero, resulting in a collector-emitter voltage, V_{CE}, of approximately V_{CC}. This gives Formula (29–13):

$$V_{CE(off)} = V_{CC} \qquad \textbf{(29–13)}$$

If a transistor is biased so that it is not operating in saturation or cutoff, it is said to be operating in the *active region.* When a transistor is operating in the active region, the following points are true:

1. $I_C = \beta_{DC} \times I_B$.
2. The collector circuit acts as a current source with high internal impedance.

Midpoint Bias

Without an AC signal applied to a transistor, specific values of I_C and V_{CE} exist. The I_C and V_{CE} values exist at a specific point on the DC load line. This is called the ***Q* point,** where Q stands for the quiescent currents and voltages with no AC input signal applied. In many cases, an amplifier is biased such that the Q point is at or near the center of the DC load line, as illustrated in Fig. 29–15. Notice that in this case, I_{CQ} equals ½ $I_{C(sat)}$ and V_{CEQ} equals $V_{CC}/2$. V_{CEQ} represents the quiescent collector-emitter voltage, and I_{CQ} represents the quiescent collector current. Centering the Q point on the load line allows optimum AC operation of the amplifier.

Figure 29–15 DC load line showing the endpoints $I_{C(sat)}$ and $V_{CE(off)}$, as well as the Q point values I_{CQ} and V_{CEQ}.

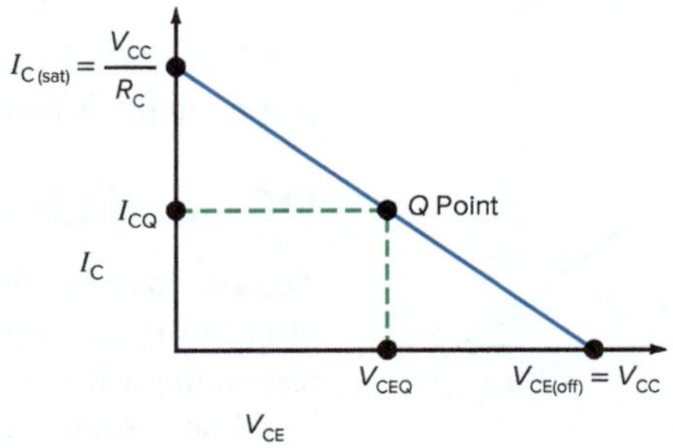

Example 29-12

In Fig. 29–16, solve for I_B, I_C, and V_{CE}. Also, construct a DC load line showing the values of $I_{C(sat)}$, $V_{CE(off)}$, I_{CQ}, and V_{CEQ}.

ANSWER Begin by using Formula (29–11) to calculate I_B:

$$I_B = \frac{V_{CC} - V_{BE}}{R_B}$$
$$= \frac{12\text{ V} - 0.7\text{ V}}{390\text{ k}\Omega}$$
$$= 28.97\ \mu\text{A}$$

Next, calculate I_C:

$$I_C = \beta_{DC} \times I_B$$
$$= 150 \times 28.97\ \mu\text{A}$$
$$= 4.35\text{ mA}$$

V_{CE} is calculated using Formula (29–10):

$$V_{CE} = V_{CC} - I_C R_C$$
$$= 12\text{ V} - (4.35\text{ mA} \times 1.5\text{ k}\Omega)$$
$$= 12\text{ V} - 6.52\text{ V}$$
$$= 5.48\text{ V}$$

The endpoints for the DC load line are calculated using Formulas (29–12) and (29–13):

$$I_{C(sat)} = \frac{V_{CC}}{R_C}$$
$$= \frac{12\text{ V}}{1.5\text{ k}\Omega}$$
$$= 8\text{ mA}$$
$$V_{CE(off)} = V_{CC}$$
$$= 12\text{ V}$$

Figure 29–16 Circuit used for Example 29–12.

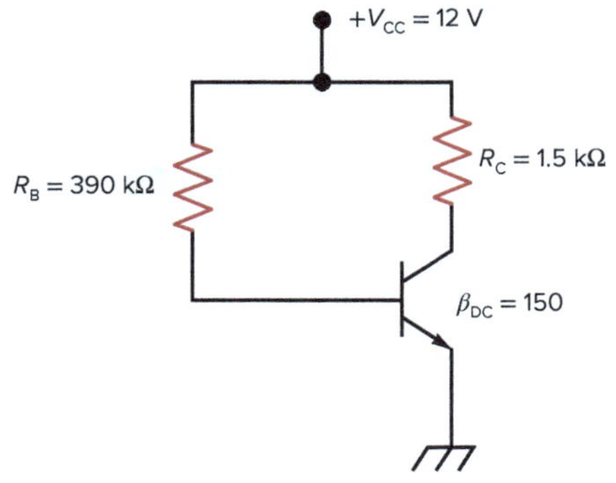

The values in Example 29–12 are shown on the DC load line in Fig. 29–17. Notice that V_{CEQ} and I_{CQ} are the same as the I_C and V_{CE} values calculated in this example, where V_{CEQ} = 5.48 V and I_{CQ} = 4.35 mA. Remember that I_{CQ} and V_{CEQ} represent the quiescent I_C and V_{CE} values without an AC input signal applied.

Figure 29–17 DC load line for base-biased transistor circuit in Fig. 29–16.

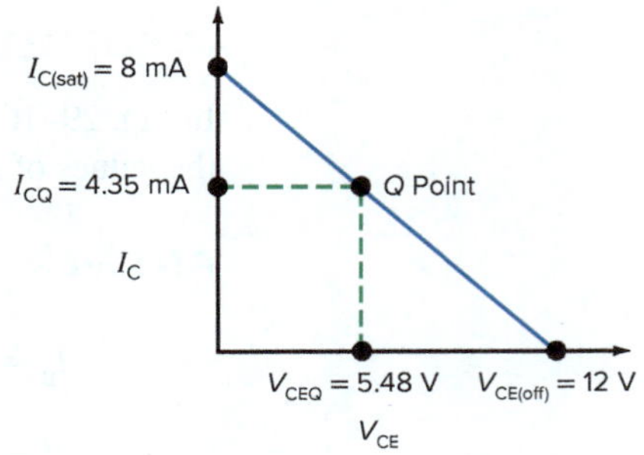

GOOD TO KNOW

Because the values of I_C and V_{CE} are dependent on the value of beta in a base-biased circuit, the circuit is said to be beta dependent.

Instability of Base Bias

Note that base bias provides a very unstable Q point because the collector current, I_C, and collector-emitter voltage, V_{CE}, are greatly affected by any change in the transistor's beta value. If the transistor is replaced with one having a significantly different value of β_{DC}, the Q point might actually shift to a point located near or at either cutoff or saturation. The beta value of a transistor also varies with temperature. Therefore, any change in the temperature of the transistor can cause the Q point to shift. The instability of base bias makes it the least popular biasing technique.

Voltage Divider Bias

The most popular way to bias a transistor is with *voltage divider bias.* The advantage lies in its stability. If designed properly, the circuit is practically immune to changes in β_{DC} caused by either transistor replacement or temperature variation. An example of voltage divider bias is shown in Fig. 29–18. Notice that V_B is the voltage measured from the base lead to ground, which is actually the voltage drop across R_2. Since the voltage divider is made up of R_1 and R_2, V_B can be calculated using the voltage divider formula shown in Formula (29–14):

$$V_B = \frac{R_2}{R_1 + R_2} \times V_{CC} \qquad \textbf{(29–14)}$$

$$= \frac{5.1\ \text{k}\Omega}{27\ \text{k}\Omega + 5.1\ \text{k}\Omega} \times 15\ \text{V}$$

$$= 2.38\ \text{V}$$

MultiSim **Figure 29–18** Voltage divider bias.

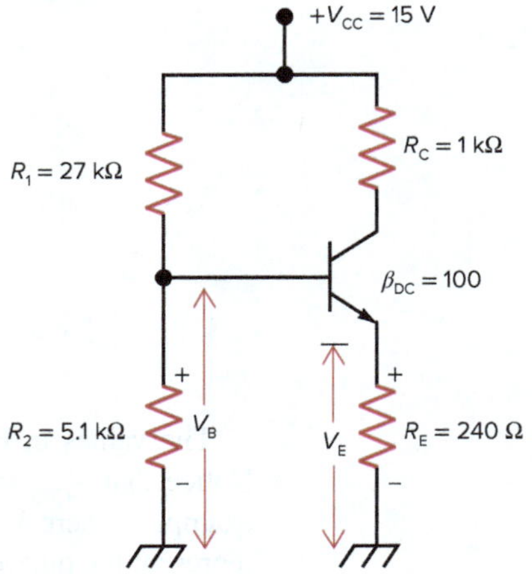

The emitter voltage, V_E, is 0.7 V less than the base voltage, V_B, assuming that the transistor is silicon. This is shown in Formula (29–15):

$$V_E = V_B - V_{BE} \qquad \textbf{(29–15)}$$
$$= 2.38\text{ V} - 0.7\text{ V}$$
$$= 1.68\text{ V}$$

Ohm's law can now be used to determine the emitter current, I_E:

$$I_E = \frac{V_E}{R_E} \qquad \textbf{(29–16)}$$
$$= \frac{1.68\text{ V}}{240\ \Omega}$$
$$= 7\text{ mA}$$

Now, because β_{DC} equals 100, assume that I_C is approximately the same as I_E. Therefore,

$$I_E \approx I_C = 7\text{ mA}$$

To calculate the collector voltage with respect to ground, use Formula (29–17):

$$V_C = V_{CC} - I_C R_C \qquad \textbf{(29–17)}$$
$$= 15\text{ V} - (7\text{ mA} \times 1\text{ k}\Omega)$$
$$= 15\text{ V} - 7\text{ V}$$
$$= 8\text{ V}$$

Finally, V_{CE} can be calculated by using Formula (29–18):

$$V_{CE} = V_{CC} - I_C\,(R_C + R_E) \qquad \textbf{(29–18)}$$
$$= 15\text{ V} - 7\text{ mA}\,(1\text{ k}\Omega + 240\ \Omega)$$
$$= 15\text{ V} - 8.68\text{ V}$$
$$= 6.32\text{ V}$$

Calculating the Endpoints for the DC Load Line

As mentioned before, when a transistor is saturated, V_{CE} is approximately zero, which is the same as saying that the collector and emitter terminals are shorted. Visualize the collector-emitter terminals shorted in Fig. 29–18. This condition would produce a voltage divider in the collector circuit whose total resistance would equal the sum of $R_C + R_E$. Thus, the saturation current, $I_{C(sat)}$, is

$$I_{C(sat)} = \frac{V_{CC}}{R_C + R_E} \qquad \textbf{(29–19)}$$

Conversely, with the transistor cut off, the collector emitter region acts like an open circuit and

$$V_{CE(off)} = V_{CC} \qquad \textbf{(29–20)}$$

Figure 29–19 shows the DC load line for the transistor circuit in Fig. 29–18. The end points of the DC load line can be calculated using Formulas (29–19) and (29–20):

$$I_{C(sat)} = \frac{V_{CC}}{R_C + R_E}$$
$$= \frac{15\text{ V}}{1\text{ k}\Omega + 240\ \Omega}$$
$$= 12.1\text{ mA}$$

$$V_{CE(off)} = V_{CC}$$
$$= 15\text{ V}$$

These values are found on the graph in Fig. 29–19.

Notice that the values of I_{CQ} and V_{CEQ} have also been included on the graph.

Figure 29–19 DC load line for voltage divider–based transistor circuit in Fig. 29–18.

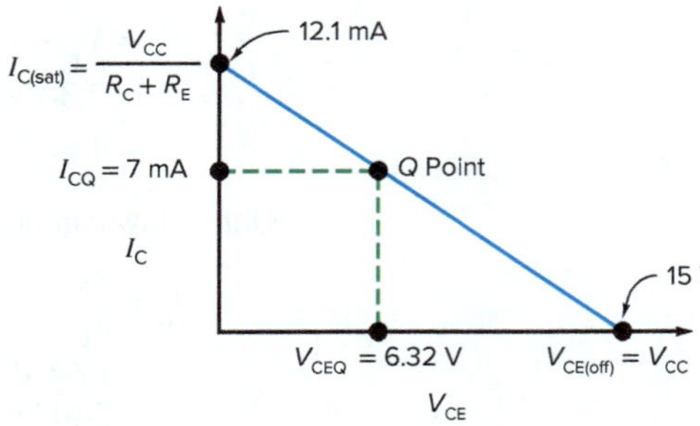

Example 29-13

For the circuit shown in Fig. 29–20, solve for V_B, V_E, I_C, V_C, and V_{CE}. Also, calculate $I_{C(sat)}$ and $V_{CE(off)}$. Finally, construct a DC load line showing the values of $I_{C(sat)}$, $V_{CE(off)}$, I_{CQ}, and V_{CEQ}.

Figure 29–20 Circuit used for Example 29–13.

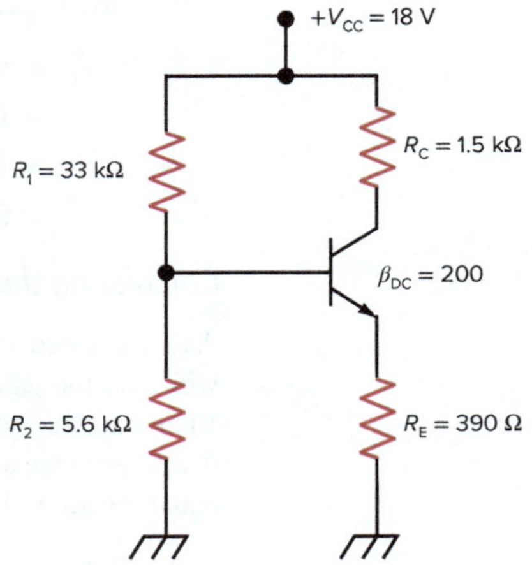

ANSWER Begin by calculating V_B. Use Formula (29–14):

$$V_B = \frac{R_2}{R_1 + R_2} \times V_{CC}$$

$$= \frac{5.6\ \text{k}\Omega}{33\ \text{k}\Omega + 5.6\ \text{k}\Omega} \times 18\ \text{V}$$

$$= 2.61\ \text{V}$$

Next, use Formula (29–15) to calculate V_E:

$$V_E = V_B - V_{BE}$$

$$= 2.61\ \text{V} - 0.7\ \text{V}$$

$$= 1.91\ \text{V}$$

$$I_E = \frac{V_E}{R_E}\ \text{(using Ohm's law)}$$

$$= \frac{1.91\ \text{V}}{390\ \Omega} = 4.9\ \text{mA}$$

since $\beta_{DC} = 200$, $I_C \cong I_E = 4.9$ mA

With I_C known, V_C and V_{CE} can be calculated using Formulas (29–17) and (29–18), respectively:

$$\begin{aligned} V_C &= V_{CC} - I_C R_C \\ &= 18\text{ V} - (4.9\text{ mA} \times 1.5\text{ k}\Omega) \\ &= 18\text{ V} - 7.35\text{ V} \\ &= 10.65\text{ V} \\ V_{CE} &= V_{CC} - I_C(R_C + R_E) \\ &= 18\text{ V} - 4.9\text{ mA}(1.5\text{ k}\Omega + 390\ \Omega) \\ &= 18\text{ V} - 9.26\text{ V} \\ &= 8.74\text{ V} \end{aligned}$$

$I_{C(sat)}$ is calculated using Formula (29–19):

$$\begin{aligned} I_{C(sat)} &= \frac{V_{CC}}{R_C + R_E} \\ &= \frac{18\text{ V}}{1.5\text{ k}\Omega + 390\ \Omega} \\ &= 9.52\text{ mA} \end{aligned}$$

Next, $V_{CE(off)} = V_{CC} = 18$ V

The DC load line is shown in Fig. 29–21. The values of I_{CQ}, V_{CEQ}, $I_{C(sat)}$, and $V_{CE(off)}$ are all included.

Figure 29–21 DC load line for voltage divider–biased transistor circuit in Fig. 29–20.

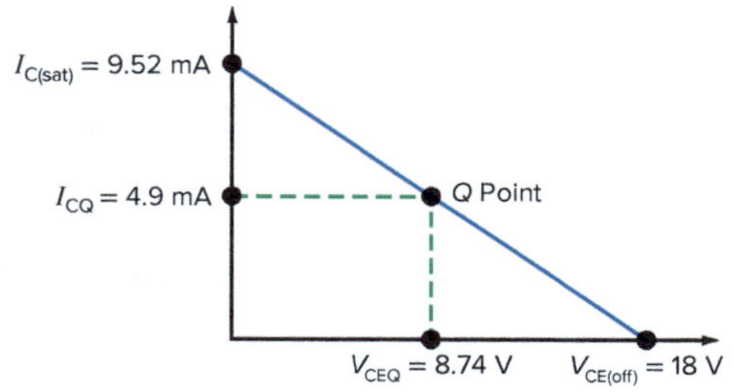

Example 29-14

For the *pnp* transistor in Fig. 29–22, solve for V_B, V_E, I_C, V_C, and V_{CE}.

ANSWER Notice that the collector supply voltage, V_{CC}, is negative. This polarity is required to bias the EB and CB junctions properly. All currents will flow in the opposite direction from the *npn* transistors used in the previous example. The calculations for all currents and voltages are similar to those of the *npn* transistor. The calculations are

$$\begin{aligned} V_B &= \frac{R_2}{R_1 + R_2} \times (-V_{CC}) \\ &= \frac{6.2\text{ k}\Omega}{6.2\text{ k}\Omega + 33\text{ k}\Omega} \times (-12\text{ V}) \\ &= -1.9\text{ V} \end{aligned}$$

Figure 29–22 Circuit used for Example 29–14.

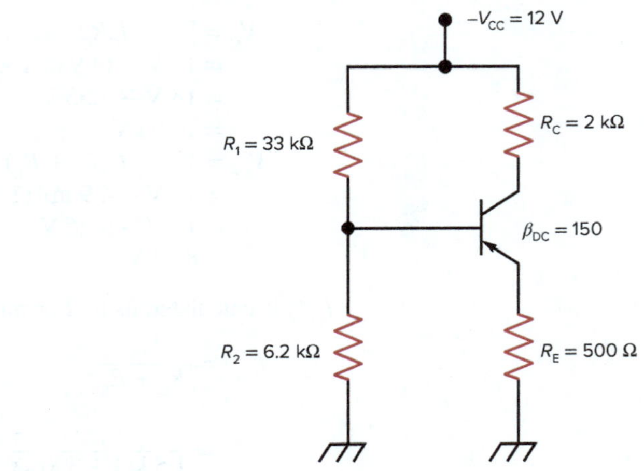

Allowing 0.7 V for V_{BE}, the emitter voltage, V_E is calculated as

$$
\begin{aligned}
V_E &= V_B - V_{BE} \\
&= -1.9\text{ V} - (-0.7\text{ V}) \\
&= -1.2\text{ V}
\end{aligned}
$$

since $I_E \cong I_C$.

$$
\begin{aligned}
\text{Next,}\quad I_C &= \frac{V_E}{R_E} \\
&= \frac{1.2\text{ V}}{500\ \Omega} \\
&= 2.4\text{ mA}
\end{aligned}
$$

V_C and V_{CE} are calculated as

$$
\begin{aligned}
V_C &= -V_{CC} + I_C R_C \\
&= -12\text{ V} + (2.4\text{ mA} \times 2\text{ k}\Omega) \\
&= -12\text{ V} + 4.8\text{ V} \\
&= -7.2\text{ V}
\end{aligned}
$$

To calculate V_{CE}, proceed as follows:

$$
\begin{aligned}
V_{CE} &= -V_{CC} + I_C (R_C + R_E) \\
&= -12\text{ V} + 2.4\text{ mA } (2\text{ k}\Omega + 500\ \Omega) \\
&= -12\text{ V} + 6\text{ V} \\
&= -6\text{ V}
\end{aligned}
$$

The values of $I_{C(sat)}$ and $V_{CE(off)}$ are determined using the same methods described earlier.

Figure 29–23 Emitter bias.

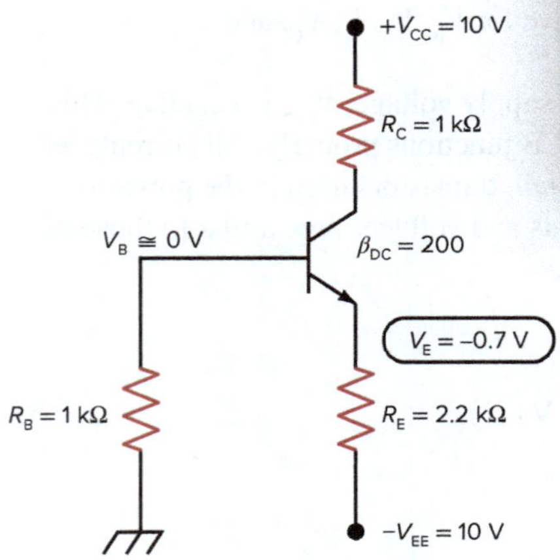

Emitter Bias

If both positive and negative power supplies are available, *emitter bias* provides a solid Q point that fluctuates very little with temperature variation and transistor replacement. An example of emitter bias is shown in Fig. 29–23. The emitter supply voltage, V_{EE}, forward-biases the emitter-base junction through the emitter resistor, R_E. To calculate the emitter current, I_E, use Formula (29–21):

$$I_E = \frac{V_{EE} - V_{BE}}{R_E} \qquad \textbf{(29–21)}$$

$$= \frac{10\text{ V} - 0.7\text{ V}}{2.2\text{ k}\Omega}$$

$$= 4.23\text{ mA}$$

Notice that R_B is ignored in the calculation for I_E. A more exact formula for I_E, however, is

$$I_E = \frac{V_{EE} - V_{BE}}{R_E + \dfrac{R_B}{\beta_{DC}}} \qquad \textbf{(29–22)}$$

Using the values from Fig. 29–23 gives

$$I_E = \frac{10\text{ V} - 0.7\text{ V}}{2.2\text{ k}\Omega + \dfrac{1\text{ k}\Omega}{200}}$$

$$= \frac{10\text{ V} - 0.7\text{ V}}{2.205\text{ k}\Omega}$$

$$= 4.22\text{ mA}$$

Notice that the difference is only 10 μA, which is small enough to be ignored.

Incidentally, notice in Fig. 29–23 that the base voltage, $V_B = 0$ V. This occurs because the I_BR_B voltage drop is very small due to the small value of base current, I_B, which is typically only a few microamperes.

To calculate V_C, proceed using Formula (29–23):

$$V_C = V_{CC} - I_CR_C$$
$$= 10\text{ V} - (4.23\text{ mA} \times 1\text{ k}\Omega)$$
$$= 10\text{ V} - 4.23\text{ V}$$
$$= 5.77\text{ V} \qquad \textbf{(29–23)}$$

GOOD TO KNOW

When transistors are biased using well-designed voltage divider bias or emitter bias configurations, they are classified as beta-independent circuits. This is because the values of I_C and V_{CE} are unaffected by changes in the transistor's beta.

Example 29-15

In Fig. 29–24, calculate I_E and V_C.

ANSWER I_E can be calculated by using either Formula (29–21) or (29–22). Here, use Formula (29–21):

$$I_E = \frac{V_{EE} - V_{BE}}{R_E}$$

$$= \frac{6\text{ V} - 0.7\text{ V}}{1\text{ k}\Omega}$$

$$= 5.3\text{ mA}$$

Next, calculate V_C using Formula (29–23):

$$V_C = V_{CC} - I_CR_C$$
$$= 15\text{ V} - (5.3\text{ mA} \times 1.5\text{ k}\Omega)$$
$$= 15\text{ V} - 7.95\text{ V}$$
$$= 7.05\text{ V}$$

Figure 29–24 Circuit used for Example 29–15.

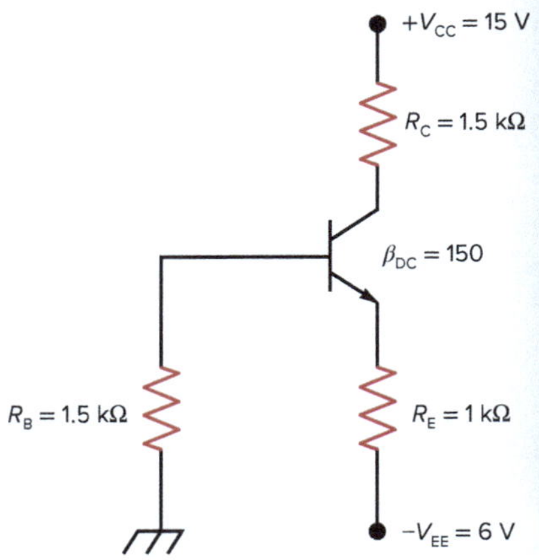

29–6 Self-Review

Answers at the end of the chapter.

a. **Base bias is the best way to bias a transistor because it provides such a stable *Q* point. (True/False)**

b. **When the collector current decreases in a transistor amplifier, the collector voltage (increases or decreases).**

c. **In a transistor circuit with voltage divider bias, the base voltage is 3.1 V. How much is the emitter voltage?**

Summary

- A transistor is made up of three doped regions: the emitter, base, and collector regions.
- The base is a very thin and lightly doped region that is sandwiched between the emitter and collector regions.
- The emitter region is the most heavily doped region in a transistor. Its function is to emit or inject current carriers into the base region.
- The collector region is moderately doped and is the largest of all three transistor regions. Most of the current carriers injected into the base are attracted into the collector region rather than flowing out from the base lead.
- In a transistor, $I_E = I_B + I_C$.
- The DC alpha (α_{DC}) is the ratio of DC collector current to DC emitter current: $\alpha_{DC} = I_C/I_E$.
- The DC beta (β_{DC}) is the ratio of DC collector current to DC base current: $\beta_{DC} = I_C/I_B$.
- A transistor has four operating regions: the breakdown region, active region, saturation region, and the cutoff region.
- When a transistor is operating in the active region, the collector acts like a current source whose value is $I_C = I_B \times \beta_{DC}$.
- The power dissipation rating of a transistor decreases for temperatures above 25°C. Manufacturers specify a derating factor in W/°C so that the transistor's power rating can be determined for any temperature.
- Transistors must be properly biased to obtain the desired circuit voltages and currents. When properly biased, the transistor can amplify AC signals.
- The most commonly used biasing techniques for transistors include base bias, voltage divider bias, and emitter bias.
- Base bias is seldom used because it has a very unstable *Q* point.
- Voltage divider bias is the most common way to bias a transistor. It provides a very stable *Q* point when designed properly.
- Emitter bias using two power supplies provides a very stable *Q* point.
- A DC load line is a graph that shows all possible operating points for I_C and V_{CE} in a transistor amplifier.
- The endpoints for the DC load line are labeled $I_{C(sat)}$ and $V_{CE(off)}$. All the points between cutoff and saturation are in the active region.

Important Terms

Active region—the region of operation where the collector of a transistor acts like a current source.

Base—a thin and very lightly doped region in a transistor. The base is sandwiched between the emitter and collector regions.

Breakdown region—the region of transistor operation where a large undesired collector current flows as a result of the collector-base diode breaking down from excessive reverse-bias voltage.

Collector—a large, moderately doped region in a transistor. The collector is the largest of all three transistor regions because it dissipates the most heat.

Cutoff—the region of transistor operation where the collector current, I_C, is zero.

DC alpha, α_{DC}—the ratio of collector current, I_C, to emitter current, I_E, in a transistor: $\alpha_{DC} = I_C/I_E$.

DC beta, β_{DC}—the ratio of collector current, I_C, to base current, I_B, in a transistor: $\beta_{DC} = I_C/I_B$.

DC load line—a graph that shows all possible values of I_C and V_{CE} for a given transistor amplifier. The end points of the DC load line are $I_{C(sat)}$ and $V_{CE(off)}$ which represent the values of I_C and V_{CE} when the transistor is in saturation and cutoff.

Derating factor—the amount that the power rating of a transistor must be reduced for each degree Celsius above 25°C. The derating factor is specified in W/°C.

Emitter—the most heavily doped region in a transistor. Its function is to inject or emit current carriers into the base region.

Midpoint bias—a bias point that is centered between cutoff and saturation on the DC load line.

***Q* point**—the values of I_C and V_{CE} in a transistor amplifier with no AC signal present.

Saturation—the region of transistor operation where the collector current no longer increases with further increases in the base current.

Transistor—a three-terminal semiconductor device that can amplify an AC signal or be used as an electronic switch.

Related Formulas

$I_E = I_B + I_C$

$I_C = I_E - I_B$

$I_B = I_E - I_C$

$\alpha_{DC} = I_C/I_E$

$\beta_{DC} = I_C/I_B$

$$\alpha_{DC} = \frac{\beta_{DC}}{1 + \beta_{DC}}$$

$$\beta_{DC} = \frac{\alpha_{DC}}{1 - \alpha_{DC}}$$

$P_d = V_{CE} \times I_C$

Base Bias (Single Supply)

$$I_B = \frac{V_{CC} - V_{BE}}{R_B}$$

$V_{CE} = V_{CC} - I_C R_C$

$$I_{C(sat)} = \frac{V_{CC}}{R_C}$$

$$V_{CE(off)} = V_{CC}$$

Voltage Divider Bias

$$V_B = \frac{R_2}{R_1 + R_2} \times V_{CC}$$

$$V_E = V_B - V_{BE}$$

$$I_E = V_E / R_E$$

$$V_C = V_{CC} - I_C R_C$$

$$V_{CE} = V_{CC} - I_C (R_C + R_E)$$

$$I_{C(sat)} = \frac{V_{CC}}{R_C + R_E}$$

$$V_{CE(off)} = V_{CC}$$

Emitter Bias

$$I_E = \frac{V_{EE} - V_{BE}}{R_E}$$

$$I_E = \frac{V_{EE} - V_{BE}}{R_E + \frac{R_B}{\beta_{DC}}} \text{ (More Accurate)}$$

$$V_C = V_{CC} - I_C R_C$$

Self-Test

Answers at the back of the book.

1. Which transistor region is very thin and lightly doped?
a. the emitter region.
b. the collector region.
c. the anode region.
d. the base region.

2. Which region in a transistor is the most heavily doped?
a. the emitter region.
b. the collector region.
c. the gate region.
d. the base region.

3. In a transistor, which is the largest of all the doped regions?
a. the emitter region.
b. the collector region.
c. the gate region.
d. the base region.

4. For a transistor to function as an amplifier,
a. both the EB and CB junctions must be forward-biased.
b. both the EB and CB junctions must be reverse-biased.
c. the EB junction must be forward-biased and the CB junction must be reverse-biased.
d. the CB junction must be forward-biased and the EB junction must be reverse-biased.

5. For a typical transistor, which two currents are nearly the same?
a. I_B and I_E.
b. I_B and I_C.
c. I_C and I_E.
d. none of the above.

6. In what operating region does the collector of a transistor act like a current source?
a. the active region.
b. the saturation region.
c. the cutoff region.
d. the breakdown region.

7. A transistor operating in the active region has a base current, I_B, of 20 μA. If $\beta_{DC} = 250$, how much is the collector current, I_C?
a. 50 mA.
b. 5 mA.
c. 12.5 mA.
d. 80 μA.

8. Which of the following biasing techniques produces the most unstable *Q* point?
a. voltage divider bias.
b. emitter bias.
c. collector bias.
d. base bias.

9. When the collector current in a transistor is zero, the transistor is
a. cut off.
b. saturated.
c. operating in the breakdown region.
d. either b or c.

10. When a transistor is in saturation,
a. $V_{CE} = V_{CC}$.
b. $I_C = 0$ A.
c. $V_{CE} = 0$ V.
d. $V_{CE} = \frac{1}{2}V_{CC}$.

11. Emitter bias with two power supplies provides a
a. very unstable *Q* point.
b. very stable *Q* point.
c. large base voltage.
d. none of the above.

12. The α_{DC} of a transistor equals
a. I_C/I_E.
b. I_B/I_C.
c. I_E/I_C.
d. I_C/I_B.

13. For a transistor operating in the active region,
a. $I_C = \beta_{DC} \times I_B$.
b. V_{CC} has little or no effect on the value of I_C.
c. I_C is controlled solely by V_{CC}.
d. both a and b.

14. In a transistor amplifier, what happens to the collector voltage, V_C, when the collector current, I_C, increases?
a. V_C increases.
b. V_C stays the same.
c. V_C decreases.
d. It cannot be determined.

15. With voltage divider bias, how much is the collector-emitter voltage, V_{CE}, when the transistor is cut off?
a. $V_{CE} = \frac{1}{2} V_{CC}$.
b. $V_{CE} = V_{CC}$.
c. $V_{CE} = 0$ V.
d. none of the above.

16. On the schematic symbol of a *pnp* transistor,
a. the arrow points out on the emitter lead.
b. the arrow points out on the collector lead.
c. the arrow points in on the base lead.
d. the arrow points in on the emitter lead.

17. What is the β_{DC} of a transistor whose α_{DC} is 0.996?

a. 249.

b. 100.

c. approximately 1.

d. It cannot be determined.

18. In a transistor, which current is the largest?

a. I_C.

b. I_B.

c. I_E.

d. I_D.

19. A bipolar junction transistor has

a. only one *p-n* junction.

b. three *p-n* junctions.

c. no *p-n* junctions.

d. two *p-n* junctions.

20. The endpoints of a DC load line are labeled

a. I_{CQ} and V_{CEQ}.

b. $I_{C(sat)}$ and $V_{CE(off)}$.

c. $I_{C(off)}$ and $V_{CE(sat)}$.

d. none of the above.

Essay Questions

1. How do the biasing polarities differ between *npn* and *pnp* transistors?

2. In a transistor, which current is called the recombination current?

3. Why does base bias produce such an unstable *Q* point?

4. When a transistor is operating in the active region, why is the collector considered a current source?

5. Derive the equation $\alpha_{DC} = \dfrac{\beta_{DC}}{1 + \beta_{DC}}$.

6. Define the following regions of operation for a transistor: a. active region b. saturation region c. cutoff region d. breakdown region.

7. What is the purpose of drawing a DC load line?

Problems

SECTION 29–1 TRANSISTOR CONSTRUCTION

29–1 Explain the characteristics and purpose of each of the following regions in a transistor:

a. emitter.

b. base.

c. collector.

29–2 In an *npn* transistor, what are the majority and minority current carriers in the

a. emitter?

b. base?

c. collector?

29–3 Repeat Prob. 29–2 for a *pnp* transistor.

29–4 What are the barrier potentials for the base-emitter (BE) and collector-base (CB) junctions in a silicon transistor?

29–5 In which direction does the arrow point on the emitter lead when viewing the schematic symbol of a(n)

a. *npn* transistor?

b. *pnp* transistor?

SECTION 29–2 PROPER TRANSISTOR BIASING

29–6 Explain how the BE and CB junctions of a transistor must be biased for a transistor to function properly as an amplifier.

29–7 In a transistor, why is the base current called recombination current?

29–8 In an *npn* transistor, explain why most of the emitter-injected electrons pass through the base region and on to the collector.

29–9 Solve for the unknown transistor current in each of the following cases:

a. $I_E = 1$ mA, $I_B = 5\ \mu$A, $I_C = ?$

b. $I_B = 50\ \mu$A, $I_C = 2.25$ mA, $I_E = ?$

c. $I_C = 40$ mA, $I_E = 40.5$ mA, $I_B = ?$

d. $I_E = 2.7$ A, $I_B = 30$ mA, $I_C = ?$

e. $I_C = 3.65$ mA, $I_E = 3.75$ mA, $I_B = ?$

f. $I_B = 90\ \mu$A, $I_C = 20.25$ mA, $I_E = ?$

29–10 Calculate the DC alpha (α_{DC}) for each set of current values listed in Prob. 29–9.

29–11 Calculate the DC beta (β_{DC}) for each set of current values listed in Prob. 29–9.

29–12 A transistor has a base current, I_B, of 15 μA. How much is the collector current, I_C, if the transistor has a β_{DC} of

a. 50?

b. 100?

c. 150?

d. 200?

29–13 A transistor has a collector current, I_C, of 10 mA. How much is the base current, I_B, if the transistor has a β_{DC} of

a. 50?

b. 100?

c. 200?

d. 250?

29–14 Calculate the DC alpha (α_{DC}) for each of the following values of β_{DC}:

a. 50.

b. 125.

c. 250.

29–15 Calculate the DC beta (β_{DC}) for each of the following values of α_{DC}:

a. 0.9875.

b. 0.996.

c. 0.9975.

SECTION 29–3 TRANSISTOR OPERATING REGIONS

29–16 When a transistor operates in the active region, does the collector current, I_C, respond to changes in

a. V_{CC}?

b. I_B?

29–17 In what region of operation does the collector of a transistor act like a current source?

29–18 In what region is a transistor operating if the collector current is zero?

29–19 When a transistor is saturated, is I_C controlled solely by I_B?

29–20 How much is V_{CE} when a transistor is saturated?

29–21 What is the ideal internal impedance of the collector current source in Fig. 29–7*b*?

SECTION 29–4 TRANSISTOR RATINGS

29–22 Calculate the power dissipation, P_d, in a transistor for each of the following values of V_{CE} and I_C:

a. $V_{CE} = 5$ V, $I_C = 20$ mA.

b. $V_{CE} = 20$ V, $I_C = 50$ mA.

c. $V_{CE} = 24$ V, $I_C = 300$ mA.

d. $V_{CE} = 30$ V, $I_C = 600$ mA.

29–23 A transistor has a power rating of 1.5 W at an ambient temperature, T_A, of 25°C. If the derate factor is 12 mW/°C what is the transistor's power rating at each of the following temperatures?

a. 50°C.

b. 75°C.

c. 100°C.

d. 125°C.

e. 150°C.

29–24 A transistor has a power rating of 2 W. Calculate the maximum allowable collector current, $I_{C(max)}$, for each of the following values of V_{CE}:

a. 5 V.

b. 12 V.

c. 25 V.

SECTION 29–5 CHECKING A TRANSISTOR WITH AN OHMMETER

29–25 When testing the BE and CB junctions of a silicon transistor with an analog ohmmeter, what should the meter show for both polarities of the meter leads if the diode is

a. good?

b. shorted?

c. open?

29–26 What should an analog ohmmeter read for both polarities of the meter leads when measuring across the collector and emitter leads of a transistor?

29–27 Why do most DMMs have a special diode range for checking diodes and transistor junctions?

SECTION 29–6 TRANSISTOR BIASING TECHNIQUES

29–28 What form of bias is shown in Fig. 29–25?

29–29 In Fig. 29–25, solve for the following:

a. I_B.

b. I_C.

c. V_{CE}.

d. $I_{C(sat)}$.

e. $V_{CE(off)}$.

Figure 29–25

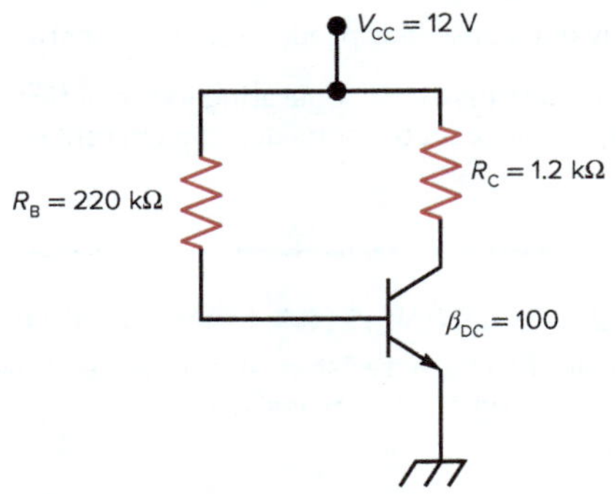

29–30 Draw a DC load line for the transistor circuit in Fig. 29–25, and indicate the values of $I_{C(sat)}$, $V_{CE(off)}$, I_{CQ}, and V_{CEQ} on the load line.

29–31 In Fig. 29–25, recalculate the values of I_B, I_C, and V_{CE} if $\beta_{DC} = 150$.

29–32 In Fig. 29–26, solve for the following:

a. I_B.

b. I_C.

c. V_{CE}.

d. $I_{C(sat)}$.

e. $V_{CE(off)}$.

Figure 29–26

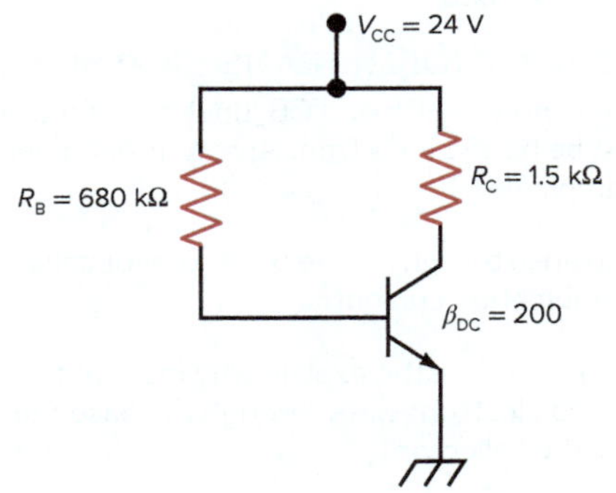

29–33 Recalculate the values in Prob. 29–32 if $R_C = 1$ kΩ.

29–34 In Fig. 29–27, what value of R_B will produce an I_{CQ} of 3.75 mA and a V_{CEQ} of 9 V?

Figure 29–27

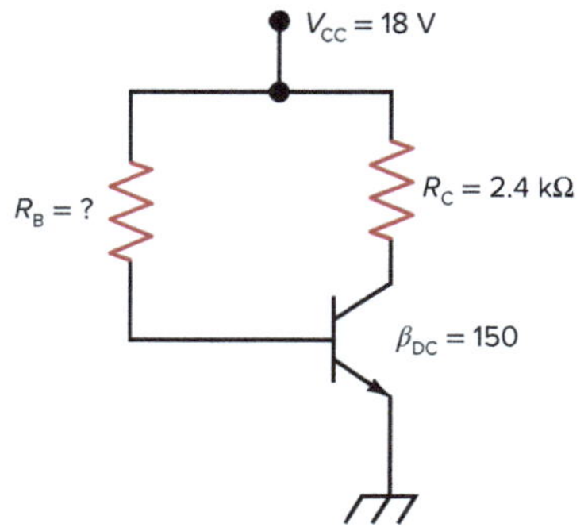

29–35 What form of bias is shown in Fig. 29–28?.

29–36 In Fig. 29–28, solve for the following:

a. V_B.
b. V_E.
c. I_E.
d. I_C.
e. V_C.
f. V_{CE}.
g. $I_{C(sat)}$.
h. $V_{CE(off)}$.

Figure 29–28

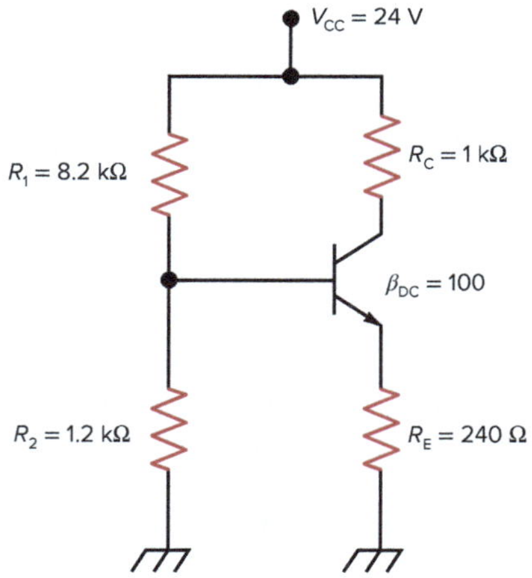

29–37 Draw a DC load line for the transistor circuit in Fig. 29–28, and indicate the values of $I_{C(sat)}$, $V_{CE(off)}$, I_{CQ}, and V_{CEQ} on the load line.

29–38 In Fig. 29–29, solve for the following:

a. V_B.
b. V_E.
c. I_C.
d. V_C.
e. V_{CE}.
f. $I_{C(sat)}$.
g. $V_{CE(off)}$.

Figure 29–29

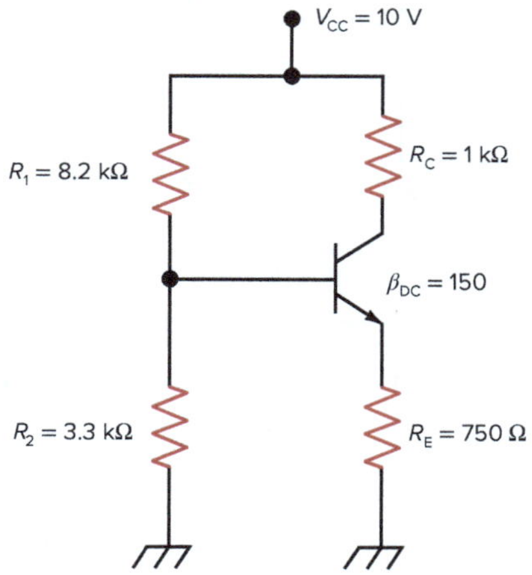

29–39 For the *pnp* transistor in Fig. 29–30, solve for the following:

a. V_B.
b. V_E.
c. I_C.
d. V_C.
e. V_{CE}.
f. $I_{C(sat)}$.
g. $V_{CE(off)}$.

Figure 29–30

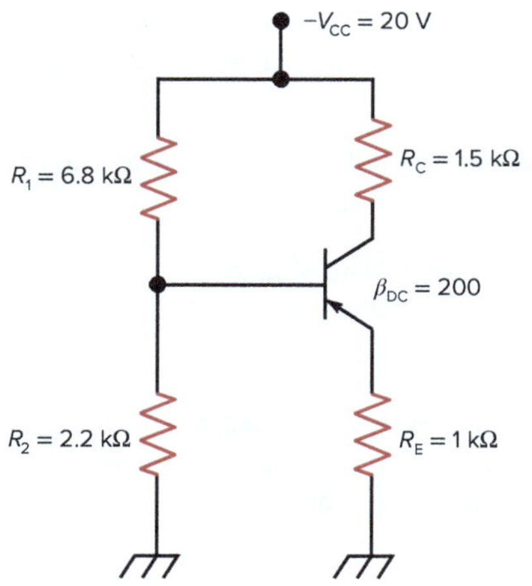

29–40 What form of bias is shown in Fig. 29–31?

29–41 In Fig. 29–31, solve for I_E and V_C.

29–42 Recalculate the values in Fig. 29–31 if $R_C = 1.5$ kΩ.

Figure 29–31

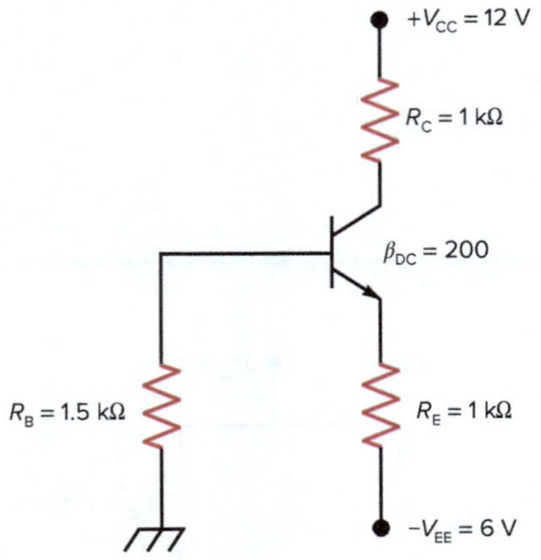

Answers to Self-Reviews

29–1 **a.** the emitter region
b. the collector region
c. the base region
d. the emitter lead

29–2 **a.** forward/reverse
b. 120 μA
c. 0.992
d. 180

29–3 **a.** the active region
b. the cutoff region
c. I_B

29–4 **a.** 4.5 W
b. 600 mW

29–5 **a.** true
b. false

29–6 **a.** false
b. increases
c. 2.4 V

Laboratory Application Assignment

In this lab application assignment, you will examine two different biasing techniques used with transistors: base bias and voltage divider bias. You will see that with base bias, I_C and V_{CE} are beta-dependent values, whereas with voltage divider bias they are not.

Equipment: Obtain the following items from your instructor.

- Two 2N2222A npn transistors or equivalent
- DMM
- Assortment of carbon-film resistors
- Variable dc power supply

Beta, β_{DC}

Many handheld and bench-top DMMs available today are capable of measuring the DC beta of a transistor. If your DMM has this capability, measure and record the DC beta of each *npn* transistor supplied to you for this experiment. Keep each transistor separate.

Q_1, β_{DC} = ____________

Q_2, β_{DC} = ____________

Base Bias

Examine the circuit in Fig. 29–32. Calculate and record the following DC values for each of the two transistor betas:

Q_1	Q_2
I_B = ____________	I_B = ____________
I_C = ____________	I_C = ____________
V_{CE} = ____________	V_{CE} = ____________

Figure 29–32

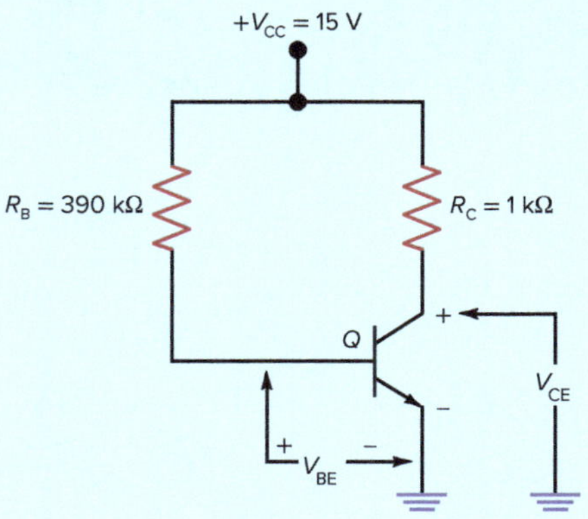

Figure 29–33

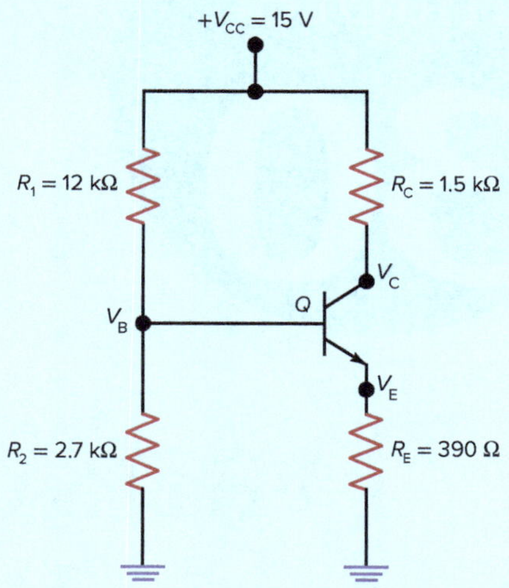

Construct the circuit in Fig. 29–32. Measure and record the following DC values for each of the two transistors:

Q_1	Q_2
I_B = ________	I_B = ________
I_C = ________	I_C = ________
V_{CE} = ________	V_{CE} = ________

Voltage Divider Bias

Examine the circuit in Fig. 29–33. Calculate and record the following circuit values:

V_B = ________ V_C = ________

V_E = ________ V_{CE} = ________

I_E = ________

Construct the circuit in Fig. 29–33. Measure the following circuit values using Q_1 as the transistor:

V_B = ________ V_C = ________

V_E = ________ V_{CE} = ________

I_E = ________

Replace Q_1 with Q_2 and repeat the measurements listed.

V_B = ________ V_C = ________

V_E = ________ V_{CE} = ________

I_E = ________

Which form of bias produces more stable results, base bias or voltage divider bias? ________ Explain your answer.

____
____
____
____
____
____
____
____

chapter

30

Transistor Amplifiers

The biasing of a transistor deals specifically with DC voltages and currents. The purpose of the DC bias is to establish the desired *Q* point so that the correct variations in base and collector currents are possible when an AC signal is applied as an input to the amplifier. The AC signal driving the amplifier forces the operating point to swing above and below the designated *Q* point, thereby producing an AC output voltage. The AC output voltage is typically much larger than the AC signal driving the input to the amplifier. The input signal applied to the amplifier should not be so large that it shifts the instantaneous operating point to either saturation or cutoff. In this chapter, you will learn about the common-emitter, common-collector, and common-base amplifiers. As you will learn, each amplifier configuration has its own unique characteristics. ■

Chapter Outline

Chapter Objectives

After studying this chapter, you should be able to

- *Calculate* the AC resistance of a diode when the DC diode current is known.
- *Calculate* the AC resistance of the emitter diode in a transistor when the DC emitter current is known.
- *Define* the term AC beta, β.
- *Explain* how a common-emitter amplifier can amplify an AC signal.
- *List* the characteristics of a common-emitter amplifier.
- *Draw* the AC equivalent circuit of a common-emitter amplifier.
- *Calculate* the voltage gain, input impedance, and output impedance of a common-emitter amplifier.
- *Explain* the effects of a swamping resistor in a common-emitter amplifier.
- *Define* a small AC signal as it relates to a transistor amplifier.
- *List* the characteristics of a common-collector amplifier.
- *Draw* the AC equivalent circuit of a common-collector amplifier.
- *Calculate* the voltage gain, input impedance, and output impedance of a common-collector amplifier.
- *Explain* the main applications of an emitter follower.
- *List* the characteristics of a common-base amplifier.
- *Draw* the AC equivalent circuit of a common-base amplifier.
- *Calculate* the voltage gain, input impedance, and output impedance of a common-base amplifier.

Important Terms

AC beta
AC equivalent circuit
common-base amplifier
common-collector amplifier
common-emitter amplifier
current gain, A_i
emitter bypass capacitor, C_E
emitter follower
input impedance, Z_{in}
output impedance, Z_{out}
power gain, A_P
small signal
swamping resistor
voltage gain, A_V

30–1 AC Resistance of a Diode

Figure 30–1*a* shows a DC source in series with an AC source. Together, both sources supply current to the diode, D_1. The DC source provides the forward bias for D_1, while the AC source produces fluctuations in the diode current. The graph in Fig. 30–1*b* illustrates how the diode current varies with the AC voltage. The AC source can produce fluctuations in diode current because its alternating voltage is, in fact, producing slight variations in the amount of forward bias for the diode, D_1. The fluctuations in diode current are usually quite small compared to the DC diode current.

For small AC signals, the diode acts like a resistance. (The term ***small signal*** is generally meant to be a signal that has a peak-to-peak current equal to or less than one-tenth the DC diode current.) The AC resistance for a diode is calculated using Formula (30–1):

$$r_{AC} = \frac{25\text{ mV}}{I_d} \qquad \textbf{(30–1)}$$

where r_{AC} represents the AC resistance of the diode to *small AC signals* and I_d represents the DC diode current. Note that as the DC diode current, I_d, increases, the AC resistance decreases.

The derivation of Formula (30–1) is quite lengthy, and because it involves the use of calculus, it is beyond the scope of this book.

Figure 30–1 Combining AC and DC voltages in a diode circuit. (*a*) V_{DC} provides a steady DC voltage that forward-biases the diode D_1. The AC voltage source produces fluctuations in the amount of forward bias. (*b*) Graph of V_F versus I_F showing AC variations.

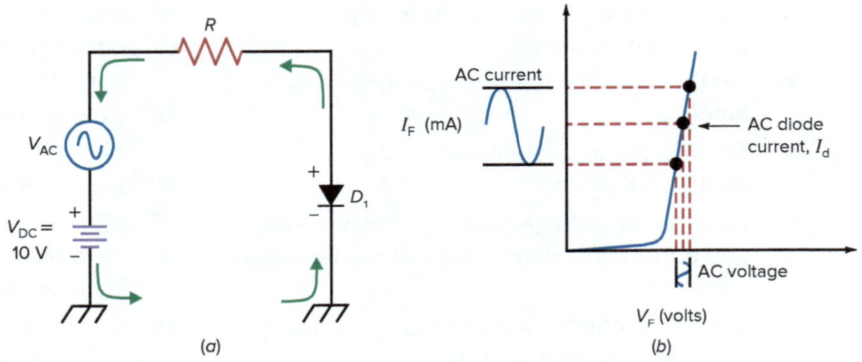

Example 30-1

For the diode circuit shown in Fig. 30–1, calculate the AC resistance, r_{AC}, for the following values of R: (*a*) 10 kΩ, (*b*) 5 kΩ, and (*c*) 1 kΩ. Use the second approximation of a diode.

ANSWER First calculate the DC diode current for each value of R:

$$\text{(a) } I_d = \frac{V_{DC} - 0.7\text{ V}}{R}$$

$$= \frac{10\text{ V} - 0.7\text{ V}}{10\text{ k}\Omega}$$

$$= \frac{9.3\ \text{V}}{10\ \text{k}\Omega}$$

$$= 0.93\ \text{mA}$$

$$\text{(b) } I_d = \frac{10\ \text{V} - 0.7\ \text{V}}{5\ \text{k}\Omega}$$

$$= \frac{9.3\ \text{V}}{5\ \text{k}\Omega}$$

$$= 1.86\ \text{mA}$$

$$\text{(c) } I_d = \frac{10\ \text{V} - 0.7\ \text{V}}{1\ \text{k}\Omega}$$

$$= \frac{9.3\ \text{V}}{1\ \text{k}\Omega}$$

$$= 9.3\ \text{mA}$$

Next, use Formula (30–1) to calculate the AC resistance, r_{AC}, for each value of DC diode current:

$$\text{(a) } r_{AC} = \frac{25\ \text{mV}}{I_d}$$

$$= \frac{25\ \text{mV}}{0.93\ \text{mA}}$$

$$= 26.88\ \Omega$$

$$\text{(b) } r_{AC} = \frac{25\ \text{mV}}{1.86\ \text{mA}}$$

$$= 13.44\ \Omega$$

$$\text{(c) } r_{AC} = \frac{25\ \text{mV}}{9.3\ \text{mA}}$$

$$= 2.69\ \Omega$$

This example confirms that the diodes' AC resistance, r_{AC}, decreases for higher values of DC diode current. The reason for the decrease in r_{AC} is that the slope of the diode curve is steeper for higher values of diode current.

AC Resistance of Emitter Diode

Since the emitter-base junction of a transistor is forward-biased, it too will act like a small AC resistance (illustrated in the diagram of Fig. 30–2). The resistance of the emitter diode is represented as r'_e, rather than r_{AC}. The DC current through the diode equals the emitter current, I_E. Use this value when calculating the resistance, r'_e, as shown in Formula (30–2):

$$r'_e = \frac{25\ \text{mV}}{I_E} \qquad \textbf{(30–2)}$$

where r'_e represents the AC resistance of the emitter diode and I_E is the DC emitter current.

When analyzing a **common-emitter amplifier**, it is common practice to represent the emitter diode as a small resistance. By doing this, important characteristics of an amplifier, such as its voltage gain and input impedance, can be calculated.

GOOD TO KNOW

In practice, the value of r'_e may range from $\frac{25\ \text{mV}}{I_E}$ to $\frac{50\ \text{mV}}{I_E}$.

GOOD TO KNOW

There are other, more accurate, transistor equivalent circuits (models) in addition to the one shown in Fig. 30–2. A highly accurate model will include something called the base spreading resistance, r'_b, and the internal resistance, r'_c, of the collector current source. These other models are used when very precise answers are required.

Figure 30–2 Equivalent circuit showing the AC resistance of an emitter diode. The AC resistance is designated r'_e.

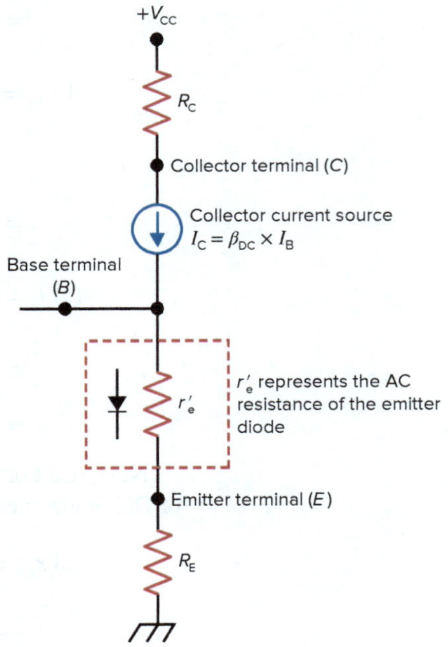

30–1 Self-Review

Answers at the end of the chapter.

a. **Does the AC resistance of a diode increase or decrease with an increase in the DC diode current?**

b. **A small AC signal is one whose peak-to-peak current is (one-tenth/ten times) that of the DC emitter current.**

c. **What is the formula for the AC resistance of the emitter diode?**

30–2 Small Signal Amplifier Operation

Figure 30–3*a* shows a common-emitter amplifier. Notice the capacitors C_{in} and C_E. C_{in} is an *input coupling capacitor* that couples the AC generator voltage to the base of the transistor. The internal resistance of the generator will not affect the DC bias of the transistor circuit because the capacitor, C_{in}, blocks DC. C_E is called an *emitter bypass capacitor*. It provides a low-impedance path for AC signals between the emitter terminal and ground.

The AC source driving the base of the transistor produces sinusoidal variations in the base current, I_B. This, in turn, provides variations in the collector current, I_C. Notice in Fig. 30–3*a* that the AC base voltage is riding on a DC axis, which is actually the DC base voltage to ground. Notice also that the bypass capacitor, C_E, makes the AC signal voltage zero at the emitter terminal. C_E and R_E have a long time constant compared to the period of the input waveform.

When the AC signal voltage driving the base goes positive, the forward bias for the transistor increases. This causes the base current, I_B, and the collector current, I_C, to increase. Likewise, when the AC signal voltage driving the base goes negative, the forward bias for the transistor decreases, causing I_B and I_C to decrease.

MultiSim **Figure 30–3** Common-emitter amplifier. (*a*) Circuit. (*b*) DC load line showing how the operating point is shifted by the AC signal.

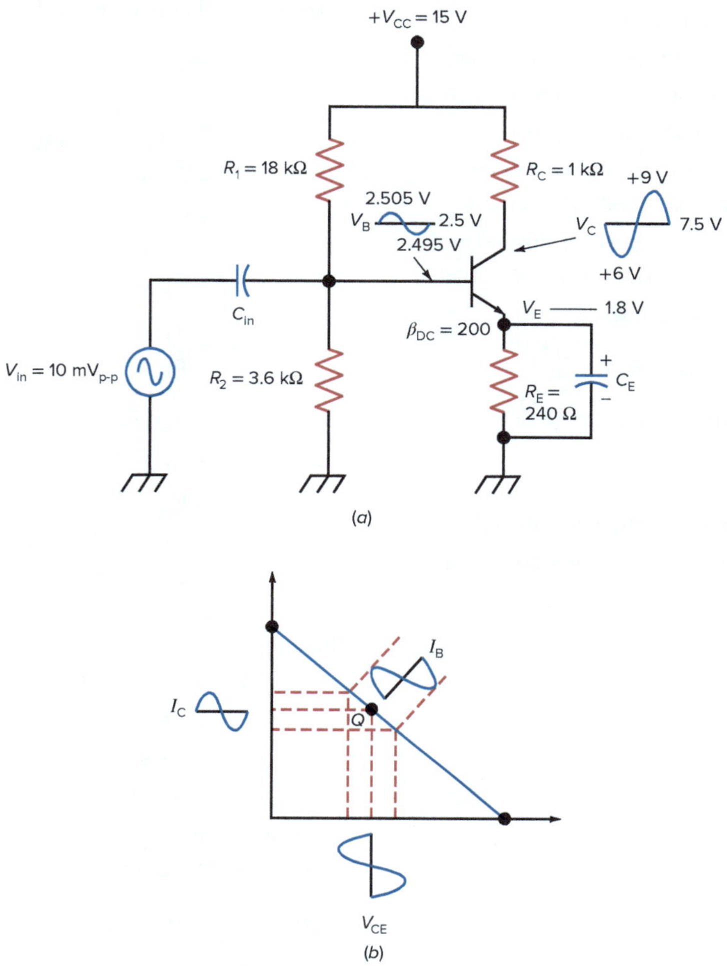

The emitter bypass capacitor holds the emitter constant, as shown in Fig. 30–3*a*. Thus, when the base voltage varies and the emitter voltage is held constant, the change is directly across the base-emitter junction of the transistor.

AC Beta

The ratio of AC collector current to AC base current is called the ***AC beta***, usually symbolized as β. This is expressed as

$$\beta = \frac{i_c}{i_b}$$

where β equals the AC beta and i_c and i_b represent the AC values of collector and base current, respectively. The AC beta is the AC current gain of a transistor. β_{DC} and β are usually quite close in value.

Figure 30–3*b* shows how the operating point moves up and down the DC load line with changes in I_B and I_C. For small signal operation, only a small portion of the DC load line is used.

Calculating the DC Quantities

Before examining exactly how the common-emitter circuit in Fig. 30–3*a* amplifies the AC signal applied to the base, first calculate the DC voltages and currents:

$$V_B = \frac{R_2}{R_1 + R_2} \times V_{CC}$$
$$= \frac{3.6\ \text{k}\Omega}{18\ \text{k}\Omega + 3.6\ \text{k}\Omega} \times 15\ \text{V}$$
$$= 2.5\ \text{V}$$

$$V_E = V_B - V_{BE}$$
$$= 2.5\ \text{V} - 0.7\ \text{V}$$
$$= 1.8\ \text{V}$$

$$I_E = \frac{V_E}{R_E}$$
$$= \frac{1.8\ \text{V}}{240\ \Omega}$$
$$= 7.5\ \text{mA}$$

$$V_C = V_{CC} - I_C R_C$$
$$= 15\ \text{V} - (7.5\ \text{mA} \times 1\ \text{k}\Omega)$$
$$= 15\ \text{V} - 7.5\ \text{V}$$
$$= 7.5\ \text{V}$$

Finally,

$$V_{CE} = V_{CC} - I_C(R_C + R_E)$$
$$= 15\ \text{V} - 7.5\ \text{mA}\ (1\ \text{k}\Omega + 240\ \Omega)$$
$$= 15\ \text{V} - 9.3\ \text{V}$$
$$= 5.7\ \text{V}$$

The voltages V_B, V_E, and V_C are shown in Fig. 30–3*a*. All voltages are specified with respect to ground.

GOOD TO KNOW

The current gain, A_i, of a common-emitter amplifier equals the ratio of output current, i_{out}, to the input current, i_{in}. The output current, however, is not i_C as you might think. The output current is the current flowing in the load, R_L. The formula for A_i is derived as follows:

$$A_i = \frac{i_{out}}{i_{in}} \quad \text{or} \quad A_i = \frac{V_{out}/R_L}{V_{in}/Z_{in}}$$

This simplifies to

$$A_i = A_V \times \frac{Z_{in}}{R_L}$$

(You will learn how to calculate Z_{in} in an upcoming section in this chapter.)

Amplifying the Input Signal

Figure 30–4 shows a graph of I_E versus V_{BE} for the transistor in the common-emitter circuit of Fig. 30–3*a*. Notice the sinusoidal variations in V_{BE} and I_E. When V_{in} is zero, $V_{BE} = 0.7$ V and $I_E = 7.5$ mA. When V_{BE} is increased to 0.705 V by the AC source, I_E increases to 9 mA. Likewise, when the AC source decreases V_{BE} to 0.695 V, I_E decreases to 6 mA. At the instant I_C equals 9 mA, V_C equals 15 V – (9 mA × 1 kΩ) = 6 V. When I_C equals 6 mA, V_C equals 15 V – (6 mA × 1 kΩ) = 9 V. The voltage and current values illustrated here are very real and do exist for the circuit shown in Fig. 30–3*a*.

Figure 30–4 Graph of I_E versus V_{BE} for the common-emitter amplifier circuit in Fig. 30–3.

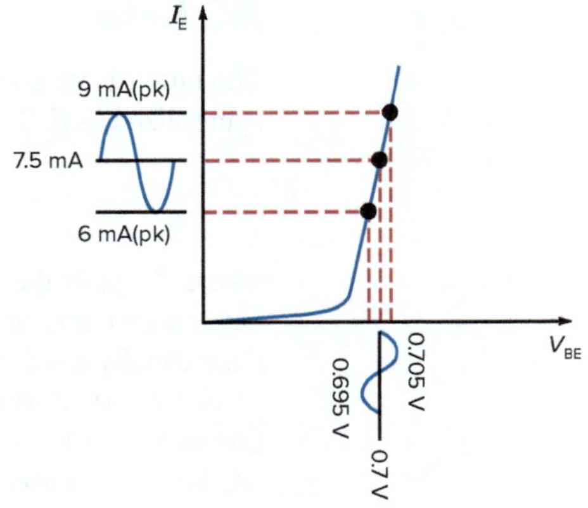

Voltage Gain, A_V

Notice how a very small AC voltage of 10 mV$_{p\text{-}p}$(0.705 V – 0.695 V) across the BE junction produces a very large change in the collector current, I_C. This in turn produces very large changes in the AC voltage at the collector.

For the circuit in Fig. 30–3*a*, the output voltage has a peak-to-peak value of 9 V – 6 V = 3 V$_{p\text{-}p}$. The input voltage is 0.705 V – 0.695 V = 10 mV$_{p\text{-}p}$. The ratio of output voltage, v_{out}, to input voltage, v_{in}, is called the voltage gain, usually designed A_V.

The **voltage gain, A_V**, is calculated as shown in Formula (30–3):

$$A_V = \frac{v_{out}}{v_{in}} \qquad \textbf{(30–3)}$$

where A_V equals voltage gain, v_{out} equals peak-to-peak output voltage, and v_{in} is the peak-to-peak input voltage. For Fig. 30–3*a*, A_V is calculated as

$$\begin{aligned} A_V &= \frac{v_{out}}{v_{in}} \\ &= \frac{3\ V_{p\text{-}p}}{10\ mV_{p\text{-}p}} \\ &= 300 \end{aligned}$$

Note one more point: In Fig. 30–4, V_{BE} has a value of 0.7 V when v_{in} is zero. This is correct, but it should be pointed out that in Fig. 30–3*a*, V_E is rock solid at +1.8 V because of the **emitter bypass capacitor, C_E**. V_B is varying ±5 mV above and below the DC base voltage of 2.5 V to provide the variation in forward bias for the transistor. This V_{BE} variation produces the variations in the collector current, I_C.

Example 30-2

A common-emitter amplifier circuit similar to Fig. 30–3*a* has an input of 25 mV$_{p\text{-}p}$ and an output of 5 V$_{p\text{-}p}$. Calculate A_V.

ANSWER Using Formula (30–3), the calculations are

$$\begin{aligned} A_V &= \frac{v_{out}}{v_{in}} \\ &= \frac{5\ V_{p\text{-}p}}{25\ mV_{p\text{-}p}} \\ &= 200 \end{aligned}$$

Example 30-3

In Fig. 30–3*a*, assume A_V still equals 300. If v_{in} = 5 mV$_{p\text{-}p}$, calculate v_{out}.

ANSWER Formula (30–3) must be rearranged to solve for v_{out}:

$$\begin{aligned} v_{out} &= A_V \times v_{in} \\ &= 300 \times 5\ mV_{p\text{-}p} \\ &= 1.5\ V_{p\text{-}p} \end{aligned}$$

Phase Inversion

In Fig. 30–3, it is important to note that the AC input and output voltages are 180° out of phase. The reason is that when v_{in} goes positive, I_C increases, causing V_{CE} to decrease. Likewise, when v_{in} decreases, I_C decreases, causing V_{CE} to increase. The common-emitter amplifier is the only transistor amplifier configuration that produces a 180° phase shift between v_{in} and v_{out}.

30–2 Self-Review

Answers at the end of the chapter.

a. In a common-emitter amplifier, what is the phase relationship between the AC input and output voltages?

b. In Fig. 30–3*a*, how much AC signal is present at the emitter terminal of the transistor?

30–3 AC Equivalent Circuit of a CE Amplifier

When analyzing transistor amplifier circuits, it is commonplace to draw the **AC equivalent circuit**. Figure 30–5 shows the AC equivalent circuit for the CE amplifier in Fig. 30–3. Notice the following points shown in the AC equivalent circuit of Fig. 30–5*a*:

1. C_{in} and C_E appear as AC short circuits because the X_C values of these capacitors is assumed to be zero for AC operation.
2. V_{CC} has been reduced to zero because it provides a very low impedance path for AC signals.
3. The emitter diode has been replaced with its equivalent AC resistance, r'_e, of 3.33 Ω equal to 25 mV/7.5 mA.
4. The biasing resistors, R_1 and R_2, are shown in parallel because V_{CC} appears as a short to AC signals.

A simplified and more condensed version of the AC equivalent circuit is shown in Fig. 30–5*b*. Notice that the input voltage v_{in} of 10 $mV_{p\text{-}p}$ appears directly across the AC resistance (r'_e) of the emitter diode. Also, notice that the output is directly across the collector resistance, R_C.

GOOD TO KNOW

The output of a power supply (such as V_{cc}) normally contains a very large electrolytic capacitor such as 1000 μF or more. This means that any AC signal will be effectively bypassed at the power supply terminals. Because of this, a power supply can be considered as an AC short when analyzing transistor amplifiers.

30–3 Self-Review

Answers at the end of the chapter.

a. In the AC equivalent circuit of a CE amplifier, all capacitors are treated as a(n) (short/open) to AC signals.

b. In the AC equivalent circuit of a CE amplifier, how is the DC supply voltage represented?

30–4 Calculating the Voltage Gain, A_V, of a CE Amplifier

The AC equivalent circuit is used to help understand the AC operation of the common-emitter amplifier circuit. To calculate the amount of voltage gain, A_V, for the CE amplifier in Fig. 30–3, use the AC equivalent circuit in Fig. 30–5*b*. Remember that the voltage gain, A_V, is expressed as

$$A_V = \frac{v_{out}}{v_{in}}$$

GOOD TO KNOW

Even though the emitter resistor, R_E, in Fig. 30–5 does not appear in the equation for A_V, its value indirectly affects the voltage gain of the circuit. For example, if the value of R_E is reduced, I_E increases, r'_e decreases, and in turn A_V increases. Conversely, if R_E is increased, I_E decreases, r'_e increases, and A_V decreases.

Figure 30–5 AC equivalent circuit for the common-emitter amplifier in Fig. 30–3. (*a*) AC equivalent circuit showing V_{CC}, C_{in}, and C_E as AC shorts. (*b*) Condensed version of AC equivalent circuit.

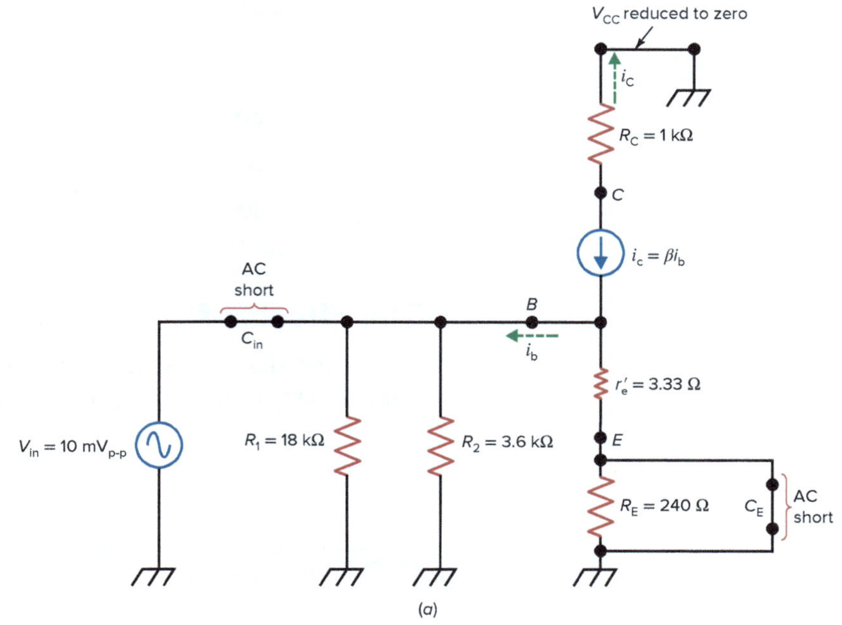

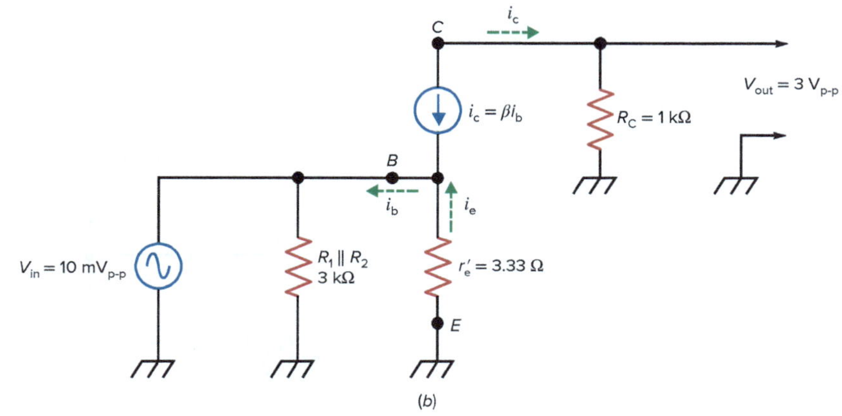

In Fig. 30–5*b*, the output equals the AC voltage across the collector resistor, R_C. This can be expressed as

$$v_{out} = i_C R_C$$

Likewise, in Fig. 30–5*b*, the input voltage, v_{in}, of 10 $mV_{p\text{-}p}$ is directly across the emitter resistance, r'_e. This is shown as

$$v_{in} = i_e r'_e$$

Since $i_e \approx i_c$,

$$v_{in} = i_C r'_e$$

A formula for A_V can now be derived:

$$A_V = \frac{v_{out}}{v_{in}} = \frac{i_C R_C}{i_C r'_e} = \frac{R_C}{r'_e} \qquad \textbf{(30–4)}$$

For the values shown in Fig. 30–5*b*,

$$A_V = \frac{R_C}{r'_e}$$
$$= \frac{1\ k\Omega}{3.33\ \Omega}$$
$$= 300$$

Since $A_V = 300$,

$$v_{out} = A_V \times v_{in}$$
$$= 300 \times 10\ mV_{p\text{-}p}$$
$$= 3\ V_{p\text{-}p}$$

The Effects of Connecting a Load Resistor, R_L

Figure 30–6 shows the effects of adding a load resistor, R_L, to the collector circuit. Since the output coupling capacitor appears as a short circuit for AC operation, R_C

MultiSim **Figure 30-6** Connecting a load resistor, R_L, to the output of a common-emitter amplifier circuit. (*a*) Original circuit. (*b*) AC equivalent circuit.

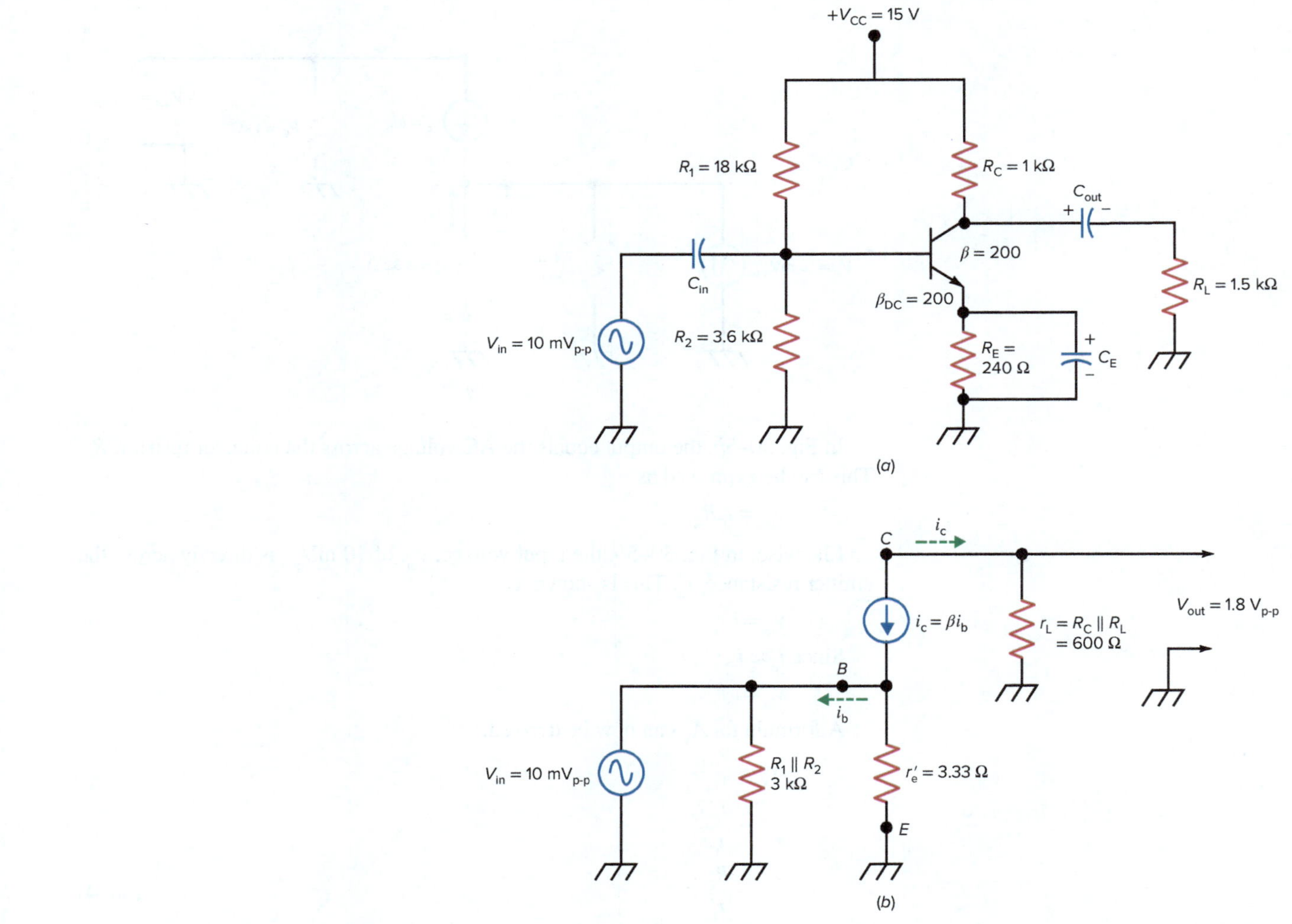

and R_L appear in parallel (see the AC equivalent circuit in Fig. 30–6*b*). The voltage gain, A_V, now is

$$A_V = \frac{i_C r_L}{i_C r'_e}$$

$$= \frac{r_L}{r'_e} \qquad \textbf{(30–5)}$$

where r_L indicates the parallel combination of R_C and R_L. In Fig. 30–6, r_L is calculated as

$$r_L = \frac{R_C \times R_L}{R_C + R_L}$$

$$= \frac{1\ \text{k}\Omega \times 1.5\ \text{k}\Omega}{1\ \text{k}\Omega + 1.5\ \text{k}\Omega}$$

$$= 600\ \Omega$$

Now calculate the value of A_V that exists with the load, R_L, connected:

$$A_V = \frac{r_L}{r'_e}$$

$$= \frac{600\ \Omega}{3.33\ \Omega}$$

$$= 180$$

The output voltage can be calculated as

$$v_{out} = A_V \times v_{in}$$

$$= 180 \times 10\ \text{mV}_{\text{p-p}}$$

$$= 1.8\ \text{V}_{\text{p-p}}$$

Notice that A_V and v_{out} are reduced when the resistance in the collector circuit is decreased.

GOOD TO KNOW

The power gain, A_P, of a CE amplifier equals $A_V \times A_i$. Since A_i can be expressed as $A_i = A_V \times \frac{Z_{in}}{R_L}$, then A_P can be expressed as

$$A_P = A_V \times A_V \times \frac{Z_{in}}{R_L}$$

$$= {A_V}^2 \times \frac{Z_{in}}{R_L}$$

Swamped Amplifier

The equivalent AC resistance, r'_e, of the emitter diode, calculated as 25 mV/I_E, is a rough approximation. The actual value of r'_e may vary with the type of transistor used, a shift in circuit bias, or fluctuations in temperature. With any change in r'_e, A_V may vary drastically. This is undesirable.

One way to greatly reduce the variations in A_V caused by changes in r'_e is to add a **swamping resistor** in the emitter circuit, as shown in Fig. 30–7*a*. It is important to note that this is still the same circuit, as shown in Fig. 30–3, but the emitter resistor is split into two values of 180 Ω and 60 Ω. Especially important is the fact that only the lower 180-Ω resistor is bypassed. This is illustrated in the AC equivalent circuit shown in Fig. 30–7*b*. With a swamping resistor in the emitter circuit, the voltage gain equals

$$A_V = \frac{v_{out}}{v_{in}}$$

$$= \frac{i_C r_L}{i_C(r'_e + r_E)}$$

which reduces to

$$A_V = \frac{r_L}{r'_e + r_E} \qquad \textbf{(30–6)}$$

For Fig. 30–7*a*, the voltage gain A_V is calculated as follows:

$$A_V = \frac{r_L}{r'_e + r_E}$$

$$= \frac{600\ \Omega}{3.33\ \Omega + 60\ \Omega}$$

$$= 9.47$$

If r_E is very large in relation to r'_e, the equation for the voltage gain, A_V, simplifies to

$$A_V = \frac{r_L}{r_E}$$

where r_E represents the unbypassed resistance in the emitter circuit.

Example 30-4

In Fig. 30–6, assume that r'_e varies from 3.33 Ω to 6.67 Ω as the temperature of the transistor changes. Calculate the variation in the voltage gain, A_V.

ANSWER We recall from our previous calculations that $A_V = 180$ when r'_e is 3.33 Ω. To calculate A_V with $r'_e = 6.67\ \Omega$, proceed as follows:

$$A_V = \frac{r_L}{r'_e}$$

$$= \frac{600\ \Omega}{6.67\ \Omega}$$

$$\approx 90$$

This is a 2:1 variation in A_V, which indicates that A_V is quite unpredictable due to possible fluctuations in r'_e.

Example 30-5

Assume that r'_e fluctuates from 3.33 Ω to 6.67 Ω in Fig. 30–7*a*. Calculate the minimum and maximum values for A_V.

ANSWER The maximum voltage gain, A_V, occurs when $r'_e = 3.33\ \Omega$. The voltage gain for this value was calculated earlier as 9.47.

To calculate the minimum voltage gain, A_V, when $r'_e = 6.67\ \Omega$, proceed as follows:

$$A_V = \frac{r_L}{r'_e + r_E}$$

$$= \frac{600}{6.67\ \Omega + 60\ \Omega}$$

$$= 9$$

Notice that the swamping resistor r_E has stabilized the voltage gain, A_V, by swamping out the effects of r'_e.

Swamping Resistor Reduces Distortion

Adding a swamping resistor to the emitter circuit also reduces the distortion significantly. Without the swamping resistor, all of the input signal would appear across the emitter diode, which has a nonlinear V_{BE} versus I_E curve. For most small signal conditions, this distortion is insignificant because only a very small portion of the diode curve is used. However, as the signal levels increase, a larger portion of the emitter diode curve is used, resulting in more distortion.

To summarize, a swamping resistor produces two very desirable effects:

1. The voltage gain, A_V, is stabilized.
2. Distortion is reduced.

30–4 Self-Review

Answers at the end of the chapter.

a. In Fig. 30–5*b*, how much is A_V if $R_C = 1.2\ k\Omega$?
b. In Fig. 30–6*a*, how much is A_V if $R_C = 1.2\ k\Omega$?
c. List two advantages of a swamping resistor in a common-emitter amplifier.

30–5 Calculating the Input and Output Impedances in a CE Amplifier

Any impedance in the emitter circuit of a CE amplifier appears greater by a factor of beta (β) when viewed from the base. The reason for this is that the base current, i_b, is smaller than the emitter current, i_e, by a factor of beta.

In Fig. 30–6*b*, the input voltage v_{in} is

$$v_{in} = i_c r'_e$$

Since the current on the input side is i_b, $i_{in} = i_b$. Therefore,

$$z_{in(base)} = \frac{v_{in}}{i_{in}} \tag{30–7}$$

$$= \frac{i_c r'_e}{i_b}$$

since $\frac{i_c}{i_b} = \beta$, then,

$$z_{in(base)} = \beta r'_e \tag{30–8}$$

The input impedance of an amplifier is the input impedance seen by the AC source driving the amplifier. Therefore, in Fig. 30–6, the biasing resistors, R_1 and R_2, are included as follows:

$$z_{in} = z_{in(base)} \parallel R_1 \parallel R_2 \tag{30–9}$$

In Fig. 30–6, $Z_{in(base)}$ and z_{in} are calculated as follows:

$$z_{in(base)} = \beta r'_e$$

$$= 200 \times 3.33\ \Omega$$

$$= 667\ \Omega$$

Z_{in} includes the effects of the biasing resistors, R_1 and R_2.

$$z_{in} = z_{in(base)} \parallel R_1 \parallel R_2$$

$$= 667\ \Omega \parallel 18\ k\Omega \parallel 3.6\ k\Omega$$

$$= 545.6\ \Omega$$

Effect of Swamping Resistance on $z_{in(base)}$ and z_{in}

The effects of the swamping resistance on z_{in} can now be determined. For the amplifier in Fig. 30–7, $z_{in(base)}$ is calculated as

$$z_{in(base)} = \beta(r'_e + r_E) \tag{30–10}$$

$$= 200\ (3.33\ \Omega + 60\ \Omega)$$

$$= 200 \times 63.33\ \Omega$$

$$= 12.67\ k\Omega$$

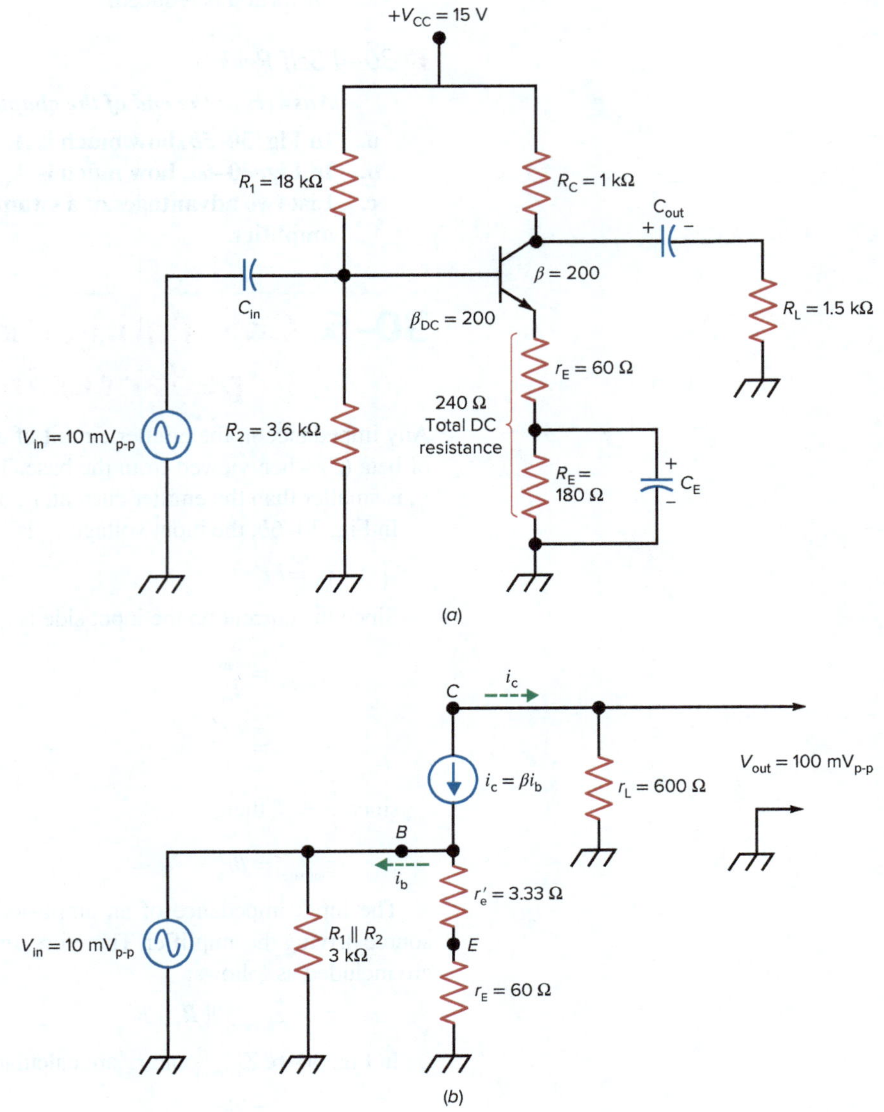

MultiSim **Figure 30–7** Adding a swamping resistor, r_E. (*a*) Original circuit. (*b*) AC equivalent circuit.

Notice that $z_{in(base)}$ has increased significantly. z_{in} is calculated as

$$z_{in} = z_{in(base)} \parallel R_1 \parallel R_2$$
$$= 12.67\ \text{k}\Omega \parallel 18\ \text{k}\Omega \parallel 3.6\ \text{k}\Omega$$
$$= 2.425\ \text{k}\Omega$$

Notice the effect of the swamping resistor, r_E, on z_{in}: r_E increases $z_{in(base)}$ and z_{in} substantially.

Output Impedance, z_{out}

The **output impedance, z_{out},** of a CE amplifier equals the value of the collector resistor, R_C, but does not include the load resistor, R_L. This is because the load, R_L, is driven by the amplifier. In Fig. 30–7, $z_{out} = R_C = 1\ \text{k}\Omega$.

Generator Impedance, R_G

In most cases, the AC source driving an amplifier will have some internal impedance that must be taken into account. The generator impedance prevents all of the

AC voltage from reaching the input of the amplifier. Actually, the generator impedance and the input impedance of the amplifier form a voltage divider, thus reducing the amount of AC voltage at the base.

30–5 Self-Review

Answers at the end of the chapter.

a. For a common-emitter amplifier, which is larger, $z_{in(base)}$ or z_{in}?
b. How does the addition of a swamping resistor affect the input impedance of a common-emitter amplifier?
c. In Fig. 30–7, what is the output impedance of the amplifier?

30–6 Common-Collector Amplifier

The **common-collector amplifier** is used to provide current gain and power gain. The voltage gain equals approximately one, or unity. As the name implies, the collector is common to both the input and output sides of the amplifier. The input signal is applied to the base, while the output is taken from the emitter. The output signal at the emitter is in phase with the input signal applied to the base. Because the output signal follows the input signal, the common-collector amplifier is usually referred to as an ***emitter follower.*** Another important characteristic of the emitter follower is that it has a high input impedance and a low output impedance, which make it ideal for impedance matching applications.

DC Analysis of an Emitter Follower

Figure 30–8*a* shows a common-collector amplifier circuit. Notice that the amplifier uses voltage divider bias and that the collector of the transistor is connected directly to V_{CC}. Because the collector is connected directly to V_{CC}, it is at AC ground.

To calculate the DC voltages and currents, proceed as follows.

The base voltage is calculated as

$$V_B = \frac{R_2}{R_1 + R_2} \times V_{CC} \qquad \textbf{(30–11)}$$

$$= \frac{5.6\text{ k}\Omega}{4.7\text{ k}\Omega + 5.6\text{ k}\Omega} \times 15\text{ V}$$

$$= 8.15\text{ V}$$

The DC emitter voltage is

$$V_E = V_B - V_{BE} \qquad \textbf{(30–12)}$$

$$= 8.15\text{ V} - 0.7\text{ V}$$

$$= 7.45\text{ V}$$

Next, calculate the DC emitter current, I_E:

$$I_E = \frac{V_E}{R_E} \qquad \textbf{(30–13)}$$

$$= \frac{7.45\text{ V}}{1\text{ k}\Omega}$$

$$= 7.45\text{ mA}$$

Since $I_E \approx I_C$, $I_C = 7.45$ mA.

Next, calculate the collector-emitter voltage, V_{CE}:

$$V_{CE} = V_{CC} - V_E \qquad \textbf{(30–14)}$$

$$= 15\text{ V} - 7.45\text{ V}$$

$$= 7.55\text{ V}$$

GOOD TO KNOW

In some emitter follower circuits, a small collector resistance is used to limit the DC collector current in case a short occurs between the emitter and ground. If a small R_C is used, the collector will also have a bypass capacitor going to ground. The small value of R_C will have only a slight bearing on the DC operation of the circuit and no bearing at all on the circuit's AC operation.

Figure 30–8 Common-collector amplifier. The circuit is also called an emitter follower. (*a*) Circuit. (*b*) DC load line showing $I_{C(sat)}$, $V_{CE(off)}$, I_C, and V_{CE}.

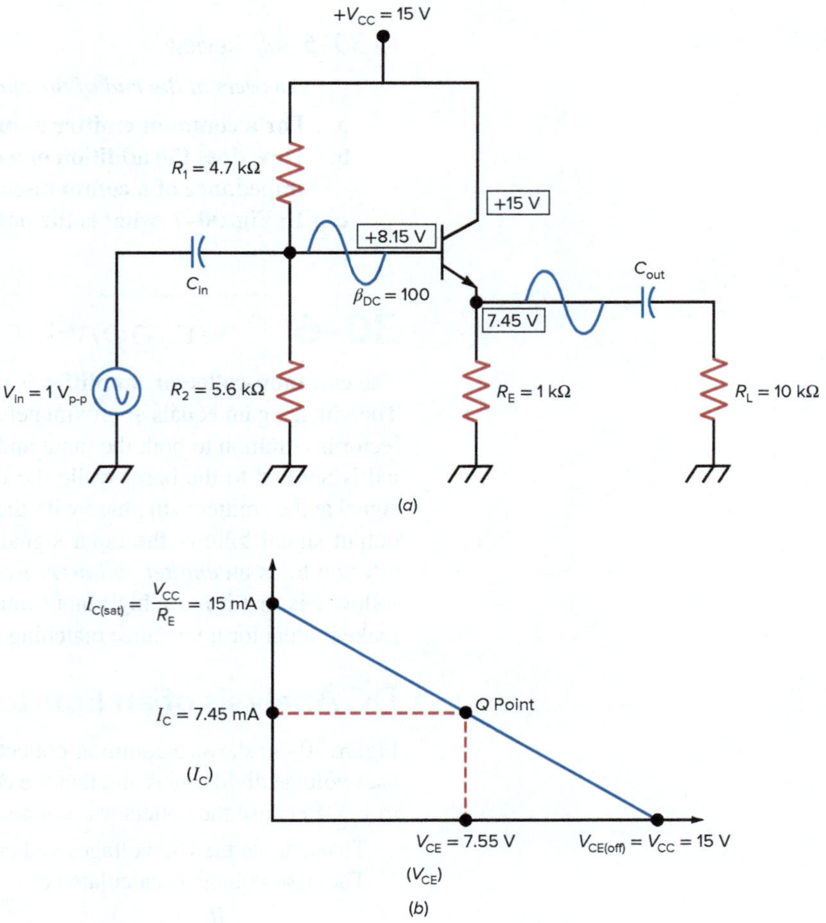

It is important to note that the collector voltage, V_C, equals +15 V with respect to ground because of its direct connection to V_{CC}. The values of V_B, V_E, and V_C are shown in the diagram of Fig. 30–8*a*.

DC Load Line

Figure 30–8*b* shows the DC load line. The endpoints are calculated using Formulas (30–15) and (30–16):

$$I_{C(sat)} = \frac{V_{CC}}{R_E} \quad \textbf{(30–15)}$$

$$V_{CE(off)} = V_{CC} \quad \textbf{(30–16)}$$

To obtain Formula (30–15), envision the collector-emitter terminals shorted; for Formula (30–16), envision the collector-emitter terminals open. For the DC load line in Fig. 30–8*b*, the endpoints of the DC load line are calculated as follows:

$$\begin{aligned} I_{C(sat)} &= \frac{V_{CC}}{R_E} \\ &= \frac{15\text{ V}}{1\text{ k}\Omega} \\ &= 15\text{ mA} \end{aligned}$$

$$\begin{aligned} V_{CE(off)} &= V_{CC} \\ &= 15\text{ V} \end{aligned}$$

The endpoints of $I_{C(sat)} = 15$ mA and $V_{CE(off)} = 15$ V are plotted on the DC load line in Fig. 30–8*b*. The calculated values for I_C and V_{CE} are also shown on the DC load line.

The values of R_1 and R_2 in Fig. 30–8 are usually chosen to produce a Q point near the center of the DC load line.

30–6 Self-Review

Answers at the end of the chapter.

a. Where is the output signal taken from in a common-collector amplifier?

b. What is another name for the common-collector amplifier?

30–7 AC Analysis of an Emitter Follower

In Fig. 30–8*a*, the input is applied to the base while the output is taken from the emitter. Because the collector is tied directly to the collector supply voltage, V_{CC}, no AC signal appears there. Also, because the emitter follower takes its output from the emitter, an emitter bypass capacitor is not used. With R_E unbypassed, the swamping is heavy, and the distortion in the output signal is extremely small.

No Phase Inversion

When the input signal voltage at the base increases, so does the output voltage at the emitter. Likewise, when the input signal voltage decreases, the output voltage decreases. Thus, the input signal and output signal are in phase with each other.

AC Equivalent Circuit

Figure 30–9 shows the AC equivalent circuit for the emitter follower circuit in Fig. 30–8*a*. Notice the AC resistance, r'_e, of the emitter diode. Its value is determined using the I_E value of 7.45 mA, as calculated earlier:

$$r'_e = \frac{25 \text{ mV}}{I_E}$$

$$= \frac{25 \text{ mV}}{7.45 \text{ mA}}$$

$$= 3.35\ \Omega$$

Figure 30–9 AC equivalent circuit of Fig. 30–8.

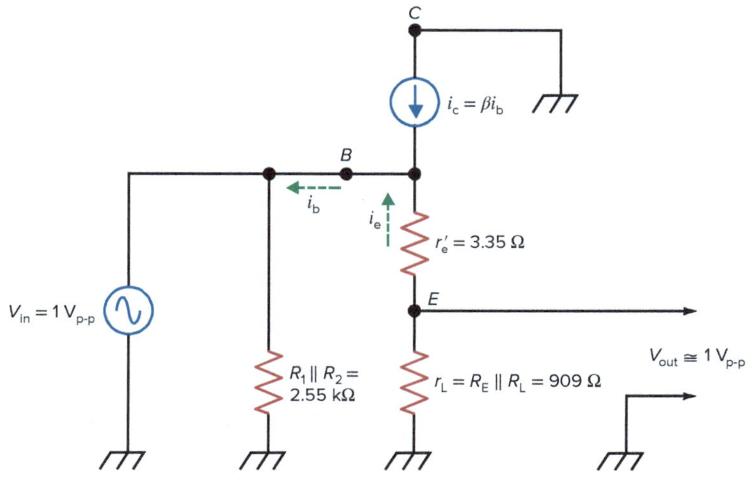

The AC load in the emitter circuit equals R_E and R_L in parallel. Its value is calculated as

$$r_L = \frac{R_E \times R_L}{R_E + R_L}$$
$$= \frac{1\ \text{k}\Omega \times 10\ \text{k}\Omega}{1\ \text{k}\Omega + 10\ \text{k}\Omega}$$
$$= 909\ \Omega$$

The biasing resistors R_1 and R_2 are in parallel for AC operation because the top of R_1 is connected to V_{CC}, which is at AC ground.

Voltage Gain, A_V

To develop a formula for the voltage gain, A_V, proceed as shown:

$$v_{out} = i_e r_L$$
$$v_{in} = i_e(r'_e + r_L)$$
$$A_V = \frac{v_{out}}{v_{in}}$$
$$= \frac{i_e r_L}{i_e(r'_e + r_L)}$$
$$= \frac{r_L}{r'_e + r_L} \qquad \textbf{(30–17)}$$

In most cases, r_L is much larger than r'_e, so Formula (30–17) simplifies to

$$A_V \approx \frac{r_L}{r_L} \approx 1 \text{ or unity}$$

Example 30-6

In Fig. 30–8*a*, use Formula (30–17) to find the exact value of A_V. Also, find v_{out}.

ANSWER

$$A_V = \frac{r_L}{r'_e + r_L}$$
$$= \frac{909\ \Omega}{3.35\ \Omega + 909\ \Omega}$$
$$= 0.996$$

With 1 $V_{p\text{-}p}$ applied as an input, v_{out} is

$$v_{out} = A_V \times v_{in}$$
$$= 0.996 \times 1\ V_{p\text{-}p}$$
$$= 0.996\ V_{p\text{-}p} \text{ or } 996\ mV_{p\text{-}p}$$

For all practical purposes, $v_{out} = v_{in}$.

Current Gain, A_i

The **current gain, A_i**, of a common-collector amplifier equals i_e/i_b. Since $i_e \approx i_c$, the current gain, A_i, is approximately equal to β.

Power Gain, A_P

The **power gain, A_P**, equals the product of the voltage and current gains.

$$A_P = A_V \times A_i \qquad \textbf{(30–18)}$$

Since $A_V \approx 1$ and $A_i \approx \beta$,

$$A_P = 1 \times \beta$$
$$= \beta$$

Calculating Input Impedance, z_{in}

To derive a formula for calculating z_{in}, use the AC equivalent circuit shown in Fig. 30–9.

The **input impedance, z_{in},** looking into the base can be expressed as

$$z_{in(base)} = \frac{v_{in}}{i_b}$$

Since $v_{in} \approx i_c(r'_e + r_L)$, then,

$$z_{in(base)} = \frac{i_c(r'_e + r_L)}{i_b}$$

and since

$$\frac{i_c}{i_b} = \beta$$

$$z_{in(base)} = \beta(r'_e + r_L) \quad \textbf{(30–19)}$$

Notice that this is the same formula used for a swamped CE amplifier. The input impedance of the circuit includes the effect of the biasing resistors, R_1 and R_2. This is shown in Formula (30–20):

$$z_{in} = z_{in(base)} \parallel R_1 \parallel R_2 \quad \textbf{(30–20)}$$

The biasing resistors, R_1 and R_2, actually lower z_{in} to a value that is not much different from that of a swamped CE amplifier. This is a disadvantage that must be overcome when using the emitter follower. In most cases, the emitter follower does not use biasing resistors because the stage driving the emitter follower provides the required DC bias. With the omission of the biasing resistors z_{in} increases substantially. How this is done will be shown in Section 30–8.

Example 30-7

Calculate Z_{in} in Fig. 30–8. (Note: $\beta = 100$)

ANSWER Referring to the AC equivalent circuit in Fig. 30–9, proceed as follows:

$$z_{in(base)} = \beta\,(r'_e + r_L)$$
$$= 100\,(3.35\ \Omega + 909\ \Omega)$$
$$= 100 \times 912.35\ \Omega$$
$$= 91.23\ \text{k}\Omega$$

Now calculate z_{in}:

$$z_{in} = z_{in(base)} \parallel R_1 \parallel R_2$$
$$= 91.23\ \text{k}\Omega \parallel 4.7\ \text{k}\Omega \parallel 5.6\ \text{k}\Omega$$
$$= 2.48\ \text{k}\Omega$$

Output Impedance, z_{out}

The output impedance, z_{out}, of an emitter follower is important because it prevents some AC signal voltage from reaching the load connected at the emitter. The output impedance of an emitter follower is usually quite low. For an emitter follower, z_{out} can be calculated by using Formula (30–21):

$$z_{out} = R_E \parallel \left(r_e' + \frac{R_G \parallel R_1 \parallel R_2}{\beta} \right) \quad \textbf{(30–21)}$$

Figure 30–10 Thevenin equivalent circuit for the output of an emitter follower circuit.

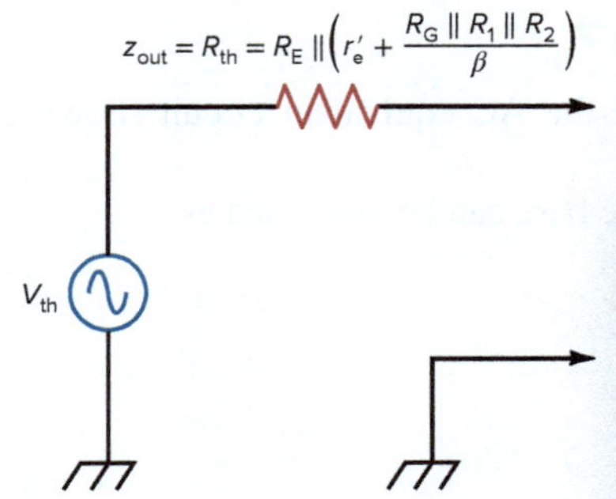

where R_G represents the internal resistance of the AC generator driving the base of the emitter follower. The derivation of this equation is quite complex and is therefore not covered here. However, Fig. 30–10 represents the Thevenin equivalent circuit for the output of an emitter follower. Since the circuit is assumed to be predominantly resistive, z_{out} can also be designated as R_{th}. A typical emitter follower has an output impedance of 10 Ω or less.

Example 30-8

In Fig. 30–11, calculate the following quantities: V_B, V_E, I_C, V_C, V_{CE}, r'_e, $z_{in(base)}$, z_{in}, A_V, v_b, and v_{out}. Also, plot the DC load line.

ANSWER Calculate all the DC quantities first:

$$V_B = \frac{R_2}{R_1 + R_2} \times V_{CC}$$
$$= \frac{18\ \text{k}\Omega}{22\ \text{k}\Omega + 18\ \text{k}\Omega} \times 20\ \text{V}$$
$$= 9.0\ \text{V}$$
$$V_E = V_B - V_{BE}$$
$$= 9.0\ \text{V} - 0.7\ \text{V}$$
$$= 8.3\ \text{V}$$

Figure 30–11 Emitter follower circuit used for Example 30–8. (*a*) Circuit. (*b*) DC load line.

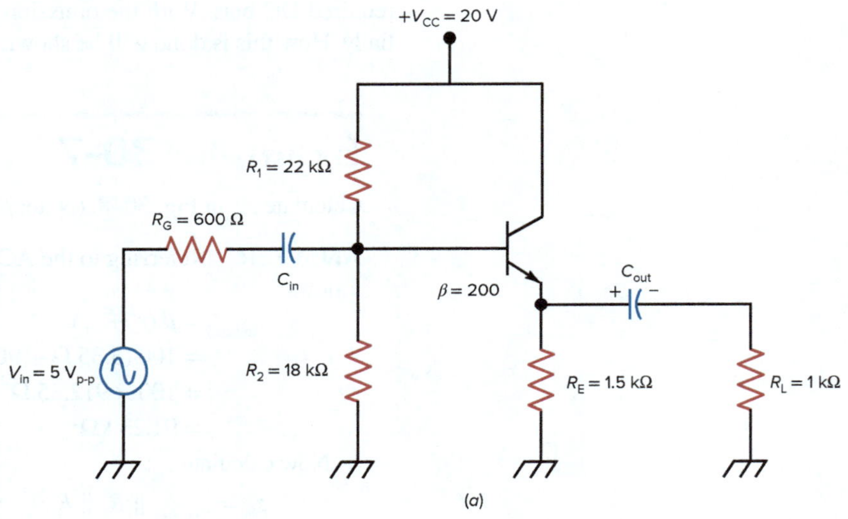

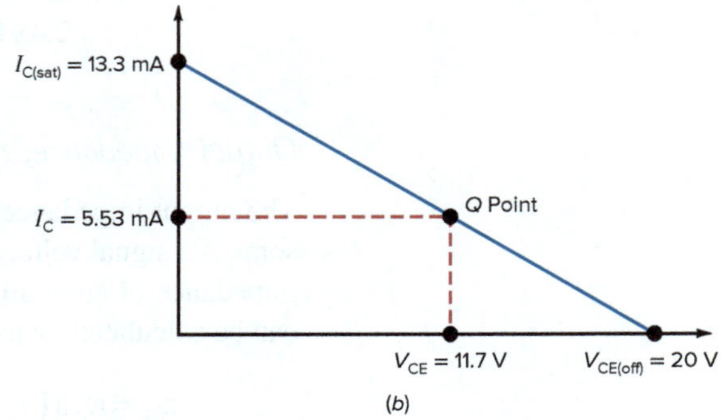

Since $I_E \approx I_C$,

$$I_C = \frac{V_E}{R_E}$$
$$= \frac{8.3\text{ V}}{1.5\text{ k}\Omega}$$
$$= 5.53\text{ mA}$$

Since the collector is tied directly to V_{CC}, $V_C = 20$ V. To calculate V_{CE}, proceed as follows:

$$V_{CE} = V_{CC} - V_E$$
$$= 20\text{ V} - 8.3\text{ V}$$
$$= 11.7\text{ V}$$

Next, calculate the endpoints for the DC load line shown in Fig. 30–11*b*:

$$I_{C(sat)} = \frac{V_{CC}}{R_E}$$
$$= \frac{20\text{ V}}{1.5\text{ k}\Omega}$$
$$= 13.3\text{ mA}$$
$$V_{CE(off)} = V_{CC}$$
$$= 20\text{ V}$$

The Q point values for I_C and V_{CE} are also shown on the DC load line in Fig. 30–11*b*.

Next, calculate the AC quantities:

$$r'_e = \frac{25\text{ mV}}{I_E}$$
$$= \frac{25\text{ mV}}{5.53\text{ mA}}$$
$$= 4.52\ \Omega$$

Now calculate the AC load resistance, r_L:

$$r_L = \frac{R_E \times R_L}{R_E + R_L}$$
$$= \frac{1.5\text{ k}\Omega \times 1\text{ k}\Omega}{1.5\text{ k}\Omega + 1\text{ k}\Omega}$$
$$= 600\ \Omega$$

Knowing r'_e and r_L, A_V can be accurately calculated:

$$A_V = \frac{r_L}{r'_e + r_L}$$
$$= \frac{600\ \Omega}{4.52\ \Omega + 600\ \Omega}$$
$$= 0.992$$

Next, calculate $z_{in(base)}$ and z_{in}:

$$z_{in(base)} = \beta(r'_e + r_L)$$
$$= 200\ (4.52\ \Omega + 600\ \Omega)$$
$$= 120.9\text{ k}\Omega$$
$$z_{in} = z_{in(base)} \parallel R_1 \parallel R_2$$
$$= 120.9\text{ k}\Omega \parallel 22\text{ k}\Omega \parallel 18\text{ k}\Omega$$
$$= 9.15\text{ k}\Omega$$

To calculate the output voltage, v_{out}, first calculate the AC base voltage, v_b. Since R_G and z_{in} form a voltage divider, the AC signal voltage at the base is calculated as

$$v_b = \frac{z_{in}}{R_G + z_{in}} \times v_{in}$$

$$= \frac{9.15\text{ k}\Omega}{600\ \Omega + 9.15\text{ k}\Omega} \times 5\text{ V}_{\text{p-p}}$$

$$= 4.69\text{ V}_{\text{p-p}}$$

Since $A_V = 0.992$, v_{out} is calculated as

$$v_{out} = A_V \times v_{in}$$

$$= 0.992 \times 4.69\text{ V}_{\text{p-p}}$$

$$= 4.65\text{ V}_{\text{p-p}}$$

Because the collector is tied to V_{CC}, the signal voltage at the collector is zero.

30–7 Self-Review

Answers at the end of the chapter.

a. What is approximate voltage gain of an emitter follower?

b. Is the output impedance of an emitter follower a high or low value?

30–8 Emitter Follower Applications

Figure 30–12 shows how an emitter follower is typically used. Notice that the common-emitter amplifier is driving the base of the emitter follower. The direct connection from the collector of Q_1 to the base of Q_2 implies that the DC collector voltage of Q_1 supplies the required bias voltage for the emitter follower circuit. This eliminates the need for the biasing resistors, which increases the input impedance of the emitter follower substantially. It should be pointed out that the DC base current of Q_2 is made very small with respect to the collector current in Q_1, so that the DC voltage at the collector of Q_1 will not be affected. The main purpose of the circuit is to use the emitter follower as a buffer to isolate the relatively low value of load resistance, R_L, from the high impedance collector of Q_1. Doing so allows the common-emitter circuit to have a much higher overall voltage gain.

Figure 30–12 Base lead of emitter follower circuit tied directly to the collector of the common-emitter amplifier. Note that the biasing resistors for the emitter follower have been omitted. This increases the z_{in} of the emitter follower.

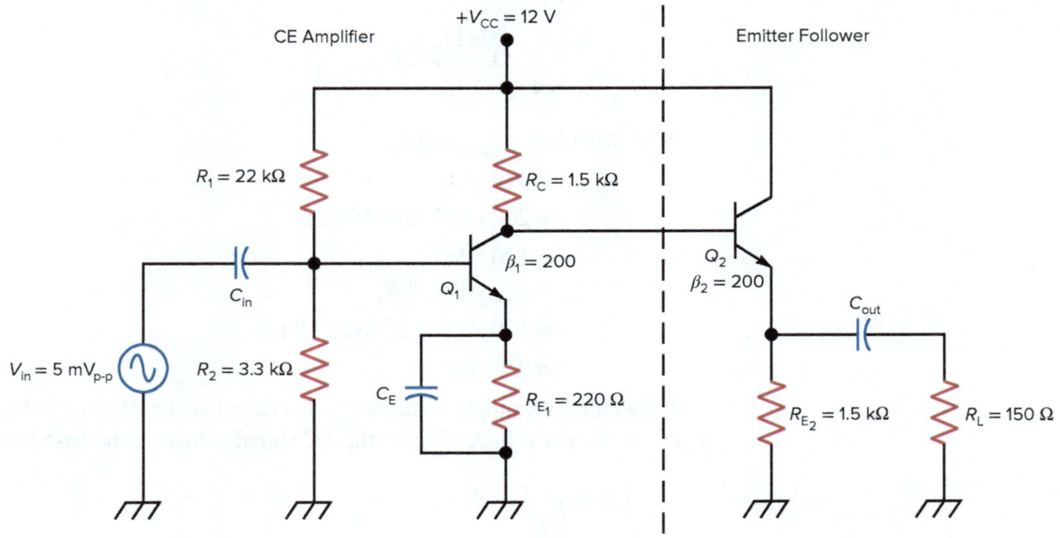

Circuit Analysis

To calculate the DC voltages and currents in the circuit, proceed as shown. Start with the CE amplifier, including the transistor, Q_1:

$$V_B = \frac{R_2}{R_1 + R_2} \times V_{CC}$$
$$= \frac{3.3\ k\Omega}{22\ k\Omega + 3.3\ k\Omega} \times 12\ V$$
$$= 1.56\ V$$
$$V_E = V_B - V_{BE}$$
$$= 1.56\ V - 0.7\ V$$
$$= 0.86\ V$$
$$I_{C_1} \approx \frac{V_E}{R_{E_1}}$$
$$= \frac{0.86\ V}{220\ \Omega}$$
$$= 3.9\ mA$$
$$V_{C_1} = V_{CC} - I_{C_1} R_C$$
$$= 12\ V - (3.9\ mA \times 1.5\ k\Omega)$$
$$= 12\ V - 5.85\ V$$
$$= 6.15\ V$$

Since the base of Q_2 is attached directly to the collector of $Q_1, V_{B_2} = 6.15$ V also. To calculate I_{E_2}, proceed as follows:

$$V_{E_2} = V_{B_2} - V_{BB}$$
$$= 6.15\ V - 0.7\ V$$
$$= 5.45\ V$$
$$I_{E_2} = \frac{V_{E2}}{R_{E_2}}$$
$$= \frac{5.45\ V}{1.5\ k\Omega}$$
$$= 3.63\ mA$$

With I_{E_2} known, r'_{e_2} can be calculated:

$$r'_{e_2} = \frac{25\ mV}{I_{E_2}}$$
$$= \frac{25\ mV}{3.63\ mA}$$
$$= 6.88\ \Omega$$

To find $z_{in(base\ Q_2)}$, the value of the AC load resistance in the emitter is needed:

$$r_L = \frac{R_{E_2} \times R_L}{R_{E_2} + R_L}$$
$$= \frac{1.5\ k\Omega \times 150\ \Omega}{1.5\ k\Omega + 150\ \Omega}$$
$$= 136.3\ \Omega$$

Now calculate $z_{in(base\ Q_2)}$:

$$z_{in(base\ Q_2)} = \beta_2(r'_{e_2} + r_L)$$
$$= 200\ (6.88\ \Omega + 136.3\ \Omega)$$
$$= 28.63\ k\Omega$$

This is the AC load in parallel with the 1.5 kΩ R_C at the collector of Q_1.

The AC load resistance in the collector of Q_1 is

$$r_{LQ_1} = z_{in(base\ Q_2)} \parallel R_C$$
$$= 28.63\ k\Omega \parallel 1.5\ k\Omega$$
$$= 1.425\ k\Omega$$

To calculate the voltage gain, A_V, in Q_1 proceed as follows:

$$A_{V_{Q_1}} = \frac{r_{L_{Q_1}}}{r'_{e1}}$$

r'_{e_1} is calculated as

$$r'_{e_1} = \frac{25\ mV}{I_{E_1}}$$
$$= \frac{25\ mV}{3.9\ mA}$$
$$= 6.41\ \Omega$$

Therefore,

$$A_{V_{Q_1}} = \frac{1.425\ k\Omega}{6.41\ \Omega}$$
$$= 222$$

This means that the AC voltage at the collector of Q_1 is

$$v_{out_{Q_1}} = A_{V_{Q_1}} \times v_{in}$$
$$= 222 \times 5\ mV_{p\text{-}p}$$
$$= 1.11\ V_{p\text{-}p}$$

Since $A_{V_{Q_2}} \approx 1$, $v_{out_{Q_2}}$ also equals approximately 1.11 $V_{p\text{-}p}$.

It must be understood that the emitter follower is used to step up the impedance of the 150-Ω load, R_L, so that the collector of Q_1 has a higher resistance and, in turn, a larger overall voltage gain.

Advantage of the Emitter Follower

Figure 30–13 shows the same CE amplifier driving the same 150-Ω load, R_L, without the emitter follower. For this connection, the voltage gain, $A_{V_{Q_1}}$, is much less than it was with the emitter follower. To prove this, calculate the AC collector load resistance:

$$r_L = R_C \parallel R_L$$
$$= 1.5\ k\Omega \parallel 150\ \Omega$$
$$= 136.3\ \Omega$$

Figure 30–13 Common-emitter circuit with R_L connected directly to the output of the common-emitter amplifier circuit.

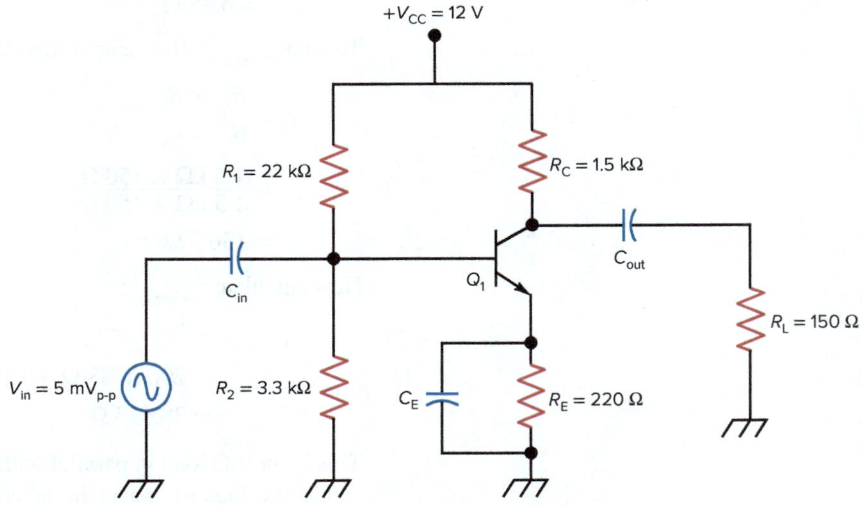

Next, calculate the voltage gain, A_V:

$$A_{V_{Q_1}} = \frac{r_L}{r'_{e_1}}$$
$$= \frac{136.3\ \Omega}{6.41\ \Omega}$$
$$= 21.3$$

For this value of A_V, v_{out} is

$$v_{out} = A_{V_{Q_1}} \times v_{in}$$
$$= 21.3 \times 5\ \text{mV}_{\text{p-p}}$$
$$= 106.5\ \text{mV}_{\text{p-p}}$$

Notice the difference from 1.11 $V_{\text{p-p}}$ obtained at the output in Fig. 30–12 with the use of the emitter follower circuit.

30–8 Self-Review

Answers at the end of the chapter.

a. **In Fig. 30–12, what is the DC voltage at the base of Q_2?**
b. **In Fig. 30–12, how much is the AC voltage at the collector of Q_2?**

30–9 Common-Base Amplifier

The **common-base amplifier** is used less often than the common-emitter or common-collector amplifier. The common-base amplifier provides a high voltage and power gain, but the current gain is less than one. Unlike the common-collector amplifier, the common-base amplifier has an extremely low input impedance, z_{in}. This is a big disadvantage. The low input impedance is the main reason the common-base amplifier is used so infrequently. The low input impedance loads down the AC signal source driving the common-base amplifier. (The AC signal source could be the output of a common-emitter amplifier circuit.)

The common-base amplifier, however, does provide some desirable features for operation at higher frequencies. It is also used in a circuit called a *differential amplifier*. Differential amplifiers are widely used in linear integrated circuits known as *op amps*.

DC Analysis of a Common-Base Amplifier

Figure 30–14*a* shows a common-base amplifier. Notice that the base is grounded. Notice also that the input signal, V_{in}, is applied to the emitter and the output is taken from the collector.

The common-base amplifier in Fig. 30–14 uses emitter bias. The emitter supply voltage, $-V_{EE}$, forward-biases the emitter-base junction, and the collector supply voltage, V_{CC}, reverse-biases the collector-base junction. The DC equivalent circuit (shown in Fig. 30–14*b*) is obtained by reducing all AC sources to zero and treating all capacitors like open circuits. Notice in Fig. 30–14*b* that the emitter diode acts like any forward-biased diode. Also, notice in Fig. 30–14*b* the collector diode acts as a current source.

To calculate the DC emitter current, use Formula (30–22):

$$I_E = \frac{V_{EE} - V_{BE}}{R_E} \qquad \textbf{(30–22)}$$

Inserting the values from Fig. 30–14 gives

$$I_E = \frac{6\ \text{V} - 0.7\ \text{V}}{3.3\ \text{k}\Omega}$$
$$= 1.6\ \text{mA}$$

GOOD TO KNOW

Although the common-base amplifier in Fig. 30–14 is biased using emitter bias, other biasing techniques such as voltage divider bias could also be used.

Figure 30–14 Common-base amplifier. (*a*) Original circuit with input applied to the emitter and the output taken from the collector. (*b*) DC equivalent circuit.

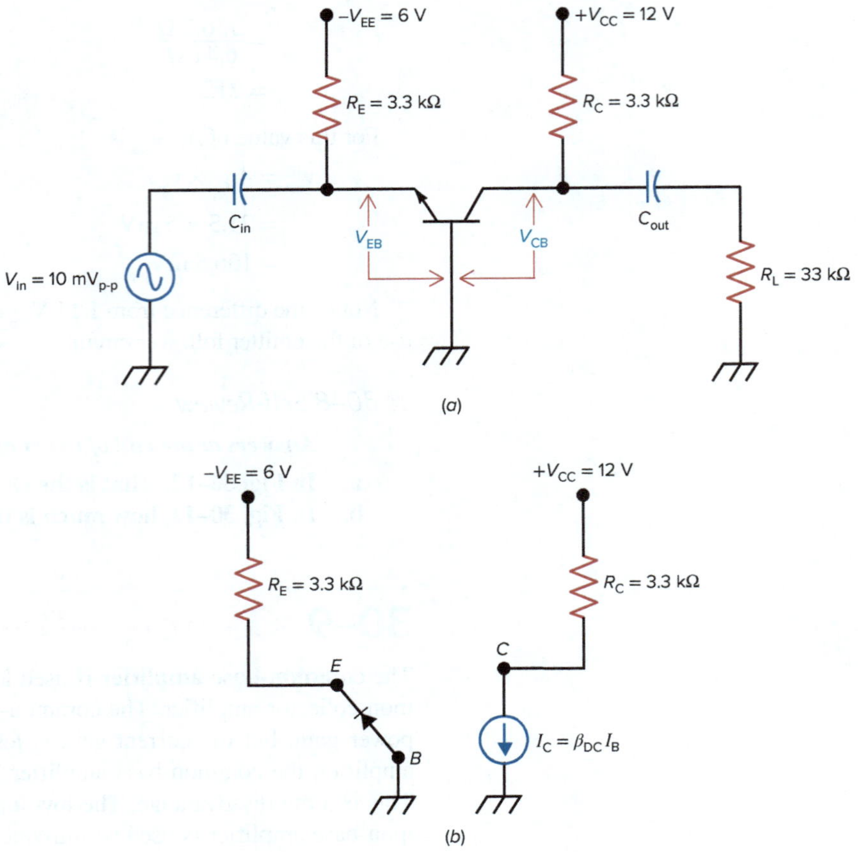

It is important to note that the DC emitter voltage equals −0.7 V with respect to ground.

Since $I_E \approx I_C$, $I_C = 1.6$ mA also.

To calculate the DC voltage at the collector with respect to the base, use Formula (30–23):

$$
\begin{aligned}
V_{CB} &= V_{CC} - I_C R_C \qquad \textbf{(30–23)}\\
&= 12\text{ V} - (1.6\text{ mA} \times 3.3\text{ k}\Omega)\\
&= 12\text{ V} - 5.28\text{ V}\\
&= 6.72\text{ V}
\end{aligned}
$$

30-9 Self-Review

Answers at the end of the chapter.

a. Where is the input signal applied in a common-base amplifier?

b. Where is the output signal taken from in a common-base amplifier?

30–10 AC Analysis of a Common-Base Amplifier

To understand how the common-base amplifier in Fig. 30–14 operates with an AC signal applied, draw the AC equivalent circuit, as shown in Fig. 30–15. Notice that both DC sources have been reduced to zero, all coupling capacitors have been treated

Figure 30–15 AC equivalent circuit of the common-base amplifier in Fig. 30–14.

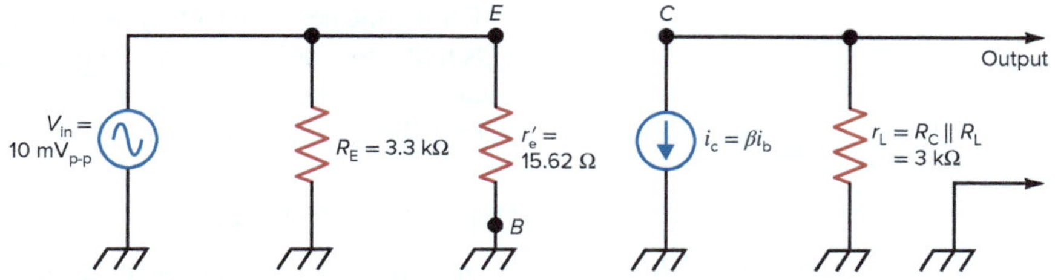

as AC shorts, and the emitter diode has been replaced with its equivalent AC resistance, designated r'_e. The AC resistance of the emitter diode is calculated as follows:

$$r'_e = \frac{25\ \text{mV}}{I_E}$$
$$= \frac{25\ \text{mV}}{1.6\ \text{mA}}$$
$$= 15.62\ \Omega$$

Notice the value of the AC load resistance, r_L, in the collector circuit. It is calculated as follows:

$$r_L = \frac{R_C \times R_L}{R_C + R_L}$$
$$= \frac{3.3\ \text{k}\Omega \times 33\ \text{k}\Omega}{3.3\ \text{k}\Omega + 33\ \text{k}\Omega}$$
$$= 3\ \text{k}\Omega$$

Voltage Gain, A_V

To calculate the voltage gain, A_V, write an expression for v_{out} and v_{in}:

$$v_{out} = i_c r_L$$
$$v_{in} = i_e r'_e$$

Since $i_e \approx i_c$, then $v_{in} = i_c r'_e$.
Next calculate A_V:

$$A_V = \frac{v_{out}}{v_{in}}$$
$$= \frac{i_c r_L}{i_c r'_e}$$
$$= \frac{r_L}{r'_e} \qquad \textbf{(30–24)}$$

In Fig. 30–14, the voltage gain, A_V, is

$$A_V = \frac{3\ \text{k}\Omega}{15.62\ \Omega}$$
$$= 192$$

Since v_{in} equals 10 $\text{mV}_{\text{p-p}}$, v_{out} is calculated as

$$v_{out} = A_V \times v_{in}$$
$$= 192 \times 10\ \text{mV}_{\text{p-p}}$$
$$= 1.92\ \text{V}_{\text{p-p}}$$

No Phase Inversion

For a common-base amplifier, the output signal voltage is in phase with the input signal voltage. In Fig. 30–14*a*, the emitter current, i_e, is decreased during the positive half-cycle of v_{in}. This occurs because the input signal voltage is opposing the forward bias from the $-V_{EE}$ supply. Because i_e decreases during the positive half-cycle of v_{in}, the voltage across the collector resistor, R_C, is also decreasing, thus causing the collector voltage to increase.

During the negative half-cycle of v_{in}, i_e increases because the input signal voltage adds to the forward bias provided by the emitter supply voltage. As i_e increases, the voltage across the collector resistor, R_C, also increases, which in turn causes the collector voltage to decrease.

GOOD TO KNOW

Common-base amplifiers are often used in high-frequency communications equipment. One reason is that they have better high-frequency response than a common-emitter amplifier. Also, in rf circuitry 50 Ω is a common impedance for matching antennas to receivers and/or transmitters. In these cases, the lower input impedance of a common-base amplifier is an advantage because it closely matches 50 Ω.

Input Impedance, z_{in}

As mentioned earlier, the main disadvantage of a common-base amplifier is its extremely low input impedance, z_{in}. In Fig. 30–15, the input voltage source, v_{in}, is connected directly across the AC resistance of the emitter diode. Therefore,

$$v_{in} = i_e r'_e$$

Also, since the input current to the transistor is i_e,

$$z_{in(emitter)} = \frac{v_{in}}{i_{in}} = \frac{i_e r'_e}{i_e} = r'_e$$

The input impedance, z_{in}, of the stage includes the emitter resistor, R_E. Therefore,

$$z_{in} = R_E \parallel r'_e \qquad \textbf{(30–25)}$$

In most cases, R_E is so much larger than r'_e that $z_{in} \approx r'_e$.

For Fig. 30–14, $z_{in(emitter)}$ and z_{in} are

$$z_{in(emitter)} = r'_e = 15.62\ \Omega$$

$$z_{in} = R_E \parallel r'_e = 3.3\ \text{k}\Omega \parallel 15.62\ \Omega = 15.54\ \Omega$$

For all practical purposes, $z_{in} = r'_e = 15.62\ \Omega$.

Example 30-9

In Fig. 30–16, calculate the following: I_E, V_{CB}, r'_e, A_V, v_{out}, and z_{in}.

Figure 30–16 Circuit used for Example 30–9.

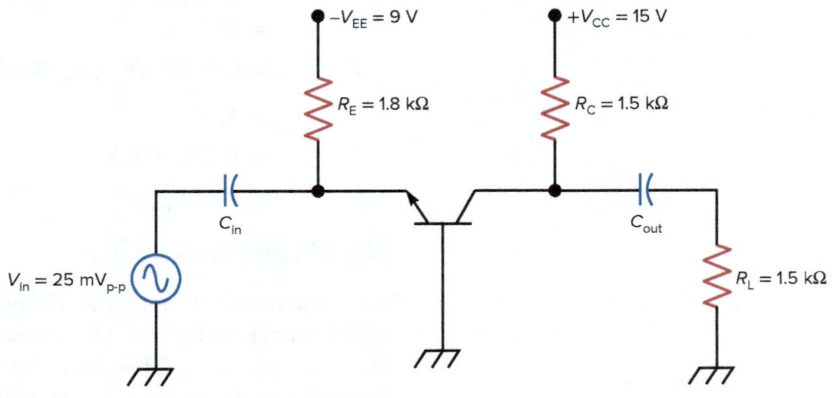

ANSWER Begin by calculating the two DC quantities. First, calculate I_E:

$$I_E = \frac{V_{EE} - V_{BE}}{R_E}$$
$$= \frac{9\text{ V} - 0.7\text{ V}}{1.8\text{ k}\Omega}$$
$$= 4.61\text{ mA}$$

Next, calculate V_{CB}:

$$V_{CB} = V_{CC} - I_C R_C$$
$$= 15\text{ V} - (4.61\text{ mA} \times 1.5\text{ k}\Omega)$$
$$= 15\text{ V} - 6.91\text{ V}$$
$$= 8.09\text{ V}$$

To find the AC quantities, first calculate r'_e and r_L:

$$r'_e = \frac{25\text{ mV}}{I_E}$$
$$= \frac{25\text{ mV}}{4.61\text{ mA}}$$
$$= 5.42\ \Omega$$
$$r_L = R_C \parallel R_L$$
$$= \frac{1.5\text{ k}\Omega}{2}$$
$$= 750\ \Omega$$

A_V can now be calculated:

$$A_V = \frac{r_L}{r'_e}$$
$$= \frac{750\ \Omega}{5.42\ \Omega}$$
$$= 138.3$$

With A_V known, v_{out} can be calculated:

$$v_{out} = A_V \times v_{in}$$
$$= 138.3 \times 25\text{ mV}_{p\text{-}p}$$
$$= 3.46\text{ V}_{p\text{-}p}$$

To calculate z_{in}, proceed as follows:

$$z_{in} = R_E \parallel r'_e$$
$$= 1.8\text{ k}\Omega \parallel 5.42\ \Omega$$
$$= 5.40\ \Omega$$

Notice how close in value r'_e and z_{in} actually are.

Example 30-10

In Fig. 30–17, calculate the AC output voltage, v_{out}.

ANSWER Begin by calculating I_E so that r'_e can be calculated:

$$I_E = \frac{V_{EE} - V_{BE}}{R_E}$$

Figure 30–17 Circuit used for Example 30–10. (*a*) Circuit. (*b*) z_{in} and R_G form a voltage divider. Because z_{in} is so small, most of v_{in} is dropped across R_G.

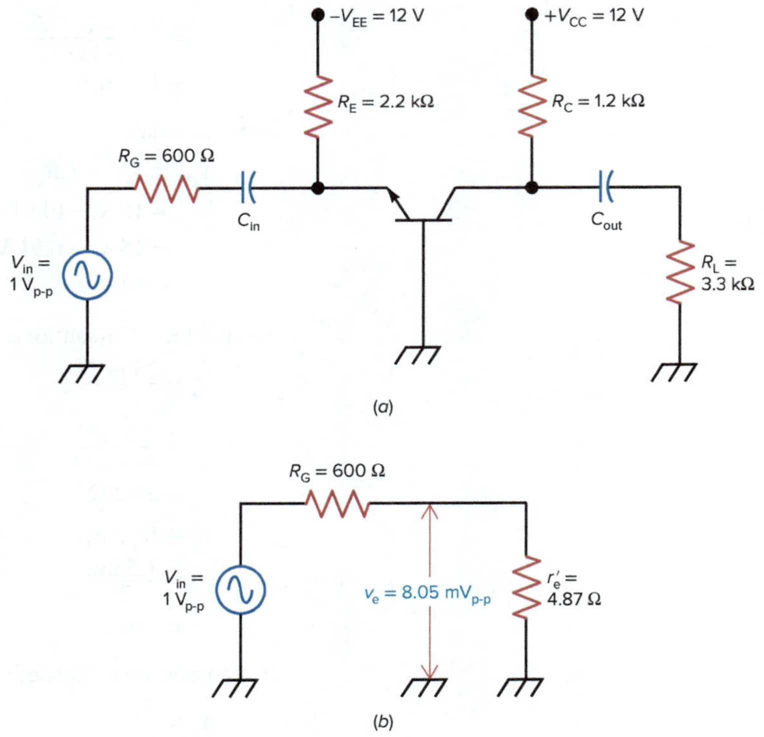

$$= \frac{12\text{ V} - 0.7\text{ V}}{2.2\text{ k}\Omega}$$

$$= 5.13\text{ mA}$$

Next, calculate r'_e:

$$r'_e = \frac{25\text{ mV}}{I_E}$$

$$= \frac{25\text{ mV}}{5.13\text{ mA}}$$

$$= 4.87\ \Omega$$

To calculate A_V, first calculate r_L:

$$r_L = R_C \parallel R_L$$

$$= \frac{1.2\text{ k}\Omega \times 3.3\text{ k}\Omega}{1.2\text{ k}\Omega + 3.3\text{ k}\Omega}$$

$$= 880\ \Omega$$

$$A_V = \frac{r_L}{r'_e}$$

$$= \frac{880\ \Omega}{4.87\ \Omega}$$

$$= 180.7$$

To calculate v_{out}, determine how much AC signal voltage is at the emitter. Figure 30–17*b* shows that the generator resistance, R_G, and the AC resistance of the emitter diode form a voltage divider. Notice how lopsided the voltage divider is. Most of the input voltage, v_{in}, will be dropped across R_G and not reach the amplifier input. To calculate the AC emitter voltage, v_e, proceed as follows:

$$
\begin{aligned}
v_e &= \frac{z_{in}}{R_G + z_{in}} \times v_{in} \\
&= \frac{4.87\ \Omega}{600\ \Omega + 4.87\ \Omega} \times 1\ V_{p\text{-}p} \\
&= 8.05\ mV_{p\text{-}p}
\end{aligned}
$$

This is the only portion of the input signal voltage that is amplified. To calculate v_{out},

$$
\begin{aligned}
v_{out} &= A_V \times v_e \\
&= 180.7 \times 8.05\ mV_{p\text{-}p} \\
&= 1.45\ V_{p\text{-}p}
\end{aligned}
$$

This example illustrates the downfall of the common-base amplifier. In real-life situations, the input voltage source could actually be the output of a CE transistor amplifier. Then the collector load resistance of the CE amplifier would be severely reduced by the low input impedance of the common-base amplifier. This in turn would drastically lower the voltage gain of the CE stage.

30–10 Self-Review

Answers at the end of the chapter.

a. What is the approximate value of Z_{in} for a common-base amplifier?

b. What is the major disadvantage of a common-base amplifier?

Summary

- The emitter diode of a transistor has an equivalent resistance for ac signals. The ac resistance of the emitter diode, designated $r'_e = 25\text{ mV}/I_E$.
- In a common-emitter amplifier, the input signal is applied to the base and the output signal is taken from the collector.
- A common-emitter amplifier provides a large voltage gain, a large current gain, and a very high power gain.
- The input and output AC voltages in a common-emitter amplifier are 180° out of phase.
- In the AC equivalent circuit of a transistor amplifier, the coupling and bypass capacitors and DC voltage sources appear as AC shorts.
- In the AC equivalent circuit of a transistor amplifier, the emitter diode is replaced with its equivalent AC resistance, r'_e.
- The voltage gain of a common-emitter amplifier equals r_L/r'_e when the dc emitter resistance is completely bypassed. The ac load resistance, designated r_L, equals the equivalent resistance of R_C and R_L in parallel.
- A swamping resistor is an unbypassed resistance in the emitter circuit of a common-emitter amplifier. The swamping resistor stabilizes the voltage gain and reduces distortion.
- The input impedance of a common-emitter amplifier equals $Z_{in(base)} \parallel R_1 \parallel R_2$. The output impedance equals R_C.
- A common-collector amplifier provides a large current gain and a large power gain, but its voltage gain is approximately one or unity.
- In a common-collector amplifier, the input signal is applied to the base, and the output signal is taken from the emitter. Since the AC signal at the emitter follows, or is in phase with, the AC signal at the base, the common-collector amplifier is also referred to as an emitter follower.
- An emitter follower has high input impedance and low output impedance. This makes it ideal for impedance-matching applications.
- In a common-base amplifier, the input signal is applied to the emitter, and the output signal is taken from the collector.
- A common-base amplifier has high voltage and power gain, but its current gain is slightly less than one.
- The main drawback of the common-base amplifier is its extremely low input impedance which is approximately equal to the low value of r'_e.

Important Terms

AC beta—the ratio of AC collector current, i_c, to AC base current, i_b: $\beta = i_c/i_b$.

AC equivalent circuit—a circuit as it appears to an AC signal. In an AC equivalent circuit, all capacitors and voltage sources appear as shorts.

AC resistance of a diode—the equivalent resistance of a forward-biased diode as it appears to small AC signals. For a standard diode, $r_{AC} = 25\text{ mV}/I_d$. For the emitter diode in a transistor, $r'_e = 25\text{ mV}/I_E$.

Common-base amplifier—a transistor amplifier whose input is applied to the emitter and whose output is taken from the collector. The common-base amplifier provides high voltage and power gain, but its current gain is less than one.

Common-collector amplifier—a transistor amplifier whose input is applied to the base and whose output is taken from the emitter. The common-collector amplifier provides high current and power gain, but its voltage gain is less than one.

Common-emitter amplifier—a transistor amplifier whose input is applied to the base and whose output is taken from the collector. The common-emitter amplifier provides high voltage and current gain and very high power gain.

Current gain, A_i—the ratio of output current to input current in a transistor amplifier.

Emitter bypass capacitor, C_E—a capacitor that bypasses the AC signal around the emitter resistor in a transistor amplifier.

Emitter follower—another name for the common-collector amplifier.

Input impedance, Z_{in}—the impedance of the input of an amplifier as seen by the AC signal source driving the amplifier.

Output impedance, Z_{out}—the impedance at the output of an amplifier as seen by the load driven by the amplifier.

Power gain, A_p—the ratio of output power to input power in a transistor amplifier. A_p can also be calculated as $A_p = A_V \times A_i$.

Small signal—a signal whose peak-to-peak current value is one-tenth or less the DC diode or DC emitter current.

Swamping resistor—an unbypassed resistor in the emitter circuit of a common-emitter amplifier. A swamping resistor stabilizes the voltage gain and reduces distortion.

Voltage gain, A_V—the ratio of output voltage to input voltage in a transistor amplifier: $A_V = V_{out}/V_{in}$.

Related Formulas

$r_{AC} = 25\text{ mV}/I_d$ (Ordinary Diode)

$r'_e = 25\text{ mV}/I_E$ (Emitter Diode)

$A_V = V_{out}/V_{in}$ (Any Amplifier)

Common-Emitter Amplifier

$A_V = R_C/r'_e$ (No Load Resistor)

$A_V = r_L/r'_e$ (With Load Resistor)

$A_V = r_L/(r'_e + r_E)$ (With Emitter Swamping Resistor)

$Z_{in(base)} = \beta r'_e$ (No Emitter Swamping Resistor)

$Z_{in(base)} = \beta(r'_e + r_E)$ (With Emitter Swamping Resistor)

$Z_{in} = Z_{in(base)} \parallel R_1 \parallel R_2$

Common-Collector Amplifier

$V_B = \frac{R_2}{R_1 + R_2} \times V_{CC}$

$V_E = V_B - V_{BE}$

$I_E = \frac{V_E}{R_E}$

$V_{CE} = V_{CC} - V_E$

$I_{C(sat)} = V_{CC}/R_E$

$V_{CE(off)} = V_{CC}$

DC Load Line Endpoints

$A_V = \frac{r_L}{r'_e + r_L}$

$A_P = A_V \times A_i$

$Z_{in(base)} = \beta(r'_e + r_L)$

$Z_{in} = Z_{in(base)} \parallel R_1 \parallel R_2$

$Z_{out} = R_E \parallel \left(r'_e + \frac{R_G \parallel R_1 \parallel R_2}{\beta}\right)$

Common-Base Amplifier

$I_E = \frac{V_{EE} - V_{BE}}{R_E}$

$V_{CB} = V_{CC} - I_C R_C$

$A_V = r_L/r'_e$

$Z_{in} = R_E \parallel r'_e$

Self-Test

Answers at the back of the book.

1. **The AC resistance, r_{AC}, of a diode**
   a. is not affected by the DC current in the diode.
   b. decreases as the DC current in the diode increases.
   c. increases as the DC current in the diode increases.
   d. decreases as the DC current in the diode decreases.

2. **If a transistor has a DC emitter current, I_E, of 6.25 mA, how much is the AC resistance, r'_e, of the emitter diode?**
   a. 0.25 Ω.
   b. 4 kΩ.
   c. 4 Ω.
   d. It cannot be determined.

3. **Which of the following transistor amplifier configurations provides a 180° phase shift between the AC input and output voltages?**
   a. the common-emitter amplifier.
   b. the emitter follower.
   c. the common-base amplifier.
   d. the common-collector amplifier.

4. **Removing the emitter bypass capacitor in a common-emitter amplifier will**
   a. decrease the voltage gain, A_V.
   b. increase the input impedance, Z_{in}.
   c. increase the voltage gain, A_V.
   d. both a and b.

5. **In a common-emitter amplifier, what happens to the voltage gain, A_V, when a load resistor, R_L, is connected to the output?**
   a. A_V decreases.
   b. A_V increases.
   c. A_V doesn't change.
   d. A_V doubles.

6. **Which type of transistor amplifier is also known as the emitter follower?**
   a. the common-base amplifier.
   b. the common-collector amplifier.
   c. the common-emitter amplifier.
   d. none of the above.

7. **Which type of transistor amplifier has a voltage gain, A_V, of approximately one, or unity?**
   a. the common-base amplifier.
   b. the common-emitter amplifier.
   c. the common-collector amplifier.
   d. none of the above.

8. **What is the biggest disadvantage of the common-base amplifier?**
   a. its high input impedance.
   b. its low voltage gain.
   c. its high output impedance.
   d. its low input impedance.

9. **Which of the following transistor amplifiers has the lowest output impedance?**
   a. the common-collector amplifier.
   b. the common-base amplifier.
   c. the common-emitter amplifier.
   d. none of the above.

10. **A swamping resistor in a common-emitter amplifier**
   a. stabilizes the voltage gain.
   b. increases the voltage gain.
   c. reduces distortion.
   d. both a and c.

11. In the AC equivalent circuit of a transistor amplifier,

a. the resistors appear as an AC short.

b. the capacitors and V_{CC} appear as an open to AC.

c. the capacitors and V_{CC} appear as AC shorts.

d. the X_C values of all capacitors are assumed to be infinite.

12. Which of the following transistor amplifier configurations is often used in impedance-matching applications?

a. the common-base amplifier.

b. the emitter follower.

c. the common-emitter amplifier.

d. the common-source amplifier.

13. In which type of amplifier is the input applied to the emitter and the output taken from the collector?

a. the common-base amplifier.

b. the common-emitter amplifier.

c. the common-collector amplifier.

d. the emitter follower.

14. The collector of an emitter follower is connected directly to the positive terminal of a 12-V_{DC} supply. If the AC input voltage applied to the base is 2 $V_{p\text{-}p}$, how much AC signal is present at the collector?

a. approximately 2 $V_{p\text{-}p}$.

b. 0 $V_{p\text{-}p}$.

c. approximately 1.98 $V_{p\text{-}p}$.

d. It cannot be determined.

15. What is the only amplifier configuration that provides both voltage and current gain?

a. the common-base amplifier.

b. the emitter follower.

c. the common-emitter amplifier.

d. the common-collector amplifier.

Essay Questions

1. In a transistor amplifier, how is a small AC signal defined?

2. Why are capacitors and voltage sources treated as short circuits when analyzing the AC operation of a transistor amplifier?

3. How does a swamping resistor in a common-emitter amplifier stabilize the voltage gain and reduce distortion?

4. How does the AC emitter resistance, r'_e, of an emitter diode vary with the DC emitter current?

5. In a common-emitter amplifier, why are the AC input and output voltages 180° out of phase?

6. Why is a common-collector amplifier often referred to as an emitter follower?

7. What is the main application of an emitter follower?

8. What is the main drawback of a common-base amplifier?

Problems

SECTION 30–1 AC RESISTANCE OF A DIODE

30–1 In Fig. 30–18, calculate the AC resistance, r_{AC}, of the diode for each of the following values of R (use the second approximation of a diode):

a. $R = 10\ \text{k}\Omega$.

b. $R = 5.6\ \text{k}\Omega$.

c. $R = 2.7\ \text{k}\Omega$.

d. $R = 1\ \text{k}\Omega$.

Figure 30–18

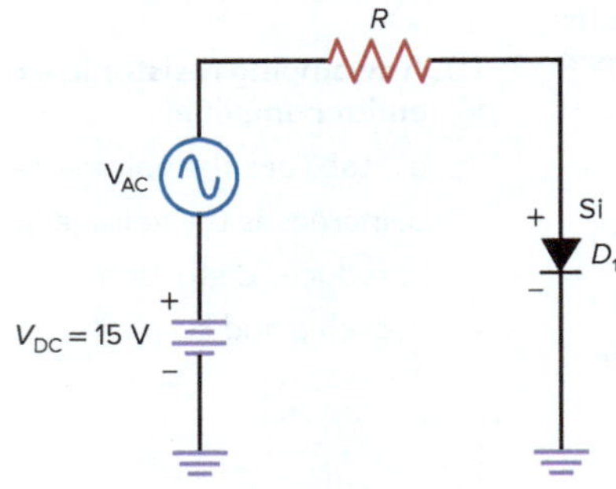

30–2 Calculate the AC resistance, r'_e, of the emitter diode in a transistor for each of the following values of DC emitter current, I_E:

a. $I_E = 1$ mA.

b. $I_E = 5$ mA.

c. $I_E = 7.5$ mA.

d. $I_E = 20$ mA.

SECTION 30–2 SMALL SIGNAL AMPLIFIER OPERATION

30–3 What type of transistor amplifier is shown in Fig. 30–19*a*?

30–4 Calculate the following DC quantities in Fig. 30–19*a*:

a. V_B.

b. V_E.

c. I_E.

d. V_C.

e. V_{CE}.

Figure 30–19

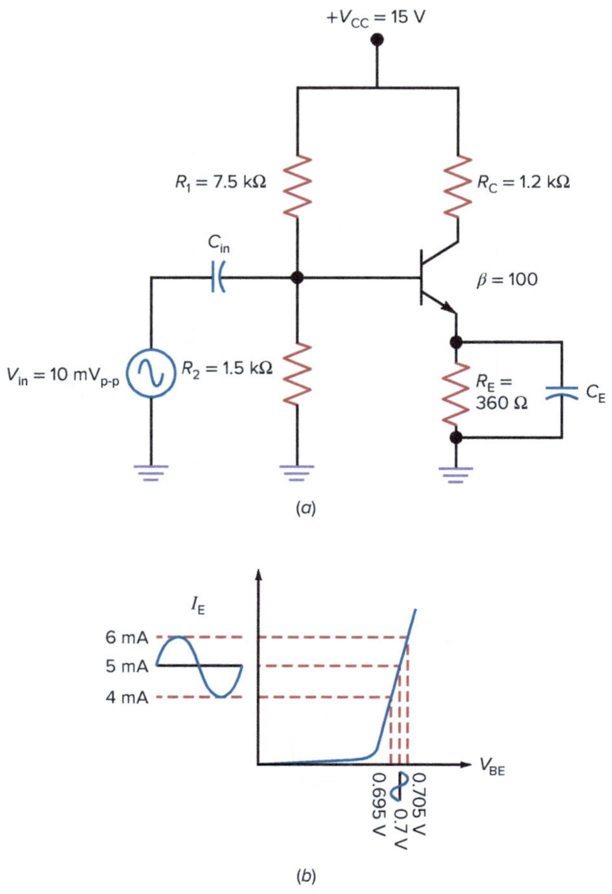

30–5 Figure 30–19*b* shows a graph of I_E versus V_{BE} for the transistor in Fig. 30–19*a*. How much is the collector voltage, V_C, when the emitter current, I_E, is

a. 4 mA?

b. 5 mA?

c. 6 mA?

30–6 Based on your answers from Prob. 30–5, how much is

a. the peak-to-peak collector voltage?

b. the voltage gain, A_V, of the amplifier?

30–7 In Fig. 30–19*a*, how much AC voltage would be measured from

a. the base to ground?

b. the emitter to ground?

c. the collector to ground?

30–8 In Fig. 30–19*a*, how much is V_{BE} when V_{in} is at

a. zero?

b. its positive peak?

c. its negative peak?

30–9 Calculate the voltage gain, A_V, of a common-emitter amplifier for each of the following values of V_{out} and V_{in}:

a. $V_{out} = 1.2\ V_{p-p}$ and $V_{in} = 4.8\ mV_{p-p}$.

b. $V_{out} = 500\ mV_{p-p}$ and $V_{in} = 25\ mV_{p-p}$.

c. $V_{out} = 10\ V_{p-p}$ and $V_{in} = 200\ mV_{p-p}$.

30–10 What is the output voltage of a common-emitter amplifier if $V_{in} = 5\ mV_{p-p}$ and A_V equals

a. 20?

b. 50?

c. 1500?

SECTION 30–3 AC EQUIVALENT CIRCUIT OF A CE AMPLIFIER

30–11 In Fig. 30–19*a*, which components appear as short circuits in the AC equivalent circuit?

30–12 Calculate the AC resistance, r'_e, of the emitter diode in Fig. 30–19*a*.

30–13 Draw the AC equivalent circuit (the condensed version) for the common-emitter amplifier in Fig. 30–19*a*. Show all values.

SECTION 30–4 CALCULATING THE VOLTAGE GAIN, A_V, OF A CE AMPLIFIER

30–14 In Fig. 30–19*a*, calculate the voltage gain, A_V, using the equation $A_V = R_C/r'_e$. After you determine A_V, calculate the output voltage, V_{out}. How do your answers compare to those calculated earlier in Prob. 30–6?

30–15 In Fig. 30–19*a*, what are the new values for A_V and V_{out} if R_C is replaced with a

a. 600-Ω value?

b. 1.5-kΩ value?

c. 2-kΩ value?

30–16 In Fig. 30–19*a*, calculate the new values for A_V and V_{out} if the following load resistors are connected via a coupling capacitor to the collector ($R_C = 1.2$ kΩ):

a. $R_L = 600$ Ω.

b. $R_L = 1.2$ kΩ.

c. $R_L = 1.5$ kΩ.

d. $R_L = 7.2$ kΩ.

30–17 In Fig. 30–20*a*, calculate the following DC quantities:

a. V_B.

b. V_E.

c. I_E.

d. V_C.

e. V_{CE}.

30–18 In Fig. 30–20*a*, calculate the following AC quantities:

a. r'_e.

b. r_L.

c. A_V.

d. V_{out}.

Figure 30–20

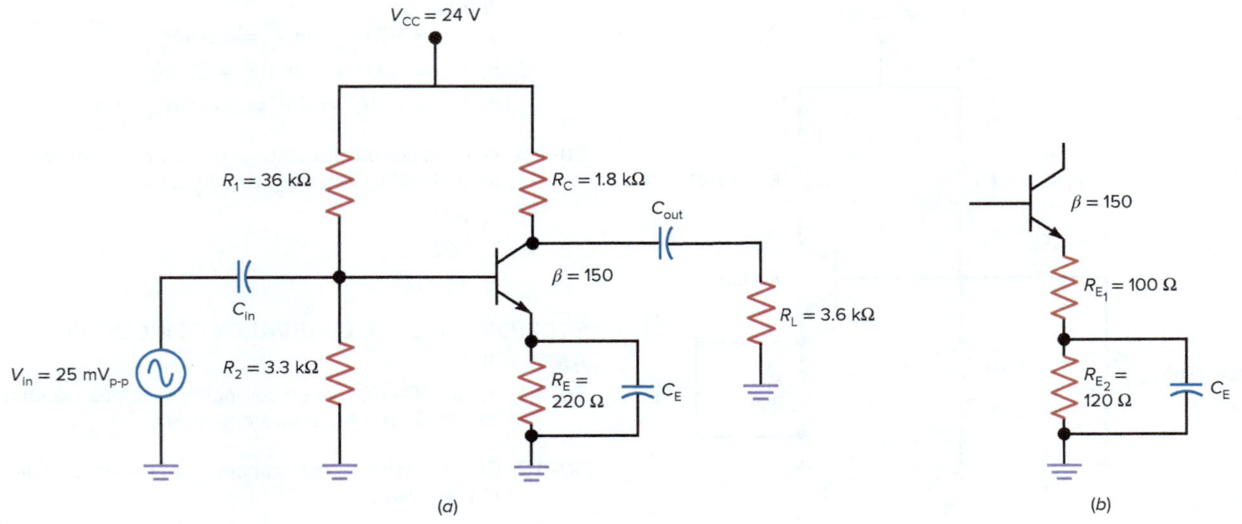

30–19 In Fig. 30–20*a*, what are the new values for A_V and V_{out} if R_L is removed?

30–20 In Fig. 30–20*a*, suppose that the emitter resistance is split into two resistors as shown in Fig. 30–20*b*. Calculate

a. A_V.

b. V_{out}.

SECTION 30–5 CALCULATING THE INPUT AND OUTPUT IMPEDANCES IN A CE AMPLIFIER

30–21 In Fig. 30–19*a*, calculate

a. $Z_{in(base)}$.

b. Z_{in}.

30–22 Repeat Prob. 30–21 for each of the following values of beta:

a. $\beta = 150$.

b. $\beta = 200$.

c. $\beta = 300$.

30–23 In Fig. 30–19*a*, assume that the emitter bypass capacitor, C_E, is removed. If $\beta = 100$, calculate

a. $Z_{in(base)}$.

b. Z_{in}.

30–24 In Fig. 30–20*a*, calculate

a. $Z_{in(base)}$.

b. Z_{in}.

c. Z_{out}.

30–25 Repeat Prob. 30–24 if the emitter circuit is modified, as shown in Fig. 30–20*b*.

SECTION 30–6 COMMON-COLLECTOR AMPLIFIER

30–26 What type of transistor amplifier is shown in Fig. 30–21?

Figure 30–21

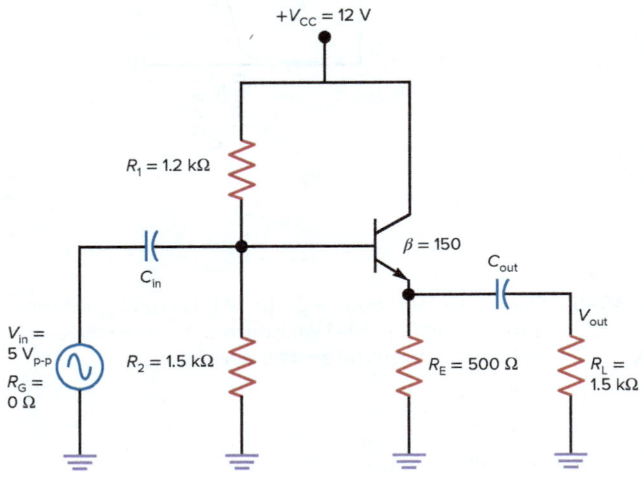

30–27 Calculate the following DC quantities in Fig. 30–21:

a. V_B.

b. V_E.

c. I_E.

d. V_C.

e. V_{CE}.

f. $I_{C(sat)}$.

g. $V_{CE(off)}$.

30–28 Draw the DC load line for the emitter follower in Fig. 30–21. Include the values of I_C and V_{CE} at the Q point.

SECTION 30–7 AC ANALYSIS OF AN EMITTER FOLLOWER

30–29 In Fig. 30–21, solve for the following AC quantities:

a. r'_e.
b. r_L.
c. A_V.
d. V_{out}.
e. $Z_{in(base)}$.
f. Z_{in}.
g. Z_{out}. (Note: Assume $R_G = 0\ \Omega$.)

30–30 In Fig. 30–21, how much AC voltage would you expect to measure at the collector with respect to ground?

30–31 What is the phase relationship between V_{in} and V_{out} in Fig. 30–21?

30–32 In Fig. 30–22, solve for the following DC and AC quantities:

a. V_B.
b. V_E.
c. I_E.
d. V_C.
e. V_{CE}.
f. r'_e.
g. r_L.
h. A_V.
i. $Z_{in(base)}$.
j. Z_{in}.
k. v_b.
l. V_{out}.
m. Z_{out}. (Note: $R_G = 600\ \Omega$.)

Figure 30–22

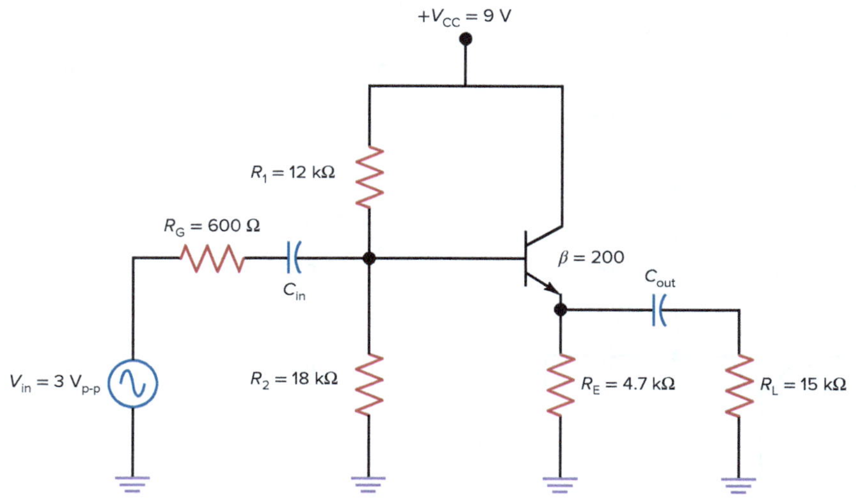

SECTION 30–8 EMITTER FOLLOWER APPLICATIONS

30–33 In Fig. 30–23, solve for the following DC quantities:

a. $V_{B(Q_1)}$.
b. $V_{E(Q_1)}$.
c. $I_{E(Q_1)}$.
d. $V_{C(Q_1)}$.
e. $V_{B(Q_2)}$.
f. $V_{E(Q_2)}$.
g. $I_{E(Q_2)}$.
h. $V_{C(Q_2)}$.

30–34 In Fig. 30–23, solve for the following AC quantities:

a. $r'_{e(Q_1)}$.
b. $r'_{e(Q_2)}$.
c. $r_{L(Q_2)}$.
d. $Z_{in(base\ Q_2)}$.
e. $r_{L(Q_1)}$.
f. $A_{V(Q_1)}$.
g. $A_{V(Q_2)}$.
h. $V_{out(Q_1)}$.
i. $V_{out(Q_2)}$.

30–35 In Fig. 30–23, suppose that the emitter follower is omitted and the output from the collector of Q_1 is capacitively coupled to the 250-Ω load, R_L. What is the output voltage across the 250-Ω load?

Figure 30–23

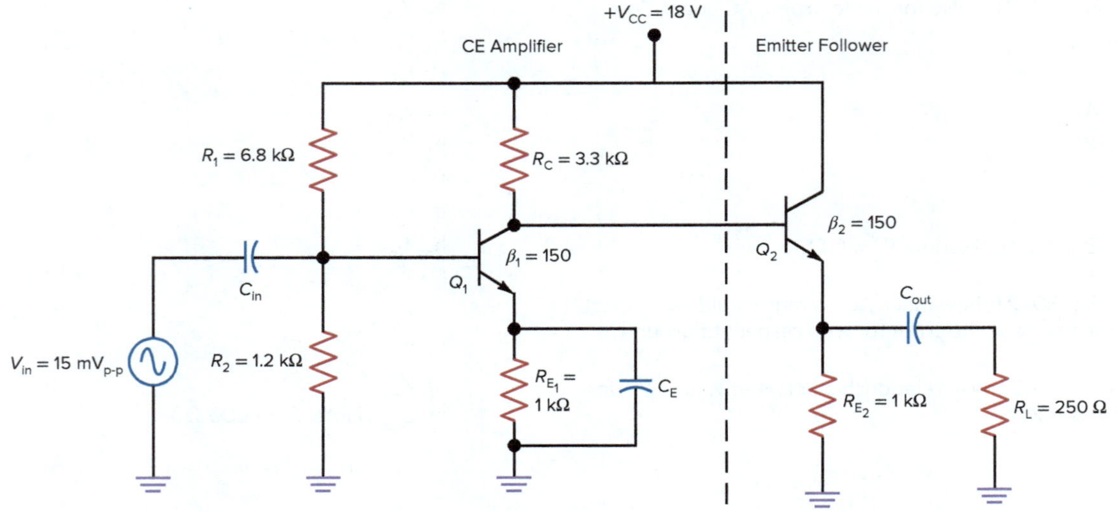

SECTION 30–9 COMMON-BASE AMPLIFIER

30–36 What type of transistor amplifier is shown in Fig. 30–24?

Figure 30–24

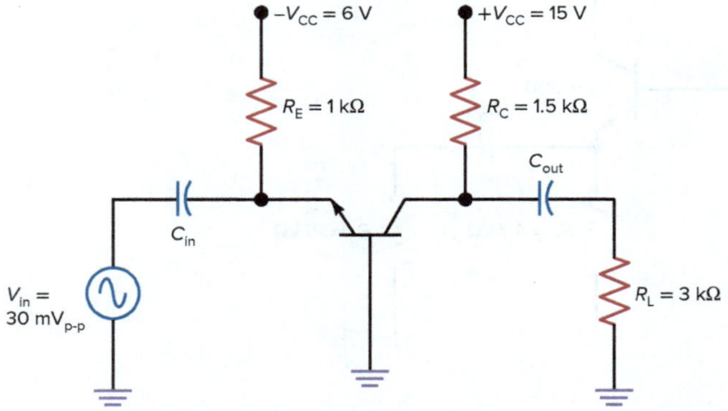

30–37 In Fig. 30–24, solve for the following DC quantities:

a. V_E.

b. I_E.

c. V_{CB}.

SECTION 30–10 AC ANALYSIS OF A COMMON-BASE AMPLIFIER

30–38 In Fig. 30–24, solve for the following AC quantities:

a. r'_e.

b. r_L.

c. A_V.

d. V_{out}.

e. Z_{in}.

30–39 In Fig. 30–24, what is the phase relationship between V_{in} and V_{out}?

30–40 In Fig. 30–25, calculate the following DC and AC quantities:

a. V_E.

b. I_E.

c. V_{CB}.

d. r'_e.

e. r_L.

f. A_V.

g. Z_{in}.

h. v_e.

i V_{out}.

Figure 30–25

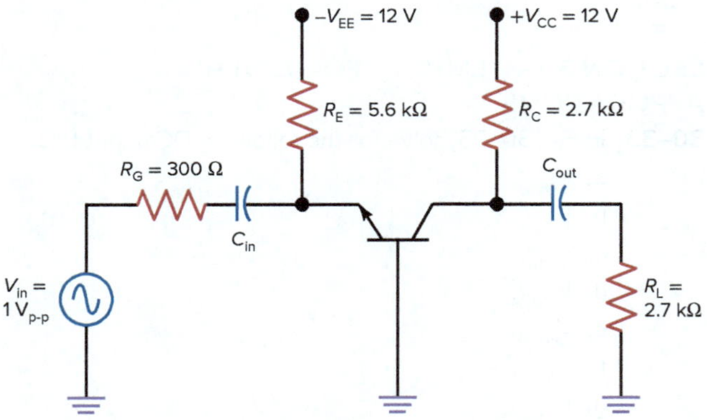

Answers to Self-Reviews

30–1 **a.** decrease
b. one-tenth
c. 25 mV/I_E

30–2 **a.** 180°
b. 0 $V_{p\text{-}p}$

30–3 **a.** short
b. as a short

30–4 **a.** 360 approx.
b. 200
c. A_V is stabilized and distortion is reduced

30–5 **a.** $Z_{in(base)}$
b. it increases Z_{in}
c. 1 kΩ

30–6 **a.** the emitter
b. emitter follower

30–7 **a.** one or unity
b. low

30–8 **a.** 6.15 V
b. 0 $V_{p\text{-}p}$

30–9 **a.** the emitter
b. the collector

30–10 **a.** $Z_{in} \approx r'_e$
b. its very low input impedance

Laboratory Application Assignment

In this lab application assignment, you will examine the ability of a transistor to amplify a small AC signal. You will build a common-emitter amplifier and measure the input and output voltages so you can determine the voltage gain, A_V. You will also see how the emitter bypass capacitor, C_E, and load resistance, R_L, affect the voltage gain, A_V.

Equipment: Obtain the following items from your instructor.

- 2N2222A *npn* transistor or equivalent
- Assortment of carbon-film resistors
- Two 100-μF electrolytic capacitors and one 220-μF electrolytic capacitor
- DMM
- Oscilloscope
- Function generator
- Variable dc power supply

Common-Emitter Amplifier Calculations

Examine the common-emitter amplifier in Fig. 30–26. Calculate and record the following DC quantities:

$V_B =$ __________, $V_E =$ __________, $I_E =$ __________,
$V_C =$ __________, $V_{CE} =$ __________

Next, calculate and record the following AC values with the emitter bypass capacitor, C_E, connected but *without* the load resistor, R_L. Note that $V_{in} = 10\ \text{mV}_{p\text{-}p}$.

$r'_e =$ __________, $A_V =$ __________, $V_{out} =$ __________

Connect the load resistor, R_L, and recalculate A_V and V_{out}.

$A_V =$ __________, $V_{out} =$ __________

Figure 30–26

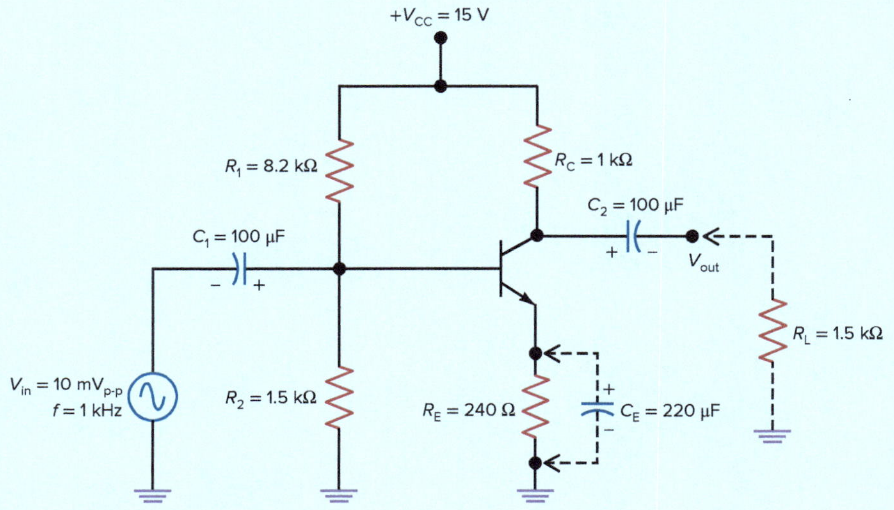

What happened to A_V and V_{out} when R_L was added?__________

With the load resistor, R_L, still connected, remove the emitter bypass capacitor, C_E, and recalculate A_V and V_{out}.

A_V = __________, V_{out} = __________

What happened to A_V and V_{out} when C_E was removed?__________

____

Common-Emitter Amplifier Measurements

Construct the common-emitter amplifier in Fig. 30–26. Adjust the input voltage, V_{in}, to exactly 10 mV_{p-p}. Measure and record the following DC quantities:

V_B = __________, V_E = __________, I_E = __________,

V_C = __________, V_{CE} = __________

Next, measure and record the output voltage, V_{out}, with the emitter bypass capacitor, C_E, connected but *without the* load resistor, R_L. Using the measured value of V_{out}, calculate the voltage gain, A_V.

V_{out} = __________, A_V = __________

Connect the load resistor, R_L, and remeasure V_{out}. Using the measured value of V_{out}, recalculate A_V.

A_V = __________, V_{out} = __________

What happened to V_{out} and A_V?

____

With the load resistor, R_L, still connected, remove the emitter bypass capacitor, C_E, and remeasure V_{out}. Using the measured value of V_{out}, recalculate A_V.

V_{out} = __________, A_V = __________

Did A_V and V_{out} decrease substantially?

__________ If yes, explain why. __________

____

____

____

Design credits: Chapter Opener Background Image, and Multisim: Stockbyte/Getty Images

chapter 31

Field Effect Transistors

The *field effect transistor (FET)* is a three-terminal device similar to the bipolar junction transistor. The FET, however, is a **unipolar** device that depends on only one type of charge carrier, either free electrons or holes. There are basically two types of FETs: the *junction field effect transistor*, abbreviated JFET, and the *metal-oxide-semiconductor field effect transistor*, abbreviated MOSFET.

Unlike bipolar transistors, which are current-controlled devices, FETs are voltage-controlled devices (i.e., an input voltage controls an output current). The input impedance is extremely high (of the order of megohms) for FETs and therefore they require very little power from the driving source. Their high input impedance is one reason that FETs are sometimes preferred over bipolar transistors.

This chapter covers JFET and MOSFET characteristics, biasing techniques, the different types of FET amplifiers, as well as MOSFET applications. ■

Chapter Outline

Chapter Objectives

After studying this chapter, you should be able to

- *Describe* the construction of a JFET.
- *Explain* how an input voltage controls the output current in a JFET.
- *Explain* why the input impedance of a JFET is so high.
- *Identify* the schematic symbols of an *n*-channel and *p*-channel JFET.
- *Explain* the ohmic and current-source regions of operation for a JFET.
- *Define* the term *gate-source cutoff voltage*.
- *Define* the term *pinch-off voltage*.
- *Explain* why JFETs are referred to as normally ON devices.
- *Calculate* the drain current in a JFET when I_{DSS}, V_{GS}, and $V_{GS(off)}$ are known.
- *Explain* the biasing techniques for JFETs.
- *Define* the term *transconductance*, g_m.
- *Explain* the operation of the common-source, common-drain, and common-gate amplifiers.
- *Explain* the differences in construction and operation of a depletion-type and enhancement-type MOSFET.
- *List* the precautions to observe when handling MOSFETs.

Important Terms

asymmetrical JFET
channel
common-drain amplifier
common-gate amplifier
common-source amplifier
current-source region
depletion mode
drain
enhancement mode
field effect transistor (FET)
gate
gate-source cutoff voltage, $V_{GS(off)}$
IGFET
JFET
MOSFET
ohmic region
pinch-off voltage, V_P
source
symmetrical JFET
threshold voltage, $V_{GS(th)}$
transconductance, g_m
unipolar

GOOD TO KNOW

JFETs are typically much smaller than bipolar transistors. This makes JFETs particularly suitable for use in ICs, where the size of each component is critical.

31–1 JFETs and Their Characteristics

Figure 31–1*a* shows the construction of an *n*-channel **JFET**. Notice there are four leads: the **drain, source**, and two **gates**. The area between the source and drain terminals is called the ***channel***. Because *n*-type semiconductor material is used for the channel, the device is called an *n*-channel JFET. Embedded on each side of the *n*-channel are two smaller *p*-type regions. Each *p* region is called a *gate*. When the manufacturer connects a separate lead to each gate, the device is called a *dual-gate* JFET. Dual-gate JFETs are most commonly used in frequency mixers, circuits that are frequently encountered in communications electronics. In most cases, the gates are internally connected and the device acts like a single-gate JFET.

A *p*-channel JFET is shown in Fig. 31–1*b*. Embedded on both sides of the *p*-channel are two *n*-type gate regions. Again, these are normally connected together to form a single gate lead.

The current flow is between the drain and source terminals in a JFET. For the *n*-channel JFET in Fig. 31–1*a*, the majority current carriers in the channel are free electrons. Conversely, for the *p*-channel JFET in Fig. 31–1*b*, the majority current carriers in the channel are holes.

Schematic Symbols

The schematic symbols for a JFET are shown in Fig. 31–2. Figure 31–2*a* is the schematic symbol for the *n*-channel JFET, and Fig. 31–2*b* shows the symbol for the *p*-channel JFET. Notice that the only difference is the direction of the arrow on the gate lead. In Fig. 31–2*a*, the arrow points in toward the *n*-type channel, whereas in Fig. 31–2*b* the arrow points outward from the *p*-type channel. In each symbol, the thin vertical line connecting the drain and source is a reminder that these terminals are connected to each end of the channel.

One more point: When the gate regions of a JFET are located in the center of the channel, the JFET is said to be *symmetrical*, meaning that the drain and source leads may be interchanged without affecting its operation. If the construction of a JFET is such that the gate regions are offset from center, the JFET is called *asymmetrical*. The drain and source leads may not be interchanged in an **asymmetrical JFET**. Figure 31–2*c* represents the schematic symbol of an asymmetrical JFET, and Fig. 31–2*a* and *b* show the schematic symbols of a **symmetrical JFET**. Note that when the gates are offset from center in an asymmetrical JFET, they are placed close to the source terminal. This is shown in the schematic symbol of Fig. 31–2*c*.

Figure 31–1 Construction of a JFET. (*a*) *n*-channel JFET. (*b*) *p*-channel JFET.

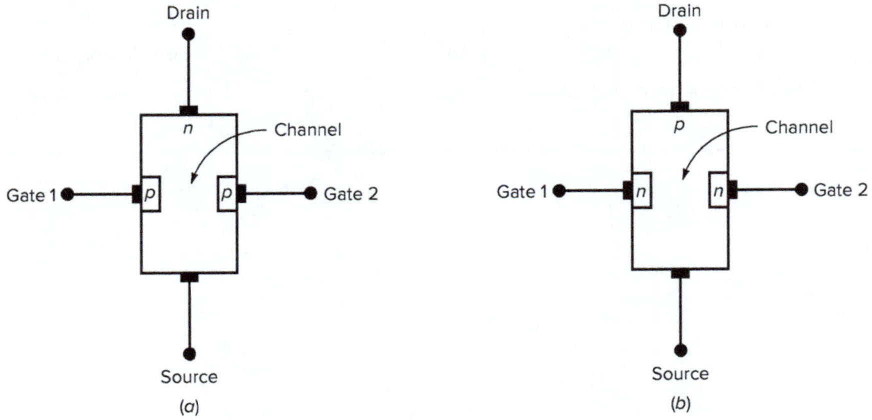

Figure 31–2 Schematic symbols for a JFET. (*a*) *n*-channel symmetrical JFET. (*b*) *p*-channel symmetrical JFET. (*c*) *n*-channel asymmetrical JFET.

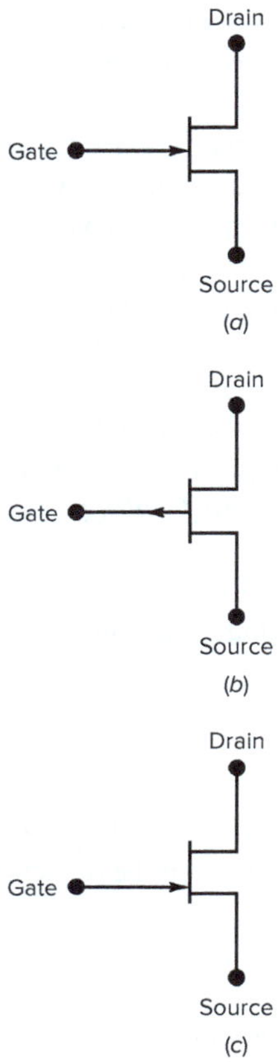

Figure 31–3 Current flow in the *n*-channel. Gates are open.

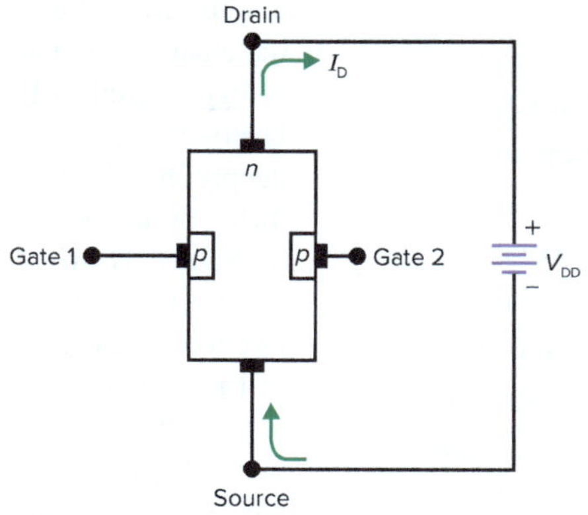

JFET Operation

Figure 31–3 illustrates the current flow in an *n*-channel JFET with the *p*-type gates left disconnected. Here the amount of current flow depends upon two factors: the value of the drain-source voltage, V_{DS}, and the drain-source resistance, designated r_{DS}. Furthermore, the ohmic value of r_{DS} is dependent on the doping level, cross-sectional area, and length of the doped semiconductor material used for the channel.

In Fig. 31–3, electrons flow in the channel between the two *p*-type gate regions. Because the drain is made positive relative to the source, electrons flow through the channel from source to drain. In a JFET, the source current, I_S, and the drain current, I_D, are the same. In most cases, therefore, the current flow in the channel of a JFET is considered to be only the drain current, I_D.

Gate Action

The gate regions in a JFET are embedded on each side of the channel to help control the amount of current flow. Figure 31–4*a* shows an *n*-channel JFET with both gates shorted to the source. The drain supply voltage, V_{DD}, reverse-biases both *p-n*

Figure 31–4 Effect of gate on drain current. (*a*) Gate leads shorted to source. (*b*) External gate bias reduces drain current.

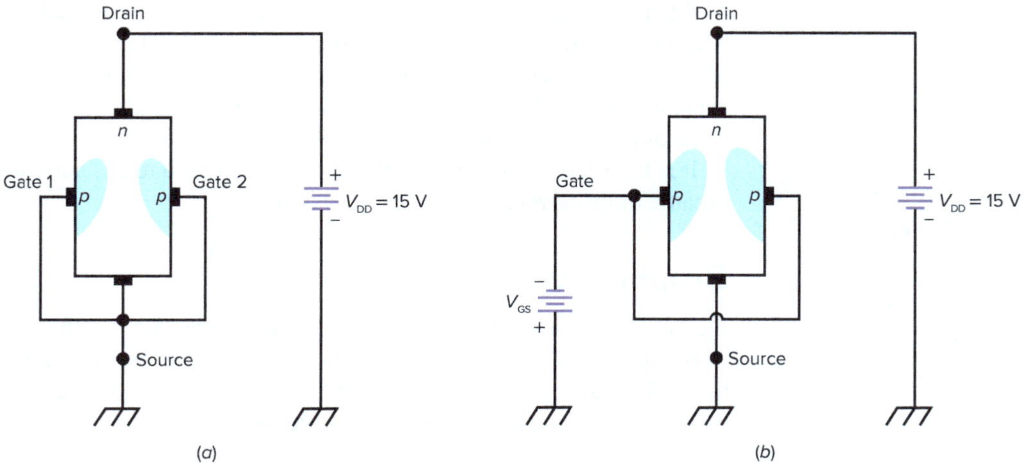

> **GOOD TO KNOW**
>
> A noteworthy static characteristic of a JFET is I_{GSS}, which is the gate current when the gate-source junction is reverse-biased. This current value allows us to determine the DC input resistance of a JFET. For example, if the spec sheet of a JFET shows that $I_{GSS} = 0.01\ \mu A$ when $V_{GS} = -15$ V, then the gate-source resistance under those circumstances is $\frac{15\text{ V}}{0.01\ \mu A} = 1.5\text{ G}\Omega$.

junctions. This results in zero gate current. If both gates are centered vertically in the channel (which is the case for a symmetrical JFET), the voltage distribution over the length of the channel makes the width of the depletion layer wider near the top of the channel and narrower at the bottom. Thus, the depletion layers are shown to be wedge-shaped in Fig. 31–4*a*. Current flows in the channel between the depletion layers and not in the depletion layers themselves. The depletion layers penetrate deeply into the *n*-channel and only slightly into the *p*-type gate regions due to the different doping levels in the *p* and *n* materials.

Figure 31–4*b* shows how an *n*-channel JFET is normally biased. Not only is the drain made positive relative to the source, but the gate is made negative relative to the source. The effect of the negative gate voltage is to expand the width of the depletion regions, which in turn narrows the channel. Since the channel is narrower, the drain current, I_D, is reduced. By varying the gate-source voltage, designated V_{GS}, the drain current, I_D, can be controlled. Notice how much narrower the channel is in Fig. 31–4*b* versus 31–4*a*. If V_{GS} is made negative enough, the depletion layers touch, which pinches off the channel. The result is zero drain current. The amount of gate-source voltage required to reduce the drain current, I_D, to zero is called the ***gate-source cutoff voltage***, designated $V_{GS(off)}$.

The polarity of the biasing voltages for a *p*-channel JFET is opposite from that of an *n*-channel JFET. For a *p*-channel JFET, the drain voltage is negative and the gate voltage is positive.

Shorted Gate-Source Junction

Figure 31–5*a* shows an *n*-channel JFET connected to the proper biasing voltages. Note that the drain is positive and the gate is negative, creating the depletion layers depicted earlier in Fig. 31–4*b*.

> **GOOD TO KNOW**
>
> Above the pinch-off voltage, V_P, it is common for the drain curves to have a slight upward slope.

When the gate supply voltage, V_{GG}, is reduced to zero in Fig. 31–5*a*, the gate is effectively shorted to the source and V_{GS} equals zero volts. Figure 31–5*b* shows the graph of I_D versus V_{DS} (drain-source voltage) for this condition. As V_{DS} is increased from zero, the drain current, I_D, increases proportionally. When the drain-source voltage, V_{DS}, reaches the pinch-off voltage, designated V_P, the drain current, I_D, levels off. In Fig. 31–5*b*, the **pinch-off voltage, V_P** = 4 V. Technically, the pinch-off voltage, V_P, is the border between the **ohmic region** and **current-source region**. The region below V_P is called the *ohmic region* because I_D increases in direct proportion to V_{DS}. Above V_P is the current-source region, where I_D is unaffected by changes in V_{DS}.

The drain current, I_D, levels off above V_P because at this point the channel resistance, r_{DS}, increases in direct proportion to V_{DS}. This results in a constant value of drain current for V_{DS} values above the pinch-off voltage, V_P.

> **GOOD TO KNOW**
>
> For every JFET, there is a value of V_{GS} near $V_{GS(off)}$ that results in a zero temperature coefficient. This means that, for some value of V_{GS} near $V_{GS(off)}$, I_D does not increase or decrease with increases in temperature.

The maximum drain current that a JFET can have under normal operating conditions occurs when V_{GS} is 0 V. This current is designated as I_{DSS}. I_{DSS} represents the drain-source current with the gate shorted. In Fig. 31–5*b*, I_{DSS} = 10 mA, a typical value for many JFETs. If V_{GS} is negative, the drain current, I_D, will be less than the value of I_{DSS}; how much less depends on the value of V_{GS}.

It is important to note that JFETs are often referred to as "normally on" devices because drain current flows when V_{GS} is 0 V.

Drain Curves

Figure 31–5*c* shows a complete set of drain curves for the JFET in Fig. 31–5*a*. Notice that as V_{GS} becomes increasingly more negative, the drain current, I_D, is reduced. Again, this is due to the fact that the channel is becoming much narrower with the increasing reverse bias in the gate regions.

Figure 31–5 JFET drain curves, (*a*) Normal biasing voltages for an *n*-channel JFET. (*b*) Drain curve with $V_{GS} = 0$ V. (*c*) Drain curves for different values of V_{GS}.

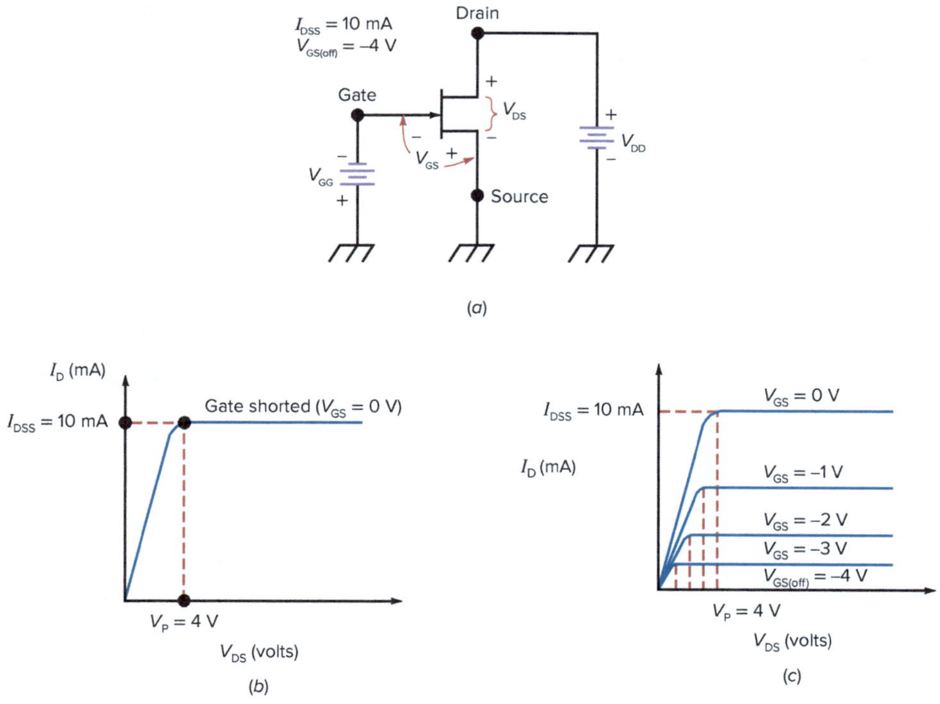

Notice the magnitude of both V_P and $V_{GS(off)}$ in Fig. 31–5*c*. It is interesting to note that for any JFET, $V_P = -V_{GS(off)}$. Most data sheets do not list V_P but almost always list $V_{GS(off)}$. The $V_{GS(off)}$ value of −4 V for the JFET in Fig. 31–5*a* is a typical value for many JFETs. In Fig. 31–5*c*, $V_P = -(-4\text{ V})$ or $V_P = +4$ V.

There are two other important points to be brought out in Fig. 31–5*c*. The first is that the slope of each separate drain curve in the ohmic region decreases as V_{GS} becomes more negative. This occurs because the channel resistance, r_{DS}, increases as V_{GS} becomes more negative. This useful feature allows using JFETs as voltage-variable resistances.

The second important feature is that the drain-source voltage, V_{DS}, at which pinch-off occurs, decreases as V_{GS} becomes more negative. Technically, the pinch-off value of V_{DS} can be specified for any value of V_{GS}. This is expressed in Formula (31–1):

$$V_{DS(P)} = V_P - V_{GS} \qquad \textbf{(31–1)}$$

where V_P is the pinch-off voltage for $V_{GS} = 0$ V and $V_{DS(P)}$ is the pinch-off voltage for any value of V_{GS}. In Formula (31–1), $V_{DS(P)}$, V_P, and V_{GS} are absolute values, that is, their polarities are ignored.

For any value of V_{GS}, $V_{DS(P)}$ is the border between the ohmic and current-source regions.

Transconductance Curve

Figure 31–6 shows a graph of I_D versus V_{GS} for the JFET in Fig. 31–5. This curve is called a *transconductance* curve. Notice that the graph is not linear because equal changes in V_{GS} do not produce equal changes in I_D.

GOOD TO KNOW

The transconductance curve of a JFET is unaffected by the circuit or biasing configuration in which the JFET is used.

Figure 31-6 Transconductance curve.

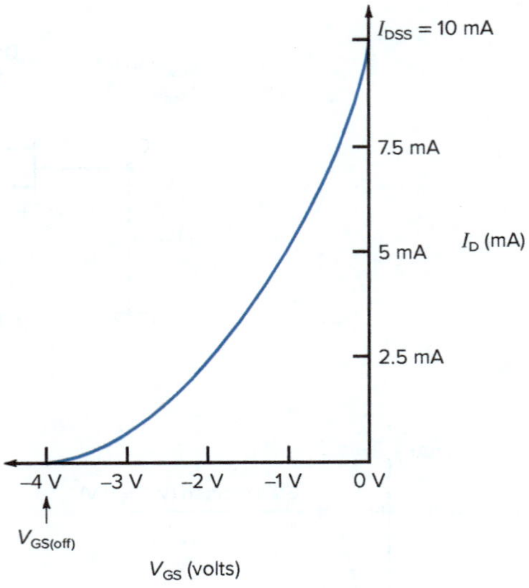

Calculating the Drain Current

When the values of I_{DSS} and $V_{GS(off)}$ are known for any JFET, the drain current, I_D, can be calculated using Formula (31–2):

$$I_D = I_{DSS}\left[1 - \frac{V_{GS}}{V_{GS(off)}}\right]^2 \qquad \textbf{(31–2)}$$

Formula (31–2) holds true only when V_{DS} is equal to or greater than $V_{DS(P)}$. Formula (31–2) can be used for any JFET (*n*-channel or *p*-channel) when I_{DSS} and $V_{GS(off)}$ are known.

Example 31-1

In Fig. 31–5*a*, calculate the drain current, I_D, for the following values of V_{GS}: (a) 0 V, (b) –0.5 V, (c) –1 V, (d) –2 V, (e) –3 V. Assume $V_{DS} \geq V_{DS(P)}$.

ANSWER Determine I_D for each value of V_{GS} by using Formula (31–2).

(a) When V_{GS} = 0 V, $I_D = I_{DSS}$, which is 10 mA in this case.

(b) When V_{GS} = –0.5 V, I_D is calculated using Formula (31–2):

$$I_D = I_{DSS}\left[1 - \frac{V_{GS}}{V_{GS(off)}}\right]^2$$

$$= 10\text{ mA}\left[1 - \frac{-0.5\text{ V}}{-4\text{ V}}\right]^2$$

$$= 10\text{ mA} \times 0.875^2$$

$$= 7.65\text{ mA}$$

The calculations for (c), (d), and (e) are

(c) $$I_D = 10\text{ mA}\left[1 - \frac{-1\text{ V}}{-4\text{ V}}\right]^2$$

$$= 10\text{ mA} \times 0.75^2$$

$$= 5.62\text{ mA}$$

(d) $I_D = 10 \text{ mA}\left[1 - \frac{-2 \text{ V}}{-4 \text{ V}}\right]^2$

$= 10 \text{ mA} \times 0.5^2$

$= 2.5 \text{ mA}$

(e) $I_D = 10 \text{ mA}\left[1 - \frac{-3 \text{ V}}{-4 \text{ V}}\right]^2$

$= 10 \text{ mA} \times 0.25^2$

$= 0.625 \text{ mA}$

Remember that Formula (31–2) applies only when V_{DS} is equal to or greater than $V_{DS(P)}$.

GOOD TO KNOW

The ohmic resistance of a JFET can be determined for any value of V_{GS} by using the following formula:

$$r_{DS} = \frac{R_{DS(on)}}{1 - \frac{V_{GS}}{V_{GS(off)}}}$$

where $R_{DS(on)}$ is the ohmic resistance when V_{DS} is small and $V_{GS} = 0$ V.

More about the Ohmic Region

When V_{DS} is below $V_{DS(P)}$, Formula (31–2) no longer applies. Instead the JFET must be considered a resistance. This resistance is designated $r_{DS(on)}$. The exact value of $r_{DS(on)}$ for a given JFET is dependent on the value of V_{GS}. When $V_{GS} = 0$ V, $r_{DS(on)}$ has its lowest value. When $V_{GS} = V_{GS(off)}$, $r_{DS(on)}$ approaches infinity. The main point is that the channel resistance, $r_{DS(on)}$, increases as V_{GS} becomes more negative. As mentioned earlier, this useful feature allows using the JFET as a voltage-variable resistance.

31–1 Self-Review

Answers at the end of the chapter.

a. Is a JFET a voltage- or current-controlled device?

b. What is a symmetrical JFET?

c. In what region does I_D increase in direct proportion to V_{DS}?

d. The pinch-off voltage, V_P, is the border between which two regions?

e. Why is a JFET considered a "normally ON" device?

31–2 JFET Biasing Techniques

Many techniques can be used to bias JFETs. In all the cases, however, the gate-source junction is reverse-biased. The most common biasing techniques are covered in this section including gate bias, self-bias, voltage divider bias, and current-source bias.

Gate Bias

Figure 31–7*a* shows an example of gate bias. The negative gate voltage is applied through a gate resistor, R_G. R_G can be any value, but it is usually 100 kΩ or larger. Since there is zero current in the gate lead of the JFET, the voltage drop across R_G is zero. The main purpose of R_G is to isolate the gate from ground for AC signals. Figure 31–7*b* shows how an AC signal is coupled to the gate of a JFET. If R_G were omitted, as shown in Fig. 31–7*c*, no AC signal would appear at the gate because V_{GG} is at ground for AC signals. R_G is usually made equal to the value desired for the input impedance, Z_{in}, of the amplifier. Since the gate-source junction of the JFET is reverse-biased, its impedance is at least several hundred megohms, and therefore $Z_{in} = R_G$. In Fig. 31–7*b*, $R_G = 1$ MΩ, so $Z_{in} = 1$ MΩ also.

Figure 31–7 Gate bias. (*a*) Negative gate voltage applied through gate resistor, R_G. (*b*) AC signal coupled to gate. (*c*) No AC signal at gate because R_G has been omitted and gate is at AC ground.

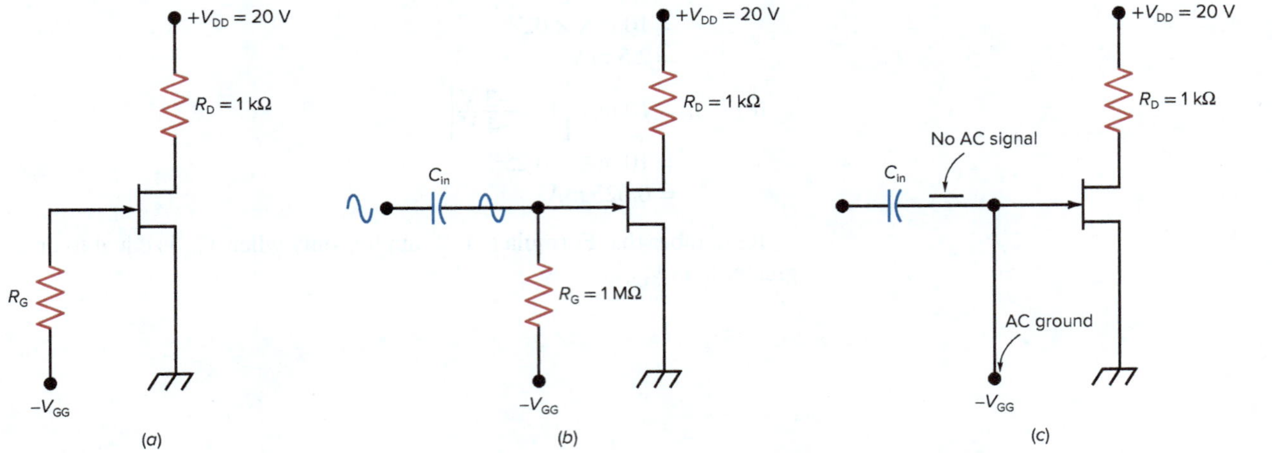

When V_{GS} is known, the value of the drain current is calculated using Formula (31–2). Then V_{DS} is calculated as

$$V_{DS} = V_{DD} - I_D R_D \tag{31–3}$$

Gate bias is seldom used with JFETs because the characteristics of the individual JFETs used in mass production may vary over a wide range. Thus, for some circuits, the amount of V_{GS} applied to the JFET may provide a very large drain current, whereas in other circuits, the same gate voltage might reduce the drain current, I_D, to nearly zero.

Example 31-2

Assume that the JFET circuit in Fig. 31–7*a* is to be mass-produced. The JFET has the following parameters:

Parameter	Minimum	Maximum
I_{DSS}	2 mA	20 mA
$V_{GS(off)}$	−2 V	−8 V

For the range of JFET parameters shown, calculate the minimum and maximum values for I_D and V_{DS} if $V_{GS} = -1.5$ V.

ANSWER Begin by calculating I_D and V_{DS} using the minimum values for I_{DSS} and $V_{GS(off)}$:

$$I_D = I_{DSS}\left[1 - \frac{V_{GS}}{V_{GS(off)}}\right]^2$$
$$= 2\text{ mA}\left[1 - \frac{-1.5\text{ V}}{-2\text{ V}}\right]^2$$
$$= 2\text{ mA} \times 0.25^2$$
$$= 125\ \mu\text{A}$$

Now calculate V_{DS}:

$$V_{DS} = V_{DD} - I_D R_D$$
$$= 20\text{ V} - (125\ \mu\text{A} \times 1\text{ k}\Omega)$$
$$= 20\text{ V} - 0.125\text{ V}$$
$$= 19.875\text{ V}$$

Next, use the maximum values of I_{DSS} and $V_{GS(off)}$ to calculate I_D and V_{DS} in Fig. 31–7*a*:

$$I_D = I_{DSS}\left[1 - \frac{V_{GS}}{V_{GS(off)}}\right]^2$$
$$= 20\text{ mA}\left[1 - \frac{-1.5\text{ V}}{-8\text{ V}}\right]^2$$
$$= 20\text{ mA} \times 0.8125^2$$
$$= 13.2\text{ mA}$$

V_{DS} is calculated next:

$$V_{DS} = V_{DD} - I_D R_D$$
$$= 20\text{ V} - (13.2\text{ mA} \times 1\text{ k}\Omega)$$
$$= 20\text{ V} - 13.2\text{ V}$$
$$= 6.8\text{ V}$$

Using Formula (31–1), the value of $V_{DS(P)}$ is

$$V_{DS(P)} = V_P - V_{GS} \qquad \text{Note: } V_P = -(-V_{GS(off)})$$
$$= 8\text{ V} - 1.5\text{ V}$$
$$= 6.5\text{ V}$$

This means that V_{DS} is still high enough to operate the JFET in the current-source region. If, however, V_{DS} is below the value of $V_{DS(P)}$, then the JFET is no longer operating in the current-source region. Instead, it is operating in the ohmic region. This means that Formula (31–2) can no longer be used to calculate the drain current, I_D. Under these circumstances, the actual value of $r_{DS(on)}$ must be known to calculate I_D and V_{DS} accurately.

This example illustrates the problem with gate bias. Even though V_{GS} remains fixed at −1.5 V in both cases, I_D and V_{DS} can vary over an extremely wide range. Then if the circuit is used as an AC amplifier, the *Q* point is very unpredictable from one circuit to the next.

Self-Bias

One of the best ways to bias a JFET is with self-bias. (See Fig. 31–8*a*.) Notice that only a single power supply is used, the drain supply voltage, V_{DD}. In this case, the voltage across the source resistor, R_S, provides the gate-to-source bias voltage. But how is this possible? Here's how.

When power is first applied, drain current flows and produces a voltage drop across the source resistor, R_S. For the direction of drain current shown, the source is positive with respect to ground. Because there is no gate current, $V_G = 0$ V. Therefore, V_{GS} is calculated as

$$V_{GS} = V_G - V_S$$
$$= 0\text{ V} - V_S$$

Also, since

$$V_S = I_D R_S$$

Figure 31–8 Self-bias. (*a*) Circuit. (*b*) Transconductance curve.

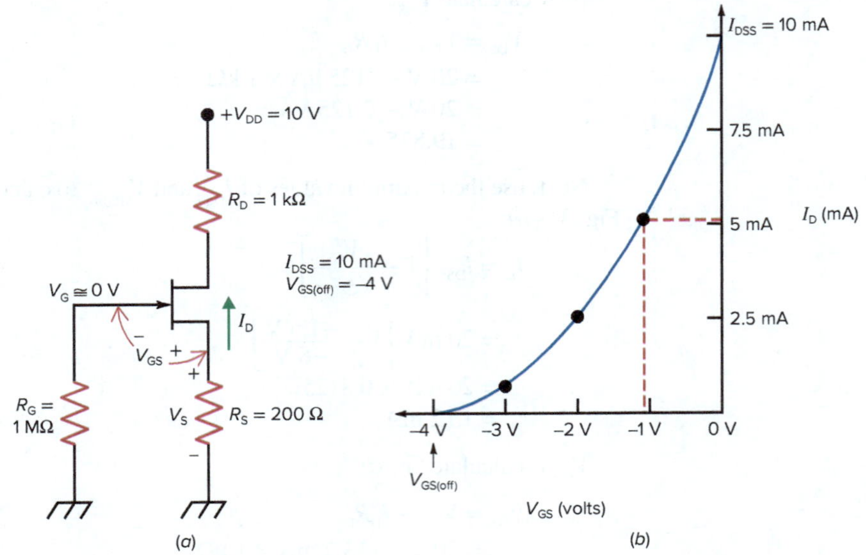

then,

$$V_{GS} = 0\text{ V} - I_D R_S$$

or

$$V_{GS} = -I_D R_S \qquad \textbf{(31–4)}$$

Self-bias is very stable because any increase in the drain current, I_D, causes V_{GS} to increase. The increase in V_{GS} causes the drain current, I_D, to decrease, thereby offsetting the original increase in drain current.

Likewise, a decrease in the drain current, I_D, causes V_{GS} to decrease. This decrease in V_{GS} causes the drain current, I_D, to increase to its original value.

Self-Bias Calculations

The source resistor, R_S, must be carefully selected for any JFET circuit using self-bias. Normally the source resistor, R_S, is chosen so that the drain current, I_D, equals approximately one-half of I_{DSS}. To do this, V_{GS} must be set equal to approximately one-fourth the value of $V_{GS(off)}$. This is a rough approximation that provides reasonably accurate results. [To prove that $I_D \approx \frac{I_{DSS}}{2}$ when $V_{GS} \approx \frac{V_{GS(off)}}{4}$, rearrange Formula (31–2).] A convenient formula for determining the source resistor, R_S, is

$$R_S = \frac{\dfrac{-V_{GS(off)}}{4}}{\dfrac{I_{DSS}}{2}}$$

which simplifies to

$$R_S = \frac{-V_{GS(off)}}{2I_{DSS}} \qquad \textbf{(31–5)}$$

Therefore, in Fig. 31–8*a*, where $V_{GS(off)} = -4$ V and $I_{DSS} = 10$ mA, $R_S = \frac{-(-4\text{ V})}{20\text{ mA}} = 200\ \Omega$.

The quiescent values of V_{GS} and I_D are shown in Fig. 31–8*b*. Notice that the values of I_D and V_{GS} on the transconductance curve are not exactly 5 mA and −1 V, respectively, because a 200-Ω source resistor, R_S, allows slightly more than 5 mA of drain current. Using too small a value for R_S makes I_D too close to the value of I_{DSS}. However, using too large a value for R_S allows too small a value for I_D.

Example 31-3

In Fig. 31–8, calculate the drain voltage, V_D.

ANSWER With V_S at approximately 1 V, I_D is calculated as

$$I_S = \frac{V_S}{R_S}$$
$$= \frac{1\ \text{V}}{200\ \Omega}$$
$$= 5\ \text{mA}$$

Since $I_S = I_D$, $I_D = 5$ mA also.

To calculate V_D, proceed as follows:

$$V_D = V_{DD} - I_D R_D$$
$$= 10\ \text{V} - (5\ \text{mA} \times 1\ \text{k}\Omega)$$
$$= 5\ \text{V}$$

Voltage Divider Bias

Figure 31–9 shows a JFET using voltage divider bias. Since the gate-source junction has extremely high resistance (several hundred megohms), the R_1-R_2 voltage divider is practically unloaded. Therefore, the gate voltage, V_G, is calculated as

$$V_G = \frac{R_2}{R_1 + R_2} \times V_{DD} \qquad \textbf{(31–6)}$$

The source voltage, V_S, is calculated as

$$V_S = V_G - V_{GS} \qquad \textbf{(31–7)}$$

Since $I_D = I_S$, the drain current is

$$I_D = \frac{V_S}{R_S} \qquad \textbf{(31–8)}$$

Also, the drain voltage, V_D, is

$$V_D = V_{DD} - I_D R_D \qquad \textbf{(31–9)}$$

Although not proven here, voltage divider bias is more stable than either gate bias or self-bias. Voltage divider bias, however, does have its drawbacks. The value of I_D for a given value of V_{GS} varies from one JFET to the next, making it difficult to predict the exact values of I_D and V_D for a given circuit.

MultiSim **Figure 31–9** Voltage divider bias.

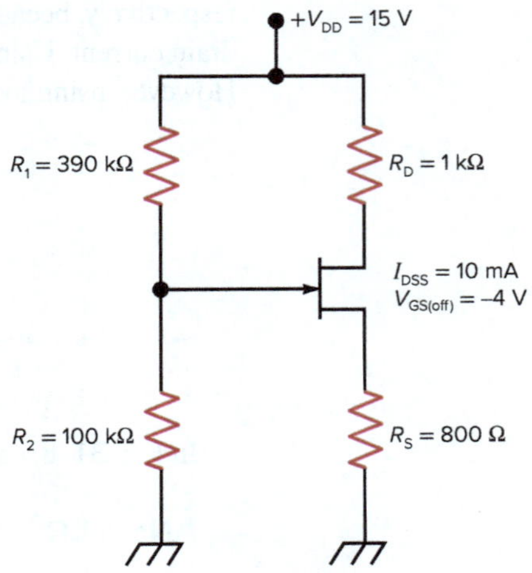

Example 31-4

In Fig. 31–9, $V_{GS} = -1$ V. Calculate V_G, V_S, I_D, and V_D.

ANSWER Begin by calculating V_G.

$$V_G = \frac{R_2}{R_1 + R_2} \times V_{DD}$$
$$= \frac{100\ \text{k}\Omega}{390\ \text{k}\Omega + 100\ \text{k}\Omega} \times 15\ \text{V}$$
$$= 3\ \text{V}$$

Next, calculate V_S:

$$V_S = V_G - V_{GS}$$
$$= 3\ \text{V} - (-1\ \text{V})$$
$$= 4\ \text{V}$$

Calculate I_D as follows:

$$I_D = \frac{V_S}{R_S}$$
$$= \frac{4\ \text{V}}{800\ \Omega}$$
$$= 5\ \text{mA}$$

Last, calculate V_D:

$$V_D = V_{DD} - I_D R_D$$
$$= 15\ \text{V} - (5\ \text{mA} \times 1\ \text{k}\Omega)$$
$$= 10\ \text{V}$$

Figure 31–10 Current-source bias.

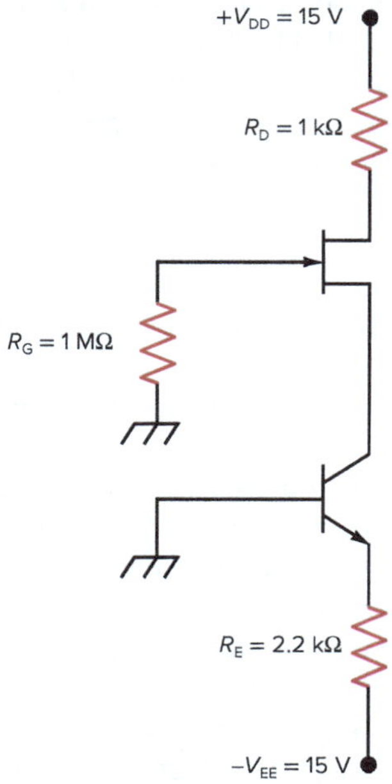

Current-Source Bias

Figure 31–10 shows one of the best ways to bias JFETs. The *npn* transistor with emitter bias acts like a current source for the JFET. The drain current, I_D, equals the collector current, I_C, which is independent of the value of V_{GS}. Therefore,

$$I_C = I_D$$

I_C is calculated as

$$I_C = \frac{V_{EE} - V_{BE}}{R_E} \qquad \textbf{(31–10)}$$

Note that the drain current, I_D, will equal the collector current I_C for all JFETs in the circuit.

Example 31-5

In Fig. 31–10, calculate the drain current, I_D, and the drain voltage, V_D.

ANSWER Begin by using Formula (31–10) to calculate the collector current, I_C:

$$I_C = \frac{V_{EE} - V_{BE}}{R_E}$$
$$= \frac{15 \text{ V} - 0.7 \text{ V}}{2.2 \text{ k}\Omega}$$
$$= 6.5 \text{ mA}$$

Since $I_C = I_D$, $I_D = 6.5$ mA also.
V_D is calculated as

$$V_D = V_{DD} - I_D R_D$$
$$= 15 \text{ V} - (6.5 \text{ mA} \times 1 \text{ k}\Omega)$$
$$= 8.5 \text{ V}$$

31–2 Self-Review

Answers at the end of the chapter.

a. Which form of JFET bias produces the most unstable *Q* point?
b. Why is the gate voltage 0 V with self-bias?
c. When biasing an *n*-channel JFET with voltage divider bias, is the source voltage less positive or more positive than the gate voltage?

31–3 JFET Amplifiers

JFETs are commonly used to amplify small AC signals. One reason for using a JFET instead of a bipolar transistor is that a very high input impedance, Z_{in}, can be obtained. A big disadvantage, however, is that the voltage gain, A_V, obtainable with a JFET is much smaller. This section analyzes the following JFET amplifier configurations: *common-source (CS)*, *common-gate (CG)*, and *common-drain (CD)*.

g_m Transconductance

Before examining the basic JFET amplifier configurations, an analysis of the JFET's transconductance curve is necessary (refer to Fig. 31–11). The transconductance curve reveals that equal changes in V_{GS} do not produce equal changes in I_D. Higher on the transconductance curve, notice that I_D is more sensitive to changes in V_{GS}.

Mathematically, **transconductance, g_m**, is defined as follows:

$$g_m = \frac{\Delta I_D}{\Delta V_{GS}} (V_{DS} > V_P) \qquad \textbf{(31–11)}$$

where ΔI_D = change in drain current, and ΔV_{GS} = change in gate-source voltage.

Therefore, the transconductance, g_m, equals the change in drain current divided by the change in gate-source voltage for a fixed value of V_{DS}.

The unit of g_m is the siemens (S). In some cases, g_m is designated in mhos. Either way, the siemens or mho units represent the conductance, g_m, which is a ratio of current to voltage. Actually, the transconductance, g_m, indicates how effective the gate-source voltage is in controlling the drain current, I_D.

To prove that g_m varies along the transconductance curve, make some calculations (refer again to Fig. 31–11). Begin by calculating g_m using the V_{GS} values of −2 V and −3 V. For $V_{GS} = -2$ V, $I_D = 2.5$ mA. For $V_{GS} = -3$ V, $I_D = 0.625$ mA. To calculate the transconductance, g_m, between these two points, proceed as shown:

$$g_m = \frac{2.5 \text{ mA} - 0.625 \text{ mA}}{-2 \text{ V} - (-3 \text{ V})}$$
$$= \frac{1.875 \text{ mA}}{1 \text{ V}}$$
$$= 1.875 \text{ mS}$$

Next, calculate g_m for the V_{GS} values of 0 V and −1 V. The graph shows that I_D = 5.625 mA when $V_{GS} = -1$ V. Of course, when $V_{GS} = 0$ V, $I_D = I_{DSS}$, which is 10 mA. The transconductance, g_m, is calculated as

$$g_m = \frac{10 \text{ mA} - 5.625 \text{ mA}}{0 \text{ V} - (-1 \text{ V})}$$
$$= \frac{4.375 \text{ mA}}{1 \text{ V}}$$
$$= 4.375 \text{ mS}$$

Figure 31–11 JFET transconductance curve.

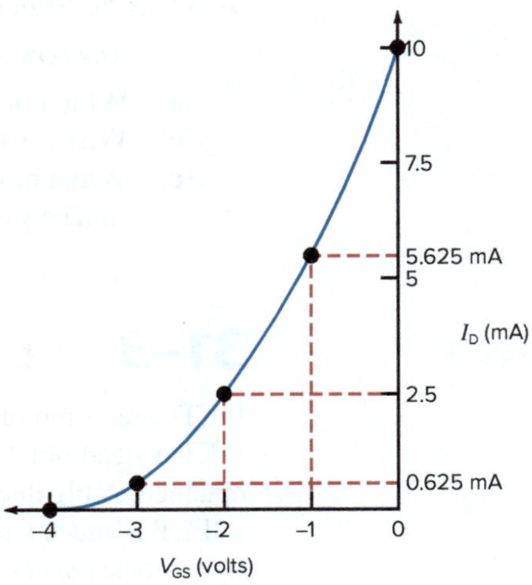

Notice how the transconductance, g_m, increases with the height of the transconductance curve.

The value of g_m can be calculated for any value of V_{GS} by using Formula (31–12):

$$g_m = g_{mo}\left[1 - \frac{V_{GS}}{V_{GS(off)}}\right] \quad \textbf{(31–12)}$$

If g_{mo} is not known, it can be calculated as

$$g_{mo} = \frac{2I_{DSS}}{-V_{GS(off)}} \quad \textbf{(31–13)}$$

where g_{mo} = transconductance when $V_{GS} = 0$ V, and g_m = transconductance for any gate-source voltage.

Common-Source (CS) Amplifier—DC Analysis

GOOD TO KNOW

Because of the extremely high input impedance of a JFET, the input current is generally assumed to be 0 μA and the current gain of a JFET amplifier is an undefined quantity.

Figure 31–12*a* shows a **common-source amplifier**. The resistor, R_S, provides self-bias. Since $V_{GS(off)} = -4$ V and $I_{DSS} = 10$ mA, the R_S value of 200 Ω is calculated by using Formula (31–5). This sets the Q point so that the drain current I_D equals approximately one-half of I_{DSS}. At this point, the transconductance, g_m, is relatively high, which is important to obtain as much voltage gain as possible. When V_S and R_S are known, the drain current, I_D, can be calculated. In Fig. 31–12*a*, I_D is calculated as

$$I_D = \frac{V_S}{R_S} = \frac{1\text{ V}}{200\ \Omega} = 5\text{ mA}$$

The drain voltage is calculated as follows:

$$\begin{aligned} V_D &= V_{DD} - I_D R_D \\ &= 15\text{ V} - (5\text{ mA} \times 1.5\text{ k}\Omega) \\ &= 15\text{ V} - 7.5\text{ V} \\ &= 7.5\text{ V} \end{aligned}$$

Common-Source Amplifier—AC Analysis

For a common-source amplifier, the input voltage is applied to the gate and the output is taken at the drain. A common-source amplifier has high input impedance and moderate voltage gain. Also, the input and output voltages are 180° out of phase.

In Fig. 31–12, the input voltage source, V_{in}, is capacitively coupled to the gate of the JFET, whereas the output is taken from the drain terminal. The AC signal voltage at the gate produces variations in the gate-source voltage. This in turn produces variations in the drain current, I_D.

GOOD TO KNOW

For any JFET amplifier, the input signal that drives the gate (or source) should never be so large that it forward-biases the gate-source junction.

Since the gate-source junction is reverse-biased, the gate-source impedance is of the order of several hundred megohms. Therefore, the input impedance, Z_{in}, of the amplifier equals R_G. In Fig. 31–12*a*, $Z_{in} = 1$ MΩ.

The source bypass capacitor, C_S, holds the source terminal of the JFET constant at 1 V. C_S also places the source at ground for AC signals. Since C_S holds the DC source voltage constant, the input voltage, V_{in}, is directly across the gate-source junction, implying that V_{in} equals v_{gs}, which is the AC gate-source voltage. Therefore, V_{in} will make the gate-source voltage, v_{gs}, vary in accordance with the input voltage, V_{in}. The output coupling capacitor, C_{out}, couples the AC signal voltage at the drain to the load resistor, R_L.

MultiSim **Figure 31–12** Common-source amplifier. (*a*) Original circuit. (*b*) AC equivalent circuit.

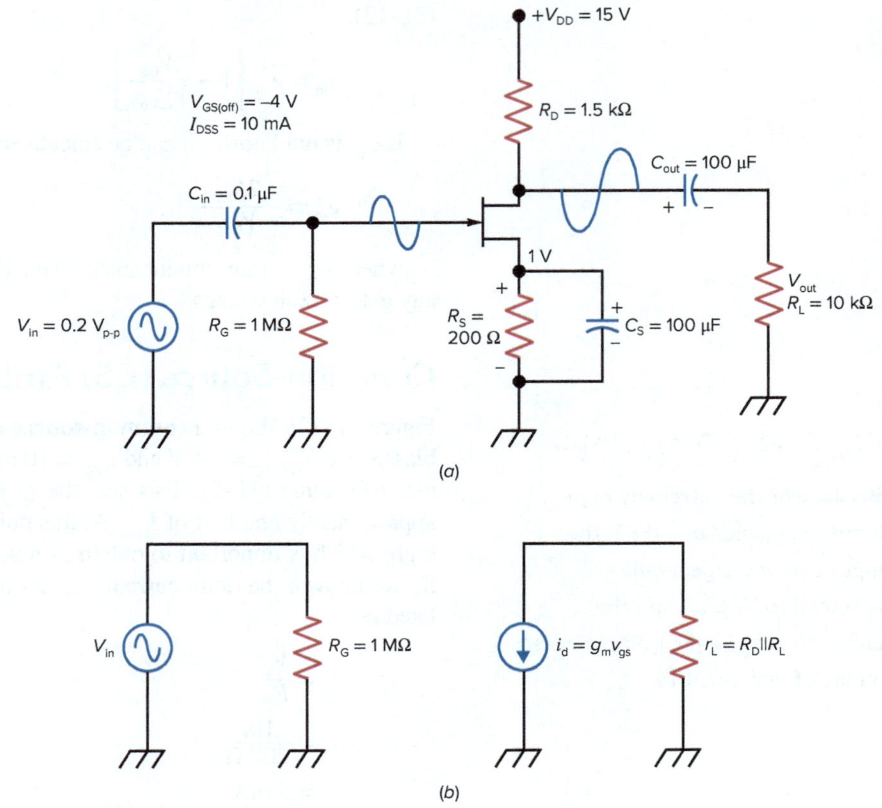

The AC equivalent circuit is shown in Fig. 31–12*b*. Notice that on the input side, $R_G = Z_{in}$, which is 1 MΩ. Again, this occurs because with practically zero gate current, the gate-source resistance, designated R_{GS}, approaches infinity.

Also, notice that the drain circuit acts like a constant current source with a value equal to $g_m v_{gs}$, where v_{gs} is the AC voltage across the gate-source junction. Because V_{DD} is at AC ground, R_D and R_L are in parallel for AC signals. The voltage gain, A_V, is calculated as follows:

$$A_V = \frac{V_{out}}{V_{in}}$$

$$= \frac{i_d r_L}{V_{in}}$$

Since $i_d = g_m v_{gs}$ and $V_{in} = v_{gs}$, then,

$$A_V = \frac{g_m v_{gs} r_L}{v_{gs}}$$

which simplifies to

$$A_V = g_m r_L \qquad \textbf{(31–14)}$$

This formula applies only when the source resistor, R_S, is bypassed. If R_S is not bypassed, A_V is calculated using Formula (31–15):

$$A_V = \frac{g_m r_L}{1 + g_m R_S} \qquad \textbf{(31–15)}$$

Notice from Formulas (31–14) and (31–15) that A_V is affected by the value of g_m. As shown earlier, g_m is not constant but is controlled by the gate-source voltage, V_{GS}.

For smaller values of V_{GS}, g_m is higher, which makes A_V larger. Likewise, A_V can be reduced by increasing V_{GS}.

Example 31-6

In Fig. 31–12, calculate the voltage gain, A_V, and the output voltage, V_{out}.

ANSWER Begin by calculating the AC load resistance in the drain circuit:

$$\begin{aligned} r_L &= \frac{R_D \times R_L}{R_D + R_L} \\ &= \frac{1.5\ \text{k}\Omega \times 10\ \text{k}\Omega}{1.5\ \text{k}\Omega + 10\ \text{k}\Omega} \\ &= 1.3\ \text{k}\Omega \end{aligned}$$

Next, calculate the transconductance, g_m, using Formulas (31–12) and (31–13). First calculate g_{mo}:

$$\begin{aligned} g_{mo} &= \frac{2I_{DSS}}{-V_{GS(off)}} \\ &= \frac{20\ \text{mA}}{4\ \text{V}} \\ &= 5\ \text{mS} \end{aligned}$$

Next, calculate g_m:

$$\begin{aligned} g_m &= g_{mo}\left[1 - \frac{V_{GS}}{V_{GS(off)}}\right] \\ &= 5\ \text{mS}\left[1 - \frac{-1}{-4}\right] \\ &= 5\ \text{mS} \times 0.75 \\ &= 3.75\ \text{mS} \end{aligned}$$

With g_m known, calculate the voltage gain, A_V:

$$\begin{aligned} A_V &= g_m r_L \\ &= 3.75\ \text{mS} \times 1.3\ \text{k}\Omega \\ &= 4.875 \end{aligned}$$

Finally, calculate the output voltage, V_{out}:

$$\begin{aligned} V_{out} &= A_V \times V_{in} \\ &= 4.875 \times 0.2\ \text{V}_{p\text{-}p} \\ &= 0.975\ \text{V}_{p\text{-}p} \end{aligned}$$

Common-Drain (CD) Amplifier

Figure 31–13*a* shows a **common-drain amplifier**, usually referred to as a source follower. This circuit is similar to the emitter follower circuit used with bipolar transistors. A source follower has high input impedance, low output impedance, and a voltage gain of less than one, or unity. In Fig. 31–13*a*, notice that the input signal is

Figure 31–13 Common-drain amplifier. (*a*) Original circuit. (*b*) AC equivalent circuit.

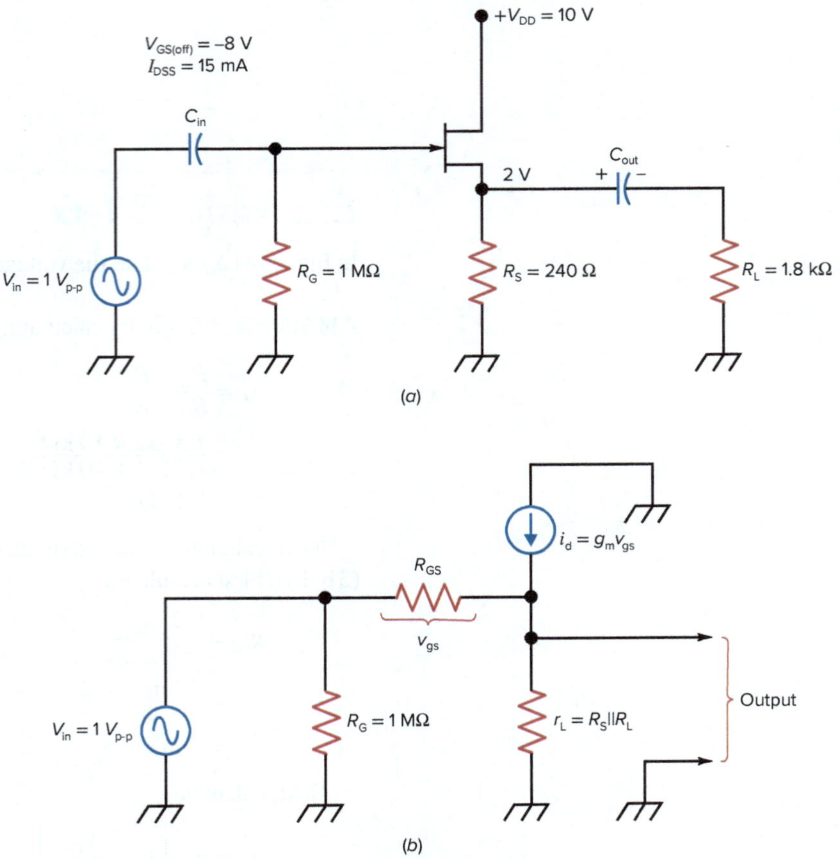

applied to the gate, whereas the output is taken from the source. Notice that there is no drain resistor. Since the drain is tied directly to V_{DD}, it is at ground for AC signals.

Figure 31–13*b* shows the AC equivalent circuit. Notice that R_S and R_L are in parallel for AC signals. R_{GS} represents the very high resistance of the reverse-biased gate-source junction.

As with the CS amplifier, the input impedance, Z_{in}, equals the gate resistor, R_G. Again, this occurs because R_{GS} approaches infinity. In this circuit, $R_G = 1\ \text{M}\Omega$, so $Z_{in} = 1\ \text{M}\Omega$ also. The equation for the voltage gain, A_V, is calculated by determining the formulas for V_{in} and V_{out}:

$$V_{out} = g_m v_{gs} r_L$$

$$V_{in} = v_{gs} + g_m v_{gs} r_L$$

Since $$A_V = \frac{V_{out}}{V_{in}}$$

then, $$A_V = \frac{g_m r_L}{1 + g_m r_L} \qquad \textbf{(31–16)}$$

Note that for a source follower the voltage gain, A_V, will always be less than one.

The output impedance, Z_{out}, of a source follower is given by Formula (31–17):

$$Z_{out} = R_S \parallel \frac{1}{g_m} \qquad \textbf{(31–17)}$$

Example 31-7

In Fig. 31–13, calculate A_V, V_{out}, and Z_{out}.

ANSWER Begin by calculating the AC load resistance in the source circuit:

$$r_L = \frac{R_S \times R_L}{R_S + R_L}$$
$$= \frac{240\ \Omega \times 1.8\ k\Omega}{240\ \Omega + 1.8\ k\Omega}$$
$$= 211.7\ \Omega$$

Next, calculate g_{mo} using Formula (31–13):

$$g_{mo} = \frac{2I_{DSS}}{-V_{GS(off)}}$$
$$= \frac{30\ mA}{8\ V}$$
$$= 3.75\ mS$$

Next, calculate g_m:

$$g_m = g_{mo}\left[1 - \frac{V_{GS}}{V_{GS(off)}}\right]$$
$$= 3.75\ mS\left[1 - \frac{-2}{-8}\right]$$
$$= 3.75\ mS \times 0.75$$
$$= 2.8\ mS$$

Now A_V can be calculated:

$$A_V = \frac{g_m r_L}{1 + g_m r_L}$$
$$= \frac{2.8\ mS \times 211.7\ \Omega}{1 + 2.8\ mS \times 211.7\ \Omega}$$
$$= 0.37$$

Notice the low value for the voltage gain, A_V. When the product $g_m r_L$ is much greater than one, A_V approaches unity.

For a source follower, 0.37 is not an uncommon value for the voltage gain, A_V.

Next, calculate V_{out}

$$V_{out} = A_V \times V_{in}$$
$$= 0.37 \times 1\ V_{p\text{-}p}$$
$$= 0.37\ V_{p\text{-}p}$$

Finally, calculate Z_{out}:

$$Z_{out} = R_S \parallel \frac{1}{g_m}$$
$$= 240\ \Omega \parallel \frac{1}{2.8\ mS}$$
$$= 143.5\ \Omega$$

Figure 31–14 Common-gate amplifier. (*a*) Original circuit. (*b*) AC equivalent circuit.

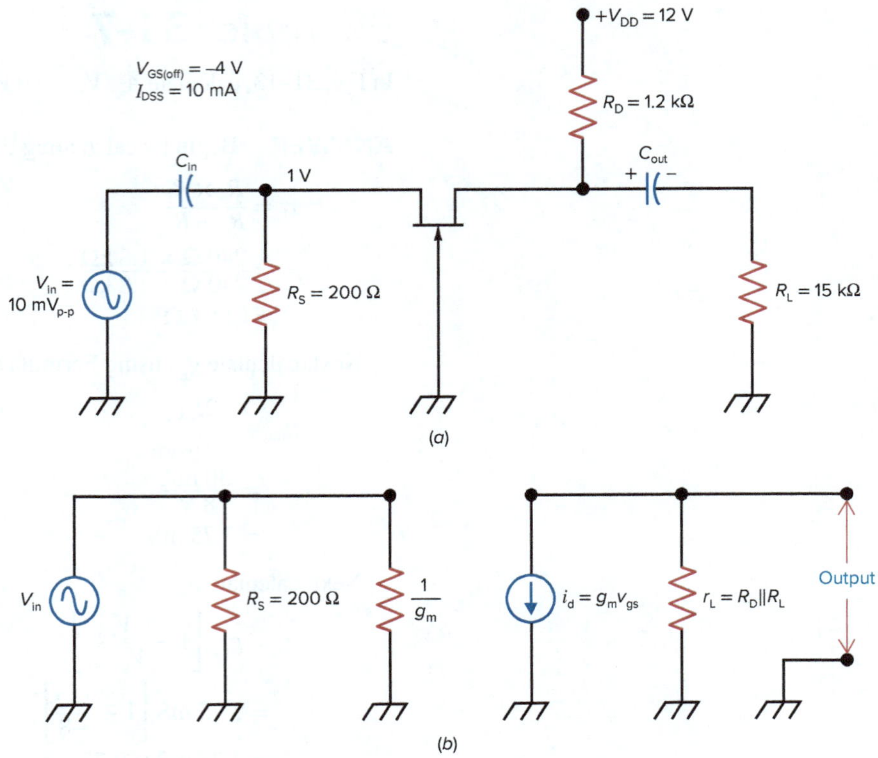

Common-Gate (CG) Amplifier

A **common-gate amplifier** has a moderate voltage gain. Its biggest drawback, however, is that Z_{in} is quite low. Figure 31–14*a* shows a CG amplifier. Resistor R_S is used to provide self-bias for the JFET. Notice that the input is applied to the source, whereas the output is taken from the drain. The AC equivalent circuit is shown in Fig. 31–14*b*.

The AC output voltage, V_{out}, can be represented as

$$V_{out} = g_m v_{gs} r_L$$

Since $V_{in} = v_{gs}$, write a formula for A_V as follows:

$$A_V = \frac{g_m v_{gs} r_L}{v_{gs}}$$

$$= g_m r_L \qquad \textbf{(31–18)}$$

The input impedance, Z_{in}, is

$$Z_{in} = R_S \parallel \frac{1}{g_m} \qquad \textbf{(31–19)}$$

The biggest disadvantage of the common-gate amplifier is its low value of input impedance, Z_{in}. Because of this undesirable characteristic, the CG amplifier is very seldom used.

Example 31-8

In Fig. 31–14, $g_m = 3.75$ mS for $V_{GS} = -1$ V. Calculate A_V, V_{out}, and Z_{in}.

ANSWER Begin by calculating r_L in the drain circuit:

$$r_L = \frac{R_D \times R_L}{R_D + R_L}$$
$$= \frac{1.2\ \text{k}\Omega \times 15\ \text{k}\Omega}{1.2\ \text{k}\Omega + 15\ \text{k}\Omega}$$
$$= 1.11\ \text{k}\Omega$$

Next, calculate A_V and V_{out}:

$$A_V = g_m r_L$$
$$= 3.75\ \text{mS} \times 1.11\ \text{k}\Omega$$
$$= 4.16$$
$$V_{out} = A_V \times V_{in}$$
$$= 4.16 \times 10\ \text{mV}_{p\text{-}p}$$
$$= 41.6\ \text{mV}_{p\text{-}p}$$

Finally, calculate Z_{in}:

$$Z_{in} = R_S \parallel \frac{1}{g_m}$$
$$= 200\ \Omega \parallel \frac{1}{3.75\ \text{mS}}$$
$$= 114\ \Omega$$

31–3 Self-Review

Answers at the end of the chapter.

a. **Does the transconductance, g_m, of a JFET vary with V_{GS}?**
b. **Which JFET amplifier is also known as the source follower?**
c. **Which JFET amplifier provides a 180° phase shift between V_{in} and V_{out}?**
d. **Which JFET amplifier has a high Z_{in}, a low Z_{out}, and a voltage gain less than one?**
e. **What is the biggest drawback of the common-gate amplifier?**

31–4 MOSFETs and Their Characteristics

The metal-oxide-semiconductor **field effect transistor** has a gate, source, and drain just like the JFET. Like a JFET, the drain current in a **MOSFET** is controlled by the gate-source voltage V_{GS}. There are two basic types of MOSFETs: the *enhancement-type* and the *depletion-type*. The enhancement-type MOSFET is usually referred to as an E-MOSFET, and the depletion-type MOSFET is referred to as a D-MOSFET.

The key difference between JFETs and MOSFETs is that the gate terminal in a MOSFET is insulated from the channel. Because of this, MOSFETs are sometimes referred to as *insulated gate* FETs or **IGFETs**. Because of the insulated gate, the input impedance of a MOSFET is many times higher than that of a JFET.

Depletion-Type MOSFET

Figure 31–15*a* shows the construction of an *n*-channel depletion-type MOSFET, and Fig. 31–15*b* shows the schematic symbol. In Fig. 31–15*a*, the drain terminal is

Figure 31–15 *n*-channel depletion-type MOSFET. (*a*) Construction. (*b*) Schematic symbol.

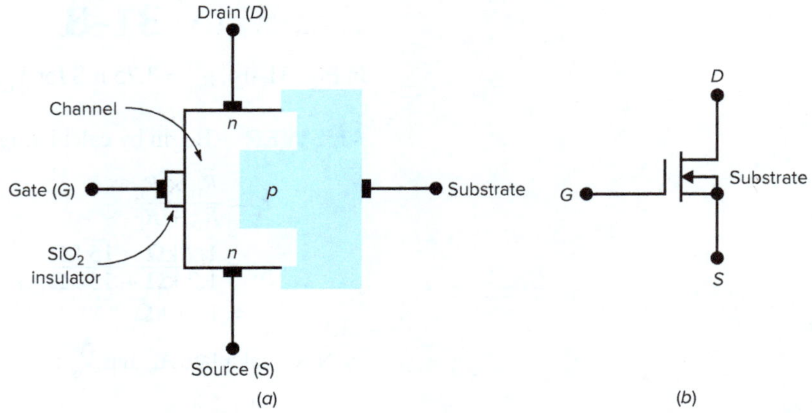

at the top of the *n*-material and the source terminal is at the bottom. The block of *p*-type material forms the substrate into which the *n*-type material is embedded. The *n*-type material forms the channel. Along the *n*-channel, a thin layer of silicon dioxide (SiO_2) is deposited to isolate the gate from the channel. From gate to channel are the metal, silicon dioxide, and *n*-type semiconductor materials, in that order, which give the MOSFET its name.

Notice in Fig. 31–15*b* that the substrate is connected to the source. This results in a three-terminal device. The solid line connecting the source and drain terminals indicates that depletion-type MOSFETs are "normally on" devices, which means that drain current flows when the gate-source voltage is zero.

Zero Gate Voltage

A depletion-type MOSFET can operate with either positive or negative gate voltages. As shown in Fig. 31–16*a*, the depletion-type MOSFET also conducts with the gate shorted to the source for $V_{GS} = 0$ V.

In Fig. 31–16*a*, notice that V_{DD} is connected between the drain and source with the drain positive relative to the source. Also, notice that the substrate is connected

Figure 31–16 Current in an *n*-channel depletion-type MOSFET. (*a*) Current flow in an *n*-channel depletion-type MOSFET with $V_{GS} = 0$ V. (*b*) Drain curves.

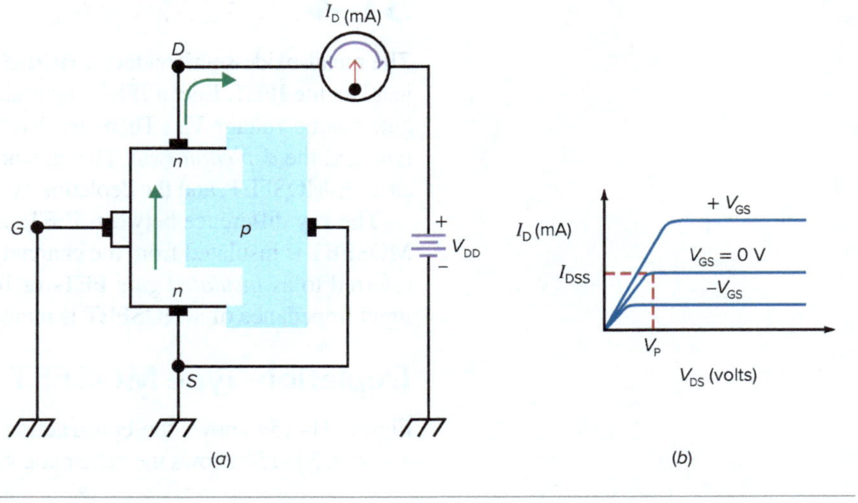

Figure 31–17 (a) Drain current increases with positive gate voltage. (b) Drain current decreases with negative gate voltage.

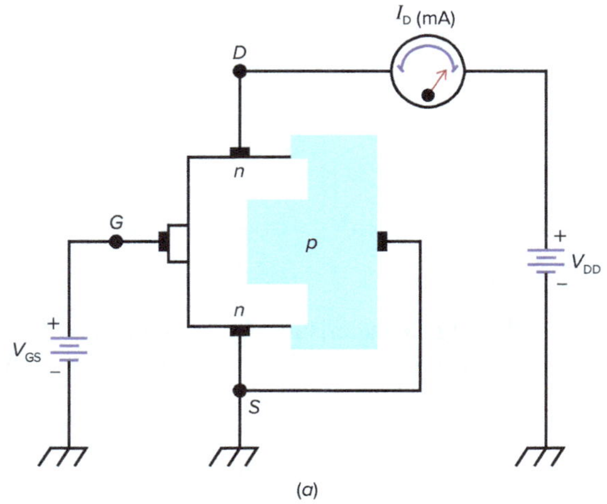

to the source. With the gate shorted to the source, drain current, I_D, will flow in the n-type channel. Because the p-type substrate is grounded, the n-channel and p-type substrate are always reverse-biased; this results in zero current in the substrate. Also, note that zero gate current flows because of the extremely high resistance of the SiO_2 insulating layer. The resistance between the gate and channel is of the order of several thousands of megohms.

A depletion-type MOSFET is similar to a JFET in its operating characteristics, as shown in Fig. 31–16*b*. Notice that for each drain curve, the drain current increases linearly until the pinch-off voltage, V_P, is reached. When V_{GS} is negative, pinch-off occurs sooner (lower values of V_{DS}), and when V_{GS} is made positive, pinch-off occurs later. Notice in Fig. 31–16*b* that the maximum drain current I_D does not exist when $V_{GS} = 0$ V. Yet I_{DSS} is still defined the same way: it is the drain current with the gate shorted.

Figure 31–17*a* shows a positive gate voltage applied to the depletion-type MOSFET. The positive gate voltage attracts free electrons into the channel from the substrate, thereby enhancing its conductivity. When the gate is made positive relative to the source, the depletion-type MOSFET is said to be operating in the ***enhancement mode***.

Figure 31–17*b* shows a negative voltage applied to the gate. The negative gate voltage sets up an electric field that repels free electrons from the channel. When the gate is made negative relative to the source, the depletion-type MOSFET is said to be operating in the ***depletion mode***. Making the gate negative enough will reduce the drain current, I_D, to zero.

Figure 31–18 shows a transconductance curve for the n-channel depletion-type MOSFET. I_{DSS} is the drain current that flows with the gate shorted to the source. It is important to note, however, that I_{DSS} is not the maximum drain current that is obtainable. When V_{GS} is positive, the depletion-type MOSFET operates in the enhancement mode, and the drain current increases beyond the value of I_{DSS}. With V_{GS} negative, the MOSFET operates in the depletion mode. If V_{GS} is made negative enough, the drain current, I_D, will be reduced to zero. As with JFETs, the value of the gate-source voltage that reduces the drain current to zero is called the gate-source cutoff voltage, designated $V_{GS(off)}$.

One more point: because there is drain current with zero gate-source voltage, the device is referred to as a "normally on" MOSFET.

Figure 31–18 Transconductance curve of an n-channel depletion-type MOSFET.

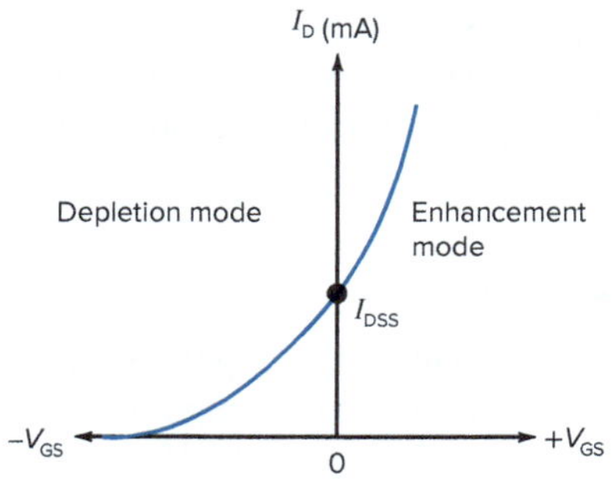

Because D-MOSFETS are normally ON devices, the drain current, I_D, can be calculated using Formula (31–2). This formula was given earlier as

$$I_D = I_{DSS}\left[1 - \frac{V_{GS}}{V_{GS(off)}}\right]^2 \quad (V_{DS} > V_P)$$

Example 31-9

A D-MOSFET has the following characteristics: $I_{DSS} = 10$ mA and $V_{GS(off)} = -4$ V. Calculate the drain current, I_D, for (a) $V_{GS} = +2$ V, (b) $V_{GS} = -2$ V, and (c) $V_{GS} = 0$ V.

ANSWER For each value of V_{GS} listed, the calculations are

a. $I_D = I_{DSS}\left[1 - \frac{V_{GS}}{V_{GS(off)}}\right]^2$

$= 10 \text{ mA}\left[1 - \frac{+2 \text{ V}}{-4 \text{ V}}\right]^2$

$= 10 \text{ mA} \times 2.25$

$= 22.5$ mA

b. $I_D = 10 \text{ mA}\left[1 - \frac{-2 \text{ V}}{-4 \text{ V}}\right]^2$

$= 10 \text{ mA}\,[1 - 0.5]^2$

$= 10 \text{ mA} \times 0.25$

$= 2.5$ mA

c. When $V_{GS} = 0$ V, then $I_D = I_{DSS}$, and therefore $I_D = 10$ mA.

p-Channel Depletion-Type MOSFET

Figure 31–19 shows the construction, schematic symbol, and transconductance curve for a *p*-channel depletion-type MOSFET.

Figure 31–19*a* shows that the channel is made of *p*-type semiconductor material and the substrate is made of *n*-type semiconductor material. Because of this, *p*-channel depletion-type MOSFETs require a negative drain voltage.

Figure 31–19*b* shows the schematic symbol. Notice that the arrow points outward away from the *p*-type channel.

Finally, the transconductance curve is shown in Fig. 31–19*c*. Compare this curve to the one in Fig. 31–18. Notice that they are opposite. As shown in Fig. 31–19*c*, the *p*-channel depletion-type MOSFET operates in the enhancement mode when V_{GS} is negative and in the depletion mode when V_{GS} is positive. Note that holes are the majority current carriers in the *p*-channel.

Depletion-Type MOSFET Applications

Depletion-type MOSFETs are frequently used as small signal amplifiers and frequency mixers. A depletion-type MOSFET used as a small signal amplifier is often biased so that the *Q* point has the following values: $I_D = I_{DSS}$ and $V_{DS} = V_{DD}/2$.

Figure 31–19 *p*-channel depletion-type MOSFET. (*a*) Construction. (*b*) Schematic symbol. (*c*) Transconductance curve.

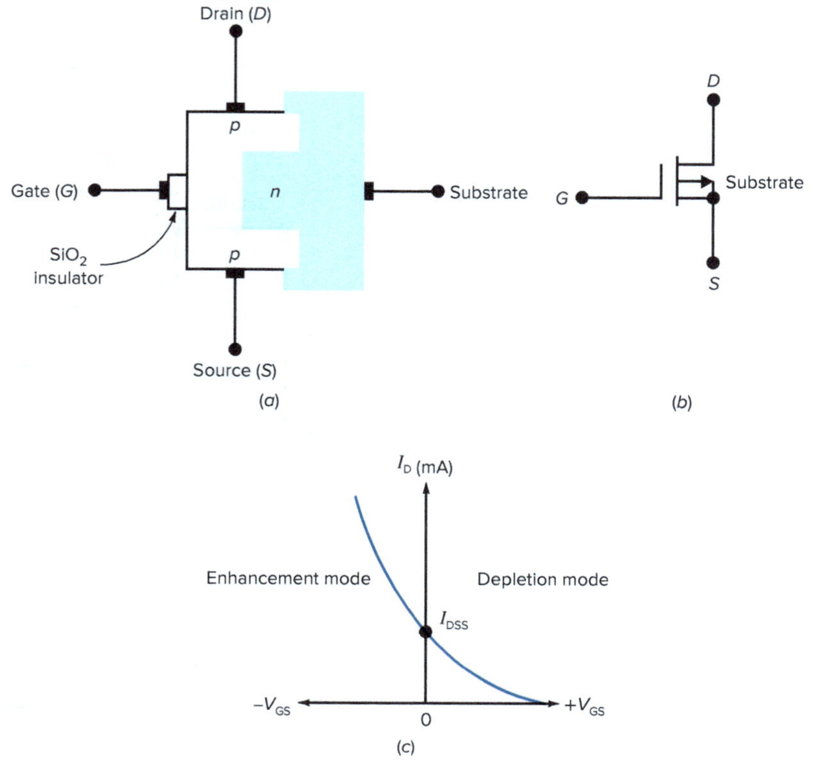

Selecting these values for the Q point allows the D-MOSFET to amplify small AC signals. Because D-MOSFETs are quite similar to JFETs, the AC analysis used with JFETs can also be used with D-MOSFETs. For example, the voltage gain A_V of a CS amplifier that uses a D-MOSFET equals $g_m r_L$.

Also, MOSFETs might be used instead of JFETs if the input resistance of the JFETs is not high enough.

Enhancement-Type MOSFETs

Figure 31–20*a* shows the construction of an *n*-channel enhancement-type MOSFET. Notice that the *p*-type substrate makes contact with the SiO_2 insulator. Because of this, there is no channel for conduction between the drain and source terminals.

Notice the polarities of the supply voltages in Fig. 31–20*a*. The drain and gate are made positive with respect to the source. With $V_{GS} = 0$ V, there is no channel between the source and drain and so the drain current, I_D, is zero.

To produce drain current, the positive gate voltage must be increased. This attracts electrons along the right edge of the SiO_2 insulator, as shown in Fig. 31–20*b*. The minimum gate-source voltage that makes drain current flow is called the ***threshold voltage***, designated $V_{GS(th)}$. When the gate voltage is less than $V_{GS(th)}$, the drain current, I_D, is zero. The value of $V_{GS(th)}$ varies from one E-MOSFET to the next.

Figure 31–20*c* shows the schematic symbol for the *n*-channel enhancement-type MOSFET. Notice the broken channel line. The broken line represents the "off" condition that exists with zero gate voltage. Because of this characteristic, enhancement-type MOSFETs are called "normally OFF" devices.

GOOD TO KNOW

With the E-MOSFET, V_{GS} has to be greater than $V_{GS(th)}$ to get any drain current at all!

Figure 31–20 *n*-channel enhancement-type MOSFET. (*a*) Construction. (*b*) Electric charges on surface of SiO_2 insulating layer. (*c*) Schematic symbol. (*d*) Drain curves. (*e*) Transconductance curve.

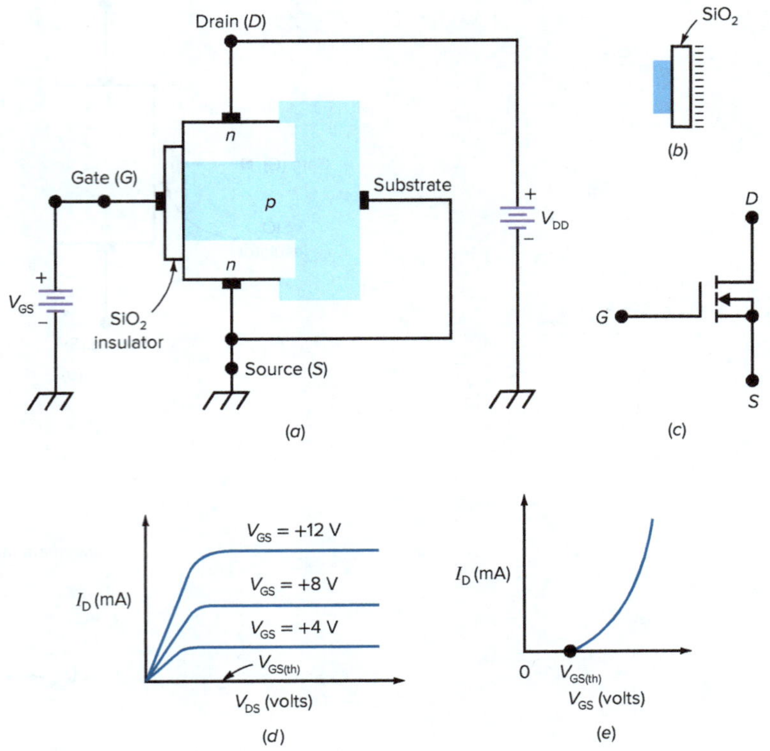

Figure 31–20*d* shows a typical set of drain curves for the *n*-channel enhancement-type MOSFET. The lowest curve is the $V_{GS(th)}$ curve. For more positive gate voltages, the drain current, I_D, increases. The transconductance curve is shown in Fig. 31–20*e*. Notice that I_D is zero when the gate-source voltage is less than $V_{GS(th)}$.

Figure 31–21*a* shows a *p*-channel enhancement-type MOSFET. Notice the arrow pointing outward away from the *p*-type channel. Also, notice the negative gate and drain voltages. These are the required polarities for biasing the *p*-channel enhancement-type MOSFET.

Figure 31–21 Biasing a *p*-channel enhancement-type MOSFET. (*a*) Proper biasing voltages. (*b*) Transconductance curve.

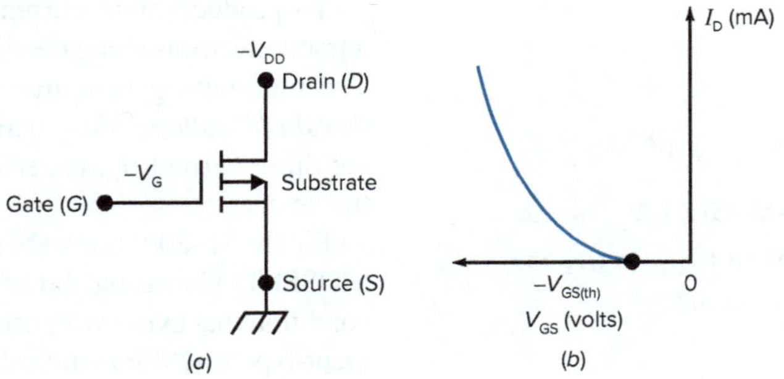

Figure 31–21*b* shows the transconductance curve for the *p*-channel enhancement-type MOSFET. Notice that I_D is zero until the gate-source voltage is more negative than $-V_{GS(th)}$.

Enhancement-Type MOSFET Application

E-MOSFETs have many applications in electronics. The most important application is in digital computer electronics. E-MOSFETs are used because they take up very little space on a chip (an integrated circuit) compared to the space used by an equivalent circuit with bipolar transistors. Thus, when packaging hundreds or even thousands of transistors onto an IC, MOSFETs are used. Another big reason that enhancement-type MOSFETs are used frequently is that they consume extremely little power.

31–4 Self-Review

Answers at the end of the chapter.

a. Is a D-MOSFET a normally ON or normally OFF device?

b. Is an E-MOSFET considered normally ON or normally OFF?

c. What are the two operating modes for a D-MOSFET?

d. What is the most common application for E-MOSFETs?

31–5 MOSFET Biasing Techniques

Zero Bias for Depletion-Type MOSFETs

Figure 31–22*a* shows a popular biasing technique that can be used only with depletion-type MOSFETs. This form of bias is called *zero bias* because the potential difference across the gate-source region is zero. With V_{GS} equal to zero, the quiescent drain current, I_D, equals I_{DSS} (see Fig. 31–22*b*). When V_{in} drives the gate positive, the drain current I_D increases. When V_{in} becomes negative, the drain current I_D decreases. During the positive alternation of V_{in}, the D-MOSFET operates in the enhancement mode. When V_{in} is negative, the *n*-channel D-MOSFET operates in the depletion mode.

Figure 31–22 Zero-biasing technique. (*a*) Common-source amplifier using zero bias. (*b*) Transconductance curve showing how I_D varies with variations in V_{GS}.

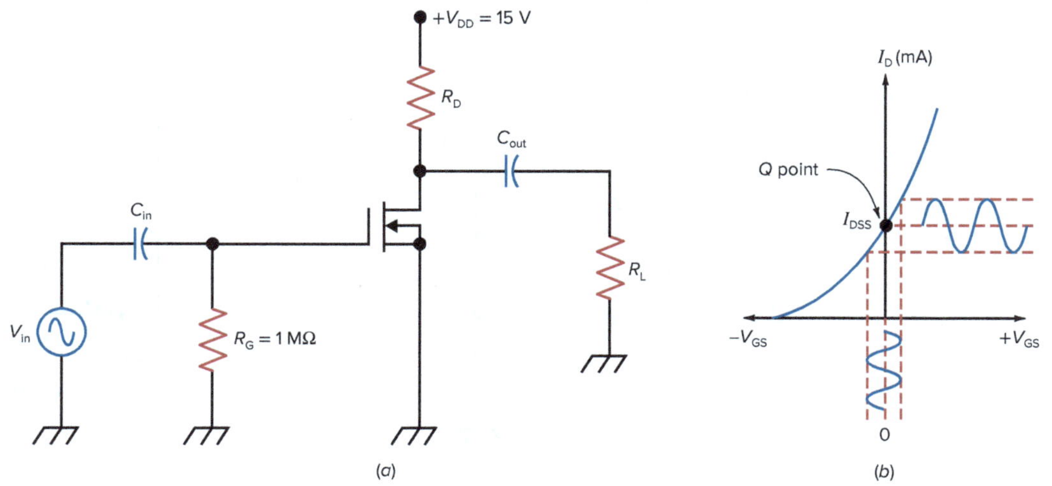

To calculate the DC voltage at the drain, use Formula (31–20):

$$V_{DS} = V_{DD} - I_{DSS}R_D \quad \textbf{(31–20)}$$

It is common to select R_D such that $V_{DS} = V_{DD}/2$. To do this, R_D is calculated by using Formula (31–21):

$$R_D = \frac{V_{DD}}{2I_{DSS}} \quad \textbf{(31–21)}$$

Because of the common-source configuration shown in Fig. 31–22, the voltage gain, A_V, is calculated the same way as for the JFET common-source amplifier. Therefore, $A_V = g_m r_L$.

Even though zero bias is the most commonly used technique for biasing depletion-type MOSFETs, other techniques can also be used. These include self-bias, voltage divider bias, and current-source bias.

Biasing Enhancement-Type MOSFETs

GOOD TO KNOW

In many cases, bipolar transistors and MOSFETs are used in the same electronic circuit.

Enhancement-type MOSFETs cannot be biased using the zero-bias technique because V_{GS} must exceed $V_{GS(th)}$ to produce any drain current at all. Because of this, zero bias, self-bias, and current-source bias cannot be used with enhancement-type MOSFETs.

Figure 31–23*a* shows one way to bias enhancement-type MOSFETs. This form of bias is called drain-feedback bias, which is similar to collector-feedback bias used with bipolar transistors. The manufacturer's data sheet for enhancement-type MOSFETs usually specifies the value of $V_{GS(th)}$ and the coordinates of one point on the transconductance curve. The quantities $I_{D(on)}$, $V_{GS(on)}$, and $V_{GS(th)}$ are the parameters that are important when biasing E-MOSFETs. (See Fig. 31–23*b*.)

The transconductance curve shown in Fig. 31–23*b* is for a Motorola 3N169 enhancement-type MOSFET. The values shown for $I_{D(on)}$, $V_{GS(on)}$, and $V_{GS(th)}$ are "typical" values.

It is somewhat unusual, but the drain resistor, R_D, must be properly selected to provide the required bias. R_D can be calculated using Formula (31–22):

$$R_D = \frac{V_{DD} - V_{GS(on)}}{I_{D(on)}} \quad \textbf{(31–22)}$$

MultiSim **Figure 31–23** Biasing an *n*-channel enhancement-type MOSFET. (*a*) Circuit using drain feedback bias. (*b*) Transconductance curve showing values of $V_{GS(th)}$, $V_{GS(on)}$, and $I_{D(on)}$.

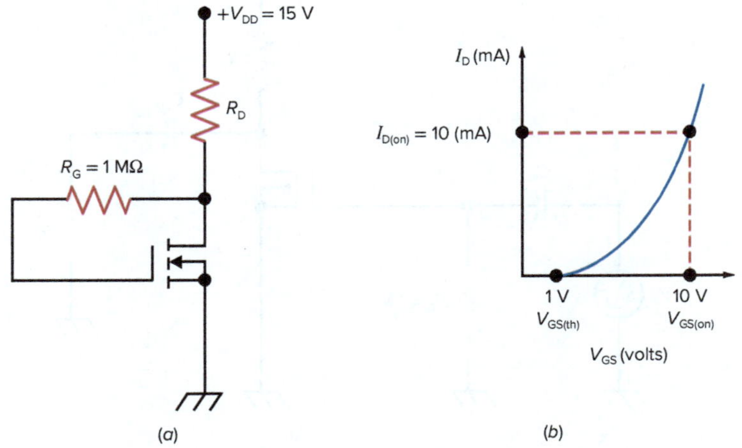

Since the gate current is zero, no voltage is dropped across the gate resistor, R_G. Therefore, $V_{GS} = V_{DS}$.

Example 31-10

In Fig. 31–23, calculate the value of R_D to provide an $I_{D(on)}$ of 10 mA.

ANSWER Using Formula (31–22) and the values from the transconductance curve in Fig. 31–23*b*, the calculations are as follows:

$$R_D = \frac{V_{DD} - V_{GS(on)}}{I_{D(on)}}$$

$$= \frac{15\ \text{V} - 10\ \text{V}}{10\ \text{mA}}$$

$$= 500\ \Omega$$

A 470-Ω resistor would provide the proper biasing voltage at the gate.

31–5 Self-Review

Answers at the end of the chapter.

a. When a D-MOSFET uses zero bias, what is the value of I_D with no signal?

b. How much drain current flows in an E-MOSFET if $V_{GS} < V_{GS(th)}$?

31–6 Handling MOSFETs

One disadvantage of MOSFET devices is their extreme sensitivity to electrostatic discharge (ESD) due to their insulated gate-source regions. The SiO_2 insulating layer is extremely thin and can be easily punctured by an electrostatic discharge.

Because MOSFETs can be easily damaged from electrostatic discharge, extreme caution is recommended when handling them. (For more information on electrostatic discharge, see Appendix G at the back of the book.) The following is a list of precautions:

1. Never insert or remove MOSFETs from a circuit with the power ON.
2. Never apply input signals when the DC power supply is OFF.
3. Wear a grounding strap on your wrist when handling MOSFET devices. This keeps the body at ground potential by bleeding off any buildup of static electric charge.
4. When storing MOSFETs, keep the device leads in contact with conductive foam, or connect a shorting ring around the leads.

It is extremely important to observe these precautions to avoid possible damage to the MOSFET device.

Many manufacturers put protective zener diodes across the gate-source region to protect against ESD (shown in Fig. 31–24). The diodes are arranged so that they will conduct for either polarity of gate-source voltage, V_{GS}. The breakdown voltage of the diodes is much higher than any voltage normally applied between the gate-source region but less than the breakdown voltage of the insulating material. One drawback of using the protective diodes is that the input impedance of the device is lowered considerably.

Figure 31–24 Protective zener diodes connected across the gate-source region to protect against ESD.

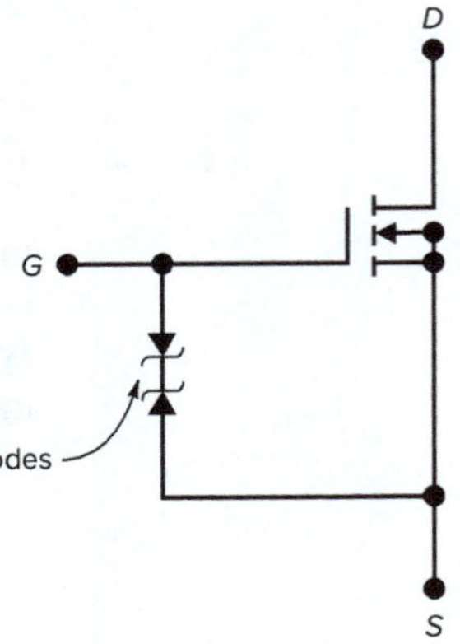

31–6 Self-Review

Answers at the end of the chapter.

a. What area of a MOSFET is easily damaged by ESD?

b. How does wearing a grounded wrist strap protect a MOSFET from damage by ESD?

Summary

- A JFET has three terminals: the gate, source, and drain.
- JFETs are voltage-controlled devices which means that the drain current, I_D, is controlled by the amount of gate-source voltage, V_{GS}.
- The input impedance of a JFET is normally very high, usually of the order of several megohms (MΩ).
- A JFET is a "normally on" device because drain current flows when $V_{GS} = 0$ V.
- The pinch-off voltage, V_P, is the drain-source voltage at which I_D levels off.
- I_{DSS} represents the drain-source current that flows when $V_{GS} = 0$ V and $V_{DS} > V_P$. I_{DSS} is the maximum possible drain current for a JFET.
- The gate-source cutoff voltage, designated $V_{GS(off)}$, is the amount of V_{GS} required to reduce the drain current, I_D, to zero. $V_P = -(-V_{GS(off)})$.
- The transconductance curve is a graph of I_D versus V_{GS}. The transconductance curve is nonlinear. The ratio of the change in drain current, ΔI_D, to the change in gate-source voltage, ΔV_{GS}, is the transconductance, g_m. The unit of transconductance is the siemens (S).
- The transconductance, g_m, tells us how effective V_{GS} is in controlling I_D.
- Gate bias, self-bias, voltage divider bias, and current-source bias are common ways of biasing a JFET. Gate bias is seldom used because its *Q* point is so unpredictable.
- JFET amplifiers provide less voltage gain than bipolar transistor amplifiers. The input impedance of a JFET amplifier is much higher, however.
- When the source resistor is bypassed, a CS amplifier has a voltage gain of $g_m r_L$. In a CS amplifier, V_{in} and V_{out} are 180° out of phase.
- A common-drain amplifier is more commonly known as a source follower. A source follower has high input impedance, low output impedance, and a voltage gain less than one.
- A CG amplifier has a voltage gain equal to $g_m r_L$ just like the CS amplifier. However, V_{in} and V_{out} are in phase. The major drawback of a CG amplifier is its low input impedance.
- There are two basic types of MOSFETs: E-MOSFETs and D-MOSFETs. Like JFETs, MOSFETs are voltage-controlled devices.
- The main difference between a JFET and a MOSFET is that the gate in the MOSFET is insulated from the channel by a thin layer of silicon dioxide (SiO_2). This makes the input impedance of a MOSFET many times higher than that of a JFET.
- D-MOSFETs are "normally on" devices because drain current flows when $V_{GS} = 0$ V. For D-MOSFETs, I_{DSS} is not the maximum possible drain current.
- A D-MOSFET can operate in either the enhancement or depletion mode.
- An E-MOSFET is a "normally off" device because there is no drain current when $V_{GS} = 0$ V. For an E-MOSFET, $V_{GS(th)}$ is the minimum gate-source voltage that produces drain current.
- One disadvantage of MOSFETs is their extreme sensitivity to damage by electrostatic discharge (ESD). When handling MOSFETs, extreme care must be taken to ensure that static electricity does not puncture the thin layer of SiO_2 separating the gate and channel.
- When handling MOSFETs, wear a grounded wrist strap!

Important Terms

Asymmetrical JFET—a JFET whose gate regions are offset from the center of the channel. The drain and source leads of an asymmetrical JFET cannot be interchanged.

Channel—the area or conducting region between the drain and source terminals of an FET. The channel can be made of either *n*-type or *p*-type semiconductor material.

Common-drain amplifier—an amplifier whose input is applied to the gate and its output is taken from the source. Another name for the common-drain amplifier is the source follower.

Common-gate amplifier—an amplifier whose input is applied to the source and its output is taken from the drain.

Common-source amplifier—an amplifier whose input is applied to the gate and its output is taken from the drain.

Current-source region—the region of operation in which the drain of a JFET acts as a current source. The current-source region of operation exists when $V_{DS} > V_P$.

Depletion mode—the mode of operation for a MOSFET in which the polarity of V_{GS} reduces the drain current as the channel becomes depleted of available charge carriers.

Drain—one of the three leads of an FET. The drain lead connects to one end of the conducting channel.

Enhancement mode—the mode of operation for a MOSFET in which the polarity of V_{GS} enhances the conductivity of the channel, thus increasing the drain current.

Field effect transistor (FET)—a unipolar device that relies on only one type of charge carrier, either free electrons or holes. FETs are voltage-controlled devices with an input voltage controlling the output current.

Gate—one of the three leads of an FET. The gate is used to control the drain current.

Gate-source cutoff voltage, $V_{GS(off)}$—the amount of gate-source voltage required to reduce the drain current, I_D, to zero.

IGFET—insulated gate field effect transistor. Another name for a MOSFET.

JFET—junction field effect transistor.

MOSFET—metal-oxide-semiconductor field effect transistor.

Ohmic region—the region of operation for a JFET where the drain current, I_D, increases in direct proportion to V_{DS}. The ohmic region of operation exists when $V_{DS} < V_P$.

Pinch-off voltage, V_P—the drain-source voltage at which the drain current, I_D, levels off. V_P is the border between the ohmic and current-source regions of operation.

Source—one of the three leads of an FET. The source lead connects to one end of the conducting channel.

Symmetrical JFET—a JFET whose gate regions are located in the center of the channel. The drain and source leads of a symmetrical JFET can be interchanged without affecting its operation.

Threshold voltage, $V_{GS(th)}$—the minimum value of V_{GS} in an enhancement-type MOSFET that makes drain current flow.

Transconductance, g_m—the ratio of the change in drain current, ΔI_D, to the change in gate-source voltage, ΔV_{GS}, for a fixed value of V_{DS}. The unit of g_m is the siemens (S).

Unipolar—a term that describes a device having only one type of charge carrier, either free electrons or holes.

Related Formulas

JFET

$$V_{DS(P)} = V_P - V_{GS}$$

$$I_D = I_{DSS}\left[1 - \frac{V_{GS}}{V_{GS(off)}}\right]^2 \quad \text{(JFETs and D-MOSFETs)}$$

Gate Bias

$$V_{DS} = V_{DD} - I_D R_D$$

Self-Bias

$$V_{GS} = -I_D R_S$$

$$R_S = \frac{-V_{GS(off)}}{2I_{DSS}}$$

Voltage Divider Bias

$$V_G = \frac{R_2}{R_1 + R_2} \times V_{DD}$$

$$V_S = V_G - V_{GS}$$

$$I_D = V_S/R_S$$

$$V_D = V_{DD} - I_D R_D$$

Current-Source Bias

$$I_C = \frac{V_{EE} - V_{BE}}{R_E}$$

JFET Amplifiers

$$g_m = \Delta I_D/\Delta V_{GS} \quad (V_{DS} > V_P)$$

$$g_m = g_{mo}\left[1 - \frac{V_{GS}}{V_{GS(off)}}\right]$$

$$g_{mo} = 2I_{DSS}/-V_{GS(off)}$$

Common-Source Amplifier

$$A_V = g_m r_L \quad (R_S \text{ bypassed})$$

$$A_V = \frac{g_m r_L}{1 + g_m R_S} \quad (R_S \text{ unbypassed})$$

Common-Drain Amplifier

$$A_V = \frac{g_m r_L}{1 + g_m r_L}$$

$$Z_{out} = R_S \,\|\, \frac{1}{g_m}$$

Common-Gate Amplifier

$$A_V = g_m r_L$$

$$Z_{in} = R_S \,\|\, \frac{1}{g_m}$$

D-MOSFET Zero Bias

$$V_{DS} = V_{DD} - I_{DSS} R_D$$

$$R_D = V_{DD}/2I_{DSS}$$

E-MOSFET Drain-Feedback Bias

$$R_D = \frac{V_{DD} - V_{GS(on)}}{I_{D(on)}}$$

Self-Test

Answers at the back of the book.

1. A JFET is a

a. unipolar device.

b. voltage-controlled device.

c. current controlled device.

d. both a and b.

2. The drain and source leads may be interchanged when using a(n)

a. asymmetrical JFET.

b. symmetrical JFET.

c. D-type MOSFET.

d. none of the above.

3. A JFET is a

a. normally ON device.

b. normally OFF device.

c. bipolar device.

d. current-controlled device.

4. **When a JFET is operating in the ohmic region,**
   a. I_D is independent of V_{DS}.
   b. I_D is independent of V_{GS}.
   c. I_D increases in direct proportion to V_{DS}.
   d. the drain acts like a current source.

5. **The value of drain to source voltage, V_{DS}, at which the drain current, I_D, levels off is called the**
   a. cutoff voltage, $V_{GS(off)}$.
   b. pinch-off voltage, V_P.
   c. breakdown voltage, V_{BR}.
   d. threshold voltage, $V_{GS(th)}$.

6. **A JFET operates in the current-source region when**
   a. $V_{DS} > V_P$.
   b. $V_{DS} < V_P$.
   c. $V_{DS} = 0$ V.
   d. $V_{GS} = 0$ V.

7. **A JFET parameter that describes how effective the gate-source voltage is in controlling the drain current is called its**
   a. gamma, γ.
   b. Beta, β.
   c. transconductance, g_m.
   d. none of the above.

8. **Which JFET amplifier is also known as a source follower?**
   a. the common-source amplifier.
   b. the common-gate amplifier.
   c. the common-channel amplifier.
   d. the common-drain amplifier.

9. **In which JFET amplifier are the AC input and output voltages 180° out of phase?**
   a. the common-gate amplifier.
   b. the common-source amplifier.
   c. the common-drain amplifier.
   d. the source follower.

10. **Which of the following JFET amplifiers has the lowest input impedance?**
   a. the common-gate amplifier.
   b. the common-source amplifier.
   c. the common-drain amplifier.
   d. the source follower.

11. **Which of the following JFET amplifiers has a high Z_{in}, a low Z_{out}, and a voltage gain less than one?**
   a. the common-gate amplifier.
   b. the common-source amplifier.
   c. the source follower.
   d. both a and b.

12. **A depletion-type MOSFET is a**
   a. normally OFF Device.
   b. normally ON Device.
   c. current-controlled device.
   d. none of the above.

13. **An enhancement-type MOSFET is a**
   a. normally OFF device.
   b. normally ON device.
   c. low input impedance device.
   d. current-controlled device.

14. **For an enhancement-type MOSFET, the threshold voltage, $V_{GS(th)}$, is the**
   a. maximum allowable gate-source voltage before breakdown.
   b. gate-source voltage that produces a leveling off of I_D.
   c. minimum gate-source voltage that makes drain current flow.
   d. none of the above.

15. **To avoid damaging MOSFETs during handling,**
   a. always wear a grounded wrist strap.
   b. never apply an input signal when the DC power supply is OFF.
   c. never insert or remove them from a circuit when the power is ON.
   d. all of the above.

16. **Which of the following types of bias produces the most unstable *Q* point in a JFET amplifier?**
   a. gate bias.
   b. current-source bias.
   c. voltage divider bias.
   d. self-bias.

17. **When an *n*-channel JFET operates in the ohmic region,**
   a. r_{DS} increases as V_{GS} becomes less negative.
   b. r_{DS} increases as V_{GS} becomes more positive.
   c. r_{DS} increases as V_{GS} becomes more negative.
   d. r_{DS} is independent of V_{GS}.

18. **In a JFET amplifier with self-bias,**
   a. $V_G = 0$ V.
   b. $V_S = I_D R_S$.
   c. $V_{GS} = -I_D R_S$.
   d. all of the above.

19. **For a depletion-type MOSFET with zero bias, the drain current, I_D, equals**
   a. zero.
   b. I_{DSS}.
   c. $\frac{1}{2} I_{DSS}$.
   d. It cannot be determined.

20. **The input impedance of a MOSFET is**
   a. higher than that of a JFET.
   b. lower than that of a JFET.
   c. no different than that of a JFET.
   d. approximately zero ohms.

Essay Questions

1. How is the gate-source junction of a JFET normally biased? How much is the gate current, I_G, under these circumstances?
2. For a JFET, what is the difference between the ohmic and current-source operating regions?
3. What does the notation I_{DSS} stand for?
4. Why are JFETs called "normally ON" devices?
5. Why is self-bias a better way to bias a JFET than gate bias?
6. How is the transconductance, g_m, of a JFET affected by V_{GS}?

7. Which JFET amplifier has a
   a. low input impedance?
   b. low output impedance?
   c. high input impedance?
   d. 180° phase difference between V_{in} and V_{out}?
   e. voltage gain less than one?

8. Is an enhancement-type MOSFET considered to be a normally on or normally off device? Why?

9. In what mode is a depletion-type MOSFET operating if I_D is
   a. greater than I_{DSS}?
   b. less than I_{DSS}?

10. Why are MOSFETs so sensitive to damage by electrostatic discharge (ESD)?

Problems

SECTION 31–1 JFETS AND THEIR CHARACTERISTICS

31–1 In what part of a JFET does current flow?

31–2 When looking at the schematic symbol of a JFET, how can you tell if it is a *p*-channel or *n*-channel JFET?

31–3 In a JFET, which two currents are identical?

31–4 For an *n*-channel JFET, what is the proper polarity for
   a. V_{GS}?
   b. V_{DS}?

31–5 For a *p*-channel JFET, what is the proper polarity for
   a. V_{GS}?
   b. V_{DS}?

31–6 Define $V_{GS(off)}$.

31–7 For a JFET, what is the pinch-off voltage, V_P?

31–8 How are V_P and $V_{GS(off)}$ related?

31–9 What happens to the pinch-off voltage of an *n*-channel JFET as V_{GS} becomes more negative?

31–10 Explain the difference between the ohmic and current-source regions of operation for a JFET.

31–11 An *n*-channel JFET has the following specifications: I_{DSS} = 15 mA and $V_{GS(off)}$ = −4 V. Calculate the drain current, I_D, for each of the following values of V_{GS} (assume $V_{DS} > V_P$):
   a. V_{GS} = 0 V.
   b. V_{GS} = −0.5 V.
   c. V_{GS} = −1 V.
   d. V_{GS} = −1.5 V.
   e. V_{GS} = −2 V.
   f. V_{GS} = −2.5 V.
   g. V_{GS} = −3 V.
   h. V_{GS} = −3.5 V.
   i. V_{GS} = −4 V.

31–12 An *n*-channel JFET has an I_{DSS} value of 8 mA and a $V_{GS(off)}$ value of −3 V. Calculate the drain current, I_D, for each of the following values of V_{GS} (assume $V_{DS} > V_P$):
   a. V_{GS} = 0 V.
   b. V_{GS} = −0.5 V.
   c. V_{GS} = −1 V.
   d. V_{GS} = −1.5 V.
   e. V_{GS} = −2.25 V.
   f. V_{GS} = −2.75 V.

31–13 A *p*-channel JFET has an I_{DSS} value of 20 mA and a $V_{GS(off)}$ value of +5 V. Calculate the drain current, I_D, for each of the following values of V_{GS} (assume $V_{DS} > V_P$):
   a. V_{GS} = 0 V.
   b. V_{GS} = 1 V.
   c. V_{GS} = 2 V.
   d. V_{GS} = 3 V.
   e. V_{GS} = 4 V.
   f. V_{GS} = 5 V.

SECTION 31–2 JFET BIASING TECHNIQUES

31–14 In Fig. 31–25, solve for I_D and V_{DS} for each of the following values of V_{GS}:
   a. V_{GS} = −1 V.
   b. V_{GS} = −1.5 V.
   c. V_{GS} = −2 V.
   d. V_{GS} = −2.5 V.

Figure 31–25

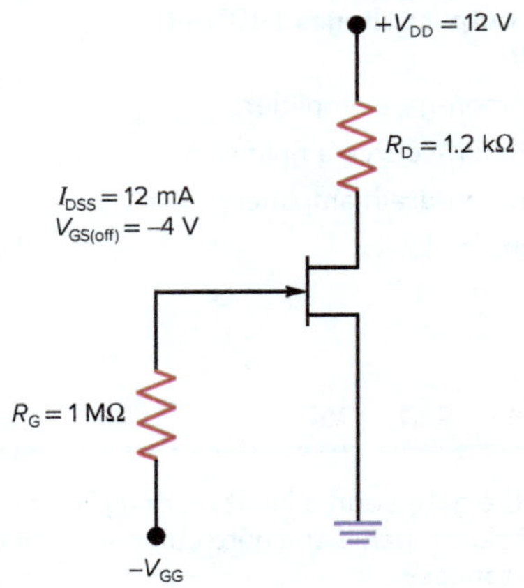

31–15 The JFET in Fig. 31–26 has a drain current, I_D, of 2.15 mA. Solve for

a. V_G.
b. V_S.
c. V_{GS}.
d. V_D.

Figure 31–26

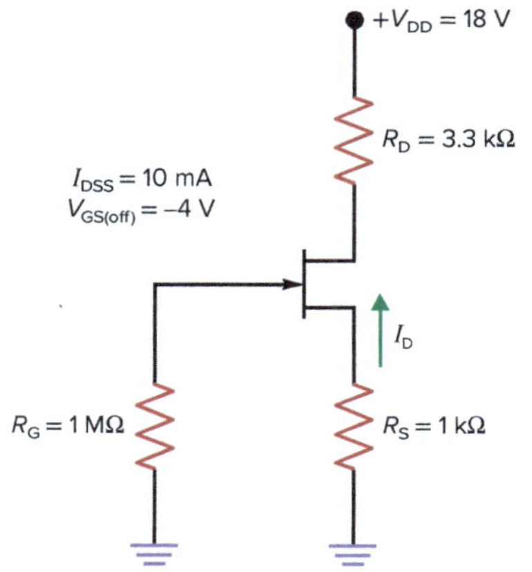

31–16 In Fig. 31–27, what value of R_S will provide a drain current, I_D, of approximately one-half I_{DSS}?

Figure 31–27

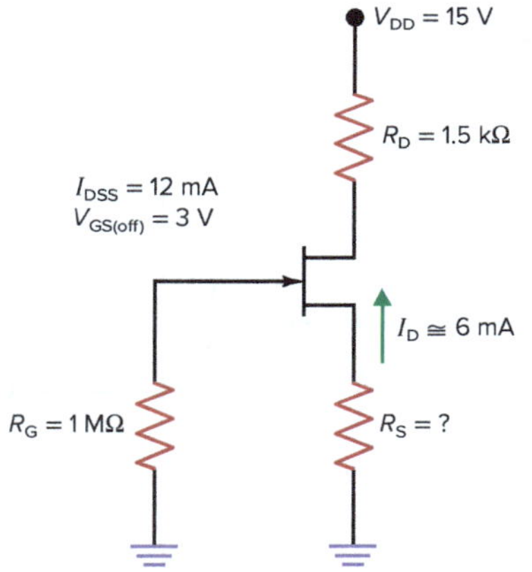

31–17 For the value of R_S calculated in Prob. 31–16, solve for the following:

a. V_G.
b. V_S.
c. V_{GS}.
d. V_D.

31–18 If $V_{GS} = -1.15$ V in Fig. 31–28, solve for the following:

a. V_G.
b. V_S.
c. I_D.
d. V_D.

Figure 31–28

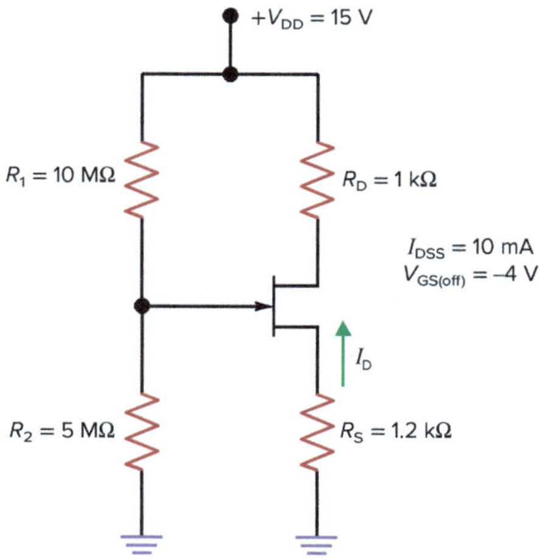

31–19 In Fig. 31–29, solve for I_D and V_D.

Figure 31–29

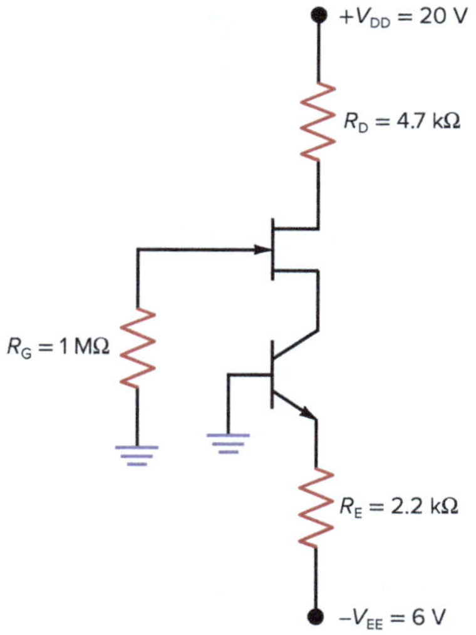

SECTION 31–3 JFET AMPLIFIERS

31–20 List one advantage and one disadvantage of a JFET amplifier versus a bipolar transistor amplifier.

31–21 What is the formula for the transconductance of a JFET, and what is its unit of measure?

31–22 A JFET has an I_{DSS} value of 12 mA and a $V_{GS(off)}$ value of −3 V. How much is g_{mo}?

31–23 For the JFET in Prob. 31–22, what is the value of g_m for each of the following values of V_{GS}?

a. $V_{GS} = 0$ V.
b. $V_{GS} = -0.5$ V.

c. $V_{GS} = -1$ V.

d. $V_{GS} = -1.5$ V.

e. $V_{GS} = -2$ V.

f. $V_{GS} = -2.5$ V.

31–24 In Fig. 31–30, solve for each of the following DC quantities:

a. V_G.

b. V_{GS}.

c. I_D.

d. V_D.

31–28 In Fig. 31–31, solve for the following AC quantities:

a. Z_{in}.

b. r_L.

c. g_{mo}.

d. g_m.

e. A_V.

f. V_{out}.

g. Z_{out}.

Figure 31–30

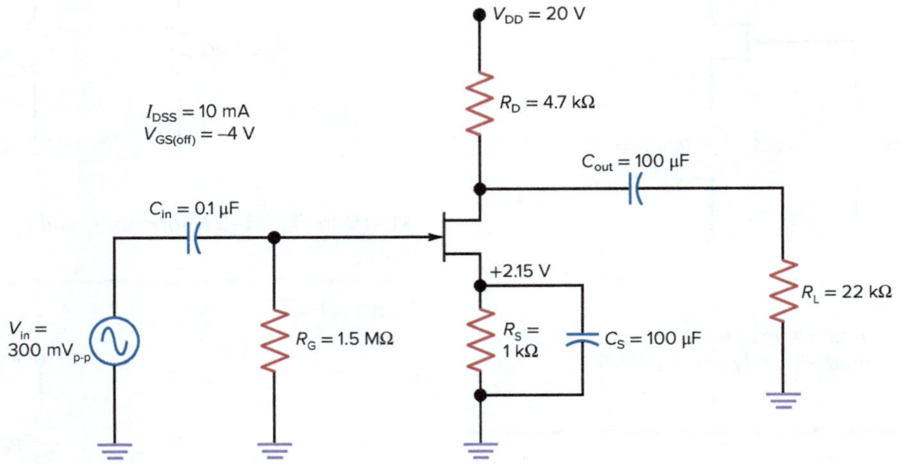

31–25 In Fig. 31–30, solve for each of the following AC quantities:

a. Z_{in}.

b. r_L.

c. g_{mo}.

d. g_m.

e. A_V.

f. V_{out}.

31–26 If the source bypass capacitor is removed in Fig. 31–30, calculate

a. A_V.

b. V_{out}.

31–27 In Fig. 31–31, solve for the following DC quantities:

a. V_G.

b. V_{GS}.

c. I_D.

d. V_D.

Figure 31–31

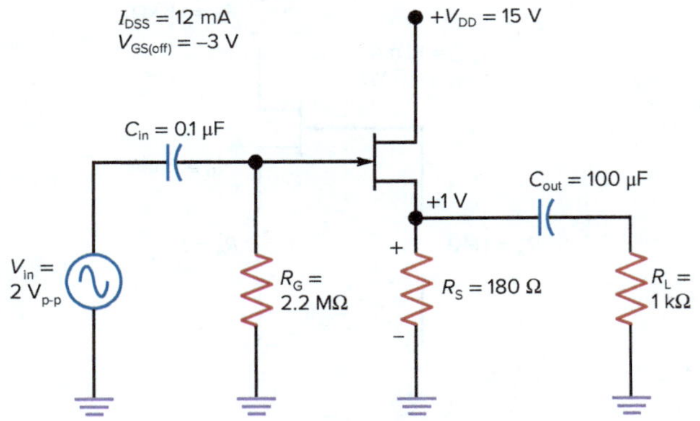

31–29 In Fig. 31–32, solve for the following AC quantities:

a. g_{mo}.

b. g_m.

c. r_L.

d. Z_{in}.

e. A_V.

f. V_{out}.

Figure 31–32

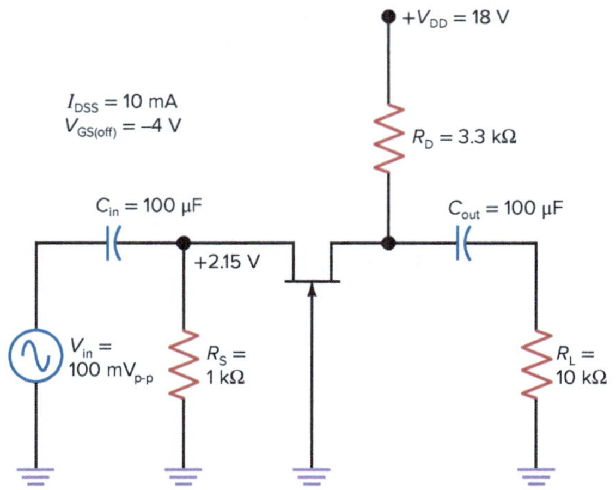

Figure 31–33

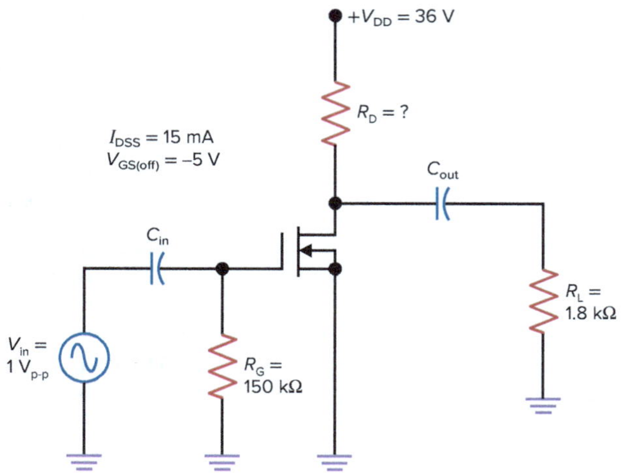

SECTION 31–4 MOSFETS AND THEIR CHARACTERISTICS

31-30 What is the key difference in the way a JFET and MOSFET are constructed?

31–31 What is another name for a MOSFET?

31–32 What are the two different types of MOSFETs?

31–33 Is I_{DSS} the maximum possible drain current for a D-MOSFET?

31–34 A D-MOSFET has an I_{DSS} value of 20 mA and a $V_{GS(off)}$ value of −5 V. Calculate the drain current, I_D, for each of the following values of V_{GS}:

a. $V_{GS} = -4$ V.
b. $V_{GS} = -3$ V.
c. $V_{GS} = -2$ V.
d. $V_{GS} = -1$ V.
e. $V_{GS} = +1$ V.
f. $V_{GS} = +2$ V.
g. $V_{GS} = +3$ V.
h. $V_{GS} = +4$ V.

31–35 How much drain current flows in an *n*-channel E-MOSFET when $V_{GS} < V_{GS(th)}$?

31–36 List two reasons why E-MOSFETs are typically used in computers.

SECTION 31–5 MOSFET BIASING TECHNIQUES

31–37 What type of bias is shown in Fig. 31–33?

31–38 In what mode is the D-MOSFET operating in Fig. 31–33 when V_{GS} is

a. positive?
b. negative?

31–39 In Fig. 31–33, how much is the DC drain current?

31–40 In Fig. 31–33, what value of R_D will produce a drain-source voltage, V_{DS}, of 18 V?

31–41 In Fig. 31–33, how much is V_{DS} if R_D equals

a. 470 Ω?
b. 820 Ω?
c. 2 kΩ?

31–42 In Fig. 31–33, calculate A_V and V_{out} using the value of R_D calculated in Prob. 31–40.

31–43 What type of biasing arrangements will not work with E-MOSFETs?

31–44 Which type of bias is shown in Fig. 31–34?

31–45 In Fig. 31–34, calculate the value of R_D that will provide an $I_{D(on)}$ of 10 mA for each of the following values of V_{DD}:

a. $V_{DD} = 12$ V.
b. $V_{DD} = 18$ V.
c. $V_{DD} = 24$ V.
d. $V_{DD} = 36$ V.

Figure 31–34

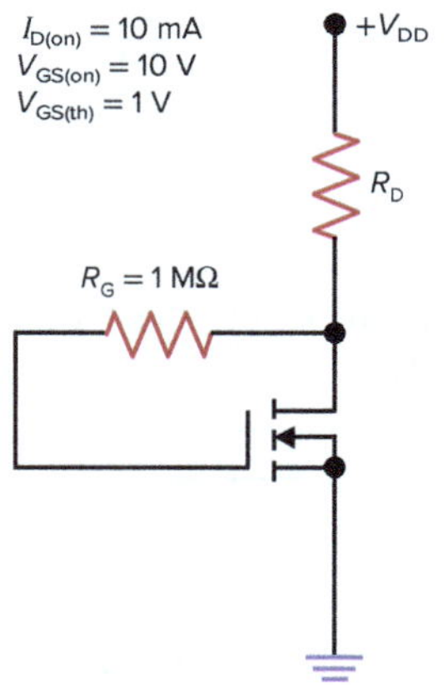

SECTION 31–6 HANDLING MOSFETS

31–46 In Fig. 31–24, what is the purpose of the protective diodes?

31–47 What is one drawback of the protective diodes in Fig. 31–24?

Answers to Self-Reviews

31–1
- **a.** voltage-controlled
- **b.** a JFET where the drain and source terminals may be interchanged
- **c.** the ohmic region
- **d.** the ohmic and current-source regions
- **e.** because drain current flows when $V_{GS} = 0$ V

31–2
- **a.** gate bias
- **b.** because the gate current, I_G, is zero
- **c.** more positive

31–3
- **a.** yes
- **b.** the common-drain amplifier
- **c.** the common-source amplifier
- **d.** the source follower (common drain)
- **e.** its low Z_{in}

31–4
- **a.** normally ON
- **b.** normally OFF
- **c.** the enhancement and depletion modes
- **d.** their use in digital computers

31–5
- **a.** I_{DSS}
- **b.** zero

31–6
- **a.** the SiO_2 insulator separating the gate from the channel
- **b.** it helps bleed off any buildup of static electric charge on the person handling the MOSFET

Laboratory Application Assignment

In this lab application assignment, you will examine a biasing technique commonly used with JFETs known as self-bias. Due to the wide range over which JFET parameters can vary, predicting exact circuit values can be difficult. Therefore, this experiment is unique in that you will not make any circuit calculations, only measurements. You will build a common-source amplifier and determine its voltage gain, A_V.

Equipment: Obtain the following items from your instructor.

- Two MPF102 *n*-channel JFETs or equivalent
- 0.1-μF capacitor and two 100-μF electrolytic capacitors
- Assortment of carbon-film resistors
- Oscilloscope
- DMM
- Variable dc power supply
- Function generator

Self-Bias

Construct the circuit in Fig. 31–35*a*. Measure and record the following DC values:

V_G = __________, I_D = __________, V_S = __________,
V_D = __________, V_{DS} = __________

Based on your measured values of V_G and V_S, calculate V_{GS}.
V_{GS} = __________.

Replace the JFET in Fig. 31–35*a* with a different MPF102, and repeat the same measurements.

V_G = __________, I_D = __________, V_S = __________,
V_D = __________, V_{DS} = __________

Based on your measured values of V_G and V_S, calculate V_{GS}.
V_{GS} = __________

Was there any difference in V_{GS} and I_D from one JFET to the next?__________

Common-Source Amplifier

Modify the circuit in Fig. 31–35*a* to that shown in Fig. 31–35*b*. (Which JFET you use doesn't matter.) Connect channel 1 of the oscilloscope to the gate and channel 2 to the load resistor, R_L. With the input voltage, V_{in}, adjusted to exactly 100 $mV_{p\text{-}p}$, measure and record the peak-to-peak output voltage. $V_{out(p\text{-}p)}$ = __________ Calculate the voltage gain, A_V, based on the measured values of V_{out} and V_{in}. A_V = __________
Remove the source bypass capacitor, C_S, and remeasure $V_{out(p\text{-}p)}$. $V_{out(p\text{-}p)}$ = __________ Also, recalculate A_V. A_V = __________
How did removing the source bypass capacitor affect the voltage gain, A_V? ______________________________
____

Measure and record the phase relationship between V_{in} and V_{out}. θ = __________
In general, is the voltage gain of a common-source amplifier less than or greater than that of a common-emitter amplifier?______
____

Figure 31-35

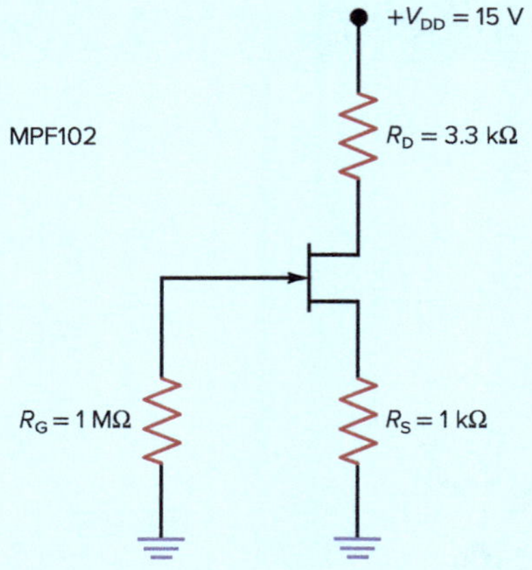

(a)

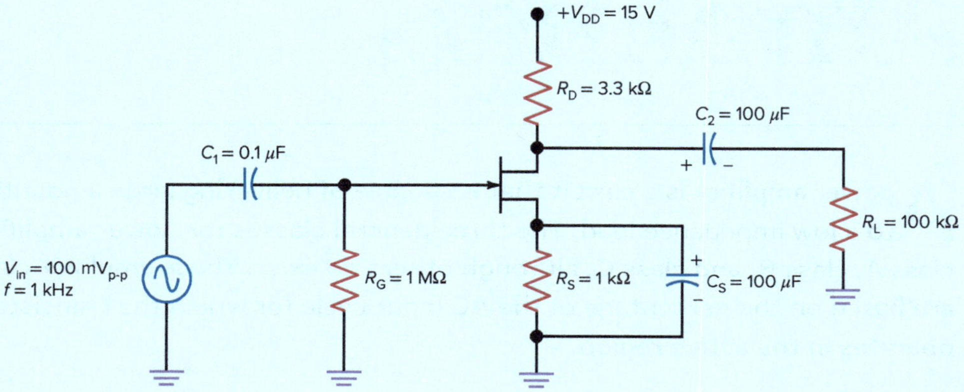

(b)

Design credits: Chapter Opener Background Image, and Multisim: Stockbyte/Getty Images

chapter 32

Power Amplifiers

A **power amplifier** is a circuit that is capable of delivering large amounts of power to a low impedance load. The three general classes for power amplifiers are class A, class B, and class C, although others do exist. These amplifier classifications are based on the percentage of the AC input cycle for which the transistor amplifier operates in the active region.

This chapter discusses how to calculate the AC load power, transistor power dissipation, DC input power, and percent efficiency for a power amplifier.

Chapter Outline

Chapter Objectives

After studying this chapter, you should be able to

- *Define* the different classes of operation for a transistor amplifier.
- *Calculate* the AC load power, DC input power, and percent efficiency of a class A amplifier.
- *Draw* the AC load line for an *RC*-coupled class A amplifier.
- *Explain* the operation of a class B push-pull amplifier.
- *Calculate* the DC quantities in a class B push-pull amplifier.
- *Explain* the advantage of using diode bias instead of standard resistor biasing in a class B push-pull amplifier.
- *Define* the term crossover distortion and *explain* why it occurs in a class B push-pull amplifier.
- *Calculate* the AC load power, DC input power, and percent efficiency of a class B push-pull amplifier.
- *Explain* the operation of a class C amplifier with a tuned *LC* tank circuit in the collector.
- *Explain* why the DC base voltage is negative for a class C amplifier using an *npn* transistor.
- *Explain* how a class C rf amplifier can be used as a frequency multiplier.

Important Terms

AC load line
AC load power, P_L
class A amplifier
class B amplifier
class C amplifier
class B push-pull amplifier
crossover distortion
DC input power, P_{CC}
diode bias
frequency multiplier
linear amplifier
percent efficiency
power amplifier

32–1 Classes of Operation

The class of operation for an amplifier is defined by the percentage of the AC input cycle that produces an output current. The class of operation for an amplifier determines its power efficiency. It also determines how much the input signal is distorted by the amplifier. Figure 32–1 shows typical input and output waveforms for class A, B, and C transistor amplifiers.

Class A Operation

The collector current, I_C, of a transistor in a **class A amplifier** flows for the full 360° of the input waveform, as shown in Fig. 32–1*a* and *b*. Figure 32–1*a* shows the AC signal voltage driving the base of the transistor, and Fig. 32–1*b* shows the resultant collector current, I_C.

A class A amplifier is one that is used as a **linear amplifier**, that is, the circuit must produce an output signal, although amplified, that is an exact replica of the input signal. The input signal must never drive the transistor into either cutoff or saturation. If it does, the output waveform will be clipped off at one or both of its peaks. For class A operation, the DC bias should provide a quiescent collector

GOOD TO KNOW

Between class A and class B is class AB operation. The collector current, I_C, of a transistor in a class AB amplifier flows for 210° of the AC input cycle.

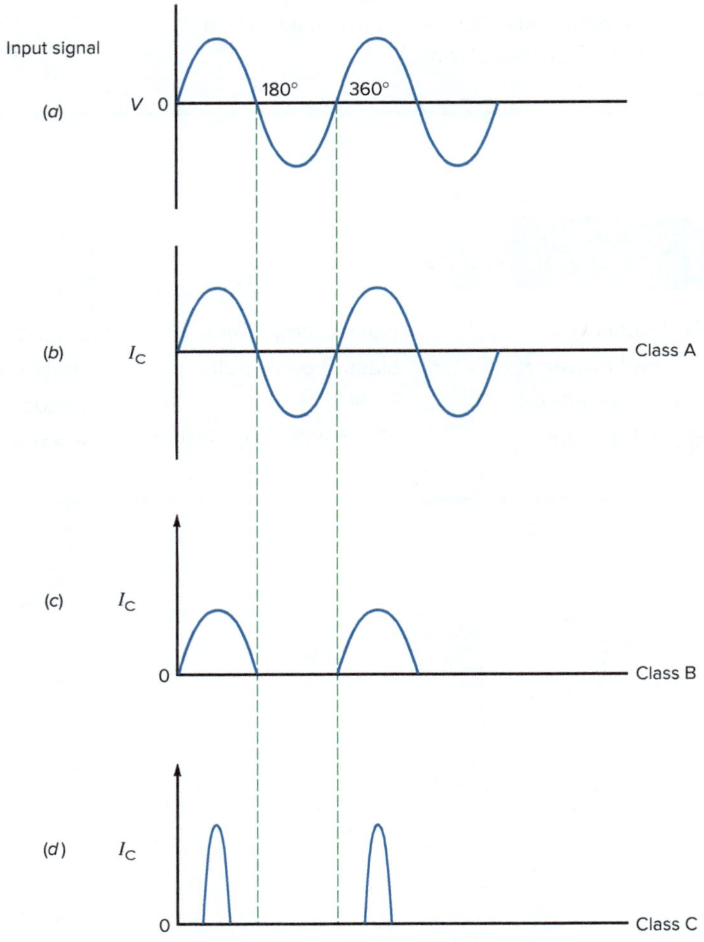

Figure 32–1 Class of operation for transistor amplifiers in terms of the conduction angle. (*a*) Sine wave of input voltage. Two full cycles are shown. (*b*) Collector current, I_C, flows for 360° of the input cycle in a class A amplifier. (*c*) Collector current, I_C, flows for 180° of the input cycle in a class B amplifier. (*d*) Collector current, I_C, flows for 120° or less of the input cycle in a class C amplifier.

current, I_C, that is approximately one-half its maximum value at saturation. Then the AC output signal can swing above and below this value without the transistor immediately reaching either cutoff or saturation. The characteristics of a class A amplifier include both low distortion and low power efficiency.

GOOD TO KNOW

Another class of operation is class D. In a class D amplifier, the transistors operate in either saturation or cutoff. Since the transistors are being used as switches, the power dissipation is nearly zero. Therefore, the efficiency of a class D amplifier is even higher than that of a class C amplifier. In a class D amplifier, 100% efficiency is almost attainable.

Class B Operation

The collector current, I_C, of a transistor in a **class B amplifier** flows for only 180° of the input waveform. Because of this, the waveform at the output of a class B amplifier is badly distorted. The input and output waveforms for a class B amplifier are shown in Fig. 32–1*a* and *c*. The collector current, I_C, flows only for 180° of the input cycle because the DC bias for the transistor produces a quiescent collector current, I_C, of zero. In other words, the transistor in a class B amplifier is biased right at cutoff. During the half-cycle when the transistor does conduct the collector current, I_C may or may not increase to its maximum value at saturation.

Class B operation with a single transistor corresponds to half-wave rectification of the input signal. When the input signal makes the transistor conduct, this half of the input is amplified linearly and is a good replica of that corresponding half of the input cycle. If the class B amplifier must yield a symmetrical output in special cases, two transistors can be used to provide opposite half-cycles of the signal at the output. The characteristics of a class B amplifier using a single transistor include medium efficiency and severe distortion.

Class C Operation

The collector current, I_C, of a transistor in a **class C amplifier** flows for less than 180° of the input waveform, which distorts the output waveform from the amplifier. The input and output waveforms of a class C amplifier are shown in Fig. 32–1*a* and *d*. The typical operation of a class C amplifier provides a collector current that flows for approximately 120° or less of the AC input cycle. In some cases, the collector current, I_C, flows in very short narrow pulses where the conduction angle of the transistor is 30° or less. The collector current, I_C, flows for less than 180° of the input cycle because the transistor is biased beyond cutoff. Thus, part of the input signal must be used to overcome the DC bias before the transistor can conduct. Class C operation is generally used for rf amplifiers with a tuned or resonant tank circuit in the output. The *LC* tank circuit is capable of reproducing the full sine-wave cycle at the output for each short pulse of collector current.

The characteristics of a class C amplifier include very high efficiency (approaching 100%) and severe distortion of the input signal. However, in the case of distortion, a tank circuit can be used to reproduce the full sine wave at the output.

32–1 Self-Review

Answers at the end of the chapter.

a. **In a class A amplifier, the collector current flows for 360° of the AC input cycle. (True/False)**
b. **The transistor in a true class B amplifier is biased right at cutoff. (True/False)**
c. **Class C amplifiers cannot be used as tuned rf amplifiers. (True/False)**

32–2 Class A Amplifiers

All of the small signal amplifiers covered so far in this text have been biased to operate as class A amplifiers. The input signal amplitude of any class A amplifier should not be large enough to drive the transistor into either cutoff or saturation. If the signal amplitude at the input is too large, either or both peaks of the output waveform will be clipped off (flattened).

Analyzing the Class A Amplifier

Figure 32–2*a* shows a common-emitter class A amplifier. For simplicity, the transistor is biased using base bias. The base resistor, R_B, is a variable resistor adjusted to a value equal to 190.67 kΩ. This provides a *Q* point located at the center of the DC load line.

To begin the analysis, calculate the DC quantities. Begin by calculating the DC base current, I_B:

$$I_B = \frac{V_{CC} - V_{BE}}{R_B}$$
$$= \frac{15\text{ V} - 0.7\text{ V}}{190.67\text{ k}\Omega}$$
$$= 75\ \mu\text{A}$$

Next, calculate the collector current, I_C:

$$I_C = I_B \times \beta_{DC}$$
$$= 75\ \mu\text{A} \times 100$$
$$= 7.5\text{ mA}$$

Finally, calculate the collector-emitter voltage, V_{CE}:

$$V_{CE} = V_{CC} - I_C R_C$$
$$= 15\text{ V} - (7.5\text{ mA} \times 1\text{ k}\Omega)$$
$$= 15\text{ V} - 7.5\text{ V}$$
$$= 7.5\text{ V}$$

The endpoints for the DC load line can be calculated as:

$$V_{CE(off)} = V_{CC}$$
$$= 15\text{ V}$$

$$I_{C(sat)} = \frac{V_{CC}}{R_C}$$
$$= \frac{15\text{ V}}{1\text{ k}\Omega}$$
$$= 15\text{ mA}$$

Figure 32–2 Common-emitter class A amplifier. (*a*) Circuit. (*b*) DC load line.

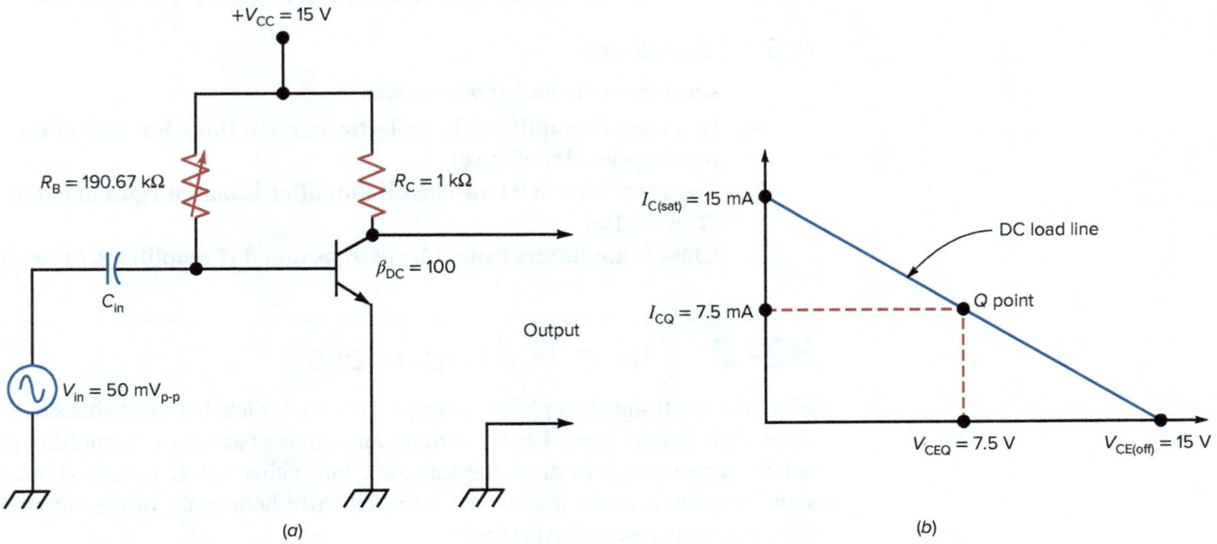

Note that when the transistor is cut off ($I_C = 0$ mA), the collector-emitter region appears open and the voltage $V_{CE} = V_{CC}$, which is 15 V in this case. Conversely, when the transistor is saturated, the collector-emitter region appears shorted ($V_{CE} = 0$ V) and only the values of V_{CC} and R_C limit I_C.

Figure 32–2*b* shows the values for $V_{CE(off)}$ and $I_{C(sat)}$. The DC load line also shows the quiescent (*Q* point) values for I_C and V_{CE}. Note that these values are designated I_{CQ} and V_{CEQ}. Notice also in Fig. 31–2*b* that the *Q* point is centered exactly on the DC load line. Because the *Q* point is exactly centered, the maximum possible peak-to-peak output voltage can be obtained from the amplifier. In this case, the collector-emitter voltage, V_{CE}, can swing ±7.5 V from its *Q* point value of 7.5 V without any clipping.

In Fig. 32–2*a*, the voltage gain, A_V, is calculated as

$$A_V = \frac{R_C}{r'_e}$$

In Fig. 32–2*a*, $I_E \approx I_C = 7.5$ mA. Therefore, r'_e is calculated as

$$\begin{aligned} r'_e &= \frac{25 \text{ mV}}{I_E} \\ &= \frac{25 \text{ mV}}{7.5 \text{ mA}} \\ &= 3.33\ \Omega \end{aligned}$$

The voltage gain, A_V, can now be calculated:

$$\begin{aligned} A_V &= \frac{R_C}{r'_e} \\ &= \frac{1\text{ k}\Omega}{3.33\ \Omega} \\ &= 300 \end{aligned}$$

With A_V known, V_{out} is calculated as

$$\begin{aligned} V_{out} &= A_V \times V_{in} \\ &= 300 \times 50 \text{ mV}_{p\text{-}p} \\ &= 15 \text{ V}_{p\text{-}p} \end{aligned}$$

With 15 $V_{p\text{-}p}$ at the output, the entire DC load line is used. On the positive alternation of output voltage, V_{CE} increases from 7.5 V to 15 V. On the negative alternation, V_{CE} decreases from 7.5 V to 0 V. Even though clipping will not occur, driving the amplifier this hard is going to cause extreme distortion of the input signal being amplified due to the nonlinearity of the emitter diode. However, for the analysis here, assume that the output of 15 $V_{p\text{-}p}$ is a pure undistorted waveform.

Transistor Power Dissipation

With no AC input signal applied to the amplifier, the transistor has a power dissipation of

$$P_d = V_{CEQ} \times I_{CQ}$$

This power dissipation must not exceed the power rating, $P_{d(max)}$, of the transistor.

In Fig. 32–2*a,* the transistor power dissipation, P_d, is

$$\begin{aligned} P_d &= V_{CEQ} \times I_{CQ} \\ &= 7.5 \text{ V} \times 7.5 \text{ mA} \\ &= 56.25 \text{ mW} \end{aligned}$$

The P_d of 56.25 mW represents the maximum power dissipation of the transistor in Fig. 32–2*a*. In a class A amplifier, the power dissipation in the transistor decreases when an AC signal is applied to the input.

AC Load Power, P_L

In Fig. 32–2*a*, the AC load power equals the power dissipated by the collector resistance, R_C. Therefore, P_L is calculated using the following formula:

$$P_L = \frac{V^2_{out(p\text{-}p)}}{8R_C} \qquad \textbf{(32–1)}$$

In Fig. 32–2*b*, P_L calculated as

$$P_L = \frac{V^2_{out(p\text{-}p)}}{8R_C}$$
$$= \frac{15\ V^2_{p\text{-}p}}{8\ k\Omega}$$
$$= 28.125\ mW$$

DC Input Power, P_{CC}

The DC power supplied to the class A amplifier in Fig. 32–2*a* is the product of V_{CC} and the total DC current drain from the power supply. The total DC current drain is designated I_{CC}. This gives the following formula:

$$P_{CC} = V_{CC} \times I_{CC} \qquad \textbf{(32–2)}$$

where P_{CC} represents the DC power supplied to the class A amplifier.

In Fig. 32–2*a*, I_B is 100 times smaller than I_C and therefore $I_{CC} \approx I_C$. Since $I_{CC} \approx I_C = 7.5$ mA in Fig. 32–2*a*, P_{CC} is calculated as

$$P_{CC} = 15\ V \times 7.5\ mA$$
$$= 112.5\ mW$$

Remember that the base current, I_B, can be ignored because I_B is 100 times smaller than the collector current, I_C.

Percent Efficiency

The **percent efficiency** of any amplifier is defined as the percentage of the DC input power (P_{CC}) that is converted to useful AC power output. This is expressed in Formula (32–3):

$$\text{Percent efficiency} = \frac{P_L}{P_{CC}} \times 100 \qquad \textbf{(32–3)}$$

In Fig. 32–2*a*, the calculations are

$$\text{Percent efficiency} = \frac{P_L}{P_{CC}} \times 100$$
$$= \frac{28.125\ mW}{112.5\ mW} \times 100$$
$$= 25\%$$

The maximum theoretical efficiency possible for a class A amplifier using a single collector resistor, R_C, is 25%.

RC-Coupled Class A Amplifier

Figure 32–3*a* shows the addition of a load resistor, R_L. This circuit is called an *RC*-coupled amplifier because the AC voltage at the collector is capacitively coupled to the load resistor R_L. There are two loads for this type of amplifier: a DC load and

Figure 32–3 *RC*-coupled class A amplifier. (*a*) Circuit. (*b*) DC and AC load lines.

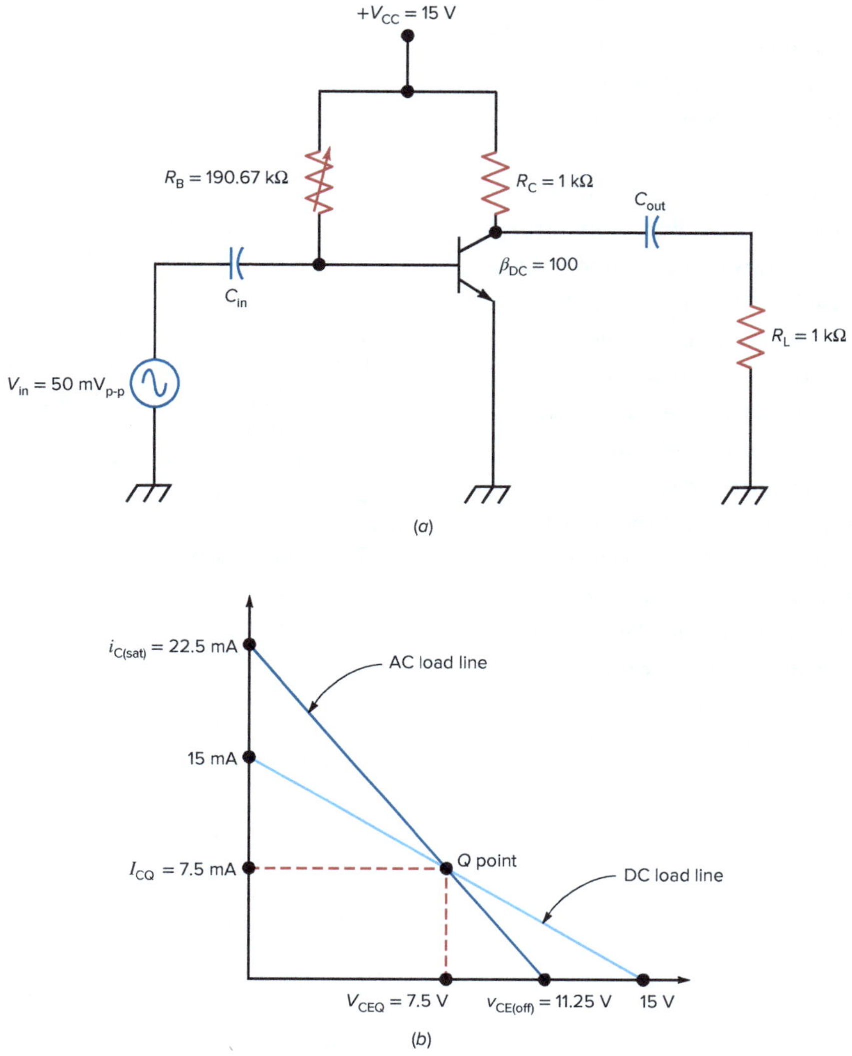

an AC load. This implies that there are two load lines: a DC load line and an **AC load line.** The DC load line shown in Fig. 32–3*b* has the same endpoints as calculated previously with Fig. 32–2 because the DC quantities do not change with the addition of R_L.

AC Load Line

When the AC load resistance is known, the endpoints for the AC load line can be determined. In Fig. 32–3*a*, the AC load resistance, $r_L = R_C \parallel R_L = 500\ \Omega$.

To calculate the endpoints of the AC load line in Fig. 32–3*b*, use Formulas (32–4) and (32–5):

$$i_{C(sat)} = I_{CQ} + \frac{V_{CEQ}}{r_L} \qquad \textbf{(32–4)}$$

$$v_{CE(off)} = V_{CEQ} + I_{CQ} r_L \qquad \textbf{(32–5)}$$

[The derivation of Formulas (32–4) and (32–5) is quite lengthy and is therefore not covered in this text.]

GOOD TO KNOW

The voltage gain and output voltage of an amplifier are affected by the following capacitances: coupling capacitors, bypass capacitors, internal transistor capacitance, and stray wiring capacitance. The coupling and bypass capacitors affect the output voltage at low frequencies because their X_C begins to increase substantially. The internal transistor capacitance and stray wiring capacitance affect the output voltage at higher frequencies because their X_C begins to decrease substantially. A graph of output voltage versus frequency for an amplifier is called its frequency response.

For Fig. 32–3*b*, $i_{C(sat)}$ and $v_{CE(off)}$ are calculated as shown:

$$\begin{aligned} i_{C(sat)} &= I_{CQ} + \frac{V_{CEQ}}{r_L} \\ &= 7.5 \text{ mA} + \frac{7.5 \text{ V}}{500 \, \Omega} \\ &= 7.5 \text{ mA} + 15 \text{ mA} \\ &= 22.5 \text{ mA} \end{aligned}$$

$$\begin{aligned} v_{CE(off)} &= V_{CEQ} + I_{CQ} r_L \\ &= 7.5 \text{ V} + (7.5 \text{ mA} \times 500 \, \Omega) \\ &= 7.5 \text{ V} + 3.75 \text{ V} \\ &= 11.25 \text{ V} \end{aligned}$$

These values for $v_{CE(off)}$ and $i_{C(sat)}$ are shown on the AC load line in Fig. 32–3*b*.

In Fig. 32–3*b*, notice that the *Q* point is centered on the DC load line but not on the AC load line. The *Q* point for the AC load line is below center. Because of this, the collector-emitter voltage, V_{CE}, can change only from 7.5 V to 11.25 V in the positive direction, which is a 3.75 V excursion. If V_{CE} tries to increase beyond 11.25 V, clipping will occur. Therefore, for the circuit shown, the maximum unclipped peak-to-peak output voltage equals $2 \times 3.75 \text{ V}_{p\text{-}p} = 7.5 \text{ V}_{p\text{-}p}$.

To calculate the voltage gain A_V with R_L connected, proceed as follows:

$$\begin{aligned} A_V &= \frac{r_L}{r'_e} \\ &= \frac{500 \, \Omega}{3.33 \, \Omega} \\ &= 150 \end{aligned}$$

Therefore,

$$\begin{aligned} V_{out} &= A_V \times V_{in} \\ &= 150 \times 50 \text{ mV}_{p\text{-}p} \\ &= 7.5 \text{ V}_{p\text{-}p} \end{aligned}$$

With $V_{out(p\text{-}p)}$ known, the AC load power can be calculated using Formula (32–6):

$$P_L = \frac{V^2_{out(p\text{-}p)}}{8 \, R_L} \qquad \textbf{(32–6)}$$

Note: R_L is the load driven by the amplifier.

In Fig. 32–3*a*, P_L is calculated as

$$\begin{aligned} P_L &= \frac{7.5 \text{ V}^2_{p\text{-}p}}{8 \text{ k}\Omega} \\ &\approx 7.031 \text{ mW} \end{aligned}$$

Since P_{CC} was calculated earlier (in Fig. 32–2) as 112.5 mW, the percent efficiency is calculated as:

$$\begin{aligned} \text{Percent efficiency} &= \frac{P_L}{P_{CC}} \times 100 \\ &= \frac{7.031 \text{ mW}}{112.5 \text{ mW}} \times 100 \\ &= 6.25\% \end{aligned}$$

Notice the significant drop in the efficiency of the amplifier. This is due to the fact that the AC output voltage has been reduced by a factor of 2, while the DC input power remains the same.

If a slight increase in the AC output power is desired, the *Q* point must be centered on the AC load line. Engineers, or technicians, will not usually worry about this unless they want to obtain the absolute maximum possible AC load power. For a small signal class A amplifier, this is usually not a major concern.

One more point: The maximum possible efficiency of an RC-coupled class A amplifier cannot exceed 8.33% no matter what is done. To obtain an efficiency of 8.33%, R_C must equal R_L and the Q point must be centered on the AC load line. Also, the power losses in the biasing resistors must be insignificant with respect to the DC power consumed in the collector circuit.

Example 32-1

In Fig. 32-4, calculate the following DC quantities: I_{CQ}, V_{CEQ}, P_d, $I_{C(sat)}$, and $V_{CE(off)}$. Also, draw the DC load line.

ANSWER Begin by calculating the DC voltage at the base and emitter terminals:

$$V_B = \frac{R_2}{R_1 + R_2} \times V_{CC}$$
$$= \frac{2.7\text{ k}\Omega}{18\text{ k}\Omega + 2.7\text{ k}\Omega} \times 20\text{ V}$$
$$= 2.6\text{ V}$$

MultiSim **Figure 32-4** Common-emitter class A amplifier used for Example 32-1. (*a*) Circuit. (*b*) DC and AC load lines.

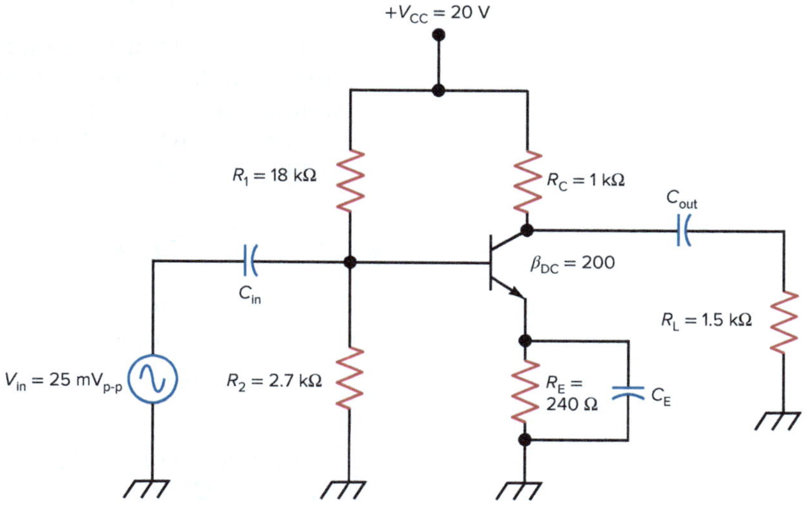

(*a*)

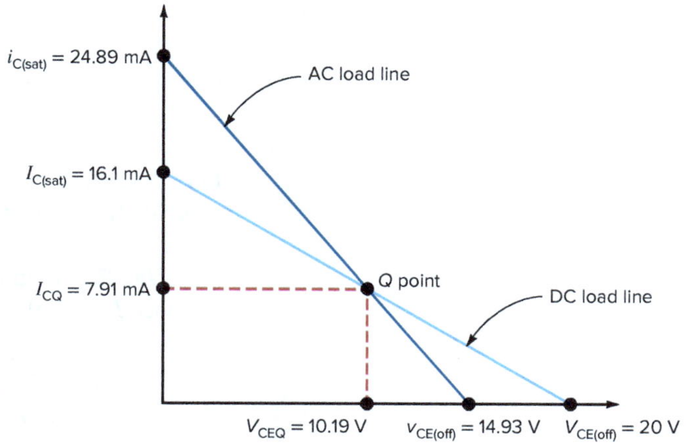

(*b*)

$$V_E = V_B - V_{BE}$$
$$= 2.6\text{ V} - 0.7\text{ V}$$
$$= 1.9\text{ V}$$

Since $I_E \approx I_C$,

$$I_{CQ} = \frac{V_E}{R_E}$$
$$= \frac{1.9\text{ V}}{240\ \Omega}$$
$$= 7.91\text{ mA}$$

V_{CEQ} is calculated as

$$V_{CEQ} = V_{CC} - I_{CQ}(R_C + R_E)$$
$$= 20\text{ V} - (7.91\text{ mA} \times 1.24\text{ k}\Omega)$$
$$= 10.19\text{ V}$$

The transistor power dissipation, P_d, is

$$P_d = V_{CEQ} \times I_{CQ}$$
$$= 10.19\text{ V} \times 7.91\text{ mA}$$
$$= 80.6\text{ mW}$$

Remember that the transistor power dissipation, P_d, for a class A amplifier is maximum when there is no input signal. As the signal voltage at the collector increases, the amount of power dissipated by the transistor decreases.

The DC load line is shown in Fig. 32–4*b*. The values for the endpoints $I_{C(sat)}$ and $V_{CE(off)}$ are calculated as

$$I_{C(sat)} = \frac{V_{CC}}{R_C + R_E}$$
$$= \frac{20\text{ V}}{1.24\text{ k}\Omega}$$
$$= 16.1\text{ mA}$$
$$V_{CE(off)} = V_{CC}$$
$$= 20\text{ V}$$

The values for $I_{C(sat)}$, $V_{CE(off)}$ as well as I_{CQ} and V_{CEQ} are shown on the DC load line in Fig. 32–4*b*.

Example 32–2

In Fig. 32–4, calculate the following AC quantities: A_V, V_{out}, P_L, P_{CC}, and percent efficiency. Also, calculate the endpoints for the AC load line.

ANSWER Begin by calculating the value for r'_e, and r_L. First, find r'_e. Since $I_{CQ} = 7.91$ mA, then,

$$r'_e = \frac{25\text{ mV}}{7.91\text{ mA}}$$
$$= 3.16\ \Omega$$

Next, calculate the AC load resistance, r_L:

$$r_L = \frac{R_C \times R_L}{R_C + R_L}$$
$$= \frac{1\ k\Omega \times 1.5\ k\Omega}{1\ k\Omega + 1.5\ k\Omega}$$
$$= 600\ \Omega$$

Knowing r_L and r'_e, now calculate the voltage gain, A_V.

$$A_V = \frac{r_L}{r'_e}$$
$$= \frac{600\ \Omega}{3.16\ \Omega}$$
$$\approx 190$$

Next, calculate V_{out}:

$$V_{out} = A_V \times V_{in}$$
$$= 190 \times 25\ mV_{p\text{-}p}$$
$$= 4.75\ V_{p\text{-}p}$$

With V_{out} known, calculate the **AC load power, P_L**:

$$P_L = \frac{V^2_{out(p\text{-}p)}}{8\ R_L}$$
$$= \frac{4.75\ V^2_{p\text{-}p}}{12\ k\Omega}$$
$$= 1.88\ mW$$

The DC input power is calculated as

$$P_{CC} = V_{CC} \times I_{CC}$$

where I_{CC} is the total DC current drain from the power supply, V_{CC}. I_{CC} equals the sum of the collector current, I_C, and the current through the base voltage divider, consisting of R_1 and R_2. I_{CC} is calculated as

$$I_{CC} = I_{V\text{-}d} + I_C$$

where

$$I_{V\text{-}d} = \frac{V_{CC}}{R_1 + R_2}$$
$$= \frac{20\ V}{18\ k\Omega + 2.7\ k\Omega}$$
$$= 966\ \mu A$$

Since $I_C = 7.91$ mA, I_{CC} is

$$I_{CC} = 966\ \mu A + 7.91\ mA$$
$$= 8.87\ mA$$

With I_{CC} known, P_{CC} is calculated as

$$P_{CC} = V_{CC} \times I_{CC}$$
$$= 20\ V \times 8.87\ mA$$
$$= 177.4\ mW$$

With P_L and P_{CC} known, the percent efficiency can be calculated:

$$\text{Percent efficiency} = \frac{P_L}{P_{CC}} \times 100$$
$$= \frac{1.88 \text{ mW}}{177.4 \text{ mW}} \times 100$$
$$\cong 1\%$$

Notice the extremely low efficiency. Remember, even under ideal conditions, the maximum theoretical efficiency of an *RC*-coupled class A amplifier is 8.33%. An efficiency of 1% means that only 1% of the DC input power, P_{CC}, is converted to useful AC power output.

Finally, calculate the endpoints for the AC load line using Formulas (32–4) and (31–5):

$$i_{C(sat)} = I_{CQ} + \frac{V_{CEQ}}{r_L}$$
$$= 7.91 \text{ mA} + \frac{10.19 \text{ V}}{600\ \Omega}$$
$$= 24.89 \text{ mA}$$

$$v_{CE(off)} = V_{CEQ} + I_{CQ}r_L$$
$$= 10.19 \text{ V} + (7.91 \text{ mA} \times 600\ \Omega)$$
$$= 14.93 \text{ V}$$

Figure 32–4*b* shows the AC load line with the calculated values for $i_{C(sat)}$ and $v_{CE(off)}$. It is important to note that the output voltage cannot increase to a peak value greater than 14.93 V, which means that the maximum positive excursion from the *Q* point is 14.93 V − 10.19 V = 4.74 V. Since V_{CE} can also decrease by the same amount without clipping, the maximum unclipped output from the circuit equals 2 × 4.74 V = 9.48 $V_{p\text{-}p}$. V_{out} cannot increase beyond this value without having the positive output peak flattened. Since V_{out} = 4.75 $V_{p\text{-}p}$, as calculated earlier, the circuit is operating without the possibility of reaching either cutoff or saturation on the AC load line.

32–2 Self-Review

Answers at the end of the chapter.

a. Class A amplifiers are nearly 100% efficient. (True/False)

b. In a class A amplifier, the power dissipation in the transistor decreases as the peak-to-peak output voltage increases. (True/False)

32–3 Class B Push-Pull Amplifiers

The collector current, I_C, of a transistor in a class B amplifier flows for 180° of the AC input cycle. For the other 180°, the transistor is cut off. A true class B amplifier is biased such that the *Q* point is located right at cutoff. The main advantage of class B operation versus class A operation is that class B operation is more efficient, that is, more AC load power can be obtained for the same amount of DC input power. The main disadvantage of class B operation, however, is that two transistors must be used to get a linear reproduction of the input waveform being amplified.

Figure 32–5*a* shows a **class B push-pull amplifier**. The transistors, Q_1 and Q_2, conduct during opposite half-cycles of the input waveform. When V_{in} is positive, Q_1

Figure 32–5 Class B push-pull amplifier. (*a*) Circuit. (*b*) DC and AC load lines. (*c*) Crossover distortion at cutoff.

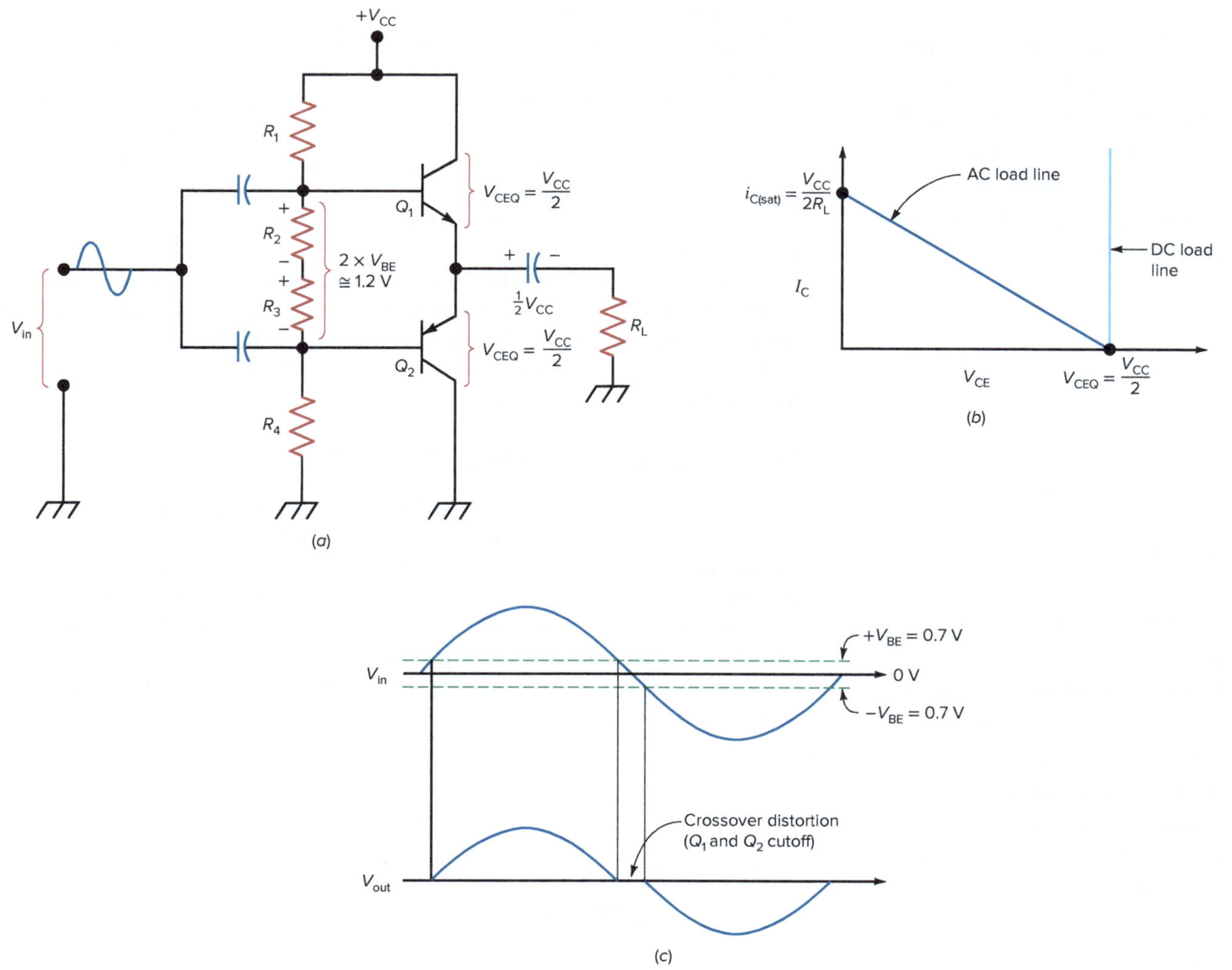

conducts and Q_2 is cut off. Conversely, when V_{in} is negative, Q_2 conducts and Q_1 is cut off. Careful examination of the circuit reveals that each transistor acts like an emitter follower for one half-cycle of the input voltage.

The biasing resistors $R_1 - R_4$ are selected to set the Q point right at cutoff. Ideally, the quiescent collector current, I_C, should be zero. Because both transistors are in series, V_{CEQ} for each transistor equals approximately one-half of V_{CC} (see Fig. 32–5*a*). Notice also that the voltage drop across R_2 and R_3 is approximately 1.2 V, which is assumed to be below the value required to turn on each transistor.

The DC and AC load lines are shown in Fig. 32–5*b*. Notice that the DC load line is perfectly vertical. With no AC input signal, both transistors, Q_1 and Q_2, are cut off, and one-half of V_{CC} appears across the collector-emitter region of each transistor. When both transistors are saturated, the collector current, I_C, increases to infinity, which is why the DC load line is shown to be perfectly vertical.

Formulas (32–4) and (32–5) still apply to the AC load line of a class B push-pull amplifier. Since $I_{CQ} \approx 0$, however, the endpoints for $i_{C(sat)}$ and $v_{CE(off)}$ can be shown as

$$i_{C(sat)} = \frac{V_{CC}}{2R_L} \tag{32–7}$$

$$v_{CE(off)} = \frac{V_{CC}}{2} \tag{32–8}$$

GOOD TO KNOW

To provide a linear output, many class B push-pull audio amplifiers are biased to operate as class AB amplifiers with a conduction angle near 210°. With class AB operation, crossover distortion is entirely eliminated.

Ideally, the maximum peak-to-peak output voltage obtainable in Fig. 32–5*a* equals the value of V_{CC}. If, for example, $V_{CC} = 15$ V, then the maximum peak-to-peak output voltage would be 15 $V_{p\text{-}p}$.

Figure 32–5*c* shows the problem with biasing the transistors exactly at cutoff. When V_{in} crosses through zero, Q_1 and Q_2 are both cut off, resulting in a time when the output voltage does not follow the input voltage because both transistors are still cut off. The effect is called ***crossover distortion***. Crossover distortion is undesirable because it produces a distortion that can be heard in the speaker output. Figure 32–5*c* is somewhat exaggerated because R_2 and R_3 bias Q_1 and Q_2 only slightly below cutoff. Therefore, the crossover distortion would not be as severe as the illustration in Fig. 32–5*c*.

The class B push-pull amplifier in Fig. 32–5 is extremely sensitive to changes in temperature. Small changes in operating temperature can produce extreme changes in the collector current, I_C, of each transistor. This is highly undesirable. In most cases, voltage divider bias is not used with class B push-pull amplifiers because thermal runaway can destroy the transistor.

Typical Class B Push-Pull Amplifier

GOOD TO KNOW

In actual designs, the compensating diodes are mounted on the case of the power transistors so that as the transistors heat up so do the diodes. The diodes are usually mounted to the power transistors with a nonconductive adhesive that has good thermal transfer characteristics. Since the diode curves match the V_{BE} versus I_E curve of each transistor, the circuit is almost immune to changes in temperature.

Figure 32–6 shows how a typical class B push-pull amplifier would be biased. This form of bias is called ***diode bias***. The diodes, D_1, and D_2, produce the required bias for the base-emitter junction of each transistor. For this bias method to work properly, the I_F versus V_F curve of each diode must match the V_{BE} versus I_E curves of each transistor. Because the series combination of D_1 and D_2 is in parallel with the emitter diodes of Q_1 and Q_2, both series combinations have the same voltage drop. Because the diode curves match the V_{BE} curves of the transistors, the diode currents and emitter currents are the same. Therefore, the collector current, I_C, in both transistors can be calculated using Formula (32–9):

$$I_{CQ} = \frac{V_{CC} - 2\,V_{BE}}{2R} \qquad \textbf{(32–9)}$$

In Fig. 32–6, I_{CQ} is

$$I_{CQ} = \frac{24\text{ V} - 1.4\text{ V}}{2 \times 2.7\text{ k}\Omega} = 4.18\text{ mA}$$

In Fig. 32–6*a*, the collector-emitter voltage, V_{CE}, of each transistor equals one-half of V_{CC}, which is 12 V in this case. Because of this, the DC voltage at the emitter junction also equals 12 V. The DC voltage at the base of Q_1 is 12 V + 0.7 V = 12.7 V, and at the base of Q_2, the DC voltage is 12 V − 0.7 V = 11.3 V.

To calculate the quiescent power dissipation in Q_1 and Q_2, proceed as follows:

$$P_{dq} = V_{CEQ} \times I_{CQ} \qquad \textbf{(32–10)}$$

$$= 12\text{ V} \times 4.18\text{ mA}$$

$$= 50.16\text{ mW}$$

Since I_{CQ} is usually quite small, the value of P_{dq} is also small. This means that a class B push-pull amplifier will run very cool when the input signal is zero. For diode bias to be insensitive to changes in temperature, the diode curves must match the emitter diode curves of the transistors. Diode bias is one of the best ways to bias a class B push-pull amplifier.

Load Current Paths

When V_{in} is positive, Q_1 conducts and Q_2 cuts off (see Fig. 32–6*b*). Since Q_1 acts like an emitter follower, the AC signal voltage at the base and emitter are the same. Notice that the output coupling capacitor, C_{out}, is charging during the positive

Figure 32–6 Class B push-pull amplifier using diode bias. (*a*) Circuit. (*b*) Load current path when V_{in} is positive. (*c*) Load current path when V_{in} is negative.

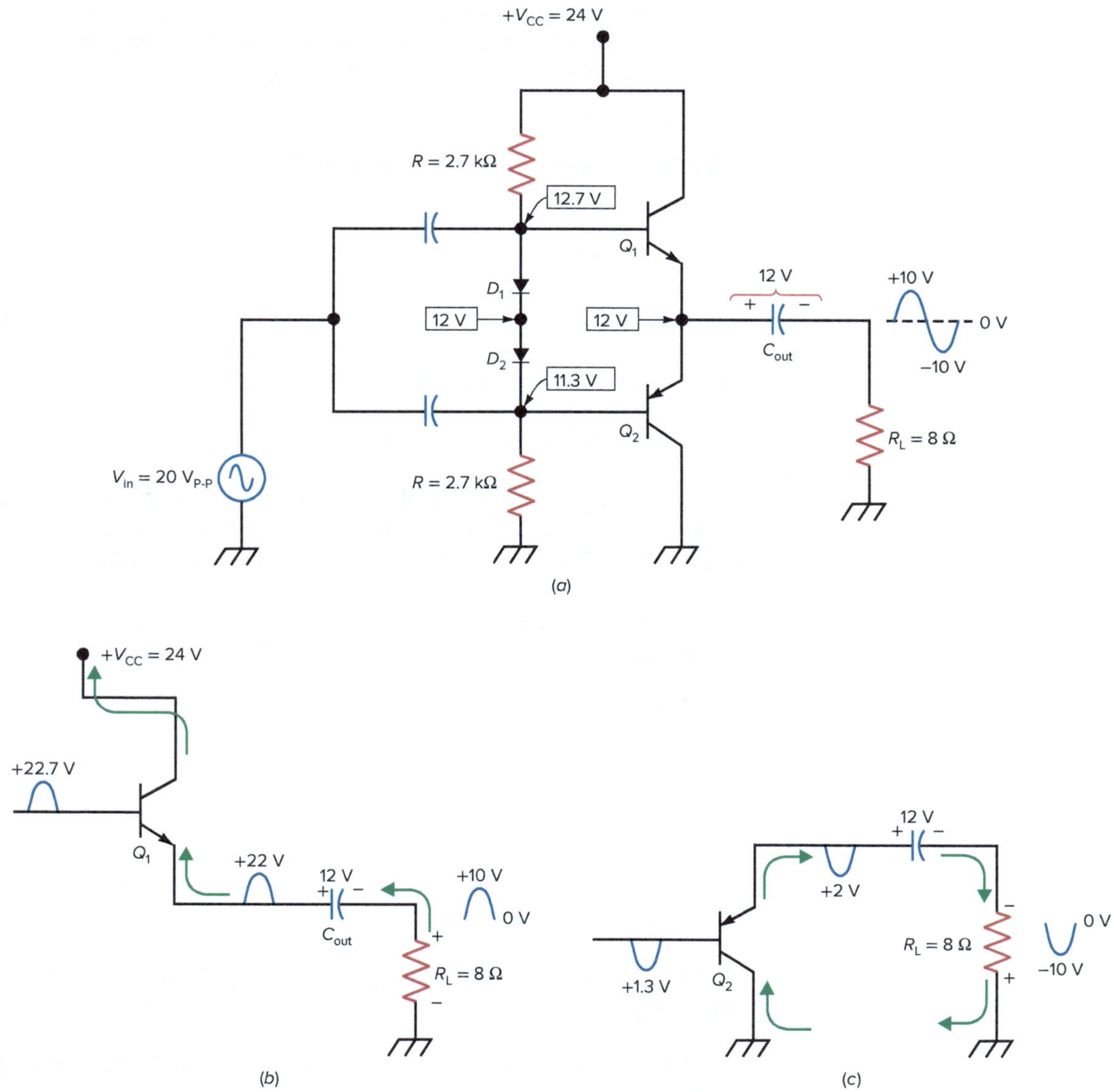

alternation of V_{in}. The charging current flows through R_L and the collector-emitter region of Q_1.

Figure 32–6*c* shows the output when V_{in} is negative; Q_2 conducts and Q_1 cuts off. Q_2 then acts like an emitter follower. Notice that Q_2 provides a discharge path for the output coupling capacitor, C_{out}. The discharge path is through R_L and the collector-emitter region of Q_2. Note that the charging and discharging time constant is made very long with respect to the period of the input waveform. The repetitive charging and discharging of the output coupling capacitor, C_{out}, produces the output signal across the load resistor, R_L.

Power Formulas

Formula (32–6) is still used to calculate the AC load power. For clarity, this equation is

$$P_L = \frac{V^2_{out(p\text{-}p)}}{8R_L}$$

To calculate the **DC input power, P_{CC},** consider the alternation during which current is drawn from the power supply, V_{CC}. This is during the positive alternation only, with the current path shown in Fig. 32–6*b*. Since current is drawn from V_{CC} only during the positive alternations of V_{in}, the waveform of power supply current is the same as that of a half-wave rectified signal. Remember from Chapter 28 that the average or DC current of a half-wave rectified signal is

$$I_{DC} = 0.318 \times I_{pk}$$

The DC input power, P_{CC}, for a class B push-pull amplifier is calculated as

$$P_{CC} = V_{CC} \times I_{CC}$$

where $I_{CC} = I_{DC}$. Or

$$P_{CC} = V_{CC} \times \frac{V_{out(pk)}}{R_L} \times 0.318 \qquad \textbf{(32–11)}$$

The percent efficiency is calculated as shown earlier:

$$\text{Percent efficiency} = \frac{P_L}{P_{CC}} \times 100$$

It is interesting to note that the efficiency of a class B push-pull amplifier varies with the amount of AC load power. In fact, the efficiency increases as the AC load power increases. The maximum obtainable efficiency for a class B push-pull amplifier is 78.6%.

One more point: The worst case power dissipation in the transistors of a class B push-pull amplifier can be found using Formula (32–12), shown here:

$$P_{d(max)} = \frac{V_{CC}^2}{40R_L} \qquad \textbf{(32–12)}$$

Example 32–3

In Fig. 32–6, calculate the following quantities: P_L, P_{CC}, $P_{d(max)}$, and percent efficiency.

ANSWER Begin by calculating the AC load power, P_L. Since $V_{in} = 20\ V_{p\text{-}p}$, then $V_{out(p\text{-}p)} = 20\ V_{p\text{-}p}$.

The calculations are

$$P_L = \frac{V_{out(p\text{-}p)}^2}{8R_L}$$

$$= \frac{20\ V_{p\text{-}p}^2}{8 \times 8\ \Omega}$$

$$= 6.25\ W$$

Next, calculate the DC input power, P_{CC}. Begin by calculating the value of I_{CC}:

$$I_{CC} = \frac{V_{out(pk)}}{R_L} \times 0.318$$

$$= \frac{10\ V}{8\ \Omega} \times 0.318$$

$$= 397.5\ mA$$

Next,

$$\begin{aligned} P_{CC} &= V_{CC} \times I_{CC} \\ &= 24\text{ V} \times 397.5\text{ mA} \\ &= 9.54\text{ W} \end{aligned}$$

The percent efficiency equals

$$\begin{aligned} \text{Percent efficiency} &= \frac{P_L}{P_{CC}} \times 100 \\ &= \frac{6.25\text{ W}}{9.54\text{ W}} \times 100 \\ &= 65.5\% \end{aligned}$$

The worst case power dissipation in the transistors is

$$\begin{aligned} P_{d(max)} &= \frac{V_{CC}^2}{40R_L} \\ &= \frac{24^2\text{ V}}{320\ \Omega} \\ &= 1.8\text{ W} \end{aligned}$$

To avoid damage to the transistors, Q_1 and Q_2 must have power ratings in excess of 1.8 W.

Using a Split Supply

To obtain a greater amount of output power, a split supply can be used (see Fig. 32–7). Notice that the collector of Q_2 connects to $-V_{CC}$ rather than to ground. As before, Q_1 conducts during the positive alternation of V_{in} and Q_2 cuts off. When V_{in} is negative, Q_2 conducts and Q_1 cuts off.

Figure 32–7 Class B push-pull amplifier using a split supply ($\pm V_{CC}$).

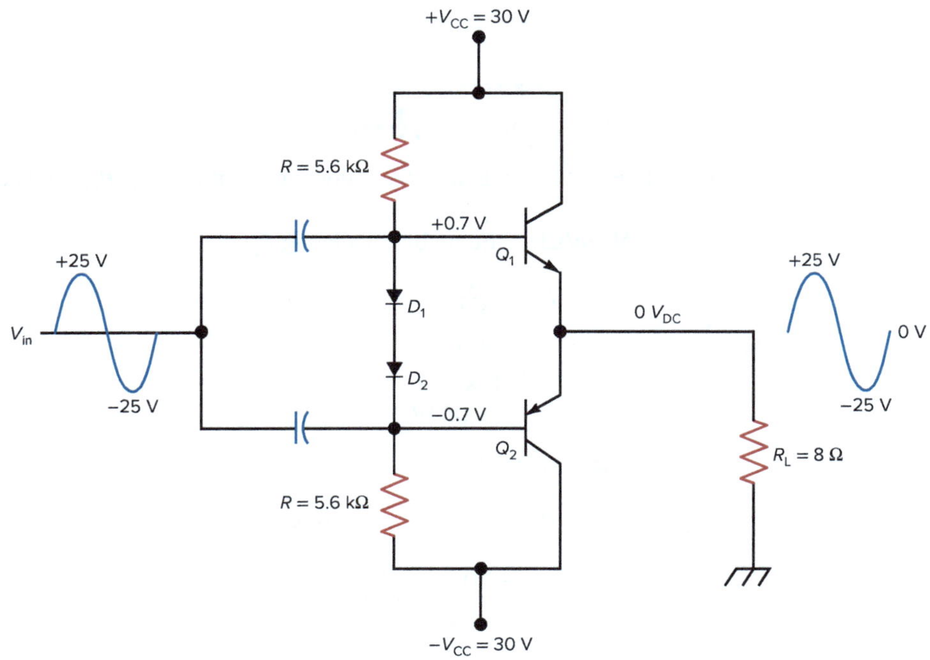

The quiescent collector current, I_{CQ}, is calculated as

$$I_{CQ} = \frac{V_{CC} - V_{BE}}{R} \qquad \textbf{(32–13)}$$

For Fig. 32–7, I_{CQ} is calculated as

$$I_{CQ} = \frac{30\text{ V} - 0.7\text{ V}}{5.6\text{ k}\Omega}$$

$$= 5.23\text{ mA}$$

Because the collectors of Q_1 and Q_2 are in series, the collector-emitter voltage, V_{CEQ}, for each transistor equals V_{CC}, or 30 V in this case. Therefore, V_{CEQ}, equals V_{CC} in a class B push-pull amplifier using a split supply. This makes the DC voltage at the emitter terminals of Q_1 and Q_2 equal to 0 V (refer to Fig. 32–7). Furthermore, the DC voltage at the base of Q_1 equals +0.7 V, while the DC voltage at the base of Q_2 equals −0.7 V. Because the emitter voltage of each transistor is at 0 V, the output coupling capacitor can be omitted. Because the DC voltage at the emitters is zero, no DC current will flow through the speaker load. Also, without the output coupling capacitor, the low-frequency response of the amplifier is greatly improved.

Power Calculations

To calculate P_{dq} proceed as follows:

$$P_{dq} = V_{CEQ} \times I_{CQ}$$
$$= 30\text{ V} \times 5.23\text{ mA}$$
$$= 156.9\text{ mW}$$

The AC load power is still given by Formula (32–6). The DC input power, P_{CC}, is now

$$P_{CC} = V_{CC} \times \frac{V_{out(pk)}}{R_L} \times 0.636 \qquad \textbf{(32–14)}$$

The factor 0.636 indicates that current is drawn from both $+V_{CC}$ and $-V_{CC}$. Current is drawn from $+V_{CC}$ when V_{in} is positive. Conversely, current is drawn from $-V_{CC}$ when V_{in} is negative.

Example 32–4

In Fig. 32–7, calculate the following quantities: P_L, P_{CC}, and percent efficiency.

ANSWER Begin by calculating P_L:

$$P_L = \frac{V^2_{out(p\text{-}p)}}{8\,R_L}$$
$$= \frac{50\text{ V}^2_{p\text{-}p}}{8 \times 8\ \Omega}$$
$$= 39.06\text{ W}$$

Next, calculate P_{CC}:

$$P_{CC} = V_{CC} \times \frac{V_{out(pk)}}{R_L} \times 0.636$$
$$= 30\text{ V} \times \frac{25\text{ V}}{8\ \Omega} \times 0.636$$
$$\approx 59.62\text{ W}$$

The percent efficiency is calculated as

$$\text{Percent efficiency} = \frac{P_L}{P_{CC}} \times 100$$
$$= \frac{39.06\text{ W}}{59.62\text{ W}} \times 100$$
$$= 65.5\%$$

A split power supply is used with the class B push-pull amplifier when it is necessary to obtain large amounts of AC load power, P_L.

32–3 Self-Review

Answers at the end of the chapter.

a. What type of distortion occurs in a class B push-pull amplifier when the transistors are biased right at cutoff?

b. Is voltage divider or diode bias the preferred way to bias a class B push-pull amplifier?

32–4 Class C Amplifiers

The collector current, I_C, of a transistor in a class C amplifier flows for 120° or less of the AC input waveform. The result is that the collector current, I_C, flows in very short, narrow pulses. Since the collector current is nonsinusoidal, it contains a large number of harmonic components and is said to be rich in harmonic content. Because of their high distortion, class C amplifiers cannot be used in audio circuitry where full reproduction of the input signal is required. Class C amplifiers, however, can be used as tuned rf amplifiers where undesired harmonic frequencies can be filtered out, passing only the fundamental frequency to the load, R_L. In some cases, however, it might be desirable to tune the *LC* tank circuit to a harmonic (multiple) of the input frequency. The tuned *LC* circuit in a class C amplifier usually has high *Q*, so that only a very narrow band of frequencies is amplified.

Class C amplifiers are much more efficient than either class A or class B amplifiers. Typical class C amplifiers have efficiencies in excess of 90%.

Class C Amplifier Circuit Analysis

Figure 32–8*a* shows a tuned class C amplifier. The input coupling capacitor, base resistor, and base-emitter junction form a negative clamper. The equivalent input circuit is shown in Fig. 32–8*b*. During the initial positive half-cycle of input voltage, the input coupling capacitor, C_{in}, charges through the low resistance of the base-emitter junction, which is forward-biased. The capacitor, C_{in}, charges to 0.8 V with the polarity shown. The DC voltage to which the capacitor charges equals $V_{in(pk)} - V_{BE}$. The negative polarity on the right plate of C_{in} will now reverse-bias the base-emitter junction. The capacitor, C_{in}, clamps or holds the DC base voltage at –0.8 V. Because of the clamping action, only the positive peaks of the input signal drive the transistor, Q_1, into conduction. The $R_B C_{in}$ time constant during discharge is made long with respect to the period of the input waveform to provide the proper clamping action. As a general rule,

$$R_B C_{in} \geq 10T \qquad \textbf{(32–15)}$$

where *T* equals the period of the input waveform.

GOOD TO KNOW

Most class C amplifiers are designed so that the peak value of input voltage is just sufficient to drive the transistor into saturation.

MultiSim **Figure 32-8** Class C rf amplifier. (*a*) Circuit. (*b*) C_{in}, R_B, and the base-emitter diode form a negative clamper. (*c*) $V_{out} = 2V_{CC(p\text{-}p)}$. (*d*) Graph of A_V versus frequency.

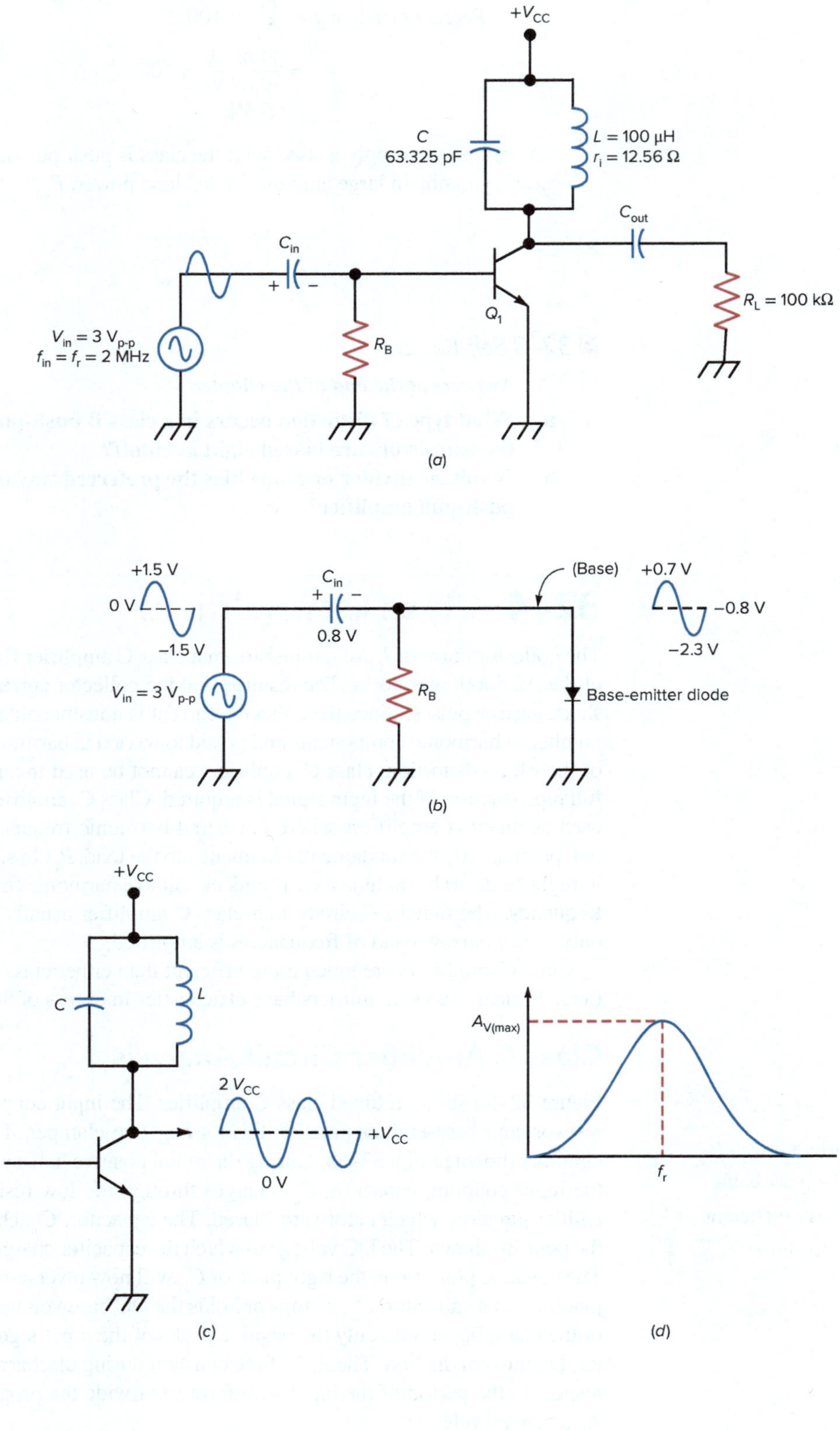

The $R_B C_{in}$ time constant affects the conduction angle of the transistor. The longer the $R_B C_{in}$ time constant with respect to the period, T, the shorter the conduction angle of the transistor.

The collector tank circuit, consisting of L and C, can be tuned to the same frequency as the input signal or some multiple of the input frequency. Every time the transistor is driven into conduction, the capacitor, C, in the tank circuit is recharged to the value of V_{CC}. When the transistor is cut off, the flywheel effect of the resonant tank circuit reproduces the full sine wave of output voltage. The resonant frequency of the tank circuit is given by

$$f_r = \frac{1}{2\pi\sqrt{LC}} \tag{32–16}$$

Figure 32–8*c* shows the peak-to-peak output voltage from the tuned class C amplifier. Notice that the minimum voltage is zero and the maximum voltage is $2V_{CC}$. Therefore, the peak-to-peak voltage available at the output equals $2V_{CC}$ because the tank voltage adds to the positive value of V_{CC} during the positive alternation of the output voltage.

Figure 32–8*d* shows a graph of frequency versus voltage gain for the tuned class C amplifier. At the resonant frequency, f_r, the impedance of the tuned LC circuit is maximum. Also, the tank impedance, Z_{tank}, is purely resistive at f_r. When the frequency of the input voltage is above or below f_r, the voltage gain, A_V, is less because the impedance of tank circuit decreases as the frequency of input voltage moves above or below f_r.

When the tank circuit is adjusted to resonance, an ammeter placed in the collector circuit will dip to its minimum value. A commonly used method of tuning a class C rf amplifier is to adjust either L or C for minimum collector current, I_C, as indicated by an ammeter.

Example 32–5

In Fig. 32–8*a*, calculate the following: (a) f_r of the LC tank circuit and (b) DC voltage at the base.

ANSWER (a) The resonant frequency, f_r, can be calculated using Formula (32–16):

$$\begin{aligned} f_r &= \frac{1}{2\pi \times \sqrt{LC}} \\ &= \frac{1}{2 \times 3.141 \times \sqrt{100\ \mu\text{H} \times 63.325\ \text{pF}}} \\ &= 2\ \text{MHz} \end{aligned}$$

At this frequency, the class C amplifier has its maximum voltage gain, A_V.

(b) The DC voltage at the base equals

$$\begin{aligned} -V_{DC} &= V_{in(pk)} - V_{BE} \\ &= 1.5\ \text{V} - 0.7\ \text{V} \\ &= 0.8\ \text{V} \end{aligned}$$

Therefore, $-V_{DC} = 0.8$ V (see Fig. 32–8*b*).

Example 32-6

Assume $C_{in} = 0.01\ \mu F$ in Fig. 32–8*a*. Calculate the minimum base resistance, R_B, necessary to provide the proper clamping action.

ANSWER Begin by calculating the period of the input waveform. Since $f_{in} = f_r$,

$$T = \frac{1}{f_{in}}$$

$$= \frac{1}{2\ \text{MHz}}$$

$$= 0.5\ \mu S$$

Therefore,

$$R_B C_{in} = 10T$$

$$= 10 \times 0.5\ \mu S$$

$$= 5\ \mu S$$

Solving for R_B,

$$R_B = \frac{5\ \mu S}{0.01\ \mu F}$$

$$= 500\ \Omega$$

Amplifier Bandwidth

As mentioned earlier, the voltage gain of a class C rf amplifier is maximum at the resonant frequency, f_r, of the tank circuit. However, frequencies close to f_r also provide a high voltage gain. Therefore, any class C rf amplifier has an associated band of frequencies at which there is a high voltage gain. The bandwidth of an amplifier refers to those frequencies in which the voltage gain, A_V, is 70.7% or more of its maximum value at resonance. The bandwidth (BW) of the tuned amplifier is affected by the Q of the circuit. Recall from basic AC circuit theory that the impedance of the tank circuit at resonance is

$$Z_{tank} = Q_{coil} \times X_L \qquad \textbf{(32–17)}$$

Remember that any shunt or parallel resistance, R_P, lowers the circuit Q. When $Z_{tank} \geq 10 \times R_p$, then $Q_{ckt} = R_P/X_L$. However, when $Z_{tank} < 10 \times R_p$ then the circuit Q, Q_{ckt}, is

$$Q_{ckt} = \frac{Z_{tank} \parallel R_P}{X_L} \qquad \textbf{(32–18)}$$

With Q_{ckt} known, the bandwidth (BW) is calculated as

$$BW = \frac{f_r}{Q_{ckt}} \qquad \textbf{(32–19)}$$

Example 32–7

In Fig. 32–8, calculate the bandwidth (BW).

ANSWER Begin by calculating X_L at f_r:

$$\begin{aligned} X_L &= 2\pi f_r L \\ &= 2 \times 3.141 \times 2.0 \text{ MHz} \times 100\ \mu\text{H} \\ &= 1.256 \text{ k}\Omega \end{aligned}$$

Next, we calculate the Q of the LC tank. The Q of the tank circuit equals the Q of the coil. Therefore,

$$\begin{aligned} Q_{coil} &= \frac{X_L}{r_i} \\ &= \frac{1.256 \text{ k}\Omega}{12.56\ \Omega} \\ &= 100 \end{aligned}$$

Now calculate the tank impedance, Z_{tank}:

$$\begin{aligned} Z_{tank} &= Q_{coil} \times X_L \\ &= 100 \times 1.256 \text{ k}\Omega \\ &= 125.6 \text{ k}\Omega \end{aligned}$$

Next, calculate Q_{ckt}:

$$Q_{ckt} = \frac{Z_{tank} \parallel R_P}{X_L}$$

where

$$\begin{aligned} Z_{tank} \parallel R_P &= \frac{125.6 \text{ k}\Omega \times 100 \text{ k}\Omega}{125.6 \text{ k}\Omega + 100 \text{ k}\Omega} \\ &= 55.67 \text{ k}\Omega \end{aligned}$$

Therefore,

$$\begin{aligned} Q_{ckt} &= \frac{55.67 \text{ k}\Omega}{1.256 \text{ k}\Omega} \\ &= 44.32 \end{aligned}$$

Finally, calculate the bandwidth (BW):

$$\begin{aligned} \text{BW} &= \frac{f_r}{Q_{ckt}} \\ &= \frac{2 \text{ MHz}}{44.32} \\ &= 45 \text{ kHz} \end{aligned}$$

Frequency Multipliers

Tuned class C amplifiers can also be used as **frequency multipliers** by tuning the *LC* tank circuit to a harmonic (multiple) of the input frequency. For the class C rf amplifier shown in Fig. 32–8*a*, the capacitor, *C*, in the tank circuit is charged once per input cycle. If the tank circuit is tuned to an f_r of 4 MHz, then the capacitor in the tank is recharged once every other cycle. If the tank is tuned to an f_r of 6 MHz, then the capacitor is recharged on every third cycle of input voltage.

Figure 32-9 Collector current in a class C rf amplifier used as a frequency multiplier. (*a*) Collector current, I_C, flows once per cycle when the tank circuit is tuned to the same frequency as f_{in}. (*b*) I_C flows once every other cycle when the tank circuit is tuned to $2f_{in}$. (*c*) I_C flows once every third cycle when the tank circuit is tuned to $3f_{in}$.

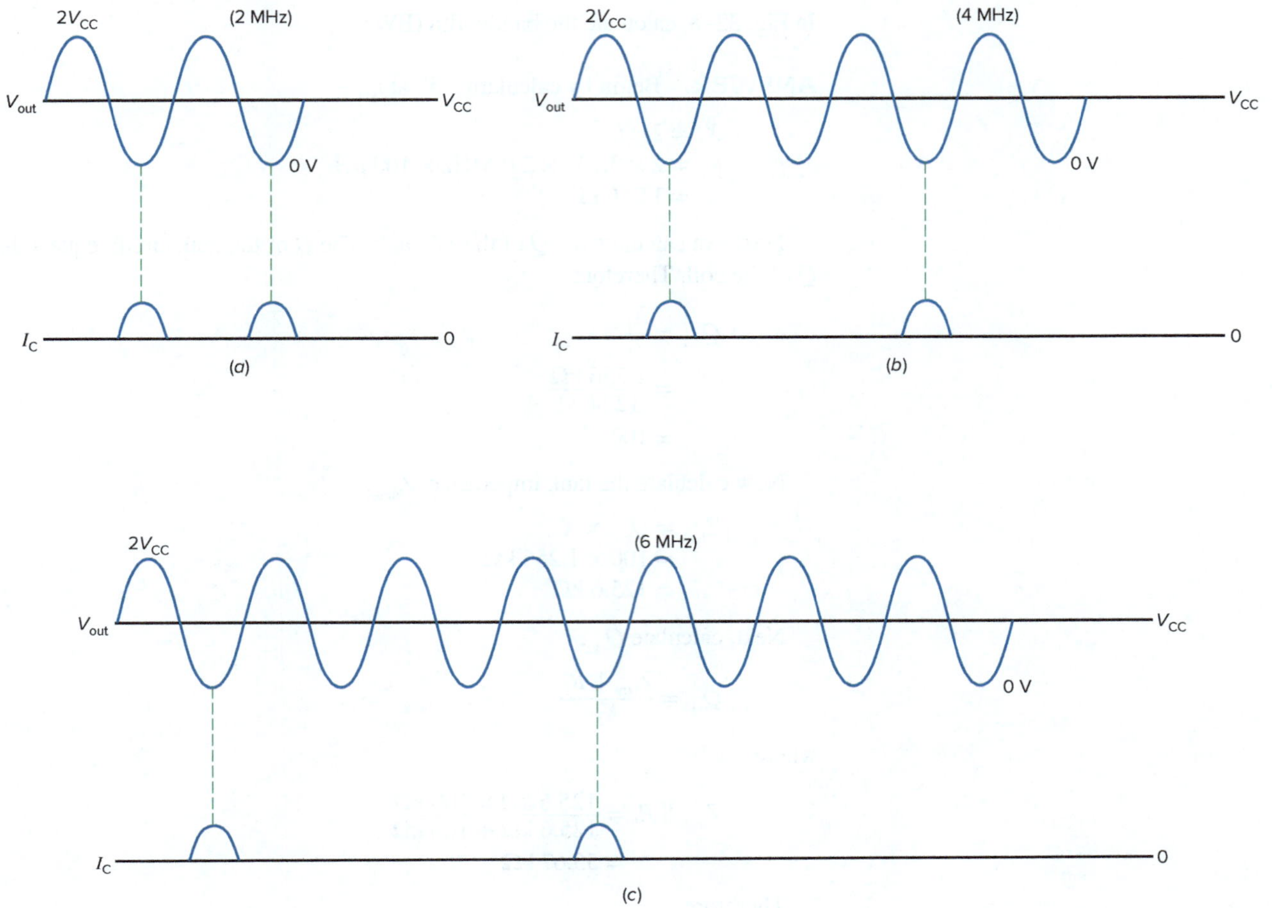

Figure 32–9 shows the collector current pulses and their relationships with respect to the output waveform. In Fig. 32–9*a*, the tank circuit is tuned to the same frequency as the input voltage applied to the class C amplifier. Therefore, the collector current, I_C, flows only once during the cycle. In Fig. 32–9*b*, the tank circuit is tuned to the second harmonic of the fundamental frequency and the collector current, I_C, flows once every other cycle. Finally, in Fig. 32–9*c*, the tank circuit is tuned to the third harmonic of the fundamental frequency and the collector current, I_C, flows once every third cycle.

Note that in Fig. 32–9 when collector current flows, the transistor is driven into saturation. This is why the collector voltage is at 0 V when the collector current reaches its peak value.

32–4 Self-Review

Answers at the end of the chapter.

a. The collector current in a class C amplifier is rich in harmonics. (True/False)

b. The peak-to-peak output voltage from a tuned class C rf amplifier is approximately two times V_{CC}. (True/False)

Summary

- The class of operation of an amplifier is defined by the percentage of the AC input cycle which produces an output current.
- In a class A amplifier, collector current flows for the full 360° of the AC input cycle.
- In a class B amplifier, collector current flows for only 180° of the AC input cycle.
- In a class C amplifier, collector current flows for 120° or less of the AC input cycle.
- A class A amplifier has very low distortion and very low power efficiency.
- A class A amplifier sees two loads, an AC load and a DC load. Therefore, there are two load lines, a DC load line and an AC load line. The AC load line is always steeper than the DC load line.
- In a class A amplifier, maximum transistor power dissipation occurs when there is no input signal applied to the amplifier.
- A class B amplifier using a single transistor has very severe distortion and medium power efficiency.
- A class B push-pull amplifier uses two transistors to get a linear reproduction of the input waveform being amplified. A class B push-pull amplifier has medium power efficiency.
- A common way to bias a class B push-pull amplifier is with diode bias. With diode bias, the diode curves must match the emitter diode curves of the transistors to obtain a stable bias.
- The transistors in a class B push-pull amplifier dissipate very little power when no AC signal is present at the input because the idling current in the transistors is quite low.
- When large amounts of AC load power are required, a split power supply is often used with class B push-pull amplifiers.
- Class C amplifiers cannot be used in audio circuitry because of their high distortion. However, class C amplifiers can be used as tuned rf amplifiers where the undesired harmonic frequencies can be filtered out.
- A class C amplifier is more efficient than either a class A or class B amplifier; its power efficiency approaches 100%.
- A tuned class C amplifier can be used as a frequency multiplier by tuning the *LC* tank circuit to a multiple of the input frequency.

Important Terms

AC load line—a graph that shows all possible values of I_C and V_{CE} when a transistor amplifier is driven by an AC signal.

AC load power, P_L—the AC power that is dissipated by the load, R_L.

Class A amplifier—an amplifier in which the collector current, I_C, flows for the full 360° of the AC input cycle.

Class B amplifier—an amplifier in which the collector current, I_C, flows for only 180° of the AC input cycle.

Class C amplifier—an amplifier in which the collector current, I_C, flows for 120° or less of the AC input cycle.

Class B push-pull amplifier—a class B amplifier that uses two transistors to reproduce the full AC cycle of input voltage. Each transistor conducts on opposite half-cycles of the input voltage.

Crossover distortion—the distortion that occurs in a class B push-pull amplifier when the transistors are biased right at cutoff. When the input voltage crosses through zero, both transistors in the push-pull amplifier are off and the output voltage cannot follow the input voltage.

DC input power, P_{CC}—the amount of DC power delivered to a transistor amplifier.

Diode bias—a form of biasing for class B push-pull amplifiers that uses diodes to provide a slight amount of forward bias for the base-emitter junctions of each transistor.

Frequency multiplier—a tuned class C amplifier whose *LC* tank circuit is tuned to a harmonic or multiple of the input frequency.

Linear amplifier—any amplifier that produces an output signal that is an exact replica of the input signal.

Percent efficiency—for an amplifier, this is the percentage of DC input power that is converted to useful ac output power.

Power amplifier—a circuit that is designed to deliver large amounts of power to a low impedance load.

Related Formulas

Class A Amplifiers

$P_L = V^2_{out(p-p)}/8R_C$ (No Load, R_L)

$P_L = V^2_{out(p-p)}/8R_L$

$P_{CC} = V_{CC} \times I_{CC}$

Percent Efficiency $= \frac{P_L}{P_{CC}} \times 100$

$i_{C(sat)} = I_{CQ} + \frac{V_{CEQ}}{r_L}$ (Endpoints for AC Load Line)

$v_{CE(off)} = V_{CEQ} + I_{CQ}r_L$

Class B Push-Pull Amplifiers

$i_{C(sat)} = V_{CC}/2R_L$ (Endpoints for AC Load Line)

$v_{CE(off)} = V_{CC}/2$

$I_{CQ} = \frac{V_{CC} - 2V_{BE}}{2R}$ (Single Supply)

$P_{dq} = V_{CEQ} \times I_{CQ}$

$P_L = V^2_{out(p-p)}/8R_L$

$P_{CC} = V_{CC} \times \frac{V_{out(pk)}}{R_L} \times 0.318$ (Single Supply)

$P_{d(max)} = V^2_{CC}/40R_L$

$I_{CQ} = \frac{V_{CC} - V_{BE}}{R}$ (Split Supply)

$P_{CC} = V_{CC} \times \frac{V_{out(pk)}}{R_L} \times 0.636$ (Split Supply)

Class C Amplifiers

$R_B C_{in} \geq 10T$

$f_r = 1/2\pi\sqrt{LC}$

$Z_{tank} = Q_{coil} \times X_L$

$Q_{ckt} = Z_{tank} \| R_P / X_L$

$BW = f_r/Q_{ckt}$

Self-Test

Answers at the back of the book.

1. In a class A amplifier, the collector current, I_C, flows for

a. 180° of the AC input cycle.

b. 360° of the AC input cycle.

c. 120° or less of the AC input cycle.

d. 90° of the AC input cycle.

2. Which of the following classes of amplifier operation has the highest power efficiency?

a. class A.

b. class B.

c. class C.

d. class AB.

3. The collector current in a class C amplifier

a. is rich in harmonics.

b. flows for 120° or less of the AC input cycle.

c. is nonsinusoidal.

d. all of the above.

4. The transistors in a class B push-pull amplifier are biased slightly above cutoff to prevent

a. crossover distortion.

b. excessive power dissipation in the transistors.

c. excessive efficiency.

d. none of the above.

5. In a class B amplifier, the collector current, I_C, flows for

a. 120° of the AC input cycle.

b. 180° of the AC input cycle.

c. 360° of the AC input cycle.

d. 60° of the AC input cycle.

6. When a class B push-pull amplifier uses a split power supply,

a. no output coupling capacitor is required.

b. a greater amount of output power can be obtained.

c. the efficiency decreases.

d. both a and b.

7. A class C amplifier is typically used as a(n)

a. audio amplifier.

b. linear amplifier.

c. tuned rf amplifier.

d. none of the above.

8. A class B push-pull amplifier uses a single DC power supply voltage of 15 V. How much voltage should exist across the collector-emitter region of each transistor?

a. 7.5 V.

b. 0 V.

c. 15 V.

d. It cannot be determined.

9. A class A amplifier should be biased

a. at cutoff.

b. midway between saturation and cutoff.

c. very near saturation.

d. none of the above.

10. A tuned class C amplifier has a power supply voltage of 12 V. What is the ideal peak-to-peak output voltage?

a. 12 V_{p-p}.

b. 48 V_{p-p}.

c. 24 V_{p-p}.

d. 6 V_{p-p}.

11. Which of the following amplifiers has the lowest efficiency under large signal conditions?

a. class B push-pull amplifier.

b. class C rf amplifier.

c. *RC*-coupled class A amplifier.

d. class B push-pull amplifier with split supplies.

12. In a class B push-pull amplifier, the transistors Q_1 and Q_2 conduct

a. on the same half-cycle of input voltage.

b. on opposite half-cycles of the input voltage.

c. only on the positive and negative peaks of the input voltage.

d. none of the above.

13. In an *RC*-coupled class A amplifier,

a. the DC and AC load lines are usually the same.

b. the DC load line is steeper than the AC load line.

c. the *Q* point should be located near cutoff on the AC load line.

d. the AC load line is steeper than the DC load line.

14. Which of the following is the best way to bias a class B push-pull amplifier?

a. diode bias.

b. voltage divider bias.

c. zero bias.

d. none of the above.

15. A power amplifier delivers 25 W of AC power to a 4-Ω speaker load. If the DC input power is 40 W, what is the efficiency of the amplifier?

a. 78.6%.

b. 25%.

c. 62.5%.

d. 160%.

Essay Questions

1. Why can't a class C amplifier be used as an audio amplifier?

2. Why are the transistors in a class B push-pull amplifier biased slightly above cutoff?

3. Why isn't it practical to use a class A amplifier if 100 W of AC load power is required?

4. How can class C rf amplifiers function as frequency multipliers?

5. What are the advantages of using a split power supply with a class B push-pull amplifier?

Problems

SECTION 32–1 CLASSES OF OPERATION

32–1 For how many degrees of the AC input cycle does collector current flow in a

a. class A amplifier?

b. class B amplifier?

c. class C amplifier?

32–2 How should a class A amplifier be biased?

32–3 List two characteristics of a class A amplifier.

32–4 How is a true class B amplifier biased?

32–5 How does a class B amplifier with a single transistor correspond to a half-wave rectifier?

32–6 List two characteristics of a class B amplifier using a single transistor.

32–7 What is the main application for class C amplifiers?

32–8 List two characteristics of a class C amplifier.

SECTION 32–2 CLASS A AMPLIFIERS

32–9 In Fig. 32–10, calculate the following DC quantities:

a. I_B.

b. I_{CQ}.

c. V_{CEQ}.

d. $V_{CE(off)}$.

e. $I_{C(sat)}$.

32–10 In Fig. 32–10, calculate the following AC quantities:

a. r'_e.

b. r_L.

Figure 32–10

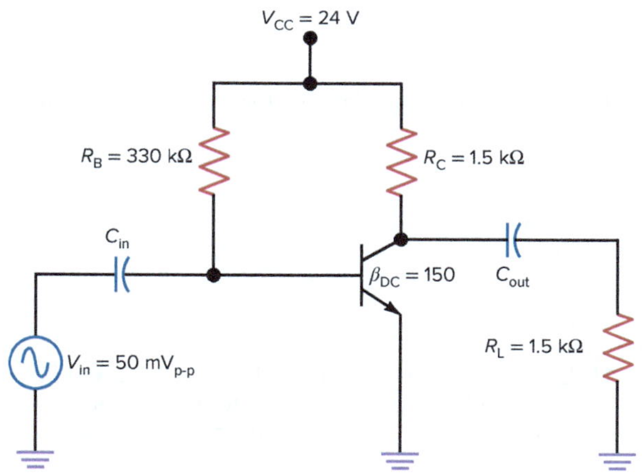

c. A_V.

d. V_{out}.

e. $v_{CE(off)}$.

f. $i_{C(sat)}$.

32–11 Using the values obtained in Probs. 32–9 and 32–10, draw the DC and AC load lines for the *RC*-coupled amplifier in Fig. 32–10. Indicate the *Q* point on the graph.

32–12 In Fig. 32–10, solve for the following:

a. P_L.

b. P_{CC}.

c. % efficiency.

32–13 In Fig. 32–11, calculate the following DC quantities:

a. V_B.
b. V_E.
c. I_{CQ}.
d. V_{CEQ}.
e. P_d.
f. $V_{CE(off)}$.
g. $I_{C(sat)}$.

Figure 32–11

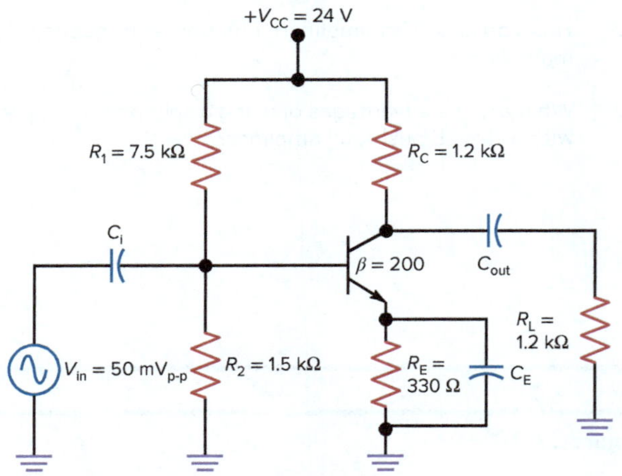

32–14 In Fig. 32–11, calculate the following AC quantities:

a. r'_e.
b. r_L.
c. A_V.
d. V_{out}.
e. $V_{CE(off)}$.
f. $i_{C(sat)}$.

32–15 Using the values obtained in Probs. 32–13 and 32–14, draw the DC and AC load lines for the *RC*-coupled amplifier in Fig. 32–11. Indicate the *Q* point on the graph.

32–16 In Fig. 32–11, solve for the following:

a. P_L.
b. P_{CC}.
c. % efficiency.

32–17 If the input voltage, V_{in}, is reduced to 25 $mV_{p\text{-}p}$ in Fig. 32–10, then recalculate the following values:

a. V_{out}.
b. P_L.
c. P_{CC}.
d. % efficiency.

32–18 Compare the % efficiency calculated in Probs. 32–12 and 32–17. Are they different? If so, why?

SECTION 32–3 CLASS B PUSH-PULL AMPLIFIERS

32–19 In Fig. 32–12, which transistor conducts during

a. the positive alternation of input voltage?
b. the negative alternation of input voltage?

Figure 32–12

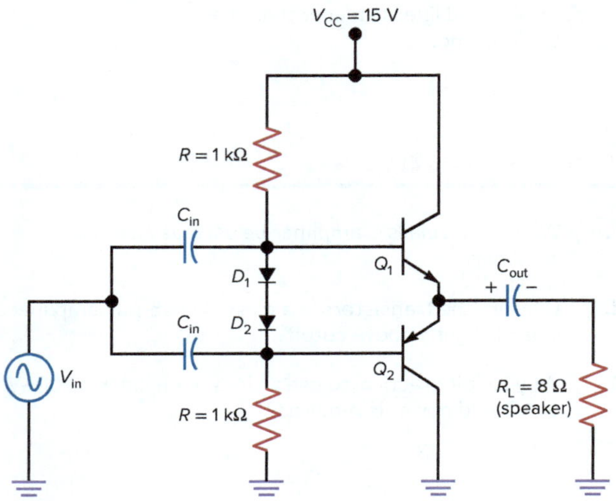

32–20 In Fig. 32–12, solve for the following DC quantities:

a. I_{CQ}.
b. V_{BQ_1}.
c. V_{BQ_2}.
d. V_{EQ_1} and V_{EQ_2}.
e. V_{CEQ_1} and V_{CEQ_2}.
f. P_{dQ}.

32–21 Draw the AC load line for Fig. 32–12 and indicate the values of $i_{C(sat)}$ and $v_{CE(off)}$.

32–22 In Fig. 32–12, solve for P_L, P_{CC}, $P_{d(max)}$, and % efficiency if $V_{in} = 12\ V_{p\text{-}p}$.

32–23 In Fig. 32–12, is C_{out} charging or discharging during the

a. positive alternation of input voltage?
b. negative alternation of input voltage?

32–24 In Fig. 32–13, solve for the following DC quantities:

a. I_{CQ}.
b. V_{BQ_1}.
c. V_{BQ_2}.
d. V_{EQ_1} and V_{EQ_2}.
e. V_{CEQ_1} and V_{CEQ_2}.
f. P_{dQ}.

32–25 If $R_L = 10\ \Omega$ in Fig. 32–13, what are the values of $i_{C(sat)}$ and $v_{CE(off)}$ on the AC load line?

32–26 In Fig. 32–13, solve for P_L, P_{CC}, and % efficiency for each of the following values of R_L ($V_{in} = 15\ V_{p\text{-}p}$):

a. $R_L = 16\ \Omega$.
b. $R_L = 8\ \Omega$.
c. $R_L = 4\ \Omega$.

Figure 32–13

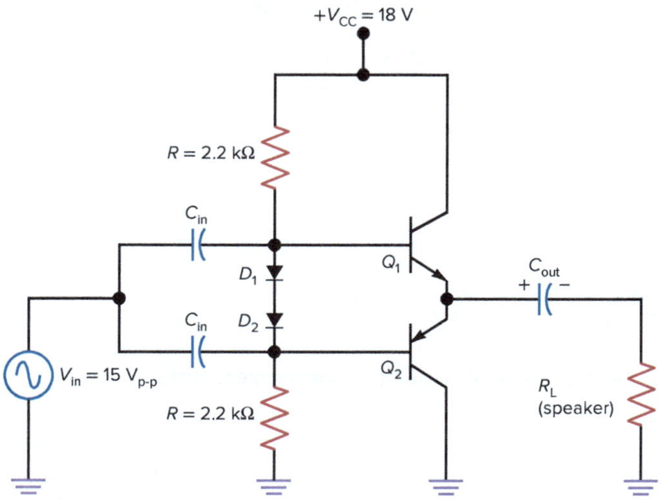

32–27 In Fig. 32–13, what is the worst case power dissipation, $P_{d(max)}$, in each transistor if $R_L = 4\ \Omega$?

32–28 In Fig. 32–13, how much is the DC voltage across

a. the output coupling capacitor, C_{out}?

b. the load, R_L?

32–29 In Fig. 32–14, solve for the following DC quantities:

a. I_{CQ}.

b. V_{BQ_1}.

c. V_{BQ_2}.

d. V_{EQ_1} and V_{EQ_2}.

e. V_{CEQ_1} and V_{CEQ_2}.

f. P_{dQ} in each transistor.

Figure 32–14

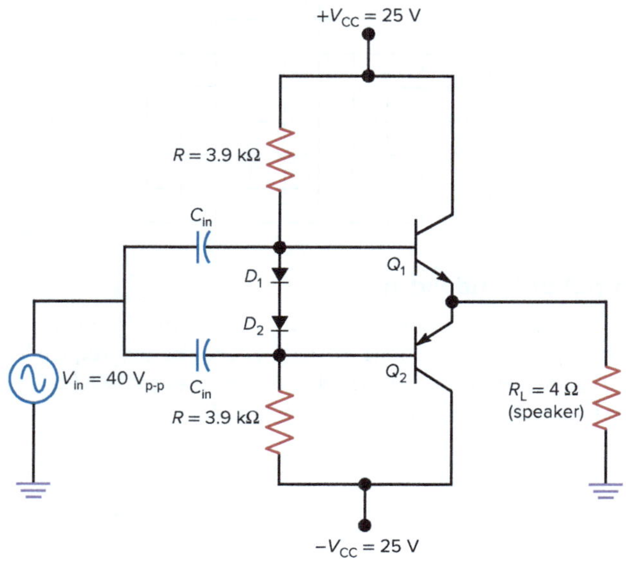

32–30 In Fig. 32–14, solve for the following quantities:

a. P_L.

b. P_{CC}.

c. % efficiency.

32–31 In Fig. 32–14, how much DC voltage is across the 4-Ω load, R_L?

SECTION 32–4 CLASS C AMPLIFIERS

32–32 In Fig. 32–15, what is the resonant frequency, f_r, of the LC tank circuit?

Figure 32–15

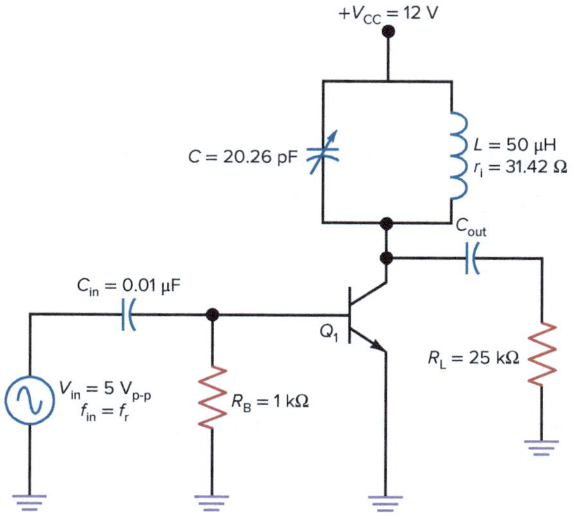

32–33 In Fig. 32–15, how much is the

a. DC voltage at the base of Q_1?

b. DC voltage at the collector of Q_1?

c. peak-to-peak voltage at the collector of Q_1?

32–34 In Fig. 32–15, calculate

a. Z_{tank}.

b. Q_{ckt}.

c. BW.

32–35 In Fig. 32–15, at what point in the AC cycle does the input voltage cause Q_1 to conduct?

32–36 At the instant V_{in} reaches its positive peak in Fig. 32–15, what is the voltage at the collector of Q_1?

32–37 If the frequency of the input voltage is changed to 2.5 MHz in Fig. 32–15, what is the frequency of the output waveform?

Answers to Self-Reviews

32–1 a. true
b. true
c. false

32–2 a. false
b. true

32–3 a. crossover distortion
b. diode bias

32–4 a. true
b. true

Laboratory Application Assignment

In this lab application assignment, you will examine a tuned class C amplifier. As you will see, the tuned class C amplifier uses a parallel resonant circuit in the collector, which results in maximum output voltage at only one frequency. The frequency at which maximum output occurs is the resonant frequency, f_r, of the *LC* tank circuit. You will also see how the tuned class C amplifier can be used as a frequency multiplier.

Equipment: Obtain the following items from your instructor.

- 2N2222A *npn* transistor or equivalent
- Two 100-kΩ carbon-film resistors
- 0.001-μF, 0.01-μF, and 0.1-μF capacitors
- 10-mH inductor
- Function generator
- Oscilloscope
- DMM
- Variable dc power supply

Tuned Class C Amplifier: Calculations and Predictions

Examine the tuned class C amplifier in Fig. 32–16. Calculate and record the resonant frequency, f_r, of the *LC* tank circuit.
f_r = ____________
With an input voltage, V_{in}, of 2 $V_{p\text{-}p}$, calculate and record the DC base voltage, V_B. $V_{B(DC)}$ = ____________
What DC voltage do you expect to measure at the collector?
$V_{C(DC)}$ = ____________

Figure 32–16

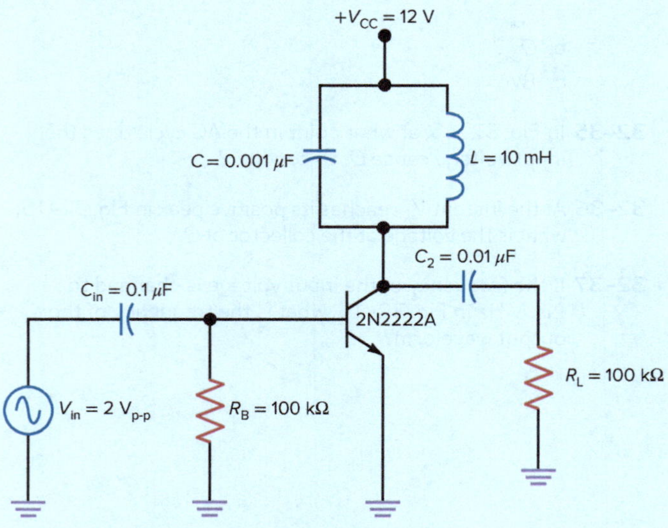

Next, predict the peak-to-peak output voltage across the load R_L if the frequency of V_{in} equals f_r. $V_{out(p\text{-}p)}$ = ____________

Tuned Class C Amplifier: Measurements

Construct the circuit in Fig. 32–16. Set V_{in} to exactly 2 $V_{p\text{-}p}$ as shown. Next, set the frequency of the function generator to the resonant frequency, f_r, calculated earlier. With channel 2 of the oscilloscope connected across the load, R_L, move the frequency dial back and forth until the output voltage is at its maximum peak-to-peak value. Measure and record the frequency where $V_{out(p\text{-}p)}$ is maximum. This frequency is the resonant frequency, f_r, of the tank circuit. f_r = ____________
Measure and record the maximum peak-to-peak output voltage.
$V_{out(p\text{-}p)}$ = ____________
Next, measure and record the DC voltage at the base and collector. $V_{B(DC)}$ = ____________, $V_{C(DC)}$ = ____________
Connect channel 1 of the oscilloscope across V_{in} and channel 2 across the load, R_L. Measure and record the phase relationship between V_{in} and V_{out} at f_r. θ = ____________
Connect channel 1 of the oscilloscope directly to the base of the transistor. Set the channel 1 volts/div. setting to 0.5 volt/div., and move the input coupling switch to DC. Draw the measured waveform, including all values, on the scope graticule in Fig. 32–17.

Figure 32–17

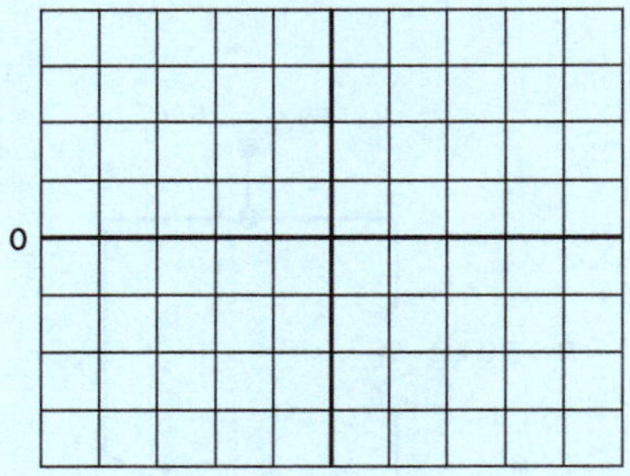

Amplifier Bandwidth

Reduce the input voltage, V_{in}, to 1.6 $V_{p\text{-}p}$. While monitoring the peak-to-peak output voltage across R_L, decrease the frequency of V_{in} below f_r until the output voltage decreases to 0.707 of its maximum value. Record this frequency as f_1.
f_1 = ____________

Now increase the frequency of V_{in} above f_r until V_{out} decreases to 0.707 of its maximum value. Record this frequency as f_2.
f_2 = ____________

Calculate the amplifier bandwidth, BW, as $f_2 - f_1$.
BW = ____________

Frequency Multiplier

Decrease the frequency of the function generator to approximately one-half of f_r. (Keep V_{in} at 1.6 $V_{p\text{-}p}$.) Move the function generator frequency dial back and forth to produce the maximum peak-to-peak output voltage. Is the frequency of the output waveform twice that of the input frequency? ______
If yes, explain how this is possible.

____
____
____

Is the peak-to-peak value of the output waveform the same for each individual cycle? ____________ If not, explain the possible cause for this. ____________________________________
____
____
____

Design credits: Chapter Opener Background Image, and Multisim: Stockbyte/Getty Images

chapter 33

Thyristors

Unlike bipolar transistors and Field Effect Transistors (FETs), thyristors *cannot be used for amplification*. Thyristors are semiconductor devices that are specifically designed for use in high-power switching applications. Thyristors can operate only in the switching mode, where they act like either an open or closed switch. Thyristors are used extensively in high-power switching applications, where the control of several hundred amperes of current is not uncommon. High-power thyristors are commonly used in the following applications: lighting systems, heaters, welders, battery chargers, DC and AC motor speed controls, voltage regulators, and more.

Chapter Outline

Chapter Objectives

After studying this chapter, you should be able to

- *Describe* what a thyristor is and *list* its main applications.
- *Explain* the basic operating mode of a thyristor.
- *Describe* the construction and operation of a diac.
- *Describe* the construction and operation of an SCR.
- *Explain* what is meant by the forward breakover voltage of a diac, SCR, or triac.
- *Define* the term *holding current*, I_H, as it relates to thyristors.
- *List* and *explain* important SCR ratings.
- *Explain* how an *RC* phase-shifting network can control the load current in an SCR circuit.
- *Explain* the differences and similarities between an SCR and a triac.
- *Explain* the different operating modes of a triac.
- *Explain* the construction, operation, and applications of a UJT.

Important Terms

bidirectional diode thyristor
diac
forward blocking current
forward breakover voltage, V_{BRF}
holding current, I_H
interbase resistance, R_{BB}
intrinsic standoff ratio, η
peak reverse voltage rating, V_{ROM}
saturation region
silicon controlled rectifier (SCR)
thyristor
triac
unijunction transistor (UJT)

Figure 33–1 Diac. (*a*) Construction. (*b*) Schematic symbol.

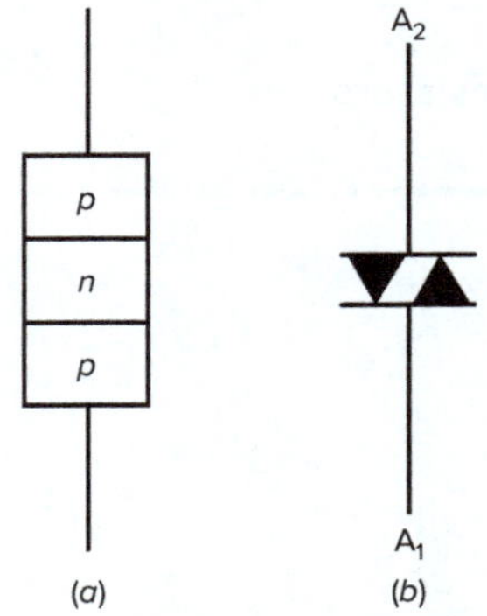

GOOD TO KNOW

Diac stands for "Diode for Alternating current."

GOOD TO KNOW

In Fig. 33–2, notice that the voltage drop across the diac decreases after it begins conducting. Also, notice that the current is increasing during this time. This region is called the negative resistance region because the voltage drop across the diac decreases as the diac current increases.

33–1 Diacs

A **diac** is a three-layer, two-junction semiconductor device that has only two leads. The diac leads are called anode (A) terminals and are labeled A_2 and A_1. In some cases, however, the leads may be labeled MT_1 and MT_2, where the letters "MT" stand for main terminal. A diac is also referred to as a ***bidirectional diode thyristor*** because it conducts current in either direction. Figure 33–1*a* shows the basic construction of a diac, and Fig. 33–1*b* shows the schematic symbol.

Current-Voltage Characteristics of a Diac

Figure 33–2 shows the current-voltage *(I-V)* characteristics of a diac. Notice that the diac does not conduct for either polarity of voltage until the breakover voltage, $\pm V_{BO}$, is reached. Notice also that the breakover voltage is the same for either polarity of voltage across its leads. In other words, the breakover voltage is symmetrical. When the breakover voltage is reached, the diac conducts and its voltage decreases to a lower value. The diac continues to conduct until its current drops below a specified value called the *holding current*, usually designated I_H. The **holding current**, I_H, is defined as the minimum amount of current required to hold the diac in its "on" or conducting state. Diacs are primarily used in power control circuits. The diac helps provide a sharp trigger current pulse that can be used to turn on another **thyristor** device known as a ***triac***. It is important to note that a diac cannot be tested with a VOM or DMM because the output voltage from either type of meter is less than the diac's breakover voltage.

33–1 Self-Review

Answers at the end of the chapter.

a. How many *p-n* junctions does a diac have?
b. Does a diac conduct for either polarity of voltage across its terminals?

33–2 SCRs and Their Characteristics

A ***silicon-controlled rectifier (SCR)*** is a four-layer *pnpn* device. Figure 33–3*a* shows the basic construction of an SCR, and Fig. 33–3*b* shows the schematic symbol. Notice that the SCR has three external leads: the anode (A), cathode (K), and gate (G). Like an ordinary rectifier diode, SCRs are unidirectional devices. An SCR differs from an ordinary rectifier diode, however, in that the SCR will remain in a

Figure 33–2 Current-voltage characteristics of a diac.

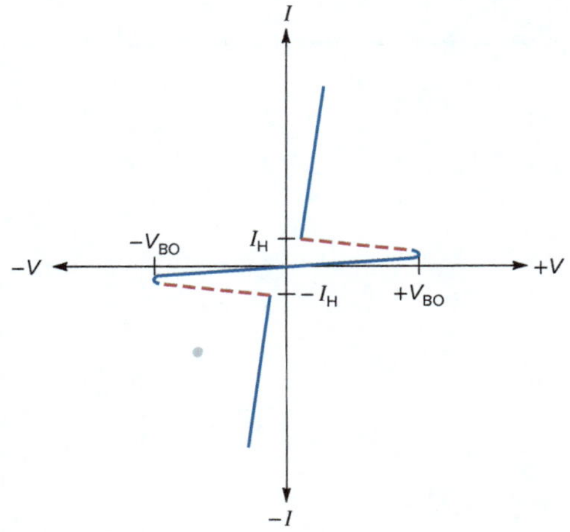

Figure 33–3 Silicon-controlled rectifier. (*a*) Construction. (*b*) Schematic symbol.

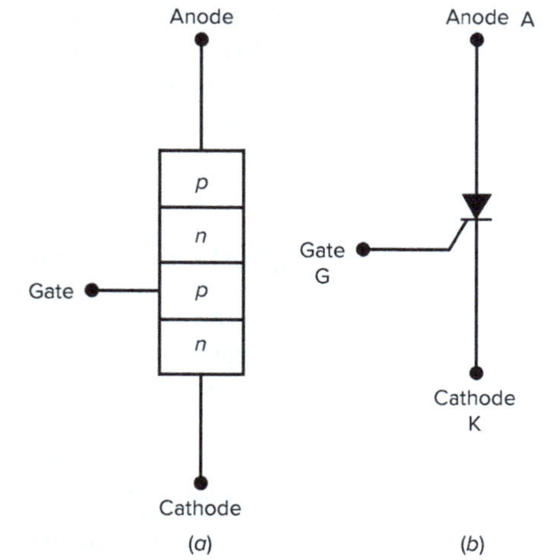

> **GOOD TO KNOW**
>
> Low-current SCRs resemble an ordinary bipolar transistor with three leads protruding from the body. High-current SCRs may be stud-mounted for proper heat sinking.

nonconducting state, although forward-biased, until the **forward breakover voltage, V_{BRF}**, is reached. Once the breakover voltage is reached, the SCR conducts and its voltage drop decreases sharply. Perhaps the most important feature of an SCR is that the forward breakover voltage, V_{BRF}, can be controlled by changing the level of the gate current, I_G.

SCRs come in many shapes and sizes. Some SCRs can safely handle anode currents of less than 1 A, and others can handle anode currents of several hundred amperes.

Figure 33–4*a* shows the current-voltage characteristics of an SCR with the gate open. Notice that when the anode-cathode circuit is reverse-biased, only a very small reverse current flows, called the *reverse blocking current*. When the anode-cathode voltage reaches the ***peak reverse voltage rating***, designated V_{ROM} on the graph, the reverse current increases sharply.

To forward-bias the SCR, the anode is made positive relative to the cathode. As shown in the graph in Fig. 33–4*a*, the forward current, I_F, remains very small until V_{BRF} is reached. The small current that flows before breakover is reached is called the ***forward blocking current***. When the breakover voltage is reached, the forward current (sometimes called the anode current) increases sharply and the voltage drop across the SCR falls to a much lower value. The SCR remains on as long as the anode current stays above the holding current, I_H.

Figure 33–4 Current-voltage characteristics of an SCR. (*a*) Gate open. (*b*) Forward breakover voltage, V_{BRF}, decreases as the gate current, I_G, increases.

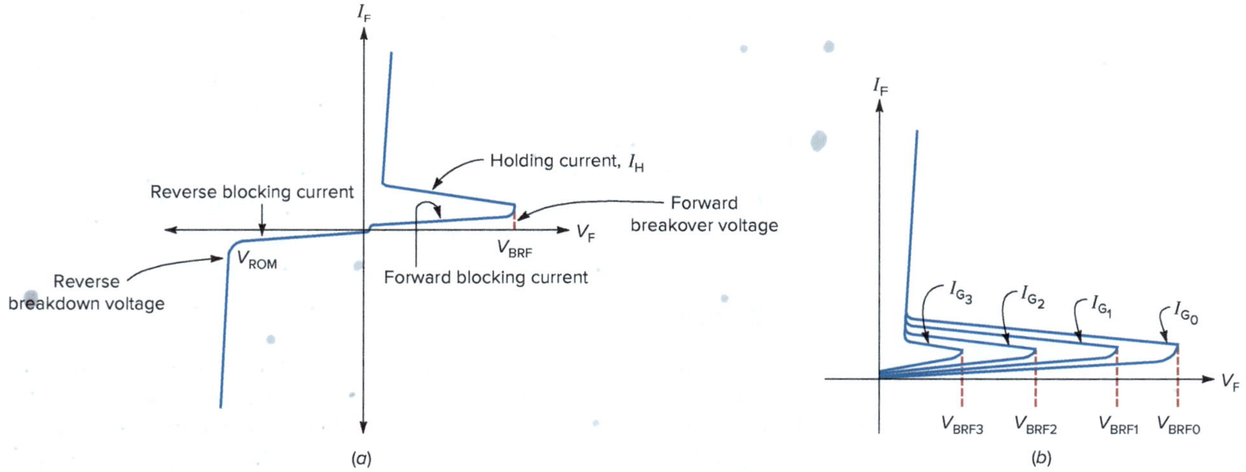

An SCR has only two distinct states of operation: ON or OFF. When the forward voltage is below the value of V_{BRF}, the SCR acts like an open switch. When the forward voltage exceeds the breakover voltage, V_{BRF}, the SCR conducts and acts like a closed switch. As a reminder, note that the SCR remains in the ON state as long as the anode current is greater than the holding current, I_H. The voltage drop across a conducting SCR is typically around 1 V. It is important to note that resistance in the external circuit limits current flow when the SCR is conducting.

Gate Current Controls Forward Breakover Voltage

Figure 33–4*b* shows how the level of the gate current, I_G, can control the forward breakover voltage, V_{BRF}. The maximum forward breakover voltage, V_{BRF}, occurs when the gate current, I_G, equals zero. When the gate-cathode junction is forward-biased, the SCR will fire at a lower anode-cathode voltage. Notice in Fig. 33–4*b* that as the gate current, I_G, is increased, the value of V_{BRF} is decreased. As the value of gate current, I_G, is increased, the SCR functions much like an ordinary rectifier diode.

An important characteristic of an SCR is that once it is turned on by gate current, the gate loses all control. The only way to turn off the SCR is to reduce the anode current below the level of holding current, I_H. Not even a negative gate voltage will turn the SCR off in this case. In most cases, the anode supply voltage is an alternating voltage. This means that the SCR will automatically turn off when the anode voltage drops to zero or goes negative. Of course, when the anode voltage is negative, the SCR is reverse-biased. The process of turning off an SCR is called *commutation*.

To see how an SCR can be tested with an analog VOM, see the "Lab Application Assignment" at the end of this chapter.

Important SCR Ratings

To ensure proper performance and long operating life, the manufacturer's maximum ratings for an SCR should never be exceeded. Exceeding any of the SCR's maximum ratings could permanently damage the device. The following is a list of some important SCR ratings:

I_F (av)	The maximum continuous average forward current. This is usually specified for a half-cycle of a sine wave at a particular frequency. This rating is sometimes referred to as the *maximum DC current rating*.
I_F (rms)	The maximum continuous rms current that the SCR can safely conduct.
I_H	The minimum anode current required to hold the SCR in its conducting or ON state. With the gate lead open, this current is specified as I_{HO}. With the gate returned to the cathode through a resistance, this current is specified as I_{HX}.
V_{ROM}	The maximum reverse-bias voltage that can be applied between the anode and cathode with the gate open.
V_{BRF}	Peak repetitive forward blocking voltage that may be applied with the gate open and the anode positive relative to the cathode.
I_{GT}	The value of gate current required to switch the SCR from its OFF state to its ON state.
dv/dt	The maximum rate of increase in anode-cathode voltage that the SCR can handle safely without false triggering.

SCR Applications

SCRs are frequently used to control the amount of DC power that is delivered to a load. Figure 33–5*a* shows a circuit where an SCR is used to control the amount of load current supplied to a lamp. The input voltage is applied across terminals *A* and *B*. In this example, the input voltage is the 120-V_{AC} power line. R_1 is a

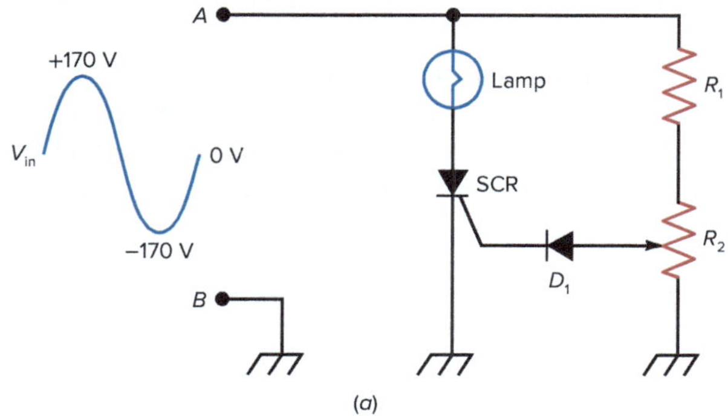

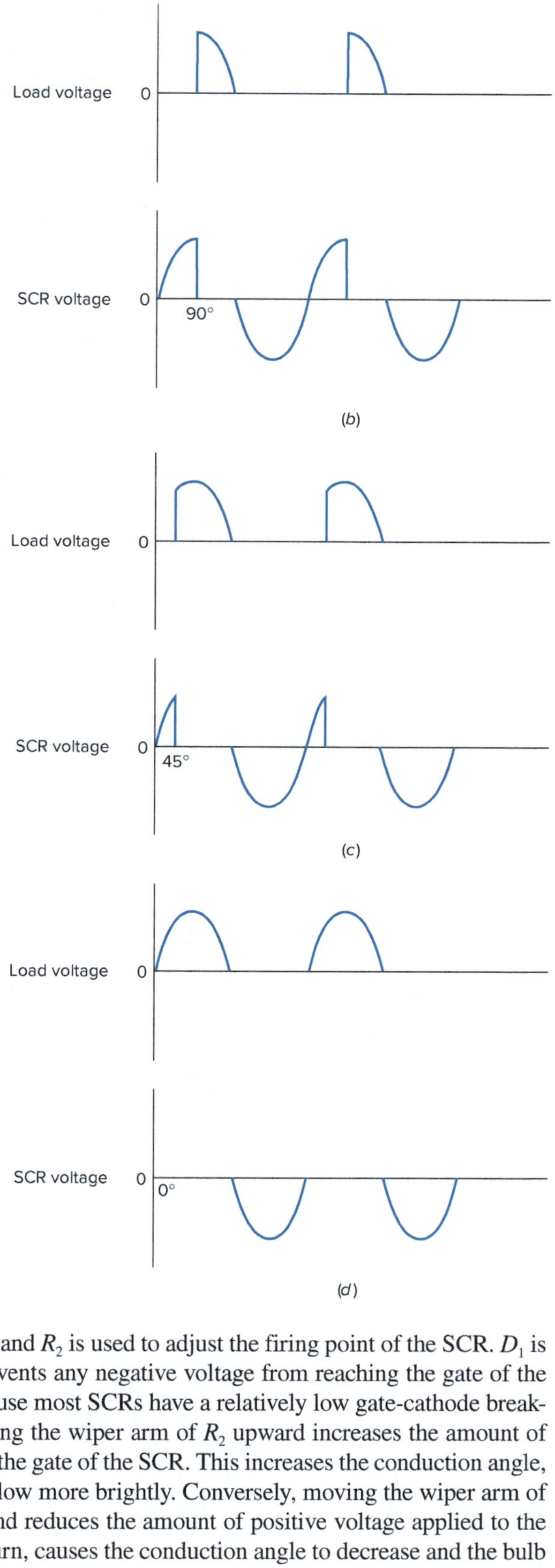

Figure 33–5 SCR used to control the current in the lamp. (*a*) Circuit. (*b*) R_2 set so that the SCR fires when the input voltage reaches its peak value at 90°. (*c*) R_2 set so that the SCR fires earlier at 45° in the input cycle. (*d*) R_2 set at its maximum value. SCR conducts near 0° and the lamp is at full brilliance.

current-limiting resistance and R_2 is used to adjust the firing point of the SCR. D_1 is a protection diode that prevents any negative voltage from reaching the gate of the SCR. D_1 is necessary because most SCRs have a relatively low gate-cathode breakdown voltage rating. Moving the wiper arm of R_2 upward increases the amount of positive voltage applied to the gate of the SCR. This increases the conduction angle, which causes the bulb to glow more brightly. Conversely, moving the wiper arm of R_2 downward toward ground reduces the amount of positive voltage applied to the gate of the SCR. This, in turn, causes the conduction angle to decrease and the bulb glows more dimly.

Figure 33–5*b*, *c*, and *d* show the load and SCR voltage waveforms for different settings of R_2. In Fig. 33–5*b*, R_2 is set so that the SCR will turn on at the peak (90°)

of the positive half-cycle of input voltage, and so the SCR will conduct for only 90° of each cycle. If R_2 were reduced below this setting, the SCR would not fire at any time during the input voltage cycle, and no power would be delivered to the lamp, which would then be dark.

In Fig. 33–5*c*, the setting of R_2 is increased so that the SCR fires at 45° on the positive half-cycle, causing the load current to increase and the bulb to glow more brightly. For this setting, the SCR conducts for 135° of the AC cycle.

Finally, in Fig. 33–5*d*, R_2 is set at its maximum value, causing the SCR to fire at a voltage just above zero on the positive half-cycle. For this setting of R_2, the SCR conducts for 180°, the load current is maximum, and the bulb has maximum brilliance. With the circuit in Fig. 33–5, the conduction angle of the SCR can only be controlled over the range of 90° to 180°.

Notice that the negative alternation appears across the SCR in all waveforms in Fig. 33–5. This occurs because the SCR acts like an open during this time. Also, when the positive alternation drops to a very low value, the anode current drops below the level of holding current, I_H, and the SCR stops conducting. It does not conduct again until the anode-cathode voltage reaches the required value during the positive alternation.

Figure 33–6*a* shows how the conduction angle of an SCR can be controlled over the range of 0° to 180° by using an *RC* phase-shifting network. Recall from basic AC circuit theory that the capacitor and resistor voltage in a series *RC* circuit are always 90° out of phase. In Fig. 33–6*a*, the voltage across the capacitor is applied to

Figure 33–6 *RC* phase-shifting network used to improve the control of load current. (*a*) Circuit. (*b*) *R* near maximum value. (*c*) *R* in middle of its range. (*d*) *R* adjusted to 0 Ω.

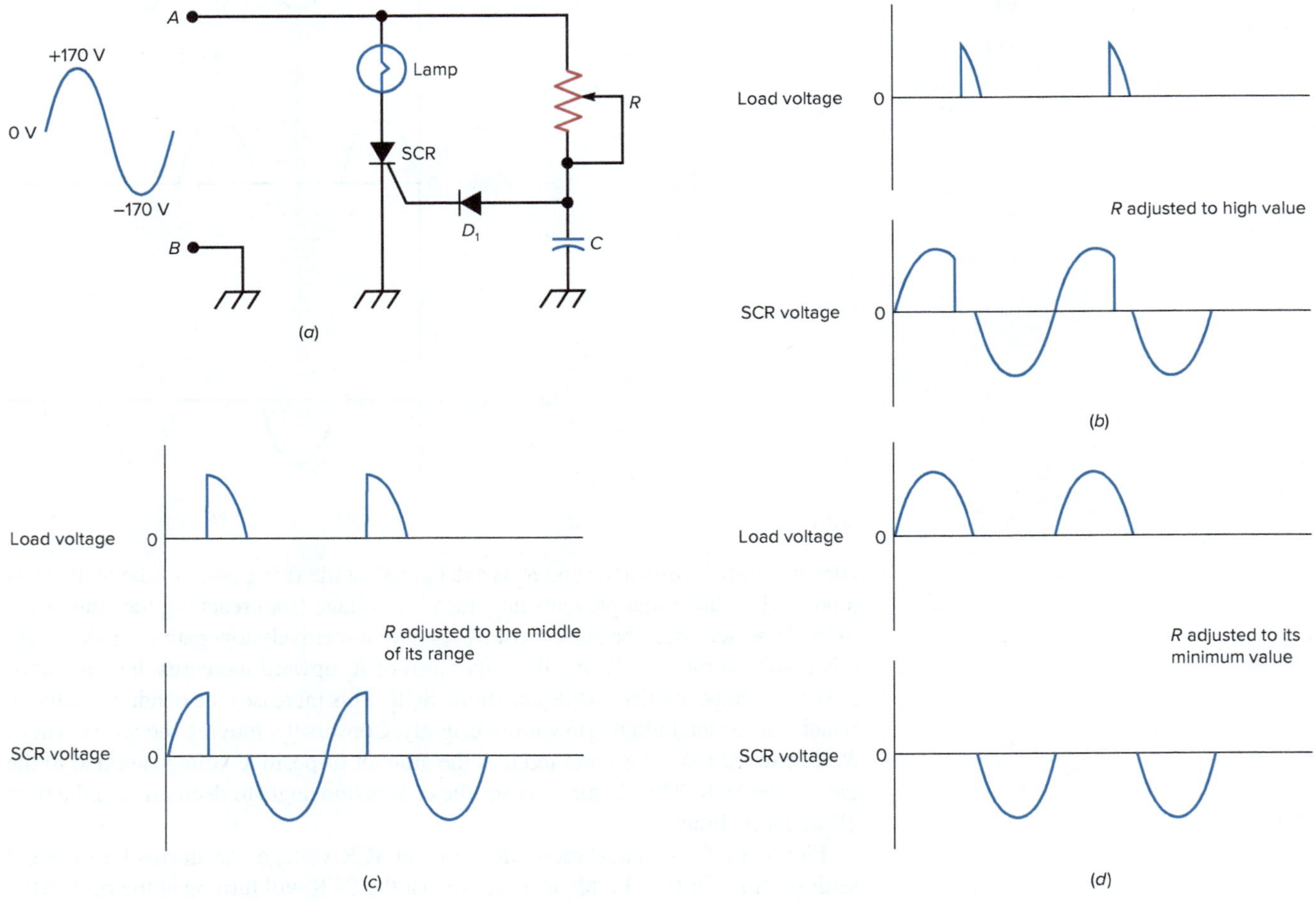

the anode side of the diode, D_1. The cathode lead of the diode connects to the gate of the SCR. Again, the purpose of using the diode is to ensure that the negative alternation of the input voltage cannot apply excessive reverse-bias voltage to the SCR's gate-cathode junction.

When R is increased to nearly its maximum value, the phase angle, θ, between V_{in} and the capacitor voltage, V_C, is approximately 90°. This means it will take longer for the voltage across C to reach the voltage required to fire the SCR. Since the RC network provides a "delay," the SCR can be triggered in the 90° to 180° portion of the input cycle, resulting in smoother control of the load current.

When R is decreased, the SCR fires with less delay, allowing the SCR to fire in the 0° to 90° portion of the input cycle. Because there is control over 180° of the input waveform, the brilliance of the bulb can be varied over a wider range compared to the circuit, as shown in Fig. 33–5*a*. In Fig. 33–6*b*, *c*, and *d*, the load and SCR voltage waveforms are shown for three different settings of resistance, R.

33–2 Self-Review

Answers at the end of the chapter.

a. What happens to the forward breakover voltage of an SCR as the gate current increases?

b. Once an SCR is conducting, can a negative gate voltage turn it off?

33–3 Triacs

GOOD TO KNOW

Triac stands for "Triode alternating current switch."

SCRs have a distinct drawback in that they can conduct current in only one direction. This is a big disadvantage if it is desired to control the power in an AC circuit. A device that can control AC power, because it can conduct in either direction, is called a *triac*. The schematic symbol for a triac is shown in Fig. 33–7*a*. Notice that there are two anode terminals, A_2 and A_1, and a gate lead. The triac is the equivalent of two SCRs connected in parallel, as shown in Fig. 33–7*b*. Notice in Fig. 33–7*b* that both gate leads are tied together. The *I-V* characteristics of a triac are shown in Fig. 33–7*c*. It operates identically to an SCR except that conduction also occurs in the negative voltage region. The curve shown in Fig. 33–7*c* indicates the forward breakover voltages $+V_{BRF}$ and $-V_{BRF}$ with the gate open. Lower values of breakover voltage occur as the gate current, I_G, is increased. As with the SCR, the holding

MultiSim **Figure 33–7** Triac. (*a*) Schematic symbol. (*b*) Equivalent circuit with two SCRs connected in parallel. (*c*) Current-voltage characteristics of a triac with gate open.

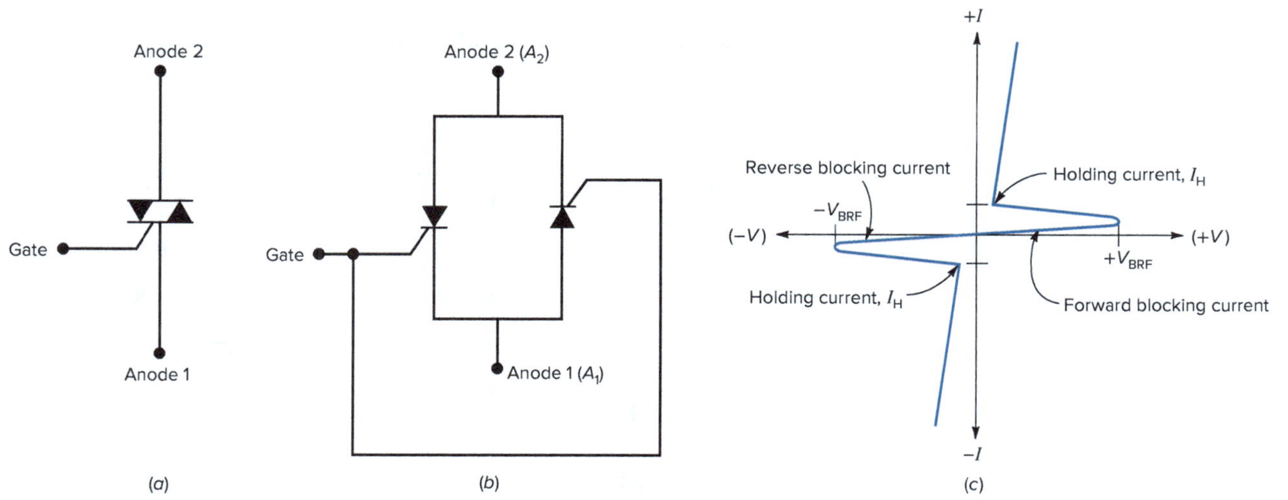

GOOD TO KNOW

The anode terminals of a triac are sometimes referred to as MT_1 and MT_2, where the letters MT stand for main terminal.

Figure 33–8 Operating modes for a triac.

A_2	Gate	Mode
+	+	1
+	−	2
−	+	3
−	−	4

GOOD TO KNOW

Some triacs have built-in diacs which reduces the number of components and connections in some cases.

current, I_H, is the minimum anode current required to keep the triac operating in the on state. The voltage drop across a conducting triac is typically around 1 V. Also, resistance in the external circuit limits current flow when the triac is conducting.

Gate Triggering Characteristics

A triac can be triggered with the gate either positive or negative relative to the anode 1 terminal. This is true regardless of the polarity of the voltage at the anode 2 terminal. (Incidentally, all voltage polarities in a triac are relative to the anode 1 terminal.) Figure 33–8 shows the four operating modes for a triac. Notice that when the anode 2 (A_2) terminal is positive, the triac can be turned on with either a positive or negative gate voltage. Likewise, when the anode 2 (A_2) terminal is negative, the triac can be turned on with either a positive or negative gate voltage.

The triac can be triggered in each mode, but some modes require more or less gate current, I_G, than others. Mode 1 is the most sensitive of all modes; it requires the least amount of gate current to fire the triac. Mode 4 is the next most sensitive mode, but it is not as sensitive as mode 1. The other modes require higher gate current. The sensitivity of each mode is affected by temperature. As the temperature increases, less gate current is required to fire the triac.

AC Control Using Triacs

As shown in Fig. 33–7*c*, a triac can operate with either positive or negative voltage across its terminals. Because of this, a triac can control the amount of current supplied to a load for both the positive and negative alternations of the input cycle. Since a triac requires different gate currents for each mode of operation, it is asymmetrical. Thus, a triac may not trigger at the same point for each alternation of the input cycle, and so a diac is often used in conjunction with a triac to ensure that the triggering time is the same for both the positive and negative alternation of the applied voltage.

Figure 33–9 shows a very effective way to provide a wide range of control over load current. R_1-C_1 and R_2-C_2 provide the required phase shift necessary for full control of the load current.

By adjusting R_1 in Fig. 33–9, the conduction angle can be varied from nearly 0° to approximately 360°. With R_1 adjusted at or near its maximum, the triac does not fire at any point during the AC cycle, resulting in zero AC power delivered to the load, R_L. As R_1 is decreased, the triac fires at the end (near 180° and 360°) of both

Figure 33–9 Triac power control circuit.

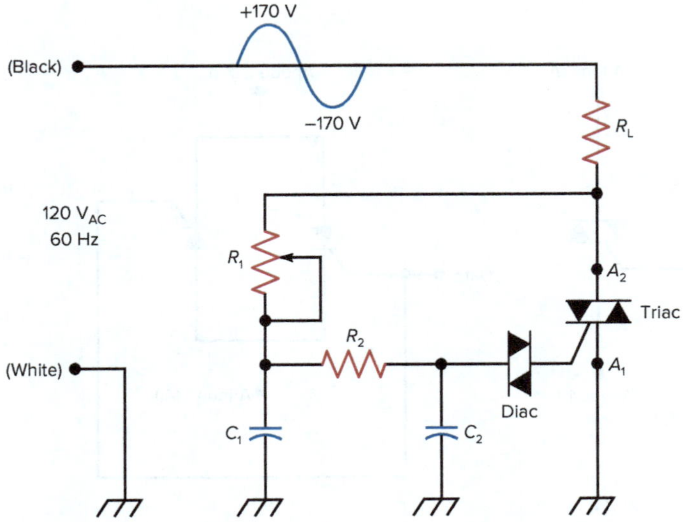

the positive and negative alternations. As a result, the conduction angle is very small, resulting in only a small amount of AC power delivered to the load, R_L. As R_1 is decreased further in value, the triac fires earlier during both alternations. This increases the conduction angle, which means that more AC power will be delivered to the load, R_L.

Note that the current in the triac reverses for each half-cycle of applied voltage. Also, note that the diac is responsible for providing symmetrical triggering during both the positive and negative alternations. It provides symmetrical triggering of the triac because the diac has the same breakover voltage, V_{BO}, for either polarity of applied voltage. Therefore, the triac will fire at the same point during both positive and negative alternations.

The load, R_L, could be a motor, lamp, heater, or any other device that operates on AC power.

33–3 Self-Review

Answers at the end of the chapter.

a. What is the equivalent circuit of a triac?

b. Is the sensitivity of a triac the same in all modes?

33–4 Unijunction Transistors

The **unijunction transistor (UJT)** is a three-terminal semiconductor device that has only one *p-n* junction. Its construction is shown in Fig. 33–10*a*. A bar of *n*-type silicon (Si) is placed on two separate pieces of ceramic. Each piece of ceramic is bonded by a gold film to each end of the *n*-type Si bar, which forms a very low resistance contact. Each end of the Si bar is called a *base*. The top base is referred to as base 2 (B_2) and the bottom base is called base 1(B_1). The *n*-type Si bar is lightly doped and has a resistance value between 5 and 10 kΩ. A *p-n* junction is formed by placing a *p*-type region in the *n*-type Si bar. The *p* material is called the emitter and is placed closer to base 2 than to base 1. Figure 33–10*b* shows the schematic symbol of a UJT. Notice the emitter (E) and the two bases, B_1, and B_2. The equivalent circuit of a UJT is shown in Fig. 33–11. The **interbase resistance, R_{BB}**, is the resistance of the *n*-type silicon bar. R_{BB} appears as two resistances, designated R_{B_1} and R_{B_2}. The value of R_{BB} is dependent upon the doping level and the physical dimensions of the Si bar.

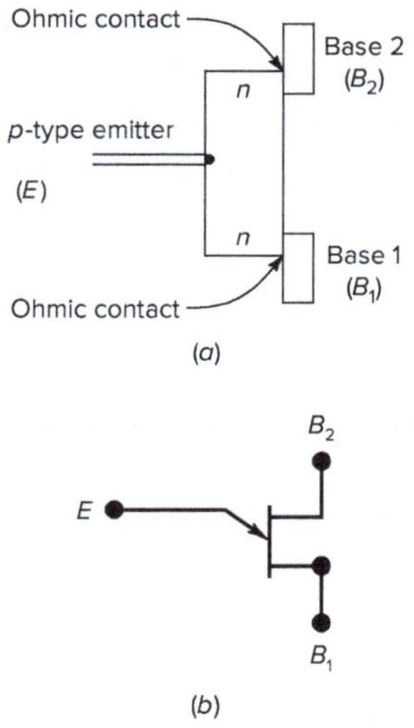

Figure 33–10 Unijunction transistor (UJT). (*a*) Construction. (*b*) Schematic symbol.

If there is zero current in the emitter circuit, R_{BB} appears as a resistive voltage divider for the base supply voltage, V_{BB}.

With zero emitter current in Fig. 33–11, the voltage across R_{B_1} is

$$V_{R_{B_1}} = \frac{R_{B_1}}{R_{B_1} + R_{B_2}} \times V_{BB} \tag{33–1}$$

where V_{BB} is the voltage applied across the interbase resistance, R_{BB}.

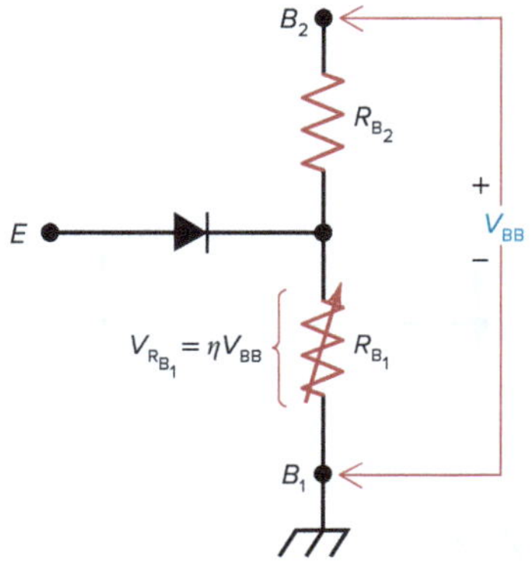

Figure 33–11 Equivalent circuit of a UJT. $R_{BB} = R_{B_1} + R_{B_2}$, when $I_E = 0$.

The ratio $\frac{R_{B_1}}{R_{B_1} + R_{B_2}}$ is called the ***intrinsic standoff ratio*** and is usually designated as η (eta). Typical values of η range from 0.47 to about 0.85.

The emitter diode does not conduct unless the emitter voltage, V_E, exceeds $\eta V_{BB} + V_D$, where V_D equals the emitter diode voltage drop. When V_E is greater than $\eta V_{BB} + V_D$, the emitter diode is forward-biased and emitter current, I_E, flows, which in turn decreases R_{B_1}. R_{B_1} decreases as I_E increases. This is a negative resistance effect and is illustrated in the emitter characteristic curve, as shown in Fig. 33–12. Notice that once V_P is reached, the emitter voltage, V_E, decreases as I_E increases. This occurs until the saturation point is reached. The region to the right of V_V is called the ***saturation region***. Beyond this point, V_E increases with increases in the emitter current, I_E. Beyond the valley points labeled I_V and V_V (in the saturation region), the UJT's resistance is positive.

Figure 33–12 Emitter characteristic curve of a UJT.

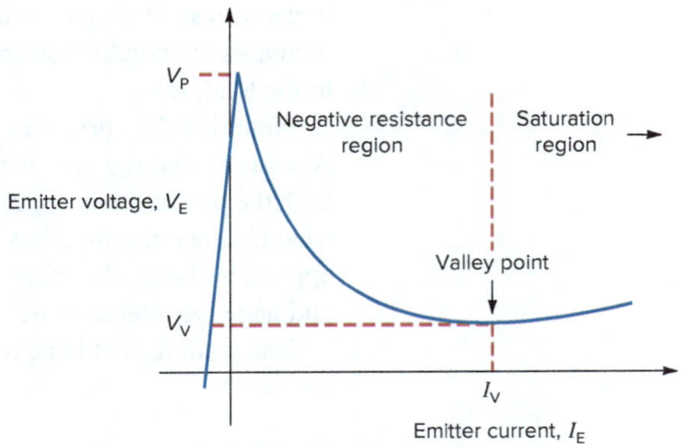

UJT Relaxation Oscillator

Figure 33–13 shows how a UJT can be used as a relaxation oscillator. (An oscillator is a circuit that produces its own output waveform without an AC input signal.) Notice the voltage waveform, V_{B_1}. Since V_{B_1} is a sharp pulse of short duration, it is the ideal gate triggering source for either an SCR or triac. The circuit operates as follows: When power (V_{BB}) is applied, C_T charges exponentially through R_T. When the voltage across C_T reaches the peak point voltage, V_P (equal to $\eta V_{BB} + 0.7$ V), the emitter diode conducts, and then C_T discharges rapidly through the base resistance, R_1, and the lowered resistance, R_{B_1}. The time required for the capacitor to reach V_p is given by Formula (33–2):

$$T = R_T C_T \ln\left(\frac{1}{1-\eta}\right) \qquad \textbf{(33–2)}$$

where T is the period of the emitter voltage waveform shown in Fig. 33–13. Increasing the value of either R_T or C_T increases the time required for the voltage across C_T to reach V_p.

As shown in Fig. 33–13, when the voltage across C has dropped to the value V_V, the UJT turns off and the cycle repeats itself.

Figure 33–13 UJT relaxation oscillator.

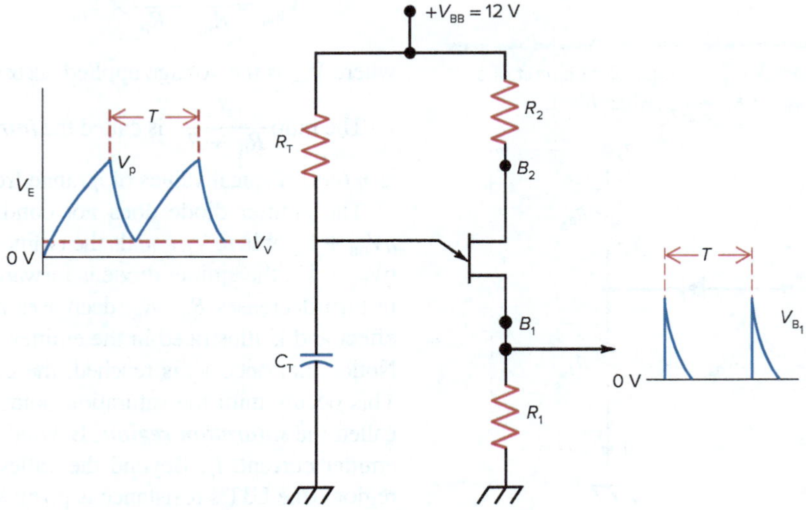

The base 1 voltage waveform, V_{B_1}, is the most important because it is used to trigger the SCR or triac. The frequency of the base 1 voltage waveform is the same as that of the emitter voltage waveform. The frequency, f, equals $1/T$, where T is given by Formula (33–2).

Example 33-1

In Fig. 33–13, $C_T = 0.1\ \mu F$ and $R_T = 220\ k\Omega$. Calculate the frequency of the emitter voltage waveform. Assume $\eta = 0.6$.

ANSWER Using Formula (33–2), the period T is calculated as

$$T = R_T C_T \ln\left(\frac{1}{1-\eta}\right)$$

$$= 220\ k\Omega \times 0.1\ \mu F \times \ln\left(\frac{1}{1-0.6}\right)$$

$$= 20.16\ ms$$

f is calculated as $\frac{1}{T}$:

$$f = \frac{1}{T}$$

$$= \frac{1}{20.16\ ms}$$

$$= 49.6\ Hz$$

UJT Phase Control Circuit

When it is necessary to control very large amounts of power, SCRs rather than triacs are used. SCRs can be designed to handle much higher load currents than triacs.

Figure 33–14 shows how the firing of an SCR can be controlled by a UJT. The circuit operates as follows: The 120 V_{AC} power-line voltage is applied to the bridge rectifier, consisting of diodes D_1 to D_4. The full-wave output from the bridge rectifier is then applied to the rest of the circuit. The full-wave output from the bridge rectifier allows twice the load current available with a sinusoidal input. The pulsating DC also allows the SCR to turn off because when the pulsating DC voltage returns to zero, the anode current drops below the level of the holding current, I_H. The zener diode, D_5, clips off the rectified signal and provides a relatively stable voltage for the UJT relaxation oscillator circuit. The variable resistance, R_T, controls the frequency of the UJT relaxation oscillator, which in turn controls the conduction angle of the SCR. R_T is adjusted to control the firing point of the SCR at different points on the pulsating input voltage waveform. Increasing R_T reduces the load current, I_L, because the SCR is fired later in the input voltage cycle. Decreasing R_T increases the load current, I_L, because the SCR is fired earlier during the input cycle.

In Fig. 33–14, the voltage source for the UJT is a series of flat-topped pulses with the same frequency as the full-wave output from the bridge rectifier. This allows the UJT oscillator to be synchronized with the full-wave output from the bridge rectifier. However, the values of R_T and C_T still affect the frequency of the UJT relaxation oscillator. Synchronization with the full-wave output from the bridge rectifier is achieved when the full-wave output drops to zero every 8.33 ms and the charge cycle starts over for the timing capacitor, C_T.

GOOD TO KNOW

In Fig. 33–14, a bridge rectifier must be used to rectify the 120-V_{AC} power-line voltage because a two-diode full-wave rectifier would require a center-tap connection from a transformer.

Figure 33–14 UJT phase control circuit.

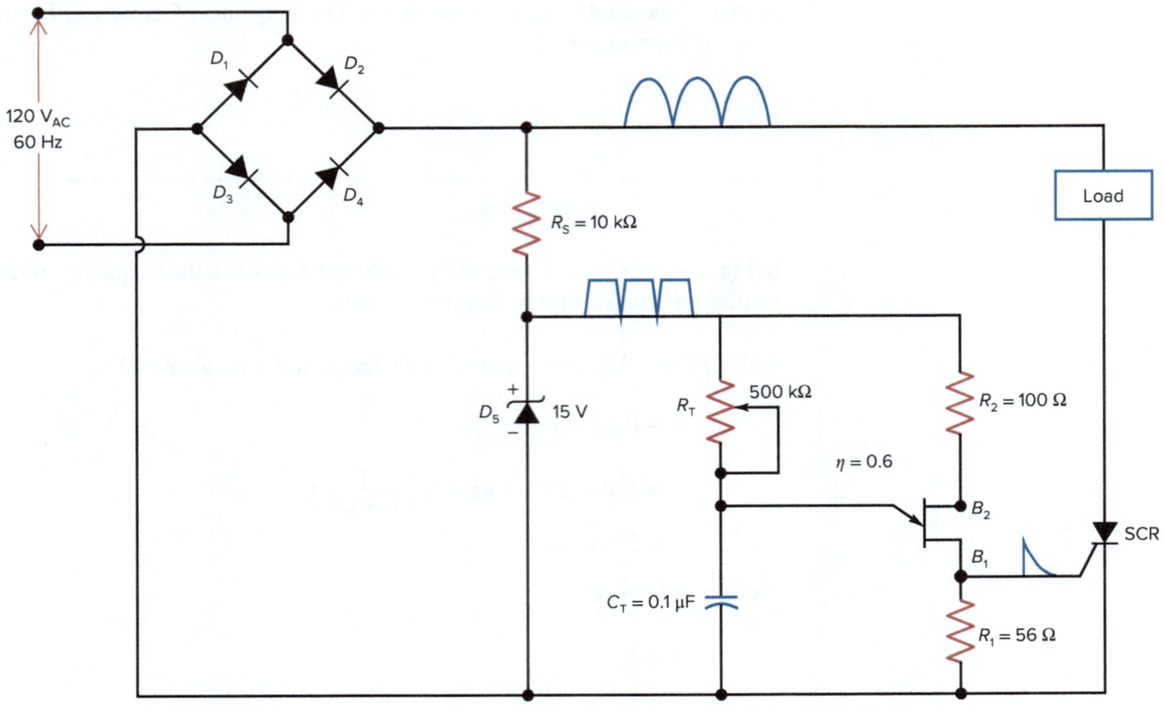

33–4 Self-Review

Answers at the end of the chapter.

a. In a UJT, what is meant by the intrinsic standoff ratio?

b. In the negative resistance region, does V_E increase or decrease as I_E increases?

Summary

- Thyristors are semiconductor devices that can operate only in the switching mode.
- Thyristors are typically used in high-power switching applications to control very large amounts of current.
- A diac is a two-terminal device that consists of three semiconductor layers and two *p-n* junctions. Diacs are often used in conjunction with triacs to provide symmetrical triggering.
- The SCR is a four-layer *pnpn* device with three leads, the anode, gate, and cathode.
- Unlike an ordinary rectifier diode, an SCR will not conduct until the forward breakover voltage is reached, even though its anode cathode is forward-biased.
- The gate current in an SCR controls the forward breakover voltage. The higher the gate current, the lower the forward breakover voltage.
- Once an SCR turns on, the gate loses all control.
- The only way to turn off an SCR is to reduce the anode current below the holding current, I_H, which is the minimum anode current required to keep the SCR conducting.
- A triac is a bidirectional thyristor used to control the power in AC circuits.
- A triac has two leads designated MT_1 and MT_2 or A_1 and A_2.
- Like an SCR, a triac has a gate lead which is used to control its conduction.
- A triac is equivalent to two SCRs in parallel.
- The UJT has two base leads, B_1 and B_2, and an emitter (E) lead.
- The interbase resistance, R_{BB}, of a UJT is the resistance of its *n*-type silicon bar. R_{BB} appears as two resistances designated R_{B_1} and R_{B_2}.
- The ratio $\frac{R_{B_1}}{R_{B_1}+R_{B_2}}$ is called the intrinsic standoff ratio, designated η.
- UJTs are used in conjunction with SCRs and triacs to control their conduction angle.

Important Terms

Bidirectional diode thyristor—another name for a diac.

Diac—a bidirectional semiconductor device that conducts when the voltage across its terminals reaches the breakover voltage, $\pm V_{BO}$. Once conducting, the voltage across the diac drops to a very low value.

Forward blocking current—the small current that flows in an SCR before breakover is reached.

Forward breakover voltage, V_{BRF}—the forward voltage across an SCR at which the SCR begins to conduct. The value of V_{BRF} is controlled by the amount of gate current, I_G.

Holding current, I_H—the minimum amount of current required to hold a thyristor (diac, SCR, or triac) in its conducting state.

Interbase resistance, R_{BB}—the resistance of the *n*-type silicon bar in a UJT. R_{BB} appears as two resistances, R_{B_1} and R_{B_2}: $R_{BB} = R_{B_1} + R_{B_2}$.

Intrinsic standoff ratio, η—the ratio of R_{B_1} to R_{BB}: $\eta = \frac{R_{B_1}}{R_{B_1}+R_{B_2}}$.

Peak reverse voltage rating, V_{ROM}—the maximum reverse-bias voltage that can be safely applied between the anode and cathode terminals of an SCR with the gate open.

Saturation—the region to the right of the valley point on the characteristic curve of a UJT.

Silicon controlled rectifier (SCR)—a unidirectional semiconductor device, like a diode, that remains in a nonconducting state, although forward-biased, until the forward breakover voltage is reached. Once conducting, the voltage across the SCR drops to a very low value.

Thyristor—a semiconductor device with alternating layers of *p* and *n* material that can only be operated in the switching mode where they act as either an open or closed switch.

Triac—a bidirectional semiconductor device that remains in a nonconducting state until the forward breakover voltage is reached. Once conducting, the voltage across the triac drops to a very low value. Like an SCR, the breakover voltage can be controlled by gate current.

Unijunction transistor (UJT)—a three-terminal semiconductor device that has only one *p-n* junction. UJTs are used to control the conduction angle of an SCR.

Related Formulas

$$\eta = \frac{R_{B_1}}{R_{B_1}+R_{B_2}}$$

$$V_{RB_1} = \frac{R_{B_1}}{R_{B_1}+R_{B_2}} \times V_{BB}$$

$$T = R_T C_T \ln\left(\frac{1}{1-\eta}\right)$$

Self-Test

Answers at the back of the book.

1. A diac is a

a. unidirectional device.

b. device with three leads.

c. bidirectional device.

d. both a and b.

2. The forward breakover voltage of an SCR

a. decreases as the gate current increases.

b. cannot be controlled by gate current.

c. increases as the gate current increases.

d. none of the above.

3. Which of the following is best suited for controlling power in AC circuits?

a. the SCR.

b. the triac.

c. an ordinary rectifier diode.

d. none of the above.

4. For a UJT, the intrinsic standoff ratio, η, equals

a. $\frac{R_{B_1}}{R_{B_2}}$.

b. $\frac{R_{B_1}}{R_{B_1} + R_{B_2}}$.

c. $\frac{R_{B_2}}{R_{B_1} + R_{B_2}}$.

d. none of the above.

5. Once an SCR is conducting,

a. its anode to cathode voltage increases substantially.

b. the only way to turn it off is with a positive gate voltage.

c. it can never be turned off.

d. the gate loses all control.

6. For an SCR, the holding current, I_H, is defined as the

a. minimum anode current required to hold the SCR in its conducting state.

b. maximum anode current that the SCR can safely handle.

c. minimum amount of anode current that will keep the SCR off.

d. none of the above.

7. An *RC* phase-shifting network is used in SCR and triac circuits to

a. control the conduction angle of the thyristor.

b. handle some of the load current.

c. vary the holding current.

d. none of the above.

8. Which is the most sensitive mode of operation for a triac?

a. mode 1.

b. mode 2.

c. mode 3.

d. mode 4.

9. Thyristors are used extensively in

a. small signal amplifiers.

b. stereo amplifiers.

c. high-power switching applications.

d. none of the above.

10. A triac is equivalent to

a. two diacs in parallel.

b. an SCR without a gate lead.

c. two ordinary diodes in parallel.

d. two SCRs in parallel.

Essay Questions

1. Why can't an SCR or triac be used to amplify an AC signal?

2. How can an SCR or triac be turned off once it is conducting?

3. What are two similarities and two differences between an SCR and a triac?

4. Why is a diac often placed in series with the gate lead of a triac?

5. What is meant by the negative resistance region of a UJT?

Problems

SECTION 33–1 DIACS

33–1 In what type of circuits are diacs primarily used?

33–2 What makes a diac stop conducting?

33–3 Under what condition will a diac conduct?

SECTION 33–2 SCRs AND THEIR CHARACTERISTICS

33–4 Name the three leads of an SCR.

33–5 How does an SCR differ from an ordinary rectifier diode?

33–6 For an SCR, define the following terms:

a. forward blocking current.

b. reverse blocking current.

33–7 What happens to the anode current in an SCR when the breakover voltage is reached?

33–8 What are the two distinct states of operation for an SCR?

33–9 How can an SCR be turned off?

33–10 How is the forward breakover voltage of an SCR affected by gate current?

33–11 If an SCR is conducting, can a control signal at the gate turn it off?

33–12 Define the following SCR ratings:

a. I_F (av).

b. I_F (rms).

c. I_H.

d. V_{BRF}.

e. dv/dt.

33–13 In Fig. 33–5, what happens to the brightness of the lightbulb as the wiper arm of R_2 is moved upward?

33–14 In Fig. 33–5, does the SCR conduct during the negative alternation of input voltage? Does the lamp light during this time?

33–15 What is the purpose of D_1 in Fig. 33–5?

33–16 What is the advantage of using an *RC* phase-shifting network to control the conduction of an SCR versus the method shown in Fig. 33–5?

SECTION 33–3 TRIACs

33–17 Name the three leads of a triac.

33–18 What is the main advantage of a triac versus an SCR?

33–19 How can the forward breakover voltage of a triac be reduced?

33–20 Can a triac be triggered with a negative gate voltage?

33–21 How can a triac be turned off?

33–22 Why is a triac said to be asymmetrical?

33–23 In Fig. 33–9, what is the purpose of the diac in the gate circuit?

33–24 In Fig. 33–9, what happens to the current in the load, R_L, as the resistance of R_1 is increased?

33–25 In Fig. 33–9, which direction does the current flow in the load, R_L?

SECTION 33–4 UNIJUNCTION TRANSISTORS

33–26 Name the three leads of a UJT.

33–27 In Fig. 33–15, what voltage does the emitter voltage, V_E, need to reach to make the UJT conduct?

Figure 33–15

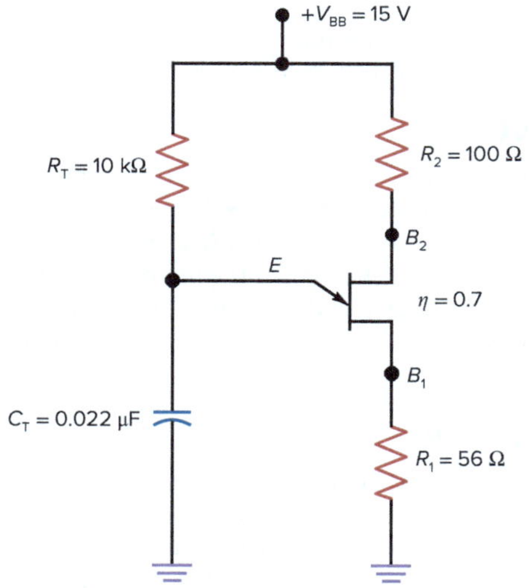

33–28 In Fig. 33–15, what is the frequency of oscillation?

33–29 If $\eta = 0.5$ in Fig. 33–15, what is the frequency of oscillation?

33–30 In Fig. 33–15, what type of waveform appears at the base 1 (B_1) terminal?

33–31 Refer to Fig. 33–14.

a. What type of waveform is produced by the bridge rectifier consisting of diodes $D_1 - D_4$?

b. What is the purpose of the zener diode, D_5

c. What happens to the conduction angle of the SCR as the value of R_T is increased?

Answers to Self-Reviews

33–1 a. two
b. yes

33–2 a it decreases
b. no

33–3 a. two SCRs in parallel
b. no

33–4 a. the ratio $\dfrac{R_{B_1}}{R_{B_1} + R_{B_2}}$
b. V_E decreases

Laboratory Application Assignment

In this lab application assignment, you will examine how to properly test an SCR with an analog VOM. You will also see that a DMM cannot properly test an SCR. Finally, you will build a simple test circuit that reinforces the basic operation of an SCR.

Equipment: Obtain the following items from your instructor.

- Two SPST switches
- Simpson 260 analog VOM or equivalent
- DMM
- 12-V incandescent lamp
- Variable dc Power Supply
- 330-Ω carbon-film resistor
- Low- or medium-current SCR

Analog VOM

Examine the analog VOM you will be using for this part of the experiment. Set the VOM to the $R \times 1$ resistance range, and short the ohmmeter leads together. Adjust the zero-ohms control for full-scale deflection of the meter's pointer. If the pointer does not deflect all the way to 0 Ω, the meter's battery needs to be replaced. If this is the case, ask your instructor for a new battery.

Testing an SCR

With the analog VOM set to the $R \times 1$ range, connect the ohmmeter leads as shown in Fig. 33–16*a*. As you can see, this connection provides a positive (+) voltage at the anode (A) of the SCR with respect to the cathode (K). Even though the SCR is forward-biased, the forward voltage applied by the VOM is much less than the SCR's forward breakover voltage, V_{BRF}. Therefore, the meter should read infinite ohms, indicating that the SCR is *not* conducting. Does your meter show infinite ohms? ______________________________

Reverse the connection of the ohmmeter leads, as shown in Fig. 33–16*b*. As you can see, this connection reverse-biases the SCR because it applies a negative (−) voltage at the anode (A) with respect to the cathode (K). As a result, the meter should still read infinite ohms. Does it? ______________

Return the ohmmeter leads to their original polarity, as shown in Fig. 33–16*c*. With the gate lead still open, the meter should read infinite ohms. Now place a jumper from the anode (A) to the gate (G) as shown. This connection provides the gate with a positive (+) voltage with respect to the cathode (K). The meter should now read a low resistance. Does it? ______________
If the meter shows a low resistance, it indicates the SCR is ON or conducting.

Remove the jumper from the anode to the gate. The meter should still show a low resistance. Does it? __________ Does this test indicate that once the SCR is conducting, the gate loses all control? ______________

Figure 33–16

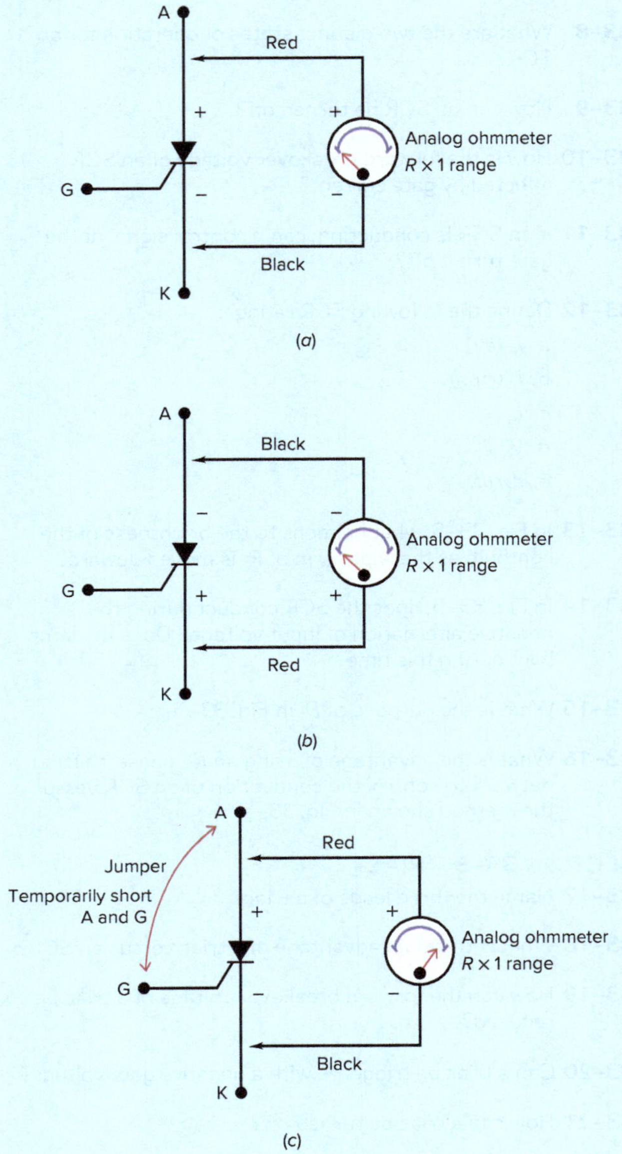

Do the results of your test indicate that the SCR is good or bad? ______________________________

Repeat this testing procedure with a DMM. Pay special attention to the step where the jumper is removed from the anode to the gate. Describe, in detail, your results below.

____
____
____

On a separate sheet of paper show how an analog VOM could be used to test a triac.

SCR Test Circuit

Connect the SCR test circuit in Fig. 33–17. Switches S_1 and S_2 should initially be in the *open* position. Indicate the state of the lamp in each step of the following procedure.

a. With both S_1 and S_2 open, is the lamp lit or is it dark?

____

b. Close S_1. Did the lamp light? ____________________

c. With S_1 still closed, now close S_2. Is the lamp lit now?

____

d. Open S_2. Did the lamp stay lit, or did it go dark?

____

e. Open S_1. What is the condition of the lamp now?

____

In step d, did the SCR turn off when S_2 was opened? ________ If not, explain why.

____

____

Imagine the DC voltage source in Fig. 33–17 was replaced with an AC voltage source whose output is 48 Vpp. Would the results of the SCR test circuit differ or be the same using the same switch sequence for S_1 and S_2?

Figure 33–17

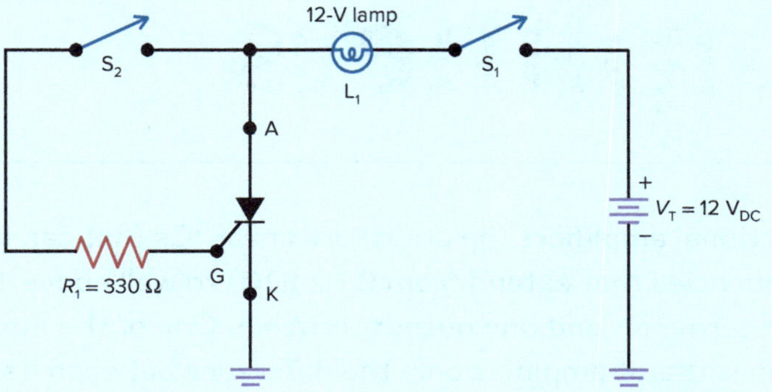

Design credits: Chapter Opener Background Image, and Multisim: Stockbyte/Getty Images

chapter 34

Operational Amplifiers

Operational amplifiers (op amps) are linear ICs that can be used to amplify signal frequencies that extend from 0 Hz (DC) to well above 1 MHz. Op amps have two input terminals and one output terminal. One of the most important qualities of an op amp is that it amplifies only the difference between its two input signals, while rejecting or severely attenuating signals common to both inputs. This allows op amps to be used in systems where a large amount of electrical noise is present. In this case, the desired signal is amplified, while the noise common to both inputs is attenuated.

The first stage of every op amp is a differential amplifier. This chapter describes the DC and AC characteristics of this circuit and then explains the op amp in greater detail. Following a discussion of op-amp theory, several op-amp circuits are introduced. ■

Chapter Outline

Chapter Objectives

After studying this chapter, you should be able to

- *Calculate* the DC values in a differential amplifier.
- *Calculate* the differential voltage gain, A_d, of a differential amplifier.
- *Calculate* the common-mode voltage gain, A_{CM}, of a differential amplifier.
- *Calculate* the voltage gain, input impedance, and output impedance in inverting and noninverting amplifiers.
- *Explain* what is meant by the gain-bandwidth product.
- *Calculate* the output voltage of a summing amplifier.
- *Calculate* the cutoff frequencies of an active low-pass and active high-pass filter.
- *Explain* the operation of a voltage-to-current and current-to-voltage converter.
- *Explain* the operation of an op-amp comparator.
- *Explain* the operation of a precision half-wave rectifier.

Important Terms

active filter

closed-loop cutoff frequency, f_{CL}

closed-loop voltage gain, A_{CL}

common-mode input

common-mode rejection ratio (CMRR)

common-mode voltage gain, A_{CM}

comparator

differential input voltage, V_{id}

differential voltage gain, A_d

f_{unity}

input bias current, I_B

input offset current, I_{os}

negative feedback

negative saturation voltage, $-V_{sat}$

open-loop cutoff frequency, f_{OL}

open-loop voltage gain, A_{VOL}

operational amplifier (op amp)

positive saturation voltage, $+V_{sat}$

power bandwidth (f_{max})

Schmitt trigger

slew rate, S_R

slew-rate distortion

summing amplifier

tail current, I_T

voltage follower

zero-crossing detector

34–1 Differential Amplifiers

Figure 34–1*a* shows the most common form of a differential amplifier. Notice that it has two inputs but only one output. The output is taken from the collector of Q_2. The base of Q_1 is called the *noninverting* input; the base of Q_2 is the *inverting* input. The voltage applied to the noninverting input is designated V_1, and the voltage applied to the inverting input is designated V_2. The output voltage for the differential amplifier in Fig. 34–1*a* equals

$$V_{out} = A_d (V_1 - V_2) \quad \textbf{(34–1)}$$

where A_d represents the differential voltage gain.

GOOD TO KNOW

An op amp that uses JFETs for the input differential amplifier and bipolar transistors for the following stages is called a biFET op amp.

If the bases of Q_1 and Q_2 are grounded, the DC output voltage equals the quiescent collector voltage, V_C.

In Fig. 34–1*b*, the base of Q_2 is grounded, and a signal is applied at the base of Q_1. Notice that V_{out} and V_1 are in phase. Figure 34–1*c* shows the other condition in which the base of Q_1 is grounded and a signal is applied to the base of Q_2.

For this condition, the input and output signals are 180° out of phase. Finally, Fig. 34–1*d* shows the condition in which two inputs are applied simultaneously. Notice that the two inputs, V_1 and V_2, are 180° out of phase. Also, note that V_{out} is in phase with V_1.

DC Analysis of a Differential Amplifier

Figure 34–2*a* shows a differential amplifier with both bases connected to ground through the base resistors, R_{B_1} and R_{B_2}. Each base must have a DC return path to ground; otherwise, the transistor with the open base will go into cutoff.

Under ideal conditions, the transistors Q_1 and Q_2 would be perfectly matched. For this analysis, assume that Q_1 and Q_2 are matched.

The DC current through the emitter resistor, R_E, is often called the *tail current* and is usually designated I_T. With Q_1 and Q_2 perfectly matched, I_T splits evenly between the emitter of each transistor. The **tail current, I_T,** is calculated using Formula (34–2):

$$I_T = \frac{V_{EE} - V_{BE}}{R_E} \quad \textbf{(34–2)}$$

where $V_{BE} = 0.7$ V at the emitter terminals. In Fig. 34–2*a*, I_T is

$$I_T = \frac{V_{EE} - V_{BE}}{R_E}$$

$$= \frac{15\text{ V} - 0.7\text{ V}}{10\text{ k}\Omega}$$

$$= 1.43\text{ mA}$$

The tail current, I_T, equals $2I_E$ or

$$I_T = 2I_E$$

where I_E is the emitter current for each transistor. Therefore, each transistor has an emitter current, I_E, of $I_T/2$ or

$$I_E = \frac{V_{EE} - V_{BE}}{2R_E} \quad \textbf{(34–3)}$$

Figure 34–2*b* shows the DC equivalent circuit where each transistor has its own separate emitter resistor of $2R_E$.

In Fig. 34–2, the DC emitter current for each transistor is

$$I_E = \frac{I_T}{2}$$

$$= \frac{1.43\text{ mA}}{2}$$

$$= 715\ \mu\text{A}$$

Figure 34–1 Differential amplifier. (*a*) Circuit showing the inverting and noninverting inputs. (*b*) V_2 grounded. Signal applied to noninverting input. V_1 and V_{out} are in phase. (*c*) V_1 grounded. Signal applied to inverting input. V_2 and V_{out} are 180° out of phase. (*d*) Signals applied to both the noninverting and inverting inputs.

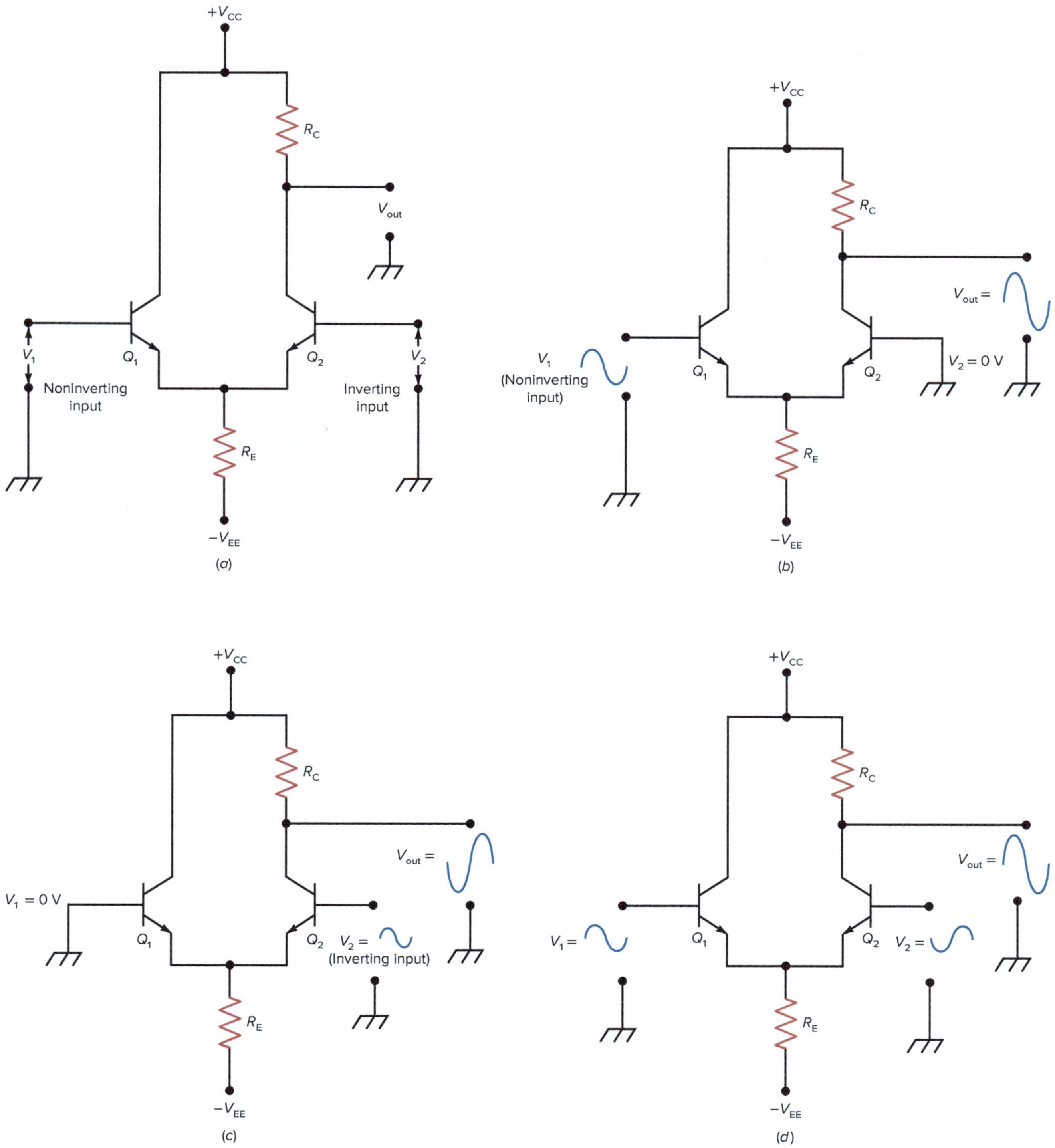

Figure 34–2 DC analysis of a differential amplifier. (*a*) Both bases grounded through the base resistors, R_{B_1} and R_{B_2}. (*b*) Equivalent circuit showing R_E split into two separate resistors of $2R_E$.

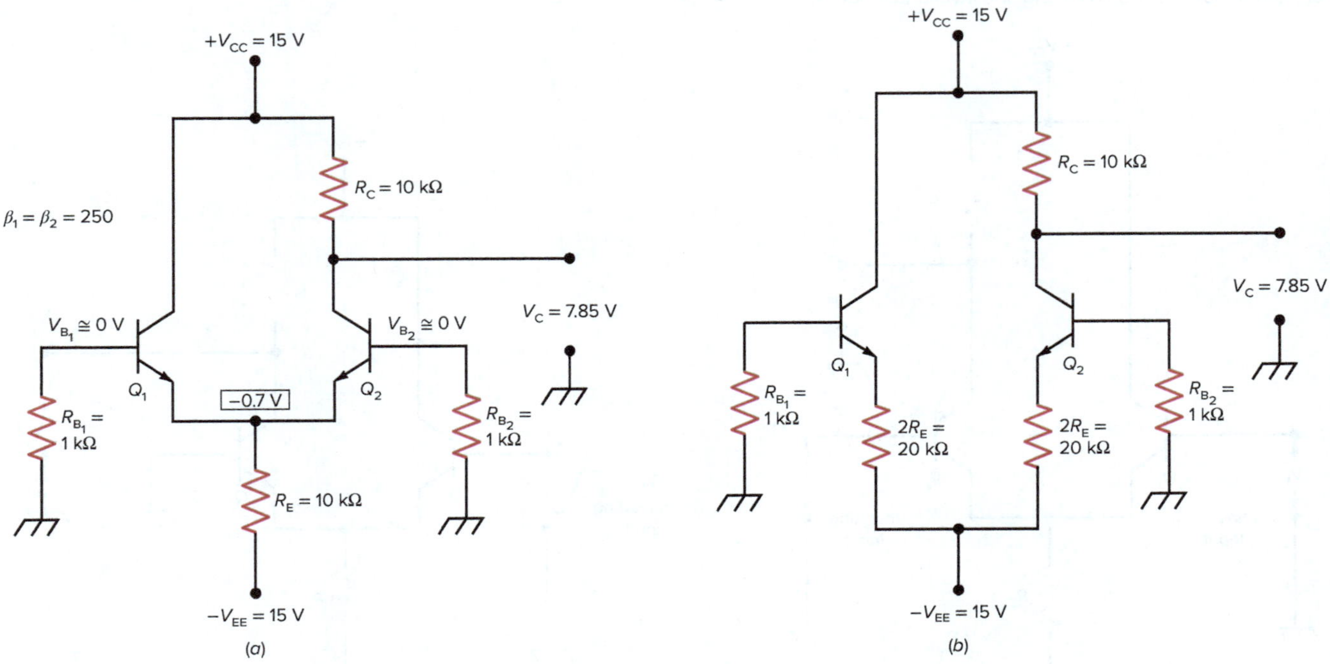

or

$$I_E = \frac{V_{EE} - V_{BE}}{2R_E}$$
$$= \frac{15\text{V} - 0.7\text{ V}}{20\text{ k}\Omega}$$
$$= 715\ \mu\text{A}$$

Either method produces the same results.

The DC collector voltage equals

$$V_C = V_{CC} - I_C R_C \tag{34–4}$$

This assumes that $I_E \approx I_C$. In Fig. 34–2, V_C is

$$V_C = V_{CC} - I_C R_C$$
$$= 15\text{ V} - (715\ \mu\text{A} \times 10\text{ k}\Omega)$$
$$= 15\text{ V} - 7.15\text{ V}$$
$$= 7.85\text{ V}$$

This DC voltage exists when both bases are grounded. This condition, of course, assumes that both transistors are identical.

The base voltages, V_{B_1} and V_{B_2}, are assumed to be approximately 0 V. As proof, assume that $B_1 = B_2 = 250$. Then, I_B is

$$I_B = \frac{I_C}{\beta}$$
$$= \frac{715\ \mu\text{A}}{250}$$
$$= 2.86\ \mu\text{A}$$

Therefore, the DC base voltage is

$$V_{B_1} = V_{B_2} = I_B R_B$$

where $R_{B_1} = R_{B_2} = 1\text{ k}\Omega$.

In Fig. 34–2, V_B is

$$V_{B_1} = V_{B_2} = 2.86\ \mu A \times 1\ k\Omega$$
$$= 2.86\ mV$$

Since this voltage is very small, it can be ignored in the calculations of the emitter current, I_E.

AC Analysis of a Differential Amplifier

Figure 34–3 shows how to analyze a differential amplifier with an AC input. In Fig. 34–3*a*, the input, V_1, is applied to the base of Q_1 while the base of Q_2 is grounded. Figure 34–3*b* shows the circuit redrawn. Q_1 acts as an emitter follower, and Q_2 acts as a common-base amplifier. Because neither transistor has phase inversion, the output signal is in phase with V_1. Hence, the base of Q_1 is called the noninverting input. Notice in Fig. 34–3*b* that V_1 divides evenly between Q_1 and Q_2. This can best be explained by examining the AC equivalent circuit in Fig. 34–3*c*. Notice that as far as the AC equivalent circuit is concerned, r'_{e1} and r'_{e2} serve as a voltage divider for the input, V_1. Since $I_{E_1} = I_{E_2}$, then $r'_{e1} = r'_{e2}$, assuming Q_1 and Q_2 are matched. Notice also in Fig. 34–3*c* that R_E is in parallel with r'_{e2}. Since $R_E \gg r'_{e2}$, the effects of R_E can be ignored. Figure 34–3*d* shows the AC equivalent circuit with R_E omitted.

In Fig. 34–3*d*, the output voltage equals $i_C R_C$. The input voltage equals $2i_C r'_e$. Therefore, the voltage gain, A_d, is

$$A_d = \frac{V_{out}}{V_{in}}$$
$$= \frac{i_C R_C}{2i_C r'_e}$$

which simplifies to

$$= \frac{R_C}{2r'_e}$$

This can be clearly stated in Formula (34–5):

$$A_d = \frac{R_C}{2r'_e} \qquad \textbf{(34–5)}$$

Inverting Input

A similar analysis can be applied when the noninverting input is grounded and a signal is applied to the inverting input. Then the formula for A_d is

$$A_d = -\frac{R_C}{2r'_e} \qquad \textbf{(34–6)}$$

where the minus (−) sign is used to indicate the 180° phase inversion.

Differential Voltage Gain, A_d

When V_1 and V_2 are applied simultaneously, the output voltage is

$$V_{out} = \frac{R_C}{2r'_e}(V_1 - V_2) \qquad \textbf{(34–7)}$$

where $\frac{R_C}{2r'_e}$ represents the **differential voltage gain, A_d.**

If V_1 and V_2 are equal, then $V_1 - V_2 = 0$ V, and the output voltage equals its quiescent value of 7.85 *V* (ideally).

MultiSim **Figure 34–3** AC analysis of a differential amplifier. (*a*) Original circuit. (*b*) Circuit redrawn. Q_1 acts as an emitter follower, and Q_2 acts as a common-base amplifier. (*c*) AC equivalent circuit showing how V_1 divides evenly between r'_{e1} and r'_{e2}. (*d*) AC equivalent circuit with R_E omitted, since $R_E \gg r'_{e2}$.

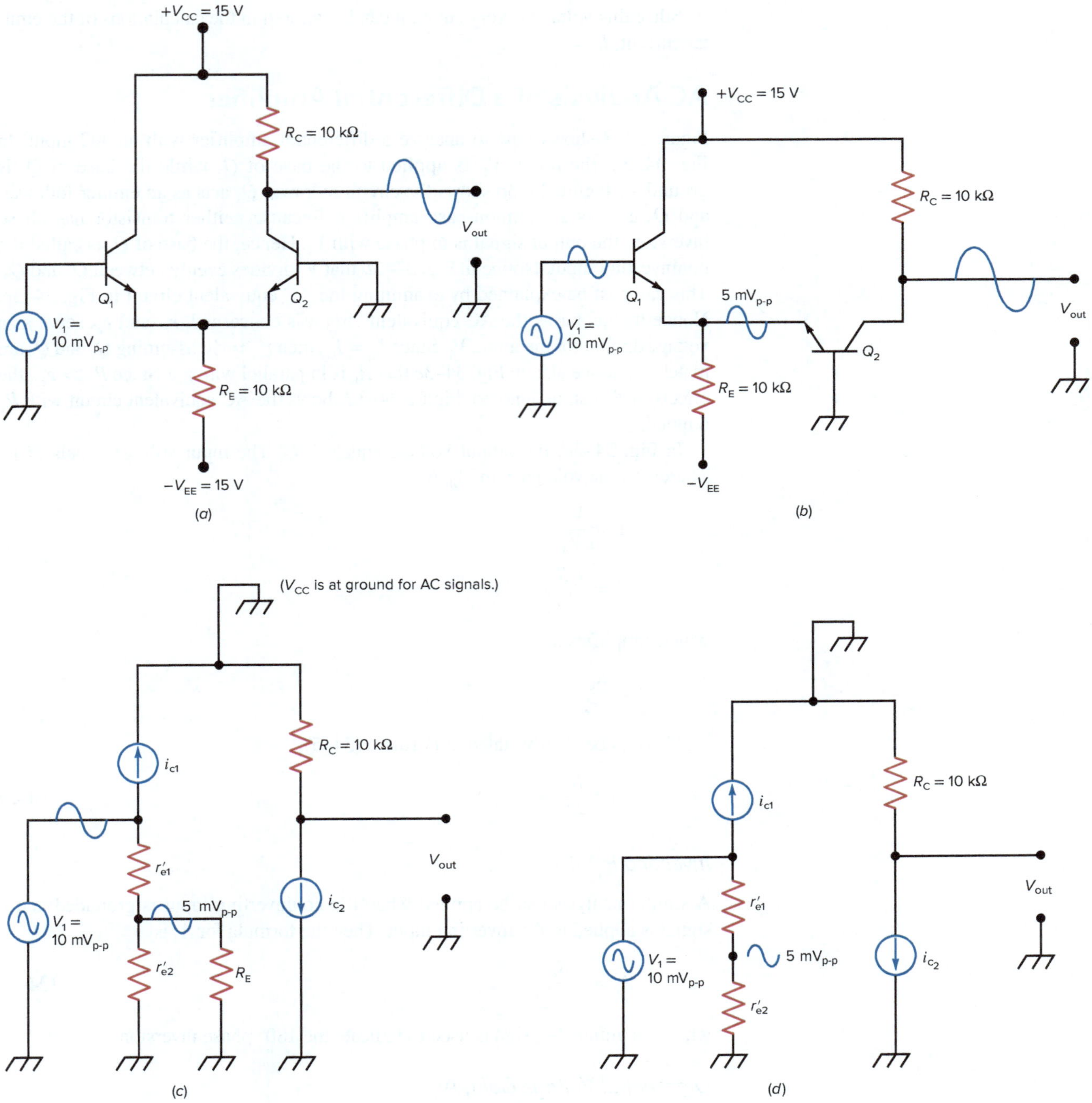

Common-Mode Voltage Gain, A_{CM}

Figure 34–4 shows how to analyze a differential amplifier when a **common-mode input** signal is applied. The original circuit is shown in Fig. 34–4*a*. Notice that the same signal is applied to each base. The signal applied to each base is assumed to have exactly the same phase and amplitude, hence the name *common-mode* input. Since the DC emitter currents are equal, R_E can be split into two separate resistances each equal to $2R_E$, as shown in Fig. 34–4*b*.

Figure 34-4 Differential amplifier with a common-mode input, $V_{in(cm)}$. (a) Original circuit. (b) Equivalent circuit. (c) AC equivalent circuit.

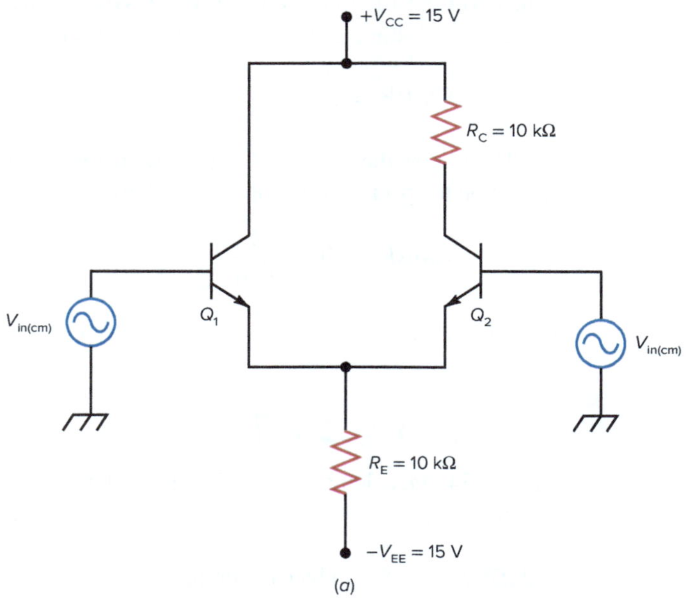

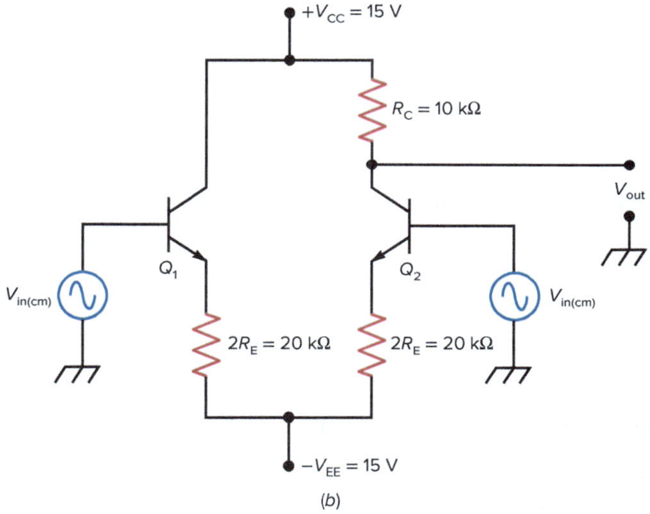

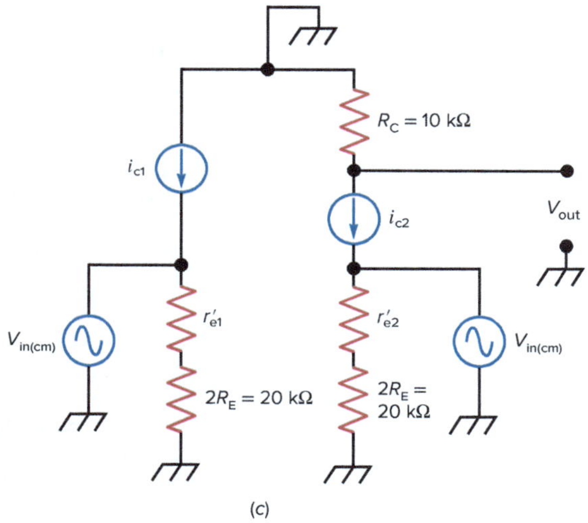

The AC equivalent circuit is shown in Fig. 34-4c. The output voltage equals

$$V_{out} = i_C R_C$$

The input voltage equals

$$V_{in(CM)} = i_C\,(r'_e + 2R_E)$$

Therefore, the **common-mode voltage gain, A_{CM},** is

$$A_{CM} = \frac{i_C R_C}{i_C(r'_e + 2R_E)}$$

which simplifies to

$$A_{CM} = \frac{R_C}{r'_e + 2R_E}$$

In most cases, $r'_e \ll 2R_E$ and the formula simplifies to

$$A_{CM} = \frac{R_C}{2R_E} \qquad \textbf{(34-8)}$$

Common-Mode Rejection Ratio (CMRR)

The **common-mode rejection ratio (CMRR)** is usually defined as the ratio of the differential voltage gain, A_d, to the common-mode voltage gain, A_{CM}:

$$\text{CMRR} = \frac{A_d}{A_{CM}} \tag{34–9}$$

The higher the value of CMRR, the better the differential amplifier. It is most common to specify the CMRR in decibels:

$$\text{CMRR} = 20 \log \frac{A_d}{A_{CM}} \tag{34–10}$$

Example 34-1

In Fig. 34–3*a*, calculate the differential voltage gain, A_d, and the AC output voltage, V_{out}.

ANSWER Begin by calculating the AC emitter resistance, r'_e. Since $I_E = 715\ \mu A$, the calculations are

$$\begin{aligned} r'_e &= \frac{25\text{ mV}}{I_E} \\ &= \frac{25\text{ mV}}{715\ \mu\text{A}} \\ &\approx 35\ \Omega \end{aligned}$$

The differential voltage gain, A_d, is calculated by Formula (34–5):

$$\begin{aligned} A_d &= \frac{R_C}{2r'_e} \\ &= \frac{10\text{ k}\Omega}{70\ \Omega} \\ &= 142.86 \end{aligned}$$

Now the AC output voltage, V_{out}, can be calculated:

$$\begin{aligned} V_{out} &= A_V \times V_{in} \\ &= 142.86 \times 10\text{ mV}_{p\text{-}p} \\ &\approx 1.43\text{ V}_{p\text{-}p} \end{aligned}$$

Example 34-2

In Fig. 34–3, calculate the common-mode voltage gain, A_{CM}, and the CMRR (dB).

ANSWER Begin by calculating A_{CM}. Use Formula (34–8):

$$\begin{aligned} A_{CM} &= \frac{R_C}{2R_E} \\ &= \frac{10\text{ k}\Omega}{20\text{ k}\Omega} \\ &= 0.5 \end{aligned}$$

A_d was calculated earlier in Example 34–1. Its value is 142.86. To calculate the common-mode rejection ratio in dB, use Formula (34–10):

$$\begin{aligned} \text{CMRR} &= 20 \log \frac{A_d}{A_{CM}} \\ &= 20 \log \frac{142.86}{0.5} \\ &= 20 \log 285.7 \\ &= 49.1 \text{ dB} \end{aligned}$$

A CMRR of 49.1 dB means that the differential input signal will appear 285.7 times larger at the output than the common-mode input signal.

34–1 Self-Review

Answers at the end of the chapter.

a. **How does the DC emitter current in each transistor of a differential amplifier relate to the tail current?**

b. **Which is higher, the differential voltage gain or common-mode voltage gain?**

c. **If a differential amplifier has an A_d of 200 and A_{CM} of 0.25, what is the CMRR in dB?**

34–2 Operational Amplifiers and Their Characteristics

Operational amplifiers (op amps) are the most commonly used type of linear integrated circuit (IC). By definition, an *op amp* is a high-gain, direct coupled, differential amplifier. An op amp referred to as the 741 has become an industry standard. This op amp, which is contained in an eight-pin IC, is made by several manufacturers. They are, however, all equivalent since the specifications are nearly identical from one manufacturer to another.

Figure 34–5*a* shows the internal diodes, transistors, resistors, and capacitors for a 741 op amp. The base leads of Q_1 and Q_2 connect to pins on the IC unit and serve as the two inputs for the op amp. Q_1 and Q_2 form a differential amplifier circuit. This circuit is used because it can amplify the difference in voltage between the two input signals.

The output of the op amp is taken at the emitters of transistors Q_8 and Q_9. These transistors are connected in a push-pull configuration. Q_8 conducts during the positive half-cycle of the output waveform and Q_9 conducts during the negative half-cycle. This push-pull configuration allows the op amp to have very low output impedance, which is analagous to a voltage source having very low internal resistance.

GOOD TO KNOW

Many general-purpose op amps are now produced with biFET technology because this provides superior performance over bipolar op amps. BiFET op amps generally have a wider bandwidth, higher slew rate (S_R), larger power output, higher input impedances, and much lower **input bias currents.**

When viewing the circuit in Fig. 34–5*a*, it is important to note that direct coupling is used between all stages. Direct coupling means that the output of one stage is connected directly to the input of the next, without using any capacitors or transformers to isolate the DC voltages in each stage. For this reason, the op amp can amplify signals all the way down to DC. Capacitor C_C affects the operation of the op amp at higher frequencies. This capacitor is called a *compensating* capacitor and has a value of about 30 pF. C_C is used to prevent undesirable oscillations within the op amp. This capacitor also produces ***slew-rate distortion***, which will be discussed later.

The schematic symbol commonly used for op amps is shown in Fig. 34–5*b*. Notice that the triangular schematic symbol shows only the pin connections to different points inside the op amp. Pin 7 connects to $+V_{CC}$, and pin 4 connects to $-V_{CC}$. Also, pins 2 and 3 connect to the op-amp inputs, and pin 6 connects to the op-amp output.

Figure 34–5 741 op amp. (*a*) Simplified schematic diagram. (*b*) Schematic symbol for op amp showing pin numbers.

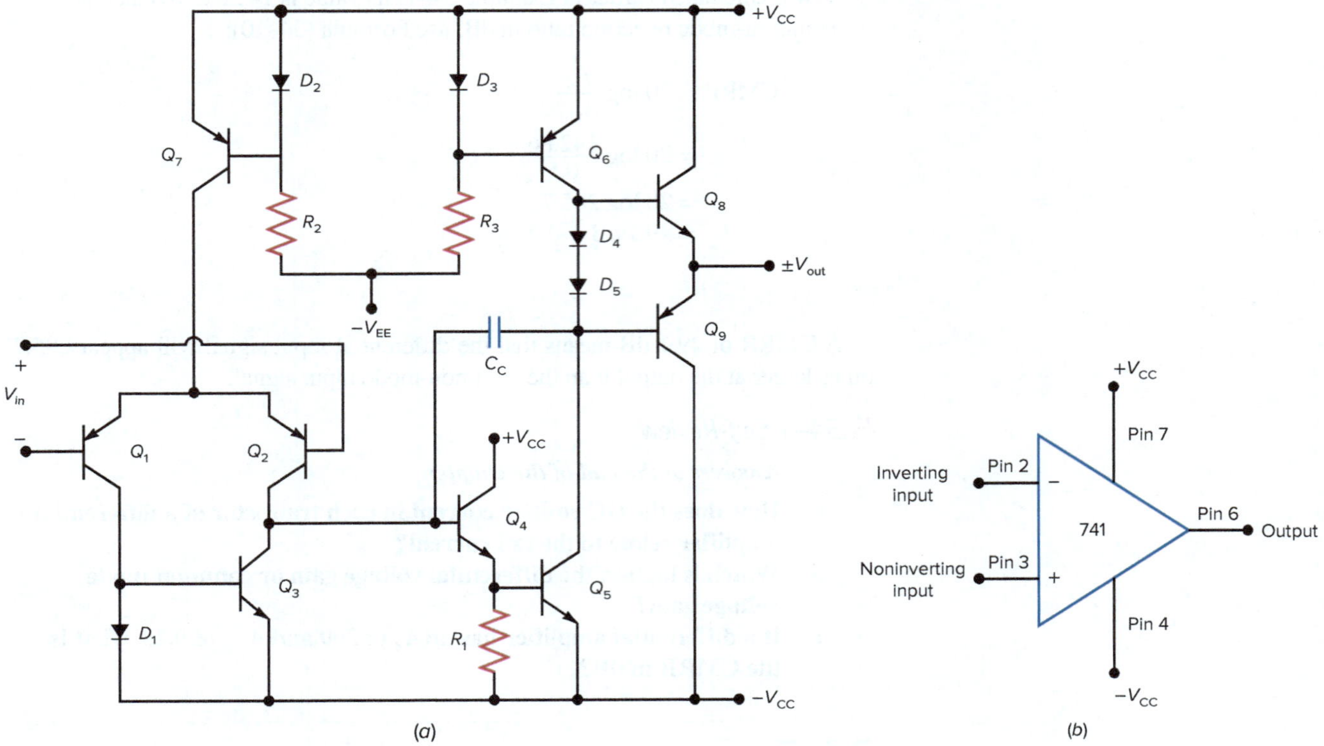

Open-Loop Voltage Gain, A_{VOL}

The **open-loop voltage gain,** A_{VOL}, of an op amp is its voltage gain when there is no **negative feedback**. The open-loop voltage gain of an op amp is the ratio of its output voltage, V_{out}, to its **differential input voltage,** V_{id}:

$$A_{VOL} = \frac{V_{out}}{V_{id}}$$

where A_{VOL} = open-loop voltage gain of op amp, V_{out} = output voltage, and V_{id} = differential input voltage. The typical value of A_{VOL} for a 741 op amp is 200,000. Figure 34–6*a* illustrates the concept. Notice that $V_{id} = V_1 - V_2$, and that $V_{out} = A_{VOL} \times V_{id}$. It is important to note that only the differential voltage, $V_1 - V_2$, is amplified, not the individual values of V_1 and V_2.

As a numerical example, assume that the differential input, V_{id}, in Fig. 34–6*b* equals ±50 μV and that A_{VOL} = 200,000. Then the following is true:

$$\begin{aligned} V_{out} &= A_{VOL} \times V_{id} \\ &= 200{,}000 \times (\pm 50\ \mu V) \\ &= \pm 10\ V \end{aligned}$$

The answer is shown as ±10 V in Fig. 34–6*b* because the polarity of V_{id} is not specified. The polarity of output voltage, V_{out}, for an op amp is determined by the following two rules:

1. When the voltage at the noninverting (+) input is made positive with respect to its inverting (−) input, the output is positive.
2. When the voltage at the noninverting (+) input is made negative with respect to its inverting (−) input, the output is negative.

Figure 34–6 Op-amp circuits used to amplify the small value of V_{id} by the high value of A_{VOL}. See text for analysis.

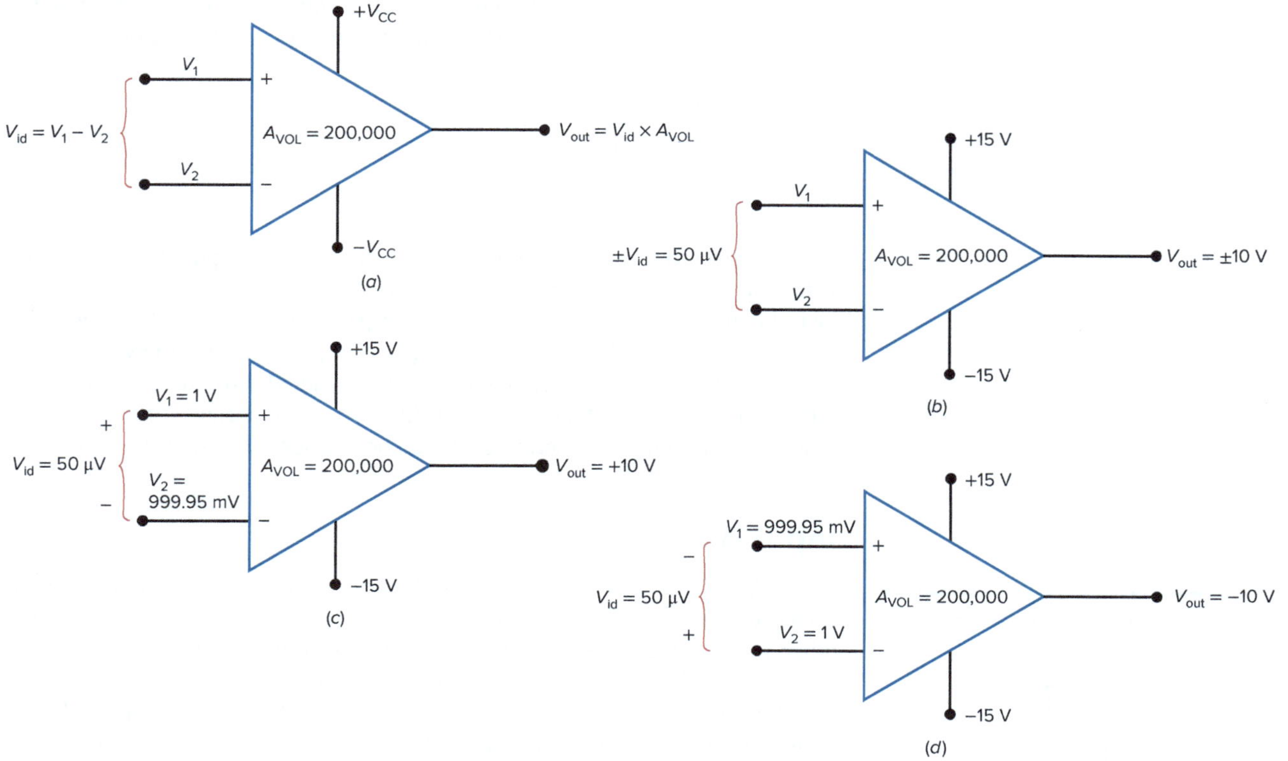

Assume in Fig. 34–6*c* that $V_1 = 1$ V and $V_2 = 999.95$ mV. What will the output be? Simply multiply V_{id} by A_{VOL}. First, find V_{id}:

$$\begin{aligned} V_{id} &= V_1 - V_2 \\ &= 1\text{ V} - 999.95\text{ mV} \\ &= 50\ \mu\text{V} \end{aligned}$$

To calculate V_{out}, proceed as follows:

$$\begin{aligned} V_{out} &= A_{VOL} \times V_{id} \\ &= 200{,}000 \times 50\ \mu\text{V} \\ &= +10\text{ V} \end{aligned}$$

Notice again that the actual values of V_1 and V_2 are not amplified, just the difference between them. Also, note that V_1 in Fig. 34–6*c* is connected to the noninverting (+) input terminal and V_2 is connected to the inverting input terminal. Because V_1 is more positive than V_2, the output is positive.

In Fig. 34–6*d*, V_1 and V_2 are reversed. Here V_1 is negative with respect to V_2. V_{id} is calculated as shown:

$$V_{id} = V_1 - V_2 = 999.95\text{ mV} - 1\text{ V} = -50\ \mu\text{V}$$

To find V_{out}, again multiply V_{id} by A_{VOL} as shown:

$$\begin{aligned} V_{out} &= A_{VOL} \times V_{id} \\ &= 200{,}000 \times (-50\ \mu\text{V}) \\ &= -10\text{ V} \end{aligned}$$

The key point from all the circuits in Fig. 34–6 is that only the differential input voltage, V_{id}, is amplified by the op amp's high value of open-loop voltage gain.

There are upper and lower limits for the output voltage, V_{out}. The upper limit of V_{out} is called the ***positive saturation* voltage**, designated $+V_{sat}$. The lower limit of V_{out}

is called the ***negative saturation* voltage**, designated $-V_{sat}$. For the 741 op amp, $\pm V_{sat}$ is usually within a couple volts of $\pm V_{CC}$. For example, if $\pm V_{CC} = \pm 15$ V, then $\pm V_{sat} = \pm 13$ V. Incidentally, the amount of differential input voltage, V_{id}, required to produce positive or negative saturation in Fig. 34–6 is found as follows:

$$\pm V_{id} = \frac{\pm V_{sat}}{A_{VOL}}$$
$$= \frac{\pm 13\text{ V}}{200{,}000}$$
$$= \pm 65\ \mu V$$

Remember that V_{out} will be positive if the noninverting input (+) is made positive with respect to the inverting (−) input. Likewise, V_{out} will be negative if the noninverting input (+) is made negative with respect to the inverting (−) input.

One more point: If the output voltage of any op amp lies between $-V_{sat}$ and $+V_{sat}$, then V_{id} will be so small that it can be considered zero. Realistically, it is very difficult to measure a V_{id} of 65 μV in the laboratory because of the presence of induced noise voltages. Therefore, V_{id} can be considered zero, or $V_{id} = 0$ V.

Input Bias Currents

Figure 34–7 Input bias currents.

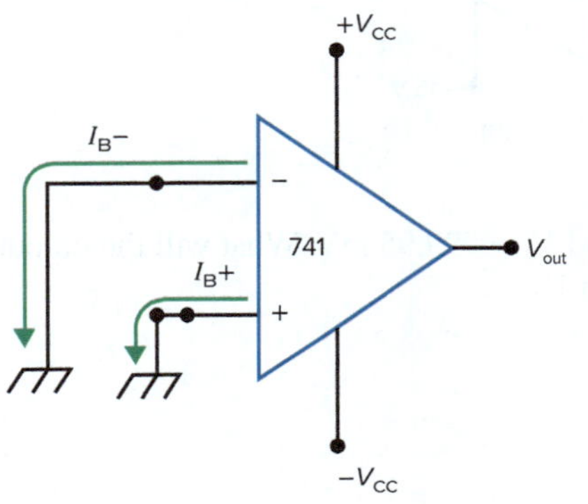

In Fig. 34–5*a*, the base leads of Q_1 and Q_2 serve as the inputs to the op amp. These transistors must be biased correctly before any signal voltage can be amplified. In other words, Q_1 and Q_2 must have external DC return paths back to the power supply ground. Figure 34–7 shows current flowing from the noninverting and inverting input terminals when they are grounded. For a 741 op amp, these currents are very, very small, usually 80 nA (80×10^{-9} A) or less. In Fig. 34–7, I_{B+} designates the current flowing from the noninverting input terminal, and I_{B-} designates the current flowing from the inverting input terminal.

Manufacturers specify I_B as the average of the two currents, I_{B+} and I_{B-}. This can be shown as

$$I_B = \frac{|I_{B+}| + |I_{B-}|}{2} \qquad \textbf{(34–11)}$$

where || means magnitude without regard to polarity. I_{B+} and I_{B-} may be different because it is difficult to match Q_1 and Q_2 exactly. The difference between these two currents is designated I_{OS}, for ***input offset current***. I_{OS} can be expressed as shown here:

$$I_{OS} = |I_{B+}| - |I_{B-}| \qquad \textbf{(34–12)}$$

For a 741 op amp, I_{OS} is typically 20 nA.

For our analysis, we will assume that the values of I_{B+} and I_{B-} are zero because they are so small. However, in some cases of high precision, their effects on circuit operation must be taken into account.

Frequency Response

Figure 34–8 shows the frequency response curve for a typical 741 op amp. Notice that at frequencies below 10 Hz, $A_{VOL} = 200{,}000$. Notice, however, that A_{VOL} is down to 70.7% of its maximum value at 10 Hz. In Fig. 34–8, $A_{VOL} = 141{,}400$ at 10 Hz. This frequency is designated f_{OL}, for *open-loop cutoff frequency*.

Beyond f_{OL}, the gain decreases by a factor of 10 for each decade increase in frequency. This is equivalent to saying that A_{VOL} decreases at the rate of 20 dB/decade above f_{OL}. This drop in A_{VOL} at higher frequencies is caused by capacitor C_C inside the op amp. The frequency where $A_{VOL} = 1$ is designated $\boldsymbol{f_{unity}}$. For a 741, f_{unity} is approximately 1 MHz.

Figure 34–8 Frequency response curve for the 741 op amp.

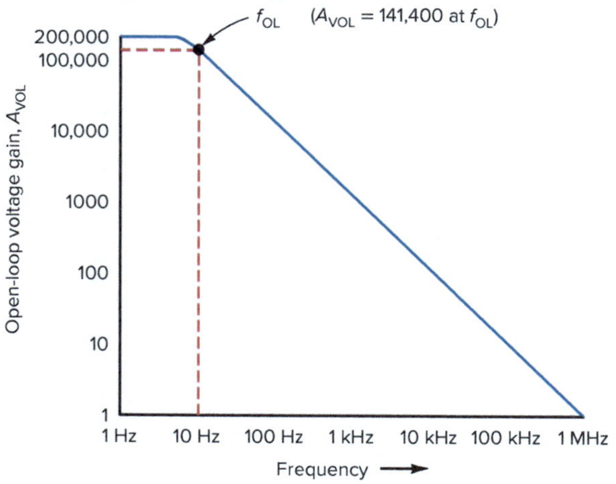

Slew Rate

> **GOOD TO KNOW**
>
> Another commonly used op amp is the LM318. It has a minimum slew rate of 50 V/μs and an f_{unity} of 15 MHz. This is a considerable improvement over the 741 op amp.

Another very important op-amp specification is its **slew rate**, usually designated S_R. The slew-rate specification of an op amp tells how fast the output voltage can change in volts per microsecond, or $V/\mu s$. For a 741 op amp, the S_R is 0.5 $V/\mu s$. This means that no matter how fast the input voltage to a 741 op amp changes, the output voltage can change only as fast as 0.5 $V/\mu s$, which is its slew rate. Figure 34–9 illustrates this concept. Here the op amp's output waveform should be an amplified version of the sinusoidal input, V_{id}. In this case, waveform *A* would be the expected output. However, if the slope of the output sine wave exceeds the S_R rating of the op amp, the waveform appears triangular. Therefore, slew-rate distortion of a sine wave produces a triangular wave, such as waveform *B* in Fig. 34–9.

Power Bandwidth

There are two ways to avoid slew-rate distortion of a sine wave: Either use an op amp with a higher slew rate or accept an output waveform with a lower peak voltage. Using an op amp with a higher slew rate seems like a logical solution because then the output waveform will be able to follow the sinusoidal input voltage, V_{id}. But why would less peak voltage for the output waveform solve the

Figure 34–9 Slew-rate distortion occurs when the initial slope of the output waveform exceeds the S_R rating of the op amp.

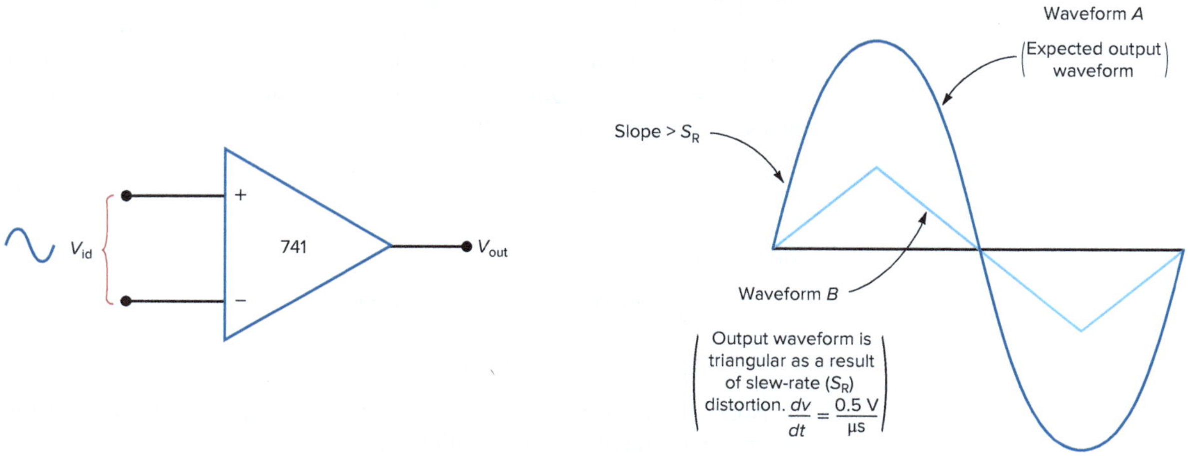

Figure 34–10 Waveforms *A* and *B* have the same frequency but different slopes during positive and negative alternations.

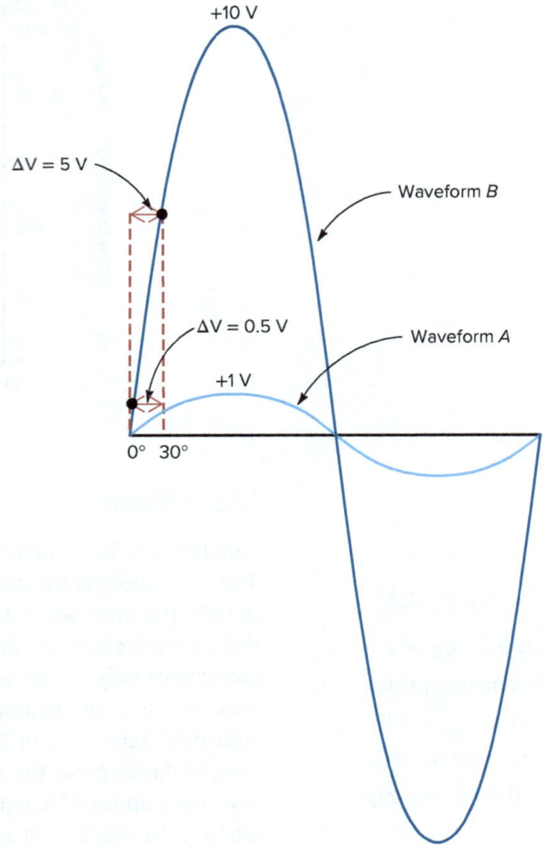

problem? The answer can best be illustrated as shown in Fig. 34–10. Here both waveforms *A* and *B* have exactly the same frequencies, but different peak values. Waveform *A* has a peak value of 1 V, whereas waveform *B* has a peak value of 10 V. Now compare the voltage change during the first 308 of each waveform. Notice that the change in voltage, ΔV, for waveform *A* is 0.5 V during the first 308, whereas ΔV for waveform *B* is 5 V during this same interval. Notice that the rate of voltage change for waveform *B* is 10 times that of waveform *A* during the same time interval, even though both waveforms have exactly the same frequency! Therefore it is true to say that two waveforms having identical frequencies but different peak values have significantly different slopes during their positive and negative alternations.

The higher the peak voltage of a sine wave for a given frequency, the greater its initial slope. If the initial slope of the output waveform exceeds the S_R rating of the op amp, slew-rate distortion will occur. The following formula gives the highest undistorted frequency out of an op amp for a given S_R and peak voltage:

$$f_{max} = \frac{S_R}{2\pi V_{pk}} \qquad \textbf{(34–13)}$$

where

f_{max} = highest undistorted frequency
S_R = slew rate
V_{pk} = peak value of output sine wave

Notice that f_{max} can be increased by using an op amp with a higher slew rate or by accepting an output waveform with lower peak voltage.

Example 34-3

Calculate f_{max} for an op amp that has an S_R of 5 V/μs and a peak output voltage of 10 V.

ANSWER

$$f_{max} = \frac{S_R}{2\pi V_{pk}}$$
$$= \frac{5\ \text{V}/\mu\text{s}}{2 \times 3.141 \times 10}$$
$$= 79.6\ \text{kHz}$$

The frequency f_{max} of 79.6 kHz is commonly called the 10-V **power bandwidth**. This means that slew-rate distortion for a 10-V peak sine wave will not occur for frequencies at or below 79.6 kHz.

Output Short-Circuit Current

An op amp such as the 741 has short-circuit output protection: Its output short-circuit current is approximately 25 mA. Thus, if the op-amp output (pin 6) is tied directly to ground, the output current cannot exceed 25 mA. Small load resistances connected to the op-amp output usually have lower amplitudes of output voltage because the output voltage cannot exceed 25 mA × R_L.

Common-Mode Rejection Ratio (CMRR)

As mentioned earlier, an op amp amplifies only the difference in voltage between its two inputs. Remember that the input stage of an op amp is a differential amplifier. Therefore, an op amp has the inherent ability to amplify the differential mode input signal and attenuate any common-mode input signal. If two identical signals are applied to the inputs of an op amp, each with exactly the same phase relationship and voltage values, the output will be zero. Such a signal is called a *common-mode* signal. Unfortunately, even with a perfect common-mode signal, the output from the op amp will not be zero because op amps are not ideal. The rejection of the common-mode signal is very high. For a typical 741 op amp, the common-mode rejection ratio (CMRR) is 90 dB, which corresponds to a ratio of about 30,000:1.

What does this mean? If two input signals, one a differential input signal and the other a common-mode input signal, are simultaneously applied to a 741 op amp, the differential input signal will appear about 30,000 times larger at the output than the common-mode input signal.

34–2 Self-Review

Answers at the end of the chapter.

a. What is the input stage of every op amp?

b. If the voltage at the noninverting input (+) of an op amp is negative relative to the inverting (−) input, what is the polarity of the output voltage?

c. Calculate f_{max} for an op amp that has an S_R of 10 V/μs and a peak output voltage of 5 V.

34–3 Op-Amp Circuits with Negative Feedback

The term *feedback* in electronics refers to sampling a portion of the output signal from amplifier and feeding it back either to aid or to oppose the input signal. Negative feedback means that the returning signal has a phase that opposes the input signal. Negative feedback can significantly improve the performance of an amplifier. Any op-amp circuit that does not use negative feedback is considered too unstable to be useful. This section describes how negative feedback can be used to stabilize the voltage gain, improve the input and output impedances, and increase the bandwidth of an amplifier.

The Inverting Amplifier

Figure 34–11*a* shows an op-amp circuit that uses negative feedback. The circuit is called an *inverting* amplifier because the input and output signals are 180° out of phase. The 180° phase inversion occurs because V_{in} is applied to the inverting (−) input terminal of the op amp.

Resistors R_F and R_i provide the negative feedback, which in turn controls the circuit's overall voltage gain. The output signal is fed back to the inverting input through resistors R_F and R_i. The voltage between the inverting input and ground is the differential input voltage, designated V_{id}. The exact value of V_{id} is determined by the values A_{VOL} and V_{out}. Even with negative feedback, the output voltage of an op amp can be found from

$$V_{out} = A_{VOL} \times V_{id}$$

MultiSim **Figure 34–11** Inverting amplifier. (*a*) Circuit. (*b*) Circuit emphasizing the concept of virtual ground.

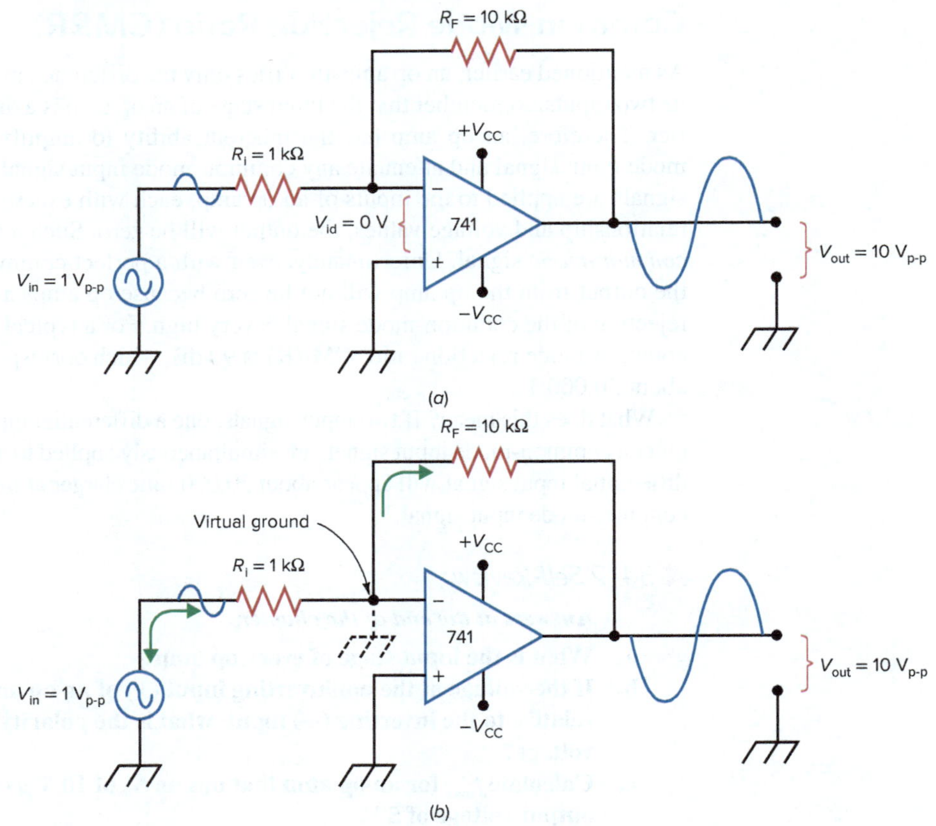

For all practical purposes, V_{id} is so small that it can be considered zero in most cases. This introduces little or no error in circuit analysis. Since V_{id} is so small (practically zero), the inverting input terminal of the op amp is said to be at virtual ground. This means that the voltage at the op amp's inverting input is at the same potential as ground, yet it can sink no current. This point of virtual ground is illustrated in Fig. 34–11*b*.

Closed-Loop Voltage Gain, A_{CL}

Because the inverting input terminal of the op amp is at virtual ground, the voltage drop across the input resistor, R_i, equals V_{in}. Therefore, the current through R_i, is

$$I = \frac{V_{in}}{R_i}$$

Because negligible current flows from the inverting input terminal, all of the current produced by V_{in} must flow through the feedback resistor, R_F. Therefore, the voltage across the feedback resistor, R_F, is

$$V_{R_F} = I \times R_F$$

where I equals the input current through R_i.

Because the inverting input is at virtual ground, the output voltage must equal the voltage across the feedback resistor, R_F. Therefore,

$$V_{out} = I \times R_F$$

By rearranging the equation, $I = V_{in}/R_i$, the voltage gain of the circuit can be calculated as

$$\begin{aligned} A_{CL} &= \frac{V_{out}}{V_{in}} \\ &= \frac{I \times R_F}{I \times R_i} \\ &= -\frac{R_F}{R_i} \end{aligned} \qquad \textbf{(34–14)}$$

where A_{CL} equals the closed-loop voltage gain, which is the gain of the amplifier with negative feedback.

The minus (−) sign in Formula (34–14) indicates that V_{in} and V_{out} are 180° out of phase.

The negative feedback keeps the overall voltage gain constant, even if the open-loop voltage gain, A_{VOL}, changes. In other words, the **closed-loop voltage gain, A_{CL}**, is independent of any changes in the op amp's open-loop voltage gain, A_{VOL}.

Example 34-4

In Fig. 34–11*a*, calculate the closed-loop voltage gain, A_{CL}, and the output voltage, V_{out}.

ANSWER Using Formula (34–14), the voltage gain is calculated as

$$\begin{aligned} A_{CL} &= -\frac{R_F}{R_i} \\ &= -\frac{10\ \text{k}\Omega}{1\ \text{k}\Omega} \\ &= -10 \end{aligned}$$

The output voltage is

$$\begin{aligned} V_{out} &= V_{in} \times A_{CL} \\ &= 1\ V_{p\text{-}p} \times 10 \\ &= 10\ V_{p\text{-}p} \end{aligned}$$

As shown in Fig. 34–11*a*, the input and output voltages are 180° out of phase.

Example 34-5

If A_{VOL} equals 100,000 in Fig. 34–11*a*, calculate the value of V_{id}.

ANSWER Rearranging the formula, $A_{VOL} = V_{out}/V_{id}$ gives

$$\begin{aligned} V_{id} &= \frac{V_{out}}{A_{VOL}} \\ &= \frac{10\ V_{p\text{-}p}}{100{,}000} \\ &= 100\ \mu V_{p\text{-}p} \end{aligned}$$

This voltage is very difficult to measure in the laboratory and therefore, as mentioned earlier, can be considered zero in most cases.

Input Impedance, Z_{in}

Since the inverting input of the op amp is at virtual ground, the voltage source, V_{in}, sees an input impedance equal to R_i. Therefore,

$$Z_{in} \approx R_i \tag{34–15}$$

The inverting input of the op amp has extremely high input impedance, but its value is not the input impedance of the circuit.

Output Impedance, Z_{out}

Because of the negative feedback in Fig. 34–11, the output impedance of the circuit is significantly less than the open-loop output impedance of the op amp. The output impedance of a circuit with negative feedback is called the closed-loop output impedance, designated $Z_{out(CL)}$, and can be calculated using Formula (34–16):

$$Z_{out(CL)} = \frac{Z_{out(OL)}}{1 + A_{VOL}\beta} \tag{34–16}$$

where $Z_{out(CL)}$ = the open-loop output impedance of the op amp, and

$$\beta = \frac{R_i}{R_i + R_F}$$

β is called the *feedback fraction* because it determines how much of the output signal is fed back to the input.

Example 34-6

In Fig. 34–11*a*, calculate Z_{in} and $Z_{out(CL)}$. Assume $A_{VOL} = 100{,}000$ and $Z_{out(OL)} = 75\ \Omega$.

ANSWER In Fig. 34–11*a*, Z_{in} is calculated as

$$\begin{aligned} Z_{in} &\approx R_i \\ &= 1\ \text{k}\Omega \end{aligned}$$

Notice how simple this is. Because $Z_{in} = R_i$ the designer can easily control the input impedance of the amplifier.

To calculate $Z_{out(CL)}$, calculate the feedback fraction, β:

$$\begin{aligned} \beta &= \frac{R_i}{R_i + R_F} \\ &= \frac{1\ \text{k}\Omega}{1\ \text{k}\Omega + 10\ \text{k}\Omega} \\ &= 0.0909 \end{aligned}$$

Next, use Formula (34–16):

$$\begin{aligned} Z_{out(CL)} &= \frac{Z_{out(OL)}}{1 + A_{VOL}\beta} \\ &= \frac{75\ \Omega}{1 + 9090} \\ &= 0.0082\ \Omega \end{aligned}$$

Example 34-7

In Fig. 34–11, calculate the 5-V power bandwidth.

ANSWER Use the formula for power bandwidth. Since the op amp used in this circuit is a 741, $S_R = 0.5\ \text{V}/\mu\text{s}$. Also, with 10 $\text{V}_{\text{p-p}}$ at the output, $V_{pk} = 10\ \text{V}_{\text{p-p}}/2 = 5\ V_{pk}$. Therefore,

$$\begin{aligned} f_{max} &= \frac{S_R}{2\pi V_{pk}} \\ &= \frac{0.5\ \text{V}/\mu\text{s}}{2 \times 3.141 \times 5\ \text{V}} \\ &= 15.915\ \text{kHz} \end{aligned}$$

To avoid slew-rate distortion of the output sine wave, the operating frequency must not exceed 15.915 kHz if a peak output voltage of 5 V is desired.

The Noninverting Amplifier

Figure 34–12 shows how to connect an op amp to work as a noninverting amplifier. Notice that the input signal, V_{in}, is applied directly to the noninverting (+) input of the op amp, so that the input and output signals will be in phase.

MultiSim **Figure 34–12** Noninverting amplifier.

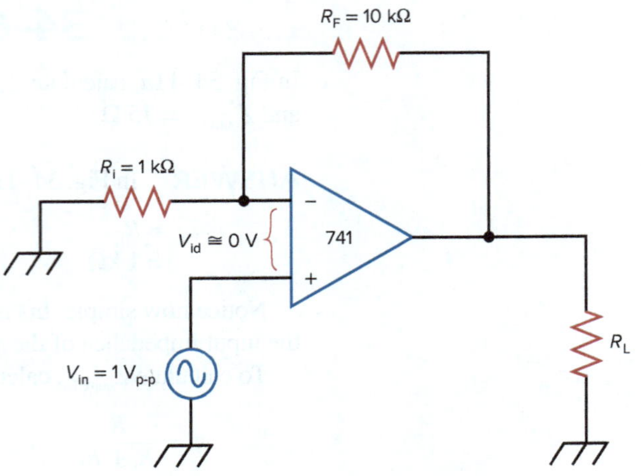

As with the inverting amplifier, the values of R_F and R_i determine the closed-loop voltage gain, A_{CL}. Because the differential input voltage, V_{id}, between the op amp input terminals is approximately zero, the current through R_i is found as follows:

$$I = \frac{V_{in}}{R_i}$$

where I is the current through R_i.

Rearranging gives $V_{in} = I \times R_i$.

Because practically no current flows from the inverting input terminal of the op amp, all of the input current must flow through the feedback resistor, R_F. Therefore,

$$V_{R_F} = I \times R_F$$

Since the output voltage, V_{out} is taken with respect to ground, V_{out} is the sum of V_{R_i} and V_{R_F}. Expressed as a formula,

$$\begin{aligned} V_{out} &= IR_F + IR_i \\ &= I(R_F + R_i) \end{aligned}$$

The closed-loop voltage gain, A_{CL}, is calculated by dividing V_{out} by V_{in}:

$$\begin{aligned} A_{CL} &= \frac{V_{out}}{V_{in}} \\ &= \frac{I(R_F + R_i)}{IR_i} \\ &= \frac{R_F + R_i}{R_i} \end{aligned}$$

or $$A_{CL} = \frac{R_F}{R_i} + 1 \qquad \textbf{(34–17)}$$

As with the inverting amplifier, the closed-loop voltage gain, A_{CL}, is independent of the changes in A_{VOL}.

Example 34-8

In Fig. 34–12, calculate the closed-loop voltage gain, A_{CL}, and the output voltage, V_{out}.

ANSWER Using Formula (34–17), the calculations are

$$A_{CL} = \frac{R_F}{R_i} + 1$$
$$= \frac{10\text{ k}\Omega}{1\text{ k}\Omega} + 1$$
$$= 11$$

The output voltage, V_{out}, is

$$V_{out} = V_{in} \times A_{CL}$$
$$= 1\text{ V}_{p\text{-}p} \times 11$$
$$= 11\text{ V}_{p\text{-}p}$$

Input Impedance, Z_{in}

Since the voltage source has to supply virtually no current to the op amp's noninverting input terminal, the voltage source, V_{in}, sees a very high input impedance, Z_{in}. Hence, there is virtually no loading of the input voltage source. To calculate Z_{in} in a noninverting amplifier, use Formula (34–18):

$$Z_{in(CL)} = R_{in}(1 + A_{VOL}\beta) \qquad \textbf{(34–18)}$$

where R_{in} represents the open-loop input resistance of the op amp and $Z_{in(CL)}$ represents the closed-loop input impedance of the circuit with negative feedback. Because $1 + A_{VOL}\beta$ is large in most cases, $Z_{in(CL)}$ approaches infinity.

Output Impedance, Z_{out}

Negative feedback also affects the output impedance of a noninverting amplifier. Here's how.

Refer to Fig. 34–12. Suppose that there is an increase in the load resistance, R_L. Then less current will flow through the output impedance of the op amp, which in turn increases the output voltage. This causes more output signal voltage to be fed back to the inverting input of the op amp. Since V_{in} remains constant, the differential input voltage, V_{id}, decreases. Furthermore, since $V_{out} = A_{VOL} \times V_{id}$, the decrease in V_{id} offsets the original increase in output voltage.

Suppose that there is a decrease in the load resistance, R_L. Then more current will flow through the output impedance of the op amp, which in turn decreases the output voltage. This causes less voltage to be fed back to the inverting input. Because V_{in} remains constant, the differential input voltage, V_{id}, increases. Since $V_{out} = A_{VOL} \times V_{id}$, the increase in V_{id} offsets the original decrease in output voltage.

In summary, negative feedback lowers the output impedance of the op amp. Just as in the inverting amplifier, $Z_{out(CL)}$ is calculated using Formula (34–16):

$$Z_{out(CL)} = \frac{Z_{out(OL)}}{1 + A_{VOL}\beta}$$

Example 34-9

In Fig. 34–12, calculate $Z_{in(CL)}$ and $Z_{out(CL)}$. Assume $R_{in} = 2\ M\Omega$, $A_{VOL} = 100{,}000$, and $Z_{out(OL)} = 75\ \Omega$.

ANSWER Begin by calculating $Z_{in(CL)}$:

$$Z_{in(CL)} = R_{in}(1 + A_{VOL}\,\beta)$$
$$= 2\ M\Omega\,(1 + 100{,}000 \times 0.0909)$$
$$\approx 18\ G\Omega$$

Wow! For all practical purposes, this can be considered infinity. Next, calculate $Z_{out(CL)}$:

$$Z_{out(CL)} = \frac{Z_{out(OL)}}{1 + A_{VOL}\beta}$$
$$= \frac{75\ \Omega}{1 + (100{,}000 \times 0.0909)}$$
$$= 0.0082\ \Omega$$

Notice how close this is to zero ohms.

The Voltage Follower

Figure 34–13 shows a very popular op-amp circuit called a ***voltage follower***. This circuit is also called a *unity gain amplifier, buffer amplifier,* or *isolation amplifier*. Notice that the input voltage, V_{in}, is applied directly to the noninverting (+) input of the op amp. Because of this, the input and output voltages are in phase. Also, notice that the output is connected directly to the inverting (−) input terminal. Since $V_{id} \approx 0$ V, then,

$$V_{out} = V_{in}$$

Therefore,

$$A_{CL} = 1$$

Because $V_{out} = V_{in}$, the output voltage must follow the input voltage; hence the name voltage follower. The voltage follower uses the maximum amount of negative

Figure 34–13 Voltage follower.

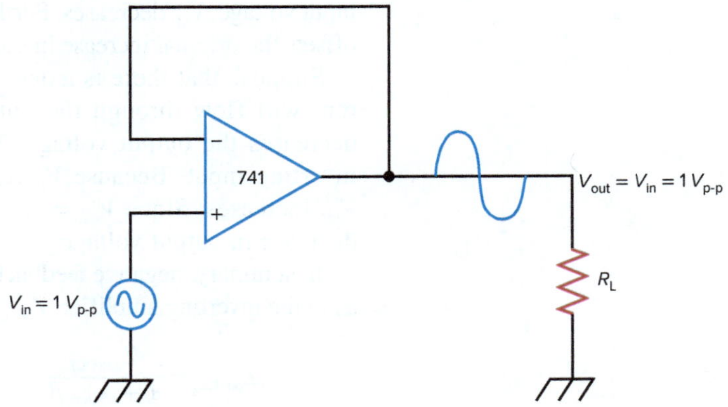

feedback possible. Because of this, $Z_{in(CL)}$ is extremely high and $Z_{out(CL)}$ is extremely low. Formulas (34–18) and (34–16) can be used to calculate $Z_{in(CL)}$ and $Z_{out(CL)}$, respectively.

GOOD TO KNOW

Using an op amp as a voltage follower (buffer amplifier) is very common in electronic circuits.

One might question why such a circuit would ever be used if the voltage gain is only one. The reason lies in the fact that the circuit can buffer or isolate a low impedance load from the voltage source, V_{in}. This means that rather than connect a relatively low value of load resistance across the terminals of V_{in}, the op amp can be used to eliminate any loading that might occur. Since V_{in} is connected to the noninverting (+) input terminal of the op amp, it has to supply virtually no current to the circuit. Thus, the source voltage, V_{in}, won't be loaded down. Also, because of the heavy negative feedback, the output impedance, $Z_{out(CL)}$, is very low. This means that the circuit acts as an ideal voltage source with nearly zero internal impedance.

Example 34-10

In Fig. 34–13, $R_{in} = 2\ \text{M}\Omega$, $Z_{out(OL)} = 75\ \Omega$, and $A_{VOL} = 100{,}000$. Calculate $Z_{in(CL)}$ and $Z_{out(CL)}$.

ANSWER Begin with $Z_{in(CL)}$:

$$Z_{in(CL)} = R_{in}\,(1 + A_{VOL}\,\beta)$$

Since $\beta = 1$,

$$\begin{aligned} Z_{in(CL)} &= R_{in}(1 + A_{VOL}) \\ &= 2\ \text{M}\Omega\,(1 + 100{,}000) \\ &\approx 200\ \text{G}\Omega \end{aligned}$$

Next calculate $Z_{out(CL)}$:

$$Z_{out(CL)} = \frac{Z_{out(OL)}}{1 + A_{VOL}\beta}$$

Since $\beta = 1$,

$$\begin{aligned} Z_{out(CL)} &= \frac{Z_{out(OL)}}{1 + A_{VOL}} \\ &= \frac{75\ \Omega}{1 + 100{,}000} \\ &\approx 0.00075\ \Omega \end{aligned}$$

For all practical purposes, Z_{in} is infinity and Z_{out} is zero ohms.

Op-Amp Bandwidth

Because the stages inside an op amp are direct coupled, there is no lower cutoff frequency. An op amp does, however, have an open-loop upper cutoff frequency, designated f_{OL}. Remember that f_{OL} for a 741 op amp is 10 Hz. At this frequency, the open-loop voltage gain, A_{VOL}, is down to 70.7% of its maximum value. With negative feedback, the upper cutoff frequency of the op-amp circuit can be extended well beyond the value of f_{OL}.

Here is why: Assume that in a noninverting amplifier (such as the one shown in Fig. 34–12), the frequency of V_{in} increases above the **open-loop cutoff frequency, f_{OL}**. This causes the open-loop voltage gain, A_{VOL}, to decrease. Because A_{VOL} decreases, however, the amount of negative feedback also decreases. Because V_{in} remains constant, the differential input voltage, V_{id}, increases. Since $V_{out} = A_{VOL} \times V_{id}$, the increase in V_{id} compensates for the reduction in A_{VOL}, which maintains the output voltage at a constant value. Therefore, the closed-loop gain, A_{CL}, also remains constant. If the frequency of V_{in} keeps increasing, the open-loop voltage gain will eventually equal the closed-loop voltage gain, A_{CL}. Then the curves for A_{VOL} and A_{CL} will superimpose and decrease together at the same rate.

The frequency at which the closed-loop gain, A_{CL}, decreases to 70.7% of its maximum value is called the **closed-loop cutoff frequency**, designated f_{CL}. f_{CL} can be calculated as

$$f_{CL} = \frac{f_{unity}}{A_{CL}} \qquad \textbf{(34–19)}$$

where f_{unity} represents the frequency at which the open-loop voltage gain of the op amp equals one, or unity. Data sheets always list the value for f_{unity}.

Formula (34–19) indicates that the closed-loop cutoff frequency is affected by the closed-loop voltage gain, A_{CL}. As A_{CL} increases, the closed-loop bandwidth, f_{CL}, decreases and vice versa.

Closed-Loop Gain-Bandwidth Product

Formula (34–19) can be rearranged as

$$A_{CL} f_{CL} = f_{unity}$$

For a particular op amp, the product of A_{CL} and f_{CL} will always equal f_{unity}. This is called the closed-loop gain-bandwidth product. For a 741 op amp, the product of A_{CL} and f_{CL} will always equal 1 MHz. This is true regardless of the values of the resistors used for negative feedback.

Figure 34–14 shows the frequency response curves for A_{VOL} and closed-loop gains of 10 and 1000. Notice that the curve for A_{VOL} is down to 70,700 at f_{OL}, which is 10 Hz. Notice that the curves for $A_{CL} = 1000$ and $A_{CL} = 10$, however, are unaffected at this frequency.

When $A_{CL} = 1000$, the closed-loop bandwidth, f_{CL}, is

$$\begin{aligned} f_{CL} &= \frac{f_{unity}}{A_{CL}} \\ &= \frac{1\ \text{MHz}}{1000} \\ &= 1\ \text{kHz} \end{aligned}$$

Figure 34–14 Frequency response curve showing A_{VOL} and closed-loop gains of 10 and 1000.

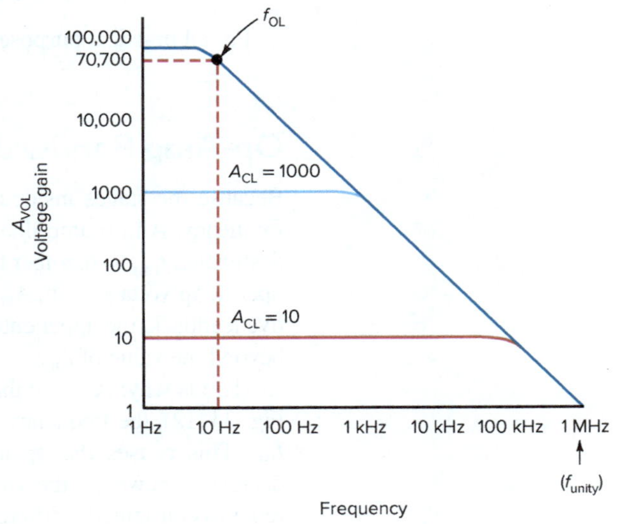

When $A_{CL} = 10$, then,

$$f_{CL} = \frac{1 \text{ MHz}}{10}$$
$$= 100 \text{ kHz}$$

Refer to Fig. 34–14. In each case, notice that when the value of $A_{VOL} = A_{CL}$, the curves superimpose and decrease to 1 or unity at 1 MHz.

Notice that there is a trade-off between gain and bandwidth. If the closed-loop voltage gain, A_{CL}, is decreased, the bandwidth will increase. Conversely, increasing the closed-loop voltage gain will decrease the bandwidth.

Single Supply Operation

Op-amp circuits can also work with a single power supply voltage, as shown in Fig. 34–15. Notice that pin 7 is connected to +15 V and pin 4 is grounded. Notice also that R_1 and R_2 supply a DC voltage to the noninverting input of the op amp. Because R_1 and R_2 are equal, the DC voltage at the noninverting input equals $V_{CC} \div 2$ or +7.5 V. Because the coupling capacitors C_{in} and C_{out} appear open to DC, the DC voltages at the inverting input and the op-amp output also equal 7.5 V. For DC, the op amp works as a voltage follower. The resistors R_i and R_F will not affect the DC voltages in the circuit because with C_{in} open to DC, R_i and R_F cannot affect the feedback fraction.

For AC operation, the input and output coupling capacitors appear as shorts. Because the AC signal is applied to the inverting input, the circuit functions as an inverting amplifier with a voltage gain of

$$A_{CL} = -\frac{R_F}{R_i}$$

The output coupling capacitor blocks the DC voltage at the op-amp output from the load, R_L. The bypass capacitor placed at the noninverting input of the op amp reduces any power supply noise at this point.

The peak positive output voltage from the op amp in Fig. 34–15 is about 2 V less than the value of V_{CC}, which is +13 V in this case. The minimum positive output voltage is usually about 2 V.

Figure 34–15 Single supply operation of an op amp.

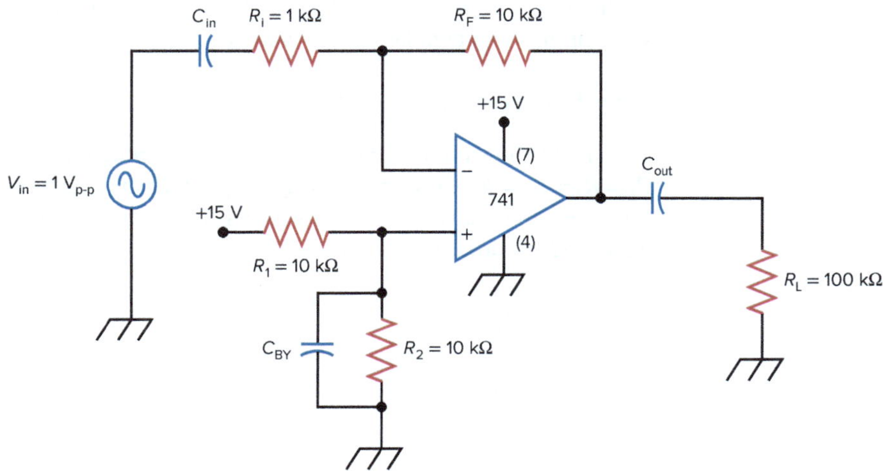

Example 34-11

In Fig. 34–15, calculate the closed-loop voltage gain, A_{CL}, and the DC voltage at the op-amp output terminal.

ANSWER The voltage gain is calculated as

$$A_{CL} = -\frac{R_F}{R_i}$$
$$= -\frac{10\text{ k}\Omega}{1\text{ k}\Omega}$$
$$= -10$$

The DC voltage at the op-amp output terminal equals the DC voltage at the noninverting input, calculated as

$$V_{(+)} = \frac{R_2}{R_1 + R_2} \times 15\text{ V}$$
$$= \frac{10\text{ k}\Omega}{10\text{ k}\Omega + 10\text{ k}\Omega} \times 15\text{ V}$$
$$= 7.5\text{ V}$$

where $V_{(+)}$ represents the DC voltage at the noninverting input terminal.

34–3 Self-Review

Answers at the end of the chapter.

a. What is the phase relationship between V_{in} and V_{out} in an inverting amplifier?
b. How does negative feedback affect the output impedance of a noninverting amplifier?
c. How is f_{CL} related to A_{CL}?

34–4 Popular Op-Amp Circuits

Op amps are used in a wide variety of applications in today's electronics industry. Because op amps are used in so many different ways, it is not possible to cover all of the different circuits in this section. However, this section describes in detail some of the more popular op-amp circuits.

The Summing Amplifier

The circuit shown in Fig. 34–16 is called a ***summing amplifier***, or *summer*. When $R_1 = R_2 = R_3 = R_F$, the output voltage, V_{out}, equals the negative sum of the input voltages. Because the right ends of resistors R_1, R_2, and R_3, are at virtual ground, the input currents are calculated as

$$I_1 = \frac{V_1}{R_1}$$
$$I_2 = \frac{V_2}{R_2}$$
$$I_3 = \frac{V_3}{R_3}$$

Figure 34–16 Summing amplifier.

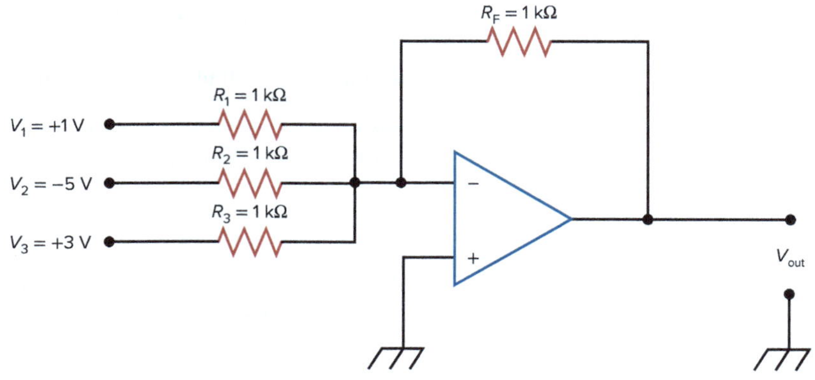

Because the inverting input has zero current, all of the input currents combine to flow through the feedback resistor, R_F. Because $R_1 = R_2 = R_3 = R_F$, the voltage gain of the circuit is one and the output voltage is

$$V_{out} = -(V_1 + V_2 + V_3) \tag{34–20}$$

Note that additional input resistors can be added to the circuit in Fig. 34–16 if necessary. It is possible to do this because the inverting input is at virtual ground. Thus, all inputs are effectively isolated from each other. Each input sees its own input resistance and nothing else.

If each input voltage is amplified by a different amount, then the output voltage will equal the negative of the amplified sum of the inputs. When the voltage gain is different for each input, the formula for the output voltage becomes

$$V_{out} = -\left[\frac{R_F}{R_1}V_1 + \frac{R_F}{R_2}V_2 + \frac{R_F}{R_3}V_3\right] \tag{34–21}$$

Technically, when each input voltage is amplified by a different factor, the circuit is called a *scaling* or *weighted amplifier*. Formula (34–21) corresponds to a circuit with only three inputs, but the formula could be expanded to handle any number of inputs.

Example 34-12

In Fig. 34–16, calculate the output voltage, V_{out}.

ANSWER Since all resistors are equal, the output voltage can be found by using Formula (34–20):

$$\begin{aligned} V_{out} &= -(V_1 + V_2 + V_3) \\ &= -(1\ V - 5\ V + 3\ V) \\ &= -(-1\ V) \\ &= 1\ V \end{aligned}$$

Example 34-13

Calculate the output voltage, V_{out}, in Fig. 34–17.

Figure 34–17 Scaling amplifier.

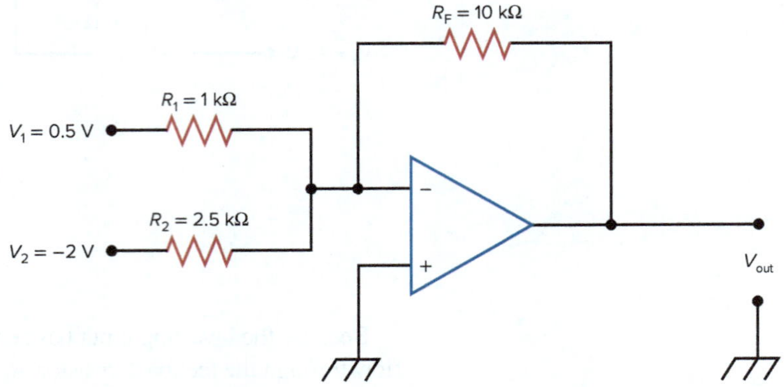

ANSWER Since the voltage gain is different for each input, use Formula (34–21). The calculations are

$$V_{out} = -\left[\frac{R_F}{R_1}V_1 + \frac{R_F}{R_2}V_2\right]$$
$$= -\left[\frac{10\text{ k}\Omega}{1\text{ k}\Omega} \times 0.5\text{ V} + \frac{10\text{ k}\Omega}{2.5\text{ k}\Omega} \times (-2\text{ V})\right]$$
$$= -(5\text{ V} - 8\text{ V})$$
$$= +3\text{ V}$$

Differential Amplifiers

Figure 34–18 shows an op-amp differential amplifier. *Differential amplifiers* are circuits that can amplify differential input signals but reject or attenuate common-mode input signals. Differential amplifiers are typically found in instrumentation and industrial applications. Differential amplifiers are often used in conjunction with resistive bridge circuits where the output from the bridge serves as the input to the op-amp differential amplifier.

To derive a formula for the output voltage, V_{out}, use the superposition theorem. Begin by shorting the input V_Y.

Then the formula for the output voltage becomes

$$V_{out} = -\frac{R_F}{R_i}V_X \quad (V_Y \text{ is shorted})$$

When the input V_X is shorted, the amplifier is noninverting, with an output voltage of

$$V_{out} = \frac{R_3 V_Y}{R_2 + R_3} \times \frac{R_F + R_1}{R_1} \quad (V_X \text{ is shorted})$$

Since $R_1 = R_2$ and $R_F = R_3$, then,

$$V_{out} = \frac{R_F V_Y}{R_1 + R_F} \times \frac{R_F + R_1}{R_1}$$

Figure 34–18 Differential amplifier.

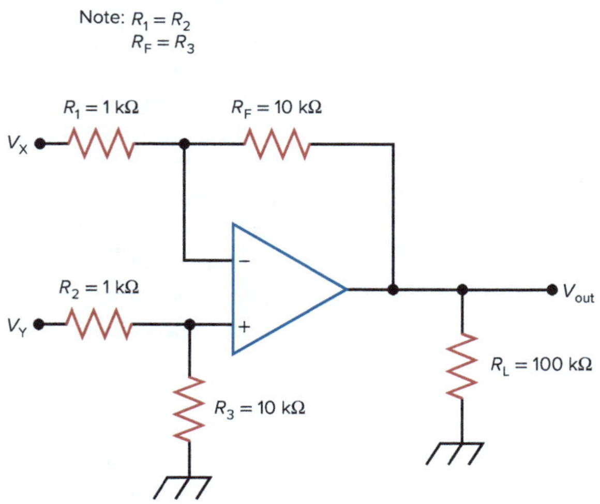

which reduces to

$$V_{out} = \frac{R_F}{R_1} \times V_Y$$

The output voltage is found by combining the results:

$$V_{out} = -\frac{R_F}{R_1} V_X + \frac{R_F}{R_1} V_Y$$

or

$$V_{out} = -\frac{R_F}{R_1}(V_X - V_Y) \qquad \textbf{(34–22)}$$

Note that if $V_X = V_Y$, then the output voltage will be zero. This circuit will amplify only the difference in voltage that exists between the inputs V_X and V_Y.

Example 34–14

In Fig. 34–18, calculate the output voltage, V_{out}, if (a) $V_X = 1\ V_{DC}$ and $V_Y = -0.25\ V_{DC}$, (b) $V_X = -0.5\ V_{DC}$ and $V_Y = +0.5\ V_{DC}$, (c) $V_X = 0.3$ V and $V_Y = 0.3$ V.

ANSWER For each case, use Formula (34–22), where

$$-\frac{R_F}{R_1} = -\frac{10\text{ k}\Omega}{1\text{ k}\Omega} = -10$$

then (a)
$$\begin{aligned} V_{out} &= -10\,(V_X - V_Y) \\ &= -10\,[1\text{ V} - (-0.25\text{ V})] \\ &= -10 \times 1.25\text{ V} \\ &= -12.5\text{ V} \end{aligned}$$

(b)
$$\begin{aligned} V_{out} &= -10\,(-0.5\text{ V} - 0.5\text{ V}) \\ &= -10\,(-1) \\ &= +10\text{ V} \end{aligned}$$

(c)
$$\begin{aligned} V_{out} &= -10\,(0.3\text{ V} - 0.3\text{ V}) \\ &= -10 \times 0 \\ &= 0\text{ V} \end{aligned}$$

Op-Amp Instrumentation Circuit

Figure 34–19 shows a differential amplifier that uses three op amps A_1, A_2, and A_3. The output voltage from the bridge is the input to the differential amplifier. Notice in Fig. 34–19 that there are three main parts to the circuit: the bridge circuit on the left, the buffer stage in the middle, and the differential amplifier on the right.

In the bridge circuit, R_D is a thermistor with a positive temperature coefficient (PTC). Its value equals 5 kΩ at room temperature, which is considered 25°C. At room temperature, R_B is adjusted to provide balance, which implies that $V_X = V_Y$, and $V_X - V_Y = 0$ V. The voltages V_X and V_Y are applied to the voltage followers in the buffer stage. Since a voltage follower has a closed-loop gain, A_{CL}, of one (or unity), the voltage at point *A* equals V_X and the voltage at point *B* equals V_Y. Because of this, the voltage output in Fig. 34–19 is

$$V_{out} = -\frac{R_F}{R_1}(V_X - V_Y)$$

The main purpose of the voltage followers A_1 and A_2 is to isolate the bridge circuit resistances from the differential amplifier input. Remember, a voltage follower has nearly infinite input impedance and almost zero output impedance.

Here is how the circuit in Fig. 34–19 works: Assume that the bridge has been balanced at room temperature. Therefore, at this temperature, $V_X - V_Y = 0$, and $V_{out} = 0\ V_{DC}$. Assume now that the ambient or surrounding temperature increases above 25°C. This causes the resistance, R_D, of the thermistor to increase above 5 kΩ, which in turn causes the voltage, V_Y, to become increasingly more positive. Since V_X remains constant, the voltage $V_X - V_Y$ becomes negative. Since the differential amplifier has a gain equal to $-R_F/R_1$, the output voltage becomes positive.

Assume now that the ambient temperature drops below 25°C. This causes the resistance of the thermistor to decrease, which means that the voltage, V_Y, becomes

Figure 34–19 Op-amp instrumentation amplifier.

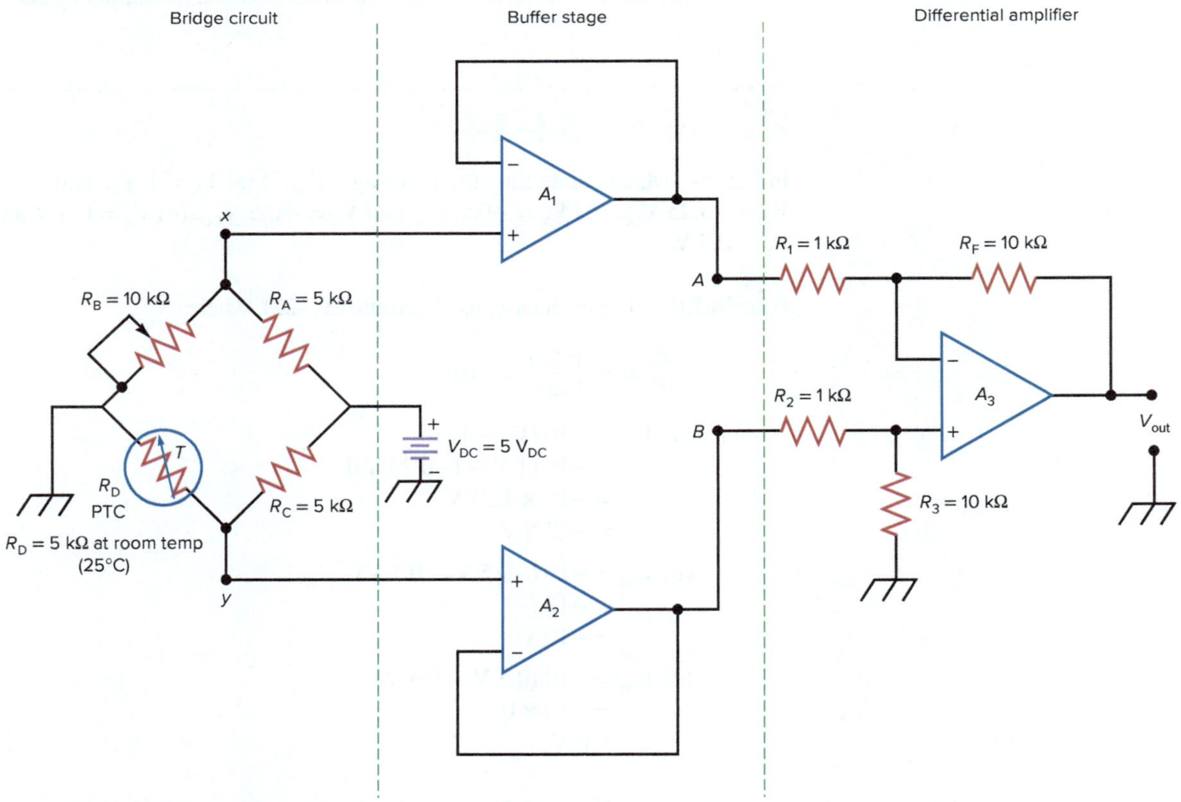

less positive. Since V_X does not change, the voltage $V_X - V_Y$ is a positive quantity. Since the gain of the differential amplifier is negative, the output voltage is negative.

Example 34–15

In Fig. 34–19, assume that R_D increases to 7.5 kΩ due to an increase in the ambient temperature. Calculate the output of the differential amplifier. Note: $R_B = 5$ kΩ.

ANSWER Begin by calculating the voltages V_X and V_Y. Since $R_A = R_B$, then V_X is

$$V_X = \frac{1}{2} \times 5\ V_{DC}$$
$$= \frac{5\ V}{2}$$
$$= 2.5\ V_{DC}$$
$$V_Y = \frac{R_D}{R_C + R_D} \times 5\ V$$
$$= \frac{7.5\ k\Omega}{5\ k\Omega + 7.5\ k\Omega} \times 5\ V$$
$$= 3\ V_{DC}$$

Next, calculate $V_X - V_Y$, which is effectively the input voltage applied to the differential amplifier, A_3:

$$V_X - V_Y = 2.5\ V - 3.0\ V$$
$$= -0.5\ V$$

Next, calculate the output voltage, V_{out}:

$$V_{out} = -\frac{R_F}{R_1}(V_X - V_Y)$$
$$= -\frac{10\ k\Omega}{1\ k\Omega}(2.5\ V - 3.0\ V)$$
$$= -10 \times (-0.5\ V)$$
$$= +5\ V$$

Active Filters

An **active filter** is one that uses active components or devices such as transistors and op amps, that can amplify. A passive filter is one that uses only passive components such as inductors, capacitors, and resistors.

A first-order filter is one that uses one resistor and one reactive component, which is either an inductor or capacitor. A second-order filter is one that uses two resistors and two reactive components.

GOOD TO KNOW

With a second-order, active, low-pass filter the rate of rolloff above the cutoff frequency is 12 dB/octave which corresponds to 40 dB/decade.

Active Low-Pass Filter

Figure 34–20*a* shows a first-order, active low-pass filter. Notice that a capacitor is placed in parallel with the feedback resistor, R_F. This capacitor is identified as C_F. At low frequencies, the capacitor has an extremely high capacitive reactance. Therefore, at low frequencies the circuit acts as an inverting amplifier with a voltage gain of $-R_F/R_i$. At higher frequencies, the capacitive reactance of C_F decreases. This causes the voltage gain to decrease due to the increased amount of negative

Figure 34–20 First-order, active low-pass filter. (*a*) Circuit. (*b*) Graph of A_{CL} versus frequency.

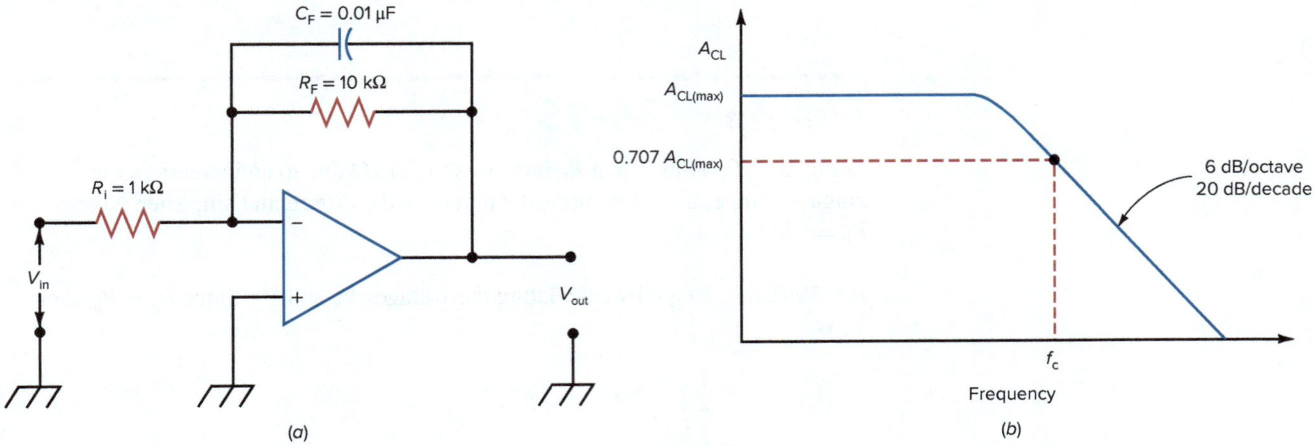

feedback. At extremely high frequencies, X_C equals almost zero ohms and the voltage gain approaches zero.

When $X_{CF} = R_F$, the voltage gain equals 70.7% of its midband value. This is the cutoff frequency, designated f_c:

$$f_c = \frac{1}{2\pi R_F C_F} \qquad \textbf{(34–23)}$$

The voltage gain for any frequency is calculated as

$$A_v = -\frac{Z_F}{R_i} \qquad \textbf{(34–24)}$$

where $$Z_F = \frac{X_{C_F} R_F}{\sqrt{R_F^{\,2} + X_{C_F}^{\,2}}}$$

Above the cutoff frequency, f_c, the voltage gain decreases at the rate of 6 dB/octave or 20 dB/decade. Figure 34–20*b* shows the frequency response curve for the circuit of Fig. 34–20*a*. In many cases, the voltage gain is specified in decibels. Then Formula (34–25) is used:

$$A_{CL(dB)} = 20 \log \frac{Z_F}{R_i} \qquad \textbf{(34–25)}$$

where $A_{CL(dB)}$ is the voltage gain in decibels.

The cutoff frequency, f_c, can be varied by adjusting the value of C_F. This, however, will not affect the passband voltage gain of the filter. The voltage gain, A_{CL}, can be adjusted by varying R_i. Adjusting the input resistor, R_i, will not affect the cutoff frequency, f_c.

Example 34–16

In Fig. 34–20, calculate the cutoff frequency, f_c.

ANSWER Using Formula (34–23), the calculations are

$$f_c = \frac{1}{2\pi R_F C_F}$$

$$= \frac{1}{2 \times 3.141 \times 10\ \text{k}\Omega \times 0.01\ \mu\text{F}}$$

$$= 1.591\ \text{kHz}$$

Example 34-17

In Fig. 34–20, calculate the voltage gain, A_{CL}, at (a) 0 Hz and (b) 1 MHz.

ANSWER (a) At 0 Hz, $X_{C_F} \approx \infty\ \Omega$. Therefore, the voltage gain is

$$A_{CL} = -\frac{R_F}{R_1}$$

$$= -\frac{10\text{ k}\Omega}{1\text{ k}\Omega}$$

$$= -10$$

(b) At 1 MHz, X_{C_F} is

$$X_{CF} = \frac{1}{2\pi f C_F}$$

$$= \frac{1}{2 \times 3.141 \times 1\text{ MHz} \times 0.01\ \mu\text{F}}$$

$$= 15.9\ \Omega$$

Next, calculate Z_F:

$$Z_F = \frac{15.9\ \Omega \times 10\text{ k}\Omega}{\sqrt{10\text{ k}\Omega^2 + 15.9\ \Omega^2}}$$

$$= \frac{159\text{ k}\Omega}{10\text{ k}\Omega}$$

$$= 15.9\ \Omega$$

Next, calculate A_{CL} using Formula (34–24):

$$A_{CL} = -\frac{Z_F}{R_1}$$

$$= -\frac{15.9\ \Omega}{1\text{ k}\Omega}$$

$$= -0.0159$$

Example 34-18

Calculate the dB voltage gain in Fig. 34–20 at (a) 0 Hz and (b) 1.591 kHz.

ANSWER (a) At 0 Hz, $X_{C_F} \approx \infty\ \Omega$, and therefore $Z_F = R_F$.

$$A_{CL(dB)} = 20 \log \frac{R_F}{R_i}$$

$$= 20 \log \frac{10\text{ k}\Omega}{1\text{ k}\Omega}$$

$$= 20 \times 1$$

$$= 20\text{ dB}$$

(b) The frequency 1.591 kHz is the cutoff frequency, f_c. At this frequency, $Z_F = 0.707$ and $R_F = 0.707 \times 10\text{ k}\Omega = 7.07\text{ k}\Omega$. Therefore,

$$A_{CL(dB)} = 20 \log \frac{Z_F}{R_i}$$

$$= 20 \log \frac{7.07\text{ k}\Omega}{1\text{ k}\Omega}$$

$$= 20 \times 0.85$$

$$= 17\text{ dB}$$

Notice that A_{CL} is down 3 dB from its passband value of 20 dB.

GOOD TO KNOW

For a second-order, high-pass filter the rate of rolloff below the cutoff frequency is 12 dB/octave which corresponds to 40 dB/decade.

Figure 34–21 First-order, active high-pass filter. (*a*) Circuit. (*b*) Graph of A_{CL} versus frequency.

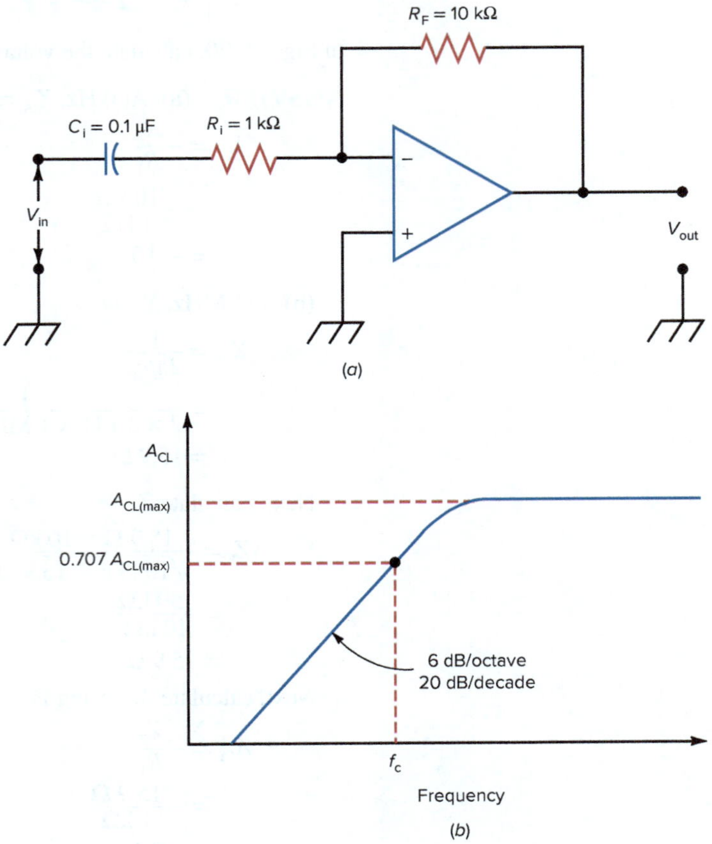

Active High-Pass Filter

Figure 34–21*a* shows a first-order, active high-pass filter, and Fig. 34–21*b* shows its frequency response curve. At high frequencies, $X_{Ci} \approx 0\ \Omega$ and the voltage gain, A_{CL}, equals $-R_F/R_i$. At very low frequencies, however, $X_{CF} \approx \infty\ \Omega$ and the voltage gain, A_{CL}, approaches zero. The cutoff frequency, f_c, is given by Formula (34–26):

$$f_C = \frac{1}{2\pi R_i C_i} \quad \textbf{(34-26)}$$

At f_c, the voltage gain, A_{CL}, is at 70.7% of its midband value.

The voltage gain, A_{CL}, for any frequency is

$$A_{CL} = -\frac{R_F}{Z_i} \quad \textbf{(34-27)}$$

where $Z_i = \sqrt{R_i^2 + X_{C_i}^2}$.

The dB voltage gain is calculated from Formula (34–28):

$$A_{CL(dB)} = 20 \log \frac{R_F}{Z_i} \quad \textbf{(34-28)}$$

The cutoff frequency, f_c, can be varied by adjusting the capacitor, C_i. However, this will have no effect on the passband voltage gain, A_{CL}. The voltage gain can be adjusted by varying R_F. This will have no effect on the cutoff frequency, f_c.

Note that the voltage gain, A_{CL}, decreases at the rate of 6 dB/octave or 20 dB/decade below the cutoff frequency, f_c.

Example 34–19

In Fig. 34–21, calculate the cutoff frequency, f_c.

ANSWER Using Formula (34–26), the calculations are

$$f_C = \frac{1}{2\pi R_i C_i}$$

$$= \frac{1}{2 \times 3.141 \times 1\ \text{k}\Omega \times 0.1\ \mu\text{F}}$$

$$= 1.591\ \text{kHz}$$

Voltage-to-Current and Current-to-Voltage Converters

Voltage-to-Current Converter

In some cases, it is necessary to have a constant load current that is not affected by changes in the load resistance, R_L. If the load doesn't have to be grounded, it can be placed in the feedback path. Figure 34–22 shows a voltage-to-current converter. Because the differential input voltage, V_{id}, equals zero, V_{in} appears across the resistor, R, connected from the op amp's inverting input to ground. The output current, I_{out}, is

$$I_{out} = \frac{V_{in}}{R} \qquad \textbf{(34-29)}$$

Notice that the output current is determined by the value of V_{in} and the resistance of R. Notice that the value of R_L *does not* affect the output current!

Because V_{in} is connected to the noninverting input of the op amp, the input impedance seen by the source, V_{in}, approaches infinity. Therefore, $Z_{in(CL)} \approx \infty\ \Omega$.

Example 34–20

In Fig. 34–22, $V_{in} = 5$ V, $R = 1$ kΩ, and $R_L = 100$ Ω. Calculate the output current, I_{out}.

ANSWER R_L will not affect the output current. To calculate I_{out}, use Formula (34–29):

$$I_{out} = \frac{V_{in}}{R}$$

$$= \frac{5\ \text{V}}{1\ \text{k}\Omega}$$

$$= 5\ \text{mA}$$

Note that I_{out} remains the same, even if R_L is reduced to zero ohms. Because I_{out} is not affected by the value of R_L, the circuit acts as a constant current source with infinite internal resistance. Therefore, $Z_{out} \approx \infty\ \Omega$.

Figure 34–22 Voltage-to-current converter.

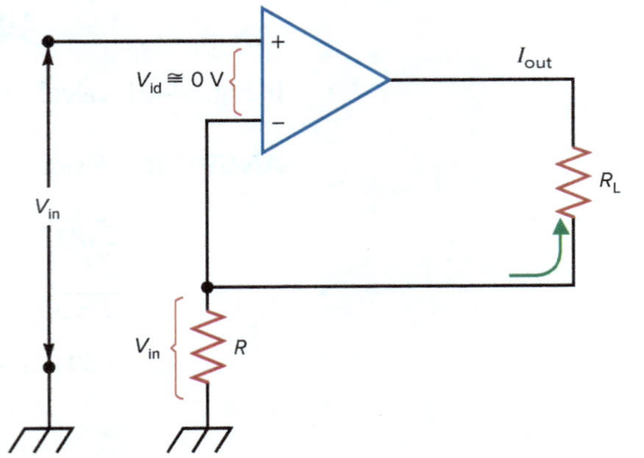

Current-to-Voltage Converter

Figure 34–23 shows a current-to-voltage converter. The circuit is driven by a current source, I_{in}. Since the inverting input of the op amp is at virtual ground, all of the input current flows through the feedback resistor, R. The output voltage V_{out} is calculated as

$$V_{out} = I_{in} \times R \qquad \textbf{(34-30)}$$

The output voltage is not affected by the value of load resistance, R_L, because the output impedance, $Z_{out(CL)}$, is approximately zero ohms.

Example 34–21

In Fig. 34–23, $I_{in} = 1.5$ mA, $R = 1$ kΩ, and $R_L = 10$ kΩ. Calculate V_{out}.

ANSWER The value of R_L will not affect the output voltage. To calculate V_{out}, use Formula (34–30):

$$\begin{aligned} V_{out} &= I_{in} \times R \\ &= 1.5 \text{ mA} \times 1 \text{ k}\Omega \\ &= 1.5 \text{ V} \end{aligned}$$

Figure 34–23 Current-to-voltage converter.

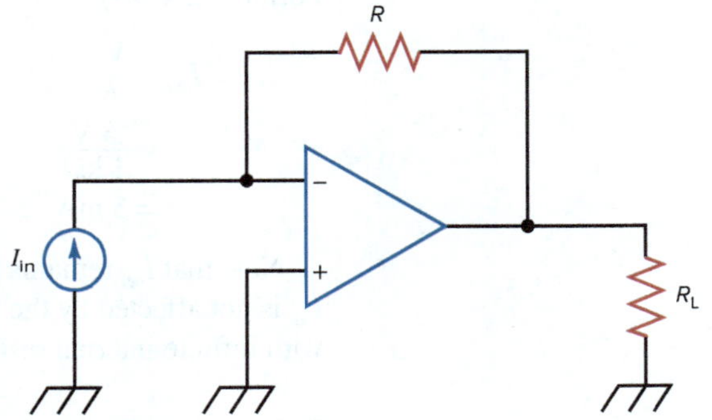

Figure 34–24 Op-amp comparator. (a) Circuit. (b) Transfer characteristic.

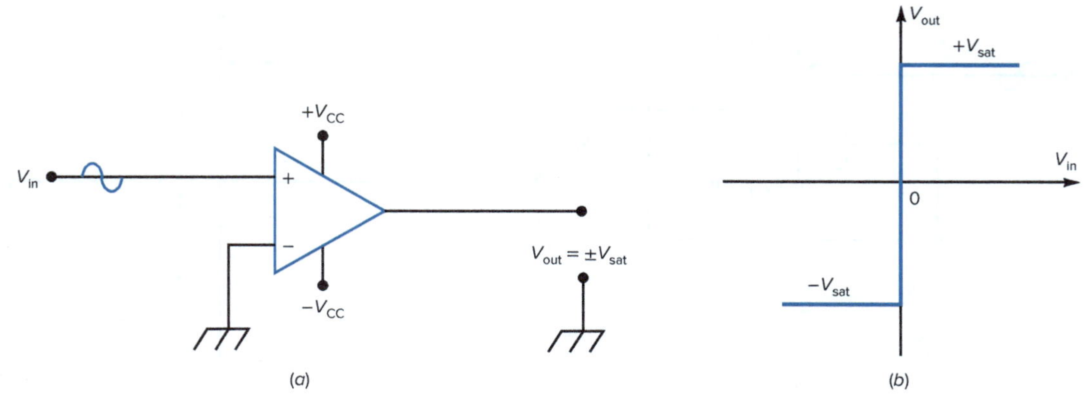

Comparators

A ***comparator*** is a circuit that compares the signal voltage on one input with a reference voltage on the other. An op-amp comparator is shown in Fig. 34–24*a*. Notice that the circuit does not use any feedback resistors. As a result, the op amp is running in the open-loop mode with a voltage gain equal to A_{VOL}.

In Fig. 34–24*a*, the inverting input of the op amp is grounded while the input signal is applied to the noninverting input. The comparator compares V_{in} to the zero volt reference on the inverting input. When V_{in} is positive, V_{out} is driven to $+V_{sat}$. When V_{in} is negative, V_{out} is driven to $-V_{sat}$. Usually, $\pm V_{sat}$ is within a couple of volts of $\pm V_{CC}$.

Because the op amp has an extremely high open-loop voltage gain (A_{VOL}), even the slightest input voltage produces an output of $\pm V_{sat}$. Figure 34–24*b* shows the transfer characteristic for the comparator. Notice that the output switches to $+V_{sat}$ if V_{in} is positive, and to $-V_{sat}$ if V_{in} is negative. Because the output voltage switches when V_{in} crosses zero, the circuit is sometimes called a ***zero-crossing detector***.

In some cases, the noninverting input of the op amp is grounded, and the signal is applied to the inverting input. If so, the output switches to $+V_{sat}$ when V_{in} is negative, and to $-V_{sat}$ when V_{in} is positive.

GOOD TO KNOW

The output of a comparator can be characterized as digital in the sense that the output is always at either $\pm V_{sat}$.

Shifting the Reference Point

Figure 34–25*a* shows a comparator that uses a reference voltage other than zero. In this case, the reference voltage, V_{ref}, equals +5 V. When V_{in} exceeds +5 V, V_{out}

Figure 34–25 Comparator with +5 V reference. (a) Circuit. (b) Transfer characteristic.

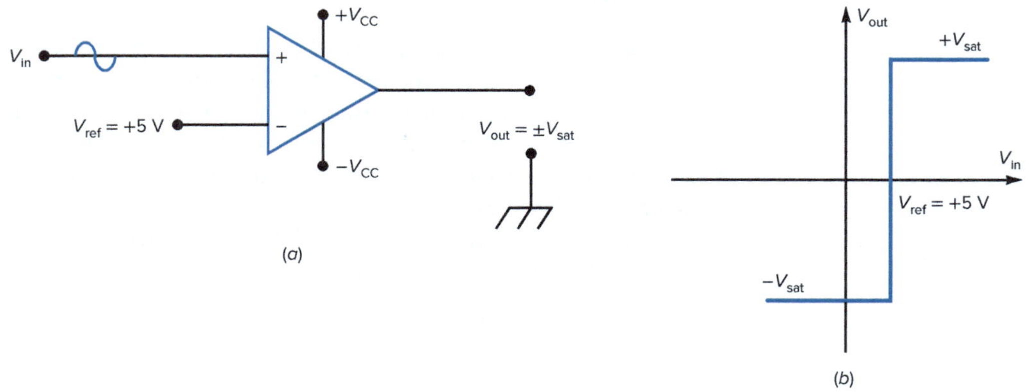

Figure 34–26 Schmitt trigger. (*a*) Circuit. (*b*) Transfer characteristic.

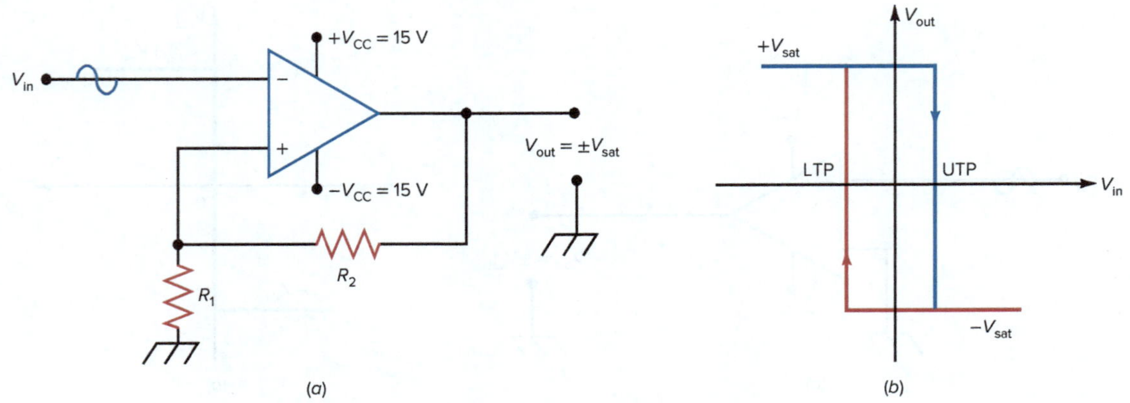

switches to $+V_{sat}$. When V_{in} drops below +5 V, V_{out} switches to $-V_{sat}$. The transfer characteristic is shown in Fig. 34–25*b*. In fact, V_{ref} can be either positive or negative. The reference voltage is usually obtained by attaching a resistive voltage divider to either $+V_{CC}$ or $-V_{CC}$. In some cases, it may even be desirable to have an adjustment that allows the reference voltage to vary.

The Schmitt Trigger

Because the op-amp comparator is so sensitive, the output may switch back and forth erratically between $+V_{sat}$ and $-V_{sat}$ when V_{in} is near the reference voltage, V_{ref}. The reason that the comparator output switches erratically between $+V_{sat}$ and $-V_{sat}$ is that the input signal usually contains some noise.

To eliminate the erratic operation, use the **Schmitt trigger**, shown in Fig. 34–26*a*. A *Schmitt trigger* is an op-amp comparator that uses positive feedback. R_1 and R_2 provide the positive feedback. When $V_{out} = +V_{sat}$, a positive voltage appears at the noninverting input. Likewise, when $V_{out} = -V_{sat}$, a negative voltage appears at the noninverting input. The feedback fraction, usually specified as β, indicates the fraction of output voltage fed back to the noninverting input. β equals

$$\beta = \frac{R_1}{R_1 + R_2} \tag{34-31}$$

The upper threshold point, designated UTP, is

$$\text{UTP} = +\beta V_{sat} \tag{34-32}$$

The lower threshold point, designated LTP, equals

$$\text{LTP} = -\beta V_{sat} \tag{34-33}$$

The output of the Schmitt trigger will remain in its present state until the input voltage exceeds the reference or threshold voltage for that state. This is best described by viewing the transfer characteristic shown in Fig. 34–26*b*. Assume that $V_{out} = +V_{sat}$. The input voltage must exceed the value of $+\beta V_{sat}$ (UTP) to switch the output to its opposite state, which would be $-V_{sat}$. The output remains at $-V_{sat}$ until V_{in} goes more negative than $-\beta V_{sat}$ (LTP).

The difference between the upper and lower threshold points is called the *hysteresis voltage*, designated V_H. V_H is calculated as

$$\begin{aligned} V_H &= \text{UTP} - \text{LTP} \\ &= +\beta V_{sat} - (-\beta V_{sat}) \\ &= 2\beta V_{sat} \end{aligned} \tag{34-34}$$

If the peak-to-peak value of noise voltage is less than the hysterisis voltage, V_H, there is no way the output can switch states. Because of this, a Schmitt trigger can be designed so that it is immune to erratic triggering caused by noise.

Example 34-22

In Fig. 34–26, $R_1 = 1\ k\Omega$ and $R_2 = 100\ k\Omega$. Calculate UTP, LTP, and V_H.

ANSWER Since $\pm V_{CC} = 15$ V, assume that $\pm V_{sat} = \pm 13$ V. To calculate the threshold voltages, UTP and LTP, calculate the feedback fraction, β:

$$\beta = \frac{R_1}{R_1 + R_2} = \frac{1\ k\Omega}{1\ k\Omega + 100\ k\Omega} = 0.0099$$

Knowing β, calculate UTP and LTP as follows:

$$\text{UTP} = +\beta V_{sat} = 0.0099 \times 13\ \text{V} = 128.7\ \text{mV}$$

$$\text{LTP} = -\beta V_{sat} = 0.0099 \times 13\ \text{V} = -128.7\ \text{mV}$$

The hysterisis voltage is calculated as

$$V_H = \text{UTP} - \text{LTP} = 128.7\ \text{mV} - (-128.7\ \text{mV}) = 257.4\ \text{mV}$$

Op-Amp Diode Circuits

Op amps can be used with diodes to rectify signals with peak values in the millivolt region. A conventional diode cannot do this by itself because it requires a larger voltage to turn on. A silicon diode requires about 0.7 V and a germanium diode requires 0.3 V.

Figure 34–27 shows a precision half-wave rectifier. This circuit is capable of rectifying signals having peak values in the millivolt region. Here is how it works:

Figure 34–27 Precision half-wave rectifier.

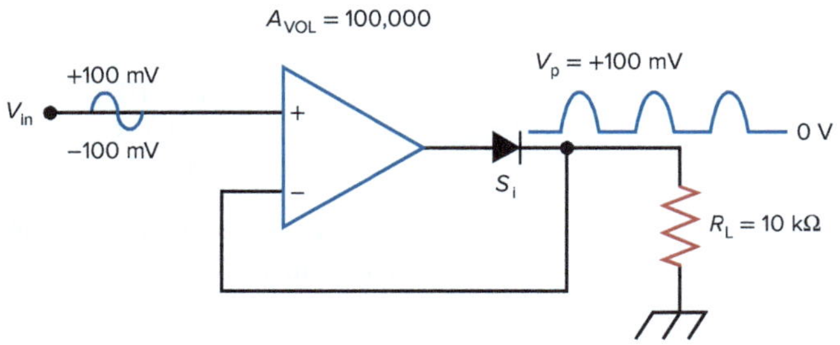

When V_{in} goes positive, the output of the op amp also goes positive. The diode will turn on when the output of the op amp equals 0.7 V. The amount of input voltage required to obtain 0.7 V at the op-amp output is

$$V_{in} = \frac{0.7\ V}{A_{VOL}}$$
$$= \frac{0.7\ V}{100{,}000}$$
$$= 7\ \mu V$$

When the input voltage exceeds 7 μV, the diode conducts and the circuit acts as a voltage follower. Then the output voltage follows the input. The op amp effectively reduces the diode turn-on voltage by a factor equal to A_{VOL}.

When V_{in} goes negative, so does the op-amp output. This turns off the diode, and so the output voltage equals zero volts. Notice that the output voltage is a series of positive pulses that have a peak value equal to $V_{in(pk)}$. Whenever it is necessary to rectify lower amplitude signals, a precision rectifier can do the job. Reversing the connection of the diode will produce a series of negative pulses at the output.

Precision Peak Detector

To peak-detect very low level signals, add a capacitor to the output of the precision half-wave rectifier. Figure 34–28 shows a precision peak detector. The output from the precision peak detector is a DC voltage whose value is equal to the positive peak of the input voltage. When V_{in} goes positive, the diode conducts and charges the capacitor at the output to the peak positive value of input voltage.

When V_{in} goes negative, the diode is off and the capacitor discharges through the load, R_L. The discharge time constant, R_LC, must be

$$R_LC \geq 10T$$

where T is the period of the input waveform. To obtain a negative output voltage, reverse the diode.

Example 34-23

In Fig. 34–28, $R_L = 1\ k\Omega$ and the frequency of the input voltage equals 100 Hz. Calculate the minimum value of C required.

ANSWER The period T equals

$$T = \frac{1}{f}$$
$$= \frac{1}{100\ Hz}$$
$$= 10\ ms$$

Next transpose the formula $R_LC = 10T$:

$$C = \frac{10T}{R_L}$$
$$= \frac{10 \times 10\ ms}{1\ k\Omega}$$
$$= 100\ \mu F$$

Using a C value larger than 100 μF improves the operation of the precision peak detector.

Figure 34–28 Precision peak detector.

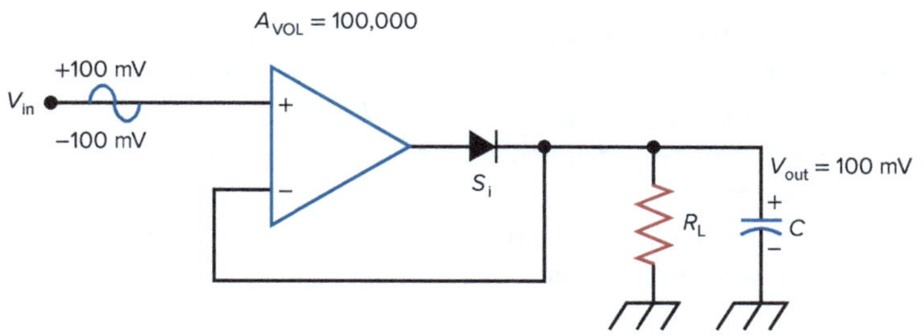

34–4 Self-Review

Answers at the end of the chapter.

a. **In a first-order, active high-pass filter, what is the rate of rolloff below the cutoff frequency?**

b. **In a summing amplifier, what isolates the inputs from each other?**

c. **What benefit does positive feedback provide in a Schmitt trigger?**

Summary

- The first stage of every op amp is a differential amplifier.
- A differential amplifier has two inputs, an inverting (−) input and a noninverting (+) input.
- A differential amplifier amplifies the difference between its two input signals.
- A differential amplifier rejects or severely attenuates signals that are common to both inputs.
- The common-mode rejection ratio (CMRR) is defined as the ratio of differential voltage gain, A_d, to common-mode voltage gain, A_{CM}.
- Op amps are high gain, direct-coupled, differential amplifiers.
- The open-loop voltage gain of an op amp is its voltage gain without negative feedback.
- When the output voltage of an op amp lies between $\pm V_{sat}$, the differential input voltage, V_{id}, is so small it can be considered zero.
- The open-loop cutoff frequency is the frequency where the open-loop voltage gain of an op amp is down to 70.7% of its maximum value at DC.
- The frequency where the open-loop voltage gain equals one is called f_{unity}.
- The slew-rate specification of an op amp indicates how fast the output voltage can change. The slew rate is specified in V/μs.
- The slew-rate distortion of a sine wave makes the output waveform appear triangular.
- Most op-amp circuits use negative feedback. With negative feedback, a portion of the output signal is fed back, 180° out of phase, to oppose the input signal.
- Negative feedback stabilizes the voltage gain of an amplifier and improves the bandwidth and input and output impedances.
- An inverting amplifier has a voltage gain, A_{CL}, of $-R_F/R_i$. The minus sign indicates that V_{in} and V_{out} are 180° out of phase.
- A noninverting amplifier has a voltage gain, A_{CL}, of $R_F/R_i + 1$.
- A voltage follower is a noninverting amplifier with a voltage gain of one or unity.
- An op-amp summing amplifier is a circuit whose output voltage is the negative sum of the input voltages.
- An op-amp differential amplifier is a circuit that amplifies the difference between two input voltages whose values may be several volts or more.
- An active filter uses an op amp to provide voltage gain in addition to filtering.
- A first-order, active filter uses only one reactive component. The output voltage of a first-order, active filter rolls off at the rate of 6 dB/octave beyond the cutoff frequency.
- An op-amp comparator is a circuit that compares the signal voltage on one input with a reference voltage on the other. A comparator uses no negative feedback and its output voltage is at either $\pm V_{sat}$.
- A Schmitt trigger is an op-amp comparator that uses positive feedback to eliminate the erratic operation caused by undesired noise.
- Op amps are often used in conjunction with diodes to rectify and filter small signals in the millivolt region.

Important Terms

Active filter—a filter that uses components or devices, such as transistors and op amps, that can amplify.

Closed-loop cutoff frequency, f_{CL}—the frequency at which the closed-loop voltage gain of an op amp decreases to 70.7% of its maximum.

Closed-loop voltage gain, A_{CL}—the voltage gain of an amplifier with negative feedback.

Common-mode input—an identical voltage appearing on both inputs of a differential amplifier.

Common-mode rejection ratio (CMRR)—the ratio of differential voltage gain, A_d, to common mode voltage gain, A_{CM}. CMRR is usually given in decibels.

Common-mode voltage gain, A_{CM}—the voltage gain of a differential amplifier for a common-mode signal.

Comparator—a circuit that compares the signal voltage on one input with a reference voltage on the other.

Differential input voltage, V_{id}—the voltage difference between the two inputs applied to a differential amplifier.

Differential voltage gain, A_d—the ratio of output voltage, V_{out}, to differential input voltage, V_{id}.

f_{unity}—the frequency where the open-loop voltage gain, A_{VOL}, of an op amp equals one or unity.

Input bias current, I_B—the average of the two op-amp input currents I_{B+} and I_{B-}.

Input offset current, I_{os}—the difference between the two input bias currents I_{B+} and I_{B-}.

Negative feedback—a form of amplifier feedback where the returning signal has a phase that opposes the input signal.

Negative saturation voltage, $-V_{sat}$—the lower limit of output voltage of an op amp.

Open-loop cutoff frequency, f_{OL}—the frequency at which the open-loop voltage gain of an op amp is down to 70.7% of its maximum value at DC.

Open-loop voltage gain, A_{VOL}—the voltage gain of an op amp without negative feedback.

Operational amplifier (op amp)—a high-gain, direct-coupled, differential amplifier.

Positive saturation voltage, $+V_{sat}$—the upper limit of output voltage of an op amp.

Power bandwidth (f_{max})—the highest undistorted frequency out of an op amp without slew-rate distortion.

Schmitt trigger—an op-amp comparator that uses positive feedback.

Slew rate, S_R—an op-amp specification indicating the maximum rate at which the output voltage can change. S_R is specified in $V/\mu s$.

Slew-rate distortion—a distortion in op amps when the rate of change in output voltage exceeds the slew-rate specification of the op amp.

Summing amplifier—an amplifier whose output voltage equals the negative sum of the input voltages.

Tail current, I_T—the DC current in the emitter resistor of a differential amplifier.

Voltage follower—an op-amp circuit with unity voltage gain. A voltage follower has very high input impedance and very low output impedance. Voltage followers are also known as unity-gain amplifiers, buffer amplifiers, and isolation amplifiers.

Zero-crossing detector—an op-amp comparator whose output voltage switches to either $\pm V_{sat}$ when the input voltage crosses through zero.

Related Formulas

Differential Amplifier

$$V_{out} = A_d(V_1 - V_2)$$

$$I_T = \frac{V_{EE} - V_{BE}}{R_E}$$

$$I_E = \frac{V_{EE} - V_{BE}}{2R_E}$$

$$V_C = V_{CC} - I_C R_C$$

$$A_d = \frac{R_C}{2r'_e} \text{ (Signal Applied to Noninverting Input)}$$

$$A_d = -R_C/2r'_e \text{ (Signal Applied to Inverting Input)}$$

$$V_{out} = \frac{R_C}{2r'_e}(V_1 - V_2)$$

$$A_{CM} = R_C/2R_E$$

$$\text{CMRR} = \frac{A_d}{A_{CM}}$$

$$\text{CMRR(dB)} = 20 \log (A_d/A_{CM})$$

Op-Amp Specifications

$$I_B = \frac{|I_{B+}| + |I_{B-}|}{2}$$

$$I_{OS} = |I_{B+}| - |I_{B-}|$$

$$f_{max} = S_R/2\pi V_{pk}$$

Inverting Amplifier

$$A_{CL} = -R_F/R_i$$

$$Z_{in} = R_i$$

$$Z_{out(CL)} = \frac{Z_{out(OL)}}{1 + A_{VOL}\beta}$$

$$\beta = \frac{R_i}{R_i + R_F}$$

Noninverting Amplifier

$$A_{CL} = \frac{R_F}{R_i} + 1$$

$$Z_{in(CL)} = R_{in}(1 + A_{VOL}\beta)$$

$$Z_{out(CL)} = \frac{Z_{out(OL)}}{1 + A_{VOL}\beta}$$

$$f_{CL} = \frac{f_{unity}}{A_{CL}}$$

Summing Amplifier

$$V_{out} = -(V_1 + V_2 + V_3) \quad (R_1 = R_2 = R_3 = R_F)$$

$$V_{out} = -\left[\frac{R_F}{R_1}V_1 + \frac{R_F}{R_2}V_2 + \frac{R_F}{R_3}V_3\right]$$

Differential Amplifier

$$V_{out} = -\frac{R_F}{R_1}(V_x - V_y)$$

Active Filters

$$f_C = 1/2\pi R_F C_F \quad \text{(Low-Pass)}$$

$$A_V = -Z_F/R_i \quad \text{(Low-Pass)}$$

$$A_{CL}\text{(dB)} = 20 \log \frac{Z_F}{R_i} \quad \text{(Low-Pass)}$$

$$f_C = 1/2\pi R_i C_i \quad \text{(High-Pass)}$$

$$A_{CL} = -R_F/Z_i \quad \text{(High-Pass)}$$

$$A_{CL}\text{(dB)} = 20 \log \frac{R_F}{Z_i} \quad \text{(High-Pass)}$$

Voltage-to-Current Converter

$$I_{out} = V_{in}/R$$

Current-to-Voltage Converter

$$V_{out} = I_{in} \times R$$

Schmitt Trigger

$$\beta = \frac{R_1}{R_1 + R_2}$$

$$\text{UTP} = +\beta V_{sat}$$

$$\text{LTP} = -\beta V_{sat}$$

$$V_H = 2\beta V_{sat}$$

Self-Test

Answers at the back of the book.

1. The input stage of every op amp is a

a. differential amplifier.
b. push-pull amplifier.
c. common-base amplifier.
d. none of the above.

2. The DC emitter current in each transistor of a differential amplifier equals

a. the tail current, I_T.
b. twice the tail current.
c. one-half the tail current.
d. zero.

3. A differential amplifier has an A_d of 100 and an A_{CM} of 0.1. What is its CMRR in dB?

a. 1000 dB.
b. 60 dB.
c. 30 dB.
d. It cannot be determined.

4. The output stage of a 741 op amp is a

a. differential amplifier.
b. common-base amplifier.
c. common-emitter amplifier.
d. push-pull amplifier.

5. A typical value of open-loop voltage gain for a 741 op amp is

a. 100.
b. 0.5.
c. 200,000.
d. none of the above.

6. When the inverting (−) input of an op amp is positive with respect to its noninverting (+) input, the output voltage is

a. negative.
b. positive.
c. zero.
d. none of the above.

7. If an op amp has an open-loop voltage gain of 100,000, what is the voltage gain at the open-loop cutoff frequency?

a. 100,000.
b. 70,700.
c. 141,400.
d. none of the above.

8. The slew-rate specification of an op amp is the

a. maximum value of positive or negative output voltage.
b. maximum rate at which its output voltage can change.
c. attenuation against a common-mode signal.
d. frequency where the voltage gain is one or unity.

9. In an inverting amplifier, the input and output voltages are

a. in phase.
b. 90° out of phase.
c. 180° out of phase.
d. 360° out of phase.

10. In an inverting amplifier, a virtual ground

a. is no different from an ordinary ground.
b. can sink a lot of current.
c. usually has a significant voltage drop.
d. has the same potential as ground, yet it can sink no current.

11. The input impedance of an inverting amplifier is approximately equal to

a. R_i.
b. zero.
c. infinity
d. R_F

12. A noninverting amplifier has a 15-kΩ R_F and a 1.2-kΩ R_i. How much is its closed-loop voltage gain, A_{CL}?

a. 12.5.
b. −12.5.
c. 13.5.
d. 9.

13. A voltage follower has a

a. high input impedance.
b. low output impedance.
c. voltage gain of one.
d. all of the above.

14. An op-amp circuit has a closed-loop voltage gain of 50. If the op amp has an f_{unity} of 15 MHz, what is the closed-loop cutoff frequency?

a. 30 kHz.
b. 300 kHz.
c. 750 MHz.
d. 750 kHz.

15. When an op-amp circuit uses a single supply voltage, the DC output voltage from the op amp should be equal to

a. ½ V_{CC}.
b. V_{CC},
c. zero.
d. 0.1 V_{CC}.

16. In an op-amp summing amplifier, the inputs are effectively isolated from each other because of the

a. low output impedance of the op amp.
b. feedback resistor.
c. virtual ground.
d. none of the above.

17. For a first-order, active low-pass filter, how fast does the output voltage roll off above the cutoff frequency?

a. 6 dB/decade.
b. 20 dB/decade.
c. 6 dB/octave.
d. both b and c.

18. An op-amp comparator that uses positive feedback is known as a

a. zero-crossinq detector.
b. Schmitt trigger.
c. peak detector.
d. voltage follower.

19. A comparator never uses

a. positive feedback.
b. an input signal.
c. negative feedback.
d. none of the above.

20. In a voltage-to-current converter, the output current is not affected by

a. the load resistance value.
b. the input voltage.
c. the resistance, R, across which the input voltage is present.
d. none of the above.

Essay Questions

1. What type of circuit is used for the input stage of an op amp?
2. What is a common-mode signal?
3. What is the common-mode rejection ratio and how is it usually specified?
4. What is slew-rate distortion and how can it be prevented?
5. What are the advantages of using negative feedback with an amplifier?
6. When and where would you use a voltage follower?
7. What is meant by the closed-loop gain-bandwidth product?
8. Explain the concept called virtual ground.
9. How does an active filter differ from a passive filter?
10. Why is a Schmitt trigger immune to erratic triggering caused by noise?

Problems

SECTION 34–1 DIFFERENTIAL AMPLIFIERS

34–1 In Fig. 34–29, solve for the following DC quantities:

a. I_T.
b. I_E for each transistor.
c. V_C at the collector of Q_2.

Figure 34–29

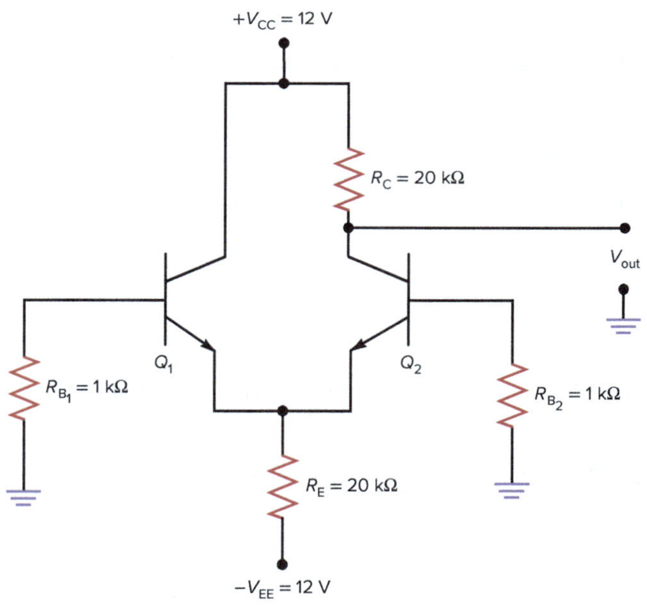

34–2 If both transistors have a β_{DC} of 200 in Fig. 34–29, then calculate

a. the base current, I_B, for each transistor.
b. the DC base voltage for each transistor.

34–3 In Fig. 34–29, which transistor base serves as the

a. inverting input?
b. noninverting input?

34–4 In Fig. 34–29, how much is the

a. differential voltage gain, A_d?
b. common-mode voltage gain, A_{CM}?
c. CMRR?
d. CMRR (dB)?

34–5 In Fig. 34–30, solve for the following DC quantities:

a. I_T.
b. I_E for each transistor.
c. V_C at the collector of Q_2.

Figure 34–30

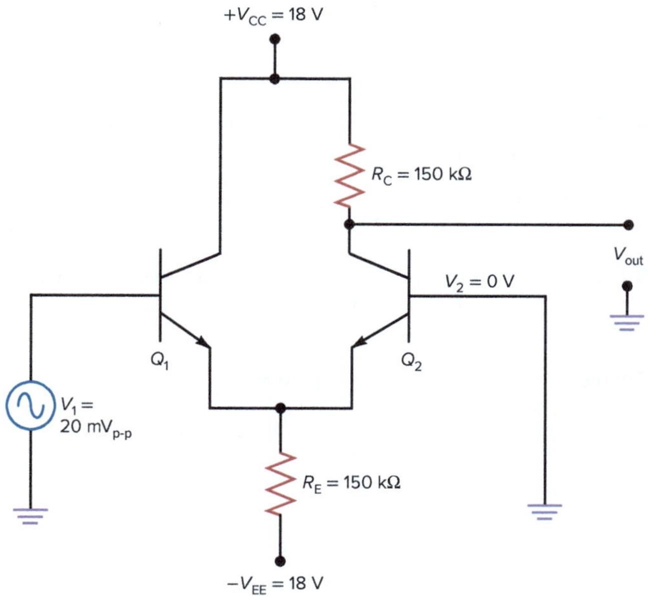

34–6 In Fig. 34–30, how much is the

a. differential voltage gain, A_d?
b. AC output voltage, V_{out}?
c. common-mode voltage gain, A_{CM}?
d. CMRR?
e. CMRR (dB)?

SECTION 34–2 OPERATIONAL AMPLIFIERS AND THEIR CHARACTERISTICS

34–7 What type of circuit is used for the input stage of the 741 op amp?

34–8 What type of circuit is used for the output stage of the 741 op amp?

34–9 In the 741 op amp, what type of coupling is used between stages? What advantage does this provide?

34–10 In Fig. 34–31, calculate the output voltage, V_{out}, for each of the following values of V_1 and V_2. Be sure to denote the proper polarity of the output voltage.

a. $V_1 = 100$ mV, $V_2 = 100.05$ mV.

b. $V_1 = 100.05$ mV, $V_2 = 100$ mV.

c. $V_1 = -75$ μV, $V_2 = -100$ μV.

d. $V_1 = 3.01$ mV, $V_2 = 3$ mV.

Figure 34–31

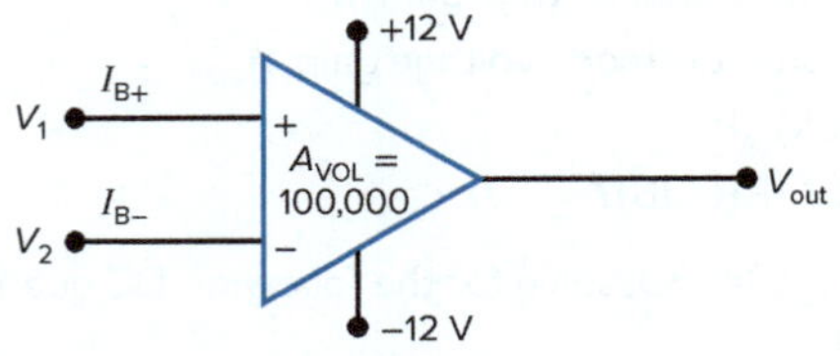

34–11 In Fig. 34–31, assume $\pm V_{sat} = \pm 10$ V. What value of V_{id} will produce positive or negative saturation?

34–12 In Fig. 34–31, assume that $I_{B+} = 105$ nA and $I_{B-} = 85$ nA. Calculate

a. I_B.

b. I_{OS}.

34–13 What is the open-loop cutoff frequency, f_{OL}, for a 741 op amp?

34–14 What is the open-loop voltage gain at f_{OL} for the op amp in Fig. 34–31?

34–15 At what frequency does the open-loop voltage gain of a 741 op amp equal one? How is this frequency designated?

34–16 What is the S_R of a 741 op amp?

34–17 Calculate f_{max} for a 741 op amp for each of the following peak output voltages:

a. $V_{pk} = 0.5$ V.

b. $V_{pk} = 1$ V.

c. $V_{pk} = 2$ V.

d. $V_{pk} = 5$ V.

34–18 What is output short-circuit current of a 741 op amp?

34–19 What is CMRR in dB for a 741 op amp?

SECTION 34–3 OP-AMP CIRCUITS WITH NEGATIVE FEEDBACK

34–20 What type of amplifier is shown in Fig. 34–32?

$S_R = 0.5$ V/μs

$A_{VOL} = 100{,}000$

$Z_{out(OL)} = 75$ Ω

$f_{unity} = 1$ MHz

Figure 34–32

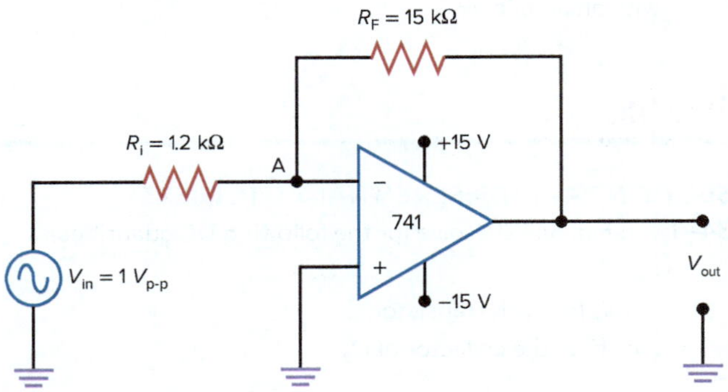

34–21 In Fig. 34–32, what is the phase relationship between V_{in} and V_{out}?

34–22 What is the approximate value of V_{id} in Fig. 34–32?

34–23 In Fig. 34–32, why is point A said to be at virtual ground?

34–24 In Fig. 34–32, solve for the following values:

a. A_{CL}.

b. V_{out}.

c. Z_{in}.

d. $Z_{out(CL)}$.

34–25 To avoid slew-rate distortion in Fig. 34–32, what is the highest allowable frequency of V_{in}?

34–26 In Fig. 34–32, calculate A_{CL} and V_{out} for each of the following combinations of values for R_F and R_i.

a. $R_F = 12$ kΩ and $R_i = 750$ Ω.

b. $R_F = 27$ kΩ and $R_i = 1.5$ kΩ.

c. $R_F = 100$ kΩ and $R_i = 20$ kΩ.

34–27 What type of amplifier is shown in Fig. 34–33?

$A_{VOL} = 100{,}000$

$R_{in} = 2$ MΩ

$Z_{out(OL)} = 75$ Ω

$f_{unity} = 1$ MHz

Figure 34–33

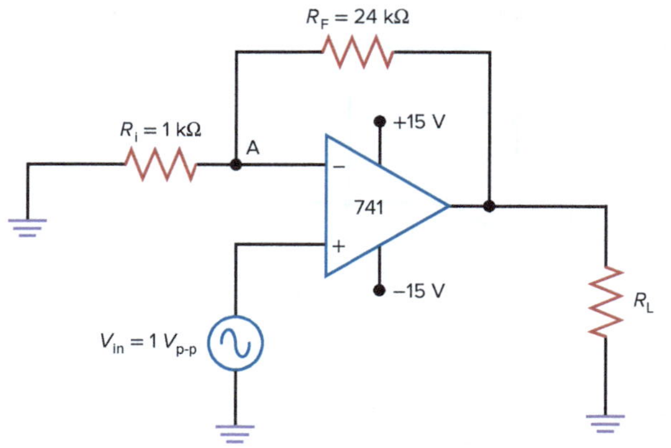

34–28 In Fig. 34–33, what is the phase relationship between V_{in} and V_{out}?

34–29 What is the approximate value for V_{id} in Fig. 34–33?

34–30 In Fig. 34–33, how much AC signal voltage would be measured at point A?

34–31 In Fig. 34–33, solve for the following:

a. A_{CL}.

b. V_{out}.

c. $Z_{in(CL)}$.

d. $Z_{out(CL)}$.

34–32 In Fig. 34–33, calculate A_{CL} and V_{out} for each of the following combinations for R_F and R_i.

a. $R_F = 15\ k\Omega$ *and* $R_i = 1\ k\Omega$.

b. $R_F = 24\ k\Omega$ *and* $R_i = 1.5\ k\Omega$.

c. $R_F = 10\ k\Omega$ *and* $R_i = 2\ k\Omega$.

34–33 What type of circuit is shown in Fig. 34–34?

$A_{VOL} = 100{,}000$

$f_{unity} = 1$ MHz

$R_{in} = 2\ M\Omega$

$Z_{out(OL)} = 75\ \Omega$

Figure 34–34

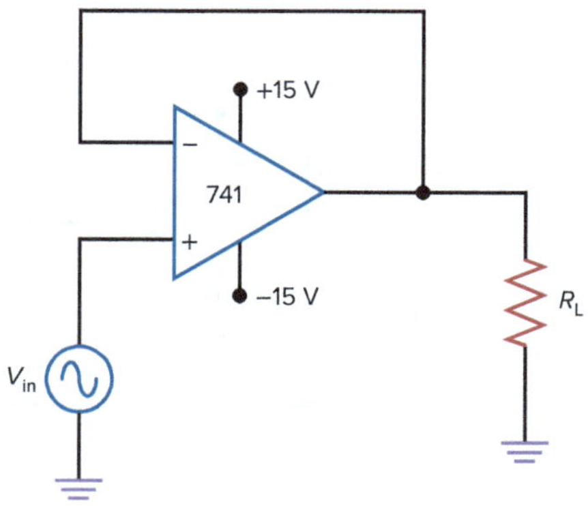

34–34 If $V_{in} = 10\ V_{p\text{-}p}$ in Fig. 34–34, what is V_{out}?

34–35 Calculate the closed-loop cutoff frequency, f_{CL}, for the values in

a. Fig. 34–32.

b. Fig. 34–33.

c. Fig. 34–34.

34–36 In Fig. 34–35, how much DC voltage exists at the

a. noninverting (+) input?

b. inverting (−) input?

c. the op-amp output?

Figure 34–35

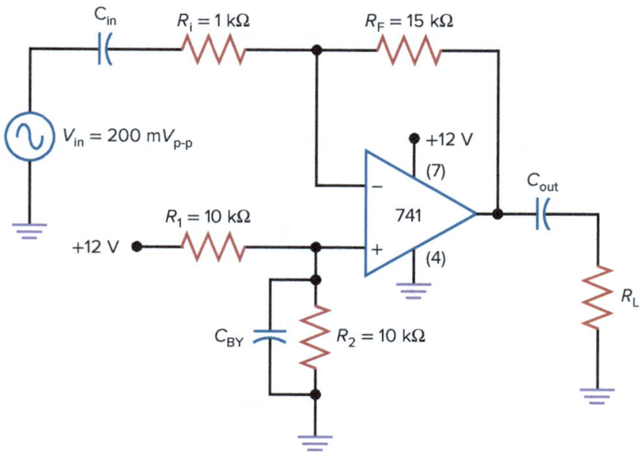

34–37 In Fig. 34–35, solve for

a. A_{CL}.

b. V_{out}.

SECTION 34–4 POPULAR OP-AMP CIRCUITS

34–38 Calculate the output voltage in

a. Fig. 34–36.

b. Fig. 34–37.

Figure 34–36

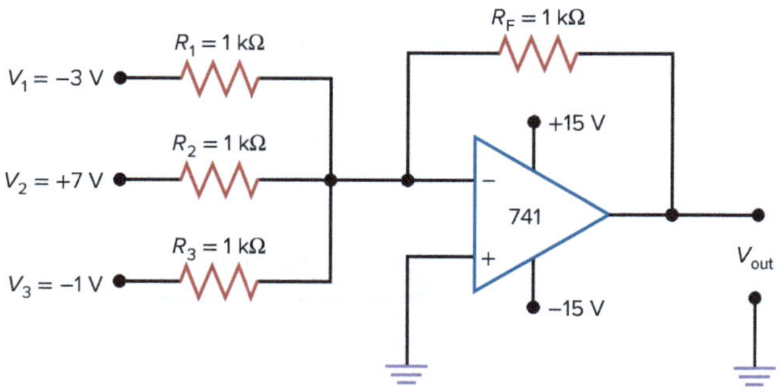

34–39 In Fig. 34–38, calculate the output voltage (including polarity) for the following values of V_X and V_Y.

a. $V_X = 1.5$ V and $V_Y = -1$ V.

b. $V_X = -2$ V and $V_Y = -1.5$ V.

c. $V_X = 8$ V and $V_Y = 10$ V.

d. $V_X = 5.5$ V and $V_Y = 6.25$ V.

Figure 34–37

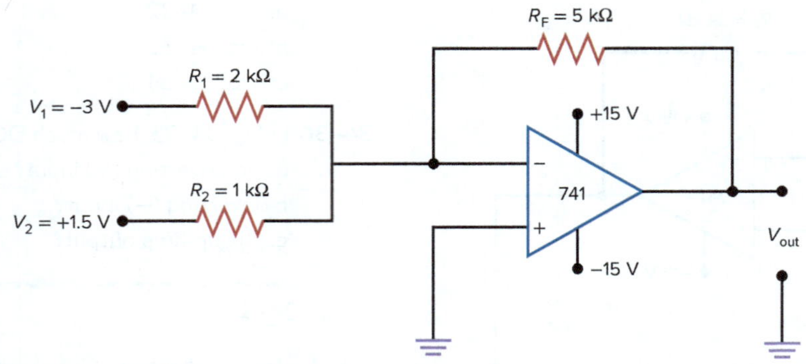

Figure 34–38

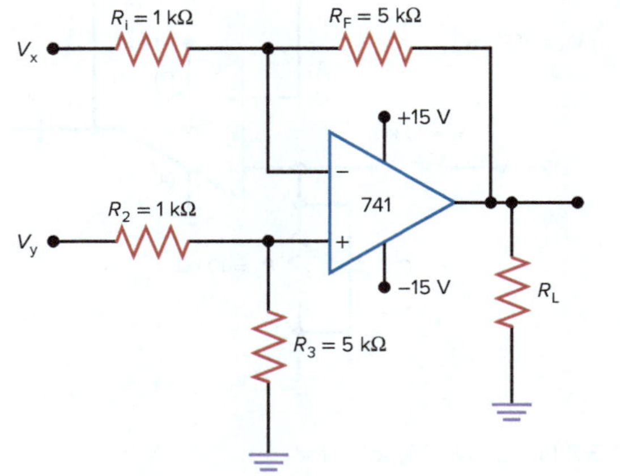

34–40 Assume that the instrumentation amplifier in Fig. 34–19 has an output of 0 V at room temperature (25°C). What happens to the output voltage when the temperature

a. increases above 25°C?

b. decreases below 25°C?

34–41 In Fig. 34–19, assume that R_D decreases to 3 kΩ as a result of a decrease in temperature. How much is the output voltage? (Note: $R_B = 5$ kΩ.)

34–42 What type of circuit is shown in Fig. 34–39?

Figure 34–39

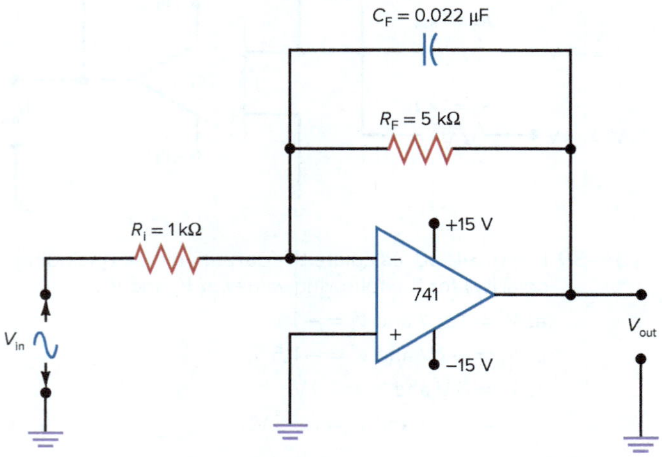

34–43 Calculate the cutoff frequency, f_C, for the circuit in Fig. 34–39.

34–44 Calculate the dB voltage gain in Fig. 34–39 for the following frequencies:

a. $f = 0$ Hz.

b. f_c.

c. $f = 10$ kHz.

34–45 What type of circuit is shown in Fig. 34–40?

Figure 34–40

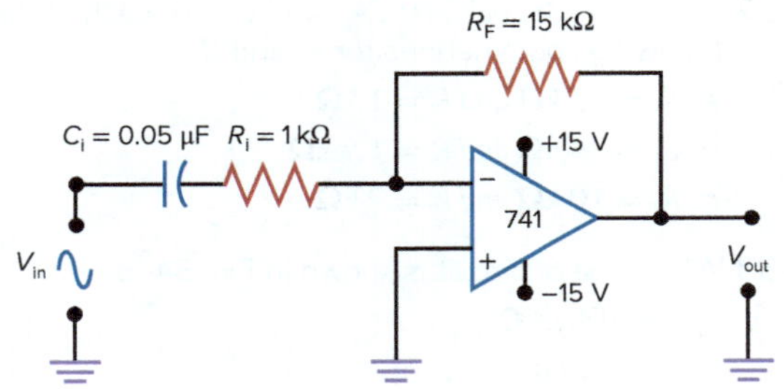

34–46 Calculate the cutoff frequency, f_c, in Fig. 34–40.

34–47 What type of circuit is shown in Fig. 34–41?

Figure 34–41

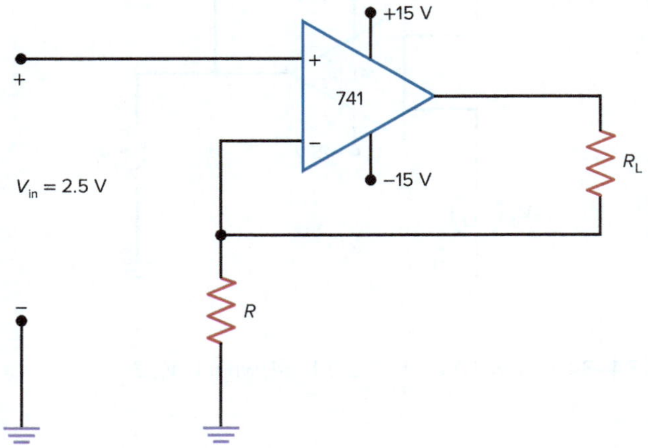

34–48 In Fig. 34–41, calculate the output current, I_{out}, for the following values of R:

a. $R = 10$ kΩ.

b. $R = 5$ kΩ.

c. $R = 2$ kΩ.

34–49 What type of circuit is shown in Fig. 34–42?

Figure 34–42

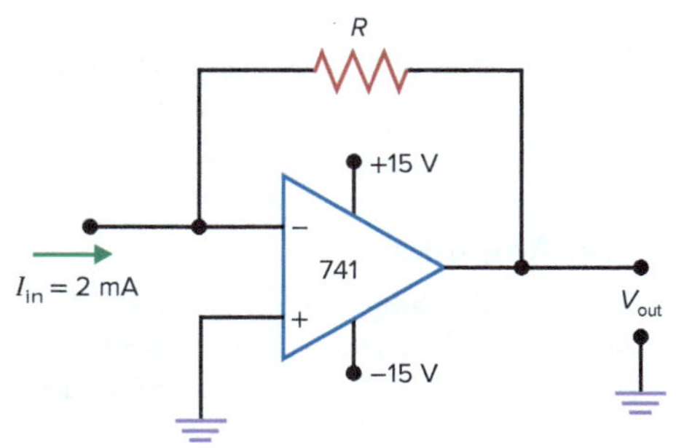

34–50 In Fig. 34–42, calculate the output voltage, V_{out}, for the following values of R:

a. $R = 1$ kΩ.

b. $R = 1.5$ kΩ.

c. $R = 5$ kΩ.

34–51 In Fig. 34–43, what value of V_{in} causes the output to be at

a. $+V_{sat}$?

b. $-V_{sat}$?

Figure 34–43

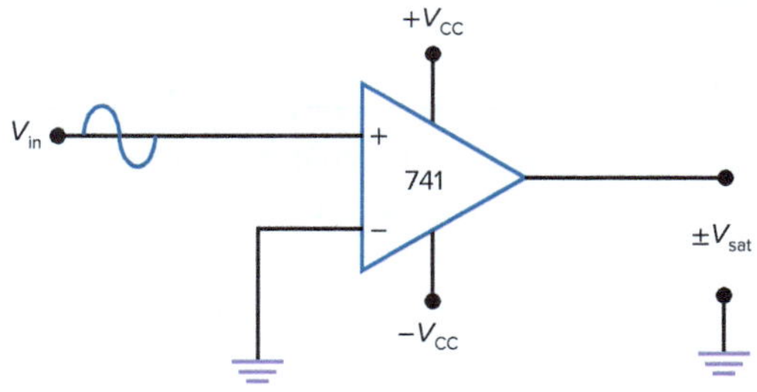

34–52 In Fig. 34–44, calculate the following (assume $\pm V_{sat} = \pm 10.2$ V):

a. UTP.

b. LTP.

c. V_H.

Figure 34–44

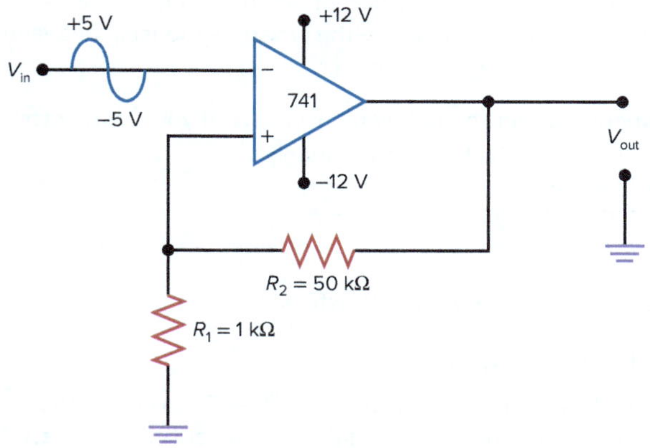

34–53 What type of circuit is shown in Fig. 34–45?

34–54 What will the output look like in Fig. 34–45?

Figure 34–45

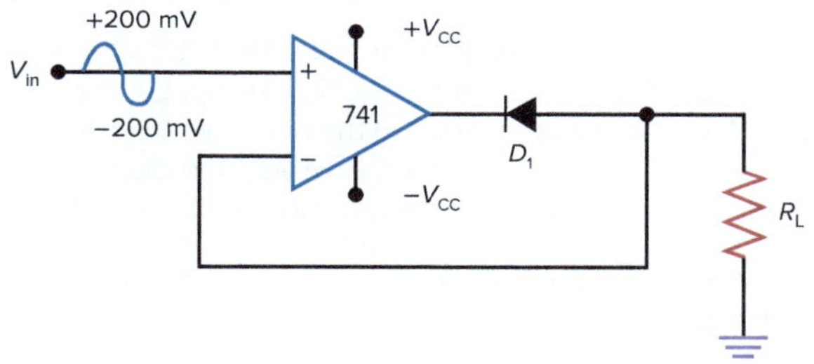

34–55 What is the output in Fig. 34–45 if a 100-μF capacitor is connected at the output?

Answers to Self-Reviews

34–1 a. the DC emitter current in each transistor is one-half the tail current
b. the differential voltage gain
c. approximately 58.1 dB

34–2 a. differential amplifier
b. negative
c. 318.3 kHz

34–3 a. 180°
b. it reduces it significantly.
c. f_{CL} is inversely proportional to A_{CL}

34–4 a. 6 dB/octave or 20 dB/decade
b. the virtual ground
c. it makes the comparator almost immune to erratic triggering from noise

Laboratory Application Assignment

In this lab application assignment, you will examine two relatively simple op-amp circuits: the inverting and noninverting amplifier. In each amplifier circuit you will measure the output voltage, V_{out}, and determine the closed-loop voltage gain, A_{CL}, with different amounts of negative feedback. You will also measure the phase relationship between V_{in} and V_{out} in each type of amplifier.

Equipment: Obtain the following items from your instructor.

- Dual-output variable dc power supply
- Oscilloscope
- Function generator
- DMM
- 741C op amp
- Assortment of carbon-film resistors

Inverting Amplifier

Examine the inverting amplifier in Fig. 34–46. Calculate and record the closed-loop voltage gain, A_{CL}, and output voltage, V_{out}, for each of the following values of feedback resistance, R_F. Note that $V_{in} = 1\ V_{p-p}$.

$R_F = 4.7\ k\Omega$ A_{CL} = ______ $V_{out(p-p)}$ = ______
$R_F = 10\ k\Omega$ A_{CL} = ______ $V_{out(p-p)}$ = ______
$R_F = 15\ k\Omega$ A_{CL} = ______ $V_{out(p-p)}$ = ______
$R_F = 22\ k\Omega$ A_{CL} = ______ $V_{out(p-p)}$ = ______

Construct the inverting amplifier in Fig. 34–46. (The IC pin numbers are shown in parentheses.) Set the input voltage, V_{in}, to exactly 1 V_{p-p}. Measure and record the output voltage, V_{out}, for each value of R_F listed below. Then from your measured values of V_{out}, calculate the closed-loop voltage gain, A_{CL}, as V_{out}/V_{in}.

$R_F = 4.7\ k\Omega$ $V_{out(p-p)}$ = ______ A_{CL} = ______
$R_F = 10\ k\Omega$ $V_{out(p-p)}$ = ______ A_{CL} = ______
$R_F = 15\ k\Omega$ $V_{out(p-p)}$ = ______ A_{CL} = ______
$R_F = 22\ k\Omega$ $V_{out(p-p)}$ = ______ A_{CL} = ______

How do your measured and calculated values compare? ______
______

With channel 1 of the oscilloscope connected to the input voltage, V_{in}, and channel 2 connected to the output of the op amp, measure and record the phase relationship between V_{in} and V_{out}. θ = ______

Measure and record the AC voltage at the inverting input (pin 2) of the op amp. $V_{(-)}$ = ______ $_{p-p}$. Explain your measurement.
______

Noninverting Amplifier

Examine the noninverting amplifier in Fig. 34–47. Calculate and record the closed-loop voltage gain, A_{CL}, and output voltage, V_{out}, for each of the following values of feedback resistance, R_F. Note that $V_{in} = 1\ V_{p-p}$.

$R_F = 1\ k\Omega$ A_{CL} = ______ $V_{out(p-p)}$ = ______
$R_F = 2\ k\Omega$ A_{CL} = ______ $V_{out(p-p)}$ = ______
$R_F = 10\ k\Omega$ A_{CL} = ______ $V_{out(p-p)}$ = ______
$R_F = 15\ k\Omega$ A_{CL} = ______ $V_{out(p-p)}$ = ______

Construct the noninverting amplifier in Fig. 34–47. (The IC pin numbers are shown in parentheses.) Set the input voltage, V_{in}, to exactly 1 V_{p-p}. Measure and record the output voltage, V_{out}, for each value of R_F listed below. Then from your measured values of V_{out}, calculate the closed-loop voltage gain, A_{CL}, as V_{out}/V_{in}.

$R_F = 1\ k\Omega$ $V_{out(p-p)}$ = ______ A_{CL} = ______
$R_F = 2\ k\Omega$ $V_{out(p-p)}$ = ______ A_{CL} = ______
$R_F = 10\ k\Omega$ $V_{out(p-p)}$ = ______ A_{CL} = ______
$R_F = 15\ k\Omega$ $V_{out(p-p)}$ = ______ A_{CL} = ______

How do your measured and calculated values compare? ______
______

Figure 34–46

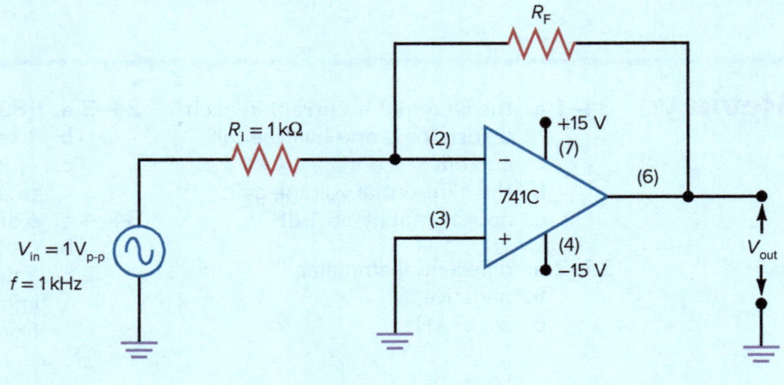

Figure 34–47

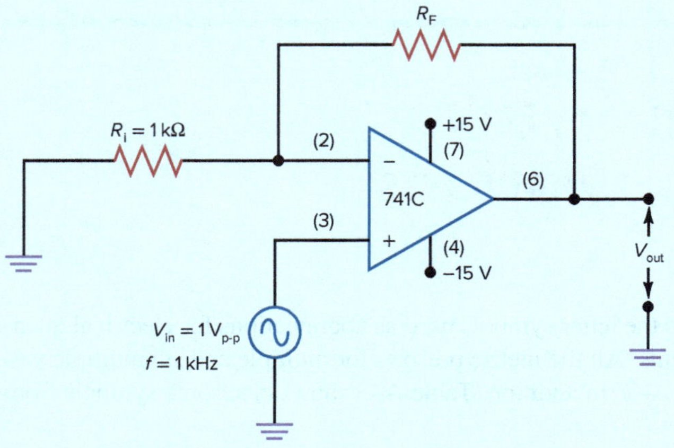

With channel 1 of the oscilloscope connected to the input voltage, V_{in}, and channel 2 connected to the output of the op amp, measure and record the phase relationship between V_{in} and V_{out}. $\theta =$ ___________

Measure and record the AC voltage at the noninverting input (pin 3) of the op amp. $V_{(+)} =$ ___________$_{p\text{-}p}$. Next, measure and record the AC voltage at the inverting input (pin 2) of the op amp. $V_{(-)} =$ ___________$_{p\text{-}p}$. Are these two values the same? ________ If yes, explain why. ______________________

____

Design credits: Chapter Opener Background Image, and Multisim: Stockbyte/Getty Images

Appendix A

Electrical Symbols and Abbreviations

Table A–1 summarizes the letter symbols used as abbreviations for electrical quantities and their basic units. All the metric prefixes for multiple and submultiple values are listed in Table A–2. In addition, Table A–3 shows electronic symbols from the Greek alphabet.

Table A–1 Electrical Quantities

Quantity	Symbol*	Basic Unit
Current	I or i	ampere (A)
Charge	Q or q	coulomb (C)
Power	P	watt (W)
Voltage	V or v	volt (V)
Resistance	R	ohm (Ω)
Reactance	X	ohm (Ω)
Impedance	Z	ohm (Ω)
Conductance	G	siemens (S)
Admittance	Y	siemens (S)
Susceptance	B	siemens (S)
Capacitance	C	farad (F)
Inductance	L	henry (H)
Frequency	f	hertz (Hz)
Period	T	second (s)

* Capital letters for I, Q, and V are generally used for peak, rms, or DC values, whereas small letters are used for instantaneous values. Small r and g are also used for internal values, such as r_i for the internal resistance of a battery and g_m for the transconductance of a JFET or MOSFET.

Table A–2 Multiples and Submultiples of Units*

Value	Prefix	Symbol	Example
1 000 000 000 000 = 10^{12}	tera	T	THz = 10^{12} Hz
1 000 000 000 = 10^{9}	giga	G	GHz = 10^{9} Hz
1 000 000 = 10^{6}	mega	M	MHz = 10^{6} Hz
1 000 = 10^{3}	kilo	k	kV = 10^{3} V
100 = 10^{2}	hecto	h	hm = 10^{2} m
10 = 10	deka	da	dam = 10 m
0.1 = 10^{-1}	deci	d	dm = 10^{-1} m
0.01 = 10^{-2}	centi	c	cm = 10^{-2} m
0.001 = 10^{-3}	milli	m	mA = 10^{-3} A
0.000 001 = 10^{-6}	micro	μ	μV = 10^{-6} V
0.000 000 001 = 10^{-9}	nano	n	ns = 10^{-9} s
0.000 000 000 001 = 10^{-12}	pico	p	pF = 10^{-12} F

* Additional prefixes are exa = 10^{8}, peta = 10^{15}, femto = 10^{-15}, and atto = 10^{-18}.

Table A–3 Greek Letter Symbols*

	LETTER		
Name	Capital	Small	Uses
Alpha	A	α	α for angles, transistor characteristic
Beta	B	β	β for angles, transistor characteristic
Gamma	Γ	γ	γ Transistor characteristic
Delta	Δ	δ	Δ Small change in value
Epsilon	E	ε	ε for permittivity; also base of natural logarithms
Zeta	Z	ζ	
Eta	H	η	η for intrinsic standoff ratio of a unijunction transistor (UJT)
Theta	Θ	θ	θ Phase angle
Iota	I	ι	
Kappa	K	κ	

* This table includes the complete Greek alphabet, although some letters are not used for electronic symbols.

Table A–3	Greek Letter Symbols* (*Continued*)		
	LETTER		
Name	Capital	Small	Uses
Lambda	Λ	λ	λ for wavelength
Mu	M	μ	μ for prefix micro-, permeability, amplification factor
Nu	N	ν	
Xi	Ξ	ξ	
Omicron	O	ο	
Pi	Π	π	π is 3.1416 for ratio of circumference to diameter of a circle
Rho	P	ρ	ρ for resistivity
Sigma	Σ	σ	Σ Summation
Tau	T	τ	τ Time constant
Upsilon	Y	υ	
Phi	Φ	ϕ	ϕ Magnetic flux, angles
Chi	X	χ	
Psi	Ψ	ψ	ψ Electric flux
Omega	Ω	ω	Ω for ohms; ω for angular velocity

* This table includes the complete Greek alphabet, although some letters are not used for electronic symbols.

Appendix B

Solder and the Soldering Process*

From Simple Task to Fine Art

Soldering is the process of joining two metals together by the use of a low-temperature melting alloy. Soldering is one of the oldest known joining techniques, first developed by the Egyptians in making weapons such as spears and swords. Since then, it has evolved into what is now used in the manufacturing of electronic assemblies. Soldering is far from the simple task it once was; it is now a fine art, one that requires care, experience, and a thorough knowledge of the fundamentals.

The importance of having high standards of workmanship cannot be overemphasized. Faulty solder joints remain a cause of equipment failure, and because of that soldering has become a *critical skill.*

The material contained in this appendix is designed to provide students with both the fundamental knowledge and the practical skills needed to perform many of the high-reliability soldering operations encountered in today's electronics.

Covered here are the fundamentals of the soldering process, the proper selection, and the use of the soldering station.

The key concept in this appendix is *high-reliability soldering*. Much of our present technology is vitally dependent on the reliability of countless, individual soldered connections. High-reliability soldering was developed in response to early failures with space equipment. Since then the concept and practice have spread into military and medical equipment. We have now come to expect it in everyday electronics as well.

The Advantage of Soldering

Soldering is the process of connecting two pieces of metal together to form a reliable electrical path. Why solder them in the first place? The two pieces of metal could be put together with nuts and bolts, or some other kind of mechanical fastening. The disadvantages of these methods are twofold. First, the reliability of the connection cannot be ensured because of vibration and shock. Second, because oxidation and corrosion are continually occurring on the metal surfaces, electrical conductivity between the two surfaces would progressively decrease.

A soldered connection does away with both of these problems. There is no movement in the joint and no interfacing surfaces to oxidize. A continuous conductive path is formed, made possible by the characteristics of the solder itself.

The Nature of Solder

Solder used in electronics is a low-temperature melting alloy made by combining various metals in different proportions. The most common types of solder are made from tin and lead. When the proportions are equal, it is known as 50/50 solder—50% tin and 50% lead. Similarly, 60/40 solder consists of 60% tin and 40% lead. The percentages are usually marked on the various types of solder available; sometimes only the

* This material is provided courtesy of PACE Worldwide, Southern Pines, North Carolina.

tin percentage is shown. The chemical symbol for tin is Sn; thus Sn 63 indicates a solder that contains 63% tin.

Pure lead (Pb) has a melting point of 327°C (621°F); pure tin, a melting point of 232°C (450°F). But when they are combined into a 60/40 solder, the melting point drops to 190°C (374°F)—lower than either of the two metals alone.

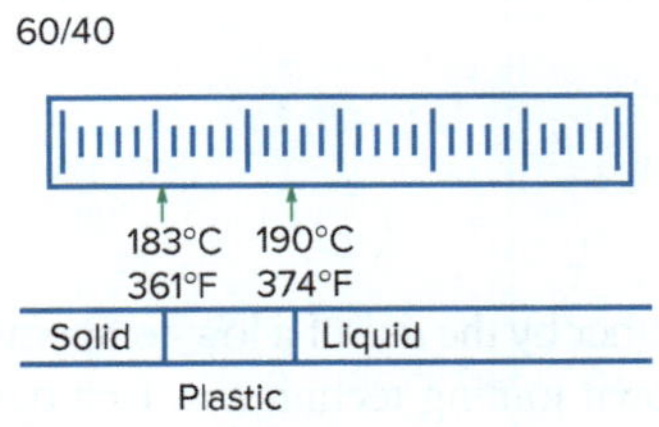

Figure B–1 Plastic range of 60/40 solder. Melt begins at 183°C (361°F) and is complete at 190°C (374°F).

Melting generally does not take place all at once. As illustrated in Fig. B–1, 60/40 solder begins to melt at 183°C (361°F), but it has not fully melted until the temperature reaches 190°C (374°F). Between these two temperatures, the solder exists in a plastic (semiliquid) state—some, but not all, of the solder has melted.

The plastic range of solder will vary, depending on the ratio of tin to lead, as shown in Fig. B–2. Various ratios of tin to lead are shown across the top of this figure. With most ratios, melting begins at 183°C (361°F), but the full melting temperatures vary dramatically. There is one ratio of tin to lead that has no plastic state. It is known as *eutectic solder*. This ratio is 63/37 (Sn 63), and it fully melts and solidifies at 183°C (361°F).

The solder most commonly used for hand soldering in electronics is the 60/40 type, but because of its plastic range, care must be taken not to move any elements of the joint during the cool-down period. Movement may cause a disturbed joint. Characteristically, this type of joint has a rough, irregular appearance and looks dull instead of bright and shiny. It is unreliable and therefore one of the types of joints that is unacceptable in high-reliability soldering.

In some situations, it is difficult to maintain a stable joint during cooling. In other cases, it may be necessary to use minimal heat to avoid damage to heat-sensitive components. In both of these situations, eutectic solder is the preferred choice because it changes from a liquid to a solid during cooling with no plastic range.

The Wetting Action

To someone watching the soldering process for the first time, it looks as though the solder simply sticks the metals together like a hot-melt glue, but what actually happens is far different.

A chemical reaction takes place when the hot solder comes into contact with the copper surface. The solder dissolves and penetrates the surface. The molecules of solder and copper blend together to form a new metal alloy, one that is part copper and part solder and that has characteristics all its own. This reaction is called *wetting* and forms the intermetallic bond between the solder and copper (Fig. B–3).

Proper wetting can occur only if the surface of the copper is free of contamination and from oxide films that form when the metal is exposed to air. Also, the solder and copper surfaces need to have reached the proper temperature.

Figure B–2 Fusion characteristics of tin/lead solders.

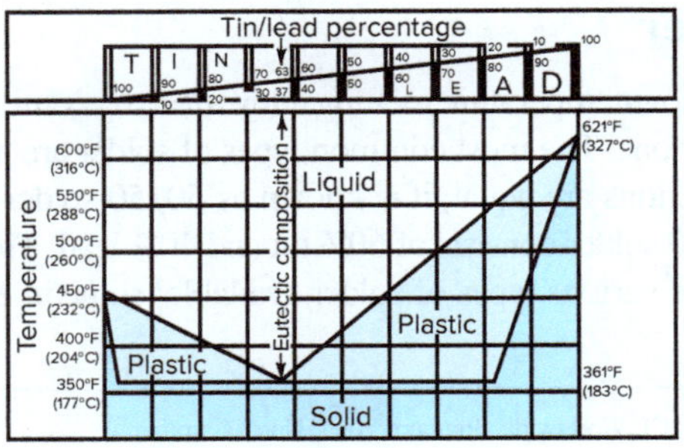

Figure B–3 The wetting action. Molten solder dissolves and penetrates a clean copper surface, forming an intermetallic bond.

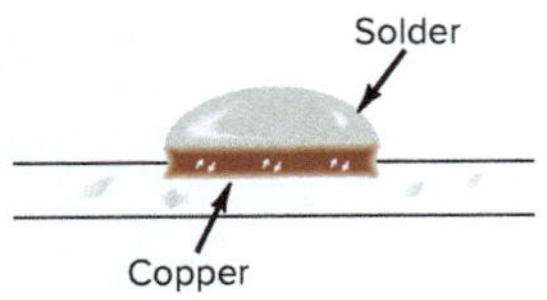

Even though the surface may look clean before soldering, there may still be a thin film of oxide covering it. When solder is applied, it acts like a drop of water on an oily surface because the oxide coating prevents the solder from coming into contact with the copper. No reaction takes place, and the solder can be easily scraped off. For a good solder bond, surface oxides must be removed during the soldering process.

The Role of Flux

Reliable solder connections can be accomplished only on clean surfaces. Some sort of cleaning process is essential in achieving successful soldered connections, but in most cases it is insufficient. This is due to the extremely rapid rate at which oxides form on the surfaces of heated metals, thus creating oxide films which prevent proper soldering. To overcome these oxide films, it is necessary to utilize materials, called *fluxes*, which consist of natural or synthetic rosins and sometimes additives called *activators*.

It is the function of flux to remove surface oxides and keep them removed during the soldering operation. This is accomplished because the flux action is very corrosive at or near solder melt temperatures and accounts for the flux's ability to rapidly remove metal oxides. It is the fluxing action of removing oxides and carrying them away, as well as preventing the formation of new oxides, that allows the solder to form the desired intermetallic bond.

Flux must activate at a temperature lower than solder so that it can do its job prior to the solder flowing. It volatilizes very rapidly; thus it is mandatory that the flux be activated to flow onto the work surface and not simply be volatilized by the hot iron tip if it is to provide the full benefit of the fluxing action.

There are varieties of fluxes available for many applications. For example, in soldering sheet metal, acid fluxes are used; silver brazing (which requires a much higher temperature for melting than that required by tin/lead alloys) uses a borax paste. Each of these fluxes removes oxides and, in many cases, serves additional purposes. The fluxes used in electronic hand soldering are the pure rosins, rosins combined with mild activators to accelerate the rosin's fluxing capability, low-residue/no-clean fluxes, or water-soluble fluxes. Acid fluxes or highly activated fluxes should never be used in electronic work. Various types of flux-cored solder are now in common use. They provide a convenient way to apply and control the amount of flux used at the joint (Fig. B–4).

Figure B–4 Types of cored solder, with varying solder-flux percentages.

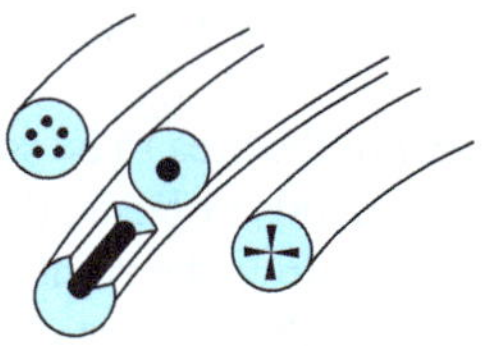

Soldering Irons

In any kind of soldering, the primary requirement, beyond the solder itself, is heat. Heat can be applied in a number of ways—conductive (e.g., soldering iron, wave, vapor phase), convective (hot air), or radiant (IR). We are mainly concerned with the conductive method, which uses a soldering iron.

Soldering stations come in a variety of sizes and shapes, but consist basically of three main elements: a resistance heating unit; a heater block, which acts as a heat reservoir; and the tip, or bit, for transferring heat to the work. The standard production station is a variable-temperature, closed-loop system with interchangeable tips and is made with Static Dissipative (ESD)-safe plastics.

Controlling Heat at the Joint

Controlling tip temperature is not the real challenge in soldering; the real challenge is to control the *heat cycle* of the work—how fast the work gets hot, how hot it gets, and how long it stays that way. This is affected by so many factors that, in reality, tip temperature is not that critical.

The first factor that needs to be considered is the *relative thermal mass* of the area to be soldered. This mass may vary over a wide range.

Consider a single land on a single-sided circuit board. There is relatively little mass, so the land heats up quickly. But on a double-sided board with plated-through holes, the mass is more than doubled. Multilayered boards may have an even greater mass, and that's before the mass of the component lead is taken into consideration. Lead mass may vary greatly, since some leads are much larger than others.

Moreover, there may be terminals (e.g., turret or bifurcated) mounted on the board. Again, the thermal mass is increased, and will further increase as connecting wires are added.

Each connection, then, has its particular thermal mass. How this combined mass compares with the mass of the iron tip, the "relative" thermal mass, determines the time and temperature rise of the work.

With a large work mass and a small iron tip, the temperature rise will be slow. With the situation reversed, using a large iron tip on a small work mass, the temperature rise of the work will be much more rapid—even though the *temperature of the tip is the same.*

Now consider the capacity of the iron itself and its ability to sustain a given flow of heat. Essentially, irons are instruments for generating and storing heat, and the reservoir is made up of both the heater block and the tip. The tip comes in various sizes and shapes; it's the *pipeline* for heat flowing into the work. For small work, a conical (pointed) tip is used, so that only a small flow of heat occurs. For large work, a large chisel tip is used, providing greater flow.

The reservoir is replenished by the heating element, but when an iron with a large tip is used to heat massive work, the reservoir may lose heat faster than it can be replenished. Thus, the *size* of the reservoir becomes important: a large heating block can sustain a larger outflow longer than a small one.

An iron's capacity can be increased by using a larger heating element, thereby increasing the wattage of the iron. These two factors, block size and wattage, are what determine the iron's recovery rate.

If a great deal of heat is needed at a particular connection, the correct temperature with the right size tip is required, as is an iron with a large enough capacity and an ability to recover fast enough. *Relative thermal mass*, then, is a major consideration for controlling the heat cycle of the work.

A second factor of importance is the *surface condition* of the area to be soldered. If there are any oxides or other contaminants covering the lands or leads, there will be a barrier to the flow of heat. Then, even though the iron tip is the right size and has the correct temperature, it may not supply enough heat to the connection to melt the solder. In soldering, a cardinal rule is that a good solder connection cannot be created on a dirty surface. Before you attempt to solder, the work should always be cleaned with an approved solvent to remove any grease or oil film from the surface. In some cases, pretinning may be required to enhance solderability and remove heavy oxidation of the surfaces prior to soldering.

A third factor to consider is *thermal linkage*—the area of contact between the iron tip and the work.

Figure B–5 Cross-sectional view (left) of iron tip on a round lead. The "X" shows point of contact. Use of a solder bridge (right) increases the linkage area and speeds the transfer of heat.

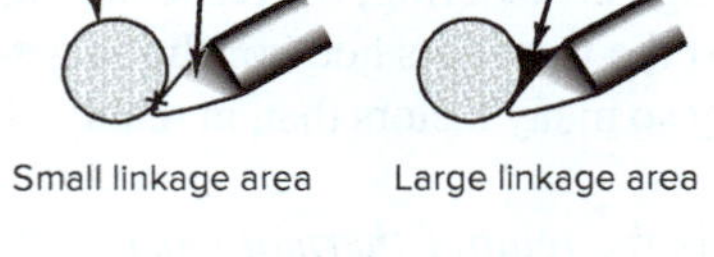

Figure B–5 shows a cross-sectional view of an iron tip touching a round lead. The contact occurs only at the point indicated by the "X," so the linkage area is very small, not much more than a straight line along the lead.

The contact area can be greatly increased by applying a small amount of solder to the point of contact between the tip and workpiece. This solder heat bridge provides the thermal linkage and ensures rapid heat transfer into the work.

From the aforementioned, it should now be apparent that there are many more factors than just the temperature of the iron tip that affect how quickly any particular connection is going to heat up. In reality, soldering is a very complex control problem, with a number of variables to it, each influencing the other. And what makes it so critical is *time*. The general rule for high-reliability soldering on printed circuit boards is to apply heat for no more than 2 s from the time solder starts to melt (wetting). Applying heat for longer than 2 s after wetting may cause damage to the component or board.

Figure B–6 Work can be viewed as a closed-loop system (left). Feedback comes from the reaction of the workpiece and is used to modify the action. Workpiece indicators (right)—changes discernible to the human senses—are the way the "work talks back to you."

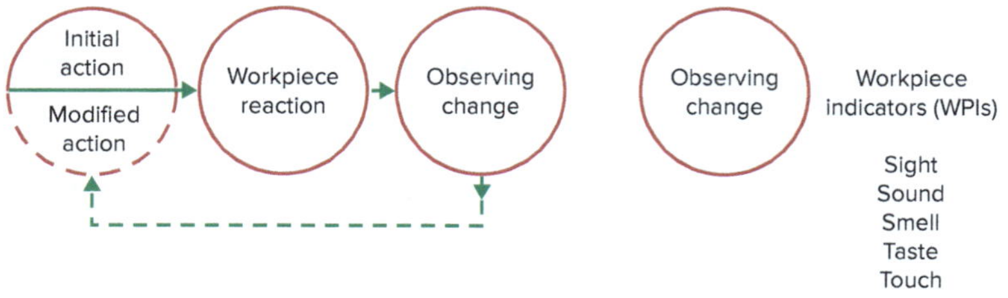

With all these factors to consider, the soldering process would appear to be too complex to accurately control in so short a time, but there is a simple solution—the *workpiece indicator* (WPI). This is defined as the reaction of the workpiece to the work being performed on it—a reaction that is discernible to the human senses of sight, touch, smell, sound, and taste.

Put simply, workpiece indicators are the way the work talks back to you—the way it tells you what effect you are having and how to control it so that you accomplish what you want.

In any kind of work, you become part of a closed-loop system. It begins when you take some action on the workpiece; then the workpiece reacts to what you did; you sense the change, and then modify your action to accomplish the result. It is in the sensing of the change, by sight, sound, smell, taste, or touch, that the workpiece indicators come in (Fig. B–6).

For soldering and desoldering, a primary workpiece indicator is *heat rate recognition*—observing how fast heat flows into the connection. In practice, this means observing the rate at which the solder melts, which should be within 1 to 2 s.

This indicator encompasses all the variables involved in making a satisfactory solder connection with minimum heating effects, including the capacity of the iron and its tip temperature, the surface conditions, the thermal linkage between tip and workpiece, and the relative thermal masses involved.

If the iron tip is too large for the work, the heating rate may be too fast to be controlled. If the tip is too small, it may produce a "mush" kind of melt; the heating rate will be too slow, even though the temperature at the tip is the same.

A general rule for preventing overheating is, "Get in and get out as fast as you can." That means using a heated iron you can react to—one giving a 1- to 2-s dwell time on the particular connection being soldered.

Selecting the Soldering Iron and Tip

A good all-around soldering station for electronic soldering is a variable-temperature, ESD-safe station with a pencil-type iron and tips that are easily interchangeable, even when hot (Fig. B–7).

The soldering iron tip should always be fully inserted into the heating element and tightened. This will allow for maximum heat transfer from the heater to the tip.

The tip should be removed daily to prevent an oxidation scale from accumulating between the heating element and the tip. A bright, thin tinned surface must be maintained on the tip's working surface to ensure proper heat transfer and to avoid contaminating the solder connection.

The plated tip is initially prepared by holding a piece of flux-cored solder to the face so that it will tin the surface when it reaches the lowest temperature at which solder will melt. Once the tip is up to operating temperature, it will usually be too hot for good tinning, because of the rapidity of oxidation at elevated temperatures.

Figure B–7 Pencil-type iron with changeable tips.

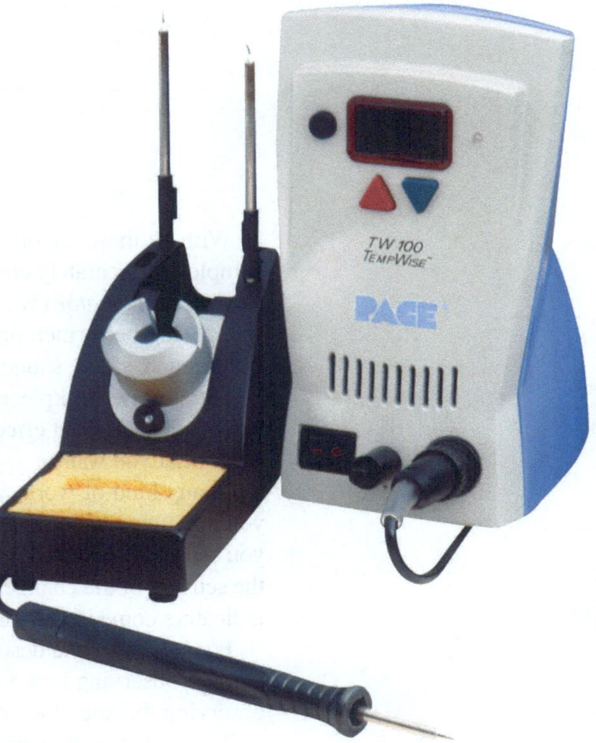

Courtesy of PACE, Inc.

The hot tinned tip is maintained by wiping it lightly on a damp sponge to shock off the oxides. When the iron is not being used, the tip should be coated with a layer of solder.

Making the Solder Connection

The soldering iron tip should be applied to the area of maximum thermal mass of the connection being made. This will permit the rapid thermal elevation of the parts being soldered. Molten solder always flows toward the heat of a properly prepared connection.

When the solder connection is heated, a small amount of solder is applied to the tip to increase the thermal linkage to the area being heated. The solder is then applied to the opposite side of the connection so that the work surfaces, not the iron, melt the solder. Never melt the solder against the iron tip and allow it to flow onto a surface cooler than the solder melting temperature.

Solder, with flux, applied to a cleaned and properly heated surface will melt and flow without direct contact with the heat source and provide a smooth, even surface, feathering out to a thin edge (Fig. B–8). Improper soldering will exhibit a built-up, irregular appearance and poor filleting. The parts being soldered must be held rigidly in place until the temperature decreases to solidify the solder. This will prevent a disturbed or fractured solder joint.

Figure B–8 Cross-sectional view of a round lead on a flat surface.

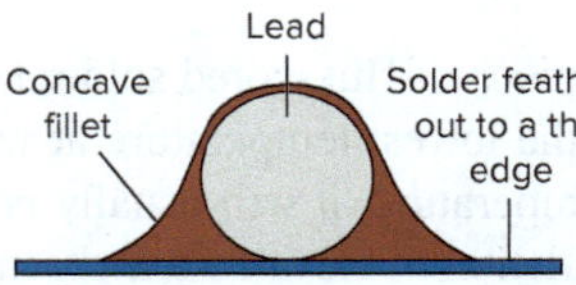

Selecting cored solder of the proper diameter will aid in controlling the amount of solder being applied to the connection (e.g., a small-gauge solder for a small connection; a large-gauge solder for a large connection).

Removal of Flux

Cleaning may be required to remove certain types of fluxes after soldering. If cleaning is required, the flux residue should be removed as soon as possible, preferably within 1 hour after soldering.

Lead-Free Solder

The use of 60/40 and eutectic (63/37) solder has significantly declined over the past several years due to the health concerns associated with lead. Although 60/40 and eutectic solder is still used for certain applications in the electronics industry, most new electronic products are assembled using lead-free solder. Most lead-free solder is made up of a mixture of copper (Cu), tin (Sn), and silver (Ag). This combination (alloy) of metals has a higher melting temperature than the traditional 60/40 and 63/37 tin/lead ratios discussed previously. One of the biggest challenges associated with lead-free solder is the growth of "tin whiskers," which can cause arcing and/or short circuits to occur in an electronic assembly. It is important to note that the soldering techniques and processes discussed earlier also apply to lead-free solder.

Appendix C

Listing of Preferred Resistance Values

Table C–1 Preferred Resistance Values for Tolerances of ±5%, ±10%, and ±20% *

+20%	+10%	+5%
10	10	10
		11
	12	12
		13
15	15	15
		16
	18	18
		20
22	22	22
		24
	27	27
		30
33	33	33
		36
	39	39
		43
47	47	47
		51
	56	56
		62
68	68	68
		75
	82	82
		91
100	100	100

* Multiple and submultiple values apply to those values which are shown.

Appendix D

Component Schematic Symbols

Voltage and Current Sources

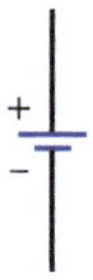

Individual voltaic cell (longer line indicates the positive electrode)

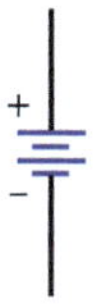

Standard symbol for a DC voltage source

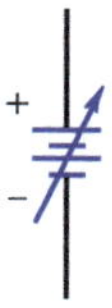

Variable DC voltage source

AC voltage source

Current source (solid arrow represents conventional current flow)

Current source (dashed arrow represents electron flow)

Ground Symbols

Earth ground

Chassis ground

Common ground

Connections

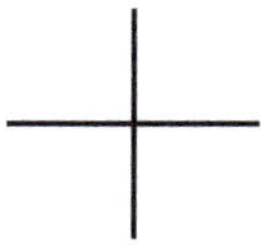

No connection

Connection

Dot indicates connection

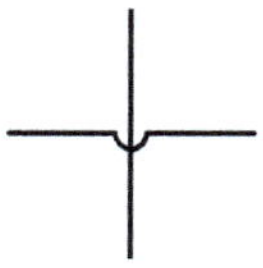

No connection

Resistors

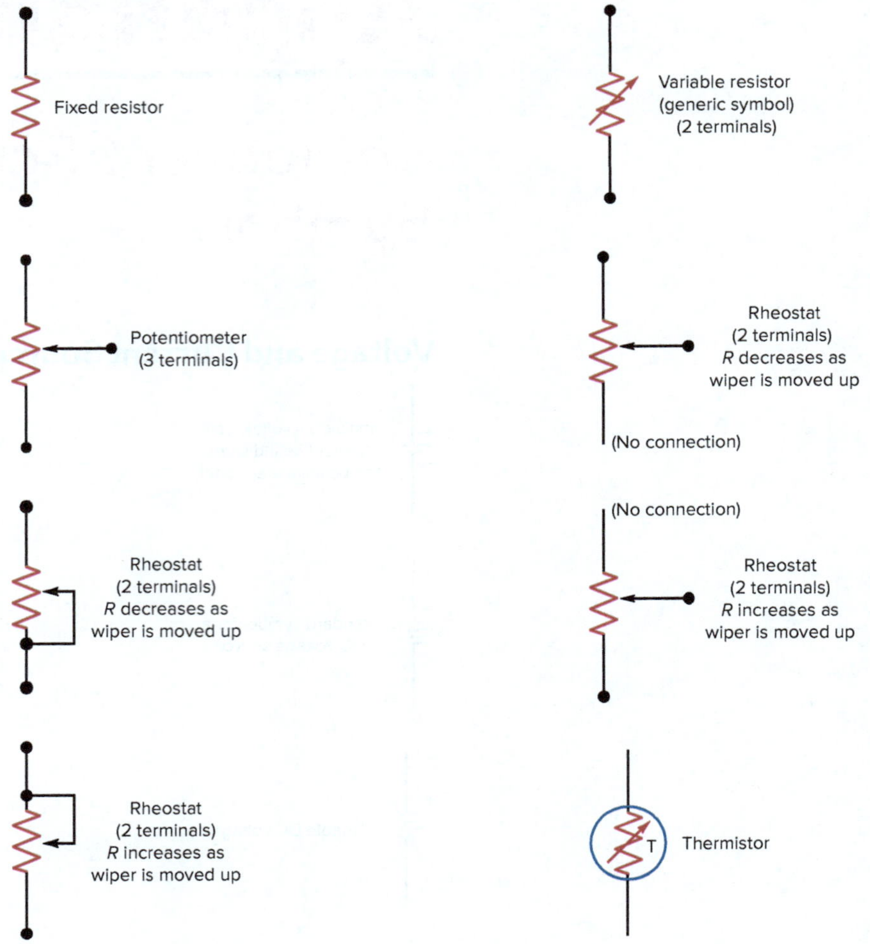

Capacitors

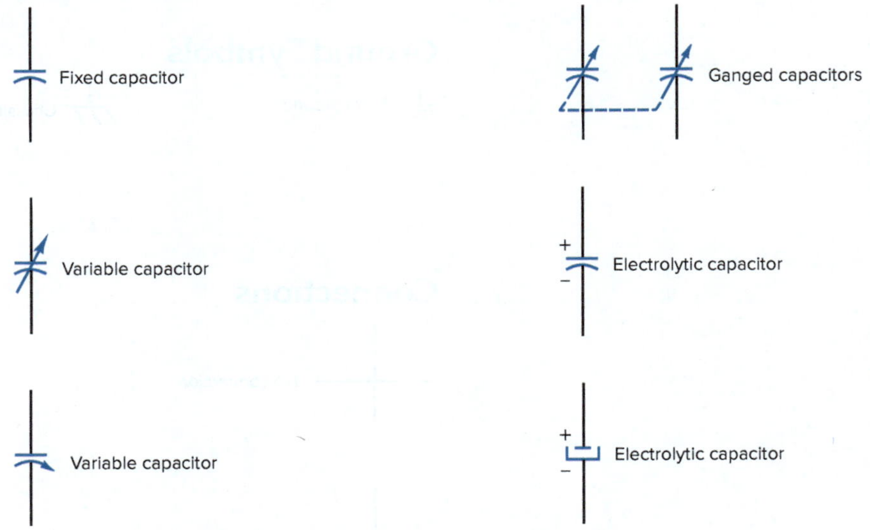

Inductors (Coils)

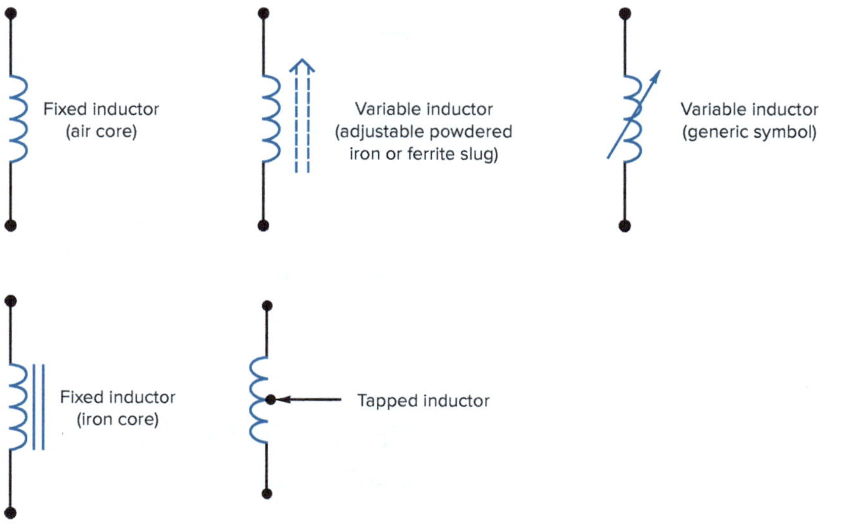

Transformers

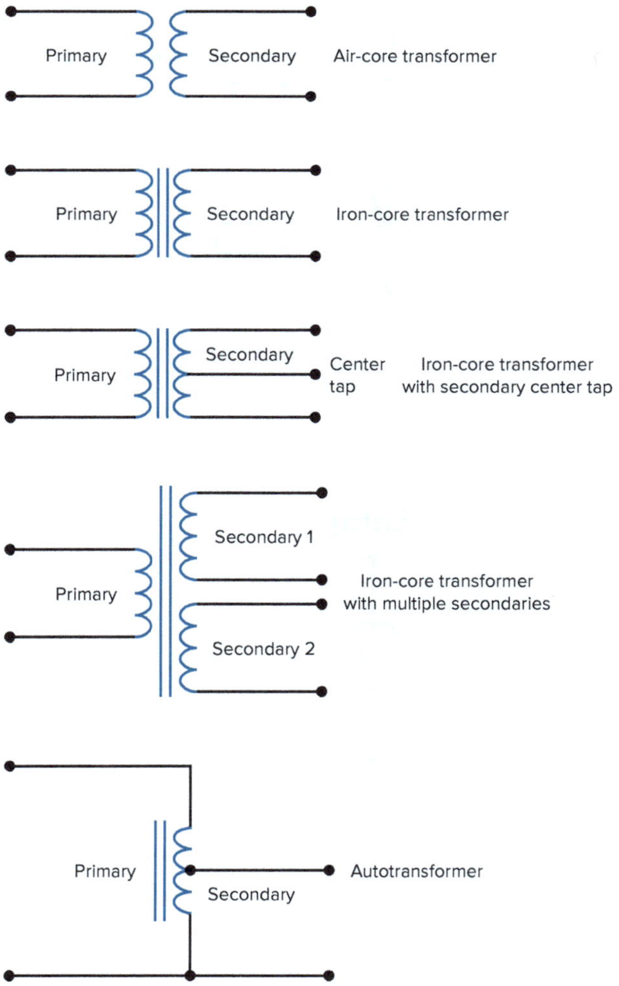

Switches

SPST

SPDT

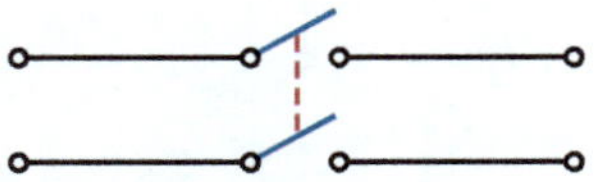

DPST

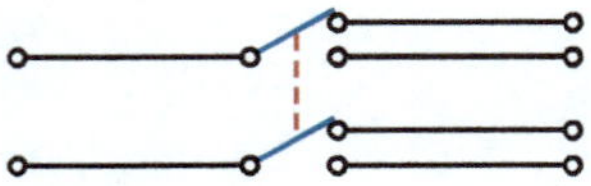

DPDT

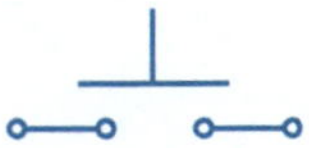

Normally open (NO) push-button switch

Normally closed (NC) push-button switch

Protective Devices

Fuse

Circuit breaker

Relays

or

Normally open (NO) relay contacts

or

Normally closed (NC) relay contacts

Relay coil

Relay coil

Lamp

Incandescent lamp

Test Instruments

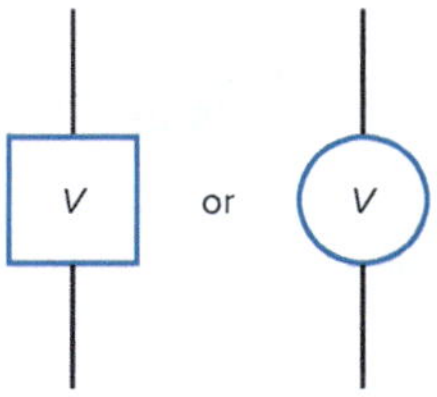

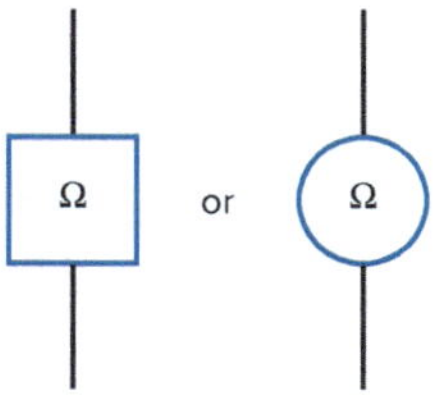

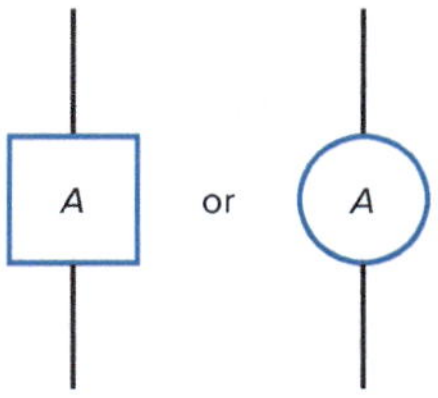

Diodes

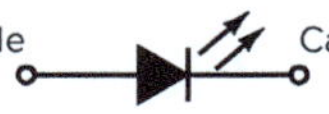

Transistors

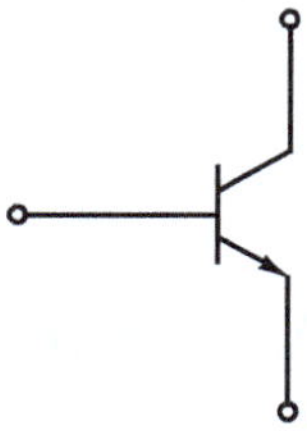

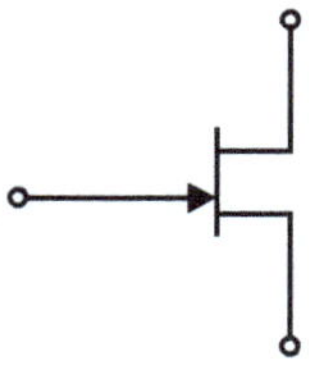

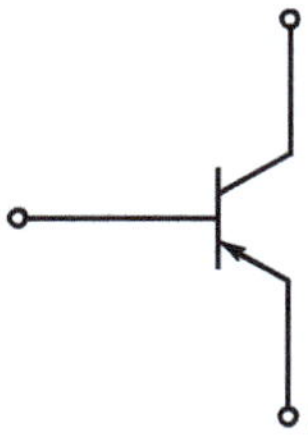

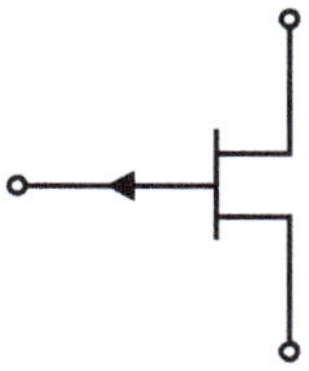

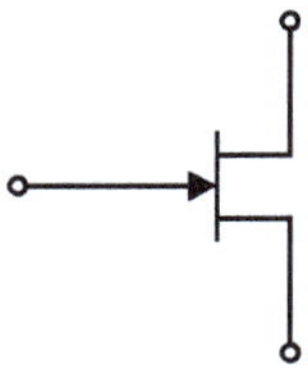

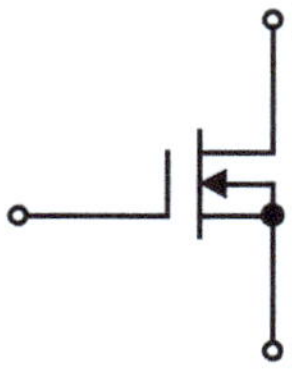

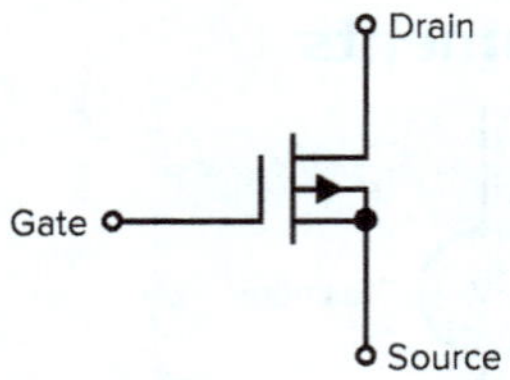

p-channel
D-MOSFET

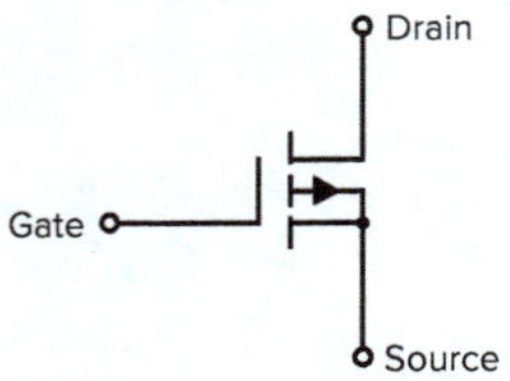

p-channel
E-MOSFET

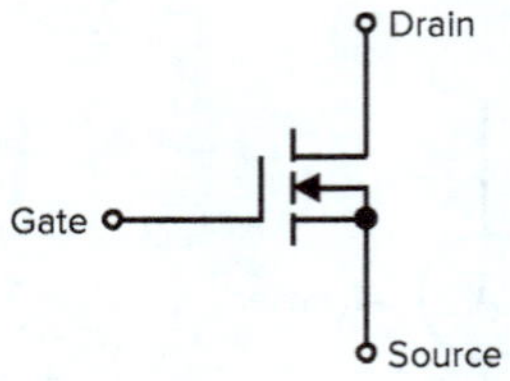

n-channel
E-MOSFET

Thyristors

Diac

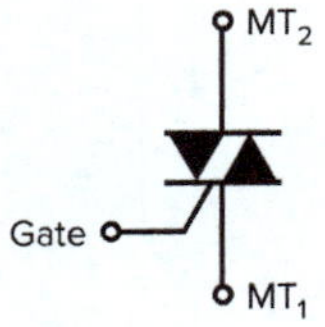

Triac

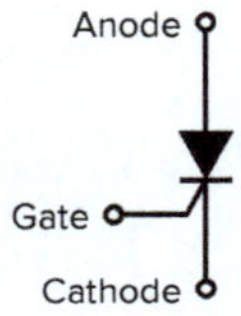

Silicon-controlled rectifier (SCR)

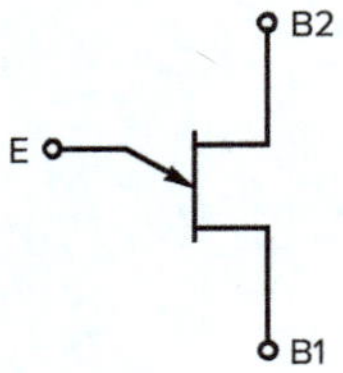

Unijunction transistor (UJT)

Operational Amplifier (Op Amp)

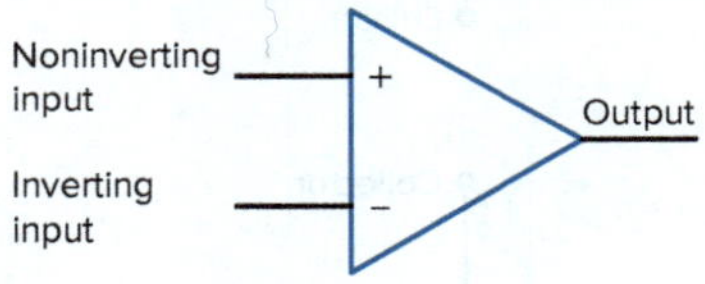

Op amp

Appendix E

Using the Oscilloscope

Basic Information

An oscilloscope or "scope," as it is commonly known, is one of the most versatile test instruments in electronics. Oscilloscopes are used in a wide variety of applications including consumer electronics repair, digital systems troubleshooting, control system design, and physics laboratories. Oscilloscopes have the ability to measure the time, frequency, and voltage level of a signal, view rapidly changing waveforms, and determine if an output signal is distorted. The technician must therefore be able to operate this instrument and understand how and where it is used.

Oscilloscopes can be classified as either analog or digital. Both types are shown in Fig. E–1. Analog oscilloscopes directly apply the voltage being measured to an electron beam moving across the oscilloscope screen. This voltage deflects the beam up, down, and across, thus tracing the waveform on the screen. Digital (also referred to as menu-driven) oscilloscopes sample the input waveform and then use an analog-to-digital converter (ADC) to change the voltage being measured into a digital format. The digital information is then used to reconstruct the waveform to be displayed on the screen.

A digital or analog oscilloscope may be used for many of the same applications. Each type of oscilloscope possesses unique characteristics and capabilities. The analog oscilloscope can display high-frequency varying signals in "real time," whereas a digital oscilloscope allows you to capture and store information which can be accessed at a later time or be interfaced to a computer.

WHAT AN OSCILLOSCOPE DOES

An analog oscilloscope displays the instantaneous amplitude of an AC voltage waveform versus time on the screen of a cathode-ray tube (CRT). Basically, the oscilloscope is a graph-displaying device. It has the ability to show how signals change over time. As shown in Fig. E–2, the vertical axis (Y) represents voltage and the horizontal axis (X) represents time. The Z axis or intensity is sometimes used in special measurement applications. Inside the cathode-ray tube is an electron gun assembly, vertical and horizontal deflection plates, and a phosphorous screen. The electron gun emits a high-velocity, low-inertia beam of electrons that strike the chemical coating on the inside face of the CRT, causing it to emit light. The brightness (called intensity) can be varied by a control located on the oscilloscope front panel. The motion of the beam over the CRT screen is controlled by the deflection voltages generated in the oscilloscope's circuits outside of the CRT and the deflection plates inside the CRT to which the deflection voltages are applied.

Figure E–3 is an elementary block diagram of an analog oscilloscope. The block diagram is composed of a CRT and four-system blocks. These blocks include the display system, vertical system, horizontal system, and trigger system. The CRT provides the screen on which waveforms of electrical signals are viewed. These signals are applied to the vertical input system. Depending on how the volts/div. control is set, the vertical attenuator—a variable voltage divider—reduces the input

Figure E–1 Oscilloscope types. (*a*) Analog oscilloscope. (*b*) Digital storage oscilloscope.

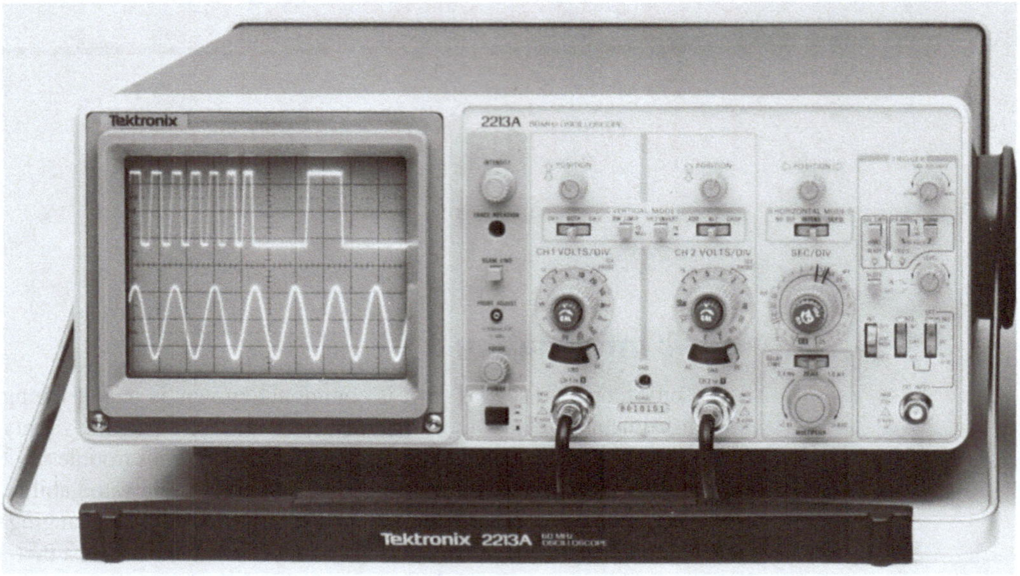

(*a*)

Reproduced by permission of Tektronix, Inc.

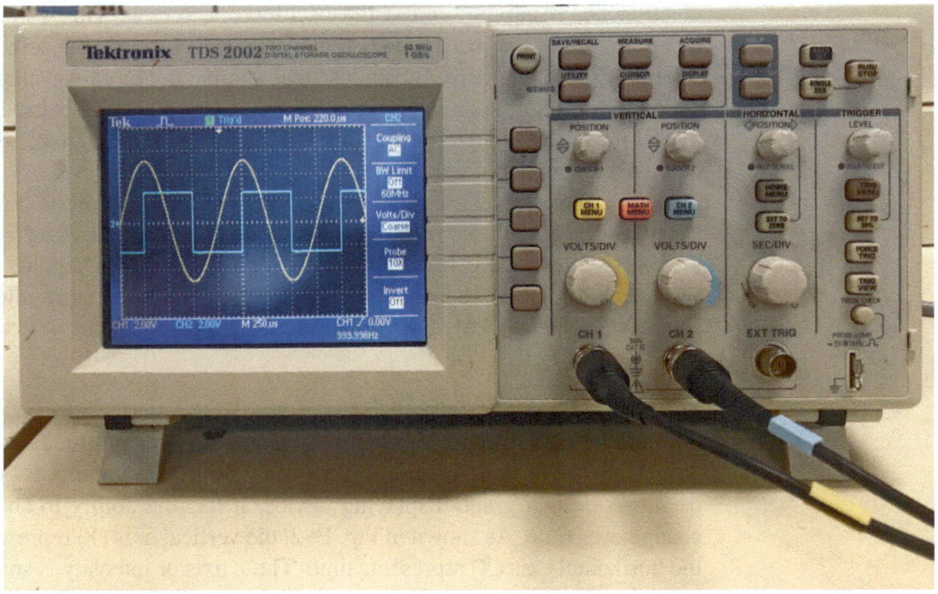

(*b*)

Sarah Schultz Photography

signal voltage to the desired signal level for the vertical amplifier. This is necessary because the oscilloscope must handle a wide range of signal-voltage amplitudes. The vertical amplifier then processes the input signal to produce the required voltage levels for the vertical deflection plates. The signal voltage applied to the vertical deflection plates causes the electron beam of the CRT to be deflected vertically. The resulting up-and-down movement of the beam on the screen, called the trace, is significant in that *the extent of vertical deflection is directly proportional to the amplitude of the signal voltage applied to the vertical, or V, input.* A portion of the input signal, from the vertical amplifier, travels to the trigger system to start or trigger a horizontal sweep. The trigger system determines *when* and *if* the sweep generator will be activated. With the proper LEVEL and SLOPE control adjustment,

Figure E–2 *X*, *Y*, and *Z* components of a displayed waveform.

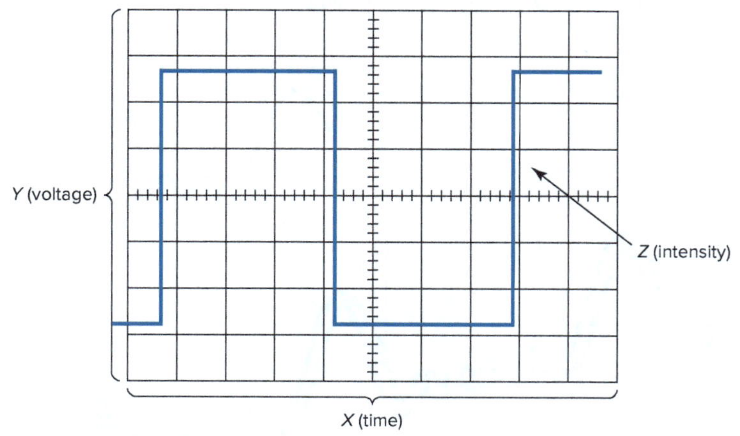

Figure E–3 Analog oscilloscope block diagram.

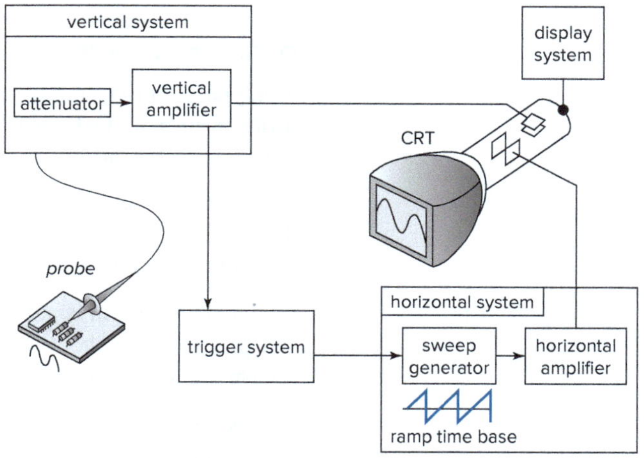

the sweep will begin at the same trigger point each time. This will produce a stable display as shown in Fig. E–4. The sweep generator produces a linear time-based deflection voltage. The resulting time-based signal is amplified by the horizontal amplifier and applied to the CRT's horizontal deflection plates. This makes it possible for the oscilloscope to graph a time-varying voltage. The sweep generator may be triggered from sources other than the vertical amplifier. External trigger input signals or internal 60-Hz (line) sources may be selected.

The display system includes the controls and circuits necessary to view the CRT signal with optimum clarity and position. Typical controls include intensity, focus, and trace rotation along with positioning controls.

DUAL-TRACE OSCILLOSCOPES

Most oscilloscopes have the ability to measure two input signals at the same time. These dual-trace oscilloscopes have two separate vertical amplifiers and an electronic switching circuit. It is then possible to observe two time-related waveforms simultaneously at different points in an electric circuit.

Figure E–4 Triggering produces a stable display because the same trigger point starts the sweep each time. The SLOPE and LEVEL controls define the trigger points on the trigger signal. The waveform on the screen is all those sweeps overlaid in what appears to be a single picture.

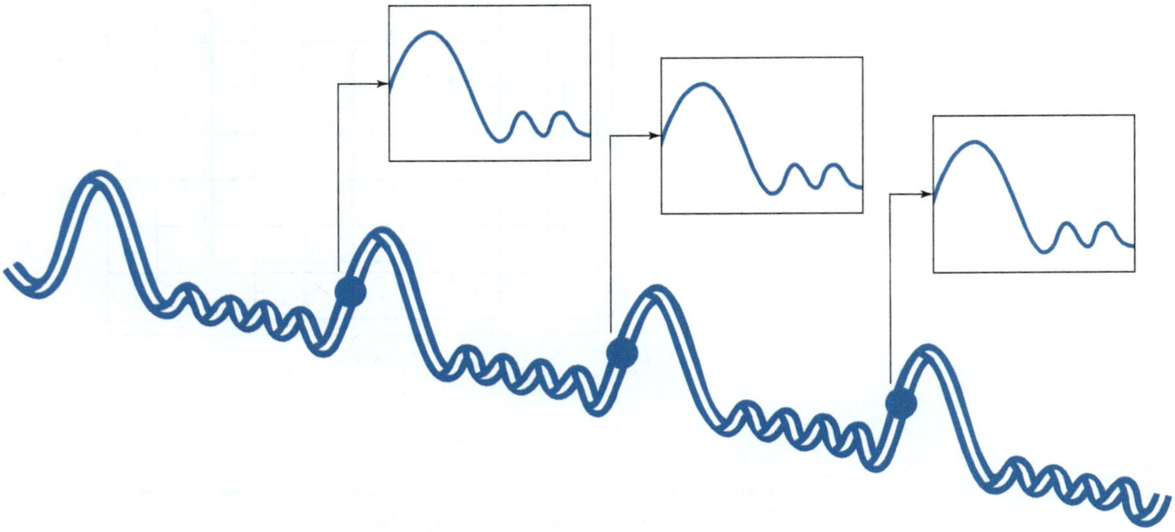

OPERATING CONTROLS OF A TRIGGERED OSCILLOSCOPE

The type, location, and function of the front panel controls of an analog oscilloscope differ from manufacturer to manufacturer and from model to model. The descriptions that follow apply to the broadest range of general-use analog scope models.

INTENSITY. This control sets the level of brightness or intensity of the light trace on the CRT. Rotation in a clockwise (CW) direction increases the brightness. Too high an intensity can damage the phosphorous coating on the inside of the CRT screen.

FOCUS. This control is adjusted in conjunction with the intensity control to give the sharpest trace on the screen. There is interaction between these two controls, so adjustment of one may require readjustment of the other.

ASTIGMATISM. This is another beam-focusing control found on older oscilloscopes that operates in conjunction with the focus control for the sharpest trace. The astigmatism control is sometimes a screwdriver adjustment rather than a manual control.

HORIZONTAL AND VERTICAL POSITIONING OR CENTERING. These are trace-positioning controls. They are adjusted so that the trace is positioned or centered both vertically and horizontally on the screen. In front of the CRT screen is a faceplate called the *graticule*, on which is etched a grid of horizontal and vertical lines. Calibration markings are sometimes placed on the center vertical and horizontal lines on this faceplate. This is shown in Fig. E–5.

VOLTS/DIV. This control attenuates the vertical input signal waveform that is to be viewed on the screen. This is frequently a click-stop control that provides step adjustment of vertical sensitivity. A separate Volts/Div. control is available for each channel of a dual-trace scope. Some scopes mark this control Volts/cm.

VARIABLE. In some scopes, this is a concentric control in the center of the Volts/Div. control. In other scopes, this is a separately located control. In either case, the functions are similar. The variable control works with the Volts/Div. control to provide a more sensitive control of the vertical height of the waveform on the screen. The variable control also has a calibrated position (CAL) either at the extreme counterclockwise or clockwise position. In the CAL position, the Volts/Div. control is

Figure E–5 An oscilloscope graticule.

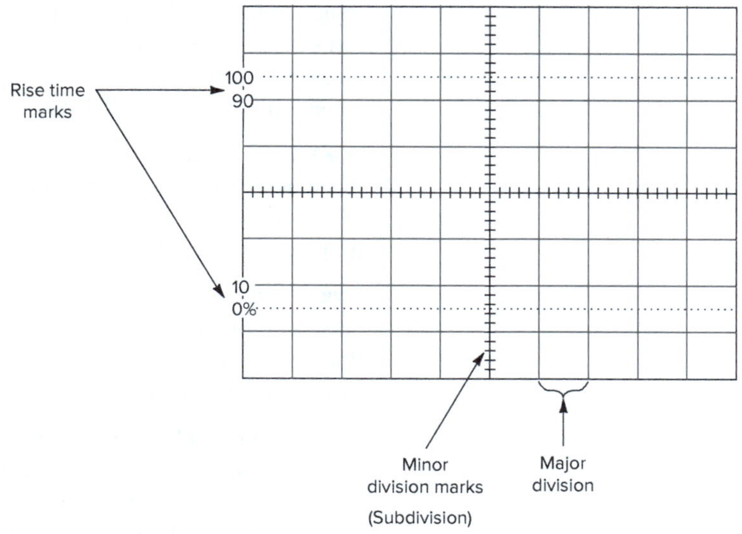

calibrated at some set value—for example, 5 mV/Div., 10 mV/Div., or 2 V/Div. This allows the scope to be used for peak-to-peak voltage measurements of the vertical input signal. Dual-trace scopes have a separate variable control for each channel.

INPUT COUPLING AC-GND-DC SWITCHES. This three-position switch selects the method of coupling the input signal into the vertical system.

AC—The input signal is capacitively coupled to the vertical amplifier. The DC component of the input signal is blocked.

GND—The vertical amplifier's input is grounded to provide a zero-volt (ground) reference point. It does not ground the input signal.

DC—This direct-coupled input position allows all signals (AC, DC, or AC-DC combinations) to be applied directly to the vertical system's input.

VERTICAL MODE SWITCHES. These switches select the mode of operation for the vertical amplifier system.

CH1—Selects only the Channel 1 input signal for display.

CH2—Selects only the Channel 2 input signal for display.

Both—Selects both Channel 1 and Channel 2 input signals for display. When in this position, ALT, CHOP, or ADD operations are enabled.

ALT—Alternatively displays Channel 1 and Channel 2 input signals. Each input is completely traced before the next input is traced. Effectively used at sweep speeds of 0.2 ms per division or faster.

CHOP—During the sweep, the display switches between Channel 1 and Channel 2 input signals. The switching rate is approximately at 500 kHz. This is useful for viewing two waveforms at slow sweep speeds of 0.5 ms per division or slower.

ADD—This mode algebraically sums the Channel 1 and Channel 2 input signals.

INVERT—This switch inverts Channel 2 (or Channel 1 on some scopes) to enable a differential measurement when in the ADD mode.

TIME/DIV. This is usually two concentric controls that affect the timing of the horizontal sweep or time-base generator. The outer control is a click-stop switch that provides step selection of the sweep rate. The center control provides a more sensitive adjustment of the sweep rate on a continuous basis. In its extreme clockwise position, usually marked CAL, the sweep rate is calibrated. Each step of the outer control is, therefore, equal to an exact time unit per scale division. Thus, the time it takes the trace to move horizontally across one division of the screen graticule is known. Dual-trace scopes generally have one Time/Div. control. Some scopes mark this control Time/cm.

X-Y SWITCH. When this switch is engaged, one channel of the dual-trace scope becomes the horizontal, or *X*, input, while the other channel becomes the vertical, or *Y*, input. In this condition, the trigger source is disabled. On some scopes, this setting occurs when the Time/Div. control is fully counterclockwise.

TRIGGERING CONTROLS. The typical dual-trace scope has a number of controls associated with the selection of the triggering source, the method by which it is coupled, the level at which the sweep is triggered, and the selection of the slope at which triggering takes place:

1. *Level Control*. This is a rotary control that determines the point on the triggering waveform where the sweep is triggered. When no triggering signal is present, no trace will appear on the screen. Associated with the level control is an Auto switch, which is often an integral part of the level rotary control or may be a separate push button. In the Auto position, the rotary control is disengaged and automatic triggering takes place. In this case, a sweep is always generated and therefore a trace will appear on the screen even in the absence of a triggering signal. When a triggering signal is present, the normal triggering process takes over.
2. *Coupling*. This control is used to select the manner in which the triggering is coupled to the signal. The types of coupling and the way they are labeled vary from one manufacturer and model to another. For example, AC coupling usually indicates the use of capacitive coupling that blocks DC; line coupling indicates the 50- or 60-Hz line voltage is the trigger. If the oscilloscope was designed for television testing, the coupling control might be marked for triggering by the horizontal or vertical sync pulses.
3. *Source*. The trigger signal may be external or internal. As already noted, the line voltage may also be used as the triggering signal.
4. *Slope*. This control determines whether triggering of the sweep occurs at the positive going or negative going portion of the triggering signal. The switch itself is usually labeled positive or negative, or simply + or −.

Oscilloscope Probes

Oscilloscope probes are the test leads used for connecting the vertical input signal to the oscilloscope. There are three types: a direct lead that is just a shielded cable, the low-capacitance probe (LCP) with a series-isolating resistor, and a demodulator probe. Figure E–6 shows a circuit for an LCP for an oscilloscope. The LCP usually has a switch to short out the isolating resistor so that the same probe can be used either as a direct lead or with low capacitance. (See S_1 in Fig. E–6.)

DIRECT PROBE

The direct probe is just a shielded wire without any isolating resistor. A shielded cable is necessary to prevent any pickup of interfering signals, especially with the high resistance at the vertical input terminals of the oscilloscope. The higher the resistance, the more voltage that can be developed by induction. Any interfering signals in the test lead produce distortion of the trace pattern. The main sources of interference are 60-Hz magnetic fields from the power line and stray rf signals.

Figure E–6 Circuit for LCP for an oscilloscope.

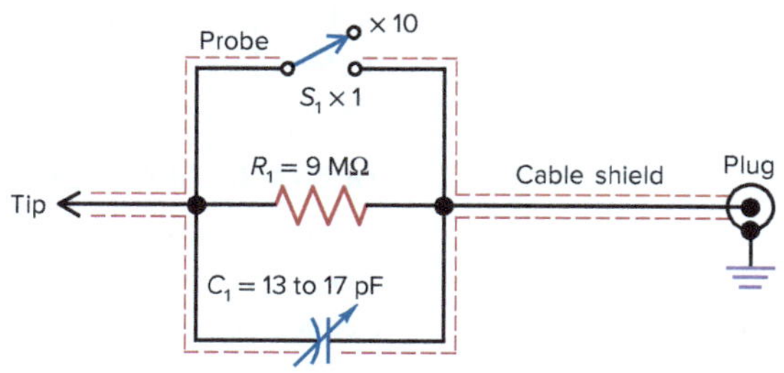

The direct probe as a shielded lead has relatively high capacitance. A typical value is 90 pF for 3 ft (0.9 m) of 50-Ω coaxial cable. Also, the vertical input terminals of the oscilloscope have a shunt capacitance of about 40 pF. The total C then is $90 + 40 = 130$ pF. This much capacitance can have a big effect on the circuit being tested. For example, it could detune a resonant circuit. Also, nonsinusoidal waveshapes are distorted. Therefore, the direct probe can be used only when the added C has little or no effect. These applications include voltages for the 60-Hz power line or sine-wave audio signals in a circuit with a relatively low resistance of several kilohms or less. The advantage of the direct probe is that it does not divide down the amount of input signal, since there is no series-isolating resistance.

LOW-CAPACITANCE PROBE (LCP)

Refer to the diagram in Fig. E–6. The 9-MΩ resistor in the probe isolates the capacitance of the cable and the oscilloscope from the circuit connected to the probe tip. With an LCP, the input capacitance of the probe is only about 10 pF. The LCP must be used for oscilloscope measurements when

1. The signal frequency is above audio frequencies.
2. The circuit being tested has R higher than about 50 kΩ.
3. The waveshape is nonsinusoidal, especially with square waves and sharp pulses.

Without the LCP, the observed waveform can be distorted. The reason is that too much capacitance changes the circuit while it is being tested.

THE 1:10 VOLTAGE DIVISION OF THE LCP

Refer to the voltage divider circuit in Fig. E–7. The 9-MΩ of R_P is a series resistor in the probe. Also, R_S of 1 MΩ is a typical value for the shunt resistance at the vertical terminals of the oscilloscope. Then $R_T = 9 + 1 = 10$ MΩ. The voltage across R_S for the scope equals R_S/R_T or 1⁄10 of the input voltage. For the example in Fig. E–7 with 10 V at the tip of the LCP, 1 V is applied to the oscilloscope.

Remember, when using the LCP, multiply by 10 for the actual signal amplitude. As an example, for a trace pattern on the screen that measures 2.4 V, the actual

Figure E–7 Voltage division of 1:10 with a low-capacitance probe.

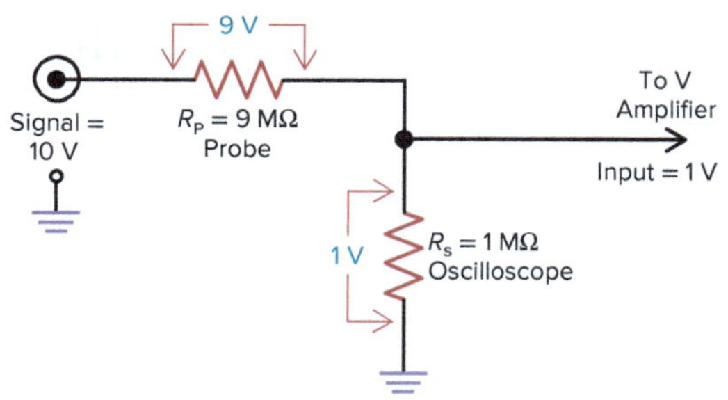

signal input at the probe is 24 V. For this reason, the LCP is generally called the "× 10" probe. Check to see whether or not the switch on the probe is in the direct or LCP position. Even though the scope trace is reduced by the factor of 1⁄10, it is preferable to use the LCP for almost all oscilloscope measurements to minimize distortion of the waveshapes.

TRIMMER CAPACITOR OF THE LCP

Referring back to Fig. E–6, note that the LCP has an internal variable capacitor C_1 across the isolating resistor R_1. The purpose of C_1 is to compensate the LCP for high frequencies. Its time constant with R_1 should equal the RC time constant of the circuit at the vertical input terminals of the oscilloscope. When necessary, C_1 is adjusted for minimum tilt on a square-wave signal.

CURRENT MEASUREMENTS WITH OSCILLOSCOPE

Although it serves as an AC voltmeter, the oscilloscope can also be used for measuring current values indirectly. The technique is to insert a low R in series where the current is to be checked. Use the oscilloscope to measure the voltage across R. Then the current is $I = V/R$. Keep the value of the inserted R much lower than the resistance of the circuit being tested to prevent any appreciable change in the actual I. Besides measuring the current this way, the waveform of V on the screen is the same as I because R does not affect the waveshape.

Voltage and Time Measurements

In general, an oscilloscope is normally used to make two basic measurements; amplitude and time. After making these two measurements, other values can be determined. Figure E–8 shows the screen of a typical oscilloscope.

As mentioned earlier, the vertical or Y axis represents values of voltage amplitude whereas the horizontal or X axis represents values of time. The volts/division control on the oscilloscope determines the amount of voltage needed at the scope input to deflect the electron beam one division vertically on the Y axis. The seconds/division control on the oscilloscope determines the time it takes for the scanning electron beam to scan one horizontal division. In Fig. E–8, note that there are 8 vertical divisions and 10 horizontal divisions.

Refer to the sine wave being displayed on the oscilloscope graticule in Fig. E–9. To calculate the peak-to-peak value of the waveform, simply count the number of

Figure E–8 Oscilloscope screen (graticule).

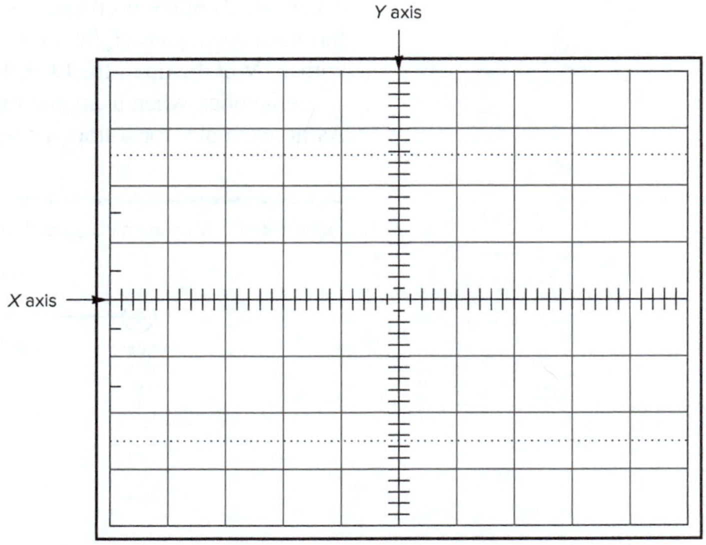

Figure E–9 Determining V_{p-p}, T, and f from the sine wave displayed on the scope graticule.

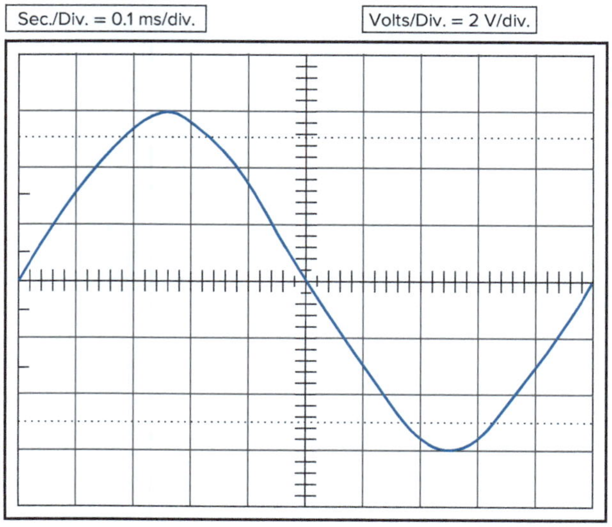

vertical divisions occupied by the waveform and then multiply this number by the volts/division setting. Expressed as a formula,

$$V_{p\text{-}p} = \#\text{ vertical divisions} \times \frac{\text{volts}}{\text{division}}\text{ setting}$$

In Fig. E–9, the sine wave occupies 6 vertical divisions. Since the Volts./Div. setting equals 2 V/division, the peak-to-peak calculations are as follows:

$$V_{p\text{-}p} = 6\text{ vertical divisions} \times \frac{2\text{ V}}{\text{division}} = 12\ V_{p\text{-}p}$$

To calculate the period, T, of the waveform, all you do is count the number of horizontal divisions occupied by one cycle. Then, simply multiply the number of horizontal divisions by the Sec./Div. setting. Expressed as a formula,

$$T = \#\text{ horizontal divisions} \times \frac{\text{sec.}}{\text{division}}\text{ setting}$$

In Fig. E–9, one cycle of the sine wave occupies exactly 10 horizontal divisions. Since the Sec./Div. setting is set to 0.1 ms/div., the calculations for T are as follows:

$$T = 10\text{ horizontal divisions} \times \frac{0.1\text{ ms}}{\text{div.}} = 1\text{ ms}$$

With the period, T, known, the frequency, f, can be found as follows:

$$\begin{aligned} f &= \frac{1}{T} \\ &= \frac{1}{1\text{ ms}} \\ &= 1\text{ kHz} \end{aligned}$$

EXAMPLE 1. In Fig. E–10, determine the peak-to-peak voltage, the period, T, and the frequency, f, of the displayed waveform.

ANSWER. Careful study of the scopes graticule reveals that the height of the waveform occupies 3.4 vertical divisions. With the Volts/Div. setting at 0.5 V/div., the peak-to-peak voltage is calculated as follows:

$$V_{p\text{-}p} = 3.4\text{ vertical divisions} \times \frac{0.5\text{ V}}{\text{div.}} = 1.7\ V_{p\text{-}p}$$

Figure E–10 Determining $V_{p\text{-}p}$, T, and f from the sine wave displayed on the scope graticule.

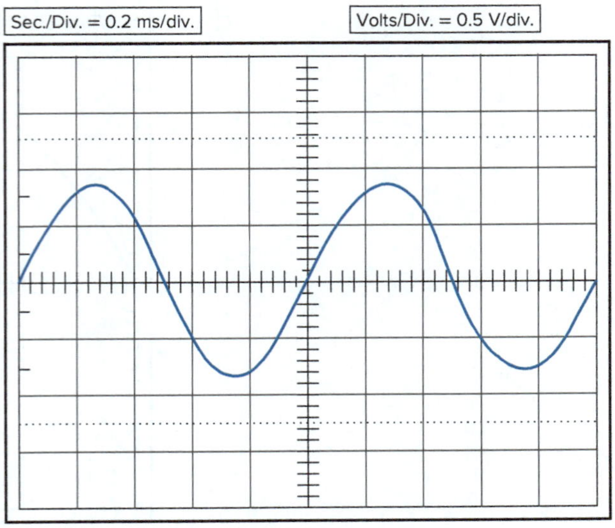

To find the period, T, of the displayed waveform, count the number of horizontal divisions occupied by just one cycle. By viewing the scopes graticule, we see that one cycle occupies 5 horizontal divisions. Since the Sec./Div. control is set to 0.2 ms/div., the period, T, is calculated as:

$$T = 5 \text{ horizontal divisions} \times \frac{0.2 \text{ ms}}{\text{div.}} = 1 \text{ ms}$$

To calculate the frequency, f, take the reciprocal of the period, T.

$$f = \frac{1}{T} = \frac{1}{1 \text{ ms}} = 1 \text{ kHz}$$

EXAMPLE 2. In Fig. E–11, determine the pulse time, *tp*, pulse repetition time, *prt*, and the peak value, V_{pk}, of the displayed waveform. Also, calculate the waveform's % duty cycle and the pulse repetition frequency, *prf*.

Figure E–11 Determining V_{pk}, *tp*, *prt*, *prf*, and % duty cycle from the rectangular wave displayed on the scope graticule.

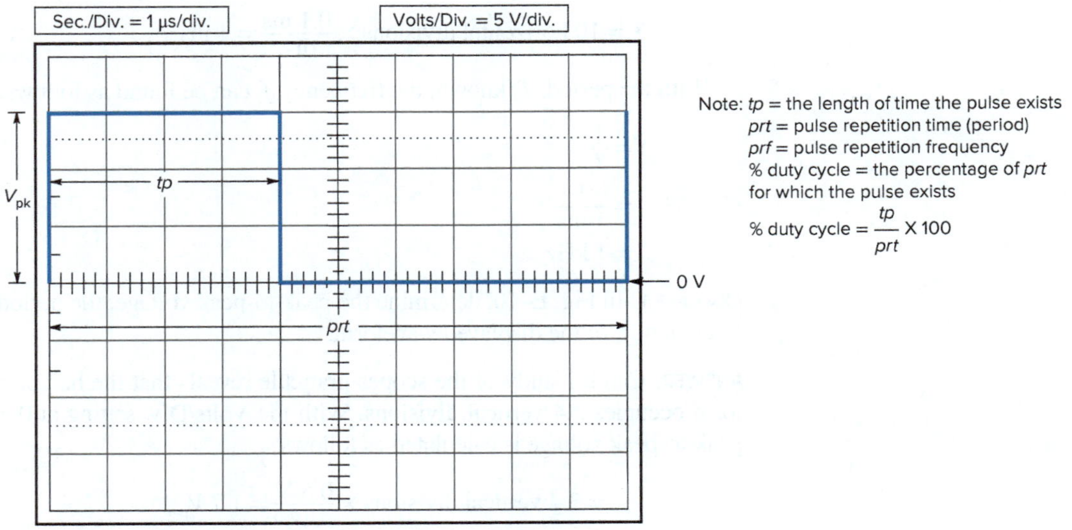

ANSWER. To find the pulse time, *tp*, count the number of horizontal divisions occupied by just the pulse. In Fig. E–11, the pulse occupies exactly 4 horizontal divisions. With the Sec./Div. control set to 1 μs/div., the pulse time, *tp*, is calculated as

$$tp = 4 \text{ horizontal divisions} \times \frac{1\ \mu\text{s}}{\text{div.}} = 4\ \mu\text{s}$$

The pulse repetition time, *prt*, is found by counting the number of horizontal divisions occupied by one cycle of the waveform. Since one cycle occupies 10 horizontal divisions, the pulse repetition time, *prt*, is calculated as follows:

$$prt = 10 \text{ horizontal divisions} \times \frac{1\ \mu\text{s}}{\text{div.}} = 10\ \mu\text{s}$$

With *tp* and *prt* known, the % duty cycle is calculated as follows:

$$\begin{aligned}\%\text{ duty cycle} &= \frac{tp}{prt} \times 100\\ &= \frac{4\ \mu\text{s}}{10\ \mu\text{s}} \times 100\\ &= 40\%\end{aligned}$$

The pulse repetition frequency, *prf*, is calculated by taking the reciprocal of *prt*.

$$\begin{aligned}prt &= \frac{1}{prt}\\ &= \frac{1}{10\ \mu\text{s}}\\ &= 100\text{ kHz}\end{aligned}$$

The peak value of the waveform is based on the fact that the baseline value of the waveform is 0 V as shown. The positive peak of the waveform is shown to be three vertical divisions above zero. Since the Volts/Div. setting of the scope is 5 V/div., the peak value of the waveform is

$$V_{\text{pk}} = 3 \text{ vertical divisions} \times \frac{5\text{ V}}{\text{div.}} = 15\text{ V}$$

Notice that the waveform shown in Fig. E–11 is entirely positive because the waveform's pulse makes a positive excursion from the zero-volt reference.

PHASE MEASUREMENT

Phase measurements can be made with a dual-trace oscilloscope when the signals are of the same frequency. To make this measurement, the following procedure can be used:

1. Preset the scope's controls, and obtain a baseline trace (the same for both channels). Set the Trigger Source to whichever input is chosen to be the reference input. Channel 1 is often used as the reference, but Channel 2 as well as External Trigger or Line could be used.
2. Set both Vertical Input Coupling switches to the same position, depending on the type of input.
3. Set the Vertical MODE to Both; then select either ALT or CHOP, depending on the input frequency.
4. Although not necessary, set both Volts/Div. and both Variable controls so that both traces are approximately the same height.
5. Adjust the TRIGGER LEVEL to obtain a stable display. Typically set so that the beginning of the reference trace begins at approximately zero volts.

Figure E–12 Oscilloscope phase shift measurement.

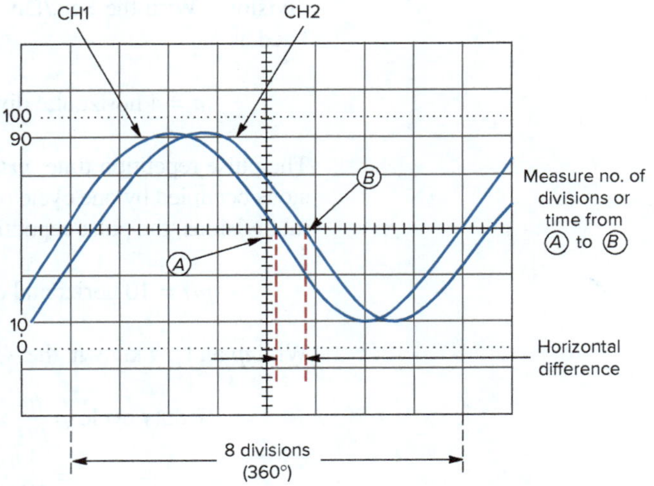

6. Set the Time/Div. switch to display about one full cycle of the reference waveform.
7. Use the Position controls, Time/Div. switch, and Variable time control so that the reference signal occupies exactly 8 horizontal divisions. The entire cycle of this waveform represents 360°, and each division of the graticule now represents 45° of the cycle.
8. Measure the horizontal difference between corresponding points of each waveform on the horizontal graticule line, as shown in Fig. E–12.
9. Calculate the phase shift by using the formula
   Phase shift = (no. of horizontal difference divisions) × (no. of degrees per division)

As an example, Fig. E–12 displays a difference of 0.6 division at 45° per division. The phase shift = (0.6 div.) × (45°/div.) = 27°.

DIGITAL (MENU-DRIVEN) OSCILLOSCOPES

Digital, menu-driven oscilloscopes have replaced analog oscilloscopes in almost all electronic industries and educational facilities. In addition to being able to make the traditional voltage, time, and phase measurements, digital or menu-driven scopes can also store a measured waveform for later viewing. Digital scopes are also much smaller and weigh less than their analog counterparts. These two advantages alone have prompted many schools and industries to make the switch from analog to digital scopes.

Like any piece of test equipment, there is a learning curve involved before you will be totally comfortable operating a digital oscilloscope. The biggest challenge facing you will be familiarizing yourself with the vast number of menus and submenus of a digital scope to access its features and functions. But it's not too bad once you sit down and start with some simple and straightforward measurements. It's always best if you can obtain the operating manual and educational materials for the digital scope you are learning to use. Keep these materials nearby so you can refer to them when you need help in making a measurement. This is not an uncommon practice, even for very experienced users of digital oscilloscopes.

Figure E–13 shows a Tektronix TDS-2002 (2-channel) digital storage oscilloscope. This scope is similar to the ones used in Appendix F. What follows is a brief explanation of the scope's vertical, horizontal, trigger, and menu and control buttons.

Figure E–13 Two-channel digital oscilloscope.

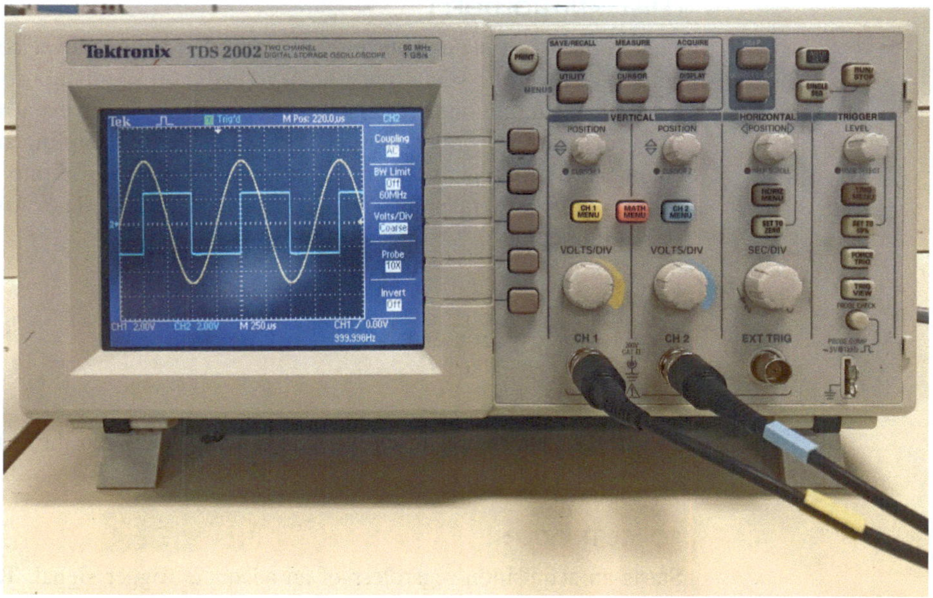

Sarah Schultz Photography

Vertical Controls (See Fig. E–13)

CHS. 1 AND 2, AND CURSOR 1 AND 2 POSITION

Positions the waveform vertically. When cursors are turned on and the cursor menu is displayed, these knobs position the cursors. (*Note:* Cursors are horizontal or vertical lines that can be moved up and down or left and right to make either voltage or time measurements.)

CHS. 1 AND 2 AND MENU

Displays the channel input menu selections and toggles the channel display on and off.

VOLTS/DIV. (CHS. 1 AND 2)

Selects calibrated scale factors also referred to as Volts/Div. settings.

MATH MENU

Displays waveform math operations menu and can also be used to toggle the math waveform on and off.

Horizontal Controls (See Fig. E–13)

POSITION

Adjusts the horizontal position of both channels and math waveforms. The resolution of this control varies with the time base.

HORIZONTAL MENU

Displays the horizontal menu.

SEC/DIV

Selects the horizontal Time/Div. setting (scale factor).

Trigger Controls (See Fig. E–13)

LEVEL (AND HOLDOFF)

This control has a dual purpose. As an edge trigger level control, it sets the amplitude level the signal must cross to cause an acquisition. As a holdoff control, it sets the amount of time before another trigger event can be accepted. (*Note:* The term *acquisition* refers to the process of sampling signals from input channels, digitizing the samples, processing the results into data points, and assembling the data points into a waveform record. The waveform record is stored in memory.)

TRIGGER MENU

Displays the trigger menu.

SET LEVEL TO 50%

The trigger level is set to the vertical midpoint between the peaks of the trigger signal.

FORCE TRIGGER

Starts an acquisition regardless of an adequate trigger signal. This button has no effect if the acquisition is already stopped.

TRIGGER VIEW

Displays the trigger waveform in place of the channel waveform while the TRIGGER VIEW button is held down. You can use this to see how the trigger settings affect the trigger signal, such as trigger coupling.

Menu and Control Buttons (See Fig. E–13)

SAVE/RECALL

Displays the save/recall menu for setups and waveforms.

MEASURE

Displays the automated measurements menu.

ACQUIRE

Displays the acquisition menu.

DISPLAY

Displays the display menu.

CURSOR

Displays the cursor menu. Vertical position controls adjust cursor position while displaying the cursor menu, and the cursors are turned on. Cursors remain displayed (unless turned off) after leaving the cursor menu but are not adjustable.

UTILITY

Displays the utility menus.

AUTOSET

Automatically sets the scopes controls to produce a usable display of the input signal.

PRINT

Starts print operations.

RUN/STOP

Starts and stops waveform acquisition.

Since the complexity of the internal operation of a digital oscilloscope is based on many advanced topics that you have not yet covered, we will provide no further explanation of digital scopes in this appendix.

Appendix F

Introduction to Multisim

Introduction

In an effort to help the reader understand the concepts presented in this textbook, key examples and problems are presented through the use of computer simulation using Multisim. Multisim is an interactive circuit simulation software package that allows the user to view their circuit in schematic form while measuring the various parameters of the circuit. The ability to quickly create a schematic and then analyze the circuit through simulation makes Multisim a wonderful tool to help students understand the concepts covered in the study of electronics. In addition, Multisim provides the opportunity to practice valuable troubleshooting skills through the use of the virtual test equipment without risking the safety of the student or damage to the equipment.

This appendix will introduce the reader to the features of Multisim that directly relate to the study of DC, AC, and semiconductor electronics. The topics covered are:

- Work Area
- Opening a File
- Running a Simulation
- Saving a File
- Components
- Sources
- Measurement Equipment
- Circuit Examples
- User Customization
- Exporting Data to Excel
- Adding Text and Graphics

Work Area

The power of this software lies in its simplicity. With just a few steps, a circuit can be either retrieved from disk or drawn from scratch and then simulated. The main screen, as shown in Fig. F–1, is divided into three areas: The drop-down menu, the tool bars, and the work area.

The drop-down menu gives the user access to all the functions of the program, including the visual appearance of the work area. The visual appearance of the work area can be modified to suit the user's needs. For example, the grid comprised of black dots can be removed and the color scheme can be modified, if the user intends to capture the schematic for use in printed documents. The Sheet Properties menu shown in Fig. F–3 is accessed by left mouse clicking on "Options" in the drop-down menu shown in Fig. F–2, or pressing <Alt><O> at the same time.

The workspace tab provides access to the grid option, as shown in Fig. F–3. Multisim provides a layout grid to help align the various components. When the box is checked, the grid is displayed as a series of black dots. When the box is left unchecked, the functionality of the grid is still present, however, the dots are not visible.

The Colors tab provides access to the various color schemes, as shown in Fig. F–4. If the schematic is going to be used in a printed document, the white background with black wires, components, and text tends to work very well.

Figure F–1 Main screen.

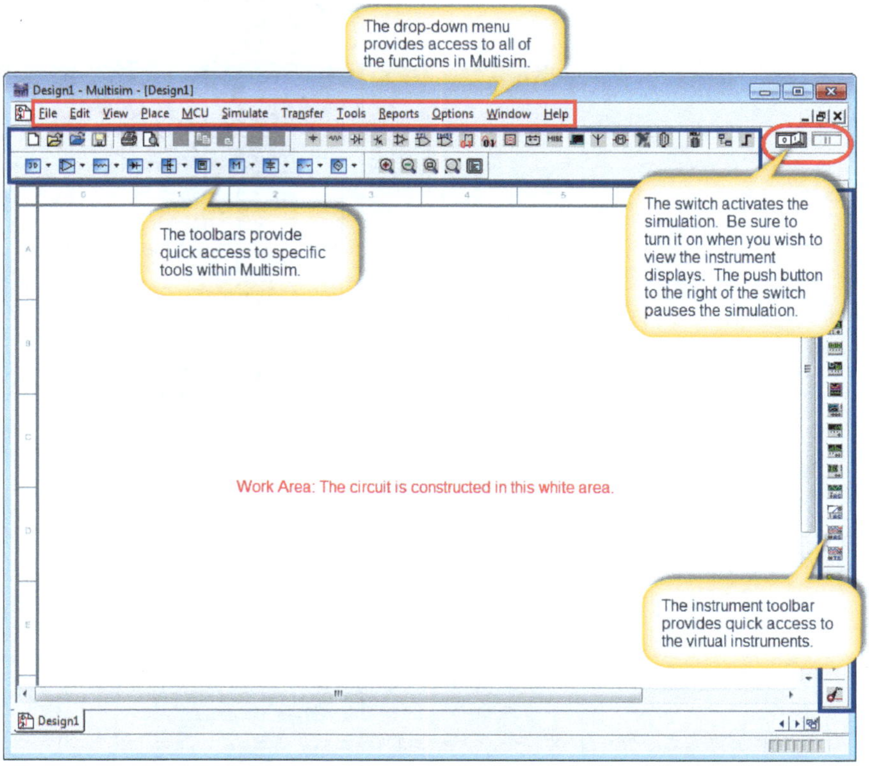

Figure F–2 Options drop-down menu.

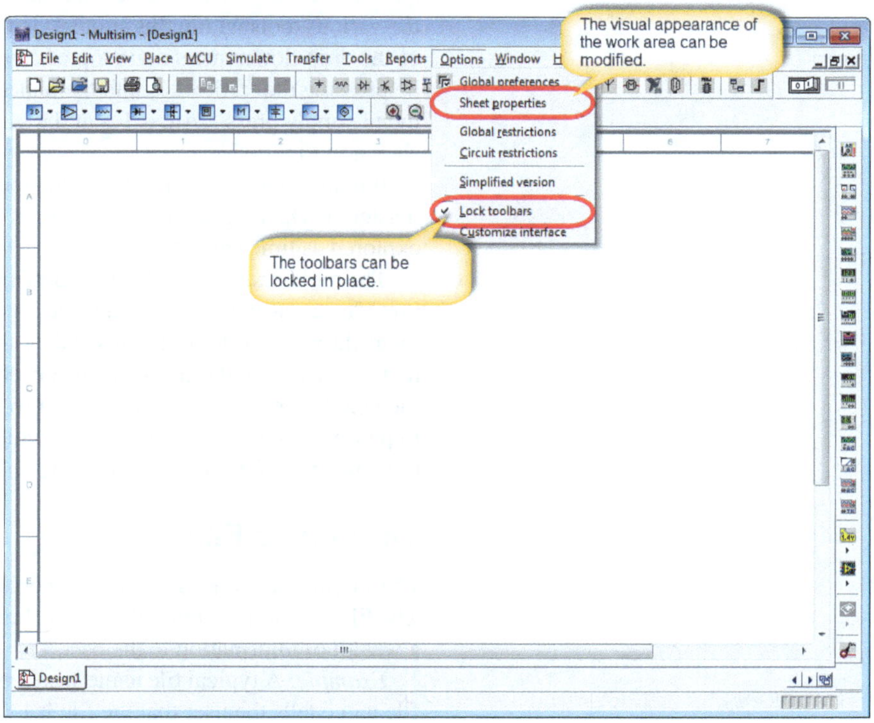

Figure F–3 Sheet Properties Workspace tab.

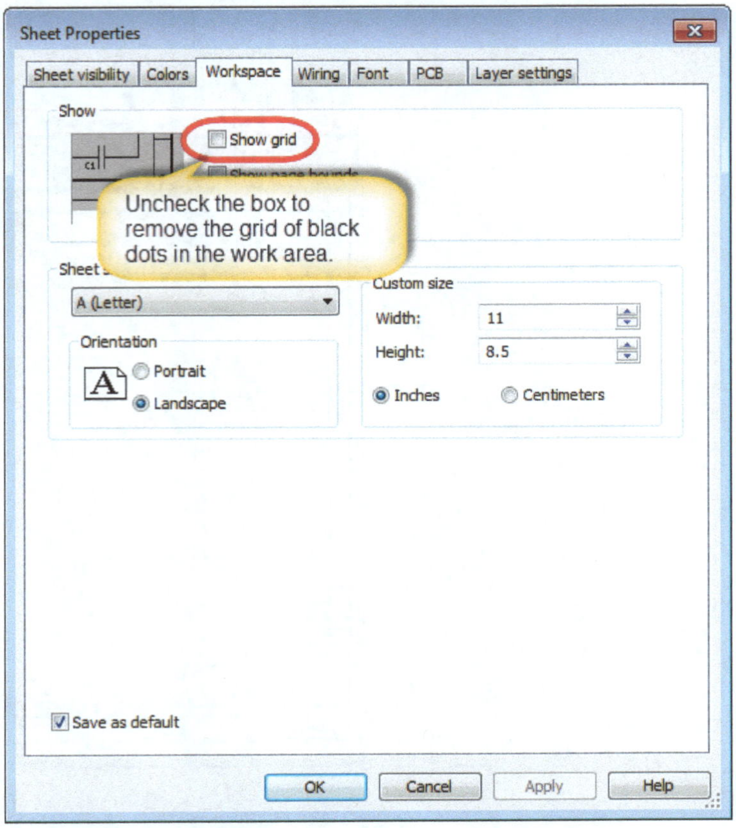

Once the visual appearance of the work area is configured, it will remain that way until changed. Once this has been done, a new user need only access a few of the drop-down menus to create a circuit, run the simulation, and save the circuit for later use. As the user becomes familiar with the software program, they can access the various tools within this simulation package through the drop-down menu. Each of the drop-down menu main topics can be accessed by either a left mouse click or by pressing the <Alt> key and the underlined letter. For example, to access the File menu simply press <Alt><F> at the same time. The File menu will drop down, as shown in Fig. F–5.

Initially, there are only two selections from the drop-down menu that need be mastered: Opening a file and saving a file. The rest of the menu options can be explored as time permits.

The tool bars beneath the drop-down menu and on the right side of the work area provide access to the commonly used menu selections. Typically, a user will access them through the tool bars instead of the drop-down menus. The most important icon in the assorted tool bars is the on-off switch. The on-off switch starts and stops the simulation. The push button next to the on-off switch will cause the simulation to pause. Pressing the Pause button while the simulation is running allows the user to view a waveform or meter reading without the display changing.

Opening a File

The circuits referenced in this textbook are available on McGraw Hill's Connect. The files are divided into folders, one for each chapter. The name of the file provides a wealth of information to the user.

Example: A typical file name would be "Ch 3 Problems 3-1." The first part of the file name tells the user that the file is located in the folder labeled "Chapter 3." The

Figure F–4 Sheet Properties Colors tab.

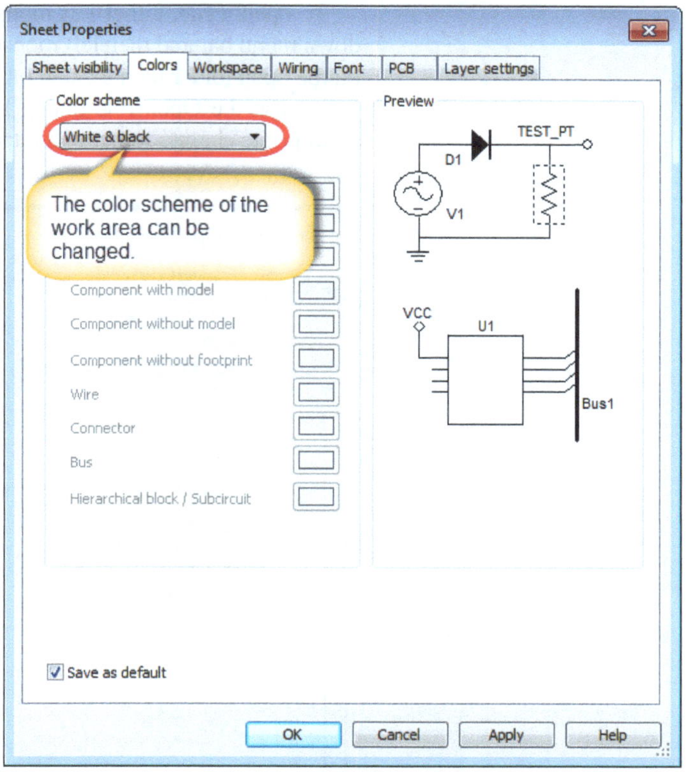

Figure F–5 File drop-down menu.

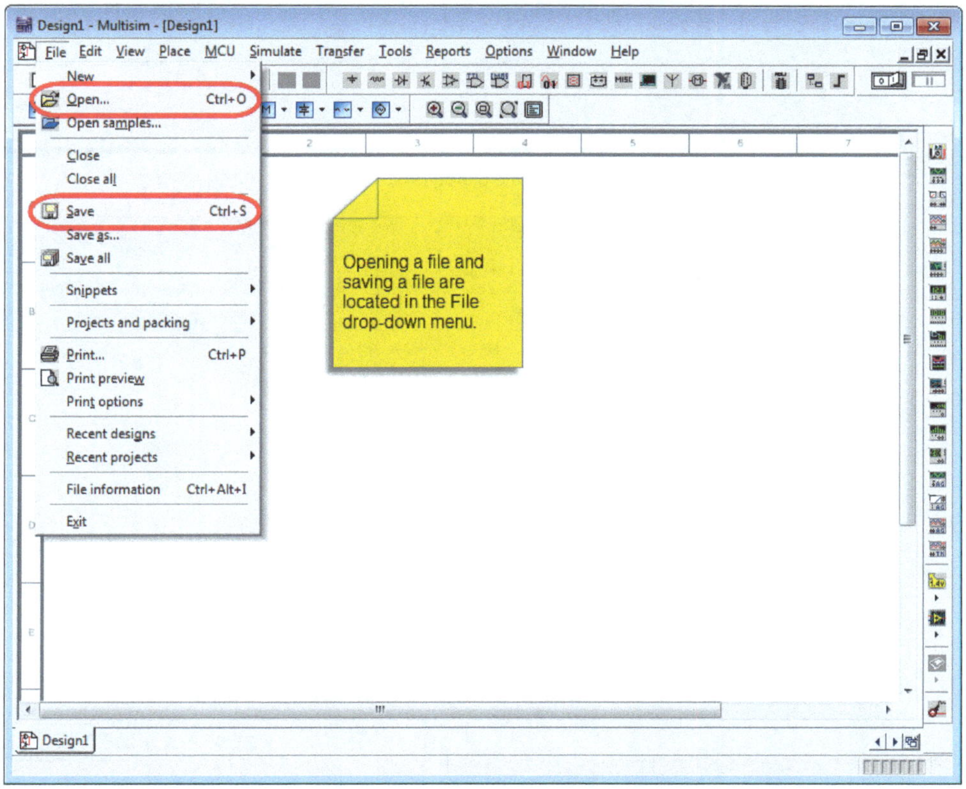

second part of the file name tells the user that it is question 1 out of the Problems section at the end of Chapter 3.

To open a file, either left mouse click on the word "File" located on the drop-down menu bar and then left mouse click on the Open command or left mouse click on the open folder icon located on the tool bar, as shown in Fig. F–6. Both methods will cause the Open File dialog box shown in Fig. F–7 to open. Navigate to the appropriate chapter folder and retrieve the file needed.

Figure F–6 Opening a file.

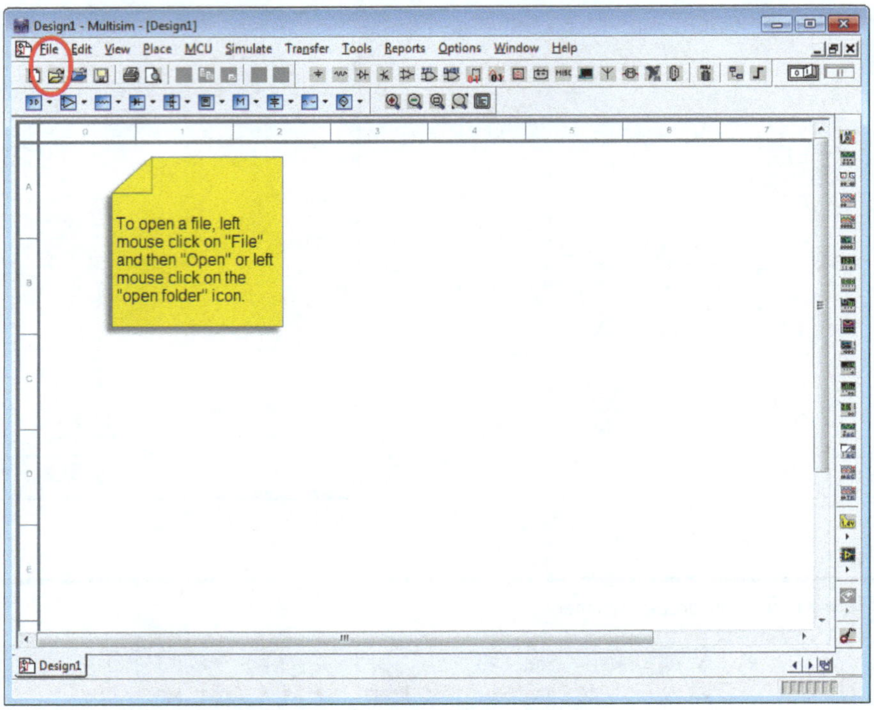

Figure F–7 Open File dialog box.

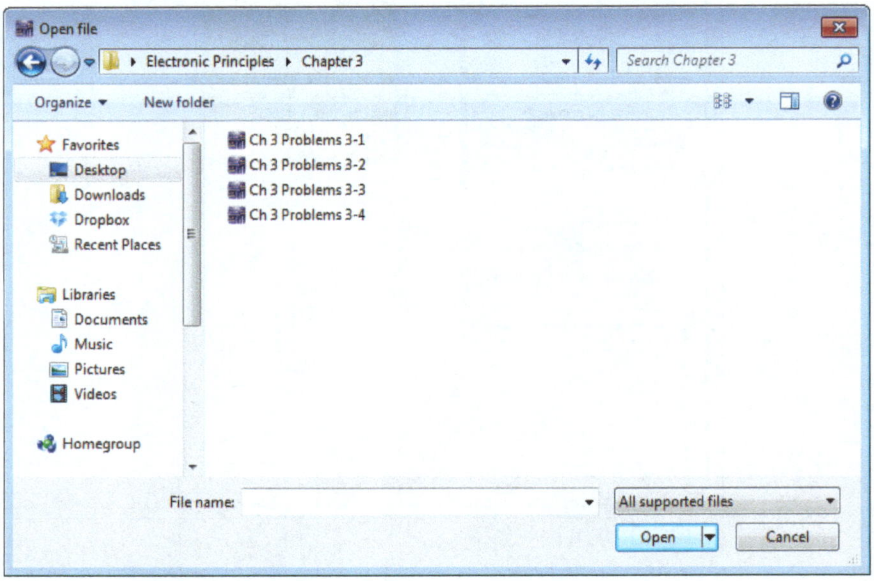

Figure F–8 Starting the simulation.

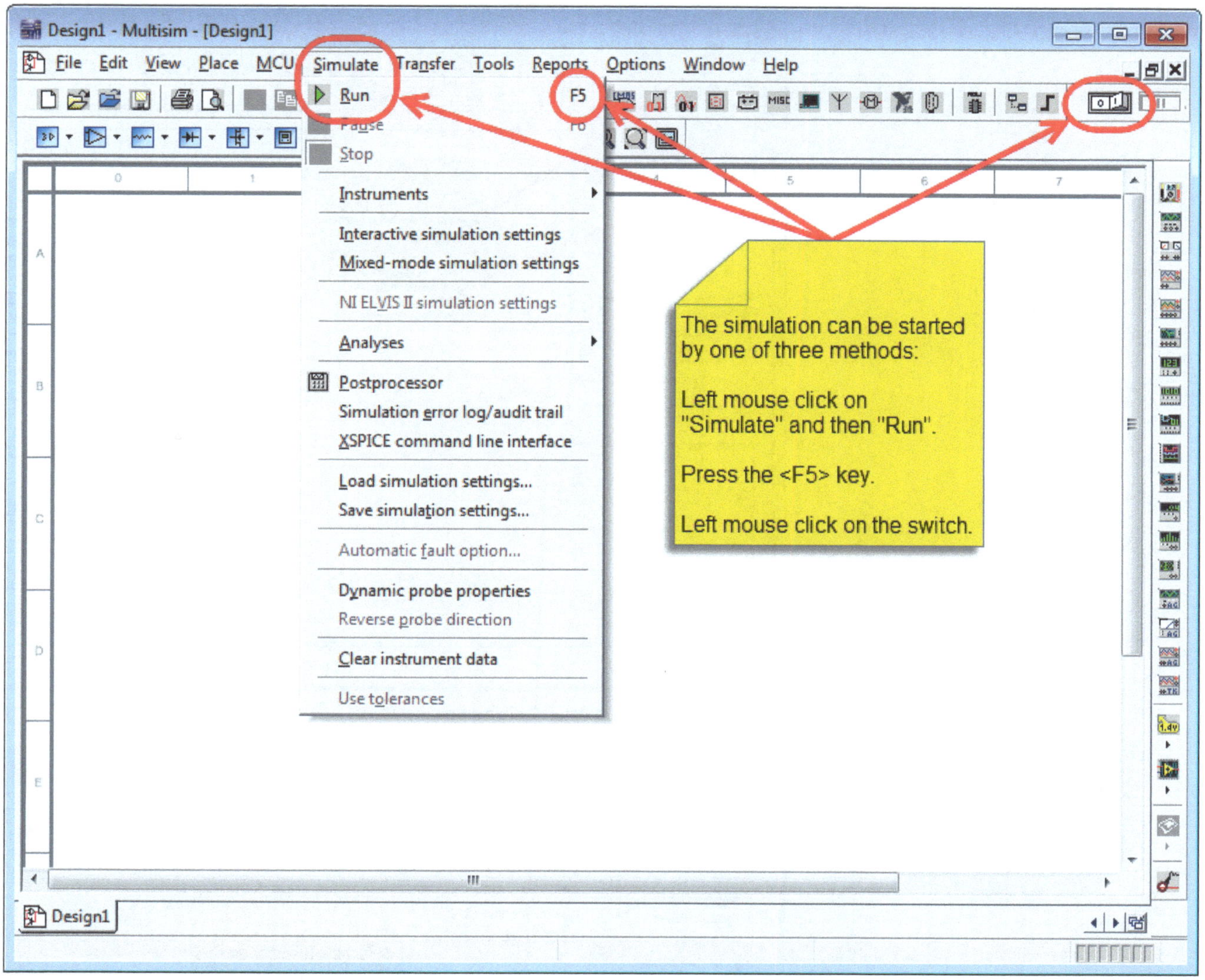

Running a Simulation

Once the circuit is constructed in the work area, the simulation can be started by one of three ways:

1. Select "Simulate" from the drop-down menu and then select "Run."
2. Press the <F5> key.
3. Press the toggle switch with a left mouse click.

All three of these methods are illustrated in Fig. F–8.

Saving a File

If the file has been modified, it needs to be saved under a new file name. As shown in Fig. F–9, select "File", located on the drop-down menu bar with a left mouse click, then select "Save As" from the drop-down menu. This will cause the "Save As" dialog box to open. Give the file a new name and press the Save button with a left mouse click. The process is demonstrated in Fig. F–10.

Figure F–9 "Save As. . ." screen.

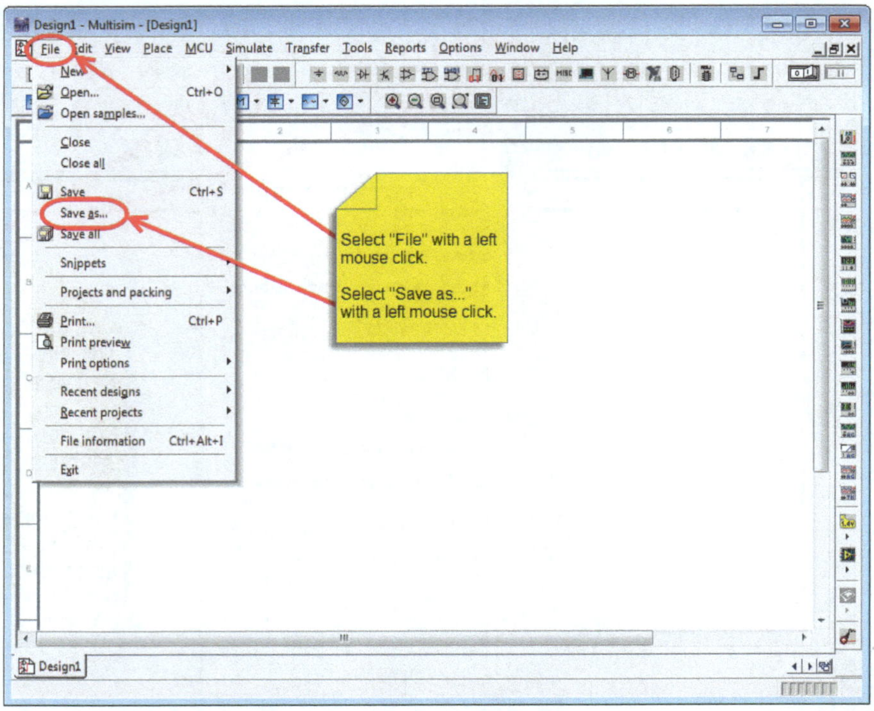

Figure F–10 "Save As . . ." dialog box.

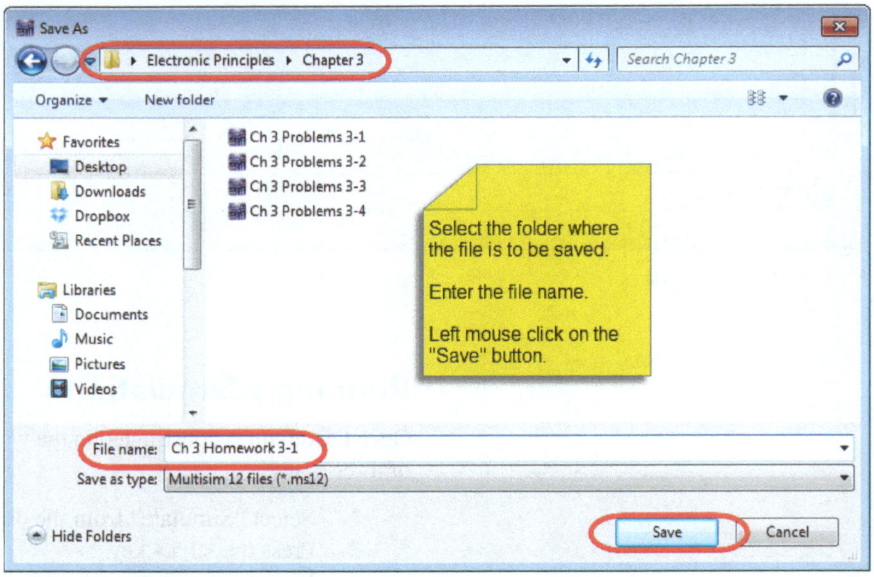

Components

There are two kinds of component models used in Multisim: Those modeled after actual components and those modeled after "ideal" components. Those modeled after ideal components are referred to as "virtual" components. There is a broad selection of virtual components available, as shown in Fig. F–11. The virtual components toolbar can be added to the top of the work area for ease of access, as shown in Fig. F–12.

The difference between the two types of components resides in their rated values. The virtual components can have any of their parameters varied, whereas those

Figure F–11 Virtual component list.

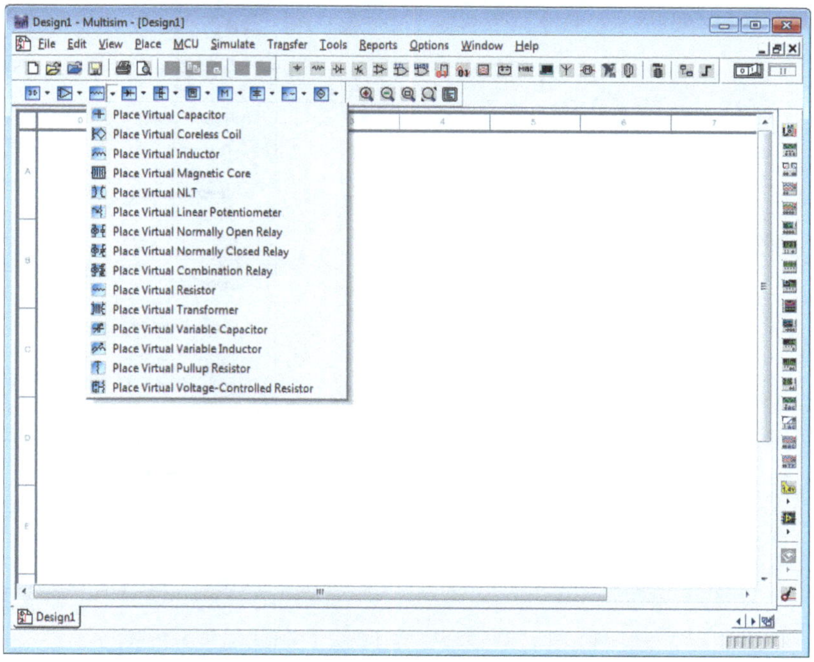

modeled after actual components are limited to real world values. For example, a virtual resistor can have any value resistance and percent tolerance, as shown in Fig. F–13. The configuration screen for each component can be opened by double left mouse click, selecting the component with a single left mouse click and then pressing <Ctrl><M> or selecting the component with a single left mouse click and then choosing "Properties" in the drop-down menu.

The models of the actual resistors are available with tolerance values of 0, 0.1, 0.5, 1, 2, 5, and 10%. The same is true for all other components modeled after real

Figure F–12 Toolbar selection.

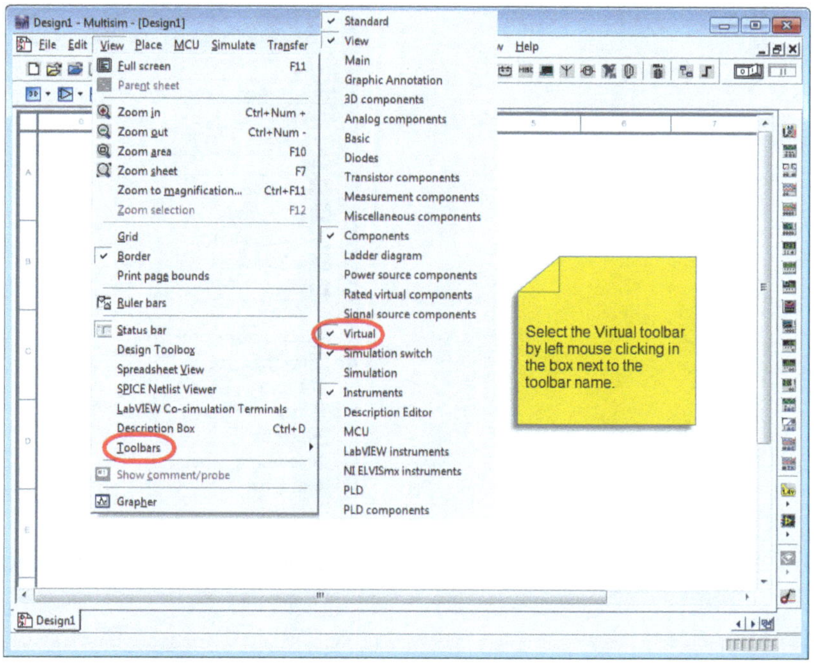

Figure F–13 Configuration screen for a virtual resistor.

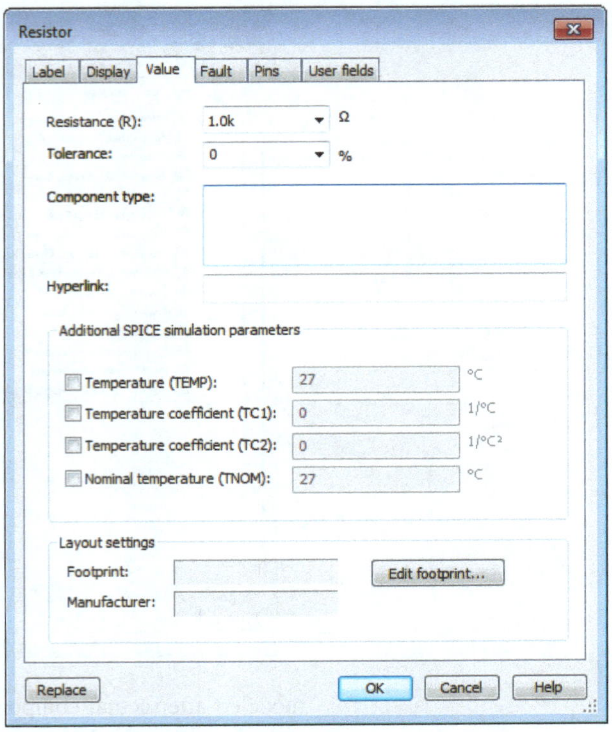

components, as shown in Fig. F–14. This is especially important when the semiconductor devices are used in a simulation. Each of the models of actual semiconductors will function in accordance with their data sheets. These components will be listed by their actual device number as identified by the manufacturers. For example, a common diode is the 1N4001. This diode, along with many others, can be found

Figure F–14 Component listing for resistors.

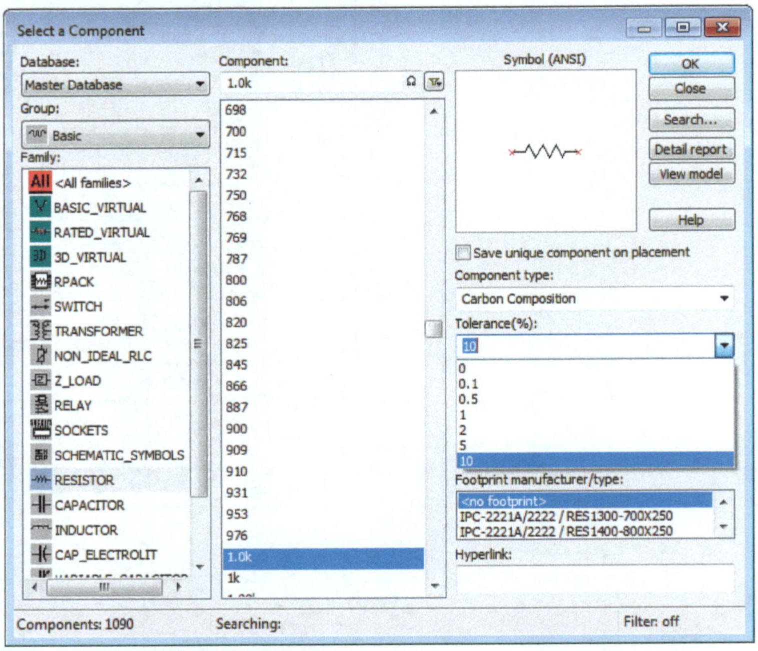

Figure F-15 Switches.

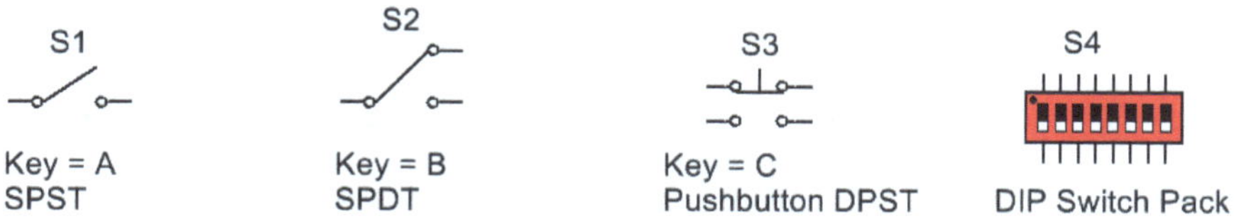

in the semiconductor library of actual components. The parameters of the actual component libraries can also be modified, but that requires an extensive understanding of component modeling and is beyond the scope of this appendix.

If actual components are selected for a circuit to be simulated, the measured value may differ slightly from the calculated values as the software will utilize the tolerances to vary the results. If precise results are required, the virtual components can be set to specific values with a zero percent tolerance.

Several of the components require interaction with the user. The two most commonly used interactive components are the switch and the potentiometer. The movement of the switch is triggered by pressing the key associated with each switch, as shown in Fig. F–15. The key is selected while in the switch configuration screen, as shown in Fig. F–16. If two switches are assigned the same key, they both will move when the key is pressed. The DIP switch packs are available in two to ten switch configurations. Each individual switch within the switch pack is activated by either pressing the associated key or by a left mouse click. When the cursor is placed over

Figure F-16 Switch configuration screen.

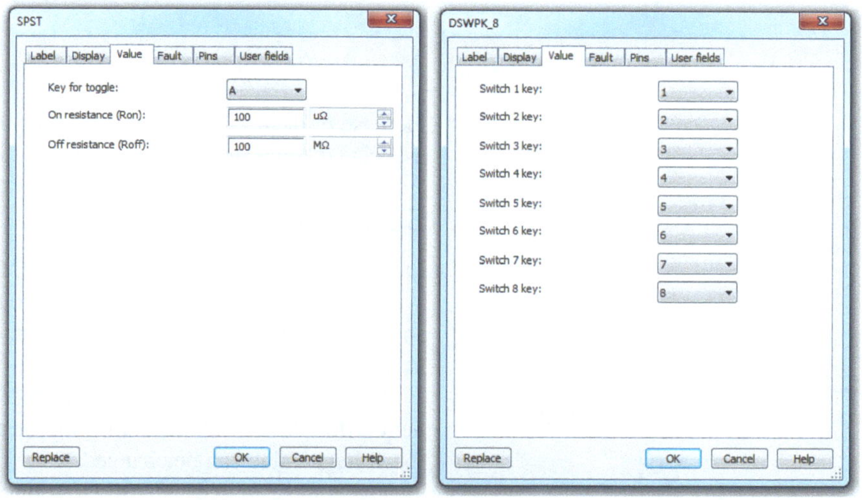

an individual switch within a switch pack, a white rectangle appears around the switch and the cursor changes into a pointing hand symbol. When the white rectangle is present and the cursor resembles a pointing hand, a single left mouse click will cause the switch to change positions. The black dot on the switch pack signifies the ON position.

Figure F-17 Potentiometer.

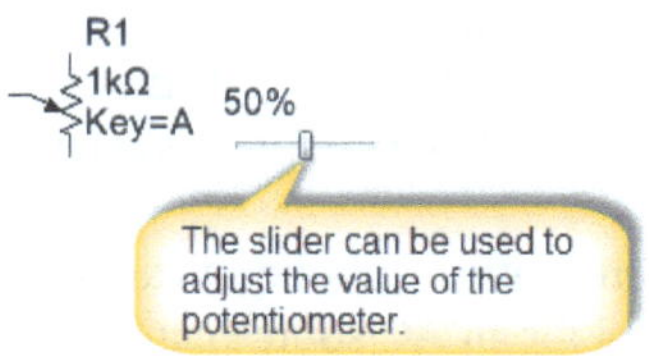

The second commonly used component that requires interaction with the user is the potentiometer. The potentiometer will vary its resistance in predetermined steps with each key press. The pressing of the associated letter on the keyboard will increase the resistance and the pressing of the <Shift> key and the letter will decrease the resistance. A "slider" located to the right of the potentiometer can also be used to adjust the resistance value by dragging it with the mouse. As shown in Fig. F–17, the percent of the total resistance is displayed next to the potentiometer. The incremental

Figure F–18 Potentiometer configuration screen.

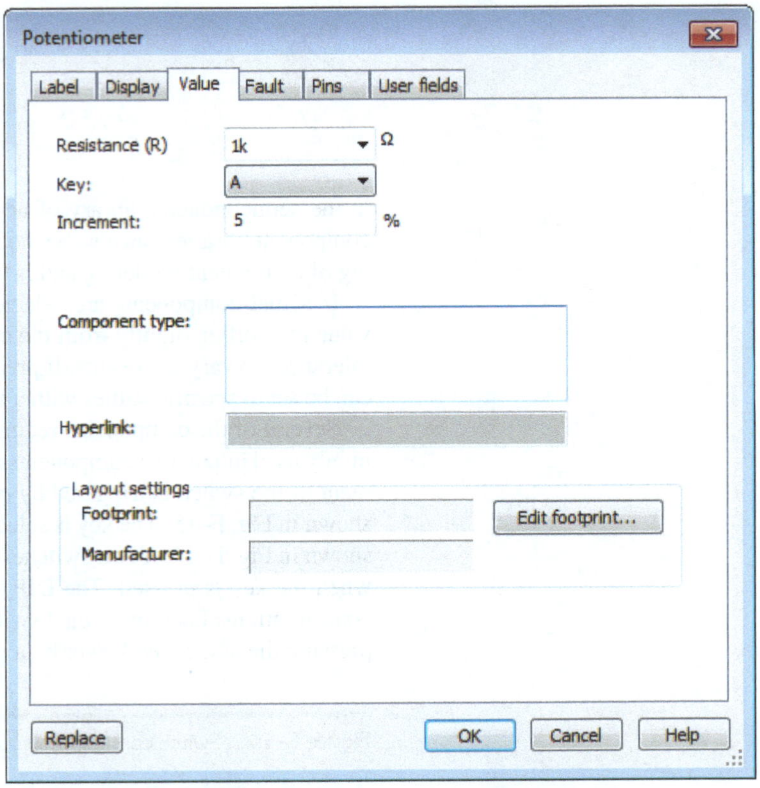

increase or decrease of resistance is set by the user in the configuration screen. The associated key is also set in the configuration screen, as shown in Fig. F–18.

Sources

In the study of DC and AC electronics, the majority of the circuits include either a voltage or current source. There are two main types of voltage sources: DC and AC sources. The DC source can be represented two ways: As a battery in Fig. F–19 and as a voltage source.

Figure F–19 DC source as a battery.

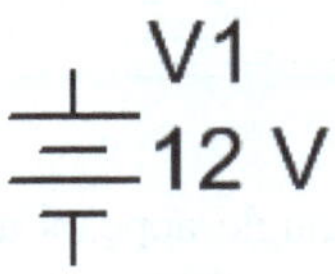

The voltage rating is fully adjustable. The default value is 12 V_{DC}. If the component is double clicked, the configuration screen shown in Fig. F–20 will open up and the voltage value can be changed.

The voltage sources are used in semiconductor circuits to represent either a positive or negative voltage source. Figure F–21 contains the $+V_{CC}$ voltage source used in transistor circuits. FET circuits will utilize the $+V_{DD}$ voltage source, as illustrated in Fig. F–22.

Figure F–23 depicts the $+V_{CC}$ and the $-V_{EE}$ voltage sources. These sources are found in operational amplifier circuits. Operational amplifiers typically have two voltage sources: A negative ($-V_{EE}$) and a positive ($+V_{CC}$) voltage source, as shown in Fig. F–24.

The voltage rating is fully adjustable for all three voltage sources. The default value is +5 V_{DC} for V_{CC} and V_{DD}. The default value for V_{EE} is −5 V_{DC}. If the component is double clicked, the configuration screen shown in Fig. F–25 will open up and the voltage value can be changed.

The AC source can be represented as either a schematic symbol or it can take the form of a function generator. The schematic symbol for an AC source can represent

Figure F-20 Configuration screen for the DC source.

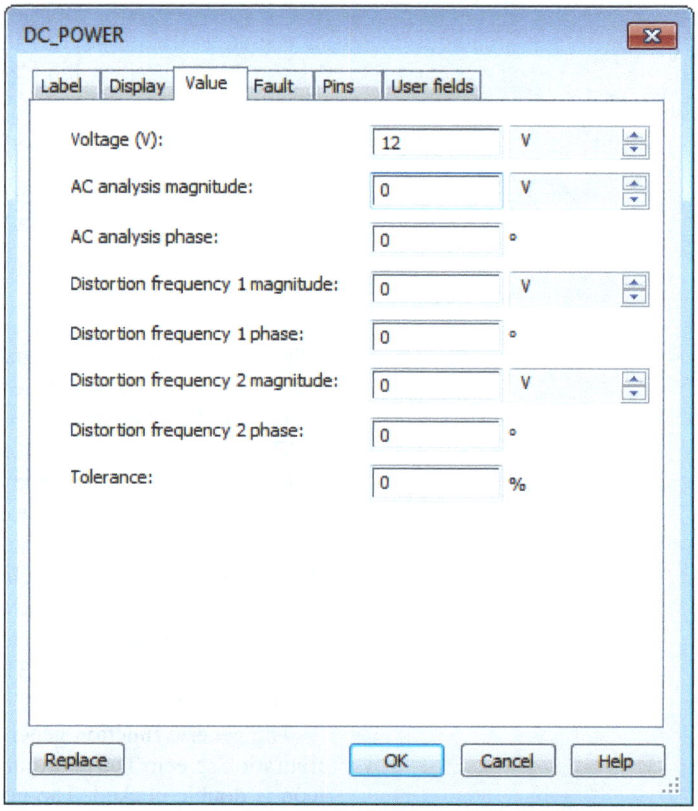

Figure F-21 V_{CC} voltage source.

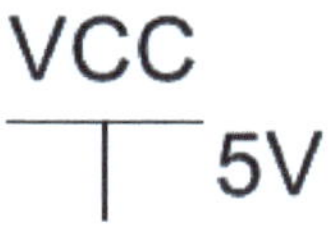

Figure F-22 V_{DD} voltage source.

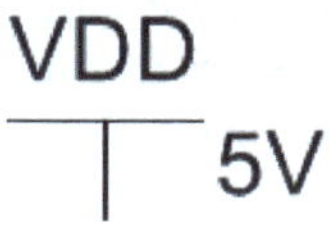

Figure F-23 V_{CC} and V_{EE} voltage sources.

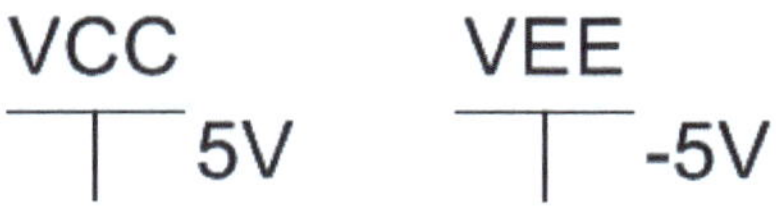

Figure F-24 V_{CC} and V_{EE} Op Amp example.

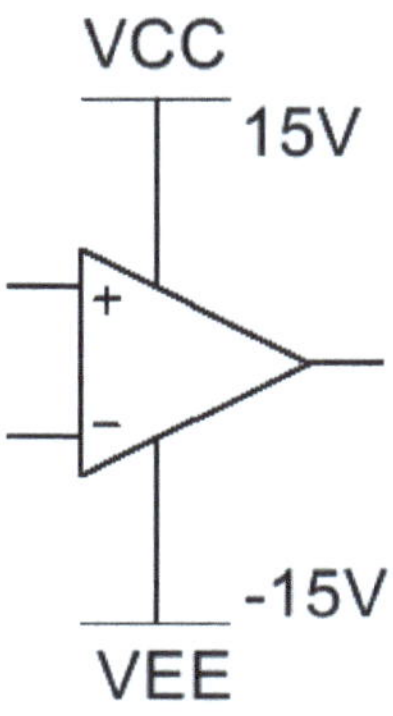

Figure F-25 V_{CC} configuration screen.

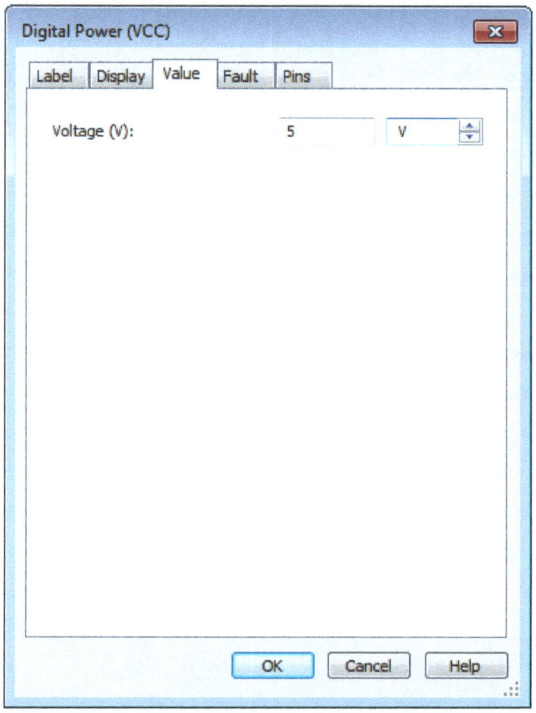

Figure F–26 AC sources.

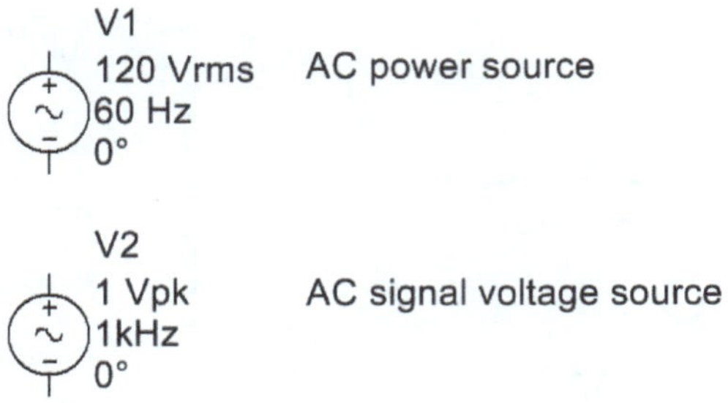

either a power source or a signal voltage source. The amplitude of the power source will be given in RMS voltage (V_{RMS}), whereas the amplitude of the signal voltage source will be given in peak voltage (V_P).

The schematic symbols for the two AC sources, as shown in Fig. F–26, will include information about the AC source. This information will include the device reference number, V_{RMS} value or V_P value, frequency, and phase shift. These values are fully adjustable. The default values for the two sources are shown in Fig. F–26. If the component is double clicked, the configuration screen will open up and the values can be changed, as shown in Fig. F–27.

Multisim provides two function generators: The generic model and the Agilent model. The Agilent model 33120A has the same functionality as the actual Agilent function generator.

The generic function generator icon is shown in Fig. F–28, along with the configuration screen. The configuration screen is displayed when the function generator icon is double clicked. The generic function generator can produce three types of waveforms: Sinusoidal wave, triangular wave, and square wave. The frequency, duty cycle, amplitude, and DC offset are all fully adjustable.

Figure F–27 Configuration screens for the two AC sources.

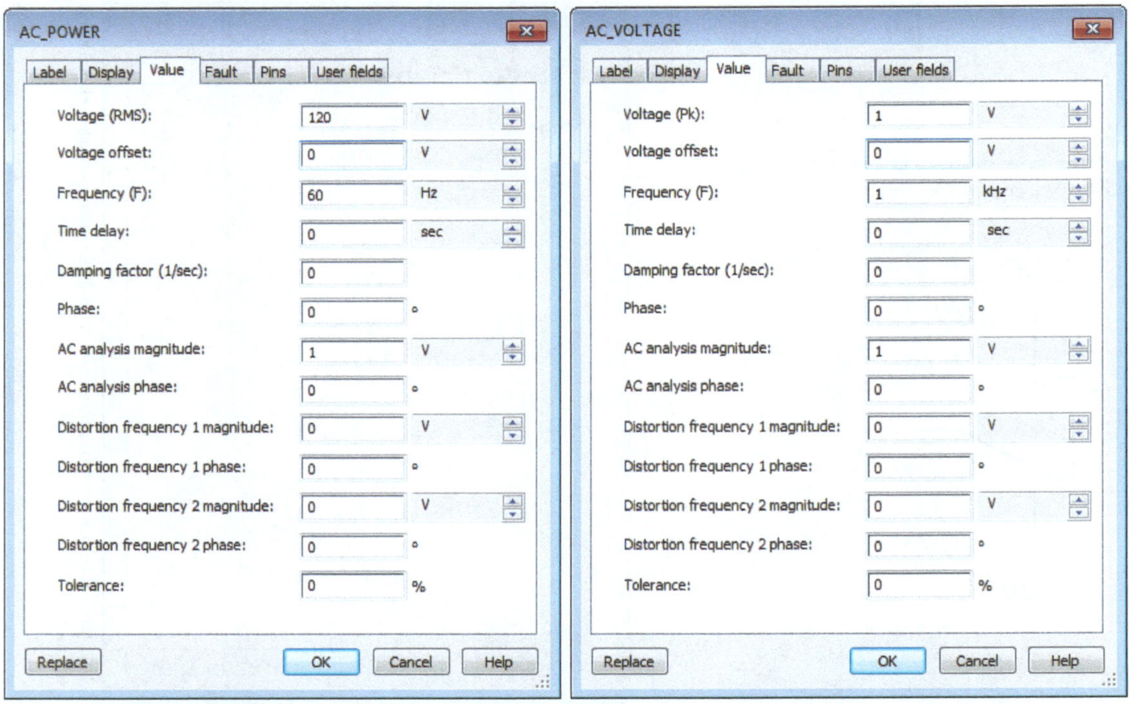

Figure F–28 Generic function generator and configuration screen.

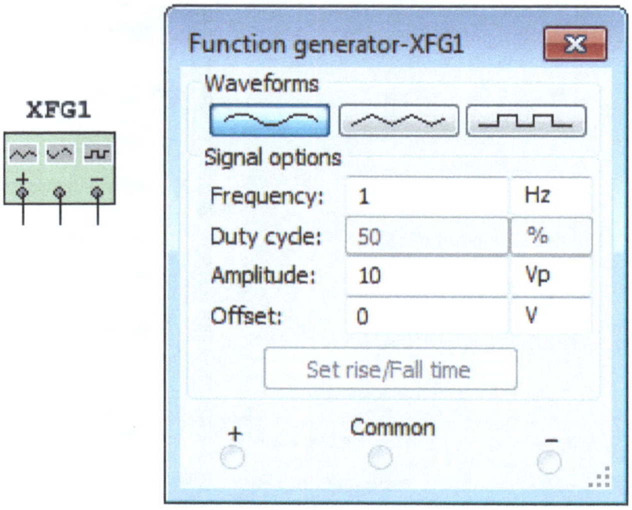

Figure F–29 Agilent function generator.

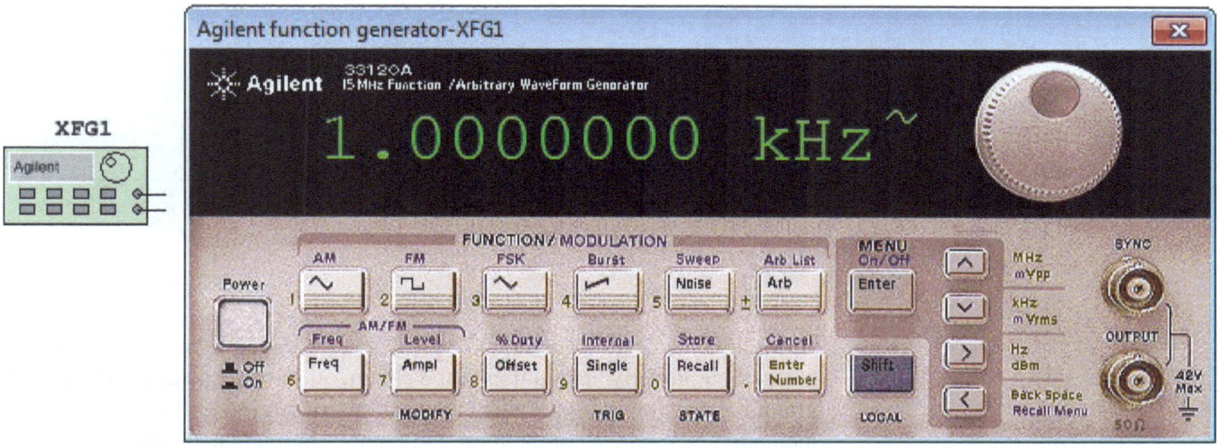

The Agilent function generator is controlled via the front panel, as shown in Fig. F–29. The buttons are "pushed" by a left mouse click. The dial can be turned by dragging the mouse over it or by placing the cursor over it and spinning the wheel on the mouse. The latter is by far the preferred method.

There are two types of current sources: DC and AC sources. The DC current source is represented as a circle with an upward pointing arrow in it. The arrow in Fig. F–30 represents the direction of current flow. The arrow can be pointed downward by rotating the symbol 180°.

The current rating is fully adjustable. The default value is 1 A. If the component is double clicked, the configuration screen in Fig. F–31 will open up and the current value can be changed.

Figure F–30 DC current source.

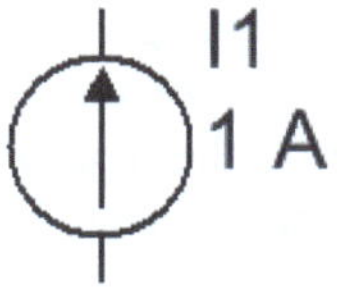

The AC current source is represented as a circle with an upward pointing arrow. There is a sine wave across the arrow. The schematic symbol in Fig. F–32 will include information about the AC current source. This information will include the device reference number, amplitude (I_{Pk}), offset, frequency, and phase shift. These values are all fully adjustable. The default values are shown in Fig. F–33. If the component is double clicked, the configuration screen will open up and the values can be changed.

Figure F–31 DC current source configuration screen.

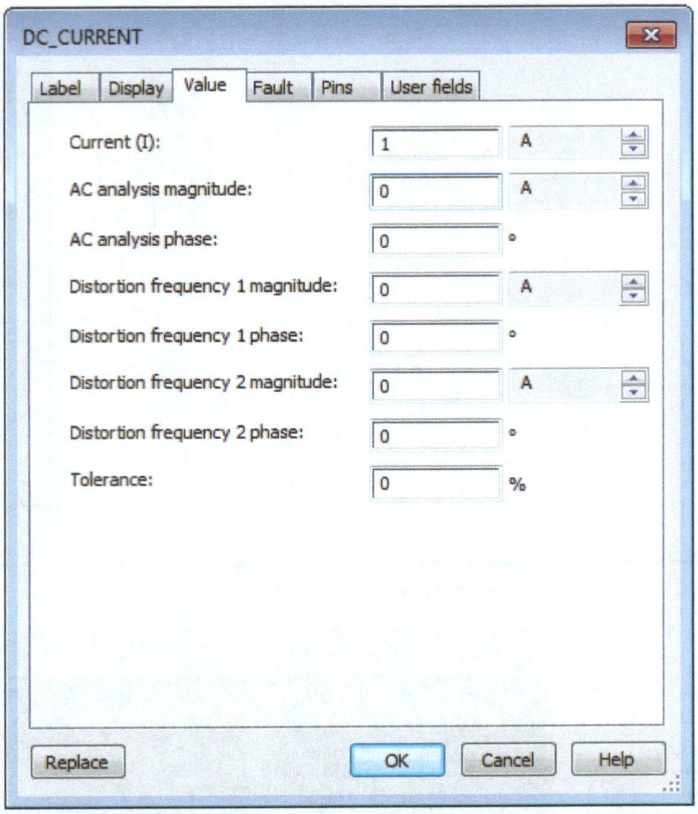

Figure F–32 AC current source.

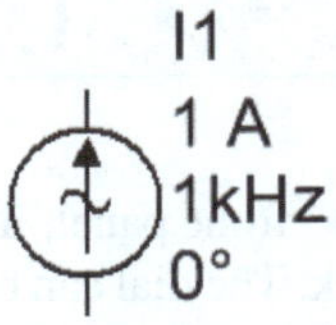

Multisim requires a ground to be present in the circuit in order for the simulation to function properly. In addition to the circuit, all instrumentation with an available ground connection must have that connection tied to ground. The circuit and instrumentation can share a common ground point or individual ground symbols may be used. The schematic symbol for ground is shown in Fig. F–34.

Measurement Equipment

Multisim provides a wide assortment of measurement equipment. In the study of DC, AC, and semiconductor electronics, the three main pieces of measurement equipment are the digital multimeter, the oscilloscope, and the Bode plotter. The first two pieces of equipment are found in test labs across the world. The Bode plotter is a virtual device that automates the task of plotting voltage gain over a wide range of frequencies. This is usually done by taking many measurements and plotting the results in a spreadsheet. The Bode plotter automates this time-consuming process.

MULTIMETERS

There are two multimeters to choose from: The generic multimeter and the Agilent multimeter. The generic multimeter can measure current, voltage, resistance, and decibels. The meter can be used for both DC and AC measurements. The different functions of the meter are selected by double clicking on the icon to the left in Fig. F–35. The double mouse click will cause the multimeter face to be displayed. The different functions on the display can be selected by pushing the different buttons via a left mouse click.

Figure F-33 AC current source configuration screen.

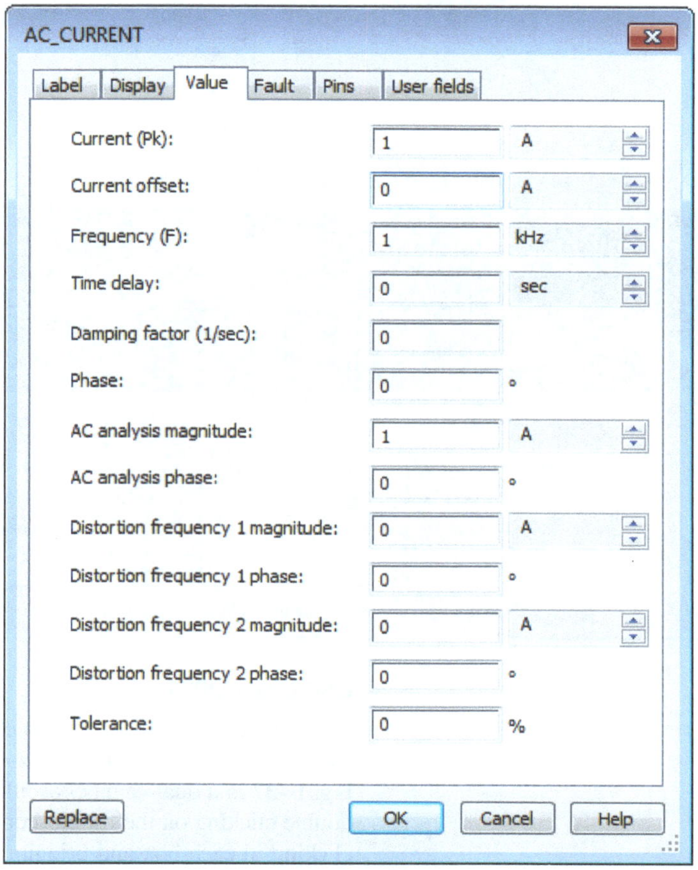

Figure F-34 Ground.

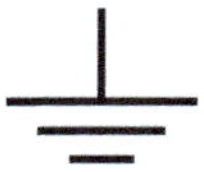

Figure F-35 Generic multimeter icon and meter display.

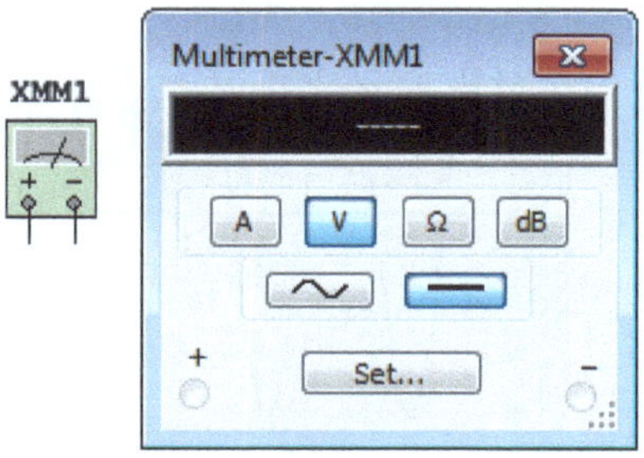

The Agilent multimeter icon and meter display are shown in Fig. F–36. The display is brought up by double clicking on the Agilent multimeter icon. This multimeter has the same functionality as the actual Agilent multimeter. The different functions are accessed by pushing the buttons. This is accomplished by left mouse clicking on the button. The input jacks on the right side of the meter display correspond to the five inputs on the icon. If something is connected to the icon, the associated jacks on the display will have a white "X" in them to show a connection.

Figure F–36 Agilent multimeter icon and meter display.

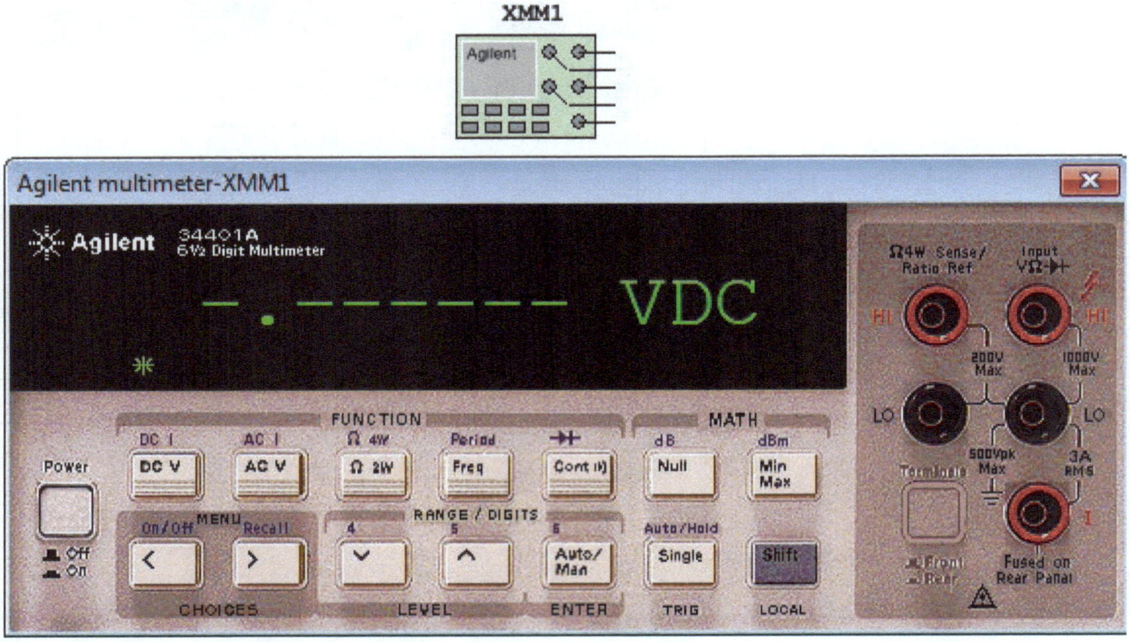

OSCILLOSCOPES

There are three oscilloscopes to choose from: The generic oscilloscope, the Agilent oscilloscope, and the Tektronix oscilloscope. The generic oscilloscope shown in Fig. F–37 is a dual channel oscilloscope. The oscilloscope display is brought up by double clicking on the oscilloscope icon. The settings can be changed by left mouse clicking in each box and bringing up the scroll arrows. The color of the traces will

Figure F–37 Generic oscilloscope icon and oscilloscope display.

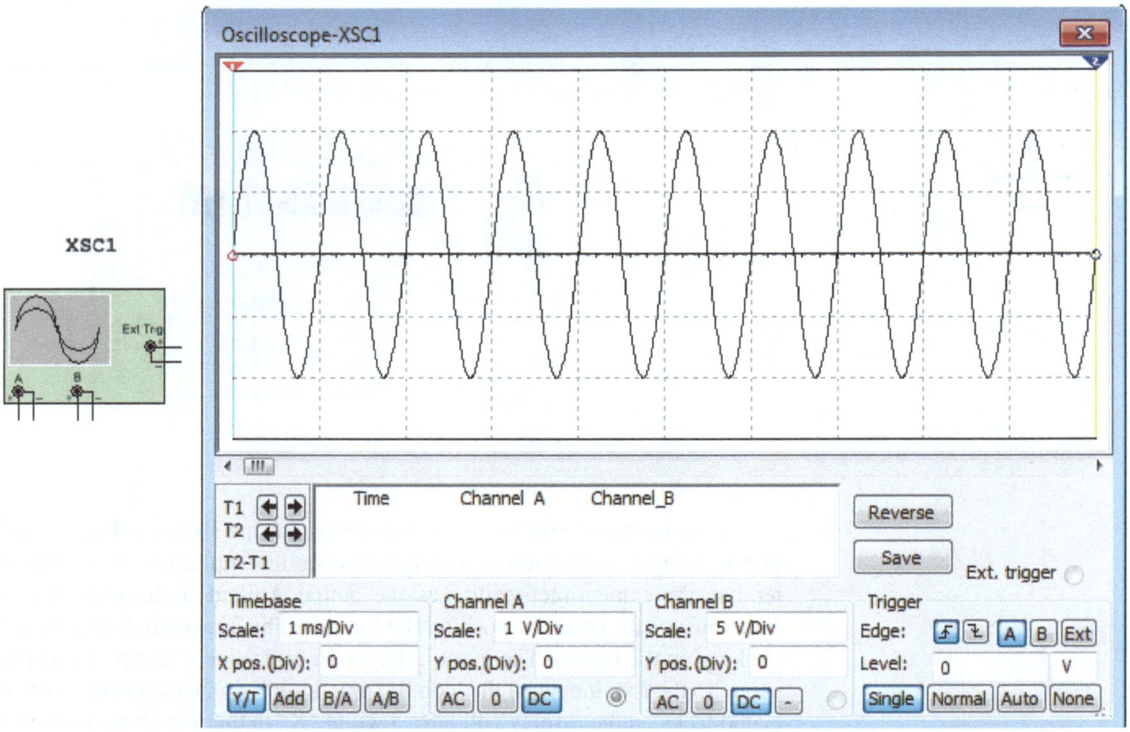

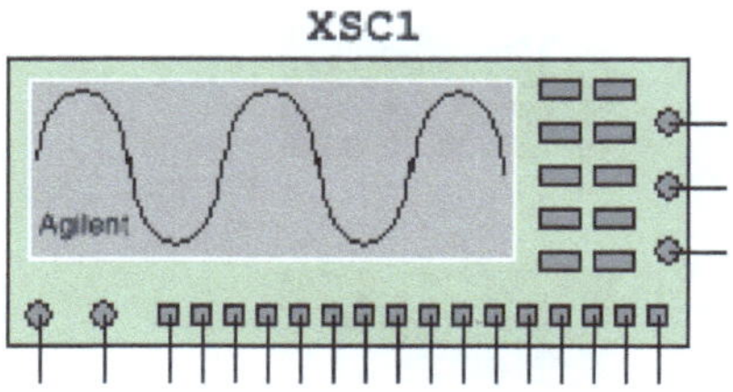

Figure F–38 Agilent oscilloscope icon.

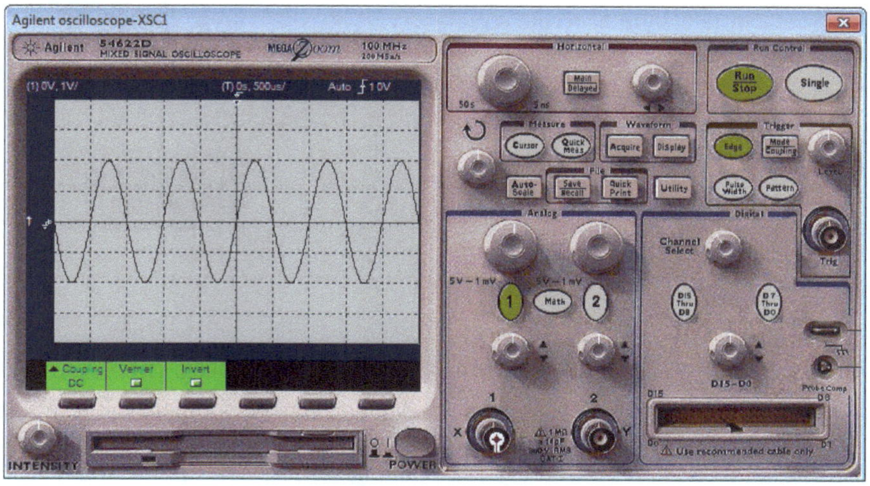

Figure F–39 Agilent oscilloscope display.

match the color of the wire segments connecting the oscilloscope to the circuit. To change the color of the trace, right mouse click on the wire segment connecting the oscilloscope to the circuit and then select "color segment" from the Properties menu. Select the color for the wire segment and corresponding trace from the color palette.

The Agilent oscilloscope icon in Fig. F–38 has all the functionality of the Model 54622D dual channel oscilloscope. The Agilent oscilloscope is controlled via the front panel, as shown in Fig. F–39. The buttons are "pushed" by a left mouse click. Each dial can be turned by dragging the mouse over it or by placing the cursor over it and spinning the wheel on the mouse. The latter is by far the preferred method.

The Tektronix oscilloscope icon shown in Fig. F–40 has all of the functionality of the Model TDS2024 four channel digital storage oscilloscope. The colors of the four channels are the same as the channel selection buttons on the display: Yellow, blue, purple, and green for channels one through four respectively. The Tektronix oscilloscope is controlled via the front panel, as seen in Fig. F–41. The buttons are "pushed" by a left mouse click. Each dial can be turned by dragging the mouse over it or by placing the cursor over it and spinning the wheel on the mouse. The latter is by far the preferred method.

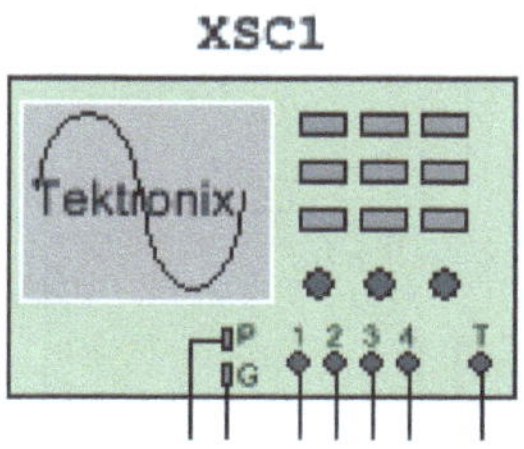

Figure F–40 Tektronix oscilloscope icon.

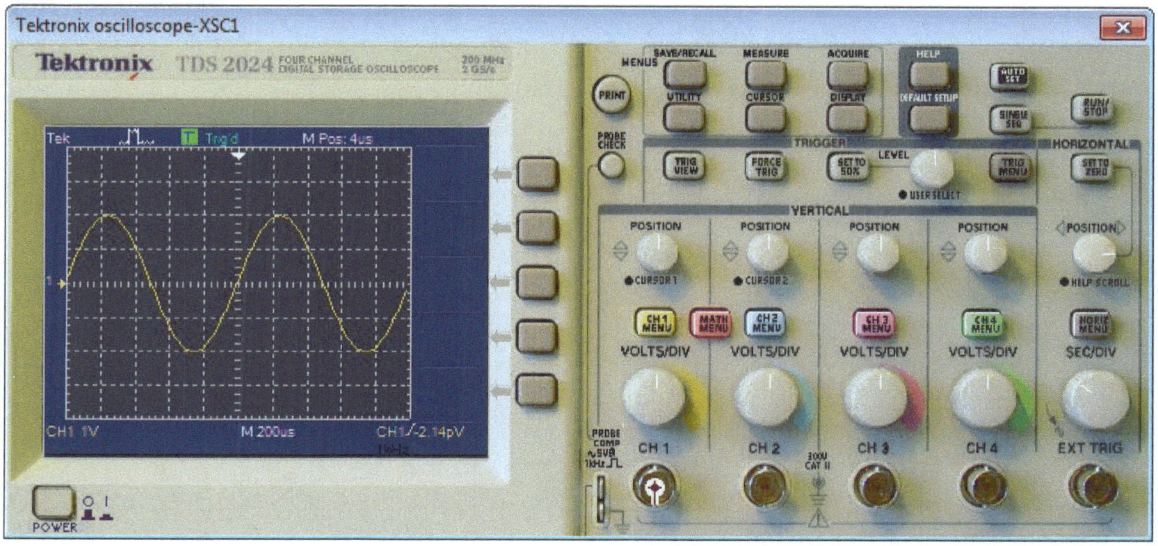

Figure F–41 Tektronix oscilloscope display.

Figure F–42 Voltage and current meters.

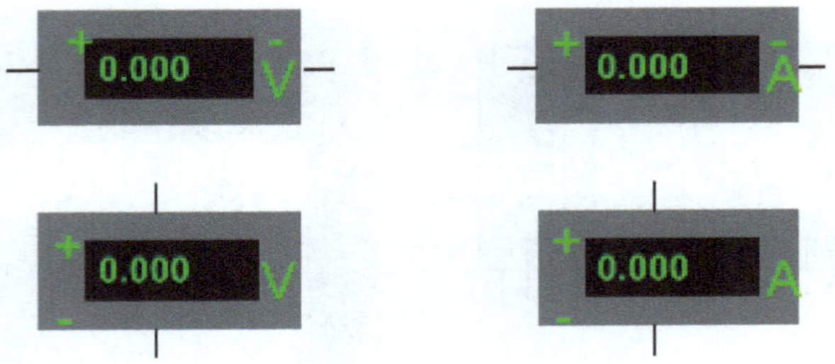

VOLTAGE AND CURRENT METERS

Multisim provides simple voltmeters and ammeters, as shown in Fig. F–42, for use when voltage or current need to be measured. These meters can be placed throughout the circuit. The meters are available in both vertical and horizontal orientation to match the layout of the circuit. The default is "DC." If the meters are to be used for AC measurement, then the configuration screen shown in Fig. F–43 must be opened and that parameter changed to reflect AC measurement. To open the configuration screen, double mouse click on the meter.

BODE PLOTTER

The Bode plotter is used to view the frequency response of a circuit. In the actual lab setting, the circuit would be operated at a base frequency and the output of the circuit measured. The frequency would be incremented by a fixed amount and the

Figure F–43 Voltmeter configuration screen.

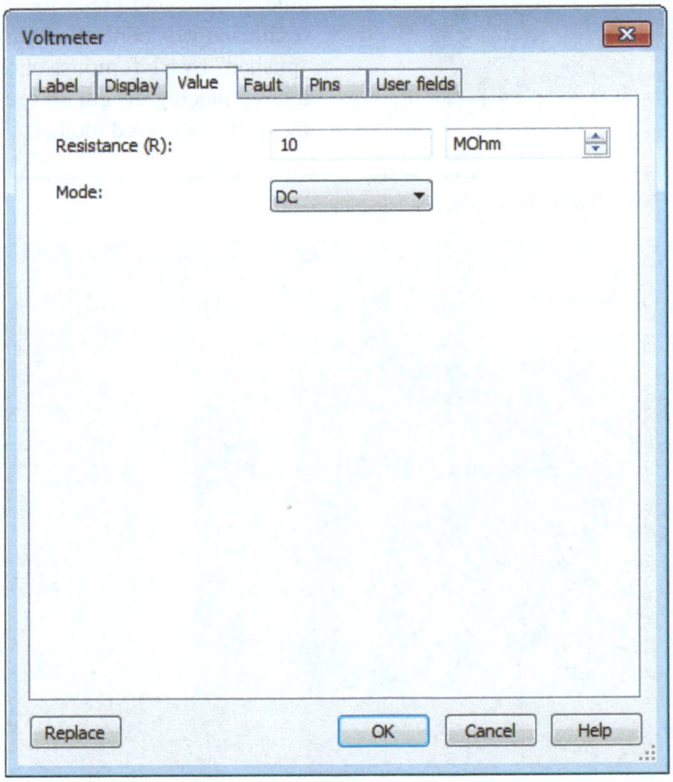

Figure F–44 Bode plotter.

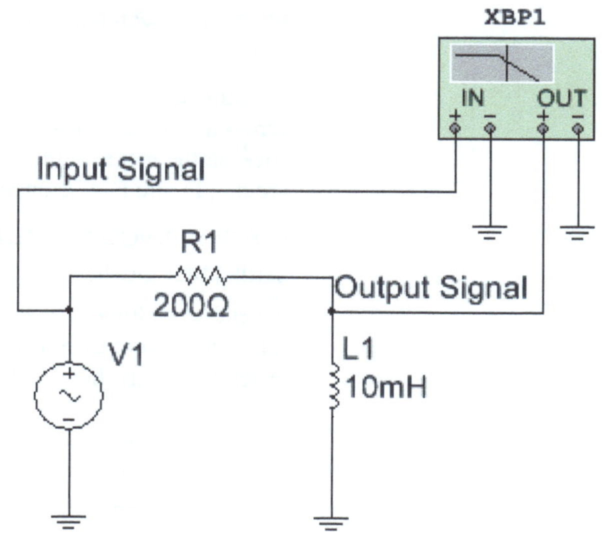

measurement repeated. After operating the circuit at a sufficient number of incremental frequencies, the data would be graphed, with independent variable "frequency" on the x-axis and dependent variable "amplitude" on the y-axis. This process can be very time-consuming. Multisim provides a simpler method of determining the frequency response of a circuit through the use of the virtual Bode plotter.

In Fig. F–44, the positive terminal of the input is connected to the applied signal source. The positive terminal of the output is connected to the output signal of the circuit. The other two terminals are connected to ground. The amplitude or frequency settings of the AC source do not matter; the AC source just needs to be in the circuit. The Bode plotter will provide the input signal, as seen in Fig. F-45.

Figure F–45 Bode plotter display.

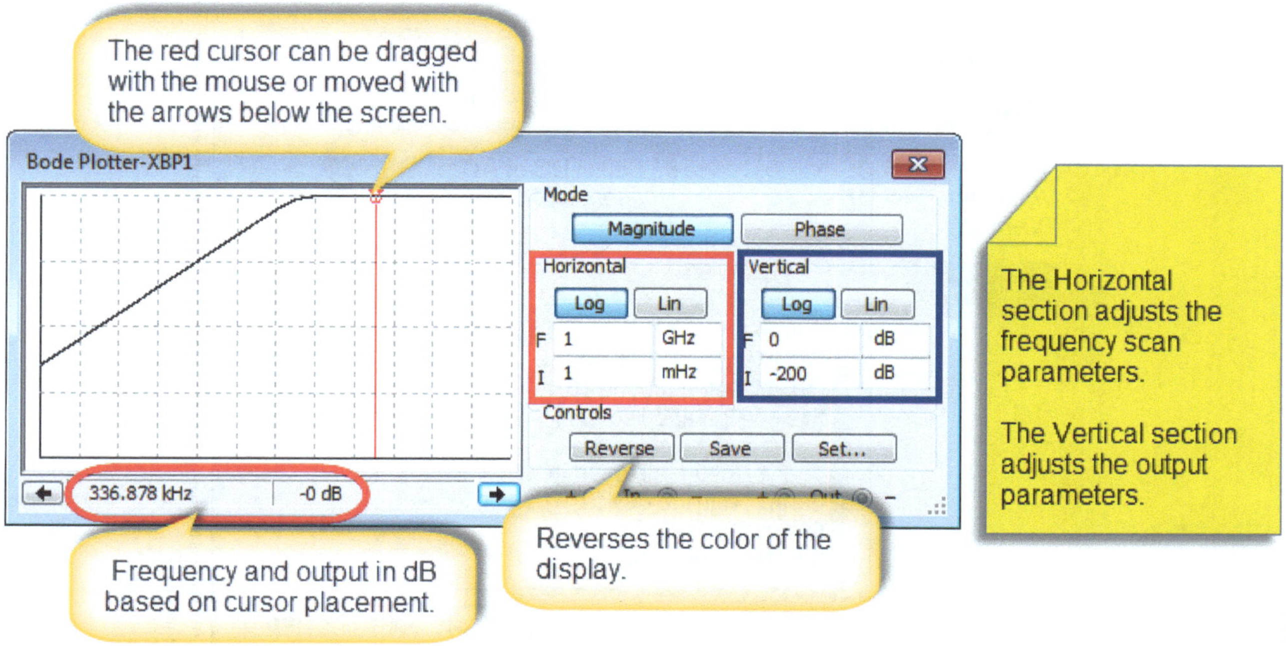

Circuit Examples

EXAMPLE 1: VOLTAGE MEASUREMENT USING A VOLTMETER IN A SERIES DC CIRCUIT.

A voltmeter in Fig. F–46 is placed in parallel with the resistor to measure the voltage across it. The default is set for "DC" measurement. If AC is required, double mouse click on the meter to bring up the configuration screen. Every circuit must have a ground. Figure F–47 contains a *Quick Hint* on the use of the voltmeter.

EXAMPLE 2: VOLTAGE MEASUREMENT USING A GENERIC MULTIMETER IN A SERIES DC CIRCUIT.

A generic multimeter is placed in parallel with the resistor to measure the voltage across it. Be sure to double mouse click the generic multimeter icon to bring up the meter display, as shown in Fig. F–48. Press the appropriate buttons for "Voltage" and

Figure F–46 DC voltage measurement with a voltmeter.

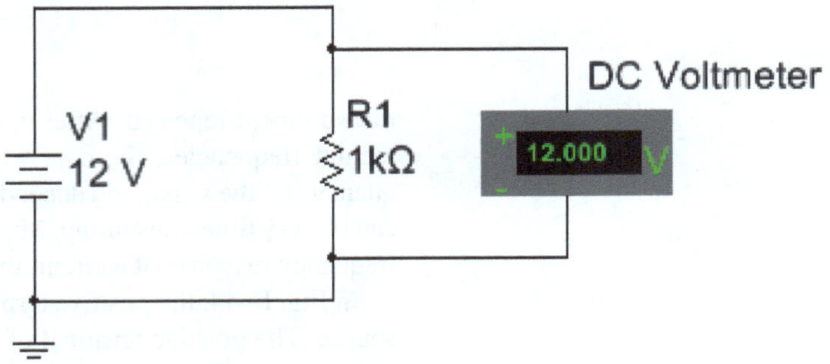

Figure F–47 Voltmeter Quick Hint.

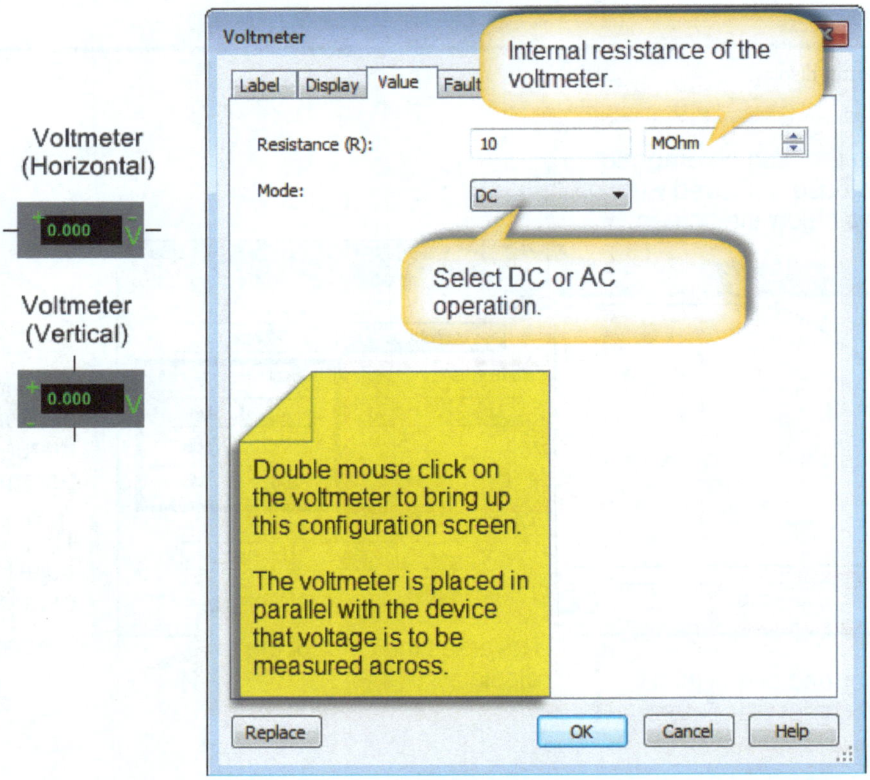

Figure F–48 DC voltage measurement with a generic multimeter.

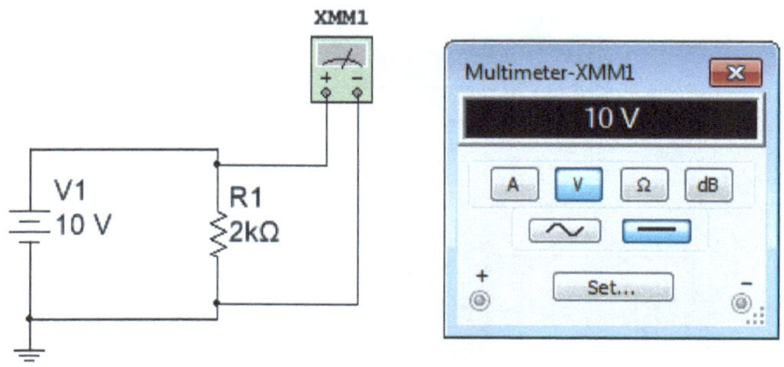

Figure F–49 Generic multimeter Quick Hint.

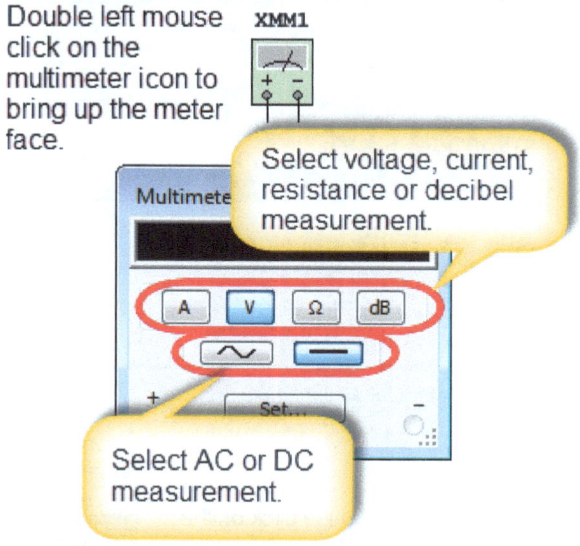

then "DC" or "AC" measurement. Every circuit must have a ground. Figure F–49 contains a Quick Hint on the use of the generic multimeter.

EXAMPLE 3: VOLTAGE MEASUREMENT USING AN AGILENT MULTIMETER IN A SERIES DC CIRCUIT.

In Fig. F–50, an Agilent multimeter is placed in parallel with the resistor to measure the voltage across it. Be sure to double click the Agilent multimeter icon to bring up the meter display. Press the appropriate buttons for "Voltage" and then "DC" or "AC" measurement. Every circuit must have a ground. Note the two white circles and black X's on the right side of the display to indicate a connection to the meter. This instrument requires that its power button be pressed to "turn on" the meter. Figure F–51 contains a *Quick Hint* on the use of the Agilent multimeter.

Figure F–50 DC voltage measurement with an Agilent multimeter.

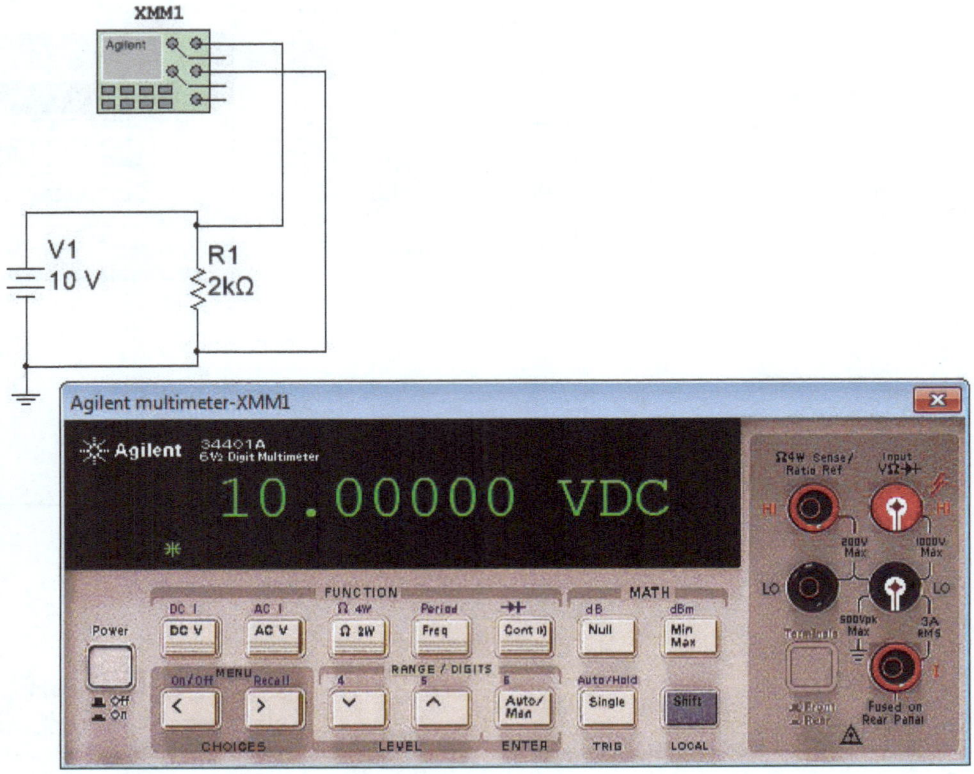

Figure F–51 Agilent multimeter Quick Hint.

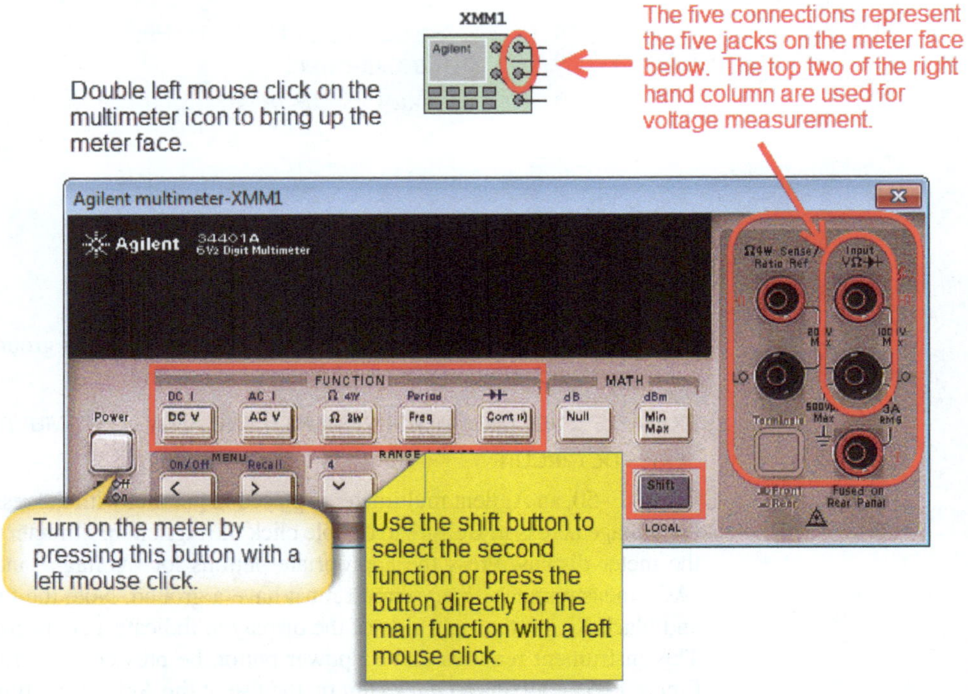

Figure F–52 DC current measurement with an ammeter.

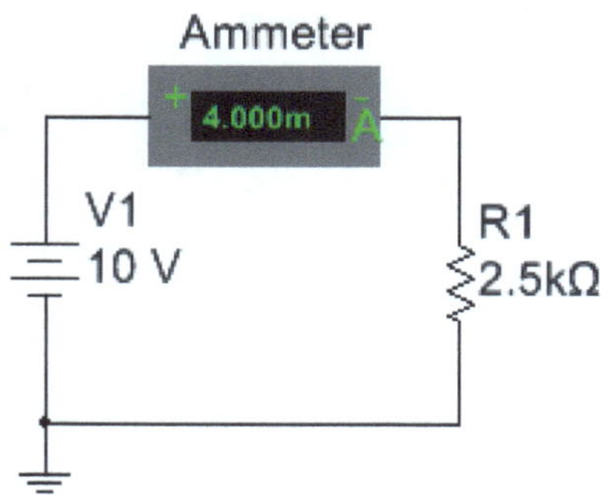

EXAMPLE 4: CURRENT MEASUREMENT USING AN AMMETER IN A SERIES DC CIRCUIT.

In Fig. F–52, an ammeter is placed in series with the resistor and DC source to measure the current flowing through the circuit. The default is set for "DC" measurement. If AC is required, double click on the meter to bring up the configuration screen. Every circuit must have a ground. Figure F–53 contains a *Quick Hint* on the use of the ammeter.

EXAMPLE 5: CURRENT MEASUREMENT USING A GENERIC MULTIMETER IN A SERIES DC CIRCUIT.

In Fig. F–54, a generic multimeter is placed in series with the resistor and DC source to measure the current flowing through the circuit. Be sure to double click the generic multimeter icon to bring up the meter display. The current function is selected by clicking on the "A" on the meter display. Since the source is DC, the DC function of the meter is also selected, as indicated by the depressed button. Every circuit must have a ground. Figure F–55 contains a *Quick Hint* on the use of the generic multimeter.

Figure F–53 Ammeter Quick Hint.

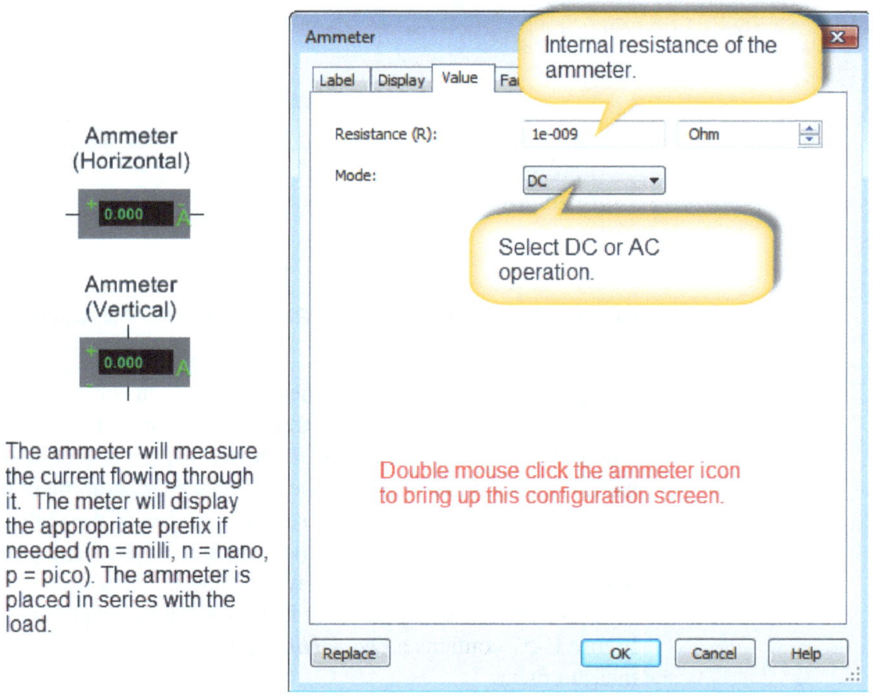

Figure F–54 DC current measurement with a generic multimeter.

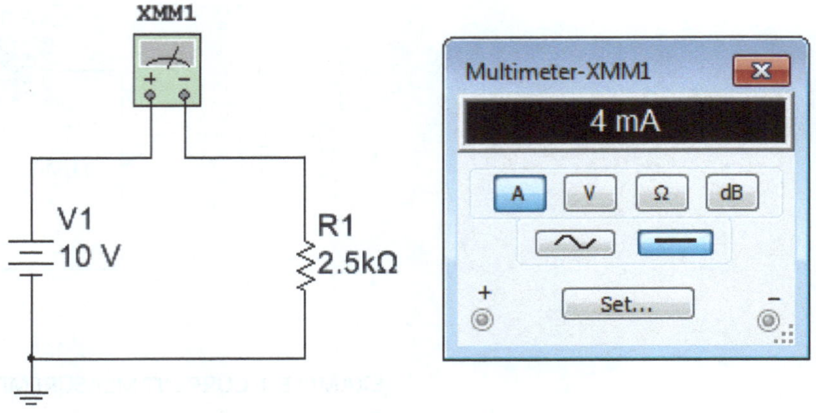

Figure F–55 Generic multimeter Quick Hint.

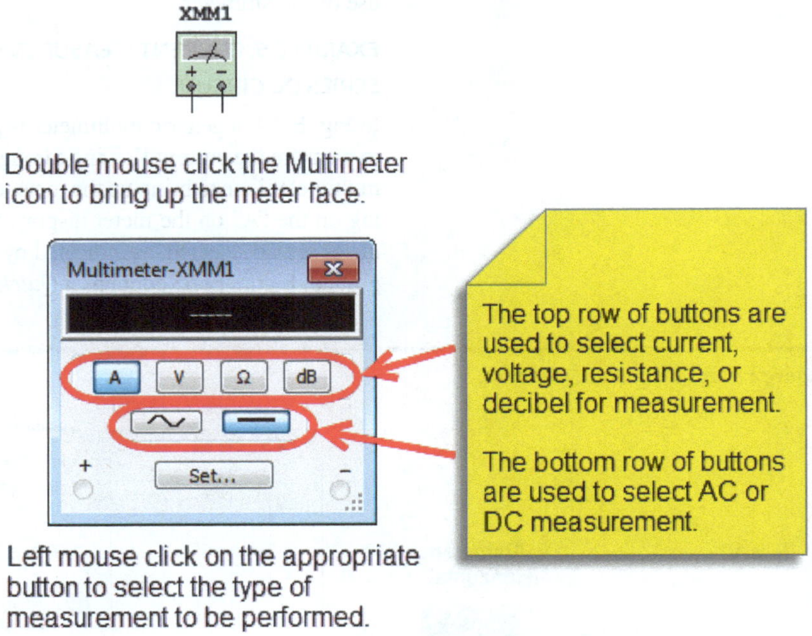

EXAMPLE 6: CURRENT MEASUREMENT USING AN AGILENT MULTIMETER IN A SERIES DC CIRCUIT.

In Fig. F–56, an Agilent multimeter is placed in series with the resistor and source to measure the current flowing through the circuit. Be sure to double click the Agilent multimeter icon to bring up the meter display. Selection of DC current measurement is the second function of the DC voltage measurement button. Be sure to press the "Shift" button to access the second function of the voltage button. Note the two white circles and black X's on the right side of the display to indicate a connection to the meter. All circuits must have a ground.

This instrument requires that its power button be pressed to "turn on" the meter. Figure F–57 contains a *Quick Hint* on the use of the Agilent multimeter for current measurement.

Figure F–56 DC current measurement with an Agilent multimeter.

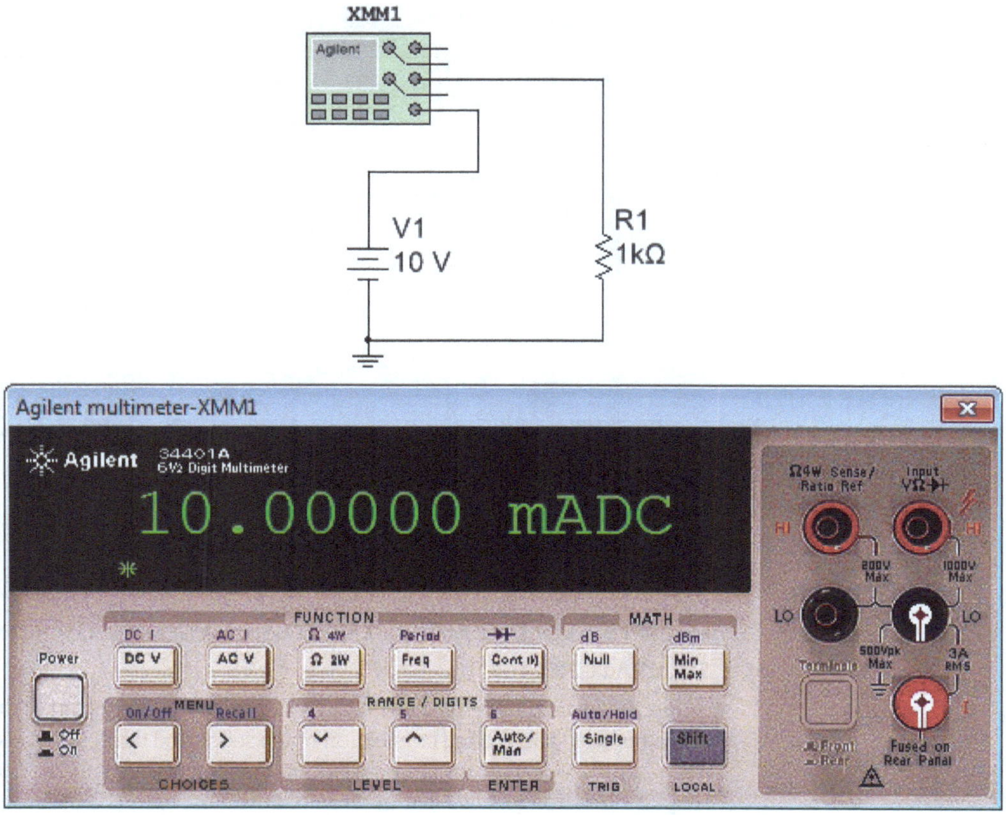

Figure F–57 Agilent multimeter Quick Hint for current measurement.

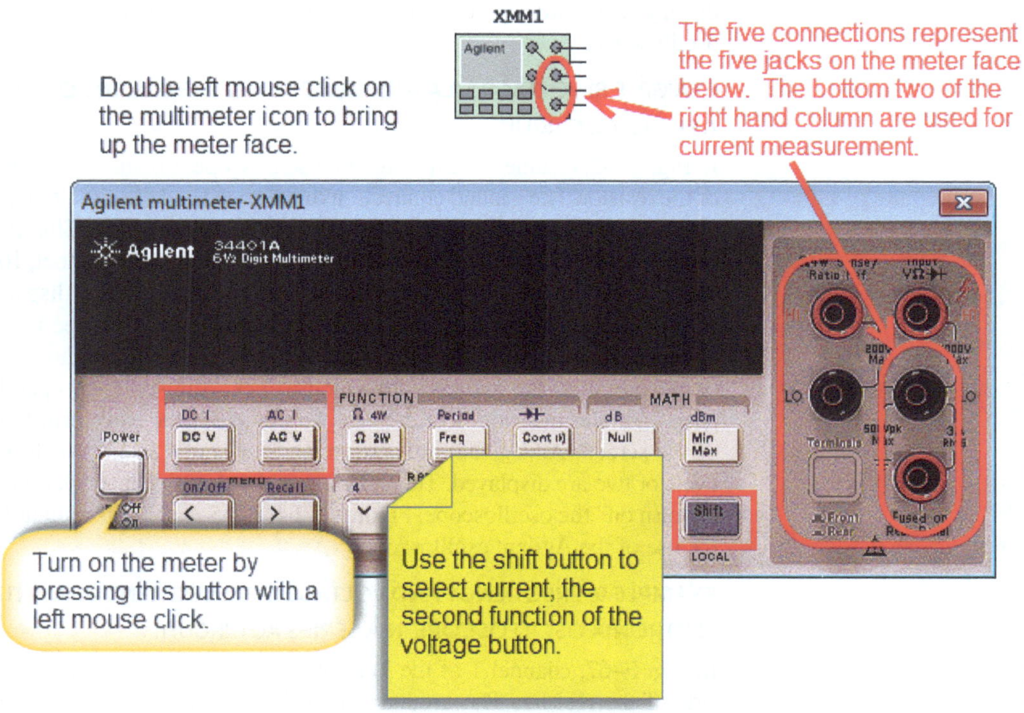

Figure F–58 Voltage measurement with a generic oscilloscope.

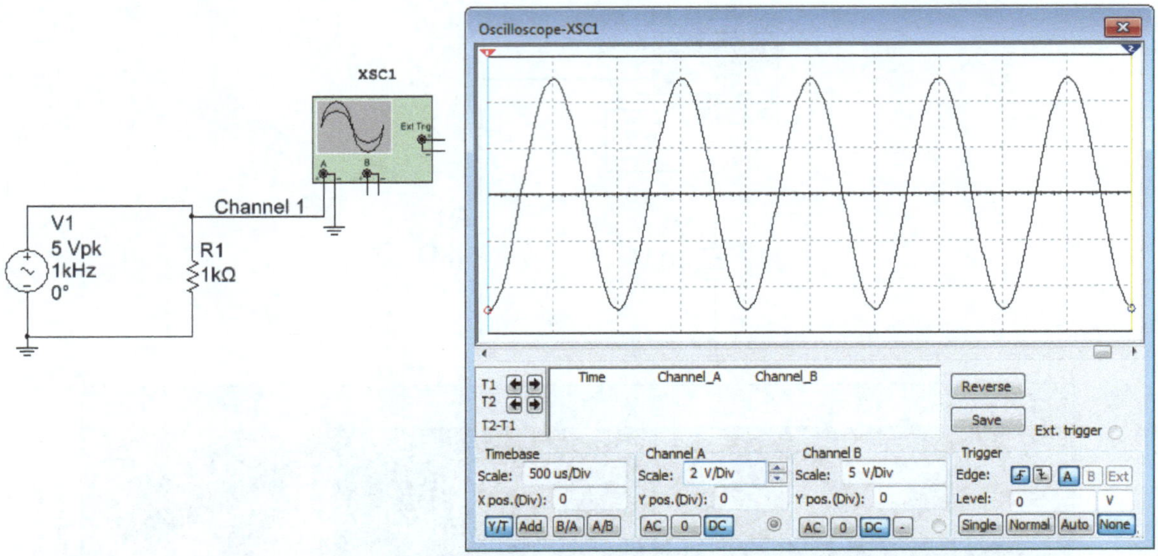

Figure F–59 Generic oscilloscope icon Quick Hint.

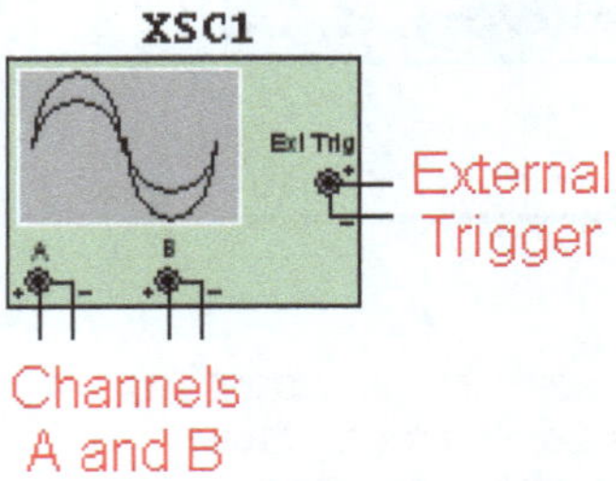

EXAMPLE 7: VOLTAGE MEASUREMENT USING A GENERIC OSCILLOSCOPE IN A SERIES AC CIRCUIT.

In Fig. F–58, channel 1 of the generic oscilloscope is connected to the positive side of the resistor. The ground connection of the scope and the circuit must be grounded. The oscilloscope will use this ground as a reference point. This generic oscilloscope's operation follows that of an actual oscilloscope. The oscilloscope display is brought up by mouse clicking on the oscilloscope icon. The settings can be changed by clicking in each box and bringing up the scroll arrows. Adjust the volts per division on the channel under measurement until the amplitude of the waveform fills the majority of the screen. Adjust the timebase such that a complete cycle or two are displayed. Figures 59, 60, and 61 contain *Quick Hints* on the use of the generic oscilloscope.

EXAMPLE 8: VOLTAGE MEASUREMENT USING THE AGILENT OSCILLOSCOPE IN A SERIES AC CIRCUIT.

In Fig. F–62, channel 1 of the Agilent oscilloscope is connected to the positive side of the resistor. The ground connection of the scope and the circuit must be grounded. The oscilloscope will use this ground as a reference point. This Agilent oscilloscope's operation follows that of a dual channel, +16 logic channel, 100-MHz bandwidth Agilent Model 54622D oscilloscope. The oscilloscope display, as shown in Fig. F–63, is brought up by mouse clicking on the oscilloscope icon. The settings can be changed by placing the mouse over the dials and spinning the mouse wheel or by "pressing" the buttons with a mouse click. Adjust the volts per division on the channel under measurement until the amplitude of the waveform fills the majority of the screen. Adjust the timebase in the Horizontal section such that a complete cycle or two are displayed. This instrument requires that its power button be pressed to "turn on" the oscilloscope. Figures F–64, F–65, and F–66 contain *Quick Hints* on the use of the Agilent oscilloscope.

EXAMPLE 9: FREQUENCY AND VOLTAGE MEASUREMENT USING THE TEKTRONIX OSCILLOSCOPE IN A SERIES AC CIRCUIT.

In Fig. F–67, channel 1 of the Tektronix oscilloscope is connected to the positive side of the resistor. The ground connection of the scope and the circuit must be grounded. The oscilloscope will use this ground as a reference point. The Tektronix

Figure F-60 Generic oscilloscope Quick Hint.

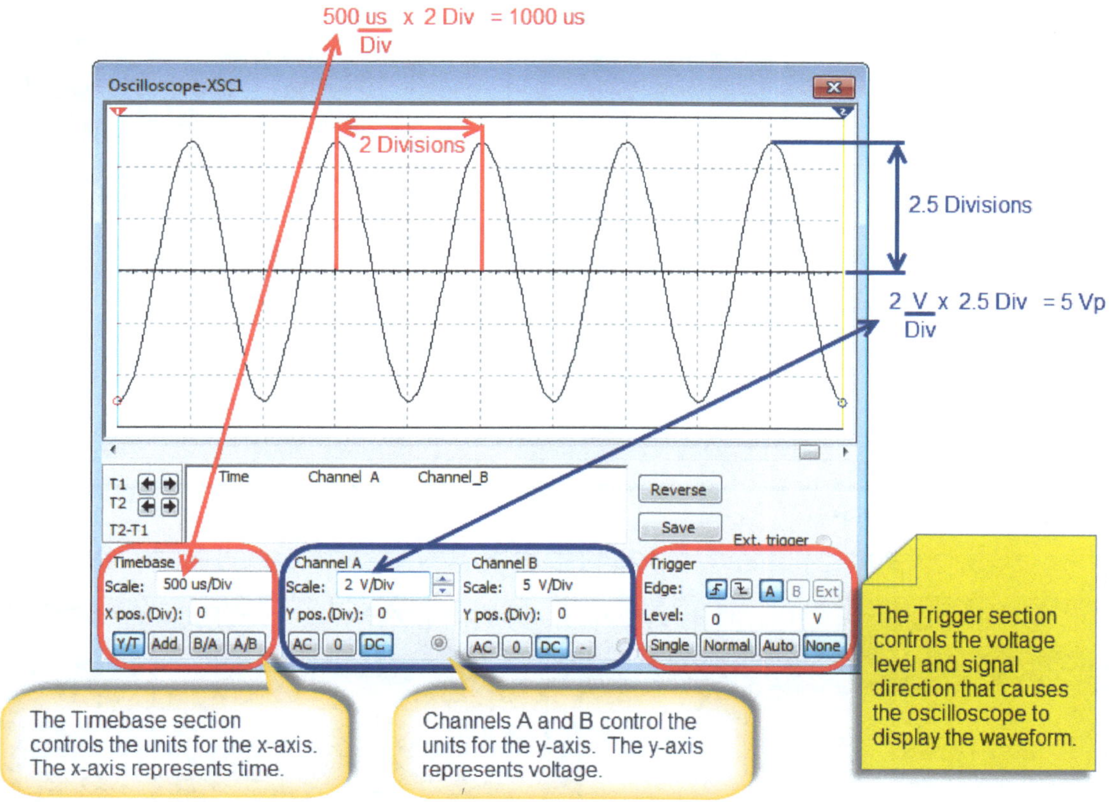

Figure F-61 Generic oscilloscope Quick Hint.

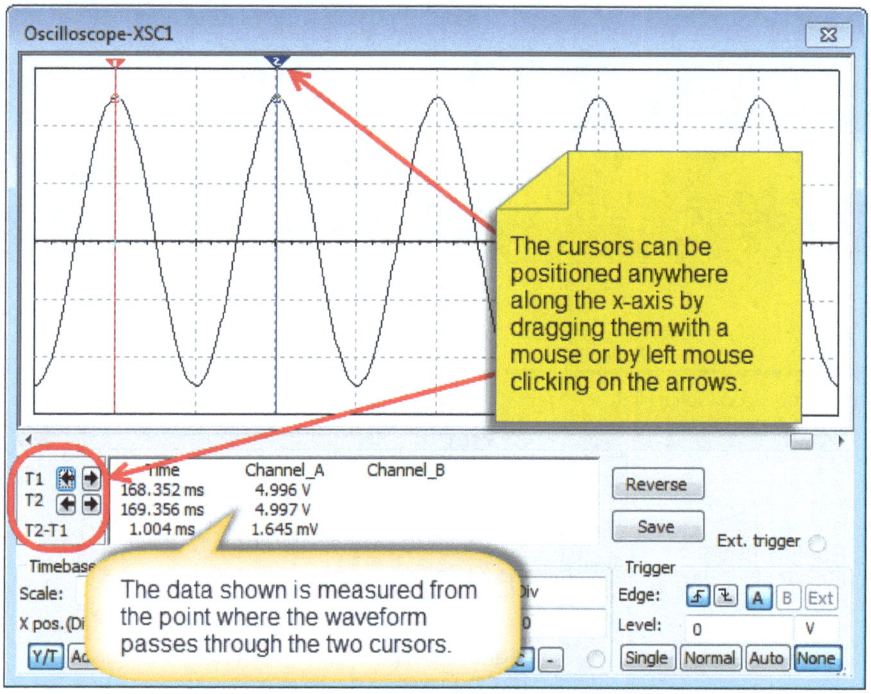

Figure F–62 Voltage measurement with an Agilent oscilloscope.

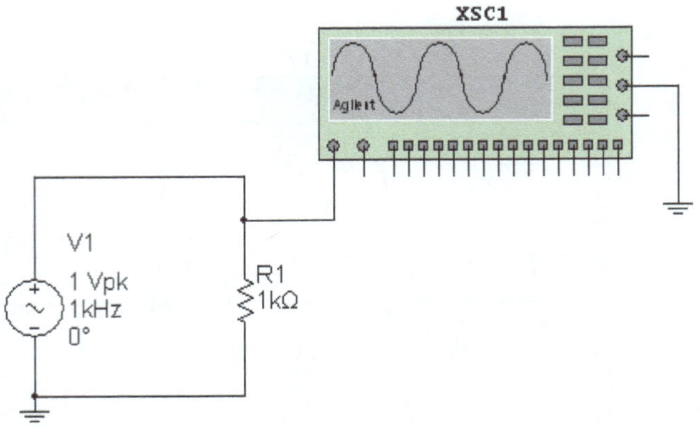

Figure F–63 Agilent oscilloscope display.

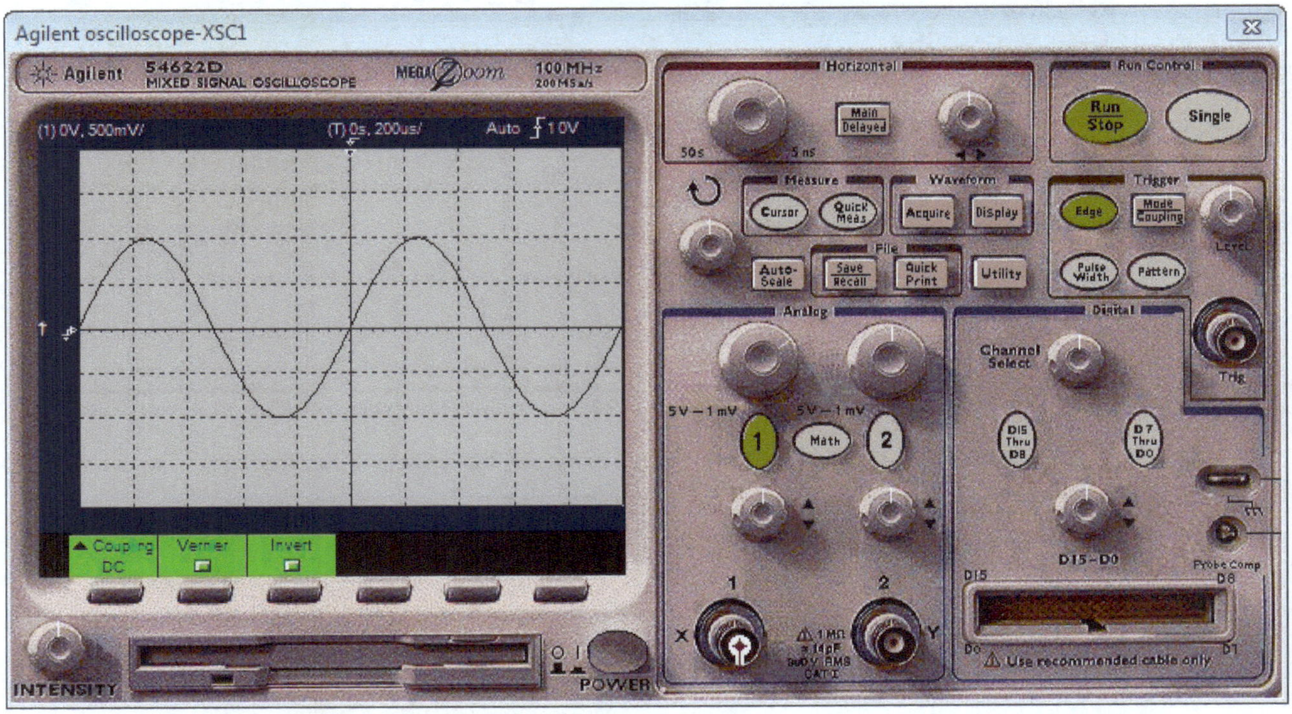

Figure F–64 Agilent oscilloscope icon Quick Hint.

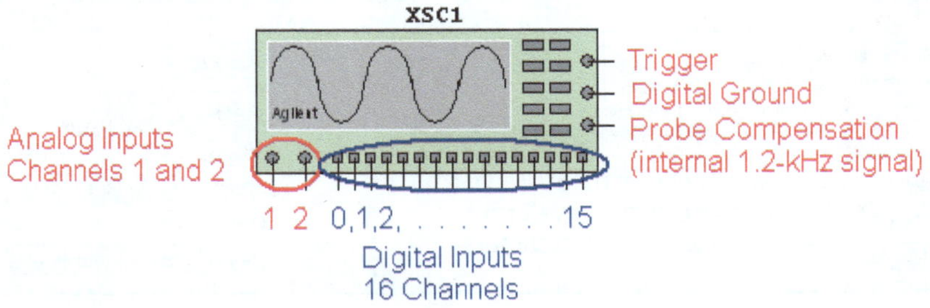

Figure F–65 Agilent oscilloscope Quick Hint.

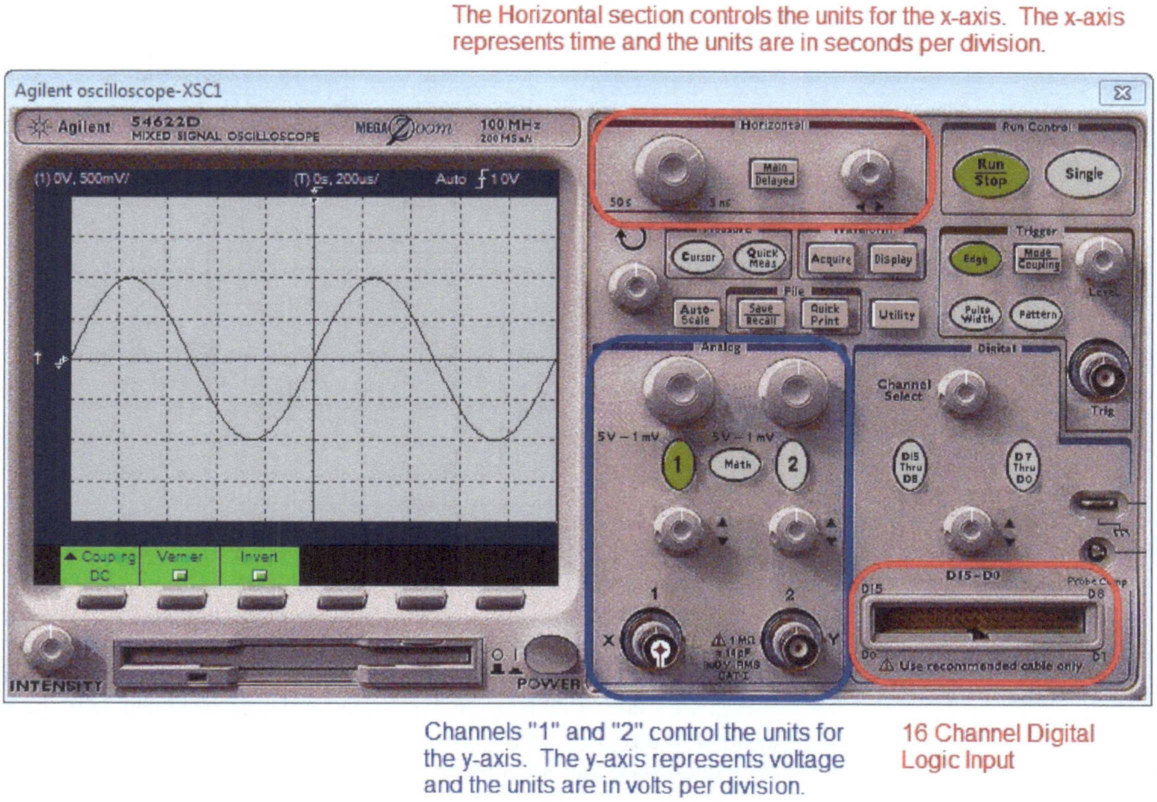

Figure F–66 Agilent oscilloscope Quick Hint.

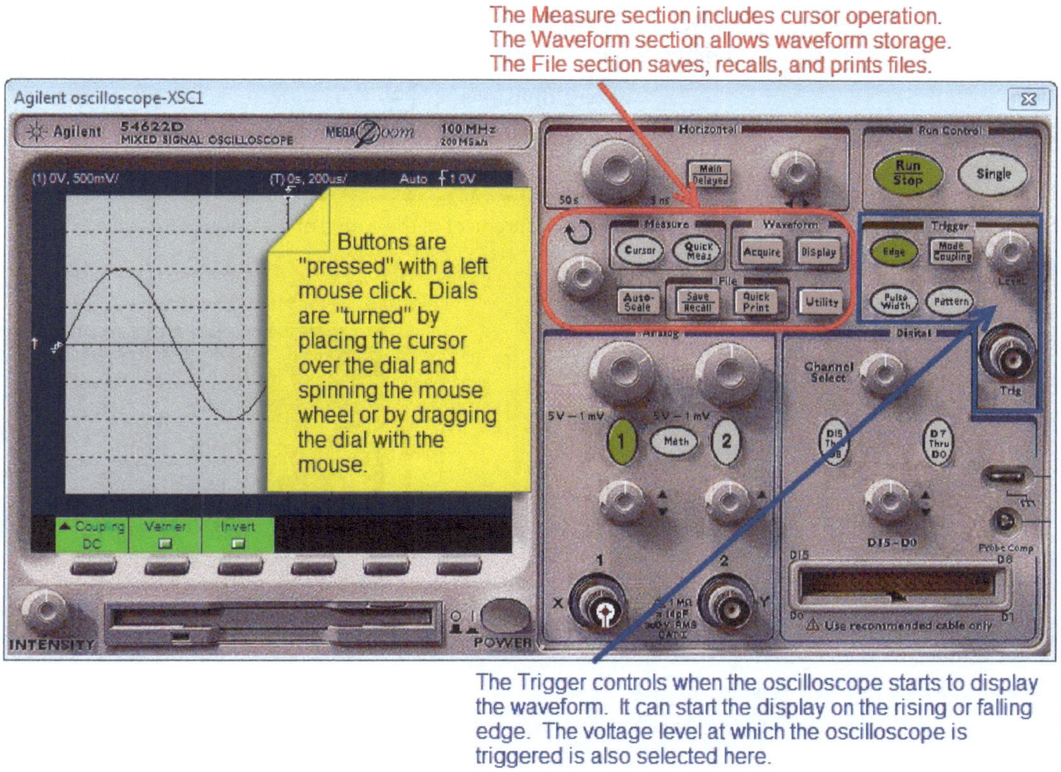

Figure F–67 Voltage and frequency measurement with a Tektronix oscilloscope.

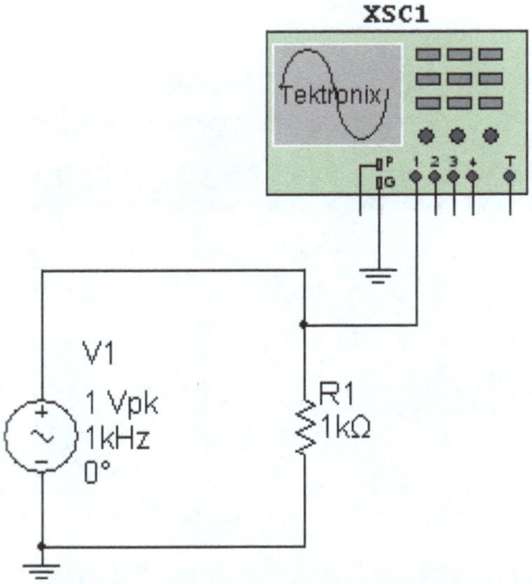

oscilloscope has all of the functionality of the Model TDS2024 four channel digital storage oscilloscope. The oscilloscope display is brought up by mouse clicking on the oscilloscope icon. The settings can be changed by placing the mouse over the dials and spinning the mouse wheel or by "pressing" the buttons with a mouse click. Adjust the volts per division on the channel under measurement until the amplitude of the waveform fills the majority of the screen. Adjust the timebase in the Horizontal section such that a complete cycle or two of the waveform is displayed. This instrument requires that its power button be pressed to "turn on" the oscilloscope. The voltage and frequency can be measured by the user or by using the "Measure" function of the oscilloscope.

Using volts per division and the seconds per division settings, the amplitude and frequency of the waveform in Fig. F–68 can be determined. The amplitude of the waveform is two divisions above zero volts. (The yellow arrow points to the zero reference point.) The volts per division setting is set to 500 mV per division.

$$V_p = 2 \text{ divisions} \times \frac{500 \text{ mV}}{\text{division}}$$

$$V_p = 1 \text{ V}$$

Figure F–68 Measurement of the period of the waveform.

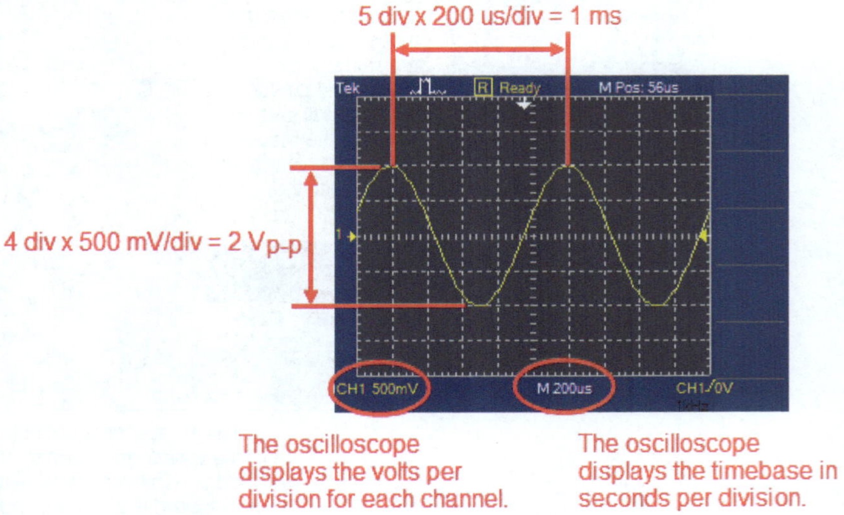

$$V_{p\text{-}p} = 4 \text{ divisions} \times \frac{500 \text{ mV}}{\text{division}}$$
$$V_{p\text{-}p} = 2 \text{ V}$$

The period of the waveform is measured to be 1 ms. Since frequency is the reciprocal of the period, the frequency can be calculated.

$$f = \frac{1}{T}$$
$$f = \frac{1}{1 \text{ ms}}$$
$$f = 1 \text{ kHz}$$

The Tektronix oscilloscope can also perform the voltage and frequency measurements automatically through the use of the "Measure" function. The four steps and the resulting display are shown in Figs. 69 and 70. The five steps to set up the oscilloscope to measure these values automatically are:

1. Press the Measure button.
2. Select the channel to be measured.

Figure F–69 Tektronix oscilloscope measurement function set-up.

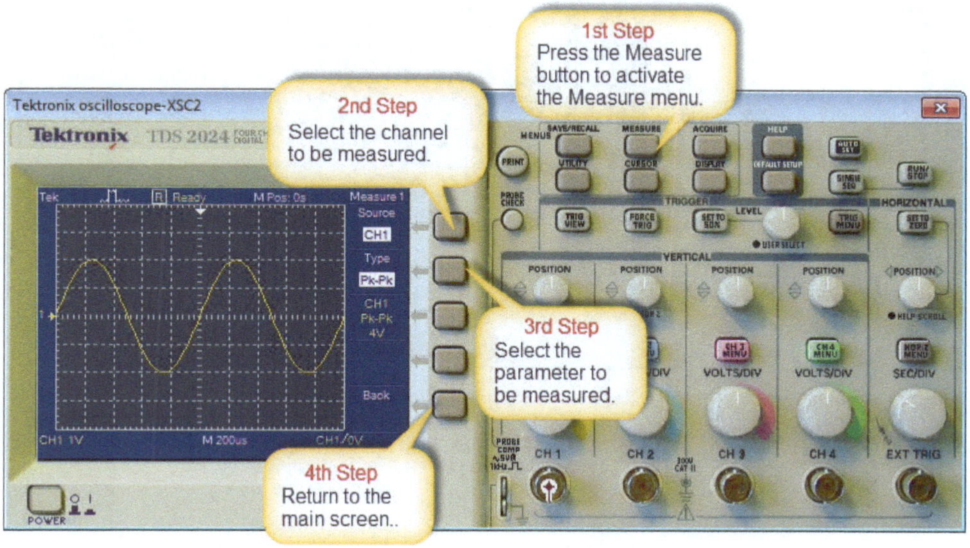

Figure F–70 Tektronix oscilloscope measurement display.

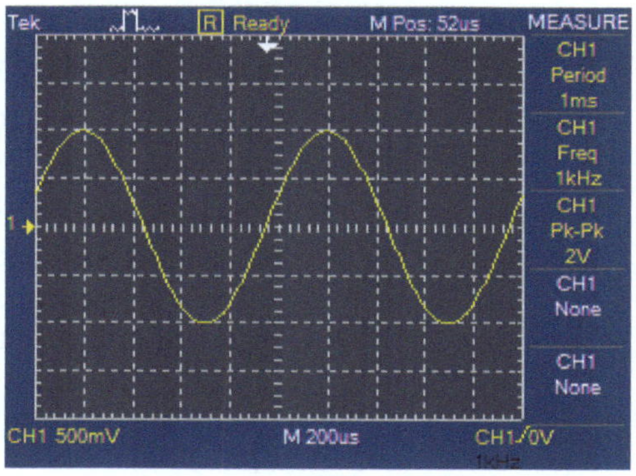

Figure F–71 Tektronix oscilloscope icon Quick Hint.

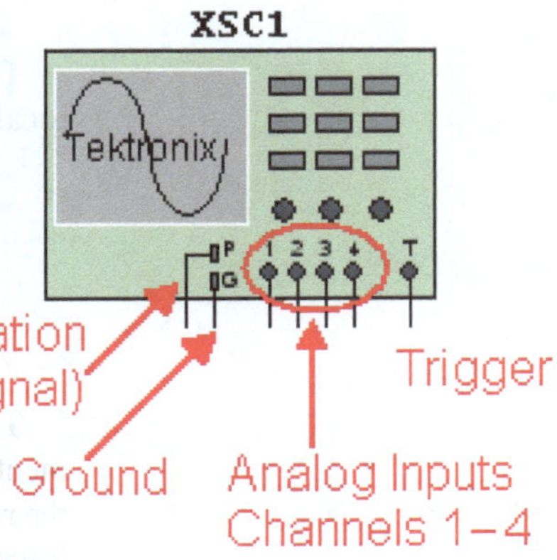

Figure F–72 Tektronix oscilloscope Quick Hint.

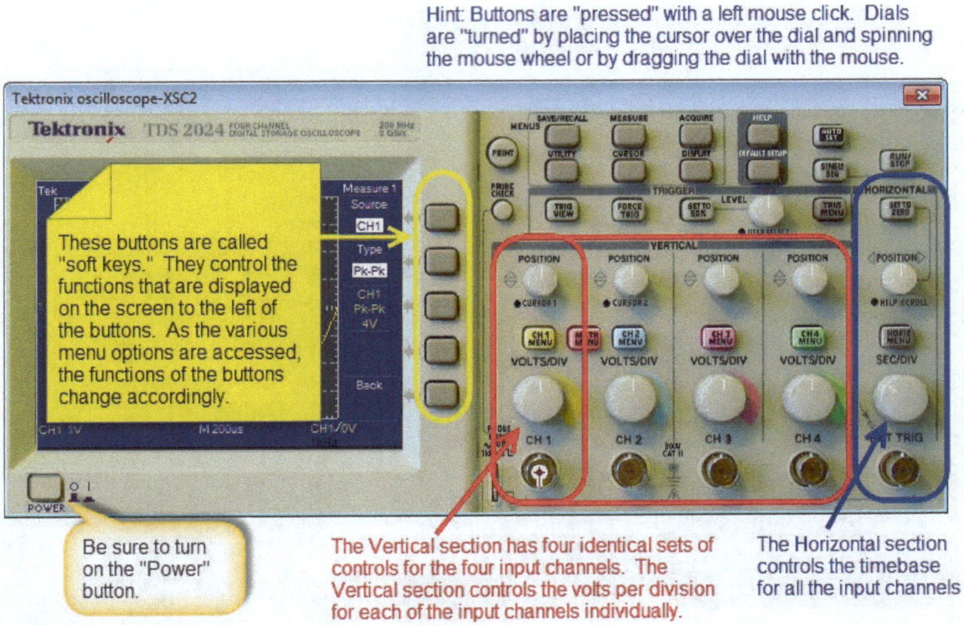

3. Select what is to be measured: $V_{p\text{-}p}$, Frequency, etc.
4. Return to the Main Screen.
5. Repeat for other channels and or values.

Figures 71, 72, and 73 contain *Quick Hints* on the use of the Tektronix oscilloscope.

User Customization

There are several changes to the base configuration of Multisim that will make it easier for you to use. Multisim uses a grid system to align the various components in the work area. When the program is first installed, the grid will be visible in the work area. It will appear as a pattern of dots, as shown in Fig. F–74.

The grid pattern can be turned off in the Sheet Properties submenu of the Options menu. The grid will still be used to align the components, however the dots will not be visible. To access the Sheet Properties submenu, either press <Alt><O> or mouse click on "Options" in the drop-down menu at the top of the screen, as shown in Fig. F–75.

Figure F–73 Tektronix oscilloscope Quick Hint.

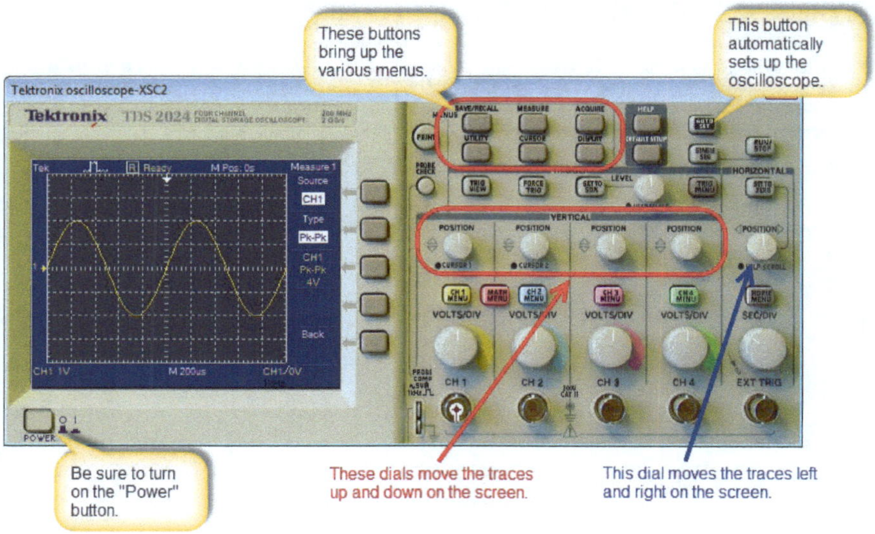

Figure F–74 The work area with the grid displayed.

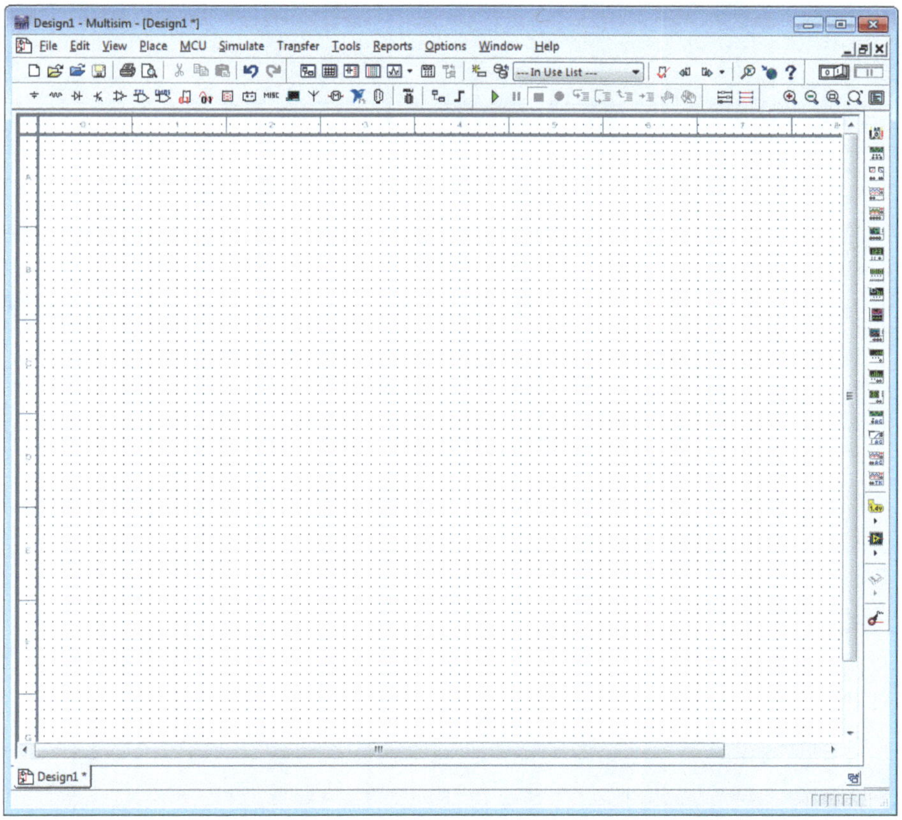

The Sheet Properties submenu shown in Fig. F–76 will allow the user to control the work area environment. The Workspace tab gives the user access to the grid controls, the sheet size, and orientation.

Figure F–75 Access to the Sheet Properties submenu.

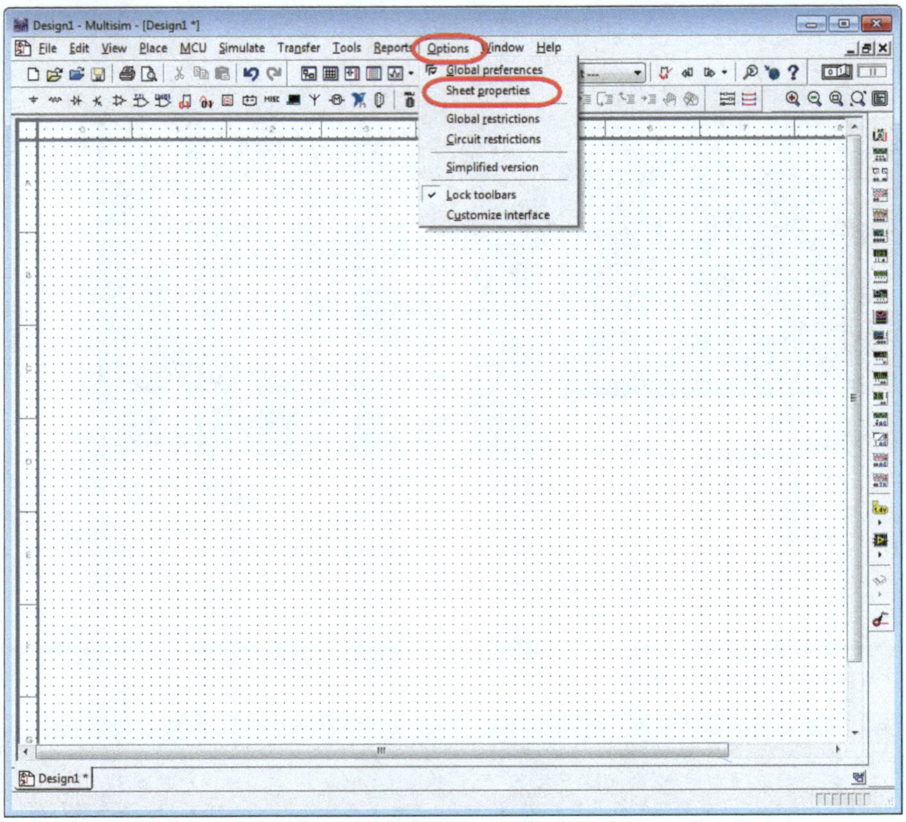

Figure F–76 Workspace tab of the Sheet Properties submenu.

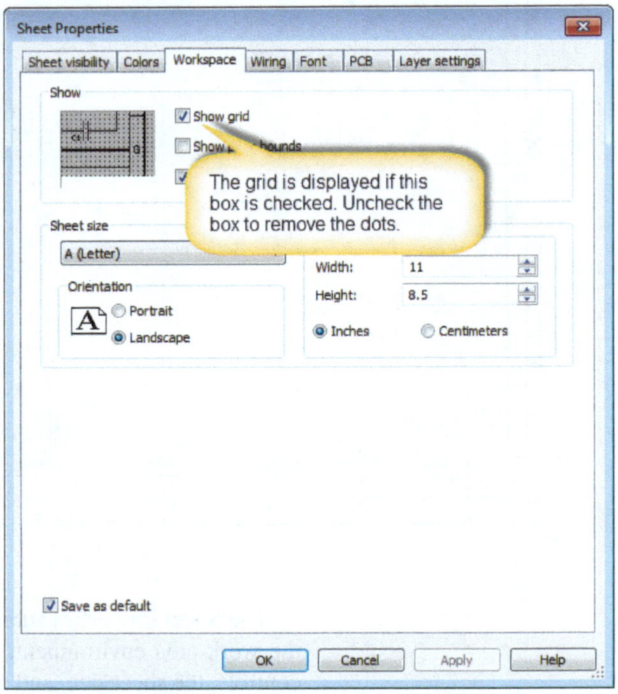

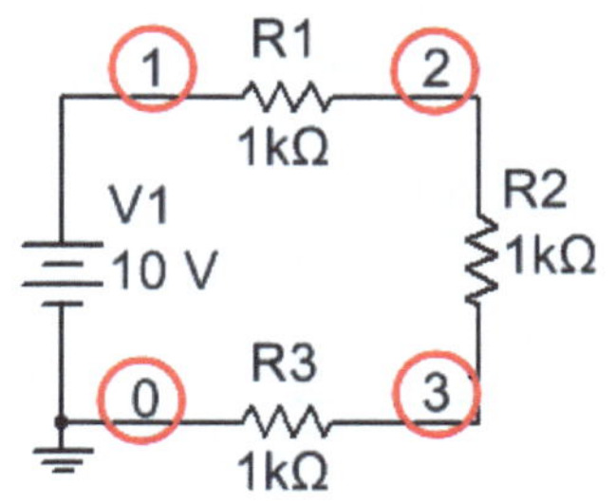

Figure F–77 Net names displayed.

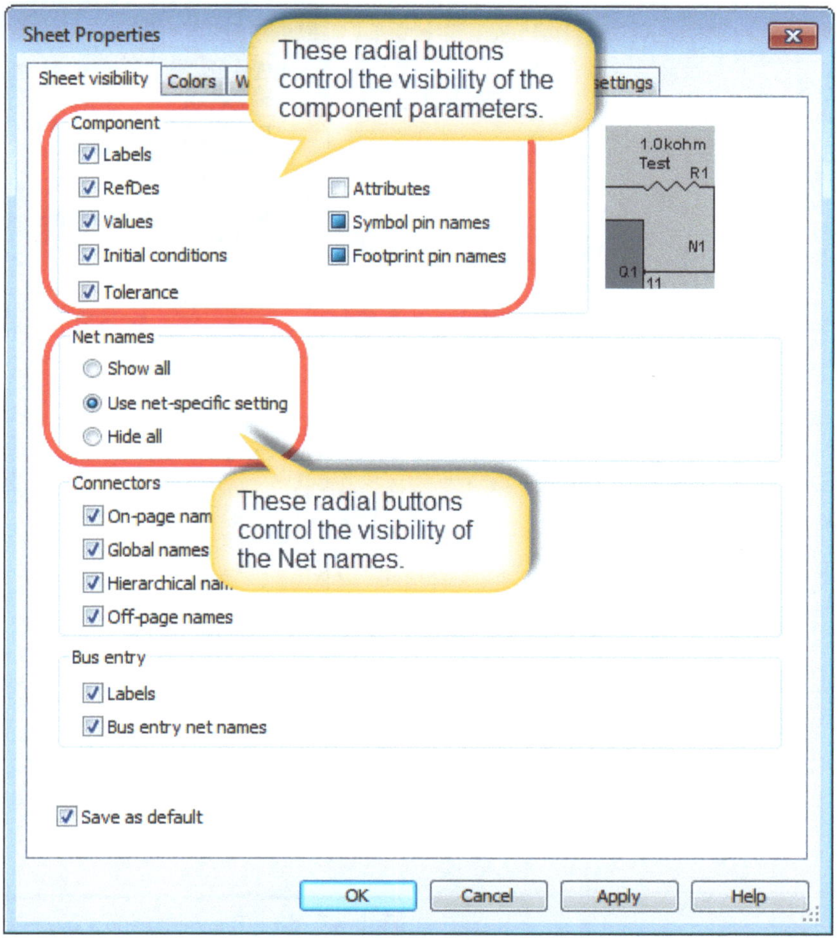

Figure F–78 The Sheet visibility tab of the Sheet Properties submenu.

The initial configuration of Multisim will display the Net names within the circuit. The Net names are the numbers circled in red in Fig. F–77.

If the Net names are a distraction, they can be turned off from within the Sheet Visibility tab of the Sheet Properties submenu, as shown in Fig. F–78. In addition to controlling the display of the Net names, this tab also gives the user access to which component parameters are displayed at the global level. The background color and the color of the various components and wiring are selected under the Colors tab of the Sheet Properties submenu, as shown in Fig. F–79. In addition to the default color combinations, a custom color scheme can be defined on this screen as well.

Exporting Data to Excel

The Bode plotter and the oscilloscope were introduced in the Measurement Equipment section. Although the displays of the Bode plotter and the oscilloscope are quite informative, it may be useful to have the data in numerical form for further analysis. After the Bode plotter simulation is run or a waveform is viewed on the oscilloscope, the data can be viewed in the Grapher. The Grapher is located in the View pull-down menu. To access the View menu, either press <Alt><V> or mouse click on "View" in the drop-down menu at the top of the screen, as shown in Fig. F–80.

If Excel is also loaded on the computer, the data used to generate the plot in the Grapher can be exported to an Excel worksheet for further analysis. The plot to be exported must be selected by mouse clicking on it. The red indicator arrow on the Grapher screen, as shown in Fig. F–81, points to the plot to be exported.

Figure F–79 The Colors tab of the Sheet Properties submenu.

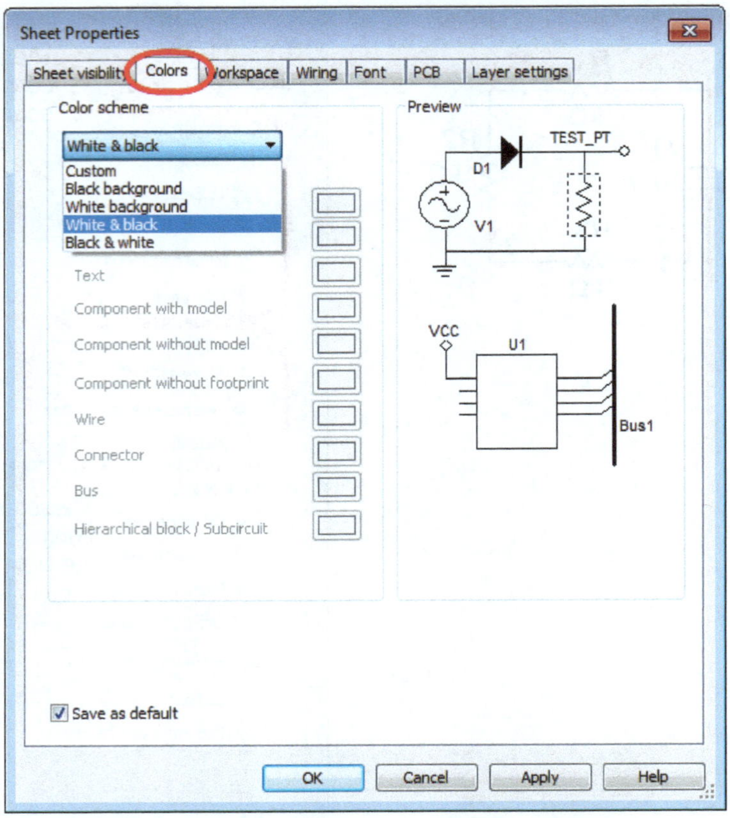

Figure F–80 The View menu provides access to the Grapher.

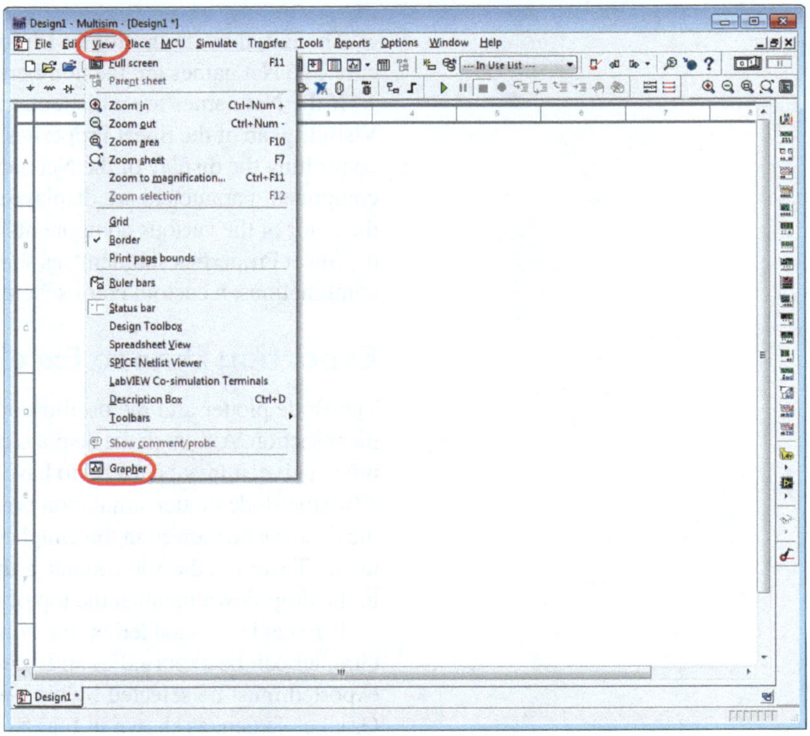

Figure F–81 The Grapher.

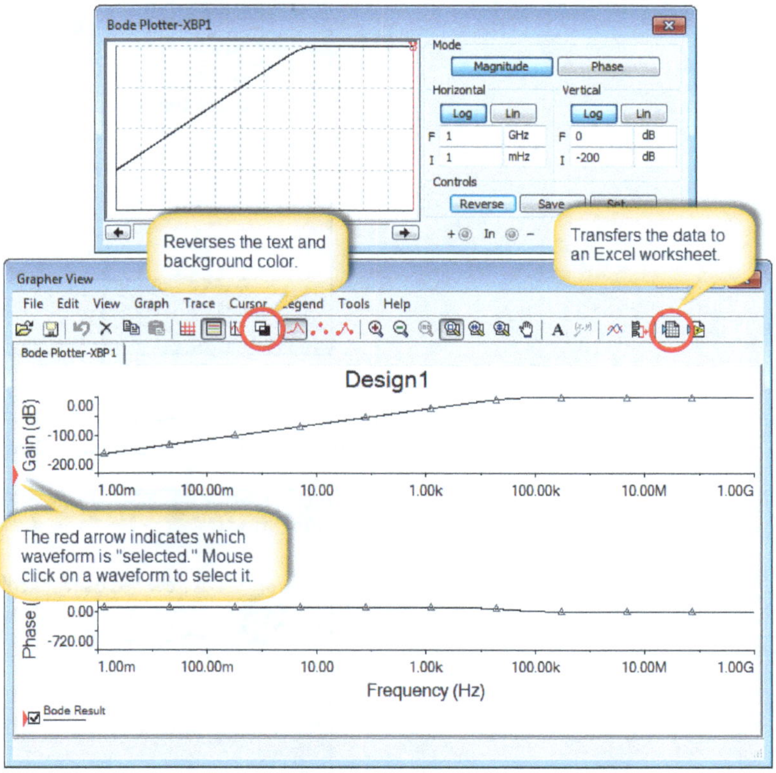

Once the plot is selected, the data can be exported to an Excel worksheet by mouse clicking on the Excel icon in the upper right hand side of the tool bar, as shown in Fig. F–81. The export feature will create a new worksheet with the X and Y data in adjacent columns. Figure F–82 shows data from the generic oscilloscope captured by the Grapher.

Figure F–82 Exporting oscilloscope data to Excel.

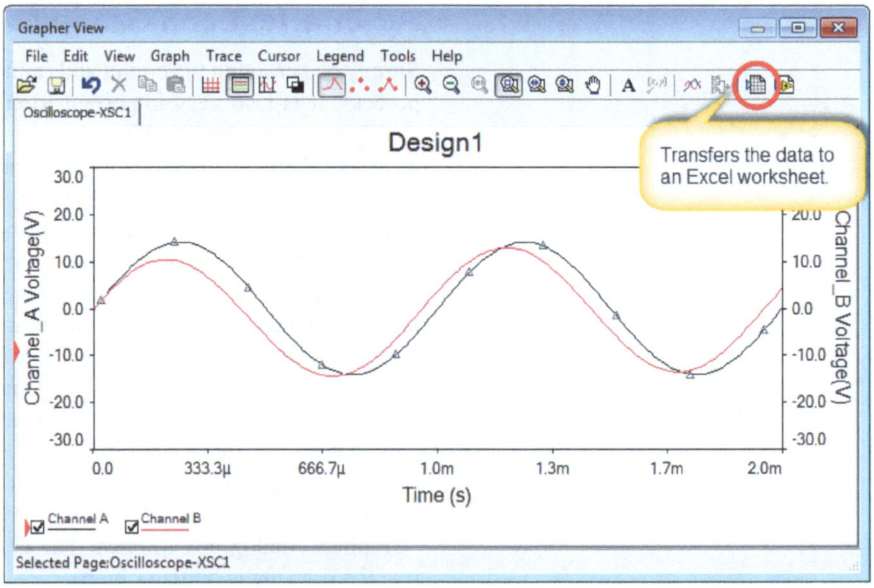

Figure F–83 Select the trace for export.

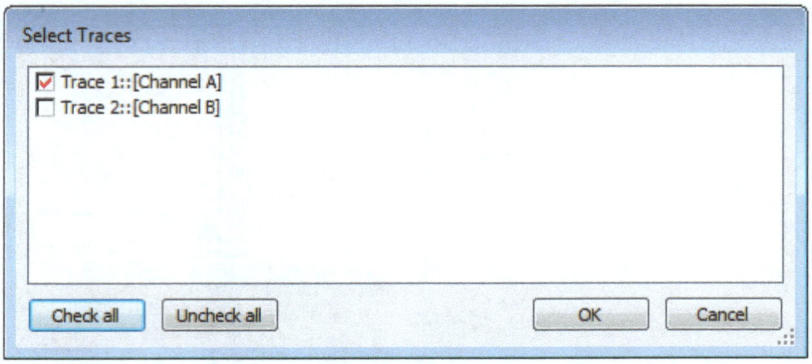

Figure F–84 A new Excel worksheet will be created.

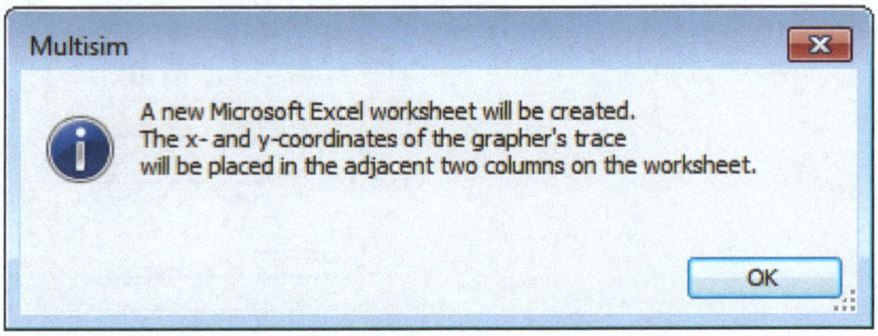

Once the "Export to Excel" icon is mouse clicked, Multisim will prompt the user for the desired trace to be exported, as shown in Fig. F–83. Once the desired trace has been selected, you will be prompted to create a new Microsoft Excel worksheet, as shown in Fig. F–84. As part of the export process, Excel will be started and a new worksheet will be created with the data from the x-axis in column A and the data from the y-axis in column B. This is shown in Fig. F–85. The data used to create the waveform within Multisim can now be examined using the analytical capabilities of Excel.

Adding Text and Graphics

Simply constructing the circuit and running the simulation is only part of the overall process. It is also very important to create circuits that include information that clearly explains what is taking place. Both annotation and graphics can be added to a circuit schematic to aid in the understanding of it by the viewer. To add text to the work area, either press < Ctrl + Alt + A > or mouse click on "Place" in the drop-down menu at the top of the screen, as shown in Fig. F–86. Move the cursor to the location within the work area where the text should begin and click the left mouse button once. To exit the "text edit" mode after the text has been entered, mouse click anywhere on the work area. The text can be edited by double mouse clicking on the text. The text can be repositioned in the work area by placing the cursor over the text, holding down the left mouse button and dragging the text its new location.

Basic drawing tools are available to annotate the work area. The option to import a picture into the work area is also accessible from the "Graphics" dialog box. As shown in Fig. F–87, select "Place" with a mouse click located on the drop-down menu bar, and then select "Graphics" from the drop-down menu. This will cause the "Graphics" dialog box to open. The default file format for pictures to be inserted into the drawing is bitmap (.bmp).

Figure F–85 The Excel worksheet with exported data.

	A	B
1	X--Trace 1::[Channel A]	Y--Trace 1::[Channel A]
2	0	0
3	0.0000001	0.008885765
4	1.08401E-07	0.009632279
5	1.25204E-07	0.011125307
6	1.58809E-07	0.014111362
7	2.26019E-07	0.020083469
8	3.60438E-07	0.032027673
9	6.29278E-07	0.055916002
10	1.16696E-06	0.103692115
11	2.24231E-06	0.199240245

Figure F–86 Placing text in the work area.

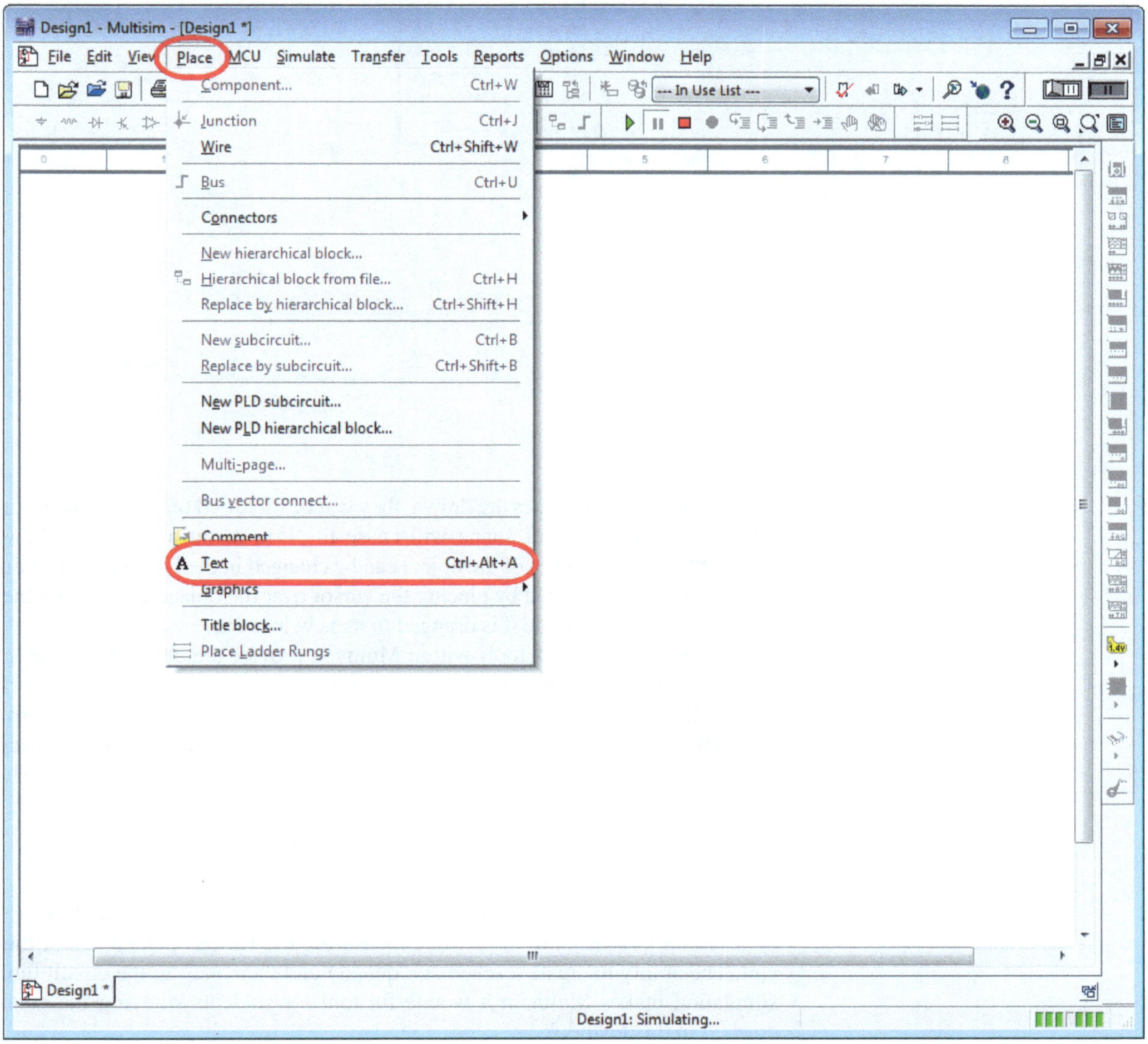

Figure F–87 Placing graphics in the work area.

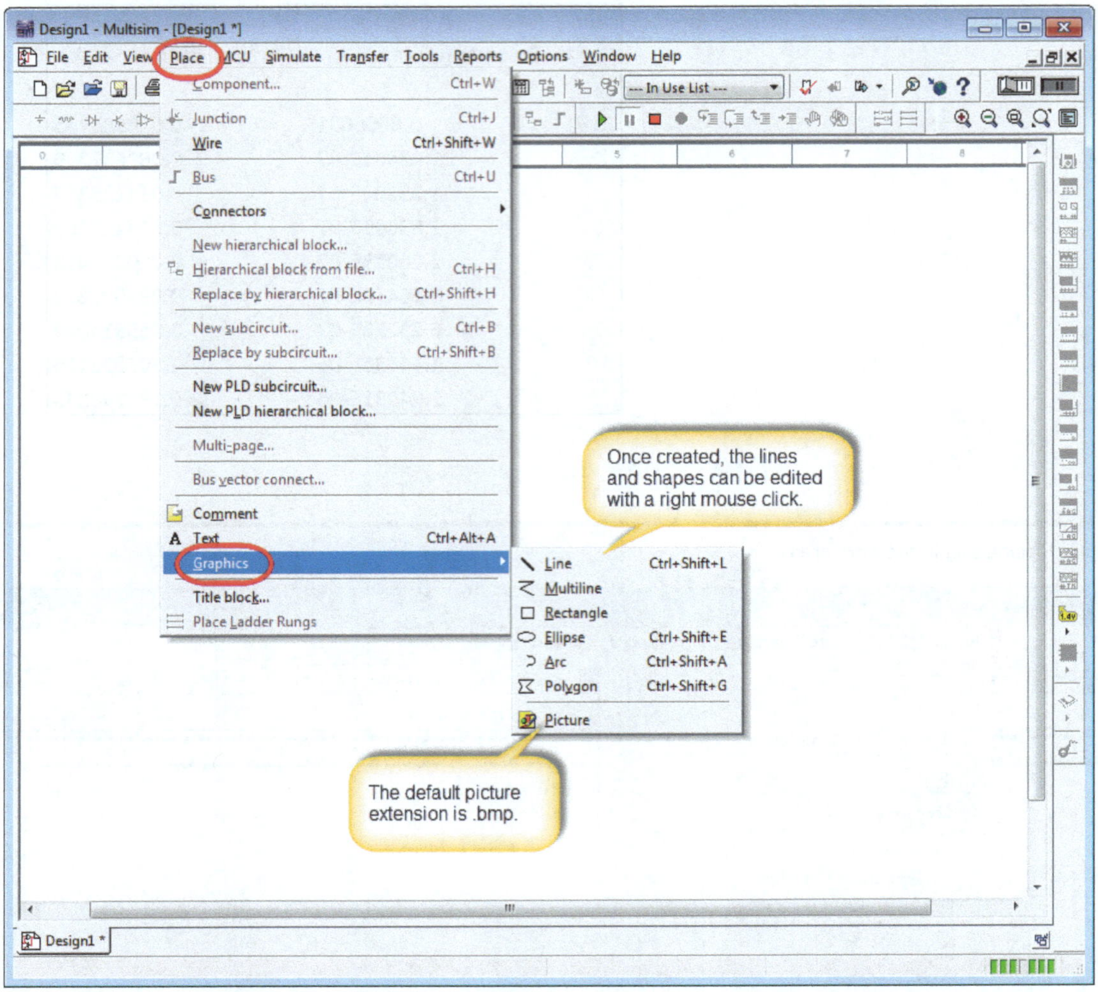

Once the lines or shapes are drawn, they can be modified or repositioned. A right mouse click on a line or shape will cause the Edit menu to open, as shown in Fig. F–88. The properties of the object can be changed in the Edit menu. The line or shape can be repositioned by placing the cursor over the object and holding the left mouse button down while it is dragged to its new location.

The text and graphics tools within Multisim provide the ability to add graphics and text to the work area. This feature enhances the readability of the schematic, especially when it is going to be used in a laboratory report. Figure F–89 provides an example of an annotated schematic that could be used within a laboratory report. Notice how the addition of the color, text, and graphics helps document what is being measured within the circuit.

Conclusion

Multisim is an interactive circuit simulation package that allows the user to view their circuit in schematic form while measuring the various parameters of the circuit. The ability to create a schematic quickly and then analyze the circuit through simulation makes Multisim a wonderful tool for students studying the concepts covered in this textbook.

Figure F–88 Modifying the graphics in the work area.

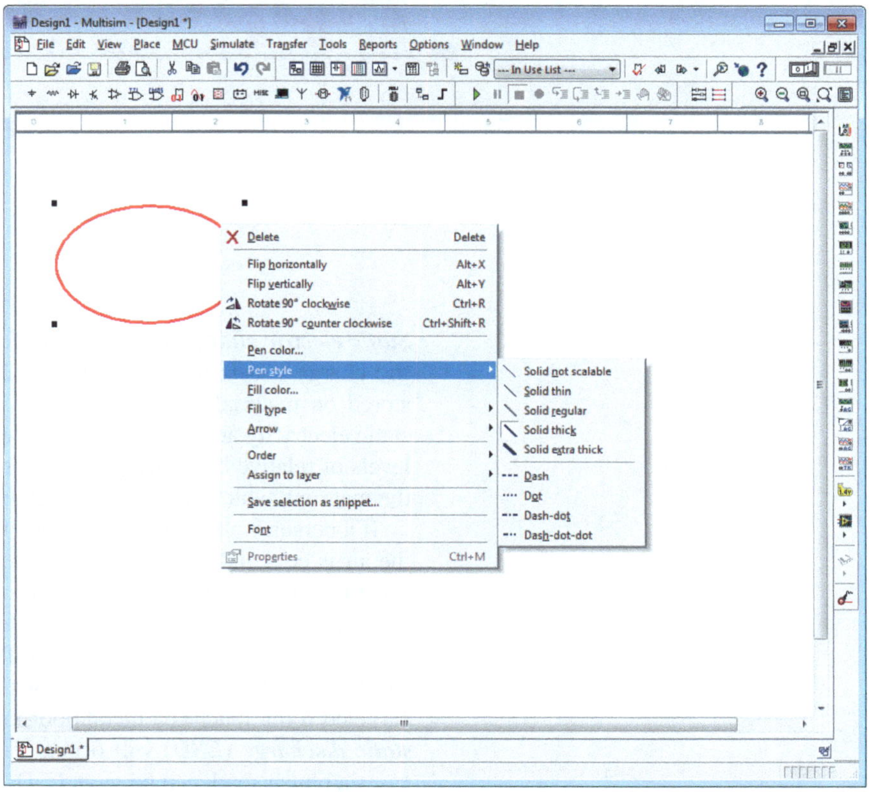

Figure F–89 The use of annotation and graphics in the work area.

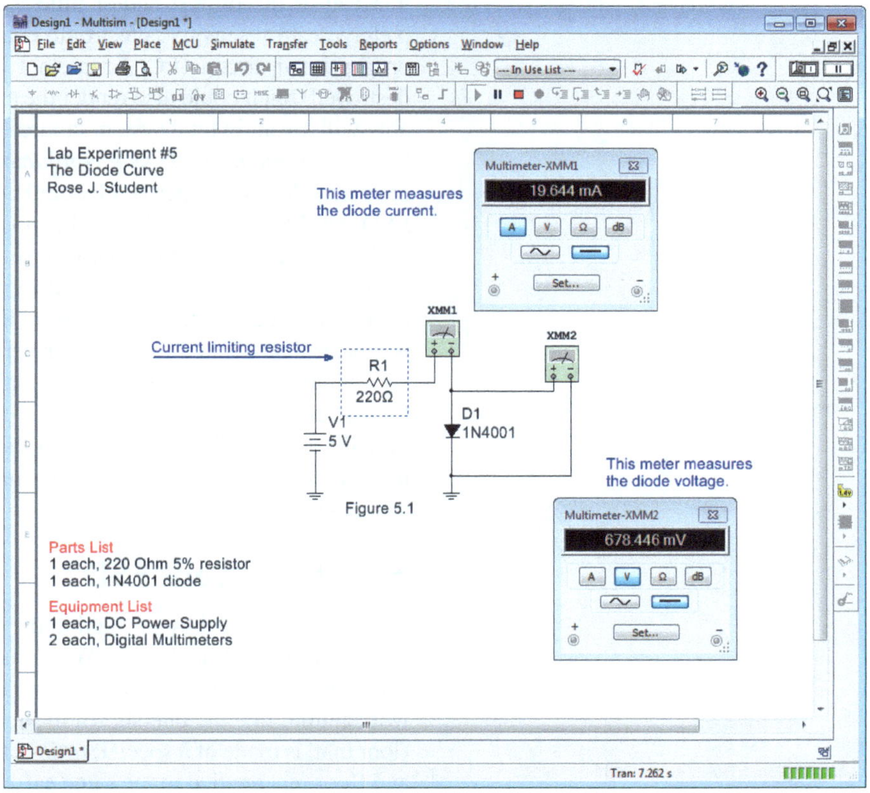

Appendix G

Electrostatic Discharge (ESD)

Static electric charge, or ***static electricity*** as it is commonly known, refers to electric charge that is motionless or stationary. Static electric charge is deposited or stored on materials with insulating or non-conductive properties. The effects of static electricity are much more prevalent when the relative humidity is low. Higher levels of relative humidity (moisture) result in a conductive layer on the surface of the material, which allows electric charge to bleed off rather than be stored.

If a person walks across a carpeted room, or similar surface, in the winter, when the air is cold and dry, the soles of their shoes sliding across the floor results in a separation of electric charge. Electrons are either added or removed from the sole of the shoe. This accumulation of static electric charge results in a build-up of potential difference or voltage. The potential difference is usually several thousand volts but can be as high as 50 to 100 kilovolts (kV). When the person walking across the carpeted room touches a metal door knob, or some other metal object, an ***electrostatic discharge (ESD)*** will occur. Usually, ESD produces a shock, and in some cases a bluish spark can be seen. ESD, which is a transfer of electric charge, is possible any time two dissimilarly charged objects are in close proximity to each other. The transfer of electric charge neutralizes both objects. Although ESD may seem harmless at first, its effect is highly undesirable. First, there is the annoying shock that causes discomfort and/or mild pain. Second, ESD can damage and/or completely destroy sensitive electronic components. If these components are mounted on a printed circuit (PC) board, not only will the device itself be damaged (or destroyed) but it could also render the entire PC board useless. This could, and probably will, result in total system failure. The component most sensitive to damage by ESD is the ***metal-oxide semiconductor field effect transistor (MOSFET)***. There are many different types of electronic components that are sensitive to damage by ESD, but MOSFETs are by far the most sensitive.

It is worth noting that many sensitive electronic components can be partially damaged or completely destroyed by voltages of 100 V or less. However, most people will not even feel a shock from ESD unless the potential difference or voltage is greater than about 2 to 3 kV. Furthermore, ESD damage to an electronic component is not something you can normally see, not even under a microscope. And finally, an ESD damaged electronic component and/or assembly may work fine initially but fail after it has been in the field for a while. Because ESD can do so much damage, special precautionary measures must be taken when handling sensitive electronic components.

How to Prevent ESD

ESD can be avoided by preventing the build-up of static electric charge on nonconductive surfaces (insulators). When handling, testing, or servicing sensitive electronic equipment and/or components, the best practice is to use an ESD-protected workstation, like the one shown in Fig. G–1. The work surface mat, as well as the floor mat, is made of a special ***static-dissipative*** material. Both the work surface mat and floor mat are tied to the same earth ground connection via a 1 MΩ resistor. The

Figure G–1 ESD-protected work station.

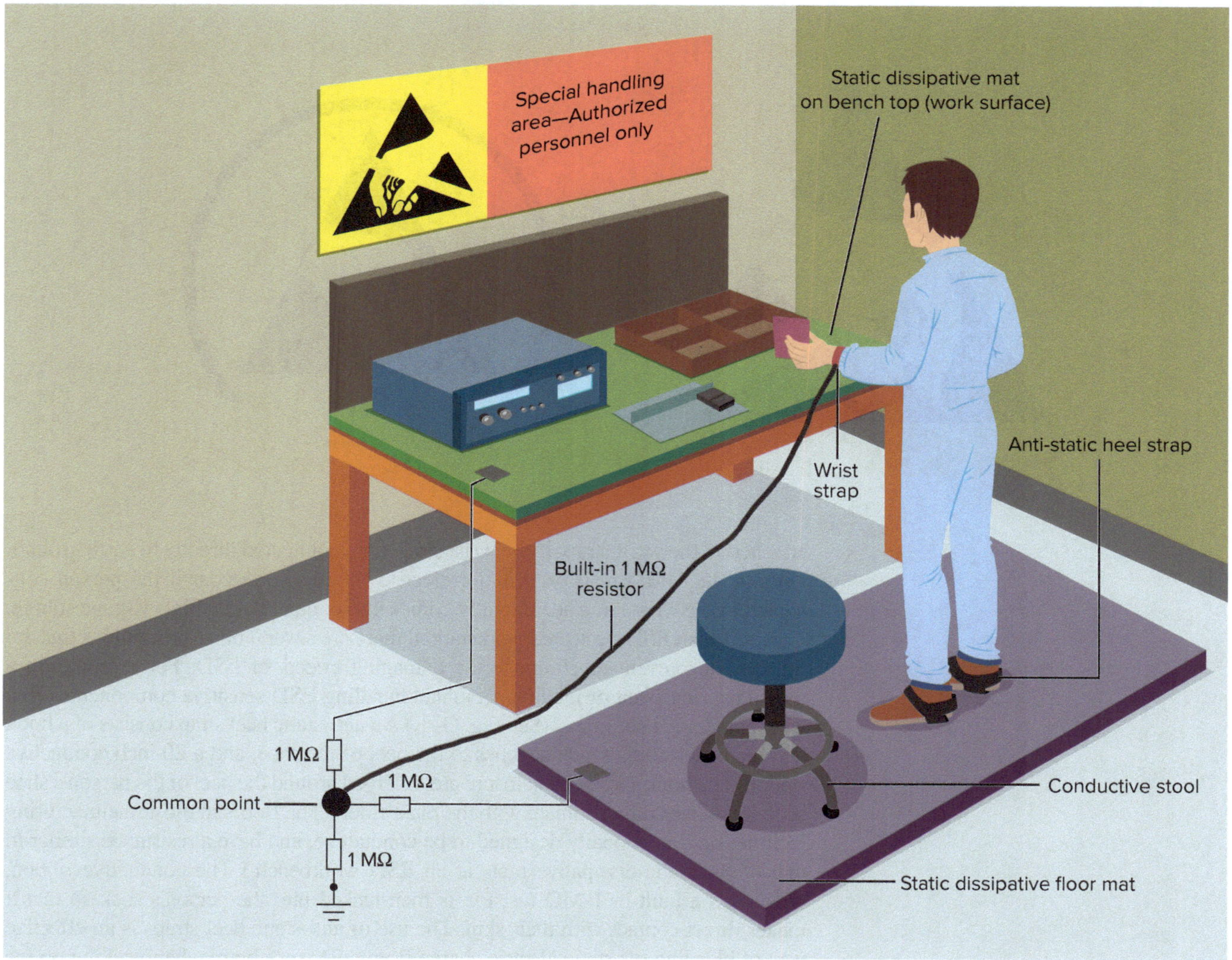

surface of the static-dissipative material has a resistance that falls somewhere within the range of $1 \times 10^5 \frac{\Omega}{sq}$ to $1 \times 10^{12} \frac{\Omega}{sq}$, where "*sq*" refers to a specific square area on the mat. This level of conductivity allows static electric charge to bleed-off slowly to ground. Also included at the ESD-protected workstation is a wrist strap. As the name implies, the wrist strap is worn around the wrist of the person at the workstation. The wrist strap itself is conductive and must be in direct contact with the skin of the person wearing it. For safety purposes, the cord of the wrist strap has a built-in 1 MΩ resistor, which is also connected to earth ground. The purpose of the grounded wrist strap is to continually bleed-off static electric charge from the body that may result from normal body motion.

The main reason the resistance of the work surface (the static-dissipative material) is very high rather than very low is for safety reasons. If high voltage is present in the electronic equipment being serviced, touching both a highly conductive work surface and a high-voltage source at the same time could be potentially dangerous or fatal. Accordingly, safety precautions necessitate that a high-resistance rather than a low-resistance work surface be used. The other reason for using a high-resistance material for the work surface is that the static electric charge will be dissipated more slowly rather than too quickly. If the discharge is too quick or sudden, it can produce a spark and thereby damage sensitive electronic components. The 1 MΩ resistor in the cord of the wrist strap provides electrical safety for the person

Figure G–2 (*a*) Symbol indicating an ESD sensitive component or assembly. (*b*) Symbol indicating a component or assembly is ESD protected.

(*a*) (*b*)

attached to the wrist strap. If the wrist strap were connected directly to earth ground, without the 1 MΩ resistor, a major electric shock could occur if the person connected to the wrist strap accidentally comes into contact with a high-voltage source.

Besides an ESD-protected workstation, there are several other precautions that are followed to prevent or eliminate the damaging effects of ESD. For example, in a manufacturing plant or facility, personnel handling ESD-sensitive components often wear anti-static heel straps. (See Fig. G–1.) An anti-static heel strap consists of a hook and loop enclosure, which attaches to the heel of the shoe, and a 20-inch conductive ribbon. The hook and loop enclosure are Velcroed around the heel of the person's shoe so that it makes direct contact with the plant floor. (The floors in most manufacturing facilities are intentionally designed to be conductive, and have a resistance similar to that of the static-dissipative mats at an ESD workbench.) The conductive ribbon, which has a built-in 1 MΩ resistor, is then tucked into the person's sock so that it makes direct contact with their skin. The use of anti-static heel straps is an effective way of bleeding off static electric charge from a person's body when walking on the plant floor. Anti-static heel straps should be worn on both shoes to provide consistent grounding while in motion. Also, special clothing, which does not generate static electricity, is also worn when working with ESD sensitive components or assemblies.

ESD Symbols

In the electronics industry, certain symbols are used to identify products that require special ESD handling precautions. The symbol shown in Fig. G–2*a*, for example, shows a slash through a hand inside of a triangle. This symbol appears on packages, boxes or other containers with ESD-sensitive devices or assemblies. When you see this symbol, you know that you must handle this package with the appropriate ESD precautions. When ESD-sensitive components or assemblies are placed in boxes for shipping, this symbol is used along with the following statement on the box: "Electrostatic sensitive devices, observe precautions when handling".

Another symbol, shown in Fig. G–2*b*, shows a hand inside of a triangle but with an arc over the top of it. The arc means "ESD protected." This symbol is often placed on chairs, mats, wrist straps and clothing, and so forth that are ESD safe.

In industry, most people working with ESD-sensitive components and assemblies undergo special training regarding how to recognize and handle ESD-sensitive devices. Without the proper training, millions of dollars of electronic components and assemblies can be permanently damaged or completely destroyed.

Appendix H

Introduction to Breadboarding

Introduction

Breadboarding is a way or means of constructing electronic circuits when performing laboratory experiments. ***Breadboards***, or ***protoboards*** as they are sometimes called, allow electronic circuits to be readily and easily assembled (or dissembled) without the need for soldering. A typical breadboard is shown in Fig. H–1. As you can see from the photo, the breadboard consists of a rectangular white-colored construction base, usually made of plastic, with holes (or sockets) arranged in columns and rows. The construction of electronic circuits is accomplished by inserting the component leads into the proper holes on the breadboard. It is important to learn how the holes are interconnected before attempting to assemble a circuit on the breadboard.

In Fig. H–2, notice the horizontal rows in the top, middle, and bottom sections of the breadboard. These rows are referred to as ***bus*** bars or ***buses*** and are typically connected to the positive (+) and negative (−) terminals of a power supply. The power supply connections are usually brought into the red, black, and yellow banana jacks shown on the left-hand side of the breadboard. Wires are then run from the base of the banana jacks to the buses located on the breadboard. As you can see from the photo, the holes in each bus bar are arranged in groups of five. The five holes in each group are electrically connected to each other using a metal strip that lies below the surface of the breadboard. Furthermore, each group of five holes is electrically connected to the others with one continuous strip of metal. The metal strips

Figure H–1 Typical breadboard used for constructing circuits.

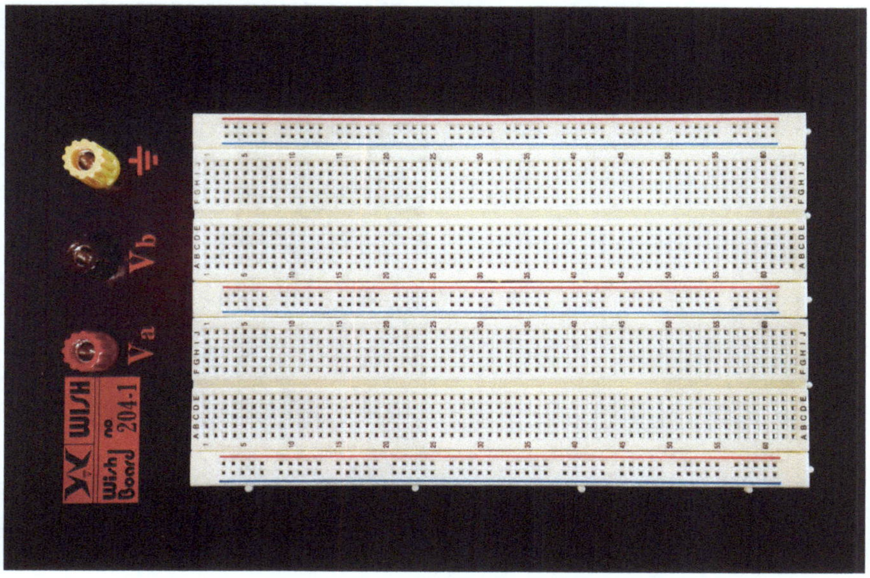

Mitchel E. Schultz

Figure H–2 Locating the bus bars on the breadboard.

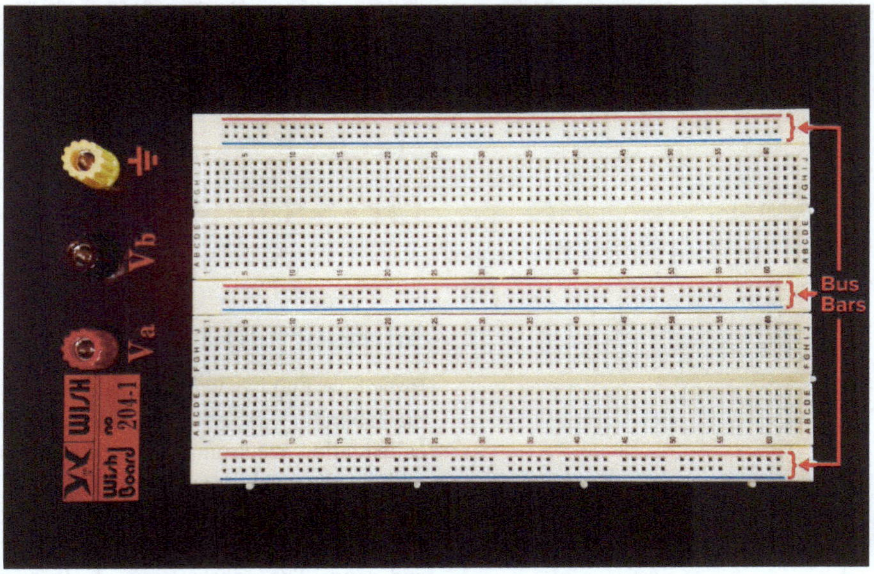

Mitchel E. Schultz

provide continuity between each hole in the row and between each group of five holes. The continuity between the holes in each bus bar is indicated by the solid red and blue lines running adjacent to the holes. It is important to note, however, that there is no continuity between the two horizontal bus bars on the breadboard.

Figure H–3 shows or highlights the main construction areas of the breadboard. The construction area is the area where circuits are constructed or assembled. On this particular breadboard, there are actually two different construction areas. One construction area lies between the top and middle bus bars, whereas the other construction area lies between the middle and bottom bus bars. Unlike the bus bars, the holes in the construction area have continuity in the vertical direction but not in the horizontal direction. Like the bus bars, metal strips lie below the surface

Figure H–3 Main construction areas on the breadboard.

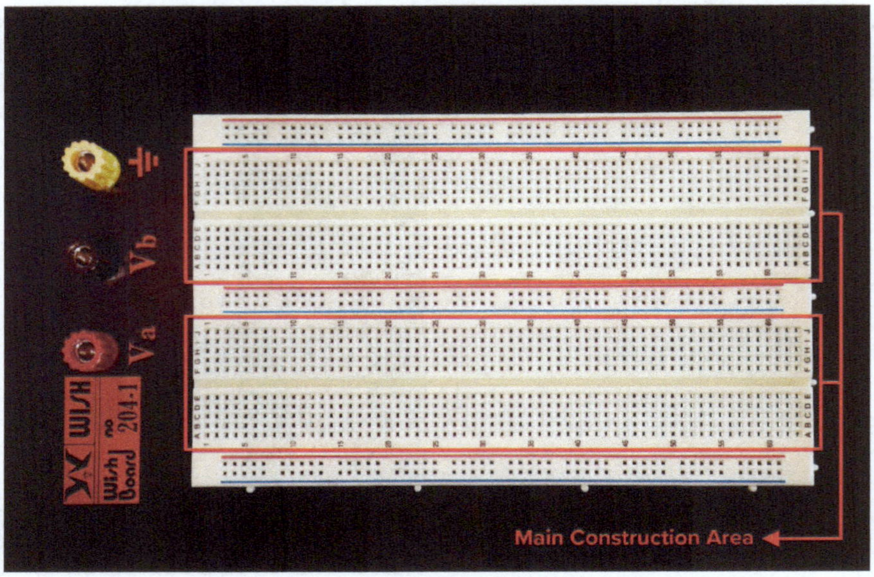

Mitchel E. Schultz

of the construction area to provide vertical continuity between the holes. Notice that each column consists of a group of five holes. The groove that exists in each construction area electrically isolates the two blocks or groups of holes from each other. In other words, there is no continuity between the vertical holes on either side of the groove.

In Fig. H–3, notice the letters and numbers written next to the holes on the breadboard. The letters **A** through **J** are used to identify each row of holes moving horizontally, whereas the numbers **1, 5, 10, 15, 20**, and so on are used to identify a specific vertical column. The columns between **1** and **5** are not numbered but are understood to be **2, 3,** and **4**, respectively. The same sequencing of numbers also applies between columns **5** and **10, 10** and **15**, and so on. The purpose of having the rows and columns labeled with letters and numbers is to identify and/or pinpoint a specific row and column when constructing a circuit on the breadboard. For example, the leads of two resistors may be inserted in rows **J** and **H** in column **15**.

Constructing a Circuit on the Breadboard

Figure H–4 shows how to construct a simple series circuit on a breadboard. The series circuit consists of three resistors which are labeled R_1, R_2, and R_3. Notice the left lead of resistor R_1 is connected in row **H** in Column **10**. The right lead of R_1 is also connected in row **H** but in column **25**. Recall that there is no horizontal continuity between the holes in the main construction area. The top lead of resistor R_2 is connected in column **25** but in row G. Since all holes in column **25** have vertical continuity, the right lead of resistor R_1 and the top lead of resistor R_2 are electrically connected at this point on the breadboard. The bottom lead of R_2 is connected in row C and column **25** but on the other side of the groove in the breadboard. The groove separates the holes in the construction area and provides electrical isolation between the vertical holes on either side of the groove. In other words, there is no continuity between the vertical holes on either side of the groove. The right lead of resistor R_3 is placed in row **B** in column **25**. And finally, the left lead of resistor R_3 is connected in row **B** in column **10**.

The power supply connections are brought into the top and middle bus bars on the breadboard as shown. The positive (+) side of the power supply is connected to the top bus bar identified with the red line above it. A jumper wire is then run from the top bus bar to column **10** in row **J**. This connects the positive (+) terminal of the power supply to the left lead of resistor R_1. Similarly, the negative (−) side of

Figure H–4 Constructing a simple series circuit on the breadboard.

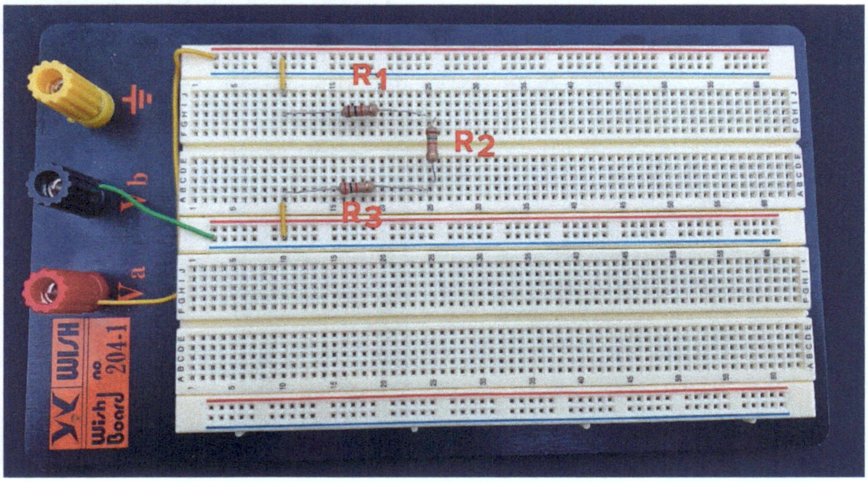

Mitchel E. Schultz

Figure H–5 Constructing a simple parallel circuit on the breadboard.

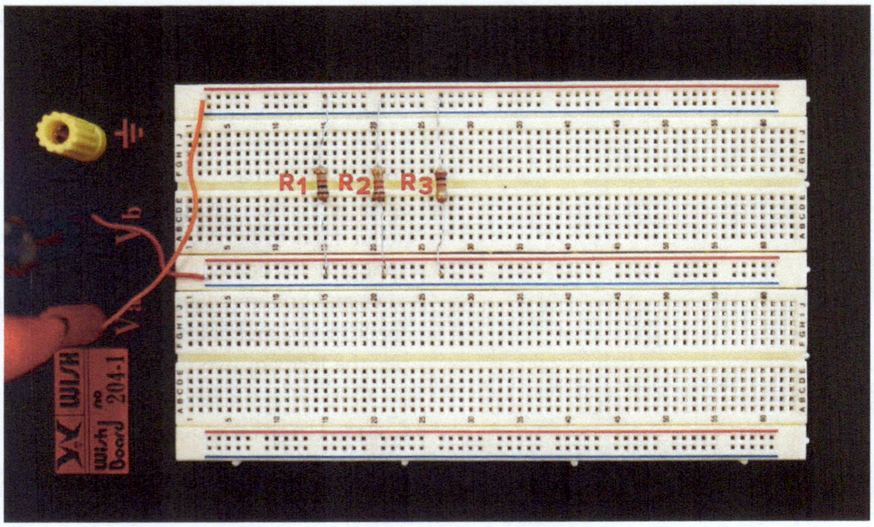

Mitchel E. Schultz

the power supply is connected to the bottom bus bar closest to the blue line in the middle of the breadboard. A jumper wire is then run from the blue bus bar to row **A** in column **10**. This connects the left lead of resistor R_3 to the negative (−) side of the power supply.

Figure H–5 shows how to connect resistors in parallel. The parallel circuit consists of three resistors labeled R_1, R_2, and R_3. Notice that the top lead of each resistor is connected to the red bus bar at the top of the breadboard, whereas the bottom lead of each resistor is connected to the blue bus bar in the middle of the breadboard. Since the bus bars have continuity in the horizontal direction, all resistor leads in each bus bar have continuity between them. The positive (+) and negative (−) power supply connections are brought into each bus bar as shown.

Figure H–6 shows how to construct a simple series-parallel (combination) circuit. The series-parallel circuit consists of four resistors, which are labeled R_1, R_2, R_3, and R_4. Notice that resistors R_2 and R_3 are connected in parallel. The top leads of R_2 and R_3 are connected together using very short jumper wires that run

Figure H–6 Constructing a simple series-parallel (combination) circuit on the breadboard.

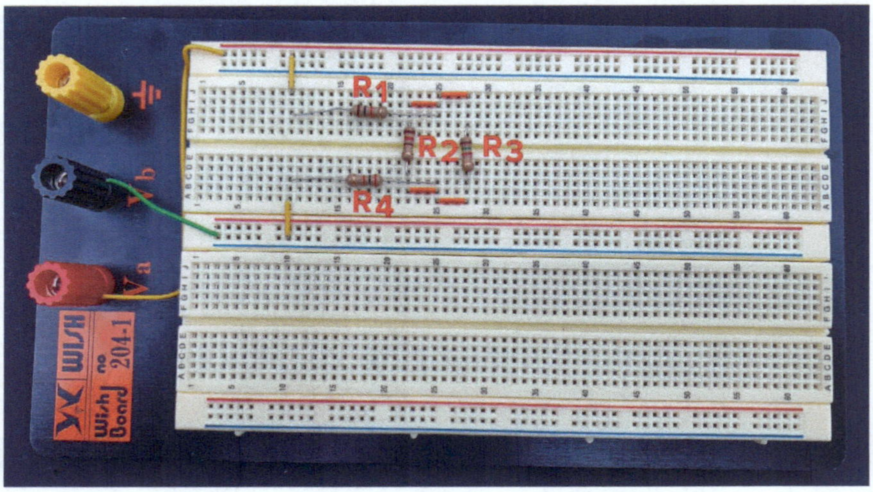

Mitchel E. Schultz

to the left and right of column **25** in rows **I** and **J**. The bottom leads of R_2 and R_3 are connected together using short jumper wires that run to the left and right of column **25** but in rows **A** and **B**. Notice the other end of the jumper wires are inserted into columns **22** and **28**. The top leads of resistors R_2 and R_3 are inserted into columns **22** and **28** in row **G**. The bottom leads of R_2 and R_3 are also inserted into columns **22** and **28** in rows **C** and **D** on the other side of the groove. The right leads of resistors R_1 and R_4 can be inserted into any of the unused holes in column **25**. The left leads of resistors R_1 and R_4 are connected in rows **H** and **C**, respectively, in column **10**. The power supply connections are brought into the bus bars as shown. Notice the short jumper wires that are running from each bus bar to column **10** on both sides of the groove. The jumper wires connect the power supply's positive (+) and negative (−) terminals to the left leads of resistors R_1 and R_4.

A word of caution. Never use jumper wires and/or components whose lead diameter is too large for the holes or sockets on the breadboard. Doing so will most definitely result in damage to the holes or sockets. Furthermore, when the correct diameter leads and/or wires are inserted into these same holes at a later point, it will most likely result in loosely connected components, and in turn, intermittent operation of the circuit being tested.

Inserting Integrated Circuits (ICs) Into the Breadboard

When inserting an IC into a breadboard, it must be done as shown in Fig. H–7. Notice the IC leads are inserted into the holes on either side of the groove in the main construction area. This is done to avoid shorting the opposite-side pins together on the IC. If the IC were inserted into the holes on just one side of the groove, the pins opposite each other would be shorted together due to the vertical continuity between the holes in each column.

Final Comments

Oftentimes, students construct multiple circuits in different areas on the breadboard. This is commonplace when students are using the same breadboard in more than one course. After constructing a few circuits, understanding how to use the breadboard will become second nature to you.

Figure H–7 Proper insertion of an integrated circuit (IC) on a breadboard.

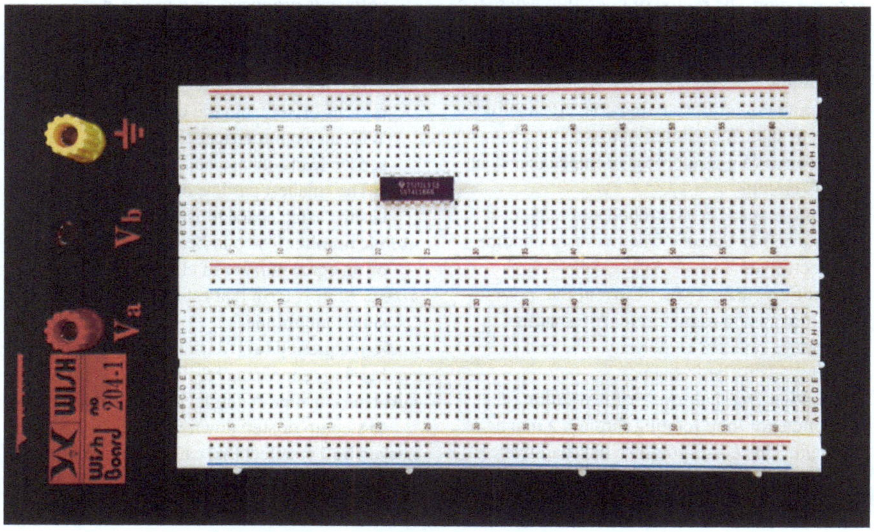

Mitchel E. Schultz

Glossary

A

AC *See* alternating current.

AC beta The ratio of AC collector current, i_c, to AC base current, i_b, or $\beta = \frac{i_c}{i_b}$.

AC effective resistance, R_e The resistance of a coil for higher-frequency alternating current. The value of R_e is more than the DC resistance of the coil because it includes the losses associated with high-frequency alternating current in a coil. These losses include skin effect, eddy currents, and hysteresis losses.

AC equivalent circuit A circuit as it appears to an AC signal. In an AC equivalent circuit, all capacitors and voltage sources appear as shorts.

AC load line A graph that shows all the possible values of I_C and V_{CE} when a transistor amplifier is driven by an AC signal.

AC load power, P_L The AC power that is dissipated by the load, R_L.

AC resistance of a diode The equivalent resistance of a forward-biased diode as it appears to small AC signals. For a standard diode, $r_{AC} = \frac{25\text{ mV}}{I_d}$. For the emitter diode in a transistor, $r'_e = \frac{25\text{ mV}}{I_E}$.

active component One that can control voltage or current. Examples are transistors and diodes.

active filter A filter that uses components or devices that have the ability to amplify such as transistors and op amps.

active region The region of operation where the collector of a transistor acts like a current source.

acute angle Less than 90°.

A/D converter A device that converts analog input signals to digital output.

admittance (*Y*) Reciprocal of impedance *Z* in AC circuits. $Y = 1/Z$.

air gap Air space between poles of a magnet.

alkaline cell or battery One that uses alkaline electrolyte.

alternating current (AC) Current that reverses direction at a regular rate. Alternating voltage reverses in polarity. The rate of reversals is the frequency.

alternation One-half cycle of revolution of a conductor loop rotating through a magnetic field. This corresponds to one-half cycle of alternating voltage or current.

alternator AC generator.

amp-clamp probe A meter that can measure AC currents, generally from the 60-Hz AC power line, without breaking open the circuit. The probe of the meter is actually a clamp that fits around the current-carrying conductor.

ampere The basic unit of current. $1\text{ A} = \frac{1\text{ C}}{1\text{ s}}$ or $1\text{ A} = \frac{1\text{ V}}{1\ \Omega}$.

ampere-hour (A·h) rating A common rating for batteries that indicates how much load current a battery can supply over a specified discharge time. For example, a battery with a 100-A·h rating can deliver 1 A for 100 h or 2 A for 50 h, 4 A for 25 h, etc.

ampere-turn (A·t) Unit of magnetizing force equal to $1\text{ A} \times 1$ turn.

ampere turns/meter $\left(\frac{\text{A·t}}{\text{m}}\right)$ The SI unit of field intensity, H.

analog multimeter A test instrument that is used to make voltage, current, and resistance measurements. An analog multimeter uses a moving pointer and a printed scale to display the value of the measured quantity.

antiresonance A term to describe the condition of unity power factor in a parallel *LC* circuit. *Antiresonance* is used to distinguish it from the case of equal X_L and X_C values in a series *LC* circuit.

apparent power The product of voltage and current *VA* when *V* and *I* are out of phase.

arctangent (arctan) An inverse trigonometric function that specifies the angle, θ, corresponding to a given tangent (tan) value.

armature The part of a generator in which the voltage is produced. In a motor, it is commonly the rotating member. Also, the movable part of a relay.

asymmetrical JFET A JFET that has its gate regions offset from the center of the channel. With an asymmetrical JFET, the drain and source leads cannot be interchanged.

atom The smallest particle of an element that still has the same characteristics as the element.

atomic number The number of protons, balanced by an equal number of electrons, in an atom.

attenuation A term that refers to a reduction in signal amplitude.

audio frequency (af) Within the range of hearing, approximately 16 to 16,000 Hz.

autotransformer A single, tapped winding used to step up or step down voltage.

avalanche The effect that causes a sharp increase in reverse current, I_R, when the reverse-bias voltage across a diode becomes excessive.

average value In sine-wave AC voltage or current, 0.637 of peak value.

B

***B-H* magnetization curve** A graph of field intensity *H* versus flux density *B*.

back-off ohmmeter scale Ohmmeter readings from right to left.

balanced bridge A circuit consisting of two series strings in parallel. The balanced condition occurs when the voltage ratio in each series string is identical. The output from the bridge is taken between the centers of each series string. When the voltage ratios in each series string are identical, the output voltage is zero and the bridge circuit is said to be balanced.

balanced load A wye (Y)- or delta-connected load in which each individual load impedance is identical.

band-pass filter Filter that allows coupling a band of frequencies to the load.

band-stop filter Filter that prevents a band of frequencies from being coupled to the load.

bandwidth A range of frequencies that has a resonant effect in *LC* circuits.

bank Components connected in parallel.

banks in series Parallel resistor banks that are connected in series with each other.

barrier potential, V_B The potential difference at the *p-n* junction of a diode. V_B exists between the wall of positive and negative ions that are created as a result of free electrons diffusing from the *n*-side of the diode to the *p*-side.

base A thin and very lightly doped region in a transistor. The base is sandwiched between the emitter and collector regions.

battery Group of cells connected in series or parallel.

bias A control voltage or current.

bidirectional diode thyristor Another name for a diac.

bilateral components Electronic components that have the same current for opposite polarities of applied voltage.

bleeder current Steady current from a source used to stabilize output voltage with changes in load current.

branch Part of a parallel circuit.

breakdown region The region of transistor operation where a large undesired collector current flows as a result of the collector-base diode breaking down due to excessive reverse-bias voltage.

breakdown voltage, V_{BR} The reverse-bias voltage at which the avalanche effect occurs. The avalanche effect causes the reverse current, I_R, to increase sharply.

brushes In a motor or generator, devices that provide stationary connections to the rotor.

bulk resistance, r_B The resistance of the *p* and *n* materials in a diode.

bypass capacitor One that has very low reactance in a parallel path.

C

C Symbol for capacitance.

C Abbreviation for *coulomb,* the unit of electric charge.

calorie Amount of heat energy needed to raise the temperature of one gram of water by 1°C.

capacitance The ability to store electric charge.

capacitive reactance, X_C A measure of a capacitor's opposition to the flow of alternating current. X_C is measured in ohms. $X_C = \frac{1}{2\pi fC}$ or $X_C = \frac{V_C}{I_C}$. X_C applies only to sine-wave AC circuits.

capacitive voltage divider A voltage divider that consists of series-connected capacitors. The amount of voltage across each capacitor is inversely proportional to its capacitance value.

capacitor Device used to store electric charge.

carbon-composition resistors Resistors made of finely divided carbon or graphite mixed with a powdered insulating material.

carbon-film resistors Resistors made by depositing a thin layer of carbon on an insulated substrate. The carbon film is cut in the form of a spiral.

Celsius scale (°C) Temperature scale that uses 0° for the freezing point of water and 100° for the boiling point. Formerly called *centigrade*.

ceramic Insulator with a high dielectric constant.

cgs Centimeter-gram-second system of units.

channel The area or conducting region between the drain and source terminals of an FET. The channel can be made of either *n*-type or *p*-type semiconductor material.

charging (1) The effect of increasing the amount of charge stored in a capacitor. The accumulation of stored charge results in a buildup of voltage across the capacitor. (2) The process of reversing the current, and thus the chemical action, in a cell or battery to re-form the electrodes.

charging current The current that flows to and from the plates of a capacitor as the charge stored by the dielectric increases.

chassis ground Common return for all electronic circuits mounted on one metal chassis or PC board. Usually connects to one side of DC supply voltage.

chip capacitor A surface-mounted capacitor.

choke Inductance with high X_L compared with the *R* of the circuit.

circuit A path for current flow.

circuit breaker A protective device that opens when excessive current flows in circuit. Can be reset.

circular mil (cmil) Cross-sectional area of round wire with diameter of 1 mil or 0.001 in.

class A amplifier An amplifier in which the collector current, I_C, flows for the full 360° of the AC input cycle.

class B amplifier An amplifier in which the collector current, I_C, flows for only 180° of the AC input cycle.

class B push-pull amplifier A class B amplifier that uses two transistors to reproduce the full AC cycle of input votlage. Each transistor conducts on opposite half-cycles of the input voltage.

class C amplifier An amplifier in which the collector current, I_C, flows for 120° or less of the AC input cycle.

closed-loop cutoff frequency, f_{CL} The frequency at which the closed-loop voltage gain decreases to 70.7% of its maximum value.

closed-loop voltage gain, A_{CL} The voltage gain of an amplifier with negative feedback.

coaxial cable An inner conductor surrounded by an outer conductor that serves as a shield.

coding of capacitors The method used to indicate the value of a capacitor.

coefficient of coupling, *k* The fraction of total flux from one coil linking another coil nearby.

coil Turns of wire conductor to concentrate a magnetic field.

collector A large, moderately doped region in a transistor. The collector is the largest of all three transistor regions because it dissipates the most heat.

color code System in which colors are used to indicate values in resistors.

common-base amplifier A transistor amplifier that has its input applied to the emitter and its output taken from the collector. The common-base amplifier provides a high voltage and power gain but its current gain is less than 1.

common-collector amplifier A transistor amplifier that has its input applied to the base and its output taken from the emitter. The common-collector amplifier provides a high current and power gain but its voltage gain is less than 1.

common-drain amplifier An amplifier that has its input applied to the gate and its output taken from the source. Another name for the common-drain amplifier is the source follower.

common-emitter amplifier A transistor amplifier that has its input applied to the base and its output taken from the collector. The common-emitter amplifier provides a high voltage and current gain and a very high power gain.

common-gate amplifier An amplifier that has its input applied to the source and its output taken from the drain.

common-mode input An identical input voltage appearing on both inputs of a differential amplifier.

common-mode rejection ratio (CMRR) The ratio of differential voltage gain, A_d, to common mode voltage gain, A_{CM}. CMRR is usually specified in decibels.

common-mode voltage gain, A_{CM} The voltage gain of a differential amplifier for a common-mode signal.

common-source amplifier An amplifier that has its input applied to the gate and its output taken from the drain.

commutator Converts reversing polarities to one polarity.

comparator An op-amp circuit that compares the signal voltage on one input with a reference voltage on the other.

complex number Has real and *j* terms; uses form $A + jB$.

compound A combination of two or more elements.

condenser Another (older) name for a capacitor.

conductance (*G*) Ability to conduct current. It is the reciprocal of resistance, $G = 1/R$. The unit is the siemens (S).

conductor Any material that allows the free movement of electric charges, such as electrons, to provide an electric current.

constant-current source A generator whose internal resistance is very high compared with the load resistance. Since its internal resistance is so high, it can supply a constant current to a load whose resistance value varies over a wide range.

constant-voltage source A generator whose internal resistance is very low compared with the load resistance. Since its internal resistance is so low, it can supply a constant voltage to a load whose resistance value varies over a wide range.

continuity Continuous path for current. Reading of zero ohms with an ohmmeter.

continuity testing A resistance measurement that determines whether there is zero ohms of resistance (approximately) between two points, such as across the ends of a wire conductor.

conventional current The direction of current flow associated with positive charges in motion. The current flow direction is from a positive to a negative potential, which is in the opposite direction of electron flow.

corona effect Effect of ionization of air around a point at high potential.

cosine A trigonometric function of an angle, equal to the ratio of the adjacent side to the hypotenuse in a right triangle.

cosine wave One whose amplitudes vary as the cosine function of an angle. It is 90° out of phase with the sine wave.

coulomb (C) Unit of electric charge. $1\ C = 6.25 \times 10^{18}$ electrons.

counter emf (cemf) A term used to describe the effect of an induced voltage in opposing a change in current.

coupling capacitor Has very low X_C in series path.
covalent bond Pairing of atoms with electron valence of ±4.
covalent bonding The sharing of valence electrons between neighboring atoms in a silicon crystal or other crystalline structure.
cps Cycles per second. Formerly used as unit of frequency. Replaced by hertz (Hz) unit, where 1 Hz = 1 cps.
crossover distortion The distortion that occurs in a class B push-pull amplifier when the transistors are biased right at cutoff. When the input voltage crosses through zero both transistors in the push-pull amplifier are off and the output voltage cannot follow the input voltage.
CRT Cathode-ray tube. A device that converts electric signals to a visual display on a fluorescent screen.
crystal filter A filter that is made up of a crystalline material such as quartz. Crystal filters are often used in place of conventional *LC* circuits because there *Q* is so much higher.
Curie temperature The temperature at which a magnetic material loses its ferromagnetic properties.
current A movement of electric charges around a closed path or circuit.
current divider A parallel circuit to provide branch *I* less than the main-line current.
current gain, A_i The ratio of output current to input current in a transistor amplifier.
current source Supplies $I = V/r_i$ to load with r_i in parallel.
current-source region The region of operation in which the drain of a JFET acts as a current source. The current-source region of operation exists when $V_{DS} > V_P$.
cutoff The region of transistor operation where the collector current, I_C, is zero.
cutoff frequency The frequency at which the attenuation of a filter reduces the output amplitude to 70.7% of its value in the passband.
cycle One complete set of values for a repetitive waveform.

D

damping Reducing the *Q* of a resonant circuit to increase the bandwidth.
D'Arsonval meter A DC analog meter movement commonly used in ammeters and voltmeters.
dB Abbreviation for *decibel*. Equals 10 times the logarithm of the ratio of two power levels.
DC *See* direct current.
DC alpha, α_{DC} The ratio of collector current, I_C, to emitter current, I_E, in a transistor; $\alpha_{DC} = \frac{I_C}{I_E}$.
DC beta, β_{DC} The ratio of collector current, I_C, to base current, I_B, in a transistor; $\beta_{DC} = \frac{I_C}{I_B}$.
DC input power, P_{CC} The amount of DC power dissipated by a transistor amplifier.
DC load line A graph that shows all of the possible values of I_C and V_{CE} for a given transistor amplifier. The endpoints of the DC load line are $I_{C(sat)}$ and $V_{CE(off)}$ which represent the values of I_C and V_{CE} when the transistor is in saturation and cutoff.
decade A 10:1 range of values.
decade resistance box A unit for providing any resistance within a wide range of values.
decibel *See* dB.
decimal notation Numbers that are written in standard form without using powers of 10 notation.
degaussing Demagnetizing by applying an AC field and gradually reducing it to zero.
delta (Δ) network Three components connected in series in a closed loop. Same as pi (π) network.
depletion mode The mode of operation for a MOSFET in which the polarity of V_{GS} causes the drain current to be reduced as the channel becomes depleted of available charge carriers.
depletion zone The area located at the *p-n* junction of a diode that is void or depleted of all charge carriers.
derating curve A graph showing how the power rating of a resistor decreases as its operating temperature increases.
derating factor The amount that the power rating of a transistor must be reduced for each degree Celsius above 25°C. The derating factor is specified in W/°C.
diac A bidirectional semiconductor device that conducts when the voltage across its terminals reaches the breakover voltage $\pm V_{BO}$. Once conducting the voltage across the diac drops to a very low value.
diamagnetic Material that can be weakly magnetized in the direction opposite from the magnetizing field.
dielectric Another name for insulator.
dielectric absorption The inability of a capacitor to completely discharge to zero. Dielectric absorption is sometimes called battery action or capacitor memory.
dielectric constant Ability to concentrate the electric field in a dielectric.
dielectric material Insulating material. It cannot conduct current but does store charge.
dielectric strength The ability of a dielectric to withstand a potential difference without internal arcing.
differential input voltage, V_{id} The voltage difference between the two inputs applied to a differential amplifier.
differential voltage gain, A_d The ratio of output voltage, V_{out}, to differential input voltage, V_{id}.
differentiator An *RC* circuit with a short time constant for pulses across *R*.
digital multimeter (DMM) A popular test instrument that is used to measure voltage, current, and resistance.
diode A unidirectional device that allows current to flow through it in only one direction.
diode bias A form of biasing for class B push-pull amplifiers that uses diodes to provide a slight amount of forward bias for the base-emitter junctions of each transistor.
direct current (DC) Current that flows in only one direction. DC voltage has a steady polarity that does not reverse.
discharge current The current that flows to and from the plates of a capacitor as the charge stored by the dielectric decreases.
discharging (1) The action of neutralizing the charge stored in a capacitor by connecting a conducting path across the capacitor leads. (2) The process of neutralizing the separated charges on the electrodes of a cell or battery as a result of supplying current to a load resistance.
DMM *See* digital multimeter.
doping The process of adding impurity atoms to a pure semiconductor material such as silicon.
double-subscript notation A notational system that identifies the points in the circuit where a voltage measurement is to be taken, i.e.; V_{AG}. The first letter in the subscript indicates the point in the circuit where the measurement is to be taken whereas the second letter indicates the point of reference.
DPDT Double-pole double-throw switch or relay contacts.
DPST Double-pole single-throw switch or relay contacts.
drain One of the three leads of an FET. The drain lead connects to one end of the conducting channel.
dynamometer Type of AC meter, generally for 60 Hz.

E

earth ground A direct connection to the earth usually made by driving copper rods into the earth and then connecting the ground wire of an electrical system to this point. The earth ground connection can serve as a common return path for the current in a circuit.
eddy current Circulating current induced in the iron core of an inductor by AC variations of magnetic flux.
effective value For sine-wave AC waveform, 0.707 of peak value. Corresponds to heating effect of same DC value. Also called *rms value*.
efficiency Ratio of power output to power input × 100%.
EIA Electronic Industries Alliance.
electric field The invisible lines of force between opposite electric charges.
electricity Dynamic electricity is the effect of voltage in producing current in conductors. Static electricity is accumulation of charge.
electrolyte Solution that forms ion charges.
electrolytic capacitor Type with very high *C* because electrolyte is used to form very thin dielectric. Must be connected with correct polarity in a circuit.

electromagnet Magnet whose magnetic field is associated with electric current in a coil.

electron Basic particle of negative charge in orbital rings around the nucleus in an atom.

electron flow Current of negative charges in motion. Direction is from the negative terminal of the voltage source, through the external circuit, and returning to the positive side of the source. Opposite to the direction of conventional current.

electron-hole pair The creation of a free electron and a hole when a valence electron gains enough energy to leave its covalent bond in a silicon crystal.

electron valence The number of electrons in an incomplete outermost shell of an atom.

electron volt Unit of energy equal to the work done in moving a charge of 1 electron through a potential difference of 1 V.

element A substance that cannot be decomposed any further by chemical action.

emf Electromotive force; voltage that produces current in a circuit.

emitter The most heavily doped region in a transistor. Its job is to inject or emit current carriers into the base region.

emitter bypass capacitor, C_E A capacitor that bypasses the AC signal around the emitter resistor in a transistor amplifier.

emitter follower Another name for the common-collector amplifier.

engineering notation A form of powers of 10 notation in which a number is expressed as a number between 1 and 1000 times a power of 10 that is a multiple of 3.

enhancement mode The mode of operation for a MOSFET in which the polarity of V_{GS} causes the conductivity of the channel to be enhanced, thus increasing the drain current.

equivalent resistance, R_{EQ} In a parallel circuit, this refers to a single resistance that would draw the same amount of current as all the parallel connected branches.

equivalent series resistance (ESR) A resistance in series with an ideal capacitor that collectively represents all of the losses in a capacitor. Ideally, the ESR of a capacitor should be zero.

extrinsic semiconductor A semiconductor that has been doped with impurity atoms to alter the characteristics of the material, mainly its conductivity.

F

F connector Solderless plug for coaxial cable.

f_{unity} The frequency where the open-loop voltage gain, A_{vol}, of an op amp equals 1 or unity.

Fahrenheit scale (°F) Temperature scale that uses 32° for the freezing point of water and 212° for the boiling point.

farad (F) Unit of capacitance. Value of one farad stores one coulomb of charge with one volt applied.

Faraday's law For magnetic induction, the generated voltage is proportional to the flux and its rate of change.

ferrite Magnetic material that is not a metal conductor.

ferrite core A type of core that has a high value of flux density, like iron, but is an insulator. A ferrite core used in a coil has minimum eddy current losses due to its high resistance.

ferromagnetic Magnetic properties of iron and other metals that can be strongly magnetized in the same direction as the magnetizing field.

field Group of lines of force; magnetic or electric field.

field effect transistor (FET) A unipolar device that relies on only one type of charge carrier, either electrons or holes. FETs are voltage-controlled devices with an input voltage controlling the output current.

field intensity (*H*) The mmf per unit of length.

field winding The part of a motor or generator that supplies the magnetic field cut by the armature.

film capacitor A capacitor that uses a plastic film for its dielectric.

filter Circuit that separates different frequencies.

float charging A method of charging in which the charger and the battery are always connected to each other for supplying current to the load. With this method, the charger provides the current for the load and the current necessary to keep the battery fully charged.

fluctuating direct current Varying voltage and current but no change in polarity.

flux (ϕ) Magnetic lines of force.

flux density (*B*) Amount of flux per unit area.

flywheel effect Ability of an *LC* circuit to continue oscillating after the energy source has been removed.

form factor The ratio of the rms to average values. For a sine wave, $\frac{\text{rms}}{\text{avg}} = 1.11$.

forward blocking current The small current that flows in an SCR before breakover is reached.

forward breakover voltage, V_{BRF} The forward voltage across an SCR at which the SCR begins to conduct. The value of V_{BRF} is controlled by the amount of gate current, I_G.

forward-bias The polarity of voltage across a diode that permits current to flow through it easily.

free electron Electron that can move freely from one atom to the next.

frequency (*f*) Number of cycles per second for a waveform with periodic variations. The unit is hertz (Hz).

frequency multiplier A tuned class C amplifier that has its *LC* tank circuit tuned to a harmonic or multiple of the input frequency.

fuel cell An electrochemical device that converts the chemicals hydrogen and oxygen into water, and in the process produces electricity. A fuel cell provides a steady DC output voltage that can power motors, lights, or other appliances. Unlike a regular battery, however, a fuel cell constantly has chemicals flowing into it so it never goes dead.

full-wave rectifier A circuit that provides an entirely positive or negative output voltage when an AC input voltage is applied. A full-wave rectifier provides an output for both the positive and negative alternations of the input voltage.

fuse Metal link that melts from excessive current and opens a circuit.

G

galvanic cell Electrochemical type of voltage source.

galvanometer Measures electric charge or current.

ganged capacitors Two or three capacitor sections on one common shaft that can be rotated.

gate One of the three leads of an FET. The gate is used to control the drain current.

gate-source cutoff voltage, $V_{GS(off)}$ The amount of gate-source voltage required to reduce the drain current, I_D, to zero.

gauss (G) Unit of flux density in cgs system equal to one magnetic line of force per square centimeter.

generator A device that produces voltage output. Is a source for either DC or AC *V* and *I*.

germanium (Ge) Semiconductor element used for transistors and diodes.

giga (G) Metric prefix for 10^9.

gilbert (Gb) Unit of magnetomotive force in cgs system. One gilbert equals 0.794 ampere-turn.

graph cycle A 10:1 range of values on logarithmic graph paper.

ground Common return to earth for AC power lines. Chassis ground in electronic equipment is the common return to one side of the internal power supply.

H

half-power points Bandwidth defined with 70.7% response for resonant *LC* circuit.

half-wave rectifier A circuit that provides an entirely positive or negative output voltage when an AC input voltage is applied. A half-wave rectifier provides an output for either the positive or negative alternation of the input voltage but not both.

Hall effect Small voltage generated by a conductor with current in an external magnetic field.

harmonic frequency Exact multiple of fundamental frequency.

henry (H) Unit of inductance. Current change of one ampere per second induces one volt across an inductance of one henry.

hertz (Hz) Unit of frequency. One hertz equals one cycle per second.

high-pass filter A filter that allows the higher-frequency components of the applied voltage to develop appreciable output voltage while at the same time attenuating or eliminating the lower-frequency components.

holding current, I_H (1) The minimum amount of current required to hold a thyristor (diac, SCR, or triac) in its conducting state. (2) The minimum amount of current required to keep a relay energized.

hole The absence of a valence electron in a covalent bond structure. The hole exhibits a positive charge.

hole current Motion of hole charges. Direction is the same as that of conventional current, opposite from electron flow.

horsepower (hp) A unit of mechanical power corresponding to 550 ft·lb/s. In terms of electric power, 1 hp = 746 W.

hot resistance The R of a component with its normal load current. Determined by V/I.

hot-wire meter Type of AC meter.

hydrometer A device used to check the state of charge of a cell within a lead-acid battery.

hypotenuse Side of a right triangle opposite the 90° angle.

hysteresis In electromagnets, the effect of magnetic induction lagging in time behind the applied magnetizing force.

Hz *See* hertz.

I

IGFET Insulated gate field effect transistor. Another name for a MOSFET.

imaginary number Value at 90°, indicated by j operator, as in the form jA.

impedance matching Occurs when a transformer is used for its impedance transformation properties. With impedance matching, maximum power is delivered to the load, R_L.

impedance, Z The total opposition to the flow of current in a sine-wave AC circuit. In an RC circuit, the impedance, Z, takes into account the 90° phase relation between X_C and R. Impedance, Z, is measured in ohms.

inductance (L) Ability to produce induced voltage when cut by magnetic flux. Unit of inductance is the henry (H).

induction Ability to generate V or I without physical contact. Electromagnetic induction by magnetic field; electrostatic induction by electric field.

inductive reactance, X_L A measure of an inductor's opposition to the flow of alternating current. X_L is measured in ohms and is calculated as $X_L = 2\pi fL$ or $X_L = \frac{V_L}{I_L}$.

inductor Coil of wire with inductance.

input bias current, I_B The average of the two op-amp input currents I_{B+} and I_{B-}.

input impedance, Z_{in} The impedance of the input of an amplifier as seen by the AC signal source driving the amplifier.

input offset current, I_{os} The difference between the two input bias currents I_{B+} and I_{B-}.

insulator A material that does not allow current to flow when voltage is applied, because of its high resistance.

integrator An RC circuit with a long time constant. Voltage output across C.

interbase resistance, R_{BB} The resistance of the n-type silicon bar in a UJT. R_{BB} appears as two resistances, R_{B1} and R_{B2}. $R_{BB} = R_{B1} + R_{B2}$.

internal resistance r_i Limits the current supplied by the voltage source to $I = V/r_i$.

intrinsic semiconductor A semiconductor material with only one type of atom.

intrinsic standoff ratio, η The ratio of R_{B1} to R_{BB} or $\eta = \frac{R_{B1}}{R_{B1} + R_{B2}}$.

inverse relation Same as reciprocal relation. As one variable increases, the other decreases.

inversely proportional The same as a reciprocal relation; as the value in the denominator increases the resultant quotient decreases. In the formula $X_C = \frac{1}{2\pi fC}$. X_C is inversely proportional to both f and C. This means that as f and C increase, X_C decreases.

ion Atom or group of atoms with net charge. Can be produced in liquids, gases, and doped semiconductors.

ionization current A current that results from the movement of ion charges in a liquid or gas.

***IR* drop** Voltage across a resistor.

iron-vane meter Type of AC meter, generally for 60 Hz.

J

***j* operator** Indicates 90° phase angle, as in $j8\ \Omega$ for X_L. Also, $-j8\ \Omega$ is at $-90°$ for X_C.

JFET Junction field effect transistor.

joule (J) Practical unit of work or energy. One joule equals one watt-second of work.

K

k Coefficient of coupling between coils.

keeper Magnetic material placed across the poles of a magnet to form a complete magnetic circuit. Used to maintain strength of magnetic field.

Kelvin (K) scale Absolute temperature scale, 273° below values on Celsius scale.

kilo (k) Metric prefix for 10^3.

kilowatt-hour A large unit of electrical energy corresponding to 1 kW·1 h.

Kirchhoff's current law (KCL) The algebraic sum of all currents into and out of any branch point in a circuit must equal zero.

Kirchhoff's voltage law (KVL) The algebraic sum of all voltages around any closed path must equal zero.

L

laminations Thin sheets of steel insulated from one another to reduce eddy-current losses in inductors, motors, etc.

leakage current (1) The current that flows through the dielectric of a capacitor when voltage is applied across the capacitor plates. (2) The very small current that flows when a diode is reverse-biased. The leakage current is mainly due to the thermally generated minority carriers that exist in both sections of the diode.

leakage flux Any magnetic field lines that do not link two coils that are in close proximity to each other.

leakage resistance A resistance in parallel with a capacitor that represents all of the leakage paths through which a capacitor can discharge.

Leclanché cell Carbon-zinc primary cell.

left-hand rule If the coil is grasped with the fingers of the left hand curled around the coil in the direction of electron flow, the thumb points to the north pole of the coil.

Lenz's law Induced current has magnetic field that opposes the change causing the induction.

light-emitting diode (LED) A diode that emits a certain color light when forward-biased. The color of light emitted by the diode is determined by the type of material used in the doping process.

line current The current flowing in the wires or lines that connect the terminals of a three-phase AC generator to the terminals of a wye- or delta-connected load.

line voltage The voltage measured across two terminals or lines (not including neutral) in a three-phase AC generator.

linear amplifier Any amplifier that produces an output signal that is an exact replica of the input signal.

linear component An electronic component whose current is proportional to the applied voltage.

linear proportion Straight-line graph between two variables. As one increases, the other increases in direct proportion.

linear resistance A resistance with a constant value of ohms.

load Takes current from the voltage source, resulting in load current.

load currents The currents drawn by the electronic devices and/or components connected as loads in a loaded voltage divider.

loaded voltage The voltage at a point in a series voltage divider where a parallel load has been connected.

loading effect Source voltage is decreased as amount of load current increases.

long time constant A long time constant can arbitrarily be defined as one that is five or more times longer than the pulse width of the applied voltage.

loop Any closed path in a circuit.

loop equation An equation that specifies the voltages around a loop.

low-pass filter A filter that allows the lower frequency components of the applied voltage to develop appreciable output voltage while at the same time attenuating or eliminating the higher-frequency components.

M

magnetic flux (ϕ) Another name used to describe magnetic field lines.
magnetic pole Concentrated point of magnetic flux.
magnetism Effects of attraction and repulsion by iron and similar materials without the need for an external force. Electromagnetism includes the effects of a magnetic field associated with an electric current.
magnetomotive force (mmf) Ability to produce magnetic lines of force. Measured in units of ampere-turns.
magnitude Value of a quantity regardless of phase angle.
main line The pair of leads connecting all of the individual branches in a parallel circuit to the terminals of the applied voltage, V_A. The main line carries the total current, I_T, flowing to and from the terminals of the voltage source.
majority current carrier The dominant type of charge carrier in a doped semiconductor material. In an *n*-type semiconductor, free electrons are the majority current carriers whereas in a *p*-type semiconductor holes are the majority current carriers.
make and break Occurs when contacts close and open.
maximum working voltage rating The maximum allowable voltage a resistor can safely withstand without internal arcing.
maxwell (Mx) Unit of magnetic flux equal to one line of force in the magnetic field.
mega (M) Metric prefix for 10^6.
mesh The simplest possible closed path within a circuit.
mesh current Assumed current in a closed path, without any current division, for application of Kirchhoff's current law.
metal-film resistors Resistors made by spraying a thin film of metal onto a ceramic substrate. The metal film is cut in the form of a spiral.
metric prefixes Letter symbols used to replace the powers of 10 that are multiples of 3.
micro (μ) Metric prefix for 10^{-6}.
microfarad A small unit of capacitance equal to 1×10^{-6} F.
midpoint bias A bias point that is centered between cutoff and saturation on the DC load line.
milli (m) Metric prefix for 10^{-3}.
Millman's theorem A theorem that provides a shortcut for finding the common voltage across any number of parallel branches with different voltage sources.
minority current carrier The type of charge carrier that appears sparsely throughout a doped semiconductor material. In an *n*-type semiconductor, holes are the minority current carriers whereas free electrons are the minority current carriers in a *p*-type semiconductor.
mks Meter-kilogram-second system of units.
molecules The smallest unit of a compound with the same chemical characteristics.
MOSFET Metal-oxide-semiconductor field effect transistor.
motor A device that produces mechanical motion from electric energy.
motor action A motion that results from the net force of two magnetic fields that can aid or cancel each other. The direction of the resultant force is always from a stronger field to a weaker field.
multiplier resistor Resistor in series with a meter movement for voltage ranges.
mutual induction (L_M) Ability of one coil to induce voltage in another coil.

N

nano (n) Metric prefix for 10^{-9}.
nanofarad A small unit of capacitance equal to 1×10^{-9} F.
NC Normally closed for relay contacts, or no connection for pinout diagrams.
negative feedback A form of amplifier feedback where the returning signal has a phase that opposes the input signal.
negative saturation voltage, $-V_{sat}$ The lower limit of output voltage for an op amp.
negative temperature coefficient (NTC) A characteristic of a thermistor indicating that its resistance decreases with an increase in operating temperature.
neutron Particle without electric charge in the nucleus of an atom.
node A common connection for two or more branch currents.
nonlinear resistance A resistance whose value changes as a result of current producing power dissipation and heat in the resistance.
nonsinusoidal waveform Any waveform that is not a sine or a cosine wave.
Norton's theorem Method of reducing a complicated network to one current source with shunt resistance.
***n*-type semiconductor** A semiconductor that has been doped with pentavalent impurity atoms. The result is a large number of free electrons throughout the material. Since the electron is the basic particle of negative charge, the material is called *n*-type semiconductor material.
nucleus The massive, stable part of the atom which contains both protons and neutrons.

O

obtuse angle More than 90°.
octave A 2:1 range of values.
oersted (Oe) Unit of magnetic field intensity; 1 Oe = 1 Gb/cm.
ohm (Ω) Unit of resistance. Value of one ohm allows current of one ampere with potential difference of one volt.
Ohm's law In electric circuits, $I = V/R$.
ohmic region The region of operation for a JFET where the drain current, I_D, increases in direct proportion to V_{DS}. The ohmic region of operation exists when $V_{DS} < V_P$.
ohms-per-volt rating Sensitivity rating for a voltmeter. High rating means less meter loading.
open circuit One that has infinitely high resistance, resulting in zero current.
open-circuit voltage The voltage present across the output terminals of a voltage source when no load is present.
open-loop cutoff frequency, f_{OL} The frequency at which the open-loop voltage gain of an op amp is down to 70.7% of its maximum value at DC.
open-loop voltage gain, A_{VOL} The voltage gain of an op amp without negative feedback.
operational amplifier (op amp) A high-gain, direct-coupled differential amplifier.
oscilloscope A piece of test equipment used to view and measure a variety of different AC waveforms.
output impedance, Z_{out} The impedance at the output of an amplifier as seen by the load being driven by the amplifier.

P

parallel bank A combination of parallel-connected branches.
parallel circuit One that has two or more branches for separate currents from one voltage source.
paramagnetic Material that can be weakly magnetized in the same direction from the magnetizing force.
passive component Components such as resistors, capacitors, and inductors. They do not generate voltage or control current.
PC board A device that has printed circuits.
peak inverse voltage (PIV) The maximum instantaneous reverse-bias voltage across a diode.
peak reverse voltage rating, V_{ROM} The maximum reverse-bias voltage that can be safely applied between the anode and cathode terminals of an SCR with the gate open.
peak-to-peak value (p-p) Amplitude between opposite peaks.
peak value Maximum amplitude, in either polarity; 1.414 times rms value for sine-wave *V* or *I*.
pentavalent atom An atom with 5 valence electrons.
percent efficiency For an amplifier, this refers to the percentage of DC input power that is converted to useful AC output power.
period (*T*) The amount of time it takes to complete one cycle of alternating voltage or current.
permanent magnet (PM) It has magnetic poles produced by internal atomic structure. No external current needed.
permeability Ability to concentrate magnetic lines of force.
phase angle θ (1) Angle between two phasors; denotes time shift. (2) The angle between the applied voltage and current in a sine-wave AC circuit.

phase current The current flowing in a coil winding of a three-phase AC generator.

phase voltage The voltage directly across a coil winding in a three-phase AC generator.

phasing dots Used on transformer windings to identify those leads having the same instantaneous polarity.

phasor A line representing magnitude and direction of a quantity, such as voltage or current, with respect to time.

phasor triangle A right triangle that represents the phasor sum of two quantities 90° out of phase with each other.

pickup current The minimum amount of current required to energize a relay.

pico (p) Metric prefix for 10^{-12}.

picofarad A small unit of capacitance equal to 1×10^{-12} F.

pinch-off voltage, V_P The drain-source voltage at which the drain current, I_D, levels off. V_P is the border between the ohmic and current-source regions of operation.

polar form Form of complex numbers that gives magnitude and phase angle in the form $A\angle\theta°$.

polarity Property of electric charge and voltage. *Negative* polarity is excess of electrons. *Positive* polarity means deficiency of electrons.

pole The number of completely isolated circuits that can be controlled by a switch.

polyphase AC circuit A circuit containing two or more AC voltage sources, each with a different phase angle.

positive saturation voltage, $+V_{sat}$ The upper limit of output voltage for an op amp.

positive temperature coefficient (PTC) A characteristic of a thermistor indicating that its resistance increases with an increase in operating temperature.

potential difference Ability of electric charge to do work in moving another charge. Measured in volt units.

potentiometer Variable resistor with three terminals connected as a voltage divider.

power (*P*) Rate of doing work. The unit of electric power is the watt.

power amplifier A circuit that is designed to deliver large amounts of power to a low impedance load.

power bandwidth (f_{max}) The highest undistorted frequency out of an op amp without slew-rate distortion.

power factor Cosine of the phase angle for a sine-wave AC circuit. Value is between 1 and 0.

power gain, A_P The ratio of output power to input power in a transistor amplifier. A_P can also be calculated as $A_P = A_V \times A_i$.

power supply A piece of test equipment used to supply DC voltage and current to electronic circuits under test.

powers of 10 A numerical representation consisting of a base of 10 and an exponent; the base 10 raised to a power.

preferred values Common values of resistors and capacitors generally available.

primary cell or battery Type that cannot be recharged.

primary winding Transformer coil connected to the source voltage.

principal node A common connection for three or more components in a circuit where currents can combine or divide.

printed wiring Conducting paths printed on plastic board.

proportional A mathematical term used to describe the relationship between two quantities. For example, in the formula $X_L = 2\pi fL$, X_L is said to be directly proportional to both the frequency, *f*, and the inductance, *L*. The term proportional means that if either *f* or *L* is doubled X_L will double. Similarly, if either *f* or *L* is reduced by one-half, X_L will be reduced by one-half. In other words, X_L will increase or decrease in direct proportion to either *f* or *L*.

proton Particle with positive charge in the nucleus of an atom.

***p*-type semiconductor** A semiconductor that has been doped with trivalent impurity atoms. The result is a large number of holes in the material. Since a hole exhibits a positive charge, the material is called *p*-type semiconductor material.

pulsating DC A DC voltage or current that varies in magnitude but does not reverse in polarity or direction. Another name for pulsating DC is fluctuating DC. Includes AC component on average DC axis.

pulse A sharp rise and decay of voltage or current of a specific peak value for a brief period of time.

Q

***Q* of a coil** The quality or figure of merit for a coil. More specifically, the *Q* of a coil can be defined as the ratio of reactive power in the inductance to the real power dissipated in the coil's resistance. $Q = \frac{X_L}{r_i}$.

***Q* point** The values of I_C and V_{CE} that exist in a transistor amplifier with no AC signal present.

quadrature phase A 90° phase angle.

R

R Symbol for resistance.

radian (rad) Angle of 57.3°. Complete circle includes 2π rad.

radio frequency (rf) A frequency high enough to be radiated efficiently as electromagnetic waves, generally above 30 kHz. Usually much higher.

ramp Sawtooth waveform with linear change in *V* or *I*.

ratio arm Accurate, stable resistors in one leg of a Wheatstone bridge or bridge circuit in general. The ratio arm fraction, $\frac{R_1}{R_2}$, can be varied in most cases, typically in multiples of 10. The ratio arm fraction in a Wheatstone bridge determines two things; the placement accuracy of the measurement of an unknown resistor, R_x, and the maximum unknown resistance, $R_{x(max)}$, that can be measured.

***RC* phase-shifter** An application of a series *RC* circuit in which the output across either *R* or *C* provides a desired phase shift with respect to the input voltage. *RC* phase-shifter circuits are commonly used to control the conduction time of semiconductors in power control circuits.

reactance Property of *L* and *C* to oppose flow of *I* that is varying. Symbol is X_C or X_L. Unit is the ohm.

real number Any positive or negative number not containing *j*. $(A + jB)$ is a complex number but *A* and *B* by themselves are real numbers.

real power The net power consumed by resistance. Measured in watts.

reciprocal relation Same as inverse relation. As one variable increases, the other decreases.

reciprocal resistance formula A formula that states that the equivalent resistance, R_{EQ}, of a parallel circuit equals the reciprocal of the sum of the reciprocals of the individual branch resistances.

rectangular form Representation of a complex number in the form $A + jB$.

reflected impedance The value of impedance reflected back into the primary from the secondary.

relative permeability (μ_r) The ability of a material to concentrate magnetic flux. Mathematically, relative permeability, designated μ_r, is a ratio of the flux density (*B*) in a material such as iron to the flux density, *B*, in air. There are no units for μ_r because it is comparison of two flux densities and the units cancel.

relative permittivity (ε_r) A factor that indicates the ability of an insulator to concentrate electric flux, also known as the dielectric constant, K_ε.

relay Automatic switch operated by current in a coil.

relay chatter The vibrating of relay contacts.

resistance (*R*) Opposition to current. Unit is the ohm (Ω).

resistance wire A conductor having a high resistance value.

resonance Condition of $X_L = X_C$ in an *LC* circuit to favor the resonant frequency for a maximum in *V*, *I*, or *Z*.

resonant frequency The frequency at which the inductive reactance, X_L and the capacitive reactance, X_C of an *LC* circuit are equal.

reverse bias The polarity of voltage across a diode that prevents the diode from conducting any current.

rheostat Variable resistor with two terminals to vary *I*.

ringing Ability of an *LC* circuit to oscillate after a sharp change in *V* or *I*.

root-mean-square (rms) value For sine-wave AC waveform, 0.707 of peak value. Also called effective value.

rotor Rotating part of generator or motor.

S

saturation The region of transistor operation where the collector current no longer increases with further increases in base current.

saturation region The region to the right of the valley point, on the characteristic curve of a UJT.

sawtooth wave One in which amplitude values have a slow linear rise or fall and a sharp change back to the starting value. Same as a *linear ramp.*

Schmitt trigger An op-amp comparator that utilizes positive feedback.

scientific notation A form of powers of 10 notation in which a number is expressed as a number between 1 and 10 times a power of 10.

secondary cell or battery Type that can be recharged.

secondary winding Transformer coil connected to the load.

self-inductance (*L*) Inductance produced in a coil by current in the coil itself.

semiconductor A material that is neither a good conductor nor a good insulator.

series-aiding A connection of coils in which the coil current produces the same direction of magnetic field for both coils.

series-aiding voltages Voltage sources that are connected so that the polarities of the individual sources aid each other in producing current in the same direction in the circuit.

series circuit One that has only one path for current.

series components Components that are connected in the same current path.

series opposing A connection of coils in which the coil current produces opposing magnetic fields for each coil.

series-opposing voltages Voltage sources that are connected so that the polarities of the individual sources will oppose each other in producing current flow in the circuit.

series string A combination of series resistances.

shield Metal enclosure preventing interference from radio waves.

short circuit Has zero resistance, resulting in excessive current.

short time constant A short time constant can arbitrarily be defined as one that is one-fifth or less the time of the pulse width of the applied voltage.

shunt resistor A parallel connection. A device to increase the range of an ammeter.

SI Abbreviation for *Système International,* a system of practical units based on the meter, kilogram, second, ampere, kelvin, mole, and candela.

siemens (S) Unit of conductance. Reciprocal of ohms unit.

silicon (Si) Semiconductor element used for transistors, diodes, and integrated circuits.

silicon controlled rectifier (SCR) A unidirectional semiconductor device, like a diode, that remains in a nonconducting state, although forward-biased, until the forward breakover voltage is reached. Once conducting, the voltage across the SCR drops to a very low value.

sine Trigonometric function of an angle, equal to the ratio of the opposite side to the hypotenuse in a right triangle.

sine wave One in which amplitudes vary in proportion to the sine function of an angle.

single-phase AC circuit A circuit containing only one AC voltage source.

skin effect A term used to describe current flowing on the outer surface of a conductor at very high frequencies. Skin effect causes the effective resistance of a coil to increase at higher frequencies because the effect is the same as reducing the cmil area of the wire.

slew rate, S_R An op-amp specification indicating the maximum rate at which the output voltage can change. S_R is specified in $\frac{V}{\mu s}$.

slew-rate distortion A distortion that occurs in op amps when the rate of change in output voltage tries to exceed the slew rate capabilities of the op amp.

slip rings In an AC generator, devices that provide connections to the rotor.

slow-blow fuse A type of fuse that can handle a temporary surge current that exceeds the current rating of the fuse. This type of fuse has an element with a coiled construction and is designed to open only on a continued overload such as short-circuit.

small signal A signal whose peak-to-peak current value is one-tenth or less the DC diode or DC emitter current.

solder Alloy of tin and lead used for fusing wire connections.

solenoid Coil used for electromagnetic devices.

source One of the three leads of an FET. The source lead connects to one end of the conducting channel.

spade lug A type of wire connector.

SPDT Single-pole double-throw switch or relay contacts.

specific gravity Ratio of weight of a substance with that of an equal volume of water.

specific resistance The *R* for a unit length, area, or volume.

SPST Single-pole single-throw switch or relay contacts.

square wave An almost instantaneous rise and decay of voltage or current in a periodic pattern with time and with a constant peak value. The *V* or *I* is ON and OFF for equal times and at constant values.

standard resistor A variable resistor in one leg of a Wheatstone bridge that is varied to provide equal voltage ratios in both series strings of the bridge. With equal voltage ratios in each series string, the bridge is said to be balanced.

static electricity Electric charges not in motion.

stator Stationary part of a generator or motor.

steady-state value The *V* or *I* produced by a source without any sudden changes. Can be DC or AC value. Final value of *V* or *I* after transient state.

storage cell or battery Type that can be recharged.

stray capacitance A very small capacitance that exists between any two conductors separated by an insulator. The capacitance can be between two wires in a wiring harness or between a single wire and a metal chassis as examples.

stray inductance The small inductance associated with any length of conductor or component lead. The effects of both stray inductance and stray capacitance are most noticeable with very high frequencies.

string Components connected in series.

strings in parallel Series resistor strings that are connected in parallel with each other.

summing amplifier An amplifier whose output voltage equals the negative sum of the input voltages.

superconductivity Very low *R* at extremely low temperatures.

superposition theorem Method of analyzing a network with multiple sources by using one at a time and combining their effects.

supersonic Frequency above the range of hearing, generally above 16,000 Hz.

surface-mount resistor Resistor made by depositing a thick carbon film on a ceramic base. Electrical connection to the resistive element is made by two leadless solder, end electrodes that are C-shaped.

surface-mount technology Components soldered directly to the copper traces of a printed circuit board. No holes need to be drilled for surface-mounted components.

susceptance (*B*) Reciprocal of reactance in sine-wave AC circuits; $B = 1/X$.

swamping resistor An unbypassed resistor in the emitter circuit of a common-emitter amplifier. A swamping resistor stabilizes the voltage gain and reduces distortion.

switch Device used to open or close connections of a voltage source to a load circuit.

switching contacts The contacts that open and close when a relay is energized.

symmetrical JFET A JFET that has its gate regions located in the center of the channel. With a symmetrical JFET the drain and source leads can be interchanged without affecting its operation.

T

tail current, I_T The DC current in the emitter resistor of a differential amplifier.

tangent (tan) Trigonometric function of an angle, equal to the ratio of the opposite side to the adjacent side in a right triangle.

tank circuit An *LC* tuned circuit. Stores energy in *L* and *C*.

tantalum Chemical element used for electrolytic capacitors.

taper How *R* of a variable resistor changes with the angle of shaft rotation.

tapered control The manner in which the resistance of a potentiometer varies with shaft rotation. For a linear taper, one-half shaft rotation corresponds to a resistance change of one-half its maximum value. For a nonlinear taper, the resistance change is more gradual at one end, with larger changes at the other end.

taut-band meter Type of construction for meter movement often used in VOM.

temperature coefficient For resistance, how R varies with a change in temperature.

tesla (T) Unit of flux density, equal to 10^8 lines of force per square meter.

thermistor A resistor whose resistance value changes with changes in its operating temperatures.

Thevenin's theorem Method of reducing a complicated network to one voltage source with series resistance.

three-phase power AC voltage generated with three components differing in phase by 120°.

threshold voltage, $V_{GS(th)}$ The minimum value of V_{GS} in an enhancement-type MOSFET which causes drain current to flow.

throw The number of closed contact positions that exist per pole on a switch.

thyristor A semiconductor device with alternating layers of p and n material that can only be operated in the switching mode where they act as either an open or closed switch.

time constant Time required to change by 63% after a sudden rise or fall in V and I. Results from the ability of L and C to store energy. Equals RC or L/R.

tolerance The maximum allowable percent difference between the measured and coded values of resistance.

toroid Electromagnet with its core in the form of a closed magnetic ring.

transconductance, g_m The ratio of the change in drain current, ΔI_D, to the change in gate-source voltage, ΔV_{GS} for a fixed value of V_{DS}. The unit of g_m is the siemens (S).

transformer A device that has two or more coil windings used to step up or step down AC voltage.

transient response Temporary value of V or I in capacitive or inductive circuits caused by abrupt change.

transistor A three-terminal semiconductor device that can amplify an AC signal or be used as an electronic switch.

triac A bidirectional semiconductor device that remains in a nonconducting state until the forward breakover voltage is reached. Once conducting, the voltage across the triac drops to a very low value. Like an SCR, the breakover voltage can be controlled by gate current.

trigonometry Analysis of angles and triangles.

trivalent atom An atom with 3 valence electrons.

troubleshooting A term that refers to the diagnosing or analyzing of a faulty electronic circuit.

tuning Varying the resonant frequency of an LC circuit.

turns ratio Comparison of turns in primary and secondary of a transformer.

twin lead Transmission line with two conductors in plastic insulator.

U

UHF Ultra high frequencies in band of 30 to 300 MHz.

unijunction transistor (UJT) A 3-terminal semiconductor device that has only 1 p-n junction. UJTs are used to control the conduction angle of an SCR.

unipolar A device having only one type of charge carrier, either electrons or holes.

universal time constant graph A graph that shows the percent change in voltage or current in an RC or RL circuit with respect to the number of time constants that have elapsed.

V

valence electrons Electrons in the outermost ring or shell of an atom.

VAR *See* volt-ampere reactive.

Variac Transformer with variable turns ratio to provide different amounts of secondary voltage.

vector A line representing magnitude and direction in space.

VHF Very high frequencies in band of 30 to 300 MHz.

volt (V) Practical unit of potential difference. One volt produces one ampere of current in a resistance of one ohm. $1\text{ V} = \frac{1\text{ J}}{1\text{ C}}$.

voltage divider A series circuit to provide V less than the source voltage.

voltage drop Voltage across each component in a series circuit. The proportional part of total applied V.

voltage follower An op-amp circuit with unity voltage gain. A voltage follower has a very high input impedance and a very low output impedance. Voltage followers are also known as unity-gain amplifiers, buffer amplifiers, and isolation amplifiers.

voltage gain, A_V The ratio of output voltage to input voltage in a transistor amplifier or $A_V = \frac{V_{out}}{V_{in}}$.

voltage polarity The positive and negative ends of a potential difference across a component such as a resistor.

voltage source Supplies potential difference across two terminals. Has internal series r_i.

voltage taps The points in a series voltage divider that provide different voltages with respect to ground.

voltaic cell A device that converts chemical energy into electric energy. The output voltage of a voltaic cell depends on the type of elements used for the electrodes.

volt-ampere (VA) Unit of apparent power, equal to $V \times I$.

volt-ampere characteristic Graph to show how I varies with V.

volt-ampere reactive (VAR) The volt-amperes at the angle of 90°.

voltmeter loading The amount of current taken by the voltmeter acting as a load. As a result, the measured voltage is less than the actual value.

VOM Volt-ohm-milliammeter.

W

watt (W) Unit of real power. Equal to I^2R or $VI \cos \theta$.

watt-hour Unit of electric energy, as power × time.

wattmeter Measures real power as instantaneous value of $V \times I$.

wavelength (λ) Distance in space between two points with the same magnitude and direction in a propagated wave.

wavetrap An LC circuit tuned to reject the resonant frequency.

weber (Wb) Unit of magnetic flux, equal to 10^8 lines of force.

Wheatstone bridge Balanced circuit used for precise measurements of resistance.

wire gage A system of wire sizes based on the diameter of the wire. Also, the tool used to measure wire size.

wire-wound resistors Resistors made with wire known as *resistance wire* that is wrapped around an insulating core.

work Corresponds to energy. Equal to power × time, as in kilowatt-hour unit. Basic unit is one joule, equal to one volt-coulomb, or one watt-second.

wye network Three components connected with one end in a common connection and the other ends to three lines. Same as T network.

X

X_C Capacitive reactance equal to $1/(2\pi fC)$.

X_L Inductive reactance equal to $2\pi fL$.

Y

Y Symbol for admittance in an AC circuit. Reciprocal of impedance Z; $Y = 1/Z$.

Y network Another way of denoting a wye network.

Z

Z Symbol for AC impedance. Includes resistance with capacitive and inductive reactance.

zener current, I_Z The name given to the reverse current in a zener diode.

zener diode A diode that has been optimized for operation in the breakdown region.

zero-crossing detector An op-amp comparator whose output voltage switches to either $\pm V_{sat}$ when the input voltage crosses through zero.

zero-ohm resistor A resistor whose value is practically 0 Ω. The 0-Ω value is denoted by a single black band around the center of the resistor body.

zero-ohms adjustment Used with ohmmeter of a VOM to set the correct reading at zero ohms.

zero-power resistance The resistance of a thermistor with zero power dissipation, designated R_0.

Answers

Self-Tests

Introduction to Powers of 10

1. d
2. a
3. b
4. c
5. c
6. a
7. a
8. d
9. b
10. c
11. c
12. b
13. a
14. d
15. c
16. b
17. c
18. a
19. b
20. d

CHAPTER 1

1. b
2. a
3. c
4. a
5. a
6. d
7. b
8. a
9. c
10. d
11. b
12. a
13. b
14. d
15. c
16. a
17. c
18. d
19. b
20. d
21. c
22. b
23. a
24. d
25. a

CHAPTER 2

1. b
2. d
3. a
4. c
5. c
6. a
7. a
8. d
9. c
10. c
11. c
12. b
13. d
14. a
15. b

CHAPTER 3

1. c
2. d
3. a
4. b
5. d
6. b
7. d
8. c
9. a
10. c
11. b
12. c
13. d
14. a
15. d
16. c
17. a
18. b
19. c
20. a

CHAPTER 4

1. c
2. b
3. a
4. a
5. c
6. c
7. a
8. b
9. a
10. c
11. b
12. c
13. d
14. b
15. a
16. c
17. b
18. d
19. b
20. c

CHAPTER 5

1. a
2. c
3. b
4. d
5. b
6. d
7. c
8. a
9. d
10. a
11. d
12. b
13. c
14. b
15. c
16. b
17. a
18. c
19. d
20. b

CHAPTER 6

1. c
2. b
3. a
4. c

5. d
6. c
7. a
8. a
9. b
10. d
11. b
12. d
13. a
14. c
15. d
16. b
17. d
18. c
19. a
20. b

REVIEW: CHAPTERS 1–6

1. a
2. c
3. b
4. c
5. c
6. c
7. b
8. c
9. b
10. d
11. b
12. a
13. c
14. b
15. a
16. a
17. a
18. a
19. b
20. a
21. b

CHAPTER 7

1. b
2. c
3. a
4. d
5. b
6. a
7. c
8. a
9. d
10. c

CHAPTER 8

1. a
2. c
3. b
4. c
5. d
6. a
7. b
8. a
9. b
10. d
11. c
12. b
13. c
14. a
15. d
16. b
17. c
18. a
19. b
20. d

REVIEW: CHAPTERS 7 AND 8

1. T
2. T
3. T
4. T
5. F
6. F
7. T
8. T
9. T
10. F
11. F
12. T

CHAPTER 9

1. b
2. a
3. c
4. d
5. c
6. a
7. c
8. b
9. c
10. d
11. d
12. a
13. c
14. a
15. b

CHAPTER 10

1. d
2. b
3. a
4. b
5. c
6. a
7. b
8. c
9. d
10. d

REVIEW: CHAPTERS 9 AND 10

1. T
2. T
3. T
4. T
5. T
6. F
7. F
8. T
9. T
10. T
11. T
12. T
13. T
14. T
15. T

CHAPTER 11

1. b
2. a
3. c
4. b
5. d
6. b
7. a
8. c
9. c
10. d
11. b
12. c
13. d
14. a
15. d

CHAPTER 12

1. d
2. b
3. d
4. a
5. c
6. b
7. a
8. c
9. b
10. c
11. d
12. a
13. a
14. c
15. d

REVIEW: CHAPTERS 11 AND 12

1. d
2. c
3. a
4. c
5. b
6. d
7. b
8. b
9. a
10. d

CHAPTER 13

1. c
2. b
3. a
4. d
5. a
6. b
7. d
8. c
9. c
10. b
11. a
12. d
13. c
14. b
15. d
16. a
17. c
18. d
19. d
20. c
21. a
22. c
23. b
24. b
25. d

CHAPTER 14

1. c
2. d
3. a
4. b
5. a
6. c
7. b
8. c
9. d
10. d
11. b
12. a
13. d
14. d
15. c
16. b
17. a
18. b
19. b
20. c

CHAPTER 15

1. d
2. c
3. a
4. d
5. b
6. c
7. a
8. b
9. c
10. a
11. a
12. c
13. b
14. c
15. b
16. d
17. a
18. b
19. a
20. b
21. a
22. d
23. c
24. d
25. c

REVIEW: CHAPTERS 13–15

1. b
2. a
3. c
4. d
5. b
6. d
7. a
8. d
9. c
10. a

CHAPTER 16

1. b
2. a
3. d
4. c
5. c
6. d
7. a
8. b
9. c
10. b
11. c
12. a
13. d
14. c
15. a
16. d
17. a
18. c
19. a
20. b
21. a
22. d
23. b
24. c
25. d

CHAPTER 17

1. a
2. c
3. b
4. d
5. b
6. c
7. a
8. d
9. b
10. c
11. b
12. d
13. a
14. b
15. a

CHAPTER 18

1. d
2. c
3. b
4. a
5. c
6. b
7. d
8. a
9. c
10. a
11. c
12. a
13. d
14. c
15. b

REVIEW: CHAPTERS 16–18

1. T
2. T
3. T
4. T
5. T

6. T
7. T
8. T
9. F
10. T
11. T
12. F
13. F
14. T
15. T
16. T
17. F
18. T
19. T
20. T
21. T
22. T
23. T
24. T
25. T
26. T
27. F
28. F
29. T
30. T
31. T
32. T
33. T
34. T
35. T
36. F
37. F
38. T
39. T
40. F

CHAPTER 19

1. a
2. d
3. b
4. c
5. b
6. a
7. b
8. d
9. c
10. a
11. b
12. d
13. a
14. c
15. d
16. c
17. b
18. a
19. b
20. d

CHAPTER 20

1. b
2. c
3. a
4. d
5. a
6. c
7. b
8. a
9. d
10. b

CHAPTER 21

1. b
2. a
3. d
4. c
5. c
6. a
7. b
8. d
9. c
10. a
11. c
12. d
13. a
14. d
15. b

CHAPTER 22

1. c
2. d
3. b
4. a
5. a
6. c
7. b
8. d
9. c
10. b
11. d
12. a
13. d
14. a
15. c
16. b
17. b
18. d
19. c
20. d
21. c
22. d
23. a
24. b
25. a

REVIEW: CHAPTERS 19–22

1. c
2. b
3. d
4. d
5. d
6. d
7. c
8. a
9. b
10. c
11. c
12. a
13. c
14. d
15. b
16. a

CHAPTER 23

1. d
2. a
3. c
4. b
5. a
6. a
7. b
8. d
9. c
10. b
11. d
12. a
13. c
14. a
15. c

CHAPTER 24

1. c
2. a
3. b
4. d
5. b
6. a
7. c
8. b
9. a
10. d
11. d
12. b
13. c
14. c
15. b

REVIEW: CHAPTERS 23 AND 24

1. 300
2. 300
3. 300
4. 250
5. 250
6. 200
7. 200
8. 14.1
9. 14.1
10. 1
11. 45°
12. –45°
13. 1
14. 1.41
15. 7.07
16. 600
17. $5.66\underline{/45^\circ}$
18. $4\underline{/10^\circ}$
19. T
20. T
21. T
22. F

CHAPTER 25

1. b
2. c
3. a
4. d
5. b
6. a
7. c
8. c
9. b
10. c
11. a
12. a
13. d
14. b
15. d
16. c
17. a
18. d
19. b
20. a

CHAPTER 26

1. c
2. a
3. b
4. c
5. d
6. b
7. a
8. d
9. c
10. b
11. c
12. a
13. d
14. a
15. b
16. a
17. b
18. c
19. d
20. b

REVIEW: CHAPTERS 25 AND 26

1. 8
2. 0.8
3. 0.4
4. 10
5. 10
6. 1
7. 5
8. 0.08
9. 40
10. 150
11. $f_c = 31.83$ kHz
12. −100 dB
13. octave, decade
14. 70.7
15. F
16. T
17. T
18. T
19. T
20. T
21. F

CHAPTER 27

1. b
2. c
3. c
4. a
5. d
6. b
7. c
8. a
9. d
10. b
11. b
12. d
13. d
14. b
15. a

CHAPTER 28

1. b
2. c
3. a
4. d
5. a
6. b
7. c
8. b
9. d
10. a
11. b
12. d
13. c
14. a
15. b
16. c
17. d
18. a
19. d
20. c

CHAPTER 29

1. d
2. a
3. b
4. c
5. c
6. a
7. b
8. d
9. a
10. c
11. b
12. a
13. d
14. c
15. b
16. d
17. a
18. c
19. d
20. b

CHAPTER 30

1. b
2. c
3. a
4. d
5. a
6. b
7. c
8. d
9. a
10. d
11. c
12. b
13. a

14. b
15. c

CHAPTER 31

1. d
2. b
3. a
4. c
5. b
6. a
7. c
8. d
9. b
10. a
11. c
12. b
13. a
14. c
15. d
16. a
17. c
18. d
19. b
20. a

CHAPTER 32

1. b
2. c
3. d
4. a
5. b
6. d
7. c
8. a
9. b
10. c
11. c
12. b
13. d
14. a
15. c

CHAPTER 33

1. c
2. a
3. b
4. b
5. d
6. a
7. a
8. a
9. c
10. d

CHAPTER 34

1. a
2. c
3. b
4. d
5. c
6. a
7. b
8. b
9. c
10. d
11. a
12. c
13. d
14. b
15. a
16. c
17. d
18. b
19. c
20. a

Answers

Odd-Numbered Problems and Critical Thinking Problems

Introduction to Powers of 10

SECTION I-1 SCIENTIFIC NOTATION

1. 3.5×10^6
3. 1.6×10^8
5. 1.5×10^{-1}
7. 2.27×10^3
9. 3.3×10^{-2}
11. 7.77×10^7
13. 8.7×10^1
15. 9.5×10^{-8}
17. 6.4×10^5
19. 1.75×10^{-9}
21. 0.000165
23. 863
25. 0.0000000017
27. 1660
29. 0.0000000000033

SECTION I-2 ENGINEERING NOTATION AND METRIC PREFIXES

31. 5.5×10^3
33. 6.2×10^6
35. 99×10^3
37. 750×10^{-6}
39. 10×10^6
41. 68×10^{-6}
43. 270×10^3
45. 450×10^{-9}
47. 2.57×10^{12}
49. 70×10^{-6}
51. 1 kW
53. 35 mV
55. 1 μF
57. 2.2 MΩ
59. 1.25 GHz
61. 250 μA
63. 500 mW
65. 180 kΩ
67. 4.7 Ω
69. 50 μW

SECTION I-3 CONVERTING BETWEEN METRIC PREFIXES

71. 55 mA
73. 0.0068 μF
75. 22 μF
77. 1500 kΩ
79. 39 kΩ
81. 7.5 mA
83. 100,000 W
85. 4.7 nF
87. 1.296 GHz
89. 7,500,000 pF

SECTION I-4 ADDITION AND SUBTRACTION INVOLVING POWERS OF 10 NOTATION

91. 7.5×10^4
93. 5.9×10^{-10}
95. 2.15×10^{-3}
97. 5.0×10^7
99. 1.45×10^{-2}
101. 2.6×10^4

SECTION I-5 MULTIPLICATION AND DIVISION INVOLVING POWERS OF 10 NOTATION

103. 1.8×10^6
105. 3.0×10^9
107. 1.0×10^{-5}
109. 2.5×10^4
111. 1.25×10^2
113. 5.0×10^7

SECTION I-6 RECIPROCALS WITH POWERS OF 10

115. 10^{-4}
117. 10^{-1}
119. 10^7
121. 10^{-15}

SECTION I-7 SQUARING NUMBERS EXPRESSED IN POWERS OF 10 NOTATION

123. 2.5×10^7
125. 8.1×10^{11}
127. 1.44×10^{-16}

SECTION I-8 SQUARE ROOTS OF NUMBERS EXPRESSED IN POWERS OF 10 NOTATION

129. 2.0×10^{-2}
131. 6.0×10^{-6}
133. 3.87×10^{-2}

SECTION I-9 THE SCIENTIFIC CALCULATOR

135. Enter the problem using the following keying sequence:
① ⑤ [EXP] [±] ③ [×] ① [·] ② [EXP] ③ [=].
The calculator will display the answer as 18.000×10^{00}.
137. Enter the problem using the following keying sequence:
① ② [÷] ① ⓪ [EXP] ③ [=].
The calculator will display the answer as 1.200×10^{-03}.
139. Enter the problem using the following keying sequence:
⑥ [·] ⑤ [EXP] ④ ⊕ ② ⑤ [EXP] ③ [=].
The calculator will display the answer as 90.000×10^{03}.

Chapter 1

SECTION 1-4 THE COULOMB UNIT OF ELECTRIC CHARGE

1. $+Q = 5$ C
3. $+Q = 2$ C
5. $-Q = 6$ C

SECTION 1-5 THE VOLT UNIT OF POTENTIAL DIFFERENCE

7. $V = 6$ V
9. $V = 1.25$ V

SECTION 1-6 CHARGE IN MOTION IS CURRENT

11. $I = 4$ A
13. $I = 500$ mA
15. $I = 10$ A
17. $Q = 1$ C

SECTION 1-7 RESISTANCE IS OPPOSITION TO CURRENT

19. **a.** $R = 1$ kΩ
b. $R = 100$ Ω
c. $R = 10$ Ω
d. $R = 1$ Ω
21. **a.** $G = 5$ mS
b. $G = 10$ mS
c. $G = 20$ mS
d. $G = 40$ mS

ANSWERS TO CRITICAL THINKING PROBLEMS

23. $Q = 1.6 \times 10^{-16}$ C
25. $I = 100$ μA

Chapter 2

SECTION 2-2 RESISTOR COLOR CODING

1. **a.** 1.5 kΩ, ±10%
b. 27 Ω, ±5%
c. 470 kΩ, ±5%
d. 6.2 Ω, ±5%
e. 91 kΩ, ±5%
f. 10 Ω, ±5%
g. 1.8 MΩ, ±10%
h. 1.5 kΩ, ±20%
i. 330 Ω, ±10%
j. 560 kΩ, ±5%
k. 2.2 kΩ, ±5%
l. 8.2 Ω, ±5%
m. 51 kΩ, ±5%
n. 680 Ω, ±5%
o. 0.12 Ω, ±5%
p. 1 kΩ, ±5%
q. 10 kΩ, ±10%
r. 4.7 kΩ, ±5%
3. **a.** 470 kΩ
b. 1.2 kΩ
c. 330 Ω
d. 10 kΩ
5. Reading from left to right the colors are:
a. Brown, black, orange, and gold
b. Red, violet, gold, and gold
c. Green, blue, red, and silver
d. Brown, green, green, and gold
e. Red, red, silver, and gold

SECTION 2-3 VARIABLE RESISTORS

7. **a.** 680,225 Ω
b. 8250 Ω
c. 18,503 Ω
d. 275,060 Ω
e. 62,984 Ω

ANSWERS TO CRITICAL THINKING PROBLEMS

9. Above 250 kΩ

Chapter 3

SECTION 3-1 THE CURRENT $I = \frac{V}{R}$

1. **a.** $I = 2$ A
b. $I = 3$ A
c. $I = 8$ A
d. $I = 4$ A
3. **a.** $I = 0.005$ A
b. $I = 0.02$ A
c. $I = 0.003$ A
d. $I = 0.015$ A
5. Yes, because the current, I, is only 15 A.

SECTION 3-2 THE VOLTAGE $V = IR$

7. **a.** $V = 50$ V
b. $V = 30$ V
c. $V = 10$ V
d. $V = 7.5$ V
9. $V = 10$ V

SECTION 3-3 THE RESISTANCE $R = \frac{V}{I}$

11. **a.** $R = 7$ Ω
b. $R = 5$ Ω
c. $R = 4$ Ω
d. $R = 6$ Ω
13. **a.** $R = 6000$ Ω
b. $R = 200$ Ω
c. $R = 2500$ Ω
d. $R = 5000$ Ω
15. $R = 8.5$ Ω

SECTION 3-5 MULTIPLE AND SUBMULTIPLE UNITS

17. **a.** $I = 80$ mA
b. $V = 19.5$ V
c. $V = 3$ V
d. $R = 33$ kΩ
19. $I = 50$ μA

SECTION 3-6 THE LINEAR PROPORTION BETWEEN *V* AND *I*

21. See Instructor's Manual.

SECTION 3-7 ELECTRIC POWER

23. **a.** $P = 1.5$ kW
b. $P = 75$ W
c. $I = 10$ A
d. $V = 12$ V
25. **a.** $I = 31.63$ mA
b. $I = 2$ mA
c. $P = 150$ μW
d. $V = 200$ V
27. $V = 15$ V
29. Cost = $7.20
31. Cost = $64.80

SECTION 3-8 POWER DISSIPATION IN RESISTANCE

33. **a.** $P = 1.98$ mW
b. $P = 675$ mW
c. $P = 24.5$ mW
d. $P = 1.28$ W
35. $P = 500$ mW
37. $P = 2.16$ W

SECTION 3-9 POWER FORMULAS

39. **a.** $I = 5$ mA
b. $R = 144$ Ω
c. $R = 312.5$ Ω
d. $V = 223.6$ V
41. **a.** $V = 44.72$ V
b. $V = 63.25$ V
c. $I = 100$ μA
d. $I = 400$ μA
43. $I = 2.38$ mA
45. $V = 100$ V
47. $R = 12$ Ω
49. $V = 50$ V
51. $R = 7.2$ Ω

SECTION 3-10 CHOOSING A RESISTOR FOR A CIRCUIT

53. $R = 1.2$ kΩ. Best choice for power rating is ¼ W.
55. $R = 2$ kΩ. Best choice for power rating is 1 W.
57. $R = 150$ Ω. Best choice for power rating is ⅛ W.
59. $R = 2.2$ MΩ. Best choice for power rating is ½ W because it has a 350-V maximum working voltage rating.

ANSWERS TO CRITICAL THINKING PROBLEMS

61. $I = 21.59$ A
63. Cost = $7.52
65. I_{max} at 120°C = 13.69 mA

Chapter 4

SECTION 4-1 WHY *I* IS THE SAME IN ALL PARTS OF A SERIES CIRCUIT

1. **a.** $I = 100$ mA
b. $I = 100$ mA
c. $I = 100$ mA
d. $I = 100$ mA
e. $I = 100$ mA
f. $I = 100$ mA

3. $I = 100$ mA

SECTION 4-2 TOTAL *R* EQUALS THE SUM OF ALL SERIES RESISTANCES

5. $R_T = 900\ \Omega$
$I = 10$ mA

7. $R_T = 6$ kΩ
$I = 4$ mA

9. $R_T = 300$ kΩ
$I = 800\ \mu$A

SECTION 4-3 SERIES *IR* VOLTAGE DROPS

11. $V_1 = 6$ V
$V_2 = 7.2$ V
$V_3 = 10.8$ V

13. $R_T = 2$ kΩ
$I = 10$ mA
$V_1 = 3.3$ V
$V_2 = 4.7$ V
$V_3 = 12$ V

15. $R_T = 16$ kΩ
$I = 1.5$ mA
$V_1 = 2.7$ V
$V_2 = 4.05$ V
$V_3 = 12.3$ V
$V_4 = 4.95$ V

SECTION 4-4 KIRCHHOFF'S VOLTAGE LAW (KVL)

17. $V_T = 15$ V

19. $V_1 = 7.2$ V
$V_2 = 8.8$ V
$V_3 = 4$ V
$V_4 = 60$ V
$V_5 = 40$ V
$V_T = 120$ V

SECTION 4-5 POLARITY OF *IR* VOLTAGE DROPS

21. **a.** $R_T = 100\ \Omega$, $I = 500$ mA, $V_1 = 5$ V, $V_2 = 19.5$ V, $V_3 = 25.5$ V
b. See Instructor's manual.
c. See Instructor's manual.
d. See Instructor's manual.

23. The polarity of the individual resistor voltage drops is opposite to that in Prob. 21. The reason is that the polarity of a resistor's voltage drop depends on the direction of current flow and reversing the polarity of V_T reverses the direction of current.

SECTION 4-6 TOTAL POWER IN A SERIES CIRCUIT

25. $P_1 = 36$ mW
$P_2 = 43.2$ mW
$P_3 = 64.8$ mW
$P_T = 144$ mW

27. $P_1 = 33$ mW
$P_2 = 47$ mW
$P_3 = 120$ mW
$P_T = 200$ mW

SECTION 4-7 SERIES-AIDING AND SERIES-OPPOSING VOLTAGES

29. **a.** $V_T = 27$ V
b. $I = 10$ mA
c. Electrons flow up through R_1.

31. **a.** $V_T = 6$ V
b. $I = 6$ mA
c. Electrons flow up through R_1.

33. **a.** $V_T = 12$ V
b. $I = 400$ mA
c. Electrons flow down through R_1 and R_2.
d. $V_1 = 4.8$ V and $V_2 = 7.2$ V

SECTION 4-8 ANALYZING SERIES CIRCUITS WITH RANDOM UNKNOWNS

35. $I = 20$ mA
$V_1 = 2.4$ V
$V_2 = 2$ V
$V_3 = 13.6$ V
$V_T = 18$ V
$R_3 = 680\ \Omega$
$P_T = 360$ mW
$P_2 = 40$ mW
$P_3 = 272$ mW

37. $I = 20$ mA
$R_T = 6$ kΩ
$V_T = 120$ V
$V_2 = 36$ V
$V_3 = 24$ V
$V_4 = 40$ V
$R_4 = 2$ kΩ
$P_1 = 400$ mW
$P_2 = 720$ mW
$P_3 = 480$ mW
$P_4 = 800$ mW

39. $R_3 = 800\ \Omega$
$I = 50$ mA
$V_T = 100$ V
$V_1 = 10$ V
$V_2 = 20$ V
$V_3 = 40$ V
$V_4 = 30$ V
$P_1 = 500$ mW
$P_3 = 2$ W
$P_4 = 1.5$ W
$P_T = 5$ W

41. $R = 1$ kΩ

43. $V_T = 25$ V

SECTION 4-9 GROUND CONNECTIONS IN ELECTRICAL AND ELECTRONIC SYSTEMS

45. $V_{AG} = 18$ V
$V_{BG} = 7.2$ V
$V_{CG} = 1.2$ V

47. $V_{AG} = 20$ V
$V_{BG} = 16.4$ V
$V_{CG} = -9.4$ V
$V_{DG} = -16$ V

SECTION 4-10 TROUBLESHOOTING: OPENS AND SHORTS IN SERIES CIRCUITS

49. $R_T = 6$ kΩ
$I = 4$ mA
$V_1 = 4$ V
$V_2 = 8$ V
$V_3 = 12$ V

51. **a.** $R_T = 3$ kΩ
b. $I = 8$ mA
c. $V_1 = 8$ V, $V_2 = 16$ V, and $V_3 = 0$ V

ANSWERS TO CRITICAL THINKING PROBLEMS

53. $R_1 = 300\ \Omega$, $R_2 = 600\ \Omega$, and $R_3 = 1.8$ kΩ

55. $I_{max} = 35.36$ mA

57. $R_1 = 250\ \Omega$ and $V_T = 1.25$ V

Answers to Troubleshooting Challenge, Table 4-1

Trouble 1: R_2 open
Trouble 3: R_4 shorted
Trouble 5: R_1 shorted
Trouble 7: R_2 shorted

Trouble 9: R_5 open
Trouble 11: R_3 decreased in value
Trouble 13: R_1 increased in value

Chapter 5

SECTION 5-1 THE APPLIED VOLTAGE V_A IS THE SAME ACROSS PARALLEL BRANCHES

1. **a.** 12 V
b. 12 V
c. 12 V
d. 12 V

3. 12 V

SECTION 5-2 EACH BRANCH I EQUALS $\frac{V_A}{R}$

5. I_2 is double I_1 because R_2 is one-half the value of R_1.

7. $I_1 = 600$ mA
$I_2 = 900$ mA
$I_3 = 300$ mA

9. $I_1 = 200$ mA
$I_2 = 15$ mA
$I_3 = 85$ mA
$I_4 = 20$ mA

SECTION 5-3 KIRCHHOFF'S CURRENT LAW (KCL)

11. $I_T = 300$ mA

13. $I_T = 1.8$ A

15. $I_T = 320$ mA

17. $I_1 = 24$ mA
$I_2 = 20$ mA
$I_3 = 16$ mA
$I_T = 60$ mA

19. **a.** 300 mA
b. 276 mA
c. 256 mA
d. 256 mA
e. 276 mA
f. 300 mA

21. **a.** 2.5 A
b. 500 mA
c. 300 mA
d. 300 mA
e. 500 mA
f. 2.5 A
g. 2 A
h. 2 A

23. $I_2 = 90$ mA

SECTION 5-4 RESISTANCES IN PARALLEL

25. $R_{EQ} = 8\ \Omega$

27. $R_{EQ} = 20\ \Omega$

29. $R_{EQ} = 318.75\ \Omega$

31. $R_{EQ} = 26.4\ \Omega$

33. $R_{EQ} = 20\ \Omega$

35. $R_{EQ} = 200\ \Omega$

37. $R_2 = 1.5$ kΩ

39. $R_{EQ} = 96\ \Omega$

41. $R_{EQ} = 112.5\ \Omega$

43. **a.** $R_{EQ} = 269.9\ \Omega$ (Ohmmeter will read 270 Ω approximately.)
b. $R_{EQ} = 256.6$ kΩ (Ohmmeter will read 257 kΩ approximately.)
c. $R_{EQ} = 559.3$ kΩ (Ohmmeter will read 559 kΩ approximately.)
d. $R_{EQ} = 1.497$ kΩ (Ohmmeter will read 1.5 kΩ approximately.)
e. $R_{EQ} = 9.868$ kΩ (Ohmmeter will read 9.87 kΩ approximately.)

SECTION 5-5 CONDUCTANCES IN PARALLEL

45. $G_1 = 2$ mS
$G_2 = 500\ \mu$S
$G_3 = 833.3\ \mu$S
$G_4 = 10$ mS
$G_T = 13.33$ mS
$R_{EQ} = 75\ \Omega$

47. $G_T = 500$ mS
$R_{EQ} = 2\ \Omega$

SECTION 5-6 TOTAL POWER IN PARALLEL CIRCUITS

49. $P_1 = 20.4$ W
$P_2 = 1.53$ W
$P_3 = 8.67$ W
$P_4 = 2.04$ W
$P_T = 32.64$ W

51. $P_1 = 13.2$ W
$P_2 = 19.8$ W
$P_3 = 132$ W
$P_T = 165$ W

SECTION 5-7 ANALYZING PARALLEL CIRCUITS WITH RANDOM UNKNOWNS

53. $V_A = 18$ V
$R_1 = 360\ \Omega$
$I_2 = 150$ mA
$R_{EQ} = 90\ \Omega$
$P_1 = 900$ mW
$P_2 = 2.7$ W
$P_T = 3.6$ W

55. $R_3 = 500\ \Omega$
$V_A = 75$ V
$I_1 = 150$ mA
$I_2 = 300$ mA
$I_T = 600$ mA
$P_1 = 11.25$ W
$P_2 = 22.5$ W
$P_3 = 11.25$ W
$P_T = 45$ W

57. $I_T = 100$ mA
$I_1 = 30$ mA
$I_2 = 20$ mA
$I_4 = 35$ mA
$R_3 = 2.4$ kΩ
$R_4 = 1.029$ kΩ
$P_1 = 1.08$ W
$P_2 = 720$ mW
$P_3 = 540$ mW
$P_4 = 1.26$ W
$P_T = 3.6$ W

59. $V_A = 24$ V
$I_1 = 20$ mA
$I_2 = 30$ mA
$I_4 = 24$ mA
$R_1 = 1.2$ kΩ
$R_3 = 4$ kΩ
$R_{EQ} = 300\ \Omega$

ANSWERS TO CRITICAL THINKING PROBLEMS

61. $I_{T(max)} = 44.55$ mA

63. $R_1 = 2$ kΩ, $R_2 = 6$ kΩ, and $R_3 = 3$ kΩ

65. $R_1 = 15$ kΩ, $R_2 = 7.5$ kΩ, $R_3 = 3.75$ kΩ, and $R_4 = 1.875$ kΩ

Answers to Troubleshooting Challenge

67. The current meter, M_3, is open; or the wire between points G and H is open. The fault could be isolated by measuring the voltage across points C and D and points G and H. The voltage will measure 36 V across the open points.

69. **a.** M_1 and M_3 will both read 0 A.
b. M_2 will read 0 V.
c. 36 V
d. The blown fuse was probably caused by a short in one of the four parallel branches.

e. With the blown fuse, F_1, still in place, open S_1. (This is an additional precaution.) Connect an ohmmeter across points B and I. The ohmmeter will probably read 0 Ω. Next, remove one branch at a time while observing the ohmmeter. When the shorted branch is removed, the ohmic value indicated by the ohmmeter will increase to a value that is normal for the circuit. Be sure the ohmmeter is set to its lowest range when following this procedure. The reason is that the equivalent resistance, R_{EQ}, of this circuit is normally quite low anyway. Setting the ohmmeter on too high of a range could result in a reading of 0 Ω even after the shorted branch has been removed.

71. 0 Ω. One way to find the shorted branch would be to disconnect all but one of the branches along the top at points B, C, D, or E. (When doing this, make certain S_1 is open if the fuse has been replaced.) Next, with F_1 replaced, close S_1. If the only remaining branch blows the fuse, then you know that's the shorted branch. If the fuse F_1 did not blow, open S_1 and reconnect the next branch. Repeat this procedure until the fuse F_1 blows. The branch that blows the fuse is the shorted branch.

73. **a.** 0 V
b. 0 V

75. **a.** M_1 will read 1.5 A and M_3 will read 0 A.
b. 36 V
c. 0 V

Chapter 6

SECTION 6-1 FINDING R_T FOR SERIES-PARALLEL RESISTANCES

1. Resistors R_1 and R_2 are in series and resistors R_3 and R_4 are in parallel. It should also be noted that the applied voltage, V_T, is in series with R_1 and R_2 because they all have the same current.

3. $I_1 = 10$ mA
$I_2 = 10$ mA
$V_1 = 2.2$ V
$V_2 = 6.8$ V
$V_3 = 6$ V
$V_4 = 6$ V
$I_3 = 6$ mA
$I_4 = 4$ mA

5. **a.** 80 Ω
b. 200 Ω
c. 60 mA
d. 60 mA

7. $P_1 = 432$ mW
$P_2 = 230.4$ mW
$P_3 = 57.6$ mW
$P_T = 720$ mW

9. **a.** 250 Ω
b. 200 Ω
c. 450 Ω
d. 40 mA
e. 40 mA

SECTION 6-2 RESISTANCE STRINGS IN PARALLEL

11. **a.** 800 Ω
b. 1.2 kΩ
c. $I_1 = 30$ mA and $I_2 = 20$ mA
d. $I_T = 50$ mA
e. $R_T = 480$ Ω
f. $V_1 = 9.9$ V, $V_2 = 14.1$ V, and $V_3 = 24$ V

13. **a.** 300 Ω
b. 900 Ω
c. $I_1 = 120$ mA and $I_2 = 40$ mA
d. $I_T = 160$ mA
e. $R_T = 225$ Ω
f. $V_1 = 12$ V, $V_2 = 24$ V, $V_3 = 27.2$ V, and $V_4 = 8.8$ V

15. **a.** $I_1 = 8$ mA, $I_2 = 24$ mA, $I_3 = 16$ mA, $I_T = 48$ mA
b. $R_T = 500$ Ω
c. $V_1 = 8$ V, $V_2 = 16$ V, $V_3 = 24$ V, $V_4 = 8$ V, and $V_5 = 16$ V

SECTION 6-3 RESISTANCE BANKS IN SERIES

17. **a.** 150 Ω
b. $R_T = 250$ Ω
c. $I_T = 100$ mA
d. $V_{AB} = 15$ V
e. $V_1 = 10$ V
f. $I_2 = 25$ mA and $I_3 = 75$ mA
g. 100 mA

19. **a.** $R_T = 1.8$ kΩ
b. $I_T = 30$ mA
c. $V_1 = 36$ V, $V_2 = 18$ V, and $V_3 = 18$ V
d. $I_2 = 12$ mA and $I_3 = 18$ mA

SECTION 6-4 RESISTANCE BANKS AND STRINGS IN SERIES-PARALLEL

21. $R_T = 500$ Ω
$I_T = 70$ mA
$V_1 = 8.4$ V
$V_2 = 14$ V
$V_3 = 5.6$ V
$V_4 = 8.4$ V
$V_5 = 12.6$ V
$I_1 = 70$ mA
$I_2 = 14$ mA
$I_3 = 56$ mA
$I_4 = 56$ mA
$I_5 = 70$ mA

23. $R_T = 4$ kΩ
$I_T = 30$ mA
$V_1 = 30$ V
$V_2 = 10$ V
$V_3 = 20$ V
$V_4 = 30$ V
$V_5 = 60$ V
$I_1 = 30$ mA
$I_2 = 10$ mA
$I_3 = 10$ mA
$I_4 = 20$ mA
$I_5 = 30$ mA

25. $R_T = 6$ kΩ
$I_T = 6$ mA
$V_1 = 36$ V
$V_2 = 5.4$ V
$V_3 = 27$ V
$V_4 = 9$ V
$V_5 = 13.5$ V
$V_6 = 4.5$ V
$V_7 = 3.6$ V
$I_1 = 2.4$ mA
$I_2 = 3.6$ mA
$I_3 = 2.7$ mA
$I_4 = 900$ μA
$I_5 = 900$ μA
$I_6 = 900$ μA
$I_7 = 3.6$ mA

27. $R_T = 200$ Ω
$I_T = 120$ mA
$V_1 = 6$ V
$V_2 = 6$ V
$V_3 = 12$ V
$V_4 = 12$ V
$V_5 = 18$ V

$V_6 = 24$ V
$I_1 = 40$ mA
$I_2 = 30$ mA
$I_3 = 10$ mA
$I_4 = 20$ mA
$I_5 = 10$ mA
$I_6 = 80$ mA

SECTION 6-5 ANALYZING SERIES-PARALLEL CIRCUITS WITH RANDOM UNKNOWNS

29. $R_T = 300\ \Omega$
$I_T = 70$ mA
$V_T = 21$ V
$V_1 = 7$ V
$V_2 = 9.24$ V
$V_4 = 4.76$ V
$I_2 = 42$ mA
$I_3 = 28$ mA

31. $R_T = 800\ \Omega$
$I_T = 30$ mA
$V_T = 24$ V
$V_1 = 5.4$ V
$V_2 = 12$ V
$V_3 = 3$ V
$V_4 = 6.6$ V
$V_5 = 2.4$ V
$V_6 = 6.6$ V
$I_2 = 10$ mA
$I_3 = 20$ mA
$I_4 = 20$ mA
$I_5 = 20$ mA

33. $I_T = 30$ mA
$R_T = 1$ kΩ
$V_2 = 10.8$ V
$V_3 = 10.8$ V
$V_4 = 10.8$ V
$V_5 = 21.6$ V
$R_2 = 600\ \Omega$
$V_6 = 3$ V
$I_2 = 18$ mA
$I_3 = 10.8$ mA
$I_4 = 7.2$ mA
$I_5 = 12$ mA
$I_6 = 30$ mA

SECTION 6-6 THE WHEATSTONE BRIDGE

35. **a.** $R_X = 6{,}816\ \Omega$
b. $V_{CB} = V_{DB} = 8.33$ V
c. $I_T = 1.91$ mA

37. **a.** $R_{X(max)} = 99.999\ \Omega$
b. $R_{X(max)} = 999.99\ \Omega$
c. $R_{X(max)} = 9999.9\ \Omega$
d. $R_{X(max)} = 99{,}999\ \Omega$
e. $R_{X(max)} = 999{,}990\ \Omega$
f. $R_{X(max)} = 9{,}999{,}900\ \Omega$

39. R_3 must be adjusted to 1 kΩ.

41. **a.** The thermistor resistance is 4250 Ω.
b. T_A has increased above 25°C.

ANSWERS TO CRITICAL THINKING PROBLEMS

43. **a.** $R_1 = 657\ \Omega$
b. Recommended wattage rating is 25 W approximately.
c. $R_T = 857\ \Omega$

45. With V_T reversed in polarity, the circuit will not operate properly. For example, if the ambient temperature increases, the voltage across points C and D becomes positive. This causes the output voltage from the amplifier to go negative, which turns on the heater. This will increase the temperature even more. Unfortunately, the heater will continue to stay on. If the temperature would have decreased initially the voltage across points C and D would have gone negative. This would make the output of the amplifier go positive, thus turning on the air conditioner, making the temperature decrease even further.

Answers to Troubleshooting Challenge, Table 6-1

Trouble 1: R_6 open
Trouble 3: R_4 open
Trouble 5: R_6 shorted
Trouble 7: R_1 open
Trouble 9: R_1 shorted
Trouble 11: R_3 shorted

Chapter 7

SECTION 7-1 SERIES VOLTAGE DIVIDERS

1. $V_1 = 3$ V
$V_2 = 6$ V
$V_3 = 9$ V

3. $V_1 = 4$ V
$V_2 = 6$ V
$V_3 = 8$ V

5. $V_1 = 2.5$ V
$V_2 = 7.5$ V
$V_3 = 15$ V

7. **a.** $V_1 = 9$ V; $V_2 = 900$ mV; $V_3 = 100$ mV
b. $V_{AG} = 10$ V; $V_{BG} = 1$ V; $V_{CG} = 100$ mV

9. $V_1 = 16$ V; $V_2 = 8$ V; $V_3 = 16$ V, $V_4 = 8$ V
$V_{AG} = 48$ V; $V_{BG} = 32$ V;
$V_{CG} = 24$ V; $V_{DG} = 8$ V

11. **a.** $R_T = 15$ kΩ
b. $I = 1.6$ mA
c. $V_1 = 16$ V, $V_{AB} = 8$ V
d. 0 to 8 V

SECTION 7-2 CURRENT DIVIDER WITH TWO PARALLEL RESISTANCES

13. $I_1 = 16$ mA
$I_2 = 8$ mA

15. $I_1 = 64$ mA
$I_2 = 16$ mA

17. $I_1 = 48$ mA
$I_2 = 72$ mA

SECTION 7-3 CURRENT DIVISION BY PARALLEL CONDUCTANCES

19. $I_1 = 3.6$ A
$I_2 = 2.4$ A
$I_3 = 3$ A

21. $I_1 = 45$ mA
$I_2 = 4.5$ mA
$I_3 = 16.5$ mA

23. $I_1 = 25\ \mu$A
$I_2 = 37.5\ \mu$A
$I_3 = 12.5\ \mu$A
$I_4 = 75\ \mu$A

SECTION 7-4 SERIES VOLTAGE DIVIDER WITH PARALLEL LOAD CURRENT

25. The voltage, V_{BG}, decreases when S_1 is closed because R_L in parallel with R_2 reduces the resistance from points B to G. This lowering of resistance changes the voltage division in the circuit. With S_1 closed, the resistance from B to G is a smaller fraction of the total resistance, which in turn means the voltage, V_{BG}, must also be less.

27. Resistor R_2

SECTION 7-5 DESIGN OF A LOADED VOLTAGE DIVIDER

29. **a.** $I_1 = 56$ mA
$I_2 = 11$ mA
$I_3 = 6$ mA
$I_T = 66$ mA

b. $V_1 = 10\text{ V}$
$V_2 = 9\text{ V}$
$V_3 = 6\text{ V}$
c. $R_1 = 178.6\ \Omega$
$R_2 = 818.2\ \Omega$
$R_3 = 1\text{ k}\Omega$
d. $P_1 = 560\text{ mW}$
$P_2 = 99\text{ mW}$
$P_3 = 36\text{ mW}$

31. **a.** $I_1 = 38\text{ mA}$
$I_2 = 18\text{ mA}$
$I_3 = 6\text{ mA}$
$I_T = 68\text{ mA}$
b. $V_1 = 9\text{ V}$
$V_2 = 6\text{ V}$
$V_3 = 9\text{ V}$
c. $R_1 = 236.8\ \Omega$
$R_2 = 333.3\ \Omega$
$R_3 = 1.5\text{ k}\Omega$
d. $P_1 = 342\text{ mW}$
$P_2 = 108\text{ mW}$
$P_3 = 54\text{ mW}$

ANSWERS TO CRITICAL THINKING PROBLEMS

33. $R_1 = 1\text{ k}\Omega$ and $R_3 = 667\ \Omega$
35. See Instructor's Manual.

Answers to Troubleshooting Challenge

TABLE 7–2

Trouble 1: R_2 open
Trouble 3: R_3 shorted
Trouble 5: R_2 shorted
Trouble 7: R_3 open

TABLE 7–3

Trouble 1: R_2 open
Trouble 3: R_3 open
Trouble 5: R_1 open
Trouble 7: R_4 or load C shorted

Chapter 8

SECTION 8-2 METER SHUNTS

1. **a.** $R_S = 50\ \Omega$
b. $R_S = 5.56\ \Omega$
c. $R_S = 2.08\ \Omega$
d. $R_S = 0.505\ \Omega$

3. **a.** $R_S = 1\text{ k}\Omega$
b. $R_S = 52.63\ \Omega$
c. $R_S = 10.1\ \Omega$
d. $R_S = 5.03\ \Omega$
e. $R_S = 1\ \Omega$
f. $R_S = 0.5\ \Omega$

5. **a.** $R_{L_1} = 111.1\ \Omega$
$R_{L_2} = 20.41\ \Omega$
$R_{L_3} = 4.02\ \Omega$
b. 1 mA range; $R_M = 100\ \Omega$
5 mA range; $R_M = 20\ \Omega$
25 mA range; $R_M = 4\ \Omega$

7. So that the current in the circuit is approximately the same with or without the meter present. If the current meter's resistance is too high, the measured value of current could be significantly less than the current without the meter present.

SECTION 8-3 VOLTMETERS

9. $\frac{\Omega}{V}\text{ rating} = \frac{1\text{ k}\Omega}{V}$

11. $\frac{\Omega}{V}\text{ rating} = \frac{50\text{ k}\Omega}{V}$

13. **a.** $R_1 = 58\text{ k}\Omega$
$R_2 = 140\text{ k}\Omega$
$R_3 = 400\text{ k}\Omega$
$R_4 = 1.4\text{ M}\Omega$
$R_5 = 4\text{ M}\Omega$
$R_6 = 14\text{ M}\Omega$
b. 3 V range; $R_V = 60\text{ k}\Omega$
10 V range; $R_V = 200\text{ k}\Omega$
30 V range; $R_V = 600\text{ k}\Omega$
100 V range; $R_V = 2\text{ M}\Omega$
300 V range; $R_V = 6\text{ M}\Omega$
1000 V range;
$R_V = 20\text{ M}\Omega$
c. $\frac{\Omega}{V}\text{ rating} = \frac{20\text{ k}\Omega}{V}$

15. **a.** $\frac{\Omega}{V}\text{ rating} = \frac{1\text{ k}\Omega}{V}$
b. $\frac{\Omega}{V}\text{ rating} = \frac{10\text{ k}\Omega}{V}$
c. $\frac{\Omega}{V}\text{ rating} = \frac{20\text{ k}\Omega}{V}$
d. $\frac{\Omega}{V}\text{ rating} = \frac{100\text{ k}\Omega}{V}$

SECTION 8-4 LOADING EFFECT OF A VOLTMETER

17. **a.** $V = 7.2\text{ V}$
b. $V = 7.16\text{ V}$
c. $V = 7.2\text{ V}$
Notice that there is little or no voltmeter loading with either meter since R_V is so much larger than the value of R_2.

19. The analog voltmeter with an R_V of 1 MΩ produced a greater loading effect. The reason is that its resistance is less than that of the DMM whose R_V is 10 MΩ.

SECTION 8-5 OHMMETERS

21. **a.** $R_X = 0\ \Omega$
b. $R_X = 250\ \Omega$
c. $R_X = 750\ \Omega$
d. $R_X = 2.25\text{ k}\Omega$
e. $R_X = \infty\ \Omega$

23. The scale would be nonlinear with values being more spread out on the right-hand side and more crowded on the left-hand side. The ohmmeter scale is nonlinear because equal increases in measured resistance do not produce equal decreases in current.

25. Because the ohms values increase from right to left as the current in the meter backs off from full-scale deflection.

27. On any range the zero-ohms adjustment control is adjusted for zero ohms with the ohmmeter leads shorted. The zero ohms control is adjusted to compensate for the slight changes in battery voltage, V_b, when changing ohmmeter ranges. Without a zero ohms adjustment control, the scale of the ohmmeter would not be properly calibrated.

SECTION 8-8 METER APPLICATIONS

29. The ohmmeter could be damaged or the meter will read an incorrect value of resistance. When measuring resistance, power must be off in the circuit being tested!

31. A current meter is connected in series to measure the current at some point in a circuit. Connecting a current meter in parallel could possibly ruin the meter due to excessive current. Remember, a current meter has a very low resistance and connecting it in parallel can effectively short-out a component.

33. **a.** 0 Ω
b. Infinite (∞) Ω

ANSWERS TO CRITICAL THINKING PROBLEMS

35. $R_1 = 40\ \Omega$, $R_2 = 8\ \Omega$, and $R_3 = 2\ \Omega$
37. 10 kΩ/V

Chapter 9

SECTION 9-1 KIRCHHOFF'S CURRENT LAW (KCL)

1. $I_3 = 15$ A
3. $I_3 = 6$ mA
5. Point X: 6 A + 11 A + 8 A − 25 A = 0
Point Y: 25 A − 2 A − 16 A − 7 A = 0
Point Z: 16 A − 5 A − 11 A = 0

SECTION 9-2 KIRCHHOFF'S VOLTAGE LAW (KVL)

7. **a.** 4.5 V + 5.4 V + 8.1 V = 18 V. This voltage is the same as the voltage V_{R_3}.
b. 6 V + 18 V + 12 V = 36 V. This voltage equals the applied voltage, V_T.
c. 6 V + 4.5 V + 5.4 V + 8.1 V + 12 V = 36 V. This voltage equals the applied voltage, V_T.
d. 18 V − 8.1 V − 5.4 V = 4.5 V. This voltage is the same as the voltage V_{R_4}.
9. $V_{AG} = +8$ V
$V_{BG} = 0$ V
11. $V_{AG} = -50$ V
$V_{BG} = -42.5$ V
$V_{CG} = -33.5$ V
$V_{DG} = -20$ V
$V_{AD} = -30$ V
13. 20 V − 2.5 V − 17.5 V = 0
15. 17.5 V + 12.5 V − 10 V − 20 V = 0

SECTION 9-3 METHOD OF BRANCH CURRENTS

17. **a.** $I_1 + I_3 - I_2 = 0$
b. $I_3 = I_2 - I_1$
c. $-V_1 - V_{R_3} + V_{R_1} = 0$ or $-24\text{ V} - V_{R_3} + V_{R_1} = 0$
d. $-V_2 + V_{R_2} + V_{R_3} = 0$ or $-12\text{ V} + V_{R_2} + V_{R_3} = 0$
e. $V_{R_1} = I_1R_1 = I_1 12\ \Omega$ or $12I_1$
$V_{R_2} = I_2R_2 = I_2 24\ \Omega$ or $24I_2$
$V_{R_3} = I_3R_3 = (I_2 - I_1)12\ \Omega$ or $12(I_2 - I_1)$
f. Loop 1:
$-24\text{ V} - 12(I_2 - I_1) + 12I_1 = 0$
g. Loop 2:
$-12\text{ V} + 24I_2 + 12(I_2 - I_1) = 0$
h. Loop 1:
$24I_1 - 12I_2 = 24$ V which can be reduced further to:
$2I_1 - I_2 = 2$ V
Loop 2:
$-12I_1 + 36I_2 = 12$ V which can be reduced further to:
$-I_1 + 3I_2 = 1$ V
i. $I_1 = 1.4$ A
$I_2 = 800$ mA
$I_3 = -600$ mA
j. No. The assumed direction for I_3 was incorrect as indicated by its negative value. I_3 actually flows downward through R_3.
k. $V_{R_1} = I_1R_1 = 1.4\text{ A} \times 12\ \Omega = 16.8$ V
$V_{R_2} = I_2R_2 = 800\text{ mA} \times 24\ \Omega = 19.2$ V
$V_{R_3} = I_3R_3 = 600\text{ mA} \times 12\ \Omega = 7.2$ V
l. Loop 1:
−24 V + 7.2 V + 16.8 V = 0
Loop 2:
12 V + 7.2 V − 19.2 V = 0
m. $I_1 - I_2 - I_3 = 0$ or 1.4 A − 800 mA − 600 mA = 0

SECTION 9-4 NODE-VOLTAGE ANALYSIS

19. **a.** $I_2 - I_1 - I_3 = 0$ or $I_2 = I_1 + I_3$
b. $\dfrac{V_{R_2}}{10\ \Omega} - \dfrac{V_{R_1}}{10\ \Omega} - \dfrac{V_N}{5\ \Omega} = 0$
or
$\dfrac{V_{R_2}}{10\ \Omega} = \dfrac{V_{R_1}}{10\ \Omega} + \dfrac{V_N}{5\ \Omega}$
c. $V_1 + V_{R_3} - V_{R_1} = 0$ or $10\text{ V} + V_{R_3} - V_{R_1} = 0$
d. $-V_2 + V_{R_3} + V_{R_2} = 0$ or $-15\text{ V} + V_{R_3} + V_{R_2} = 0$
e. $V_{R_1} = V_1 + V_N$ or $V_{R_1} = 10\text{ V} + V_N$
$V_{R_2} = V_2 - V_N$ or $V_{R_2} = 15\text{ V} - V_N$
f. $\dfrac{15\text{ V} - V_N}{10\ \Omega} = \dfrac{10\text{ V} + V_N}{10\ \Omega} + \dfrac{V_N}{5\ \Omega}$
g. $V_N = 1.25$ V
h. $V_{R_1} = 10\text{ V} + 1.25\text{ V} = 11.25$ V
$V_{R_2} = 15\text{ V} - 1.25\text{ V} = 13.75$ V
i. Yes. Because the solutions for V_{R_1} and V_{R_2} were both positive.
j. $I_1 = 1.125$ A
$I_2 = 1.375$ A
$I_3 = 250$ mA
k. For the loop with V_1 we have 10 V + 1.25 V − 11.25 V = 0 going CCW from the positive (+) terminal of V_1.
For the loop with V_2 we have −15 V + 1.25 V + 13.75 V = 0 going CW from the negative (−) terminal of V_2.
l. $I_2 - I_1 - I_3 = 0$ or 1.375 A − 1.125 A − 250 mA = 0

SECTION 9-5 METHOD OF MESH CURRENTS

21. **a.** V_1, R_1 and R_3
b. V_2, R_2 and R_3
c. R_3
d. $20I_A - 10I_B = -40$ V
e. $-10I_A + 25I_B = -20$ V
f. $I_A = -3$ A
$I_B = -2$ A
g. $I_1 = I_A = -3$ A
$I_2 = I_B = -2$ A
$I_3 = 1$ A
h. No. Because the answers for the mesh currents I_A and I_B were negative.
i. I_3 flows in the same direction as I_A or up through R_3.
j. $V_{R_1} = I_1R_1 = 3\text{ A} \times 10\ \Omega = 30$ V
$V_{R_2} = I_2R_2 = 2\text{ A} \times 15\ \Omega = 30$ V
$V_{R_3} = I_3R_3 = 1\text{ A} \times 10\ \Omega = 10$ V
k. 30 V + 10 V − 40 V = 0
l. −20 V − 10 V + 30 V = 0
m. $I_2 + I_3 - I_1 = 0$ or 2 A + 1 A − 3 A = 0

ANSWERS TO CRITICAL THINKING PROBLEMS

23. $I_A = 413.3$ mA, $I_B = 40$ mA, and $I_C = 253.3$ mA
$I_1 = 413.3$ mA
$I_2 = 373.3$ mA
$I_3 = 413.3$ mA
$I_4 = 40$ mA
$I_5 = 293.3$ mA
$I_6 = 40$ mA

$I_7 = 253.3$ mA
$I_8 = 253.3$ mA

Chapter 10

SECTION 10-1 SUPERPOSITION THEOREM

1. $V_P = -6$ V
3. $V_P = 10$ V
5. $V_{AB} = 0.2$ V

SECTION 10-2 THEVENIN'S THEOREM

7. When $R_L = 3\ \Omega$, $I_L = 2.5$ A and $V_L = 7.5$ V
   When $R_L = 6\ \Omega$, $I_L = 1.67$ A and $V_L = 10$ V
   When $R_L = 12\ \Omega$, $I_L = 1$ A and $V_L = 12$ V
9. When $R_L = 100\ \Omega$, $I_L = 48$ mA and $V_L = 4.8$ V
   When $R_L = 1$ kΩ, $I_L = 30$ mA and $V_L = 30$ V
   When $R_L = 5.6$ kΩ, $I_L =$ 10.29 mA and $V_L = 57.6$ V
11. When $R_L = 200\ \Omega$, $I_L = 45$ mA and $V_L = 9$ V
   When $R_L = 1.2$ kΩ, $I_L = 20$ mA and $V_L = 24$ V
   When $R_L = 1.8$ kΩ, $I_L = 15$ mA and $V_L = 27$ V
13. $I_L = 20$ mA and $V_L = 24$ V

SECTION 10-3 THEVENIZING A CIRCUIT WITH TWO VOLTAGE SOURCES

15. $I_3 = 208.3$ mA and $V_{R_3} = 3.75$ V
17. $I_3 = 37.5$ mA and $V_{R_3} = 2.1$ V

SECTION 10-4 THEVENIZING A BRIDGE CIRCUIT

19. $V_{TH} = 10$ V and $R_{TH} = 150\ \Omega$
21. $V_{TH} = 9$ V and $R_{TH} = 200\ \Omega$

SECTION 10-5 NORTON'S THEOREM

23. $I_N = 5$ A and $R_N = 6\ \Omega$
25. $I_N = 1.5$ A and $R_N = 10\ \Omega$
27. $I_N = 500$ mA and $R_N = 30\ \Omega$
   $I_L = 333.3$ mA
   $V_L = 5$ V

SECTION 10-6 THEVENIN-NORTON CONVERSIONS

29. $V_{TH} = 24$ V
   $R_{TH} = 1.2$ kΩ
31. $I_N = 30$ mA
   $R_{TH} = 1.2$ kΩ

SECTION 10-7 CONVERSION OF VOLTAGE AND CURRENT SOURCES

33. **a.** See Instructor's Manual.
   b. $V_T = 30$ V and $R = 24\ \Omega$
   c. $I_3 = 1$ A
   $V_{R_3} = 6$ V

SECTION 10-8 MILLMAN'S THEOREM

35. $V_{XY} = -28$ V
37. $V_{XY} = 0$ V

SECTION 10-9 T OR Y AND π OR Δ CONNECTIONS

39. $R_A = 17.44\ \Omega$, $R_B = 19.63\ \Omega$, and $R_C = 31.4\ \Omega$
41. $R_T = 7\ \Omega$
   $I_T = 3$ A

ANSWERS TO CRITICAL THINKING PROBLEMS

43. $I_L = 0$ A and $V_L = 0$ V
45. $V_{TH} = 21.6$ V
   $R_{TH} = 120\ \Omega$
   $I_2 = 67.5$ mA
   $V_2 = 13.5$ V

Chapter 11

SECTION 11-1 FUNCTION OF THE CONDUCTOR

1. **a.** 100 ft
   b. $R_T = 8.16\ \Omega$
   c. $I = 14.71$ A
   d. 1.18 V
   e. 117.7 V
   f. 17.31 W
   g. 1.731 kW
   h. 1.765 kW
   i. 98.1%

SECTION 11-2 STANDARD WIRE GAGE SIZES

3. **a.** 25 cmils
   b. 441 cmils
   c. 1024 cmils
   d. 2500 cmils
   e. 10,000 cmils
   f. 40,000 cmils
5. **a.** $R = 1.018\ \Omega$
   b. $R = 2.042\ \Omega$
   c. $R = 4.094\ \Omega$
   d. $R = 26.17\ \Omega$
7. A 1000-ft length of No. 23 gage copper wire

SECTION 11-3 TYPES OF WIRE CONDUCTORS

9. No. 10 gage
11. No. 19 gage

SECTION 11-6 SWITCHES

13. **a.** 6.3 V
   b. 0 V
   c. No
   d. 0 A
15. **a.** See Instructor's Manual.
   b. See Instructor's Manual.

SECTION 11-8 WIRE RESISTANCE

17. **a.** $R = 2.54\ \Omega$
   b. $R = 4.16\ \Omega$
   c. $R = 24.46\ \Omega$
19. **a.** $R = 0.32\ \Omega$
   b. $R = 0.64\ \Omega$
21. $R = 0.127\ \Omega$ (approx.)
23. No. 16 gage

SECTION 11-9 TEMPERATURE COEFFICIENT OF RESISTANCE

25. $R = 12.4\ \Omega$
27. $R = 140\ \Omega$
29. $R = 8.5\ \Omega$

ANSWERS TO CRITICAL THINKING PROBLEMS

31. See Instructor's Manual.

Chapter 12

SECTION 12-6 SERIES-CONNECTED AND PARALLEL-CONNECTED CELLS

1. $V_L = 3$ V, $I_L = 30$ mA, the current in each cell equals 30 mA.
3. $V_L = 1.25$ V, $I_L = 50$ mA, the current in each cell equals 25 mA.
5. $V_L = 3$ V, $I_L = 300$ mA, the current in each cell equals 100 mA.

SECTION 12-8 INTERNAL RESISTANCE OF A GENERATOR

7. $r_i = 2\ \Omega$
9. $r_i = 6\ \Omega$
11. $r_i = 15\ \Omega$

SECTION 12-9 CONSTANT-VOLTAGE AND CONSTANT-CURRENT SOURCES

13. **a.** $I_L = 1\ \mu$A; $V_L = 0$ V
   b. $I_L \approx 1\ \mu$A; $V_L \approx 100\ \mu$V
   c. $I_L \approx 1\ \mu$A; $V_L \approx 1$ mV
   d. $I_L = 0.99\ \mu$A; $V_L = 99$ mV

SECTION 12-10 MATCHING A LOAD RESISTANCE TO THE GENERATOR r_i

15. a. $I_L = 1.67$ A
$V_L = 16.67$ V
$P_L = 27.79$ W
$P_T = 167$ W
% Efficiency = 16.64%
b. $I_L = 1.33$ A
$V_L = 33.3$ V
$P_L = 44.44$ W
$P_T = 133$ W
% Efficiency = 33.4%
c. $I_L = 1$ A
$V_L = 50$ V
$P_L = 50$ W
$P_T = 100$ W
% Efficiency = 50%
d. $I_L = 800$ mA
$V_L = 60$ V
$P_L = 48$ W
$P_T = 80$ W
% Efficiency = 60%
e. $I_L = 667$ mA
$V_L = 66.67$ V
$P_L = 44.44$ W
$P_T = 66.7$ W
% Efficiency = 66.67%
17. a. $R_L = 8\ \Omega$
b. $P_L = 78.125$ W
c. % Efficiency = 50%

ANSWERS TO CRITICAL THINKING PROBLEMS

19. a. $R_L = 30\ \Omega$
b. $P_{L(max)} = 2.7$ W

Chapter 13

SECTION 13-2 MAGNETIC FLUX, ϕ

1. a. 1 Mx = 1 magnetic field line
b. 1 Wb = 1×10^8 Mx or 1×10^8 magnetic field lines
3. a. 1×10^{-5} Wb
b. 1×10^{-4} Wb
c. 1×10^{-8} Wb
d. 1×10^{-6} Wb or 1 μWb
5. a. 4000 Mx
b. 2.25×10^{-6} Wb
c. 8×10^{-4} Wb
d. 6.5×10^4 Wb
7. 9×10^4 magnetic field lines

SECTION 13-3 FLUX DENSITY, B

9. a. $1\text{G} = \dfrac{1\text{ Mx}}{\text{cm}^2}$
b. $1\text{T} = \dfrac{1\text{ Wb}}{\text{m}^2}$
11. a. 0.4 T
b. 80 T
c. 0.06 T
d. 1 T
13. a. 905 G
b. 1×10^6 G
c. 7.5 T
d. 175 T
15. 0.05 T
17. 4000 G or 4 kG
19. 0.133 T
21. 240,000 Mx
23. 0.02 Wb
25. 1×10^6 magnetic field lines

ANSWERS TO CRITICAL THINKING PROBLEMS

27. a. 40,300 Mx
b. 403 μWb

Chapter 14

SECTION 14-1 AMPERE-TURNS OF MAGNETOMOTIVE FORCE (MMF)

1. a. Gilbert (Gb)
b. Ampere-turn (A·t)
3. a. 20 A·t
b. 2 A·t
c. 10 A·t
d. 180 A·t
5. a. $N = 1000$ turns
b. $N = 4000$ turns
c. $N = 2500$ turns
d. $N = 50$ turns
7. a. 100 A·t
b. 30 A·t
c. 500 A·t

SECTION 14-2 FIELD INTENSITY (H)

9. a. $500\ \dfrac{\text{A·t}}{\text{m}}$
b. $100\ \dfrac{\text{A·t}}{\text{m}}$
c. $50\ \dfrac{\text{A·t}}{\text{m}}$
d. $200\ \dfrac{\text{A·t}}{\text{m}}$
11. a. 0.63 Oersteds
b. 1.89 Oersteds
13. a. 12.6×10^{-6}
b. 63×10^{-6}
c. 126×10^{-6}
d. 630×10^{-6}
e. 1.26×10^{-3}
15. $\mu_r = 133.3$

SECTION 14-3 B-H MAGNETIZATION CURVE

17. a. 126×10^{-6}
b. 88.2×10^{-6}

SECTION 14-9 GENERATING AN INDUCED VOLTAGE

19. $v_{ind} = 8$ V
21. $v_{ind} = 2$ kV

ANSWERS TO CRITICAL THINKING PROBLEMS

23. See Instructor's Manual.
25. a. $R_W = 1.593\ \Omega$
b. $R_T = 17.593\ \Omega$
c. $V_L = 218.3$ V
d. I^2R power loss = 296.5 W
e. $P_L = 2.98$ kW
f. $P_T = 3.27$ kW
g. % Efficiency = 91.1%
27. With a relay, the 1000-ft length of wire does not carry the load current, I_L, and thus the circuit losses are reduced significantly.

Chapter 15

SECTION 15-2 ALTERNATING-VOLTAGE GENERATOR

1. a. 90°
b. 180°
c. 270°
d. 360°
3. a. at 90°
b. at 270°
c. 0°, 180°, and 360°

SECTION 15-3 THE SINE WAVE

5. a. $v = 10$ V
b. $v = 14.14$ V
c. $v = 17.32$ V
d. $v = 19.32$ V
e. $v = 17.32$ V
f. $v = -10$ V
g. $v = -17.32$ V
7. $V_{pk} = 51.96$ V

SECTION 15-4 ALTERNATING CURRENT

9 a. Counterclockwise
b. Clockwise

SECTION 15-5 VOLTAGE AND CURRENT VALUES FOR A SINE WAVE

11. a. 100 V peak-to-peak
b. 35.35 V rms
c. 31.85 V average

13. a. 56.56 V peak
b. 113.12 V peak-to-peak
c. 36 V average

15. a. 47.13 mA rms
b. 66.7 mA peak
c. 133.3 mA peak-to-peak
d. 42.47 mA average

17. a. 16.97 V peak
b. 113 V peak
c. 25 V peak
d. 1.06 V peak

SECTION 15-6 FREQUENCY

19. a. 2000 cps
b. 15,000,000 cps
c. 10,000 cps
d. 5,000,000,000 cps

SECTION 15-7 PERIOD

21. a. $T = 500\ \mu s$
b. $T = 250\ \mu s$
c. $T = 5\ \mu s$
d. $T = 0.5\ \mu s$

23. a. $f = 200$ Hz
b. $f = 100$ kHz
c. $f = 2$ MHz
d. $f = 30$ kHz

SECTION 15-8 WAVELENGTH

25. a. 186,000 mi/s
b. 3×10^{10} cm/s
c. 3×10^{8} m/s

27. a. 8000 cm
b. 4000 cm
c. 2000 cm
d. 1500 cm

29. 2 m

31. a. $f = 1.875$ MHz
b. $f = 30$ MHz
c. $f = 17.65$ MHz
d. $f = 27.3$ MHz

SECTION 15-9 PHASE ANGLE

33. A sine wave has its maximum values at 90° and 270° whereas a cosine wave has its maximum values at 0° and 180°.

SECTION 15-10 THE TIME FACTOR IN FREQUENCY AND PHASE

35. a. $t = 83.3\ \mu s$
b. $t = 125\ \mu s$
c. $t = 166.7\ \mu s$
d. $t = 250\ \mu s$

SECTION 15-11 ALTERNATING CURRENT CIRCUITS WITH RESISTANCE

37. $R_T = 250\ \Omega$
$I = 40$ mA
$V_1 = 4$ V
$V_2 = 6$ V
$P_1 = 160$ mW
$P_2 = 240$ mW
$P_T = 400$ mW

39. $R_T = 900\ \Omega$
$I_T = 40$ mA
$V_1 = 7.2$ V
$V_2 = 28.8$ V
$V_3 = 28.8$ V
$I_2 = 24$ mA
$I_3 = 16$ mA
$P_1 = 288$ mW
$P_2 = 691.2$ mW
$P_3 = 460.8$ mW
$P_T = 1.44$ W

SECTION 15-12 NONSINUSOIDAL AC WAVEFORMS

41. a. $V = 100$ V peak-to-peak
$f = 20$ kHz
b. $V = 30$ V peak-to-peak
$f = 500$ Hz
c. $V = 100$ V peak-to-peak
$f = 2.5$ kHz

SECTION 15-13 HARMONIC FREQUENCIES

43. 1 kHz First odd harmonic
2 kHz First even harmonic
3 kHz Second odd harmonic
4 kHz Second even harmonic
5 kHz Third odd harmonic
6 kHz Third even harmonic
7 kHz Fourth odd harmonic

45. 750 Hz

47. 100 kHz

SECTION 15-14 THE 60-Hz AC POWER LINE

49. Transformer

ANSWERS TO CRITICAL THINKING PROBLEMS

51. a. 68.3 ft
b. 2083 cm

53. $f = 4.1$ MHz

Chapter 16

SECTION 16-3 THE FARAD UNIT OF CAPACITANCE

1. a. $Q = 50\ \mu C$
b. $Q = 25\ \mu C$
c. $Q = 1.5\ \mu C$
d. $Q = 11\ \mu C$
e. $Q = 136$ nC
f. $Q = 141$ nC

3. a. $V = 2.5$ V
b. $V = 6.25$ V
c. $V = 12.5$ V
d. $V = 50$ V
e. $V = 75$ V

5. a. $C = 15\ \mu F$
b. $C = 0.5\ \mu F$ or 500 nF
c. $C = 4\ \mu F$
d. $C = 0.22\ \mu F$ or 220 nF
e. $C = 0.001\ \mu F$
f. $C = 0.04\ \mu F$ or 40 nF

7. a. $C = 1.77$ pF
b. $C = 2.213$ nF
c. $C = 44.25$ nF
d. $C = 106.2$ nF

SECTION 16-6 CAPACITOR CODING

9. a. $C = 0.0033\ \mu F$; +80%, −20%
b. $C = 0.022\ \mu F$; +100%, −0%
c. $C = 1800$ pF; ±10%
d. $C = 0.0047\ \mu F$; +80%, −20%
e. $C = 100{,}000$ pF; ±5%
f. $C = 0.15\ \mu F$; ±20%

11. a. $C = 56$ pF
b. $C = 12{,}000$ pF
c. $C = 560{,}000$ pF
d. $C = 22$ pF

13. a. $C = 0.47\ \mu F$, ±10%
b. $C = 6.2\ \mu F$, ±5%
c. $C = 15\ \mu F$, ±10%
d. $C = 820\ \mu F$, ±5%

15. See Instructor's Manual.

SECTION 16-7 PARALLEL CAPACITANCES

17. $C_T = 0.38\ \mu F$

19. a. $V = 10$ V
b. $Q_1 = 1$ mC
c. $Q_2 = 2.2$ mC
d. $Q_3 = 6.8$ mC
e. $Q_7 = 10$ mC
f. $Q_9 = 1000\ \mu F$

SECTION 16-8 SERIES CAPACITANCES

21. $C_{EQ} = 600$ pF

23. a. $C_{EQ} = 5\ \mu F$
b. $Q_1 = Q_2 = Q_3 = 180\ \mu C$
c. $V_{C1} = 18$ V, $V_{C2} = 6$ V, $V_{C3} = 12$ V
d. 180 μC

25. 600 pF

SECTION 16-9 ENERGY STORED IN ELECTROSTATIC FIELD OF CAPACITANCE

27. **a.** $\mathscr{E} = 5.4\ \mu J$
b. $\mathscr{E} = 135\ \mu J$
c. $\mathscr{E} = 3.375$ mJ

SECTION 16-10 MEASURING AND TESTING CAPACITORS

29. **a.** 47,000 pF
b. 1,500 pF
c. 0.39 μF
d. 0.001 μF

31. Yes

33. No

SECTION 16-11 TROUBLES IN CAPACITORS

35. The ohmmeter needle will deflect all the way to the right and then back off to infinity as the capacitor charges.

ANSWERS TO CRITICAL THINKING PROBLEMS

37. See Instructor's Manual.

Chapter 17

SECTION 17-1 ALTERNATING CURRENT IN A CAPACITIVE CIRCUIT

1. **a.** $I = 0$ A
b. $V_{\text{lamp}} = 0$ V
c. $V_C = 12$ V

3. **a.** $I = 400$ mA
b. $I = 400$ mA
c. $I = 400$ mA
d. $I = 400$ mA
e. $I = 0$ A

5. The amplitude of the applied voltage, the frequency of the applied voltage, and the amount of capacitance.

SECTION 17-2 THE AMOUNT OF X_C EQUALS $\frac{1}{2\pi fC}$

7. **a.** $X_C = 265.26\ \Omega$
b. $X_C = 132.63\ \Omega$
c. $X_C = 31.83\ \Omega$
d. $X_C = 15.92\ \Omega$

9. **a.** $f = 33.86$ Hz
b. $f = 677.26$ Hz
c. $f = 2.26$ kHz
d. $f = 67.73$ kHz

11. $f = 776.37$ kHz

13. **a.** $X_C = 20\ k\Omega$
b. $X_C = 5\ k\Omega$
c. $X_C = 2.5\ k\Omega$
d. $X_C = 1\ k\Omega$

15. **a.** $C = 0.05\ \mu F$
b. $C = 0.0125\ \mu F$
c. $C = 0.004\ \mu F$
d. $C = 1.592$ nF

SECTION 17-3 SERIES OR PARALLEL CAPACITIVE REACTANCES

17. **a.** $X_{CT} = 5\ k\Omega$
b. $X_{CT} = 3\ k\Omega$
c. $X_{CT} = 150\ k\Omega$
d. $X_{CT} = 3\ k\Omega$

SECTION 17-4 OHM'S LAW APPLIED TO X_C

19. $I = 50$ mA

21. **a.** $X_{CT} = 2.4\ k\Omega$
b. $I = 15$ mA
c. $V_{C_1} = 6$ V, $V_{C_2} = 12$ V, and $V_{C_3} = 18$ V

23. $C_1 = 1.25\ \mu F$, $C_2 = 0.625\ \mu F$, $C_3 = 0.417\ \mu F$, $C_{EQ} = 0.208\ \mu F$

25. **a.** $X_{C_1} = 400\ \Omega$, $X_{C_2} = 320\ \Omega$ and $X_{C_3} = 80\ \Omega$
b. $I_{C_1} = 60$ mA, $I_{C_2} = 75$ mA and $I_{C_3} = 300$ mA
c. $I_T = 435$ mA
d. $X_{CEQ} = 55.17\ \Omega$
e. $C_T = 1.45\ \mu F$

SECTION 17-5 APPLICATIONS OF CAPACITIVE REACTANCE

27. $C = 3.183\ \mu F$
$C = 159$ nF
$C = 6.37$ nF
$C = 31.83$ pF

SECTION 17-6 SINE-WAVE CHARGE AND DISCHARGE CURRENT

29. **a.** $i_C = 1\ \mu A$
b. $i_C = 1$mA
c. $i_C = 500$ mA

31. For any capacitor, i_C and V_C are 90° out of phase with each other, with i_C reaching its maximum value 90° ahead of V_C. The reason that i_C leads V_C by 90° is that the value of i_C depends on the rate of voltage change across the capacitor plates rather than on the actual value of voltage itself.

33. $\frac{dv}{dt} = 2.5$ MV/s

ANSWERS TO CRITICAL THINKING PROBLEMS

35. $X_{C_T} = 625\ \Omega$
$X_{C_1} = 500\ \Omega$
$X_{C_2} = 500\ \Omega$
$C_1 = 0.01\ \mu F$
$C_3 = 0.03\ \mu F$
$V_{C_1} = 20$ V
$V_{C_2} = V_{C_3} = 5$ V
$I_2 = 10$ mA
$I_3 = 30$ mA

Chapter 18

SECTION 18-1 SINE WAVE V_C LAGS i_C BY 90°

1. **a.** 10 V
b. 10 mA
c. 10 kHz
d. 90° (i_C leads V_C by 90°)

3. **a.** See Instructor's Manual.
b. See Instructor's Manual.

SECTION 18-2 X_C AND R IN SERIES

5. **a.** I and V_R are in phase
b. V_C lags I by 90°
c. V_C lags V_R by 90°

7. See Instructor's Manual.

9. **a.** $V_R = 7.07$ V
b. $V_C = 7.07$ V
c. $V_T = 10$ V

SECTION 18-3 IMPEDANCE Z TRIANGLE

11. $Z_T = 25\ \Omega$
$I = 4$ A
$V_C = 80$ V
$V_R = 60$ V
$\theta_Z = -53.13°$

13. $Z_T = 21.63\ \Omega$
$I = 2.31$ A
$V_C = 41.58$ V
$V_R = 27.72$ V
$\theta_Z = -56.31°$

15. $Z_T = 10.44\ k\Omega$
$I = 2.3$ mA
$V_C = 6.9$ V
$V_R = 23$ V
$\theta_Z = -16.7°$

17. $X_C = 5\ k\Omega$
$Z_T = 6.34\ k\Omega$
$I = 5.68$ mA
$V_R = 22.15$ V
$V_C = 28.4$ V
$\theta_Z = -52°$

19. **a.** X_C increases.
b. Z_T increases.
c. I decreases.
d. V_C increases.
e. V_R decreases.
f. θ_Z increases (becomes more negative).

SECTION 18-4 *RC* PHASE-SHIFTER CIRCUIT

21. **a.** V_R leads V_T by 28°.
b. V_C lags V_T by 62°.

23. **a.** $Z_T = 26.55\ \text{k}\Omega$
$I = 4.52$ mA
$V_C = 119.9$ V
$V_R = 4.52$ V
$\theta_Z = -87.84°$
b. V_R leads V_T by 87.84°.
c. V_C lags V_T by 2.16°.

SECTION 18-5 X_C AND *R* IN PARALLEL

25. **a.** 120 V
b. 120 V

27. $I_R = 3$ A
$I_C = 4$ A
$I_T = 5$ A
$Z_{EQ} = 24\ \Omega$
$\theta_I = 53.13°$

29. $I_R = 2$ A
$I_C = 4$ A
$I_T = 4.47$ A
$Z_{EQ} = 22.37\ \Omega$
$\theta_I = 63.4°$

31. $I_R = 200$ mA
$I_C = 200$ mA
$I_T = 282.8$ mA
$Z_{EQ} = 63.65\ \Omega$
$\theta_I = 45°$

33. **a.** $I_R = 1$ A
$I_C = 1$ A
$I_T = 1.414$ A
$Z_{EQ} = 35.36\ \Omega$
$\theta_I = 45°$
b. $I_R = 2$ A
$I_C = 200$ mA
$I_T = 2.01$ A
$Z_{EQ} = 9.95\ \Omega$
$\theta_I = 5.7°$
c. $I_R = 200$ mA
$I_C = 2$ A
$I_T = 2.01$ A
$Z_{EQ} = 9.95\ \Omega$
$\theta_I = 84.3°$

35. $X_C = 500\ \Omega$
$I_R = 20$ mA
$I_C = 48$ mA
$I_T = 52$ mA
$Z_{EQ} = 461.54\ \Omega$
$\theta_I = 67.38°$

37. **a.** I_R stays the same.
b. I_C decreases.
c. I_T decreases.
d. Z_{EQ} increases.
e. θ_I decreases.

SECTION 18-6 RF AND AF COUPLING CAPACITORS

39. $f = 33.86$ kHz
$\theta_Z = -5.71°$

SECTION 18-7 CAPACITIVE VOLTAGE DIVIDERS

41. $V_{C_1} = 50$ V
$V_{C_2} = 20$ V
$V_{C_3} = 10$ V

SECTION 18-8 THE GENERAL CASE OF CAPACITIVE CURRENT i_C

43. See Instructor's Manual.

ANSWERS TO CRITICAL THINKING PROBLEMS

45. $I_C = 400$ mA
$I_R = 300$ mA
$V_A = 36$ V
$X_C = 90\ \Omega$
$C = 5.56\ \mu$F
$Z_{EQ} = 72\ \Omega$

Chapter 19

SECTION 19-1 INDUCTION BY ALTERNATING CURRENT

1. A small current change of 1 to 2 mA
3. A high-frequency alternating current

SECTION 19-2 SELF-INDUCTANCE *L*

5. **a.** $L = 10$ H
b. $L = 1.5$ mH
c. $L = 1.5$ H
d. $L = 6$ mH
e. $L = 3$ mH
f. $L = 375\ \mu$H
g. $L = 15$ H

7. $L = 50$ mH
9. $L = 2.53\ \mu$H

SECTION 19-3 SELF-INDUCED VOLTAGE, v_L

11. $v_L = 500$ V

13. **a.** $v_L = 10$ V
b. $v_L = 20$ V
c. $v_L = 5$ V
d. $v_L = 100$ V

SECTION 19-5 MUTUAL INDUCTANCE L_M

15. $k = 0.75$
17. $L_M = 61.24$ mH

SECTION 19-6 TRANSFORMERS

19. **a.** $V_S = 24\ V_{AC}$
b. $I_S = 2$ A
c. $P_{sec} = 48$ W
d. $P_{pri} = 48$ W
e. $I_P = 400$ mA

21. **a.** $V_{S_1} = 120\ V_{AC}$
b. $V_{S_2} = 24\ V_{AC}$
c. $I_{S_1} = 50$ mA
d. $I_{S_2} = 1$ A
e. $P_{sec1} = 6$ W
f. $P_{sec2} = 24$ W
g. $P_{pri} = 30$ W
h. $I_P = 250$ mA

23 **a.** $\dfrac{N_P}{N_S} = \dfrac{3}{1}$
b. $I_S = 2.5$ A
c. $I_P = 833.3$ mA

25. % Efficiency = 80%

SECTION 19-7 TRANSFORMER RATINGS

27. The power rating of a transformer is specified in volt-amperes (VA), which is the unit of apparent power.

29. To identify those transformer leads with the same instantaneous polarity

31. **a.** $V_{sec1} = 32\ V_{AC}$
b. $V_{sec2} = 60\ V_{AC}$
c. $I_{S1(max)} = 1.875$ A
d. $I_{S2(max)} = 1.67$ A
e. $I_{P(max)} = 1.33$ A

33. $I_P = 210$ mA

SECTION 19-8 IMPEDANCE TRANSFORMATION

35. **a.** $Z_P = 200\ \Omega$
b. $Z_P = 12.5\ \Omega$
c. $Z_P = 6.25\ \text{k}\Omega$
d. $Z_P = 5\ \text{k}\Omega$
e. $Z_P = 5\ \Omega$

37. $\dfrac{N_P}{N_S} = 11.18{:}1$

SECTION 19-12 INDUCTANCES IN SERIES OR PARALLEL

39. **a.** $L_T = 20$ mH
b. $L_T = 18$ mH

c. $L_T = 1$ mH
d. $L_T = 10$ mH

41. $L_T = 660$ mH

43. a. $L_T = 82.63$ mH
b. $L_T = 37.37$ mH

SECTION 19-13 ENERGY IN MAGNETIC FIELD OF INDUCTANCE

45. Energy = 243 μJ

47. Energy = 675 mJ

ANSWERS TO CRITICAL THINKING PROBLEMS

49. $Z_P = 36.36\ \Omega$

51. $I_P = 312.5$ mA

Chapter 20

SECTION 20-1 HOW X_L REDUCES THE AMOUNT OF I

1. $X_L = 0\ \Omega$ at DC

3. $I_{DC} = 2.5$ A

5. a. Because with S_1 in position 2 the inductor has an inductive reactance, X_L, in addition to the DC resistance, r_i, to limit the circuit's current flow. With S_1 in position 1 only the DC resistance of the coil limits current flow since there is no X_L for direct current.
b. $X_L = 4$ kΩ

SECTION 20-2 $X_L = 2\pi fL$

7. a. $X_L = 37.7\ \Omega$
b. $X_L = 75.4\ \Omega$
c. $X_L = 1$ kΩ
d. $X_L = 6.28$ kΩ

9. a. $L = 500$ mH
b. $L = 100$ mH
c. $L = 31.83$ mH
d. $L = 25$ mH

11. $L = 254.65$ mH

13. a. $I = 1$ mA
b. $I = 4$ mA
c. $I = 1$ mA
d. $I = 4$ mA

15. a. $X_L = 2.64$ kΩ
b. $X_L = 1.1$ kΩ
c. $X_L = 1$ kΩ
d. $X_L = 1$ kΩ

17. a. $f = 1.99$ kHz
b. $f = 530.52$ kHz
c. $f = 3.183$ kHz
d. $f = 7.96$ kHz

SECTION 20-3 SERIES OR PARALLEL INDUCTIVE REACTANCES

19. a. $X_{LEQ} = 720\ \Omega$
b. $X_{LEQ} = 600\ \Omega$
c. $X_{LEQ} = 150\ \Omega$
d. $X_{LEQ} = 133.3\ \Omega$

SECTION 20-4 OHM'S LAW APPLIED TO X_L

21. a. I increases.
b. I decreases.

23. $L_1 = 10$ mH, $L_2 = 12$ mH, $L_3 = 18$ mH, and $L_T = 40$ mH

25. a. $I_{L_1} = 600$ mA, $I_{L_2} = 200$ mA, $I_{L_3} = 800$ mA
b. $I_T = 1.6$ A
c. $X_{LEQ} = 75\ \Omega$

27. a. $X_{L_1} = 1.6$ kΩ, $X_{L_2} = 6.4$ kΩ and $X_{L_3} = 1.28$ kΩ
b. $I_{L_1} = 20$ mA, $I_{L_2} = 5$ mA, $I_{L_3} = 25$ mA
c. $I_T = 50$ mA
d. $X_{LEQ} = 640\ \Omega$
e. $L_{EQ} = 16$ mH

SECTION 20-6 WAVESHAPE OF V_L INDUCED BY SINE-WAVE CURRENT

29. V_L leads i_L by a phase angle of 90°. This 90° phase relationship exists because V_L depends on the rate of current change rather than on the actual value of current itself.

ANSWERS TO CRITICAL THINKING PROBLEMS

31. $L_1 = 60$ mH
$L_2 = 40$ mH
$L_3 = 120$ mH
$L_T = 90$ mH
$X_{L_1} = 1.2$ kΩ
$X_{L_2} = 800\ \Omega$
$X_{L_T} = 1.8$ kΩ
$V_{L_1} = 24$ V
$V_{L_3} = 12$ V
$I_{L_2} = 15$ mA
$I_{L_3} = 5$ mA

33. $L_1 = 10$ mH, $L_2 = 120$ mH, and $L_3 = 40$ mH

Chapter 21

SECTION 21-1 SINE WAVE i_L LAGS v_L BY 90°

1. a. 10 V
b. 10 mA
c. 10 kHz
d. 90°

3. a. See Instructor's Manual.
b. See Instructor's Manual.

SECTION 21-2 X_L AND R IN SERIES

5. a. 0°
b. 90°
c. 90°

7. See Instructor's Manual.

9. a. $V_R = 7.07$ V
b. $V_L = 7.07$ V
c. $V_T = 10$ V

SECTION 21-3 IMPEDANCE Z TRIANGLE

11. $Z_T = 125\ \Omega$
$I = 288$ mA
$V_L = 21.6$ V
$V_R = 28.8$ V
$\theta_Z = 36.87°$

13. $Z_T = 11.18$ kΩ
$I = 10.73$ mA
$V_L = 107.3$ V
$V_R = 53.67$ V
$\theta_Z = 63.44°$

15. $Z_T = 42.43\ \Omega$
$I = 1.18$ A
$V_L = 35.35$ V
$V_R = 35.35$ V
$\theta_Z = 45°$

17. $X_L = 1.8$ kΩ
$Z_T = 3.25$ kΩ
$I = 30.77$ mA
$V_R = 83.1$ V
$V_L = 55.4$ V
$\theta_Z = 33.7°$

19. a. X_L decreases.
b. Z_T decreases.
c. I increases.
d. V_R increases.
e. V_L decreases.
f. θ_Z decreases.

SECTION 21-4 X_L AND R IN PARALLEL

21. a. 0°
b. I_L lags V_A by 90°.
c. I_L lags I_R by 90°.

23. See Instructor's Manual.

25. $I_R = 3$ A
$I_L = 2$ A
$I_T = 3.61$ A
$Z_{EQ} = 33.24\ \Omega$
$\theta_I = -33.7°$

27. $I_R = 4.8$ mA
$I_L = 2$ mA
$I_T = 5.2$ mA
$Z_{EQ} = 4.62$ kΩ
$\theta_I = -22.62°$

29. $Z_{EQ} = 192\ \Omega$

31. a. I_R stays the same.
b. I_L decreases.
c. I_T decreases.
d. Z_{EQ} increases.
e. θ_I becomes less negative.

SECTION 21-5 *Q* OF A COIL

33. a. $Q = 3.14$
b. $Q = 6.28$
c. $Q = 10$
d. $Q = 62.83$

35. $R_e = 94.25\ \Omega$

SECTION 21-6 AF AND *RF* CHOKES

37. a. $L = 4.78$ H
b. $L = 954.9$ mH
c. $L = 11.94$ mH
d. $L = 2.39$ mH

39. a. $V_{out} = 9.95\ V_{p\text{-}p}$
b. $V_{out} = 7.07\ V_{p\text{-}p}$
c. $V_{out} = 995\ mV_{p\text{-}p}$

SECTION 21-7 THE GENERAL CASE OF INDUCTIVE VOLTAGE

41. See Instructor's Manual.

ANSWERS TO CRITICAL THINKING PROBLEMS

43. $I_T = 6$ mA
$I_R = 3$ mA
$I_L = 5.2$ mA
$X_L = 2.31$ kΩ
$R = 4$ kΩ
$L = 36.77$ mH

Chapter 22

SECTION 22-1 RESPONSE OF RESISTANCE ALONE

1. The current, I, reaches its steady-state value immediately because a resistor does not provide any reaction to a change in either voltage or current.

3. The resistor provides 2 Ω of resistance to oppose current from the 12 V source but it does not provide any reaction to the closing or opening of the switch, S_1.

SECTION 22-2 $\frac{L}{R}$ TIME CONSTANT

5. a. $T = 200\ \mu s$
b. 240 mA
c. 0 mA
d. Approximately 151.7 mA
e. 1 ms

7. a. Either increase L or decrease R
b. Either decrease L or increase R

SECTION 22-3 HIGH VOLTAGE PRODUCED BY OPENING AN *RL* CIRCUIT

9. Without a resistor across S_1 there is no way to determine the time constant of the circuit with S_1 open. This is because there is no way of knowing what the resistance of the open switch is. We do know, however, that the time constant will be very short with S_1 open. This short time constant will result in a very large $\frac{di}{dt}$ value which in turn will produce a very large induced voltage across the open contacts of the switch. This will most likely produce internal arcing across the open switch contacts.

SECTION 22-4 *RC* TIME CONSTANT

11. a. $V_C = 31.6$ V
b. $V_C = 50$ V
c. $V_C = 50$ V

13. a. $T = 1$s
b. $T = 1.5\ \mu s$
c. $T = 89.1\ \mu s$
d. $T = 200$ ms

15. a. 25 V
b. $V_C = 40.8$ V
c. $V_C = 50$ V

SECTION 22-5 *RC* CHARGE AND DISCHARGE CURVES

17. a. 500 μA
b. Zero
c. $V_R = 18.4$ V
d. 184 μA

SECTION 22-6 HIGH CURRENT PRODUCED BY SHORT-CIRCUITING AN *RC* CIRCUIT

19. a. $T = 100$ ms
b. $T = 250\ \mu s$

21. a. $V_C = 0$ V
b. $V_R = 3$ V
c. $I = 30$ mA

23. $\varepsilon = 4.5$ mJ

SECTION 22-8 LONG AND SHORT TIME CONSTANTS

25. a. Long
b. Short

27. a. The output is taken across the capacitor.
b. Long

SECTION 22-10 LONG TIME CONSTANT FOR RC COUPLING CIRCUIT

29. a. $T = 1$ ms
b. $\frac{tp}{RC} = \frac{1}{10}$
c. See Instructor's Manual.

SECTION 22-11 ADVANCED TIME CONSTANT ANALYSIS

31. a. $V_C = 0$ V
b. $V_C = 151$ V
c. $V_C = 189.6$ V
d. $V_C = 233.1$ V
e. $V_C = 259.4$ V
f. $V_C = 275.4$ V
g. $V_C = 290.9$ V

33. a. $t = 356.7$ ms
b. $t = 693.1$ ms
c. $t = 1.1$ s
d. $t = 1.61$ s
e. $t = 2.3$ s

35. $T = 7.5$ ms

37. a. $V_R = 24$ V
b. $V_R = 13.17$ V
c. $V_R = 6.33$ V
d. $V_R = 3.25$ V
e. $V_R = 856.5$ mV

SECTION 22-12 COMPARISON OF REACTANCE AND TIME CONSTANT

39. Reactance

41. Long

ANSWERS TO CRITICAL THINKING PROBLEMS

43. a. 3 ms
b. $V_C = 24.35$ V
c. $V_C = 15$ V
d. $V_C = 27.54$ V

Chapter 23

SECTION 23-1 AC CIRCUITS WITH RESISTANCE BUT NO REACTANCE

1. $R_T = 30\ \Omega$
$I = 500$ mA
$V_1 = 6$ V
$V_2 = 9$ V

3. $I_1 = 3$ A
$I_2 = 2$ A
$I_T = 5$ A
$R_{EQ} = 7.2\ \Omega$

SECTION 23-2 CIRCUITS WITH X_L ALONE

5. $X_{LT} = 250\ \Omega$
$I = 480$ mA
$V_1 = 48$ V
$V_2 = 72$ V

7. $I_1 = 1.2$ A
$I_2 = 300$ mA
$I_T = 1.5$ A
$X_{LEQ} = 80\ \Omega$

SECTION 23-3 CIRCUITS WITH X_C ALONE

9. $X_{CT} = 900\ \Omega$
$I = 20$ mA
$V_1 = 4.4$ V
$V_2 = 13.6$ V

11. $I_1 = 100$ mA
$I_2 = 400$ mA
$I_T = 500$ mA
$X_{CEQ} = 20\ \Omega$

SECTION 23-4 OPPOSITE REACTANCES CANCEL

13. **a.** net $X = X_L = 60\ \Omega$
b. $I = 400$ mA
c. $V_L = 72$ V
d. $V_C = 48$ V

15. **a.** $I_L = 300$ mA
b. $I_C = 200$ mA
c. $I_T = I_L = 100$ mA
d. $X = X_L = 180\ \Omega$

SECTION 23-5 SERIES REACTANCE AND RESISTANCE

17. **a.** $X = X_C = 75\ \Omega$
b. $Z_T = 125\ \Omega$
c. $I = 1$ A
d. $V_R = 100$ V
e. $V_L = 50$ V
f. $V_C = 125$ V
g. $\theta_Z = -36.87°$

19. **a.** $X = X_L = 600\ \Omega$
b. $Z_T = 750\ \Omega$
c. $I = 20$ mA
d. $V_R = 9$ V
e. $V_L = 36$ V
f. $V_C = 24$ V
g. $\theta_Z = 53.13°$

SECTION 23-6 PARALLEL REACTANCE AND RESISTANCE

21. **a.** $I_R = 150$ mA
b. $I_C = 600$ mA
c. $I_L = 400$ mA
d. $I_X = I_C = 200$ mA
e. $I_T = 250$ mA
f. $Z_{EQ} = 144\ \Omega$
g. $\theta_I = 53.13°$

23. **a.** $I_R = 120$ mA
b. $I_C = 80$ mA
c. $I_L = 120$ mA
d. $I_X = I_L = 40$ mA
e. $I_T = 126.5$ mA
f. $Z_{EQ} = 94.86\ \Omega$
g. $\theta_I = -18.44°$

SECTION 23-7 SERIES-PARALLEL REACTANCE AND RESISTANCE

25. **a.** $Z_T = 50\ \Omega$
b. $I_T = 2$ A
c. $V_{R_1} = 60$ V
d. $V_{C_1} = 24$ V, $V_{C_2} = 240$ V, and $V_{C_3} = 96$ V
e. $V_{L_1} = 144$ V and $V_{L_2} = 40$ V
f. $\theta_Z = -53.13°$

SECTION 23-8 REAL POWER

27. **a.** Real power = 100 W
Apparent power = 125 VA
PF = 0.8
b. Real power = 180 mW
Apparent power = 300 mVA
PF = 0.6
c. Real power = 5.4 W
Apparent power = 9 VA
PF = 0.6
d. Real power = 1.44 W
Apparent power = 1.52 VA
PF = 0.947

ANSWERS TO CRITICAL THINKING PROBLEMS

29. $C = 6.63\ \mu$F or $46.4\ \mu$F

Chapter 24

SECTION 24-1 POSITIVE AND NEGATIVE NUMBERS

1. **a.** 0°
b. 180°

SECTION 24-2 THE *j* OPERATOR

3. The *j* axis

5. **a.** Real numbers
b. Imaginary numbers

7. **a.** 25 units with a leading phase angle of +90°
b. 36 units with a lagging phase angle of −90°

SECTION 24-3 DEFINITION OF A COMPLEX NUMBER

9. Rectangular form

11. **a.** The phase angle is greater than 45°.
b. The phase angle is less than 45°.
c. The phase angle is −45°.
d. The phase angle is more negative than −45°.
e. The phase angle is less than 45°.

SECTION 24-4 HOW COMPLEX NUMBERS ARE APPLIED TO AC CIRCUITS

13. 0°

SECTION 24-5 IMPEDANCE IN COMPLEX FORM

15. **a.** $10\ \Omega + j20\ \Omega$
b. $15\ \Omega + j10\ \Omega$
c. $0\ \Omega - j1$ kΩ
d. 1.5 k$\Omega - j2$ kΩ
e. $150\ \Omega \pm j0\ \Omega$
f. $75\ \Omega - j75\ \Omega$

SECTION 24-6 OPERATIONS WITH COMPLEX NUMBERS

17. **a.** $15 + j15$
b. $40 - j20$
c. $200 + j150$
d. $90 - j50$
e. $36 - j48$

19. **a.** −72
b. 60
c. −28
d. −24
e. 2
f. −12.5
g. 25
h. −25

21. **a.** $1.19 - j0.776$
b. $0.188 + j0.188$
c. $0.461 + j0.194$
d. $1 - j0.5$

SECTION 24-8 POLAR FORM OF COMPLEX NUMBERS

23. **a.** $14.14\angle 45°$
b. $12.81\angle -51.34°$
c. $21.63\angle 56.3°$
d. $150.4\angle -21.45°$

25. **a.** $3\angle -125°$
b. $5\angle 150°$
c. $4\angle 0°$
d. $4.67\angle 66°$
e. $30\angle 75°$
f. $25\angle 80°$
g. $12.5\angle 50°$

SECTION 24-10 COMPLEX NUMBERS IN SERIES AC CIRCUITS

27. **a.** $Z_T = 30\ \Omega + j40\ \Omega$
b. $Z_T = 50\angle 53.13°\ \Omega$
c. $I = 2\angle -53.13°$ A
d. $V_R = 60\angle -53.13°$ V
e. $V_L = 140\angle 36.87°$ V
f. $V_C = 60\angle -143.13°$ V

SECTION 24-11 COMPLEX NUMBERS IN PARALLEL AC CIRCUITS

29. $Z_T = 33.3\angle 33.69°\ \Omega$ (polar form)
$Z_T = 27.7\ \Omega + j18.47\ \Omega$ (rectangular form)

31. $Y_T = 20$ mS $+ j6.67$ mS (rectangular form)
$Y_T = 21.08\angle 18.44°$ mS (polar form)
$Z_T = 47.44\angle -18.44°\ \Omega$ (polar form)

33. **a.** See Instructor's Manual.
b. See Instructor's Manual.

SECTION 24-12 COMBINING TWO COMPLEX BRANCH IMPEDANCES

35. **a.** $Z_1 = 30\ \Omega - j40\ \Omega = 50\angle -53.13°\ \Omega$
b. $Z_2 = 20\ \Omega + j15\ \Omega = 25\angle 36.87°\ \Omega$
c. $Z_T = 22\ \Omega + j4\ \Omega = 22.4\angle 10.3°\ \Omega$

SECTION 24-13 COMBINING COMPLEX BRANCH CURRENTS

37. **a.** $I_1 = 1\angle 53.13°$ A $= 600$ mA $+ j800$ mA
b. $I_2 = 2\angle -36.87°$ A $= 1.6$ A $- j1.2$ A
c. $I_T = 2.236\angle -10.3°$ A $= 2.2$ A $- j400$ mA

ANSWERS TO CRITICAL THINKING PROBLEMS

39. $V_{in} = 24\ 0°/V$

Chapter 25

SECTION 25-1 THE RESONANCE EFFECT

1. The condition of equal and opposite reactances in an *LC* circuit. Resonance occurs at only one particular frequency, known as the resonant frequency.

3. $X_L = X_C = 1\ k\Omega$

SECTION 25-2 SERIES RESONANCE

5. **a.** $X = 0\ \Omega$
b. $Z_T = 40\ \Omega$
c. $I = 25\ \mu A$
d. $\theta = 0°$
e. $V_L = 50$ mV
f. $V_C = 50$ mV
g. $V_{rs} = 1$ mV

7. Because at f_r the total impedance, Z_T is purely resistive.

SECTION 25-3 PARALLEL RESONANCE

9. **a.** Z_{EQ} is maximum.
b. I_T is minimum.
c. $\theta = 0°$

11. The resistance, r_s

SECTION 25-4 RESONANT FREQUENCY $f_r = \frac{1}{2\pi\sqrt{LC}}$

13. **a.** $f_r = 2.5$ MHz
b. $f_r = 400$ kHz
c. $f_r = 3$ MHz
d. $f_r = 5$ MHz

15. $C = 70.17$ pF

17. **a.** $f_r = 5$ MHz
b. $X_L = X_C = 628.3\ \Omega$
c. $Z_T = r_s = 12.56\ \Omega$
d. $I = 796.2\ \mu A$
e. $V_L = V_C = 500$ mV
f. $\theta_Z = 0°$

19. With *C* set to 360 pF, $f_r = 1.875$ MHz. To double f_r *C* must be reduced to 90 pF.

SECTION 25-5 *Q* MAGNIFICATION FACTOR OF A RESONANT CIRCUIT

21. **a.** $f_r = 1$ MHz
b. $Q = 100$
c. $V_L = V_C = 1$ V

23. $Q = 300$

25. **a.** $f_r = 1.25$ MHz
b. $X_L = X_C = 785.4\ \Omega$
c. $I_L = I_C = 12.73$ mA
d. $Q = 100$
e. $Z_{EQ} = 78.54\ k\Omega$
f. $I_T = 127.3\ \mu A$

27. $Q = 191$

SECTION 25-6 BANDWIDTH OF A RESONANT CIRCUIT

29. **a.** $\Delta f = 12.5$ kHz
b. $f_1 = 1.24375$ MHz (exactly) and $f_2 = 1.25625$ MHz (exactly)
c. $Z_{EQ} = 78.54\ k\Omega$ at f_r, Z_{EQ} at f_1 and $f_2 = 55.53\ k\Omega$

31. **a.** $f_r = 3$ MHz
b. $X_L = X_C = 942.5\ \Omega$
c. $Z_T = 18.85\ \Omega$
d. $I = 2.65\ \mu A$
e. $Q = 50$
f. $V_L = V_C = 2.5$ mV
g. $\theta = 0°$
h. $\Delta f = 60$ kHz, $f_1 = 2.97$ MHz and $f_2 = 3.03$ MHz
i. $I = 1.87\ \mu A$

33. At f_1 *I* is approximately 70.7% of *I* at f_r. This is because at f_1, Z_T is approximately 1.41 times the value of Z_T at f_r.

35. At f_1 and f_2 $Z_{EQ} = 138.8\ k\Omega$ and I_T is 14.41 μA.

SECTION 25-7 TUNING

37. No, because as *C* is varied to provide different resonant frequencies the *Q* of the circuit varies. Recall that $V_C = Q \times V_{in}$ at f_r. (This assumes that V_{in} remains the same for all frequencies.)

SECTION 25-8 MISTUNING

39. **a.** The circuit appears inductive with a lagging phase angle because $I_L > I_C$.
b. The circuit appears capacitive with a leading phase angle because $I_C > I_L$.

SECTION 25-9 ANALYSIS OF PARALLEL RESONANT CIRCUITS

41. At f_r, $Q = 125$ which is considered a high Q.

43. $I_L = 1.24$ mA and $I_C = 1.27$ mA. I_L is less than I_C at f_r because the impedance of the inductive branch is greater than X_C or X_L alone.

45. Z_{EQ} is maximum below f_r because this will cause X_C to increase and the impedance of the inductive branch to decrease. At some frequency below f_r the impedance of the inductive branch will equal X_C and Z_{EQ} will be maximum.

SECTION 25-10 DAMPING OF PARALLEL RESONANT CIRCUITS

47. **a.** $Q = 114$
b. $\Delta f = 8.77$ kHz

49. $R_P = 196.4$ kΩ

ANSWERS TO CRITICAL THINKING PROBLEMS

51. $Q = 2\pi f_r L/r_s$

$Q_{r_s} = 2\pi f_r L$

$Q_{r_s} = 2\pi L \times 1/2\pi\sqrt{LC}$

$Q_{r_s} = L/\sqrt{LC}$

$Q^2 r_s^2 = L^2/LC$

$Q^2 r_s^2 = L/C$

$\dfrac{X_L^2}{r_s^2} \times r_s^2 = L/C$

$X_L^2 = L/C$

$X_L = \sqrt{L/C}$

Chapter 26

SECTION 26-1 EXAMPLES OF FILTERING

1. **a.** A low-pass filter allows the lower frequency signals to pass from its input to its output with little or no attenuation while at the same time severely attenuating or eliminating the higher frequency signals.
b. A high-pass filter does just the opposite of a low-pass filter.

SECTION 26-2 DIRECT CURRENT COMBINED WITH ALTERNATING CURRENT

3. **a.** 10 V_{DC}
b. 5 mA

5. See Instructor's Manual.

SECTION 26-3 TRANSFORMER COUPLING

7. See Instructor's Manual.

SECTION 26-4 CAPACITIVE COUPLING

9. **a.** 159.2 Ω
b. 15 V
c. 15 V
d. 0 V
e. 10 $V_{p\text{-}p}$
f. 0 $V_{p\text{-}p}$
g. 10 $V_{p\text{-}p}$
h. 3.53 V

11. **a.** C_C charges
b. C_C discharges

13. $f = 1.59$ kHz

SECTION 26-5 BYPASS CAPACITORS

15. **a.** $X_{C_1} = 159.2$ Ω
b. 20 V
c. 8 V
d. 12 V
e. 12 V
f. 15 $V_{p\text{-}p}$
g. 0 $V_{p\text{-}p}$
h. 15 $V_{p\text{-}p}$

17. $C = 6.37$ μF

SECTION 26-6 FILTER CIRCUITS

19. **a.** Low-pass
b. High-pass

SECTION 26-7 LOW-PASS FILTERS

21. **a.** The term *passband* refers to frequencies below the cutoff frequency of a low-pass filter. Signal frequencies in the passband are allowed to pass from the input to the output of the filter with little or no attenuation.
b. The term *stopband* refers to frequencies above the cutoff frequency of a low-pass filter. Signal frequencies in the stopband are severely attenuated as they pass through the filter from input to output.

SECTION 26-8 HIGH-PASS FILTERS

23. Yes, except that for a high-pass filter the passband is above the cutoff frequency and the stopband is below the cutoff frequency.

SECTION 26-9 ANALYZING FILTER CIRCUITS

25. **a.** Low-pass
b. High-pass
c. Low-pass
d. High-pass

27. **a.** $V_{out} = 49.99$ mV and $\theta = -0.87°$
b. $V_{out} = 49.9$ mV and $\theta = -3.48°$
c. $V_{out} = 47.84$ mV and $\theta = -16.91°$
d. $V_{out} = 35.35$ mV and $\theta = -45°$
e. $V_{out} = 15.63$ mV and $\theta = -71.79°$
f. $V_{out} = 8.12$ mV and $\theta = -80.66°$
g. $V_{out} = 1.64$ mV and $\theta = -88.12°$

29. **a.** $V_{out} = 99.98$ mV and $\theta = -1.08°$
b. $V_{out} = 99.56$ mV and $\theta = -5.38°$
c. $V_{out} = 93.57$ mV and $\theta = -20.66°$
d. $V_{out} = 70.71$ mV and $\theta = -45°$
e. $V_{out} = 33.34$ mV and $\theta = -70.5°$
f. $V_{out} = 17.41$ mV and $\theta = -80°$
g. $V_{out} = 5.3$ mV and $\theta = -87°$

31. 0.707

33. **a.** 0°
b. −90°

35. Bandpass filter

37. **a.** $f_{C_1} = 1.06$ kHz
b. $f_{C_2} = 10.26$ kHz
c. BW = 9.2 kHz

39. $f_N = 4.42$ kHz

SECTION 26-10 DECIBELS AND FREQUENCY RESPONSE CURVES

41. **a.** $N_{dB} = -3$ dB
b. $N_{dB} = -10$ dB
c. $N_{dB} = -60$ dB
d. $N_{dB} = -20$ dB

43. **a.** $N_{dB} = -30.6$ dB
b. $N_{dB} = -16.72$ dB
c. $N_{dB} = -10.97$ dB
d. $N_{dB} = -3$ dB
e. $N_{dB} = -0.491$ dB
f. $N_{dB} = 0$ dB
g. $N_{dB} = 0$ dB

SECTION 26-11 RESONANT FILTERS

45. The circuit Q

SECTION 26-12 INTERFERENCE FILTERS

47. A low-pass filter with a cutoff frequency around 30 MHz

ANSWERS TO CRITICAL THINKING PROBLEMS

49. a. $f_c = 965$ Hz
    b. $V_{out} = 3.535\ V_{p\text{-}p}$
    c. $V_{out} = 68.2\ mV_{p\text{-}p}$
51. $L = 191\ \mu H$
    $C = 132.63$ pF

Chapter 27

SECTION 27-1 THREE-PHASE AC GENERATORS

1. The wye (Y) or star connection
3. $V_L = \sqrt{3}V_\theta$
5. The delta (Δ) connection
7. a. 120°
   b. 120°

SECTION 27-2 THE WYE (Y)-CONNECTED THREE-PHASE GENERATOR

7. $V_{NA} = 60\angle 150°$
11. $V_{NA} = 25\angle 240°$
13. a. 77.94 V
    b. 173.2 V
    c. 3.464 kV

SECTION 27-3 THE DELTA (Δ)-CONNECTED THREE-PHASE GENERATOR

15. a. 5.196 A
    b. 50.23 A
    c. 12.99 A

SECTION 27-4 THREE-PHASE SOURCE/LOAD CONFIGURATIONS

17. a. $V_{AN} = 480\angle 0°$; $V_{BN} = 480\angle -120°$ V; $V_{CN} = 480\angle 120°$ V
    b. $V_{AB} = 831.38\angle 30°$ V; $V_{BC} = 831.38\angle -90°$ V; $V_{CA} = 831.38\angle 150°$ V
    c. $I_{Z1} = 50\angle 0°$ A; $I_{Z2} = 50\angle -120°$ A; $I_{Z3} = 50\angle 120°$ A
    d. $I_{Aa} = I_{Z1} = 50\angle 0°$ A; $I_{Bb} = I_{Z2} = 50\angle -120°$ A; $I_{Cc} = I_{Z3} = 50\angle 120°$ A
    e. $I_{\Theta A} = I_{Aa} = I_{Z1} = 50\angle 0°$ A; $I_{\Theta B} = I_{Bb} = I_{Z2} = 50\angle -120°$ A; $I_{\Theta C} = I_{Cc} = I_{Z3} = 50\angle 120°$ A
    f. $I_N = 0$ A
19. a. $V_{Z1} = 280\angle 0°$ V; $V_{Z2} = 280\angle 120°$ V; $V_{Z3} = 280\angle -120°$ V
    b. $I_{Z1} = 6.22\angle -45°$ A; $I_{Z2} = 6.22\angle 75°$ A; $I_{Z3} = 6.22\angle -165°$ A
    c. $I_{Aa} = 10.77\angle -75°$ A; $I_{Bb} = 10.77\angle 165°$ A; $I_{Cc} = 10.77\angle 45°$ A
21. a. Each load voltage, V_Z = 277.13 V
    b. Each load current, I_Z = 13.86 A
    c. Each line current, I_L = 13.86 A
    d. $I_N = 0$ A

SECTION 27-5 THREE-PHASE AC POWER CALCULATIONS

a. $P_{Z1} = P_{Z2} = P_{Z3} = 24$ kW
b. $P_{Z1} = P_{Z2} = P_{Z3} = 14.463$ kW
c. $P_{Z1} = P_{Z2} = P_{Z3} = 7.92$ kW
d. $P_{Z1} = P_{Z2} = P_{Z3} = 7.58$ kW
e. $P_{Z1} = P_{Z2} = P_{Z3} = 3.84$ kW
f. $P_{Z1} = P_{Z2} = P_{Z3} = 3.6$ kW

ANSWERS TO CRITICAL THINKING PROBLEMS

25. a. $I_{Z1} = 48\angle 0°$ A $= 48 + j0$ A
    $I_{Z2} = 20\angle -60°$
    A $= 10 - j17.321$ A
    $I_{Z3} = 32\angle 60°$
    A $= 16 + j27.713$ A
    b. $I_{Aa} = I_{Z1} = 48\angle 0°$
    A $= 48 + j0$ A
    $I_{Bb} = I_{Z2} = 20\angle -60°$
    A $= 10 - j17.321$ A
    $I_{Cc} = I_{Z3} = 32\angle 60°$
    A $= 16 + j27.713$ A
    c. $I_N = 74 + j10.392$
    A $= 74.726\angle 8°$ A
    d. $P_T = 35.52$ kW

Chapter 28

SECTION 28-1 SEMICONDUCTOR MATERIALS

1. Four
3. a. A pure semiconductor that has only one type of atom.
   b. A semiconductor that has been doped with impurity atoms, which means that other atoms have been mixed in.
5. a. A pentavalent impurity atom (one with 5 valence electrons).
   b. A trivalent impurity atom (one with 3 valence electrons).

SECTION 28-2 THE *p-n* JUNCTION DIODE

7. Because it only allows current to flow through it in one direction.
9. a. Approximately 0.3 V
   b. Approximately 0.7 V
11. a. The anode or *p*-side of the diode must be positive with respect to the cathode.
    b. The anode or *p*-side of the diode must be negative with respect to the cathode.
13. An open switch

SECTION 28-3 VOLT-AMPERE CHARACTERISTIC CURVE

15. The forward voltage at which the diode current increases sharply. For silicon diodes, the diode current starts to increase sharply at a forward voltage of about 0.6 V and for germanium at about 0.3 V.
17. The breakdown voltage, V_{BR}, is the reverse-bias voltage at which the reverse current, I_R, increases sharply.
19. a. $R_F = 10\ k\Omega$
    b. $R_F = 1.1\ k\Omega$
    c. $R_F = 600\ \Omega$
    d. $R_F = 125\ \Omega$
    e. $R_F = 43.3\ \Omega$
    f. $R_F = 17\ \Omega$
    g. $R_F = 10\ \Omega$
21. a. The meter should read a high resistance for one polarity of the meter leads and a low resistance for the opposite polarity. For a silicon diode the ratio $\frac{R_R}{R_F}$ should be at least $\frac{1000}{1}$.
    b. A low resistance for both polarities of the meter leads
    c. A high or infinite resistance for both polarities of the meter leads
23. No, because most DMMs do not provide enough voltage and current on the resistance ranges to properly forward-bias a diode.

SECTION 28-4 DIODE APPROXIMATIONS

25. a. The second approximation
b. The first approximation
c. The third approximation

27. The first approximation.

29. a. $I_L = 200$ mA and $V_L = 6$ V
b. $I_L = 176.7$ mA and $V_L = 5.3$ V
c. $I_L = 165.6$ mA and $V_L = 4.97$ V

31. a. $I_L = 80$ mA and $V_L = 120$ V
b. $I_L = 79.53$ mA and $V_L = 119.3$ V
c. $I_L = 79.38$ mA and $V_L = 119.1$ V

SECTION 28-5 DIODE RATINGS

33. The breakdown voltage rating, V_{BR}

35. $R_R = 20$ GΩ

SECTION 28-6 RECTIFIER CIRCUITS

37. a. $V_S = 20$ V
b. $V_{out(pk)} = 27.58$ V
c. $V_{DC} = 8.77$ V
d. $I_L = 175.4$ mA
e. $I_{diode} = 175.4$ mA
f. PIV = 28.28 V
g. $f_{out} = 60$ Hz

39. A full-wave rectifier

41. a. $V_{out(pk)} = 33.24$ V
b. $V_{DC} = 21.14$ V
c. $I_L = 422.8$ mA
d. $I_{diode} = 211.4$ mA
e. PIV = 67.17 V
f. $f_{out} = 120$ Hz

43. a. $V_{out(pk)} = 26.88$ V
b. $V_{DC} = 17.1$ V
c. $I_L = 85.5$ mA
d. $I_{diode} = 42.75$ mA
e. PIV = 27.58 V
f. $f_{out} = 120$ Hz

45. a. $V_{ripple} = 3.88\ V_{p\text{-}p}$
b. $V_{DC} = 25.64$ V
c. $I_L = 512.8$ mA
d. PIV = 55.86 V

47. a. $V_{ripple} = 1.6$ V
b. $V_{DC} = 26.08$ V
c. $I_L = 130.4$ mA
d. PIV = 27.58 V

SECTION 28-7 SPECIAL DIODES

49. $R_S = 650\ \Omega$

51. a. $I_S = 50$ mA
b. $I_L = 30$ mA
c. $I_Z = 20$ mA

53. $I_Z = 76$ mA

Chapter 29

SECTION 29-1 TRANSISTOR CONSTRUCTION

1. a. The emitter (E) is the most heavily doped region in a transistor. Its job is to inject an abundance of current carriers (either free electrons or holes) into the base region.
b. The base (B) is a very thin and lightly doped region. It is sandwiched between the larger emitter and collector regions. Most of the current carriers injected into the base from the emitter flow on through to the collector.
c. The collector (C) is moderately doped and is the largest region in a transistor since it must dissipate the bulk of the heat. The main job of the collector is to attract current carriers from the base region.

3. a. Holes are the majority current carriers and electrons are the minority current carriers.
b. Electrons are the majority current carriers and holes are the minority current carriers.
c. Holes are the majority current carriers and electrons are the minority current carriers.

5. a. Outward
b. Inward

SECTION 29-2 PROPER TRANSISTOR BIASING

7 Because the only current that flows out of the base lead is a result of free electrons and holes recombining in the base region.

9. a. $I_C = 0.995$ mA
b. $I_E = 2.3$ mA
c. $I_B = 500\ \mu$A
d. $I_C = 2.67$ A
e. $I_B = 100\ \mu$A
f. $I_E = 20.34$ mA

11. a. $\beta_{DC} = 199$
b. $\beta_{DC} = 45$
c. $\beta_{DC} = 80$
d. $\beta_{DC} = 89$
e. $\beta_{DC} = 36.5$
f. $\beta_{DC} = 225$

13. a. $I_B = 200\ \mu$A
b. $I_B = 100\ \mu$A
c. $I_B = 50\ \mu$A
d. $I_B = 40\ \mu$A

15. a. $\beta_{DC} = 79$
b. $\beta_{DC} = 249$
c. $\beta_{DC} = 399$

SECTION 29-3 TRANSISTOR OPERATING REGIONS

17. The active region

19. No, I_C is controlled by other external parameters besides I_B.

21. Infinity

SECTION 29-4 TRANSISTOR RATINGS

23. a. $P_d = 1.2$ W
b. $P_d = 900$ mW
c. $P_d = 600$ mW
d. $P_d = 300$ mW
e. $P_d = 0$ W

SECTION 29-5 CHECKING A TRANSISTOR WITH AN OHMMETER

25. a. A high resistance for one polarity of the meter leads and a low resistance for the opposite polarity
b. A low resistance for both polarities of the meter leads
c. A high or infinite resistance for both polarities of the meter leads

27. Because the ohmmeter ranges of a typical DMM do not supply enough voltage and current to forward-bias the PN junction being tested.

SECTION 29-6 TRANSISTOR BIASING TECHNIQUES

29. a. $I_B = 51.4\ \mu$A
b. $I_C = 5.14$ mA
c. $V_{CE} = 5.83$ V
d. $I_{C(sat)} = 10$ mA
e. $V_{CE(off)} = 12$ V

31. $I_B = 51.4\ \mu$A
$I_C = 7.71$ mA
$V_{CE} = 2.75$ V

33. a. $I_B = 34.3\ \mu A$
b. $I_C = 6.86$ mA
c. $V_{CE} = 17.14$ V
d. $I_{C(sat)} = 24$ mA
e. $V_{CE(off)} = 24$ V
35. Voltage-divider bias
37. See Instructor's Manual.
39. a. $V_B = -4.89$ V
b. $V_E = -4.19$ V
c. $I_C = 4.19$ mA
d. $V_C = 13.71$ V
e. $V_{CE} = 9.52$ V
f. $I_{C(sat)} = 8$ mA
g. $V_{CE(off)} = -20$ V
41. $I_E = 5.3$ mA and $V_C = 6.7$ V

Chapter 30

SECTION 30-1 AC RESISTANCE OF A DIODE

1. a. $r_{AC} = 17.5\ \Omega$
b. $r_{AC} = 9.8\ \Omega$
c. $r_{AC} = 4.72\ \Omega$
d. $r_{AC} = 1.75\ \Omega$

SECTION 30-2 SMALL SIGNAL AMPLIFIER OPERATION

3 A common-emitter amplifier
5. a. $V_C = 10.2$ V
b. $V_C = 9$ V
c. $V_C = 7.8$ V
7. a. 10 $mV_{p\text{-}p}$
b. 0 $V_{p\text{-}p}$
c. 2.4 $V_{p\text{-}p}$
9. a. $A_V = 250$
b. $A_V = 20$
c. $A_V = 50$

SECTION 30-3 AC EQUIVALENT CIRCUIT OF A CE AMPLIFIER

11. C_{in}, C_E, and V_{CC}
13. See Instructor's Manual.

SECTION 30-4 CALCULATING THE VOLTAGE GAIN, A_V, OF A CE AMPLIFIER

15. a. $A_V = 120$, $V_{out} = 1.2\ V_{p\text{-}p}$
b. $A_V = 300$, $V_{out} = 3\ V_{p\text{-}p}$
c. $A_V = 480$, $V_{out} = 4.8\ V_{p\text{-}p}$
17. a. $V_B = 2.02$ V
b. $V_E = 1.32$ V
c. $I_E = 6$ mA
d. $V_C = 13.2$ V
e. $V_{CE} = 11.88$ V
19. $A_V = 432$ and $V_{out} = 10.8\ V_{p\text{-}p}$

SECTION 30-5 CALCULATING THE INPUT AND OUTPUT IMPEDANCES IN A CE AMPLIFIER

21. a. $Z_{in(base)} = 500\ \Omega$
b. $Z_{in} = 357\ \Omega$
23. a. $Z_{in(base)} = 36.5\ k\Omega$
b. $Z_{in} = 1.21\ k\Omega$
25. a. $Z_{in(base)} = 15.6\ k\Omega$
b. $Z_{in} = 2.53\ k\Omega$
c. $Z_{out} = 1.8\ k\Omega$

SECTION 30-6 THE COMMON-COLLECTOR AMPLIFIER

27. a. $V_B = 6.67$ V
b. $V_E = 5.97$ V
c. $I_E = 11.94$ mA
d. $V_C = 12$ V
e. $V_{CE} = 6.03$ V
f. $I_{C(sat)} = 24$ mA
g. $V_{CE(off)} = 12$ V

SECTION 30-7 AC ANALYSIS OF AN EMITTER FOLLOWER

29. a. $r'_e = 2.1\ \Omega$
b. $r_L = 375\ \Omega$
c. $A_V = 0.994$
d. $V_{out} = 4.97\ V_{p\text{-}p}$
e. $Z_{in(base)} = 56.6\ k\Omega$
f. $Z_{in} = 659\ \Omega$
g. $Z_{out} = 2.1\ \Omega$
31. 0° (V_{out} and V_{in} are in-phase)

SECTION 30-8 EMITTER FOLLOWER APPLICATIONS

33. a. $V_{B(Q_1)} = 2.7$ V
b. $V_{E(Q_1)} = 2$ V
c. $I_{E(Q_1)} = 2$ mA
d. $V_{C(Q_1)} = 11.4$ V
e. $V_{B(Q_2)} = 11.4$ V
f. $V_{E(Q_2)} = 10.7$ V
g. $I_{E(Q_2)} = 10.7$ mA
h. $V_{C(Q_2)} = 18$ V
35. $V_{out} = 279$ mVpp. Notice how much less the output voltage is without the emitter follower buffering the low impedance load from the collector of Q_1.

SECTION 30-9 COMMON-BASE AMPLIFIER

37. a. $V_E = -0.7$ V
b. $I_E = 5.3$ mA
c. $V_{CB} = 7.05$ V

SECTION 30-10 AC ANALYSIS OF A COMMON-BASE AMPLIFIER

39. 0° (V_{in} and V_{out} are in phase)

Chapter 31

SECTION 31-1 JFETs AND THEIR CHARACTERISTICS

1. In the channel
3. The source current, I_S, and the drain current, I_D
5. a. V_{GS} is made positive.
b. V_{DS} is made negative.
7. V_P is the drain-source voltage at which the drain current, I_D, levels off when $V_{GS} = 0$ V.
9. The pinchoff voltage decreases by the same amount that V_{GS} increases.
11. a. $I_D = 15$ mA
b. $I_D = 11.5$ mA
c. $I_D = 8.44$ mA
d. $I_D = 5.86$ mA
e. $I_D = 3.75$ mA
f. $I_D = 2.11$ mA
g. $I_D = 938\ \mu A$
h. $I_D = 234\ \mu A$
i. $I_D = 0$ mA
13. a. $I_D = 20$ mA
b. $I_D = 12.8$ mA
c. $I_D = 7.2$ mA
d. $I_D = 3.2$ mA
e. $I_D = 800\ \mu A$
f. $I_D = 0$ mA

SECTION 31-2 JFET BIASING TECHNIQUES

15. a. $V_G = 0$ V
b. $V_S = 2.15$ V
c. $V_{GS} = -2.15$ V
d. $V_D = 10.91$ V
17. a. $V_G = 0$ V
b. $V_S = 0.75$ V
c. $V_{GS} = -0.75$ V
d. $V_D = 6$ V
19. $I_D = 2.41$ mA and $I_D = 8.67$ V

SECTION 31-3 JFET AMPLIFIERS

21. $g_m = \dfrac{\Delta I_D}{\Delta V_{GS}}$ (V_{DS} constant)
The unit is the Siemen (S).
23. a. $g_m = 8$ mS
b. $g_m = 6.67$ mS
c. $g_m = 5.33$ mS
d. $g_m = 4$ mS
e. $g_m = 2.67$ mS
f. $g_m = 1.33$ mS
25. a. $Z_{in} = 1.5\ M\Omega$
b. $r_L = 3.87\ k\Omega$
c. $g_{mo} = 5$ mS
d. $g_m = 2.31$ mS

e. $A_V = 8.94$
f. $V_{out} = 2.68\ V_{p-p}$

27. a. $V_G = 0$ V
b. $V_{GS} = -1$ V
c. $I_D = 5.56$ mA
d. $V_D = 15$ V

29. a. $g_{mo} = 5$ mS
b. $g_m = 2.31$ mS
c. $r_L = 2.48\ k\Omega$
d. $Z_{in} = 302\ \Omega$
e. $A_V = 5.73$
f. $V_{out} = 573\ mV_{p-p}$

SECTION 31-4 MOSFETs AND THEIR CHARACTERISTICS

31. Insulated gate field effect transistor (IGFET)

33. No

35. Zero

SECTION 31-5 MOSFET BIASING TECHNIQUES

37. Zero-bias

39. $I_D = 15$ mA

41. a. $V_{DS} = 29$ V
b. $V_{DS} = 23.7$ V
c. $V_{DS} = 3$ V

43. Zero-bias, self-bias, and current-source bias

45. a. $R_D = 200\ \Omega$
b. $R_D = 800\ \Omega$
c. $R_D = 1.4\ k\Omega$
d. $R_D = 2.6\ k\Omega$

SECTION 31-6 HANDLING MOSFETs

47. They lower the input impedance.

Chapter 32

SECTION 32-1 CLASSES OF OPERATION

1. a. 360°
b. 180°
c. 120° or less

3. Low distortion and low power efficiency

5. It only conducts during the positive or negative alternation of the AC input voltage but not both.

7. Tuned rf amplifiers

SECTION 32-2 CLASS A AMPLIFIERS

9. a. $I_B = 70.6\ \mu A$
b. $I_{CQ} = 10.6$ mA
c. $V_{CEQ} = 8.1$ V
d. $V_{CE(off)} = 24$ V
e. $I_{C(sat)} = 16$ mA

11. See Instructor's Manual.

13. a. $V_B = 4$ V
b. $V_E = 3.3$ V
c. $I_{CQ} = 10$ mA
d. $V_{CEQ} = 8.7$ V
e. $P_d = 87$ mW
f. $V_{CE(off)} = 24$ V
g. $I_{C(sat)} = 15.7$ mA

15. See Instructor's Manual.

17. a. $V_{out} = 7.95\ V_{p-p}$
b. $P_L = 5.27$ mW
c. $P_{CC} = 254$ mW
d. % Efficiency = 2.08%

SECTION 32-3 CLASS B PUSH-PULL AMPLIFIERS

19. a. Q_1
b. Q_2

21. See Instructor's Manual.

23. a. Charging
b. Discharging

25. $i_{C(sat)} = 900$ mA
$v_{ce(off)} = 9$ V

27. $P_{d(max)} = 2.03$ W

29. a. $I_{CQ} = 6.23$ mA
b. $V_{BQ_1} = 0.7$ V
c. $V_{BQ_2} = -0.7$ V
d. V_{EQ_1} and $V_{EQ_2} = 0$ V
e. V_{CEQ_1} and $V_{CEQ_2} = 25$ V
f. $P_{dQ} = 156$ mW

31. 0 V

SECTION 32-4 CLASS C AMPLIFIERS

33. $V_B = -1.8$ V
$V_C = 12$ V
$v_C = 24\ V_{p-p}$

35. At or near the positive peak

37. 5 MHz

Chapter 33

SECTION 33-1 DIACs

1. Power control circuits

3. When the voltage across the diac (irregardless of polarity) reaches or exceeds the breakover voltage, $\pm V_{BO}$

SECTION 33-2 SCRs AND THEIR CHARACTERISTICS

5. Even though an SCR is forward-biased, it will not conduct until the forward breakover voltage is reached.

7. It increases sharply.

9. By reducing the anode current below the level of holding current, I_H

11. No! Once an SCR fires, the gate loses all control!

13. It gets brighter.

15. To ensure that the negative alternation of voltage cannot apply excessive reverse-bias voltage to the SCRs gate-cathode junction

SECTION 33-3 TRIACs

17. Anode 2 (A_2), Anode 1 (A_1), and the gate (*G*).

19. By increasing the gate current

21. By reducing the anode current below the holding current, I_H

23. To provide symmetrical triggering of the triac

25. In both directions

SECTION 33-4 UNIJUNCTION TRANSISTORS

27. V_E must reach 11.2 V

29. $f = 6.56$ kHz

31. a. A pulsating DC voltage
b. To provide a relatively stable voltage for the UJT circuit
c. The conduction angle decreases

Chapter 34

SECTION 34-1 DIFFERENTIAL AMPLIFIERS

1. a. $I_T = 565\ \mu A$
b. $I_E = 282.5\ \mu A$
c. $V_C = 6.35$ V

3. a. The base of Q_2
b. The base of Q_1

5. a. $I_T = 115.3\ \mu A$
b. $I_E = 57.7\ \mu A$
c. $V_C = 9.35$ V

SECTION 34-2 OPERATIONAL AMPLIFIERS AND THEIR CHARACTERISTICS

7. A differential amplifier

9. Direct coupling. The advantage of direct coupling is that a DC input can also be amplified.

11. $V_{id} = \pm 100\ \mu V$

13. $f_{OL} = 10$ Hz

15. 1 MHz, f_{unity}

17. a. $f_{max} = 159.2$ kHz
    b. $f_{max} = 79.58$ kHz
    c. $f_{max} = 39.79$ kHz
    d. $f_{max} = 15.92$ kHz

19. CMRR(dB) = 90 dB

SECTION 34-3 OP-AMP CIRCUITS WITH NEGATIVE FEEDBACK

21. 180°

23. Since it has the same potential as ground yet it can sink no current.

25. $f_{max} = 12.73$ kHz

27. Noninverting amplifier

29. 0 V

31. $A_{CL} = 25$
    $V_{out} = 25\ V_{p\text{-}p}$
    $Z_{in} = 8$ GΩ
    $Z_{out(CL)} = 18.75$ mΩ

33. Voltage follower

35. a. $f_{CL} = 80$ kHz
    b. $f_{CL} = 40$ kHz
    c. $f_{CL} = 1$ MHz

37. a. $A_{CL} = -15$
    b. $V_{out} = 3\ V_{p\text{-}p}$

SECTION 34-4 POPULAR OP-AMP CIRCUITS

39. a. $V_{out} = -12.5$ V
    b. $V_{out} = 2.5$ V
    c. $V_{out} = 10$ V
    d. $V_{out} = 3.75$ V

41. $V_{out} = -6.25$ V

43. $f_C = 1.45$ kHz

45. An active high-pass filter

47. A voltage to current converter

49. A current to voltage converter

51. a. Any voltage that is even slightly positive.
    b. Any voltage that is even slightly negative.

53. A precision half-wave rectifier

55. −200 mV

Index

C

D

E

F

G

H

I

J

N

O

P

S

T

U

V